The Greek Alphabet

Alpha	A	α	Iota	I				
Beta	B	β	Kappa	K				
Gamma	Γ	γ	Lambda	Λ	λ	Tau	T	τ
Delta	Δ	δ	Mu	M	μ	Upsilon	Υ	υ
Epsilon	E	ϵ	Nu	N	ν	Phi	Φ	ϕ
Zeta	Z	ζ	Xi	Ξ	ξ	Chi	X	χ
Eta	H	η	Omicron	O	o	Psi	Ψ	ψ
Theta	Θ	θ	Pi	Π	π	Omega	Ω	ω

Abbreviations for Units

A	ampere
Å	angstrom (10^{-10} m)
atm	atmosphere
Btu	British thermal unit
Bq	becquerel
C	coulomb
°C	degree Celsius
cal	calorie
Ci	curie
cm	centimeter
dyn	dyne
eV	electron volt
°F	degree Fahrenheit
fm	femtometer, fermi (10^{-15} m)
ft	foot
Gm	gigameter (10^{9} m)
G	gauss
Gy	gray
g	gram
H	henry
h	hour
Hz	hertz
in	inch
J	joule
K	kelvin
kg	kilogram
km	kilometer
keV	kilo-electron volts
lb	pound
L	liter
m	meter
MeV	mega-electron volts
Mm	megameter (10^{6} m)
mi	mile
min	minute
mm	millimeter
ms	millisecond
N	newton
nm	nanometer (10^{-9} m)
pt	pint
qt	quart
rev	revolution
R	roentgen
Sv	seivert
s	second
T	tesla
u	unified mass unit
V	volt
W	watt
Wb	weber
y	year
yd	yard
μm	micrometer (10^{-6} m)
μs	microsecond
μC	microcoulomb
Ω	ohm

Third Edition

Volume 2

Physics

For Scientists and Engineers

Paul A. Tipler

Worth Publishers

For Claudia

Physics for Scientists and Engineers, Third Edition, Volume 2

Paul A. Tipler

Printed in the United States of America

Library of Congress Catalog Card Number: 89-52166

Extended Version (Chapters 1–42) ISBN: 0-87901-432-6

Standard Version (Chapters 1–35) ISBN: 0-87901-430-X

Volume 1 (Chapters 1–17) ISBN: 0-87901-433-4

Volume 2 (Chapters 18–42) ISBN: 0-87901-434-2

Printing: 5 6 7 Year: 99 98 97

Development Editors: Valerie Neal and Steven Tenney

Design: Malcolm Grear Designers

Art Director: George Touloumes

Production Editor: Elizabeth Mastalski

Production Supervisor: Sarah Segal

Layout: Patricia Lawson

Photographs: Steven Tenney, John Schultz of PAR/NYC, and Lana Berkovich

Line Art: York Graphic Services and Demetrios Zangos

Composition: York Graphic Services

Printing and binding: R. R. Donnelley and Sons

Cover: Supersonic Candlelight. A stroboscopic color schlieren or shadow picture taken at one-third microsecond exposure shows a supersonic bullet passing through the hot air rising above a candle. Schlieren pictures make visible the regions of nonuniform density in air. Estate of Harold E. Edgerton/ Courtesy of Palm Press.

Illustration credits begin on p. **IC-1** which constitute an extension of the copyright page.

Worth Publishers

33 Irving Place

New York, NY 10003

Preface

This third edition of *Physics*, now titled *Physics for Scientists and Engineers*, is a textbook for the standard two- or three-semester introductory physics course for engineering and science majors. It is assumed that the student has taken, or is concurrently taking, calculus. The book is divided into six parts: Mechanics, Oscillations and Waves, Thermodynamics, Electricity and Magnetism, Optics, and Modern Physics. It is available both in the standard version of 35 chapters, including two on modern physics (Relativity and The Origins of Quantum Theory), and an extended version of 42 chapters (available in one volume or two), which has seven additional chapters on modern physics (Quantum Mechanics, Atoms, Molecules, Solids, Nuclei, Elementary Particles, and Astrophysics and Cosmology), of which six are new.

SI units are used throughout the text. Except for a few problems on the conversion of force units (for example, pounds to newtons) in Chapter 4 and in the discussion of *R* factors for insulation materials in Chapter 16, all the worked examples, exercises, and problems are given in SI units.

My primary goals in writing this text have been

1. To provide a balanced introduction to the most important concepts and phenomena in classical and modern physics in a way that reflects the beauty and excitement of physics and also provides a solid foundation for further study.
2. To present physics in a logical and coherent manner that is interesting and accessible to all students.
3. To help students build self-confidence in their understanding of physics and in their problem-solving skills.
4. To stimulate students by exposing them to some of the many contemporary applications of and developments in physics in everyday life, in today's technology, and in the study of the cosmos.

Some of the features of this new edition are described below.

Streamlining and Consolidation

Every chapter has been extensively revised to make the presentation more concrete and to provide clear, logical, and succinct introductions to the central ideas of physics. For example,

> Work and energy are treated in a single chapter (Chapter 6) that discusses both the application of the work–energy theorem and conservation of mechanical energy to the solving of mechanics problems, and the generalized law of the conservation of energy. The difficulties that arise in the application of the work–energy theorem to work done on extended objects is discussed fully, as is the conversion of internal energy to mechanical energy. The concept of pseudowork is presented in Chapter 7 after the center of mass has been introduced. The treatment of escape speed has been moved to Chapter 10 (Gravity).

The discussion of rotational motion is now covered in a single chapter (Chapter 8) and has been reorganized so that angular momentum and rolling bodies can be discussed before the full vector treatment of rotation.

The chapter on oscillations (Chapter 12) has been moved so that it immediately precedes the chapters on mechanical waves. Care has been taken in the discussion of resonance to match the equations in it to those in Chapter 28 (Alternating-Current Circuits) describing resonance in *LRC* circuits. Many of the concepts of waves are introduced in Chapter 13 (Waves on Strings), with Chapter 14 being devoted entirely to sound waves. Reflection, refraction, interference, and diffraction are discussed qualitatively as applying to all types of waves, but all detailed calculations are deferred to the chapters on optics (Chapters 30 through 33).

The thermodynamics unit has been rewritten and is now organized into three chapters (Chapters 15 through 17). The introduction to temperature has been made more physical, and the section on the kinetic-theory interpretation of temperature has been expanded by the inclusion of the Maxwell–Boltzmann distribution.

The unit on electricity and magnetism (Chapters 18 through 29) has been extensively revised with particular attention to enhancing its accessibility. Electric charge, electric fields due to discrete charges, and the behavior of point charges and dipoles in an electric field are covered in Chapter 18 (Electric Fields I). In Chapter 19 (Electric Fields II), the fields due to continuous charge distributions are calculated using Coulomb's law, Gauss's law, or both. Expanded coverage of the classical model of conduction, together with a qualitative discussion of contemporary theory in Chapter 22 (Electric Current), complements the more detailed introduction to the band theory of solids in Chapter 39 of the extended edition.

In Chapter 25 (Sources of the Magnetic Field), the Biot–Savart law is given in terms of the field of a moving charge as well as that of a current element. The field is then calculated for a current loop, a solenoid, and a straight wire before Ampère's law is discussed. The material on magnetic flux has been moved to Chapter 26 (Magnetic Induction), the coverage of the magnetization of a bar magnet has been moved to Chapter 27 (Magnetism in Matter), and the discussion of Maxwell's displacement current has been moved to Chapter 29 (Maxwell's Equations and Electromagnetic Waves).

In the optics unit (Chapters 30 through 33), the material on the eye, microscopes, and telescopes is now in a short, new chapter on optical instruments (Chapter 32), which also includes a new section on the camera. A discussion of rainbows has been added to Chapter 30 (Light). The diffraction of light is first covered qualitatively and is then given a full mathematical treatment using phasors in Chapter 33 (Interference and Diffraction).

Because of this streamlining and consolidation, the length of the book has not increased significantly despite the substantial number of new worked examples, problems, photographs, and figures.

Modern Physics and Contemporary Applications

In addition to the separate chapters on modern physics (two in the standard version and nine in the extended version), the coverage of modern and applied physics has been greatly expanded throughout this edition. A section on numerical methods to support computer-based problem solving has been written for Chapter 5 (Newton's Laws II). A set of computer problems that supplement the end-of-chapter problems is available.

A solid foundation for the study of modern physics has been provided in the presentation of classical physics. For example, wave packets and group velocity are introduced in Chapter 14 (Sound) so that these concepts will be available for the study of electron waves in Chapter 36 (Quantum Mechanics) in the extended edition. Similarly, the Maxwell–Boltzmann energy distribution is given in Chapter 15 (Temperature) so that it can be used in the discussion of the electrical properties of solids in Chapter 39, and vacuum-tube diodes and triodes are discussed in Chapter 28 (Alternating-Current Circuits) to provide a basis for the treatment of semiconductor devices, also in Chapter 39.

The nine chapters on modern physics in the extended version of this edition provide an introduction to quantum physics and to the physics of atoms, molecules, solids, nuclei, and elementary particles, plus a concluding chapter on astrophysics and cosmology. Although the material in these chapters is necessarily more descriptive than that in the previous chapters, the important ideas of modern physics are often introduced by the study of specific, simple problems. For example, the concepts of quantum numbers and degeneracy arise naturally in the study of the three-dimensional infinite square well in Chapter 36. Similarly, the Pauli exclusion principle arises from symmetry considerations in the study of two identical particles in a one-dimensional infinite square well in the same chapter. Later in Chapter 39, the infinite square well is again used to introduce the concept of Fermi energy needed to understand the electrical and thermal properties of solids.

New to this edition are hundreds of photographs, most of them in color, that complement and amplify the text and provide effective examples of contemporary applications of physics.

Seventeen guest essays, of which fourteen are new, have been written for the enjoyment and enlightenment of students and instructors. In Volume 2 these include:

Examples, Exercises, and Problems

The understanding of physics and the development of problem-solving skills are enhanced by the extensive and integrated use of examples, in-text exercises, and graded sets of problems. Of the 434 worked examples in the 42-chapter extended version, 290 are new. Nearly all of the examples are numerical and have been written to ensure correspondence between the examples and the end-of-chapter problems (especially those at the intermediate level). Worked examples are often paired with a numerical in-text exercise (with its answer given immediately) that asks the student to perform a simple calculation to extend the example and reinforce understanding. Many of the simple, single-step examples in the second edition have been converted into in-text exercises, and many of the nonnumerical examples have been converted into text material under an appropriate subhead. Despite these conversions, the total number of worked examples has been increased by about 50 percent.

Problems at the end of each chapter are grouped into three levels of difficulty. Level I problems are relatively easy, single-step problems. They are keyed to the appropriate sections of the chapter so that the student can quickly find help if needed. Level II problems require a more sophisticated understanding and are not divided by section. Great effort has been invested in providing an extensive selection of these intermediate level problems. Level III problems are the most challenging and will be of value for more advanced students. A typical chapter has about 35 problems at Level I, 20 at Level II, and 12 at Level III. The number of intermediate level problems has been increased by more than 40 percent.

Mathematics

The teaching of introductory physics is complicated by the wide range of student backgrounds and abilities. Some have had both physics and calculus courses in high school; others have had no high school physics and are just beginning their first calculus course. To accommodate this diversity of backgrounds, a general review of mathematics (including algebra, geometry, trigonometry, complex numbers, and calculus) is provided in Appendix A.

In the text itself, the mathematical sophistication required increases gradually so that those students who are studying calculus concurrently will not be at a disadvantage. New mathematical methods and concepts are introduced as dictated by the physics, with the introduction of a new method or concept immediately followed by an example. The derivative and integral are presented in Chapter 2 (Motion in One Dimension) as an intrinsic component of the discussion of instantaneous velocity and displacement. Vector dot products are introduced in Chapter 6, where they are needed for the discussion of work and energy, and cross products are introduced in Chapter 8 in conjunction with the discussion of torque and rotational motion. A few sections that are more mathematically sophisticated are optional and are located at the ends of chapters, where they do not interrupt the flow of ideas and can easily be omitted.

The gradual increase in mathematical level through the text, combined with the relatively easy in-text exercises and numerous Level I problems, should help the least prepared students to build confidence. At the same time, the broad spectrum of complexity in the problem sets, culminating in the Level III problems, ensures that all students can be challenged and stimulated at a level appropriate to their abilities.

Ease of Review

Several pedagogical features will aid students as they review the material just covered. Important equations, laws, and tables are highlighted by a color screen. Margin heads are provided for quick reference. Key terms are introduced in boldface type, defined in the text, and listed in the review section of each chapter. Thought questions immediately follow some sections within each chapter. These include routine questions that can be easily answered from the preceding text as well as open-ended questions that can serve as a basis for classroom discussion.

Each chapter concludes with a summary, which lists the important laws and results that were discussed in the chapter, along with the equations that will be most useful in solving problems. Following each chapter is a list of suggested further readings, a review section, and the graded sets of problems. The review section contains a list of learning objectives, a list of key terms that the student should be able to identify and define, and a set of true–false questions.

Use of Color

Color has been used to improve the pedagogical effectiveness of the illustrations, most of which have been redrawn to enhance their clarity and dimensionality. Students will appreciate, for example, the way color is used to distinguish between vectors and their components, between force and velocity vectors in the same diagram, or between conductor surfaces and gaussian surfaces. The conventions used in the illustrations, such as the placement of force vectors in a diagram, have been chosen with pedagogical effectiveness as the primary goal.

Comparison with the Second Edition

Those instructors familiar with my previous edition should be reassured that the features that contributed to its usefulness have not been diminished in this extensive revision. All of the many changes have been made to enhance and complement what was already sound. A few of the changes are given here to expedite familiarization with this edition for those who have taught from the second edition:

The list of learning objectives, formerly at the beginning of each chapter, now appears in the review section following the summary.

The reviews of algebra and trigonometry have been moved from Chapter 1 to Appendix A.

Relative velocity is presented in Chapter 3 (Motion in Two and Three Dimensions).

Each of Newton's three laws of motion is discussed in a separate section in Chapter 4 (Newton's Laws I). The basic interactions are then listed with a brief qualitative discussion. In Chapter 4, applications are limited to problems involving a single particle. A brief discussion of rolling friction has been included in Chapter 5 (Newton's Laws II).

Static equilibrium is now treated separately in a new chapter (Chapter 9).

Discussions of the Maxwell–Boltzmann distribution and the heat pump have been added to the thermodynamics unit, while the Clausius inequality and the proof of the existence of entropy have been deleted.

Chapter 29 (Maxwell's Equations and Electromagnetic Waves) now includes material on Maxwell's displacement current, energy and momentum in electromagnetic waves, and the electromagnetic spectrum.

The Lorentz transformation is treated early in Chapter 34 and is used to simplify the discussion of clock synchronization and simultaneity.

Seven new chapters on modern physics have been added to the two in the standard version to complete the modern physics unit in the extended version. This material can easily serve as the basis of a one-quarter or one-semester course in modern physics.

A number of deletions have been made to streamline the text or to make space for the coverage of contemporary topics. These include reaction thresholds, the vector nature of angular displacement, gravitational field lines, the buoyant force in accelerated reference frames, the capacitance of an isolated sphere, the betatron, the Wheatstone bridge, and the magnetic vector **H**.

Acknowledgements

Many people have contributed to this edition. I would like to thank everyone who used the first or second edition and offered comments and suggestions. Your letters have been valuable and welcome.

Ralph Llewellyn (University of Central Florida) wrote the exciting and informative chapter on astrophysics and cosmology (Chapter 42) in the extended edition.

Many new and interesting end-of-chapter problems were provided by Howard Miles (Washington State University), Lawrence B. Golden and James Klein (Pennsylvania State University), Robert Rundel (Mississippi State University), and William E. Fasnacht (United States Naval Academy). Among them, they have also provided one of the independently worked sets of solutions for all of the problems in the text.

Lawrence Golden and James Klein also wrote the Computer Investigations supplement.

James Walker (Washington State University) prepared the answers listing at the end of the text, produced the elegant solutions that are published in the accompanying complete *Solutions Manual* (for instructors), and also offered many valuable suggestions for improving and clarifying the end-of-chapter problems.

Robin Macqueen (University of British Columbia) contributed the Suggestions for Further Reading for each chapter.

The accuracy of the numerical calculations in the examples and exercises has been expertly checked by Robert Weidman (Michigan Technological University), Chelcie Liu (City College of San Francisco), and Edward Brown (Manhattan College). Professor Brown also offered many helpful suggestions in his reviews of the end-of-chapter problems.

Gene Mosca (United States Naval Academy) has applied his profound understanding of the teaching of physics to the revision of the *Study Guide* originally written by Granvil C. Kyker.

David Mills (College of the Redwoods) has produced a test bank of about 3000 conceptual and numerical multiple-choice questions.

The *Instructors Resource Manual* has been prepared by Robert Allen (Inver Hills Community College), John Davis (University of Washington), John Risley (North Carolina State University), James Walker (Washington State University), Vicki Williams (Pennsylvania State University), and myself. It provides a comprehensive selection of demonstrations, listings of commercially available and public-domain software programs suitable for use in teaching introductory physics, a film and video guide, suggested homework assignments, critical-thinking questions, and a complete answers listing for all end-of-chapter problems.

Roger Clapp (University of South Florida), Manuel Gómez-Rodríguez (University of Puerto Rico, Río Piedras), John Russell (Southeastern Massachusetts University), and Jim Smith (University of Illinois, Champaign–Urbana) reviewed the entire second edition; John Russell also reviewed parts of the second draft manuscript, and Jim Smith reviewed the entire first draft for the third edition. Brooke Pridmore (Clayton State College) reviewed the second draft manuscript; Grant Hart (Brigham Young University) reviewed major portions of the second edition, as well as both drafts of the manuscript; and James Brown (The Colorado School of Mines) reviewed major portions of the second edition and the first draft of the manuscript. Their substantive and constructive comments and suggestions were a continual source of inspiration over the many months of this project.

Many other instructors have provided extensive and invaluable reviews. They have all made a deeply appreciated and fundamental contribution to the quality of this revision, and I therefore wish to thank:

Walter Borst, *Texas Technological University*

Edward Brown, *Manhattan College*

James Brown, *The Colorado School of Mines*

Christopher Cameron, *University of Southern Mississippi*

Roger Clapp, *University of South Florida*

Bob Coakley, *University of Southern Maine*

Andrew Coates, *University College, London, United Kingdom*

Miles Dresser, *Washington State University*

Manuel Gómez-Rodríguez, *University of Puerto Rico, Río Piedras*

Allin Gould, *John Abbott College C.E.G.E.P. Canada*

Dennis Hall, *University of Rochester*

Grant Hart, *Brigham Young University*

Jerold Izatt, *University of Alabama*

Alvin Jenkins, *North Carolina State University*

Lorella Jones, *University of Illinois, Champaign–Urbana*

Michael Kambour, *Miami-Dade Junior College*

Patrick Kenealy, *California State University at Long Beach*

Doug Kurtze, *Clarkson University*

Lui Lam, *San Jose State University*

Chelcie Liu, *City College of San Francisco*

Robert Luke, *Boise State University*

Stefan Machlup, *Case Western Reserve University*

Eric Matthews, *Wake Forest University*

Konrad Mauersberger, *University of Minnesota, Minneapolis*

Duncan Moore, *University of Rochester*

Gene Mosca, *United States Naval Academy*

Elizabeth Nickles, *Albany College of Pharmacy*

Harry Otteson, *Utah State University*

Jack Overley, *University of Oregon*

Larry Panek, *Widener University*

Malcolm Perry, *Cambridge University, United Kingdom*

Brooke Pridmore, *Clayton State College*

Arthur Quinton, *University of Massachusetts, Amherst*

John Risley, *North Carolina State University*

Robert Rundel, *Mississippi State University*

John Russell, *Southeastern Massachusetts University*

Michael Simon, *Housatonic Community College*

Jim Smith, *University of Illinois, Champaign-Urbana*

Richard Smith, *Montana State University*

Larry Sorenson, *University of Washington*

Thor Stromberg, *New Mexico State University*

Edward Thomas, *Georgia Institute of Technology*

Colin Thomson, *Queens University, Canada*

Gianfranco Vidali, *Syracuse University*

Brian Watson, *St. Lawrence University*

Robert Weidman, *Michigan Technological University*

Stan Williams, *Iowa State University*

Thad Zaleskiewicz, *University of Pittsburgh, Greensburg*

George Zimmerman, *Boston University*

Finally, I would like to thank everyone at Worth Publishers for their help and encouragement, and particularly Steven Tenney, Valerie Neal, Betsy Mastalski, Wendy Schechter, Anne Vinnicombe, George Touloumes, and Sarah Segal.

Berkeley, California
June 1991

Paul Tipler

Contents in Brief

Volume 2

Supplements

For Students

Study Guide

Volume 1 (Chapters 1–17) and
Volume 2 (Chapters 18–42)

Each chapter contains:

Key ideas and equations

Possible pitfalls

Questions and answers

True/False questions

Problems with detailed solutions, followed by a similar problem with the answer only

Student's Solutions Manual

Contains worked solutions to selected odd-numbered problems, totaling about 25% of the text problems.

Computer Investigations

An introduction to the application of computers to physics. Spreadsheets and programs, including 25 that are supplied for the student, are used to model, visualize, and calculate. No previous computer literacy is assumed. Explanatory text describes spreadsheets, rudiments of BASIC, numerical methods, and elements of equation-solver software. The approximately 110 problems that follow are divided into three levels: Level I problems supply a spreadsheet or program and require variation of input parameters; Level II problems indicate a prototype spreadsheet or program to be modified by the student; Level III problems require the student to create the needed spreadsheet or program. An accompanying IBM or Mac disk contains spreadsheets and programs. Answers to selected problems are supplied.

For Instructors

Solutions Manual

Volume 1 (Chapters 1–17) and
Volume 2 (Chapters 18–42)

Contains complete worked-out solutions to all problems in the text.

Test Bank

Volume 1 (Chapters 1–17) and
Volume 2 (Chapters 18–42)

Approximately 3000 multiple-choice questions spanning all sections of the text. Each question is identified by topic and noted as factual, conceptual, or numerical.

Computerized Test-Generation System

(IBM and Macintosh)

A data base comprising the questions in the Test Bank.

Instructor's Resource Manual

Contains:

Demonstrations

Film and videocassette guide

Software guide

Critical thinking questions

Answers to all end-of-chapter problems

Transparencies

Volume 1 (Chapters 1–17) and
Volume 2 (Chapters 18–42)

Approximately 150 full-color acetates of figures and tables from the text, with type enlarged for projection.

Contents

Part 4 Electricity and Magnetism 597

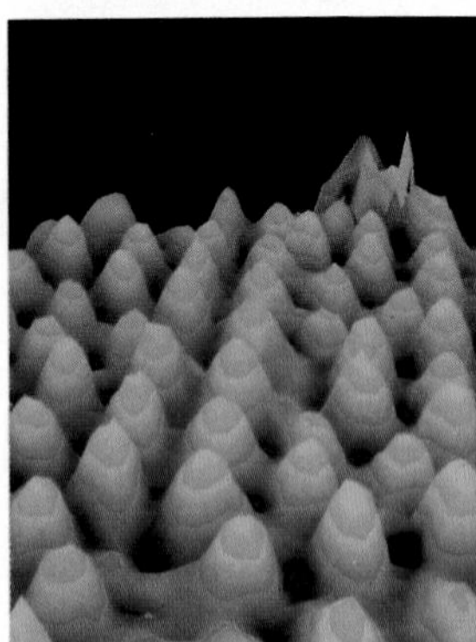

Part 4 Electricity and Magnetism

A small, cubicle permanent magnet levitates above a disk of the superconductor yttrium-barium-copper oxide, cooled by liquid nitrogen to 77 K. At temperatures below 92 K, the disk becomes superconducting. The magnetic field of the cube sets up circulating electric currents in the superconducting disk, such that the resultant magnetic field in the superconductor is zero. These currents produce a magnetic field that repels the cube.

Chapter 18

The Electric Field I: Discrete Charge Distributions

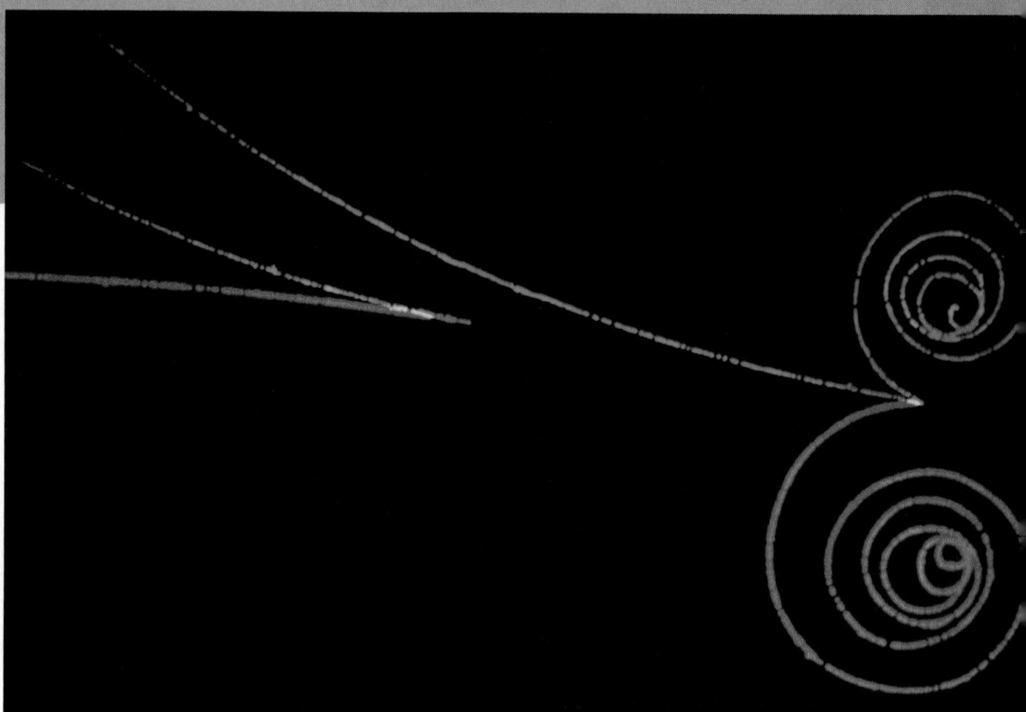

Pair production. An electron of charge $-e$ and a positron of charge $+e$ are created by the interaction of electromagnetic radiation with matter. The paths of the oppositely charged particles, made visible in a bubble chamber, are bent in opposite directions by a magnetic field.

Electricity is in such common use today that we normally give it little thought. Yet just a century ago, there were few electric lights and no electric heaters, motors, radios, or television sets. Although the practical use of electricity has been developed mostly in the twentieth century, the study of electricity has a long history. Observations of electrical attraction can be traced as far back as the time of the ancient Greeks. They noticed that after amber has been rubbed, it attracts small objects such as straw or feathers. Indeed, the word "electric" comes from the Greek word for amber, *elektron.*

In this chapter, we will begin our study of electricity with a short discussion of the concept of electric charge, followed by a brief look at conductors and insulators and how conductors can be given a charge. We will then study Coulomb's law, which describes the force exerted by one electric charge on another. Next, we will introduce the electric field and show how it can be described by electric-field lines, which indicate the magnitude and direction of the field. Last, we will discuss the behavior of point charges and electric dipoles in electric fields.

18-1 Electric Charge

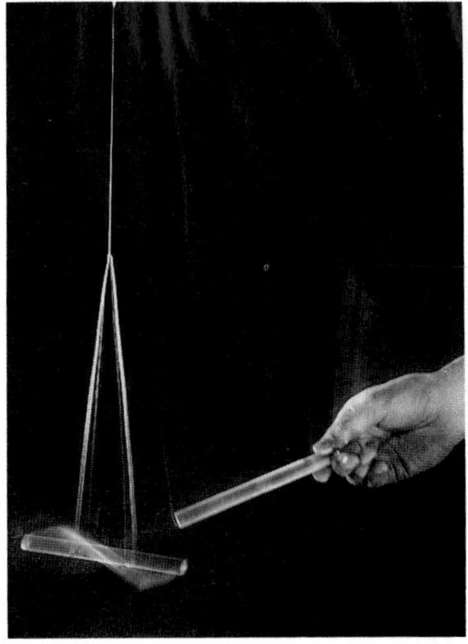

Figure 18-1 Two plastic rods that have been rubbed with fur repel each other.

Let us begin by considering a simple experiment involving electrical attraction. Suppose that we rub a plastic rod with fur and suspend the rod from a string such that it is free to rotate. If we then approach this rod with a second plastic rod that has also been rubbed with fur, we find that the rods repel each other (Figure 18-1). We can repeat the experiment and achieve the same results if we use two glass rods that have been rubbed with a piece of silk. However, if we use a plastic rod that has been rubbed with fur and a glass rod that has been rubbed with silk, we find that the rods attract each other.

When we rub the plastic rod with fur or the glass rod with silk, we cause the rod to become "electrified" or "charged." If we repeat our experiment with various types of materials, we find that we can classify all charged objects into just two groups—those that are charged like the plastic rod rubbed with fur and those that are charged like the glass rod rubbed with silk. The great American statesman and scientist Benjamin Franklin proposed a model explaining why this is so. He suggested that every object has a "normal" amount of electricity and that, when two objects are rubbed together, some of this electricity can be transferred from one of the objects to the other. This leaves one with an excess amount of electricity and the other with an equal deficiency. Franklin described the resulting charges with plus and minus signs. He chose the type of charge acquired by a glass rod when it is rubbed with a piece of silk to be positive, which meant that the piece of silk acquired a negative charge of equal magnitude. Based on Franklin's convention, then, plastic rubbed with fur acquires a negative charge and the fur acquires a positive charge of equal magnitude. As we saw in our experiment, two objects that carry the same type of charge—that is, two objects that are both positive or both negative—repel each other, and two objects that carry opposite charges attract each other (Figure 18-2).

Today, we know that when glass is rubbed with silk, electrons are transferred from the glass to the silk, leaving the silk with an excess number of electrons and the rod with a deficiency. According to Franklin's classification, which we still use, the silk is negatively charged, so the electrons are said to carry a negative charge.

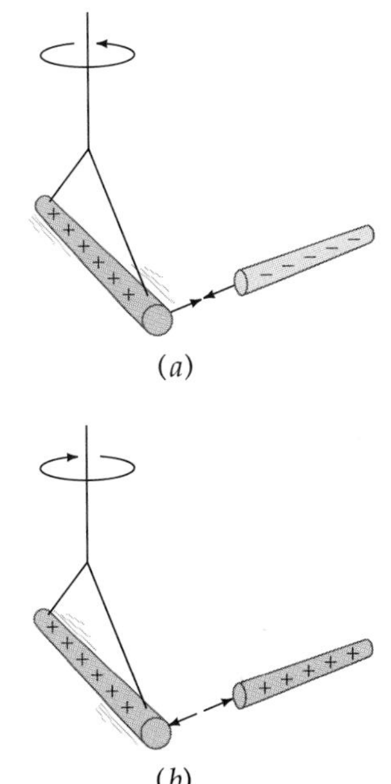

Figure 18-2 (*a*) Objects carrying charges of opposite sign attract each other. (*b*) Objects carrying charges of the same sign repel each other.

We now know also that matter consists of atoms that are electrically neutral. Each atom has a tiny but massive nucleus that contains protons, each of which carries a positive charge, and neutrons, which have no charge. The number of protons in the nucleus is the atomic number Z of the element. Surrounding the nucleus are an equal number of negatively charged electrons. The electron and the proton are very different particles. For one thing, the proton is about 2000 times more massive than the electron. Yet their charges are exactly equal but opposite in sign. The charge of the proton is e and that of the electron is $-e$, where e is called the **fundamental unit of charge.** All charges occur in integral amounts of the fundamental unit of charge e. That is, **charge is quantized.** Any charge Q occurring in nature can be written $Q = \pm Ne$, where N is an integer.* The quantization of electric charge is usually not noticed because N is usually very large. For example, charging a plastic rod by rubbing it with a piece of fur typically transfers about 10^{10} electrons to the rod.

*In the quark model of elementary particles, protons, neutrons, and some other elementary particles are thought to be made up of particles called quarks that carry charges of $\pm\frac{1}{3}e$ or $\pm\frac{2}{3}e$. Apparently, quarks cannot be observed individually but only in combinations that result in a net charge of $\pm Ne$ or 0.

Quantized charge. These individual mercury ions are trapped in an electric field configuration called a Paul trap. In this false-color photograph, ions are preferentially located in the red areas. Neighboring ions are separated by several micrometers.

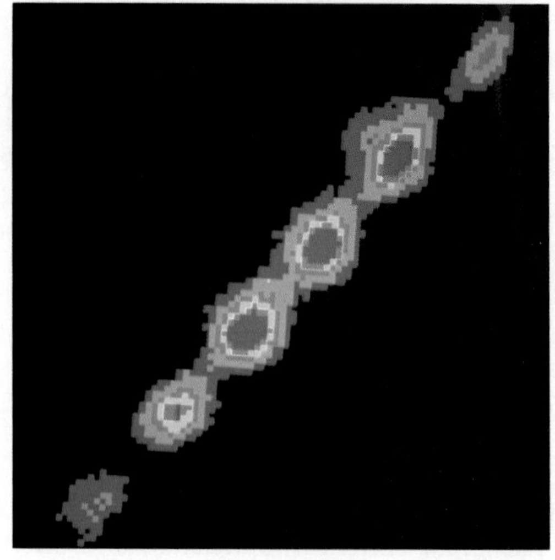

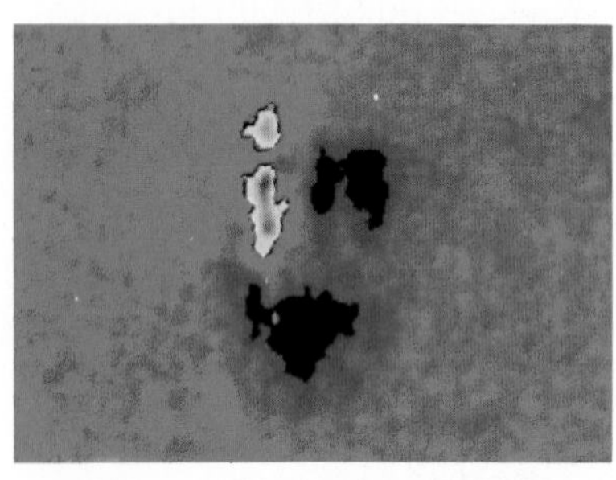

Charging by contact. A sample of plastic about 0.02 mm wide that was charged by contact with a piece of nickel. Although the plastic carries a net positive charge, regions of negative charge (dark) as well as regions of positive charge (yellow) are indicated. The photograph was taken by sweeping a charged needle of width 10^{-7} m over the sample and recording the electrostatic force on the needle.

When objects are in very close contact, as when they are rubbed against each other, electrons are transferred from one object to the other. One object is left with an excess number of electrons and is therefore negatively charged, and the other object is left lacking electrons and is therefore positively charged. In this process, charge is not created but is merely transferred. The net charge of the two objects taken together does not change. That is, *charge is conserved*. The **law of conservation of charge** is a fundamental law of nature. In certain interactions among elementary particles, charged particles such as electrons are created or annihilated. However, in all these processes, equal amounts of positive and negative charge are produced or destroyed, so the net charge of the universe is unchanged. For instance, whenever an electron of charge $-e$ is created, a particle of charge $+e$ called a *positron* is created simultaneously. (This process is called *pair production*.)

The SI unit of charge is the coulomb, which is defined in terms of the unit of electric current, the *ampere*. (The ampere is defined in terms of a magnetic-force measurement, which we will discuss in Chapter 25. It is the unit of current used in everyday electrical work.) The **coulomb** (C) is the amount of charge flowing through a cross-sectional area of a wire in one second when the current in the wire is one ampere. The fundamental unit of electric charge e is related to the coulomb by

Fundamental unit of charge

$$e = 1.60 \times 10^{-19}\ \text{C} \tag{18-1}$$

Charges from about 10 nC (1 nC = 10^{-9} C) to about 0.1 μC (1 μC = 10^{-6} C) can be produced in the laboratory by putting certain objects in intimate contact, often by simply rubbing their surfaces together. Such procedures involve the transfer of many electrons.

Example 18-1

A copper penny has a mass of 3 g. The atomic number of copper is $Z = 29$ and the atomic mass is 63.5 g/mol. What is the total charge of all the electrons in the penny?

We must first find the number of atoms in 3 g of copper. Since 1 mol of copper contains Avogadro's number of atoms and has a mass of 63.5 g, the number of atoms in 3 g of copper is

$$N = (3\text{ g})\frac{6.02 \times 10^{23}\text{ atoms/mol}}{63.5\text{ g/mol}} = 2.84 \times 10^{22}\text{ atoms}$$

Each atom contains $Z = 29$ electrons, so the total charge Q is

$$Q = (2.84 \times 10^{22}\text{atoms})(29\text{electrons/atom})(-1.60 \times 10^{-19}\text{ C/electron})$$

$$= -1.32 \times 10^{5}\text{ C}$$

Question

1. After you stroke your cat, do you think the cat is positively or negatively charged?

18-2 Conductors and Insulators and Charging by Induction

In many materials, such as copper and other metals, some of the electrons are free to move about the entire material. Such materials are called **conductors.** In other materials, such as wood or glass, all the electrons are bound to nearby atoms and none can move freely. These materials are called **insulators.**

In a single atom of copper, 29 electrons are bound to the nucleus by the electrostatic attraction between the negatively charged electrons and the positively charged nucleus. The outer electrons are more weakly bound than the inner electrons because of their greater distance from the nucleus and because of the repulsion of the inner electrons. When a large number of copper atoms are combined in a piece of metallic copper, the binding of the electrons of each individual atom is changed by interactions with neighboring atoms. One or more of the outer electrons in each atom is no longer bound but is free to move throughout the whole piece of metal, much as a gas molecule is free to move about in a box. The number of free electrons depends on the particular metal, but it is typically about one per atom. The copper atom minus one of its outer electrons carries a positive charge and is called a *positive ion.* In metallic copper, the copper ions are arranged in a regular array called a *lattice.* Normally, a conductor is electrically neutral because there is a lattice ion carrying a positive charge $+e$ for each free electron carrying a negative charge $-e$. A conductor can be given a net charge by adding or removing free electrons.

Figure 18-3 An electroscope. The two gold leaves are connected to a metal rod that has a metal ball on top. When a negative charge is placed on the metal ball, the charge is conducted to the leaves and they repel each other.

Figure 18-3 shows an **electroscope,** which is a device for detecting electric charge. Two gold leaves are attached to a conducting post that has a conducting ball on top and are otherwise insulated from the container. When uncharged, the leaves hang together vertically. When the ball is touched by a negatively charged plastic rod, some of the negative charge from the rod is transferred to the ball and moves to the gold leaves, which then spread apart because of the electrical repulsion of their negative charges. Touching the ball with a positively charged glass rod also causes the leaves to spread apart. In this case, the positively charged glass rod attracts electrons from the metal ball, leaving the leaves charged positively.

In Figure 18-4, a long metal rod is in contact with the ball of an electroscope. When the far end of the rod is touched by a charged plastic rod, the leaves of the electroscope spread apart because the electrons from the plastic rod are conducted along the metal rod to the electroscope. If the metal rod is replaced by a wooden rod and the far end of the wooden rod is touched by a charged plastic rod, nothing happens. The wooden rod is an insulator, which does not conduct electricity.

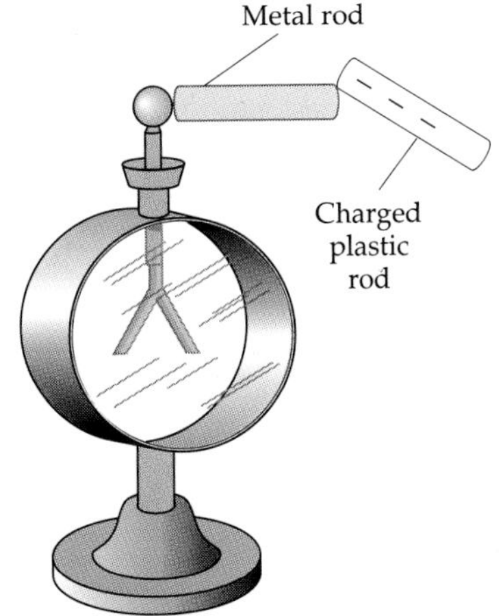

Figure 18-4 A metal rod is in contact with the metal ball on the electroscope. When the far end of the metal rod is touched by a negatively charged plastic rod, some of the charge conducts along the metal rod to the electroscope, as evidenced by the spreading of the gold leaves.

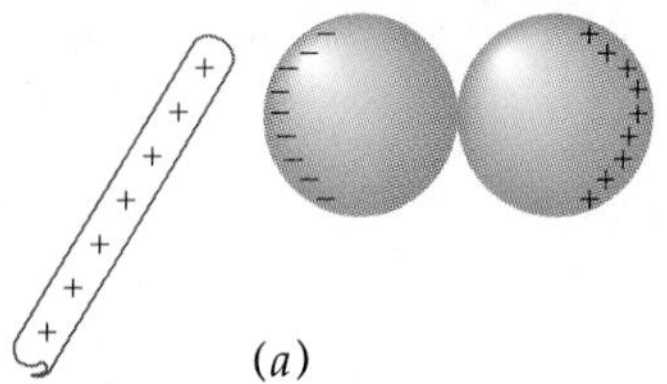

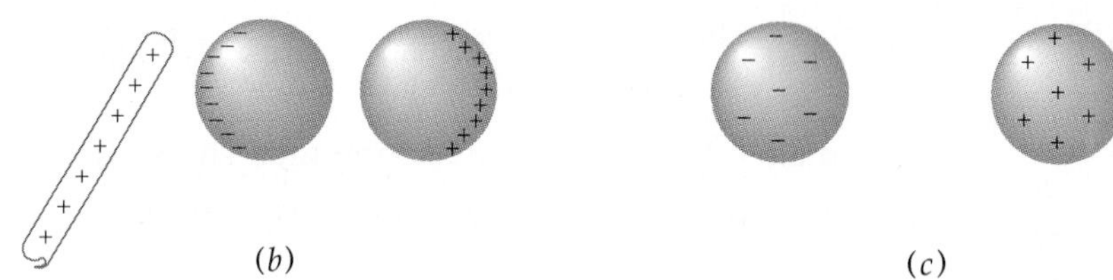

Figure 18-5 Charging by induction. (*a*) The two spherical conductors in contact become oppositely charged because the positively charged rod attracts electrons to the left sphere, leaving the right sphere positively charged. (*b*) If the spheres are now separated while the rod is in place, they retain their equal and opposite charges. (*c*) When the rod is removed and the spheres are far apart, the spheres are uniformly charged with equal and opposite charges.

A simple and practical method of charging a conductor makes use of the ready movement of free electrons in a conductor. In Figure 18-5, two uncharged metal spheres are in contact. When a charged rod is brought near one of the spheres, free electrons flow from one sphere to the other. If the rod is positively charged, it attracts the negatively charged electrons, and the sphere nearest the rod acquires electrons from the other. This leaves the near sphere with a net negative charge and the far sphere with an equal net positive charge (Figure 18-5*a*). If the spheres are separated before the rod is removed (Figure 18-5*b*), they will have equal and opposite charges (Figure 18-5*c*). A similar result is obtained with a negatively charged rod, which drives electrons from the near sphere to the far sphere. In either case, the spheres are charged without being touched by the rod, and the charge on the rod is undisturbed. This process is called **electrostatic induction** or **charging by induction.** If a charged spherical conductor is brought into contact with an identical but uncharged sphere, the charge on the first sphere will be shared equally by both conductors. If the spheres are then separated, each sphere will have half the excess charge originally on the first sphere.

A convenient large conductor is the earth itself, which for most purposes can be considered to be an infinitely large conductor. When a conductor is connected to the earth, it is said to be **grounded.** This is indicated schematically by showing a connecting wire ending in parallel horizontal lines as in Figure 18-6*b*. We can use the earth to charge a single conductor by induction. In Figure 18-6*a*, a positively charged rod is brought near an uncharged conducting sphere. Free electrons are attracted to the side near the positive rod, leaving the other side with a positive charge. If we ground the sphere while the charged rod is still present, the sphere acquires a charge opposite to that of the rod because electrons from the earth travel along the connecting wire to the sphere and neutralize the positive charge on the right side of the sphere (Figure 18-6*b*). The connection to ground is broken before the rod is removed to complete the charging by induction (Figure 18-6*c*). When the charged rod is removed, the sphere has a uniform negative charge as shown in Figure 18-6*d*.

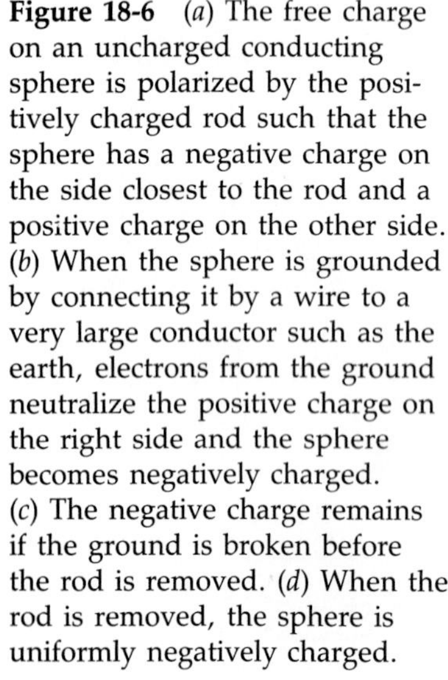

Figure 18-6 (*a*) The free charge on an uncharged conducting sphere is polarized by the positively charged rod such that the sphere has a negative charge on the side closest to the rod and a positive charge on the other side. (*b*) When the sphere is grounded by connecting it by a wire to a very large conductor such as the earth, electrons from the ground neutralize the positive charge on the right side and the sphere becomes negatively charged. (*c*) The negative charge remains if the ground is broken before the rod is removed. (*d*) When the rod is removed, the sphere is uniformly negatively charged.

Question

2. Can insulators as well as conductors be charged by induction?

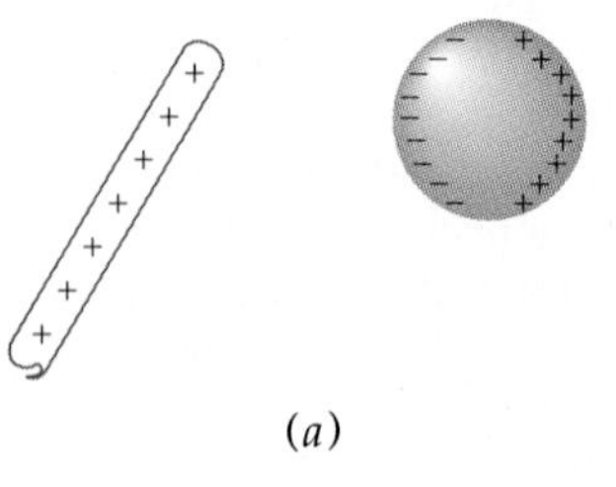

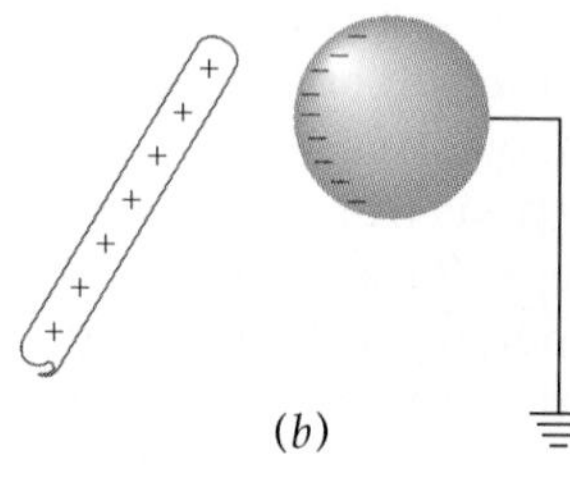

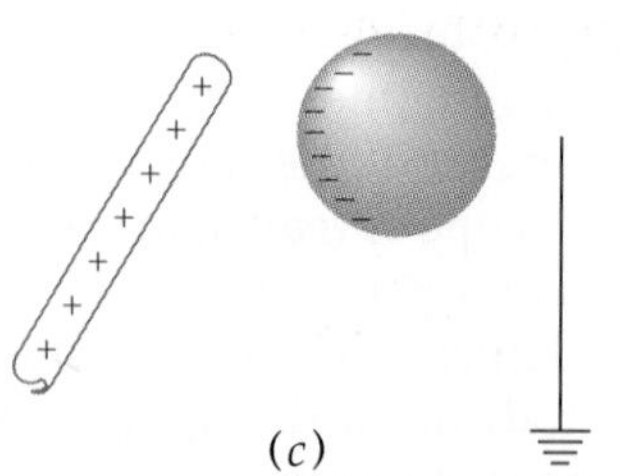

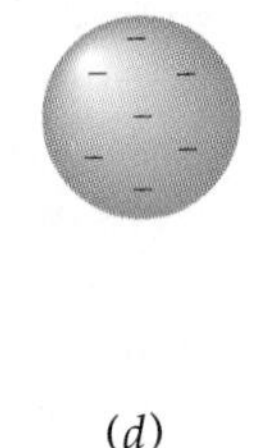

(*Left*) The lightning rod on this building is grounded to guide the electric current from a lightning discharge to the ground. (*Right*) These fashionable ladies are wearing hats with metal chains that drag along the ground supposedly to protect them from lightning.

18-3 Coulomb's Law

The force exerted by one charge on another was studied by Charles Coulomb (1736–1806) using a torsion balance of his own invention. Coulomb's experimental apparatus was essentially the same as that described for the Cavendish experiment (Chapter 10), with the masses replaced by small charged spheres. For the magnitudes of charges easily transferred by rubbing, the gravitational attraction of the spheres is completely negligible compared with their electric attraction or repulsion. In Coulomb's experiment, the charged spheres were much smaller than the distance between them so that the charges could be considered to be point charges. Coulomb used induction to produce equally charged spheres and to vary the amount of charge on the spheres. For example, beginning with charge q_0 on each sphere, he could reduce the charge to $\frac{1}{2}q_0$ by temporarily grounding one sphere to discharge it and then placing the two spheres in contact. The results of the experiments of Coulomb and others on the forces exerted by one point charge on another are summarized in **Coulomb's law:**

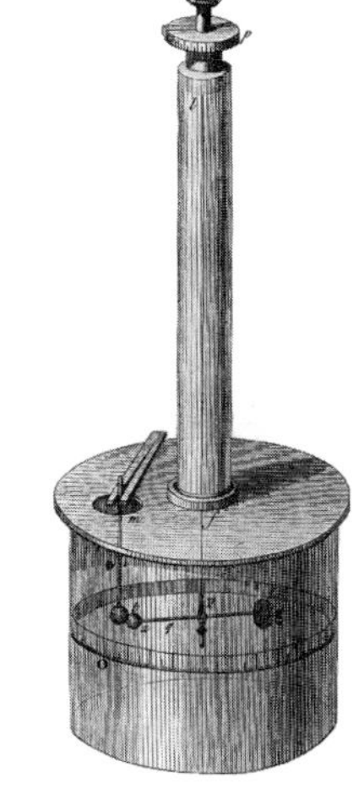

Coulomb's torsion balance.

> The force exerted by one point charge on another acts along the line joining the charges. It varies inversely as the square of the distance separating the charges and is proportional to the product of the charges. The force is repulsive if the charges have the same sign and attractive if the charges have opposite signs.

Figure 18-7 shows the forces exerted between two charges with the same sign and between two charges of opposite sign.

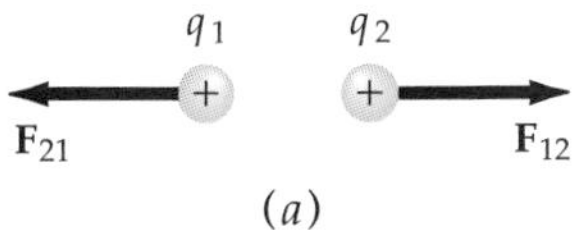

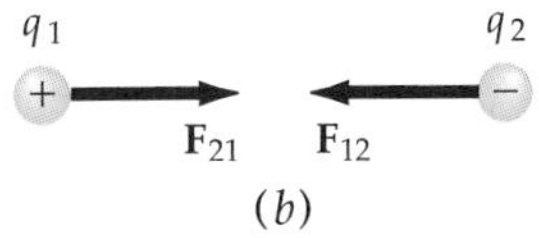

Figure 18-7 (*a*) Like charges repel, whereas (*b*) unlike charges attract.

Coulomb's law can be stated more simply using a mathematical expression. Let q_1 and q_2 be two point charges that are separated by a distance r_{12}, which is the magnitude of the vector $\mathbf{r}_{12}$ pointing from charge q_1 to q_2 (Figure 18-8). The force $\mathbf{F}_{12}$ exerted by charge q_1 on q_2 is then

Coulomb's law

$$\mathbf{F}_{12} = \frac{kq_1q_2}{r_{12}^2}\hat{\mathbf{r}}_{12} \qquad 18\text{-}2$$

where $\hat{\mathbf{r}}_{12} = \mathbf{r}_{12}/r_{12}$ is a unit vector pointing from q_1 to q_2 and k is the **Coulomb constant,** which has the value

$$k = 8.99 \times 10^9 \text{ N·m}^2/\text{C}^2 \qquad 18\text{-}3$$

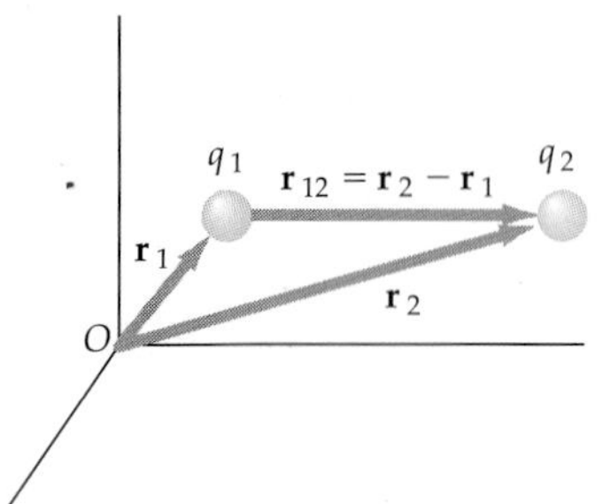

Figure 18-8 Charge q_1 at position $\mathbf{r}_1$ and charge q_2 at $\mathbf{r}_2$ relative to origin O. The force exerted by q_1 on q_2 is in the direction of the vector $\mathbf{r}_{12} = \mathbf{r}_2 - \mathbf{r}_1$ if both charges have the same sign and in the opposite direction if they have opposite signs.

By Newton's third law, the force $\mathbf{F}_{21}$ exerted by q_2 on q_1 is the negative of $\mathbf{F}_{12}$. That is, $\mathbf{F}_{21}$ is equal in magnitude to $\mathbf{F}_{12}$ but opposite in direction. The *magnitude* of the electric force exerted by a charge q_1 on another charge q_2 a distance r away is thus given by

$$F = \frac{kq_1q_2}{r^2} \qquad 18\text{-}4$$

If both charges have the same sign, that is, if both are positive or both are negative, the force is repulsive. If the two charges have opposite signs, the force is attractive. Note the similarity between Coulomb's law and Newton's law of gravity (Equation 10-2). Both are inverse-square laws. However, the gravitational force between two particles is proportional to the masses of the particles and is always attractive, whereas the electric force is proportional to the charges of the particles and may be attractive or repulsive.

Example 18-2

Two point charges of 0.05 μC each are separated by 10 cm. Find (*a*) the magnitude of the force exerted by one charge on the other and (*b*) the number of fundamental units of charge in each.

(*a*) From Coulomb's law, the magnitude of the force is

$$F = \frac{kq_1q_2}{r^2}$$

$$= \frac{(8.99 \times 10^9 \text{ N·m}^2/\text{C}^2)(0.05 \times 10^{-6}\text{ C})(0.05 \times 10^{-6}\text{ C})}{(0.1\text{ m})^2}$$

$$= 2.25 \times 10^{-3}\text{ N}$$

(*b*) The number of electrons required to produce a charge of 0.05 μC is found from

$$q = Ne$$

$$N = \frac{q}{e} = \frac{0.05 \times 10^{-6}\text{ C}}{1.6 \times 10^{-19}\text{ C}} = 3.12 \times 10^{11}$$

A charge of this size does not reveal that electric charge is quantized. A million electrons could be added to or subtracted from this charge without being detected by ordinary instruments.

Since both the electric force and the gravitational force between any two particles vary inversely with the square of their separation, the ratio of these forces is independent of their separation. We can therefore compare the relative strengths of these forces for elementary particles such as two protons, two electrons, or an electron and a proton.

Example 18-3

Compute the ratio of the electric force to the gravitational force exerted by one proton on another.

Since each proton has charge $+e$, the electric force is repulsive and has the magnitude

$$F_e = \frac{ke^2}{r^2}$$

The gravitational force, given by Newton's law of gravity, is attractive and has the magnitude

$$F_g = \frac{Gm_p^2}{r^2}$$

where m_p is the mass of a proton. The ratio of these forces is independent of the separation distance r:

$$\frac{F_e}{F_g} = \frac{ke^2}{Gm_p^2}$$

Substituting the values $k = 8.99 \times 10^9\ \text{N·m}^2/\text{C}^2$, $e = 1.60 \times 10^{-19}$ C, $G = 6.67 \times 10^{-11}\ \text{N·m}^2/\text{kg}^2$, and $m_p = 1.67 \times 10^{-27}$ kg, we obtain

$$\frac{F_e}{F_g} = \frac{(8.99 \times 10^9\ \text{N·m}^2/\text{C}^2)(1.60 \times 10^{-19}\ \text{C})^2}{(6.67 \times 10^{-11}\ \text{N·m}^2/\text{kg}^2)(1.67 \times 10^{-27}\ \text{kg})^2} = 1.24 \times 10^{36}$$

Exercise

In the hydrogen atom, the electron is separated from the proton by a distance of about 5.3×10^{-11} m on the average. Calculate the magnitude of the electrostatic force exerted by the proton on the electron. (Answer: 8.2×10^{-8} N)

We can see from Example 18-3 that the gravitational force between two elementary particles is so much smaller than the electric force between them (assuming they are charged) that it can always be neglected in describing interactions between them. It is only because large masses such as the earth contain almost exactly equal numbers of positive and negative charges that the gravitational force is important. If the positive and negative charges in such objects did not cancel each other, the electric forces between them would be much greater than the gravitational forces.

In a system of charges, each charge exerts a force given by Equation 18-2 on every other charge. Thus, the net force on any charge is the vector sum of the individual forces exerted on that charge by all the other charges in the system.

Example 18-4

Three point charges lie on the x axis; $q_1 = 25$ nC is at the origin, $q_2 = -10$ nC is at $x = 2$ m, and $q_0 = 20$ nC is at $x = 3.5$ m (Figure 18-9). Find the net force on q_0 due to q_1 and q_2.

The force on q_0 due to q_1, which is 3.5 m away, is given by

$$\mathbf{F}_{10} = \frac{kq_1q_0}{r_{10}^2}\,\hat{\mathbf{r}}_{10}$$

$$= \frac{(8.99 \times 10^9\ \text{N·m}^2/\text{C}^2)(25 \times 10^{-9}\ \text{C})(20 \times 10^{-9}\ \text{C})}{(3.5\ \text{m})^2}\,\mathbf{i}$$

$$= (0.367\ \mu\text{N})\mathbf{i}$$

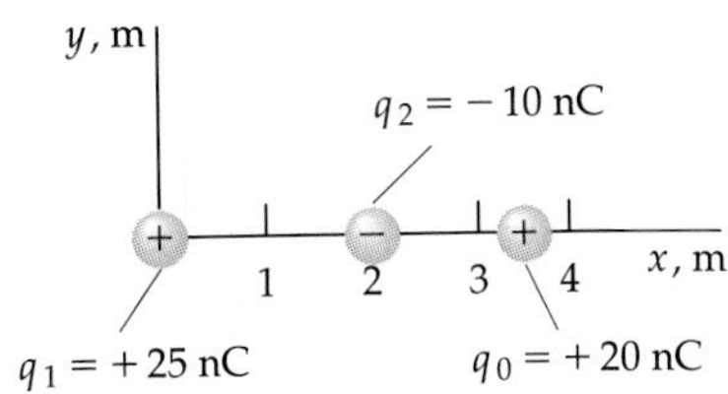

Figure 18-9 Point charges on the x axis for Example 18-4.

where we have used **i** to designate the unit vector $\hat{\mathbf{r}}_{10}$ from charge q_1 to q_0, which points in the x direction. The force on q_0 due to q_2, which is 1.5 m away, is

$$\mathbf{F}_{20} = \frac{kq_2q_0}{r_{20}^2}\hat{\mathbf{r}}_{20}$$

$$= \frac{(8.99 \times 10^9\ \mathrm{N{\cdot}m^2/C^2})(-10 \times 10^{-9}\ \mathrm{C})(20 \times 10^{-9}\ \mathrm{C})}{(1.5\ \mathrm{m})^2}\mathbf{i}$$

$$= (-0.799\ \mu\mathrm{N})\mathbf{i}$$

The net force on the charge q_0 due to the charges q_1 and q_2 is

$$\mathbf{F}_{\text{net}} = \mathbf{F}_{10} + \mathbf{F}_{20}$$

$$= (0.367\ \mu\mathrm{N})\mathbf{i} - (0.799\ \mu\mathrm{N})\mathbf{i} = (-0.432\ \mu\mathrm{N})\mathbf{i}$$

Note that in Example 18-4, the charge q_2, which is between q_1 and q_0, has no effect on the force $\mathbf{F}_{10}$ exerted by q_1 on q_0, just as the charge q_1 has no effect on the force exerted by q_2 on q_0. The net force exerted on a charge (such as q_0 in this example) by a system of charges is found by the simple superposition of the separate forces exerted by each charge in the system. This **principle of superposition of electric forces** has been verified by experiment.

If a system of charges is to remain stationary, there must be other, nonelectrical forces acting on the charges so that the net force from all sources acting on each charge is zero. In the preceding example and those that follow, we assume that there are such forces so that all the charges remain stationary.

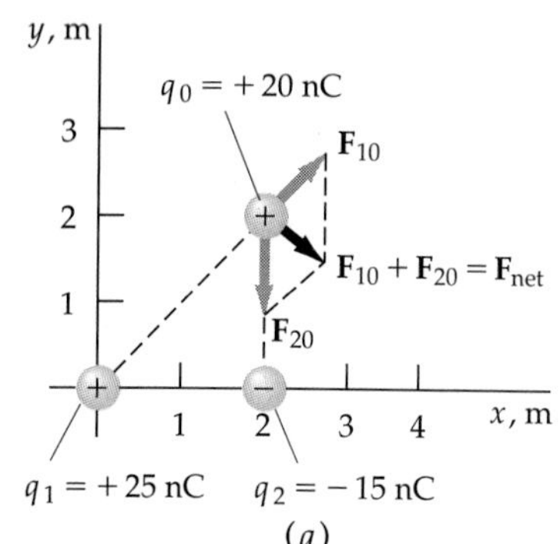

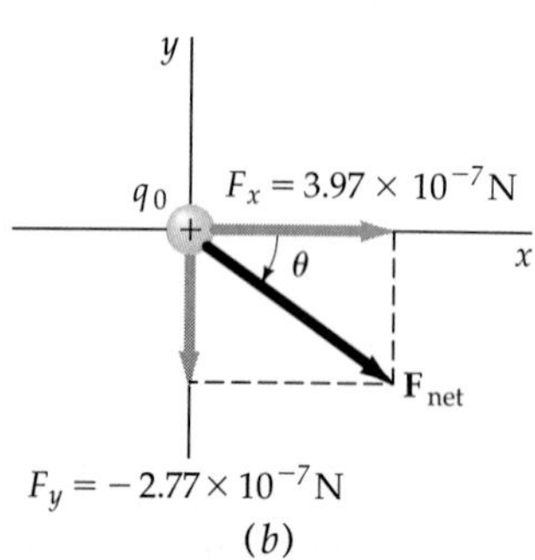

Figure 18-10 (*a*) Force diagram for Example 18-5. The net force on charge q_0 is the vector sum of the forces $\mathbf{F}_{10}$ due to q_1 and $\mathbf{F}_{20}$ due to q_2. (*b*) Diagram showing net force in (*a*) and its x and y components.

Example 18-5

Charge $q_1 = +25$ nC is at the origin, charge $q_2 = -15$ nC is on the x axis at $x = 2$ m, and charge $q_0 = +20$ nC is at the point $x = 2$ m, $y = 2$ m as shown in Figure 18-10*a*. Find the force on q_0.

Since q_2 and q_0 have opposite signs, the force exerted by q_2 on q_0 is attractive and in the negative y direction as shown in the figure. It is given by

$$\mathbf{F}_{20} = \frac{kq_2q_0}{r_{20}^2}\hat{\mathbf{r}}_{20}$$

$$= \frac{(8.99 \times 10^9\ \mathrm{N{\cdot}m^2/C^2})(-15 \times 10^{-9}\ \mathrm{C})(20 \times 10^{-9}\ \mathrm{C})}{(2\ \mathrm{m})^2}\mathbf{j}$$

$$= (-6.74 \times 10^{-7}\ \mathrm{N})\mathbf{j}$$

The distance between q_1 and q_0 is $2\sqrt{2}$ m. The force exerted by q_1 on q_0 is

$$\mathbf{F}_{10} = \frac{kq_1q_0}{r_{10}^2}\hat{\mathbf{r}}_{10}$$

$$= \frac{(8.99 \times 10^9\ \mathrm{N{\cdot}m^2/C^2})(25 \times 10^{-9}\ \mathrm{C})(20 \times 10^{-9}\ \mathrm{C})}{(2\sqrt{2}\ \mathrm{m})^2}\hat{\mathbf{r}}_{10}$$

$$= (5.62 \times 10^{-7}\ \mathrm{N})\,\hat{\mathbf{r}}_{10}$$

where $\hat{\mathbf{r}}_{10}$ is the unit vector that is directed along the line from q_1 to q_0.

The vector sum of these two forces is most easily found by first writing the forces in terms of their rectangular components. Since $\mathbf{F}_{10}$ makes an angle of 45° with the x and y axes, its x and y components are equal to each other and to $F_{10}/\sqrt{2}$:

$$F_{10x} = F_{10y} = \frac{5.62 \times 10^{-7}\ \text{N}}{\sqrt{2}} = 3.97 \times 10^{-7}\ \text{N}$$

The x and y components of the net force are therefore

$$F_x = F_{10x} + F_{20x} = (3.97 \times 10^{-7}\ \text{N}) + 0 = 3.97 \times 10^{-7}\ \text{N}$$

$$F_y = F_{10y} + F_{20y} = (3.97 \times 10^{-7}\ \text{N}) + (-6.74 \times 10^{-7}\ \text{N})$$

$$= -2.77 \times 10^{-7}\ \text{N}$$

The magnitude of the net force is

$$F_{\text{net}} = \sqrt{F_x^2 + F_y^2} = \sqrt{(3.97 \times 10^{-7}\ \text{N})^2 + (-2.77 \times 10^{-7}\ \text{N})^2}$$

$$= 4.84 \times 10^{-7}\ \text{N}$$

The net force points to the right and downward as shown in Figure 18-10b, making an angle θ with the x axis given by

$$\tan\theta = \frac{F_y}{F_x} = \frac{-2.77}{3.97} = -0.698$$

$$\theta = -34.9°$$

Questions

3. Discuss the similarities and differences in the properties of electric charge and gravitational mass.
4. If the sign convention for charge were changed so that the charge on the electron were positive and the charge on the proton were negative, would Coulomb's law still be written the same?

18-4 The Electric Field

The electric force exerted by one charge on another is an example of an action-at-a-distance force that is similar to the gravitational force exerted by one mass on another. To avoid the problem of action at a distance, we introduce the concept of the **electric field E.** One charge produces an electric field **E** everywhere in space, and this field exerts the force on the other charge. The force is thus exerted by the field at the position of the second charge, rather than by the first charge itself, which is some distance away.

Figure 18-11 shows a set of point charges, q_1, q_2, and q_3, arbitrarily arranged in space. If we place a charge q_0 at some point near this system of charges, there will be a force exerted on q_0 due to the other charges. The presence of the charge q_0 will generally change the original distribution of the other charges, particularly if the charges are on conductors. However, we may choose q_0 to be small enough so that its effect on the original charge distribution is negligible. We call such a small charge a **test charge** because we use it to test the field of other charges without disturbing them. The net force exerted on q_0 is the vector sum of the individual forces exerted on q_0 by each of the other charges in the system. By Coulomb's law, each of these forces is proportional to q_0, so the net force will be proportional to q_0. The electric field **E** at a point is defined as the net force on a positive test charge q_0 divided by q_0:

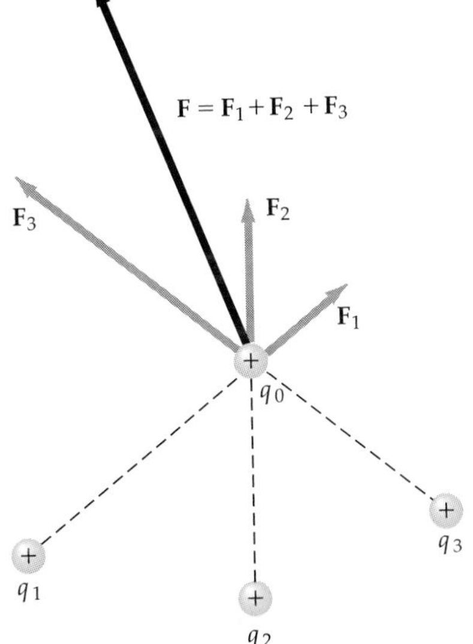

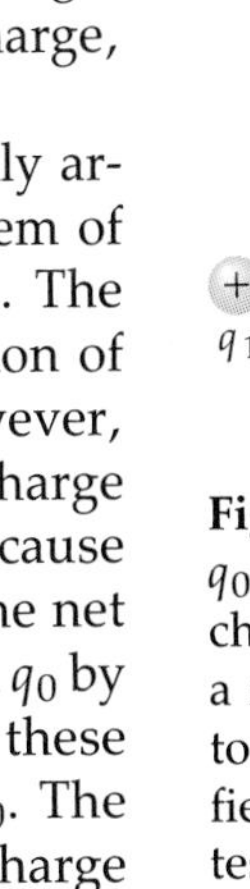

Figure 18-11 A small test charge q_0 in the vicinity of a system of charges, q_1, q_2, and q_3, experiences a force **F** that is proportional to q_0. The ratio $\mathbf{F}/q_0$ is the electric field **E** at the position of the test charge.

$$\mathbf{E} = \frac{\mathbf{F}}{q_0} \quad (q_0 \text{ small}) \qquad 18\text{-}5$$

Electric field defined

This definition is similar to that for the gravitational field of the earth, which was defined in Section 4-3 as the force per unit mass exerted by the earth on an object. The gravitational field of the earth **g** describes the property of the space around the earth such that when a mass m is placed at some point, the force exerted by the earth is $m\mathbf{g}$.

The SI unit of electric field is the newton per coulomb (N/C). Table 18-1 lists the magnitudes of some of the electric fields found in nature.

We note that the electric field is a vector and that it obeys the superposition principle. That is, the net electric field due to a system of charges can be found by computing the electric field due to each charge in the system separately and then adding these vectors to obtain the net electric field.

Table 18-1 **Some Electric Fields in Nature**

	E, N/C
In household wires	10^{-2}
In radio waves	10^{-1}
In the atmosphere	10^{2}
In sunlight	10^{3}
Under a thundercloud	10^{4}
In a lightning bolt	10^{4}
In an x-ray tube	10^{6}
At the electron in a hydrogen atom	6×10^{11}
At the surface of a uranium nucleus	2×10^{21}

The electric field **E** is a vector that describes the condition in space set up by the system of point charges. By moving a test charge q_0 from point to point, we can find **E** at all points in space (except one occupied by a charge q). The electric field **E** is thus a vector function of position.

The force exerted on a test charge q_0 at any point is related to the electric field at that point by

$$\mathbf{F} = q_0\mathbf{E} \qquad 18\text{-}6$$

Example 18-6

When a 5-nC test charge is placed at a certain point, it experiences a force of 2×10^{-4} N in the x direction. What is the electric field **E** at that point?

Since the force on the positive test charge is in the x direction, the electric field vector is also in the x direction. From its definition (Equation 18-5), the electric field is

$$\mathbf{E} = \frac{\mathbf{F}}{q_0} = \frac{(2 \times 10^{-4}\ \text{N})\mathbf{i}}{5 \times 10^{-9}\ \text{C}} = (4 \times 10^{4}\ \text{N/C})\mathbf{i}$$

Exercise

What is the force on an electron placed at the point in Example 18-6 where the electric field is $\mathbf{E} = (4 \times 10^{4}\ \text{N/C})\mathbf{i}$ [Answer: $(-6.4 \times 10^{-15}\ \text{N})\mathbf{i}$]

The electric field due to a single point charge q_i at a position $\mathbf{r}_i$ can be calculated from Coulomb's law. If we place a small, positive test charge q_0 at some point P a distance r_{i0} away, the force on it is

$$\mathbf{F}_{i0} = \frac{kq_iq_0}{r_{i0}^2}\,\hat{\mathbf{r}}_{i0}$$

where $\hat{\mathbf{r}}_{i0}$ is the unit vector that points from q_i to q_0. The electric field at point P due to charge q_i is thus

$$\mathbf{E}_i = \frac{kq_i}{r_{i0}^2}\hat{\mathbf{r}}_{i0} \tag{18-7}$$

Coulomb's law for **E** *due to a point charge*

where r_{i0} is the distance from the charge to point P called the **field point,** and $\hat{\mathbf{r}}_{i0}$ is a unit vector pointing from the charge to P. We will refer to this equation, which follows directly from Coulomb's law, as Coulomb's law for the electric field due to a single point charge. The net electric field due to a distribution of point charges is found by summing the fields due to each charge separately:

$$\mathbf{E} = \sum_i \mathbf{E}_i = \sum_i \frac{kq_i}{r_{i0}^2}\hat{\mathbf{r}}_{i0} \tag{18-8}$$

Electric field due to a system of point charges

Example 18-7

A positive charge $q_1 = +8$ nC is at the origin, and a second positive charge $q_2 = +12$ nC is on the x axis at $a = 4$ m (Figure 18-12). Find the net electric field (a) at point P_1 on the x axis at $x = 7$ m and (b) at point P_2 on the x axis at $x = 3$ m.

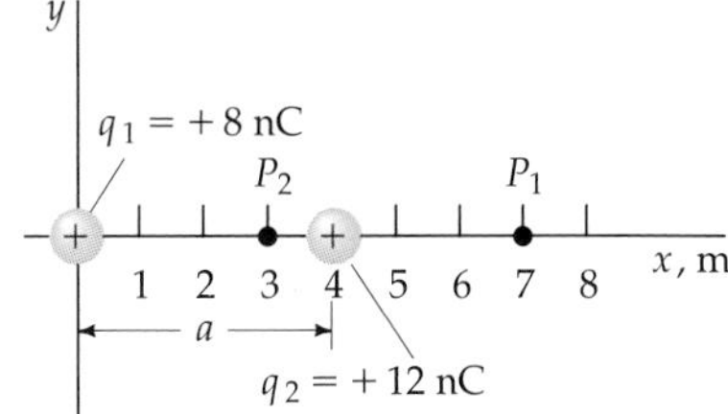

Figure 18-12 Two point charges on the x axis for Example 18-7. The net electric field is directed to the right at point P_1 and to the left at point P_2.

(a) The point P_1 at $x = 7$ m is to the right of both charges. The electric field at P_1 due to each charge is in the positive x direction. The distance from the field point to charge q_1 is $x = 7$ m and that to charge q_2 is $x - a = 7\text{ m} - 4\text{ m} = 3$ m. The net electric field at P_1 is thus

$$\begin{aligned}\mathbf{E} &= \frac{kq_1}{x^2}\mathbf{i} + \frac{kq_2}{(x-a)^2}\mathbf{i}\\ &= \frac{(8.99\times10^9\text{ N·m}^2/\text{C}^2)(8\times10^{-9}\text{ C})}{(7\text{ m})^2}\mathbf{i}\\ &\quad + \frac{(8.99\times10^9\text{ N·m}^2/\text{C}^2)(12\times10^{-9}\text{ C})}{(3\text{ m})^2}\mathbf{i}\\ &= (1.47\text{ N/C})\mathbf{i} + (12.0\text{ N/C})\mathbf{i} = (13.5\text{ N/C})\mathbf{i}\end{aligned}$$

(b) The point P_2 at $x = 3$ m is between the charges. A positive test charge placed at P_2 would experience a repulsive force to the right due to the +8-nC charge and a repulsive force to the left due to the +12-nC charge. The distance to the +8-nC charge is $x = 3$ m and that to the +12-nC charge is $a - x = 4\text{ m} - 3\text{ m} = 1$ m. The net electric field at P_2 is thus

$$\begin{aligned}\mathbf{E} &= \frac{kq_1}{x^2}\mathbf{i} - \frac{kq_2}{(a-x)^2}\mathbf{i}\\ &= \frac{(8.99\times10^9\text{ N·m}^2/\text{C}^2)(8\times10^{-9}\text{ C})}{(3\text{ m})^2}\mathbf{i}\\ &\quad - \frac{(8.99\times10^9\text{ N·m}^2/\text{C}^2)(12\times10^{-9}\text{ C})}{(1\text{ m})^2}\mathbf{i}\\ &= (7.99\text{ N/C})\mathbf{i} - (108\text{ N/C})\mathbf{i} = (-100\text{ N/C})\mathbf{i}\end{aligned}$$

The electric field at point P_2 is in the negative x direction because the contribution to the field due to the +12-nC charge, which is 1 m away, is larger than that due to the +8-nC charge, which is 3 m away. As we move toward the +8-nC charge at the origin, the magnitude of the field due to the +8-nC charge increases and that due to the +12-nC charge decreases. There is one point between the charges where the net electric field due to the two charges is zero. At this point a positive test charge would experience no net force because the repulsive force to the right due to the +8-nC charge would just balance the repulsive force to the left due to the +12-nC charge. At points closer to the +8-nC charge at the origin, the electric field points in the positive x direction.

Exercise

Find the point on the x axis in Figure 18-12 where the electric field is zero. (Answer: $x = 1.80$ m)

Example 18-8

Find the electric field at a point P_3 on the y axis at $y = 3$ m for the charges in Example 18-7.

The fields due to each charge at point P_3 on the y axis are shown in Figure 18-13*a*. The field $\mathbf{E}_1$ due to the +8-nC charge is in the positive y direction and has a magnitude

$$E_1 = \frac{kq_1}{y^2} = \frac{(8.99 \times 10^9\ \text{N}\cdot\text{m}^2/\text{C}^2)(8 \times 10^{-9}\ \text{C})}{(3\ \text{m})^2} = 7.99\ \text{N/C}$$

The field $\mathbf{E}_2$ due to the +12-nC charge is in the direction of the line from that charge to point P_3. The distance from the +12-nC charge to point P_3 is 5 m, which is found using the pythagorean theorem. The magnitude of $\mathbf{E}_2$ is

$$E_2 = \frac{(8.99 \times 10^9\ \text{N}\cdot\text{m}^2/\text{C}^2)(12 \times 10^{-9}\ \text{C})}{(5\ \text{m})^2} = 4.32\ \text{N/C}$$

The field $\mathbf{E}_2$ has a component in the positive y direction equal to $E_2 \cos\theta$ and a component in the negative x direction equal to $-E_2 \sin\theta$. From the triangle in Figure 18-13*a*, we can see that $\cos\theta = \frac{3}{5} = 0.6$ and $\sin\theta = \frac{4}{5} = 0.8$. The x and y components of E_2 are thus

$$E_{2x} = -E_2 \sin\theta = -(4.32\ \text{N/C})(0.8) = -3.46\ \text{N/C}$$

and

$$E_{2y} = E_2 \cos\theta = (4.32\ \text{N/C})(0.6) = 2.59\ \text{N/C}$$

We obtain the x and y components of the net electric field $\mathbf{E}$ from

$$E_x = E_{1x} + E_{2x} = 0 + (-3.46\ \text{N/C}) = -3.46\ \text{N/C}$$

and

$$E_y = E_{1y} + E_{2y} = 7.99\ \text{N/C} + 2.59\ \text{N/C} = 10.6\ \text{N/C}$$

Figure 18-13 Example 18-8. (*a*) On the y axis, the electric field $\mathbf{E}_1$ due to charge q_1 is directed along the y axis, and the field $\mathbf{E}_2$ due to charge q_2 makes an angle θ with the y axis. The net electric field is the vector sum $\mathbf{E} = \mathbf{E}_1 + \mathbf{E}_2$. (*b*) The net electric field and its x and y components.

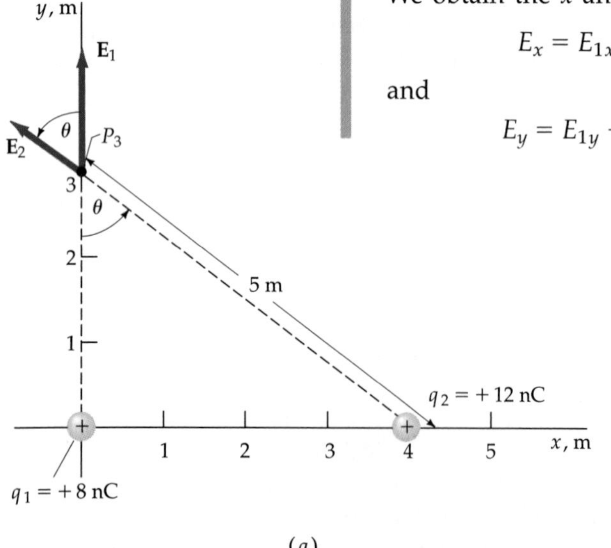

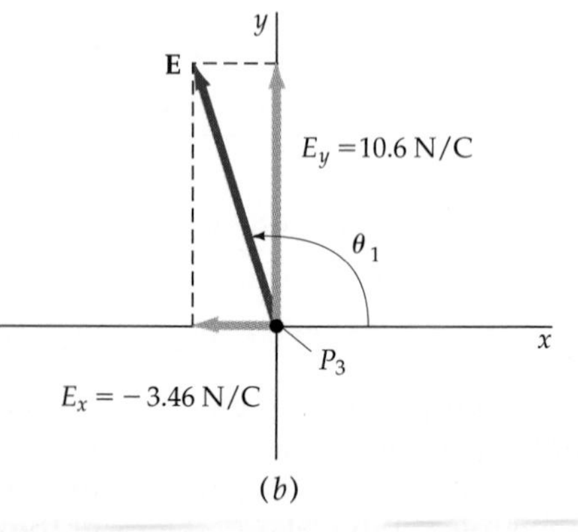

The magnitude of the net electric field is

$$E = \sqrt{E_x^2 + E_y^2} = \sqrt{(-3.46\ \text{N/C})^2 + (10.6\ \text{N/C})^2} = 11.2\ \text{N/C}$$

The net electric field **E** makes an angle θ_1 with the x axis (see Figure 18-13*b*) given by

$$\tan\theta_1 = \frac{E_y}{E_x} = \frac{10.6\ \text{N/C}}{-3.46\ \text{N/C}} = -3.06$$

$$\theta_1 = 108°$$

Example 18-9

A charge $+q$ is at $x = a$ and a second charge $-q$ is at $x = -a$ (Figure 18-14). Find the electric field on the x axis at a field point P, which is a large distance away compared to the separation of the charges.

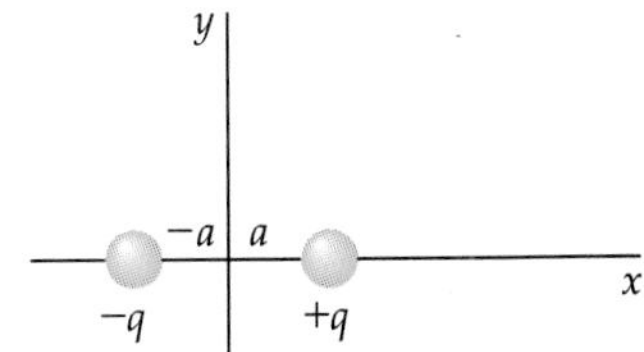

Figure 18-14 A point charge $+q$ at $x = a$ and a second point charge $-q$ at $x = -a$ for Example 18-9. This charge distribution is called an electric dipole.

The point P on the x axis is a distance $x - a$ from the positive charge and a distance $x + a$ from the negative charge. The electric field at P due to these two charges is thus

$$\mathbf{E} = \frac{kq}{(x-a)^2}\,\mathbf{i} + \frac{k(-q)}{(x+a)^2}\,\mathbf{i} = kq\mathbf{i}\left[\frac{1}{(x-a)^2} - \frac{1}{(x+a)^2}\right]$$

Putting the terms in brackets over a common denominator, we get

$$\frac{1}{(x-a)^2} - \frac{1}{(x+a)^2} = \frac{(x+a)^2 - (x-a)^2}{(x+a)^2(x-a)^2} = \frac{4ax}{(x^2-a^2)^2}$$

For $x \gg a$, we can neglect a^2 as compared with x^2 in the denominator. Therefore,

$$\frac{4ax}{(x^2-a^2)^2} \approx \frac{4ax}{x^4} = \frac{4a}{x^3}$$

Thus, the electric field at P is approximately

$$\mathbf{E} = \frac{4kqa}{x^3}\,\mathbf{i}$$

A system of two equal and opposite charges q separated by a small distance L is called an **electric dipole.** An electric dipole is characterized by the **electric dipole moment p,** which is a vector that points from the negative charge to the positive charge and has a magnitude that is the product of the charge q times the separation L (Figure 18-15). If **L** is the displacement vector of the positive charge from the negative charge, the dipole moment is

−q L +q

$\mathbf{p} = q\mathbf{L}$

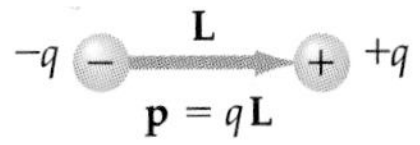

Figure 18-15 A dipole consists of two equal and opposite charges separated by some distance L. The dipole moment points from the negative charge to the positive charge and has the magnitude $p = qL$.

$$\mathbf{p} = q\mathbf{L} \qquad 18\text{-}9$$

Electric dipole moment defined

For the dipole shown in Figure 18-14, the displacement of the positive charge is $\mathbf{L} = 2a\mathbf{i}$, and the electric dipole moment is

$$\mathbf{p} = 2aq\mathbf{i}$$

In terms of the dipole moment, the electric field on the axis of the dipole at a point a great distance x away has the magnitude

$$E = \frac{2kp}{x^3} \qquad 18\text{-}10$$

Thus, the electric field far from a dipole is proportional to the dipole moment and decreases with the cube of the distance.

18-5 Electric-Field Lines

It is convenient to picture the electric field by drawing lines to indicate the direction of the field at any point. The field vector **E** is tangent to the line at each point and indicates the direction of the electric field at that point. Electric-field lines are also called lines of force because they show the direction of the force exerted on a positive test charge.

At any point near a positive charge, the electric field points radially away from the charge. The electric-field lines therefore diverge from a point occupied by a positive charge. Similarly, the electric field near a negative point charge points inward toward the charge, so the electric-field lines point toward a negative charge.

Figure 18-16 shows the electric-field lines of a single, positive point charge. As we move away from the charge, the electric field becomes weaker and the lines become farther apart. There is a connection between the spacing of the lines and the strength of the electric field. Consider a spherical surface of radius r with its center at the charge. We are interested in the number of lines per unit area of the sphere, which we will call the density of the lines. If we make r larger, the area of the sphere increases, but the same number of lines passes through it. The number of lines per unit area thus decreases as r increases. The area of the sphere is given by $A = 4\pi r^2$. The number of lines per unit area of the sphere thus decreases inversely with the square of the distance from the point charge. But the strength of the electric field, $E = kq/r^2$, also decreases inversely with the square of this distance. Thus, if we adopt the convention of drawing a fixed number of lines from a point charge, the number being proportional to the charge q, and if we draw the lines symmetrically about the point charge, the field strength is indicated by the density of the lines. The more closely spaced the lines, the stronger the electric field.

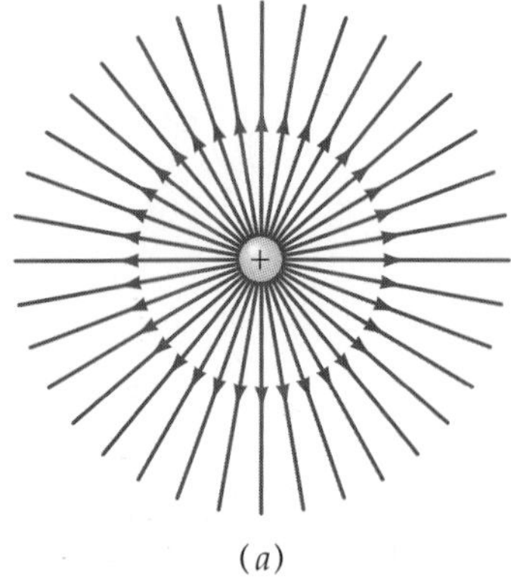
(a)

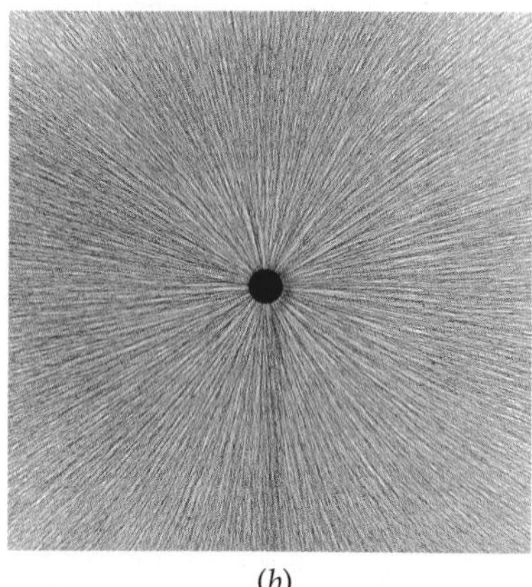
(b)

Figure 18-16 (*a*) Electric-field lines of a single positive point charge. If the charge were negative, the arrows would be reversed. (*b*) The same electric-field lines shown by bits of thread suspended in oil. The electric field of the charged object in the center induces opposite charges on the ends of each bit of thread, causing the threads to align themselves parallel to the field.

Figure 18-17 shows the electric-field lines for two equal positive point charges q separated by a distance a. We can sketch this pattern without calculating the field at each point. We again use the fact that the contribution to the field due to each of the charges varies as $1/r^2$, where r is the distance from that charge. At a point near one of the charges, the field is approximately due to that charge alone because the other charge is far enough away that we can ignore its contribution to the field. Thus, on a sphere of very small radius about either of the charges, the field lines are radial and equally spaced. Since the charges are equal, we draw an equal number of lines originating from each charge. At very large distances from the charges, the details of the system are not important. For example, if the two charges were 1 mm apart and we were looking at them from a point 100 km away, they

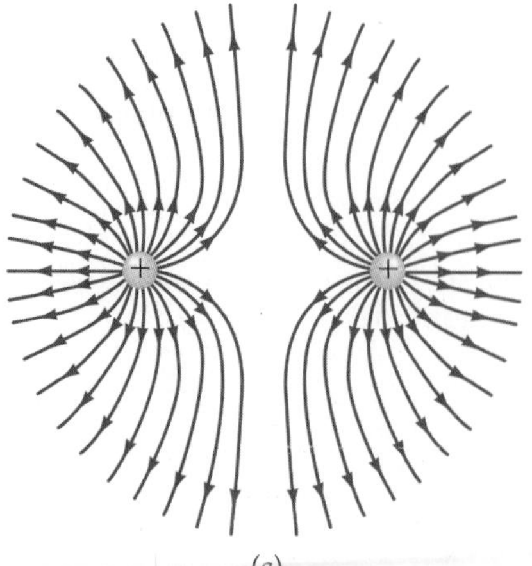
(a)

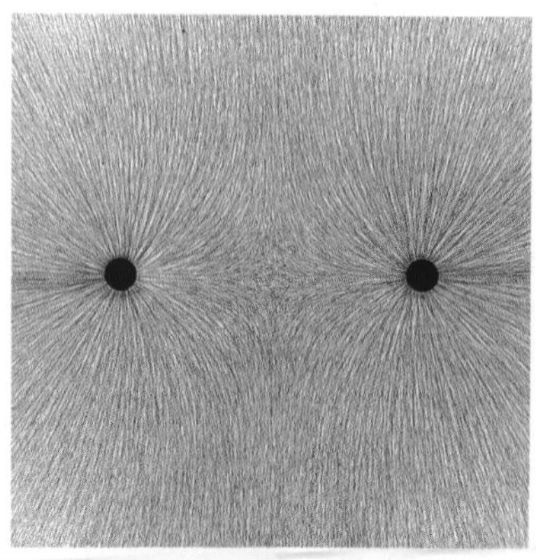
(b)

Figure 18-17 (*a*) Electric-field lines due to two positive point charges. The arrows would be reversed if both charges were negative. (*b*) The same electric-field lines shown by bits of thread in oil.

would look like a single charge. So on a sphere of radius r, where r is much greater than a, the field is approximately the same as that due to a single point charge of magnitude $2q$, and the lines are approximately equally spaced. We can see by merely looking at the figure that the electric field in the space between the charges is weak because there are few lines in this region compared with the region just to the right or left of the charges where the lines are more closely spaced. This information can, of course, also be obtained by direct calculation of the field at points in these regions.

We can apply the reasoning used in the preceding examples in drawing the electric-field lines for any system of point charges. Very near each charge, the field lines are equally spaced and leave or enter the charge radially, depending on the sign of the charge. Very far from all the charges, the detailed structure of the system is not important, so the field lines are just like those of a single point charge carrying the net charge of the system. For future reference, the rules for drawing electric-field lines can be summarized as follows:

Rules for drawing electric-field lines

1. Electric-field lines begin on positive charges and end on negative charges (or at infinity).
2. The lines are drawn symmetrically leaving or entering a charge.
3. The number of lines leaving a positive charge or entering a negative charge is proportional to the charge.
4. The density of the lines (the number of lines per unit area perpendicular to the lines) at some point is proportional to the magnitude of the field at that point.
5. At large distances from a system of charges, the field lines are equally spaced and radial as if they came from a single point charge equal to the net charge of the system.
6. No two field lines can cross.

Rule 6 follows from the fact that **E** has a unique direction at any point in space (except at a point occupied by a point charge or where $\mathbf{E} = 0$). If two lines crossed, two directions would be indicated for **E** at the point of intersection.

Figure 18-18 shows the electric-field lines due to an electric dipole. Very near the positive charge, the lines are directed radially outward. Very near the negative charge, the lines are directed radially inward. Since the charges have equal magnitudes, the number of lines that begin at the positive charge equals the number that end at the negative charge. In this case, the field is strong in the region between the charges, as indicated by the high density of field lines in this region in the figure.

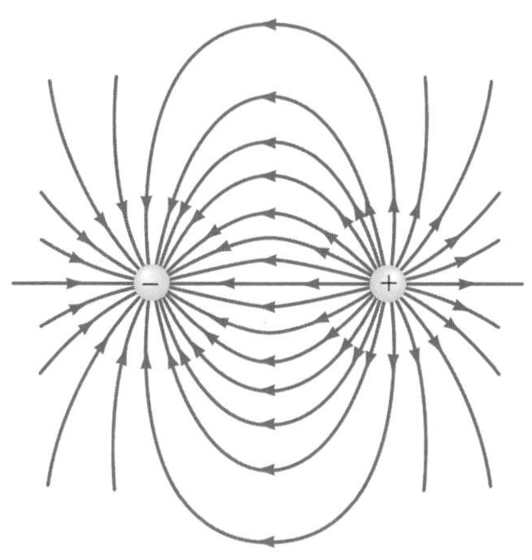

(a)

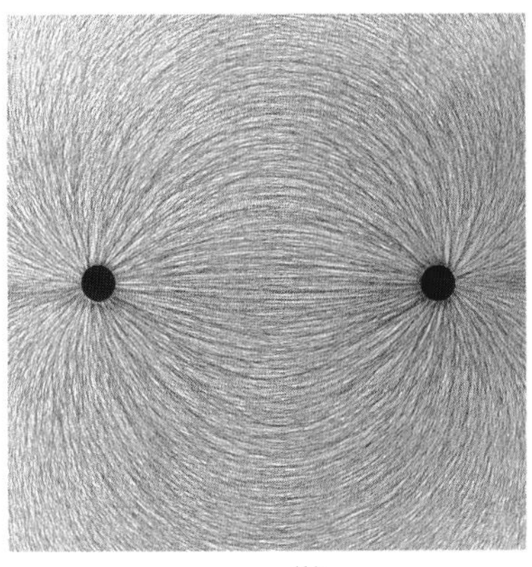

(b)

Figure 18-18 (*a*) Electric-field lines for an electric dipole. (*b*) The same field lines shown by bits of thread in oil.

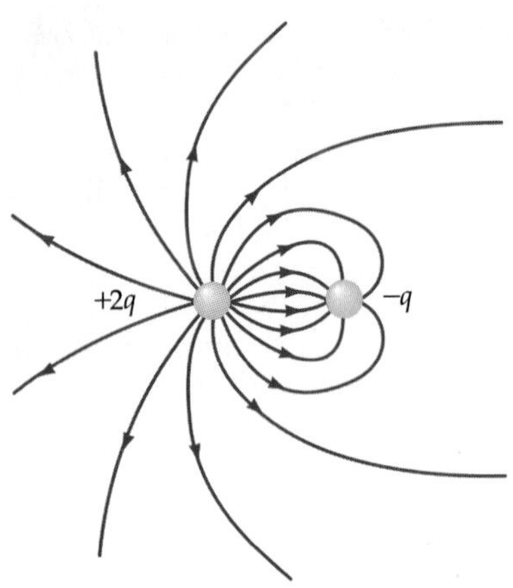

Figure 18-19 Electric-field lines for a point charge $+2q$ and a second point charge $-q$. At great distances from the charges, the lines are the same as those for a single charge $+q$.

Figure 18-19 shows the electric-field lines for a negative charge $-q$ at a distance a from a positive charge $+2q$. Since the positive charge has twice the magnitude of the negative charge, twice as many lines leave the positive charge as enter the negative charge. That is, half the lines beginning on the positive charge $+2q$ enter the negative charge $-q$, and the other half leave the system. On a sphere of radius r, where r is much larger than the separation of the charges, the lines leaving the system are approximately symmetrically spaced and point radially outward, just as they would for a single positive point charge $+q$. Thus at great distances from the charges, the system looks like a single charge $+q$. At a great distance from a system of charges, only the net charge is important. The convention indicating the electric field strength by the electric-field lines works because the electric field varies inversely as the square of the distance from a point charge.

Since the gravitational field of a point mass also varies inversely as the square of the distance, the concept of field lines is also useful for picturing the gravitational field. Near a point mass, the gravitational field lines converge toward the mass just as electric-field lines converge toward a negative charge. However, there are no points in space where gravitational field lines diverge as electric-field lines do near a positive charge because the gravitational force is always attractive, never repulsive.

18-6 Motion of Point Charges in Electric Fields

When a particle with a charge q is placed in an electric field **E**, it experiences a force $q\mathbf{E}$. As we have seen, the gravitational forces acting on a particle are usually negligible in comparison with the electric forces. If the electric force is the only significant force acting on the particle, the particle has an acceleration

$$\mathbf{a} = \frac{q}{m}\mathbf{E}$$

where m is the mass of the particle.* If the electric field is known, the charge-to-mass ratio of the particle can be determined from the measured acceleration. For example, in a uniform electric field, the path of the particle is a parabola, similar to that of a projectile in a uniform gravitational field. The deflection of electrons in a uniform electric field was used by J. J. Thomson in 1897 to demonstrate the existence of electrons and to measure their charge-to-mass ratio. Familiar examples of devices that rely on the motion of electrons in electric fields are the oscilloscope and the television picture tube.

We will now look at some examples involving the motion of electrons in constant electric fields. Problems of this type can be worked using the constant-acceleration equations from Chapter 2 or the equations for projectile motion from Chapter 3.

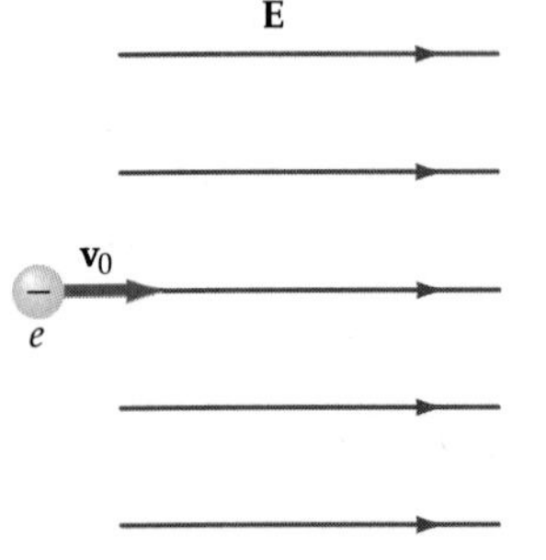

Figure 18-20 An electron projected into a uniform electric field with an initial velocity parallel to the field for Example 18-10.

Example 18-10

An electron is projected into a uniform electric field $\mathbf{E} = (1000\ \mathrm{N/C})\mathbf{i}$ with an initial velocity $\mathbf{v}_0 = (2 \times 10^6\ \mathrm{m/s})\mathbf{i}$ in the direction of the field (Figure 18-20). How far does the electron travel before it is brought momentarily to rest?

*Often the speed of an electron in an electric field is a significant fraction of the speed of light; in which case, Newton's laws of motion must be modified by Einstein's special theory of relativity.

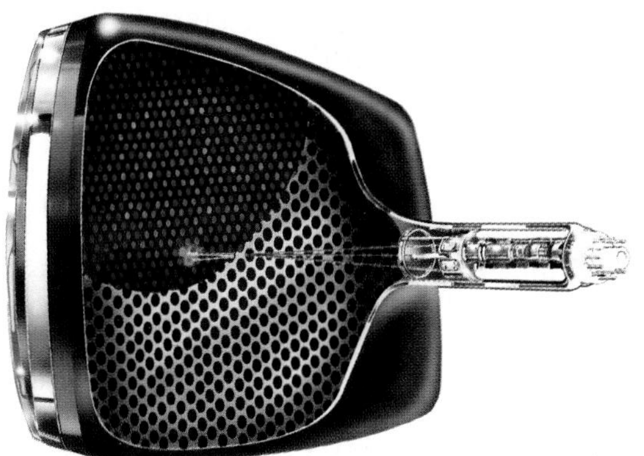

Schematic drawing of a cathode ray tube used for color television. The beams of electrons from the electron gun on the right activate phosphors on the screen at the left, giving rise to a bright spot whose color depends on the relative intensity of each beam. Electric fields between deflection plates in the gun (or magnetic fields from coils within the gun) deflect the beams. The beams sweep across the screen in a horizontal line, are then deflected downward, and sweep across again. The entire screen is covered once each 1/30 s.

Since the charge of the electron is negative, the force $-e\mathbf{E}$ acting on it is in the direction opposite that of the field. We thus have a constant-acceleration problem in which the acceleration of a particle is opposite to its initial velocity, and we are asked to find the distance the particle travels in its original direction. We can use the constant-acceleration equation relating distance to velocity:

$$v^2 = v_0^2 + 2a(x - x_0)$$

Using $x_0 = 0$, $v = 0$, $v_0 = 2 \times 10^6$ m/s, and $a = -eE/m$, we obtain

$$x = \frac{mv_0^2}{2eE} = \frac{(9.11 \times 10^{-31}\ \text{kg})(2 \times 10^6\ \text{m/s})^2}{2(1.6 \times 10^{-19}\ \text{C})(1000\ \text{N/C})} = 1.14 \times 10^{-2}\ \text{m}$$

Example 18-11

An electron is projected into a uniform electric field $\mathbf{E} = (-2000\ \text{N/C})\mathbf{j}$ with an initial velocity $\mathbf{v}_0 = (10^6\ \text{m/s})\mathbf{i}$ perpendicular to the field (Figure 18-21). (*a*) Compare the gravitational force acting on the electron to the electric force acting on it. (*b*) By how much has the electron been deflected after it has traveled 1 cm in the x direction?

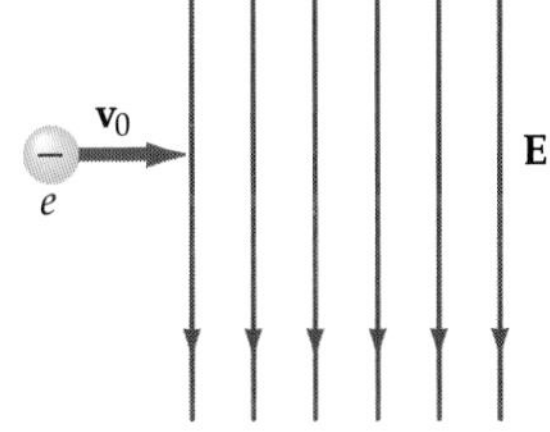

Figure 18-21 An electron projected into a uniform electric field with an initial velocity perpendicular to the field for Example 18-11.

(*a*) The electric force on the electron is $-e\mathbf{E}$, and the gravitational force on it is $m\mathbf{g}$. Since the electric field is downward, the electric force on the negative electron is upward. The gravitational force is, of course, downward. The ratio of their magnitudes is

$$\frac{F_e}{F_g} = \frac{eE}{mg} = \frac{(1.6 \times 10^{-19}\ \text{C})(2000\ \text{N/C})}{(9.1 \times 10^{-31}\ \text{kg})(9.8\ \text{N/kg})} = 3.6 \times 10^{13}$$

As is most commonly the case, the electric force is huge compared with the gravitational force, so the gravitational force is wholly negligible.

(*b*) The time it takes the electron to travel a distance of 1 cm in the x direction is

$$t = \frac{x}{v_0} = \frac{10^{-2}\ \text{m}}{10^6\ \text{m/s}} = 10^{-8}\ \text{s}$$

In this time, the electron is deflected upward, antiparallel to the field, a distance y given by

$$y = \frac{1}{2}at^2 = \frac{1}{2}\frac{eE}{m}t^2$$

Substituting in the known values of e, m, E, and t gives

$$y = 1.76 \times 10^{-2}\ \text{m} = 1.76\ \text{cm}$$

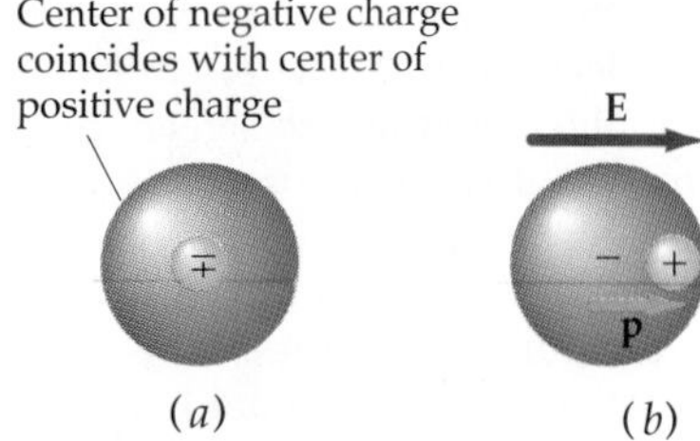

Figure 18-22 Schematic diagrams of the charge distributions of an atom or nonpolar molecule. (*a*) In the absence of an external electric field, the center of positive charge coincides with the center of negative charge. (*b*) In the presence of an external electric field, the centers of positive and negative charge are displaced, producing an induced dipole moment in the direction of the external field.

18-7 Electric Dipoles in Electric Fields

Although atoms and molecules are electrically neutral, they are affected by electric fields because they contain positive and negative charges. We can think of an atom as consisting of a very small, positively charged nucleus surrounded by a negatively charged electron cloud. Since the radius of the nucleus is about 100,000 times smaller than that of the electron cloud, we can consider it to be a point charge. In some atoms and molecules, the electron cloud is spherically symmetric, so its "center of charge" is at the center of the atom or molecule, coinciding with the positive charge. Such an atom or molecule is said to be **nonpolar.** However, in the presence of an external electric field, the center of positive charge does not coincide with the center of negative charge. The electric field exerts a force on the positively charged nucleus in the direction of the field and a force on the negatively charged electron cloud in the opposite direction. The positive and negative charges separate until the attractive force they exert on each other balances the forces due to the external electric field (Figure 18-22). Such a charge distribution behaves like an electric dipole.

The dipole moment of an atom or nonpolar molecule in an external electric field is called an **induced dipole moment.** It has the same direction as the electric field. If the electric field is uniform, there is no net force on the dipole because the forces on the positive and negative charges are equal and opposite. However, if the electric field is not uniform, there will be a net external force acting on the dipole. Figure 18-23 shows a nonpolar molecule in the electric field of a positive point charge q. The induced dipole moment is parallel to $\mathbf{E}$ in the radial direction from the point charge. The field is stronger at the center of negative charge because it is closer to the point charge, so the net force on the dipole is toward the point charge, and the dipole is attracted toward the point charge. If the point charge were negative, the induced dipole would be in the opposite direction, and the dipole would again be attracted to the point charge. The force produced by a nonuniform electric field on an electrically neutral particle is responsible for the familiar attraction of a charged comb for uncharged bits of paper. It is also responsible for the forces that hold an electrostatically charged balloon against a wall or ceiling. In this case, the charge on the balloon provides the nonuniform electric field that polarizes (that is, induces dipole moments in) the molecules of the wall or ceiling and then attracts them.

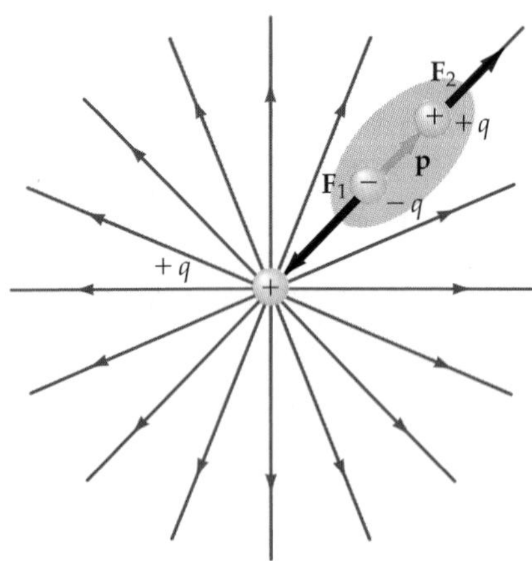

Figure 18-23 Nonpolar molecule in the nonuniform electric field of a positive point charge. The induced electric dipole moment **p** is parallel to the field of the point charge. Since the point charge is closer to the center of negative charge than to the center of positive charge, there is a net force of attraction between the dipole and the point charge.

In some molecules, the center of positive charge does not coincide with the center of negative charge even in the absence of an external electric field. These **polar molecules** have a permanent electric dipole moment. When a polar molecule is placed in a uniform electric field, there is no net force on it, but there is a torque that tends to rotate the molecule so that the dipole lines up with the field. Figure 18-24 shows the forces exerted on a dipole of moment $\mathbf{p} = q\mathbf{L}$ in a uniform electric field $\mathbf{E}$. We found in Section 9-4 that the torque produced by two equal and opposite forces, called a couple, is the same about any point in space. From the figure, we see that the torque about the negative charge has the magnitude $F_1L \sin\theta = qEL \sin\theta = pE \sin\theta$. The direction of the torque is into the paper such that it rotates the dipole moment $\mathbf{p}$ into the direction of the electric field $\mathbf{E}$. The torque can be conveniently written as the cross product of the dipole moment $\mathbf{p}$ and the electric field $\mathbf{E}$:

$$\boldsymbol{\tau} = \mathbf{p} \times \mathbf{E} \qquad 18\text{-}11$$

When the dipole rotates through an angle $d\theta$, the electric field does work

$$dW = -\tau\, d\theta = -pE \sin\theta\, d\theta$$

The minus sign arises because the torque tends to decrease θ. Setting this work equal to the decrease in potential energy, we have

$$dU = -dW = +pE \sin \theta \, d\theta$$

Integrating, we obtain

$$U = -pE \cos \theta + U_0$$

It is customary to choose the potential energy to be zero when the dipole is perpendicular to the electric field, that is, when $\theta = 90°$. Then $U_0 = 0$, and the potential energy of the dipole is

$$U = -pE \cos \theta = -\mathbf{p} \cdot \mathbf{E} \qquad 18\text{-}12$$

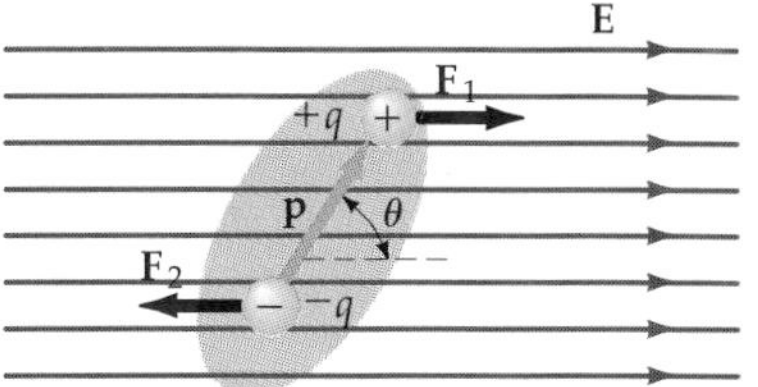

Figure 18-24 A dipole in a uniform electric field experiences equal and opposite forces that tend to rotate the dipole so that its dipole moment is aligned with the electric field.

In a nonuniform electric field, a polar molecule experiences a net force because the electric field has different magnitudes at the centers of positive and negative charge. An example of a polar molecule is HCl, which is essentially a positive hydrogen ion of charge $+e$ combined with a negative chlorine ion of charge $-e$. Another example of a polar molecule is water (Figure 18-25). The dipole moment of the water molecule is mainly responsible for the energy absorption of food in a microwave oven. Like all electromagnetic waves, microwaves have an oscillating electric field that can cause electric dipoles to vibrate. The vibration of the electric dipole moment of the water molecule in resonance with the oscillating electric field of the microwaves leads to the absorption of energy from the microwaves.

The diameter of an atom or molecule is of the order of 10^{-10} m = 0.1 nm. A convenient unit for the electric dipole moment of atoms and molecules is the fundamental electronic charge e times the distance 1 nm. For example, the dipole moment of NaCl in these units has a magnitude of about 0.2 e·nm.

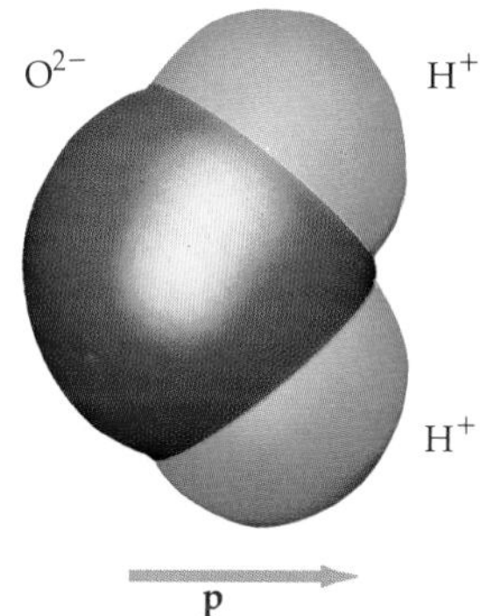

Figure 18-25 A computer-generated three-dimensional model of an H_2O molecule, consisting of oxygen ion of charge $-2e$ and two hydrogen ions of charge $+e$ each. This molecule has a permanent electric dipole moment in the direction shown.

Example 18-12

A dipole with a moment of magnitude 0.02 e·nm makes an angle of 20° with a uniform electric field of magnitude 3×10^3 N/C. Find (*a*) the magnitude of the torque on the dipole and (*b*) the potential energy of the system.

(*a*) The magnitude of the torque is

$$\begin{aligned}\tau &= |\mathbf{p} \times \mathbf{E}| = pE \sin \theta \\ &= (0.02)(1.60 \times 10^{-19}\ \text{C})(10^{-9}\ \text{m})(3 \times 10^3\ \text{N/C})(\sin 20°) \\ &= 3.28 \times 10^{-27}\ \text{N·m}\end{aligned}$$

(*b*) The potential energy of the system is

$$\begin{aligned}U &= -\mathbf{p} \cdot \mathbf{E} = -pE \cos \theta \\ &= -(0.02)(1.60 \times 10^{-19}\ \text{C})(10^{-9}\ \text{m})(3 \times 10^3\ \text{N/C})(\cos 20°) \\ &= -9.02 \times 10^{-27}\ \text{J}\end{aligned}$$

Question

5. A small, nonconducting ball with no net electric charge is suspended from a thread. When a positive charge is brought near the ball, the ball is attracted toward the charge. How does this come about? How would the situation be different if the charge brought near the ball were negative instead of positive?

Summary

1. There are two kinds of electric charge, labeled positive and negative. Electric charge always occurs in integral multiples of the fundamental unit of charge e. The charge of the electron is $-e$ and that of the proton is $+e$. Objects become charged by the transfer of electric charge from one object to another, usually in the form of electrons. Charge is conserved. It is neither created nor destroyed in the charging process, but is merely transferred.

2. The force exerted by one charge on another acts along the line joining the charges. It is proportional to the product of the charges and inversely proportional to the square of the separation. The force is repulsive if the charges have the same sign and attractive if they have opposite signs. This result is known as Coulomb's law:

$$\mathbf{F}_{12} = \frac{kq_1q_2}{r_{12}^2}\,\hat{\mathbf{r}}_{12}$$

where k is the Coulomb constant, which has the value

$$k = 8.99 \times 10^9\ \mathrm{N{\cdot}m^2/C^2}$$

3. The electric field due to a system of charges at a point is defined as the net force exerted by those charges on a positive test charge q_0 divided by q_0:

$$\mathbf{E} = \frac{\mathbf{F}}{q_0}$$

4. The electric field at some point P due to a single point charge q_i at a position $\mathbf{r}_i$ is

$$\mathbf{E}_i = \frac{kq_i}{r_{i0}^2}\,\hat{\mathbf{r}}_{i0}$$

where r_{i0} is the distance from charge q_i to the field point P and $\hat{\mathbf{r}}_{i0}$ is the unit vector pointing from q_i to P. The electric field due to several charges is the vector sum of the fields due to the individual charges:

$$\mathbf{E} = \sum_i \mathbf{E}_i = \sum_i \frac{kq_i}{r_{i0}^2}\,\hat{\mathbf{r}}_{i0}$$

5. The electric field can be represented by electric-field lines that originate on positive charges and end on negative charges. The strength of the electric field is indicated by the density of the electric-field lines.

6. An electric dipole is a system of two equal but opposite charges separated by a small distance. The dipole moment $\mathbf{p}$ is a vector with a magnitude equal to the charge times the separation that points in the direction from the negative charge to the positive charge:

$$\mathbf{p} = q\mathbf{L}$$

The electric field far from a dipole is proportional to the dipole moment and decreases with the cube of the distance.

7. In a uniform electric field, the net force on a dipole is zero, but there is a torque $\boldsymbol{\tau}$ given by

$$\boldsymbol{\tau} = \mathbf{p} \times \mathbf{E}$$

that tends to align the dipole in the direction of the field. The potential energy of a dipole in an electric field is given by

$$U = -\mathbf{p}\cdot\mathbf{E}$$

where the potential energy is taken to be zero when the dipole is perpendicular to the electric field. In a nonuniform electric field, there is a net force on a dipole.

8. Polar molecules, such as H_2O, have permanent dipole moments because their centers of positive and negative charge do not coincide. They behave like simple dipoles in an electric field. Nonpolar molecules do not have permanent dipole moments, but they acquire induced dipole moments in the presence of an electric field.

Suggestions for Further Reading

Bordeau, Sanford P.: *Volts to Hertz . . . The Rise of Electricity,* Burgess Publishing Company, Minneapolis, 1982.

This book presents a history of electrical science through discussion of the experiments and lives of the men after whom electrical and magnetic units are named. Illustrated with photographs and original engravings.

Cohn, Sherrye: "Painting the Fields of Faraday: Physics Inspired America's First Abstract Artist," *The Sciences,* November/December 1985, p. 44.

The painting "Rise of the Full Moon" by Arthur Dove, 1937, is reproduced here and discussed with reference to the field concept.

Goldhaber, Alfred Scharff, and Michael Martin Nieto: "The Mass of the Photon," *Scientific American,* May 1976, p. 86.

Strange as it may seem, tests of Coulomb's law provide an upper limit on a possible mass for the photon. This article describes the history of such tests, which began before Coulomb started his investigations and continue today.

Kevles, Daniel J.: "Robert Millikan," *Scientific American,* January 1979, p. 142.

The life and work of the second American scientist to win the Nobel prize in physics: this, in part, for his beautiful demonstration that all electrons carry the same charge, and his measurement of that charge.

Shamos, Morris H.: "The Laws of Electric and Magnetic Force—Charles Coulomb," in *Great Experiments in Physics,* Henry Holt and Co., New York, 1959. Reprinted by Dover, 1987.

Coulomb's description of his experiments in English translation, with editorial annotations for clarity and a biographical sketch.

Walker, Jearl: "The Amateur Scientist: How to Map Electronically Charged Patches with Parsley, Sage, Rosemary and Thyme," *Scientific American,* April 1988, p. 92.

Colorful experiments with charged surfaces.

Walker, Jearl: "The Amateur Scientist: The Secret of a Microwave Oven's Rapid Cooking Action Is Disclosed," *Scientific American,* February 1987, p. 134.

This article describes several mechanisms that have been suggested to explain the fact that microwaves heat water. All depend on the fact that the water molecule is polar.

Williams, Earle R.: "The Electrification of Thunderstorms," *Scientific American,* November 1988, p. 88.

The mechanisms of lightning, one of the most spectacular of natural electrical phenomena, are still in dispute.

Review

A. Objectives: After studying this chapter, you should:

1. Be able to state Coulomb's law and use it to find the force exerted by one point charge on another.

2. Know the value of the Coulomb constant in SI units.

3. Know the magnitude of the fundamental unit of electric charge e in coulombs.

4. Be able to use Coulomb's law to calculate the electric field due to a system of point charges.

5. Be able to draw the electric-field lines for simple charge distributions and to obtain information about the direction and strength of an electric field from such a diagram.

6. Be able to state the difference between a polar and a nonpolar molecule, and describe the behavior of each in uniform and nonuniform electric fields.

7. Be able to explain why bits of paper are attracted to a charged comb, and why an electrostatically charged balloon will stick to a wall.

B. Define, explain, or otherwise identify:

Charge quantization
Fundamental unit of charge
Law of conservation of charge
Coulomb
Conductors
Insulators
Electroscope

Electrostatic induction
Charging by induction
Grounded
Coulomb's law
Coulomb constant
Principle of superposition of electric forces
Electric field
Test charge
Field point
Electric dipole
Electric dipole moment
Electric-field lines
Nonpolar molecule
Induced dipole moment
Polar molecule

C. True or false: If the statement is true, explain why it is true. If it is false, give a counterexample.

1. The electric field of a point charge always points away from the charge.

2. The charge of the electron is the smallest unit of charge found.

3. Electric field lines never diverge from a point in space.

4. Electric field lines never cross at a point in space.

5. All molecules have electric dipole moments in the presence of an external electric field.

Problems

Level I

18-1 Electric Charge

1. A plastic rod is rubbed against a wool shirt, thereby acquiring a charge of $-0.8\ \mu C$. How many electrons are transferred from the wool shirt to the plastic rod?

2. A charge equal to the charge of Avogadro's number of protons ($N_A = 6.02 \times 10^{23}$) is called a *faraday*. Calculate the number of coulombs in a faraday.

3. How many coulombs of positive charge are there in 1 kg of carbon? Twelve grams of carbon contain Avogadro's number of atoms, with each atom having six protons and six electrons.

18-2 Conductors and Insulators and Charging by Induction

4. Explain, giving each step, how a positively charged insulating rod can be used to give a metal sphere (*a*) a negative charge and (*b*) a positive charge. (*c*) Can the same rod be used to simultaneously give one sphere a positive charge and another sphere a negative charge without the rod having to be recharged?

5. Two uncharged conducting spheres with their conducting surfaces in contact are supported on a large wooden table by insulated stands. A positively charged rod is brought up close to the surface of one of the spheres on the side opposite its point of contact with the other sphere. (*a*) Describe the induced charges on the two conducting spheres, and sketch the charge distributions on them. (*b*) The two spheres are separated far apart and the charged rod is removed. Sketch the charge distributions on the separated spheres.

18-3 Coulomb's Law

6. A charge $q_1 = 4.0\ \mu C$ is at the origin, and charge $q_2 = 6.0\ \mu C$ is on the x axis at $x = 3.0$ m. (*a*) Find the force on charge q_2. (*b*) Find the force on q_1. (*c*) How would your answers for parts (*a*) and (*b*) differ if q_2 were $-6.0\ \mu C$?

7. Three point charges are on the x axis; $q_1 = -6.0\ \mu C$ is at $x = -3.0$ m, $q_2 = 4.0\ \mu C$ is at the origin, and $q_3 = -6.0\ \mu C$ is at $x = 3.0$ m. Find the force on q_1.

8. Two equal charges of $3.0\ \mu C$ are on the y axis, one at the origin and the other at $y = 6$ m. A third charge $q_3 = 2\ \mu C$ is on the x axis at $x = 8$ m. Find the force on q_3.

9. Three charges, each of magnitude 3 nC, are at the corners of a square of side 5 cm. The two charges at the opposite corners are positive, and the other is negative. Find the force exerted by these charges on a fourth charge $q = +3$ nC at the remaining corner.

10. A charge of $5\ \mu C$ is on the y axis at $y = 3$ cm, and a second charge of $-5\ \mu C$ is on the y axis at $y = -3$ cm. Find the force on a charge of $2\ \mu C$ on the x axis at $x = 8$ cm.

18-4 The Electric Field

11. A charge of $4.0\ \mu C$ is at the origin. What is the magnitude and direction of the electric field on the x axis at (*a*) $x = 6$ m and (*b*) $x = -10$ m? (*c*) Sketch the function E_x versus x for both positive and negative values of x. (Remember that E_x is negative when $\mathbf{E}$ points in the negative x direction.)

12. Two charges, each $+4\ \mu C$, are on the x axis, one at the origin and the other at $x = 8$ m. Find the electric field on the x axis at (*a*) $x = -2$ m, (*b*) $x = 2$ m, (*c*) $x = 6$ m, and (*d*) $x = 10$ m. (*e*) At what point on the x axis is the electric field zero? (*f*) Sketch E_x versus x.

13. Two equal positive charges of magnitude $q_1 = q_2 = 6.0$ nC are on the y axis at $y_1 = +3$ cm and $y_2 = -3$ cm. (*a*) What is the magnitude and direction of the electric field on the x axis at $x = 4$ cm? (*b*) What is the force exerted on a test charge $q_0 = 2$ nC placed on the x axis at $x = 4$ cm?

14. When a test charge $q_0 = 2$ nC is placed at the origin, it experiences a force of 8.0×10^{-4} N in the positive y direction. (*a*) What is the electric field at the origin? (*b*) What would be the force on a charge of -4 nC placed at the origin? (*c*) If this force is due to a charge on the y axis at $y = 3$ cm, what is the value of that charge?

15. An oil drop has a mass of 4×10^{-14} kg and a net charge of 4.8×10^{-19} C. An upward electric force just balances the downward force of gravity so that the oil drop is stationary. What is the direction and magnitude of the electric field?

16. The electric field near the surface of the earth points downward and has a magnitude of 150 N/C. (*a*) Compare the upward electric force on an electron with the downward gravitational force. (*b*) What charge should be placed on a penny of mass 3 g so that the electric force balances the weight of the penny near the earth's surface?

18-5 Electric-Field Lines

17. Figure 18-26 shows the electric-field lines for a system of two point charges. (*a*) What are the relative magnitudes of the charges? (*b*) What are the signs of the charges? (*c*) In what regions of space is the electric field strong? In what regions is it weak?

Figure 18-26 Electric-field lines for Problem 17.

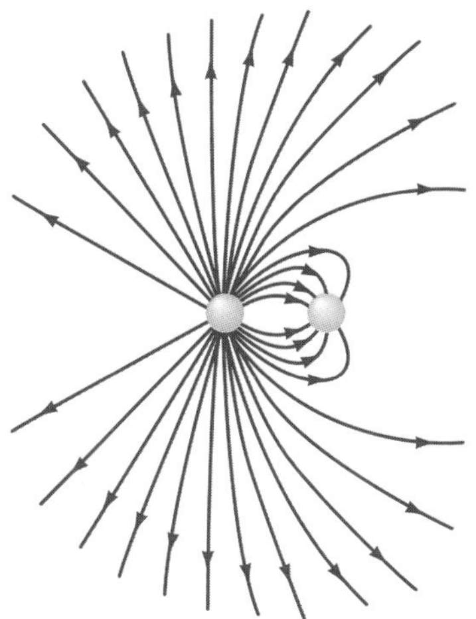

18. Two charges $+q$ and $-3q$ are separated by a small distance. Draw the electric-field lines for this system.

19. Three equal positive point charges are situated at the corners of an equilateral triangle. Sketch the electric-field lines in the plane of the triangle.

20. Two conducting spheres, each with a net positive charge, are held close together so that the electric-field lines are as shown in Figure 18-27. What is the relative charge on the small sphere compared to the large sphere?

Figure 18-27 Problem 20.

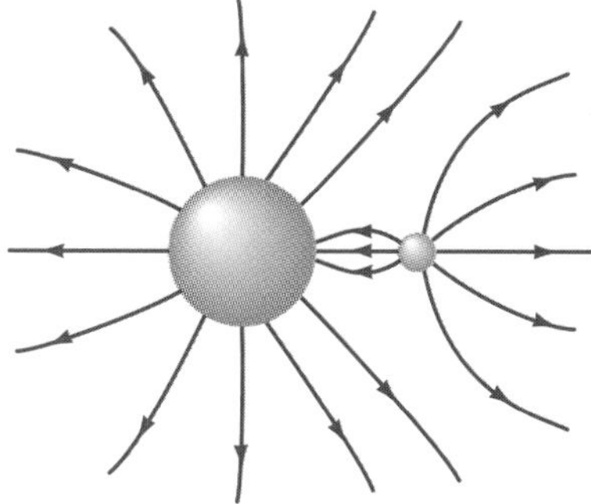

18-6 Motion of Point Charges in Electric Fields

21. In finding the acceleration of an electron or some other charged particle, the ratio of the charge to the mass of the particle is important. (*a*) Compute e/m for an electron. (*b*) What is the magnitude and direction of the acceleration of an electron in a uniform electric field with a magnitude of 100 N/C? (*c*) Nonrelativistic mechanics can be used only if the speed of the electron is significantly less than the speed of light c. Compute the time it takes for an electron placed at rest in an electric field with a magnitude of 100 N/C to reach a speed of $0.01c$. (*d*) How far does the electron travel in that time?

22. (*a*) Compute e/m for a proton, and find its acceleration in a uniform electric field with a magnitude of 100 N/C. (*b*) Find the time it takes for a proton initially at rest in such a field to reach a speed of $0.01c$ (where c is the speed of light).

23. An electron has an initial velocity of 2×10^6 m/s in the x direction. It enters a uniform electric field $\mathbf{E} = (400\ \text{N/C})\mathbf{j}$, which is in the y direction. (*a*) Find the acceleration of the electron. (*b*) How long does it take for the electron to travel 10 cm in the x direction in the field? (*c*) By how much and in what direction is the electron deflected after traveling 10 cm in the x direction in the field?

24. An electron moves in a circular orbit about a stationary proton. The centripetal force is provided by the electrostatic force of attraction between the proton and the electron. The electron has a kinetic energy of 2.18×10^{-18} J. (*a*) What is the speed of the electron? (*b*) What is the radius of the orbit of the electron?

18-7 Electric Dipoles in Electric Fields

25. Two point charges, $q_1 = 2.0$ pC and $q_2 = -2.0$ pC, are separated by 4 μm. (*a*) What is the dipole moment of this pair of charges? (*b*) Sketch the pair, and show the direction of the dipole moment.

26. A dipole of moment 0.5 $e\cdot$nm is placed in a uniform electric field with a magnitude of 4.0×10^4 N/C. What is the magnitude of the torque on the dipole when (*a*) the dipole is parallel to the electric field, (*b*) the dipole is perpendicular to the electric field, and (*c*) the dipole makes an angle of 30° with the electric field? (*d*) Find the potential energy of the dipole in the electric field for each case.

Level II

27. In copper, about one electron per atom is free to move about. A copper penny has a mass of 3 g. (*a*) What percentage of the free charge would have to be removed to give the penny a charge of 15 μC? (See Example 18-1.) (*b*) What would be the force of repulsion between two pennies carrying this charge if they were 25 cm apart? Assume that the pennies are point charges.

28. A point charge of $-5\ \mu$C is located at $x = 4$ m, $y = -2$ m. A second point charge of 12 μC is located at $x = 1$ m, $y = 2$ m. (*a*) Find the magnitude and direction of the electric field at $x = -1$ m, $y = 0$. (*b*) Calculate the magnitude and direction of the force on an electron at $x = -1$ m, $y = 0$.

29. A 5-μC point charge is located at $x = 1$ m, $y = 3$ m, and a -4-μC point charge is located at $x = 2$ m, $y = -2$ m. (*a*) Find the magnitude and direction of the electric field at $x = -3$ m, $y = 1$ m. (*b*) Find the magnitude and direction of the force on a proton at $x = -3$ m, $y = 1$ m.

30. A point charge of $-2.5\ \mu$C is located at the origin. A second point charge of $6\ \mu$C is at $x = 1$ m, $y = 0.5$ m. Find the x and y coordinates of the position at which an electron would be in equilibrium.

31. A particle leaves the origin with a speed of 3×10^6 m/s at 35° to the x axis. It moves in a constant electric field $\mathbf{E} = E_y\mathbf{j}$. Find E_y such that the particle will cross the x axis at $x = 1.5$ cm if the particle is (*a*) an electron and (*b*) a proton.

32. An electron starts at the position shown in Figure 18-28 with an initial speed $v_0 = 5 \times 10^6$ m/s at 45° to the x axis. The electric field is in the positive y direction and has a magnitude of 3.5×10^3 N/C. On which plate and at what location will the electron strike?

Figure 18-28 An electron moving in a uniform electric field for Problem 32.

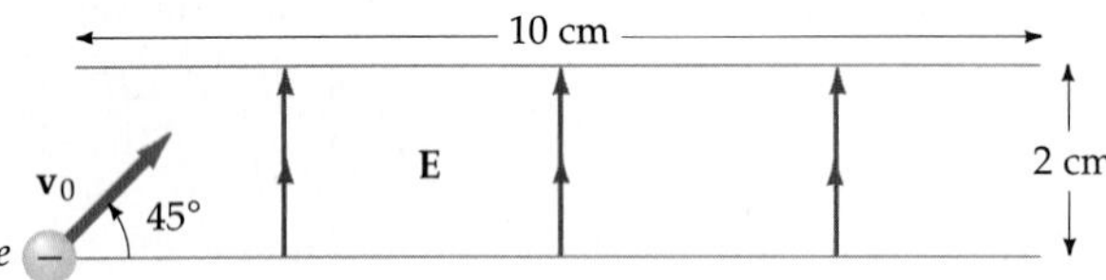

33. An electron with kinetic energy of 2×10^{-16} J is moving to the right along the axis of a cathode ray tube as shown in Figure 18-29. There is an electric field $\mathbf{E} = (2 \times 10^4\ \text{N/C})\mathbf{j}$ in the region between the deflection plates. Everywhere else, $\mathbf{E} = 0$. (*a*) How far is the electron from the axis of the tube when it reaches the end of the plates? (*b*) At what angle is the electron moving with respect to the axis? (*c*) How far from the axis will the electron be when it strikes the fluorescent screen?

Figure 18-29 An electron in a cathode ray tube for Problem 33.

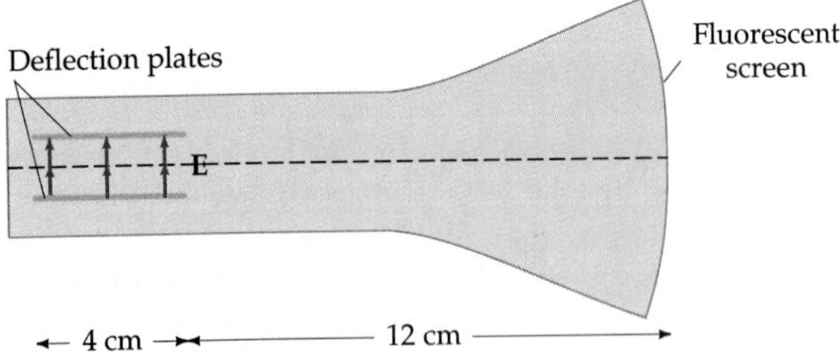

34. Four charges of equal magnitude are arranged at the corners of a square of side L as shown in Figure 18-30. (*a*) Find the magnitude and direction of the force exerted on the charge in the lower left corner by the other charges. (*b*) Show that the electric field at the midpoint of one of the sides of the square is directed along that side toward the negative charge and has a magnitude E given by

$$E = k\frac{8q}{L^2}\left(1 - \frac{\sqrt{5}}{25}\right)$$

Figure 18-30 Problem 34.

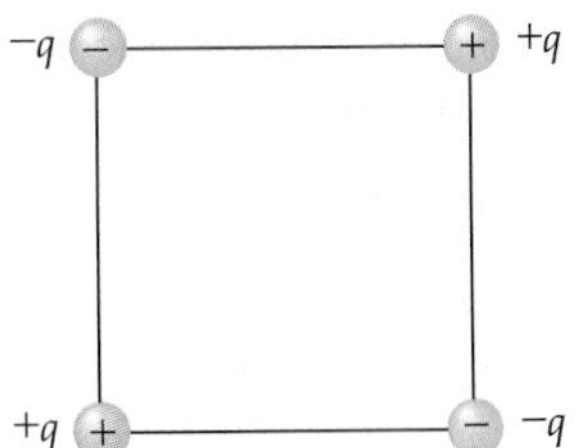

35. Two charges q_1 and q_2 have a total charge of $6\ \mu$C. When they are separated by 3 m, the force exerted by one charge on the other has a magnitude of 8 mN. Find q_1 and q_2 if (*a*) both are positive so that they repel each other and (*b*) one is positive and the other is negative so that they attract each other.

36. A positive charge Q is to be divided into two positive charges q_1 and q_2. Show that, for a given separation D, the force exerted by one charge on the other is greatest if $q_1 = q_2 = \frac{1}{2}Q$.

37. Two equal positive charges q are on the y axis, one at $y = a$ and the other at $y = -a$. (*a*) Show that the electric field on the x axis is along the x axis with $E_x = 2kqx(x^2 + a^2)^{-3/2}$. (*b*) Show that near the origin, when x is much smaller than a, E_x is approximately $2kqx/a^3$. (*c*) Show that for values of x much larger than a, E_x is approximately $2kq/x^2$. Explain why you would expect this result even before calculating it.

38. (*a*) Show that the electric field for the charge distribution in Problem 37 has its greatest magnitude at the points $x = a/\sqrt{2}$ and $x = -a\sqrt{2}$ by computing dE_x/dx and setting the derivative equal to zero. (*b*) Sketch the function E_x versus x using your results for part (*a*) of this problem and parts (*b*) and (*c*) of Problem 37.

39. An electric dipole consists of a positive charge q on the x axis at $x = a$ and a negative charge $-q$ on the x axis at $x = -a$. Find the magnitude and direction of the electric field at a point y on the y axis, and show that for $y \gg a$, the field is approximately $\mathbf{E} = -(kp/y^3)\mathbf{i}$, where p is the magnitude of the dipole moment.

40. Five equal charges Q are equally spaced on a semicircle of radius R as shown in Figure 18-31. Find the force on a charge q located at the center of the semicircle.

Figure 18-31 Problem 40.

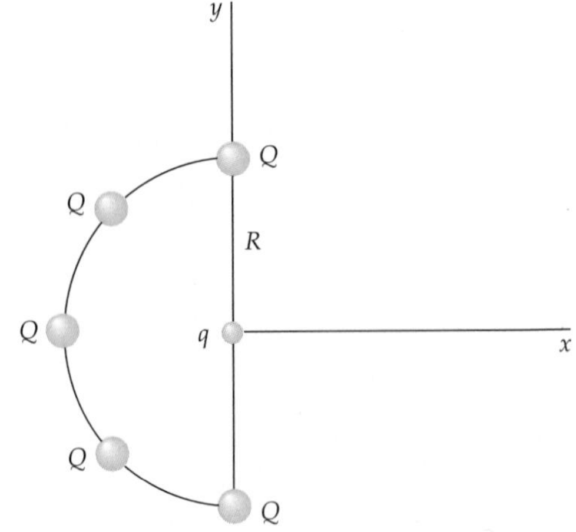

41. Two small spheres of mass m are suspended from a common point by threads of length L. When each sphere carries a charge q, each thread makes an angle θ with the vertical as shown in Figure 18-32. (*a*) Show that the charge q is given by

$$q = 2L \sin \theta \sqrt{\frac{mg \tan \theta}{k}}$$

where k is the Coulomb constant. (*b*) Find q if $m = 10$ g, $L = 50$ cm, and $\theta = 10°$.

Figure 18-32 Problem 41.

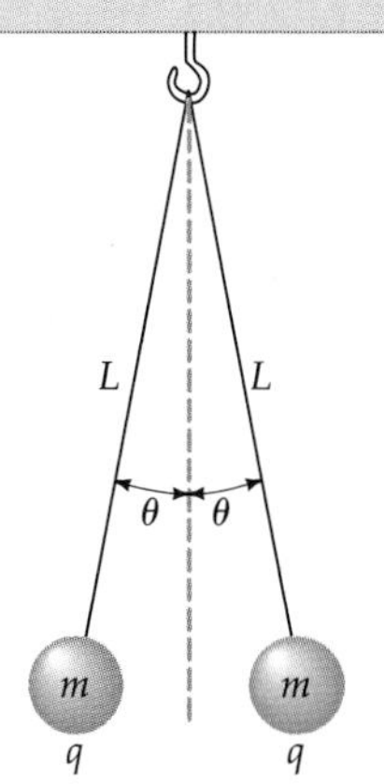

42. A water molecule has its oxygen atom at the origin, one hydrogen nucleus at $x = 0.077$ nm, $y = 0.058$ nm and the other hydrogen nucleus at $x = -0.077$ nm, $y = 0.058$ nm. If the hydrogen electrons are transferred completely to the oxygen atom so that it has a charge of $-2e$, what is the dipole moment of the water molecule? This characterization of the chemical bonds of water as being totally ionic overestimates the dipole moment of a water molecule.

Level III

43. For the charge distribution in Problem 37, the electric field at the origin is zero. A test charge q_0 placed at the origin will therefore be in equilibrium. (*a*) Discuss the stability of the equilibrium for a positive test charge by considering small displacements from equilibrium along the x axis and small displacements along the y axis. (*b*) Repeat part (*a*) for a negative test charge. (*c*) Find the magnitude and sign of a charge q_0 that when placed at the origin results in a net force of zero on each of the three charges. (*d*) What will happen if any of the charges is displaced slightly from equilibrium?

44. Two positive point charges $+q$ are on the y axis at $y = +a$ and $y = -a$ as in Problem 37. A bead of mass m carrying a negative charge $-q$ slides along a thread that runs along the x axis. (*a*) Show that for small displacements of $x \ll a$, the bead experiences a restoring force that is proportional to x and therefore undergoes simple harmonic motion. (*b*) Find the period of the motion.

45. An electric dipole consists of two charges $+q$ and $-q$ separated by a very small distance $2a$. Its center is on the x axis at $x = x_1$, and it points along the x axis in the positive x direction. The dipole is in a nonuniform electric field, which is also in the x direction, given by $\mathbf{E} = Cx\mathbf{i}$, where C is a constant. (*a*) Find the force on the positive charge and that on the negative charge, and show that the net force on the dipole is $Cp\mathbf{i}$. (*b*) Show that, in general, if a dipole of moment $\mathbf{p}$ lies along the x axis in an electric field in the x direction, the net force on the dipole is given approximately by $(dE_x/dx)p\mathbf{i}$.

46. A positive point charge $+Q$ is at the origin, and a dipole of moment $\mathbf{p}$ is a distance r away and in the radial direction as in Figure 18-23. (*a*) Show that the force exerted by the electric field of the point charge on the dipole is attractive and has a magnitude of approximately $2kQp/r^3$ (see Problem 45). (*b*) Now assume that the dipole is centered at the origin and that a point charge Q is a distance r away along the line of the dipole. From your result for part (*a*) and Newton's third law, show that the magnitude of the electric field of the dipole along the line of the dipole a distance r away is approximately $2kp/r^3$.

47. A quadrupole consists of two dipoles that are close together as shown in Figure 18-33. The effective charge at the origin is $-2q$ and the other charges on the y axis at $y = a$ and $y = -a$ are each $+q$. (*a*) Find the electric field at a point on the x axis far away so that $x \gg a$. (*b*) Find the electric field on the y axis far away so that $y \gg a$.

Figure 18-33 Problem 47.

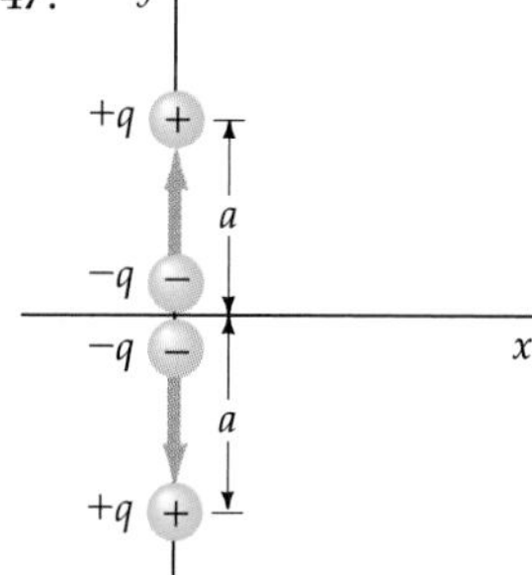

Chapter 19

The Electric Field II: Continuous Charge Distributions

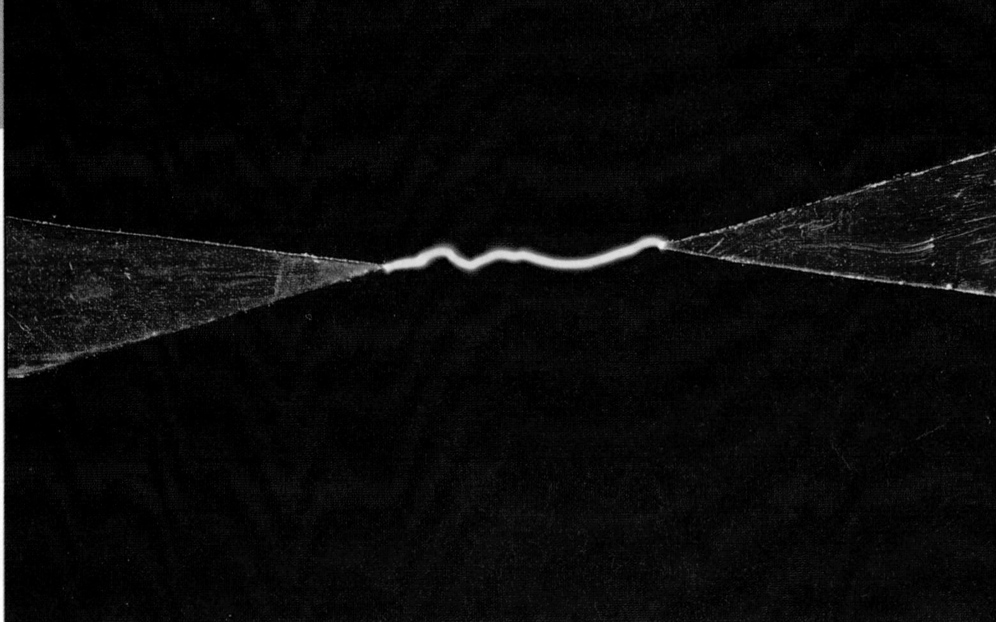

Electrical discharge between two charged conductors. The electric field near the sharp points is strong enough to strip the electrons from nearby air molecules, thus ionizing them and causing the air to conduct.

On a microscopic scale, electric charge is quantized. However, there are often situations in which many charges are so close together that the total charge can be considered to be continuously distributed in space. The use of a continuous charge density to describe a distribution of a large number of discrete charges is similar to the use of a continuous mass density to describe air, which actually consists of a large number of discrete molecules. In either case, it is usually easy to find a volume element ΔV that is large enough to contain many (billions of) individual charges or molecules and yet is small enough that replacing ΔV by a differential dV and using calculus introduces negligible error. We describe the charge per unit volume by the **volume charge density** ρ:

$$\rho = \frac{\Delta Q}{\Delta V} \qquad 19\text{-}1$$

Often charge is distributed in a thin layer on the surface of an object. In such cases, we define the **surface charge density** σ as the charge per unit area:

$$\sigma = \frac{\Delta Q}{\Delta A} \qquad 19\text{-}2$$

Similarly, we sometimes encounter a charge distributed along a line in space. In such cases, we define the **linear charge density** λ as the charge per unit length:

$$\lambda = \frac{\Delta Q}{\Delta L} \qquad 19\text{-}3$$

In this chapter, we will give examples of using Coulomb's law to calculate the electric field due to various types of continuous charge distributions. We will then discuss Gauss's law, which relates the electric field on a closed surface to the net charge within the surface, and use this relation to calculate the electric field for certain charge distributions that have a high degree of symmetry.

19-1 Calculation of the Electric Field from Coulomb's Law

The electric field produced by a given charge distribution can be calculated in a straightforward way using Coulomb's law. In Figure 19-1, we have chosen an element of charge $dq = \rho\, dV$ that is small enough that we may consider it to be a point charge. The electric field $d\mathbf{E}$ at a field point P due to this charge element is given by Coulomb's law:

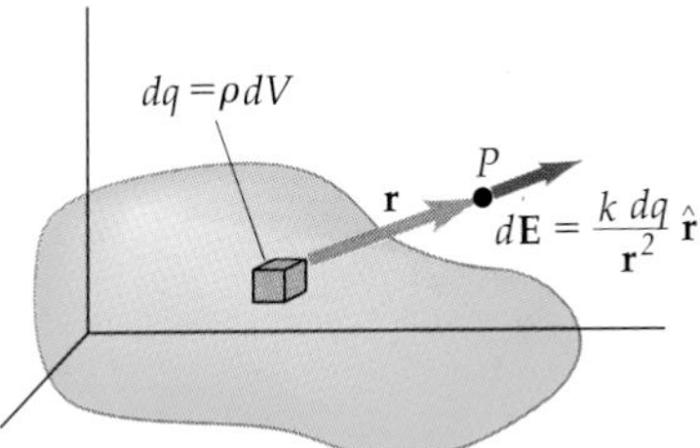

Figure 19-1 An element of charge dq produces a field $d\mathbf{E} = (k\,dq/r^2)\,\hat{\mathbf{r}}$ at point P. The field at P due to the total charge is found by integrating over the entire charge distribution.

$$d\mathbf{E} = \frac{k\,dq}{r^2}\,\hat{\mathbf{r}}$$

where r is the distance from the charge element to the field point P, and $\hat{\mathbf{r}}$ is a unit vector that points from the element to the field point. The total field at P is found by integrating this expression over the entire charge distribution, which we consider to occupy some volume V:

$$\mathbf{E} = \int_V \frac{k\,dq}{r^2}\,\hat{\mathbf{r}} \qquad 19\text{-}4$$

Electric field due to a continuous charge distribution

where $dq = \rho\, dV$. If the charge is distributed on a surface, we use $dq = \sigma\, dA$ and integrate over the surface. If the charge is along a line, we use $dq = \lambda\, dL$ and integrate over the line.

E on the Axis of a Finite Line Charge

A uniform charge Q lies along the x axis from $x = 0$ to $x = L$ as shown in Figure 19-2. The linear charge density for this charge is $\lambda = Q/L$. We wish to find the electric field produced by this line charge at some point P on the x axis at $x = x_0$, for $x_0 > L$. In the figure, we have chosen a small differential element dx at a distance x from the origin. The field point P is at a distance $r = x_0 - x$ from this charge element. The electric field due to this element of charge is directed along the x axis and has the magnitude

$$dE_x = \frac{k\,dq}{(x_0 - x)^2} = \frac{k\lambda\,dx}{(x_0 - x)^2}$$

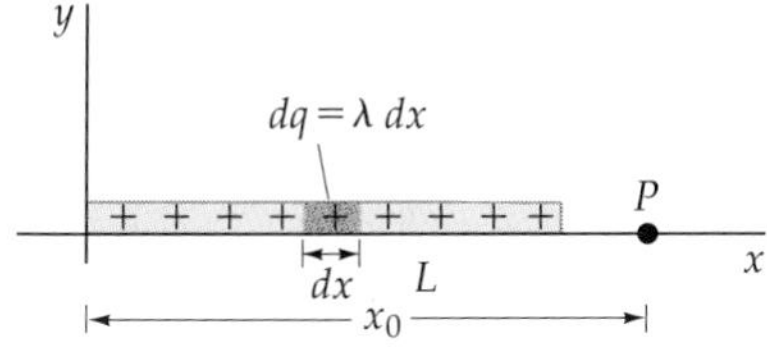

Figure 19-2 Geometry for the calculation of the electric field on the axis of a uniform line charge of charge Q, length L, and linear charge density $\lambda = Q/L$. An element $dq = \lambda\,dx$ of the line charge is treated as a point charge. The field due to this element is found from Coulomb's law, and the total field is then found by integrating from $x = 0$ to $x = L$.

We find the total field by integrating over the entire line charge from $x = 0$ to $x = L$:

$$E_x = k\lambda \int_0^L \frac{dx}{(x_0 - x)^2} = k\lambda \left[\frac{1}{x_0 - x}\right]_0^L$$

$$= k\lambda \left\{\frac{1}{x_0 - L} - \frac{1}{x_0}\right\} = k\lambda \left\{\frac{L}{x_0(x_0 - L)}\right\}$$

Using $\lambda = Q/L$, we obtain

$$E_x = \frac{kQ}{x_0(x_0 - L)} \qquad 19\text{-}5$$

We can see that if L is much smaller than x_0, the electric field at x_0 is approximately kQ/x_0^2. That is, if we are sufficiently far away from the line charge, it looks like a point charge.

E on the Perpendicular Bisector of a Finite Line Charge

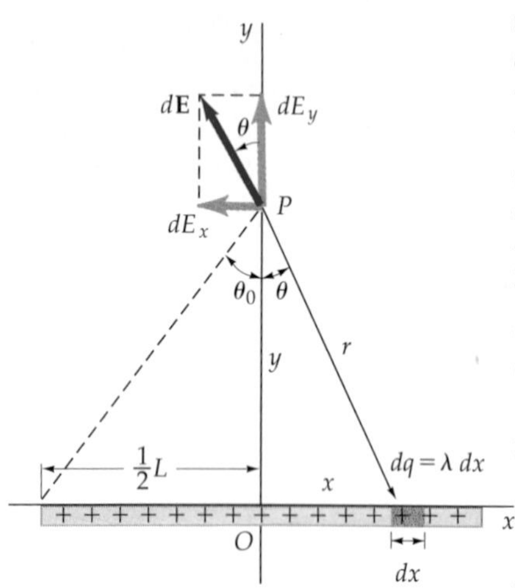

Figure 19-3 Geometry for the calculation of the electric field at a point on the perpendicular bisector of a uniform finite line charge. By symmetry, the net electric field is perpendicular to the line charge. The net field is found by integrating the expression for the perpendicular component from $\theta = 0$ to $\theta = \theta_0$ and multiplying by 2.

We will now find the electric field due to a uniform line charge of length L and total charge Q at a point P on the perpendicular bisector of the line as shown in Figure 19-3. In the figure, we have chosen a coordinate system such that the origin is at the center of the line charge, the charge is on the x axis, and the field point P is on the y axis. The charge element $dq = \lambda\, dx$ and the field $d\mathbf{E}$ that it produces are shown. The field has a component parallel to the line charge and one perpendicular to it. However, we can see from the symmetry of the charge distribution that for each charge element to the right of the origin (such as the one shown) there is one to the left that produces a parallel component of $d\mathbf{E}$ that is equal and opposite to the one shown. When we sum over all the charge elements in the line, the parallel components will sum to zero. We therefore need to calculate only the component of **E** perpendicular to the line charge.

The magnitude of the field produced by an element of charge $dq = \lambda\, dx$ is

$$|d\mathbf{E}| = \frac{k\, dq}{r^2} = \frac{k\lambda\, dx}{r^2}$$

The perpendicular component (in this case, the y component) is

$$dE_y = \frac{k\lambda\, dx}{r^2} \cos\theta \qquad 19\text{-}6$$

The total field E_y is computed by integrating from $x = -\frac{1}{2}L$ to $x = +\frac{1}{2}L$. Because of the symmetry of the charge distribution, the contribution of each half of the line charge to the total field is the same, so we can get the same result by integrating from $x = 0$ to $x = \frac{1}{2}L$ and multiplying by 2. That is,

$$E_y = \int_{x=-\frac{1}{2}L}^{x=+\frac{1}{2}L} dE_y = 2\int_{x=0}^{x=\frac{1}{2}L} dE_y \qquad 19\text{-}7$$

This integration can be simplified by rewriting it as an integration over the variable θ instead of the variable x. We can see from Figure 19-3 that x and θ are related by

$$x = y \tan\theta \qquad 19\text{-}8$$

where y is the perpendicular distance from the line charge to the field point, which does not vary during the integration. Then

$$\frac{dx}{d\theta} = y \sec^2\theta = y\left(\frac{r}{y}\right)^2$$

The increment dx is thus related to the increment $d\theta$ by

$$dx = \frac{r^2}{y}\,d\theta$$

Substituting $r^2\,d\theta/y$ for dx in Equation 19-6, we obtain

$$dE_y = \frac{k\lambda}{y}\cos\theta\,d\theta \qquad \text{19-9}$$

From Equation 19-8, we can see that $\theta = 0$ when $x = 0$, and that $\theta = \theta_0$ when $x = \frac{1}{2}L$, which is given by

$$\tan\theta_0 = \frac{\frac{1}{2}L}{y}$$

The total y component of the field is obtained by integrating Equation 19-9 from $\theta = 0$ to $\theta = \theta_0$ and multiplying by 2:

$$E_y = 2\int_{\theta=0}^{\theta=\theta_0} dE_y = \frac{2k\lambda}{y}\int_0^{\theta_0}\cos\theta\,d\theta$$

or

$$E_y = \frac{2k\lambda}{y}\sin\theta_0 = \frac{2k\lambda}{y}\,\frac{\frac{1}{2}L}{\sqrt{(\frac{1}{2}L)^2 + y^2}} \qquad \text{19-10}$$

E on the perpendicular bisector of a finite line charge

where (from Figure 19-3) $\sin\theta_0$ is related to L and y by

$$\sin\theta_0 = \frac{\frac{1}{2}L}{\sqrt{(\frac{1}{2}L)^2 + y^2}}$$

When y is much greater than L, $\sin\theta_0$ is given approximately by

$$\sin\theta_0 \approx \frac{\frac{1}{2}L}{y} \qquad y \gg L$$

and E_y is approximately

$$E_y \approx \frac{k\lambda L}{y^2} = \frac{kQ}{y^2}$$

where $Q = \lambda L$ is the total charge. As expected, when we are very far away, a finite line charge looks like a point charge.

E near an Infinite Line Charge

When we are very close to a line charge or, alternatively, when the line charge is very long, so that $y \ll L$, the angle θ_0 in Figure 19-3 is approximately 90°. Substituting $\theta_0 = 90°$ into Equation 19-10 gives

$$E_y = \frac{2k\lambda}{y} \qquad \text{19-11}$$

E at a distance y from an infinite line charge

Thus, as the distance y of a point from an infinite line charge increases, the electric field decreases as $1/y$.

Exercise

Show that Equation 19-11 has the correct units for the electric field.

Example 19-1

An infinite line charge of linear charge density $\lambda = 0.6\ \mu\text{C/m}$ lies along the z axis, and a point charge $q = 8\ \mu\text{C}$ lies on the y axis at $y = 3$ m as shown in Figure 19-4. Find the electric field at the point P on the x axis at $x = 4$ m.

The electric field at any point in space is found from the superposition of the electric field due to the line charge and that due to the point charge. At the field point P on the x axis at $x = 4$ m, the electric field due to the line charge is in the x direction and is given by

$$\mathbf{E}_\text{L} = \frac{2k\lambda}{x}\,\mathbf{i} = \frac{2(8.99 \times 10^9\ \text{N·m}^2/\text{C}^2)(0.6 \times 10^{-6}\ \text{C/m})}{4\ \text{m}}\,\mathbf{i}$$

$$= 2.70\ \text{kN/C}\ \mathbf{i}$$

The electric field due to the point charge, which is $\sqrt{(4\ \text{m})^2 + (3\ \text{m})^2} = 5$ m away, is

$$\mathbf{E}_\text{P} = \frac{kq}{r^2}\,\hat{\mathbf{r}} = \frac{(8.99 \times 10^9\ \text{N·m}^2/\text{C}^2)(8 \times 10^{-6}\ \text{C})}{(5\ \text{m})^2}\,\hat{\mathbf{r}} = (2.88\ \text{kN/C})\hat{\mathbf{r}}$$

where $\hat{\mathbf{r}}$ is a unit vector that points from q to the field point P. The field $\mathbf{E}_\text{P}$ makes an angle θ with the x axis as shown in the figure. The x and y components of $\mathbf{E}_\text{P}$ are

$$E_{\text{P}x} = E_\text{P} \cos\theta = (2.88\ \text{kN/C})(\tfrac{4}{5}) = 2.30\ \text{kN/C}$$

and

$$E_{\text{P}y} = -E_\text{P} \sin\theta = -(2.88\ \text{kN/C})(\tfrac{3}{5}) = -1.73\ \text{kN/C}$$

The net electric field at P has x and y components given by

$$E_x = E_{\text{L}x} + E_{\text{P}x} = 2.70\ \text{kN/C} + 2.30\ \text{kN/C} = 5.00\ \text{kN/C}$$

and

$$E_y = E_{\text{L}y} + E_{\text{P}y} = 0 + (-1.73\ \text{kN/C}) = -1.73\ \text{kN/C}$$

The net electric field therefore has a magnitude of

$$E = \sqrt{E_x^2 + E_y^2} = \sqrt{(5.00)^2 + (-1.73)^2}\ \text{kN/C} = 5.29\ \text{kN/C}$$

It is directed at an angle ϕ below the x axis given by

$$\phi = \tan^{-1}\left(-\frac{1.73}{5.00}\right) = -19.1°$$

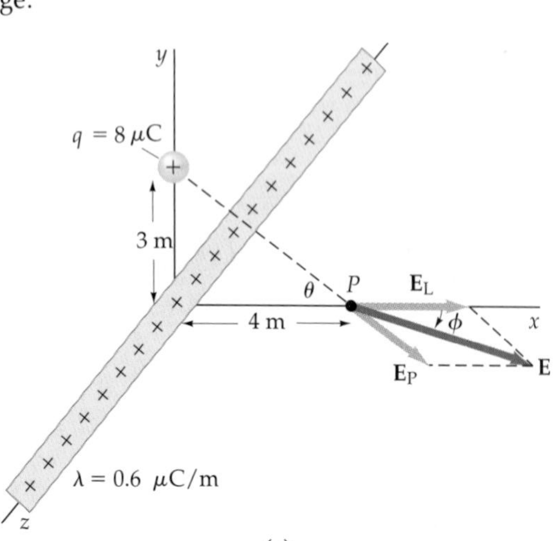

Figure 19-4 (*a*) An infinite line charge along the z axis and a point charge q on the y axis for Example 19-1. The electric field **E** at a point P on the x axis is found from the superposition of the fields due to the line charge and the point charge. The field due to the point charge is directed along the line from the point charge to the field point. (*b*) Electric-field lines near a long wire. The electric field near a high-voltage power line can be large enough to strip the electrons from air molecules, thus ionizing them and making the air a conductor. The glow resulting from the recombination of free electrons with the ions is called corona discharge.

E on the Axis of a Ring Charge

Figure 19-5 shows a uniform ring charge of radius a and total charge Q. We wish to find the electric field at a point P on the axis of the ring at a distance x from the center of the ring. The field $d\mathbf{E}$ due to the charge element dq is shown in the figure. This field has a component dE_x directed along the axis of the ring and a component $dE_\perp$ directed perpendicular to the axis. From the symmetry of the charge distribution, we can see that the net field due to the entire ring must lie along the axis of the ring; that is, the perpendicular components will sum to zero. In particular, the perpendicular component shown will be canceled by that due to the charge element on the ring directly opposite the one shown. The axial component of the field due to the charge element shown is

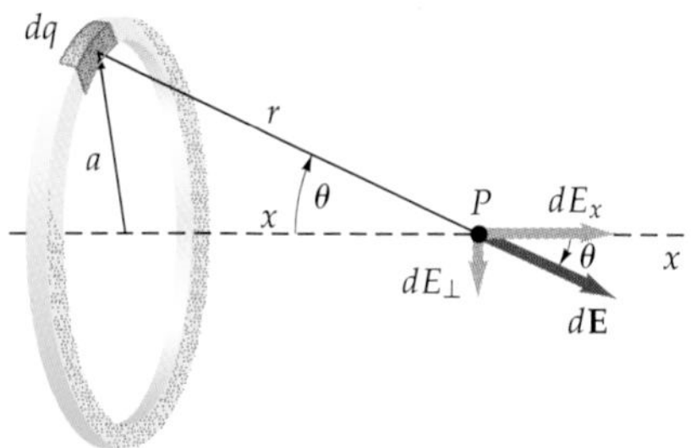

Figure 19-5 A ring charge of radius a. The electric field at point P on the x axis due to the charge element dq shown has a component along the x axis and one perpendicular to the x axis. When summed over the total ring, the perpendicular components cancel, so the net field is along the x axis.

$$dE_x = \frac{k\,dq}{r^2}\cos\theta = \frac{k\,dq}{r^2}\frac{x}{r} = \frac{k\,dq\,x}{(x^2+a^2)^{3/2}}$$

where

$$r^2 = x^2 + a^2$$

and

$$\cos\theta = \frac{x}{r} = \frac{x}{\sqrt{x^2+a^2}}$$

The field due to the entire ring of charge is

$$E_x = \int \frac{kx\,dq}{(x^2+a^2)^{3/2}}$$

Since x does not vary as we integrate over the elements of charge, we can remove it from the integral. Then

$$E_x = \frac{kx}{(x^2+a^2)^{3/2}}\int dq$$

or

$$E_x = \frac{kQx}{(x^2+a^2)^{3/2}} \qquad \text{19-12}$$

E on the axis of a ring charge

A useful check of our result is to examine it at the extreme values of x. At $x = 0$, we obtain $E_x = 0$, which is what we should expect since for each element on the ring the field at the center is canceled by that due to the element directly opposite on the other side of the ring. When x is much larger than a, we can neglect a^2 compared with x^2 in the denominator of Equation 19-12. We then obtain $E_x \approx kQ/x^2$. That is, far from the ring, the ring looks like a point charge as expected.

E on the Axis of a Uniformly Charged Disk

Figure 19-6 shows a uniformly charged disk of radius R and total charge Q. We wish to find the electric field on the axis of the disk. Since the area of the disk is πR^2, the charge per unit area is $\sigma = Q/\pi R^2$. The electric field on the axis of the disk will be parallel to the axis. We can calculate the field by treating the disk as a set of concentric ring charges. Consider a ring of radius a and width da as shown in the figure. The area of this ring is $dA = 2\pi a\,da$, and its charge is $dq = \sigma\,dA = 2\pi\sigma a\,da$. The field produced by this ring is given by Equation 19-12 if we replace Q with $dq = 2\pi\sigma a\,da$. Thus

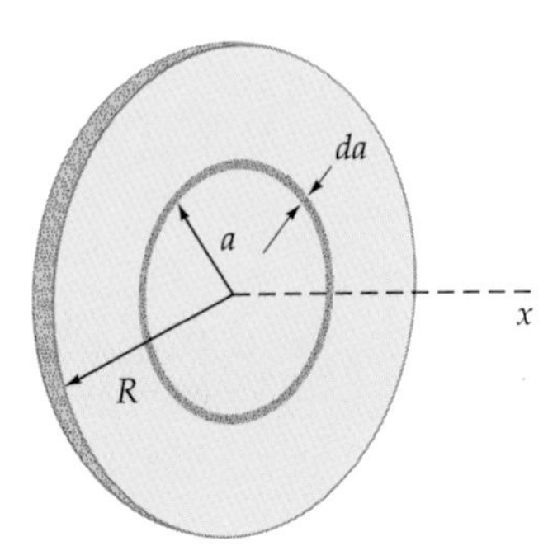

Figure 19-6 A uniform disk of charge can be treated as a set of ring charges, each of radius a and thickness da carrying a charge $dq = \sigma\,dA = (Q/\pi R^2)2\pi a\,da$.

$$dE_x = \frac{kx2\pi\sigma a\,da}{(x^2+a^2)^{3/2}}$$

The total field produced by the disk is found by integrating this expression from $a = 0$ to $a = R$:

$$E_x = \int_0^R \frac{kx2\pi\sigma a\,da}{(x^2+a^2)^{3/2}} = kx\pi\sigma \int_0^R (x^2+a^2)^{-3/2}\,2a\,da$$

This integral is of the form $\int u^n\,du$, with $u = x^2 + a^2$ and $n = -\frac{3}{2}$. The integration thus gives

$$E_x = kx\pi\sigma\left[\frac{(x^2+a^2)^{-1/2}}{-\frac{1}{2}}\right]_0^R$$

$$= -2kx\pi\sigma\left(\frac{1}{\sqrt{x^2+R^2}} - \frac{1}{x}\right)$$

or

E on the axis of a disk charge

$$E_x = 2\pi k\sigma\left(1 - \frac{x}{\sqrt{x^2+R^2}}\right) \qquad 19\text{-}13$$

When we are very far from the disk, we expect the disk to look like a point charge. For $x \gg R$, we can find an approximation for the second term in Equation 19-13 using the binomial expansion, $(1+\epsilon)^n \approx 1 + n\epsilon$, for $\epsilon \ll 1$. This gives

$$\frac{x}{\sqrt{x^2+R^2}} = \frac{x}{x(1+R^2/x^2)^{1/2}} = \left(1 + \frac{R^2}{x^2}\right)^{-1/2} \approx 1 - \frac{R^2}{2x^2} + \cdots$$

Equation 19-13 then becomes

$$E_x \approx 2\pi k\sigma\left(1 - 1 + \frac{R^2}{2x^2} + \cdots\right) = \frac{k\pi R^2\sigma}{x^2} = \frac{kQ}{x^2}$$

where $Q = \sigma\pi R^2$ is the total charge on the disk.

E near an Infinite Plane of Charge

The interesting and important result for the field near an infinite plane of charge can be obtained from Equation 19-13 by letting either R go to infinity or x go to zero. Then

E near an infinite plane of charge

$$E_x = 2\pi k\sigma \qquad x > 0 \qquad 19\text{-}14a$$

Thus, the field due to an infinite-plane charge distribution is uniform; that is, the field does not depend on x. On the other side of the infinite plane, for negative values of x, the field points in the negative x direction, so

$$E_x = -2\pi k\sigma \qquad x < 0 \qquad 19\text{-}14b$$

As we move along the x axis, the electric field jumps from $-2\pi k\sigma\mathbf{i}$ to $+2\pi k\sigma\mathbf{i}$ when we pass through an infinite plane of charge.

Example 19-2

A disk of radius 5 cm carries a uniform surface charge density of 4 $\mu C/m^2$. Using reasonable approximations, find the electric field on the axis of the disk at distances of (*a*) 0.01 cm, (*b*) 0.03 cm, (*c*) 6 m, and (*d*) 6 cm.

(*a*) Since 0.01 cm is much less than the radius of the disk, we can approximate the disk as an infinite plane of charge and use Equation 19-14*a*. The electric field is then

$$\begin{aligned} E_x &= 2\pi k\sigma \\ &= 2\pi(8.99 \times 10^9\ \text{N}\cdot\text{m}^2/\text{C}^2)(4 \times 10^{-6}\ \text{C/m}^2) \\ &= 226\ \text{kN/C} \end{aligned}$$

(*b*) Again, 0.03 cm is much less than the 5-cm radius of the disk, so the disk looks like an infinite plane from this point and the electric field is $2\pi k\sigma = 226$ kN/C.

(*c*) Since 6 m is much greater than the radius of the disk, we may treat the disk as a point charge $Q = \sigma\pi r^2 = (4\ \mu\text{C/m}^2)\pi(0.05\ \text{m})^2 = 31.4$ nC. The electric field at a distance of 6 m from such a point charge is

$$\begin{aligned} E_x &= \frac{kQ}{x^2} = \frac{(8.99 \times 10^9\ \text{N}\cdot\text{m}^2/\text{C}^2)(31.4 \times 10^{-9}\ \text{C})}{(6\ \text{m})^2} \\ &= 7.84\ \text{N/C} \end{aligned}$$

(*d*) Since 6 cm is neither much less than nor much greater than the radius of 5 cm, we use the exact expression given by Equation 19-13:

$$\begin{aligned} E_x &= 2\pi k\sigma\left(1 - \frac{x}{\sqrt{(x^2 + R^2)}}\right) \\ &= (226\ \text{kN/C})\left(1 - \frac{6\ \text{cm}}{\sqrt{(6\ \text{cm})^2 + (5\ \text{cm})^2}}\right) \\ &= (226\ \text{kN/C})(1 - 0.768) = 52.4\ \text{kN/C} \end{aligned}$$

Note that we did not need to convert from centimeters to meters to find $x/\sqrt{x^2 + R^2}$ because the units cancel.

Exercise

Calculate the electric field to four significant figures for parts (*a*) and (*b*) of Example 19-2, and compare your results with the approximate results found in that example. [Answers: (*b*) $E_x = 225.9$ kN/C, which differs from 226 kN/C by about 0.04 percent.]

Question

1. Since electric charge is quantized, how is it possible to talk about continuous charge distributions?

19-2 Gauss's Law

The qualitative description of the electric field using electric-field lines discussed in Chapter 18 is related to a mathematical equation known as Gauss's law, which relates the electric field on a closed surface to the net charge within the surface. Electric fields arising from some symmetrical charge distributions, such as a spherical shell of charge or an infinite line of charge, can be easily calculated using Gauss's law. In this section, we give a plausible argument for Gauss's law based on the properties of electric-field lines. A rigorous derivation of Gauss's law is given in Section 19-5.

Figure 19-7 shows a surface of arbitrary shape enclosing a dipole. The number of electric-field lines coming from the positive charge, crossing the surface, and going outside of the volume enclosed by the surface depends on where the surface is drawn, but it is exactly equal to the number of lines entering the volume enclosed by the surface and ending on the negative charge. If we count the number of lines leaving as positive and the number entering as negative, the net number leaving and entering is zero. For surfaces enclosing other types of charge distributions, such as that shown in Figure 19-8, the net number of lines leaving any surface enclosing the charges is proportional to the net charge enclosed by the surface. This is a qualitative statement of Gauss's law.

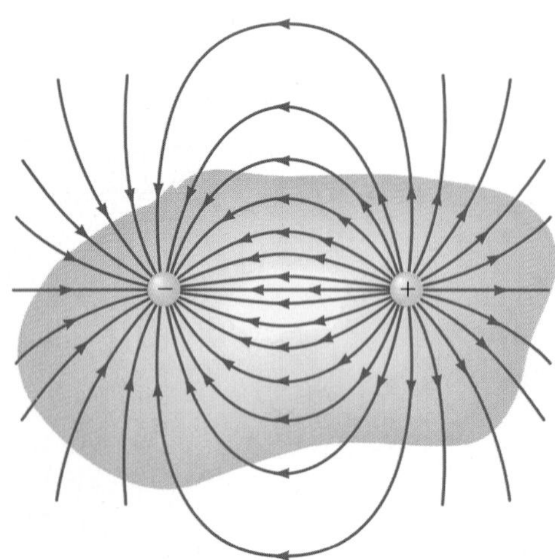

Figure 19-7 A surface of arbitrary shape enclosing an electric dipole. As long as the surface encloses both charges, the number of lines leaving the surface is exactly equal to the number of lines entering the surface no matter where the surface is drawn.

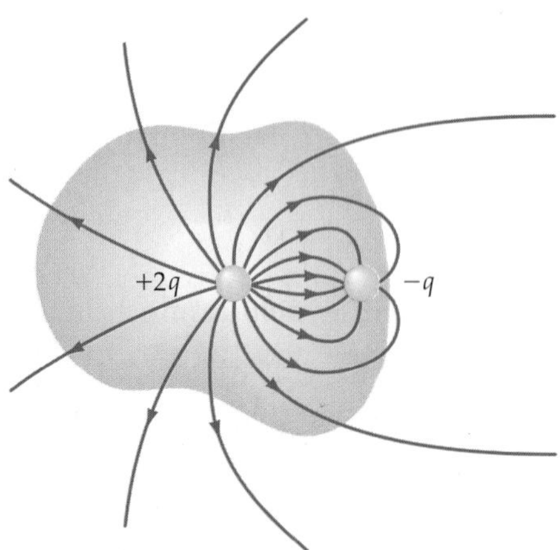

Figure 19-8 A surface of arbitrary shape enclosing the charges $+2q$ and $-q$. The field lines that end on $-q$ either do not pass through the surface or they exit once and enter once. The net number that exit is the same as that for a single charge equal to the net charge within the surface.

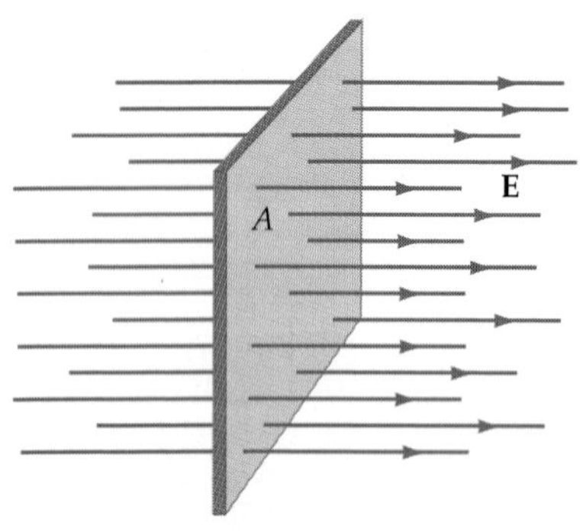

Figure 19-9 Electric-field lines of a uniform field crossing an area A that is perpendicular to the field. The product EA is the electric flux ϕ through the area.

The mathematical quantity related to the number of field lines crossing a surface is called the electric flux. Figure 19-9 shows an area A that is perpendicular to a uniform electric field. The **electric flux** ϕ through a surface of area A that is perpendicular to the field is defined as the product of the field **E** and the area A:

$$\phi = EA$$

The units of flux are newton-meters squared per coulomb ($\mathrm{N{\cdot}m^2/C}$). Since the electric field is proportional to the number of lines per unit area, the flux is proportional to the number of field lines through the area.

In Figure 19-10, the surface of area A_2 is not perpendicular to the electric field **E**. The number of lines that cross area A_2 is the same as the number that cross area A_1. These areas are related by

$$A_2 \cos \theta = A_1 \qquad \text{19-15}$$

where θ is the angle between **E** and the unit vector $\hat{\mathbf{n}}$ that is perpendicular to the surface A_2 as shown in the figure. The flux through a surface that is not perpendicular to **E** is defined to be

$$\phi = \mathbf{E}\cdot\hat{\mathbf{n}}A = EA \cos \theta = E_n A$$

where $E_n = \mathbf{E}\cdot\hat{\mathbf{n}}$ is the component of the electric-field vector that is perpendicular, or normal, to the surface.

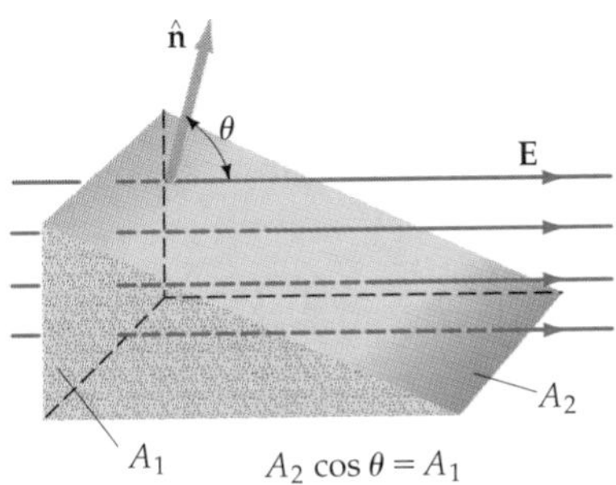

Figure 19-10 Electric-field lines of a uniform electric field that is perpendicular to the area A_1 but makes an angle θ with the unit vector $\hat{\mathbf{n}}$ that is normal to the area A_2. When **E** is not perpendicular to the area, the flux is $E_n A$, where $E_n = E \cos\theta$ is the component of **E** that is perpendicular to the area. The flux through A_2 is the same as that through A_1.

We can generalize our definition of electric flux to a curved surface over which the electric field may vary in magnitude or direction or both by dividing the surface into a large number of very small elements. If each element is small enough, it can be considered to be a plane, and the variation of the electric field across the element can be neglected. Let $\hat{\mathbf{n}}_i$ be the unit vector perpendicular to the ith element and ΔA_i be its area (Figure 19-11). (If the surface is curved, the unit vectors for different elements will have different directions.) The flux of the electric field through this element is

$$\Delta\phi_i = \mathbf{E}\cdot\hat{\mathbf{n}}_i \, \Delta A_i$$

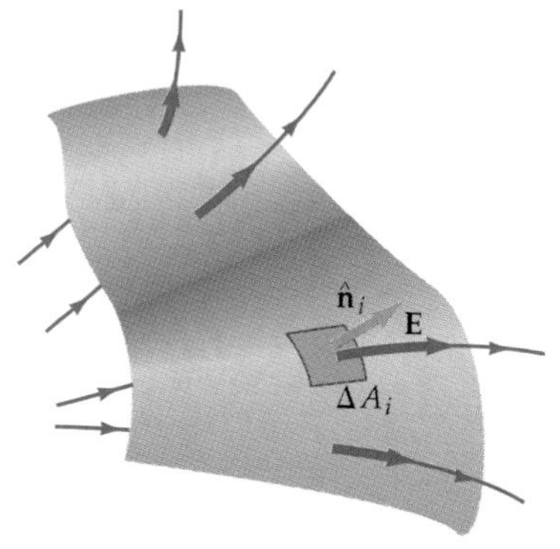

Figure 19-11 When **E** varies in either magnitude or direction, the area of the surface is divided into small elements ΔA_i. The flux through the area is computed by summing $\mathbf{E}\cdot\hat{\mathbf{n}}_i \, \Delta A_i$ over all the area elements.

The total flux through the surface is the sum of $\Delta\phi_i$ over all the elements. In the limit as the number of elements approaches infinity and the area of each element approaches zero, this sum becomes an integral. The general definition of electric flux is thus

$$\phi = \lim_{\Delta A_i \to 0} \sum_i \mathbf{E}\cdot\hat{\mathbf{n}}_i \, \Delta A_i = \int \mathbf{E}\cdot\hat{\mathbf{n}} \, dA \qquad \text{19-16}$$

Electric flux defined

We are often interested in the flux of an electric field through a closed surface. On a closed surface, the normal unit vector $\hat{\mathbf{n}}$ is defined to be outward at each point. At a point where an electric-field line leaves the surface, **E** is directed outward and ϕ is positive; at a point where a field line enters the surface, **E** is directed inward and ϕ is negative. The total or net flux ϕ_{net} through the closed surface is positive or negative depending on whether **E** is predominantly outward or inward at the surface. Since the flux through any part of the surface is proportional to the number of lines passing through the surface, the net flux is proportional to the net number of electric-field lines leaving the surface, that is, the number going out minus the number going in. The integral over a closed surface is indicated by the symbol $\oint$. The net flux through a closed surface is therefore written

$$\phi_{net} = \oint_S \mathbf{E}\cdot\hat{\mathbf{n}} \, dA = \oint_S E_n \, dA \qquad \text{19-17}$$

Figure 19-12 shows a spherical surface of radius R with its center at a point charge Q. The electric field everywhere on this surface is perpendicular to the surface and has the magnitude

$$E_n = \frac{kQ}{R^2}$$

The net flux through this spherical surface is

$$\phi_{net} = \oint_S E_n \, dA = E_n \oint_S dA$$

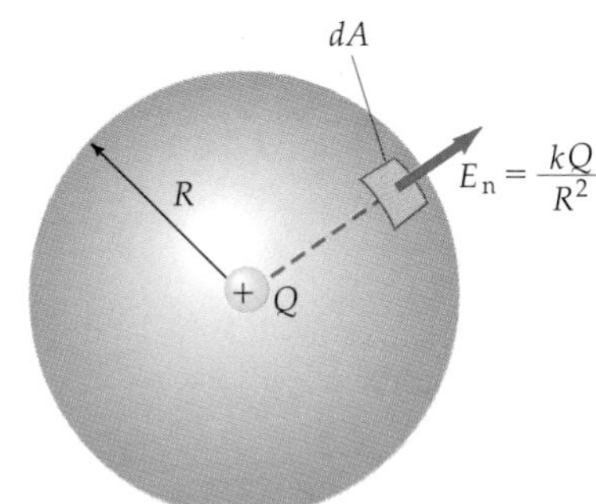

Figure 19-12 A spherical surface enclosing the point charge Q. The same number of electric-field lines that passes through this surface passes through any surface that encloses Q. The flux is easily calculated for a spherical surface. It equals E_n times the surface area $4\pi R^2$.

where we have taken E_n out of the integral because it is constant everywhere on the surface. The integral of dA over the surface is just the total area of the surface, which equals $4\pi R^2$. Using this and substituting kQ/R^2 for E_n, we obtain

$$\phi_{\text{net}} = \frac{kQ}{R^2} 4\pi R^2 = 4\pi kQ \quad \text{19-18}$$

Thus, the net flux through a spherical surface with a point charge at its center is independent of the radius of the sphere and is equal to $4\pi k$ times the magnitude of the point charge. This is consistent with our previous observation that the net number of lines going out of a surface is proportional to the net charge inside the surface. This number of lines is the same for all surfaces surrounding the charge, independent of the shape of the surface. Since the number of lines and the flux are proportional to each other, it follows that Equation 19-18 holds for the flux through any surface enclosing the point charge Q. Thus, the net flux through any surface surrounding a point charge Q equals $4\pi kQ$.

We can extend this result to systems consisting of more than one point charge. In Figure 19-13, the surface encloses two point charges q_1 and q_2, and there is a third point charge q_3 outside the surface. Since the electric field at any point on the surface is the vector sum of the electric fields produced by each of the three charges, the net flux $\phi_{\text{net}} = \oint \mathbf{E}\cdot\hat{\mathbf{n}}\, dA$ through the surface is just the sum of the fluxes due to the individual charges. The flux due to charge q_3, which is outside the surface, is zero because every field line from q_3 that enters the surface at one point leaves the surface at some other point. The net number of lines passing through a surface from a charge outside the surface is zero. The flux through the surface due to charge q_1 is $4\pi kq_1$ and that due to charge q_2 is $4\pi kq_2$. The net flux through the surface therefore equals $4\pi k(q_1 + q_2)$, which may be positive, negative, or zero depending on the signs and magnitudes of the two charges.

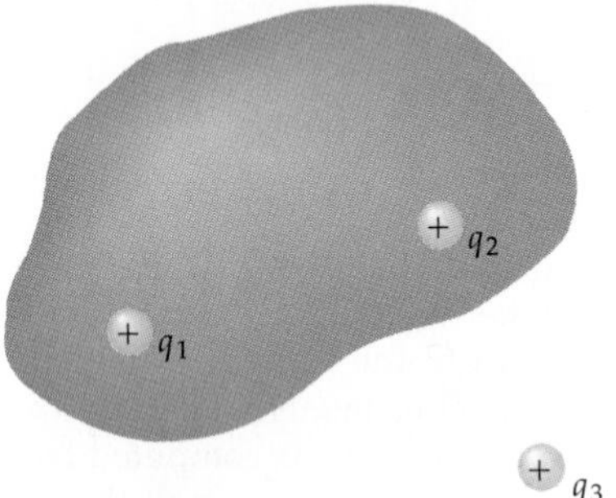

Figure 19-13 A surface enclosing point charges q_1 and q_2 but not q_3. The net flux through this surface is $4\pi k(q_1 + q_2)$.

The net flux through any surface equals $4\pi k$ times the net charge inside the surface:

Gauss's law

$$\phi_{\text{net}} = \oint_S E_n\, dA = 4\pi kQ_{\text{inside}} \quad \text{19-19}$$

This is **Gauss's law.** Its validity depends on the fact that the electric field due to a single point charge varies inversely with the square of the distance from the charge. It was this property of the electric field that made it possible to draw a fixed number of electric-field lines from a charge and have the density of lines be proportional to the field strength.

It is customary to write the Coulomb constant k in terms of another constant ϵ_0, which is called the **permittivity of free space:**

$$k = \frac{1}{4\pi\epsilon_0} \quad \text{19-20}$$

Using this notation, Coulomb's law is written

Coulomb's law in terms of ϵ_0

$$\mathbf{F}_{12} = \frac{1}{4\pi\epsilon_0}\frac{q_1 q_2}{r_{12}^2}\hat{\mathbf{r}}_{12} \quad \text{19-21}$$

and Gauss's law is written

$$\phi_{\text{net}} = \oint_S E_n \, dA = \frac{1}{\epsilon_0} Q_{\text{inside}} \qquad \text{19-22}$$

Gauss's law in terms of ϵ_0

The value of ϵ_0 in SI units is

$$\epsilon_0 = \frac{1}{4\pi k} = \frac{1}{4\pi(8.99 \times 10^9 \text{ N·m}^2/\text{C}^2)} = 8.85 \times 10^{-12} \text{ C}^2/\text{N·m}^2$$

Gauss's law is valid for all surfaces and all charge distributions. For some special charge distributions that have high degrees of symmetry, it can be used to calculate the electric field as we will illustrate in the next section. The real power of Gauss's law is theoretical. For electric fields arising from static or slowly moving charges, Gauss's law and Coulomb's law are equivalent. However, Gauss's law is more general in that it also applies to electric fields arising from rapidly moving charges and accelerating charges.

Questions

2. If the electric field **E** is zero everywhere on a closed surface, is the net flux though the surface necessarily zero? What, then, is the net charge inside the surface?
3. If the net flux through a closed surface is zero, does it follow that the electric field **E** is zero everywhere on the surface? Does it follow that the net charge inside the surface is zero?
4. Is the electric field **E** in Gauss's law only that part of the electric field due to the charge inside a surface, or is it the total electric field due to all charges both inside and outside the surface?

19-3 Calculation of the Electric Field from Gauss's Law

For some highly symmetrical charge distributions, such as a uniformly charged sphere or an infinite line of charge, we can find a mathematical surface on which we know from symmetry that the electric field is constant and perpendicular to the surface. We can then easily evaluate the electric flux through this surface and use Gauss's law to relate the electric field to the charge inside the surface. A surface used to calculate the electric field from Gauss's law is called a **gaussian surface.** In this section, we will use this method to calculate the electric field due to a number of symmetrical charge distributions.

E near a Point Charge

We first use Gauss's law to find the electric field at a distance r from a point charge q. Let the point charge be at the origin. By symmetry, **E** must be radial, and its magnitude can depend only on the distance from the charge. For our gaussian surface, we choose a spherical surface of radius r centered at the charge. The normal component of **E**, $E_n = \mathbf{E}\cdot\hat{\mathbf{n}} = E_r$, has the same value everywhere on the spherical surface. The net flux through this surface is thus

$$\phi_{\text{net}} = \oint \mathbf{E}\cdot\hat{\mathbf{n}} \, dA = \oint E_r \, dA = E_r \oint dA$$

But $\oint dA = 4\pi r^2$, the total area of the spherical surface. Since the total charge inside the surface is just the point charge q, Gauss's law gives

$$E_r 4\pi r^2 = \frac{q}{\epsilon_0}$$

and

$$E_r = \frac{1}{4\pi\epsilon_0}\frac{q}{r^2}$$

We have thus derived Coulomb's law from Gauss's law. Since we originally obtained Gauss's law from Coulomb's law, we have shown that the two laws are equivalent for static charges.

E near an Infinite Plane of Charge

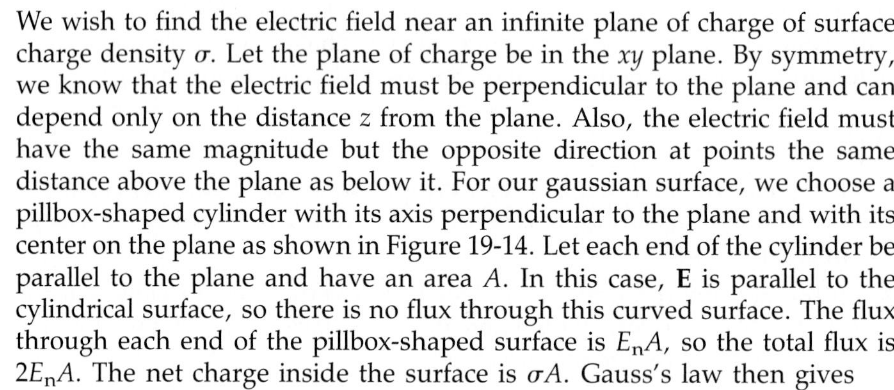

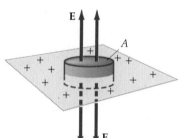

Figure 19-14 Gaussian surface for the calculation of the electric field due to an infinite plane of charge. On the upper and lower faces of this pillbox surface, **E** is perpendicular to the surface and constant in magnitude. The flux through the surface is $2E_nA$, where A is the area of each face.

We wish to find the electric field near an infinite plane of charge of surface charge density σ. Let the plane of charge be in the xy plane. By symmetry, we know that the electric field must be perpendicular to the plane and can depend only on the distance z from the plane. Also, the electric field must have the same magnitude but the opposite direction at points the same distance above the plane as below it. For our gaussian surface, we choose a pillbox-shaped cylinder with its axis perpendicular to the plane and with its center on the plane as shown in Figure 19-14. Let each end of the cylinder be parallel to the plane and have an area A. In this case, **E** is parallel to the cylindrical surface, so there is no flux through this curved surface. The flux through each end of the pillbox-shaped surface is E_nA, so the total flux is $2E_nA$. The net charge inside the surface is σA. Gauss's law then gives

$$\phi_{\text{net}} = \oint E_n\, dA = \frac{1}{\epsilon_0} Q_{\text{inside}}$$

$$2E_nA = \frac{1}{\epsilon_0}\sigma A$$

or

E near an infinite plane of charge

$$E_n = \frac{\sigma}{2\epsilon_0} = 2\pi k\sigma \qquad 19\text{-}23$$

This result agrees with that we obtained by direct integration for **E** near a disk charge in the limit of infinite radius (Equation 19-14*a*).

Example 19-3

An infinite plane of surface charge density $\sigma = +4\ \text{nC/m}^2$ lies in the yz plane at the origin, and a second infinite plane of surface charge density $\sigma = -4\ \text{nC/m}^2$ lies in a plane parallel to the yz plane at $x = 2$ m. Find the electric field at (*a*) $x = 1.8$ m and (*b*) $x = 5$ m.

(*a*) The magnitude of the electric field due to either charge distribution is constant and equal to

$$E_1 = E_2 = \frac{\sigma}{2\epsilon_0} = \frac{4\times 10^{-9}\ \text{C/m}^2}{2(8.85\times 10^{-12}\ \text{C}^2/\text{N·m}^2)} = 226\ \text{N/C}$$

The electric field due to the positive charge on the yz plane points away from the yz plane. Similarly, the electric field due to the negative charge on the plane at $x = 2$ m points toward that plane. Thus, between the planes, the magnitudes of the fields add, whereas to the right of both planes or to the left of both planes, the magnitudes subtract. The electric field at $x = 1.8$ m, which is between the planes, is therefore

$$E_x = E_1 + E_2 = 226\text{ N/C} + 226\text{ N/C} = 452\text{ N/C}$$

(*b*) Since the point $x = 5$ m is to the right of both planes, the magnitudes of the fields subtract, so the net electric field is 0. The electric field lines for this charge distribution are shown in Figure 19-15.

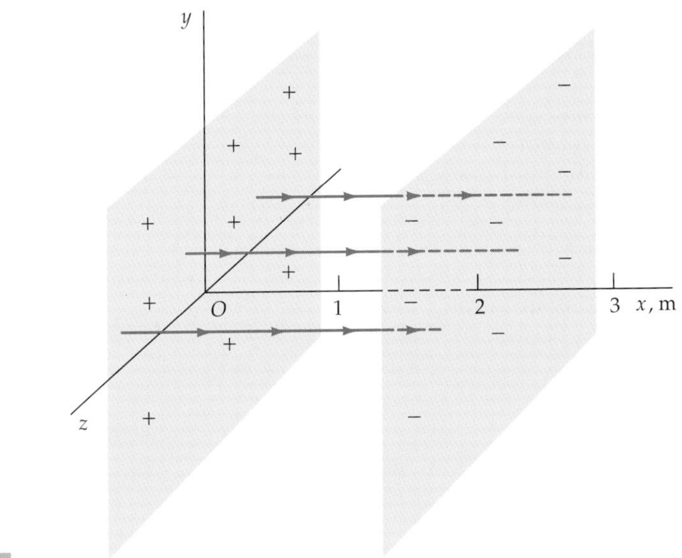

Figure 19-15 An infinite plane of charge with a positive surface charge density in the *yz* plane at the origin and a parallel infinite plane of charge with a negative surface charge density of equal magnitude at $x = 2$ m. The electric field is zero except in the region between the planes. The electric field lines begin on the positively charged plane and end on the negatively charged plane.

E near an Infinite Line Charge

We next consider the electric field at a distance r from a very long line charge of uniform linear charge density λ. In Figure 19-16, a cylindrical surface of length L and radius r has been drawn around the line. By symmetry, at points far from the ends of the line, the electric-field lines radiate out from the line uniformly (if the line charge is positive). The electric field is thus perpendicular to the cylindrical surface and has the same value E_r everywhere on the surface. The electric flux is then just the product of the electric field and the area of the cylindrical surface. There is no flux through the flat surfaces at the ends of the cylinder because $\mathbf{E}\cdot\hat{\mathbf{n}} = 0$ on these surfaces. The net charge inside this surface is the charge per unit length λ times the length L. Gauss's law then gives

$$\phi_{\text{net}} = \oint E_{\text{n}}\, dA = \frac{1}{\epsilon_0} Q_{\text{inside}}$$

$$\oint E_{\text{n}}\, dA = E_r \oint dA = \frac{\lambda L}{\epsilon_0}$$

Since the area of the cylindrical surface is $2\pi rL$, we have

$$E_r 2\pi rL = \frac{\lambda L}{\epsilon_0}$$

or

$$E_r = \frac{1}{2\pi\epsilon_0}\frac{\lambda}{r} = 2k\frac{\lambda}{r} \qquad 19\text{-}24$$

E a distance r from an infinite line charge

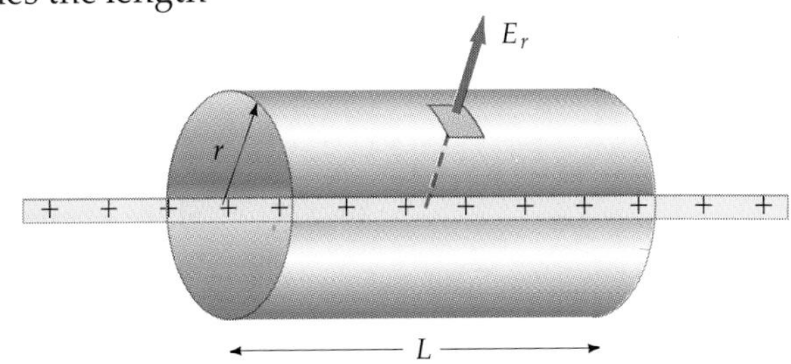

Figure 19-16 A very long, uniform line charge with a cylindrical surface enclosing part of the charge. The flux through the surface is E_r times the surface area $2\pi rL$.

This result is the same as Equation 19-11, which we obtained by direct integration over the line charge.

It is important to realize that a high degree of symmetry is needed to use Gauss's law. In the preceding calculation, we needed to assume that the field point was very far from the ends of the line charge so that E_n would be constant everywhere on the cylindrical gaussian surface. This is equivalent to assuming that, at the distance r from the line, the line charge appears to be infinitely long. If we have a line charge of finite length, we cannot assume that E_n is constant everywhere on our cylindrical surface, so we cannot use Gauss's law to calculate the electric field.

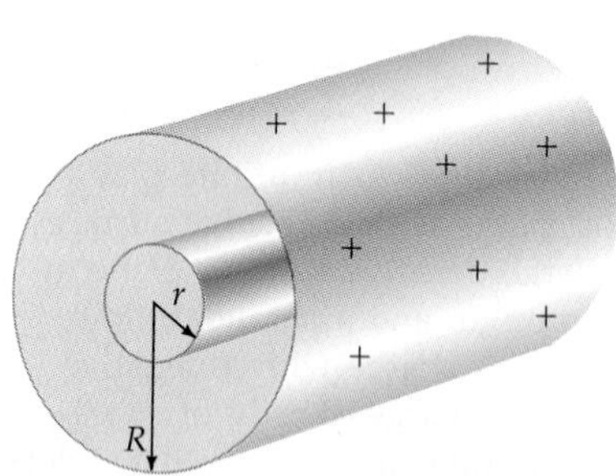

Figure 19-17 A cylindrical shell of radius R carrying a uniform surface charge density σ. To find the electric field inside the shell, we construct a cylindrical gaussian surface concentric with the shell of radius $r < R$ as shown. Since there is no charge inside the gaussian surface, the net flux through this surface is zero.

E Inside and Outside a Cylindrical Shell of Charge

We will now calculate the electric field both inside and outside a cylindrical shell of radius R carrying a uniform surface charge density σ. To calculate the field inside the shell, we construct a cylindrical gaussian surface of length L and radius $r < R$ that is concentric with the shell as shown in Figure 19-17. By symmetry, the electric field is perpendicular to this gaussian surface and its magnitude E_r is constant everywhere on the surface. The flux of **E** through the gaussian surface is then

$$\phi_{\text{net}} = \oint E_n\, dA = E_r \oint dA = E_r 2\pi r L$$

where $2\pi rL$ is the area of the gaussian surface. Since the total charge inside this surface is zero, Gauss's law gives

$$\phi_{\text{net}} = E_r 2\pi r L = 0$$

Therefore,

E inside a cylindrical shell of charge

$$E_r = 0 \qquad r < R \qquad \text{19-25}a$$

Thus, the electric field everywhere inside a cylindrical shell of charge is zero.

To find the electric field outside the cylindrical shell, we construct a cylindrical gaussian surface of radius $r > R$. Again, by symmetry, the electric field is perpendicular to this gaussian surface and its magnitude E_r is constant everywhere on the surface. The flux is again $E_r 2\pi rL$, but this time the total charge inside the surface is $\sigma 2\pi RL$. Gauss's law then gives

$$\phi_{\text{net}} = E_r 2\pi r L = \frac{\sigma 2\pi R L}{\epsilon_0}$$

Therefore,

$$E_r = \frac{\sigma R}{\epsilon_0 r}$$

Since a length L of the cylindrical shell carries a charge of $\sigma 2\pi RL$, the charge per unit length of the shell is $\lambda = \sigma 2\pi R$. Substituting $\lambda/2\pi R$ for σ in the previous equation, we obtain

E outside a cylindrical shell of charge

$$E_r = \frac{\sigma R}{\epsilon_0 r} = \frac{1}{2\pi\epsilon_0}\frac{\lambda}{r} \qquad r > R \qquad \text{19-25}b$$

which is the same as Equation 19-24 for **E** a distance r from an infinite line charge. Thus, the field outside a cylindrical shell of charge is the same as if

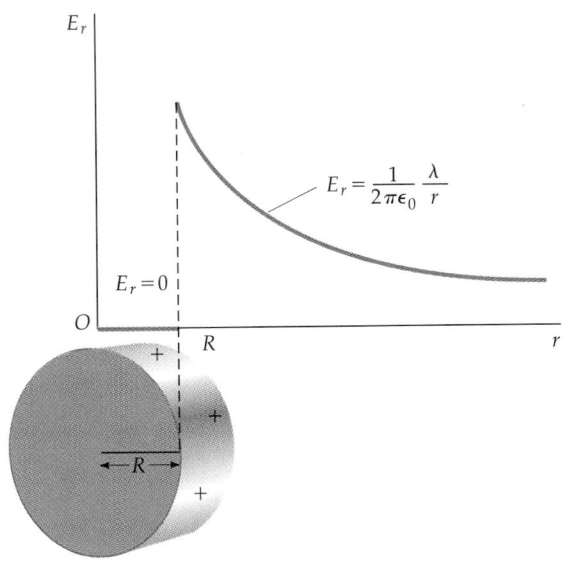

Figure 19-18 A plot of E_r versus r for a cylindrical-shell charge distribution. The electric field is discontinuous at $r = R$, where there is a surface charge of density σ. Just inside the shell the field is zero, whereas just outside the shell it has the magnitude σ/ϵ_0.

all the charge were on the axis of the cylinder. Figure 19-18 shows E_r versus r for this charge distribution. Just outside the shell at $r \approx R$, the electric field is $E_r = \sigma/\epsilon_0$. Since the field just inside the shell is zero, the electric field is discontinuous by the amount σ/ϵ_0 as we pass through the shell. This is the same result that we found for an infinite plane of charge, where the electric field is $-\sigma/2\epsilon_0$ on one side of the plane and $+\sigma/2\epsilon_0$ on the other side. It is a general result, which we will derive at the end of this section.

E Inside and Outside an Infinitely Long Solid Cylinder of Charge

Figure 19-19 shows a solid cylinder of radius R carrying a charge that is uniformly distributed throughout the volume of the cylinder with charge density ρ. As was the case with the cylindrical shell of charge, the flux through a cylindrical gaussian surface of radius r and length L is

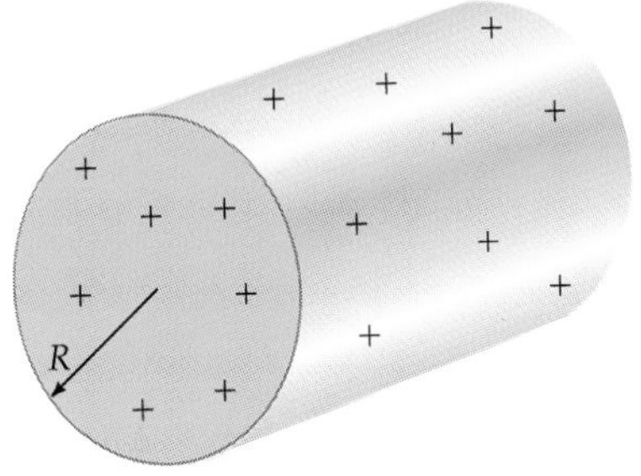

Figure 19-19 A solid cylinder carrying a uniform volume charge density ρ.

$$\phi_{\text{net}} = E_r 2\pi r L$$

If the gaussian surface is outside of the cylinder, that is, if $r > R$, the total charge inside the surface is ρ times the volume of the solid cylinder, which is $\pi R^2 L$. Gauss's law then gives

$$E_r 2\pi r L = \frac{\rho \pi R^2 L}{\epsilon_0}$$

$$E_r = \frac{\rho R^2}{2\epsilon_0 r}$$

Again, we may write this in terms of the charge per unit length along the cylinder, which is $\lambda = (\rho\pi R^2 L)/L = \rho\pi R^2$. Substituting $\lambda/\pi R^2$ for ρ in the previous equation, we obtain

$$E_r = \frac{\rho R^2}{2\epsilon_0 r} = \frac{1}{2\pi\epsilon_0}\frac{\lambda}{r} \qquad r \geq R \qquad \text{19-26}a$$

E outside a solid cylinder of charge

which is the same as Equations 19-24 and 19-25*b*. Thus, the electric field outside a solid cylinder of charge is the same as if all the charge were on the axis of the cylinder.

If the gaussian surface is chosen to be inside the cylinder so that $r < R$, the total charge inside the surface is $\rho V'$, where $V' = \pi r^2 L$ is the volume inside the gaussian surface. Then, for the electric field inside the solid cylinder of charge, Gauss's law gives

$$\phi_{\text{net}} = \frac{1}{\epsilon_0} Q_{\text{inside}}$$

$$E_r 2\pi r L = \frac{1}{\epsilon_0}\rho V' = \frac{1}{\epsilon_0}\rho \pi r^2 L$$

or

E inside a solid cylinder of charge

$$E_r = \frac{\rho}{2\epsilon_0} r = \frac{\lambda}{2\pi\epsilon_0 R^2} r \qquad r \le R \qquad \text{19-26}b$$

Thus, the electric field inside a solid cylinder of charge increases with r. Figure 19-20 shows a plot of E_r versus r for this charge distribution. Note that E_r is continuous at $r = R$.

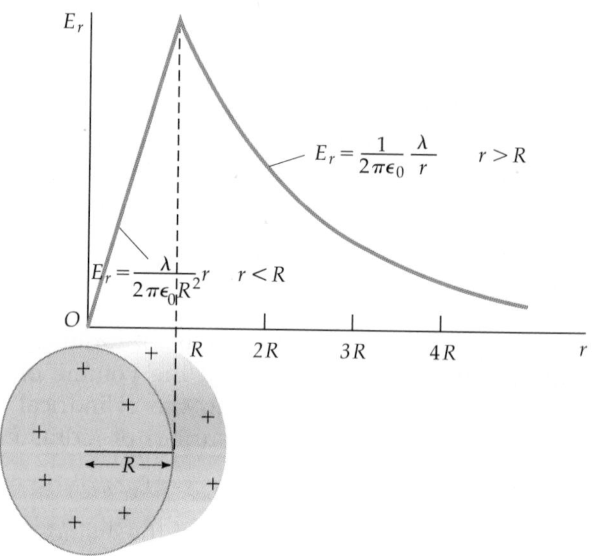

Figure 19-20 Plot of the electric field E_r due to a solid cylinder of charge of radius R versus the distance r from the axis of the cylinder. The field E_r is proportional to r for $0 < r < R$ and decreases as $1/r$ for $r > R$. The field is continuous at $r = R$.

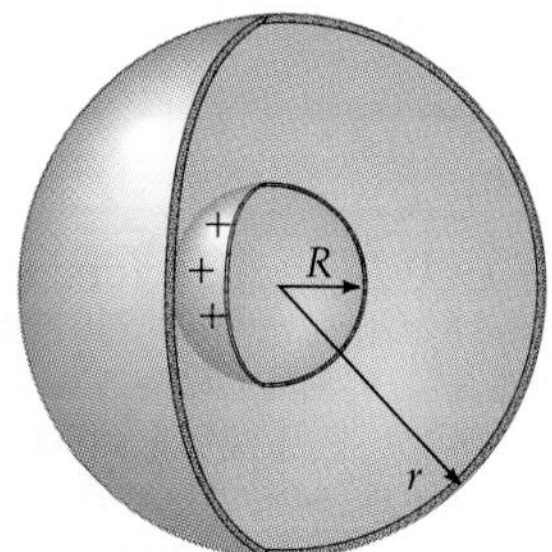

Figure 19-21 Spherical gaussian surface of radius $r > R$ for the calculation of the electric field outside a uniformly charged spherical shell of radius R. The total flux through this surface is $E_r 4\pi r^2$, and the total charge inside the surface is the total charge of the shell Q. The field is the same as if all the charge were at the center of the shell.

E Inside and Outside a Spherical Shell of Charge

We wish to find the electric field inside and outside a uniformly charged spherical shell of radius R and total charge Q. By symmetry, **E** must be radial, and its magnitude can depend only on the distance r from the center of the sphere. In Figure 19-21, we have chosen a spherical gaussian surface of radius $r > R$. Since **E** is perpendicular to this surface and constant in magnitude everywhere on it, the flux through the surface is

$$\phi_{\text{net}} = \oint E_r\, dA = E_r 4\pi r^2$$

Since the total charge inside the gaussian surface is the total charge on the shell, Q, Gauss's law gives

$$E_r 4\pi r^2 = \frac{Q}{\epsilon_0}$$

or

$$E_r = \frac{1}{4\pi\epsilon_0}\frac{Q}{r^2} \qquad r > R \tag{19-27a}$$

E outside a spherical shell of charge

Thus, the electric field outside a uniformly charged spherical shell is the same as if all the charge were at the center of the shell.

If we choose a spherical gaussian surface inside the shell, where $r < R$, the net flux is again $E_r 4\pi r^2$, but the total charge inside the surface is zero. Therefore, for $r < R$, Gauss's law gives

$$\phi_{\text{net}} = E_r 4\pi r^2 = 0$$

and

$$E_r = 0 \qquad r < R \tag{19-27b}$$

E inside a spherical shell of charge

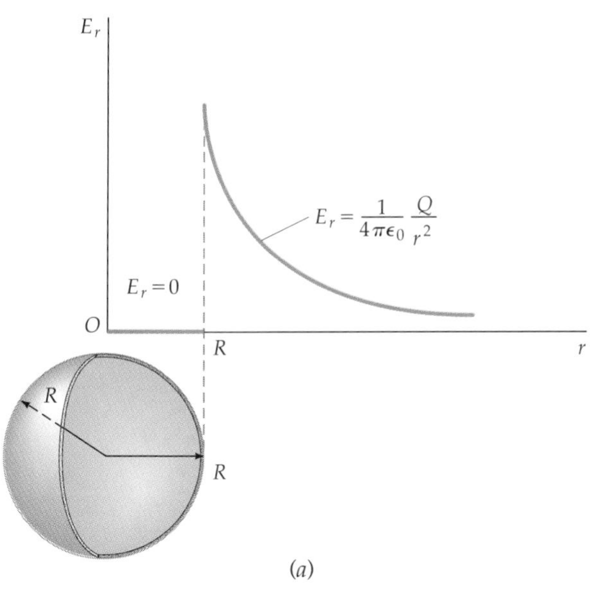

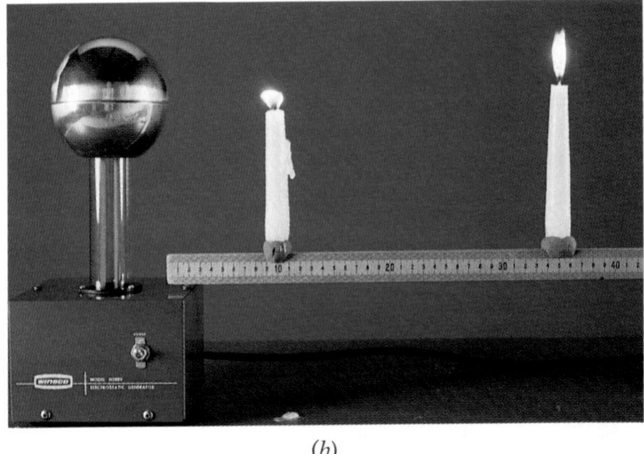

(*b*)

Figure 19-22 (*a*) A plot of E_r versus r for a spherical-shell charge distribution. The electric field is discontinuous at $r = R$, where there is a surface charge of density σ. Just inside the shell the field is zero, whereas just outside the shell it has the magnitude σ/ϵ_0. (*b*) The decrease in E_r over distance due to a charged spherical shell is evident by the effect of the field on the flames of these two candles. The spherical shell on the van de Graaff generator at the left carries a large negative charge that attracts the positive ions in the nearby candle flame. The flame at right, which is much farther away, is not affected.

Note the similarity of these results with those we obtained when we calculated the gravitational field due to a spherical shell of mass in Section 10-7. We could obtain these results by the direct integration of Coulomb's law, but that calculation is much more difficult. Figure 19-22 shows E_r versus r for this charge distribution. Again, note that the electric field is discontinuous at $r = R$, where the surface charge density is σ. Just inside the shell $E_r = 0$, whereas just outside the shell $E_r = Q/4\pi\epsilon_0 R^2 = \sigma/\epsilon_0$ since $\sigma = Q/4\pi R^2$.

Example 19-4

A spherical shell of radius $R = 3$ m has its center at the origin and carries a surface charge density $\sigma = 3\ \text{nC/m}^2$. A point charge $q = 250$ nC is on the y axis at $y = 2$ m. Find the electric field on the x axis at (*a*) $x = 2$ m and (*b*) $x = 4$ m.

(*a*) The point on the x axis at $x = 2$ m is inside the spherical shell, so the field due to the shell is zero. The electric field at this point is due only to the point charge, which is a distance $r_1 = \sqrt{(2\text{ m})^2 + (2\text{ m})^2} = \sqrt{8}$ m away. The field makes an angle of $-45°$ with the x axis and has a magnitude of

$$E = \frac{kq}{r_1^2} = \frac{(8.99 \times 10^9\text{ N·m}^2/\text{C}^2)(250 \times 10^{-9}\text{ C})}{(\sqrt{8}\text{ m})^2} = 281\text{ N/C}$$

(*b*) At the point $x = 4$ m, which is outside the spherical shell, the shell can be treated as a point charge at the origin with a magnitude $Q = \sigma 4\pi R^2 = (3\text{ nC/m}^2)4\pi(3\text{ m})^2 = 339$ nC. The electric field due to the shell at $x = 4$ m is in the x direction and has the magnitude

$$E_1 = E_{1x} = \frac{(8.99 \times 10^9\text{ N·m}^2/\text{C}^2)(339 \times 10^{-9}\text{ C})}{(4\text{ m})^2} = 190\text{ N/C}$$

The distance from the point charge q on the y axis to the field point at $x = 4$ m is $r_2 = \sqrt{(2\text{ m})^2 + (4\text{ m})^2} = \sqrt{20}$ m. The magnitude of the electric field at $x = 4$ m due to q is

$$E_2 = \frac{(8.99 \times 10^9\text{ N·m}^2/\text{C}^2)(250 \times 10^{-9}\text{ C})}{(\sqrt{20}\text{ m})^2} = 112\text{ N/C}$$

This field makes an angle θ with the x axis, where $\cos\theta = 4/\sqrt{20}$ and $\sin\theta = -2/\sqrt{20}$. The x and y components of this field are thus

$$E_{2x} = E_2\cos\theta = (112\text{ N/C})\left(\frac{4}{\sqrt{20}}\right) = 100\text{ N/C}$$

and

$$E_{2y} = E_2\sin\theta = (112\text{ N/C})\left(-\frac{2}{\sqrt{20}}\right) = -50\text{ N/C}$$

The x and y components of the net electric field are

$$E_x = E_{1x} + E_{2x} = 190\text{ N/C} + 100\text{ N/C} = 290\text{ N/C}$$

and

$$E_y = E_{1y} + E_{2y} = 0 - 50\text{ N/C} = -50\text{ N/C}$$

The magnitude and direction of the net field can be found from $E = \sqrt{E_x^2 + E_y^2}$, and $\tan\theta' = E_y/E_x$.

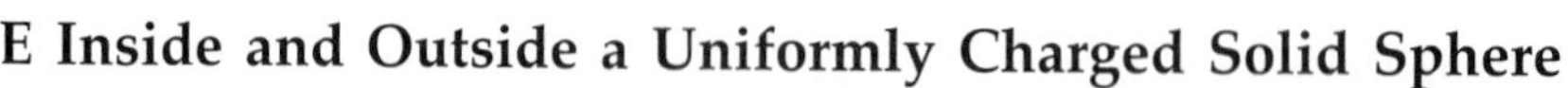

E Inside and Outside a Uniformly Charged Solid Sphere

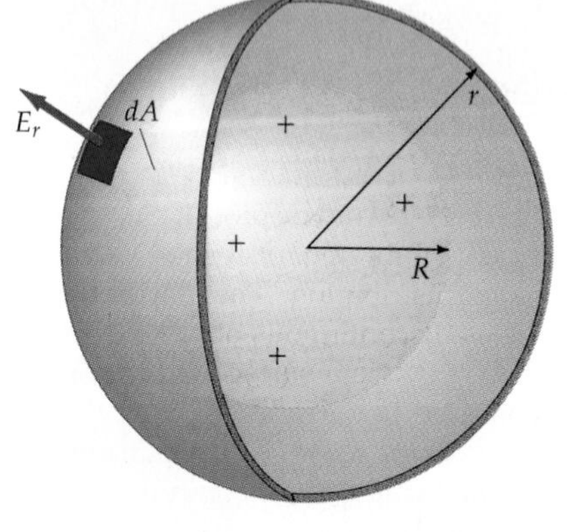

Figure 19-23 Spherical gaussian surface for the calculation of the electric field outside a uniformly charged solid sphere. The total flux through this surface is $E_r 4\pi r^2$, and the total charge inside the surface is the total charge of the sphere Q. The field is the same as if all the charge were at the center of the sphere.

We will now calculate the electric field inside and outside a uniformly charged solid sphere of radius R carrying a total charge Q that is uniformly distributed throughout the volume of the sphere with charge density $\rho = Q/V$, where $V = \frac{4}{3}\pi R^3$ is the volume of the sphere. As was the case with the spherical shell of charge, the flux through a gaussian surface of radius r is

$$\phi_{\text{net}} = E_r 4\pi r^2$$

If the gaussian surface is outside of the sphere, as in Figure 19-23, the total charge inside the surface is Q, and Gauss's law gives

E outside a solid sphere of charge

$$E_r = \frac{1}{4\pi\epsilon_0}\frac{Q}{r^2} \qquad r \geq R \tag{19-28a}$$

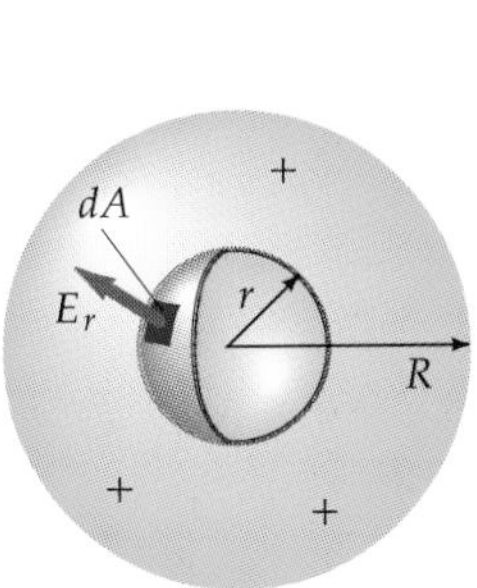

Figure 19-24 Spherical gaussian surface for the calculation of the electric field inside a uniformly charged solid sphere. The flux through the surface is again $E_r 4\pi r^2$. The total charge inside the gaussian surface is $Q(r^3/R^3)$.

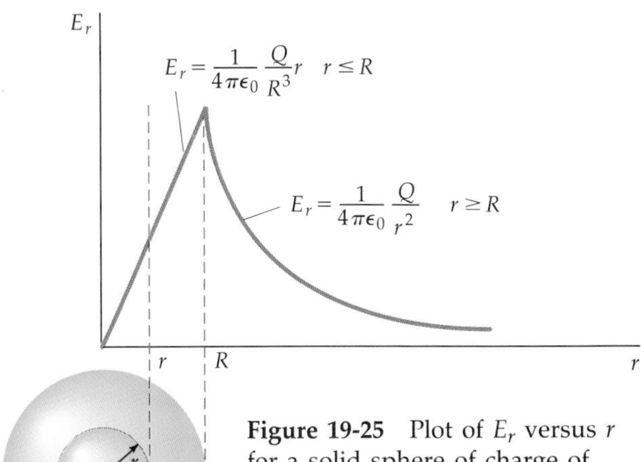

Figure 19-25 Plot of E_r versus r for a solid sphere of charge of radius R. For $r < R$, the field increases linearly with r. Outside the sphere, the electric field is the same as that due to a point charge. The field is continuous at $r = R$.

If the gaussian surface is chosen to be inside the sphere, which means that $r < R$ (Figure 19-24), the total charge inside the surface is $\rho V'$, where $V' = \frac{4}{3}\pi r^3$ is the volume inside the gaussian surface:

$$Q_{\text{inside}} = \rho V' = \frac{Q}{V}V' = \left(\frac{Q}{\frac{4}{3}\pi R^3}\right)\left(\frac{4}{3}\pi r^3\right) = Q\frac{r^3}{R^3}$$

For the electric field inside the sphere, Gauss's law then gives

$$E_r 4\pi r^2 = \frac{1}{\epsilon_0}Q\frac{r^3}{R^3}$$

or

$$E_r = \frac{1}{4\pi\epsilon_0}\frac{Q}{R^3}r \qquad r \le R \tag{19-28b}$$

E inside a solid sphere of charge

Thus, the electric field inside a solid sphere of charge increases with r. Figure 19-25 shows a plot of E_r versus r for this charge distribution. Note that E_r is continuous at $r = R$. This function is sometimes used to describe the electric field of an atomic nucleus, which can be considered to be approximately a uniform solid sphere of charge.

Discontinuity of E_n

We have seen that the electric field for an infinite plane of charge, a cylindrical shell of charge, and a spherical shell of charge is discontinuous by the amount σ/ϵ_0 at a point where there is a surface charge density σ. We will now show that this is a general result for the component of the electric field that is perpendicular to a surface carrying a charge density of σ. Figure 19-26 shows a small pillbox-shaped gaussian surface with faces of area A on each side of an arbitrary surface carrying a surface charge density σ. Let the normal component of the electric field be E_{n2} on one side of the surface and E_{n1} on the other side, as shown in the figure. If we make the length of the pillbox very small compared with the radius of the faces, we can neglect the flux through the cylindrical area $2\pi RL$ compared with the flux through the

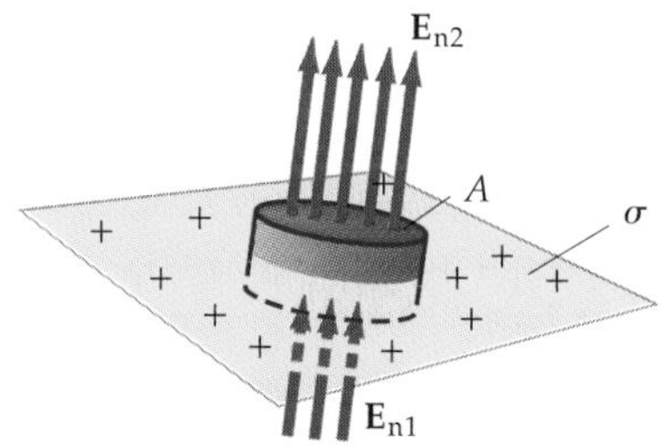

Figure 19-26 A gaussian, pillbox-shaped surface with faces of area A on each side of a surface carrying a charge density of σ. The net flux through the pillbox is $(E_{n2} - E_{n1})A$. The electric field E_{n2} on one side is greater than the electric field E_{n1} on the other side by the amount σ/ϵ_0.

faces of area πR^2. The net flux through the gaussian surface is then $E_{n2}A - E_{n1}A$, and the charge inside the surface is σA. Gauss's law gives

$$E_{n2}A - E_{n1}A = \frac{\sigma A}{\epsilon_0}$$

or

Discontinuity of E_n

$$E_{n2} - E_{n1} = \frac{\sigma}{\epsilon_0} \qquad 19\text{-}29$$

which is the result we wished to prove. Note that the electric field is *not* discontinuous at points of discontinuity of a volume charge density. The electric field just inside a solid cylinder of charge or a solid sphere of charge is the same as it is just outside such a charge distribution, as can be seen from Figures 19-20 and 19-25.

Questions

5. What information in addition to the total charge inside a surface is needed to use Gauss's law to find the electric field?
6. Explain why the electric field increases with r rather than decreasing as $1/r^2$ as one moves out from the center inside a spherical charge distribution of constant volume charge density.
7. Equation 19-10 for the electric field on the perpendicular bisector of a finite line charge is different from Equation 19-11 or 19-24 for the electric field near an infinite line charge, yet Gauss's law would seemingly give the same result for these two cases. Explain.

19-4 Charge and Field at Conductor Surfaces

The property of a conductor that is important in studying electrostatic fields is the availability of charge that is free to move about inside the conductor. In the presence of an external electric field, the free charge in a conductor moves about the conductor until it is so distributed that it creates an electric field that cancels the external field inside the conductor.* The conductor is then said to be in **electrostatic equilibrium.** Consider a charge q inside a conductor. If there is a field $\mathbf{E}$ inside the conductor, there will be a force $q\mathbf{E}$ on this charge. Therefore, if the charge is free to move—that is, if it is not bound to an atom or molecule by a stronger force—it will accelerate. Thus electrostatic equilibrium is impossible in a conductor unless the electric field is zero everywhere inside the conductor. At the surface of a conductor in equilibrium, the electric field must be perpendicular to the surface. If there were a tangential component of $\mathbf{E}$, the free charge in the conductor would move until this component became zero.

Figure 19-27 shows a conducting slab in an external electric field $\mathbf{E}_0$. The free electrons are originally distributed uniformly throughout the slab. Since the slab is made up of neutral atoms, it is electrically neutral (provided that no extra charge has been placed on it). If the external electric field is to the

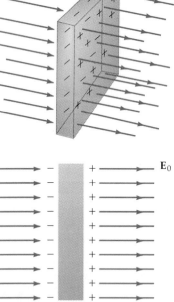

Figure 19-27 Two views of a conducting slab in an external electric field $\mathbf{E}_0$. A positive charge is induced on the right face and a negative charge is induced on the left face such that the net electric field inside the conductor is zero. The electric-field lines then end on the left face and begin again on the right face.

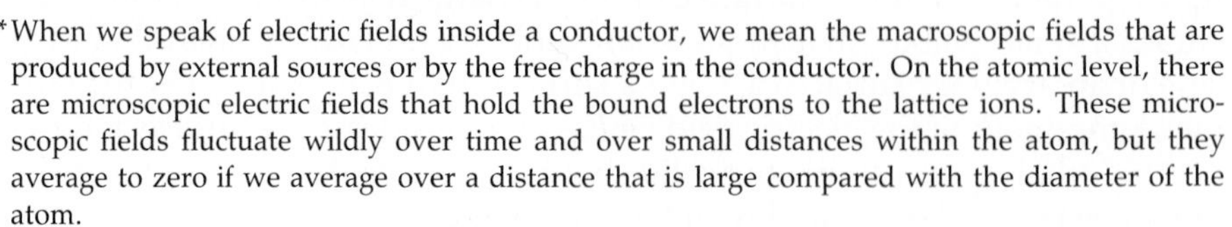

*When we speak of electric fields inside a conductor, we mean the macroscopic fields that are produced by external sources or by the free charge in the conductor. On the atomic level, there are microscopic electric fields that hold the bound electrons to the lattice ions. These microscopic fields fluctuate wildly over time and over small distances within the atom, but they average to zero if we average over a distance that is large compared with the diameter of the atom.

right, there will be a force on each electron $\mathbf{F} = -e\mathbf{E}_0$ to the left because the electron has a negative charge, so the free electrons accordingly accelerate to the left. At the surface of the conductor, the conductor exerts forces on these electrons that keep them bound to the conductor. (If the external field is very strong, the electrons can be stripped off the surface of the conductor. In electronics this is called **field emission.** We assume here that the external field is not strong enough to overcome the forces binding the electrons to the surface.) The result is an induced negative surface charge density on the left side of the slab and an induced positive surface charge density on the right side of equal magnitude because of the movement of some of the free electrons from the right side to the left side of the slab. Together, these induced charge densities produce an electric field inside the slab that is opposite to the external field. When the induced and external fields cancel, electrostatic equilibrium is established everywhere inside the conductor so that there is no unbalanced force on the free electrons.

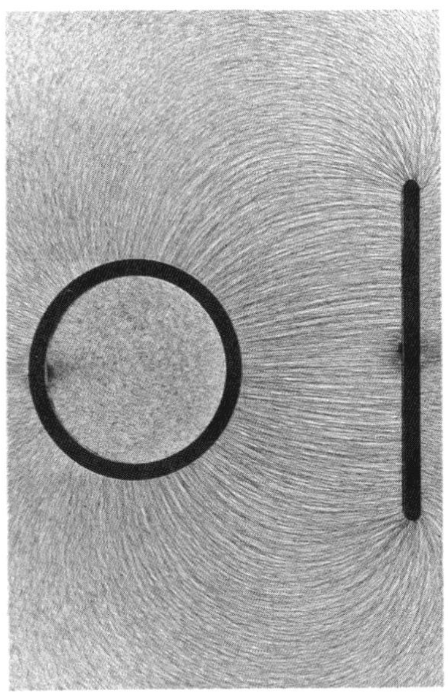

Lines of force for an oppositely charged cylinder and plate, shown by bits of fine thread suspended in oil. Note that the field lines are perpendicular to the conductors and that there are no lines inside the cylinder.

The behavior of the free charge in a conductor placed in an external electric field is the same no matter what the shape of the conductor may be. When an external field is applied, the free charge quickly moves until an equilibrium distribution is achieved such that the net electric field is zero everywhere inside the conductor. The time it takes to reach equilibrium depends on the conductor. For copper and other good conductors, it is so small that, for all practical purposes, electrostatic equilibrium is reached instantaneously.

In this section, we will use Gauss's law to show that the following results occur for conductors in electrostatic equilibrium:

1. Any net electric charge on a conductor resides on the surface of the conductor.
2. The electric field just outside the surface of a conductor is perpendicular to the surface and has the magnitude σ/ϵ_0, where σ is the surface charge density at that point on the conductor.

To obtain the first result, we consider a gaussian surface just inside the actual surface of a conductor in electrostatic equilibrium, as shown in Figure 19-28. Since the electric field is zero everywhere inside the conductor, it is zero everywhere on the gaussian surface, which is chosen to be completely within the conductor. Since $E_n = 0$ at all points on the gaussian surface, the net flux through the surface must be zero. By Gauss's law, the flux equals $1/\epsilon_0$ times the net charge inside a surface. Since the net flux is zero, there can be no net charge inside any surface lying completely within the conductor. If there is any net charge on the conductor, it must be on the conductor's surface.

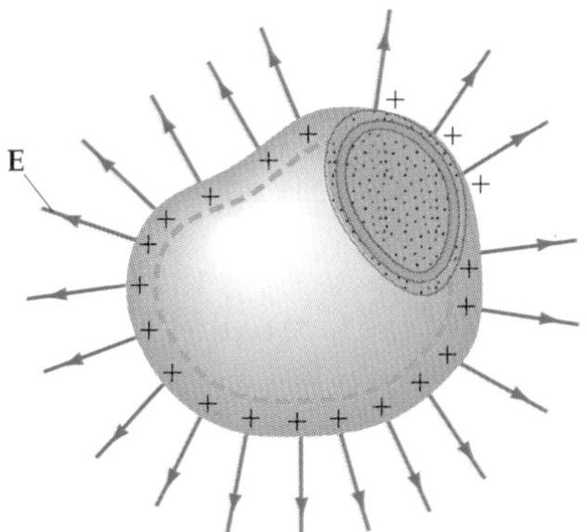

Figure 19-28 A gaussian surface (the dashed line) just inside the surface of a conductor. Since the electric field is zero inside a conductor in electrostatic equilibrium, the net flux through this surface must also be zero. Therefore, the net charge inside this surface must be zero. If there is any charge on the conductor, it must reside on the surface of the conductor. Here, the conductor carries a net positive charge.

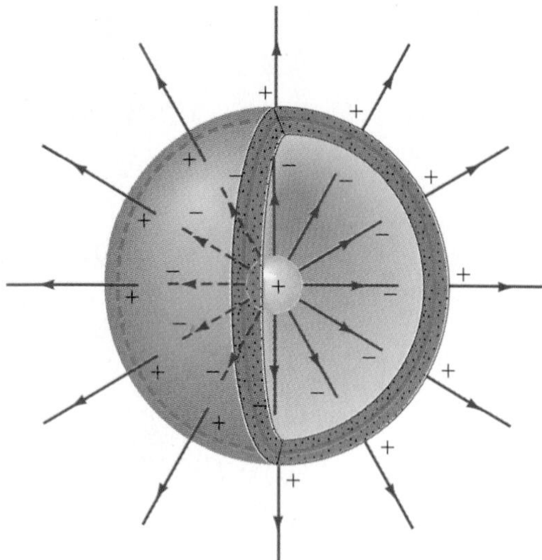

Figure 19-29 A thick spherical conducting shell with a point charge q in the cavity at the center. Since $\mathbf{E} = 0$ inside the conductor, there can be no net flux through any surface within the conductor, such as the gaussian surface indicated in blue. A surface charge $-q$ is induced on the inner surface of the shell, and the electric-field lines that begin on the point charge end on that surface. Since the conductor is neutral, an equal but opposite charge $+q$ is induced on the outer surface of the shell. The electric-field lines begin on the point charge, end on the inner surface, and begin again on the outer surface.

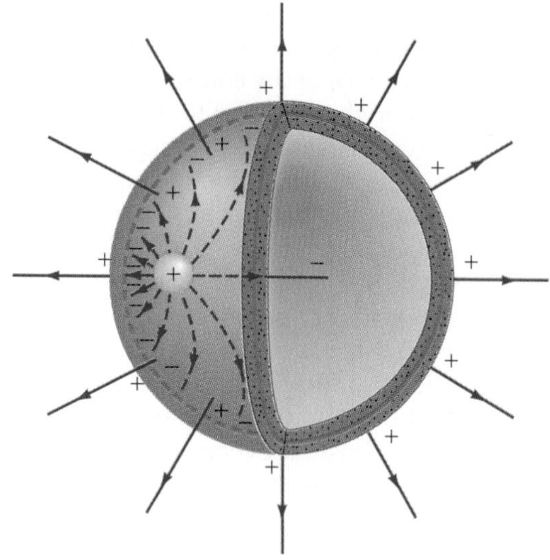

Figure 19-30 The same conductor as in Figure 19-29 with the point charge not at the center of the sphere. The charge on the outer surface and the electric-field lines outside the sphere are not affected.

Figure 19-29 shows a spherical conductor with a spherical cavity inside. At the center of the cavity is a positive point charge q. If we draw a gaussian surface within the conductor, **E** is zero everywhere on the surface, so Gauss's law tells us that there can be no net charge within the surface. A negative charge $-q$ must therefore be induced on the inside surface of the conductor. All the electric-field lines from the point charge must end on the negative charge induced on the inside surface. If the conductor has no net charge, there will be an equal induced charge $+q$ on the outside surface of the conductor. In Figure 19-30, the point charge has been moved so that it is no longer at the center of the cavity. The field lines in the cavity are altered, and the surface charge density of the induced negative charge on the inner surface is no longer uniform. However, all of the field lines still end on the induced negative charge. The positive surface charge density on the outside surface is not disturbed because it is shielded from the cavity by the conductor.

To find the electric field just outside the surface of a conductor, we consider a portion of the conductor's surface small enough to be considered flat with a surface charge density σ that is uniform over the area. We construct a pillbox-shaped gaussian surface (Figure 19-31) with one face of the cylinder just outside the conductor and parallel to its surface and the other face just inside the conductor. The electric field at the surface of the conductor in equilibrium must be perpendicular to the surface. Thus, we can take **E** to be perpendicular to the face of the pillbox. The other face of the pillbox is inside the conductor, where **E** is zero. There is no flux through the cylindrical surface of the pillbox because **E** is parallel to that surface. The flux through the pillbox is therefore $E_n A$, where E_n is the field just outside the conductor surface, and A is the area of the face of the pillbox. The net charge inside the gaussian surface is σA. Gauss's law gives

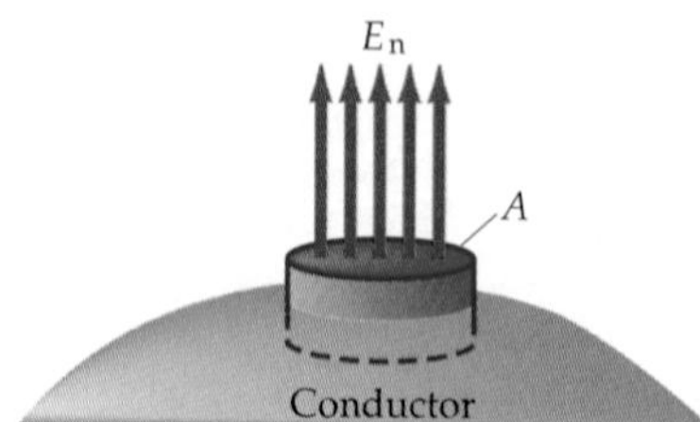

Figure 19-31 Pillbox-shaped gaussian surface for the calculation of the electric field at the surface of a conductor. There is no flux through the inside face of the pillbox because the electric field inside the conductor is zero. The flux through the outside face of the pillbox is $E_n A$.

$$\phi_{\text{net}} = \oint E_n \, dA = \frac{1}{\epsilon_0} Q_{\text{inside}}$$

$$\oint E_n \, dA = E_n \oint dA = E_n A = \frac{1}{\epsilon_0} \sigma A$$

or

$$E_n = \frac{\sigma}{\epsilon_0} \qquad \text{19-30}$$

E_n just outside conductor surface

This result is exactly twice the field produced by an infinite plane of charge. We can understand this result from Figure 19-32, which shows a large charged conducting slab carrying a surface charge density σ on each face. Near the conducting slab, each of its faces can be considered to be an infinite plane of charge. Let E_L be the electric field due to the left face and E_R be that due to the right face. Each of these fields has the magnitude $\sigma/2\epsilon_0$. Inside the conductor, these fields are oppositely directed, so they cancel. Outside the conductor, the fields add to give a total electric field of magnitude $E = E_L + E_R = \sigma/2\epsilon_0 + \sigma/2\epsilon_0 = \sigma/\epsilon_0$. Similar but slightly more complicated reasoning can be applied to a conductor of arbitrary shape, as shown in Figure 19-33. We can consider the charge on the surface of the conductor to consist of two parts: (1) the charge in the immediate neighborhood of point P and (2) all the rest of the charge, which we shall call the distant charge. Since point P is just outside the surface, the charge in its immediate neighborhood looks like an infinite plane of charge. It produces a field of magnitude $\sigma/2\epsilon_0$ at P and a field of equal magnitude just inside the conducting surface pointing away from the surface. The rest of the charge on the conductor (or elsewhere) must produce a field $\sigma/2\epsilon_0$ inside the conductor pointing toward the surface, so that the net field inside the conductor is zero. The field due to the distant charge has the same magnitude and direction at points just inside and just outside the surface. Just inside the conductor surface, the field due to the distant charge cancels that due to the neighboring charge; but outside, the fields are in the same direction and add, giving a net field $\sigma/2\epsilon_0 + \sigma/2\epsilon_0 = \sigma/\epsilon_0$ just outside the conductor.

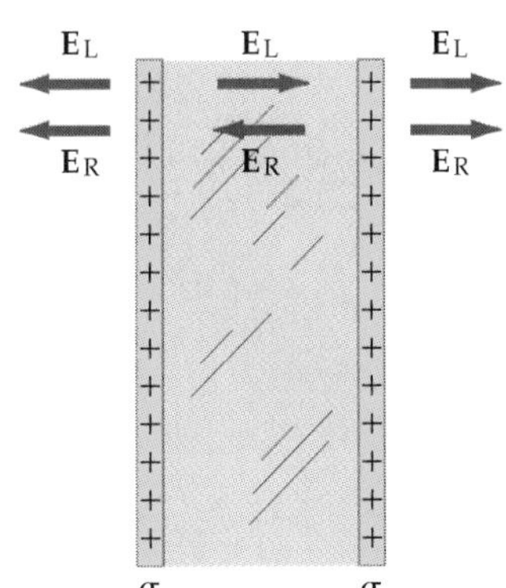

Figure 19-32 A conducting slab carrying a uniform surface charge density σ. The electric field due to the surface charge density on the left face is $\mathbf{E}_L$ and that due to the charge on the right face is $\mathbf{E}_R$. Each of these fields has a magnitude of $\sigma/2\epsilon_0$. Inside the conductor, these fields are oppositely directed and cancel. Outside the conductor, these fields are in the same direction and add to produce a field of magnitude σ/ϵ_0.

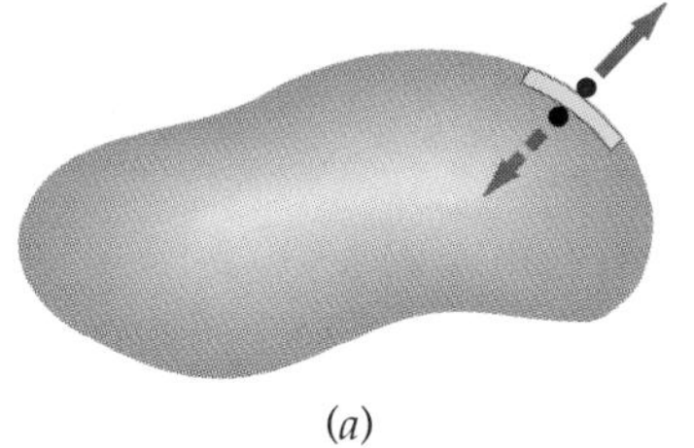

(*a*)

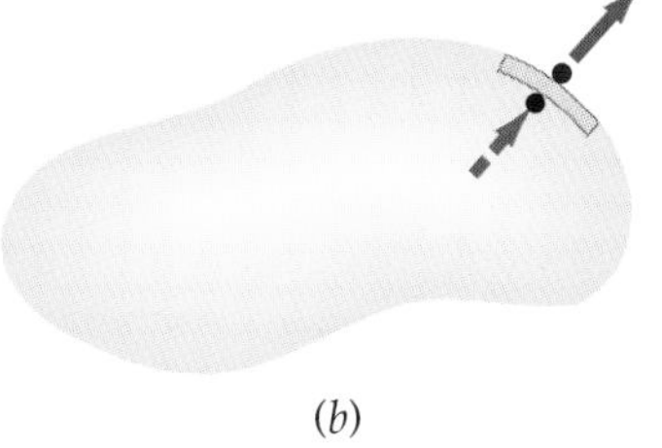

(*b*)

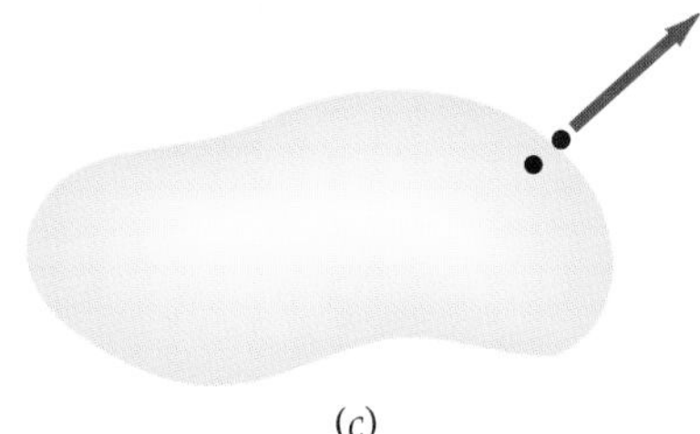

(*c*)

Figure 19-33 An arbitrarily shaped conductor carrying a charge on its surface. (*a*) The charge in the vicinity of point P near the surface looks like an infinite plane of charge, giving an electric field of magnitude $\sigma/2\epsilon_0$ pointing away from the surface both inside and outside the surface. (*b*) Since the net field inside the surface must be zero, the rest of the charge must produce a field of equal magnitude. (*c*) Inside the surface the fields shown in parts (*a*) and (*b*) cancel, but outside at point P they add to give $E_n = \sigma/\epsilon_0$.

Example 19-5

A square conducting slab of negligible thickness and with 4-m sides is placed in an external uniform field $\mathbf{E} = (450\ \text{kN/C})\mathbf{i}$ that is perpendicular to the faces of the slab. (*a*) Find the charge density on each face of the slab. (*b*) A net charge of 96 μC is placed on the slab. Find the new charge density on each face and the electric field near each face but far from the edges of the slab.

(*a*) Just outside the right face of the slab, the electric field is $E_n =$ 450 kN/C, so the charge density on this face is

$$\sigma_R = \epsilon_0 E_n = (8.85 \times 10^{-12}\ \text{C}^2/\text{N}\cdot\text{m}^2)(450\ \text{kN/C})$$
$$= 3.98 \times 10^{-6}\ \text{C/m}^2$$
$$= 3.98\ \mu\text{C/m}^2$$

At the left face, the electric field points toward the slab, so $E_n = -450$ kN/C. The charge density on this face is

$$\sigma_L = \epsilon_0 E_n$$
$$= (8.85 \times 10^{-12}\ \text{C}^2/\text{N}\cdot\text{m}^2)(-450\ \text{kN/C})$$
$$= -3.98\ \mu\text{C/m}^2$$

(*b*) The total charge of 96 μC must be distributed equally on each face of the slab so that the electric field inside the slab remains zero. Since each face has an area of 16 m² and carries a net charge of (96 μC)/2 = 48 μC, the additional charge density on each face is $\sigma_a = 48\ \mu\text{C}/16\ \text{m}^2 = 3.0\ \mu\text{C/m}^2$. The net charge density on each face will therefore be

$$\sigma_R = 3.98\ \mu\text{C/m}^2 + 3.0\ \mu\text{C/m}^2 = 6.98\ \mu\text{C/m}^2$$

and

$$\sigma_L = -3.98\ \mu\text{C/m}^2 + 3.0\ \mu\text{C/m}^2 = -0.98\ \mu\text{C/m}^2$$

The net electric field just to the right of the slab is

$$E_{nR} = \sigma_R/\epsilon_0$$
$$= (6.98\ \mu\text{C/m}^2)/(8.85 \times 10^{-12}\ \text{C}^2/\text{N}\cdot\text{m}^2)$$
$$= 789\ \text{kN/C}$$

Since the normal to the slab is in the positive x direction at the right face, the electric field just to the right of the slab is

$$\mathbf{E}_R = 789\ \text{kN/C}\ \mathbf{i}$$

Just to the left of the slab the electric field is

$$E_{nL} = \sigma_R/\epsilon_0$$
$$= (-0.98\ \mu\text{C/m}^2)/(8.85 \times 10^{-12}\ \text{C}^2/\text{N}\cdot\text{m}^2)$$
$$= -111\ \text{kN/m}^2$$

Since the normal to the slab points in the negative x direction on the left face, the electric field at this face is to the right:

$$\mathbf{E}_L = 111\ \text{kN/C}\ \mathbf{i}$$

We can look at part (*b*) in another way. The positive charge density of 3.0 μC/m² added to each face is equivalent to adding two planes of positive charge. Outside the slab these planes produce an electric field of magnitude

$$E = \tfrac{1}{2}\sigma/\epsilon_0 + \tfrac{1}{2}\sigma/\epsilon_0 = \sigma/\epsilon_0$$
$$= (3.0\ \mu\text{C/m}^2)/(8.85 \times 10^{-12}\ \text{C}^2/\text{N}\cdot\text{m}^2)$$
$$= 339\ \text{kN/C}.$$

To the right of the slab, this field adds to the original field giving

$$E_R = 450\ \text{kN/C} + 339\ \text{kN/C} = 789\ \text{kN/C}$$

To the left of the slab, this field subtracts from the original field giving

$$E_L = 450\ \text{kN/C} - 339\ \text{kN/C} = 111\ \text{kN/C}$$

Exercise

The electric field just outside the surface of a conductor points away from the conductor and has the magnitude of 2000 N/C. What is the surface charge density on the surface of the conductor?
(Answer: 17.7 nC/m²)

19-5 Mathematical Derivation of Gauss's Law (Optional)

Gauss's law can be derived mathematically using the concept of the **solid angle.** Consider an area element ΔA on a spherical surface. The solid angle $\Delta\Omega$ subtended by ΔA at the center of the sphere is defined to be

$$\Delta\Omega = \frac{\Delta A}{r^2}$$

where r is the radius of the sphere. Since ΔA and r^2 both have dimensions of length squared, the solid angle is dimensionless. The unit of solid angle is the **steradian** (sr). Since the total area of a sphere is $4\pi r^2$, the total solid angle subtended by a sphere is

$$\frac{4\pi r^2}{r^2} = 4\pi \text{ steradians}$$

There is a close analogy between the solid angle and the ordinary plane angle, which is defined to be the ratio of an element of arc length of a circle Δs to the radius of the circle:

$$\Delta\theta = \frac{\Delta s}{r} \text{ radians}$$

The total plane angle subtended by a circle is 2π radians.

In Figure 19-34, the area element ΔA is not perpendicular to the radial lines from point O. The unit vector $\hat{\mathbf{n}}$ normal to the area element makes an angle θ with the radial unit vector $\hat{\mathbf{r}}$. In this case, the solid angle subtended by ΔA at point O is defined to be

$$\Delta\Omega = \frac{\Delta A\ \hat{\mathbf{n}}\cdot\hat{\mathbf{r}}}{r^2} = \frac{\Delta A \cos\theta}{r^2} \qquad 19\text{-}31$$

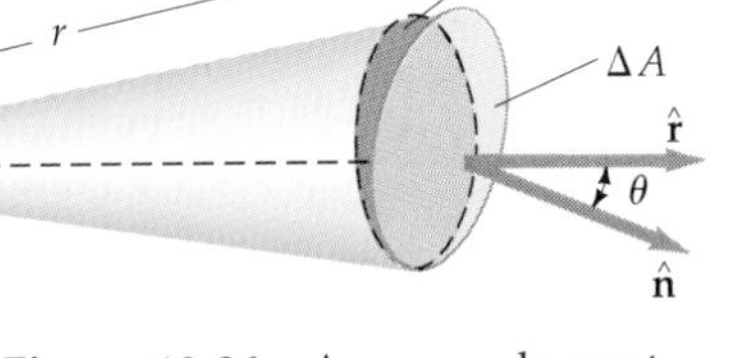

Figure 19-34 An area element ΔA, whose normal is not parallel to the radial line from O to the center of the element. The solid angle subtended by this element at O is defined to be $(\Delta A \cos\theta)/r^2$.

Figure 19-35 shows a point charge q surrounded by a surface S of arbitrary shape. To calculate the flux through this surface, we want to find $\mathbf{E}\cdot\hat{\mathbf{n}}\ \Delta A$ for each element of area on the surface and sum over the entire surface. The flux through the area element shown is

$$\Delta\phi = \mathbf{E}\cdot\hat{\mathbf{n}}\ \Delta A = \frac{kq}{r^2}\ \hat{\mathbf{r}}\cdot\hat{\mathbf{n}}\ \Delta A = kq\ \Delta\Omega$$

The solid angle $\Delta\Omega$ is the same as that subtended by the corresponding area element of a spherical surface of any radius. The sum of the flux through the entire surface is kq times the total solid angle subtended by the closed surface, which is 4π steradians:

$$\phi_{\text{net}} = \oint \mathbf{E}\cdot\hat{\mathbf{n}}\ dA = kq \oint d\Omega = 4\pi kq = \frac{q}{\epsilon_0}$$

which is Gauss's law.

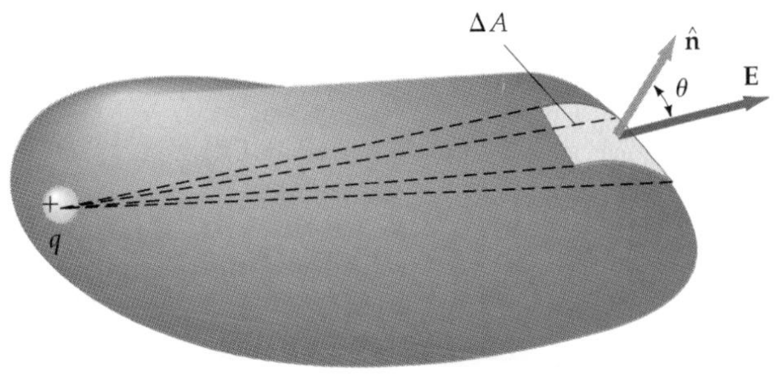

Figure 19-35 A point charge q enclosed by an arbitrary surface S. The flux through an area element ΔA is proportional to the solid angle subtended by the area element at the charge. The net flux through the surface found by summing over all the area elements is proportional to the total solid angle 4π at the charge, which is independent of the shape of the surface.

Summary

1. The electric field for continuous charge distributions can be calculated directly from Coulomb's law in the form

$$\mathbf{E} = \int_V \frac{k\,dq}{r^2}\,\hat{\mathbf{r}}$$

where $dq = \rho\,dV$ for a charge distributed throughout a volume, $dq = \sigma\,dA$ for a charge distributed on a surface, and $dq = \lambda\,dL$ for a charge distributed along a line.

2. Electric flux ϕ of a constant electric field through an area A is the product of the component of the electric field perpendicular to the area times the area:

$$\phi = \mathbf{E}\cdot\hat{\mathbf{n}}A = EA\cos\theta = E_{\mathrm{n}}A$$

For a general electric field, which may vary in space, the flux through an element of area dA is

$$d\phi = \mathbf{E}\cdot\hat{\mathbf{n}}\,dA = E\cos\theta\,dA = E_{\mathrm{n}}\,dA$$

3. The net flux through a closed surface equals $4\pi k$ times the net charge within the surface, a result known as Gauss's law:

$$\phi_{\mathrm{net}} = \oint_S E_{\mathrm{n}}\,dA = 4\pi k Q_{\mathrm{inside}}$$

Gauss's law can be used to find the electric field for highly symmetric charge distributions.

4. The Coulomb constant k is often written in terms of the permittivity of free space ϵ_0:

$$k = \frac{1}{4\pi\epsilon_0}$$

In terms of this constant, Coulomb's law is written

$$\mathbf{F}_{12} = \frac{1}{4\pi\epsilon_0}\frac{q_1q_2}{r_{12}^2}\,\hat{\mathbf{r}}_{12}$$

and Gauss's law is written

$$\phi_{\mathrm{net}} = \oint_S E_{\mathrm{n}}\,dA = \frac{1}{\epsilon_0}Q_{\mathrm{inside}}$$

5. The electric field for various charge distributions are as follows:

$E_r = \dfrac{2k\lambda}{r}\sin\theta_0$	on bisector of a finite line charge
$E_r = \dfrac{1}{2\pi\epsilon_0}\dfrac{\lambda}{r} = 2k\dfrac{\lambda}{r}$	near an infinite line charge
$E_x = \dfrac{kQx}{(x^2+a^2)^{3/2}}$	on axis of a ring charge
$E_x = 2\pi k\sigma\left(1 - \dfrac{x}{\sqrt{x^2+R^2}}\right)$	on axis of a disk charge
$E_{\mathrm{n}} = \dfrac{\sigma}{2\epsilon_0} = 2\pi k\sigma$	near an infinite plane of charge
$E_r = 0 \quad r < R$	inside a cylindrical shell of charge

$$E_r = \frac{\sigma R}{\epsilon_0 r} = \frac{1}{2\pi\epsilon_0}\frac{\lambda}{r} \qquad r > R \qquad \text{outside a cylindrical shell of charge}$$

$$E_r = \frac{\rho R^2}{2\epsilon_0 r} = \frac{1}{2\pi\epsilon_0}\frac{\lambda}{r} \qquad r \geq R \qquad \text{outside a solid cylinder of charge}$$

$$E_r = \frac{\rho}{2\epsilon_0} r = \frac{\lambda}{2\pi\epsilon_0 R^2} r \qquad r \leq R \qquad \text{inside a solid cylinder of charge}$$

$$E_r = \frac{1}{4\pi\epsilon_0}\frac{Q}{r^2} \qquad r > R \qquad \text{outside a spherical shell of charge}$$

$$E_r = 0 \qquad r < R \qquad \text{inside a spherical shell of charge}$$

$$E_r = \frac{1}{4\pi\epsilon_0}\frac{Q}{r^2} \qquad r \geq R \qquad \text{outside a solid sphere of charge}$$

$$E_r = \frac{1}{4\pi\epsilon_0}\frac{Q}{R^3} r \qquad r \leq R \qquad \text{inside a solid sphere of charge}$$

6. At a surface carrying a surface charge density σ, the component of the electric field perpendicular to the surface is discontinuous by σ/ϵ_0:

$$E_{n2} - E_{n1} = \frac{\sigma}{\epsilon_0}$$

7. In electrostatic equilibrium, the net electric charge on a conductor resides on the surface of the conductor. The electric field just outside the surface of a conductor is perpendicular to the surface and has the magnitude σ/ϵ_0, where σ is the local surface charge density at that point on the conductor.

Suggestion for Further Reading

Reichardt, Hans: "Gauss," *The New Encylcopaedia Britannica*, 1968, vol. 19, p. 698.

A two-page biography of the German mathematician, one of the greatest mathematicians of all time, who elegantly solved problems in astronomy and geodesy—as well as electromagnetism.

Review

A. Objectives: After studying this chapter, you should:

1. Be able to use Coulomb's law to calculate the electric field due to a line charge, a ring charge, and a disk charge.

2. Be able to state Gauss's law and use it to find the electric field due to an infinite plane of charge, a spherically symmetric charge distribution, and a cylindrically symmetric charge distribution.

3. Be able to show that the electric field just outside a conducting surface is σ/ϵ_0.

B. Define, explain, or otherwise identify:

Volume charge density
Surface charge density
Linear charge density
Electric flux
Gauss's law
Permittivity of free space
Gaussian surface
Electrostatic equilibrium
Field emission
Solid angle
Steradian

C. True or false: If the statement is true, explain why it is true. If it is false, give a counterexample.

1. If there is no charge in a region of space, the electric field on a surface surrounding the region must be zero everywhere.

2. Gauss's law holds only for symmetric charge distributions.

3. The electric field inside a uniformly charged spherical shell is zero.

4. In electrostatic equilibrium, the electric field inside a conductor is zero.

5. The result that $\mathbf{E} = 0$ inside a conductor can be derived from Gauss's law.

6. If the net charge on a conductor is zero, the charge density must be zero at every point on the surface of the conductor.

7. The electric field is discontinuous at all points at which the charge density is discontinuous.

8. Half of the electric field at a point just outside the surface of a conductor is due to the charge on the surface in the immediate vicinity of that point.

Problems

Level I

19-1 Calculation of the Electric Field from Coulomb's Law

1. A uniform line charge of linear charge density $\lambda =$ 3.5 nC/m extends from $x = 0$ to $x = 5$ m. (*a*) What is the total charge? Find the electric field on the x axis at (*b*) $x =$ 6 m, (*c*) $x = 9$ m, and (*d*) $x = 250$ m. (*e*) Find the field at $x = 250$ m, using the approximation that the charge is a point charge at the origin, and compare your result with that for the exact calculation in part (*d*).

2. Two infinite vertical planes of charge are parallel to each other and are separated by a distance $d = 4$ m. Find the electric field to the left of the planes, to the right of the planes, and between the planes when (*a*) each plane has a uniform surface charge density $\sigma = +3\ \mu\text{C/m}^2$ and (*b*) the left plane has a uniform surface charge density $\sigma =$ $+3\ \mu\text{C/m}^2$ and that of the right plane is $\sigma = -3\ \mu\text{C/m}^2$. Draw the electric-field lines for each case.

3. A 2.75-μC charge is uniformly distributed on a ring of radius 8.5 cm. Find the electric field on the axis at (*a*) 1.2 cm, (*b*) 3.6 cm, and (*c*) 4.0 m from the center of the ring. (*d*) Find the field at 4.0 m using the approximation that the ring is a point charge at the origin, and compare your results with that for part (*c*).

4. A disk of radius 2.5 cm carries a uniform surface charge density of 3.6 $\mu\text{C/m}^2$. Using reasonable approximations, find the electric field on the axis at distances of (*a*) 0.01 cm, (*b*) 0.04 cm, (*c*) 5 m, and (*d*) 5 cm.

5. For the disk charge of Problem 4, calculate exactly the electric field on the axis at distances of (*a*) 0.04 cm and (*b*) 5 m, and compare your results with those for parts (*b*) and (*c*) of Problem 4.

6. A uniform line charge extends from $x = -2.5$ cm to $x =$ $+2.5$ cm and has a linear charge density $\lambda = 4.5$ nC/m. (*a*) Find the total charge. Find the electric field on the y axis at (*b*) $y = 4$ cm, (*c*) $y = 12$ cm, and (*d*) $y = 4.5$ m. (*e*) Find the field at $y = 4.5$ m assuming the charge to be a point charge, and compare your result with that for part (*d*).

7. A disk of radius a lies in the yz plane with its axis along the x axis and carries a uniform surface charge density σ. Find the value of x for which $E_x = \frac{1}{2}\sigma/2\epsilon_0$.

8. A ring of radius a with its center at the origin and its axis along the x axis carries a total charge Q. Find E_x at (*a*) $x = 0.2a$, (*b*) $x = 0.5a$, (*c*) $x = 0.7a$, (*d*) $x = a$, and (*e*) $x =$ $2a$. (*f*) Use your results to plot E_x versus x for both positive and negative values of x.

9. Repeat Problem 8 for a disk of uniform surface charge density σ.

19-2 Gauss's Law

10. Consider a uniform electric field $\mathbf{E} = 2\ \text{kN/C}\ \mathbf{i}$. (*a*) What is the flux of this field through a square of side 10 cm in a plane parallel to the yz plane? (*b*) What is the flux through the same square if the normal to its plane makes a 30° angle with the x axis?

11. A single point charge $q = +2\ \mu$C is at the origin. A spherical surface of radius 3.0 m has its center on the x axis at $x = 5$ m. (*a*) Sketch electric-field lines for the point charge. Do any lines enter the spherical surface? (*b*) What is the net number of lines that leave the spherical surface, counting those that enter as negative? (*c*) What is the net flux of the electric field due to the point charge through the spherical surface?

12. An electric field is $\mathbf{E} = 200\ \text{N/C}\ \mathbf{i}$ for $x > 0$ and $\mathbf{E} =$ $-200\ \text{N/C}\ \mathbf{i}$ for $x < 0$. A cylinder of length 20 cm and radius 5 cm has its center at the origin and its axis along the x axis such that one end is at $x = +10$ cm and the other is at $x = -10$ cm. (*a*) What is the flux through each end? (*b*) What is the flux through the curved surface of the cylinder? (*c*) What is the net outward flux through the entire cylindrical surface? (*d*) What is the net charge inside the cylinder?

13. A positive point charge q is at the center of a cube of side L. A large number N of electric-field lines are drawn from the point charge. (*a*) How many of the field lines pass through the surface of the cube? (*b*) How many lines pass through each face, assuming that none pass through the edges or corners? (*c*) What is the net outward flux of the electric field through the cubic surface? (*d*) Use symmetry arguments to find the flux of the electric field through one face of the cube. (*e*) Which, if any, of your answers would change if the charge were inside the cube but not at its center?

14. Careful measurement of the electric field at the surface of a black box indicates that the net outward flux through the surface of the box is 6.0 kN·m^2/C. (*a*) What is the net charge inside the box? (*b*) If the net outward flux through the surface of the box were zero, could you conclude there were no charges inside the box? Why or why not?

15. A point charge $q = +2\ \mu$C is at the center of a sphere of radius 0.5 m. (*a*) Find the surface area of the sphere. (*b*) Find the magnitude of the electric field at points on the surface of the sphere. (*c*) What is the flux of the electric field due to the point charge through the surface of the sphere? (*d*) Would your answer to part (*c*) change if the point charge were moved so that it was inside the sphere but not at the center? (*e*) What is the net flux through a cube of side 1 m that encloses the sphere?

16. Since Newton's law of gravity and Coulomb's law have the same inverse-square dependence on distance, an expression analogous in form to Gauss's law can be found for gravity. The gravitational field **g** is the force per unit mass on a test mass m_0. Then for a point mass m at the origin, the gravitational field **g** at some position **r** is

$$\mathbf{g} = -\frac{Gm}{r^2}\,\hat{\mathbf{r}}$$

Compute the flux of the gravitational field through a spherical surface of radius r centered at the origin, and show that the gravitational analog of Gauss's law is $\phi_{\text{net}} = -4\pi G m_{\text{inside}}$.

19-3 Calculation of the Electric Field from Gauss's Law

17. A spherical shell of radius 6 cm carries a uniform surface charge density $\sigma = 9$ nC/m^2. (*a*) What is the total charge on the shell? Find the electric field at (*b*) $r = 2$ cm, (*c*) $r = 5.9$ cm, (*d*) $r = 6.1$ cm, and (*e*) $r = 10$ cm.

18. A sphere of radius 6 cm carries a uniform volume charge density $\rho = 450$ nC/m^3. (*a*) What is the total charge of the sphere? Find the electric field at (*b*) $r = 2$ cm, (*c*) $r = 5.9$ cm, (*d*) $r = 6.1$ cm, and (*e*) $r = 10$ cm.

19. A cylindrical shell of length 12 m and radius 6 cm carries a uniform surface charge density $\sigma = 9$ nC/m^2. (*a*) What is the total charge on the shell? Find the electric field at (*b*) $r = 2$ cm, (*c*) $r = 5.9$ cm, (*d*) $r = 6.1$ cm, and (*e*) $r = 10$ cm.

20. A cylinder of length 12 m and radius 6 cm carries a uniform volume charge density $\rho = 300$ nC/m^3. (*a*) What is the total charge of the cylinder? Find the electric field at (*b*) $r = 2$ cm, (*c*) $r = 5.9$ cm, (*d*) $r = 6.1$ cm, and (*e*) $r =$ 10 cm.

21. A spherical shell of radius R_1 carries a total charge q_1 that is uniformly distributed on its surface. A second, larger spherical shell of radius R_2 that is concentric with the first carries a charge q_2 that is uniformly distributed on its surface. (*a*) Use Gauss's law to find the electric field in the regions $r < R_1$, $R_1 < r < R_2$, and $r > R_2$. (*b*) What should the ratio of the charges q_1/q_2 and their relative signs be for the electric field to be zero for $r > R_2$? (*c*) Sketch the electric-field lines for the situation in part (*b*).

22. Consider two infinitely long, concentric cylindrical shells. The inner shell has a radius R_1 and carries a uniform surface charge density σ_1, and the outer shell has a radius R_2 and carries a uniform surface charge density σ_2. (*a*) Use Gauss's law to find the electric field in the regions $r < R_1$, $R_1 < r < R_2$, and $r > R_2$. (*b*) What should the ratio of the surface charge densities σ_2/σ_1 and their relative signs be for the electric field to be zero at $r > R_2$? What would the electric field between the shells be in this case? (*c*) Sketch the electric-field lines for the situation in part (*b*).

23. A nonuniform surface charge lies in the yz plane. At the origin, the surface charge density is $\sigma = 3.10\ \mu$C/m^2. There are various other charge distributions in space. Just to the right of the origin, the x component of the electric field is $E_x = 4.65 \times 10^5$ N/C. What is E_x just to the left of the origin?

19-4 Charge and Field at Conductor Surfaces

24. A penny is in an external electric field of magnitude 1.6 kN/C directed perpendicular to its faces. (*a*) Find the charge density on each face of the penny, assuming the faces are planes. (*b*) If the radius of the penny is 1 cm, find the total charge on one face.

25. An uncharged metal slab has square faces with 12-cm sides. It is placed in an external electric field that is perpendicular to its faces. The total charge induced on one of the faces is 1.2 nC. What is the magnitude of the electric field?

26. A charge of 6 nC is placed uniformly on a square sheet of nonconducting material of side 20 cm in the yz plane. (*a*) What is the surface charge density σ? (*b*) What is the magnitude of the electric field just to the right and just to the left of the sheet? (*c*) The same charge is placed on a square conducting slab of side 20 cm and thickness 1 mm. What is the surface charge density σ? (Assume that the charge distributes itself uniformly on the large square surfaces.) (*d*) What is the magnitude of the electric field just to the right and just to the left of each face of the slab?

27. A spherical conducting shell with zero net charge has an inner radius a and an outer radius b. A point charge q is placed at the center of the shell. (*a*) Use Gauss's law and the properties of conductors in equilibrium to find the electric field in the regions $r < a$, $a < r < b$, and $b < r$. (*b*) Draw the electric-field lines for this situation. (*c*) Find the charge density on the inner surface ($r = a$) and on the outer surface ($r = b$) of the shell.

19-5 Mathematical Derivation of Gauss's Law (Optional)

There are no problems for this section.

Level II

28. The electric field just above the surface of the earth has been measured to be 150 N/C downward. What total charge on the earth is implied by this measurement?

29. In a particular region of the earth's atmosphere, the electric field above the earth's surface has been measured to be 150 N/C downward at an altitude of 250 m and 170 N/C downward at an altitude of 400 m. Calculate the volume charge density of the atmosphere assuming it to be uniform between 250 and 400 m. (You may neglect the curvature of the earth. Why?)

30. An infinite line charge of uniform linear charge density $\lambda = -1.5\ \mu\text{C/m}$ lies parallel to the y axis at $x = -2$ m. A point charge of 1.3 μC is located at $x = 1$ m, $y = 2$ m. Find the electric field at $x = 2$ m, $y = 1.5$ m.

31. A solid sphere 1.2 m in diameter with its center on the x axis at $x = 4$ m carries a uniform volume charge of density $\rho = 5\ \mu\text{C/m}^3$. A spherical shell concentric with the sphere has a diameter of 2.4 m and a uniform surface charge density $\sigma = -1.5\ \mu\text{C/m}^2$. Calculate the magnitude and direction of the electric field at (*a*) $x = 4.5$ m, $y = 0$; (*b*) $x = 4.0$ m, $y = 1.1$ m; and (*c*) $x = 2.0$ m, $y = 3.0$ m.

32. Two infinite planes of charge lie parallel to each other and to the yz plane. One is at $x = -2$ m and has a surface charge density $\sigma = -3.5\ \mu\text{C/m}^2$. The other is at $x = 2$ m and has a surface charge density $\sigma = 6.0\ \mu\text{C/m}^2$. Find the electric field for (*a*) $x < -2$ m, (*b*) $-2\text{ m} < x < 2$ m, and (*c*) $x > 2$ m.

33. A model of an atom has a positive nuclear point charge $+Ze$ embedded in a rigid electron sphere of radius R containing a total charge of $-Ze$ uniformly distributed throughout the sphere. (*a*) If there is no external electric field, where is the equilibrium position of the nuclear point charge? (*b*) In an external electric field $\mathbf{E}_0$, where is the equilibrium position of the nuclear point charge with respect to the center of the negatively charged electron sphere? (*c*) What is the electric dipole moment induced by the field $\mathbf{E}_0$ for this atomic model?

34. Show that E_x on the axis of a ring charge of radius a has its maximum and minimum values at $x = +a/\sqrt{2}$ and $x = -a/\sqrt{2}$. Sketch E_x versus x for both positive and negative values of x.

35. A positive point charge of magnitude 2.5 μC is at the center of an uncharged spherical conducting shell of inner radius 60 cm and outer radius 90 cm. (*a*) Find the charge densities on the inner and outer surfaces of the shell and the total charge on each surface. (*b*) Find the electric field everywhere. (*c*) Repeat (*a*) and (*b*) if a net charge of +3.5 μC is placed on the shell.

36. A square conducting slab with 5-m sides carries a net charge of 80 μC. (*a*) Find the charge density on each face of the slab and the electric field just outside one face of the slab. (*b*) The slab is placed to the right of an infinite charged nonconducting plane with charge density $2.0\ \mu\text{C/m}^2$ so that the faces of the slab are parallel to the plane. Find the electric field on each side of the slab far from its edges and the charge density on each face.

37. (*a*) A finite line charge of uniform linear charge density λ lies on the x axis from $x = 0$ to $x = a$. Show that the y component of the electric field at a point on the y axis is given by

$$E_y = \frac{\lambda}{4\pi\epsilon_0 y}\sin\theta_1 = \frac{\lambda}{4\pi\epsilon_0 y}\frac{a}{\sqrt{y^2 + a^2}}$$

where θ_1 is the angle subtended by the line charge at the field point. (*b*) Show that if the line charge extends from $x = -b$ to $x = a$, the y component of the electric field at a point on the y axis is given by

$$E_y = \frac{\lambda}{4\pi\epsilon_0 y}(\sin\theta_1 + \sin\theta_2)$$

where $\sin\theta_2 = b/\sqrt{y^2 + b^2}$.

38. Imagine that a small hole has been punched through the wall of a thin, uniformly charged spherical shell whose surface charge density is σ. Find the electric field near the center of the hole.

39. An infinite plane of charge with surface charge density $\sigma_1 = 3\ \mu\text{C/m}^2$ is parallel to the xz plane at $y = -0.6$ m. A second infinite plane of charge with surface charge density $\sigma_2 = -2\ \mu\text{C/m}^2$ is parallel to the yz plane at $x = 1$ m. A sphere of radius 1 m with its center in the xy plane at the intersection of the two charged planes ($x = 1$ m, $y = -0.6$ m) has a surface charge density $\sigma_3 = -3\ \mu\text{C/m}^2$. Find the magnitude and direction of the electric field on the x axis at (*a*) $x = 0.4$ m and (*b*) $x = 2.5$ m.

40. An infinitely long cylindrical shell is coaxial with the y axis and has a radius of 15 cm. It carries a uniform surface charge density $\sigma = 6\ \mu\text{C/m}^2$. A spherical shell of radius 25 cm is centered on the x axis at $x = 50$ cm and carries a uniform surface charge density $\sigma = -12\ \mu\text{C/m}^2$. Calculate the magnitude and direction of the electric field at (*a*) the origin; (*b*) $x = 20$ cm, $y = 10$ cm; and (*c*) $x = 50$ cm, $y = 20$ cm.

41. A thick, nonconducting spherical shell of inner radius a and outer radius b has a uniform volume charge density ρ. Find the total charge and the electric field everywhere.

42. An infinite plane in the xz plane carries a uniform surface charge density $\sigma_1 = 65\ \text{nC/m}^2$. A second infinite plane carrying a uniform charge density $\sigma_2 = 45\ \text{nC/m}^2$ intersects the xz plane at the z axis and makes an angle of 30° with the xz plane as shown in Figure 19-36. Find the electric field in the xy plane at (*a*) $x = 6$ m, $y = 2$ m and (*b*) $x = 6$ m, $y = 5$ m.

Figure 19-36 Uniform surface charges on the xz plane and on a plane making an angle of 30° with the xz plane for Problem 42.

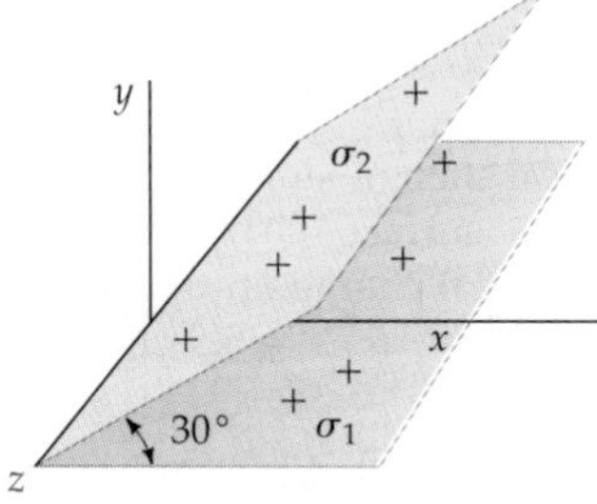

43. A ring of radius R carries a uniform, positive, linear charge density λ. Figure 19-37 shows a point P in the plane of the ring but not at the center. Consider the two elements of the ring of lengths s_1 and s_2 shown in the figure at distances r_1 and r_2, respectively, from point P. (*a*) What is the ratio of the charges of these elements? Which produces the greater field at point P? (*b*) What is the direction of the field at point P due to each element? What is the direction of the total electric field at point P? (*c*) Suppose that the electric field due to a point charge varied as $1/r$ rather than $1/r^2$. What would the electric field be at point P due to the elements shown? (*d*) How would your answers to parts (*a*), (*b*), and (*c*) differ if point P were inside a spherical shell of uniform charge and the elements were of areas s_1 and s_2?

Figure 19-37 Problem 43.

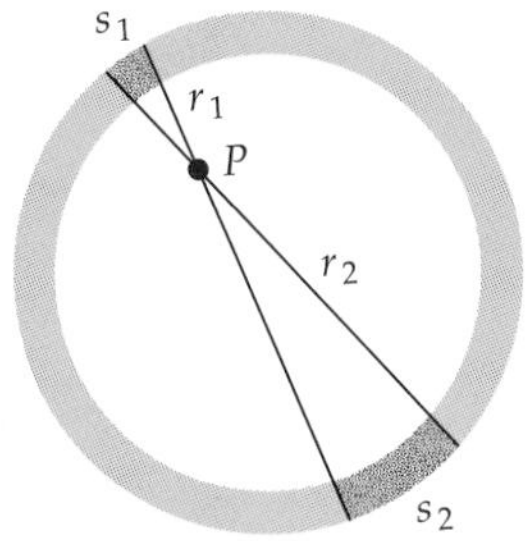

44. A disk of radius 30 cm carries a uniform charge density σ. (*a*) Compare the approximation $E = \sigma/2\epsilon_0$ with the exact expression for the electric field on the axis of the disk by computing the neglected term as a percentage of $\sigma/2\epsilon_0$ for the distances $x = 0.1$, $x = 0.2$, and $x = 3$ cm. (*b*) At what distance is the neglected term 1 percent of $\sigma/2\epsilon_0$?

45. A semi-infinite line charge of uniform linear charge density λ lies along the x axis from $x = 0$ to $x = \infty$. Find both E_x and E_y at a point on the y axis.

46. An infinite plane lies parallel to the yz plane at $x = 2$ m and carries a uniform surface charge density $\sigma = 2\ \mu C/m^2$. An infinite line charge of uniform linear charge, density $\lambda = 4\ \mu C/m$ passes through the origin at 45° to the x axis in the xy plane. A sphere of volume charge density $\rho = -6\ \mu C/m^3$ and radius 0.8 m is centered on the x axis at $x = 1$ m. Calculate the magnitude and direction of the electric field in the xy plane at $x = 1.5$ m, $y = 0.5$ m.

Level III

47. An infinitely long, thick, nonconducting cylindrical shell of inner radius a and outer radius b has a uniform volume charge density ρ. Find the electric field everywhere.

48. A nonconducting solid sphere of radius R carries a volume charge density that is proportional to the distance from the center: $\rho = Ar$ for $r \leq R$, where A is a constant; $\rho = 0$ for $r > R$. (*a*) Find the total charge in the sphere by summing the charges on shells of thickness dr and volume $4\pi r^2\, dr$. (*b*) Find the electric field E_r both inside and outside the charge distribution, and sketch E_r versus r.

49. Repeat Problem 48 for a solid sphere with volume charge density $\rho = B/r$ for $r < R$ and $\rho = 0$ for $r > R$.

50. Repeat Problem 48 for a solid sphere with volume charge density $\rho = C/r^2$ for $r < R$ and $\rho = 0$ for $r > R$.

51. A uniformly charged sphere of radius R is centered at the origin with a charge of Q. Find the force on a uniformly charged line oriented radially having a total charge q with its ends at $r = R$ and $r = R + d$.

52. Two equal uniform line charges of length L lie on the x axis a distance d apart as shown in Figure 19-38. (*a*) What is the force that one line charge exerts on the other line charge? (*b*) Show that when $d \gg L$, the force tends toward the expected result of $k(\lambda L)^2/d^2$.

Figure 19-38 Problem 52.

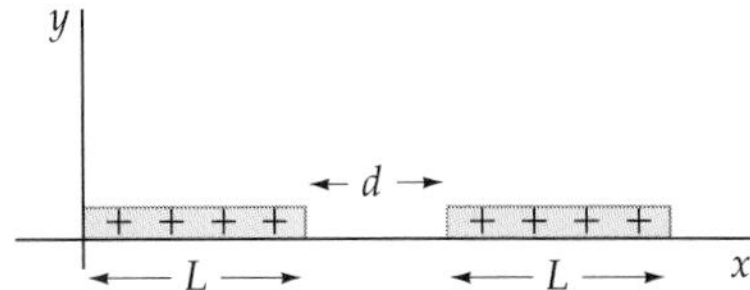

53. A line charge of linear charge density λ with the shape of a square of side L lies in the yz plane with its center at the origin. Find the electric field on the x axis at an arbitrary distance x, and compare your result to that for the field on the axis of a charged ring of radius $r = \frac{1}{2}L$ with its center at the origin and carrying the same total charge. *Hint:* Use Equation 19-10 for the field due to each segment of the square.

54. A nonconducting solid sphere of radius a with its center at the origin has a spherical cavity of radius b with its center at the point $x = b$, $y = 0$, $z = 0$ as shown in Figure 19-39. The sphere has a uniform volume charge density ρ. Show that the electric field in the cavity is uniform and is given by $E_x = \rho b/3\epsilon_0$, $E_y = E_z = 0$. *Hint:* Replace the cavity with spheres of equal positive and negative charge densities.

55. The electrostatic force on a charge at some point is the product of the charge and the electric field due to all other charges. Consider a small charge on the surface of a conductor $\Delta q = \sigma\, \Delta A$. (*a*) Show that the electrostatic force on the charge is $\sigma^2\, \Delta A/2\epsilon_0$. (*b*) Explain why this is just half of $\Delta q\, E$, where $E = \sigma/\epsilon_0$ is the electric field just outside the conductor at that point. (*c*) The force per unit area is called the electrostatic stress. Find the electrostatic stress when a charge of 2 μC is placed on a conducting sphere of radius 10 cm.

Figure 19-39 Problem 54.

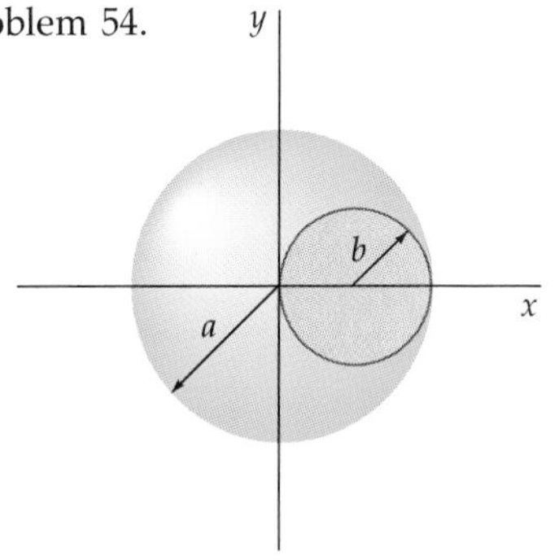

Chapter 20

Electric Potential

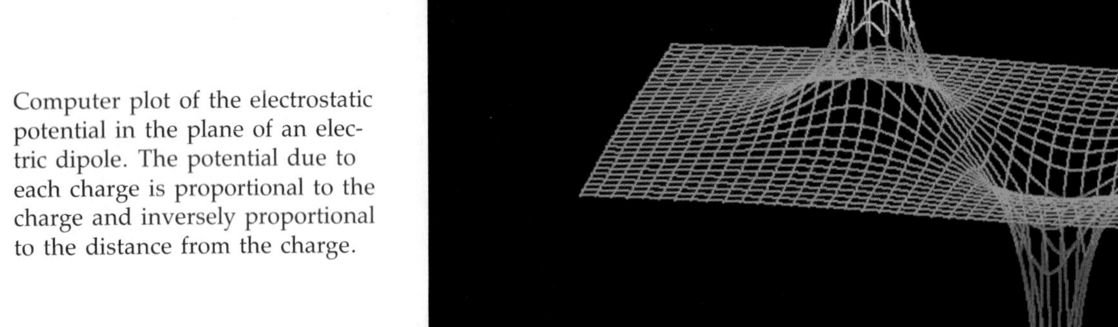

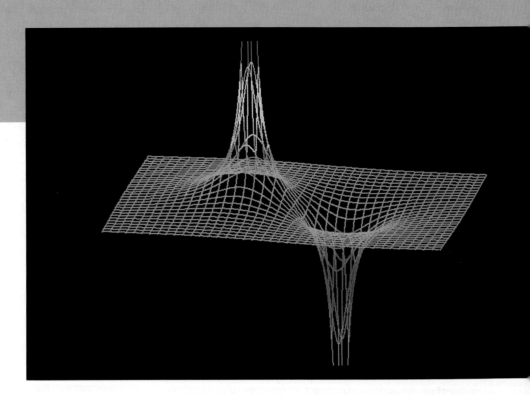

Computer plot of the electrostatic potential in the plane of an electric dipole. The potential due to each charge is proportional to the charge and inversely proportional to the distance from the charge.

In our study of mechanics, we found the concept of potential energy to be very useful. When we lift an object of mass *m* a vertical distance *h* near the earth's surface, the work we do goes into potential energy *mgh* of the earth-mass system. If we then drop the object, this potential energy is converted into kinetic energy. The electric force between two charges is directed along the line of the charges and depends on the inverse square of their separation, the same as the gravitational force between two masses. Like the gravitational force, the electric force is conservative. There is thus a potential-energy function associated with the electric force. As we will see, the potential energy associated with a particle in an electric field is proportional to the charge. The potential energy per unit charge is called the electric potential. Electric potential is measured in volts and is commonly called voltage. In this chapter, we will define the electric potential function *V* and show how to calculate the potential from a given charge distribution or from a given electric field. We will then show how the potential is related to the electric field **E** and to electrostatic potential energy. Finally, we will show that the electric potential is constant everywhere inside a conductor in an electrostatic field.

20-1 Electric Potential and Potential Difference

In general, when a conservative force **F** acts on a particle that undergoes a displacement $d\boldsymbol{\ell}$, the change in the potential-energy function dU is defined by (Equation 6-17)

$$dU = -\mathbf{F}\cdot d\boldsymbol{\ell}$$

The work done by a conservative force decreases the potential energy (Figure 20-1). The force exerted by an electric field **E** on a point charge q_0 is

$$\mathbf{F} = q_0\,\mathbf{E}$$

When the charge undergoes a displacement $d\boldsymbol{\ell}$ in an electric field **E**, the change in the electrostatic potential energy is

$$dU = -q_0\,\mathbf{E}\cdot d\boldsymbol{\ell} \qquad \text{20-1}$$

If the charge is moved from some initial point a to some final point b, the change in its electrostatic potential energy is

$$\Delta U = U_b - U_a = \int_a^b dU = -\int_a^b q_0\,\mathbf{E}\cdot d\boldsymbol{\ell} \qquad \text{20-2}$$

The potential-energy change is proportional to the test charge q_0. The potential-energy change per unit charge is called the **potential difference** dV:

$$dV = \frac{dU}{q_0} = -\mathbf{E}\cdot d\boldsymbol{\ell} \qquad \text{20-3}a$$

Potential difference defined

For a finite displacement from point a to point b, the change in potential is

$$\Delta V = V_b - V_a = \frac{\Delta U}{q_0} = -\int_a^b \mathbf{E}\cdot d\boldsymbol{\ell} \qquad \text{20-3}b$$

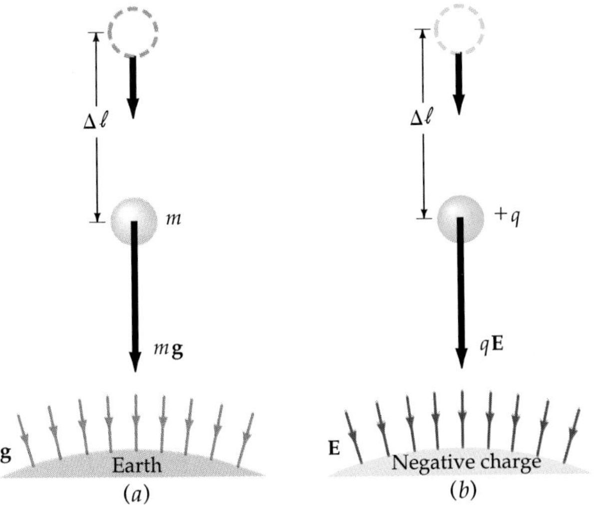

Figure 20-1 (*a*) The work done by the gravitational field on a mass decreases the gravitational potential energy. (*b*) The work done by the electric field on a positive charge $+q$ decreases the electrostatic potential energy.

> The potential difference $V_b - V_a$ is the negative of the work per unit charge done by the electric field on a positive test charge when it moves from point a to point b.

Equation 20-3 defines the change in the function V, which is called the **electric potential** (or sometimes just the **potential**). As with potential energy U, only *changes* in the potential V are important. We are free to choose the potential energy or the potential to be zero at any convenient point, just as we were for mechanical potential energy. (For example, in the expression for the gravitational potential energy near the earth's surface, mgh, we can choose h to be zero at any convenient point, such as the floor or at the top of a table. For two point masses, we found that it was usually convenient to choose the gravitational potential energy to be zero when their separation was infinite.)

Since electric potential is the electrostatic potential energy per unit charge, the SI unit for potential and potential difference is the joule per coulomb, called the **volt** (V):

$$1 \text{ V} = 1 \text{ J/C} \qquad \text{20-4}$$

Because potential difference is measured in volts, it is sometimes called **voltage.** In a 12-V car battery, the positive terminal has a potential 12 V higher than the negative terminal. If an external circuit is attached to the battery and one coulomb of charge is transferred from the positive terminal through the circuit to the negative terminal, the potential energy of the charge is decreased by $Q\,\Delta V = (1 \text{ C})(12 \text{ V}) = 12 \text{ J}$. This energy usually appears as thermal energy in the circuit.

From Equation 20-3, we can see that the dimensions of potential are also those of electric field times distance. Thus, the unit of electric field E, the newton per coulomb, is also equal to a volt per meter:

$$1 \text{ N/C} = 1 \text{ V/m} \qquad \text{20-5}$$

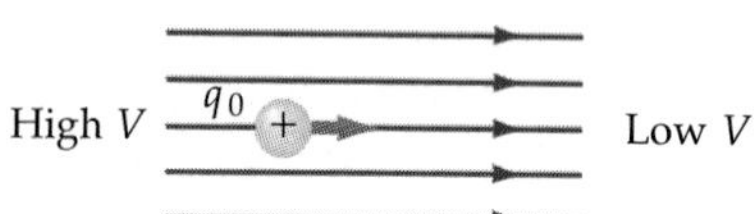

Figure 20-2 Electric-field lines point in the direction of decreasing potential. When a positive test charge q_0 is placed in an electric field, it accelerates in the direction of the field. Its kinetic energy increases and its potential energy decreases.

If we place a positive test charge q_0 in an electric field **E** and release the charge, it experiences a force in the direction of the field and will accelerate in the direction of **E** along a field line. As the kinetic energy of the charge increases, its potential energy decreases. Thus, the charge moves toward a region of lower potential energy, just as a massive body falls toward a region of lower gravitational potential energy. For a positive test charge, a region of lower potential energy is one of lower potential. Thus, as illustrated in Figure 20-2,

> Electric-field lines point in the direction of decreasing electric potential.

Example 20-1

An electric field points in the positive x direction and has a constant magnitude of 10 N/C = 10 V/m. Find the potential as a function of x, assuming that $V = 0$ at $x = 0$.

The electric-field vector is given by $\mathbf{E} = (10 \text{ N/C})\mathbf{i} = (10 \text{ V/m})\mathbf{i}$. For a general displacement $d\boldsymbol{\ell}$, the change in potential is given by Equation 20-3a:

$$dV = -\mathbf{E}\cdot d\boldsymbol{\ell} = -(10 \text{ V/m})\mathbf{i}\cdot(dx\,\mathbf{i} + dy\,\mathbf{j} + dz\,\mathbf{k})$$
$$= -(10 \text{ V/m})\,dx$$

Integrating from point x_1 to x_2, we obtain the potential difference $V(x_2) - V(x_1)$:

$$V(x_2) - V(x_1) = \int_{x_1}^{x_2} dV = \int_{x_1}^{x_2} -(10\ \text{V/m})\ dx$$
$$= -(10\ \text{V/m})(x_2 - x_1) = (10\ \text{V/m})(x_1 - x_2)$$

Since we are given that the potential is zero at $x = 0$, we have $V(x_1) = 0$ at $x_1 = 0$. Then the potential at x_2 relative to $V = 0$ at $x = 0$ is given by

$$V(x_2) - 0 = (10\ \text{V/m})(0 - x_2)$$

or

$$V(x_2) = -(10\ \text{V/m})x_2$$

At a general point x, the potential is

$$V(x) = -(10\ \text{V/m})x$$

The potential is zero at $x = 0$ and decreases by 10 V/m in the x direction.

An ordinary wall outlet. The power company maintains an electric potential difference between the slotted openings which varies sinusoidally with a frequency of 60 Hz. The effective average value of the potential difference (the rms value, see pages 900–902) is 120 V. The rounded opening is connected to the ground.

Example 20-2

A proton of mass 1.67×10^{-27} kg and charge 1.6×10^{-19} C is placed in a uniform electric field $\mathbf{E} = (5.0\ \text{N/C})\mathbf{i} = (5.0\ \text{V/m})\mathbf{i}$ and released from rest. After traveling 4 cm, how fast is it moving?

As the proton travels down the electric-field line, its potential energy decreases and its kinetic energy increases by an equal amount. According to Equation 20-3, the change in the electric potential for $\Delta x = 4$ cm $=$ 0.04 m is

$$dV = -\mathbf{E}\cdot d\boldsymbol{\ell} = -(5.0\ \text{V/m}\ \mathbf{i})\cdot(dx\ \mathbf{i}) = -(5.0\ \text{V/m})\ dx$$
$$\Delta V = -(5.0\ \text{V/m})(0.04\ \text{m}) = -0.20\ \text{V}$$

The change in the potential energy of the proton is the product of its charge times the change in its potential (Equation 20-3):

$$\Delta U = q\ \Delta V = (1.6 \times 10^{-19}\ \text{C})(-0.20\ \text{V}) = -3.2 \times 10^{-20}\ \text{J}$$

By conservation of energy, the loss in potential energy equals the gain in kinetic energy. Since the proton starts from rest, its gain in kinetic energy is just $\frac{1}{2}mv^2$, where v is its speed after traveling the 4 cm. We therefore have

$$\Delta K + \Delta U = 0$$
$$\Delta K = -\Delta U = -(-3.2 \times 10^{-20}\ \text{J})$$
$$\tfrac{1}{2}mv^2 = 3.2 \times 10^{-20}\ \text{J}$$
$$v^2 = \frac{(2)(3.2 \times 10^{-20}\ \text{J})}{1.67 \times 10^{-27}\ \text{kg}} = 3.83 \times 10^7\ \text{J/kg}$$
$$v = \sqrt{3.83 \times 10^7\ \text{J/kg}} = 6.19 \times 10^3\ \text{m/s}$$

In atomic and nuclear physics, we often have elementary particles, such as electrons and protons, with charges of magnitude e moving through potential differences of several to thousands or even millions of volts. Since energy has dimensions of electric charge times electric potential, a convenient unit of energy is the product of the electron charge e times a volt. This unit is called an **electron volt** (eV). The conversion between electron volts and joules is obtained by expressing the electronic charge in coulombs:

$$1\text{ eV} = 1.6 \times 10^{-19}\text{ C}\cdot\text{V} = 1.6 \times 10^{-19}\text{ J} \qquad 20\text{-}6$$

In Example 20-2, the change in the potential energy of the proton after traveling 4 cm is

$$\Delta U = q\,\Delta V = e(-0.20\text{ V}) = -0.20\text{ eV}$$

Questions

1. Explain in your own words the distinction between electric potential and electrostatic potential energy.
2. If a test charge moves through a small distance in the direction of an electric field, does its electrostatic potential energy increase or decrease? Does your answer depend on the sign of the charge? Does the change in potential depend on the sign of the test charge?
3. In what direction can you move relative to an electric field so that the electric potential does not change?
4. A positive charge is released from rest in an electric field. Will it move toward a region of greater or smaller electric potential?

20-2 Potential Due to a System of Point Charges

The electric potential due to a point charge q at the origin can be calculated from the electric field, which is given by

$$\mathbf{E} = \frac{kq}{r^2}\,\hat{\mathbf{r}}$$

If a test charge q_0 at distance r is given a displacement $d\boldsymbol{\ell} = dr\,\hat{\mathbf{r}}$, the change in its potential energy is $dU = -q_0\,\mathbf{E}\cdot d\boldsymbol{\ell}$, and the change in the electric potential is

$$dV = -\mathbf{E}\cdot d\boldsymbol{\ell} = -\frac{kq}{r^2}\,\hat{\mathbf{r}}\cdot dr\,\hat{\mathbf{r}} = -\frac{kq}{r^2}\,dr \qquad 20\text{-}7$$

Integrating, we obtain

Potential due to a point charge

$$V = +\frac{kq}{r} + V_0 \qquad 20\text{-}8$$

where V_0 is a constant of integration.

It is customary to define the potential to be zero at an infinite distance from the point charge (that is, at $r = \infty$). Then the constant V_0 is zero, and the potential at a distance r from the point charge is

Potential due to a point charge with $V = 0$ at $r = \infty$

$$V = \frac{kq}{r} \qquad V = 0 \text{ at } r = \infty \qquad 20\text{-}9$$

The potential is positive or negative depending on the sign of the charge q.

If a test charge q_0 is released from a point a distance r from a point charge q that is held fixed at the origin, the test charge will be accelerated outward in the direction of the electric field. The work done by the electric field as the test charge moves from r to ∞ is

$$W = \int_r^\infty q_0 \,\mathbf{E}\cdot d\boldsymbol{\ell} = q_0 \int_r^\infty E_r\, dr = q_0 \int_r^\infty \frac{kq}{r^2}\, dr = \frac{kqq_0}{r}$$

This work is the electrostatic potential energy of the two-charge system:

$$U = \frac{kqq_0}{r} = q_0\, V$$

The potential energy is thus the work done by the electric field as the test charge moves from r to ∞. Alternatively, we may think of the potential energy as the work that must be done by an applied force $\mathbf{F}_{\text{app}} = -q_0\,\mathbf{E}$ to bring a positive test charge q_0 from an infinite distance away to a distance r from a point charge q (Figure 20-3).

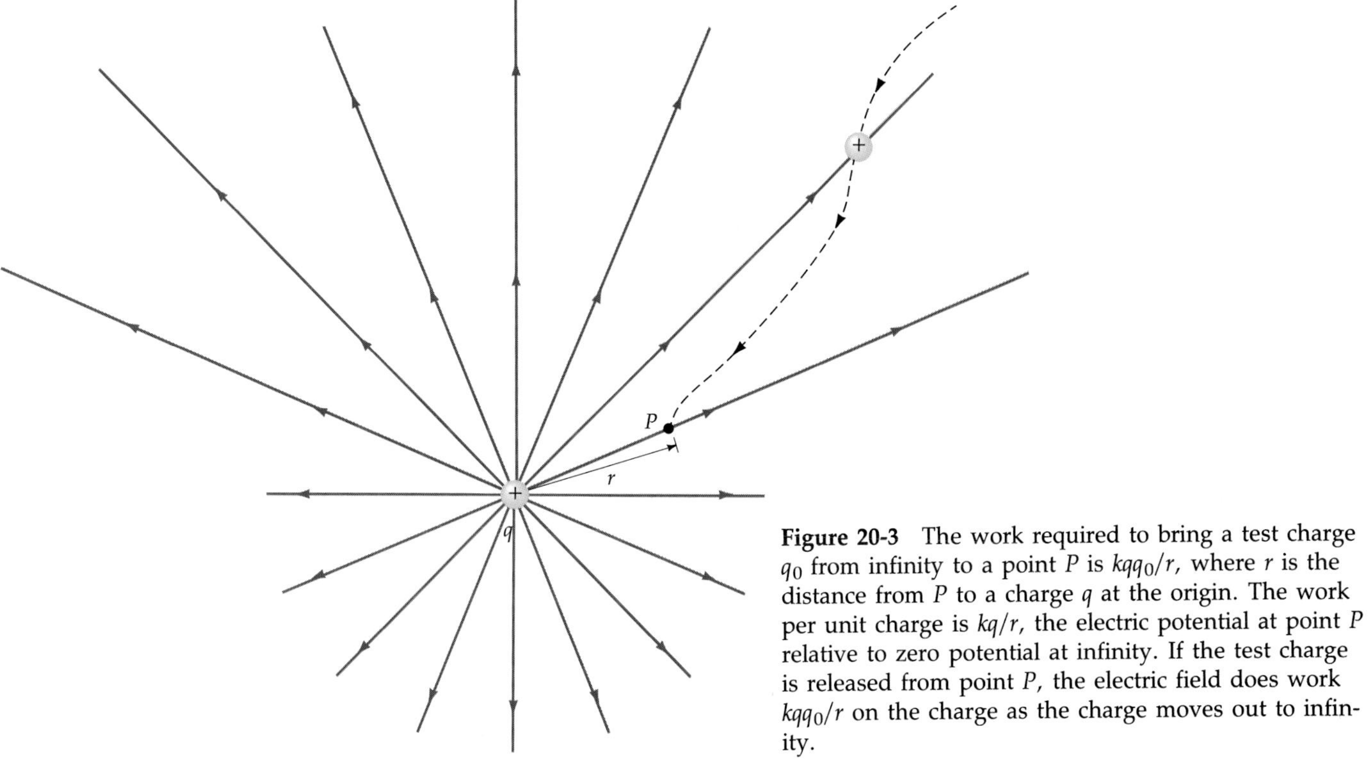

Figure 20-3 The work required to bring a test charge q_0 from infinity to a point P is kqq_0/r, where r is the distance from P to a charge q at the origin. The work per unit charge is kq/r, the electric potential at point P relative to zero potential at infinity. If the test charge is released from point P, the electric field does work kqq_0/r on the charge as the charge moves out to infinity.

The choice that the electric potential is zero at an infinite distance from a point charge is made merely for convenience. The potential energy of two charges is then zero when they are infinitely far apart. This choice is similar to that we made in our study of gravity in Chapter 10 when we chose the gravitational potential energy of two point masses to be zero when the masses were very far apart. We can also make this choice of zero potential for a system of charges as long as the system is finite, that is, as long as there are no charges an infinite distance from other charges in the system. At sufficiently great distances from any finite charge distribution, the charge distribution looks like a point charge, and the potential function V approaches that given by Equation 20-9, where q is the net charge of the distribution.

Example 20-3

(*a*) What is the electric potential at a distance $r = 0.529 \times 10^{-10}$ m from a proton? (This is the average distance between the proton and electron in the hydrogen atom.) (*b*) What is the potential energy of the electron and the proton at this separation?

(*a*) The charge of the proton is $q = 1.6 \times 10^{-19}$ C. Equation 20-9 gives

$$V = \frac{kq}{r} = \frac{(8.99 \times 10^9\ \text{N·m}^2/\text{C}^2)(1.6 \times 10^{-19}\ \text{C})}{0.529 \times 10^{-10}\ \text{m}}$$

$$= 27.2\ \text{J/C} = 27.2\ \text{V}$$

(*b*) The charge of the electron is $-e = -1.6 \times 10^{-19}$ C. In electron volts, the potential energy of the electron and proton separated by a distance of 0.529×10^{-10} m is

$$U = qV = -e(27.2\ \text{V}) = -27.2\ \text{eV}$$

In SI units, the potential energy is

$$U = qV = (-1.6 \times 10^{-19}\ \text{C})(27.2\ \text{V}) = -4.35 \times 10^{-18}\ \text{J}$$

To find the potential at some point due to several point charges, we find the potential at that point due to each charge separately and sum. This follows from the superposition principle for the electric field. If $\mathbf{E}_i$ is the electric field at some point due to charge q_i, the net field at that point due to all the charges is

$$\mathbf{E} = \mathbf{E}_1 + \mathbf{E}_2 + \cdots = \sum_i \mathbf{E}_i$$

Then, from the definition of potential difference (Equation 20-3), we have for a displacement $d\boldsymbol{\ell}$

$$dV = -\mathbf{E}\cdot d\boldsymbol{\ell} = -\mathbf{E}_1\cdot d\boldsymbol{\ell} - \mathbf{E}_2\cdot d\boldsymbol{\ell} - \cdots = dV_1 + dV_2 + \cdots$$

If the charge distribution is finite, that is, if there are no charges at infinity, we may choose the potential to be zero at infinity and use Equation 20-9 for the potential due to each point charge. The potential due to a system of point charges q_i is then given by

$$V = \sum_i \frac{kq_i}{r_{i0}} \qquad 20\text{-}10$$

where the sum is taken over all the charges and r_{i0} is the distance from the ith charge to the point P at which the potential is to be found.

Example 20-4

Two equal positive point charges of magnitude +5 nC are on the x axis. One is at the origin and the other at $x = 8$ cm as shown in Figure 20-4. Find the potential at (*a*) point P_1 on the x axis at $x = 4$ cm and (*b*) point P_2 on the y axis at $y = 6$ cm.

(*a*) Point P_1 is 4 cm from each charge. Using Equation 20-10 with $q_1 = q_2 = 5$ nC and $r_{10} = r_{20} = 0.04$ m, we find for the potential at that point

$$V = \sum_i \frac{kq_i}{r_{i0}} = \frac{kq_1}{r_{10}} + \frac{kq_2}{r_{20}}$$

$$= 2 \times \frac{(8.99 \times 10^9\ \text{N·m}^2/\text{C}^2)(5 \times 10^{-9}\ \text{C})}{0.04\ \text{m}} = 2250\ \text{V}$$

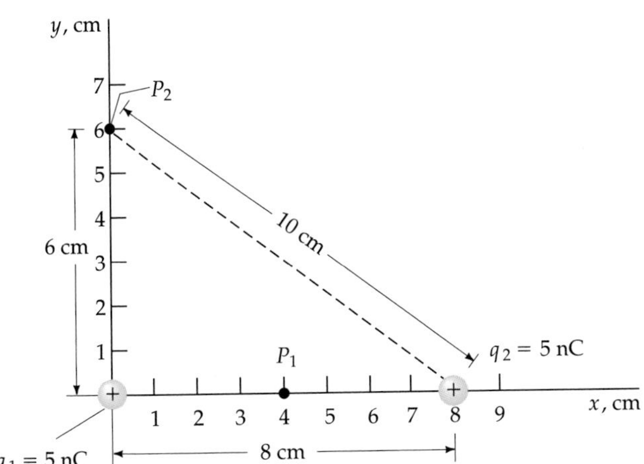

Figure 20-4 Two positive point charges on the x axis for Example 20-4. The potential is to be found at points P_1 and P_2.

Note that the electric field is zero at this point midway between the charges but the potential is not. To bring in a test charge from a long distance away to this point requires work because the electric field is zero only at the final position.

(*b*) Point P_2 is 6 cm from one charge and 10 cm from the other. The potential at this point is therefore

$$V = \frac{(8.99 \times 10^9\ \text{N·m}^2/\text{C}^2)(5 \times 10^{-9}\ \text{C})}{0.06\ \text{m}} + \frac{(8.99 \times 10^9\ \text{N·m}^2/\text{C}^2)(5 \times 10^{-9}\ \text{C})}{0.10\ \text{m}}$$

$$= 749\ \text{V} + 450\ \text{V} \approx 1200\ \text{V}$$

Example 20-5

A point charge q_1 is at the origin, and a second point charge q_2 is on the x axis at $x = a$ as shown in Figure 20-5. Find the potential everywhere on the x axis.

We need to divide the x axis into three regions: to the right of both charges $x > a$, between the two charges $0 < x < a$, and to the left of both charges $x < 0$. At a point P_1 to the right of both charges, the distance to q_1 is x and the distance to q_2 is $x - a$. The potential in this region is therefore

$$V = \frac{kq_1}{x} + \frac{kq_2}{x - a} \qquad x > a$$

At a point P_2 between the charges, the distance to q_1 is again x, but the distance to q_2 is $a - x$. The potential on the axis between the charges is therefore

$$V = \frac{kq_1}{x} + \frac{kq_2}{a - x} \qquad 0 < x < a$$

At a point P_3 to the left of both charges, the distance to q_1 is $-x$ (because x is negative) and the distance to q_2 is $a - x$. The potential on the axis to the left of the charges is therefore

$$V = \frac{kq_1}{-x} + \frac{kq_2}{a - x} \qquad x < 0$$

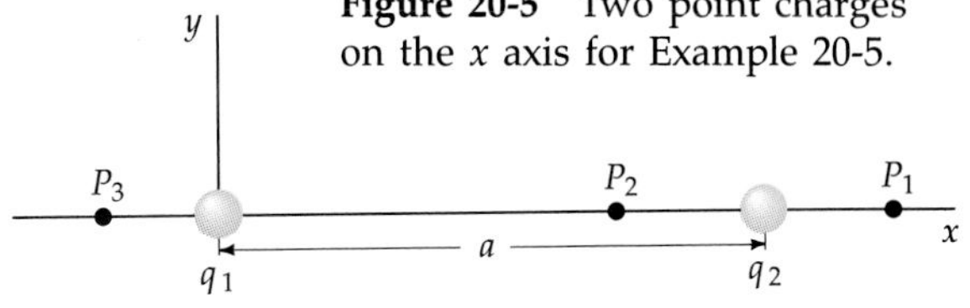

Figure 20-5 Two point charges on the x axis for Example 20-5.

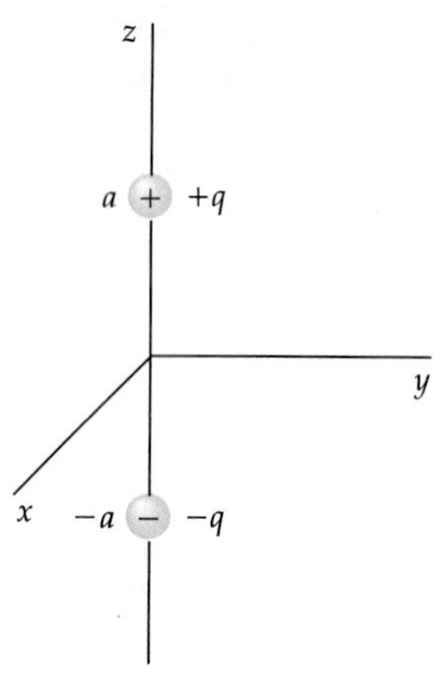

Figure 20-6 Electric dipole on the z axis for Example 20-6.

Example 20-6

An electric dipole consists of a positive charge $+q$ on the z axis at $z = +a$ and a negative charge $-q$ on the z axis at $z = -a$ (Figure 20-6). Find the potential on the z axis a great distance from the dipole.

From Equation 20-10, we have

$$V = \frac{kq}{z-a} + \frac{k(-q)}{z+a} = \frac{2kqa}{z^2 - a^2}$$

For $z \gg a$, we can neglect a^2 compared with z^2 in the denominator. We have then

$$V \approx \frac{2kqa}{z^2} = \frac{kp}{z^2} \qquad z \gg a \tag{20-11}$$

where $p = 2qa$ is the magnitude of the dipole moment.

20-3 Electrostatic Potential Energy

If we have a point charge q_1, the potential at a distance r_{12} away is given by

$$V = \frac{kq_1}{r_{12}}$$

The work required to bring up a second point charge q_2 from an infinite distance away to a distance r_{12} is $W_2 = q_2V = kq_1q_2/r_{12}$. To bring up a third charge, work must be done against the electric field produced by both q_1 and q_2. The work required to bring up a third charge q_3 to a distance r_{13} from q_1 and r_{23} from q_2 is $W_3 = kq_3q_1/r_{13} + kq_3q_2/r_{23}$. The total work required to assemble the three charges is therefore

$$W = \frac{kq_1q_2}{r_{12}} + \frac{kq_1q_3}{r_{13}} + \frac{kq_2q_3}{r_{23}}$$

This work is the **electrostatic potential energy** of the system of three point charges. It is independent of the order in which the charges are brought to their final positions. In general,

> The electrostatic potential energy of a system of point charges is the work needed to bring the charges from an infinite separation to their final positions.

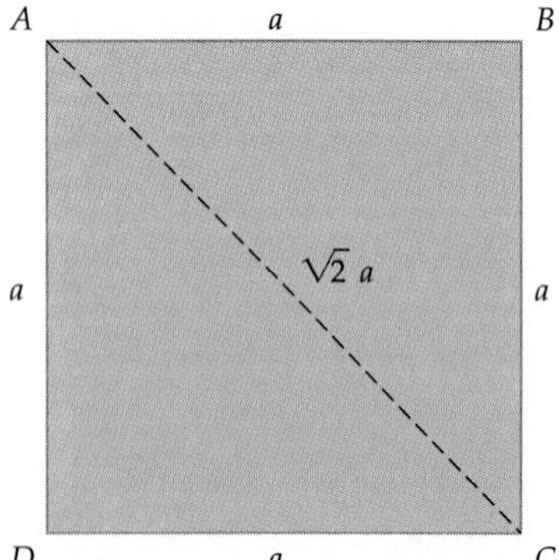

Figure 20-7 A square of side a for Example 20-7.

Example 20-7

Points A, B, C, and D are at the corners of a square of side a as shown in Figure 20-7. How much work is required to place a positive charge q at each corner of the square?

No work is needed to place the first charge at point A because the potential is zero there when the other three charges are an infinite distance away. To bring up a second charge to point B a distance a away requires work

$$W_2 = \frac{kqq}{a}$$

Point C is a distance a from point B and $\sqrt{2}a$ from point A. The potential at point C due to the charges at A and B is

$$V_C = \frac{kq}{a} + \frac{kq}{\sqrt{2}a}$$

The work required to bring a third charge q to point C is thus

$$W_3 = qV_C = \frac{kqq}{a} + \frac{kqq}{\sqrt{2}a}$$

Finally, the work needed to bring the fourth charge to point D when the other three charges are already in place is

$$W_4 = \frac{kqq}{a} + \frac{kqq}{a} + \frac{kqq}{\sqrt{2}a}$$

The total work required to assemble the four charges is

$$W_{\text{total}} = W_2 + W_3 + W_4 = \frac{4kqq}{a} + \frac{2kqq}{\sqrt{2}a} = \frac{(8 + 2\sqrt{2})kqq}{2a}$$

This work is the total electrostatic energy of the charge distribution.

20-4 Calculation of Electric Potential for Continuous Charge Distributions

In this section, we will calculate V for some important continuous charge distributions. The potential due to a continuous distribution of charge can be calculated from Equation 20-3 if the electric field is known, or it can be calculated from Equation 20-10 by choosing an element of charge dq, which we treat as a point charge, and changing the sum in Equation 20-10 to an integral:

$$V = \int \frac{k\,dq}{r} \qquad \text{20-12}$$

Potential due to a continuous charge distribution

We illustrate the application of Equation 20-12 by using it to calculate the electric potential on the axis of a uniformly charged ring and on the axis of a uniformly charged disk.

V on the Axis of a Charged Ring

Consider a uniformly charged ring of radius a and charge Q shown in Figure 20-8. In the figure, an element of charge dq is shown. The distance from this charge element to the field point P on the axis of the ring is $r = \sqrt{x^2 + a^2}$. Since this distance is the same for all elements of charge on the ring, we can remove this term from the integral in Equation 20-12. The potential at point P due to the ring is thus

$$V = \int \frac{k\,dq}{r} = \int \frac{k\,dq}{\sqrt{x^2 + a^2}}$$

$$= \frac{k}{\sqrt{x^2 + a^2}} \int dq = \frac{kQ}{\sqrt{x^2 + a^2}} \qquad \text{20-13}$$

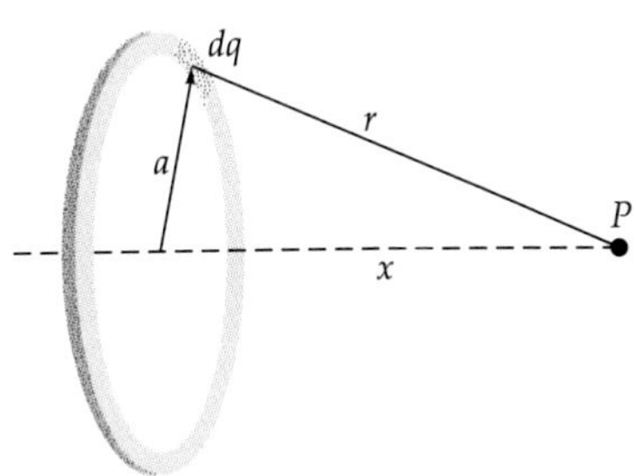

Figure 20-8 Geometry for the calculation of the electric potential at a point on the axis of a uniformly charged ring of radius a.

Example 20-8

A ring of radius 4 cm carries a uniform charge of 8 nC. A small particle of mass $m = 6$ mg $= 6 \times 10^{-6}$ kg and charge $q_0 = 5$ nC is placed at $x = 3$ cm and released. Find the speed of the charge when it is a great distance from the ring.

The potential energy of the charge q_0 at $x = 3$ cm is

$$U = q_0 V = \frac{kQq_0}{\sqrt{x^2 + a^2}}$$

$$= \frac{(8.99 \times 10^9 \text{ N·m}^2/\text{C}^2)(8 \times 10^{-9}\text{ C})(5 \times 10^{-9}\text{ C})}{\sqrt{(0.03\text{ m})^2 + (0.04\text{ m})^2}}$$

$$= 7.19 \times 10^{-6}\text{ J}$$

As the particle moves along the x axis away from the ring, its potential energy decreases and its kinetic energy increases. When the particle is very far from the ring, its potential energy is zero and its kinetic energy is 7.19×10^{-6} J. Its speed is then given by

$$\tfrac{1}{2}mv^2 = 7.19 \times 10^{-6}\text{ J}$$

$$v = \sqrt{\frac{2(7.19 \times 10^{-6}\text{ J})}{6 \times 10^{-6}\text{ kg}}} = 1.55\text{ m/s}$$

Exercise

What is the potential energy of the particle in Example 20-8 when it is at $x = 9$ cm? (Answer: 3.65×10^{-6} J)

V on the Axis of a Uniformly Charged Disk

We will now use Equation 20-13 to calculate the potential on the axis of a uniformly charged disk. Let the disk have a radius R and carry a total charge Q. The surface charge density on the disk is then $\sigma = Q/\pi R^2$. We take the x axis to be the axis of the disk and treat the disk as a set of ring charges. Figure 20-9 shows a ring of radius a and width da. The area of this ring is $2\pi a\, da$, and its charge is $dq = \sigma\, dA = \sigma 2\pi a\, da$. The potential at some point P on the x axis due to this charged ring element is given by Equation 20-13:

$$dV = \frac{k\,dq}{(x^2 + a^2)^{1/2}} = \frac{k2\pi\sigma a\, da}{(x^2 + a^2)^{1/2}}$$

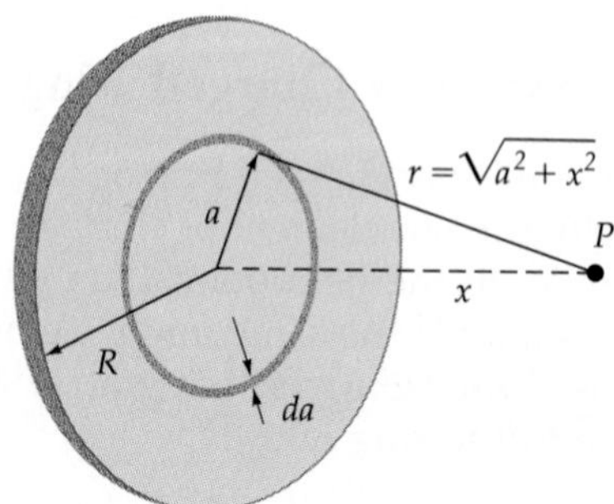

Figure 20-9 Geometry for the calculation of the electric potential at a point on the axis of a uniformly charged disk of radius R. The disk is divided into concentric rings of radius a and thickness da, each carrying charge $dq = \sigma\, dA = (Q/\pi R^2)\, 2\pi a\, da$.

The potential on the axis of the disk is found by integrating from $a = 0$ to $a = R$:

$$V = \int_0^R \frac{k2\pi\sigma a\, da}{(x^2 + a^2)^{1/2}} = k\sigma\pi \int_0^R (x^2 + a^2)^{-1/2}\, 2a\, da$$

This integral is of the form $\int u^n\,du$, with $u = x^2 + a^2$ and $n = -\frac{1}{2}$. The integration thus gives

$$V = k\sigma\pi \frac{(x^2 + a^2)^{+1/2}}{\frac{1}{2}}\Bigg|_{a=0}^{a=R} = 2\pi k\sigma[(x^2 + R^2)^{1/2} - x] \qquad 20\text{-}14$$

V near an Infinite Plane of Charge: Continuity of V

If we let R become very large, our disk approaches an infinite plane. As R becomes infinite the potential function (Equation 20-14) becomes infinite. We cannot apply Equation 20-12 to charge distributions that extend to infinity, such as an infinite plane of charge or an infinite line charge, because the potential cannot be chosen to be zero at infinity. For such cases, we first find the electric field **E** (by direct integration, or from Gauss's law) and then calculate the potential from its definition, Equation 20-3. For an infinite plane of charge of density σ in the yz plane, the electric field for positive x is given by

$$\mathbf{E} = \frac{\sigma}{2\epsilon_0}\mathbf{i}$$

The potential is calculated from its definition (Equation 20-3). If the potential at the yz plane where $x = 0$ is V_0, the potential at some arbitrary positive value of x is

$$V(x) - V_0 = -\int_0^x \mathbf{E}\cdot d\boldsymbol{\ell} = -\int_0^x \frac{\sigma}{2\epsilon_0}\mathbf{i}\cdot dx\,\mathbf{i} = -\frac{\sigma}{2\epsilon_0}\int_0^x dx = -\frac{\sigma}{2\epsilon_0}x$$

or

$$V(x) = V_0 - \frac{\sigma}{2\epsilon_0}x \qquad x > 0 \qquad 20\text{-}15a$$

For positive x, the potential has its maximum value V_0 at $x = 0$ and decreases linearly with distance from the plane. Since it does not approach any limiting value as x approaches infinity, we cannot choose the potential to be zero at $x = \infty$. We can, however, choose V to be zero at $x = 0$ or at any other point. For $x < 0$, the electric field is in the negative x direction and is given by

$$\mathbf{E} = -\frac{\sigma}{2\epsilon_0}\mathbf{i}$$

Repeating the calculation for the potential, using this function for the electric field, gives

$$V(x) = V_0 + \frac{\sigma}{2\epsilon_0}x \qquad x < 0 \qquad 20\text{-}15b$$

Since x is negative in this equation, the potential has its maximum value of V_0 at $x = 0$ and again decreases linearly with distance from the plane. Figure 20-10 shows a plot of V versus x. Note that this function is continuous at $x = 0$, even though the electric field E_x is discontinuous there. In Chapter 19, we saw that the electric field is discontinuous by σ/ϵ_0 at a point where there is a surface charge density σ. The potential function, on the other hand, is continuous everywhere in space. We can see this from its definition. Consider two nearby points x_1 and x_2. If V_1 is the potential at x_1 and V_2 is that at x_2, the potential difference can be written

$$\Delta V = (E_x)_{\text{av}}\,\Delta x = (E_x)_{\text{av}}\,(x_2 - x_1)$$

where $(E_x)_{\text{av}}$ is the average value of the electric field between the points. As x_2 approaches x_1, the potential difference ΔV approaches zero as long as $(E_x)_{\text{av}}$ is not infinite. Physically, if a test charge is moved a distance Δx, the work done by the field approaches zero as Δx approaches zero, as long as the electric field is not infinite.

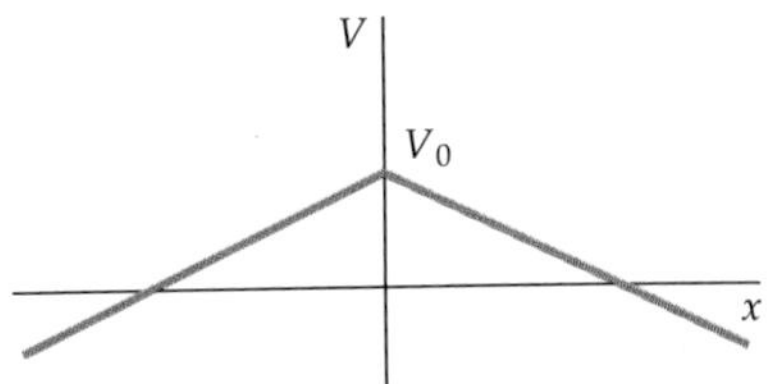

Figure 20-10 Plot of V versus x for an infinite plane of charge in the yz plane. The potential is continuous at $x = 0$ even though the electric field is not.

Example 20-9

An infinite plane of charge density σ is parallel to the yz plane at $x = -a$, and a point charge q is at the origin as shown in Figure 20-11. Find the potential at some point P a distance r from the point charge, for $x > -a$ (that is, to the right of the plane charge).

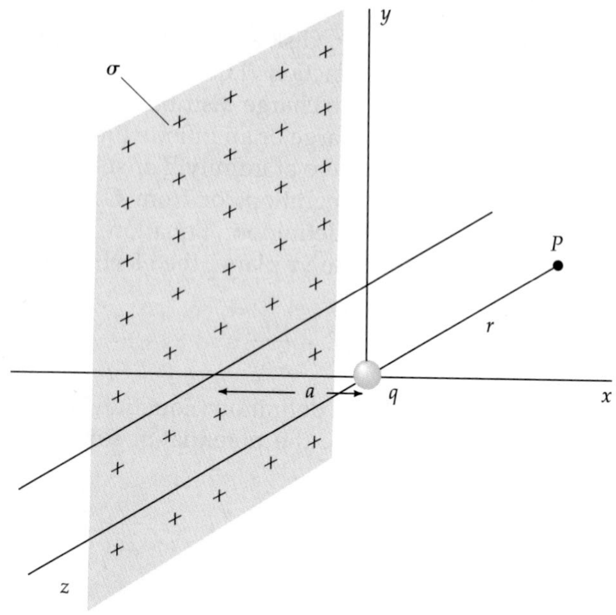

Figure 20-11 A point charge at the origin and an infinite plane of charge at $x = -a$ for Example 20-9.

The potential V_{plane} at a distance x' from the infinite plane due to the charge on the plane is given by Equation 20-15a with x' replacing x:

$$V_{\text{plane}} = A - \frac{\sigma}{2\epsilon_0}x'$$

where $x' = x + a$ and A is a constant that depends on the choice of zero potential. Because we cannot choose the potential to be zero at $r = \infty$, we use Equation 20-8 for the potential due to the point charge at the origin:

$$V_q = \frac{kq}{r} + B$$

where B is a constant that depends on the choice of zero potential. The potential due to both the infinite-plane charge and the point charge is

$$V = V_{\text{plane}} + V_q$$

$$= A - \frac{\sigma}{2\epsilon_0}x' + \frac{kq}{r} + B$$

$$= \frac{kq}{r} - \frac{\sigma}{2\epsilon_0}x' + C \qquad \text{20-16}$$

where we have combined the constants $A + B = C$. Let us choose the potential to be zero at the point at which the x axis intersects the infinite plane of charge. The coordinates of this point are $x = -a$, $y = 0$, $z = 0$. At this point $x' = 0$, and $r = a$. Equation 20-16 then gives

$$V = +\frac{kq}{a} + C = 0$$

$$C = -\frac{kq}{a}$$

which determines the constant C. At any general point, the potential is then given by

$$V = \frac{kq}{r} - \frac{kq}{a} - \frac{\sigma}{2\epsilon_0}x' = \frac{kq}{r} - \frac{kq}{a} - \frac{\sigma}{2\epsilon_0}(x + a)$$

In rectangular coordinates, $r = (x^2 + y^2 + z^2)^{1/2}$ and V is given by

$$V = \frac{kq}{(x^2 + y^2 + z^2)^{1/2}} - \frac{\sigma}{2\epsilon_0}(x + a) - \frac{kq}{a}$$

V Inside and Outside a Spherical Shell of Charge

We next find the potential due to a spherical shell of radius R with charge Q uniformly distributed on its surface. We are interested in the potential at all points inside and outside the shell. Since this shell is of finite extent, we could calculate the potential by direct integration of Equation 20-12, but this integration is somewhat difficult. Since the electric field for this charge distribution is easily obtained from Gauss's law, it is easiest to use Equation 20-3 to find the potential from the known electric field.*

Outside the spherical shell, the electric field is radial and is the same as if all the charge were at the origin:

$$\mathbf{E} = \frac{kQ}{r^2}\hat{\mathbf{r}}$$

The change in the potential for some displacement $d\boldsymbol{\ell} = dr\,\hat{\mathbf{r}}$ outside the shell is then

$$dV = -\mathbf{E}\cdot d\boldsymbol{\ell} = -\frac{kQ}{r^2}\hat{\mathbf{r}}\cdot dr\,\hat{\mathbf{r}} = -\frac{kQ}{r^2}\,dr$$

This is the same as Equation 20-7 for a point charge at the origin. Integrating, we obtain

$$V = \frac{kQ}{r} + V_0$$

where V_0 is the potential at $r = \infty$. Choosing the potential to be zero at $r = \infty$, we obtain

$$V = \frac{kQ}{r} \qquad r > R$$

Inside the spherical shell, the electric field is zero. The change in potential for any displacement inside the shell is therefore zero. Thus, the potential must be constant everywhere inside the shell. As r approaches R from outside the shell, the potential approaches kQ/R. Hence, the constant value of V inside must be kQ/R to make V continuous. Thus,

$$V = \begin{cases} \dfrac{kQ}{R} & r \leq R \\ \dfrac{kQ}{r} & r \geq R \end{cases} \qquad \text{20-17}$$

Potential due to a spherical shell

*The calculation of the electric field **E** for a uniformly charged spherical shell by direct integration of Coulomb's law is even more difficult than the integration to find V, because V is a scalar whereas **E** is a vector. The direct calculation of **E** is similar to the calculation of the gravitational field due to a spherical shell as discussed in Section 10-7.

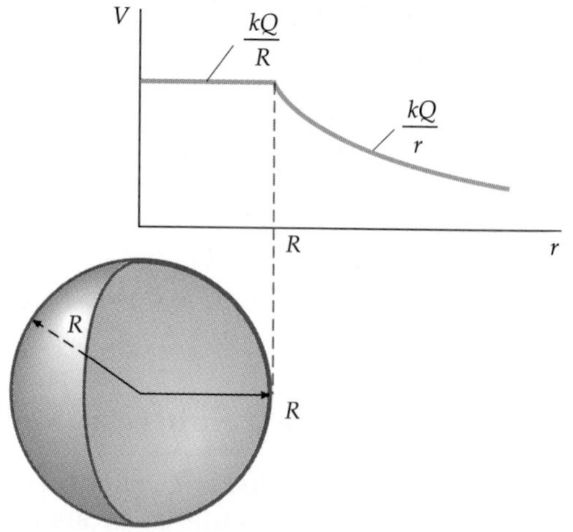

Figure 20-12 Electric potential of a uniformly charged spherical shell of radius R as a function of the distance r from the center of the shell. Inside the shell, the potential has the constant value kQ/R. Outside the shell the potential is the same as that due to a point charge at the center of the sphere.

This potential function is plotted in Figure 20-12.

A common mistake is to think that the potential must be zero inside a spherical shell because the electric field is zero. Actually, zero electric field merely implies that the potential does not change. Consider a spherical shell with a small hole so that we can move a test charge in and out of the shell. If we move the test charge from an infinite distance to the shell, the work per charge we must do is kQ/R. Inside the shell, there is no electric field, so it takes no work to move the test charge inside the shell. The total amount of work per charge it takes to bring the test charge from infinity to any point inside the shell is just the work per charge it takes to bring it up to the shell radius R, which is kQ/R. The potential is therefore kQ/R everywhere inside the shell.

Exercise

What is the potential of a spherical shell of radius 10 cm carrying a charge of 6 μC? (Answer: 5.39×10^5 V = 539 kV)

V near an Infinite Line Charge

We will now calculate the potential due to a uniform infinite line charge. Let the charge per unit length be λ. Since this charge distribution extends to infinity, we cannot use Equation 20-12 to find the potential. In Chapter 19, we found that the electric field produced by an infinite line charge points away from the line (if λ is positive) and is given by $E_r = 2k\lambda/r$. Equation 20-3 then gives for the change in potential

$$dV = -\mathbf{E}\cdot d\boldsymbol{\ell} = -E_r\, dr = -\frac{2k\lambda}{r}\, dr$$

Integrating, we obtain

$$V = V_0 - 2k\lambda \ln r \qquad \text{20-18}$$

For a positive line charge, the electric-field lines point away from the line, and the potential decreases with increasing distance from the line charge. At large values of r, the potential decreases without limit. The potential therefore cannot be chosen to be zero at $r = \infty$. (It cannot be chosen to be zero at $r = 0$ either, because $\ln r$ approaches $-\infty$ as r approaches zero.) Instead, we choose V to be zero at some distance $r = a$. Substituting $r = a$ into Equation 20-18 and setting $V = 0$, we obtain

$$V = 0 = V_0 - 2k\lambda \ln a$$

or

$$V_0 = 2k\lambda \ln a$$

Then Equation 20-18 is

$$V = 2k\lambda \ln a - 2k\lambda \ln r$$

or

$$V = -2k\lambda \ln \frac{r}{a} \qquad \text{20-19}$$

Potential due to a line charge with $V = 0$ at $r = a$

Questions

5. In the calculation of V for a ring of charge, does it matter whether the charge Q is uniformly distributed around the ring? Would either V or E_x be different if it were not?
6. If the electric potential is constant throughout a region of space, what can you say about the electric field in that region?

20-5 Electric Field and Potential

The electric-field lines point in the direction of decreasing potential. If the potential is known, it can be used to calculate the electric field. Consider a small displacement $d\boldsymbol{\ell}$ in an arbitrary electric field **E**. The change in potential is

$$dV = -\mathbf{E}\cdot d\boldsymbol{\ell} = -E_\ell \, d\ell \qquad \text{20-20}$$

where E_ℓ is the component of **E** parallel to the displacement. If we divide by $d\ell$, we have

$$E_\ell = -\frac{dV}{d\ell} \qquad \text{20-21}$$

If the displacement $d\boldsymbol{\ell}$ is perpendicular to the electric field, the potential does not change. The greatest change in V occurs when the displacement $d\boldsymbol{\ell}$ is parallel or antiparallel to **E**. A vector that points in the direction of the greatest change in a scalar function and has a magnitude equal to the derivative of that function with respect to the distance in that direction is called the **gradient** of the function. The electric field **E** is the negative gradient of the potential V. The field lines point in the direction of the greatest decrease in the potential function. In vector notation, the gradient of V is written ∇V. Thus,

$$\mathbf{E} = -\nabla V \qquad \text{20-22}$$

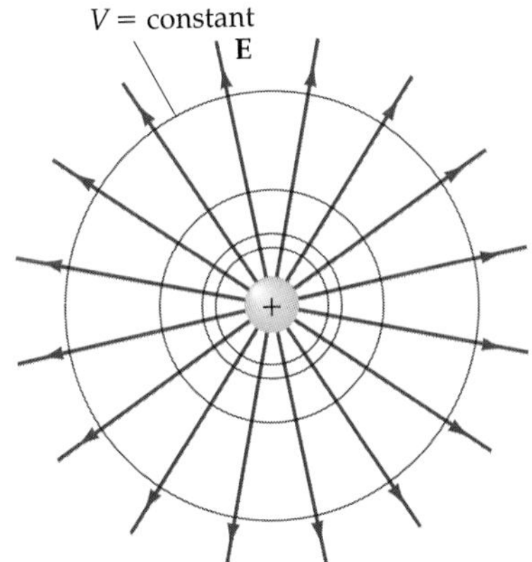

Figure 20-13 Equipotential surfaces and electric-field lines outside a point charge q. The field lines are radial and the equipotential surfaces are spherical. The electric-field lines are always perpendicular to an equipotential surface.

Figure 20-13 shows the electric-field lines due to a point charge q at the origin. If we move a test charge perpendicular to these lines, no work is done and the potential does not change. A surface on which the electric potential is constant is called an **equipotential surface.** For the potential $V = kq/r$ produced by a point charge at the origin, the equipotential surfaces are spherical surfaces defined by r = constant. Later in this chapter, we will see that the surface of any conductor in electrostatic equilibrium is an equipotential surface. The electric-field lines are always perpendicular to an equipotential surface. For a point charge at the origin, the electric-field lines are radial lines, and the equipotential surfaces are spheres. A displacement parallel to a radial electric field is written $d\boldsymbol{\ell} = dr\,\hat{\mathbf{r}}$. Equation 20-20 is then

$$dV = -\mathbf{E}\cdot d\boldsymbol{\ell} = -\mathbf{E}\cdot dr\,\hat{\mathbf{r}} = -E_r \, dr$$

and

$$E_r = -\frac{dV}{dr} \qquad \text{20-23}$$

For any spherically symmetric charge distribution, the potential varies only with r, and the electric field is related to the potential by

$$\mathbf{E} = -\nabla V = -\frac{dV}{dr}\,\hat{\mathbf{r}} \qquad \text{20-24}$$

For a uniform electric field in the x direction, for example, that is produced by an infinite plane charge in the yz plane, the electric-field lines are parallel lines in the x direction, and the equipotential surfaces are planes parallel to the yz plane. Then, the potential function V can depend only on x. The displacement vector that is parallel to this field is given by

$$d\boldsymbol{\ell} = dx\,\mathbf{i} \qquad \text{20-25}$$

For this case, Equation 20-21 is

$$E_x = -\frac{dV}{dx}$$

and the electric field is

$$\mathbf{E} = -\frac{dV}{dx}\,\mathbf{i} \qquad \text{20-26}$$

In general, the potential function can depend on x, y, and z. The rectangular components of the electric field are related to the partial derivatives of the potential with respect to x, y, or z while the other variables are held constant. For example, the x component of the electric field is given by

$$E_x = -\frac{\partial V}{\partial x} \qquad \text{20-27}a$$

Similarly, the y and z components of the electric field are related to the potential by

$$E_y = -\frac{\partial V}{\partial y}$$

and

$$E_z = -\frac{\partial V}{\partial z} \qquad \text{20-27}b$$

Thus, Equation 20-22 in rectangular coordinates is

$$\mathbf{E} = -\nabla V = -\left(\frac{\partial V}{\partial x}\,\mathbf{i} + \frac{\partial V}{\partial y}\,\mathbf{j} + \frac{\partial V}{\partial z}\,\mathbf{k}\right) \qquad \text{20-28}$$

Example 20-10

Find the electric field for the electric potential function $V(x)$ given by $V(x) = 100\text{ V} - (25\text{ V/m})x$.

This potential function depends only on x. The electric field is found from Equation 20-26:

$$\mathbf{E} = -\frac{dV}{dx}\,\mathbf{i} = +25\text{ V/m}\,\mathbf{i}$$

This electric field is uniform and in the x direction. Note that the constant 100 V in the expression for $V(x)$ has no effect on the electric field. The electric field does not depend on the choice of zero for the potential function.

Exercise

(*a*) At what point is $V = 0$ in Example 20-10? (*b*) Write the potential function corresponding to the same electric field with $V = 0$ at $x = 0$. [Answers: (*a*) $x = 4$ m, (*b*) $V = -(25\ \text{V/m})\ x$]

Example 20-11

Find the electric field for the dipole charge distribution of Example 20-6.

In that example, we found the potential on the z axis at a great distance from the dipole to be

$$V = \frac{kp}{z^2}$$

where $p = 2qa$ is the magnitude of the dipole moment. The electric field at some point on the z axis is then given by

$$\mathbf{E} = -\frac{dV}{dz}\mathbf{k}$$

$$= -(-2)\frac{kp}{z^3}\mathbf{k} = \frac{2kp}{z^3}\mathbf{k}$$

which is the same as we found directly from Coulomb's law (Equation 18-10).

Example 20-12

Use the potential functions found in the previous section for the potential on the axis of a uniformly charged ring and of a uniformly charged disk to find the electric field on the axis of these charge distributions.

The potential on the axis of a uniformly charged ring of total charge Q is given by Equation 20-13:

$$V = \frac{kQ}{\sqrt{x^2 + a^2}} = kQ(x^2 + a^2)^{-1/2}$$

The electric field is then

$$\mathbf{E} = -\frac{dV}{dx}\mathbf{i} = -\left(-\frac{1}{2}\right)kQ(x^2 + a^2)^{-3/2}(2x)\mathbf{i}$$

$$= \frac{kQx}{(x^2 + a^2)^{3/2}}\mathbf{i}$$

This is the same as Equation 19-12, which we found directly from Coulomb's law.

The potential on the axis of a uniformly charged disk is given by Equation 20-14:

$$V = 2\pi k\sigma[(x^2 + R^2)^{1/2} - x]$$

Again, the electric field is found by taking the gradient of this expression:

$$\mathbf{E} = -\frac{dV}{dx}\mathbf{i} = 2\pi k\sigma\left(1 - \frac{x}{\sqrt{x^2 + R^2}}\right)\mathbf{i}$$

This is the same as Equation 19-13, which we obtained by direct calculation from Coulomb's law.

20-6 Equipotential Surfaces, Charge Sharing, and Dielectric Breakdown

We have seen that there is no electric field inside a conductor in static equilibrium. Thus, there is no force on a test charge, and no work is done on a test charge as it moves about inside a conductor. The electric potential is thus the same throughout the conductor, that is, the volume occupied by the conductor is an equipotential volume. A surface on which the potential is constant is called an equipotential surface. The surface of a conductor is an equipotential surface. If a test charge is given a displacement $d\boldsymbol{\ell}$ parallel to an equipotential surface, $dV = -\mathbf{E} \cdot d\boldsymbol{\ell} = 0$ so the electric field lines are perpendicular to an equipotential surface. Figures 20-14 and 20-15 show equipotential surfaces near a spherical conductor and near a nonspherical conductor. Note that the field lines are everywhere perpendicular to these surfaces. If we move a short distance $d\ell$ along a field line from one equipotential surface to another, the potential changes by $dV = -\mathbf{E} \cdot d\boldsymbol{\ell} = -E\, d\ell$. If E is large, equipotential surfaces with a fixed potential difference between them are closely spaced.

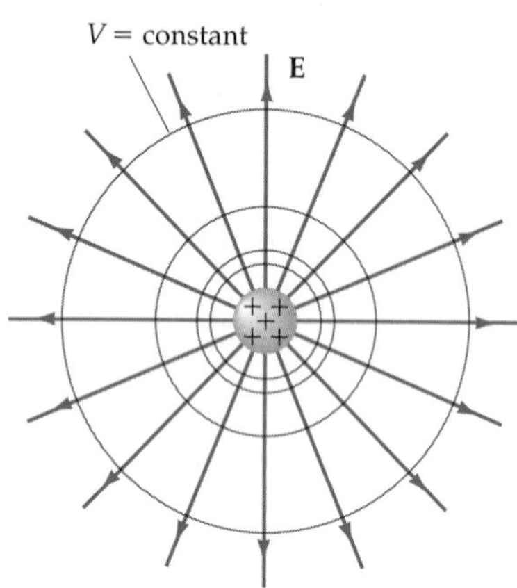

Figure 20-14 Equipotential surfaces and electric-field lines outside a uniformly charged spherical conductor. The equipotential surfaces are spherical and the field lines are radial and perpendicular to the equipotential surfaces.

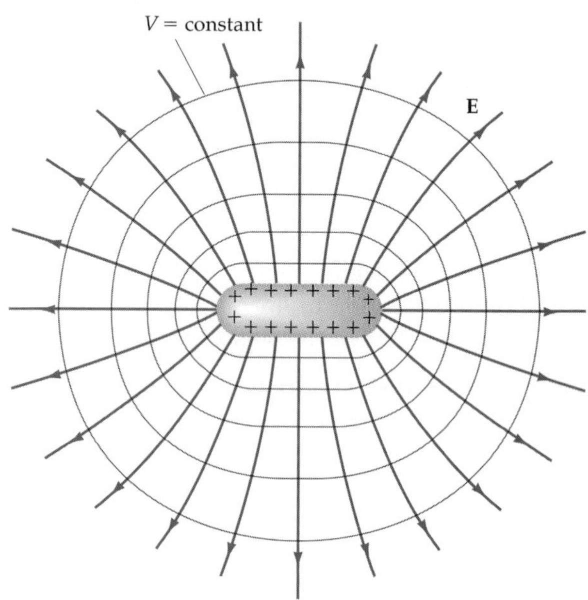

Figure 20-15 Equipotential surfaces and electric-field lines outside a nonspherical conductor. Electric-field lines are always perpendicular to equipotential surfaces.

Example 20-13

A hollow spherical conductor that is uncharged has inner radius a and outer radius b. A positive point charge $+q$ is in the cavity at the center of the sphere. Find the potential $V(r)$ everywhere, assuming that $V = 0$ at $r = \infty$.

As discussed in Chapter 19, the electric-field lines from the point charge end on the inner surface of the shell at $r = a$, where there is an induced charge $-q$ that is uniformly distributed on the inner surface. Since the conducting shell is uncharged, there is a positive charge $+q$ uniformly distributed on the outer surface at $r = b$. We thus have three charges: a point charge q at the origin, a spherical shell of total charge $-q$ and radius a, and a second spherical shell of total charge $+q$ and radius b. Outside the sphere, the electric field is the same as if the shell were not there and is given by $E_r = kq/r^2$. The electric potential outside the shell is then given by

$$V = \frac{kq}{r} \qquad r > b$$

Since the electric potential must be continuous everywhere, the potential at $r = b$ is $V = kq/b$. This is the potential everywhere inside the conducting material since the conductor is an equipotential volume. Thus,

$$V = \frac{kq}{b} \qquad a \leq r \leq b$$

Inside the cavity, the electric field is again $E_r = kq/r^2$. The potential for $r < a$ is therefore given by

$$V = \frac{kq}{r} + V_0$$

where V_0 is a constant. This constant is not determined by the condition $V = 0$ at $r = \infty$, because r does not go to infinity inside the cavity. Instead, we determine the constant V_0 by the condition that V must be continuous at $r = a$. Since $V = kq/b$ everywhere inside the conducting material, it must have this value at $r = a$. Thus, at $r = a$, we have

$$V = \frac{kq}{a} + V_0 = \frac{kq}{b}$$

which means that V_0 is given by

$$V_0 = \frac{kq}{b} - \frac{kq}{a}$$

Inside the cavity, the potential is thus given by

$$V = \frac{kq}{r} + \frac{kq}{b} - \frac{kq}{a} \qquad r < a$$

Figure 20-16 shows a plot of V versus r.

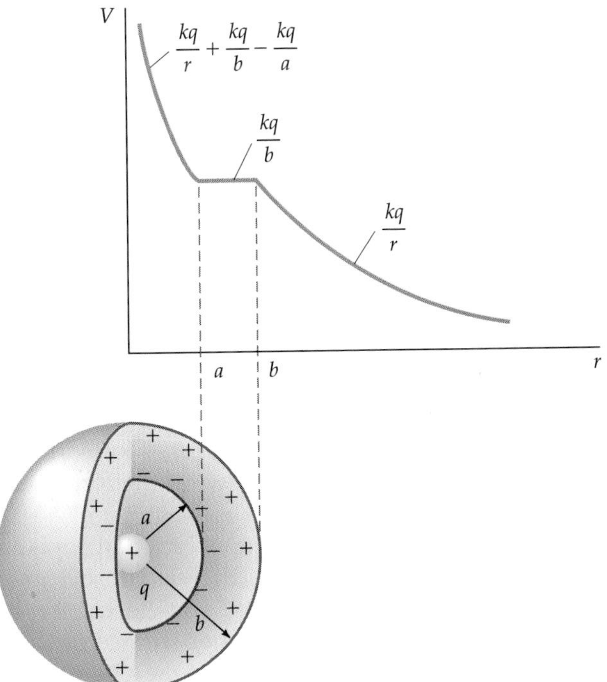

Figure 20-16 Plot of the electric potential V for a point charge at the center of an uncharged, hollow spherical conductor (Example 20-13) as a function of the distance r from the center of the cavity. Inside the conducting material, where $a \leq r \leq b$, the potential has the constant value kq/b. Outside the shell the potential is the same as that of a point charge.

In general, two conductors that are separated in space will not be at the same potential. The potential difference between the conductors depends on their geometrical shapes, their separation, and the net charge on each. When two conductors are brought into contact, the charge on the conductors distributes itself so that electrostatic equilibrium is established and the electric field is zero inside both conductors. While in contact, the two conductors may be considered to be a single conductor with a single equipotential surface. The transfer of charge from one conductor to another is called **charge sharing.**

Consider a spherical conductor of radius R carrying a charge $+Q$. The electric-field lines outside the conductor point radially outward, and the potential of the conductor relative to infinity is kQ/R. If we bring up a second, uncharged conductor, the potential and field lines will change. Electrons on the uncharged conductor will be attracted by the positive charge Q, leaving the near side of the uncharged conductor with a negative charge and the far side with a positive charge (Figure 20-17). This charge separation on the neutral conductor will affect the originally uniform charge distribution on the charged conductor. Although the detailed calculation of the charge distributions and potential for this case is quite complicated, we can see that some of the field lines leaving the positive conductor end on the negative charge on the near side of the neutral conductor, and an equal number of lines leave the far side of that conductor. Since electric-field lines point toward regions of lower potential, the positively charged conductor must be at a greater potential than the neutral conductor.

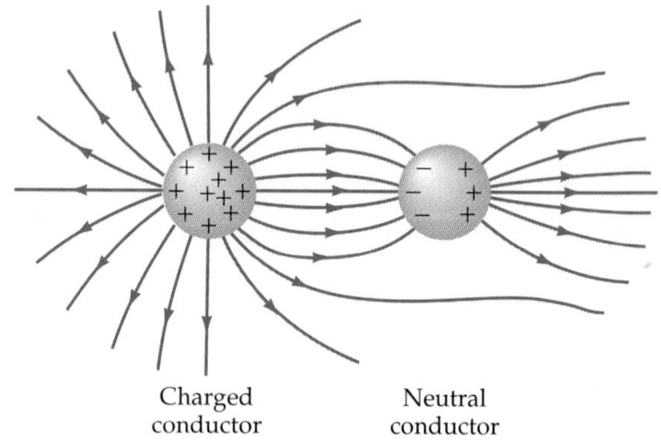

Figure 20-17 Electric-field lines for a charged spherical conductor near an uncharged spherical conductor. Some of the field lines leaving the charged conductor will end on the induced negative charge on the neutral conductor. Since electric-field lines point from regions of high potential to regions of low potential, the neutral conductor must be at a lower potential than the charged conductor.

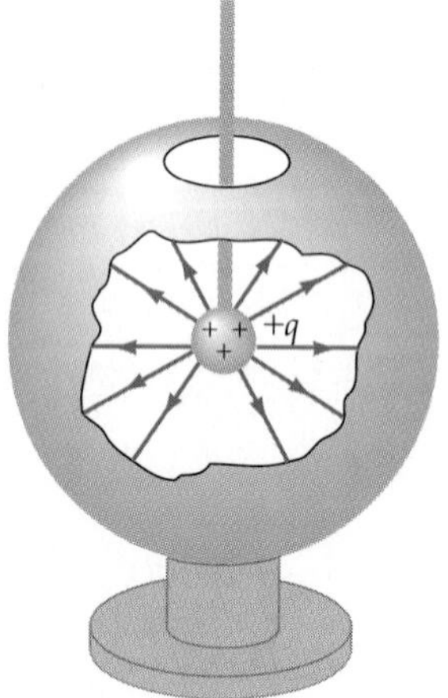

Figure 20-18 Small conductor carrying a positive charge inside a larger conductor.

If we put the two conductors in contact, positive charge will flow to the neutral conductor until both conductors are at the same potential. (Actually, negative electrons flow from the neutral conductor to the positive conductor, but it is slightly more convenient to think of this as a flow of positive charge in the opposite direction.) If the conductors are identical, they will share the original charge equally. If the conductors are now separated, each will carry charge $\frac{1}{2}Q$, and both will be at the same potential. Coulomb used this method of charge sharing to produce various charges of known ratios to some original charge in his experiment to find his law for the force between two small (point) charges.

In Figure 20-18, a small conductor carrying a positive charge q is inside the cavity of a larger conductor. In equilibrium, the electric field is zero inside the conducting material of both conductors. The electric-field lines that leave the positive charge q must end on the inner surface of the large conductor. This must occur no matter what the charge may be on the outside surface of the large conductor. Regardless of the charge on the large conductor, the small conductor in the cavity is at a greater potential because the electric-field lines go from this conductor to the larger conductor. If the conductors are now connected, say, with a fine conducting wire, *all* the charge originally on the smaller conductor will flow to the larger one. When the connection is broken, there is no charge on the small conductor in the cavity, and there are no field lines anywhere within the outer surface of the large conductor. The positive charge transferred from the smaller conductor resides completely on the outside surface of the larger conductor. If we put more positive charge on the small conductor in the cavity and again connect the conductors with a fine wire, all of the charge on the inner conductor will again flow to the outer conductor. This procedure can be repeated indefinitely. This method is used to produce large potentials in the Van de Graaff generator, in which the charge is brought to the inner surface of a larger spherical conductor by a continuous charged belt (see Figure 20-19). Work must be done by the motor driving the belt to bring the charge to the outer sphere, which is at a high potential. The greater the net charge on the outer conductor, the greater its potential.

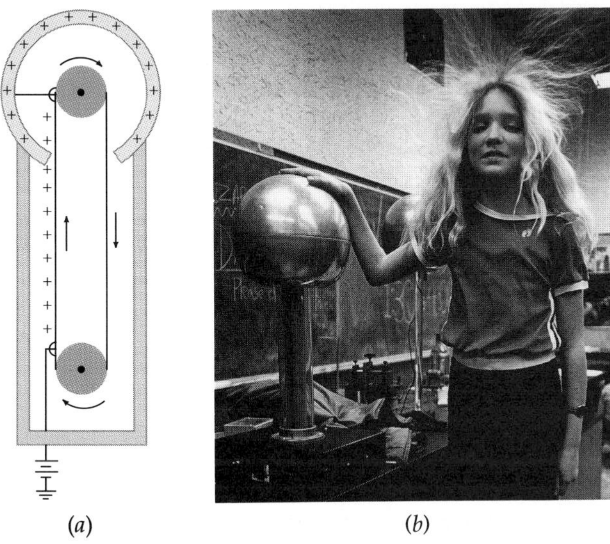

(*a*) (*b*)

(*c*)

Figure 20-19 (*a*) Schematic diagram of a Van de Graaff generator. Charge leaks off the pointed conductor near the bottom onto the belt. Near the top, the charge leaks off of the belt onto the pointed conductors attached to the large spherical conductor. (*b*) This girl has been charged to a very high potential through contact with a demonstration Van de Graaff generator while standing on an insulating block. Her hair has acquired sufficient charge to show electrostatic repulsion. Some care must be taken to acquire the charge gradually and to avoid rapid discharge to prevent a painful shock. (*c*) These large demonstration Van de Graaff generators in the Boston science museum are discharging to the grounded wire cage housing the operator.

The maximum potential obtainable in this way is limited only by the fact that air molecules become ionized in very high electric fields and the air becomes a conductor. This phenomenon, which is called **dielectric breakdown,** occurs in air at electric field strengths of $E_{\text{max}} \approx 3 \times 10^6$ V/m = 3 MV/m. The electric field strength for which dielectric breakdown occurs in a material is called the **dielectric strength** of that material. The dielectric strength of air is thus about 3 MV/m. The discharge through the conducting

(*a*) An electrical tree produced by arc discharge in a piece of plastic. The plastic is charged by an electron beam that penetrates about 0.5 cm. After the beam is turned off, the block is given a tap with a metal punch. The electrons suddenly shoot out leaving a pattern of tracks in the plastic. (*b*) A similar electrical tree produced by lightning in air.

air resulting from dielectric breakdown is called **arc discharge.** The electric shocks you receive when you touch a metal door knob after walking across a rug on a dry day is a familiar example of arc discharge. (This occurs more often on dry days because moist air will conduct away some of the charge you acquire by walking on the rug before you accumulate enough charge to reach a high potential.) Lightning is another familiar example of arc discharge.

Example 20-14

A spherical conductor has a radius of 2 m. (*a*) What maximum charge can be placed on the sphere before dielectric breakdown? (*b*) What is the maximum potential of the sphere?

(*a*) The electric field just outside the conductor carrying a surface charge σ is

$$E = \frac{\sigma}{\epsilon_0}$$

Setting this equal to the maximum electric field in air, we obtain for σ_{max}

$$E_{max} = 3 \times 10^6 \text{ N/C} = \frac{\sigma_{max}}{\epsilon_0}$$

The maximum charge on the sphere is then

$$\begin{aligned} Q &= 4\pi R^2 \sigma_{max} \\ &= 4\pi R^2(\epsilon_0 E_{max}) = 4\pi(2\text{ m})^2(8.85 \times 10^{-12}\text{ C}^2/\text{N}\cdot\text{m}^2)(3 \times 10^6\text{ N/C}) \\ &= 1.33 \times 10^{-3}\text{ C} \end{aligned}$$

(*b*) The maximum potential of the sphere carrying this charge is

$$\begin{aligned} V_{max} &= \frac{kQ}{R} = \frac{(8.99 \times 10^9\text{ N}\cdot\text{m}^2/\text{C}^2)(1.33 \times 10^{-3}\text{ C})}{2\text{ m}} \\ &= 5.98 \times 10^6\text{ V} \end{aligned}$$

When a charge is placed on a conductor of nonspherical shape, like that in Figure 20-20*b*, the surface of the conductor will be an equipotential surface, but the surface charge density (the charge per unit area) and the electric field just outside the conductor will vary from point to point. Near a point where the radius of curvature is small, such as point A in the figure, the surface charge density and electric field will be large; whereas near a point where the radius of curvature is large, such as point B in the figure, they will be small. We can understand this qualitatively by considering the ends of the conductor to be spheres of different radii. Let σ be the surface charge density. The potential of a sphere of radius r is

$$V = \frac{kq}{r} = \frac{1}{4\pi\epsilon_0}\frac{q}{r} \qquad \text{20-29}$$

Since the area of a sphere is $4\pi r^2$, the charge on a sphere is related to the charge density by

$$q = 4\pi r^2 \sigma$$

Substituting this expression for q into Equation 20-29, we have

$$V = \frac{1}{4\pi\epsilon_0}\frac{4\pi r^2 \sigma}{r} = \frac{r\sigma}{\epsilon_0}$$

Solving for σ, we obtain

$$\sigma = \frac{\epsilon_0 V}{r} \qquad \text{20-30}$$

Since both "spheres" are at the same potential, the one with the smaller radius must have the greater surface charge density. And since the electric field at the surface of a conductor is proportional to the surface charge density σ, the electric field is greatest at points on the conductor where the radius of curvature is least.

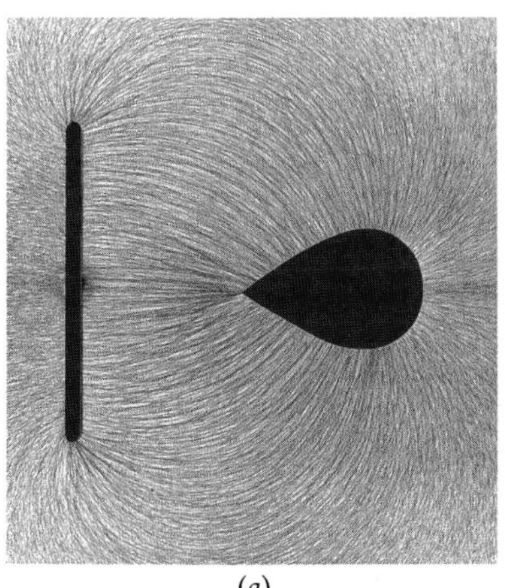

(*a*)

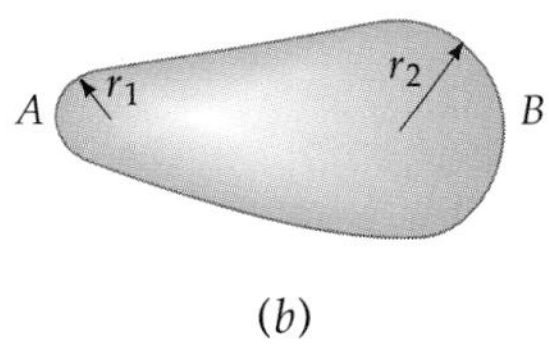

(*b*)

Figure 20-20 (*a*) Electric-field lines near a nonspherical conductor and plate carrying equal and opposite charges. The lines are shown by small bits of thread suspended in oil. The electric field is strongest near points of small radius of curvature, such as at the ends of the plate and at the pointed left side of the conductor. (*b*) A nonspherical conductor. If a charge is placed on such a conductor, it will produce an electric field that is stronger near point A, where the radius of curvature is small, than near point B, where the radius of curvature is large.

For an arbitrarily shaped conductor, the potential at which dielectric breakdown occurs depends on the smallest radius of curvature of any part of the conductor. If the conductor has sharp points of very small radius of curvature, dielectric breakdown will occur at relatively low potentials. In the Van de Graaff generator, the charge is transferred onto the belt by sharp-edged conductors near the bottom of the belt. The charge is removed from the belt by sharp-edged conductors near the top of the belt (Figure 20-19). Lightning rods at the top of a tall building draw the charge off a nearby cloud before the potential of the cloud can build up to a very large value.

Question

7. When you touch a friend after walking across a rug on a dry day, you typically draw a spark of about 2 mm. Estimate the potential difference between you and your friend before the spark.

Summary

1. Potential difference $V_b - V_a$ is defined as the negative of the work per unit charge done by the electric field when a test charge moves from point a to point b:

$$\Delta V = V_b - V_a = -\int_a^b \mathbf{E} \cdot d\boldsymbol{\ell}$$

For infinitesimal displacements this is written

$$dV = -\mathbf{E} \cdot d\boldsymbol{\ell}$$

Since only differences in electric potential are important, we can choose the potential to be zero at any convenient point. The potential at any point is the potential energy of a charge divided by the charge:

$$V = \frac{U}{q_0}$$

The SI unit of potential and potential difference is the volt (V):

$$1\ \text{V} = 1\ \text{J/C}$$

In terms of this unit, the unit for the electric field can be expressed:

$$1\ \text{N/C} = 1\ \text{V/m}$$

2. A convenient unit of energy in atomic and nuclear physics is the electron volt (eV), which is the potential energy of a particle of charge e at a point where the potential is 1 volt. The electron volt is related to the joule by

$$1\ \text{eV} = 1.6 \times 10^{-19}\ \text{J}$$

3. The electric potential at a distance r from a point charge q at the origin is given by

$$V = \frac{kq}{r} + V_0$$

where V_0 is the potential at an infinite distance from the charge. When the potential is chosen to be zero at infinity, the potential due to the point charge is

$$V = \frac{kq}{r}$$

For a system of point charges, the potential is given by

$$V = \sum_i \frac{kq_i}{r_{i0}}$$

where the sum is taken over all the charges and r_{i0} is the distance from the ith charge to the point P at which the potential is to be found.

4. The electrostatic potential energy of a system of point charges is the work needed to bring the charges from an infinite separation to their final positions.

5. For a continuous distribution of charge, the potential is found by integration over the charge distribution:

$$V = \int \frac{k\,dq}{r}$$

This expression can be used only if the charge distribution is contained in a finite volume so that the potential can be chosen to be zero at infinity.

6. The electric field points in the direction of the greatest decrease in the potential. The component of **E** in the direction of a displacement $d\boldsymbol{\ell}$ is related to the potential by

$$E_\ell = -\frac{dV}{d\ell}$$

A vector that points in the direction of the greatest change in a scalar function and has a magnitude equal to the derivative of that function with respect to the distance in that direction is called the gradient of the function. The electric field **E** is the negative gradient of the potential V. In vector notation, the gradient of V is written $\boldsymbol{\nabla}V$. Thus,

$$\mathbf{E} = -\boldsymbol{\nabla}V$$

For any spherically symmetric charge distribution, the potential varies only with r, and the electric field is related to the potential by

$$\mathbf{E} = -\boldsymbol{\nabla}V = -\frac{dV}{dr}\,\hat{\mathbf{r}}$$

In rectangular coordinates, the electric field is related to the potential by

$$\mathbf{E} = -\boldsymbol{\nabla}V = -\left(\frac{\partial V}{\partial x}\,\mathbf{i} + \frac{\partial V}{\partial y}\,\mathbf{j} + \frac{\partial V}{\partial z}\,\mathbf{k}\right)$$

7. On a conductor of arbitrary shape, the surface charge density σ is greatest at points where the radius of curvature is smallest.

8. The amount of charge that can be placed on a conductor is limited by the fact that air molecules become ionized in very high electric fields, and the air becomes a conductor—a phenomenon called dielectric breakdown, which occurs in air at electric field strengths of $E_{max} \approx 3 \times 10^6$ V/m = 3 MV/m. The electric field strength for which dielectric breakdown occurs in a material is called the dielectric strength of that material. The resulting discharge through the conducting air is called arc discharge.

Electrostatics and Xerography

Richard Zallen
Virginia Polytechnic Institute and State University

There are many important and beneficial technological applications that could be included in a discussion of uses of electrostatic phenomena. For example, a powerful air-pollution preventer is the electrostatic precipitator, which years ago made life livable near cement mills and ore-processing plants and which is currently credited with extracting better than 99 percent of the ash and dust from the gases about to issue from chimneys of coal-burning power plants. The basic idea of this very effective antipollution technique is shown in Figure 1. The outer wall of a vertical metal duct is grounded, while a wire running down the center of the duct is kept at a very large negative voltage. In this concentric geometry, a very nonuniform electric field is set up, with lines of force directed radially inward toward the negative wire electrode. Close to the wire the field attains enormous values, large enough to produce an electrical breakdown of air, and the normally placid mixture of neutral gas molecules is replaced by a turmoil of free electrons and positive ions. The electrons from this corona discharge are driven outward from the wire by the electric field. Most of them quickly become attached to oxygen molecules to produce negative O_2^- ions, which are also accelerated outward. As this stream of ions passes across the hot waste gas rising in the duct, small particles carried by the gas become charged by capturing ions and are pulled by the field to the outer wall. If the noxious particles are solid, they are periodically shaken down off the duct into a hopper; if they are liquid, the residue simply runs down the wall and is collected below.

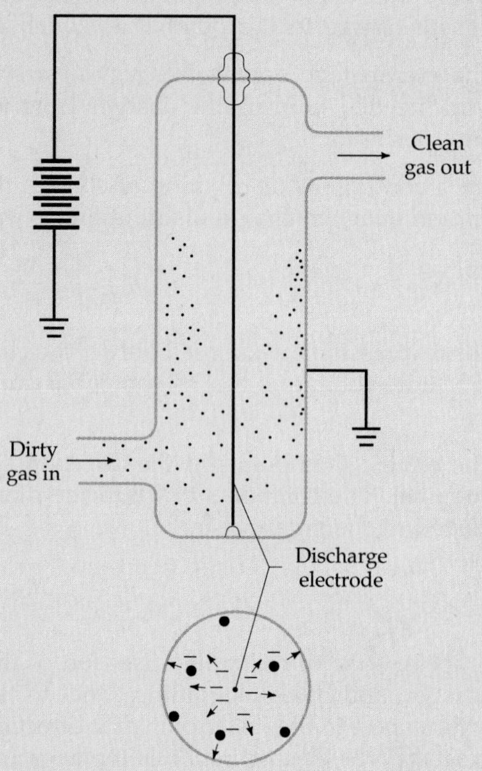

Figure 1 Schematic diagram of the use of a corona discharge in an electrostatic precipitator.

Richard Zallen received his education at Madison H.S. in Brooklyn, Rensselaer (B.S.), and Harvard (Ph.D.). He was elected a fellow of the American Physical Society in 1976. Before joining Virginia Tech in 1983, he worked for seventeen years at the Xerox Research Laboratories in Rochester, New York, which of course is where he learned about "Electrostatics and Xerography." He is married and has two children.

Throughout his career as a physicist, Professor Zallen has been involved with experimental studies of the interaction of light with solids such as semiconductors, molecular crystals, and amorphous solids. Most recently, he has been working on sol-gel systems and ion-bombarded semiconductors.

Professor Zallen is probably best known for his book *The Physics of Amorphous Solids,* Wiley, New York, 1983.

Beside electrostatic precipitation, other technological examples include electrocoating with spray paints and the electrostatic separation of granular mixtures used for the removal of rock particles from minerals, garlic seeds from wheat, even rodent excreta from rice. However, the application that is the main focus of this essay is xerography, the most widely used form of electrostatic imaging, or electrophotography. This is the most familiar use of

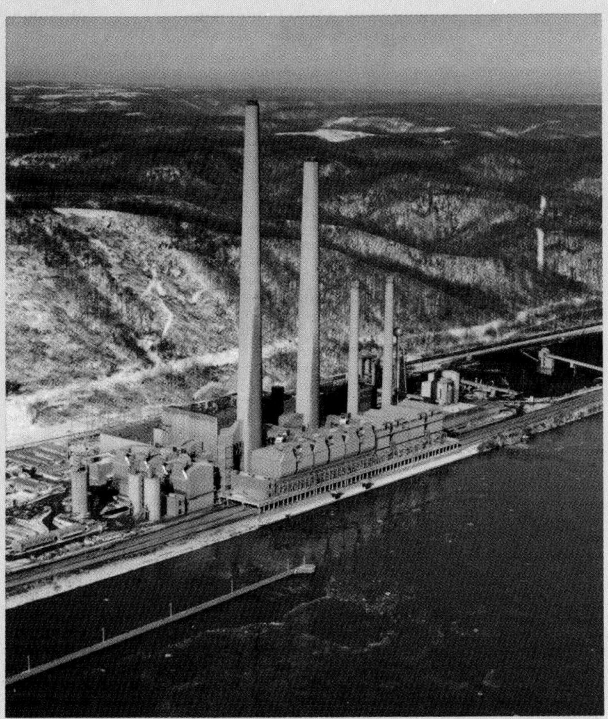

Electrostatic precipitators are housed in the gray box-like structures at the base of these smokestacks.

electrostatics in terms of the number of people who have occasion to use plain-paper copying machines in offices, libraries, and schools, and it also provides a fine example of a process utilizing a sequence of distinct electrostatic events.

The xerographic process was invented in 1937 by Chester Carlson. The term xerography, literally "dry writing," was actually adopted a bit later to emphasize the distinction from wet chemical processes. Carlson's innovative concept did not find early acceptance, and a practical realization of his idea became available only after a small company (in a famous entrepreneurial success story) risked its future in its intensive efforts to develop the process.

Four of the main steps involved in xerography are illustrated in Figure 2. In the interest of clarity the process has been oversimplified, and several subtleties (as well as gaps in our understanding) have been suppressed. Electrostatic imaging takes place on a large thin plate of a photoconducting material supported by a grounded metal backing. A photoconductor is a solid that is a good insulator *in the dark* but becomes capable of conducting electric current when exposed to light. In the dark, a uniform electrostatic charge is laid down on the surface of the photoconductor. This charging step (Figure 2*a*) is accomplished by means of a positive corona discharge surrounding a fine wire held at about +5000 V. This corona (a miniature version of, and opposite in sign to, the intense precipitator corona of Figure 1) is passed over the photoconductor surface, spraying positive ions onto it and charging it to a potential of the order of +1000 V. Since charge is free to flow within the grounded metal backing, an equal and opposite induced charge develops at the metal-photoconductor interface. In the dark the photoconductor contains no mobile charge, and the large potential difference persists across this dielectric layer, which is only 0.005 cm thick.

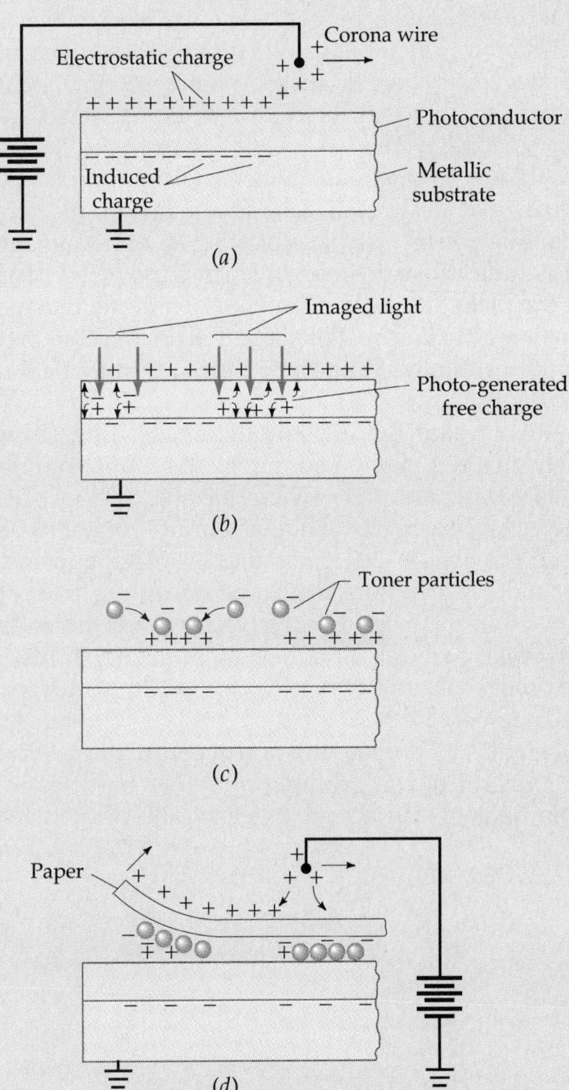

Figure 2 Steps in the xerographic process: (*a*) charging, (*b*) exposure, (*c*) development, and (*d*) transfer.

Continued

The photoconductor plate is next exposed to light in the form of an image reflected from the document being copied. What happens now is indicated in Figure 2*b*. Where light strikes the photoconductor, light quanta (photons) are absorbed, and pairs of mobile charges are created. Each photogenerated pair consists of a negative charge (an electron) and a positive charge (a hole; crudely, a missing electron). Photogeneration of this free charge depends not only on the photoconductor used and on the wavelength and intensity of the incident light but also on the electric field present. This large field (1000 V/0.005 cm $= 2 \times 10^5$ V/cm $=$ 2×10^7 V/m) helps to pull apart the mutually attracting electron-hole pairs so that they are free to move separately. The electrons then move under the influence of the field to the surface, where they neutralize positive charges, while the holes move to the photoconductor-substrate interface and neutralize negative charges there. Where intense light strikes the photoconductor, the charging step is totally undone; where weak light strikes it, the charge is partially reduced; and where no light strikes it, the original electrostatic charge remains on the surface. The critical task of converting an optical image into an electrostatic image, which is now recorded on the plate, has been completed. This latent image consists of an electrostatic potential distribution that replicates the light and dark pattern of the original document.

To develop the electrostatic image, fine negatively charged pigmented particles are brought into contact with the plate. These *toner particles* are attracted to positively charged surface regions, as shown in Figure 2*c*, and a visible image appears. The toner is then transferred (Figure 2*d*) to a sheet of paper that has been positively charged in order to attract the particles. Brief heating of the paper fuses the toner to it and produces a permanent photocopy ready for use.

Finally, to prepare the photoconductor plate for a repetition of the process, any toner particles remaining on its surface are mechanically cleaned off, and the residual electrostatic image is erased, that is, discharged, by flooding with light. The photoconductor is now ready for a new cycle, starting with the charging step. In high-speed duplicators, the photoconductor layer is often in the form of a moving continuous drum or belt, around the perimeter of which are located stations for performing the various functions of Figure 2. The speed of xerographic printing technology is presently on the order of a few copies per second.*

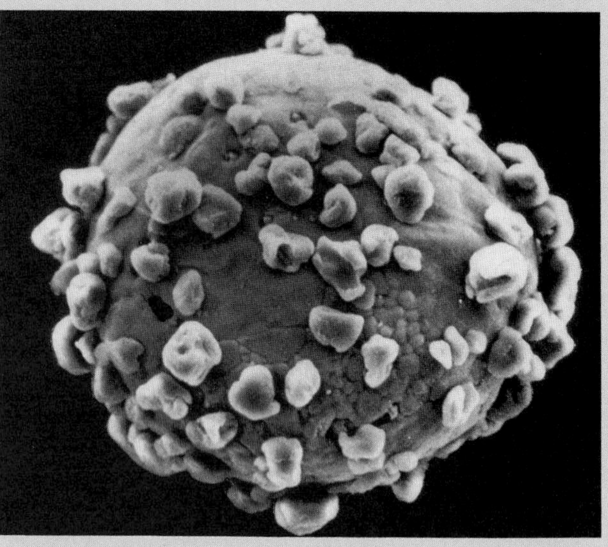

Toner particles, electrostatically attracted to a larger carrier particle.

*For further information on electrostatics in xerography, consult J. H. Dessauer and H. E. Clark (eds.), *Xerography and Related Processes*, Focal Press, New York, 1965, and R. M. Schaffert, *Electrophotography*, rev. ed., Focal Press, New York, 1973. Other applications of electrostatics are discussed in A. D. Moore, *Scientific American*, March 1972.

Suggestions for Further Reading

Moore, A. D.: "Electrostatics," *Scientific American*, March 1972, p. 46.

This article describes some modern uses of electrostatics, including precipitation of airborne industrial wastes, separation of granular solids such as minerals, efficient paint-spraying, and xerographic copying.

Rose, Peter H., and Andrew B. Wittkower: "Tandem Van de Graaff Accelerators," *Scientific American*, August 1970, p. 24.

These machines accelerate charged particles between terminals maintained at potential differences of millions of volts.

Review

A. Objectives: After studying this chapter, you should:

1. Be able to give a definition of electric potential and discuss its relation to the electric field.

2. Be able to calculate the potential difference between two points, given the electric field in the region.

3. Be able to sketch equipotential surfaces, given a pattern of electric-field lines.

4. Be able to calculate the electric potential for various charge distributions.

5. Be able to calculate the electric field from the electric-potential function.

6. Be able to discuss the phenomena of charge sharing and dielectric breakdown.

B. Define, explain, or otherwise identify:

Potential difference
Electric potential
Volt
Voltage
Electron volt
Electrostatic potential energy
Gradient
Equipotential surface
Charge sharing
Dielectric breakdown
Dielectric strength
Arc discharge

C. True or false: If the statement is true, explain why it is true. If it is false, give a counterexample.

1. If the electric field is zero in some region of space, the electric potential must also be zero in that region.

2. If the electric potential is zero in some region of space, the electric field must also be zero in that region.

3. If the electric potential is zero at a point, the electric field must also be zero at that point.

4. Electric-field lines always point toward regions of lower potential.

5. The value of the electric potential can be chosen to be zero at any convenient point.

6. In electrostatics, the surface of a conductor is an equipotential surface.

7. Dielectric breakdown occurs in air when the potential is 3×10^6 V.

Problems

Level I

20-1 Electric Potential and Potential Difference

1. A uniform electric field of 2 kN/C is in the x direction. A point charge $Q = 3\ \mu\text{C}$ is released from rest at the origin. (*a*) What is the kinetic energy of the charge when it is at $x = 4$ m? (*b*) What is the change in the potential energy of the charge from $x = 0$ to $x = 4$ m? (*c*) What is the potential difference $V(4\text{ m}) - V(0)$? Find the potential $V(x)$ if $V(x)$ is chosen to be (*d*) zero at $x = 0$, (*e*) 4 kV at $x = 0$, and (*f*) zero at $x = 1$ m.

2. An infinite plane of surface charge density $\sigma = +2.5\ \mu\text{C/m}^2$ is in the yz plane. (*a*) What is the magnitude of the electric field in newtons per coulomb? In volts per meter? What is the direction of **E** for positive values of x? (*b*) What is the potential difference $V_b - V_a$ when point b is at $x = 20$ cm and point a is at $x = 50$ cm? (*c*) How much work is required by an outside agent to move a test charge $q_0 = +1.5$ nC from point a to point b?

3. A uniform electric field is in the negative x direction. Points a and b are on the x axis, a at $x = 2$ m and b at $x = 6$ m. (*a*) Is the potential difference $V_b - V_a$ positive or negative? (*b*) If the magnitude of $V_b - V_a$ is 10^5 V, what is the magnitude E of the electric field?

4. Two parallel conducting plates carry equal and opposite surface charge densities such that the electric field between them is uniform. The difference in potential between the plates is 500 V, and they are separated by 10 cm. An electron is released from rest at the negative plate. (*a*) What is the magnitude of the electric field between the plates? Is the positive or negative plate at the higher potential? (*b*) Find the work done by the electric field on the electron as the electron moves from the negative plate to the positive plate. Express your answers in both electron volts and joules. (*c*) What is the change in potential energy of the electron when it moves from the negative plate to the positive plate? What is its kinetic energy when it reaches the positive plate?

5. An electric field is given by $\mathbf{E} = ax\,\mathbf{i}$, where $\mathbf{E}$ is in newtons per coulomb, x is in meters, and a is a positive constant. (*a*) What are the SI units of a? (*b*) How much work is done by this field on a positive point charge q_0 when the charge moves from the origin to some point x? (*c*) Find the potential function $V(x)$ such that $V = 0$ at $x = 0$.

20-2 Potential Due to a System of Point Charges

6. Four 2-μC point charges are at the corners of a square of side 4 m. Find the potential at the center of the square (relative to zero potential at infinity) if (*a*) all the charges are positive, (*b*) three of the charges are positive and one is negative, and (*c*) two are positive and two are negative.

7. Three point charges are on the x axis: q_1 at the origin, q_2 at $x = 3$ m, and q_3 at $x = 6$ m. Find the potential at the point $x = 0$, $y = 3$ m if (*a*) $q_1 = q_2 = q_3 = 2\ \mu$C, (*b*) $q_1 = q_2 = 2\ \mu$C and $q_3 = -2\ \mu$C, and (*c*) $q_1 = q_3 = 2\ \mu$C and $q_2 = -2\ \mu$C.

8. Points A, B, and C are at the corners of an equilateral triangle of side 3 m. Equal positive charges of 2 μC are at A and B. (*a*) What is the potential at point C? (*b*) How much work is required to bring a positive charge of 5 μC from infinity to point C if the other charges are held fixed? (*c*) Answer parts (*a*) and (*b*) if the charge at B is replaced by a charge of $-2\ \mu$C.

9. A sphere with radius 60 cm has its center at the origin. Equal charges of 3 μC are placed at 60° intervals along the equator of the sphere. (*a*) What is the electric potential at the origin? (*b*) What is the electric potential at the north pole?

20-3 Electrostatic Potential Energy

10. A positive charge of magnitude 2 μC is at the origin. (*a*) What is the electric potential V at a point 4 m from the origin relative to $V = 0$ at infinity? (*b*) How much work must be done by an outside agent to bring a 3-μC charge from infinity to $r = 4$ m, assuming that the 2-μC charge is held fixed at the origin? (*c*) How much work must be done by an outside agent to bring the 2-μC charge from infinity to the origin if the 3-μC charge is first placed at $r = 4$ m and is then held fixed?

11. Find the electrostatic potential energy for the charge distribution described in (*a*) Problem 6*a*, (*b*) Problem 6*b*, (*c*) Problem 6*c* with like charges at opposite corners, and (*d*) Problem 6*c* with unlike charges at opposite corners.

12. Find the electrostatic potential energy for the charge distributions described in Problem 7.

13. Point charges q_1, q_2, and q_3 are at the corners of an equilateral triangle of side 2.5 m. Find the electrostatic potential energy of this charge distribution if (*a*) $q_1 = q_2 = q_3 = 4.2\ \mu$C, (*b*) $q_1 = q_2 = 4.2\ \mu$C and $q_3 = -4.2\ \mu$C, (*c*) $q_1 = q_2 = -4.2\ \mu$C and $q_3 = +4.2\ \mu$C.

20-4 Calculation of Electric Potential for Continuous Charge Distributions

14. (*a*) Sketch $V(x)$ versus x for the uniformly charged ring in the yz plane given by Equation 20-13. (*b*) At what point is $V(x)$ a maximum? (*c*) What is E_x at this point?

15. A charge of $q = +10^{-8}$ C is uniformly distributed on a spherical shell of radius 12 cm. (*a*) What is the magnitude of the electric field just outside and just inside the shell? (*b*) What is the magnitude of the electric potential just outside and just inside the shell? (*c*) What is the electric potential at the center of the shell? What is the electric field at that point?

16. A disk of radius 6.25 cm carries a uniform surface charge density $\sigma = 7.5\ \text{nC/m}^2$. Find the potential on the axis of the disk at a distance (*a*) 0.5 cm, (*b*) 3.0 cm, and (*c*) 6.25 cm from the disk.

17. An infinite line charge of linear charge density $\lambda = 1.5\ \mu$C/m lies on the z axis. Find the potential at distances of (*a*) 2.0 m, (*b*) 4.0 m, and (*c*) 12 m from the line assuming that $V = 0$ at 2.5 m.

20-5 Electric Field and Potential

18. Two positive charges $+q$ are on the y axis at $y = +a$ and $y = -a$. (*a*) Find the potential V for any point on the x axis. (*b*) Use your result in part (*a*) to find the electric field at any point on the x axis.

19. A point charge $q = 3.00\ \mu$C is at the origin. (*a*) Find the potential V on the x axis at $x = 3.00$ m and at $x = 3.01$ m. (*b*) Does the potential increase or decrease as x increases? Compute $-\Delta V/\Delta x$, where ΔV is the change in potential from $x = 3.00$ m to $x = 3.01$ m and $\Delta x = 0.01$ m. (*c*) Find the electric field at $x = 3.00$ m, and compare its magnitude with $-\Delta V/\Delta x$ found in part (*b*). (*d*) Find the potential (to three significant figures) at the point $x = 3.00$ m, $y = 0.01$ m, and compare your result with the potential on the x axis at $x = 3.00$ m. Discuss the significance of this result.

20. A charge of $+3.00\ \mu$C is at the origin, and a charge of $-3.00\ \mu$C is on the x axis at $x = 6.00$ m. (*a*) Find the potential on the x axis at $x = 3.00$ m. (*b*) Find the electric field on the x axis at $x = 3.00$ m. (*c*) Find the potential on the x axis at $x = 3.01$ m, and compute $-\Delta V/\Delta x$, where ΔV is the change in potential from $x = 3.00$ m to $x = 3.01$ m and $\Delta x = 0.01$ m. Compare your result with your answer to part (*b*).

21. In the following, V is in volts and x is in meters. Find E_x when (*a*) $V(x) = 2000 + 3000x$; (*b*) $V(x) = 4000 + 3000x$; (*c*) $V(x) = 2000 - 3000x$; and (*d*) $V(x) = -2000$, independent of x.

22. The electric potential in some region of space is given by $V(x) = C_1 + C_2x^2$, where V is in volts, x is in meters, and C_1 and C_2 are positive constants. Find the electric field $\mathbf{E}$ in this region. In what direction is $\mathbf{E}$?

23. An infinite plane of charge has surface charge density $3.5\ \mu\text{C/m}^2$. How far apart are the equipotential surfaces whose potentials differ by 100 V?

24. A point charge $q = +\frac{1}{9} \times 10^{-8}$ C is at the origin. Taking the potential to be zero at $r = \infty$, locate the equipotential surfaces at 20-V intervals from 20 to 100 V, and sketch them to scale. Are these surfaces equally spaced?

20-6 Equipotential Surfaces, Charge Sharing, and Dielectric Breakdown

25. (*a*) Find the maximum net charge that can be placed on a spherical conductor of radius 16 cm before dielectric breakdown of the air occurs. (*b*) What is the potential of the sphere when it carries this maximum charge?

26. Sketch the electric-field lines and the equipotential surfaces both near to and far from the conductor shown in Figure 20-20*b*, assuming that the conductor carries some charge Q.

27. Find the greatest surface charge density σ_{max} that can exist on a conductor before dielectric breakdown of the air occurs.

28. If a conducting sphere is to be charged to a potential of 10,000 V, what is the smallest possible radius of the sphere such that the electric field will not exceed the dielectric strength of air?

Level II

29. A Van de Graaff generator has a potential difference of 1.25 MV between the belt and the outer shell. Charge is supplied at the rate of 200 μC/s. What minimum power is needed to drive the moving belt?

30. A uniformly charged sphere has a potential on its surface of 450 V. At a radial distance of 20 cm from this surface, the potential is 150 V. What is the radius of the sphere, and what is the charge of the sphere?

31. Four charges are at the corners of a square centered at the origin as follows: q at $(-a, +a)$; $2q$ at (a, a); $-3q$ at $(a, -a)$; and $6q$ at $(-a, -a)$. Find (*a*) the electric field at the origin and (*b*) the potential at the origin. (*c*) A fifth charge $+q$ is placed at the origin and released from rest. Find its speed when it is a great distance from the origin.

32. Two positive charges $+q$ are on the x axis at $x = +a$ and $x = -a$. (*a*) Find the potential $V(x)$ as a function of x for points on the x axis. (*b*) Sketch $V(x)$ versus x. (*c*) What is the significance of the minimum in your curve?

33. An electric field is given by $E_x = 2.0x^3$ kN/C. Find the potential difference between the points on the x axis at $x = 1$ m and $x = 2$ m.

34. Consider two infinite parallel planes of charge, one in the yz plane and the other at distance $x = a$. (*a*) Find the potential everywhere in space when $V = 0$ at $x = 0$ if the planes carry equal positive charge densities $+\sigma$. (*b*) Do the same if the charge densities are equal and opposite, with the charge in the yz plane positive.

35. In a Van de Graaff accelerator, protons are released from rest at a potential of 5 MV and travel through a vacuum to a region at zero potential. (*a*) Find the speed of the 5-MeV protons. (*b*) If the potential change occurs uniformly over a distance of 2.0 m, find the accelerating electric field.

36. Two equal positive charges are separated by a small distance. Sketch the electric-field lines and the equipotential surfaces for this system.

37. When uranium ^{235}U captures a neutron, it fissions (splits) into two nuclei (and emits several neutrons, which can cause other uranium nuclei to split). Assume that the fission products are equally charged nuclei with charge $+46e$ and that these nuclei are at rest just after fission and separated by twice their radius $2R \approx 1.3 \times 10^{-14}$ m. (*a*) Using $U = kq_1q_2/2R$, calculate the electrostatic potential energy of the fission fragments. This is approximately the energy released per fission. (*b*) About how many fissions per second are needed to produce 1 MW of power in a reactor?

38. A radioactive ^{210}Po nucleus emits an alpha particle of charge $+2e$ and energy 5.30 MeV. Assume that just after the alpha particle is formed and escapes from the nucleus, it is a distance R from the center of the daughter nucleus ^{206}Pb which has a charge $+82e$. Calculate R by setting the electrostatic potential energy of the two particles at this separation equal to 5.30 MeV.

39. An electron gun fires electrons at the screen of a television tube. The electrons start from rest and are accelerated through a potential difference of 30,000 V. What is the energy of the electrons when they hit the screen (*a*) in electron volts and (*b*) in joules. (*c*) What is the speed of impact of electrons with the screen of the picture tube?

40. Two large, parallel, nonconducting planes carry equal and opposite charge densities of magnitude σ. The planes have area A and are separated by a distance d. (*a*) Find the potential difference between the planes. (*b*) A conducting slab having thickness a and area A, the same area as the planes, is inserted between the original two planes. The slab carries no net charge. Find the potential difference between the original two planes and sketch the lines of **E** in the region between the original two planes.

41. Two concentric spherical shell conductors carry equal and opposite charges. The inner shell has radius a and charge $+q$; the outer shell has radius b and charge $-q$. Find the potential difference between the shells, $V_a - V_b$.

42. A conducting spherical shell of inner radius b and outer radius c is concentric with a small metal sphere of radius $a < b$. The metal sphere has a positive charge Q. The total charge on the conducting spherical shell is $-Q$. (*a*) What is the potential of the spherical shell? (*b*) What is the potential of the metal sphere?

43. Two very long, coaxial cylindrical shell conductors carry equal and opposite charges. The inner shell has radius a and charge $+q$; the other shell has radius b and charge $-q$. The length of each cylindrical shell is L. Find the potential difference between the shells.

44. The centers of two metal spheres of radius 10 cm are 50 cm apart on the x axis. The spheres are initially neutral, but a charge Q is transferred from one sphere to the other, creating a potential difference between the spheres of 100 V. A proton is released from rest at the surface of the positively charged sphere and travels to the negatively charged sphere. At what speed does it strike the negatively charged sphere?

45. A point charge of $+3e$ is at the origin and a second point of $-2e$ is on the x axis at $x = a$. (*a*) Sketch the potential function $V(x)$ versus x for all x. (*b*) At what point or points is $V(x)$ zero? (c) How much work is needed to bring a third charge $+e$ to the point $x = \frac{1}{2}a$ on the x axis?

46. Three equal charges lie in the xy plane. Two are on the y axis at $y = -a$ and $y = +a$, and the third is on the x axis at $x = a$. (a) What is the potential $V(x)$ due to these charges at a point on the x axis? (b) Find E_x along the x axis from the potential function $V(x)$. Evaluate your answers to (a) and (b) at the origin and at $x = \infty$ to see if they yield the expected results.

47. A charge q is at $x = 0$ and a charge $-3q$ is at $x = 1$ m. (a) Find $V(x)$ for a general point on the x axis. (b) Find the points on the x axis where the potential is zero. (c) What is the electric field at these points? (d) Sketch $V(x)$ versus x.

48. A rod of length L has a charge Q uniformly distributed along its length. The rod lies along the x axis with its center at the origin. (a) What is the electric potential as a function of position along the x axis for $x > L/2$? (b) Show that for $x \gg L/2$ your result reduces to that due to a point charge Q.

49. A charge of 2 nC is uniformly distributed around a ring of radius 10 cm that has its center at the origin and its axis along the x axis. A point charge of 1 nC is located at $x = 50$ cm. Find the work required to move the point charge to the origin in joules and electron volts.

50. A uniformly charged ring with a total charge of 100 μC and a radius of 0.1 m lies in the yz plane with its center at the origin. A meterstick has a point charge of 10 μC on the end marked 0 and a point charge of 20 μC on the end marked 100 cm. How much work does it take to bring the meterstick from a long distance away to a position along the x axis with the end marked 0 at $x =$ 0.2 m and the other end at $x = 1.2$ m.

51. Four equal charges Q are at the corners of a square of side L. The charges are released one at a time proceeding clockwise around the square. Each charge is allowed to reach its final speed a long distance from the square before the next charge is released. What is the final kinetic energy of (a) the first charge released, (b) the second charge released, (c) the third charge released, and (d) the fourth charge released?

52. Two identical uncharged metal spheres are connected by a wire as shown in Figure 20-21a. Two similar conducting spheres with equal and opposite charges are brought to the positions shown in Figure 20-21b. (a) Sketch the electric-field lines between spheres 1 and 3 and between spheres 2 and 4. (b) What can be said about the potentials V_1, V_2, V_3, and V_4 of the spheres? (c) If spheres 3 and 4 are connected by a wire, prove that the final charge on each must be zero.

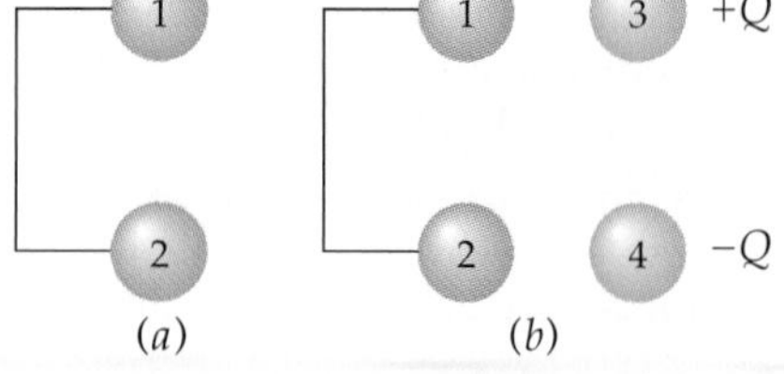

Figure 20-21 Problem 52.

53. Three large conducting plates are parallel to one another with the outer plates connected by a wire. The inner plate is isolated and carries a charge density σ_t on the upper surface and σ_b on the lower surface, where $\sigma_t + \sigma_b = 12\ \mu\text{C/m}^2$. The inner plate is 1 mm from the top plate and 3 mm from the bottom plate. Find the surface charge densities σ_t and σ_b.

54. Show that when R is much smaller than x, the potential on the axis of a disk charge approaches kQ/x, where $Q = \sigma\pi R^2$ is the total charge on the disk. *Hint:* Write $(x^2 + R^2)^{1/2} = x(1 + R^2/x^2)^{1/2}$ and use the binomial expression.

55. A uniformly charged ring of radius a and charge Q lies in the yz plane with its axis along the x axis. A point charge Q' is placed on the x axis at $x = 2a$. (a) Find the potential at any point on the x axis due to the total charge $Q + Q'$. (b) Find the electric field for any point on the x axis.

Level III

56. A potential is given by

$$V(x, y, z) = \frac{kQ}{\sqrt{(x-a)^2 + y^2 + z^2}}$$

(a) Find the components E_x, E_y, and E_z of the electric field by differentiating this potential function. (b) What simple charge distribution might be responsible for this potential?

57. The electric potential in a region of space is given by

$$V = (2\ \text{V/m}^2)x^2 + (1\ \text{V/m}^2)yz$$

Find the electric field at the point $x = 2$ m, $y = 1$ m, $z =$ 2 m.

58. A point charge q_1 is at the origin and a second point charge q_2 is on the x axis at $x = a$ as in Example 20-5. (a) Calculate the electric field everywhere on the x axis from the potential function given in that example. (b) Find the potential at a general point on the y axis. (c) Use your result from (b) to calculate the y component of the electric field on the y axis. Compare your result with that obtained directly from Coulomb's law.

59. Consider a ball of charge of uniform volume charge density with a radius R and total charge Q. (This is a model of a proton.) The center of the ball is at the origin. Use the radial component of the electric field E_r found from Gauss's law to calculate the potential $V(r)$, assuming $V = 0$ at $r = \infty$ for (a) any point outside the charge $r \geq R$ and (b) any point inside the charge $r \leq R$. (Remember that V must be continuous at $r = R$.) (c) What is the potential at the origin? (d) Sketch V versus r.

60. In the Bohr model of the hydrogen atom, the electron moves in a circular orbit of radius r around the proton. (a) Find an expression for the kinetic energy of the electron as a function of r by setting the force on the electron (given by Coulomb's law) equal to ma, where a is the centripetal acceleration. Show that at any distance r the kinetic energy is half the magnitude of the potential energy. (b) Evaluate $\frac{1}{2}mv^2$, U, and the total energy $W = \frac{1}{2}mv^2 + U$ in electron volts for $r = 0.529 \times 10^{-10}$ m, the radius of the electron's orbit in hydrogen. The energy $|W|$ that must be supplied to the hydrogen atom to remove the electron is called the **ionization energy.**

61. (*a*) For the dipole of Example 20-6, show that the potential at a point off-axis a great distance r from the origin (Figure 20-22) is given approximately by

$$V = \frac{2kqa\cos\theta}{r^2} = \frac{kp\cos\theta}{r^2} = \frac{kpz}{r^3}$$

Hint: Show that $r_+^{-1} - r_-^{-1} \approx \Delta r/r^2$ where $\Delta r = r_+ - r_- \approx 2a\cos\theta$. (*b*) Find the x, y, and z components of the electric field at a point off-axis.

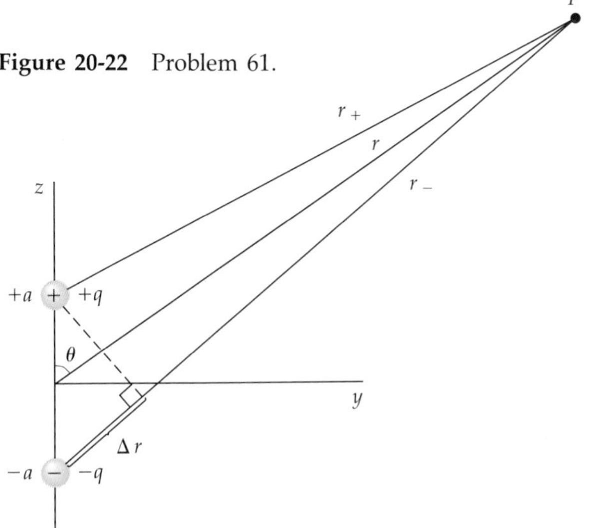

Figure 20-22 Problem 61.

62. Consider two concentric spherical metal shells of radii a and b where $b > a$. The outer shell has a charge Q, but the inner shell is grounded. This means that the inner shell is at zero potential and that electric-field lines leave the outer shell and go to infinity but other electric-field lines leave the outer shell and end on the inner shell. Find the charge on the inner shell.

63. Three concentric conducting spherical shells have radii a, b, and c such that $a < b < c$. Initially, the inner shell is uncharged, the middle shell has a positive charge Q, and the outer shell has a negative charge $-Q$. (*a*) Find the electric potential of the three shells. (*b*) If the inner and outer shells are now connected by a wire that is insulated as it passes through the middle shell, what is the electric potential of each of the three shells, and what is the final charge on each shell?

64. A nonconducting sphere of radius R has a volume charge density $\rho = \rho_0 r/R$, where ρ_0 is a constant. (*a*) Show that the total charge is $Q = \pi R^3 \rho_0$. (*b*) Show that the total charge inside a sphere of radius $r < R$ is $q = Qr^4/R^4$. (*c*) Use Gauss's law to find the electric field E_r everywhere. (*d*) Use $dV = -E_r\,dr$ to find the potential V everywhere, assuming that $V = 0$ at $r = \infty$. (Remember that V is continuous at $r = R$.)

65. A particle of mass m and charge Q is located on the x axis at $x = +a$, while a second particle of equal mass and charge $-Q$ is located on the x axis at $x = -a$. They are both released at $t = 0$. (*a*) Find the velocity of the positively charged particle as a function of its position x. (*b*) Integrate the velocity equation to find the time for the collision of the two charges.

Chapter 21

Capacitance, Dielectrics, and Electrostatic Energy

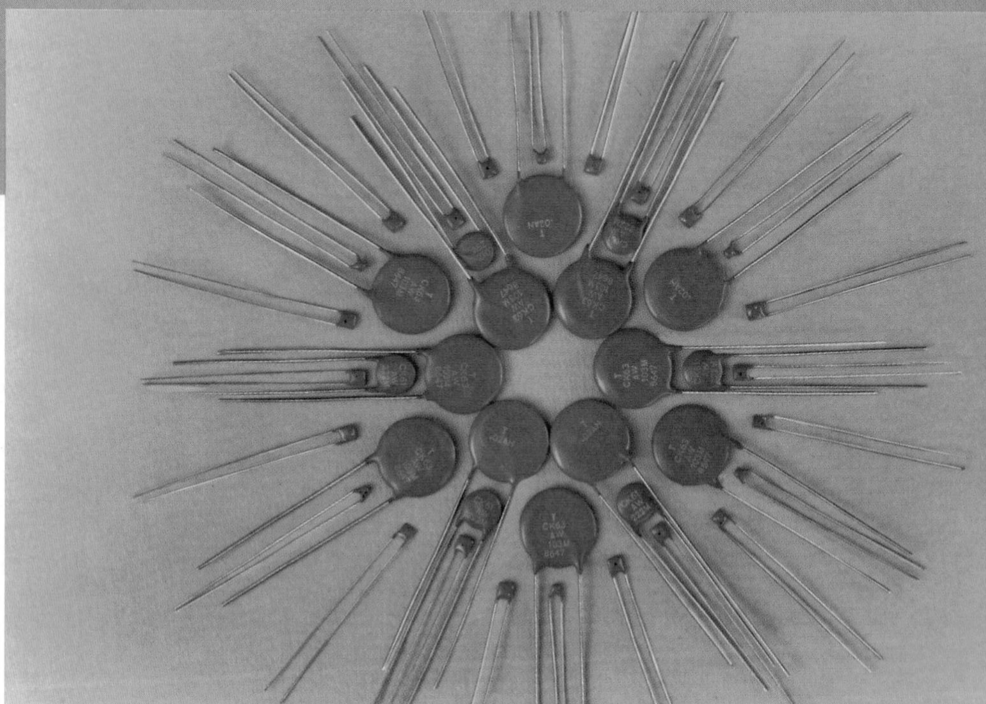

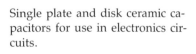

Single plate and disk ceramic capacitors for use in electronics circuits.

A **capacitor** is a useful device for storing charge and energy. It consists of two conductors, closely spaced but insulated from each other, that carry equal and opposite charges. Capacitors have many uses. The flash attachment for your camera uses a capacitor to store the energy needed to provide the sudden flash of light. Capacitors are also used to smooth out the ripples that arise when alternating current (the type of current that comes from a wall outlet) is converted into direct current for a power supply, such as that used to power your calculator or radio when the batteries are low.

The first capacitor for storing large electric charges was a jar with gold foil inside and out called a Leyden jar. It was invented in eighteenth-century Leyden (Holland) by experimenters who, while studying the effects of electric charges on people and animals, got the idea of trying to store a large amount of charge in a bottle of water. An experimenter held up a jar of water in one hand while charge was conducted to the water by a chain from a static electric generator. When he reached over to lift the chain out of the water with his other hand, he was knocked unconscious. After many experiments, it was discovered that the hand holding the jar could be replaced by metal foil on the surfaces of the jar. Benjamin Franklin realized that the device for storing charge did not have to be jar-shaped and used foil-covered window glass, called Franklin panes. With several of these connected in parallel, he stored a large charge and tried to kill a turkey with it. Instead, he knocked himself out. "I tried to kill a turkey but nearly succeeded in killing a goose," he wrote.

21-1 The Parallel-Plate Capacitor

A common capacitor is the **parallel-plate capacitor,** which utilizes two large parallel conducting plates. In practice, the plates may be thin metallic foils that are separated and insulated from one another by a thin sheet of paper. This "paper sandwich" is then rolled up to save space. When the plates are connected to a charging device, for example, a battery,* as in Figure 21-1, charge is transferred from one conductor to the other until the potential difference between the conductors due to their equal and opposite charges equals the potential difference between the battery terminals. The amount of charge on the plates depends on the potential difference and on the geometry of the capacitor, for example, on the area and separation of the plates in a parallel-plate capacitor. Let Q be the magnitude of the charge on either plate and V be the potential difference between the plates.** (When we speak of the charge on a capacitor, we mean the magnitude of the charge on either plate.) The ratio Q/V is called the **capacitance** C:

$$C = \frac{Q}{V} \tag{21-1}$$

Capacitance defined

Capacitance is a measure of the "capacity" to store charge for a given potential difference. The SI unit of capacitance is the coulomb per volt, which is called a **farad** (F) after the great English experimentalist Michael Faraday:

$$1\ \text{F} = 1\ \text{C/V} \tag{21-2}$$

Since the farad is a rather large unit, submultiples such as the microfarad ($1\ \mu\text{F} = 10^{-6}$ F) or the picofarad ($1\ \text{pF} = 10^{-12}$ F) are often used.

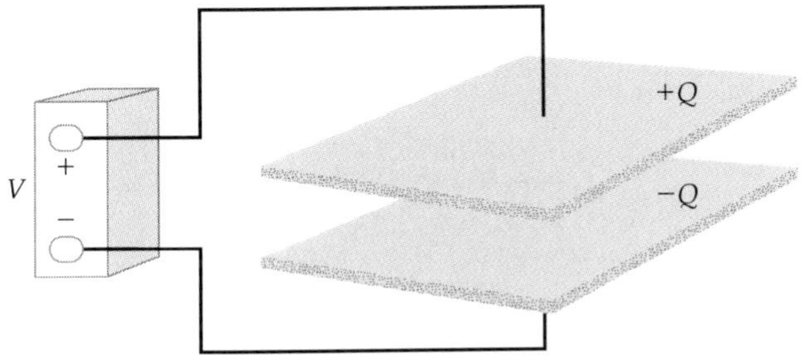

Figure 21-1 A capacitor consisting of two closely spaced parallel-plate conductors. When the conductors are connected to the terminals of a battery, the battery transfers charge from one conductor to the other until the potential difference between the conductors equals that between the battery terminals. The amount of charge transferred is proportional to the potential difference.

To calculate the capacitance of any capacitor, we first place some charge $+Q$ on one conductor and an equal and opposite charge $-Q$ on the other and then find the electric field between the conductors. We then integrate the field from one conductor to the other to find the potential difference V between the conductors. Since the potential difference is proportional to the charge Q, the capacitance $C = Q/V$ does not depend on either Q or V.

*We will discuss batteries more fully in Chapter 22. Here, all we need to know is that a battery is a device that stores and supplies electrical energy and maintains a constant potential difference V between its terminals.

**The use of V rather than ΔV for the potential difference between the plates is standard and simplifies many of the equations.

Let us consider a parallel-plate capacitor consisting of two plates of the same area A separated by a distance s, which is small compared to the length and width of the plates. We place a charge $+Q$ on one plate and $-Q$ on the other. Since the plates are close together, the electric field at any point between the plates (excluding points near the edges) is approximately equal to the field due to two equal and opposite infinite planes of charge. Each plate contributes a uniform field of magnitude $\sigma/2\epsilon_0$ (Equation 19-23), giving a total field $E = \sigma/\epsilon_0$, where $\sigma = Q/A$ is the charge per unit area on either plate. Since the field between the plates of our capacitor is uniform (Figure 21-2), the potential difference between the plates equals the field times the plate separation s:

$$V = Es = \frac{\sigma}{\epsilon_0}s = \frac{Qs}{\epsilon_0 A} \qquad 21\text{-}3$$

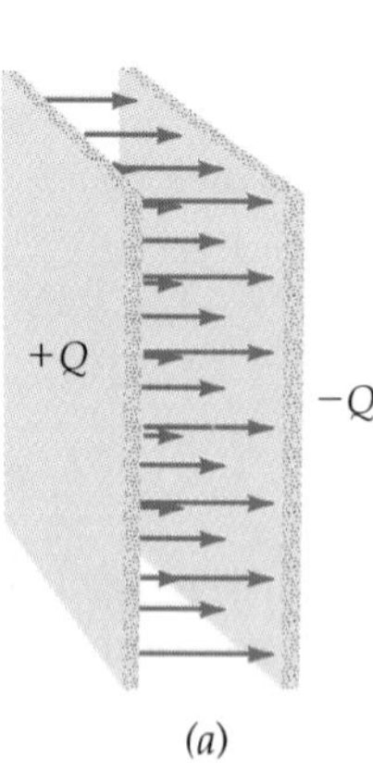

(a)

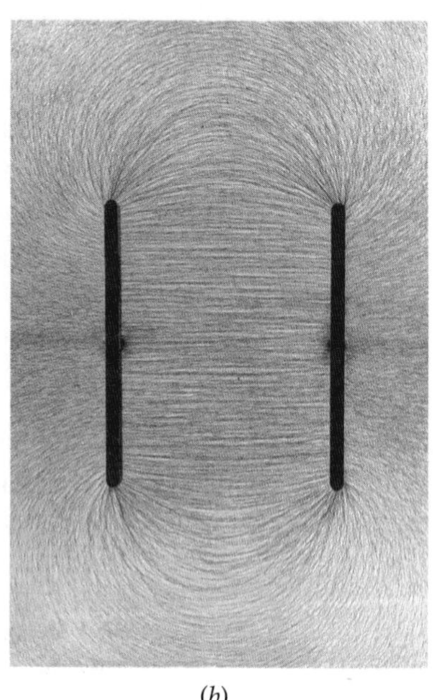

(b)

Figure 21-2 (*a*) The electric-field lines between the plates of a parallel-plate capacitor are equally spaced, indicating that the electric field there is uniform. (*b*) The electric-field lines between the plates of a parallel-plate capacitor shown by small bits of thread suspended in oil.

The capacitance of the parallel-plate capacitor is thus

Capacitance of a parallel-plate capacitor

$$C = \frac{Q}{V} = \frac{\epsilon_0 A}{s} \qquad 21\text{-}4$$

Note that since V is proportional to Q, the capacitance does not depend on either the charge or the voltage of the capacitor but only on geometric factors. For a parallel-plate capacitor, the capacitance is proportional to the area of the plates and is inversely proportional to the separation distance. In general, capacitance depends on the size, shape, and geometrical arrangement of the conductors. Since capacitance is in farads and A/s is in meters, we can see from Equation 21-4 that the SI unit for the permittivity of free space ϵ_0 can also be written as a farad per meter:

$$\epsilon_0 = 8.85 \times 10^{-12}\ \text{F/m} = 8.85\ \text{pF/m} \qquad 21\text{-}5$$

A numerical calculation will illustrate how large a unit of capacitance the farad is.

Example 21-1

A parallel-plate capacitor has square plates of side 10 cm separated by 1 mm. (*a*) Calculate its capacitance. (*b*) If this capacitor is charged to 12 V, how much charge is transferred from one plate to another?

(*a*) Using Equation 21-4, we obtain for the capacitance

$$C = \frac{\epsilon_0 A}{s} = \frac{(8.85\ \text{pF/m})(0.1\ \text{m})^2}{0.001\ \text{m}} = 8.85 \times 10^{-11}\ \text{F}$$

$$= 88.5\ \text{pF}$$

(*b*) From the definition of capacitance (Equation 21-1), the charge transferred is

$$Q = CV = (88.5 \times 10^{-12}\ \text{F})(12\ \text{V}) = 1.06 \times 10^{-9}\ \text{C} = 1.06\ \text{nC}$$

This is the magnitude of the charge on either plate.

21-2 The Cylindrical Capacitor

A cylindrical capacitor consists of a small conducting cylinder or wire of radius a and a larger, concentric cylindrical conducting shell of radius b. A coaxial cable, such as that used for cable television, can be thought of as a cylindrical capacitor. The capacitance per unit length of a coaxial cable is important in determining the transmission characteristics of the cable. Let L be the length of the capacitor, which carries a charge $+Q$ on the inner conductor and a charge $-Q$ on the outer conductor. In Chapter 19, we found the electric field outside a long wire or cylinder of charge Q (Equation 19-24 or 19-26*a*) to be

$$E_r = \frac{1}{2\pi\epsilon_0}\frac{\lambda}{r} = \frac{Q}{2\pi\epsilon_0 Lr} \qquad \text{21-6}$$

where $\lambda = Q/L$ is the linear charge density. The field due to the charge $-Q$ on the outer cylindrical shell is zero inside the shell, as was shown using Gauss's law in Chapter 19 (Equation 19-25*a*).

The potential difference V between the conductors is found from Equation 20-3*b*. Let V_a be the potential of the inner conductor and V_b be that of the outer conductor. Then

$$V_b - V_a = -\int_a^b E_r\, dr = -\frac{Q}{2\pi\epsilon_0 L}\int_a^b \frac{dr}{r} = -\frac{Q}{2\pi\epsilon_0 L}\ln\frac{b}{a} \qquad \text{21-7}$$

A coaxial cable is a long cylindrical capacitor with a solid wire for the inner conductor and a braided-wire shield for the outer conductor. The outer rubber coating has been pealed back here to show the conductors and the white plastic insulator that separates the conductors.

The potential is, of course, greater on the inner conductor, which carries the positive charge, since the electric-field lines point from this conductor to the outer conductor. The magnitude of this potential difference is

$$V = V_a - V_b = \frac{Q\ln(b/a)}{2\pi\epsilon_0 L}$$

and the capacitance is

$$C = \frac{Q}{V} = \frac{2\pi\epsilon_0 L}{\ln(b/a)} \qquad \text{21-8}$$

Thus, the capacitance is proportional to the length of the conductors. The greater the length, the greater the amount of charge that can be put on the conductors for a given potential difference, because the electric field, and therefore the potential difference, depends only on the charge per unit length.

Example 21-2

A coaxial cable consists of a wire of radius 0.5 mm and an outer conducting shell of radius 1.5 mm. Find its capacitance per unit length.

Using Equation 21-8, we obtain

$$\frac{C}{L} = \frac{2\pi\epsilon_0}{\ln (b/a)} = \frac{2\pi(8.85 \text{ pF/m})}{\ln (1.5 \text{ mm}/0.5 \text{ mm})} = 50.6 \text{ pF/m}$$

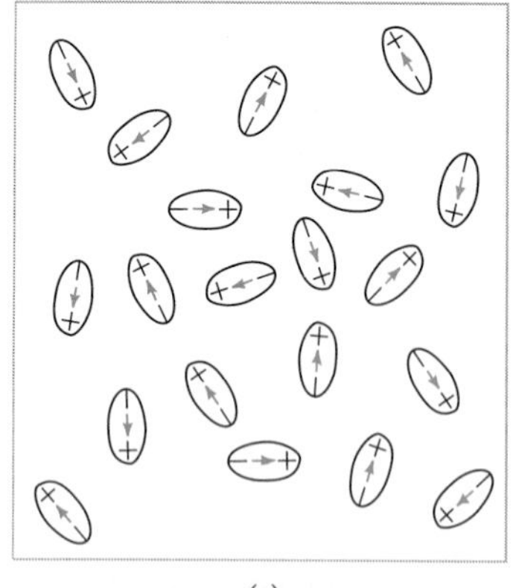

(a)

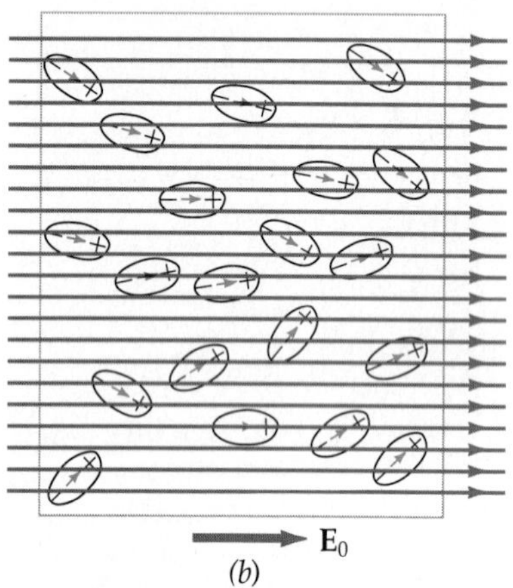

(b)

Figure 21-3 (*a*) The randomly oriented electric dipoles of a polar dielectric in the absence of an external electric field. (*b*) In the presence of an external electric field, the dipoles are partially aligned parallel to the field.

21-3 Dielectrics

A nonconducting material, such as glass, paper, or wood, is called a **dielectric.** When the space between the two conductors of a capacitor is occupied by a dielectric, the capacitance is increased by a factor κ that is characteristic of the dielectric and is called the **dielectric constant.** This was discovered experimentally by Michael Faraday. The reason for this increase is that the electric field between the plates of a capacitor is weakened by the dielectric. Thus, for a given charge on the plates, the potential difference is reduced and the ratio Q/V is increased.

A dielectric weakens the electric field between the plates of a capacitor because, in the presence of an external electric field, the molecules in the dielectric produce an additional electric field in a direction opposite to that of the external field. If the molecules of the dielectric are polar and so have permanent dipole moments, these dipole moments are normally randomly oriented (Figure 21-3*a*). In the presence of the field between the capacitor plates, however, these dipole moments experience a torque that tends to align them in the direction of the field (Figure 21-3*b*). The extent of alignment depends on the strength of the field and the temperature. At high temperatures, the random thermal motion of the molecules tends to counteract their alignment. If the molecules of the dielectric are nonpolar, they will have induced dipole moments in the presence of an external electric field in the direction of the field. A dielectric that has electric dipole moments that are predominantly in the direction of the external field is said to be polarized by the field, whether the polarization is due to the alignment of permanent dipole moments of polar molecules or to the creation of induced dipole moments in nonpolar molecules. In either case, the molecular dipoles produce an additional electric field that is in a direction opposite to that of the original field, thus weakening the original field.

The net effect of the polarization of a homogeneous dielectric is the creation of a surface charge on the dielectric faces near the plates as shown in Figure 21-4. It is this surface charge, which is bound to the dielectric, that produces an electric field opposite the direction of that due to the free charge

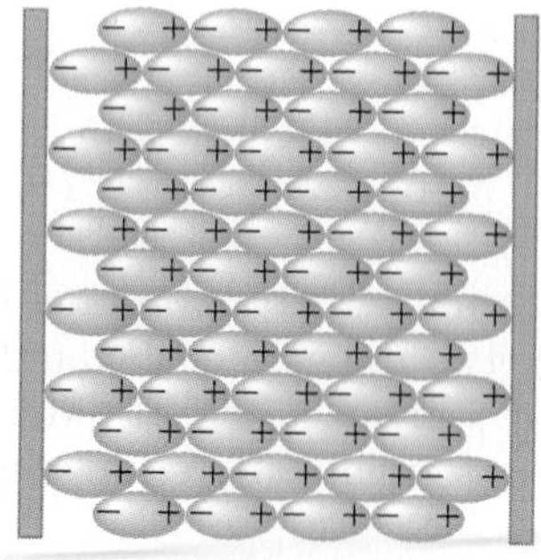

Figure 21-4 When a dielectric is placed between the plates of a capacitor, the electric field of the capacitor polarizes the molecules of the dielectric. The result is a bound charge on the surface of the dielectric that produces its own electric field that opposes the external field. The electric field between the plates is thus weakened by the dielectric.

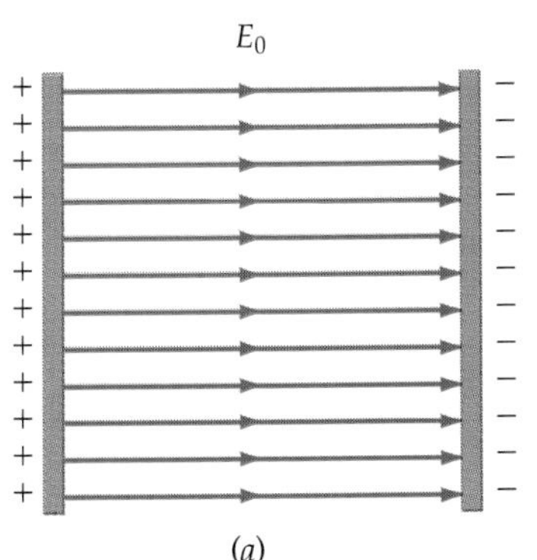

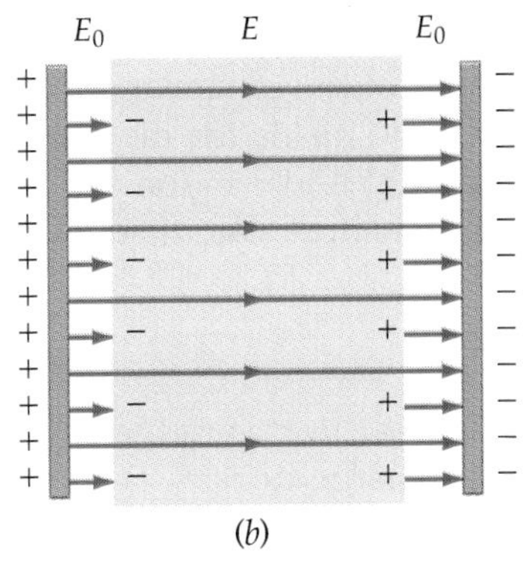

Figure 21-5 The electric field between the plates of a capacitor (*a*) with no dielectric and (*b*) with a dielectric. The surface charge on the dielectric weakens the original field between the plates.

on the conductors. Thus, the electric field between the plates is weakened as illustrated in Figure 21-5.

If the original electric field between the plates of a capacitor without a dielectric is E_0, the field in the dielectric is

$$E = \frac{E_0}{\kappa} \tag{21-9}$$

Electric field inside a dielectric

where κ is the dielectric constant. For a parallel-plate capacitor of separation s, the potential difference between the plates is

$$V = Es = \frac{E_0 s}{\kappa} = \frac{V_0}{\kappa}$$

where V is the potential difference with the dielectric and $V_0 = E_0 s$ is the original potential difference without the dielectric. The new capacitance is

$$C = \frac{Q}{V} = \frac{Q}{V_0/\kappa} = \kappa\frac{Q}{V_0}$$

or

$$C = \kappa C_0 \tag{21-10}$$

where $C_0 = Q/V_0$ is the original capacitance. The capacitance of a parallel-plate capacitor filled with a dielectric of constant κ is thus

$$C = \frac{\kappa\epsilon_0 A}{s} = \frac{\epsilon A}{s} \tag{21-11}$$

where

$$\epsilon = \kappa\epsilon_0 \tag{21-12}$$

is called the **permittivity** of the dielectric.

The charge densities on the faces of the dielectric are due to the displacement of positive and negative molecular charges near the faces caused by the external electric field of the capacitor. The charge on the dielectric is called a **bound charge** because it is bound to the molecules of the dielectric. That is, it cannot move about like the free charge on the conducting capacitor plates. Although it disappears when the external electric field disappears, it produces an electric field just like any other charge.

We will now see how the bound charge density σ_b on the surfaces of the dielectric is related to the dielectric constant κ and to the surface charge density on the plates of the capacitor, which we call the free charge density σ_f because it is free to move about the conductor. Consider a dielectric slab between the plates of a parallel-plate capacitor as shown in Figure 21-6. If

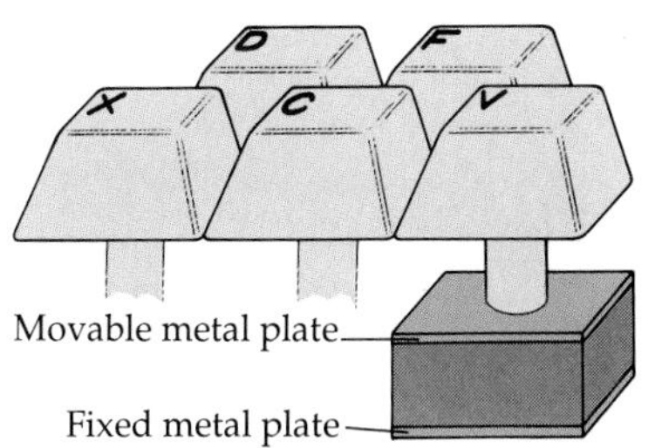

Some keyboards in computers use capacitance switching. A metal plate, mounted on a plunger attached to the key, acts as the top plate of a capacitor. When the key is depressed, the separation between the top plate and bottom plate is changed from about 5 mm to about 0.3 mm and the capacitance increases. The change in capacitance triggers the electronic circuitry to enter the information into the computer.

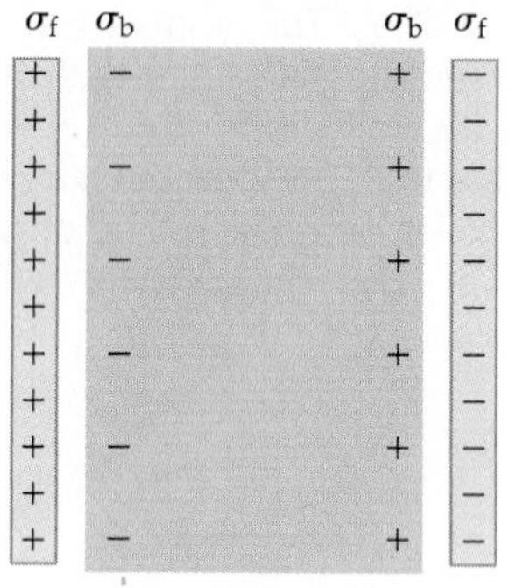

Figure 21-6 A parallel-plate capacitor with a dielectric slab between the plates. If the plates are closely spaced, each of the surface charges can be considered to be an infinite plane charge. The electric field due to the free charge on the plates is directed to the right and has a magnitude $E_0 = \sigma_f/\epsilon_0$. That due to the bound charge is directed to the left and has a magnitude $E_b = \sigma_b/\epsilon_0$.

the plates of the capacitor are close together so that the slab is very thin, the electric field inside the dielectric slab due to the bound charge densities $+\sigma_b$ on the right and $-\sigma_b$ on the left is just the field due to two infinite-plane charge densities. The field E_b thus has the magnitude

$$E_b = \frac{\sigma_b}{\epsilon_0} \qquad 21\text{-}13$$

This field is directed to the left and subtracts from the electric field E_0 due to the free charge density on the capacitor plates. E_0 has the magnitude

$$E_0 = \frac{\sigma_f}{\epsilon_0} \qquad 21\text{-}14$$

The magnitude of the net field E is the difference between these magnitudes. It also equals E_0/κ:

$$E = E_0 - E_b = \frac{E_0}{\kappa}$$

or

$$E_b = E_0\left(1 - \frac{1}{\kappa}\right) = \frac{\kappa - 1}{\kappa}E_0$$

Writing σ_b/ϵ_0 for E_b and σ_f/ϵ_0 for E_0, we obtain

$$\sigma_b = \frac{\kappa - 1}{\kappa}\sigma_f \qquad 21\text{-}15$$

The bound charge density σ_b is always less than the free charge density σ_f on the capacitor plates, and it is zero if $\kappa = 1$, which is the case when there is no dielectric.

In the preceding discussion, we assumed that the charge on the plates of the capacitor did not change when the dielectric was inserted. This would be true if the capacitor were charged and then removed from the charging source (the battery) before the insertion of the dielectric. If the dielectric is inserted while the battery is still connected, the battery will supply more charge to maintain the original potential difference. The total charge on the plates is then $Q = \kappa Q_0$. In either case, however, the capacitance is increased by the factor κ.

Exercise

The capacitor of Example 21-1 is filled with a dielectric of constant $\kappa = 2$. (*a*) Find the new capacitance. (*b*) Find the charge on the capacitor with the dielectric if the capacitor is attached to a 12-V battery. [Answers: (*a*) 177 pF, (*b*) 2.12 nC]

Exercise

The capacitor in the previous exercise is charged to 12 V without the dielectric and is then disconnected from the battery. The dielectric of constant $\kappa = 2$ is then inserted. Find the new values of the (*a*) charge Q, (*b*) the voltage V, and (*c*) the capacitance C. [Answers: (*a*) $Q = 1.06$ nC, which is unchanged; (*b*) $V = 6$ V; (*c*) $C = 177$ pF]

In addition to increasing the capacitance, a dielectric has two other functions in a capacitor. First, it provides a physical means of separating the two conductors, which must be very close together for a large capacitance since the capacitance varies inversely with the separation. Second, it increases the dielectric strength of the capacitor because the dielectric strength of a dielectric is usually greater than that of air.

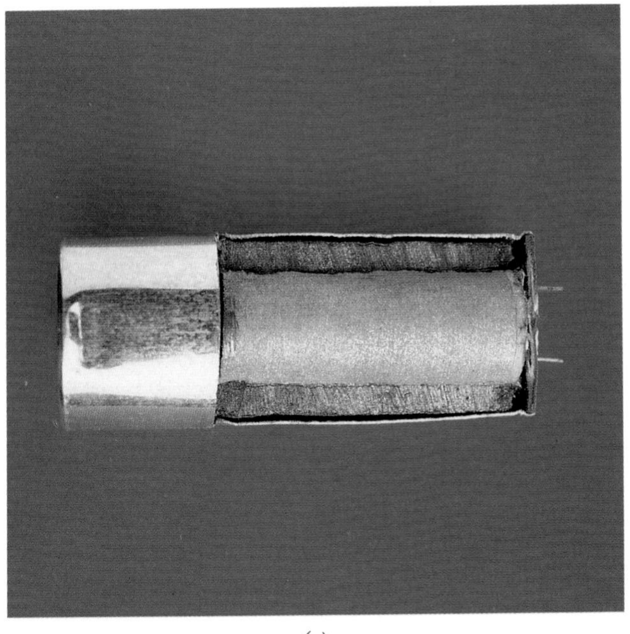

(*a*)

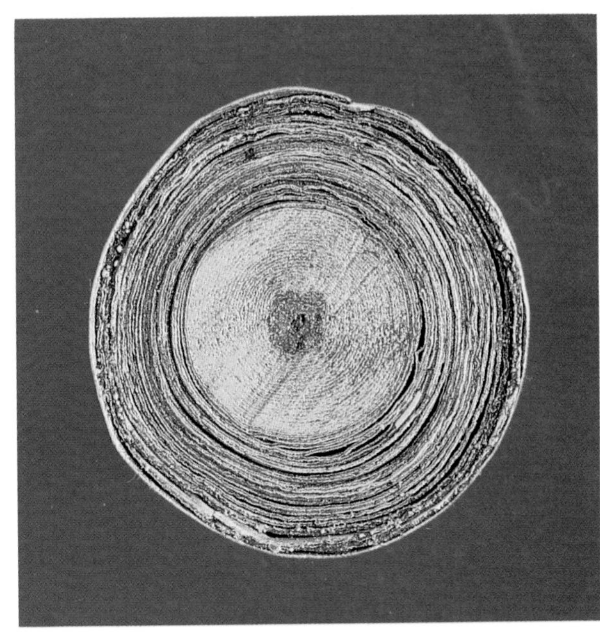

(*b*)

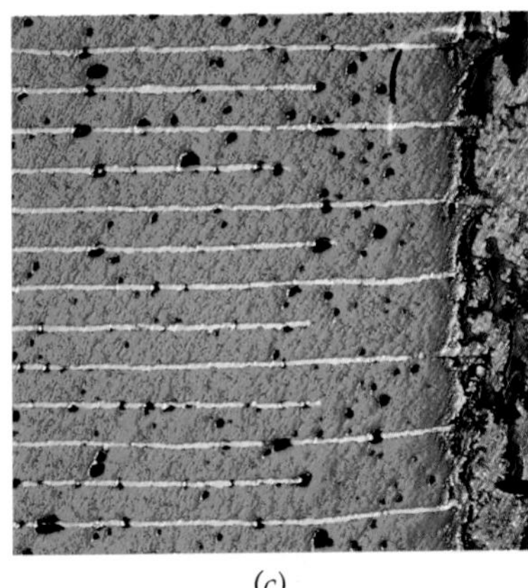

(*c*)

(*a*) A 200-μF capacitor used in an electronic strobe light. (*b*) Cross section of a foil-wound capacitor. (*c*) A cut section of a multilayer ceramic capacitor. The bright lines are the edges of the conducting plate.

We have already seen in Chapter 20 that the dielectric strength of air is about 3 MV/m = 3 kV/mm. Fields with a magnitude greater than this cannot be maintained in air because of dielectric breakdown; that is, the air becomes ionized and conducts. Many materials have dielectric strengths greater than that of air and so allow greater potential differences between the conducting plates of a capacitor.

Examples of the three dielectric functions are provided by a parallel-plate capacitor made from two sheets of metal foil of large area (to increase the capacitance) that are separated by a thin sheet of paper. The paper increases the capacitance because of its polarization; that is, κ is greater than 1. It also provides physical separation so that the sheets can be very close together without being in electrical contact. Finally, the dielectric strength of paper is greater than that of air, so greater potential differences can be attained without dielectric breakdown. Table 21-1 lists the dielectric constants and dielectric strengths of some dielectrics. Note that for air, $\kappa \approx 1$, so for most situations we do not need to distinguish between air and a vacuum.

Table 21-1 **Dielectric Constants and Strengths of Various Materials**

Material	Dielectric constant κ	Dielectric strength, kV/mm
Air	1.00059	3
Bakelite	4.9	24
Glass (Pyrex)	5.6	14
Mica	5.4	10–100
Neoprene	6.9	12
Paper	3.7	16
Paraffin	2.1–2.5	10
Plexiglas	3.4	40
Polystyrene	2.55	24
Porcelain	7	5.7
Transformer oil	2.24	12
Water (20°C)	80	

Example 21-3

A parallel-plate capacitor has square plates of side 10 cm and a separation of 4 mm. A dielectric slab of constant $\kappa = 2$ has the same area as the plates but has a thickness of 3 mm. What is the capacitance (*a*) without the dielectric and (*b*) with the dielectric?

(*a*) This capacitor is the same as the one in Example 21-1 except that the plate spacing is 4 mm rather then 1 mm. Since the capacitance varies inversely with the plate separation, the capacitance without the dielectric is one-fourth the value found in Example 21-1, or $C_0 = \frac{1}{4}(88.5\text{ pF}) = 22.1\text{ pF}$.

(*b*) To find the value of the capacitance with the dielectric inserted, we place charges $+Q$ on one plate and $-Q$ on the other plate and find the electric field between the plates. We then calculate the potential difference between the plates.

In the space between the plates without the dielectric, the electric field is $E_0 = Q/\epsilon_0 A$ because the field due to the bound charges is zero. Inside the dielectric, the electric field is $E = E_0/\kappa$. The potential difference between the plates is the integral of the field over distance from one plate to the other. If s is the total separation of the plates, the thickness of the dielectric is $\frac{3}{4}s$ and that of the free space is $\frac{1}{4}s$. The potential difference between the plates is thus

$$V = E_0\left(\frac{1}{4}\right)s + \frac{E_0}{\kappa}\left(\frac{3}{4}s\right) = E_0 s\left(\frac{1}{4} + \frac{3}{4\kappa}\right) = V_0\left(\frac{\kappa + 3}{4\kappa}\right)$$

where we have used $E_0 s = V_0$, the original potential difference with no dielectric. Using $\kappa = 2$, we find the potential difference with the dielectric to be

$$V = \tfrac{5}{8}V_0$$

The new capacitance is thus

$$C = \frac{Q}{V} = \frac{Q}{\frac{5}{8}V_0} = \frac{8}{5}\frac{Q}{V_0} = \frac{8}{5}C_0$$

$$= \frac{8}{5}(22.1\text{ pF}) = 35.4\text{ pF}$$

21-4 The Storage of Electrical Energy

While a capacitor is being charged, a positive charge is transferred from the negatively charged conductor to the positively charged conductor. Since the positive conductor is at a greater potential than the negative conductor, the potential energy of the charge being transferred is increased. For example, if a small amount of charge q is transferred through a potential difference V, the potential energy of the charge is increased by the amount qV. (Remember that, by definition, potential difference is the difference in potential energy per unit charge.) Work must therefore be done to charge a capacitor. Some of this work is stored as electrostatic potential energy. At the beginning of the charging process, neither conductor is charged. There is no electric field, and both conductors are at the same potential. After the charging process, a charge Q has been transferred from one conductor to the other, and the potential difference is $V = Q/C$, where C is the capacitance.

Let q be the charge that has been transferred at some time during the process. The potential difference is then $V = q/C$. If a small amount of additional charge dq is now transferred from the negative conductor at zero potential to the positive conductor at a potential V (Figure 21-7), the potential energy of the charge is increased by

$$dU = V\,dq = \frac{q}{C}\,dq$$

The total increase in potential energy U is the sum or integral of these charges dU as q increases from zero to its final value Q (Figure 21-8):

$$U = \int dU = \int_0^Q \frac{q}{C}\,dq = \frac{1}{2}\frac{Q^2}{C}$$

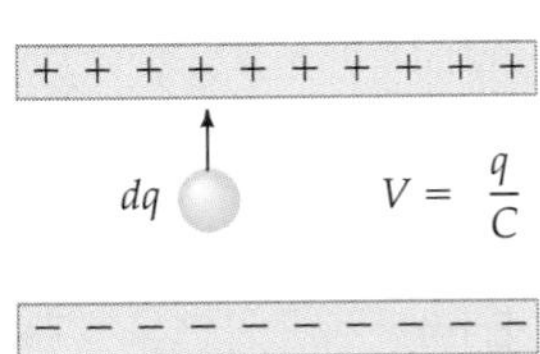

Figure 21-7 When a small amount of charge dq is moved from the negative conductor to the positive conductor, its potential energy is increased by $dU = V\,dq$, where V is the potential difference between the conductors.

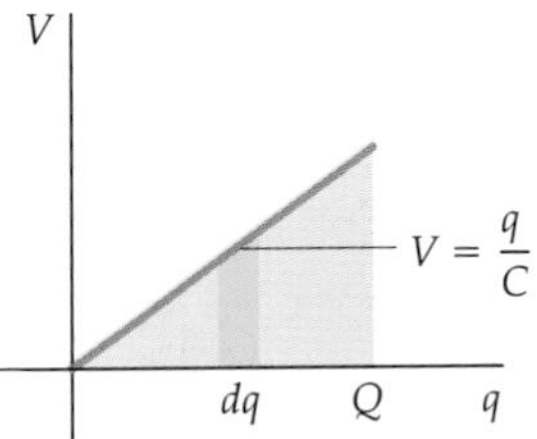

Figure 21-8 The work needed to charge a capacitor is the integral of $V\,dq$ from the original charge of $q = 0$ to the final charge of $q = Q$. This work is the area under the curve $\frac{1}{2}Q(Q/C)$.

This potential energy is the energy stored in the capacitor. Using $C = Q/V$, we can express this energy in a variety of ways:

$$U = \frac{1}{2}\frac{Q^2}{C} = \frac{1}{2}QV = \frac{1}{2}CV^2 \qquad \text{21-16}$$

Energy stored in a capacitor

Equation 21-16 is a general expression for the energy stored in a charged capacitor as electrostatic potential energy.

Exercise

A 15-μF capacitor is charged to 60 V. How much energy is stored in the capacitor? (Answer: 0.027 J)

Example 21-4

A 60-μF capacitor is charged to 12 V. The capacitor is removed from the battery and the plate separation is increased from 2.0 mm to 3.5 mm. (*a*) What is the charge on the capacitor? (*b*) How much energy was originally stored in the capacitor? (*c*) By how much is the energy increased when the plate separation is changed?

(*a*) From the definition of capacitance (Equation 21-1) the charge on the capacitor is

$$Q = CV = (60\ \mu\text{F})(12\ \text{V}) = 720\ \mu\text{C}$$

(*b*) The energy originally stored is

$$W = \frac{1}{2}QV = \frac{1}{2}(720\ \mu\text{C})(12\ \text{V}) = 4320\ \mu\text{J}$$

We could have found the energy without first finding the charge from

$$W = \frac{1}{2}CV^2 = \frac{1}{2}(60\ \mu\text{F})(12\ \text{V})^2 = 4320\ \mu\text{J}$$

(*c*) After the capacitor is removed from the battery, the charge on the plates must remain constant. When the plate separation is increased, the voltage between the plates is increased and the capacitance is decreased. We can find the increase in energy stored by either finding the increase in the voltage and using $W = \frac{1}{2}QV$ or by finding the decrease in the capacitance and using $W = \frac{1}{2}Q^2/C$. The potential difference between the plates is related to the plate separation s and the electric field E by

$$V = Es$$

Capacitor bank for energy storage for the pulsed Nova laser used at Lawrence Livermore Laboratories to study fusion. Storage capacitors at the Nova laser have a capacitance of 14.5, 25, and 52.5 μF.

The field does not change because the charge remains the same. Since the potential difference is 12 V when the plate separation is 2.0 mm, the potential difference when the separation is 3.5 mm is

$$V = (12\ \text{V})\frac{3.5\ \text{mm}}{2.0\ \text{mm}} = 21\ \text{V}$$

When the plate separation is 3.5 mm, the energy stored is

$$W = \frac{1}{2}QV = \frac{1}{2}(720\ \mu\text{C})(21\ \text{V}) = 7560\ \mu\text{J}$$

The increase in the energy stored is therefore 7560 μJ − 4320 μJ = 3240 μJ.

It is illuminating to work part *b* in another way. Because the plates of a capacitor are oppositely charged, they exert attractive forces on one another. To increase the plate separation, work must be done against these forces. Let us assume that the lower plate is held fixed and the upper plate is moved. The force on the upper plate is the charge on the plate Q times the electric field *due to the lower plate*. This field is half the total field between the plates because the charge on the upper plate also contributes equally to the field. When the potential difference is 12 V and the separation is 2 mm, the total field between the plates is

$$E = \frac{V}{s} = \frac{12\ \text{V}}{2\ \text{mm}} = 6\ \text{V/mm} = 6\ \text{kV/m}$$

The electric field due to just the charge on the bottom plate is then

$$E' = \frac{1}{2}E = 3\ \text{kV/m}$$

The force exerted on the upper plate by the bottom plate is thus

$$F = QE' = (720\ \mu\text{C})(3\ \text{kV/m}) = 2.16\ \text{N}$$

The work that must be done to move the upper plate a distance of $\Delta s = 1.5$ mm is then

$$W = F\ \Delta s = (2.16\ \text{N})(1.5\ \text{mm}) = 3.24 \times 10^{-3}\ \text{J} = 3240\ \mu\text{J}$$

This work equals the increase in the energy stored.

In the process of charging a capacitor, an electric field is produced between the plates. The work required to charge the capacitor can be thought of as the work required to create the electric field. That is, we can think of the energy stored in a capacitor as energy stored in the electric field, called **electrostatic field energy.** We can see this in the case of a parallel-plate capacitor that is filled with a dielectric of constant κ. Let $+Q$ be the charge on one of the plates of the capacitor. The potential difference between the plates is $V = Es$, where s is the plate separation and E is the electric field between the plates, which is related to the charge on the plates by

$$E = \frac{E_0}{\kappa} = \frac{\sigma}{\kappa\epsilon_0} = \frac{Q}{\epsilon A}$$

Substituting $Q = \epsilon AE$ and $V = Es$ into Equation 21-16, we obtain the potential energy U associated with an electric field:

$$U = \frac{1}{2}QV = \frac{1}{2}(\epsilon AE)(Es)$$
$$= \frac{1}{2}\epsilon E^2(As)$$

The quantity As is the volume of the space between the plates of the capacitor containing the electric field. The energy per unit volume is called the **energy density** η. The energy density in an electric field E is thus

$$\eta = \frac{\text{energy}}{\text{volume}} = \frac{1}{2}\epsilon E^2 \qquad 21\text{-}17$$

Energy density of an electrostatic field

Thus, the energy per unit volume of the electrostatic field is proportional to the square of the electric field. Although we obtained Equation 21-17 by considering the electric field between the plates of a parallel-plate capacitor, the result applies to any electric field. Whenever there is an electric field in space, the electrostatic energy per unit volume is given by Equation 21-17.

We can see the generality of Equation 21-17 by calculating the electrostatic field energy for a case that does not involve a capacitor, and in which the electric field is not constant. Let us consider the electrostatic potential energy of a spherical conductor of radius R that carries a charge Q. We will calculate the work needed to bring a charge from a great distance away to the conductor in the same way that we calculated the work needed to transfer a charge from one plate of a capacitor to the other. When the sphere carries a charge q, its potential relative to $V = 0$ at infinity is

$$V = \frac{1}{4\pi\epsilon_0}\frac{q}{R}$$

The work needed to bring an additional amount of charge dq from infinity to the conductor is $V\,dq$, which equals the increase in the potential energy of the conductor:

$$dU = V\,dq = \frac{1}{4\pi\epsilon_0 R}q\,dq$$

The total increase in potential energy U is the integral of dU as q increases from zero to its final value Q. Integrating from $q = 0$ to $q = Q$, we obtain

$$U = \frac{1}{4\pi\epsilon_0 R}\frac{Q^2}{2} = \frac{1}{2}QV \qquad 21\text{-}18$$

This is the electrostatic potential energy of the spherical conductor.

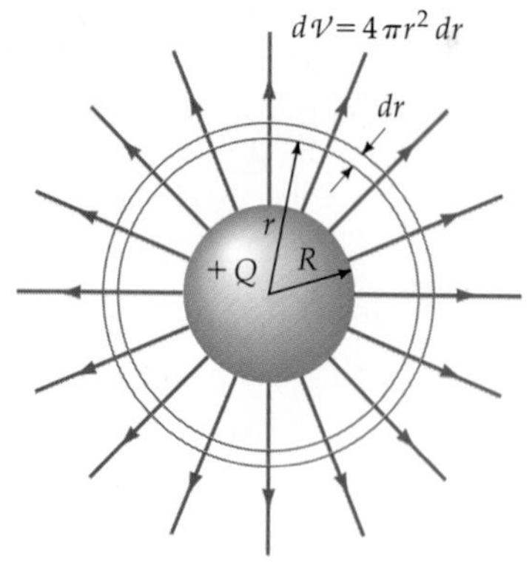

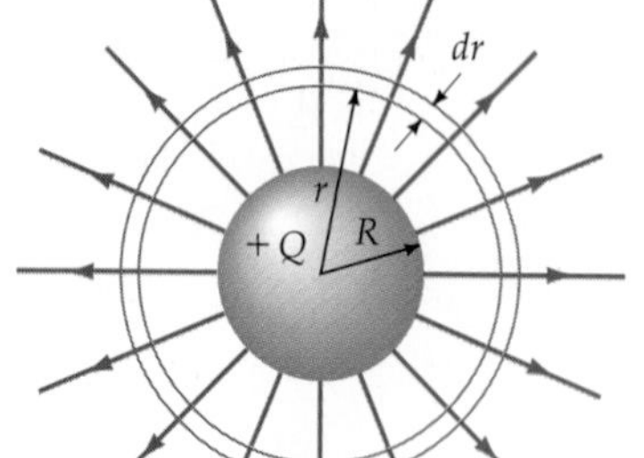

Figure 21-9 Geometry for the calculation of the electrostatic energy of a spherical conductor carrying a charge Q. The volume of the space between r and $r + dr$ is $d\mathcal{V} = 4\pi r^2\,dr$. The electrostatic field energy in this volume element is $\eta\,d\mathcal{V}$, where $\eta = \frac{1}{2}\epsilon_0 E^2$ is the energy density.

We will now obtain this same result by considering the energy density of an electric field given by Equation 21-17 with $\epsilon = \epsilon_0$, the permittivity of free space. When the conductor carries a charge Q, the electric field is radial and is given by

$$E_r = 0 \qquad r < R \text{ (inside the conductor)}$$

$$E_r = \frac{1}{4\pi\epsilon_0}\frac{Q}{r^2} \qquad r > R \text{ (outside the conductor)}$$

Since the electric field is spherically symmetric, we choose a spherical shell for our volume element. If the radius of the shell is r and its thickness is dr, the volume is $d\mathcal{V} = 4\pi r^2\,dr$ (Figure 21-9). The energy dU in this volume element is

$$dU = \eta\,d\mathcal{V} = \tfrac{1}{2}(\epsilon_0 E^2)4\pi r^2\,dr$$

$$= \frac{1}{2}\epsilon_0\left(\frac{Q}{4\pi\epsilon_0 r^2}\right)^2(4\pi r^2\,dr) = \frac{Q^2}{8\pi\epsilon_0}\frac{dr}{r^2}$$

Since the electric field is zero for $r < R$, we obtain the total energy in the electric field by integrating from $r = R$ to $r = \infty$:

$$U = \int_R^\infty \frac{Q^2}{8\pi\epsilon_0}\frac{dr}{r^2} = \frac{1}{2}\frac{Q^2}{4\pi\epsilon_0 R} = \frac{1}{2}QV \qquad 21\text{-}19$$

which is the same as Equation 21-18.

Questions

1. If the potential difference of a capacitor is doubled, by what factor does its stored electric energy change?
2. Half the charge is removed from a capacitor. What fraction of its stored energy is removed along with the charge?

21-5 Combinations of Capacitors

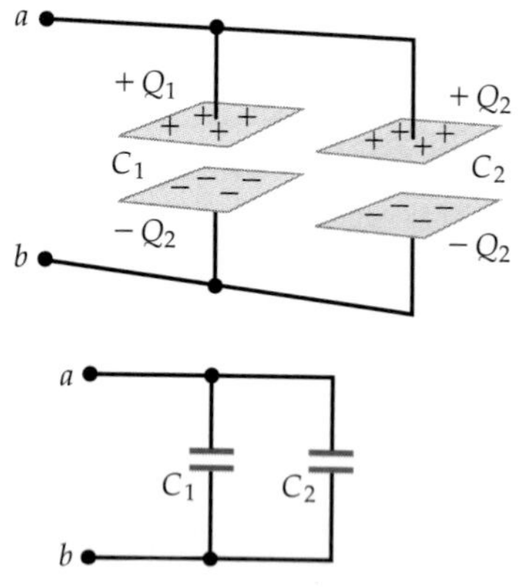

Figure 21-10 Two capacitors in parallel. The potential difference is the same across each capacitor.

Two or more capacitors are often used in combination. Figure 21-10 shows two **capacitors in parallel.** (In electric circuits, a capacitor is indicated by the symbol ⊣ ⊢.) The upper plates of the two capacitors are connected by a conducting wire and are therefore at the same potential V_a. The lower plates are also connected and are at a common potential V_b. Points a and b are connected to a battery or some other device that maintains a potential difference $V = V_a - V_b$, which is the potential difference between the plates of each capacitor. The effect of adding a second capacitor connected in this way is an increase in the capacitance. The area is essentially increased, allowing more charge to be stored for the same potential difference. If the capacitances are C_1 and C_2, the charges Q_1 and Q_2 stored on the plates are given by

$$Q_1 = C_1 V$$

and

$$Q_2 = C_2 V$$

The total charge stored is

$$Q = Q_1 + Q_2 = C_1 V + C_2 V = (C_1 + C_2)V$$

The **equivalent capacitance** is the capacitance of a single capacitor that could replace some combination of capacitors in a circuit and store the same

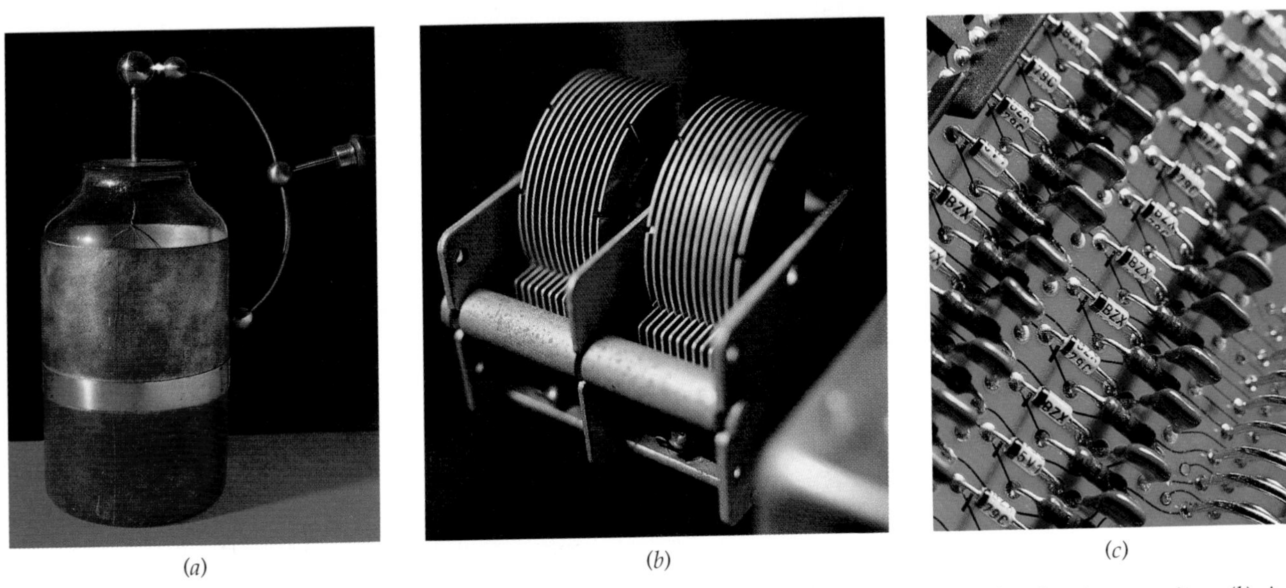

(a) (b) (c)

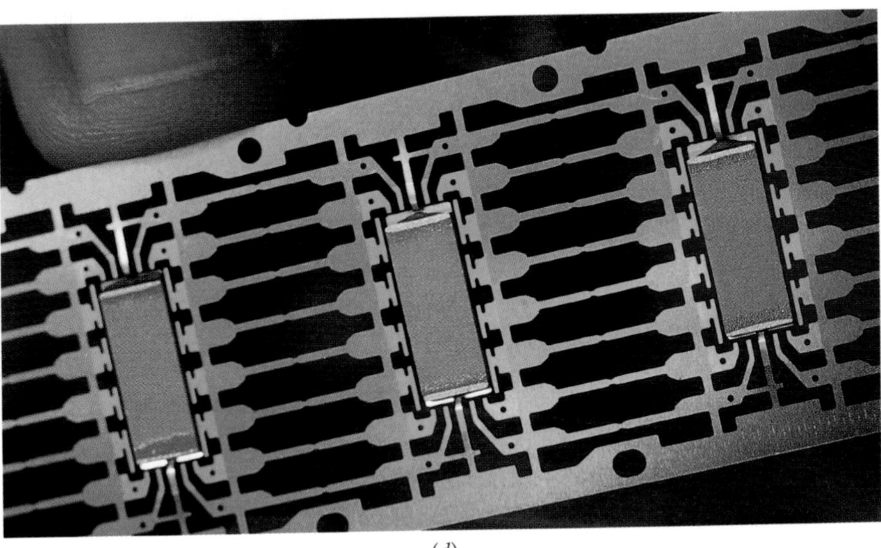

(d)

(*a*) A Leyden jar capacitor. (*b*) A variable air-gap capacitor like those that were used in the tuning circuits of old radios. (*c*) Capacitors wired into a circuit board. (*d*) Capacitors in a printed circuit.

amount of charge for a given potential difference. The equivalent capacitance of two capacitors in parallel is the ratio of the total charge stored to the potential difference:

$$C_{eq} = \frac{Q}{V} = C_1 + C_2 \qquad \text{21-20}$$

Thus, the equivalent capacitance of two capacitors in parallel equals the sum of the individual capacitances. The same reasoning can be extended to three or more capacitors connected in parallel, as in Figure 21-11:

$$C_{eq} = C_1 + C_2 + C_3 + \cdots \qquad \text{21-21}$$

Equivalent capacitance for capacitors in parallel

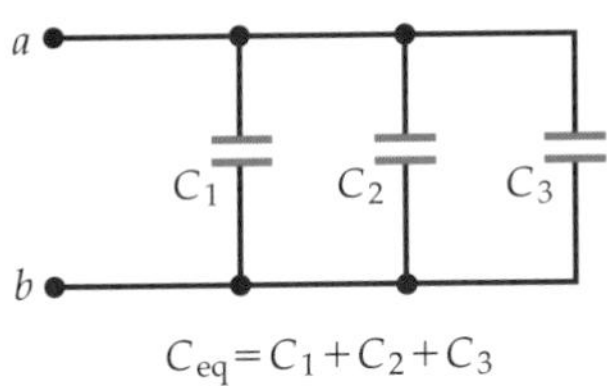

Figure 21-11 Three capacitors in parallel. The effect of adding a parallel capacitor to a circuit is an increase in the equivalent capacitance.

Figure 21-12 shows two **capacitors in series.** When points a and b are connected to the terminals of a battery, there is a potential difference $V = V_a - V_b$ across the two capacitors, but the potential difference across one of the capacitors is not necessarily the same as that across the other. If a charge $+Q$ is placed on the upper plate of the first capacitor, the electric field produced by that charge will induce an equal negative charge $-Q$ on its lower plate. This charge comes from electrons drawn from the upper plate of the second capacitor. Thus, there will be an equal charge $+Q$ on the upper plate of the second capacitor and a corresponding charge $-Q$ on its lower plate. The potential difference across the first capacitor is

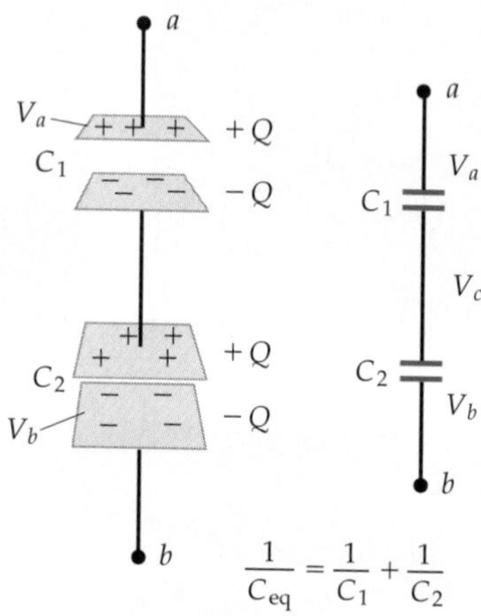

Figure 21-12 Two capacitors in series. The charge is the same on each capacitor. The potential difference across the series combination is the sum of the potential differences across the individual capacitors.

$$V_1 = V_a - V_c = \frac{Q}{C_1}$$

Similarly, the potential difference across the second capacitor is

$$V_2 = V_c - V_b = \frac{Q}{C_2}$$

The potential difference across the two capacitors in series is the sum of these potential differences:

$$V = V_a - V_b = (V_a - V_c) + (V_c - V_b)$$
$$= V_1 + V_2 = \frac{Q}{C_1} + \frac{Q}{C_2}$$

Thus,

$$V = \frac{Q}{C_1} + \frac{Q}{C_2}$$
$$= Q\left(\frac{1}{C_1} + \frac{1}{C_2}\right) \qquad \text{21-22}$$

The equivalent capacitance of two capacitors in series is that of a single capacitor that could replace the two capacitors and give the potential difference V for the same charge Q. Thus,

$$C_{eq} = \frac{Q}{V} \qquad \text{21-23}$$

Comparing Equations 21-22 and 21-23, we have

$$\frac{1}{C_{eq}} = \frac{1}{C_1} + \frac{1}{C_2} \qquad \text{21-24}$$

Equation 21-24 can be generalized to three or more capacitors connected in series:

$$\frac{1}{C_{eq}} = \frac{1}{C_1} + \frac{1}{C_2} + \frac{1}{C_3} + \cdots \qquad \text{21-25}$$

Equivalent capacitance for capacitors in series

The potential difference across a set of capacitors in series is the sum of the potential differences across the individual capacitors. Note that the addition of a capacitor in series increases $1/C_{eq}$, which means that the equivalent capacitance C_{eq} decreases.

Exercise

Two capacitors have capacitances of 20 μF and 30 μF. Find the equivalent capacitance if the capacitors are connected (*a*) in parallel, (*b*) in series. [Answers: (*a*) 50 μF, (*b*) 12 μF]

Note that in the preceding exercise the equivalent capacitance of the two capacitances in series is less than the capacitance of either capacitor.

There is a maximum voltage that a given capacitor can withstand before breakdown occurs. Suppose we have a power supply of 100 V and two identical capacitors each of which breaks down if the potential across it exceeds 60 V. Either capacitor will break down if connected across the power supply. Breakdown will also occur if we connect the two capacitors in parallel because the potential across each will be 100 V. However, if we connect the two capacitors in series, the potential across each capacitor will be only 50 V, and they will not break down.

Example 21-5

Find the equivalent capacitance of the network of the three capacitors shown in Figure 21-13.

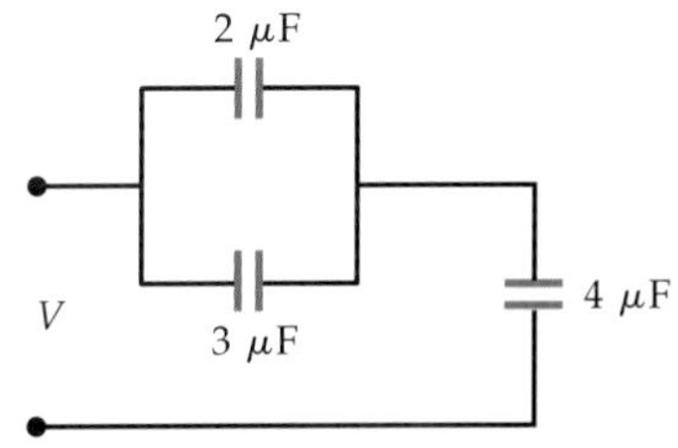

Figure 21-13 Capacitor network for Example 21-5.

In this network, the 2-μF and 3-μF capacitors are connected in parallel and this parallel combination is connected in series with the 4-μF capacitor. The equivalent capacitance of the two capacitors in parallel is

$$C_{eq} = C_1 + C_2 = 2\ \mu\text{F} + 3\ \mu\text{F} = 5\ \mu\text{F}$$

If we replace the two capacitors in parallel with a single 5-μF capacitor, we have a 5-μF capacitor in series with a 4-μF capacitor. The equivalent capacitance of this series combination is found from

$$\frac{1}{C_{eq}} = \frac{1}{C_1} + \frac{1}{C_2} = \frac{1}{5\ \mu\text{F}} + \frac{1}{4\ \mu\text{F}} = \frac{9}{20\ \mu\text{F}}$$

The equivalent capacitance of the three-capacitor network is thus

$$C_{eq} = \frac{20\ \mu\text{F}}{9} = 2.22\ \mu\text{F}$$

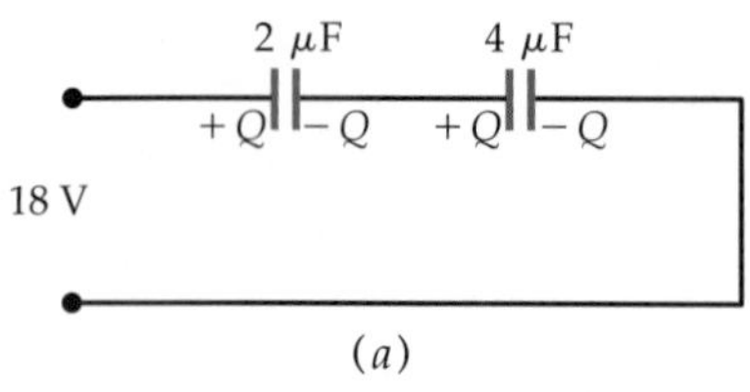

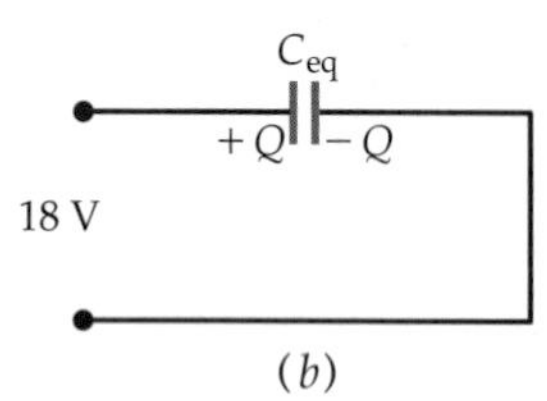

Figure 21-14 (*a*) Two capacitors connected in series across an 18-V battery for Example 21-6. (*b*) The two capacitors in (*a*) can be replaced by an equivalent capacitor.

Example 21-6

A 2-μF capacitor and a 4-μF capacitor are connected in series across an 18-V battery as shown in Figure 21-14*a*. Find the charge on the capacitors and the potential difference across each.

In Figure 21-14*b*, the two capacitors have been replaced by one capacitor with an equivalent capacitance C_{eq}, which is found from

$$\frac{1}{C_{eq}} = \frac{1}{C_1} + \frac{1}{C_2} = \frac{1}{2\ \mu\text{F}} + \frac{1}{4\ \mu\text{F}} = \frac{3}{4\ \mu\text{F}}$$

$$C_{eq} = \tfrac{4}{3}\ \mu\text{F}$$

The charge on either plate of the equivalent capacitor in Figure 21-14*b* is

$$Q = C_{eq}V = (\tfrac{4}{3}\ \mu\text{F})(18\ \text{V}) = 24\ \mu\text{C}$$

This is the magnitude of the charge on each plate of the two original capacitors. The potential difference across the 2-μF capacitor is thus

$$V_1 = \frac{Q}{C_1} = \frac{24\ \mu\text{C}}{2\ \mu\text{F}} = 12\ \text{V}$$

and that across the 4-μF capacitor is

$$V_2 = \frac{Q}{C_2} = \frac{24\ \mu\text{C}}{4\ \mu\text{F}} = 6\ \text{V}$$

Note that the sum of these potential differences is 18 V, as required.

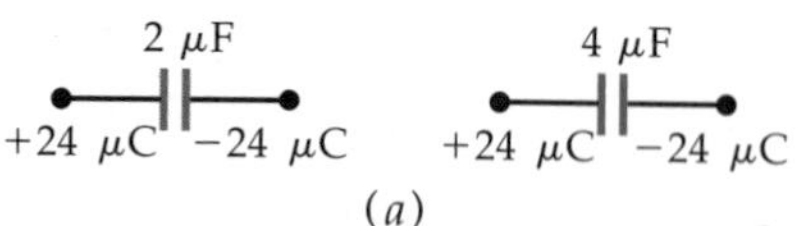

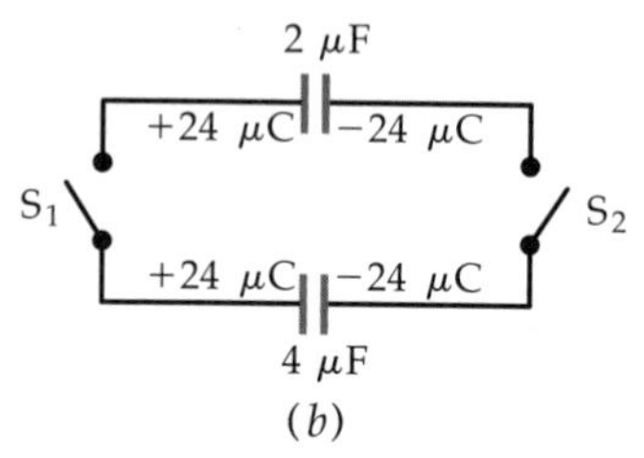

Figure 21-15 (*a*) The two capacitors in Figure 21-14*a* just after they are disconnected from the battery. (*b*) After the switches S_1 and S_2 are closed, the two capacitors are reconnected, positive plate to positive plate and negative plate to negative plate.

Example 21-7

The two capacitors in Example 21-6 are removed from the battery and are carefully disconnected from each other so that the charge on the plates is not disturbed as shown in Figure 21-15*a*. They are then reconnected with positive plate connected to positive plate and negative plate connected to negative plate as shown in Figure 21-15*b*. Find the potential difference across the capacitors and the charge on each capacitor.

After the capacitors are reconnected, the two positive plates form a single conductor as do the two negative plates. The total charge on the positive plates is +48 μC, and that on the negative plates is −48 μC. Furthermore, the potential difference across both capacitors must now be the same. The capacitors are therefore connected in parallel, so the equivalent capacitance is

$$C_{eq} = C_1 + C_2 = 2\ \mu\text{F} + 4\ \mu\text{F} = 6\ \mu\text{F}$$

Since the total charge is 48 μC, the potential difference across the parallel combination is

$$V = \frac{Q}{C_{eq}} = \frac{48\ \mu\text{C}}{6\ \mu\text{F}} = 8\ \text{V}$$

The charge on each capacitor is

$$Q_1 = C_1V = (2\ \mu\text{F})(8\ \text{V}) = 16\ \mu\text{C}$$

and

$$Q_2 = C_2V = (4\ \mu\text{F})(8\ \text{V}) = 32\ \mu\text{C}$$

The total charge adds up to 48 μC, as required.

Example 21-8

Two parallel-plate capacitors, each having a capacitance of 2 μF, are connected in parallel across a 12-V battery. Find (*a*) the charge on each capacitor and (*b*) the total energy stored in the capacitors. The capacitors are disconnected from the battery and a dielectric of constant $\kappa = 3$ is inserted between the plates of one of the capacitors. After the dielectric is inserted, find (*c*) the potential difference across each capacitor, (*d*) the charge on each capacitor, and (*e*) the total energy stored in the capacitors.

(*a*) The original charge on each capacitor is

$$Q = CV = (2\ \mu\text{F})(12\ \text{V}) = 24\ \mu\text{C}$$

(*b*) The energy stored in each capacitor is

$$U = \frac{1}{2}QV = \frac{1}{2}(24\ \mu\text{C})(12\ \text{V}) = 144\ \mu\text{J}$$

so the total energy stored is $2 \times 144\ \mu\text{J} = 288\ \mu\text{J}$.

(*c*) The capacitance of the capacitor with the dielectric is now

$$C' = \kappa C = 3(2\ \mu\text{F}) = 6\ \mu\text{F}$$

Since the capacitors are in parallel, the equivalent capacitance is

$$C_{\text{eq}} = C_1 + C_2 = 6\ \mu\text{F} + 2\ \mu\text{F} = 8\ \mu\text{F}$$

The total charge, which is 48 μC, must remain the same because the battery is disconnected. The potential difference across the parallel combination, which is the potential difference across each capacitor, is

$$V = \frac{Q}{C_{\text{eq}}} = \frac{48\ \mu\text{C}}{8\ \mu\text{F}} = 6\ \text{V}$$

(*d*) When the dielectric is inserted into one of the capacitors, the field is weakened and the potential difference is lowered. Since the two capacitors are connected in parallel, charge must flow from the other capacitor so that the potential difference is the same across both capacitors. Let Q_1 be the charge on the capacitor with the dielectric and Q_2 be that on the other capacitor. Then

$$Q_1 = C_1V = (6\ \mu\text{F})(6\ \text{V}) = 36\ \mu\text{C}$$

and

$$Q_2 = C_2V = (2\ \mu\text{F})(6\ \text{V}) = 12\ \mu\text{C}$$

The total charge is 48 μC as before.

(*e*) The energy in the capacitor with the dielectric is

$$U_1 = \frac{1}{2}Q_1V = \frac{1}{2}(36\ \mu\text{C})(6\ \text{V}) = 108\ \mu\text{J}$$

and the energy in the other capacitor is

$$U_2 = \frac{1}{2}Q_2V = \frac{1}{2}(12\ \mu\text{C})(6\ \text{V}) = 36\ \mu\text{J}$$

The total energy is $U_1 + U_2 = 144\ \mu\text{J}$. This is half the original energy of 288 μJ. Removing the dielectric from the capacitor requires 144 μJ of work which is then stored as electrostatic potential energy.

A mechanical hand is used at Lockheed to install capacitors in a circuit.

Example 21-9

Work parts *c*, *d*, and *e* of Example 21-8, if the dielectric is inserted into one of the capacitors while the battery is still connected.

(*c*) Since the battery is still connected, the potential difference across the capacitors remains 12 V.

(*d*) When the dielectric is inserted into one of the capacitors, additional charge is supplied by the battery to maintain the potential difference of 12 V. Since the new capacitance of the capacitor with the dielectric is 6 μF, the charge on that capacitor is

$$Q_1 = C_1V = (6\ \mu\text{F})(12\ \text{V}) = 72\ \mu\text{C}$$

and the charge on the other capacitor remains the same as before.

(*e*) The energy in the capacitor with the dielectric is

$$U_1 = \frac{1}{2}Q_1V = \frac{1}{2}(72\ \mu\text{C})(12\ \text{V}) = 432\ \mu\text{J}$$

The capacitor without the dielectric has the same charge and potential difference, so it has the same energy:

$$U = \frac{1}{2}QV = \frac{1}{2}(24\ \mu\text{C})(12\ \text{V}) = 144\ \mu\text{J}$$

Therefore, the total energy is 432 μJ + 144 μJ = 576 μJ. In this case, the total energy is increased because the battery supplies more charge.

Summary

1. A capacitor is a device for storing charge and energy. It consists of two conductors, closely spaced but insulated from each other, carrying equal and opposite charges. The capacitance is the ratio of the magnitude of the charge on either conductor Q to the potential difference between the conductors V:

$$C = \frac{Q}{V}$$

Capacitance depends only on the geometrical arrangement of the conductors and not on the charge or the potential difference.

2. The capacitance of a parallel-plate capacitor is proportional to the area of the plates and is inversely proportional to the separation distance:

$$C = \frac{\epsilon_0 A}{s}$$

The capacitance of a cylindrical capacitor is given by

$$C = \frac{2\pi\epsilon_0 L}{\ln (b/a)}$$

where L is the length of the capacitor and a and b are the radii of the inner and outer conductors, respectively.

3. A nonconducting material is called a dielectric. When a dielectric is inserted between the plates of a capacitor, the molecules in the dielectric become polarized and the electric field within the dielectric is weakened. If the field is E_0 without the dielectric, with the dielectric it is

$$E = \frac{E_0}{\kappa}$$

where κ is the dielectric constant. This decrease in electric field leads to an increase in the capacitance by the factor κ:

$$C = \kappa C_0$$

where C_0 is the capacitance without the dielectric. The permittivity of a dielectric ϵ is defined as

$$\epsilon = \kappa \epsilon_0$$

Dielectrics also provide a physical means of separating the plates of a capacitor, and they increase the voltage that can be applied to a capacitor before dielectric breakdown occurs.

4. The electrostatic energy stored in a capacitor of charge Q, potential difference V, and capacitance C is

$$U = \frac{1}{2}\frac{Q^2}{C} = \frac{1}{2}QV = \frac{1}{2}CV^2$$

This energy can be considered to be stored in the electric field between the plates. The energy per unit volume in an electric field E is given by

$$\eta = \frac{\text{energy}}{\text{volume}} = \frac{1}{2}\epsilon E^2$$

5. When two or more capacitors are connected in parallel, the equivalent capacitance of the combination is the sum of the individual capacitances:

$$C_{eq} = C_1 + C_2 + C_3 + \cdots \qquad \text{capacitors in parallel}$$

When two or more capacitors are connected in series, the reciprocal of the equivalent capacitance is found by adding the reciprocals of the individual capacitances:

$$\frac{1}{C_{eq}} = \frac{1}{C_1} + \frac{1}{C_2} + \frac{1}{C_3} + \cdots \qquad \text{capacitors in series}$$

Suggestion for Further Reading

Trotter, Donald M., Jr.: "Capacitors," *Scientific American*, July 1988, p. 86.

Marvelous advances in capacitor miniaturization have been achieved over the past thirty years; without this, the advantages of miniaturizing integrated circuits for radios, computers, and other electronic equipment could not have been realized.

Review

A. Objectives: After studying this chapter, you should:

1. Be able to define capacitance and calculate it for a parallel-plate capacitor and a cylindrical capacitor.

2. Be able to discuss the effects of a dielectric on the capacitance, charge, potential difference, and electric field of a parallel-plate capacitor.

3. Know what is meant by bound charge in a dielectric, and be able to discuss how bound charge arises and what effect it has.

4. Be able to derive the expression $U = \frac{1}{2}QV$ for the energy stored in a charged capacitor.

5. Be able to discuss the concept of electrostatic field energy.

6. Be able to calculate the equivalent capacitance of parallel and series combinations of capacitors.

B. Define, explain, or otherwise identify:

Capacitor
Capacitance
Farad
Dielectric
Dielectric constant
Permittivity of a dielectric
Bound charge
Electrostatic field energy
Energy density
Capacitors in parallel
Equivalent capacitance
Capacitors in series

C. True or false: If the statement is true, explain why it is true. If it is false, give a counterexample.

1. The capacitance of a capacitor is defined as the total amount of charge it can hold.

2. The capacitance of a parallel-plate capacitor depends on the voltage difference between the plates.

3. The capacitance of a parallel-plate capacitor is proportional to the charge on its plates.

4. The equivalent capacitance of two capacitors in parallel equals the sum of the individual capacitances.

5. The equivalent capacitance of two capacitors in series is less than the capacitance of either capacitor alone.

6. A dielectric inserted into a capacitor increases the capacitance.

7. The electrostatic energy per unit volume at some point is proportional to the square of the electric field at that point.

Problems

Level I

21-1 The Parallel-Plate Capacitor

1. (*a*) If a parallel-plate capacitor has a 0.15-mm separation, what must its area be for it to have a capacitance of 1 F? (*b*) If the plates are square, what is the length of their sides?

2. A parallel-plate capacitor has a capacitance of 2.0 μF and a plate separation of 1.6 mm. (*a*) What is the maximum potential difference between the plates such that dielectric breakdown of the air between the plates does not occur? (E_{max} = 3 MV/m) (*b*) How much charge is stored at this maximum potential difference?

3. A parallel-plate capacitor has a charge of 40 μC. The potential difference across the plates is 500 V. What is the capacitance?

4. An electric field of 3×10^6 V/m exists between the plates of a circular parallel-plate capacitor that has a plate separation of 2 mm. (*a*) What is the voltage across the capacitor? (*b*) What plate radius is required if the stored charge is 10 μC?

21-2 The Cylindrical Capacitor

5. A coaxial cable between two cities has an inner radius of 0.8 mm and an outer radius of 6 mm. Its length is 8×10^5 m (about 500 mi). Treat this cable as a cylindrical capacitor and calculate its capacitance.

6. A Geiger tube consists of a wire of radius 0.2 mm and length 12 cm and a coaxial cylindrical shell conductor of the same length and a radius of 1.5 cm. (*a*) Find the capacitance, assuming that the gas in the tube has a dielectric constant of 1. (*b*) Find the charge per unit length on the wire when the potential difference between the wire and shell is 1.2 kV.

21-3 Dielectrics

7. A parallel-plate capacitor is made by placing polyethylene (κ = 2.3) between two sheets of aluminum foil. The area of each sheet is 400 cm^2, and the thickness of the polyethylene is 0.3 mm. Find the capacitance.

8. What is the dielectric constant of a dielectric on which the induced bound charge density is (*a*) 80 percent of the free charge density on the plates of a capacitor filled by the dielectric, (*b*) 20 percent of the free charge density, and (*c*) 98 percent of the free charge density?

9. Two parallel plates have charges Q and $-Q$. When the space between the plates is devoid of matter, the electric field is 2.5×10^5 V/m. When the space is filled with a certain dielectric, the field is reduced to 1.2×10^5 V/m. (*a*) What is the dielectric constant of the dielectric? (*b*) If Q = 10 nC, what is the area of the plates? (*c*) What is the total induced charge on either face of the dielectric?

21-4 The Storage of Electrical Energy

10. (*a*) A 3-μF capacitor is charged to 100 V. How much energy is stored in the capacitor? (*b*) How much additional energy is required to charge the capacitor from 100 to 200 V?

11. How much energy is stored when an isolated spherical conductor of radius 10 cm is charged to 2 kV?

12. A 10-μF capacitor is charged to $Q = 4\ \mu$C. (*a*) How much energy is stored in the capacitor? (*b*) If half the charge is removed, how much energy remains?

13. (*a*) Find the energy stored in a 20-pF capacitor when it is charged to 5 μC. (*b*) How much additional energy is required to increase the charge from 5 to 10 μC?

14. Find the energy per unit volume in an electric field that is equal to the dielectric strength of air (3 MV/m).

15. A parallel-plate capacitor with a plate area of 2 m^2 and a separation of 1.0 mm is charged to 100 V. (*a*) What is the electric field between the plates? (*b*) What is the energy per unit volume in the space between the plates? (*c*) Find the total energy by multiplying your answer to part (*b*) by the total volume between the plates. (*d*) Find the capacitance *C*. (*e*) Calculate the total energy from $U = \frac{1}{2}CV^2$, and compare your answer with your result for part (*c*).

21-5 Combinations of Capacitors

16. A 10.0-μF capacitor is connected in series with a 20.0-μF capacitor across a 6.0-V battery. (*a*) What is the equivalent capacitance of this combination? (*b*) Find the charge on each capacitor. (*c*) Find the potential difference across each capacitor.

17. A 10.0-μF capacitor and a 20.0-μF capacitor are connected in parallel across a 6.0-V battery. (*a*) What is the equivalent capacitance of this combination? (*b*) What is the potential difference across each capacitor? (*c*) Find the charge on each capacitor.

18. Three capacitors have capacitances of 2.0, 4.0, and 8.0 μF. Find the equivalent capacitance (*a*) if the capacitors are connected in parallel and (*b*) if they are connected in series.

19. A 2.0-μF capacitor is charged to a potential difference of 12.0 V and is then disconnected from the battery. (*a*) How much charge is on the plates? (*b*) When a second capacitor that is initially uncharged is connected in parallel with the first capacitor, the potential difference drops to 4.0 V. What is the capacitance of the second capacitor?

20. (*a*) How many 1.0-μF capacitors connected in parallel would it take to store a total charge of 1 mC with a potential difference of 10 V across each capacitor? (*b*) What would be the potential difference across the combination? (*c*) If the number of 1.0-μF capacitors found in part (*a*) is connected in series and the potential difference across each is 10 V, find the charge on each and the potential difference across the combination.

21. A 1.0-μF capacitor is connected in parallel with a 2.0-μF capacitor, and the combination is connected in series with a 6.0-μF capacitor. What is the equivalent capacitance of this combination?

22. A 3.0-μF capacitor and a 6.0-μF capacitor are connected in series, and the combination is connected in parallel with an 8.0-μF capacitor. What is the equivalent capacitance of this combination?

23. Three capacitors are connected in a triangular network as shown in Figure 21-16. Find the equivalent capacitance across terminals *a* and *c*.

Figure 21-16 Problem 23.

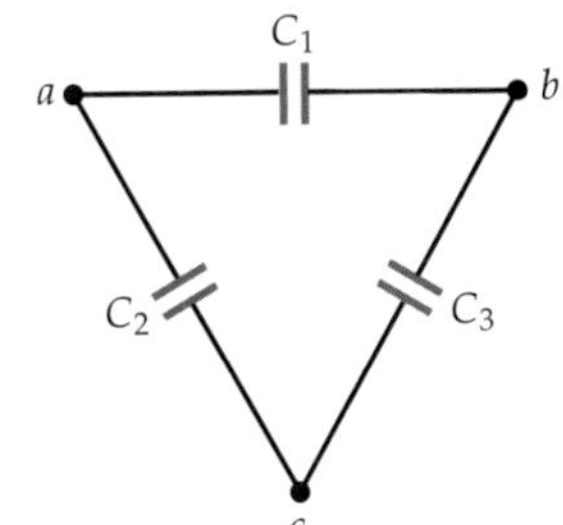

Level II

24. A parallel-plate capacitor has plates of area 600 cm^2 and a separation of 4 mm. It is charged to 100 V and is then disconnected from the battery. (*a*) Find the electric field E_0, the charge density σ, and the electrostatic potential energy *U*. A dielectric of constant $\kappa = 4$ is then inserted, completely filling the space between the plates. Find (*b*) the new electric field *E*, (*c*) the potential difference *V*, and (*d*) the bound charge density.

25. A certain dielectric with a dielectric constant $\kappa = 24$ can withstand an electric field of 4×10^7 V/m. Suppose we want to use this dielectric to construct a 0.1-μF capacitor that can withstand a potential difference of 2000 V. (*a*) What is the minimum plate separation? (*b*) What must the area of the plates be?

26. A parallel-plate capacitor of area *A* and separation *d* is charged to a potential difference *V* and is then disconnected from the charging source. The plates are then pulled apart until the separation is 2*d*. Find expressions in terms of *A*, *d*, and *V* for (*a*) the new capacitance, (*b*) the new potential difference, and (*c*) the new stored energy. (*d*) How much work was required to change the plate separation from *d* to 2*d*?

27. A parallel-plate, air-gap capacitor has a capacitance of 0.14 μF. The plates are 0.5 mm apart. (*a*) What is the area of each plate? (*b*) What is the potential difference if there is a charge of 3.2 μC on one plate and a charge of $-3.2\ \mu$C on the other? (*c*) What is the stored energy? (*d*) How much charge can the capacitor carry before dielectric breakdown of the air between the plates occurs?

28. For the circuit shown in Figure 21-17, find (*a*) the total equivalent capacitance between the terminals, (*b*) the charge stored on each capacitor, and (*c*) the total stored energy.

Figure 21-17 Problem 28.

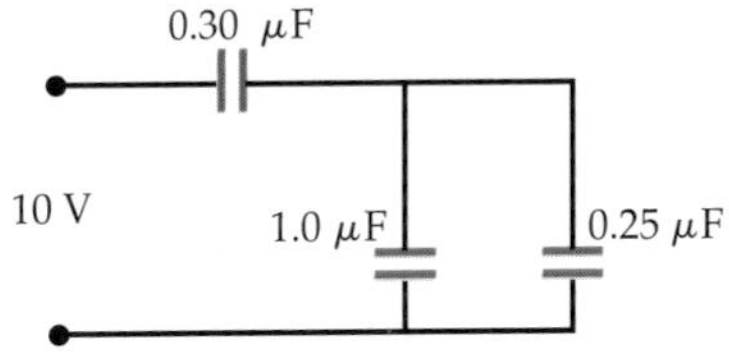

29. For the circuit shown in Figure 21-18, find (*a*) the total equivalent capacitance between the terminals, (*b*) the charge stored on each capacitor, and (*c*) the total stored energy.

Figure 21-18 Problem 29.

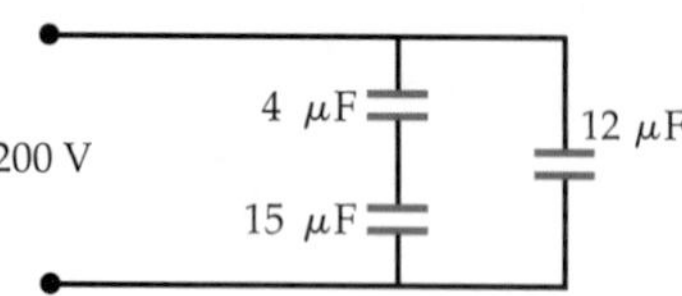

30. (*a*) Show that the equivalent capacitance of two capacitors in series can be written

$$C_{eq} = \frac{C_1 C_2}{C_1 + C_2}$$

(*b*) Use this expression to show that $C_{eq} < C_1$ and $C_{eq} < C_2$. (*c*) Show that the correct expression for the equivalent capacitance of three capacitors in series is

$$C_{eq} = \frac{C_1 C_2 C_3}{C_1 C_2 + C_2 C_3 + C_1 C_3}$$

31. A 20-pF capacitor is charged to 3.0 kV and is then removed from the battery and connected to an uncharged 50-pF capacitor. (*a*) What is the new charge on each capacitor? (*b*) Find the initial energy stored in the 20-pF capacitor and the final energy stored in the two capacitors. Is energy gained or lost when the two capacitors are connected?

32. Find all the different possible equivalent capacitances that can be obtained using a 1.0-, a 2.0-, and a 4.0-μF capacitor in any combination that includes all three or any two of the capacitors.

33. Three identical capacitors are connected so that their maximum equivalent capacitance is 15 μF. (*a*) Describe this combination. (*b*) Find the three other possible combinations using all three capacitors and their equivalent capacitances.

34. Two capacitors $C_1 = 4$ μF and $C_2 = 12$ μF are connected in series across a 12-V battery. They are carefully disconnected so that they are not discharged and are reconnected to each other with positive plate to positive plate and negative plate to negative plate. (*a*) Find the potential difference across each capacitor after they are connected. (*b*) Find the initial and final energy stored in the capacitors.

35. Work Problem 34 if the two capacitors are first connected in parallel across the 12-V battery and are then connected, with the positive plate of each capacitor connected to the negative plate of the other.

36. A 100-pF capacitor and a 400-pF capacitor are both charged to 2.0 kV. They are then disconnected from the voltage source and are connected together, positive plate to positive plate and negative plate to negative plate. (*a*) Find the resulting potential difference across each capacitor. (*b*) Find the energy lost when the connections are made.

37. Work Problem 36 if the capacitors are connected with the positive plate of each connected to the negative plate of the other, after they have been charged to 2.0 kV.

38. A parallel-plate capacitor has a capacitance C_0 and a plate separation d. Two dielectric slabs of constants κ_1 and κ_2, each of thickness $\frac{1}{2}d$ and having the same area as the plates, are inserted between the plates as shown in Figure 21-19. When the free charge on the plates is Q, find (*a*) the electric field in each dielectric and (*b*) the potential difference between the plates. (*c*) Show that the new capacitance is given by

$$C = \frac{2\kappa_1\kappa_2}{\kappa_1 + \kappa_2} C_0$$

(*d*) Show that this system can be considered to be two capacitors of thickness $\frac{1}{2}d$ connected in series.

Figure 21-19 Problem 38.

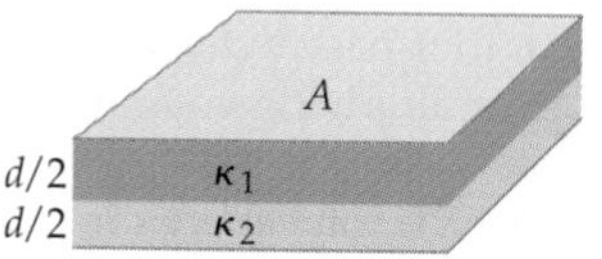

39. The membrane of the axon of a nerve cell is a thin cylindrical shell of radius $r = 10^{-5}$ m, length $L = 0.1$ m, and thickness $d = 10^{-8}$ m. The membrane has a positive charge on one side and a negative charge on the other, and acts as a parallel-plate capacitor of area $A = 2\pi rL$ and separation d. Its dielectric constant is about $\kappa = 3$. (*a*) Find the capacitance of the membrane. If the potential difference across the membrane is 70 mV, find (*b*) the charge on each side of the membrane, and (*c*) the electric field through the membrane.

40. A parallel-plate capacitor has a plate area A and a separation d. A metal slab of thickness t and area A is inserted between the plates. (*a*) Show that the capacitance is given by $C = \epsilon_0 A/(d - t)$, regardless of where the metal slab is placed. (*b*) Show that this arrangement can be considered to be a capacitor of separation a in series with one of separation b, where $a + b + t = d$.

41. In Figure 21-20, $C_1 = 2$ μF, $C_2 = 6$ μF, and $C_3 = 3.5$ μF. (*a*) Find the equivalent capacitance of this combination. (*b*) If the breakdown voltages of the individual capacitors are $V_1 = 100$ V, $V_2 = 50$ V, and $V_3 = 400$ V, what maximum voltage can be placed across points a and b?

Figure 21-20 Problem 41.

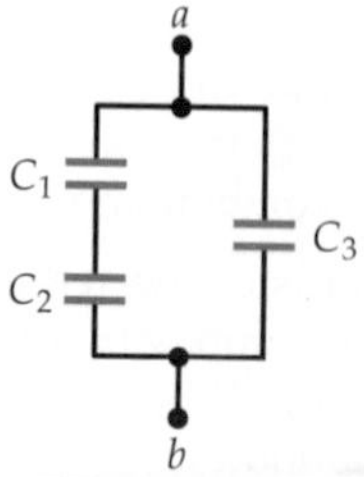

42. A parallel-plate capacitor is filled with two dielectrics of equal size as shown in Figure 21-21. Show (*a*) that this system can be considered to be two capacitors of area $\frac{1}{2}A$ connected in parallel and (*b*) that the capacitance is increased by the factor $(\kappa_1 + \kappa_2)/2$.

Figure 21-21 Problem 42.

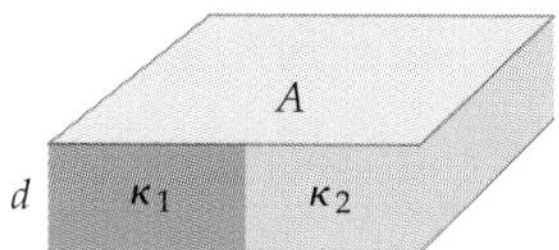

43. A Leyden jar, the earliest type of capacitor, is a glass jar coated inside and outside with metal foil. Suppose that a Leyden jar is a cylinder 40 cm high with 2.0-mm-thick walls and an inner diameter of 8 cm. Ignore any field fringing. (*a*) Find the capacitance of this Leyden jar if the dielectric constant of the glass is 5.0. (*b*) If the dielectric strength of the glass is 15 MV/m, what maximum charge can the Leyden jar carry without undergoing dielectric breakdown?

44. A parallel-plate capacitor of plate area A and separation x is given a charge Q and is then removed from the charging source. (*a*) Find the stored electrostatic energy as a function of x. (*b*) Find the increase in energy dU due to an increase in plate separation dx from $dU = (dU/dx)\,dx$. (*c*) If F is the force exerted by one plate on the other, the work needed to move one plate a distance dx is $F\,dx = dU$. Show that $F = Q^2/2\epsilon_0 A$. (*d*) Show that the force in part (*c*) equals $\frac{1}{2}EQ$, where Q is the charge on one plate and E is the electric field between the plates. Discuss the reason for the factor $\frac{1}{2}$ in this result.

45. Design a network of capacitors that has a capacitance of 2 μF and breakdown voltage of 400 V using as many 2-μF capacitors that have individual breakdown voltages of 100 V as needed.

46. A 1.2-μF capacitor is charged to 30 V. After charging, the capacitor is disconnected from the voltage source and is connected to another uncharged capacitor. The final voltage is 10 V. (*a*) What is the capacitance of the other capacitor? (*b*) How much energy was lost when the connection was made?

47. A rectangular parallel-plate capacitor of length a and width b has a dielectric of width b partially inserted a distance x between the plates as shown in Figure 21-22. (*a*) Find the capacitance as a function of x. Neglect edge effects. (*b*) Show that your answer gives the expected results for $x = 0$ and $x = a$.

Figure 21-22 Problem 47.

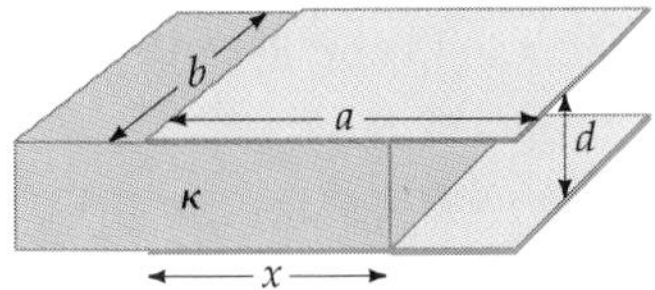

48. Determine the capacitance of each of the networks shown in Figure 21-23.

Figure 21-23 Problem 48.

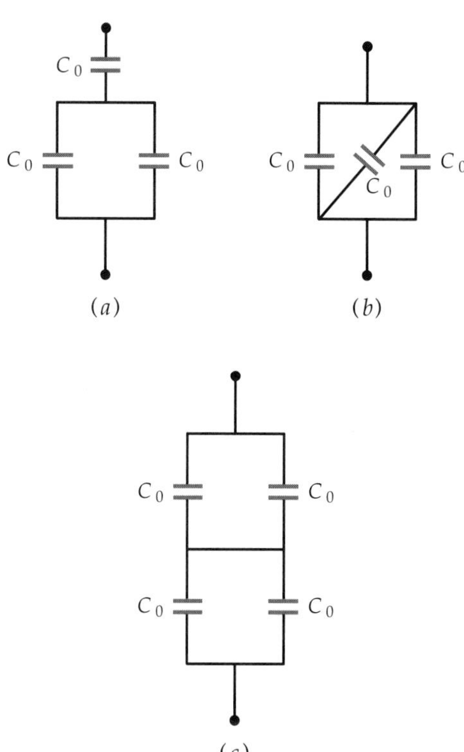

49. Five identical capacitors of capacitance C_0 are connected in a bridge network as shown in Figure 21-24. (*a*) What is the equivalent capacitance between points a and b? (*b*) Find the equivalent capacitance if the capacitance between a and b is changed to $10C_0$.

Figure 21-24 Problem 49.

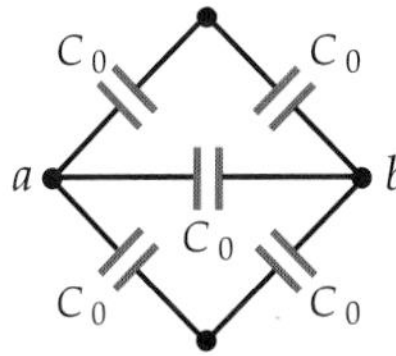

50. A parallel-plate capacitor with plates of area 500 cm^2 is charged to a potential difference V and is then disconnected from the voltage source. When the plates are moved 0.4 cm farther apart, the voltage between the plates increased by 100 V. (*a*) What is the charge Q on the positive plate of the capacitor? (*b*) How much does the energy stored in the capacitor increase due to the movement of the plates?

51. Design a 0.1-μF parallel-plate capacitor with air between the plates that can be charged to a maximum potential difference of 1000 V. (*a*) What is the minimum possible separation between the plates? (*b*) What minimum area must the plates of the capacitor have?

52. Three capacitors, $C_1 = 2\ \mu F$, $C_2 = 4\ \mu F$, and $C_3 = 6\ \mu F$, are connected in parallel and are charged with a 200-V source. They are then disconnected from the source and reconnected positive plates to negative plates as shown in Figure 21-25. (*a*) What is the voltage across each capacitor with switches S_1 and S_2 closed but with switch S_3 open? (*b*) After switch S_3 is closed, what is the final charge on each capacitor? (*c*) Give the voltage across each capacitor after switch S_3 is closed.

Figure 21-25 Problem 52.

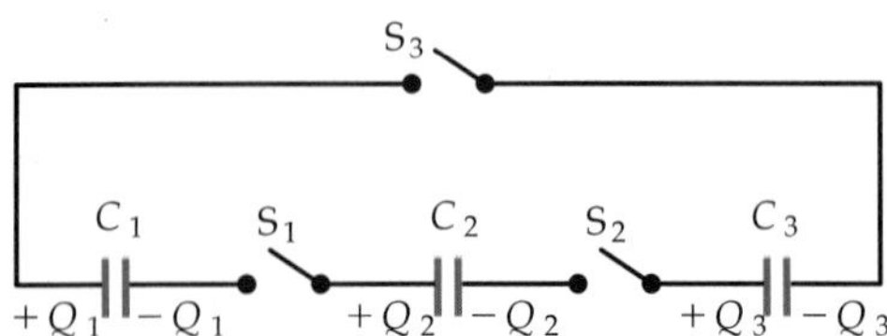

53. A parallel-plate capacitor is constructed from a layer of silicon dioxide of thickness 5×10^{-6} m between two conducting films. The dielectric constant of silicon dioxide is 3.8 and its dielectric strength is 8×10^6 V/m. (*a*) What voltage can be applied across this capacitor without dielectric breakdown? (*b*) What should the surface area of the layer of silicon dioxide be for a 100-pF capacitor? (*c*) Estimate the number of these capacitors that can fit into a square 1 cm by 1 cm.

54. Estimate the electrical energy stored in the atmosphere if the earth's electric field extends upward for 1000 m and has an average magnitude of 200 V/m. *Hint:* You may treat the atmosphere as a rectangular layer with an area equal to the surface area of the earth. Why?

55. You are asked to construct a parallel-plate, air-gap capacitor that will store 100 kJ of energy. (*a*) What minimum volume is required between the plates of the capacitor? (*b*) Suppose you have developed a dielectric that can withstand 3×10^8 V/m and has a dielectric constant of 5. What volume of this dielectric between the plates of the capacitor is required for it to be able to store 100 kJ of energy?

56. Two identical, 4-μF parallel-plate capacitors are connected in series across a 24-V battery. (*a*) What is the charge on each capacitor? (*b*) What is the total stored energy of the capacitors? A dielectric having a dielectric constant of 4.2 is inserted between the plates of one of the capacitors while the battery is still connected. After the dielectric is inserted, (*c*) what is the charge on each capacitor? (*d*) What is the potential difference across each capacitor? (*e*) What is the total stored energy of the capacitors?

57. Consider two parallel-plate capacitors, C_1 and C_2, that are connected in parallel. The capacitors are identical except that C_2 has a dielectric inserted between its plates. A potential difference V is connected across the capacitors to charge them and is then disconnected. (*a*) What is the charge on each capacitor? (*b*) What is the total stored energy of the capacitors? (*c*) The dielectric is removed from C_2. What is the final stored energy of the capacitors? (*d*) What is the final voltage across the two capacitors?

58. Find the capacitance of the parallel-plate capacitor shown in Figure 21-26.

Figure 21-26 Problem 58.

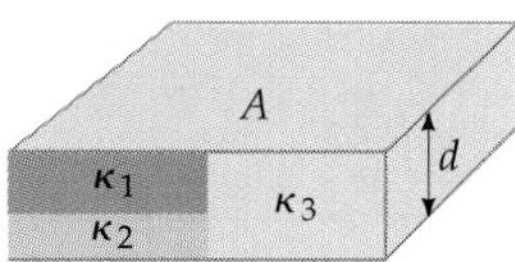

Level III

59. Two parallel-plate capacitors have the same separation and plate area. The capacitance of each is initially 10 μF. When a dielectric is inserted such that it completely fills the space between the plates of one of the capacitors, the capacitance of that capacitor increases to 35 μF. The 35-μF and 10-μF capacitors are connected in parallel and are charged to a potential difference of 100 V. The voltage source is then disconnected. (*a*) What is the stored energy of this system? (*b*) What are the charges on the two capacitors? (*c*) The dielectric is removed from the capacitor. What are the new charges on the plates of the capacitors? (*d*) What is the final stored energy of the system?

60. A parallel-plate capacitor of area A and separation d is charged to a potential difference V and is then removed from the charging source. A dielectric slab of constant $\kappa = 2$, thickness d, and area $\frac{1}{2}A$ is inserted as shown in Figure 21-27. Let σ_1 be the free charge density at the conductor–dielectric surface and σ_2 be the free charge density at the conductor–air surface. (*a*) Why must the electric field have the same value inside the dielectric as in the free space between the plates? (*b*) Show that $\sigma_1 = 2\sigma_2$. (*c*) Show that the new capacitance is $3\epsilon_0 A/2d$ and that the new potential difference is $\frac{2}{3}V$.

Figure 21-27 Problem 60.

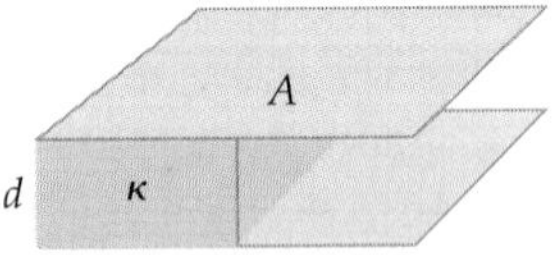

61. Two identical, 10-μF parallel-plate capacitors are given equal charges of 100 μC each and are then removed from the charging source. The charged capacitors are connected by a wire between their positive plates and another wire between their negative plates. (*a*) What is the stored energy of the system? A dielectric having a dielectric constant of 3.2 is inserted between the plates of one of the capacitors such that it completely fills the region between the plates. (*b*) What is the final charge on each capacitor? (*c*) What is the final stored energy of the system?

62. A spherical capacitor consists of two thin concentric spherical shells of radii R_1 and R_2. (*a*) Show that the capacitance is given by

$$C = 4\pi\epsilon_0 R_1 R_2/(R_2 - R_1)$$

(*b*) Show that when the radii of the shells are nearly equal, the capacitance is given approximately by the expression for the capacitance of a parallel-plate capacitor, $C = \epsilon_0 A/d$, where A is the area of the sphere and $d = R_2 - R_1$.

63. A parallel-plate capacitor of plate area 1.0 m^2 and plate separation distance 0.5 cm has a glass plate of the same area that completely fills the space between its plates. The glass has a dielectric constant of 5.0. The capacitor is charged to a potential difference of 12.0 V and is then removed from its charging source. How much work is required to pull the glass plate out of the capacitor?

64. A spherical capacitor has an inner sphere of radius R_1 with a charge of $+Q$ and an outer concentric spherical shell of radius R_2 with a charge of $-Q$. (*a*) Find the electric field and the energy density at any point in space. (*b*) How much energy is there in a spherical shell of radius r, thickness dr, and volume $4\pi r^2\,dr$ between the conductors? (*c*) Integrate your expression from part (*b*) to find the total energy stored in the capacitor, and compare your result with that obtained using $U = \frac{1}{2}QV$.

65. A cylindrical capacitor consists of a long wire of radius R_1 and length L with a charge of $+Q$ and a concentric outer cylindrical shell of radius R_2, length L, and charge $-Q$. (*a*) Find the electric field and the energy density at any point in space. (*b*) How much energy is there in a cylindrical shell of radius r, thickness dr, and volume $2\pi rL\,dr$ between the conductors? (*c*) Integrate your expression from part (*b*) to find the total energy stored in the capacitor, and compare your result with that obtained using $U = \frac{1}{2}CV^2$.

66. A ball of charge of radius R has a uniform charge density ρ and a total charge $Q = \frac{4}{3}\pi R^3 \rho$. (*a*) Find the electrostatic energy density at distance r from the center of the ball for $r < R$ and for $r > R$. (*b*) Find the energy in a spherical shell of volume $4\pi r^2\,dr$ for both $r < R$ and $r > R$. (*c*) Compute the total electrostatic energy by integrating your expressions from part (*b*), and show that your result can be written $U = \frac{3}{5}kQ^2/R$. Explain why this result is greater than that for a spherical conductor of radius R carrying a total charge Q.

67. A capacitor is constructed of two concentric cylinders of radii a and b ($b > a$) having a length $L \gg b$. A charge of $+Q$ is on the inner cylinder and a charge of $-Q$ is on the outer cylinder. The region between the two cylinders is filled with a dielectric having a dielectric constant κ. (*a*) Find the potential difference between the cylinders. Find the density of the free charge σ_f on (*b*) the inner cylinder and (*c*) the outer cylinder. Find the bound charge density σ_b on (*d*) the inner cylindrical surface of the dielectric and (*e*) the outer surface of the dielectric. (*f*) Find the total stored electrostatic energy. (*g*) If the dielectric will move without friction, how much mechanical work is required to remove the dielectric cylindrical shell?

68. A conducting sphere of radius R_1 is given a free charge Q. The sphere is surrounded by an uncharged concentric spherical dielectric shell having an inner radius R_1, an outer radius R_2, and a dielectric constant κ. The system is far removed from other objects. (*a*) Find the electric field everywhere in space. (*b*) What is the potential of the conducting sphere relative to $V = 0$ at infinity? (*c*) Find the total electrostatic energy of the system.

69. A parallel-plate capacitor is constructed using a variable dielectric. Let A be the area of the plates and y_0 be their separation. The dielectric constant is given as a function of y according to

$$\kappa = 1 + \frac{3}{y_0}y$$

The bottom plate is at $y = 0$ and the top plate is at $y = y_0$. (*a*) What is the capacitance? (*b*) Find the induced charge density on the surfaces of the dielectric. (*c*) Use Gauss's law to find the induced volume charge density $\rho(y)$ within this dielectric. (*d*) Integrate the expression for the volume charge density found in (*c*) over the dielectric, and show that the total induced bound charge, including that on the surfaces, is zero.

70. A capacitor has rectangular plates of length a and width b. The top plate is inclined at a small angle as shown in Figure 21-28. The plate separation varies from $s = y_0$ at the left to $s = 2y_0$ at the right, where y_0 is much less than a or b. Calculate the capacitance using strips of width dx and length b to approximate differential capacitors of area $b\,dx$ and separation $s = y_0 + (y_0/a)x$ that are connected in parallel.

Figure 21-28 Problem 70.

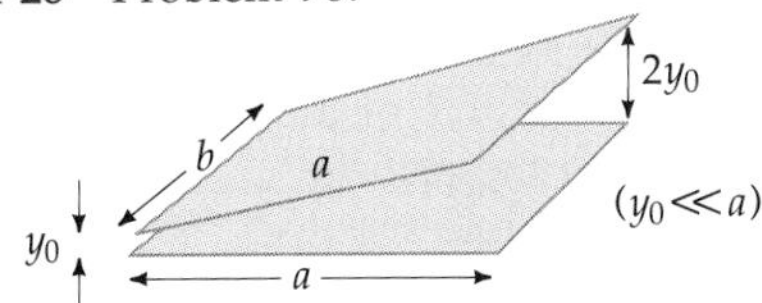

Chapter 22

Electric Current

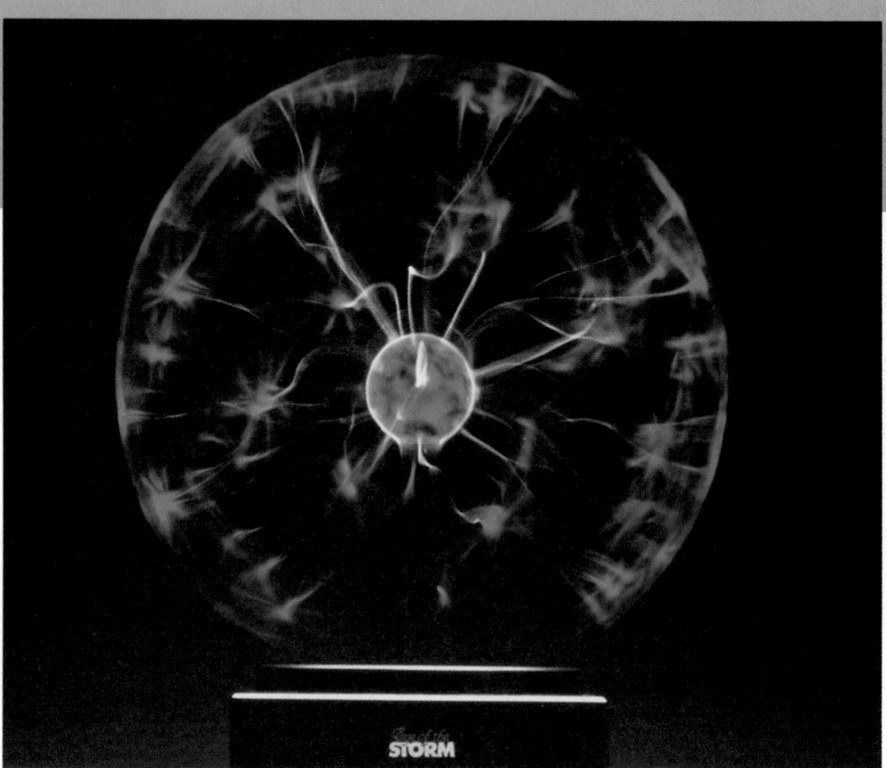

The plasma globe Eye of the Storm. In plasma globes, voltages ranging from 3 kV to 8 kV, at frequencies of between 20 kHz and 50 kHz, are applied between the sphere at the center of the globe and the outer shell, which is at ground. The region in between is filled with a mixture of inert gases. The applied voltage partially ionizes the gases, creating plasma filaments that conduct current. The colored displays are paths along which ionization is occurring.

Inert gases are more easily ionized than many noninert gases—it is easier to achieve dielectric breakdowns in plasma globes than, for instance, in the earth's nitrogen–oxygen atmosphere. The ease of ionization at the voltages applied in Eye of the Storm is apparent in the globe's ability to support multiple simultaneous paths of ionization. Because each of the discharge paths has the same instantaneous polarity, the paths repel one another.

When you touch the outer shell, regions of electrical discharge are drawn to the vicinity of your hand because your body is a much better conductor than the glass outer shell, so it becomes the path of least resistance to the earth for the charge flowing onto the shell. The current passing through your body is low enough to be unnoticeable.

When we turn on a light, we connect the wire filament in the light bulb across a potential difference that causes electric charge to flow through the wire, much like a pressure difference in a garden hose causes water to flow through the hose. The flow of electric charge constitutes an electric current. We usually think of currents as being in conducting wires, but an electric current results from any flow of charge. An example of a current that is not in a conducting wire is the electron beam in a cathode ray tube, such as that in your video monitor, or a beam of charged ions from a particle accelerator. In this chapter, we will define electric current and relate it to the motion of charged particles. After a discussion of electrical resistance and Ohm's law, we will consider the energy aspects of electric currents. Resistors in parallel and in series are then discussed, and the chapter concludes with a brief discussion of the classical microscopic model of electrical conduction.

22-1 Current and Motion of Charges

Electric **current** is defined as the rate of flow of electric charge through a cross-sectional area. Figure 22-1 shows a segment of a current-carrying wire in which charge carriers are moving with some small average velocity. If ΔQ

is the charge that flows through the cross-sectional area A in time Δt, the current is

$$I = \frac{\Delta Q}{\Delta t} \tag{22-1}$$

Electric current

The SI unit of current is the **ampere** (A):

$$1\ \text{A} = 1\ \text{C/s} \tag{22-2}$$

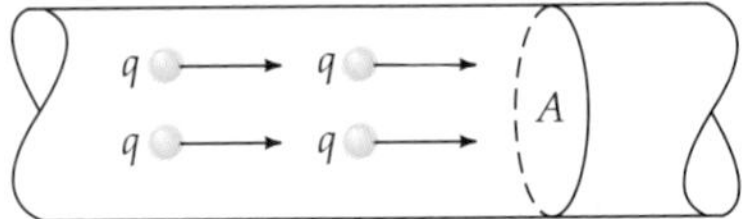

Figure 22-1 A segment of a current-carrying wire. If ΔQ is the amount of charge that flows through the cross-sectional area A in time Δt, the current is $I = \Delta Q/\Delta t$.

By convention, the direction of current is considered to be the direction of flow of positive charge. This convention was established before it was known that free electrons, which are negatively charged, are the particles that are actually moving and thus producing the current in a conducting wire. The motion of negatively charged electrons in one direction is equivalent to a flow of positive charge in the opposite direction. Thus, the electrons move in the direction opposite to the direction of the current. However, not all currents are produced by electrons flowing through a wire. In an accelerator that produces a proton beam, the direction of motion of the positively charged protons is in the direction of the current. In electrolysis, the current is produced by the flow of positive ions in the direction of the current plus the flow of negative ions and electrons opposite the direction of the current. The movement of negative particles in one direction and that of positive particles in the opposite direction both contribute to a current in the same direction. In nearly all applications, the motion of negative charges in one direction is indistinguishable from the motion of positive charges in the opposite direction. We can always think of current as motion of positive charges in the direction of the current and remember (if we need to) that in conducting wires, for example, the electrons are moving in the direction opposite to the current.

The actual motion of the free electrons in a conducting wire is quite complicated. When there is no electric field in the wire, these electrons move in random directions with relatively large speeds due to their thermal energy. Since the velocity vectors of the electrons are randomly oriented, the average velocity due to this thermal energy is zero. When an electric field is applied, for example, by connecting the wire to a battery that applies a potential difference along the wire, the free electrons experience a momentary acceleration due to the force $-e\mathbf{E}$. The electrons acquire a small velocity in the direction opposite the field, but the kinetic energy acquired is quickly dissipated by collisions with the fixed ions in the wire. The electrons are then again accelerated by the field. The net result of this repeated acceleration and dissipation of energy is that the electrons have a small **drift velocity** opposite to the electric field superimposed on their large, random, thermal velocity. The motion of the free electrons in a metal is somewhat similar to that of the molecules of a gas, such as air. In still air, the molecules move with large instantaneous velocities between collisions, but the average velocity is zero. When there is a breeze, the air molecules have a small drift velocity in the direction of the breeze superimposed on their much larger instantaneous velocities. Similarly, when there is no current in a conductor the electrons move about in random directions with very high speeds because of thermal energy. When there is a current, the electrons have a small drift velocity superimposed on their much larger, but random, thermal velocities.

Let us consider a current in a conducting wire of cross-sectional area A. Let n be the number of free charge-carrying particles per unit volume. We will assume that each particle carries a charge q and moves with a drift

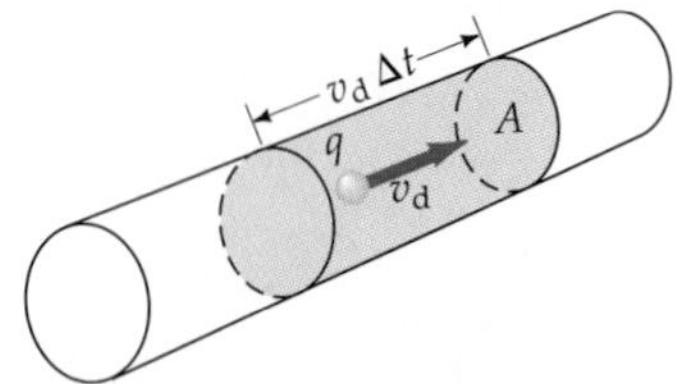

Figure 22-2 In time Δt, all the charges in the shaded volume pass through A. If there are n charge carriers per unit volume, each with charge q, the total charge in this volume is $\Delta Q = nqv_dA\,\Delta t$, where v_d is the drift velocity of the charge carriers. The total current is then $I = \Delta Q/\Delta t = nqv_dA$.

velocity v_d. In a time Δt, all the particles in the volume $Av_d\,\Delta t$, shaded in Figure 22-2, pass through the area element. The number of particles in this volume is $nAv_d\,\Delta t$, and the total charge is

$$\Delta Q = qnAv_d\,\Delta t$$

The current is thus

$$I = \frac{\Delta Q}{\Delta t} = nqAv_d \qquad 22\text{-}3$$

Equation 22-3 can be used to find the current due to the flow of any type of charged particle, simply by replacing the drift velocity v_d with the velocity of the particle.

We can get an idea of the order of magnitude of the drift velocity for electrons in a conducting wire by putting typical magnitudes into Equation 22-3.

Example 22-1

What is the drift velocity of electrons in a typical (14-gauge) copper wire of radius 0.815 mm carrying a current of 1 A?

If we assume one free electron per copper atom, the density of free electrons is the same as the density of atoms n_a, which is related to the mass density ρ, Avogadro's number N_A, and the molar mass M by

$$n_a = \frac{\rho N_A}{M}$$

For copper $\rho = 8.93\ \text{g/cm}^3$ and $M = 63.5\ \text{g/mol}$. Then

$$n_a = \frac{(8.93\ \text{g/cm}^3)(6.02 \times 10^{23}\ \text{atoms/mol})}{63.5\ \text{g/mol}}$$

$$= 8.47 \times 10^{22}\ \text{atoms/cm}^3$$

The density of electrons is then

$$n = 8.47 \times 10^{22}\ \text{electrons/cm}^3 = 8.47 \times 10^{28}\ \text{electrons/m}^3$$

The drift velocity is therefore

$$v_d = \frac{I}{Ane} = \frac{1\ \text{C/s}}{\pi(0.000815\ \text{m})^2(8.47 \times 10^{28}\ \text{m}^{-3})(1.6 \times 10^{-19}\ \text{C})}$$

$$\approx 3.54 \times 10^{-5}\ \text{m/s}$$

We see that typical drift velocities are of the order of 0.01 mm/s, which is quite small.

Exercise

How long does it take for an electron to drift a distance of 1 m if its drift velocity is 3.54×10^{-5} m/s? (Answer: 7.85 h)

At first, it may seem surprising that an electric light comes on immediately when the switch is thrown since the electrons drift down the wire at such low speeds that it should take them hours to travel from the switch to the light. An analogy with water in a hose may prove useful. When you turn on water into a long, empty hose, it takes quite a few seconds for the water to travel from the faucet to the nozzle. However, if the hose is already full of water, the water emerges from the nozzle almost instantaneously. Because of the water pressure at the faucet, the segment of water there pushes on the water immediately next to it, which pushes on the next segment of water and so on, until the last segment of water is pushed out the nozzle. This pressure wave moves down the hose with the speed of sound in the water (if the hose is rigid), and the water quickly reaches a steady flow rate. The density of the water during a steady flow remains constant in time. The water that flows out of a segment of hose is replaced by an equal amount of water flowing into the segment at the other end. The behavior of wires full of free electrons is similar. When the light switch is turned on, an electric field is propagated down the wire with nearly the speed of light, and the free electrons throughout the wire acquire their drift velocity almost immediately. The charge density within a current-carrying wire remains constant in time. The charge that flows out of a segment of wire is replaced by an equal amount of charge that flows into the segment at the other end. Thus charge starts moving through the filament almost immediately after the light switch is thrown. The transport of a significant amount of charge in a wire is accomplished not by a few charges moving rapidly down the wire, but by a very large number of charges slowly drifting down the wire.

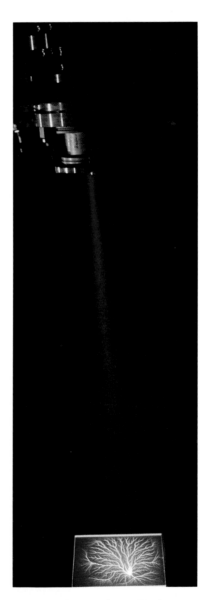

A 1000-A beam of 3-MeV electrons hits a lucite plate. The plate acquires a charge and discharges, producing the beautiful pattern shown. The electron beam, which lasts for about 1 μs, ionizes the air molecules, which give off a bluish glow when they recombine.

Example 22-2

In a certain particle accelerator, a current of 0.5 mA is carried by a 5-MeV proton beam that has a radius of 1.5 mm. (*a*) Find the number of protons per unit volume in the beam. (*b*) If the beam hits a target, how many protons hit the target in one second?

(*a*) From Equation 22-3, we have

$$n = \frac{I}{qAv}$$

where q is the charge on each proton, v is the speed of the protons, and A is the cross-sectional area of the beam. The kinetic energy of each proton is 5 MeV:

$$K_k = \frac{1}{2}mv^2 = 5\text{ MeV} = 5 \times 10^6\text{ eV} \times \frac{1.6 \times 10^{-19}\text{ J}}{1\text{ eV}} = 8 \times 10^{-13}\text{ J}$$

Using $m = 1.67 \times 10^{-27}$ kg for the mass of a proton, we have for the speed of each proton

$$v = \sqrt{\frac{2K_k}{m}} = \sqrt{\frac{(2)(8 \times 10^{-13}\text{ J})}{1.67 \times 10^{-27}\text{ kg}}} = 3.10 \times 10^7\text{ m/s}$$

The number of protons per unit volume of the beam is then

$$\begin{aligned} n &= \frac{I}{qAv} \\ &= \frac{0.5 \times 10^{-3}\text{ A}}{(1.6 \times 10^{-19}\text{ C/proton})\pi(1.5 \times 10^{-3}\text{ m})^2(3.10 \times 10^7\text{ m/s})} \\ &= 1.43 \times 10^{13}\text{ protons/m}^3 \end{aligned}$$

(*b*) The number that strikes a target in time Δt is the number in the volume $Av\,\Delta t$, which is $nAv\,\Delta t$. For $\Delta t = 1$ s, this number is

$$\begin{aligned} N &= nAv\,\Delta t \\ &= (1.43 \times 10^{13}\text{ protons/m}^3)\pi(1.5 \times 10^{-3}\text{ m})^2(3.10 \times 10^7\text{ m/s})(1\text{ s}) \\ &= 3.13 \times 10^{15}\text{ protons} \end{aligned}$$

We can check this result by noting that, from Equation 22-3,

$$nAv\ \Delta t = \frac{I\ \Delta t}{q} = \frac{Q}{q}$$

where $Q = I\ \Delta t$ is the total charge that strikes the target. Since the current is 0.5 mA, the total charge that strikes the target in 1 s is 0.5 mC. The number of protons is then

$$N = \frac{Q}{q} = \frac{0.5 \times 10^{-3}\ \text{C}}{1.6 \times 10^{-19}\ \text{C/proton}} = 3.13 \times 10^{15}\ \text{protons}$$

22-2 Resistance and Ohm's Law

In our study of conductors in electrostatics, we argued that the electric field inside a conductor in electrostatic equilibrium must be zero. If this were not so, the free charges inside the conductor would move about. We are now considering situations in which the free charge *does* move in a conductor. That is, the conductors are not in electrostatic equilibrium. The current in a conductor is produced by an electric field inside the conductor that exerts a force on the free charges. Since the field **E** is in the direction of the force on a positive charge, and since the direction of the current is the direction of the flow of positive charge, the current is in the direction of the electric field. Figure 22-3 shows a segment of a wire of length ΔL and cross-sectional area A carrying a current I. Since the electric field points from regions of higher potential towards regions of lower potential, the potential at point a is greater than that at point b. Assuming that ΔL is small enough so that we may consider the electric field E to be constant across the segment, the potential difference V between points a and b is*

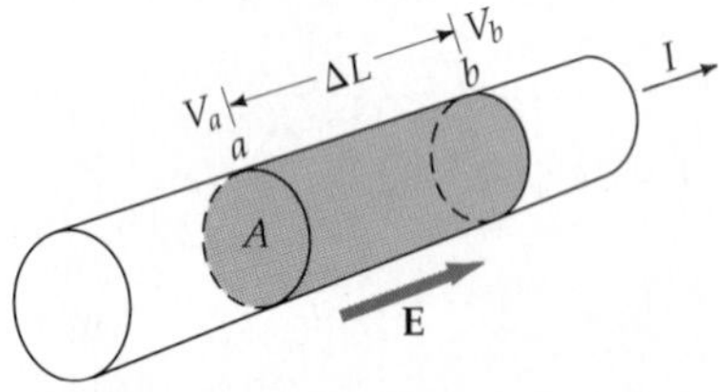

Figure 22-3 A segment of wire carrying a current I. The potential difference is related to the electric field by $V_a - V_b = E\ \Delta L$.

$$V = V_a - V_b = E\ \Delta L \qquad 22\text{-}4$$

For most materials,

> The current in a wire segment is proportional to the potential difference across the segment.

This experimental result is known as **Ohm's law.** The constant of proportionality is written $1/R$, where R is called the **resistance:**

$$I = \left(\frac{1}{R}\right) V$$

or

Resistance defined

$$R = \frac{V}{I} \qquad 22\text{-}5$$

Equation 22-5 gives a general definition of the resistance between two points in terms of the potential drop V between the points. The SI unit of resistance, the volt per ampere, is called an ohm (Ω):

$$1\ \Omega = 1\ \text{V/A} \qquad 22\text{-}6$$

The resistance of a material depends on its length, its cross-sectional area, the type of material, and its temperature. For materials obeying Ohm's law,

*Again we use V rather than ΔV for the potential difference (which in this case is a potential *decrease*) to simplify the notation.

the resistance does not depend on the current I; that is, the ratio V/I is independent of I. Such materials, which include most metals, are called **ohmic materials.** For ohmic materials, the potential drop across a segment is proportional to the current:

Ohm's law

$$V = IR \qquad R \text{ constant} \tag{22-7}$$

Equation 22-7, with the qualification that R is constant, gives a mathematical statement of Ohm's law.

For **nonohmic materials,** the ratio V/I depends on the current I, so the current is not proportional to the potential difference. For nonohmic materials, the resistance R, as defined by Equation 22-5, depends on the current I. Figure 22-4 shows the potential difference V versus the current I for ohmic and nonohmic materials. For ohmic materials (the bottom curve), the relation is linear, so $R = V/I$ does not depend on I; but for nonohmic materials (the top curve), the relation is not linear, and $R = V/I$ *does* depend on I. Ohm's law is not a fundamental law of nature like Newton's laws or the laws of thermodynamics but rather is an empirical description of a property shared by many materials.

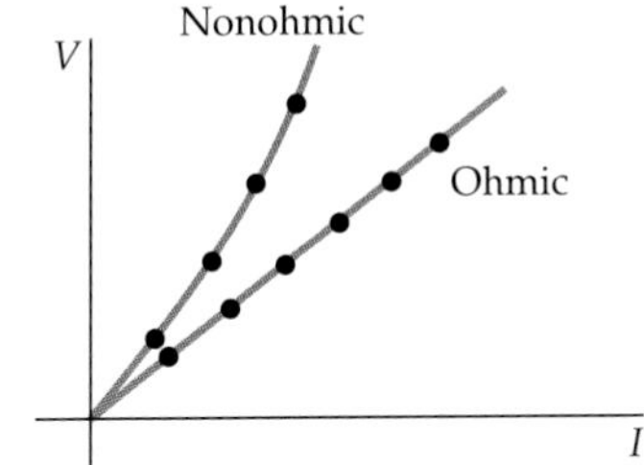

Figure 22-4 Plots of V versus I for ohmic and nonohmic materials. The resistance $R = V/I$ is independent of I for ohmic materials, as is indicated by the constant slope of the line.

Exercise

A wire of resistance 3 Ω carries a current of 1.5 A. What is the potential drop across the wire? (Answer: 4.5 V)

The resistance of a conducting wire is found to be proportional to the length of the wire and inversely proportional to its cross-sectional area:

$$R = \rho \frac{L}{A} \tag{22-8}$$

where the proportionality constant ρ is called the **resistivity** of the conducting material. The unit of resistivity is the ohm-meter (Ω·m).

Example 22-3

A nichrome wire (resistivity 10^{-6} Ω·m) has a radius of 0.65 mm. What length of wire is needed to obtain a resistance of 2.0 Ω?

The cross-sectional area of this wire is

$$A = \pi r^2 = (3.14)(6.5 \times 10^{-4}\ \text{m})^2 = 1.33 \times 10^{-6}\ \text{m}^2$$

From Equation 22-8, we have

$$L = \frac{RA}{\rho} = \frac{(2\ \Omega)(1.33 \times 10^{-6}\ \text{m}^2)}{10^{-6}\ \Omega\cdot\text{m}} = 2.66\ \text{m}$$

We sometimes refer to a wire as a conductor, and at other times we call it a resistor, depending on which property we wish to emphasize. The reciprocal of the resistivity is called the **conductivity** σ*:

$$\sigma = \frac{1}{\rho} \tag{22-9}$$

*The symbols ρ and σ, used here for the resistivity and conductivity, were used in previous chapters for volume charge density and surface charge density. Thus, care must be taken to distinguish what quantities are being referred to when these symbols are used. Usually this will be clear from the context.

Equation 22-8 may be written in terms of the conductivity instead of the resistivity:

$$R = \frac{L}{\sigma A} \qquad 22\text{-}10$$

Note that Equations 22-7 and 22-10 for electrical conduction and electrical resistance are of the same form as Equations 16-13 ($\Delta T = IR$) and 16-14 ($R = \Delta x/kA$) for thermal conduction and thermal resistance. For electrical equations, the potential difference V replaces the temperature difference ΔT and the electrical conductivity σ replaces the thermal conductivity k. Ohm was, in fact, led to his law by the similarity between the conduction of electricity and the conduction of heat.

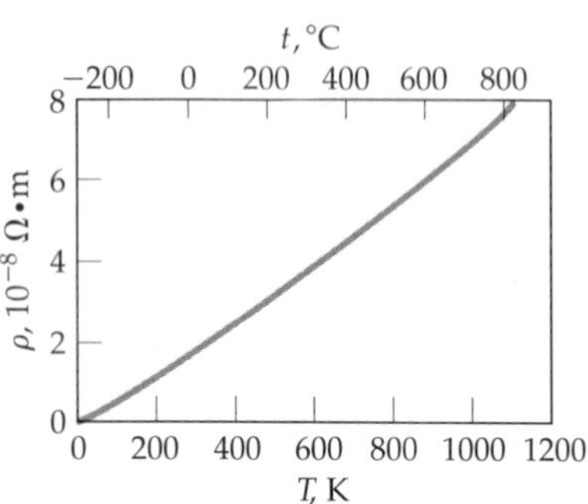

Figure 22-5 Plot of resistivity ρ versus temperature for copper.

The resistivity (and conductivity) of any given metal depends on the temperature. Figure 22-5 shows the temperature dependence of the resistivity of copper. This graph is nearly a straight line, which means that the resistivity varies nearly linearly with temperature. (There is a breakdown in this linearity at very low temperatures, which is not shown on the graph.) In tables, the resistivity is usually given in terms of its value at 20°C (ρ_{20}) along with the **temperature coefficient of resistivity** α, which is proportional to the slope of the ρ-versus-T curve. The resistivity at some other Celsius temperature t_C is then given by

$$\rho = \rho_{20}[1 + \alpha(t_C - 20°C)] \qquad 22\text{-}11$$

(Since the Celsius and absolute temperatures differ only in the choice of zero, the resistivity has the same slope whether it is plotted against t_C or T.) Table 22-1 gives the resistivity at 20°C and temperature coefficient α for various materials. Note the tremendous difference between the values of ρ for conductors (the metals listed) and those for nonconductors.

Wires used to carry electric current are manufactured in standard sizes. The diameter of the circular cross section is indicated by a "gauge number," with higher numbers corresponding to smaller diameters. Table 22-2 gives the diameters and cross-sectional areas for some common wire gauges. Handbooks typically give ρ/A-R/L in ohms per centimeter or ohms per foot.

Table 22-1 **Resistivities and Temperature Coefficients**

Material	Resistivity ρ at 20°C, Ω·m	Temperature coefficient α at 20°C, K^{-1}
Silver	1.6×10^{-8}	3.8×10^{-3}
Copper	1.7×10^{-8}	3.9×10^{-3}
Aluminum	2.8×10^{-8}	3.9×10^{-3}
Tungsten	5.5×10^{-8}	4.5×10^{-3}
Iron	10×10^{-8}	5.0×10^{-3}
Lead	22×10^{-8}	4.3×10^{-3}
Mercury	96×10^{-8}	0.9×10^{-3}
Nichrome	100×10^{-8}	0.4×10^{-3}
Carbon	3500×10^{-8}	-0.5×10^{-3}
Germanium	0.45	-4.8×10^{-2}
Silicon	640	-7.5×10^{-2}
Wood	$10^{8} - 10^{14}$	
Glass	$10^{10} - 10^{14}$	
Hard rubber	$10^{13} - 10^{16}$	
Amber	5×10^{14}	
Sulfur	1×10^{15}	

Table 22-2 **Wire Diameters and Cross-sectional Areas for Commonly Used Copper Wires**

Gauge number	Diameter at 20°C, mm	Area, mm^2
4	5.189	21.15
6	4.115	13.30
8	3.264	8.366
10	2.588	5.261
12	2.053	3.309
14	1.628	2.081
16	1.291	1.309
18	1.024	0.8235
20	0.8118	0.5176
22	0.6438	0.3255

Example 22-4

Calculate ρ/A in ohms per meter for 14-gauge copper wire, which has a diameter $d = 1.63$ mm.

From Table 22-1, the resistivity of copper is

$$\rho = 1.7 \times 10^{-8}\ \Omega\cdot\text{m}$$

The cross-sectional area of 14-gauge wire is

$$A = \frac{\pi d^2}{4} = \frac{\pi(0.00163\ \text{m})^2}{4} = 2.1 \times 10^{-6}\ \text{m}^2$$

Thus

$$\frac{\rho}{A} = \frac{1.7 \times 10^{-8}\ \Omega\cdot\text{m}}{2.1 \times 10^{-6}\ \text{m}^2} = 8.1 \times 10^{-3}\ \Omega/\text{m}$$

This example shows that the copper connecting wires used in the laboratory have very small resistances.

Resistors for use in the laboratory are often made by winding a fine wire around an insulating tube to get a long wire in a short space. Carbon, which has a relatively high resistivity, is usually used for resistors in electronic equipment. Such a resistor is often painted with colored stripes to indicate the value of its resistance.

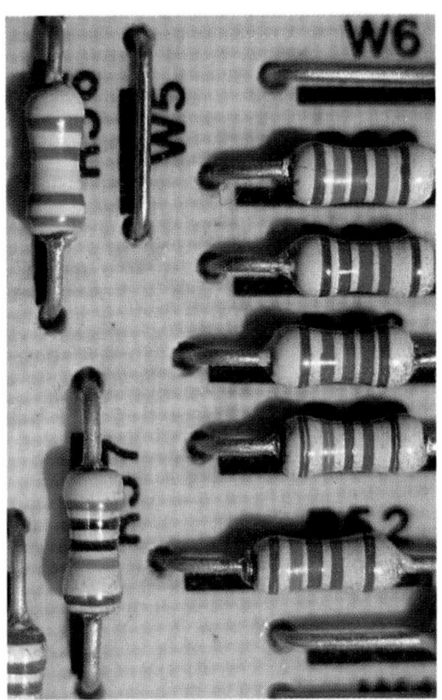

Color-coded carbon resistors on a circuit board.

Example 22-5

Assuming the electric field to be uniform, find its magnitude in a 14-gauge copper wire that is carrying a current of 1 A.

According to Example 22-4, the resistance of a 1-m length of 14-gauge copper wire is $8.1 \times 10^{-3}\ \Omega$. From Ohm's law, the voltage drop across 1 m of this wire is

$$V = IR = (1\ \text{A})(8.1 \times 10^{-3}\ \Omega) = 8.1 \times 10^{-3}\ \text{V}$$

so the electric field is

$$E = \frac{V}{\Delta L} = \frac{8.1 \times 10^{-3}\ \text{V}}{1\ \text{m}} = 8.1 \times 10^{-3}\ \text{V/m}$$

Note that the electric field in a conducting wire is very small.

Example 22-6

By what percentage does the resistance of a copper wire increase when its temperature increases from 20 to 30°C?

From Equation 22-11, the fractional change in the resistivity is

$$\frac{\rho - \rho_{20}}{\rho_{20}} = \alpha(t_C - 20°\text{C})$$

Using $\alpha = 3.9 \times 10^{-3}/\text{K}$ for copper from Table 22-1 and $(t_C - 20°\text{C}) = 10\ \text{C}° = 10\ \text{K}$, we obtain

$$\frac{\rho - \rho_{20}}{\rho_{20}} = (3.9 \times 10^{-3}/\text{K})(10\ \text{K}) = 3.9 \times 10^{-2}$$

The percentage change is thus 3.9%.

Exercise:

What temperature change will produce a 10% increase in the resistance of an iron bar? (Answer: $\Delta T = 20$ K)

Superconductivity

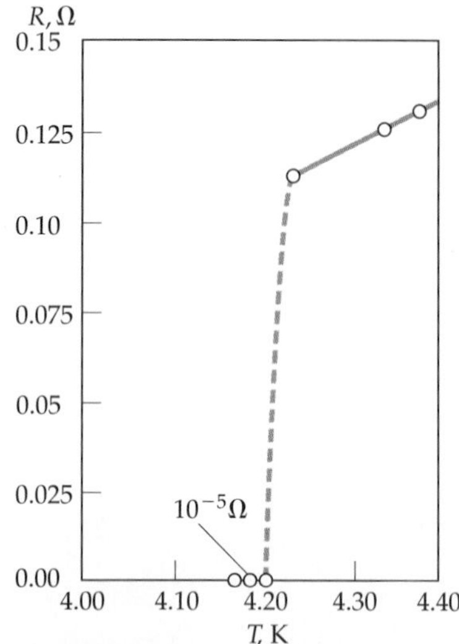

Figure 22-6 Plot by Kamerlingh Onnes of the resistance of mercury versus temperature, showing sudden decrease at the critical temperature $T = 4.2$ K.

There are some materials for which the resistivity is zero below a certain temperature, called the **critical temperature** T_c. This phenomenon, called **superconductivity,** was discovered in 1911 by the Dutch physicist H. Kamerlingh Onnes. Figure 22-6 shows his plot of the resistance of mercury versus temperature. The critical temperature for mercury is 4.2 K. Critical temperatures for other superconducting elements range from less than 0.1 K for hafnium and iridium to 9.2 K for niobium. Many metallic compounds are also superconductors. For example, the superconducting alloy Nb_3Ge, discovered in 1973, has a critical temperature of 23.2 K, which was the highest known until 1986. Despite the cost and inconvenience of refrigeration with expensive liquid helium, which boils at 4.2 K, many superconducting magnets were built using such materials, because such magnets produce no heat. In late 1986 and early 1987, it was discovered that certain ceramic oxides become superconducting at much higher temperatures. For example, the critical temperature for yttrium-barium-copper oxide ($YBa_2Cu_3O_7$) is about 92 K. These discoveries have revolutionized the science of superconductivity because relatively inexpensive liquid nitrogen, which boils at 77 K, can be used to cool them. However, there are many problems, such as the brittleness of ceramics, that make these new superconductors difficult to use. The search continues for new materials that will superconduct at even higher temperatures.

The conductivity of a superconductor cannot be defined since its resistance is zero. There can be a current in a superconductor even when the electric field in the superconductor is zero. Indeed, steady currents have been observed to persist for years without apparent loss in superconducting rings in which there was no electric field. The phenomenon of superconductivity cannot be understood in terms of classical physics. Instead, quantum mechanics, developed in the twentieth century, is needed. We will discuss some of the ideas of quantum mechanics in the latter chapters of this book. The first successful theory of superconductivity was published by John Bardeen, Leon Cooper, and J. Robert Schrieffer in 1957 and is known as the BCS theory. These physicists were awarded the Nobel Prize in Physics in 1972 for their accomplishment. The BCS theory describes the older superconductors well, but it is apparently not sufficient for understanding the newer, higher-temperature superconductors. (We will discuss superconductivity and the BCS theory in more detail in Chapter 39 of the extended version of this book.)

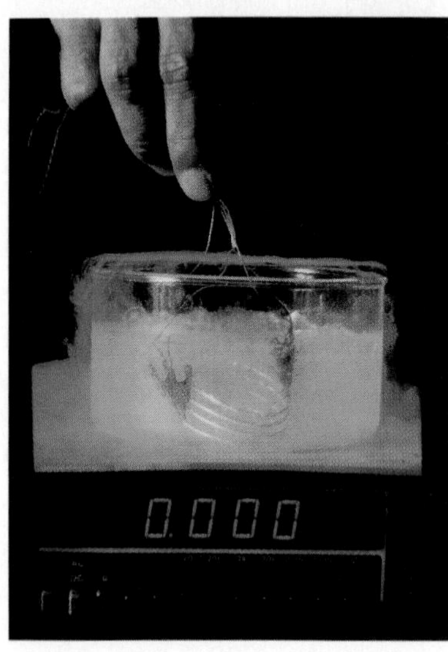

An ohmmeter measures zero resistance across this superconducting coil, which is made of a barium-yttrium-copper-oxide compound and cooled in liquid nitrogen.

Questions

1. Wire *a* and wire *b* have the same electric resistance and are made of the same material. Wire *a* has twice the diameter of wire *b*. How do the lengths of the wires compare?
2. In our study of electrostatics, we concluded that there is no electric field within a conductor in electrostatic equilibrium. How is it that we can now discuss electric fields inside a conductor?

22-3 Energy in Electric Circuits

When there is electric current in a conductor, electrical energy is continually being converted into thermal energy within the conductor. The electric field in the conductor accelerates each free electron for a short time, giving it an increased kinetic energy; but this additional energy is quickly transferred into thermal energy of the conductor by collisions between the electron and the lattice ions of the conductor. Thus, though the electrons continually gain energy from the electric field, this energy is immediately transferred to thermal energy of the conductor, and the electrons maintain a steady drift velocity.

When positive charge flows in a conductor, it flows from high potential to low potential in the direction of the electric field. (The negatively charged electrons move in the opposite direction.) The charge thus loses potential energy. The loss of potential energy appears as kinetic energy of the charge carriers only momentarily before it is transferred to the conducting material by collisions with the fixed ions. The loss of potential energy therefore goes into the increased thermal energy of the conductor.

Consider a segment of wire of length ΔL and cross-sectional area A shown in Figure 22-7. During a time interval Δt, an amount of charge ΔQ passes through area A_1 and enters the segment. If the potential at that point is V_1, the charge has potential energy equal to $\Delta Q\ V_1$. During that time interval, an equal amount of charge leaves the segment passing through area A_2, where the potential is V_2. It has potential energy $\Delta Q\ V_2$, which is less than $\Delta Q\ V_1$. The effect is the same as if the same charge ΔQ entered the segment at a high potential V_1 and left it at a low potential V_2, thereby losing potential energy in the segment given by

$$\Delta U = \Delta Q\ (V_2 - V_1) = \Delta Q(-V)$$

where $V = V_1 - V_2$ is the potential *decrease* across the segment. The energy lost in this segment of the wire is thus

$$-\Delta U = (\Delta Q)V$$

The rate of energy loss is

$$-\frac{\Delta U}{\Delta t} = \frac{\Delta Q}{\Delta t}V = IV$$

where $I = \Delta Q/\Delta t$ is the current. The energy loss per unit time is the power P dissipated in this conducting segment:

$$P = IV \qquad 22\text{-}12$$

If I is in amperes and V is in volts, the power is in watts. This expression for electric power can be remembered by recalling the definitions of V and I. The voltage drop is the decrease in potential energy per unit charge, and the current is the charge flowing per unit time. The product IV is thus the energy loss per unit time or the power put into the conductor. As we have seen, this power goes into thermal energy in the conductor. Using the definition of resistance, $R = V/I$, we can write Equation 22-12 in other useful forms by eliminating either V or I:

$$P = (IR)I = I^2R \qquad 22\text{-}13$$

or

$$P = \frac{V}{R}V = \frac{V^2}{R} \qquad 22\text{-}14$$

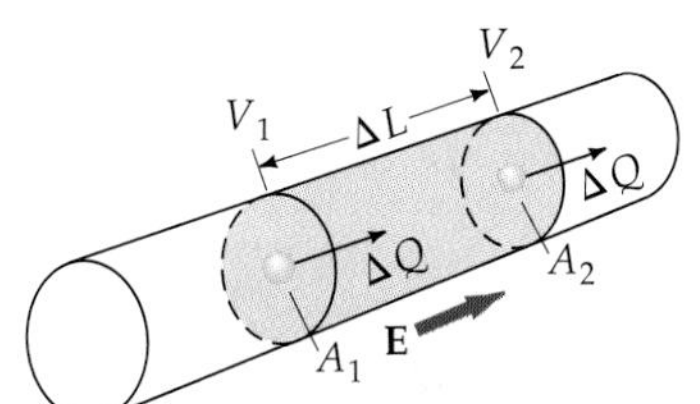

Figure 22-7 During a time Δt, an amount of charge ΔQ passes through area A_1, where the potential is V_1. During the same time interval an equal amount of charge leaves the segment passing through area A_2, where the potential is V_2. The effect is the same as if the same charge ΔQ entered the segment at a high potential V_1 and left it at a low potential V_2, thereby losing potential energy in the segment.

Equations 22-12, 22-13, and 22-14 all contain the same information. The choice of which to use depends on the particular problem. The energy put into a conductor is called **Joule heat.**

(*a*)

(*b*)

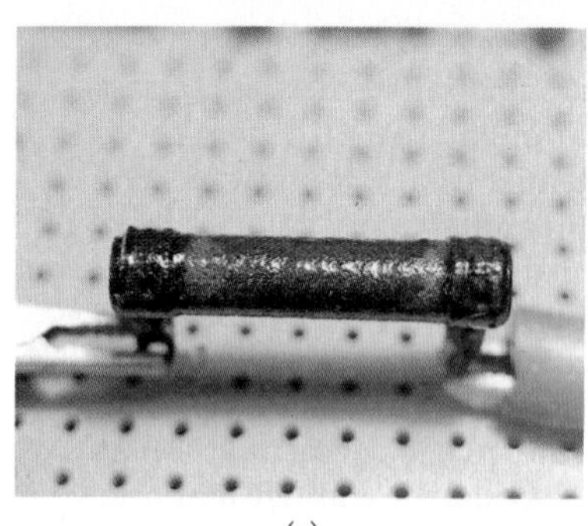

(*c*)

(*a*) Much of the electrical energy dissipated in the filament of this light bulb appears as light energy. (*b*) These rather elaborate cooling fins are used to dissipate the Joule heat developed by transistors and other semiconductor devices. (*c*) The thermal profile of this high-wattage resistor is indicated by the color of the liquid crystal coating.

Example 22-7

A 12-Ω resistor carries a current of 3 A. Find the power dissipated in this resistor.

Since we are given the current and the resistance but not the potential drop, Equation 22-13 is the most convenient to use. We have

$$P = I^2R = (3\text{ A})^2(12\ \Omega) = 108\text{ W}$$

Alternatively, we could have first found the potential drop across the resistor from $V = IR = (3\text{ A})(12\ \Omega) = 36\text{ V}$, and then used Equation 22-12 to find the power dissipated:

$$P = IV = (3\text{ A})(36\text{ V}) = 108\text{ W}$$

Exercise

A wire of resistance 5 Ω carries a current of 3 A for 6 s. (*a*) How much power is put into the wire? (*b*) How much heat is produced? [Answers: (*a*) 45 W, (*b*) 270 J]

EMF and Batteries

To have a steady current in a conductor, we have to have a constant supply of electrical energy. A device that supplies electrical energy is called a source of **electromotive force** or simply a source of **emf.** It converts chemical, mechanical, or other forms of energy into electrical energy. Examples are a battery, which converts chemical energy into electrical energy, and a generator, which converts mechanical energy into electrical energy. A source of emf does work on the charge passing through it, raising the potential energy of the charge. The work per unit charge is called the emf, $\mathcal{E}$, of the source. When a charge ΔQ flows through a source of emf, its potential energy is increased by the amount $\Delta Q\ \mathcal{E}$. The unit of emf is the volt, the same as the unit of potential difference. An **ideal battery** is a source of emf that maintains a constant potential difference between its two terminals, independent of the rate of flow of charge between them. The potential difference between the terminals of an ideal battery is equal in magnitude to the emf of the battery.

Figure 22-8 shows a simple circuit consisting of a resistance R connected to an ideal battery. In such diagrams a battery is denoted by the symbol $\dashv\vdash$, the longer line indicating the terminal with the higher potential. The resistance is indicated by the symbol ⩘. The straight lines indicate connecting wires of negligible resistance. Since we are interested only in the potential differences between various points in the circuit, we can choose any point we wish to have zero potential. The source of emf maintains a constant potential difference equal to $\mathcal{E}$ between points a and b, with point a being at the higher potential. There is negligible potential difference between points a and c or between points d and b because the connecting wire is assumed to have negligible resistance. The potential difference between points c and d is therefore equal in magnitude to the emf $\mathcal{E}$, and the current in the resistor is given by $I = \mathcal{E}/R$. The direction of the current in this circuit is clockwise, as shown in the figure. Note that *inside* the source of emf the charge flows from a region of low potential to a region of high potential, so it gains potential energy.* When charge ΔQ flows through the source of emf $\mathcal{E}$, its potential energy is increased by the amount $\Delta Q\,\mathcal{E}$. The charge then flows through the resistor, where this potential energy is converted into thermal energy. The rate at which energy is supplied by the source of emf is the power output:

$$P = \frac{\Delta Q\,\mathcal{E}}{\Delta t} = \mathcal{E}I \qquad 22\text{-}15$$

In the simple circuit of Figure 22-8, the power put out by the source of emf equals that dissipated in the resistor.

A source of emf can be thought of as a sort of charge pump that pumps the charge from a region of low electrical potential energy to a region of high electrical potential energy, much like a water pump that pumps water from low to high regions of gravitational potential energy. Figure 22-9 shows a mechanical analog to the simple electric circuit just discussed in which marbles of mass m roll down an inclined board with many nails in it. The marbles start at some height h above the bottom and are accelerated between collisions with the nails by the gravitational field. The nails are analogous to the lattice ions in the resistor. During the collisions the marbles transfer the kinetic energy they obtained between collisions to the nails. Because of the many collisions, the marbles move with a small drift velocity toward the bottom. When they reach the bottom, a child picks them up, lifts them to their original height h, and starts them again. The child, who does work mgh on each marble, is analogous to the source of emf. The work per mass is gh, which is analogous to the work per charge done by the source of emf. The energy source in this case is the internal chemical energy of the child.

In a **real battery** the potential difference across the battery terminals, called the **terminal voltage,** is not simply equal to the value of the emf of the battery. Consider the simple circuit consisting of a real battery and a resistor in Figure 22-8. If the current is varied by varying the resistance R and the terminal voltage is measured, the terminal voltage is found to decrease slightly as the current increases, just as if there were a small resistance

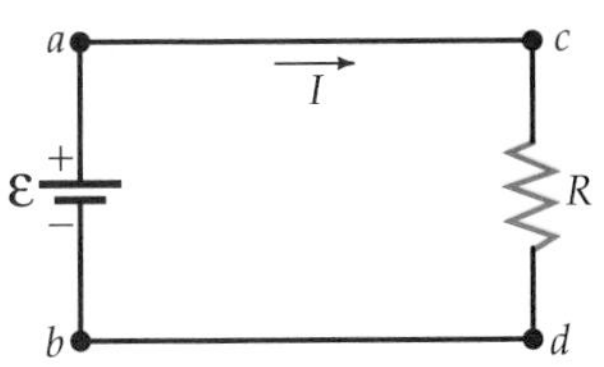

Figure 22-8 A simple circuit consisting of an ideal battery of emf $\mathcal{E}$, a resistance R, and connecting wires that are assumed to be without resistance.

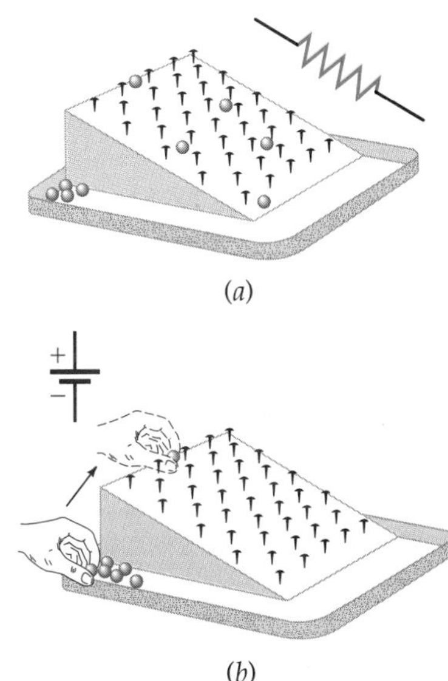

Figure 22-9 A mechanical analogy to a simple circuit consisting of a resistance and source of emf. (*a*) As the marbles roll down the incline, their potential energy is converted into kinetic energy that is quickly converted into heat because of collisions with the nails in the board. (*b*) A child lifts the marbles up from the bottom, where their potential energy is low, to the top, where their potential energy is high, thereby converting her internal chemical energy into potential energy of the marbles.

*When a battery is being charged by a generator or by another battery, the charge flows from a high potential to a low potential region within the battery being charged, thus losing electrostatic potential energy. The energy lost is stored in the battery being charged.

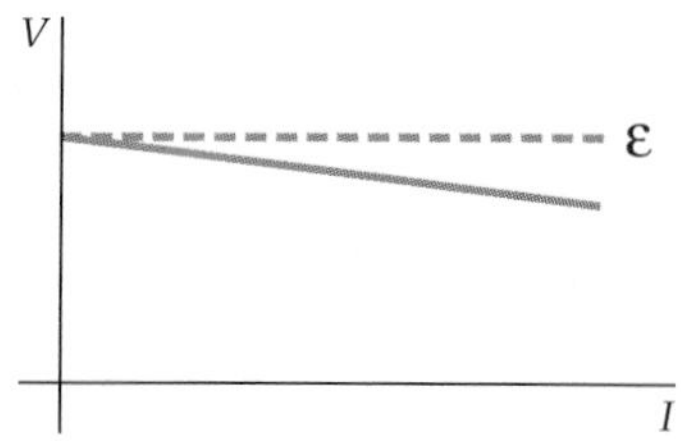

Figure 22-10 Terminal voltage V versus I for a real battery. The dashed line shows the terminal voltage of an ideal battery, which has the same magnitude as $\mathcal{E}$.

within the battery. This is shown graphically in Figure 22-10. Thus we can consider a real battery to consist of an ideal battery of emf $\mathcal{E}$ plus a small resistance r, called the **internal resistance** of the battery. Figure 22-11 shows a simple circuit consisting of a real battery, a resistor, and connecting wires. As before, we can ignore any resistance in the connecting wires. The circuit diagram for this circuit is shown in Figure 22-12. As charge passes from point b to point a, its potential energy is first increased as it passes through the source of emf and is then decreased slightly as it passes through the internal resistance of the battery. (In the actual battery, these energy changes take place concurrently.) If the current in the circuit is I, the potential at point a is related to that at point b by

$$V_a = V_b + \mathcal{E} - Ir$$

The terminal voltage is thus

$$V_a - V_b = \mathcal{E} - Ir \qquad 22\text{-}16$$

The terminal voltage of the battery decreases linearly with current, as we saw in Figure 22-10. The potential drop across the resistor R is IR and is equal to the terminal voltage:

$$IR = V_a - V_b = \mathcal{E} - Ir$$

Solving for the current I, we obtain

$$IR + Ir = \mathcal{E}$$

or

$$I = \frac{\mathcal{E}}{R + r} \qquad 22\text{-}17$$

Figure 22-11 A photograph of a simple circuit consisting of a real battery, a resistor, and connecting wires.

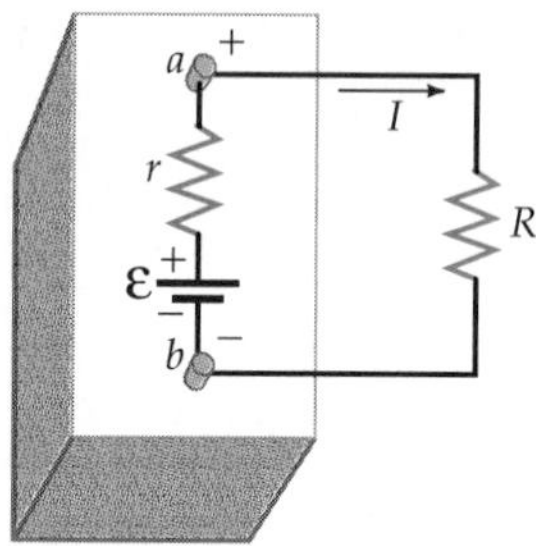

Figure 22-12 Circuit diagram for the circuit in Figure 22-11. A real battery can be represented by an ideal battery of emf $\mathcal{E}$ and a small resistance r.

The terminal voltage given by Equation 22-16 is less than the emf of the battery because of the potential drop across the internal resistance of the battery. Real batteries such as a good car battery usually have an internal resistance of the order of a few hundredths of an ohm, so the terminal voltage is nearly equal to the emf unless the current is very large. One sign of a bad battery is an unusually high internal resistance. If you suspect that your 12-V car battery is bad, and you check the terminal voltage with a voltmeter (discussed in Section 23-3), which draws very little current, the voltmeter may read nearly 12 V, as if the battery were good. However, if you check the terminal voltage while current is being drawn from the battery, such as with the car lights on or while you are trying to start your car, the terminal voltage may drop considerably below 12 V, thereby indicating a high internal resistance and a bad battery.

Example 22-8

An 11-Ω resistance is connected across a battery of emf 6 V and internal resistance 1 Ω.* Find (*a*) the current, (*b*) the terminal voltage of the battery, (*c*) the power delivered by the emf, and (*d*) the power delivered to the external resistance.

(*a*) From Equation 22-17, the current is

$$I = \frac{\mathcal{E}}{R + r} = \frac{6\text{ V}}{11\ \Omega + 1\ \Omega} = 0.5\text{ A}$$

(*b*) The terminal voltage of the battery is

$$V_a - V_b = \mathcal{E} - Ir = 6\text{ V} - (0.5\text{ A})(1\ \Omega) = 5.5\text{ V}$$

(*c*) The power delivered by the source of emf is

$$P = \mathcal{E}I = (6\text{ V})(0.5\text{ A}) = 3\text{ W}$$

(*d*) The power delivered to the external resistance is

$$I^2R = (0.5\text{ A})^2(11\ \Omega) = 2.75\text{ W}$$

This is dissipated as Joule heat. The other 0.25 W of power is dissipated as Joule heat in the internal resistance of the battery.

Example 22-9

For a battery of given emf and internal resistance r, what value of external resistance R should be placed across the terminals to obtain the greatest Joule heating in R?

The external resistance R is sometimes called the load resistance. The power input to R is I^2R, where I is given by Equation 22-17. Thus, the power input is

$$P = I^2R = \frac{\mathcal{E}^2}{(r + R)^2}R = \mathcal{E}^2R(r + R)^{-2}$$

Figure 22-13 shows a sketch of P versus R. At the value of R for which P is maximum, the slope of the curve P versus r is zero. We find this value of R by setting dP/dR equal to zero. We have

$$\frac{dP}{dR} = \mathcal{E}^2(r + R)^{-2} + \mathcal{E}^2R\,(-2)(r + R)^{-3} = 0$$

Multiplying each term by $(r + R)^3/\mathcal{E}^2$, we obtain

$$r + R = 2R$$

or

$$R = r$$

The maximum value of P occurs when $R = r$, that is, when the load resistance equals the internal resistance. A similar result also occurs for more complicated ac circuits and is known as impedance matching.

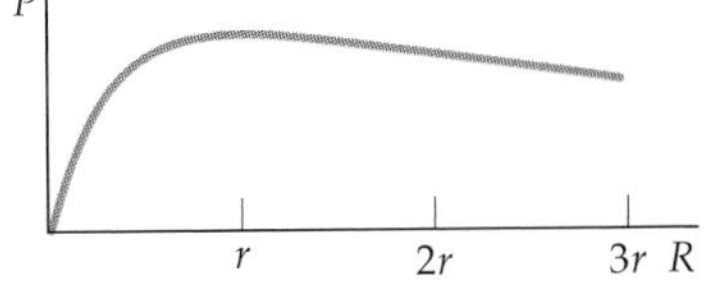

Figure 22-13 Plot of the power input to the external resistor versus R. The power is a maximum when the load resistance R equals the internal resistance of the battery.

*The value of the internal resistance is exaggerated in this example to simplify calculations. In other examples, we may simply ignore the internal resistance.

(a)

(b)

(c)

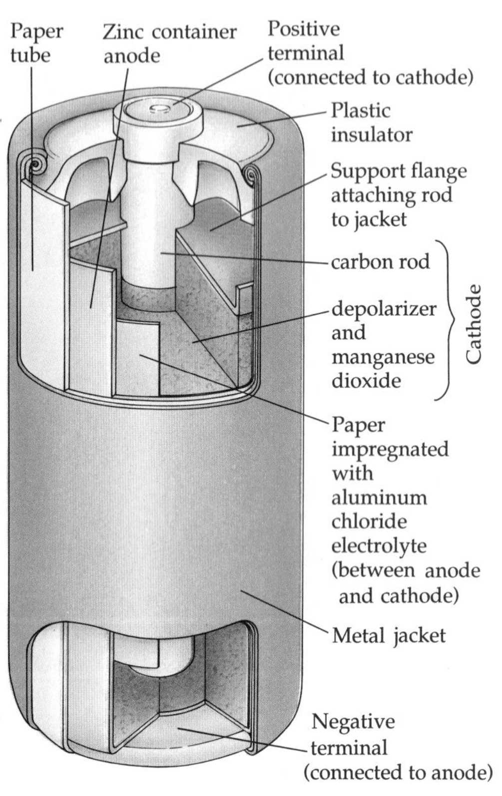

(d)

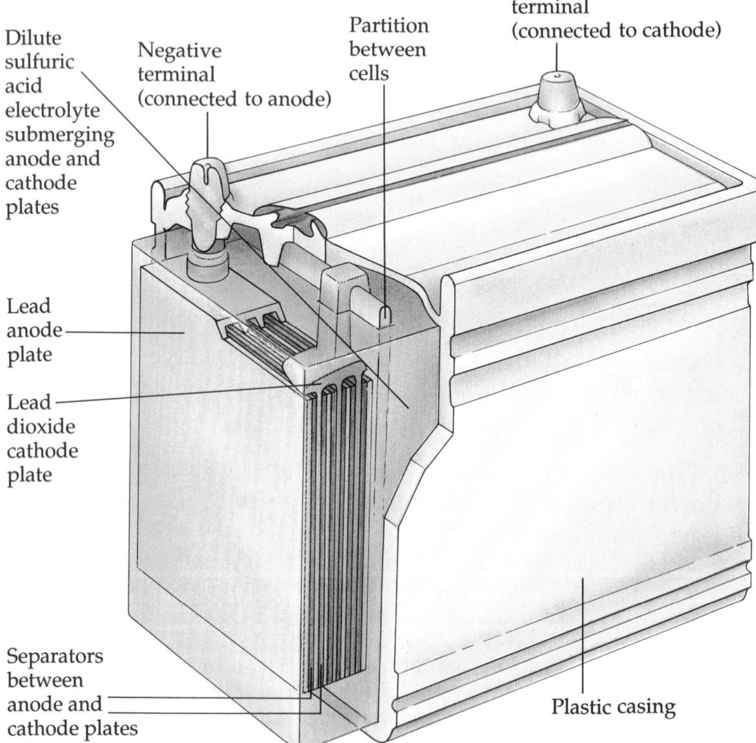

(e)

A battery is a set of chemical cells each of which consists of two metal electrodes immersed in a conducting solution called an electrolyte. Because of chemical reactions between the conductors and the electrolyte, one electrode, the anode, becomes positively charged and the other, the cathode, becomes negatively charged. (*a*) Coin-sized lithium–polymer cells are used in computers to prevent data loss when the power is removed. (*b*) An assembly of high duty silver–zinc cells that is used in satellites. (*c*) This electric car uses eight storage batteries for its power. (*d*) In a dry cell, the electrolyte is a paste of ammonium chloride and other additives. It develops an emf of 1.5 V. (*e*) A 12-V storage battery consists of six cells, which are rechargeable. Each cell has a lead anode, a lead dioxide cathode, and a water solution of sulfuric acid for its electrolyte. (*f*) This giant battery consisting of 200 cells was built in 1807 in the basement of the Royal Institution, London by Humphry Davy.

(f)

Batteries are often rated in ampere-hours (A·h), which is the total charge they can deliver. Since one ampere is one coulomb per second, and one hour contains 3600 seconds, an ampere-hour is equal to 3600 coulombs:

$$1\ \text{A·h} = 1\frac{\text{C}}{\text{s}}(3600\ \text{s}) = 3600\ \text{C}$$

Questions

3. Name several common sources of emf. What sort of energy is converted into electrical energy in each?
4. In a simple electric circuit like that shown in Figure 22-12, the charge outside the emf flows from positive voltage toward negative voltage, whereas inside the emf it flows from negative voltage to positive voltage. Explain how this is possible.
5. Figure 22-9 illustrates a mechanical analog to the simple electric circuit. Devise another mechanical analog in which the current is represented by a flow of water instead of marbles.
6. A skier is towed up a hill and then skis down with a speed that is constant because of friction. How is this analogous to a simple electric circuit?

22-4 Combinations of Resistors

Series Resistors

Two or more resistors that are connected such that the same charge must flow through each are said to be connected in series. Resistors R_1 and R_2 in Figure 22-14*a* are examples of resistors in series. Since charge does not collect at any point in a wire carrying a steady current, if a charge ΔQ flows into R_1 during some time interval, an equal amount of charge ΔQ must flow out of R_2 during the same interval. The two resistors must therefore carry the same current I. We can often simplify the analysis of a circuit with resistors in series by replacing such resistors with a single equivalent resistance R_{eq} that gives the same total potential drop V when carrying the same current I (see Figure 22-14*b*). The potential drop across R_1 is IR_1 and that across R_2 is IR_2. The potential drop across the two resistors is the sum of the potential drops across the individual resistors:

$$V = IR_1 + IR_2 = I(R_1 + R_2) \qquad 22\text{-}18$$

Setting this potential drop equal to IR_{eq}, we obtain

$$R_{eq} = R_1 + R_2$$

Figure 22-14 (*a*) Two resistors in series carry the same current. (*b*) The resistors in (*a*) can be replaced by a single equivalent resistance $R_{eq} = R_1 + R_2$ that gives the same total potential drop when carrying the same current as in (*a*).

Thus, the equivalent resistance for resistors in series is the sum of the original resistances. When there are more than two resistors in series, the equivalent resistance is

$$R_{eq} = R_1 + R_2 + R_3 + \cdots \qquad 22\text{-}19$$

Resistors in series

Parallel Resistors

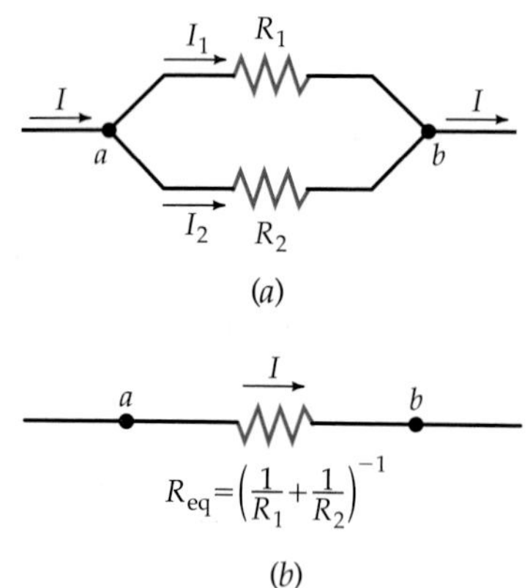

Two resistors that are connected as in Figure 22-15*a* such that they have the same potential difference across them are said to be connected in parallel. Note that the resistors are connected at both ends by wires. Let I be the current from point a to point b. At point a the current splits into two parts, I_1 in resistor R_1 and I_2 in R_2. The total current is the sum of the individual currents:

$$I = I_1 + I_2 \tag{22-20}$$

Let $V = V_a - V_b$ be the potential drop across either resistor. In terms of the currents and resistances,

$$V = I_1R_1 = I_2R_2 \tag{22-21}$$

The equivalent resistance of a combination of parallel resistors is defined to be that resistance R_{eq} for which the same total current I produces the potential drop V (Figure 22-15*b*):

$$R_{eq} = \frac{V}{I}$$

Figure 22-15 (*a*) Two resistors are in parallel when they are connected together at both ends so that the potential drop is the same across each. (*b*) The two resistors in (*a*) can be replaced by an equivalent resistance R_{eq} that is related to R_1 and R_2 by $1/R_{eq} = 1/R_1 + 1/R_2$.

Solving this equation for I and using $I = I_1 + I_2$, we have

$$I = \frac{V}{R_{eq}} = I_1 + I_2 \tag{22-22}$$

But according to Equation 22-21, $I_1 = V/R_1$ and $I_2 = V/R_2$. Equation 22-22 can thus be written

$$I = \frac{V}{R_{eq}} = \frac{V}{R_1} + \frac{V}{R_2}$$

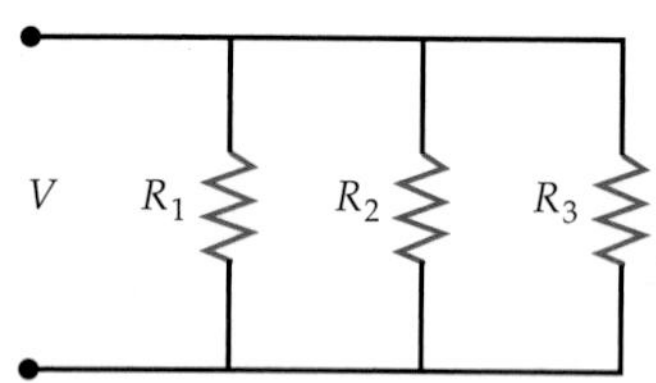

The equivalent resistance for two resistors in parallel is therefore given by

$$\frac{1}{R_{eq}} = \frac{1}{R_1} + \frac{1}{R_2}$$

Figure 22-16 Three resistors in parallel.

This result can be generalized for combinations, such as that in Figure 22-16, in which three or more resistors are connected in parallel:

Resistors in parallel

$$\frac{1}{R_{eq}} = \frac{1}{R_1} + \frac{1}{R_2} + \frac{1}{R_3} + \cdots \tag{22-23}$$

Exercise

A 2-Ω and a 4-Ω resistor are connected (*a*) in series and (*b*) in parallel. Find the equivalent resistances. [Answers: (*a*) 6 Ω, (*b*) 1.33 Ω]

Example 22-10

A 4-Ω and a 6-Ω resistor are connected in parallel as shown in Figure 22-17, and a potential difference of 12 V is applied across the combination. Find (*a*) the equivalent resistance, (*b*) the total current, (*c*) the current in each resistor, and (*d*) the power dissipated in each resistor.

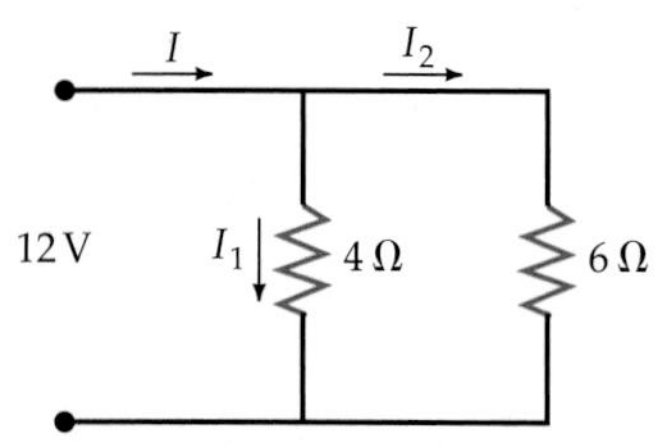

Figure 22-17 Two resistors in parallel across a potential difference of 12 V for Example 22-10.

(*a*) We first calculate the equivalent resistance from Equation 22-23:

$$\frac{1}{R_{eq}} = \frac{1}{4\ \Omega} + \frac{1}{6\ \Omega} = \frac{3}{12\ \Omega} + \frac{2}{12\ \Omega} = \frac{5}{12\ \Omega}$$

or

$$R_{eq} = \frac{12\ \Omega}{5} = 2.4\ \Omega$$

(*b*) The total current is therefore

$$I = \frac{V}{R_{eq}} = \frac{12\text{ V}}{2.4\ \Omega} = 5\text{ A}$$

(*c*) We obtain the current in each resistor from the fact that the potential drop is 12 V across each resistor (Equation 22-21). Calling the current in the 4-Ω resistor I_1 and that in the 6-Ω resistor I_2, we have

$$V = I_1R_1 = I_1(4\ \Omega) = 12\text{ V}$$

$$I_1 = \frac{12\text{ V}}{4\ \Omega} = 3.0\text{ A}$$

and

$$I_2 = \frac{12\text{ V}}{6\ \Omega} = 2.0\text{ A}$$

(*d*) The power dissipated in the 4-Ω resistor is

$$P = I_1^2R_1 = (3.0\text{ A})^2(4\ \Omega) = 36\text{ W}$$

The power dissipated in the 6-Ω resistor is

$$P = (2.0\text{ A})^2(6\ \Omega) = 24\text{ W}$$

This power comes from the source of emf that maintains the 12-V potential difference across the combination of resistors. The power required to deliver 5.0 A at 12 V is

$$P = IV = (5.0\text{ A})(12\text{ V}) = 60\text{ W}$$

which equals the total power dissipated in the two resistors.

Note from Example 22-10 that the equivalent resistance of two parallel resistances is less than the resistance of either resistor alone. This is a general result. Suppose we have a single resistor R_1 carrying current I_1 with potential drop $V = I_1R_1$. We now add a second resistor R_2 in parallel. If the potential drop is to remain the same, the second resistor must carry additional current $I_2 = V/R_2$ without affecting the original current I_1. The parallel combination thus carries more total current $I = I_1 + I_2$ for the same potential drop, so the ratio of the potential drop to the total current is less. Note also from Example 22-10 that the ratio of the currents in the two parallel resistors equals the inverse ratio of the resistances. This general result follows from Equation 22-21:

$$\frac{I_1}{I_2} = \frac{R_2}{R_1} \quad \text{parallel resistors} \qquad 22\text{-}24$$

Example 22-11

For the circuit shown in Figure 22-18, find (*a*) the equivalent resistance of the parallel combination of resistors, (*b*) the total current in the source of emf and the current carried by each resistor.

(*a*) The equivalent resistance of the 6- and 12-Ω resistors in parallel is found from

$$\frac{1}{R_{eq}} = \frac{1}{6\ \Omega} + \frac{1}{12\ \Omega} = \frac{3}{12\ \Omega} = \frac{1}{4\ \Omega}$$

$$R_{eq} = 4\ \Omega$$

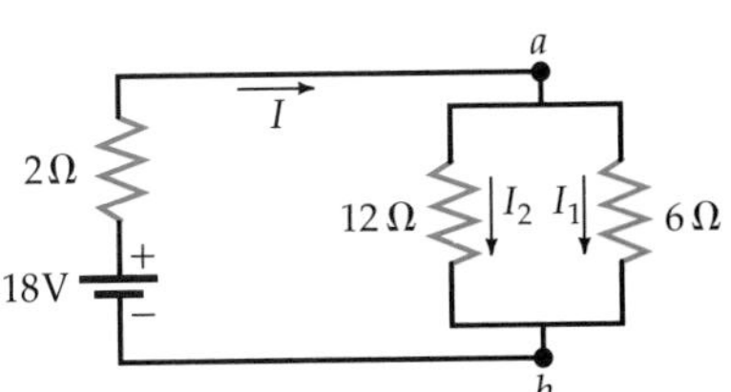

Figure 22-18 Circuit for Example 22-11. The 12- and 6-Ω resistors are in parallel, and this parallel combination is in series with the 2-Ω resistor.

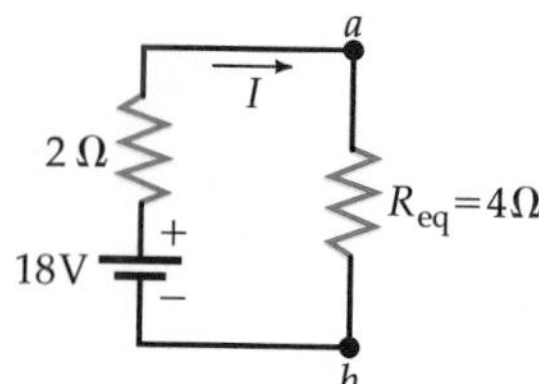

Figure 22-19 The circuit of Figure 22-18 has been simplified by replacing the two parallel resistors with their equivalent resistance.

(*b*) Figure 22-19 shows the circuit with R_{eq} replacing the parallel combination. The resistances $R_{eq} = 4\ \Omega$ and $R = 2\ \Omega$ are in series. The equivalent resistance of this series combination is $R'_{eq} = R_{eq} + R = 6\ \Omega$. The current I in the circuit is therefore

$$I = \frac{\mathcal{E}}{R'_{eq}} = \frac{18\ \text{V}}{6\ \Omega} = 3\ \text{A}$$

This is the total current in the source of emf. The potential drop from a to b across R_{eq} in Figure 22-19 is $V = IR_{eq} = (3\ \text{A})(4\ \Omega) = 12\ \text{V}$. The current in the 6-Ω resistor is thus

$$I_1 = \frac{12\ \text{V}}{6\ \Omega} = 2\ \text{A}$$

and that in the 12-Ω resistor is

$$I_2 = \frac{12\ \text{V}}{12\ \Omega} = 1\ \text{A}$$

Note that the current in the 6-Ω resistor is twice that in the 12-Ω resistor, as we should expect.

Example 22-12

Find the equivalent resistance between points a and b for the combination of resistors shown in Figure 22-20.

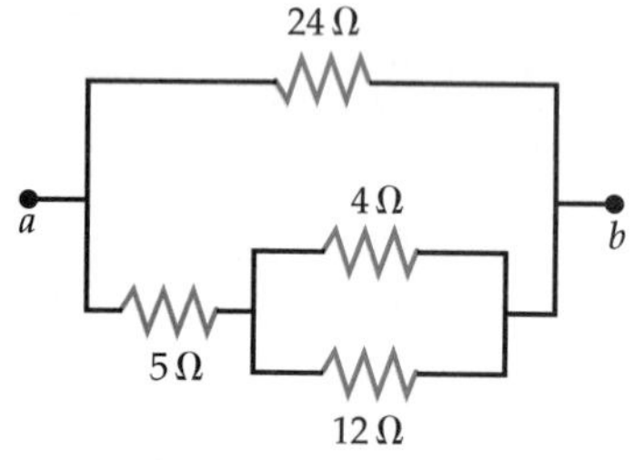

Figure 22-20 Resistor network for Example 22-12.

This combination of resistors may look complicated, but it can be analyzed step by step. The only pair of resistors that are either in series or in parallel are the 4-Ω and 12-Ω resistors, which are in parallel. We therefore first find the equivalent resistance of these resistors. From Equation 22-23, we obtain

$$\frac{1}{R_{eq}} = \frac{1}{4\ \Omega} + \frac{1}{12\ \Omega} = \frac{4}{12\ \Omega} = \frac{1}{3\ \Omega}$$

or

$$R_{eq} = 3\ \Omega$$

In Figure 22-21, the 4-Ω and 12-Ω resistors have been replaced by their equivalent, a 3-Ω resistor. Since this 3-Ω resistor is in series with the 5-Ω resistor, the equivalent resistance of the bottom branch of this combination is 8 Ω. We are now left with an 8-Ω resistor in parallel with a 24-Ω resistor (Figure 22-22). The equivalent resistance of these two parallel resistors is again found from Equation 22-23:

$$\frac{1}{R_{eq}} = \frac{1}{24\ \Omega} + \frac{1}{8\ \Omega} = \frac{4}{24\ \Omega} = \frac{1}{6\ \Omega}$$

$$R_{eq} = 6\ \Omega$$

Thus the equivalent resistance between points a and b is 6 Ω.

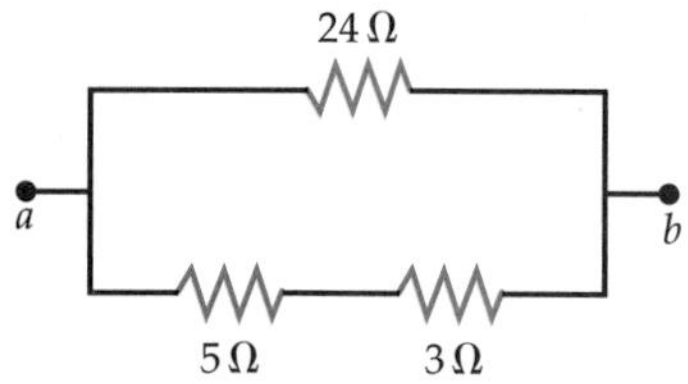

Figure 22-21 Simplification of the resistor network of Figure 22-20. The parallel 4- and 12-Ω resistors in that figure have been replaced by their equivalent resistance of 3 Ω.

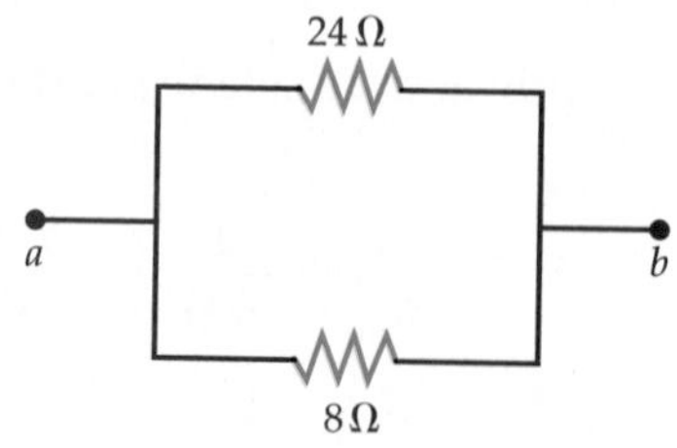

Figure 22-22 Further simplification of the resistor network of Figures 22-20 and 22-21. The 5- and 3-Ω resistors in series in Figure 22-21 are replaced by their equivalent resistance of 8 Ω. The network is now reduced to 24- and 8-Ω resistors in parallel.

Questions

7. Which will produce more heat—a small resistance or a large resistance connected across a source of emf that remains constant?
8. After the introduction of electric lighting, some people were careful to keep bulbs in all the sockets so that the electricity would not leak out. Why is this not necessary?

22-5 A Microscopic Picture of Conduction

A microscopic model of electric conduction was first proposed by P. Drude in 1900 and was developed by Hendrik A. Lorentz about 1909. This model, now called the **classical model of electric conduction,** successfully predicts Ohm's law and relates the conductivity and resistivity of conductors to the motion of free electrons within them. In this model, a metal is pictured as a regular, three-dimensional lattice of ions with a large number of electrons that are free to move throughout the metal. In the absence of an electric field, the free electrons move about the metal much like gas molecules move about in a container. The free electrons are in thermal equilibrium with the lattice ions with which they make collisions. The mean speed of the electrons can be calculated from the equipartition theorem. The result is the same as that for ideal-gas molecules, except that the electron mass m_e is used instead of the mass of a molecule in Equation 15-28. For example, at $T =$ 300 K, the rms speed (which is slightly greater than the mean speed) is

This scanning electron microscope photograph of a 21-μm–wide aluminum conducting strip such as used in computer chips shows voids and clusters of atoms caused by electromigration of aluminum atoms as a result of a current of 0.5 A. The nature of this effect is not well understood, but it adversely affects the reliability of computer chips, which use strips as narrow as 4 μm.

$$v_{\text{rms}} = \sqrt{\frac{3kT}{m_e}}$$

$$= \sqrt{\frac{3(1.38 \times 10^{-23}\ \text{J/K})(300\ \text{K})}{9.11 \times 10^{-31}\ \text{kg}}}$$

$$= 1.17 \times 10^5\ \text{m/s} \qquad 22\text{-}25$$

This is much greater than the drift velocity that we calculated in Example 22-1.

According to Ohm's law, the current in a conducting wire segment is proportional to the voltage drop across the segment:

$$I = \frac{V}{R}$$

The resistance R is proportional to the length of the wire segment L and inversely proportional to the cross-sectional area A:

$$R = \rho\frac{L}{A}$$

For a uniform electric field E, the voltage across a segment of length L is $V = EI$. Substituting $\rho L/A$ for R, and EL for V, we can write Ohm's law

$$I = \frac{EL}{\rho L/A} = \frac{1}{\rho}EA \qquad 22\text{-}26$$

The objective of the classical theory of conduction is to find an expression for ρ in terms of the properties of metals. The current in a wire is related to the number of electrons per unit volume n, the drift velocity v_d, the charge on the electron e, and the cross-sectional area A by Equation 22-3:

$$I = neAv_d$$

In the presence of an electric field, a free electron experiences a force of magnitude eE. If this were the only force acting on the electron, the electron would have an acceleration eE/m_e, and its velocity would steadily increase. However, Ohm's law implies that there is a steady-state situation in which the average velocity of the electron is proportional to the field E, since the current I is proportional to E and also to v_d. In the classical model, it is assumed that a free electron is accelerated for a short time and then makes a collision with a lattice ion. After the collision, the velocity of the electron is completely unrelated to that before the collision. The justification for this assumption is that the drift velocity is very small compared with the random thermal velocity.

Let τ be the average time before an electron, picked at random, makes its next collision. Because the collisions are random, this time does not depend on the time elapsed since the electron's last collision. If we look at an electron immediately after it makes a collision, the average time before its next collision will be τ. Thus τ, called the **collision time**, is the average time between collisions. It is also the average time since the *last* collision of an electron picked at random.*

The drift velocity is the average velocity of an electron picked at random. Since the acceleration is eE/m_e, the drift velocity is

$$v_d = \frac{eE}{m_e}\tau \qquad \text{22-27}$$

Using this result in Equation 22-3, we obtain

$$I = neAv_d = \frac{ne^2\tau}{m_e}EA \qquad \text{22-28}$$

Using $\rho = EA/I$ from Equation 22-26, we have for the resistivity

$$\rho = \frac{m_e}{ne^2\tau} \qquad \text{22-29}$$

The average distance the electron travels between collisions is called the **mean free path** λ. It is the product of the mean speed v_{av} and the mean time between collisions τ:

$$\lambda = v_{av}\tau \qquad \text{22-30}$$

In terms of the mean free path and the mean speed, the resistivity is

$$\rho = \frac{m_e v_{av}}{ne^2\lambda} \qquad \text{22-31}$$

We can relate the mean free path to the size of the copper ions. Consider one electron moving with speed v through a region of stationary ions (Figure 22-23). If the size of the electron is negligible, it will collide with an ion if it comes within a distance r from the center of the ion, where r is the radius of the ion. In some time t, the electron moves a distance vt and collides with every ion in the cylindrical volume $\pi r^2 vt$. The number of ions in this volume is $n\pi r^2 vt$, where n is the number of ions per unit volume. (After each colli-

*It is tempting but incorrect to think that if τ is the average time between collisions, the average time since its last collision is $\frac{1}{2}\tau$ rather than τ. If you find this confusing, you may take comfort in the fact that Drude used the incorrect result $\frac{1}{2}\tau$ in his original work.

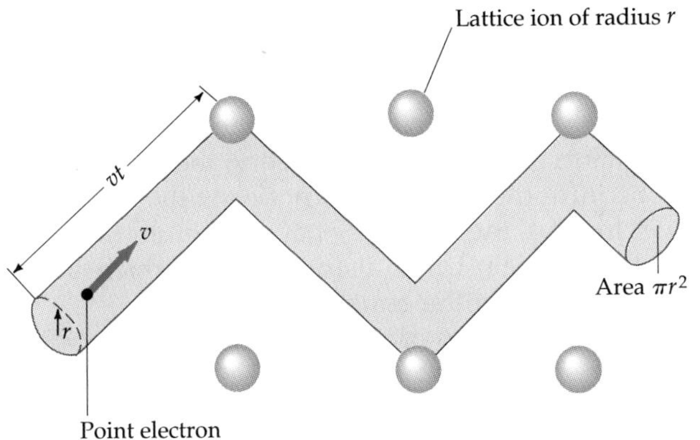

Figure 22-23 Model of an electron moving through the lattice ions in copper. The electron, which is considered to be a point, collides with a lattice ion if it comes within a distance r of the center of the ion, where r is the radius of the ion. If the electron has speed v, it collides with all the ions in the cylindrical volume $\pi r^2 vt$ in time t.

sion, the direction of the electron changes, so the path is really a zigzag one.) The total path length divided by the number of collisions is the mean free path:

$$\lambda = \frac{vt}{n\pi r^2 vt} = \frac{1}{n\pi r^2} \qquad \text{22-32}$$

Example 22-13

Estimate the mean free path for electrons in copper.

The number of copper ions per unit volume was calculated in Example 22-1 to be 8.47×10^{22} ions/cm^3. Using $r \approx 10^{-10}$ m $= 10^{-8}$ cm for the radius of a copper ion, we obtain for an estimate of the mean free path of the electrons in copper

$$\lambda = \frac{1}{(8.47 \times 10^{22}\ \text{cm}^{-3})\pi(10^{-8}\ \text{cm})^2} \approx 4 \times 10^{-8}\ \text{cm}$$
$$= 4 \times 10^{-10}\ \text{m} = 0.4\ \text{nm}$$

We can use the result of Example 22-13 and $v_{\text{av}} \sim 10^5$ m/s from Equation 22-25 to obtain for an estimate of the collision time

$$\tau = \frac{\lambda}{v_{\text{av}}} \sim \frac{4 \times 10^{-10}\ \text{m}}{10^5\ \text{m/s}} = 4 \times 10^{-15}\ \text{s}$$

According to Ohm's law, the resistivity is independent of the electric field E. The quantities in Equation 22-31 that might depend on the electric field are the mean speed v_{av} and the mean free path λ. As we have seen, the drift velocity is very much smaller than the mean speed of the electrons in thermal equilibrium with the lattice ions. Thus, the electric field has essentially no effect on the mean speed of the electrons. The mean free path of the electrons depends on the size of the lattice ions and on the density of the ions, neither of which depends on the electric field E. Thus, this model predicts Ohm's law, with the resistivity as given by Equation 22-31.

Although successful in predicting Ohm's law, the classical theory of conduction has several defects. Classical methods for finding the mean free path and the mean speed give a numerical magnitude of the resistivity calculated from Equation 22-31 about six times the measured value at $T = 300$ K, and the temperature dependence is not correct. The temperature dependence of resistivity in Equation 22-31 is given completely by the mean speed v_{av}, which is proportional to $\sqrt{T}$. Thus, this calculation does not give a linear dependence on temperature. Finally, the classical model says nothing about why some materials are conductors, others insulators, and still others semiconductors.

In the quantum-mechanical theory of electrical conduction, which is discussed in the extended version of this book (Chapter 39), the resistivity is again given by Equation 22-31, but the mean speed and the mean free path are interpreted in terms of quantum theory. In this theory, the mean speed is not proportional to $\sqrt{T}$ because the electrons do not obey the Maxwell–Boltzmann distribution law. Instead, the electrons obey a quantum-mechanical distribution law called the Fermi–Dirac distribution in which mean speed is approximately independent of the temperature.

In the quantum-mechanical calculation of the mean free path, the wave nature of the electron (Chapter 35) is important. The collision of an electron with a lattice ion is not similar to the collision of a baseball and a tree, but instead involves the scattering of an electron wave by a regularly spaced lattice. A detailed calculation of the scattering of electron waves in a *perfectly* ordered crystal gives the result that *there is no scattering, and the mean free path is infinite*. Thus, the mean free path is not determined by the size of the lattice ions. At very low temperatures, the ions look like points as far as the scattering of electrons is concerned. The scattering of electron waves arises when the crystal lattice is not perfectly periodic. There are two common causes of deviations from perfect periodicity in a lattice. One is the displacement of the lattice ions due to thermal vibrations. This effect is dominant at ordinary temperatures. The effective area that an ion presents to an electron is proportional to the square of the amplitude of vibration, which is proportional to the energy of vibration, which in turn is proportional to the absolute temperature T. The mean free path therefore varies inversely with T, and ρ is proportional to T as observed experimentally. The other cause of deviations from perfect periodicity is impurities. For example, if some zinc is introduced into pure copper, the previously perfect periodicity is destroyed. At very low temperatures, the resistance of a metal is primarily due to impurities.

Summary

1. Electric current is the rate of flow of charge through a cross-sectional area. By convention, its direction is considered to be that of the flow of positive charge. In a conducting wire, electric current is the result of the slow drift of negatively charged electrons that are accelerated by an electric field in the wire and then quickly collide with the conductor atoms. Typical drift velocities of electrons in wires are of the order of 0.01 mm/s.

2. The resistance of a segment of wire is defined as the ratio of the voltage drop across the segment to the current. In ohmic materials, which include most metals, the resistance is independent of the current, an experimental result known as Ohm's law. For all materials, the potential difference, current, and resistance are related by

$$V = IR$$

3. The resistance of a wire is proportional to its length and inversely proportional to its cross-sectional area:

$$R = \rho\frac{L}{A}$$

where ρ is the resistivity of the material, which depends on its temperature. The reciprocal of the resistivity is called the conductivity σ:

$$\sigma = \frac{1}{\rho}$$

4. The power supplied to a segment of a circuit equals the product of the current and the voltage drop across the segment:

$$P = IV$$

A device that supplies energy to a circuit is called a source of emf. The power supplied by a source of emf is the product of the emf and the current:

$$P = \mathcal{E}I$$

The power dissipated in a resistor is given by

$$P = IV = I^2R = \frac{V^2}{R}$$

An ideal battery is a source of emf that maintains a constant potential difference across its terminals independent of the current. A real battery can be considered to be an ideal battery in series with a small resistance called its internal resistance.

5. The equivalent resistance of a set of resistors in series equals the sum of the resistances:

$$R_{eq} = R_1 + R_2 + R_3 + \cdots \qquad \text{resistors in series}$$

For a set of resistors in parallel, the reciprocal of the equivalent resistance equals the sum of the reciprocals of the individual resistances:

$$\frac{1}{R_{eq}} = \frac{1}{R_1} + \frac{1}{R_2} + \frac{1}{R_3} + \cdots \qquad \text{resistors in parallel}$$

6. In the microscopic model of electrical conduction, the free electrons in a metal are accelerated by the electric field but quickly lose their additional energy in collisions with the lattice ions of the metal. Their drift velocity is proportional to the electric field. The resistivity ρ is related to the mean speed v_{av} and to the mean free path λ (the average distance traveled between collisions) by

$$\rho = \frac{m_e v_{av}}{ne^2\lambda}$$

In the classical model, the mean speed is obtained from the Maxwell–Boltzman speed distribution and is proportional to $\sqrt{T}$, and the mean free path depends on the size of the lattice ions. This model predicts Ohm's law but gives incorrect numerical magnitudes for ρ and an incorrect temperature dependence of ρ. In the modern, quantum-mechanical theory of conductivity, the average speed, which is found from the quantum-mechanical speed distribution, is independent of temperature. The wave nature of electrons is used to determine the mean free path, which turns out to be infinite for a perfectly periodic lattice. Deviations from perfect periodicity arise from impurities and from thermal vibrations of the lattice ions.

Conduction in Nerve Cells

Elizabeth Pflegl Nickles
The Albany College of Pharmacy

During a thunderstorm in 1786, Luigi Galvani touched the muscles of a frog's leg with a metal instrument and noticed that the muscles twitched. He concluded that electricity generated by the storm had been conducted through the frog's nerves, causing the muscles to contract. Nerves *do* transmit impulses from one part of the body to another, but by a mechanism very different from the transmission of electricity through a metallic conductor. For a long time, it was thought that a nerve impulse consisted of a flow of ions (that is, an electric current) along the nerve cell, in a manner analogous to the flow of electrons in a wire. However, the electrical properties of a nerve cell are very different from those of a metallic conductor; for example, neural conduction is much slower and does not vary in strength (it is all-or-nothing conduction).

A nerve cell (neuron) is the basic building block of the nervous system and is specialized to transmit information. It consists of a cell body and one or more branchlike fibers (Figure 1). The fibers are of two types, the dendrites, which along with the cell body receive information in the form of stimuli from sensory receptors or other nerve cells, and the axon, which transmits information to other nerve cells. If the stimuli received on the dendrites or the cell body at any one time are at or above a particular threshold of intensity, a nerve impulse is initiated and propagated along the axon. It flows along the axon away from the cell body toward the terminal branches. The axon is a long, thin cellular extension bounded by a membrane and filled with a viscous intracellular fluid called the axoplasm (Figure 2). Once a nerve impulse reaches the nerve cell's terminal branches, neurotransmitter substances are released that convey the impulse to receptors on the next cell.

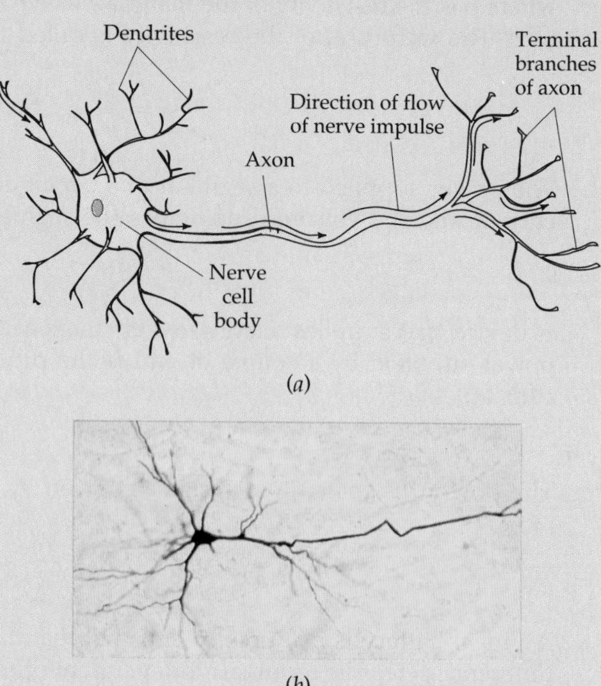

Figure 1 (*a*) A diagram of the structure of a nerve cell. (*b*) A micrograph of a nerve cell from the cerebrum of a cat.

Elizabeth Pflegl Nickles was born in Catskill, New York and raised on a dairy farm. She earned a B.S. in physics and an M.S. in biology at SUNY Albany, where she is now completing an M.S. in physics. She has taught at SUNY Brockport and at Hudson Valley Community College, and has also held the position of Senior Biophysicist with the New York State Department of Health. Currently, in addition to being a lecturer in physics at the Albany College of Pharmacy, she does research in biological materials using x-ray and backscatter spectra obtained with a scanning microbeam from a linear accelerator.

The small cross-sectional area of an axon and the high resistivity ($R = \rho L/A$, Equation 22-8) of the axoplasm combine to yield an extremely high resistance. A piece of nerve axon 1 cm in length has an electrical resistance of about $2.5 \times 10^8\ \Omega$ (comparable to that of wood; see Table 22-1). To understand how nerve impulse transmission is possible through such a seemingly unlikely medium, one must first learn something about the physiology of a nerve cell before it is stimulated, when it is in its "resting state."

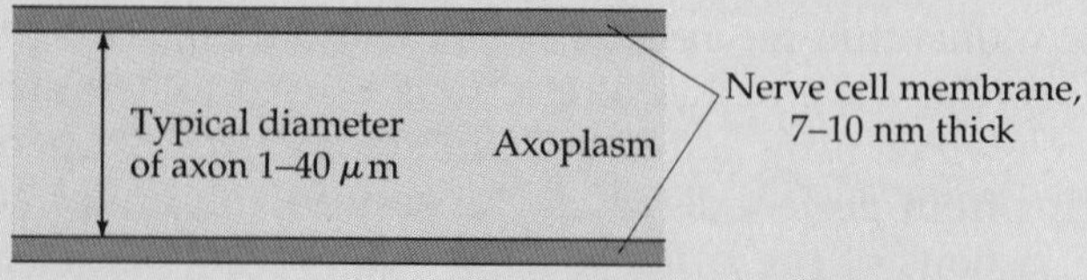

Figure 2 A longitudinal section through an axon.

The Resting Nerve Cell

In nerve cells and other living cells, the cell membrane maintains intracellular conditions that differ from those of the extracellular environment. Critical to the function of the nerve cell is the buildup of a slight excess of negative ions just inside and a slight excess of positive ions just outside the cell membrane (Figure 3).

Electrochemical gradients across the nerve cell membrane are the key to understanding nerve impulse transmission. The concentration of potassium ions (K^+) is 30 times greater in the fluid inside the cell than outside; the concentration of sodium ions (Na^+) is nearly 10 times greater in the fluid outside the cell than inside (see Table 1). Note that anions (particularly chloride, Cl^-) are also unevenly distributed. Like all living cells, nerve cells use both passive diffusion and active transport to maintain these gradients across their cell membranes. The unequal distribution of Na^+ and K^+ is established by an energy-dependent Na^+–K^+ "pump," which moves Na^+ out of the cell and K^+ into the cell. There are also specialized proteins embedded in the nerve cell membrane that function as voltage-dependent channels through which Na^+ and K^+ pass during nerve impulse transmission.

When the nerve cell is in its resting state, the voltage-dependent Na^+ channels are closed, thus maintaining the unequal distribution of Na^+. The resting nerve cell membrane is not permeable to large anions (or to other large negatively charged species, such as proteins) so a slight excess of negative charge builds up immediately inside the nerve cell membrane. The potential difference (Section 20-1) across the membrane is about 70 mV; taking the electric potential outside the cell to be zero, the electric potential inside is therefore −70 mV. This is the nerve cell's *resting potential* (Figure 3).

The arrangement of charge on each surface of the nerve cell membrane resembles that of a charged capacitor. The electric field across a parallel-plate capacitor is uniform (see Section 21-1), so we can calculate the electric field across a cell membrane of 7 nm in thickness:

Table 1 **Ion Concentrations Inside and Outside a Typical Resting Nerve Cell**

	Concentration (mmol/L)	
	Inside	Outside
Na^+	15	145
K^+	150	5
Cl^-	9	120
Other	156	30

$$E = \frac{-dV}{dL} = \frac{-(-70 \times 10^{-3}\ \text{V})}{7.0 \times 10^{-9}\ \text{m}}$$

$$= 1.0 \times 10^{7}\ \text{V/m (inward)}$$

The inward force due to this field on a positive ion would be

$$F = qE = (1.6 \times 10^{-19}\ \text{C})(1.0 \times 10^{7}\ \text{V/m})$$

$$= 1.6 \times 10^{-12}\ \text{N}$$

This force competes against the concentration gradient for K^+, while it supports the effect of the concentration gradient for Na^+

The Stimulated Nerve Cell

The resting potential of a nerve cell can be disturbed by physical or chemical stimuli. A disturbance may only slightly affect the membrane potential at the point of stimulation. In this case, the membrane potential quickly returns to its resting value of −70 mV. The effect of such a subthreshold stimulus, denoted s_1, is illustrated in Figure 4.

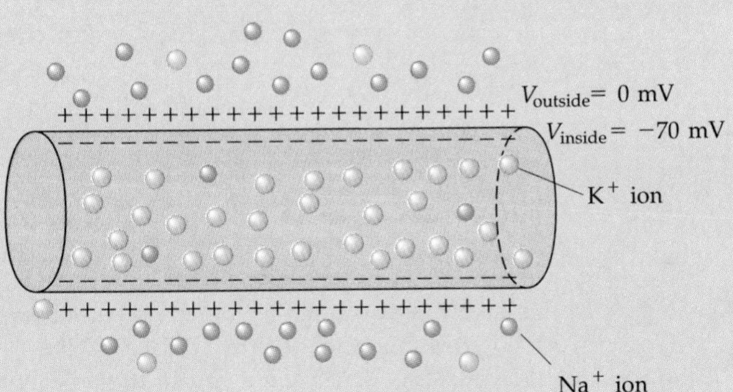

Figure 3 The electric charge distribution along the membrane of a nerve cell in its resting state.

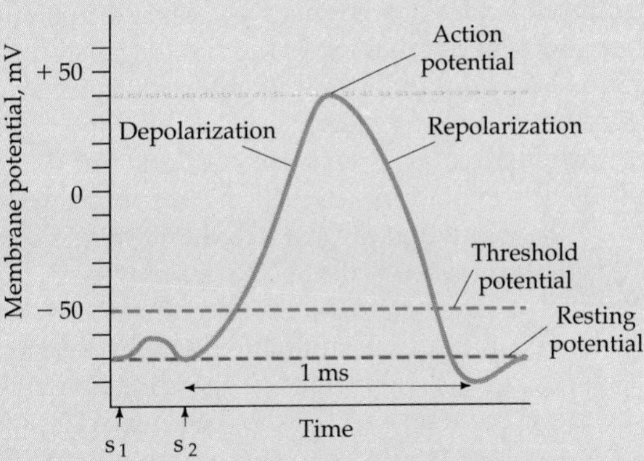

Figure 4 An action potential pulse. s_1 is a subthreshold stimulus; s_2 is a threshold stimulus.

Continued

If, however, the stimulus is strong enough to cause depolarization from the resting potential of −70 mV to around −50 mV, the voltage-dependent Na^+ transmembrane channels open. Favored by both the concentration gradient (see Table 1) and the electric gradient (see the charge distribution in Figure 3), Na^+ ions flow into the cell, creating an electric current ($I = \Delta Q/\Delta t$; see Equation 22-3). The influx of Na^+ causes a local reversal of the electric polarity of the membrane, changing the electric potential to about +40 mV (a swing of 110 mV from the resting potential). The effect of such a threshold stimulus, denoted s_2, is also shown in Figure 4.

When the effect of the Na^+ concentration gradient (outside > inside) becomes balanced by that of the electric gradient (due to the membrane potential now having become positive on the *inside*), depolarization is complete at the site of the original stimulus. The Na^+ channels then close again. The K^+ channels respond to the changes in membrane polarity somewhat after the opening of the Na^+ channels, sending K^+ ions flowing out of the cell while the Na^+ ions are flowing in. The movement of the K^+ ions and the slower action of the Na^+–K^+ pump soon restore the concentration gradients and electric gradient to those of the resting state (see repolarization in Figure 4).

After depolarization, the Na^+ channels remain closed for a brief period (of a few milliseconds) during which that portion of the nerve cell membrane cannot again be stimulated. This is called the *refractory period.* The transient change in the electric potential across the membrane is referred to as the *action potential.*

Nerve Impulse Propagation

The preceding paragraphs described a single short electric pulse at the point of stimulation. How does this impulse travel down the axon?

The local depolarization at the site of the original stimulus causes movement—passive diffusion—of ions into areas adjacent to the site of the stimulus (Figure 5). The membrane potential in an adjacent area soon reaches threshold level of −50 mV, so *its* permeability to Na^+ is, in turn, suddenly increased, and the action potential of +40 mV is propagated in a wavelike manner along the length of the nerve cell (see Figure 5). Because of the refractory period, during which that portion of the membrane recently depolarized cannot be stimulated again, the nerve impulse can be propagated in one direction only, away from the nerve cell body. The nerve impulse continues along the axon to the terminal branches, where it causes the release of neurotransmitter substances from the nerve cell membrane. These cross the gap (synapse) to the next nerve cell, allowing the process to be repeated.

Note that, during the transmission of a nerve impulse, electric currents flow in and out through the cell membrane *perpendicular* to the direction traveled by the nerve impulse. Furthermore, no matter how long the axon, the impulse never needs amplification; it continues down the axon with each pulse obtaining the same height as that induced by the original stimulus.

Some axons are surrounded by a discontinuous multilayered sheath, the myelin sheath. This is formed when Schwann cells wrap around the axon (Figure 6). Gaps about 1 μm wide in the myelin sheath called nodes of Ranvier occur at regular intervals of 1 to 2 mm along the length of the axon. Propagation of a nerve impulse along a myelinated axon differs somewhat from that seen in unmyelinated axons.

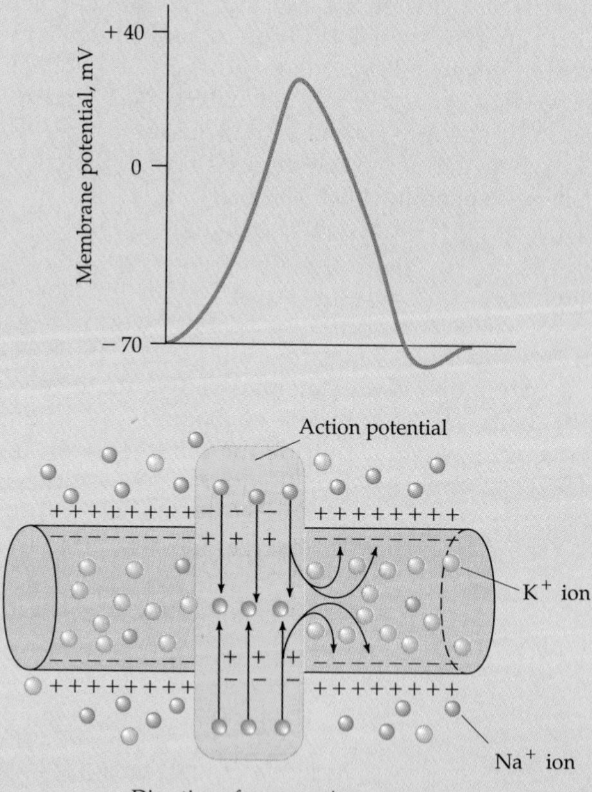

Figure 5 The propagation of an action potential pulse. In advance of the action potential pulse, a small segment of the membrane becomes slightly depolarized owing to the lateral flow of ions inside the membrane. Once this depolarization reaches threshold level, the action potential pulse is generated again in the adjacent segment of the axon. Not shown is the lateral flow of ions to the left, which, because of the refractory period, does not trigger an action potential.

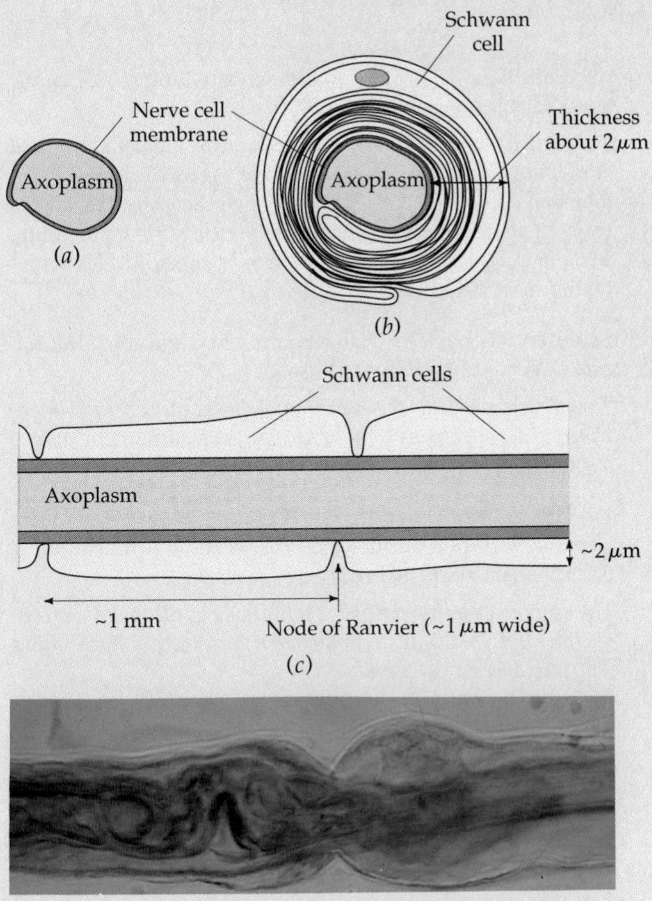

(d)

Figure 6 Transverse sections through *(a)* a "bare" (nonmyelinated) nerve axon and *(b)* a nerve axon that has been surrounded by a sheath (myelinated). *(c)* Longitudinal section through the axon of a myelinated nerve cell. *(d)* A photomicrograph of a myelinated axon shows the thickness of the myelin sheath as well as a node of Ranvier in the center of the photo.

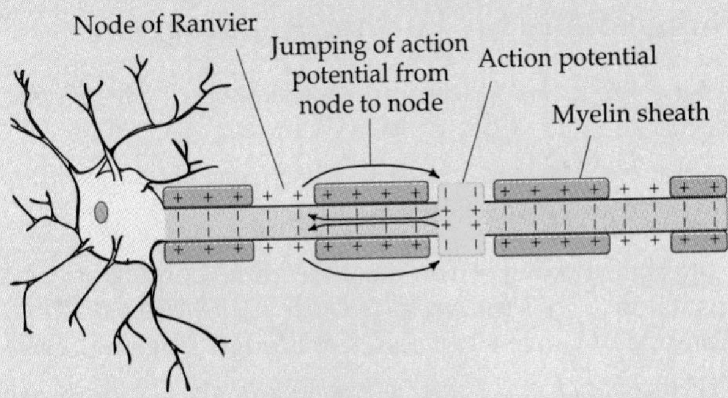

Figure 7 During the propagation of an action potential pulse along a myelinated nerve axon, the impulse jumps from node to node, greatly accelerating nerve impulse conduction. This contrasts with the continuous travel of the action potential along a nonmyelinated axon.

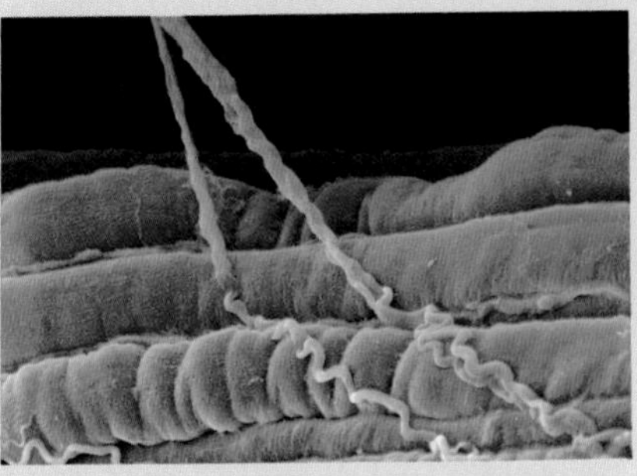

Scanning electron micrograph showing the axons of several nerve cells coming from the upper left of the photo and leading to skeletal muscle fibers lying horizontally in the photo. Electrical impulses traveling down these fibers will result in the release of acetylcholine, a neurotransmitter substance that initiates muscle contraction.

The myelin sheath is a good insulator, so ions cannot flow through it. Electric activity in myelinated nerve cells is confined to the nodes of Ranvier, where there are dense concentrations of voltage-dependent ion channels. Action potentials can be generated only at the nodes of Ranvier and "jump" rapidly from one node to the next along the nerve axon (Figure 7) due to the rapid diffusion of ions through the axoplasm and extracellular fluid. The conduction velocity in a typical myelinated nerve axon is 12 m/s.

Conduction velocity depends on resistivity of the axoplasm and on the membrane capacitance. As resistance is inversely proportional to the cross-sectional area, an axon with a large diameter has a lower resistance and a higher conduction velocity. As was demonstrated for a parallel-plate capacitor (Section 21-1), capacitance is inversely proportional to the separation of the plates. It follows that myelinated axons have a lower capacitance than unmyelinated axons. The lower the membrane capacitance, the smaller the charge, and the less time the membrane will take both to depolarize and to repolarize. This is one explanation for the higher conduction velocity observed in myelinated axons. Measurements of conduction velocities in a wide range of nerve cells have been shown to correlate closely to their calculated resistances and capacitances.

The mechanism of electrical impulse transmission in nerve cells is certainly quite different from electric conduction in metals, but physics also has an important role to play in the understanding of this process.

Suggestions for Further Reading

Allen, Philip B. "Electrical Conductivity," *The Physics Teacher*, vol. 17, 1979, p. 362.

A brief but advanced treatment of the classical and quantum theories of electrical conduction.

Cotterhill, Rodney: "The Busy Electron: Conductors and Insulators," in *The Cambridge Guide to the Material World*, Cambridge University Press, Cambridge, England, 1985.

The distribution of the outer electrons of the atoms making up a material determine whether the material is a conductor, semiconductor, or insulator. This article explains how, with diagrams and without mathematics.

Grundfest, Harry: "Electric Fishes," *Scientific American*, January 1965, p. 82.

Members of many families of fishes are able to produce appreciable voltages outside their bodies using specialized organs containing arrays of "electroplaque membranes" in series or in parallel.

Hamakawa, Yoshihiro: "Photovoltaic Power," *Scientific American*, April 1987, p. 86.

How advances in solar cell design and fabrication are bringing down the cost of photovoltaic electricity.

de Santillana, Giorgio: "Alessandro Volta," *Scientific American*, January 1965, p. 82.

This article describes the argument between Luigi Galvani and Alessandro Volta over whether or not electricity was a living force and a key to the mystery of life or a phenomenon that could manifest itself without requiring the presence of a living being. Volta won the argument in 1800 when he announced his invention of what we would now call a battery.

Schluter, Michael A.: "Unaccountable Conduct," *The Sciences*, May/June 1989, p. 44.

This article presents the background to the discovery of high-temperature superconductivity in 1986, and current attempts to account theoretically for the phenomenon.

Wolsky, Alan M., Robert F. Giese, and Edward J. Daniels: "The New Superconductors: Prospects for Applications," *Scientific American*, February 1989, p. 60.

The authors examine possible applications of the new superconductors and predict that development for practical applications will take at least a decade.

Review

A. Objectives: After studying this chapter you should:

1. Be able to define and discuss the concepts of electric current, drift velocity, resistance, and emf.

2. Be able to state Ohm's law and distinguish between it and the definition of resistance.

3. Be able to define resistivity and describe its temperature dependence.

4. Be able to discuss the simple model of a real battery in terms of a source of emf and an internal resistance, and find the terminal voltage of a battery when it delivers a current I.

5. Be able to give the general relationship between potential difference, current, and power.

6. Be able to determine the equivalent resistances of resistors in series or parallel.

7. Be able to discuss the microscopic model of electrical conduction.

B. Define, explain, or otherwise identify:

Current
Ampere
Drift velocity
Ohm's law
Resistance
Ohmic materials
Nonohmic materials
Resistivity
Conductivity
Temperature coefficient of resistivity
Critical temperature
Superconductivity
Joule heat
Electromotive force
Emf
Ideal battery
Real battery
Terminal voltage
Internal resistance
Resistors in series
Resistors in parallel
Classical model of electric conduction
Collision time
Mean free path

C. True or false: If the statement is true, explain why it is true. If it is false, give a counterexample.

1. Ohm's law is $R = V/I$.

2. Electrons drift in the direction of the current.

3. A source of emf supplies power to an electrical circuit.

4. When the potential drops by V in a segment of a circuit, the power supplied to that segment is IV.

5. The equivalent resistance of two resistors in parallel is always less than the resistance of either resistor alone.

6. The terminal voltage of a battery is the same as its emf.

Problems

Level I

22-1 Current and Motion of Charges

1. A wire carries a steady current of 2.0 A. (*a*) How much charge flows through a cross-sectional area of the wire in 5.0 min? (*b*) How many electrons flow through the cross-sectional area in this time?

2. A 10-gauge copper wire carries a current of 20 A. Assuming one free electron per copper atom, calculate the drift velocity of the electrons.

3. In a fluorescent tube of diameter 3.0 cm, 2.0×10^{18} electrons and 0.5×10^{18} positive ions (with a charge of $+e$) flow through a cross-sectional area each second. What is the current in the tube?

4. In a certain electron beam, there are 5.0×10^6 electrons per cubic centimeter. Suppose the kinetic energy of each electron is 10.0 keV, and the beam is cylindrical, with a diameter of 1.00 mm. (*a*) What is the velocity of an electron? (*b*) Find the beam current.

5. A charge $+q$ moves in a circle of radius r with speed v. (*a*) Express the frequency f with which the charge passes a particular point in terms of r and v. (*b*) Show that the average current is qf and express it in terms of v and r.

6. A ring of radius R with a linear charge density λ rotates about its axis with angular velocity ω. Find an expression for the current.

7. In a certain particle accelerator, a proton beam with a diameter of 2.0 mm constitutes a current of 1.0 mA. The kinetic energy of each proton is 20 MeV. The beam strikes a metal target and is absorbed by it. (*a*) What is the number n of protons per unit volume in the beam? (*b*) How many protons strike the target in 1.0 min? (*c*) If the target is initially uncharged, express the charge of the target as a function of time.

8. A 10-gauge copper wire and a 14-gauge copper wire are welded together end to end. The wires carry a current of 15 A. If there is one free electron per copper atom, find the drift velocity of the electrons in each wire.

22-2 Resistance and Ohm's Law

9. A 10-m–long wire of resistance 0.2 Ω carries a current of 5 A. (*a*) What is the potential difference across the wire? (*b*) What is the magnitude of the electric field in the wire?

10. A potential difference of 100 V produces a current of 3 A in a certain resistor. (*a*) What is its resistance? (*b*) What is the current when the potential difference is 25 V?

11. A copper wire and an iron wire with the same length and diameter carry the same current I. (*a*) Find the potential drop across each wire and the ratio of these drops. (*b*) In which wire is the electric field greater?

12. A block of carbon is 3.0 cm long and has a square cross-sectional area with sides of 0.5 cm. A potential difference of 8.4 V is maintained across its length. (*a*) What is the resistance of the block? (*b*) What is the current in this resistor?

13. A tungsten rod is 50 cm long and has a square cross-sectional area with sides of 1.0 mm. (*a*) What is its resistance at 20°C? (*b*) What is its resistance at 40°C?

14. A carbon rod with a radius of 0.1 mm is used to make a resistor. The resistivity of this material is 3.5×10^{-5} Ω·m. What length of the carbon rod will make a 10-Ω resistor?

15. The third (current-carrying) rail of a subway track is made of steel and has a cross-sectional area of about 55 cm^2. What is the resistance of 10 km of this track?

16. What is the potential difference across one wire of a 30-m extension cord made of 16-gauge copper wire carrying a current of 5.0 A?

17. At what temperature will the resistance of a copper wire be 10 percent greater than it is at 20°C?

18. How long is a 14-gauge copper wire that has a resistance of 2 Ω?

22-3 Energy in Electric Circuits

19. What is the power dissipated in a 10.0-Ω resistor if the potential difference across it is 50 V?

20. Find the power dissipated in a resistor connected across a constant potential difference of 120 V if its resistance is (*a*) 5 Ω and (*b*) 10 Ω.

21. A 10,000-Ω carbon resistor used in electronic circuits is rated at 0.25 W. (*a*) What maximum current can this resistor carry? (*b*) What maximum voltage can be placed across this resistor?

22. A 1-kW heater is designed to operate at 240 V. (*a*) What is its resistance, and what current does it draw? (*b*) What is the power dissipated in this resistor if it operates at 120 V? Assume that its resistance is constant.

23. A 10.0-Ω resistor is rated as being capable of dissipating 5.0 W of power. (*a*) What maximum current can this resistor tolerate? (*b*) What voltage across this resistor will produce this current?

24. If energy costs 9 cents per kilowatt-hour (*a*) how much does it cost to operate an electric toaster for 4 min if the toaster has resistance 11.0 Ω and is connected across 120 V? (*b*) How much does it cost to operate a heater of resistance 5.0 Ω connected across 120 V for 8 h?

25. A battery has an emf of 12.0 V. How much work does it do in 5 s if it delivers a current of 3 A?

26. A battery with 12-V emf has a terminal voltage of 11.4 V when it delivers a current of 20 A to the starter of a car. What is the internal resistance r of the battery?

27. (*a*) How much power is delivered by the emf of the battery in Problem 26 when it delivers a current of 20 A? (*b*) How much of this power is delivered to the starter? (*c*) By how much does the chemical energy of the battery decrease when it delivers a current of 20 A to the starter for 3 min? (*d*) How much heat is developed in the battery when it delivers a current of 20 A for 3 min?

28. A 12-V car battery has an internal resistance of 0.4 Ω. (*a*) What is the current if the battery is shorted momentarily? (*b*) What is the terminal voltage when the battery delivers a current of 20 A to start the car?

29. A physics student runs a 1200-W electric heater constantly in her basement bedroom during the winter time. If electric energy costs 9 cents per kilowatt-hour, how much does this electric heating cost per 30-day month?

22-4 Combinations of Resistors

30. (*a*) Find the equivalent resistance between points *a* and *b* in Figure 22-24. (*b*) If the potential drop between *a* and *b* is 12 V, find the current in each resistor.

Figure 22-24 Problem 30.

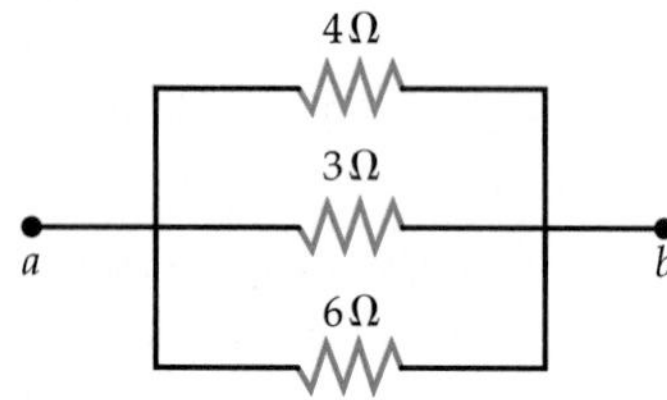

31. Repeat Problem 30 for the resistor network shown in Figure 22-25.

Figure 22-25 Problem 31.

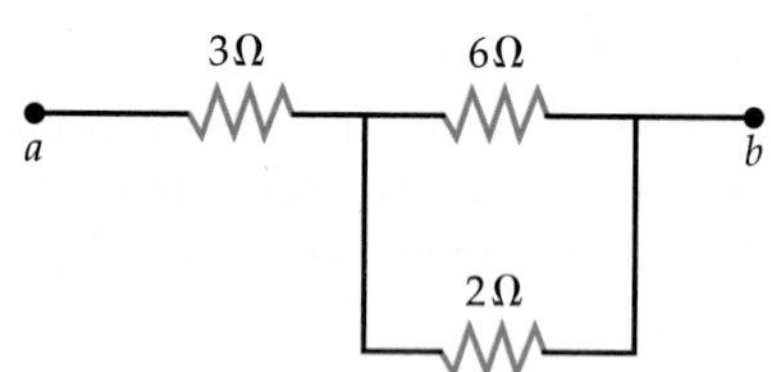

32. Repeat Problem 30 for the resistor network shown in Figure 22-26.

Figure 22-26 Problems 32 and 34.

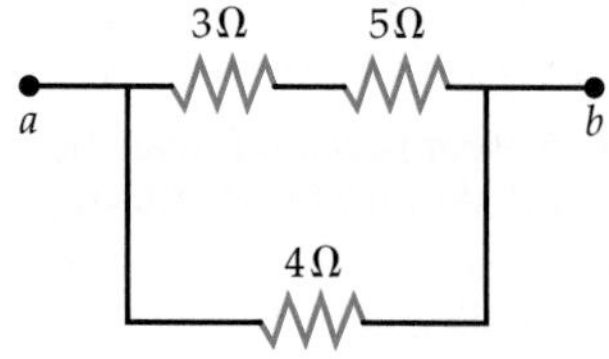

33. Repeat Problem 30 for the resistor network shown in Figure 22-27.

Figure 22-27 Problem 33.

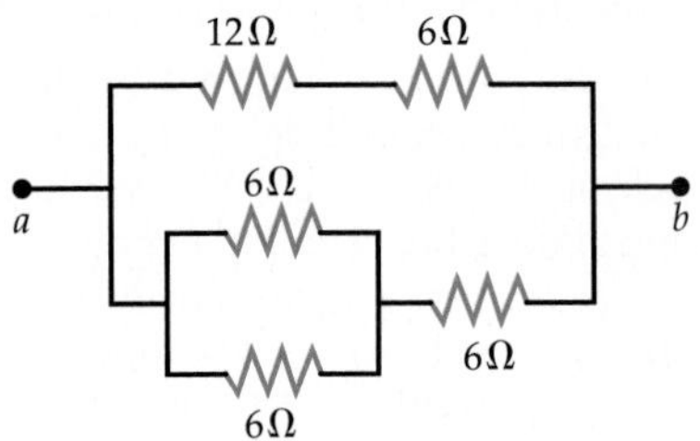

34. In Figure 22-26, the current in the 4-Ω resistor is 4 A. (*a*) What is the potential drop between *a* and *b*? (*b*) What is the current in the 3-Ω resistor?

35. (*a*) Show that the equivalent resistance between points *a* and *b* in Figure 22-28 is *R*. (*b*) What would be the effect of adding a resistance *R* between points *c* and *d*?

Figure 22-28 Problem 35.

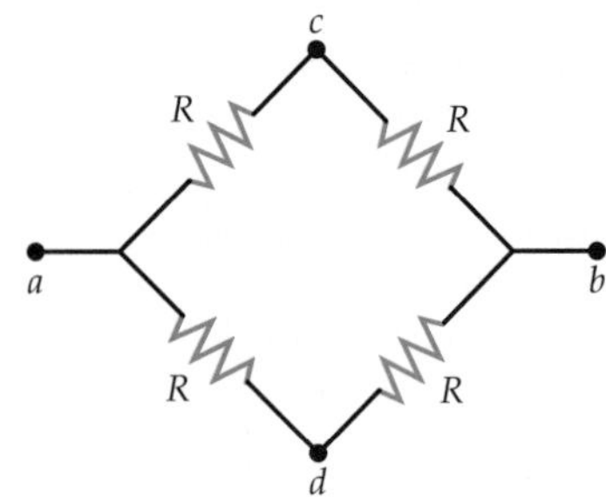

36. Repeat Problem 30 for the resistor network shown in Figure 22-29.

Figure 22-29 Problem 36.

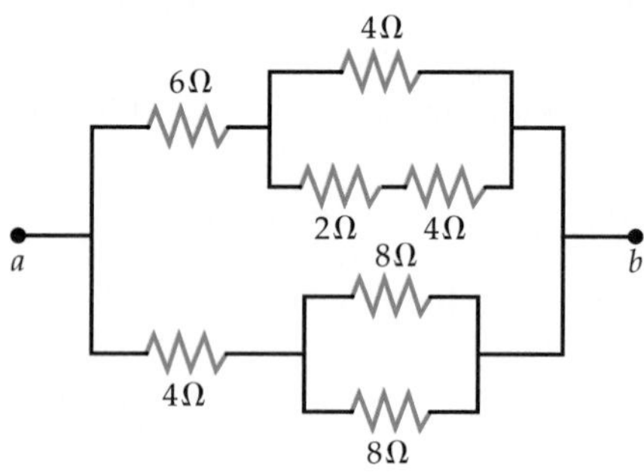

37. The battery in Figure 22-30 has negligible internal resistance. Find (*a*) the current in each resistor and (*b*) the power delivered by the battery.

Figure 22-30 Problem 37.

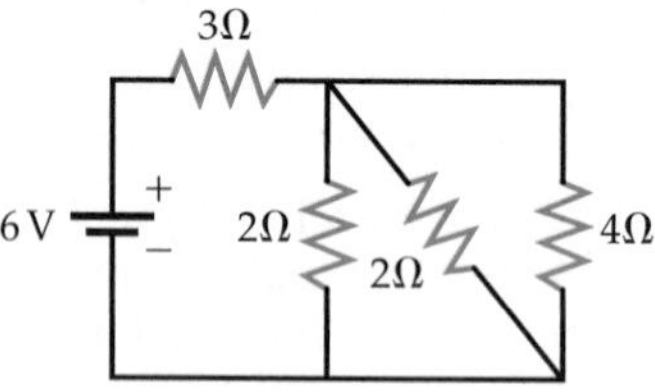

22-5 A Microscopic Picture of Conduction

There are no problems for this section.

Level II

38. A wire of length 1 m has a resistance of 0.3 Ω. It is uniformly stretched to a length of 2 m. What is its new resistance?

39. A cube of copper has sides of 2.0 cm. If it is drawn out to form a 14-gauge wire, what will its resistance be?

40. A 200-W heater is used to heat water in a cup. Assume that 90 percent of the energy produced by the heater goes into heating the water. (*a*) How long does it take to heat 0.25 kg of water from 15 to 100°C? (*b*) How long does it take to boil this water away after it reaches 100°C?

41. Consider the equivalent resistance of two resistors R_1 and R_2 connected in parallel as a function of the ratio $x = R_2/R_1$. (*a*) Show that $R_{eq} = R_1x/(1 + x)$. (*b*) Sketch a plot of R_{eq} as a function of x.

42. A 16-gauge copper wire insulated with rubber can safely carry a maximum current of 6 A. (*a*) How great a potential difference can be applied across 40 m of this wire? (*b*) Find the electric field in the wire when it carries a current of 6 A. (*c*) Find the power dissipated in the wire when it carries a current of 6 A.

43. The belt of a Van de Graaff generator carries a surface charge density of 5 mC/m^2. The belt is 0.5 m wide and moves at 20 m/s. (*a*) What current does it carry? (*b*) If this charge is raised to a potential of 100 kV, what is the minimum power of the motor needed to drive the belt?

44. A coil of nichrome wire is to be used as the heating element in a water boiler that is required to generate 8.0 g of steam per second. The wire has a diameter of 1.80 mm and is connected to a 120-V power supply. Find the length of wire required.

45. An 80.0-m copper wire 1.0 mm in diameter is joined end to end with a 49.0-m iron wire of the same diameter. The current in each is 2.0 A. (*a*) Find the electric field in each wire. (*b*) Find the potential difference across each wire. (*c*) Find the equivalent resistance that would carry a current of 2.0 A at a potential difference equal to the sum of that across the two, and compare it with the sum of the resistances of the two.

46. The current in a wire varies with time according to the relation $I = 20 + 3t^2$, where I is in amperes and t is in seconds. (*a*) How many coulombs are transported by the wire between $t = 0$ s and $t = 10$ s? (*b*) What constant current would transport the same charge in the same time interval?

47. A toaster with a nichrome heating element has a resistance of 80 Ω at 0°C and an initial current of 1.5 A. When the heating element reaches its final temperature, the current is 1.3 A. What is the final temperature of the heating element?

48. A cylinder of glass 1 cm long has a resistivity of 10^{12} Ω·m. How long would a copper wire of the same cross-sectional area need to be to have the same resistance as the glass cylinder?

49. Compact fluorescent light bulbs cost \$20 each and have an expected lifetime of 8000 h. These bulbs consume 20 W of power, but produce the illumination equivalent to 75-W incandescent bulbs. Incandescent bulbs cost about 50¢ each and have an expected lifetime of 1200 h. (*a*) If the average household has, on the average, six 75-W incandescent light bulbs on constantly, and if energy costs 9 cents per kilowatt-hour, how much money would a consumer save each year by installing the energy-efficient fluorescent light bulbs? (*b*) At what cost per kilowatt-hour would the total cost of using either type of bulb be the same? (See *Scientific American*, April 1988, p. 56.)

50. The wires in a house must be large enough in diameter so that they do not get hot enough to start a fire. Suppose a certain wire is to carry a current of 20 A, and it is determined that the joule heating of the wire should not exceed 2 W/m. What diameter must a copper wire have to be "safe" for this current?

51. A 12-V automobile battery can deliver a total charge of 160 A·h. (*a*) What is the total stored energy in the battery? (*b*) How long could this battery provide 150 W to a pair of headlights?

52. A lightweight electric car is powered by ten 12-V batteries. At a speed of 80 km/h the average frictional force is 1200 N. (*a*) What must be the power of the electric motor if the car is to travel at a speed of 80 km/h? (*b*) If each battery can deliver a total charge of 160 A·h before recharging, what is the total charge in coulombs that can be delivered by the 10 batteries before charging? (*c*) What is the total electrical energy delivered by the 10 batteries before recharging? (*d*) How far can the car travel at 80 km/h before the batteries must be recharged? (*e*) What is the cost per kilometer if the cost of recharging the batteries is 9 cents per kilowatt-hour?

53. Suppose the bulb in a two-cell flashlight draws 4 W of power. The batteries go dead in 45 min and cost \$7.99. (*a*) How many kilowatt-hours of energy can be supplied by the two batteries? (*b*) What is the cost per kilowatt-hour of energy if the batteries cannot be recharged? (*c*) If the batteries can be recharged at a cost of 9 cents per kilowatt-hour, what is the cost of recharging them?

54. In a proton supercollider, the protons in a 5-mA beam move with nearly the speed of light. (*a*) How many protons are there per meter of the beam? (*b*) If the cross-sectional area of the beam is 10^{-6} m^2, what is the average separation of the protons?

55. A 10-Ω resistor is wired into an electronic circuit by copper wire of length 50 cm and diameter 0.6 mm. (*a*) What additional resistance does the wire introduce? (*b*) What percentage error is produced by neglecting the resistance of the wiring? (*c*) If the resistor is made from nichrome wire, what change in its temperature would produce a change in its resistance equal to the resistance of the wiring?

56. The filament of a certain lamp has a resistance that increases linearly with temperature. When a constant voltage is switched on, the initial current decreases until the filament reaches its steady-state temperature. The temperature coefficient of resistivity of the filament is 4×10^{-3} K^{-1}. The final current through the filament is one-eighth the initial current. What is the change in temperature of the filament?

57. A rubber tube 1 m long with an inside diameter of 4 mm is filled with a salt solution that has a resistivity of 10^{-3} Ω·m. Metal plugs form electrodes at the ends of the tube. (*a*) What is the resistance of the filled tube? (*b*) What is the resistance of the filled tube if it is uniformly stretched to a length of 2 m?

58. A space heater in an old home draws a 15-A current. A pair of 12-gauge copper wires carries the current from the fuse box to the wall outlet, a distance of 30 m. The voltage at the fuse box is exactly 120 V. (*a*) What is the voltage delivered to the space heater? (*b*) If the fuse will blow at a current of 20 A, how many 60-W bulbs can be supplied by this line when the space heater is on? (Assume that the wires from the wall to the space heater and to the light fixtures have negligible resistance.)

59. An electric space heater has a nichrome-wire heating element with a resistance of 8 Ω at 0°C. When 120 V are applied, the electric current heats the nichrome wire to 1000°C. (*a*) What is the initial current drawn by the cold heating element? (*b*) What is the resistance of the heating element at 1000°C? (*c*) What is the operating wattage of this heater?

60. Currents up to 30 A can be carried by 10-gauge copper wire. (*a*) What is the resistance of 100 m of 10-gauge copper wire? (*b*) What is the electric field in the wire when the current is 30 A? (*c*) How long does it take for an electron to travel 100 m in the wire when the current is 30 A?

61. An automobile jumper cable 3 m long is made of three strands of 12-gauge copper wire twisted together. (*a*) What is the resistance of the jumper cable? (*b*) When the cable is used to start a car, it carries a current of 90 A. What is the potential drop that occurs across the jumper cable? (*c*) How much joule heating occurs in the jumper cable?

Level III

62. A linear accelerator produces a pulsed beam of electrons. The current is 1.6 A for the 0.1-μs duration of each pulse. (*a*) How many electrons are accelerated in each pulse? (*b*) What is the average current of the beam if there are 1000 pulses per second? (*c*) If each electron acquires an energy of 400 MeV, what is the average power output of the accelerator? (*d*) What is the peak power output? (*e*) What fraction of the time is the accelerator actually accelerating electrons? (This is called the *duty factor* of the accelerator.)

63. A wire of cross-sectional area A, length L_1, resistivity ρ_1, and temperature coefficient α_1 is connected end to end to a second wire of the same cross-sectional area, length L_2, resistivity ρ_2, and temperature coefficient α_2, so that the wires carry the same current. (*a*) Show that if $\rho_1 L_1 \alpha_1 + \rho_2 L_2 \alpha_2 = 0$, the total resistance R is independent of temperature for small temperature changes. (*b*) If one wire is made of carbon and the other is copper, find the ratio of their lengths for which R is approximately independent of temperature.

64. A 100-W heater is designed to operate across 120 V. (*a*) What is its resistance, and what current does it draw? (*b*) Show that if the potential difference across the heater changes by a small amount ΔV, the power changes by a small amount ΔP, where $\Delta P/P \approx 2\,\Delta V/V$. *Hint:* Approximate the changes with differentials. (*c*) Find the approximate power dissipated in the heater if the potential difference is decreased to 115 V.

65. Find the resistance between the ends of the half ring shown in Figure 22-31. The resistivity of the material of the ring is ρ.

Figure 22-31 Problem 65.

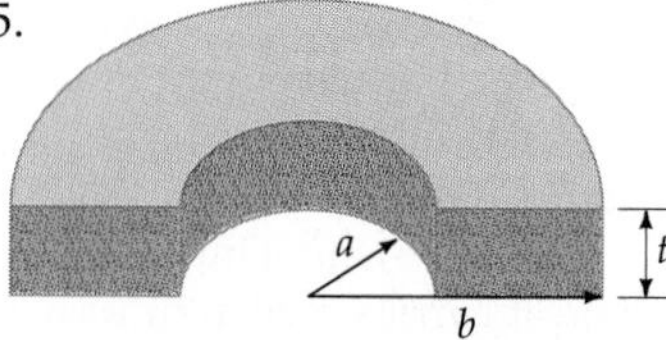

66. The space between two spherical-shell conductors is filled with a material that has a resistivity of 10^9 Ω·m. If the inner shell has a radius of 1.5 cm and the outer shell has a radius of 5 cm, what is the resistance between the conductors? *Hint:* Find the resistance of a spherical-shell element of the material of area $4\pi r^2$ and length dr, and integrate to find the total resistance of the set of shells in series.

67. The space between two metallic coaxial cylinders of length L and radii a and b is completely filled with a material having a resistivity ρ. (*a*) What is the resistance between the two cylinders? (See the hint in Problem 66.) (*b*) Find the current between the two cylinders if $\rho = 30$ Ω·m, $a = 1.5$ cm, $b = 2.5$ cm, $L = 50$ cm, and a potential difference of 10 V is maintained between the two cylinders.

68. A semiconducting diode is a nonlinear device whose current I is related to the voltage V across the diode by

$$I = I_0(e^{eV/kT} - 1)$$

where k is Boltzmann's constant, e is the charge on an electron, and T is the absolute temperature. (*a*) What is the resistance of the diode for $V = 0.5$ V if $I_0 = 10^{-9}$ A? (*b*) What is the resistance for $V = 0.6$ V?

69. The radius of a wire of length L increases linearly along its length according to

$$r = a + \frac{b-a}{L}x$$

where x is the distance from the small end of radius a. What is the resistance of this wire in terms of its resistivity ρ, length L, radius a, and radius b?

Chapter 23

Direct-Current Circuits

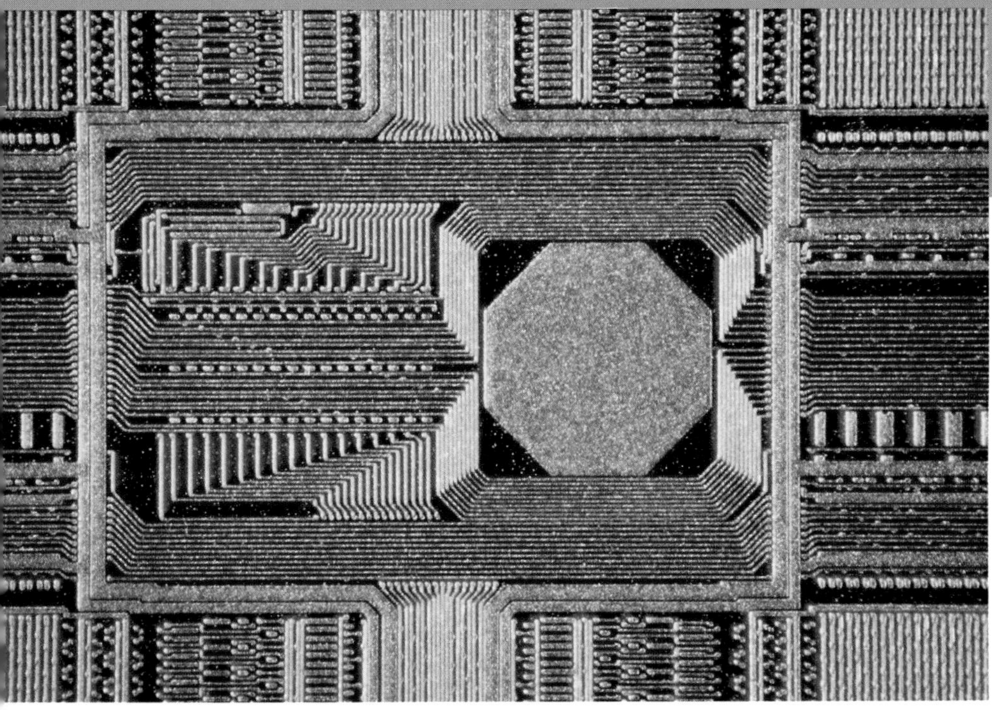

A photomicrograph (about 1000X magnification) of the center portion of a 256K dynamic random-access memory (DRAM) chip that can store 256,000 bits of data. The chip is smaller than a postage stamp and contains over 600,000 electronic components—resistors, capacitors, diodes, and transistors. These components are constructed on a single slice of silicon crystal by layering precisely defined regions of the crystal with impurities. The layering process that constructs the components also creates the wires (100 times finer than a human hair) connecting them. This kind of circuit, called an integrated circuit (IC), functions the same as circuits assembled from discrete components. But ICs operate faster and use less power because, being smaller, currents in them travel shorter distances.

In this chapter, we will analyze some simple circuits consisting of batteries, resistors, and capacitors in various combinations; that is, we will find the values of V and I and quantities derived from them at various points in the circuits. These circuits are called direct-current (dc) circuits because the current in any part of the circuit is always in the same direction. In Chapter 28, we will discuss alternating-current (ac) circuits, which are circuits in which the current in any part alternates in direction.

When a switch is thrown to turn on a circuit, an electric field propagates along the elements of the circuit, causing many complicated changes to occur in the circuit as the current builds up and charge accumulates at various points. However, since the electric field propagates at nearly the speed of light, these changes occur rapidly and equilibrium is quickly established. The time for equilibrium to be established depends on the conductivity of the elements in the circuit, but it is practically instantaneous for most purposes. In equilibrium, charge no longer accumulates at points along the circuit and the current is steady. For circuits containing capacitors, the current may increase or decrease slowly, but appreciable changes occur only over times much longer than the time needed to reach equilibrium. Thus, in circuits of this type each successive value of the current may be considered a quasi-equilibrium state.

23-1 Kirchhoff's Rules

The methods discussed in Chapter 22 for replacing combinations of resistors in series or parallel with their equivalent resistances are very useful for simplifying many combinations of resistors. However, they are not sufficient for the analysis of many simple circuits. Figure 23-1 gives an example of such a circuit. The two resistors R_1 and R_2 in this circuit look as if they might be in parallel, but they are not. The potential drop is not the same across both resistors because of the presence of the emf $\mathcal{E}_2$ in series with R_2. Also, R_1 and R_2 do not carry the same current, so they are not in series.

There are two rules, called **Kirchhoff's rules,** that apply to any circuit in the steady state:

Kirchhoff's rules

1. When any closed circuit loop is traversed, the algebraic sum of the changes in potential must equal zero.
2. At any junction point in a circuit where the current can divide, the sum of the currents into the junction must equal the sum of the currents out of the junction.

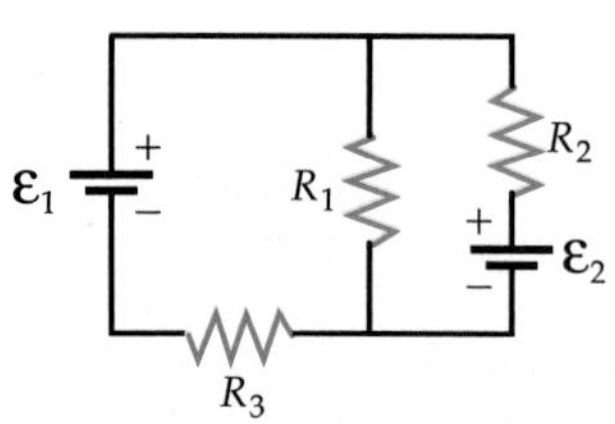

Figure 23-1 An example of a simple circuit that cannot be analyzed by replacing combinations of resistors in series or parallel with their equivalent resistances. The potential drops across R_1 and R_2 are not equal because of the emf $\mathcal{E}_2$, so these resistors are not in parallel. Note also that these resistors are not connected together at both ends. The resistors do not carry the same current, so they are not in series.

Kirchhoff's first rule, called the **loop rule,** follows from the fact that the potential difference between any two points in a circuit in the steady state is constant. In the steady state, the electric field at any point (outside a source of emf) is due to charge that has accumulated on the surfaces of battery terminals, resistors, wires, or other elements of the circuit. Since the electric field is conservative, a potential function exists at any point in space (except inside a source of emf). As we move around a loop of the circuit, the potential may decrease or increase as we pass through a resistor or battery, but when we have completely traversed the loop and have arrived back at our starting point, *the net change in the potential must be zero.* This rule is a direct result of the conservation of energy. If we have a charge q at some point where the potential is V, the potential energy of the charge is qV. As the charge traverses a loop in a circuit, it loses or gains energy as it passes through resistors, batteries, or other devices, but when it arrives back at its starting point, its energy must again be qV.

Exercise

Draw a rectangle representing some arbitrary circuit, and label any six points (such as the four corners and two other points) with the letters a, b, c, d, e, and f. Assign any number you like to each point on your diagram. Begin at point a and add the differences between each pair of numbers as you traverse the circuit, keeping track of the signs of the differences. What is the algebraic sum of these differences for one complete traversal of the circuit? (Answer: 0)

Kirchhoff's second rule, called the **junction rule,** follows from the conservation of charge. It is needed for multiloop circuits containing points where the current can divide. In the steady state, there is no further accumulation of electric charge at any point in a circuit so the amount of charge entering any point must equal the amount of charge leaving that point. Figure 23-2 shows the junction of three wires carrying currents I_1, I_2, and I_3. During a time interval Δt, charge $I_1\,\Delta t$ flows into the junction from the left. During the same time interval, charges $I_2\,\Delta t$ and $I_3\,\Delta t$ flow out of the junction to the right. Since charge does not originate at this point nor continues to accumulate there in the steady state, the conservation of charge implies the junction rule, which for this case gives

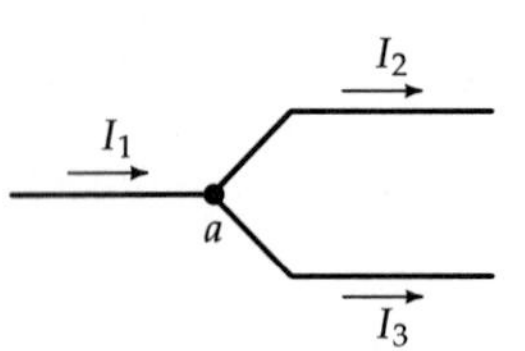

Figure 23-2 Illustration of Kirchhoff's junction rule. The current I_1 into point a equals the sum $I_2 + I_3$ of the currents out of point a.

$$I_1 = I_2 + I_3 \tag{23-1}$$

Figure 23-3 shows a circuit containing two batteries with internal resistances r_1 and r_2 and three external resistors. We wish to find the current in terms of the emfs and resistances, which are assumed to be known. We cannot predict the direction of the current unless we know which battery has the greater emf, but we do not have to know the direction of the current to analyze the circuit. We can assume either direction and solve the problem based on that assumption. If the assumption is incorrect, we will get a negative value for the current, indicating that its actual direction is opposite the direction assumed. Let us assume that I is clockwise, as indicated in the figure, and apply Kirchhoff's loop rule as we traverse the circuit in the assumed direction of the current, beginning at point a. The high- and low-potential sides of the resistors for this choice of current direction are indicated by the plus and minus signs in the figure. The potential decreases and increases are given in Table 23-1. Note that we encounter a potential drop as we traverse the source of emf between points c and d and a potential increase as we traverse the source of emf between f and g. Beginning at point a, we obtain from Kirchhoff's loop rule

$$-IR_1 - IR_2 - \mathcal{E}_2 - Ir_2 - IR_3 + \mathcal{E}_1 - Ir_1 = 0 \qquad 23\text{-}2$$

Solving for the current I, we obtain

$$I = \frac{\mathcal{E}_1 - \mathcal{E}_2}{R_1 + R_2 + R_3 + r_1 + r_2} \qquad 23\text{-}3$$

Note that if $\mathcal{E}_2$ is greater than $\mathcal{E}_1$, we will get a negative value for the current I, indicating that we have assumed the wrong direction for I. That is, if $\mathcal{E}_2$ is greater than $\mathcal{E}_1$, the current will be in the counterclockwise direction. On the other hand, if $\mathcal{E}_1$ is the greater emf, we will get a positive value for I, which means that the direction we have assumed is correct. Let us assume for this example that $\mathcal{E}_1$ is the greater emf. In battery 2, the charge flows from high potential to low potential. Therefore, a charge ΔQ moving through battery 2 from point c to point d loses energy $\mathcal{E}_2\,\Delta Q$. This electrical energy is converted into chemical energy and stored in the battery, which means that battery 2 is *charging*.

We can account for the energy balance in this circuit by rearranging Equation 23-2 and multiplying each term by the current I:

$$\mathcal{E}_1 I = \mathcal{E}_2 I + I^2R_1 + I^2R_2 + I^2R_3 + I^2r_2 + I^2r_1 \qquad 23\text{-}4$$

The term $\mathcal{E}_1 I$ is the rate at which battery 1 puts energy into the circuit. This energy comes from the internal chemical energy of the battery. The term $\mathcal{E}_2 I$ is the rate at which electric energy is converted into chemical energy in battery 2. The term I^2R_1 is the rate at which Joule heat is produced in resistor R_1. Similarly, the terms for the other resistances give the rate of Joule heating in them.

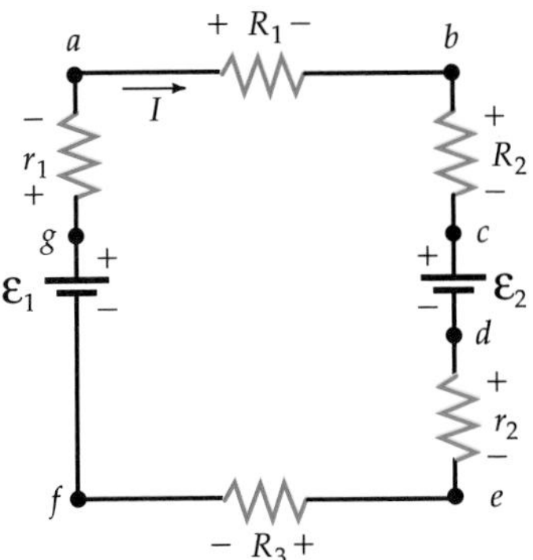

Figure 23-3 Circuit containing two batteries and three external resistors. The plus and minus signs on the resistors are there to help us remember which side of each resistor is at the higher potential for the current direction we have assumed.

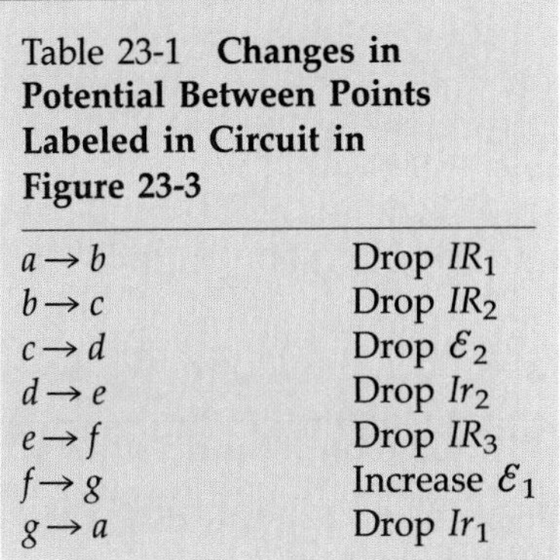

Table 23-1 **Changes in Potential Between Points Labeled in Circuit in Figure 23-3**

$a \to b$	Drop IR_1
$b \to c$	Drop IR_2
$c \to d$	Drop $\mathcal{E}_2$
$d \to e$	Drop Ir_2
$e \to f$	Drop IR_3
$f \to g$	Increase $\mathcal{E}_1$
$g \to a$	Drop Ir_1

Example 23-1

The elements in the circuit in Figure 23-3 have the values $\mathcal{E}_1 = 12$ V, $\mathcal{E}_2 = 4$ V, $r_1 = r_2 = 1\ \Omega$, $R_1 = R_2 = 5\ \Omega$, and $R_3 = 4\ \Omega$ as shown in Figure 23-4. Find the potentials at points a through g in the figure, assuming that the potential at point f is zero, and discuss the energy balance in the circuit.

The analysis of a circuit is usually simplified if we choose one point to be at zero potential and then find the potentials of the other points relative to it. Since only potential differences are important, any point in a circuit can be chosen to have zero potential. In this example, we will choose point f to be at zero potential. This is indicated by the ground symbol ⏚ at point f. As we saw in Section 18-2, the earth can be consid-

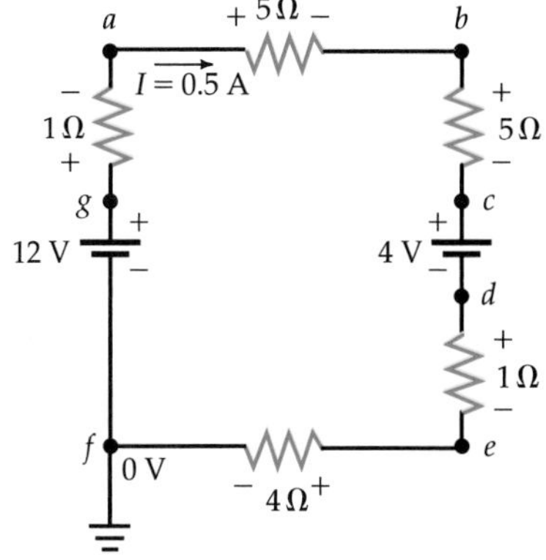

Figure 23-4 The circuit in Figure 23-3 with sample values for Example 23-1. The potential is chosen to be zero at point f. The three horizontal lines near point f indicate that it is grounded.

ered to be a very large conductor with a nearly unlimited supply of charge, which means that the potential of the earth remains essentially constant. Therefore, it is often chosen to be at zero potential. In practice, electrical circuits are often grounded by connecting one point to the earth. The outside metal case of a washing machine, for example, is usually grounded by connecting it by a wire to a water pipe that is in contact with the earth.

We first find the current in the circuit. From Equation 23-3, we have

$$I = \frac{12\text{ V} - 4\text{ V}}{5\ \Omega + 5\ \Omega + 4\ \Omega + 1\ \Omega + 1\ \Omega} = \frac{8\text{ V}}{16\ \Omega} = 0.5\text{ A}$$

We can now find the potentials at points a through g relative to zero potential at point f. Since, by definition, battery 1 maintains a constant potential difference $\mathcal{E}_1 = 12$ V between point g and point f, the potential at point g is 12 V. The potential at point a is less than that at g by the potential drop $Ir_1 = (0.5\text{ A})(1\ \Omega) = 0.5$ V. Thus, the potential at point a is 12 V − 0.5 V = 11.5 V. Similarly, the potential drops across the 5-Ω resistors R_1 and R_2 are each $IR_1 = (0.5\text{ A})(5\ \Omega) = 2.5$ V. The potential at point b is therefore 11.5 V − 2.5 V = 9 V, and that at c is 6.5 V. The potential drop across $\mathcal{E}_2$ is 4 V, so point d is at a potential of 2.5 V. Since the drop across the 1-Ω resistance r_2 is 0.5 V, the potential at e is 2 V. The potential drop across the 4-Ω resistance R_3 is $IR_3 = 2$ V. This gives zero for the potential at f, which is consistent with our original assumption. Figure 23-5 shows the potentials at all the labeled points, beginning and ending with point f.

The power delivered by the emf $\mathcal{E}_1$ is

$$P_{\mathcal{E}_1} = \mathcal{E}_1 I = (12\text{ V})(0.5\text{ A}) = 6.0\text{ W}$$

The power dissipated in the internal resistance of battery 1 is

$$P_{r_1} = I^2 r_1 = (0.5\text{ A})^2(1\ \Omega) = 0.25\text{ W}$$

Thus, the power delivered by battery 1 to the external circuit is 6.0 W − 0.25 W = 5.75 W; this also equals $V_1 I$, where $V_1 = V_a - V_f = 11.5$ V is the terminal voltage of battery 1. The total power dissipated in the external resistances in the circuit is

$$P_R = (0.5\text{ A})^2(5\ \Omega + 5\ \Omega + 4\ \Omega) = 3.5\text{ W}$$

The power going into battery 2, which is being charged, is $(V_c - V_e)I = (6.5\text{ V} - 2\text{ V})(0.5\text{ A}) = 2.25$ W. Part of this power, $P_{r_2} = I^2 r_2 = 0.25$ W, is dissipated in the internal resistance r_2 and part, $P_{\mathcal{E}_2} = \mathcal{E}_2 I = 2$ W, is the rate at which energy is being stored in that battery.

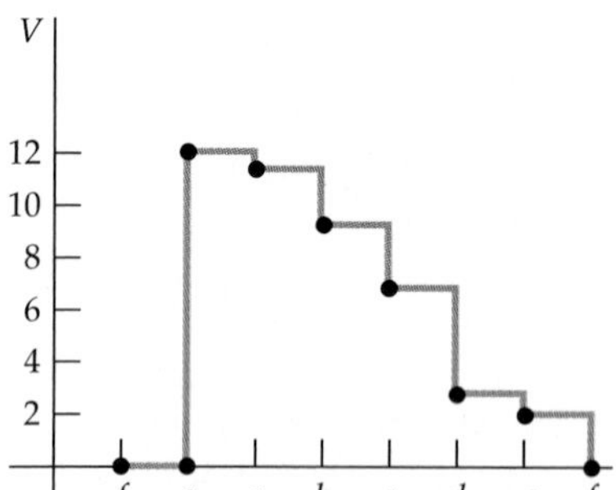

Figure 23-5 The potential at the labeled points of the circuit in Figure 23-4. The potential is zero at f and jumps to 12 V at g. It decreases by IR as we pass through each resistor in the direction of the current. When we get back to f, the potential is again zero.

Note that the terminal voltage of the battery that is being charged in Example 23-1 is $V_c - V_e = 4.5$ V, which is greater than the emf of the battery. Because of its internal resistance, a battery is not completely reversible. If the same 4-V battery were to deliver 0.5 A to an external circuit, its terminal voltage would be 3.5 V (again assuming that its internal resistance is 1 Ω). If the internal resistance is very small, the terminal voltage of a battery is nearly equal to its emf, whether the battery is delivering current to an external circuit or is being charged. Some real batteries, such as those used in automobiles, are nearly reversible and can easily be recharged. Other types of batteries are not reversible. If you attempt to recharge one of these by driving current from its positive to its negative terminal, most, if not all, of the energy will go into heat rather than into the chemical energy of the battery, and the battery may explode.

Example 23-2

A good car battery is to be connected by jumper cables to a weak car battery to charge it. (*a*) To which terminal of the weak battery should the positive terminal of the good battery be connected? (*b*) Assume that the good battery has an emf $\mathcal{E}_1 = 12$ V and the weak battery has an emf $\mathcal{E}_2 = 11$ V, that the internal resistances of the batteries are $r_1 = r_2 = 0.02\ \Omega$, and that the resistance of the jumper cables is $R = 0.01\ \Omega$, all of which are typical values. What will the charging current be? (*c*) What will the current be if the batteries are connected incorrectly?

(*a*) Since the weak battery is to be charged, we connect the positive terminal of the good battery to the positive terminal of the weak battery (and the negative terminal of the good battery to the negative terminal of the weak battery) so as to drive charge through the weak battery from the positive terminal to the negative terminal. Figure 23-6 shows the circuit diagram for this process.

(*b*) The charging current is given by

$$I = \frac{\mathcal{E}_1 - \mathcal{E}_2}{R + r_1 + r_2} = \frac{12\text{ V} - 11\text{ V}}{0.05\ \Omega} = 20\text{ A}$$

(*c*) Figure 23-7 shows the circuit diagram for the case in which the batteries are connected incorrectly, positive terminals to negative terminals. In this case, the current would be

$$I = \frac{\mathcal{E}_1 + \mathcal{E}_2}{R + r_1 + r_2} = \frac{12\text{ V} + 11\text{ V}}{0.05\ \Omega} = 460\text{ A}$$

If the batteries are connected in this way, both may explode in a shower of boiling battery acid.

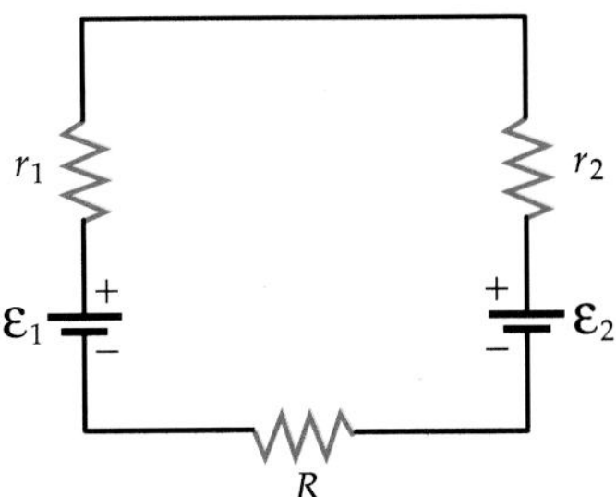

Figure 23-6 Two batteries that are connected such that one is charging the other. Since the sum of the internal resistances of the batteries and the resistance R of the connecting cables is very small, relatively large currents are obtained even when the emfs are nearly equal.

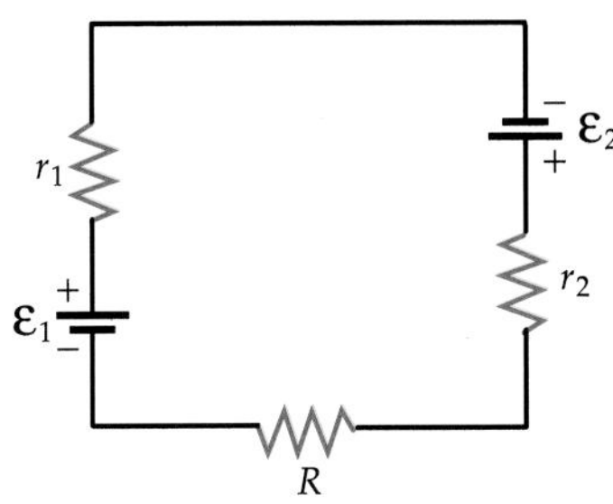

Figure 23-7 Two batteries that are *not* connected correctly for charging. Since the total resistance of the circuit is of the order of hundredths of an ohm, the current is very large, and the batteries could explode.

Multiloop Circuits

We will now consider circuits containing more than one loop. To analyze such circuits, we need to apply Kirchhoff's junction rule at points where the current splits into two or more parts.

Example 23-3

(*a*) Find the current in each part of the circuit shown in Figure 23-8. (*b*) Find the energy dissipated in the 4-Ω resistor in 3 s.

(*a*) This circuit is the same as that in Figure 23-1 with $\mathcal{E}_1 = 12$ V, $\mathcal{E}_2 = 5$ V, $R_1 = 4\ \Omega$, $R_2 = 2\ \Omega$, and $R_3 = 3\ \Omega$. Let I be the current through the 12-V battery in the direction shown in the figure. At point b, this current divides into currents I_1 and I_2. The current directions shown for I_1 and I_2 are merely guesses because we cannot be sure of their directions until after we have analyzed the circuit. For example, we need to know whether point b or point e is at the higher potential in order to know the direction of the current through the 4-Ω resistor. Applying the junction rule to point b, we obtain

$$I = I_1 + I_2$$

Applying the junction rule at point e gives us the same information since the currents I_1 and I_2 join there to form the current I directed towards point f. There are three loops to which the loop rule can be applied, the interior loops *abef* and *bcde* and the outer loop *abcdef*. We need only two more equations to determine the three unknown currents, so equations

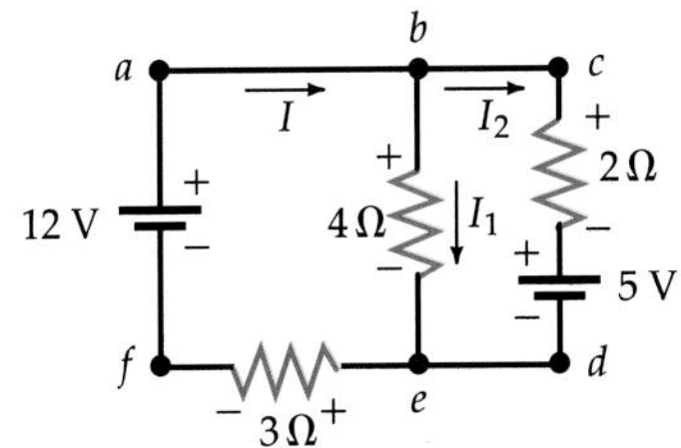

Figure 23-8 Circuit for Example 23-3. The direction of the current I_1 from b to e is not known before the circuit is analyzed. The plus and minus signs on the 4-Ω resistor are for the assumed direction of I_1 from b to e.

for any two of the loops will be sufficient. (The equation for the third loop will give redundant information.) Replacing I with $I_1 + I_2$ and applying the loop rule to the outer loop (*abcdef*), we get

$$12\ \text{V} - (2\ \Omega)I_2 - 5\ \text{V} - (3\ \Omega)(I_1 + I_2) = 0$$

Simplifying this equation and dividing it by 1 Ω, recalling that 1 V/1 Ω = 1 A, we obtain

$$7\ \text{A} - 3I_1 - 5I_2 = 0 \qquad \text{23-5}$$

Similarly, applying the loop rule to the loop on the left (*abef*) gives

$$12\ \text{V} - (4\ \Omega)I_1 - (3\ \Omega)(I_1 + I_2) = 0$$

or

$$12\ \text{A} - 7I_1 - 3I_2 = 0 \qquad \text{23-6}$$

Equations 23-5 and 23-6 can now be solved for the currents I_1 and I_2. To eliminate I_2, we multiply each term in Equation 23-5 by 3 and each term in Equation 23-6 by 5 and obtain

$$21\ \text{A} - 9I_1 - 15I_2 = 0 \qquad \text{23-5}a$$

and

$$60\ \text{A} - 35I_1 - 15I_2 = 0 \qquad \text{23-6}a$$

Then, subtracting Equation 23-5*a* from Equation 23-6*a*, we obtain

$$39\ \text{A} - 26I_1 = 0$$

or

$$I_1 = \frac{39\ \text{A}}{26} = 1.5\ \text{A}$$

Substituting this value for I_1 into Equation 23-5, we obtain

$$7\ \text{A} - 3(1.5\ \text{A}) - 5I_2 = 0$$

$$I_2 = \frac{2.5\ \text{A}}{5} = 0.5\ \text{A}$$

The total current through the 12-V battery is therefore

$$I = I_1 + I_2 = 1.5\ \text{A} + 0.5\ \text{A} = 2.0\ \text{A}$$

(*b*) Since the current in the 4-Ω resistor is 1.5 A, the power dissipated in this resistor is

$$P = I_1^2R = (1.5\ \text{A})^2(4\ \Omega) = 9\ \text{W}$$

The total energy dissipated in the 4-Ω resistor in 3 s is then

$$W = Pt = (9\ \text{W})(3\ \text{s}) = 27\ \text{J}$$

Example 23-3 illustrates some general features of the analysis of multi-loop circuits. Consider a general circuit containing 1 outer loop and n interior loops. There will be $n - 1$ junction points where the current divides and another $n - 1$ junction points where the currents join. We can solve for the currents by applying the loop rule to any n loops and the junction rule to the $n - 1$ junction points where the current divides. Further application of the junction rule or loop rule will merely lead to redundant information. In Example 23-3, we had two interior loops, so we applied the loop rule to two loops and the junction rule to the one junction point (point b) where the current divided.

Example 23-4

Find the current in each part of the circuit shown in Figure 23-9.

We first note that the 3-Ω and 6-Ω resistors in the loop at the right are in parallel. Our work will be easier if we replace these resistors with the equivalent resistance R_{eq} found from

$$\frac{1}{R_{eq}} = \frac{1}{3\ \Omega} + \frac{1}{6\ \Omega} = \frac{3}{6\ \Omega} = \frac{1}{2\ \Omega}$$

or

$$R_{eq} = 2\ \Omega$$

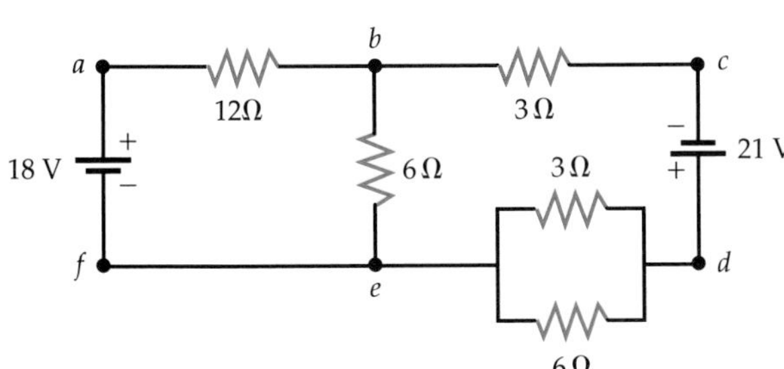

Figure 23-9 Circuit for Example 23-4.

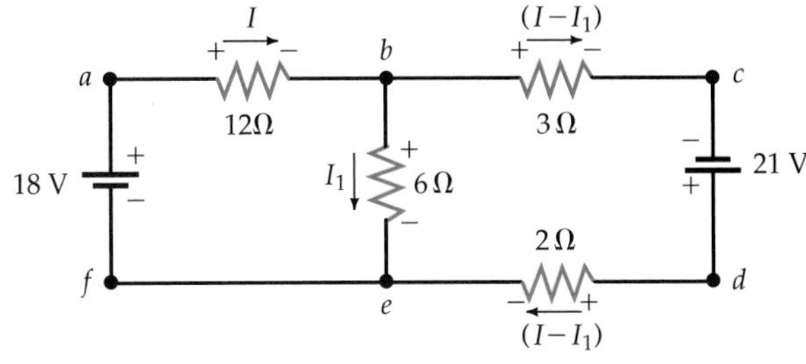

Figure 23-10 Circuit of Figure 23-9 with $R_{eq} = 2\ \Omega$ replacing the parallel combination. The plus and minus signs on the 6-Ω resistor are for the assumed direction of I_1. The junction rule is applied immediately at point b by using $(I - I_1)$ for the current from b to c.

In Figure 23-10, we have made this replacement, and we have chosen directions for the currents. We have called the current through the 18-V battery I and that through the 6-Ω resistor I_1, which we have assumed to be downward. Note that we have applied the junction rule to point b immediately by using $I - I_1$ for the current from point b to point c. Applying Kirchhoff's loop rule in the clockwise direction to the loop *abef* gives

$$18\ \text{V} - (12\ \Omega)I - (6\ \Omega)I_1 = 0$$

Simplifying this equation by dividing each term by 6 Ω and transposing the term (18 V)/(6 Ω) = 3 A gives

$$2I + I_1 = 3\ \text{A} \qquad \text{23-7}$$

Applying the loop rule to loop *bcde* gives us another equation:

$$-(3\ \Omega)(I - I_1) + 21\ \text{V} - (2\ \Omega)(I - I_1) + (6\ \Omega)I_1 = 0$$

Note that in moving from e to b we encounter a voltage increase because the assumed direction of current I_1 is opposite the direction in which we are traversing the loop. Combining terms, rearranging, and dividing the terms by 1 Ω, we obtain

$$5I - 11I_1 = 21\ \text{A} \qquad \text{23-8}$$

Solving Equations 23-7 and 23-8 for the unknown currents I and I_1, we obtain

$$I = 2\ \text{A}$$

and

$$I_1 = -1\ \text{A}$$

The negative value for I_1 shows that our original assumption about the direction of I_1 was incorrect. That is, the current through the 6-Ω resistor is actually in the direction from point e to point b. The current through the 21-V battery is therefore

$$I - I_1 = 2\text{ A} - (-1\text{ A}) = 3\text{ A}$$

In the original circuit of Figure 23-9, this current splits just after point d, with 2 A going through the 3-Ω resistor and 1 A going through the 6-Ω resistor. Figure 23-11 shows the complete circuit with the correct magnitudes and directions for the currents. A good way to check the solution of a circuit problem is to assign a potential of 0 to one point in the circuit and use the values of the currents found to determine the potential at the other points. In Figure 23-11 we have chosen point c to be at 0 V. Then point d is at +21 V, point e is at 15 V, point a is at 33 V and point b is at 9 V as indicated.

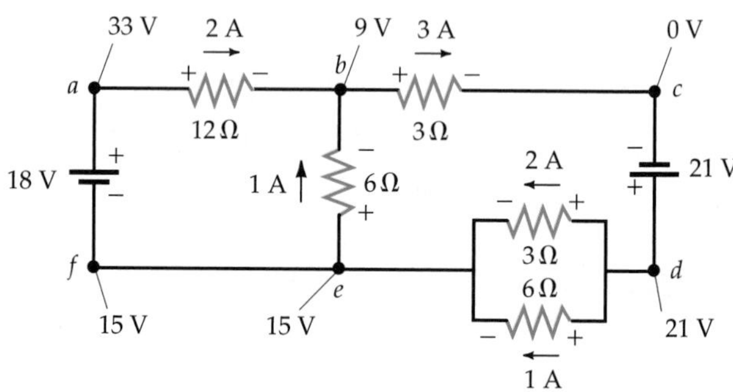

Figure 23-11 Circuit of Figure 23-9 with the correct magnitudes and directions for the currents found in Example 23-4.

We can summarize the general method for solving multiloop circuits:

General methods for solving multiloop circuits

1. Replace any combinations of resistors in series or parallel with their equivalent resistances.
2. Choose a direction for the current in each branch of the circuit, and label the currents in a circuit diagram. Add plus and minus signs to indicate the high and low potential sides of each source of emf, resistor, or capacitor.
3. Apply the junction rule to each junction where the current divides.
4. In a circuit containing n interior loops, apply the loop rule to any n loops.
5. Solve the equations to obtain the values of the unknowns.
6. Check your results by assigning a potential of zero to one point in the circuit and use the values of the currents found to determine the potentials at other points in the circuit.

Example 23-5

Find the current in each part of the circuit shown in Figure 23-12a.

This circuit contains three interior loops, so we will need to apply the loop rule 3 times. In Figure 23-12a, we have assumed a current I to the right through the 3-Ω resistor and have applied the junction rule at the two junctions c and e where the current divides. The current from c to d is labeled I_1, so the current from c to h is $I - I_1$. Similarly, the current from e to f is labeled I_2, so the current from e to h is $I_1 - I_2$. Applying

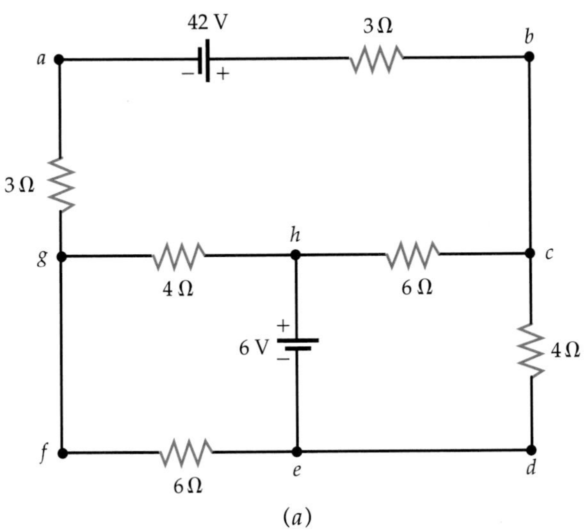

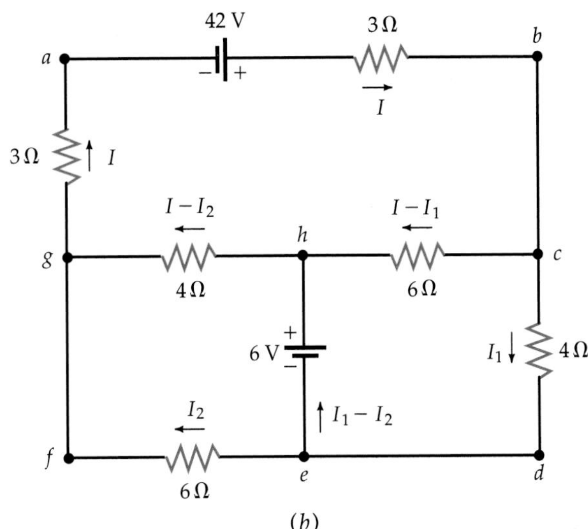

Figure 23-12 (*a*) Circuit containing three interior loops for Example 23-5. (*b*) Circuit diagram with assumed currents. The junction rule has been applied at points *c* and *e* where the current divides.

Kirchhoff's loop rule to the outer loop beginning at point *a*, we have

$$+42\text{ V} - (3\ \Omega)I - (4\ \Omega)I_1 - (6\ \Omega)I_2 - (3\ \Omega)I = 0$$

which can be simplified to

$$21\text{ A} - 3I - 2I_1 - 3I_2 = 0 \qquad \text{23-9}$$

Similarly, applying the loop rule to the upper loop (*abchga*) gives

$$+42\text{ V} - (3\ \Omega)I - (6\ \Omega)(I - I_1) - (4\ \Omega)(I - I_2) - (3\ \Omega)I = 0$$

which simplifies to

$$21\text{ A} - 8I + 3I_1 + 2I_2 = 0 \qquad \text{23-10}$$

For our third loop, we choose the lower left loop (*efghe*). We obtain

$$-(6\ \Omega)I_2 + (4\ \Omega)(I - I_2) - 6\text{ V} = 0$$

which simplifies to

$$2I - 5I_2 - 3\text{ A} = 0 \qquad \text{23-11}$$

We can eliminate I_1 by multiplying each term in Equation 23-9 by 3 and each term in Equation 23-10 by 2 to obtain

$$63\text{ A} - 9I - 6I_1 - 9I_2 = 0$$

$$42\text{ A} - 16I + 6I_1 + 4I_2 = 0$$

Adding the resulting equations gives

$$105\text{ A} - 25I - 5I_2 = 0 \qquad \text{23-12}$$

Substituting $5I_2 = 2I - 3$ A from Equation 23-11 into Equation 23-12, we obtain

$$105\text{ A} - 25I - (2I - 3\text{ A}) = 0$$

$$I = \frac{108\text{ A}}{27} = 4\text{ A}$$

Then from Equation 23-11,

$$5I_2 = 2I - 3\text{ A} = 2(4\text{ A}) - 3\text{ A} = 5\text{ A}$$

$$I_2 = 1\text{ A}$$

and from Equation 23-9,

$$21\text{ A} - 3(4\text{ A}) - 2I_1 - 3(1\text{ A}) = 0$$

$$I_1 = \frac{21\text{ A} - 12\text{ A} - 3\text{ A}}{2} = 3\text{ A}$$

In Figure 23-13, we have chosen the potential to be zero at point a and indicated the values of the currents found in our solution. The potential at point b is 42 V − (4 A)(3 Ω) = 30 V. The potentials at other points in the circuit are found by similar reasoning.

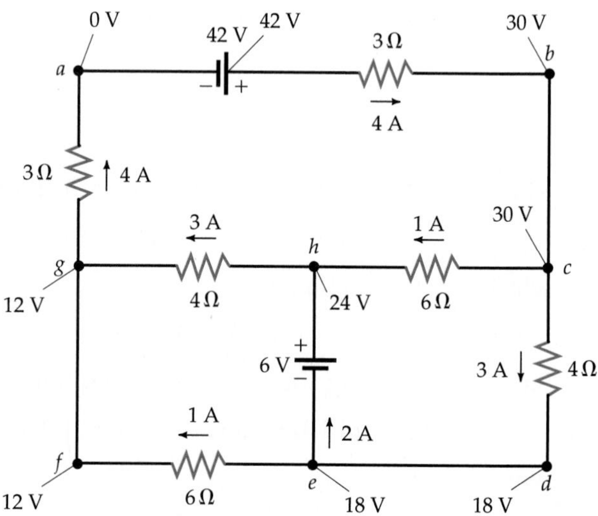

Figure 23-13 Circuit diagram of Figure 23-12 with the values of the potentials calculated from the values of the currents found in Example 23-5, assuming $V = 0$ at point a.

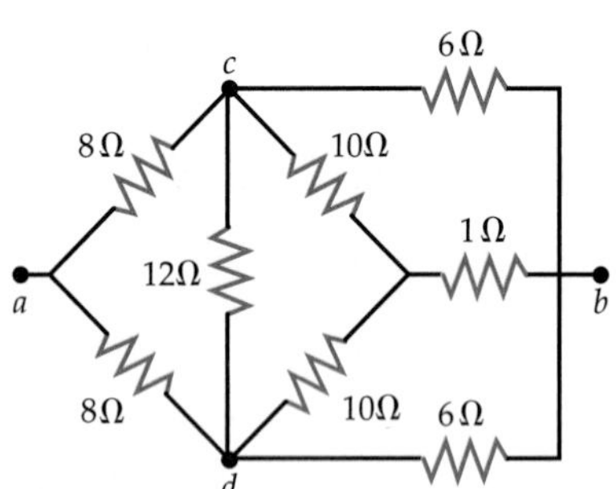

Figure 23-14 A complex, multibranch circuit. The circuit can be simplified by noting that by symmetry points c and d must be at the same potential. Because there is no current through the 12-Ω resistor between c and d, that resistor can be removed without affecting the current or potential anywhere in the circuit. Thus, points c and d can be considered a single point cd.

Analysis of Circuits by Symmetry

Some complicated circuits or networks of resistors can be more easily analyzed by using the junction rule and symmetry considerations rather than the loop rule. If two points in a circuit are at the same potential, they can be connected with a wire without causing any changes in the currents or potentials at other points in the circuit. Such points can often be identified by considering the symmetry of the circuit. The circuit can then be simplified by connecting these points and drawing a new circuit diagram.

Consider the circuit in Figure 23-14. We wish to find the current in each part of the circuit when a potential difference V_{ab} is applied between points a and b. Since this circuit has four interior loops, analyzing it by applying the loop rule to each loop becomes quite complicated. However, by symmetry, we can see that the points labeled c and d must be at the same potential, so there is no current in the 12-Ω resistor connecting these points. Thus, if we connect these points by a wire, there will be no current in the wire. In Figure 23-15 the circuit has been redrawn with points c and d connected. We can now find the equivalent resistance R_{eq} between points a and b using the methods for analyzing combinations of resistors in series and parallel discussed in Chapter 22. The total current from a to b is then V_{ab}/R_{eq}, and the current in each part of the circuit can be found by the methods discussed previously in this section.

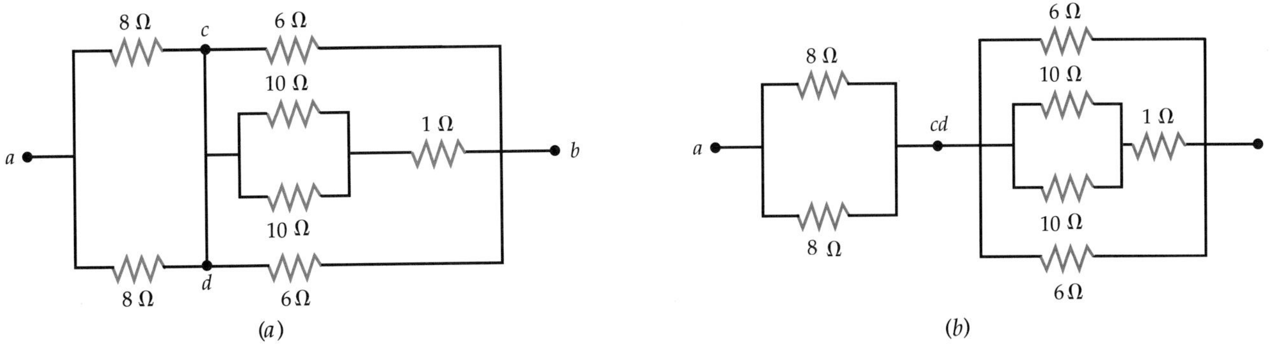

Figure 23-15 (*a*) Simplified diagram for the circuit of Figure 23-14 with points *c* and *d* connected. The two 8-Ω resistors are in parallel, and there are three parallel paths from the point *cd* to point *b*. (*b*) Alternative diagram in which points *c* and *d* are considered to be the same point.

Exercise

Find (*a*) the equivalent resistance between points *a* and *b* in the circuit in Figure 23-15 and (*b*) the current in the 10-Ω resistors if $V_{ab} = 12$ V. [Answer: (*a*) $R_{eq} = 6\ \Omega$, (*b*) $I_{10} = \frac{1}{3}$ A.]

Figure 23-16 shows 12 equal resistors that comprise the edges of a cube. We wish to find the equivalent resistance between the opposite corners of the cube labeled points *a* and *g* in the figure. Analyzing this network without using symmetry would clearly be difficult. By symmetry, we can see that if we apply a potential difference V_{ag} between points *a* and *g*, the points *b*, *d*, and *e* must all be at the same potential. If we connect these points by a wire, we have three equal resistors in parallel between point *a* and the common point *bde*. We can also see that points *c*, *f*, and *h* are at the same potential and can therefore be connected by a wire. There are six parallel paths, each of resistance *R*, between point *bde* and point *cfh* and three parallel paths from point *cfh* to point *g*. The simplified diagram for this circuit is shown in Figure 23-17. The equivalent resistance between *a* and *g* is thus

$$R_{eq} = \tfrac{1}{3}R + \tfrac{1}{6}R + \tfrac{1}{3}R = \tfrac{5}{6}R$$

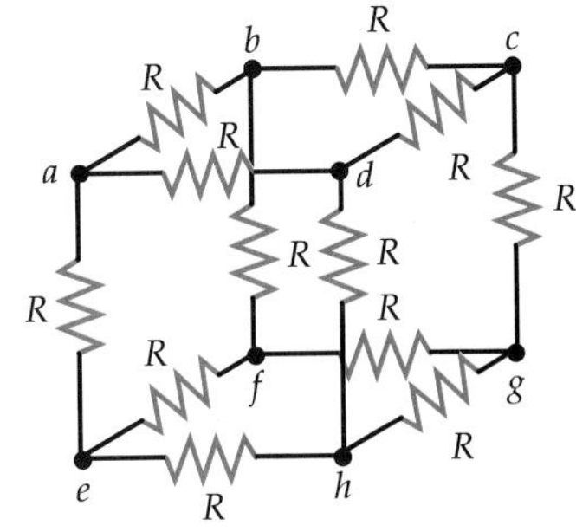

Figure 23-16 Twelve equal resistors comprising the edges of a cube. The circuit diagram can be simplified by noting from the symmetry of the figure that points *b*, *d*, and *e* must be at the same potential and points *c*, *f*, and *h* must be at the same potential.

There is another way to solve this problem. Let *I* be the current from point *a* to point *g* when the potential difference is V_{ag}. By symmetry, the current splits into three equal parts at point *a*, so the current from *a* to *b* is $\frac{1}{3}I$. At point *b*, the current splits into two equal parts since points *f* and *c* are at the same potential. Thus, the current from *b* to *c* is $\frac{1}{6}I$. Finally, since we have three symmetrical paths into point *g*, the current from *c* to *g* is $\frac{1}{3}I$. Therefore, if we move from point *a* to *b* to *c* to *g*, the potential drops are

$$V_{ag} = \tfrac{1}{3}IR + \tfrac{1}{6}IR + \tfrac{1}{3}IR = \tfrac{5}{6}IR = IR_{eq}$$

or

$$R_{eq} = \tfrac{5}{6}R$$

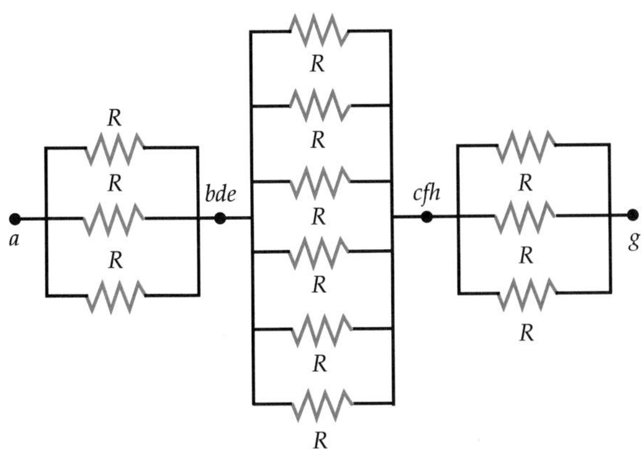

Figure 23-17 Simplified diagram for the circuit in Figure 23-16 with points *b*, *d*, and *e* connected and points *c*, *f*, and *h* connected. The six parallel paths between point *bde* and point *cfh* are the six edges of the cube *bc*, *bf*, *dc*, *dh*, *ef*, and *eh*.

23-2 *RC* Circuits

A circuit containing a resistor and capacitor is called an ***RC* circuit.** In such a circuit, the current is not steady but varies with time. Practical examples of *RC* circuits are the circuits in the flash attachment of a camera. Before a flash photograph is taken, a battery in the flash attachment charges the capacitor through a resistor. When this is accomplished, the flash is ready. When the picture is taken, the capacitor discharges through the flash bulb. The capacitor is then recharged by the battery, and a short time later the flash is ready for another picture. Using Kirchhoff's rules, we can obtain equations for the charge Q and the current I as functions of time for both the charging and discharging of a capacitor through a resistor.

Discharging a Capacitor

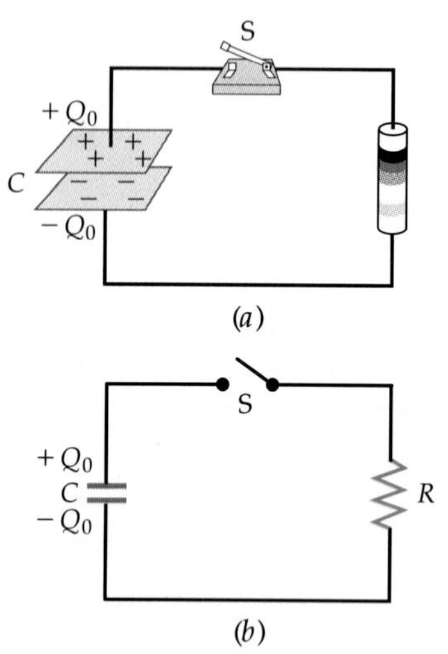

Figure 23-18 (*a*) A parallel-plate capacitor in series with a switch and a resistor R. (*b*) Circuit diagram for (*a*).

Figure 23-18 shows a capacitor with initial charges of $+Q_0$ on the upper plate and $-Q_0$ on the lower plate. It is connected to a resistor R and a switch S, which is open to prevent the charge from flowing through the resistor. The potential difference across the capacitor is initially $V_0 = Q_0/C$, where C is the capacitance. Since there is no current when the switch is open, there is no potential drop across the resistor. Thus, there is also a potential difference V_0 across the switch.

We close the switch at time $t = 0$. Since there is now a potential difference across the resistor, there must be a current in it. The initial current is

$$I_0 = \frac{V_0}{R} = \frac{Q_0}{RC} \qquad \text{23-13}$$

The current is due to the flow of charge from the positive plate of the capacitor to the negative plate through the resistor. After a time, the charge on the capacitor is reduced. Since the charge on the capacitor is *decreasing,* and we are taking the clockwise current to be positive, the current equals the rate of *decrease* of that charge. If Q is the charge on the capacitor at any time, the current at that time is

$$I = -\frac{dQ}{dt} \qquad \text{23-14}$$

Traversing the circuit in the direction of the current, we encounter a potential drop IR across the resistor and a potential increase Q/C across the capacitor. Kirchhoff's loop rule gives

$$\frac{Q}{C} - IR = 0 \qquad \text{23-15}$$

where both Q and I are functions of time and are related by Equation 23-14. Substituting $-dQ/dt$ for I in Equation 23-15, we have

$$\frac{Q}{C} + R\frac{dQ}{dt} = 0$$

or

$$\frac{dQ}{dt} = -\frac{1}{RC}Q \qquad \text{23-16}$$

Equation 23-16 states that the rate of change of the function $Q(t)$ is proportional to the function $Q(t)$. To solve this equation, we first separate the variables Q and t. Multiplying both sides of the equation by dt/Q, we obtain

$$\frac{dQ}{Q} = -\frac{dt}{RC} \qquad \text{23-17}$$

Integrating, we obtain

$$\ln Q = -\frac{t}{RC} + A$$

where A is an arbitrary constant of integration that is determined by the initial conditions. Taking the exponential of both sides of this equation, we obtain

$$e^{\ln Q} = e^{-t/RC+A} = e^{A}e^{-t/RC}$$

or

$$Q = Be^{-t/RC}$$

We obtain the constant $B = e^{A}$ from the initial condition that $Q = Q_0$ at $t = 0$. Then

$$Q(t) = Q_0e^{-t/RC} = Q_0e^{-t/\tau} \qquad 23\text{-}18$$

where τ, which is called the **time constant,** is the time it takes for the charge to decrease to $1/e$ of its original value:

$$\tau = RC \qquad 23\text{-}19$$

Time constant

Figure 23-19 shows the charge on the capacitor in the circuit of Figure 23-18 as a function of time. The dashed line is the initial slope of the charge-versus-time function. If the charge continued to decrease at a constant rate equal to this initial rate, it would reach zero in a time equal to the time constant τ. However, the actual rate of decrease $-dQ/dt$ is not constant but also decreases with time. This is evident from Equation 23-16, which shows that the rate of decrease of the charge is proportional to the charge itself. After a time $t = \tau$, the charge is $Q = Q_0e^{-1} = 0.37Q_0$. After a time $t = 2\tau$, the charge is $Q = Q_0e^{-2} = 0.135Q_0$, and so forth. After a time equal to several time constants, the charge on the capacitor is negligible. This type of decrease, which is called an **exponential decrease,** is very common in nature. It occurs whenever the rate at which a quantity decreases is proportional to the quantity itself. (We encountered exponential decreases in Chapter 11 when we studied the decrease in air pressure with altitude and again in Chapter 12 when we studied the decrease in the energy of a damped oscillator with time.) The decrease in the charge on a capacitor can be likened to the decrease in the amount of water in a bucket that has a small hole in the bottom. The rate at which the water flows out of the bucket is proportional to the pressure of the water, which is in turn proportional to the amount of water still in the bucket.

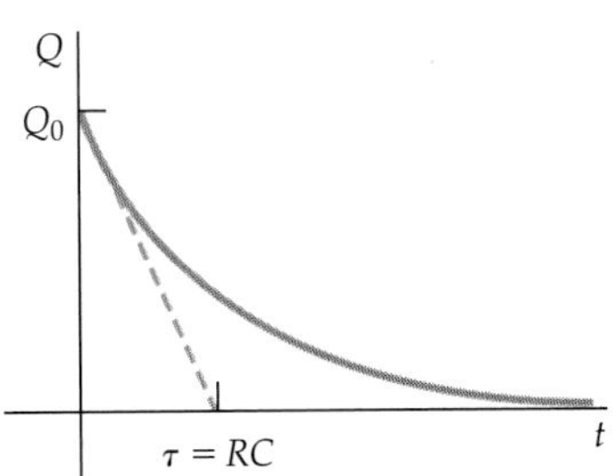

Figure 23-19 Plot of the charge on the capacitor versus time for the circuit in Figure 23-18 when the switch is closed at time $t = 0$. The time constant $\tau = RC$ is the time it takes for the charge to decrease to $e^{-1}Q_0$. After a time equal to two time constants, the charge is $e^{-2}Q_0$. This is an example of an exponential decrease. The time constant is also the time it would take the capacitor to discharge fully if its discharge rate were constant, as indicated by the dashed line.

The current is obtained from Equation 23-14 by differentiating Equation 23-18:

$$I = -\frac{dQ}{dt} = \frac{Q_0}{RC}e^{-t/RC}$$

or

$$I = \frac{V_0}{R}e^{-t/RC} = I_0e^{-t/\tau} \qquad 23\text{-}20$$

where $I_0 = Q_0/RC = V_0/R$ is the initial current. The current as a function of

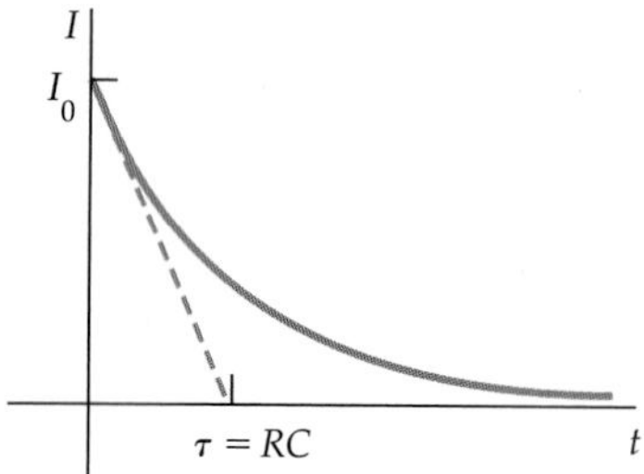

Figure 23-20 Plot of the current versus time for the circuit in Figure 23-18. The curve has the same shape as that in Figure 23-19.

time is shown in Figure 23-20. The current also decreases exponentially with time and falls to $1/e$ of its initial value after a time $t = \tau = RC$.

Example 23-6

A 4-μF capacitor is charged to 24 V and then connected across a 200-Ω resistor. Find (*a*) the initial charge on the capacitor, (*b*) the initial current through the 200-Ω resistor, (*c*) the time constant, and (*d*) the charge on the capacitor after 4 ms.

(*a*) The initial charge is $Q_0 = CV = (4\ \mu\text{F})(24\ \text{V}) = 96\ \mu\text{C}$.
(*b*) The initial current is $I_0 = V_0/R = (24\ \text{V})/(200\ \Omega) = 0.12\ \text{A}$.
(*c*) The time constant is $\tau = RC = (200\ \Omega)(4\ \mu\text{F}) = 800\ \mu\text{s} = 0.8\ \text{ms}$.
(*d*) At $t = 4$ ms, the charge on the capacitor is

$$Q = Q_0 e^{-t/\tau} = (96\ \mu\text{C})e^{-(4\ \text{ms})/(0.8\ \text{ms})} = (96\ \mu\text{C})e^{-5} = 0.647\ \mu\text{C}$$

Exercise

Find the current through the 200-Ω resistor at $t = 4$ ms. (Answer: 0.809 mA)

Charging a Capacitor

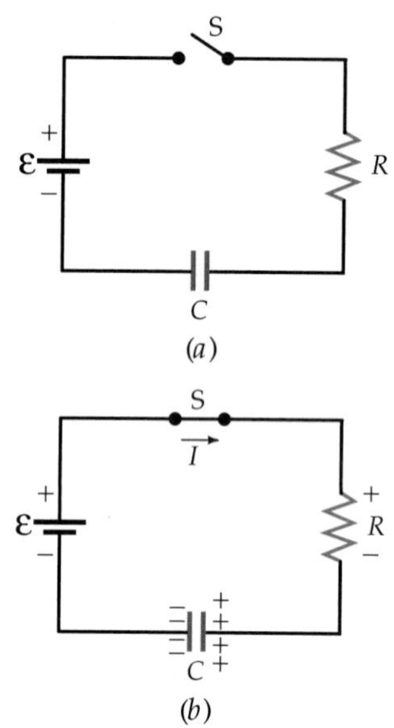

Figure 23-21 (*a*) Circuit for charging a capacitor to a potential difference $\mathcal{E}$. (*b*) After the switch is closed, there is a potential drop across the resistor and a charge on the capacitor.

Figure 23-21*a* shows a circuit for charging a capacitor, which we will assume to be initially uncharged. The switch, originally open, is closed at time $t = 0$. Charge immediately begins to flow through the resistor and onto the positive plate of the capacitor (Figure 23-21*b*). If the charge on the capacitor at some time is Q and the current in the circuit is I, Kirchhoff's loop rule gives

$$\mathcal{E} - V_R - V_C = 0$$

or

$$\mathcal{E} - IR - \frac{Q}{C} = 0 \qquad \text{23-21}$$

In this circuit, the current equals the rate at which the charge on the capacitor is *increasing:*

$$I = +\frac{dQ}{dt}$$

Substituting $+dQ/dt$ for I in Equation 23-21 gives

$$\mathcal{E} = R\frac{dQ}{dt} + \frac{Q}{C} \qquad \text{23-22}$$

At time $t = 0$, the charge on the capacitor is zero and the current is $I_0 = \mathcal{E}/R$. The charge then increases and the current decreases, as can be seen from Equation 23-21. The charge reaches a maximum value of $Q_f = C\mathcal{E}$ when the current I equals zero, as can also be seen from Equation 23-21.

Equation 23-22 is slightly more difficult to solve than was Equation 23-16. Multiplying each term by C and rearranging, we can write Equation 23-22 as

$$RC\frac{dQ}{dt} = C\mathcal{E} - Q$$

We can now separate the variables Q and t by multiplying each side by dt/RC and dividing by $C\mathcal{E} - Q$:

$$\frac{dQ}{C\mathcal{E} - Q} = \frac{dt}{RC} \qquad \text{23-23}$$

Integrating each side of Equation 23-23, we obtain

$$-\ln (C\mathcal{E} - Q) = t/RC + A \qquad 23\text{-}24$$

where A is an arbitrary constant of integration. Taking the exponent of each side of Equation 23-24, we obtain

$$C\mathcal{E} - Q = e^{-A}e^{-t/RC} = Be^{-t/RC}$$

or

$$Q = C\mathcal{E} - Be^{-t/RC} \qquad 23\text{-}25$$

where $B = e^{-A}$ is another constant. The value of B is determined by the initial condition $Q = 0$ at $t = 0$. Setting $t = 0$ and $Q = 0$ in Equation 23-25 gives

$$0 = C\mathcal{E} - B$$

or

$$B = C\mathcal{E}$$

Substituting $B = C\mathcal{E}$ into Equation 23-25, we obtain for the charge

$$Q = C\mathcal{E}(1 - e^{-t/RC}) = Q_f(1 - e^{-t/\tau}) \qquad 23\text{-}26$$

where $Q_f = QE$ is the final charge. The current is obtained by differentiating Equation 23-26:

$$I = \frac{dQ}{dt} = -C\mathcal{E}e^{-t/RC}(-1/RC)$$

or

$$I = \frac{\mathcal{E}}{R}e^{-t/RC} = I_0e^{-t/\tau} \qquad 23\text{-}27$$

Figures 23-22 and 23-23 show the charge and the current as functions of time. Note from Figure 23-22 that the time constant τ is also the time in which the capacitor would become fully charged if the current remained constant at its initial value.

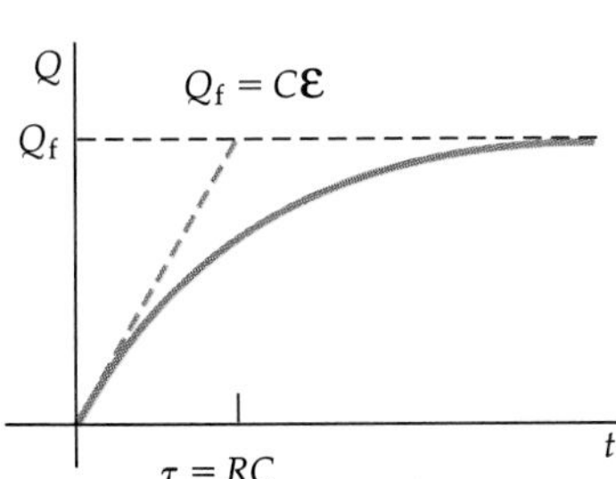

Figure 23-22 Plot of the charge on the capacitor versus time for the charging circuit of Figure 23-21 after the switch is closed at $t = 0$. After a time $t = \tau = RC$, the charge on the capacitor is $0.63C\mathcal{E}$, where $C\mathcal{E}$ is its final charge. If the charging rate were constant, the capacitor would be fully charged after a time $t = \tau$.

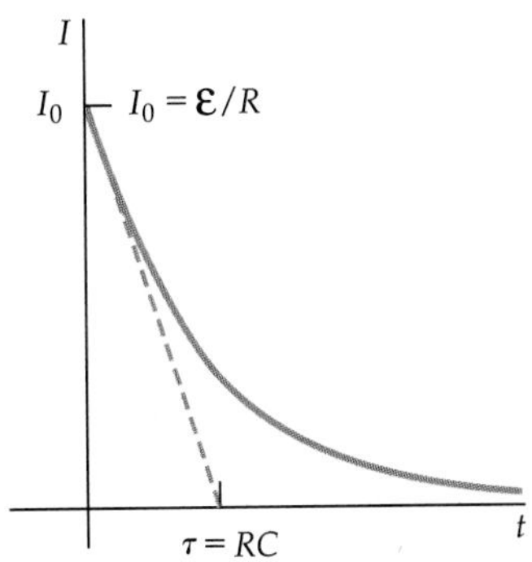

Figure 23-23 Plot of the current versus time for the charging circuit of Figure 23-21. The current is initially $\mathcal{E}/R$, and it decreases exponentially with time.

Exercise

Show that Equation 23-26 does indeed satisfy Equation 23-22 by substituting $Q(t)$ and dQ/dt into Equation 23-22.

Exercise

What fraction of the maximum charge is on the charging capacitor after a time $t = \tau$? (Answer: 0.63)

Example 23-7

A 6-V battery of negligible internal resistance is used to charge a 2-μF capacitor through a 100-Ω resistor. Find (*a*) the initial current, (*b*) the final charge on the capacitor, and (*c*) the time required for the charge to reach 90 percent of its final value.

(*a*) The initial current is

$$I_0 = \mathcal{E}/R = (6\text{ V})/(100\ \Omega) = 0.06\text{ A}$$

(*b*) The final charge on the capacitor is

$$Q_f = \mathcal{E}C = (6\text{ V})(2\ \mu\text{F}) = 12\ \mu\text{C}$$

(*c*) The time constant for this circuit is $\tau = RC = (100\ \Omega)(2\ \mu\text{F}) = 200\ \mu\text{s}$. We should expect the charge to reach 90 percent of its final value in a time of the order of several time constants. We can find the exact solution from Equation 23-26, using $Q = 0.9\mathcal{E}C$:

$$Q = 0.9\mathcal{E}C = \mathcal{E}C(1 - e^{-t/RC})$$

$$0.9 = 1 - e^{-t/RC}$$

$$e^{-t/RC} = 1 - 0.9 = 0.1$$

$$\ln e^{-t/RC} = -\frac{t}{RC} = \ln 0.1 = -2.3$$

Thus,

$$t = 2.3RC = 2.3(200\ \mu\text{s}) = 460\ \mu\text{s}$$

Example 23-8

The capacitor in the circuit shown in Figure 23-24*a* is initially uncharged. Find the current through the battery (*a*) immediately after the switch is closed and (*b*) a long time after the switch is closed.

(*a*) Since the capacitor is initially uncharged, the potential at point *d* is the same as that at point *c* just after the switch is closed. There is thus no initial current through the 8-Ω resistor between *b* and *e*. Applying the loop rule to the outer loop (*abcdefa*), we obtain

$$12\text{ V} - (4\ \Omega)I_0 = 0$$

$$I_0 = 3\text{ A}$$

(*b*) While the capacitor is charging, the current divides at point *b* and charge flows onto the upper plate and off of the lower plate. After a long time, the capacitor becomes fully charged, and no more charge flows onto or off of the plates. Applying the loop rule to the left loop (*abefa*), we obtain

$$12\text{ V} - (4\ \Omega)I_f - (8\ \Omega)I_f = 0$$

$$I_f = 1\text{ A}$$

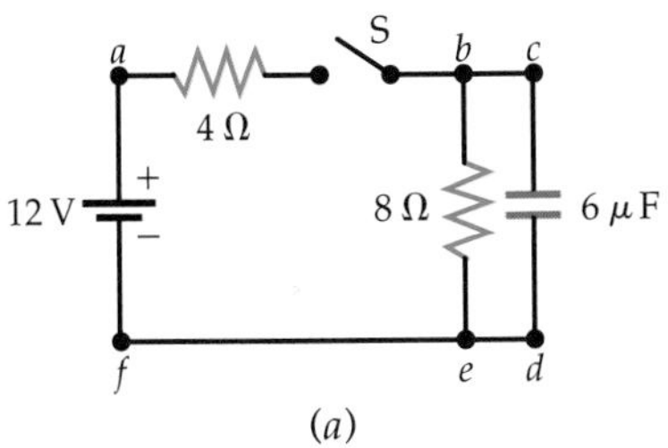

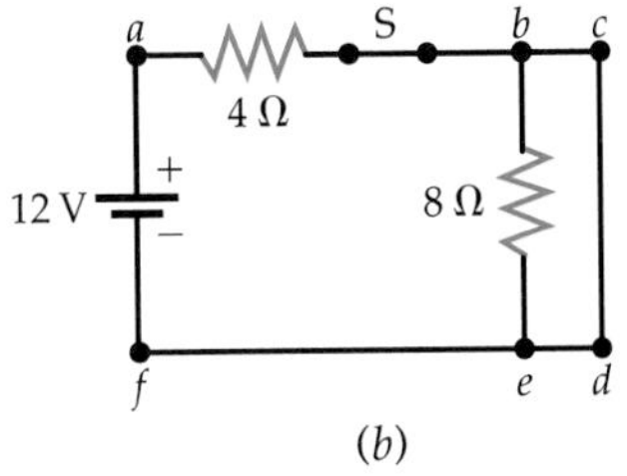

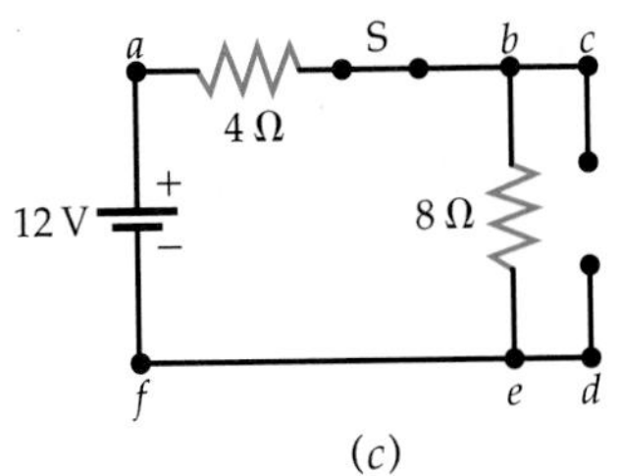

Figure 23-24 (*a*) A capacitor in parallel with a resistor across an emf. The capacitor is assumed to be uncharged before the switch is closed. (*b*) Immediately after the switch is closed, there is no potential drop across the capacitor, so the capacitor can be replaced with a wire of zero resistance. (*c*) A long time after the switch has been closed, the capacitor is fully charged and all the charge entering point *b* flows through the 8-Ω resistor. The capacitor can thus be replaced by an open circuit.

We see that the analysis of this circuit at the extreme times, when the capacitor is either uncharged or fully charged, is simple. When the capacitor is uncharged, it acts like a short circuit between points *c* and *d*, that is, the circuit is the same as shown in Figure 23-24*b*, where we have replaced the capacitor by a wire of zero resistance. When the capacitor is fully charged, it acts like an open circuit as shown in Figure 23-24*c*.

During the charging process, a total charge $Q_f = \mathcal{E}C$ flows through the battery. The battery therefore does work

$$W = Q_f\mathcal{E} = \mathcal{E}^2 C$$

Half of this work is stored as energy in the capacitor. From Equation 21-16,

$$U = \tfrac{1}{2}QV = \tfrac{1}{2}Q_f\mathcal{E} = \tfrac{1}{2}\mathcal{E}^2 C$$

We now show that the other half of the energy provided by the battery goes into Joule heat in the resistor. The rate of energy put into the resistor is

$$\frac{dW_R}{dt} = I^2 R$$

Using Equation 23-27 for the current, we have

$$\frac{dW_R}{dt} = \left(\frac{\mathcal{E}}{R}e^{-t/RC}\right)^2 R = \frac{\mathcal{E}^2}{R}e^{-2t/RC}$$

We find the total Joule heat produced by integrating from $t = 0$ to $t = \infty$:

$$W_R = \int_0^\infty \frac{\mathcal{E}^2}{R}e^{-2t/RC}\,dt$$

The integration can be done by substituting $x = 2t/RC$. Then,

$$dt = \frac{RC}{2}\,dx$$

and

$$W_R = \frac{\mathcal{E}^2}{R}\frac{RC}{2}\int_0^\infty e^{-x}\,dx = \frac{1}{2}\mathcal{E}^2 C$$

since the integral is 1. This result is independent of the resistance R. Thus, when a capacitor is charged by a battery with a constant emf, half the energy provided by the battery is stored in the capacitor and half goes into heat, independent of the resistance. The energy that goes into heat includes the energy that goes into the internal resistance of the battery.

Example 23-9

For the discharging capacitor in Example 23-6, find (*a*) the initial energy stored in the capacitor, (*b*) the initial power input to the resistor, and (*c*) the energy stored at $t = 0.8\text{ ms} = 1\tau$.

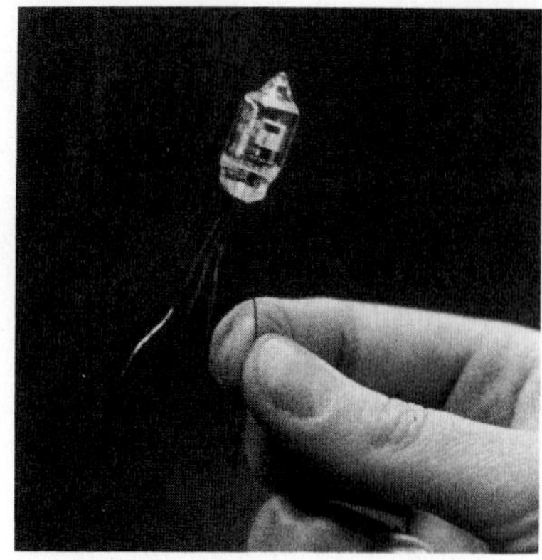

A krytron, shown here, is a high-speed triggering device used to discharge capacitors in the detonation of an atomic bomb, which consists of a uranium core surrounded by a shell of explosives. The capacitor, triggered by the krytron, discharges its energy to detonators that set off the explosive shell. The explosion rapidly and symmetrically compresses the uranium core, which reaches critical mass, undergoes runaway nuclear fission, and explodes. The sale of krytrons and certain specialized capacitors is controlled. In 1985 and 1990, they were the subject of publicized unsuccessful smuggling operations.

(*a*) In Example 23-6, we found the initial charge on the capacitor to be 96 μC. The initial energy stored is therefore

$$U = \frac{1}{2}\frac{Q^2}{C} = \frac{1}{2}\frac{(96\ \mu\text{C})^2}{4\ \mu\text{F}} = 1.152\ \text{mJ}$$

We could also have obtained this from $U = \frac{1}{2}QV = \frac{1}{2}(96\ \mu\text{C})(24\ \text{V}) = 1.152$ mJ.

(*b*) The initial power input to the resistor is

$$P_0 = I_0^2R = (0.12\ \text{A})^2(200\ \Omega) = 2.88\ \text{W}$$

where we have used $I_0 = 0.12$ A from Example 23-6.

(*c*) The charge on the capacitor after $t = 0.8$ ms $= 1\tau$ is

$$Q = Q_0e^{-t/\tau} = (96\ \mu\text{C})e^{-1} = 35.3\ \mu\text{C}$$

The energy stored at this time is thus

$$U = \frac{1}{2}\frac{Q^2}{C} = \frac{1}{2}\frac{(35.3\ \mu\text{C})^2}{4\ \mu\text{F}} = 0.156\ \text{mJ}$$

Note that we could not have used $U = \frac{1}{2}QV$ unless we first found the potential V, which also decreases as the capacitor discharges.

Example 23-10

Show that the decrease in the energy stored in the capacitor in Example 23-9 from $t = 0$ to $t = 1\tau$ equals the joule heat dissipated in the resistor.

In Example 23-9, we found the energy stored in the capacitor to be 1.152 mJ at $t = 0$ and 0.156 mJ at $t = 1\tau$. The decrease in the energy stored is thus

$$-\Delta U = 1.152\ \text{mJ} - 0.156\ \text{mJ} = 0.996\ \text{mJ}$$

The power input into the resistor is I^2R where I is given by Equation 23-20. Since this power is not constant in time, we must integrate from $t = 0$ to $t = 1\tau$ to find the total energy dissipated in the resistor. We have

$$W = \int_0^\tau I^2R\ dt = \int_0^\tau (I_0e^{-t/\tau})^2R\ dt$$
$$= I_0^2R\int_0^\tau e^{-2t/\tau}\ dt = I_0^2R\left(\frac{\tau}{-2}\right)e^{-2t/\tau}\Big|_0^\tau$$
$$= I_0^2R\left(\frac{\tau}{2}\right)(1 - e^{-2})$$

where we have used

$$\int e^{ax}\ dx = \frac{1}{a}e^{ax}$$

from integral tables. Then, using $I_0^2R = 2.88$ W from Example 23-9 and $\tau = 0.8$ ms from Example 23-6, we have

$$W = (2.88\ \text{W})(0.4\ \text{ms})(1 - e^{-2}) = 0.996\ \text{mJ}$$

which equals the decrease in the energy stored.

Question

1. A swimming pool is filled by siphoning water from a nearby lake. How is this analogous to the charging of a capacitor?

23-3 Ammeters, Voltmeters, and Ohmmeters

We turn now to the consideration of the measurement of electrical quantities in dc circuits. The devices that measure current, potential difference, and resistance are called **ammeters, voltmeters,** and **ohmmeters,** respectively. Often, all three of these meters are included in a single "multimeter" that can be switched from one use to another. You might use a voltmeter to measure the terminal voltage of your car battery and an ohmmeter to measure the resistance between two points in some electrical device at home (such as a toaster) where you suspect a short circuit or a broken wire. Therefore, some knowledge of the basic operation of these devices might prove very useful.

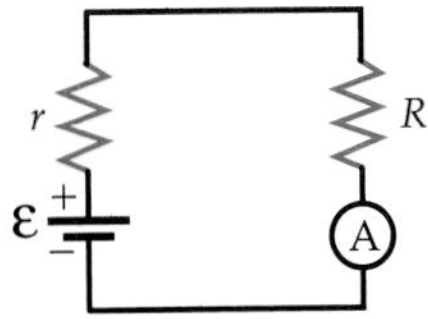

Figure 23-25 To measure the current in a resistor R, an ammeter —Ⓐ— is placed in series with the resistor so that it carries the same current as the resistor.

To measure the current through a resistor in a simple circuit, we place an ammeter in series with the resistor, as shown in Figure 23-25, so that the ammeter and the resistor carry the same current. Since the ammeter has some resistance, the current in the circuit decreases slightly when the ammeter is inserted. Ideally, the ammeter should have a very small resistance so that only a small change is caused in the current to be measured.

The potential difference across a resistor is measured by placing a voltmeter across the resistor in parallel with it, as shown in Figure 23-26, so that the potential drop across the voltmeter is the same as that across the resistor. The voltmeter reduces the resistance between points a and b, thus increasing the total current in the circuit and changing the potential drop across the resistor. A good voltmeter has a very large resistance so that its effect on the circuit is minimal.

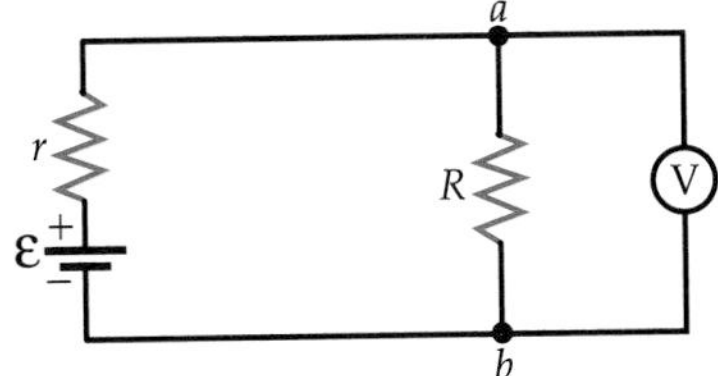

Figure 23-26 To measure the voltage drop across a resistor, a voltmeter —Ⓥ— is placed in parallel with the resistor so that the potential drops across the voltmeter and the resistor are the same.

The principal component of an ammeter and a voltmeter is a **galvanometer,** which is a device that detects a small current that passes through it. The galvanometer is designed so that the scale reading is proportional to the current passing through it. Many meters today have a digital readout rather than an indicator and a scale, but the basic way they operate is similar to that discussed here.

There are two properties of a galvanometer that are important for its use in an ammeter or a voltmeter. These are the resistance of the galvanometer R_g and the current needed to produce a full-scale deflection I_g. Typical values of these for a portable laboratory galvanometer are $R_g = 20\ \Omega$ and $I_g = 0.5$ mA. The voltage drop across such a galvanometer for full-scale deflection is thus

$$V = I_g R_g = (20\ \Omega)(5.0 \times 10^{-4}\ \text{A}) = 10^{-2}\ \text{V}$$

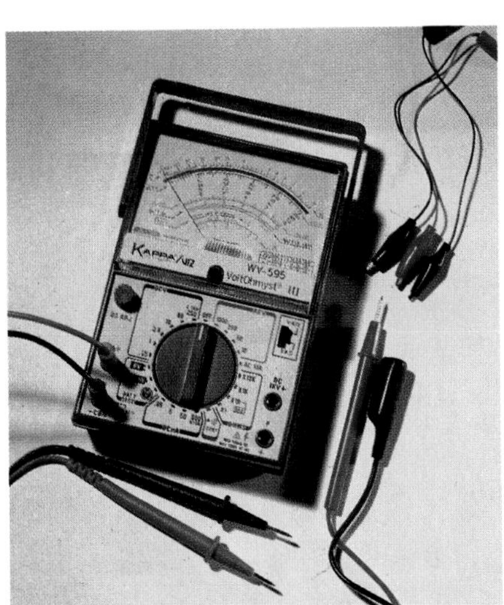

(a)

(b)

(a) An analog multimeter. (b) A digital multimeter. Digital multimeters are generally more accurate and more expensive than their analog counterparts.

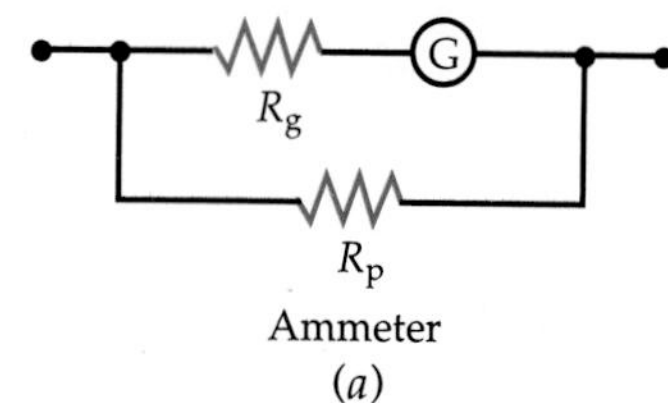

Ammeter
(a)

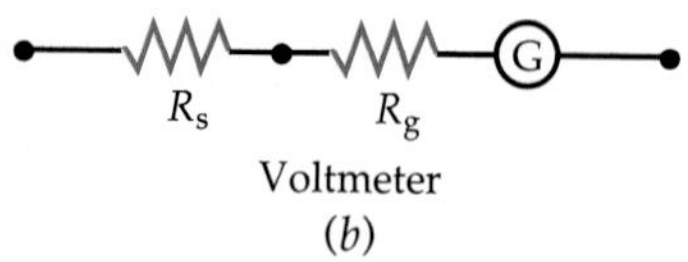

Voltmeter
(b)

Figure 23-27. (a) An ammeter consists of a galvanometer —Ⓖ— whose resistance is R_g and a small parallel resistance R_p. (b) A voltmeter consists of a galvanometer —Ⓖ— and a large series resistance R_s. In these diagrams, the galvanometer's resistance is shown next to the symbol for the galvanometer.

To construct an ammeter from a galvanometer, we place a small resistor called a **shunt resistor** in parallel with the galvanometer. The shunt resistance is usually much smaller than the resistance of the galvanometer so that most of the current is carried by the shunt resistor and the equivalent resistance of the ammeter is much smaller than that of the galvanometer alone. In fact, the equivalent resistance of the ammeter is approximately equal to the shunt resistance. To construct a voltmeter, we place a resistor with a large resistance in series with the galvanometer so that the equivalent resistance of the voltmeter is much larger than that of the galvanometer alone. Figure 23-27 illustrates the construction of an ammeter and voltmeter from a galvanometer. The resistance of the galvanometer R_g is shown separately in these schematic drawings, but it is actually part of the galvanometer since it is essentially due to the resistance of the galvanometer coil. The choice of appropriate resistors for the construction of an ammeter or voltmeter from a galvanometer is best illustrated by example.

Example 23-11

Using a galvanometer with a resistance of 20 Ω, for which a current of 5×10^{-4} A gives a full-scale deflection, design an ammeter that will give a full-scale deflection when the current is 5 A.

Since the total current through the ammeter must be 5 A when the current through the galvanometer is just 5×10^{-4} A, most of the current must go through the shunt resistor. Let R_p be the shunt resistance and I_p be the current through it. Since the galvanometer and shunt resistor are in parallel, we have

$$I_g R_g = I_p R_p$$

and

$$I_p + I_g = 5 \text{ A}$$

or

$$I_p = 5 \text{ A} - I_g = (5 \text{ A}) - (5 \times 10^{-4} \text{ A}) \approx 5 \text{ A}$$

Thus, the value of the shunt resistor should be

$$R_p = \frac{I_g}{I_p} R_g = \frac{5 \times 10^{-4} \text{ A}}{5 \text{ A}} (20 \ \Omega) = 2 \times 10^{-3} \ \Omega$$

Since the resistance of the shunt resistor is so much smaller than the resistance of the galvanometer, the effective resistance of the parallel combination is approximately equal to the shunt resistance.

Example 23-12

Using the same galvanometer as in Example 23-11, design a voltmeter that will give a full-scale deflection for a potential difference of 10 V.

Let R_s be the resistance of the resistor in series with the galvanometer. We want to choose R_s so that a current of $I_g = 5 \times 10^{-4}$ A gives a potential drop of 10 V. Thus,

$$I_g(R_s + R_g) = 10 \text{ V}$$

$$R_s + R_g = \frac{10 \text{ V}}{5 \times 10^{-4} \text{ A}} = 2 \times 10^4 \ \Omega$$

$$R_s = 2 \times 10^4 \ \Omega - R_g = (2 \times 10^4 \ \Omega) - 20 \ \Omega$$

$$= 19{,}980 \ \Omega \approx 20 \text{ k}\Omega$$

Example 23-13

The resistance of a 100-Ω resistor is to be measured using the circuit shown in Figure 23-28. The resistance of the voltmeter is 2000 Ω and that of the ammeter is 0.002 Ω. What error is made in calculating R from V/I where V is the reading of the voltmeter and I is the reading of the ammeter?

In the circuit shown, the voltmeter measures the voltage drop across the resistor, but the ammeter measures the total current in the circuit, including the current through the voltmeter. The equivalent resistance R'_{eq} of the voltmeter in parallel with the resistor is

$$R'_{eq} = \left(\frac{1}{100\ \Omega} + \frac{1}{2000\ \Omega}\right)^{-1} = 95.238\ \Omega$$

and the equivalent resistance of the entire circuit is

$$R_{eq} = R_a + R'_{eq} = 0.002\ \Omega + 95.238\ \Omega = 95.240\ \Omega$$

The current in the ammeter is

$$I = \frac{12\ \text{V}}{R_{eq}} = \frac{12\ \text{V}}{95.240\ \Omega} = 0.126\ \text{A}$$

If I_1 is the current through the 100-Ω resistor and I_2 is the current through the voltmeter, we have $100I_1 = 2000I_2$ or $I_2 = 0.05I_1$. The current through the 100-Ω resistor is thus

$$I_1 = I - I_2 = I - 0.05I_1$$

or

$$I_1 = \frac{I}{1.05} = \frac{0.126\ \text{A}}{1.05} = 0.120\ \text{A}$$

The voltage drop across the 100-Ω resistor as measured by the voltmeter is thus $RI_1 = (100\ \Omega)(0.120\ \text{A}) = 12.0\ \text{V}$, and the measured value of the resistance is

$$R = \frac{V}{I} = \frac{12.0\ \text{V}}{0.126\ \text{A}} = 95.2\ \Omega$$

which differs from the true value by about 5 percent. This is what is to be expected because the resistance of the voltmeter is only 20 times that of the resistor, so putting the voltmeter and resistor in parallel increases the total current by about 5 percent.

Figure 23-29 shows a better circuit for measuring this resistance with these meters. In this circuit, the ammeter measures the true current in the resistor, but the voltmeter measures the total potential drop across the resistor plus the ammeter. Since the resistance of the ammeter is only 0.002 percent of that of the resistor, the error made with this circuit will be only about 0.002 percent. If the resistance of the resistor were 0.1 Ω instead of 100 Ω, the circuit in Figure 23-28 would be better.

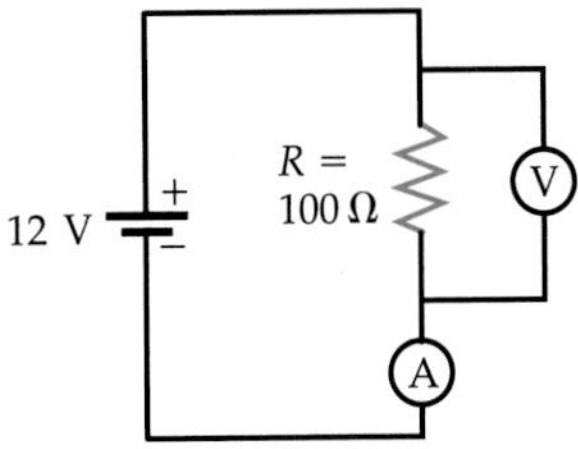

Figure 23-28 A possible circuit for measuring the resistance of a 100-Ω resistor of Example 23-13.

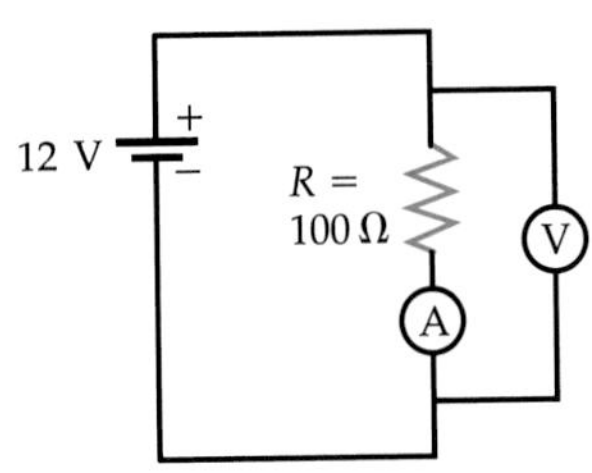

Figure 23-29 A better alternative for measuring the resistance of the 100-Ω resistor with the meters of Example 23-13.

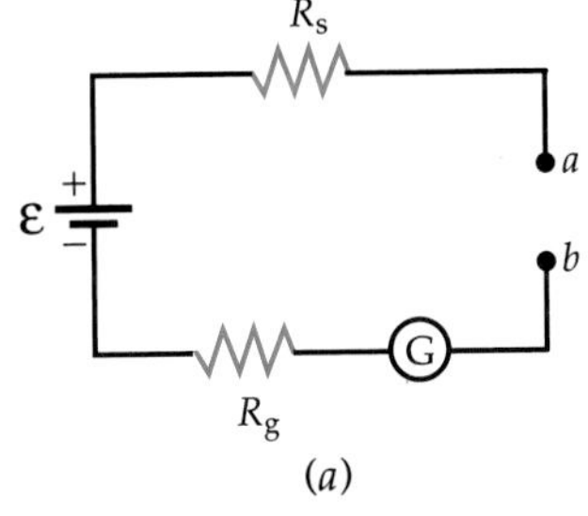

(a)

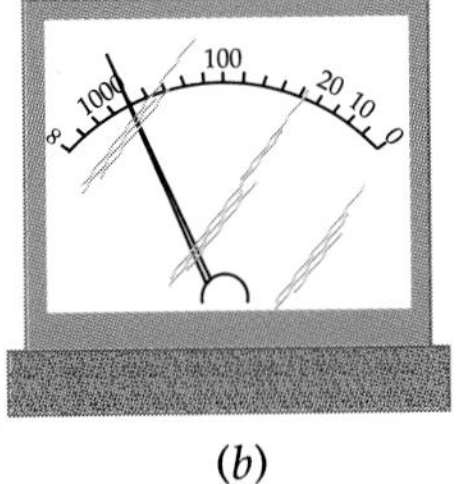

(b)

Figure 23-30 (a) An ohmmeter consists of a battery in series with a galvanometer and a resistor R_s, which is chosen such that the galvanometer gives full-scale deflection when points a and b are shorted. (b) The galvanometer scale calibrated to give a readout in ohms.

A simple ohmmeter for measuring resistance consists of a battery connected in series with a galvanometer and a resistor, as shown in Figure 23-30a. The resistance R_s is chosen such that when the terminals a and b are shorted (touched together), which means that there is no resistance between them, the current through the galvanometer is I_g. This is the current for which the galvanometer gives a full-scale deflection. Thus, a full-scale deflection indicates no resistance between terminals a and b, and a zero deflection indicates an infinite resistance between the terminals. When the termi-

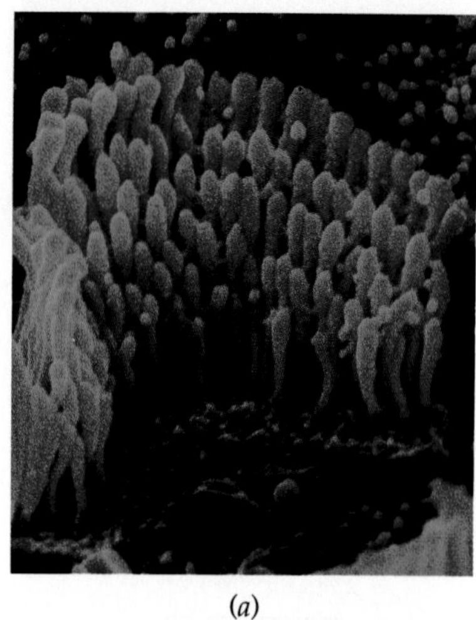

(*a*)

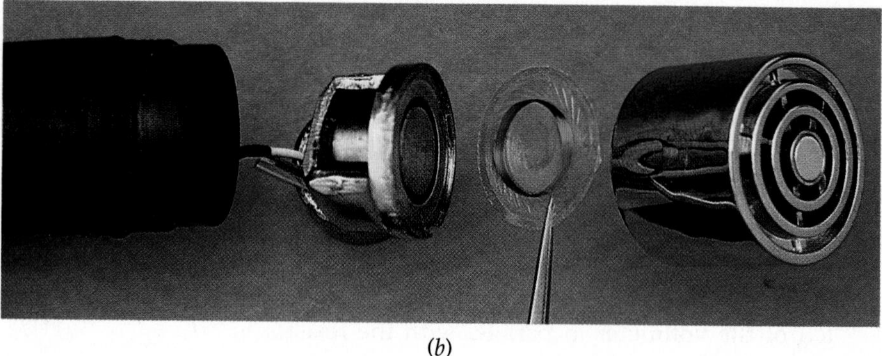

(*b*)

Devices that convert one form of energy to another are called transducers. Those shown here convert nonelectrical energy to electrical energy. (*a*) Sound waves transmitted to the spiral chamber of the inner ear causes watery fluid there to vibrate. Sensory receptor cells (shown here at about 3500X magnification) are attached to the walls of the chamber. The receptor cells are stimulated by the vibrating fluid and cause neurons, connected to their base, to transmit electrical impulses. The impulses, traveling along a chain of neurons, eventually register in the brain as the sensation "sound." Because it converts sound energy to electrical energy, this system is a biological counterpart to a microphone. (*b*) A microphone converts sound energy to electrical energy. In the kind shown here, a copper ring is attached to a thin plastic membrane. Sound waves hitting the membrane cause it and the ring to vibrate. The ring is mounted in the field of a permanent magnet. Motion back and forth across the field, caused by the vibration, induces an alternating current in the ring. This current causes a secondary alternating magnetic field to arise, which in turn creates a secondary alternating current, this time in a wire coil positioned behind the ring. (*c*) Photovoltaic cells convert light energy to electrical energy. The cells are composed of two semiconductors. At the junction where they meet, the semiconductors interact to form an electric dipole. Light absorbed near the

nals are connected across an unknown resistance R, the current through the galvanometer is less than I_g. Specifically, it is

$$I = \frac{\mathcal{E}}{R + R_s + R_g} \qquad \text{23-28}$$

Since this current depends on R, the scale can be calibrated to give a direct reading of R, as shown in Figure 23-30*b*. Note that this scale is far from linear. Because the accuracy of the reading depends on the constancy of the emf of the battery, this type of simple ohmmeter is not a highly accurate instrument. However, it is quite useful for making a quick, rough determination of an unknown resistance.

Because an ohmmeter sends a current through the resistance to be measured, some caution must be exercised when using this instrument. For example, consider an ohmmeter constructed from a 1.5-V battery and a galvanometer similar to that in Examples 23-11 and 23-12. The series resistance R_s needed is found from

$$I_g(R_s + R_g) = 1.5 \text{ V}$$

or

$$R_s = \frac{1.5 \text{ V}}{5 \times 10^{-4} \text{ A}} - R_g = 3000 \ \Omega - 20 \ \Omega = 2980 \ \Omega$$

Suppose we use this ohmmeter to measure the resistance of a very sensitive laboratory galvanometer that gives a full-scale reading when the current through it is 10^{-5} A and has a resistance of about 20 Ω. When the terminals a and b are placed across this sensitive galvanometer, the current sent through it will be just slightly less than 5×10^{-4} A because the total resistance is 3020 Ω, which is just slightly more than 3000 Ω. This current is about 50 times that needed to produce a full-scale deflection. The likely results will be some popping sounds, a puff of smoke, one less sensitive galvanometer, and probably some unkind words from the laboratory instructor.

Questions

2. Under what conditions might it be advantageous to use a galvanometer that is less sensitive than the one discussed in Examples 23-11 and 23-12, that is, one that requires a greater current I_g for a full-scale deflection?
3. When the series resistance R_s is properly chosen for the emf of a particular ohmmeter, any resistance from zero to infinity can be measured. Why, then, do practical ohmmeters have different scales for measuring different ranges of resistance?
4. A none-too-bright student decides to measure the internal resistance of his car battery with an ohmmeter borrowed from his physics laboratory. Why is this a bad idea?

(c)

(d)

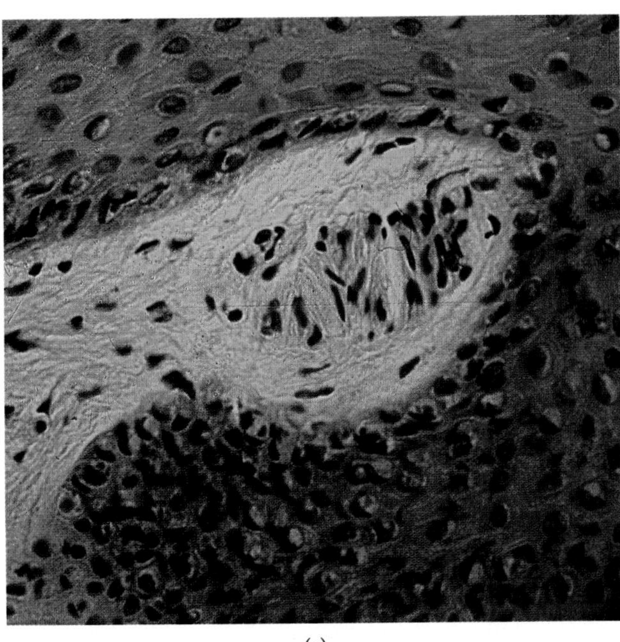

(e)

(f)

junction creates negative and positive charges which, if they drift into the junction, are swept in opposite directions by the dipole field. This separation of charge creates a voltage difference between the two semiconductors. At present, the best photovoltaic cells convert light energy to electrical energy with an efficiency of between 20 and 30 percent. (d) Piezoelectric crystals such as quartz, tourmaline, and topaz become electrically polarized when they are compressed, bent, or otherwise distorted. The polarization creates a voltage drop across the stressed crystal, which can be used to produce an electric current. The process is reversible: voltage applied across a piezoelectric will cause it to distort. Quartz piezoelectrics (shown here) are commonly introduced into oscillating electrical circuits having nearly the same frequency as the natural frequency of vibration of the crystal. The result is that the crystal and the electrical circuit both vibrate at the natural frequency of the crystal. This effect can be used to stabilize the frequency of radio oscillators and to make clocks with errors of less than 0.1 s per year. (e) Our sense of touch arises from encapsulated nerve fibers called Meissner's corpuscles that are located directly under the outer layer of skin. When skin overlying them is touched, the corpuscles are deformed, triggering electrical impulses in the nerve fibers. A stronger touch produces greater deformation and increases the frequency of impulses. The system is in some respects a biological counterpart to a strain gauge: in both transducers, the energy of mechanical stress is used to modulate changes in electrical conduction. (f) A strain gauge consists of a grid of very fine wire or foil of a substance, such as carbon, which changes its electrical resistance when mechanically stressed. The wire is bonded to a thin insulating backing, which is attached by adhesive to an object. Stresses that distort the object deform the attached strain gauge as well. The degree of deformation is measured by the change in resistance of the gauge. If a fixed voltage is applied across the ends of the gauge wire, a varying resistance will produce variations in the current.

Summary

1. Kirchhoff's rules are
   1. When any closed circuit loop is traversed, the algebraic sum of the changes in potential must equal zero.
   2. At any junction in a circuit where the current can divide, the sum of the currents into the junction must equal the sum of the currents out of the junction.
2. The general methods for analyzing multiloop circuits can be summarized as follows:
   1. Replace any combinations of resistors in series or parallel with their equivalent resistances.
   2. Choose a direction for the current in each branch of the circuit, and label the currents in a circuit diagram. Add plus and minus signs to indicate the high- and low-potential sides of each source of emf, resistor, or capacitor.
   3. Apply the junction rule to each junction where the current divides.
   4. In a circuit containing n interior loops, apply the loop rule to any n loops.
   5. Solve the equations to obtain the values of the unknowns.
   6. Check your results by assigning a potential of zero to one point in the circuit and using the values of the currents found to determine the potentials at other points in the circuit.
3. Complex circuits or networks of resistors can often be simplified by considering their symmetry. If the potential of two points is equal, the points can be connected with a wire, and a simplified circuit diagram can be drawn.
4. When a capacitor is discharged through a resistor, the charge on the capacitor and the current both decrease exponentially with time. The time constant $\tau = RC$ is the time it takes for either to decrease to $e^{-1} = 0.37$ times its original value. When a capacitor is charged through a resistor, the rate of charging, which equals the current, decreases exponentially with time. After a time $\tau = RC$, the charge on the capacitor has reached 63 percent of its final charge.
5. A galvanometer is a device that detects a small current that passes through it and gives a scale deflection that is proportional to the current. An ammeter is a device for measuring current. It consists of a galvanometer plus a parallel resistor called a shunt resistor. To measure the current through a resistor an ammeter is inserted in series with the resistor. The ammeter has a very small resistance so that it has little effect on the current to be measured. A voltmeter measures potential difference. It consists of a galvanometer plus a large series resistor. To measure the potential drop across a resistor, a voltmeter is placed in parallel with the resistor. The voltmeter has a very large resistance so that it has little effect on the potential drop to be measured. An ohmmeter is a device for measuring resistance. It consists of a galvanometer, a source of emf, and a resistor.

Suggestion for Further Reading

Rosenfeld, L.: "Gustav Robert Kirchhoff," *The Dictionary of Scientific Biography*, vol. 7, Charles C. Gillespie (ed.), Charles Scribner's Sons, New York, 1973, p. 379.

Kirchhoff's famous rules for electrical circuits were only his first of a number of important discoveries.

Review

A. Objectives: After studying this chapter, you should:

1. Be able to state Kirchhoff's rules and use them to analyze various dc circuits.

2. Be able to use symmetry to find the equivalent resistance of complex but symmetrical networks of resistors.

3. Be able to sketch both the charge Q on a capacitor and the current I as functions of time for charging and discharging a capacitor.

4. Be able to calculate the proper series or shunt resistors needed to make an ammeter, voltmeter, or ohmmeter from a given galvanometer and draw the circuit diagrams for these devices.

B. Define, explain, or otherwise identify:

Kirchhoff's rules	Voltmeter
Loop rule	Ammeter
Junction rule	Ohmmeter
RC circuit	Galvanometer
Time constant	Shunt resistor
Exponential decrease	

C. True or false: If the statement is true, explain why it is true. If it is false, give a counterexample.

1. The net change in the potential around a complete circuit in the steady state is zero.

2. The time constant of an *RC* circuit is the time needed to completely discharge the capacitor.

3. To measure the potential drop across a resistor, a voltmeter is placed in series with the resistor.

Problems

Level I

23-1 Kirchhoff's Rules

1. A battery with an emf of 6 V and an internal resistance of 0.3 Ω is connected to a variable resistance R. Find the current and power delivered by the battery when R is (*a*) 5 Ω and (*b*) 10 Ω.

2. A variable resistance R is connected across a potential difference V that remains constant independent of R. When $R = R_1$, the current is 6.0 A. When R is increased to $R_2 = R_1 + 10.0\ \Omega$, the current drops to 2.0 A. Find (*a*) R_1 and (*b*) V.

3. A battery has an emf $\mathcal{E}$ and an internal resistance r. When a 5.0-Ω resistor is connected across the terminals, the current is 0.5 A. When this resistor is replaced by an 11.0-Ω resistor, the current is 0.25 A. Find (*a*) the emf $\mathcal{E}$ and (*b*) the internal resistance r.

4. In Figure 23-31, the emf is 6 V and $R = 0.5\ \Omega$. The rate of joule heating in R is 8 W. (*a*) What is the current in the circuit? (*b*) What is the potential difference across R? (*c*) What is r?

5. For the circuit in Figure 23-32, find (*a*) the current, (*b*) the power delivered or absorbed by each emf, and (*c*) the rate of Joule heating in each resistor. (Assume that the batteries have negligible internal resistance.)

Figure 23-31 Problem 4. **Figure 23-32** Problem 5.

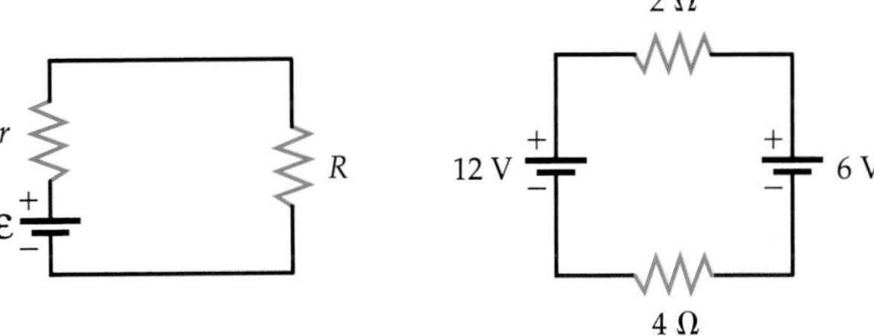

6. In the circuit in Figure 23-33, the batteries have negligible internal resistance, and the ammeter has negligible resistance. (*a*) Find the current through the ammeter. (*b*) Find the energy delivered by the 12-V battery in 3 s. (*c*) Find the total Joule heat produced in 3 s. (*d*) Account for the difference in your answers to parts (*b*) and (*c*).

Figure 23-33 Problem 6.

7. In the circuit in Figure 23-34, the batteries have negligible internal resistance. Find (*a*) the current in each resistor, (*b*) the potential difference between points *a* and *b*, and (*c*) the power supplied by each battery.

Figure 23-34 Problem 7.

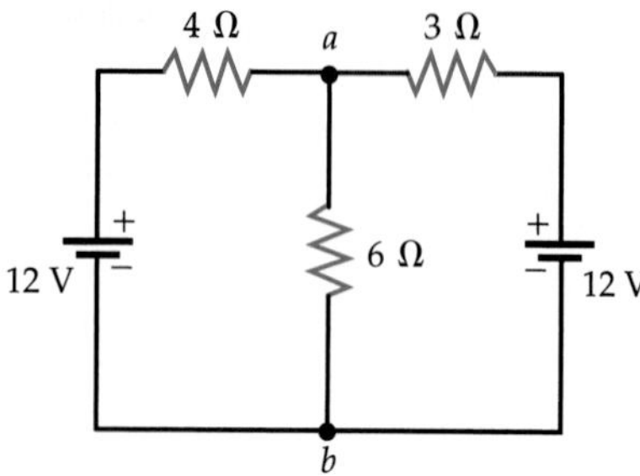

8. Repeat Problem 7 for the circuit in Figure 23-35.

Figure 23-35 Problem 8.

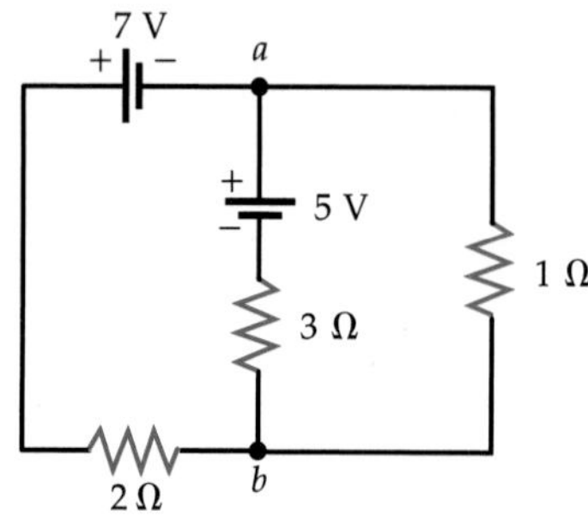

23-2 *RC* Circuits

9. A 6-μF capacitor is charged to 100 V and is then connected across a 500-Ω resistor. (*a*) What is the initial charge on the capacitor? (*b*) What is the initial current just after the capacitor is connected to the resistor? (*c*) What is the time constant of this circuit? (*d*) How much charge is on the capacitor after 6 ms?

10. (*a*) Find the initial energy stored in the capacitor of Problem 9. (*b*) Show that the energy stored in the capacitor is given by $U = U_0 e^{-2t/\tau}$, where U_0 is the initial energy and $\tau = RC$ is the time constant. (*c*) Sketch a plot of the energy U in the capacitor versus time t.

11. A 0.12-μF capacitor is given a charge Q_0. After 4 s, its charge is $\frac{1}{2}Q_0$. What is the effective resistance across this capacitor?

12. A 1.6-μF capacitor, initially uncharged, is connected in series with a 10-kΩ resistor and a 5.0-V battery of negligible internal resistance. (*a*) What is the charge on the capacitor after a very long time? (*b*) How long does it take the capacitor to reach 99 percent of its final charge?

13. A 2-MΩ resistor is connected in series with a 1.5-μF capacitor and a 6.0-V battery of negligible internal resistance. The capacitor is initially uncharged. After a time $t = \tau = RC$, find (*a*) the charge on the capacitor, (*b*) the rate at which the charge is increasing, (*c*) the current, (*d*) the power supplied by the battery, (*e*) the power dissipated in the resistor, and (*f*) the rate at which the energy stored in the capacitor is increasing.

14. Repeat Problem 13 for the time $t = 2\tau$.

23-3 Ammeters, Voltmeters, and Ohmmeters

15. A galvanometer has a resistance of 140 Ω. It requires 1.2 mA to give a full-scale deflection. (*a*) What resistance should be placed in parallel with the galvanometer to make an ammeter that gives a full-scale deflection for a current of 2 A? (*b*) What resistance should be placed in series with the galvanometer to make a voltmeter that gives a full-scale deflection for a potential difference of 5 V?

16. Sensitive galvanometers can detect currents as small as 1 pA. How many electrons per second produce this current?

17. A sensitive galvanometer has a resistance of 120 Ω and requires 1.4 μA of current to produce a full-scale deflection. (*a*) Find the shunt resistance needed to construct an ammeter that gives a full-scale deflection for a current of 1.0 mA. (*b*) What is the resistance of the ammeter? (*c*) What resistance would be required to construct a voltmeter that gives a full-scale deflection for a potential difference of 3.0 V?

18. A galvanometer with a resistance of 90 Ω gives a full-scale deflection when its current is 1.5 mA. It is used to construct an ammeter that gives a full-scale reading for a current of 200 A. (*a*) Find the shunt resistance needed. (*b*) What is the resistance of the ammeter? (*c*) If the shunt resistor consists of a piece of 10-gauge copper wire (diameter 2.59 mm), what should its length be?

19. The galvanometer in Problem 18 is used with a 1.5-V battery of negligible internal resistance to make an ohmmeter. (*a*) What resistance R_s should be placed in series with the galvanometer? (*b*) What resistance R will give a half-scale deflection? (*c*) What resistance R will give a deflection that is one-tenth full scale?

20. For the ohmmeter in Problem 19, show how the galvanometer scale should be calibrated by representing the scale by a straight line of length L, where the end of the line ($x = L$) represents a full-scale reading for $R = 0$. Divide the line into 10 equal parts and give the value of the resistance at each division.

21. A galvanometer with a resistance of 110 Ω gives a full-scale reading when its current is 0.13 mA. It is to be used in the multirange voltmeter shown in Figure 23-36, where the full-scale readings for the ranges are shown by the indicated connections. Determine R_1, R_2, and R_3.

Figure 23-36 Problem 21.

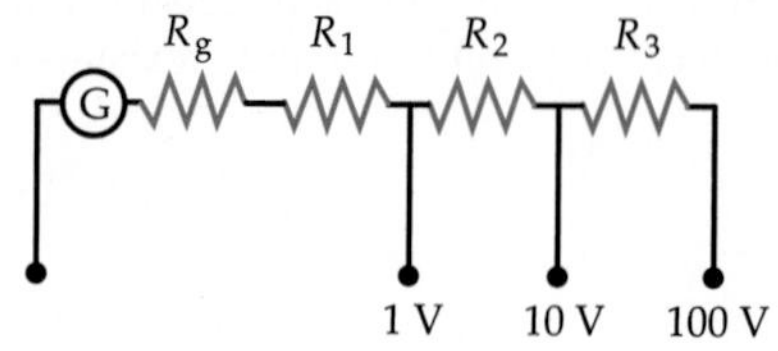

22. The galvanometer of Problem 21 is to be used in the multirange ammeter shown in Figure 23-37. Again, the full-scale reading for the ranges are shown by the indicated connections. Determine R_1, R_2, and R_3.

Figure 23-37 Problem 22.

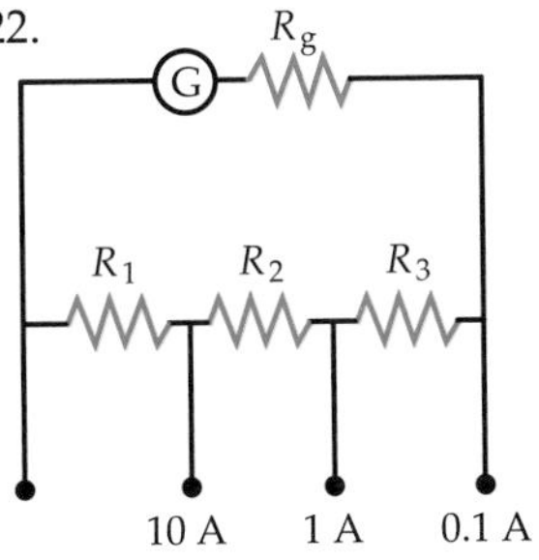

Level II

23. Two identical batteries, each with an emf $\mathcal{E}$ and an internal resistance r, can be connected across a resistance R either in series or in parallel. For which type of connection is the power supplied to R greater when (*a*) $R < r$ and (*b*) $R > r$?

24. A sick car battery with an emf of 11.4 V and an internal resistance of 0.01 Ω is connected to a load of 2.0 Ω. To help the ailing battery, a second battery with an emf of 12.6 V and an internal resistance of 0.01 Ω is connected by jumper cables to the terminals of the first battery. (*a*) Draw a diagram for this circuit. (*b*) Find the current in each part of the circuit. (*c*) Find the power delivered by the second battery and discuss where this power goes, assuming that the emfs and internal resistances of both batteries remain constant.

25. For the circuit in Figure 23-38, find (*a*) the current in each resistor, (*b*) the power supplied by each emf, and (*c*) the power dissipated in each resistor.

Figure 23-38 Problem 25.

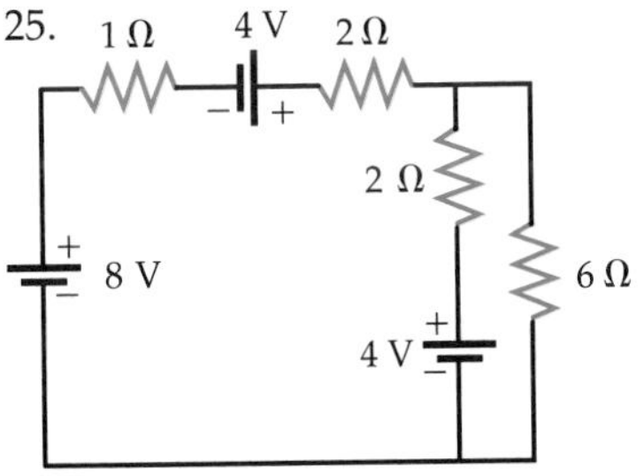

26. For the circuit in Figure 23-39, find the potential difference between points a and b.

Figure 23-39 Problem 26.

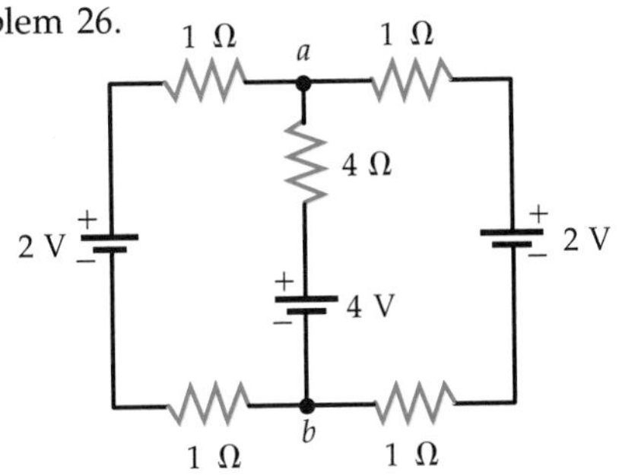

27. The space between the plates of a parallel-plate capacitor is filled with a dielectric of constant κ and resistivity ρ. (*a*) Show that the time constant for the decrease of charge on the plates is $\tau = \varepsilon_0 \kappa \rho$. (*b*) If the dielectric is mica, for which $\kappa = 5.0$ and $\rho = 9 \times 10^{13}$ Ω·m, find the time it takes for the charge to decrease to $1/e^2 \approx 14$ percent of its initial value.

28. The battery in the circuit in Figure 23-40 has an internal resistance of 0.01 Ω. (*a*) An ammeter with a resistance of 0.01 Ω is inserted at point a. What is the reading of the ammeter? (*b*) By what percentage is the current changed because of the ammeter? (*c*) The ammeter is removed and a voltmeter with a resistance of 1 kΩ is connected from a to b. What is the reading of the voltmeter? (*d*) By what percentage is the voltage drop from a to b changed by the presence of the voltmeter?

Figure 23-40 Problem 28.

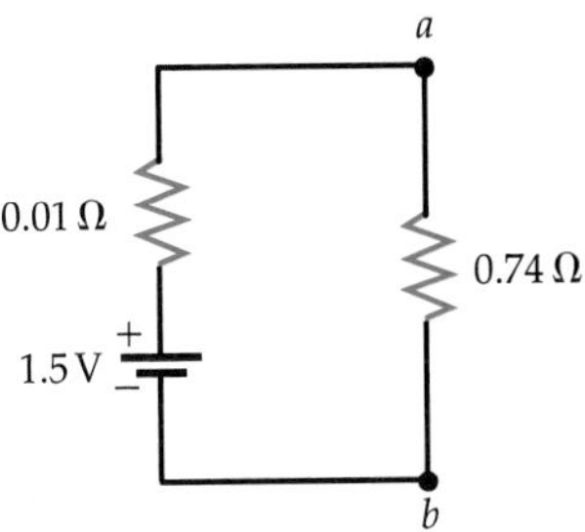

29. You have two batteries, one with $\mathcal{E} = 9.0$ V and $r = 0.8$ Ω and the other with $\mathcal{E} = 3.0$ V and $r = 0.4$ Ω. (*a*) Show how you would connect the batteries to give the largest current through a resistor R. Find the current for (*b*) $R = 0.2$ Ω, (*c*) $R = 0.6$ Ω, (*d*) $R = 1.0$ Ω, and (*e*) $R = 1.5$ Ω.

30. In the circuit in Figure 23-41, the reading of the ammeter is the same with both switches open and both closed. Find the resistance R.

Figure 23-41 Problem 30.

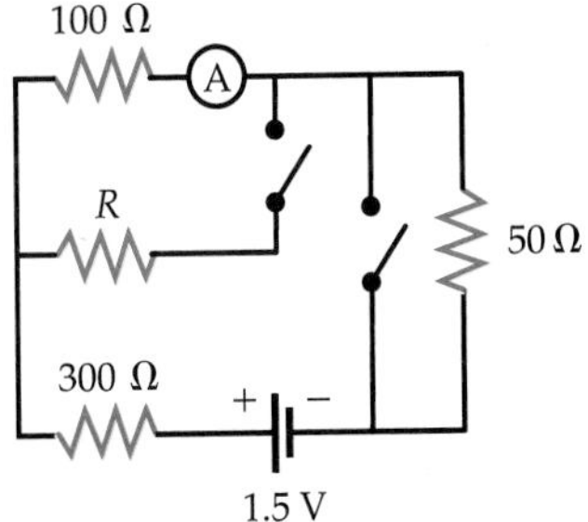

31. A galvanometer gives a full-scale deflection when the voltage across it is 10 mV and the current through it is 50 μA. (*a*) Design a voltmeter that gives a full-scale reading for a potential difference of 50 V using this galvanometer. (*b*) Design an ammeter that gives a full-scale reading for a current of 10 A using this galvanometer.

32. (*a*) Find the current in each part of the circuit shown in Figure 23-42. (*b*) Use your results of (*a*) to assign a potential at each indicated point assuming the potential at point *a* is zero.

Figure 23-42 Problem 32.

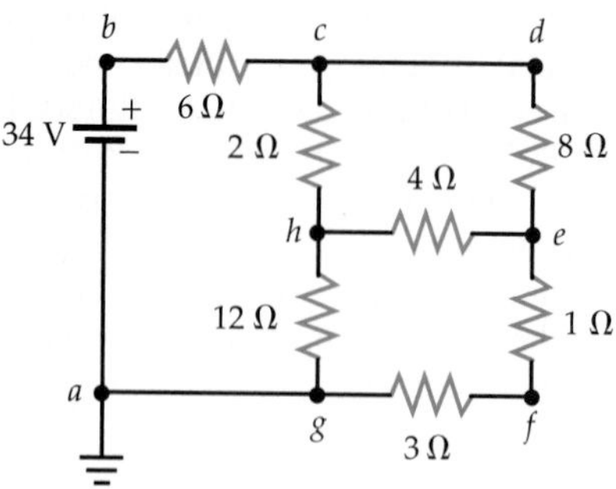

33. (*a*) Find the current in each part of the circuit shown in Figure 23-43. (*b*) Use your results of (*a*) to assign a potential at each indicated point assuming the potential at point *a* is zero.

Figure 23-43 Problem 33.

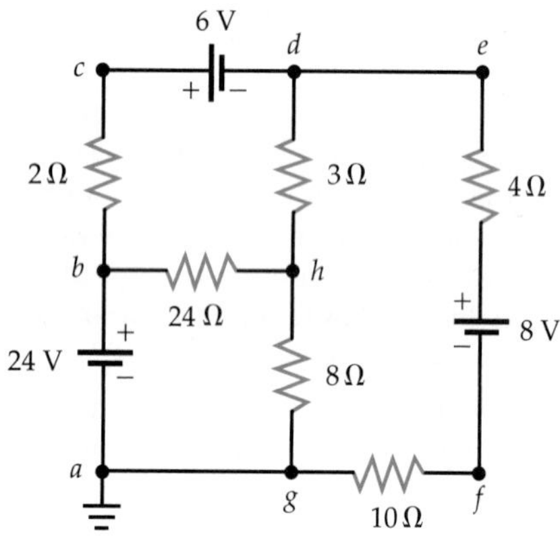

34. (*a*) Use symmetry to find the equivalent resistance of the network in Figure 23-44. (*b*) What is the current in each resistor if *R* is 10 Ω and a potential difference of 80 V is applied between *a* and *b*?

Figure 23-44 Problem 34.

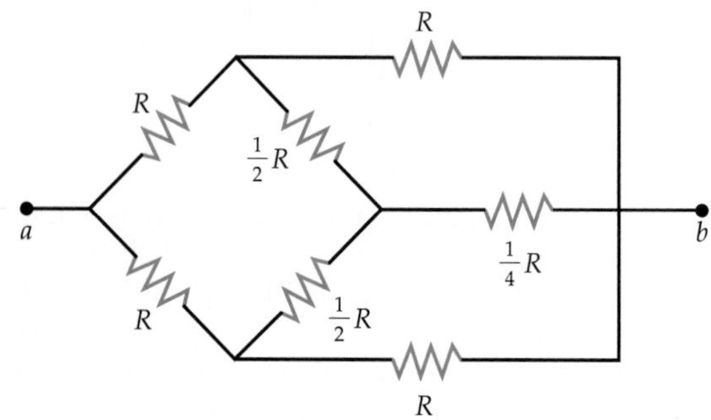

35. Nine 10-Ω resistors are connected as shown in Figure 23-45, and a potential difference of 20 V is applied between points *a* and *b*. (*a*) What is the equivalent resistance of this network? (*b*) Find the current in each of the nine resistors.

Figure 23-45 Problem 35.

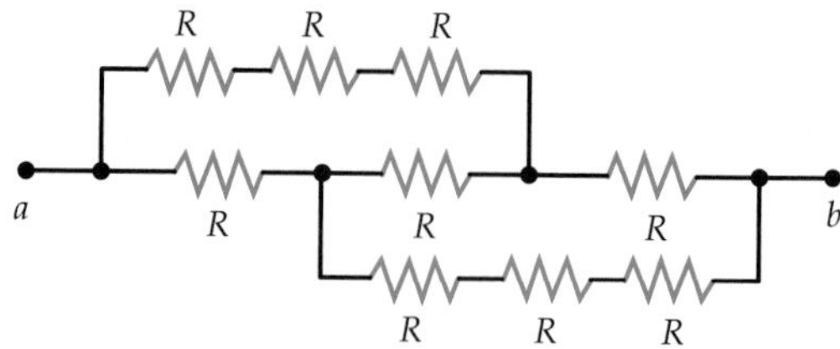

36. A parallel combination of an 8-Ω resistor and an unknown resistor *R* is connected in series with a 16-Ω resistor and a battery. The three resistors are then connected in series with each other and the same battery. In both arrangements, the current through the 8-Ω resistor is the same. What is the unknown resistance *R*?

37. A closed box has two metal terminals *a* and *b*. The inside of the box contains an unknown emf $\mathcal{E}$ in series with a resistance *R*. When a potential difference of 21 V is maintained between *a* and *b*, a current of 1 A enters the box at *a* and exits at *b*. If this potential difference is reversed, a current of 2 A in the reverse direction is observed. Find $\mathcal{E}$ and *R*.

38. A voltmeter has a resistance of 10^5 Ω. A 60-V battery with a 10-Ω internal resistance is connected across a 68-kΩ and a 56-kΩ resistor connected in series. (*a*) What is the voltmeter reading across the 68-kΩ resistor? (*b*) What is the voltmeter reading across the 56-kΩ resistor? (*c*) What is the voltmeter reading across the battery? (*d*) Give the percentage error of each voltage measurement.

39. The capacitors in the circuit in Figure 23-46 are initially uncharged. (*a*) What is the initial value of the battery current when switch S is closed? (*b*) What is the battery current after a long time? (*c*) What are the final charges on the capacitors?

Figure 23-46 Problem 39.

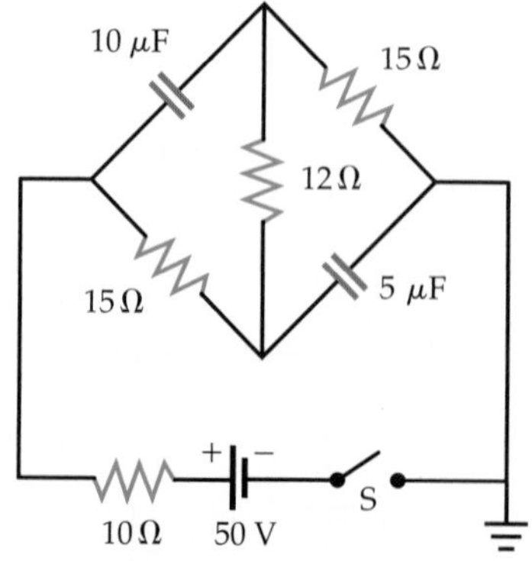

40. In the steady state, the charge on the 5-μF capacitor in the circuit in Figure 23-47 is 1000 μC. (*a*) Find the battery current. (*b*) Find the resistances R_1, R_2, and R_3.

Figure 23-47 Problem 40.

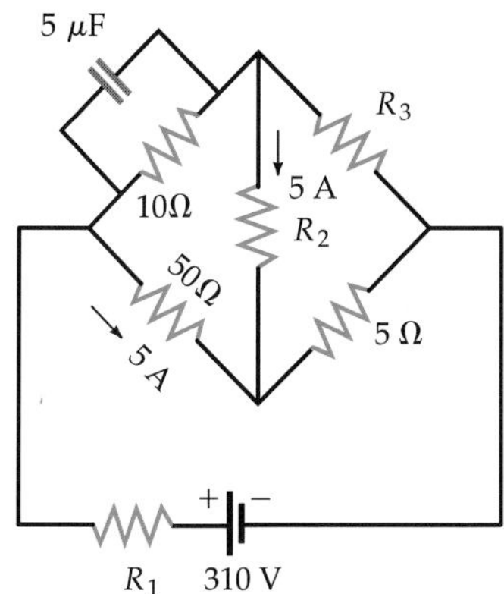

41. Consider the circuit shown in Figure 23-48. From your knowledge of how capacitors behave in circuits find (*a*) the initial current through the battery just after the switch is closed, (*b*) the steady-state current through the battery when the switch has been closed for a long time, and (*c*) the maximum voltage across the capacitor.

Figure 23-48 Problem 41.

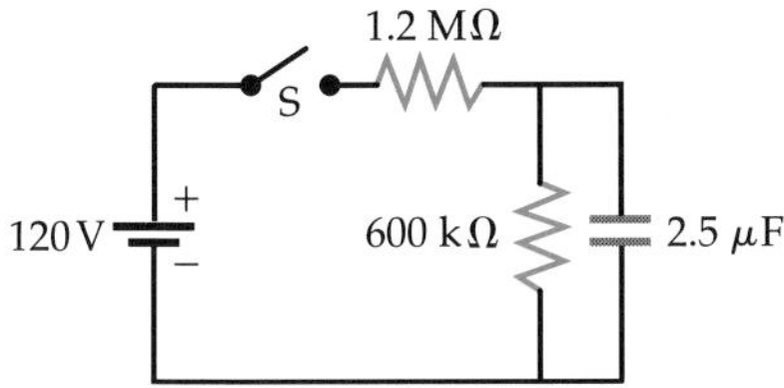

42. (*a*) What is the voltage across the capacitor in the circuit in Figure 23-49? (*b*) If the battery is disconnected, give the capacitor current as a function of time. (*c*) How long does it take the capacitor to discharge until the potential difference across it is 1 V?

Figure 23-49 Problems 42 and 59.

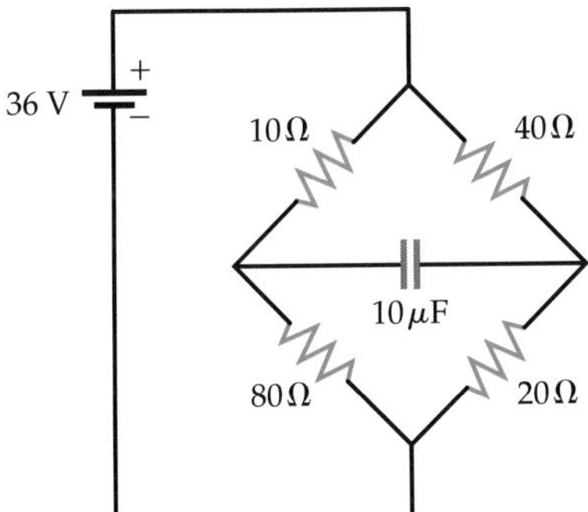

43. The circuit in Figure 23-50 is a slide-type *Wheatstone bridge*. It is used for determining an unknown resistance R_x in terms of the known resistances R_1, R_2, and R_0. The resistances R_1 and R_2 comprise a wire 1 m long. Point *a* is a sliding contact that is moved along the wire to vary these resistances. Resistance R_1 is proportional to the distance from the left end of the wire (0 cm) to point *a*, and R_2 is proportional to the distance from point *a* to the right end of the wire (100 cm). The sum of R_1 and R_2 remains constant. When points *a* and *b* are at the same potential, there is no current in the galvanometer and the bridge is said to be balanced. (Since the galvanometer is used to detect the absence of a current, it is called a *null detector*.) If the fixed resistance $R_0 = 200\ \Omega$, find the unknown resistance R_x if (*a*) the bridge balances at the 18-cm mark, (*b*) the bridge balances at the 60-cm mark, and (*c*) the bridge balances at the 95-cm mark.

Figure 23-50 Problem 43.

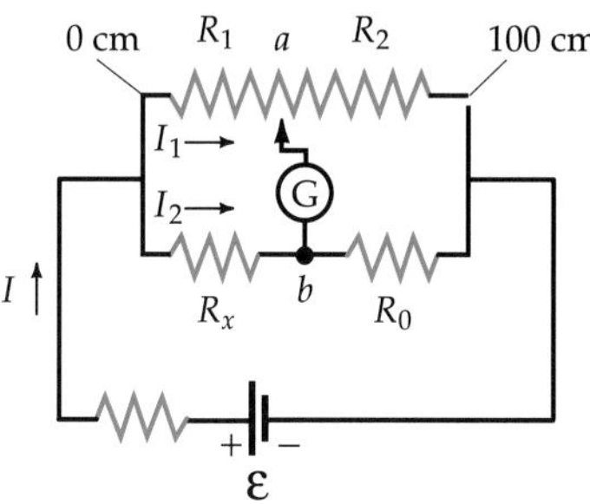

44. For the Wheatstone bridge of Problem 43, the bridge balances at the 98-cm mark when $R_0 = 200\ \Omega$. (*a*) What is the unknown resistance? (*b*) What effect would an error of 2 mm in the location of the balance point have on the measured value of the unknown resistance? (*c*) How should R_0 be changed so that the balance point for this unknown resistor will be nearer the 50-cm mark?

Level III

45. In Problem 24, assume that the emf of the first battery increases at a constant rate of 0.2 V/h while the emf of the second battery and the internal resistances remain constant. (*a*) Find the current in each part of the circuit as a function of time. (*b*) Sketch a graph of the power delivered to the first battery as a function of time.

46. In the ammeter in Figure 23-51, a galvanometer with a resistance of 10 Ω is connected across a 90-Ω resistor. Different ranges of current can be measured by choosing connections *ab*, *ac*, *ad*, or *ae*. (*a*) How should the 90-Ω resistor be divided so that the current that causes full-scale deflection decreases by a factor of 10 for each successive connection? (*b*) What should be the value of the current I_g for which the galvanometer gives a full-scale deflection so that the ranges of this ammeter are 1 A, 100 mA, 10 mA, and 1.0 mA?

Figure 23-51 Problem 46.

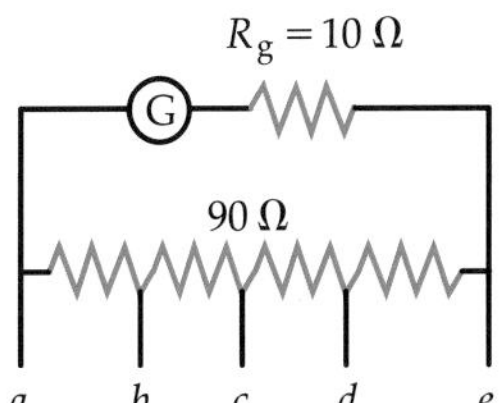

47. Figure 23-52 shows two ways a voltmeter and an ammeter can be used to measure an unknown resistance R. Assume that the internal resistance of the battery is negligible and that the resistance of the voltmeter is 1000 times that of the ammeter, that is, $R_v = 1000R_a$. The calculated value of R is taken to be $R_c = V/I$, where V and I are the readings of the voltmeter and the ammeter. (a) Which circuit in the figure should be preferred for values of R in the range from $10R_a$ to $0.9R_v$? Why? Find R_c for each circuit if $R_a = 0.1\ \Omega$, $R_v = 100\ \Omega$, and (b) $R = 0.5\ \Omega$, (c) $R = 3\ \Omega$, and (d) $R = 80\ \Omega$.

Figure 23-52 Problems 47 and 48.

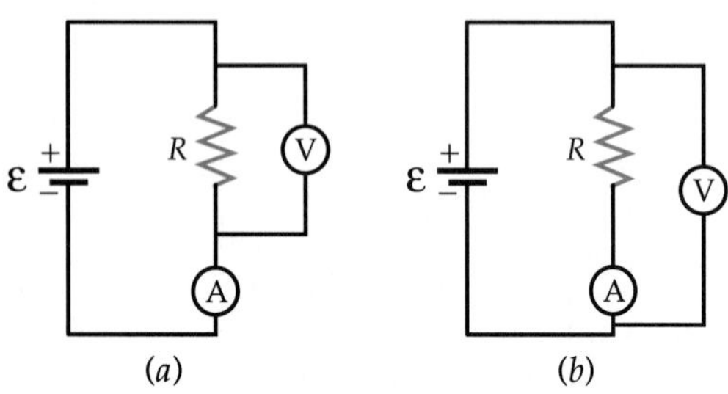

48. (a) For the circuits in Figure 23-52, show that $R_c = V/I$ is related to the true value R by $1/R_c = 1/R + 1/R_v$ for circuit a and $R_c = R + R_a$ for circuit b. (See Problem 47.) If $\mathcal{E} = 1.5$ V, $R_a = 0.01\ \Omega$, and $R_v = 10$ kΩ, for what range of values of R is R_c within 5 percent of R using (b) circuit a and (c) circuit b?

49. In the circuit in Figure 23-53, r is the internal resistance of the source of emf and R_a is the resistance of the ammeter. (a) Show that the ammeter reading is given by

$$\mathcal{E}\left(R_2 + R_a + r + \frac{R_2 + R_a}{R_1}r\right)^{-1}$$

(b) Show that if the ammeter and source of emf are interchanged, the ammeter reading will be

$$\mathcal{E}\left(R_2 + R_a + r + \frac{R_2 + r}{R_1}R_a\right)^{-1}$$

Note that if $R_a = r$ or if both are negligible, the ammeter reading is the same in both cases. (When R_a and r can be neglected, this symmetry can be very useful in analyzing circuits that have only one source of emf. It is not valid for circuits that have more than one source of emf, however.)

Figure 23-53 Problem 49.

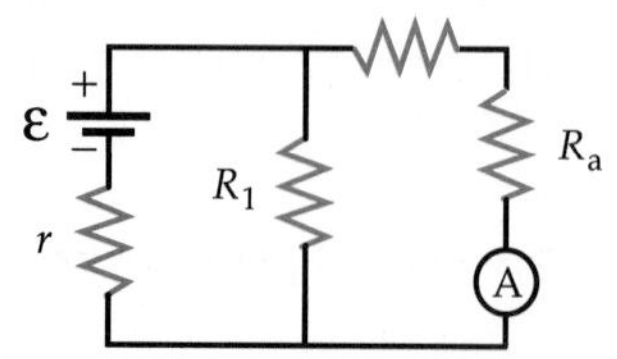

50. In the circuit in Figure 23-54 the capacitor is initially uncharged and the switch is open. At time $t = 0$, the switch is closed. (a) What current is supplied by the emf just after the switch is closed? (b) What current is supplied by the emf a long time after the switch is closed? (c) Derive an expression for the current through the emf at any time after the switch is closed. (d) After a long time t', the switch is opened. How long does it take for the charge on the capacitor to decrease to 10 percent of its value at $t = t'$ if $R_1 = R_2 = 5$ kΩ and $C = 1.0\ \mu$F?

Figure 23-54 Problem 50.

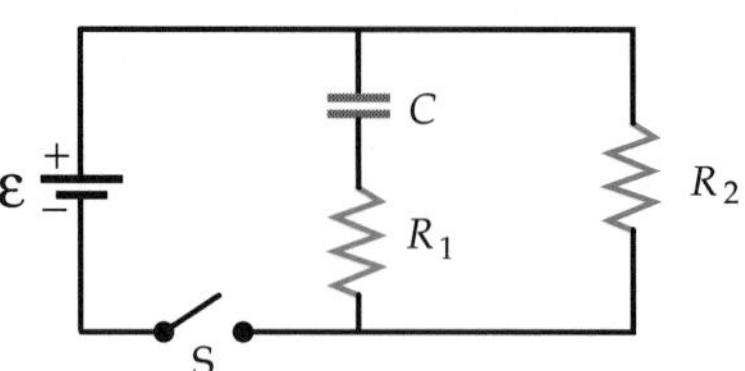

51. Two batteries with emfs $\mathcal{E}_1$ and $\mathcal{E}_2$ and internal resistances r_1 and r_2 are connected in parallel. Prove that the optimal load resistance R (for the delivery of maximum power) connected in parallel with this combination is $R = r_1r_2/(r_1 + r_2)$.

52. Figure 23-55 shows an infinite, two-dimensional, planar network of equal resistors. If the resistance of each resistor is R, find the equivalent resistance between points a and b.

Figure 23-55 Problem 52.

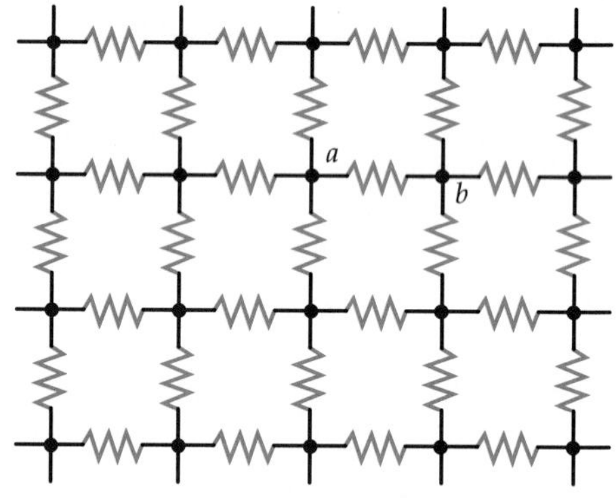

53. Consider an infinite, two-dimensional, periodic, triangular lattice of resistors. If R is the resistance of each resistor, what is the equivalent resistance across any resistor?

54. Consider a three-dimensional, periodic, cubic lattice of resistors that extends to infinity in all directions. If R is the resistance of each resistor, what is the equivalent resistance across any resistor?

55. Each of the six terminals a, b, c, d, e, and f in Figure 23-56 is connected to each other terminal by a wire of resistance R. The wires are insulated so that they make electrical contact only with the terminals. Use symmetry to find the resistance between any two terminals.

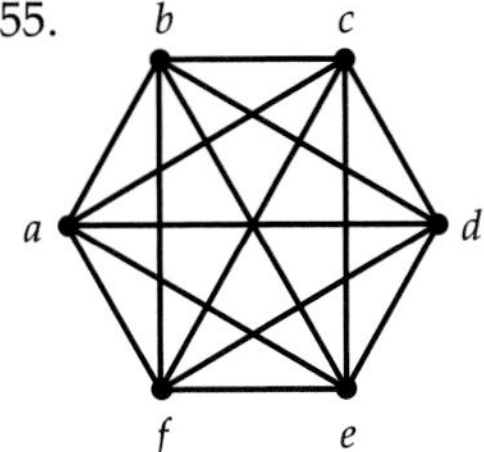

Figure 23-56 Problem 55.

56. (*a*) Find the equivalent resistance between points a and b for the 12-resistor cube in Figure 23-16. (*b*) What is the equivalent resistance between a and b if the resistor directly between these two points is removed?

57. An infinite chain of resistors is shown in Figure 23-57. Find the equivalent resistance between points a and b. *Hint:* The resistance R_{ab} is the same as $R_{a'b'}$ if the section to the left of $a'b'$ is removed. The network remains an infinite chain with the same structure.

Figure 23-57 Problem 57.

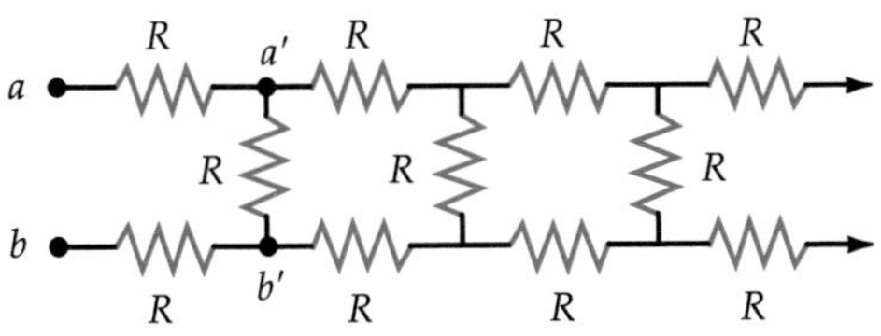

58. An infinite chain of resistors is shown in Figure 23-58. Find the input resistance. (See Problem 57.)

Figure 23-58 Problem 58.

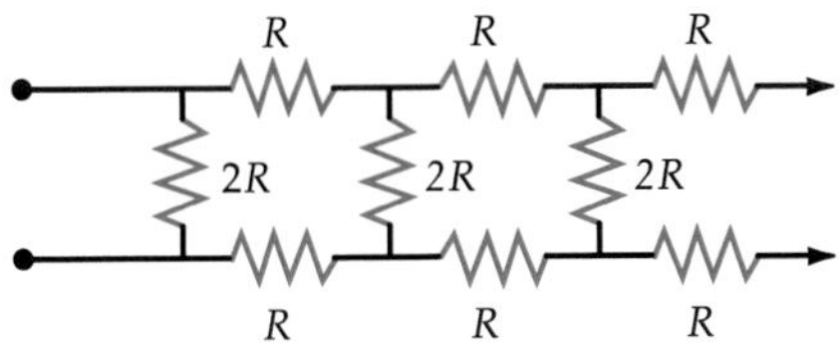

59. If the capacitor in the circuit in Figure 23-49 is replaced by a 30-Ω resistor, what currents flow through the resistors?

60. For the circuit in Figure 23-59, (*a*) what is the initial battery current immediately after switch S is closed? (*b*) What is the battery current a long time after switch S is closed? (*c*) What is the current in the 600-Ω resistor as a function of time?

Figure 23-59 Problem 60.

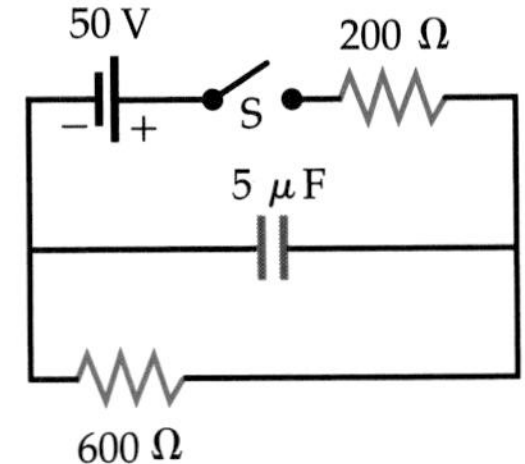

61. For the circuit in Figure 23-60, (*a*) what is the initial battery current immediately after switch S is closed? (*b*) What is the battery current a long time after switch S is closed? (*c*) If the switch has been closed for a long time and is then opened, find the current through the 600-kΩ resistor as a function of time.

Figure 23-60 Problem 61.

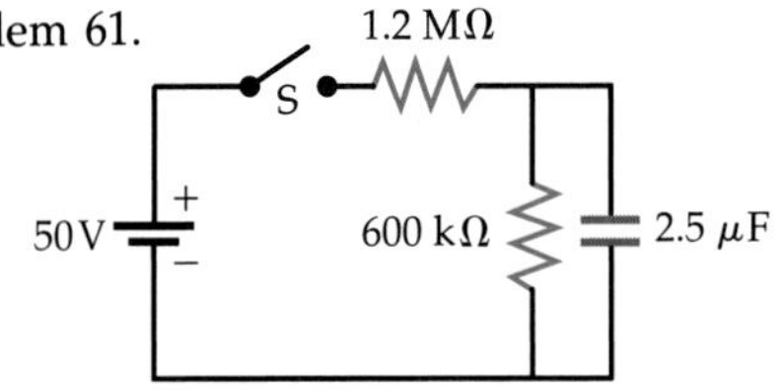

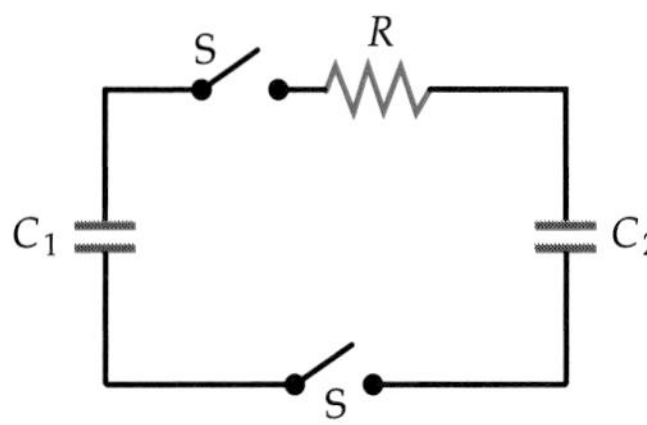

62. Capacitors C_1 and C_2 are connected in parallel by a resistor and two switches as shown in Figure 23-61. Capacitor C_1 is initially charged to a voltage V_0, and capacitor C_2 is uncharged. The switches S are then closed. (*a*) What are the final charges on C_1 and C_2? (*b*) Compare the initial and final stored energies of the system. (*c*) What caused the decrease in the capacitor-stored energy?

Figure 23-61 Problem 62.

63. (*a*) In Problem 62, find the current through R after the switches S are closed as a function of time. (*b*) Find the energy dissipated in the resistor as a function of time. (*c*) Find the total energy dissipated in the resistor and compare it with the loss of stored energy found in part (*b*) of Problem 62.

64. In the circuit in Figure 23-62, the capacitors are initially uncharged. Switch S_2 is closed and then switch S_1 is closed. (*a*) What is the battery current immediately after S_1 is closed? (*b*) What is the battery current a long time after both switches are closed? (*c*) What is the final voltage across C_1? (*d*) What is the final voltage across C_2? (*e*) Switch S_2 is opened again after a long time. Give the current in the 150-Ω resistor as a function of time.

Figure 23-62 Problem 64.

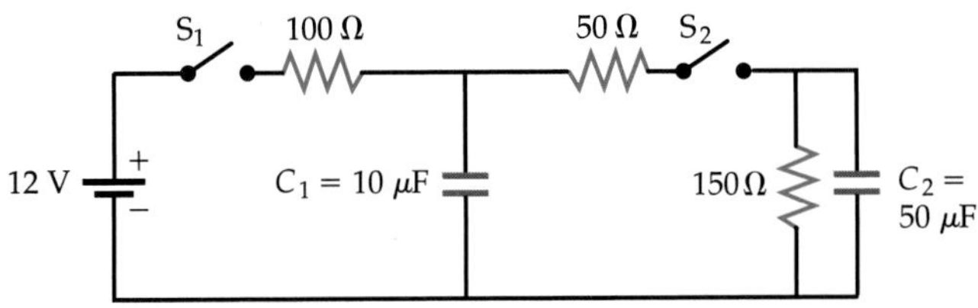

65. In the *RC* circuit in Figure 23-63 the capacitor is initially uncharged and the switch is closed at time $t = 0$. (*a*) What is the power supplied by the battery as a function of time? (*b*) What is the power dissipated in the resistor as a function of time? (*c*) What is the rate at which energy is stored in the capacitor as a function of time? Plot your answers to parts (*a*), (*b*), and (*c*) versus time on the same graph. (*d*) Find the maximum rate at which energy is stored in the capacitor as a function of the battery voltage $\mathcal{E}$ and the resistance *R*. At what time does this maximum occur?

Figure 23-63 Problem 65.

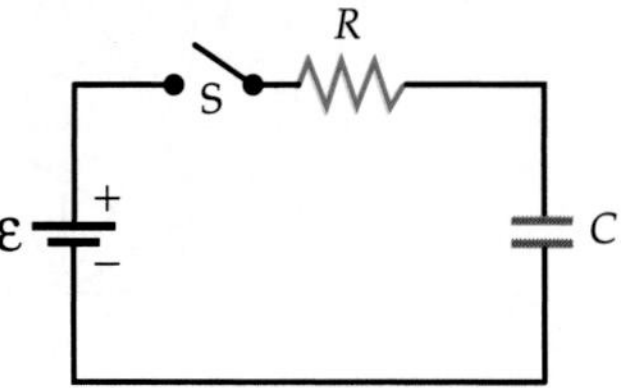

Chapter 24

The Magnetic Field

This experimental Maglev train using magnetic repulsion for levitation, guidance, and propulsion has achieved speeds greater than 300 km/h.

When the existence of magnetism was first noted is not known. However, more than 2000 years ago, the Greeks were aware that a certain type of stone (now called magnetite) attracts pieces of iron, and there are written references to the use of magnets for navigation dating from the twelfth century.

In 1269, Pierre de Maricourt discovered that a needle laid at various positions on a spherical natural magnet orients itself along lines that pass through points at the opposite ends of the sphere. He called these points the poles of the magnet. Subsequently, many experimenters noted that every magnet of whatever shape has two poles, a north pole and a south pole, where the force exerted by the magnet is strongest. It was also noted that the like poles of two magnets repel each other and the unlike poles attract each other.

In 1600, William Gilbert discovered that the earth itself is a natural magnet with magnetic poles near the north and south geographic poles. (Since the north pole of a compass needle points north, what we call the north magnetic pole is actually a south pole, as illustrated in Figure 24-1.) Around 1750, John Michell did a quantitative study of the attraction and repulsion of magnetic poles using a torsion balance. He discovered that the force exerted

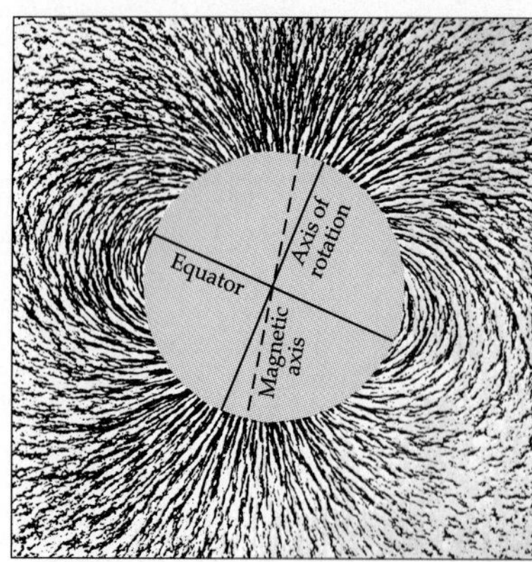

Figure 24-1 Magnetic-field lines of the earth indicated by iron filings around a uniformly magnetized sphere. The lines are somewhat similar to but not exactly the same as those of a bar magnet.

by one pole on another varies inversely with the square of the distance between the poles. These results were confirmed by Coulomb shortly thereafter.

Although the force between two magnetic poles is similar to that between two electric charges, there is an important difference between electric charges and magnetic poles, namely, magnetic poles always occur in pairs. If a magnet is broken in half, there will be equal and opposite poles at either side of the break point; that is, there will be two magnets, each with a north and south pole. There has been much speculation throughout the years as to the existence of an isolated magnetic pole, and in recent years considerable experimental effort has been made to find such an object. Thus far, there seems to be no conclusive evidence that an isolated magnetic pole exists.

The connection between electricity and magnetism was not known until the nineteenth century, when Hans Christian Oersted discovered that an electric current affects the orientation of a compass needle. Subsequent experiments by André-Marie Ampère and others showed that electric currents attract bits of iron and that parallel currents attract each other. Ampère proposed a theoretical model of magnetism that still serves as the basis of the modern theory of magnetism. He speculated that the fundamental source of magnetism is not a magnetic pole but rather an electric current. He further proposed that the magnetism of a permanent magnet is due to the alignment of molecular current loops within the material. Today, we know that these current loops result partly from the motion of electrons within the atom and partly from electron spin, a quantum-mechanical property of the electron. The basic magnetic interaction is the magnetic force one moving charge exerts on another moving charge. This force is in addition to the electric force between the two charges. As with the electric force, we consider the magnetic force to be transmitted by another agent, the magnetic field. The moving charge produces a **magnetic field,** and the field, in turn, exerts a force on the other moving charge. Since a moving charge constitutes an electric current, the magnetic interaction can also be thought of as an interaction between two currents.

In the early 1830s, Michael Faraday and Joseph Henry demonstrated in independent experiments that a changing magnetic field produces an electric field. Some years later (about 1860), James Clerk Maxwell developed a complete theory of electricity and magnetism that showed that a changing electric field produces a magnetic field.

In this chapter, we will consider only the effects of a given magnetic field on moving charges and on wires carrying currents. The sources of magnetic fields will be discussed in Chapter 28.

24-1 The Force Exerted by a Magnetic Field

The existence of a magnetic field* **B** at some point in space can be demonstrated in a straightforward way. We simply place a compass needle at that point and see if it tends to align in a particular direction. If there are no magnets or electric currents nearby, the needle will point in the direction of the magnetic field of the earth. If there are magnets or electric currents nearby, the needle will point in the direction of the net magnetic field due to the earth and the magnets or currents.

*For historical reasons, the magnetic field **B** is sometimes called the *magnetic-induction vector* or the *magnetic flux density*. We will refer to it as the magnetic field.

It is observed experimentally that, when a charge q has velocity **v** in a magnetic field, there is a force on it that depends on q and on the magnitude and direction of the velocity. Let us assume that we know the direction of the magnetic field **B** at a point in space from a measurement with a compass. Experiments with various charges moving with various velocities at such a point give the following results for the magnetic force:

1. The force is proportional to the charge q. The force on a negative charge is in the direction opposite that on a positive charge with the same velocity.
2. The force is proportional to the speed v.
3. The force is perpendicular to both the magnetic field and the velocity.
4. The force is proportional to $\sin\theta$, where θ is the angle between the velocity **v** and the magnetic field **B**. If **v** is parallel or antiparallel to **B**, the force is zero.

These experimental results can be summarized as follows. When a charge q moves with velocity **v** in a magnetic field **B**, the magnetic force **F** on the charge is

$$\mathbf{F} = q\mathbf{v} \times \mathbf{B} \qquad 24\text{-}1$$

Magnetic force on a moving charge

Since **F** is perpendicular to both **v** and **B**, it is perpendicular to the plane defined by these two vectors. The direction of **F** is given by the right-hand rule as **v** is rotated into **B**, as illustrated in Figure 24-2.

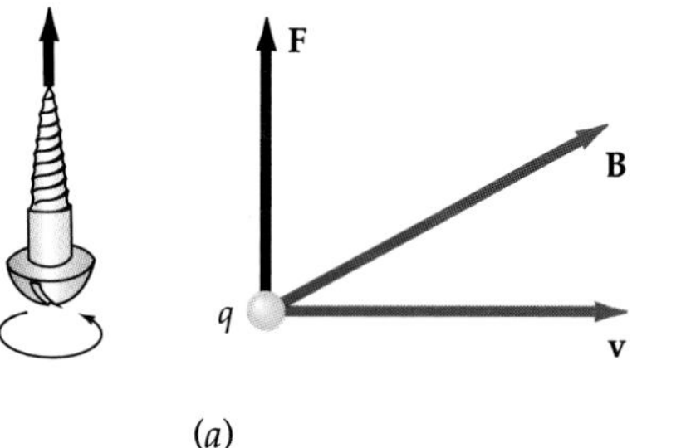

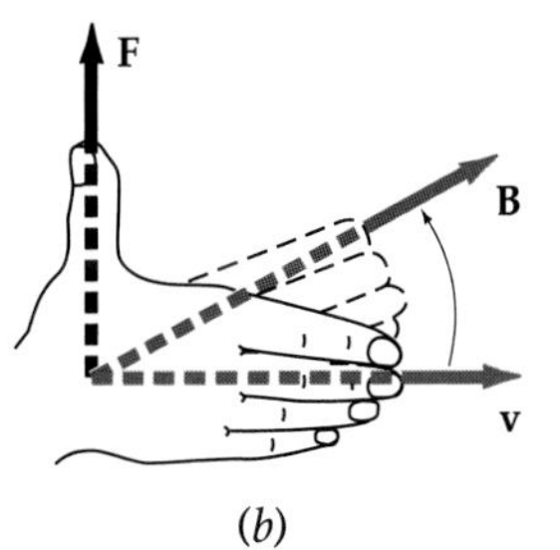

Figure 24-2 Right-hand rule for determining the direction of the magnetic force exerted on a charge moving in a magnetic field. (*a*) The force is perpendicular to both **v** and **B** and is in the direction in which a right-hand-threaded screw will advance if it is turned in the direction that will rotate **v** into **B** through the smaller of the two possible angles. (*b*) If the fingers of the right hand are in the direction of **v** such that they can be curled into **B**, the thumb points in the direction of **F**.

Examples of the direction of the forces exerted on moving charges when the magnetic-field vector **B** is in the vertical direction are given in Figure 24-3. Note that the direction of any particular magnetic field **B** can be found experimentally by measuring **F** and **v** and then applying Equation 24-1.

Equation 24-1 defines the **magnetic field B** in terms of the force exerted on a moving charge. The SI unit of magnetic field is the **tesla** (T). A charge of

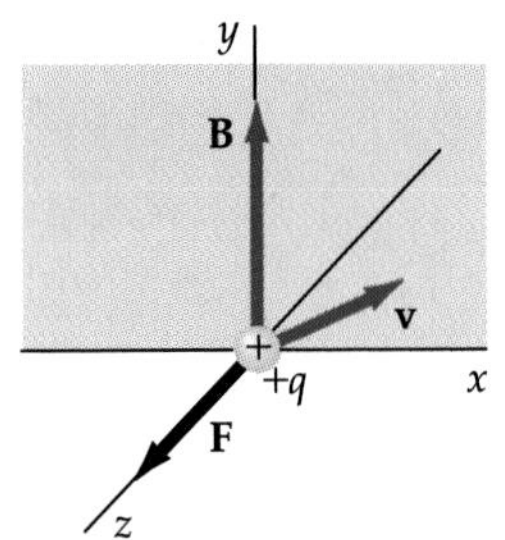

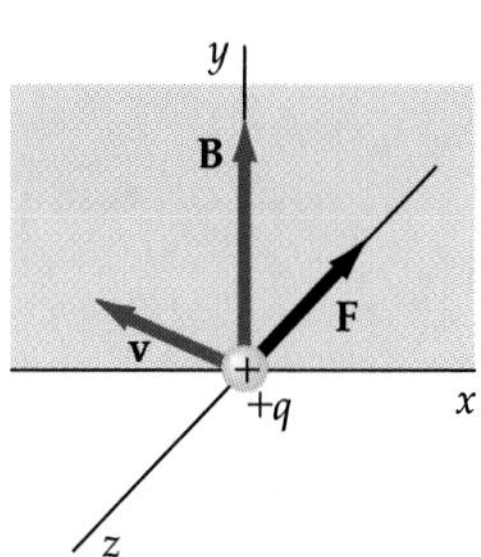

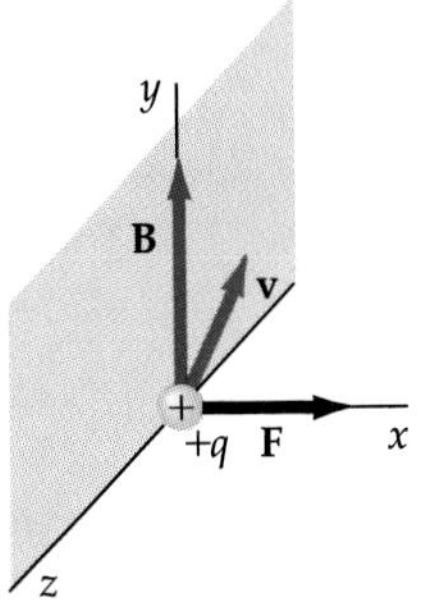

Figure 24-3 Direction of the magnetic force on a charged particle moving with velocity **v** in a magnetic field **B**. The shading indicates the plane of **v** and **B**.

one coulomb moving with a velocity of one meter per second perpendicular to a magnetic field of one tesla experiences a force of one newton:

$$1\ \mathrm{T} = 1\ \frac{\mathrm{N/C}}{\mathrm{m/s}} = 1\ \mathrm{N/A{\cdot}m} \qquad 24\text{-}2$$

This unit is rather large. The magnetic field of the earth is slightly less than 10^{-4} T. The magnetic fields near powerful permanent magnets are about 0.1 to 0.5 T, and powerful laboratory and industrial electromagnets produce fields of 1 to 2 T. Fields greater than 10 T are quite difficult to produce because the resulting magnetic forces will tear the magnets apart or crush them. A commonly used unit, derived from the cgs system, is the **gauss** (G), which is related to the tesla as follows:

$$1\ \mathrm{T} = 10^4\ \mathrm{G} \qquad 24\text{-}3$$

Since magnetic fields are often given in gauss, which is not an SI unit, it is important to remember to convert from gauss to teslas when making calculations.

Example 24-1

The magnetic field of the earth has a magnitude of 0.6 G and is directed downward and northward, making an angle of about 70° with the horizontal. (The magnitude and direction of the earth's magnetic field vary from place to place. These data are approximately correct for the central United States.) A proton of charge $q = 1.6 \times 10^{-19}$ C is moving horizontally in the northward direction with speed $v = 10^7$ m/s. Calculate the magnetic force on the proton.

Figure 24-4 shows the directions of the magnetic field **B** and the proton's velocity **v**. The angle between them is $\theta = 70°$. The magnetic force is parallel to **v × B**, which is west for a proton moving north. The magnitude of the magnetic force is

$$F = qvB \sin \theta$$
$$= (1.6 \times 10^{-19}\ \mathrm{C})(10^7\ \mathrm{m/s})(0.6 \times 10^{-4}\ \mathrm{T})(0.94) = 9.02 \times 10^{-17}\ \mathrm{N}$$

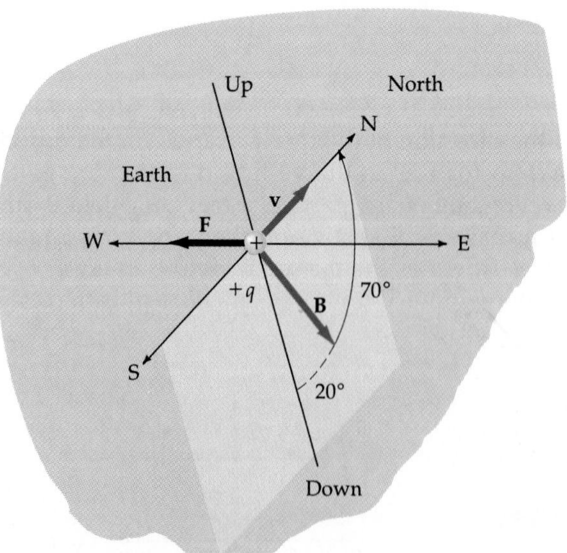

Figure 24-4 Magnetic force on a proton moving north in the magnetic field of the earth, which dips below the horizontal north direction at an angle of 70°, for Example 24-1. The force is directed toward the west.

It is instructive to do this example using unit vectors. We choose the x and y directions to be east and north, respectively, and the z direction to be upwards as shown in Figure 24-5. Then the velocity vector is in the y

direction, and the magnetic field of the earth has the components $B_x = 0$, $B_y = B \cos 70° = (0.6 \times 10^{-4}\ \text{T})(0.342) = 2.05 \times 10^{-5}\ \text{T}$, and $B_z = -B \sin 70° = (-0.6 \times 10^{-4}\ \text{T})(0.940) = -5.64 \times 10^{-5}\ \text{T}$. The magnetic-field vector is thus given by

$$\mathbf{B} = 0\ \mathbf{i} + 2.05 \times 10^{-5}\ \text{T}\ \mathbf{j} - 5.64 \times 10^{-5}\ \text{T}\ \mathbf{k}$$

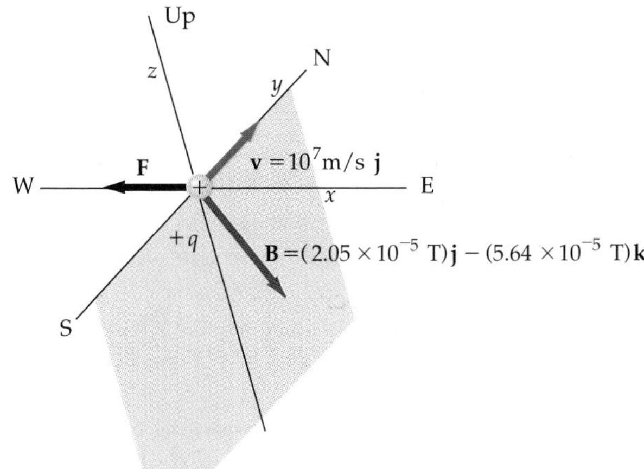

Figure 24-5 Coordinate system for expressing the velocity **v**, magnetic field **B**, and force **F** in terms of the unit vectors **i**, **j**, and **k**.

and the magnetic force on the proton is

$$\begin{aligned}\mathbf{F} &= q\mathbf{v} \times \mathbf{B} \\ &= (1.6 \times 10^{-19}\ \text{C})(10^7\ \text{m/s}\ \mathbf{j}) \\ &\quad \times (0\ \mathbf{i} + 2.05 \times 10^{-5}\ \text{T}\ \mathbf{j} - 5.64 \times 10^{-5}\ \text{T}\ \mathbf{k})\end{aligned}$$

Since $\mathbf{j} \times \mathbf{j} = 0$ and $\mathbf{j} \times \mathbf{k} = \mathbf{i}$, we have

$$\begin{aligned}\mathbf{F} &= (1.6 \times 10^{-19}\ \text{C})(10^7\ \text{m/s}\ \mathbf{j}) \times (-5.64 \times 10^{-5}\ \text{T}\ \mathbf{k}) \\ &= -9.02 \times 10^{-17}\ \text{N}\ \mathbf{i}\end{aligned}$$

Exercise

Find the force on a proton moving with velocity $\mathbf{v} = 4 \times 10^6$ m/s **i** in a magnetic field $\mathbf{B} = 2.0$ T **k**. (Answer: -1.28×10^{-12} N **j**)

When a wire carries a current in a magnetic field, there is a force on the wire that is equal to the sum of the magnetic forces on the charged particles whose motion produces the current. Figure 24-6 shows a short segment of wire of cross-sectional area A and length ℓ carrying a current I. If the wire is in a magnetic field **B**, the magnetic force on each charge is $q\mathbf{v}_d \times \mathbf{B}$, where $\mathbf{v}_d$ is the drift velocity of the charge carriers. The number of charges in the wire segment is the number n per unit volume times the volume $A\ell$. Thus the total force on the wire segment is

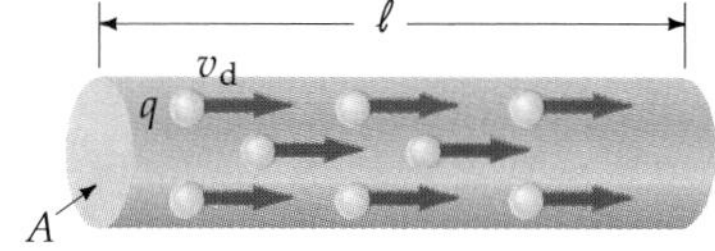

Figure 24-6 Wire segment of length ℓ carrying a current I. If the wire is in a magnetic field, there will be a force on each charge carrier resulting in a force on the wire.

$$\mathbf{F} = (q\mathbf{v}_d \times \mathbf{B})nA\ell$$

From Equation 22-3, the current in the wire is

$$I = nqv_dA$$

Hence the force can be written

$$\mathbf{F} = I\boldsymbol{\ell} \times \mathbf{B} \qquad 24\text{-}4$$

Magnetic force on a segment of current-carrying wire

where $\boldsymbol{\ell}$ is a vector whose magnitude is the length of the wire and whose direction is parallel to $q\mathbf{v}_d$, the direction of the current I. For the current in the positive x direction and the magnetic field in the xy plane shown in Figure 24-7, the force on the wire is directed along the positive z axis, as indicated. In Equation 24-4 it is assumed that the wire segment is straight and that the magnetic field does not vary over its length. The equation can be generalized for an arbitrarily shaped wire in any magnetic field. If we choose a very small wire segment $d\boldsymbol{\ell}$ and write the force on this segment as $d\mathbf{F}$, we have

Magnetic force on a current element

$$d\mathbf{F} = I\, d\boldsymbol{\ell} \times \mathbf{B} \qquad 24\text{-}5$$

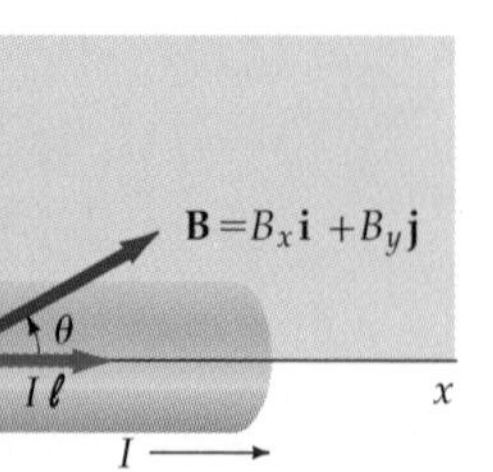

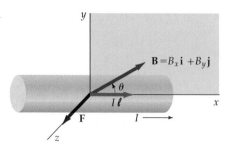

Figure 24-7 Magnetic force on a current-carrying segment of wire in a magnetic field. The current is in the x direction. The magnetic field is in the xy plane and makes an angle θ with the x axis. The force **F** is in the z direction, perpendicular to both **B** and $I\boldsymbol{\ell}$. Its magnitude is $I\ell B \sin\theta$.

where **B** is the magnetic-field vector at the segment. The quantity $I\, d\boldsymbol{\ell}$ is called a **current element.** We find the total force on the wire by summing (integrating) over all the current elements in the wire, using the appropriate field **B** at each element.

Equation 24-5 is the same as Equation 24-1 with the current element $I\, d\boldsymbol{\ell}$ replacing $q\mathbf{v}$. It defines the magnetic field **B** in terms of the force extended on a current element.

Just as the electric field **E** can be represented by electric-field lines, the magnetic field **B** can be represented by **magnetic-field lines.** In both cases, the direction of the field is indicated by the direction of the field lines and the magnitude of the field is indicated by their density.

There are, however, two important differences between electric-field lines and magnetic-field lines. The first has to do with the direction of the force the field exerts on a charge. The electric force on a positive charge is in the direction of the electric field and thus of the electric-field lines. The magnetic force on a moving charge, however, is perpendicular to the magnetic field, so magnetic-field lines are *not* in the direction of the magnetic force on a moving charge.

The second difference is that electric-field lines begin on positive charges and end on negative charges. However, because isolated magnetic poles apparently do not exist, there are no points in space where magnetic-field lines begin or end. Instead, they form closed loops. Figure 24-8 shows the magnetic-field lines both inside and outside a bar magnet.

Figure 24-8 (*a*) Magnetic-field lines inside and outside a bar magnet. The lines emerge from the north pole and enter the south pole, but they have no beginning or end. Instead, they form closed loops. (*b*) Magnetic-field lines outside a bar magnet, as indicated by iron filings.

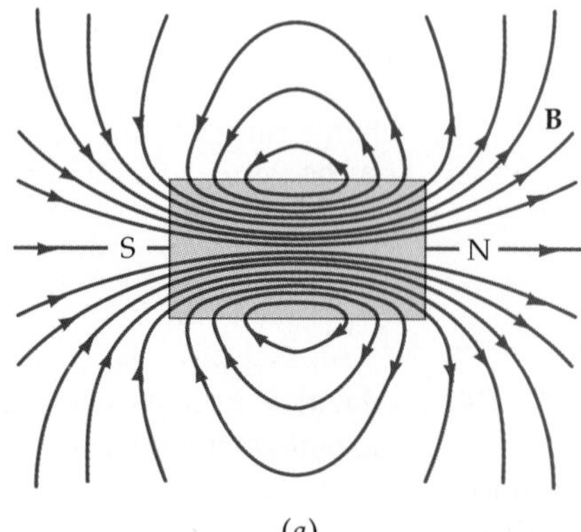

(*a*)

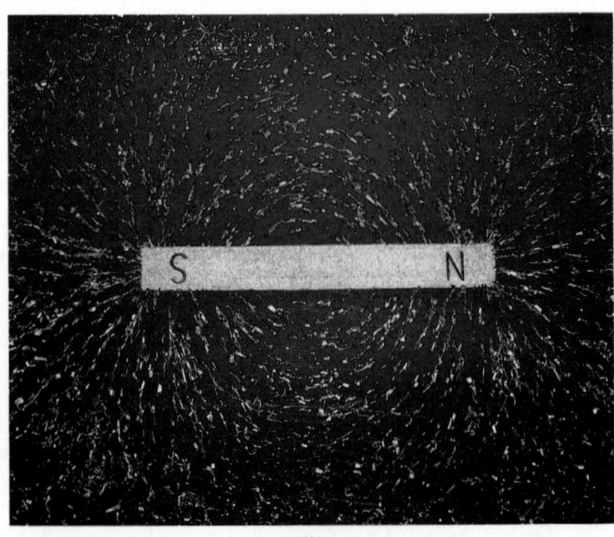

(*b*)

Example 24-2

A wire segment 3 mm long carries a current of 3 A in the x direction. It lies in a magnetic field of magnitude 0.02 T that is in the xy plane and makes an angle of 30° with the x axis as shown in Figure 24-7. What is the magnetic force exerted on the wire segment?

From the figure, we can see that the magnetic force is in the z direction. Its magnitude is given by Equation 24-4:

$$\mathbf{F} = I\boldsymbol{\ell} \times \mathbf{B} = I\ell B \sin 30° \ \mathbf{k}$$

$$= (3.0\ \text{A})(0.003\ \text{m})(0.02\ \text{T})(\sin 30°)\ \mathbf{k}$$

$$= 9 \times 10^{-5}\ \text{N}\ \mathbf{k}$$

The total force on a current-carrying wire is found by adding the forces on each of the segments.

Questions

1. Charge q moves with velocity **v** through a magnetic field **B**. At a particular instant, it experiences a magnetic force **F**. How would the force differ if the charge had the opposite sign? If the velocity were in the opposite direction? If the magnetic field were in the opposite direction?
2. For what angle between **B** and **v** is the magnetic force on q greatest? The least?
3. A moving electric charge may experience both electric and magnetic forces. How could you determine whether a force causing a charge to deviate from a straight path is an electric or a magnetic force?
4. How can a charge move through a magnetic field without ever experiencing any magnetic force?
5. Show that the force on a current element is the same in direction and magnitude regardless of whether positive charges, negative charges, or a mixture of positive and negative charges create the current.
6. A current-carrying wire is in a magnetic field, but the wire does not experience any magnetic force. How is this possible?
7. How are magnetic-field lines similar to electric-field lines? How are they different?

24-2 Motion of a Point Charge in a Magnetic Field

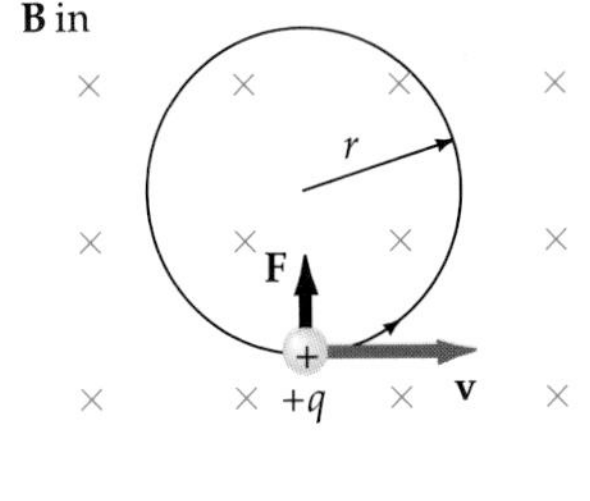

Figure 24-9 Charged particle moving in a plane perpendicular to a uniform magnetic field. The magnetic field is into the page, as indicated by the crosses. (A field out of the plane of the page would be indicated by dots.) The magnetic force is perpendicular to the velocity of the particle, causing it to move in a circular orbit.

An important characteristic of the magnetic force on a charged particle moving through a magnetic field is that the force is always perpendicular to the velocity of the particle. The magnetic force thus changes the direction of the velocity but not its magnitude. It therefore does no work on the particle and does not affect the kinetic energy of the particle.

In the special case where the velocity of a particle is perpendicular to a uniform magnetic field, as shown in Figure 24-9, the particle moves in a circular orbit. The magnetic force provides the centripetal force necessary for circular motion. We can relate the radius of the circle r to the magnetic field B and the speed v of the particle by setting the net force equal to the mass m of the particle times the centripetal acceleration v^2/r in accordance with

Newton's second law. The net force in this case is qvB since **v** and **B** are perpendicular. Thus, Newton's second law gives

$$F = ma$$

$$qvB = \frac{mv^2}{r}$$

or

$$r = \frac{mv}{qB} \tag{24-6}$$

The period of the circular motion is the time it takes the particle to travel once around the circumference of the circle. From Equation 3-23, the period is related to the speed by

$$T = \frac{2\pi r}{v}$$

(a)

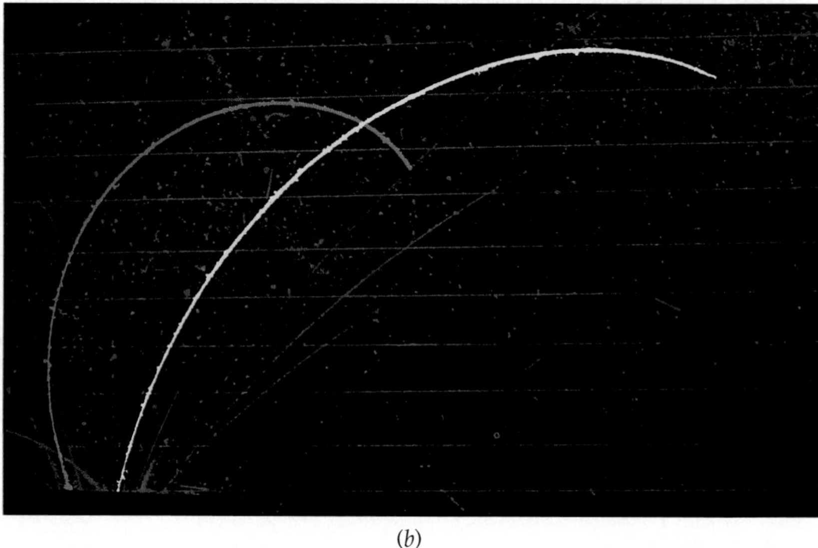

(b)

(*a*) Circular path of electrons moving in the magnetic field produced by two large coils. The electrons ionized the gas in the tube causing it to give off a bluish glow that indicates the path of the beam. (*b*) False-color photograph showing tracks of a 1.6-MeV proton (red) and a 7-MeV α particle (yellow) in a cloud chamber. The radius of curvature is proportional to the momentum and inversely proportional to the charge of the particle. For these energies, the momentum of the α particle, which has twice the charge of the proton, is about four times that of the proton and so its radius of curvature is greater.

Substituting in $r = mv/qB$ from Equation 24-6, we obtain,

$$T = \frac{2\pi(mv/qB)}{v} = \frac{2\pi m}{qB} \tag{24-7}$$

The frequency of the circular motion is the reciprocal of the period.

Cyclotron frequency

$$f = \frac{1}{T} = \frac{qB}{2\pi m} \tag{24-8}$$

Note that the period and the frequency given by Equations 24-7 and 24-8 do not depend on the radius of the orbit or the velocity of the particle. This period is called the **cyclotron period,** and the frequency the **cyclotron frequency.** Two of the many interesting applications of the circular motion of charged particles in a uniform magnetic field, the mass spectrometer and the cyclotron, will be discussed later in this section.

Example 24-3

A proton of mass $m = 1.67 \times 10^{-27}$ kg and charge $q = e = 1.6 \times 10^{-19}$ C moves in a circle of radius 21 cm perpendicular to a magnetic field $B =$ 4000 G. Find (*a*) the period of the motion and (*b*) the speed of the proton.

(*a*) We do not need to know the radius of the circle to find the period. Converting the magnetic field to SI units (4000 G = 0.4 T) and using Equation 24-7, we have

$$T = \frac{2\pi m}{qB} = \frac{2\pi(1.67 \times 10^{-27}\ \text{kg})}{(1.6 \times 10^{-19}\ \text{C})(0.4\ \text{T})}$$

$$= 1.64 \times 10^{-7}\ \text{s}$$

(*b*) The speed v of the proton is related to the radius of the circle by Equation 24-6:

$$v = \frac{rqB}{m} = \frac{(0.21\ \text{m})(1.6 \times 10^{-19}\ \text{C})(0.4\ \text{T})}{1.67 \times 10^{-27}\ \text{m}}$$

$$= 8.05 \times 10^{6}\ \text{m/s}$$

We can check our results by noting that the product of the speed v and the period T is the circumference of the circle $2\pi r$. Then

$$r = \frac{vT}{2\pi} = \frac{(8.05 \times 10^{6}\ \text{m/s})(1.64 \times 10^{-7}\ \text{s})}{2\pi}$$

$$= 0.21\ \text{m} = 21\ \text{cm}$$

In these calculations, we put each quantity in SI units so that the results will be in SI units, namely, seconds for the period and meters per second for the speed.

Note from Equation 24-6 that the radius of the circular motion is proportional to the speed. If we double the speed of the proton in this example, the radius would double, but the period and the frequency would remain unchanged.

Suppose that a charged particle enters a uniform magnetic field with a velocity that is not perpendicular to **B**. We can resolve the velocity of the particle into components $v_{\parallel}$ parallel to **B** and $v_{\perp}$ perpendicular to **B**. The motion due to the perpendicular component is the same as that just discussed. The component of the velocity parallel to **B** is not affected by the magnetic field. It therefore remains constant. The path of the particle is thus a helix, as shown in Figure 24-10.

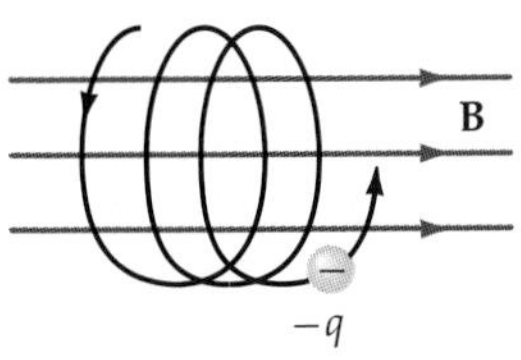

(*a*)

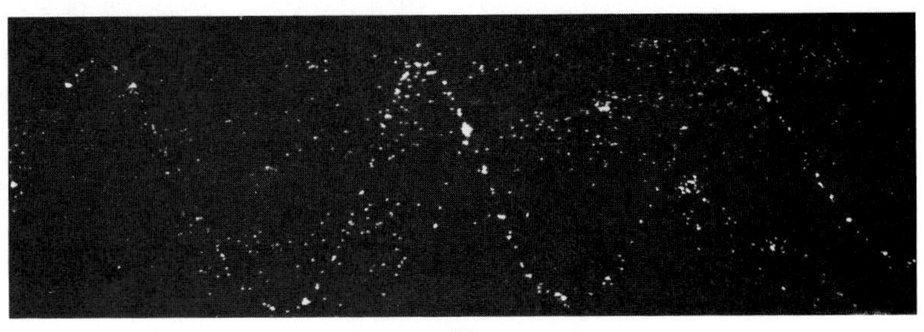

(*b*)

Figure 24-10 (*a*) When a charged particle has a velocity component that is parallel to a magnetic field as well as one that is perpendicular, it moves in a helical path around the field lines. (*b*) Cloud-chamber photograph of the helical path of an electron moving in a magnetic field. The path of the electrons is made visible by the condensation of water droplets in the cloud chamber.

The motion of charged particles in nonuniform magnetic fields is quite complicated. Figure 24-11 shows a **magnetic bottle,** an interesting magnetic-field configuration in which the field is weak at the center and strong at both ends. A detailed analysis of the motion of a charged particle in such a field shows that the particle will spiral around the field lines and become trapped, oscillating back and forth between points P_1 and P_2 in the figure. Such magnetic-field configurations are used to confine dense beams of charged parti-

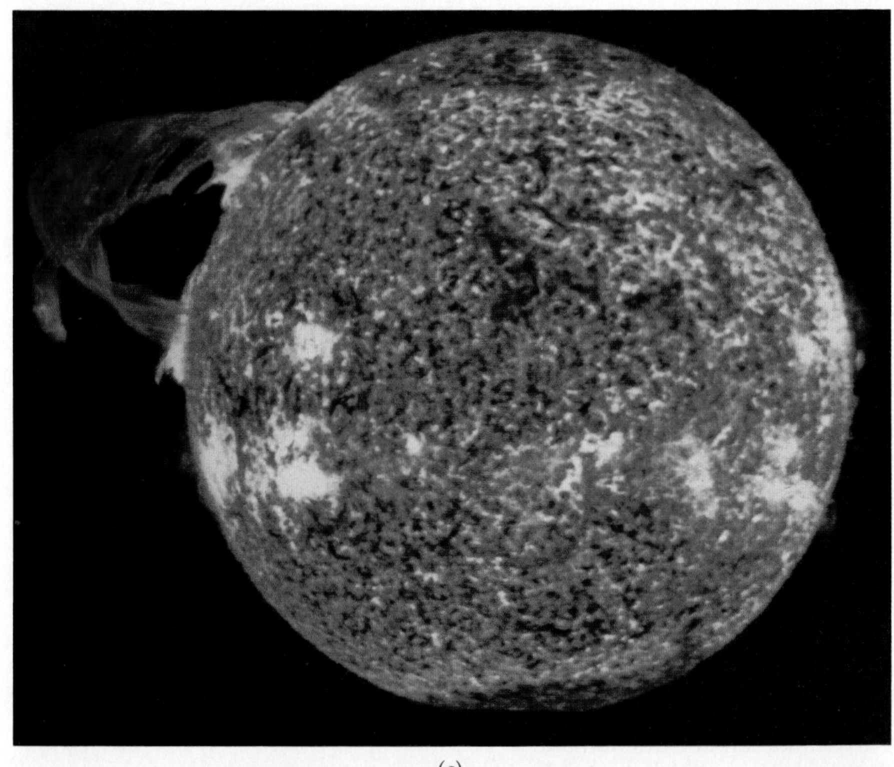

(*a*)

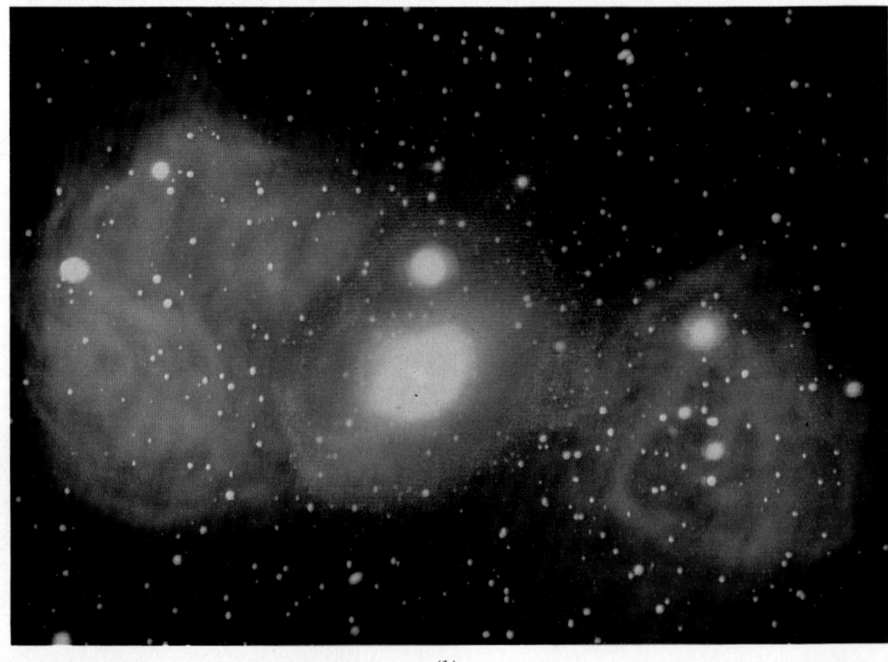

(*b*)

(*a*) The solar flare shown on the left of this skylab-4 photograph consists of charged particles confined by the magnetic field of the sun. (*b*) Interstellar dust grains, aligned by the magnetic fields of distant galaxies, act like polarizing filters. This map of the magnetic field of the galaxy NGC 1316 obtained with a radio telescope shows double-lobed magnetic jets in red. The magnetic map is superimposed on a visible-light photograph of the galaxy.

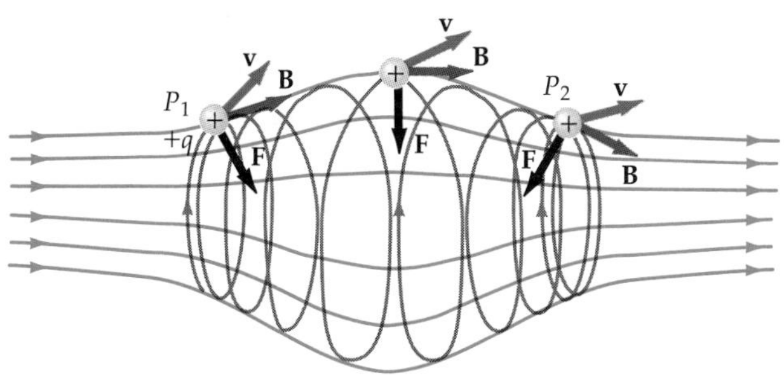

Figure 24-11 Magnetic bottle. When a charged particle moves in such a field, which is strong at both ends and weak in the middle, the particle becomes trapped and moves back and forth, spiraling around the field lines.

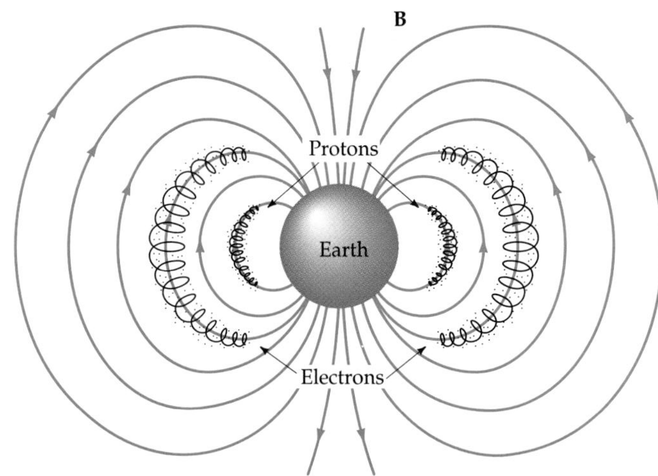

Figure 24-12 Van Allen belts. Protons (the inner belts) and electrons (the outer belts) are trapped in the earth's magnetic field and travel in helical paths around the field lines between the north and south poles.

cles, called *plasmas,* in nuclear fusion research. A similar phenomenon is the oscillation of ions back and forth between the earth's magnetic poles in the Van Allen belts (Figure 24-12).

The Velocity Selector

The magnetic force on a charged particle moving in a uniform magnetic field can be balanced by an electric force if the magnitudes and directions of the magnetic and electric fields are properly chosen. Since the electric force is in the direction of the electric field (for positive particles) and the magnetic force is perpendicular to the magnetic field, the electric and magnetic fields in the region through which the particle is moving must be perpendicular to each other if the forces are to balance. Such a region is said to have **crossed fields.** Figure 24-13 shows a region of space between the plates of a capacitor where there is an electric field and a perpendicular magnetic field (produced by a magnet not shown). Consider a particle of charge q entering this space from the left. If q is positive, the electric force of magnitude qE is down and the magnetic force of magnitude qvB is up. If the charge is negative, each of these forces is reversed. The two forces will balance if $qE = qvB$ or

$$v = \frac{E}{B} \tag{24-9}$$

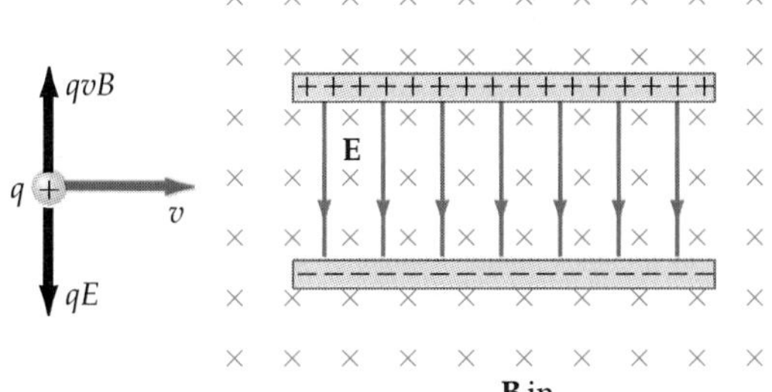

Figure 24-13 Crossed electric and magnetic fields. When a positive particle moves to the right, it experiences a downward electric force qE and an upward magnetic force qvB. These forces balance if the speed of the particle is related to the magnitudes of the fields by $v = E/B$.

For given magnitudes of the electric and magnetic fields, the forces will balance only for particles with the speed given by Equation 24-9. Any particle with this speed, no matter what its mass or charge, will traverse the space undeflected. A particle with a greater speed will be deflected in the direction of the magnetic force, and one with less speed will be deflected in the direction of the electric force. Because only a particle with a particular speed can traverse this arrangement of fields, it is called a **velocity selector.**

Exercise

A proton is moving in the x direction in a region of crossed fields where $\mathbf{E} = 2 \times 10^5$ N/C $\mathbf{k}$ and $\mathbf{B} = -3000$ G $\mathbf{j}$. (*a*) What is the speed of the proton if it is not deflected? (*b*) If the proton moves with twice this speed, in which direction will it be deflected? [Answers: (*a*) 667 km/s; (*b*) in the negative z direction]

Thomson's Measurement of q/m for Electrons

An example of the use of a velocity selector is the famous experiment performed by J. J. Thomson in 1897 in which he showed that the rays of a cathode-ray tube can be deflected by electric and magnetic fields and therefore consist of charged particles. By measuring the deflections of these particles caused by various combinations of electric and magnetic fields, Thomson showed that all the particles have the same charge-to-mass ratio q/m. He also showed that particles with this charge-to-mass ratio can be obtained using any material for the cathode, which means that these particles, now called electrons, are a fundamental constituent of all matter.

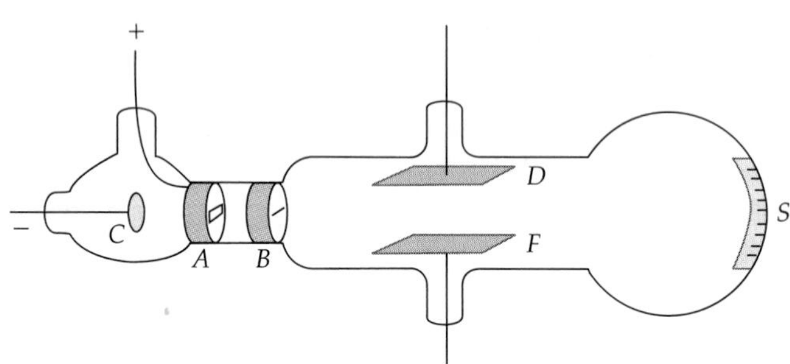

(*a*)

(*b*)

Figure 24-14 (*a*) A schematic diagram of the cathode-ray tube Thomson used to measure q/m for the particles that comprise cathode rays (electrons). Electrons from the cathode C pass through the slits at A and B and strike a phosphorescent screen S. The beam can be deflected by an electric field between plates D and F or by a magnetic field (not shown). (*b*) J. J. Thomson in his laboratory.

Figure 24-14 shows a schematic diagram of the cathode-ray tube Thomson used. Electrons are emitted from the cathode C, which is at a negative potential relative to the slits A and B. An electric field in the direction from A to C accelerates the electrons, and they pass through slits A and B into a field-free region. The electrons then enter the electric field between the capacitor plates D and F that is perpendicular to the velocity of the electrons. Because of the acceleration produced by this electric field, the velocity of the electrons has a vertical component when they leave the region between the plates. They strike the phosphorescent screen S at the far right side of the tube at some deflection Δy from the point at which they strike when there is no field between the plates D and F. The screen glows where the electrons strike it, indicating the location of the beam. The deflection Δy consists of

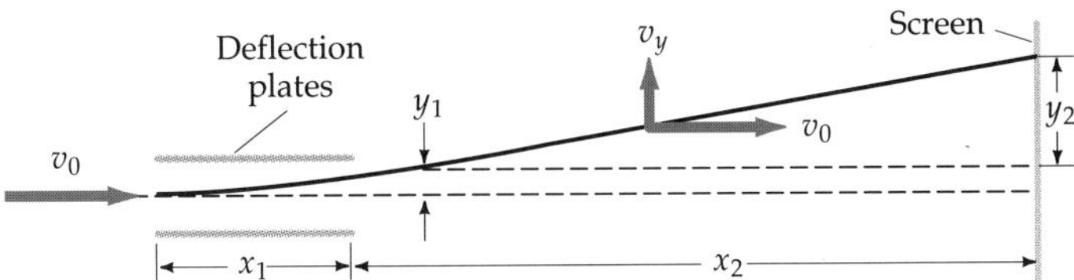

Figure 24-15 The total deflection of the beam in the J. J. Thomson experiments consists of the deflection while the electrons are between the plates, y_1, plus the deflection y_2 that occurs in the field-free region between the plates and the screen.

two parts: the deflection Δy_1, while the electrons are between the plates, and the deflection Δy_2, which occurs after the electrons leave the region between the plates (Figure 24-15).

Let x_1 be the horizontal distance across the deflection plates D and F. If the electron is moving horizontally with speed v_0 when it enters the plates, the time spent between the plates is $t_1 = x_1/v_0$, and the vertical velocity when it leaves the plates is

$$v_y = at_1 = \frac{qE}{m}t_1 = \frac{qE}{m}\frac{x_1}{v_0}$$

where E is the electric field between the plates. The deflection in this region will be

$$\Delta y_1 = \frac{1}{2}at_1^2 = \frac{1}{2}\frac{qE}{m}\left(\frac{x_1}{v_0}\right)^2$$

The electron then travels an additional horizontal distance x_2 in the field-free region from the deflection plates to the screen. Since the velocity of the electron is constant in this region, the time to reach the screen is $t_2 = x_2/v_0$, and the additional vertical deflection is

$$\Delta y_2 = v_y t_2 = \frac{qE}{m}\frac{x_1}{v_0}\frac{x_2}{v_0}$$

The total deflection at the screen is therefore

$$\Delta y = \Delta y_1 + \Delta y_2 = \frac{1}{2}\frac{qE}{m}\left(\frac{x_1}{v_0}\right)^2 + \frac{qE}{m}\frac{x_1 x_2}{v_0^2} \qquad \text{24-10}$$

The initial speed v_0 is determined by introducing a magnetic field **B** between the plates in a direction that is perpendicular to both the electric field and the initial velocity of the electrons. The magnitude of **B** is then adjusted until the beam is not deflected. The speed is then found from Equation 24-9, and the measured deflection Δy is used to determine the charge-to-mass ratio, q/m, from Equation 24-10.

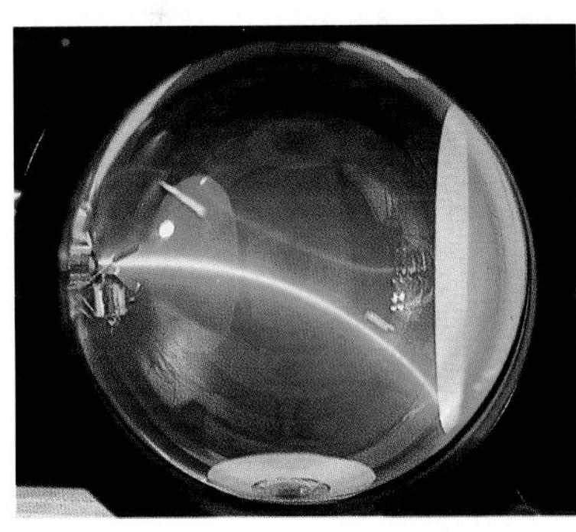

An electron beam moving from left to right is deflected downward by a magnetic field produced by coils. In what direction is the magnetic field?

Example 24-4

Electrons pass undeflected through the plates of Thomson's apparatus when the electric field is 3000 V/m and there is a crossed magnetic field of 1.40 G. If the plates are 4 cm long and the end of the plates is 30 cm from the screen, find the deflection on the screen when the magnetic field is turned off.

In this example, we use the known charge $e = 1.6 \times 10^{-19}$ C and mass $m = 9.11 \times 10^{-31}$ kg of the electron to find the deflection Δy. The initial speed of the electrons is found from Equation 24-9:

$$v_0 = \frac{E}{B} = \frac{3000\ \text{V/m}}{1.40 \times 10^{-4}\ \text{T}} = 2.14 \times 10^7\ \text{m/s}$$

Using Equation 24-10 with $v_0 = 2.14 \times 10^7$ m/s, $x_1 = 4$ cm, and $x_2 =$ 30 cm, we obtain for the deflection

$$\Delta y = \frac{1}{2}\frac{(1.6 \times 10^{-19}\text{ C})(3000\text{ V/m})}{9.11 \times 10^{-31}\text{ kg}}\left(\frac{0.04\text{ m}}{2.14 \times 10^7\text{ m/s}}\right)^2 + \frac{(1.6 \times 10^{-19}\text{ C})(3000\text{ V/m})}{9.11 \times 10^{-31}\text{ kg}}\frac{(0.04\text{ m})(0.30\text{ m})}{(2.14 \times 10^7\text{ m/s})^2}$$

$$= 9.20 \times 10^{-4}\text{ m} + 1.38 \times 10^{-2}\text{ m}$$

$$= 0.92\text{ mm} + 13.8\text{ mm} = 14.7\text{ mm}$$

The Mass Spectrometer

The **mass spectrometer,** first designed by Francis William Aston in 1919 and later improved by Kenneth Bainbridge and others, was developed as a means of measuring the masses of isotopes. Such measurements are an important way of determining both the existence of isotopes and their abundance in nature. For example, natural magnesium has been found to consist of 78.7 percent ^{24}Mg, 10.1 percent ^{25}Mg, and 11.2 percent ^{26}Mg. These isotopes have masses in the approximate ratio 24:25:26.

The mass spectrometer is used to find the mass-to-charge ratio of ions of known charge by measuring the radius of their circular orbits in a uniform magnetic field. Equation 24-6, $r = mv/qB$, gives the radius r for the circular orbit of a particle of mass m and charge q moving with speed v in a magnetic field B that is perpendicular to the velocity of the particle. Figure 24-16 shows a simple schematic drawing of a mass spectrometer. Ions from an ion source are accelerated by an electric field and enter a uniform magnetic field produced by an electromagnet. If the ions start from rest and move through a potential drop ΔV, their kinetic energy when they enter the magnetic field equals their loss in potential energy, $q\,\Delta V$:

$$\tfrac{1}{2}mv^2 = q\,\Delta V \qquad 24\text{-}11$$

The ions move in a semicircle of radius r given by Equation 24-6 and strike a photographic plate at point P_2, a distance $2r$ from the point where they enter the magnet. The speed v can be eliminated from Equations 24-6 and 24-11 to

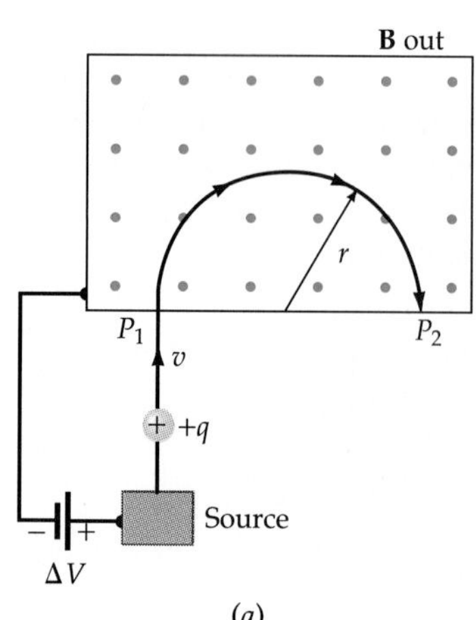

(a)

(b)

Figure 24-16 (a) Schematic drawing of a mass spectrometer. Ions from an ion source are accelerated though a potential difference ΔV and enter a uniform magnetic field. The magnetic field is out of the plane of the page as indicated by the dots. The ions are bent into circular arcs and strike a photographic plate at P_2. The radius of the circle is proportional to the mass of the ion. (b) A modern mass spectrometer used for research at Rockefeller University.

find m/q in terms of the known quantities ΔV, B, and r. We first solve Equation 24-6 for v and square each term, which gives

$$v^2 = \frac{r^2q^2B^2}{m^2}$$

Substituting this expression for v^2 into Equation 24-11, we obtain

$$\frac{1}{2}m\left(\frac{r^2q^2B^2}{m^2}\right) = q\,\Delta V$$

Simplifying this equation and solving it for m/q, we obtain

$$\frac{m}{q} = \frac{B^2r^2}{2\,\Delta V} \qquad 24\text{-}12$$

In Aston's original mass spectrometer, mass differences could be measured to a precision of about 1 part in 10,000. The precision has been improved by introducing a velocity selector between the ion source and the magnet, which makes it possible to limit the range of velocities of the incoming ions and to determine the velocities of the ions more accurately.

Example 24-5

A ^{58}Ni ion of charge $+e$ and mass 9.62×10^{-26} kg is accelerated through a potential difference of 3 kV and deflected in a magnetic field of 0.12 T. (*a*) Find the radius of curvature of the orbit of the ion. (*b*) Find the difference in the radii of curvature of ^{58}Ni ions and ^{60}Ni ions. (Assume that the mass ratio is 58/60.)

(*a*) Using Equation 24-12, we obtain

$$r^2 = \frac{2m\,\Delta V}{qB^2} = \frac{2(9.62 \times 10^{-26}\text{ kg})(3000\text{ V})}{(1.6 \times 10^{-19}\text{ C})(0.12\text{ T})^2} = 0.251\text{ m}^2$$

$$r = \sqrt{0.251\text{ m}^2} = 0.501\text{ m}$$

(*b*) The radius of the orbit of an ion in a given magnetic field is proportional to the square root of its mass, for a given accelerating voltage. If r_1 is the radius of the orbit of the ^{58}Ni ion and r_2 is that of the ^{60}Ni ion, the ratio of the radii is

$$\frac{r_2}{r_1} = \sqrt{\frac{m_2}{m_1}} = \sqrt{\frac{60}{58}} = 1.017$$

Then, the radius of the orbit of the ^{60}Ni ion is

$$r_2 = 1.017\,r_1 = (1.017)(0.501\text{ m}) = 0.510\text{ m}$$

The difference in the radii of the orbits is thus

$$r_2 - r_1 = 0.510\text{ m} - 0.501\text{ m} = 0.009\text{ m} = 9\text{ mm}$$

The Cyclotron

The **cyclotron** was invented by E. O. Lawrence and M. S. Livingston in 1932 to accelerate particles such as protons or deuterons to high kinetic energies. (A deuteron is the nucleus of heavy hydrogen, ^{2}H, which consists of a proton and neutron tightly bound together.) The high-energy particles are then used to bombard atomic nuclei to cause nuclear reactions that are then studied to obtain information about the nucleus. High-energy protons and deuterons are also used to produce radioactive materials and for medical purposes.

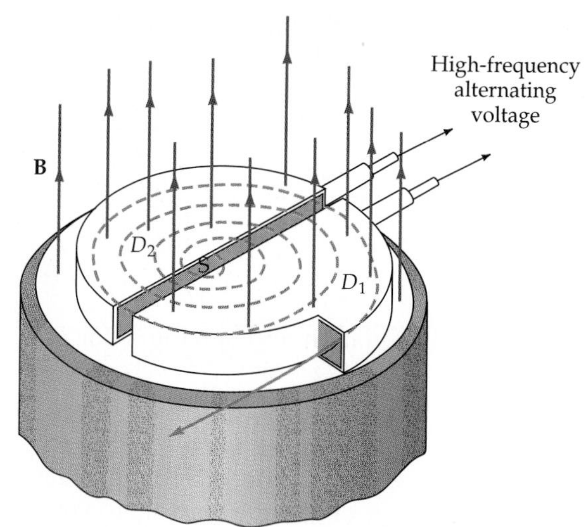

Figure 24-17 Schematic drawing of a cyclotron. The upper portion of the magnet has been omitted. Charged particles such as protons from a source S at the center are accelerated by the potential difference across the gap between the dees. The potential difference across the gap alternates with the cyclotron period of the particle, which is independent of the radius of the circle. Thus, when the particles arrive at the gap again, the potential difference has changed sign, so they are again accelerated across the gap and move in a larger circle.

The operation of the cyclotron is based on the fact that the period of the motion of a charged particle in a uniform magnetic field is independent of the velocity of the particle, as can be seen from Equation 24-7:

$$T = \frac{2\pi m}{qB}$$

Figure 24-17 is a schematic drawing of a cyclotron. The particles move in two semicircular metal containers called *dees* (because of their shape). The dees are housed in a vacuum chamber that is in a uniform magnetic field provided by an electromagnet. (The region in which the particles move must be evacuated so that the particles will not lose energy and be scattered in collisions with air molecules.) Between the dees there is maintained a potential difference ΔV that alternates in time with a period T, which is chosen to be equal to the cyclotron period given by Equation 24-7. This potential difference creates an electric field across the gap between the dees. At the same time, there is no electric field within each dee because of the shielding of the metal dees.

The charged particles are initially injected into dee 1 with a small velocity from an ion source S near the center of the dees. They move in a semicircle in dee 1 and arrive at the gap between dee 1 and dee 2 after a time $\frac{1}{2}T$, where T is the cyclotron period and is also the period with which the potential across the dees is alternated. The alternation of the potential is adjusted

(a) A cyclotron constructed by John R. Dunning in the late 1930's at Columbia University. It was used in 1939 to demonstrate uranium fission. The cyclotron is in the collection of the National Museum of American History in Washington, D.C. *(b)* A modern cyclotron used to accelerate protons or deuterons for making short-lived radioactive isotopes for medical diagnosis.

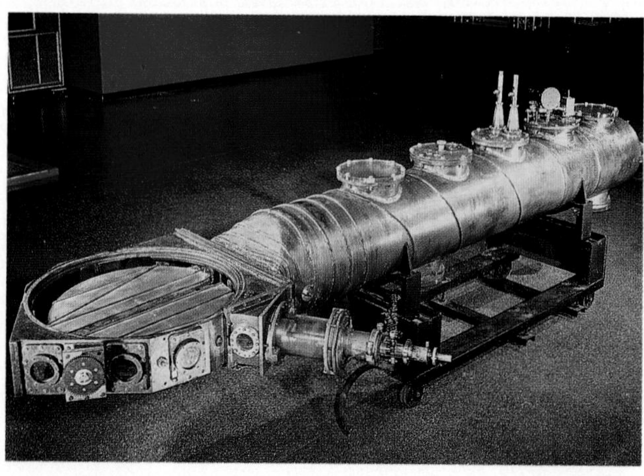

(a)

(b)

so that dee 1 is at a higher potential than dee 2 when the particles arrive at the gap between them. Each particle is therefore accelerated across the gap by the electric field across the gap and gains kinetic energy equal to $q\ \Delta V$. Because it has more kinetic energy, the particle moves in a semicircle of larger radius in dee 2, and again arrives at the gap after a time $\frac{1}{2}T$. By this time the potential between the dees has been reversed so that dee 2 is now at the higher potential. Once more the particle is accelerated across the gap and gains additional kinetic energy equal to $q\ \Delta V$. Each time the particle arrives at the gap, it is accelerated and gains kinetic energy equal to $q\ \Delta V$. Thus, it moves in larger and larger semicircular orbits until it eventually leaves the magnetic field. In the typical cyclotron, each particle may make 50 to 100 revolutions and exit with energies of up to several hundred MeV.

This kinetic energy of a particle leaving a cyclotron can be calculated by setting r in Equation 24-6 equal to the maximum radius of the dees and solving the equation for v:

$$r = \frac{mv}{qB}$$

$$v = \frac{qBr}{m}$$

Then,

$$K = \frac{1}{2}mv^2 = \frac{1}{2}\left(\frac{q^2B^2}{m}\right)r^2 \qquad 24\text{-}13$$

Example 24-6

A cyclotron for accelerating protons has a magnetic field of 1.5 T and a maximum radius of 0.5 m. (*a*) What is the cyclotron frequency? (*b*) Find the kinetic energy of the protons when they emerge.

(*a*) The cyclotron frequency is given by Equation 24-8:

$$f = \frac{qB}{2\pi m} = \frac{(1.6 \times 10^{-19}\ \text{C})(1.5\ \text{T})}{2\pi(1.67 \times 10^{-27}\ \text{kg})} = 2.29 \times 10^7\ \text{Hz} = 22.9\ \text{MHz}$$

(*b*) The kinetic energy of the emerging protons is given by Equation 24-13:

$$K = \frac{1}{2}\left[\frac{(1.6 \times 10^{-19}\ \text{C})^2(1.5\ \text{T})^2}{1.67 \times 10^{-27}\ \text{kg}}\right](0.5\ \text{m})^2$$

$$= 4.31 \times 10^{-12}\ \text{J}$$

The energies of protons and other elementary particles are usually expressed in electron volts. Since 1 eV $= 1.6 \times 10^{-19}$ J, we have

$$K = 4.31 \times 10^{-12}\ \text{J} \times \frac{1\ \text{eV}}{1.6 \times 10^{-19}\ \text{J}} = 26.9\ \text{MeV}$$

Questions

8. How can you determine by observing the path of a deflected particle whether the particle is being deflected by a magnetic field or an electric field?

9. A beam of positively charged particles passes undeflected from left to right through a velocity selector in which the electric field is up. The beam is then reversed so that it travels from right to left. Will the beam now be deflected in the velocity selector? If so, in which direction?

24-3 Torques on Current Loops and Magnets

Figure 24-18 shows a rectangular wire loop of length a and width b carrying a current I in a uniform external magnetic field $\mathbf{B}$ that is parallel to the plane of the loop. The forces on each segment of the loop are shown in the figure. There are no forces on the top or bottom of the loop because the current in those segments is parallel or antiparallel to the magnetic field $\mathbf{B}$, so $I\, d\boldsymbol{\ell} \times \mathbf{B}$ is zero. The forces on the sides of the loop have the magnitude

$$F_1 = F_2 = IaB$$

Since these forces are equal and opposite, they form a couple. The net force is therefore zero, and the torque about any point is independent of the location of the point. Point P is a convenient point about which to compute the torque. The magnitude of the torque is

$$\tau = F_1 b = IabB = IAB$$

where $A = ab$ is the area of the loop. The torque is therefore the product of the current, the area of the loop, and the magnetic field B. This torque tends to twist the loop so that its plane is perpendicular to $\mathbf{B}$.

The orientation of the loop can be described conveniently by a unit vector $\hat{\mathbf{n}}$ that is perpendicular to the plane of the loop. The sense of $\hat{\mathbf{n}}$ is chosen to be that given by the right-hand rule applied to the circulating current, as illustrated in Figure 24-19. The torque tends to rotate $\hat{\mathbf{n}}$ into the direction of $\mathbf{B}$.

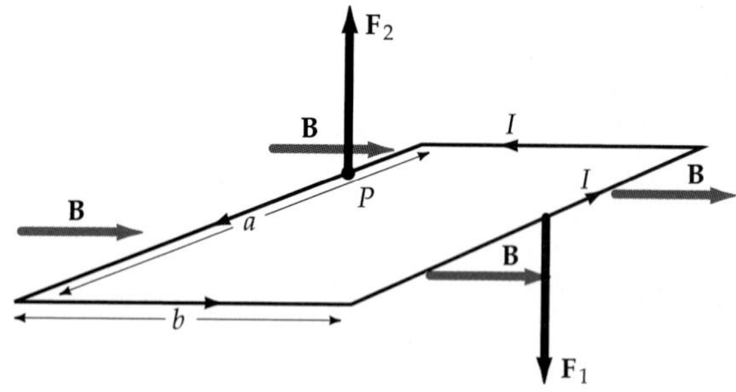

Figure 24-18 Forces exerted on a rectangular current loop in a uniform magnetic field **B** that is parallel to the plane of the loop. The forces produce a torque that tends to twist the loop so that its plane is perpendicular to **B**.

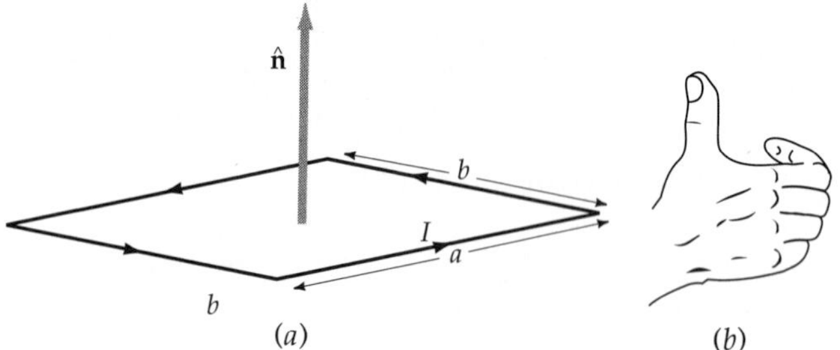

Figure 24-19 (*a*) The orientation of a current loop is described by the unit vector $\hat{\mathbf{n}}$ that is perpendicular to the plane of the loop. (*b*) Right-hand rule for determining the sense of $\hat{\mathbf{n}}$. When the fingers of the right hand curl around the loop, with the fingers pointing in the direction of the current, the thumb points in the direction of $\hat{\mathbf{n}}$.

Figure 24-20 shows the forces exerted by a uniform magnetic field on a rectangular loop whose normal unit vector $\hat{\mathbf{n}}$ makes an angle θ with the magnetic field $\mathbf{B}$. Again, the net force on the loop is zero. The torque about any point is the product of the force and the lever arm. For example, the torque about point P is the force $F_2 = IaB$ times the lever arm $b \sin \theta$. The torque thus has the magnitude

$$\tau = IaBb \sin \theta = IAB \sin \theta$$

where again $A = ab$ is the area of the loop. For a loop with N turns, the torque has the magnitude

$$\tau = NIAB \sin \theta$$

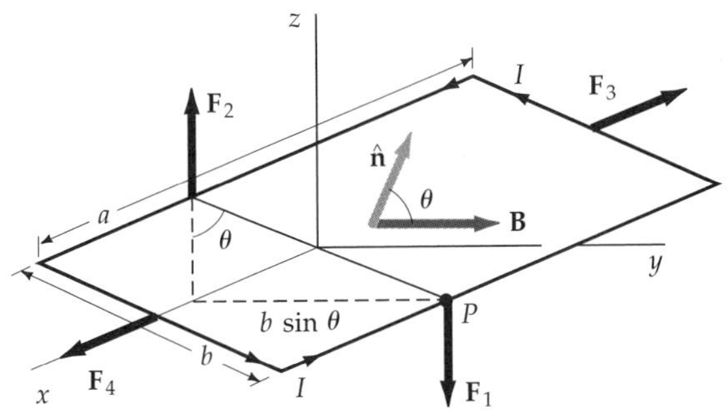

Figure 24-20 Rectangular current loop whose unit normal $\hat{\mathbf{n}}$ makes an angle θ with a uniform magnetic field $\mathbf{B}$. The torque on the loop has magnitude $IAB \sin \theta$ and is in the direction such that $\hat{\mathbf{n}}$ tends to rotate into $\mathbf{B}$. The torque can be written $\boldsymbol{\tau} = \mathbf{m} \times \mathbf{B}$, where $\mathbf{m} = IA\hat{\mathbf{n}}$ is the magnetic moment of the loop.

The torque can be written conveniently in terms of the **magnetic dipole moment m** (also referred to simply as the **magnetic moment**) of the current loop, which is defined as

$$\mathbf{m} = NIA\,\hat{\mathbf{n}} \tag{24-14}$$

Magnetic dipole moment of a current loop

The SI unit of magnetic moment is the ampere-meter^2 ($A{\cdot}m^2$). In terms of the magnetic dipole moment, the torque on the current loop is given by

$$\boldsymbol{\tau} = \mathbf{m} \times \mathbf{B} \tag{24-15}$$

Torque on a current loop

Equation 24-15, which we have derived for a rectangular loop, holds in general for a loop of any shape. The torque on any loop is the cross product of the magnetic moment **m** of the loop and the magnetic field **B**, where the magnetic moment is defined to be a vector that is perpendicular to the area of the loop (Figure 24-21) and has magnitude equal to NIA. Comparing Equation 24-15 with Equation 18-11 for the torque on an electric dipole, we see that a current loop in a magnetic field acts the same as does an electric dipole in an electric field.

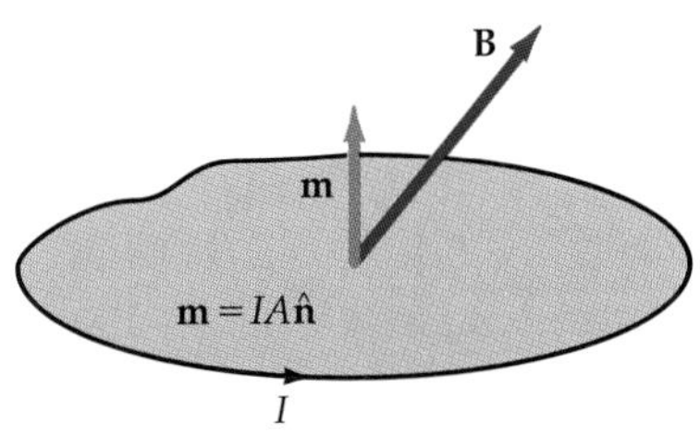

Figure 24-21 The magnetic moment of a current loop of arbitrary shape is $\mathbf{m} = IA\hat{\mathbf{n}}$. In a magnetic field **B**, the loop experiences a torque $\mathbf{m} \times \mathbf{B}$.

When a small permanent magnet such as a compass needle is placed in a magnetic field **B**, it tends to orient itself so that the north pole points in the direction of **B**. This effect also occurs with previously unmagnetized iron filings, which become magnetized in the presence of a **B** field. Figure 24-22 shows a small magnet that makes an angle θ with a magnetic field **B**. There is a force $\mathbf{F}_1$ on the north pole in the direction of **B** and an equal but opposite force $\mathbf{F}_2$ on the south pole. These two forces produce no translational motion because they are equal and opposite, but they do produce a torque that tends to rotate the magnet so that it lines up with the field. A small bar magnet thus behaves like a current loop. The fact that a small magnet behaves like a current loop is not accidental. The origin of the magnetic moment of a bar magnet is, in fact, microscopic current loops that result from the motion of electrons in the atoms of the magnet.

We can use the experimentally observed forces and torque on a bar magnet to define the pole strength and magnetic moment of the magnet. We

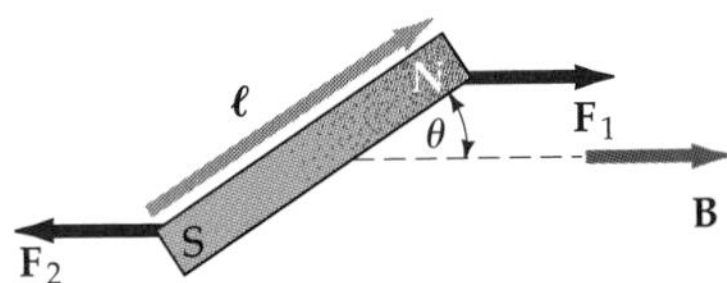

Figure 24-22 A small magnet in a uniform magnetic field experiences a torque that tends to rotate the magnet into the direction of the field. The magnetic moment of the magnet is in the direction of the vector $\boldsymbol{\ell}$ from the south pole to the north pole.

define the pole strength of a magnet q_m so that the force exerted on the pole in a magnetic field **B** is given by*

Magnetic pole strength defined

$$\mathbf{F} = q_m\mathbf{B} \tag{24-16}$$

The pole strength is positive for a north pole and negative for a south pole. The magnetic moment **m** of a magnet is then defined as

$$\mathbf{m} = |q_m|\boldsymbol{\ell} \tag{24-17}$$

where $\boldsymbol{\ell}$ is the vector from the south pole to the north pole. The torque exerted on a bar magnet in a magnetic field is then given by $\boldsymbol{\tau} = \mathbf{m} \times \mathbf{B}$, the same as Equation 24-15. Although we have defined the magnetic pole strength to be analogous to electric charge, we should remember that magnetic poles always come in pairs; that is, the fundamental unit of magnetism is the magnetic dipole. Experimentally, it is the magnetic dipole moment **m** of a magnet that is easily measured by placing the magnet in a magnetic field of known strength and measuring the torque. The magnetic pole strength is then found from Equation 24-17 by dividing the magnitude of the magnetic moment by the length of the magnet.

Example 24-7

A circular loop of radius 2 cm has 10 turns of wire and carries a current of 3 A. The axis of the loop makes an angle of 30° with a magnetic field of 8000 G. Find the torque on the loop.

The magnitude of the magnetic moment of the loop is

$$m = NIA = (10)(3\text{ A})\pi(0.02\text{ m})^2 = 3.77 \times 10^{-2}\text{ A·m}^2$$

The magnitude of the torque is then

$$\tau = mB\sin\theta = (3.77 \times 10^{-2}\text{ A·m}^2)(0.8\text{ T})(\sin 30°)$$
$$= 1.51 \times 10^{-2}\text{ N·m}$$

where we have used 8000 G = 0.8 T and 1 T = 1 N/A·m.

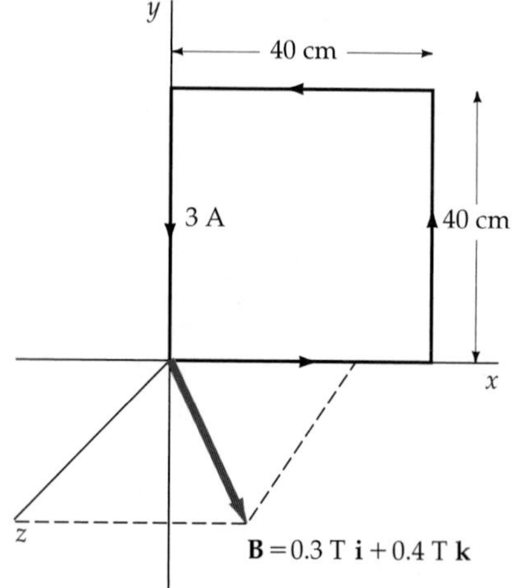

Figure 24-23 Square current loop in the xy plane in a magnetic field $\mathbf{B} = 0.3\text{ T }\mathbf{i} + 0.4\text{ T }\mathbf{k}$ for Example 24-8.

Example 24-8

A square 12-turn coil with sides of length 40 cm carries a current of 3 A. It lies in the xy plane as shown in Figure 24-23 in a uniform magnetic field $\mathbf{B} = 0.3\text{ T }\mathbf{i} + 0.4\text{ T }\mathbf{k}$. Find (*a*) the magnetic moment of the coil and (*b*) the torque exerted on the coil. (*c*) What is the pole strength and orientation of a bar magnet of length 8 cm that has a magnetic moment equal to that of the coil?

(*a*) From Figure 24-23, we see that the magnetic moment of the loop is in the positive z direction. Its magnitude is $m = NIA = (12)(3\text{ A})(0.40\text{ m})^2 = 5.76\text{ A·m}^2$. The magnetic moment is thus

$$\mathbf{m} = 5.76\text{ A·m}^2\ \mathbf{k}$$

(*b*) The torque on the current loop is given by Equation 24-15:

$$\boldsymbol{\tau} = \mathbf{m} \times \mathbf{B} = (5.76\text{ A·m}^2\ \mathbf{k}) \times (0.3\text{ T }\mathbf{i} + 0.4\text{ T }\mathbf{k}) = 1.73\text{ N·m }\mathbf{j}$$

where we have used $\mathbf{k} \times \mathbf{k} = 0$ and $\mathbf{k} \times \mathbf{i} = \mathbf{j}$.

*The notation for magnetic pole strength q_m is used so that the magnetic equations resemble the corresponding equations for electric charges in electric fields. The subscript m reminds us that q_m denotes a magnetic pole and not an electric charge.

(*c*) A bar magnet with a magnetic moment in the positive z direction must lie along the z axis or be parallel to it with the vector $\boldsymbol{\ell}$ from south to north pole in the positive z direction. For $\ell = 8$ cm $= 0.08$ m and $m = 5.76\ \text{A}\cdot\text{m}^2$, the pole strength q_m is

$$q_m = \frac{m}{\ell} = \frac{5.76\ \text{A}\cdot\text{m}^2}{0.08\ \text{m}} = 72\ \text{A}\cdot\text{m} = 72\ \text{N/T}$$

Question

10. The magnetic moment of a current loop is antiparallel to a uniform magnetic field **B**. What is the torque on the loop? Is this equilibrium stable or unstable?

24-4 The Hall Effect

In Section 24-1, we calculated the force exerted by a magnetic field on a current-carrying wire. This force is transferred to the wire by the forces that bind the electrons to the wire at the surface. Since the charge carriers themselves experience the magnetic force when a current-carrying wire is in an external magnetic field, the carriers are accelerated toward one side of the wire. This results in a separation of charge in the wire called the **Hall effect.** This phenomenon allows us to determine the sign of the charge on the charge carriers and the number of charge carriers per unit volume n in a conductor. It also provides a convenient method for measuring magnetic fields.

Figure 24-24 shows two conducting strips each of which carries a current I to the right because the left sides of the strips are connected to the positive terminal of a battery, and the right sides are connected to the negative terminal. The strips are in a magnetic field that is directed into the paper. Let us assume for the moment that the current consists of positively charged particles moving to the right as shown in Figure 24-24*a*. The magnetic force on these particles is $q\mathbf{v}_d \times \mathbf{B}$ (where $\mathbf{v}_d$ is the drift velocity of the charge carriers). This force is directed upward. The positive particles therefore move up to the top of the strip, leaving the bottom of the strip with an excess negative charge. This separation of charge produces an electrostatic field in the strip that opposes the magnetic force on the charge carriers. When the electrostatic and magnetic forces balance, the charge carriers will no longer move upward. In this equilibrium situation, the upper part of the strip is positively charged, so it is at a greater potential than the negatively charged lower part. If the current consists of negatively charged particles, as shown in Figure 24-24*b*, the charge carriers must move to the left (since the current is still to the right). The magnetic force $q\mathbf{v}_d \times \mathbf{B}$ is again up because the signs of both q

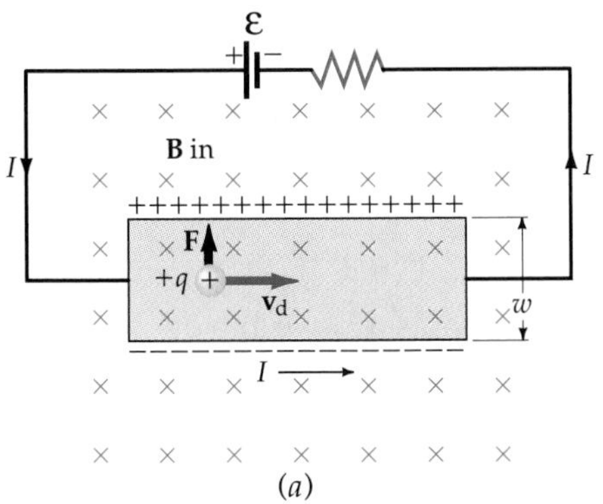

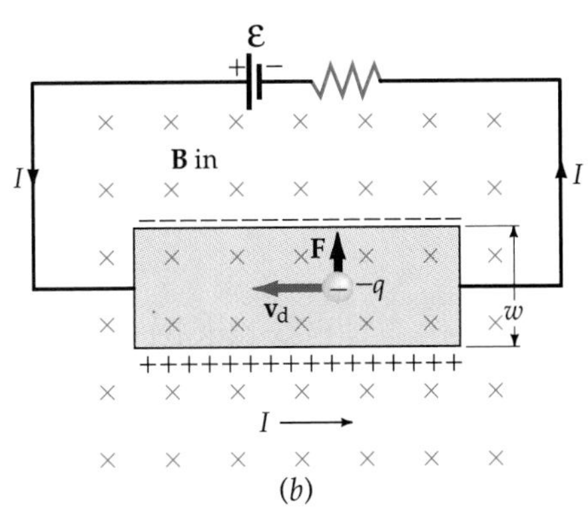

Figure 24-24 The Hall effect. The magnetic field is directed into the plane of the page as indicated by the crosses. The magnetic force on a charged particle is upward for a current to the right whether the current is due to (*a*) positive particles moving to the right or (*b*) negative particles moving to the left.

and $\mathbf{v}_d$ have been changed. Again the carriers are forced to the upper part of the strip, but the upper part of the strip now carries a negative charge (because the charge carriers are negative) and the lower part carries a positive charge.

A measurement of the sign of the potential difference between the upper and lower part of the strip tells us the sign of the charge carriers. For a normal metallic conductor, we find that the upper part of the strip in Figure 24-24 is at a lower potential than the lower part—which means that the upper part must carry a negative charge. It was this type of experiment that led to the discovery that the charge carriers in metallic conductors are negative. Thus Figure 24-24*b* is the correct illustration of the current in a normal conductor.

If we connect the upper and lower portions of the strip with a wire of resistance R, the negative electrons will flow from the upper part of the strip through the wire to the lower part. As electrons leave the upper part of the strip and enter the lower part, the magnitude of the charge separation across the strip is momentarily reduced. As a result, the electrostatic force on the electrons in the strip is momentarily weakened such that it no longer balances the magnetic force on them. The magnetic force will therefore drive more electrons toward the top of the strip, thereby maintaining the potential difference across the strip. The strip is thus a source of emf. The potential difference between the top and bottom of the strip is called the **Hall voltage.**

The magnitude of the Hall voltage is not hard to calculate. The magnitude of the magnetic force on the charge carriers in the strip is qv_dB. This magnetic force is balanced by the electrostatic force of magnitude qE, where E is the electric field due to the charge separation. Thus we have $E = v_dB$. If the width of the strip is w, the potential difference is Ew. The Hall voltage is therefore

$$V_H = Ew = v_dBw \tag{24-18}$$

Exercise

A conducting strip of width $w = 2.0$ cm is placed in a magnetic field of 8000 G. Calculate the Hall voltage if the drift velocity is 4.0×10^{-5} m/s. (Answer: 0.64 μV)

Since the drift velocity for ordinary currents is very small, we can see from Equation 24-18 that the Hall voltage is very small for ordinary-sized strips and magnetic fields. From measurements of the Hall voltage for a strip of a given size carrying a known current in a known magnetic field, we can determine the number of charge carriers per unit volume in the strip. By Equation 22-3 the current is

$$I = nqv_dA$$

where A is the cross-sectional area of the strip. For a strip of width w and thickness t, the cross-sectional area is $A = wt$. Since the charge carriers are electrons, the quantity q is the charge on one electron e. The number density of charge carriers n is thus given by

$$n = \frac{I}{Aqv_d} = \frac{I}{wtev_d} \tag{24-19}$$

Substituting $v_dw = V_H/B$ from Equation 24-18, we have

$$n = \frac{IB}{etV_H} \tag{24-20}$$

Example 24-9

When a silver slab of thickness 1 mm and width 1.5 cm carries a current of 2.5 A in a region in which there is a magnetic field of magnitude 1.25 T perpendicular to the slab, the Hall voltage is measured to be 0.334 μV. (*a*) Calculate the number density of the charge carriers. (*b*) Compare your answer in part (*a*) to the number density of atoms in silver, which has mass density $\rho = 10.5\ \text{g/cm}^3$ and molar mass $M = 107.9$ g/mol.

(*a*) From Equation 24-20, we have

$$n = \frac{(2.5\ \text{A})(1.25\ \text{T})}{(1.6 \times 10^{-19}\ \text{C})(0.001\ \text{m})(3.34 \times 10^{-7}\ \text{V})}$$

$$= 5.85 \times 10^{28}\ \text{electrons/m}^3$$

(*b*) The number of atoms per unit volume n_a is given by

$$n_a = \frac{N_A \rho}{M} = \frac{(6.02 \times 10^{23}\ \text{atoms/mol})(10.5\ \text{g/cm}^3)}{107.9\ \text{g/mol}}$$

$$= 5.86 \times 10^{22}\ \text{atoms/cm}^3 = 5.86 \times 10^{28}\ \text{atoms/m}^3$$

These results indicate that the number of charge carriers in silver is very nearly one per atom.

Although the Hall voltage is ordinarily very small, it provides a convenient method for measuring magnetic fields. If we rearrange Equation 24-20, we can write for the Hall voltage

$$V_H = \frac{I}{net} B \qquad 24\text{-}21$$

A given strip can be calibrated by measuring the Hall voltage for a given current in a known magnetic field. The strength of the magnetic field B of an unknown field can then be measured by placing the strip in the unknown field, sending a current through the strip, and measuring V_H.

The Quantum Hall Effect

According to Equation 24-21, the Hall voltage should increase linearly with magnetic field B for a given current in a given slab. In 1980, while studying the Hall effect in semiconductors at very low temperatures and very large magnetic fields, the German physicist Klaus von Klitzing discovered that a plot of V_H versus B resulted in a series of plateaus, as shown in Figure 24-25,

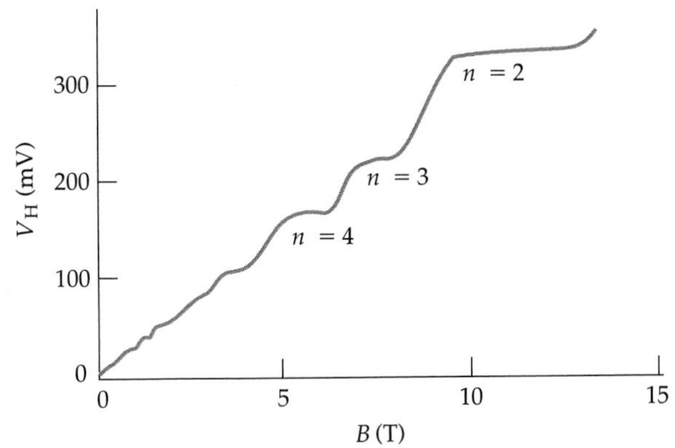

Figure 24-25 A plot of the Hall voltage versus applied magnetic field shows plateaus, indicating that the Hall voltage is quantized. These data were taken at a temperature of 1.39 K with the current I held fixed at 25.52 μA.

rather than a straight line. That is, the Hall voltage is quantized. For this discovery of the **quantum Hall effect,** von Klitzing won the Nobel Prize in Physics in 1985. According to the theory of the quantum Hall effect, the Hall resistance, defined as $R_H = V_H/I$, can take on only the values

$$R_H = \frac{V_H}{I} = \frac{R_K}{n} \qquad n = 1, 2, 3, \ldots \tag{24-22}$$

where n is an integer and R_K, called the **von Klitzing constant,** is related to the fundamental electronic charge e and Planck's constant h by

$$R_K = \frac{h}{e^2} = \frac{6.626 \times 10^{-34}\ \text{J}\cdot\text{s}}{(1.602 \times 10^{-19}\ \text{C})^2} = 25\ 813\ \Omega \tag{24-23}$$

Because the von Klitzing constant can be measured to an accuracy of a few parts in 10^9, the quantum Hall effect is now used to define a standard of resistance. As of January 1990, the ohm is now defined so that R_K has the value 25 812.807 Ω exactly.

Recent experiments have shown that under certain special conditions the Hall resistance is given by Equation 24-22 with the integer n replaced by a rational fraction f. At present, the theory of this **fractional quantum Hall effect** is incomplete.

Summary

1. Moving charges interact with each other via the magnetic force. Since electric currents consist of moving charges, they also exert magnetic forces on each other. This force is described by saying that one moving charge or current creates a magnetic field that, in turn, exerts a force on the other moving charge or current. Ultimately, all magnetic fields are caused by charges in motion.

2. When a charge q moves with velocity $\mathbf{v}$ in a magnetic field $\mathbf{B}$, it experiences a force

$$\mathbf{F} = q\mathbf{v} \times \mathbf{B}$$

The force on a current element is given by

$$d\mathbf{F} = I\, d\boldsymbol{\ell} \times \mathbf{B}$$

The SI unit of magnetic field is the tesla (T). A commonly used unit is the gauss (G), which is related to the tesla by

$$1\ \text{T} = 10^4\ \text{G}$$

3. A particle of mass m and charge q moving with speed v in a plane perpendicular to a magnetic field moves in a circular orbit of radius r given by

$$r = \frac{mv}{qB}$$

The period and frequency of this circular motion are independent of the radius of the orbit or of the speed of the particle. The period, called the cyclotron period, is given by

$$T = \frac{2\pi m}{qB}$$

The frequency, called the cyclotron frequency, is given by

$$f = \frac{1}{T} = \frac{qB}{2\pi m}$$

4. A velocity selector consists of crossed electric and magnetic fields such that the electric and magnetic forces balance for a particle whose speed is given by $v = E/B$.

5. The mass-to-charge ratio of an ion of known speed can be determined by measuring the radius of the circular path taken by the ion in a known magnetic field in a mass spectrometer.

6. A current loop in a uniform magnetic field behaves as a magnetic dipole with magnetic moment **m** given by

$$\mathbf{m} = NIA\,\hat{\mathbf{n}}$$

where N is the number of turns, A is the area of the loop, I is the current, and $\hat{\mathbf{n}}$ is a unit vector perpendicular to the plane of the loop in the direction given by the right-hand rule. When a magnetic dipole is in a magnetic field, it experiences a torque given by

$$\boldsymbol{\tau} = \mathbf{m} \times \mathbf{B}$$

tending to align the magnetic moment of the current loop with the external field. The net force on a current loop in a uniform magnetic field is zero.

7. A bar magnet also experiences a torque in a magnetic field. The experimentally measured torque can be used to define the magnetic moment of the bar magnet from $\boldsymbol{\tau} = \mathbf{m} \times \mathbf{B}$. The pole strength of a bar magnet q_m can be defined by writing the force exerted on the pole as $\mathbf{F} = q_m\mathbf{B}$. The north magnetic pole has a positive pole strength and the south pole has a negative pole strength. In terms of the pole strength, the magnetic moment of a bar magnet is $\mathbf{m} = |q_m|\boldsymbol{\ell}$, where $\boldsymbol{\ell}$ is the vector from the south pole to the north pole.

8. When a conducting strip carrying a current is placed in a magnetic field, the magnetic force on the charge carriers causes a separation of charge called the Hall effect. This results in a voltage V_H, called the Hall voltage, which is given by

$$V_H = v_d B w = \frac{I}{nqt}B$$

where v_d is the drift velocity, B is the magnetic field, w is the width of the strip, t is the thickness of the strip, n is the number density of charge carriers, and q is the charge of each carrier. The sign of the charge carriers can be determined from a measurement of the sign of the Hall voltage, and the number of carriers per unit volume can be determined from the magnitude of V_H. Measurements at very low temperatures and very large magnetic fields indicate that the Hall resistance $R_H = V_H/I$ is quantized and can take on values given by

$$R_H = \frac{V_H}{I} = \frac{R_K}{n}$$

where n is an integer and R_K is the von Klitzing constant, which has the value

$$R_K = \frac{h}{e^2} \approx 25\,813\ \Omega$$

Suggestions for Further Reading

Akasofu, Syun-Ichi: "The Dynamic Aurora," *Scientific American*, May 1989, p. 90.

How the "solar wind" of charged particles interacts with the earth's magnetic field to produce the aurora, why the aurora appears to move and change, and where its power comes from.

Nier, Alfred O. C.: "The Mass Spectrometer," *Scientific American*, March 1953, p. 68.

This device, which has made possible great advances in chemistry and other sciences, allows the determination of the composition of a substance based on the principle of the deflection of a beam of charged particles in a magnetic field.

Shamos, Morris H.: "The Electron—J. J. Thomson," in *Great Experiments in Physics*, Henry Holt and Co., New York, 1959. Reprinted by Dover, 1987.

Thomson's account of his discovery of the electron, with editorial annotations for clarity, and a biographical sketch.

Van Allen, James A.: "Interplanetary Particles and Fields," *Scientific American*, September 1975, p. 160.

This article, written by the man after whom the Van Allen radiation belts are named, describes the deflection of the "solar wind" in the earth's magnetic field.

Review

A. Objectives: After studying this chapter, you should:

1. Be able to calculate the magnetic force on a current element and on a moving charge in a given magnetic field.

2. Be able to calculate the magnetic dipole moment of a current loop and the torque exerted on a current loop in a magnetic field.

3. Be able to discuss the experiment in which Thomson measured q/m for electrons.

4. Be able to describe a velocity selector, a mass spectrometer, and a cyclotron.

5. Be able to discuss the Hall effect.

B. Define, explain, or otherwise identify:

Magnetic field
Tesla
Gauss
Current element
Magnetic-field lines
Cyclotron period
Cyclotron frequency
Magnetic bottle
Crossed fields
Velocity selector
Mass spectrometer
Cyclotron
Magnetic dipole moment
Hall effect
Hall voltage
Quantum Hall effect
von Klitzing constant

C. True or false: If the statement is true, explain why it is true. If it is false, give a counterexample.

1. The magnetic force on a moving charged particle is always perpendicular to the velocity of the particle.

2. The torque on a magnet tends to align the magnetic moment in the direction of the magnetic field.

3. A current loop in a uniform magnetic field behaves like a small magnet.

4. The period of a particle moving in a circle in a magnetic field is proportional to the radius of the circle.

5. The drift velocity of electrons in a wire can be determined from the Hall effect.

Problems

Level I

24-1 The Force Exerted by a Magnetic Field

1. Find the magnetic force on a proton moving with velocity 4.46 Mm/s in the positive x direction in a magnetic field of 1.75 T in the positive z direction.

2. A charge $q = -2.64$ nC moves with a velocity of 2.75×10^6 m/s **i**. Find the force on the charge if the magnetic field is (*a*) $\mathbf{B} = 0.48$ T **j**, (*b*) $\mathbf{B} = 0.65$ T **i** + 0.65 T **j**, (*c*) $\mathbf{B} = 0.75$ T **i**, (*d*) $\mathbf{B} = 0.65$ T **i** + 0.65 T **k**.

3. A uniform magnetic field of magnitude 1.28 T is in the positive z direction. Find the force on a proton if its velocity is (*a*) $\mathbf{v} = 3.5$ Mm/s **i**, (*b*) $\mathbf{v} = 2.5$ Mm/s **j**, (*c*) $\mathbf{v} = 6.5$ Mm/s **k**, and (*d*) $\mathbf{v} = 3.0$ Mm/s **i** + 4.0 Mm/s **j**.

4. An electron moves with a velocity of 3.75 Mm/s in the xy plane at an angle of 60° to the x axis and 30° to the y axis. A magnetic field of 0.85 T is in the positive y direction. Find the force on the electron.

5. A straight wire segment 2 m long makes an angle of 30° with a uniform magnetic field of 0.5 T. Find the magnitude of the force on the wire if it carries a current of 2 A.

6. A straight wire segment $I\ell = (2.5\ \text{A})(3\ \text{cm}\ \mathbf{i} + 4\ \text{cm}\ \mathbf{j})$ is in a uniform magnetic field $\mathbf{B} = 1.5$ T **i**. Find the force on the wire.

7. A long wire parallel to the x axis carries a current of 8.5 A in the positive x direction. There is a uniform magnetic field $\mathbf{B} = 1.65$ T **j**. Find the force per unit length on the wire.

24-2 Motion of a Point Charge in a Magnetic Field

8. A proton moves in a circular orbit of radius 65 cm perpendicular to a uniform magnetic field of magnitude 0.75 T. (*a*) What is the period for this motion? Find (*b*) the speed of the proton and (*c*) the kinetic energy of the proton.

9. An electron of kinetic energy 45 keV moves in a circular orbit perpendicular to a magnetic field of 0.325 T. (*a*) Find the radius of the orbit. (*b*) Find the frequency and the period of the motion.

10. An alpha particle (charge $+2e$) travels in a circular path of radius 0.5 m in a magnetic field of 1.0 T. Find the (*a*) period, (*b*) speed, and (*c*) kinetic energy (in electron volts) of the alpha particle. Take $m = 6.65 \times 10^{-27}$ kg for the mass of the alpha particle.

11. A beam of protons moves along the x axis in the positive x direction with a speed of 12.4 km/s through a region of crossed fields balanced for zero deflection. (*a*) If there is a magnetic field of magnitude 0.85 T in the positive y direction, find the magnitude and direction of the electric field. (*b*) Would electrons of the same velocity be deflected by these fields? If so, in what direction?

12. A velocity selector has a magnetic field of magnitude 0.28 T perpendicular to an electric field of magnitude 0.46 MV/m. (*a*) What must the speed of a particle be for it to pass through undeflected? What energy must (*b*) protons and (*c*) electrons have to pass through undeflected?

13. An electron from the sun with a speed of 1×10^8 m/s enters the earth's magnetic field high above the equator where the magnetic field is 4×10^{-7} T. The electron moves nearly in a circle except for a small drift along the direction of the earth's magnetic field that will take it toward the north pole. (*a*) What is the radius of the circular motion? (*b*) What is the radius of the circular motion near the north pole where the magnetic field is 2×10^{-5} T?

14. A singly ionized ^{24}Mg ion (mass 3.983×10^{-26} kg) is accelerated through a 2.5-kV potential difference and deflected in a magnetic field of 55.7 mT in a mass spectrometer. (*a*) Find the radius of curvature of the orbit for the ion. (*b*) What is the difference in radius for ^{26}Mg and ^{24}Mg ions? (Assume that their mass ratio is 26/24.)

15. A cyclotron for accelerating protons has a magnetic field of 1.4 T and a radius of 0.7 m. (*a*) What is the cyclotron frequency? (*b*) Find the maximum energy of the protons when they emerge. (*c*) How will your answers change if deuterons, which have the same charge but twice the mass, are used instead of protons?

16. A certain cyclotron with magnetic field 1.8 T is designed to accelerate protons to 25 MeV. (*a*) What is the cyclotron frequency? (*b*) What must the minimum radius of the magnet be to achieve a 25-MeV emergence energy? (*c*) If the alternating potential applied to the dees has a maximum value of 50 kV, how many revolutions must the protons make before emerging with an energy of 25 MeV?

24-3 Torques on Current Loops and Magnets

17. A small circular coil of 20 turns of wire lies on a uniform magnetic field of 0.5 T such that the normal to the plane of the coil makes an angle of 60° with the direction of **B**. The radius of the coil is 4 cm, and it carries a current of 3 A. (*a*) What is the magnitude of the magnetic moment of the coil? (*b*) What is the magnitude of the torque exerted on the coil?

18. The SI unit for the magnetic moment of a current loop is A·m^2. Use this to show that 1 T = 1 N/A·m.

19. What is the maximum torque on a 400-turn circular coil of radius 0.75 cm that carries a current of 1.6 mA and resides in a uniform magnetic field of 0.25 T?

20. The unit of magnetic pole strength, as defined by Equation 24-16, is a newton/tesla (N/T). Show that this unit is also equal to an ampere-meter (A·m).

21. A current-carrying wire is bent into the shape of a square of sides $L = 6$ cm and is placed in the xy plane. It carries a current $I = 2.5$ A. What is the torque on the wire if there is a uniform magnetic field of 0.3 T (*a*) in the z direction, (*b*) in the x direction?

22. Repeat Problem 21 if the wire is bent into an equilateral triangle of sides 8 cm.

23. A small bar magnet of length 8.5 cm and pole strength 25 N/T lies along the x axis in a uniform magnetic field $\mathbf{B} = 1.5\text{ T}\,\mathbf{i} + 2.5\text{ T}\,\mathbf{j} + 1.6\text{ T}\,\mathbf{k}$. (*a*) What is the magnetic moment of the magnet? (*b*) Find the torque exerted on the magnet.

24. A small magnet of length 6.8 cm is placed at an angle of 60° to the direction of a uniform magnetic field of magnitude 0.04 T. The observed torque has the magnitude of 0.10 N·m. (*a*) Find the magnetic moment of the magnet. (*b*) Find the pole strength q_m.

24-4 The Hall Effect

25. A metal strip 2.0 cm wide and 0.1 cm thick carries a current of 20 A in a uniform magnetic field of 2.0 T, as shown in Figure 24-26. The Hall emf is measured to be 4.27 μV. (*a*) Calculate the drift velocity of the electrons in the strip. (*b*) Find the number density of the charge carriers in the strip.

26. (*a*) In Figure 24-26, is point a or b at the higher potential? (*b*) If the metal strip is replaced by a p-type semiconductor in which the charge carriers are positive, which point will be at the higher potential?

Figure 24-26 Problems 25, 26, and 27.

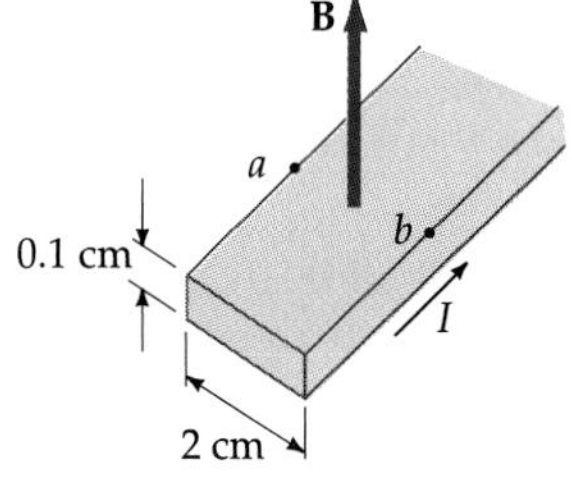

27. The number density of free electrons in copper is 8.47×10^{22} electrons per cubic centimeter. If the metal strip in Figure 24-26 is copper and the current is 10 A, find (*a*) the drift velocity v_d and (*b*) the Hall voltage. (Assume that the magnetic field is 2.0 T.)

28. A copper strip ($n = 8.47 \times 10^{22}$ electrons per cubic centimeter) 2 cm wide and 0.1 cm thick is used to measure the magnitudes of unknown magnetic fields which are perpendicular to the strip. Find the magnitude of B when $I = 20$ A and the Hall voltage is (*a*) 2.00 μV, (*b*) 5.25 μV, and (*c*) 8.00 μV.

29. Blood contains charged ions so that moving blood develops a Hall voltage across the diameter of an artery. A large artery with a diameter of 0.85 cm has a flow speed of 0.6 m/s. If a section of this artery is in a magnetic field of 0.2 T, what is the potential difference across the diameter of the artery?

Level II

30. A beam of ^{6}Li and ^{7}Li ions passes through a velocity selector and enters a magnetic spectrometer. If the diameter of the orbit of the ^{6}Li ions is 15 cm, what is the diameter of the orbit of the ^{7}Li ions?

31. The wire segment in Figure 24-27 carries a current of 1.8 A from a to b. There is a magnetic field $\mathbf{B} = 1.2$ T $\mathbf{k}$. Find the total force on the wire and show that it is the same as if the wire were a straight segment from a to b.

Figure 24-27 Problem 31.

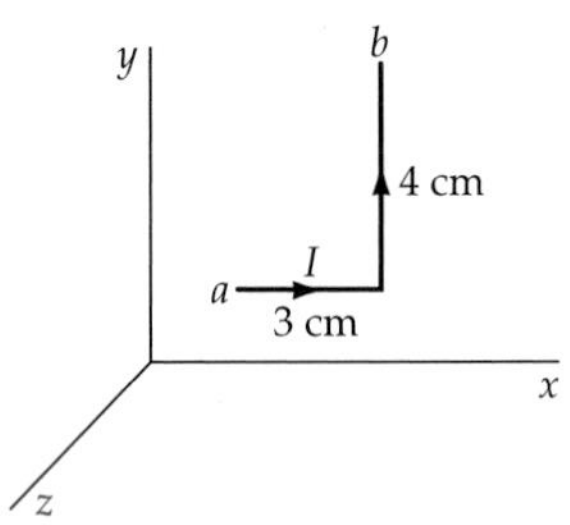

32. A straight, stiff, horizontal wire of length 25 cm and mass 50 g is connected to a source of emf by light, flexible leads. A magnetic field of 1.33 T is horizontal and perpendicular to the wire. Find the current necessary to float the wire, that is, the current such that the magnetic force balances the weight of the wire.

33. The plates of a Thomson q/m apparatus are 6.0 cm long and are separated by 1.2 cm. The end of the plates is 30.0 cm from the tube screen. The kinetic energy of the electrons is 2.8 keV. (*a*) If a potential of 25.0 V is applied across the deflection plates, by how much will the beam deflect? (*b*) Find the magnitude of the crossed magnetic field that will allow the beam to pass through undeflected.

34. A simple gaussmeter for measuring horizontal magnetic fields consists of a stiff 50-cm wire that hangs from a conducting pivot so that its free end makes contact with a pool of mercury in a dish below. The wire has a mass of 5 g and conducts a current downward. (*a*) What is the equilibrium angular displacement of the wire from vertical if the horizontal magnetic field is 0.04 T and the current is 0.20 A? (*b*) If the current is 20 A and a displacement from vertical of 0.5 mm can be detected for the free end, what is the horizontal magnetic field sensitivity of this gaussmeter?

35. A rectangular, 50-turn coil has sides 6.0 and 8.0 cm long and carries a current of 1.75 A. It is oriented as shown in Figure 24-28 and pivoted about the z axis. (*a*) If the wire in the xy plane makes an angle of 37° with the y axis as shown, what angle does the unit normal $\hat{\mathbf{n}}$ make with the x axis? (*b*) Write an expression for $\hat{\mathbf{n}}$ in terms of the unit vectors $\mathbf{i}$ and $\mathbf{j}$. (*c*) What is the magnetic moment of the coil? (*d*) Find the torque on the coil when there is a uniform magnetic field $\mathbf{B} = 1.5$ T $\mathbf{j}$.

Figure 24-28 Problems 35 and 36.

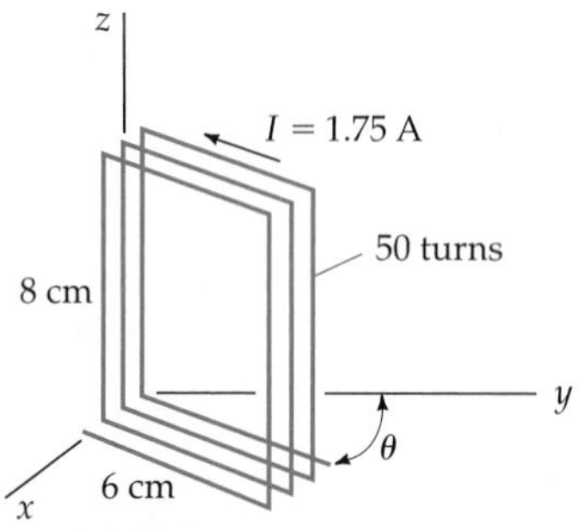

36. The coil in Problem 35 is pivoted about the z axis and held at various positions in a uniform magnetic field $\mathbf{B} = 2.0$ T $\mathbf{j}$. Sketch the position of the coil and find the torque exerted when the unit normal is (*a*) $\hat{\mathbf{n}} = \mathbf{i}$, (*b*) $\hat{\mathbf{n}} = \mathbf{j}$, (*c*) $\hat{\mathbf{n}} = -\mathbf{j}$, and (*d*) $\hat{\mathbf{n}} = (\mathbf{i} + \mathbf{j})/\sqrt{2}$.

37. A particle of charge q and mass M moves in a circle of radius r and with angular velocity ω. (*a*) Show that the average current is $I = q\omega/2\pi$ and that the magnetic moment has the magnitude $m = \frac{1}{2}q\omega r^2$. (*b*) Show that the angular momentum of this particle has the magnitude $L = Mr^2\omega$ and that the magnetic moment and angular momentum vectors are related by $\mathbf{m} = (q/2M)\mathbf{L}$.

38. A particle of charge q and mass m has momentum $p = mv$ and kinetic energy $K = \frac{1}{2}mv^2 = p^2/2m$. If it moves in a circular orbit of radius r perpendicular to a uniform magnetic field B, show that (*a*) $p = Bqr$ and (*b*) $K = B^2q^2r^2/2m$.

39. Protons, deuterons (each with charge $+e$), and alpha particles (with charge $+2e$) of the same kinetic energy enter a uniform magnetic field $\mathbf{B}$ that is perpendicular to their velocities. Let r_p, r_d, and r_α be the radii of their circular orbits. Find the ratios r_d/r_p and r_α/r_p. Assume that $m_\alpha = 2m_\text{d} = 4m_\text{p}$.

40. Show that the cyclotron frequencies of deuterons and alpha particles are the same and are half that of a proton in the same magnetic field. (See Problem 39.)

41. A proton and an alpha particle move in a uniform magnetic field in circles of the same radii. Compare (*a*) their velocities, (*b*) their kinetic energies, and (*c*) their angular momenta. (See Problem 39.)

42. Beryllium has a density of 1.83 g/cm^3 and a molar mass of 9.01 g/mol. A slab of beryllium of thickness 1.4 mm and width 1.2 cm carries a current of 3.75 A in a region in which there is a magnetic field of magnitude 1.88 T perpendicular to the slab. The Hall voltage is measured to be 0.130 μV. (*a*) Calculate the number density of the charge carriers. (*b*) Calculate the number density of atoms in beryllium. (*c*) How many free electrons are there per atom of beryllium?

43. A rigid, circular loop of radius R and mass M lies in the xy plane on a rough, flat table. The magnetic field is $\mathbf{B} = B_x\,\mathbf{i} + B_y\,\mathbf{j}$. How large must the current I be before one edge of the loop will lift off the table?

44. A moving-coil galvanometer consists of a coil of wire suspended by a thin, highly flexible fiber in a magnetic field **B**, which is in the plane of the coil. As a current I is passed through the coil, a torque is produced that tends to rotate the coil. The fiber, in turn, supplies a restoring torque $\tau = k\theta$ that is proportional to the angle θ through which it has been twisted. The constant k is called the torsion constant. Show that $I = k\theta/(NAB)$, where N is the number of turns of wire in the coil and A is the area of the coil.

45. A wire of length ℓ is wound into a circular coil of N loops. Show that when this coil carries a current I, its magnetic moment has the magnitude $I\ell^2/4\pi N$.

46. A metal disk of radius 6 cm is mounted on a frictionless axle. Current can flow through the axle out along the disk to a sliding contact at the rim of the disk. A uniform magnetic field $B = 1.25$ T is parallel to the axis of the disk. When the current is 3 A, the disk rotates with constant angular velocity. What is the frictional force at the rim between the stationary electrical contact and the rotating rim?

47. A particle of mass m and charge q enters a region where there is a uniform magnetic field B along the x axis. The initial velocity of the particle is $\mathbf{v} = v_{0x}\,\mathbf{i} + v_{0y}\,\mathbf{j}$, so the particle moves in a helix. (*a*) Show that the radius of the helix is $r = mv_{0y}/qB$. (*b*) Show that the particle takes a time $t = 2\pi m/qB$ to make one orbit around the helix.

48. Before entering a mass spectrometer, ions pass through a velocity selector consisting of parallel plates separated by 2.0 mm and having a potential difference of 160 V. The magnetic field between the plates is 0.42 T. The magnetic field in the mass spectrometer is 1.2 T. Find (*a*) the speed of the ions entering the mass spectrometer and (*b*) the difference in the diameters of the orbits of singly ionized ^{238}U and ^{235}U. (The mass of a ^{235}U ion is 3.903×10^{-25} kg.)

49. A conducting wire is parallel to the y axis. It moves in the positive x direction with a speed of 20 m/s in a magnetic field $\mathbf{B} = 0.5\text{ T}\,\mathbf{k}$. (*a*) What are the magnitude and direction of the magnetic force on an electron in the conductor? (*b*) Because of this magnetic force, electrons move to one end of the wire leaving the other end positively charged, until the electric field due to this charge separation exerts a force on the electrons that balances the magnetic force. Find the magnitude and direction of this electric field in the steady state. (*c*) Suppose the moving wire is 2 meters long. What is the potential difference between its two ends due to this electric field?

50. A metal crossbar of mass M rides on a pair of long, horizontal conducting rails separated by a distance ℓ and connected to a device that supplies constant current I to the circuit, as shown in Figure 24-29. A uniform magnetic field B is established as shown. (*a*) If there is no friction and the bar starts from rest at $t = 0$, show that at time t the bar has velocity $v = (BI\ell/M)t$. (*b*) In which direction will the bar move? (*c*) If the coefficient of static friction is μ_s, find the minimum field B necessary to start the bar moving.

Figure 24-29 Problems 50 and 51.

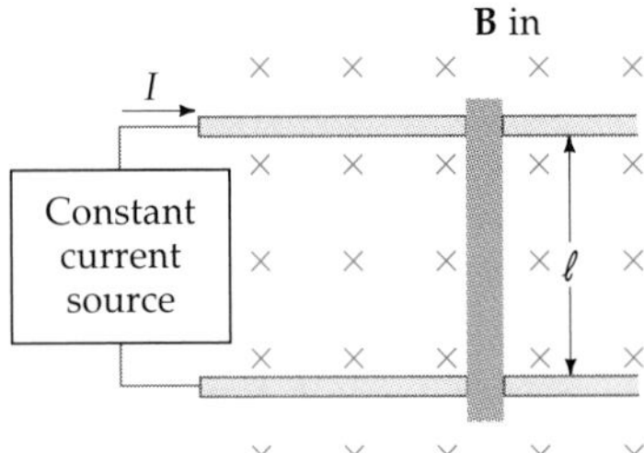

51. Assume that the rails in Figure 24-29 are frictionless but tilted upward so that they make an angle θ with the horizontal. (*a*) What vertical magnetic field B is needed to keep the bar from sliding down the rails? (*b*) What is the acceleration of the bar if B has twice the value found in part (*a*)?

Level III

52. A stiff, straight, horizontal wire of length 25 cm and mass 20 g is supported by electrical contacts at its ends, but is otherwise free to move vertically upward. The wire is in a uniform, horizontal magnetic field of magnitude 0.4 T perpendicular to the wire. A switch connecting the wire to a battery is closed and the wire is shot upward, rising to a maximum height h. The battery delivers a total charge of 2 C during the short time it makes contact with the wire. Find the height h.

53. A circular loop of wire with mass M carries a current I in a uniform magnetic field. It is initially in equilibrium with its magnetic moment vector aligned with the magnetic field. The loop is given a small twist about a diameter and then released. What is the period of the motion? (Assume that the only torque exerted on the loop is due to the magnetic field.)

54. A current-carrying wire is bent into a semicircular loop of radius R, which lies in the xy plane. There is a uniform magnetic field $\mathbf{B} = B\,\mathbf{k}$ perpendicular to the plane of the loop (Figure 24-30). Show that the force acting on the loop is $\mathbf{F} = 2IRB\,\mathbf{j}$.

Figure 24-30 Problem 54.

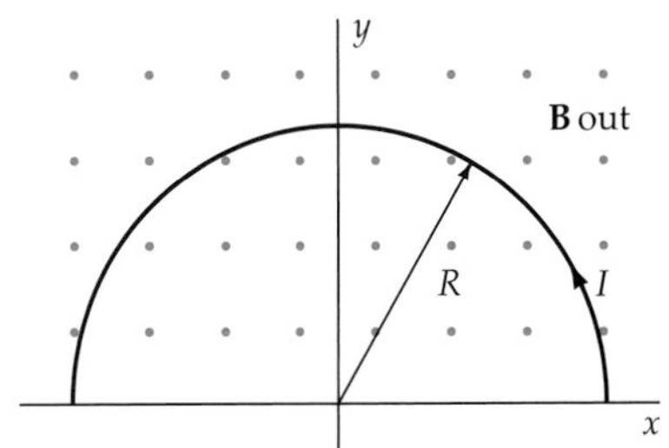

55. Show that the radius of the orbit of a charged particle in a cyclotron is proportional to the square root of the number of orbits completed.

56. A wire bent in some arbitrary shape carries a current I in a uniform magnetic field **B**. Show that the total force on the part of the wire from some point a to some point b is $\mathbf{F} = I\boldsymbol{\ell} \times \mathbf{B}$, where $\boldsymbol{\ell}$ is the vector from a to b.

57. If you make a coil of N turns from a wire of fixed length ℓ, the larger the number of turns, the smaller the area enclosed by the wire. Show that for a wire of given length carrying a current I, the maximum magnetic moment is achieved with a coil of just one turn and the magnitude of the magnetic moment is $I\ell^2/4\pi$. (You need to consider only circular coils. Why?)

58. A nonconducting rod of mass M and length ℓ has a uniform charge per unit length λ and rotates with angular velocity ω about an axis through one end and perpendicular to the rod. (*a*) Consider a small segment of the rod of length dx and charge $dq = \lambda\, dx$ at a distance x from the pivot. Show that the magnetic moment of this segment is $\frac{1}{2}\lambda\omega x^2\, dx$. (*b*) Integrate your result to show that the total magnetic moment of the rod is $m = \frac{1}{6}\lambda\omega\ell^3$. (*c*) Show that the magnetic moment **m** and angular momentum **L** are related by $\mathbf{m} = (Q/2M)\mathbf{L}$, where Q is the total charge on the rod.

59. A nonconducting disk of mass M and radius R has a surface charge density σ and rotates with angular velocity ω about its axis. (*a*) Consider a ring of radius r and thickness dr. Show that the total current in this ring is $dI = (\omega/2\pi)\, dq = \omega\sigma r\, dr$. (*b*) Show that the magnetic moment of the ring is $dm = \pi\omega\sigma r^3\, dr$. (*c*) Integrate your result for part (*b*) to show that the total magnetic moment of the disk is $m = \frac{1}{4}\pi\omega\sigma R^4$. (*d*) Show that the magnetic moment **m** and angular momentum **L** are related by $\mathbf{m} = (Q/2M)\mathbf{L}$, where Q is the total charge on the disk.

60. A small magnet of moment **m** makes an angle θ with a uniform magnetic field **B**. (*a*) How much work must be done by an external torque to twist the magnet by a small amount $d\theta$? (*b*) Show that the work required to rotate the magnet until it is perpendicular to the field is $W = mB \cos\theta$. (*c*) Use your result for part (*b*) to show that if the potential energy of the magnet is chosen to be zero when the magnet is perpendicular to the field, the potential energy at angle θ is $U(\theta) = -\mathbf{m}\cdot\mathbf{B}$. (*d*) Would any part of this problem be different if the magnet were replaced by a current-carrying coil with magnetic moment **m**?

61. A beam of particles with velocity **v** enters a region of uniform magnetic field **B** that makes a small angle θ with **v**. Show that after a particle moves a distance $2\pi(m/qB)v \cos\theta$ measured along the direction of **B** the velocity of the particle is in the same direction as it was when it entered the field.

62. A small bar magnet has a magnetic moment **m** that makes an angle θ with the x axis and lies in a nonuniform magnetic field given by $\mathbf{B} = B_x(x)\,\mathbf{i} + B_y(y)\,\mathbf{j}$. Show that there is a net force on the magnet that is given approximately by

$$\mathbf{F} \approx m_x \frac{\partial B_x}{\partial x}\mathbf{i} + m_y \frac{\partial B_y}{\partial y}\mathbf{j} \qquad \text{24-24}$$

Chapter 25

Sources of the Magnetic Field

These coils at the Kettering Magnetics Laboratory at Oakland University are called Helmholtz coils. They are used to cancel the earth's magnetic field and to provide a uniform magnetic field in a small region of space for studying the magnetic properties of matter.

We now turn to a consideration of the origins of the magnetic field **B**. The earliest known sources of magnetism were permanent magnets. One month after Oersted announced his discovery that a compass needle is deflected by an electric current, Jean Baptiste Biot and Felix Savart announced the results of their measurements of the force on a magnet near a long, current-carrying wire and analyzed these results in terms of the magnetic field produced by each element of the current. André-Marie Ampère extended these experiments and showed that current elements also experience a force in the presence of a magnetic field and that two currents exert forces on each other.

We begin by considering the magnetic field produced by a single moving charge and by the charges in a current element. We then calculate the magnetic fields produced by some common current configurations such as a straight wire segment, a long straight wire, a current loop, and a solenoid. Finally, we discuss Ampère's law, which relates the line integral of the magnetic field around a closed loop to the total current that passes through the loop.

25-1 The Magnetic Field of Moving Point Charges

When a point charge q moves with velocity $\mathbf{v}$, it produces a magnetic field $\mathbf{B}$ in space given by

Magnetic field of a moving charge

$$\mathbf{B} = \frac{\mu_0}{4\pi}\frac{q\mathbf{v} \times \hat{\mathbf{r}}}{r^2} \qquad 25\text{-}1$$

where $\hat{\mathbf{r}}$ is a unit vector that points from the charge q to the field point P (Figure 25-1) and μ_0 is a constant of proportionality called the **permeability of free space,** which has the value

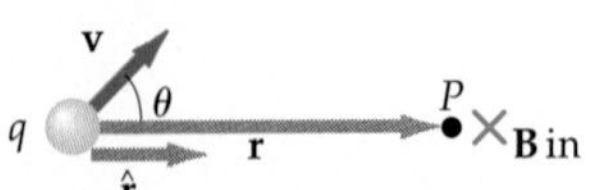

Figure 25-1 A point charge q moving with velocity $\mathbf{v}$ produces a magnetic field $\mathbf{B}$ at a field point P that is in the direction $\mathbf{v} \times \hat{\mathbf{r}}$, where $\hat{\mathbf{r}}$ is the unit vector pointing from the charge to the field point. (The blue x at the field point indicates that the direction of the field is into the page.)

$$\mu_0 = 4\pi \times 10^{-7}\ \text{T·m/A} = 4\pi \times 10^{-7}\ \text{N/A}^2 \qquad 25\text{-}2$$

The units of μ_0 are such that B is in teslas when q is in coulombs, v is in meters per second, and r is in meters. The unit N/A^2 comes from the fact that 1 T = 1 N/A·m. The constant $1/4\pi$ is arbitrarily included in Equation 25-1 so that the factor 4π will not appear in Ampere's law (Equation 25-15), which we will study in Section 25-4. Equation 25-1 for the magnetic field due to a moving point charge is analogous to Coulomb's law for the electric field due to a point charge:

$$\mathbf{E} = \frac{1}{4\pi\epsilon_0}\frac{kq}{r^2}\hat{\mathbf{r}}$$

We see from Equation 25-1 that the magnetic field of a moving point charge has the following characteristics:

1. The magnitude of $\mathbf{B}$ is proportional to the charge q and to the speed v and varies inversely with the square of the distance from the charge.
2. The magnetic field is zero along the line of motion of the charge. At other points in space, it is proportional to $\sin\theta$, where θ is the angle between the velocity $\mathbf{v}$ and the vector $\mathbf{r}$ from the charge to the field point.
3. The direction of $\mathbf{B}$ is perpendicular to both the velocity $\mathbf{v}$ and the vector $\mathbf{r}$. It is in the direction given by the right-hand rule as $\mathbf{v}$ is rotated into $\mathbf{r}$.

Example 25-1

A point charge of magnitude $q_1 = 4.5$ nC is moving with speed 3.6×10^7 m/s parallel to the x axis along the line $y = 3$ m. Find the magnetic field produced by this charge at the origin when the charge is at the point $x = -4$ m, $y = 3$ m, as shown in Figure 25-2.

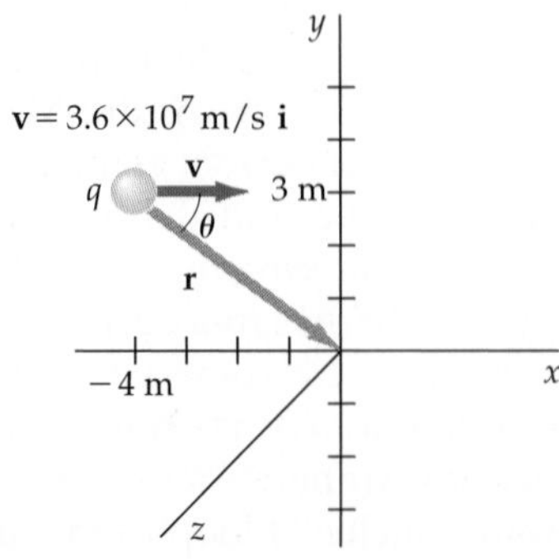

Figure 25-2 Charged particle moving parallel to the x axis for Example 25-1.

The velocity of the charge is $\mathbf{v} = v\mathbf{i} = 3.6 \times 10^7$ m/s $\mathbf{i}$, and the vector from the charge to the origin is given by $\mathbf{r} = 4$ m $\mathbf{i} - 3$ m $\mathbf{j}$. Then $r = 5$ m and the unit vector $\hat{\mathbf{r}}$ is

$$\hat{\mathbf{r}} = \frac{\mathbf{r}}{r} = \frac{4 \text{ m } \mathbf{i} - 3 \text{ m } \mathbf{j}}{5 \text{ m}}$$

$$= 0.8\mathbf{i} - 0.6\mathbf{j}$$

Then

$$\mathbf{v} \times \hat{\mathbf{r}} = (v\mathbf{i}) \times (0.8\mathbf{i} - 0.6\mathbf{j})$$

$$= -0.6\, v\mathbf{k}$$

and Equation 25-1 gives

$$\mathbf{B} = \frac{\mu_0}{4\pi} \frac{q\mathbf{v} \times \hat{\mathbf{r}}}{r^2}$$

$$= \frac{\mu_0}{4\pi} \frac{q(-0.6v\mathbf{k})}{r^2}$$

$$= -(10^{-7}\text{T}\cdot\text{m/A}) \frac{(4.5 \times 10^{-9}\text{ C})(0.6)(3.6 \times 10^7\text{ m/s})}{(5\text{ m})^2}\mathbf{k}$$

$$= -3.89 \times 10^{-10}\text{ T }\mathbf{k}$$

We can also find the magnetic field without explicitly finding the unit vector $\hat{\mathbf{r}}$. From the figure we see that $\mathbf{v} \times \hat{\mathbf{r}}$ is in the negative z direction. Since the magnitude of $\mathbf{v} \times \hat{\mathbf{r}}$ is $v \sin \theta$, where $\sin \theta = 3$ m/5 m $= 0.6$, we have

$$\mathbf{v} \times \hat{\mathbf{r}} = v \sin \theta\, (-\mathbf{k}) = -v(0.6)\mathbf{k}$$

which is the same as we found above. We note from this example that the magnetic field due to a moving charge is very small. For comparison, the magnitude of the magnetic field of the earth near its surface is about 10^{-4} T.

The Magnetic Force and Conservation of Momentum

The magnetic force exerted by one moving charge on another is found by combining Equation 24-1 for the force on a moving charge in a magnetic field and Equation 25-1 for the magnetic field of a charge. The force $\mathbf{F}_{12}$ exerted by a charge q_1 moving with velocity $\mathbf{v}_1$ on a charge q_2 moving with velocity $\mathbf{v}_2$ is given by

$$\mathbf{F}_{12} = q_2\mathbf{v}_2 \times \mathbf{B}_1 = q_2\mathbf{v}_2 \times \left(\frac{\mu_0}{4\pi} \frac{q_1\mathbf{v}_1 \times \hat{\mathbf{r}}_{12}}{r_{12}^2}\right) \qquad \text{25-3}a$$

where $\mathbf{B}_1$ is the magnetic field at the position of charge q_2 due to charge q_1, and $\hat{\mathbf{r}}_{12}$ is the unit vector pointing from q_1 to q_2. Similarly, the force $\mathbf{F}_{21}$ exerted by a charge q_2 moving with velocity $\mathbf{v}_2$ on a charge q_1 moving with velocity $\mathbf{v}_1$ is given by

$$\mathbf{F}_{21} = q_1\mathbf{v}_1 \times \mathbf{B}_2 = q_1\mathbf{v}_1 \times \left(\frac{\mu_0}{4\pi} \frac{q_2\mathbf{v}_2 \times \hat{\mathbf{r}}_{21}}{r_{21}^2}\right) \qquad \text{25-3}b$$

These relations are remarkable in that the force exerted by charge q_1 on charge q_2 is generally not equal and opposite to that exerted by charge q_2 on charge q_1. That is, these forces do not obey Newton's third law, as can be

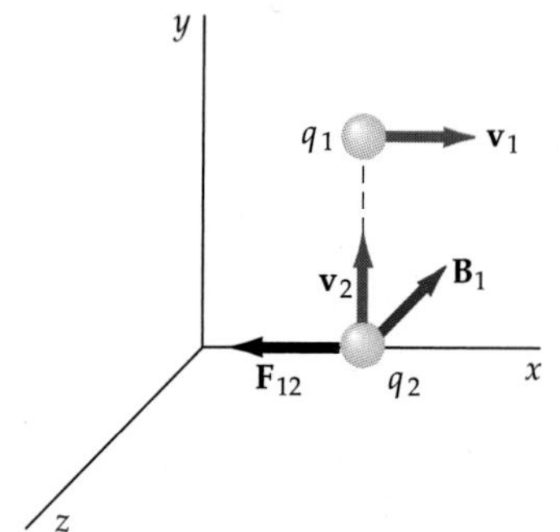

Figure 25-3 The forces exerted by moving charges on each other are not equal and opposite. The magnetic field $\mathbf{B}_1$ at charge q_2 due to charge q_1 is in the negative z direction, so it exerts a force $\mathbf{F}_{12}$ on q_2 to the left in the negative x direction. However, $\mathbf{B}_2$ at charge q_1 due to charge q_2 is zero, so there is no force on q_1.

demonstrated by considering the special case illustrated in Figure 25-3. Here, the magnetic field $\mathbf{B}_1$ due to charge q_1 at charge q_2 is in the negative z direction, and the force on q_2 is to the left in the negative x direction. However, the magnetic field $\mathbf{B}_2$ due to q_2 at q_1 is zero because q_1 lies along the line of motion of q_2. Thus, there is no magnetic force exerted by q_2 on q_1. There is thus a net force $\mathbf{F}_{12}$ acting on the two-charge system. The system will accelerate in the direction of this force, and linear momentum will not be conserved. This apparent violation of the law of conservation of linear momentum results from our treating the force exerted by one charge on another as an action-at-a-distance force and neglecting the momentum carried by the electric and magnetic fields of the moving charges. We saw in Chapter 21 that there is energy associated with an electric field, and we will see later that there is also energy associated with a magnetic field. Advanced treatments of the electric and magnetic fields of moving charges show that there is also momentum associated with these fields. When the charges move, as in Figure 25-3, the linear momentum produced when the system accelerates to the left is balanced by momentum in the opposite direction carried by the fields. Thus, when we include the momentum of the fields, the total momentum of the system is conserved.

Example 25-2

A point charge q_1 is at the point $\mathbf{R} = x\mathbf{i} + y\mathbf{j}$ and is moving parallel to the x axis with velocity $\mathbf{v}_1 = v_1\mathbf{i}$. A second point charge q_2 is at the origin and moving along the x axis with velocity $\mathbf{v}_2 = v_2\mathbf{i}$, as shown in Figure 25-4a. Find the magnetic force exerted by each charge on the other.

We first find the force exerted on charge q_1. We note that the vector $\mathbf{r}_{21}$ from charge q_2 to charge q_1 is just $\mathbf{R}$. Using $\hat{\mathbf{r}}_{21} = \mathbf{r}_{21}/r_{21} = \mathbf{R}/R$, we have

$$\frac{\mathbf{v}_2 \times \hat{\mathbf{r}}_{21}}{r_{21}^2} = \frac{\mathbf{v}_2 \times \mathbf{R}}{R^3} = \frac{v_2\mathbf{i} \times (x\mathbf{i} + y\mathbf{j})}{R^3} = \frac{yv_2}{R^3}\mathbf{k}$$

so the magnetic field produced by charge q_2 at the position of charge q_1 is

$$\mathbf{B}_2 = \frac{\mu_0}{4\pi}\frac{q_2 y v_2}{R^3}\mathbf{k}$$

The magnetic force exerted by charge q_2 on charge q_1 is then

$$\mathbf{F}_{21} = q_1\mathbf{v}_1 \times \mathbf{B}_2 = q_1(v_1\mathbf{i}) \times \left(\frac{\mu_0}{4\pi}\frac{q_2 y v_2}{R^3}\mathbf{k}\right)$$

$$= -\frac{\mu_0}{4\pi}\frac{q_1 q_2 v_1 v_2 y}{R^3}\mathbf{j}$$

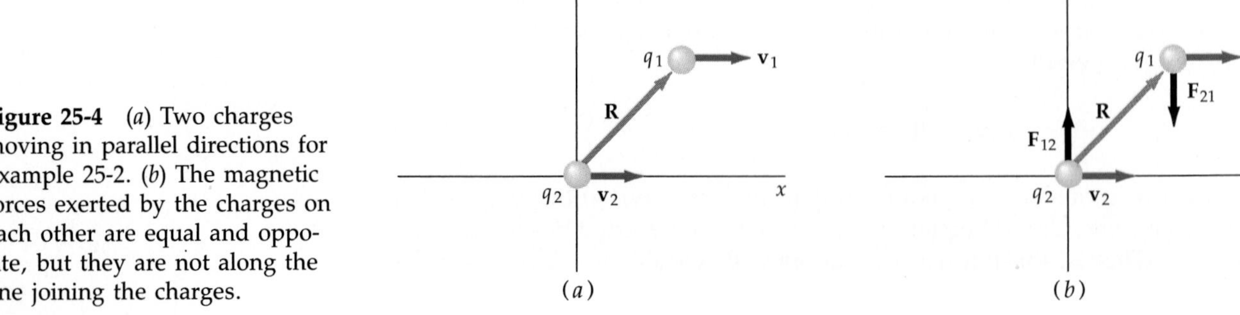

Figure 25-4 (a) Two charges moving in parallel directions for Example 25-2. (b) The magnetic forces exerted by the charges on each other are equal and opposite, but they are not along the line joining the charges.

To find the magnetic force exerted by charge q_1 on charge q_2, we note that the vector $\mathbf{r}_{12}$ from charge q_1 to charge q_2 is $-\mathbf{R}$. Then,

$$\frac{\mathbf{v}_1 \times \hat{\mathbf{r}}_{12}}{r_{12}^2} = \frac{\mathbf{v}_1 \times (-\mathbf{R})}{R^3} = \frac{v_1\mathbf{i} \times (-x\mathbf{i} - y\mathbf{j})}{R^3} = -\frac{yv_1}{R^3}\mathbf{k}$$

The magnetic force exerted by charge q_1 on charge q_2 is then

$$\mathbf{F}_{12} = q_2\mathbf{v}_2 \times \mathbf{B}_1 = q_2(v_2\mathbf{i}) \times \left(-\frac{\mu_0}{4\pi}\frac{q_1yv_1}{R^3}\mathbf{k}\right) = +\frac{\mu_0}{4\pi}\frac{q_1q_2v_1v_2y}{R^3}\mathbf{j}$$

In this case, the forces are equal and opposite as shown in Figure 25-4*b*, but they are not along the line joining the two particles. The magnetic forces thus exert a torque on the two-particle system. Here, the apparent lack of conservation of *angular* momentum implied by the existence of this torque is resolved by the consideration of the *angular* momentum carried by the electromagnetic field.

25-2 The Magnetic Field of Currents: The Biot–Savart Law

As we did in the previous chapter when we found the force exerted on charges and current elements, we can find the magnetic field $d\mathbf{B}$ produced by a current element $I\,d\boldsymbol{\ell}$ by replacing $q\mathbf{v}$ in Equation 25-1 with the current element $I\,d\boldsymbol{\ell}$. We then have

$$d\mathbf{B} = \frac{\mu_0}{4\pi}\frac{I\,d\boldsymbol{\ell} \times \hat{\mathbf{r}}}{r^2} \qquad \text{25-4}$$

Biot–Savart law

Equation 25-4, known as the **Biot–Savart law,** was also deduced by Ampère. The Biot–Savart law, like Equation 25-1, is analogous to Coulomb's law for the electric field of a point charge. The source of the magnetic field is a moving charge $q\mathbf{v}$ or a current element $I\,d\boldsymbol{\ell}$, just as the charge q is the source of the electrostatic field. The magnetic field decreases with the square of the distance from the moving charge or current element, just as the electric field decreases with the square of the distance from a point charge. However, the directional aspects of the electric and magnetic fields are quite different.

(*a*)

(*b*)

Oersted's experiment. (*a*) With no current in the wire, the compass needle points north. (*b*) When the wire carries a current, the needle is deflected in the direction of the resultant magnetic field. The current in the wire is directed upwards from left to right. The insulation has been stripped from the wire to improve the contrast of the photograph.

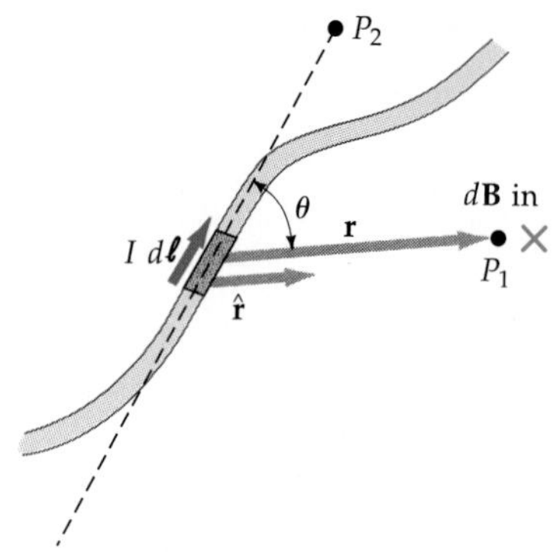

Figure 25-5 The current element $I\,d\boldsymbol{\ell}$ produces a magnetic field at point P_1 that is perpendicular to both $I\,d\boldsymbol{\ell}$ and $\hat{\mathbf{r}}$. It produces no magnetic field at point P_2, which is along the line of $I\,d\boldsymbol{\ell}$.

Whereas the electric field points in the radial direction **r** from the point charge to the field point (for a positive charge), the magnetic field is perpendicular both to **r** and to the direction of motion of the charges **v**, which is along the direction of the current element. At a point along the line of a current element, such as point P_2 in Figure 25-5, the magnetic field due to that element is zero, because the angle θ between $I\,d\boldsymbol{\ell}$ and the vector **r** to that point is zero. The magnetic field due to the total current in a circuit can be calculated by using the Biot–Savart law to find the field due to each current element and then summing (integrating) over all the current elements in the circuit. This calculation is very difficult for all but the most simple circuit geometries.

B Due to a Current Loop

One calculation that is relatively straightforward is to find the magnetic field at the center of a circular loop. Figure 25-6 shows a current element $I\,d\boldsymbol{\ell}$ of a current loop of radius R and the unit vector $\hat{\mathbf{r}}$ that is directed from the element to the center of the loop. The magnetic field at the center of the loop due to this element is directed along the axis of the loop, and its magnitude is given by

$$dB = \frac{\mu_0}{4\pi}\frac{I\,d\ell\,\sin\theta}{R^2}$$

where θ is the angle between $I\,d\boldsymbol{\ell}$ and $\hat{\mathbf{r}}$, which is 90° for each current element, so $\sin\theta = 1$. The magnetic field due to the entire current is found by integrating over all the current elements in the loop. Since R is the same for all elements, we obtain

$$B = \oint dB = \frac{\mu_0}{4\pi}\frac{I}{R^2}\oint d\ell$$

The integral of $d\ell$ around the complete loop gives the total length $2\pi R$, the circumference of the loop. The magnetic field due to the entire loop is thus

$$B = \frac{\mu_0}{4\pi}\frac{I\,2\pi R}{R^2} = \frac{\mu_0 I}{2R} \qquad \text{(at center of loop)} \qquad 25\text{-}5$$

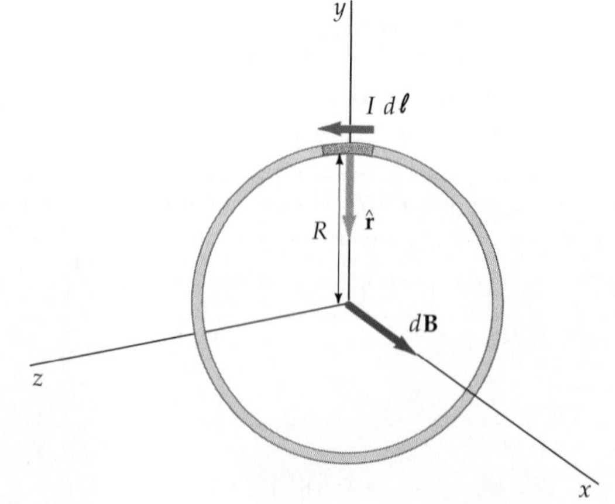

Figure 25-6 Current element for calculating the magnetic field at the center of a circular current loop. Each element produces a magnetic field that is directed along the axis of the loop.

Exercise

Find the current in a circular loop of radius 8 cm that will give a magnetic field of 2 G at the center of a loop. (Answer: 25.5 A)

Figure 25-7 shows the geometry for calculating the magnetic field at a point on the axis of a circular current loop a distance x from its center. We first consider the current element at the top of the loop. Here, as everywhere around the loop, $I\,d\boldsymbol{\ell}$ is tangent to the loop and perpendicular to the vector $\mathbf{r}$ from the current element to the field point P. The magnetic field $d\mathbf{B}$ due to this element is in the direction shown in the figure, perpendicular to $\mathbf{r}$ and also perpendicular to $I\,d\boldsymbol{\ell}$. The magnitude of $d\mathbf{B}$ is

$$|d\mathbf{B}| = \frac{\mu_0}{4\pi}\frac{I|d\boldsymbol{\ell} \times \hat{\mathbf{r}}|}{r^2} = \frac{\mu_0}{4\pi}\frac{I\,d\ell}{x^2 + R^2}$$

where we have used the facts that $r^2 = x^2 + R^2$ and that $d\boldsymbol{\ell}$ and $\hat{\mathbf{r}}$ are perpendicular, so $|d\boldsymbol{\ell} \times \hat{\mathbf{r}}| = d\ell$.

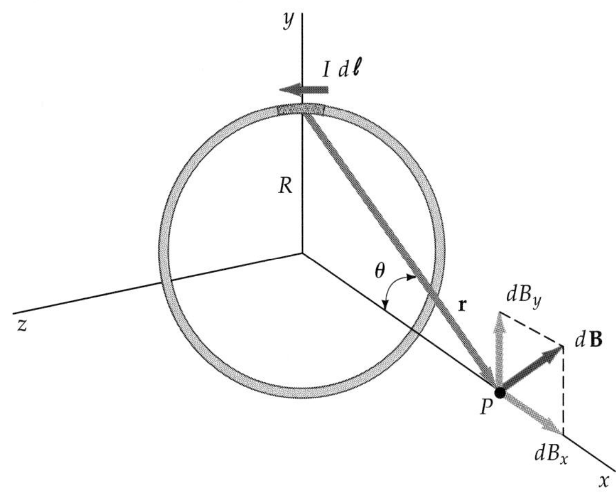

Figure 25-7 Geometry for calculating the magnetic field at a point on the axis of a circular current loop.

When we sum around all the current elements in the loop, the components of $d\mathbf{B}$ perpendicular to the axis of the loop, such as dB_y in the figure, sum to zero, leaving only the components dB_x that are parallel to the axis. We thus compute only the x component of the field. From the figure, we have

$$dB_x = dB \sin\theta = dB\left(\frac{R}{\sqrt{x^2 + R^2}}\right) = \frac{\mu_0}{4\pi}\frac{I\,d\ell}{x^2 + R^2}\frac{R}{\sqrt{x^2 + R^2}}$$

To find the field due to the entire loop of current, we integrate dB_x around the loop:

$$B_x = \oint dB_x = \oint \frac{\mu_0}{4\pi}\frac{IR}{(x^2 + R^2)^{3/2}}\,d\ell$$

Since neither x nor R varies as we sum over the elements in the loop, we can remove these quantities from the integral. Then,

$$B_x = \frac{\mu_0 IR}{4\pi(x^2 + R^2)^{3/2}}\oint d\ell$$

The integral of $d\ell$ around the loop gives $2\pi R$. Thus,

$$B_x = \frac{\mu_0}{4\pi}\frac{IR(2\pi R)}{(x^2 + R^2)^{3/2}} = \frac{\mu_0}{4\pi}\frac{2\pi R^2 I}{(x^2 + R^2)^{3/2}} \qquad 25\text{-}6$$

B on the axis of a current loop

Exercise

Show that Equation 25-6 reduces to Equation 25-5 at the center of the loop.

At great distances from the loop, x is much greater than R, so $(x^2 + R^2)^{3/2} \approx (x^2)^{3/2} = x^3$. Then,

$$B_x \longrightarrow \frac{\mu_0}{4\pi} \frac{2I\pi R^2}{x^3}$$

or

Magnetic-dipole field on the axis of the dipole

$$B_x = \frac{\mu_0}{4\pi} \frac{2m}{x^3} \tag{25-7}$$

where $m = I\pi R^2$ is the magnitude of the magnetic moment of the loop. Note the similarity of this expression to Equation 18-10 for the electric field on the axis of an electric dipole of moment p:

$$E_x \longrightarrow \frac{1}{4\pi\epsilon_0} \frac{2p}{x^3}$$

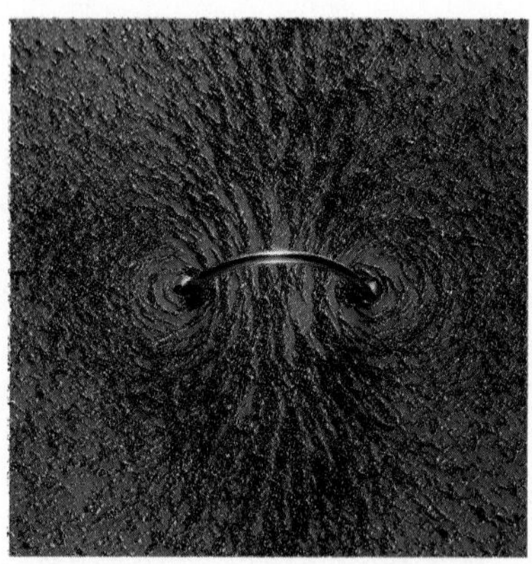

Figure 25-8 The magnetic-field lines of a circular current loop indicated by iron filings.

Although it has not been demonstrated, our result that a current loop produces a magnetic dipole field far away from it holds in general for any point whether it is on or off of the axis of the loop. Thus, a current loop behaves as a magnetic dipole both in that it experiences a torque $\mathbf{m} \times \mathbf{B}$ when placed in an external magnetic field (as was shown in Chapter 24) and in that it produces a magnetic dipole field at a great distance from it. Figure 25-8 shows the magnetic-field lines for a current loop.

Example 25-3

A circular loop of radius 5.0 cm has 12 turns and lies in the xy plane. It carries a current of 4 A in the direction such that the magnetic moment of the loop is along the x axis. Find the magnetic field on the x axis at (*a*) $x = 15$ cm and (*b*) $x = 3$ m.

(*a*) Equation 25-6 gives the magnetic field due to the current in a single turn. The magnetic field due to a loop with N turns is N times that due to a single turn. The magnetic field at $x = 15$ cm is thus

$$\begin{aligned} B_x &= \frac{\mu_0}{4\pi} \frac{2\pi R^2 NI}{(x^2 + R^2)^{3/2}} \\ &= (10^{-7}\ \text{T}\cdot\text{m/A}) \frac{2\pi(0.05\ \text{m})^2(12)(4\ \text{A})}{[(0.15\ \text{m})^2 + (0.05\ \text{m})^2]^{3/2}} \\ &= 1.91 \times 10^{-5}\ \text{T} \end{aligned}$$

Note that the magnetic field is considerably larger than that due to the single moving charge calculated in Example 25-1.

(*b*) Since 3 m is much greater than the radius 0.05 m, we can use Equation 25-7 for the magnetic field far from the loop. The magnitude of the magnetic dipole moment of a loop with N turns is

$$\begin{aligned} m = NIA &= (12)(4\ \text{A})\pi(0.05\ \text{m})^2 \\ &= 0.377\ \text{A}\cdot\text{m}^2 \end{aligned}$$

The magnetic field at $x = 3$ m is then

$$B_x = \frac{\mu_0}{4\pi} \frac{2m}{x^3} = (10^{-7}\ \text{T}\cdot\text{m/A}) \frac{2(0.377\ \text{A}\cdot\text{m}^2)}{(3\ \text{m})^3} = 2.79 \times 10^{-9}\ \text{T}$$

B Due to a Current in a Solenoid

We now use Equation 25-6 to calculate the magnetic field of a **solenoid,** which is a wire tightly wound into a helix of closely spaced turns, as illustrated in Figure 25-9. A solenoid is used to produce a strong, uniform magnetic field in the region surrounded by its loops. Its role in magnetism is analogous to that of the parallel-plate capacitor in electrostatics, in that the capacitor produces a strong, uniform electric field between its plates. The magnetic field of a solenoid is essentially that of a set of N identical current loops placed side by side. Figure 25-10 shows the magnetic-field lines for two such loops. In the region between the loops near their axis, the fields of the individual loops are in the same direction and their magnitudes add, whereas in the region between the loops but at distances from the axis that are large compared with the radius of the loops, the fields tend to cancel. Figure 25-11 shows the magnetic-field lines for a long, tightly wound solenoid. Inside the solenoid, the field lines are approximately parallel to the axis and are closely and uniformly spaced, indicating a strong, uniform magnetic field. Outside the solenoid, the lines are much less dense. They diverge from one end and converge at the other end. Comparing this figure with Figure 24-8, we see that the field lines of a solenoid, both inside and outside, are the same as those of a bar magnet of the same shape as the solenoid.

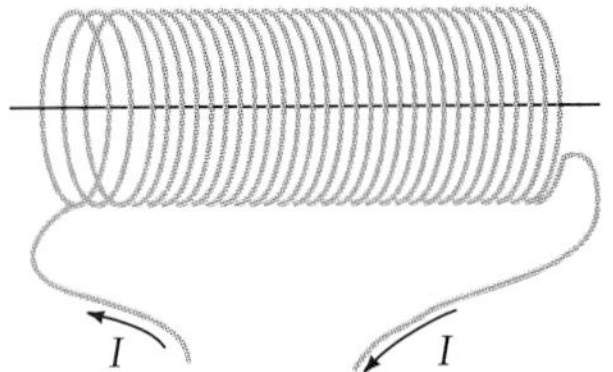

Figure 25-9 A tightly wound solenoid can be considered to be a set of circular current loops placed side by side that carry the same current. It produces a uniform magnetic field inside it.

We will calculate the magnetic field only at a point on the axis of the solenoid and between its ends. We consider a solenoid of length L consisting of N turns of wire carrying a current I. We choose the axis of the solenoid to

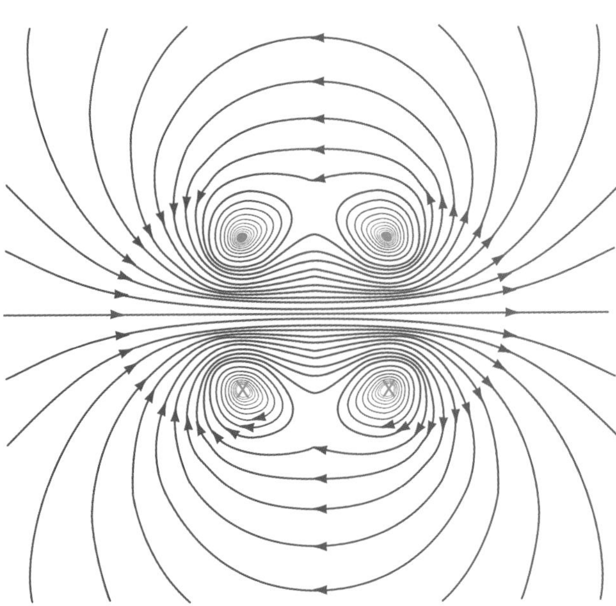

Figure 25-10 Magnetic-field lines due to two loops carrying the same current in the same sense. The points where the loops intersect the plane of the page are marked by a blue x where the current enters and a blue · where the current emerges. In the region between the loops, the magnetic fields of the individual loops add so the resultant field is strong, whereas in the regions away from the loops, the fields subtract so the resultant field is weak.

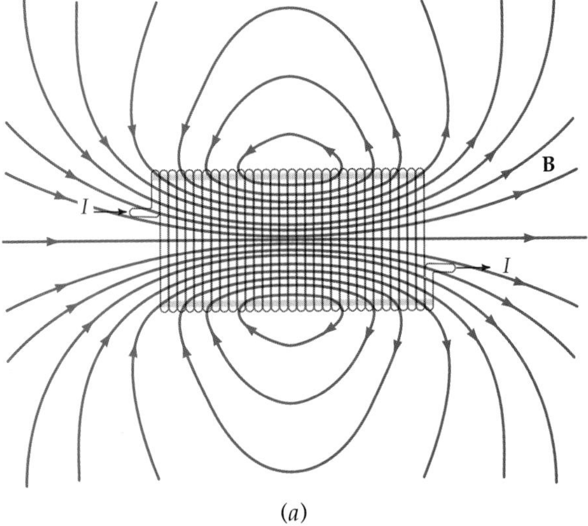

(*a*)

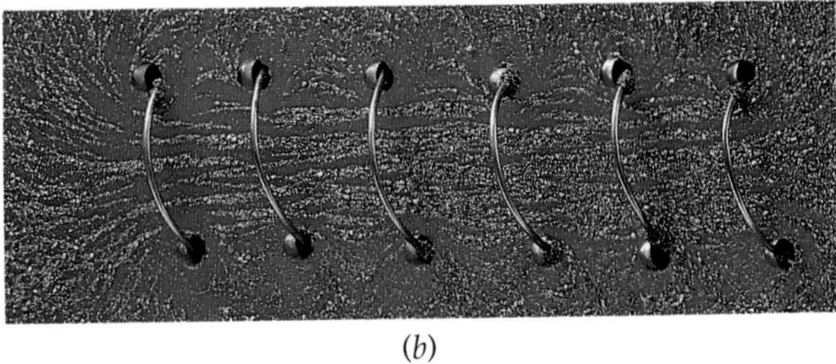

(*b*)

Figure 25-11 (*a*) Magnetic-field lines of a solenoid. The lines are identical to those of a bar magnet of the same shape (Figure 24-8). (*b*) Magnetic-field lines of a solenoid shown by iron filings.

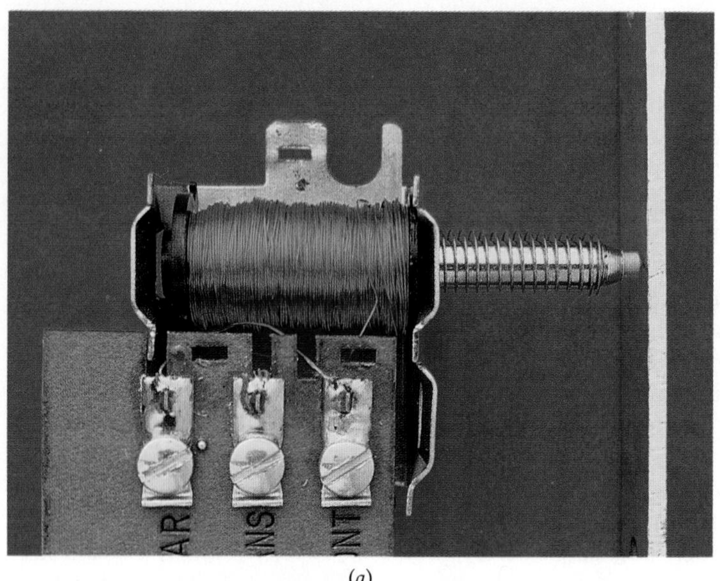

(a)

(b)

(*a*) A cross section of a doorbell. When the outer solenoid is energized, its magnetic field causes the inner plunger to strike the bell. (*b*) A cross section of a speaker. The speaker cone is attached to a coil, which is in the magnetic field of a permanent magnet. An alternating current in the coil, such as that from the output of a radio, causes the cone to vibrate resulting in the emission of sound waves.

be along the x axis, with the left end at $x = -a$ and the right end at $x = +b$ as shown in Figure 25-12. We will calculate the magnetic field at the origin. The figure shows an element of the solenoid of length dx at a distance x from the origin. If $n = N/L$ is the number of turns per unit length, there are $n\ dx$ turns of wire in this element, with each turn carrying a current I. The element is thus equivalent to a single loop carrying a current $di = nI\ dx$. The magnetic field at a point x on the axis due to a loop at the origin carrying a current $nI\ dx$ is given by Equation 25-6 with I replaced by $nI\ dx$:

$$dB_x = \frac{\mu_0}{4\pi}\frac{2\pi nIR^2\ dx}{(x^2+R^2)^{3/2}}$$

This expression also gives the magnetic field at the origin due to a current loop at x. We find the magnetic field due to the entire solenoid by integrating this expression from $x = -a$ to $x = b$:

$$B_x = \frac{\mu_0}{4\pi}\,2\pi nIR^2\int_{-a}^{b}\frac{dx}{(x^2+R^2)^{3/2}} \tag{25-8}$$

The integral in Equation 25-8 can be found in standard tables of integrals. Its value is

$$\int_{-a}^{b}\frac{dx}{(x^2+R^2)^{3/2}} = \left.\frac{x}{R^2\sqrt{x^2+R^2}}\right|_{-a}^{b} = \frac{b}{R^2\sqrt{b^2+R^2}} + \frac{a}{R^2\sqrt{a^2+R^2}}$$

Substituting this into Equation 25-8, we obtain

$$B = \frac{1}{2}\mu_0 nI\left(\frac{b}{\sqrt{b^2+R^2}} + \frac{a}{\sqrt{a^2+R^2}}\right) \tag{25-9}$$

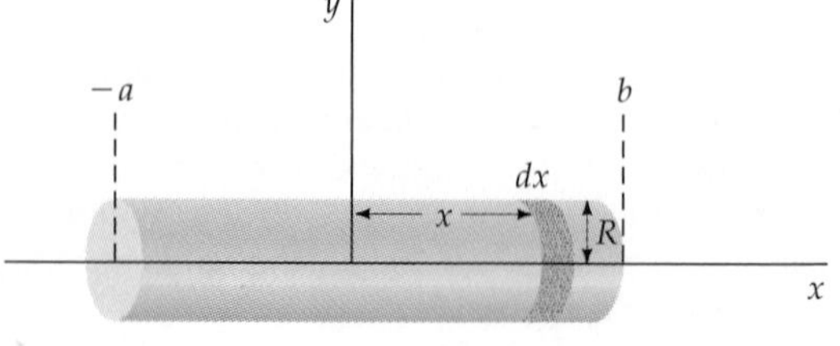

Figure 25-12 Geometry for calculating the magnetic field inside a solenoid on the axis. The number of turns in the element dx is $n\ dx$, where $n = N/L$ is the number of turns per unit length. The element dx is treated as a current loop carrying a current $di = nI\ dx$.

For a long solenoid for which a and b are much larger than R, the two terms in the parentheses each tend toward 1. For this approximation, the magnetic field is

$$B = \mu_0 nI \qquad 25\text{-}10$$

B inside a long solenoid

If the origin is at one end of the solenoid, either a or b is zero. Then, if the other end is far away compared with the radius, one of the terms in the parentheses of Equation 25-9 is zero and the other is 1, so $B \approx \frac{1}{2}\mu_0 nI$. Thus, the magnitude of **B** at a point near either end of a long solenoid is about half that at points within the solenoid away from the ends. Figure 25-13 gives a plot of the magnetic field on the axis of a solenoid versus position (with the origin at the center of the solenoid). The approximation that the field is constant independent of the position along the axis is quite good except for very near the ends.

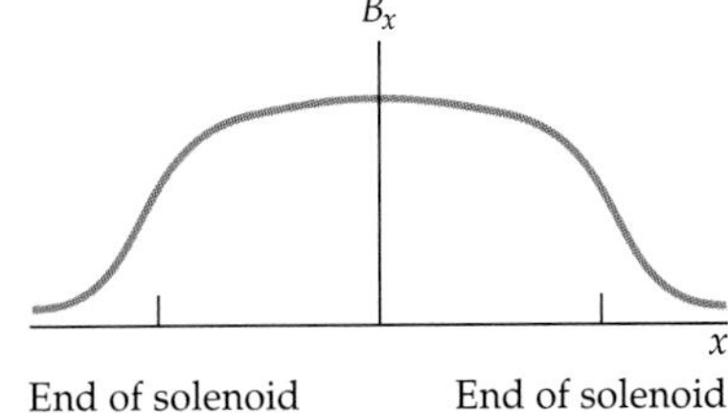

Figure 25-13 Graph of the magnetic field on the axis inside a solenoid versus the position x on the axis. The field inside the solenoid is nearly constant except near the ends.

Example 25-4

Find the magnetic field at the center of a solenoid of length 20 cm, radius 1.4 cm, and 600 turns that carries a current of 4 A.

We will calculate the field exactly using Equation 25-9. For a point at the center of the solenoid, $a = b = 10$ cm. Thus, each term in the parentheses in Equation 25-9 has the value

$$\frac{a}{\sqrt{a^2 + R^2}} = \frac{b}{\sqrt{b^2 + R^2}} = \frac{10\text{ cm}}{\sqrt{(10\text{ cm})^2 + (1.4\text{ cm})^2}} = 0.990$$

Then, from Equation 25-9, the magnetic field at the center of the solenoid is

$$B = \tfrac{1}{2}\mu_0 nI\left(\frac{b}{\sqrt{b^2 + R^2}} + \frac{a}{\sqrt{a^2 + R^2}}\right)$$

$$= (0.5)(4\pi \times 10^{-7}\text{ T·m/A})(600\text{ turns}/0.2\text{ m})(4\text{ A})(0.990 + 0.990)$$

$$= 1.50 \times 10^{-2}\text{ T}$$

Note that the approximation obtained using Equation 25-10 amounts to replacing 0.99 by 1.00, which differs by only 1 percent. Note also that the magnitude of the magnetic field inside this solenoid is fairly large—about 100 times the magnetic field of the earth.

(*a*)

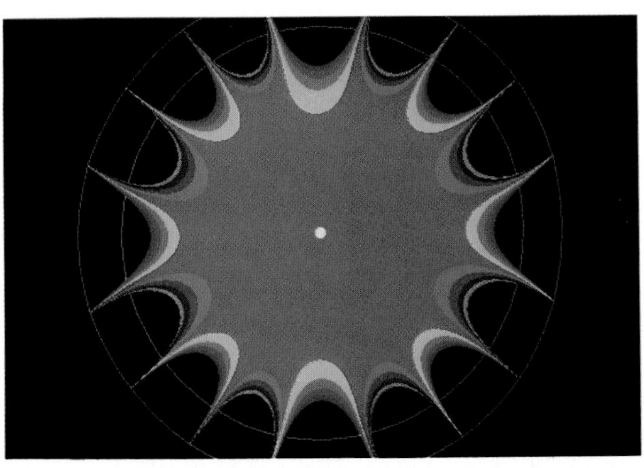

(*b*)

(*a*) A sextupole magnet used for focusing beams of electrons and positrons in the LEP collider at CERN. (*b*) Computer graphic showing the magnetic field distribution in a superconducting magnet used at Brookhaven National Laboratory. The contours show equal deviations of the field from that at the center of the magnet.

B Due to a Current in a Straight Wire

Figure 25-14 shows the geometry for calculating the magnetic field **B** at a point P due to the current in the straight wire segment shown. We choose the wire to be the x axis and point P to be on the y axis. Note that because of the symmetry in this problem, any direction perpendicular to the wire could be chosen for the y axis. A typical current element $I\,d\boldsymbol{\ell}$ at a distance x from the origin is shown. The vector **r** points from the element to the field point P.

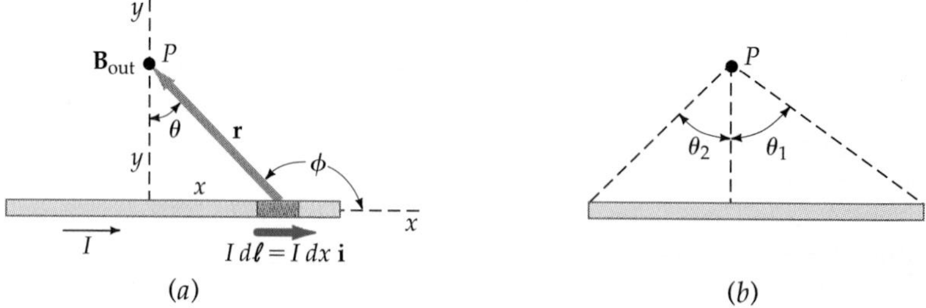

Figure 25-14 (*a*) Geometry for calculating the magnetic field at point P due to a straight current segment. Each element of the segment contributes to the total magnetic field at point P, which is directed out of the paper. (*b*) The result is expressed in terms of the angles θ_1 and θ_2.

The direction of the magnetic field at P due to this element is the direction of $I\,d\boldsymbol{\ell} \times \mathbf{r}$, which is out of the paper. Note that the magnetic fields due to all the current elements of the wire are in this same direction. Thus, we only need to compute the magnitude of the field. The field due to the current element shown has the magnitude

$$dB = \frac{\mu_0}{4\pi}\frac{I\,dx}{r^2}\sin\phi$$

It is more convenient to write this in terms of θ rather than ϕ:

$$dB = \frac{\mu_0}{4\pi}\frac{I\,dx}{r^2}\cos\theta \qquad 25\text{-}11$$

To sum over all the current elements, we need to relate the variables θ, r, and x. It turns out to be easiest to express x and r in terms of θ. We have

$$x = y\tan\theta$$

Then,

$$dx = y\sec^2\theta\,d\theta = y\frac{r^2}{y^2}\,d\theta = \frac{r^2}{y}\,d\theta$$

where we have used $\sec\theta = r/y$. Substituting this expression for dx into Equation 25-11, we obtain

$$dB = \frac{\mu_0}{4\pi}\frac{I}{r^2}\frac{r^2\,d\theta}{y}\cos\theta = \frac{\mu_0}{4\pi}\frac{I}{y}\cos\theta\,d\theta$$

Let us first calculate the contribution from the current elements to the right of the point $x = 0$. We sum over these elements by integrating from $\theta = 0$ to $\theta = \theta_1$, where θ_1 is the angle between the line perpendicular to the wire and the line from P to the right end of the wire, as shown in Figure 25-14*b*. For this contribution, we have

$$B_1 = \int_0^{\theta_1}\frac{\mu_0}{4\pi}\frac{I}{y}\cos\theta\,d\theta$$

$$= \frac{\mu_0}{4\pi}\frac{I}{y}\int_0^{\theta_1}\cos\theta\,d\theta = \frac{\mu_0}{4\pi}\frac{I}{y}\sin\theta_1$$

Similarly, the contribution from elements to the left of $x = 0$ is

$$B_2 = \frac{\mu_0}{4\pi}\frac{I}{y}\sin\theta_2$$

The total magnetic field due to the wire segment is the sum of B_1 and B_2. Writing R instead of y for the perpendicular distance from the wire segment to the field point, we obtain

$$B = \frac{\mu_0}{4\pi}\frac{I}{R}(\sin\theta_1 + \sin\theta_2) \qquad 25\text{-}12$$

B due to a straight wire segment

This result gives the magnetic field due to any wire segment in terms of the perpendicular distance R and the angles subtended at the field point by the ends of the wire. If the wire is very long, these angles are nearly 90°. The result for a very long wire is obtained from Equation 25-12 by setting $\theta_1 = \theta_2 = 90°$:

$$B = \frac{\mu_0 I}{2\pi R} = \frac{\mu_0}{4\pi}\frac{2I}{R} \qquad 25\text{-}13$$

B due to a long, straight wire

At any point in space, the magnetic field lines of a long, straight, current-carrying wire is tangent to a circle of radius R about the wire, where R is the perpendicular distance from the wire to the field point. The direction of **B** can be determined by applying the right-hand rule as shown in Figure 25-15*a*. The magnetic-field lines thus encircle the wire as shown in Figure 25-15*b*.

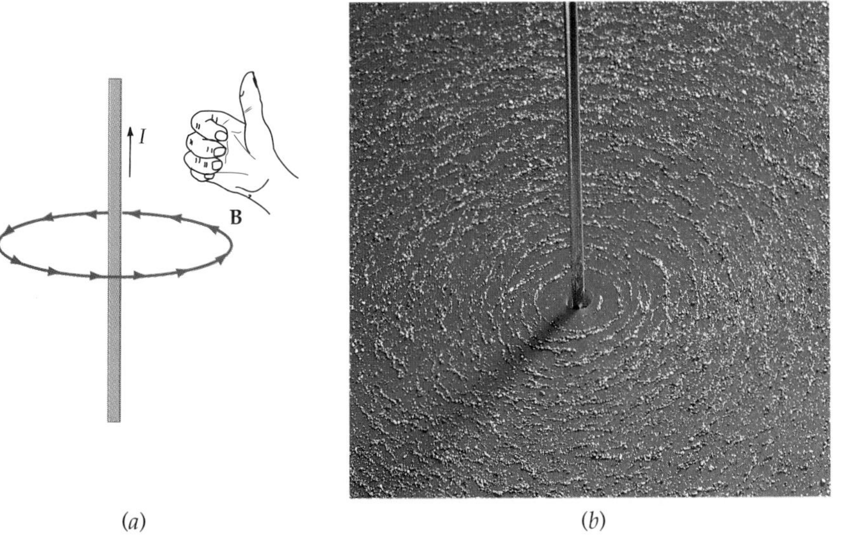

Figure 25-15 (*a*) Right-hand rule for determining the direction of the magnetic field due to a long, straight, current-carrying wire. The magnetic-field lines encircle the wire in the direction of the fingers of the right hand when the thumb points in the direction of the current. (*b*) Magnetic-field lines due to a long wire indicated by iron filings.

The result expressed by Equation 25-13 was found experimentally by Biot and Savart in 1820. From an analysis of it, they were able to discover the expression for the magnetic field due to a current element given in Equation 25-4.

Example 25-5

Find the magnetic field at the center of a square current loop of side $\ell = 50$ cm carrying current 1.5 A (Figure 25-16).

From Figure 25-16, we see that each side of the loop contributes a field in the direction out of the paper. Because of the symmetry of this figure, we need only calculate the field due to one side of the loop and multiply by 4. The distance between one side and the field point is $R = \frac{1}{2}\ell = 0.25$ m. Thus, from Equation 25-12, the field is

$$B = 4\left(\frac{\mu_0}{4\pi}\right)\frac{I}{\frac{1}{2}\ell}(\sin 45^\circ + \sin 45^\circ) = (4 \times 10^{-7}\ \text{T·m/A})\frac{1.5\ \text{A}}{0.25\ \text{m}}(2 \sin 45^\circ)$$

$$= 3.39 \times 10^{-6}\ \text{T}$$

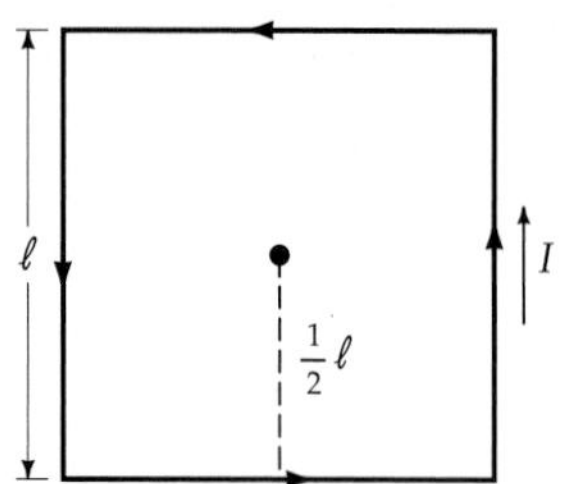

Figure 25-16 Square current loop for Example 25-5.

Exercise

Compare the magnetic field at the center of a circular current loop of radius R with that at the center of a square current loop of side $\ell = 2R$. Which is larger? (Answer: B is larger for the circle by about 10 percent)

Example 25-6

Find the magnetic field at a distance of 20 cm from a long, straight wire carrying a current of 5 A.

From Equation 25-13, we have

$$B = \frac{\mu_0}{4\pi}\frac{2I}{y} = (10^{-7}\ \text{T·m/A})\frac{2(5\ \text{A})}{0.2\ \text{m}} = 5.00 \times 10^{-6}\ \text{T}$$

We note from this example that the magnetic field near a wire carrying a current of ordinary size is small. In this case, it is only about 1 percent of the magnetic field due to the earth.

Example 25-7

A long, straight wire carrying a current of 1.7 A in the positive z direction lies along the line $x = -3$ cm. A similar wire carrying a current of 1.7 A in the positive z direction lies along the line $x = +3$ cm as shown in Figure 25-17*a*. Find the magnetic field at a point on the y axis at $y = 6$ cm.

In Figure 25-17*b*, we have drawn the xy plane in the plane of the page and indicated the directions of the magnetic field $\mathbf{B}_L$ due to the current

Figure 25-17 (*a*) Two parallel wires carrying currents in the same sense for Example 25-7. (*b*) Geometry for calculating the resultant magnetic field due to the two currents.

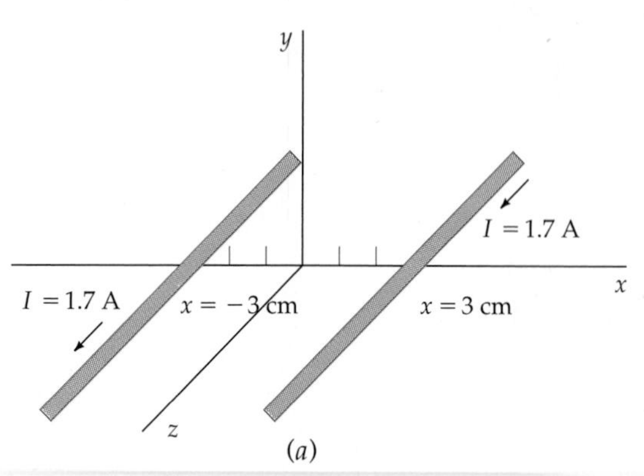

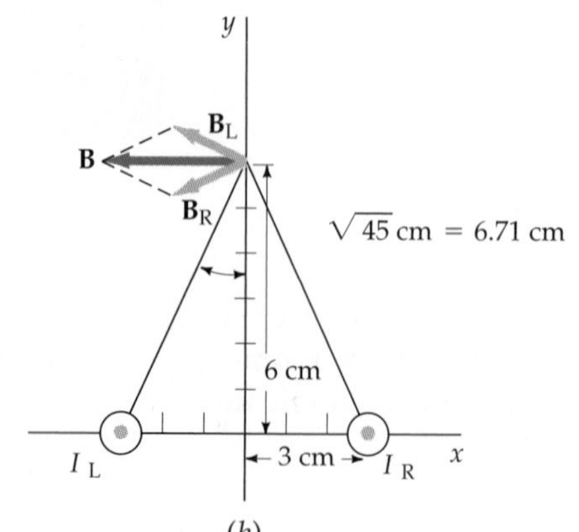

A current gun for measuring electric current. The jaws of the current gun clamp around a current-carrying wire without touching the wire. The magnetic field produced by the wire is measured with a Hall-effect device mounted in the current gun. The Hall-effect device puts out a voltage proportional to the magnetic field, which in turn is proportional to the current in the wire.

on the left and $\mathbf{B}_R$ due to the current on the right. Since the currents have equal magnitudes and are each at a distance $R = \sqrt{(3\text{ cm})^2 + (6\text{ cm})^2} = 6.71$ cm from the field point, these fields have equal magnitudes given by

$$B_R = B_L = \frac{\mu_0}{4\pi}\frac{2I}{R} = (10^{-7}\text{ T·m/A})\frac{2(1.7\text{ A})}{0.0671\text{ m}} = 5.07 \times 10^{-6}\text{ T}$$

From Figure 25-17*b*, we see that the resultant magnetic field is in the negative x direction and has the magnitude $2B_L \cos\theta$, where $\cos\theta = (6\text{ cm})/(6.71\text{ cm}) = 0.894$. The resultant magnetic field is thus

$$\mathbf{B} = -2B_L \cos\theta\, \mathbf{i} = -2(5.07 \times 10^{-6}\text{ T})(0.894)\mathbf{i} = 9.07 \times 10^{-6}\text{ T}\, \mathbf{i}$$

Example 25-8

An infinitely long wire carrying a current of 4.5 A is bent as shown in Figure 25-18. Find the magnetic field at the point $x = 3$ cm, $y = 2$ cm.

We first note that the magnetic field due to each part of the wire is in the positive z direction. The magnitude of the field due to the wire segment along the y axis B_1 is given by Equation 25-12, with $R = 3$ cm, $\theta_1 = 90°$, and $\theta_2 = \alpha = \tan^{-1}(\frac{2}{3}) = 33.7°$:

$$B_1 = \frac{\mu_0}{4\pi}\frac{I}{R}(\sin\theta_1 + \sin\theta_2)$$

$$= (10^{-7}\text{ T·m/A})\frac{4.5\text{ A}}{0.03\text{ m}}(\sin 90° + \sin 33.7°) = 2.33 \times 10^{-5}\text{ T}$$

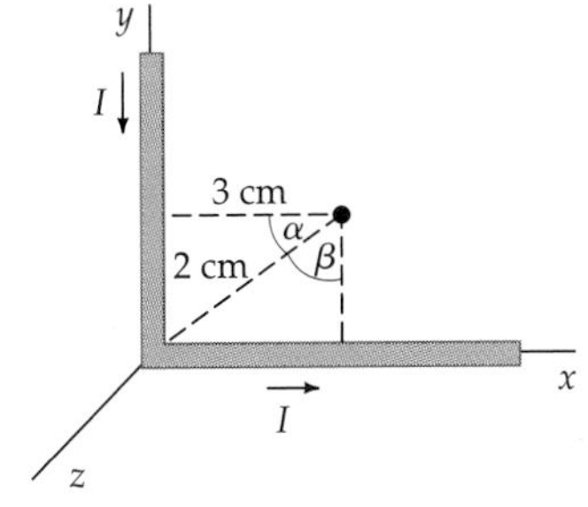

Figure 25-18 Current-carrying wire for Example 25-8.

Similarly, the magnitude of the field B_2 due to the wire segment along the x axis is given by Equation 25-12 with $R = 2$ cm, $\theta_1 = \beta = 90° - 33.7° = 56.3°$, and $\theta_2 = 90°$:

$$B_2 = \frac{\mu_0}{4\pi}\frac{I}{R}(\sin\theta_1 + \sin\theta_2)$$

$$= (10^{-7}\text{ T·m/A})\frac{4.5\text{ A}}{0.02\text{ m}}(\sin 56.3° + \sin 90°) = 4.12 \times 10^{-5}\text{ T}$$

The resultant magnetic field is thus

$$\mathbf{B} = (B_1 + B_2)\mathbf{k} = (2.33 \times 10^{-5}\text{ T} + 4.12 \times 10^{-5}\text{ T})\mathbf{k} = 6.45 \times 10^{-5}\text{ T}\,\mathbf{k}$$

25-3 Definition of the Ampere

We can use Equation 25-13 for the magnetic field due to a long, straight, current-carrying wire and Equation 24-4 for the force exerted by a magnetic field on a segment of a current-carrying wire to find the force exerted by one long, straight current on another. Figure 25-19 shows two long, parallel wires carrying currents in the same direction. We consider the force on a segment $\Delta\ell_2$ carrying current I_2 as shown. The magnetic field $\mathbf{B}_1$ at this segment due to current I_1 is perpendicular to the segment $I_2\,\Delta\ell_2$ as shown. This is true for all current elements along the wire. The magnetic force on current segment $I_2\,\Delta\ell_2$ is directed toward current I_1. Similarly, a current segment $I_1\,\Delta\ell_1$ will experience a magnetic force directed toward current I_2 due to a magnetic field arising from current I_2. Thus, two parallel currents attract each other. If one of the currents is reversed, the force will be reversed. Thus, two antiparallel currents will repel each other. The attraction or repulsion of parallel or antiparallel currents was discovered experimentally by Ampère one week after he heard of Oersted's discovery of the effect of a current on a compass needle.

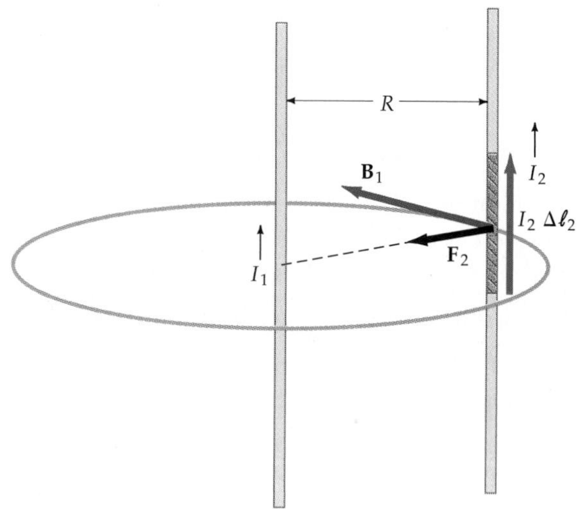

Figure 25-19 Two long, straight wires carrying parallel currents. The magnetic field $\mathbf{B}_1$ due to current I_1 is perpendicular to current I_2. The force on current I_2 is toward current I_1. There is an equal and opposite force exerted by current I_2 on I_1. The currents thus attract each other.

The magnitude of the magnetic force on the segment $I_2\,\Delta\ell_2$ is

$$F_2 = |I_2\,\Delta\boldsymbol{\ell}_2 \times \mathbf{B}_1|$$

Since the magnetic field at segment $I_2\,\Delta\ell_2$ is perpendicular to the current segment, we have

$$F_2 = I_2\,\Delta\ell_2\,B_1$$

If the distance R between the wires is much less than their length, the field at $I_2\,\Delta\ell_2$ due to current I_1 will approximate the field due to an infinitely long, current-carrying wire given by Equation 25-13. The magnitude of the force on the segment $I_2\,\Delta\ell_2$ is therefore

$$F_2 = I_2\,\Delta\ell_2\,\frac{\mu_0 I_1}{2\pi R}$$

The force per unit length is

$$\frac{F_2}{\Delta\ell_2} = \frac{\mu_0}{2\pi}\frac{I_1 I_2}{R} = 2\frac{\mu_0}{4\pi}\frac{I_1 I_2}{R} \qquad 25\text{-}14$$

In Chapter 18, the coulomb was defined in terms of the ampere, but the definition of the ampere was deferred. The ampere is defined as follows:

> If two very long parallel wires one meter apart carry equal currents, the current in each is defined to be one ampere when the force per unit length on each wire is 2×10^{-7} N/m.

Ampere defined

This definition of the ampere makes the permeability of free space μ_0 equal to exactly $4\pi \times 10^{-7}$ N/A^2. It also allows the unit of current (and therefore the unit of electric charge) to be determined by a mechanical measurement. In practice, currents much closer together than 1 m are used so that the wires need not be so long and the force is large enough to measure accurately. Figure 25-20 shows a **current balance,** which is a device that can be used to calibrate an ammeter from the definition of the ampere. The upper conductor is free to rotate about the knife edges and is balanced so that the wires (or conducting rods) are a small distance apart. The conductors are connected in series to carry the same current but in opposite directions so that they will repel each other. Weights are placed on the upper conductor until it balances again at the original separation. The force of repulsion is thus determined by measuring the total weight needed to balance the upper conductor.

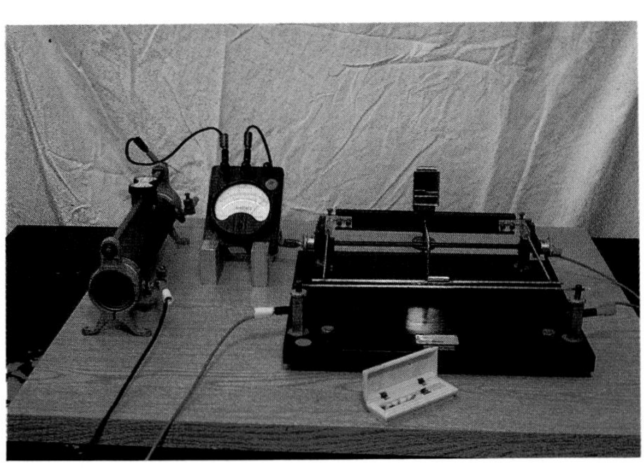

Figure 25-20 Current balance used in an elementary physics laboratory to calibrate an ammeter. The two parallel rods in front carry equal but oppositely directed currents and therefore repel each other. The force of repulsion is balanced by weights placed on the upper rod, which is part of a rectangle that is balanced on knife edges at the back. The mirror on top is used to reflect a beam of laser light for accurately determining the position of the upper rod.

Example 25-9

Two straight rods 50 cm long and 1.5 mm apart in a current balance carry currents of 15 A each in opposite directions. What mass must be placed on the upper rod to balance the magnetic force of repulsion?

The force exerted by the lower rod on the upper rod of length ℓ has the magnitude

$$F = \frac{\mu_0}{2\pi}\frac{I_1 I_2}{R}\ell$$

$$= (2 \times 10^{-7}\ \text{N/A}^2)\,\frac{(15\ \text{A})(15\ \text{A})}{0.0015\ \text{m}}\,(0.5\ \text{m}) = 1.5 \times 10^{-2}\ \text{N}$$

This force can be balanced by a weight mg:

$$mg = 1.5 \times 10^{-2}\ \text{N}$$

$$m = \frac{1.5 \times 10^{-2}\ \text{N}}{9.81\ \text{N/kg}} = 1.53 \times 10^{-3}\ \text{kg} = 1.53\ \text{g}$$

We see from this example that the magnetic force between two current-carrying wires is small even for currents as large as 15 A.

25-4 Ampère's Law

In Chapter 24, we noted that no fundamental magnetic charges or "poles" playing a role similar to that of electric charges have ever been observed. Instead, the fundamental source of magnetic fields is electric current. The magnetic fields arising from currents do not originate or end at points in space, but instead form closed loops encircling the current. For instance, we saw in Section 25-2 that the magnetic-field lines due to a long, straight, current-carrying wire encircle the wire. There is thus a definite difference between the characteristic shapes of electric-field lines and magnetic-field lines since electric-field lines begin and end on electric charges.

In Chapter 19, we studied Gauss's law, which relates the normal component of the electric field summed over a closed surface to the net charge inside the surface. There is an analogous equation for the magnetic field called Ampère's law that relates the tangential component of **B** summed around a closed curve C to the current I_C that passes through the curve. In mathematical form, **Ampère's law** is

Ampère's law

$$\oint_C \mathbf{B}\cdot d\boldsymbol{\ell} = \mu_0 I_C \qquad C, \text{ any closed curve} \tag{25-15}$$

where I_C is the net current that penetrates the area bounded by the curve C. Ampère's law holds for any curve C as long as the currents are continuous, that is, they do not begin or end at any finite point. Like Gauss's law, Ampère's law can be used to obtain an expression for the magnetic field in situations that have a high degree of symmetry. If the symmetry is great enough, the line integral $\oint \mathbf{B}\cdot d\boldsymbol{\ell}$ can be written as the product of B and some distance. Then if I_C is known, B can be determined. Also, like Gauss's law, Ampère's law is of no use in finding an expression for the magnetic field if there is no symmetry. It is, however, of considerable theoretical importance.

The simplest application of Ampère's law is to find the magnetic field of an infinitely long, straight, current-carrying wire. Figure 25-21 shows a circular curve around a point on a long wire with its center at the wire. If we assume that we are far from the ends of the wire, we can use symmetry to rule out the possibility of any component of **B** parallel to the wire. We may then assume that the magnetic field is tangent to this circle and has the same magnitude B at any point on the circle. Ampère's law then gives

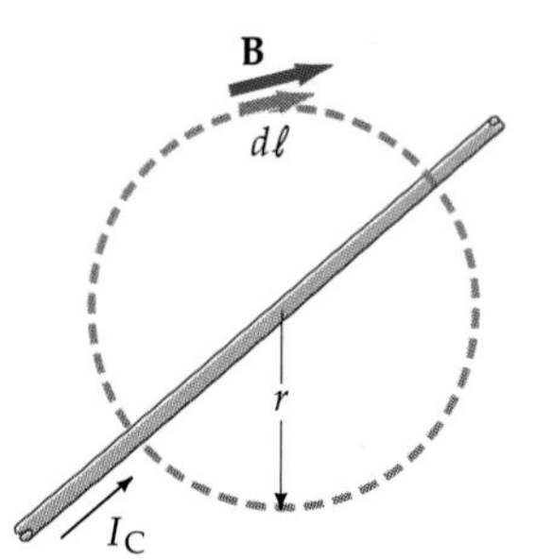

Figure 25-21 Geometry for calculating the magnetic field of a long, straight, current-carrying wire using Ampère's law. On a circle around the wire, the magnetic field is constant and tangent to the circle.

$$\oint_C \mathbf{B}\cdot d\boldsymbol{\ell} = B \oint_C d\ell = \mu_0 I_C$$

where we have taken B out of the integral because it has the same value everywhere on the circle. The integral of $d\ell$ around the circle equals $2\pi r$, the circumference of the circle. The current I_C is the current I in the wire. We thus obtain

$$B(2\pi r) = \mu_0 I$$

$$B = \frac{\mu_0}{2\pi}\frac{I}{r}$$

which is Equation 25-13.

Example 25-10

A long, straight wire of radius a carries a current I that is uniformly distributed over the cross-sectional area of the wire (Figure 25-22). Find the magnetic field both inside and outside the wire.

We can use Ampère's law to calculate **B** because of the high degree of symmetry. At a distance r, we know that **B** is tangent to the circle of radius r about the wire and constant in magnitude everywhere on the circle. Thus,

$$\oint_C \mathbf{B}\cdot d\boldsymbol{\ell} = B \oint_C d\ell = B2\pi r$$

The current through C depends on whether r is less than or greater than the radius of the wire a. For r greater than a, the total current I crosses the area bounded by C, and we obtain Equation 25-13 for the magnetic field due to a long, straight, current-carrying wire. Inside the wire, we consider a circle of radius $r < a$. The current passing through this circle is

$$I_C = \frac{\pi r^2}{\pi a^2} I$$

Ampère's law then gives

$$\oint_C \mathbf{B}\cdot d\boldsymbol{\ell} = B2\pi r = \mu_0 \frac{r^2}{a^2} I$$

$$B = \frac{\mu_0}{2\pi} \frac{I}{a^2} r \qquad r < a \qquad \text{25-16}$$

Figure 25-23 gives a graph of B versus r for this example.

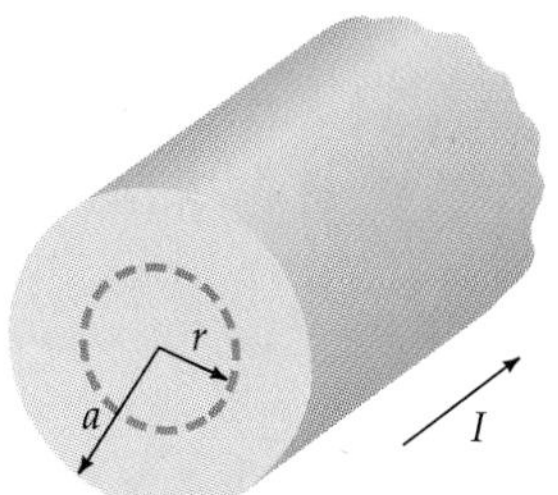

Figure 25-22 Long wire of radius a carrying a current I that is uniformly distributed over its cross-sectional area. Because of the symmetry of the wires, the magnetic field at any distance r can be calculated by applying Ampère's law to a circle of radius r.

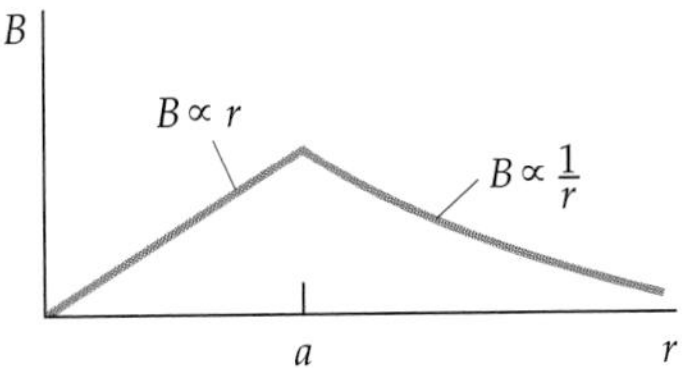

Figure 25-23 Graph of B versus r for a wire of radius a carrying a current uniformly distributed over its cross-sectional area. Inside the wire the magnetic field is proportional to the distance r from the axis of the wire.

For our next application of Ampère's law, we will calculate the magnetic field of a tightly wound **toroid,** which consists of loops of wire wound around a doughnut-shaped form as shown in Figure 25-24. There are N turns of wire, each carrying a current I. To calculate B, we evaluate the line integral $\oint \mathbf{B}\cdot d\boldsymbol{\ell}$ around a circle of radius r centered at the center of the toroid. By symmetry, **B** is tangent to this circle and constant in magnitude at every point on the circle. Then,

$$\oint \mathbf{B}\cdot d\boldsymbol{\ell} = B2\pi r = \mu_0 I_C$$

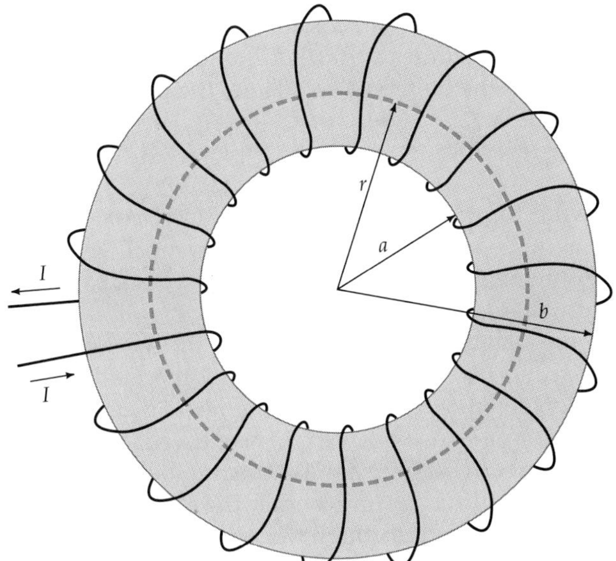

Figure 25-24 A toroid consists of loops of wire wound around a doughnut-shaped form. The magnetic field at any distance r can be found by applying Ampère's law to the circle of radius r.

(a)

(b)

(a) The Tokamak fusion test reactor is a large toroid that produces a magnetic field for confining charged particles. Coils containing over 10 km of water-cooled copper wire carry a pulsed current, which has a peak value of 73,000 A and produces a magnetic field of 5.2 T for about 3 s. (b) Inspection of the assembly of the Tokamak from inside the toroid.

Let a and b be the inner and outer radii of the toroid, respectively. The total current through the circle of radius r for $a < r < b$ is NI. Ampère's law then gives

$$\oint \mathbf{B} \cdot d\boldsymbol{\ell} = B2\pi r = \mu_0 I_C = \mu_0 NI$$

or

B inside a tightly wound toroid

$$B = \frac{\mu_0 NI}{2\pi r} \qquad a < r < b \tag{25-17}$$

If r is less than a, there is no current through the circle of radius r. If r is greater than b, the total current through r is zero because for each current I into the page in Figure 25-24, at the inner surface of the toroid, there is an equal current I out of the page at the outer surface. Thus, the magnetic field is zero for both $r < a$ and $r > b$:

$$B = 0 \qquad r < a \text{ or } r > b$$

The magnetic field inside the toroid is not uniform but decreases with r. However, if the diameter of the loops of the toroid, $b - a$, is much less than the radius of the doughnut, the variation in r from $r = a$ to $r = b$ is small, and B is approximately uniform as it is in a solenoid.

We can also use Ampère's law to find an expression for the magnetic field inside a long, tightly wound solenoid, assuming that the field is uniform inside the solenoid and zero outside. We chose the rectangle of sides a and b shown in Figure 25-25 for our closed curve C. The current that passes through this curve is the current I in each turn times the number of turns in the length a. If the solenoid has n turns per unit length, the number of turns in the length a will be na, and the current through the rectangular curve will be $I_C = naI$. The only contribution to the sum of $\oint \mathbf{B} \cdot d\boldsymbol{\ell}$ for this curve is along

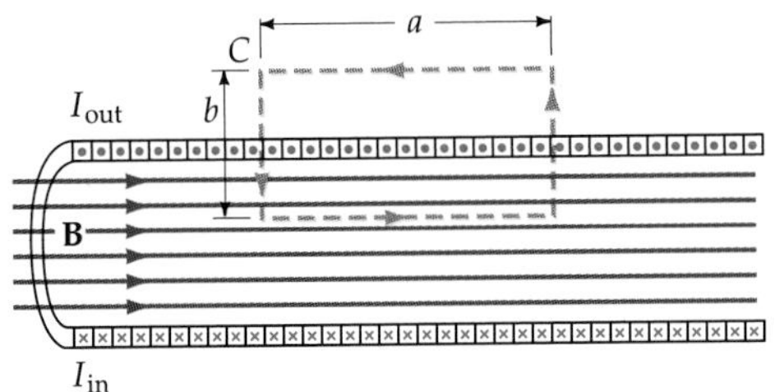

Figure 25-25 The magnetic field inside a solenoid can be calculated by applying Ampère's law to the rectangular curve C. If $\mathbf{B}$ is assumed to be uniform inside the solenoid and zero outside it, the line integral $\oint \mathbf{B}\cdot d\boldsymbol{\ell}$ around the curve C is just Ba.

the long side of the rectangle inside the solenoid, which gives Ba. Ampère's law thus gives

$$\oint \mathbf{B}\cdot d\boldsymbol{\ell} = Ba = \mu_0 I_C = \mu_0 naI$$

The magnetic field inside the solenoid is thus

$$B = \mu_0 nI$$

in agreement with Equation 25-10, which we obtained using the Biot–Savart law.

Limitations of Ampère's law

For an example of a case for which Ampère's law is not useful for calculating the magnetic field, consider the current loop shown in Figure 25-26. We have already calculated the field on the axis of such a loop using the Biot–Savart law. According to Ampère's law, the line integral $\oint \mathbf{B}\cdot d\boldsymbol{\ell}$ around a curve such as curve C in the figure equals μ_0 times the current I in the loop. Although Ampère's law is valid for this curve, the magnetic field $\mathbf{B}$ is not constant along any curve encircling the current, nor is it everywhere tangent to any such curve. Thus, there is not enough symmetry in this situation to allow us to calculate B using Ampère's law.

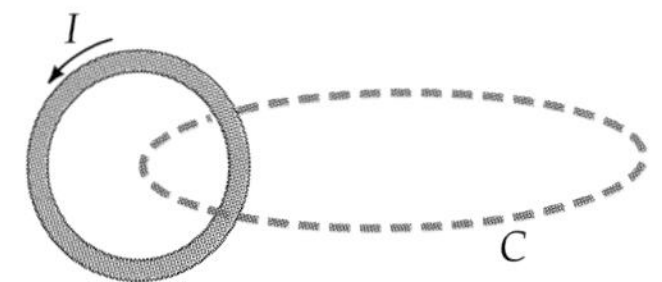

Figure 25-26 Ampère's law holds for the curve C encircling the current in the circular loop, but it is not useful in finding $\mathbf{B}$ because $\mathbf{B}$ is neither constant along the curve nor tangent to it.

Figure 25-27 shows a finite current segment of length ℓ. We wish to find the magnetic field at point P, which is equidistant from the ends of the segment and at a distance r from the center of the segment. A direct application of Ampère's law gives

$$B = \frac{\mu_0}{2\pi}\frac{I}{r}$$

This result is the same as for an infinitely long wire, since we have the same symmetry arguments. It does not agree with the result obtained from the Biot–Savart law, which depends on the length of the current segment and which agrees with experiment. If the current segment is just one part of a continuous circuit carrying a current, as shown in Figure 25-28, Ampère's law for curve C is valid, but it cannot be used to find the magnetic field at point P because there is no symmetry.

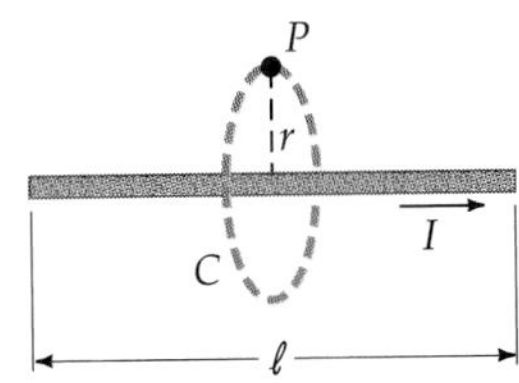

Figure 25-27 The application of Ampère's law to find the magnetic field on the bisector of a finite current segment gives an incorrect result.

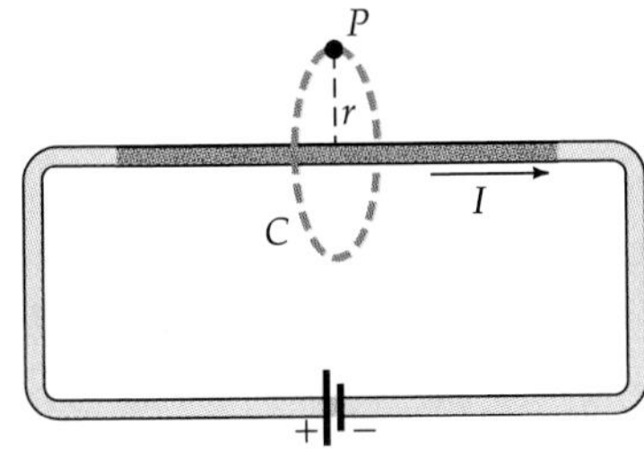

Figure 25-28 If the current segment in Figure 25-27 is part of a complete circuit, Ampère's law for the curve C is valid, but there is not enough symmetry to use it to find the magnetic field at point P.

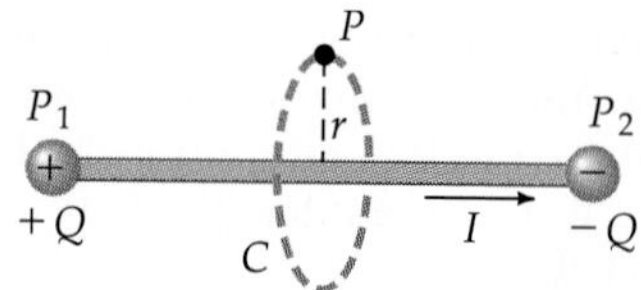

Figure 25-29 If the current segment in Figure 25-27 is due to a momentary flow of charge from a small conductor on the left to the one at the right, there is enough symmetry to use Ampère's law to compute the magnetic field at P, but Ampère's law is not valid because the current is not continuous in space.

In Figure 25-29, the current in the segment arises from a small spherical conductor with initial charge $+Q$ at the left of the segment and another one at the right with charge $-Q$. When they are connected, a current $I = -dQ/dt$ exists in the segment for a short time, until the spheres are uncharged. For this case, we *do* have the symmetry needed to assume that **B** is tangential to the curve and constant in magnitude along the curve. For a situation like this, in which the current is discontinuous in space, Ampère's law is not valid. In Chapter 29, we will see how Maxwell was able to modify Ampère's law so that it holds for all currents. When Maxwell's generalized form of Ampère's law is used to calculate the magnetic field for a current segment, such as that shown in Figure 25-29, the result agrees with that found from the Biot–Savart law.

Summary

1. The magnetic field produced by a point charge q moving with velocity $\mathbf{v}$ at a point a distance r away is given by

$$\mathbf{B} = \frac{\mu_0}{4\pi}\frac{q\mathbf{v}\times\hat{\mathbf{r}}}{r^2}$$

where $\hat{\mathbf{r}}$ is a unit vector that points from the charge to the field point and μ_0 is a constant, called the permeability of free space, that has the magnitude

$$\mu_0 = 4\pi\times 10^{-7}\ \mathrm{T{\cdot}m/A} = 4\pi\times 10^{-7}\ \mathrm{N/A^2}$$

2. The magnetic field $d\mathbf{B}$ at a distance r from a current element $I\,d\boldsymbol{\ell}$ is

$$d\mathbf{B} = \frac{\mu_0}{4\pi}\frac{I\,d\boldsymbol{\ell}\times\hat{\mathbf{r}}}{r^2}$$

which is known as the Biot–Savart law. The magnetic field is perpendicular to both the current element and to the vector $\mathbf{r}$ from the current element to the field point.

3. The magnetic forces between two moving charges do not obey Newton's third law of action and reaction, implying that linear momentum of the two-charge system is not conserved. However, when the momentum associated with t the electromagnetic field is included, the total linear momentum of the two-charge system plus field is conserved.

4. The magnetic field on the axis of a current loop is given by

$$\mathbf{B} = \frac{\mu_0}{4\pi}\frac{2\pi R^2 I}{(x^2+R^2)^{3/2}}\mathbf{i}$$

where $\mathbf{i}$ is a unit vector along the axis of the loop. At great distances from the loop, the field is a dipole field:

$$\mathbf{B} = \frac{\mu_0}{4\pi}\frac{2\mathbf{m}}{x^3}$$

where $\mathbf{m}$ is the dipole moment of the loop whose magnitude is the product of the current and the area of the loop and whose direction points in the direction perpendicular to the loop given by the right-hand rule.

5. Inside a solenoid and far from the ends, the magnetic field is uniform and has the magnitude

$$B = \mu_0 nI$$

where n is the number of turns per unit length of the solenoid.

6. The magnetic field of a straight, current-carrying wire segment is

$$B = \frac{\mu_0}{4\pi}\frac{I}{R}(\sin\theta_1 + \sin\theta_2)$$

where R is the perpendicular distance to the wire and θ_1 and θ_2 are the angles subtended at the field point by the ends of the wire. If the wire is very long, or the field point is very close to the wire, the magnetic field is approximately

$$B = \frac{\mu_0}{4\pi}\frac{2I}{R}$$

The direction of **B** is such that the lines of **B** encircle the wire in the direction of the fingers of the right hand if the thumb points in the direction of the current.

7. The magnetic field inside a tightly wound toroid is given by

$$B = \frac{\mu_0 NI}{2\pi r}$$

where r is the distance from the center of the toroid.

8. The ampere is defined such that two long parallel wires each carrying a current of 1 A and separated by 1 m exert a force of exactly 2×10^{-7} N/m on each other.

9. Ampère's law relates the integral of the tangential component of the magnetic field around a closed curve to the total current I_C passing through the area bounded by the curve:

$$\oint_C \mathbf{B}\cdot d\boldsymbol{\ell} = \mu_0 I_C \qquad C\text{, any closed curve}$$

Ampère's law is valid only if the currents are continuous. It can be used to derive expressions for the magnetic field for situations with a high degree of symmetry, such as a long, straight, current-carrying wire; a tightly wound toroid; and a long, tightly wound solenoid.

Suggestions for Further Reading

Banerjee, Subir K.: "Polar Flip-Flop," *The Sciences*, November/December 1984, p. 24.

The first half of this article is a fine short history of our understanding of the earth's magnetic field. Following this, the modern theory of earth magnetism is discussed, as well as evidence that the field may undergo a north–south reversal over the next 2000 years.

Carrigan, Charles R., and David Gubbins: "The Source of the Earth's Magnetic Field," *Scientific American*, February 1979, p. 118.

The earth may act as a huge dynamo in which electric currents in the molten and flowing metallic core maintain themselves by producing the magnetic field that deflects compass needles on the surface.

Shamos, Morris H.: "Electromagnetism—Hans Christian Oersted," in *Great Experiments in Physics*, Henry Holt and Co., New York, 1959. Reprinted by Dover, 1987.

Oersted's account, in translation, of his discovery of the magnetic field accompanying an electric current, with editorial annotations for clarity and a biographical sketch.

Williams, L. Pearce: "André-Marie Ampère," *Scientific American*, January 1989, p. 90.

A short biography of Ampère, describing his experiments and discoveries in electricity and magnetism, and the rationale behind his methods of investigation.

Review

A. Objectives: After studying this chapter, you should:

1. Be able to state the Biot–Savart law and use it to calculate the magnetic field **B** due to a straight, current-carrying wire and on the axis of a circular current loop.

2. Be able to sketch the magnetic-field lines for a long, straight current; a circular current loop; a solenoid; and a toroid.

3. Be able to state Ampère's law and discuss its uses and limitations.

4. Be able to use Ampère's law to derive expressions for **B** due to an infinite straight current; a toroid; and a long, tightly wound solenoid.

B. Define, explain, or otherwise identify:

Permeability of free space	Current balance
Biot–Savart law	Ampère's law
Solenoid	Toroid

C. True or false: If the statement is true, explain why it is true. If it is false, give a counterexample.

1. The magnetic field due to a current element is parallel to the current element.

2. The magnetic field due to a current element varies inversely with the square of the distance from the element.

3. The magnetic field due to a long wire varies inversely with the square of the distance from the wire.

4. Ampère's law is valid only if there is a high degree of symmetry.

5. Ampère's law is valid only for continuous currents.

Problems

Level I

25-1 The Magnetic Field of Moving Point Charges

1. At a certain instant of time a particle with charge $q = 12\ \mu\text{C}$ is located at $x = 0$, $y = 2$ m; its velocity at that time is $\mathbf{v} = 30$ m/s $\mathbf{i}$. Find the magnetic field at (*a*) the origin; (*b*) $x = 0$, $y = 1$ m; (*c*) $x = 0$, $y = 3$ m; and (*d*) $x = 0$, $y = 4$ m.

2. For the particle in Problem 1, find the magnetic field at (*a*) $x = 1$ m, $y = 3$ m; (*b*) $x = 2$ m, $y = 2$ m; and (*c*) $x = 2$ m, $y = 3$ m.

3. A proton (charge $+e$) traveling with a velocity of $\mathbf{v} = 1 \times 10^4$ m/s $\mathbf{i} + 2 \times 10^4$ m/s $\mathbf{j}$ is located at $x = 3$ m, $y = 4$ m at some time. Find the magnetic field at the following positions: (*a*) $x = 2$ m, $y = 2$ m; (*b*) $x = 6$ m, $y = 4$ m; and (*c*) $x = 3$ m, $y = 6$ m.

4. Two equal charges q located at $(0, 0, 0)$ and $(0, b, 0)$ at time zero are moving with speed v in the positive x direction. Find the ratio of the magnitudes of the magnetic and electrostatic force on each.

5. An electron orbits a proton at a radius of 5.29×10^{-11} m. What is the magnetic field at the proton due to the orbital motion of the electron?

Section 25-2 The Magnetic Field of Currents: The Biot–Savart Law

6. A small current element $I\,d\boldsymbol{\ell}$, with $d\boldsymbol{\ell} = 2$ mm $\mathbf{k}$ and $I = 2$ A is centered at the origin. Find the magnetic field $d\mathbf{B}$ at the following points: (*a*) on the x axis at $x = 3$ m, (*b*) on the x axis at $x = -6$ m, (*c*) on the z axis at $z = 3$ m, and (*d*) on the y axis at $y = 3$ m.

7. For the current element in Problem 6, find the magnitude and direction of $d\mathbf{B}$ at $x = 0$, $y = 3$ m, $z = 4$ m.

8. For the current element in Problem 6, find the magnitude of $d\mathbf{B}$ and indicate its direction on a diagram at (*a*) $x = 2$ m, $y = 4$ m, $z = 0$ and (*b*) $x = 2$ m, $y = 0$, $z = 4$ m.

9. A single-turn, circular loop of radius 10.0 cm is to produce a field at its center that will just cancel the earth's magnetic field at the equator, which is 0.7 G directed north. Find the current in the loop and make a sketch showing the orientation of the loop and the current.

10. A solenoid with length 30 cm, radius 1.2 cm, and 300 turns carries a current of 2.6 A. Find B on the axis of the solenoid (*a*) at the center, (*b*) inside the solenoid at a point 10 cm from one end, and (*c*) at one end.

11. A solenoid 2.7 m long has a radius of 0.85 cm and 600 turns. It carries a current I of 2.5 A. What is the approximate magnetic field B on the axis of the solenoid?

12. A single loop of wire of radius 3 cm carries a current of 2.6 A. What is the magnitude of **B** on the axis of the loop at (*a*) the center of the loop, (*b*) 1 cm from the center, (*c*) 2 cm from the center, and (*d*) 35 cm from the center?

13. For the loop of wire in Problem 12, at what point along the axis of the loop is the magnetic field (*a*) 10 percent of the field at the center, (*b*) 1 percent of the field at the center, and (*c*) 0.1 percent of the field at the center?

14. A long, straight wire carries a current of 10 A. Find the magnitude of **B** at (*a*) 10 cm, (*b*) 50 cm, and (*c*) 2 m from the center of the wire.

Problems 15 to 20 refer to Figure 25-30, which shows two long, straight wires in the xy plane and parallel to the x axis. One wire is at $y = -6$ cm and the other is at $y = +6$ cm. The current in each wire is 20 A.

Figure 25-30 Problems 15 through 20.

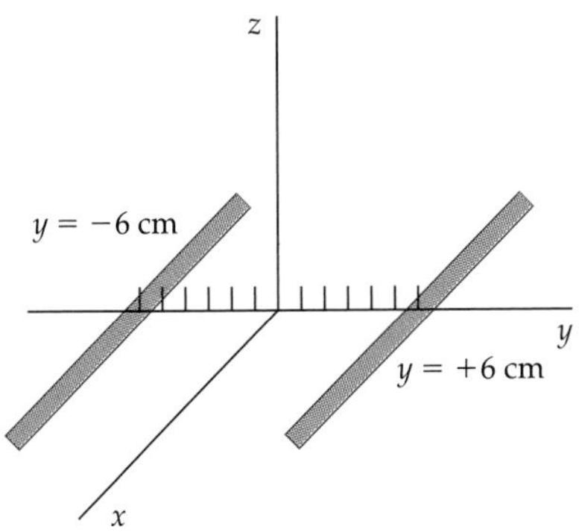

15. If the currents in Figure 25-30 are in the negative x direction, find **B** at the points on the y axis at (*a*) $y = -3$ cm, (*b*) $y = 0$, (*c*) $y = +3$ cm, and (*d*) $y = +9$ cm.

16. Sketch B_z versus y for points on the y axis when both currents are in the negative x direction.

17. Find **B** at points on the y axis as in Problem 15 when the current in the wire at $y = -6$ cm is in the negative x direction and the current in the wire at $y = +6$ cm is in the positive x direction.

18. Sketch B_z versus y for points on the y axis when the directions of the currents are opposite to those in Problem 17.

19. Find **B** on the z axis at $z = +8$ cm if (*a*) the currents are parallel, as in Problem 15 and (*b*) the currents are antiparallel, as in Problem 17.

20. Find the magnitude of the force per unit length exerted by one wire on the other.

21. The current in the wire of Figure 25-31 is 8.0 A. Find **B** at point P due to each wire segment and sum to find the resultant **B**.

Figure 25-31 Problem 21.

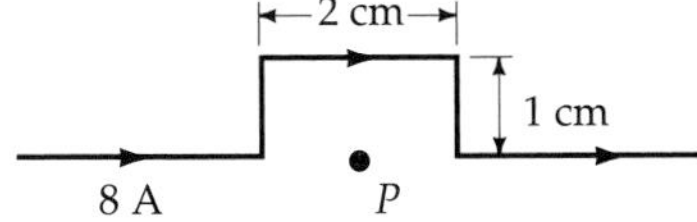

22. Find the magnetic field at point P in Figure 25-32.

Figure 25-32 Problem 22.

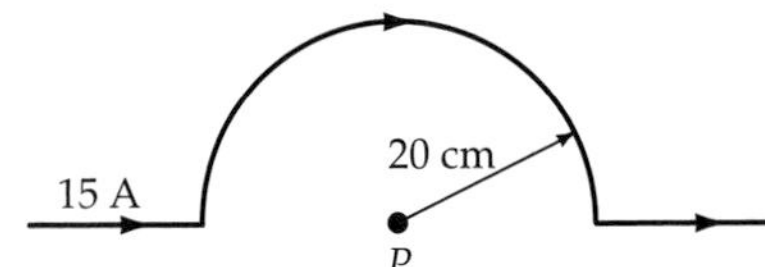

23. A single-turn circular loop of radius 8.5 cm is to produce a field at its center that will just cancel the earth's field of magnitude 0.7 G directed at 70° below the horizontal north direction. Find the current in the loop and make a sketch showing the orientation of the loop and the current.

24. In Figure 25-33, find the magnetic field at point P, which is at the common center of the two semicircular arcs.

Figure 25-33 Problem 24.

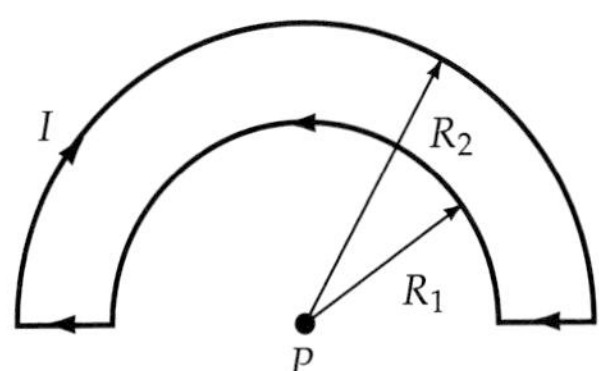

25-3 Definition of the Ampere

25. Two long, straight, parallel wires 8.6 cm apart carry currents of equal magnitude I. They repel each other with a force per unit length of 3.6 nN/m. (*a*) Are the currents parallel or antiparallel? (*b*) Find I.

26. A wire of length 16 cm is suspended by flexible leads above a long, straight wire. Equal but opposite currents are established in the wires such that the 16-cm wire floats 1.5 mm above the long wire with no tension in its suspension leads. If the mass of the 16-cm wire is 14 g, what is the current?

27. In an experiment with a current balance, the upper wire of length 30 cm is pivoted such that with no current it balances at 2 mm above a fixed, parallel wire also 30 cm long. When the wires carry equal but opposite currents I, the upper wire balances at its original position when a mass of 2.4 g is placed on it. What is the current I?

28. Three long, parallel, straight wires pass through the corners of an equilateral triangle of sides 10 cm as shown in Figure 25-34, where a dot means that the current is out of the paper and a cross means that it is into the paper. If each current is 15.0 A, find (*a*) the force per unit length on the upper wire and (*b*) the magnetic field **B** at the upper wire due to the two lower wires. *Hint:* It is easier to find the force per unit length directly from Equation 25-14 and use your result to find **B** than to find **B** and use it to find the force.

Figure 25-34 Problems 28 and 29.

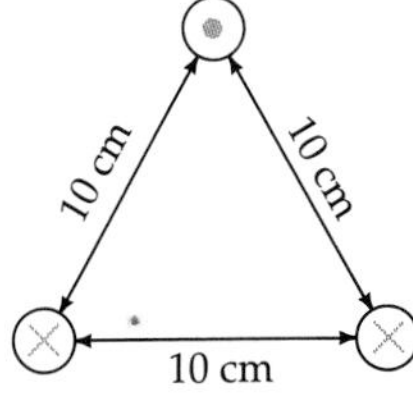

29. Work Problem 28 with the current in the lower right corner of Figure 25-34 reversed.

25-4 Ampère's Law

30. A long, straight, thin-walled, cylindrical shell of radius R carries a current I. Find **B** inside and outside the cylinder.

31. In Figure 25-35, one current is 8 A into the paper; the other current is 8 A out of the paper; and each curve is a circular path. (*a*) Find $\oint \mathbf{B}\cdot d\boldsymbol{\ell}$ for each path indicated. (*b*) Which path, if any, can be used to find **B** at some point due to these currents?

Figure 25-35 Problem 31.

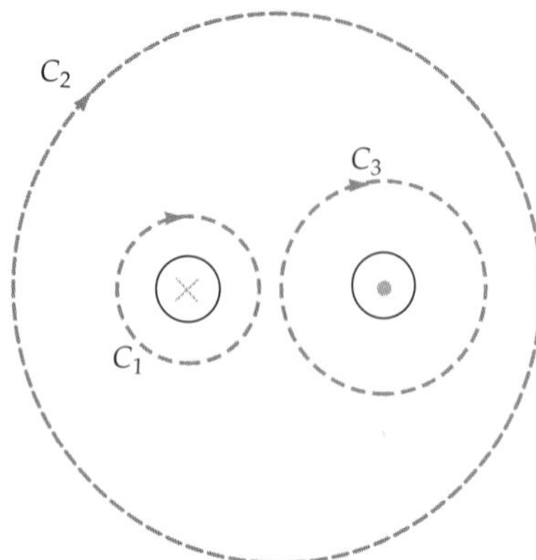

32. A very long coaxial cable consists of an inner wire and a concentric outer cylindrical conducting shell of radius R. At one end, the wire is connected to the shell. At the other end, the wire and shell are connected to opposite terminals of a battery, so there is a current down the wire and back up the shell. Assume that the cable is straight. Find **B** (*a*) at points between the wire and the shell far from the ends, and (*b*) outside the cable.

33. A wire of radius 0.5 cm carries a current of 100 A that is uniformly distributed over its cross-sectional area. Find B (*a*) 0.1 cm from the center of the wire, (*b*) at the surface of the wire, and (*c*) at a point outside the wire 0.2 cm from the surface of the wire. (*d*) Sketch a graph of B versus the distance from the center of the wire.

34. Show that a uniform magnetic field with no fringing field, such as that shown in Figure 25-36, is impossible because it violates Ampère's law. Do this by applying Ampère's law to the rectangular curve shown by the dashed lines.

Figure 25-36 Problem 34.

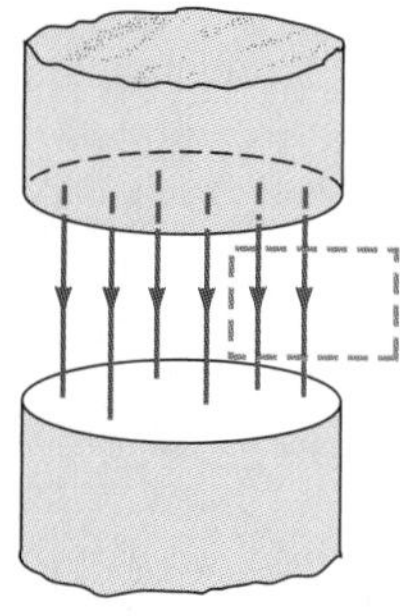

35. A tightly wound toroid of inner radius 1 cm and outer radius 2 cm has 1000 turns of wire and carries a current of 1.5 A. (*a*) What is the magnetic field at a distance of 1.1 cm from the center? (*b*) What is the field 1.5 cm from the center?

Level II

36. A wire of length ℓ is wound into a circular coil of N loops and carries a current I. Show that the magnetic field at the center of the coil is given by $B = \mu_0 \pi N^2 I/\ell$.

37. A very long, straight wire carries a current of 20.0 A. An electron 1.0 cm from the center of the wire is moving with a speed of 5.0×10^6 m/s. Find the force on the electron when it moves (*a*) directly away from the wire, (*b*) parallel to the wire in the direction of the current, and (*c*) perpendicular to the wire and tangent to a circle around the wire.

38. A very long wire carrying a current I is bent into the shape shown in Figure 25-37. Find the magnetic field at point P.

Figure 25-37 Problem 38.

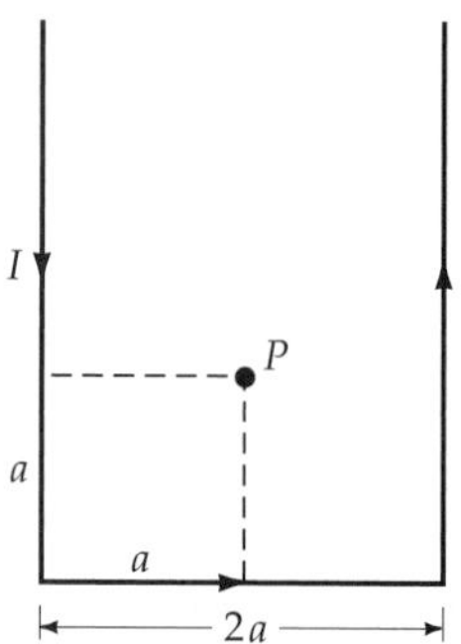

39. A loop of wire of length ℓ carries a current I. Compare the magnetic fields at the center of the loop when it is (*a*) a circle, (*b*) a square, and (*c*) an equilateral triangle. Which field is largest?

40. A power cable carrying 50.0 A is 2.0 m below the earth's surface, but its direction and precise position are unknown. Show how you could locate the cable using a compass. Assume that you are at the equator, where the earth's magnetic field is 0.7 G north.

41. Four long, straight, parallel wires each carry current I. In a plane perpendicular to the wires, the wires are at the corners of a square of side a. Find the force per unit length on one of the wires if (*a*) all the currents are in the same direction and (*b*) the currents in the wires at adjacent corners are oppositely directed.

42. A current I is uniformly distributed over the cross section of a long, straight wire of radius 1.40 mm. At the surface of the wire, the magnitude of the magnetic field is $B = 2.46$ mT. Find the magnitude of the magnetic field at (*a*) 2.10 mm from the axis and (*b*) 0.60 mm from the axis. (*c*) Find the current I.

43. A coaxial cable consists of a solid inner cylindrical conductor of radius 1.00 mm and an outer cylindrical shell conductor of inner radius 2.00 mm and outer radius 3.00 mm. There is a current of 18 A down the inner wire and an equal return current in the outer conductor. The currents are uniform over the cross section of each conductor. Find the numerical value of $\oint \mathbf{B}\cdot d\boldsymbol{\ell}$ for a closed circular path (centered on the axis of the cable and in a plane perpendicular to the axis) which has a radius r for (*a*) $r = 1.50$ mm, (*b*) $r = 2.50$ mm, and (*c*) $r = 3.50$ mm.

44. An infinitely long, insulated wire lies along the x axis and carries current I in the positive x direction. A second infinitely long, insulated wire lies along the y axis and carries current $I/4$ in the positive y direction. Where in the xy plane is the resultant magnetic field zero?

45. A large, 50-turn circular coil of radius 10.0 cm carries a current of 4.0 A. At the center of the large coil is a small 20-turn coil of radius 0.5 cm carrying a current of 1.0 A. The planes of the two coils are perpendicular. Find the torque exerted by the large coil on the small coil. (Neglect any variation in **B** due to the large coil over the region occupied by the small coil.)

46. A relatively inexpensive ammeter called a *tangent galvanometer* can be made using the earth's field. A plane circular coil of N turns and radius R is oriented such that the field B_c it produces in the center of the coil is either east or west. A compass is placed at the center of the coil. When there is no current in the coil, the compass needle points north. When there is a current I, the compass needle points in the direction of the resultant magnetic field **B** at an angle θ to the north. Show that the current I is related to θ and the horizontal component of the earth's field B_e by

$$I = \frac{2RB_e}{\mu_0 N} \tan \theta$$

47. An infinitely long, straight wire is bent as shown in Figure 25-38. The circular portion has a radius of 10 cm with its center a distance r from the straight part. Find r such that the magnetic field at the center of the circular portion is zero.

Figure 25-38 Problem 47.

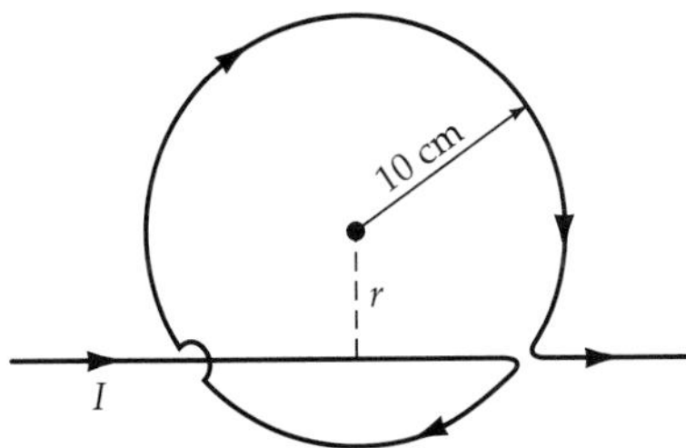

48. An infinitely long, nonconducting cylinder of radius R lies along the z axis. Five long conducting wires are parallel to the cylinder and spaced equally on the upper half of its surface. Each wire carries a current I in the positive z direction. Find the magnetic field on the z axis.

49. An infinitely long wire lies along the z axis and carries a current of 20 A in the positive z direction. A second infinitely long wire is parallel to the z axis at $x = 10$ cm. (*a*) Find the current in the second wire if the magnetic field at $x = 2$ cm is zero. (*b*) What is the magnetic field at $x = 5$ cm?

50. Three very long, parallel wires are at the corners of a square, as shown in Figure 25-39. Find the magnetic field **B** at the unoccupied corner of the square when (*a*) all the currents are into the paper, (*b*) I_1 and I_3 are in and I_2 is out, and (*c*) I_1 and I_2 are in and I_3 is out.

Figure 25-39 Problem 50.

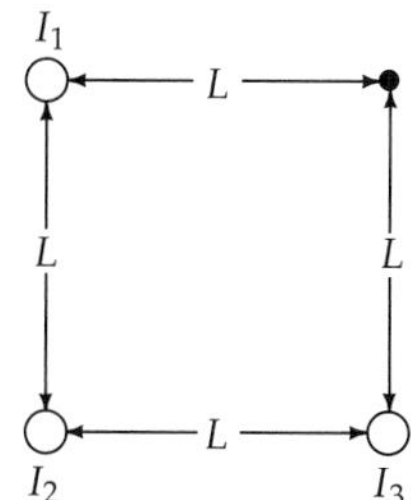

51. (*a*) Find the magnetic field at point P for the wire carrying current I shown in Figure 25-40. (*b*) Use your result from (*a*) to find the field at the center of a polygon of N sides. Show that when N is very large, your result approaches that for the magnetic field at the center of a circle.

Figure 25-40 Problem 51.

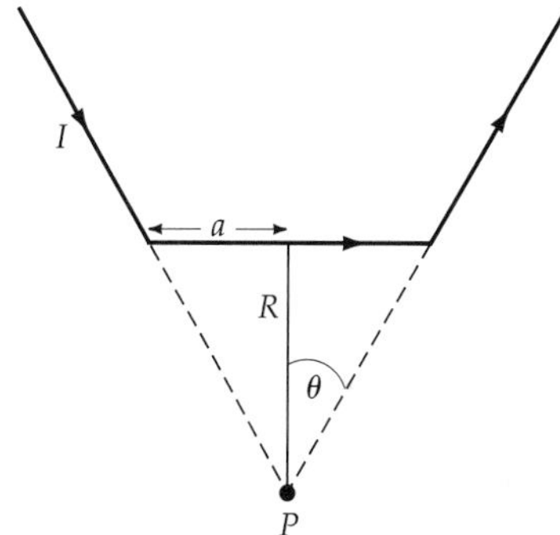

52. A circular current loop of radius R carrying a current I is centered at the origin with its axis along the x axis. Its current is such that it produces a magnetic field in the positive x direction. (*a*) Sketch a graph of B_x versus x for points on the x axis. Include both positive and negative values of x. Compare this graph with that for E_x due to a charged ring of the same size. (*b*) A second, identical current loop, carrying an equal current in the same sense, is in a plane parallel to the yz plane with its center at $x = d$. Sketch graphs of the magnetic field on the x axis due to each loop separately and the resultant field due to the two loops. Show from your sketch that dB_x/dx is zero midway between the two loops.

53. Two coils which are separated by a distance equal to their radius and carry equal currents such that their axial fields add are called **Helmholtz coils.** A feature of Helmholtz coils is that the resultant magnetic field between the coils is very uniform. Let $R = 10$ cm, $I = 20$ A, and $N = 300$ turns for each coil. Place one coil in the yz plane with its center at the origin and the other in a parallel plane at $x = 10$ cm. (*a*) Calculate the resultant field B_x at $x = 5$ cm,

$x = 7$ cm, $x = 9$ cm, and $x = 11$ cm. (*b*) Use your results and the fact that B_x is symmetric about the midpoint of the coils to sketch B_x versus x. (See also Problem 62.)

54. An infinitely long, thick cylindrical shell of inner radius a and outer radius b carries a current I uniformly distributed across a cross section of the shell. Find the magnetic field for (*a*) $r < a$, (*b*) $a < r < b$, and (*c*) $r > b$.

55. A long, straight wire carries a current of 20 A as shown in Figure 25-41. A rectangular coil with two sides parallel to the straight wire has sides 5 cm and 10 cm with the near side a distance 2 cm from the wire. The coil carries a current of 5 A. (*a*) Find the force on each segment of the rectangular coil. (*b*) What is the net force on the coil?

Figure 25-41 Problem 55.

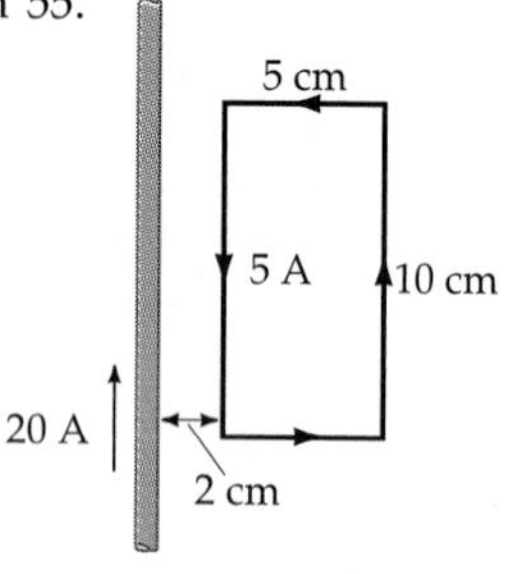

56. The xz plane contains an infinite sheet of current in the positive z direction. The current per unit length (along the x direction) is λ. Figure 25-42*a* shows a point P above the sheet ($y > 0$) and two portions of the current sheet labeled I_1 and I_2. (*a*) What is the direction of the magnetic field **B** at P due to the two portions of the current shown? (*b*) What is the direction of the magnetic field **B** at P due to the entire sheet? (*c*) What is the direction of **B** at a point below the sheet ($y < 0$)? (*d*) Apply Ampère's law to the rectangular curve shown in Figure 25-42*b* to show that the magnetic field at any point above the sheet is given by

$$\mathbf{B} = -\tfrac{1}{2}\mu_0\lambda\mathbf{i}$$

Figure 25-42 Problem 56.

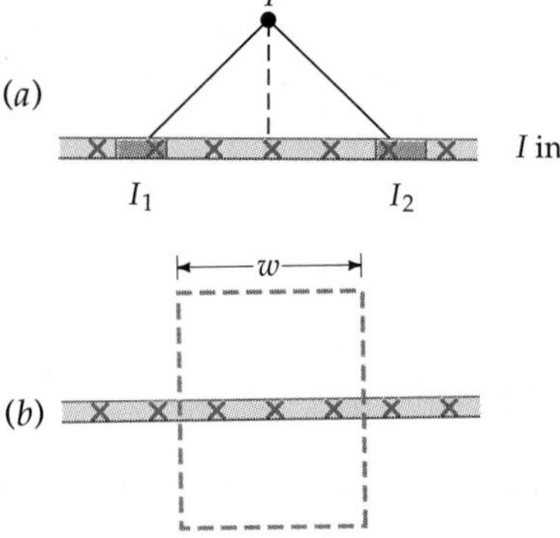

Level III

57. A square loop of side ℓ lies in the yz plane with its center at the origin. It carries a current I. Find the magnetic field **B** at any point on the x axis and show from your expression that for x much larger than ℓ

$$\mathbf{B} \approx \frac{\mu_0}{4\pi}\frac{2\mathbf{m}}{x^3}$$

where $m = I\ell^2$ is the magnetic moment of the loop.

58. A circular loop carrying current I lies in the yz plane with its axis along the x axis. (*a*) Evaluate the line integral $\int \mathbf{B}\cdot d\boldsymbol{\ell}$ along the axis of the loop from $x = -\ell_1$ to $x = +\ell_1$. (*b*) Show that when $\ell_1 \to \infty$, the line integral approaches $\mu_0 I$. This result can be related to Ampère's law by closing the curve of integration with a semicircle of radius ℓ_1 on which $B \approx 0$ for ℓ_1 very large.

59. A very long, straight conductor with a circular cross section of radius R carries a current I. Inside the conductor, there is a cylindrical hole of radius a whose axis is parallel to the axis of the conductor a distance b from it (Figure 25-43). Let the z axis be the axis of the conductor, and let the axis of the hole be at $x = b$. Find the magnetic field **B** at the point (*a*) on the x axis at $x = 2R$, and (*b*) on the y axis at $y = 2R$. *Hint*: Consider a uniform current distribution throughout the cylinder of radius R plus a current in the opposite direction in the hole.

60. For the cylinder with the hole in Problem 59, show that the magnetic field inside the hole is uniform, and find its magnitude and direction.

Figure 25-43 Problems 59 and 60.

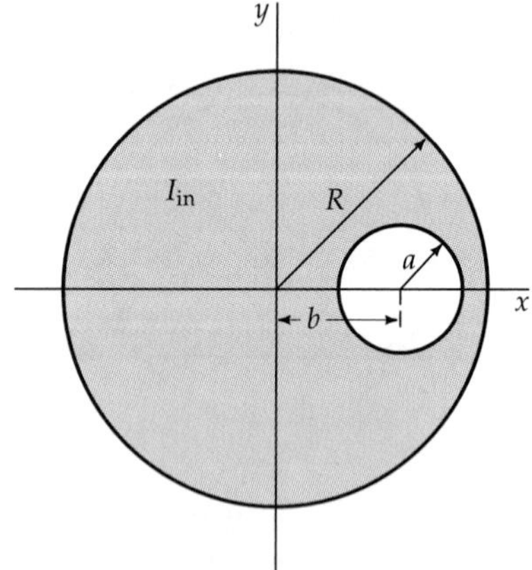

61. A disk of radius R carries a fixed charge density σ and rotates with angular velocity ω. (*a*) Consider a circular strip of radius r and width dr with charge dq. Show that the current produced by this strip $dI = (\omega/2\pi)\,dq = \omega\sigma r\,dr$. (*b*) Use your result from part (*a*) to show that the magnetic field at the center of the disk is $B = \frac{1}{2}\mu_0\sigma\omega R$. (*c*) Use your result from part (*a*) to find the magnetic field at a point on the axis of the disk a distance x from the center.

62. Two Helmholtz coils have radii R and their axes along the x axis (see Problem 53). One coil is in the yz plane and the other is in a parallel plane at $x = R$. Show that at the midpoint of the coils ($x = \frac{1}{2}R$), $dB_x/dx = 0$, $d^2B_x/dx^2 = 0$, and $d^3B_x/dx^3 = 0$. This shows that the magnetic field at points near the midpoint is approximately equal to that at the midpoint.

63. A solenoid has n turns per unit length and radius R and carries a current I. Its axis is along the x axis with one end at $x = -\frac{1}{2}\ell$ and the other end at $x = +\frac{1}{2}\ell$, where ℓ is the total length of the solenoid. Show that the magnetic field B at a point on the axis outside the solenoid is given by

$$B = \tfrac{1}{2}\mu_0 nI(\cos\theta_1 - \cos\theta_2) \qquad \text{25-18}$$

where

$$\cos\theta_1 = \frac{x + \frac{1}{2}\ell}{[R^2 + (x + \frac{1}{2}\ell)^2]^{1/2}}$$

and

$$\cos\theta_2 = \frac{x - \frac{1}{2}\ell}{[R^2 + (x - \frac{1}{2}\ell)^2]^{1/2}}$$

64. In Problem 63, a formula for the magnetic field along the axis of a solenoid is given. For $x \gg \ell$ and $\ell > R$, the angles θ_1 and θ_2 in Equation 25-18 are very small so the small-angle approximation $\cos\theta \approx 1 - \theta^2/2$ is valid. (*a*) Draw a diagram and show that

$$\theta_1 \approx \frac{R}{x + \frac{1}{2}\ell}$$

and

$$\theta_2 \approx \frac{R}{x - \frac{1}{2}\ell}$$

(*b*) Show that the magnetic field at a point far from either end of the solenoid can be written

$$B = \frac{\mu_0}{4\pi}\left(\frac{q_m}{r_1^2} - \frac{q_m}{r_2^2}\right) \qquad \text{25-19}$$

where $r_1 = x - \frac{1}{2}\ell$ is the distance to the near end of the solenoid, $r_2 = x + \frac{1}{2}\ell$ is the distance to the far end, and $q_m = nI\pi R^2 = m/\ell$, where $m = NI\pi R^2$ is the magnetic moment of the solenoid.

65. In this problem, you will derive Equation 25-19 by another method. Consider a long, tightly wound solenoid of length ℓ and radius $R \ll \ell$ lying along the x axis with its center at the origin. It has N turns and carries a current I. Consider an element of the solenoid of length dx. (*a*) What is the magnetic moment of this element? (*b*) Show that the magnetic field dB due to this element at a point on the x axis x_0 far from the element is given by

$$dB = \frac{\mu_0}{2\pi} nIA \frac{dx}{x'^3}$$

where $A = \pi R^2$ and $x' = x_0 - x$ is the distance from the element to the field point. (*c*) Integrate this expression from $x = -\frac{1}{2}\ell$ to $x = +\frac{1}{2}\ell$ to obtain Equation 25-19.

Chapter 26

Magnetic Induction

A cross section of a pickup for an electric guitar. The pickup consists of a cylindrical permanent magnet wrapped with a wire coil. The guitar string is magnetized by the magnet below it and produces an oscillating magnetic flux through the coil of the pickup when it vibrates. A current of frequency equal to that of the vibrating string is thereby induced in the coil. The current is amplified and sent to a speaker. (For this photo, the guitar string has been placed parallel to six pairs of pickups. When mounted for playing, the strings are perpendicular to the direction shown here with each string crossing over one pair of pickups.)

In the previous chapter, we learned that a current in a wire creates a magnetic field. In the early 1830s, Michael Faraday in England and Joseph Henry in America independently discovered that a magnetic field induces a current in a wire but that this happens only when the magnetic field is *changing*. When you pull the plug of an electric cord from its socket, you sometimes observe a small spark. Before the cord is disconnected, it carries a current, which, as we have seen, produces a magnetic field encircling the current. When the cord is disconnected, the current abruptly ceases and the magnetic field around it collapses. The changing magnetic field produces an emf that tries to maintain the original current, resulting in a spark across the plug. Once the magnetic field has reached zero and is therefore no longer changing, the emf is zero. The emfs and currents caused by changing magnetic fields are called **induced emfs** and **induced currents.** The process itself is referred to as **magnetic induction.**

In the case of an electric cord being disconnected from its power source, the changing magnetic field is caused by a change in the electric currents. However, a changing magnetic field can also be produced by moving a magnet. Figure 26-1 illustrates a simple classroom demonstration of induced currents. The ends of a coil are attached to a galvanometer and a strong magnet is moved toward or away from the coil. The momentary deflection shown by the galvanometer *during* the motion indicates that there is an induced electric current in the coil–galvanometer circuit. A current is also induced if the coil is moved toward or away from the magnet, or if the coil is rotated in a fixed magnetic field. A coil rotating in a magnetic field is the basic element of a generator, which converts mechanical or thermal energy into electrical energy. In a hydrostatic power plant, a river is dammed up and the water is released in a controlled way to turn the coils. In a steam-based power plant, water is heated and converted to steam by energy from burning coal or from nuclear fission. The pressure of the steam is then used to turn the coils.

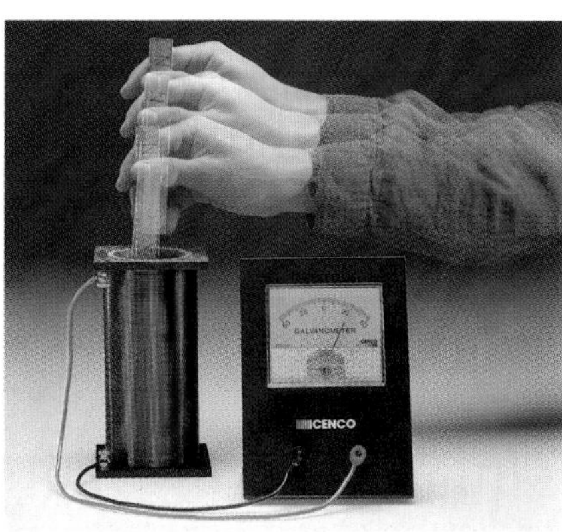

Figure 26-1 Demonstration of induced emf. When the magnet is moving away from the coil, an emf is induced in the coil as shown by the galvanometer's deflection. No deflection is observed when the magnet is stationary.

All of the various methods of magnetic induction can be summarized by a single relation known as Faraday's law, which relates the induced emf in a circuit to the change in magnetic flux through the circuit.

26-1 Magnetic Flux

Magnetic flux is the magnetic analog to electric flux, which we learned about in Section 19-2. It is related to the number of magnetic-field lines that pass through a given area. In Figure 26-2, the magnetic field is perpendicular to the area bounded by a simple circuit consisting of one turn of wire. In this case, the **magnetic flux** ϕ_m is defined as the product of the magnetic field **B** and the area A bounded by the circuit:

$$\phi_m = BA$$

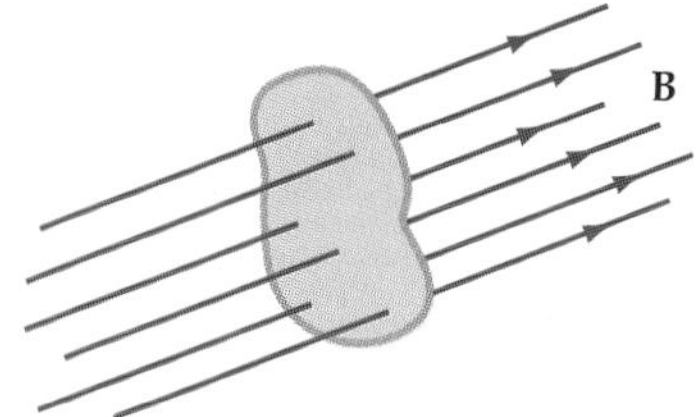

Figure 26-2 When the magnetic field **B** is perpendicular to the area enclosed by a loop, the magnetic flux through the loop is BA.

The unit of magnetic flux is that of magnetic field times area, tesla-meter squared, which is called a **weber** (Wb):

$$1\ \text{Wb} = 1\ \text{T}\cdot\text{m}^2 \qquad 26\text{-}1$$

Since the magnetic field is proportional to the number of magnetic-field lines per unit area, the magnetic flux is proportional to the number of lines through the area.

If the magnetic field is not perpendicular to the surface, as in Figure 26-3, the flux ϕ_m is defined as

$$\phi_m = \mathbf{B}\cdot\hat{\mathbf{n}}A = BA\cos\theta = B_nA \qquad 26\text{-}2$$

where $B_n = \mathbf{B}\cdot\hat{\mathbf{n}}$ is the component of the magnetic-field vector that is perpendicular or normal to the surface.

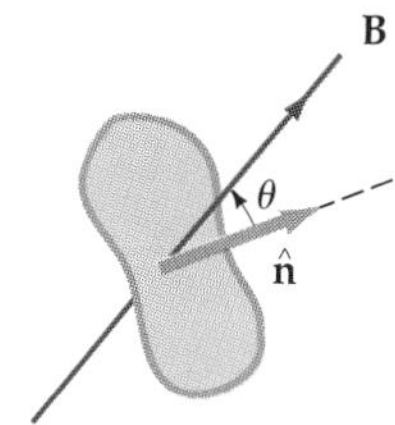

Figure 26-3 When **B** makes an angle θ with the normal to the area of a loop, the flux through the loop is $B\cos\theta\ A$.

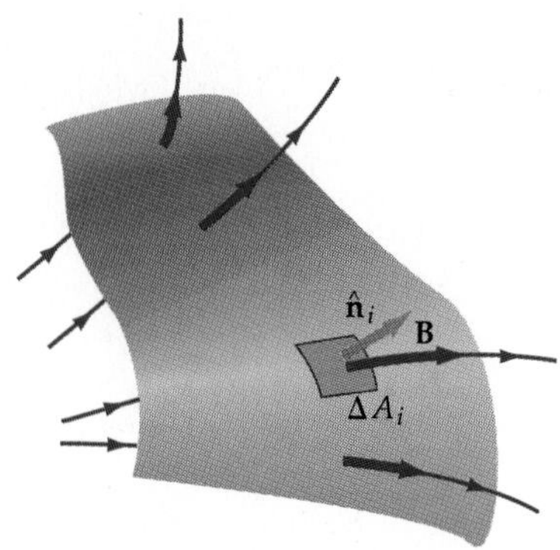

Figure 26-4 When **B** varies in either magnitude or direction, the area is divided into small elements ΔA_i. The flux through the area is computed by summing $B_n \, \Delta A_i$ over all the area elements.

We can generalize our definition of magnetic flux to curved surfaces over which the magnetic field may vary in magnitude or direction or both by dividing the surface up into a large number of very small area elements. If each element is small enough, it can be considered to be a plane, and the variation of the magnetic field across the element can be neglected. Let $\hat{\mathbf{n}}_i$ be the unit vector perpendicular to such an element and ΔA_i be its area (Figure 26-4). The flux of the magnetic field through this element is

$$\Delta\phi_{mi} = \mathbf{B}\cdot\hat{\mathbf{n}}_i \, \Delta A_i$$

The total flux through the surface is the sum of $\Delta\phi_{mi}$ over all the elements. In the limit as the number of elements approaches infinity and the area of each element approaches zero, this sum becomes an integral. The general definition of magnetic flux is then

$$\phi_m = \lim_{\Delta A_i \to 0} \sum_i \mathbf{B}\cdot\hat{\mathbf{n}}_i \, \Delta A_i = \int_S \mathbf{B}\cdot\hat{\mathbf{n}} \, dA$$

For a coil of N turns, the flux through the coil is N times the flux through a single turn:

$$\phi_m = NBA \cos\theta \qquad 26\text{-}3$$

For the general case in which **B** is not necessarily constant over the area, the flux is

Magnetic flux defined

$$\phi_m = \int_S N\mathbf{B}\cdot\hat{\mathbf{n}} \, dA = \int_S NB_n \, dA \qquad 26\text{-}4$$

Example 26-1

A uniform magnetic field of magnitude 2000 G makes an angle of 30° with the axis of a circular coil of 300 turns and a radius of 4 cm. Find the magnetic flux through the coil.

Since 1 G = 10^{-4} T, this magnetic field in SI units is 0.2 T. The area of the coil is

$$A = \pi r^2 = (3.14)(0.04 \text{ m})^2 = 0.00502 \text{ m}^2$$

The flux through the coil is thus

$$\phi_m = NBA \cos\theta = (300)(0.2 \text{ T})(0.00502 \text{ m}^2)(0.866) = 0.26 \text{ Wb}$$

Example 26-2

Find the magnetic flux through a solenoid that is 40 cm long, has a radius of 2.5 cm, has 600 turns, and carries a current of 7.5 A.

The magnetic field inside the solenoid is given by Equation 25-10:

$$B = \mu_0 nI = (4\pi \times 10^{-7} \text{ T·m/A})(600 \text{ turns}/0.40 \text{ m})(7.5 \text{ A})$$
$$= 1.41 \times 10^{-2} \text{ T}$$

Since the magnetic field is essentially constant across the cross-sectional area of the coil the magnetic flux is

$$\phi_m = NBA = (600)(1.41 \times 10^{-2} \text{ T})\pi(0.025 \text{ m})^2 = 1.66 \times 10^{-2} \text{ Wb}$$

Note that since $\phi_m = NBA$ and B is proportional to the number of turns N, the flux is proportional to N^2.

26-2 Induced Emf and Faraday's Law

Michael Faraday (1791–1867).

The experiments of Faraday, Henry, and others have shown that if the magnetic flux through a circuit is changed by any means, an emf equal in magnitude to the rate of change of the flux is induced in the circuit. The emf is usually detected by observing a current in the circuit, but it is present even if the circuit is incomplete (not closed) so that no current exists. In our previous discussions, the emf in a circuit has been localized in a specific region of the circuit, such as between the terminals of the battery. However, the emf induced by a changing magnetic flux can be considered to be distributed throughout the circuit.

Consider a single loop of wire in a magnetic field as shown in Figure 26-5. If the flux through the loop is changing, there is an emf induced in the loop. Since emf is the work done per unit charge, there must be a force exerted on the charge associated with the emf. The force per unit charge is the electric field **E**, which in this case is induced by the changing flux. The line integral of the electric field around a complete circuit equals the work done per unit charge, which, by definition, is the emf in the circuit:

$$\mathcal{E} = \oint_C \mathbf{E}\cdot d\boldsymbol{\ell} \qquad \text{26-5}$$

Emf defined

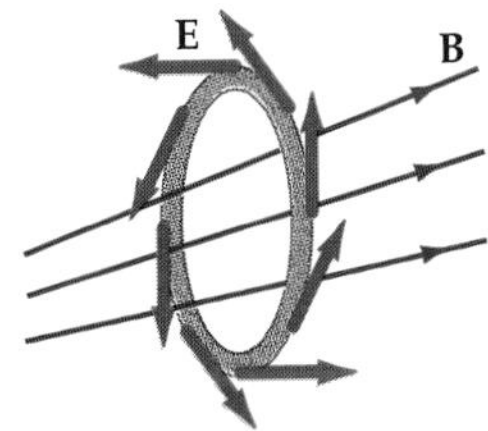

Figure 26-5 When the magnetic flux through the wire loop is changing, an emf is induced in the loop. The emf is distributed throughout the loop and is equivalent to a nonconservative electric field **E** parallel to the wire. In this figure, the direction of **E** corresponds to the case in which the flux through the loop is increasing.

The electric fields that we studied previously resulted from static electric charges. These fields are conservative, meaning that the line integral of the electrostatic field around a closed curve is zero. However, the electric field resulting from a changing magnetic flux is not conservative. The line integral around a closed curve equals the induced emf, which equals the rate of change of the magnetic flux:

$$\mathcal{E} = \oint_C \mathbf{E}\cdot d\boldsymbol{\ell} = -\frac{d\phi_{\mathrm{m}}}{dt} \qquad \text{26-6}$$

Faraday's law

This result is known as **Faraday's law.** The negative sign in Faraday's law has to do with the direction of the induced emf, which will be discussed shortly.

Exercise

Show that a weber per second is a volt.

The magnetic flux through a circuit can be changed in many different ways. The current producing the magnetic field may be increased or decreased, permanent magnets may be moved toward the circuit or away from it, the circuit itself may be moved toward or away from the source of the flux, the orientation of the circuit may be changed, or the area of the circuit in a fixed magnetic field may be increased or decreased. In every case, an emf is induced in the circuit that is equal in magnitude to the rate of change of the magnetic flux.

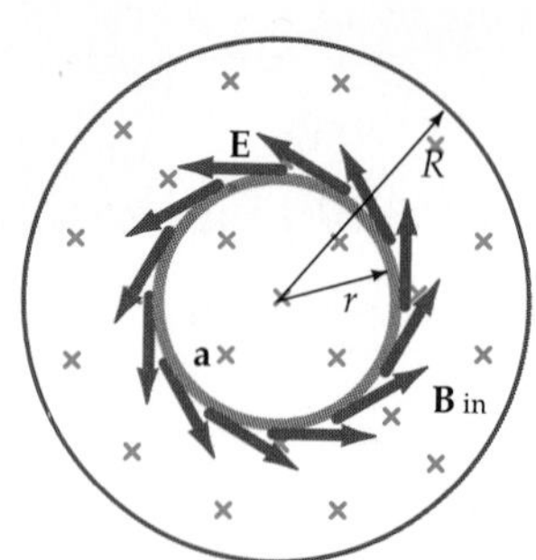

Figure 26-6 Diagram for Example 26-3. The magnetic field **B** is into the page and uniform over a circular region of radius R. When **B** changes, the magnetic flux changes and an emf $\mathcal{E} = \oint \mathbf{E} \cdot d\boldsymbol{\ell}$ is induced around any curve enclosing the flux. The induced electric field **E** at a distance r is tangent to the circle of radius r and is in the direction shown if **B** is increasing.

Example 26-3

A magnetic field **B** is perpendicular to the plane of the page and uniform in a circular region of radius R as shown in Figure 26-6. Outside of the circular region, $\mathbf{B} = 0$. The rate of change of the magnitude of **B** is dB/dt. What is the magnitude of the induced electric field in the plane of the page at a distance r from the center of the circular region?

According to Equation 26-6, the line integral of **E** around a closed curve equals the negative of the rate of change of the magnetic flux through the curve. Since we are interested only in magnitudes, we neglect the negative sign in this equation. Then

$$\oint_C \mathbf{E}\cdot d\boldsymbol{\ell} = \frac{d\phi_m}{dt}$$

In the figure, we have chosen a circular curve of radius $r < R$ to compute the line integral $\oint \mathbf{E}\cdot d\boldsymbol{\ell}$. By symmetry, **E** is tangent to this curve and has the same value at any point on it. Then

$$\oint_C \mathbf{E}\cdot d\boldsymbol{\ell} = E(2\pi r)$$

Since **B** is perpendicular to the plane of this curve, the flux through the curve is

$$\phi_m = BA = B\pi r^2$$

and the rate of change of the flux is

$$\frac{d\phi_m}{dt} = \pi r^2 \frac{dB}{dt}$$

Equation 26-6 then gives

$$2\pi r E = \pi r^2 \frac{dB}{dt}$$

or

$$E = \frac{r}{2}\frac{dB}{dt} \qquad r < R$$

For a circle with $r > R$, the line integral $\oint \mathbf{E}\cdot d\boldsymbol{\ell}$ again gives $2\pi r E$. However, since $\mathbf{B} = 0$ for $r > R$, the flux is now $\pi R^2 B$. For this case, Equation 26-6 gives

$$2\pi r E = \pi R^2 \frac{dB}{dt}$$

$$E = \frac{R^2}{2r}\frac{dB}{dt}$$

This example shows that Faraday's law implies that a changing magnetic field produces an electric field.

Example 26-4

An 80-turn coil has a radius of 5.0 cm and a resistance of 30 Ω. At what rate must a perpendicular magnetic field change to produce a current of 4.0 A in the coil?

The emf in the coil must equal the potential drop across its resistance:

$$\mathcal{E} = IR = (4.0\text{ A})(30\ \Omega) = 120\text{ V}$$

Since the plane of the coil is perpendicular to the field, the flux is

$$\phi_m = NBA = NB\pi r^2$$

According to Faraday's law, the magnitude of the induced emf equals the rate of change of this flux. Thus,

$$\mathcal{E} = 120\text{ V} = \frac{d\phi_m}{dt} = N\pi r^2 \frac{dB}{dt}$$

$$\frac{dB}{dt} = \frac{120\text{ V}}{(80)\pi(0.05\text{ m})^2} = 191\text{ T/s}$$

Example 26-5

A small coil of N turns has its plane perpendicular to a uniform magnetic field **B** as shown in Figure 26-7. The coil is connected to a current integrator ©, a device designed to measure the total charge passing through it. Find the charge passing through the coil if the coil is rotated through 180° about its diameter.

The flux through the coil is

$$\phi_m = NBA$$

where N is the number of turns and A is the area of the coil. If the coil is rotated through 180°, the flux reverses, so the total change in flux has the magnitude $2NBA$. While the flux is changing, there is an emf and therefore a current in the coil. The current is

$$I = \frac{\mathcal{E}}{R} = \frac{1}{R}\frac{d\phi_m}{dt}$$

where R is the total resistance of the coil. The total charge that passes through the coil is

$$Q = \int I\, dt = \frac{1}{R}\int d\phi_m$$

$$= \frac{\Delta\phi_m}{R} = \frac{2NBA}{R}$$

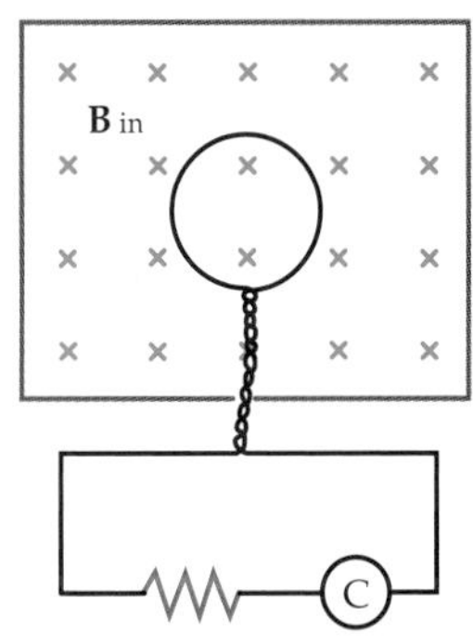

Figure 26-7 Flip-coil circuit for measuring the magnetic field **B**. When the coil is flipped over, the total charge that flows through the current integrator © is proportional to B.

The coil described in Example 26-5 is called a **flip coil.** It is used to measure magnetic fields. If a total charge Q passes through the coil when it is flipped 180°, the magnetic field B can be found from

$$B = \frac{RQ}{2NA} \qquad 26\text{-}7$$

Exercise

A flip coil of 40 turns has a radius of 3 cm and a resistance of 16 Ω. If the coil is turned through 180° in a magnetic field of 5000 G, how much charge passes through it? (Answer: 7.07 mC)

26-3 Lenz's Law

The negative sign in Faraday's law has to do with the direction of the induced emf. The direction of the induced emf and the induced current can be found from a general physical principle known as **Lenz's law:**

Lenz's law

> The induced emf and induced current are in such a direction as to oppose the change that produces them.

This statement of Lenz's law does not specify just what kind of change it is that causes the induced emf and current. It was purposefully left vague to cover a variety of conditions. A few illustrations will clarify this point.

Figure 26-8 shows a bar magnet moving toward a loop that has a resistance R. Since the magnetic field from the bar magnet is to the right, out of the north pole of the magnet, the movement of the magnet toward the loop tends to increase the flux through the loop to the right. (The magnetic field at the loop is stronger when the magnet is closer.) The induced current in the loop produces a magnetic field of its own. This induced current is in the direction shown, so the magnetic flux it produces is opposite that of the magnet. The induced magnetic field tends to *decrease* the flux through the loop. If the magnet were moved away from the loop, which would decrease the flux through the loop due to the magnet, the induced current would be in the opposite direction from that in Figure 26-8. In that case, the current would produce a magnetic field to the right, which would tend to increase the flux through the loop. As we might expect, moving the loop toward or away from the magnet has the same effect as moving the magnet. Only the relative motion is important.

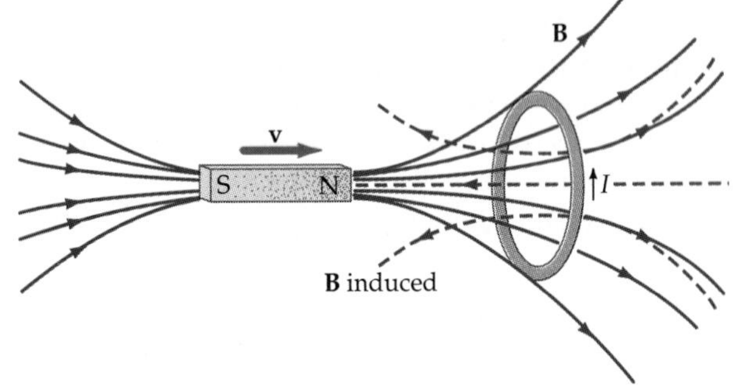

Figure 26-8 When the bar magnet is moving toward the loop, the emf induced in the loop produces a current in the direction shown. The magnetic field due to the induced current in the loop (indicated by the dashed lines) produces a flux that opposes the increase in flux through the loop due to the motion of the magnet.

Figure 26-9 shows the induced magnetic moment of the current loop when the magnet is moving toward it as in Figure 26-8. The loop acts like a small magnet with its north pole to the left and its south pole to the right. Since opposite poles attract and like poles repel, the induced magnetic moment of the loop exerts a force on the bar magnet to the left to oppose its motion toward the loop. Thus, we can express Lenz's law in terms of forces rather than flux. If the bar magnet is moved toward the loop, the induced current must produce a magnetic moment to oppose this change.

Note that Lenz's law is required by the law of conservation of energy. If the current in the loop in Figure 26-9 were opposite the direction shown, the induced magnetic moment of the loop would attract the magnet when it is moving toward the loop and cause it to accelerate toward the loop. If we begin with the magnet a great distance from the loop and give it a very slight push toward the loop, the force due to the induced current would be toward the loop, which would increase the velocity of the magnet. As the speed of the magnet increases, the rate of change of the flux would increase, thereby increasing the induced current. This would further increase the force on the magnet. Hence, the kinetic energy of the magnet and the rate at which Joule heat is produced in the loop (I^2R) would both increase with no source of energy. This would violate the law of conservation of energy.

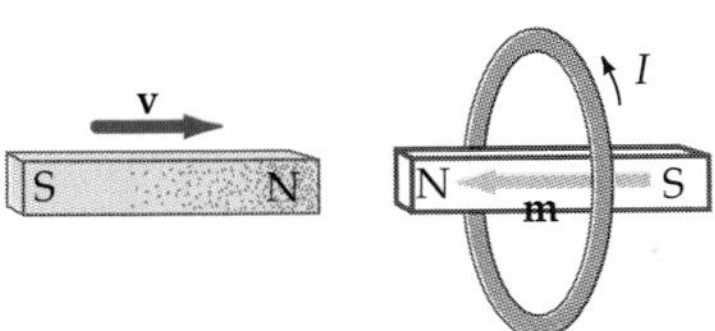

Figure 26-9 The magnetic moment of the loop (indicated by the outline magnet) due to the induced current is such as to oppose the motion of the bar magnet. Here the bar magnet is moving toward the loop, so the induced magnetic moment repels the bar magnet.

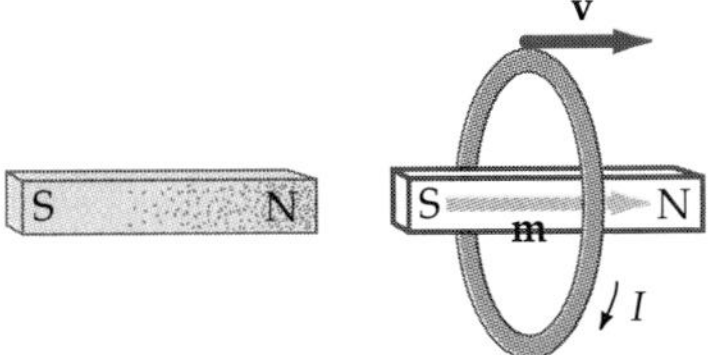

Figure 26-10 When the loop is moving away from the stationary bar magnet, the induced magnetic moment in the loop attracts the bar magnet, again opposing the relative motion.

In Figure 26-10, the bar magnet is at rest, and the loop is moving away from it. The induced current and magnetic moment are shown in the figure. In this case, the magnetic moment of the loop attracts the bar magnet, thus opposing the motion of the loop as required by Lenz's law.

In Figure 26-11, when the current in circuit 1 is changing, there is a changing flux through circuit 2. Suppose that the switch S in circuit 1 is initially open so that there is no current in the circuit (Figure 26-11*a*). When the switch is closed (Figure 26-11*b*), the current in circuit 1 does not reach its steady value $\mathcal{E}_1/R_1$ instantaneously but takes some time to change from zero to this value. During this time, while the current is increasing, the flux through circuit 2 is changing, and there is an induced current in that circuit in the direction shown. When the current in circuit 1 reaches its steady value, the flux through circuit 2 is no longer changing, so there is no induced current in circuit 2. An induced current in circuit 2 in the opposite direction appears momentarily when the switch in circuit 1 is opened (Figure 26-11*c*) and the current is decreasing to zero. It is important to understand that there is an induced emf *only while the flux is changing*. The emf does not depend on the magnitude of the flux, only on its rate of change. If there is a large, steady flux through a circuit, there is no induced emf.

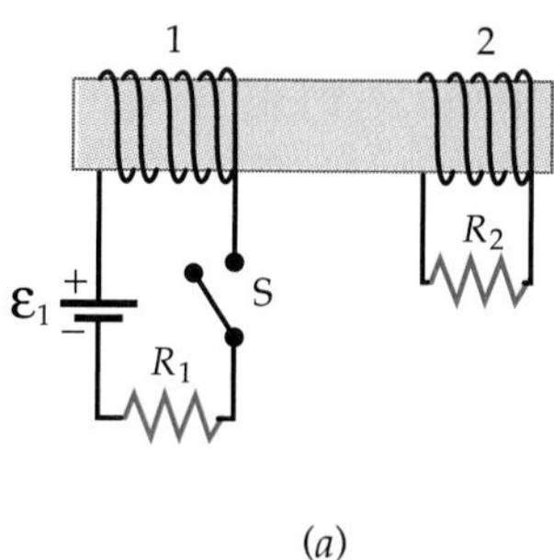

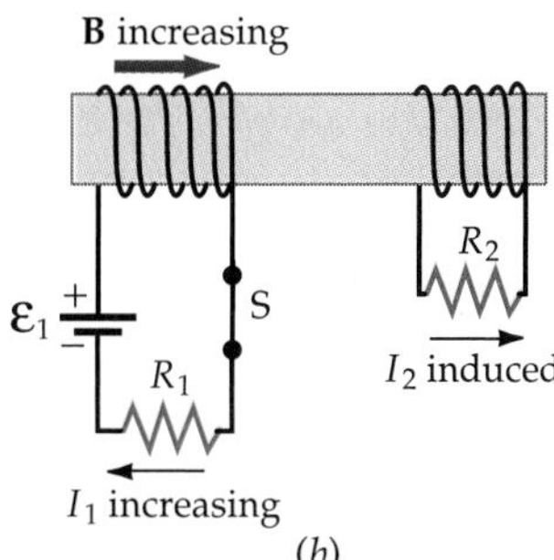

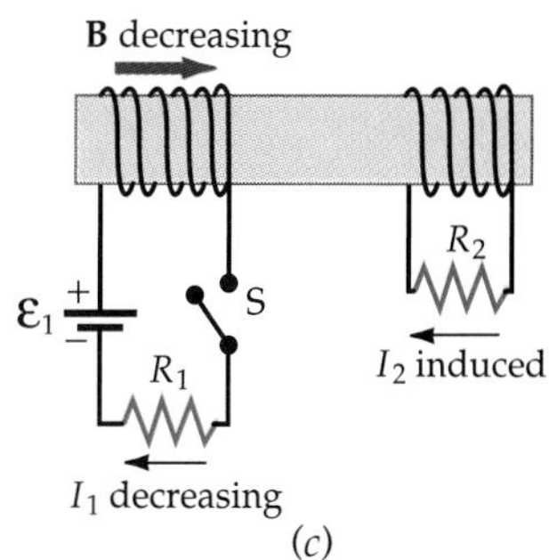

Figure 26-11 (*a*) Two adjacent circuits. (*b*) Just after the switch is closed, I_1 is increasing in the direction shown. The changing flux in circuit 2 induces the current I_2. The flux due to I_2 opposes the increase in flux due to I_1. (*c*) As the switch is opened, I_1 decreases and B decreases. The induced current I_2 then tends to maintain the flux in the circuit, opposing the change.

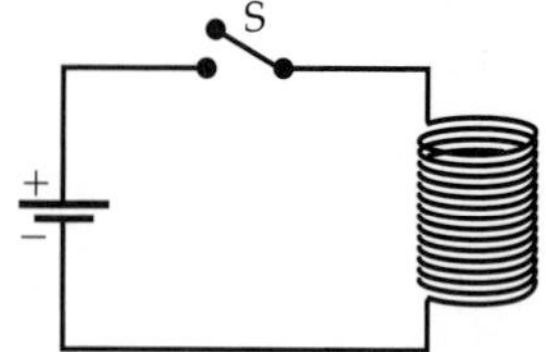

Figure 26-12 The coil with many turns of wire gives a large flux for a given current in the circuit. When the current changes, there is a large emf induced in the coil opposing the change.

For our next example, we consider the single, isolated circuit shown in Figure 26-12. When there is a current in the circuit, there is a magnetic flux through the coil due to its own current. When the current is changing, the flux in the coil is changing and there is an induced emf in the circuit. This *self-induced* emf opposes the change in the current. It is therefore called a **back emf.** Because of this self-induced emf, the current in a circuit cannot jump instantaneously from zero to some finite value or from some finite value to zero. Henry first noticed this effect when he was experimenting with a circuit consisting of many turns of a wire like that in Figure 26-12. This arrangement gives a large flux through the circuit for even a small current. Henry noticed a spark across the switch when he tried to break the circuit. Such a spark is due to the large induced emf that occurs when the current varies rapidly, as during the opening of the switch. In this case, the induced emf tries to maintain the original current. The large induced emf produces a large voltage drop across the switch as it is opened. The electric field between the contacts of the switch is large enough to tear electrons from the air molecules, causing dielectric breakdown. When the molecules in the air dielectric are ionized, the air conducts electric current in the form of a spark.

Question

1. Figure 26-13*a* shows a rectangular loop in a uniform magnetic field into the paper. Indicate the direction of the current induced in loop as it is twisted into the position shown in Figure 26-13*b*.

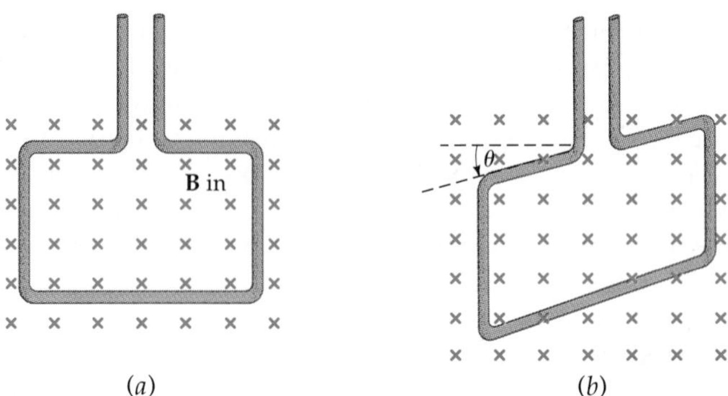

Figure 26-13 (*a*) A rectangular loop whose plane is perpendicular to a magnetic field **B**. (*b*) When the loop is twisted, the flux through it is changed and an emf is induced in it.

26-4 Motional Emf

Figure 26-14 shows a conducting rod sliding along conducting rails that are connected by a resistor. A uniform magnetic field **B** is directed into the paper. Since the area of the circuit increases as the rod moves to the right, the magnetic flux through the circuit is increasing. An emf is therefore induced in the circuit. Let ℓ be the separation of the rails and x be the distance from the left end of the rails to the rod at some time. The area enclosed by the circuit is then ℓx, and the magnetic flux through the circuit at this time is

$$\phi_m = BA = B\ell x$$

When the rod moves through a distance dx, the area enclosed by the circuit changes by $dA = \ell\, dx$ and the flux changes by $d\phi_m = B\ell\, dx$. The rate of change of the flux is

$$\frac{d\phi_m}{dt} = B\ell\frac{dx}{dt} = B\ell v$$

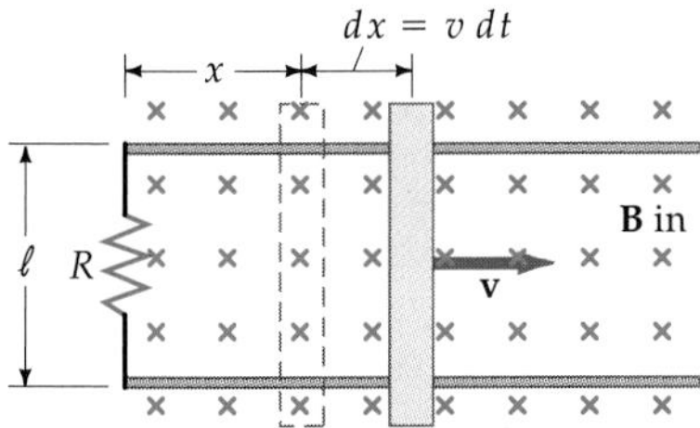

Figure 26-14 A conducting rod sliding on conducting rails in a magnetic field. As the rod moves to the right, the area of the circuit increases, so the magnetic flux through the circuit into the paper increases. An emf of magnitude $B\ell v$ is induced in the circuit, producing a counterclockwise current that produces flux out of the paper opposing the change.

where $v = dx/dt$ is the speed of the rod. The magnitude of the emf induced in this circuit is therefore

$$|\mathcal{E}| = \frac{d\phi_m}{dt} = B\ell v$$

The direction of the emf in this case is such as to produce a current in the counterclockwise sense. The flux produced by this induced current is out of the paper, opposing the increase in flux due to the motion of the rod. Because of the induced current, which is upward in the rod, there is a magnetic force on the rod of magnitude $I\ell B$. The direction of this force, obtained from the right-hand rule, is to the left, opposing the motion of the rod. If the rod is given some initial velocity **v** to the right and is then released, the force due to the induced current slows the rod down until it stops. To maintain the motion of the rod, an external force must be exerted on it to the right.

The emf in this case is called **motional emf.** More generally,

Motional emf defined

> Motional emf is any emf induced by the relative motion of a magnetic field and a current path.

Motional emf is induced in a conducting rod or wire moving in a magnetic field even when there is no complete circuit and thus no current.

Figure 26-15 shows an electron in a conducting rod that is moving through a uniform magnetic field directed into the paper. Because the electron is moving horizontally with the rod, there is a magnetic force on the electron that has a downward component of magnitude qvB. Because of this magnetic force, free electrons in the rod move downward, producing a net negative charge at the bottom and leaving a net positive charge at the top. The electrons continue to move down until the electric field produced by the separated charges exerts an upward force of magnitude qE on the electrons that balances the magnetic force qvB. In equilibrium, the electric field in the rod is thus

$$E = vB$$

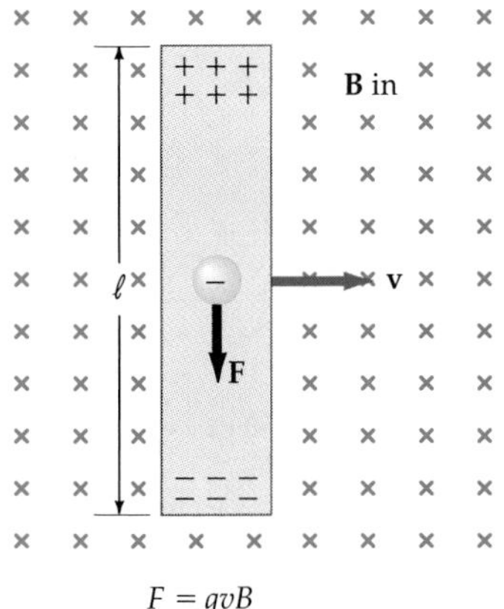

Figure 26-15 An electron in a conducting rod that is moving through a magnetic field experiences a magnetic force that has a downward component. Electrons move to the bottom of the rod, leaving the top of the rod positive. The charge separation produces an electric field of magnitude $E = vB$. The potential at the top of the rod is greater than that at the bottom by $E\ell = vB\ell$.

The potential difference across the rod is

$$\Delta V = E\ell = vB\ell$$

This potential difference equals the magnitude of the induced emf, that is, the motional emf:

Motional emf

$$|\mathcal{E}| = vB\ell \tag{26-8}$$

Motional emf is an example of Faraday's law in which we can understand the origin of the emf by considering the known forces acting on the electrons in the circuit. Figure 26-16 shows a typical electron in the conducting bar that is moving to the right in a magnetic field directed into the paper. The velocity of a typical electron $\mathbf{v}_e$ makes an angle θ with the rod as shown in the figure. It has a downward vertical component, $v_d = v_e \cos\theta$, the drift velocity, and a horizontal component $v = v_e \sin\theta$ equal to the speed of the rod. The magnetic force $\mathbf{f}_m = -e\mathbf{v}_e \times \mathbf{B}$ is in the plane of the figure and perpendicular to $\mathbf{v}_e$ as shown. It has the magnitude

$$f_m = ev_eB \tag{26-9}$$

If $\mathbf{f}_m$ were the only force acting on the electron, the electron could not stay in the rod as the rod moves to the right. The rod exerts a horizontal force $\mathbf{f}_r$ on the electron to balance the horizontal component of $\mathbf{f}_m$, which is $f_m \cos\theta$:

$$f_r = f_m \cos\theta \tag{26-10}$$

Since the magnetic force $\mathbf{f}_m$ is perpendicular to the motion of the electron, it does no work. The work done on the electron is done by the force $\mathbf{f}_r$. As the electron moves down the rod, the rod moves to the right so the electron moves along a diagonal path of length $S = \ell/\cos\theta$ shown in the figure. Since the component of $\mathbf{f}_r$ in the direction of the motion is $f_r \sin\theta$, the work done on the electron as it moves down the complete length of the rod is

$$W = f_r \sin\theta\ S = (f_m \cos\theta) \sin\theta\ S = f_m \sin\theta\ \ell$$

where $\ell = S\cos\theta$ is the length of the rod. Substituting ev_eB for f_m, we obtain

$$W = ev_eB \sin\theta\ \ell$$

But $v_e \sin\theta$ is the velocity of the rod v. The work done on the electron is thus

$$W = eBv\ell \tag{26-11}$$

The work per unit charge is the emf $B\ell v$, in agreement with our result obtained from Faraday's law.

Since the rod exerts a force f_r on each electron, each electron exerts an

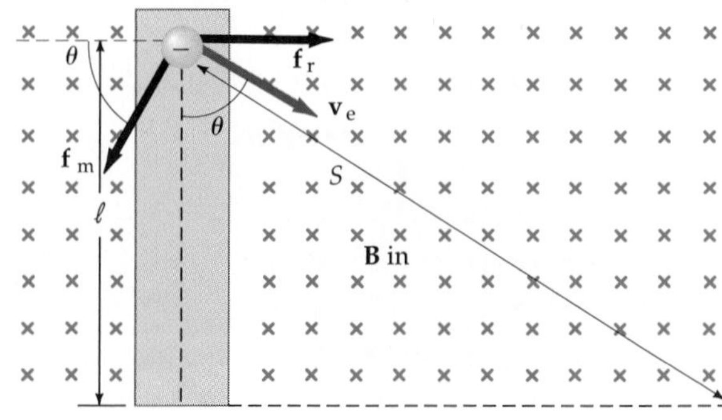

Figure 26-16 Forces on an electron in the rod of Figure 26-14. The electron's velocity $\mathbf{v}_e$ has a horizontal component v, the speed of the rod, and a vertical component v_d, its drift velocity along the rod. The magnetic force $\mathbf{f}_m$ is perpendicular to $\mathbf{v}_e$ and does no work. The rod exerts a horizontal force $\mathbf{f}_r$ on the electron with magnitude $f_m \cos\theta$. This force has a component in the direction of motion of the electron and therefore does work on it. The work per charge is equal to $B\ell v$.

equal and opposite force on the rod. This force is exerted to the left in Figure 26-16. If the rod has a cross-sectional area A and there are n free electrons per unit volume in the rod, the total number of electrons in the rod is $nA\ell$. The total force exerted on the rod is thus

$$F = nA\ell f_r = nA\ell f_m \cos\theta = nA\ell e v_e B \cos\theta \qquad \text{26-12}$$

But $v_e \cos\theta = v_d$, the drift velocity of the electrons, and $nAev_d = I$, the total current in the rod. Substituting I for $nAev_e \cos\theta$ in Equation 26-12, we obtain for the total force exerted by the electrons on the rod

$$F = I\ell B \qquad \text{26-13}$$

which is the same as Equation 24-4 for the magnetic force exerted on a current-carrying segment. To keep the rod moving with a constant speed v, an external force of magnitude $F = I\ell B$ must be exerted to the right. The power input of this force is the force times the speed:

$$P = Fv = I\ell Bv$$

Setting the power equal to the rate of production of Joule heat in the resistor I^2R, we obtain

$$IB\ell v = I^2R$$

or

$$B\ell v = IR$$

Thus, the induced emf $\mathcal{E} = B\ell v$ equals the potential drop across the resistor $\Delta V = IR$.

Exercise

A rod 40 cm long moves at 12 m/s in a plane perpendicular to a magnetic field of 3000 G. Its velocity is perpendicular to its length. Find the emf induced in the rod. (Answer: 1.44 V)

Example 26-6

In Figure 26-14, let $B = 0.6$ T, $v = 8$ m/s, $\ell = 15$ cm, and $R = 25\ \Omega$, and assume that the resistance of the rods and rails is negligible. Find (*a*) the induced emf in the circuit, (*b*) the current in the circuit, (*c*) the force needed to move the rod with constant velocity, and (*d*) the power dissipated in the resistor.

(*a*) The emf is given by Equation 26-8:

$$\mathcal{E} = Bv\ell = (0.6\text{ T})(8\text{ m/s})(0.15\text{ m}) = 0.72\text{ V}$$

(*b*) Since the total resistance in the circuit is 25 Ω, the current is

$$I = \frac{\mathcal{E}}{R} = \frac{0.72\text{ V}}{25\ \Omega} = 28.8\text{ mA}$$

(*c*) The force needed to move the rod with constant velocity is equal and opposite to the force exerted by the magnetic field on the rod. The magnitude of this force is

$$F = IB\ell = (0.0288\text{ A})(0.6\text{ T})(0.15\text{ m}) = 2.59\text{ mN}$$

(*d*) The power dissipated in the resistor is

$$P = I^2R = (0.0288\text{ A})^2(25\ \Omega) = 20.7\text{ mW}$$

We can check our answer in (*d*) by computing the power input of the force found in (*c*):

$$P = Fv = (2.59 \times 10^{-3}\text{ N})(8\text{ m/s}) = 2.07 \times 10^{-2}\text{ W} = 20.7\text{ mW}$$

Example 26-7

In Figure 26-14, the rod has a mass m. At time $t = 0$, the rod is moving with initial speed v_0, and the external force acting on it is removed. Find the speed of the rod as a function of time.

The current induced in the circuit is $\mathcal{E}/R$, where $\mathcal{E} = B\ell v$ is the induced emf. The magnitude of the magnetic force acting on the rod is therefore

$$F = IB\ell = \frac{\mathcal{E}}{R}B\ell$$

$$= \frac{B\ell v}{R}B\ell = \frac{B^2\ell^2 v}{R}$$

This force is directed opposite the direction of motion. If we take the positive direction as the direction of the initial velocity, the force is negative and Newton's second law for the rod is

$$F = ma = m\frac{dv}{dt}$$

$$-\frac{B^2\ell^2 v}{R} = m\frac{dv}{dt}$$

Separating the variables and integrating, we obtain

$$\frac{dv}{v} = -\frac{B^2\ell^2}{mR}dt$$

$$\ln v = -\frac{B^2\ell^2}{mR}t + C$$

where C is an arbitrary constant of integration. Then

$$v = e^C e^{-(B^2\ell^2/mR)t} = v_0 e^{-(B^2\ell^2/mR)t}$$

where $v_0 = e^C$ is the velocity at time $t = 0$.

26-5 Eddy Currents

In the examples we have discussed, the currents produced by a changing magnetic flux were set up in clearly defined circuits. Often, though, a changing flux sets up circulating currents, called **eddy currents,** in a piece of bulk metal such as the core of a transformer. (We will discuss transformers in Section 28-6.) The heat produced by eddy currents constitutes a power loss in the transformer.

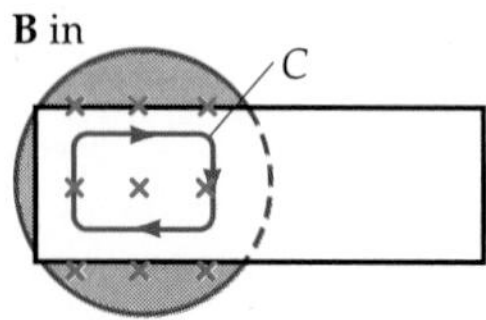

Figure 26-17 Eddy currents. When the magnetic field through a metal slab is changing, an emf is induced in any closed loop in the metal such as loop C. The induced emf causes a current in the loop.

Consider a conducting slab between the pole faces of an electromagnet as shown in Figure 26-17. If the magnetic field **B** between the pole faces is changing with time (as it will be if the current in the magnet windings is alternating current), the flux through any closed loop in the slab will be changing. For example, the flux through the closed loop C in the figure is the product of the magnetic field B and the area enclosed by the loop. If B varies, the flux will vary, and there will be an induced emf around the loop C. Since loop C is in a conductor, there will be a current along it equal to the emf divided by the resistance of the loop. In the figure, just one of the many closed loops that will contain currents if the magnetic field between the pole faces varies has been indicated.

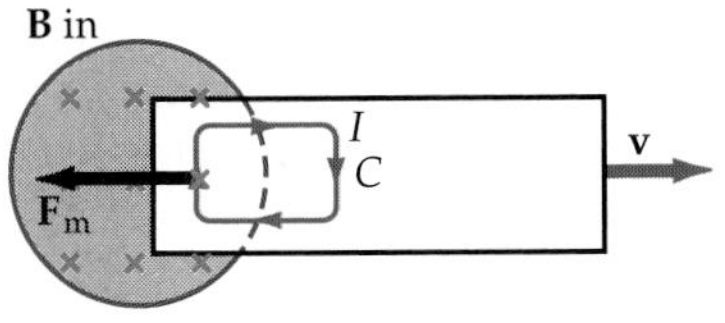

Figure 26-18 Demonstration of eddy currents. When the metal sheet is pulled to the right, there is a magnetic force to the left on the induced current opposing the motion.

The existence of eddy currents can be demonstrated by pulling a copper or aluminum sheet from between the poles of a strong magnet (Figure 26-18). Part of the area enclosed by loop C in this figure is in the magnetic field, and part is outside the field. As the sheet is pulled to the right, the flux through loop C decreases (assuming that the flux into the paper is positive). According to Faraday's law and Lenz's law, a clockwise current will be induced around the loop. Since this current is directed upward in the region between the pole faces, the magnetic field exerts a force on the current to the left, opposing motion of the sheet. You can feel this force on a conducting sheet if you try to pull it quickly through a strong magnetic field.

Eddy currents are usually unwanted because the heat they produce is a power loss. In addition, this heat must be dissipated. The power loss can be reduced by increasing the resistance of the possible paths of the eddy currents. In Figure 26-19, for instance, the conducting slab in Figure 26-17 is laminated; that is, it is made up of small strips of metal glued together. Because of the insulating glue between the strips, the eddy currents are essentially confined to the strips. Thus, the large eddy-current loops are broken up, and the power loss is greatly reduced. Similarly, if the sheet in Figure 26-19 has slots cut in it, as in Figure 26-20, the eddy currents are lessened and the force is greatly reduced. Figure 26-21 shows a common lecture demonstration of a magnetic brake. The apparatus consists of a sheet of metal fixed to the end of a pivoted rod that swings from the pivot like a pendulum. When the sheet swings through the gap between two poles of a magnet, the oscillation rapidly damps out. If the magnetic field is strong enough, the motion is stopped suddenly during the first encounter of the sheet with the magnetic field. If slots are cut in the sheet, as in Figure 26-21*c*, the damping is considerably reduced.

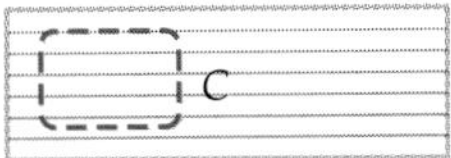

Figure 26-19 Eddy currents in a metal slab can be reduced by constructing the slab from small strips of metal glued together. The resistance of the closed loop C is now larger because of the insulating glue between the strips.

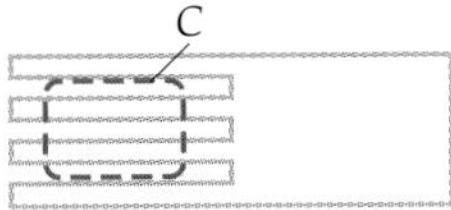

Figure 26-20 If the metal sheet of Figure 26-17 has slots cut, the eddy currents are greatly reduced because of the lack of good conducting paths.

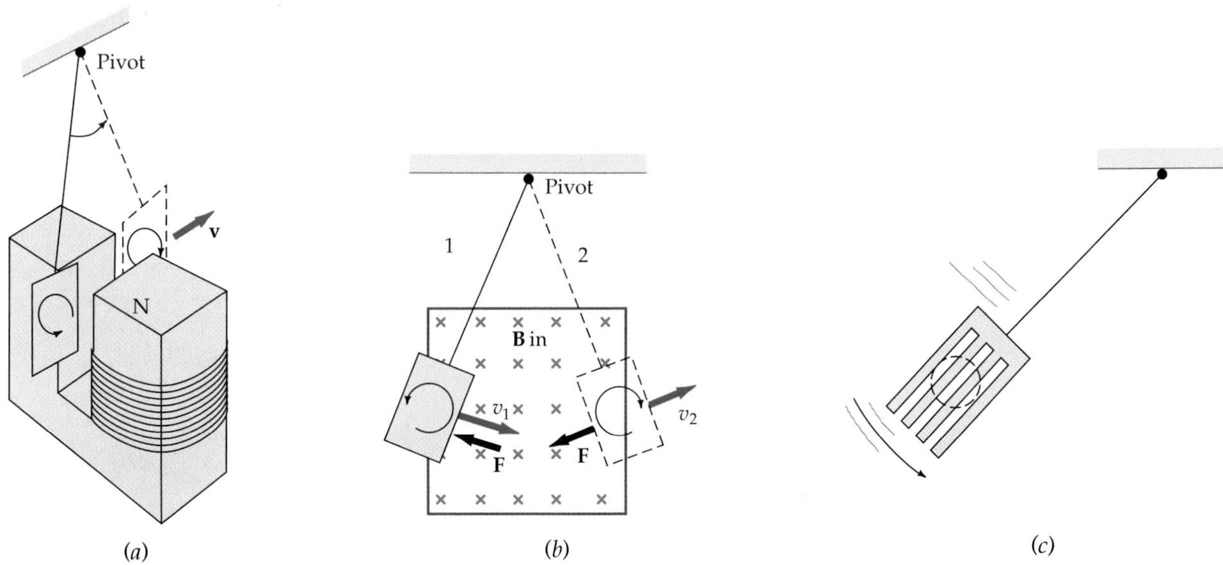

Figure 26-21 Lecture demonstration of a magnetic brake. (*a*) A pendulum with a metallic sheet for a bob is released from a large angle and swings between the poles of an electromagnet. (*b*) The bob is slowed considerably by the magnetic force on the induced eddy currents as the sheet enters or leaves the field. If the magnetic field is strong enough, the bob is stopped between the poles of the magnet. (*c*) If the sheet has slots cut in it, the eddy currents are greatly reduced and the bob swings through the magnetic field with little slowing.

Eddy currents are not always undesired. They are often used to damp unwanted oscillations. For example, sensitive mechanical balance scales used to weigh small masses tend to oscillate back and forth around their equilibrium reading many times. Therefore, such scales are usually designed with a small piece of metal that moves between the poles of a magnet as the scales oscillate. The resulting eddy currents dampen the oscillations so that equilibrium is reached more quickly. Another use of eddy currents is to provide *magnetic braking* for rapid transit cars. A large electromagnet is positioned in the car over the rails. When the magnet is energized by a current in its windings, eddy currents are induced in the rails by the motion of the magnet, and the resulting magnetic forces provide a drag force on the car that stops it.

Question

2. A bar magnet is dropped inside a long vertical tube. If the tube is made of metal, the magnet quickly reaches a terminal velocity, but if the tube is made of cardboard, it does not. Explain.

26-6 Generators and Motors

Most electrical energy used today is produced by electric generators in the form of alternating current (ac). A simple **generator** of alternating current is a coil rotating in a uniform magnetic field as shown in Figure 26-22. The ends of the coil are connected to rings called slip rings that rotate with the coil. Electrical contact is made with the coil by stationary graphite brushes in contact with the rings. When the line perpendicular to the plane of the coil makes an angle θ with a uniform magnetic field **B**, as shown in the figure, the magnetic flux through the coil is

$$\phi_{\mathrm{m}} = NBA \cos \theta \tag{26-14}$$

where N is the number of turns in the coil and A is the area of the coil. When the coil is mechanically rotated, the flux through it will change, and an emf will be induced in the coil according to Faraday's law. If the initial angle is δ, the angle at some later time t is given by

$$\theta = \omega t + \delta$$

where ω is the angular frequency of rotation. Substituting this expression for θ into Equation 26-14, we obtain

$$\phi_m = NBA \cos (\omega t + \delta) = NBA \cos (2\pi f t + \delta)$$

The emf in the coil will then be

$$\mathcal{E} = -\frac{d\phi_{\mathrm{m}}}{dt} = -NBA \frac{d}{dt} \cos (\omega t + \delta) = +NBA\omega \sin (\omega t + \delta) \tag{26-15}$$

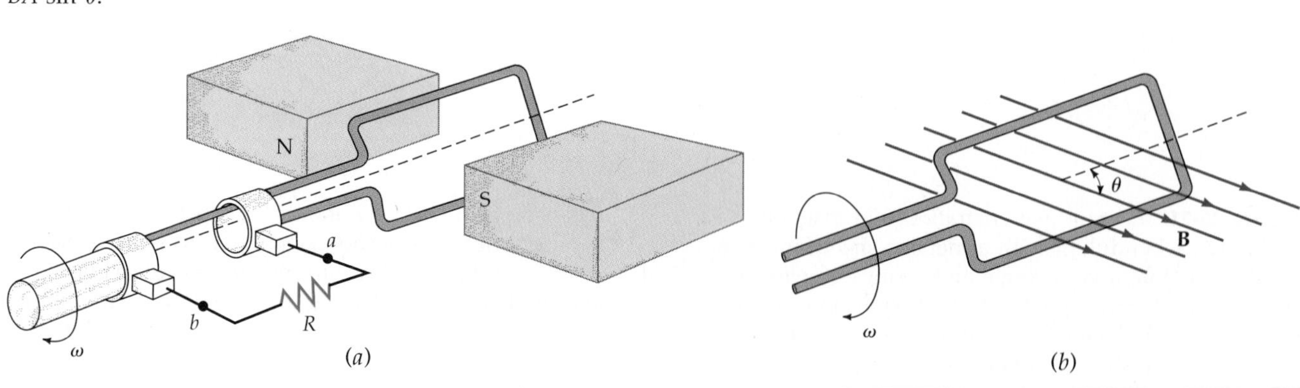

Figure 26-22 (*a*) An ac generator. A coil rotating with constant angular frequency ω in a magnetic field **B** generates a sinusoidal emf. Energy from a waterfall or a steam turbine is used to rotate the coil to produce electrical energy. The emf is supplied to an external circuit by the brushes in contact with the rings. (*b*) At this instant, the normal to the plane of the coil makes an angle θ with the magnetic field and the flux is $BA \sin \theta$.

A generator under construction in China. Shown here is the rotor of Turbo Generator #1 at Dahua Power Station being lowered into place.

This can be written

$$\mathcal{E} = \mathcal{E}_{max} \sin (\omega t + \delta) \quad 26\text{-}16$$

where

$$\mathcal{E}_{max} = NBA\omega \quad 26\text{-}17$$

is the maximum value of the emf. We can thus produce a sinusoidal emf in a coil by rotating it with constant frequency in a magnetic field. In this source of emf, the mechanical energy of the rotating coil is converted into electric energy. The mechanical energy usually comes from a waterfall or a steam turbine. Although practical generators are considerably more complicated, they work on the same principle that an alternating emf is produced in a coil rotating in a magnetic field, and they are designed so that the emf produced is sinusoidal.

The same coil in a magnetic field that can be used to generate an alternating emf can also be used as an ac **motor.** Instead of mechanically rotating the coil to generate an emf, we apply an alternating current to the coil from another ac generator as shown in Figure 26-23. (In circuit diagrams, an ac generator is represented by the symbol ⊙̃.) We learned in Chapter 24 that a current loop in a magnetic field experiences a torque that tends to rotate the loop such that its magnetic moment points in the direction of **B** and the plane of the loop is perpendicular to **B**. If direct current were supplied to the coil in Figure 26-23, the torque on the coil would change directions when the coil rotates past its equilibrium position, which is when the plane of the coil is vertical in the figure. The coil would then oscillate about its equilibrium position, eventually coming to rest there with its plane vertical. However, if the direction of the current is reversed just as the coil passes the vertical position, the torque does not change direction but continues to rotate the coil in the same direction. As the coil rotates in the magnetic field, a back emf is generated that tends to counter the emf that supplies the current. When the motor is first turned on, there is no back emf and the current is very large, being limited only by the resistance in the circuit. As the motor begins to rotate, the back emf increases and the current decreases.

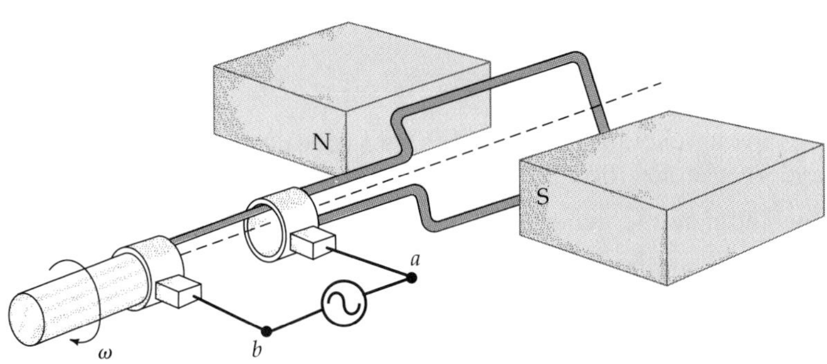

Figure 26-23 When alternating current is supplied to the coil of Figure 26-22, the coil becomes a motor. As the coil rotates, a back emf is generated, limiting the current.

(a)

(b)

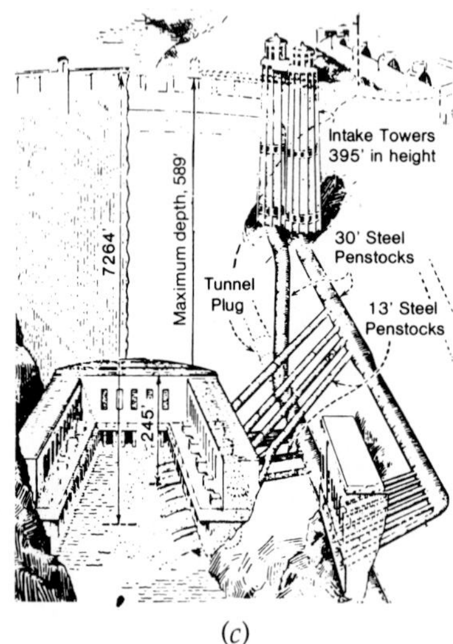

(c)

(a) River-level view of Hoover Dam with the Nevada wing of its power plant on the left and the Arizona wing on the right. The mechanical energy of falling water drives the turbines shown (b) for the generation of electricity.
(c) Schematic drawing of Hoover Dam showing the intake towers and pipes (penstocks) that carry the water to the generators below.

Example 26-8

A 250-turn coil has an area of 3 cm^2. If it rotates in a magnetic field of 0.4 T at 60 Hz, what is $\mathcal{E}_{max}$?

From Equation 26-17, we have

$$\mathcal{E}_{max} = NBA\omega = NBA(2\pi f) = (250)(0.4\text{ T})(3 \times 10^{-4}\text{ m}^2)(2\pi)(60\text{ Hz})$$
$$= 11.3\text{ V}$$

Example 26-9

The windings of a dc motor have a resistance of 1.5 Ω. When the motor is connected across 40 V and running at full speed, the current in the windings is 2.0 A. (*a*) What is the back emf when the motor is running at full speed? (*b*) What is the initial current in the windings at start-up when the back emf is negligible?

(*a*) The potential drop across the windings is

$$V = IR = (2.0\text{ A})(1.5\ \Omega) = 3\text{ V}$$

Since the total potential drop across the motor is 40 V, the back emf is 40 V − 3 V = 37 V.

(*b*) When the motor is first turned on, the back emf is negligible. Since the potential drop across the motor is still 40 V, the current is

$$I = \frac{40\text{ V}}{1.5\ \Omega} = 26.7\text{ A}$$

Questions

3. Does the sinusoidal nature of the emf produced by an ac generator depend on the size or shape of the coil?
4. How could an ac generator be used to generate a nonsinusoidal emf?
5. When a generator delivers electric energy to a circuit, where does the energy come from?
6. A motor sometimes burns out when its load is suddenly increased. Why?

26-7 Inductance

Self Inductance

The magnetic flux through a circuit can be related to the current in that circuit and the currents in other, nearby circuits. (We will assume that there are no permanent magnets around.) Consider a coil carrying a current I. The current produces a magnetic field that could, in principle, be calculated from the Biot–Savart law. Since the magnetic field at every point in the neighborhood of the coil is proportional to I, the magnetic flux through the coil is also proportional to I:

$$\phi_m = LI \tag{26-18}$$

Self inductance defined

where L is a constant called the **self inductance** of the coil. The self inductance depends on the geometric shape of the coil. The SI unit of inductance is the **henry** (H). From Equation 26-18, we can see that the unit of inductance equals the unit of flux, the weber, divided by the unit of current, the ampere:

$$1\ \text{H} = 1\ \frac{\text{Wb}}{\text{A}} = 1\frac{\text{T}\cdot\text{m}^2}{\text{A}}$$

In principle, the self inductance of any coil or circuit can be calculated by assuming a current I, finding the flux ϕ_m, and using $L = \phi_m/I$. In actual practice, the calculation is very difficult. However, there is one case, that of a tightly wound solenoid, for which the self inductance can be calculated directly. The magnetic field inside a tightly wound solenoid of length ℓ carrying a current I is given by Equation 25-10:

$$B = \mu_0 nI$$

where $n = N/\ell$ is the number of turns per unit length. If the solenoid has a cross-sectional area A, the flux through the N turns is

$$\phi_m = NBA = n\ell BA = \mu_0 n^2 A\ell I$$

As expected, the flux is proportional to the current I. The proportionality constant is the self inductance:

$$L = \frac{\phi_m}{I} = \mu_0 n^2 A\ell \tag{26-19}$$

Self inductance of a solenoid

The self inductance is proportional to the square of the number of turns per unit length n and to the volume $A\ell$. Thus, like capacitance, self inductance depends only on geometric factors. From the dimensions of Equation 26-19, we can see that μ_0 can be expressed in henrys per meter:

$$\mu_0 = 4\pi \times 10^{-7}\ \text{H/m}$$

Example 26-10

Find the self inductance of a solenoid of length 10 cm, area 5 cm^2, and 100 turns.

We can calculate the self inductance in henrys from Equation 26-19 if we put all the quantities in SI units. Using $\ell = 0.1$ m, $A = 5 \times 10^{-4}$ m^2, $n = N/\ell = (100\ \text{turns})/(0.1\ \text{m}) = 1000$ turns/m, and $\mu_0 = 4\pi \times 10^{-7}$ H/m, we obtain

$$L = \mu_0 n^2 A\ell = (4\pi \times 10^{-7}\ \text{H/m})(10^3\ \text{turns/m})^2(5 \times 10^{-4}\ \text{m}^2)(0.1\ \text{m})$$
$$= 6.28 \times 10^{-5}\ \text{H}$$

When the current in a circuit is changing, the magnetic flux due to the current is also changing, so an emf is induced in the circuit. Since the self inductance of a circuit is constant, the change in flux is related to the change in current by

$$\frac{d\phi_{\mathrm{m}}}{dt} = \frac{d(LI)}{dt} = L\frac{dI}{dt}$$

According to Faraday's law, we have

$$\mathcal{E} = -\frac{d\phi_{\mathrm{m}}}{dt} = -L\frac{dI}{dt} \qquad \text{26-20}$$

Thus, the self-induced emf is proportional to the rate of change of the current.

Exercise

At what rate must the current in the solenoid of Example 26-10 change to induce an emf of 20 V? (Answer: 3.18×10^5 A/s)

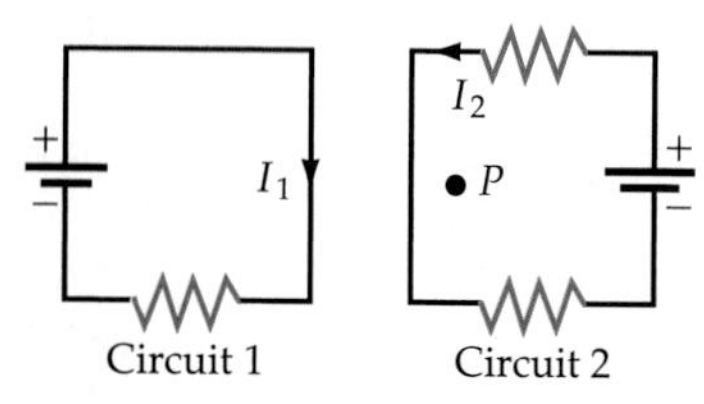

Figure 26-24 Two adjacent circuits. The magnetic field at point P is partly due to current I_1 and partly due to I_2. The flux through either circuit is the sum of two terms, one proportional to I_1 and the other to I_2.

Mutual Inductance

When two or more circuits are close to each other, as in Figure 26-24, the magnetic flux through one circuit depends not only on the current in that circuit but also on the current in the nearby circuits. Let I_1 be the current in circuit 1 on the left in Figure 26-24 and I_2 be that in circuit 2 on the right. The magnetic field at some point P has a part due to I_1 and a part due to I_2. These fields are proportional to the currents that produce them. We can therefore write the flux through circuit 2, $\phi_{\mathrm{m}2}$, as the sum of two parts, one proportional to the current I_1 and the other proportional to the current I_2:

Mutual inductance defined

$$\phi_{\mathrm{m}2} = L_2 I_2 + M_{12} I_1 \qquad \text{26-21}a$$

where L_2 is the self inductance of circuit 2 and M_{12} is called the **mutual inductance** of the two circuits. The mutual inductance depends on the geometrical arrangement of the two circuits. For instance, if the circuits are far apart, the flux through circuit 2 due to the current I_1 will be small and the mutual inductance will be small. An equation similar to Equation 26-21*a* can be written for the flux through circuit 1:

$$\phi_{\mathrm{m}1} = L_1 I_1 + M_{21} I_2 \qquad \text{26-21}b$$

where L_1 is the self inductance of circuit 1.

Figure 26-25 shows a long, narrow, tightly wound solenoid inside another tightly wound solenoid of equal length but larger radius. For this situation we can actually calculate the mutual inductance of the two solenoids. Let ℓ be the length of both solenoids, and let the inner solenoid have N_1 turns and radius r_1 and the outer solenoid have N_2 turns and radius r_2. We will first calculate the mutual inductance M_{12} by assuming that the inner solenoid carries a current I_1 and finding the magnetic flux $\phi_{\mathrm{m}2}$ due to this current through the outer solenoid. The magnetic field due to the current in the inner solenoid is constant in the space within the solenoid and has magnitude

$$B_1 = \mu_0 n_1 I_1 \qquad \text{26-22}$$

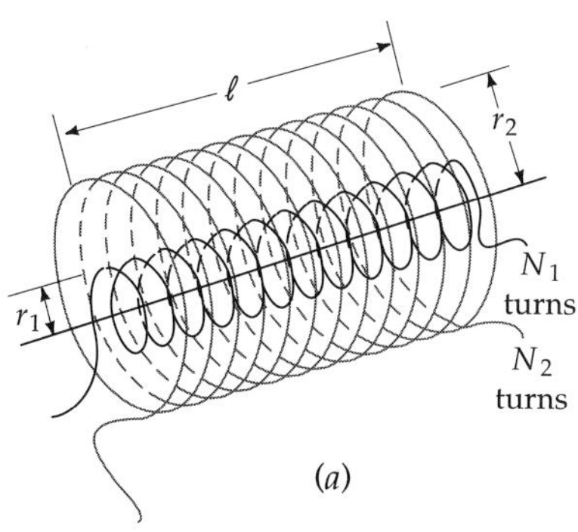

(*b*)

Figure 26-25 (*a*) A long, narrow solenoid inside a second solenoid of the same length. A current in either solenoid produces magnetic flux in the other. (*b*) Tesla coil illustrating the geometry of the wires in part (*a*). Such a device functions as a transformer (Chapter 28). Here, low-voltage alternating current in the outer winding is transformed into a higher-voltage alternating current in the inner winding. Induced alternating voltage from the changing fields is great enough to light the bulb above the coil.

Outside the inner solenoid, the magnetic field is zero. The flux through the outer solenoid due to this magnetic field is therefore

$$\phi_{m2} = N_2B_1(\pi r_1^2) = n_2\ell B_1(\pi r_1^2) = \mu_0 n_2 n_1 \ell(\pi r_1^2)I_1$$

Note that the area used to compute the flux through the outer solenoid is not the area of that solenoid, πr_2^2, but rather is the area of the inner solenoid, πr_1^2, because there is no magnetic field outside the inner solenoid. The mutual inductance M_{12} is thus

$$M_{12} = \frac{\phi_{m2}}{I_1} = \mu_0 n_2 n_1 \ell \pi r_1^2 \qquad \text{26-23}$$

We will now calculate M_{21} by finding the magnetic flux through the inner solenoid due to a current I_2 in the outer solenoid. When the outer solenoid carries a current I_2, there is a uniform magnetic field B_2 inside that solenoid given by Equation 26-22 with I_2 replacing I_1 and n_2 replacing n_1:

$$B_2 = \mu_0 n_2 I_2$$

The magnetic flux through the inner solenoid is then

$$\phi_{m1} = N_1B_2(\pi r_1^2) = n_1\ell B_2(\pi r_1^2) = \mu_0 n_1 n_2 \ell(\pi r_1^2)I_2$$

The area used here is also πr_1^2 because it is the cross-sectional area of the inner solenoid, and the magnetic field is uniform everywhere inside that solenoid. The mutual inductance M_{21} is thus

$$M_{21} = \frac{\phi_{m1}}{I_2} = \mu_0 n_1 n_2 \ell(\pi r_1^2) \qquad \text{26-24}$$

Note that Equations 26-23 and 26-24 are the same; that is, $M_{21} = M_{12}$. It can be shown that this is a general result. We will therefore drop the subscripts for mutual inductance and simply write M.

Question

7. How would the self inductance of a solenoid be changed if the same length of wire were wound onto a cylinder of the same diameter but twice the length? If twice as much wire were wound onto the same cylinder?

26-8 *LR* Circuits

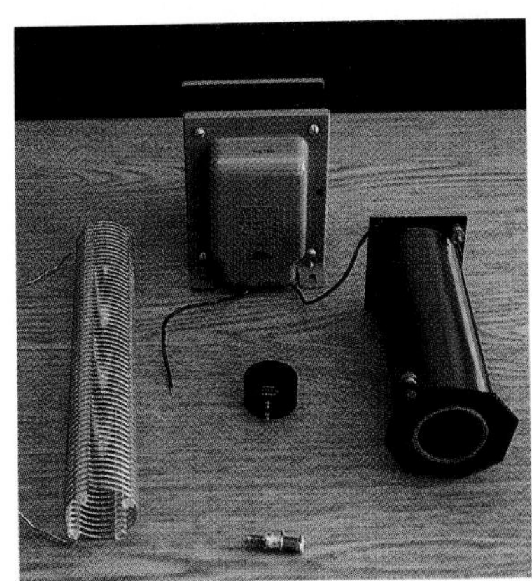
Various inductors.

As we have seen, the self inductance of a circuit prevents the current from rising or falling instantaneously. Circuits containing coils or solenoids of many turns have a large self inductance. Such a coil or solenoid is called an **inductor.** The symbol for an inductor is —ꝏ—. We can often neglect the self inductance of the rest of the circuit compared with that of an inductor.

A circuit containing a resistor and an inductor is called an ***LR* circuit.** Since all circuits have resistance and self inductance, the analysis of an *LR* circuit can be applied to some extent to all circuits. All circuits also have some capacitance between parts of the circuits at different potentials. We will consider the effects of capacitance in Chapter 28 when we study ac circuits. Here we will neglect capacitance to simplify the analysis and to focus on the effects of inductance.

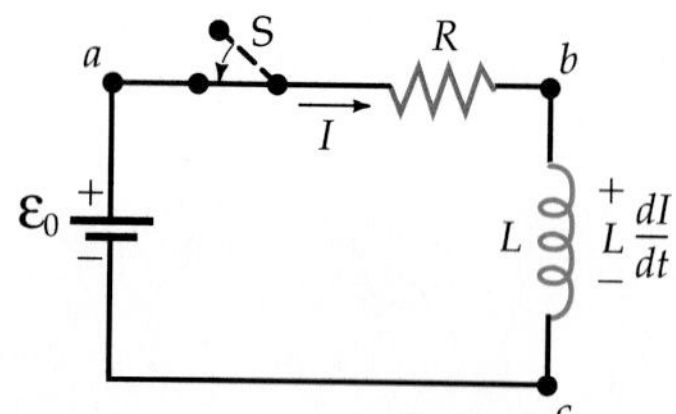

Figure 26-26 A typical *LR* circuit. Just after the switch S is closed, the current begins to increase in the circuit and a back emf of magnitude $L\,dI/dt$ is generated in the inductor. The potential drop across the resistor IR plus the potential drop across the inductor equals the emf of the battery.

Figure 26-26 shows an *LR* circuit in which an inductance L and a resistance R are in series with a battery of emf $\mathcal{E}_0$ and a switch S. We will assume that the resistance R includes the resistance of the inductor coil and that the inductance of the rest of the circuit is negligible compared with that of the inductor. The switch is initially open, so there is no current in the circuit. Just after the switch is closed, the current is still zero, but it is changing at a rate dI/dt, and there is a back emf of magnitude $L\,dI/dt$ in the inductor. In the circuit diagram, plus and minus signs have been placed on the inductor to indicate the direction of the emf when the current is increasing, that is, when dI/dt is positive. A short time after the switch is closed, there is a current I in the circuit and a potential drop IR across the resistor. Applying Kirchhoff's loop rule to this circuit gives

$$\mathcal{E}_0 - IR - L\frac{dI}{dt} = 0 \qquad \text{26-25}$$

We can understand many of the features of the current in this circuit from this equation without solving the equation. Initially (just after the switch is closed) the current is zero and the back emf $L\,dI/dt$ equals the emf of the battery $\mathcal{E}_0$. The initial rate of change of the current is, from Equation 26-25,

$$\left(\frac{dI}{dt}\right)_0 = \frac{\mathcal{E}_0}{L} \qquad \text{26-26}$$

As the current increases, the potential drop IR increases, and the rate of change of the current decreases. After a short time, the current has reached a positive value I, and the rate of change of the current is

$$\frac{dI}{dt} = \frac{\mathcal{E}_0}{L} - \frac{IR}{L}$$

At this time the current is still increasing, but its rate of increase is less than at $t = 0$. The final value of the current can be obtained by setting dI/dt equal to zero. The final value of the current is thus

$$I_f = \frac{\mathcal{E}_0}{R} \qquad \text{26-27}$$

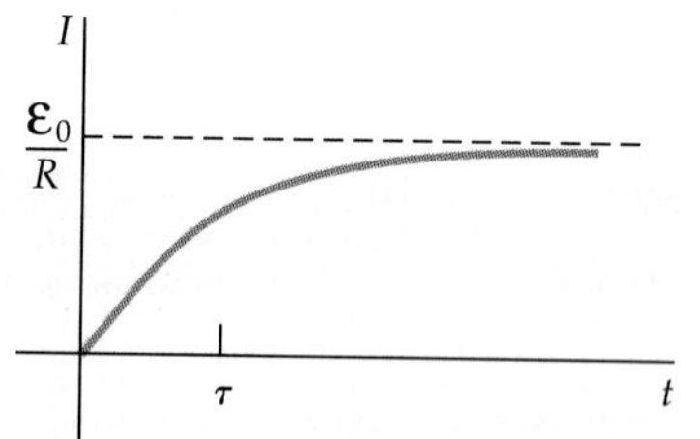

Figure 26-27 Current versus time in an *LR* circuit. At a time $t = \tau = L/R$, the current is at 63% of its maximum value $\mathcal{E}_0/R$.

Figure 26-27 shows the current in this circuit as a function of time. This figure is similar to that for the charge on a capacitor when the capacitor is charged in an *RC* circuit (Figure 23-22).

Equation 26-25 is of the same form as Equation 23-22 for the charging of a capacitor and can be solved in the same way. The result is

$$I = \frac{\mathcal{E}_0}{R}(1 - e^{-Rt/L}) = \frac{\mathcal{E}_0}{R}(1 - e^{-t/\tau}) = I_f(1 - e^{-t/\tau}) \qquad \text{26-28}$$

where

$$\tau = \frac{L}{R} \qquad \text{26-29}$$

is the **time constant** of the circuit. The larger the self inductance L or the smaller the resistance R, the longer it takes for the current to build up. Note that the product of the time constant L/R and the initial slope $\mathcal{E}_0/L$ equals the final current $I_f = \mathcal{E}_0/R$. If the rate of increase of the current were constant at its original rate, the current would reach its maximum value in a time $t = \tau$. However, the rate of increase of the current is not constant, but decreases with time. At a time τ, the current is $0.63I_f$.

Example 26-11

A coil of self inductance 5.0 mH and a resistance of 15.0 Ω is placed across the terminals of a 12-V battery of negligible internal resistance. (*a*) What is the final current? (*b*) What is the current after 100 μs?

(*a*) The final current is

$$I_f = \frac{\mathcal{E}_0}{R} = \frac{12\text{ V}}{15\ \Omega} = 0.800\text{ A}$$

(*b*) The time constant for this circuit is

$$\tau = \frac{L}{R} = \frac{5 \times 10^{-3}\text{ H}}{15\ \Omega} = 333\ \mu\text{s}$$

The current after 100 μs is given by Equation 26-28:

$$I = \frac{\mathcal{E}_0}{R}(1 - e^{-t/\tau}) = (0.800\ A)(1 - e^{-100/333}) = (0.800\ A)(1 - 0.741)$$
$$= 0.207\text{ A}$$

In Figure 26-28, the circuit has an additional switch that allows us to remove the battery, and an additional resistor R_1 to protect the battery so that it is not shorted when both switches are momentarily closed. If both switches are initially open, and we close switch S_1, the current builds up in the circuit just as discussed, except that the total resistance is now $R_1 + R$ and the final current is $\mathcal{E}_0/(R + R_1)$. Suppose that this switch has been closed for a long time compared with the time constant so that the current is approximately steady at its final value, which we will call I_0. Switch S_2 is then closed and switch S_1 is opened to remove the battery from consideration completely. Let us choose the time when switch S_2 is closed to be $t = 0$. We now have a circuit with just a resistor and an inductor (loop *abcd*) carrying an initial current I_0. Applying Kirchhoff's loop rule to this circuit gives

$$-IR - L\frac{dI}{dt} = 0$$

or

$$\frac{dI}{dt} = -\frac{R}{L}I \qquad 26\text{-}30$$

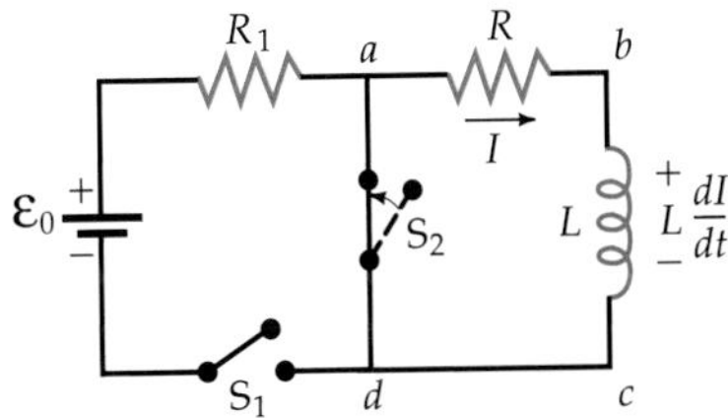

Figure 26-28 An *LR* circuit with two switches so that the battery can be removed from the circuit. After the current in the inductor reaches its maximum value with S_1 closed, S_2 is closed and S_1 is opened. The current then decreases exponentially with time.

Note that to apply Kirchhoff's loop rule, we have assumed a direction for I and assigned plus and minus signs to the inductor in Figure 26-28 for positive dI/dt, as usual. The solution of the equation then gives us the correct algebraic signs for these quantities, which indicate whether or not our assumption was correct. In this case, the direction of I is already known from the initial conditions. Equation 26-30 then tells us that, since I is positive, dI/dt is negative, indicating that the current is decreasing. For a negative dI/dt, the induced emf, $-L\,dI/dt$, is in the direction of the current and opposes the decrease in current in agreement with Lenz's law. Equation 26-30

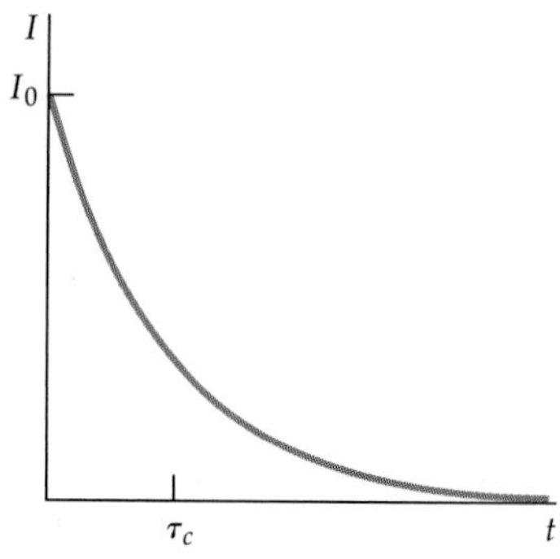

Figure 26-29 Current versus time for the circuit in Figure 26-28. The current decreases exponentially with time.

is of the same form as Equation 23-16 for the discharge of a capacitor. It can be solved by direct integration. We will omit the details and merely state the solution. The current I is given by

$$I = I_0 e^{-Rt/L} = I_0 e^{-t/\tau} \qquad \text{26-31}$$

where $\tau = L/R$ is the time constant. Figure 26-29 shows the current as a function of time.

Exercise

What is the time constant of a circuit of resistance 85 Ω and inductance 6 mH? (Answer: 70.6 μs)

Example 26-12

Find the total heat produced in the resistor R in Figure 26-28 when the current in the inductor decreases from its initial value of I_0 to 0.

The rate of heat production is

$$P = \frac{dW}{dt} = I^2R$$

where I is given by Equation 26-31. In a time dt, the heat produced is

$$dW = I^2R\,dt$$

The total energy dissipated as heat in the resistor is thus

$$W = \int_0^\infty I^2R\,dt = \int_0^\infty I_0^2 e^{-2Rt/L} R\,dt = I_0^2 R \int_0^\infty e^{-2Rt/L}\,dt$$

The integration can be done by substituting $x = 2Rt/L$. Then

$$dt = \frac{L}{2R}\,dx$$

and

$$W = I_0^2 R \frac{L}{2R} \int_0^\infty e^{-x}\,dx = \frac{1}{2} L I_0^2$$

since the value of the integral is 1. This energy was originally stored in the inductor. In the next section, we will see that, in general, the energy stored in an inductor carrying a current I is $\frac{1}{2}LI^2$.

26-9 Magnetic Energy

In Section 21-4, we saw that it takes work to charge a capacitor and that a charged capacitor stores energy given by

$$U = \frac{1}{2}QV = \frac{1}{2}CV^2 = \frac{1}{2}\frac{Q^2}{C}$$

where Q is the charge on either plate, V is the potential difference between the plates, and C is the capacitance. We also saw that this energy can be considered to be stored in the electric field between the plates and found that, in general, when there is an electric field E in space, the electric energy density (the electric energy per unit volume) is

$$\eta = \tfrac{1}{2}\epsilon_0 E^2$$

There is a similar expression for the energy in a magnetic field. It takes work to produce a current in an inductor. We can see this by multiplying each term in Equation 26-25 by the current I and then rearranging:

$$\mathcal{E}_0 I = I^2 R + LI\frac{dI}{dt} \qquad 26\text{-}32$$

The term $\mathcal{E}_0 I$ is the power output of the battery. The term I^2R is the power dissipated as heat in the resistance of the circuit. The term $LI\,dI/dt$ is the rate at which energy is put into the inductor. If U_m is the energy in the inductor, then

$$\frac{dU_m}{dt} = LI\frac{dI}{dt}$$

We can find the total energy in the inductor by integrating this equation from time $t = 0$, when the current is zero, to $t = \infty$, when the current has reached its final value I_f:

$$U_m = \int dU_m = \int_0^{I_f} LI\,dI = \tfrac{1}{2}LI_f^2$$

The energy stored in an inductor carrying a current I is thus given by

$$U_m = \tfrac{1}{2}LI^2 \qquad 26\text{-}33$$

Energy stored in an inductor

This is in agreement with the result calculated in Example 26-12 that the heat produced in the resistor is $\tfrac{1}{2}LI^2$ when the current in the inductor decreases from I to 0.

Exercise

How much energy is stored in the inductor of Example 26-11 when the final current has been attained? (Answer: 1.6×10^{-3} J)

In the process of producing a current in an inductor, a magnetic field is created in the space within the inductor coil. The work done in producing a current in an inductor can be thought of as the work required to create a magnetic field. That is, we can think of the energy stored in an inductor as energy stored in the magnetic field of the inductor. For the special case of a solenoid, the magnetic field is related to the current I and the number of turns per unit length n by

$$B = \mu_0 n I$$

and the self inductance is given by Equation 26-19:

$$L = \mu_0 n^2 A\ell$$

where A is the cross-sectional area and ℓ is the length. Substituting $B/\mu_0 n$ for I and $\mu_0 n^2 A\ell$ for L in Equation 26-33, we obtain

$$U_m = \frac{1}{2}LI^2 = \frac{1}{2}\mu_0 n^2 \ell A\left(\frac{B}{\mu_0 n}\right)^2 = \frac{B^2}{2\mu_0}\ell A$$

The quantity $A\ell$ is the volume of the space within the solenoid containing the magnetic field.

The energy per unit volume is the **magnetic energy density** η_m:

$$\eta_m = \frac{B^2}{2\mu_0} \qquad 26\text{-}34$$

Magnetic energy density

Although we have obtained Equation 26-34 by considering the special case of the magnetic field in a solenoid, the result is general. That is, whenever there is a magnetic field in space, the magnetic energy per unit volume is given by Equation 26-34.

Example 26-13

A certain region of space contains a magnetic field of 200 G and an electric field of 2.5×10^6 N/C. Find (*a*) the total energy density and (*b*) the energy in a cubical box of side 12 cm.

(*a*) The electrical energy density is

$$\eta_e = \tfrac{1}{2}\varepsilon_0 E^2 = (0.5)(8.85 \times 10^{-12}\ \text{C}^2/\text{N}\cdot\text{m}^2)(2.5 \times 10^6\ \text{N/C})^2 = 27.7\ \text{J/m}^3$$

and the magnetic energy density is

$$\eta_m = \frac{B^2}{2\mu_0} = \frac{(0.02\ \text{T})^2}{2(4\pi \times 10^{-7}\ \text{N/A}^2)} = 159\ \text{J/m}^3$$

The total energy density is therefore

$$\eta = \eta_e + \eta_m = 27.7\ \text{J/m}^3 + 159\ \text{J/m}^3 = 187\ \text{J/m}^3$$

(*b*) The volume in a cube of side 12 cm is

$$V = (0.12\ \text{m})^3 = 1.73 \times 10^{-3}\ \text{m}^3$$

The total energy in this volume is then

$$U = \eta V = (187\ \text{J/m}^3)(1.73 \times 10^{-3}\ \text{m}^3) = 0.324\ \text{J}$$

Summary

1. For a magnetic field that is constant in space, the magnetic flux through a coil is the product of the component of the magnetic field that is perpendicular to the plane of the coil and the area of the coil. In general, for a coil of N turns, the magnetic flux through the coil is

$$\phi_m = \int NB_n\, dA$$

The SI unit of magnetic flux is the weber:

$$1\ \text{Wb} = 1\ \text{T}\cdot\text{m}^2$$

2. When the magnetic flux through a circuit changes, there is an emf induced in the circuit given by Faraday's law

$$\mathcal{E} = \oint_C \mathbf{E}\cdot d\boldsymbol{\ell} = -\frac{d\phi_m}{dt}$$

The induced emf and induced current are in such a direction as to oppose the change that produces them. This is known as Lenz's law.

3. The emf induced in a conducting wire or rod of length ℓ moving with velocity **v** perpendicular to a magnetic field **B** is called motional emf. Its magnitude is

$$|\mathcal{E}| = \frac{d\phi_m}{dt} = B\ell v$$

4. Circulating currents that are set up in bulk metal by a changing magnetic flux are called eddy currents.

5. A coil rotating with angular frequency ω in a magnetic field generates an alternating emf given by

$$\mathcal{E} = \mathcal{E}_{\text{max}} \sin(\omega t + \delta)$$

where $\mathcal{E}_{\text{max}} = NBA\omega$ is the maximum value of the emf.

6. The magnetic flux through a circuit is related to the current in the circuit by

$$\phi_{\text{m}} = LI$$

where L is the self inductance of the circuit, which depends on the geometrical arrangement of the circuit. The SI unit of inductance is the henry (H):

$$1\ \text{H} = 1\ \text{Wb/A} = 1\ \text{T}\cdot\text{m}^2/\text{A}$$

The self inductance of a tightly wound solenoid of length ℓ and area A with n turns per unit length is given by

$$L = \frac{\phi_{\text{m}}}{I} = \mu_0 n^2 A\ell$$

If there is another circuit nearby carrying current I_2, there is additional flux through the first circuit

$$\phi_{\text{m}} = MI_2$$

where M is the mutual inductance, which depends on the geometrical arrangement of the two circuits.

7. When the current in an inductor is changing, the emf induced in the inductor is given by

$$\mathcal{E} = -\frac{d\phi_{\text{m}}}{dt} = -L\frac{dI}{dt}$$

8. In an LR circuit, which consists of a resistance R, an inductance L, and a battery of emf $\mathcal{E}_0$ in series, the current does not reach its maximum value I instantaneously but rather takes some time to build up. If the current is initially zero, its value at some later time t is given by

$$I = \frac{\mathcal{E}_0}{R}(1 - e^{-Rt/L}) = \frac{\mathcal{E}_0}{R}(1 - e^{-t/\tau})$$

where $\tau = L/R$ is the time constant of the circuit.

9. The energy stored in an inductor carrying a current I is

$$U_{\text{m}} = \tfrac{1}{2}LI^2$$

This energy can be considered to be stored in the magnetic field inside the inductor. In general, the magnetic energy density (the energy per unit volume) of a magnetic field B is given by

$$\eta_{\text{m}} = \frac{B^2}{2\mu_0}$$

The Aurora

Syun-Ichi Akasofu
Geophysical Institute, University of Alaska, Fairbanks

The aurora appears as an awesome but silent light display in the far northern and far southern night skies (Figure 1). It often appears as a faint greenish-white arc, but it is actually a long, shimmering, undulating curtain of glowing bands and rays in several colors. The intensity of auroral light varies. At its brightest, the colors can be dramatic and beautiful. The lower edge of the auroral curtain is located at an altitude of 100 km, and the upper edge can extend to as high as 1000 km above the surface of the earth within two ring-shaped zones lying between 60 and 75 degrees of latitude (Figures 2*a*, 2*b*), each centered on one of the earth's magnetic poles. These ring-shaped belts are called the auroral ovals.

Auroral light was once thought to be sunlight reflected by ice crystals in the sky. However, in 1888 Anders Jonas Ångström showed that auroral light differs from sunlight; many of the wavelengths present in sunlight are absent in auroral light (Figure 3). A spectrum analogous to that of auroral light can result from applying a high voltage to electrodes inserted into a vacuum glass tube containing a gas such as neon. The electrons flow from the negative electrode to the positive one. They collide with neon atoms, exciting them and ultimately causing them to emit light. Similarly, auroral light is the result of an electrical discharge process and is emitted by atoms and molecules in the upper atmosphere (Figure 4) as they are stuck by high-speed electrons.

Syun-Ichi Akasofu, Director of the Geophysical Institute of the University of Alaska, Fairbanks, is the author of hundreds of publications on the aurora. His research work on the aurora has earned him national and international recognition and honors. Since becoming Director of the Geophysical Institute, Professor Akasofu has concentrated his efforts on establishing it as the key research center in the Arctic.

Figure 1 The aurora borealis.

(*a*)

Figure 2 (*a*) The southern aurora, as photographed by astronaut Robert Overmyer using a 35-mm camera.

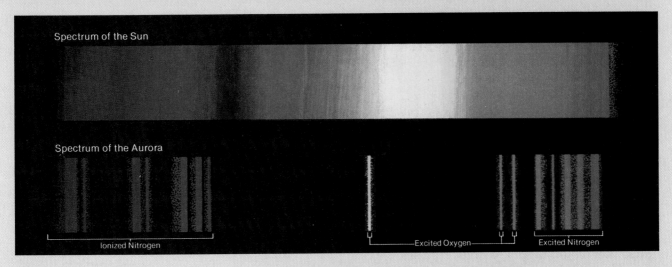

Figure 3 A comparison of the visible light spectra of the sun and the aurora.

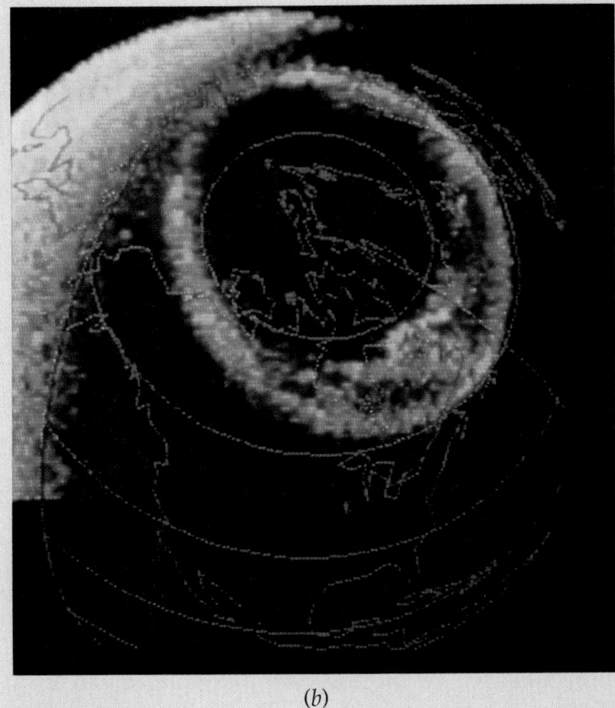

(*b*)

(*b*) A computer-enhanced image of the northern aurora, taken by satellite at a distance of 3 earth radii.

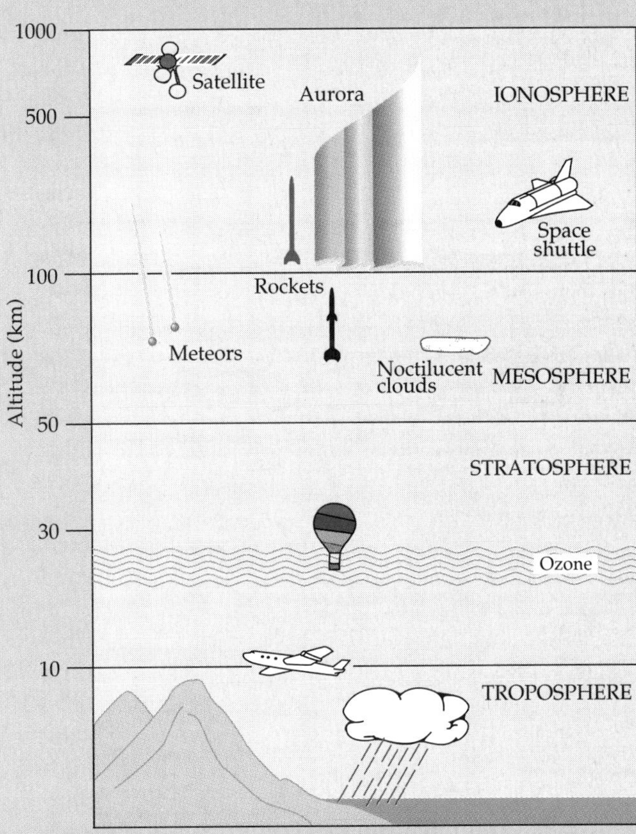

Figure 4 A schematic representation of the earth's atmosphere showing some artificial and some natural features and their altitudes. Auroras form in and above the ionosphere, the layer of the atmosphere that contains many free electrons and ions that have been created by the effects of solar ultraviolet and x-ray emissions.

Continued

In trying to understand the process that powers the auroral discharge, it is useful to consider an analogy with an electric generator. In a generator, an electrical current is produced in a conductor by moving it through a magnetic field. Auroral power is generated in a similar way: A stream of charged particles flowing outward from the sun (called the "solar wind") functions as a conductor, and the earth itself provides the magnetic field (see Figure 5).

The outer layer of the sun's atmosphere, the corona, consists of gas (primarily hydrogen) that is so hot that electrically neutral atoms break down into positive ions (primarily protons) and electrons. The solar wind, streaming out from the corona, is a tenuous hot plasma of these charged particles. Moving at a speed between 300 and 1000 km/s, it travels away from the sun in all directions to the limit of the solar system. Magnetic field lines in the solar wind behave like elastic strings. When the solar wind blows outward, it carries the sun's magnetic field lines and stretches them out. On its way, it confines the earth's magnetic field into a comet-shaped cavity, which is called the magnetosphere (Figure 6). The outer boundary of this cavity is called the magnetopause.

At distances of about 10 earth radii from the surface of our planet, the strength of the earth's magnetic field (30×10^{-5} G) is equal to that of the sun's magnetic field stretched out by the solar wind. Both magnetic fields are interconnected on the boundary of the comet-shaped magnetosphere. Here charged particles in the solar wind blow across the interconnected field. This motion is equivalent to motion of an electrical conductor through a magnetic field. Looking at the earth from the sun, one would see the protons in the solar wind deflected (by the $e\mathbf{v} \times \mathbf{B}$ force) toward the left and the electrons deflected toward the right, creating the positive and negative terminals of the auroral generator (Figure 7*a*). The magnetosphere is filled with tenuous plasma. This enables current to flow between the terminals. The current flows from the positive terminal, spiraling down magnetic field lines, into the ionosphere (the atmosphere's electrically conducting layer), through the ionosphere across the polar region, and up magnetic field lines from the ionosphere to the negative terminal. This is the primary electrical discharge circuit.

On what is called the "morning side" of the magnetosphere, current flows into the inner edge of the auroral oval as part of the primary discharge circuit and is conducted to the outer edge of the oval. Since the region outside the oval is not very conductive, some current finds its way back out along magnetic field lines, giving rise to a parallel, secondary circuit. A corresponding process occurs on the "evening side" (Figure 7*b*). Thus there are a pair of electrical currents (upward and downward) flowing along magnetic field lines in both the morning and evening sides of the magnetosphere. The upward current in both sectors is carried by downward-flowing electrons, which collide with atoms and molecules in the atmosphere, exciting them and causing them to emit light. This is the part of the discharge circuit that produces auroral light (exactly as occurs in a neon tube, as described earlier).

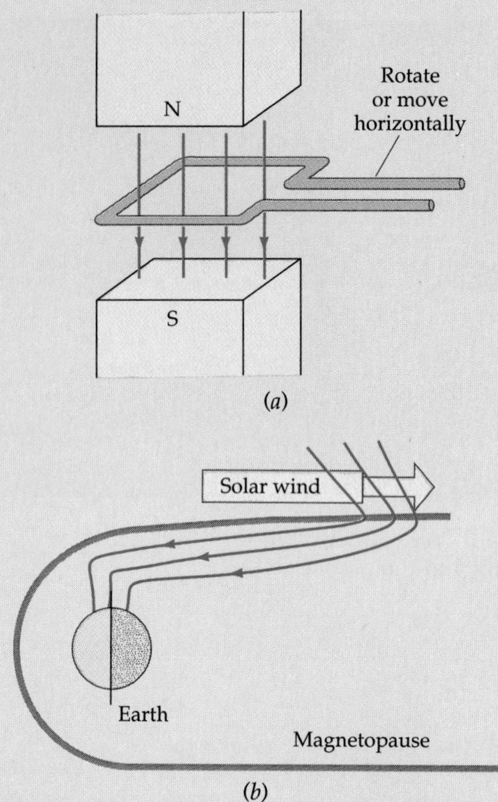

Figure 5 (*a*) Schematic of a conventional generator. (*b*) The interaction of the solar wind with the earth's magnetic field creates a naturally occurring generator.

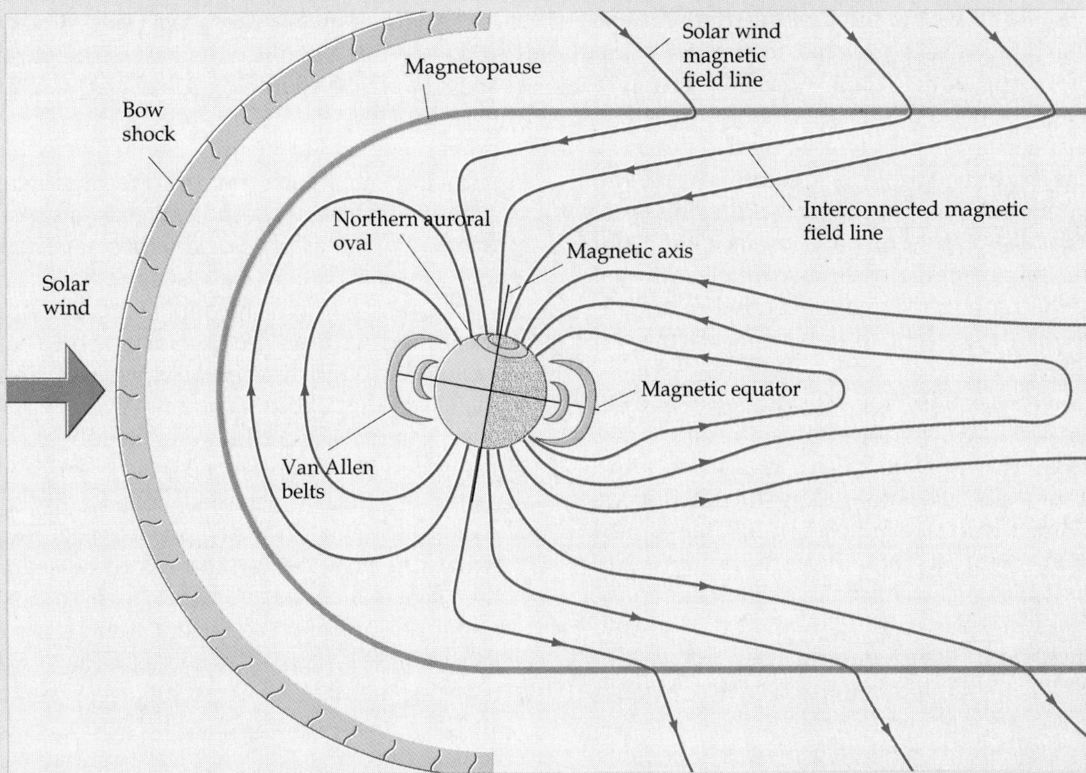

Figure 6 The earth's magnetosphere. The solar wind confines the earth's magnetic field to a comet-shaped zone that has our planet as its nucleus. The distance between the earth and the sunward side of the magnetosphere is about 10 earth radii. The magnetosphere is drawn out into a very long tail (not shown) that stretches for more than 1000 earth radii away from the sun (to the right of this illustration).

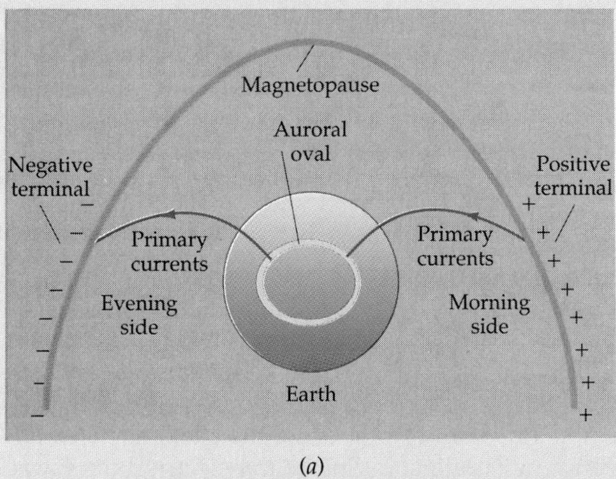

(*a*)

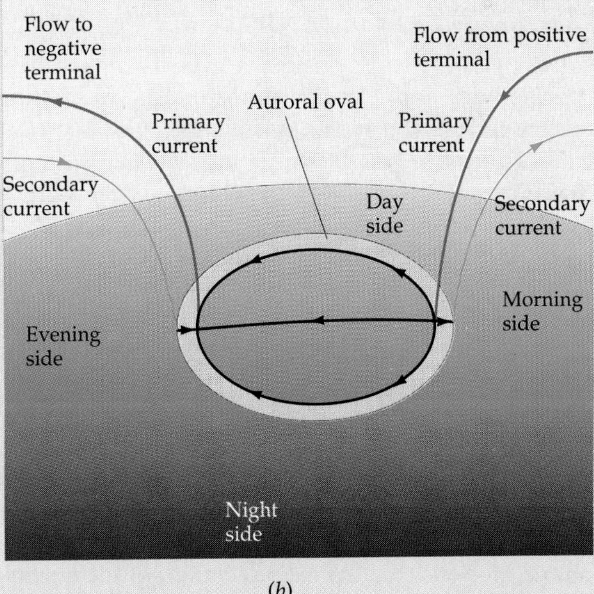

(*b*)

Figure 7 (*a*) View of the magnetosphere and the earth from above. The positive and negative terminals (morning and evening side magnetopause, respectively), together with the primary currents, are shown. (*b*) The primary and secondary electrical discharge circuits. The currents across the polar cap and along the auroral oval depend on the atmosphere's conductivity.

Continued

What is the basis for the curtain shape of auroras? It is thought to be related to the very thin sheetlike forms in which these electrons flow into the upper atmosphere, but the mechanism by which these thin sheets arise is not yet known. The lower limit of the auroral curtain is determined by the depth of penetration of the current-carrying electrons. Collisions with atoms and molecules in the upper atmosphere deplete much of the electrons' energy by the time they have descended to an altitude of about 100 km above the earth, so few descend farther.

Two factors explain the variations in auroral color. First, the color produced by an electrical discharge process varies from gas to gas and with the energy of the electrons producing the excitation. Second, the chemical composition of the atmosphere differs with height. These factors together explain the variations in auroral color. In the ionosphere, the atmosphere consists mainly of atomic oxygen, produced when energy from ultraviolet sunlight splits O_2 molecules apart. When oxygen atoms are excited, a greenish-white light is emitted (the most common auroral color). More energetic electrons penetrating farther into the atmosphere collide with neutral nitrogen molecules, producing auroras with red-violet or pink borders and rippled edges. Ionized nitrogen molecules produce a blue-violet light. Visible light is only a small portion of auroral emissions; x-rays, ultraviolet, and infrared radiation are also produced by the aurora.

To understand the movement observed in most auroral displays, consider an analogy with the image produced on the screen of a cathode-ray tube, such as that in a television set. The screen corresponds here to the upper atmosphere. The coating on the back of the screen emits light when the electron beam from the electron gun strikes it; this light is seen from the front of the screen as the image. Similarly, as just described, the ionosphere fluoresces when it is struck by the current-carrying electron sheets. Just as the impact point of an electron beam in a cathode-ray tube can change, causing movement of the image on the screen, so too can the auroral electron sheet rapidly shift, causing the auroral curtain to move, often violently. In the case of both the cathode-ray tube and the aurora, changes in a magnetic field and/or an electric field modulate the behavior of the electron beam. So, it is changes in the magnetic field, rather than atmospheric motions, that cause movement in the auroral curtain.

A large power-generating plant produces about 1000 MW on a continuous basis. The aurora generates about 1 to 10 million MW (1 to 10 TW), equivalent to 1000 to 10,000 large power plants. This power, generated by the interaction of solar wind with the earth's magnetosphere, fluctuates, sometimes considerably, because the strength of the solar wind and its magnetic field varies depending on the level of solar flare activity. A solar flare is associated with an eruption in the sun's corona that in turn is associated with a "gusty" solar wind that radiates rapidly through interplanetary space and reaches the earth after about 40 hours. As this "gusty" solar wind interacts with the magnetosphere, the power generated can be enhanced a thousandfold. In these instances, the ring-shaped belts of the aurora expand from the polar regions toward the equator, making it possible for auroras to be seen south of the U.S.–Canadian border. Auroral displays are also much brighter than usual after a solar flare, and the top of the auroral curtain extends to greater altitudes, sometimes allowing the upper portion of the aurora borealis to be seen as far south as Mexico and central Europe.

The enhanced electrical discharge currents associated with the gusty solar wind produce intensely fluctuating magnetic fields. When such fields are recorded, we say that a magnetic storm is in progress. The electric currents heat the upper atmosphere, causing upward motion of the lower, denser atmosphere and thus increasing the density at higher altitudes. This, in turn, increases the friction between orbiting satellites and the atmosphere, causing reductions in orbital altitudes. Satellites have even been known to fall out of orbit after major magnetic storms.

We now have a partial understanding of auroras: of the origin of the ring-shaped auroral belts around the geomagnetic poles; of the processes that power the gigantic electrical discharges that create them; of the amount of power generated during a display; of the reasons for the fluctuations in that power, and of the relationship between auroral and solar activity that is manifested by various transient solar processes including solar flares. As the twentieth century draws to a close, the challenge is to advance our understanding of the electrical discharge process that underlies this beautiful phenomenon, a powerful natural generator.

Suggestions for Further Reading

Akasofu, Syun-Ichi: "The Dynamic Aurora," *Scientific American*, May 1989, p. 90.

How the "solar wind" of charged particles interacts with the earth's magnetic field to produce the aurora, why the aurora appears to move and change, and where its power comes from.

Kondo, Herbert: "Michael Faraday," *Scientific American*, October 1953, p. 90.

This article describes Faraday's experiments and his revolutionary concept of the electromagnetic field.

Shamos, Morris H.: "Electromagnetic Induction and Laws of Electrolysis—Michael Faraday," in *Great Experiments in Physics*, Henry Holt and Company, New York, 1959. Reprinted by Dover, 1987.

This is Faraday's account of his discovery of electromagnetic induction, with editorial annotations for clarity and a biographical sketch.

Shamos, Morris H.: "Lenz's Law—Heinrich Lenz," in *Great Experiments in Physics*, Henry Holt and Company, New York, 1959. Reprinted by Dover, 1987.

This is Lenz's account of some experiments on electromagnetic induction resulting in a rule for determining the direction of an induced current, with editorial annotations and a biographical sketch.

Review

A. Objectives: After studying this chapter, you should:

1. Be able to state Faraday's law and use it to find the emf induced by a changing magnetic flux.

2. Be able to state Lenz's law and use it to find the direction of the induced current in various applications of Faraday's law.

3. Be able to discuss eddy currents.

4. Be able to discuss how simple ac generators and motors work.

5. Be able to sketch a graph of current versus time in an *LR* circuit.

B. Define, explain, or otherwise identify:

Magnetic induction	Lenz's law	Motor	Inductor
Magnetic flux	Back emf	Self inductance	*LR* circuit
Weber	Motional emf	Henry	Time constant of an *LR* circuit
Faraday's law	Eddy currents	Mutual inductance	Magnetic energy density
Flip coil	Generator		

C. True or false: If the statement is true, explain why it is true. If it is false, give a counterexample.

1. The induced emf in a circuit is proportional to the magnetic flux through the circuit.

2. There can be an induced emf at an instant when the flux through the circuit is zero.

3. Lenz's law is related to the conservation of energy.

4. The inductance of a solenoid is proportional to the rate of change of the current in it.

5. The magnetic energy density at some point in space is proportional to the square of the magnetic field at that point.

Problems

Level I

26-1 Magnetic Flux

1. A uniform magnetic field of magnitude 2000 G is parallel to the x axis. A square coil of side 5 cm has a single turn and makes an angle θ with the z axis as shown in Figure 26-30. Find the magnetic flux through the coil when (*a*) $\theta = 0$, (*b*) $\theta = 30°$, (*c*) $\theta = 60°$, and (*d*) $\theta = 90°$.

Figure 26-30 Problem 1.

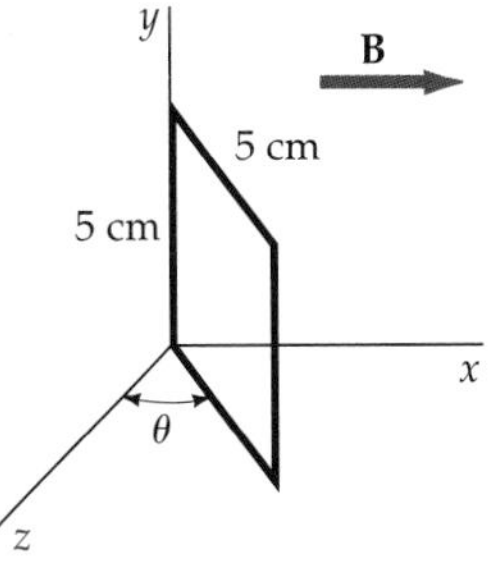

2. A circular coil has 25 turns and a radius of 5 cm. It is at the equator, where the earth's magnetic field is 0.7 G north. Find the magnetic flux through the coil when (*a*) its plane is horizontal, (*b*) its plane is vertical and its axis points north, (*c*) its plane is vertical and its axis points east, and (*d*) its plane is vertical and its axis makes an angle of 30° with north.

3. Find the magnetic flux through a solenoid of length 25 cm, radius 1 cm, and 400 turns that carries a current of 3 A.

4. Work Problem 3 for a solenoid of length 30 cm, radius 2 cm, and 800 turns that carries a current of 2 A.

5. A circular coil of radius 3.0 cm has 6 turns. A magnetic field $B = 5000$ G is perpendicular to the coil. (*a*) Find the magnetic flux through the coil. (*b*) Find the magnetic flux through the coil if the coil makes an angle of 20° with the magnetic field.

6. A magnetic field of 1.2 T is perpendicular to a square coil of 14 turns. The length of each side of the coil is 5 cm. (*a*) Find the magnetic flux through the coil. (*b*) Find the magnetic flux through the coil if the magnetic field makes an angle of 60° with the normal to the plane of the coil.

7. A circular coil of radius 3.0 cm has its plane perpendicular to a magnetic field of 400 G. (*a*) What is the magnetic flux through the coil if it has 75 turns? (*b*) How many turns must the coil have for the flux to be 0.015 Wb?

26-2 Induced Emf and Faraday's Law

8. A uniform magnetic field **B** is established perpendicular to the plane of a loop of radius 5.0 cm, resistance 0.4 Ω, and negligible self inductance. The magnitude of **B** is increasing at a rate of 40 mT/s. Find (*a*) the induced emf in the loop, (*b*) the induced current in the loop, and (*c*) the rate of Joule heating in the loop.

9. A 100-turn coil has a radius of 4.0 cm and a resistance of 25 Ω. At what rate must a perpendicular magnetic field change to produce a current of 4.0 A in the coil?

10. The flux through a loop is given by $\phi_m = (t^2 - 4t) \times 10^{-1}$ T·m², where t is in seconds. (*a*) Find the induced emf $\mathcal{E}$ as a function of time. (*b*) Find both ϕ_m and $\mathcal{E}$ at $t = 0$, $t = 2$ s, $t = 4$ s, and $t = 6$ s.

11. (*a*) For the flux given in Problem 10, sketch graphs of ϕ_m and $\mathcal{E}$ versus t. (*b*) At what time is the flux maximum? What is the emf at this time? (*c*) At what times is the flux zero? What is the emf at these times?

12. A 100-turn circular coil has a diameter of 2.0 cm and resistance of 50 Ω. The plane of the coil is perpendicular to a uniform magnetic field of magnitude 1.0 T. The direction of the field is suddenly reversed. (*a*) Find the total charge that passes through the coil. If the reversal takes 0.1 s, find (*b*) the average current in the coil and (*c*) the average emf in the coil.

13. At the equator, a 1000-turn coil with a cross-sectional area of 300 cm² and a resistance of 15.0 Ω is aligned with its plane perpendicular to the earth's magnetic field of 0.7 G. If the coil is flipped over, how much charge flows through it?

14. A circular coil of 300 turns and radius 5.0 cm is connected to a current integrator. The total resistance of the circuit is 20 Ω. The plane of the coil is originally aligned perpendicular to the earth's magnetic field at some point. When the coil is rotated through 90°, the charge that passes through the current integrator is measured to be 9.4 μC. Calculate the magnitude of the earth's magnetic field at that point.

15. The magnetic field in Problem 5 is steadily reduced to zero in 1.2 s. Find the emf induced in the coil when (*a*) the magnetic field is perpendicular to the coil and (*b*) the magnetic field makes an angle of 20° with the normal to the coil.

16. The magnetic field in Problem 7 is steadily reduced to zero in 0.8 s. What is the magnitude of the emf induced in the coil of part (*b*)?

17. A solenoid of length 25 cm and radius 0.8 cm with 400 turns is in an external magnetic field of 600 G that makes an angle of 50° with the axis of the solenoid. (*a*) Find the magnetic flux through the solenoid. (*b*) Find the magnitude of the emf induced in the solenoid if the external magnetic field is reduced to zero in 1.4 s.

26-3 Lenz's Law

18. The two loops in Figure 26-31 have their planes parallel to each other. As viewed from A toward B, there is a counterclockwise current in loop A. Give the direction of the current in loop B and state whether the loops attract or repel each other if the current in loop A is (*a*) increasing and (*b*) decreasing.

Figure 26-31 Problem 18.

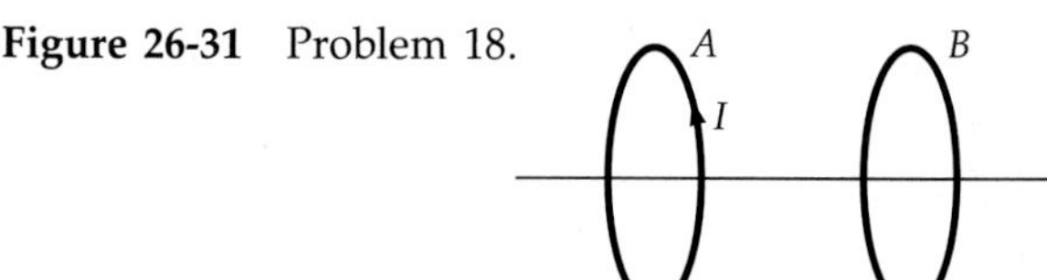

19. A bar magnet moves with constant velocity along the axis of a loop as shown in Figure 26-32. (*a*) Make a qualitative graph of the flux ϕ_m through the loop as a function of time. Indicate the time t_1 when the magnet is halfway through the loop. (*b*) Sketch a graph of the current I in the loop versus time, choosing I to be positive when it is counterclockwise as viewed from the left.

Figure 26-32 Problem 19.

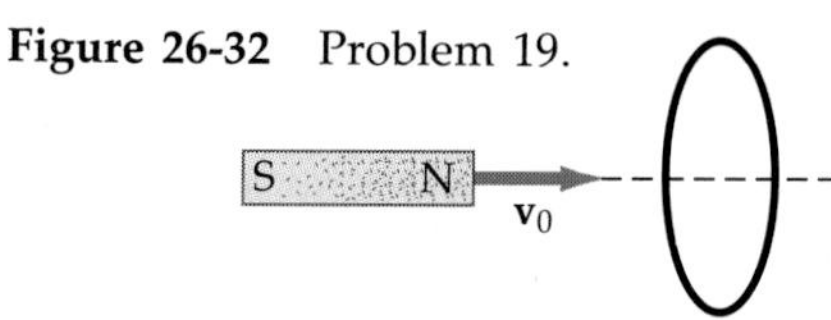

20. Give the direction of the induced current in the circuit on the right in Figure 26-33 when the resistance in the circuit on the left is suddenly (*a*) increased and (*b*) decreased.

Figure 26-33 Problem 20.

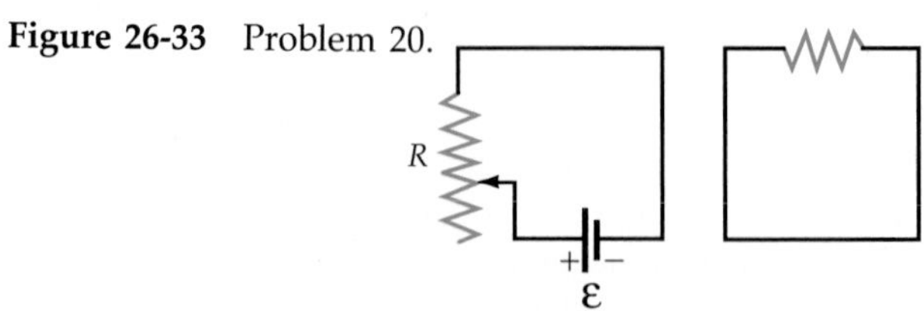

21. A bar magnet is mounted on the end of a coiled spring in such a way that it moves with simple harmonic motion along the axis of a loop as shown in Figure 26-34. (*a*) Make a qualitative graph of the flux ϕ_m through the loop as a function of time. Indicate the time t_1 when the magnet is halfway through the loop. (*b*) Sketch the current I in the loop versus time, choosing I to be positive when it is counterclockwise as viewed from above.

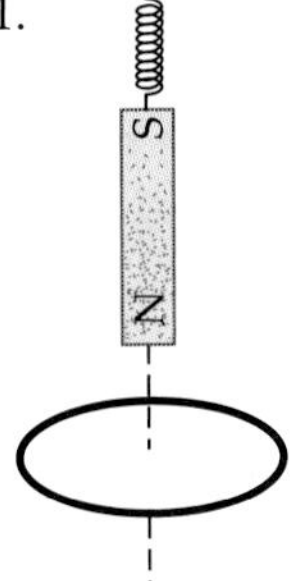

Figure 26-34 Problem 21.

26-4 Motional Emf

22. A rod 30 cm long moves at 8 m/s in a plane perpendicular to a magnetic field of 500 G. The velocity of the rod is perpendicular to its length. Find (*a*) the magnetic force on an electron in the rod, (*b*) the electrostatic field **E** in the rod, and (*c*) the potential difference V between the ends of the rod.

23. Find the speed of the rod in Problem 22 if the potential difference between the ends is 6 V.

24. In Figure 26-14, let B be 0.8 T, $v = 10.0$ m/s, $\ell =$ 20 cm, and $R = 2\ \Omega$. Find (*a*) the induced emf in the circuit, (*b*) the current in the circuit, and (*c*) the force needed to move the rod with constant velocity assuming negligible friction. Find (*d*) the power input by the force found in part (*c*) and (*e*) the rate of Joule heat production I^2R.

25. Work Problem 24 for $B = 1.5$ T, $v = 6$ m/s, $\ell = 40$ cm, and $R = 1.2\ \Omega$.

26-5 Eddy Currents

There are no problems for this section.

26-6 Generators and Motors

26. A 200-turn coil has an area of 4 cm^2. It rotates in a magnetic field of 0.5 T. (*a*) What should the frequency be to generate a maximum emf of 10 V? (*b*) If the coil rotates at 60 Hz, what is the maximum emf?

27. In what magnetic field must the coil of Problem 26 be rotating to generate a maximum emf of 10 V at 60 Hz?

28. A 2-cm by 1.5-cm rectangular coil has 300 turns and rotates in a magnetic field of 4000 G. (*a*) What is the maximum emf generated when the coil rotates at 60 Hz? (*b*) What must its frequency be to generate a maximum emf of 110 V?

29. The coil of Problem 28 rotates at 60 Hz in a magnetic field B. What must the value of B be so that the maximum emf generated is 24 V?

26-7 Inductance

30. A coil with a self inductance of 8.0 H carries a current of 3 A that is changing at a rate of 200 A/s. Find (*a*) the magnetic flux through the coil and (*b*) the induced emf in the coil.

31. A coil with self inductance L carries a current I, given by $I = I_0 \sin 2\pi ft$. Find and graph the flux ϕ_m and the self-induced emf as functions of time.

32. A solenoid has a length of 25 cm, a radius of 1 cm, and 400 turns and carries a 3-A current. Find (*a*) B on the axis at the center of the solenoid; (*b*) the flux through the solenoid, assuming B to be uniform; (*c*) the self inductance of the solenoid; and (*d*) the induced emf in the solenoid when the current changes at 150 A/s.

33. Two solenoids of radii 2 cm and 5 cm are coaxial. They are each 25 cm long and have 300 and 1000 turns, respectively. Find their mutual inductance.

26-8 *LR* Circuits

34. The current in an LR circuit is zero at time $t = 0$ and increases to half its final value in 4.0 s. (*a*) What is the time constant of this circuit? (*b*) If the total resistance is 5 Ω, what is the self inductance?

35. A coil of resistance 8.0 Ω and self inductance 4.0 H is suddenly connected across a constant potential difference of 100 V. Let $t = 0$ be the time of connection, at which the current is zero. Find the current I and its rate of change dI/dt at times (*a*) $t = 0$, (*b*) $t = 0.1$ s, (*c*) $t = 0.5$ s, and (*d*) $t = 1.0$ s.

36. How many time constants must elapse before the current in an LR circuit that is initially zero reaches (*a*) 90 percent, (*b*) 99 percent, and (*c*) 99.9 percent of its final value?

37. The current in a coil with a self inductance of 1 mH is 2.0 A at $t = 0$, when the coil is shorted through a resistor. The total resistance of the coil plus the resistor is 10.0 Ω. Find the current after (*a*) 0.5 ms and (*b*) 10 ms.

26-9 Magnetic Energy

38. In the circuit of Figure 26-26, let $\mathcal{E}_0 = 12.0$ V, $R =$ 3.0 Ω, and $L = 0.6$ H. The switch is closed at time $t = 0$. At time $t = 0.5$ s, find (*a*) the rate at which the battery supplies energy, (*b*) the rate of Joule heating, and (*c*) the rate at which energy is being stored in the inductor.

39. Do Problem 38 for the times $t = 1$ s and $t = 100$ s.

40. A coil with a self inductance of 2.0 H and a resistance of 12.0 Ω is connected across a 24-V battery of negligible internal resistance. (*a*) What is the final current? (*b*) How much energy is stored in the inductor when the final current is attained?

41. Find (*a*) the magnetic energy, (*b*) the electric energy, and (*c*) the total energy in a volume of 1.0 m^3 in which there is an electric field of 10^4 V/m and a magnetic field of 1 T.

Level II

42. A circular coil of radius 4 cm with 15 turns is in a uniform magnetic field of 4000 G in the positive x direction. Find the flux through the coil when the unit normal vector to the plane of the coil is (*a*) $\hat{\mathbf{n}} = \mathbf{i}$, (*b*) $\hat{\mathbf{n}} = \mathbf{j}$, (*c*) $\hat{\mathbf{n}} = (\mathbf{i} + \mathbf{j})/\sqrt{2}$, (*d*) $\hat{\mathbf{n}} = \mathbf{k}$, and (*e*) $\hat{\mathbf{n}} = 0.6\mathbf{i} + 0.8\mathbf{j}$.

43. A uniform magnetic field $\mathbf{B}$ is perpendicular to the base of a hemisphere of radius R. Calculate the magnetic flux through the spherical surface of the hemisphere.

44. An elastic circular conducting loop is expanding at a constant rate so that its radius is given by $R = R_0 + vt$. The loop is in a region of constant magnetic field perpendicular to the loop. What is the emf generated in the expanding loop? Neglect possible effects of self inductance.

45. A solenoid has n turns per unit length, radius R_1, and carries a current I. (*a*) A large circular loop of radius $R_2 > R_1$ and N turns encircles the solenoid at a point far away from the ends of the solenoid. Find the magnetic flux through the loop. (*b*) A small circular loop of N turns and radius $R_3 < R_1$ is completely inside the solenoid, far from its ends, with its axis parallel to that of the solenoid. Find the magnetic flux through the loop.

46. Show that if the flux through each turn of an N-turn coil of resistance R changes from ϕ_{m1} to ϕ_{m2} in any manner, the total charge passing through the coil is given by $Q = N(\phi_{m1} - \phi_{m2})/R$.

47. An ac generator's rectangular loop of dimensions a and b has N turns. The loop is connected to slip rings (Figure 26-35) and rotates with an angular velocity ω in a uniform magnetic field $\mathbf{B}$. (*a*) Show that the potential difference between the two slip rings is $\mathcal{E} = NBab\omega \sin \omega t$. (*b*) If $a = 1.0$ cm, $b = 2.0$ cm, $N = 1000$, and $B = 2$ T, at what angular frequency ω must the coil rotate to generate an emf whose maximum value is 110 V?

Figure 26-35 Problem 47.

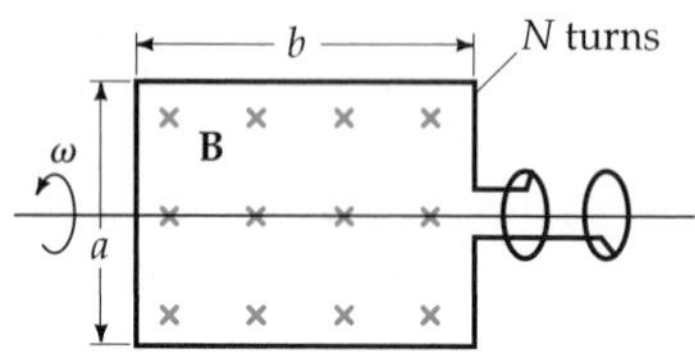

48. A dc motor has coils of resistance 5.5 Ω. When connected to dc power supply of 120 V, it draws 6 A. (*a*) How large is the back emf? (*b*) What is the initial current drawn before it starts to rotate?

49. To limit the current drawn by a motor as it starts up, a resistance is sometimes placed in series with the motor. The resistance is removed when the motor attains operating speed. (*a*) How much resistance should be placed in series with a motor that has a resistance of 0.75 Ω and draws 8 A when operated at 220 V if the current is not to exceed 15 A? (*b*) What is the back emf of this motor when it attains operating speed and the resistance is removed?

50. Compute the initial slope dI/dt at $t = 0$ from Equation 26-31, and show that if the current decreased steadily at this rate, it would be zero after one time constant.

51. An inductance L and resistance R are connected in series with a battery as in Figure 26-28. A long time after switch S_1 is closed, the current is 2.5 A. When the battery is switched out of the circuit by opening switch S_1 and closing S_2, the current drops to 1.5 A in 45 ms. (*a*) What is the time constant for this circuit? (*b*) If $R = 0.4$ Ω, what is L?

52. A coil with inductance 4 mH and resistance 150 Ω is connected across a battery of emf 12 V and negligible internal resistance. (*a*) What is the initial rate of increase of the current? (*b*) What is the rate of increase when the current is half its final value? (*c*) What is the final current? (*d*) How long does it take for the current to reach 99 percent of its final value?

53. A large electromagnet has an inductance of 50 H and a resistance of 8.0 Ω. It is connected to a dc power source of 250 V. Find the time for the current to reach (*a*) 10 A and (*b*) 30 A.

54. When the current in a certain coil is 5.0 A and is increasing at the rate of 10.0 A/s, the potential difference across the coil is 140 V. When the current is 5.0 A and is decreasing at the rate of 10.0 A/s, the potential difference is 60 V. Find the resistance and self inductance of the coil.

55. In a plane electromagnetic wave such as a light wave, the magnitudes of the electric and magnetic fields are related by $E = cB$, where $c = 1/\sqrt{\epsilon_0\mu_0}$ is the speed of light. Show that in this case the electric and the magnetic energy densities are equal.

56. Show that the effective inductance for two inductors L_1 and L_2 connected in series such that none of the flux from either passes through the other is given by $L_{\text{eff}} = L_1 + L_2$.

57. Given the circuit shown in Figure 26-36, assume that switch S has been closed for a long time so that steady currents exist in the circuit and that the inductor L is made of superconducting wire so that its resistance may be considered to be zero. (*a*) Find the battery current, the current in the 100-Ω resistor, and the current through the inductor. (*b*) Find the initial voltage across the inductor when switch S is opened. (*c*) Give the current in the inductor as a function of time measured from the instant of opening switch S.

Figure 26-36 Problem 57.

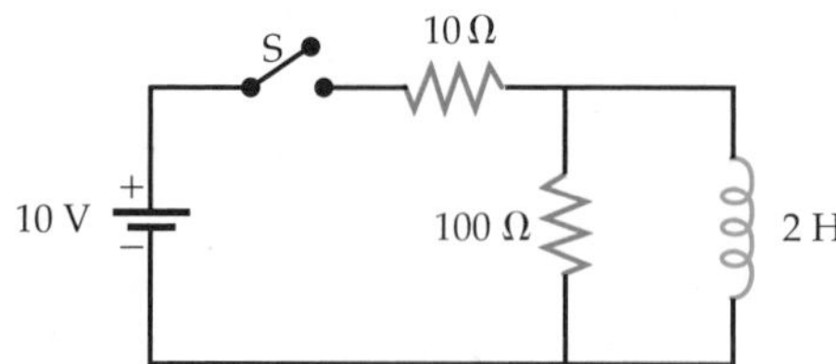

58. Show that the effective inductance for two inductors L_1 and L_2 connected in parallel such that none of the flux from either passes through the other is given by

$$1/L_{\text{eff}} = 1/L_1 + 1/L_2$$

59. For the circuit of Figure 26-37, (*a*) find the rate of change of current in each inductor and in the resistor just after the switch is closed. (*b*) What is the final current? (See Problem 58.)

Figure 26-37 Problem 59.

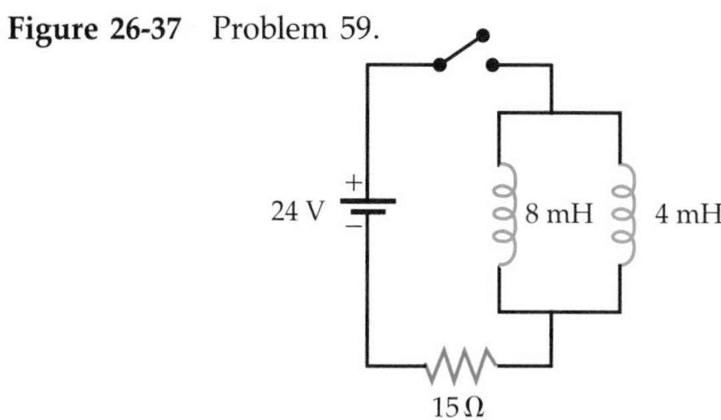

60. For the circuit shown in Figure 26-38, find the currents I_1, I_2, and I_3 (*a*) immediately after switch S is closed and (*b*) a long time after switch S has been closed. After the switch has been closed for a long time, it is opened. Find the three currents (*c*) immediately after switch S is opened and (*d*) a long time after switch S was opened.

Figure 26-38 Problem 60.

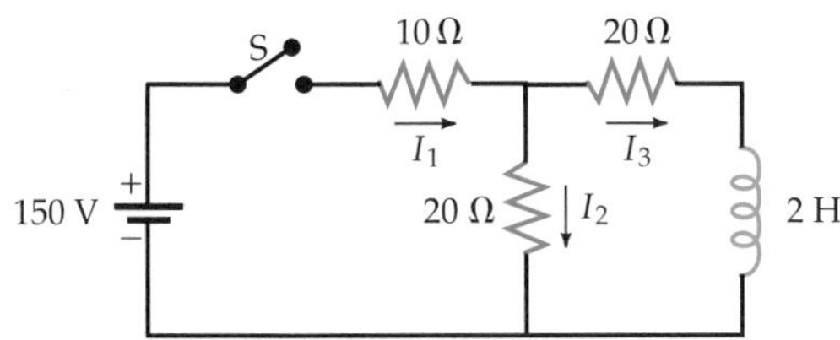

61. A solenoid of 2000 turns, area 4 cm^2, and length 30 cm carries a current of 4.0 A. (*a*) Calculate the magnetic energy stored in the solenoid from $\frac{1}{2}LI^2$. (*b*) Divide your answer in part (*a*) by the volume of the solenoid to find the magnetic energy per unit volume in the solenoid. (*c*) Find B in the solenoid. (*d*) Compute the magnetic energy density from $\eta_m = B^2/2\mu_0$, and compare your answer with your result for part (*b*).

62. A toroid of mean radius 25 cm and a coil radius of 2 cm is wound with a superconducting wire of length 1000 m that carries a current of 400 A. (*a*) What is the number of turns on the coil? (*b*) What is the magnetic field at the mean radius? (*c*) Assuming that B is constant over the area of the coil, calculate the magnetic energy density and the total energy stored in the toroid.

63. A long solenoid has n turns per unit length and carries a current given by $I = I_0 \sin \omega t$. The solenoid has a circular cross section of radius R. Find the induced electric field at a radius r from the axis of the solenoid for (*a*) $r < R$ and (*b*) $r > R$.

64. A uniform magnetic field of magnitude 1.2 T is in the z direction. A conducting rod of length 15 cm lies parallel to the y axis and oscillates in the x direction with displacement given by $x = (2\text{ cm}) \cos 120\,\pi t$. What is the emf induced in the rod?

65. A 10-cm by 5-cm rectangular loop with resistance 2.5 Ω is pulled through a region of uniform magnetic field $B = 1.7$ T (Figure 26-39) with constant speed $v = 2.4$ cm/s. The front of the loop enters the region of the magnetic field at time $t = 0$. (*a*) Find and graph the flux through the loop as a function of time. (*b*) Find and graph the induced emf and the current in the loop as functions

Figure 26-39 Problem 65.

20 cm
10 cm
B
5 cm
v
x

of time. Neglect any self inductance of the loop and extend your graphs from $t = 0$ to $t = 16$ s.

66. In Example 26-7, find the total energy dissipated in the resistance and show that it is equal to $\frac{1}{2}mv_0^2$.

67. In Figure 26-40, the rod has a resistance R and the rails have negligible resistance. A battery of emf $\mathcal{E}$ and negligible internal resistance is connected between points a and b such that the current in the rod is downward. The rod is placed at rest at $t = 0$. (*a*) Find the force on the rod as a function of the speed v and write Newton's second law for the rod when it has speed v. (*b*) Show that the rod reaches a terminal velocity and find an expression for it. (*c*) What is the current when the rod reaches its terminal velocity?

Figure 26-40 Problems 67 and 68.

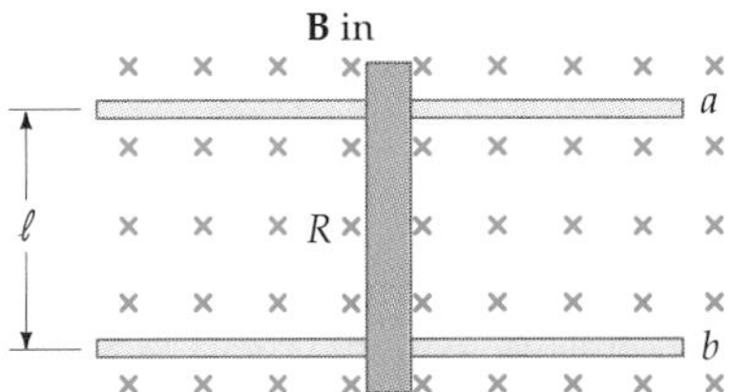

68. In Figure 26-40, the rod has a resistance R and the rails have negligible resistance. A capacitor with charge Q_0 and capacitance C is connected between points a and b such that the current in the rod is downward. The rod is placed at rest at $t = 0$. (*a*) Write the equation of motion for the rod on the rails. (*b*) Show that the terminal speed of the rod down the rails is related to the final charge on the capacitor.

69. A conducting rod of mass m and resistance R is free to slide without friction along two parallel rails of negligible resistance connected by a wire of negligible resistance. The rails are separated by a distance ℓ and inclined at an angle θ to the horizontal. There is a magnetic field B directed upward. (*a*) Show that there is a retarding force directed up the incline given by $F = (B^2\ell^2 v \cos^2 \theta)/R$.

(*b*) Show that the terminal speed of the rod is

$$v_t = (mgR \sin \theta)/(B^2\ell^2 \cos^2 \theta)$$

70. A simple pendulum has a wire of length ℓ supporting a metal ball of mass m. The wire has negligible mass and moves in a uniform horizontal magnetic field B. This pendulum executes simple harmonic motion with angular amplitude θ_0, in a plane perpendicular to **B**. What is the emf generated along the wire?

71. A wire lies along the z axis and carries current $I = 20$ A in the positive z direction. A small conducting sphere of radius $R = 2$ cm is initially at rest on the y axis at a distance $h = 45$ m above the wire. The sphere is dropped at time $t = 0$. (*a*) What is the electric field at the center of the sphere at $t = 3$ s? Assume that the only magnetic field is that produced by the wire. (*b*) What is the voltage across the sphere at $t = 3$ s?

Level III

72. A long straight wire carries a current I. A rectangular coil with two sides parallel to the straight wire has sides a and b with its near side a distance of d from the wire as shown in Figure 26-41. (*a*) Compute the magnetic flux through the rectangular coil. *Hint:* Calculate the flux through a strip of area $dA = b\,dx$ and integrate from $x = d$ to $x = d + a$. (*b*) Evaluate your answer for $a = 5$ cm, $b = 10$ cm, $d = 2$ cm, and $I = 20$ A.

Figure 26-41 Problems 72 and 74.

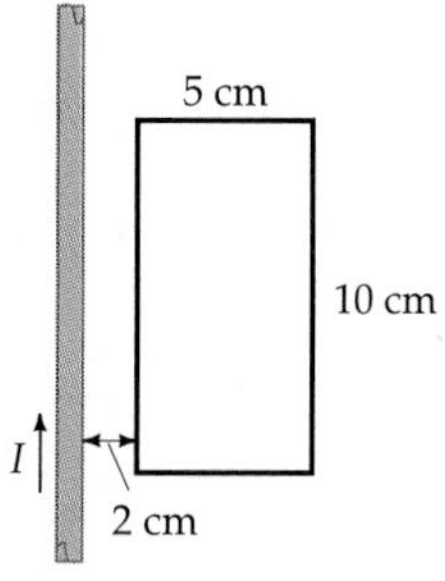

73. A rod of length ℓ lies with its length perpendicular to a long wire carrying current I, as shown in Figure 26-42. The near end of the rod is a distance d away from the wire. The rod moves with a speed v in the direction of the current I. (*a*) Show that the potential difference between the ends of the rod is given by

$$V = \frac{\mu_0 I}{2\pi} v \ln \frac{d + \ell}{d}$$

(*b*) Use Faraday's law to obtain this result by considering the flux through a rectangular area $A = \ell v t$ swept out by the rod.

Figure 26-42 Problem 73.

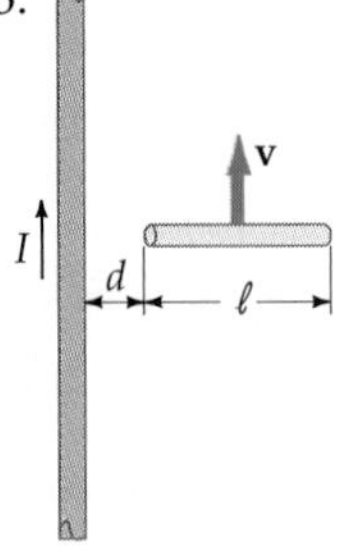

74. The loop in Problem 72 moves away from the wire with a constant speed v. At time $t = 0$, the left side of the loop is a distance d from the long straight wire. (*a*) Compute the emf in the loop by computing the motional emf in each segment of the loop that is parallel to the long wire. Explain why you can neglect the emf in the segments that are perpendicular to the wire. (*b*) Compute the emf in the loop by first computing the flux through the loop as a function of time and then using $\mathcal{E} = -d\phi_m/dt$ and compare your answer with that obtained in part (*a*).

75. A thin-walled hollow wire of radius a lies with its axis along the z axis and carries current I in the positive z direction. A second identical wire is parallel to it with its axis along the line $x = d$. It carries current I in the negative z direction. (*a*) Find the magnetic flux per unit length through the space in the xz plane between the wires. (*b*) If the far ends of the wires are connected together so that the parallel wires form two sides of a loop, find the self inductance per unit length of the loop.

76. A long cylindrical conductor of radius R carries a current I that is uniformly distributed over its cross-sectional area. Find the magnetic flux per unit length through the area indicated in Figure 26-43.

Figure 26-43 Problem 76.

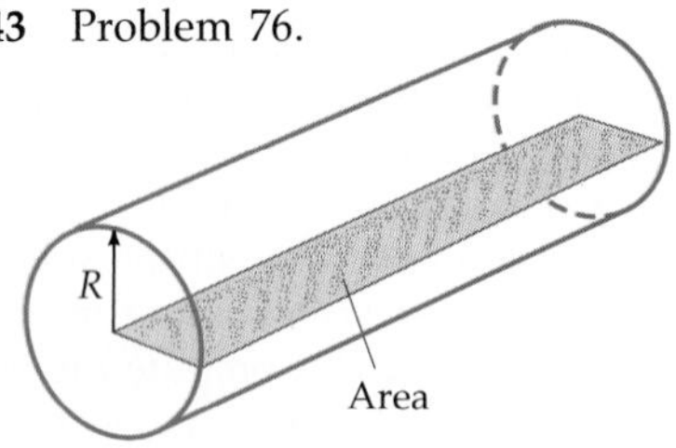

77. A conducting rod of length ℓ rotates at constant angular velocity about one end, in a plane perpendicular to a uniform magnetic field B (Figure 26-44). (*a*) Show that the magnetic force on a charge q at a distance r from the pivot is $Bqr\omega$. (*b*) Show that the potential difference between the ends of the rod is $V = \frac{1}{2}B\omega\ell^2$. (*c*) Draw any radial line in the plane from which to measure $\theta = \omega t$. Show that the area of the pie-shaped region between the reference line and the rod is $A = \frac{1}{2}\ell^2\theta$. Compute the flux through this area, and show that $\mathcal{E} = \frac{1}{2}B\omega\ell^2$ follows when Faraday's law is applied to this area.

Figure 26-44 Problem 77.

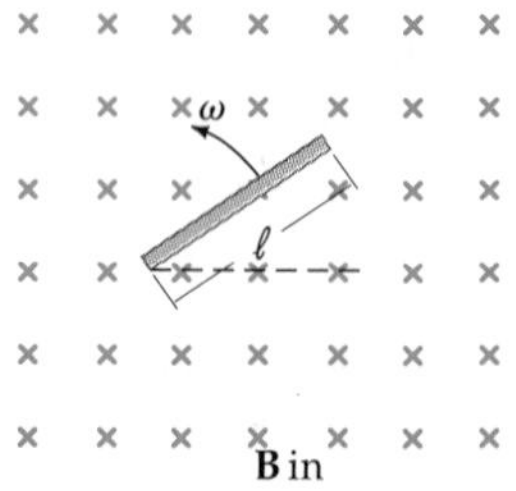

78. In the circuit of Figure 26-26, let $\mathcal{E}_0 = 12.0$ V, $R = 3.0\ \Omega$, and $L = 0.6$ H. The switch is closed at time $t = 0$. From time $t = 0$ to $t = \tau$, find (*a*) the total energy that has been supplied by the battery, (*b*) the total energy that has been dissipated in the resistor, and (*c*) the energy that has been stored in the inductor. *Hint:* Find the rates as functions of time and integrate from $t = 0$ to $t = \tau = L/R$.

79. In Figure 26-45, circuit 2 has a total resistance of 300 Ω. A total charge of 2×10^{-4} C flows through the galvanometer in circuit 2 when switch S in circuit 1 is closed. After a long time, the current in circuit 1 is 5 A. What is the mutual inductance between the two coils?

Figure 26-45 Problem 79.

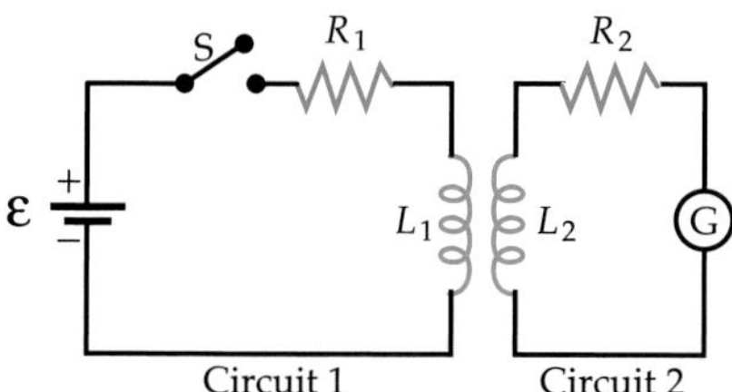

80. A coaxial cable consists of two very thin-walled conducting cylinders of radii r_1 and r_2 (Figure 26-46). Current I goes in one direction down the inner cylinder and in the opposite direction in the outer cylinder. (*a*) Use Ampere's law to find B. Show that $B = 0$ except in the region between the conductors. (*b*) Show that the magnetic energy density in the region between the cylinders is

$$\eta_m = \frac{\mu_0 I^2}{8\pi^2 r^2}$$

(*c*) Find the magnetic energy in a cylindrical shell volume element of length ℓ and volume $d\mathcal{V} = \ell 2\pi r\, dr$, and integrate your result to show that the total magnetic energy in the volume of length ℓ is

$$U_m = \frac{\mu_0}{4\pi} I^2 \ell \ln\frac{r_2}{r_1}$$

(*d*) Use the result in part (*c*) and $U_m = \frac{1}{2}LI^2$ to show that the self inductance per unit length is

$$\frac{L}{\ell} = \frac{\mu_0}{2\pi} \ln\frac{r_2}{r_1}$$

Figure 26-46 Problems 80 and 81.

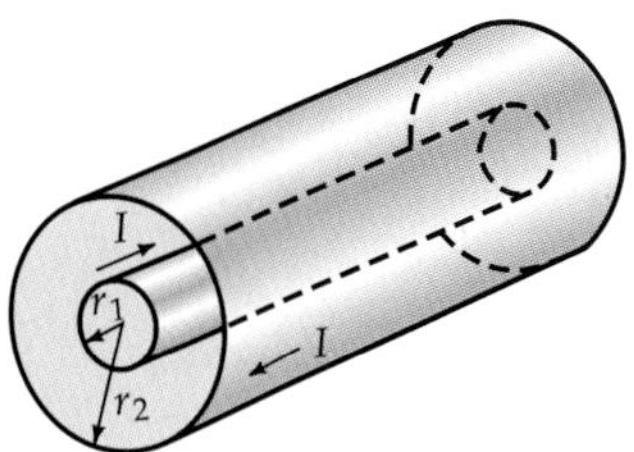

81. In Figure 26-46, compute the flux through a rectangular area of sides ℓ and $r_2 - r_1$ between the conductors. Show that the self inductance per unit length can be found from $\phi_m = LI$ [see part (*d*) of Problem 80].

82. Show that the inductance of a toroid of rectangular cross section as shown in Figure 26-47 is given by

$$L = \frac{\mu_0 N^2 h \ln (b/a)}{2\pi}$$

where N is the total number of turns, a is the inside radius, b is the outside radius, and h is the height of the toroid.

Figure 26-47 Problem 82.

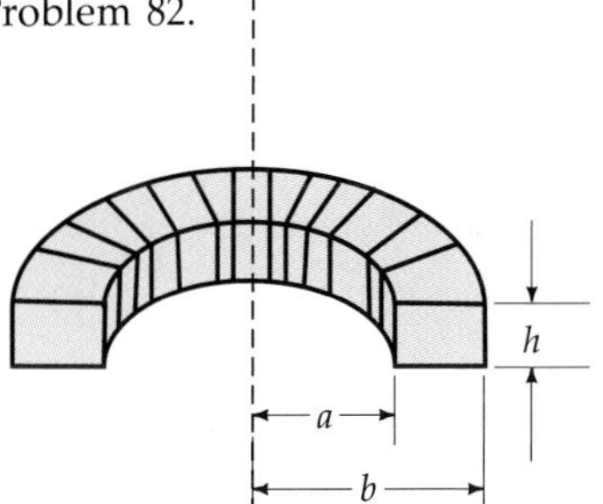

Chapter 27

Magnetism in Matter

Magnetic domains on the surface of an Fe-3%Si crystal—observed when using a scanning electron microscope with polarization analysis. The four colors indicate four possible domain orientations.

In studying electric fields in matter, we found that the electric field is affected by the presence of electric dipoles. Polar molecules have permanent electric dipole moments that are partially aligned by the electric field in the direction of the field. In nonpolar molecules, electric dipole moments are induced by the electric field in the direction of the field. In both cases, the alignment of the dipole moments parallel to the external electric field tends to weaken the field.

Similar but more complicated effects occur in magnetism. Atoms have magnetic dipole moments due to the motion of their electrons. In addition, each electron has an intrinsic magnetic dipole moment associated with its spin. The net magnetic moment of an atom depends on the arrangement of the electrons in the atom. Unlike the situation with electric dipoles, the alignment of magnetic dipoles parallel to an external magnetic field tends to *increase* the field. We can see this difference by comparing the electric-field lines of an electric dipole with the magnetic-field lines of a magnetic dipole, such as a small current loop, as in Figure 27-1. Far from the dipoles, the field lines are identical. However, between the charges of the electric dipole, the electric-field lines are opposite the direction of the dipole moment, whereas inside the current loop, the magnetic-field lines are parallel to the magnetic dipole moment. Thus, inside an electrically polarized material, the electric dipoles create an electric field that is *antiparallel* to their dipole-moment vectors, whereas inside a magnetically polarized material, the magnetic dipoles create a magnetic field that is *parallel* to the magnetic-dipole-moment vectors.

Materials fall into three categories—paramagnetic, diamagnetic, and ferromagnetic—according to the behavior of their molecules in an external magnetic field. Paramagnetic and ferromagnetic materials have molecules with permanent magnetic dipole moments. In paramagnetic materials, the magnetic dipoles do not interact strongly with each other and are normally randomly oriented. In the presence of an external magnetic field, the dipoles are partially aligned in the direction of the field, thereby increasing the field. However, in external magnetic fields of ordinary strength at ordinary temperatures, only a very small fraction of the molecules are aligned because thermal motion tends to randomize their orientation. The increase in the total magnetic field is therefore very small. Ferromagnetism is much more complicated. Because of a strong interaction between neighboring magnetic dipoles, a high degree of alignment occurs even in weak external magnetic fields, causing a very large increase in the total field. Even when there is no external magnetic field, a ferromagnetic material may have its magnetic dipoles aligned, as in permanent magnets. Diamagnetism is observed in materials whose molecules have no permanent magnetic moments. It is the result of an induced magnetic dipole moment opposite in direction to the external field. The induced dipoles thus decrease the total magnetic field. This effect actually occurs in all materials, but because it is very small, it is masked by paramagnetic or ferromagnetic effects when the individual molecules of the material have permanent magnetic dipole moments.

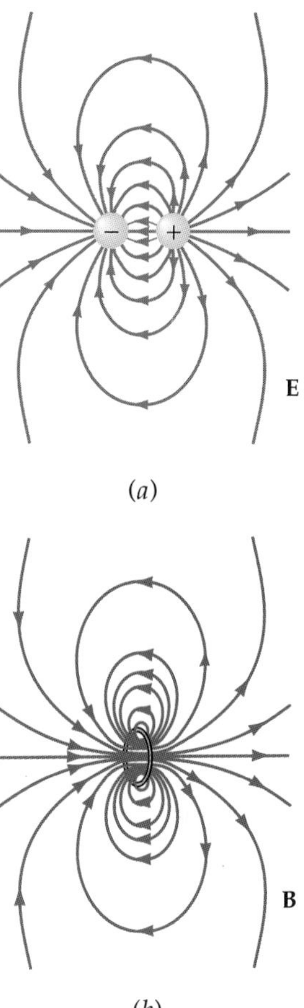

Figure 27-1 (*a*) Electric-field lines of an electric dipole. (*b*) Magnetic-field lines of a magnetic dipole. Far from the dipoles, the field lines are identical. In the region between the charges in (*a*), the electric field is opposite the dipole moment, whereas inside the loop in (*b*), the magnetic field is parallel to the dipole moment.

27-1 Magnetization and Magnetic Susceptibility

In our study of the electrical effects of matter, we placed a nonconducting material (dielectric) in a strong, uniform electric field between the plates of a parallel-plate capacitor. We found that the applied field tends to align the electric dipole moments (either permanent or induced). Similarly, when some material is placed in a strong magnetic field, such as that of a solenoid, the magnetic field of the solenoid tends to align the magnetic dipole moments (either permanent or induced) inside the material, and the material is said to be magnetized. We describe a magnetized material by its **magnetization M,** which is defined as the net magnetic dipole moment per unit volume of the material:

$$\mathbf{M} = \frac{d\mathbf{m}}{d\mathcal{V}} \qquad 27\text{-}1$$

Long before we had any understanding of atomic or molecular structure, Ampère proposed a model of magnetism in which the magnetization of materials is due to microscopic current loops inside the magnetized material. We now know that these current loops are the result of the intrinsic motion of atomic charges. Although these motions are very complicated, for Ampère's model, we need only assume that the motions are equivalent to closed-circuit loops. Let us assume that our magnetized material is a cylinder. Figure 27-2 shows atomic current loops in the cylinder aligned with their magnetic moments along the axis of the cylinder. If the material is homogeneous, the net current at any point inside the material is zero because of cancellation of neighboring current loops. However, since there is no cancellation on the surface of the material, the result of these current

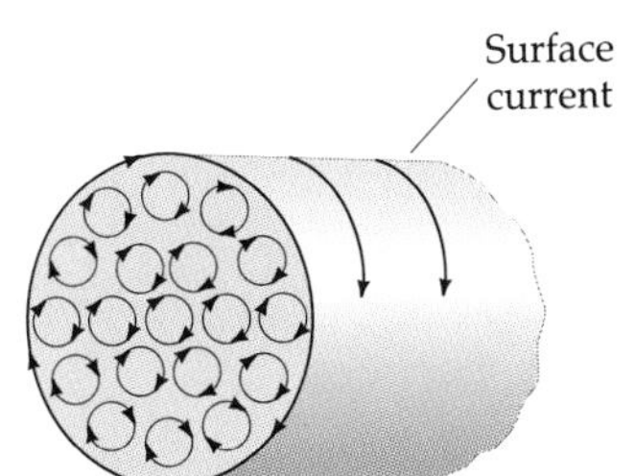

Figure 27-2 A model of atomic current loops in which all the atomic dipoles are parallel to the axis of the cylinder. The net current at any point inside the material is zero due to cancellation of neighboring atoms. The result is a surface current similar to that of a solenoid.

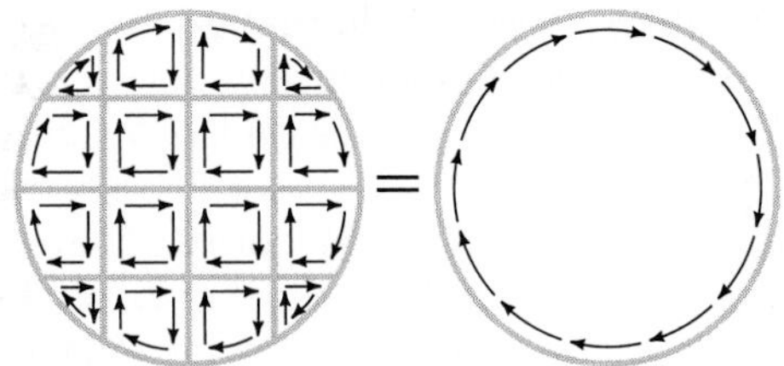

Figure 27-3 The currents in the adjacent current loops in the interior of a uniformly magnetized material cancel leaving only a surface current. This cancellation occurs at every interior point independent of the shape of the loops.

loops is equivalent to a current on the surface of the material (Figure 27-3). This surface current, called an **amperian current,** is similar to the real current in the windings of the solenoid.

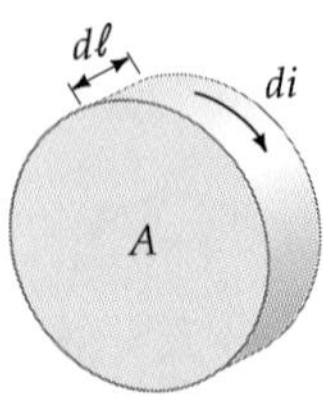

Figure 27-4 Disk element for relating the magnetization M to the surface current per unit length.

Figure 27-4 shows a small disk-shaped section of the cylinder. The disk has a cross-sectional area A, length $d\ell$, and volume $d\mathcal{V} = A\,d\ell$. Let di be the amperian current on the surface of the disk. The magnitude of the magnetic dipole moment dm of the disk is the same as that of a current loop of area A carrying a current di:

$$dm = A\,di$$

The magnetization M of the disk is the magnetic moment per unit volume:

$$M = \frac{dm}{d\mathcal{V}} = \frac{A\,di}{A\,d\ell} = \frac{di}{d\ell} \qquad 27\text{-}2$$

Thus the magnitude of the magnetization vector is the amperian current per unit length along the surface of the magnetized material. We see from this result that the units of M are amperes per meter.

Consider a cylinder that has a uniform magnetization **M** parallel to its axis. As we have just seen, the effect of the magnetization is the same as if the cylinder carried a surface current per unit length of magnitude M. This current is similar to the current carried by a tightly wound solenoid. For a solenoid, the current per unit length is nI where n is the number of turns per unit length and I is the current in each turn. We can calculate the magnetic field produced by the magnetized cylinder in the same way that we calculated the field produced by a tightly wound solenoid. If the cylinder is the same shape as the solenoid and if $M = nI$, the magnetic field produced by the cylinder at any point is exactly the same as that produced by the solenoid. In particular, the magnetic field B inside a solenoid and far from its ends is given by

$$B = \mu_0 nI$$

Then the magnetic field B_m inside and far from the ends of a cylinder with a uniform magnetization M is given by

$$B_m = \mu_0 M \qquad 27\text{-}3$$

Example 27-1

A small, cylindrical bar magnet of radius 0.5 cm and length 12 cm has a magnetic dipole moment of magnitude $m = 1.5\ \text{A}\cdot\text{m}^2$. (*a*) Find the magnetization M, assuming it to be uniform in the magnet. Find the magnetic field (*b*) at the center of the magnet and (*c*) just outside one end of the magnet. (*d*) Find the pole strength q_m of the magnet.

(*a*) The volume of the magnet is $\mathcal{V} = \pi r^2\ell = \pi(0.005\ \text{m})^2(0.12\ \text{m}) = 9.42 \times 10^{-6}\ \text{m}^3$. The magnetization is the magnetic moment per unit volume

$$M = \frac{m}{\mathcal{V}} = \frac{1.5\ \text{A}\cdot\text{m}^2}{9.42 \times 10^{-6}\ \text{m}^3} = 1.59 \times 10^5\ \text{A/m}$$

(*b*) The magnetic field inside a cylindrical magnet is the same as that inside a solenoid with nI replaced by M—the amperian current per unit length on the surface of the magnet. If end effects are neglected, the magnetic field at the center of the magnet is

$$B = \mu_0 M$$
$$= (4\pi \times 10^{-7}\ \text{T·m/A})(1.59 \times 10^5\ \text{A/m}) = 0.200\ \text{T}$$

(*c*) In Chapter 25, we saw that the magnetic field near the end of a solenoid is half that at the center of the solenoid. Thus, the magnetic field near the end of the cylinder is

$$B = \tfrac{1}{2}\mu_0 M = 0.100\ \text{T}$$

(*d*) The magnetic pole strength of the bar magnet equals the magnitude of the magnetic dipole moment divided by length. The pole strength is thus

$$q_m = \frac{m}{\ell} = \frac{1.5\ \text{A·m}^2}{0.12\ \text{m}} = 12.5\ \text{A·m}$$

Consider a long solenoid with n turns per unit length that carries a current I. We will call the magnetic field due to the current in the solenoid the applied field $\mathbf{B}_{app}$. We now place a cylinder of material inside the solenoid. The applied field of the solenoid magnetizes the material so that it has a magnetization **M**. The resultant magnetic field at a point inside the solenoid and far from its ends due to the current in the solenoid plus the magnetized material is

$$\mathbf{B} = \mathbf{B}_{app} + \mu_0\mathbf{M} \qquad 27\text{-}4$$

For paramagnetic and ferromagnetic materials, **M** is in the same direction as $\mathbf{B}_{app}$; for diamagnetic materials, **M** is opposite to $\mathbf{B}_{app}$. For paramagnetic and diamagnetic materials, the magnetization is found to be proportional to the applied magnetic field that produces the alignment of the magnetic dipoles in the material. We can thus write

$$\mathbf{M} = \chi_m\left(\frac{\mathbf{B}_{app}}{\mu_0}\right) \qquad 27\text{-}5$$

where χ_m is a dimensionless number called the **magnetic susceptibility.** Equation 27-4 is then

$$\mathbf{B} = \mathbf{B}_{app} + \mu_0\mathbf{M} = \mathbf{B}_{app}\,(1 + \chi_m) \qquad 27\text{-}6$$

For paramagnetic materials, χ_m is a small positive number that depends on temperature. For diamagnetic materials, it is a small negative constant independent of temperature. Table 27-1 lists the magnetic susceptibility of various paramagnetic and diamagnetic materials. We see that the magnetic susceptibility for the solids listed is of the order of 10^{-5}.

Equations 27-5 and 27-6 are not very useful for ferromagnetic materials because χ_m depends on B_{app} and on the previous state of magnetization of the material.

Table 27-1 **Magnetic Susceptibility of Various Materials at 20°C**

Material	χ_m
Aluminum	2.3×10^{-5}
Bismuth	-1.66×10^{-5}
Copper	-0.98×10^{-5}
Diamond	-2.2×10^{-5}
Gold	-3.6×10^{-5}
Magnesium	1.2×10^{-5}
Mercury	-3.2×10^{-5}
Silver	-2.6×10^{-5}
Sodium	-0.24×10^{-5}
Titanium	7.06×10^{-5}
Tungsten	6.8×10^{-5}
Hydrogen (1 atm)	-9.9×10^{-9}
Carbon dioxide (1 atm)	-2.3×10^{-9}
Nitrogen (1 atm)	-5.0×10^{-9}
Oxygen (1 atm)	2090×10^{-9}

Question

1. Why are some values of χ_m in Table 27-1 positive and others negative?

27-2 Atomic Magnetic Moments

The magnetization of a paramagnetic or ferromagnetic material can be related to the permanent magnetic moments of the individual atoms of the material. In general, the magnetic moment of an atom is related to its angular momentum. We can illustrate this by considering a particle of mass m_q and charge q moving with speed v in a circle of radius r as shown in Figure 27-5. (We use m_q here for the mass to avoid confusion with m for the magnetic moment.) The angular momentum of the particle is

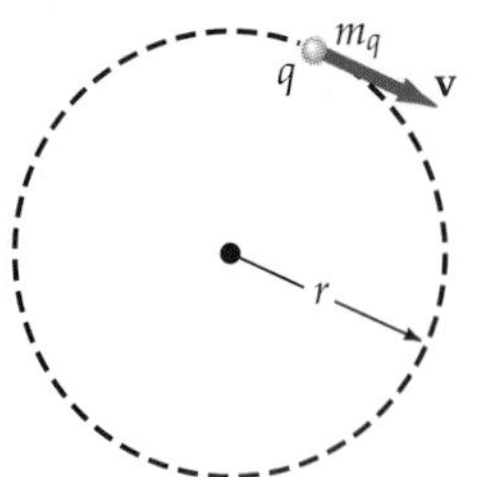

Figure 27-5 Particle of charge q and mass m_q moving in a circle of radius r. The angular momentum is into the paper and has a magnitude $m_q vr$, and the magnetic moment is into the paper (if q is positive) and has a magnitude $\frac{1}{2}qvr$.

$$L = m_q vr \qquad \text{27-7}$$

The magnitude of the magnetic moment is the product of the current and the area of the circle:

$$m = IA = I\pi r^2$$

For a charge moving in a circle, the current equals the charge times the frequency:

$$I = qf = \frac{q}{T}$$

where T is the period of the motion. Since the speed times the period equals the circumference of the circle, the period is given by

$$T = \frac{2\pi r}{v}$$

The current is thus

$$I = \frac{q}{T} = \frac{qv}{2\pi r}$$

and the magnetic moment is

$$m = IA \qquad \text{27-8}$$

$$= \frac{qv}{2\pi r}\pi r^2 = \frac{1}{2}qvr$$

Using $vr = L/m_q$ from Equation 27-7, we have for the magnetic moment

$$m = \frac{q}{2m_q}L$$

If the charge q is positive, the angular momentum and magnetic moment are in the same direction. We can therefore write

Magnetic moment and angular momentum

$$\mathbf{m} = \frac{q}{2m_q}\mathbf{L} \qquad \text{27-9}$$

Equation 27-9 is the general classical relation between magnetic moment and angular momentum. It also holds in the quantum theory of the atom for orbital angular momentum, but not for the intrinsic spin angular momentum of the electron. For electron spin, the magnetic moment is twice that predicted by this equation. The extra factor of 2 is a result from quantum theory that has no analog in classical mechanics.

In the quantum theory of the atom, angular momentum is quantized; that is, it can have only a certain discrete set of values. The orbital angular momentum must be an integral multiple of $h/2\pi$, and spin angular momentum must be a half-integral multiple of $h/2\pi$, where h is a fundamental constant called Planck's constant* and has the value

$$h = 6.67 \times 10^{-34}\ \text{J·s}$$

The combination $h/2\pi$ occurs often and is designated by $\hbar$ (read "h bar"):

$$\hbar = \frac{h}{2\pi} = 1.05 \times 10^{-34}\ \text{J·s}$$

The magnetic moment of an atom is therefore also quantized. It is convenient to write Equation 27-9 for the magnetic moment as

$$\mathbf{m} = \frac{q\hbar}{2m_q}\frac{\mathbf{L}}{\hbar}$$

For an electron, $m_q = m_e$ and $q = -e$, so the magnetic moment of the electron is

$$\mathbf{m} = -\frac{e\hbar}{2m_e}\frac{\mathbf{L}}{\hbar} = -m_B\frac{\mathbf{L}}{\hbar} \qquad 27\text{-}10$$

where

$$m_B = \frac{e\hbar}{2m_e} = 9.27 \times 10^{-24}\ \text{A·m}^2 = 9.27 \times 10^{-24}\ \text{J/T} \qquad 27\text{-}11$$

Bohr magneton

is called a **Bohr magneton.** The magnetic moment of an electron due to its intrinsic spin angular momentum has a magnitude of 1 Bohr magneton. Although the calculation of the magnetic moment of any atom is a complicated problem in quantum theory, the result for all atoms, according to both theory and experiment, is that the magnetic moment is of the order of a few Bohr magnetons (or zero for atoms with closed-shell electronic structures that have zero angular momentum). See Section 37-6 in the *Extended Version* for discussion of the shell structure of atoms.

If all the atoms or molecules in some material have their magnetic moments aligned, the magnetic moment per unit volume of the material is the product of the number of molecules per unit volume n and the magnetic moment m of each molecule. For this extreme case, the **saturation magnetization** M_s is

$$M_s = nm \qquad 27\text{-}12$$

The number of molecules per unit volume can be found from the molecular mass $\mathcal{M}$, the density of the material ρ, and Avogadro's number N_A:

$$n = \frac{N_A(\text{atoms/mol})}{\mathcal{M}(\text{kg/mol})}\rho(\text{kg/m}^3)$$

Example 27-2

Find the saturation magnetization and the magnetic field it produces for iron, assuming that each iron atom has a magnetic moment of 1 Bohr magneton.

The density of iron is 7.9×10^3 kg/m^3, and its molecular mass is 55.8×10^{-3} kg/mol. The number of iron molecules (atoms) per unit volume is then

$$n = \frac{6.02 \times 10^{23}\ \text{atoms/mol}}{55.8 \times 10^{-3}\ \text{kg/mol}}(7.9 \times 10^3\ \text{kg/m}^3)$$
$$= 8.52 \times 10^{28}\ \text{atoms/m}^3$$

*Quantization and Planck's constant will be discussed in Chapter 35.

Taking the magnetic moment of each atom to be 1 Bohr magneton, we have for the saturation magnetization

$$M_s = (8.52 \times 10^{28}\ \text{atoms/m}^3)(9.27 \times 10^{-24}\ \text{A·m}^2)$$
$$= 7.90 \times 10^5\ \text{A/m}$$

The magnetic field on the axis inside a long iron cylinder resulting from this maximum magnetization is then (Equation 27-3)

$$B = \mu_0 M_s = (4\pi \times 10^{-7}\ \text{T·m/A})(7.90 \times 10^5\ \text{A/m})$$
$$= 0.993\ \text{T} \approx 1\ \text{T}$$

The measured saturation magnetic field of annealed iron is about 2.16 T, indicating that the magnetic moment of an iron atom is slightly greater than 2 Bohr magnetons. This magnetic moment is due mainly to the spins of two unpaired electrons in the iron atom.

Questions

2. Can a particle have angular momentum and not have a magnetic moment?
3. Can a particle have a magnetic moment and not have angular momentum?
4. A circular loop of wire carries a current I. Is there angular momentum associated with the magnetic moment of the loop? If so, why is it not noticed?

27-3 Paramagnetism

Paramagnetic materials are those having a very small, positive magnetic susceptibility χ_m. Paramagnetism occurs in materials whose atoms have permanent magnetic moments that interact with each other only very weakly. When there is no external magnetic field, these magnetic moments are randomly oriented. In the presence of an external magnetic field, they tend to line up parallel to the field, but this is counteracted by the tendency for the moments to be randomly oriented due to thermal motion. The fraction of the moments that line up with the field depends on the strength of the field and on the temperature. In strong external magnetic fields at very low temperatures, nearly all the moments are aligned with the field. In this situation, the contribution to the total magnetic field due to the material is very large, as indicated in the numerical estimate in Example 27-2. However, even with the strongest magnetic field obtainable in the laboratory, the temperature must be as low as a few kelvins to obtain a high degree of alignment. At higher temperatures only a small fraction of the moments are aligned with the external field, and the contribution of the material to the total magnetic field is very small. We can state this more quantitatively by comparing the energy of a magnetic moment in an external magnetic field with the thermal energy of an atom of the material, which is of the order of kT, where k is Boltzmann's constant and T is the absolute temperature.

In Chapter 18, we saw that the potential energy of an electric dipole of moment **p** in an electric field **E** is given by (Equation 18-12)

$$U = -pE\cos\theta = -\mathbf{p}\cdot\mathbf{E}$$

Liquid oxygen, which is paramagnetic, is attracted by the magnetic field of a permanent magnet. In a uniform magnetic field, a magnetic dipole experiences a torque but no net force. In a nonuniform field a dipole experiences a net force that depends on the spatial rate of change or gradient of the field. Here the liquid oxygen collects near the edges of the closely spaced region where the field gradient is maximum.

The potential energy of a magnetic dipole of moment **m** in an external magnetic field **B** is given by a similar equation:

$$U = -mB\cos\theta = -\mathbf{m}\cdot\mathbf{B} \qquad 27\text{-}13$$

The potential energy when the moment is parallel with the field ($\theta = 0$) is thus lower than when it is antiparallel ($\theta = 180°$) by the amount $2mB$ (Figure 27-6). For a typical magnetic moment of 1 Bohr magneton and a typical strong magnetic field of 1 T, the difference in potential energy is

$$\Delta U = 2m_{\text{B}}B = 2(9.27 \times 10^{-24}\ \text{J/T})(1\ \text{T}) = 1.85 \times 10^{-23}\ \text{J}$$

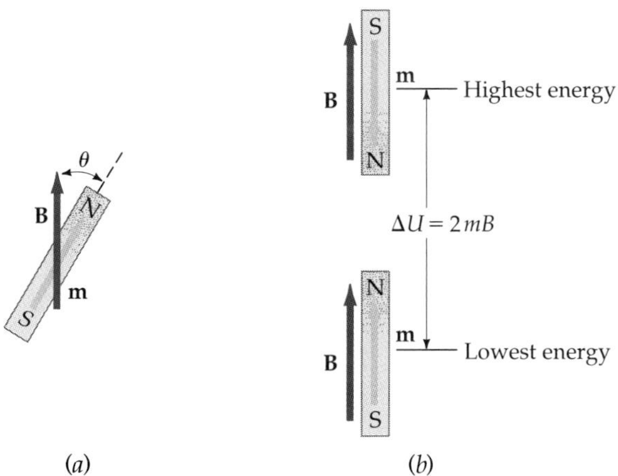

Figure 27-6 (*a*) A magnetic moment that makes an angle θ with a magnetic field **B**. (*b*) The potential energy is greatest when the moment is aligned with the field and least when it is antiparallel to the field. The difference in energy is $2mB$.

At a normal temperature of $T = 300$ K, the typical thermal energy kT is

$$kT = (1.38 \times 10^{-23}\ \text{J/K})(300\ \text{K}) = 4.14 \times 10^{-21}\ \text{J}$$

which is about 200 times greater than $2m_{\text{B}}B$. Thus, even in a strong magnetic field of 1 T, most of the magnetic moments will be randomly oriented because of thermal motions.

Figure 27-7 shows a plot of the magnetization M versus an applied external magnetic field B_{app} at a given temperature. In very strong fields, nearly all the magnetic moments are aligned with the field and $M \approx M_s$. (For magnetic fields attainable in the laboratory, this can only occur for very low temperatures.) When $B_{\text{app}} = 0$, $M = 0$, indicating that the orientation of the moments is completely random. In weak fields, the magnetization is approximately proportional to the applied field, as indicated by the orange dashed line in the figure. In this region, the magnetization is given by

Curie's law

$$M = \frac{1}{3}\frac{mB_{\text{app}}}{kT}M_s \tag{27-14}$$

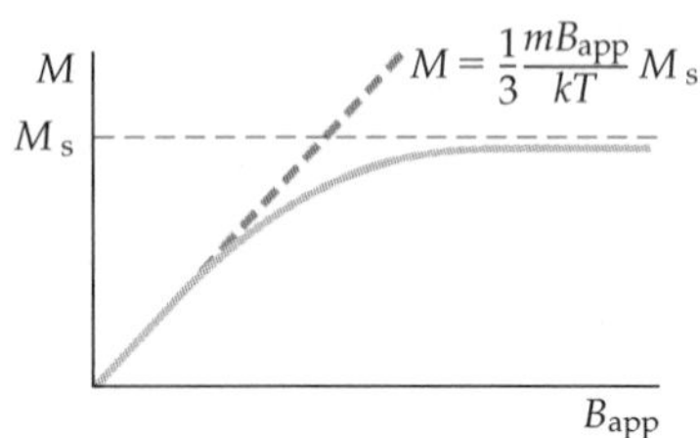

Figure 27-7 Plot of magnetization M versus applied field B_{app}. In very strong fields, the magnetization approaches the saturation value M_s. This can be achieved only at very low temperatures. In weak fields, the magnetization is approximately proportional to B_{app}, a result known as Curie's law.

Note that (mB_{app}/kT) is the ratio of the maximum energy of a dipole in the magnetic field to the characteristic thermal energy and is, therefore, a dimensionless number. The result that the magnetization varies inversely with the absolute temperature was discovered experimentally by Pierre Curie and is known as **Curie's law.**

Example 27-3

If $m = m_B$, at what temperature will the magnetization be 1 percent of the saturation magnetization in an applied magnetic field of 1 T?

From Curie's law, we have

$$M = \frac{1}{3}\frac{mB_{\text{app}}}{kT}M_s = 0.01\ M_s$$

Then,

$$T = \frac{mB_{\text{app}}}{0.03k} = \frac{(9.27 \times 10^{-24}\ \text{J/T})(1\ \text{T})}{(0.03)(1.38 \times 10^{-23}\ \text{J/K})} = 22.4\ \text{K}$$

From this example, we see that even in a strong applied magnetic field of 1 T the magnetization is less than 1 percent of saturation at temperatures above 22.4 K.

Exercise

If $m = m_B$, what fraction of the saturation magnetization is M at 300 K for an external magnetic field of 15,000 G? (Answer: $M/M_s = 1.12 \times 10^{-3}$)

27-4 Ferromagnetism

A chunk of magnetite (loadstone) attracts the needle of a compass.

Ferromagnetic materials are those having very large, positive values of magnetic susceptibility χ_m (as measured under conditions described below). Ferromagnetism occurs in pure iron, cobalt, and nickel and in alloys of these metals with each other. It also occurs in gadolinium, dysprosium, and a few compounds. In these substances, a small external magnetic field can produce a very large degree of alignment of the atomic magnetic dipole moments. In some cases, this alignment can persist even when the external magnetizing field is removed. This occurs because the magnetic dipole moments of the atoms of these substances exert strong forces on their neighbors so that over a small region of space the moments are aligned with each other even when there is no external field. The region of space over which the magnetic dipole moments are aligned is called a **magnetic domain.** The size of a domain is usually microscopic. Within the domain, all the magnetic moments are aligned, but the direction of alignment varies from domain to domain so that the net magnetic moment of a macroscopic piece of ferromagnetic material is zero in the normal state. Figure 27-8 illustrates this situation. The dipole forces that produce this alignment are predicted by quantum theory but cannot be explained with classical physics. At temperatures above a critical temperature, called the **Curie temperature,** thermal agitation is great enough to break up this alignment, and ferromagnetic materials become paramagnetic.

When an external magnetic field is applied, the boundaries of the domains may shift or the direction of alignment within a domain may change so that there is a net macroscopic magnetic moment in the direction of the applied field. Since the degree of alignment is large for even a small external field, the magnetic field produced in the material by the dipoles is often much greater than the external field.

Let us consider what happens when we magnetize a long iron rod by placing it inside a solenoid and gradually increase the current in the solenoid windings. We assume that the rod and the solenoid are long enough to permit us to neglect end effects. The magnetic field at the center of the rod is given by Equation 27-4:

$$\mathbf{B} = \mathbf{B}_{\text{app}} + \mu_0\mathbf{M} \qquad 27\text{-}15$$

where

$$B_{\text{app}} = \mu_0 nI$$

In ferromagnetic materials, the magnetic field $\mu_0 M$ due to the magnetic moments is often greater than the magnetizing field B_{app} by a factor of several thousand.

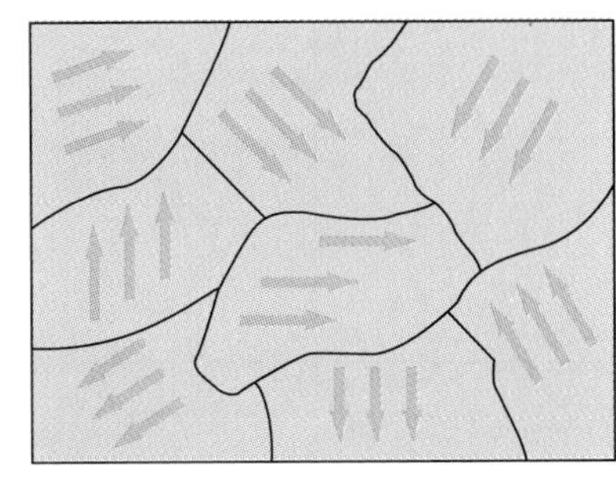

Figure 27-8 Schematic illustration of ferromagnetic domains. Within a domain, the magnetic dipoles are aligned, but the direction of alignment varies from domain to domain so that the net magnetic moment is zero. A small external magnetic field may cause the enlargement of those domains that are aligned parallel to the field, or it may cause the alignment within a domain to rotate. In either case, the result is a net magnetic moment parallel to the field.

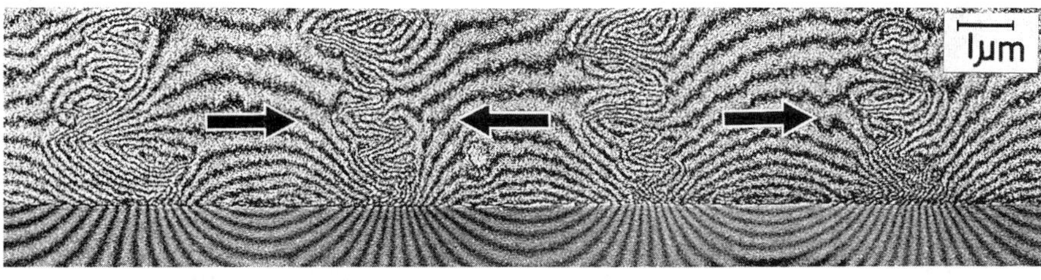

(*a*)

(*b*)

(*c*)

(*a*) Magnetic field lines on a cobalt magnetic recording tape. The solid arrows indicate the encoded magnetic bits. (*b*) Cross section of a magnetic tape recording head. Current from an audio amplifier is sent to wires around a magnetic core in the recording head where it produces a magnetic field. When the tape passes over a gap in the core of the recording head, the fringing magnetic field encodes information on the tape. The information is retrieved when the tape passes over a reading head shown in cross section in (*c*). In this case, the change in flux due to the magnetized tape induces currents in the wires around the core of the reading head.

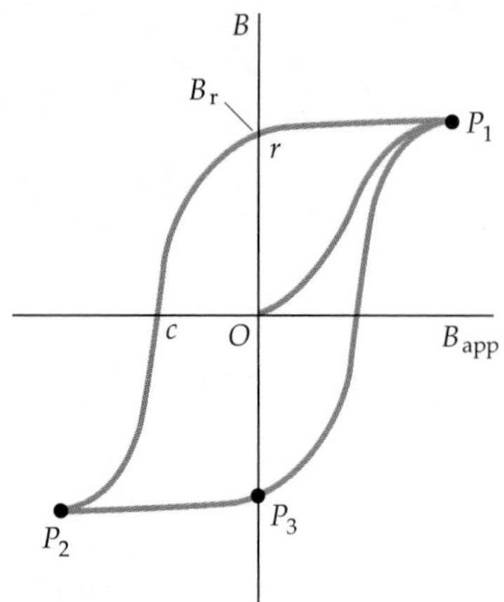

Figure 27-9 Plot of B versus the applied field B_{app}. The outer curve is called a hysteresis curve. The field B_r is called the remnant field. It remains when the applied field returns to zero.

Figure 27-9 shows a plot of B versus the magnetizing field B_{app}. As the current is gradually increased from zero, B increases from zero along the part of the curve from the origin O to point P_1. The flattening of this curve near point P_1 indicates that the magnetization M is approaching its saturation value M_s, at which all the atomic magnetic moments are aligned. Above saturation, B increases only because the magnetizing field $B_{app} = \mu_0 nI$ increases. When B_{app} is gradually decreased from point P_1, there is not a corresponding decrease in the magnetization. The shift of the domains in a ferromagnetic material is not completely reversible, and some magnetization remains even when B_{app} is reduced to zero, as indicated in the figure. This effect is called **hysteresis,** from the Greek word *hysteros* meaning later or behind, and the curve in Figure 27-9 is called a **hysteresis curve.** The value of the magnetic field at point r when B_{app} is zero is called the **remnant field** B_r. At this point, the iron rod is a permanent magnet. If the current in the solenoid is now reversed so that B_{app} is in the opposite direction, the magnetic field B is gradually brought to zero at point c. The remaining part of the hysteresis curve is obtained by further increasing the current in the opposite direction until point P_2 is reached, which corresponds to saturation in the opposite direction, and then decreasing the current to zero at point P_3 and increasing it again in its original direction.

Since the magnetization M depends on the previous history of the material, and it can have a large value even when the applied field is zero, it is not simply related to the applied field B_{app}. However, if we confined ourselves to that part of the magnetization curve from the origin to point P_1 in Figure 27-9, **M** and $\mathbf{B}_{app}$ are parallel and M is zero when B_{app} is zero. We can then define the magnetic susceptibility as in Equation 27-5,

$$M = \chi_m(B_{app}/\mu_0)$$

and

$$B = B_{app} + \mu_0 M = B_{app}(1 + \chi_m) = \mu_0 nI(1 + \chi_m) = \mu nI \qquad 27\text{-}16$$

where

$$\mu = (1 + \chi_m)\mu_0 \qquad 27\text{-}17$$

is called the **permeability** of the material. (The permeability is defined in the same way for paramagnetic and diamagnetic materials, but since χ_m is much less than 1 for these materials, the permeability μ and the permeability of free space μ_0 are very nearly equal.) The **relative permeability** K_m is a dimensionless number that is defined by

$$K_m = \frac{\mu}{\mu_0} = 1 + \chi_m = \frac{B}{B_{app}} \qquad 27\text{-}18$$

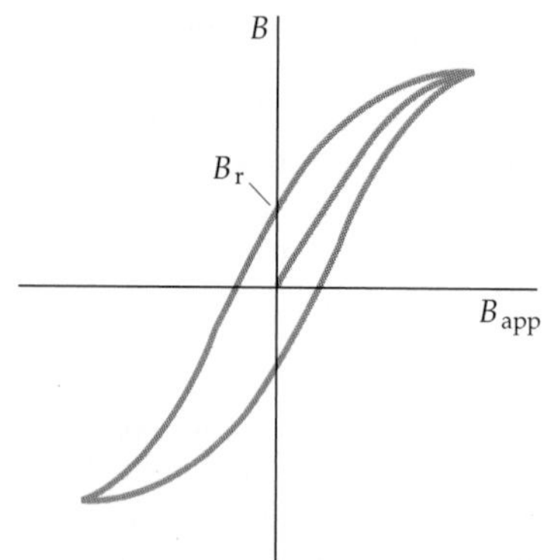

Figure 27-10 Hysteresis curve for a magnetically soft material. The remnant field is very small compared with that for a magnetically hard material such as that in Figure 27-9.

Since B does not vary linearly with B_{app}, as can be seen from Figure 27-9, the relative permeability is not constant. The maximum value of K_m occurs at a magnetization that is considerably less than the saturation magnetization. Table 27-2 lists the saturation magnetic field $\mu_0 M_s$ and the maximum values of K_m for some ferromagnetic materials. Note that the maximum values of K_m are much greater than 1.

The area enclosed by the hysteresis curve is proportional to the energy dissipated as heat in the irreversible process of magnetizing and demagnetizing. If the hysteresis effect is small, so that the area is small indicating a small energy loss, the material is called **magnetically soft.** Soft iron is an example. The hysteresis curve for a magnetically soft material is shown in Figure 27-10. Here the remnant field B_r is nearly zero, and the energy loss per cycle is small. Magnetically soft materials are used for transformer cores to allow the magnetic field B to change without incurring large energy losses

Table 27-2 **Maximum Values of $\mu_0 M$ and K_m for Some Ferromagnetic Materials**

Material	$\mu_0 M_s$, T	K_m
Iron (annealed)	2.16	5,500
Iron-silicon (96% Fe, 4% Si)	1.95	7,000
Permalloy (55% Fe, 45% Ni)	1.60	25,000
Mu-metal (77% Ni, 16% Fe, 5% Cu, 2% Cr)	0.65	100,000

as the field alternates. On the other hand, a large remnant field is desirable in a permanent magnet. **Magnetically hard** materials, such as carbon steel and the alloy Alnico 5, are used for permanent magnets.

(*a*)

(*b*) 10μm

(*a*) Computer hard disk drive for magnetic storage of information. (*b*) A magnetic test pattern on a hard disk, magnified 2400 times. The light and dark regions correspond to oppositely directed magnetic fields. The smooth region just outside the pattern is a region of the disk that has been erased just prior to writing.

Example 27-4

A long solenoid with 12 turns per centimeter has a core of annealed iron. When the current is 0.50 A, the magnetic field inside the iron core is 1.36 T. Find (*a*) the applied field B_{app}, (*b*) the relative permeability K_m, and (*c*) the magnetization M.

(*a*) The applied field is

$$B_{app} = \mu_0 nI = (4\pi \times 10^{-7}\ \text{T·m/A})(1200\ \text{turns/m})(0.50\ \text{A})$$
$$= 7.54 \times 10^{-4}\ \text{T}$$

Note that the total magnetic field is 1.36 T, so this applied magnetic field is a negligible fraction of the total field.

(*b*) From Equation 27-18, the relative permeability is

$$K_m = \frac{B}{B_{app}} = \frac{1.36\ \text{T}}{7.54 \times 10^{-4}\ \text{T}} = 1.80 \times 10^3 = 1800$$

Note that this is considerably smaller than the maximum value of K_m which is about 5500 (Table 27-2). To the three-place accuracy with which we have calculated K_m, the susceptibility χ_m is equal to the relative permeability:

$$\chi_m = K_m - 1 \approx K_m = 1800$$

(*c*) We can find the magnetization from Equation 27-3 or from Equation 27-6. Using Equation 27-6, we have

$$\mu_0 M = B - B_{app} = 1.36\ \text{T} - 7.54 \times 10^{-4}\ \text{T} \approx B = 1.36\ \text{T}$$

Then

$$M = \frac{B}{\mu_0} = \frac{1.36\ \text{T}}{4\pi \times 10^{-7}\ \text{T·m/A}} = 1.08 \times 10^6\ \text{A/m}$$

Questions

5. In a common lecture demonstration, a long iron rod is held such that its axis is along the magnetic field of the earth and is then struck with a hammer. This makes the rod a permanent magnet. It can be demagnetized if it is held perpendicular to the earth's field and struck again. Explain what happens in the rod.

6. A permanent magnet may lose much of its magnetization if it is dropped or banged against something. Why?

27-5 Diamagnetism

Diamagnetic materials are those having very small, negative values of magnetic susceptibility χ_m. Diamagnetism was discovered by Faraday in 1846 when he found that a piece of bismuth is repelled by either pole of a magnet, indicating that the external field of the magnet induces a magnetic moment in bismuth in the direction opposite the field. We can understand this effect qualitatively using Lenz's law. Figure 27-11 shows two positive charges moving in circular orbits with the same speed but in opposite directions. Their magnetic moments are in opposite directions and therefore cancel. (It is simpler to consider positive charges even though it is the negatively charged electrons that provide the magnetic moments in matter.) Consider now what happens when an external magnetic field **B** is turned on in the direction into the paper. According to Lenz's law, current will be induced to oppose the change in flux. If we assume that the radius of the circle does not change, the charge on the left will be speeded up to increase its flux out of the page, and that on the right will be slowed down to decrease its flux into the page. In each case, the *change* in the magnetic moment of the charges will be in the direction out of the page, opposite that of the external applied field. Since the permanent magnetic moments of the two charges are equal and oppositely directed, they add to zero, leaving only the induced magnetic moments, which are both opposite the direction of the applied magnetic field.

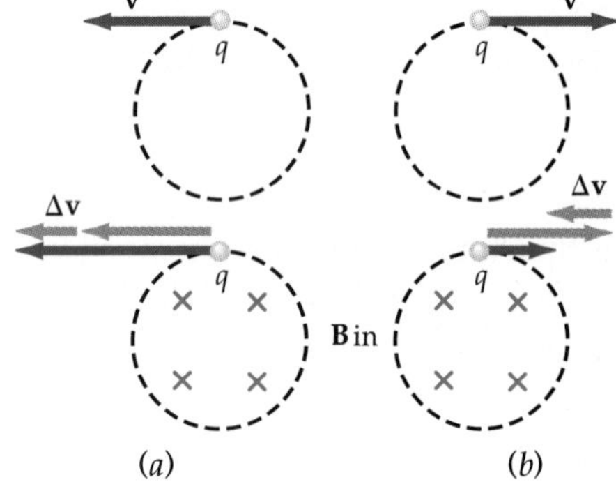

Figure 27-11 (*a*) A positive charge moving counterclockwise in a circle has its magnetic moment outward. When an external, inward magnetic field is turned on, the speed of the particle will increase to oppose the change in flux. The change in the magnetic moment is outward. (*b*) A positive charge moving clockwise in a circle has its magnetic moment inward. When an external, inward magnetic field is turned on, the speed of the particle will decrease to oppose the change in flux. As in (*a*), the change in the magnetic moment is outward.

Atoms that have a closed-shell electronic structure have a net angular momentum of zero and therefore no net permanent magnetic moment. Materials that have such atoms—bismuth, for example—are diamagnetic. As we will see below, the induced magnetic moments that cause diamagnetism have magnitudes of the order of 10^{-5} Bohr magnetons. Since this is much smaller than the permanent magnetic moments of the atoms of paramagnetic or ferromagnetic materials, which do not have closed-shell structure, the diamagnetic effect in these atoms is masked by the alignment of their permanent magnetic moments. However, since this alignment decreases with temperature, all materials are theoretically diamagnetic at sufficiently high temperatures.

A superconductor is a perfect diamagnet, that is, it has a magnetic susceptibility of -1. When a superconductor is placed in an external magnetic field, electric currents are induced on its surface so that the net magnetic field in the superconductor is zero. Consider a superconducting rod inside a solenoid of n turns per unit length. When the solenoid is connected to a source of emf so that it carries a current I, the magnetic field due to the solenoid is $\mu_0 nI$. A surface current of $-nI$ per unit length is induced on the

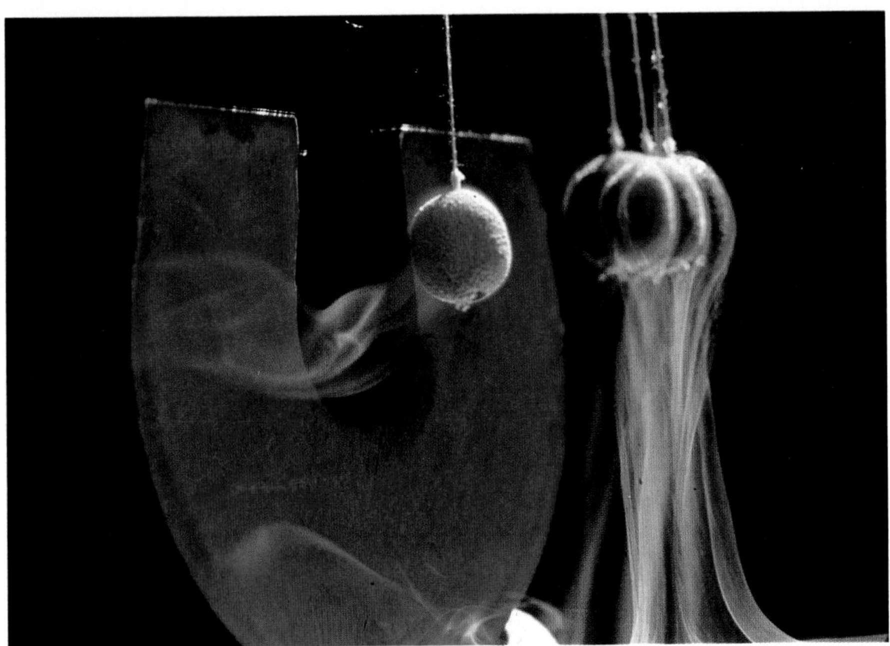

A superconductor is a perfect diamagnet. Here, the superconducting pendulum bob is repelled by the permanent magnet.

superconducting rod that cancels out the field due to the solenoid so that the net field inside the superconductor is zero. From Equation 27-6,

$$\mathbf{B} = \mathbf{B}_{\text{app}}(1 + \chi_{\text{m}}) = 0$$

so

$$\chi_{\text{m}} = -1$$

Estimating Induced Magnetic Moments

We can estimate the magnitude of the induced magnetic moments in diamagnetic materials by relating the change in the speeds of the electrons to the change in the centripetal force due to the external magnetic field. We assume that the radius of the orbit does not change and that the change in speed is small compared with the original speed. Both of these assumptions can be justified. The original centripetal force is provided by the electrostatic force of attraction F of the electron to the nucleus. Setting this force equal to the mass times the acceleration, we obtain

$$F = \frac{m_{\text{e}}v^2}{r} \qquad \text{27-19}$$

where m_{e} is the mass of the electron. In the presence of an external magnetic field $\mathbf{B}$, there is an additional force $q\mathbf{v} \times \mathbf{B}$ on each particle. For the particle on the left in Figure 27-11, this force is directed inward (for a positively charged particle), thus increasing the net inward force, which is necessary because this particle speeds up when the magnetic field is turned on. Similarly, for the particle on the right, the magnetic force is outward, which reduces the net inward force. Again, this is in the correct direction since the particle on the right slows down when the field is turned on.

Since the change in the net inward force is small, we can approximate it by a differential. Differentiating Equation 27-19, we obtain

$$dF = \frac{2m_{\text{e}}v}{r}\,dv \approx \Delta F$$

Setting this change in force equal to the magnetic force qvB, we have

$$qvB = \frac{2m_e v}{r}\,dv$$

or

$$dv = \frac{qrB}{2m_e}$$

This change in speed causes a small change in the magnetic moment of the charge. Using Equation 27-8, we have for the magnetic moment

$$m = \tfrac{1}{2}qvr$$

Then

$$\Delta m \approx dm = \frac{1}{2}qr\,dv = \frac{1}{2}qr\left(\frac{qrB}{2m_e}\right) = \frac{q^2r^2}{4m_e}\,B \qquad 27\text{-}20$$

Using $r = 10^{-10}$ m for a typical atomic radius, $B = 1$ T for a typical strong magnetic field, and the values of q and m_e for the electron, we obtain for the order of magnitude of the induced magnetic moment

$$\Delta m \approx \frac{(1.60 \times 10^{-19}\ \text{C})^2(10^{-10}\ \text{m})^2}{4(9.11 \times 10^{-31}\ \text{kg})}(1\ \text{T}) \approx 7 \times 10^{-29}\ \text{A·m}^2 \sim 10^{-28}\ \text{A·m}^2$$

Comparing this induced magnetic moment with the Bohr magneton ($m_B = 9.27 \times 10^{-24}$ A·m$^2 \sim 10^{-23}$ A·m^2), we see that the induced magnetic moments are about 100,000 times smaller than the magnetic moment of an electron due to its intrinsic spin or to its orbital motion in an atom. Thus, materials that have atoms with permanent magnetic dipole moments, which are of the order of a Bohr magneton, are paramagnetic (or ferromagnetic). The resultant magnetic moment in the direction of the applied magnetic field that results from even a very small partial alignment of these permanent magnetic moments at ordinary temperatures is much greater than the induced magnetic dipole moment in the direction opposite the applied magnetic field.

Question

7. Why might you expect a heavy diamagnetic element to have a susceptibility of greater magnitude than a light diamagnetic element?

Summary

1. All materials can be classified as either paramagnetic, ferromagnetic, or diamagnetic.

2. A magnetized material is described by its magnetization vector **M**, which is defined to be the net magnetic dipole moment per unit volume of the material:

$$\mathbf{M} = \frac{d\mathbf{m}}{dV}$$

The magnetic field due to a uniformly magnetized cylinder is the same as if the cylinder carried a current per unit length of magnitude M on its surface. This current, which is due to the intrinsic motion of the atomic charges in the cylinder, is called an amperian current.

3. If a long cylinder of magnetic material is placed inside a solenoid of n turns per unit length carrying a current I, the resultant magnetic field at

a point inside the solenoid and far from its ends due to the current in the solenoid plus the magnetized material is

$$\mathbf{B} = \mathbf{B}_{\text{app}} + \mu_0\mathbf{M}$$

where the applied magnetic field has the magnitude

$$B_{\text{app}} = \mu_0 nI$$

For paramagnetic and ferromagnetic materials, **M** is in the same direction as $\mathbf{B}_{\text{app}}$; for diamagnetic materials, **M** is opposite to $\mathbf{B}_{\text{app}}$.

4. In paramagnetic and diamagnetic materials the magnetization is proportional to the magnetizing field B_{app}:

$$M = \chi_{\text{m}}(B_{\text{app}}/\mu_0)$$

where χ_{m} is called the magnetic susceptibility. For paramagnetic materials, χ_{m} is a small positive number that depends on temperature. For diamagnetic materials (other than superconductors), it is a small negative constant independent of temperature. For superconductors, $\chi_{\text{m}} = -1$. For ferromagnetic materials, the magnetization depends not only on the magnetizing current but also on the past history of the material.

5. The magnetic moment of a particle of charge q and mass m_q is related to its angular momentum **L** by

$$\mathbf{m} = \frac{q}{2m_q}\mathbf{L} = \frac{q\hbar}{2m_q}\frac{\mathbf{L}}{\hbar}$$

where

$$\hbar = \frac{h}{2\pi} = 1.05 \times 10^{-34}\ \text{J·s}$$

is a convenient unit for expressing the angular momentum of electrons and atoms and

$$h = 6.67 \times 10^{-34}\ \text{J·s}$$

is a fundamental constant called Planck's constant. The magnetic moments of electrons and atoms are conveniently expressed in units of the Bohr magneton m_{B}:

$$m_{\text{B}} = \frac{e\hbar}{2m_{\text{e}}} = 9.27 \times 10^{-24}\ \text{A·m}^2 = 9.27 \times 10^{-24}\ \text{J/T}$$

The magnetic moment associated with the spin angular momentum of the electron is 1 Bohr magneton, and the magnetic moment of an atom is of the order of a few Bohr magnetons.

6. Paramagnetic materials have permanent atomic magnetic moments that have random directions in the absence of an external magnetic field. In an external field, some of these dipoles are aligned producing a small contribution to the total field that adds to the external field. The degree of alignment is small except in very strong fields and at very low temperatures. At ordinary temperatures, thermal motion tends to maintain the random directions of the magnetic moments. At low fields, the magnetization is approximately proportional to the external field and is given by Curie's law:

$$M = \frac{1}{3}\frac{mB_{\text{app}}}{kT}M_{\text{s}}$$

where M_{s} is the saturation magnetization that results when all the magnetic dipole moments are aligned.

7. Ferromagnetic materials have small regions of space called magnetic domains in which the permanent atomic magnetic moments are aligned.

When unmagnetized, the direction of alignment in one domain is independent of that in another so that no net magnetic field is produced. When magnetized, the domains of a ferromagnetic material are aligned, producing a very strong contribution to the magnetic field. This alignment can persist even when the external field is removed, thus leading to permanent magnetism.

8. A plot of the magnetic field in a ferromagnetic material versus magnetizing field is called a hysteresis curve. In the upper right quadrant of this curve **M** and $\mathbf{B}_{app}$ are in the same direction and a magnetic susceptibility χ_m can be defined for ferromagnetic materials in the same way as it is defined for paramagnetic and diamagnetic materials. The magnetic field inside a ferromagnetic material in a solenoid carrying current I is then related to the applied field by

$$B = B_{app} + \mu_0 M = B_{app}(1 + \chi_m) = \mu_0 nI(1 + \chi_m) = \mu nI$$

where

$$\mu = (1 + \chi_m)\mu_0$$

is the permeability of the material. The relative permeability K_m is a dimensionless number, which is defined as the ratio of the permeability to the permeability of free space:

$$K_m = \frac{\mu}{\mu_0} = 1 + \chi_m = \frac{B}{B_{app}}$$

The maximum value of K_m is much greater than 1 for ferromagnetic materials.

9. Diamagnetic materials are those in which the magnetic moments of all electrons in each atom cancel, leaving each atom with zero magnetic moment in the absence of an external field. In an external field, a very small magnetic moment is induced that tends to weaken the field. This effect is independent of temperature. Superconductors are diamagnetic with susceptibility equal to −1.

Suggestion for Further Reading

Becker, Joseph J.: "Permanent Magnets," *Scientific American*, December 1970, p. 92.

Magnets made from the new alloys described in this article can be made many times stronger than those made from conventional metals.

Review

A. Objectives: After studying this chapter, you should:

1. Be able to list the three types of magnetism and discuss the origins, directions, and strengths of the magnetic effects in each.

2. Be able to derive the relation between magnetic moment and angular momentum for a charged particle moving in a circle.

3. Know the sign and order of magnitude of the magnetic susceptibility of paramagnetic and diamagnetic materials.

4. Be able to describe the general temperature dependence of the magnetization in paramagnetic materials and explain its origin.

5. Be able to sketch B versus B_{app} for ferromagnetic materials.

B. Define, explain, or otherwise identify:

Magnetization
Amperian current
Magnetic susceptibility
Bohr magneton

Saturation magnetization
Paramagnetic materials
Curie's law
Ferromagnetic materials
Magnetic domain
Curie temperature
Hysteresis
Hysteresis curve
Remnant field
Permeability
Relative permeability
Magnetically soft
Magnetically hard
Diamagnetic materials

C. True or false: If the statement is true, explain why it is true. If it is false, give a counterexample.

1. Diamagnetism occurs in all materials.

2. Diamagnetism is the result of induced magnetic dipole moments.

3. Paramagnetism is the result of the partial alignment of permanent magnetic dipole moments.

4. Hysteresis is associated with a loss in electromagnetic energy.

5. Magnetically hard materials are good for transformer cores.

Problems

Level I

27-1 Magnetization and Magnetic Susceptibility

1. A tightly wound solenoid 20 cm long has 400 turns and carries a current of 4 A such that its axial field is in the z direction. Neglecting end effects, find B and B_{app} at the center when (*a*) there is no core in the solenoid and (*b*) there is an iron core with a magnetization $M = 1.2 \times 10^6$ A/m.

2. Which of the four gases listed in Table 27-1 are diamagnetic and which are paramagnetic?

3. If the solenoid of Problem 1 has an aluminum core, find B_{app}, M, and B at the center, neglecting end effects.

4. Repeat Problem 3 for a tungsten core.

5. A long solenoid is wound around a tungsten core and carries a current. (*a*) If the core is removed while the current is held constant, does the magnetic field inside the solenoid decrease or increase? By what percentage? (*b*) Does the self inductance of the solenoid decrease or increase? By what percentage?

6. When a sample of liquid is inserted into a solenoid carrying a constant current, the magnetic field inside the solenoid decreases by 0.004 percent. What is the magnetic susceptibility of the liquid?

7. A long solenoid carrying a current of 10 A has 50 turns/cm. What is the magnetic field in the interior of the solenoid when the interior is (*a*) a vacuum, (*b*) filled with aluminum, and (*c*) filled with silver?

27-2 Atomic Magnetic Moments

8. Nickel has a density of 8.7 g/cm^3 and molecular mass of 58.7 g/mol. Its saturation magnetization is given by $\mu_0 M_s = 0.61$ T. Calculate the magnetic moment of a nickel atom in Bohr magnetons.

9. Repeat Problem 8 for cobalt, which has a density of 8.9 g/cm^3, a molecular mass of 58.9 g/mol, and a saturation magnetization given by $\mu_0 M_s = 1.79$ T.

27-3 Paramagnetism

10. Show that Curie's law predicts that the magnetic susceptibility of a paramagnetic substance is $\chi_m = m\mu_0 M_s / 3kT$.

11. Assume that the magnetic moment of an aluminum atom is 1 Bohr magneton. The density of aluminum is 2.7 g/cm^3, and its molecular mass is 27 g/mol. (*a*) Calculate M_s and $\mu_0 M_s$ for aluminum. (*b*) Use the results of Problem 10 to calculate χ_m at $T = 300$ K. (*c*) Explain why the result for part (*b*) is larger than the value listed in Table 27-1.

27-4 Ferromagnetism

12. The saturation magnetization for annealed iron occurs when $B_{app} = 0.201$ T. Find the permeability μ and the relative permeability K_m of annealed iron at saturation. (See Table 27-2.)

13. For annealed iron, the relative permeability K_m has its maximum value of about 5500 at $B_{app} = 1.57 \times 10^{-4}$ T. Find M and B when K_m is maximum.

14. The coercive force is defined to be the applied magnetic field needed to bring B back to zero along the hysteresis curve (point c in Figure 27-9). For a certain permanent bar magnet, $B_{app} = 5.53 \times 10^{-2}$ T. The bar magnet is to be demagnetized by placing it inside a 15-cm-long solenoid with 600 turns. What minimum current is needed in the solenoid to demagnetize the magnet?

15. A long solenoid with 50 turns/cm carries a current of 2 A. The solenoid is filled with iron, and B is measured to be 1.72 T. (*a*) Neglecting end effects, what is B_{app}? (*b*) What is M? (*c*) What is the relative permeability K_m?

16. When the current in Problem 15 is 0.2 A, the magnetic field is measured to be 1.58 T. (*a*) Neglecting end effects, what is B_{app}? (*b*) What is M? (*c*) What is the relative permeability K_m?

27-5 Diamagnetism

There are no problems for this section.

Level II

17. A long, iron-core solenoid with 2000 turns/m carries a current of 20 mA. At this current, the relative permeability of the iron core is 1200. (*a*) What is the magnetic field within the solenoid? (*b*) With the iron core removed, find the current necessary to produce the same field within the solenoid.

18. The magnetic dipole moment of an iron atom is 2.219 m_B. (*a*) If all the atoms in an iron bar of length 20 cm and cross-sectional area 2 cm^2 have their dipole moments aligned, what is the dipole moment of the bar? (*b*) What torque must be supplied to hold the iron bar perpendicular to a magnetic field of 0.25 T?

19. A small magnetic sample is in the form of a disk. It has a radius of 1.4 cm, a thickness of 0.3 cm, and a uniform magnetization throughout its volume. The magnetic moment of the sample is 1.5×10^{-2} A·m^2. (*a*) What is the magnetization **M** of the sample? (*b*) If this magnetization is due to the alignment of N electrons each with a magnetic moment of 1 m_B, what is N? (*c*) If the magnetization is along the axis of the disk, what is the magnitude of the amperian surface current?

20. A very long solenoid of length ℓ and cross-sectional area A has n turns per unit length and carries a current I. It is filled with iron of relative permeability K_m. (*a*) Find the self inductance of the solenoid. (*b*) Use $U_m = \frac{1}{2}LI^2$ to find the magnetic energy stored in the solenoid in terms of the magnetic field B. (*c*) Show that the energy density in the solenoid is $\eta_m = B^2/(2K_m\mu_0) = B^2/2\mu$.

21. The magnetic moment of the earth is about 9×10^{22} A·m^2. (*a*) If the magnetization of the earth's core were 1.5×10^9 A/m, what is the core volume? (*b*) What is the radius of such a core if it were spherical and centered with the earth?

22. In a simple model of paramagnetism, we can consider that some fraction f of the molecules have their magnetic moments aligned with the external magnetic field and that the rest of the molecules are randomly oriented and so do not contribute to the magnetic field. (*a*) Use this model and Curie's law to show that at temperature T and external magnetic field B the fraction of aligned molecules is $f = mB/3kT$. (*b*) Calculate this fraction for $T = 300$ K, $B = 1$ T, assuming m to be 1 Bohr magneton.

23. It is desired to fill a solenoid with a mixture of oxygen and nitrogen at room temperataure and 1 atmosphere pressure such that K_m is exactly 1. Assume that the magnetic dipole moments of the gas molecules are all aligned and that the susceptibility of a gas is proportional to the number density of its molecules. What should the ratio of the number densities of oxygen to nitrogen molecules be so that $K_m = 1$?

24. A cylinder of magnetic material is placed in a long solenoid of n turns per unit length and current I. Table 27-3 gives the magnetic field B versus nI. Use these values to plot B versus B_{app} and K_m versus nI.

Table 27-3 **For Problem 24**

nI, A/m	B, T
0	0
50	0.04
100	0.67
150	1.00
200	1.2
500	1.4
1,000	1.6
10,000	1.7

25. A toroid with N turns carrying a current I has mean radius R and cross-sectional radius r, where $r < R$ (Figure 27-12). When the toroid is filled with material, it is called a *Rowland ring*. Find $\mathbf{B}_{app}$ and $\mathbf{B}$ in such a ring, assuming a magnetization **M** everywhere parallel to $\mathbf{B}_{app}$.

Figure 27-12 Problem 25.

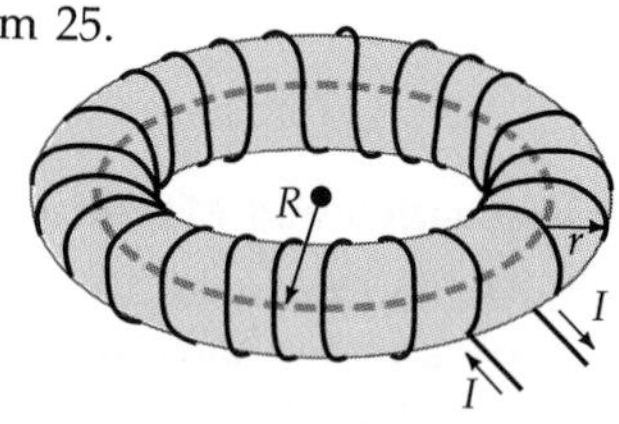

26. A toroid is filled with liquid oxygen that has a susceptibility of 4×10^{-3}. The toroid has 2000 turns and carries a current of 15 A. Its mean radius is 20 cm, and the radius of its cross section is 0.8 cm. (*a*) What is the magnetization M? (*b*) What is the magnetic field B? (*c*) What is the percentage increase in B produced by the liquid oxygen?

27. The toroid of Problem 26 has its core filled with iron. When the current is 10 A, the magnetic field in the toroid is 1.8 T. (*a*) What is the magnetization M? (*b*) Find the values for K_m, μ, and χ_m for the iron sample.

28. A toroid has average radius 14 cm and a cross-sectional area of 3 cm^2. It is wound with fine wire with 60 turns/cm measured along its mean circumference that carries a current of 4 A. The core is filled with a paramagnetic material of magnetic susceptibility 2.9×10^{-4}. (*a*) What is the magnitude of the magnetic field within the substance? (*b*) What is the magnitude of the magnetization? (*c*) What would the magnitude of the magnetic field be if there were no paramagnetic core present?

29. What would the result be for Problem 28 if soft iron, having a relative permeability of 500, were substituted for the paramagnetic core?

30. Two long, straight wires 4.0 cm apart are embedded in a uniform insulator having relative permeability of $K_m = 120$. The wires carry 40 A in opposite directions. (*a*) What is the magnetic field at the midpoint of the plane of the wires? (*b*) What is the force per unit length on the wires?

31. A long, narrow bar magnet that has magnetic moment **m** parallel to its long axis is suspended at its center as a frictionless compass needle. When placed in a magnetic field **B**, the needle lines up with the field. If it is displaced by a small angle θ, show that the needle will oscillate about its equilibrium position with frequency $f = (1/2\pi)\sqrt{mB/I}$, where I is the moment of inertia about the point of suspension.

32. Suppose the needle in Problem 31 is a uniformly magnetized iron rod that is 8 cm long and has a cross-sectional area of 3 mm^2. Assume that the magnetic dipole moment for each iron atom is 2.2 m_B and that all the iron atoms have their dipole moments aligned. Calculate the frequency of small oscillations about the equilibrium position when the magnetic field is 0.5 G.

33. The needle of a magnetic compass has a length of 3 cm, a radius of 0.85 mm, and a density of 7.96×10^3 kg/m^3. It is free to rotate in a horizontal plane, where the horizontal component of the earth's magnetic field is 0.6 G. When disturbed slightly, the compass executes simple harmonic motion about its midpoint with a frequency of 1.4 Hz. (*a*) What is the magnetic dipole moment of the needle? (*b*) What is the magnetization M? (*c*) What is the amperian current on the surface of the needle? (See Problem 31.)

34. A long, straight wire with a radius of 1.0 mm is coated with an insulating ferromagnetic material that has a thickness of 3.0 mm and a relative magnetic permeability of $K_m = 400$. The coated wire is in air and the wire itself is nonmagnetic. The wire carries a current of 40 A. (*a*) Find the magnetic field inside the wire as a function of radius r. (*b*) Find the magnetic field inside the ferromagnetic material as a function of radius r. (*c*) Find the magnetic field outside the ferromagnetic material as a function of r. (*d*) What must the magnitudes and directions of the amperian currents be on the surfaces of the ferromagnetic material to account for the magnetic fields observed?

35. In Section 27-5, the change in speed of an electron in an atom is found for the case in which a magnetic field is impressed on the atom. From this result, show that the change in angular frequency is $\Delta\omega = eB/2m$. This frequency is called the *Larmor frequency*.

36. An iron bar of length 1.4 m has a diameter of 2 cm and a uniform magnetization of 1.72×10^6 A/m directed along the bar's length. The bar is stationary in space and is suddenly demagnetized so that its magnetization disappears. What is the rotational angular velocity of the bar if its angular momentum is conserved? (Assume that Equation 27-9 holds where m_q is the mass of an electron and $q = -e$.)

37. A bar magnet has a diameter of 2 cm and has a magnetic field in its center of 0.1 T. If the magnet cracks apart in its middle, the magnetic attraction holds the two pieces together. (*a*) Show that if the pieces move apart a small distance dx, the additional magnetic energy is $dU_m = (B^2/2\mu_0)A\,dx$, where A is the cross-sectional area of the magnet and B is the magnetic field in the gap, which is assumed to be the same as the field inside the magnet. (*b*) Estimate the required force needed to pull apart the two pieces by computing the work done to separate them by a distance dx.

Level III

38. A proton has a magnetic moment **m**, which is parallel to its angular momentum **L**. It is in a uniform magnetic field **B** that makes an angle θ with **m** and **L**. Show that the magnetic moment vector precesses about the magnetic field, and find the angular frequency of precession.

39. Two long conducting strips are each 20 m wide and 0.3 mm thick. The strips are in parallel planes, which are separated by a 4.0-cm ferromagnetic spacer that has a relative permeability of $K_m = 400$. The conducting strips carry a uniform current of 4800 A in opposite directions. In the space between the strips far from their edges find (*a*) B_{app}, (*b*) B, and (*c*) the magnetic energy per unit volume.

40. In our derivation of the magnetic moment induced in an atom, we assumed that the radius of the electron's orbit did not change in the presence of an external magnetic field. In this problem you will show that this assumption of constant radius is justified by showing that, when **B** is applied, there is an impulse that increases or decreases the speed of the electron by just the correct amount, which is given by $\Delta v = qrB/2m$. (*a*) Use Faraday's law to show that the induced electric field is related to the rate of change of the magnetic field by $E = \frac{1}{2}r\,dB/dt$, assuming r to be constant. (*b*) Use Newton's second law to show that the change in speed of the electron dv is related to the change in B by $dv = (qr/2m)\,dB$. Integrate this expression to find Δv.

41. Equation 27-20 gives the induced magnetic moment for a single electron in an orbit that has its plane perpendicular to **B**. If an atom has Z electrons, a reasonable simplifying assumption is that, on the average, one-third have their planes perpendicular to **B**. Show that the diamagnetic susceptibility obtained from Equation 27-20 is then

$$\chi_m = \frac{-nZq^2r^2}{12m_e}\mu_0$$

where n is the number of atoms per unit volume. Use $n \approx 6 \times 10^{28}$ atoms/m^3 and $r \approx 5 \times 10^{-11}$ m to estimate χ_m for $Z \approx 50$.

Chapter 28

Alternating-Current Circuits

Most high-voltage long-distance power transmission is presently accomplished using alternating current. Historically, the difficulty and expense in converting ac to dc at the sending end (such as a hydroelectric power plant) and from dc back to ac at the consuming end have been major disadvantages in transmitting power via high-voltage direct current (HVDC). Because of recent advances in technology, interest in HVDC has been revived. Shown here is a dc to ac conversion station near Boston, linked by HVDC lines to a 2000-MW hydroelectric generator unit in James Bay, Quebec. Converter valves called thyristors are linked via Δ-Y connections to bushings that lead out of the station. This particular station is used to convert direct current to alternating current; but, like many others, it can be run in reverse.

Toward the end of the nineteenth century, there was a heated debate as to whether direct or alternating current should be used to deliver electrical energy to consumers in the United States. Thomas Edison lobbied for the use of direct current while Nikola Tesla and George Westinghouse were proponents of the use of alternating current. In 1893, alternating current was chosen to light the World Columbian Exposition in Chicago, and a contract was awarded to Westinghouse to deliver alternating current generated at Niagara Falls to American homes and factories.

Alternating current has the great advantage that electrical energy can be transported over long distances at very high voltages and low currents to reduce energy losses in the form of Joule heat. It can then be transformed, with almost no energy loss, to lower and safer voltages and correspondingly higher currents for everyday use. The transformer that accomplishes this change in voltage and current works on the basis of magnetic induction. We will study the transformer in Section 28-6.

More than 99 percent of the electrical energy used today is produced by electrical generators in the form of alternating current. In North America, power is delivered by a sinusoidal current of frequency 60 Hz. Devices such as radios, television sets and microwave ovens detect or generate alternating currents of much greater frequencies. Alternating current is easily produced by magnetic induction in an ac generator, as we discussed in Chapter 26. Although practical generators are considerably more complicated than the simple device we studied in Section 26-6, they are designed to put out a sinusoidal emf. We will see that when the generator output is sinusoidal, the current in an inductor, a capacitor, or a resistor is also sinusoidal, though it is generally not in phase with the generator's emf. When the emf and current are both sinusoidal, their maximum values can be simply related. The study of sinusoidal currents is important because even currents that are not sinusoidal can be analyzed in terms of sinusoidal components using Fourier analysis.

We will first look at the behavior of alternating current in resistors, inductors, and capacitors and in some simple circuits containing these elements.

28-1 Alternating Current in a Resistor

In our study of dc circuits in Chapter 23, we noted that Kirchhoff's rules apply to any circuit in a steady state. We also noted that steady states are reached in circuit elements almost immediately after a change in voltage or current is introduced. Since the time it takes to reach steady state is much smaller than the period of oscillation of ac circuits, we can apply Kirchhoff's rules to alternating-current circuits as well as direct-current circuits.

Figure 28-1 shows a simple ac circuit consisting of a generator and a resistor. In the figure, plus and minus signs indicate the higher-potential and lower-potential sides of the source of emf for the assumed direction of the current. Plus and minus signs have also been placed on the resistor to indicate the direction of the potential drop for the assumed direction of the current. Note that the point at which current enters the resistor is at a higher potential than the point at which the current leaves. The voltage drop across the resistor V_R is given by

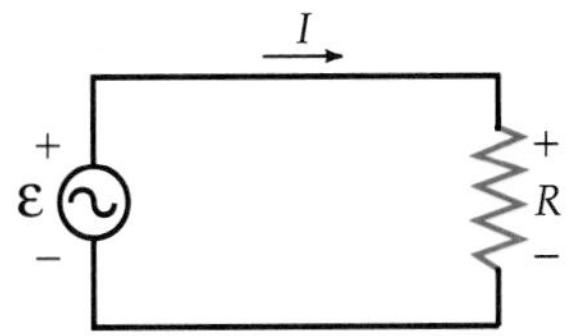

Figure 28-1 An ac generator in series with a resistor R.

$$V_R = V_+ - V_- = IR \tag{28-1}$$

If $\mathcal{E}$ is the emf supplied by the generator, applying Kirchhoff's loop rule to this circuit gives

$$\mathcal{E} - V_R = 0$$

If the generator produces an emf* given by

$$\mathcal{E} = \mathcal{E}_{max} \cos \omega t$$

we have

$$\mathcal{E}_{max} \cos \omega t - IR = 0 \tag{28-2}$$

The current in the resistor is

$$I = \frac{\mathcal{E}_{max}}{R} \cos \omega t \tag{28-3}$$

The maximum value of I occurs when $\cos \omega t$ has its maximum value of 1, in which case

$$I_{max} = \frac{\mathcal{E}_{max}}{R} \tag{28-4}$$

We may thus write Equation 28-3 as

$$I = I_{max} \cos \omega t \tag{28-5}$$

*The general equation for the emf of a generator was found in Section 26-6 to be $\mathcal{E} = \mathcal{E}_{max} \sin (\omega t + \delta)$. We are free to choose any convenient phase constant δ since it depends merely on the choice of zero time. For simplicity, we choose $\delta = \pi/2$ so that $\mathcal{E} = \mathcal{E}_{max} \sin (\omega t + \pi/2) = \mathcal{E}_{max} \cos \omega t$.

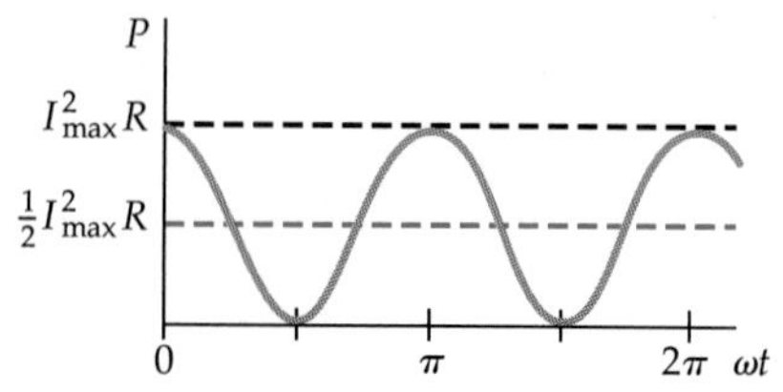

Figure 28-2 Plot of the power dissipated in the resistor in Figure 28-1 versus time. The power varies from zero to a maximum value $I^2_{max}R$. The average power is half the maximum power.

Note that the current through the resistor is in phase with the voltage across the resistor.

The power dissipated in the resistor varies with time. Its instantaneous value is

$$P = I^2R = (I_{max} \cos \omega t)^2R = I^2_{max}R \cos^2 \omega t \qquad 28\text{-}6$$

Figure 28-2 shows the power as a function of time. It varies from zero to its maximum value $I^2_{max}R$ as shown. We are usually interested in the average power over one or more cycles. The energy W_T delivered during the time of one period ($t = T = 2\pi/\omega$) is

$$W_T = \int_0^T P\,dt = \int_0^T I^2_{max}R \cos^2 \omega t\,dt$$

Substituting $\theta = \omega t$, we obtain

$$W_T = \frac{I^2_{max}R}{\omega} \int_0^{2\pi} \cos^2 \theta\,d\theta$$

The integral in this expression can be found in tables. It has the value π. The average power delivered to the resistor during one period is this energy divided by T:

$$P_{av} = \frac{W_T}{T} = \frac{(\pi I^2_{max}R)/\omega}{2\pi/\omega} = \frac{1}{2}I^2_{max}R$$

We could also obtain this result directly from Equation 28-6 by noting that the average value of $\cos^2 \omega t$ over one or more periods is $\frac{1}{2}$. This can be seen from the identity $\cos^2 \omega t + \sin^2 \omega t = 1$. A plot of $\sin^2 \omega t$ looks the same as one of $\cos^2 \omega t$ except that it is shifted by 90°. Both have the same average value over one or more periods, and since their sum is 1, the average value of each must be $\frac{1}{2}$. The average power dissipated in the resistor is thus

$$P_{av} = (I^2R)_{av} = \tfrac{1}{2}I^2_{max}R \qquad 28\text{-}7$$

rms Values

Most ac ammeters and voltmeters are designed to measure **root-mean-square (rms) values** of current and voltage rather than the maximum or peak values. The **rms** value of a current I_{rms} is defined by

rms current defined

$$I_{rms} = \sqrt{(I^2)_{av}} \qquad 28\text{-}8$$

The average value of I^2 is

$$(I^2)_{av} = [(I_{max} \cos \omega t)^2]_{av} = \tfrac{1}{2}I^2_{max}$$

where we have used $(\cos^2 \omega t)_{av} = \frac{1}{2}$. Substituting $\frac{1}{2}I^2_{max}$ for $(I^2)_{av}$ in Equation 28-8, we obtain

$$I_{rms} = \frac{1}{\sqrt{2}}I_{max} \qquad 28\text{-}9$$

The rms value of any quantity that varies sinusoidally equals the maximum value of that quantity divided by $\sqrt{2}$.

Substituting I^2_{rms} for $\frac{1}{2}I^2_{max}$ in Equation 28-7, we obtain for the average power dissipated in the resistor

$$P_{av} = I_{rms}^2 R \qquad \text{28-10}$$

We can see from Equation 28-10 that the rms current equals the steady dc current that would produce the same Joule heating as the ac current of Equation 28-5.

For this simple circuit in Figure 28-1, the average power delivered by the generator is equal to that dissipated in the resistor:

$$P_{av} = (\mathcal{E}I)_{av} = [(\mathcal{E}_{max} \cos \omega t)(I_{max} \cos \omega t)]_{av} = \mathcal{E}_{max} I_{max} (\cos^2 \omega t)_{av}$$

or

$$P_{av} = \tfrac{1}{2}\mathcal{E}_{max} I_{max}$$

Using $I_{rms} = I_{max}/\sqrt{2}$ and $\mathcal{E}_{rms} = \mathcal{E}_{max}/\sqrt{2}$, this can be written

$$P_{av} = \mathcal{E}_{rms} I_{rms} \qquad \text{28-11}$$

Average power delivered by a generator

The rms current is related to the rms emf in the same way that the maximum current is related to the maximum emf. We can see this by dividing each side of Equation 28-4 by $\sqrt{2}$ and using $I_{rms} = I_{max}/\sqrt{2}$ and $\mathcal{E}_{rms} = \mathcal{E}_{max}/\sqrt{2}$:

$$I_{rms} = \frac{\mathcal{E}_{rms}}{R} \qquad \text{28-12}$$

Equations 28-10, 28-11, and 28-12 are of the same form as the corresponding equations for direct-current circuits with I replaced by I_{rms} and $\mathcal{E}$ replaced by $\mathcal{E}_{rms}$. Thus, if we use rms values for the current and emf, we can calculate the power input and the heat generated using the same equations as we did for direct current.

Exercise

A 12-Ω resistor is connected across a sinusoidal emf that has a peak value of 48 V. Find (*a*) the rms current, (*b*) the average power, and (*c*) the maximum power. [Answer: (*a*) 2.83 A, (*b*) 96 W, (*c*) 192 W]

The ac power supplied to your house by the power company has a frequency of 60 Hz and a voltage of 120 V rms. (For some high-power appliances, such as an electric clothes dryer or an oven, separate lines carrying power at 240 V are often supplied. For a given power requirement, only half as much current is required at 240 V as at 120 V, but 240 V is much more dangerous than 120 V. A shock at 240 V is much more likely to be fatal than one at 120 V.) If you plug in a 1600-W heater, it will draw a current of

$$I_{rms} = \frac{P_{av}}{\mathcal{E}_{rms}} = \frac{1600\text{ W}}{120\text{ V}} = 13.3\text{ A}$$

The voltage across the outlets is maintained at 120 V, which is independent of the current drawn. Thus, all appliances plugged into the outlets of a single circuit are essentially in parallel. Therefore, if you plug a 500-W toaster into another outlet of the circuit supplying the heater, it will draw a current of 500 W/120 V = 4.17 A, so the total current through the circuit will be more than 17 A. Most household wiring is rated at 15 or 20 A. A current greater than this will overheat the wiring and create a fire hazard. Each circuit is therefore equipped with a circuit breaker (or a fuse in older houses). For a 20-A circuit, the circuit breaker trips (or the fuse blows), opening the circuit, when the current exceeds 20 A. The maximum power load that can be handled by a circuit with a 20-A circuit breaker is

$$P_{av} = \mathcal{E}_{rms} I_{rms} = (120\ \text{V})(20\ \text{A}) = 2.4\ \text{kW}$$

Since most modern houses require considerably more than 2.4 kW of power, many circuits are supplied, each with its own circuit breaker and each having several outlets.

Example 28-1

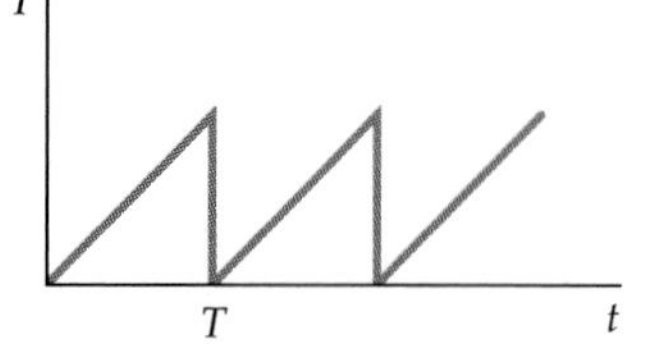

Figure 28-3 Saw-tooth waveform for the current in Example 28-1.

A current with a sawtooth waveform as shown in Figure 28-3 occurs sometimes in electronics. In the region $0 < t < T$, the current is given by $I = (I_0/T)t$. Find (*a*) the average current and (*b*) the rms current for this waveform.

(*a*) The average of any quantity over some interval T is the integral of the quantity over the interval divided by T. The average current is thus

$$I_{av} = \frac{1}{T}\int_0^T I\,dt = \frac{1}{T}\int_0^T (I_0/T)t\,dt = \frac{I_0}{T^2}\frac{T^2}{2} = \frac{1}{2}I_0$$

The average current is half the maximum current as we would expect.

(*b*) The average squared current is

$$(I^2)_{av} = \frac{1}{T}\int_0^T I^2\,dt = \frac{1}{T}\int_0^1 (I_0/T)^2t^2\,dt = \frac{I_0^2}{T^3}\frac{T^3}{3} = \frac{1}{3}I_0^2$$

The rms current is therefore $I_{rms} = I_0/\sqrt{3}$.

Questions

1. What is the average current in the resistor in Figure 28-1?
2. Is the instantaneous power in the resistor in Figure 28-1 ever negative?

28-2 Alternating Current in Inductors and Capacitors

The behavior of alternating current in inductors and capacitors is very different from that of direct current. For example, when a capacitor is in series in a dc circuit, the current stops completely when the capacitor becomes fully charged. But if the current alternates, charge continually flows onto or off of the plates of the capacitor. We will see that if the frequency of the alternating current is great, a capacitor hardly impedes the current at all. Conversely, an inductor coil usually has a very small resistance and therefore has little effect on direct current. But when the current is changing in an inductor, a back emf is generated that is proportional to the rate of change of the current. The greater the frequency of alternating current in an inductor, the greater the rate of change of current and, therefore, the greater the back emf. Thus, an inductor has just the opposite effect on alternating current as a capacitor. At very low frequencies, an inductor hardly impedes the current at all, but at high frequencies it impedes the current flow greatly because of its back emf.

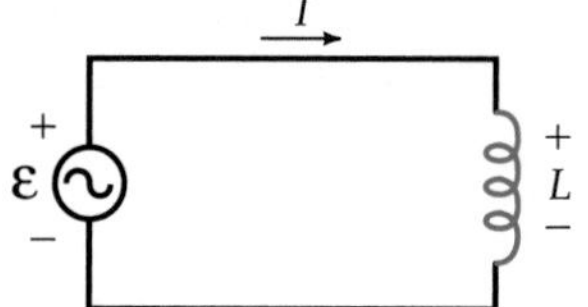

Figure 28-4 An ac generator in series with an inductor L.

Inductors

Figure 28-4 shows an inductor coil across the terminals of an ac generator. When the current increases in the inductor, a back emf of magnitude $L\,dI/dt$ is generated due to the changing flux (Equation 26-20). Usually, the potential drop across the inductor due to this back emf is much greater than the drop IR due to the resistance of the coil. Hence, we can normally neglect any resistance in the coil. In the figure, plus and minus signs have been placed on the inductor to indicate the direction of the potential drop when dI/dt is positive for the assumed direction of the current. Note that for a positive dI/dt, the point at which the current enters the inductor is at a higher potential than the point at which the current leaves. The voltage drop across the inductor V_L is then given by

$$V_L = V_+ - V_- = L\frac{dI}{dt} \qquad \text{28-13}$$

Applying Kirchhoff's loop rule to this circuit, we obtain

$$\mathcal{E} - V_L = 0$$

where $V_L = L\,dI/dt$ is the magnitude of the emf of the inductor. Setting the emf of the generator equal to $\mathcal{E}_{max} \cos \omega t$, we obtain

$$\mathcal{E} = L\frac{dI}{dt} = \mathcal{E}_{max} \cos \omega t \qquad \text{28-14}$$

Multiplying both sides by dt and dividing by L, we obtain

$$dI = \frac{\mathcal{E}_{max}}{L} \cos \omega t \, dt \qquad \text{28-15}$$

We solve for the current I by integrating both sides of Equation 28-15:

$$I = \frac{\mathcal{E}_{max}}{L}\int \cos \omega t \, dt = \frac{\mathcal{E}_{max}}{\omega L} \sin \omega t + C \qquad \text{28-16}$$

where C is a constant of integration. The constant is the average value of the current, since the average of $\sin \omega t$ over one or more cycles is zero. Choosing the dc component of the current to be zero, we have

$$I = \frac{\mathcal{E}_{max}}{\omega L} \sin \omega t = I_{max} \sin \omega t \qquad \text{28-17}$$

where

$$I_{max} = \frac{\mathcal{E}_{max}}{\omega L} \qquad \text{28-18}$$

is the maximum value of the current.

Figure 28-5 shows the current I and the voltage drop across the inductor V_L as functions of time. The current is not in phase with the voltage drop across the inductor, which equals the generator voltage. From the figure, we can see that the maximum value of the voltage occurs 90° or one-fourth period before the corresponding maximum value of the current. Thus, *the voltage drop across an inductor is said to lead the current by 90°*. We can understand this physically. When the current is zero but increasing, its rate of change is at its maximum, so the back emf induced in the inductor is at its maximum. One-quarter cycle later, the current is at its maximum. At this time, dI/dt is zero, so V_L is zero. Using the trigonometric identity $\sin \omega t = \cos(\omega t - \pi/2)$, Equation 28-17 for the current can be written

$$I = I_{max} \cos(\omega t - \pi/2) \qquad \text{28-19}$$

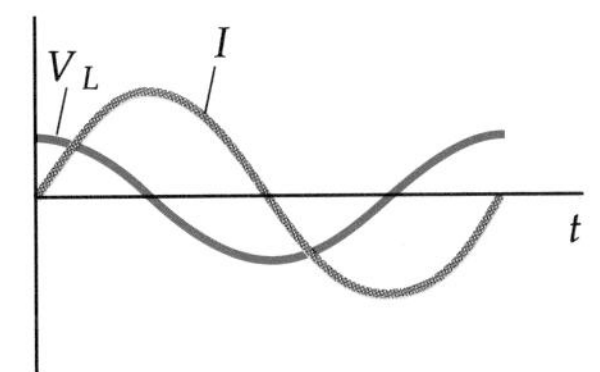

Figure 28-5 Current and voltage across the inductor in Figure 28-4 as functions of time. The maximum voltage occurs one-fourth period before the maximum current. Thus, the voltage is said to lead the current by one-fourth period or 90°.

The relation between the maximum (or rms) current and the maximum (or rms) voltage for an inductor can be written in a form similar to Equation 28-4 for a resistor. From Equation 28-18, we have

$$I_{max} = \frac{\mathcal{E}_{max}}{\omega L} = \frac{\mathcal{E}_{max}}{X_L} \qquad \text{28-20}$$

where

Inductive reactance

$$X_L = \omega L \tag{28-21}$$

is called the **inductive reactance.** Since $I_{\rm rms} = I_{\rm max}/\sqrt{2}$ and $\mathcal{E}_{\rm rms} = \mathcal{E}_{\rm max}/\sqrt{2}$, the rms current is given by

$$I_{\rm rms} = \frac{\mathcal{E}_{\rm rms}}{\omega L} = \frac{\mathcal{E}_{\rm rms}}{X_L} \tag{28-22}$$

Like resistance, inductive reactance has units of ohms. As we can see from Equation 28-20, the larger the reactance for a given emf, the smaller the current. Unlike resistance, the inductive reactance depends on frequency of the current—the greater the frequency, the greater the reactance.

The instantaneous power input to the inductor from the generator is

$$P = \mathcal{E}I = (\mathcal{E}_{\rm max} \cos \omega t)(I_{\rm max} \sin \omega t)$$
$$= \mathcal{E}_{\rm max} I_{\rm max} \cos \omega t \sin \omega t$$

The average power into the inductor is zero. We can see this by using

$$\cos \omega t \sin \omega t = \tfrac{1}{2} \sin 2\omega t$$

The value of this term oscillates twice during each cycle and is negative as often as it is positive. Thus, no energy is dissipated in an inductor. (This is true only if the resistance of the inductor can be neglected.)

Example 28-2

A 40-mH inductor is placed across an ac generator that has a maximum emf of 120 V. Find the inductive reactance and the maximum current when the frequency is 60 Hz and when it is 2000 Hz.

The inductive reactance at 60 Hz is

$$X_{L1} = \omega_1 L = 2\pi f_1 L = (2\pi)(60\ \text{Hz})(40 \times 10^{-3}\ \text{H}) = 15.1\ \Omega$$

and that at 2000 Hz is

$$X_{L2} = \omega_2 L = 2\pi f_2 L = (2\pi)(2000\ \text{Hz})(40 \times 10^{-3}\ \text{H}) = 503\ \Omega$$

The maximum value of the currents for these frequencies are

$$I_{1,\rm max} = \frac{\mathcal{E}_{\rm max}}{X_{L1}} = \frac{120\ \text{V}}{15.1\ \Omega} = 7.95\ \text{A}$$

$$I_{2,\rm max} = \frac{120\ \text{V}}{503\ \Omega} = 0.239\ \text{A}$$

Capacitors

Figure 28-6 shows a capacitor connected across the terminals of a generator. For the current direction shown, the current is related to the charge by

$$I = \frac{dQ}{dt}$$

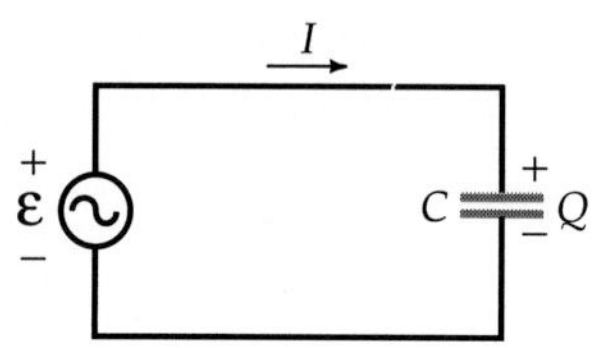

Figure 28-6 An ac generator in series with a capacitor C.

Again plus and minus signs have been placed on the capacitor plates showing a positive charge on the plate where the current enters and a negative charge on the plate where the current leaves. The voltage drop across the capacitor is

$$V_C = V_+ - V_- = \frac{Q}{C} \qquad \text{28-23}$$

From Kirchhoff's loop rule, we have

$$\mathcal{E} - V_C = 0$$

or

$$\mathcal{E} = \mathcal{E}_{max} \cos \omega t = \frac{Q}{C}$$

Thus

$$Q = \mathcal{E}_{max} C \cos \omega t$$

The current is

$$I = \frac{dQ}{dt} = -\omega \mathcal{E}_{max} C \sin \omega t$$

The maximum value of I occurs when $\sin \omega t = -1$, in which case

$$I_{max} = \omega \mathcal{E}_{max} C \qquad \text{28-24}$$

The current can thus be written

$$I = -\omega \mathcal{E}_{max} C \sin \omega t = -I_{max} \sin \omega t$$

Using the trigonometric identity $\sin \omega t = -\cos(\omega t + \pi/2)$, we obtain

$$I = -\omega C \mathcal{E}_{max} \sin \omega t = I_{max} \cos(\omega t + \pi/2) \qquad \text{28-25}$$

As with the inductor, the capacitor current is not in phase with the voltage drop across the capacitor, which equals the generator voltage. In Figure 28-7, the current I and the voltage drop across the capacitor $V_C = Q/C$ are plotted as functions of time. We can see that the maximum value of the voltage occurs 90° or one-fourth period *after* the maximum value of the current. Thus, *the voltage drop across a capacitor lags the current by 90°*. Again, we can understand this physically. At $\omega t = 3\pi/2$ in Figure 28-7, the current I is maximum. The maximum rate of charge buildup dQ/dt must occur when the charge Q is zero and, therefore, V_C is zero. As the charge on the capacitor plate increases, the current decreases until the charge is a maximum (so that V_C is a maximum) and the current is zero. The current then becomes negative as the charge flows back in the opposite direction, off the capacitor.

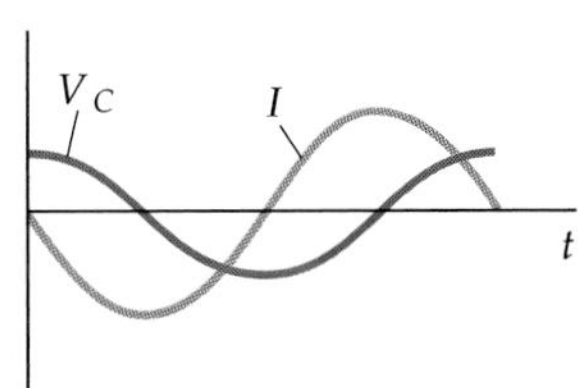

Figure 28-7 Current and voltage across the capacitor in Figure 28-6 versus time. The maximum voltage occurs one-fourth period after the maximum current. Thus, the voltage is said to lag the current by 90°.

Again, the relation between the maximum (or rms) current and the maximum (or rms) voltage for a capacitor can be written in a form similar to Equation 28-4 for a resistor. From Equation 28-24, we have

$$I_{max} = \omega C \mathcal{E}_{max} = \frac{\mathcal{E}_{max}}{1/\omega C} = \frac{\mathcal{E}_{max}}{X_C}$$

and, similarly,

$$I_{rms} = \frac{\mathcal{E}_{rms}}{1/\omega C} = \frac{\mathcal{E}_{rms}}{X_C} \qquad \text{28-26}$$

where

$$X_C = \frac{1}{\omega C} \qquad \text{28-27}$$

Capacitive reactance

is called the **capacitive reactance** of the circuit. Like resistance and inductive reactance, capacitive reactance has units of ohms, and like inductive reactance, capacitive reactance depends on the frequency of the current. In this case, the greater the frequency, the smaller the reactance. As for an inductor, the average power input to a capacitor from an ac generator is zero. This is because the emf is proportional to cos ωt and the current is proportional to sin ωt and $(\cos \omega t \sin \omega t)_{av} = 0$. Thus, like inductors, ideal capacitors dissipate no energy.

Since charge cannot pass across the space between the plates of a capacitor, it may seem strange that there is a continuing alternating current in the circuit of Figure 28-6. Recall, though, that when an uncharged capacitor is placed across the terminals of a dc voltage source (such as a battery), there is a current that decreases exponentially with time until the plates are charged to the same potential as the battery. Consider an initially uncharged capacitor across a source of emf, with the upper plate attached to the positive terminal. Initially, positive charge flows to the upper plate and away from the lower plate. (Of course, it is actually the negative electrons flowing in the opposite direction.) The effect is the same as if the charge actually flows *across* the space between the plates. If the source of emf is an ac generator, the potential difference changes sign every half period, as shown in Figure 28-7. Let us keep the generator emf constant but increase the frequency. During each half cycle, the same charge $\Delta Q = 2C\mathcal{E}_{max}$ is transferred into or out of the capacitor, but the number of cycles per second increases, so the current "through" the capacitor increases in proportion to the frequency. Hence, the greater the frequency, the less the capacitor impedes the flow of charge.

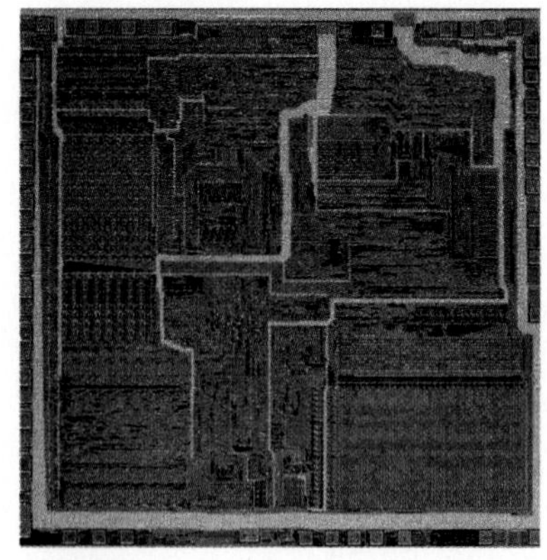

(*a*)

(*b*)

(*a*) The 8086 microprocessor is an integrated circuit containing 29,000 transistors used primarily in personal computers. Much of the 8086 design was drawn by hand on paper, which was used to make cutouts from rubyliths, red cellophanelike sheets as shown in (*b*). Today computers are used to draw microcircuit designs.

Example 28-3

A 20-μF capacitor is placed across a generator that has a maximum emf of 100 V. Find the capacitive reactance and the maximum current when the frequency is 60 Hz and when it is 5000 Hz.

The capacitive reactance at 60 Hz is

$$X_{C1} = \frac{1}{\omega_1 C} = \frac{1}{2\pi f_1 C}$$
$$= [2\pi(60\text{ Hz})(20 \times 10^{-6}\text{ F})]^{-1} = 133\ \Omega$$

and that at 5000 Hz is

$$X_{C2} = \frac{1}{\omega_2 C} = \frac{1}{2\pi f_2 C}$$
$$= [2\pi(5000\text{ Hz})(20 \times 10^{-6}\text{ F})]^{-1} = 1.59\ \Omega$$

The maximum currents are then

$$I_{1,max} = \frac{\mathcal{E}_{max}}{X_{C1}} = \frac{100\text{ V}}{133\ \Omega} = 0.754\text{ A}$$

and

$$I_{2,max} = \frac{100\text{ V}}{1.59\ \Omega} = 62.8\text{ A}$$

The circuits of Figures 28-4 and 28-6 contain only a generator and an inductor or capacitor. In them, the voltage drop across the inductor or capacitor equals the voltage of the generator. In more complicated circuits containing three or more elements, the voltage drop across each element is usually not equal to the generator voltage. It is useful, therefore, to write Equations 28-22 and 28-26 in terms of the voltage drops across the inductor and capacitor, respectively. If $V_{L,\text{rms}}$ is the rms voltage drop across an inductor, the rms current in the inductor is given by

$$I_{\text{rms}} = \frac{V_{L,\text{rms}}}{\omega L} = \frac{V_{L,\text{rms}}}{X_L} \qquad 28\text{-}28$$

The voltage drop across the inductor leads the current by 90°. Similarly, if $V_{C,\text{rms}}$ is the rms voltage across a capacitor, the rms current in the capacitor is given by

$$I_{\text{rms}} = \frac{V_{C,\text{rms}}}{1/\omega C} = \frac{V_{C,\text{rms}}}{X_C} \qquad 28\text{-}29$$

The voltage drop across the capacitor lags the current by 90°. Equations 28-28 and 28-29 can also be written in terms of the maximum voltages and maximum currents.

Questions

3. In a circuit consisting of a generator and an inductor, are there any times when the inductor absorbs power from the generator? Are there any times when the inductor supplies power to the generator?
4. In a circuit consisting of a generator and a capacitor, are there any times when the capacitor absorbs power from the generator? Are there any times when the capacitor supplies power to the generator?

28-3 Phasors

In the previous sections, we saw that the voltage across a resistor is in phase with the current, whereas the voltage across an inductor leads the current by 90° and the voltage across a capacitor lags the current by 90°. These phase relations can be represented by two-dimensional vectors called **phasors.** In Figure 28-8, the voltage across a resistor V_R is represented by a vector $\mathbf{V}_R$ that has magnitude $I_{\text{max}}R$ and makes an angle θ with the x axis. This voltage is in phase with the current. In general, a steady-state current in an ac circuit varies with time as

$$I = I_{\text{max}} \cos \theta = I_{\text{max}} \cos (\omega t - \delta) \qquad 28\text{-}30$$

where ω is the angular frequency and δ is some phase constant. The voltage drop across a resistor is then given by

$$V_R = IR = I_{\text{max}}R \cos (\omega t - \delta) \qquad 28\text{-}31$$

The instantaneous value of the voltage drop across a resistor is thus equal to the x component of the phasor vector $\mathbf{V}_R$, which rotates counterclockwise with an angular frequency ω. The current I may be written as the x component of a phasor $\mathbf{I}$ having the same orientation as $\mathbf{V}_R$.

When several components are connected together in a series circuit, their voltages add. When they are connected in parallel, their currents add. Adding sines or cosines of different amplitudes and phases algebraically is awkward. It is much easier to do this by vector addition.

Phasors are used as follows. Any ac voltage or current is written in the form $A \cos (\omega t - \delta)$, which in turn is treated as the x component A_x of a

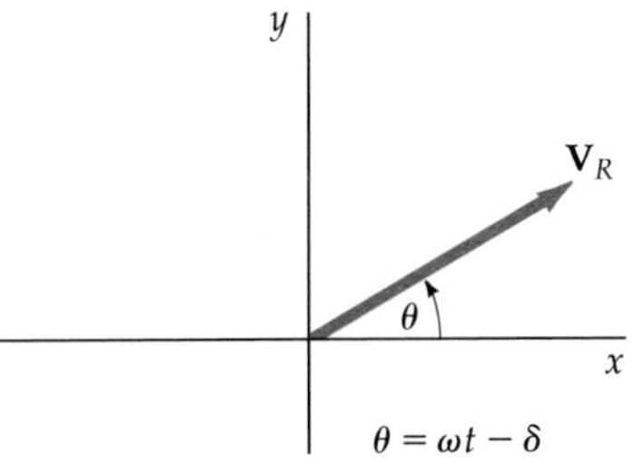

Figure 28-8 The voltage across a resistor can be represented by a vector $\mathbf{V}_R$, called a phasor, that has magnitude $I_{\text{max}}R$ and makes an angle $\theta = \omega t - \delta$ with the x axis. The phasor rotates with an angular frequency ω. The voltage $V_R = IR$ is the x component of $\mathbf{V}_R$.

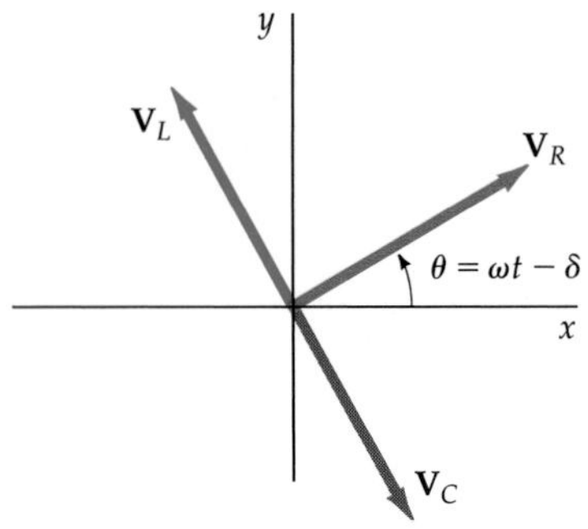

Figure 28-9 Phasor representations of the voltages V_R, V_L, and V_C. Each vector rotates in the counterclockwise direction with an angular frequency ω. At any instant, the voltage across an element equals the x component of the corresponding phasor, and the sum of the voltages equals the x component of the vector sum $\mathbf{V}_R + \mathbf{V}_L + \mathbf{V}_C$.

phasor **A** that makes an angle $(\omega t - \delta)$ with the x axis. Instead of adding two voltages or currents algebraically as $A\cos(\omega t - \delta_1) + B\cos(\omega t - \delta_2)$, we represent these quantities as phasors **A** and **B** and find the phasor sum $\mathbf{C} = \mathbf{A} + \mathbf{B}$ geometrically. The resultant voltage or current is then the x component of the resultant phasor, $C_x = A_x + B_x$. The geometric representation conveniently shows the relative amplitudes and phases of the phasors.

Consider a circuit containing an inductor L, a capacitor C, and a resistor R all connected in series. They all carry the same current, which is represented as the x component of the current phasor **I**. The voltage across the inductor V_L is represented by a phasor $\mathbf{V}_L$ that has magnitude $I_{\max}X_L$ and leads the current phasor **I** by 90°. Similarly, the voltage across the capacitor V_C is represented by a phasor $\mathbf{V}_C$ that has magnitude $I_{\max}X_C$ and lags the current by 90°. Figure 28-9 shows the three phasors $\mathbf{V}_R$, $\mathbf{V}_L$, and $\mathbf{V}_C$. As time goes on, the three phasors rotate counterclockwise with an angular frequency ω, so the relative positions of the vectors do not change. At any time, the instantaneous value of the voltage drop across any of these elements equals the x component of the corresponding phasor.

28-4 *LC* and *LCR* Circuits without a Generator

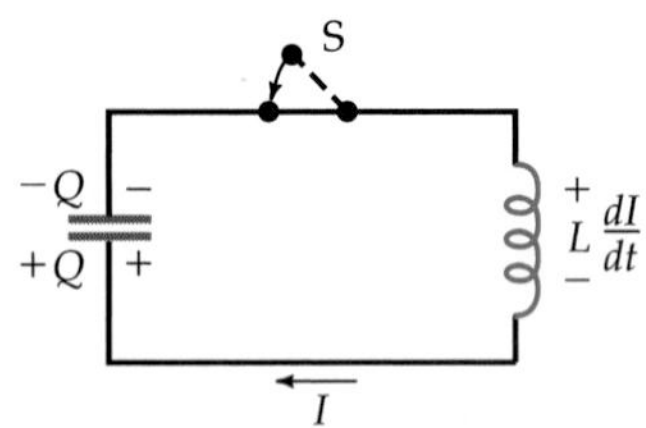

Figure 28-10 An *LC* circuit. When the switch is closed, the initially charged capacitor discharges through the inductor, producing a back emf.

In this section, we will study some simple circuits containing inductance, capacitance, and resistance but no generator. We will add a generator to these circuits in the next section. We first look at a simple circuit with inductance and capacitance but no resistance as shown in Figure 28-10. We assume that the capacitor carries an initial charge Q_0 and that the switch is initially open. After the switch is closed at $t = 0$, the charge begins to flow through the inductor. In the figure, the signs of Q on the capacitor and the direction of the current I have been chosen such that

$$I = \frac{dQ}{dt}$$

Applying Kirchhoff's loop rule to the circuit for the assumed signs of Q and I, we have

$$L\frac{dI}{dt} + \frac{Q}{C} = 0 \qquad 28\text{-}32$$

Substituting dQ/dt for I in this equation, we obtain

$$L\frac{d^2Q}{dt^2} + \frac{Q}{C} = 0 \qquad 28\text{-}33$$

Equation 28-33 is of the same form as the equation for the acceleration of a mass on a spring:

$$m\frac{d^2x}{dt^2} + kx = 0 \qquad 28\text{-}34$$

The behavior of an *LC* circuit is thus analogous to that of a mass on a spring, with L analogous to the mass m, Q analogous to the position x, and $1/C$ analogous to the spring constant k. Also, the current I is analogous to the velocity v, since $v = dx/dt$ and $I = dQ/dt$. In mechanics, the mass of an object describes the inertia of the object. The greater the mass, the more difficult it is to change the velocity of the object. Similarly, the inductance L can be thought of as the inertia of an ac circuit. The greater the inductance, the more difficult it is to change the current I.

If we divide each term in Equation 28-33 by L and rearrange, we obtain

$$\frac{d^2Q}{dt^2} = -\frac{1}{LC}Q \qquad 28\text{-}35$$

which is analogous to

$$\frac{d^2x}{dt^2} = -\frac{k}{m}x = -\omega^2 x \qquad 28\text{-}36$$

where $\omega^2 = k/m$. In Chapter 12, we found that we could write the solution of Equation 28-36 for simple harmonic motion in the form

$$x = A\cos(\omega t - \delta)$$

where $\omega = \sqrt{k/m}$ is the angular frequency, A is the amplitude, and δ is the phase constant, which depends on the initial conditions. We can put Equation 28-35 in this same form by writing ω^2 for $1/LC$. Then

$$\frac{d^2Q}{dt^2} = -\omega^2 Q \qquad 28\text{-}37$$

$$\omega = \frac{1}{\sqrt{LC}} \qquad 28\text{-}38$$

The solution of Equation 28-38 is

$$Q = A\cos(\omega t - \delta)$$

The current is found by differentiating this solution:

$$I = \frac{dQ}{dt} = -\omega A\sin(\omega t - \delta)$$

If we choose our initial conditions to be $Q = Q_0$ and $I = 0$ at $t = 0$, the phase constant δ is zero and $A = Q_0$. Our solutions are then

$$Q = Q_0\cos\omega t \qquad 28\text{-}39$$

and

$$I = -\omega Q_0\sin\omega t = -I_{\max}\sin\omega t \qquad 28\text{-}40$$

where $I_{\max} = \omega Q_0$.

Figure 28-11 shows graphs of Q and I versus time. The charge oscillates between the values $+Q_0$ and $-Q_0$ with angular frequency $\omega = \sqrt{1/LC}$. The current oscillates between $+\omega Q_0$ and $-\omega Q_0$ with the same frequency and is 90° out of phase with the charge. The current is maximum when the charge is zero and zero when the charge is maximum.

In our study of the oscillations of a mass on a spring, we found that the total energy is constant but oscillates between potential and kinetic energy. In our *LC* circuit, we also have two kinds of energy, electric energy and magnetic energy. The electric energy stored in the capacitor is

$$U_e = \frac{1}{2}QV_C = \frac{1}{2}\frac{Q^2}{C}$$

Substituting $Q_0\cos\omega t$ for Q, we have for the electric energy

$$U_e = \frac{Q_0^2}{2C}\cos^2\omega t \qquad 28\text{-}41$$

The electric energy oscillates between its maximum value $Q_0^2/2C$ and zero. The magnetic energy stored in the inductor is

$$U_m = \tfrac{1}{2}LI^2 \qquad 28\text{-}42$$

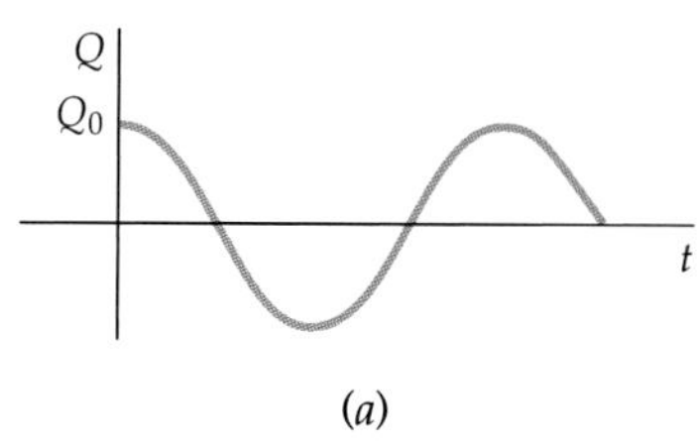

(*a*)

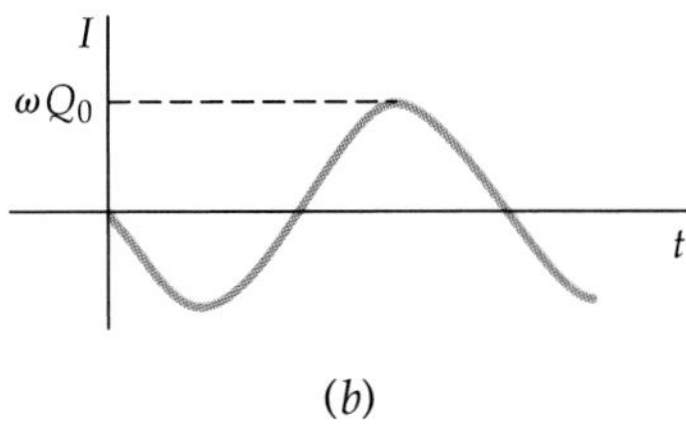

(*b*)

Figure 28-11 Graphs of (*a*) Q versus t and (*b*) I versus t for the *LC* circuit of Figure 28-10.

Substituting in the value for the current from Equation 28-30, we obtain

$$U_m = \frac{1}{2}LI_{max}^2 \sin^2 \omega t = \frac{1}{2}L\omega^2 Q_0^2 \sin^2 \omega t = \frac{Q_0^2}{2C} \sin^2 \omega t \qquad 28\text{-}43$$

where we have used the fact that $\omega^2 = 1/LC$. The magnetic energy also oscillates between its maximum value of $Q_0^2/2C$ and zero. The sum of the electrostatic and magnetic energies is the total energy, which is constant in time:

$$U_{total} = U_e + U_m = \frac{Q_0^2}{2C} \cos^2 \omega t + \frac{Q_0^2}{2C} \sin^2 \omega t = \frac{Q_0^2}{2C}$$

which is the energy initially stored on the capacitor.

Example 28-4

A 2-μF capacitor is charged to 20 V and is then connected across a 6-μH inductor. (*a*) What is the frequency of oscillation? (*b*) What is the maximum value of the current?

(*a*) The frequency of oscillation depends only on the values of the capacitance and inductance:

$$f = \frac{\omega}{2\pi} = \frac{1}{2\pi\sqrt{LC}}$$

$$= \frac{1}{2\pi\sqrt{(6 \times 10^{-6}\ \text{H})(2 \times 10^{-6}\ \text{F})}} = 4.59 \times 10^4\ \text{Hz}$$

(*b*) According to Equation 28-40, the maximum value of the current is related to the maximum value of the charge by

$$I_{max} = \omega Q_0 = \frac{Q_0}{\sqrt{LC}}$$

The initial charge on the capacitor is

$$Q_0 = CV_0 = (2\ \mu\text{F})(20\ \text{V}) = 40\ \mu\text{C}$$

Thus,

$$I_{max} = \frac{40\ \mu\text{C}}{\sqrt{(6\ \mu\text{H})(2\ \mu\text{F})}} = 11.5\ \text{A}$$

Exercise

A 5-μF capacitor is charged and is then discharged through an inductor. What should the value of the inductance be so that the current oscillates with frequency 8 kHz? (Answer: 79.2 μH)

In Figure 28-12, a resistor is included in series with the capacitor and inductor. Again we assume that the switch is initially open, that the capacitor carries an initial charge Q_0, and that we close the switch at $t = 0$. Since there is now a drop IR across the resistance, Kirchhoff's loop rule gives

$$L\frac{dI}{dt} + \frac{Q}{C} + IR = 0 \qquad 28\text{-}44a$$

or

$$L\frac{d^2Q}{dt^2} + \frac{Q}{C} + R\frac{dQ}{dt} = 0 \qquad 28\text{-}44b$$

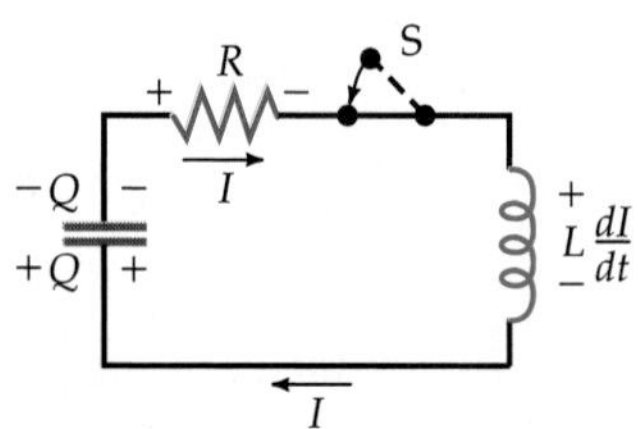

Figure 28-12 An *LCR* circuit.

where we have used $I = dQ/dt$ as before. Equations 28-44*a* and *b* are analogous to the equation for a damped harmonic oscillator (see Equation 12-46):

$$m\frac{d^2x}{dt^2} + kx + b\frac{dx}{dt} = 0$$

The first term, $L\,dI/dt = L\,d^2Q/dt^2$, is analogous to the mass times the acceleration, $m\,dv/dt = m\,d^2x/dt^2$; the second term, Q/C, is analogous to the restoring force kx; and the third term, $IR = R\,dQ/dt$, is analogous to the damping term, $bv = b\,dx/dt$. In the oscillation of a mass on a spring, the damping constant b leads to a dissipation of mechanical energy as heat. In an *LCR* circuit, the resistance R is analogous to the damping constant b and leads to a dissipation of electrical energy as Joule heat.

If the resistance is small, the charge and current oscillate with a frequency that is very nearly equal to $1/\sqrt{LC}$, but the oscillations are damped; that is, the maximum values of the charge and current decrease with each oscillation. We can understand this qualitatively from energy considerations. If we multiply each term in Equation 28-44*a* by the current I, we obtain

$$IL\frac{dI}{dt} + I\frac{Q}{C} + I^2R = 0 \qquad 28\text{-}45$$

The first term in this equation is the current times the voltage across the inductor. This is the rate at which energy is put into the inductor or taken out of it; that is, it is the rate of change of the magnetic energy, $d(\frac{1}{2}LI^2)/dt$, which is positive or negative depending on whether I and dI/dt have the same sign or different signs. Similarly, the second term is the current times the voltage across the capacitor. This is the rate of change of the energy of the capacitor, which may be positive or negative. The last term, I^2R, is the rate at which energy is dissipated in the resistor as Joule heat. This term is always positive. The sum of the electric and magnetic energies is not constant for this circuit because energy is continually dissipated in the resistor. Figure 28-13 shows graphs of Q versus t and I versus t for a small resistance R. If we increase R, the oscillations become more heavily damped until a critical value of R is reached for which there is not even one oscillation. Figure 28-14 shows Q versus t when the value of R is greater than the critical damping value.

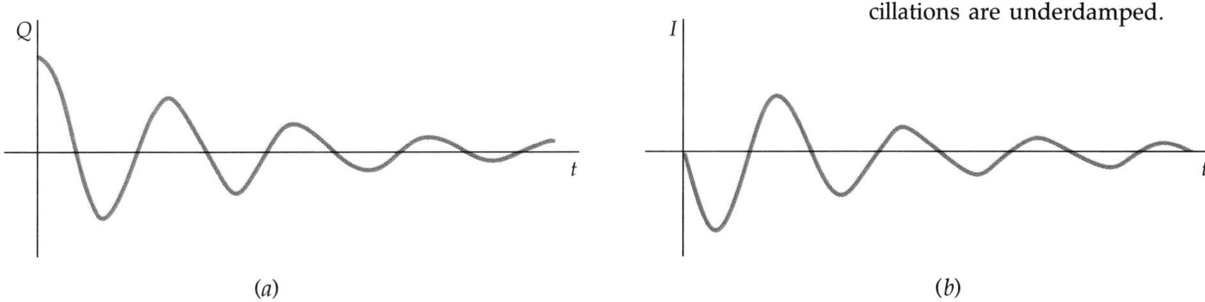

Figure 28-13 Graphs of (*a*) Q versus t and (*b*) I versus t for the *LCR* circuit of Figure 28-12 when R is small enough so that the oscillations are underdamped.

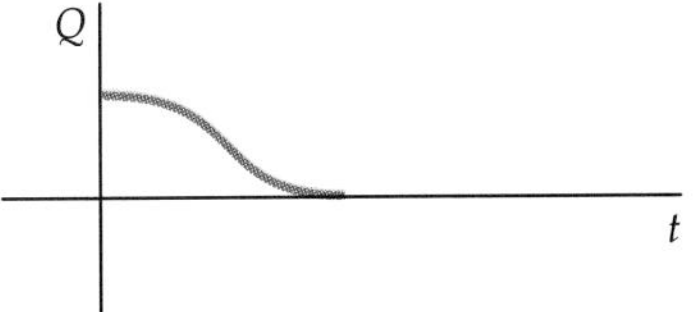

Figure 28-14 Graph of Q versus t for the *LCR* circuit of Figure 28-12 when R is so large that the oscillations are overdamped.

Question

5. It is easy to make *LC* circuits that have frequencies of oscillation of thousands of hertz or more, but it is difficult to make *LC* circuits that have small frequencies. Why?

28-5 *LCR* Circuits with a Generator

Series

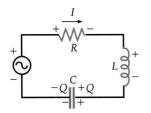

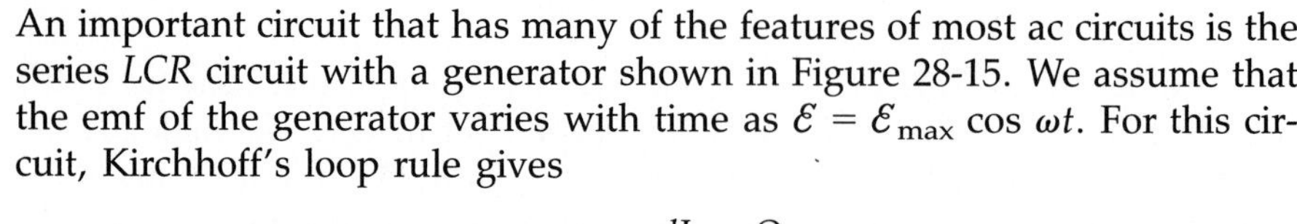

An important circuit that has many of the features of most ac circuits is the series *LCR* circuit with a generator shown in Figure 28-15. We assume that the emf of the generator varies with time as $\mathcal{E} = \mathcal{E}_{max} \cos \omega t$. For this circuit, Kirchhoff's loop rule gives

$$\mathcal{E}_{max} \cos \omega t - L\frac{dI}{dt} - \frac{Q}{C} - IR = 0$$

Figure 28-15 A series *LCR* circuit with an ac generator.

Using $I = dQ/dt$ and rearranging, we obtain

$$L\frac{d^2Q}{dt^2} + R\frac{dQ}{dt} + \frac{Q}{C} = \mathcal{E}_{max} \cos \omega t \qquad \text{28-46}$$

This equation is analogous to Equation 12-63 for the forced oscillation of a mass on a spring:

$$m\frac{d^2x}{dt^2} + b\frac{dx}{dt} + m\omega_0^2 x = F_0 \cos \omega t$$

(In Equation 12-63, the force constant k was written in terms of the mass m and the natural angular frequency ω_0 using $k = m\omega_0^2$. In Equation 28-46, the capacitance could be similarly written in terms of L and the natural angular frequency using $1/C = L\omega_0^2$.)

Rather than solving Equation 28-46, we will discuss its solution qualitatively as we did with Equation 12-63 for the forced oscillator in Chapter 12. The current in the circuit consists of two parts, a transient current and a steady-state current. The transient current depends on the initial conditions, such as the initial phase of the generator and the initial charge on the capacitor. The steady-state current, on the other hand, is independent of the initial conditions. The transient current decreases exponentially with time and is eventually negligible compared with the steady-state current. We will ignore the transient current and concentrate on the steady-state current.

The steady-state current obtained by solving Equation 28-46 is

$$I = I_{max} \cos (\omega t - \delta) \qquad \text{28-47}$$

where the phase angle δ is given by

$$\tan \delta = \frac{X_L - X_C}{R} \qquad \text{28-48}$$

The maximum current is

Impedance of series LCR circuit

$$I_{max} = \frac{\mathcal{E}_{max}}{\sqrt{R^2 + (X_L - X_C)^2}} = \frac{\mathcal{E}_{max}}{Z} \qquad \text{28-49}$$

where

$$Z = \sqrt{R^2 + (X_L - X_C)^2} \qquad \text{28-50}$$

The quantity $X_L - X_C$ is called the **total reactance,** and the quantity Z is called the **impedance.** Combining these results, we have

$$I = \frac{\mathcal{E}_{\max}}{Z} \cos(\omega t - \delta) \qquad 28\text{-}51$$

Equation 28-51 can also be obtained from a simple diagram using the phasor representations discussed in Section 28-3. Figure 28-16 shows the phasors representing the voltage drops across the resistance, the inductance, and the capacitance. The x component of each of these vectors equals the instantaneous voltage drop across the corresponding element. Since the sum of the x components equals the x component of the sum of the vectors, the sum of the x components equals the sum of the voltage drops across these elements, which by Kirchhoff's loop rule equals the instantaneous emf. If we represent the applied emf, $\mathcal{E}_{\max} \cos \omega t$, as a phasor $\boldsymbol{\mathcal{E}}$ that has the magnitude $\mathcal{E}_{\max}$, we have

$$\boldsymbol{\mathcal{E}} = \mathbf{V}_R + \mathbf{V}_L + \mathbf{V}_C \qquad 28\text{-}52$$

In terms of the magnitudes,

$$\mathcal{E}_{\max} = |\mathbf{V}_R + \mathbf{V}_L + \mathbf{V}_C| = \sqrt{V_{R,\max}^2 + (V_{L,\max} - V_{C,\max})^2}$$

But $V_R = I_{\max}R$, $V_L = I_{\max}X_L$, and $V_C = I_{\max}X_C$. Thus,

$$\mathcal{E}_{\max} = I_{\max}\sqrt{R^2 + (X_L - X_C)^2} = I_{\max}Z$$

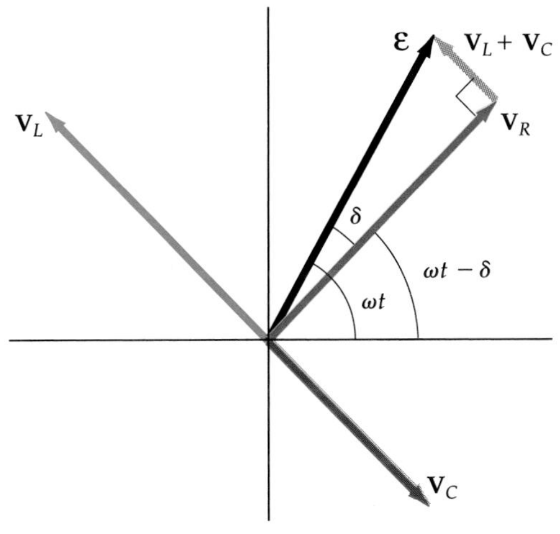

Figure 28-16 Phase relations among voltages in a series *LCR* circuit. The voltage across the resistor is in phase with the current. The voltage across the inductor V_L leads the current by 90°. The voltage across the capacitor lags the current by 90°. The sum of the vectors representing these voltages gives a vector at an angle δ with the current representing the applied emf. For the case shown here, V_L is greater than V_C and the current lags the emf by δ.

The phasor $\boldsymbol{\mathcal{E}}$ makes an angle δ with $\mathbf{V}_R$ as shown in Figure 28-16. From the figure, we can see that

$$\tan \delta = \frac{|\mathbf{V}_L + \mathbf{V}_C|}{|\mathbf{V}_R|} = \frac{I_{\max}X_L - I_{\max}X_C}{I_{\max}R} = \frac{X_L - X_C}{R}$$

in agreement with Equation 28-48. Since $\boldsymbol{\mathcal{E}}$ makes an angle ωt with the x axis, $\mathbf{V}_R$ makes an angle $\omega t - \delta$ with the x axis. This voltage is in phase with the current, which is therefore given by

$$I = I_{\max} \cos(\omega t - \delta) = \frac{\mathcal{E}_{\max}}{Z} \cos(\omega t - \delta)$$

This is Equation 28-51. The relation between the impedance Z and the resistance R and the total reactance $X_L - X_C$ can be remembered using the right triangle shown in Figure 28-17.

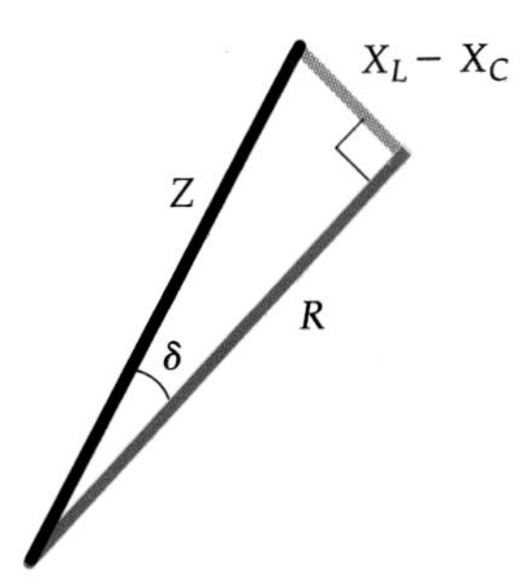

Figure 28-17 Triangle relating capacitive and inductive reactance, resistance, impedance, and the phase angle in an *LCR* circuit.

Resonance

Although Equations 28-50 and 28-51 appear to be complicated, we can use them to learn some simple and important features of the behavior of the circuit in Figure 28-15. Since both the inductive reactance $X_L = \omega L$ and the capacitive reactance $X_C = 1/\omega C$ depend on the frequency of the applied emf, the impedance Z and the maximum current I_{max} do also. At very low frequencies, $X_C = 1/\omega C$ is much greater than $X_L = \omega L$, so the impedance is large and I_{max} is small. The phase angle δ is negative, which means that the current leads the generator voltage. As we increase ω, the inductive reactance increases and the capacitive reactance decreases. When X_L and X_C are equal, the impedance Z has its smallest value, equal to R, and I_{max} has its greatest value. At $X_L = X_C$ the phase angle δ is zero, which means that the current is in phase with the applied emf. As we increase ω further, X_L becomes greater than X_C. The impedance increases and the maximum current decreases. The phase angle is positive, meaning that the current lags the generator voltage.

The value of ω for which X_L and X_C are equal is obtained from

$$X_L = X_C$$

$$\omega L = \frac{1}{\omega C}$$

or

$$\omega = \frac{1}{\sqrt{LC}} = \omega_0 = 2\pi f_0$$

The frequency f_0 (or the angular frequency ω_0) is called the natural frequency or **resonance frequency** of the circuit. The impedance is smallest and the maximum value of the current is greatest when the frequency of the emf equals the natural frequency. At this frequency, the circuit is said to be at **resonance.** At resonance, the current is in phase with the generator voltage. This resonance condition in a driven *LCR* circuit is similar to that in a driven simple harmonic oscillator.

We noted previously that neither an inductor nor a capacitor dissipates energy. The average power delivered to a series *LCR* circuit therefore equals the average power supplied to the resistor. The instantaneous power supplied to the resistor is

$$P = I^2R = [I_{max} \cos (\omega t - \delta)]^2R$$

Averaging over one or more cycles and using $(\cos^2 \theta)_{av} = \frac{1}{2}$, we obtain for the average power

$$P_{av} = \tfrac{1}{2}I_{max}^2R$$

which is the same as Equation 28-7. Using $V_R = I_{max}R$, this can be written

$$P_{av} = \tfrac{1}{2}I_{max}V_R$$

From Figure 28-16, we can see that $V_R = \mathcal{E}_{max} \cos \delta$. The average power supplied to the circuit can thus be written

$$P_{av} = \tfrac{1}{2}I_{max}\mathcal{E}_{max} \cos \delta$$

In terms of the rms values, $I_{rms} = I_{max}/\sqrt{2}$ and $\mathcal{E}_{rms} = \mathcal{E}_{max}/\sqrt{2}$, the average power is

$$P_{av} = \mathcal{E}_{rms}I_{rms} \cos \delta \qquad \text{28-53}$$

The quantity $\cos \delta$ is called the **power factor** of the *LCR* circuit. At resonance, δ is zero, and the power factor is 1.

The power can be expressed as a function of the angular frequency ω. From the triangle in Figure 28-17, we have

$$\cos\delta = \frac{R}{Z} \qquad 28\text{-}54$$

Using this result and $I_{\text{rms}} = \mathcal{E}_{\text{rms}}/Z$, we have for the average power

$$P_{\text{av}} = \mathcal{E}_{\text{rms}}^2 \frac{R}{Z^2}$$

From the definition of impedance Z, we have

$$Z^2 = (X_L - X_C)^2 + R^2 = \left(\omega L - \frac{1}{\omega C}\right)^2 + R^2$$

$$= \frac{L^2}{\omega^2}\left(\omega^2 - \frac{1}{LC}\right)^2 + R^2$$

$$= \frac{L^2}{\omega^2}(\omega^2 - \omega_0^2)^2 + R^2$$

where we have used $\omega_0^2 = 1/LC$. Using this expression for Z^2, we obtain the average power as a function of ω:

$$P_{\text{av}} = \frac{\mathcal{E}_{\text{rms}}^2 R\omega^2}{L^2(\omega^2 - \omega_0^2)^2 + \omega^2 R^2} \qquad 28\text{-}55$$

Figure 28-18 shows the average power supplied by the generator to the circuit as a function of generator frequency for two different values of the resistance R. These curves, called **resonance curves,** are the same as the power-versus-frequency curves for a driven damped oscillator (see Section 12-8). The average power is maximum when the generator frequency equals the resonance frequency. When the resistance is small, the resonance curve is narrow; when it is large, the curve is broad. A resonance curve can be characterized by the **resonance width** Δf shown in the figure. This width is the frequency difference between the two points on the curve where the power is half its maximum value. When the width is small compared with the resonance frequency, the resonance is sharp, that is, the resonance curve is narrow.

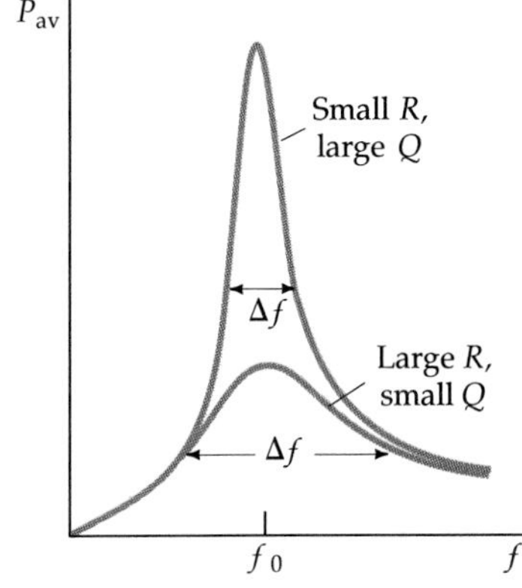

Figure 28-18 Plot of average power versus frequency for a series *LCR* circuit. The power is maximum when the frequency of the generator f equals the natural frequency of the circuit $f_0 = (1/2\pi\sqrt{LC})$. If the resistance is small, the Q factor is large and the resonance is sharp. The resonance width Δf of the curves is measured between points where the power is half its maximum value.

In Chapter 12, the Q factor for a mechanical oscillator was defined as $Q = 2\pi E/|\Delta E|$ (Equation 12-54), where E is the total energy of the system and ΔE is the energy lost in one cycle. We saw that $Q = 2\pi m/bT$, where m is the mass, b is the damping constant, and T is the period. Since $\omega_0 = 2\pi/T$, the Q factor for the damped driven mechanical oscillator is given by

$$Q = \frac{\omega_0 m}{b} \qquad 28\text{-}56$$

The **Q factor** for an *LCR* circuit can be defined in a similar way. Since L is analogous to the mass m and R is analogous to the damping constant b, the Q factor for an *LCR* circuit is given by

$$Q = \frac{2\pi E}{|\Delta E|} = \frac{\omega_0 L}{R} \qquad 28\text{-}57$$

When the resonance is reasonably narrow (that is, when Q is greater than about 2 or 3), the Q factor can be approximated by

$$Q = \frac{\omega_0}{\Delta\omega} = \frac{f_0}{\Delta f} \qquad 28\text{-}58$$

Q factor for an LCR circuit

A shipboard radio, circa 1920. Exposed at the operator's left are the inductance coils and capacitor plates of the tuning circuit.

Resonance circuits are used in radio receivers, where the resonance frequency of the circuit is varied by varying the capacitance. Resonance occurs when the natural frequency of the circuit equals one of the frequencies of the radio waves picked up at the antenna. At resonance, there is a relatively large current in the antenna circuit. If the Q factor of the circuit is sufficiently high, currents due to other station frequencies off resonance will be negligible compared with those due to the station frequency to which the circuit is tuned.

Example 28-5

A series LCR circuit with $L = 2$ H, $C = 2\ \mu$F, and $R = 20\ \Omega$ is driven by a generator with a maximum emf of 100 V and a variable frequency. (*a*) Find the resonance frequency f_0. Find (*b*) the maximum current I_{max} and (*c*) the phase angle δ when the generator frequency is $f = 60$ Hz.

(*a*) The resonance frequency is

$$f_0 = \frac{\omega_0}{2\pi} = \frac{1}{2\pi\sqrt{LC}} = \frac{1}{2\pi\sqrt{(2\text{ H})(2 \times 10^{-6}\text{ F})}} = 79.6\text{ Hz}$$

(*b*) When the generator frequency is 60 Hz, it is well below the resonance frequency. The capacitive and inductive reactances at 60 Hz are

$$X_C = \frac{1}{\omega C} = \frac{1}{(2\pi)(60\text{ Hz})(2 \times 10^{-6}\text{ F})} = 1326\ \Omega$$

and

$$X_L = \omega L = (2\pi)(60\text{ Hz})(2\text{ H}) = 754\ \Omega$$

The total reactance is $X_L - X_C = 754\ \Omega - 1326\ \Omega = -572\ \Omega$. This is of much greater magnitude than the resistance, a result that always holds far from resonance. The total impedance is

$$Z = \sqrt{R^2 + (X_L - X_C)^2} = \sqrt{(20\ \Omega)^2 + (-572\ \Omega)^2} \approx 572\ \Omega$$

since $(20)^2$ is negligible compared with $(572)^2$. The maximum current is then

$$I_{max} = \frac{\mathcal{E}_{max}}{Z} = \frac{100\text{ V}}{572\ \Omega} = 0.175\text{ A}$$

This is small compared with I_{max} at resonance, which is (100 V)/(20 Ω) = 5 A.

(*c*) The phase angle δ is given by

$$\tan\delta = \frac{X_L - X_C}{R} = \frac{-572\ \Omega}{20\ \Omega} = -28.6$$

$$\delta = -88°$$

From Equation 28-51 (or from Figure 28-16), we can see that a negative phase angle means that the current leads the generator voltage.

Example 28-6

Find the average power delivered by the generator in Example 28-5 at 60 Hz.

Since we are given the maximum emf and have calculated the maximum current in Example 28-5, it is convenient to write the average power in terms of these quantities. We have

$$P_{av} = \mathcal{E}_{rms}I_{rms}\cos\delta = \tfrac{1}{2}\mathcal{E}_{max}I_{max}\cos\delta$$

$$= \tfrac{1}{2}(100\ \text{V})(0.175\ \text{A})[\cos(-88°)] = 0.306\ \text{W}$$

As we have noted, this power goes into Joule heat in the resistor. We could also have calculated the average power from

$$P_{av} = I_{rms}^2R = \tfrac{1}{2}I_{max}^2R = \tfrac{1}{2}(0.175\ \text{A})^2(20\ \Omega) = 0.306\ \text{W}$$

Example 28-7

Find (*a*) the Q value and (*b*) the resonance width of the circuit in Example 28-5.

(*a*) In Example 28-5, we found the resonance frequency to be $f_0 =$ 79.6 Hz. The Q value is therefore

$$Q = \frac{\omega_0 L}{R} = \frac{2\pi(79.6\ \text{Hz})(2\ \text{H})}{20\ \Omega} = 50$$

(*b*) The width of the resonance is

$$\Delta f = \frac{f_0}{Q} = \frac{79.6\ \text{Hz}}{50} = 1.6\ \text{Hz}$$

This is a very sharp resonance. The width is only 1.6 Hz at the resonance frequency of 79.6 Hz.

Example 28-8

Find the maximum voltage across the resistor, the inductor, and the capacitor at resonance for the circuit in Example 28-5.

At resonance, the impedance is just the resistance $R = 20\ \Omega$. Since the maximum emf is 100 V, the maximum current is

$$I_{max} = \frac{\mathcal{E}_{max}}{Z} = \frac{100\ \text{V}}{20\ \Omega} = 5\ \text{A}$$

The maximum voltage across the resistor is thus

$$V_{R,max} = I_{max}R = (5\ \text{A})(20\ \Omega) = 100\ \text{V}$$

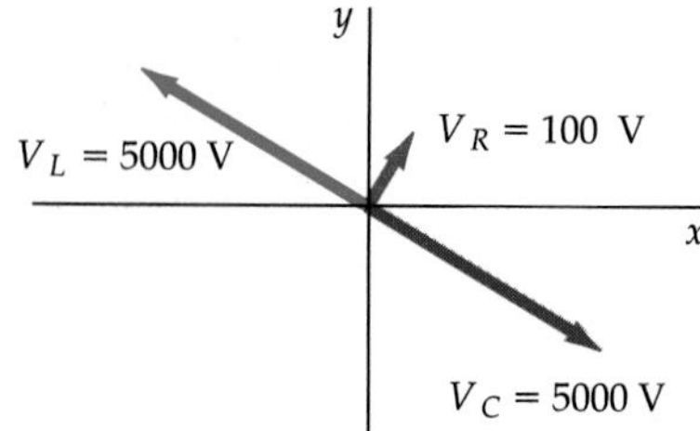

Figure 28-19 Voltages in Example 28-8. The voltages across the inductor and capacitor in a series *LCR* circuit are always 180° out of phase with each other. At resonance, they are equal in magnitude, so they sum to zero, leaving the sum of the voltages across all three elements equal to V_R. In this example, the maximum voltage drop across the resistor is 100 V, whereas the maximum drops across the inductor and the capacitor are 5000 V.

The resonance frequency found in Example 28-4 was $f_0 = 79.6$ Hz. The inductive and capacitive reactances at resonance are

$$X_L = \omega_0 L = (2\pi)(79.6 \text{ Hz})(2 \text{ H}) = 1000 \ \Omega$$

and

$$X_C = \frac{1}{\omega_0 C} = \frac{1}{(2\pi)(79.6 \text{ Hz})(2 \times 10^{-6} \text{ F})} = 1000 \ \Omega$$

The inductive and capacitive reactances are equal, as we would expect since we found the resonance frequency by setting them equal. The maximum voltage across the inductor is then

$$V_{L,\text{max}} = I_{\text{max}} X_L = (5 \text{ A})(1000 \ \Omega) = 5000 \text{ V}$$

and that across the capacitor is

$$V_{C,\text{max}} = I_{\text{max}} X_C = (5 \text{ A})(1000 \ \Omega) = 5000 \text{ V}$$

Figure 28-19 shows the phasor diagram for these voltages. The maximum voltage across the resistor is a relatively safe 100 V, equal to the maximum emf of the generator. However, the maximum voltages across the inductor and the capacitor are a dangerously high 5000 V. These voltages are 180° out of phase. At resonance, the voltage across the inductor at any instant is the negative of that across the capacitor, so they always sum to zero, leaving the voltage across the resistor equal to the emf in the circuit.

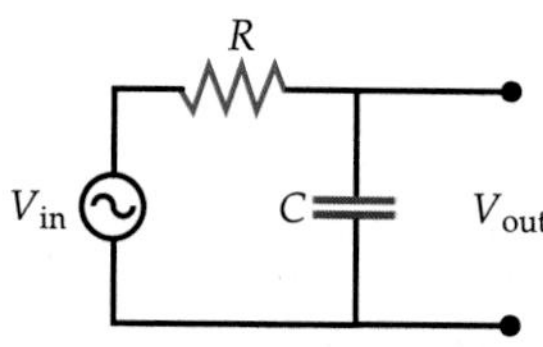

Figure 28-20 Circuit for Example 28-9.

Example 28-9

A resistor R and capacitor C are in series with a generator, which has a voltage given by $V_{\text{in}} = V_0 \cos \omega t$ as shown in Figure 28-20. Find the voltage across the capacitor as a function of frequency ω.

This circuit is simpler than the ones we have been discussing because there is no inductance. In the figure, we have labeled the generator voltage V_{in} (the input voltage) and the voltage across the capacitor V_{out} (the output voltage). Figure 28-21 shows the phasors representing the voltage drops across the resistance and capacitance. The total impedance in the circuit is

$$Z = \sqrt{R^2 + X_C^2}$$

where $X_C = 1/\omega C$. The rms current is then

$$I_{\text{rms}} = \frac{V_{\text{in rms}}}{Z} = \frac{V_{\text{in rms}}}{\sqrt{R^2 + X_C^2}}$$

The rms output voltage across the capacitor is

$$V_{\text{out rms}} = I_{\text{rms}} X_C = \frac{X_C V_{\text{in rms}}}{\sqrt{R^2 + X_C^2}}$$

$$= \frac{(1/\omega C) V_{\text{in rms}}}{\sqrt{R^2 + (1/\omega C)^2}} = \frac{V_{\text{in rms}}}{\sqrt{\omega^2 C^2 R^2 + 1}}$$

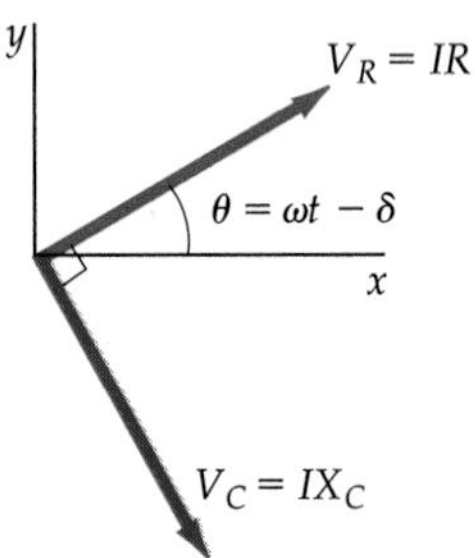

Figure 28-21 Phasor diagram for the voltages across the resistor and capacitor of Figure 28-20.

Figure 28-22 shows the ratio of the output voltage to the input voltage as a function of frequency ω. This circuit is called an *RC* **low-pass filter** because low input frequencies are transmitted with greater amplitude than high input frequencies.

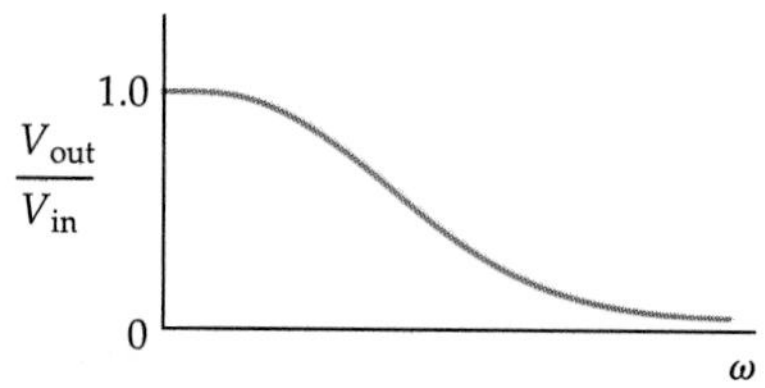

Figure 28-22 Graph of the ratio of output voltage to input voltage for the low-pass filter circuit of Example 28-9.

Parallel

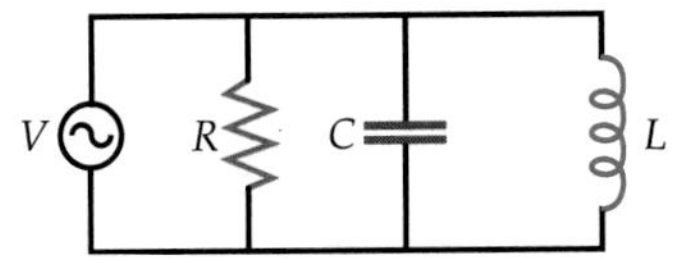

Figure 28-23 A parallel *LCR* circuit.

Figure 28-23 shows a resistor R, capacitor C, and an inductor L connected in parallel across an ac generator. The total current I from the generator divides into three currents, the current I_R in the resistor, the current I_C in the capacitor, and the current I_L in the inductor. The instantaneous voltage V is the same across each element. The current in the resistor is in phase with the voltage and has amplitude V/R. Since the voltage drop across an inductor *leads* the current in the inductor by 90°, the current in the inductor *lags* the voltage by 90° and has magnitude V/X_L. Similarly, the current in the capacitor leads the voltage by 90° and has magnitude V/X_C. These currents are represented by phasors in Figure 28-24. The total current I is the x component of the vector sum of the individual currents as shown in the figure. The magnitude of the total current is

$$I = \sqrt{I_R^2 + (I_L - I_C)^2} = \sqrt{\left(\frac{V}{R}\right)^2 + \left(\frac{V}{X_L} - \frac{V}{X_C}\right)^2} = \frac{V}{Z} \qquad 28\text{-}59$$

where the impedance Z is related to the resistance and the capacitive and inductive reactances by

$$\frac{1}{Z} = \sqrt{\left(\frac{1}{R}\right)^2 + \left(\frac{1}{X_L} - \frac{1}{X_C}\right)^2} \qquad 28\text{-}60$$

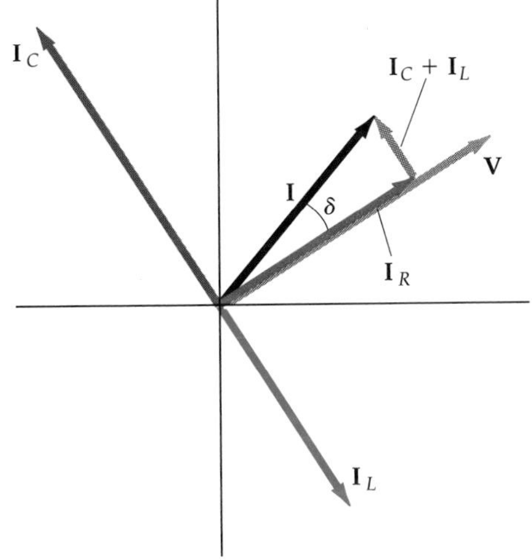

Figure 28-24 Phasor diagram for the voltage and currents in the parallel *LCR* circuit of Figure 28-23. The voltage is the same across each element. The current in the resistor is in phase with the voltage. The current in the capacitor leads the voltage by 90° and that in the inductor lags the voltage by 90°. The phase difference δ between the total current and the voltage depends on the relative magnitudes of the currents, which depend on the values of the resistance and of the capacitive and inductive reactances.

At resonance, the generator frequency ω equals the natural frequency $\omega_0 = 1/\sqrt{LC}$, and the inductive and capacitive reactances are equal. Then from Equation 28-60, we see that $1/Z$ has its minimum value $1/R$, so the impedance Z is maximum and the total current is minimum. We can understand this if we note that at resonance $X_C = X_L$, and the currents in the inductor and capacitor are equal but 180° out of phase, so the total current is just the current in the resistor.

Questions

6. Does the power factor depend on the frequency?
7. Are there any disadvantages in a radio tuning circuit having an extremely large Q factor?
8. What is the power factor for a circuit that has inductance and capacitance but no resistance?

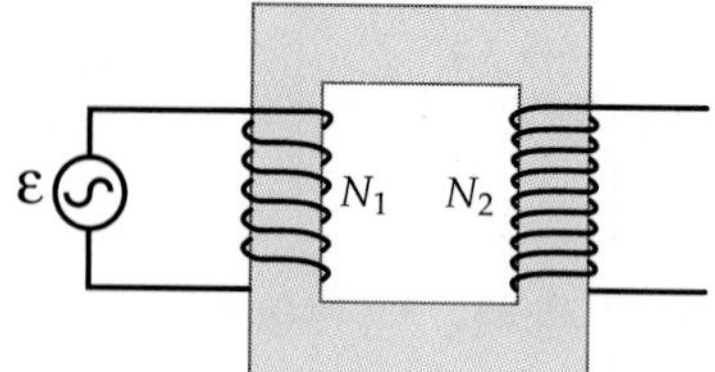

Figure 28-25 Transformer with N_1 turns in the primary and N_2 turns in the secondary.

28-6 The Transformer

A **transformer** is a device for changing alternating voltage and current without an appreciable loss in power. Its operation is based on the fact that an alternating current in one circuit will induce an alternating emf in a nearby circuit because of the mutual inductance of the two circuits. Figure 28-25 shows a simple transformer consisting of two coils of wire around a common iron core. The coil carrying the input power is called the **primary,** and the other coil is called the **secondary.** Either coil of a transformer can be used for the primary or secondary. The function of the iron core is to increase the magnetic field for a given current and to guide it so that nearly all the magnetic flux through one coil goes through the other coil. The iron core is laminated to reduce eddy-current losses. Other power losses arise because of the Joule heating in the small resistances in both coils and from hysteresis in the iron cores. We will neglect these losses and consider an ideal transformer of 100 percent efficiency, for which all of the power supplied to the primary coil appears in the secondary coil. Actual transformers are often 90 to 95 percent efficient.

Consider a transformer with an ac generator of emf $\mathcal{E}$ across the primary of N_1 turns and the circuit of the secondary coil of N_2 turns open. Because of the iron core, there is a large flux through each coil even when the **magnetizing current** I_{m} in the primary circuit is very small. We can neglect the resistances of the coils compared with their inductive reactances. The primary is then a simple circuit consisting of an ac generator and a pure inductance like that discussed in Section 28-2. The (magnetizing) current and the voltage in the primary are out of phase by 90°, and the average power dissipated in the primary coil is zero. If ϕ_{turn} is the magnetic flux in one turn of the primary coil, the voltage drop across the primary coil is $V_{L1} = N_1\, d\phi_{\mathrm{turn}}/dt$. Applying Kirchhoff's loop rule to the primary circuit then gives

$$\mathcal{E} - N_1 \frac{d\phi_{\mathrm{turn}}}{dt} = 0$$

or

$$\mathcal{E} = N_1 \frac{d\phi_{\mathrm{turn}}}{dt} \qquad 28\text{-}61$$

If there is no flux leakage out of the iron core, the flux through each turn is the same for both coils. Thus, the total flux through the secondary coil is $N_2\, \phi_{\mathrm{turn}}$, and the voltage across the secondary coil is

$$V_2 = -N_2 \frac{d\phi_{\mathrm{turn}}}{dt} \qquad 28\text{-}62$$

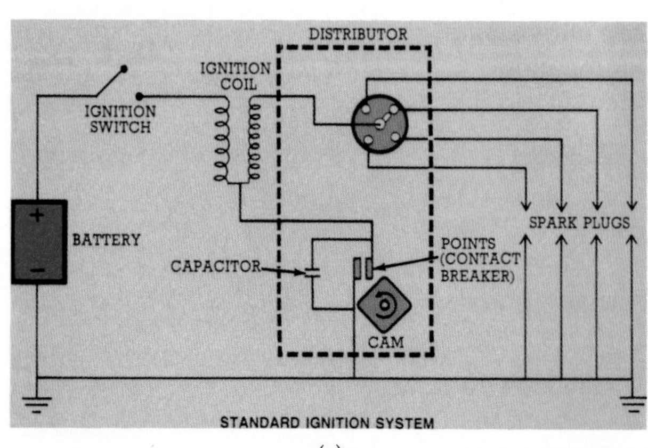

(*a*)

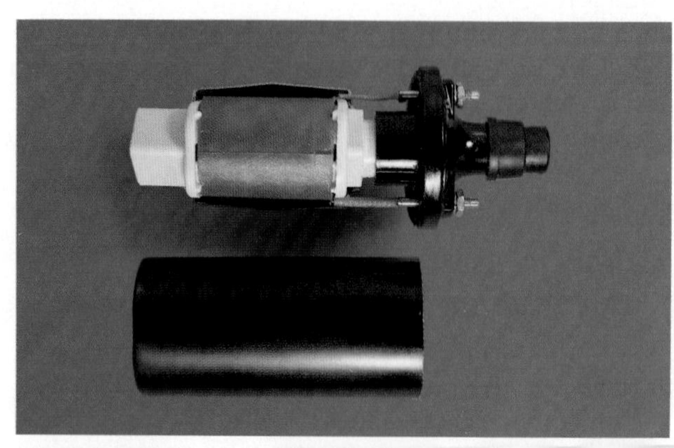

(*b*)

(*a*) Circuit for automobile ignition. The ignition shown in the photo in (*b*) is a transformer with a primary winding of heavy wire and a secondary winding of finer wire. With the ignition switch on and contact points closed, there is a current through the primary of the ignition coil. The rotating cam opens or closes the contact points, resulting in a rapid change in the primary current that produces a voltage of 30 to 40 kV in the secondary coil. This high voltage is distributed through the rotor to the spark plugs at precisely timed intervals, resulting in a spark across the gap in the plugs. The spark ignites the air-fuel mixture in the cylinders.

Comparing these equations, we can see that

$$V_2 = -\frac{N_2}{N_1}\mathcal{E} \qquad 28\text{-}63$$

If N_2 is greater than N_1, the voltage in the secondary coil is greater than that in the primary coil, and the transformer is called a **step-up transformer.** If N_2 is less than N_1, the voltage in the secondary coil is less than that in the primary coil, and the transformer is called a **step-down transformer.**

Consider now what happens when we put a resistance R called a **load resistance** across the secondary coil. There will then be a current I_2 in the secondary circuit that is in phase with the voltage V_2 across the resistance. This current will set up an additional flux ϕ'_{turn} through each turn that is proportional to N_2I_2. This flux opposes the original flux set up by the original magnetizing current I_m in the primary. However, the voltage across the primary coil is determined by the generator emf, which is unaffected by the secondary circuit. According to Equation 28-61, the flux in the iron core must change at the original rate; that is, the total flux in the iron core must be the same as when there is no load across the secondary. The primary coil thus draws an additional current I_1 to maintain the original flux ϕ_{turn}. The flux through each turn produced by this additional current is proportional to N_1I_1. Since this flux equals $-\phi'_{\text{turn}}$, the additional current I_1 in the primary is related to the current I_2 in the secondary by

$$N_1I_1 = -N_2I_2 \qquad 28\text{-}64$$

These currents are 180° out of phase and produce counteracting fluxes. Since I_2 is in phase with V_2, the additional current I_1 is in phase with the applied emf. The power input from the generator is $\mathcal{E}_{\text{rms}}I_{1,\text{rms}}$, and the power output is $V_{2,\text{rms}}I_{2,\text{rms}}$. (The magnetizing current does not contribute to the power input because it is 90° out of phase with the generator voltage.) If there are no losses,

$$\mathcal{E}_{\text{rms}}I_{1,\text{rms}} = V_{2,\text{rms}}I_{2,\text{rms}} \qquad 28\text{-}65$$

In most cases the additional current in the primary I_1 is much greater than the original magnetizing current I_m that is drawn from the generator when there is no load. This can be demonstrated by putting a light bulb in series with the primary coil. The bulb is much brighter when there is a load across the secondary than when the secondary circuit is open. If I_m can be neglected, Equation 28-64 relates the total currents in the primary and secondary circuits.

(*a*)

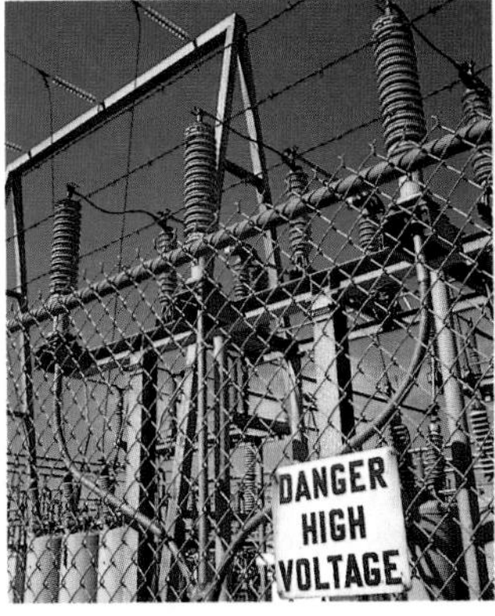

(*b*)

(*a*) A power box with transformer for stepping down voltage for distribution to homes. (*b*) A suburban power substation where transformers step down voltage from high-voltage transmission lines.

Example 28-10

A doorbell requires 0.4 A at 6 V. It is connected to a transformer whose primary contains 2000 turns that is connected to 120 V ac. (*a*) How many turns should there be in the secondary? (*b*) What is the current in the primary?

(*a*) Since the input voltage is 120 V and the output is 6 V, the turns ratio can be obtained from Equation 28-63:

$$\frac{N_2}{N_1} = \frac{V_2}{\mathcal{E}} = \frac{6\text{ V}}{120\text{ V}}$$

The number of turns in the secondary is thus

$$N_2 = \frac{6}{120}(2000\text{ turns}) = 100\text{ turns}$$

(*b*) Since we are assuming 100 percent efficiency in power transmission, the input and output currents are related by

$$V_2 I_2 = \mathcal{E} I_1$$

The current in the primary is thus

$$I_1 = \frac{V_2}{\mathcal{E}} I_2 = \frac{6}{120}(0.4\text{ A}) = 0.02\text{ A}$$

An important use of transformers is in the transport of electrical power. To minimize the I^2R heat loss in transmission lines, it is economical to use a high voltage and a low current. On the other hand, safety and other considerations, such as insulation, make it necessary to use power at lower voltage and higher current to run motors and other electrical appliances. Suppose, for example, that each person in a city with a population of 50,000 uses 1.2 kW of electric power. (The per capita consumption of power in the United States is actually somewhat higher than this.) At 120 V, the current required for each person would be

$$I = \frac{1200\text{ W}}{120\text{ V}} = 10\text{ A}$$

The total current for 50,000 people would then be 500,000 A. The transport of such a current from a power-plant generator to a city many kilometers away would require wires of enormous size (actually, the wires would probably have to be large copper cylinders), and the I^2R power loss would be substantial. Rather than transmit the power at 120 V, step-up transformers are used at the power plant to step up the voltage to some very large value, such as 600,000 V. The current needed is thus reduced to

$$I = \frac{120\text{ V}}{600{,}000\text{ V}}(500{,}000\text{ A}) = 100\text{ A}$$

To reduce the voltage to a safer level for transport within a city, power stations are located just outside the city to step down the voltage to a safer value, such as 10,000 V. Transformers in boxes attached to the power poles outside each house again step down the voltage to 120 V (or 240 V) for distribution to the house. It is because of the ease of stepping the voltage up or down with transformers that alternating current rather than direct current is in common use.

Example 28-11

A transmission line has a resistance of 0.02 Ω/km. Calculate the I^2R power loss if 200 kW of power is transmitted from a power generator to a city 10 km away at (*a*) 240 V and (*b*) 4.4 kV.

(*a*) The total resistance of 10 km of wire is $R = (0.02\ \Omega/\text{km})(10\ \text{km}) = 0.20\ \Omega$. The current required to transmit 200 kW at 240 V is

$$I = \frac{200{,}000\ \text{W}}{240\ \text{V}} = 833\ \text{A}$$

The power loss is then

$$I^2R = (833\ \text{A})^2(0.20\ \Omega) = 139\ \text{kW}$$

Thus, about 70 percent of the power is wasted through heat loss.

(*b*) If the transmission voltage is 4.4 kV, the current is

$$I = \frac{200\ \text{kW}}{4.4\ \text{kV}} = 45.5\ \text{A}$$

The power loss is then

$$I^2R = (45\ \text{A})^2(0.20\ \Omega) = 414\ \text{W}$$

The energy loss is considerably less.

28-7 Rectification and Amplification

Although alternating current is readily available, direct current is often needed to power such devices as portable radios or calculators. These devices often come with batteries and with ac–dc converters to save the batteries when ac power is available. These converters contain a transformer for stepping down the voltage from 120 V to whatever voltage is needed (typically 9 V) and a circuit for converting from ac to dc. The process of converting alternating current to direct current is called **rectification.** The principal element in a rectifier circuit is a **diode.** The first diodes, developed by John Fleming in 1904, were vacuum tubes containing two main elements, a cathode that emits electrons and an anode, called the plate, that collects them. The important feature of a diode is that it conducts current in one direction and not the other. Most diodes in use today are semiconductor devices, which will be discussed in Chapter 39 of the extended version of this book. The symbol for a diode as a circuit element is →|. The arrow indicates the direction in which current can pass through the diode. (This is opposite the direction of the motion of the electrons.)

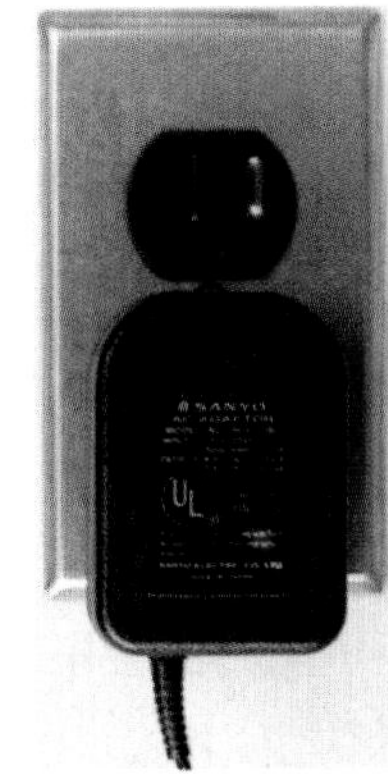

A transformer and rectifier for converting 120 V ac from a wall outlet to 9 V dc for use as a battery saver to power a radio or calculator.

Figure 28-26 shows a vacuum-tube diode. When the cathode is heated (by a heating element in a separate circuit), it emits electrons, a process called **thermionic emission** discovered by Thomas Edison in 1883. If the

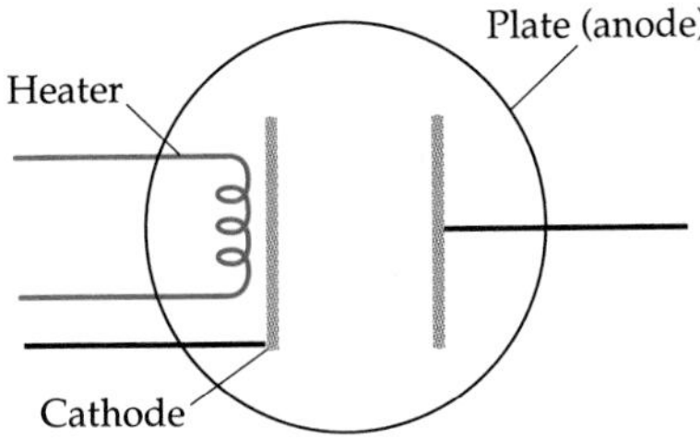

Figure 28-26 Vacuum-tube diode. When the cathode is heated, it emits electrons. The electrons are drawn to the plate when it is at a higher potential than the cathode.

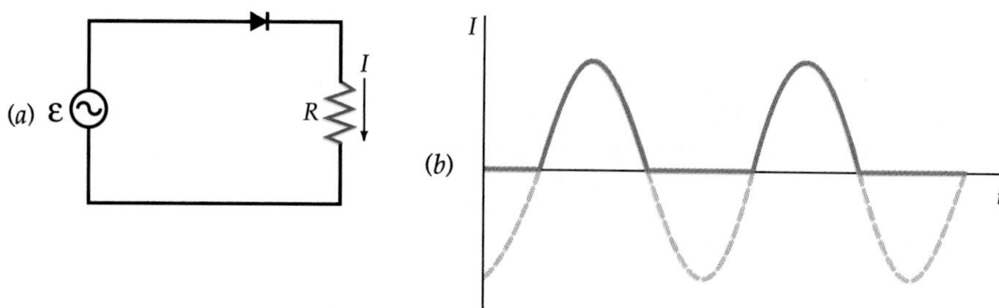

Figure 28-27 (*a*) Simple circuit containing an ac generator, a diode, and a resistor. (*b*) Current versus time in the resistor in the circuit in (*a*). The negative current indicated by the dashed lines does not get through the diode.

plate is at a higher potential than the cathode, it attracts the electrons and the tube conducts a current. The current is called the **plate current.** If the plate is at a lower potential than the cathode, the electrons are repelled and there is no current through the tube.

Figure 28-27*a* shows a simple circuit containing an ac generator, a diode, and a resistor. The current in the resistor is shown in Figure 28-27*b*. The diode is said to be a half-wave rectifier because there is current in the resistor for only half of each cycle of the ac generator. Figure 28-28 shows a circuit that gives full-wave rectification. In it, two diodes are connected to the terminals *a* and *b* of a transformer. The outputs of the diodes are connected together and to a resistor. The other side of the resistor is connected to the midpoint *c* of the transformer. When point *a* is at a higher potential than point *c*, diode 1 conducts the current I_1 to the resistor. One-half cycle later, point *b* is at a higher potential than point *c* and diode 2 conducts the current I_2 to the resistor. The current $I = I_1 + I_2$ through the resistor is shown in Figure 28-29*d*. The undesirable variations in rectifier output are called **ripple.**

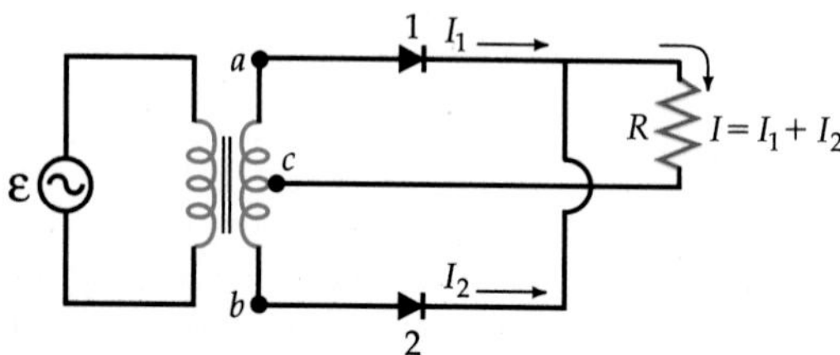

Figure 28-28 Full-wave rectifying circuit. When the potential of point *a* is greater than that of point *c*, the current I_1 passes through diode 1. One-half cycle later, the potential of point *b* is greater than that of point *c*, and the current I_2 passes through diode 2.

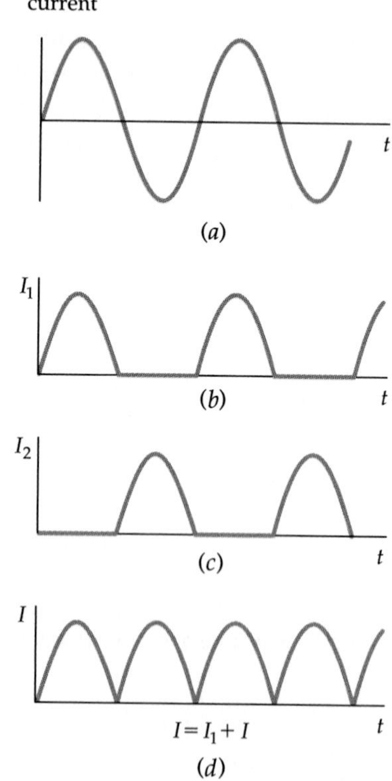

Figure 28-29 (*a*) Input current to the transformer in the circuit shown in Figure 28-28. (*b*) Current I_1 through diode 1. (*c*) Current I_2 through diode 2. (*d*) The total current $I = I_1 + I_2$ through the resistor in Figure 28-28.

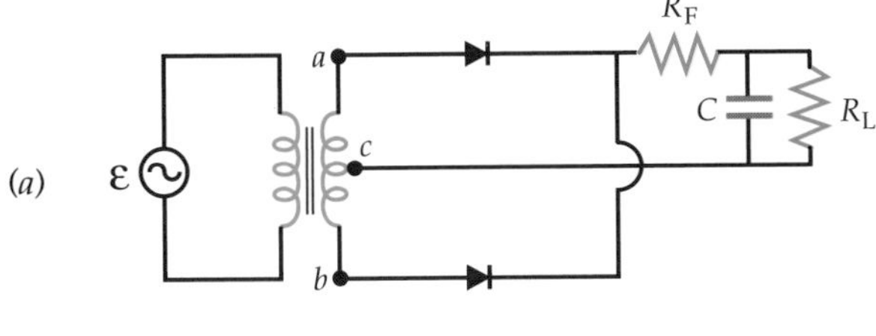

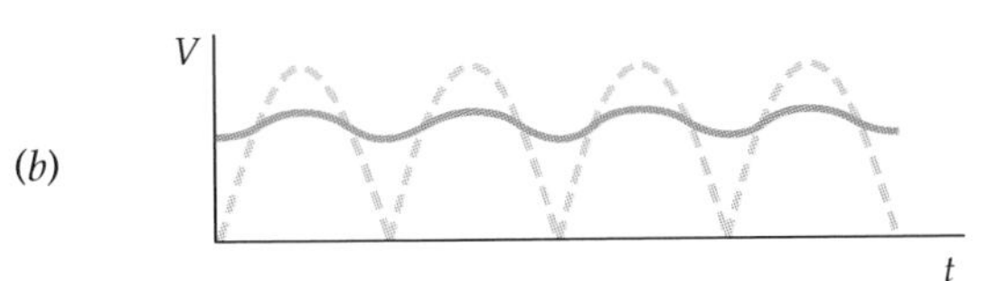

Figure 28-30 (*a*) Full-wave rectifying circuit of Figure 28-28 with a low-pass filter to smooth out some of the ripples in the rectified voltage. (*b*) Input voltage (dashed) and output voltage (solid) of the low-pass filter.

In Figure 28-30*a* a low-pass filter consisting of a resistor R_F and a capacitor C have been added between the rectifier and the load resistor R_L. (This filter was analyzed above in Example 28-9.) The resistance R_F is chosen to be much smaller than R_L, so that the dc voltage drop (and hence the power loss) across the filter resistor is small compared to that across the load resistor. The capacitance C is large so that the filter has a time constant R_FC much larger than the period between ripple cycles, so the change of the capacitor charge (and voltage) due to the ripple is very small compared to what it could become if the ripple varied slowly. The waveforms for input and output voltage for the filter are plotted in Figure 28-30*b*.

In 1907, Lee de Forest discovered that the plate current could be greatly modified by small voltage changes on a third electrode inserted between the cathode and the plate. A vacuum-tube **triode** is shown in Figure 28-31. The third electrode is a fine wire mesh called the **grid.** As with the diode, the cathode of the triode is heated and emits electrons that are collected by the plate, which is at a higher potential (typically 100 to 200 V) than the cathode. Because the grid is closer to the cathode than the plate, the potential of the grid relative to the cathode has a large effect on the plate current.

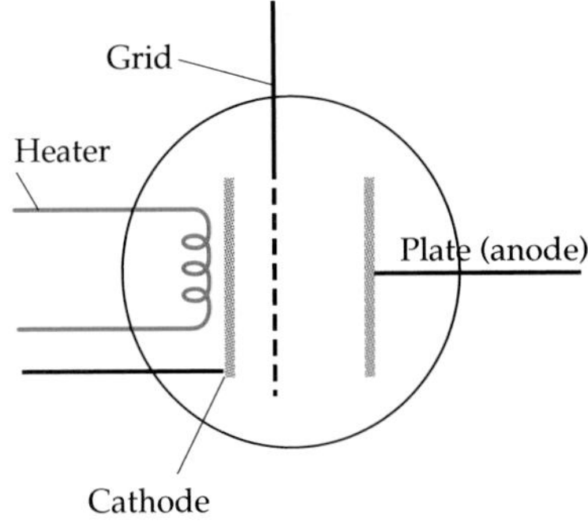

Figure 28-31 Vacuum-tube triode. The grid placed near the cathode controls the plate current. When the grid is negative relative to the cathode, it repels the electrons emitted by the cathode and diminishes the plate current. When the grid is positive relative to the cathode, it attracts the electrons emitted by the cathode and increases the plate current.

(*a*)

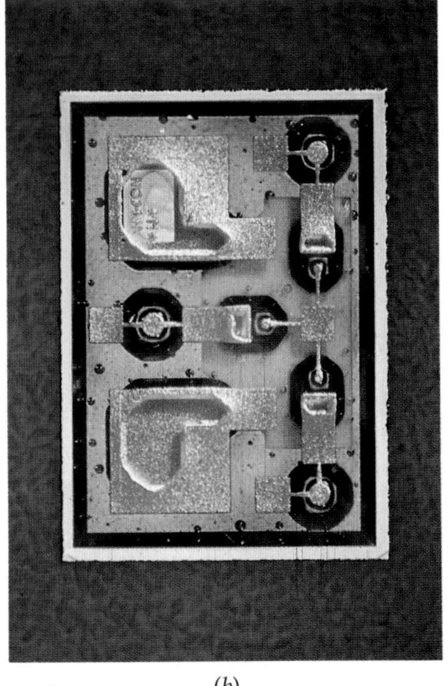

(*b*)

(*a*) The triode vacuum tube, invented by Lee De Forest in 1907. On either side of the cathode (which is hidden from view at the center) is the partially visible control grid, two zigzagging nickel wires. A pair of nickel plates surrounding the grid serves as the anode. The tube (not counting its ceramic base) is 9 cm tall. (*b*) A silicon chip containing six PIN diodes (dark, octagonal elements). PIN diodes function as a resistance that varies according to the voltage applied across them. They are used to switch microwave signals on and off by short-circuiting the waveguides that transmit them.

Figure 28-32 Amplification by a triode. A small sinusoidal signal applied to the grid results in a large sinusoidal signal across the resistance R.

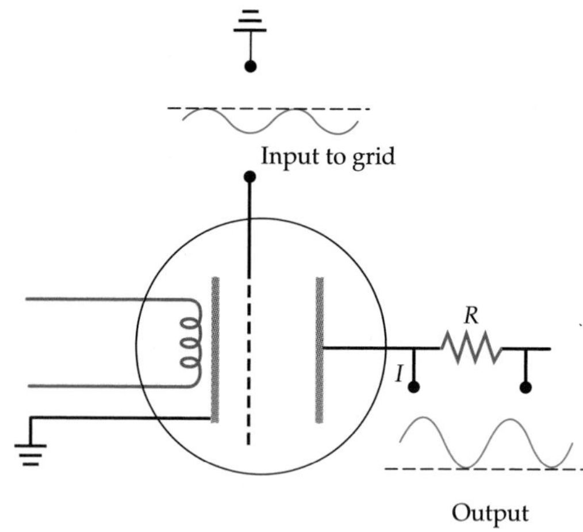

When the grid is at the same potential as the cathode, the plate current is essentially unaffected by the grid. When the grid is negative relative to the cathode, the electrons emitted by the cathode are repelled by the grid, and the plate current is greatly diminished. Figure 28-32 shows how a triode can be used as an **amplifier.** The input signal is a small sinusoidal voltage that is applied between the grid, which is negatively biased, and the cathode. The output signal is the voltage across the resistance R. The output signal is considerably larger than the input signal because small voltage changes on the grid produce large changes in the plate current. Today, vacuum-tube triodes have been largely replaced by transistors, which will be discussed in Chapter 38 of the extended version of this book.

Question

9. Explain why the rms current in a half-wave rectifier circuit is half that in a full-wave rectifier circuit.

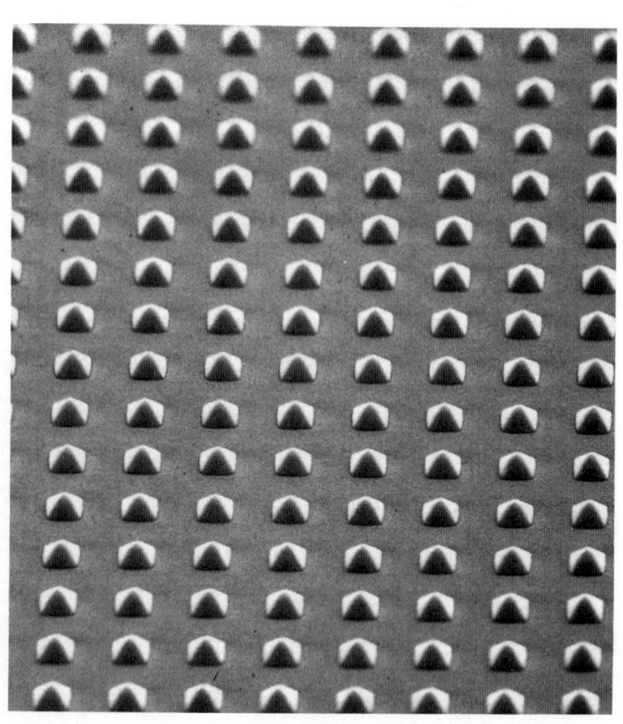

Vacuum tubes can now be constructed in miniaturized forms known as microelectric vacuum devices. Such tubes, presently in the research state, may someday replace the cathode-ray tubes in television sets, making much thinner sets possible. This array of tungsten-clad pyramids etched in a single silicon crystal functions as a replacement for a cathode. The pyramids are 3 μm high and separated by 10 μm. Instead of heat, a strong electric field is applied above the pyramids so that they emit electrons. (At a sufficiently strong field, electrons overcome the forces that bind them to the tungsten surface.) Unlike transistors, microvacuum tubes are insensitive to heat and radiation. Furthermore, at equivalent sizes, microvacuum tubes operate more rapidly than transistors because electrons travel through them without collisions.

Summary

1. The root-mean-square (rms) value of alternating current, I_{rms}, is defined as

$$I_{rms} = \sqrt{(I^2)_{av}}$$

It is related to the maximum current by

$$I_{rms} = \frac{I_{max}}{\sqrt{2}}$$

The average power dissipated in a resistor carrying a sinusoidal current is

$$P_{av} = \tfrac{1}{2}\mathcal{E}_{max}I_{max} = \mathcal{E}_{rms}I_{rms} = I_{rms}^2 R$$

2. The voltage across an inductor leads the current by 90°. The rms or maximum current is related to the rms or maximum voltage by

$$I = \frac{V_L}{X_L}$$

where

$$X_L = \omega L$$

is the inductive reactance of the inductor. The average power dissipated in an inductor is zero.

The voltage across a capacitor lags the current by 90°. The rms or maximum current is related to the rms or maximum voltage by

$$I = \frac{V_C}{X_C}$$

where

$$X_C = \frac{1}{\omega C}$$

is the capacitive reactance. The average power dissipated in a capacitor is zero. Like resistance, inductive and capacitive reactances have units of ohms.

3. The phase relations among the voltages across a resistor, a capacitor, and an inductor in an ac circuit can be described graphically by representing the voltages with rotating, two-dimensional vectors called phasors. These phasors rotate in the counterclockwise direction with an angular frequency ω that is equal to the angular frequency of the current. The phasor **I** represents the current. The phasor $\mathbf{V}_R$ representing the voltage across a resistor is in phase with the current. The phasor $\mathbf{V}_L$ representing the voltage across an inductor leads the current by 90°. The phasor $\mathbf{V}_C$ representing the voltage across a capacitor lags the current by 90°. The x component of each phasor equals the magnitude of the current or the corresponding voltage drop at any instant.

4. If a capacitor is discharged through an inductor, the charge and voltage on the capacitor oscillate with angular frequency

$$\omega_0 = \frac{1}{\sqrt{LC}}$$

The current in the inductor oscillates with the same frequency, but it is out of phase with the charge by 90°. The energy oscillates between electric energy in the capacitor and magnetic energy in the inductor. If the circuit also has resistance, the oscillations are damped because energy is dissipated in the resistor.

5. The current in a series *LCR* circuit driven by an ac generator is given by

$$I = \frac{\mathcal{E}_{max}}{Z} \cos(\omega t - \delta)$$

where the impedance Z is

$$Z = \sqrt{R^2 + (X_L - X_C)^2}$$

and the phase angle δ is found from

$$\tan \delta = \frac{X_L - X_C}{R}$$

The average power input to such a circuit depends on the frequency and is given by

$$P_{av} = \mathcal{E}_{rms} I_{rms} \cos \delta$$

where $\cos \delta$ is called the power factor. The average power is maximum at the resonance frequency, which is given by

$$f_0 = \frac{1}{2\pi\sqrt{LC}}$$

At the resonance frequency, the phase angle δ is zero, the power factor is 1, the inductive and capacitive reactances are equal, and the impedance Z equals the resistance R.

6. The sharpness of the resonance is described by the Q factor, which is defined by

$$Q = \frac{\omega_0 L}{R}$$

When the resonance is reasonably narrow, the Q factor can be approximated by

$$Q = \frac{\omega_0}{\Delta\omega} = \frac{f_0}{\Delta f}$$

where Δf is the width of the resonance curve.

7. A transformer is a device for changing alternating voltage and current without an appreciable loss in power. For a transformer with N_1 turns in the primary and N_2 turns in the secondary, the voltage across the secondary coil is related to the generator emf across the primary coil by

$$V_2 = -\frac{N_2}{N_1}\mathcal{E}$$

A transformer is called a step-up transformer if N_2 is greater than N_1 so that the output voltage is greater than the input voltage. If N_2 is less than N_1, it is called a step-down transformer.

8. A diode is a device that conducts current in one direction only. It can be used to convert alternating current to direct current, a process called rectification.

9. Small changes in the voltage of the grid in a triode produce large changes in the plate current, an effect that can be used to amplify ac signals.

Electric Motors

John Dentler
United States Naval Academy

The development of a variety of electric motors has revolutionized our society. Early in the twentieth century, the machines in most large factories were driven by one or two large steam engines via belts and pulleys. Automobiles were started with hand cranks, refrigerators used large blocks of ice for cooling, and sewing machines were powered by foot treadles. Today, electric motors perform these tasks.

The wide variety of applications for electric motors requires many different designs. Electric-clock motors must operate at a precise speed. Automobile starters must deliver a tremendous torque from a standstill. A hand-held hair dryer must be lightweight and operate at several different speeds. Engineers design motors for various applications using models derived from the physical principles discussed in this text. These models are equations that predict a motor's performance for a set of specific applications or loads.

John Dentler graduated from the United States Naval Academy in 1972 with a B.S. in Physics. After completing the Navy Nuclear Power Training Program, he was assigned as Reactor Controls Division Officer on the USS Nimitz. He returned to the Nuclear Power Prototype as an instructor and department head. Following three shipboard tours in Engineering, Weapons and Operations, he attended the Naval Postgraduate School, earning a Master of Science in Electrical Engineering (concentrating in electro-optic, electronic warfare, and radar systems). Assigned as an instructor at the United States Naval Academy, he taught Fundamental Electrical Engineering for all midshipmen, and the Energy Conversion Course in the Electrical Engineering Major Curriculum. He served as Associate Chairman of the Electrical Engineering Department. He is currently assigned as a Power Systems Engineer at the Navy's David Taylor Research Center.

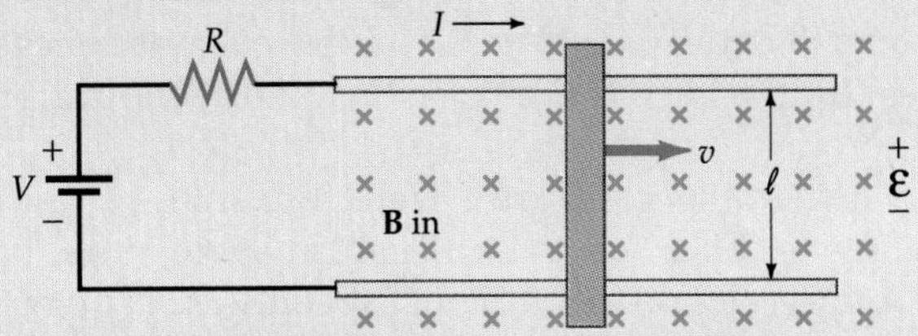

Figure 1 A simple electric motor constructed by adding a battery to the linear machine shown in Figures 26-14 and 26-16.

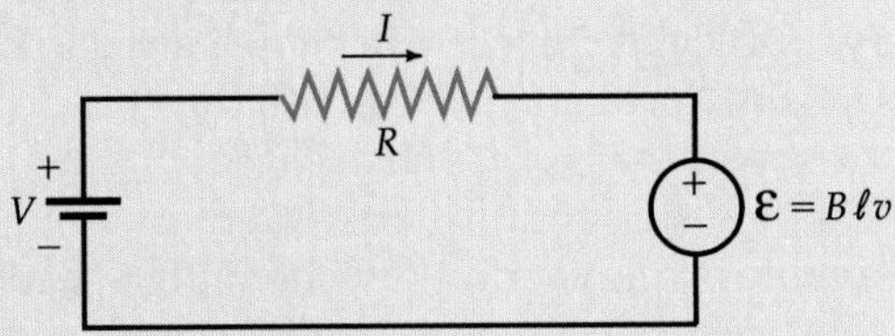

Figure 2 Circuit for the linear electric motor in Figure 1.

The simplest motor to model is derived from the linear machine introduced in Section 26-4. Figure 1 is similar to Figures 26-14 and 26-16 except a battery of voltage V is inserted with its voltage in the same sense as the emf developed across the moving bar. In general, the rails and the bar have resistance and inductance and the battery has internal resistance. For simplicity, we assume that the inductance is negligible and that the total resistance of the system can be considered to be that of a single resistor R. The bar can be modeled as an ideal source of emf with $\mathcal{E} = B\ell v$. The circuit that models this simple linear motor is shown in Figure 2.

Applying Kirchhoff's loop rule to this circuit yields

$$V - IR - \mathcal{E} = 0 \tag{1}$$

Substituting $\mathcal{E} = B\ell v$, we obtain

$$V - IR - B\ell v = 0 \tag{2}$$

The current is thus given by

$$I = -\frac{B\ell}{R}v + \frac{V}{R} \tag{3}$$

The current is thus a linear function of velocity provided that V, B, ℓ, and R are constant. For low velocities, the emf is small and the current is positive (down the bar). For high velocities, the emf is larger

Continued

than the battery voltage V and the current is negative (up the bar). If the velocity is $V/B\ell$, the current is zero.

If the rails in Figure 1 were frictionless and the magnetic field were sufficiently broad, the bar would accelerate to a terminal speed $V/B\ell$, at which time the force, $F = IB\ell$, is zero because the current is zero. If the bar is accelerated to the right by an external force, the current charges the battery as the generator charges the battery in an automobile. If the bar is slowed by an external force directed to the left, the current is driven through the bar by the battery. Sufficient current will be delivered to match the opposing force and the bar will travel at an equilibrium speed that is less than $V/B\ell$.

A motor designer is interested in predicting how a motor will respond (by changing speed) to a load. For this linear motor, the load is an external force on the bar. The equal and opposite force delivered by the motor at the equilibrium speed is $F = IB\ell$. Equation 3 can be rearranged to show the velocity as a linear function of the current:

$$v = -\frac{R}{B\ell}I + \frac{V}{B\ell} \tag{4}$$

Substituting $F/B\ell$ for the current, we obtain

$$v = -\frac{R}{(B\ell)^2}F + \frac{V}{B\ell} \tag{5}$$

Equation 5 relating the velocity to the load is called the *motor performance characteristic*. Figure 3 shows a graph of the velocity versus force. Line 1 represents v versus F for typical values of the battery voltage V and the magnetic field B. Line 2 shows the effect of raising the voltage V. Line 3 shows the effect of lowering the magnetic field B. Line 4 represents a typical (for instance, frictional) load that increases proportionately with speed. The motor operates at the intersection of the load line and the motor performance characteristic. Thus, the operating speed of the motor shown in Figure 1 can be controlled by changing either the voltage or the magnetic field.

The linear motor in Figure 1 is not practical for most applications. Instead, a rotating motor is more appropriate. Figure 4 shows most of the parts of a simple rotating electric motor. Although this motor appears to be quite different from the linear one, the operation of the two motors is similar.

Like the linear motor, the rotating motor has current-carrying conductors that react with an external field. The field, called the *stator field*, is created

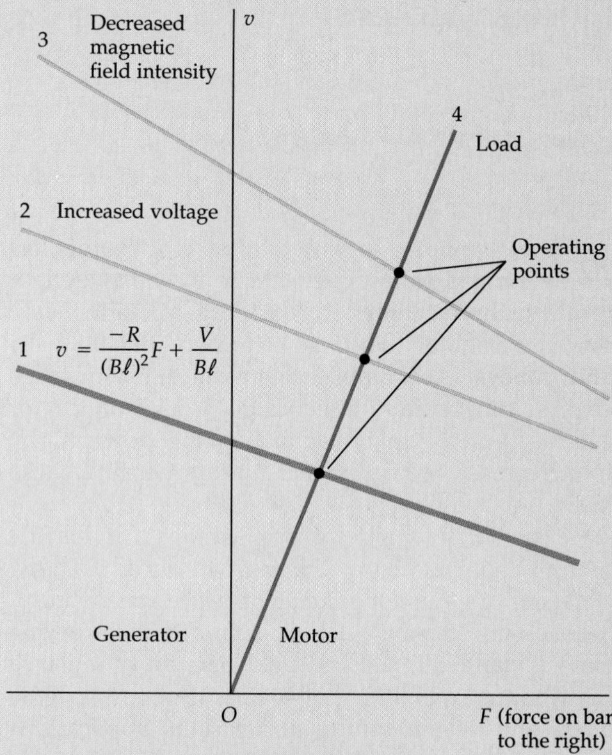

Figure 3 Graph of different values of the motor performance characteristic of the linear electric motor in Figure 1 and a typical load. The load line indicated is typical for a frictional load.

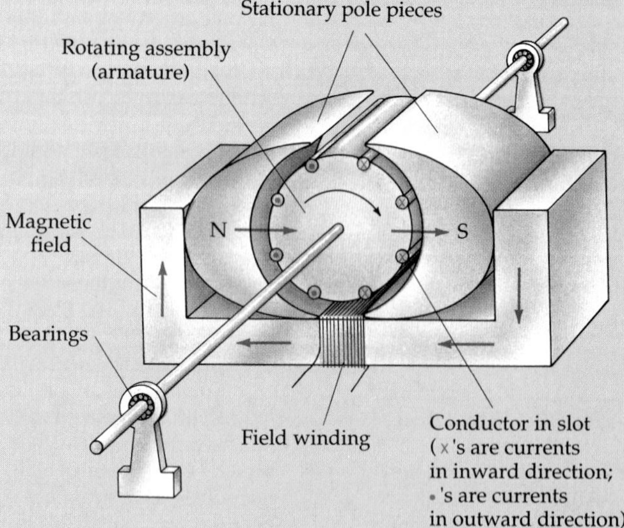

Figure 4 A simple rotating electric motor.

and controlled by the coil of wire at the bottom of Figure 4. Flux from the wire coil passes through the core, creating a north pole on the left and a south pole on the right of the rotating element. The rotating assembly is supported by two bearings, one at the front and one at the back of the motor. The rotating assembly, called the *armature,* consists of an iron cylinder with eight slots. The slots contain conductors, which are similar to the bar in the linear motor. If current can be driven through these conductors in the direction shown (front to back near the south pole and back to front near the north pole), then a net clockwise torque (down on the south-pole side and up on the north-pole side) will be developed to turn the armature. The torque developed in the rotating motor is analogous to the force developed in the linear motor.

Constructing a device that maintains the proper current direction in each conductor as the armature turns is a complicated task. Such devices are called *commutator brush assemblies.* Figure 5 shows the connections of such an assembly for the motor in Figure 4. Figure 6 is a photo of an actual commutator brush assembly used in a car starter motor. The photo shows an armature with many slots and a commutator with many segments. Figure 7 is a photo of a slot-car motor with three discrete slots,

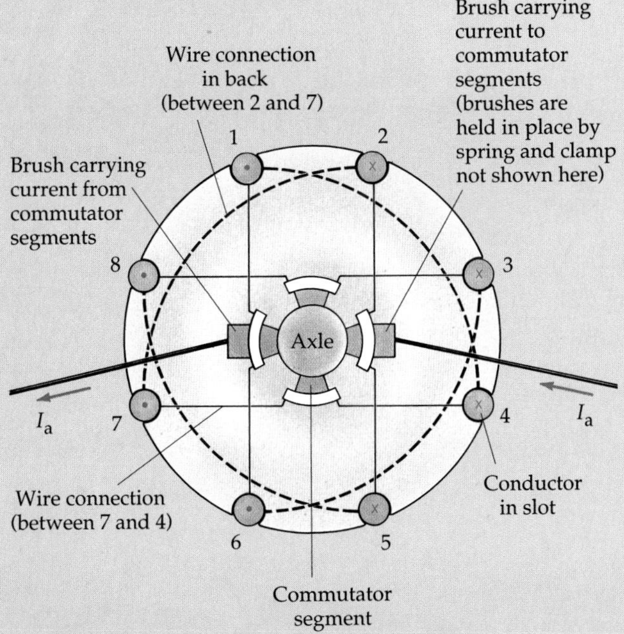

Figure 5 A commutator brush assembly for the motor in Figure 4. A commutator, in its most general sense, is a switching device. The device shown switches current direction through the armature to maintain the clockwise rotation.

each containing many windings and a three-segment commutator (of which only two are visible).

The commutator brush assembly shown in Figure 5 consists of four segments that protrude along the motor shaft and two brushes that conduct current from a source to the segments. Each segment is connected to two conductors, which run through the slots of the rotating assembly. The conductors are interconnected through wires in the rear of the rotating assembly and by the commutator segments in the front of the assembly. This method of connection results in two parallel paths between the brushes; thus all of the conductors are used all of the time.

In the commutator shown in Figure 5, current is delivered from the brush on the right. It then follows one of the two parallel paths through the armature. The conductors in slots 2 and 5 both carry the current from the front to the back of the armature. Conductors 2 and 5 are connected to conductors 7 and 8 via wires in the rear of the armature. The current returns to the front through slots 7 and 8, which are connected to slots 3 and 4 via the common connection on the commutator segments. The current is carried to the back along slots 3 and 4 and is then returned to the front through slots 1 and 6, where it is picked off by the brush on the left. The commutator assembly rotates with the armature and moves under the brushes. The brushes are stationary and will contact different commutator segments when the armature has moved 90 degrees. Since the armature is symmetrically wound, the slots on the right will always carry current from front to back and the slots on the left will always carry current from back to front, thus maintaining the clockwise torque.

The total torque turning the motor is the sum of the torques exerted by the conductors in each slot. In any position there are four armature conductors acting on the right and four on the left; therefore, the torque is approximately constant. Similarly, the total emf developed between the brushes is the sum of the emfs on each conductor. At any position there are two parallel paths, each consisting of four conductors. From Section 26-6, the emf developed across a single length of wire can be shown to be

$$\mathcal{E} = B\ell r\omega \sin(\omega t + \delta) \tag{6}$$

where ℓ is the length of the armature (front to back) and r is the radius of the rotor. The total emf across the armature will be the average of that developed

Continued

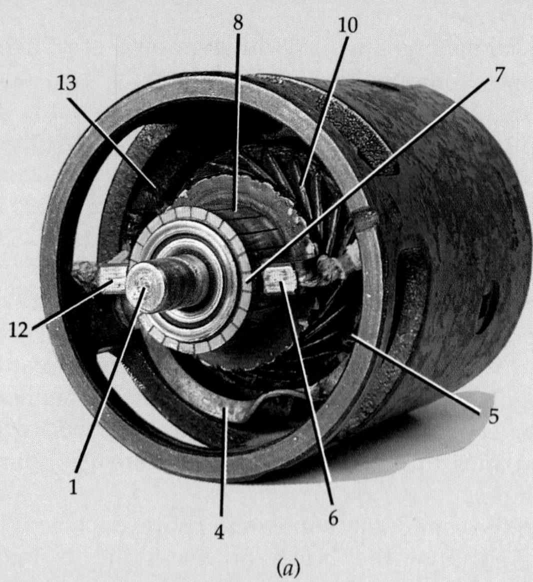

(a)

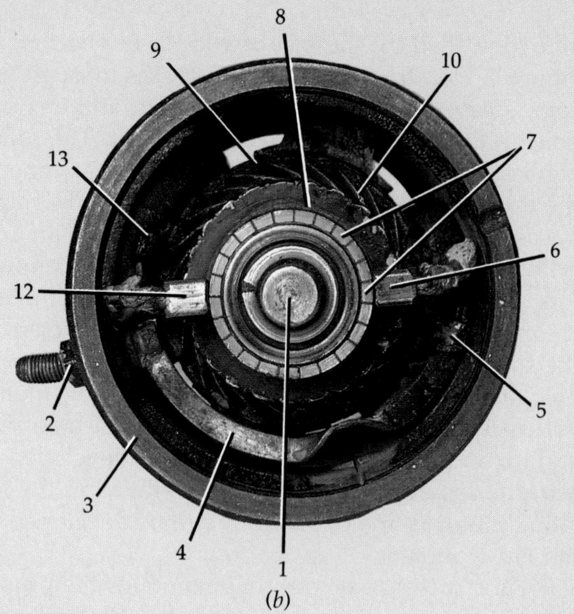

(b)

Figure 6 An automobile starter motor shown from several views. The shaft **(1)** is supported by bearings normally contained in the end bells of the motor housing. (The end bells have been removed.) Current from the battery enters the motor from a wire attached to the bolt **(2)** on the side of the motor. The bolt is insulated from the motor housing **(3)** and is connected through the housing to the field via the metal strip **(4)** running under the shaft. (This extra length is for thermal expansion.) The strip is coiled on the right side to form the winding for the right side of the stator pole **(5)**. The strip conductor is then connected to the wire and the soft metal brush **(6)** on the right side. The brush is normally held tight against the commutator **(7)** by a spring clip. (The spring clip has been removed.) Current is conducted through one of the 23 commutator segments **(8)** to the armature. The armature (Figure 6*d*) has 23 slots **(9)** each containing a pair of conductors for a total of 46 conductors **(10)** (one of each pair is close to the axle and hidden, and one is close to the outer circumference of the armature and can be seen). In this motor, the armature is the rotating assembly. These conductors are interconnected in the rear of the motor **(11)** forming two parallel paths, each 23 conductors long, between the brushes. Current is conducted from a commutator segment on the left to the brush on the left **(12)**, and to the metal strip that is coiled to form the left side of the stator pole **(13)**. Current is conducted to the automobile's "common" or ground via a hard connection between the left stator pole winding and the motor housing. This connection is deep inside the motor and cannot be seen. The magnetic flux path for the stator field runs from the right side, through the armature, to the left side and is completed through the motor housing.

(c)

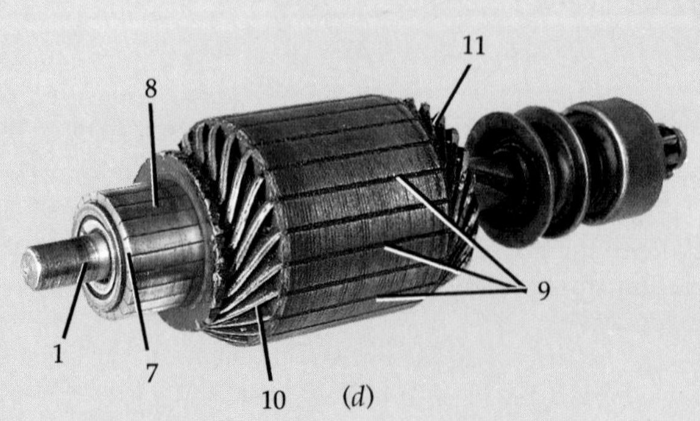

(d)

Figure 7 A common slot car motor with the top half of the housing cut away. The shaft **(1)** is supported by bearings **(2)** on the right and the left. The armature is made of 12 thin plates **(3)** of magnetically permeable material, laminated together, forming three slots and three salient poles. The slots in this armature are very large and contain many turns of wire. The set of turns around one pole is called a winding. The windings are soldered together at their ends **(4)** then connected to the commutator segments **(5)**. The armature is surrounded by black permanent magnets **(6)**, creating a stationary field that reacts with the currents and fields of the armature. A battery connected to the external connections **(7)** drives current through the top hairpin brush **(8)**, to the commutator segment on the top, through the interconnected windings, and back to the battery through the commutator segment on the bottom **(9)**. Notice how the commutator segments are separated **(10)**. The operation of this motor can be thought of in two different ways. Either the current through the windings acts with the stationary field to cause a torque, or current in the windings around the salient armature pieces form poles that alternately repel and attract the stationary pole pieces. Very careful analysis will show that these two approaches are really exactly the same.

across the two parallel paths described above. The slots are separated by only 45 degrees, so the variation in the total emf across the armature as the armature turns will be relatively small. Therefore, the time-varying term of Equation 6, sin $(\omega t + \delta)$, can be discarded and the total emf can be expressed as

$$\mathcal{E}_{\text{total}} = BK\omega \tag{7}$$

where the constant K, the motor constant, includes ℓ and r and the results of the summing and averaging of the total emf across the armature. The validity of Equation 7 improves as more slots and commutator segments are added to the armature.

Equation 7 is similar to $\mathcal{E} = B\ell v$ with K replacing ℓ and ω replacing v. The power delivered to the armature is the product of the emf and the armature current I_a. For a rotating motor, the load is a torque τ applied to the shaft opposing the direction of rotation. The mechanical power delivered to the load is the product of the torque and the angular velocity. At equilibrium, the driving torque from the motor is equal and opposite to the load torque. Thus

$$P = \mathcal{E}I_a = \tau\omega \tag{8}$$

Substituting $BK\omega$ for the emf from Equation 7, we obtain

$$P = BKI_a\omega = \tau\omega \tag{9}$$

Applying the same logic used to develop the model for the linear motor, we can represent the armature by a simple voltage source with an external resistance R_a. The field winding connections of the coil shown at the bottom of Figure 4 can be connected either in series or in parallel (shunt) with the armature. These two methods of connection yield motors with extremely different characteristics.

Parallel or Shunt Connection

Figure 8 shows the circuit for the shunt or parallel field connection. A variable resistance called a rheostat is included to control the field and thereby control the speed of the motor. Applying Kirchhoff's loop rule to this circuit yields

$$V - I_aR_a - BK\omega = 0 \tag{10}$$

which can be rearranged to express the rotational speed ω in terms of the armature current I_a:

$$\omega = -\frac{R_a}{BK}I_a + \frac{V}{BK} \tag{11}$$

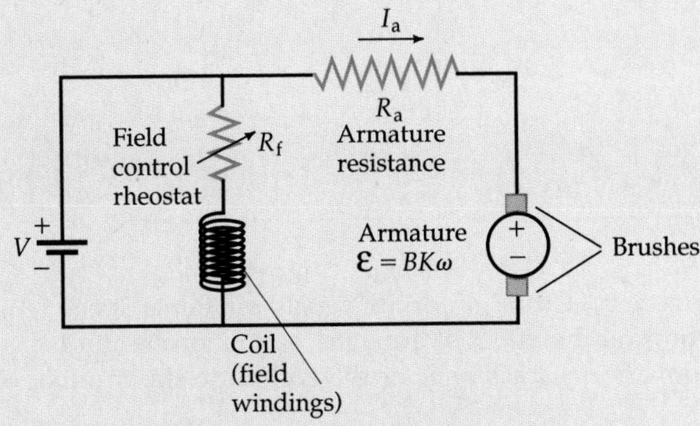

Figure 8 Circuit for a typical dc shunt motor.

Continued

If we substitute τ/BK for the current from Equation 9, the rotational speed is

$$\omega = -\frac{R_a}{(BK)^2}\tau + \frac{V}{BK} \qquad (12)$$

Equation 12 is a linear equation relating the rotational speed to the load. It is analogous to the performance characteristic of the linear motor. The speed can be controlled either by varying the voltage V or, more commonly, by varying the current into the coil with the rheostat.

At high armature currents, the armature core saturates, the voltage drop due to the armature inductance becomes significant, and the relationship between the torque and speed becomes nonlinear. However, for normal loads, Equation 12 accurately describes the motor's operation. Figure 9 shows how the performance characteristic is affected by armature saturation.

Series Connection

In the motor circuit of Figure 10, the coil is connected in series with the armature, so the field strength is a function of the armature current. If the armature current is small and the field does not saturate, the product of the field strength and the motor constant K can be expressed as a linear function of the armature current:

$$BK = CI$$

where C is some constant. Substituting this value of BK in the expressions for the armature emf, power, and torque yields

$$\mathcal{E} = CI\omega \qquad P = CI^2\omega \qquad \text{and} \qquad \tau = CI^2$$

The Kirchhoff loop rule then gives

$$V - IR - CI\omega = 0$$

where R represents the total resistance of the coil and armature, and I represents the only current in the circuit. This gives the following speed-versus-current equation:

$$\omega = \frac{V}{CI} - \frac{R}{C}$$

Substituting $\sqrt{\tau/C}$ for I, we obtain the equation for the speed versus torque, which is the performance characteristic for the series motor:

$$\omega = \frac{V}{(C\tau)^{1/2}} - \frac{R}{C}$$

Figure 11 shows the performance characteristic for the series motor. Comparing this performance characteristic with that for the shunt motor reveals striking differences. At low torques, the series motor runs very fast, almost without limit (the only load is the friction of its bearings and the air around the rotor) whereas the shunt motor is regulated to run close to the speed V/BK. At high torques, the speed of the shunt motor tapers off and the motor stalls, but the series motor delivers its greatest torque when the motor is stopped. A series motor is therefore the best choice to start a car engine, which re-

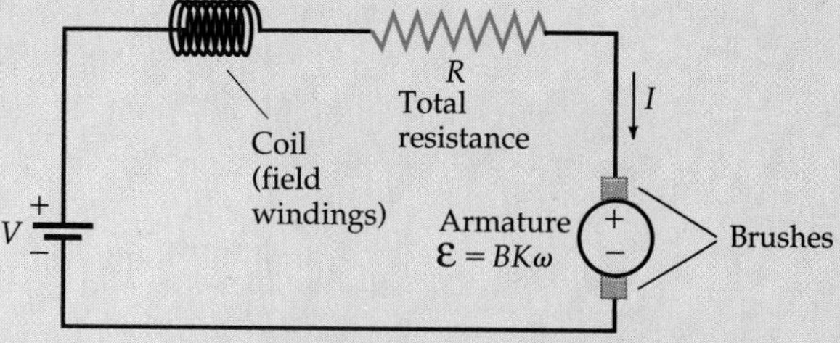

Figure 10 Circuit for a typical dc series motor.

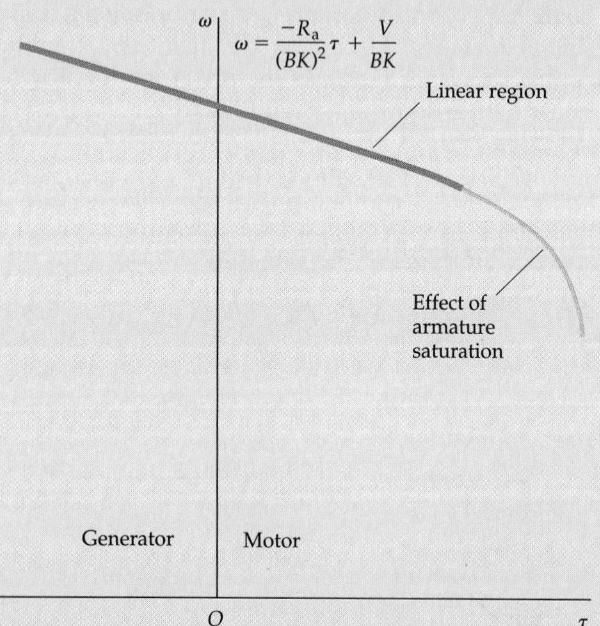

Figure 9 Graph of torque versus rotational speed showing the effect of armature saturation on the performance characteristic of the typical dc shunt motor.

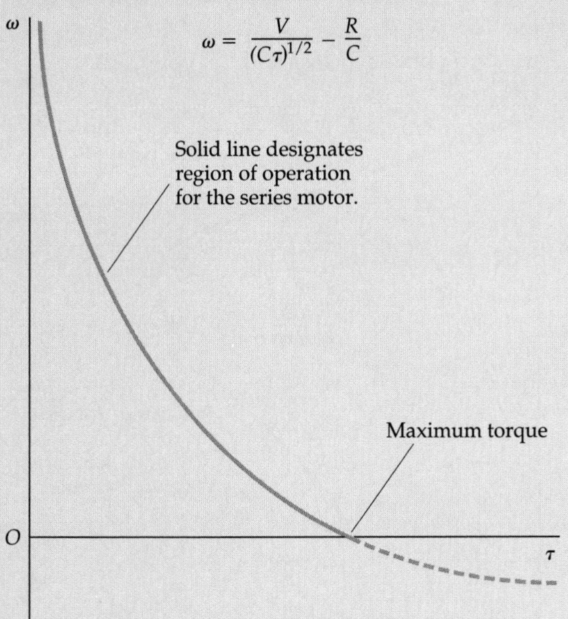

Figure 11 Graph of torque versus rotational speed showing the performance characteristic of the typical dc series motor.

quires a high torque at $\omega = 0$. On the other hand, a shunt motor is the better choice to drive a speed-sensitive load, like a tape recorder.

The linear, shunt, and series motors discussed so far all run on direct current whereas power companies deliver alternating current. With only minor modifications, the principles of dc motor construction and operation also apply to ac motors.

The torque of a series motor is proportional to I^2 and is thus independent of current direction. This is because the same current is in both the stationary field and rotating armature. With this cursory examination, it might be concluded that any series dc motor would run on alternating current. However, an assumption made to simplify the analysis of the dc motor was that inductance could be ignored. Inductance cannot be ignored when driving a motor from an ac source. Inductance has two effects: (1) it acts as a throttle limiting the amount of ac current for a given input voltage and (2) it changes the phase relationship of the current and voltage.

A dc shunt motor typically has field windings with high resistance and an armature with high inductance. Applying alternating current to such a motor would create a phase difference between the field and the armature currents resulting in unsatisfactory performance.

A series motor, like the car starter in Figure 6, has a very tight magnetic circuit with close tolerances to develop a very high torque in a small package. Such a device has a high inductance, thus limiting the ac current drawn by the motor. A series motor designed to run on alternating current must have a relatively low inductance. The low inductance is achieved by limiting the amount of iron used in the pole pieces and the armature. Such a motor is called a *universal motor*. By its nature, it is both lightweight and limited to driving devices with relatively light loads such as vacuum cleaners, food blenders, hair dryers, and sewing machines. Its performance characteristic is similar to that for the dc series motor shown in Figure 11. From this figure, can you explain why a vacuum cleaner motor speeds up when the suction is blocked? (Hint: The load on the vacuum cleaner motor is the air moving through it. If air flow decreases, the load decreases.)

The most common ac motor is the *induction motor*. This motor has a rotating assembly like the one shown in Figure 4, but unlike the dc motor, the commutator and interconnecting wires are replaced with shorting plates, connecting all the slotted conductors, mounted on the front and back. Such a design greatly simplifies construction. The challenge becomes how to make the shorted rotor rotate. The solution is to make the field from the stator appear to rotate. If the field rotates, there will be a relative velocity between the rotor and the stator field. An emf develops across the shorted rotor, driving current through the conductors in the slots. The rotating stator field produces a torque on the induced current in the rotor. Recall that the linear motor tended to move just fast enough to develop an emf that matched the source voltage. The induction motor responds in the same way, but the source voltage to the rotor is zero. To keep the emf close to zero, the rotor moves such as to minimize the relative motion between it and the field. Thus, the rotor turns almost as fast as the rotating stator field.

There are many schemes for creating apparent field rotation. The one shown in Figure 12 is known as the *shaded pole*. The motor is identical to the one in Figure 4 with the exception that the rotor is shorted on either end and the stator pole pieces have been sliced, with a conducting band around the small pieces of each pole. This construction allows the magnetic field to be quickly established through the faces of the large pole pieces and to be delayed through the small faces by the inductance of the conducting band. The phase delay between the field

Continued

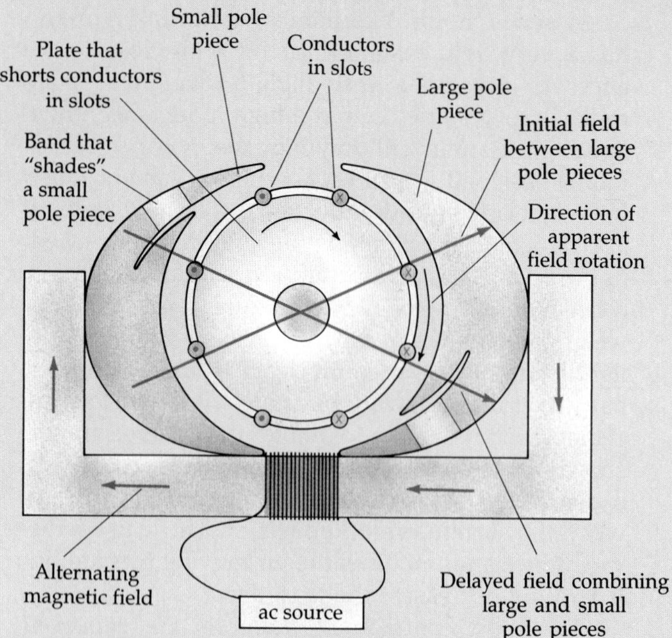

Figure 12 A shaded-pole induction motor.

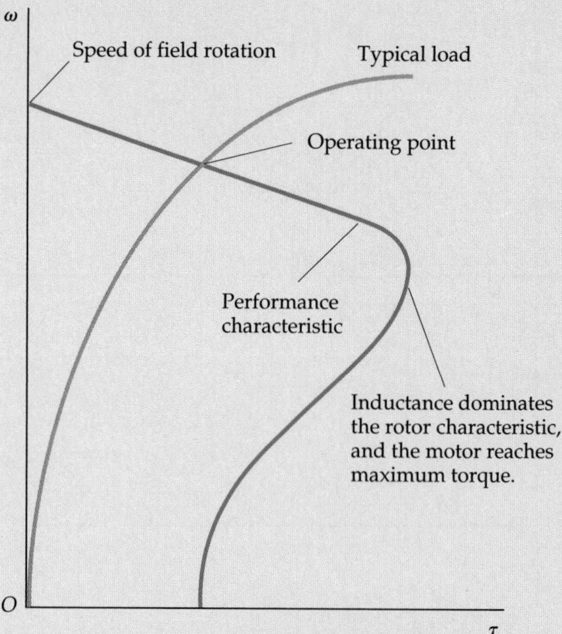

Figure 13 Graph of torque versus speed showing the performance characteristic of a typical induction motor. The load line indicated is typical for a centrifugal pump.

through the large pole faces and the field through the small pole faces creates the appearance of a rotating field.

The performance characteristic of a typical induction motor is shown in Figure 13. Normal operation is close to the speed of field rotation. If the motor in Figure 12 were connected to a 60-Hz source, the rotational speed would be somewhat less than 60 rev/s.

A rotor with conductors surrounding a metal, magnetically permeable core has characteristics of inductance and resistance. The effect of inductance within the rotor is proportional to the frequency of the rotor currents, which in turn is proportional to the difference between the field rotation and the rotor rotation. The maximum torque shown on the performance characteristic curve occurs where the difference between the rotor speed and the field rotation speed is large enough for the effects of rotor inductance to significantly delay rotor currents. The delayed rotor currents cannot interact with the stator field, and so the motor stalls if the load is increased.

Shaded-pole motors are used in devices with light loads such as cooling fans in electrical equipment. More complex schemes for creating field rotation are used in the induction motors for refrigerators and air conditioners. Large industrial induction motors use three-phase electricity to rotate the field. No matter what the size of the motor or the method used to create field rotation, the principles and fundamental operating characteristics for all induction motors are the same.

Whether run off direct or alternating current, all electric motors are based upon the fundamental principles discussed in Chapters 26 through 28 of this text. There is an extraordinary opportunity for engineers to creatively combine these fundamental principles with advances in related fields, such as superconductivity, and design motors meeting the needs of the twenty-first century. Development and improvement of electric cars, trains, and satellites offer just a few of the challenges facing the motor designer.

There are many excellent electrical engineering textbooks devoted to the subject of motors or machines. Two of my favorites are *Electric Machines and Power Systems* by Vincent Deltoro (Prentice Hall, Englewood Cliffs, New Jersey, 1985) and *Electromechanical Motion Devices* by Paul Krause and Oleg Wasynczuk. (McGraw-Hill, New York, 1989).

Suggestion for Further Reading

Coltman, John W.: "The Transformer," *Scientific American*, January 1988, p. 86.

This article describes the nineteenth-century struggle between proponents of alternating current and those of direct current and how the transformer helped to settle the matter. More recent advances in transformer technology are also discussed.

Review

A. Objectives: After studying this chapter, you should:

1. Be able to define rms current and relate it to the maximum current in an ac circuit.

2. Be able to define capacitive reactance, inductive reactance, and impedance.

3. Be able to give the phase relations between the current and the voltage across a resistor, inductor, or capacitor.

4. Be able to draw a phasor diagram for a series *LCR* circuit and from it relate the phase angle δ to the capacitive reactance, inductive reactance, and resistance.

5. Be able to define *Q* factor and discuss its significance.

6. Be able to state the resonance condition for a series *LCR* circuit with a generator and sketch the power versus frequency for both a high-*Q* and a low-*Q* circuit.

7. Be able to describe a step-up and a step-down transformer.

B. Define, explain, or otherwise identify:

Root-mean-square (rms) values
Inductive reactance
Capacitive reactance
Phasors
Total reactance
Impedance
Resonance frequency
Resonance
Power factor
Resonance curves
Resonance width
Q factor
Low-pass filter
Transformer
Primary
Secondary
Magnetizing current
Step-up transformer
Step-down transformer
Load resistance
Rectification
Diode
Thermionic emission
Plate current
Ripple
Triode
Grid
Amplifier

C. True or false: If the statement is true, explain why it is true. If it is false, give a counterexample.

1. Alternating current in a resistance dissipates no power because the current is negative as often as it is positive.

2. At very high frequencies, a capacitor acts like a short circuit.

3. An *LCR* circuit with a high *Q* factor has a narrow resonance curve.

4. At resonance, the impedance of an *LCR* circuit equals the resistance *R*.

5. At resonance, the current and generator voltage are in phase.

6. If a transformer increases the current, it must decrease the voltage.

Problems

Level I

28-1 Alternating Current in a Resistor

1. A 100-W light bulb is plugged into a standard 120-V (rms) outlet. Find (*a*) I_{rms}, (*b*) I_{max}, and (*c*) the maximum power.

2. A 3-Ω resistor is placed across a generator having a frequency of 60 Hz and a maximum emf of 12.0 V. (*a*) What is the angular frequency ω of the current? (*b*) Find I_{max} and I_{rms}. What is (*c*) the maximum power into the resistor, (*d*) the minimum power, and (*e*) the average power?

3. A 5.0-kW electric clothes dryer runs on 240 V rms. Find (*a*) I_{rms} and (*b*) I_{max}. (*c*) Find the same quantities for a dryer of the same power that operates at 120 V rms.

4. A circuit breaker is rated for a current of 15 A rms at a voltage of 120 V rms. (*a*) What is the largest value of I_{max} that the breaker can carry? (*b*) What average power can be supplied by this circuit?

28-2 Alternating Current in Inductors and Capacitors

5. What is the reactance of a 1.0-mH inductor at (*a*) 60 Hz, (*b*) 600 Hz, and (*c*) 6 kHz?

6. An inductor has a reactance of 100 Ω at 80 Hz. (*a*) What is its inductance? (*b*) What is its reactance at 160 Hz?

7. At what frequency would the reactance of a 10.0-μF capacitor equal that of a 1.0-mH inductor?

8. Sketch a graph of X_L versus f for $L = 3$ mH.

9. What is the reactance of a 1.0-nF capacitor at (*a*) 60 Hz, (*b*) 6 kHz, and (*c*) 6 MHz?

10. Find the reactance of a 10.0-μF capacitor at (*a*) 60 Hz, (*b*) 6 kHz, and (*c*) 6 MHz.

11. Sketch a graph of X_C versus f for $C = 100$ μF.

12. An emf of 10.0 V maximum and frequency 20 Hz is applied to a 20-μF capacitor. Find (*a*) I_{max} and (*b*) I_{rms}.

13. At what frequency is the reactance of a 10-μF capacitor (*a*) 1 Ω, (*b*) 100 Ω, and (*c*) 0.01 Ω?

28-3 Phasors

14. Draw the resultant phasor diagram for a series *LCR* circuit when $V_L < V_C$. Show on your diagram that the emf will lag the current by the phase angle δ given by

$$\tan \delta = \frac{V_C - V_L}{V_R}$$

28-4 *LC* and *LCR* Circuits without a Generator

15. Show from the definitions of the henry and the farad that $1/\sqrt{LC}$ has the unit s^{-1}.

16. What is the period of oscillation of an *LC* circuit consisting of a 2-mH coil and a 20-μF capacitor?

17. What inductance is needed with an 80-μF capacitor to construct an *LC* circuit that oscillates with a frequency of 60 Hz?

18. An *LC* circuit has capacitance C_1 and inductance L_1. A second circuit has $C_2 = \frac{1}{2}C_1$ and $L_2 = 2L_1$, and a third circuit has $C_3 = 2C_1$ and $L_3 = \frac{1}{2}L_1$. (*a*) Show that each circuit oscillates with the same frequency. (*b*) In which circuit would the maximum current be the greatest if the capacitor in each were charged to the same potential *V*?

19. A 5-μF capacitor is charged to 30 V and is then connected across a 10-mH inductor. (*a*) How much energy is stored in the system? (*b*) What is the frequency of oscillation of the circuit? (*c*) What is the maximum current in the circuit?

28-5 *LCR* Circuits with a Generator

20. A series *LCR* circuit with $L = 10$ mH, $C = 2$ μF, and $R = 5$ Ω is driven by a generator with a maximum emf of 100 V and a variable angular frequency ω. Find (*a*) the resonant frequency ω_0 and (*b*) I_{rms} at resonance. When $\omega = 8000$ rad/s, find (*c*) X_C and X_L, (*d*) *Z* and I_{rms}, and (*e*) the phase angle δ.

21. For the circuit of Problem 20, let the generator frequency be $f = \omega/2\pi = 1$ kHz. Find (*a*) the resonance frequency $f_0 = \omega_0/2\pi$, (*b*) X_C and X_L, (*c*) the total impedance *Z* and I_{rms}, and (*d*) the phase angle δ.

22. A series *LCR* circuit in a radio receiver is tuned by a variable capacitor so that it can resonate at frequencies from 500 to 1600 kHz. If $L = 1.0$ μH, find the range of capacitances necessary to cover this range of frequencies.

23. FM radio stations have carrier frequencies that are separated by 0.20 MHz. When the radio is tuned to a station, such as 100.1 MHz, the resonance width of the receiver circuit should be much smaller than 0.2 MHz so that adjacent stations are not received. If $f_0 = 100.1$ MHz and $\Delta f = 0.05$ MHz, what is the *Q* factor for the circuit?

24. (*a*) Find the power factor for the circuit in Example 28-5 when $\omega = 400$ rad/s. (*b*) At what angular frequency is the power factor 0.5?

25. Find (*a*) the *Q* factor and (*b*) the resonance width for the circuit in Problem 20. (*c*) What is the power factor when $\omega = 8000$ rad/s?

26. An ac generator with a maximum emf of 20 V is connected in series with a 20-μF capacitor and an 80-Ω resistor. There is no inductance in the circuit. Find (*a*) the power factor, (*b*) the rms current, and (*c*) the average power if the angular frequency of the generator is 400 rad/s.

27. A coil can be considered to be a resistance and an inductance in series. Assume that $R = 100$ Ω and $L = 0.4$ H. The coil is connected across a 120-V-rms, 60-Hz line. Find (*a*) the power factor, (*b*) the rms current, and (*c*) the average power supplied.

28. Find the power factor and the phase angle δ for the circuit in Problem 20 when the generator frequency is (*a*) 900 Hz, (*b*) 1.1 kHz, and (*c*) 1.3 kHz.

28-6 The Transformer

29. A transformer has 400 turns in the primary and 8 turns in the secondary. (*a*) Is this a step-up or step-down transformer? (*b*) If the primary is connected across 120 V rms, what is the open-circuit voltage across the secondary? (*c*) If the primary current is 0.1 A, what is the secondary current, assuming negligible magnetization current and no power loss?

30. The primary of a step-down transformer has 250 turns and is connected to a 120-V-rms line. The secondary is to supply 20 A at 9 V. Find (*a*) the current in the primary and (*b*) the number of turns in the secondary, assuming 100 percent efficiency.

31. A transformer has 500 turns in its primary, which is connected to 120 V rms. Its secondary coil is tapped at three places to give outputs of 2.5, 7.5, and 9 V. How many turns are needed for each part of the secondary coil?

32. The distribution circuit of a residential power line is operated at 2000 V rms. This voltage must be reduced to 240 V rms for use within the residences. If the secondary side of the transformer has 400 turns, how many turns are in the primary?

28-7 Rectification and Amplification

33. The maximum output current in a half-wave rectifier circuit is 3.5 A. (*a*) Find the rms current. (*b*) Find the rms current if the maximum output current is 3.5 A in a full-wave rectifier circuit.

34. Sketch a graph of the current versus time if a low-pass filter such as that in Figure 28-30*a* is inserted before the load resistor in Figure 28-27.

Level II

35. Figure 28-33 shows the voltage V versus time t for a "square wave" voltage. If $V_0 = 12$ V, (*a*) what is the rms voltage of this waveform? (*b*) If this alternating waveform is rectified so that only the positive voltages remain, what now is the rms voltage of the rectified waveform?

Figure 28-33 Problem 35.

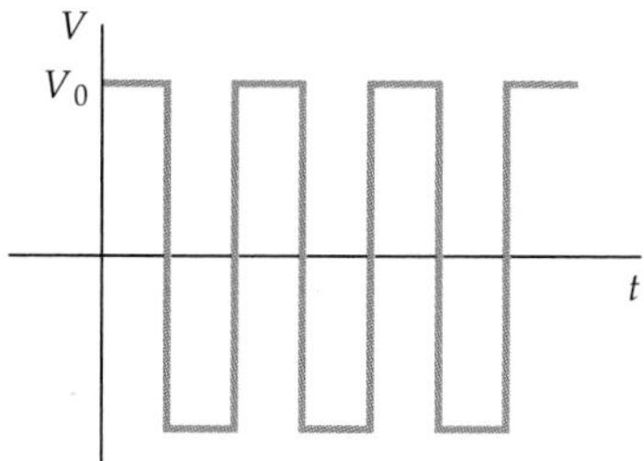

36. A pulsed current has a constant value of 15 A for the first 0.1 s of each second and is then 0 for the next 0.9 s of each second. (*a*) What is the rms value for this current waveform? (*b*) Each current pulse is generated by a 100-V pulse. What is the average power delivered by the pulse generator?

37. A 100-V-rms voltage is applied to a series *RC* circuit. The rms voltage across the capacitor is 80 V. What is the rms voltage across the resistor?

38. Show that the formula $P_{av} = \mathcal{E}^2_{rms}R/Z^2$ gives the correct result for a circuit containing a generator and only (*a*) a resistor, (*b*) a capacitor, and (*c*) an inductor.

39. Sketch the impedance Z versus ω for (*a*) a series *LR* circuit, (*b*) a series *RC* circuit, and (*c*) a series *LCR* circuit.

40. The charge on the capacitor of a series *LC* circuit is given by $Q = (15\ \mu\text{C}) \cos (1250\ t + \pi/4)$ where t is in seconds. (*a*) Find the current as a function of time. (*b*) Find C if $L = 28$ mH. (*c*) Write expressions for the electrical energy U_e, the magnetic energy U_m, and the total energy U.

41. A resistance R and a 1.4-H inductance are in series across a 60-Hz ac voltage. The voltage across the resistor is 30 V and the voltage across the inductor is 40 V. (*a*) What is the resistance R? (*b*) What is the ac input voltage?

42. A resistance R carries a current $I = (5.0\text{ A}) \sin 120\pi t + (7.0\text{ A}) \sin 240\pi t$. (*a*) What is the rms current? (*b*) If the resistance R is 12 Ω, what is the power dissipated in the resistor? (*c*) What is the rms voltage across the resistor?

43. A coil has a dc resistance of 80 Ω and an impedance of 200 Ω at a frequency of 1 kHz. One may neglect the wiring capacitance of the coil at this frequency. What is the inductance of the coil?

44. Two ac voltage sources are connected in series with a resistor $R = 25$ Ω. One source is given by $V_1 = (5.0\text{ V}) \cos (\omega t - \alpha)$, and the other source is $V_2 = (5.0\text{ V}) \cos (\omega t + \alpha)$, with $\alpha = \pi/6$. (*a*) Find the current in R using a trigonometric identity for the sum of two cosines. (*b*) Use phasor diagrams to find the current in R. (*c*) Find the current in R if $\alpha = \pi/4$ and the amplitude of V_2 is increased from 5.0 V to 7.0 V.

45. Given the circuit shown in Figure 28-34, (*a*) find the power loss in the inductor. (*b*) Find the resistance r of the inductor. (*c*) Find the inductance L.

Figure 28-34 Problem 45.

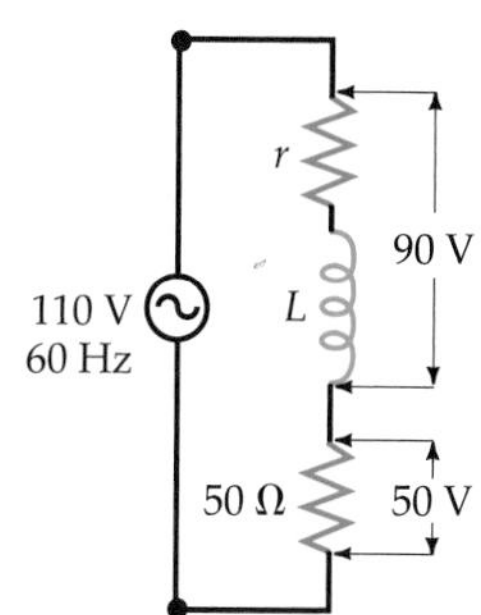

46. A coil of resistance R, inductance L, and negligible capacitance has a power factor of 0.866 at a frequency of 60 Hz. What is the power factor for a frequency of 240 Hz?

47. A coil draws 15 A when connected to a 220-V ac 60-Hz line. When it is in series with a 4-Ω resistor and the combination is connected to a 100-V battery, the battery current after a long time is observed to be 10 A. (*a*) What is the resistance in the coil? (*b*) What is the inductance of the coil?

48. A coil is connected to a 60-Hz, 100-V ac generator. At this frequency the coil has an impedance of 10 Ω and a reactance of 8 Ω. (*a*) What is the current in the coil? (*b*) What is the phase angle between the current and the applied voltage? (*c*) What series capacitance is required so that the current and voltage are in phase? (*d*) What then is the voltage measured across the capacitor?

49. A 0.25-H inductor and a capacitor C are connected in series with a 60-Hz ac generator. An ac voltmeter is used to measure the rms voltages across the inductor and capacitor separately. The rms voltage across the capacitor is 75 V and that across the inductor is 50 V. (*a*) Find the capacitance C and the rms current in the circuit. (*b*) What would be the measured rms voltage across both the capacitor and inductor together?

50. Show that Equation 28-49 can be written as

$$I_{max} = \frac{\omega\mathcal{E}_{max}}{\sqrt{L^2(\omega^2 - \omega_0^2)^2 + \omega^2R^2}}$$

51. (*a*) Show that Equation 28-48 can be written as

$$\tan \delta = L(\omega^2 - \omega_0^2)/\omega R$$

Find δ approximately at (*b*) very low frequencies and (*c*) very high frequencies.

52. (*a*) Show that in a series *RC* circuit with no inductance, the power factor is given by

$$\cos \delta = \frac{RC\omega}{\sqrt{1 + (RC\omega)^2}}$$

(*b*) Sketch a graph of the power factor versus ω.

53. In the circuit in Figure 28-35, the ac generator produces an rms voltage of 115 V when operated at 60 Hz. What is the rms voltage across points (*a*) *AB*, (*b*) *BC*, (*c*) *CD*, (*d*) *AC*, and (*e*) *BD*?

Figure 28-35 Problem 53.

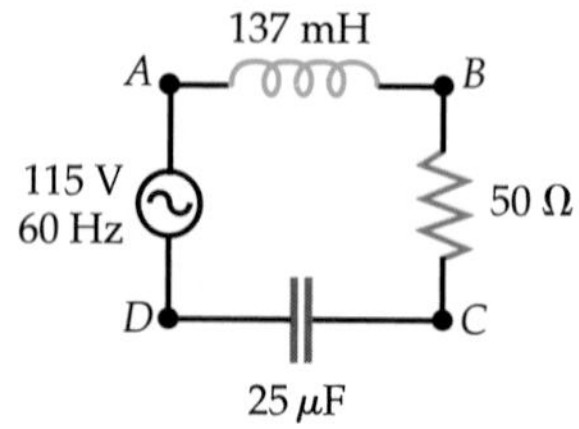

54. A variable-frequency ac generator is connected to a series *LCR* circuit for which $R = 1\ \text{k}\Omega$, $L = 50$ mH, and $C = 2.5\ \mu$F. (*a*) What is the resonance frequency of the circuit? (*b*) What is the *Q* value? (*c*) At what frequencies is the value of the average power delivered by the generator half of its maximum value?

55. A series *LCR* circuit is driven at a frequency of 500 Hz. The phase angle between the applied voltage and current is determined from an oscilloscope measurement to be $\delta = 75°$. If the total resistance is known to be 35 Ω and the inductance is 0.15 H, what is the capacitance of the circuit?

56. A series *LCR* circuit with $R = 400\ \Omega$, $L = 0.35$ H, and $C = 5\ \mu$F is driven by a generator of variable frequency *f*. (*a*) What is the resonance frequency f_0? Find *f* and f/f_0 when the phase angle δ is (*b*) 60°, and (*c*) −60°.

57. An experimental physicist wishes to design a series *LCR* circuit with a *Q* value of 10 and a resonance frequency of 33 kHz. She has as a 45-mH inductor with negligible resistance. What values for the resistance *R* and capacitance *C* should she use?

58. The generator voltage in Figure 28-36 is given by $V =$ (100 V) $\cos \omega t$. (*a*) For each branch, what is the current amplitude and its phase relative to the applied voltage? (*b*) What is the angular frequency ω such that the current in the generator vanishes? (*c*) At this resonance, what is the current in the inductor? What is the current in the capacitor? (*d*) Draw a phasor diagram showing the general relationships between the applied voltage, the generator current, the capacitor current, and the inductor current for the case where the inductive reactance is larger than the capacitive reactance.

Figure 28-36 Problem 58.

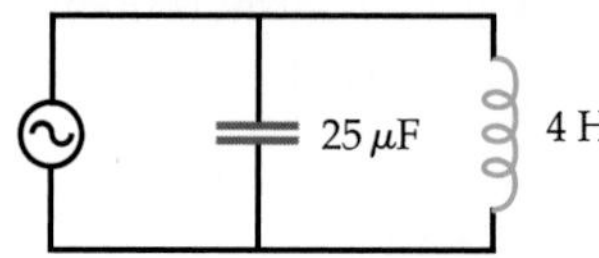

59. The circuit shown in Figure 28-37 is called an *RC* high-pass filter because high input frequencies are transmitted with greater amplitude than low input frequencies. (*a*) If the input voltage is $V_{in} = V_0 \cos \omega t$, show that the output voltage is

$$V_{out} = \frac{V_0}{\sqrt{(1/\omega RC)^2 + 1}}$$

(*b*) At what angular frequency is the output voltage half the input voltage? (*c*) Sketch a graph of V_{out}/V_0 as a function of ω.

Figure 28-37 Problem 59.

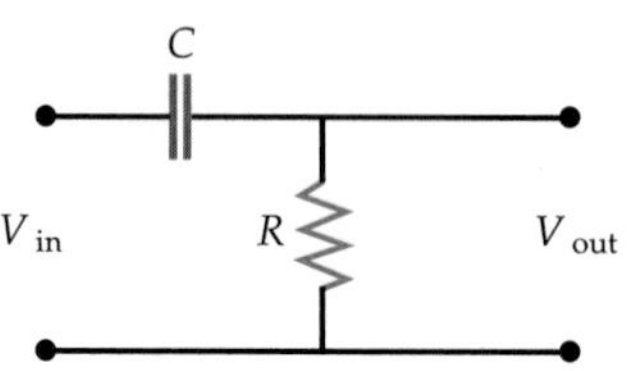

60. A circuit consists of two capacitors, a 24-V battery, and an ac voltage connected as shown in Figure 28-38. The ac voltage is given by (20 V) $\cos 120\pi t$, where *t* is in seconds. (*a*) Find the charge on each capacitor as a function of time. Assume transient effects have had sufficient time to decay. (*b*) What is the steady-state current? (*c*) What is the maximum energy stored in the capacitors? (*d*) What is the minimum energy stored in the capacitors?

Figure 28-38 Problem 60.

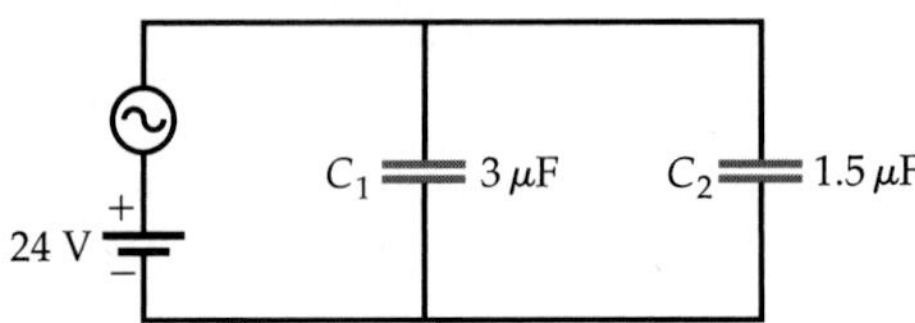

61. A single transmission line carries two voltage signals given by $V_1 =$ (10 V) $\cos 100t$ and $V_2 =$ (10 V) $\cos 10{,}000t$, where *t* is in seconds. A series inductor of 1 H and a shunting resistor of 1 kΩ is inserted into the transmission line as indicated in Figure 28-39. (*a*) What is the voltage signal observed at the output side of the transmission line? (*b*) What is the ratio of the low-frequency amplitude to the high-frequency amplitude?

Figure 28-39 Problem 61.

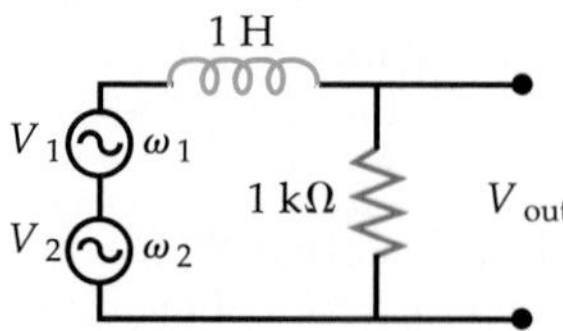

62. A coil with resistance and inductance is connected to a 120-V-rms, 60-Hz line. The average power supplied to the coil is 60 W, and the rms current is 1.5 A. Find (*a*) the power factor, (*b*) the resistance of the coil, and (*c*) the inductance of the coil. (*d*) Does the current lag or lead the voltage? What is the phase angle δ?

63. In a series *LCR* circuit $X_C = 16\ \Omega$ and $X_L = 4\ \Omega$ at some frequency. The resonance frequency is $\omega_0 = 10^4$ rad/s. (*a*) Find L and C. If $R = 5\ \Omega$ and $\mathcal{E}_{max} = 26$ V, find (*b*) the Q factor and (*c*) the maximum current.

64. In a series *LCR* circuit connected to an ac generator whose maximum emf is 200 V, the resistance is 60 Ω and the capacitance is 8.0 μF. The inductance can be varied from 8.0 to 40.0 mH by the insertion of an iron core in the solenoid. The angular frequency of the generator is 2500 rad/s. If the capacitor voltage is not to exceed 150 V, find (*a*) the maximum current and (*b*) the range of inductance that is safe to use.

65. When an *LCR* series circuit is connected to a 120-V-rms, 60-Hz line, the current is $I_{rms} = 11.0$ A and the current leads the voltage by 45°. (*a*) Find the power supplied to the circuit. (*b*) What is the resistance? (*c*) If the inductance $L = 0.05$ H, find the capacitance C. (*d*) What capacitance or inductance would you add to make the power factor 1?

Level III

66. Consider the parallel circuit shown in Figure 28-40. (*a*) What is the impedance of each branch? (*b*) For each branch, what is the current amplitude and its phase relative to the applied voltage? (*c*) Give the current phasor diagram, and use it to find the total current and its phase relative to the applied voltage.

Figure 28-40 Problem 66.

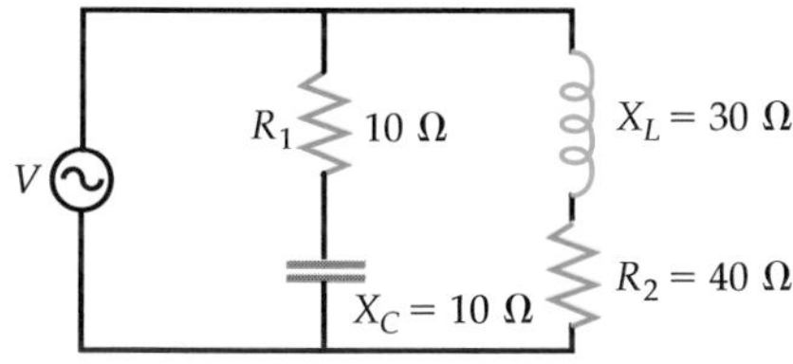

67. (*a*) Show that Equation 28-48 can be written as

$$\tan\delta = \frac{Q(\omega^2 - \omega_0^2)}{\omega\omega_0}$$

(*b*) Show that near resonance

$$\tan\delta \approx \frac{2Q(\omega - \omega_0)}{\omega}$$

(*c*) Sketch a plot of δ versus x, where $x = \omega/\omega_0$, for a circuit with high Q and for one with low Q.

68. Show by direct substitution that the current given by Equation 28-47 with δ and I_{max} given by Equations 28-48 and 28-49, respectively, satisfies Equation 28-46. *Hint:* Use trigonometric identities for the sine and cosine of the sum of two angles, and write the equation in the form

$$A \sin\omega t + B\cos\omega t = 0$$

Since this equation must hold for all times, $A = 0$ and $B = 0$.

69. A certain electrical device draws 10 A rms and has an average power of 720 W when connected to a 120-V-rms, 60-Hz power line. (*a*) What is the impedance of the device? (*b*) What series combination of resistance and reactance is this device equivalent to? (*c*) If the current leads the emf, is the reactance inductive or capacitive?

70. An ac generator is in series with a capacitor and an inductor in a circuit with negligible resistance. (*a*) Show that the charge on the capacitor obeys the equation

$$L\frac{d^2Q}{dt^2} + \frac{Q}{C} = \mathcal{E}_{max}\cos\omega t$$

(*b*) Show by direct substitution that this equation is satisfied by $Q = Q_{max}\cos\omega t$ if

$$Q_{max} = -\frac{\mathcal{E}_{max}}{L(\omega^2 - \omega_0^2)}$$

(*c*) Show that the current can be written as $I = I_{max}\cos(\omega t - \delta)$, where

$$I_{max} = \frac{\omega\mathcal{E}_{max}}{L|\omega^2 - \omega_0^2|} = \frac{\mathcal{E}_{max}}{|X_L - X_C|}$$

and $\delta = -90°$ for $\omega < \omega_0$ and $\delta = 90°$ for $\omega > \omega_0$.

71. A method for measuring inductance is to connect the inductor in series with a known capacitance, a known resistance, an ac ammeter, and a variable-frequency signal generator. The frequency of the signal generator is varied and the emf is kept constant until the current is maximum. (*a*) If $C = 10$ μF, $\mathcal{E}_{max} = 10$ V, $R = 100\ \Omega$, and I is maximum at $\omega = 5000$ rad/s, what is L? (*b*) What is I_{max}?

72. A resistor and an inductor are connected in parallel across an emf $\mathcal{E} = \mathcal{E}_{max}\cos\omega t$ as shown in Figure 28-41. Show that (*a*) the current in the resistor is $I_R = (\mathcal{E}_{max}/R)\cos\omega t$, (*b*) the current in the inductor is $I_L = (\mathcal{E}_{max}/X_L)\cos(\omega t - 90°)$, and (*c*) $I = I_R + I_L = I_{max}\cos(\omega t - \delta)$, where $\tan\delta = R/X_L$ and $I_{max} = \mathcal{E}_{max}/Z$ with $Z^{-2} = R^{-2} + X_L^{-2}$.

Figure 28-41 Problem 72.

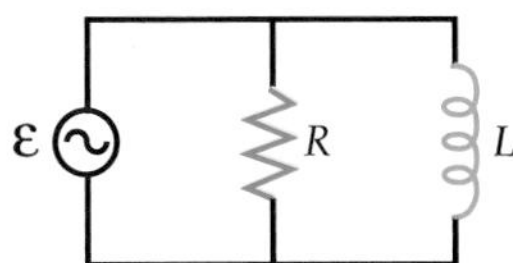

73. A resistor and a capacitor are connected in parallel across a sinusoidal emf $\mathcal{E} = \mathcal{E}_{max}\cos\omega t$ as shown in Figure 28-42. (*a*) Show that the current in the resistor is $I_R = (\mathcal{E}_{max}/R)\cos\omega t$. (*b*) Show that the current in the capacitor branch is $I_C = (\mathcal{E}_{max}/X_C)\cos(\omega t + 90°)$. (*c*) Show that the total current is given by $I = I_R + I_C = I_{max}\cos(\omega t + \delta)$, where $\tan\delta = R/X_C$ and $I_{max} = \mathcal{E}_{max}/Z$ with $Z^{-2} = R^{-2} + X_C^{-2}$.

Figure 28-42 Problem 73.

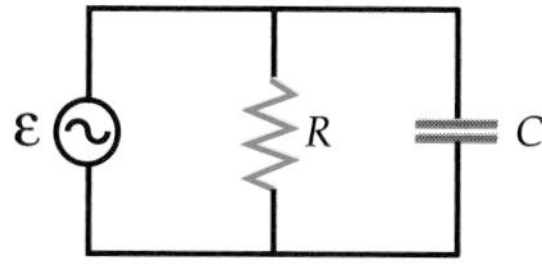

74. Figure 28-18 shows a plot of average power P_{av} versus generator frequency f for an LCR circuit with a generator. The average power P_{av} is given by Equation 28-55, where $\omega = 2\pi f$. The "full width at half maximum" Δf is the width of the resonance curve between the two points where P_{av} is one-half its maximum value. Show that, for a sharply peaked resonance, $\Delta f \approx R/2\pi L = \Delta\omega/2\pi$ and, hence, that $Q \approx \omega_0/\Delta\omega = f_0/\Delta f$ in this case (Equation 28-58). *Hint:* At resonance, the denominator of the expression on the right of Equation 28-55 is $\omega^2 R^2$. The half-power points will occur when the denominator is twice the value near resonance, that is, when $L^2(\omega^2 - \omega_0^2)^2 = \omega^2 R^2 \approx \omega_0^2 R^2$. Let ω_1 and ω_2 be the solutions of this equation. For a sharply peaked resonance, $\omega_1 \approx \omega_0$ and $\omega_2 \approx \omega_0$. Then, using the fact that $\omega + \omega_0 \approx 2\omega_0$, one finds that $\Delta\omega = \omega_2 - \omega_1 \approx R/L$.

75. One use of a transformer is for *impedance matching*. For example, the output impedance of a stereo amplifier is matched to the impedance of a speaker by a transformer. In Equation 28-65, the currents I_1 and I_2 can be related to the impedance Z in the secondary since $I_2 = V_2/Z$. Using Equations 28-63 and 28-64, show that

$$I_1 = \frac{\mathcal{E}}{(N_1/N_2)^2 Z}$$

and, therefore, $Z_{eff} = (N_1/N_2)^2 Z$.

76. Show by direct substitution that Equation 28-44*b* is satisfied by

$$Q = Q_0 e^{-Rt/2L} \cos(\omega' t)$$

where

$$\omega' = \sqrt{(1/LC) - (R/2L)^2}$$

and Q_0 is the charge on the capacitor at $t = 0$.

77. (*a*) Compute the current $I = dQ/dt$ from the solution of Equation 28-44*b* given in Problem 76, and show that

$$I = -I_0\left(\sin \omega' t + \frac{R}{2L\omega'} \cos \omega' t\right)e^{-Rt/2L}$$

where $I_0 = \omega' Q_0$. (*b*) Show that this can be written

$$I = -\frac{I_0}{\cos\delta}(\cos\delta \sin \omega' t + \sin\delta \cos \omega' t)e^{-Rt/2L}$$

$$= -\frac{I_0}{\cos\delta} \sin(\omega' t + \delta)\, e^{-Rt/2L}$$

where $\tan\delta = R/2L\omega'$. When $R/2L\omega'$ is small, $\cos\delta \approx 1$, and

$$I \approx I_0 \sin(\omega' t + \delta) e^{-Rt/2L}.$$

Chapter 29

Maxwell's Equations and Electromagnetic Waves

A multiple-exposure view showing the 26-m tracking antenna at Wallops Station, Virginia, and a total solar eclipse. Electromagnetic radiation at radio wavelengths, like that at optical wavelengths, is not readily absorbed by the earth's atmosphere—making it a viable means of communication between two distant points on the ground or between a point on the ground and a plane, satellite, or spacecraft. Objects are tracked by aiming a continuous radar beam at them and receiving the reflected beam.

About 1860, the great Scottish physicist James Clerk Maxwell found that the experimental laws of electricity and magnetism—the laws of Coulomb, Gauss, Biot–Savart, Ampère, and Faraday, which we have studied in Chapters 18 through 28—could be summarized in a concise mathematical form now known as Maxwell's equations. One of the laws, Ampère's law, contained an inconsistency, which Maxwell was able to remove with the invention of the displacement current (Section 29-1). The new consistent set of equations predicts the possibility of electromagnetic waves.

Maxwell's equations relate the electric- and magnetic-field vectors **E** and **B** to their sources, which are electric charges, currents, and changing fields. These equations play a role in classical electromagnetism analogous to that of Newton's laws in classical mechanics. In principle, all problems in classical electricity and magnetism can be solved using Maxwell's equations, just as all problems in classical mechanics can be solved using Newton's laws. Maxwell's equations are considerably more complicated than Newton's laws, however, and their application to most problems involves mathematics beyond the scope of this book. Nevertheless, Maxwell's equations are of great theoretical importance.

Maxwell showed that these equations could be combined to yield a wave equation for the electric- and magnetic-field vectors **E** and **B**. Such electromagnetic waves are caused by accelerating charges, for example, the charges in an alternating current in an antenna. They were first produced in the laboratory by Heinrich Hertz in 1887. Maxwell showed that the speed of electromagnetic waves in free space should be

$$c = \frac{1}{\sqrt{\mu_0 \epsilon_0}} \qquad 29\text{-}1$$

where ϵ_0, the permittivity of free space, is the constant appearing in Coulomb's and Gauss's laws and μ_0, the permeability of free space, is that in the Biot–Savart law and Ampère's law. When the measured value of ϵ_0 and the defined value of μ_0 are put into Equation 29-1, the speed of electromagnetic waves is found to be about 3×10^8 m/s, the same as the measured speed of light. Maxwell noted this "coincidence" with great excitement and correctly surmised that light itself is an electromagnetic wave.

In this chapter we begin by showing that Ampère's law as stated in Chapter 25 does not hold for discontinuous currents. We then show how Maxwell generalized Ampère's law by adding a term now called Maxwell's displacement current. After stating Maxwell's equations and relating them to the laws of electricity and magnetism that we have already studied, we will show that these equations imply that electric and magnetic field vectors obey a wave equation that describes waves that propagate through free space with speed $c = 1/\sqrt{\mu_0 \epsilon_0}$. Finally, we will illustrate how electromagnetic waves carry energy and momentum, and discuss the electromagnetic spectrum.

29-1 Maxwell's Displacement Current

As we studied in Chapter 25, Ampère's law (Equation 25-15) relates the line integral of the magnetic field around some closed curve C to the current that passes through any area bounded by that curve:

$$\oint_C \mathbf{B}\cdot d\boldsymbol{\ell} = \mu_0 I \quad \text{for any closed curve } C \qquad 29\text{-}2$$

We noted that this equation holds only for continuous currents. We can see that it does not hold for discontinuous currents by considering the charging of a capacitor (Figure 29-1). According to Ampère's law, the line integral of the magnetic field **B** around a closed curve equals μ_0 times the total current through any surface bounded by the curve. Such a surface need not be a plane. Two surfaces bounded by the curve C are indicated in Figure 29-1.

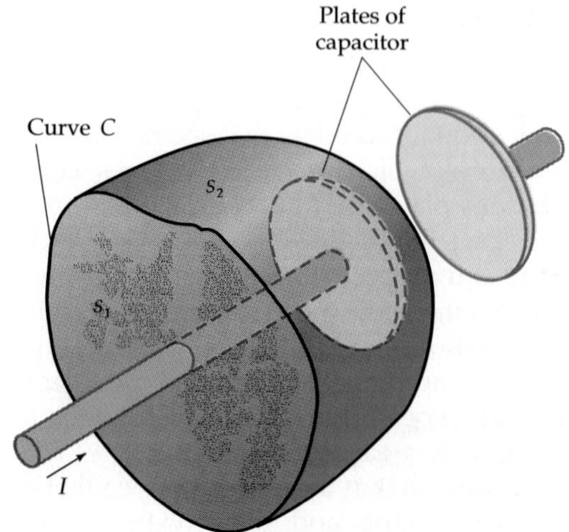

Figure 29-1 Two surfaces S_1 and S_2 bounded by the same curve C. The current I passes through surface S_1 but not S_2. Ampère's law, which relates the line integral of the magnetic field **B** around the curve C to the total current passing through any surface bounded by C, is not valid when the current is not continuous, as when it stops at the capacitor plate here.

The current through surface S_1 is I. There is no current through surface S_2 because the charge stops on the capacitor plate. There is thus ambiguity in the phrase "the current through any surface bounded by the curve." For continuous currents, we get the same current no matter which surface we choose.

Maxwell recognized this flaw in Ampère's law and showed that the law can be generalized to include all situations if the current I in the equation is replaced by the sum of the conduction current I and another term I_d, called **Maxwell's displacement current,** defined as

$$I_d = \epsilon_0 \frac{d\phi_e}{dt} \qquad 29\text{-}3$$

where ϕ_e is the flux of the electric field through the same surface bounded by the curve C. The generalized form of Ampère's law is then

$$\oint_C \mathbf{B}\cdot d\boldsymbol{\ell} = \mu_0(I + I_d) = \mu_0 I + \mu_0\epsilon_0 \frac{d\phi_e}{dt} \qquad 29\text{-}4$$

We can understand this generalization by considering Figure 29-1 again. Let us call the sum $I + I_d$ the generalized current. According to the argument just stated, the same generalized current must cross any area bounded by the curve C. Thus, there can be no net generalized current into or out of the closed volume. If there is a net true current I into the volume, there must be an equal net displacement current I_d out of the volume. In the volume in the figure, there is a net conduction current I into the volume that increases the charge within the volume:

$$I = \frac{dQ}{dt}$$

The flux of the electric field out of the volume is related to the charge by Gauss's law

$$\phi_{e,net} = \oint_S E_n\, dA = \frac{1}{\epsilon_0} Q_{inside}$$

The rate of increase of the charge is thus proportional to the rate of increase of the net flux out of the volume:

$$\epsilon_0 \frac{d\phi_{e,net}}{dt} = \frac{dQ}{dt} = I_d$$

Thus, the net conduction current into the volume equals the net displacement current out of the volume. The generalized current is always continuous.

It is interesting to compare Equation 29-4 to Faraday's law (Equation 26-6)

$$\mathcal{E} = \oint_C \mathbf{E}\cdot d\boldsymbol{\ell} = -\frac{d\phi_m}{dt} \qquad 29\text{-}5$$

where $\mathcal{E}$ is the induced emf in a circuit and ϕ_m is the magnetic flux through the circuit. According to Faraday's law, a changing magnetic flux produces an electric field whose line integral around a curve is proportional to the rate of change of magnetic flux through the curve. Maxwell's modification of Ampère's law shows that a changing electric flux produces a magnetic field whose line integral around a curve is proportional to the rate of change of the electric flux. We thus have the interesting reciprocal result that a changing magnetic field produces an electric field (Faraday's law) and a changing electric field produces a magnetic field (generalized form of Ampère's law.) Note that there is no magnetic analogue of a conduction current I.

Example 29-1

A parallel-plate capacitor has closely spaced circular plates of radius R. Charge is flowing onto the positive plate and off of the negative plate at the rate $I = dQ/dt = 2.5$ A. Compute the displacement current between the plates.

Since the plates are closely spaced, the electric field between them is uniform in the direction from the positive plate to the negative plate with magnitude $E = \sigma/\epsilon_0$, where σ is the magnitude of the charge per unit area on either plate. Consider any plane between the plates and parallel to them. Since **E** is perpendicular to the plates and therefore to the plane, and is uniform between the plates and zero outside the plates, the electric flux through the plane is

$$\phi_e = \pi R^2 E = (\pi R^2)(\sigma/\epsilon_0) = Q/\epsilon_0$$

where $Q = \pi R^2 \sigma$ is the magnitude of the total charge on either plate. The displacement current is then

$$I_d = \epsilon_0 d\phi_e/dt = dQ/dt = 2.5 \text{ A}$$

Example 29-2

The circular plates in Example 29-1 have a radius of $R = 3.0$ cm. Find the magnetic field at a point between the plates at a distance $r = 2.0$ cm from the axis of the plates when the current into the positive plate is 2.5 A.

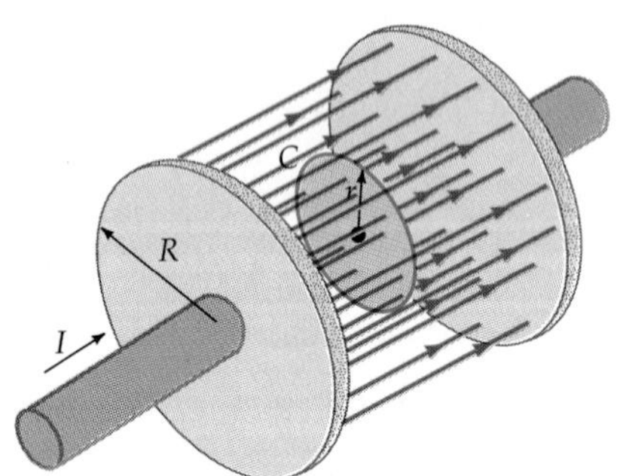

Figure 29-2 Curve C for computing the displacement current in Example 29-2.

We find B from the generalized form of Ampère's law (Equation 29-4). In Figure 29-2, we have chosen a circular path of radius $r = 2.0$ cm about the center line joining the plates to compute $\oint \mathbf{B}\cdot d\boldsymbol{\ell}$. By symmetry, **B** is tangent to this circle and has the same magnitude everywhere on it. Then

$$\oint \mathbf{B}\cdot d\boldsymbol{\ell} = B(2\pi r)$$

The electric flux through the area bounded by this curve is

$$\phi_e = \pi r^2 E = (\pi r^2)\frac{\sigma}{\epsilon_0}$$

$$= (\pi r^2)\frac{Q}{\pi R^2 \epsilon_0} = \frac{r^2 Q}{R^2 \epsilon_0}$$

where we have used $\sigma = Q/\pi R^2$. Since there is no conduction current between the plates of the capacitor, the generalized current is just the displacement current

$$I_d = \epsilon_0 \frac{d\phi_e}{dt} = \frac{r^2}{R^2}\frac{dQ}{dt}$$

$$\oint_C \mathbf{B}\cdot d\boldsymbol{\ell} = \mu_0(I + I_d) = \mu_0 I_d = \mu_0 \epsilon_0 \frac{d\phi_e}{dt}$$

$$B(2\pi r) = \mu_0 \frac{r^2}{R^2}\frac{dQ}{dt}$$

$$B = \frac{\mu_0}{2\pi}\frac{r}{R^2}\frac{dQ}{dt} = (2 \times 10^{-7} \text{ T}\cdot\text{m/A})\frac{0.02 \text{ m}}{(0.03 \text{ m})^2}(2.5 \text{ A}) = 1.11 \times 10^{-5} \text{ T}$$

29-2 Maxwell's Equations

Maxwell's equations are

$$\oint_S E_n \, dA = \frac{1}{\epsilon_0} Q_{\text{inside}} \qquad \text{29-6}a$$

$$\oint_S B_n \, dA = 0 \qquad \text{29-6}b$$

$$\oint_C \mathbf{E} \cdot d\boldsymbol{\ell} = -\frac{d}{dt} \int_S B_n \, dA \qquad \text{29-6}c$$

$$\oint_C \mathbf{B} \cdot d\boldsymbol{\ell} = \mu_0 I + \mu_0 \epsilon_0 \frac{d}{dt} \int_S E_n \, dA \qquad \text{29-6}d$$

Maxwell's equations

Equation 29-6*a* is Gauss's law; it states that the flux of the electric field through any closed surface equals $1/\epsilon_0$ times the net charge inside the surface. As discussed in Chapter 19, Gauss's law implies that the electric field due to a point charge varies inversely as the square of the distance from the charge. This law describes how electric-field lines diverge from a positive charge and converge on a negative charge. Its experimental basis is Coulomb's law.

Equation 29-6*b*, sometimes called Gauss's law for magnetism, states that the flux of the magnetic-field vector **B** is zero through any closed surface. This equation describes the experimental observation that magnetic-field lines do not diverge from any point in space or converge on any point; that is, it implies that isolated magnetic poles do not exist.

Equation 29-6*c* is Faraday's law; it states that the integral of the electric field around any closed curve C, which is the emf, equals the (negative) rate of change of the magnetic flux through any surface *S* bounded by the curve. (This is not a closed surface, so the magnetic flux through *S* is not necessarily zero.) Faraday's law describes how electric-field lines encircle any area through which the magnetic flux is changing, and it relates the electric-field vector **E** to the rate of change of the magnetic-field vector **B**.

Equation 29-6*d*, Ampère's law with Maxwell's displacement-current modification, states that the line integral of the magnetic field **B** around any closed curve C equals μ_0 times the current through any surface bounded by the curve plus $\mu_0\epsilon_0$ times the rate of change of the electric flux through the surface. This law describes how the magnetic-field lines encircle an area through which a current is passing or the electric flux is changing.

29-3 The Wave Equation for Electromagnetic Waves (Optional)

In Section 13-8, we saw that the harmonic wave functions for waves on a string obey a partial differential equation called the **wave equation:**

$$\frac{\partial^2 y(x, t)}{\partial x^2} = \frac{1}{v^2} \frac{\partial^2 y(x, t)}{\partial t^2} \qquad \text{29-7}$$

In this equation, $y(x, t)$ is the wave function, which for string waves is the displacement of the string. The derivatives are partial derivatives because the wave function depends on both x and t. The quantity v is the velocity of the wave, which depends on the medium (and on the frequency if the medium is dispersive). We also saw that the wave equation for string waves can be derived by applying Newton's laws of motion to a string under tension, and we found that the velocity of the waves is $\sqrt{F/\mu}$, where F is the tension and μ the linear mass density. The solutions of this equation are harmonic wave functions of the form

$$y(x, t) = y_0 \sin (kx - \omega t)$$

where $k = 2\pi/\lambda$ is the wave number and $\omega = 2\pi f$ is the angular frequency.

A double rainbow over the radio telescope at Socorro, New Mexico. The telescope consists of a Very Large Array (VLA) of antenna dishes. The direction of incoming radio waves from distant galaxies can be determined by the interference of signals detected in the array.

In this section, we will use Maxwell's equations to derive the wave equation for electromagnetic waves. We will not consider how such waves arise from the motion of charges but merely show that the laws of electricity and magnetism imply a wave equation, which in turn implies the existence of electric and magnetic fields **E** and **B** that propagate through space with the velocity of light c. We will consider only free space, in which there are no charges or currents. We will assume that the electric and magnetic fields **E** and **B** are functions of time and one space coordinate only, which we will take to be the x coordinate. Such a wave is called a **plane wave,** because field quantities are constant across any plane perpendicular to the x axis.

To obtain the wave equation relating the time and space derivatives of either the electric field **E** or the magnetic field **B**, we first relate the space derivative of one of the field vectors to the time derivative of the other. We do this by applying Equations 29-6c and 29-6d to appropriately chosen curves in space. We first relate the space derivative of E_y to the time derivative of B_z by applying Equation 29-6c (which is Faraday's law) to the rectangular curve of sides Δx and Δy lying in the xy plane shown in Figure 29-3. If Δx and Δy are very small, the line integral of **E** around this curve is approximately

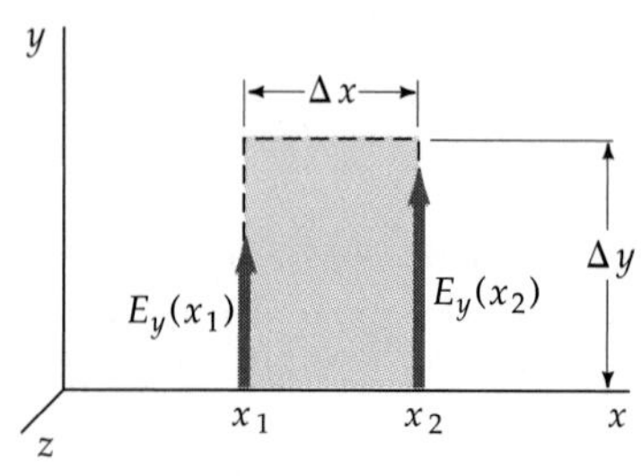

Figure 29-3 A rectangular curve in the xy plane for the derivation of Equation 29-8.

$$\oint \mathbf{E} \cdot d\boldsymbol{\ell} = E_y(x_2)\, \Delta y - E_y(x_1)\, \Delta y$$

where $E_y(x_1)$ is the value of E_y at the point x_1 and $E_y(x_2)$ is the value of E_y at the point x_2. The contributions of the type $E_x\, \Delta x$ from the top and bottom of this curve cancel because we have assumed that **E** does not depend on y (or

z). Because Δx is very small, we can approximate the difference in E_y at the points x_1 and x_2 by

$$E_y(x_2) - E_y(x_1) = \Delta E \approx \frac{\partial E_y}{\partial x} \Delta x$$

Then

$$\oint \mathbf{E} \cdot d\boldsymbol{\ell} \approx \frac{\partial E_y}{\partial x} \Delta x \, \Delta y$$

The flux of the magnetic field through this curve is approximately

$$\int_S B_n \, dA = B_z \, \Delta x \, \Delta y$$

Faraday's law then gives

$$\frac{\partial E_y}{\partial x} \Delta x \, \Delta y = -\frac{\partial B_z}{\partial t} \Delta x \, \Delta y$$

or

$$\frac{\partial E_y}{\partial x} = -\frac{\partial B_z}{\partial t} \qquad \text{29-8}$$

Equation 29-8 implies that if there is a component of the electric field E_y that depends on x, there must be a component of the magnetic field B_z that depends on time or, conversely, that if there is a component of the magnetic field B_z that depends on time, there must be a component of the electric field E_y that depends on x. We can get a similar equation relating the space derivative of the magnetic field B_z to the time derivative of the electric field E_y by applying Equation 29-6*d* to the curve of sides Δx and Δz in the xz plane shown in Figure 29-4. For the case of no conduction currents, Equation 29-6*d* is

$$\oint \mathbf{B} \cdot d\boldsymbol{\ell} = \mu_0 \epsilon_0 \frac{d}{dt} \int_S E_n \, dA$$

The details of this calculation are similar to those for Equation 29-8. The result is

$$\frac{\partial B_z}{\partial x} = -\mu_0 \epsilon_0 \frac{\partial E_y}{\partial t} \qquad \text{29-9}$$

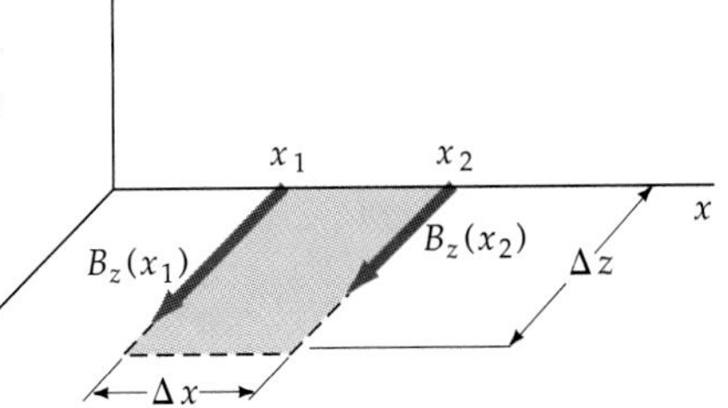

Figure 29-4 A rectangular curve in the xz plane for the derivation of Equation 29-9.

We can eliminate either B_z or E_y from Equations 29-8 and 29-9 by differentiating either equation with respect to x or t. If we differentiate both sides of Equation 29-8 with respect to x, we obtain

$$\frac{\partial}{\partial x}\left(\frac{\partial E_y}{\partial x}\right) = -\frac{\partial}{\partial x}\left(\frac{\partial B_z}{\partial t}\right)$$

or

$$\frac{\partial^2 E_y}{\partial x^2} = -\frac{\partial}{\partial t}\left(\frac{\partial B_z}{\partial x}\right)$$

where the order of the time and space derivatives on the right side have been interchanged. We now use Equation 29-9 for $\partial B_z/\partial x$:

$$\frac{\partial^2 E_y}{\partial x^2} = -\frac{\partial}{\partial t}\left(-\mu_0 \epsilon_0 \frac{\partial E_y}{\partial t}\right)$$

which yields the wave equation

$$\frac{\partial^2 E_y}{\partial x^2} = \mu_0 \epsilon_0 \frac{\partial^2 E_y}{\partial t^2} \qquad 29\text{-}10$$

Comparing this equation with Equation 29-7, we see that E_y obeys a wave equation for waves with speed

$$c = \frac{1}{\sqrt{\mu_0 \epsilon_0}}$$

which is Equation 29-1.

If we had instead chosen to eliminate E_y from Equations 29-8 and 29-9 (by differentiating Equation 29-8 with respect to t, for example), we would have obtained an equation identical to Equation 29-10 except with B_z replacing E_y. We can thus see that both the electric field E_y and the magnetic field B_z obey a wave equation for waves traveling with the velocity $1/\sqrt{\mu_0 \epsilon_0}$, which is the velocity of light.

By following the same line of reasoning as used above, we can readily show (as in Problem 29-49) that if Equation 29-6c (Faraday's law) is applied to the curve in the xz plane shown in Figure 29-4, the spatial variation in E_z is related to the time variation in B_y by

$$\frac{\partial E_z}{\partial x} = \frac{\partial B_y}{\partial t} \qquad 29\text{-}11$$

Similarly, the application of Equation 26-6d to the curve in the xy plane of Figure 29-3 gives

$$\frac{\partial B_y}{\partial x} = \mu_0 \epsilon_0 \frac{\partial E_z}{\partial t} \qquad 29\text{-}12$$

We can use these results to show that, for a wave propagating in the x direction, the components E_z and B_y also obey the wave equation.

So far we have considered only the y and z components of the electric and magnetic fields. The same type of analysis may be applied to a rectangular loop in the yz plane, similar to the loops in Figure 29-3 and 29-4, to obtain equations analogous to Equations 29-8, 29-9, 29-11, and 29-12 in which the time derivatives of E_x and B_x are proportional to the y and z derivatives of the field quantities. In the plane wave under consideration, the y and z derivatives of the field quantities are zero, so the time derivatives of E_x and B_x must be zero. In an electromagnetic wave, we are interested only in time-varying fields, and so we may subtract out any x components of the fields that are constant in time.

We have shown that in any plane electromagnetic wave traveling parallel to the x axis the x components of the fields are zero, so the vectors **E** and **B** are perpendicular to the x axis. They are also perpendicular to each other and each obeys the wave equation:

Wave equation for **E**

$$\frac{\partial^2 \mathbf{E}}{\partial x^2} = \frac{1}{c^2} \frac{\partial^2 \mathbf{E}}{\partial t^2} \qquad 29\text{-}13a$$

Wave equation for **B**

$$\frac{\partial^2 \mathbf{B}}{\partial x^2} = \frac{1}{c^2} \frac{\partial^2 \mathbf{B}}{\partial t^2} \qquad 29\text{-}13b$$

As we noted in discussing harmonic waves, a particularly important solution to a wave equation like Equation 29-10 is the harmonic wave function of the form

$$E_y = E_{y0} \sin (kx - \omega t) \tag{29-14}$$

If we substitute this solution into either Equation 29-8 or 29-9, we can see that the magnetic field B_z is in phase with the electric field E_y. From Equation 29-8, we have

$$\frac{\partial B_z}{\partial t} = -\frac{\partial E_y}{\partial x} = -kE_{y0} \cos (kx - \omega t)$$

Solving for B_z gives

$$B_z = \frac{k}{\omega} E_{y0} \sin (kx - \omega t)$$

$$= B_{z0} \sin (kx - \omega t) \tag{29-15}$$

where

$$B_{z0} = \frac{k}{\omega} E_{y0} = \frac{E_{y0}}{c}$$

and $c = \omega/k$ is the velocity of the wave.* Since the electric and magnetic fields oscillate in phase with the same frequency, we have the general result that the magnitude of the electric field is c times the magnitude of the magnetic field for an electromagnetic wave:

$$E = cB \tag{29-16}$$

Suppose the electric field vector **E** is confined to the y direction, as exemplified by Equation 29-14. Then $E_z = 0$, and, according to Equation 29-11, $dB_y/dt = 0$. Thus, if **E** is in the y direction, then the time-varying part (the only part we are interested in) of **B** is in the z direction, as shown in Figure 29-5. Such a wave is said to be **linearly polarized,** because if we plot **E** (or **B**) as a function of time in any plane perpendicular to the x axis, we obtain a straight line.

We see that Maxwell's equations imply wave equations 29-13*a* and *b* for the electric and magnetic fields; and that if E_y varies harmonically, as in Equation 29-14, the magnetic field B_z is in phase with E_y and has an amplitude related to the amplitude of E_y by Equation 29-16. The electric and magnetic fields are perpendicular to each other and to the direction of the wave propagation, as shown in Figure 29-5. In general, the direction of propagation of an electromagnetic wave is the direction of the cross product **E × B**.

*In obtaining Equation 29-15 by the integration of the previous equation, an arbitrary constant of integration arises. We have omitted this constant magnetic field from Equation 29-15 because it plays no part in the electromagnetic waves we are interested in. Note that if any constant electric field is added to Equation 29-14, the new electric field still satisfies the wave equation.

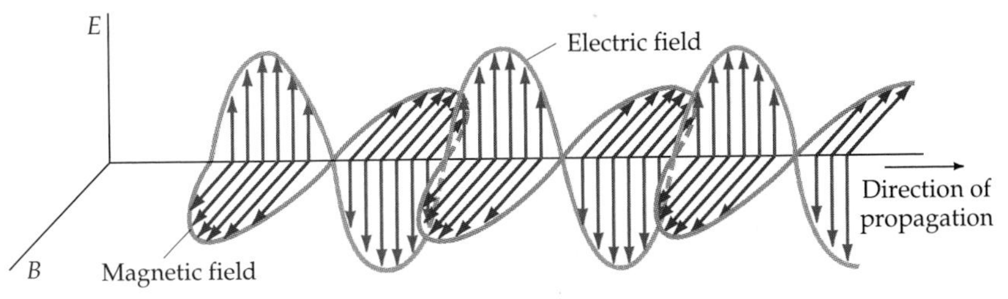

Figure 29-5 The electric- and magnetic-field vectors in a plane-polarized electromagnetic wave. The fields are in phase, perpendicular to each other, and perpendicular to the direction of propagation of the wave.

A radar image of the south coast of New Guinea. A radar system operates by transmitting radio waves toward objects, sensing the echo of waves reflected back, and determining the distance to the object from the intervening time interval.

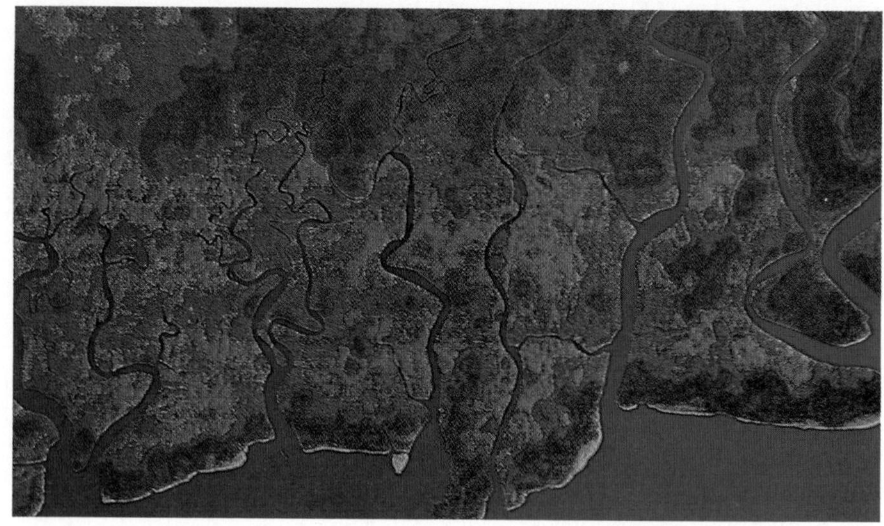

Example 29-3

The electric field vector of an electromagnetic wave is given by $\mathbf{E}(x, t) = E_0 \sin(kx - \omega t)\,\mathbf{j} + E_0 \cos(kx - \omega t)\,\mathbf{k}$. (*a*) Find the corresponding magnetic field. (*b*) Compute $\mathbf{E}\cdot\mathbf{B}$ and $\mathbf{E} \times \mathbf{B}$.

(*a*) We can use either Equation 29-11 or 29-12 to find B_y. From Equation 29-11, we obtain

$$\frac{\partial B_y}{\partial t} = \frac{\partial E_z}{\partial x} = \frac{\partial}{\partial x}[E_0 \cos(kx - \omega t)] = -kE_0 \sin(kx - \omega t)$$

Then, neglecting the arbitrary constant of integration, we obtain

$$B_y = [kE_0 \cos(kx - \omega t)](-1/\omega) = -B_0 \cos(kx - \omega t)$$

where $B_0 = kE_0/\omega = E_0/c$. We can find B_z from either Equation 29-8 or 29-9. Using Equation 29-8, we obtain

$$\frac{\partial B_z}{\partial t} = -\frac{\partial E_y}{\partial x} = -\frac{\partial}{\partial x}[E_0 \sin(kx - \omega t)] = -kE_0 \cos(kx - \omega t)$$

and

$$B_z = [-kE_0 \sin(kx - \omega t)](-1/\omega) = B_0 \sin(kx - \omega t)$$

where again $B_0 = kE_0/\omega = E_0/c$. The magnetic field is thus given by

$$\mathbf{B}(x, t) = -B_0 \cos(kx - \omega t)\,\mathbf{j} + B_0 \sin(kx - \omega t)\,\mathbf{k}.$$

This type of electromagnetic wave is said to be **circularly polarized.** Both **E** and **B** are constant in magnitude, as can be seen by computing $\mathbf{E}\cdot\mathbf{E}$ or $\mathbf{B}\cdot\mathbf{B}$. For example, $\mathbf{E}\cdot\mathbf{E} = E_x^2 + E_y^2 = E_0^2 \sin^2(kx - \omega t) + E_0^2 \cos^2(kx - \omega t) = E_0^2$. At a fixed point x, both vectors rotate in a circle in the plane perpendicular to x with angular frequency ω.

(*b*) Computing $\mathbf{E}\cdot\mathbf{B}$, with $\theta = kx - \omega t$ to simplify the notation, we obtain

$$\begin{aligned}\mathbf{E}\cdot\mathbf{B} &= [E_0 \sin\theta\,\mathbf{j} + E_0 \cos\theta\,\mathbf{k}]\cdot[-B_0 \cos\theta\,\mathbf{j} + B_0 \sin\theta\,\mathbf{k}] \\ &= -E_0B_0 \sin\theta \cos\theta\,\mathbf{j}\cdot\mathbf{j} + E_0B_0 \sin^2\theta\,\mathbf{j}\cdot\mathbf{k} \\ &\qquad - E_0B_0 \cos^2\theta\,\mathbf{k}\cdot\mathbf{j} + E_0B_0 \cos\theta \sin\theta\,\mathbf{k}\cdot\mathbf{k} \\ &= -E_0B_0 \sin\theta \cos\theta + 0 - 0 + E_0B_0 \cos\theta \sin\theta = 0\end{aligned}$$

The electric and magnetic fields are perpendicular to each other as well as to the direction of propagation. Computing $\mathbf{E} \times \mathbf{B}$ and using $\mathbf{j} \times \mathbf{j} = \mathbf{k} \times \mathbf{k} = 0$, $\mathbf{j} \times \mathbf{k} = \mathbf{i}$, and $\mathbf{k} \times \mathbf{j} = -\mathbf{i}$, we obtain

$$\begin{aligned}\mathbf{E} \times \mathbf{B} &= [E_0 \sin \theta\, \mathbf{j} + E_0 \cos \theta\, \mathbf{k}] \times [-B_0 \cos \theta\, \mathbf{j} + B_0 \sin \theta\, \mathbf{k}] \\ &= E_0 B_0 \sin^2 \theta\, \mathbf{j} \times \mathbf{k} + (-E_0 B_0 \cos^2 \theta\, \mathbf{k} \times \mathbf{j}) \\ &= E_0 B_0 \sin^2 \theta\, \mathbf{i} + E_0 B_0 \cos^2 \theta\, \mathbf{i} = E_0 B_0 \mathbf{i}\end{aligned}$$

We note that $\mathbf{E} \times \mathbf{B}$ is in the direction of propagation of the wave.

29-4 Energy and Momentum in an Electromagnetic Wave

In our discussion of the transport of energy by a wave of any kind, we saw that the intensity of the wave (the average energy per unit time per unit area) is equal to the product of the average energy density (the energy per unit volume) and the speed of the wave. The energy density stored in the electric field is (Equation 21-17)

$$\eta_e = \tfrac{1}{2}\epsilon_0 E^2$$

and the energy density stored in the magnetic field is (Equation 26-34)

$$\eta_m = \frac{B^2}{2\mu_0}$$

In an electromagnetic wave in free space, $E = cB$, so we can express the magnetic energy density in terms of the electric field:

$$\eta_m = \frac{B^2}{2\mu_0} = \frac{(E/c)^2}{2\mu_0} = \frac{E^2}{2\mu_0 c^2} = \frac{1}{2}\epsilon_0 E^2$$

where we have used $c^2 = 1/\epsilon_0\mu_0$. Thus, the electric and magnetic energy densities are equal. The total energy density η in the wave is the sum of the electric and magnetic energy densities. Using $E = cB$, we may express the total energy density in several useful ways:

$$\eta = \eta_e + \eta_m = \frac{1}{2}\epsilon_0 E^2 + \frac{1}{2}\epsilon_0 E^2 = \epsilon_0 E^2 = \frac{B^2}{\mu_0} = \frac{EB}{\mu_0 c} \qquad 29\text{-}17$$

Energy density in an electromagnetic wave

In Section 14-3, we saw that the intensity of a wave (the average power flowing across an area per unit area) is equal to the product of the average energy density and the speed of the wave. The instantaneous intensity is the instantaneous power flowing across an area per unit area. It equals the product of the instantaneous energy density and the wave speed. For an electromagnetic wave in free space, the instantaneous intensity is therefore

$$I_{\text{instantaneous}} = \eta c = c\epsilon_0 E^2 = c\frac{B^2}{\mu_0} = \frac{EB}{\mu_0} \qquad 29\text{-}18$$

Equation 29-18 can be generalized to a vector expression:

$$\mathbf{S} = \frac{\mathbf{E} \times \mathbf{B}}{\mu_0} \qquad 29\text{-}19$$

The vector **S** is called the **Poynting vector** after its discoverer Sir John Poynting. Since **E** and **B** are perpendicular in an electromagnetic wave, the magnitude of **S** is the instantaneous intensity of the wave and the direction of **S** is the direction of propagation of the wave.

In a harmonic plane wave of angular frequency ω and wave number k, the instantaneous electric and magnetic fields are given by

$$E = E_0 \sin(kx - \omega t) \qquad \text{and} \qquad B = B_0 \sin(kx - \omega t)$$

Using these results for E and B in Equation 29-17, we obtain for the instantaneous energy density

$$\eta = \frac{EB}{\mu_0 c} = \frac{E_0 B_0 \sin^2(kx - \omega t)}{\mu_0 c}$$

When we average the sine squared function over space or time, we obtain a factor of $\frac{1}{2}$. The average energy density is therefore

$$\eta_{av} = \frac{1}{2}\frac{E_0 B_0}{\mu_0 c} = \frac{E_{rms} B_{rms}}{\mu_0 c} \qquad \text{29-20}$$

where we have used $E_{rms} = E_0/\sqrt{2}$ and $B_{rms} = B_0/\sqrt{2}$. The intensity is thus

Intensity of an electromagnetic wave

$$I = \eta_{av} c = \frac{1}{2}\frac{E_0 B_0}{\mu_0} = \frac{E_{rms} B_{rms}}{\mu_0} = |\mathbf{S}|_{av} \qquad \text{29-21}$$

We will now show by a simple example that an electromagnetic wave carries momentum. In this example, we will calculate the momentum and energy absorbed from the wave by a free charged particle. Consider an electromagnetic wave moving along the x axis with the electric field in the y direction and the magnetic field in the z direction that is incident on a stationary charge on the x axis as shown in Figure 29-6. For simplicity, we will neglect the time dependence of the electric and magnetic fields. The particle experiences a force $q\mathbf{E}$ in the y direction and is thus accelerated by the electric field. At any time t, the velocity in the y direction is

$$v_y = at = \frac{qE}{m}t$$

After a short time t_1 the charge has acquired a velocity in the y direction given by

$$v_y = at_1 = \frac{qE}{m}t_1$$

The energy acquired by the charge after time t_1 is

$$K = \frac{1}{2}mv_y^2 = \frac{1}{2}\frac{mq^2E^2t_1^2}{m^2} = \frac{1}{2}\frac{q^2E^2}{m}t_1^2 \qquad \text{29-22}$$

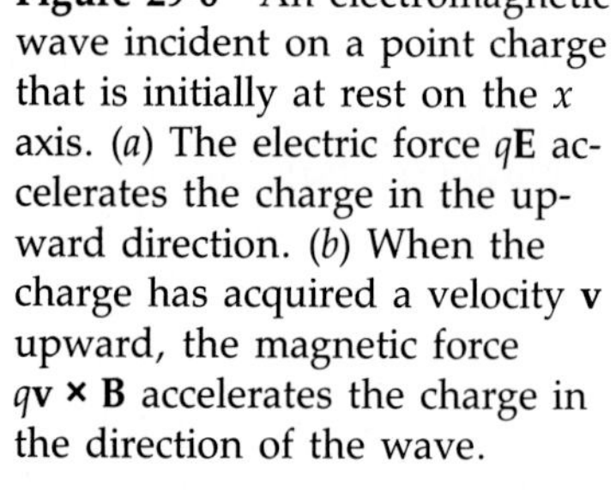

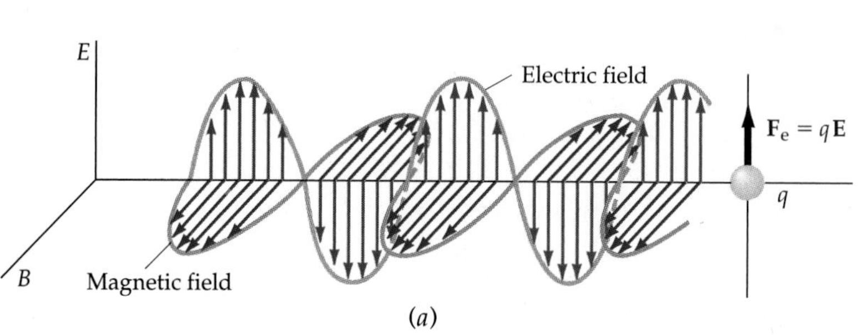

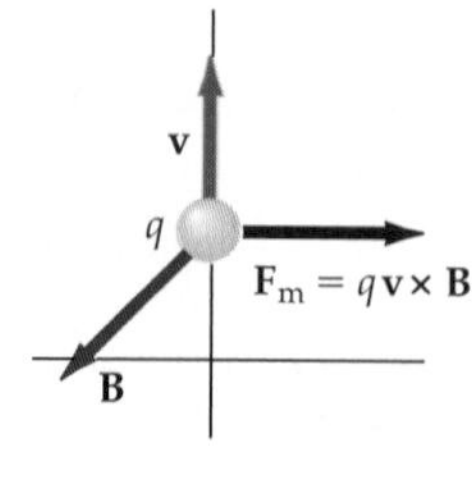

Figure 29-6 An electromagnetic wave incident on a point charge that is initially at rest on the x axis. (*a*) The electric force $q\mathbf{E}$ accelerates the charge in the upward direction. (*b*) When the charge has acquired a velocity **v** upward, the magnetic force $q\mathbf{v} \times \mathbf{B}$ accelerates the charge in the direction of the wave.

When the charge is moving in the y direction, it experiences a magnetic force $q\mathbf{v} \times \mathbf{B}$, which is in the positive x direction (the direction of propagation of the wave) when $\mathbf{B}$ is in the z direction. The magnetic force at any time t is

$$F_x = qv_yB = \frac{q^2EB}{m}t$$

The impulse of this force equals the momentum transferred by the wave to the particle. Setting the impulse equal to the momentum p_x, we obtain

$$p_x = \int_0^{t_1} F_x\, dt$$

$$= \int_0^{t_1} \frac{q^2EB}{m}t\, dt = \frac{1}{2}\frac{q^2EB}{m}t_1^2$$

If we use $B = E/c$, this becomes

$$p_x = \frac{1}{c}\left(\frac{1}{2}\frac{q^2E^2}{m}t_1^2\right) \qquad \text{29-23}$$

Comparing Equations 29-22 and 29-23, we see that the momentum acquired by the charge in the direction of the wave is $1/c$ times the energy. Although our simple calculation was not rigorous, the results are correct. In general,

The magnitude of the momentum carried by an electromagnetic wave is $1/c$ times the energy carried by the wave:

$$p = \frac{U}{c} \qquad \text{29-24}$$

Momentum and energy in an electromagnetic wave

Since the intensity of a wave is the energy per unit time per unit area, the intensity divided by c is the momentum carried by the wave per unit time per unit area. The momentum carried per unit time is a force. The intensity divided by c is thus a force per unit area, which is a pressure. This pressure is called **radiation pressure** P_r:

$$P_r = \frac{I}{c} \qquad \text{29-25}$$

We can relate the radiation pressure to the electric or magnetic fields by using Equations 29-21 and 29-16:

$$P_r = \frac{I}{c} = \frac{E_0B_0}{2\mu_0c} = \frac{E_{rms}B_{rms}}{\mu_0c} = \frac{E_0^2}{2\mu_0c^2} = \frac{B_0^2}{2\mu_0} \qquad \text{29-26}$$

Radiation pressure

Consider an electromagnetic wave incident normally on some surface. If the surface absorbs energy U from the electromagnetic wave, it also absorbs momentum p given by Equation 29-24, and the pressure exerted on the surface equals the radiation pressure. If the wave is reflected, the momentum transferred is twice the energy incident on the surface because the wave now carries momentum in the opposite direction. The pressure exerted on the surface by the wave is then twice the radiation pressure.

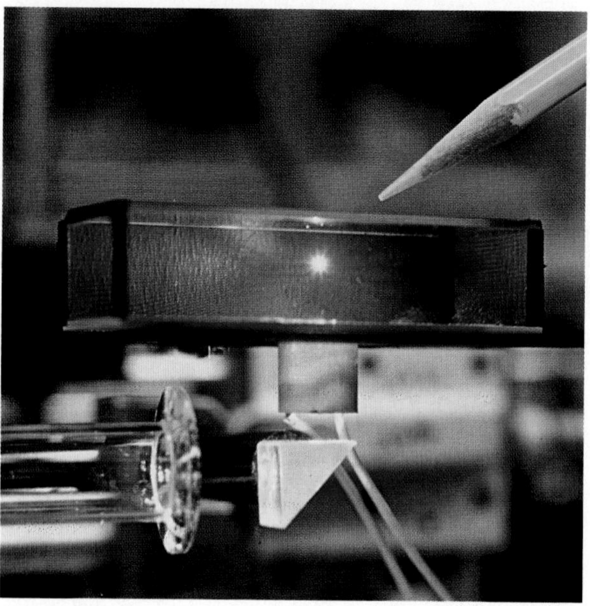

(a)

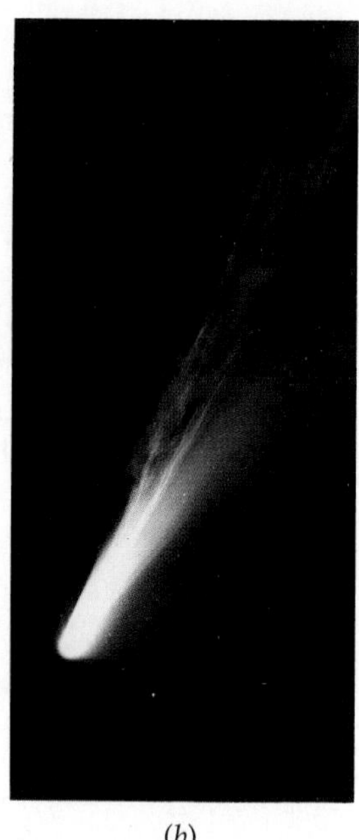

(b)

(a) A transparent glass sphere, about 25 μm in diameter (and visible here as a starlike speck) is suspended by the radiation pressure of an upward-directed 250-mW laser beam. (b) The comet Mrkos photographed in August 1957. The tail is pushed away from the sun and split into two parts by solar radiation pressure and by the solar wind, which is a stream of charged particles emitted by the sun. The splitting occurs because lighter particles in the tail are more readily deflected than heavier ones.

Example 29-4

A 100-W light bulb emits spherical electromagnetic waves uniformly in all directions. Find the intensity, the radiation pressure, and the electric and magnetic fields at a distance of 3 m from the bulb, assuming that 50 W goes into electromagnetic radiation.

At a distance r from the bulb, the energy is spread uniformly over an area $4\pi r^2$. The intensity is therefore

$$I = \frac{50\ \text{W}}{4\pi r^2}$$

At $r = 3$ m, the intensity is

$$I = \frac{50\ \text{W}}{4\pi(3\ \text{m})^2} = 0.442\ \text{W/m}^2$$

The radiation pressure is the intensity divided by the speed of light:

$$P_r = \frac{I}{c} = \frac{0.442\ \text{W/m}^2}{3 \times 10^8\ \text{m/s}} = 1.47 \times 10^{-9}\ \text{Pa}$$

This is a very small pressure compared with atmospheric pressure of the order of 10^5 Pa.

The maximum value of the magnetic field is, from Equation 29-26,

$$\begin{aligned} B_0 &= (2\mu_0 P_r)^{1/2} \\ &= [2(4\pi \times 10^{-7})(1.47 \times 10^{-9})]^{1/2} \\ &= 6.08 \times 10^{-8}\ \text{T} \end{aligned}$$

The maximum value of the electric field is the speed of light times B_0:

$$E_0 = cB_0 = 18.2\ \text{V/m}$$

The electric and magnetic fields are of the form $E = E_0 \sin(kx - \omega t)$ and $B = B_0 \sin(kx - \omega t)$ with $E_0 = 18.2$ V/m and $B_0 = 6.08 \times 10^{-8}$ T.

Example 29-5

An astronaut is stranded in space a distance of 20 m from her spaceship. She carries a 1-kW laser gun. If the total mass of the astronaut, space suit, and laser gun is 95 kg, how long will it take her to reach the ship if she points the laser gun directly away from it?

The power of the laser gun is the rate at which it emits energy. If we divide the power by c, we obtain the rate at which momentum is emitted. If P is the power of the laser gun, the rate of change of momentum of the system due to the momentum carried away by the radiation is

$$\frac{dp}{dt} = \frac{P}{c}$$

The astronaut must then have an equal and opposite rate of change of momentum. The force on the astronaut is thus

$$F = \frac{dp}{dt} = \frac{P}{c} = ma$$

If the power of the laser gun is constant, the acceleration of the astronaut will be constant and given by

$$a = \frac{P}{mc} = \frac{1000 \text{ W}}{(95 \text{ kg})(3 \times 10^8 \text{ m/s}^2)}$$

$$= 3.51 \times 10^{-8} \text{ m/s}^2$$

This is a very small acceleration. The time required for the astronaut to travel 20 m is found from the constant-acceleration formula:

$$x = \tfrac{1}{2}at^2$$

$$t = \sqrt{\frac{2x}{a}} = \sqrt{\frac{2(20 \text{ m})}{3.51 \times 10^{-8} \text{ m/s}^2}}$$

$$= 3.38 \times 10^4 \text{ s} = 9.38 \text{ h}$$

Dave Cooke and his colleagues at the University of Chicago use a 40.6-cm diameter silver-coated telescope mirror and a tiny cone-shaped sapphire crystal to concentrate sunlight to intensities that are 84,000 times that received normally at the surface of the earth. The world record 72 MW/m^2 exceeds the intensity of light at the surface of the sun itself (63 MW/m^2)—making it the highest concentration of sunlight produced in the solar system. The light is focused to a 1-cm diameter spot by the mirror and is then compressed to a 1-mm diameter spot by the nonimaging sapphire concentrator. The sapphire was chosen for its low absorption and high index of refraction. Sunlight hits the base of the cone, is funnelled through the sapphire, and exits at its tip. The specific shape of the crystal maximizes the concentration of light leaving the tip—rather than providing an accurate image of the source of light (as in conventional lenses). Despite its low absorptivity, the sapphire crystal is often heated by the concentrated sunlight to the point of explosion.

29-5 The Electromagnetic Spectrum

Electromagnetic waves include light, radio waves, x rays, gamma rays, microwaves, and others. The various types of electromagnetic waves differ only in wavelength and frequency, which is related to the wavelength in the usual way:

$$f = \frac{c}{\lambda}$$

Table 29-1 gives the **electromagnetic spectrum** and the names usually associated with the various frequency and wavelength ranges. These ranges are often not well defined and sometimes overlap. For example, electromagnetic waves with wavelengths of about 0.1 nm are usually called x rays, but if they originate from nuclear radioactivity, they are called gamma rays.

The human eye is sensitive to electromagnetic radiation with wavelengths from about 400 to 700 nm, the range called **visible light.** The shortest wavelengths in the visible spectrum correspond to violet light and the longest to red light, with all the colors of the rainbow falling between these extremes. Electromagnetic waves with wavelengths slightly less than those of visible light are called **ultraviolet rays,** and those with wavelengths

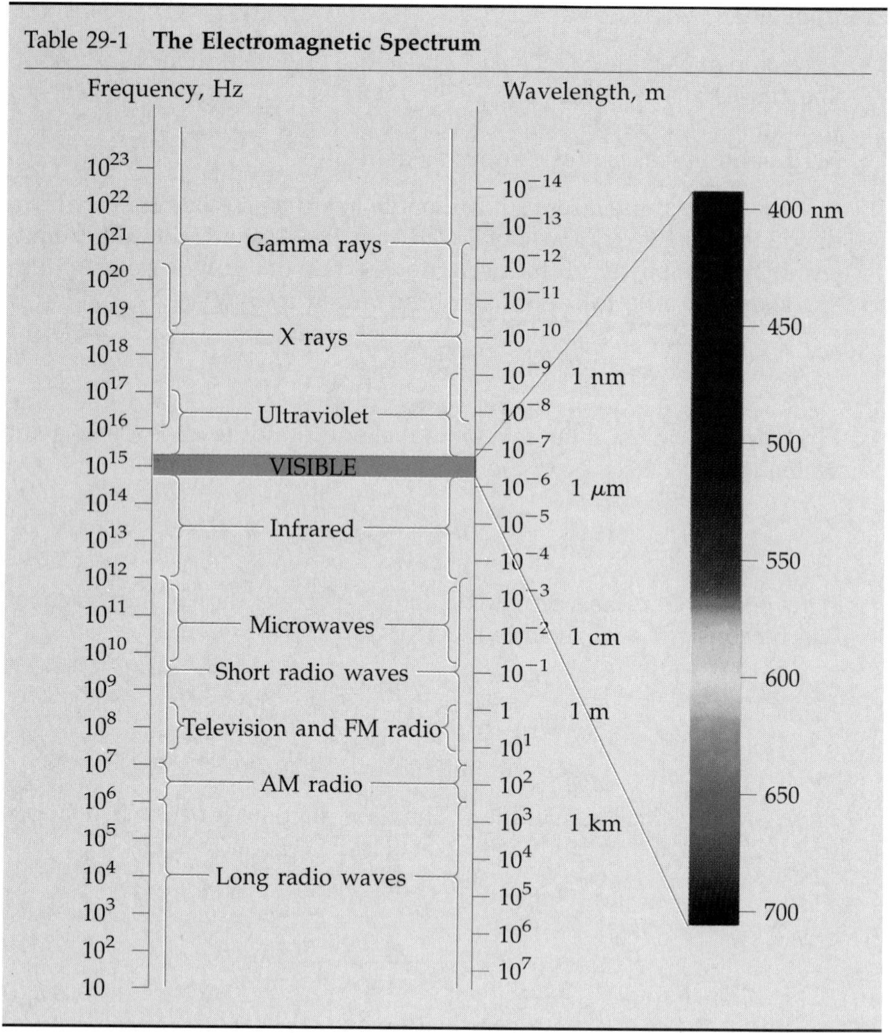

Table 29-1 **The Electromagnetic Spectrum**

slightly greater than those of visible light are called **infrared waves.** Heat radiation given off by bodies at ordinary temperatures is in the infrared region of the electromagnetic spectrum. There are no limits on the wavelengths of electromagnetic radiation; that is, all wavelengths (or frequencies) are theoretically possible.

The differences in wavelengths of the various kinds of electromagnetic waves are very important. As we know, the behavior of waves depends strongly on the relative sizes of the wavelengths and the physical objects or apertures the waves encounter. Since the wavelengths of light are in the rather narrow range from about 400 to 700 nm, they are much smaller than most obstacles, so the ray approximation is often valid. The wavelength and frequency are also important in determining the kinds of interactions between electromagnetic waves and matter. X rays, for example, which have very short wavelengths and high frequencies, easily penetrate many materials that are opaque to lower-frequency light waves, which are absorbed by the materials. Microwaves have wavelengths of the order of a few centimeters and frequencies that are close to the natural resonance frequencies of water molecules in solids and liquids. Microwaves are therefore readily absorbed by the water molecules in foods, which is the mechanism for heating in microwave ovens.

(a)

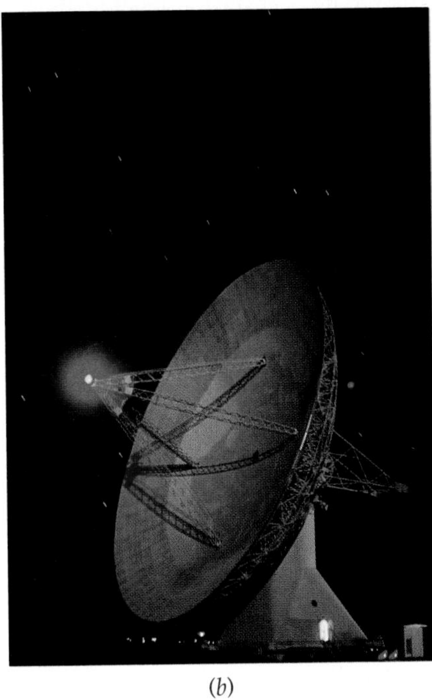

(b)

(*a*) Television antennas, shown here, operate at radio frequencies. Messages are transmitted by encoding them as modulations of frequency (FM) or amplitude (AM). The same antenna will function as a transmitter or receiver. However, antennas used for transmission operate at significantly higher power levels than those used for reception. (*b*) The Caltech radio telescope in Owens Valley, California. Stars, galaxies, quasars, and pulsars are all sources of radio waves. Nearly one-fifth of cosmic radio sources are presently "unidentified", that is, are not correlated with any source that has been seen optically.

Electromagnetic waves are produced when electric charges accelerate. When electric charges oscillate they radiate electromagnetic waves whose frequency equals the frequency of oscillation. The wavelength of the waves emitted is therefore determined by the frequency of oscillation of the charges. Radio waves, which have frequencies from about 550 to 1600 kHz for AM radio waves and from about 88 to 108 MHz for FM radio waves, are produced by macroscopic electric currents oscillating in radio antennas. Light waves, which have frequencies of the order of 10^{14} Hz, originate from the motion of atomic charges.

Figure 29-7 is a schematic drawing of an **electric-dipole antenna,** which consists of two bent conducting rods fed by an alternating-current generator. At the time $t = 0$, shown in Figure 29-6*a*, the ends of the rods are charged, and there is an electric field near the rods parallel to the rods. There is also a magnetic field (not shown) encircling the rods due to the current in the rods. The magnetic field is perpendicular to the page. These fields move outward from the rods with the speed of light. After one-fourth period, at $t = \frac{1}{4}T$, the rods are uncharged, and the electric field near the rods is zero, as shown in Figure 29-7*b*. At $t = \frac{1}{2}T$, the rods are again charged, but the charges are opposite to those at $t = 0$, as shown in Figure 29-7*c*. The electric and

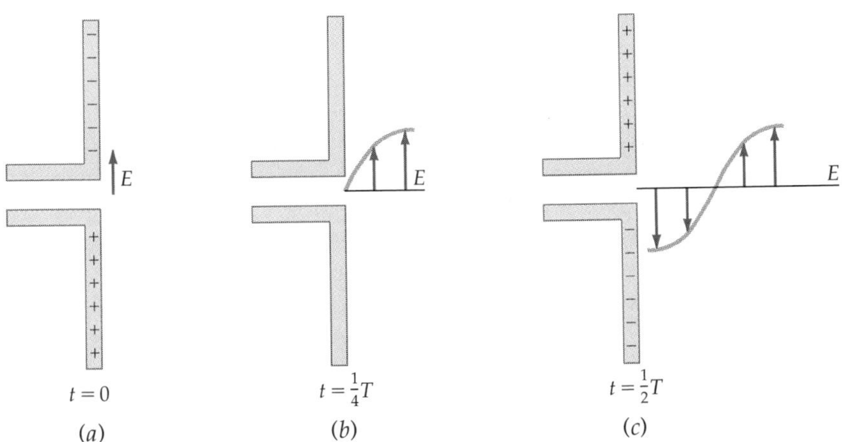

Figure 29-7 An electric-dipole antenna. Alternating current is supplied to the antenna by a generator (not shown). The electric field due to the charges in the antenna propagates outward at the speed of light. There is also a propagating magnetic field (not shown) perpendicular to the paper due to the current in the antenna.

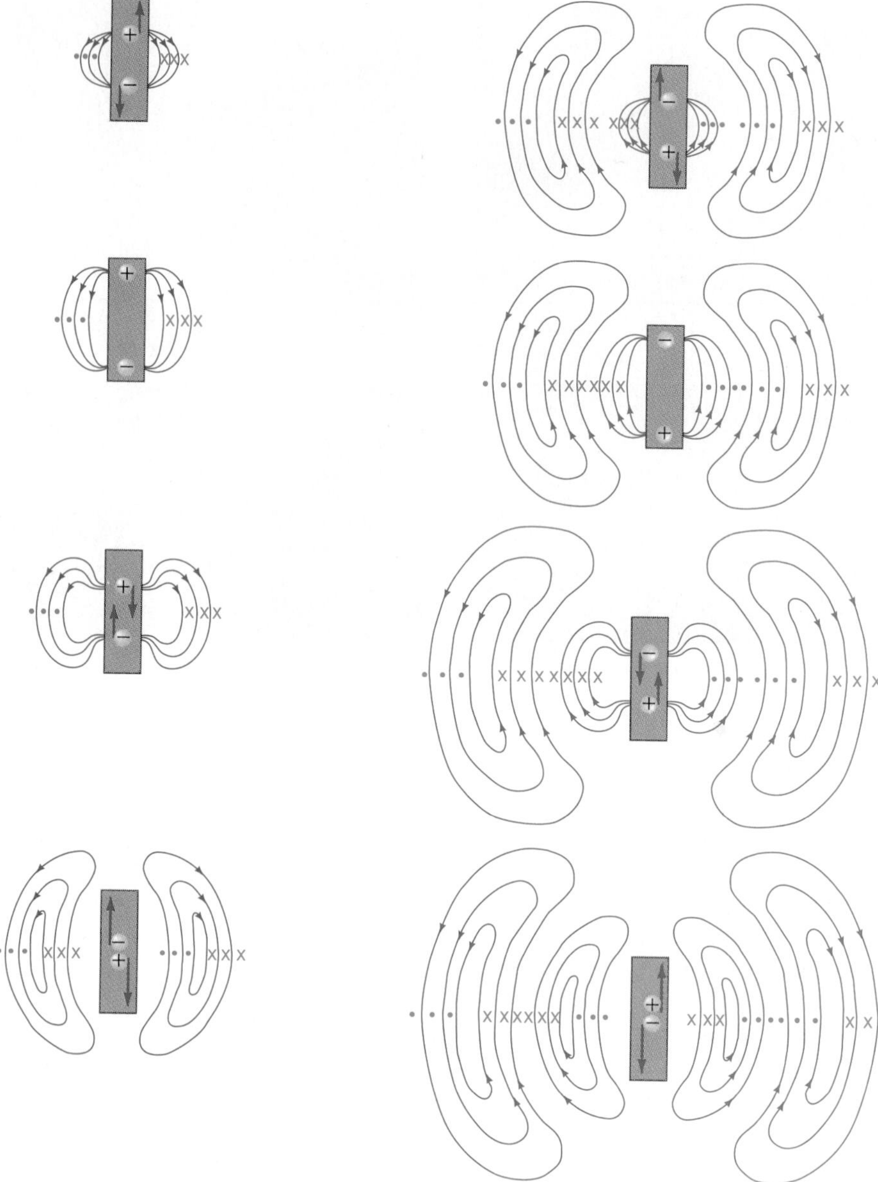

Figure 29-8 Electric- and magnetic-field lines produced by an oscillating electric dipole.

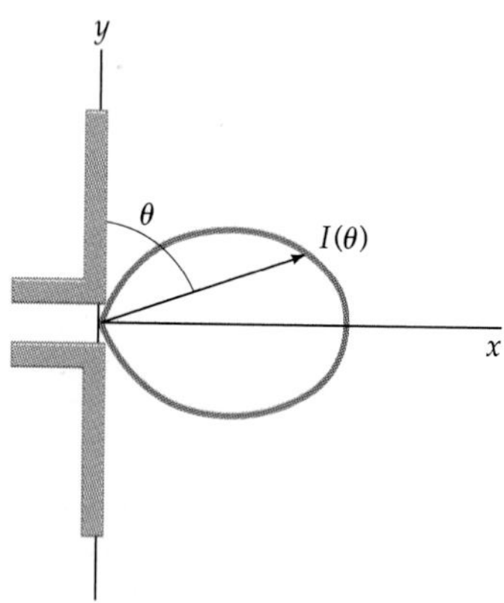

Figure 29-9 Polar plot of the intensity of electromagnetic radiation from an electric-dipole antenna versus angle. The intensity $I(\theta)$ is proportional the length of the arrow. The intensity is maximum perpendicular to the antenna at $\theta = 90°$ and minimum along the antenna at $\theta = 0°$ or $\theta = 180°$.

magnetic fields at a great distance from this transmitting antenna are quite different from the fields near the antenna. Far from the antenna, the electric and magnetic fields oscillate in phase with simple harmonic motion, perpendicular to each other and to the direction of propagation of the wave. The wave is thus a transverse wave. Figure 29-8 shows the electric and magnetic fields far from an electric-dipole antenna.

The radiation from an electric-dipole antenna such as that in Figure 29-7 is called **electric-dipole radiation.** Many electromagnetic waves exhibit the characteristics of electric-dipole radiation. A characteristic that will be important in our discussion of light in Chapter 30 is that the intensity of the electromagnetic waves radiated by a dipole antenna is zero along the axis of the antenna and maximum in the directions perpendicular to the axis. That is, if the dipole is in the z direction with center at the origin, the intensity is zero along the z axis and maximum in the xy plane. In the direction of a line making an angle θ with the z axis, as in Figure 29-9, the intensity is proportional to $\sin^2 \theta$. Far from the antenna, the electric field is parallel to the dipole.

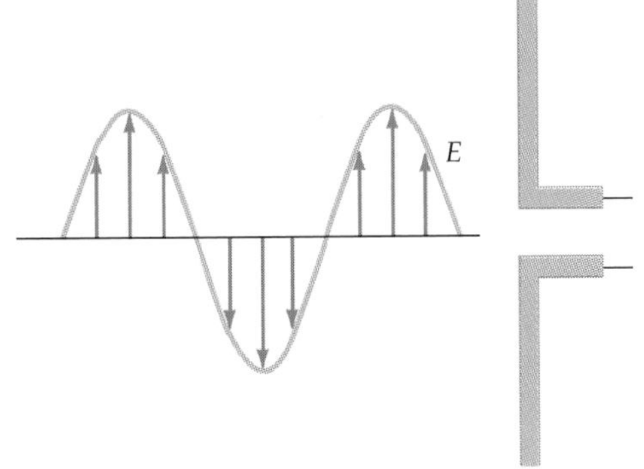

Figure 29-10 An electric-dipole antenna for detecting electromagnetic radiation. The alternating electric field of the radiation produces an alternating current in the antenna.

Electromagnetic waves of radio or television frequencies can be detected by a dipole receiving antenna oriented parallel to the electric field so that it induces an alternating current in the antenna (Figure 29-10). They can also be detected by a loop antenna oriented perpendicular to the magnetic field so that the changing magnetic flux through the loop induces a current in the loop (Figure 29-11). Electromagnetic waves of frequencies in the visible light range can be detected by the eye or by photographic film, both of which are mainly sensitive to the electric field.

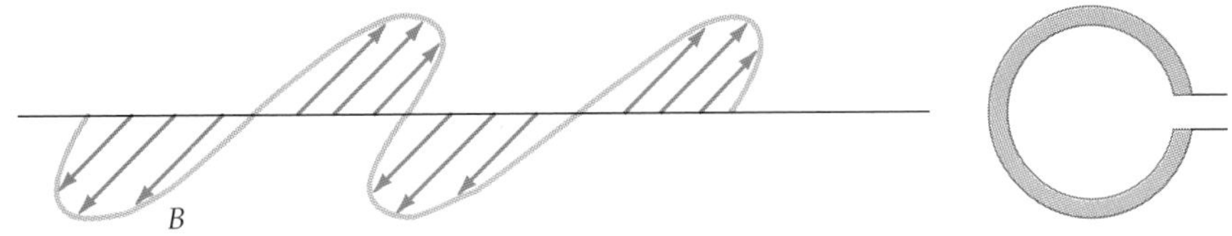

Figure 29-11 A loop antenna for detecting electromagnetic radiation. The alternating magnetic flux in the loop due to the magnetic field of the radiation induces an alternating current in the loop.

Example 29-6

An antenna consisting of a single loop of wire of radius 10 cm is used to detect electromagnetic waves for which $E_{rms} = 0.15$ V/m. Find the rms induced emf in the loop if the wave frequency is (*a*) 600 kHz and (*b*) 600 MHz.

(*a*) From Faraday's law the magnitude of the induced emf is

$$|\mathcal{E}| = \frac{d\phi_m}{dt} = \pi r^2 \frac{dB}{dt}$$

Then $\mathcal{E}_{rms} = \pi r^2 (dB/dt)_{rms}$. If $B = B_0 \sin(kx - \omega t)$, $dB/dt = -\omega B_0 \cos(kx - \omega t)$ and $(dB/dt)_{rms} = \omega B_{rms} = \omega E_{rms}/c = (2\pi f/c)E_{rms}$. The rms induced emf is then

$$\mathcal{E}_{rms} = \pi r^2 (2\pi f/c) E_{rms}$$

At $f = 600$ kHz, the emf is

$$\mathcal{E}_{rms} = \pi(0.1\text{ m})^2 2\pi(6 \times 10^5\text{ Hz})(3 \times 10^8\text{ m/s})^{-1}(0.15\text{ V/m})$$
$$= 5.92 \times 10^{-5}\text{ V}$$

(*b*) The induced emf is proportional to the frequency, so at 600 MHz it will be 1000 times greater than at 600 kHz. Then $\mathcal{E}_{rms} = (10^3)(5.92 \times 10^{-5}\text{ V}) = 0.0592$ V.

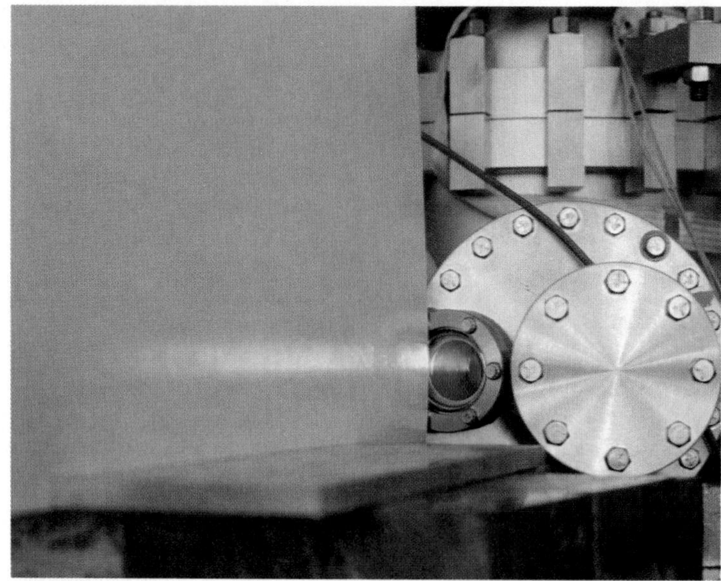

(*Left*) Magnetrons are used to generate the radar beams in tracking antennas and, as in the one shown above, to power microwave ovens. The central horizontal cylinder is a cathode that is heated and emits electrons. (The thin strips attached to the central cylinder are cooling fins.) Two disk-shaped magnets at either end provide an axial magnetic field. The emitted electrons are accelerated, creating oscillating electric fields that generate oscillating magnetic fields. The anode is configured so that electromagnetic oscillations at microwave frequencies can be sustained with little energy loss—that is, the magnetron acts as a resonant cavity for standing electromagnetic waves that have wavelengths of the order of a few centimeters. The microwaves exit from their standing pattern via the tube at the right, much as sound waves exit from a clarinet.

(*Right*) A beam of visible light emitted by electrons undergoing acceleration in a synchrotron. Electrons oscillating continuously in a typical radio antenna radiate sinusoidal electromagnetic fields. In a synchrotron, bunches of electrons move in circular paths at speeds near the speed of light and emit radiation that is largely compressed into brief pulses in the direction of motion of the electrons.

Questions

1. Which waves have greater frequencies, light waves or x rays?
2. Are the frequencies of ultraviolet radiation greater or less than those of infrared radiation?
3. What kind of waves have wavelengths of the order of a few meters?

Summary

1. Ampère's law can be generalized to apply to discontinuous currents if the conduction current I is replaced by $I + I_d$, where I_d is called Maxwell's displacement current:

$$I_d = \epsilon_0 \frac{d\phi_e}{dt}$$

2. The laws of electricity and magnetism are summarized by Maxwell's equations, which are

$$\oint_S E_n\, dA = \frac{1}{\epsilon_0} Q_{\text{inside}}$$ Gauss's law

$$\oint_S B_n\, dA = 0$$ Gauss's law for magnetism (isolated magnetic poles do not exist)

$$\oint_C \mathbf{E}\cdot d\boldsymbol{\ell} = -\frac{d}{dt}\int_S B_n\, dA$$ Faraday's law

$$\oint_C \mathbf{B}\cdot d\boldsymbol{\ell} = \mu_0 I + \mu_0\epsilon_0 \frac{d}{dt}\int_S E_n\, dA$$ Ampère's law modified

3. Maxwell's equations imply that the electric- and magnetic-field vectors in free space obey a wave equation of the form

$$\frac{\partial^2 \mathbf{E}}{\partial x^2} = \frac{1}{c^2}\frac{\partial^2 \mathbf{E}}{\partial t^2}$$

where

$$c = \frac{1}{\sqrt{\mu_0 \epsilon_0}}$$

is the wave speed. The fact that this speed equals the speed of light led Maxwell to surmise correctly that light is an electromagnetic wave.

4. In an electromagnetic wave, the electric and magnetic field vectors are perpendicular to each other and to the direction of propagation. Their magnitudes are related by

$$E = cB$$

5. Electromagnetic waves carry energy and momentum. The average energy density of an electromagnetic wave is

$$\eta_{av} = \frac{1}{2}\frac{E_0 B_0}{\mu_0 c} = \frac{E_{rms} B_{rms}}{\mu_0 c}$$

The intensity of an electromagnetic wave is given by

$$I = \eta_{av} c = \frac{1}{2}\frac{E_0 B_0}{\mu_0} = \frac{1}{2}\frac{E_0^2}{\mu_0 c} = \frac{1}{2}\frac{c B_0^2}{\mu_0} = |\mathbf{S}|_{av}$$

where $\mathbf{S}$, called the Poynting vector, describes the transport of electromagnetic energy:

$$\mathbf{S} = \frac{\mathbf{E} \times \mathbf{B}}{\mu_0}$$

6. An electromagnetic wave carries momentum that is equal to $1/c$ times the energy carried by the wave:

$$p = \frac{U}{c}$$

The intensity of an electromagnetic wave divided by c is the momentum carried by the wave per unit time per unit area, which is called the radiation pressure of the wave:

$$P_r = \frac{I}{c}$$

If the wave is incident normally on a surface and completely absorbed, it exerts a pressure equal to its radiation pressure. If it is incident normally and reflected, the pressure exerted is twice the radiation pressure.

7. Electromagnetic waves include light, radio waves, x rays, gamma rays, microwaves, and others. The various types of electromagnetic waves differ only in wavelength and frequency, which is related to the wavelength in the usual way:

$$f = \frac{c}{\lambda}$$

8. Electromagnetic waves are produced when electric charges accelerate. Oscillating charges in an electric-dipole antenna radiate electromagnetic waves with an intensity that is maximum in directions perpendicular to the antenna and zero along the axis of the antenna. Perpendicular to the antenna and far away from it, the electric field of the electromagnetic wave is parallel to the antenna.

James Clerk Maxwell (1831–1879)

C. W. F. Everitt
Stanford University

One day in 1877 a young Scottish undergraduate named Donald MacAlister, afterwards a distinguished physician and academic statesman, wrote home from Cambridge University that he had just had dinner with a professor who was "one of the best of our men, and a thorough old Scotch laird in ways and speech." This description of James Clerk Maxwell was accurate. He was wealthy, an expert swimmer and horseman, proprietor of an estate of 2,000 acres in the southwest of Scotland—few would suspect that this man who seemed to MacAlister charmingly old-fashioned in 1877 was also a scientist whose writings remain astonishingly up to date in the 1990s; that he was the greatest mathematical physicist since Newton; that he had created the electromagnetic theory of light and predicted the existence of radio waves; that he had written the first significant paper on control theory; that he was just then writing a profound article on statistical mechanics, a science he and Ludwig Boltzmann jointly invented; that he had performed with his wife's aid a brilliant series of experiments on color vision and had taken the first color photograph; and that in the remaining 2 years until his death due to cancer in 1879, at the age of 48, he would lay the foundations of another new subject that was to reach fruition in the twentieth century, rarefied gas dynamics.

C.W.F. (Francis) Everitt obtained his doctorate in 1959 from Imperial College, London, in the then new field of paleomagnetic studies of plate tectonics. He decided that, as much as he liked geology, his true métier was physics. So he moved in 1960 to the University of Pennsylvania where he investigated "third sound," a peculiar kind of surface wave propagating in thin films of superfluid helium. He then transferred to Stanford University and with others initiated research on a long-range NASA program (Gravity Probe B) to test Einstein's general theory of relativity using precise gyroscopes in Earth orbit. Everitt's interest in history goes back to his high school days. He has written on the history of precise measurement; on scientific creativity; on the relationship between theory and experiment in physics; on spectroscopy; on science, history and religion; on the dynamics of "big physics"; and on several physicists, including three books about Maxwell.

Maxwell's undergraduate career was unusually protracted. He spent 3 years at Edinburgh University and another $3\frac{1}{4}$ at Cambridge. His student friendships were more with classical scholars than scientists. Unlike Einstein, he enjoyed student life and was fortunate to gain the attention of some outstanding teachers. At Edinburgh he was influenced by two powerful and sharply contrasted men, James David Forbes and Sir William Hamilton, the metaphysician. Forbes was an experimentalist, who had invented the seismometer and done important work on the polarization of infrared radiation and the study of the motion of glaciers. He gave Maxwell the run of his laboratory and with his help began the experiments on color vision that led eventually to Maxwell's own work on the subject. Hamilton, who had a genius for inspiring youth, imparted to Maxwell the ranging philosophic vision that can be seen in the many interesting metaphysical asides in his papers.

In 1850 Maxwell went to Cambridge. By then his mathematical bent was clear and, like many another clever undergraduate before and since, Maxwell worked hard while pretending not to. His tutor was William Hopkins, the founder of modern geophysics and arguably the greatest teacher Cambridge has ever produced. Others who influenced him were G. G. Stokes, the mathematical physicist who held the chair Newton had occupied, and William Whewell.

Maxwell's electromagnetic theory of light was rooted in the work of two men, Michael Faraday and William Thomson. Faraday's invention of the electric motor and his researches on electromagnetic induction, electrochemistry, dielectric and diamagnetic action, and magneto-optical rotation made him in Maxwell's words "the nucleus of everything electric since 1830." His contributions to theory lay in his progressively advancing ideas about lines of electric and magnetic force, in particular the geometrical relations governing electromagnetic phenomena and the idea that magnetic forces might be accounted for not by direct attractions and repulsions between elements of current but by attributing

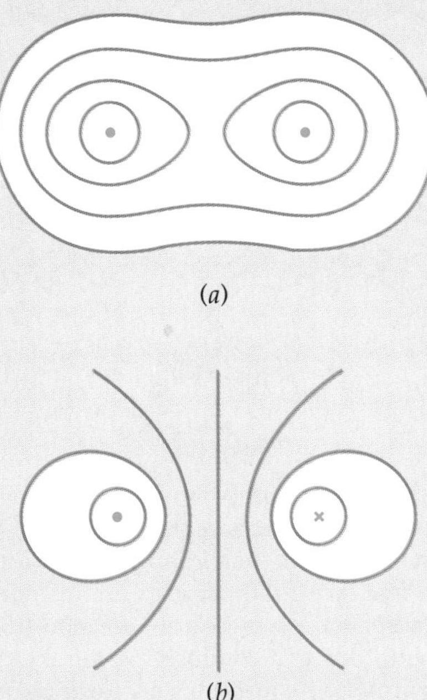

Figure 1 Faraday's explanation of forces between current-carrying wires. The two diagrams show the lines of force observed when currents are flowing in parallel wires. Faraday assumed that the lines of force tend to shorten and repel each other sideways. (*a*) For wires with currents flowing in the same direction, the lines of force pull the two wires together. (*b*) For wires with currents flowing in opposite directions, the lines of force push the wires apart.

to lines of force the property of shortening themselves and repelling each other sideways (Figure 1). Thomson's role was to relate lines of force to existing theories in electrostatics and magnetostatics, to invent a number of highly ingenious analytical techniques for solving electrical problems, and to emphasize the cardinal importance of energy principles in electromagnetism. Maxwell then introduced a series of new concepts: the *electrotonic function* (vector potential), the energy density of the field, the displacement current (see Equation 29-4), and the significance of the operation *curl* in the field equations; he organized the subject into a coherent structure and made in 1861 the momentous discovery of the equivalence between light and electromagnetic waves.

The first part of Maxwell's paper "On Faraday's Lines of Force" (1855–1856) developed an analogy, due in essence to Thomson, between lines of electric and magnetic force and streamlines in a moving incompressible fluid. Maxwell applied this to interpret many of Faraday's observations, prefacing his paper with a luminous discussion of the significance of analogies in physics. Next, still building on Faraday and Thomson, Maxwell extended the discussion to electromagnetism. He formulated a group of equations summarizing the relations of the electric and magnetic fields to the charges and currents producing them—the beginnings of what we now call Maxwell's equations (see Section 29-2). They described the phenomena with elegant precision from a point of view completely different from the then-popular action-at-a-distance theories of André-Marie Ampère and Wilhelm Weber. The central theorem in all this work was one which, following Maxwell, we now call *Stokes' theorem*, which had been published as a question by Stokes in one of the examinations taken by Maxwell in January 1854, while he was still an undergraduate.

After such a brilliant start one might have expected a rush of papers following up the new ideas. But other physicists ignored them, and Maxwell had the habit of investigating different subjects in turn, often with long intervals between successive papers in the same field. Six years elapsed before the appearance of his next paper, "On Physical Lines of Force," published in four parts in 1861–1862. During the interval Maxwell made brilliant contributions to three distinct subjects before returning to electromagnetism: color vision, the theory of Saturn's rings, and the kinetic theory of gases. He left Cambridge, became a professor at Marischal College in Aberdeen, married the daughter of the Principal of the College, and then found himself in the odd position of being forced to retire at age 29 with a life pension after his chair had been abolished when the two universities in Aberdeen were united by Act of Parliament. Fortunately, a suitable post (the chair at King's College in London) had just fallen vacant, and so he went there.

"On Physical Lines of Force" contained Maxwell's extraordinary molecular-vortex model of the electromagnetic field. In order to account for the pattern of stresses associated with lines of force by Faraday, Maxwell investigated the properties of a medium occupying all space in which tiny molecular vortices rotate with their axes parallel to the lines of force. The closer together the lines are, the faster the rotation of the vortices. In a medium of this kind the lines of force do tend to shorten themselves and repel each other sideways, yielding the right forces between currents and magnets. The question is: what makes the vortices rotate? Here Maxwell put forward an idea as ingenious as it was weird. He postulated that an electric current consists in the motion of tiny particles that mesh like gear wheels with the vortices, and that the medium is filled with similar particles between the vortices. Figure 2 gives the picture. Maxwell remarks:

> I do not bring [this hypothesis] forward as a mode of connexion existing in nature . . . [but] I venture to say that anyone who understands [its] provisional and temporary character . . . will find himself helped rather than hindered by it in his search for the true interpretation of [electromagnetic] phenomena.

The question then was how to fit electrostatic phenomena into the model. Maxwell made the medium an elastic one. Thus magnetic forces were accounted for by rotations in the medium, and electric forces by its elastic distortion. Any elastic medium will transmit waves. In Maxwell's medium the velocity of the waves turned out to be related to the ratio of electric to magnetic forces. Putting in numbers from an experiment of 1856 by G. Kohlrausch and W. Weber, Maxwell found to his astonishment that the propagation velocity was equal to the velocity of light. With excitement he wrote "we can scarcely avoid the inference that *light consists in the transverse undulation of the same medium which is the cause of electric and magnetic phenomena*" (see Section 29-3).

Having made the great discovery, Maxwell promptly jettisoned his model. Instead of attempting a more refined mechanical explanation of the phenomena, he formulated a system of electromagnetic equations from which he deduced that waves of electric and magnetic force would propagate through space with the velocity of light. That is why his is called an *electromagnetic* theory of light, in contrast to the theories of the mechanical ether that preceded it. The theory appeared in two papers of 1865 and 1868, and in its most general form in the great *Treatise on Electricity and Magnetism,* published in 1873. This was a work of such scope that Robert Andrews Millikan, author of the famous oil-drop experiment to measure the charge on the electron, ranked it with Newton's *Principia* in considering them the two most influential books in the history of physics, "the one creating our modern mechanical world and the other our modern electrical world."

Equally profound were Maxwell's contributions to statistical and molecular physics. They began with a paper in 1859 on the kinetic theory of gases, in which Maxwell introduced the velocity distribution function and enunciated the *equipartition theorem* (Section 16-7), which in its original form stated that the average translational and rotational energies of large numbers of colliding molecules, whether of the same or different species, are equal. One surprising conclusion, afterwards confirmed experimentally by Maxwell and his wife, was that the viscosity of a gas should over a wide range be independent of its pressure. Another result was Maxwell's estimate of the mean free path of a gas molecule, which Loschmidt in 1865 applied to make the first serious estimates of the diameters of molecules. Later, Maxwell developed the general theory of transport phenomena, from which the Boltzmann equation is derived; invented the concept of ensemble averaging; created rarefied-gas dynamics; and conceived that "very small BUT lively being" the Maxwell demon.

The demon, so named by Kelvin, is one of the earliest examples in physics of a "thought experiment." Maxwell imagined two chambers of gas, A and B, separated by a wall in which there was a trap door guarded by a tiny being with eyesight so acute that it could discern the motion of individual molecules. By opening the door when a fast molecule approached from chamber A or as a slow molecule approached from chamber B, the demon could redistribute the velocities to make B hotter than A without doing any work and thereby defeat the sec-

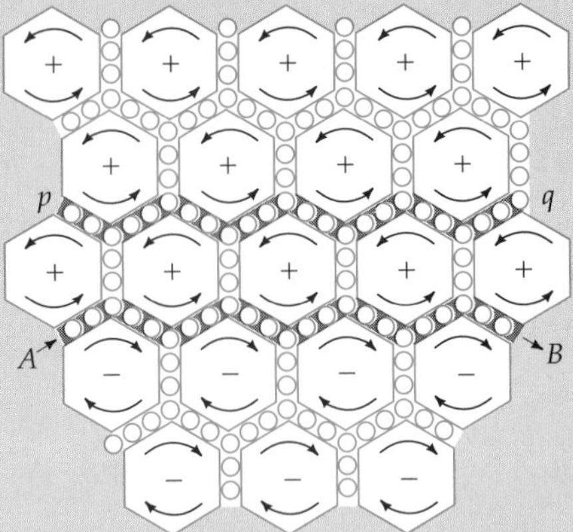

Figure 2 Maxwell's vortex model of the magnetic field. The rotating vortices represent lines of magnetic force. They mesh with small particles that act like gear wheels. In free space the particles are restrained from moving, except for a small elastic reaction (the displacement current), but in a conducting wire they are free to move. Their motion constitutes an electric current, which in turn sets the vortices in rotation, creating the magnetic field around the wire. *A* and *B* represent current through a wire, and *p* and *q* represent an induced current in an adjacent wire. (Redrawn from The Scientific Papers of James Clerk Maxwell, Vol. I, fig. 2 after p. 488.)

James Clerk Maxwell (1831–1878) with his wife, Katherine Mary, and dog.

ond law of thermodynamics. Maxwell's point was to demonstrate that the second law of thermodynamics is inherently a statistical law and not a dynamical one.

The work by Maxwell and Boltzmann on statistical mechanics had profound implications for modern physics. Brilliant as its successes were, the failures were—as Maxwell saw—in some ways even more striking. The equipartition theorem gave an answer for the ratio of the specific heats of gases that disagreed with experiment, while some of Boltzmann's theorems "proved too much" because they would apply to the properties of solids and liquids as well as gases. These questions remained shrouded in mystery until the emergence in 1900 of Planck's quantum hypothesis (see Section 35-1). Writing about them in 1877, Maxwell confessed his bewilderment and stated that nothing remained but to adopt the attitude of "thoroughly conscious ignorance that is the prelude to every real advance in knowledge."

Maxwell was an unusually sensitive man, with strong religious feeling and a fascinating and astonishing sense of humor. Many of his letters reveal a delightfully sly irony. He also had some talent for writing poetry, usually light, but occasionally touching a deeper note. The last stanza of one poem to his wife, written in 1867, was

All powers of mind, all force of will
  May lie in dust when we are dead,
But love is ours, and shall be still
  When earth and seas are fled.

Suggestions for Further Reading

Campbell, L., and W. Garnett: *The Life of James Clerk Maxwell*, Johnson Reprint Corporation, Harcourt, Brace & Jovanovich, New York, 1970 (reprint of Oxford 1882 edition).

Everitt, C. W. F.: *James Clerk Maxwell: Physicist and Natural Philosopher*, Scribner, New York, 1975.

These books offer more details on Maxwell's fascinating life and work than found in this chapter's essay.

Mulligan, Joseph F.: "Heinrich Hertz and the Development of Physics," *Physics Today*, March 1989, p. 50.

This article describes Hertz's life and work, including not only his experiments on electromagnetism, but also his discovery of the photoelectric effect and his work with cathode rays.

Shamos, Morris H: "The Electromagnetic Field—James Clerk Maxwell," in *Great Experiments in Physics*, Henry Holt and Company, New York, 1959. Reprinted by Dover, 1987.

In annotated excerpts from his paper of 1865, "A Dynamical Theory of the Electromagnetic Field," Maxwell discusses his conception of the electromagnetic field and introduces his equations. The excerpts are introduced by a brief biography.

Shamos, Morris H.: "Electromagnetic Waves—Heinrich Hertz," in *Great Experiments in Physics*, Henry Holt and Company, New York, 1959. Reprinted by Dover, 1987.

Hertz's account of the set of experiments in which he showed the existence of electromagnetic waves of much greater wavelength than light waves. He was able to demonstrate in these experiments on reflection, refraction, and polarization that these waves behave the same as light. The chapter also includes a brief biography.

Review

A. Objectives: After studying this chapter, you should:

1. Be able to write Maxwell's equations and discuss the experimental basis of each.

2. Be able to state the expression for the speed of an electromagnetic wave in terms of the fundamental constants μ_0 and ϵ_0.

3. Be able to give the expression for the Poynting vector and discuss its significance.

4. Be able to state the relationships between the Poynting vector, the intensity of an electromagnetic wave, and radiation pressure.

5. Be able to calculate the radiation pressure and the maximum values of E and B from the intensity of an electromagnetic wave.

B. Define, explain, or otherwise identify:

Maxwell's displacement current
Maxwell's equations
Wave equation
Plane wave
Linearly polarized wave
Circularly polarized wave
Poynting vector
Radiation pressure
Electromagnetic spectrum
Visible light
Ultraviolet rays
Infrared waves
Electric-dipole antenna
Electric-dipole radiation

C. True or false: If the statement is true, explain why it is true. If it is false, give a counterexample.

1. Maxwell's equations apply only to fields that are constant in time.

2. The wave equation can be derived from Maxwell's equations.

3. Electromagnetic waves are transverse waves.

4. In an electromagnetic wave, the electric and magnetic fields are in phase.

5. In an electromagnetic wave, the electric- and magnetic-field vectors **E** and **B** are equal in magnitude.

6. In an electromagnetic wave, the electric and magnetic energy densities are equal.

Problems

Level I

29-1 Maxwell's Displacement Current

1. A parallel-plate capacitor in air has circular plates of radius 2.3 cm separated by 1.1 mm. Charge is flowing onto the upper plate and off the lower plate at a rate of 5 A. (*a*) Find the time rate of change of the electric field between the plates. (*b*) Compute the displacement current between the plates and show that it equals 5 A.

2. In a region of space, the electric field varies according to

$$E = (0.05\ \text{N/C})\sin 2000t$$

where t is in seconds. Find the maximum displacement-current through a 1 m^2 area perpendicular to **E**.

29-2 Maxwell's Equations

There are no problems for this section.

29-3 The Wave Equation for Electromagnetic Waves

3. Show by direct substitution that the wave function

$$E_y = E_0 \sin(kx - \omega t) = E_0 \sin k(x - ct)$$

where $c = \omega/k$, satisfies Equation 29-10.

4. Use the known values of μ_0 and ϵ_0 in SI units to compute

$$c = \frac{1}{\sqrt{\mu_0 \epsilon_0}}$$

and show that it is approximately 3×10^8 m/s.

29-4 Energy and Momentum in an Electromagnetic Wave

5. An electromagnetic wave has an intensity of 100 W/m^2. Find (*a*) the radiation pressure P_r, (*b*) E_{rms}, and (*c*) B_{rms}.

6. The amplitude of an electromagnetic wave is $E_0 =$ 400 V/m. Find (*a*) E_{rms}, (*b*) B_{rms}, (*c*) the intensity I, and (*d*) the radiation pressure P_r.

7. (*a*) Show that if E is in volts per meter and B is in teslas, the units of the Poynting vector $\mathbf{S} = (\mathbf{E} \times \mathbf{B})/\mu_0$ are watts per square meter. (*b*) Show that if the intensity I is in watts per square meter, the units of radiation pressure $P_r = I/c$ are newtons per square meter.

8. (*a*) An electromagnetic wave of intensity 200 W/m^2 is incident normally on a rectangular black card with sides of 20 and 30 cm that absorbs all the radiation. Find the force exerted on the card by the radiation. (*b*) Find the force exerted by the same wave if the card reflects all the radiation incident on it.

9. Find the force exerted by the electromagnetic wave on the reflecting card in part (*b*) of Problem 8 if the radiation is incident at an angle of 30° to the normal.

10. The rms value of the electric field in an electromagnetic wave is $E_{rms} = 400$ V/m. (*a*) Find B_{rms}, (*b*) the average energy density, and (*c*) the intensity.

11. Show that the units of $E = cB$ are consistent; that is, show that when B is in teslas and c is in meters per second, the units of cB are volts per meter or newtons per coulomb.

12. The root-mean-square value of the magnitude of the magnetic field in an electromagnetic wave is $B_{rms} = 0.245\ \mu T$. Find (*a*) E_{rms}, (*b*) the average energy density, and (*c*) the intensity.

29-5 The Electromagnetic Spectrum

13. Find the wavelength for (*a*) a typical AM radio wave with a frequency of 1000 kHz and (*b*) a typical FM radio wave of 100 MHz.

14. What is the frequency of a 3-cm microwave?

15. What is the frequency of an x ray with a wavelength of 0.1 nm?

Level II

16. For Problem 1, show that at a distance r from the axis of the plates the magnetic field between the plates is given by $B = (1.89 \times 10^{-3}\ T/m)r$ if r is less than the radius of the plates.

17. (*a*) Show that for a parallel-plate capacitor the displacement current is given by $I_d = C\, dV/dt$, where C is the capacitance and V the voltage across the capacitor. (*b*) A parallel plate capacitor $C = 5$ nF is connected to an emf $\mathcal{E} = \mathcal{E}_0 \cos \omega t$, where $\mathcal{E}_0 = 3$ V and $\omega = 500\ \pi$. Find the displacement current between the plates as a function of time. Neglect any resistance in the circuit.

18. Current of 10 A flows into a capacitor having plates with areas of 0.5 m^2. (*a*) What is the displacement current between the plates? (*b*) What is dE/dt between the plates for this current? (*c*) What is the line integral of $\mathbf{B} \cdot d\boldsymbol{\ell}$ around a circle of radius 10 cm that lies within and parallel to the plates?

19. The intensity of radiation from an electric dipole is proportional to $(\sin^2 \theta)/r^2$, where θ is the angle between the electric dipole moment and the position vector $\mathbf{r}$. A radiating electric dipole lies along the z axis (its dipole moment is in the z direction). Let I be the intensity of the radiation at a distance $r = 10$ m and at angle $\theta = 90°$. Find the intensity (in terms of I) at (*a*) $r = 30$ m, $\theta = 90°$; (*b*) $r = 10$ m, $\theta = 45°$; and (*c*) $r = 20$ m, $\theta = 30°$.

20. (*a*) For the situation described in Problem 19, at what angle is the intensity at $r = 5$ m equal to I? (*b*) At what distance is the intensity equal to I at $\theta = 45°$?

21. A typical AM radio station radiates an isotropic sinusoidal wave with an average power of 50 kW. What are the amplitudes of E_{max} and B_{max} at a distance of (*a*) 500 m, (*b*) 5 km, and (*c*) 50 km?

22. The intensity of sunlight striking the earth's upper atmosphere is called the solar constant and is 1.35 kW/m^2. (*a*) Find E_{rms} and B_{rms} due to the sun at the upper atmosphere of the earth. (*b*) Find the average power output of the sun. (*c*) Find the intensity and the radiation pressure at the surface of the sun.

23. At the surface of the earth, there is an approximate average solar flux of 0.75 kW/m^2. A family wishes to construct a solar energy conversion system to power their home. If the conversion system is 30 percent efficient and the family needs a maximum of 25 kW, what effective surface area is needed for perfectly absorbing collectors?

24. Instead of sending power by a 750-kV, 1000-A transmission line, one desires to beam this energy via an electromagnetic wave. The beam has a uniform intensity within a cross-sectional area of 50 m^2. What are the rms values of the electric and the magnetic fields?

25. A demonstration laser has an average output power of 0.9 mW and a beam diameter of 1.2 mm. What is the force exerted by the laser beam on (*a*) a 100 percent absorbing black surface? (*b*) a 100 percent reflecting surface?

26. A laser beam has a diameter of 1.0 mm and average power of 1.5 mW. Find (*a*) the intensity of the beam, (*b*) E_{rms}, (*c*) B_{rms}, and (*d*) the radiation pressure.

27. A laser pulse has an energy of 20 J and a beam radius of 2 mm. The pulse duration is 10 ns and the energy density is constant within the pulse. (*a*) What is the spatial length of the pulse? (*b*) What is the energy density within the pulse? (*c*) Find the electric and magnetic amplitudes of the laser pulse.

28. An electromagnetic wave has a frequency of 100 MHz and is traveling in a vacuum. The magnetic field is given by

$$\mathbf{B}(z, t) = (10^{-8}\ \mathrm{T}) \cos (kz - \omega t)\ \mathbf{i}$$

(*a*) Find the frequency, wavelength, and the direction of propagation of this wave. (*b*) Find the electric vector $\mathbf{E}(z, t)$. (*c*) Give Poynting's vector, and find the intensity of this wave.

29. The electric field of an electromagnetic wave oscillates in the y direction and the Poynting vector is given by

$$\mathbf{S}(x, t) = (100\ \mathrm{W/m^2}) \cos^2 [10x - (3 \times 10^9)t]\ \mathbf{i}$$

where x is in meters and t is in seconds. (*a*) What is the direction of propagation of the wave? (*b*) Find the wavelength and the frequency. (*c*) Find the electric and magnetic fields.

30. A pulsed laser fires a 1000-MW pulse of 200 ns duration at a small object of mass 10 mg suspended by a fine fiber 4 cm long. If the radiation is completely absorbed without other effects, what is the maximum angle of deflection of this pendulum?

31. A very long wire of radius 4 mm is heated to 1000 K. The surface of the wire is an ideal blackbody radiator. (*a*) What is the total power radiated per unit length? Find (*b*) the Poynting vector S, (*c*) E_{rms}, and (*d*) B_{rms} at a distance of 25 cm from the wire.

32. A blackbody sphere of radius R is a distance 2×10^{11} m from the sun. The effective area of the body for absorption of energy from the sun is πR^2, but the area for radiation by the object is $4\pi R^2$. The power output of the sun is 3.83×10^{26} W. What is the temperature of the sphere?

33. (*a*) If the earth were an ideal blackbody with infinite thermal conductivity and no atmosphere, what would be the temperature of the earth? (*b*) If 40 percent of the incident sun's energy were reflected, what then would be the temperature of the earth? (See Problem 32.)

34. Two harmonic waves of angular frequency ω_1 and ω_2 have their electric fields given by $\mathbf{E}_1 = E_{10} \cos (k_1x - \omega_1t)\,\mathbf{j}$ and $\mathbf{E}_2 = E_{20} \cos (k_2x - \omega_2t + \delta)\,\mathbf{j}$. Find (*a*) the instantaneous Poynting vector for the resultant wave motion and (*b*) the time-average Poynting vector. If $\mathbf{E}_2 = E_{20} \cos (k_2x + \omega_2t + \delta)\,\mathbf{j}$ find (*c*) the instantaneous Poynting vector for the resultant wave motion and (*d*) the time-average Poynting vector.

35. A 10- by 15-cm card has a mass of 2 g and is perfectly reflecting. The card hangs in a vertical plane and is free to rotate about a horizontal axis through one edge. The card is illuminated uniformly by an intense light that causes the card to make an angle of 1° with the vertical. Find the intensity of the light.

36. A valuable 0.08-kg diamond gem and a 105-kg spaceperson are separated by 95 m. Both objects are initially at rest. The spaceperson has a 1.5-kW laser that can be used as a photon rocket motor to propel the person towards the diamond. How long would it take the spaceperson to move 95 m using the laser rocket propulsion?

37. A circular loop of wire can be used to detect electromagnetic waves. Suppose a 100-MHz FM station radiates 50 kW uniformly in all directions. What is the maximum rms voltage induced in a loop of radius 30 cm at a distance of 10^5 m from the station?

38. Suppose one has an excellent radio capable of detecting a signal as weak as 10^{-14} W/m^2. This radio has a 2000-turn coil antenna having a radius of 1 cm wound on an iron core of permeability 200. The ratio frequency is 140 KHz. (*a*) What is the amplitude of the magnetic field in this wave? (*b*) What is the emf induced in the antenna? (*c*) What would be the emf induced in a 2-m wire oriented in the direction of the electric field?

39. The electric field from a radio station some distance from the transmitter is given by

$$E = (10^{-4} \text{ N/C}) \cos 10^6 t$$

where t is in seconds. (*a*) What voltage is picked up on a 50-cm wire oriented along the electric field direction? (*b*) What voltage can be induced in a loop of radius 20 cm?

40. A parallel-plate capacitor with circular plates is given a charge Q_0. Between the plates is a leaky dielectric having a dielectric constant of κ and a resistivity ρ. (*a*) Find the conduction current between the plates as a function of time. (*b*) Find the displacement current between the plates as a function of time. What is the total (conduction plus displacement) current? (*c*) Find the magnetic field as a function of time between the plates produced by the leakage discharge current. (*d*) Find the magnetic field as a function of time between the plates produced by the displacement current. (*e*) What is the total magnetic field between the plates during discharge of the capacitor?

41. The leaky capacitor of Problem 40 is charged such that the voltage across the capacitor is given by $V(t) = 10^{-2}t$. (*a*) Find the conduction current as a function of time. (*b*) Find the displacement current. (*c*) Find the time for which the displacement current is equal to the conduction current.

42. The space between the plates of a capacitor is filled with a material of resistivity ρ of 10^4 Ω·m and dielectric constant $\kappa = 2.5$. The parallel plates are circular with a radius of 20 cm and are separated by 1 mm. The voltage across the plates is given by $V_0 \cos \omega t$, with $V_0 = 40$ V and $\omega = 120\pi$ rad/s. (*a*) What is the displacement current? (*b*) What is the conduction current between the plates? (*c*) At what angular frequency is the total current 45° out of phase with the applied voltage?

43. A circular capacitor of area A has a wire of resistance R connecting the centers of the two plates. A voltage $V_0 \sin \omega t$ is applied between the plates. (*a*) What is the current drawn by this capacitor? (*b*) What is the magnetic field as a function of radial distance r from the centerline within the plates of this capacitor? (*c*) What is the phase angle between current and applied voltage?

44. Show that the normal component of the magnetic field $\mathbf{B}$ is continuous across a surface. Do this by applying Gauss's law for $\mathbf{B}$ ($\oint B_n\, dA = 0$) to a pillbox gaussian surface that has a face on each side of the surface.

Level III

The following two problems do not concern waves but illustrate the use of the Poynting vector to describe the flow of electromagnetic energy.

45. A long, cylindrical conductor of length L, radius a, and resitivity ρ carries a steady current I that is uniformly distributed over its cross-sectional area. (*a*) Use Ohm's law to relate the electric field E in the conductor to I, ρ, and a. (*b*) Find the magnetic field B just outside the conductor. (*c*) Use the results for parts (*a*) and (*b*) to compute the Poynting vector $\mathbf{S} = \mathbf{E} \times \mathbf{B}/\mu_0$ at $r = a$ (the edge of the conductor). In what direction is $\mathbf{S}$? (*d*) Find the flux $\oint S_n\, dA$ through the surface of the conductor into the conductor, and show that the rate of energy flow into the conductor equals I^2R, where R is the resistance. (Here S_n is the *inward* component of $\mathbf{S}$ perpendicular to the surface of the conductor.)

46. A long solenoid of n turns per unit length has a current that slowly increases with time. The solenoid has radius R, and the current in the windings has the form $I(t) = at$. (*a*) Find the induced electric field at a distance $r < R$ from the solenoid axis. (*b*) Find the magnitude and direction of the Poynting vector $\mathbf{S}$ at the cylindrical surface $r = R$ just inside the solenoid windings. (*c*) Calculate the flux $\oint S_n\, dA$ into the solenoid, and show that it equals the rate of increase of the magnetic energy inside the solenoid. (Here S_n is the *inward* component of $\mathbf{S}$ perpendicular to the surface of the solenoid.)

47. Small particles might be blown out of solar systems by the radiation pressure of sunlight. Assume the particles are spherical with a radius r and a density of 1 g/cm^3 and absorb all the radiation in a cross-sectional area of πr^2. They are a distance R from the sun, which has a power output of 3.83×10^{26} W. What is the radius r for which the radiation force of repulsion just balances the gravitational force of attraction to the sun?

48. In this problem, you are to show that the generalized form of Ampère's law (Equation 29-4) and the Biot–Savart law give the same result in a situation in which they both can be used. Figure 29-12 shows two charges $+Q$ and $-Q$

Figure 29-12 Problem 48.

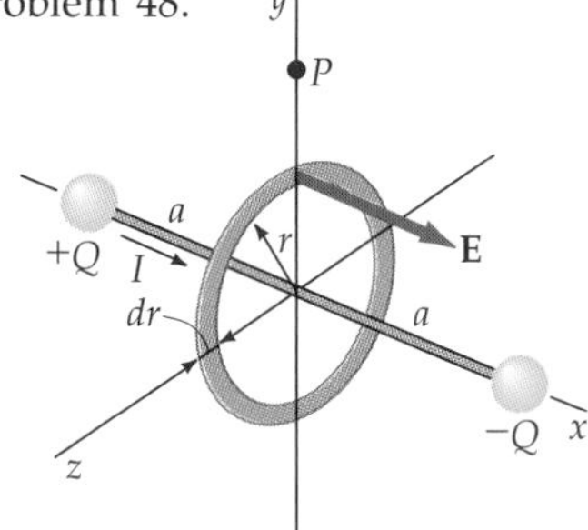

on the x axis at $x = -a$ and $x = +a$, with a current $I = -dQ/dt$ along the line between them. Point P is on the y axis at $y = R$. (*a*) Use Equation 25-12, obtained from the Biot–Savart law, to show that the magnitude of B at point P is

$$B = \frac{\mu_0 Ia}{2\pi R}\frac{1}{\sqrt{R^2 + a^2}}$$

(*b*) Consider a circular strip of radius r and width dr in the yz plane with its center at the origin. Show that the flux of the electric field through this strip is

$$E_x\, dA = (Q/\epsilon_0)a(r^2 + a^2)^{-3/2}\, r\, dr$$

(*c*) Use your result for part (*b*) to find the total flux ϕ_e through a circular area of radius R. Show that

$$\epsilon_0\phi_e = Q(1 - a/\sqrt{a^2 + R^2})$$

(*d*) Find the displacement current I_d, and show that

$$I + I_d = I\frac{a}{\sqrt{a^2 + R^2}}$$

(*e*) Then show that Equation 29-4 gives the same result for B as that found in part (*a*).

49. (*a*) Using arguments similar to those given in the text, show that for a plane wave, in which **E** and **B** are independent of y and z,

$$\frac{\partial E_z}{\partial x} = \frac{\partial B_y}{\partial t}$$

and

$$\frac{\partial B_y}{\partial x} = \mu_0\epsilon_0\frac{\partial E_z}{\partial t}$$

(*b*) Show that E_z and B_y also satisfy the wave equation.

50. Some science fiction writers have used solar sails to propel interstellar spaceships. Imagine a giant sail erected on a spacecraft subjected to the solar radiation pressure. (*a*) Show that the spacecraft's acceleration is given by

$$a = \frac{P_s A}{4\pi r^2 cm}$$

where P_s is the power output of the sun and is equal to 3.8×10^{26} W, A is the surface area of the sail, m is the total mass of the spacecraft, r is the distance from the sun, and c is the speed of light. (*b*) Show that the velocity of the spacecraft at a distance r from the sun is found from

$$v^2 = v_0^2 + \left(\frac{P_s A}{2\pi mc}\right)\left(\frac{1}{r_0} - \frac{1}{r}\right)$$

where v_0 is the initial velocity at r_0. (*c*) Compare the relative accelerations due to the radiation pressure and the gravitational force. Use reasonable values for A and m. Will such a system work?

51. Novelty stores sell a device called a radiometer shown in Figure 29-13, in which a balanced vane spins rapidly. A card is mounted on each arm of the vane. One side of each card is white and the other is black. Assume that the mass of each card is 2 g, that the light-collecting area for each card is 1 cm^2, and that each arm of the vane has a length of 2 cm. (*a*) If a 100-W light bulb produces 50 W of electromagnetic energy and the bulb is 50 cm from the radiometer, find the maximum angular acceleration of the vane. (Estimate the moment of inertia of the vane by assuming all the mass of each card is at the end of the arms.) (*b*) How long will it take for the vane to accelerate to 10 rev/min if it starts from rest and is subject to the maximum angular acceleration at all times? (*c*) Can the radiation pressure account for the rapid motion of the radiometer? (The radiometer actually spins in the opposite direction from what would be expected if the force were due to radiation pressure. The reason is that the air near the black side is warmer than that near the white side, so the air molecules hitting the black side have greater energy than those hitting the white side.)

Figure 29-13 A radiometer. See Problem 51.

Part 5 Optics

Stress patterns around a crack in a sheet of transparent plastic are revealed by polarized light. The stress is perpendicular to the crack. Two smaller cracks have propagated from the lower end of the large one, creating additional patterns of stress. Smaller circular flaws surround the upper end of the large crack.

Chapter 30

Light

A bright primary rainbow and the fainter secondary rainbow in a sheet of rain over Lake Michigan. The primary bow is formed by light rays that enter spherical drops of water and are reflected once internally before leaving the drops. The secondary bow results from rays that experience two internal reflections before leaving the drops.

Light has intrigued mankind for centuries. Early theories considered light to be something that emanated *from* the eye. Later it was realized that light must come from the objects seen and enter the eye causing the sensation of vision. The question of whether light consists of a beam of particles or is some kind of wave motion is one of the most interesting in the history of science. The most influential proponent of the **particle theory of light** was Newton. Using it, he was able to explain the laws of reflection and refraction. However, his derivation of the law of refraction depends on the assumption that light travels faster in water or glass than in air, an assumption later shown to be false. The chief proponents of the **wave theory of light** were Christian Huygens and Robert Hooke. Using his own theory of wave propagation, Huygens was able to explain reflection and refraction by assuming that light travels more slowly in glass or water than in air. Newton understood the virtues of the wave theory of light, particularly as it explained the colors formed by thin films, which he studied extensively. However, he rejected the wave theory because of the apparent rectilinear propagation of light. Diffraction, the bending of a light beam around obstacles, had not been observed in his time. Because of Newton's great reputation and authority, his reluctant rejection of the wave theory of light was strictly adhered to by his followers. Even after evidence of diffraction was available,

Newton's followers sought to explain it as the scattering of light particles from the edges of slits.

Newton's particle theory of light was accepted for more than a century. Then, in 1801, Thomas Young revived the wave theory of light. He was one of the first to introduce the idea of interference as a wave phenomenon that occurs in both light and sound. His observation of interference with light was a clear demonstration of the wave nature of light. However, Young's work went unnoticed by the scientific community for more than a decade. Perhaps the greatest advance in the general acceptance of the wave theory of light was due to the French physicist Augustin Fresnel (1788–1827), who performed extensive experiments on interference and diffraction and put the wave theory on a mathematical basis. He showed, for example, that the observed rectilinear propagation of light is a result of the very short wavelengths of visible light. In 1850, Jean Foucault measured the speed of light in water and showed that it is less than that in air, thus ruling out Newton's particle theory. In 1860, James Clerk Maxwell published his mathematical theory of electromagnetism, which predicted the existence of electromagnetic waves that propagate with a speed calculated from the laws of electricity and magnetism to be 3×10^8 m/s, the same as the speed of light. Maxwell's theory was confirmed in 1887 by Hertz, who used a tuned electrical circuit to generate the waves and another, similar circuit to detect them. In the latter half of the nineteenth century, Kirchhoff and others applied Maxwell's equations to explain the interference and diffraction of light and other electromagnetic waves and put Huygens' empirical methods of construction on a firm mathematical basis.

Although the wave theory is generally correct in describing the propagation of light (and other electromagnetic waves), it fails to account for all properties of light, particularly the interaction of light with matter. In his famous experiment of 1887 that confirmed Maxwell's wave theory, Hertz also discovered the photoelectric effect, which will be discussed in detail in Chapter 35. This effect can be explained only by a particle model of light, as Einstein showed only a few years later. Thus, a particle model of light was reintroduced. Light particles are called **photons.** The energy of a photon E is related to the frequency f of the associated light wave by the famous Einstein relation $E = hf$, where h is a constant called *Planck's constant*. Complete understanding of the dual nature of light did not come until the 1920s, when experiments by C. J. Davisson and L. Germer and by G. P. Thompson showed that electrons (and other "particles") also have a dual nature and exhibit the wave properties of interference and diffraction in addition to their well-known particle properties. (We will discuss the dual nature of light and electrons in Chapter 35.)

The development of the quantum theory of atoms and molecules by Rutherford, Bohr, Schrödinger, and others in the twentieth century led to an understanding of the emission and absorption of light by matter. Light emitted or absorbed by atoms is now known to be the result of energy changes of the outermost electrons in the atom. Because these energy changes are quantized rather than being continuous, the photons emitted have discrete energies resulting in light waves with a discrete set of frequencies and wavelengths similar to the set of frequencies and wavelengths observed in standing sound waves. Viewed through a spectroscope with a narrow-slit aperture, the light emitted by an atom appears as a discrete set of lines of different colors or wavelengths, with the spacing and intensities of the lines being characteristic of the element.

The technological developments in the second half of the twentieth century have led to a renewed interest in both theoretical and applied optics. The advent of high-speed computers has brought vast improvement in the design of complex optical systems. Optical fibers are rapidly replacing elec-

(a)

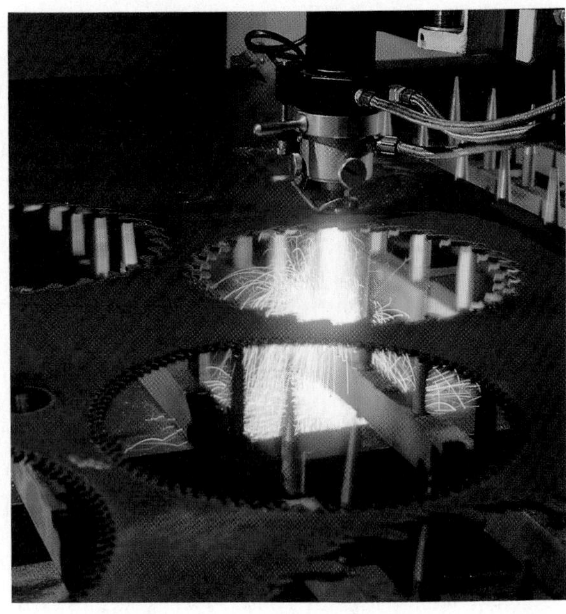

(b)

(a) Since its invention in 1960, the laser has found many uses. (b) A carbon dioxide laser takes just two minutes to cut out a steel saw blade. (c) The supermarket laser scanner is now a familiar sight. (d) A helium-neon laser mounted on a truck is used to precisely measure the clearance in a tunnel. (e) A liquid dye laser beam transmitted by an optical fiber is used here in eye surgery.

trical wires for the transmission of data. The laser, invented in 1960, has led to the discovery of a number of new optical effects. Today, lasers are used to scan labels in supermarkets, to perform surgery in hospitals, to set type for newspapers, and to read compact discs for home audio systems. The wavefront reconstruction technique known as holography, developed in the late 1940s, is now used in nondestructive testing and in data storage.

In this chapter, we will begin by discussing some of the early measurements of the speed of light. We will then discuss the basic phenomena of reflection, refraction, dispersion, and polarization. These can all be adequately understood by using rays to describe the straight-line propagation of light and neglecting interference and diffraction effects. As discussed in Chapter 14, the ray approximation is valid for the propagation of any wave motion if the wavelength is small compared with any apertures or obstacles. This is often the situation in optics because the wavelengths of visible light range from about 400 nm (1 nm = 10^{-9} m) for violet light to about 700 nm for red light.

30-1 The Speed of Light

The first effort to measure the speed of light was made by Galileo. He and a partner stood on hilltops about a kilometer apart, each with a lantern and a shutter to cover it. Galileo proposed to measure the time it took for light to travel back and forth between the experimenters. First, one would uncover his lantern, and when the other saw the light, he would uncover his. The time between the first partner's uncovering his lantern and his seeing the light from the other lantern would be the time it took for light to travel back and forth between the experimenters. Though this method is sound in principle, the speed of light is so great that the time interval to be measured is much smaller than fluctuations in human response time, so Galileo was unable to obtain any value for the speed of light.

The first indication of the true magnitude of the speed of light came from astronomical observations of the period of Io, one of the moons of Jupiter.

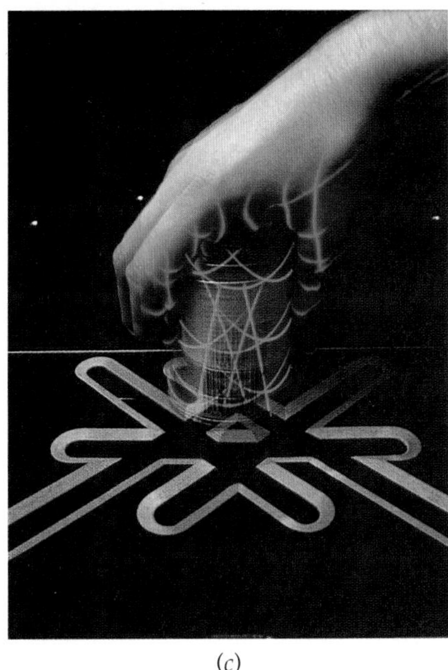

(c)

(d)

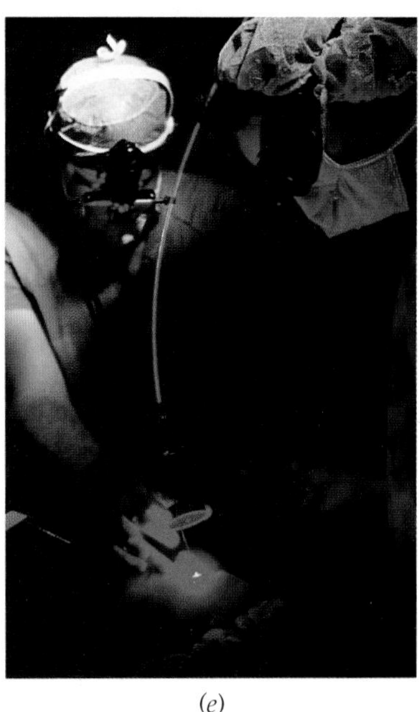

(e)

This period is determined by measuring the time between eclipses (when the moon Io disappears behind Jupiter). The eclipse period is about 42.5 hours, but measurements made when the earth is moving away from Jupiter along path *ABC* in Figure 30-1 give a greater time for this period than do measurements made when the earth is moving toward Jupiter along path *CDA* in the figure. Since these measurements differ from the average value by only about 15 seconds, the discrepancies were difficult to measure accurately. In 1675, the astronomer Ole Römer attributed these discrepancies to the fact that the speed of light is not infinite. During the 42.5 hours between eclipses of Jupiter's moon, the distance between the earth and Jupiter changes, making the path for the light longer or shorter. Römer devised the following method for measuring the cumulative effect of these discrepancies. Because Jupiter moves much more slowly than earth, we can neglect its motion. When the earth is at point *A*, nearest to Jupiter, the distance between the earth and Jupiter is changing negligibly. The period of Io's eclipse is measured, providing the time between the beginnings of successive eclipses. Based on this measurement, the number of eclipses in 6 months is computed, and the time when an eclipse should begin a half-year later when the earth is at point *C* is predicted. When the earth is actually at *C*, the observed beginning of the eclipse is about 16.6 minutes later than predicted. This is the time it takes light to travel a distance equal to the diameter of the earth's orbit.

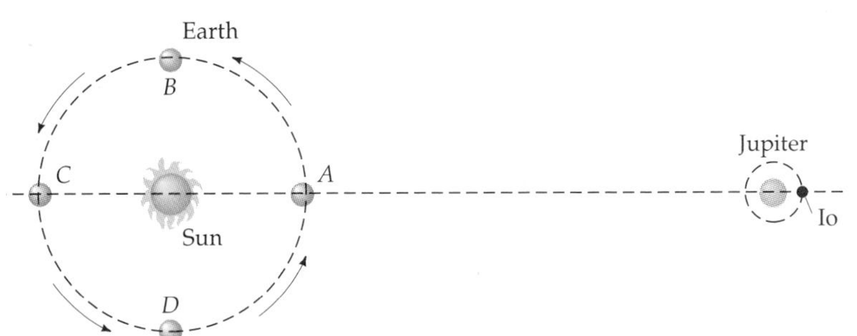

Figure 30-1 Römer's method of measuring the speed of light. The time between eclipses of Jupiter's moon Io appears to be greater when the earth is moving along path *ABC* than when it is moving along path *CDA*. The difference is due to the time it takes light to travel the distance traveled by the earth along the line of sight during one period of Io. (The distance traveled by Jupiter in one earth year is negligible.)

Example 30-1

The diameter of the earth's orbit is 3.00×10^{11} m. If light takes 16.6 min to travel this distance, what is the speed of light?

The number of seconds in 16.6 min is (16.6 min) × (60 s/min) = 996 s. The measured speed of light is thus

$$c = \frac{\Delta x}{\Delta t} = \frac{3.00 \times 10^{11}\ \text{m}}{996\ \text{s}} = 3.01 \times 10^{8}\ \text{m/s}$$

Römer obtained a considerably smaller value for c because he used 22 min for Δt.

The first nonastronomical measurement of the speed of light was made by the French physicist Fizeau in 1849. On a hill in Paris, Fizeau placed a light source and a system of lenses arranged such that the light reflected from a semitransparent mirror was focused on a gap in a toothed wheel as shown in Figure 30-2. On a distant hill (about 8.63 km away), he placed a mirror to reflect the light back to be viewed by an observer as shown. The toothed wheel was rotated, and the speed of rotation was varied. At low speeds of rotation, no light was visible because the reflected light was obstructed by the teeth of the rotating wheel. The speed of rotation was then increased. The light suddenly became visible when the rotation speed was such that the reflected light passed through the next gap in the wheel.

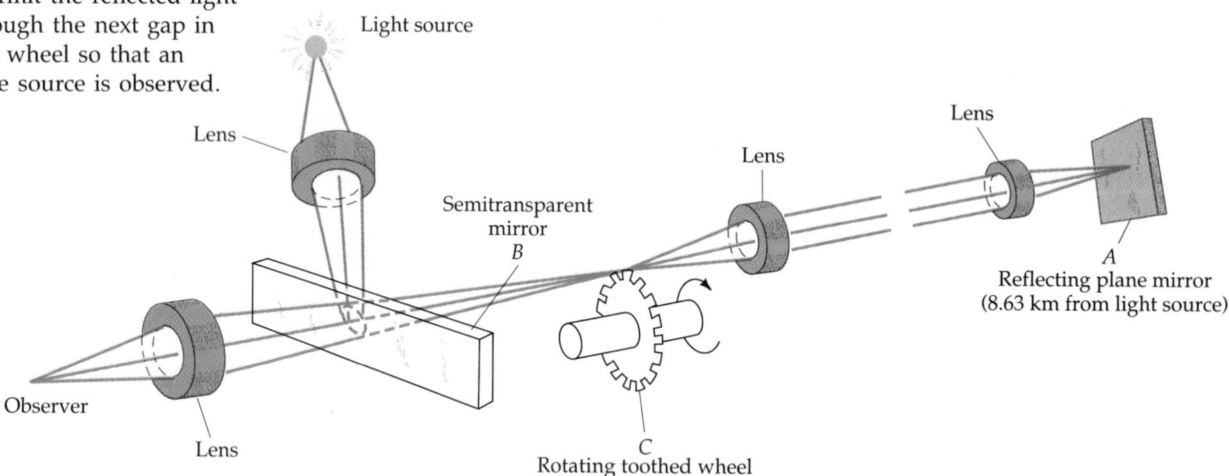

Figure 30-2 Fizeau's method of measuring the speed of light. Light from the source is reflected by mirror B and is transmitted through a gap in the toothed wheel to mirror A. The speed of light is determined by measuring the angular speed of the wheel that will permit the reflected light to pass through the next gap in the toothed wheel so that an image of the source is observed.

Fizeau's method was improved upon by Foucault, who replaced the toothed wheel with an eight-sided rotating mirror as shown in Figure 30-3. Light strikes one face of the mirror and is reflected from a distant fixed mirror to another face and then to an observing telescope. (The actual experimental arrangement was somewhat more complicated than the simple diagram shown in Figure 30-3.) When the mirror rotates through one-eighth of a turn (or $n/8$ turns, where n is an integer) another face of the mirror is in the right position for the reflected light to enter the telescope. In about 1850, Foucault measured the speed of light in air and in water and showed that it is less in water. Using essentially the same method, the American physicist A. A. Michelson made precise measurements of the speed of light from 1880 to 1930.

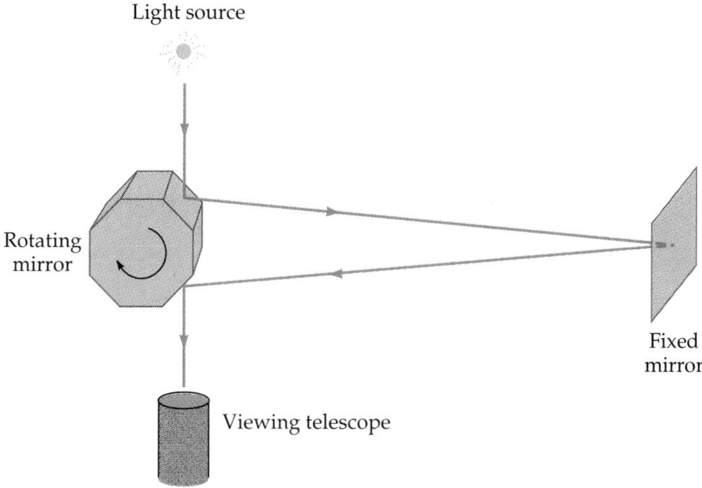

Figure 30-3 Simplified drawing of Foucault's method of measuring the speed of light. Essentially, Fizeau's rotating toothed wheel is replaced by a rotating octagonal mirror. When the mirror makes an eighth of a revolution during the time it takes for the light to travel to the fixed mirror and back, the next face of the mirror is in the proper position to reflect the light into the telescope.

Another method of determining the speed of light involves the measurement of the electrical constant ϵ_0 to determine c from

$$c = \frac{1}{\sqrt{\epsilon_0 \mu_0}}$$

The constant ϵ_0 can be obtained by measuring the capacitance of a parallel-plate capacitor. The constant μ_0 is defined in terms of the definition of the ampere, which in turn determines the coulomb.

The various methods we have discussed for measuring the speed of light are all in general agreement. Today, the speed of light is defined to be exactly

$$c = 299\ 792\ 457 \text{ m/s}$$

and the standard unit of length, the meter, is defined in terms of this speed. A measurement of the speed of light is therefore now a measurement of the size of the meter, which is the distance light travels in (1/299 792 457) s. The value 3×10^8 m/s for the speed of light is accurate enough for nearly all calculations. The speed of radio waves and all other electromagnetic waves (in a vacuum) is the same as the speed of light.

Example 30-2

In Fizeau's experiment, the wheel had 720 teeth, and light was observed when the wheel rotated at 25.2 revolutions per second. If the distance from the wheel to the distant mirror was 8.63 km, what was Fizeau's value for the speed of light?

The total distance traveled by the light from the gap in the wheel to the mirror and back was 2×8.63 km = 17.3 km. The reflected light passed through the next gap in the wheel, so the wheel made 1/720 revolution. Since the wheel made 25.3 revolutions in 1 s, the time it took to make 1/720 revolution was

$$\Delta t = \frac{1 \text{ s}}{25.3 \text{ rev}}\left(\frac{1}{720} \text{ rev}\right) = 5.49 \times 10^{-5} \text{ s}$$

The value for the speed of light from this experiment is then

$$c = \frac{\Delta x}{\Delta t} = \frac{17.3 \times 10^3 \text{ m}}{5.49 \times 10^{-5} \text{ s}} = 3.15 \times 10^8 \text{ m/s}$$

which is about 5 percent too high.

Example 30-3

Space travelers on the moon use electromagnetic waves to communicate with the space control center on earth. What is the time delay for their signal to reach the earth, which is 3.84×10^8 m away?

Using 3×10^8 m/s for the speed of light and other electromagnetic waves, we find the time it takes for a signal to travel from the moon to the earth to be

$$\Delta t = \frac{\Delta x}{c} = \frac{3.84 \times 10^8 \text{ m}}{3 \times 10^8 \text{ m/s}} = 1.28 \text{ s}$$

The time delay is thus 1.28 s each way.

Example 30-4

The sun is 1.50×10^{11} m from the earth. How long does it take for the sun's light to reach the earth?

$$\Delta t = \frac{\Delta x}{c} = \frac{1.50 \times 10^{11} \text{ m}}{3 \times 10^8 \text{ m/s}} = 500 \text{ s} = 8.33 \text{ min}$$

Large distances are often given in terms of the time it takes for light to travel those distances. For example, the distance to the sun is 8.33 light-minutes, written 8.33 c-min. A light-year is the distance light travels in one year. We can easily find a conversion factor between light-years and meters. The number of seconds in one year is

$$1 \text{ y} = 1 \text{ y} \times \frac{365.24 \text{ d}}{1 \text{ y}} \times \frac{24 \text{ h}}{1 \text{ d}} \times \frac{3600 \text{ s}}{1 \text{ h}} = 3.156 \times 10^7 \text{ s}$$

The number of meters in one light-year is thus

$$1 \text{ } c\text{-year} = (2.998 \times 10^8 \text{ m/s})(3.156 \times 10^7 \text{ s}) = 9.46 \times 10^{15} \text{ m}$$

Question

1. Estimate the time required for light to make the round trip in Galileo's experiment to determine the speed of light.

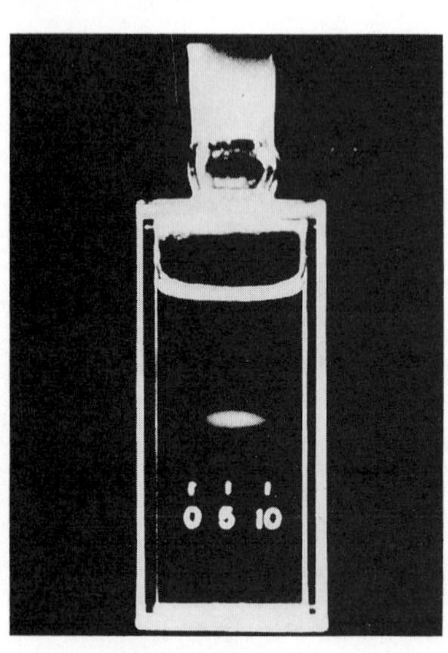

(*a*)

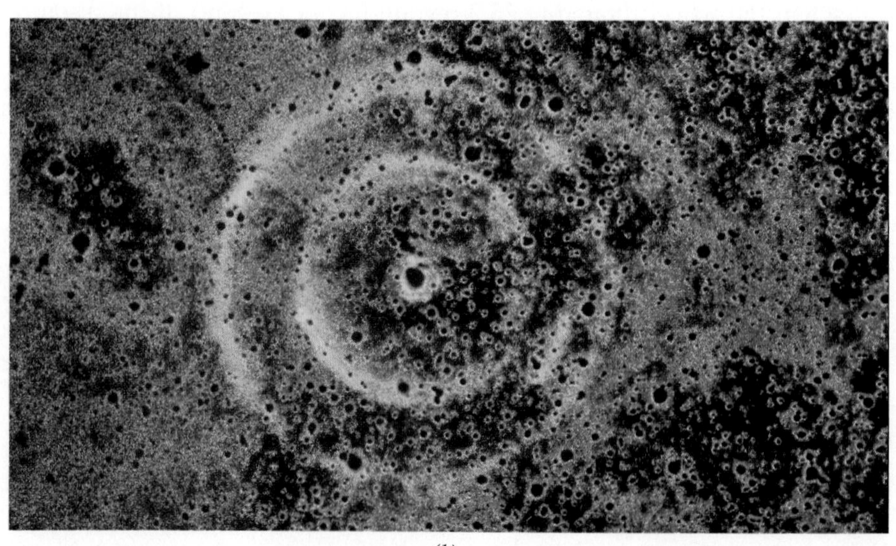

(*b*)

(*a*) A 10-ps pulse of light from a neodymium-doped glass laser passing through a water cell calibrated in millimeters. The pulse length is about 2.2 mm. (*b*) Light echoes from a supernova explosion that occurred 170,000 years ago. Direct light from the explosion was first seen on earth in February 1987. Light reflected from dust clouds near the explosion arrives at the earth months or years later. Comparison of photograph plates exposed in 1987 and 1989 show the outer rings that result from reflection from dust clouds.

30-2 The Propagation of Light: Huygens' Principle

Figure 30-4 shows a portion of a spherical wavefront emanating from a point source. The wavefront is the locus of points of constant phase. If the radius of the wavefront is r at time t, its radius at time $t + \Delta t$ is $r + c\,\Delta t$, where c is the speed of the wave. However, if a part of the wave is blocked by some obstacle or if the wave passes through a different medium, as in Figure 30-5, the determination of the new wavefront at time $t + \Delta t$ is much more difficult.

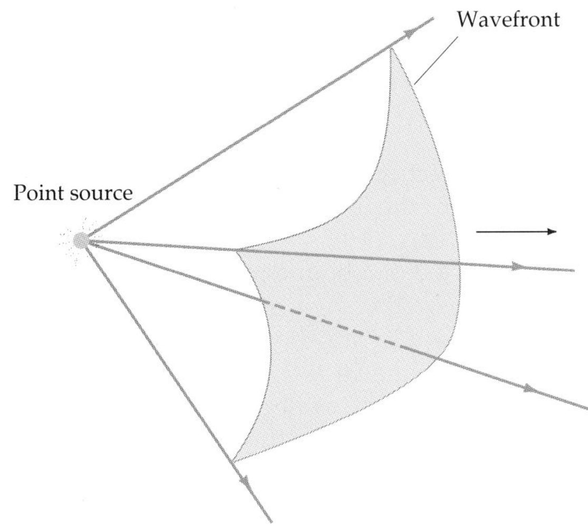

Figure 30-4 Spherical wavefront from a point source.

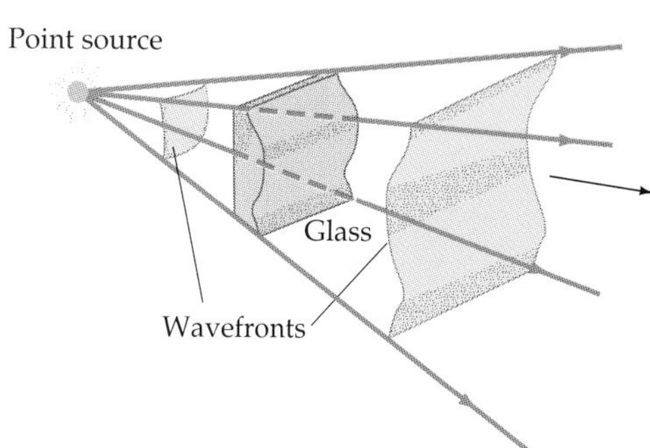

Figure 30-5 Wavefront from a point source before and after passing through an irregularly shaped piece of glass.

The propagation of any wave through space can be described using a geometric method discovered by Christian Huygens about 1678 that is now known as **Huygens' principle** or **Huygens' construction:**

> Each point on a primary wavefront serves as the source of spherical secondary wavelets that advance with a speed and frequency equal to those of the primary wave. The primary wavefront at some later time is the envelope of these wavelets.

Figure 30-6 shows the application of Huygens' principle to the propagation of a plane wave and of a spherical wave. Of course, if each point on a wavefront were really a point source, there would be waves in the backward direction as well. Huygens ignored these back waves.

Huygens' principle was later modified by Fresnel so that the new wavefront was calculated from the old wavefront by superposition of the wavelets considering their relative amplitudes and phases. Kirchhoff later showed that the Huygens–Fresnel principle was a consequence of the wave equation, thus putting it on a firm mathematical basis. Kirchhoff showed that the intensity of the wavelets depends on the angle and is zero in the backward direction.

In this chapter, we will use Huygens' principle to derive the laws of reflection and refraction. In Chapter 33, we will apply Huygens' principle with Fresnel's modification to calculate the diffraction pattern of a single slit.

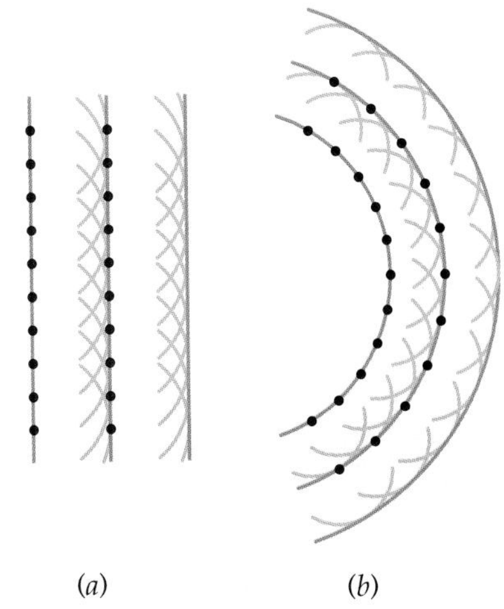

Figure 30-6 Huygens' construction for the propagation to the right of (*a*) a plane wave and (*b*) an outgoing spherical, or circular, wave.

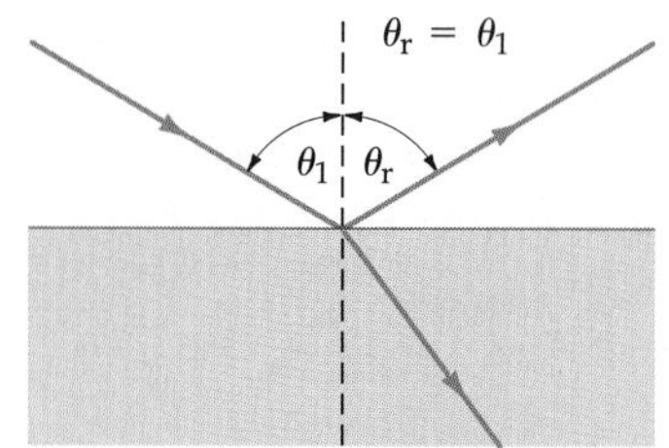

Figure 30-7 The angle of reflection θ_r equals the angle of incidence θ_1.

30-3 Reflection

When waves of any type strike a plane barrier such as a mirror, new waves are generated that move away from the barrier. This phenomenon is called **reflection.** Reflection occurs at a boundary between two different media such as an air–glass surface, in which case part of the incident energy is reflected and part is transmitted. Figure 30-7 shows a light ray striking a smooth air–glass surface. The angle θ_1 between the incident ray and the normal (the line perpendicular to the surface) is called the **angle of incidence,** and the plane defined by these two lines is called the **plane of incidence.** The reflected ray lies in the plane of incidence and makes an angle θ_r with the normal that is equal to the angle of incidence as shown in the figure:

Law of reflection

$$\theta_r = \theta_1 \tag{30-1}$$

This result is known as the **law of reflection.** The law of reflection holds for any type of wave. Figure 30-8 illustrates the law of reflection for rays of light and for wavefronts of ultrasonic waves.

The fraction of light energy reflected at a boundary such as an air–glass surface depends in a complicated way on the angle of incidence, the orientation of the electric-field vector associated with the wave, and the relative speed of light in the first medium (air) and in the second medium (glass). The speed of light in a medium such as glass, water, or air is characterized by the **index of refraction** n, which is defined as the ratio of the speed of light in a vacuum c to the speed in the medium v:

Index of refraction

$$n = \frac{c}{v} \tag{30-2}$$

For the special case of normal incidence ($\theta_1 = \theta_r = 0°$), the reflected intensity can be shown to be

$$I = \left(\frac{n_1 - n_2}{n_1 + n_2}\right)^2 I_0 \tag{30-3}$$

where I_0 is the incident intensity and n_1 and n_2 are the indexes of refraction of the two media. For a typical case of reflection from an air–glass surface for

Figure 30-8 (*a*) Ultrasonic plane waves in water reflecting from a steel plate. (*b*) Light rays reflecting from an air–glass surface showing equal angles of incidence and reflection.

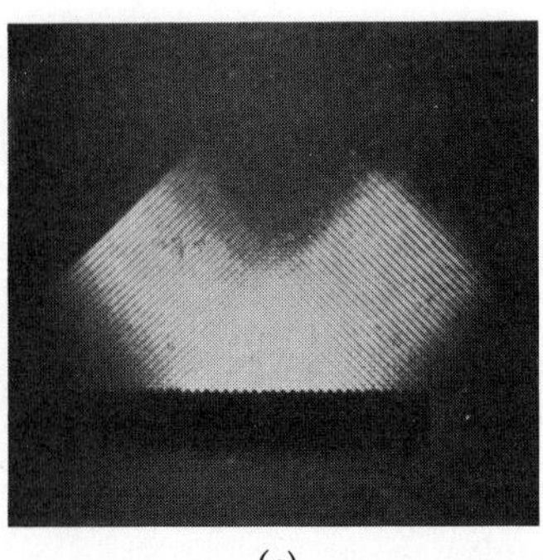

(*a*)

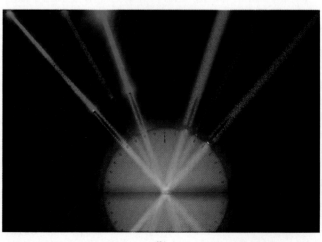

(*b*)

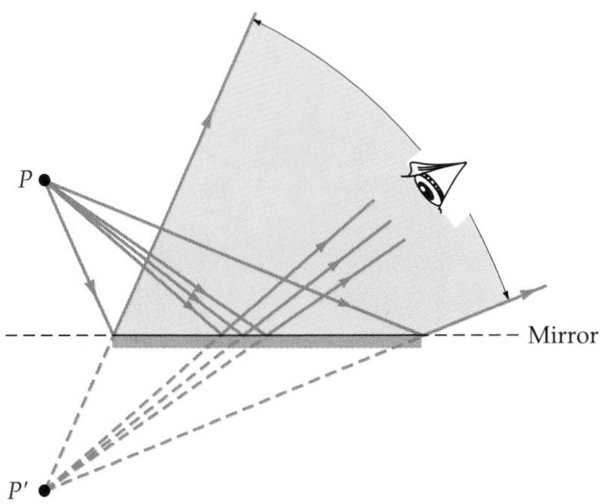

Figure 30-9 Rays from a source P reflected by a mirror into the eye appear to come from the image point P' behind the mirror. The image can be seen when the eye is anywhere in the shaded region.

which $n_1 = 1$ and $n_2 = 1.5$, Equation 30-3 gives $I = I_0/25$. That is, only about 4 percent of the energy is reflected; the rest is transmitted.

Figure 30-9 shows a narrow bundle of light rays from a point source P that are reflected from a flat surface. After reflection, the rays diverge exactly as if they came from a point P' behind the surface. The point P' is called the **image** of point P. When these rays enter the eye, they cannot be distinguished from rays diverging from a source at P' as though no reflecting surface were present. (We will study the formation of images by reflecting and refracting surfaces in the next chapter.) Reflection from a smooth surface is called **specular reflection.** It differs from **diffuse reflection,** which is illustrated in Figure 30-10. Here, because the surface is rough, the rays enter the eye after reflecting from many different points on the surface, so there is no image. The reflection of light from the page of this book is diffuse reflection. Ground glass is sometimes used in a picture frame to give diffuse reflection and thereby cut out the glare from the light used to illuminate the picture. It is diffuse reflection from the surface of the road that allows you to see the road when you are driving at night because some of the light from your headlights reflects back toward you.

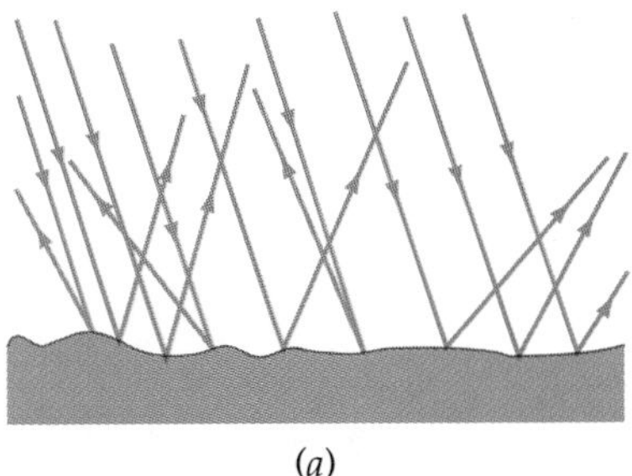

(*a*)

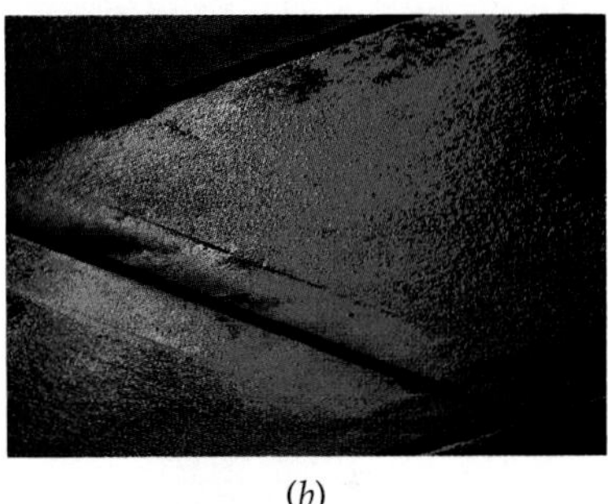

(*b*)

Figure 30-10 (*a*) Diffuse reflection from a rough surface. (*b*) Photograph of diffuse reflection of colored lights from a sidewalk.

The physical mechanism of the reflection of light can be understood in terms of the absorption and reradiation of the light by the atoms in the reflecting medium. When light traveling in air strikes a glass surface, the atoms in the glass absorb the light and reradiate it at the same frequency in all directions. The waves radiated backwards by the glass atoms interfere constructively at an angle equal to the angle of incidence to produce the reflected wave.

The law of reflection can be derived from Huygens' principle. Figure 30-11 shows a plane wavefront AA' striking a mirror at point A. As can be seen from the figure, the angle ϕ_1 between the wavefront and the mirror is the same as the angle of incidence θ_1, which is the angle between the per-

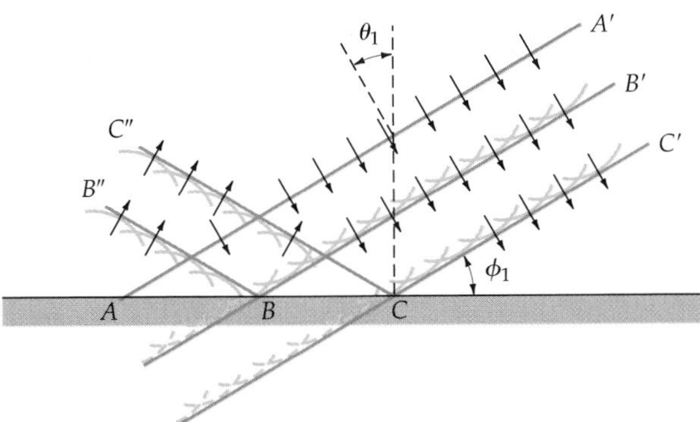

Figure 30-11 Plane wave reflected at a plane mirror. The angle θ_1 between the incident ray and the normal to the mirror is the angle of incidence. It is equal to the angle ϕ_1 between the incident wavefront and the mirror.

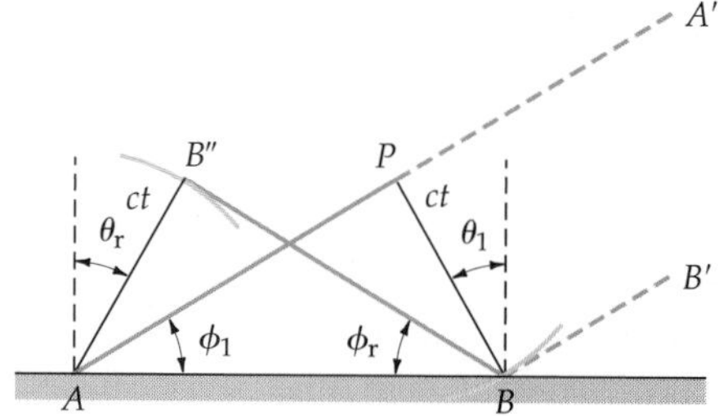

Figure 30-12 Geometry of Huygens' construction for the calculation of the law of reflection. The wavefront AP initially strikes the mirror at point A. After a time t, the Huygens' wavelet from P strikes the mirror at point B, and the one from A reaches point B''.

pendicular to the mirror and the rays that are perpendicular to the wavefronts. According to Huygens' principle, each point on a given wavefront can be considered to be a point source of secondary wavelets. The position of the wavefront after a time t is found by constructing wavelets of radius ct with their centers on the wavefront AA'. Wavelets that do not strike the mirror form the portion of the new wavefront BB'. Wavelets that do strike the mirror are reflected and form the portion of the new wavefront BB''. By a similar construction, the wavefront $C''CC'$ is obtained from the Huygens' wavelets originating on the wavefront $B''BB'$. Figure 30-12 is an enlargement of a portion of Figure 30-11 showing the part of the original wavefront AP that strikes the mirror during time t. In this time, the wavelet from point P reaches the mirror at point B, and the wavelet from point A reaches point B''. The reflected wave BB'' makes an angle ϕ_r with the mirror that is equal to the angle of reflection θ_r between the reflected ray and the normal to the mirror. The triangles ABP and BAB'' are both right triangles with a common side AB and equal sides $AB'' = BP = ct$. Hence, these triangles are congruent, and the angles ϕ_1 and ϕ_r are equal, implying that the angle of reflection θ_r equals the angle of incidence θ_1.

Question

2. How does a thin layer of water on the road affect the light you see reflected off the road from your own headlights? How does it affect the light you see reflected from the headlights of an oncoming car?

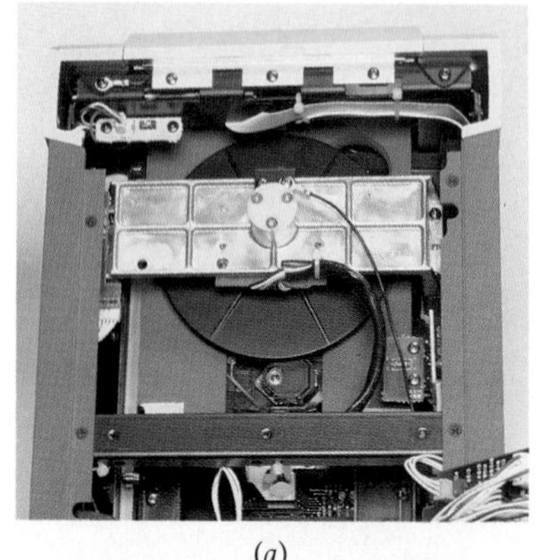

(*a*)

(*b*)

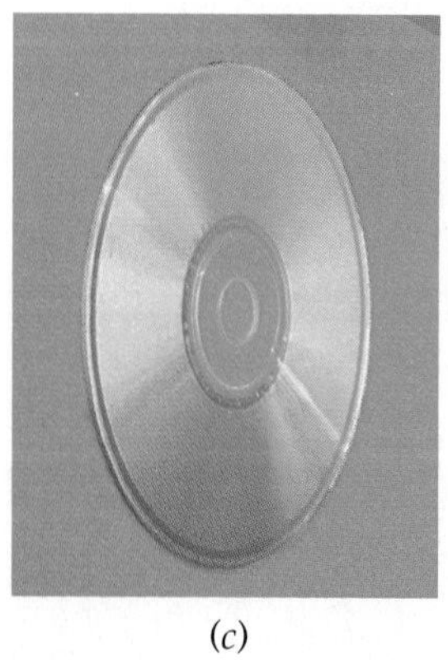

(*c*)

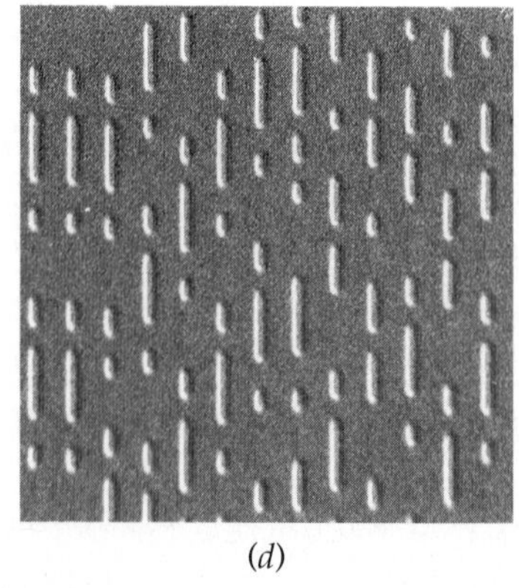

(*d*)

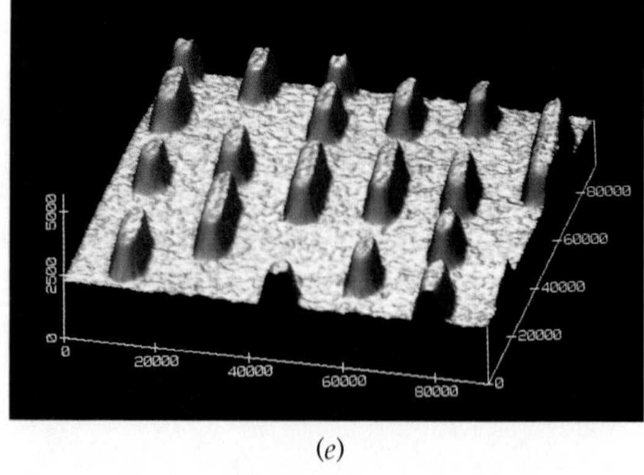

(*e*)

The reflection of laser light is used to read compact discs. (*a*) In a compact disc player, the laser is less than 2 mm from the bottom of the disc as it spins. (*b*) The laser head used to read a compact disc. (*c*) A compact disc has a diameter of 12 cm and is 1.2 mm thick. (*d*) Information is stored digitally on a compact disc in these pits, which are 0.12 μm deep, 0.6 μm wide, and 0.9 to 3.3 μm long. The leading and trailing edges of a pit represent a binary 1, whereas intervening areas represent binary 0's. (*e*) Stamper used to produce the pits in (*d*).

30-4 Refraction

When a beam of light strikes a boundary surface separating two different media, such as an air–glass surface, part of the light energy is reflected and part enters the second medium. The change in direction of the transmitted ray is called **refraction.**

The transmitted wave is the result of the interference of the incident wave and the wave produced by the absorption and reradiation of light energy by the atoms in the medium. For light entering glass from air, there is a phase lag between the reradiated wave and the incident wave. There is therefore also a phase lag between the resultant wave and the incident wave. This phase lag means that the position of a wave crest of the transmitted wave is retarded relative to the position of a wave crest of the incident wave in the medium. Therefore, in a given time, the transmitted wave does not travel as far in the medium as the original incident wave; that is, the velocity of the transmitted wave is less than that of the incident wave. The index of refraction, which is the ratio of the speed of light in a vacuum to that in the medium, is therefore greater than 1. For example, the speed of light in glass is about two-thirds of the speed of light in free space. The index of refraction of glass is therefore about $n = c/v = 3/2$.

Since the frequency of the light in the second medium is the same as that of the incident light—the atoms absorb and reradiate the light at the same frequency—but the wave speed is different, the wavelength of the transmitted light is different from that of the incident light. If λ is the wavelength of light in a vacuum, the wavelength λ' in a medium of index of refraction n is

$$\lambda' = \frac{v}{f} = \frac{c/n}{f} = \frac{\lambda}{n} \qquad 30\text{-}4$$

Exercise

Sodium light has a wavelength of 589 nm in a vacuum. Find the wavelength of sodium light (*a*) in water, for which $n = 1.33$, and (*b*) in glass, for which $n = 1.50$. [Answers: (*a*) 443 nm, (*b*) 393 nm]

Figure 30-13 shows light striking a flat air–glass surface. The ray that enters the glass is called the refracted ray, and the angle θ_2 is called the angle of refraction. The angle of refraction is less than the angle of incidence θ_1, as shown in the figure; that is, the refracted ray is bent toward the normal. If, on the other hand, the light beam originates in the glass and is refracted into

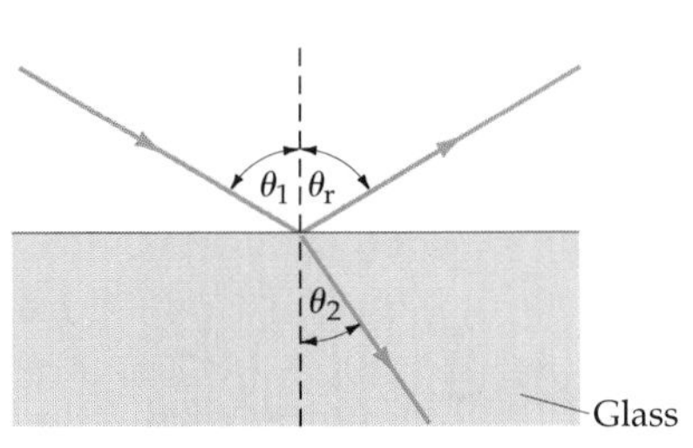

(*a*)

(*b*)

Figure 30-13 (*a*) Incident, reflected, and refracted rays for light striking an air–glass surface. The angle of refraction θ_2 is less than the angle of incidence θ_1. (*b*) Reflection and refraction of a beam of light incident on a glass slab. The refracted beam is partially reflected and partially refracted at the bottom glass–air surface.

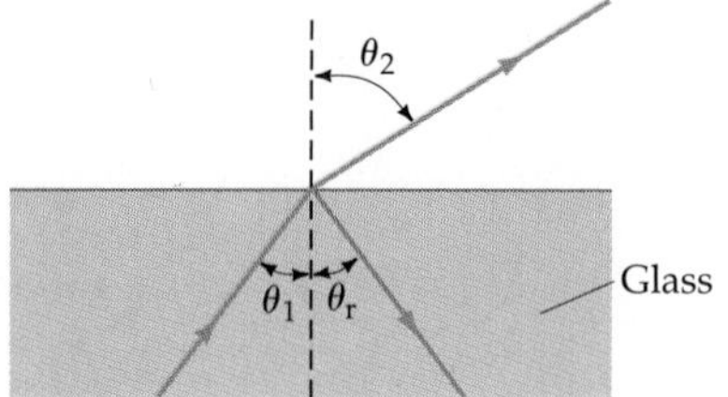

Figure 30-14 Refraction from a dense medium to a less dense medium. Here the angle of refraction is greater than the angle of incidence. The light ray is bent away from the normal.

the air, the angle of refraction is greater than the angle of incidence, and the refracted ray is bent away from the normal as shown in Figure 30-14. We can relate the angle of refraction θ_2 to the indexes of refraction of the two media n_1 and n_2 and to the angle of incidence θ_1 using Huygens' principle. Figure 30-15 shows a plane wave incident on an air–glass surface. We apply Huygens' construction to find the wavefront of the transmitted wave. Line AP indicates a portion of the wavefront in medium 1 that strikes the glass surface at an angle of incidence θ_1. In time t the wavelet from P travels the distance v_1t and reaches the point B on the line AB separating the two media, while the wavelet from point A travels a shorter distance v_2t into the second medium. The new wavefront BB' is not parallel to the original wavefront AP because the speeds v_1 and v_2 are different. From the triangle APB,

$$\sin \phi_1 = \frac{v_1t}{AB}$$

or

$$AB = \frac{v_1t}{\sin \phi_1} = \frac{v_1t}{\sin \theta_1}$$

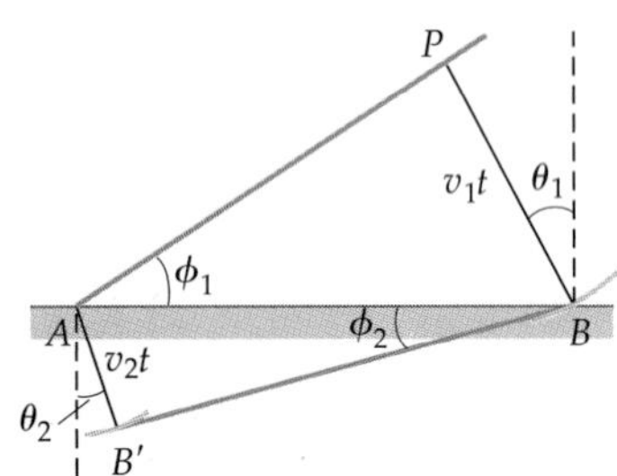

Figure 30-15 Application of Huygens' principle to the refraction of plane waves at the surface separating a medium in which the wave speed is v_1 from a medium in which the wave speed v_2 is less than v_1. The angle of refraction in this case is less than the angle of incidence.

using the fact that the angle ϕ_1 equals the angle of incidence θ_1. Similarly, from triangle $AB'B$,

$$\sin \phi_2 = \frac{v_2t}{AB}$$

or

$$AB = \frac{v_2t}{\sin \phi_2} = \frac{v_2t}{\sin \theta_2}$$

where $\theta_2 = \phi_2$ is the angle of refraction. Equating the two values for AB, we obtain

$$\frac{\sin \theta_1}{v_1} = \frac{\sin \theta_2}{v_2} \qquad 30\text{-}5$$

Substituting $v_1 = c/n_1$ and $v_2 = c/n_2$ in this equation and multiplying by c, we obtain

Snell's law of refraction

$$n_1 \sin \theta_1 = n_2 \sin \theta_2 \qquad 30\text{-}6$$

This result was discovered experimentally in 1621 by Willebrod Snell, a Dutch scientist, and is known as **Snell's law** or the **law of refraction.** It was independently discovered a few years later by René Descartes. Equation 30-6 holds for the refraction of any kind of wave incident on a boundary surface separating two media. Figure 30-16 shows the refraction of plane water waves at a boundary at which the wave speed changes because the depth of the water changes.

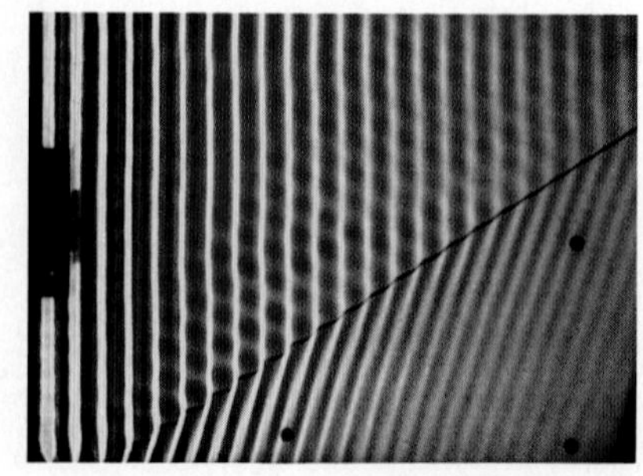

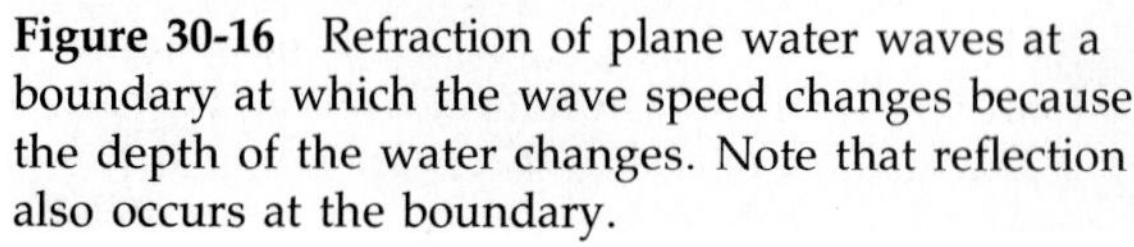

Figure 30-16 Refraction of plane water waves at a boundary at which the wave speed changes because the depth of the water changes. Note that reflection also occurs at the boundary.

Example 30-5

Light traveling in air enters water with an angle of incidence of 45°. If the index of refraction of water is 1.33, what is the angle of refraction?

Taking $n = 1$ for air, we obtain from Equation 30-6

$$(1.00)\sin 45° = (1.33)\sin\theta_2$$

$$\sin\theta_2 = \frac{(1.00)\sin 45°}{1.33} = \frac{(1.00)(0.707)}{1.33} = 0.53$$

The angle whose sine is 0.53 is 32°.

Figure 30-17 shows a point source in glass with rays striking the glass–air surface at various angles. All the rays are bent away from the normal. As the angle of incidence is increased, the angle of refraction increases until a critical angle of incidence θ_c is reached for which the angle of refraction is 90°. For incident angles greater than this critical angle, there is no refracted ray. All the energy is reflected. This phenomenon is called **total internal reflection.** The critical angle can be found in terms of the indexes of refraction of the two media by solving Equation 30-6 for $\sin\theta_1$ and setting θ_2 equal to 90°:

$$\sin\theta_1 = (n_2/n_1)\sin\theta_2$$

Setting $\theta_2 = 90°$, we obtain

$$\sin\theta_c = \frac{n_2}{n_1} \qquad 30\text{-}7$$

Critical angle for total internal reflection

Note that total internal refraction occurs only when the light is originally in the medium with the higher index of refraction. Mathematically, if n_2 is greater than n_1, Equation 30-7 cannot be satisfied because there is no real angle whose sine is greater than 1.

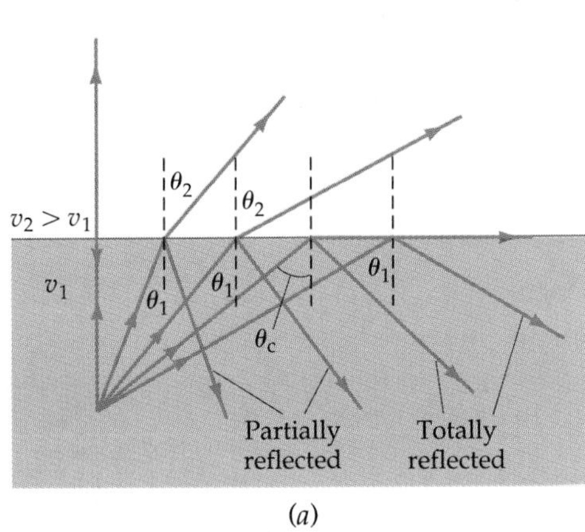

(*a*)

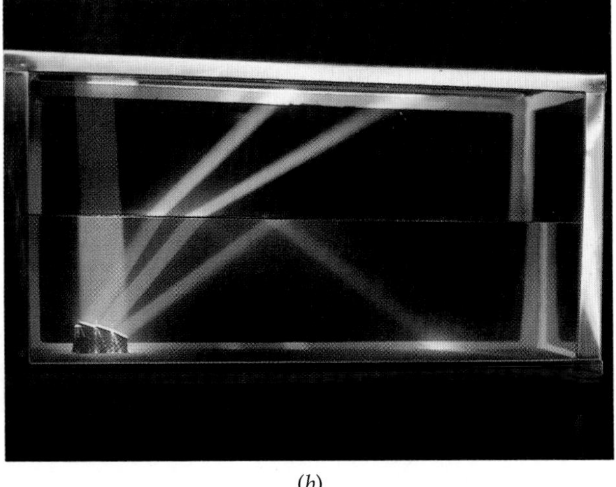

(*b*)

Figure 30-17 (*a*) Total internal reflection. For refraction from glass or water to air, the refracted ray is bent away from the normal. As the angle of incidence is increased, the angle of refraction is increased until, at a critical angle of incidence θ_c, the angle of refraction is 90°. For angles of incidence greater than the critical angle, there is no refracted ray. All the energy is reflected. (*b*) Photograph of refraction and total internal reflection from a water–air surface.

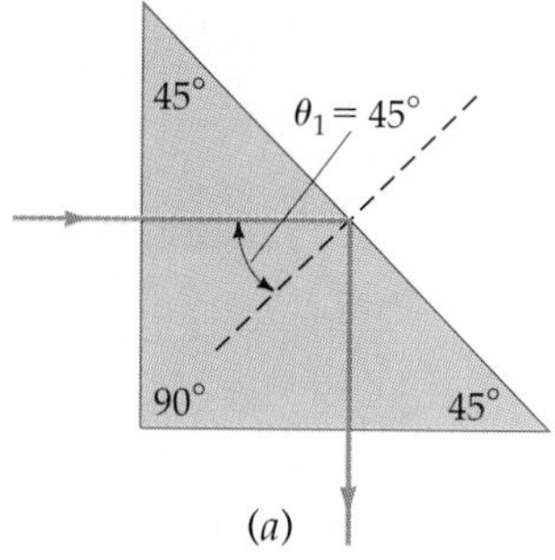

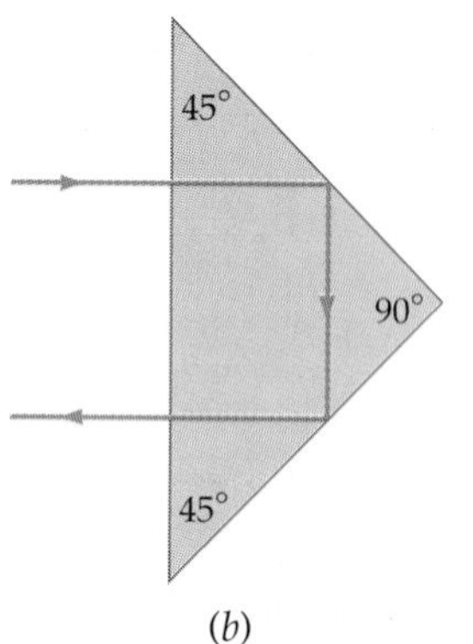

Figure 30-18 (*a*) Light entering through one of the short sides of a 45-45-90° glass prism is totally reflected in the prism and emerges through the other short side at 90° to the direction of the incident light. (*b*) Light entering through the long side of the prism is totally reflected twice and emerges in the direction opposite to that of the incident light.

Example 30-6

A particular glass has an index of refraction of $n = 1.50$. What is the critical angle for total internal reflection for light leaving this glass and entering air, for which $n = 1.00$?

From Equation 30-7, we obtain for the critical angle

$$\sin \theta_c = \frac{1.00}{1.50} = 0.667$$

The angle whose sine is 0.667 is 42°.

Figure 30-18*a* shows light incident normally on one of the short sides of a 45-45-90° glass prism. If the index of refraction of the prism is 1.5, the critical angle for total internal reflection is 42°, as we found in Example 30-6. Since the angle of incidence of the ray on the glass–air surface is 45°, the light will be totally reflected and will exit perpendicular to the other face of the prism as shown. In Figure 30-18*b*, the light is incident perpendicular to the hypotenuse of the prism and is totally reflected twice such that it emerges at 180° to its original direction. Prisms are used to change the direction of a light ray. In binoculars, four prisms are used to reinvert the image that was inverted by the binocular lens. Diamonds have a very high index of refraction ($n \approx 2.4$), so nearly all the light that enters a diamond is eventually reflected back out, giving the diamond its sparkle.

An interesting application of total internal reflection is the transmission of a beam of light down a long, narrow, transparent glass fiber (Figure 30-19). If the beam begins approximately parallel to the axis of the fiber, it will strike the walls of the fiber at angles greater than the critical angle (if the bends in the fiber are not too sharp) and no light energy will be lost through the walls of the fiber.

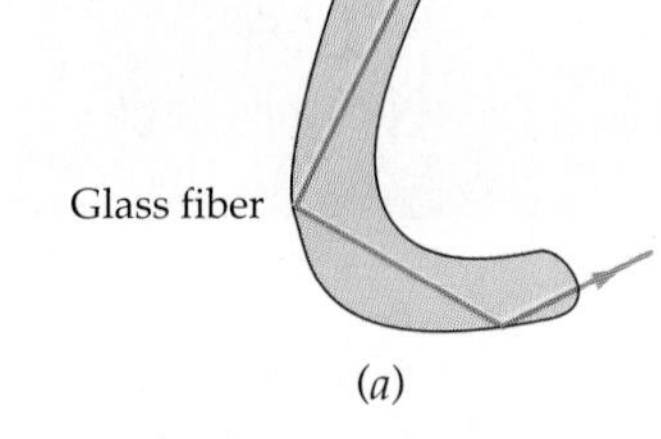

(*a*)

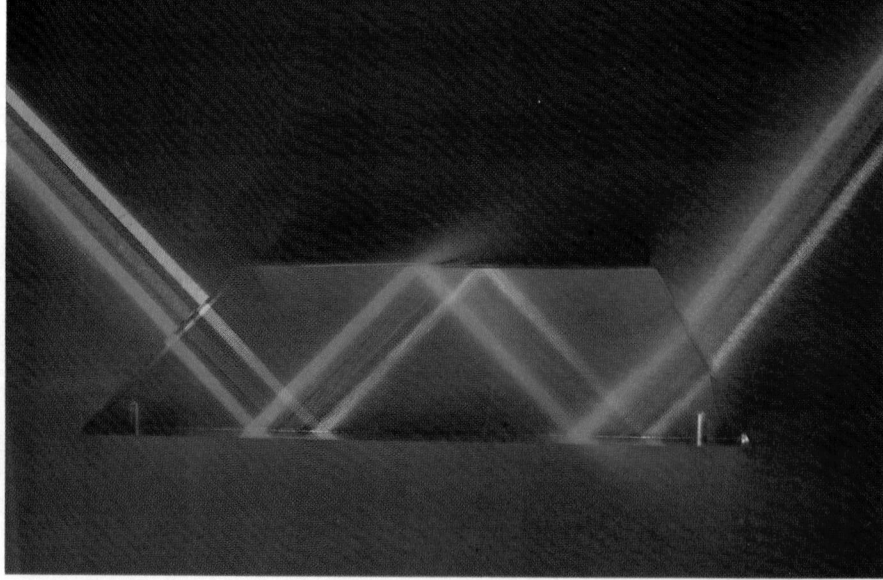

(*b*)

Figure 30-19 (*a*) A light pipe. Light inside the pipe is always incident at an angle greater than the critical angle, so no light escapes the pipe by refraction. (*b*) These laser beams are totally reflected three times at the sides of the glass bar before exiting the end.

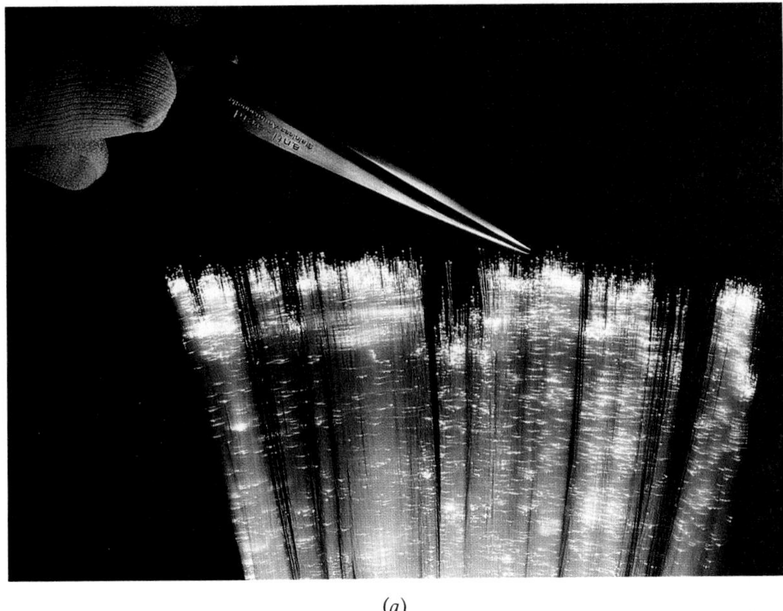

(a)

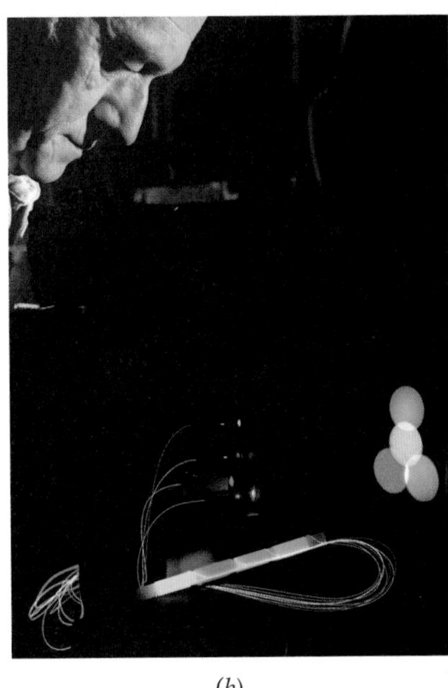

(b)

(a) Light emerging from a loose bundle of glass fibers. (b) In this demonstration at the Naval Research Laboratory, a combination of laser sources generates different colors that excite adjacent fiber sensor elements leading to a separation of the information as indicated by the separation of the colors.

A bundle of such fibers can be used for imaging, as illustrated in Figure 30-20. Fiber optics has many applications in medicine and in communications. In medicine, tiny bundles of fibers are used as probes to examine various internal organs without surgery. In communications, the rate at which information can be transmitted is related to the signal frequency. A transmission system using light of frequencies of the order of 10^{14} Hz can transmit information at a much greater rate than one using radio waves, which have frequencies of the order of 10^6 Hz. In telecommunication systems, a single glass fiber the size of a human hair can now transmit audio or video information equivalent to 25,000 voices speaking simultaneously.

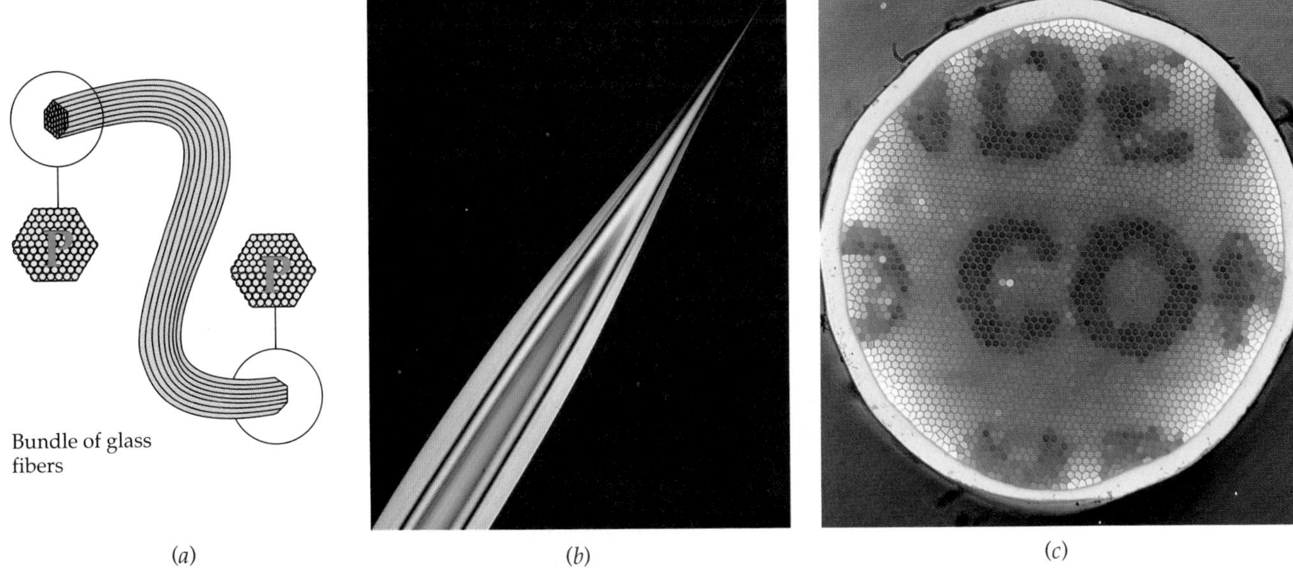

Figure 30-20 (a) Light from the object is transported by a bundle of glass fibers to form an image of the object at the other end of the pipe. (b) The tip of a lightguide preform is softened by heat and drawn into a long tiny fiber. The colors in the preform indicate a layered structure of differing compositions, which is retained in the fiber. (c) An image transmitted through a bundle of fibers.

When the index of refraction of a medium changes gradually, the refraction is continuous, leading to a gradual bending of the light. An interesting example of this is the formation of a mirage. On a hot day, there is often a layer of very hot air near the ground. This air is warmer and is therefore less dense than the air just above it. The speed of light is slightly greater in this less dense layer, so a light beam passing from the cooler layer into the warmer layer is bent. Figure 30-21*a* shows the light from a tree when all the air is at the same temperature. The wavefronts are spherical, and the rays are straight lines. In Figure 30-21*b*, the air near the ground is warmer, resulting in a greater speed of light there. The portions of the wavefronts near the ground travel faster and get ahead of the higher portions, creating a nonspherical wavefront and causing a curving of the rays. Thus, the ray shown initially heading for the ground is bent upwards. As a result, the viewer sees an image of the tree and thinks the light has been reflected from the ground.

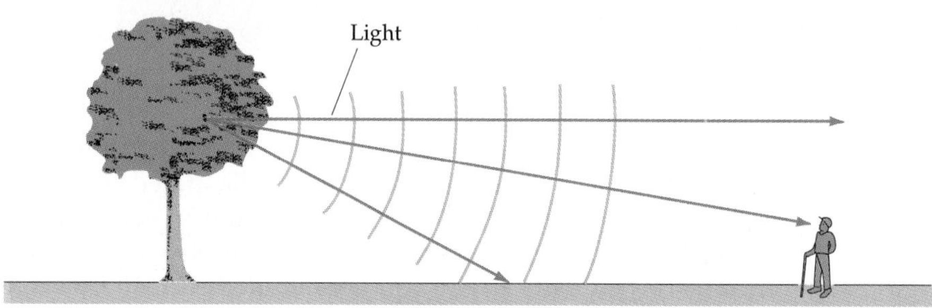

(*c*)

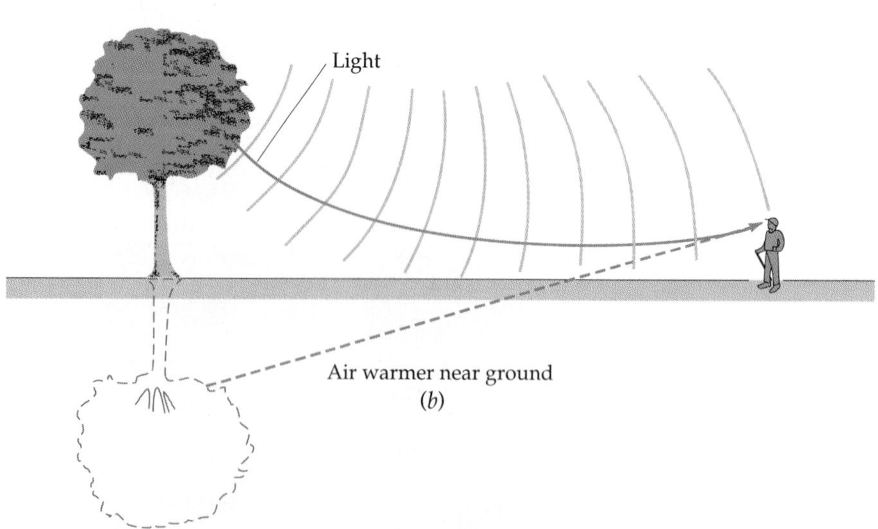

Figure 30-21 A mirage. (*a*) When the air is at a uniform temperature, the wavefronts of the light from the tree are spherical. (*b*) When the air near the ground is warmer, the wavefronts are not spherical and the light from the tree is continuously refracted into a curved path. Because an image of the tree is seen, the viewer may think that there is a reflecting body of water in front of the tree. (*c*) Photograph of apparent reflections of motorcycles and cars on a hot road.

The viewer often attributes this reflection to a water surface near the tree. When driving on a very hot day, you may have noticed apparent wet spots on the highway that disappear when you get to them. This is due to the refraction of light from a hot air layer near the pavement.

Table 30-1 **Index of Refraction for Yellow Sodium Light (λ = 589 nm)**

Substance	Index of refraction	Substance	Index of refraction
Solids		Liquids at 20°C	
Ice (H_2O)	1.309	Methyl alcohol (CH_3OH)	1.329
Fluorite (CaF_2)	1.434	Water (H_2O)	1.333
Rock salt (NaCl)	1.544	Ethyl alcohol (C_2H_5OH)	1.36
Quartz (SiO_2)	1.544	Carbon tetra-chloride (CCl_4)	1.460
Zircon ($ZrO_2 \cdot SiO_2$)	1.923	Turpentine	1.472
Diamond (C)	2.417	Glycerine	1.473
Glasses (typical values)		Benzene	1.501
Crown	1.52	Carbon disulfide (CS_2)	1.628
Light flint	1.58		
Medium flint	1.62		
Dense flint	1.66		

Dispersion

Table 30-1 lists the indexes of refraction for sodium light of wavelength 589 nm for various transparent materials. The index of refraction of a material has a slight dependence on wavelength. Figure 30-22 shows this dependence for several materials. We can see that the indexes of refraction for these materials decrease slightly as the wavelength increases. This dependence of the index of refraction on wavelength (and therefore on frequency) is called **dispersion.** When a beam of white light is incident at some angle on the surface of a glass prism, the angle of refraction for the shorter wavelengths toward the violet end of the visible spectrum is slightly larger than that for the longer wavelengths toward the red end of the spectrum. The light of shorter wavelength is therefore bent more than that of longer wavelength. The beam of white light is thus spread out or dispersed into its component colors or wavelengths (Figure 30-23).

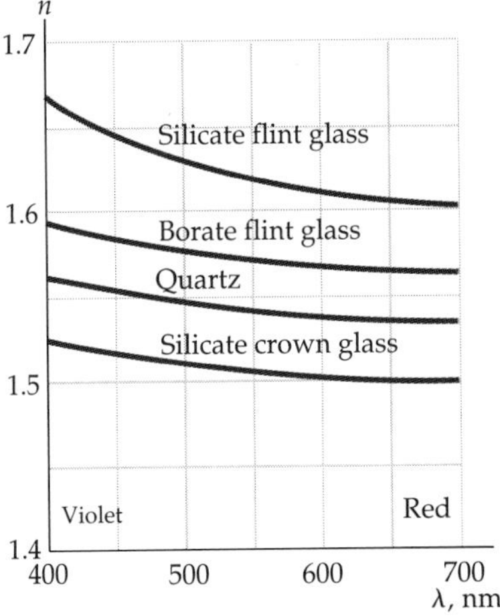

Figure 30-22 The graph of the index of refraction versus wavelength for various materials.

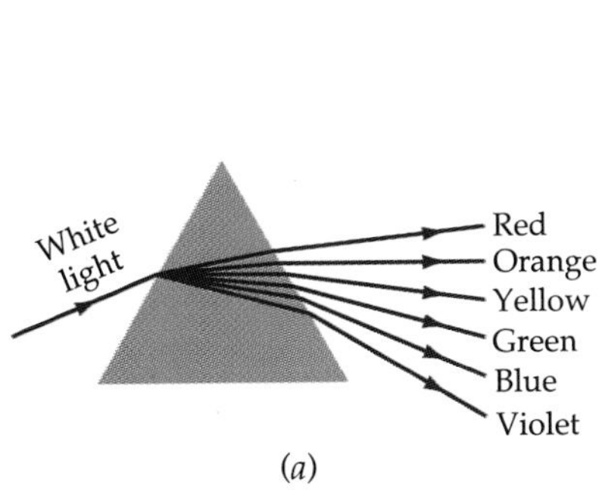

(*a*)

(*b*)

Figure 30-23 (*a*) A beam of white light incident on a glass prism is dispersed into its component colors. The index of refraction decreases as the wavelength increases so that the longer wavelengths (red) are bent less than the shorter wavelengths (blue). (*b*) Photograph of the dispersion of light by a glass prism.

Rainbows

The formation of a rainbow is a familiar example of the dispersion of sunlight by refraction in water drops. Figure 30-24 is a diagram originally drawn by Descartes showing parallel rays of light from the sun entering a spherical water drop. First, the rays are refracted as they enter the drop. They are then reflected from the back water–air surface and finally are refracted again as they leave the drop. Ray 1 enters the drop along a diameter (with an angle of incidence of zero) and is reflected back along its incident path. Ray 2 enters slightly above the diameter and emerges below the diameter at a small angle with it. The rays entering farther and farther away from the diameter emerge at greater and greater angles up to ray 7, shown as the heavy line.

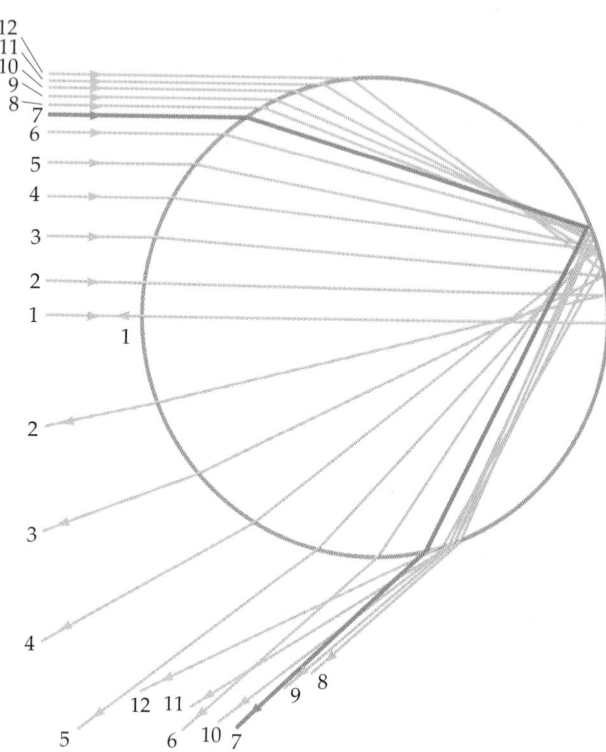

Figure 30-24 Descartes' construction of parallel rays of light entering a spherical water drop. The rays are refracted at the first surface, reflected from the back surface, and refracted again as they leave the drop. The angle between the emerging ray and the diameter increases as we move to rays further away from the diameter up to ray number 7, which emerges at the maximum angle. The concentration of rays emerging at approximately the maximum angle gives rise to the rainbow.

Rays entering above ray 7 emerge at smaller and smaller angles with the diameter. We can see from the diagram that a concentration of rays emerges at angles near the maximum angle. This concentration of rays near the maximum angle gives rise to the rainbow. By construction (using the law of refraction), Descartes showed that the maximum angle is about 42°. To observe a rainbow, we must therefore look at the water drops at an angle of 42° relative to the line back to the sun as shown in Figure 30-25. The angular radius of the rainbow is therefore 42°.

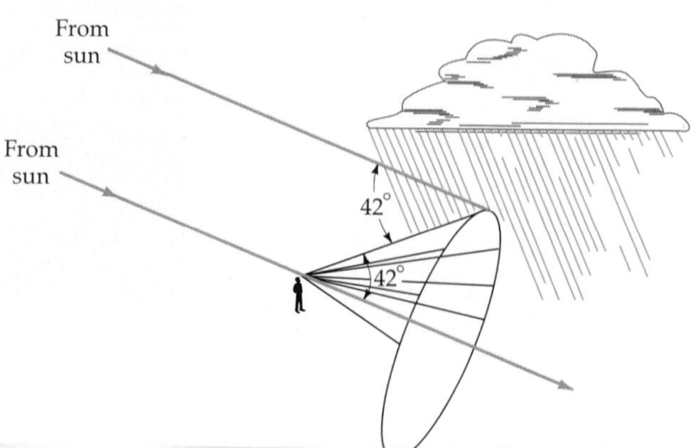

Figure 30-25 A rainbow is viewed at an angle of 42° from the line to the sun, as predicted by Descartes' construction in Figure 30-24.

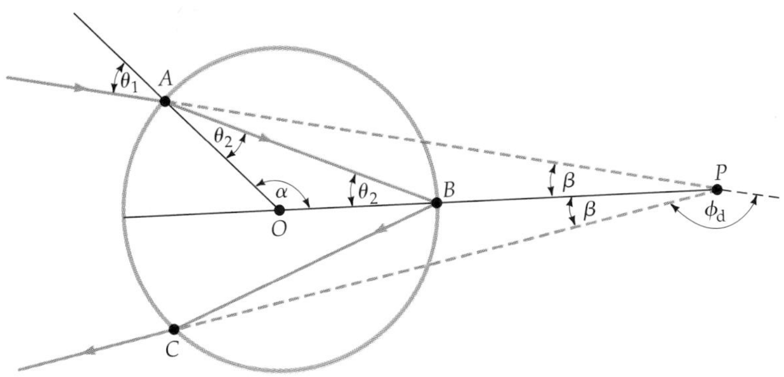

Figure 30-26 Light ray incident on a spherical water drop. The ray is refracted at point A, reflected at point B, and again refracted at point C, where it leaves the drop. The line of the incident ray intersects that of the emerging ray at point P. The angle ϕ_d is called the angle of deviation of the ray.

We can calculate the angular radius of the rainbow from the laws of reflection and refraction. Figure 30-26 shows a ray of light incident on a spherical water droplet at point A. The angle of refraction θ_2 is related to the angle of incidence θ_1 by Snell's law:

$$n_{\text{air}} \sin \theta_1 = n_{\text{water}} \sin \theta_2 \qquad \text{30-8}$$

The refracted ray strikes the back of the water droplet at point B. It makes an angle θ_2 with the radial line OB and is reflected at an equal angle. The ray is refracted again at point C, where it leaves the droplet. Point P is the intersection of the line of the incident ray and the line of the emerging ray. The angle ϕ_d is called the angle of deviation of the ray. It is related to the angle β by

$$\phi_d + 2\beta = \pi \qquad \text{30-9}$$

The angle 2β is the angular radius of the rainbow. We wish to relate the angle of deviation ϕ_d to the angle of incidence θ_1. From the triangle AOB, we have

$$2\theta_2 + \alpha = \pi \qquad \text{30-10}$$

Similarly, from the triangle AOP, we have

$$\theta_1 + \beta + \alpha = \pi \qquad \text{30-11}$$

Eliminating α from Equations 30-10 and 30-11 and solving for β, we obtain

$$\beta = \pi - \theta_1 - \alpha = \pi - \theta_1 - (\pi - 2\theta_2) = 2\theta_2 - \theta_1$$

Substituting this value for β into Equation 30-9, we obtain for the angle of deviation

$$\phi_d = \pi - 2\beta = \pi - 4\theta_2 + 2\theta_1 \qquad \text{30-12}$$

Equation 30-12 can be combined with Snell's law (Equation 30-8) to eliminate θ_2 and give the angle of deviation ϕ_d in terms of the angle of incidence θ_1:

$$\phi_d = \pi + 2\theta_1 - 4 \arcsin\left(\frac{n_{\text{air}} \sin \theta_1}{n_{\text{water}}}\right) \qquad \text{30-13}$$

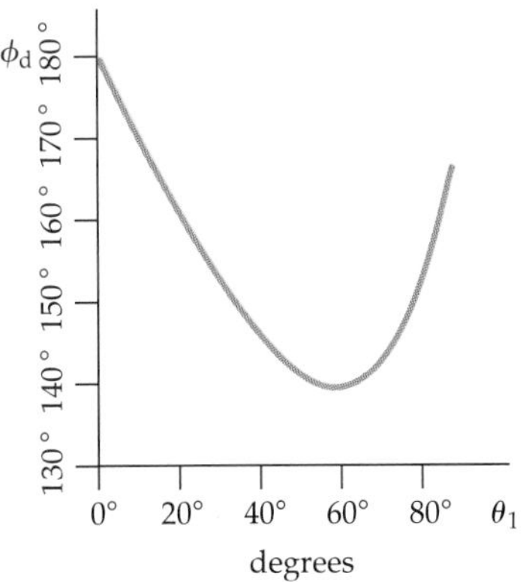

Figure 30-27 Plot of the angle of deviation ϕ_d as a function of incident angle θ_1. The angle of deviation has its minimum value of 138° when the angle of incidence is 60°. Since $d\phi_d/d\theta_1 = 0$ at minimum deviation, the deviation of rays with incident angles slightly less or slightly greater than 60° will be approximately the same.

(The arcsine of a quantity is the angle whose sine is that quantity, that is, the arcsin x is the angle whose sine is x.) Figure 30-27 shows a plot of ϕ_d versus θ_1. The angle of deviation ϕ_d has its minimum value when $\theta_1 = 60°$. At this angle of incidence, the angle of deviation is $\phi_d = 138°$. This angle is called the **angle of minimum deviation.** At incident angles that are slightly greater or slightly smaller than 60°, the angle of deviation is approximately the same. Therefore, the light reflected by the water droplet will be concentrated near the angle of minimum deviation. The angular radius of the rainbow is thus

$$2\beta = \pi - \phi_d = 180° - 138° = 42°$$

(a)

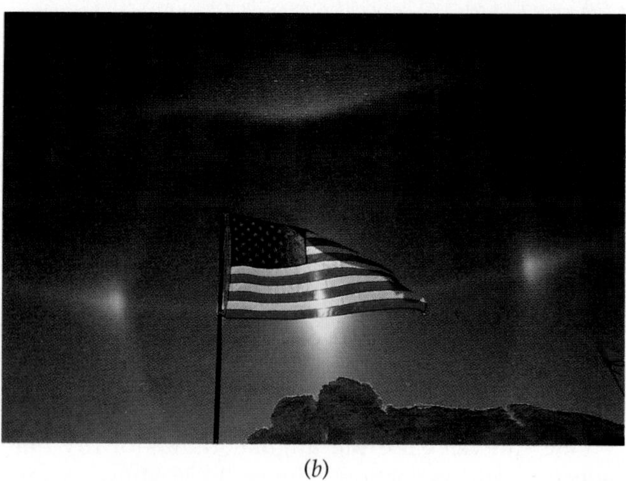

(b)

(a) This 22° halo around the sun results from reflection and refraction from hexagonal ice crystals that are randomly oriented in the atmosphere. (b) When the ice crystals are not randomly oriented but are falling with their flat bases horizontal, only parts of the halo on each side of the sun, called "sun dogs," are seen.

The separation of the colors in the rainbow results from the fact that the index of refraction of water depends slightly on the wavelength of light. The angle of minimum deviation and the angular radius of the bow will therefore depend slightly on the wavelength of the light. The observed rainbow is made up of light rays from many different droplets of water as shown in Figure 30-28. The color seen at a particular angular radius corresponds to the wavelength of light that has an angle of minimum deviation that allows the light to reach the eye from the droplets at that angular radius. For example, since n_{water} is smaller for red light than for blue light, the arcsine in Equation 30-13 is greater for red light than for blue light. Then ϕ_d is smaller for red light than for blue light and β is greater for red light than for blue light. Consequently, the red part of the rainbow is at a slightly greater angular radius than the blue part of the rainbow.

When a light ray strikes a surface separating water and air, part of the light is reflected and part is refracted. A secondary rainbow results from the light rays that are reflected twice within a droplet (Figure 30-29). The secondary bow has an angular radius of 51°, and its color sequence is the reverse of that of the primary bow; that is, the violet is on the outside in the secondary bow. Because of the small fraction of light reflected from a water–air surface, the secondary bow is considerably fainter than the primary bow.

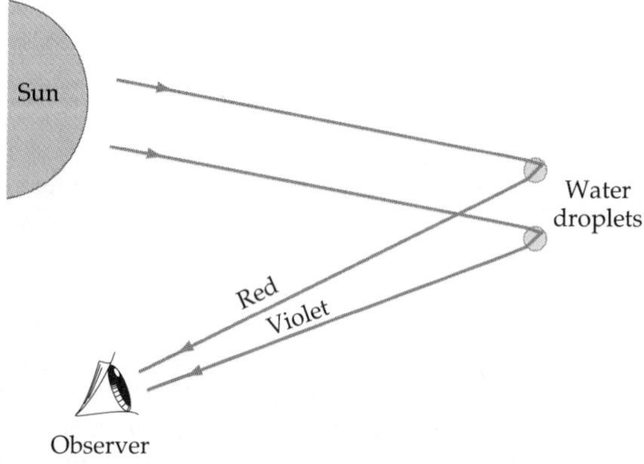

Figure 30-28 The rainbow results from light from many different water droplets.

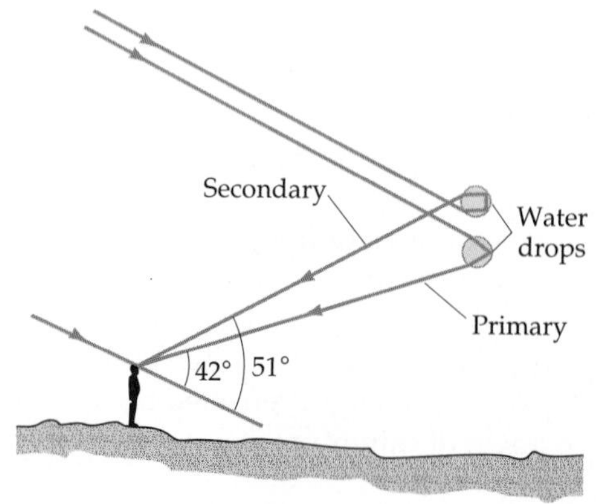

Figure 30-29 The secondary rainbow results from light rays that are reflected twice within a water droplet.

30-5 Fermat's Principle

We have seen that the propagation of light and other waves can be described by Huygens' principle. The propagation of light can also be described by **Fermat's principle** first enunciated in the seventeenth century by the French mathematician Pierre de Fermat:

> The path taken by light in traveling from one point to another is such that the time of travel is a minimum.

Fermat's principle

This statement of Fermat's principle does not cover all cases. The time of travel for the path taken by light is not always a minimum. Sometimes it is a maximum. A more complete and general formulation of Fermat's principle is

> The path taken by light in traveling from one point to another is such that the time of travel is stationary with respect to variations in that path.

Fermat's principle; general form

If the time t is expressed in terms of some parameter x, the path taken by light will be such that $dt/dx = 0$; that is, t is either a minimum, a maximum, or a point of inflection in the t-versus-x curve. The important characteristic of a stationary path is that the time taken along nearby paths will be approximately the same as that along the true path.

In this section, we will use Fermat's principle for alternative derivations of the laws of reflection and refraction.

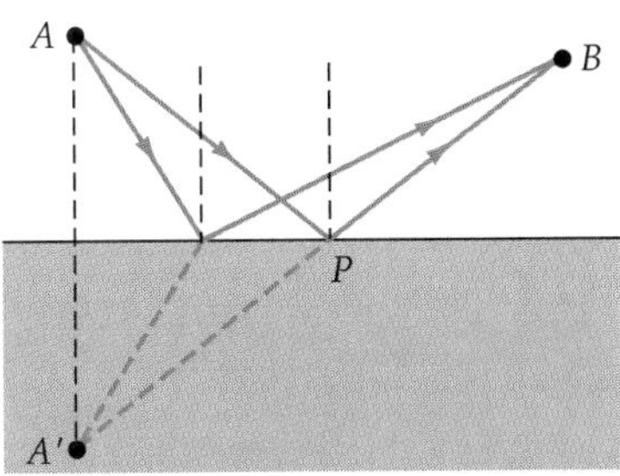

Figure 30-30 Geometry for deriving the law of reflection from Fermat's principle. The time it takes for the light to travel from point A to point B is a minimum when the light strikes the surface at point P.

Reflection

In Figure 30-30, we assume that light leaves point A, strikes the plane surface, which we can consider to be a mirror, and travels to point B. We wish to find the path taken by the light. The problem for the application of Fermat's principle to reflection can be stated as follows: At what point P in Figure 30-30 must the light strike the mirror so that it will travel from point A to point B in the least time? Since the light is traveling in the same medium for this problem, the time will be minimum when the distance is minimum. In Figure 30-30, the distance APB is the same as the distance $A'PB$, where A' is the image of the source A. Point A' lies along the perpendicular from A to the mirror and is equidistant behind the mirror. Obviously, as we vary point P, the distance $A'PB$ is least when the points A', P, and B lie on a straight line. We can see from the figure that this occurs when the angle of incidence equals the angle of reflection.

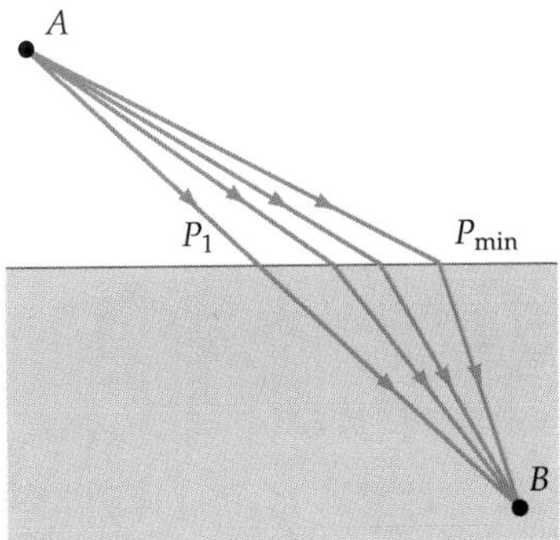

Figure 30-31 Geometry for deriving Snell's law of refraction from Fermat's principle. The point P_{min} is the point at which light must strike the glass for the travel time from A to B to be a minimum.

Refraction

The derivation of Snell's law of refraction from Fermat's principle is more complicated than that for the law of reflection. Figure 30-31 shows the possible paths for light traveling from point A in air to point B in glass. Point P_1 is on the straight line between A and B, but this path is not the one for the shortest travel time because light travels with a smaller speed in the glass. If we move slightly to the right of P_1, the total path length is greater, but the distance traveled in the slower medium is less than for the path through P_1. It is not apparent from the figure which path is that of least time, but it is not surprising that a path slightly to the right of the straight-line path takes less time because the time gained by traveling a shorter distance in the glass more than compensates for the time lost traveling a longer distance in the air. As we move the point of intersection of the possible path to the right of point P_1, the total time of travel from A to B decreases until we reach a minimum at point P_{min}. Beyond this point, the time saved by traveling a shorter distance in the glass does not compensate for the greater time required for the greater distance traveled in the air.

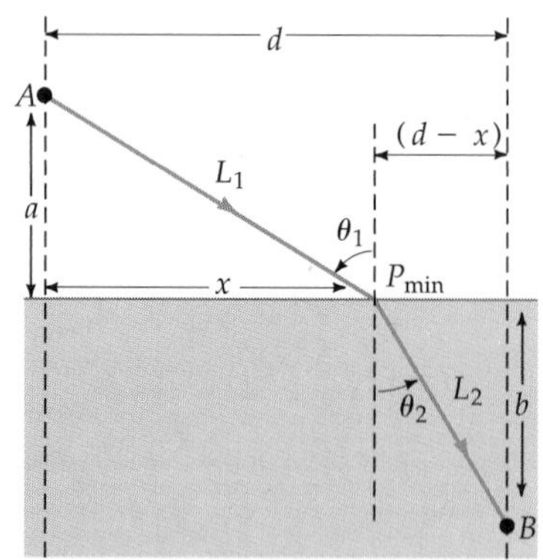

Figure 30-32 Geometry for calculating the minimum time in the derivation of Snell's law from Fermat's principle.

Figure 30-32 shows the geometry for finding the path of least time. If L_1 is the distance traveled in medium 1 with index of refraction n_1 and L_2 is the distance traveled in medium 2 with index of refraction n_2, the time for light to travel the total path AB is

$$t = \frac{L_1}{v_1} + \frac{L_2}{v_2} = \frac{L_1}{c/n_1} + \frac{L_2}{c/n_2} = \frac{n_1 L_1}{c} + \frac{n_2 L_2}{c} \qquad \text{30-14}$$

We wish to find the point P_{min} for which this time is a minimum. We do this by expressing the time in terms of a single parameter indicating the position of point P_{min}. In terms of the distance x in Figure 30-32, we have

$$L_1^2 = a^2 + x^2 \qquad \text{and} \qquad L_2^2 = b^2 + (d - x)^2 \qquad \text{30-15}$$

Figure 30-33 shows the time t as a function of x. At the value of x for which the time is a minimum, the slope of this graph is zero:

$$\frac{dt}{dx} = 0$$

Differentiating each term in Equation 30-14 with respect to x, we obtain

$$\frac{dt}{dx} = \frac{1}{c}\left(n_1 \frac{dL_1}{dx} + n_2 \frac{dL_2}{dx}\right)$$

Setting $dt/dx = 0$, we obtain

$$n_1 \frac{dL_1}{dx} + n_2 \frac{dL_2}{dx} = 0 \qquad \text{30-16}$$

We can compute these derivatives from Equations 30-15. We have

$$2L_1 \frac{dL_1}{dx} = 2x$$

or

$$\frac{dL_1}{dx} = \frac{x}{L_1}$$

But x/L_1 is just $\sin\theta_1$, where θ_1 is the angle of incidence. Thus,

$$\frac{dL_1}{dx} = \sin\theta_1$$

Similarly,

$$2L_2 \frac{dL_2}{dx} = 2(d - x)(-1)$$

or

$$\frac{dL_2}{dx} = -\frac{d - x}{L_2} = -\sin\theta_2$$

where θ_2 is the angle of refraction. Hence, Equation 30-16 is

$$n_1 \sin\theta_1 + n_2(-\sin\theta_2) = 0$$

or

$$n_1 \sin\theta_1 = n_2 \sin\theta_2$$

which is Snell's law.

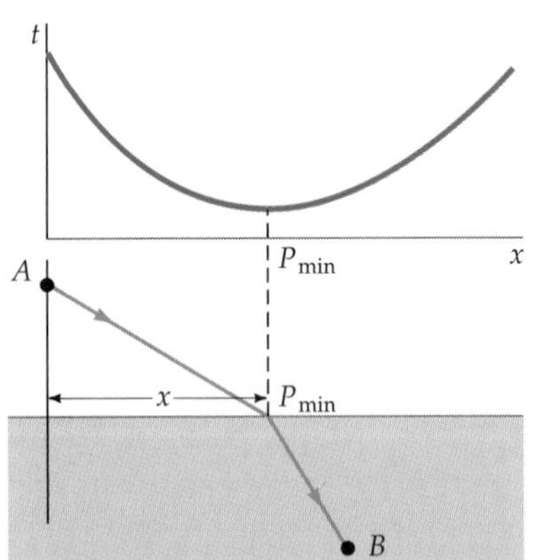

Figure 30-33 Graph of the time it takes for light to travel from A to B versus x, measured along the refracting surface. The time is a minimum at the point at which the angles of incidence and refraction obey Snell's law.

30-6 Polarization

In any transverse wave, the vibration is perpendicular to the direction of propagation of the wave. For example, in waves moving down a string, the elements of the string move in a plane perpendicular to the string. Similarly, in a light wave traveling in the z direction, the electric field is perpendicular to the z direction. (The magnetic field of a light wave is also perpendicular to the z direction.) If the vibration of a transverse wave remains parallel to a fixed line in space, the wave is said to be **linearly polarized.** We can visualize polarization most easily by considering mechanical waves on a string. If one end is moved up and down, the resulting waves on the string are linearly polarized with each element of the string vibrating in the vertical direction. Similarly, if one end is moved along a horizontal line (perpendicular to the string), the displacements of the string are linearly polarized in the horizontal direction. If one end of the string is moved with constant speed in a circle, the resulting wave is said to be **circularly polarized.** In this case, each element of the string moves in a circle. Unpolarized waves can be produced by moving the end of the string vertically and horizontally in a random way. Then, if the string itself is in the z direction, the vibrations will have both x and y components that vary in a random way.

Most waves produced by a single source are polarized. For example, string waves produced by the regular vibration of one end of a string or electromagnetic waves produced by a single atom or by a single antenna are polarized. Waves produced by many sources are usually unpolarized. A typical light source, for example, contains millions of atoms acting independently. The electric field for such a wave can be resolved into x and y components that vary randomly because there is no correlation between the individual atoms producing the light.

There are four phenomena that produce polarized light from unpolarized light: (1) absorption, (2) scattering, (3) reflection, and (4) birefringence (also called double refraction).

Polarization by Absorption

Several naturally occurring crystals, when cut into appropriate shapes, absorb and transmit light differently depending on the polarization of the light. These crystals can be used to produce linearly polarized light. In 1938, E. H. Land invented a simple commercial polarizing film called Polaroid. This material contains long-chain hydrocarbon molecules that are aligned when the sheet is stretched in one direction during the manufacturing process. These chains become conducting at optical frequencies when the sheet is dipped in a solution containing iodine. When light is incident with its electric-field vector parallel to the chains, electric currents are set up along the chains, and the light energy is absorbed. If the electric field is perpendicular to the chains, the light is transmitted. The direction perpendicular to the chains is called the **transmission axis.** We will make the simplifying assumption that all the light is transmitted when the electric field is parallel to the transmission axis and all is absorbed when it is perpendicular to the transmission axis.

Consider an unpolarized light beam traveling in the z direction incident on a polarizing film with its transmission axis in the y direction. On the average, half of the incident light has its electric field in the y direction and half has it in the x direction. Thus, half the intensity is transmitted, and the transmitted light is linearly polarized with its electric field in the y direction.

Suppose we have a second polarizing film whose transmission axis makes an angle θ with that of the first as shown in Figure 30-34. If $\mathbf{E}$ is the

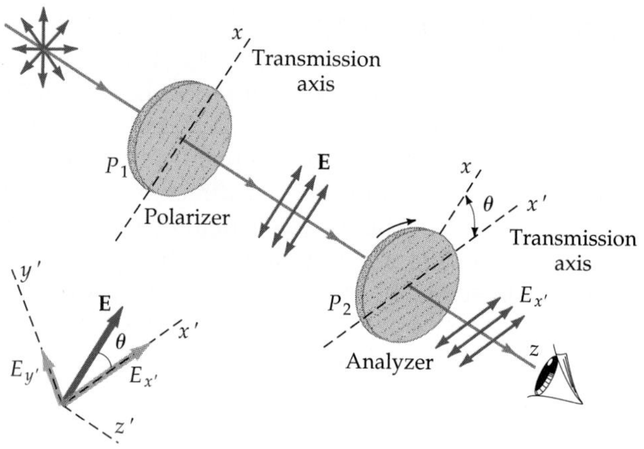

Figure 30-34 Two polarizing films with their transmission axes making an angle θ with each other. Only the component $E \cos \theta$ is transmitted through the second film. If the intensity between the films is I_0, that transmitted by both films is $I_0 \cos^2 \theta$.

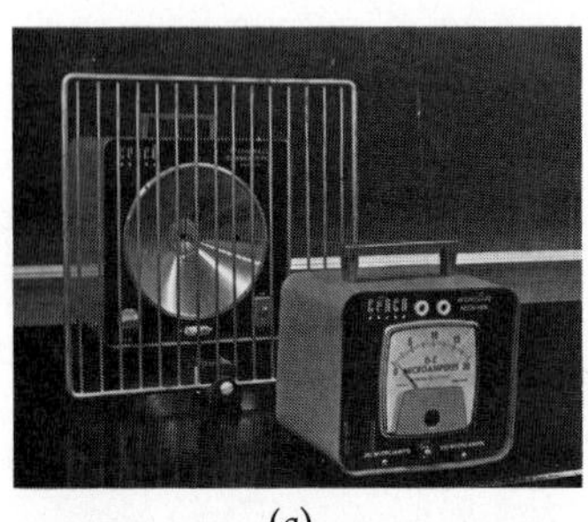

(*a*)

(*b*)

Figure 30-35 Demonstration showing the polarization of microwaves. The electric field of the microwaves is vertical, parallel to the vertical dipole antenna. (*a*) When the metal wires of the absorber are vertical, electric currents are set up in the wires and energy is absorbed, as indicated by the low reading on the detector. (*b*) When the wires are horizontal, no currents are set up, and the microwaves are transmitted, as indicated by the high reading on the detector. The transmission axis for the wire screen is perpendicular to the wires.

electric field between the films, its component along the direction of the transmission axis of the second film is $E \cos \theta$. Since the intensity of light is proportional to E^2, the intensity of light transmitted by both films will be given by

$$I = I_0 \cos^2 \theta \tag{30-17}$$

where I_0 is the intensity incident on the second film and is, of course, half the intensity incident on the first film. When two polarizing elements are placed in succession in a beam of light as described here, the first is called the **polarizer** and the second is called the **analyzer.** If the polarizer and analyzer are crossed, that is, if their transmission axes are perpendicular to each other, no light gets through. Equation 30-17 is known as **Malus's law** after its discoverer E. L. Malus (1775–1812). It applies to any two polarizing elements whose transmission axes make an angle θ with each other.

The polarization of electromagnetic waves can be demonstrated with microwaves, which have wavelengths on the order of centimeters. In a typical microwave generator, polarized waves are radiated by a dipole antenna. In Figure 30-35, the dipole antenna is vertical, so the electric-field vector **E** of the radiated waves is vertical. An absorber can be made of a screen of parallel straight wires. When the wires are vertical, as in Figure 30-35*a*, the electric field parallel to the wires sets up currents in the wires and energy is absorbed. When the wires are horizontal and so perpendicular to **E**, as in Figure 30-35*b*, no currents are set up and the waves are transmitted. The transmission axis of the wire screen is thus perpendicular to the wires.

Example 30-7

Unpolarized light of intensity 3.0 W/m^2 is incident on two polarizing films whose transmission axes make an angle of 60°. What is the intensity of light transmitted by both films?

The intensity of the light that is transmitted by the first film is half the incident intensity, or 1.5 W/m^2. Calling this intensity I_0, that transmitted by the second film is

$$I = I_0 \cos^2 \theta = (1.5\ \text{W/m}^2) \cos^2 60° = (1.5\ \text{W/m}^2)(0.500)^2 = 0.375\ \text{W/m}^2$$

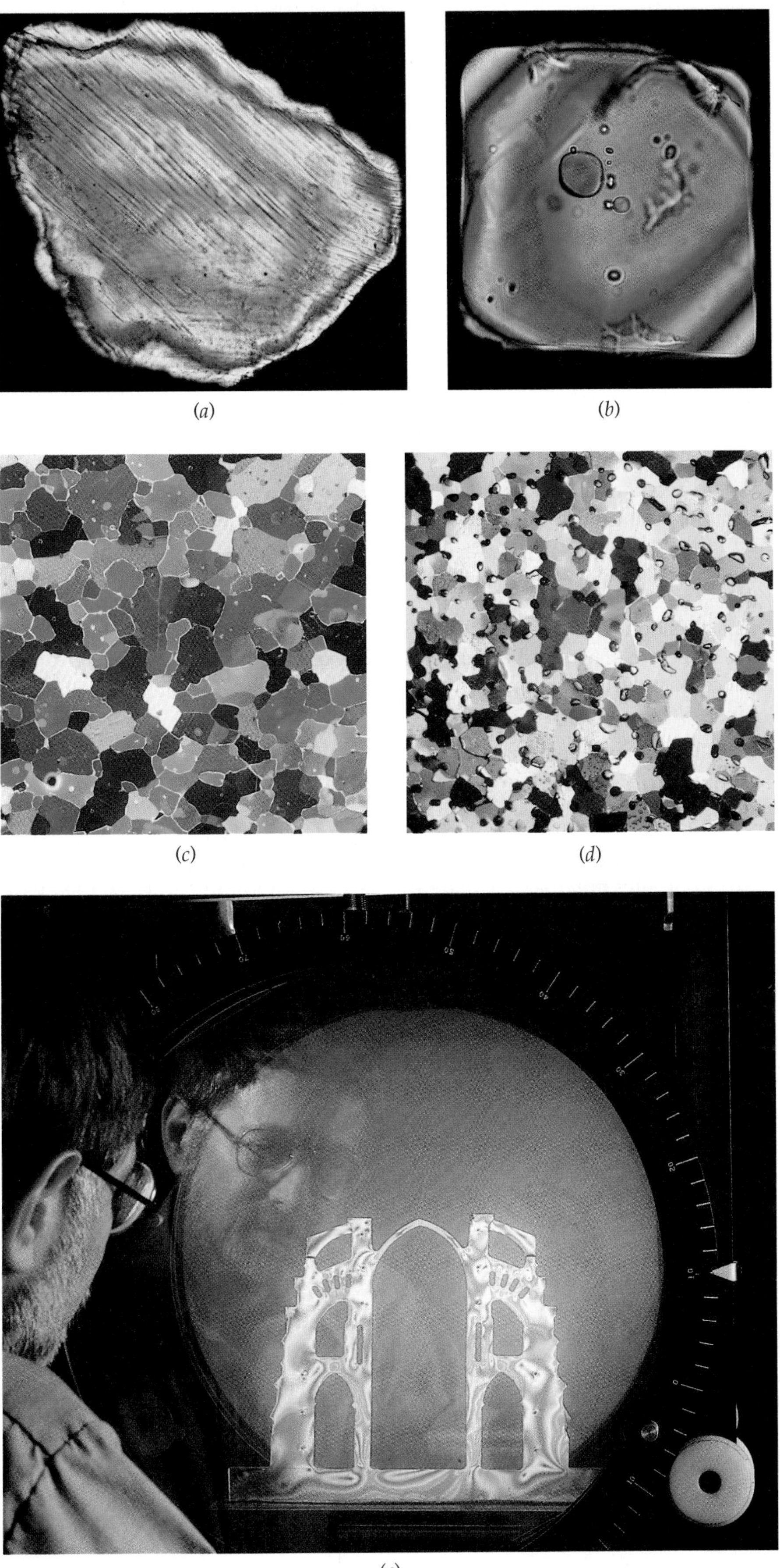

(*a*) (*b*) (*c*) (*d*) (*e*)

When the transmission axes of two polarizing films are perpendicular, the polarizers are said to be crossed and no light is transmitted. However, many materials are birefringent (page 1001) or become so under stress. Such materials rotate the direction of polarization of the light so that light of a particular wavelength is transmitted through both polarizers. When a birefringent material is viewed between crossed polarizers, information about its internal structure is revealed. (*a*) A shocked quartz grain from the site of a meteorite crater. The layered structure, evidenced by the parallel lines, arises from the shock of the impact of the meteor. (*b*) A grain of quartz typically found in silicic volcanic rocks. No shock lines are seen. (*c*) Thin sections of ice core from the antarctic ice sheet reveal bubbles of trapped CO_2, which appear amber-colored. This sample was taken from a depth of 194 meters corresponding to air trapped 1600 years ago, whereas that in (*d*) is from a depth of 56 meters corresponding to air trapped 450 years ago. Ice core measurements have replaced the less reliable technique of analyzing carbon in tree rings to compare current atmospheric CO_2 levels with those of the recent past. (*e*) Robert Mark of the Princeton School of Architecture examines the stress patterns in a plastic model of the nave structure of Chartres Cathedral.

Polarization by Scattering

The phenomenon of absorption and reradiation is called **scattering.** Scattering can be demonstrated by passing a light beam through a container of water to which a small amount of powdered milk has been added. The milk particles absorb light and reradiate it, making the light beam visible. Similarly, laser beams can be made visible by introducing chalk or smoke particles into the air to scatter the light. A familiar example of light scattering is that from air molecules, which tend to scatter short wavelengths more than long wavelengths, thereby giving the sky its blue color.

We can understand polarization by scattering if we think of an absorbing molecule as an electric-dipole antenna that radiates waves with a maximum intensity in the direction perpendicular to the antenna with the electric-field vector parallel to the antenna and zero intensity in the direction along the antenna. Figure 30-36 shows a beam of unpolarized light initially traveling along the z axis striking a scattering center at the origin. The electric field in the light beam has components in both the x and y directions perpendicular to the direction of motion of the light beam. These fields set up oscillations of the scattering center in both the x and y directions, but there is no oscillation in the z direction. The oscillation of the scattering center in the x direction produces light along the y axis but not along the x axis, which is along the line of oscillation. The light radiated along the y axis is thus polarized in the x direction. Similarly, the light radiated along the x axis is polarized in the y direction. This can be seen easily by examining the scattered light with a piece of polarizing film.

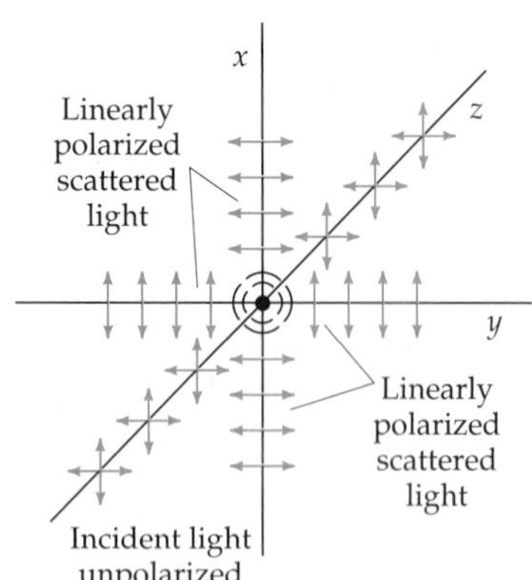

Figure 30-36 Polarization by scattering. Unpolarized light propagating in the z direction is incident on a scattering center at the origin. The light scattered in the x direction is polarized in the y direction and that scattered in the y direction is polarized in the x direction.

Polarization by Reflection

When unpolarized light is reflected from a plane surface boundary between two transparent media, such as air and glass or air and water, the reflected light is partially polarized. The degree of polarization depends on the angle of incidence and the indexes of refraction of the two media. When the angle of incidence is such that the reflected and refracted rays are perpendicular to each other, the reflected light is completely polarized. This result was discovered experimentally by Sir David Brewster in 1812.

Figure 30-37 shows light incident at the polarizing angle θ_p for which the reflected light is completely polarized. The electric field of the incident light can be resolved into components parallel and perpendicular to the plane of incidence. The reflected light is completely polarized with its electric field perpendicular to the plane of incidence. We can relate the polarizing angle to the indexes of refraction of the media using Snell's law. If n_1 is the index of refraction of the first medium and n_2 is that of the second medium, Snell's law gives

$$n_1 \sin \theta_p = n_2 \sin \theta_2$$

where θ_2 is the angle of refraction. From Figure 30-37, we can see that the sum of the angle of reflection and the angle of refraction is 90°. Since the angle of reflection equals the angle of incidence, we have

$$\theta_2 = 90° - \theta_p$$

Then

$$n_1 \sin \theta_p = n_2 \sin (90° - \theta_p) = n_2 \cos \theta_p$$

or

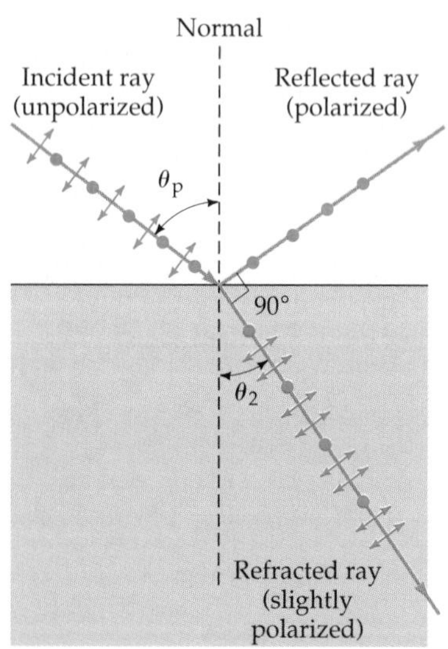

Figure 30-37 Polarization by reflection. The incident wave is unpolarized and has components of the electric field parallel to the plane of incidence (arrows) and components perpendicular to this plane (dots). For incidence at the polarizing angle, the reflected wave is completely polarized, with its electric field perpendicular to the plane of incidence.

$$\tan \theta_p = \frac{n_2}{n_1} \qquad 30\text{-}18$$

Brewster's law

Equation 30-18 is known as **Brewster's law.** Although the reflected light is completely polarized when the incident angle is θ_p, the transmitted light is only partially polarized because only a small fraction of the incident light is reflected. If the incident light itself is polarized with **E** in the plane of incidence, there is no reflected light when the angle of incidence is θ_p. We can understand this qualitatively from Figure 30-38. If we consider the molecules in the second medium to be oscillating parallel to the electric field of the refracted ray, there can be no reflected ray because no energy is radiated along the line of oscillation.

Because of the polarization of reflected light, sunglasses made of polarizing material can be very effective in cutting out glare. If light is reflected from a horizontal surface such as a lake or snow on the ground, the plane of incidence will be vertical and the electric field of the reflected light will be predominantly horizontal. Polarized sunglasses with a vertical transmission axis will then reduce glare by absorbing much of the reflected light. If you have polarized sunglasses, you can observe this by looking through them at such reflected light and then rotating the glasses 90°, so that much more of the light is transmitted.

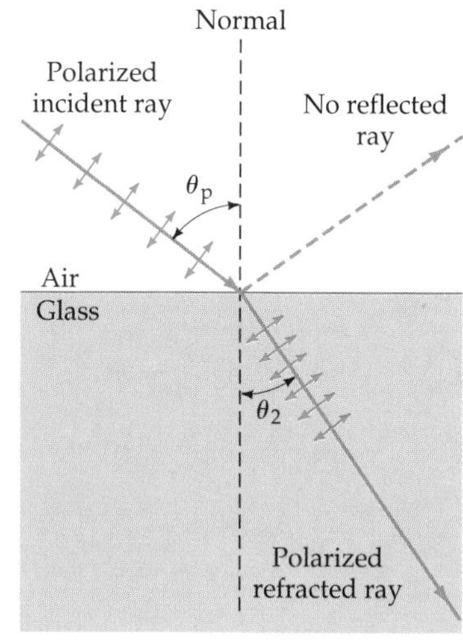

Figure 30-38 Polarized light incident at the polarizing angle. When the light is polarized with **E** in the plane of incidence there is no reflected ray.

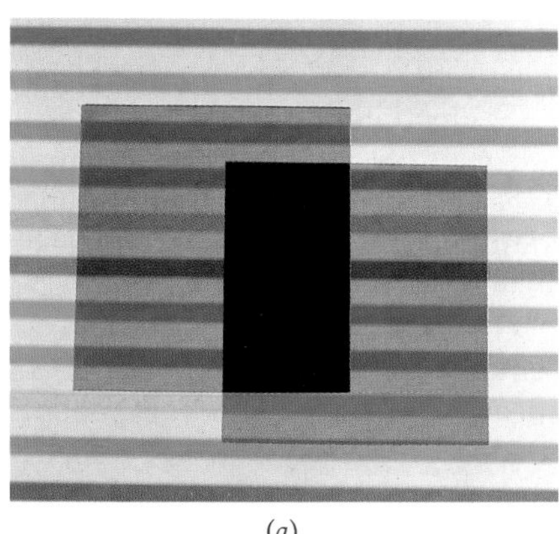

(*a*)

(*b*)

(*a*) Cross polarizers block out all of the light. (*b*) In a liquid crystal display, the crystal is between crossed polarizers. Light incident on the crystal is transmitted because the crystal rotates the direction of polarization of the light 90°. The light is reflected back out through the crystal by a mirror behind the crystal, and a uniform background is seen. When a voltage is applied across a small segment of the crystal, the polarization is not rotated, so no light is transmitted and the segment appears black.

Polarization by Birefringence

Birefringence, or **double refraction,** is a complicated phenomenon that occurs in calcite and other noncubic crystals and in some stressed plastics such as cellophane. In most materials, the speed of light is the same in all directions. Such materials are **isotropic.** Because of their atomic structure, birefringent materials are **anisotropic.** The speed of light depends on its direction of propagation through the material. When a light ray is incident on such materials, it may be separated into two rays called the *ordinary ray* and the *extraordinary ray.* These rays are polarized in mutually perpendicular directions, and they travel with different speeds. Depending on the relative orientation of the material and the incident light, the rays may also travel in different directions.

There is one particular direction in a birefringent material in which both rays propagate with the same speed. This direction is called the **optic axis** of the material. (The optic axis is actually a *direction* rather than a line in the material.) Nothing unusual happens when light travels along the optic axis.

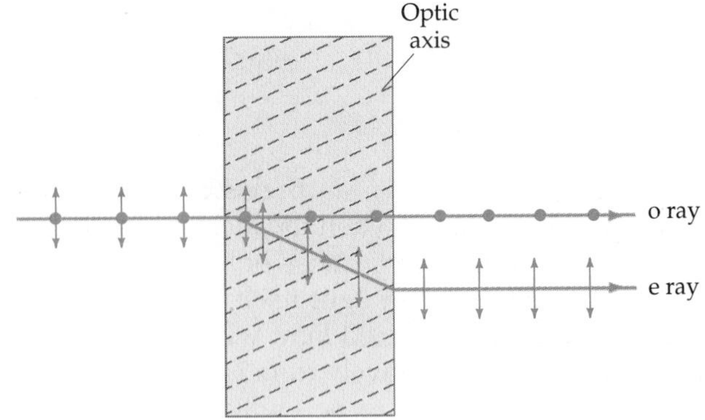

Figure 30-39 A narrow beam of light incident on a birefringent crystal such as calcite is split into two beams, called the ordinary ray (o ray) and the extraordinary ray (e ray) that have mutually perpendicular polarizations. If the crystal is rotated, the extraordinary ray rotates in space.

However, when light is incident at an angle to the optic axis as shown in Figure 30-39, the rays travel in different directions and emerge separated in space. If the material is rotated, the extraordinary ray (the e ray in the figure) rotates in space.

If light is incident on a birefringent plate perpendicular to its crystal face and perpendicular to the optic axis, the two rays travel in the same direction but at different speeds. The number of wavelengths in the two rays in the plate is different because the wavelengths ($\lambda = v/f$) of the rays differ. The rays emerge with a phase difference that depends on the thickness of the plate and on the wavelength of the incident light. In a **quarter-wave plate,** the thickness is such that there is a 90° phase difference between the waves of a particular wavelength when they emerge. In a **half-wave plate,** the rays emerge with a phase difference of 180°.

Suppose that the incident light is linearly polarized such that the electric-field vector is at 45° to the optic axis, as illustrated in Figure 30-40. The ordinary and extraordinary rays start out in phase and have equal amplitudes. With a quarter-wave plate, the waves emerge with a phase difference of 90°, so the resultant electric field has components $E_x = E_0 \sin \omega t$ and $E_y = E_0 \sin (\omega t + 90°) = E_0 \cos \omega t$. The electric-field vector thus rotates in a circle and the wave is circularly polarized.

With a half-wave plate, the waves emerge with a phase difference of 180°, so the resultant electric field is linearly polarized with components $E_x = E_0 \sin \omega t$ and $E_y = E_0 \sin (\omega t + 180°) = -E_0 \sin \omega t$. The net effect is that the direction of polarization of the wave is rotated by 90° relative to that of the incident light as shown in Figure 30-41.

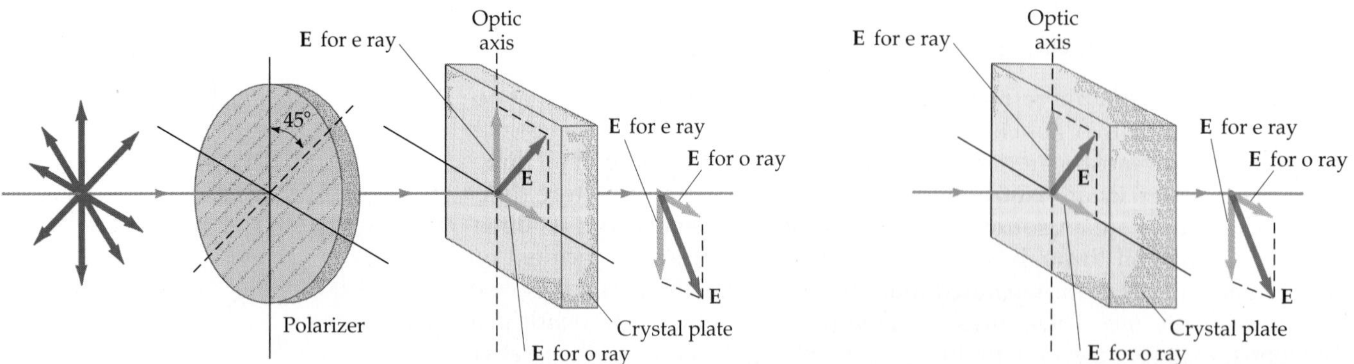

Figure 30-40 Polarized light emerging from the polarizer is incident on a birefringent crystal such that the electric-field vector makes a 45° angle with the optic axis, which is perpendicular to the light beam. The ordinary and extraordinary rays travel in the same direction but at different speeds.

Figure 30-41 When the birefringent crystal in Figure 30-40 is a half-wave plate, the emerging light has its direction of polarization rotated.

Interesting and beautiful patterns, like those on page 999, can be observed by placing birefringent materials, such as cellophane or a piece of stressed plastic, between two polarizing sheets with their transmission axes perpendicular to each other (crossed Polaroids). Ordinarily, no light is transmitted through crossed polarizing sheets because the polarization direction of light transmitted by the first sheet is perpendicular to the axis of the second. However, if we place a birefringent material between the crossed Polaroids, the material acts as a half-wave plate for light of a certain color depending on the material's thickness. The direction of polarization is rotated and some light gets through both films. Various glasses and plastics become birefringent when under stress. The stress patterns can be observed when the material is placed between crossed polarizing sheets.

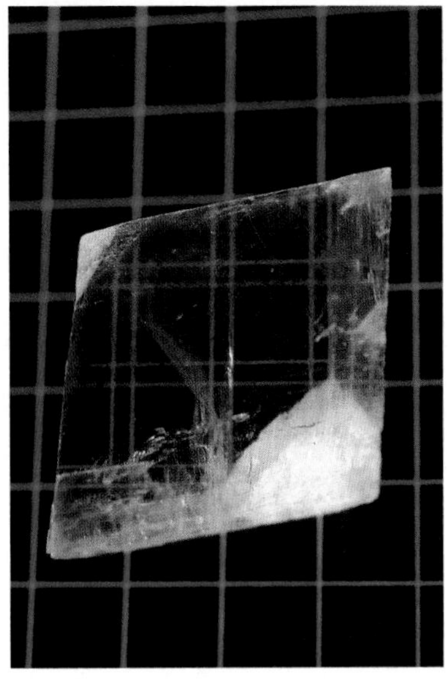

A double image of the cross hatching is produced by this birefringent crystal of calcium carbonate.

Summary

1. When light is incident on a surface separating two media in which the speed of light differs, part of the light energy is transmitted and part is reflected. The angle of reflection equals the angle of incidence:

$$\theta_r = \theta_1$$

The angle of refraction depends on the angle of incidence and on the indexes of refraction of the two media and is given by Snell's law of refraction:

$$n_1 \sin \theta_1 = n_2 \sin \theta_2$$

where the index of refraction of a medium n is the ratio of the speed of light in a vacuum c to that in the medium v:

$$n = \frac{c}{v}$$

2. When light is traveling in a medium with an index of refraction n_1 and is incident on the boundary of a second medium with a lower index of refraction $n_2 < n_1$, the light is totally reflected if the angle of incidence is greater than the critical angle θ_c given by

$$\sin \theta_c = \frac{n_2}{n_1}$$

3. The speed of light in a medium and therefore the index of refraction of that medium depends on the wavelength of light, a phenomenon known as dispersion. Because of dispersion, a beam of white light incident on a refracting prism is dispersed into its component colors. Similarly, the reflection and refraction of sunlight by raindrops produces a rainbow.

4. When two polarizers have their transmission axes at an angle θ, the intensity transmitted by the second polarizer is reduced by the factor $\cos^2 \theta$, a result known as Malus's law. If I_0 is the intensity of the light between the polarizers, the intensity transmitted by the second polarizer is

$$I = I_0 \cos^2 \theta$$

5. The four phenomena that produce polarized light from unpolarized light are (1) absorption, (2) scattering, (3) reflection, and (4) birefringence.

Beyond the (Visible) Rainbow*

Robert Greenler
University of Wisconsin, Milwaukee

Sometimes science is a very personal activity. Ever since I was a small boy, I have been excited by the beauty and grandeur of the rainbow. This was my reaction long before I had acquired the tools of the scientist with which to understand the origin of this marvelous arch of color.

A professional interest in which I have invested a considerable amount of energy over the past three decades is the understanding of the structure of molecules that become attached (adsorbed) to the surface of a solid material. This understanding is important for such a diverse assortment of phenomena as the functioning of catalysts, the electrical properties of small integrated circuits, the separation of ores, and the processes that take place within a fusion reactor.

It might seem that this interest would have nothing to do with rainbows, but not so. I have developed a technique for deducing the structure of molecules adsorbed on a metal surface using infrared radiation. So, some understanding of the nature of infrared radiation is one of the tools of my scientific trade.

These two different strands of my experience came together one day while I was sitting at my desk woolgathering rather than addressing the task at hand. The question occurred to me: I wonder if there is an infrared rainbow in the sky?

How does one explore such a question? Here is the process I went through. For there to be an infrared rainbow, a number of conditions must be met. *First,* the source of light must emit infrared radiation (the sun emits light over the entire electromagnetic spectrum from x-rays to radio waves; see Section 29-5). *Second,* the infrared radiation must pass through the Earth's atmosphere (water vapor and carbon dioxide in the atmosphere absorb some infrared wavelengths, as discussed in the essay on global warming, but others pass through unimpeded). The rainbow is caused by light rays that enter a droplet of water and are reflected internally before emerging from the drop (see Figure 30-26 and Section 30-4). For there to be an infrared rainbow, the *third* requirement is that the infrared rays would have to pass through a water droplet. This is a serious consideration. Just because a droplet of water appears transparent in visible light, we cannot assume that it is transparent to infrared "light"; indeed, liquid water does absorb over a broad range of infrared wavelengths. However, the measured spectral transmittance of water shows that water drops should be quite transparent from the visible region out to an infrared wavelength of about 1300 nm. *Finally,* after emerging from the raindrop, the infrared rays that have survived all these losses must again pass through air to the (unseeing) eye of the would-be observer.

The Search

This line of reasoning produced a tentative answer to the question that prompted the speculation: Yes, there should be an infrared rainbow in the sky and it should lie in a band just outside of the red of the visible rainbow.

I decided to try to photograph this invisible bow using a film that is sensitive to a portion of the infrared spectrum. Figure 1 shows the curve of the sensitivity of the film. The figure also shows a curve of the sensitivity of the human eye, as a way of defining the limits of the visible spectral region (extending from about 400 nm at the violet end of the spectrum to 700 nm at the red end). Note that the infrared film has a sensitivity extending out to about 930 nm.

*This essay is adapted from an article that first appeared in *Optic News,* published by the Optical Society of America, in November 1988.

Robert Greenler has been professor of physics at the University of Wisconsin–Milwaukee since 1962, where he has been instrumental in the development of the Laboratory for Surface Studies at Milwaukee. His research interests include the study of the structure of molecules adsorbed on solid surfaces, optical effects of the sky, and the understanding of the iridescent colors seen in many biological organisms.

The pursuit of rainbows (and other scientific interests) has taken him to the University of East Anglia in Norwich, England; the Fritz Haber Institute in West Berlin, Germany; the Institute Teknologi MARA in Shah Alam, Malaysia; and the U.S. Antarctic research station at the South Pole.

He served as President of the Optical Society of America in 1987 and, in 1988, he received the Millikan Lecture Award of the American Association of Physics Teachers for "notable and creative contributions to the teaching of physics."

The problem in using this film to record an infrared scene is that the film is not only sensitive to the infrared but throughout the visible region (it is *very* sensitive to blue light). If we were to look at the black-and-white image produced by such a film, we would have no way of knowing which parts of the image resulted from exposure to infrared and which parts from exposure to visible radiation. This problem was solved by using a filter that appears to be an opaque sheet of black plastic. The material is opaque to visible light and transmits only wavelengths longer than about 800 nm. As can be seen from Figure 1, this combination of film and filter will permit the recording of only those wavelengths in a band between 800 and 930 nm, well removed from the visible spectral region.

The Capture

Anyone who has tried to photograph rainbows knows that they usually occur when a camera is not at hand and fade just before one is located. I decided to first try an easier subject—that of a rainbow in a water spray that I could turn on at my convenience in my backyard. Figure 2 shows one of the first photographic results. A garden hose with many holes was wrapped back and forth across a board resting on top of the ladder. And in the spray of the hose—the infrared rainbow! You can also see the fainter, secondary rainbow outside the brighter, primary bow. This corresponds to the secondary bow seen rather commonly in visible light and it results from rays that enter a water droplet and experience two internal reflections before leaving the drop (see Figure 30-29).

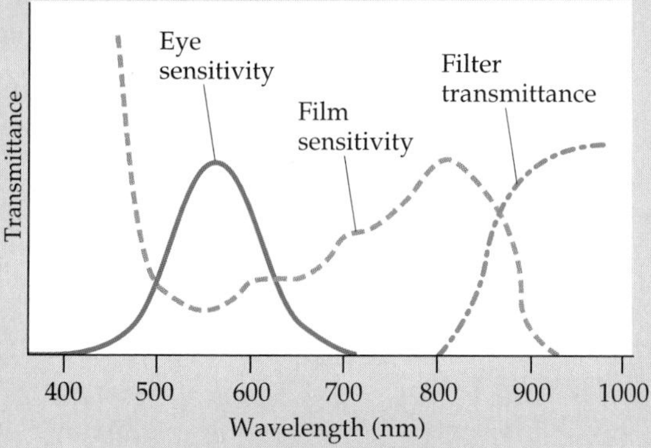

Figure 1 The infrared film (Eastman Kodak infrared film IR 135) has a sensitivity extending throughout the visible and into the near infrared region. The filter (Eastman Kodak 87C infrared transmitting filter) is opaque to visible light but transmits in the infrared for wavelengths longer than 800 nm. The combination of film and filter records an image with wavelengths between 800 and 930 nm, well outside the visible spectrum.

Figure 2 An infrared rainbow photographed in the water spray of a garden hose. The fainter secondary bow is shown outside (to the left of) the primary bow. The fringes seen inside (to the right of) the primary bow are caused by interference effects. A rare interference fringe outside the secondary bow is visible on the original photograph but may be difficult to pick out on the reproduction.

There is another interesting feature in this infrared photograph: immediately *inside* (to the right side of) the bright primary bow there is another bright band—or perhaps two bands. Such fringes, sometimes seen inside a visible bow, are called supernumerary bows and result from the interference of light waves (see Chapter 33).

Close inspection of the negative of Figure 2 reveals yet another feature, which is difficult to reproduce in a printed picture. There is a faint fringe just *outside* the secondary bow. A process similar to the one that produces supernumerary bows inside the primary should, in theory, produce a similar set of fringes outside the secondary. I have never seen any of those fringes associated with any rainbow or with any rainbow photograph, but they are visible in the original of this photograph. They are even visible when I project the slide onto a screen. For the first attempt, that was quite an exciting collection of effects.

Other Effects in the Infrared Photos

Some other features of these infrared photographs are worth considering. If the only radiation that produced these photographic images is invisible, infrared radiation, is it surprising that we can see the ladder, trees, and grass? One should not be too surprised. These objects absorb some wavelengths and reflect or scatter others. Objects that absorb the infrared appear dark in the photos and those that scatter it strongly appear bright. To make it clear just what these photos show, we need to understand the difference between reflected (or scattered) radiation and emitted radiation.

Continued

Normally when you look at objects in your landscape, you see them only by the light they scatter. However, if the temperature of an object is high enough, it emits light (see discussion of radiation in Section 16-3). If it is very hot—you might call it "white hot"—it emits a broad spectrum of wavelengths with the peak of the emission curve in the visible spectrum. If the object cools down a bit, the peak in the emission curve moves to longer wavelengths. The result is that there is more red light than blue light being emitted, and the appropriate description for its temperature is "red hot." At a lower temperature, you might see a dull red glow. At this point, the peak of emission curve is in the infrared with just a small amount of emission in the red end of the visible region. At a slightly lower temperature, the object appears dark; the emission peak has moved further into the infrared, so no visible radiation can be seen. If the object cools to where it is warm to the touch, its emission peak is far out in the infrared—perhaps at 10,000 nm—and it is emitting almost nothing in the visible or in the near infrared region to which photographic films are sensitive.

If, however, you could produce a picture with 10,000-nm radiation, objects slightly warmer than their surroundings would appear to be bright—they would be glowing with emitted infrared radiation. There *are* ways to produce such pictures; they are used to show sources of heat loss in homes or to record relatively warm spots on a human body (thermograms) that may indicate the site of some physical disorder. These pictures are usually described as infrared pictures, but they are quite different from the photographs taken with infrared-sensitive film. This film is sensitive only to the near infrared, but the "heat pictures" result from emitted radiation in the far infrared. So the infrared photographs shown here show only the infrared radiation from the sun that is scattered by the leaves or ladder or transformed by raindrop spheres into an invisible rainbow.

Another interesting feature of the photographs is the darkness of the clear sky background. We see light in the clear, clean sky, away from the sun, due to scattering from the molecules of gases in the air. These small scattering particles (much smaller than the wavelength of the light) scatter the shorter waves more effectively than the longer waves. Thus, more blue light is scattered than red light, giving the sky its blue color. This same effect, which makes the sky darker in red light than in blue, makes it even darker in the infrared light sampled by these photographs.

After taking the initial photographs, made with the leaky hose, I waited to capture Nature's own natural, infrared rainbow. It was four years before I saw a natural rainbow when I had at hand my camera, infrared film, the filter, *and* time enough to put them together for the photographs shown in Figures 3 and 4.

Figure 3 A natural infrared rainbow. This photograph shows the clouds to be brighter inside the bow than outside, a common feature of visible rainbows.

Figure 4 A photograph of Nature's own invisible rainbow, showing the primary bow, secondary bow, and a series of interference fringes (supernumerary bows) inside the primary.

Public Response

I received an interesting collection of letters in response to a brief published note describing the infrared rainbow. Some were from people who had a "scientific interest" in the matter; others from friends, with whom ties had been stretched by distance and neglect, saying: I'm glad to see you're still at it. Other letters represented unique interests, such as the psychologist studying color blindness, wondering whether such a disability might be a reason for a person named Greenler to be interested in invisible light—or the person from Belgium Television wanting photographs of the infrared rainbow for a show they were producing, insisting that they be *in color*. But most of the letters were from people who shared with me the fascination of "seeing" for the first time this bow, whose undetected presence in the sky predated that of a human consciousness on this planet.

Suggestions for Further Reading

Boyle, W. S.: "Light-Wave Communications," *Scientific American*, August 1977, p. 40.

The physics and technology of a telephone system that transmits signals via light pulses carried along optical fibers are discussed in this article.

Greenler, Robert: *Rainbows, Halos, and Glories*, Cambridge University Press, Cambridge, 1980.

In this book, the author of this chapter's essay ["Beyond the (Visible) Rainbow"] discusses rainbows as well as reflection and refraction effects due to atmospheric ice crystals and refraction through a nonuniform atmosphere.

Katzir, Abraham: "Optical Fibers in Medicine," *Scientific American*, May 1989, p. 120.

Optical fibers can carry light into and out of the body for diagnostic purposes (imaging, blood flow measurement) and treatment (laser surgery). This article examines instruments presently used and those likely to be developed in the future.

Können, G. P.: *Polarized Light in Nature*, Cambridge University Press, Cambridge, 1985.

This unusual book is a kind of field guide for observing the polarization of light from diverse objects, including the sky, clouds, rainbows, plants, ice sheets, beetles, and minerals. It includes many color photographs and diagrams.

"Light," *Scientific American* special issue, September 1968.

How light interacts with both living and nonliving matter, how images are formed, vision, and laser light are some of the topics covered in this issue.

Sobel, Michael I: *Light*, University of Chicago Press, Chicago, 1987.

This book presents light as a central concept of the natural sciences, one that ties together nonmathematical discussions of x rays and radio waves, cosmic background radiation, fiber-optic communication, light-sensitive pigments in the eye, and much more.

Walker, Jearl: "The Amateur Scientist: Studying Polarized Light with Quarter-Wave and Half-Wave Plates of One's Own Making," *Scientific American*, December 1977, p. 172.

Walker, Jearl: "The Amateur Scientist: More about Polarizers and How to Use Them, Particularly for Studying Polarized Sky Light," *Scientific American*, January 1978, p. 132.

These two articles are instructive even if one doesn't choose to repeat the experiments.

Wehner, Rüdiger: "Polarized-Light Navigation by Insects," *Scientific American*, July 1976, p. 106.

This article describes how ants and bees are able to use the natural polarization of sky light as an aid to navigation.

Review

A. Objectives: After studying this chapter, you should:

1. Be able to state Huygens' principle and use it to derive the law of reflection and Snell's law of refraction.

2. Be able to state Fermat's principle and use it to derive the law of reflection and Snell's law of refraction.

3. Be able to derive an expression relating the critical angle for total internal reflection to the index of refraction of a substance.

4. Be able to describe how a rainbow is formed, and explain qualitatively why the primary bow is seen at an angular radius of 42°.

5. Be able to list the four means of producing polarized light from unpolarized light.

6. Be able to state Malus's law and use it in problems involving the transmission of light through a polarizer.

7. Be able to derive Brewster's law using Snell's law and the fact that at the polarizing angle the reflected and refracted rays are perpendicular.

B. Define, explain, or otherwise identify:

Particle theory of light
Wave theory of light
Photons
Huygens' principle
Huygens' construction
Reflection
Angle of incidence
Plane of incidence
Law of reflection
Index of refraction
Image
Specular reflection
Diffuse reflection
Refraction
Snell's law of refraction
Total internal reflection
Dispersion
Angle of minimum deviation
Fermat's principle
Polarization
Linearly polarized
Circularly polarized
Transmission axis
Polarizer
Analyzer
Malus's law
Scattering
Brewster's law
Birefringence
Double refraction
Isotropic materials
Anisotropic materials
Optic axis
Quarter-wave plate
Half-wave plate

C. True or false: If the statement is true, explain why it is true. If it is false, give a counterexample.

1. Light and radio waves travel with the same speed through a vacuum.

2. Most of the light incident normally on an air–glass surface is reflected.

3. The angle of refraction of light is always less than the angle of incidence.

4. The index of refraction of water is the same for all wavelengths in the visible spectrum.

5. Longitudinal waves cannot be polarized.

Problems

Level I

30-1 The Speed of Light

1. The spiral galaxy in the Andromeda constellation is about 2×10^{19} km away from us. How many light-years is this?

2. On a spacecraft sent to Mars to take pictures, the camera is triggered by radio waves, which like all electromagnetic waves travel with the speed of light. What is the time delay between sending the signal from the earth and receiving it at Mars? (Take the distance to Mars to be 9.7×10^{10} m.)

3. The distance from a point on the surface of the earth to one on the surface of the moon is measured by aiming a laser light beam at a reflector on the surface of the moon and measuring the time required for the light to make a round trip. The uncertainty in the measured distance Δx is related to the uncertainty in the time Δt by $\Delta x = c\,\Delta t$. If the time intervals can be measured to ± 1.0 ns, find the uncertainty in the distance in meters.

30-2 The Propagation of Light: Huygens' Principle

There are no problems for this section.

30-3 Reflection

4. Calculate the fraction of light energy reflected from an air–water surface at normal incidence ($n = 1.33$ for water).

5. Light is incident normally on a slab of glass with an index of refraction $n = 1.5$. Reflection occurs at both surfaces of the slab. About what percentage of the incident light energy is transmitted by the slab?

30-4 Refraction

6. The index of refraction of water is 1.33. Find the angle of refraction of a beam of light in air that hits a water surface at an angle with the normal of (*a*) 20°, (*b*) 30°, (*c*) 45°, and (*d*) 60°. Show these rays on a diagram.

7. Repeat Problem 6 for a beam of light initially in water that is incident on a water–air surface.

8. What is the critical angle for total internal reflection for light traveling initially in water ($n = 1.33$) that is incident on a water–air surface?

9. Find the speed of light in water ($n = 1.33$) and in glass ($n = 1.5$).

10. A beam of monochromatic red light with a wavelength of 700 nm in air travels in water. (*a*) What is the wavelength in water? (*b*) Does a swimmer underwater observe the same color or a different color for this light?

11. A slab of glass with an index of refraction of 1.5 is submerged in water with an index of refraction of 1.33. Light in the water is incident on the glass. Find the angle of refraction if the angle of incidence is (*a*) 60°, (*b*) 45°, and (*c*) 30°.

12. Repeat Problem 11 for a beam of light initially in the glass that is incident on the glass–water surface at the same angles.

13. A glass surface ($n = 1.50$) has a layer of water ($n = 1.33$) on it. Light in the glass is incident on the glass–water surface. Find the critical angle for total internal reflection.

14. The index of refraction for silicate flint glass is 1.66 for light with a wavelength of 400 nm and 1.61 for light with a

wavelength of 700 nm. Find the angles of refraction for light of these wavelengths that is incident on this glass at an angle of 45°.

30-5 Fermat's Principle

15. A physics student playing pocket billiards wants to strike her cue ball such that it hits a cushion and then hits the eight ball squarely. She chooses several points on the cushion and for each point measures the distance from it to the cue ball and to the eight ball. She aims at the point for which the sum of these distances is least. (*a*) Will her cue ball hit the eight ball? (*b*) How is her method related to Fermat's principle?

16. A swimmer at S in Figure 30-42 develops a leg cramp while swimming near the shore of a calm lake and calls for help. A lifeguard at L hears the call. The lifeguard can run 9 m/s and swim 3 m/s. He knows physics and chooses a path that will take the least time to reach the swimmer. Which of the paths shown in Figure 30-42 does he take?

Figure 30-42 Problem 16.

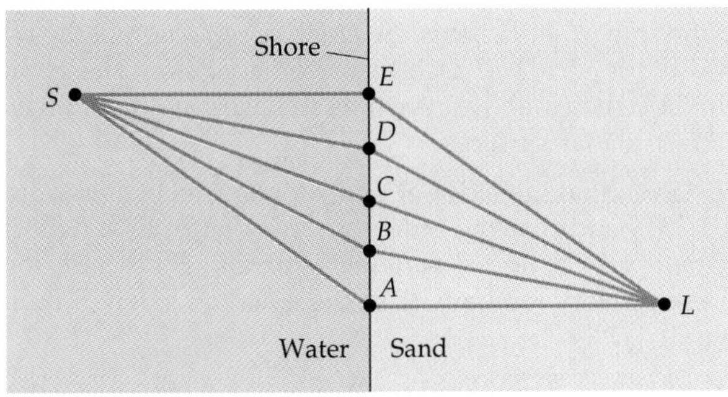

30-6 Polarization

17. Two polarizing sheets have their transmission axes crossed so that no light gets through. A third sheet is inserted between the first two such that its transmission axis makes an angle θ with that of the first sheet. Unpolarized light of intensity I_0 is incident on the first sheet. Find the intensity of the light transmitted through all three sheets if (*a*) $\theta = 45°$ and (*b*) $\theta = 30°$.

18. The polarizing angle for a certain substance is 60°. (*a*) What is the angle of refraction of light incident at this angle? (*b*) What is the index of refraction of this substance?

19. The critical angle for total internal reflection for a substance is 45°. What is the polarizing angle for this substance?

20. What is the polarizing angle for (*a*) water with $n = 1.33$ and (*b*) glass with $n = 1.5$?

Level II

21. A point source of light is 5 cm above a plane reflecting surface (such as a mirror). Draw a ray from the source that strikes the surface at an angle of incidence of 45° and two more rays that strike the surface at angles slightly less than 45°, and draw the reflected ray for each. The reflected rays appear to diverge from a point called the image of the light source. Draw dotted lines extending the reflected rays back until they meet at a point behind the surface to locate the image point.

22. A silver coin sits on the bottom of a swimming pool that is 4 m deep. A beam of light reflected from the coin emerges from the pool making an angle of 20° with respect to the water's surface and enters the eye of an observer. Draw a ray from the coin to the eye of the observer. Extend this ray, which goes from the water–air surface to the eye, straight back until it intersects with the vertical line drawn through the coin. What is the apparent depth of the swimming pool to this observer?

23. Two affluent students decide to improve on Galileo's experiment to measure the speed of light. One student goes to London and calls the other in New York on the telephone. The telephone signals are transmitted by reflecting electromagnetic waves from a satellite that is 37.9 Mm above the earth's surface. If the distance between London and New York is neglected, the distance traveled is twice this distance. One student claps his hands, and when the other student hears the sound over the phone, she claps her hands. The first student measures the time between his clap and his hearing the second one. Calculate this time lapse, neglecting the students' response times. Do you think this experiment would be successful? What improvements for measuring this time interval would you suggest? (Time delays in the electronic circuits that are greater than those due to the light traveling to the satellite and back make this experiment not feasible.)

24. In Galileo's attempt to determine the speed of light, he and his assistant were located on hilltops about 3 km apart. Galileo flashed a light and received a return flash from his assistant. (*a*) If his assistant had an instant reaction, what time difference would Galileo need to be able to measure for this method to be successful? (*b*) How does this time compare with human reaction time, which is about 0.2 s?

25. A point source of light is located 5 m below the surface of a large pool of water. Find the area of the largest circle on the pool's surface through which light coming directly from the source can emerge.

26. A swimmer at the bottom of a pool 3 m deep looks up and sees a circle of light. If the index of refraction of the water in the pool is 1.33, find the radius of the circle.

27. Show that when a mirror is rotated through an angle θ, the reflected beam of light is rotated through 2θ.

28. Light is incident normally on the largest face of an isosceles–right-triangle prism. What is the speed of light in this prism if the prism is just barely able to produce total internal reflection?

29. Show that the transmitted intensity through a glass slab with an index of refraction of n for normally incident light is approximately given by

$$I_T = I_0 \left[\frac{4n}{(n+1)^2} \right]^2$$

30. A ray of light begins at the point $x = -2$ m, $y = 2$ m, strikes a mirror in the xz plane at some point x, and reflects through the point $x = 2$ m, $y = 6$ m. (*a*) Find the value of x that makes the total distance traveled by the ray a minimum. (*b*) What is the angle of incidence on the reflecting plane? What is the angle of reflection?

31. Light passes symmetrically through a prism having an apex angle of α as shown in Figure 30-43. (*a*) Show that the angle of deviation δ is given by

$$\sin \frac{\alpha + \delta}{2} = n \sin \frac{\alpha}{2}$$

(*b*) If the refractive index for red light is 1.48 and for violet light is 1.52, what is the angular separation of visible light for a prism with an apex angle of 60°?

32. A beam of light strikes the plane surface of silicate flint glass at an angle of incidence of 45°. The index of refraction of the glass varies with wavelength as shown in the graph in Figure 30-22. How much smaller is the angle of refraction for violet light of wavelength 400 nm than that for red light of wavelength 700 nm?

33. Repeat Problem 32 for quartz.

34. Use Figure 30-22 to calculate the critical angles for total internal reflection for light initially in silicate flint glass that is incident on a glass–air surface if the light is (*a*) violet light of wavelength 400 nm, and (*b*) red light of wavelength 700 nm.

35. (*a*) For a light ray inside a transparent medium having a planar interface with a vacuum, show that the polarizing angle and the critical angle for internal reflection satisfy $\tan \theta_p = \sin \theta_c$. (*b*) Which angle is the larger?

36. Light is incident from air on a transparent substance at an angle of 58.0° with the normal. The reflected and refracted rays are observed to be mutually perpendicular. (*a*) What is the index of refraction of the transparent substance? (*b*) What is the critical angle for total internal reflection in this substance?

37. Two polarizing sheets have their transmission axes crossed and a third sheet is inserted so that its transmission axis makes an angle θ with that of the first sheet as in Problem 17. Show that the intensity transmitted through all three sheets is maximum when $\theta = 45°$.

38. If the middle polarizing sheet in Problem 37 is rotating at an angular velocity ω about an axis parallel to the light beam, find the intensity transmitted through all three sheets as a function of time. Assume that $\theta = 0$ at time $t = 0$.

39. Given a stack of $N + 1$ ideal polarizing sheets with each sheet rotated by an angle of $\pi/2N$ rad with respect to the preceding sheet. A plane linearly polarized light wave of intensity I_0 is incident normally on the stack. The incident light is polarized along the transmission axis of the first sheet and therefore normal to the transmission axis of the last sheet in the stack. (*a*) What is the transmitted intensity through the stack? (*b*) For 3 sheets ($N = 2$), what is the transmitted intensity? (*c*) For 101 sheets, what is the transmitted intensity? (*d*) What is the direction of polarization of the transmitted beam in each case?

40. A point source of light is located at the bottom of a steel tank, and an opaque circular card of radius 6.0 cm is placed over it. A transparent fluid is gently added to the tank such that the card floats on the surface with its center directly above the light source. No light is seen by an observer above the surface until the fluid is 5 cm deep. What is the index of refraction of the fluid?

41. A light ray in dense flint glass with index of refraction 1.655 is incident on the glass interface. An unknown liquid condenses on the surface of the glass. Total internal reflection on the glass–liquid surface occurs for an angle of incidence on the glass–liquid interface of 53.7°. (*a*) What is the refractive index of the unknown liquid? (*b*) If the liquid is removed, what is the angle of incidence for total internal reflection? (*c*) For the angle of incidence found in part (*b*), what is the angle of refraction of the ray into the liquid film? Does a ray emerge from the liquid film into the air above? Assume the glass and liquid have perfect planar surfaces.

42. Given that the index of refraction for red light in water is 1.3318 and that the index of refraction of blue light in water is 1.3435, find the angular separation of these colors in the primary rainbow. (Use the equation given in Problem 51.)

43. In Figure 30-44, light is initially in a medium (such as air) of index of refraction n_1. It is incident at angle θ_1 on the surface of a liquid (such as water) of index of refraction n_2. The light passes through the layer of water and enters glass of index of refraction n_3. If θ_3 is the angle of refraction in the glass, show that $n_1 \sin \theta_1 = n_3 \sin \theta_3$. That is, show that the second medium can be neglected when finding the angle of refraction in the third medium.

44. A ray of light falls on a rectangular glass block ($n = 1.5$) that is almost completely submerged in water ($n = 1.33$) as shown in Figure 30-45. (*a*) Find the angle θ for which total internal reflection just occurs at point P. (*b*) Would total internal reflection occur at point P for the value of θ found in part (*a*) if the water were removed? Explain.

Figure 30-43 Problem 31.

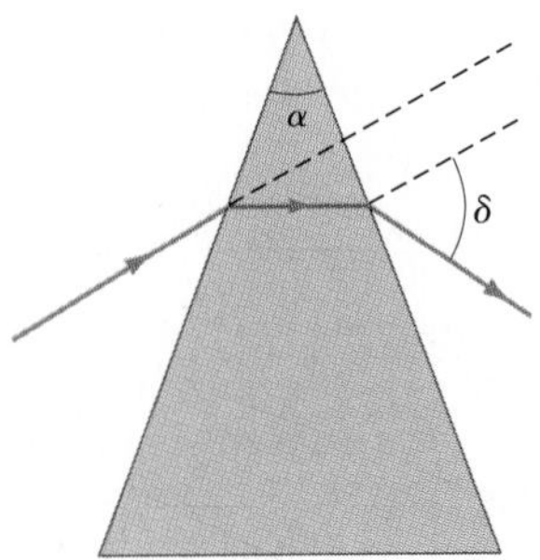

Figure 30-44 Problem 43.

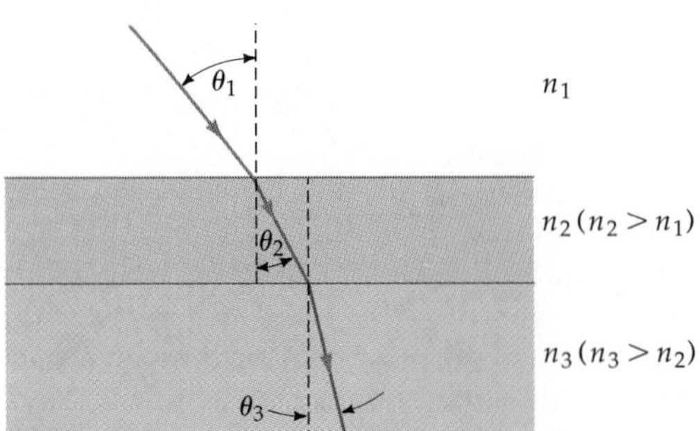

Figure 30-45 Problem 44.

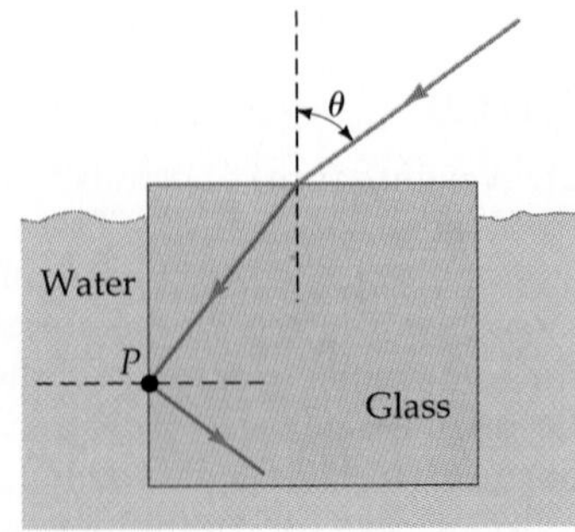

45. Light of wavelength λ in air is incident on a slab of calcite such that the ordinary and extraordinary rays travel in the same direction as shown in Figure 30-46. Show that the phase difference between these rays after they traverse a thickness t of calcite is

$$\delta = \frac{2\pi}{\lambda}(n_o - n_e)t$$

46. (*a*) Use the result for Problem 29 to find the ratio of the transmitted intensity to the incident intensity through N parallel slabs of glass for light of normal incidence. (*b*) Find this ratio for three slabs of glass with $n = 1.5$. (*c*) How many slabs of glass with $n = 1.5$ will reduce the intensity to 10 percent of the incident intensity?

47. Light is incident on a slab of transparent material at an angle θ_1 as shown in Figure 30-47. The slab has a thickness t and an index of refraction n. Show that

$$n = \frac{\sin\theta_1}{\sin[\arctan(d/t)]}$$

where d is the distance shown in the figure and $\arctan(d/t)$ is the angle whose tangent is d/t.

48. Suppose rain falls vertically from a stationary cloud 10,000 m above a confused marathoner running in a circle with constant speed of 4 m/s. The rain has a terminal speed of 9 m/s. (*a*) What is the angle that the rain appears to make with the vertical to the marathoner? (*b*) What is the apparent motion of the cloud as observed by the marathoner? (*c*) A star on the axis of the earth's orbit appears to have a circular orbit of angular diameter of 41.2 seconds of arc. How is this angle related to the earth's speed in its orbit and the velocity of photons falling from this distant star? (*d*) What is the speed of light using this method?

49. This problem is a refraction analogy. A band is marching down a football field with a constant speed v_1. About midfield, the band comes to a section of muddy ground that has a sharp boundary making an angle of 30° with the 50-yd line as shown in Figure 30-48. In the mud, the marchers move with speed $v_2 = \frac{1}{2}v_1$. Diagram how each line of marchers is bent as it encounters the muddy section of the field so that the band is eventually marching in a different direction. Indicate the original direction by a ray, the final direction by a second ray, and find the angles between the rays and the line perpendicular to the boundary. Is their direction of motion bent toward the perpendicular to the boundary or away from it?

Level III

50. Light is incident normally upon one face of a prism of glass with an index of refraction n (Figure 30-49). The light is totally reflected at the right side. (*a*) What is the minimum value n can have? (*b*) When the prism is immersed in a liquid whose index of refraction is 1.15, there is still total reflection, but when it is immersed in water whose index of refraction is 1.33, there is no longer total reflection. Use this information to establish limits for possible values of n.

51. Equation 30-13 gives the relation between the angle of deviation ϕ_d of a light ray incident on a spherical drop of water in terms of the incident angle θ_1 and the index of refraction of water. (*a*) Assume that $n_{air} = 1$, and differentiate ϕ_d with respect to θ_1. [Hint: If $y = \arcsin x$, $dy/dx = (1 - x^2)^{-1/2}$.] (*b*) Set $d\phi_d/d\theta_1 = 0$ and show that the angle of incidence θ_{1m} for minimum deviation is given by

$$\cos\theta_{1m} = \sqrt{\frac{n^2 - 1}{3}}$$

and find θ_{1m} for water, where the index of refraction for water is 1.33.

52. Investigate how a thin film of water on a glass surface affects the critical angle for total reflection. Take $n = 1.5$ for glass and $n = 1.33$ for water. (*a*) What is the critical angle for total internal reflection at the glass–water surface? (*b*) Is there any range of incident angles that are greater than θ_c for glass-to-air refraction, and for which light rays will leave the glass and the water and pass into the air?

53. A laser beam is incident on a plate of glass of thickness 3 cm. The glass has an index of refraction of 1.5 and the angle of incidence is 40°. The top and bottom surfaces of the glass are parallel and both produce reflected beams of nearly the same intensity. What is the perpendicular distance d between the two adjacent reflected beams?

54. (*a*) Show that a light ray transmitted through a glass slab emerges parallel to the incident ray but displaced from it. (*b*) For an incident angle of 60°, glass of index of refraction $n = 1.5$, and a slab of thickness 10 cm, find the displacement measured perpendicularly from the incident ray.

55. An isotropic point source of light is placed below the surface of a large pool of liquid having an index of refraction n. What fraction of light energy leaves the surface directly?

Figure 30-46 Problem 45. **Figure 30-47** Problem 47.

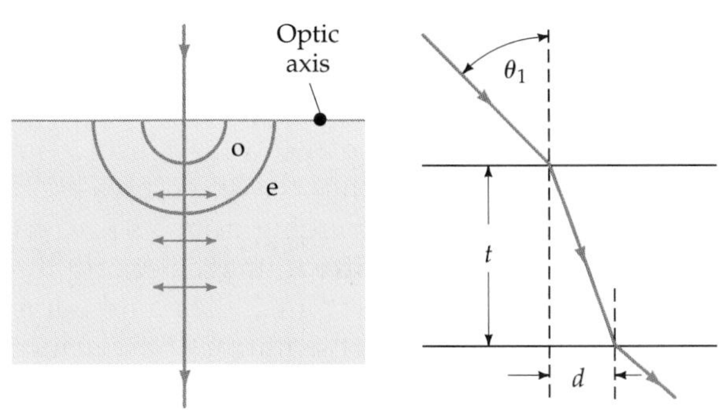

Figure 30-48 Problem 49. **Figure 30-49** Problem 50.

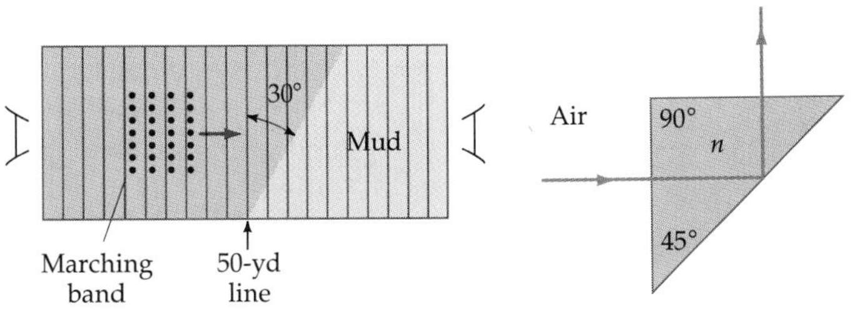

Chapter 31

Geometrical Optics

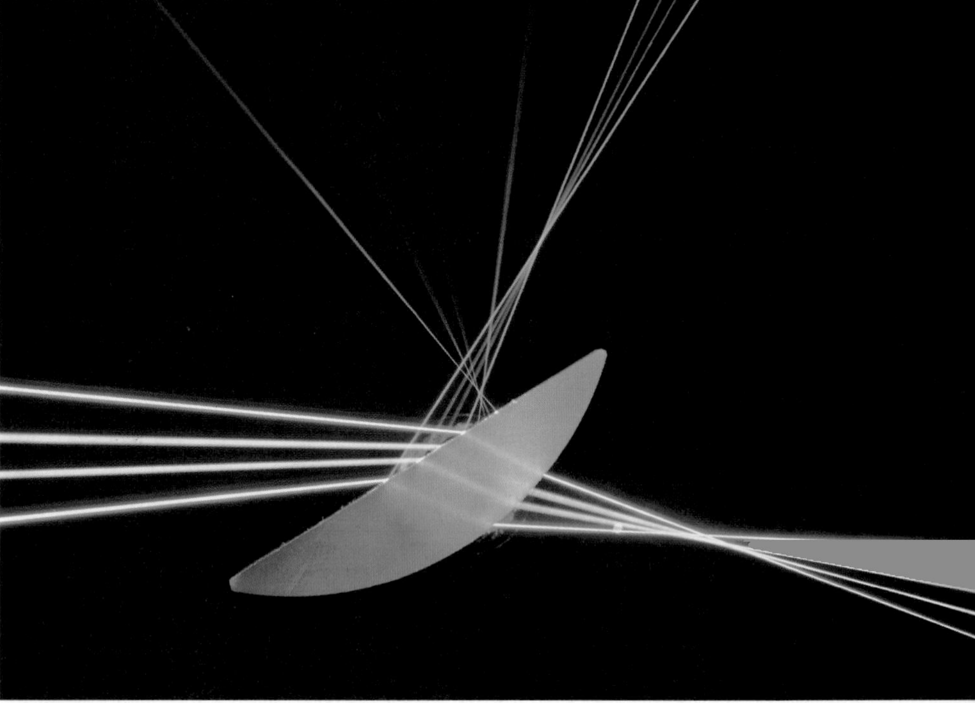

The focusing of rays by reflection and by refraction is illustrated by these laser beams incident on a glass lens.

The wavelength of light is very small compared with most obstacles and openings. Because of this, diffraction—the bending of waves around corners—is often negligible, and the ray approximation, in which waves are considered to propagate in straight lines, is valid. **Geometrical optics** is the study of those phenomena in which the ray approximation is valid. In this chapter, we will apply the laws of reflection and refraction to the formation of images by mirrors and lenses.

31-1 Plane Mirrors

Figure 31-1 shows a narrow bundle of light rays emanating from a point source P and reflected from a plane mirror. After reflection, the rays diverge exactly as if they came from a point P' behind the plane of the mirror. The point P' is called the **image** of the object P. When these reflected rays enter the eye, they cannot be distinguished from rays diverging from a source at P' with no mirror present. This image is called a **virtual image** because the light does not actually emanate from it. The image point P' lies on the line through the object P perpendicular to the plane of the mirror at a distance behind the plane equal to that from the plane to the object as shown in the

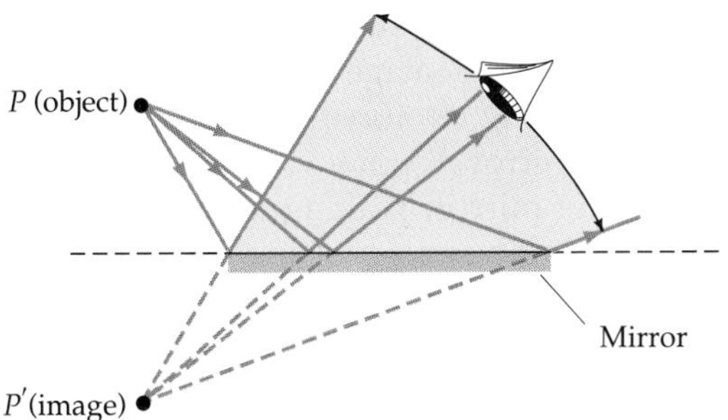

Figure 31-1 Image formed by a plane mirror. The rays from point P that strike the mirror and enter the eye appear to come from the image point P' behind the mirror. The image can be seen by the eye anywhere in the shaded region.

figure. (Figure 31-1 was produced by geometric construction using the law of reflection.) The image can be seen by an eye anywhere in the shaded region indicated in which a straight line from the image to the eye passes through the mirror. The object need not be directly in front of the mirror. The image can be seen as long as the object is not behind the plane of the mirror.

The image you see if you hold the palm of your right hand up to a plane mirror is shown in Figure 31-2. The image is the same size as the object, but the image is not the same as what would be seen by someone facing you or what you would see if you looked at the palm of your right hand. The image of a right hand in a mirror is a left hand. This right-to-left inversion is a result of **depth inversion;** that is, the hand is transformed from a right hand to a left hand because the front and back of the hand are reversed by the mirror. Depth inversion is also illustrated in Figure 31-3, which shows a person lying down with his feet touching a plane mirror. The image of a simple rectangular coordinate system that has its x and y axes parallel to the plane of the mirror is shown in Figure 31-4. The images of the arrows along the x and y axes are parallel to the object arrows. But the image of the arrow along the z axis is directed opposite to the object arrow along the z axis. The mirror transforms a right-handed coordinate system for which $\mathbf{i} \times \mathbf{j} = \mathbf{k}$, where $\mathbf{i}$, $\mathbf{j}$, and $\mathbf{k}$ are the unit vectors along the x, y, and z axes, respectively, into a left-handed coordinate system for which $\mathbf{i} \times \mathbf{j} = -\mathbf{k}$.

Figure 31-2 The image of a right hand in a plane mirror is a left hand. This right-to-left reversal is a result of depth inversion.

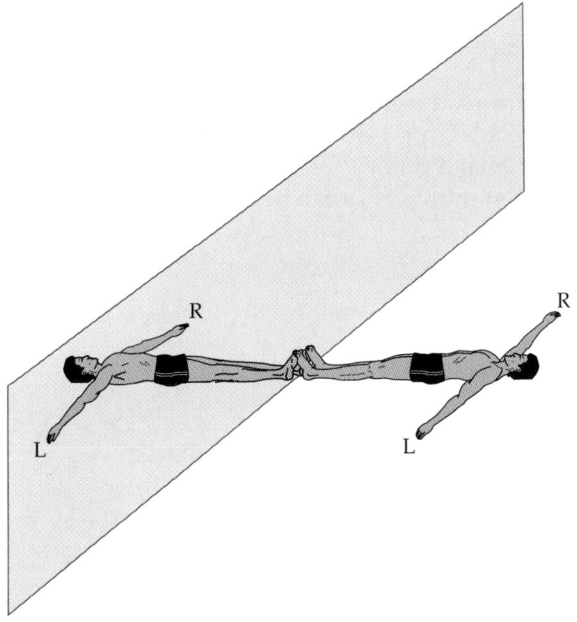

Figure 31-3 A person lying down with his feet against the mirror. The image is depth inverted.

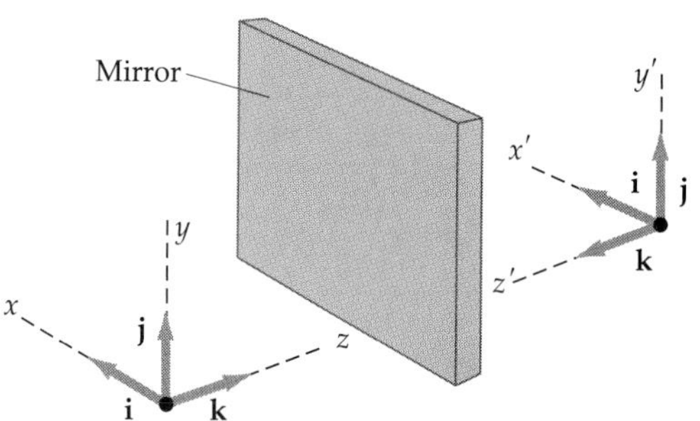

Figure 31-4 Image of a rectangular coordinate system in a plane mirror. The arrows along the x and y axes, which are parallel to the plane of the mirror, are in the same directions in the image as in the object. The direction of the arrow along the z axis is reversed in the image. The image of the original right-handed coordinate system, for which $\mathbf{i} \times \mathbf{j} = \mathbf{k}$, is a left-handed coordinate system, for which $\mathbf{i} \times \mathbf{j} = -\mathbf{k}$.

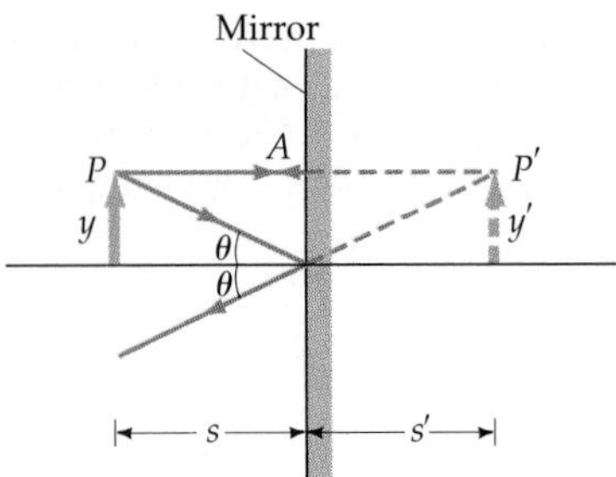

Figure 31-5 Ray diagram for locating the image of an arrow in a plane mirror.

Figure 31-5 shows an arrow of height y standing parallel to a plane mirror a distance s from it. We can locate the image of the arrowhead (and of any other point on the arrow) by drawing two rays. One ray is drawn perpendicular to the mirror. It hits the mirror at point A and is reflected back onto itself. The other ray strikes the mirror, making an angle θ with the normal to the mirror. It is reflected, making an equal angle θ with the x axis. The extension of these two rays back behind the mirror locates the image of the arrowhead, as shown by the dashed lines in the figure. We can see from this figure that the image is the same distance behind the mirror as the object is in front of the mirror and that the image is erect and the same size as the object.

The formation of multiple images by two plane mirrors making an angle with each other is illustrated in Figure 31-6. We frequently see this phenomenon in clothing stores that provide adjacent mirrors. Light reflected from mirror 1 strikes mirror 2 just as if it came from the image point P_1'. The image P_1' is called the object for mirror 2. Its image is at point $P_{1,2}''$. This image will be formed whenever the image point P_1' is in front of the plane of mirror 2.

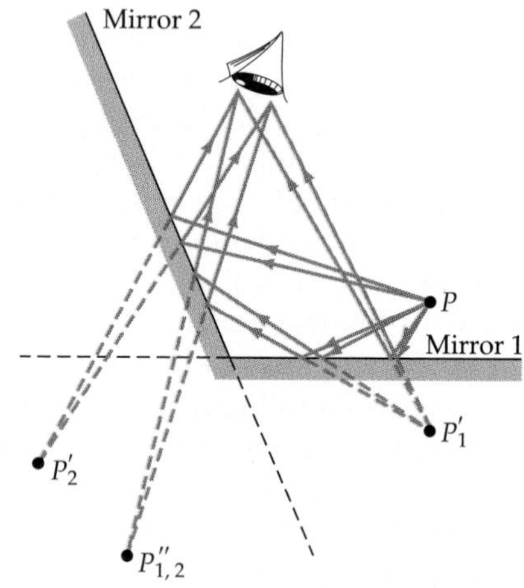

Figure 31-6 Images formed by two plane mirrors. P_1' is the image of the object P in mirror 1, and P_2' is the image of the object in mirror 2. Point $P_{1,2}''$ is the image of P_1' in mirror 2 seen when light rays from the object reflect first from mirror 1 and then from mirror 2. The image P_2' does not have an image in mirror 1 because it is behind that mirror.

The image at point P_2' is due to rays from the object that reflect directly from mirror 2. Since P_2' is behind the plane of mirror 1, it cannot serve as an object point for a further image in mirror 1. The number of multiple images formed by two mirrors depends on the angle between the mirrors and the position of the object.

Figure 31-7 shows two mirrors at right angles to each other. Rays from the object to the eye that strike mirror 1 and then mirror 2 are shown in Figure 31-7a. In this case, the image point $P_{1,2}''$ is the same as that for rays that strike mirror 2 first and then mirror 1, as can be seen from Figure 31-7b.

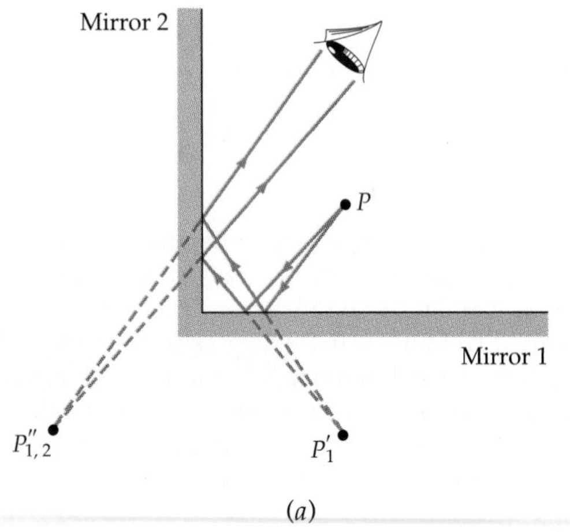

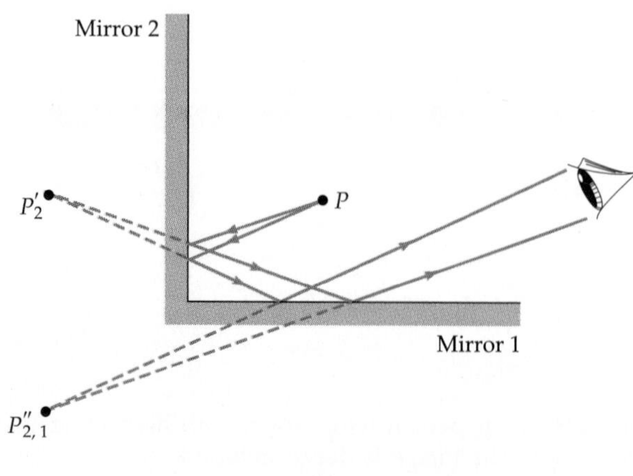

Figure 31-7 Two plane mirrors at right angles to each other. (a) Rays that strike mirror 1 first and then mirror 2. The image of P_1' in mirror 2 is $P_{1,2}''$. (b) Rays that strike mirror 2 first and then mirror 1. The image of P_2' in mirror 1 is $P_{2,1}''$, which coincides with $P_{1,2}''$ for perpendicular mirrors.

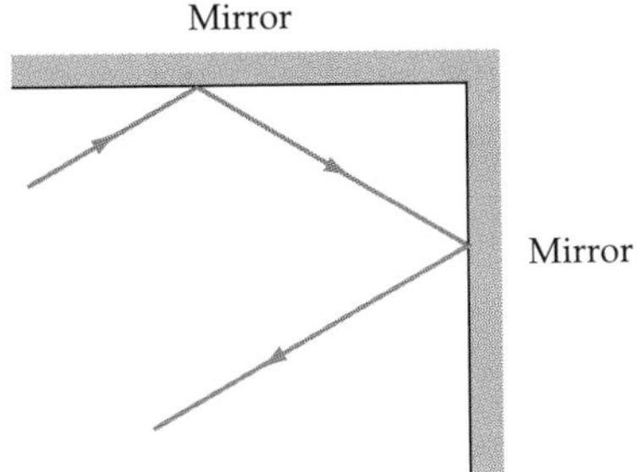

Figure 31-8 A ray striking one of two perpendicular plane mirrors is reflected from the second mirror in the direction opposite the original direction for any angle of incidence.

If you stand in front of two vertical mirrors that are perpendicular to each other, such as in the corner of a room, the image you see *is* the same as that seen by others who are facing you because depth inversion occurs twice, once in each mirror. Figure 31-8 illustrates the fact that a horizontal ray reflected from two perpendicular vertical mirrors is exactly reversed in direction no matter what angle the ray makes with the mirrors. If three mirrors are placed perpendicular to each other like the sides of an inside corner of a box, any ray incident on any of the mirrors from any direction is exactly reversed. A set of mirrors of this type (actually a set of reflecting prisms) was placed on the moon facing the earth. A laser beam from earth directed at the mirrors is reflected back to the same place on the earth. Such a beam has been used to measure the distance to the mirrors to within a few centimeters by measuring the time it takes for the light to reach the mirrors and return.

This reflecting pool serves as a plane mirror producing a virtual image of the Taj Mahal.

Questions

1. Can a virtual image be photographed?
2. Suppose each axis of a coordinate system like the one in Figure 31-4 is painted a different color. One photograph is take of the coordinate system and another is taken of its image in a plane mirror. Is it possible to tell that one of the photographs is of a mirror image rather than both being photographs of the real coordinate system from different angles?

31-2 Spherical Mirrors

Figure 31-9 shows a bundle of rays from a point source P on the axis of a concave spherical mirror reflecting from the mirror and converging at point P'. The rays then diverge from this point just as if there were an object at that point. This image is called a **real image** because light actually does emanate from the image point. The image can be seen by an eye at the left of the image looking into the mirror. It could also be observed on a ground-glass viewing screen or photographic film placed at the image point. A virtual image, such as that formed by a plane mirror as discussed in the previous section, cannot be observed on a screen at the image point because there is no light there. Despite this distinction between real and virtual images,

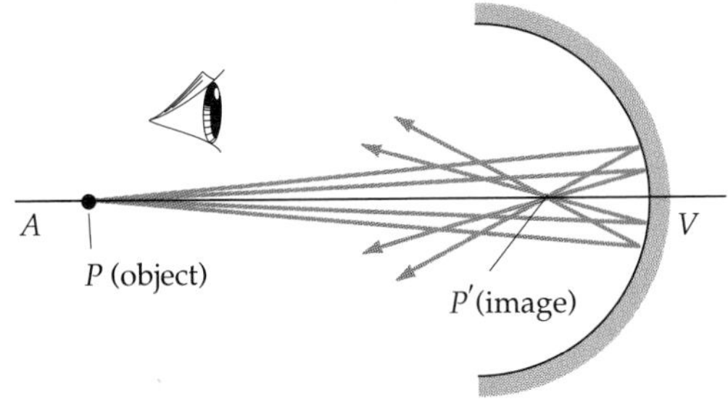

Figure 31-9 Rays from a point object P on the axis AV of a concave spherical mirror form an image at P'. The image is sharp if the rays strike the mirror near the axis.

the light rays diverging from a real image and those appearing to diverge from a virtual image are identical, so no distinction is made by the eye between viewing a real or a virtual image.

From Figure 31-10, we can see that only rays that strike the spherical mirror at points near the axis AV are reflected through the image point. Such rays are called **paraxial rays.** Rays that strike the mirror at points far from the axis, called *nonparaxial rays,* converge to different points near the image point. Such rays cause the image to appear blurred, an effect called **spherical aberration.** The image can be sharpened by blocking off all but the central part of the mirror so that nonparaxial rays do not strike it. Although the image is then sharper, its brightness is reduced because less light is reflected to the image point.

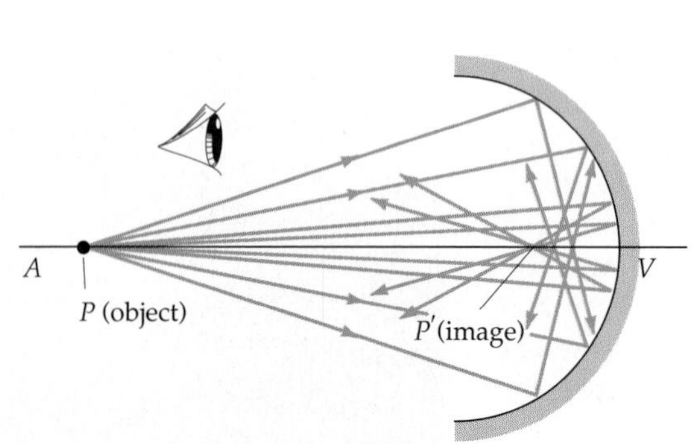

Figure 31-10 Spherical aberration. Nonparaxial rays that strike the mirror at points far from the axis AV are not reflected through the image point P'. These rays blur the image.

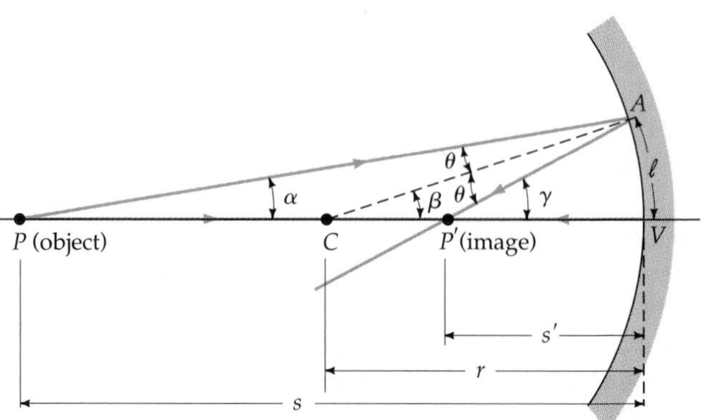

Figure 31-11 Geometry for calculating the image distance s' from the object distance s and the radius of curvature r.

The image distance from the vertex of the mirror V to P' can be related to the object distance from the vertex V to point P and the radius of curvature of the mirror by elementary geometry. Figure 31-11 shows a ray from an object point P reflecting off the mirror and passing through the image point P'. Point C is the center of curvature of the mirror. The incident and reflected rays make equal angles with the radial line CA, which is perpendicular to the surface of the mirror. Let s be the object distance, s' be the image distance, and r be the radius of curvature of the mirror. The angle β is an exterior angle to the triangle PAC and is therefore equal to $\alpha + \theta$:

$$\beta = \alpha + \theta \qquad 31\text{-}1$$

Similarly, from the triangle PAP',

$$\gamma = \alpha + 2\theta \qquad 31\text{-}2$$

Eliminating θ from these equations gives

$$2\theta = \gamma - \alpha = 2\beta - 2\alpha$$

or

$$2\beta = \alpha + \gamma \qquad 31\text{-}3$$

Using the small angle approximations $\alpha \approx \ell/s$, $\beta \approx \ell/r$, and $\gamma \approx \ell/s'$, we have

$$\frac{1}{s} + \frac{1}{s'} = \frac{2}{r} \qquad 31\text{-}4$$

The derivation of this equation assumes that angles made by the incident and reflected rays with the axis are small. This is equivalent to assuming that the rays are paraxial.

When the object distance is large compared with the radius of curvature of the mirror, the term $1/s$ in Equation 31-4 is much smaller than $2/r$ and can be neglected. For $s = \infty$, the image distance is $s' = \frac{1}{2}r$. This distance is called the **focal length** f of the mirror:

$$f = \tfrac{1}{2}r \qquad \text{31-5}$$

Focal length for mirror

In terms of the focal length f, the mirror equation is

$$\frac{1}{s} + \frac{1}{s'} = \frac{1}{f} \qquad \text{31-6}$$

Mirror equation

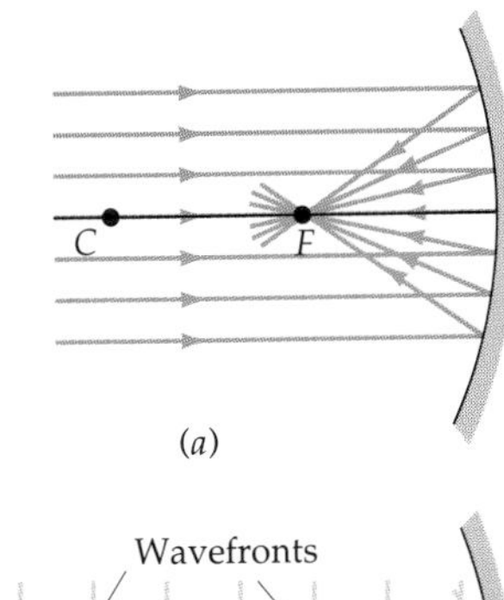

(a)

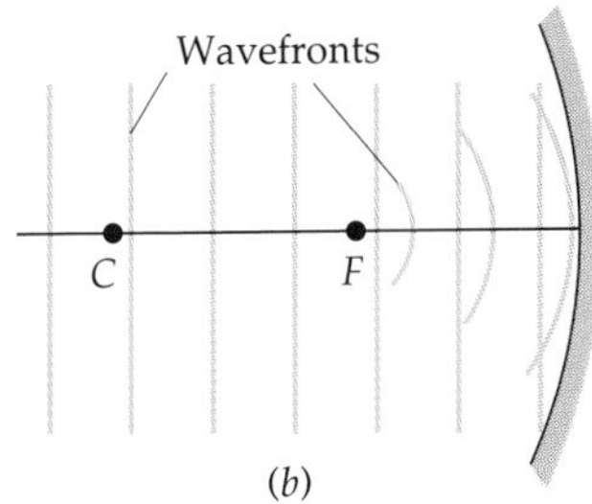

(b)

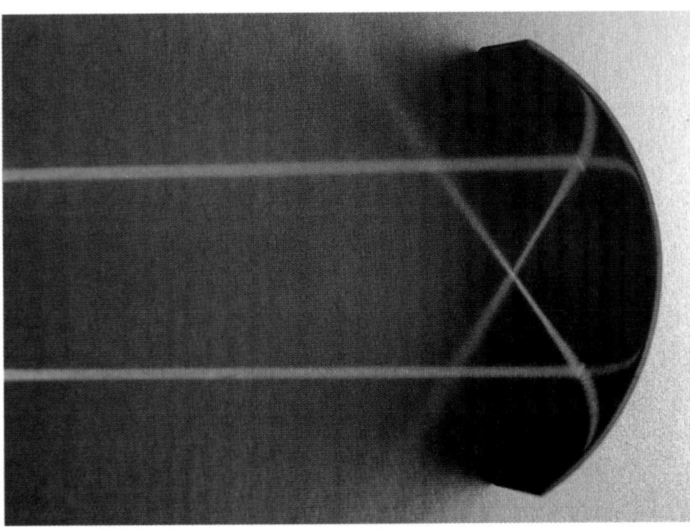

(c)

Figure 31-12 *(a)* Parallel rays strike a concave mirror and are reflected through the focal point F at a distance $r/2$. *(b)* The incoming wavefronts are plane waves; upon reflection they become spherical waves that converge at the focal point. *(c)* Photograph of parallel rays focused by a concave mirror.

The **focal point** F (the image point) is the point at which parallel rays incident on the mirror are focused, as illustrated in Figure 31-12*a*. (Again, only paraxial rays are focused at a single point.)

When an object is very far from the mirror, the wavefronts are approximately planes, as shown in Figure 31-12*b*, and the rays are parallel. In Figure 31-12*b*, note how the edges of the wavefront hit the concave mirror surface before the central portion near the axis, resulting in a spherical wavefront upon reflection. Figure 31-13 shows the wavefronts and rays for plane waves striking a convex mirror. In this case, the central part of the wavefront strikes the mirror first, and the reflected waves appear to come from the focal point behind the mirror.

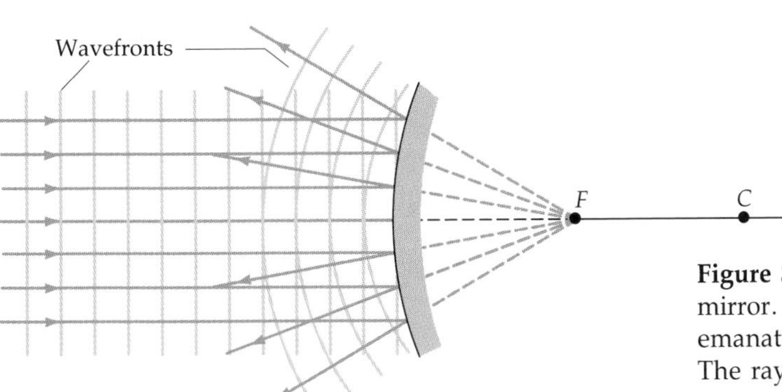

Figure 31-13 Reflection of plane waves from a convex mirror. The outgoing wavefronts are spherical as if emanating from the focal point F behind the mirror. The rays are perpendicular to the wavefronts, and appear to diverge from F.

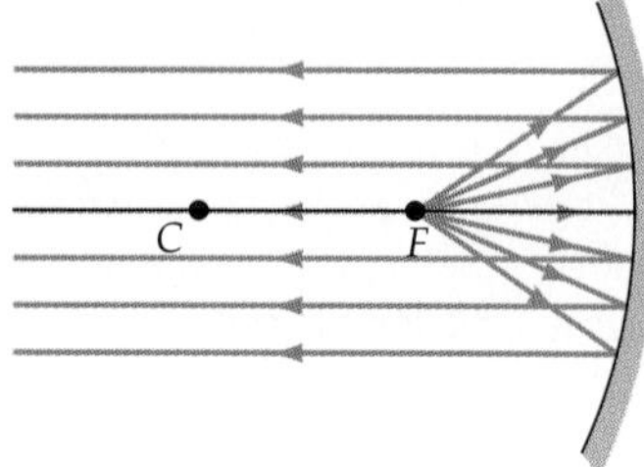

Figure 31-14 Illustration of reversibility. Rays diverging from a point source at the focal point of a concave mirror are reflected from the mirror as parallel rays. The rays are the same as in Figure 31-12*a* but in the reverse direction.

In Figure 31-14, rays from a point source at the focal point strike a concave mirror and are reflected parallel to the axis. This illustrates a property of waves called **reversibility.** If we reverse the direction of a reflected ray, the law of reflection assures us that the reflected ray will be along the original incoming ray but in the opposite direction. (Reversibility holds also for refracted rays, which are discussed in later sections.) Thus, if we have a real image of an object formed by a reflecting (or refracting) surface, we can place an object at the image point and a new image will be formed at the position of the original object.

Example 31-1

An object is 12 cm from a concave mirror with a radius of curvature of 6 cm. Find the focal length of the mirror and the image distance.

From Equation 31-5, the focal length is

$$f = \tfrac{1}{2}r = \tfrac{1}{2}(6 \text{ cm}) = 3 \text{ cm}$$

Equation 31-6 then gives

$$\frac{1}{12 \text{ cm}} + \frac{1}{s'} = \frac{1}{3 \text{ cm}}$$

$$\frac{1}{s'} = \frac{4}{12 \text{ cm}} - \frac{1}{12 \text{ cm}} = \frac{3}{12 \text{ cm}}$$

$$s' = 4 \text{ cm}$$

Ray Diagrams for Mirrors

A useful method for locating images is by geometric construction of a **ray diagram.** This is illustrated in Figure 31-15, where the object is a human figure perpendicular to the axis a distance s from the mirror. By the judicious choice of rays from the head of the figure, we can quickly locate the image. There are four **principal rays** that are convenient to use:

Principal rays for a mirror

1. The **parallel ray,** drawn parallel to the axis. This ray is reflected through the focal point.
2. The **focal ray,** drawn through the focal point. This ray is reflected parallel to the axis.
3. The **radial ray,** drawn through the center of curvature. This ray strikes the mirror perpendicular to its surface and is thus reflected back on itself.
4. The **central ray,** drawn to the vertex of the mirror. It reflects at an equal angle to the axis.

The first three of these rays are shown in Figure 31-15. The intersection of any two rays locates the image point of the head. The third ray can be used to provide a check.

We can see from Figure 31-15 that the image is inverted and is not the same size as the object. The ratio of the image size to the object size is defined as the **lateral magnification** of the image. In Figure 31-16 we have drawn the central ray from the top of the object to the center of the mirror. This ray makes an angle θ with the axis. The reflected ray to the top of the image makes an equal angle with the axis. A comparison of the triangle formed by the incident ray, the axis, and the object with that formed by the reflected ray, the axis, and the image shows that the lateral magnification

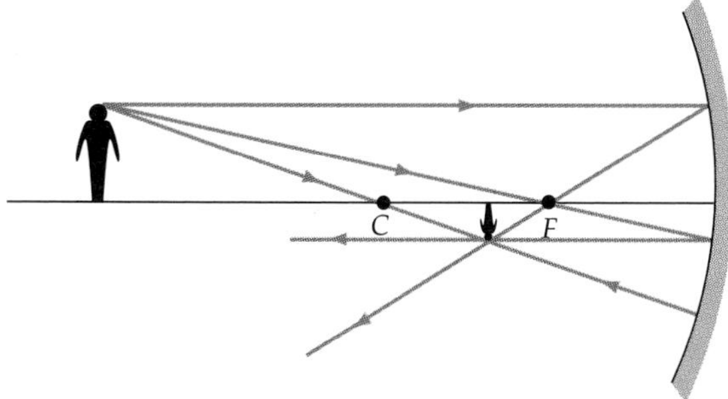

Figure 31-15 Ray diagram for the location of the image by geometric construction.

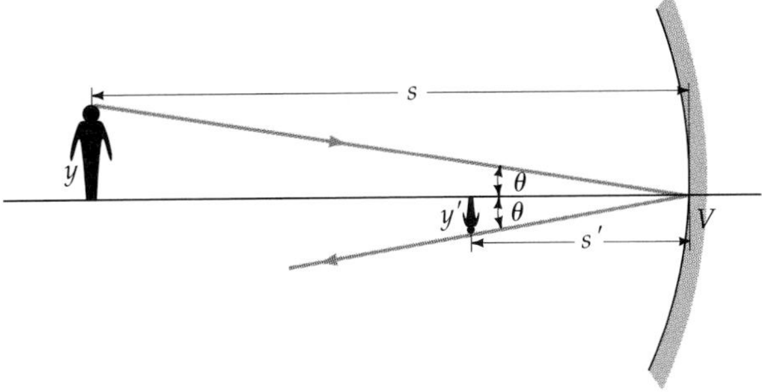

Figure 31-16 Geometry for finding the magnification of a concave spherical mirror.

y'/y equals the ratio of the distances s'/s. Ray diagrams are easier to draw if the mirror is replaced by a straight line that extends as far as necessary to intercept the rays as shown in Figure 31-17.

When the object is between the mirror and its focal point, the rays reflected from the mirror do not converge but appear to diverge from a point behind the mirror as illustrated in Figure 31-18. In this case the image is virtual and erect. ("Erect" means not inverted relative to the object.) For an object between the mirror and the focal point, s is less than $\frac{1}{2}r$, so the image distance s' calculated from Equation 31-4 turns out to be negative. We can apply Equations 31-4, 31-5 and 31-6 to this case and to convex mirrors if we adopt a convenient sign convention. Whether the mirror is convex or concave, real images can be formed only in front of the mirror, that is, on the same side of the mirror as the object. Virtual images are formed behind the mirror where there are no actual light rays. Our sign convention is as follows:

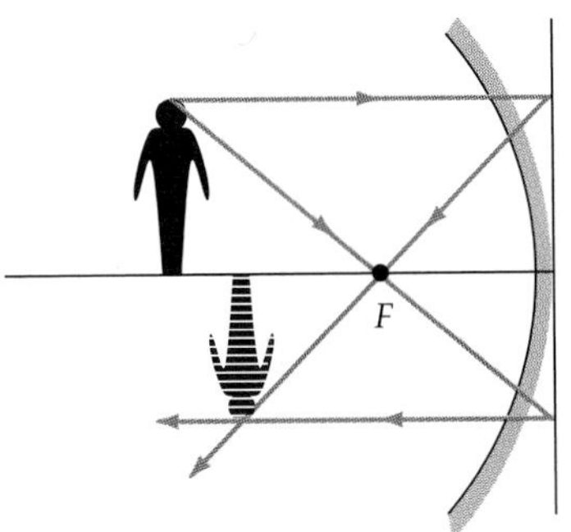

Figure 31-17 Ray diagrams are easier to construct if the curved surface is replaced by a plane surface.

Sign conventions for reflection

s	+ if the object is in front of the mirror (real object)
	− if the object is behind the mirror (virtual object)*
s'	+ if the image is in front of the mirror (real image)
	− if the image is behind the mirror (virtual image)
r,f	+ if the center of curvature is in front of the mirror (concave mirror)
	− if the center of curvature is behind the mirror (convex mirror)

*You may wonder how an object can be behind a mirror. This occurs when there is a lens in front of the mirror and the rays to the image of the lens are intercepted by the mirror. The image of the lens is then never formed, but the distance to the unformed image behind the mirror is taken as the object distance for the mirror, and the object is called a virtual object. We will discuss examples of this in Section 31-4 when we discuss lenses.

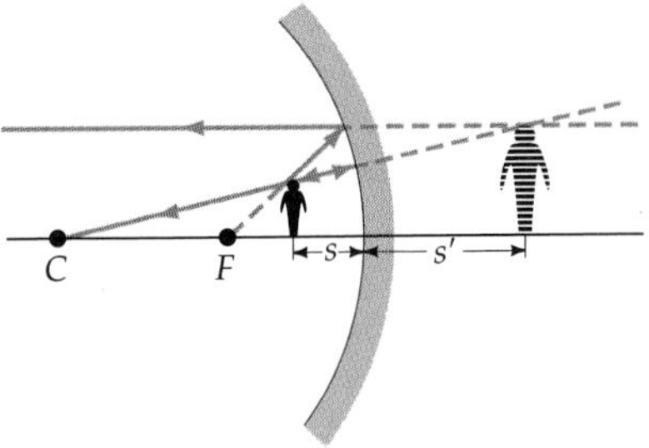

Figure 31-18 A virtual image formed by a concave mirror. The image is located by the radial ray, which is reflected back on itself, and the focal ray, which is reflected parallel to the axis. These two rays appear to diverge from a point behind the mirror found by extending them. A third ray (not shown) could be drawn from the object parallel to the axis. It would be reflected through the focal point, and its extension would intersect the other two rays at the image point.

With these sign conventions, Equations 31-4, 31-5, and 31-6 can be used for all situations with any type of mirror. The lateral magnification of the image is then given by

Lateral magnification

$$m = \frac{y'}{y} = -\frac{s'}{s} \qquad 31\text{-}7$$

(a)

(b)

Reflection in *(a)* a concave mirror and *(b)* a convex mirror.

A negative magnification, which occurs when both s and s' are positive, indicates that the image is inverted.

Exercise

A concave mirror has a focal length of 4 cm. (*a*) What is its radius of curvature? (*b*) Find the image distance for an object 2 cm from the mirror. Draw a ray diagram for this situation. Is the image erect or inverted? [Answers: (*a*) 8 cm; (*b*) $s' = -4$ cm, erect]

For plane mirrors, discussed in the previous section, the radius of curvature is infinite. The focal length given by Equation 31-5 is then also infinite. Equation 31-6 then gives $s' = -s$, indicating that the image is behind the mirror at a distance equal to the object distance. The magnification given by Equation 31-7 is then +1, indicating that the image is erect and the same size as the object.

Although the preceding equations coupled with our sign conventions are relatively easy to use, we often need to know only whether an image is real or virtual, whether it is erect or inverted, and its approximate location and magnification. This knowledge is usually easiest to obtain by constructing a ray diagram. However, it is always a good idea to use both the graphical method and the algebraic method to locate an image so that one method serves as a check on the results of the other.

Convex Mirrors

Figure 31-19 shows a ray diagram for an object in front of a convex mirror. The central ray heading toward the center of curvature C is perpendicular to the mirror and is reflected back on itself. The parallel ray is reflected as if it came from the focal point *F* behind the mirror. The focal ray (not shown) would be drawn towards the focal point and would be reflected parallel to the axis. We can see from the figure that the image is behind the mirror and is therefore virtual. It is also erect and smaller than the object.

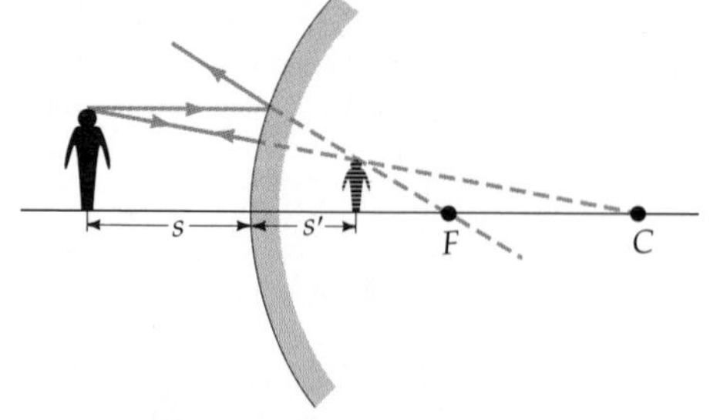

Figure 31-19 Ray diagram for a convex mirror. The parallel ray is reflected as if it came from the focal point behind the mirror, and the radial ray is reflected back on itself. These rays appear to diverge from a point behind the mirror. A third ray (not shown) could be drawn toward the focal point. It would be reflected parallel to the axis and its extension would intersect the other two rays at the image point.

Example 31-2

An object 2 cm high is 10 cm from a convex mirror with a radius of curvature of 10 cm. Locate the image and find its height.

Since the center of curvature of a convex mirror is behind the mirror, the radius and the focal length are negative:

$$f = \tfrac{1}{2}r = \tfrac{1}{2}(-10 \text{ cm}) = -5 \text{ cm}$$

Using Equation 31-6 to find the image distance, we obtain

$$\frac{1}{10 \text{ cm}} + \frac{1}{s'} = \frac{1}{f} = -\frac{1}{5 \text{ cm}}$$

$$\frac{1}{s'} = -\frac{2}{10 \text{ cm}} - \frac{1}{10 \text{ cm}} = -\frac{3}{10 \text{ cm}}$$

$$s' = -3.33 \text{ cm}$$

The image distance is negative, indicating a virtual image behind the mirror. The magnification is

$$m = -\frac{s'}{s} = -\frac{-3.33 \text{ cm}}{10 \text{ cm}} = +0.333$$

Thus, the image is erect and is one-third the size of the object. Since the object height is 2 cm, the image height is 2/3 cm. The ray diagram for this example is similar to Figure 31-19.

Exercise

Find the image distance and magnification for an object 5 cm away from the mirror in Example 31-2. Draw a ray diagram. (Answers: $s' = -2.5$ cm, $m = +0.5$)

Questions

3. Under what condition will a concave mirror produce an erect image? A virtual image? An image smaller than the object? An image larger than the object?
4. Answer Question 3 for a convex mirror.
5. Convex mirrors are often used for rear-view mirrors on cars and trucks to give a wide-angle view. Below the mirror is written, "Warning, objects are closer than they appear." Yet, according to a ray diagram such as Figure 31-19, the image distance for distant objects is much smaller than the object distance. Why then do they appear more distant?

(a)

(b)

(a) A convex mirror resting on paper with equally spaced parallel stripes. Note the large number of lines imaged in a small space and the reduction in size and distortion in shape of the image. *(b)* A convex mirror is used for security in a store.

31-3 Images Formed by Refraction

The formation of an image by refraction at a spherical surface separating two media with indexes of refraction n_1 and n_2 is illustrated in Figure 31-20. In this figure, n_2 is greater than n_1, so the waves travel more slowly in the second medium. Again, only paraxial rays converge to one point. An equation relating the image distance to the object distance, the radius of curvature, and the indexes of refraction can be derived by applying Snell's law of refraction to these rays and using small-angle approximations. The geometry is shown in Figure 31-21. The angles θ_1 and θ_2 are related by Snell's law:

$$n_1 \sin \theta_1 = n_2 \sin \theta_2$$

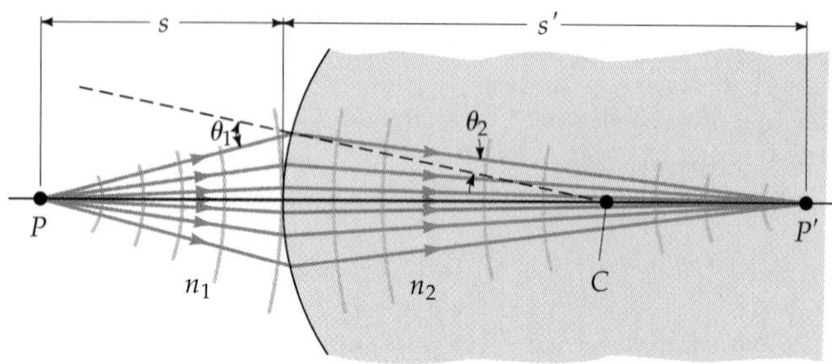

Figure 31-20 Image formed by refraction at a spherical surface between two media where the waves move slower in the second medium.

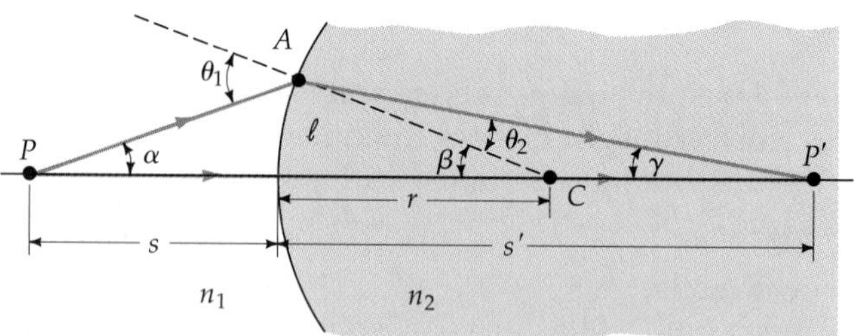

Figure 31-21 Geometry for relating the image position to the object position for refraction at a single spherical surface. Snell's law is applied to the ray incident at point A, and the small-angle approximation is used.

Using the small-angle approximation $\sin \theta \approx \theta$, we have

$$n_1\theta_1 = n_2\theta_2 \qquad \text{31-8}$$

From triangle ACP', we have

$$\beta = \theta_2 + \gamma = \frac{n_1}{n_2}\theta_1 + \gamma \qquad \text{31-9}$$

We can obtain another relation for θ_1 from triangle PAC:

$$\theta_1 = \alpha + \beta \qquad \text{31-10}$$

Eliminating θ_1 from Equations 31-9 and 31-10, we obtain

$$n_1\alpha + n_1\beta + n_2\gamma = n_2\beta$$

or

$$n_1\alpha + n_2\gamma = (n_2 - n_1)\beta \qquad \text{31-11}$$

Using the small-angle approximations $\alpha \approx \ell/s$, $\beta \approx \ell/r$, and $\gamma \approx \ell/s'$, we obtain

$$\frac{n_1}{s} + \frac{n_2}{s'} = \frac{n_2 - n_1}{r} \qquad \text{31-12}$$

In refraction, real images are formed in back of the surface, which we will call the transmission side, whereas virtual images occur on the incident side in front of the surface. The sign conventions we use for refraction are similar to those for reflection:

Sign conventions for refraction

s	+ (real object) for objects in front of the surface (incident side) – (virtual object) for objects in back of the surface (transmission side)
s'	+ (real image) for images in back of the surface (transmission side) – (virtual image) for images in front of the surface (incident side)
r,f	+ if the center of curvature is on the transmission side – if the center of curvature is on the incident side

If we compare these sign conventions with those for reflection, we see that s' is positive and the image is real when the image is on the side of the surface traversed by the reflected or refracted light. For reflection, this side is in front of the mirror, whereas for refraction, it is behind the refracting surface. Similarly, r and f are positive when the center of curvature is on the side traversed by the reflected or refracted light.

We can obtain an expression for the magnification of an image formed by a refracting surface by considering Figure 31-22, which shows a ray from the top of the object to the top of the image. The ray is bent toward the normal as it crosses the surface, so θ_2 is less than θ_1. These angles are related by Snell's law:

$$n_1 \sin \theta_1 = n_2 \sin \theta_2$$

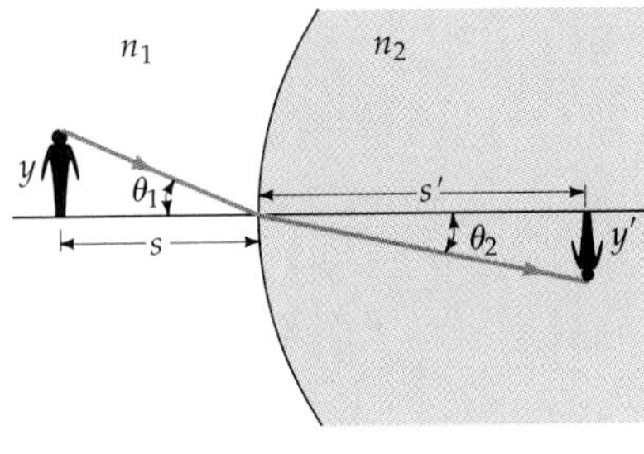

Figure 31-22 Geometry for finding the lateral magnification of an image formed by refraction from a single spherical surface.

The object and image sizes are related to the angles by

$$\tan \theta_1 = \frac{y}{s}$$

$$\tan \theta_2 = -\frac{y'}{s'}$$

where the minus sign arises because y' is negative. Since we are considering only paraxial rays for which the angles are small, the sine of an angle is approximately equal to its tangent. With this approximation, Snell's law becomes

$$n_1 \frac{y}{s} = n_2 \frac{-y'}{s'}$$

The magnification is thus

$$m = \frac{y'}{y} = -\frac{n_1 s'}{n_2 s} \qquad 31\text{-}13$$

Example 31-3

A fish is in a spherical bowl of water of index of refraction 1.33. The radius of the bowl is 15 cm. The fish looks through the bowl and sees a cat sitting on the table with its nose 10 cm from the bowl. Where is the image of the cat's nose, and what is its magnification? Neglect any effect of the thin glass wall of the bowl.

The object distance between the cat and the bowl is 10 cm. The indexes of refraction are $n_1 = 1$ and $n_2 = 1.33$. The radius of curvature is +15 cm. Equation 31-12 then gives for the image distance

$$\frac{1.00}{10\text{ cm}} + \frac{1.33}{s'} = \frac{1.33 - 1.00}{15\text{ cm}}$$

Solving for s', we obtain

$$s' = -17.1\text{ cm}$$

The negative image distance means that the image is virtual and in front

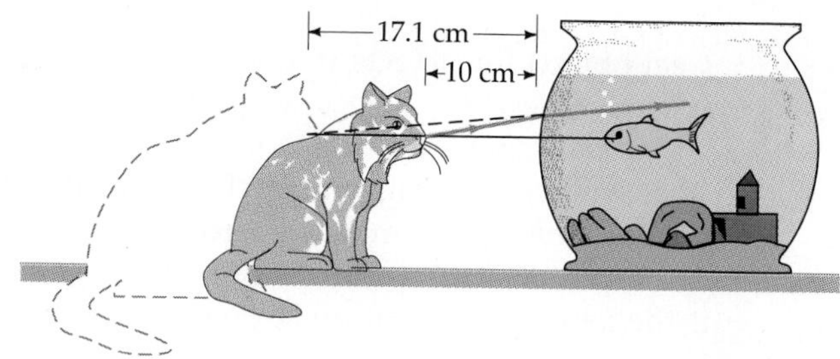

Figure 31-23 Fish looking at cat for Example 31-3. Because of refraction at the spherical surface, the cat appears farther away and slightly larger.

of the refracting surface, on the same side as the object, as shown in Figure 31-23. The magnification of the image is

$$m = -\frac{n_1 s'}{n_2 s} = -\frac{-17.1 \text{ cm}}{1.33(10 \text{ cm})} = 1.29$$

Thus, the cat appears to be farther away and slightly larger.

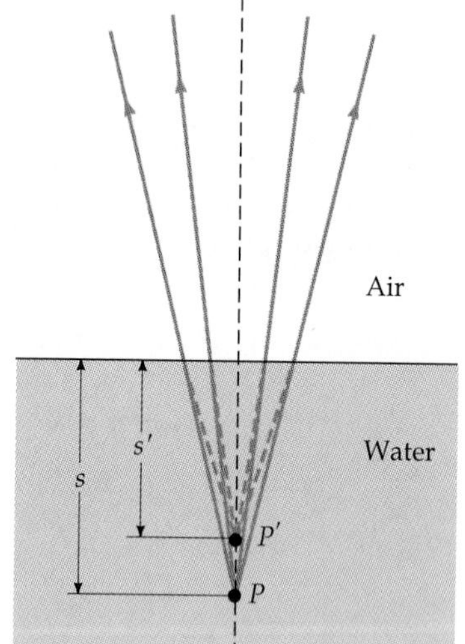

Figure 31-24 Ray diagram for the image of an object in water as viewed from directly overhead. The depth of the image is less than the depth of the object. The apparent depth equals the real depth divided by the index of refraction of water.

We can use Equation 31-12 to find the **apparent depth** of an object under water when it is viewed from directly overhead. For this case, the surface is a plane surface, so the radius of curvature is infinite. The image and object distances are related by

$$\frac{n_1}{s} + \frac{n_2}{s'} = 0$$

where n_1 is the index of refraction of the first medium (water) and n_2 is that of the second medium (air). The apparent depth is therefore

$$s' = -\frac{n_2}{n_1} s \qquad 31\text{-}14$$

The negative sign indicates that the image is virtual and on the same side of the refracting surface as the object, as shown in the ray diagram in Figure 31-24. The magnification is

$$m = -\frac{n_1 s'}{n_2 s} = +1$$

Since $n_2 = 1$ for air, we see from Equation 31-14 that the apparent depth equals the real depth divided by the index of refraction of water.

Example 31-4

Find the apparent depth of a fish resting 1 m below the surface of water that has an index of refraction $n = 4/3$.

Using $n_1 = 4/3$ and $n_2 = 1$ in Equation 31-14, we obtain

$$s' = -\frac{1}{(4/3)}(1 \text{ m}) = -\frac{3}{4}(1 \text{ m}) = -0.75 \text{ m}$$

The apparent depth is three-fourths the actual depth, so the fish appears to be 75 cm below the surface. Note that this result holds only when the object is viewed from directly overhead so that the rays are paraxial.

Questions

6. If a fish under water is viewed from a point not directly above it, is its apparent depth greater or less than three-fourths its actual depth? (Draw rays from the fish to the eye for large angles to help answer this question.)

7. A bird on a limb above the water is viewed by a scuba diver submerged beneath the water's surface. Does the bird appear to the diver to be closer to or father from the surface than it actually is?

31-4 Thin Lenses

The most important application of Equation 31-12 is finding the position of the image formed by a lens. This is done by considering the refraction at each surface separately to derive an equation relating the image distance to the object distance, the radius of curvature of each surface of the lens, and the index of refraction of the lens.

We will consider a very thin lens of index of refraction n with air on both sides. Let the radii of curvature of the surfaces of the lens be r_1 and r_2. If an object is at a distance s from the first surface (and therefore from the lens), the distance s'_1 of the image due to refraction at the first surface can be found using Equation 31-12:

$$\frac{1}{s} + \frac{n}{s'_1} = \frac{n-1}{r_1} \qquad 31\text{-}15$$

Because of refraction, the apparent depth of the submerged portion of the straw is less than the real depth. Consequently, the straw appears to be bent. A reflected image of the straw is also seen.

This image is not formed because the light is again refracted at the second surface. Figure 31-25 shows the case when the image distance s'_1 for the first surface is negative, indicating a virtual image to the left of the surface. Rays in the glass refracted from the first surface diverge as if they came from the image point P'_1. They strike the second surface at the same angles as if there were an object at this image point. The image for the first surface therefore becomes the object for the second surface. Since the lens is of negligible thickness, the object distance is equal in magnitude to s'_1, but since object distances in front of the surface are positive whereas image distances are negative there, the object distance for the second surface is $s_2 = -s'_1$. (If s'_1 were positive, the rays would converge as they strike the second surface. The object for the second surface would then be to the right of the surface. This object would be a **virtual object.** Again, $s_2 = -s'_1$.) We now write Equation 31-12 for the second surface with $n_1 = n$, $n_2 = 1$, and $s = -s'_1$. The image distance for the second surface is the final image distance s' for the lens.

$$\frac{n}{-s'_1} + \frac{1}{s'} = \frac{1-n}{r_2} \qquad 31\text{-}16$$

We can eliminate the image distance for the first surface s'_1 by adding Equations 31-15 and 31-16. We obtain

$$\frac{1}{s} + \frac{1}{s'} = (n-1)\left(\frac{1}{r_1} - \frac{1}{r_2}\right) \qquad 31\text{-}17$$

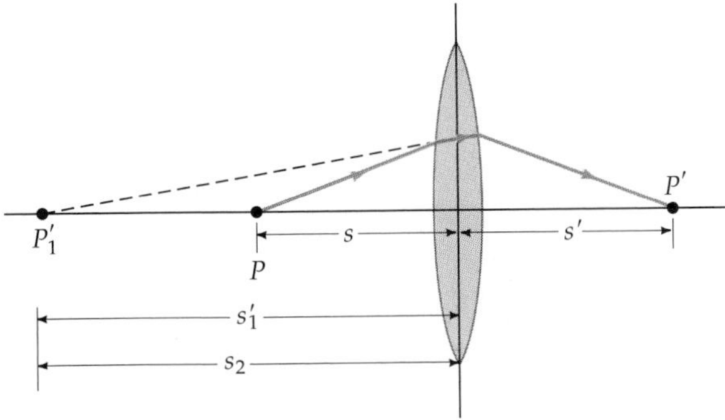

Figure 31-25 Refraction occurs at both surfaces of a lens. Here, the refraction at the first surface leads to a virtual image at P'_1. The rays strike the second surface as if they came from P'_1. Since image distances are negative when the image is on the incident side of the surface whereas object distances are positive for objects there, $s_2 = -s'_1$ is the object distance for the second surface of the lens.

Equation 31-17 gives the image distance s' in terms of the object distance s and the properties of the thin lens—r_1, r_2, and the index of refraction n. As with mirrors, the focal length of a thin lens is defined as the image distance when the object distance is infinite. Setting s equal to infinity and writing f for the image distance s', we obtain

Lens-maker's equation

$$\frac{1}{f} = (n - 1)\left(\frac{1}{r_1} - \frac{1}{r_2}\right) \qquad 31\text{-}18$$

Equation 31-18 is called the **lens-makers' equation;** it gives the focal length of a thin lens in terms of the properties of the lens. Substituting $1/f$ for the right side of Equation 31-17, we obtain

Thin-lens equation

$$\frac{1}{s} + \frac{1}{s'} = \frac{1}{f} \qquad 31\text{-}19$$

Equation 31-19 is called the **thin-lens equation.** Note that it is the same as the mirror equation (Equation 31-6). Recall, however, that the sign conventions for refraction are somewhat different from those for reflection. For lenses, the image distance s' is positive when the image is on transmission side of the lens, that is, when it is on the side opposite the side on which light is incident. The sign convention for r in Equation 31-18 is the same as that for refraction at a single surface. The radius is positive if the center of curvature is on the transmission side of the lens and negative if it is on the incident side.

Figure 31-26*a* shows the wavefronts of plane waves incident on a double convex lens. The central part of the wavefront strikes the lens first. Since the wave speed in the lens is less than that in air (assuming $n > 1$), the central

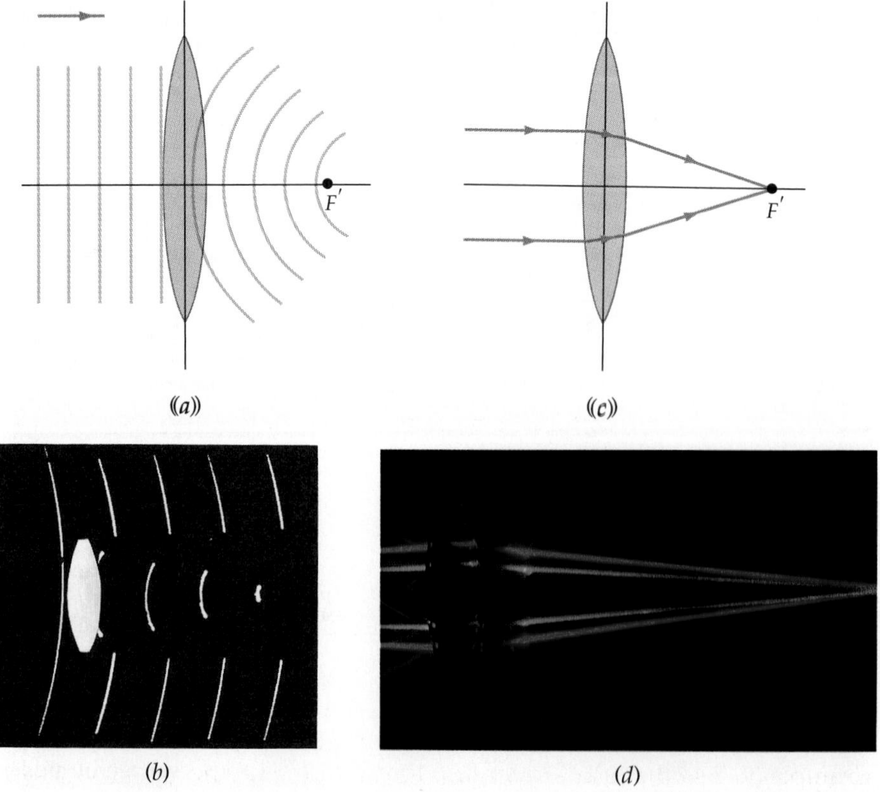

Figure 31-26 *(a)* Wavefronts for plane waves striking a converging lens. The central part of the wavefront is retarded more by the lens than the outer part, resulting in a spherical wave that converges at the focal point F'. *(b)* Wavefronts passing through a lens, shown by a photographic technique called *light-in-flight-recording* that uses a pulsed laser to make a hologram of the wavefronts of light. *(c)* Rays for plane waves striking a converging lens. The rays are bent at each surface and converge at the focal point. *(d)* Photograph of rays focused by a converging lens.

part of the wavefront lags behind the outer parts, resulting in a spherical wave that converges at the focal point F'. The rays for this situation are shown in Figure 31-26*c*. Such a lens is called a **converging lens.** Since its focal length as calculated from Equation 31-18 is positive, it is also called a **positive lens.** Any lens that is thicker in the middle than at the edges is a converging lens (providing that the index of refraction of the lens is greater than that of the surrounding medium). Figures 31-27*a* and 31-27*b* show the wavefronts and rays for plane waves incident on a double concave lens. In this case, the outer part of the wavefronts lag behind the central parts, resulting in outgoing spherical waves that diverge from a focal point on the incident side of the lens. The focal length of this lens is negative. Any lens (with index of refraction greater than that of the surrounding medium) that is thinner in the middle than at the edges is a **diverging,** or **negative, lens.**

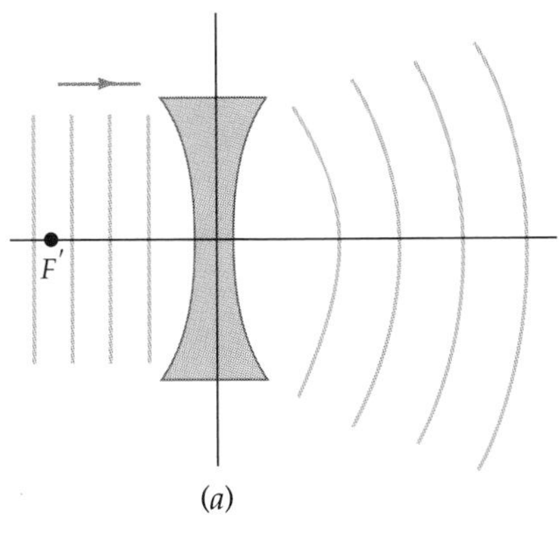

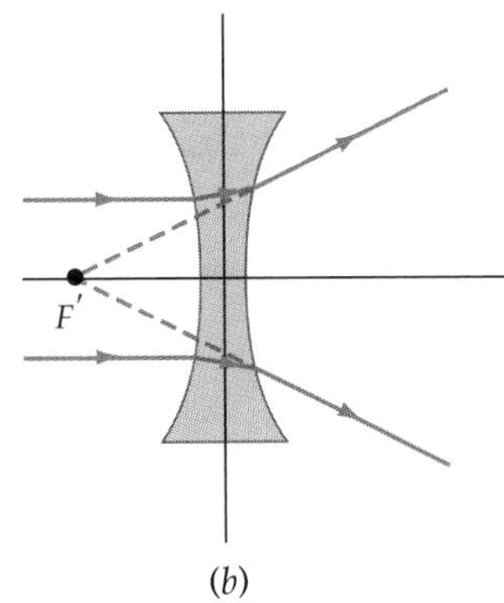

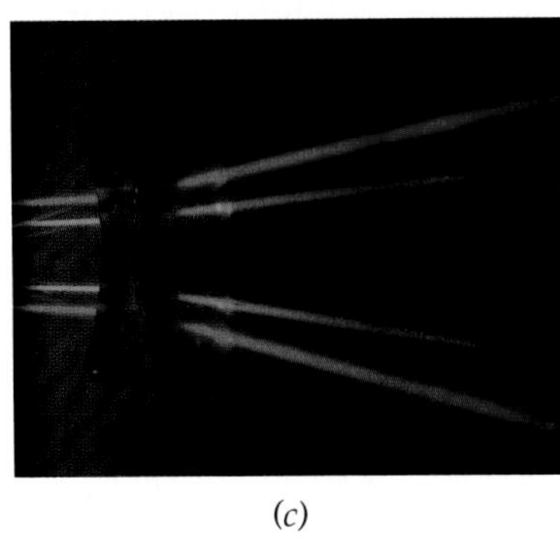

Figure 31-27 (*a*) Wavefronts for plane waves striking a diverging lens. Here, the outer parts of the wavefronts are retarded more than the central part, resulting in a spherical wave that diverges as it moves out as if it came from the focal point F' in front of the lens. (*b*) Rays for plane waves striking the same diverging lens. The rays are bent outward and diverge as if they came from the focal point F'. (*c*) Photograph of rays passing through a diverging lens.

Example 31-5

A double convex thin lens made of glass of index of refraction $n = 1.5$ has radii of curvature of magnitude 10 cm and 15 cm as shown in Figure 31-28. Find its focal length.

We assume that the light is incident on the surface with the smaller radius of curvature (the left surface in Figure 31-28). The center of curvature of the first surface, C_1, is on the transmission side of the lens, so r_1 is positive and equal to +10 cm. The center of curvature of the second surface, C_2, is on the incident side, so r_2 is negative and equal to −15 cm. Equation 31-18 is then

$$\frac{1}{f} = (1.5 - 1)\left(\frac{1}{+10\text{ cm}} - \frac{1}{-15\text{ cm}}\right)$$

$$= 0.5\left(\frac{3}{30\text{ cm}} + \frac{2}{30\text{ cm}}\right) = 0.5\left(\frac{1}{6\text{ cm}}\right)$$

$$f = 12\text{ cm}$$

Exercise

A double convex thin lens has an index of refraction $n = 1.6$ and radii of curvature of equal magnitude. If its focal length is 15 cm, what is the magnitude of the radius of curvature of each surface? (Answer: 18 cm)

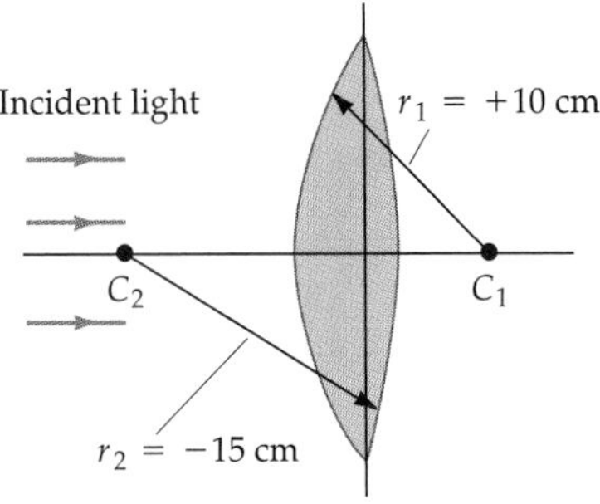

Figure 31-28 Double convex lens with radii of curvature of magnitude 15 cm and 10 cm for Example 31-5. The center of curvature of the first surface is on the transmission side of the lens, so r is positive for this surface. The center of curvature for the second surface is on the incident side of the lens, so r for it is negative. Both surfaces tend to converge the light rays and contribute to a positive focal length for the lens.

Note that if we reverse the direction of the incoming light for the lens in Example 31-5 so that it is incident on the surface with the greater radius of

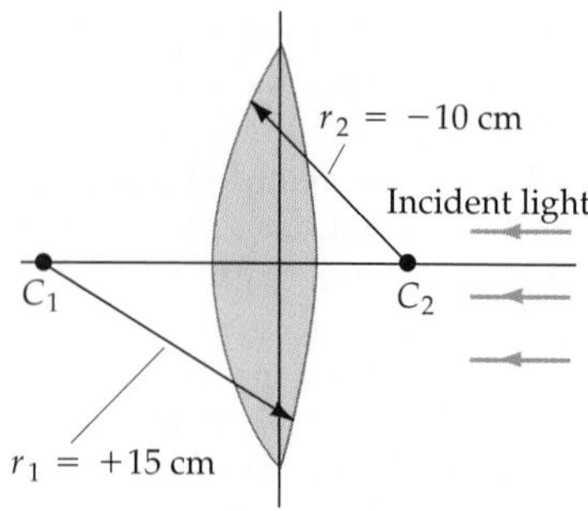

Figure 31-29 The same lens as in Figure 31-28 with the light incident from the other side. The order of the surfaces and the signs of the radii of curvature are interchanged, but the focal length is the same.

curvature (from the right as in Figure 31-29), the order of the surfaces is interchanged. The radius of the first surface has a magnitude of 15 cm and is positive because the center of curvature of that surface is on the transmission side, whereas the center of curvature of the surface with the 10-cm radius is on the incident side, so $r_2 = -10$ cm. Using these values in Equation 31-18 gives the same result, $f = 12$ cm, for the focal length. Thus, the focal length of a lens is the same for light incident on either side. If parallel light strikes the lens of Example 31-5 from the left, it is focused at a point 12 cm to the right of the lens, whereas if parallel light strikes the lens from the right, it is focused at 12 cm to the left of the lens. Both of these points are focal points of the lens. Using the reversibility property of light rays, we can see that light diverging from a focal point and striking the lens will leave the lens as a parallel beam as shown in Figure 31-30. In a particular lens problem in which the direction of the incident light is specified, the object point for which light emerges as a parallel beam is called the **first focal point** F and the point at which parallel light is focused is called the **second focal point** F'. For a positive lens, the first focal point is on the incident side and the second focal point is on the transmission side. If parallel light is incident on the lens at a small angle with the axis as in Figure 31-31, it is focused at a point in the **focal plane** a distance f from the lens.

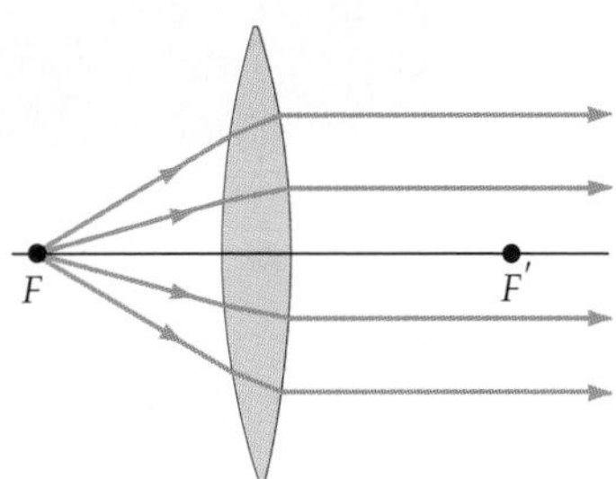

Figure 31-30 Light rays diverging from the focal point of a positive lens emerge parallel to the axis. This point is called the first focal point F. The point at which parallel light is converged by the lens is called the second focal point F'.

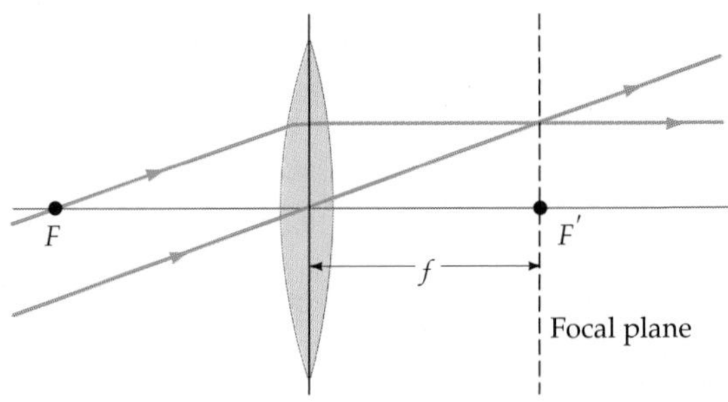

Figure 31-31 Parallel rays incident on the lens at an angle to its axis are focused at a point in the focal plane of the lens.

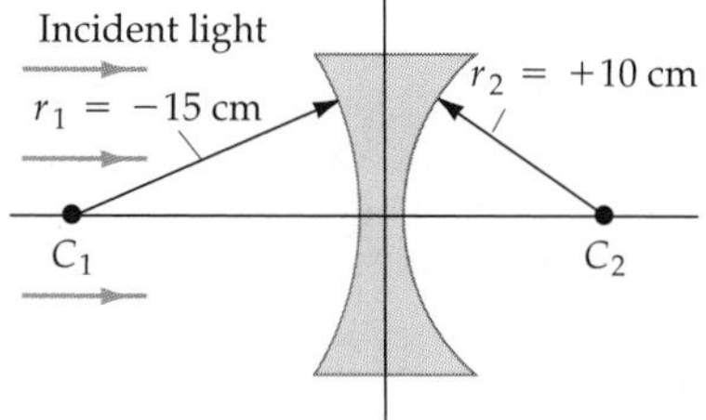

Figure 31-32 Double concave lens with radii of magnitude 15 cm and 10 cm. The center of curvature of the first surface is on the incident side of the lens and that of the second surface is on the transmission side, so r_1 is negative and r_2 is positive. Both surfaces tend to diverge light rays and contribute to a negative focal length.

Example 31-6

A double concave lens has an index of refraction of 1.5 and radii of curvature of magnitude 10 cm and 15 cm. Find its focal length.

For the orientation of the lens relative to the incident light shown in Figure 31-32, the radius of curvature of the first surface is $r_1 = -15$ cm and that of the second surface is $r_2 = +10$ cm. The lens-maker's equation (Equation 31-18) then gives

$$\frac{1}{f} = (1.5 - 1.0)\left(\frac{1}{-15\text{ cm}} - \frac{1}{+10\text{ cm}}\right)$$

Solving for f, we obtain $f = -12$ cm. Again, we obtain the same result no matter which surface the light strikes first.

In laboratory experiments involving lenses, it is usually much easier to measure the focal length rather than to calculate it from the radii of curvature of the surfaces.

Ray Diagrams for Lenses

As with images formed by mirrors, it is convenient to locate the images of lenses by graphical methods. Figure 31-33 illustrates the graphical method for a converging lens. We use three principal rays. For the sake of simplicity, we consider the rays to bend at the plane through the center of the lens. For a positive lens, the principal rays are

Principal rays for a positive lens

1. The **parallel ray,** drawn parallel to the axis. This ray is bent through the second focal point of the lens.
2. The **central ray,** drawn through the center (the vertex) of the lens. This ray is undeflected. (The faces of the lens are parallel at this point, so the ray emerges in the same direction but displaced slightly. Since the lens is thin, the displacement is negligible.)
3. The **focal ray,** drawn through the first focal point. This ray emerges parallel to the axis.

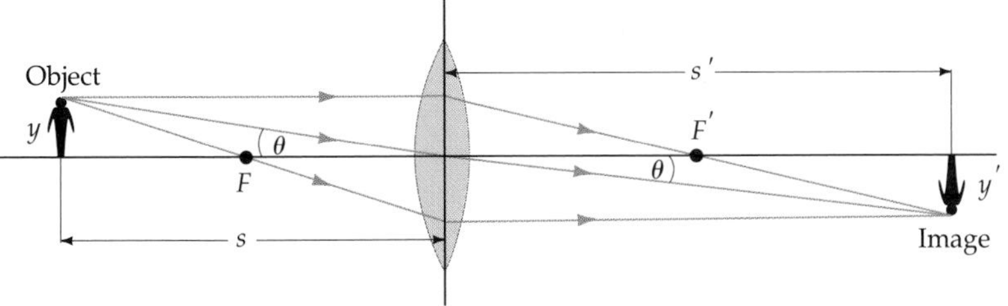

Figure 31-33 Ray diagram for a thin converging lens. For the sake of simplicity, we assume that all the bending takes place at the central plane. The ray through the center is undeflected because the lens surfaces there are parallel and are close together.

These three rays converge to the image point, as shown in the figure. In this case, the image is real and inverted. From Figure 31-33, we have $\tan\theta = y/s = -y'/s'$. The lateral magnification is then

$$m = \frac{y'}{y} = -\frac{s'}{s}$$

This expression is the same as that for mirrors. Again, a negative magnification indicates that the image is inverted.

The principal rays for a negative, or diverging, lens are

Principal rays for a negative lens

1. The **parallel ray,** drawn parallel to the axis. This ray diverges from the lens as if it came from the second focal point.
2. The **central ray,** drawn through the center (the vertex) of the lens. This ray is undeflected.
3. The **focal ray,** drawn toward the first focal point. This ray emerges parallel to the axis.

The ray diagram for a diverging lens is shown in Figure 31-34.

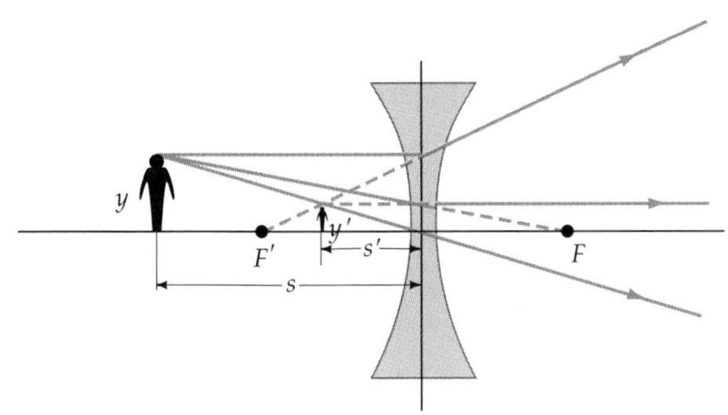

Figure 31-34 Ray diagram for a diverging lens. The parallel ray is bent away from the axis as if it came from the second focal point F'. The ray toward the first focal point F emerges parallel to the axis. The central ray is undeflected. The three rays appear to diverge from the image point.

Example 31-7

An object 1.2 cm high is placed 4 cm from the double convex lens of Example 31-5. Locate the image, state whether it is real or virtual, and find its height.

In Example 31-5, the focal length for this lens was found to be $f =$ 12 cm. Figure 31-35 shows the ray diagram for an object placed 4 cm in front of a positive lens of focal length 12 cm. The parallel ray is bent through the second focal point, and the central ray is undeflected. These rays are diverging on the transmission side of the lens. The image is located by extending the rays back until they meet. These two rays are sufficient to locate the image. (As a check, we could draw the third ray, the focal ray, along the line from the first focal point F on the incident side of the lens. This ray would leave the lens parallel to the axis.) We can see immediately from the figure that the image is virtual, erect, and enlarged. It is on the same side of the lens as the object and is slightly farther away from the lens. Since it is quite easy to make an error when calculating the image distance using Equation 31-19, it is always a good idea to check your result with a ray diagram.

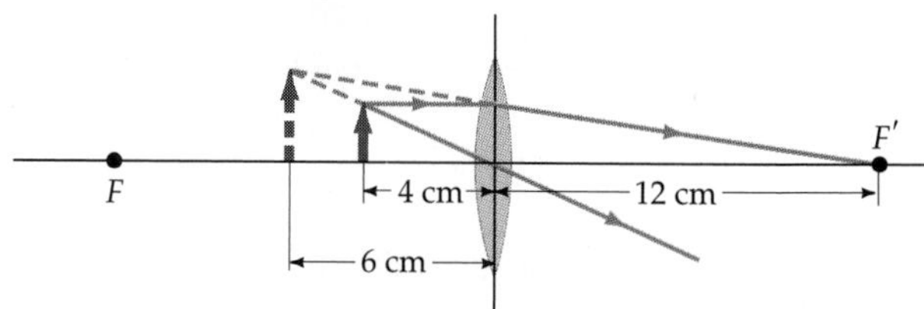

Figure 31-35 Ray diagram for Example 31-7. When the object is between the first focal point and the converging lens, the image is virtual and erect.

The image distance is found algebraically using Equation 31-19:

$$\frac{1}{4\text{ cm}} + \frac{1}{s'} = \frac{1}{12\text{ cm}}$$

$$\frac{1}{s'} = \frac{1}{12\text{ cm}} - \frac{1}{4\text{ cm}} = \frac{1}{12\text{ cm}} - \frac{3}{12\text{ cm}} = -\frac{1}{6\text{ cm}}$$

$$s' = -6\text{ cm}$$

The image distance is negative, indicating that the image is virtual and on the incident side of the lens. The magnification is

$$m = -\frac{s'}{s} = -\frac{-6\text{ cm}}{4\text{ cm}} = +1.5$$

The image is thus 1.5 times as large as the object and is erect. Since the height of the object is 1.2 cm, the height of the image is 1.8 cm.

Exercise

An object is placed 15 cm from a double convex lens of focal length 10 cm. Find the image distance and the magnification. Draw a ray diagram. Is the image real or virtual? Erect or inverted? (Answers: $s' =$ 30 cm, $m = -2$, real, inverted)

Exercise

Work the previous exercise for an object placed 5 cm from a lens with a focal length of 10 cm. (Answers: $s' = -10$ cm, $m = 1$, virtual, erect)

Multiple Lenses

If we have two or more thin lenses, we can find the final image produced by the system by finding the image distance for the first lens and using it along with the distance between lenses to find the object distance for the second lens. That is, we consider each image, whether it is real or virtual and whether it is formed or not, as the object for the next lens.

Example 31-8

A second lens of focal length +6 cm is placed 12 cm to the right of the lens in Example 31-7. Locate the final image.

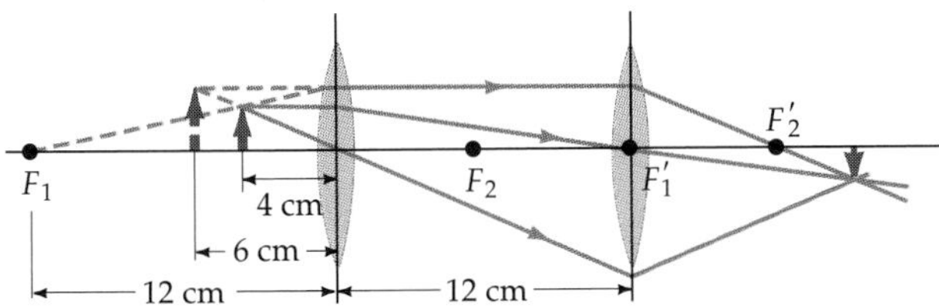

Figure 31-36 Ray diagram for Example 31-8. The image of the first lens acts as the object for the second lens. The final image is located by drawing two rays from the first image through the second lens. In this case, one of the rays used to locate the first image happens to be the central ray for the second lens. A second ray parallel to the axis from the first image locates the final image.

Figure 31-36 shows the ray diagram for this example. The rays used to locate the image of the first lens will not necessarily be the principal rays for the second lens. If they are not, we merely draw additional rays from the first image that are principal rays for the second lens, such as a ray from the image parallel to the axis and one from the image through the first focal point of the second lens or one through the vertex of the second lens. In this case, two of the principal rays for the first lens are also principal rays for the second lens. The parallel ray for the first lens turns out to be the central ray for the second lens. Also, the focal ray for the first lens emerges parallel to the axis and is therefore refracted through the focal point of the second lens. (In the figure, we have extended the central ray for the first lens so that it passes through the image found from the other two rays.) We can see that the final image is real, inverted, and just outside the second focal point of the second lens. We can locate its position algebraically by noting that the virtual image of the first lens is 6 cm to the left of that lens and is therefore 18 cm to the left of the second lens. Using $s_2 = 18$ cm and $f_2 = 6$ cm, we have

$$\frac{1}{s_2} + \frac{1}{s_2'} = \frac{1}{f_2}$$

$$\frac{1}{18\text{ cm}} + \frac{1}{s_2'} = \frac{1}{6\text{ cm}}$$

giving

$$s_2' = 9\text{ cm}$$

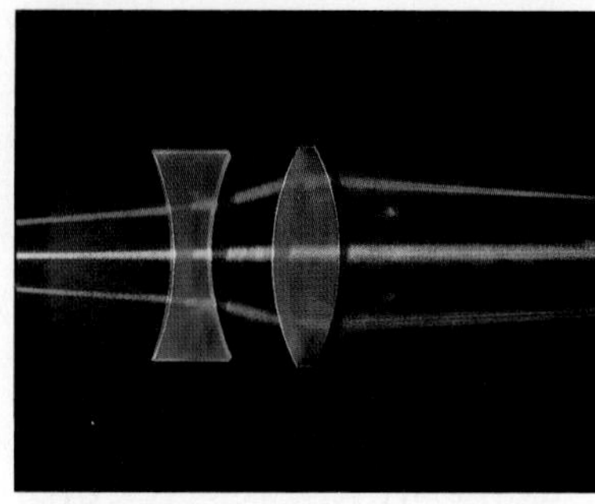

Light rays passing through a diverging lens followed by a converging lens, which is placed at the first focal point of the diverging lens.

Example 31-9

Two lenses, each of focal length 10 cm, are 15 cm apart. Find the final image of an object 15 cm from one of the lenses.

In the ray diagram of Figure 31-37, the image of the first lens would be 30 cm to the right of the lens if the second lens were not there. We calculate this using $s_1 = 15$ cm and $f_1 = 10$ cm in the thin-lens equation:

$$\frac{1}{s_1} + \frac{1}{s_1'} = \frac{1}{f_1}$$

$$\frac{1}{15\text{ cm}} + \frac{1}{s_1'} = \frac{1}{10\text{ cm}}$$

Solving for s_1', we obtain

$$s_1' = 30\text{ cm}$$

This image is not formed because the light rays strike the second lens before they reach the image position. We can locate the final image graphically by choosing rays that are heading toward the unformed image when they strike the lens. These rays need not be the principal rays for the first lens. Any ray that leaves the object and strikes the first lens is directed toward the image of the first lens. We choose a ray that leaves the first lens parallel to the axis (the bottom ray in the figure) and one that goes through the center of the second lens (the top ray in the figure). We see that the final image is between the second lens and its focal point. (In the figure, we have extended the middle two rays used to locate the first, unformed image so that they pass through the final image.) The final image can be located algebraically by using the first image as the object for the second lens. Since this unformed image is on the transmission side of the second lens, it is a virtual object. Since it is 15 cm from the second lens, the object distance is $s_2 = -15$ cm. Then

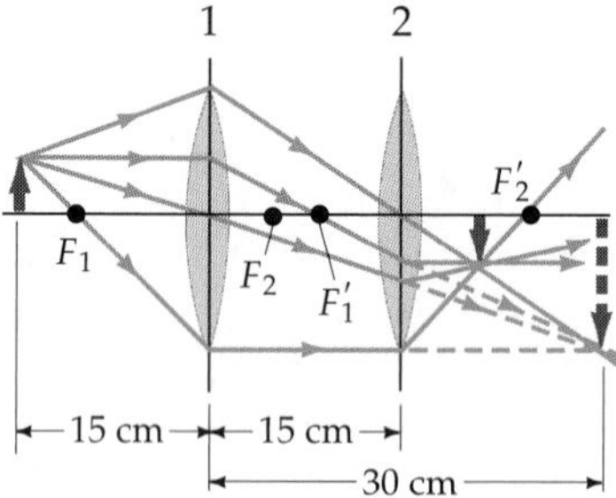

Figure 31-37 Ray diagram for Example 31-9. The image of the first lens is to the right of the second lens. This image is not formed because the rays are refracted by the second lens before they get to the first image. Nevertheless, this unformed image acts as a virtual object for the second lens. The final image is found by drawing rays toward the first image as shown. A ray through the center of the second lens and a ray parallel to the axis as it strikes the second lens are used.

$$\frac{1}{-15\text{ cm}} + \frac{1}{s_2'} = \frac{1}{f_2} = \frac{1}{10\text{ cm}}$$

Solving for s_2', we obtain

$$s_2' = 6\text{ cm}$$

Example 31-10

Two thin lenses of focal lengths f_1 and f_2 are placed together. Show that the equivalent focal length of the combination f is given by

$$\frac{1}{f} = \frac{1}{f_1} + \frac{1}{f_2} \tag{31-20}$$

Let s be the object distance for the first lens (and, therefore, for the lens combination) and s'_1 be the image distance. Applying the thin-lens equation to the first lens, we have

$$\frac{1}{s} + \frac{1}{s'_1} = \frac{1}{f_1}$$

Since the lenses are together, the object distance for the second lens is the negative of the image distance for the first lens, so $s_2 = -s'_1$. Calling the final image distance s', we have for the second lens

$$\frac{1}{-s'_1} + \frac{1}{s'} = \frac{1}{f_2}$$

Adding these two equations to eliminate s'_1, we obtain

$$\frac{1}{s} + \frac{1}{s'} = \frac{1}{f_1} + \frac{1}{f_2} = \frac{1}{f}$$

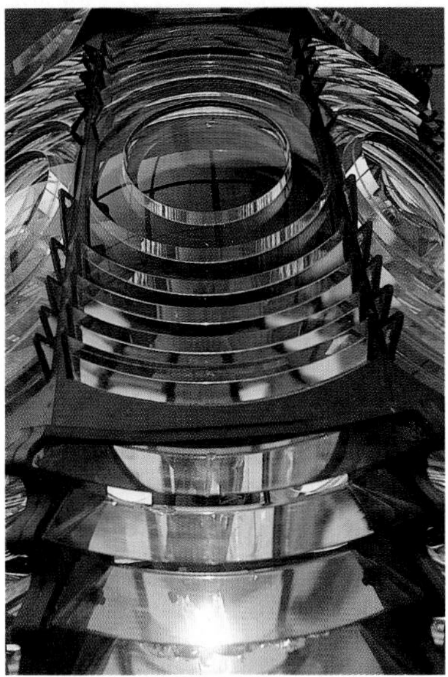

The weight and bulk of a large diameter lens can be reduced by constructing the lens from annular segments at different angles such that light from a point is refracted by the segments into a parallel beam. Such an arrangement is called a Fresnel lens. Several Fresnel lenses are used in this lighthouse to produce intense parallel beams of light from a source at the focal point of the lenses.

Example 31-10 gives us the important result that when two lens are placed in contact (or very close together), the reciprocals of their focal lengths add. The reciprocal of the focal length is called the **power of a lens.** When the focal length is expressed in meters, the power is given in reciprocal meters called **diopters** (D):

$$P = \frac{1}{f} \quad \text{diopters} \qquad 31\text{-}21$$

The power of a lens measures its ability to focus parallel light at a short distance from the lens. The shorter the focal length, the greater the power. For example, a lens with a focal length of 25 cm = 0.25 m has a power of 4.0 diopters. A lens with a focal length of 10 cm = 0.10 m has a power of 10 diopters. Since the focal length of a diverging lens is negative, its power is also negative.

Example 31-11

A lens has a power of −2.5 diopters. What is its focal length?

Solving Equation 31-21 for the focal length, we obtain

$$f = \frac{1}{P} = \frac{1}{-2.5\ \text{D}} = -0.40\ \text{m} = -40\ \text{cm}$$

where we have used the fact that a diopter is the same as a reciprocal meter, that is, $1\ \text{D} = 1\ \text{m}^{-1}$.

The results of Example 31-10 can be simply stated in terms of the power of a lens. When two lenses are in contact, the power of the combination equals the sum of the powers of the lenses:

$$P = P_1 + P_2 \quad \text{two lenses in contact} \qquad 31\text{-}22$$

Questions

8. Under what conditions will the focal length of a thin lens be positive? Negative?
9. The focal length of a simple lens is different for different colors of light. Why?

31-5 Aberrations

When all the rays from a point object are not focused at a single image point, the resulting blurring of the image is called **aberration.** Figure 31-38 shows rays from a point source on the axis traversing a thin lens with spherical surfaces. Rays that strike the lens far from the axis are bent much more than those near the axis, with the result that not all the rays are focused at a single point. Instead, the image appears as a circular disk. The **circle of least confusion** is at point C, where the diameter is minimum. This type of aberration is called **spherical aberration.** It is the same as the spherical aberration of mirrors discussed in Section 31-2. Similar but more complicated aberrations called *coma* (for the comet-shaped image) and *astigmatism* occur when objects are off axis. The aberration in the shape of the image of an extended object due to the fact that the magnification depends on the distance of the object point from the axis is called **distortion.** We will not discuss these aberrations further except to point out that they do not arise from any defect in the lens or mirror but instead result from the application of the laws of refraction and reflection to spherical surfaces. They are not evident in our simple equations because we used small-angle approximations in the derivation of these equations.

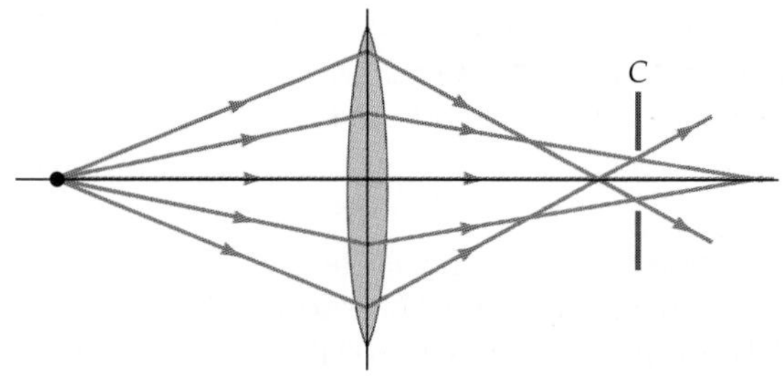

Figure 31-38 Spherical aberration. Rays from a point object on the axis are not focused at a point. The image is thus a circle about the axis rather than a point. At point C, the circle has its least diameter. This is called the circle of least confusion. Spherical aberration can be reduced by blocking off the outer parts of the lens and thereby reducing the diameter of the circle of least confusion, but this also reduces the amount of light reaching the image.

Some aberrations can be eliminated or partially corrected by using nonspherical surfaces for mirrors or lenses, but nonspherical surfaces are usually much more difficult and costly to produce than spherical surfaces. One example of a nonspherical reflecting surface is the parabolic mirror illustrated in Figure 31-39. Rays that are parallel to the axis of a parabolic surface are reflected and focused at a common point no matter how far they are from the axis. Parabolic reflecting surfaces are commonly used in large astronomical telescopes, which need a large reflecting surface to gather as much light as possible to make the image as intense as possible. A parabolic surface can also be used in a searchlight to produce a parallel beam of light from a small source placed at the focal point of the surface.

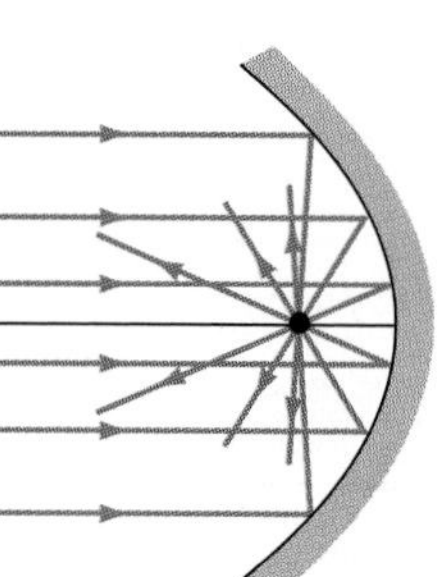

Figure 31-39 A parabolic mirror focuses all rays parallel to the axis to a single point with no spherical aberration. Parabolic surfaces are more costly to produce than spherical surfaces.

An important aberration found with lenses but not with mirrors is **chromatic aberration,** which is due to variations in the index of refraction with wavelength. From Equation 31-18, we can see that the focal length of a lens depends on its index of refraction and is therefore different for different wavelengths. Since n is slightly greater for blue light than for red light (see Figure 30-22), the focal length for blue light will be shorter than that for red light. Since chromatic aberration does not occur for mirrors, many large telescopes use mirrors rather than lenses.

Chromatic and other aberrations can be partially corrected by using combinations of lenses instead of a single lens. For example, a positive lens and a negative lens of greater focal length can be used together to produce a converging lens system that has much less chromatic aberration than a single lens of the same focal length. The lens of a good camera typically contains six elements to correct for the various aberrations that are present.

Summary

1. The image formed by a spherical mirror or by a lens is at a distance s', which is related to the object distance s by

$$\frac{1}{s} + \frac{1}{s'} = \frac{1}{f}$$

where f is the focal length, which is the image distance when $s = \infty$. For a mirror, the focal length is equal to half the radius of curvature. For a thin lens in the air, the focal length is related to the index of refraction n and radii of curvature of the two sides r_1 and r_2 by

$$\frac{1}{f} = (n - 1)\left(\frac{1}{r_1} - \frac{1}{r_2}\right)$$

In these equations s, s', r, r_1, and r_2 are taken to be positive when the object, image, or center of curvature lies on the real side of the element. For mirrors, the real side is the incident side. For lenses, the real side is the incident side for objects and the transmission side for images and centers of curvature. When s' is positive, the image is real, meaning that light rays actually diverge from the image point. Real images can be seen on a ground-glass viewing screen or photographic film placed at the image point. When s' is negative, the image is virtual, meaning that no light actually diverges from the image point.

2. The lateral magnification of the image is given by

$$m = \frac{y'}{y} = \frac{-s'}{s}$$

where y is the object size and y' is the image size. A negative magnification means that the image is inverted.

3. For a plane mirror, r and f are infinite, $s' = -s$, and the image is virtual, erect, and the same size as the object.

4. Images can be conveniently located by a ray diagram using any two principal rays. The point from which these rays diverge or appear to diverge is the image point. For spherical mirrors, there are four principal rays: the parallel ray, parallel to the axis; the focal ray, through the focal point; the radial ray, through the center of curvature of the mirror; and the central ray, toward the vertex of the mirror. For a lens, there are three principal rays: the parallel ray, parallel to the axis; the focal ray, through the second focal point; and the central ray, through the center of the lens.

5. A positive or converging lens is one that is thicker at the middle than at the edges. Parallel light incident on a positive lens is focused at the second focal point, which is on the transmission side of the lens. A negative or diverging lens is one that is thicker at the edges than at the

middle. Parallel light incident on a negative lens emerges as if it originated from the second focal point, which is on the incident side of the lens.

6. The power of a lens equals the reciprocal of the focal length. When the focal length is in meters, the power is in diopters (D):

$$P = \frac{1}{f} \quad \text{diopters}$$

$$1\ \text{D} = 1\ \text{m}^{-1}$$

7. The image distance s' for refraction at a single spherical surface of radius r is related to the object distance s and the radius of curvature of the surface r by

$$\frac{n_1}{s} + \frac{n_2}{s'} = \frac{n_2 - n_1}{r}$$

where n_1 is the index of refraction of the medium on the incident side of the surface and n_2 is the index of refraction of the medium on the transmission side. The magnification of the image due to refraction at a single surface is

$$m = -\frac{n_1 s'}{n_2 s}$$

8. The blurring of the image of a single object point is known as aberration. Spherical aberration results from the fact that a spherical surface focuses only paraxial rays (those that travel close to the axis) at a single point. Nonparaxial rays are focused at nearby points depending on the angle made with the axis. Spherical aberration can be reduced by reducing the size of the spherical surface, which also reduces the amount of light reaching the image. Chromatic aberration, which occurs with lenses but not mirrors, results from the variation in the index of refraction with wavelength. Lens aberrations are most commonly reduced by using a series of lens elements.

Review

A. Objectives: After studying this chapter, you should:

1. Be able to draw simple ray diagrams for mirrors and lenses to locate images and determine whether they are real or virtual, erect or inverted, and enlarged or reduced.

2. Be able to determine algebraically the location of the image formed by a mirror or by a thin lens and calculate the magnification of the image.

3. Be able to use the lens-maker's equation to determine the focal length of a lens from the radii of curvature of the surfaces.

4. Be able to discuss spherical aberration and chromatic aberration.

B. Define, explain, or otherwise identify:

Geometrical optics
Image
Virtual image
Depth inversion
Real image
Paraxial ray
Spherical aberration
Focal length
Focal point
Reversibility
Ray diagram
Principal ray
Lateral magnification
Apparent depth
Virtual object
Lens-maker's equation
Converging lens
Positive lens
Diverging lens
Negative lens
Focal plane
Virtual object
Power of a lens
Diopter
Chromatic aberration

C. True or false: If the statement is true, explain why it is true. If it false, give a counterexample.

1. A virtual image cannot be displayed on a screen.

2. Aberrations occur only for real images.

3. A negative image distance implies that the image is virtual.

4. All rays parallel to the axis of a spherical mirror are reflected through a single point.

5. A diverging lens cannot form a real image from a real object.

6. The image distance for a positive lens is always positive.

7. Chromatic aberration does not occur with mirrors.

Suggestions for Further Reading

Walker, Jearl: "The Amateur Scientist: The Kaleidoscope Now Comes Equipped with Flashing Diodes and Focusing Lenses," *Scientific American,* December 1985, p. 134.

Multiple reflections created by arrangements of different numbers of plane mirrors are investigated.

Walker, Jearl: "The Amateur Scientist: What Is a Fish's View of a Fisherman and the Fly He Has Cast on the Water?" *Scientific American,* March 1984, p. 138.

Refraction causes the world above water to appear to a fish the way things appear to us when viewed through a "fish-eye" lens.

Walker, Jearl: "The Amateur Scientist: Shadows Cast on the Bottom of a Pool Are Not Like Other Shadows. Why?" *Scientific American,* July 1988, p. 116.

This article describes simple experiments in refraction from curved surfaces to try the next time you take a bath!

Problems

Use $n = 1.33$ for the index of refraction of water unless otherwise specified.

Level I

31-1 Plane Mirrors

1. The image of the point object P in Figure 31-40 is viewed by an eye as shown. Draw a bundle of rays from the object that reflect from the mirror and enter the eye. For this object position and mirror, indicate the region of space in which the eye can see the image.

Figure 31-40 Problem 1.

2. When two plane mirrors are parallel, such as on opposite walls in a barber shop, multiple images arise because each image in one mirror serves as an object for the other mirror. A point object is placed between parallel mirrors separated by 30 cm. The object is 10 cm in front of the left mirror and 20 cm in front of the right mirror. (*a*) Find the distance from the left mirror to the first four images in that mirror. (*b*) Find the distance from the right mirror to the first four images in that mirror.

3. A person 1.62 m tall wants to be able to see her full image in a plane mirror. (*a*) What must be the minimum height of the mirror? (*b*) How far above the floor should it be placed, assuming that the top of the person's head is 15 cm above her eye level? Draw a ray diagram.

4. Two plane mirrors make an angle of 90°. Show by considering various object positions that there are three images for any position of an object. Draw appropriate bundles of rays from the object to the eye for viewing each image.

5. (*a*) Two plane mirrors make an angle of 60° with each other. Show on a sketch the location of all the images formed of a point object on the bisector of the angle between the mirrors. (*b*) Repeat for an angle of 120°.

31-2 Spherical Mirrors

6. A concave spherical mirror has a radius of curvature of 40 cm. Draw ray diagrams to locate the image (if one is formed) for an object at a distance of (*a*) 100 cm, (*b*) 40 cm, (*c*) 20 cm, and (*d*) 10 cm from the mirror. For each case, state whether the image is real or virtual; erect or inverted; and enlarged, reduced, or the same size as the object.

7. Use the mirror equation to locate and describe the images for the object distances and mirror of Problem 6.

8. Repeat Problem 6 for a convex mirror with the same radius of curvature.

9. Repeat Problem 7 for the convex mirror in Problem 8.

10. Show that a convex mirror cannot form a real image of a real object, no matter where the object is placed, by showing that s' is always negative for a positive s.

11. Convex mirrors are used in stores to provide a wide angle of surveillance for a reasonable mirror size. The mirror shown in Figure 31-41 allows a clerk 5 m away from the mirror to survey the entire store. It has a radius of curvature of 1.2 m. (*a*) If a customer is 10 m from the mirror, how far from the mirror surface is his image? (*b*) Is the image in front of or behind the mirror? (*c*) If the customer is 2 m tall, how high is his image?

Figure 31-41 Problem 11.

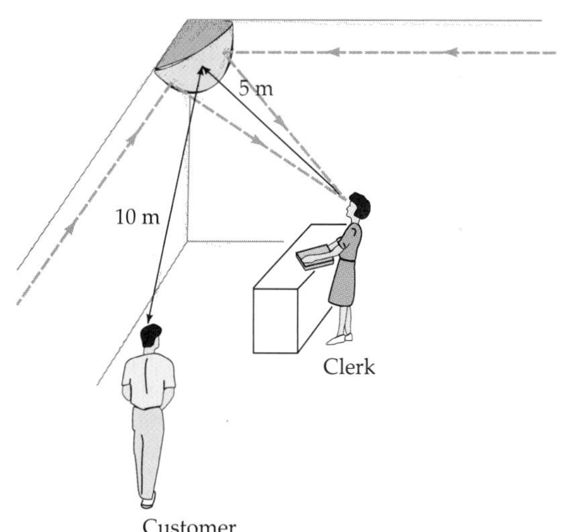

12. A certain telescope uses a concave spherical mirror of radius 8 m. Find the location and diameter of the image of the moon formed by this mirror. The moon has a diameter of 3.5×10^6 m and is 3.8×10^8 m from the earth.

13. A dentist wants a small mirror that will produce an upright image with a magnification of 5.5 when the mirror is located 2.1 cm from a tooth. (*a*) What should the radius of the mirror be? (*b*) Should it be concave or convex?

31-3 Images Formed by Refraction

14. A sheet of paper with writing on it is protected by a thick glass plate having an index of refraction of 1.5. If the plate is 2 cm thick, at what distance beneath the top of the plate does the writing appear when it is viewed from directly overhead?

15. A very long glass rod has one end ground to a convex hemispherical surface of radius 5 cm. Its index of refraction is 1.5. (*a*) A point object in air is on the axis of the rod 20 cm from the surface. Find the image and state whether it is real or virtual. Repeat for (*b*) an object 5 cm from the surface and (*c*) an object very far from the surface. Draw a ray diagram for each case.

16. At what distance from the rod of Problem 15 should the object be placed so that the light rays in the rod are parallel? Draw a ray diagram for this situation.

17. Repeat Problem 15 for a glass rod with a concave hemispherical surface of radius −5 cm.

18. Repeat Problem 15 when the glass rod and objects are immersed in water.

19. Repeat Problem 15 for a glass rod with a concave hemispherical surface of radius −5 cm when it and the objects are immersed in water.

20. A fish is 10 cm from the front surface of a fish bowl of radius 20 cm. (*a*) Where does the fish appear to be to someone in air viewing it from in front of the bowl? (*b*) Where does the fish appear to be when it is 30 cm from the front surface of the bowl?

21. A scuba diver wears a diving mask with a face plate that bulges outward with a radius of curvature of 0.5 m. There is thus a convex spherical surface between the water and the air in the mask. A fish is 2.5 m in front of the diving mask. (*a*) Where does the fish appear to be? (*b*) What is the magnification of the image of the fish?

31-4 Thin Lenses

22. The following thin lenses are made of glass with an index of refraction of 1.5. Make a sketch of each lens, and find its focal length in air: (*a*) double convex, $r_1 = 10$ cm and $r_2 = -21$ cm; (*b*) plano-convex, $r_1 = \infty$ and $r_2 = -10$ cm; (*c*) double concave, $r_1 = -10$ cm and $r_2 = +10$ cm; (*d*) plano-concave, $r_1 = \infty$ and $r_2 = +20$ cm.

23. Glass with an index of refraction of 1.6 is used to make a thin lens that has radii of equal magnitude. Find the radii of curvature and make a sketch of the lens if the focal length in air is (*a*) +5 cm and (*b*) −5 cm.

24. Find the focal length of a glass lens of index of refraction 1.62 that has a concave surface with radius of magnitude 100 cm and a convex surface with a radius of magnitude 40 cm.

25. A double-concave lens of index of refraction 1.45 has radii of magnitude 30 cm and 25 cm. An object is located 80 cm to the left of the lens. Find (*a*) the focal length of the lens, (*b*) the location of the image, and (*c*) the magnification of the image. (*d*) Is the image real or virtual? Upright or inverted?

26. A converging lens made of polystyrene (index of refraction, 1.59) has a focal length of 50 cm. One surface is convex with radius of magnitude 50 cm. Find the radius of the second surface. Is it convex or concave?

27. The following thin lenses are made of glass of index of refraction 1.6. Make a sketch of each lens, and find its focal length in air: (*a*) $r_1 = 20$ cm, $r_2 = 10$ cm; (*b*) $r_1 = 10$ cm, $r_2 = 20$ cm; (*c*) $r_1 = -10$ cm, $r_2 = -20$ cm.

28. Show that a diverging lens can never form a real image from a real object. (*Hint:* Show that s' is always negative.)

29. For the following object distances and focal lengths of thin lenses in air, find the image distance and the magnification and state whether the image is real or virtual and erect or inverted: (*a*) $s = 40$ cm, $f = 20$ cm; (*b*) $s = 10$ cm, $f = 20$ cm; (*c*) $s = 40$ cm, $f = -30$ cm; (*d*) $s = 10$ cm, $f = -30$ cm.

30. An object 3.0 cm high is placed 20 cm in front of a thin lens of power 20 diopters. Draw a precise ray diagram to find the position and size of the image and check your results using the thin-lens equation.

31. Repeat Problem 30 for an object 1.0 cm high placed 10 cm in front of a thin lens of power 20 diopters.

32. Repeat Problem 30 for an object 1.0 cm high placed 10 cm in front of a thin lens whose power is −20 diopters.

33. A thin converging lens of focal length 10 cm is used to obtain an image that is twice as large as a small object. Find the object and image distances if (*a*) the image is to be erect and (*b*) the image is to be inverted. Draw a ray diagram for each case.

34. (*a*) What is meant by a negative object distance? How can it occur? Find the image distance and magnification and state whether the image is virtual or real and erect or inverted for a thin lens in air when (*b*) $s = -20$ cm, $f = +20$ cm and (*c*) $s = -10$ cm, $f = -30$ cm. Draw a ray diagram for each of these cases.

35. Two converging lenses, each of focal length 10 cm, are separated by 35 cm. An object is 20 cm to the left of the first lens. (*a*) Find the position of the final image using both a ray diagram and the thin-lens equation. (*b*) Is the image real or virtual? Erect or inverted? (*c*) What is the overall lateral magnification of the image?

36. Work Problem 35 for a second lens that is a diverging lens of focal length −15 cm.

31-5 Aberrations

37. A double-convex lens of radii $r_1 = +10$ cm and $r_2 = -10$ cm is made from glass with indexes of refraction of 1.53 for blue light and 1.47 for red light. Find the focal length of this lens for (*a*) red light and (*b*) blue light.

Level II

38. A concave spherical mirror has a radius of curvature of 6.0 cm. A point object is on the axis 9 cm from the mirror. Construct a precise ray diagram showing rays from the object that make angles of 5°, 10°, 30°, and 60° with the axis, strike the mirror, and are reflected back across the axis. (Use a compass to draw the mirror, and use a protractor to measure the angles needed to find the reflected rays.) What is the spread along the axis of the image for these rays?

39. An object placed 8 cm from a concave spherical mirror produces a virtual image 10 cm behind the mirror. (*a*) If the object is moved back to 25 cm from the mirror, where is the image located? (*b*) Is it real or virtual?

40. A concave mirror has a radius of curvature 6.0 cm. Draw rays parallel to the axis at 0.5, 1.0, 2.0, and 4.0 cm above the axis and find the points at which the reflected rays cross the axis. (Use a compass to draw the mirror and a protractor to find the angle of reflection for each ray.) (*a*) What is the spread Δx of the points where these rays cross the axis? (*b*) By what percentage could this spread be reduced if the edge of the mirror were blocked off so that parallel rays more than 2.0 cm from the axis could not strike the mirror?

41. An object located 100 cm from a concave mirror forms a real image 75 cm from the mirror. The mirror is then turned around so that its convex side faces the object. The mirror is moved so that the image is now 35 cm behind the mirror. How far should the mirror be moved? Should it be moved toward or away from the object?

42. (*a*) Show that if f is the focal length of a thin lens in air, its focal length in water is f' given by

$$f' = \frac{n_w(n-1)}{n-n_w}f$$

where n_w is the index of refraction of water and n is that of the lens. (*b*) Calculate the focal length in air and in water of a double-concave lens of index of refraction $n =$ 1.5 that has radii of magnitude 30 cm and 35 cm.

43. You wish to see an image of your face for applying makeup or shaving. If you want the image to be upright, virtual, and magnified 1.5 times when your face is 30 cm from the mirror, what kind of mirror should you use, convex or concave, and what should its focal length be?

44. A glass rod 96 cm long with an index of refraction of 1.6 has its ends ground to convex spherical surfaces of radii 8 cm and 16 cm. A point object is in air on the axis of the rod 20 cm from the end with the 8-cm radius. (*a*) Find the image distance due to refraction at the first surface. (*b*) Find the final image due to refraction at both surfaces. (*c*) Is the final image real or virtual?

45. Repeat Problem 44 for a point object in air on the axis of the rod 20 cm from the end with the 16-cm radius.

46. (*a*) Find the focal length of a *thick* double-convex lens with an index of refraction of 1.5, a thickness of 4 cm, and radii of +20 cm and −20 cm. (*b*) Find the focal length of this lens in water.

47. A thin lens of index of refraction 1.5 has one convex side with a radius of magnitude 20 cm. When an object 1-cm in height is placed 50 cm from this lens, an upright image 2.15 cm in height is formed. (*a*) Calculate the radius of the second side of the lens. Is it concave or convex? (*b*) Draw a sketch of the lens.

48. (*a*) Show that to obtain a magnification of magnitude m with a converging thin lens of focal length f, the object distance must be given by

$$s = \frac{m-1}{m}f$$

(*b*) A camera lens with 50-mm focal length is used to take a picture of a person 1.75 m tall. How far from the camera should the person stand so that the image size is 24 mm?

49. A 2-cm thick layer of water (n = 1.33) floats on top of a 4-cm thick layer of carbon tetrachloride (n = 1.46) in a tank. How far below the top surface of the water does the bottom of the tank appear to be to an observer looking from above at normal incidence?

50. While sitting in your car, you see a jogger in your side mirror, which is convex with a radius of curvature of magnitude 2 m. The jogger is 5 m from the mirror and is approaching at 3.5 m/s. How fast does the jogger appear to be running when viewed in the mirror?

51. Parallel light from a distant object strikes the large mirror in Figure 31-42 (r = 5 m) and is reflected by the small mirror that is 2 m from the large mirror and is actually spherical, not planar as shown. The light is focused at the vertex of the large mirror. (*a*) What is the radius of curvature of the small mirror? (*b*) Is it convex or concave?

Figure 31-42 Problem 51.

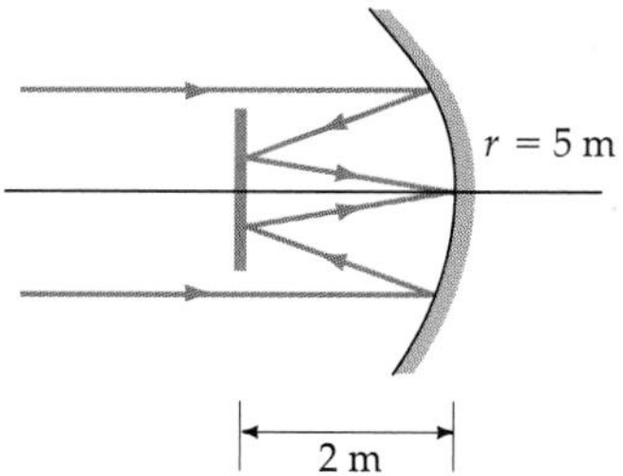

52. A small object is 20 cm from a thin positive lens of focal length 10 cm. To the right of the lens is a plane mirror that crosses the axis at the second focal point of the lens and is tilted so that the reflected rays do not go back through the lens (see Figure 31-43). (*a*) Find the position of the final image. (*b*) Is this image real or virtual? (*c*) Sketch a ray diagram showing the final image.

Figure 31-43 Problem 52.

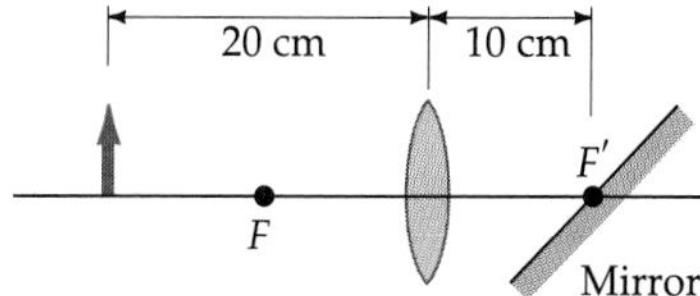

53. An object is placed 12 cm to the left of a lens of focal length 10 cm. A second lens of focal length 12.5 cm is placed 20 cm to the right of the first lens. (*a*) Find the position of the final image. (*b*) What is the magnification of the image? (*c*) Sketch a ray diagram showing the final image.

54. An object is 15 cm in front of a positive lens of focal length 15 cm. A second positive lens of focal length 15 cm is 20 cm from the first lens. Find the final image and draw a ray diagram.

55. Work Problem 54 for a second lens with a focal length of −15 cm.

56. In the seventeenth century, Antonie van Leeuwenhoek, the first great microscopist, used simple spherical lenses made first of water droplets and then of glass for his first instruments. He made staggering discoveries with these simple lenses. Consider a glass sphere of radius 2.0 mm with an index of refraction of 1.50. Find the focal length of this lens. *Hint:* Use the equation for refraction at a single spherical surface to find the image distance for an infinite object distance for the first surface. Then use this image point as the object point for the second surface.

Level III

57. An object is 15 cm to the left of a thin convex lens of focal length 10 cm. A concave mirror of radius 10 cm is 25 cm to the right of the lens. (*a*) Find the position of the final image formed by the mirror and lens. (*b*) Is the image real or virtual? Erect or inverted? (*c*) Show on a diagram where your eye must be to see this image.

58. Find the final image for the situation in Problem 52 when the mirror is not tilted. Assume that the image is viewed by an eye to the left of the object looking through the lens into the mirror.

59. A women uses a concave makeup mirror of radius 1.5 m. How far from the mirror should her face be for the image to be 80 cm from her face?

60. When a bright light source is placed 30 cm in front of a lens, there is an erect image 7.5 cm from the lens. There is also a faint inverted image 6 cm in front of the lens due to reflection from the front surface of the lens. When the lens is turned around, this weaker, inverted image is 10 cm in front of the lens. Find the index of refraction of the lens.

61. A horizontal concave mirror with radius of curvature of 50 cm holds a layer of water with an index of refraction of 1.33 and a maximum depth of 1 cm. At what height above the mirror must an object be placed so that its image is at the same position as the object?

62. A lens with one concave side with a radius of magnitude 17 cm and one convex side with a radius of magnitude 8 cm has a focal length in air of 27.5 cm. When placed in a liquid with an unknown index of refraction, the focal length increases to 109 cm. What is the index of refraction of the liquid?

Figure 31-44 Problem 63.

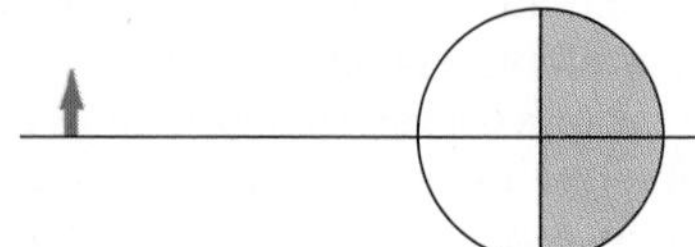

63. A glass ball of radius 10 cm has an index of refraction of 1.5. The back half of the ball is silvered so that it acts as a concave mirror (Figure 31-44). Find the position of the final image seen by an eye to the left of the object and ball for an object at (*a*) 30 cm and (*b*) 20 cm to the left of the front surface of the ball.

64. In a convenient form of the thin-lens equation used by Newton, the object and image distances are measured from the focal points. Show that if $x = s - f$ and $x' = s' - f$, the thin-lens equation can be written as $xx' = f^2$, and the lateral magnification is given by $m = -x'/f = -f/x$. Indicate x and x' on a sketch of a lens.

65. An object is placed 2.4 m from a screen, and a lens of focal length f is placed between the object and the screen so that a real image of the object is formed on the screen. When the lens is moved 1.2 m toward the screen, another real image of the object is formed on the screen. (*a*) Where was the lens located before it was moved? (*b*) What is the focal length of the lens?

66. An object is 17.5 cm to the left of a lens of focal length 8.5 cm. A second lens of focal length −30 cm is 5 cm to the right of the first lens. (*a*) Find the distance between the object and the final image formed by the second lens. (*b*) What is the overall magnification? (*c*) Is the final image real or virtual? Upright or inverted?

67. (*a*) Show that a small change dn in the index of refraction of a lens material produces a small change in the focal length df given approximately by

$$\frac{df}{f} = \frac{-dn}{n-1}$$

(*b*) Use this result to find the focal length of a thin lens for blue light, for which $n = 1.53$, if the focal length for red light, for which $n = 1.47$, is 20 cm.

68. The lateral magnification of a spherical mirror or a thin lens is given by $m = -s'/s$. Show that for objects of small horizontal extent, the longitudinal magnification is approximately $-m^2$. *Hint:* Show that $ds'/ds = s'^2/s^2$.

69. A thin double-convex lens has radii r_1 and r_2 and an index of refraction n_L. The surface of radius r_1 is in contact with a liquid of index of refraction n_1, and the surface of radius r_2 is in contact with a liquid of index of refraction n_2. Show that the thin-lens equation for this situation can be expressed as

$$\frac{n_1}{s} + \frac{n_2}{s'} = \frac{n_2}{f}$$

where the focal length is given by

$$\frac{1}{f} = \frac{n_L - n_1}{n_2 r_1} - \frac{n_L - n_2}{n_2 r_2}$$

Chapter 32

Optical Instruments

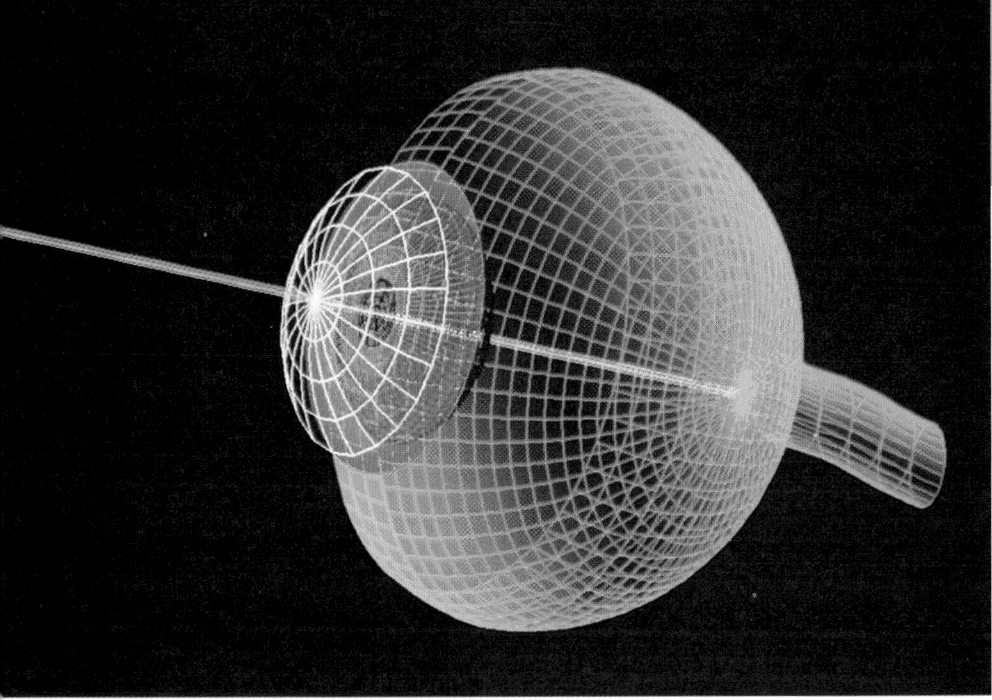

A computer-modeled image of the human eye.

In this chapter, we will use what we have learned about mirrors and lenses to examine the workings of various optical instruments, such as the camera, the simple magnifier, the microscope, and the telescope. The most important optical instrument is the eye, which we will study first. Many optical instruments used today are quite complicated. The basic principles behind their workings are often quite simple, but imaginative applications of these principles have revolutionized our capacity to see and understand the world around us.

32-1 The Eye

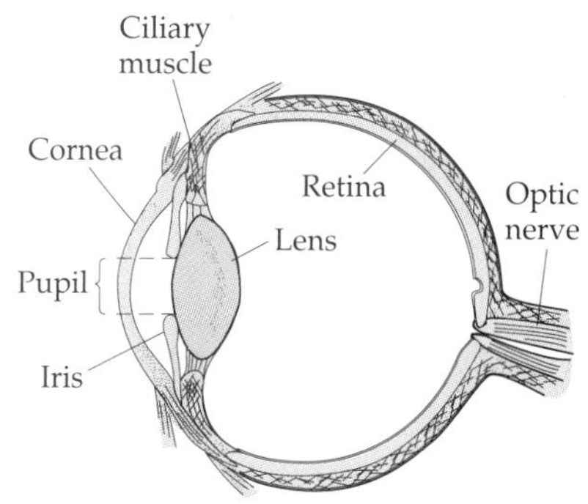

Figure 32-1 Cutaway view of the human eye. The amount of light entering the eye is controlled by the iris, which regulates the size of the pupil. The lens thickness is controlled by the ciliary muscle.

The optical system of prime importance is the eye, shown in Figure 32-1. Light enters the eye through a variable aperture, the *pupil,* and is focused by the *cornea–lens system* on the *retina,* a film of nerve fibers covering the back surface. The retina contains tiny light-sensing structures called *rods* and *cones* that receive the image and transmit the information along the optic nerve to the brain. The shape of the crystalline lens can be altered slightly by the action of the *ciliary muscle.* When the eye is focused on an object far away, the muscle is relaxed and the cornea–lens system has its maximum focal length, about 2.5 cm, the distance from the cornea to the retina. When the object is brought closer to the eye, the ciliary muscle increases the curvature of the lens slightly, thereby decreasing its focal length so that the image is again focused on the retina. This process is called **accommodation.** If the object is too close to the eye, the lens cannot focus the light on the retina and the image is blurred. The closest point for which the lens can focus the image on the retina is called the **near point.** The distance from the eye to the near point varies greatly from one person to another and changes with age. At the age of 10 years, the near point may be as close as 7 cm, whereas by 60 years it may have receded to 200 cm because of the loss of flexibility of the lens. The standard value taken for the near point is 25 cm.

If the eye underconverges, resulting in the images being focused behind the retina, the person is said to be **farsighted.** A farsighted person can see distant objects, for which little convergence is required, but has trouble seeing close objects clearly. Farsightedness is corrected with a converging (positive) lens (Figure 32-2).

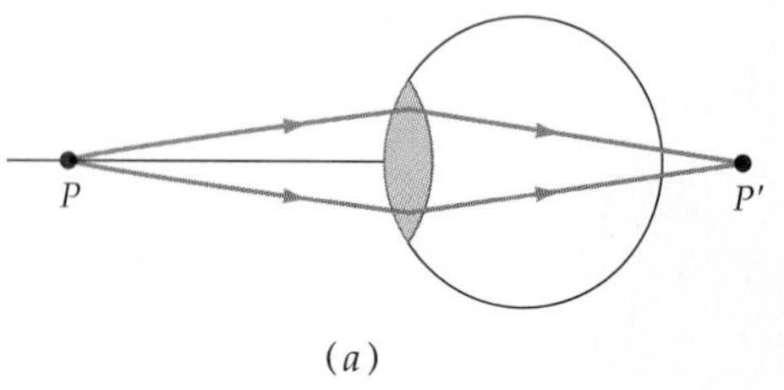

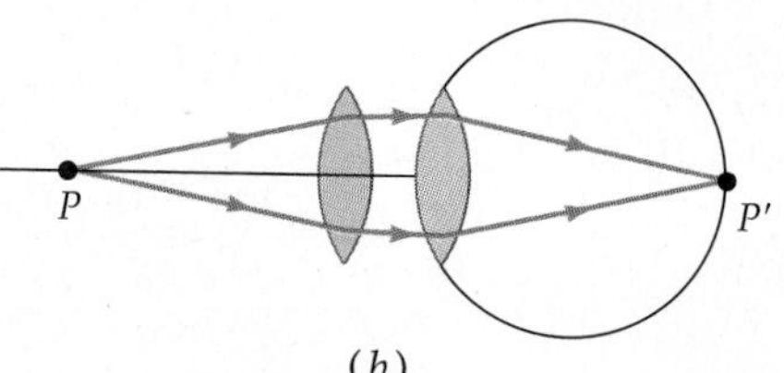

Figure 32-2 (*a*) A farsighted eye focuses rays from a nearby object P to a point P' behind the retina. (*b*) A converging lens corrects this defect by bringing the image onto the retina. These diagrams and those following are drawn as if all the focusing of the eye is done at the lens; in fact, the cornea–lens system acts more like a spherical refracting surface than a thin lens.

On the other hand, the eye of a **nearsighted** person overconverges and focuses light from distant objects in front of the retina. A nearsighted person can see nearby objects, for which the widely diverging incident rays can be focused on the retina, but has trouble focusing on distant objects. Nearsightedness is corrected with a diverging (negative) lens (Figure 32-3).

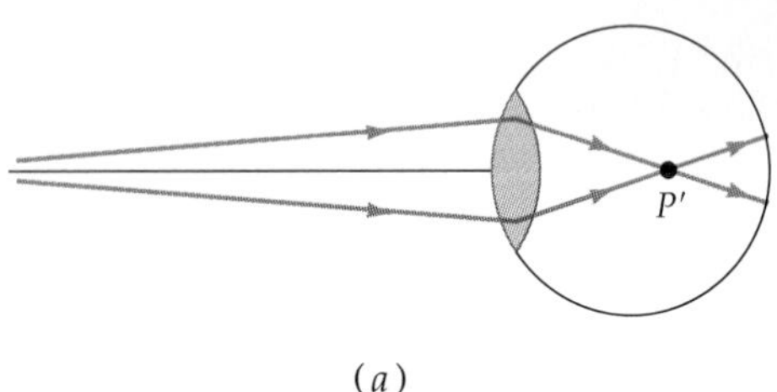

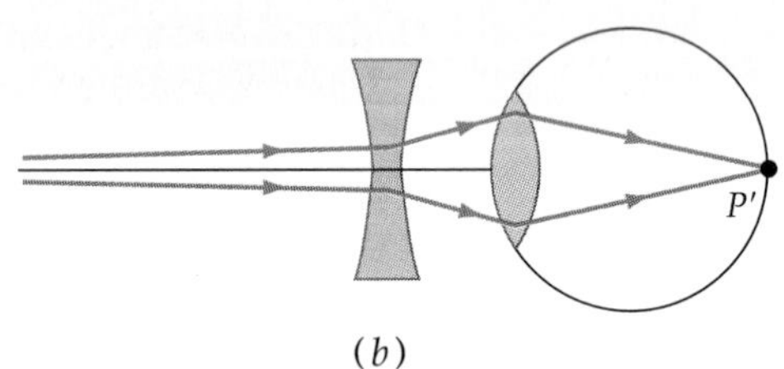

Figure 32-3 (*a*) A nearsighted eye focuses rays from a distant object to a point P' in front of the retina. (*b*) A diverging lens corrects this defect.

Another common defect of vision is *astigmatism,* which is caused by the cornea being not quite spherical but having a different curvature in one plane than in another. This results in the blurring of the image of a point object into a short line. Astigmatism is corrected by glasses with lenses of cylindrical rather than spherical shape.

Example 32-1

By how much must the focal length of the cornea–lens system of the eye change when an object is moved from infinity to the near point at 25 cm? Assume that the distance from the cornea to the retina is 2.5 cm.

When the object is at infinity, the rays from the object are parallel and are focused by the eye on the retina, giving a focal length for the cornea–lens system of 2.5 cm. When the object is at 25 cm, the focal length f must be such that the image distance is 2.5 cm. Using $s = 25$ cm for the object distance and $s' = 2.5$ cm for the image distance in the thin-lens equation (Equation 31-19), we have

$$\frac{1}{25\text{ cm}} + \frac{1}{2.5\text{ cm}} = \frac{1}{f}$$

$$\frac{1}{f} = \frac{1}{25\text{ cm}} + \frac{10}{25\text{ cm}} = \frac{11}{25\text{ cm}}$$

$$f = \frac{25\text{ cm}}{11} = 2.27\text{ cm}$$

The focal length must therefore decrease by 0.23 cm. In terms of the power of the cornea–lens system, when the focal length is 2.5 cm = 0.025 m for distant objects, the power is $P = 1/f = 40$ diopters. When the focal length is 2.27 cm, the power is 44 diopters.

Exercise

Find the change in the focal length of the eye when an object originally at 4 m is brought to 40 cm from the eye. (Assume the distance from the cornea to the retina is 2.5 cm.) (Answer: 0.13 cm)

The apparent size of an object is determined by the size of the image on the retina. The larger the image on the retina, the greater the number of rods and cones activated. From Figures 32-4*a* and *b*, we can see that the size of the image on the retina is greater when the object is close than it is when the object is far away. Thus, even though the actual size of the object does not change, its apparent size is greater when it is brought closer to the eye. A convenient measure of the size of the image on the retina is the angle θ subtended by the object at the eye as shown in Figure 32-4. From Figure 32-4*c*, we can see that the angle θ is related to the image size y' by

$$\theta = \frac{y'}{2.5\text{ cm}} \qquad 32\text{-}1$$

The image size is therefore directly proportional to the angle subtended by the object. From Figure 32-4*a* or *b*, we can see that the angle θ is related to the object size y and object distance s by

$$\tan\theta = \frac{y}{s}$$

For small angles, we can use the approximation $\tan\theta \approx \theta$ and write

$$\theta \approx \frac{y}{s} \qquad 32\text{-}2$$

Combining Equations 32-1 and 32-2, we obtain

$$y' = (2.5\text{ cm})\theta \approx (2.5\text{ cm})\frac{y}{s} \qquad 32\text{-}3$$

Thus, the size of the image on the retina is proportional to the size of the object and inversely proportional to the distance between the object and the eye.

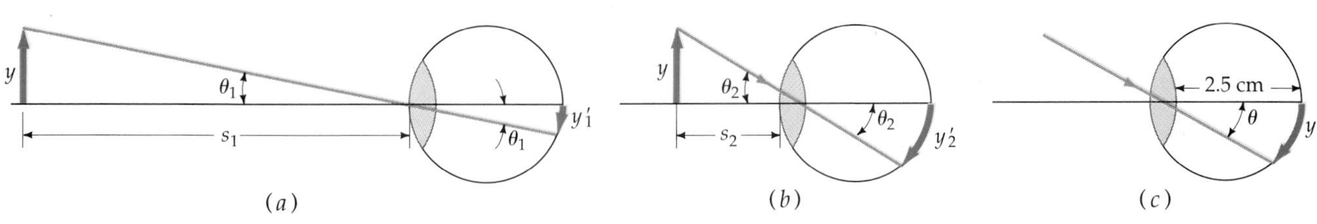

Figure 32-4 (*a*) A distant object of height y looks small because the image on the retina is small. (*b*) When the same object is closer, it looks larger because the image on the retina is larger. The size of the image on the retina is proportional to the angle θ subtended by the object, which in turn is inversely proportional to the object distance. (*c*) The angle subtended is $\theta = y'/(2.5\text{ cm})$.

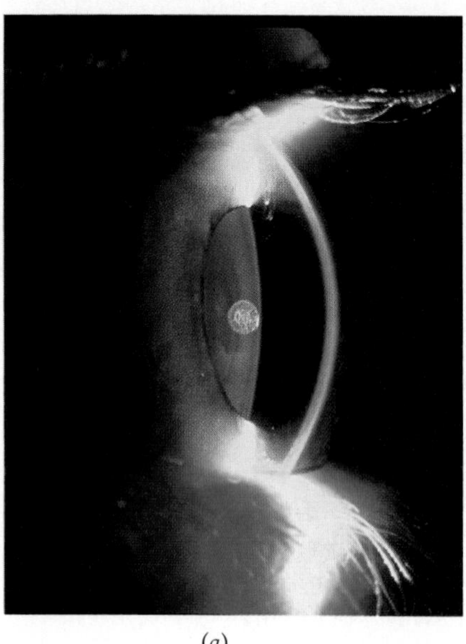

(a)

(b)

(a) The human eye in profile. (b) The lens of the eye is kept in place by ciliary muscle fibers shown here in the upper left. When the fibers contract, the tension on the lens is reduced and the lens, which is made of elastic tissue, tends to balloon outward. The greater lens curvature enables the eye to focus on nearby objects. (c) Some of the 120 million rods and 7 million cones in the eye magnified approximately 5000 times. The rods (the more slender of the two) are more sensitive in dim light, whereas the cones are more sensitive to color. The rods and cones form the bottom layer of the retina and are covered by nerve cells, blood vessels, and supporting cells. Most of the light entering the eye is reflected or absorbed before reaching the rods and cones. The light that does reach them triggers electrical impulses along nerve fibers that ultimately reach the brain. (d) A neural net used in the vision system of certain robots. Loosely modeled on the human eye, it contains 1920 sensors.

Example 32-2

Assume that the near point of your eye is 75 cm. What power reading glasses should you use to bring your near point to 25 cm?

If your near point is 75 cm, you are farsighted. To read a book you must hold the book at least 75 cm from your eye so that you can focus on the print. The image of the print on your retina is then very small. A converging lens, which is used in reading glasses, allows the book to be brought closer to the eye so that the image of the print is larger. When the book is 25 cm from the eye, we want the image formed by the converging lens to be 75 cm from the eye. Recall that a converging lens forms a virtual, erect image when the object is between the lens and the focal point. We therefore expect the focal length of the lens to be greater than 25 cm.

Figure 32-5 shows a diagram of an object 25 cm from a converging lens that produces virtual, erect image at $s' = -75$ cm. Using the thin-lens equation with $s = 25$ cm and $s' = -75$ cm, we obtain

$$\frac{1}{25 \text{ cm}} + \frac{1}{-75 \text{ cm}} = \frac{1}{f}$$

$$\frac{1}{f} = \frac{2}{75 \text{ cm}} = \frac{1}{0.375 \text{ m}} = 2.67 \text{ diopters}$$

$$f = \frac{1}{2.67 \text{ m}^{-1}} = 0.375 \text{ m} = 37.5 \text{ cm}$$

Thus, the power of your reading glasses should be 2.67 diopters, which is obtained when the focal length of the glasses is 37.5 cm.

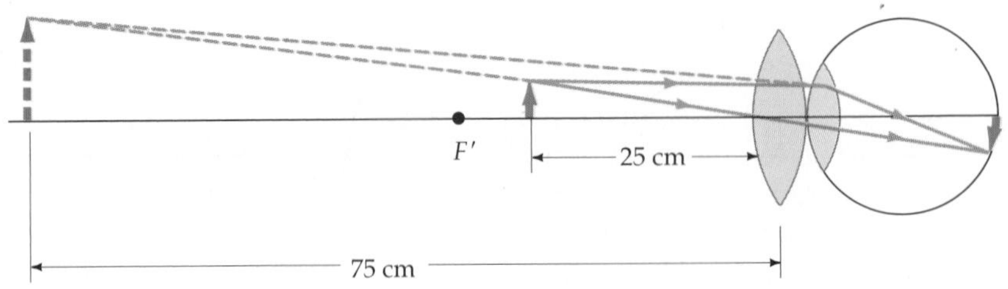

Figure 32-5 Ray diagram for Example 32-2. When the object is placed just inside the first focal point of a converging lens, the image is virtual, erect, enlarged, and far from the lens. In this example, the image distance is chosen to be 75 cm, the near point of a farsighted eye, and the object distance is chosen to be 25 cm. The focal length of the lens for these choices is then calculated from the thin-lens equation.

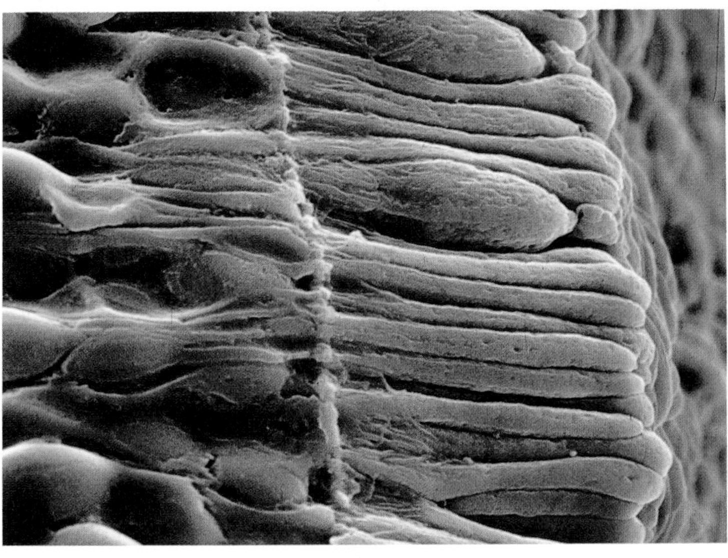

(c)

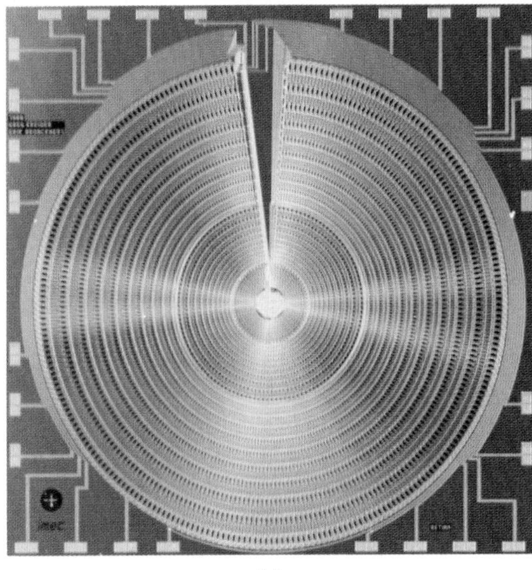

(d)

There is an alternative way to solve this problem that uses the result we found in Section 31-4 that the power of two lenses in contact is the sum of the powers of the individual lenses. Without reading glasses, when the object is at 75 cm = 0.75 m from the eye, the image is at 2.5 cm = 0.025 m, the distance from the eye lens to the retina. The focal length of the eye can then be determined from the thin-lens equation:

$$\frac{1}{0.75\ \text{m}} + \frac{1}{0.025\ \text{m}} = \frac{1}{f_e}$$

The power of the eye lens is then

$$P_e = \frac{1}{f_e} = 1.33\ \text{m}^{-1} + 40.00\ \text{m}^{-1} = 41.33\ \text{diopters}$$

When the reading-glasses lens is used, the image distance of the combination should be 0.025 m when the object is at a distance of 25 cm. If f_c is the focal length of the combination and $P_c = 1/f_c$ is the power of the combination, we have

$$\frac{1}{0.25\ \text{m}} + \frac{1}{0.025\ \text{m}} = \frac{1}{f_c} = P_c$$

or

$$P_c = 4.00\ \text{m}^{-1} + 40\ \text{m}^{-1} = 44.0\ \text{diopters}$$

The power of the combination equals the sum of the power of the eye lens and that of reading-glasses lens:

$$P_c = P_e + P_g$$

The power of the reading-glasses lens is thus

$$P_g = P_c - P_e = 44.0\ \text{diopters} - 41.33\ \text{diopters} = 2.67\ \text{diopters}$$

which is in agreement with our first calculation. In both calculations, we have assumed that the lens of the reading glasses is in contact with that of the eye. Our results therefore apply for contact lenses. For reading glasses that are a short distance in front of the eye, the results are somewhat different.

Question

1. Glasses with a power of −2 diopters are prescribed for a certain person. Is that person nearsighted or farsighted?

32-2 The Simple Magnifier

We saw in Example 32-2 that the apparent size of an object can be increased by using a converging lens to allow the object to be brought closer to the eye and thereby increase the size of the image on the retina. Such a converging lens is called a **simple magnifier.** In Figure 32-6*a*, a small object of height y is at the near point of the eye at a distance x_{np}. As we discussed previously, the size of the image on the retina is proportional to the angle θ_o subtended by the object at the eye. In this case, θ_o is given approximately by

$$\theta_o = \frac{y}{x_{np}}$$

In Figure 32-6*b*, a converging lens of focal length f, which is less than x_{np}, has been placed in front of the eye, and the object has been placed at the focal point of the lens. The rays emerge from the lens parallel, indicating that the image is an infinite distance in front of the lens. The parallel rays are focused by the relaxed eye on the retina. If the lens is in close contact with the eye, the angle subtended by the object is now approximately

$$\theta = \frac{y}{f}$$

Figure 32-6 (*a*) An object at the near point subtends an angle θ_o at the eye. (*b*) When the object is at the focal point of the converging lens, the rays emerge from the lens parallel and enter the eye as if they came from an object a very large distance away. The image can thus be viewed at infinity by the relaxed eye. When f is less than the near point, the converging lens allows the object to be brought closer to the eye, increasing the angle subtended by the object to θ and thereby increasing the size of the image on the retina.

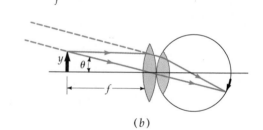

The ratio θ/θ_o is called the **angular magnification** or **magnifying power** M of the lens:

Magnifying power of a lens

$$M = \frac{\theta}{\theta_o} = \frac{x_{np}}{f} \qquad 32\text{-}4$$

Example 32-3

A person with a near point of 25 cm uses a 40-diopter lens as a simple magnifier. What magnification is obtained?

The focal length of a 40-diopter lens is

$$f = \frac{1}{P} = \frac{1}{40\ \text{m}^{-1}} = 0.025\ \text{m} = 2.5\ \text{cm}$$

Using $x_{np} = 25$ cm and $f = 2.5$ cm in Equation 32-4, we obtain for the magnification

$$M = \frac{25\ \text{cm}}{2.5\ \text{cm}} = 10$$

The object looks 10 times larger because it can be placed at 2.5 cm rather than at 25 cm from the eye, so the size of the image on the retina is increased tenfold. The additional converging power of the magnifier is used to focus the very divergent rays from the very close object.

Exercise

What is the magnification in Example 32-3 if the near point of the person is 30 cm rather than 25 cm? (Answer: $M = 12$)

It is possible to increase this magnification slightly by moving the object closer to the magnifier. When the object is inside the focal point of the magnifier, the image is virtual and erect as in Figure 32-5. As the object is moved toward the magnifier, the image moves closer to the eye, and the angle subtended increases slightly. The largest usable magnification occurs when the image is at the near point of the eye, as it was in Example 32-2. Calculation shows that the magnification in this case is just 1 greater than that with the image at infinity (see Problem 26). For example, if the near point distance is 25 cm, a lens with focal length of 2.5 cm would give a magnification of 11 with the image at 25 cm rather than a magnification of 10 with the image at infinity. In Example 32-2, we found that a lens of focal length 37.5 cm produces an image at 75 cm from the eye when the object at 25 cm from the eye. Since the angle subtended by the image on the retina varies inversely as the distance of the object (or image) from the eye, the magnification was (75 cm)/(25 cm) = 3 in that example. If we were to move the object back from 25 cm to the focal point of the lens at 37.5 cm, the image could be viewed with a relaxed eye and the magnification would be 2 instead of 3. In practice, the gain in magnification obtained by viewing the image at the near point, rather than at infinity with a relaxed eye, is not worth the additional eye strain, so we will use Equation 32-4 for the magnification of a simple magnifier.

Simple magnifiers are used as eyepieces (called *oculars*) in compound microscopes and telescopes to view the image formed by another lens or lens system. To correct aberrations, combinations of lenses with a resulting short positive focal length are often used instead of a single lens, but the principle is the same as that of the simple magnifier.

32-3 The Camera

The basic camera consists of a positive lens, a variable aperture, a shutter that can be opened for a short time that can be varied, a light-tight box, and a film (Figure 32-7). Unlike the eye, which has a lens of variable focal length, the focal length of a camera lens is fixed. Typically, the focal length for the lens in a 35-mm camera is 50 mm. (The 35-mm refers to the width of film.) Focusing is accomplished by varying the distance from the lens to the film by moving the lens closer to or farther from the film.

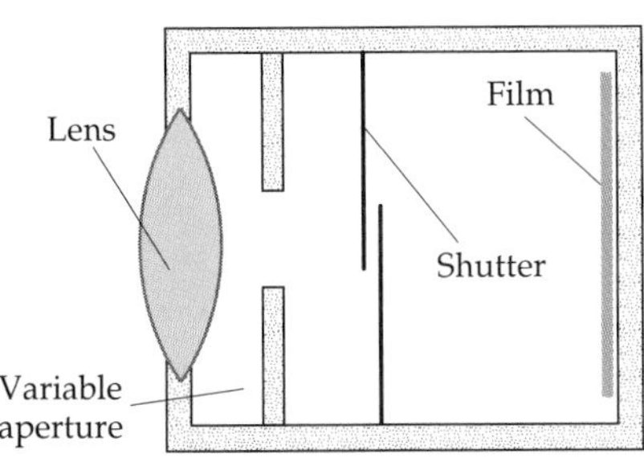

Figure 32-7 Schematic diagram of a camera. The positive lens focuses the light on the film. The variable aperture limits the amount of light entering the camera and the area of the lens used. The shutter speed can be varied to vary the exposure time.

Example 32-4

The focal length of a camera lens is 50 mm. By how much must the lens be moved to change from focusing on an object far away to one 2 m away?

When the object is far away, the image of the lens is at the focal length of the lens, so the film should be 50 mm from the lens. When the object is 2 m away, the image distance is s', given by

$$\frac{1}{2\text{ m}} + \frac{1}{s'} = \frac{1}{f} = \frac{1}{50\text{ mm}}$$

$$\frac{1}{s'} = \frac{1}{50\text{ mm}} - \frac{1}{2000\text{ mm}}$$

$$= \frac{40}{2000\text{ mm}} - \frac{1}{2000\text{ mm}}$$

$$s' = \frac{2000\text{ mm}}{39} = 51.3\text{ mm}$$

The lens must therefore be moved 1.3 mm farther away from the film.

Exercise

In Example 32-4, how far must the lens be moved to change from focusing on an object far away to one that is 1 m away? (Answer: 2.6 mm)

The amount of light that strikes the film can be controlled by varying the time that the shutter is open and by varying the size of the aperture. For a given film type, there is an optimum amount of light that will give a good picture of proper contrast. Too little light results in a dark picture. Too much light results in a washed-out picture with too little contrast. The amount of light needed for proper contrast is related to the film's "speed," which is rated by an ASA number. The higher the ASA number, the faster the film and the smaller the amount of light needed. A film with a high ASA number, such as ASA 400 or ASA 1000, is good for taking pictures indoors where there is little available light. With high-speed films, some reduction in picture quality (the sharpness of the image or the trueness of the color reproduction) usually occurs; so for outdoor photographs taken with plenty of available light, it is usually preferable to use a lower-speed film, such as ASA 100 or ASA 64. Lower-speed film may also be needed in bright-light situations if the shutter speed of the camera has limited variability. Shutter speeds on good cameras can often be varied from exposures of several seconds for low-light photography to 1/1000 of a second for stop-action photography. For a hand-held camera, exposure times of more than about 1/60 of a second often result in a blurring of the image because of camera motion.

The maximum size of the aperture is limited by the size of the lens, which is in turn limited by the various lens aberrations we discussed in Section 31-5. (Although we treat the camera lens as a single positive lens, the optical systems in good cameras are combinations of lenses designed to reduce chromatic, spherical, and other aberrations.) The size of the aperture is given by the ***f*-number,** which is the ratio of the focal length to the diameter of the aperture:

f-number

$$f\text{-number} = \frac{f}{D} \qquad 32\text{-}5$$

Cutaway view of a 35-mm camera.

The maximum aperture is the f-number of the lens. For example, an $f/2.8$ lens with a focal length of 50 mm has a maximum usable diameter given by

$$D = \frac{f}{f\text{-number}} = \frac{50 \text{ mm}}{2.8} = 17.9 \text{ mm}$$

Lenses with large diameters (small f-numbers) are costly to make because of the expense of correcting for aberrations. Aperture settings on a camera are usually marked $f/22$, $f/16$, $f/11$, $f/8$, $f/5.6$, $f/4$, $f/2.8$, $f/2.0$, $f/1.4$, $f/1.0$ and on down to the lowest f-number, which corresponds to the largest usable diameter of the lens. Note that for each successive setting, called an f-stop, the diameter of the aperture is $\sqrt{2} = 1.4$ times that of the previous one. Since the amount of light that enters the camera is proportional to the area of the lens, which is in turn proportional to the square of the diameter, opening up the aperture by one f-stop, for example, from $f/2.0$ to $f/1.4$, increases the area and therefore the amount of light entering in a given time by a factor of 2. Increasing the aperture, however, reduces the depth of focus—the range of distances to objects that are in sharp focus on the film.

Exercise

What is the f-number of a lens whose focal length is 50 mm and whose maximum usable diameter is 8.93 mm? (Answer: $f/5.6$)

Example 32-5

The instructions for a certain film say to set the aperture at $f/11$ and the shutter speed at 1/250 s to take pictures on bright sunny days. What should the shutter speed be if the aperture is set at $f/5.6$?

Since $11^2 = 121$ and $5.6^2 = 31.4$, the area of the aperture at $f/5.6$ is $121/(31.4) = 3.85$ or approximately 4 times greater than that at $f/11$. Thus, to get the desired amount of light, you should use a shutter speed 4 times faster or 1/1000 s.

Wide-angle lens photograph of the skating rink at Rockefeller Center in New York. Note the distortion at the top of the photograph.

A 35-mm camera with a lens having a focal length of 50 mm gives a field of view that is approximately the same as that of ordinary vision, about 45°.

To increase the field of view, a wide-angle lens with a smaller focal length, for example, 24 mm, is used. When the object distance is much greater than the focal length, which is usually the case for a camera, the image distance s' is approximately equal to f. Since the lateral magnification of a lens is $m = (-)s'/s$, the size of the image on the film is approximately proportional to the focal length. A wide-angle lens thus gives a smaller image on the film than does a normal lens for a given object size. A telephoto lens has a large focal length to increase the size of the image on the film and thus make the object seem closer. A telephoto lens with a focal length of 200 mm would give a magnification approximately 4 times that of an ordinary lens with a focal length of 50 mm.

Questions

2. What are the advantages of having a camera with a fast shutter?
3. Why is an $f/1.0$ lens more expensive than an $f/2.8$ lens?

32-4 The Compound Microscope

The compound microscope (Figure 32-8) is used to look at very small objects at short distances. In its simplest form, it consists of two converging lenses. The lens nearest the object, called the **objective,** forms a real image of the object. This image is enlarged and inverted. The lens nearest the eye, called the **eyepiece** or **ocular,** is used as a simple magnifier to view the image formed by the objective. The eyepiece is placed such that the image formed by the objective falls at the first focal point of the eyepiece. The light thus emerges from the eyepiece as a parallel beam as if it were coming from a point a great distance in front of the lens. (This is commonly called "viewing the image at infinity.") As we discussed in Section 32-2, the function of a simple magnifier (the eyepiece in this case) is to allow the object (the image formed by the objective in this case) to be brought closer to the eye than the near point. Since a simple magnifier produces a virtual image that is erect, the final image produced by the two lenses is inverted.

(a)

Objective

Eyepiece

y β β y' f_o L f_e

(b)

Figure 32-8 (a) William Marin at the Brookhaven National Laboratory chooses an objective lens for viewing a sample of radioactive waste with a modern optical microscope. (b) Schematic diagram of a compound microscope consisting of two positive lenses, the objective of focal length f_o and the ocular, or eyepiece, of focal length f_e. The real image of the object formed by the objective is viewed by the eyepiece, which acts as a simple magnifier. The final image is at infinity.

The distance between the second focal point of the objective and the first focal point of the eyepiece is called the **tube length** L. It is typically fixed at 16 cm. The object is placed just outside the first focal point of the objective so that an enlarged image is formed at the first focal point of the eyepiece a distance $L + f_o$ from the objective, where f_o is the focal length of the objective. From Figure 32-8, $\tan \beta = y/f_o = -y'/L$. The lateral magnification of the objective is therefore

$$m_o = \frac{y'}{y} = -\frac{L}{f_o} \qquad \text{32-6}$$

The angular magnification of the eyepiece is

$$M_e = \frac{x_{np}}{f_e}$$

where x_{np} is the near point of the viewer and f_e is the focal length of the eyepiece. As we discussed in Section 32-2, a slightly greater angular magnification can be obtained by placing the object (image formed by the objective) at a point just inside the first focal point of the eyepiece so that the final image is at the near point. The slight gain in the angular magnification of the eyepiece is usually not worth the eye strain caused by viewing the image at the near point rather than viewing it at infinity with a relaxed eye. The magnifying power of the compound microscope is the product of the lateral magnification of the objective and the angular magnification of the eyepiece:

$$M = m_o M_e = -\frac{L}{f_o}\frac{x_{np}}{f_e} \qquad 32\text{-}7$$

Magnifying power of a microscope

Example 32-6

A microscope has an objective lens of focal length 1.2 cm and an eyepiece of focal length 2.0 cm separated by 20 cm. (*a*) Find the magnifying power if the near point of the viewer is 25 cm. (*b*) Where should the object be placed if the final image is to be viewed at infinity?

(*a*) The distance between the second focal point of the objective and the first focal point of the eyepiece is 20 cm − 2 cm − 1.2 cm = 16.8 cm. The magnifying power is given by Equation 32-7 with $L = 16.8$ cm, $f_o = 1.2$ cm, $f_e = 2.0$ cm, and $x_{np} = 25$ cm:

$$M = -\frac{16.8\text{ cm}}{1.2\text{ cm}}\frac{25\text{ cm}}{2\text{ cm}} = -175$$

The negative sign indicates that the final image is inverted.

(*b*) We can calculate the object distance between the original object and the objective from the thin-lens equation. From Figure 32-8, we can see that the image distance is

$$s' = f_o + L = 1.2\text{ cm} + 16.8\text{ cm} = 18\text{ cm}$$

The object distance is then found from

$$\frac{1}{s} + \frac{1}{s'} = \frac{1}{f}$$

$$\frac{1}{s} + \frac{1}{18\text{ cm}} = \frac{1}{1.2\text{ cm}}$$

Solving for s, we obtain $s = 1.29$ cm. The object should thus be placed at 1.29 cm from the objective or 0.09 cm outside its first focal point.

The shorter the wavelength of light used to image a specimen, the smaller the specimen that can be imaged. Since shorter wavelength light carries more energy, there is a trade-off between resolving power of the scope and potential damage to the illuminated specimen. (*a*) A state-of-the-art near-field microscope. (*b*) The aperture of the microscope is a 50-nm wide hole at the tip of a glass pipette. The specimen is positioned so near the aperture that light encounters it before the light has a chance to diverge. The beam is swept over the specimen's surface, partially transmitted through the specimen, and collected by a photodetector. Line by line an image is eventually built up on a video monitor. Near-field microscopes using greenish-yellow light of 500-nm wavelength have achieved a resolution of about 40 nm. This resolution is an order of magnitude better than that possible with a conventional optical microscope.

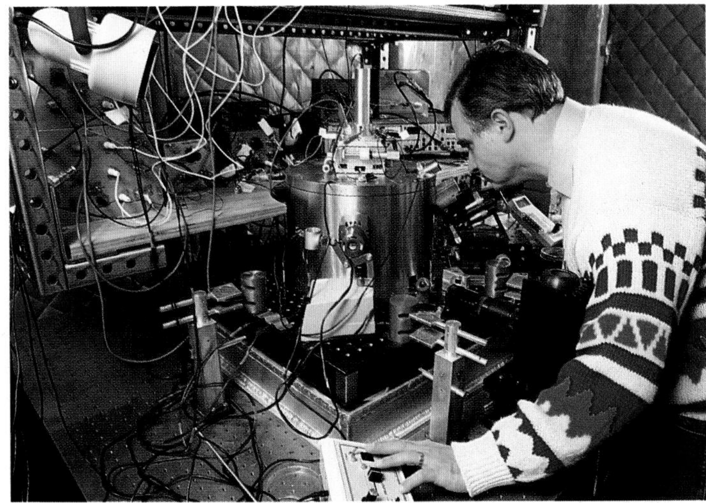

(*a*)

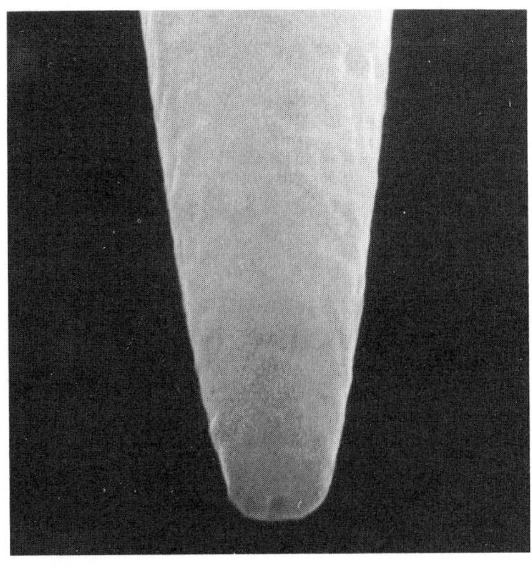

(*b*)

32-5 The Telescope

A telescope is used to view objects that are far away and often large. Its purpose is to bring the image of the object closer, that is, to increase the angle subtended by the image so that the object appears larger. The astronomical telescope, illustrated schematically in Figure 32-9, consists of two positive lenses—an objective lens that forms a real, inverted image and an eyepiece that is used as a simple magnifier to view that image. Since the object is very far away, the image of the objective lies at the focal point of the objective, and the image distance equals the focal length f_o. Since the object distance is much larger than the focal length of the objective, the image formed by the objective is much smaller than the object. For example, if we are looking at the moon, the image of the moon formed by the objective is much smaller than the moon itself. The purpose of the objective is not to magnify the object, but to produce an image that is close so it can be viewed by the eyepiece. Since this image is at the second focal point of the objective and at the first focal point of the ocular, the objective and ocular must be separated by the sum of the focal lengths of the objective and eyepiece, $f_o + f_e$, where f_e is the focal length of the eyepiece.

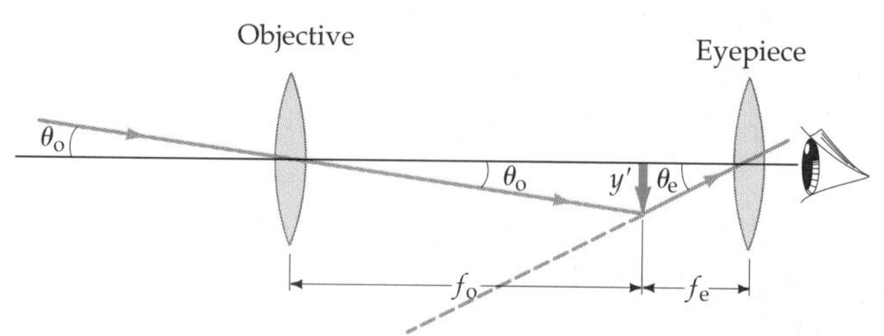

Figure 32-9 Schematic diagram of an astronomical telescope. The objective forms a real image of a distant object near its second focal point, which coincides with the first focal point of the eyepiece. The eyepiece serves as a simple magnifier for viewing the image.

The magnifying power of the telescope is the angular magnification θ_e/θ_o, where θ_e is the angle subtended by the final image as viewed through the eyepiece and θ_o is the angle subtended by the object when it is viewed directly by the unaided eye. The angle θ_o is the same as that subtended by the object at the objective shown in Figure 32-9. (The distance from a distant object, such as the moon, to the objective is essentially the same as the distance to the eye.) From this figure, we can see that

$$\tan \theta_o = -\frac{y'}{f_o} \approx \theta_o$$

where we have used the small-angle approximation $\tan \theta \approx \theta$ and have introduced a negative sign to make θ_o positive when y' is negative. The angle θ_e in the figure is that subtended by the final image:

$$\tan \theta_e = \frac{y'}{f_e} \approx \theta_e$$

Since y' is negative, θ_e is negative, indicating that the image is inverted. The magnifying power of the telescope is then

Magnifying power of a telescope

$$M = \frac{\theta_e}{\theta_o} = -\frac{f_o}{f_e} \qquad 32\text{-}8$$

From Equation 32-8, we can see that a large magnifying power is obtained with an objective of large focal length and an eyepiece of short focal length.

(*a*) (*b*) (*c*)

(*d*) (*e*)

Astronomy at optical wavelengths began with Galileo approximately 400 years ago. In this century, astronomers have begun to explore the electromagnetic spectrum at other wavelengths beginning with radio astronomy in the 1940s, satellite-based x-ray astronomy in the early 1960s, and more recently ultraviolet, infrared, and gamma-ray astronomy. (*a*) Galileo's seventeenth-century telescope with which he discovered mountains on the moon, sunspots, Saturn's rings, and the bands and moons of Jupiter. (*b*) An engraving of the reflector telescope built in the 1780s, which was used by the great astronomer Hershel who was the first to observe galaxies outside our own. (*c*) Because of the difficulty of making large, flaw-free lenses, refractor telescopes as this 91.4-cm refractor telescope at Lick Observatory, have been superseded in light-gathering power by reflector telescopes. (*d*) The great astronomer Hubble, who discovered the apparent expansion of the universe, is shown seated in the observer's cage of the 5.08-m Hale reflecting telescope, which is large enough for the observer to sit at the prime focus itself. (*e*) This 10-meter optical reflector at the Whipple Observatory in southern Arizona is the largest instrument designed exclusively for use in gamma-ray astronomy. High-energy gamma rays of unknown origin strike the upper atmosphere and create cascades of particles, such as high-energy electrons which emit Čherenkov radiation that can be observed from the ground. According to one hypothesis, high-energy gamma rays are emitted when matter is accelerated towards ultradense, rotating stars called pulsars. (For an image of Čherenkov radiation, see p. 469; for an image of a pulsar, see p. 237.)

(*a*)

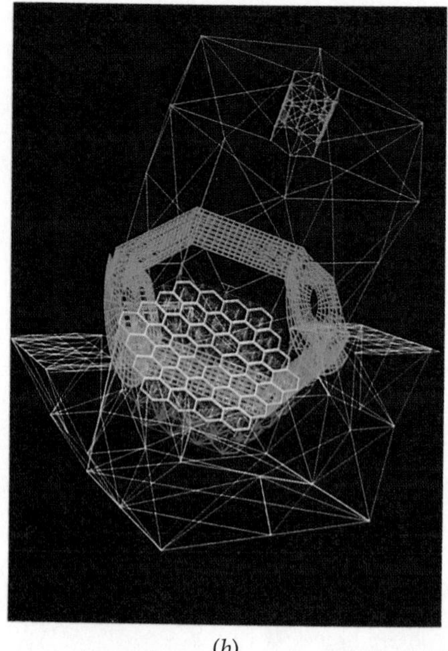

(*b*)

(*c*)

(*d*)

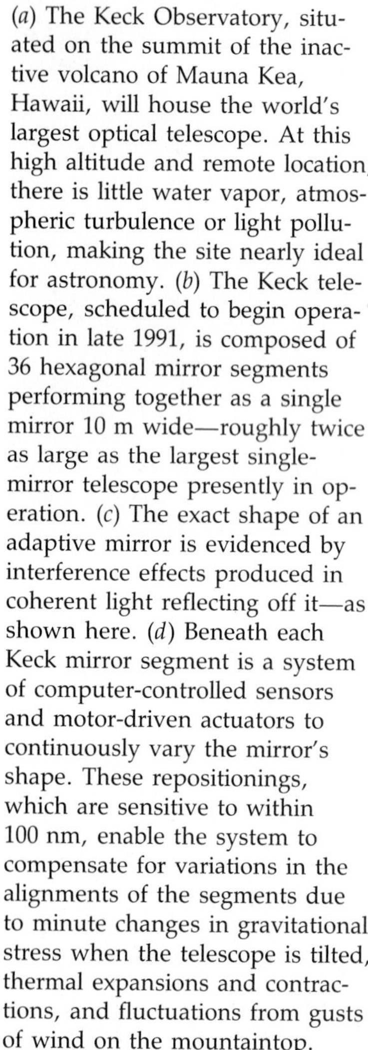

(*a*) The Keck Observatory, situated on the summit of the inactive volcano of Mauna Kea, Hawaii, will house the world's largest optical telescope. At this high altitude and remote location, there is little water vapor, atmospheric turbulence or light pollution, making the site nearly ideal for astronomy. (*b*) The Keck telescope, scheduled to begin operation in late 1991, is composed of 36 hexagonal mirror segments performing together as a single mirror 10 m wide—roughly twice as large as the largest single-mirror telescope presently in operation. (*c*) The exact shape of an adaptive mirror is evidenced by interference effects produced in coherent light reflecting off it—as shown here. (*d*) Beneath each Keck mirror segment is a system of computer-controlled sensors and motor-driven actuators to continuously vary the mirror's shape. These repositionings, which are sensitive to within 100 nm, enable the system to compensate for variations in the alignments of the segments due to minute changes in gravitational stress when the telescope is tilted, thermal expansions and contractions, and fluctuations from gusts of wind on the mountaintop.

Example 32-7

The world's largest refracting telescope is at the Yerkes Observatory of the University of Chicago at Williams Bay, Wisconsin. The objective has a diameter of 102 cm and a focal length of 19.5 m. The focal length of the eyepiece is 10 cm. What is the magnifying power of this telescope?

From Equation 32-8, we have

$$M = -\frac{f_o}{f_e} = -\frac{19.5\text{ m}}{0.10\text{ m}} = -195$$

The main consideration with an astronomical telescope is not its magnifying power but its light-gathering power, which depends on the size of the objective. The larger the objective, the brighter the image. Very large lenses without aberrations are difficult to produce. In addition, there are mechanical problems in supporting very large, heavy lenses by their edges. A reflecting telescope uses a concave mirror instead of a lens for its objective. This offers several advantages. For one, a mirror does not produce chromatic aberration. In addition, mechanical support is much simpler since the mirror weighs far less than a lens of equivalent optical quality and can be supported over its entire back surface.

One problem with the reflecting telescope is that the image of the objective mirror must be viewed in the region of the incoming rays (Figure 32-10). In very large reflecting telescopes, such as the 200-in (5.1 m) diameter telescope at Mt. Palomar, California, the viewer sits in an observer's cage near the focal point of the mirror. To obstruct as little light as possible, the cage is very small and cramped, so there is little space for auxiliary instruments such as spectrographs. In smaller telescopes, the fraction of light obstructed by such an arrangement would be too great. One method of reducing the amount of light obstructed is to use a second, smaller mirror to reflect the rays through a small hole in the center of the objective as shown in Figure 32-11. This has the further advantages of making the viewing area more accessible and providing more room for auxiliary instruments.

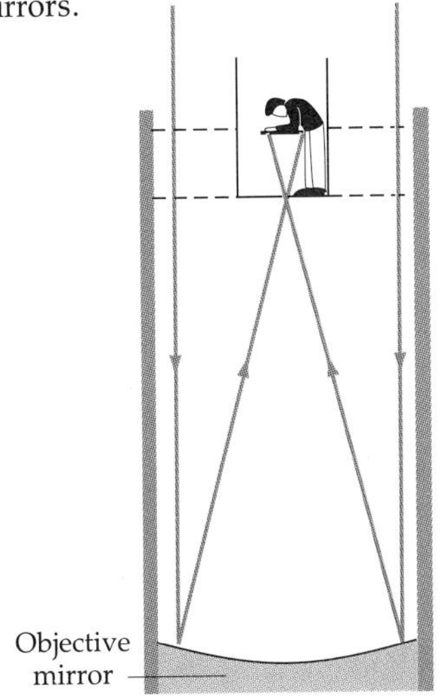

Figure 32-10 A reflecting telescope uses a mirror for its objective. Because the viewer compartment blocks off some of the incoming light, the arrangement shown here is used only in telescopes with very large objective mirrors.

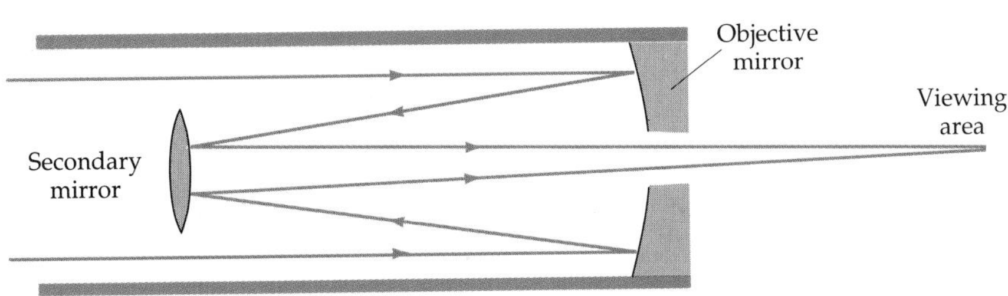

Figure 32-11 Reflecting telescope with secondary mirror to redirect the light through a small hole in the objective mirror. This arrangement has a further advantage over that of Figure 32-10 in that there is more room for auxiliary instruments in the viewing region.

The fact that the final image is inverted in a simple telescope is not a disadvantage when viewing astronomical objects such as stars and planets, but it is when viewing terrestrial objects. Binoculars use two 45-45-90° prisms in each side to provide a second inversion of the image so that the final image is upright. Figure 32-12*a* shows a 45-45-90° prism with its hypotenuse horizontal and its reflecting sides vertical. Light entering through the long face is reflected twice and emerges back through the long face in the opposite direction. Horizontal images are inverted, but vertical ones are not. In Figure 32-12*b*, a second prism with its hypotenuse vertical redirects the light back into its original direction and inverts vertical images without changing horizontal ones. The multiple reflections of the prisms also in-

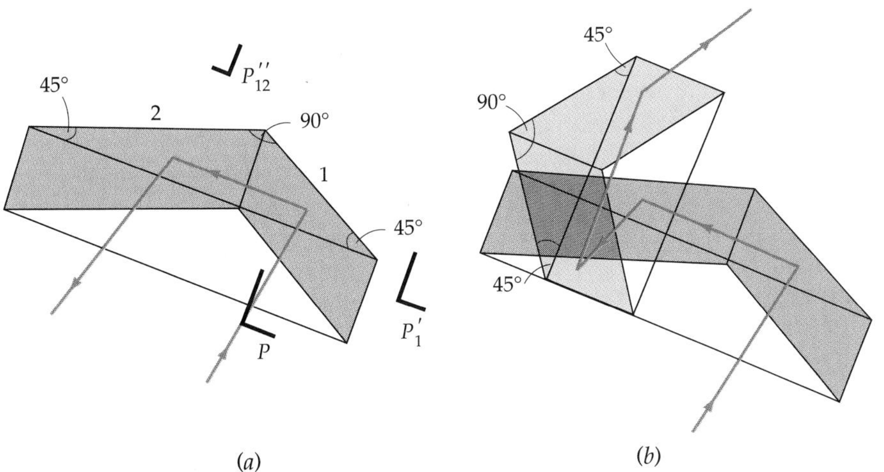

Figure 32-12 (*a*) A 45-45-90° prism with its hypotenuse horizontal and the reflecting faces 1 and 2 vertical. The image P''_{12} due to reflection from surfaces 1 and 2 is inverted in the horizontal direction but not in the vertical direction. (*b*) If the light enters a second, identical prism oriented with its hypotenuse vertical, the image will be inverted in the vertical direction. After passing through both prisms, the light emerges in its original direction with the image inverted in both the vertical and horizontal directions.

(a)

(b)

The Hubble Space Telescope (*a*) during and (*b*) after its deployment from the cargo bay of the space shuttle using a long mechanical arm. The telescope orbits nearly 615 km above the earth's surface, high above atmospheric turbulence that limits the ability of ground-based telescopes to resolve images at optical wavelengths. Due to an error in the device used to measure the shape of the Hubble's main reflecting mirror, the telescope suffers from spherical aberration. At present it is not clear to what extent this can be remedied. (*c*) A false-color image of 30 Doradus, a star cluster in the Large Magellanic Cloud from a 2.2-m ground-based telescope. (*d*) The same region as imaged by the Hubble with a sixfold improvement in resolution. (*e*) A computer-processed version of the Hubble image, removing the haze attributed to spherical aberrations. Such processing may work for bright sources, but is unlikely to enable the Hubble to image dim ones from very distant galaxies as originally hoped.

crease the path length of the light so that a relative long focal length for the objective can be used in a relatively short space.

Summary

1. The cornea–lens system of the eye focuses light on the retina, where it is sensed by the rods and cones, which transmit the information along the optic nerve to the brain. When the eye is relaxed, the focal length of the cornea–lens system is about 2.5 cm, the distance from the cornea to the retina. When objects are brought near the eye, the shape of the lens changes slightly to decrease the overall focal length so that the image is again focused on the retina. The closest distance for which the lens can focus the image on the retina is called the near point, which is typically about 25 cm but varies with age and from person to person. The apparent size of an object depends on the size of the image on the retina, which in turn depends on the distance from the object to the eye. The closer the object, the larger the image on the retina and, therefore, the larger the apparent size of the object.

2. A simple magnifier consists of a lens with a positive focal length that is smaller than the near point distance. The angular magnification of a simple magnifier is the ratio of the near-point distance to the focal length of the lens:

$$M = \frac{x_{\text{np}}}{f}$$

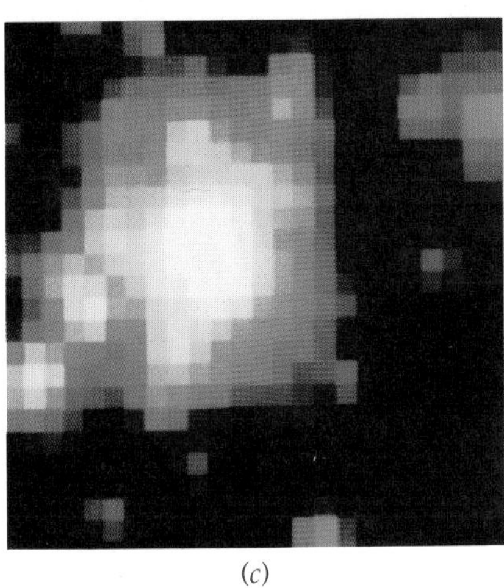
(c)

(d)

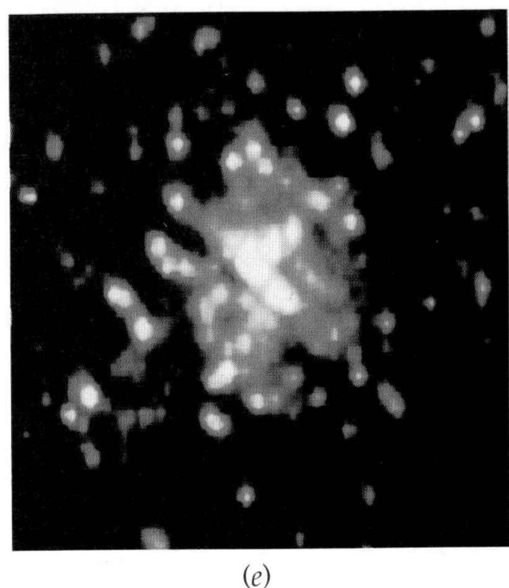
(e)

3. A basic camera consists of a lens, a variable aperture, a shutter, a light-tight box, and a film. Since the focal length of the lens is fixed, focusing is accomplished by moving the lens toward or away from the film. The *f*-number of the aperture is the ratio of the focal length to the diameter of the lens:

$$f\text{-number} = \frac{f}{D}$$

The focal length of a typical lens in a 35-mm camera is about 50 mm. A telephoto lens has a larger focal length, which gives a larger image on the film but a narrower field of view. A wide-angle lens has a smaller focal length, which gives a smaller image on the film but a wider field of view.

4. The compound microscope is used to look at very small objects at short distances. In its simplest form, it consists of two lenses—an objective and an ocular or eyepiece. The object to be viewed is placed just outside the focal point of the objective, which forms an enlarged image of the object at the focal point of the eyepiece. The eyepiece acts as a simple magnifier for viewing the final image. The magnifying power of the microscope is the product of the lateral magnification of the objective and the angular magnification of the eyepiece:

$$M = m_o M_e = -\frac{L}{f_o}\frac{x_{np}}{f_e}$$

where L is the tube length, which is the distance between the second focal point of the objective and the first focal point of the eyepiece.

5. The telescope is used to view objects that are far away. The objective of the telescope forms a real image that is much smaller than the object but much closer. The eyepiece is then used as a simple magnifier for viewing the image. A reflecting telescope uses a mirror for its objective. The magnifying power of a telescope equals the (negative) ratio of the focal length of the objective to the focal length of the eyepiece:

$$M = -\frac{f_o}{f_e}$$

The most important feature of an astronomical telescope is its light-gathering power, which is proportional to the area of the objective.

Suggestions for Further Reading

Everhart, Thomas E., and Thomas L. Hayes: "The Scanning Electron Microscope," *Scientific American*, January 1972, p. 54.

This article describes how the interaction between a beam of high-energy electrons and matter is used by the scanning electron microscope to create an image of three-dimensional appearance.

Koretz, Jane F., and George H. Handelman: "How the Human Eye Focuses," *Scientific American*, July 1988, p. 92.

Throughout a person's lifetime, the shortest distance at which the eye can focus gradually increases. This article describes measurements and analyses that suggest that changes in the lens, including increasing thickness and decreasing index of refraction, are responsible.

Land, Michael F.: "Animal Eyes with Mirror Optics," *Scientific American*, December 1978, p. 126.

Several sea creatures, described in this article, have been found to use reflection rather than refraction to form images of their surroundings.

Price, William H.: "The Photographic Lens," *Scientific American*, August 1976, p. 72.

A history of lens design, including a description of the defects and aberrations one wishes to minimize in a lens, and a discussion of modern computer-aided design are presented in this article.

Review

A. Objectives: After studying this chapter, you should:

1. Be able to discuss how the eye works.

2. Be able to show with a simple diagram why an object appears larger when it is brought closer to the eye.

3. Be able to describe how a simple magnifier works and calculate its angular magnification.

4. Be able to discuss how a camera works.

5. Be able to describe with diagrams and equations how a microscope and a telescope work.

B. Define, explain, or otherwise identify:

Accommodation	Magnifying power
Near point	*f*-number
Farsighted	Objective
Nearsighted	Eyepiece
Simple magnifier	Ocular
Angular magnification	Tube length

C. True or false: If the statement is true, explain why it is true. If it is false, give a counterexample.

1. The lens of the eye forms a real image.

2. A simple magnifier should have a short focal length.

3. A simple magnifier forms a virtual image.

4. The lens of a camera forms a real image.

5. The area of a camera's aperture is proportional to the *f*-number.

6. The focal length of a telephoto lens is greater than that of a wide-angle lens.

7. The image formed by the objective of a microscope is inverted and larger than the object.

8. The image formed by the objective of a telescope is inverted and larger than the object.

9. A reflecting telescope uses a mirror for its ocular.

Problems

Level I

32-1 The Eye

In the following problems, take the distance from the cornea–lens system of the eye to the retina to be 2.5 cm and assume that the lenses of corrective glasses are in contact with the eye unless otherwise stated.

1. Suppose the eye were designed like a camera with a lens of fixed focal length $f = 2.5$ cm that could move toward or away from the retina. Approximately how far would the lens have to move to focus the image of an object 25 cm from the eye onto the retina? *Hint:* Find the distance from the retina to the image behind it for an object at 25 cm.

2. Find the change in the focal length of the eye when an object originally at 3 m is brought to 30 cm from the eye.

3. A farsighted person needs to read from a computer screen that is 45 cm from her eye. Her near point is at 80 cm. (*a*) Find the focal length of the lenses in reading glasses that will produce an image of the screen at 80 cm from her eye. (*b*) What is the power of the lenses?

4. Find (*a*) the focal length and (*b*) the power of a lens that will produce an image at 80 cm from the eye of a book that is 30 cm from the eye.

5. A farsighted person requires lenses with a power of 1.75 diopters to read comfortably from a book that is 25 cm from his eye. What is his near point without the lenses?

6. A nearsighted person cannot focus clearly on objects more distant that 225 cm from her eye. What power lenses are required for her to see distant objects clearly?

7. Since the index of refraction of the lens of the eye is not very different from that of the surrounding material, most of the refraction takes place at the cornea, where n changes abruptly from 1.0 in air to about 1.4. Assuming the cornea to be a homogeneous sphere with an index of refraction of 1.4, calculate its radius if it focuses parallel light on the retina a distance 2.5 cm away. Do you expect your result to be larger or smaller than the actual radius of the cornea?

8. If two point objects that are close together are to be seen as two distinct objects, their images must fall on the retina on two different cones that are not adjacent. That is, there must be an unactivated cone between them. The separation of the cones is about 1 μm. (*a*) What is the smallest angle the two points can subtend? (See Figure 32-13.) (*b*) How close together can two points be if they are 20 m from the eye?

Figure 32-13 Problem 8. The two points will look like two separate points only if their images fall on two different, nonadjacent cones of the retina.

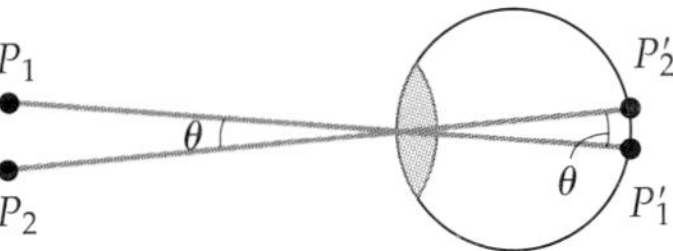

32-2 The Simple Magnifier

9. A person with a near-point distance of 30 cm uses a simple magnifier of power 20 diopters. What is the magnification obtained if the final image is at infinity?

10. A person with a near-point distance of 25 cm wishes to obtain a magnifying power of 5 with a simple magnifier. What should be the focal length of the lens used?

11. What is the magnifying power of a lens of focal length 7 cm when the image is viewed at infinity by a person whose near point is at 35 cm?

12. A lens of focal length 6 cm is used as a simple magnifier with the image at infinity by one person whose near point is at 25 cm and by another whose near point is at 40 cm. (*a*) What is the effective magnifying power of the lens for each person? (*b*) Compare the size of the image on the retina when each looks at the same object with the magnifier.

32-3 The Camera

13. What is the diameter of an $f/1.4$ lens if its focal length is 50 mm?

14. A lens has a usable diameter of 2.5 cm. What is its f-number if its focal length is 50 mm?

15. A telephoto lens has a focal length of 200 mm. By how much must it be moved to change from focusing on an object at infinity to one at a distance of 30 m?

16. A wide-angle lens has a focal length of 28 mm. By how much must it be moved to change from focusing on an object at infinity to one at a distance of 5 m?

17. A camera produces a proper exposure for an aperture stop of $f/16$ at 1/30 s. What shutter speed should be used at (*a*) $f/11$, (*b*) $f/8$, (*c*) $f/5.6$, (*d*) $f/4$, and (*e*) $f/2.8$?

18. Light conditions on a certain day for a certain film call for an aperture stop of $f/8$ at 1/250 s. (*a*) If you wish to take a picture of a humming bird at 1/1000 s, what f/number should you use? (*b*) If you set the aperture at $f/22$ for another picture, what shutter speed should you use?

32-4 The Compound Microscope

19. A microscope objective has a focal length of 0.5 cm. It forms an image at 16 cm from its second focal point. What is the magnifying power for a person whose near point is at 25 cm if the focal length of the eyepiece is 3 cm?

20. A microscope has an objective of focal length 16 mm and an eyepiece that gives an angular magnification of 5 for a person whose near point is 25 cm. The tube length is 18 cm. (*a*) What is the lateral magnification of the objective? (*b*) What is the magnifying power of the microscope?

21. A crude, symmetric, hand-held microscope consists of two converging 20-diopter lenses fastened in the ends of a tube 30 cm long. (*a*) What is the tube length of this microscope? (*b*) What is the lateral magnification of the objective? (*c*) What is the magnifying power of the microscope? (*d*) How far from the objective should the object be placed?

22. Repeat Problem 21 for the same two lenses separated by 40 cm.

32-5 The Telescope

23. A simple telescope has an objective of focal length 100 cm and an eyepiece of focal length 5 cm. It is used to look at the moon, which subtends an angle of about 0.009 rad. (*a*) What is the diameter of the image formed by the objective? (*b*) What angle is subtended by the final image at infinity? (*c*) What is the magnifying power of the telescope?

24. The objective lens of the refracting telescope at the Yerkes Observatory has a focal length of 19.5 m. It is used to look at the moon, which subtends an angle of about 0.009 rad. What is the diameter of the image of the moon formed by the objective?

25. The 200-in reflecting telescope at Mt. Palomar has a mirror with a diameter of 200 in = 5.1 m and a focal length of 1.68 m. (*a*) By what factor is the light-gathering power increased over the 40-in (1.02-m) diameter refract-

ing lens of the telescope at the Yerkes Observatory? (*b*) If the focal length of the eyepiece is 1.25 cm, what is the magnifying power of this telescope?

Level II

26. (*a*) Show that if the final image of a simple magnifier is at the near point of the eye rather than at infinity, the angular magnification is given by

$$M = \frac{x_{np}}{f} + 1$$

(*b*) Find the magnification of a 20-diopter lens for a person with a near point of 30 cm if the final image is at the near point. Draw a ray diagram for this situation.

27. A botanist examines a leaf using a convex lens of power 12 diopters as a simple magnifier. What is the expected angular magnification (*a*) if the final image is at infinity and (*b*) if the final image is at 25 cm?

28. Show that when the image of a simple magnifier is viewed at the near point, the lateral and angular magnification of the magnifier are equal.

29. A 35-mm camera has a picture size of 24 mm by 36 mm. It is to be used to take a picture of a person 175 cm tall in which the image of the person just fills the height (24 mm) of the film. How far should the person stand from the camera if the focal length of the lens is 50 mm?

30. A 35-mm camera with interchangeable lenses is used to take a picture of a hawk that has a wingspan of 2 m. The hawk is 30 m away. What would be the ideal focal length for the lens so that the image of the wings just fills the width of the film, which is 36 mm?

31. An astronomical telescope has a magnifying power of 7. The two lenses are 32 cm apart. Find the focal length of each lens.

32. The near point of a certain person is 80 cm. Reading glasses are prescribed so that he can read a book at 25 cm from his eye. The glasses are 2 cm from the eye. What power lenses should be used in the glasses?

33. A disadvantage of the astronomical telescope for terrestrial use (for example, at a football game) is that the image is inverted. A galilean telescope uses a converging lens as its objective but a diverging lens as its eyepiece. The image formed by the objective is behind the eyepiece at its focal point so that the final image is virtual, erect, and at infinity. (*a*) Show that the magnifying power is $M = -f_o/f_e$, where f_o is the focal length of the objective and f_e is that of the eyepiece (which is negative). (*b*) Draw a ray diagram to show that the final image is indeed virtual, erect, and at infinity.

34. A galilean telescope (see Problem 33) is designed so that the final image is at the near point, which is at 25 cm, rather than at infinity. The focal length of the objective is 100 cm and that of the eyepiece is −5 cm. (*a*) If the object distance is 30 m, where is the image of the objective? (*b*) What is the object distance for the eyepiece for which the final image is at the near point? (*c*) How far apart are the lenses? (*d*) If the object height is 1.5 m, what is the height of the final image? (*e*) What is the angular magnification of the image?

35. A compound microscope has an objective with a power of 45 diopter and an eyepiece with a power of 80 diopter. The lenses are separated by 28 cm. Assuming that the final image is formed 25 cm from the eye, what is the magnifying power?

Level III

36. At age 45, a person was fitted for reading glasses of power 2.1 diopters so she could read a newspaper at 25 cm. Now at age 55, she has discovered she must hold her newspaper at a distance of 40 cm in order to see it clearly with her glasses on. (*a*) Where was her near point at age 45? (*b*) Where is her near point now? (*c*) What power is now required for the lenses of her reading glasses so that she can again read at 25 cm? (Assume the glasses are 2.2 cm from her eyes.)

37. If you look into the wrong end of a telescope, that is, into the objective, you will see a distant object that is reduced in size. For a refracting telescope with an objective of focal length 2.25 m and an eyepiece of focal length 1.5 cm, by what factor is the angular size of the object reduced?

38. An aging physics professor discovers that he can see objects clearly only between 0.75 m and 2.5 m, so he decides he needs bifocals. The upper part of the lens allows him to see objects clearly at infinity, and the lower part allows him to see objects clearly at 25 cm. Assume that the lens is 2 cm from his eye. (*a*) Calculate the power of the lens required for the upper part his bifocals. (*b*) Calculate the power of the lens required for the lower part of his bifocals. (*c*) Is there a range of distances over which he cannot see objects clearly no matter which part of the bifocals he looks through? If so, what is that range? (*d*) Is there a range of distances over which he cannot see objects clearly whether or not he is wearing his bifocals? If so, what is that range?

39. A microscope has a magnifying power of −600 and an eyepiece with an angular magnification of 15. The objective lens is 22 cm from the eyepiece. Without making any approximations, calculate (*a*) the focal length of the eyepiece, (*b*) the location of the object at which it is in focus for a normal relaxed eye, and (*c*) the focal length of the objective lens.

40. A hunter lost in the mountains tries to make a telescope from two lenses, one of power 2.0 diopters and the other of power 6.5 diopters, and a cardboard tube. (*a*) What is the maximum possible magnifying power? (*b*) How long must the tube be? (*c*) Which lens should be used as the eyepiece? Why?

Chapter 33

Interference and Diffraction

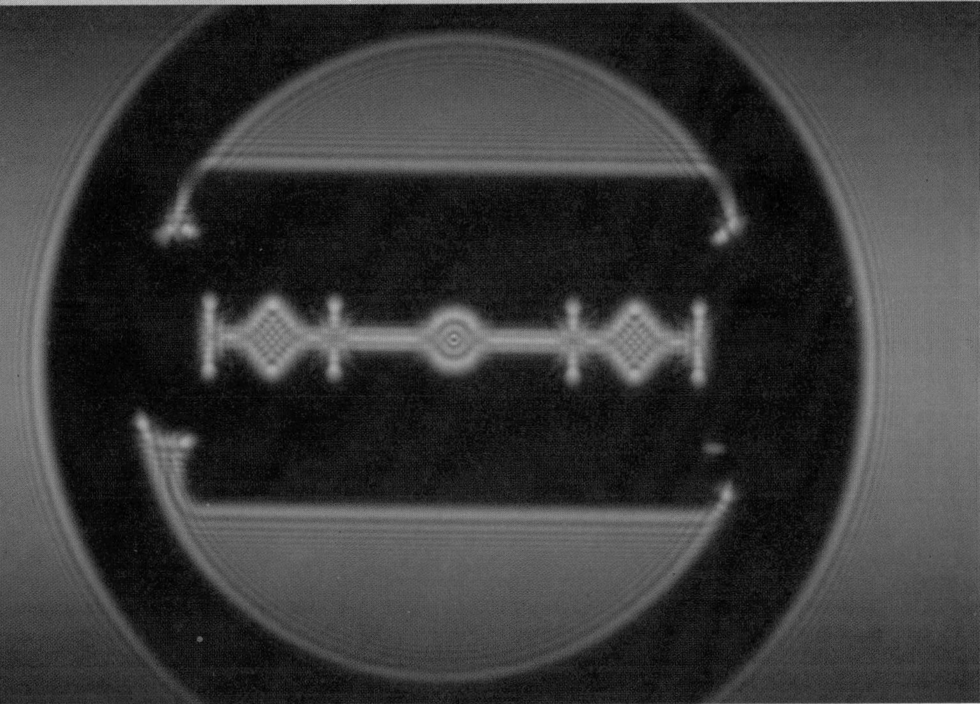

The diffraction of light incident on a razor blade.

Interference and diffraction are the important phenomena that distinguish waves from particles. Interference is the combining by superposition of two or more waves that meet at one point in space. Diffraction is the bending of waves around corners that occurs when a portion of a wavefront is cut off by a barrier or obstacle. The pattern of the resulting wave can be calculated by treating each point on the original wavefront as a point source according to Huygens' principle and calculating the interference pattern resulting from these sources.

In Chapter 14, we discussed the interference of sound waves from two point sources and the diffraction of sound qualitatively. Since the analytical treatment of interference and diffraction is the same for all waves whether they are sound waves, waves on strings, water waves, or electromagnetic waves, you should review Chapter 14 before you begin this chapter.

33-1 Phase Difference and Coherence

When two harmonic waves of the same frequency and wavelength but differing in phase combine, the resultant wave is a harmonic wave whose amplitude depends on the phase difference. If the phase difference is 0 or an integer times 360°, the waves are in phase and interfere constructively. The resultant amplitude equals the sum of the individual amplitudes, and the intensity (which is proportional to the square of the amplitude) is maximum. If the phase difference is 180° (π radians) or any odd integer times 180°, the waves are out of phase and interfere destructively. The resultant amplitude is then the difference between the individual amplitudes, and the intensity is a minimum. If the amplitudes are equal, the maximum intensity is 4 times that of either source and the minimum intensity is zero.

A phase difference between two waves is often the result of a difference in path length traveled by the two waves. A path difference of one wavelength produces a phase difference of 360°, which is equivalent to no phase difference at all. A path difference of one-half wavelength produces a 180° phase difference. In general, a path difference of Δr contributes a phase difference δ given by

$$\delta = \frac{\Delta r}{\lambda} 2\pi = \frac{\Delta r}{\lambda} 360° \qquad 33\text{-}1$$

Another cause of phase difference is the 180° phase change a wave sometimes undergoes upon reflection from a boundary surface. This phase change is analogous to the inversion of a pulse on a string when it reflects from a point where the density suddenly increases, such as when a light string is attached to a heavier string or rope. The inversion of the reflected pulse is equivalent to a phase change of 180° for a sinusoidal wave, which can be thought of as a series of pulses. When light traveling in air strikes the surface of a medium in which light travels more slowly, such as glass or water, there is a 180° phase change in the reflected light. When light is originally traveling in glass or water, there is no phase change in the light reflected from the glass–air or water–air surface. This is analogous to the reflection without inversion of a pulse on a heavy string at a point where the heavy string is attached to a lighter string.

As we saw in Chapter 14, interference of waves from two sources is not observed unless the sources are coherent, that is, unless the phase difference between the waves is constant in time. Because a light beam is usually the result of millions of atoms radiating independently, two light sources are usually not coherent. Indeed, the phase difference between the waves from such sources fluctuates randomly many times per second. Coherence in optics is often achieved by splitting the light beam from a single source into two or more beams, which can then be combined to produce an interference pattern. This splitting can be achieved by reflecting the light from the two closely spaced surfaces of a thin film (Section 33-2); the simultaneous reflection from and transmission through a half-silvered mirror, as in the Michelson interferometer (Section 33-3); or the diffraction of the beam by two small openings or slits in an opaque barrier (Section 33-4). Coherent sources can also be obtained by using a single point source and its image in a plane mirror for the two sources, an arrangement called *Lloyd's mirror*. Today, lasers are the most important sources of coherent light in the laboratory. Lasers have the property that all the atoms within a laser radiate in phase with one another, which leads to strong collimation of the light radiated.

Example 33-1

(*a*) What is the minimum path difference that will produce a phase difference of 180° for light of wavelength 800 nm? (*b*) What phase difference will that path difference produce in light of wavelength 700 nm?

(*a*) From Equation 33-1, we have

$$\delta = \frac{\Delta r}{\lambda}360^\circ = 180^\circ$$

$$\Delta r = \tfrac{1}{2}\lambda = \tfrac{1}{2}(800\text{ nm}) = 400\text{ nm}$$

(*b*) For $\lambda = 700$ nm and $\Delta r = 400$ nm, we have

$$\delta = \frac{\Delta r}{\lambda}360^\circ = \frac{400\text{ nm}}{700\text{ nm}}360^\circ = 206^\circ = 3.59\text{ rad}$$

33-2 Interference in Thin Films

Interference of light rays from the front and back surface of a thin soap film. At the top where the film is very thin, the rays from the front surface of the film (that undergo a 180° phase change) and the rays from the back surface of the film (that do not change phase) interfere destructively and the film appears dark. At other parts of the film, the interference is destructive or constructive depending on the wavelength and on the film thickness.

Undoubtedly, you have noticed the colored bands in a soap bubble or in an oil film on a water-covered street. These bands are due to the interference of light reflected from top and bottom surfaces of the film. The different colors arise because of variations in the thickness of the film, causing interference for different wavelengths at different points.

Consider a thin film of water (such as a small section of a soap bubble) of uniform thickness viewed at small angles with the normal as shown in Figure 33-1. Part of the light is reflected from the upper, air–water surface. Since light travels more slowly in water than in air, there is a 180° phase change in this reflected light. Some of the light enters the film and is partially reflected by the bottom water–air surface. There is no phase change in this reflected light. If the light is nearly perpendicular to the surfaces, both the ray reflected from the top surface and the one reflected from the bottom surface can enter the eye at point P in the figure. The path difference between these two rays is $2t$, where t is the thickness of the film. This path difference produces a phase difference of $(2t/\lambda')360^\circ$, where λ' is the wavelength of the light in the film. The wavelength in the film is related to the wavelength in air λ by $\lambda' = \lambda/n$ (Equation 30-4), where n is the index of refraction of the film. The total phase difference between these two rays is thus 180° plus that due to the path difference. Destructive interference occurs when the path difference $2t$ is zero or a whole number of wavelengths λ' (in the film). Constructive interference occurs when the path difference is an odd number of half wavelengths.

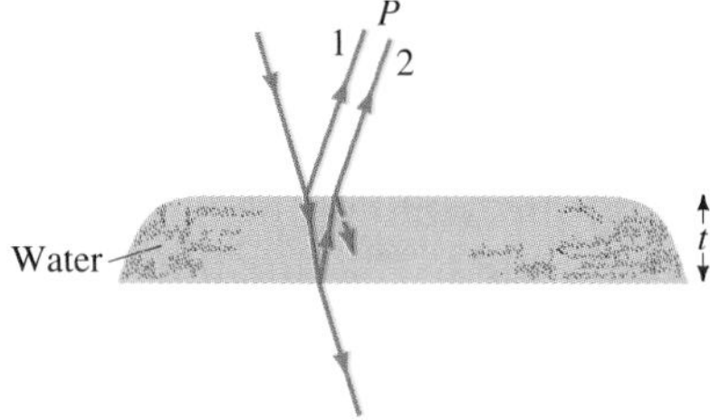

Figure 33-1 Light rays reflected from the top and bottom surfaces of a thin film are coherent because both rays come from the same source. If the light is incident nearly normally, the two reflected rays will be very close to each other and will produce interference.

We can express these conditions mathematically. When there is one phase change of 180° due to reflection, the conditions for interference are

$$\frac{2t}{\lambda'} = m \qquad m = 0, 1, 2, 3, \ldots \quad \text{(destructive)} \tag{33-2a}$$

$$\frac{2t}{\lambda'} = m + \frac{1}{2} \qquad m = 0, 1, 2, 3, \ldots \quad \text{(constructive)} \tag{33-2b}$$

Conditions for interference with one 180° phase change

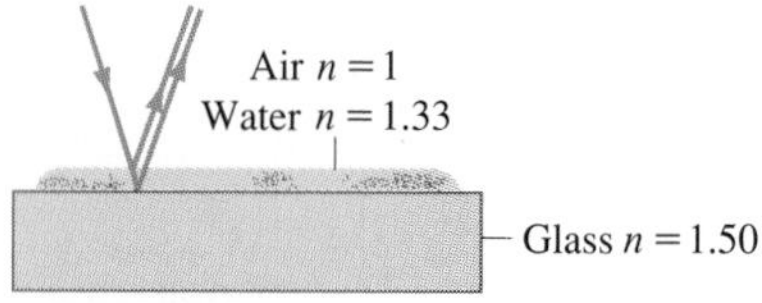

Figure 33-2 Interference of light reflected from a thin film of water resting on a glass surface. In this case, both rays undergo a change in phase of 180° upon reflection.

When a thin water film lies on a glass surface as in Figure 33-2, the ray that reflects from the lower, water–glass surface also undergoes a 180° phase change because the index of refraction of glass (about 1.5) is greater than that of water (about 1.33). Thus, both the rays shown in the figure have undergone a 180° phase change upon reflection. The phase difference between these rays is due solely to the path difference and is given by $\delta = (2t/\lambda')360°$. When there are two 180° phase changes upon reflection (or if there are no phase changes), the conditions for interference are

Conditions for interference with two 180° phase changes

$$\frac{2t}{\lambda'} = m \qquad m = 0, 1, 2, 3, \ldots \quad \text{(constructive)} \qquad \text{33-3}a$$

$$\frac{2t}{\lambda'} = m + \frac{1}{2} \qquad m = 0, 1, 2, 3, \ldots \quad \text{(destructive)} \qquad \text{33-3}b$$

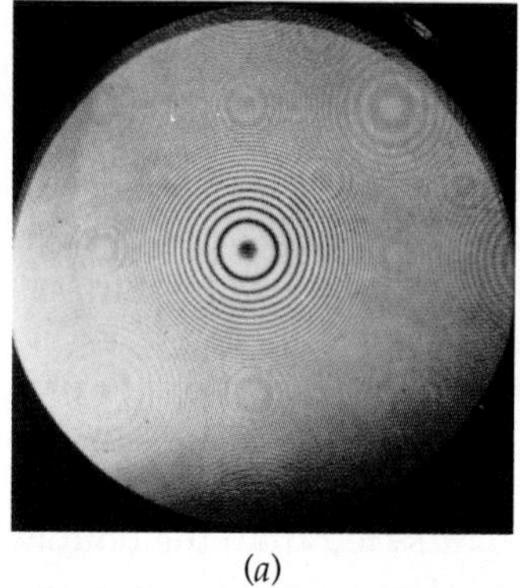

(*a*)

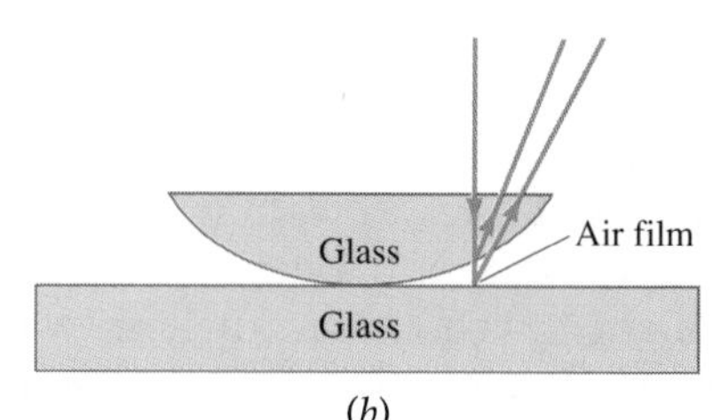

(*b*)

Figure 33-3 (*a*) Newton's rings observed with light reflected from a thin film of air between a plane glass surface and a spherical glass surface. At the center, the thickness of the air film is negligible and the interference is destructive because of the phase change of one of the rays. (*b*) Glass surfaces for the observation of Newton's rings shown in part (*a*). The thin film in this case is the film of air between the glass surfaces.

When a thin film of varying thickness is viewed with monochromatic light, such as the yellow light from a sodium lamp, alternating bright and dark bands or lines called **fringes** are observed. The distance between a bright fringe and a dark fringe is that distance over which the film's thickness changes such that the path difference $2t$ is $\lambda'/2$. Figure 33-3*a* shows the interference pattern observed when light is reflected from an air film between a spherical glass surface and a plane glass surface in contact. These circular interference fringes are known as **Newton's rings.** Typical rays reflected at the top and bottom of the air film are shown in Figure 33-3*b*. Near the point of contact of the surfaces, where the path difference between the ray reflected from the upper, glass–air surface and the ray reflected from the lower, air–glass surface is essentially zero or is at least small compared with the wavelength of light, the interference is perfectly destructive because of the 180° phase shift of the ray reflected from the lower air–glass surface. This central region in Figure 33-3*a* is therefore dark. The first bright fringe occurs at the radius at which the path difference is $\lambda/2$, which contributes a phase difference of 180°. This adds to the phase shift due to reflection to produce a total phase difference of 360°, which is equivalent to a zero phase difference. The second dark region occurs at the radius at which the path difference is λ, and so on.

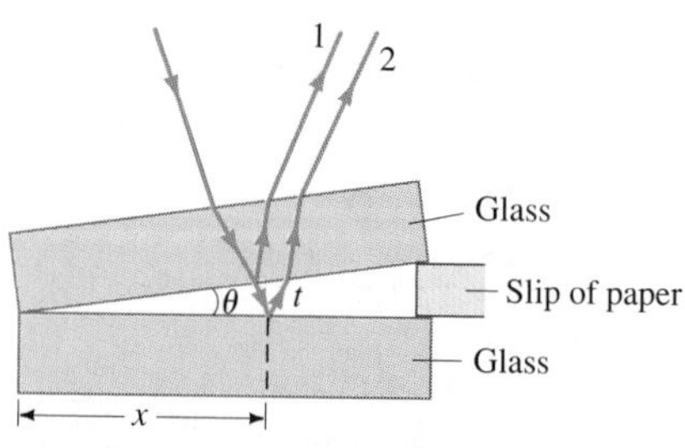

Figure 33-4 Light incident normally on a wedge-shaped film of air between two glass plates. The path difference $2t$ is proportional to x. When viewed from above, alternate bright and dark bands or fringes are seen.

Example 33-2

A wedge-shaped film of air is made by placing a small slip of paper between the edges of two flat pieces of glass as shown in Figure 33-4. Light of wavelength 500 nm is incident normally on the glass plates, and interference fringes are observed by reflection. If the angle θ made by the plates is 3×10^{-4} rad, how many interference fringes per centimeter are observed?

Because of the 180° phase shift in the ray reflected from the bottom plate, the first fringe near the point of contact (where the path difference is zero) will be dark. Let x be the horizontal distance to the mth dark fringe, where the plate separation is t as shown in the figure. Since the angle θ is very small, it is given approximately by

$$\theta = \frac{t}{x}$$

Using Equation 33-2a for m, we have

$$m = \frac{2t}{\lambda'} = \frac{2t}{\lambda}$$

since the film is an air film. Substituting $t = x\theta$ gives

$$m = \frac{2x\theta}{\lambda}$$

or

$$\frac{m}{x} = \frac{2\theta}{\lambda} = \frac{2(3 \times 10^{-4})}{5 \times 10^{-5}\ \text{cm}} = 12\ \text{cm}^{-1}$$

where we have used $\lambda = 5 \times 10^{-7}$ m $= 5 \times 10^{-5}$ cm. We therefore observe 12 dark fringes per centimeter. In practice, the number of fringes per centimeter, which is easy to count, can be used to determine the angle.

Note that if the angle of the wedge is increased, the fringes become more closely spaced. The distance along the glass between adjacent dark (or adjacent bright) fringes is that distance that results in an additional path difference equal to the wavelength of the light in the film. If the angle of the wedge is increased, this distance is decreased.

Exercise

How many fringes per centimeter are observed in Example 33-2 if light of wavelength 650 nm is used? (Answer: $9.2\ \text{cm}^{-1}$)

Figure 33-5a shows interference fringes produced by a wedge-shaped air film between two flat glass plates as in Example 33-2. The straightness of the fringes indicates the flatness of the glass plates. Such plates are said to be **optically flat.** A similar wedge-shaped air film formed by two ordinary glass plates yields the irregular fringe pattern in Figure 33-5b, which indicates that these plates are not optically flat.

Figure 33-5 (a) Straight-line fringes from a wedge-shaped film of air like that in Figure 33-4. The straightness of the fringes indicates the flatness of the glass plates. (b) Fringes from a wedge-shaped film of air between glass plates that are not optically flat.

(a)

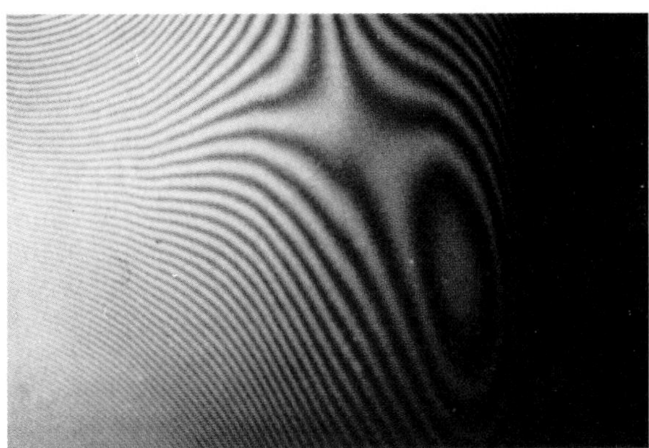

(b)

(a)

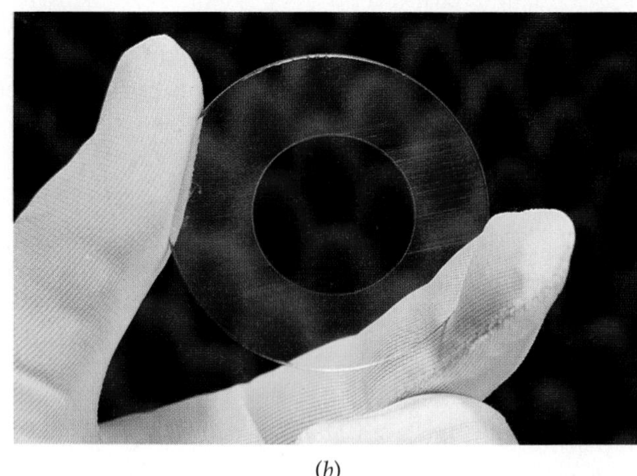

(b)

(*a*) A narrow-angle lens (*left*) and a wide-angle lens (*right*) prior to assembly in the cameras of Voyager 2. Each lens has two types of thin-film coatings: a low thermal emittance coating that reflects thermal wavelengths to prevent heat loss and keep the telescope elements warm and an antireflecting coating at optical wavelengths. (*b*) The antireflecting film used on the lenses in (*a*) is demonstrated here on the central region of a clear, polycarbonate disk.

A nonreflecting lens is made by covering the lens with a thin film of a material that has an index of refraction of about 1.22, which is between that of glass and that of air, so that the intensities of the light reflected from the top and bottom surfaces of the film are approximately equal. Since both rays undergo a 180° phase change, there is no phase difference between the rays due to reflection. The thickness of the film is chosen to be $\lambda'/4$, where $\lambda' = \lambda/n$ and λ is in the middle of the visible spectrum, so that there is a phase change of 180° due to the path difference of $\lambda'/2$. Reflection from the coated surface is thus minimized.

Questions

1. Why must a film used to observe interference colors be thin?
2. If the angle of a wedge-shaped air film such as that in Example 33-2 is too large, fringes are not observed. Why?
3. The spacing between Newton's rings decreases rapidly as the diameter of the rings increases. Explain qualitatively why this occurs.

33-3 The Michelson Interferometer

An **interferometer** is a device that uses interference fringes to make precise measurements of distance. Figure 33-6 is a schematic diagram of a Michelson interferometer. Light from a broad source strikes plate A, a beam splitter that is partially silvered so that the light is partially reflected and partially transmitted. The reflected beam travels to mirror M_2 and is reflected back toward the eye at O. The transmitted beam travels through a compensating plate B, which is of the same thickness as plate A, to mirror M_1 and is reflected back to plate A and then to the eye at O. The purpose of the compensating plate B is to make both beams pass through the same thickness of glass. Mirror M_1 is fixed, but mirror M_2 can be moved back and forth with fine and accurately calibrated screw adjustments. The two beams combine at O and form an interference pattern. This pattern is most easily understood by considering the mirror M_2 and the image of mirror M_1 produced by the mirror in the beam splitter A. This image is labeled M_1' in the diagram. If the mirrors M_1 and M_2 are exactly perpendicular to each other and are equidistant from the beam splitter, the image M_1' will coincide with M_2. If not, M_1' will be slightly displaced and will make a small angle with M_2, as shown in the diagram. The interference pattern at O will then be that of a thin, wedge-shaped film of air between M_1' and M_2 similar to that discussed in Example 33-2. If mirror M_2 is now moved, the fringe pattern will shift. If, for

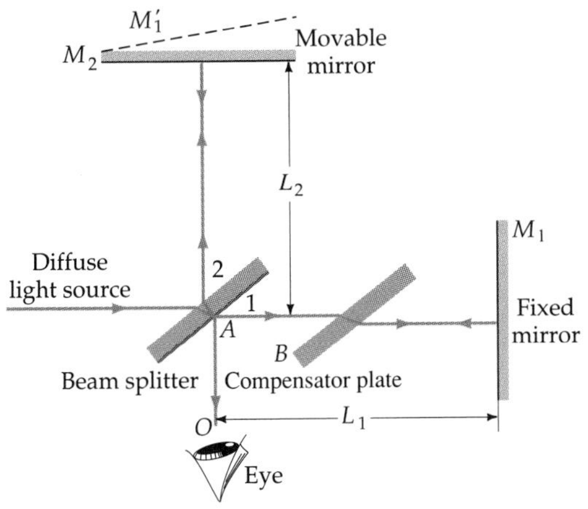

Figure 33-6 Michelson interferometer. The dashed line M_1' is the image of mirror M_1 in the mirror A. The interference fringes observed are those of a small wedge-shaped film of air formed by the sources M_1' and M_2. As M_2 is moved, the fringes move across the field of view.

example, the mirror M_2 is moved toward the splitter by a distance equal to $\frac{1}{4}\lambda$, the thickness of the wedge will increase by $\frac{1}{4}\lambda$ at each point. This will introduce an additional path difference of $\frac{1}{2}\lambda$ everywhere in the wedge (because the light traverses the wedge twice). The fringe pattern will move over by one-half fringe; that is, a previously dark fringe will now be a bright fringe, and so forth. If the distance mirror M_2 is moved is known, the wavelength of the light can be determined. Michelson used such an interferometer to measure the wavelength of a spectral line of light emitted by krypton 86 in terms of the standard meter bar. This measurement was then used to redefine the standard meter in terms of this wavelength. (The definition of the standard meter has since been changed. It is now defined in terms of the speed of light.)

Another use for the Michelson interferometer is to measure the index of refraction of air (or of some other gas). One of the beams from plate A is made to pass through a container that can be evacuated. The wavelength of the light in air λ' is related to that in a vacuum by $\lambda' = \lambda/n$, where n is the index of refraction of air (about 1.0003). When the container is evacuated, the wavelength of the light passing through it increases so that there are fewer waves in the length of the container. This causes a shift in the fringe pattern. By measuring the shift, the index of refraction can be determined (see Problem 10).

Michelson also used his interferometer in a famous experiment with Edward W. Morley in 1887 in which they attempted to measure the difference between the speed of light relative to the earth in the direction of motion of the earth and in a perpendicular direction. This experiment will be described in some detail in the next chapter.

Example 33-3

A film of index of refraction 1.33 and thickness 12 μm is inserted in one arm of a Michelson interferometer. The light used has a wavelength of 589 nm in air. By how many fringes is the interference pattern shifted?

The number of waves in the thin film N_f is the thickness $2t$ divided by the wavelength $\lambda' = \lambda/n$:

$$N_f = \frac{2t}{\lambda'} = \frac{2nt}{\lambda} = \frac{2(1.33)(12 \times 10^{-6}\text{ m})}{589 \times 10^{-9}\text{ m}} = 54.2 \text{ waves}$$

The number of waves originally in this space when it was occupied by air is

$$N_a = \frac{2t}{\lambda} = \frac{2(12 \times 10^{-6}\text{ m})}{589 \times 10^{-9}\text{ m}} = 40.8 \text{ waves}$$

There are thus $54.2 - 40.8 = 13.4$ more waves in one arm of the interferometer, so the interference pattern will be shifted by 13.4 fringes.

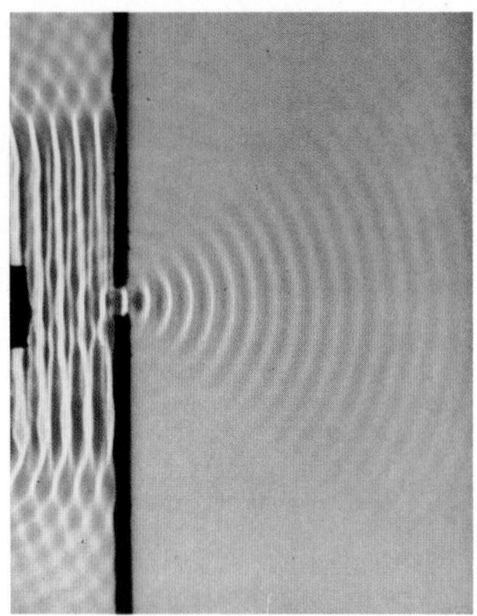

Figure 33-7 Plane water waves in a ripple tank encountering a barrier with a small opening. The waves to the right of the barrier are circular waves that are concentric about the opening just as if there were a point source at the opening.

33-4 The Two-Slit Interference Pattern

Interference patterns of light from two or more sources can be observed only if sources are coherent, that is, only if they are in phase or have a phase difference that is constant in time. We have noted that the randomness of the emissions of light by atoms means that two different light sources are generally incoherent. The interference in thin films discussed previously can be observed because the two beams come from the same light source but are separated by reflection.

In the famous experiment devised by Thomas Young in 1801 in which he demonstrated the wave nature of light, two coherent light sources are produced by illuminating two parallel slits with a single source. We assume here that each slit is very narrow. (We will treat the general case in Section 33-8.) We saw in Chapter 14 that when a wave encounters a barrier with a very small opening, the opening acts as a point source of waves (Figure 33-7). In Young's experiment each slit acts as a line source, which is equivalent to a point source in two dimensions. The interference pattern is observed on a screen far from the slits, which are separated by a distance d. At very large distances from the slits, the lines from the two slits to some point P on the screen are approximately parallel, and the path difference is approximately $d \sin \theta$, as shown in Figure 33-8c. We thus have interference maxima at an angle given by

Two-slit interference maxima

$$d \sin \theta = m\lambda \qquad m = 0, 1, 2, \ldots \tag{33-4}$$

The interference minima occur at

Two-slit interference minima

$$d \sin \theta = (m + \tfrac{1}{2})\lambda \qquad m = 0, 1, 2, \ldots \tag{33-5}$$

The phase difference δ at a point P is $2\pi/\lambda$ times the path difference $d \sin \theta$:

$$\delta = \frac{2\pi}{\lambda} d \sin \theta \tag{33-6}$$

The distance y_m measured along the screen from the central point to the mth bright fringe (see Figure 33-8b) is related to the angle θ by

$$\tan \theta = \frac{y_m}{L}$$

where L is the distance from the slits to the screen. For small θ, we have

$$\sin \theta \approx \tan \theta = \frac{y_m}{L}$$

so $d \sin \theta$ is given approximately by

$$d \sin \theta \approx d\frac{y_m}{L}$$

Substituting this into Equation 33-4, we obtain

$$d\frac{y_m}{L} = m\lambda$$

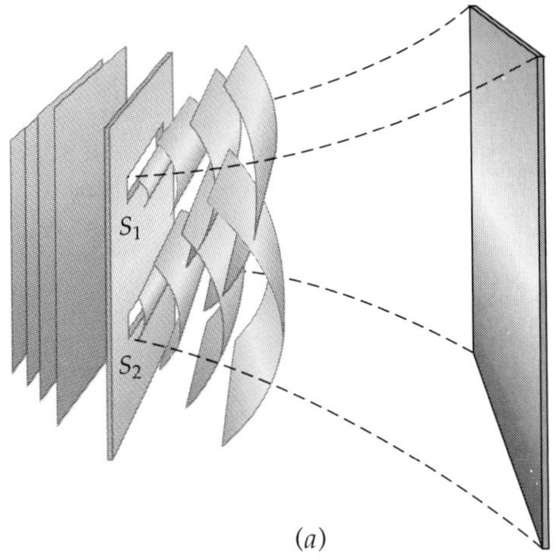

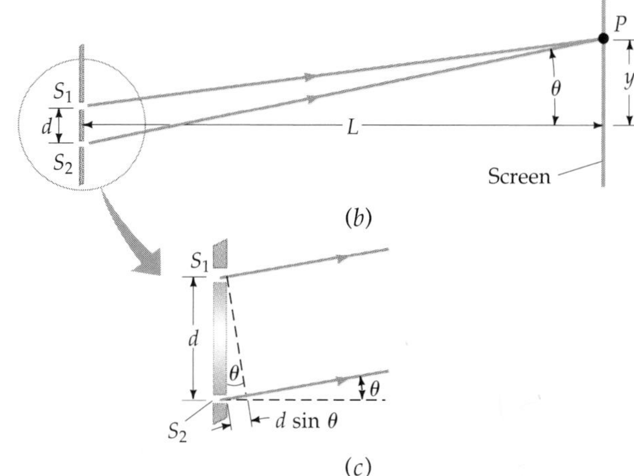

Figure 33-8 (*a*) Two slits act as coherent sources of light for the observation of interference in Young's experiment. Cylindrical waves from the slits overlap and produce an interference pattern on a screen far away. (*b*) Geometry for relating the distance y measured along the screen to L and θ. (*c*) Because the screen is very far away compared with the slit separation, the rays from the slits to a point on the screen are approximately parallel, and the path difference between the two rays is $d \sin \theta$.

Thus, for small angles (which is nearly always the case), the distance measured along the screen to the mth bright fringe is given by

$$y_m = m\frac{\lambda L}{d} \qquad \text{33-7}$$

Note from this result that the fringes are equally spaced on the screen, with the distance between two successive bright fringes given by

$$\Delta y = \frac{\lambda L}{d}$$

To calculate the intensity of the light on the screen at a general point P, we need to add two harmonic wave functions that differ in phase, as we did in Chapter 14 where we discussed variations in intensity due to the interference of sound waves. The wave functions for electromagnetic waves are the electric-field vectors. Let E_1 be the electric field at some point P on the screen due to the waves from slit 1, and let E_2 be the electric field at that point due to waves from slit 2. Since the angles of interest are small, we can assume that these fields are parallel and consider only their magnitudes. Both electric fields oscillate with the same frequency (since they result from a single source that illuminates both slits) and they have the same amplitude. (The path difference is only of the order of a few wavelengths of light at most.) They have a phase difference δ given by Equation 33-6. If we represent these wave functions by

$$E_1 = A_0 \sin \omega t \qquad \text{and} \qquad E_2 = A_0 \sin (\omega t + \delta)$$

the resultant wave function is

$$E = E_1 + E_2 = A_0 \sin \omega t + A_0 \sin (\omega t + \delta) \qquad \text{33-8}$$

We use the following trigonometric identity for the two sine functions:

$$\sin \alpha + \sin \beta = 2 \cos \tfrac{1}{2}(\alpha - \beta) \sin \tfrac{1}{2}(\alpha + \beta) \qquad \text{33-9}$$

Equation 33-8 then becomes

$$E = 2A_0 \cos \tfrac{1}{2}\delta \sin (\omega t + \tfrac{1}{2}\delta) \qquad \text{33-10}$$

The amplitude of the resultant wave is thus $2A_0 \cos \frac{1}{2}\delta$. It has its maximum value of $2A_0$ when the waves are in phase ($\delta = 0$ or an integer times 2π) and is zero when they are 180° out of phase ($\delta = \pi$ or an odd integer times π).

Since the intensity is proportional to the square of the amplitude, the intensity at any point P is

$$I = 4I_0 \cos^2 \tfrac{1}{2}\delta \qquad 33\text{-}11$$

where I_0 is the intensity of the light on the screen from either slit separately. The phase angle δ is related to the position on the screen by Equation 33-6. For small angles, $d \sin \theta \approx dy_m/L$ and the phase angle is related to y_m by

$$\delta = \frac{2\pi}{\lambda} d \sin \theta \approx \frac{2\pi}{\lambda}\frac{y_m d}{L} \qquad 33\text{-}12$$

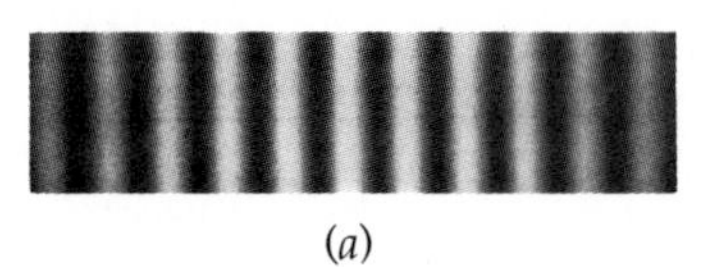

(a)

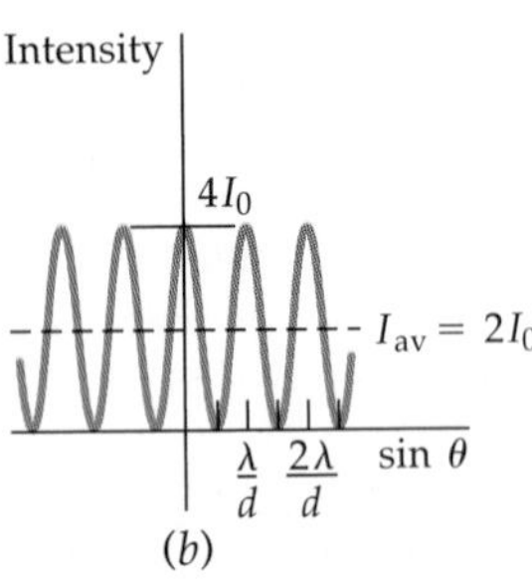

(b)

Figure 33-9 (a) The interference pattern observed on a screen far away from the two slits of Figure 33-8. (b) Plot of intensity versus $\sin \theta$. The maximum intensity is $4I_0$, where I_0 is the intensity due to each slit separately. The average intensity (dashed line) is $2I_0$. For small θ, this is also a plot of intensity versus the distance y measured along the screen because $y = L \tan \theta \approx L \sin \theta$.

Figure 33-9*a* shows the intensity pattern as seen on a screen. A graph of the intensity as a function of $\sin \theta$ is shown in Figure 33-9*b*. For small θ, this is equivalent to a plot of intensity versus y since $y \approx L \sin \theta$. The intensity I_0 is that from each slit separately. The dashed line shows the average intensity $2I_0$, which is the result of averaging over many interference maxima and minima. This is the intensity that would arise from the two sources if they acted independently without interference. In other words, it is the intensity we would observe if the sources were incoherent because there would then be an additional phase difference between them that would fluctuate randomly so that only the average intensity would be observed.

Example 33-4

Two narrow slits separated by 1.5 mm are illuminated by sodium light of wavelength 589 nm. Interference fringes are observed on a screen 3 m away. Find the spacing of the fringes on the screen.

The distance y_m measured along the screen to the mth bright fringe is given by Equation 33-7, with $L = 3$ m, $d = 1.5$ mm, and $\lambda = 589$ nm. The spacing of the fringes is this distance divided by the number of fringes, or y_m/m. Solving Equation 33-7 for y_m/m and substituting in the given values, we obtain

$$\frac{y_m}{m} = \lambda\frac{L}{d} = \frac{(589 \times 10^{-9}\text{ m})(3\text{ m})}{0.0015\text{ m}} = 1.18 \times 10^{-3}\text{ m} = 1.18\text{ mm}$$

The fringes are thus 1.18 mm apart.

Figure 33-10 shows another method of producing the two-slit interference pattern, an arrangement known as **Lloyd's mirror.** A single slit is placed at a distance $\frac{1}{2}d$ above the plane of a mirror. Light striking the screen directly from the source interferes with that reflected from the mirror. The reflected light can be considered to come from the virtual image of the slit formed by the mirror. Because of the 180° change in phase upon reflection at the mirror, the interference pattern is that of two coherent line sources that differ in phase by 180°. The pattern is the same as that shown in Figure 33-9 for two slits except that the maxima and minima are interchanged. The central fringe just above the mirror at a point equidistant from the sources is

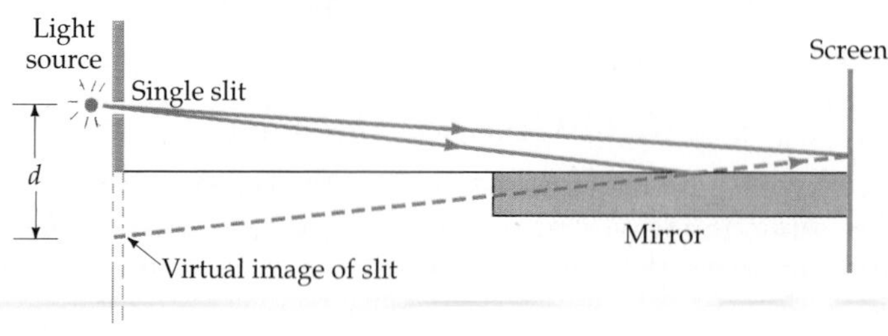

Figure 33-10 Lloyd's mirror for producing a two-slit interference pattern. The two sources (the slit and its image) are coherent and are 180° out of phase. The central interference band at the point equidistant from the sources is dark.

dark. Constructive interference occurs at points for which the path difference is a half wavelength or any odd number of half wavelengths. At these points, the 180° phase difference due to the path difference combines with the 180° phase difference of the sources to produce constructive interference.

Exercise

A point source of light ($\lambda = 589$ nm) is placed 0.4 mm above a mirror. Interference fringes are observed on a screen 6 m away. Find the spacing of the fringes on the screen. (Answer: 4.42 mm)

Question

4. When destructive interference occurs, what happens to the energy in the light waves?

33-5 The Addition of Harmonic Waves Using Phasors

To calculate the interference pattern produced by three, four, or more coherent light sources and to calculate the diffraction pattern of a single slit, we need to combine several harmonic waves of the same frequency that differ in phase. A simple geometric interpretation of harmonic wave functions leads to a method of adding harmonic waves of the same frequency by geometric construction. This method lets us find the sum of two or more harmonic waves geometrically without having to remember the trigonometric identity given in Equation 33-9. It is useful even if the amplitudes of the waves are different. The method is based on the fact that the y (or x) component of the resultant of two vectors equals the sum of the y (or x) components of the vectors.

Let

$$E_1 = A_1 \sin \omega t \qquad \text{and} \qquad E_2 = A_2 \sin (\omega t + \delta)$$

be the wave functions of the two waves at some point. (We have chosen our time t so that $E_1 = 0$ at $t = 0$.) We can simplify our notation by writing α for the quantity ωt. Our problem is then to find the sum

$$E_1 + E_2 = A_1 \sin \alpha + A_2 \sin (\alpha + \delta)$$

Consider a vector in the xy plane of magnitude A_1 that makes an angle α with the x axis (Figure 33-11). The y component of this vector is $A_1 \sin \alpha$, which is the wave function E_1. As the time varies, this vector rotates in the xy plane with angular frequency ω. Such a vector is called a **phasor.** We encountered phasors in our study of ac circuits in Section 28-3. The wave function $E_2 = A_2 \sin (\alpha + \delta)$ is the y component of a phasor of magnitude A_2 that makes an angle $\alpha + \delta$ with the x axis. By the laws of vector addition, the sum of these components equals the y component of the resultant vector $\mathbf{A}'$, as shown in Figure 33-11. The y component of the resultant vector, $A' \sin (\alpha + \delta')$, is a harmonic wave function that is the sum of the two original wave functions:

$$A_1 \sin \alpha + A_2 \sin (\alpha + \delta) = A' \sin (\alpha + \delta') \qquad 33\text{-}13$$

where A' (the amplitude of the resultant wave) and δ' (the phase of the resultant wave relative to the first wave) are found by adding the phasors representing the waves as in Figure 33-11. As time varies, α varies. The phasors representing the two wave functions and the resultant phasor representing the resultant wave function rotate in space, but their relative positions do not change because they all rotate with the same angular velocity ω.

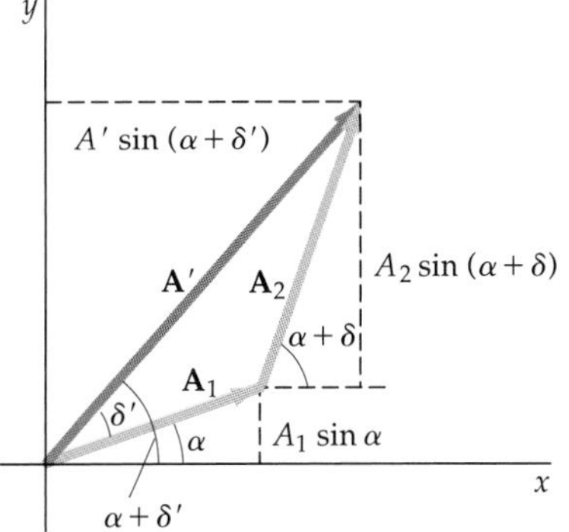

Figure 33-11 The wave function $A_1 \sin \alpha$ is the y component of the vector $\mathbf{A}_1$, which makes an angle α with the x axis. The wave function $A_2 \sin (\alpha + \delta)$ is the y component of the vector $\mathbf{A}_2$, which makes an angle $\alpha + \delta$ with the x axis. The sum of these wave functions is $A' \sin (\alpha + \delta')$, which is the y component of the resultant vector $\mathbf{A}' = \mathbf{A}_1 + \mathbf{A}_2$.

Example 33-5

Use the phasor method of addition to derive Equation 33-10 for the superposition of two waves of the same amplitude.

Figure 33-12 shows the phasors representing two waves of equal amplitude A_0 and the resultant wave of amplitude A'. These three phasors form an isosceles triangle in which the two equal angles are δ'. Since the sum of these angles equals the exterior angle δ, we have

$$\delta' = \tfrac{1}{2}\delta$$

The amplitude A' can be found from the right triangles formed by bisecting the resultant phasor, as shown in Figure 33-12*b*. From these triangles, we have

$$\cos \tfrac{1}{2}\delta = \frac{\tfrac{1}{2}A'}{A_0}$$

Therefore, the amplitude is given by $A' = 2A_0 \cos \frac{1}{2}\delta$, and the resultant wave is

$$A' \sin (\alpha + \delta') = 2A_0 \cos \tfrac{1}{2}\delta \sin (\alpha + \tfrac{1}{2}\delta)$$

which is in agreement with Equation 33-10 when $\alpha = \omega t$.

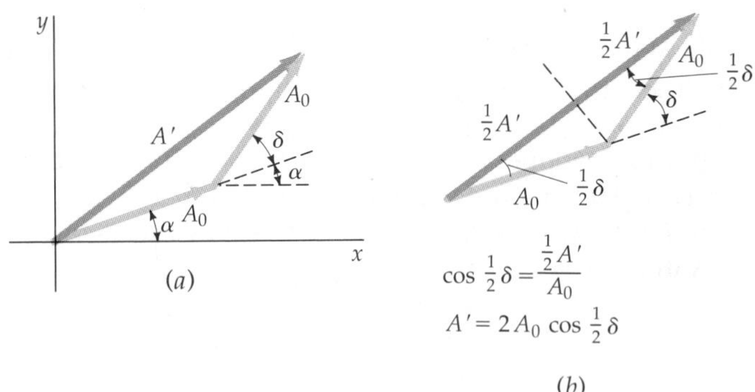

Figure 33-12 The phasor addition of two waves having equal amplitudes A_0 and a phase difference of δ. (*a*) The phasors at a particular time at which $\alpha = \omega t$. (*b*) Geometric construction for finding the amplitude A' of the resultant wave.

Example 33-6

Find the resultant of the two waves

$$E_1 = 4 \sin (\omega t) \qquad \text{and} \qquad E_2 = 3 \sin (\omega t + 90^\circ)$$

Figure 33-13 shows the phasor diagram for this addition. The phasors make an angle of 90° with each other. The resultant of these two phasors has a magnitude of 5 and makes an angle of 37° with the first phasor, as shown in the figure. The sum of these two waves is therefore

$$E_1 + E_2 = 5 \sin (\omega t + 37^\circ)$$

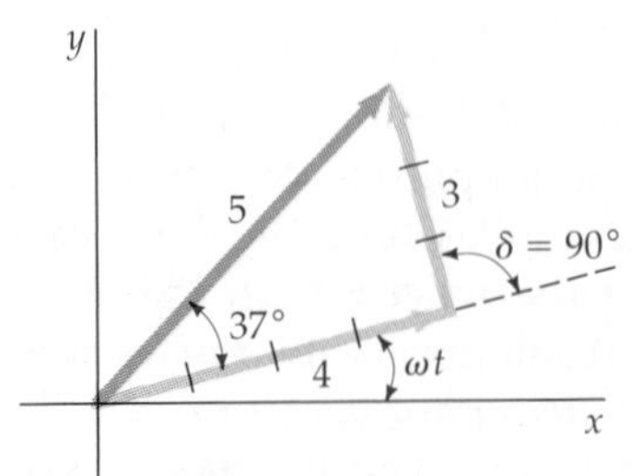

Figure 33-13. Phasor diagram for the addition of the two waves in Example 33-6.

33-6 Interference Pattern of Three or More Equally Spaced Sources

If we have three or more sources that are equally spaced and in phase with each other, the intensity pattern on a screen far away is similar to that due to two sources, but there are important differences. The positions on the screen of the intensity maxima are the same no matter how many sources we have, but these maxima have much greater intensities and are much sharper if there are many sources. We can calculate the intensity pattern for the interference of three or more equally spaced sources using the phasor method of adding harmonic waves discussed in the previous section. We will be most interested in the points of perfectly constructive interference and perfectly destructive interference, that is, in the interference maxima and minima.

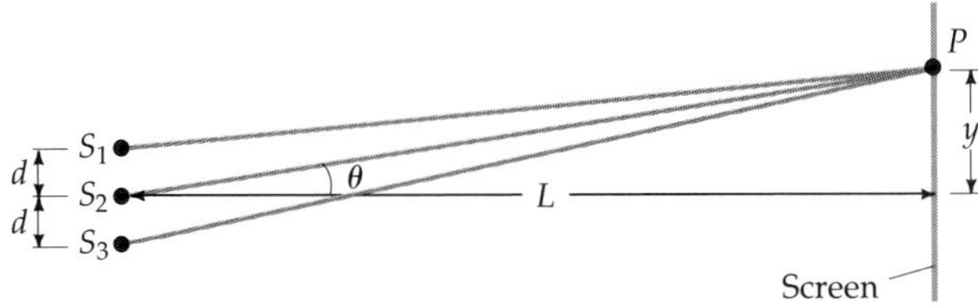

Figure 33-14 Geometry for calculating the intensity pattern far from three equally spaced sources that are in phase.

Figure 33-14 shows the situation for the case of three sources. The geometry is the same as for two sources. At a great distance from the sources, the rays from the sources to a point P on the screen are approximately parallel. The path difference between the first and second source is then $d \sin \theta$, as before, and that between the first and third source is $2d \sin \theta$. The wave at point P is the sum of three waves. Let $\alpha = \omega t$ be the phase of the first wave at point P. We thus have the problem of adding three waves of the form

$$E_1 = A_0 \sin \alpha$$

$$E_2 = A_0 \sin (\alpha + \delta) \qquad 33\text{-}14$$

$$E_3 = A_0 \sin (\alpha + 2\delta)$$

where

$$\delta = \frac{2\pi}{\lambda} d \sin \theta \approx \frac{2\pi}{\lambda} \frac{yd}{L} \qquad 33\text{-}15$$

as in the two-slit problem.

It is easiest to analyze the resulting pattern in terms of the phase angle δ between the first and second sources or between the second and third sources instead of directly in terms of the space angle θ. If we know the resultant amplitude due to the three waves at some point P corresponding to a particular phase angle δ, we can relate this phase angle to the space angle θ by using Equation 33-15.

At $\theta = 0$, the phase angle δ is zero; that is, all the waves are in phase. The amplitude of the resultant wave is 3 times that of each individual wave. Since the intensity is proportional to the square of the amplitude, the intensity at $\delta = 0$ is 9 times that due to each source acting separately. As the angle δ increases from $\theta = 0$, the phase angle δ increases and the intensity decreases. The position $\theta = 0$ is thus a position of maximum intensity.

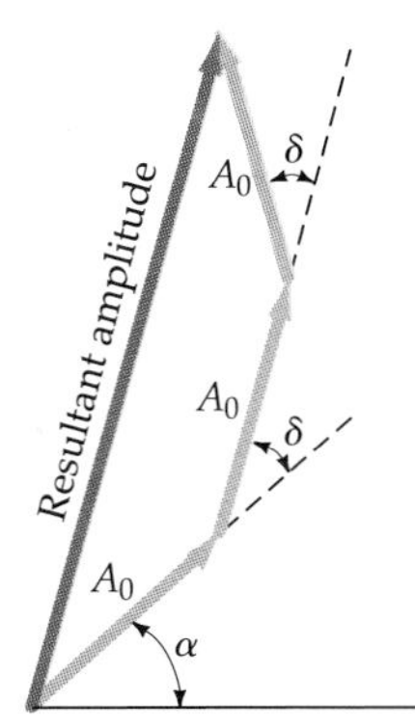

Figure 33-15 Phasor diagram for determining the resultant amplitude due to three waves, each of amplitude A_0, that have phase differences of δ and 2δ due to path differences of $d \sin \theta$ and $2d \sin \theta$. The angle $\alpha = \omega t$ varies with time but does not affect the calculation of the resultant amplitude.

Figure 33-15 shows the phasor addition of three waves for a phase angle δ of about $30° = \pi/6$ rad. (This corresponds to a point P on the screen for which θ is given by $\sin \theta = \lambda\delta/2\pi d = \lambda/12d$.) The resultant amplitude is considerably less than 3 times that of each source. As the phase angle δ

increases, the resultant amplitude decreases until the amplitude is zero at $\delta = 120°$. For this phase difference, the three phasors form an equilateral triangle (Figure 33-16). This first interference minimum for three sources occurs at a smaller phase angle (and therefore at a smaller space angle θ) than it does for only two sources (for which the first minimum occurs at $\delta = 180°$). As δ increases from 120°, the resultant amplitude increases, reaching a **secondary maximum** near $\delta = 180°$. At the phase angle $\delta = 180°$, the amplitude is the same as that from a single source since the waves from the first two sources cancel each other, leaving only the third. The intensity of the secondary maximum is one-ninth that of the maximum at $\theta = 0$. As δ increases beyond 180°, the amplitude again decreases and is zero at $\delta = 180° + 60° = 240°$. For δ greater than 240°, the amplitude increases and is again 3 times that of each source when $\delta = 360°$. This phase angle corresponds to a path difference of 1 wavelength for the waves from the first two sources and 2 wavelengths for the waves from the first and third sources. Hence, the three waves are in phase at this point. The largest maxima, called the **principal maxima,** are at the same positions as for just two sources, which are those points corresponding to the angles θ given by

Principal interference maxima

$$d \sin \theta = m\lambda \qquad m = 0, 1, 2, \ldots \tag{33-16}$$

These maxima are stronger and narrower than those for two sources. They occur at points for which the path difference between adjacent sources is zero or an integral number of wavelengths.

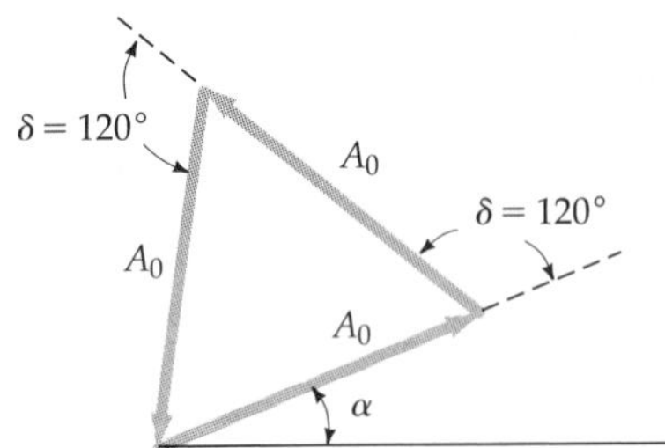

Figure 33-16 The resultant amplitude for the waves from three sources is zero when δ is 120°. This interference minimum occurs at a smaller angle θ than does the first minimum for two sources, which occurs when δ is 180°.

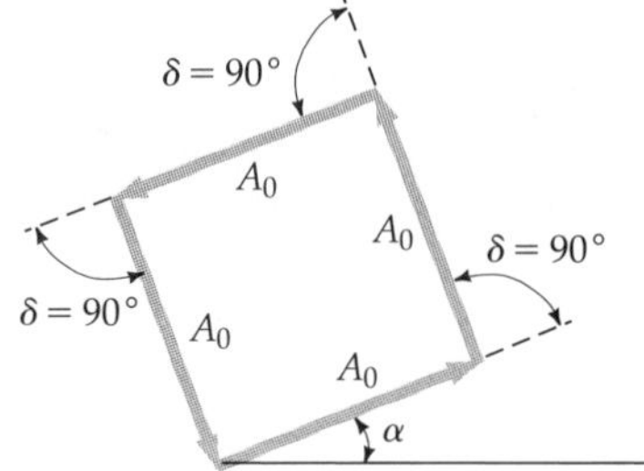

Figure 33-17 Phasor diagram for the first minimum for four equally spaced in-phase sources. The amplitude is zero when the phase difference of the waves from adjacent sources is 90°.

These results can be generalized to more than three sources. For example, if we have four equally spaced sources that are in phase, the principal interference maxima are again given by Equation 33-16, but these maxima are even more intense and narrower and there are two small secondary maxima between each pair of principal maxima. At $\theta = 0$, the intensity is 16 times that due to a single source. The first interference minimum occurs when δ is 90°, as can be seen from the phasor diagram of Figure 33-17. The first secondary maximum is near $\delta = 120°$, where the waves from three of the sources cancel, leaving only the wave from the fourth source. The intensity of the secondary maximum is approximately one-sixteenth that of the central maximum. There is another minimum at $\delta = 180°$, another secondary maximum near $\delta = 240°$, and another minimum at $\delta = 270°$ before the next principal maximum at $\delta = 360°$.

Figures 33-18*a*–*c* show the intensity patterns for two equally spaced sources, three equally spaced sources, and four equally spaced sources. In

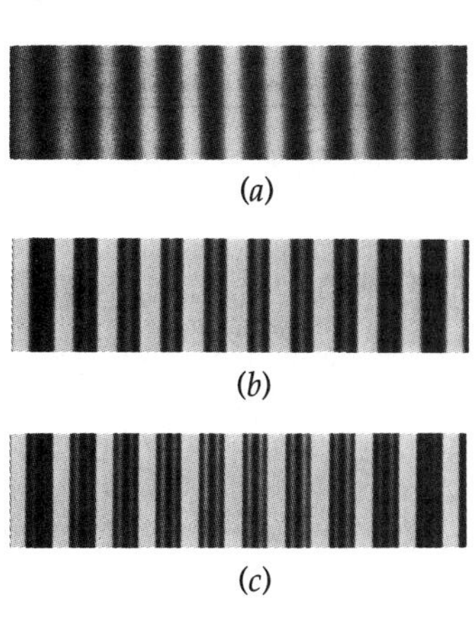

(a)

(b)

(c)

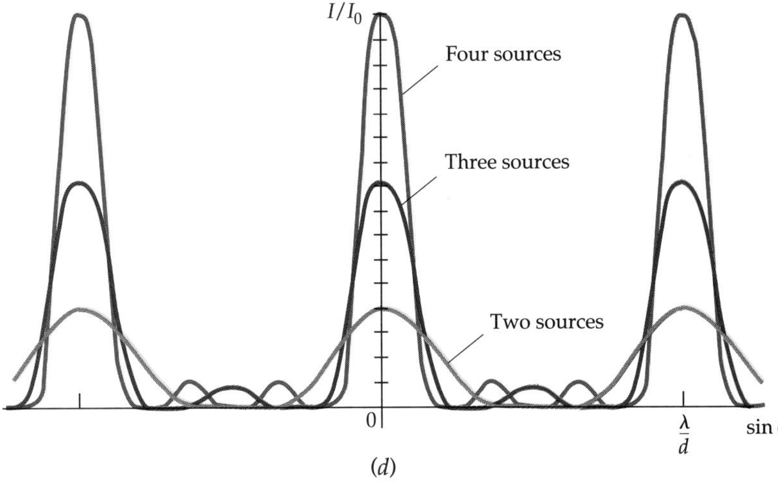

(d)

Figure 33-18 Intensity patterns for (*a*) two, (*b*) three, and (*c*) four equally spaced coherent sources. There is a secondary maximum between each pair of principal maxima for three sources, and two secondary maxima for four sources. (*d*) Plot of intensity versus sin θ for two, three, and four equally spaced coherent sources.

Figure 33-18*d*, I_0 is the intensity due to each source acting separately. For three sources, there is a very small secondary maximum between each pair of principal maxima, and the principal maxima are sharper and more intense than those due to just two sources. For four sources, there are two small secondary maxima between each pair of principal maxima, and the principal maxima are even more narrow and intense.

From this discussion, we can see that as we increase the number of sources, the intensity becomes more and more concentrated in the principal maxima given by Equation 33-16, and these maxima become narrower. For N sources, the intensity of the principal maxima is N^2 times that due to a single source. The first minimum occurs at a phase angle of $\delta = 360°/N$, for which the N phasors form a closed polygon of N sides. There are $N - 2$ secondary maxima between each pair of principal maxima. These secondary maxima are very weak compared with the principal maxima. As the number of sources is increased, the principal maxima become sharper and more intense, and the intensities of the secondary maxima become negligible compared to those of the principal maxima.

Example 33-7

Four equally spaced coherent light sources with a wavelength of 500 nm are separated by a distance $d = 0.1$ mm. The interference pattern is viewed on a screen at a distance of 1.4 m. Find the positions of the principal interference maxima and compare their width with that for just two sources with the same spacing.

According to Equation 33-16, the maxima are at angles given by

$$\sin\theta = m\frac{\lambda}{d} = m\frac{5 \times 10^{-7}\text{ m}}{1 \times 10^{-4}\text{ m}} = (5 \times 10^{-3})m$$

where $m = 0, 1, 2, 3, \ldots$. Since θ is small, we can approximate $\sin\theta \approx \tan\theta \approx \theta$. The distance y measured along the screen from the central maximum is related to θ by

$$y = L\tan\theta \approx L\theta$$

The position of the mth principal maximum is thus

$$y_m = L\theta_m = m(1.4\text{ m})(5 \times 10^{-3}) = m(7.0\text{ mm})$$

The principal maxima are thus separated by 7.0 mm on the screen.

The first minimum occurs when the phase difference between two adjacent sources is $\delta = 90° = \pi/2$. This corresponds to a path difference of $\lambda/4$. The angle θ of this minimum is given by $d \sin\theta = \lambda/4$, or

$$\sin\theta = \frac{\lambda}{4d} = \frac{5\times10^{-7}\text{ m}}{4\times10^{-4}\text{ m}} = 1.25\times10^{-3}$$

The position y of this minimum is

$$y = L\theta = (1.4\text{ m})(1.25\times10^{-3}) = 1.75\text{ mm}$$

The distance between the first minima on either side of the maximum at $\theta = 0$ is $2y = 3.5$ mm. This is a measure of the width of the principal maxima. If we had only two sources with the same spacing, the principal maxima would be at the same points but the first minimum would be at an angle θ corresponding to a path difference of $\lambda/2$. The width of these maxima would therefore be twice as great as when there are four sources, as is shown in Figure 33-18.

Question

5. How many secondary maxima would there be between the main maxima in the interference pattern produced by five equally spaced sources? Why would they be difficult to see?

33-7 Diffraction Pattern of a Single Slit

Qualitative

In our discussion of the interference patterns produced by two or more slits, we assumed the slits were very narrow so that we could consider them to be line sources of cylindrical waves, which in our two-dimensional diagrams are point sources of circular waves. We could therefore assume that the intensity due to one slit acting alone was the same (I_0) at any point P on the screen independent of the angle θ made between the ray to point P and the normal line between the slit and the screen. When the slit is not narrow, the intensity on a screen far away is not independent of angle but decreases as the angle increases. Let us consider a slit of width a. Figure 33-19 shows

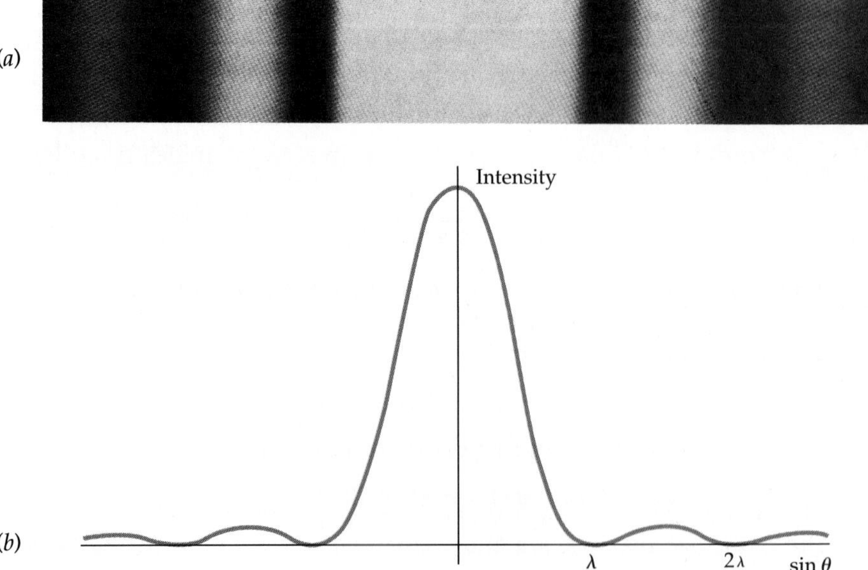

Figure 33-19 (*a*) Diffraction pattern of a single slit as observed on a screen far away. (*b*) Plot of intensity versus sin θ for the pattern in (*a*).

the intensity pattern on a screen far away from the slit of width a as a function of $\sin\theta$. We can see that the intensity is maximum in the forward direction ($\sin\theta = 0$) and decreases to zero at an angle that depends on the slit width a and the wavelength λ. Most of the light intensity is concentrated in the broad **central diffraction maximum,** though there are minor secondary maxima bands on either side of the central maximum. The first zeroes in the intensity occur at angles given by

$$\sin\theta = \lambda/a \qquad 33\text{-}17$$

Note that for a given wavelength λ, the width of the central maximum varies inversely with the width of the slit. That is, if we *increase* the slit width a, the angle θ at which the intensity first becomes zero *decreases,* giving a more narrow central diffraction maximum. Conversely, if we *decrease* the slit width, the angle of the first zero *increases,* giving a wider central diffraction maximum. When a is very small, there are no points of zero intensity in the pattern, and the slit acts as a line source (a point source in two dimensions), radiating light energy essentially equally in all directions.

We can write Equation 33-17 slightly differently. Multiplying both sides by a, we obtain

$$a\sin\theta = \lambda \qquad 33\text{-}18$$

The quantity $a\sin\theta$ is the path difference between a light ray leaving the top of the slit and one leaving the bottom of the slit. We see that the first diffraction minimum occurs when these two rays are in phase, that is, when their path difference equals 1 wavelength. We can understand this result by considering each point on a wavefront to be a point source of light in accordance with Huygens' principle. In Figure 33-20, we have placed a line of dots on the wavefront at the slit to represent these point sources schematically. Suppose, for example, that we have 100 such dots and that we look at an angle θ for which $a\sin\theta = \lambda$, that is, the angle for which the waves from the top and bottom of the slit are in phase. Let us consider the slit to be divided into two regions, with the first 50 sources in the first, upper region and sources 51 through 100 in the second, lower region. When the path difference between the top and bottom of the slit equals one wavelength, the path difference between source 1 (the first source in the upper region) and source 51 (the first source in the lower region) is $\frac{1}{2}$ wavelength. The waves from these two sources will be out of phase by 180° and will thus cancel. Similarly, waves from the second source in each region (source 2 and source 52) will cancel. Continuing this argument, we can see that the waves from each pair of sources separated by $a/2$ will cancel. Thus, there will be no light energy at this angle. We can extend this argument to the second and third minima in the diffraction pattern of Figure 33-19. At an angle such that $a\sin\theta = 2\lambda$, we can divide the slit into four regions, two for the top half and two for the bottom half. Using this same argument, the light intensity from the top half is zero because of the cancellation of pairs of sources, and, similarly, that from the bottom half is zero. The general expression for the points of zero intensity in the diffraction pattern of a single slit is thus

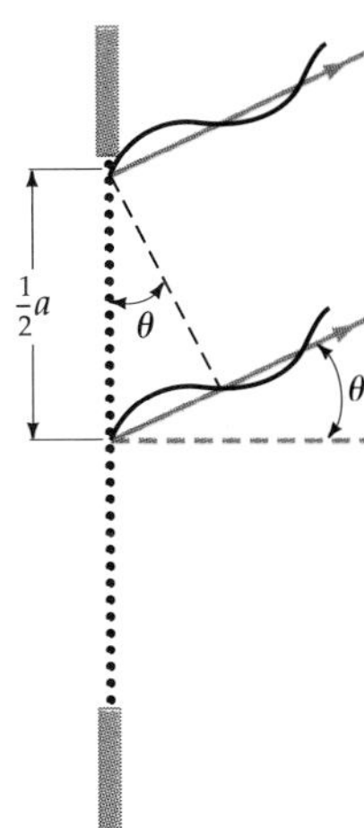

Figure 33-20 A single slit is represented by a large number of point sources of equal amplitude. At the first diffraction minimum of a single slit, the waves from the source near the top and those from the source just below the middle of the slit are 180° out of phase and cancel, as do all other pairs of sources.

$$a\sin\theta = m\lambda \qquad m = 1, 2, 3, \ldots \qquad 33\text{-}19$$

Points of zero intensity for a single-slit diffraction pattern

Usually, we are just interested in the first occurrence of a minimum in the light intensity because nearly all of the light energy is contained in the central diffraction maximum.

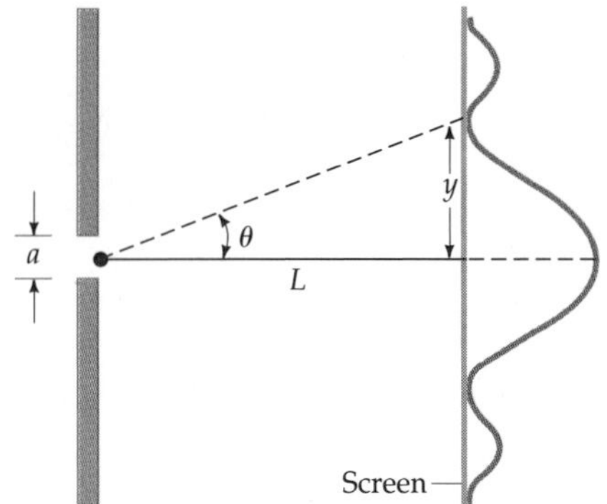

Figure 33-21 The distance y measured along the screen from the central maximum to the first diffraction minimum is related to the angle θ by $\tan\theta = y/L$, where L is the distance to the screen. Since the angle is very small, $\tan\theta \approx \sin\theta$. Then $y = L\tan\theta \approx L\sin\theta = L\lambda/a$.

In Figure 33-21, the distance y from the central maximum to the first diffraction minimum is related to the angle θ and the distance L from the slit to the screen by

$$\tan\theta = \frac{y}{L}$$

Since this angle is very small, $\tan\theta \approx \sin\theta$. Then, according to Equation 33-17, we have

$$\sin\theta = \frac{\lambda}{a} \approx \frac{y}{L}$$

or

$$y = \frac{L\lambda}{a} \qquad 33\text{-}20$$

Example 33-8

In a lecture demonstration of single-slit diffraction, a laser beam of wavelength 700 nm passes through a vertical slit 0.2 mm wide and hits a screen 6 m away. Find the width of the central diffraction maximum on the screen, that is, the distance between the first minimum on the left and the first minimum on the right of the central maximum.

The width of the central diffraction maximum in Figure 33-21 is $2y$. From Equation 33-20, we have

$$y = \frac{L\lambda}{a} = \frac{(6\text{ m})(700 \times 10^{-9}\text{ m})}{0.0002\text{ m}} = 2.1 \times 10^{-2}\text{ m} = 2.1\text{ cm}$$

The width of the central maximum is thus $2y = 4.2$ cm.

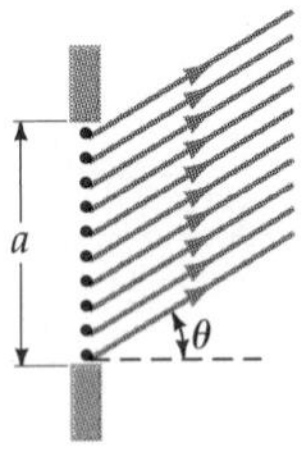

Figure 33-22 Diagram for calculating the diffraction pattern far away from a narrow slit. The slit width a is assumed to contain a large number of in-phase point sources separated by a distance d. The rays from these sources to a point very far away are approximately parallel. The path difference for the waves from adjacent sources is $d\sin\theta$.

Quantitative

We will now use the phasor method of addition of harmonic waves to calculate the intensity pattern shown in Figure 33-19. We assume that the slit of width a is divided into N equal intervals and that there is a point source of waves at the midpoint of each interval. If d is the distance between two adjacent sources and a is the width of the opening, we have $d = a/N$. Since the screen on which we are calculating the intensity is very far from the sources, the rays from the sources to a point P on the screen are approximately parallel. The path difference between any two adjacent sources is $d\sin\theta$ (Figure 33-22), and the phase difference is

$$\delta = \frac{2\pi}{\lambda} d\sin\theta$$

If A_0 is the amplitude due to a single source, the amplitude at the central maximum, where $\theta = 0$ and all the waves are in phase, is $A_{max} = NA_0$ (Figure 33-23). We can find the amplitude at some other point at an angle θ by using the phasor method for the addition of harmonic waves. As in the addition of two, three, or four waves, the intensity is zero at any point

Figure 33-23 A single slit is represented by N sources, each of amplitude A_0. At the central maximum point at $\theta = 0$, the waves from the sources add in phase, giving a resultant amplitude $A_{max} = NA_0$.

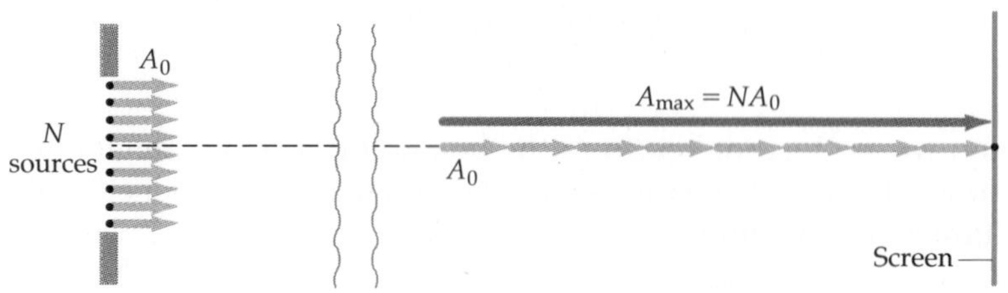

where the phasors representing the waves form a closed polygon. In this case the polygon has N sides (Figure 33-24). At the first minimum, the wave from the first source just below the top of the opening and that from the source just below the middle of the opening are 180° out of phase. In this case, the waves from the source near the top of the opening differ from those from the bottom by nearly 360°. (The phase difference is, in fact, $360° - (360°)/N$.) Thus, if the number of sources is very large, we get complete cancellation when the waves from the first and last sources are out of phase by 360°, corresponding to a path difference of 1 wavelength in agreement with Equation 33-18.

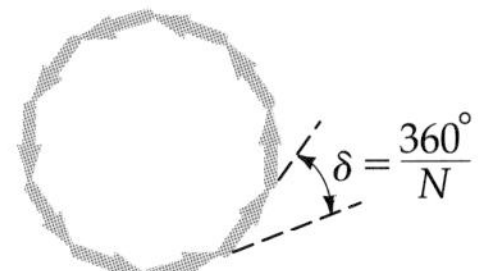

Figure 33-24 Phasor diagram for calculating the first minimum in the single-slit diffraction pattern. When the waves from the N sources completely cancel, the N phasors form a closed polygon. The phase difference between the waves from adjacent sources is then $\delta = 360°/N$. When N is very large, the waves from the first and last sources are approximately in phase.

We will now calculate the amplitude at a general point at which the waves from two adjacent sources differ in phase by δ. Figure 33-25 shows the phasor diagram for the addition of N waves where the subsequent waves differ in phase from the first wave by $\delta, 2\delta, \ldots, (N-1)\delta$. When N is very large and δ is very small, the phasor diagram approximates the arc of a circle. The resultant amplitude A is the length of the chord of this arc. We will calculate this resultant amplitude in terms of the phase difference ϕ between the first wave and the last wave. From Figure 33-25, we have

$$\sin \tfrac{1}{2}\phi = \frac{A/2}{r}$$

or

$$A = 2r \sin \tfrac{1}{2}\phi \qquad \text{33-21}$$

where r is the radius of the arc. Since the length of the arc is $A_{max} = NA_0$ and the angle subtended is ϕ, we have

$$\phi = \frac{A_{max}}{r} \qquad \text{33-22}$$

or

$$r = \frac{A_{max}}{\phi}$$

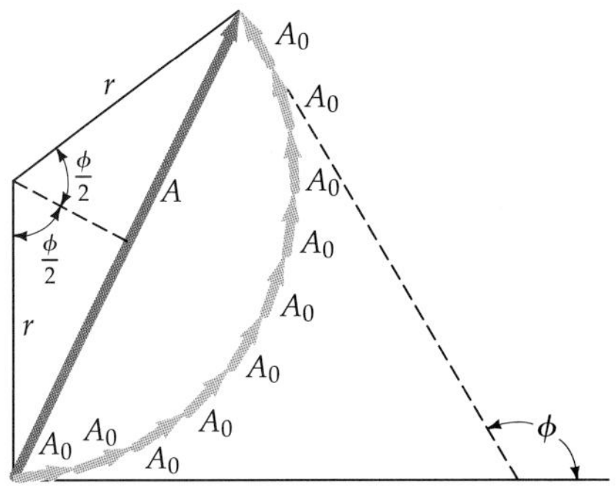

Figure 33-25 Phasor diagram for calculating the resultant amplitude due to the waves from N sources in terms of the phase difference ϕ between the wave from the first source just below the top of the slit and that from the last source just above the bottom of the slit. When N is very large, the resultant amplitude A is the chord of a circular arc of length $NA_0 = A_{max}$.

Substituting this into Equation 33-21 gives

$$A = \frac{2A_{max}}{\phi} \sin \tfrac{1}{2}\phi = A_{max}\frac{\sin \frac{1}{2}\phi}{\frac{1}{2}\phi}$$

Since the amplitude at the center of the central maximum ($\theta = 0$) is A_{max}, the ratio of the intensity at any other point to that at the center of the central maximum is

$$\frac{I}{I_0} = \frac{A^2}{A_{max}^2} = \left(\frac{\sin \frac{1}{2}\phi}{\frac{1}{2}\phi}\right)^2$$

or

$$I = I_0\left(\frac{\sin \frac{1}{2}\phi}{\frac{1}{2}\phi}\right)^2 \qquad \text{33-23}$$

Intensity for a single-slit diffraction pattern

The phase difference ϕ between the first and last waves is $2\pi/\lambda$ times the path difference $a \sin \theta$ between the top and bottom of the opening:

$$\phi = \frac{2\pi}{\lambda} a \sin \theta \qquad \text{33-24}$$

Equations 33-23 and 33-24 describe the intensity pattern shown in Figure 33-19. The first minimum occurs at $a \sin \theta = \lambda$, the point where the waves from the top and bottom of the opening have a path difference of λ and are in phase. The second minimum occurs at $a \sin \theta = 2\lambda$, where the waves from the top and bottom of the opening have a path difference of 2λ.

There is a secondary maximum approximately midway between the first and second minima at $a \sin \theta \approx \frac{3}{2}\lambda$. Figure 33-26 shows the phasor diagram for determining the approximate intensity of this secondary maximum. The phase difference between the first and last waves is approximately 360° + 180°. The phasors thus complete $1\frac{1}{2}$ circles. The resultant amplitude is the diameter of a circle with a circumference that is two-thirds the total length A_{max}. If $C = \frac{2}{3}A_{max}$ is the circumference, the diameter A is

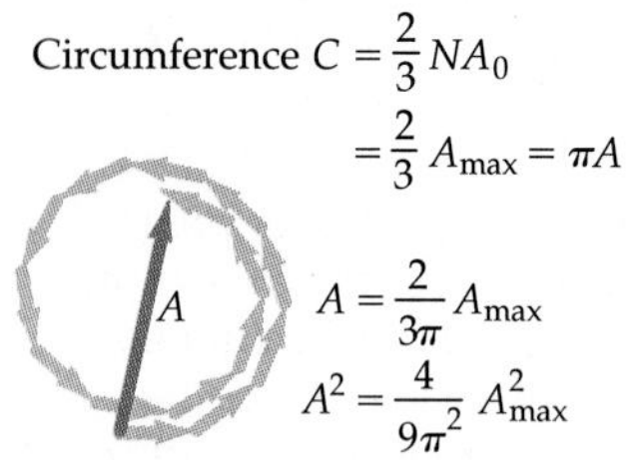

Circumference $C = \frac{2}{3}NA_0 = \frac{2}{3}A_{max} = \pi A$

$A = \frac{2}{3\pi}A_{max}$

$A^2 = \frac{4}{9\pi^2}A^2_{max}$

Figure 33-26 Phasor diagram for calculating the approximate amplitude of the first secondary maximum of the single-slit diffraction pattern. This secondary maximum occurs near the midpoint between the first and second minima when the N phasors complete $1\frac{1}{2}$ circles.

$$A = \frac{C}{\pi} = \frac{\frac{2}{3}A_{max}}{\pi} = \frac{2}{3\pi}A_{max}$$

and

$$A^2 = \frac{4}{9\pi^2}A^2_{max}$$

The intensity at this point is

$$I = \frac{4}{9\pi^2}I_0 = \frac{1}{22.2}I_0 \qquad 33\text{-}25$$

Question

6. As the width of a slit producing a single-slit diffraction pattern is slowly and steadily reduced, how will the diffraction pattern change?

33-8 Interference–Diffraction Pattern of Two Slits

When there are two or more slits, the intensity pattern on a screen far away is a combination of the single-slit diffraction pattern and the multiple-slit interference pattern we have studied. Figure 33-27 shows the intensity pattern on a screen far from two slits whose separation d is 10 times the width a of each slit. The pattern is the same as the two-slit pattern with very narrow slits (Figure 33-9) except that it is modulated by the single-slit diffraction pattern; that is, the intensity due to each slit separately is now not constant but decreases with angle as shown in Figure 33-27*b*. The intensity can be calculated from the two-slit pattern (Equation 33-11) with the intensity of each slit (I_0 in that equation) replaced by the diffraction-pattern intensity due to each slit, I, given by Equation 33-23. The intensity for the two-slit interference–diffraction pattern is thus

Interference–diffraction intensity for two slits

$$I = 4I_0\left(\frac{\sin \frac{1}{2}\phi}{\frac{1}{2}\phi}\right)^2 \cos^2 \tfrac{1}{2}\delta \qquad 33\text{-}26$$

where ϕ is the difference in phase between rays from the top and bottom of each slit, which is related to the width of each slit by

$$\phi = \frac{2\pi}{\lambda} a \sin \theta$$

and δ is the difference in phase between rays from the centers of two adjacent slits, which is related to the slit separation by

$$\delta = \frac{2\pi}{\lambda} d \sin \theta$$

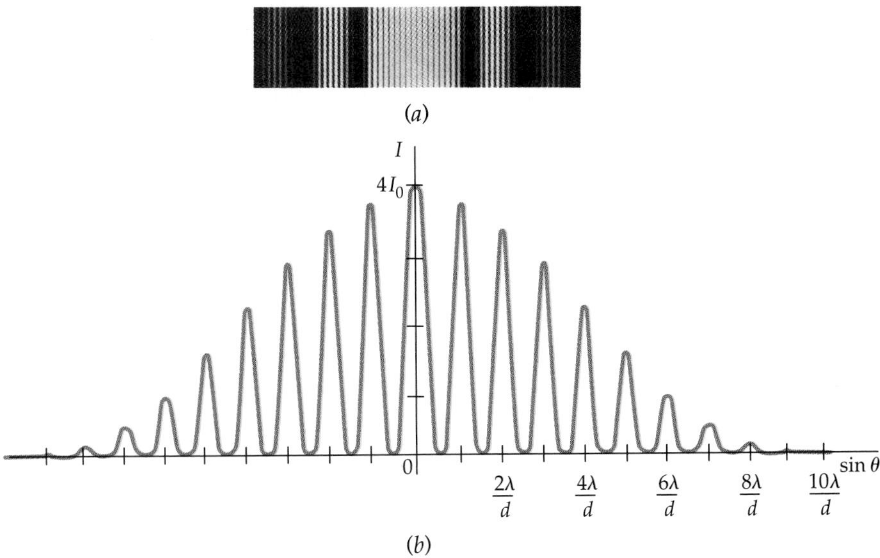

Figure 33-27 (*a*) Interference–diffraction pattern for two slits whose separation d is equal to 10 times their width a. The tenth interference maximum on either side of the central interference maximum is missing because it falls at the first diffraction minimum. (*b*) Plot of intensity versus sin θ for the central band of the pattern in (*a*).

In Equation 33-26, the intensity I_0 is the intensity at $\theta = 0$ due to one slit alone. Note that in Figure 33-27 the central diffraction maximum contains 19 interference maxima—the central interference maximum and 9 maxima on either side. The tenth interference maximum on either side of the central one is at the angle θ given by $\sin\theta = 10\lambda/d = \lambda/a$ since $d = 10a$. This coincides with the position of the first diffraction minimum, so this interference maximum is not seen. At these points, the light from the two slits would be in phase and would interfere constructively, but there is no light from either slit because the points are diffraction minima.

Example 33-9

Two slits of width $a = 0.015$ mm are separated by a distance $d = 0.06$ mm and illuminated by light of wavelength $\lambda = 650$ nm. How many bright fringes are seen in the central diffraction maximum?

The number of bright fringes in the central diffraction maximum does not depend on the wavelength of light, but only on the ratio of the slit separation to the slit width:

$$\frac{d}{a} = \frac{0.06\text{ mm}}{0.015\text{ mm}} = 4$$

The angle of the first diffraction minimum is given by

$$\sin\theta = \lambda/a$$

Because $a = d/4$, this can be written

$$\sin\theta = 4\lambda/d$$

Thus the position of the fourth interference maximum coincides with the position of the first diffraction minimum, so there will be 3 interference maxima on either side of the central interference maximum for total of 7 bright fringes in the central diffraction maximum.

Question

7. How many interference maxima will be contained in the central diffraction maximum in the diffraction–interference pattern of two slits if the separation of the slits d is 5 times their width a? How many will there be if $d = Na$ for any value of N?

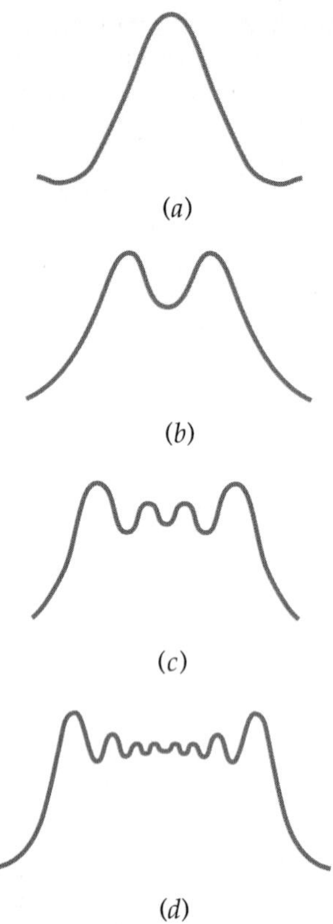

Figure 33-28 Diffraction patterns for a single slit at various screen distances. As the screen is moved closer to the slit, the Fraunhofer pattern (*a*) observed far from the slit gradually changes into the Fresnel pattern (*d*) observed near the slit.

33-9 Fraunhofer and Fresnel Diffraction

The assumptions made in deriving Equation 33-23 describing the diffraction pattern for a single slit were

1. Plane waves are incident normally on the slit. (We assumed that the amplitudes and phases of the many Huygens sources are equal.)
2. The pattern is observed at a great distance from the slit compared with the size of the openings. (We assumed that the rays from the sources to a point on the screen are approximately parallel to simplify the geometry.)

Diffraction patterns that are observed at points for which the rays from an aperture or obstacle are nearly parallel are called **Fraunhofer diffraction patterns.** The intensity pattern of Figure 33-19 is thus a Fraunhofer diffraction pattern of a single slit. Fraunhofer patterns can be observed at great distances from the obstacle or aperture so that the rays reaching any point are approximately parallel, or they can be observed using a lens to focus parallel rays on a viewing screen placed in the focal plane of the lens. If a slit is many wavelengths wide, the Fraunhofer pattern will not be observed because the angle of the first minimum will be very small. For example, if $a = 1000\lambda$, the first minimum will occur at an angle θ given by $\sin\theta = 1/1000 \approx \theta$. This small angle is not much different from the angle made by the rays from the top and bottom of the slit to the central maximum, rays which we assumed to be parallel in our derivation.

When the diffraction pattern is observed near an aperture or obstacle, it is called a **Fresnel diffraction pattern.** Because of the geometry, this pattern is much more difficult to analyze. Figure 33-28 illustrates the difference between the Fresnel and Fraunhofer patterns for a single slit.*

Figure 33-29*a* shows the Fresnel diffraction pattern of an opaque disk illuminated by light from a source on its axis. Note the bright spot at the center of the pattern caused by the constructive interference of the light waves diffracted from edge of the disk. This pattern is of some historical interest. In an attempt to discredit Fresnel's wave theory of light, Poisson applied it to this situation and considered the prediction of the bright spot at the center of the shadow to be a ridiculous contradiction of fact. However, Fresnel immediately demonstrated experimentally that such a spot does, in fact, exist. This demonstration convinced many doubters of the validity of the wave theory of light. The Fresnel diffraction pattern of a circular aperture is shown in Figure 33-29*b*. Comparing this with the pattern of the opaque disk in Figure 33-29*a*, we can see that the two patterns are complements of each other.

*See Richard E. Haskel, "A Simple Experiment on Fresnel Diffraction," *American Journal of Physics*, vol. 38, 1970, p. 1039.

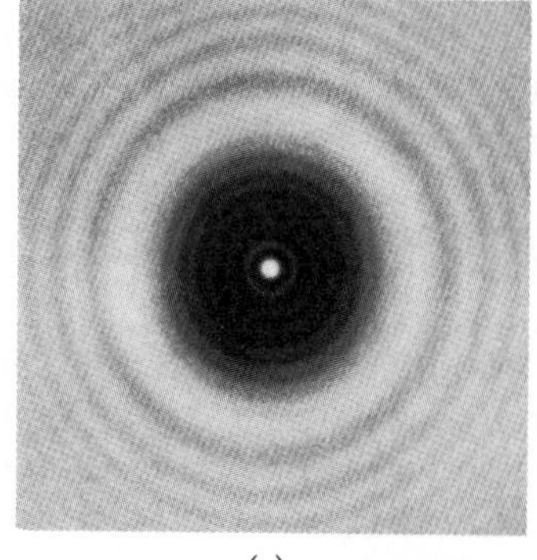

(*a*)

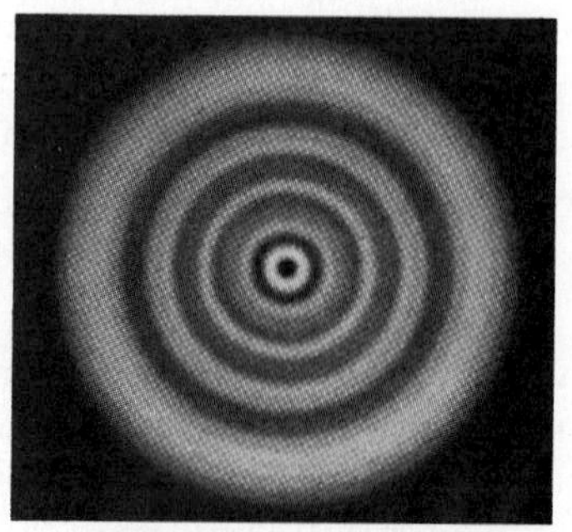

(*b*)

Figure 33-29 (*a*) Fresnel diffraction pattern of an opaque disk. At the center of the shadow, the light waves diffracted from the edge of the disk are in phase and produce a bright spot called the *Poisson spot.* (*b*) Fresnel diffraction pattern of a circular aperture. Compare this with part (*a*).

(a)

Intensity

Geometric shadow

Edge

Distance

(b)

Figure 33-30 (*a*) Fresnel diffraction of a straightedge. (*b*) Intensity versus distance along a line perpendicular to the edge.

Figure 33-30*a* shows the Fresnel diffraction pattern of a straightedge illuminated by light from a point source. A graph of the intensity versus distance (measured along a line perpendicular to the edge) is shown in Figure 33-30*b*. The light intensity does not fall abruptly to zero in the geometric shadow, but it decreases rapidly and is negligible within a few wavelengths of the edge. The Fresnel diffraction pattern of a rectangular aperture is shown in Figure 33-31. These patterns cannot be seen with broad light sources like an ordinary light bulb because the dark fringes of the pattern produced by light from one point on the source overlap the bright fringes of the pattern produced by light from another point.

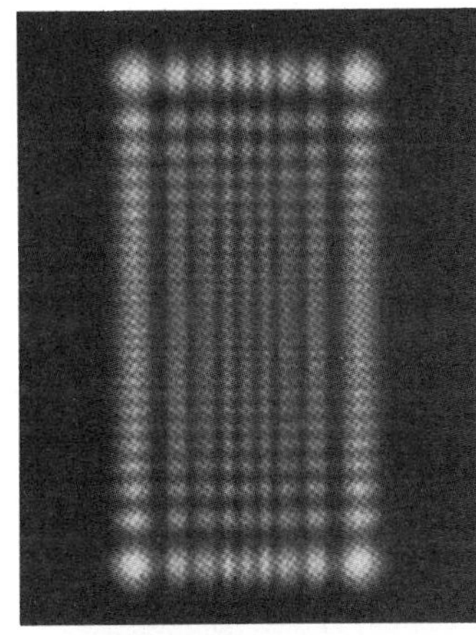

Figure 33-31 Fresnel diffraction of a rectangular aperture.

33-10 Diffraction and Resolution

Figure 33-32 shows the Fraunhofer diffraction pattern of a circular aperture. This pattern has important applications to the resolution of many optical instruments. The angle θ subtended by the first diffraction minimum is related to the wavelength and the diameter of the opening D by

$$\sin \theta = 1.22\frac{\lambda}{D} \qquad \text{33-27}$$

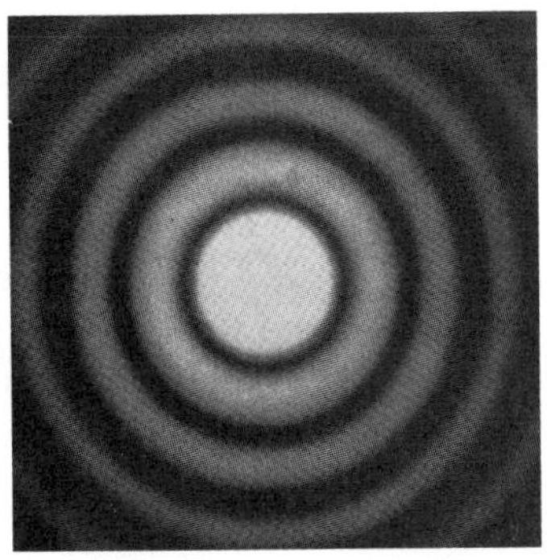

Figure 33-32 Fraunhofer diffraction pattern of a circular aperture.

Equation 33-27 is similar to Equation 33-17 except for the factor 1.22. This factor arises from the mathematical analysis, which is similar to that for a single slit but more complicated because of the circular geometry. In many applications, the angle θ is small, so $\sin\theta$ can be replaced by θ. The first diffraction minimum is then at an angle θ given by

$$\theta \approx 1.22\frac{\lambda}{D} \qquad 33\text{-}28$$

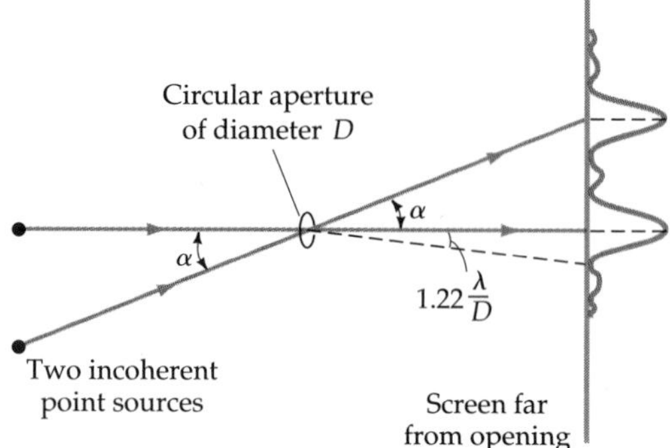

Figure 33-33 Two distant sources that subtend an angle α. If α is much greater than 1.22 λ/D, where λ is the wavelength of light and D is the diameter of the aperture, the diffraction patterns have little overlap and the sources are easily seen as two sources. If α is not much greater than 1.22 λ/D, the overlap of the diffraction patterns makes it difficult to distinguish two sources from one.

Figure 33-33 shows two point sources that subtend an angle α at a circular aperture far from the sources. The Fraunhofer diffraction patterns are also included in this figure. If α is much greater than $1.22\lambda/D$, they will be seen as two sources. However, as α is decreased, the overlap of the diffraction patterns increases, and it becomes difficult to distinguish the two sources from one source. At the critical angular separation α_c given by

$$\alpha_c = 1.22\frac{\lambda}{D} \qquad 33\text{-}29$$

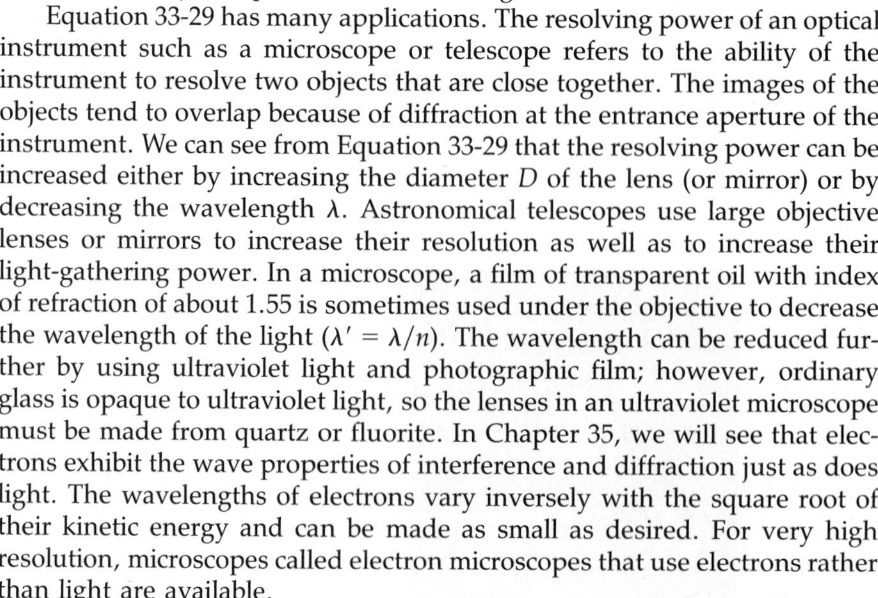

the first minimum of the diffraction pattern of one source falls at the central maximum of the other source. These objects are said to be just resolved by **Rayleigh's criterion for resolution.** Figure 33-34 shows the diffraction patterns for two sources when α is greater than the critical angle for resolution and when α is just equal to the critical angle for resolution.

Equation 33-29 has many applications. The resolving power of an optical instrument such as a microscope or telescope refers to the ability of the instrument to resolve two objects that are close together. The images of the objects tend to overlap because of diffraction at the entrance aperture of the instrument. We can see from Equation 33-29 that the resolving power can be increased either by increasing the diameter D of the lens (or mirror) or by decreasing the wavelength λ. Astronomical telescopes use large objective lenses or mirrors to increase their resolution as well as to increase their light-gathering power. In a microscope, a film of transparent oil with index of refraction of about 1.55 is sometimes used under the objective to decrease the wavelength of the light ($\lambda' = \lambda/n$). The wavelength can be reduced further by using ultraviolet light and photographic film; however, ordinary glass is opaque to ultraviolet light, so the lenses in an ultraviolet microscope must be made from quartz or fluorite. In Chapter 35, we will see that electrons exhibit the wave properties of interference and diffraction just as does light. The wavelengths of electrons vary inversely with the square root of their kinetic energy and can be made as small as desired. For very high resolution, microscopes called electron microscopes that use electrons rather than light are available.

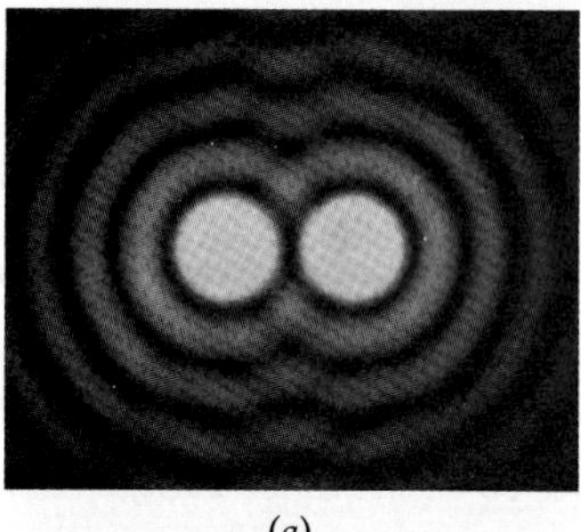

(*a*)

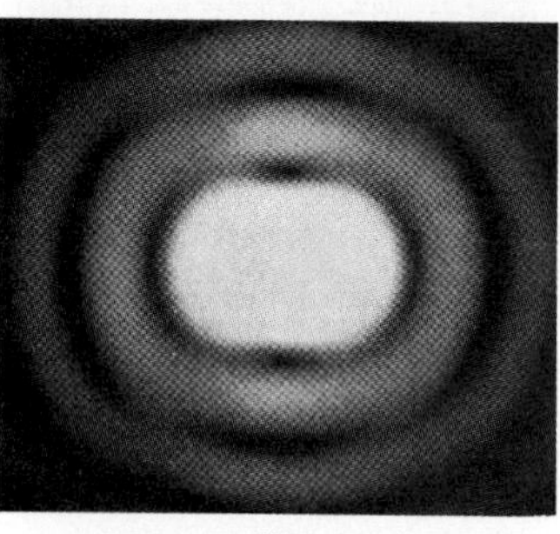

(*b*)

Figure 33-34 Diffraction patterns for a circular aperture and two incoherent point sources when (*a*) α is much greater than 1.22 λ/D and (*b*) when α is at the limit of resolution, $\alpha_c = 1.22\ \lambda/D$.

Example 33-10

What minimum angular separation must two point objects have if they are to be just resolved by the eye? How far apart must they be if they are 100 m away? Assume that the diameter of the pupil of the eye is 5 mm and that the wavelength of the light is 600 nm.

Using Equation 33-29 with $D = 5$ mm and $\lambda = 600$ nm, we have for the minimum angular separation

$$\alpha_c = 1.22\frac{6 \times 10^{-7}\text{ m}}{5 \times 10^{-3}\text{ m}} = 1.46 \times 10^{-4}\text{ rad}$$

If the objects are separated by a distance y and are 100 m away, they will be just barely resolved if $\tan \alpha_c = y/(100\text{ m})$. Then

$$y = (100\text{ m}) \tan \alpha_c \approx (100\text{ m})\, \alpha_c = 1.46 \times 10^{-2}\text{ m} = 1.46\text{ cm}$$

where we have used the small-angle approximation $\tan \alpha_c \approx \alpha_c$.

It is instructive to compare the limitation on resolution of the eye due to diffraction as seen in Example 33-10 with that due to the separation of the receptors (cones) on the retina. To be seen as two distinct objects, the images of the objects must fall on the retina on two nonadjacent cones. (See Problem 8 in Chapter 32.) Because the retina is about 2.5 cm from the eye lens, the distance y on the retina corresponding to an angular separation of 1.5×10^{-4} rad is found from

$$\alpha_c = 1.5 \times 10^{-4}\text{ rad} = \frac{y}{2.5\text{ cm}}$$

or

$$y \approx 4 \times 10^{-4}\text{ cm} = 4 \times 10^{-6}\text{ m} = 4\ \mu\text{m}$$

The actual separation of the cones in the fovea centralis, where the cones are the most tightly packed, is about 1 μm. Outside this region, they are about 3 to 5 μm apart.

Exercise

Two objects are 4 cm apart. How far away from them can you be and still resolve them with your eye if $\lambda = 600$ nm and the diameter of the pupil of your eye is 5 mm? (Answer: 274 m)

33-11 Diffraction Gratings

A useful tool for measuring the wavelength of light is the **diffraction grating,** which consists of a large number of equally spaced lines or slits on a flat surface. Such a grating can be made by cutting parallel, equally spaced grooves on a glass or metal plate with a precision ruling machine. With a reflection grating, light is reflected from the ridges between the lines. A phonograph record exhibits some of the properties of a reflection grating. In a transmission grating, the light passes through the clear gaps between the rulings. Inexpensive plastic gratings with 10,000 or more slits per centimeter are not uncommon. The spacing of the slits in a grating with 10,000 slits per centimeter is $d = (1\text{ cm})/10{,}000 = 10^{-4}$ cm.

Consider a plane light wave incident normally on a transmission grating (Figure 33-35) and assume that the width of each slit is very small so that each slit produces a widely diffracted beam. The interference pattern produced on a screen a large distance from the grating is that due to a large

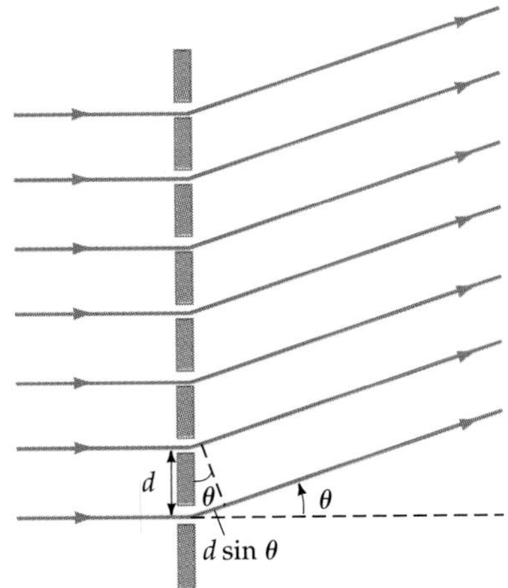

Figure 33-35 Light incident normally on a diffraction grating. At an angle θ, the path difference between rays from adjacent slits is $d \sin \theta$.

number of equally spaced light sources. The interference maxima are at angles θ given by

$$d \sin \theta = m\lambda \qquad m = 0, 1, 2, \ldots \tag{33-30}$$

where m is called the **order number.** The position of an interference maximum does not depend on the number of sources, but the more sources there are, the sharper and more intense the maximum will be, as was illustrated in Figure 33-18.

Figure 33-36*b* shows a typical spectroscope, which uses a diffraction grating to analyze light from a source such as a tube containing atoms of a gas, for example, helium or sodium vapor. The gas atoms are excited because of bombardment by electrons that are accelerated by the high voltage across the tube. The light emitted by such a source does not consist of a continuous spectrum. Instead the spectrum contains only certain wavelengths that are characteristic of the atoms in the source. Light from the source passes through a narrow collimating slit and is made parallel by a lens. Parallel light from the lens is incident on the grating. Instead of falling on a screen a large distance away, the parallel light from the grating is focused by a telescope and viewed by the eye. The telescope is mounted on a rotating platform that has been calibrated so that the angle θ can be measured. In the forward direction ($\theta = 0$), the central maximum for all wavelengths is seen. If light of a particular wavelength λ is emitted by the source, the first interference maximum is seen at the angle θ given by Equation 33-30 with $m = 1$. Each wavelength emitted by the source produces a separate image of the collimating slit in the spectroscope called a **spectral line.** The set of lines corresponding to $m = 1$ is called the **first-order spectrum.** The **second-order spectrum** corresponds to $m = 2$ for each wavelength. Higher orders may be seen if the angle θ given by Equation 33-30 is less than 90°. Depending on the wavelengths and the spacing of the slits in the grating, the orders may be mixed; that is, the third-order line for one wavelength may occur before the second-order line for another wavelength. If the spacing of the slits in the grating is known, the wavelengths emitted by the source can be determined by measuring the angle θ.

Figure 33-36 (*a*) A late nineteenth century spectroscope belonging to Gustav Kirchhoff used a prism rather than a diffraction grating to disperse the light. (*b*) Typical student spectroscope. Light from a collimating slit near the source is made parallel by a lens and falls on a grating. The diffracted light is viewed with a telescope at an angle that can be accurately measured.

Example 33-11

Sodium light is incident on a diffraction grating with 10,000 lines per centimeter. At what angles will the two yellow lines of wavelengths 589.00 nm and 589.59 nm be seen in the first order?

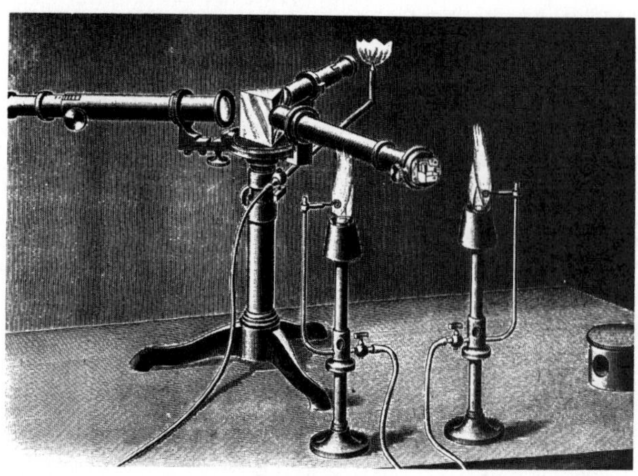

(*a*)

(*b*)

Using $m = 1$ and $d = 10^{-4}$ cm $= 10^{-6}$ m in Equation 33-30, we have for $\lambda = 589 \times 10^{-9}$ m

$$\sin\theta = \frac{\lambda}{d} = \frac{589 \times 10^{-9}\ \text{m}}{10^{-6}\ \text{m}} = 0.589$$

$$\theta = 36.09°$$

For $\lambda = 589.59$ nm, a similar calculation gives $\sin\theta = 0.58959$, resulting in $\theta = 36.13°$.

An important feature of a spectroscope is its ability to measure light of two nearly equal wavelengths λ_1 and λ_2. For example, the two prominent yellow lines in the spectrum of sodium have wavelengths 589.00 and 589.59 nm, which can be seen as two separate wavelengths if their interference maxima do not overlap. According to Rayleigh's criterion for resolution, these wavelengths are resolved if the angular separation of their interference maxima is greater than the angular separation between one interference maximum and the first interference minimum on either side of it. The **resolving power** of a diffraction grating is defined to be $\lambda/|\Delta\lambda|$, where $|\Delta\lambda|$ is the smallest difference between two nearby wavelengths, each approximately equal to λ, that may be resolved. The resolving power is proportional to the number of slits illuminated because the more slits illuminated, the sharper the interference maxima. The resolving power R can be shown to be

$$R = \frac{\lambda}{|\Delta\lambda|} = mN \qquad 33\text{-}31$$

where N is the number of slits and m is the order number (see Problem 73). We can see from Equation 33-31 that to resolve the two yellow lines in the sodium spectrum the resolving power must be

$$R = \frac{589.00\ \text{nm}}{589.59 - 589.00\ \text{nm}} = 998$$

Thus, to resolve the two yellow sodium lines in the first order ($m = 1$), we need a grating containing about 1000 slits in the area illuminated by the light.

(c) Photomicrograph of the ridged pattern on the surface of a diffraction grating. (d) Aerial view of the very large array (VLA) radio telescope in New Mexico. Radio signals from distant galaxies add constructively when Equation 33-30 is satisfied, where d is the distance between two adjacent telescopes.

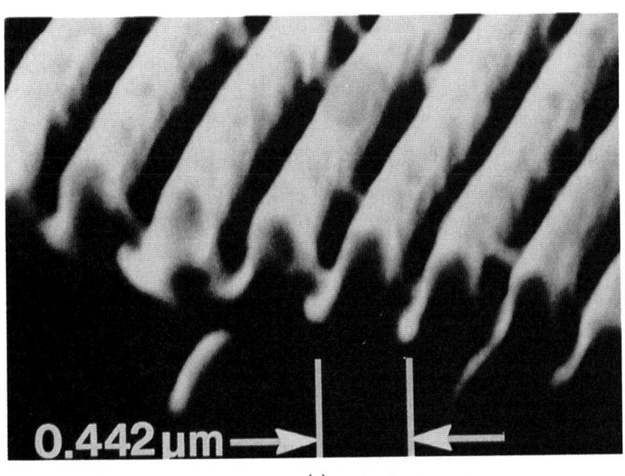

(c)

(d)

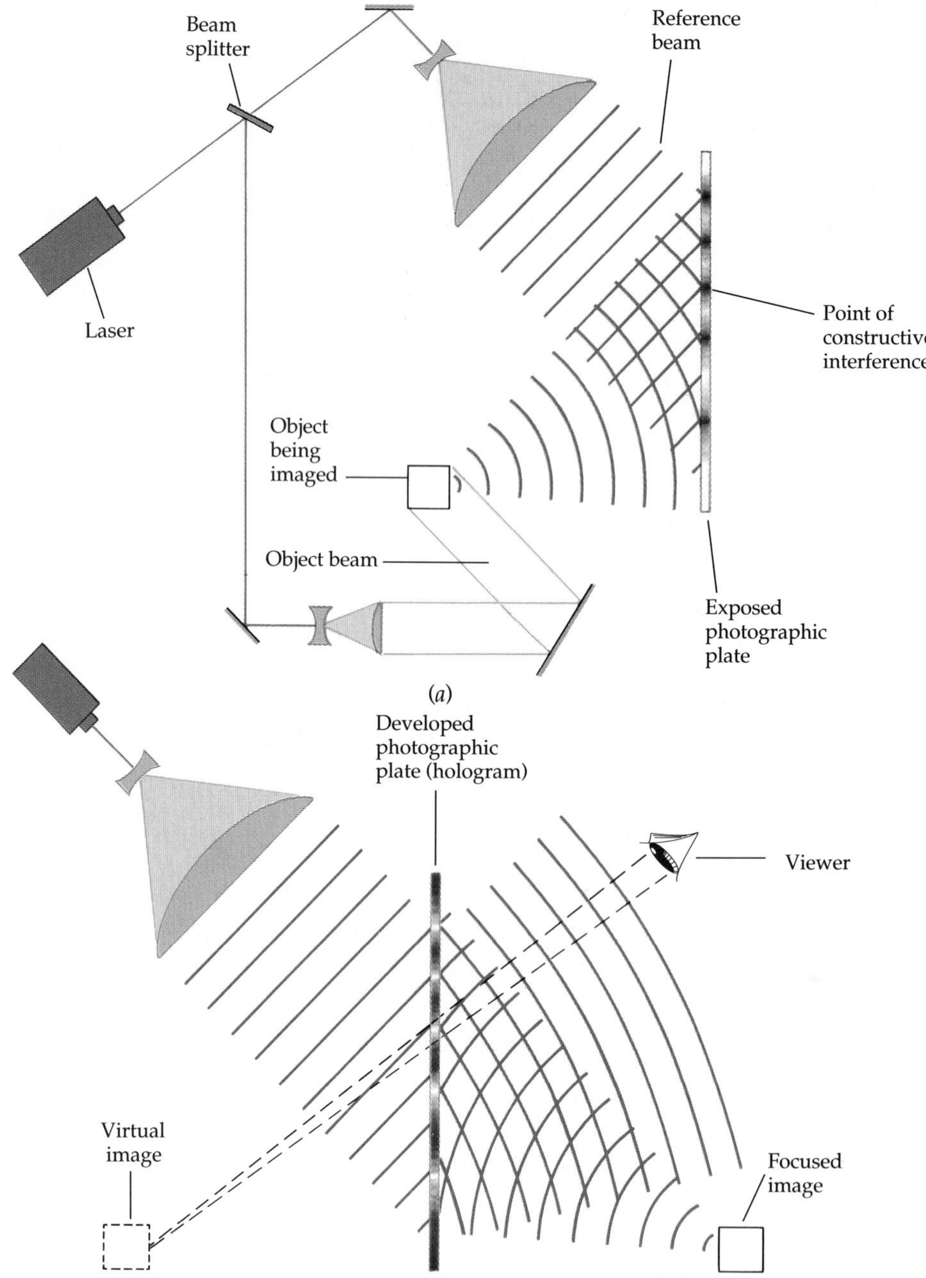

(*a*) The production of a hologram. The interference pattern produced by the reference beam and object beam is recorded on a photographic film. (*b*) When the film is developed and illuminated by coherent laser light, a three-dimensional image is seen. Holograms that you see on credit cards or postage stamps, called rainbow holograms, are more complicated. A horizontal strip of the original hologram is used to make a second hologram. The three-dimensional image can be seen as the viewer moves from side to side, but if viewed with laser light, the image disappears when the viewer's eyes move above or below the slit image. When viewed with white light, the image is seen in different colors as the viewer moves in the vertical direction.

Holograms

An interesting application of diffraction gratings is the production of a three-dimensional photograph called a **hologram.** In an ordinary photograph, the intensity of reflected light from an object is recorded on a film. When the film is viewed by transmitted light, a two-dimensional image is produced. In a hologram, a beam from a laser is split into two beams, a reference beam and an object beam. The object beam reflects from the object to be photographed and the interference pattern between it and the reference beam is recorded on a photographic film. This can be done because the laser beam is coherent so that the relative phase difference between the reference beam and object beam can be kept constant during the exposure. The interference fringes on the film act as a diffraction grating. When the film is illuminated with a laser, a three-dimensional replica of the object is produced.

(*a*)

(*a*) A technician produces a hologram of a statuette at the University of Strasbourg. When the glass plate is later illuminated by laser light, the statuette appears as a three-dimensional image. (*b*) and (*c*) Two views of the hologram "Digital." Note that different parts of the circuit board appear behind the front magnifying lens. (*d*) A holograph emulsion magnified 100 times. (*e*) A "head-up display" uses a holographic projection of important information from the airplane's control panel, so the pilot can view the runway and the control panel at the same time.

(*b*)

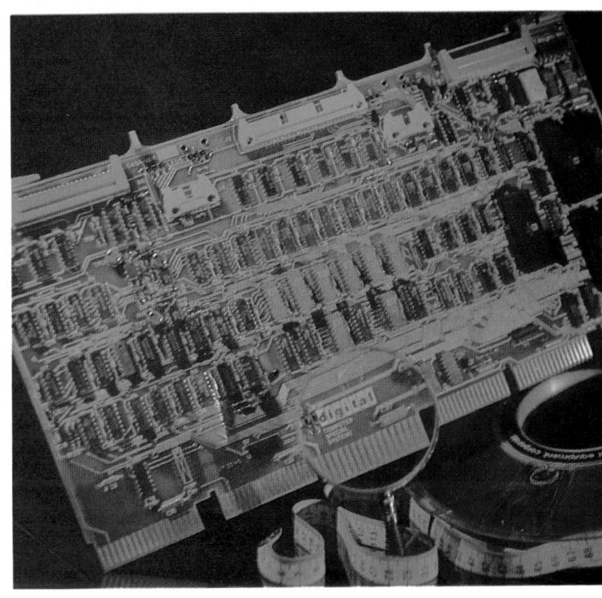

(*c*)

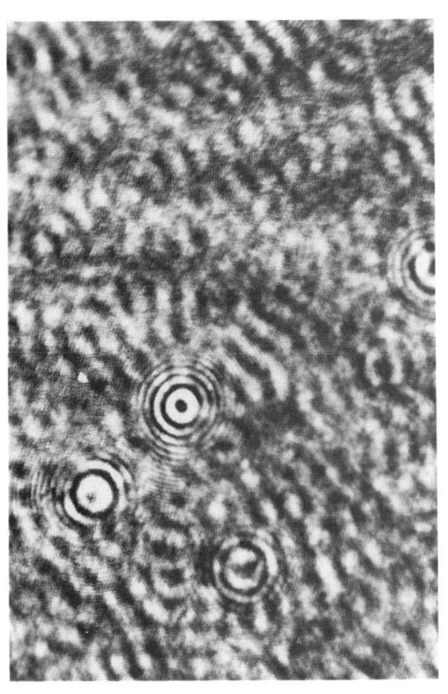

(*d*)

(*e*)

Summary

1. Two light rays interfere constructively if their phase difference is zero or an integer times 360°. They interfere destructively if their phase difference is 180° or an odd integer times 180°. A common source of phase difference is a path difference. A path difference Δr introduces a phase difference δ given by

$$\delta = \frac{\Delta r}{\lambda} 2\pi = \frac{\Delta r}{\lambda} 360°$$

A phase difference of 180° is introduced when a light wave is reflected from a boundary between two media for which the wave speed is greater in the original medium, such as the one between air and glass.

2. The interference of light rays reflected from the top and bottom surfaces of a thin film produces colored bands or fringes commonly observed in soap films or oil films. The difference in phase between the two rays results from the path difference of twice the thickness of the film plus any phase change due to reflection of one or both of the rays.

3. The Michelson interferometer uses interference to measure small distances such as the wavelength of light, or to measure a small difference in the index of refraction such as that between air and a vacuum.

4. The path difference at an angle θ on a screen far away from two narrow slits separated by a distance d is $d \sin \theta$. When this path difference is an integer times the wavelength, the interference is constructive and the intensity is maximum. When the path difference is an odd integer times $\lambda/2$, the interference is destructive, resulting in minimum intensity.

$$d \sin \theta = m\lambda \qquad m = 0, 1, 2, \ldots \qquad \text{maxima}$$

$$d \sin \theta = (m + \tfrac{1}{2})\lambda \qquad m = 0, 1, 2, \ldots \qquad \text{minima}$$

If the intensity due to each slit separately is I_0, the intensity at points of constructive interference is $4I_0$ and that at points of destructive interference is 0. When there are many equally spaced slits, the principal interference maxima occur at the same points as for two slits, but the maxima are much more intense and much narrower. For N slits, the intensity of the principal maxima is N^2I_0, and there are $N - 2$ secondary maxima between each pair of principal maxima.

5. Diffraction occurs whenever a portion of a wavefront is limited by an obstacle or aperture. The intensity of light at any point in space can be computed using Huygens' principle by taking each point on the wavefront to be a point source and computing the resulting interference pattern. Fraunhofer patterns are observed at great distances from the obstacle or aperture so that the rays reaching any point are approximately parallel, or they can be observed using a lens to focus parallel rays on a viewing screen placed in the focal plane of the lens. Fresnel patterns are observed at points close to the source. Diffraction of light is often difficult to observe because the wavelength is so small or because the light intensity is not great enough. Except for the Fraunhofer pattern of a long narrow slit, diffraction patterns are usually difficult to analyse.

6. When light is incident on a single slit of width a, the intensity pattern on a screen far away shows a broad central diffraction maximum that de-

creases to zero at an angle θ given by

$$a \sin \theta = \lambda$$

The width of the central maximum is inversely proportional to the width of the slit. Other zeros in the single-slit diffraction pattern occur at angles given by

$$\sin \theta = m\frac{\lambda}{a} \qquad m = 1, 2, 3 \ldots$$

On each side of the central maximum are secondary maxima of much smaller intensity.

7. The Fraunhofer interference–diffraction pattern of two slits is the same as the interference pattern for two narrow slits modulated by the single-slit diffraction pattern.

8. When light from two point sources that are close together passes through an aperture, the diffraction patterns of the sources may overlap. If the overlap is too great the two sources cannot be resolved as two separate sources. When the central diffraction maximum of one source falls at the diffraction minimum of the other source, the two sources are said to be just resolved by Rayleigh's criterion for resolution. For a circular aperture, the critical angular separation of two sources for resolution by Rayleigh's criterion is

$$\alpha_c = 1.22\frac{\lambda}{D}$$

where D is the diameter of the aperture.

9. A diffraction grating consisting of a large number of closely spaced lines or slits is used to measure the wavelength of light emitted by a source. The positions of the interference maxima from a grating are at angles given by

$$d \sin \theta = m\lambda \qquad m = 0, 1, 2, \ldots$$

where m is the order number. The resolving power of a grating is

$$R = \frac{\lambda}{|\Delta\lambda|} = mN$$

where N is the number of slits of the grating that are illuminated and m is the order number.

Suggestions for Further Reading

Baumeister, Philip, and Gerald Pincus: "Optical Interference Coatings," *Scientific American,* December 1970, p. 58.

Color television cameras, lasers, projector bulbs, and lenses of various types employ thin films to reflect or transmit light of certain wavelengths.

Nassau, Kurt: "Color Involving Geometrical and Physical Optics," *The Physics and Chemistry of Color: The Fifteen Causes of Color,* Part IV, John Wiley & Sons, New York, 1983.

Contains a good discussion of color production by thin films.

Walker, Jearl: "The Amateur Scientist: A Ball Bearing Aids in the Study of Light and Also Serves as a Lens," *Scientific American,* November 1984, p. 186.

This is a report of an unusual investigation into the properties of the diffraction pattern of a ball bearing placed in the beam of a laser.

Review

A. Objectives: After studying this chapter, you should:

1. Be able to work problems involving interference in thin films.

2. Be able to describe the Michelson interferometer.

3. Be able to sketch the two-slit interference intensity pattern and calculate the positions of the interference maxima and minima.

4. Be able to use the phasor method to find the sum of several harmonic waves.

5. Be able to sketch the interference pattern of three or more equally spaced slits.

6. Be able to sketch the single-slit diffraction pattern and calculate the position of the first diffraction minimum.

7. Be able to sketch the combined interference–diffraction pattern for several slits.

8. Be able to state Rayleigh's criterion for resolution and use it to investigate the conditions for the resolution of two nearby objects.

9. Be able to discuss the use of diffraction gratings and find the resolving power of a grating.

B. Define, explain, or otherwise identify:

Fringes
Newton's rings
Optically flat
Interferometer
Lloyd's mirror
Phasor
Secondary maxima
Principal maxima
Central diffraction maximum
Fraunhofer diffraction pattern
Fresnel diffraction pattern
Rayleigh's criterion for resolution
Diffraction grating
Order number
Spectral line
Resolving power
Hologram

C. True or false: If the statement is true, explain why it is true. If it is false, give a counterexample.

1. When waves interfere destructively, the energy is converted into heat energy.

2. Interference is observed only for waves from coherent sources.

3. In the Fraunhofer diffraction pattern for a single slit, the narrower the slit, the wider the central maximum of the diffraction pattern.

4. A circular aperture can produce both a Fraunhofer and a Fresnel diffraction pattern.

5. The ability to resolve two point sources depends on the wavelength of the light.

Problems

Level I

33-1 Phase Difference and Coherence

1. Which of the following pairs of light sources are coherent? (*a*) Two candles. (*b*) One point source and its image in a plane mirror. (*c*) Two pinholes uniformly illuminated by the same point source. (*d*) Two headlights of a car. (*e*) Two images of a point source due to reflection from the front and back surfaces of a soap film.

2. (*a*) What minimum path difference is needed to introduce a phase shift of 180° in light of wavelength 600 nm? (*b*) What phase shift will that path difference introduce in light of wavelength 800 nm?

3. Two coherent microwave sources that produce waves of wavelength 1.5 cm are in the xy plane, one on the y axis at $y = 15$ cm and the other at $x = 3$ cm, $y = 14$ cm. If the sources are in phase, find the difference in phase between the two waves from these sources at the origin.

4. Light of wavelength 500 nm is incident normally on a film of water 10^{-4} cm thick. The index of refraction of water is 1.33. (*a*) What is the wavelength of the light in the water? (*b*) How many wavelengths are contained in the distance $2t$, where t is the thickness of the film? (*c*) What is the phase difference between the wave reflected from the top of the film and the one reflected from the bottom after it has traveled this distance?

33-2 Interference in Thin Films

5. A loop of wire is dipped in soapy water and held so that the soap film is vertical. (*a*) Viewed by reflection with white light, the top of the film appears black. Explain why. (*b*) Below the black region are colored bands. Is the first band red or violet? (*c*) Describe the appearance of the film when it is viewed by *transmitted* light.

6. A wedge-shaped film of air is made by placing a small slip of paper between the edges of two flat plates of glass. Light of wavelength 700 nm is incident normally on the glass plates, and interference bands are observed by reflection. (*a*) Is the first band near the point of contact of the plates dark or bright? Why? (*b*) If there are five dark bands per centimeter, what is the angle of the wedge?

7. A thin layer of a transparent material with an index of refraction of 1.30 is used as a nonreflective coating on the surface of glass with an index of refraction of 1.50. What should the thickness of the material be for it to be nonreflecting for light of wavelength 600 nm (in a vacuum)?

8. The diameters of fine wires can be accurately measured using interference patterns. Two optically flat pieces of glass of length L are arranged with the wire between them as shown in Figure 33-37. The setup is illuminated by monochromatic light, and the resulting interference fringes are detected. Suppose $L = 20$ cm and yellow sodium light ($\lambda \approx 590$ nm) is used for illumination. If 19 bright fringes are seen along this 20 cm distance, what are the limits on the diameter of the wire? *Hint:* The nineteenth fringe might not be right at the end, but you do not see a twentieth fringe at all.

Figure 33-37 Problem 8.

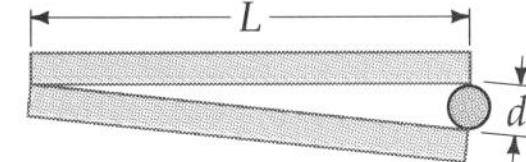

33-3 The Michelson Interferometer

9. A thin film of index of refraction $n = 1.5$ for light of wavelength 600 nm is inserted in one arm of a Michelson interferometer. (*a*) If a fringe shift of 12 fringes occurs, what is the thickness of this film? (*b*) If the illuminating light is changed to 400 nm, the fringe shift when this film is inserted is 16 fringes. What is the index of refraction of this film for light of wavelength 400 nm?

10. A hollow cell of length 5 cm with glass windows is inserted into one arm of a Michelson interferometer. The air is pumped out of the cell and the mirrors are adjusted to give a bright fringe at the center. As the air is gradually let back into the cell, there is a shift of 49.6 fringes when light of wavelength 589.29 nm is used. (*a*) How many waves are there in the 5.0 cm long cell when it is evacuated? (*b*) How many waves are there in the cell when it contains air? (*c*) What is the index of refraction of air as determined by this experiment?

33-4 The Two-Slit Interference Pattern

11. Two narrow slits separated by 1 mm are illuminated by light of wavelength 600 nm, and the interference pattern is viewed on a screen 2 m away. Calculate the number of bright fringes per centimeter on the screen.

12. Using a conventional two-slit apparatus with light of wavelength 589 nm, 28 bright fringes per centimeter are observed on a screen 3 m away. What is the slit separation?

13. Two narrow slits are separated by a distance d. Their interference pattern is to be observed on a screen a large distance L away. (*a*) Calculate the spacing y of the maxima on the screen for light of wavelength 500 nm when $L =$ 1 m and $d = 1$ cm. (*b*) Would you expect to observe the interference of light on the screen for this situation? (*c*) How close together should the slits be placed for the maxima to be separated by 1 mm for this wavelength and screen distance?

14. A long, narrow, horizontal slit lies 1 mm above a plane mirror, which is in the horizontal plane. The interference pattern produced by the slit and its image is viewed on a screen 1 m from the slit. The wavelength of the light is 600 nm. (*a*) Find the distance from the mirror to the first maximum. (*b*) How many dark bands per centimeter are seen on the screen?

15. In a lecture demonstration, laser light is used to illuminate two slits separated by 0.5 mm, and the interference pattern is observed on a screen 5 m away. The distance on the screen from the centerline to the thirty-seventh bright fringe is 25.7 cm. What is the wavelength of the light?

16. Light of wavelength 633 nm from a helium–neon laser is shone normally on a plane containing two slits. The first interference maximum is 82 cm from the central maximum on a screen 12 m away. (*a*) Find the separation of the slits. (*b*) How many interference maxima can be observed?

33-5 The Addition of Harmonic Waves Using Phasors

17. Find the resultant of the two waves

$$E_1 = 2 \sin \omega t \qquad \text{and} \qquad E_2 = 3 \sin (\omega t + 270°)$$

18. Find the resultant of the two waves

$$E_1 = 4 \sin \omega t \qquad \text{and} \qquad E_2 = 3 \sin (\omega t + 60°)$$

33-6 Interference Pattern of Three or More Equally Spaced Sources

19. Three waves have electric fields given by

$$E_1 = E_0 \sin \omega t \; E_2 = E_0 \sin (\omega t + \delta) \; E_3 = E_0 \sin (\omega t + 2\delta)$$

Using the phasor method, draw the appropriate phasors. Calculate the amplitude (in terms of E_0) and phase of the resultant wave for (*a*) $\delta = 30°$, (*b*) $\delta = 60°$, (*c*) $\delta = 90°$, and (*d*) $\delta = 240°$.

20. Three equally spaced slits separated by 0.1 mm are uniformly illuminated by light of wavelength 600 nm. The interference pattern is viewed on a screen 2 m away. Find the positions of the interference maxima and minima.

21. Five equally spaced slits are uniformly illuminated, and the slit spacing is large enough that the small angle approximation $\sin \theta \approx \theta$ is good for the first few maxima. (*a*) Find the angle θ_1 between the first principal interference maximum and the central maximum ($\theta = 0$), and compare this angle with the angle to the first minimum. (*b*) Sketch the interference pattern.

22. Repeat Problem 21 for six equally spaced slits.

33-7 Diffraction Pattern of a Single Slit

23. Equation 33-4, $d \sin \theta = m\lambda$, and Equation 33-19, $a \sin \theta = m\lambda$, are sometimes confused. For each equation, define the symbols and explain the equation's application.

24. Light of wavelength 600 nm is incident on a long, narrow slit. Find the angle of the first diffraction minimum if the width of the slit is (*a*) 1 mm, (*b*) 0.1 mm, and (*c*) 0.01 mm.

25. The single-slit diffraction pattern of light is observed on a screen a large distance L from the slit. Note from Equation 33-20 that the width $2y$ of the central maximum varies inversely with the width a of the slit. Calculate the width $2y$ for $L = 2$ m, $\lambda = 500$ nm, and (*a*) $a = 0.1$ mm, (*b*) $a = 0.01$ mm, and (*c*) $a = 0.001$ mm.

26. In a lecture demonstration of diffraction, a laser beam of wavelength 700 nm passes through a vertical slit 0.5 mm wide and hits a screen 6 m away. Find the horizontal length of the principal diffraction maximum on the screen; that is, find the distance between the first minimum on the left and the first minimum on the right of the central maximum.

27. Plane microwaves are incident on a long, narrow metal slit of width 5 cm. The first diffraction minimum is observed at $\theta = 37°$. What is the wavelength of the microwaves?

33-8 Interference–Diffraction Pattern of Two Slits

28. A two-slit Fraunhofer interference–diffraction pattern is observed with light of wavelength 500 nm. The slits have a separation of 0.1 mm and a width of a. (*a*) Find the width a if the fifth interference maximum is at the same angle as the first diffraction minimum. (*b*) For this case, how many bright interference fringes will be seen in the central diffraction maximum?

29. A two-slit Fraunhofer interference-diffraction pattern is observed with light of wavelength 700 nm. The slits have widths of 0.01 mm and are separated by 0.2 mm. How many bright fringes will be seen in the central diffraction maximum?

30. Suppose that the *central* diffraction maximum for two slits contains 17 interference fringes for some wavelength of light. How many interference fringes would you expect in the first *secondary* diffraction maximum?

33-9 Fraunhofer and Fresnel Diffraction

There are no problems for this section.

33-10 Diffraction and Resolution

31. Light of wavelength 700 nm is incident on a pinhole of diameter 0.1 mm. (*a*) What is the angle between the central maximum and the first diffraction minimum for a Fraunhofer diffraction pattern? (*b*) What is the distance between the central maximum and the first diffraction minimum on a screen 8 m away?

32. Two sources of light of wavelength 700 nm are 10 m away from the pinhole of Problem 31. How far apart must the sources be for their diffraction patterns to be resolved by Rayleigh's criterion?

33. (*a*) How far apart must two objects be on the moon to be resolved by the eye? Take the diameter of the pupil of the eye to be 5 mm, the wavelength of the light to be 600 nm, and the distance to the moon to be 380,000 km. (*b*) How far apart must the objects on the moon be to be resolved by a telescope that has a 5-m-diameter mirror?

34. Two sources of light of wavelength 700 nm are separated by a horizontal distance x. They are 5 m from a vertical slit of width 0.5 mm. What is the least value of x for which the diffraction pattern of the sources can be resolved by Rayleigh's criterion?

35. What minimum aperture, in millimeters, is required for opera glasses (binoculars) if an observer is to be able to distinguish the soprano's individual eyelashes (separated by 0.5 mm) at an observation distance of 25 m? Assume the effective wavelength of the light to be 550 nm.

36. The headlights on a small car are separated by 112 cm. At what maximum distance could you resolve them if the diameter of your pupil is 5 mm and the effective wavelength of the light is 550 nm?

37. You are told not to shoot until you see the whites of their eyes. If their eyes are separated by 6.5 cm and the diameter of your pupil is 5 mm, at what distance can you resolve the two eyes using light of wavelength 550 nm?

33-11 Diffraction Gratings

38. A diffraction grating with 2000 slits per centimeter is used to measure the wavelengths emitted by hydrogen gas. At what angles θ in the first-order spectrum would you expect to find the two violet lines of wavelengths 434 and 410 nm?

39. With the grating used in Problem 38, two other lines in the first-order hydrogen spectrum are found at angles $\theta_1 = 9.72 \times 10^{-2}$ rad and $\theta_2 = 1.32 \times 10^{-1}$ rad. Find the wavelengths of these lines.

40. Repeat Problem 38 for a diffraction grating with 15,000 slits per centimeter.

41. A diffraction grating of 2000 slits per centimeter is used to analyze the spectrum of mercury. (*a*) Find the angular separation in the first-order spectrum of the two lines of wavelength 579.0 and 577.0 nm. (*b*) How wide must the beam on the grating be for these lines to be resolved?

42. What is the longest wavelength that can be observed in the fifth-order spectrum using a diffraction grating with 4000 slits per centimeter?

Level II

43. Laser light falls normally on three evenly spaced, very narrow slits. When one of the side slits is covered, the first-order maximum is at 0.60° from the normal. If the center slit is covered and the other two are open, find (*a*) the angle of the first-order maximum and (*b*) the order number of the maximum that now occurs at the same angle as the fourth-order maximum did before.

44. The ceiling of your lecture hall is probably covered with acoustic tile, which has small holes separated by about 6 mm. (*a*) Using light with a wavelength of 500 nm, how far could you be from this tile and still resolve these holes? The diameter of the pupil of your eye is about 5 mm. (*b*) Could you resolve these holes better with red light or with violet light?

45. The telescope on Mount Palomar has a diameter of 200 inches. Suppose a double star were 4 light-years away. Under ideal conditions, what must be the minimum separation of the two stars for their images to be resolved?

46. A mica sheet 1.2 μm thick is suspended in air. In re-

flected light, there are gaps in the visible spectrum at 421, 474, 542, and 633 nm. Find the index of refraction of the mica sheet.

47. A thin film having an index of refraction of 1.5 is surrounded by air. It is illuminated normally by white light and is viewed by reflection. Analysis of the resulting reflected light shows that the wavelengths 360, 450, and 602 nm are the only missing wavelengths in or near the visible portion of the spectrum. That is, for these wavelengths, there is destructive interference. (*a*) What is the thickness of the film? (*b*) What visible wavelengths are brightest in the reflected interference pattern? (*c*) If this film were resting on glass with an index of refraction of 1.6, what wavelengths in the visible spectrum would be missing from the reflected light?

48. For a ruby laser of wavelength 694 nm, the end of the ruby crystal is the aperture that determines the diameter of the light beam emitted. If the diameter is 2 cm and the laser is aimed at the moon, 380,000 km away, find the approximate diameter of the light beam when it reaches the moon, assuming the spread is due solely to diffraction.

49. Sodium light of wavelength 589 nm falls normally on a 2-cm-square diffraction grating ruled with 4000 lines per centimeter. The Fraunhofer diffraction pattern is projected onto a screen at 1.5 m by a lens of focal length 1.5 m placed immediately in front of the grating. Find (*a*) the positions of the first two intensity maxima on one side of the central maximum, (*b*) the width of the central maximum, and (*c*) the resolution in the first order.

50. At the second secondary maximum of the diffraction pattern of a single slit, the phase difference between the waves from the top and bottom of the slit is approximately 5π. The phasors used to calculate the amplitude at this point complete 2.5 circles. If I_0 is the intensity at the central maximum, find the intensity I at this second secondary maximum.

51. A camera lens is made of glass with an index of refraction of 1.6. This lens is coated with a magnesium fluoride film ($n = 1.38$) to enhance its light transmission. This film is to produce zero reflection for light of wavelength 540 nm. Treat the lens surface as a flat plane and the film as a uniformly thick flat film. (*a*) How thick must the film be to accomplish its objective in the first order? (*b*) Would there be destructive interference for any other visible wavelengths? (*c*) By what factor would the reflection for light of wavelengths 400 and 700 nm be reduced by this film? Neglect the variation in the reflected light amplitudes from the two surfaces.

52. (*a*) Show that the positions of the interference minima on a screen a large distance L away from three equally spaced sources (spacing d, with $d >> \lambda$) are given approximately by

$$y = \frac{n\lambda L}{3d} \qquad \text{where } n = 1, 2, 4, 5, 7, 8, 10, \ldots$$

that is, n is not a multiple of 3. (*b*) For $L = 1$ m, $\lambda = 5 \times 10^{-7}$ m, and $d = 0.1$ mm, calculate the width of the principal interference maxima (the distance between successive minima) for three sources.

53. (*a*) Show that the positions of the interference minima on a screen a large distance L away from four equally spaced sources (spacing d, with $d >> \lambda$) are given approximately by

$$y = \frac{n\lambda L}{4d} \qquad \text{where } n = 1, 2, 3, 5, 6, 7, 9, 10, \ldots$$

that is, n is not a multiple of 4. (*b*) For $L = 2$ m, $\lambda = 6 \times 10^{-7}$ m, and $d = 0.1$ mm, calculate the width of the principal interference maxima (the distance between successive minima) for four sources. Compare this width with that for two sources with the same spacing.

54. In a pinhole camera, the image is fuzzy because of geometrical optics effects of the pinhole and because of diffraction. As the pinhole is made smaller, the fuzziness due to its size (that is, due to rays arriving at the film through different parts of the pinhole) is reduced, but the fuzziness due to diffraction is increased. The optimum size of the pinhole for the sharpest possible image occurs when the spread due to diffraction equals that due to geometrical optics effects of the pinhole. Estimate the optimum size of the pinhole if the distance from it to the film is 10 cm and the wavelength of the light is 550 nm.

55. Light of wavelength 480 nm falls normally on four slits. Each slit is 2 μm wide and is separated from the next by 6 μm. (*a*) Find the angle from the center to the first point of zero intensity of the single-slit diffraction pattern on a distant screen. (*b*) Find the angles of any bright interference maxima that lie inside the central diffraction maximum. (*c*) Find the angular spread between the central interference maximum and the first interference minimum on either side of it. (*d*) Sketch the intensity as a function of angle.

56. A drop of oil ($n = 1.22$) floats on water ($n = 1.33$). When reflected light is observed from above as shown in Figure 33-38, what is the thickness of the drop at the point where the second red fringe, counting from the edge of the drop, is observed? Assume red light has a wavelength of 650 nm.

Figure 33-38 Problem 56.

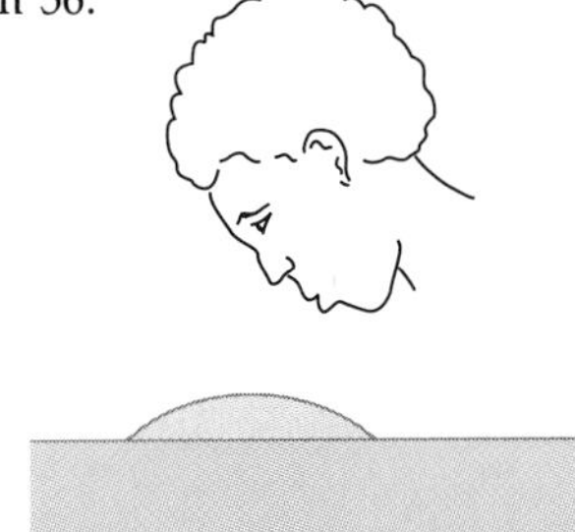

57. The Impressionist painter Georges Seurat used a technique called "pointillism," in which his paintings are composed of small, closely spaced dots of pure color, each about 2 mm in diameter. The illusion of the colors blending together smoothly is produced in the eye of the viewer by diffraction effects. Calculate the minimum viewing distance for this effect to work properly. Use the wavelength of visible light that requires the *greatest* distance, so that you're sure the effect will work for *all* visible wavelengths. Assume the pupil of the eye has a diameter of 5 mm.

58. Light of wavelength 600 nm is used to illuminate normally two glass plates 22 cm in length that touch at one end and are separated at the other end by a wire of radius 0.025 mm. How many bright fringes appear along the total length of the plates?

59. A square diffraction grating with an area of 25 cm^2 has a resolution of 22,000 in the fourth order. At what angle should you look to see a wavelength of 510 nm in the fourth order?

60. Light of wavelength 550 nm illuminates two slits of width 0.03 mm and separation 0.15 mm. (*a*) How many interference maxima fall within the full width of the central diffraction maximum? (*b*) What is the ratio of the intensity of the third interference maximum to the side of the centerline (not counting the center interference maximum) to the intensity of the center interference maximum?

61. Light is incident at an angle ϕ with the normal to a vertical plane containing two slits of separation d (Figure 33-39). Show that the interference maxima are located at angles θ given by $\sin\theta + \sin\phi = m\lambda/d$.

Figure 33-39 Problem 61.

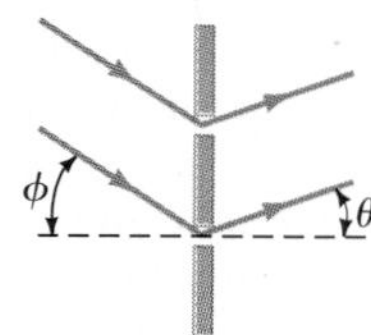

62. White light falls at an angle of 30° to the normal of a plane containing a pair of slits separated by 2.5 μm. What visible wavelengths give a bright interference maximum in the transmitted light in the direction normal to the plane? (See Problem 61.)

Level III

63. A Newton's-ring apparatus consists of a glass lens with radius of curvature R that rests on a flat glass plate as shown in Figure 33-40. The thin film is air of variable thickness. The pattern is viewed by reflected light. (*a*) Show that for a thickness t the condition for a bright (constructive) interference ring is

$$t = (m + \tfrac{1}{2})\frac{\lambda}{2} \qquad m = 0, 1, 2, \ldots$$

Figure 33-40 Problem 63.

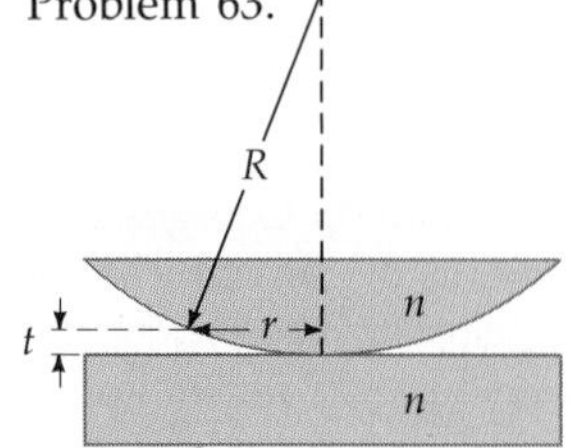

(*b*) Show that as long as $t/R << 1$, the radius r of a bright circular fringe is given by

$$r = \sqrt{(m + \tfrac{1}{2})\lambda R} \qquad m = 0, 1, 2, \ldots$$

(*c*) How would the transmitted pattern look in comparison with the reflected one? (*d*) Use $R = 10$ m and a diameter of 4 cm for the lens. How many bright fringes would you see if the apparatus were illuminated by yellow sodium light ($\lambda \approx 590$ nm) and were viewed by reflection? (*e*) What would be the diameter of the sixth bright fringe? (*f*) If the glass used in the apparatus has an index of refraction $n =$ 1.5 and water is placed between the two pieces of glass, what change will take place in the bright fringes?

64. A *Jamin refractometer* is a device for measuring or comparing the indexes of refraction of fluids. A beam of monochromatic light is split into two parts, each of which is directed along the axis of a separate cylindrical tube before being recombined into a single beam that is viewed through a telescope. Suppose that each tube is 0.4 m long and that sodium light of wavelength 589 nm is used. Both tubes are initially evacuated, and constructive interference is observed in the center of the field of view. As air is slowly allowed to enter one of the tubes, the central field of view changes to dark and back to bright a total of 198 times. (*a*) What is the index of refraction of air? (*b*) If the fringes can be counted to ±0.25 fringe, where one fringe is equivalent to one complete cycle of intensity variation at the center of the field of view, to what accuracy can the index of refraction of air be determined by this experiment?

65. With a diffraction grating, we are interested not only in its resolving power R, which is the ability of the grating to separate two close wavelengths, but also in the dispersion D of the grating. This is defined by $D = \Delta\theta_m/\Delta\lambda$ in the mth order. (*a*) Show that D can be written

$$D = \frac{m}{\sqrt{d^2 - m^2\lambda^2}}$$

where d is the slit spacing. (*b*) If a diffraction grating with 2000 slits per centimeter is to resolve the two yellow sodium lines in the second order (wavelengths 589.0 and 589.6 nm), how many slits must be illuminated by the beam? (*c*) What would the separation be between these resolved yellow lines if the pattern were viewed on a screen 4 m from the grating?

66. Light of wavelength λ is diffracted through a single slit of width a, and the resulting pattern is viewed on a screen a long distance L away from the slit. (*a*) Show that the width of the central maximum on the screen is approximately $2L\lambda/a$. (*b*) If a slit of width $2L\lambda/a$ is cut in the screen and is illuminated, show that the width of its central diffraction maximum at the same distance L is a to the same approximation.

67. A double-slit experiment uses a helium–neon laser with a wavelength of 633 nm and a slit separation of 0.12 mm. When a thin sheet of plastic is placed in front of one of the slits, the interference pattern shifts by 5.5 fringes. When the experiment is repeated under water, the shift is 3.5 fringes. Calculate (*a*) the thickness of the plastic sheet and (*b*) the index of refraction of the plastic sheet.

68. Three slits, each separated from its neighbor by 0.06 mm, are illuminated by a coherent light source of wavelength 550 nm. The slits are extremely narrow, so you may ignore the effects of diffraction. A screen is located 2.5 m from the slits. The intensity on the centerline is 0.05 W/m^2. Consider a location 1.72 cm from the centerline. (*a*) Draw the phasors, according to the phasor model for the addition of harmonic waves, appropriate for this location. (*b*) From the phasor diagram, calculate the intensity of light at this location.

69. Two coherent sources are located on the y axis at $+\lambda/4$ and $-\lambda/4$. They emit waves of wavelength λ and intensity I_0. (*a*) Calculate the net intensity I as a function of the angle θ measured from the $+x$ axis. (*b*) Make a polar plot of $I(\theta)$.

70. Repeat Problem 69 for four sources that are located on the y axis at $+3\lambda/4$, $+\lambda/4$, $-\lambda/4$, and $-3\lambda/4$.

71. For single-slit diffraction, calculate the first three values of ϕ (the total phase difference between rays from each edge of the slit) that produce subsidiary maxima by (*a*) using the phasor model and (*b*) by setting $dI/d\phi = 0$, where I is given by Equation 33-23.

72. For a diffraction grating in which all the surfaces are normal to the incident radiation, most of the energy goes into the zeroth order, which is useless from a spectroscopic point of view since the various wavelengths are not dispersed in angle. Therefore, modern gratings have shaped, or *blazed,* grooves as shown in Figure 33-41. This shifts the specular reflection, which contains most of the energy, from the zeroth order to some higher order. (*a*) Calculate the blaze angle ϕ in terms of a (the groove separation), λ, (the wavelength), and m (the order in which specular reflection is to occur). (*b*) Calculate the proper blaze angle for the specular reflection to occur in the second order for light of wavelength 450 nm incident on a grating with 10,000 lines per centimeter.

73. In this problem you will derive Equation 33-31 for the resolving power of a diffraction grating containing N slits separated by a distance d. To do this you will calculate the angular separation between the maximum and minimum for some wavelength λ and set it equal to the angular separation of the mth-order maximum for two nearby wavelengths. (*a*) Show that the phase difference ϕ between the light from two adjacent slits is given by

$$\phi = \frac{2\pi d}{\lambda} \sin\theta$$

(*b*) Differentiate this expression to show that a small change in angle $d\theta$ results in a change in phase of $d\phi$ given by

$$d\phi = \frac{2\pi d}{\lambda} \cos\theta \, d\theta$$

(*c*) For N slits, the angular separation between an interference maximum and interference minimum corresponds to a phase change of $d\phi = 2\pi/N$. Use this to show that the angular separation $d\theta$ between the maximum and minimum for some wavelength λ is given by

$$d\theta = \frac{\lambda}{Nd\cos\theta} \qquad \text{33-32}$$

(*d*) The angle of the mth-order interference maximum for wavelength λ is given by Equation 33-30. Compute the differential of each side of this equation to show that angular separation of the mth-order maximum for two nearly equal wavelengths differing by $d\lambda$ is given by

$$d\theta \approx \frac{m\, d\lambda}{d\cos\theta} \qquad \text{33-33}$$

(*e*) According to Rayleigh's criterion, two wavelengths will be resolved in the mth-order if the angular separation of the wavelengths given by Equation 33-33 equals the angular separation of the interference maximum and interference minimum given by Equation 33-32. Use this to derive Equation 33-31 for the resolving power of a grating.

Figure 33-41 Problem 72.

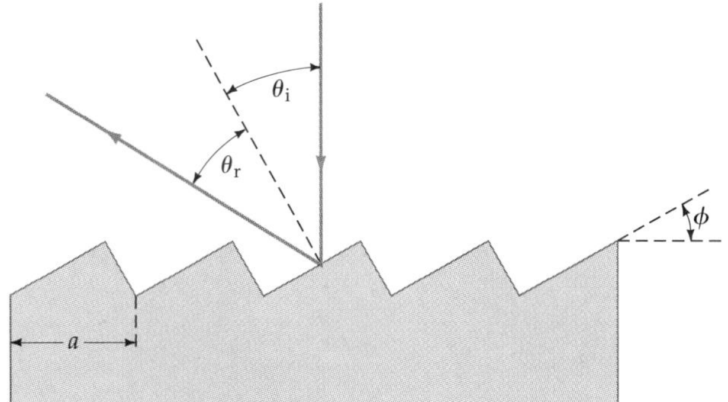

Part 6 Modern Physics

The surface of silicon, an important ingredient in many semiconductor devices, is shown here at a magnification of about 1,000,000,000. Individual atoms of a silicon crystal are seen as hills in this micrograph, which was obtained with a scanning tunneling microscope. The digital information collected by the microscope is plotted by a computer, which assigns false colors to accentuate the crystalline structure.

Chapter 34

Relativity

Albert Einstein in 1916.

Near the end of the nineteenth century, many physicists thought that all the important laws of physics had been discovered and that there was little left for them to do other than work out the remaining details. Newton's laws of motion and gravity seemed to describe all known motion on earth as well as that of the planets and other heavenly bodies, whereas Maxwell's equations of electricity and magnetism seemed to give a complete description of electromagnetic phenomena. Even as evidence of the microscopic world of molecules and atoms began to accumulate, it was assumed that these new phenomena would be adequately described by the theories of Newton and Maxwell. However, the discovery of radioactivity by Becquerel in 1896, the theoretical papers of Planck in 1897 and Einstein in 1905, and the work of Rutherford, Millikan, Bohr, de Broglie, Schrödinger, Heisenberg, and others in the early twentieth century led to two completely new theories: relativity and quantum mechanics. These theories revolutionized the world of science and became the foundation for new technologies that have changed the face of civilization.

In this chapter we will study relativity. The theory of relativity consists of two rather different theories, the special theory and the general theory.

The special theory, developed by Einstein and others in 1905, concerns the comparison of measurements made in different inertial reference frames moving with constant velocity relative to one another. Its consequences, which can be derived with a minimum of mathematics, are applicable in a wide variety of situations encountered in physics and engineering. On the other hand, the general theory, also developed by Einstein and others around 1916, is concerned with accelerated reference frames and gravity. A thorough understanding of the general theory requires sophisticated mathematics, and the applications of this theory are chiefly in the area of gravitation. It is of great importance in cosmology, but it is rarely encountered in other areas of physics or in engineering. We will therefore concentrate on the special theory (often referred to as *special relativity*) and discuss the general theory only briefly in the last section of this chapter.

34-1 Newtonian Relativity

Newton's first law does not distinguish between a particle at rest and one moving with constant velocity. If there is no net external force acting, the particle will remain in its initial state—either at rest or moving with its initial velocity. Consider a particle at rest relative to you with no forces acting on it. According to Newton's first law, it will remain at rest. Now consider the same particle from the point of view of a second observer who is moving with constant velocity relative to you. From this observer's "frame of reference," both you and the particle are moving with constant velocity. Newton's first law holds also for him. (Note that if the second observer were accelerating relative to you, he would see the particle accelerating relative to him with no external forces acting on it. Thus Newton's first law would not hold for him.) How might we distinguish whether you and the particle are at rest and the second observer is moving with constant velocity, or the second observer is at rest and you and the particle are moving?

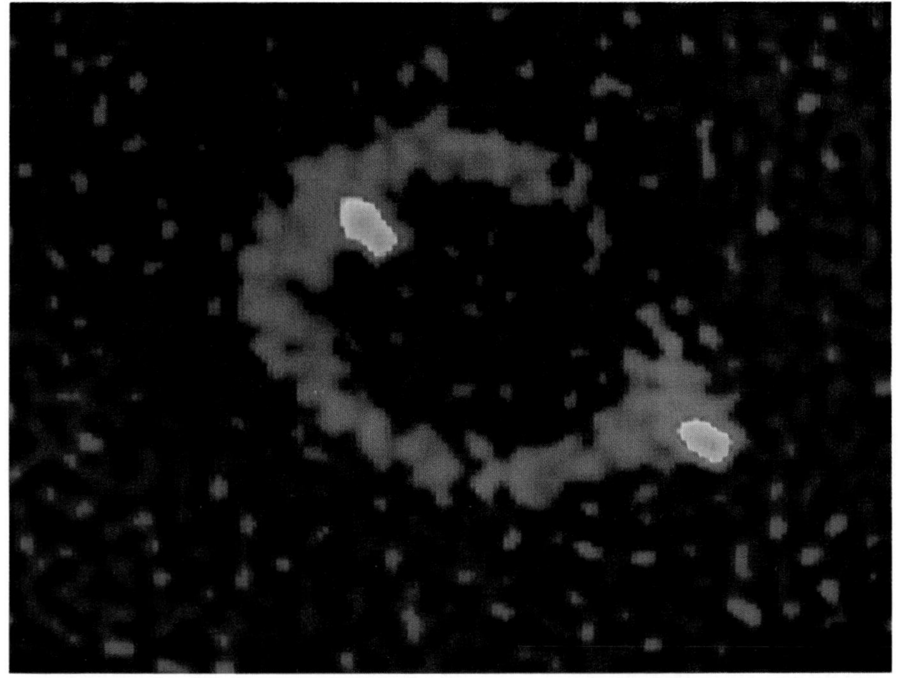

This ringlike structure of the radio source MG1131 + 0456 is thought to be due to "gravitational lensing," first proposed by Einstein in 1936, in which a source is imaged into a ring by a large massive object in the foreground.

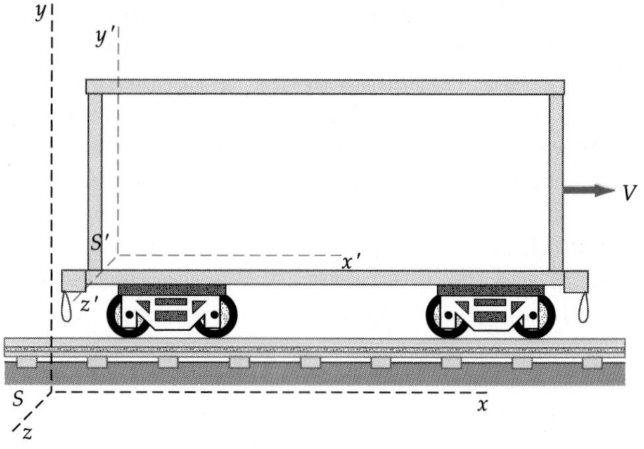

Figure 34-1 A boxcar moving with constant velocity along a straight track. The reference frame S' is at rest relative to the car and is moving with speed V relative to S, which is at rest relative to the track. It is impossible to tell by doing mechanics experiments inside the car whether the car is moving to the right with speed V or the track is moving to the left with speed V.

Let us consider some simple experiments. Suppose we have a train moving along a straight, flat track with a constant velocity V. (We assume there are no bumps or shakes in the motion.) Let us choose a coordinate system xyz with its x axis along the track as shown in Figure 34-1. It doesn't matter where along the track we choose the origin. For different choices, the position (relative to the origin) of the train and its contents will differ, but their velocities will be the same. A set of coordinate systems at rest relative to each other is called a **reference frame.** We will call the reference frame at rest relative to the track frame S. We now consider doing various mechanics experiments in a closed boxcar of the train. For these we choose a coordinate system at rest relative to the train. This coordinate system is in reference frame S', which is moving to the right with speed V relative to frame S. We note that a ball at rest in the train remains at rest. If we drop the ball, it falls straight down in frame S' with an acceleration g due to gravity. (Of course, when viewed in frame S, the ball moves along a parabolic path because it has an initial velocity V to the right.) No mechanics experiment that we can do—measuring the period of a pendulum or a body on a spring, observing the collisions between two bodies, or whatever—will tell us whether the train is moving and the track is at rest or the track is moving and the train is at rest. Newton's laws hold for reference frame S' as well as reference frame S.

A reference frame in which Newton's laws hold is called an **inertial reference frame.**

All reference frames moving at constant velocity relative to an inertial reference frame are also inertial reference frames.

If we have two inertial reference frames moving with constant velocity relative to each other, such as S and S', there are no mechanics experiments that can tell us which is at rest and which is moving or if they are both moving. This result is known as the principle of **newtonian relativity:**

Absolute motion cannot be detected.

This principle was well known by Galileo, Newton, and others in the seventeenth century. By the late nineteenth century, however, this view had changed. It was then generally thought that newtonian relativity was not valid and that absolute motion could be detected in principle by a measurement of the speed of light.

34-2 The Michelson–Morley Experiment

From our study of wave motion, we know that all mechanical waves require a medium for their propagation, and that the speed of such waves depends only on the properties of the medium. For example, the speed of sound waves in air depends on the temperature of the air. This speed is relative to still air. Motion relative to still air can indeed be detected. If you are moving relative to still air, you feel a wind.

It was therefore natural to expect that some kind of medium supports the propagation of light and other electromagnetic waves. This proposed medium was called the **ether.** As proposed, the ether had to have unusual properties. For example, it had to have great rigidity to support waves of such high velocity. (Recall that the velocity of waves on a string depends on the tension of the string and that of longitudinal sound waves in a solid depends on the bulk modulus of the solid.) Yet the ether could introduce no drag force on the planets, as their motion is fully accounted for by the law of gravitation. It was suspected that the ether was at rest relative to the distant stars, but this was considered to be an open question. It was therefore of considerable interest to determine the velocity of the earth relative to the ether. Experiments to do this were undertaken by Albert Michelson, first in 1881 and then again with Edward Morley in 1887 with greater precision. It was thought that a measurement of the speed of light relative to some reference frame moving through the ether would yield a result greater or less than c by an amount that depended on the speed of the frame relative to the ether and the direction of motion relative to the direction of the light beam. Thus, in 1881 Michelson set out to measure the speed of light relative to the earth and from this measurement to determine the velocity of the earth relative to the ether.

According to Maxwell's theory of electromagnetism, the speed of light and other electromagnetic waves is

$$c = \frac{1}{\sqrt{\epsilon_0 \mu_0}} = 3 \times 10^8 \text{ m/s}$$

where ϵ_0 and μ_0 are, respectively, the permittivity and permeability of free space. There is nothing in Maxwell's equations that tells us in what reference frame the speed of light will have this value, but the expectation was that this was the speed of light relative to its natural medium, the ether.

In the usual measurements of the speed of light (Section 30-1), the time it takes for a light pulse to travel to and from a mirror is determined. Figure 34-2 shows a light source and a mirror a distance L apart. If we assume that both are moving with speed v through the ether, classical theory predicts that the light will travel toward the mirror with speed $c - v$ and back with speed $c + v$ (both speeds being relative to the mirror and the light source). The time for the total trip will be

$$t_1 = \frac{L}{c - v} + \frac{L}{c + v} = 2c\,\frac{L}{c^2 - v^2} = \frac{2L}{c}\left(1 - \frac{v^2}{c^2}\right)^{-1} \qquad 34\text{-}1$$

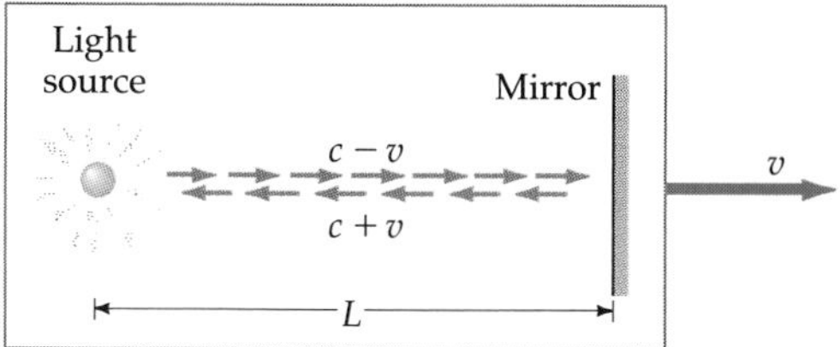

Figure 34-2 A light source and mirror moving with speed v relative to the "ether." According to classical theory, the speed of light relative to the source and mirror is $c - v$ toward the mirror and $c + v$ away from the mirror.

We can see that this differs from the time $2L/c$ by the factor $(1 - v^2/c^2)^{-1}$, which is very nearly equal to 1 if v is much less than c. We can simplify this expression for small values of v/c by using the binomial expansion

$$(1 + x)^n = 1 + nx + n(n - 1)\frac{x^2}{2} + \cdots \simeq 1 + nx \tag{34-2}$$

when x is much less than 1. Using $n = -1$ and $x = v^2/c^2$, Equation 34-1 becomes

$$t_1 \simeq \frac{2L}{c}\left(1 + \frac{v^2}{c^2}\right) \tag{34-3}$$

The orbital speed of the earth about the sun is about 3×10^4 m/s. If we take this for an estimate of v we have $v = 3 \times 10^4$ m/s, $v/c = (3 \times 10^4 \text{ m/s})/(3 \times 10^8 \text{ m/s}) = 10^{-4}$, and $v^2/c^2 = 10^{-8}$. Thus the correction for the earth's motion is small indeed.

Michelson realized that although this effect is too small to be measured directly, it should be possible to determine v^2/c^2 by a difference measurement. For this measurement, he used the Michelson interferometer, which we discussed in Section 33-3. In this experiment, one beam of light moves along the direction of the earth's motion and another moves perpendicular to that direction (Figure 34-3). The difference between the round-trip times of these beams depends on the speed of the earth and can be determined by an interference measurement. Let us assume that the interferometer is oriented such that the beam that strikes mirror M_1 is in the direction of the assumed motion of the earth. Equation 34-3 then gives the classical result for the round-trip time t_1 for the transmitted beam. The beam that reflects from the beam splitter and strikes mirror M_2 travels with some velocity **u** (relative to the earth) perpendicular to the earth's velocity. Relative to the ether, it travels with velocity **c** as shown in Figure 34-4. The velocity **u** (according to classical theory) is then the vector difference $\mathbf{u} = \mathbf{c} - \mathbf{v}$, as shown in the figure. The magnitude of **u** is $\sqrt{c^2 - v^2}$, so the round-trip time t_2 for this beam is

$$t_2 = \frac{2L}{\sqrt{c^2 - v^2}} = \frac{2L}{c}(1 - v^2/c^2)^{-1/2} \tag{34-4}$$

Again using the binomial expansion, we obtain

$$t_2 \simeq \frac{2L}{c}\left(1 + \frac{1}{2}\frac{v^2}{c^2}\right) \tag{34-5}$$

This expression is slightly different from that given for t_1 in Equation 34-3.

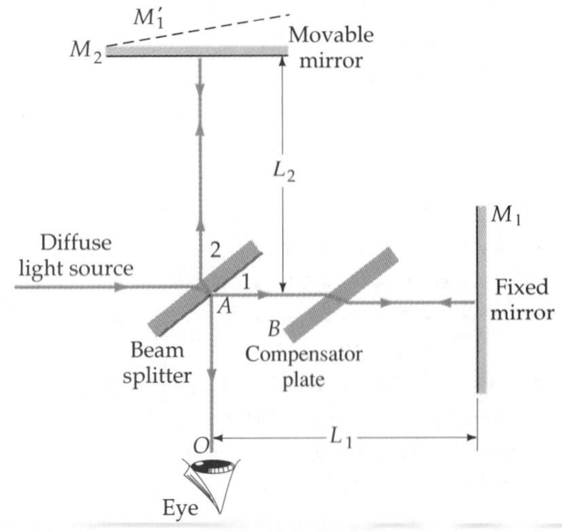

Figure 34-3 Michelson interferometer. The dashed line M_1' is the image of mirror M_1 in mirror A. The interference fringes observed are those of a small wedge-shaped film of air formed between the sources M_2 and M_1'. Assume that the light beam reflecting off mirror M_1 is parallel to the motion of the earth and that reflecting off mirror M_2 is perpendicular to the earth's motion. The interference of the two beams depends on the relative number of waves in each path, which depends in turn on the speed of the light beams relative to the earth. If the speed of light along the parallel path is different from that along the perpendicular path, the interference fringe pattern will shift when the interferometer is rotated through 90°.

The difference in these two times is

$$\Delta t = t_1 - t_2 \approx \frac{L}{c}\frac{v^2}{c^2} \qquad 34\text{-}6$$

This time difference is to be detected by observing the interference of the two beams of light.

Because of the difficulty of making the two paths of equal length to the precision required, the interference pattern of the two beams is observed and the whole apparatus is then rotated 90°. The rotation produces a time difference given by Equation 34-6 for each beam. The total time difference of 2 Δt results in a phase difference $\Delta\phi$ between the two beams, where

$$\Delta\phi = 2\pi\,\frac{2c\,\Delta t}{\lambda}$$

and λ is the wavelength of the light. The interference fringes observed in the first orientation should thus shift by a number of fringes ΔN given by

$$\Delta N = \frac{\Delta\phi}{2\pi} = \frac{2c\,\Delta t}{\lambda} = \frac{2L}{\lambda}\frac{v^2}{c^2} \qquad 34\text{-}7$$

In Michelson's first attempt in 1881, L was about 1.2 m and λ was 590 nm. For $v^2/c^2 = 10^{-8}$, ΔN was expected to be 0.04 fringes. However, no shift was observed. In case the earth just happened to be at rest relative to the ether when the experiment was performed, the experiment was repeated six months later when the motion of the earth relative to the sun was in the opposite direction. Even though the experimental uncertainties were estimated to be of about the same magnitude as the expected fringe shift, Michelson reported the observation of no fringe shift as evidence that the earth did not move relative to the ether. In 1887, when he repeated the experiment with Edward W. Morley, he used an improved system for rotating the apparatus without introducing a fringe shift due to mechanical strains, and he increased the effective path length L to about 11 m by a series of multiple reflections. Figure 34-5 shows the configuration of the Michelson–Morley apparatus. For this attempt, ΔN was expected to be 0.4 fringes, about 20 to 40 times the minimum that could be observed. Once again, no shift was observed. The experiment has since been repeated under various conditions by a number of people, and no shift has ever been found.

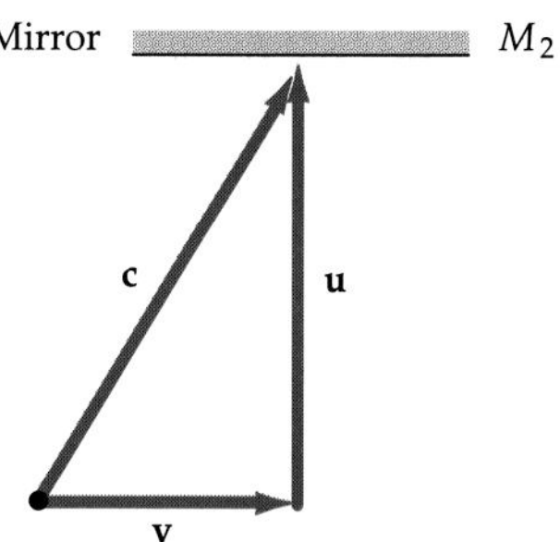

Figure 34-4 The light beam reflected from the beam splitter in a Michelson interferometer. The interferometer is moving to the right relative to the ether with velocity **v**, and the light beam moves perpendicular to mirror M_2 with velocity **u**. The velocity of light is **c** in the frame of the ether. Relative to the earth, in which the interferometer is fixed, the velocity of light is $\mathbf{u} = \mathbf{c} - \mathbf{v}$. The speed of light relative to the earth according to classical theory is then $u = (c^2 - v^2)^{1/2} = c(1 - v^2/c^2)^{1/2}$.

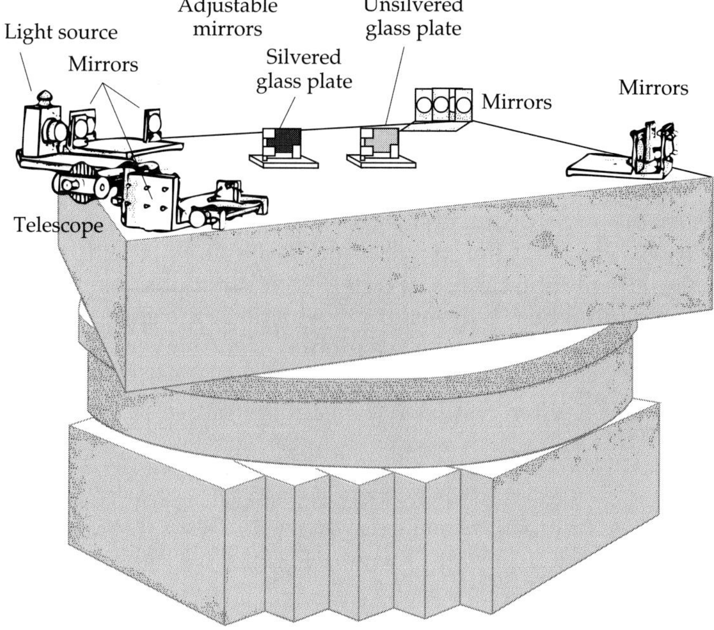

Figure 34-5 Drawing of Michelson and Morley's apparatus for their 1887 experiment. The optical parts were mounted on a sandstone slab 5 ft square, which was floated in mercury to reduce the strains and vibrations that had affected the earlier experiments. Observations could be made in all directions by rotating the apparatus in the horizontal plane.

In 1905, at the age of 26, Albert Einstein published a paper on the electrodynamics of moving bodies.* In this paper, he postulated that absolute motion cannot be detected by any experiment. (We will discuss the Einstein postulates in detail in the next section.) The null result of the Michelson–Morley experiment is therefore expected. We can consider the whole apparatus and the earth to be at rest. Thus no fringe shift is expected when the apparatus is rotated 90° since all directions are equivalent. Einstein did not set out to explain the results of the Michelson–Morley experiment. His theory arose from his considerations of the theory of electricity and magnetism and the unusual property of electromagnetic waves that they propagate in a vacuum. In his first paper, which contains the complete theory of special relativity, he made only a passing reference to the Michelson–Morley experiment, and in later years he could not recall whether he was aware of the details of this experiment before he published his theory.

34-3 Einstein's Postulates

The theory of special relativity can be derived from two postulates proposed by Einstein in his original paper in 1905. Simply stated, these postulates are

Einstein's postulates

> Postulate 1. Absolute, uniform motion cannot be detected.
>
> Postulate 2. The speed of light is independent of the motion of the source.

Postulate 1 is merely an extension of the newtonian principle of relativity to include all types of physical measurements (not just those that are mechanical). Postulate 2 describes a common property of all waves. For example, the speed of sound waves does not depend on the motion of the sound source. When an approaching car sounds its horn, the frequency heard increases according to the doppler effect we studied in Section 14-6, but the speed of the waves traveling through the air does not depend on the speed of the car. The speed of the waves depends only on the properties of the air, such as its temperature.

Although each postulate seems quite reasonable, many of the implications of the two together are quite surprising and contradict what is often called common sense. For example, one important implication of these postulates is that every observer measures the same value for the speed of light independent of the relative motion of the source and the observer. Consider a light source S and two observers, R_1 at rest relative to S and R_2 moving toward S with speed v, as shown in Figure 34-6*a*. The speed of light measured by R_1 is $c = 3 \times 10^8$ m/s. What is the speed measured by R_2? The answer is *not* $c + v$. By postulate 1, Figure 34-6*a* is equivalent to Figure 34-6*b*, in which R_2 is at rest and the source S and R_1 are moving with speed v. That is, since absolute motion cannot be detected, it is not possible to say which is really moving and which is at rest. By postulate 2, the speed of light from a moving source is independent of the motion of the source. Thus, looking at Figure 34-6*b*, we see that R_2 measures the speed of light to be c, just as R_1 does. This result is often considered as an alternative to Einstein's second postulate:

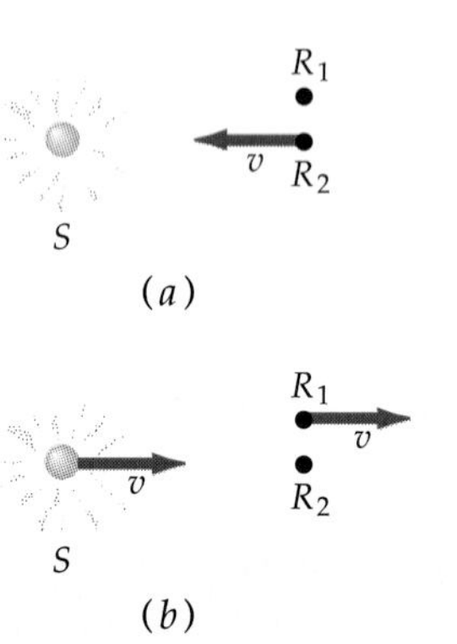

Figure 34-6 (*a*) A stationary light source S and a stationary observer R_1, with a second observer R_2 moving toward the source with speed v. (*b*) In the reference frame in which the observer R_2 is at rest, the light source S and observer R_1 move to the right with speed v. If absolute motion cannot be detected, the two views are equivalent. Since the speed of light does not depend on the motion of the source, observer R_2 measures the same value for that speed as observer R_1.

* *Annalen der Physik*, vol. 17, 1905, p. 841. For a translation from the original German, see W. Perrett and G.B. Jeffery (trans.), *The Principle of Relativity: A collection of Original Memoirs on the Special and General Theory of Relativity* by H. A. Lorentz, A. Einstein, H. Minkowski, and W. Weyl, Dover, New York, 1923.

Postulate 2 (Alternate). Every observer measures the same value c for the speed of light.

This result contradicts our intuitive ideas about relative velocities. If a car moves at 50 km/h away from an observer and another car moves at 80 km/h in the same direction, the velocity of the second car relative to the first car is 30 km/h. This result is easily measured and conforms to our intuition. However, according to Einstein's postulates, if a light beam is moving in the direction of the cars, observers in both cars will measure the same speed for the light beam. Our intuitive ideas about the combination of velocities are approximations that hold only when the speeds are very small compared with the speed of light. Even in an airplane moving with the speed of sound, it is not possible to measure the speed of light accurately enough to distinguish the difference between the results c and $c + v$, where v is the speed of the plane. In order to make such a distinction, we must either move with a very great velocity (much greater than that of sound) or make extremely accurate measurements, as in the Michelson–Morley experiment.

34-4 The Lorentz Transformation

Einstein's postulates have important consequences for measuring time intervals and space intervals as well as relative velocities. Throughout this chapter we will be comparing measurements of the positions and times of events (such as lightning flashes) made by observers who are moving relative to each other. We will use a rectangular coordinate system xyz with origin O, called the S reference frame, and another system $x'y'z'$ with origin O', called the S' frame, that is moving with a constant velocity $\mathbf{V}$ relative to the S frame. Relative to the S' frame, the S frame is moving with a constant velocity $-\mathbf{V}$. For simplicity, we will consider the S' frame to be moving with speed V along the x axis in the positive x direction relative to S. Then, relative to S', the S frame is moving with speed V in the negative x' direction along the x' axis. In each frame, we will assume that there are as many observers as are needed who are equipped with measuring devices, such as clocks and metersticks, that are identical when compared at rest (see Figure 34-7).

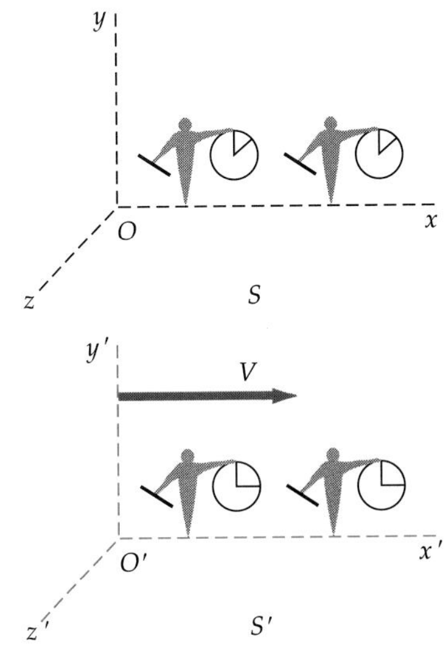

Figure 34-7 Coordinate reference frames S and S' moving with relative speed V. In each frame, there are observers with metersticks and clocks that are identical when compared at rest.

We need many observers, for example, to determine the times of events. If one observer is distant from an event, then his time observation can be thrown off by the time it takes for the information about the event to travel to his location (such as the transit time for a light pulse). The observer can avoid such problems by recording only events *local* to him, and leaving other events to other observers at those locations. It's like having one official at the beginning of a racetrack and another at the end.

We will use Einstein's postulates to find the general relation between the coordinates x, y, and z and the time t of an event as seen in reference frame S and the coordinates x', y', and z' and the time t' of the same event as seen in reference frame S', which is moving with uniform velocity relative to S. We will consider only the simple case in which the origins are coincident at time $t = t' = 0$. The classical relation, called the **galilean transformation,** is

$$x = x' + Vt' \qquad y = y' \qquad z = z' \qquad t = t' \tag{34-8a}$$

Galilean transformation

The inverse transformation is

$$x' = x - Vt \qquad y' = y \qquad z' = z \qquad t' = t \tag{34-8b}$$

These equations are consistent with experimental observations as long as V is much less than c. They lead to the familiar classical addition law for velocities. If a particle has velocity $u_x = dx/dt$ in frame S, its velocity in frame S' is

$$u'_x = \frac{dx'}{dt'} = \frac{dx'}{dt} = \frac{dx}{dt} - V = u_x - V \qquad \text{34-9}$$

If we differentiate this equation again, we find that the acceleration of the particle is the same in both frames:

$$a_x = du_x/dt = du'_x/dt' = a'_x$$

It should be clear that the galilean transformation is not consistent with Einstein's postulates of special relativity. If light moves along the x axis with speed c in S, these equations imply that the speed in S' is $u'_x = c - V$; rather than $u'_x = c$, which is consistent with Einstein's postulates and with experiment. The classical transformation equations must therefore be modified to make them consistent with Einstein's postulates. We will give a brief outline of one method of obtaining the relativistic transformation.

We assume that the relativistic transformation equation for x is the same as the classical equation (Equation 34-8*a*) except for a constant multiplier on the right side. That is, we assume the equation is of the form

$$x = \gamma(x' + Vt') \qquad \text{34-10}$$

where γ is a constant that can depend on V and c but not on the coordinates. The inverse transformation must look the same except for the sign of the velocity:

$$x' = \gamma(x - Vt) \qquad \text{34-11}$$

Let us consider a light pulse that starts at the origin of S at $t = 0$. Since we have assumed that the origins are coincident at $t = t' = 0$, the pulse also starts at the origin of S' at $t' = 0$. Einstein's postulates require that the equation for the x component of the wavefront of the light pulse is $x = ct$ in frame S and $x' = ct'$ in frame S'. Substituting ct for x and ct' for x' in Equations 34-10 and 34-11, we obtain

$$ct = \gamma(ct' + Vt') = \gamma(c + V)t' \qquad \text{34-12}$$

and

$$ct' = \gamma(ct - Vt) = \gamma(c - V)t \qquad \text{34-13}$$

We can eliminate either t' or t from these two equations and determine γ. We get

$$\gamma^2 = (1 - V^2/c^2)^{-1}$$

$$\gamma = \frac{1}{\sqrt{1 - V^2/c^2}} \qquad \text{34-14}$$

(It is important to note that γ is always greater than 1 and that when V is much less than c, $\gamma \approx 1$.) The relativistic transformation for x and x' is therefore given by Equations 34-10 and 34-11 with γ given by Equation 34-14. We can obtain equations for t and t' by combining Equation 34-10 with the inverse transformation given by Equation 34-11. Substituting $x = \gamma(x' + Vt')$ for x in Equation 34-11, we obtain

$$x' = \gamma[\gamma(x' + Vt') - Vt] \qquad \text{34-15}$$

which can be solved for t in terms of x' and t'. The complete relativistic transformation is

Lorentz transformation

$$x = \gamma(x' + Vt') \qquad y = y' \qquad z = z' \tag{34-16}$$

$$t = \gamma\left(t' + \frac{Vx'}{c^2}\right) \tag{34-17}$$

The inverse is

$$x' = \gamma(x - Vt) \qquad y' = y \qquad z' = z \tag{34-18}$$

$$t' = \gamma\left(t - \frac{Vx}{c^2}\right) \tag{34-19}$$

The transformation described by Equations 34-16 through 34-19 is called the **Lorentz transformation.** It relates the space and time coordinates x, y, z, and t of an event in frame S to the coordinates x', y', z', and t' of the same event as seen in frame S', which is moving along the x axis with speed V relative to frame S.

We will now look at some applications of the Lorentz transformation.

Time Dilation

An important consequence of Einstein's postulates and the Lorentz transformation is that the time interval between two events that occur at the same place in some reference frame is always less than the time interval between the same events that is measured in another reference frame in which the events occur at different places. Consider two events that occur at x'_0 at times t'_1 and t'_2 in frame S'. We can find the times t_1 and t_2 for these events in S from Equation 34-17. We have

$$t_1 = \gamma\left(t'_1 + \frac{Vx'_0}{c^2}\right)$$

and

$$t_2 = \gamma\left(t'_2 + \frac{Vx'_0}{c^2}\right)$$

so

$$t_2 - t_1 = \gamma(t'_2 - t'_1)$$

The time between events that happen at the *same place* in a reference frame is called **proper time** t_p. In this case, the time interval $\Delta t_p = t'_2 - t'_1$ measured in frame S' is proper time. The time interval Δt measured in any other reference frame is always longer than the proper time. This expansion is called **time dilation:**

Time dilation

$$\Delta t = \gamma\, \Delta t_p \tag{34-20}$$

Example 34-1

Two events occur at the same point x'_0 at times t'_1 and t'_2 in frame S', which is traveling at speed V relative to frame S. What is the spatial separation of these events in frame S?

From Equation 34-16, we have

$$x_1 = \gamma(x'_0 + Vt'_1)$$

and

$$x_2 = \gamma(x'_0 + Vt'_2)$$

Then

$$x_2 - x_1 = \gamma V(t'_2 - t'_1)$$
$$= V(t_2 - t_1)$$

The spatial separation of these events in S is the distance a single point, such as x'_0 in S', moves in S during the time interval between the events.

We can understand time dilation directly from Einstein's postulates without using the Lorentz transformation. Figure 34-8*a* shows an observer A' a distance D from a mirror. The observer and the mirror are in a spaceship that is at rest in frame S'. He explodes a flash gun and measures the time interval $\Delta t'$ between the original flash and his seeing the return flash from the mirror. Since light travels with speed c, this time is

$$\Delta t' = \frac{2D}{c}$$

We now consider these same two events, the original flash of light and the receiving of the return flash, as observed in reference frame S, in which observer A' and the mirror are moving to the right with speed V as shown in Figure 34-8*b*. The events happen at two different places x_1 and x_2 in frame S. During the time interval Δt (as measured in S) between the original flash and the return flash, observer A' and his spaceship have moved a horizontal distance $V\,\Delta t$. In Figure 34-8*b*, we can see that the path traveled by the light is longer in S than in S'. However, by Einstein's postulates, light travels with the same speed c in frame S as it does in frame S'. Since it travels farther in S at the same speed, it takes longer in S to reach the mirror and return. The time interval in S is thus longer than it is in S'. From the triangle in Figure 34-8*c*, we have

$$\left(\frac{c\,\Delta t}{2}\right)^2 = D^2 + \left(\frac{V\,\Delta t}{2}\right)^2$$

or

$$\Delta t = \frac{2D}{\sqrt{c^2 - V^2}} = \frac{2D}{c}\,\frac{1}{\sqrt{1 - V^2/c^2}}$$

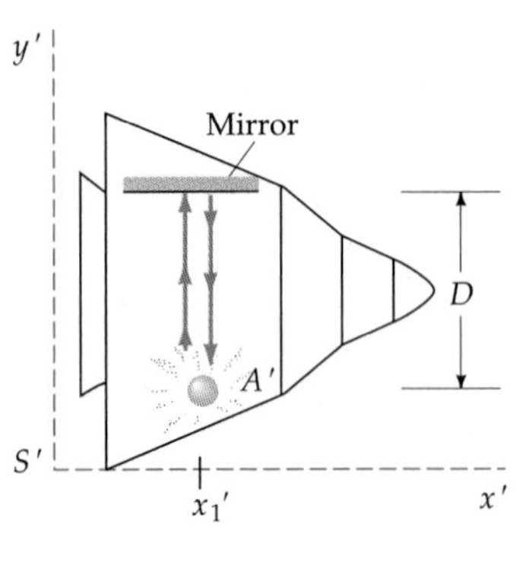

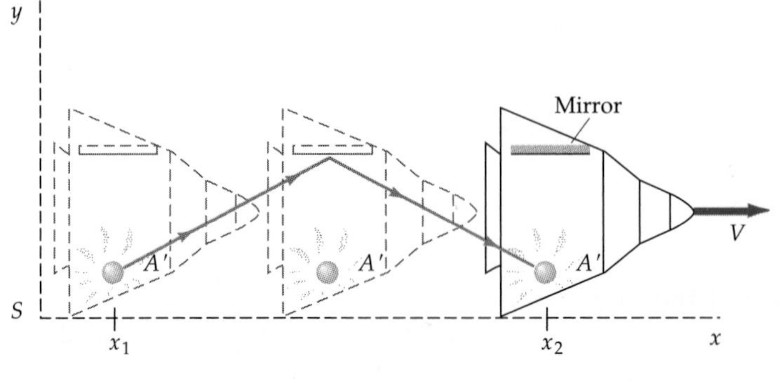

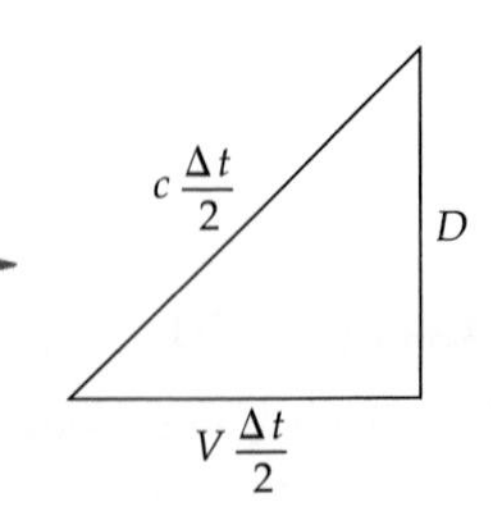

Figure 34-8 (*a*) Observer A' and the mirror are in a spaceship at rest in frame S'. The time it takes for the light pulse to reach the mirror and return is measured by A' to be $2D/c$. (*b*) In frame S, the spaceship is moving to the right with speed V. If the speed of light is the same in both frames, the time it takes for the light to reach the mirror and return is longer than $2D/c$ in S because the distance traveled is greater than $2D$. (*c*) A right triangle for computing the time Δt in frame S.

Using $\Delta t' = 2D/c$, we obtain

$$\Delta t = \frac{\Delta t'}{\sqrt{1 - V^2/c^2}} = \gamma\, \Delta t'$$

Example 34-2

Astronauts in a spaceship traveling away from the earth at $V = 0.6c$ sign off from space control, saying that they are going to nap for 1 hour and then call back. How long does their nap last as measured on earth?

Since the astronauts go to sleep and wake up at the same place in their reference frame, the time interval for their nap of 1 hour as measured by them is proper time. In the earth's reference frame, they move a considerable distance between these two events. The time interval measured in the earth's frame (using two clocks located at those events) is longer by the factor γ. With $V = 0.6c$, we have

$$1 - \frac{V^2}{c^2} = 1 - (0.6)^2 = 0.64$$

Then γ is

$$\gamma = \frac{1}{\sqrt{1 - V^2/c^2}} = \frac{1}{\sqrt{0.64}} = \frac{1}{0.8} = 1.25$$

The nap thus lasts for 1.25 hours as measured on earth.

Exercise

If the spaceship in Example 34-2 is moving at $V = 0.8c$, how long would a 1-hour nap last as measured on earth? (Answer: 1.67 h)

Length Contraction

A phenomenon closely related to time dilation is **length contraction.** The length of an object measured in the reference frame in which the object is at rest is called its **proper length** L_p. In a reference frame in which the object is moving, the measured length is shorter than its proper length. Consider a rod at rest in frame S' with one end at x_2' and the other end at x_1'. The length of the rod in this frame is its proper length $L_p = x_2' - x_1'$. Some care must be taken to find the length of the rod in frame S. In this frame, the rod is moving to the right with speed V, the speed of frame S'. The length of the rod in frame S is *defined* as $L = x_2 - x_1$, where x_2 is the position of one end at some time t_2, and x_1 is the position of the other end *at the same time* $t_1 = t_2$ as measured in frame S. Equation 34-18 is convenient to use to calculate $x_2 - x_1$ at some time t because it relates x, x' and t, whereas Equation 34-16 is not convenient because it relates x, x', and t':

$$x_2' = \gamma(x_2 - Vt_2)$$

and

$$x_1' = \gamma(x_1 - Vt_1)$$

Since $t_2 = t_1$, we obtain

$$x_2' - x_1' = \gamma(x_2 - x_1)$$

$$x_2 - x_1 = \frac{1}{\gamma}(x_2' - x_1') = \sqrt{1 - V^2/c^2}\,(x_2' - x_1')$$

or

Length contraction

$$L = \frac{1}{\gamma} L_p = \sqrt{1 - V^2/c^2}\, L_p \qquad 34\text{-}21$$

Thus the length of a rod is smaller when it is measured in a frame in which it is moving. Before Einstein's paper was published, Lorentz and FitzGerald tried to explain the null result of the Michelson–Morley experiment by assuming that distances in the direction of motion contracted by the amount given in Equation 34-21. This contraction is now known as the **Lorentz–FitzGerald contraction.**

Example 34-3

A stick that has a proper length of 1 m moves in a direction along its length with speed V relative to you. The length of the stick as measured by you is 0.914 m. What is the speed V?

The length of the stick measured in a frame in which it is moving with speed V is related to its proper length by Equation 34-21:

$$L = \frac{L_p}{\gamma}$$

Then

$$\gamma = \frac{L_p}{L} = \frac{1 \text{ m}}{0.914 \text{ m}} = \frac{1}{\sqrt{1 - V^2/c^2}} = 1.094$$

$$\sqrt{1 - V^2/c^2} = 0.914$$

$$1 - \frac{V^2}{c^2} = (0.914)^2 = 0.835$$

$$\frac{V^2}{c^2} = 1 - 0.835 = 0.165$$

$$V = 0.406c$$

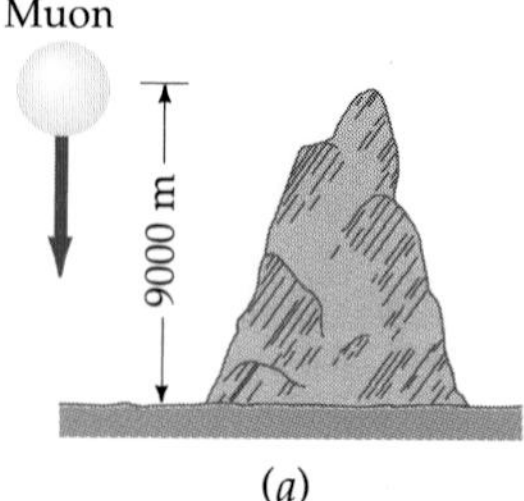

(a)

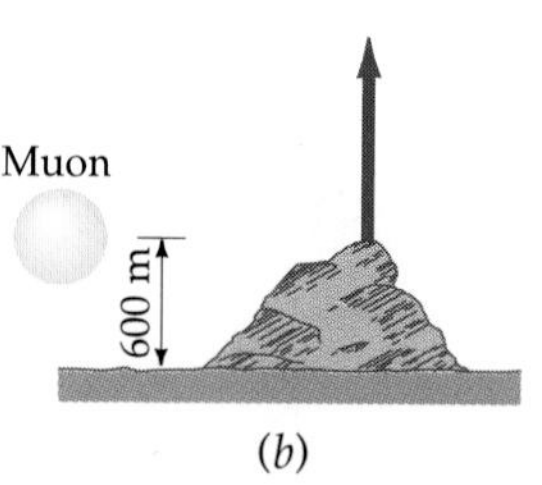

(b)

Figure 34-9 Although muons are created high above the earth and their mean lifetime is only about 2 μs when at rest, many appear at the earth's surface. (*a*) In the earth's reference frame, a typical muon moving at $0.998c$ has a mean lifetime of 30 μs and travels 9000 m in this time. (*b*) In the reference frame of the muon, the distance traveled by the earth is only 600 m in the muon's lifetime of 2 μs.

An interesting example of time dilation or length contraction is afforded by the appearance of muons as secondary radiation from cosmic rays. Muons decay according to the statistical law of radioactivity:

$$N(t) = N_0\, e^{-t/\tau} \qquad 34\text{-}22$$

where N_0 is the original number of muons at time $t = 0$, $N(t)$ is the number remaining at time t, and τ is the mean lifetime, which is about 2 μs for muons at rest. Since muons are created (from the decay of pions) high in the atmosphere, usually several thousand meters above sea level, few muons should reach sea level. A typical muon moving with speed $0.998c$ would travel only about 600 m in 2 μs. However, the lifetime of the muon measured in the earth's reference frame is increased by the factor $1/\sqrt{1 - V^2/c^2}$, which is 15 for this particular speed. The mean lifetime measured in the earth's reference frame is therefore 30 μs, and a muon with speed $0.998c$ travels about 9000 m in this time. From the muon's point of view, it lives only 2 μs, but the atmosphere is rushing past it with a speed of $0.998c$. The distance of 9000 m in the earth's frame is thus contracted to only 600 m in the muon's frame as indicated in Figure 34-9.

It is easy to distinguish experimentally between the classical and relativistic predictions of the observation of muons at sea level. Suppose that we observe 10^8 muons at an altitude of 9000 m in some time interval with a muon detector. How many would we expect to observe at sea level in the same time interval? According to the nonrelativistic prediction, the time it takes for these muons to travel 9000 m is $(9000 \text{ m})/0.998c \approx 30\ \mu\text{s}$, which is 15 lifetimes. Substituting $N_0 = 10^8$ and $t = 15\tau$ into Equation 34-22, we obtain

$$N = 10^8\, e^{-15} = 30.6$$

We would thus expect all but about 31 of the original 100 million muons to decay before reaching sea level.

According to the relativistic prediction, the earth must travel only the contracted distance of 600 m in the rest frame of the muon. This takes only $2\ \mu\text{s} = 1\tau$. Therefore the number of muons expected at sea level is

$$N = 10^8\, e^{-1} = 3.68 \times 10^7$$

Thus relativity predicts that we would observe 36.8 million muons in the same time interval. Experiments of this type have confirmed the relativistic predictions.

Question

1. You are standing on a corner and a friend is driving past in an automobile. Both of you note the times when the car passes two different intersections and determine from your watch readings the time that elapses between the two events. Which of you has determined the proper time interval?

34-5 Clock Synchronization and Simultaneity

We saw in Section 34-4 that proper time is the time interval between two events that occur at the same point in some reference frame. It can therefore be measured on a single clock. However, in another reference frame moving relative to the first, the same two events occur at different places, so two clocks are needed to record the times. The time of each event is measured on a different clock, and the interval is found by subtraction. This procedure requires that the clocks be **synchronized.** We will show in this section that

> Two clocks that are synchronized in one reference frame are not synchronized in any other frame moving relative to the first frame.

A corollary to this result is

> Two events that are simultaneous in one reference frame are not simultaneous in another frame moving relative to the first.

(This is true unless the events and clocks are in the same plane and are perpendicular to the relative motion). Comprehension of these facts usually resolves all relativity paradoxes. Unfortunately, the intuitive (and incorrect) belief that simultaneity is an absolute relation is difficult to overcome.

Suppose we have two clocks at rest at points A and B a distance L apart in frame S. How can we synchronize these two clocks? If an observer at A looks at the clock at B and sets her clock to read the same time, the clocks will not be synchronized because of the time L/c it takes light to travel from one clock to another. To synchronize the clocks, the observer at A must set her clock ahead by the time L/c. Then she will see that the clock at B reads a time that is L/c behind the time on her clock, but she will calculate that the clocks are synchronized when she allows for the time L/c for the light to reach her. Any other observers except those equidistant from the clocks will see the clocks reading different times, but they will also calculate that the clocks are synchronized when they correct for the time it takes the light to reach them. An equivalent method for synchronizing two clocks would be for an observer C at a point midway between the clocks to send a light signal and for observers at A and B to set their clocks to some prearranged time when they receive the signal.

We now examine the question of **simultaneity.** Suppose A and B agree to explode flashguns at t_0 (having previously synchronized their clocks). Observer C will see the light from the two flashes at the same time, and since he is equidistant from A and B, he will conclude that the flashes were simultaneous. Other observers in frame S will see the light from A or B first, depending on their location, but after correcting for the time the light takes to reach them, they also will conclude that the flashes were simultaneous. We can thus define simultaneity as follows:

> Two events in a reference frame are simultaneous if light signals from the events reach an observer halfway between the events at the same time.

To show that two events that are simultaneous in frame S are not simultaneous in another frame S' moving relative to S, we will use an example introduced by Einstein. A train is moving with speed V past a station platform. We will consider the train to be at rest in S' and the platform to be at rest in S. We have observers A', B', and C' at the front, back, and middle of the train. We now suppose that the train and platform are struck by lightning at the front and back of the train and that the lightning bolts are simultaneous in the frame of the platform (S) (Figure 34-10). That is, an observer C on the platform halfway between the positions A and B, where the lightning strikes, sees the two flashes at the same time. It is convenient to suppose that the lightning scorches the train and platform so that the events can be easily located in each reference frame. Since C' is in the middle of the train, halfway between the places on the train that are scorched, the events are simultaneous in S' only if C' sees the flashes at the same time. However, the flash from the front of the train is seen by C' before the flash from the back of the train. We can understand this by considering the motion of C' as seen in frame S (Figure 34-11). By the time the light from the front flash reaches C', C' has moved some distance toward the front flash and some distance away from the back flash. Thus the light from the back flash has not yet reached C' as indicated in the figure. Observer C' must therefore conclude that the events are not simultaneous and that the front of the train was struck before the back. Furthermore, all observers in S' on the train will agree with C' when they have corrected for the time it takes the light to reach them.

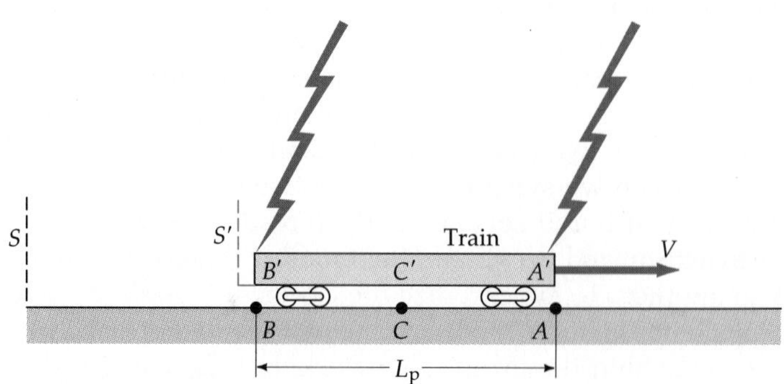

Figure 34-10 Simultaneous lightning bolts strike the ends of a train traveling with speed V in frame S attached to the platform. The light from these simultaneous events reaches observer C midway between the events at the same time. The distance between the bolts is $L_{p,\text{platform}}$.

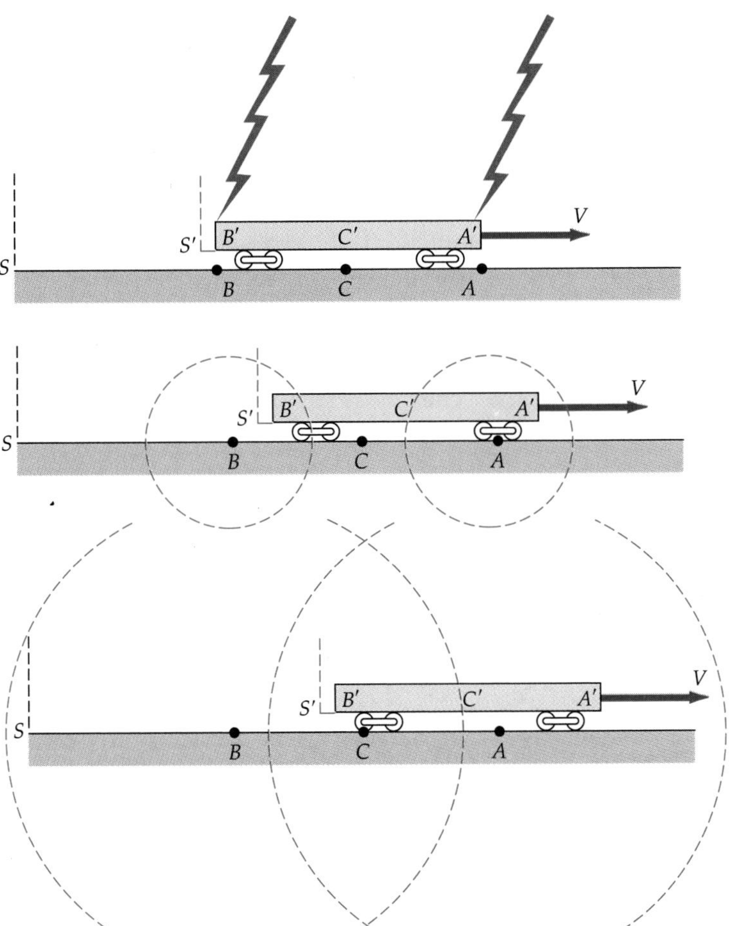

Figure 34-11 The light from the lightning bolt at the front of the train reaches observer C' at the middle of the train before that from the bolt at the back of the train. Since C' is midway between the events (which occur at the front and rear of the train), these events are not simultaneous for him.

Let $L_{\text{p,train}}$ be the proper length of the train, that is, the length of the train as measured in S' in which it is at rest. Also, let $L_{\text{p,platform}}$ be the proper length of the platform, that is, the distance between the scorch marks as seen in S. Since the scorch marks on the platform coincide with the front and back of the train at the instant (in S) that the lightning bolts strike, the distance between the scorch marks $L_{\text{p,platform}}$ equals the length of the train L_{T} as measured in frame S in which it is moving. This length is smaller than the proper length of the train because of length contraction; that is, $L_{\text{T}} = L_{\text{p,platform}} < L_{\text{p,train}}$.

Figure 34-12 shows the events of the lightning bolts as seen in the reference frame of the train (S') in which the train is at rest and the platform is moving. In this frame, the distance between the burns on the platform is contracted, so the platform is shorter than it is in S, and the train is at rest, so the train is longer than it is in S. When the lightning bolt strikes the front of the train at A', the front of the train is at point A, and the back of the train has not yet reached point B. Later, when the lightning bolt strikes the back of the train at B', the back has reached point B on the platform.

Figure 34-12 The lightning bolts of Figure 34-10 as seen in frame S' of the train. In this frame, the distance between A and B on the platform is less than $L_{\text{p,platform}}$, and the proper length of the train $L_{\text{p,train}}$ is longer than $L_{\text{p,platform}}$. The first lightning bolt strikes the front of the train when A' and A are coincident. The second bolt strikes the rear of the train when B' and B are coincident.

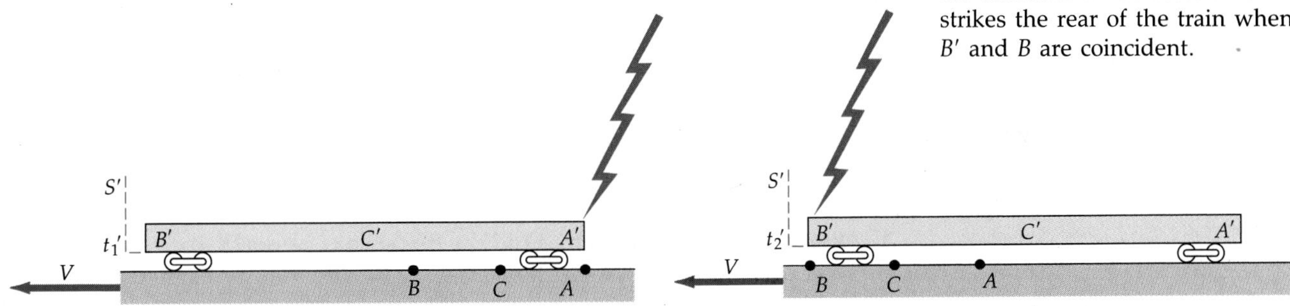

In reference frame S, the lightning bolts strike A and B simultaneously. Suppose there are clocks on the platform at A and B that are synchronized in frame S. From the point of view of frame S' attached to the train, the clocks and the platform are moving past the train. Lightning first strikes the front of the train, which is at point A, and some time later, lightning strikes the back of the train, which is now at point B. The moving clocks are thus not synchronized as seen from frame S'. If the clock at A reads 12:00 noon when the lightning bolt strikes A, the clock at B must read some time before 12:00 noon at that time. The clock at B reads 12:00 noon later when it reaches the back of the train and the lightning bolt strikes at B. Another way of saying this is that the clock at A leads the clock at B as seen in S'. In frame S', we will call the clock at A the "chasing" clock because in that frame the two clocks are moving in the negative x' direction with clock A at x_2 following, or chasing, clock B at x_1.

The time discrepancy of two clocks that are synchronized in frame S as seen in frame S' can be found from the Lorentz transformation equations. Suppose we have clocks at points x_1 and x_2 that are synchronized in S. What are the times t_1 and t_2 on these clocks as observed from frame S' at a time t_0'? From Equation 34-19, we have

$$t_0' = \gamma\left(t_1 - \frac{Vx_1}{c^2}\right)$$

and

$$t_0' = \gamma\left(t_2 - \frac{Vx_2}{c^2}\right)$$

Then

$$t_2 - t_1 = \frac{V}{c^2}(x_2 - x_1) \qquad 34\text{-}23$$

Note that the chasing clock (at x_2) leads the other (at x_1) by an amount that is proportional to their proper separation $x_2 - x_1$.

> If two clocks are synchronized in the frame in which they are at rest, they will be out of synchronization in another frame. In the frame in which they are moving, the chasing clock leads (shows a later time) by an amount
>
> $$\Delta t_s = L_p \frac{V}{c^2}$$
>
> where L_p is the proper distance between the clocks.

A numerical example should help clarify time dilation, clock synchronization, and the internal consistency of these results.

Example 34-4

An observer in a spaceship has a flash gun and a mirror (as in our time dilation example in Figure 34-8). The distance from the gun to the mirror is 15 light-minutes (written 15c·min) and the spaceship travels with speed $V = 0.8c$. The spaceship travels past a very long space platform that has two synchronized clocks, one at the position of the spaceship when the observer explodes the flash gun and the other at the position of the spaceship when the light returns to the gun from the mirror. Find the time intervals between the events (exploding the flash gun and receiving the return flash from the mirror) in the frame of the ship and in the frame of the platform. Find the distance traveled by the ship, and the amount by which the clocks on the platform are out of synchronization as viewed by the ship.

We will call the frame of the spaceship S' and that of the platform S. In the spaceship, the light travels from the gun to the mirror and back, a total distance $D = 30c\cdot\text{min}$. The time it takes for light to travel $30c\cdot\text{min}$ is

$$\Delta t' = \frac{D}{c} = \frac{(30c\cdot\text{min})}{c} = 30\text{ min}$$

Since these events happen at the same place in the spaceship, the time interval is proper time:

$$\Delta t_{\text{p}} = D/c = 30\text{ min}$$

During this time, the platform travels backwards past the ship a distance equal to the distance L' between the platform clocks measured in frame S':

$$L' = \Delta x' = V\,\Delta t' = (0.8c)(30\text{ min}) = 24c\cdot\text{min}$$

In frame S, the time between the events is longer by the factor γ. Since $V/c = 0.8$, $1 - V^2/c^2 = 1 - 0.64 = 0.36$. The factor γ is thus

$$\gamma = \frac{1}{\sqrt{1 - V^2/c^2}} = \frac{1}{\sqrt{0.36}} = \frac{1}{0.6} = \frac{5}{3}$$

The time between the events as observed in frame S is therefore

$$\Delta t = \gamma\,\Delta t_{\text{p}} = \frac{5}{3}(30\text{ min}) = 50\text{ min}$$

During this time, the spaceship travels a distance in frame S equal to the proper distance between the platform clocks:

$$L_{\text{p}} = \Delta x = V\,\Delta t = (0.8c)(50\text{ min}) = 40c\cdot\text{min}$$

Note that this distance is longer than the contracted distance between the clocks that is measured by observers in frame S' of the spaceship.

Observers on the platform would say that the spaceship's clock is running slow because it records a time of only 30 min between the events, whereas the time measured on the platform is 50 min.

Figure 34-13 shows the situation viewed from the spaceship in S'. The platform is traveling past the ship with speed $0.8c$. There is a clock at point x_1, which coincides with the ship when the flash gun is exploded, and another at point x_2, which coincides with the ship when the return flash is received from the mirror. We assume that the clock at x_1 reads 12:00 noon at the time of the light flash. The clocks at x_1 and x_2 are synchronized in S but not in S'. In S', the clock at x_2, which is chasing the one at x_1, leads by

$$\frac{L_{\text{p}}V}{c^2} = \frac{(40c\cdot\text{min})(0.8c)}{c^2} = 32\text{ min}$$

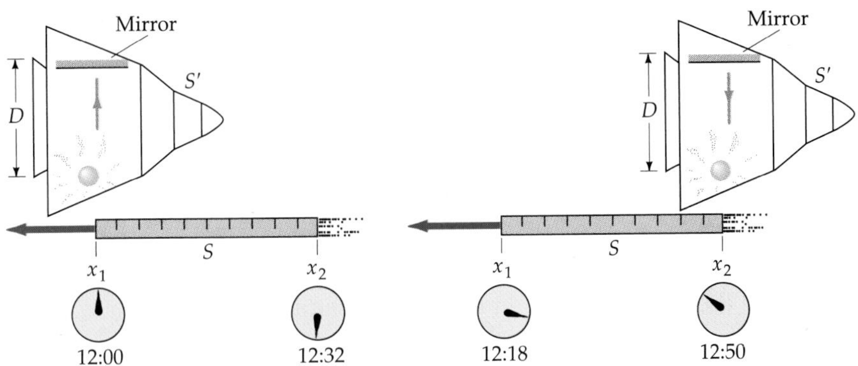

Figure 34-13 Example 34-4. Clocks on a platform as observed from the spaceship's frame of reference S'. During the time $\Delta t' =$ 30 min it takes for the platform to pass the spaceship, the clocks on the platform run slow and tick off $(30\text{ min})/\gamma = 18$ min. But the clocks are unsynchronized, with the chasing clock leading by $L_{\text{p}}V/c^2$, which for this case is 32 min. The time it takes for the spaceship to pass as measured on the platform is therefore 32 min + 18 min = 50 min.

When the spaceship coincides with x_2, the clock there reads 12:50. The time between the events is therefore 50 min in S. Note that according to observers in S', this clock ticks off 50 min − 32 min = 18 min for a trip that takes 30 min in S'. Thus, observers in S' see this clock run slow by the factor 30/18 = 5/3.

Every observer in one frame sees the clocks in the other frame run slow. According to observers in S, who measure 50 min for the time interval, the time interval in S' (30 min) is too small, so they see the single clock in S' run too slow by the factor 5/3. According to the observers in S', the observers in S measure a time that is too *long* despite the fact that their clocks run too slow because the clocks in S are out of synchronization. The clocks tick off only 18 min, but the second one leads the first by 32 min, so the time interval is 50 min.

Questions

2. Two observers are in relative motion. In what circumstances can they agree on the simultaneity of two different events?

3. If event A occurs before event B in some frame, might it be possible for there to be a reference frame in which event B occurs before event A?

4. Two events are simultaneous in a frame in which they also occur at the same point in space. Are they simultaneous in other reference frames?

34-6 The Doppler Effect

In our discussion of the doppler effect for sound (Section 14-9), we found that the change in frequency for a given velocity V depends on whether it is the source or the receiver that is moving with that speed. Such a distinction is possible for sound because there is a medium (the air) relative to which the motion takes place, so it is not surprising that the motion of the source or the receiver relative to the still air can be distinguished. Such a distinction between motion of the source or receiver cannot be made for light or other electromagnetic waves in a vacuum. Therefore, the expressions we have derived for the doppler effect cannot be correct for light. We will now derive the relativistic doppler-effect equations that are correct for light.

We will consider a source moving toward a receiver with velocity V, and we will work in the frame of the receiver. Let the source emit N electromagnetic waves. The first wave will travel a distance $c\,\Delta t_R$ and the source will travel a distance $V\,\Delta t_R$ in the time Δt_R measured in the frame of the receiver. The wavelength will be

$$\lambda' = \frac{(c\,\Delta t_R - V\,\Delta t_R)}{N}$$

The frequency f' observed by the receiver will therefore be

$$f' = \frac{c}{\lambda'} = \frac{c}{c - V}\frac{N}{\Delta t_R}$$

$$= \frac{1}{1 - V/c}\frac{N}{\Delta t_R}$$

If the frequency of the source is f_0, it will emit $N = f_0\,\Delta t_S$ waves in the time Δt_S measured by the source. Here Δt_S is the proper time interval (the first wave and the Nth wave are emitted at the same place in the source's reference frame). Times Δt_S and Δt_R are related by Equation 34-20 for time dilation, so $\Delta t_R = \gamma\, \Delta t_S$. Thus, when the source and receiver are moving toward one another we obtain

$$f' = \frac{1}{1 - V/c} \frac{f_0\, \Delta t_S}{\Delta t_R} = \frac{f_0}{1 - V/c} \frac{1}{\gamma}$$

or

$$f' = \frac{\sqrt{1 - V^2/c^2}}{1 - V/c} f_0 = \sqrt{\frac{1 + V/c}{1 - V/c}}\, f_0 \qquad \text{approaching} \qquad 34\text{-}24a$$

This differs from our classical equation only in the time-dilation factor. When the source and receiver are moving away from one another, the same analysis shows that the observed frequency is given by

$$f' = \frac{\sqrt{1 - V^2/c^2}}{1 + V/c} f_0 = \sqrt{\frac{1 - V/c}{1 + V/c}}\, f_0 \qquad \text{receding} \qquad 34\text{-}24b$$

It is left as a problem (Problem 34-64) for you to show that the same results are obtained if the calculations are done in the reference frame of the source.

Example 34-5

The longest wavelength of light emitted by hydrogen in the Balmer series (see Chapter 35) has a wavelength of $\lambda_0 = 656$ nm. In light from a distant galaxy, this wavelength is measured to be $\lambda' = 1458$ nm. Find the speed at which the distant galaxy is receding from the earth.

If we substitute $f' = c/\lambda'$ and $f_0 = c/\lambda_0$ into Equation 34-24b, we obtain

$$\sqrt{\frac{1 - V/c}{1 + V/c}} = \frac{f'}{f_0} = \frac{\lambda_0}{\lambda'}$$

This equation is somewhat simplified if we use $\beta = V/c$. Then squaring the above equation and taking the reciprocal of each side, we obtain

$$\frac{1 + \beta}{1 - \beta} = \left(\frac{\lambda'}{\lambda_0}\right)^2 = \left(\frac{1458 \text{ nm}}{656 \text{ nm}}\right)^2 = 4.94$$

so

$$1 + \beta = 4.94 - 4.94\,\beta$$

$$\beta = \frac{4.94 - 1}{4.94 + 1} = 0.663 = \frac{V}{c}$$

The galaxy is thus receding at a speed of $V = 0.663c$. The shift towards longer wavelengths of light from distant galaxies that are receding from us is called the **redshift.**

34-7 The Twin Paradox

Homer and Ulysses are identical twins. Ulysses travels at high speed to a planet beyond the solar system and returns while Homer remains at home. When they are together again, which twin is older, or are they the same age? The correct answer is that Homer, the twin who stays at home, is older. This problem, with variations, has been subject of spirited debate for decades, though there are very few who disagree with the answer.* The problem is a paradox because of the seemingly symmetric roles played by the twins with the asymmetric result in their aging. The paradox is resolved when the asymmetry of the twins' roles is noted. The relativistic result conflicts with common sense based on our strong but incorrect belief in absolute simultaneity. We will consider a particular case with some numerical magnitudes that, though impractical, make the calculations easy.

Let planet P and Homer on earth be at rest in reference frame S a distance L_p apart, as illustrated in Figure 34-14. We neglect the motion of the earth. Reference frames S' and S'' are moving with speed V toward and away from the planet, respectively. Ulysses quickly accelerates to speed V, then coasts in S' until he reaches the planet, where he stops and is momentarily at rest in S. To return he quickly accelerates to speed V toward earth and then coasts in S'' until he reaches earth, where he stops. We can assume that the acceleration times are negligible compared with the coasting times. We use the following values for illustration: $L_p = 8$ light-years and $V = 0.8c$. Then $\sqrt{1 - V^2/c^2} = 3/5$ and $\gamma = 5/3$.

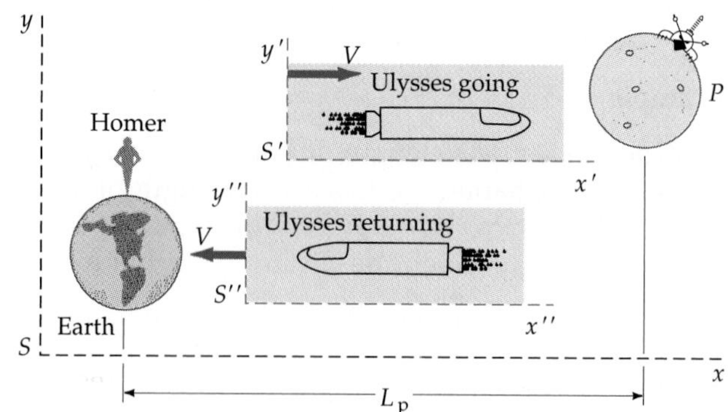

Figure 34-14 The twin paradox. The earth and a distant planet are fixed in frame S. Ulysses coasts in frame S' to the planet and then coasts back in frame S''. His twin Homer stays on earth. When Ulysses returns, he is younger than his twin. The roles played by the twins are not symmetric. Homer remains in one inertial reference frame, but Ulysses must accelerate if he is to return home.

It is easy to analyze the problem from Homer's point of view on earth. According to Homer's clock, Ulysses coasts in S' for a time $L_p/V = 10$ y and in S'' for an equal time. Thus Homer is 20 y older when Ulysses returns. The time interval in S' between Ulysses' leaving earth and his arriving at the planet is shorter because it is proper time. The time it takes to reach the planet by Ulysses' clock is

$$\Delta t' = \frac{\Delta t}{\gamma} = \frac{10 \text{ y}}{5/3} = 6 \text{ y}$$

Since the same time is required for the return trip, Ulysses will have recorded 12 y for the round trip and will be 8 y younger than Homer upon his return.

From Ulysses' point of view, the distance from the earth to the planet is contracted and is only

$$L' = \frac{L_p}{\gamma} = \frac{8 \text{ light-years}}{5/3} = 4.8 \text{ light-years}$$

At $V = 0.8c$, it takes only 6 y each way.

The real difficulty in this problem is for Ulysses to understand why his twin aged 20 y during his absence. If we consider Ulysses as being at rest and Homer as moving away, Homer's clock should run slow and measure

*A collection of some important papers concerning this paradox can be found in *Special Relativity Theory, Selected Reprints*, American Association of Physics Teachers, New York, 1963.

only 3/5(6) = 3.6 y. Then why shouldn't Homer age only 7.2 y during the round trip? This, of course, is the paradox. The difficulty with the analysis from the point of view of Ulysses is that he does not remain in an inertial frame. What happens while Ulysses is stopping and starting? To investigate this problem in detail, we would need to treat accelerated reference frames, a subject dealt with in the study of general relativity and beyond the scope of this book. However, we can get some insight into the problem by having the twins send regular signals to each other so that they can record the other's age continuously. If they arrange to send a signal once a year, each can determine the age of the other merely by counting the signals received. The arrival frequency of the signals will not be 1 per year because of the doppler shift. The frequency observed will be given by Equations 34-24*a* and 34-24*b*. Using $V/c = 0.8$ and $V^2/c^2 = 0.64$, we have for the case in which the twins are receding from each other

$$f' = \frac{\sqrt{1 - V^2/c^2}}{1 + V/c} f_0 = \frac{\sqrt{1 - 0.64}}{1 + 0.8} f_0 = \frac{1}{3} f_0$$

When they are approaching, Equation 34-24 gives $f' = 3f_0$.

Consider the situation first from the point of view of Ulysses. During the 6 y it takes him to reach the planet (remember that the distance is contracted in his frame), he receives signals at the rate of $\frac{1}{3}$ signal per year, and so he receives 2 signals. As soon as he turns around and starts back to earth, he begins to receive 3 signals per year. In the 6 y it takes him to return he receives 18 signals, giving a total of 20 for the trip. He accordingly expects his twin to have aged 20 years.

We now consider the situation from Homer's point of view. He receives signals at the rate of $\frac{1}{3}$ signal per year not only for the 10 y it takes Ulysses to reach the planet but also for the time it takes for the last signal sent by Ulysses before he turns around to get back to earth. (He cannot know that Ulysses has turned around until the signals begin reaching him with increased frequency.) Since the planet is 8 light-years away, there is an additional 8 y of receiving signals at the rate of $\frac{1}{3}$ signal per year. During the first 18 y, Homer receives 6 signals. In the final 2 y before Ulysses arrives, Homer receives 6 signals, or 3 per year. (The first signal sent after Ulysses turns around takes 8 y to reach earth, whereas Ulysses, traveling at $0.8c$, takes 10 y to return and therefore arrives just 2 y after Homer begins to receive signals at the faster rate.) Thus Homer expects Ulysses to have aged 12 y. In this analysis, the asymmetry of the twins' roles is apparent. When they are together again, both twins agree that the one who has been accelerated will be younger than the one who stayed home.

The predictions of the special theory of relativity concerning the twin paradox have been tested using small particles that can be accelerated to such large speeds that γ is appreciably greater than 1. Unstable particles can be accelerated and trapped in circular orbits in a magnetic field, for example, and their lifetimes can then be compared with those of identical particles at rest. In all such experiments, the accelerated particles live longer on the average than those at rest, as predicted. These predictions have also been confirmed by the results of an experiment in which high-precision atomic clocks were flown around the world in commercial airplanes, but the analysis of this experiment is complicated due to the necessity of including gravitational effects treated in the general theory of relativity.*

*The details of this experiment can be found in J. C. Hafele and Richard E. Keating, "Around-the-world Atomic Clocks: Predicted Relativistic Time Gains" and "Around-the-world Atomic Clocks: Observed Relativistic Time Gains," *Science,* July 14, 1972, p. 166.

34-8 The Velocity Transformation

We can find how velocities transform from one reference frame to another by differentiating the Lorentz transformation equations. Suppose a particle has velocity $u'_x = dx'/dt'$ in frame S', which is moving to the right with speed V relative to frame S. Its velocity in frame S is

$$u_x = \frac{dx}{dt}$$

From the Lorentz transformation equations (Equations 34-16 and 34-17), we have

$$dx = \gamma(dx' + V\,dt')$$

and

$$dt = \gamma\left(dt' + \frac{V\,dx'}{c^2}\right)$$

The velocity in S is thus

$$u_x = \frac{dx}{dt} = \frac{\gamma(dx' + V\,dt')}{\gamma\left(dt' + \frac{V\,dx'}{c^2}\right)} = \frac{\frac{dx'}{dt'} + V}{1 + \frac{V}{c^2}\frac{dx'}{dt'}} = \frac{u'_x + V}{1 + Vu'_x/c^2}$$

If a particle has components of velocity along the y or z axes, we can use the same relation between dt and dt', with $dy = dy'$ and $dz = dz'$, to obtain

$$u_y = \frac{dy}{dt} = \frac{dy'}{\gamma\left(dt' + \frac{V\,dx'}{c^2}\right)} = \frac{dy'/dt'}{\gamma\left(1 + \frac{V}{c^2}\frac{dx'}{dt'}\right)} = \frac{u'_y}{\gamma\left(1 + \frac{Vu'_x}{c^2}\right)}$$

and

$$u_z = \frac{u'_z}{\gamma\left(1 + \frac{Vu'_x}{c^2}\right)}$$

The complete relativistic velocity transformation is

Relativistic velocity transformation

$$u_x = \frac{u'_x + V}{1 + Vu'_x/c^2} \qquad \text{34-25}a$$

$$u_y = \frac{u'_y}{\gamma(1 + Vu'_x/c^2)} \qquad \text{34-25}b$$

$$u_z = \frac{u'_z}{\gamma(1 + Vu'_x/c^2)} \qquad \text{34-25}c$$

The inverse velocity transformation equations are

$$u'_x = \frac{u_x - V}{1 - Vu_x/c^2} \qquad \text{34-26}a$$

$$u'_y = \frac{u_y}{\gamma(1 - Vu_x/c^2)} \qquad \text{34-26}b$$

$$u'_z = \frac{u_z}{\gamma(1 - Vu_x/c^2)} \qquad \text{34-26}c$$

These equations differ from the classical and intuitive result $u_x = u'_x + V$, $u_y = u'_y$ and $u_z = u'_z$ because the denominators in Equations 34-25 and 34-26 are not equal to 1. When V and u'_x are small compared with the speed of light c, $\gamma \approx 1$ and $Vu'_x/c^2 \ll 1$. Then the relativistic and classical expressions are the same.

Example 34-6

A supersonic plane moves with speed 1000 m/s (about 3 times the speed of sound) along the x axis relative to you. Another plane moves along the x axis at speed 500 m/s relative to the first plane. How fast is the second plane moving relative to you?

According to the classical formula for combining velocities, the speed of the second plane relative to you is 1000 m/s + 500 m/s = 1500 m/s. If we assume that you are at rest in the S reference frame and that the first plane is at rest in the S' frame, which is moving at $V = 1000$ m/s relative to S, the second plane has velocity $u'_x = 500$ m/s in S'. The correction term for u_x in the denominator of Equation 34-25a is then

$$\frac{Vu'_x}{c^2} = \frac{(1000)(500)}{(3 \times 10^8)^2} \approx 5 \times 10^{-12}$$

This correction term is so small that the classical and relativistic results are essentially the same.

Example 34-7

Work Example 34-6 if the first plane moves with speed $V = 0.8c$ relative to you and the second plane moves with the same speed $0.8c$ relative to the first plane.

In this case, the correction term is

$$\frac{Vu'_x}{c^2} = \frac{(0.8c)(0.8c)}{c^2} = 0.64$$

The speed of the second plane in frame S is then

$$u_x = \frac{0.8c + 0.8c}{1 + 0.64} = 0.98c$$

This is quite different from the classically expected result of $0.8c + 0.8c = 1.6c$. In fact, it can be shown from Equation 34-25 that if the speed of an object is less than c in one frame, it is less than c in all other frames moving relative to that frame with a speed less than c. We will see in Section 34-10 that it takes an infinite amount of energy to accelerate a particle to the speed of light. The speed of light c is thus an upper, unattainable limit for the speed of a particle having mass. (Massless particles, such as photons, always move at the speed of light.)

Example 34-8

Light moves along the x axis with speed $u_x = c$. What is its speed in S'?

From Equation 34-26a, we have

$$u'_x = \frac{c - V}{1 - Vc/c^2} = \frac{c(1 - V/c)}{1 - V/c} = c$$

as required by Einstein's postulates.

Question

5. The Lorentz transformation for y and z is the same as the classical result: $y = y'$ and $z = z'$. Yet the relativistic velocity transformation does not give the classical result $u_y = u'_y$ and $u_z = u'_z$. Explain.

34-9 Relativistic Momentum

We have seen in previous sections that Einstein's postulates require important modifications in our ideas of simultaneity and in our measurements of time and length. Perhaps more importantly, they also require modifications in our concepts of mass, momentum, and energy. In classical mechanics, the momentum of a particle is defined as the product of its mass and its velocity, $\mathbf{p} = m\mathbf{u}$, where $\mathbf{u}$ is the velocity. In an isolated system of particles, with no net force acting on the system, the total momentum of the system remains constant.

In this section we will see from a simple thought experiment that the classical expression for momentum, $\mathbf{p} = m\mathbf{u}$, is just an approximation. That is, this quantity is not conserved in an isolated system. We consider two observers: observer A in reference frame S and observer B in frame S'. Each observer has a ball of mass m. The two balls are identical when compared at rest. Each observer throws his ball vertically with a speed u_0 such that it travels a distance L, makes an elastic collision with the other ball, and returns. Figure 34-15 shows how the collision looks in each reference frame. Classically, each ball has vertical momentum of magnitude mu_0. Since the vertical components of the momenta are equal and opposite, the total vertical component of momentum is zero before the collision. The collision merely reverses the momentum of each ball, so the total vertical momentum is zero after the collision.

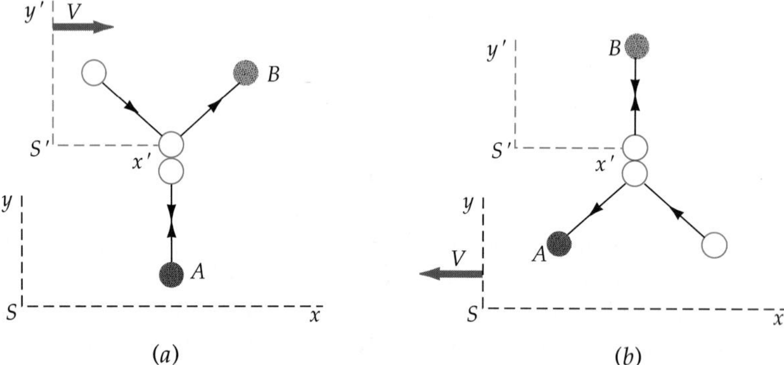

Figure 34-15 (*a*) Elastic collision of two identical balls as seen in frame S. The vertical component of the velocity of ball B is u_0/γ in S if it is u_0 in S'. (*b*) The same collision as seen in S'. In this frame, ball A has vertical component of velocity equal to u_0/γ.

Relativistically, however, the vertical components of the velocities of the two balls as seen by either observer are not equal and opposite. Thus, when they are reversed by the collision, classical momentum is not conserved. Consider the collision as seen by A in frame S. The velocity of his ball is $u_{Ay} = +u_0$. Since the velocity of B's ball in frame S' is $u'_{Bx} = 0$, $u'_{By} = -u_0$, the y component of the velocity of B's ball in frame S is (Equation 34-25*b*) $u_{By} = -u_0/\gamma$. Thus, if the classical expression for momentum $\mathbf{p} = m\mathbf{u}$ is used, the vertical components of momentum of the two balls are not equal and opposite as seen by observer A. Since the balls are reversed by the collision, momentum is not conserved. Of course, the same result is observed by B. In the classical limit, when u is much less than c, γ is approximately 1, and the momentum of the system is conserved as seen by either observer.

The reason that the total momentum of a system is important in classical mechanics is that it is conserved when there are no external forces acting on the system, as is the case in collisions. We now see that the quantity $\Sigma m\mathbf{u}$ is conserved only in the approximation that $u \ll c$. We will define the relativistic momentum $\mathbf{p}$ of a particle to have the following properties:

1. In collisions, $\mathbf{p}$ is conserved.
2. As u/c approaches zero, $\mathbf{p}$ approaches $m\mathbf{u}$.

We will now show that the quantity

$$\mathbf{p} = \frac{m\mathbf{u}}{\sqrt{1 - u^2/c^2}} \qquad 34\text{-}27$$

is conserved in the elastic collision shown in Figure 34-15. Since this quantity also approaches $m\mathbf{u}$ as u/c approaches zero, we take this equation for the definition of the **relativistic momentum** of a particle.

We will compute the y component of the relativistic momentum of each particle in reference frame S and show that the y component of the total relativistic momentum is zero. The speed of ball A in S is u_0, so the y component of its relativistic momentum is

$$p_{Ay} = \frac{mu_0}{\sqrt{1 - u_0^2/c^2}}$$

The speed of ball B in S is more complicated. Its x component is V and its y component is $-u_0/\gamma$. Thus

$$u_B^2 = u_{Bx}^2 + u_{By}^2 = V^2 + (-u_0\sqrt{1 - V^2/c^2})^2 = V^2 + u_0^2 - \frac{u_0^2 V^2}{c^2}$$

Using this result to compute $\sqrt{1 - u_B^2/c^2}$, we obtain

$$1 - \frac{u_B^2}{c^2} = 1 - \frac{V^2}{c^2} - \frac{u_0^2}{c^2} + \frac{u_0^2 V^2}{c^4} = (1 - V^2/c^2)(1 - u_0^2/c^2)$$

and

$$\sqrt{1 - u_B^2/c^2} = \sqrt{1 - V^2/c^2}\,\sqrt{1 - u_0^2/c^2} = \frac{1}{\gamma}\sqrt{1 - u_0^2/c^2}$$

The y component of the relativistic momentum of ball B as seen in S is therefore

$$p_{By} = \frac{mu_{By}}{\sqrt{1 - u_B^2/c^2}} = \frac{-mu_0/\gamma}{(1/\gamma)\sqrt{1 - u_0^2/c^2}} = \frac{-mu_0}{\sqrt{1 - u_0^2/c^2}}$$

Since $p_{By} = -p_{Ay}$, the y component of the total momentum of the two balls is zero. If the speed of each ball is reversed by the collision, the total momentum will remain zero and momentum will be conserved.

One interpretation of Equation 34-27 is that the mass of an object increases with speed. The quantity $m/\sqrt{1 - u^2/c^2}$ is called the **relativistic mass** of a particle. The mass of a particle when it is at rest in some reference frame is called its **rest mass** m_0. The mass thus increases from m_0 at rest to $m_r = m_0/\sqrt{1 - u^2/c^2}$ when it is moving at speed u. To avoid confusion, we will label the rest mass m_0 and use $m_0/\sqrt{1 - u^2/c^2}$ for the relativistic mass in this chapter. The rest mass of a particle is the same in all reference frames. Using this notation, the relativistic momentum of a particle is then

$$\mathbf{p} = \frac{m_0\mathbf{u}}{\sqrt{1 - u^2/c^2}} \qquad 34\text{-}28$$

Relativistic momentum

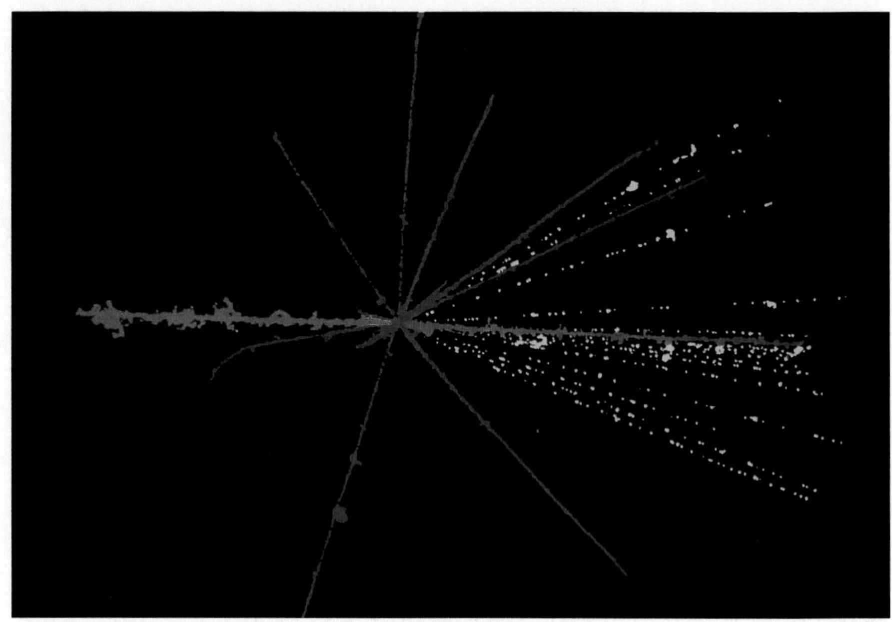

The creation of elementary particles demonstrates the conversion of kinetic energy to rest energy. In this 1950 photograph of a cosmic ray shower, a high-energy sulfur nucleus (red) collides in a photographic emulsion and produces a spray of particles, including a fluorine nucleus (green), other nuclear fragments (blue), and about 16 pions (yellow).

34-10 Relativistic Energy

In classical mechanics, the work done by an unbalanced force acting on a particle equals the change in the kinetic energy of the particle. In relativistic mechanics, we equate the unbalanced force to the rate of change of the relativistic momentum. The work done by such a force can then be calculated and set equal to the change in kinetic energy. As in classical mechanics, we will define kinetic energy as the work done by an unbalanced force in accelerating a particle from rest to some velocity. Considering one dimension only, we have

$$K = \int_{u=0}^{u} \sum F\, ds = \int_0^u \frac{dp}{dt}\, ds = \int_0^u u\, dp = \int_0^u u\, d\left(\frac{m_0 u}{\sqrt{1 - u^2/c^2}}\right) \qquad 34\text{-}29$$

where we have used $u = ds/dt$. It is left as a problem (Problem 70) for you to show that

$$d\left(\frac{m_0 u}{\sqrt{1 - u^2/c^2}}\right) = m_0\left(1 - \frac{u^2}{c^2}\right)^{-3/2} du$$

If we substitute this expression into the integrand in Equation 34-29, we obtain

$$K = \int_0^u u\, d\left(\frac{m_0 u}{\sqrt{1 - u^2/c^2}}\right) = \int_0^u m_0 \left(1 - \frac{u^2}{c^2}\right)^{-3/2} u\, du$$

$$= m_0 c^2\left(\frac{1}{\sqrt{1 - u^2/c^2}} - 1\right)$$

or

$$K = \frac{m_0 c^2}{\sqrt{1 - u^2/c^2}} - m_0 c^2 \qquad 34\text{-}30$$

The expression for kinetic energy consists of two terms. The first term depends on the speed of the particle. The second, m_0c^2, is independent of the speed. The quantity m_0c^2 is called the **rest energy** of the particle E_0. The rest energy is the product of the rest mass and c^2:

$$E_0 = m_0c^2 \qquad \text{34-31}$$

Rest energy

The total **relativistic energy** E is then defined to be the sum of the kinetic energy and the rest energy:

$$E = K + m_0c^2 = \frac{m_0c^2}{\sqrt{1 - u^2/c^2}} \qquad \text{34-32}$$

Relativistic energy

Thus, the work done by an unbalanced force increases the energy from the rest energy m_0c^2 to the final energy $m_0c^2/\sqrt{1 - u^2/c^2} = m_rc^2$, where $m_r = m_0/\sqrt{1 - u^2/c^2}$ is the relativistic mass. We can obtain a useful expression for the velocity of a particle by multiplying Equation 34-28 for the relativistic momentum by c^2 and comparing the result with Equation 34-32 for the relativistic energy. We have

$$pc^2 = \frac{m_0c^2u}{\sqrt{1 - u^2/c^2}} = Eu$$

or

$$\frac{u}{c} = \frac{pc}{E} \qquad \text{34-33}$$

Example 34-9

An electron with rest energy 0.511 MeV moves with speed $u = 0.8c$. Find its total energy, kinetic energy, and momentum.

We first calculate the factor $1/\sqrt{1 - u^2/c^2}$.

$$\frac{1}{\sqrt{1 - u^2/c^2}} = \frac{1}{\sqrt{1 - 0.64}} = \frac{5}{3} = 1.67$$

The total energy is then

$$E = \frac{m_0c^2}{\sqrt{1 - u^2/c^2}} = 1.67(0.511 \text{ MeV}) = 0.853 \text{ MeV}$$

The kinetic energy is the total energy minus the rest energy:

$$K = E - m_0c^2 = 0.853 \text{ MeV} - 0.511 \text{ MeV} = 0.342 \text{ MeV}$$

The magnitude of the momentum is

$$p = \frac{m_0u}{\sqrt{1 - u^2/c^2}} = (1.67)m_0(0.8c) = \frac{1.33m_0c^2}{c}$$

$$= \frac{(1.33)(0.511 \text{ MeV})}{c} = 0.680 \text{ MeV}/c$$

The unit MeV/c is a convenient unit for momentum.

The expression for kinetic energy given by Equation 34-30 doesn't look much like the classical expression $\frac{1}{2}m_0u^2$. However, when u is much less than c, we can approximate $1/\sqrt{1-u^2/c^2}$ using the binomial expansion (Equation 34-2):

$$\frac{1}{\sqrt{1-u^2/c^2}} = \left(1-\frac{u^2}{c^2}\right)^{-1/2}$$

$$\approx 1 + \frac{1}{2}\frac{u^2}{c^2}$$

From this result, when u is much less than c, the expression for relativistic kinetic energy becomes

$$K = m_0c^2\left(\frac{1}{\sqrt{1-u^2/c^2}} - 1\right)$$

$$\approx m_0c^2\left(1 + \frac{1}{2}\frac{u^2}{c^2} - 1\right)$$

$$= \frac{1}{2}m_0u^2$$

Thus at low speeds, the relativistic expression is the same as the classical expression.

We note from Equation 34-32 that as the speed u approaches the speed of light c, the energy of the particle becomes very large because $1/\sqrt{1-u^2/c^2}$ becomes very large. At $u = c$, the energy becomes infinite. For u greater than c, $\sqrt{1-u^2/c^2}$ is the square root of a negative number and is therefore imaginary. A simple interpretation of the result that it takes an infinite amount of energy to accelerate a particle to the speed of light is that no particle that is ever at rest in any inertial reference frame can travel as fast or faster than the speed of light c. As we noted in Example 34-7, if the speed of a particle is less than c in one reference frame, it is less than c in all other reference frames moving relative to that frame at speeds less than c.

In practical applications, the momentum or energy of a particle is often known rather than the speed. Equation 34-28 for the relativistic momentum and Equation 34-32 for the relativistic energy can be combined to eliminate the speed u. (See Problem 48.) The result is

Relation for total energy, momentum, and rest energy

$$E^2 = p^2c^2 + (m_0c^2)^2 \qquad 34\text{-}34$$

This useful equation can be conveniently remembered from the right triangle shown in Figure 34-16. If the energy of a particle is much greater than its rest energy mc^2, the second term on the right of Equation 34-34 can be neglected, giving the useful approximation

$$E \approx pc \qquad \text{for } E \gg m_0c^2 \qquad 34\text{-}35$$

Equation 34-35 is an exact relation between energy and momentum for particles with no rest mass, such as photons and neutrinos.

$E^2 = (pc)^2 + (m_0c^2)^2$
E
pc
m_0c^2

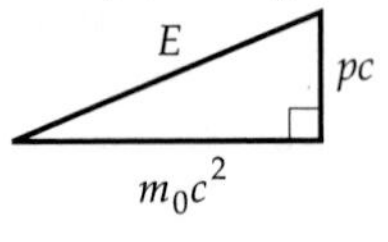

Figure 34-16 Right triangle for remembering Equation 34-34.

Exercise

A proton with a rest mass of 938 MeV/c^2 has a total energy of 1400 MeV. Find (*a*) $1/\sqrt{1-u^2/c^2}$, (*b*) its momentum, and (*c*) its speed u. [Answers: (*a*) 1.49, (*b*) p = 1040 MeV/c, (*c*) $u = 0.74c$]

The identification of the term m_0c^2 as rest energy is not merely a convenience. The conversion of rest energy to kinetic energy with a corresponding loss in rest mass is a common occurrence in radioactive decay and nuclear reactions, including nuclear fission and nuclear fusion. We will give some examples of this in this section. Einstein considered Equation 34-31 relating the energy of a particle to its mass to be the most significant result of the theory of relativity. Energy and inertia, which were formerly two distinct concepts, are related through this famous equation.

To illustrate the interrelation of rest mass and energy, we consider a perfectly inelastic collision of two particles. Classically, kinetic energy is lost in such a collision. For example, in the zero-momentum reference frame, the particles are moving toward one another with equal and opposite momenta and are at rest after the collision. In this frame, the total kinetic energy of the system before the collision is lost. In any other reference frame, the particles move with the velocity of the center of mass, but the amount of kinetic energy lost is the same. We will now see that if we assume that the total relativistic energy is conserved, the loss in kinetic energy equals the gain in rest energy of the system. Consider a particle of rest mass m_{10} moving with initial speed u_1 that collides with a particle of rest mass m_{20} moving with initial speed u_2. The particles collide and stick together, forming a particle of rest mass M_0 that moves with speed u_f, as shown in Figure 34-17. Let E_1 be the initial total energy and K_1 be the initial kinetic energy of particle 1 and E_2 be the initial total energy and K_2 be the initial kinetic energy of particle 2. The initial total energy of the system is

$$E_i = E_1 + E_2$$

and the initial kinetic energy of the system is

$$K_i = K_1 + K_2 = (E_1 - m_{10}c^2) + (E_2 - m_{20}c^2)$$

After the collision, the composite particle has a rest mass M_0, total energy E_f, and kinetic energy $K_f = E_f - M_0c^2$. The loss in kinetic energy of the system is thus

$$K_i - K_f = (E_1 + E_2 - m_{10}c^2 - m_{20}c^2) - (E_f - M_0c^2) \qquad 34\text{-}36$$

If we assume the conservation of total energy, we have $E_f = E_i = E_1 + E_2$. Substituting $E_1 + E_2 - E_f = 0$ in Equation 34-36 and rearranging, we obtain

$$K_i - K_f = [M_0 - (m_{10} + m_{20})]c^2 = (\Delta m_0)c^2 \qquad 34\text{-}37$$

where $\Delta m_0 = M_0 - (m_{10} + m_{20})$ is the increase in rest mass of the system.

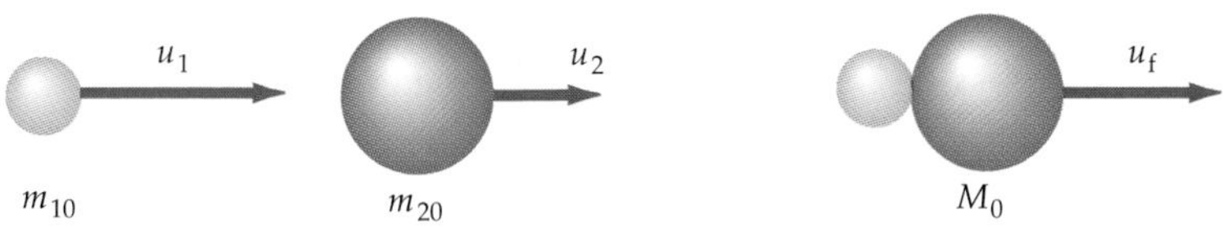

Figure 34-17 A perfectly inelastic collision between two particles. One particle of rest mass m_{10} collides with another particle of rest mass m_{20}. After the collision, the particles stick together, forming a composite particle of rest mass M_0 that moves with speed u_f such that relativistic momentum is conserved. Kinetic energy is lost in this process. If we assume that the total energy is conserved, the loss in kinetic energy must equal c^2 times the increase in the rest mass of the system.

Some numerical examples from atomic and nuclear physics will illustrate changes in rest mass and rest energy. Energies in atomic and nuclear physics are usually expressed in units of electron volts (eV) or mega electron volts (MeV):

$$1\text{ eV} = 1.6 \times 10^{-19}\text{ J}$$

A convenient unit for the masses of atomic particles is eV/c^2 or MeV/c^2, which is just the rest energy of the particle divided by c^2. The rest masses and rest energies of some elementary particles and light nuclei are given in Table 34-1, from which we can see that the mass of a nucleus is not the same as the sum of the masses of its parts.

Table 34-1 **Rest Energies of Some Elementary Particles and Light Nuclei**

Particle	Symbol	Rest energy, MeV
Photon	γ	0
Electron (positron)	e or e^- (e^+)	0.5110
Muon	$\mu^\pm$	105.7
Pion	π^0	135
	$\pi^\pm$	139.6
Proton	p	938.280
Neutron	n	939.573
Deuteron	^{2}H or d	1875.628
Triton	^{3}H or t	2808.944
Alpha particle	^{4}He or α	3727.409

Example 34-10

A deuteron consists of a proton and neutron bound together. It is the nucleus of the deuterium atom, which is an isotope of hydrogen called heavy hydrogen and written ^{2}H. How much energy is required to separate the proton from the neutron in the deuteron?

From Table 34-1, we can see that the rest energy of the deuteron is 1875.63 MeV. The rest energy of the proton is 938.28 Mev, and that of the neutron is 939.57 MeV. The sum of the rest energies of the proton and neutron is 938.28 MeV + 939.57 MeV = 1877.85 MeV. This is greater than the rest energy of the deuteron by 1877.85 − 1875.63 = 2.22 MeV. The energy needed to break up a nucleus into its constituent parts is called the **binding energy** of the nucleus. The binding energy of the deuteron is 2.22 MeV. This is the energy that must be added to the deuteron to break it up into a proton plus a neutron. This can be done by bombarding deuterons with energetic particles or with electromagnetic radiation with energy of at least 2.22 MeV.

When a deuteron is formed by the combination of a neutron and proton, energy must be released. When neutrons from a reactor collide with protons, some neutrons are captured to form deuterons. In the capture process, 2.22 MeV of energy is released, usually in the form of electromagnetic radiation.

Example 34-10 illustrates an important property of atoms and nuclei. Any stable composite particle, such as a deuteron or a helium nucleus (2 neutrons plus 2 protons), that is made up of other particles has a rest energy that is less than the sum of the rest energies of its parts. The difference is the binding energy of the composite particle. The binding energies of atoms and molecules are of the order of a few electron volts, which leads to a negligible difference in mass between the composite particle and its parts. The binding energies of nuclei are of the order of several MeV, which leads to a noticeable difference in mass. Some very heavy nuclei, such as radium, are radioactive and decay into a lighter nucleus plus an alpha particle. In this case, the original nucleus has a rest energy greater than that of the decay particles. The excess energy appears as the kinetic energy of the decay products.

Example 34-11

In a typical nuclear fusion reaction, a tritium nucleus (^{3}H) and a deuterium nucleus (^{2}H) fuse together to form a helium nucleus (^{4}He) plus a neutron. How much energy is released in this fusion reaction?

The reaction is written

$$^2\text{H} + {}^3\text{H} \longrightarrow {}^4\text{He} + n$$

From Table 34-1, we see that the rest energy of the deuterium plus tritium nuclei is 1875.628 MeV + 2808.944 MeV = 4684.572 MeV. The rest energy of the helium nucleus plus the neutron is 3727.409 + 939.573 = 4666.982 MeV. This is less than that of the deuterium plus tritium by 4684.572 − 4666.982 = 17.59 MeV. The energy released in this reaction is thus 17.59 MeV. This and other fusion reactions occur in the sun and are responsible for the energy supplied to the earth. As the sun gives off energy, its rest mass continually decreases.

Example 34-12

A hydrogen atom consisting of a proton and an electron has a binding energy of 13.6 eV. By what percentage is the mass of the proton plus the electron greater than that of the hydrogen atom?

The rest energy of a proton plus that of an electron is 938.28 MeV + 0.511 eV = 938.791 MeV. The sum of the masses of these two particles is 938.791 MeV/c^2. The mass of the hydrogen atom is less than this by 13.6 eV/c^2. The percentage difference is

$$\frac{13.6 \text{ eV}/c^2}{938.791 \times 10^6 \text{ eV}/c^2} = 1.45 \times 10^{-8} = 1.45 \times 10^{-6}\%$$

This mass difference is so small as to be hardly measurable.

Example 34-13

A particle of rest mass 2 MeV/c^2 and kinetic energy 3 MeV collides with a stationary particle of rest mass 4 MeV/c^2. After the collision, the two particles stick together. Find (*a*) the initial momentum of the system, (*b*) the final velocity of the two-particle system, and (*c*) the rest mass of the two-particle system.

(*a*) Since the moving particle has kinetic energy of 3 MeV and rest energy of 2 MeV, its total energy is E_1 = 5 MeV. We obtain its momentum from Equation 34-34,

$$pc = \sqrt{E_1^2 - (m_0^2c^2)^2} = \sqrt{(5 \text{ MeV})^2 - (2 \text{ MeV})^2} = \sqrt{21} \text{ MeV}$$

or

$$p = 4.58 \text{ MeV}/c$$

Since the other particle is rest, this is the total momentum of the system.

(*b*) We can find the final velocity of the two-particle system from its total energy E and its momentum p using Equation 34-33. By the conservation of total energy, the final energy of the system equals the initial total energy of the two particles:

$$E_f = E_i = E_1 + E_2 = 5\text{ MeV} + 4\text{ MeV} = 9\text{ MeV}$$

By the conservation of momentum, the final momentum of the two-particle system equals the initial momentum, $p = 4.58\text{ MeV}/c$. The velocity of the two-particle system is thus given by

$$\frac{u}{c} = \frac{pc}{E} = \frac{4.58\text{ MeV}}{9\text{ MeV}} = 0.509$$

(*c*) We can find the rest mass of the final two-particle system from Equation 34-34 using $pc = 4.58$ MeV and $E = 9$ MeV. We have

$$E^2 = (pc)^2 + (M_0c^2)^2$$

$$(9\text{ MeV})^2 = (4.58\text{ MeV})^2 + (M_0c^2)^2$$

$$M_0c^2 = \sqrt{81 - 21}\text{ MeV} = 7.75\text{ MeV}$$

$$M_0 = 7.75\text{ MeV}/c^2$$

It is instructive to check our answers by computing the initial and final kinetic energies. The initial kinetic energy is $K_i = 3$ MeV. The final kinetic energy is

$$K_f = E - M_0c^2 = 9\text{ MeV} - 7.75\text{ MeV} = 1.25\text{ MeV}$$

The loss in kinetic energy is

$$K_i - K_f = 3\text{ MeV} - 1.25\text{ MeV} = 1.75\text{ MeV}$$

Since the initial rest energy is 2 MeV + 4 MeV = 6 MeV and the final rest energy is $M_0c^2 = 7.75$ MeV, the gain in rest energy is 7.75 MeV − 6 MeV = 1.75 MeV.

34-11 General Relativity

The generalization of the theory of relativity to noninertial reference frames by Einstein in 1916 is known as the general theory of relativity. It is much more difficult mathematically than the special theory of relativity, and there are fewer situations in which it can be tested. Nevertheless, its importance calls for a brief qualitative discussion.

The basis of the general theory of relativity is the **principle of equivalence:**

> A homogeneous gravitational field is completely equivalent to a uniformly accelerated reference frame.

This principle arises in newtonian mechanics because of the apparent identity of gravitational mass and inertial mass. In a uniform gravitational field, all objects fall with the same acceleration **g** independent of their mass because the gravitational force is proportional to the (gravitational) mass, whereas the acceleration varies inversely with the (inertial) mass. Consider a compartment in space far from any matter and undergoing a uniform acceleration **a**, as shown in Figure 34-18*a*. No mechanics experiment can be performed *inside* the compartment that will distinguish whether the compartment is actually accelerating in space or is at rest (or is moving with uniform

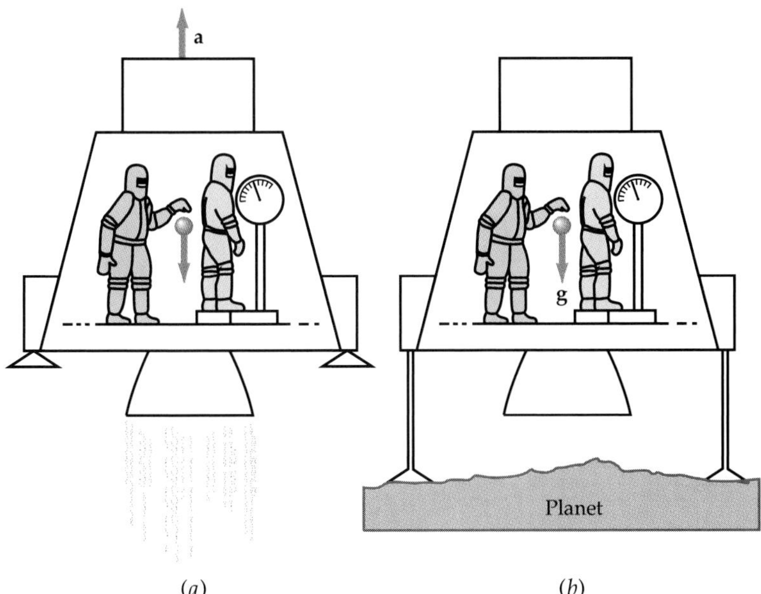

Figure 34-18 The results of experiments in a uniformly accelerated reference frame (*a*) cannot be distinguished from those in a uniform gravitational field (*b*) if the acceleration **a** and the gravitational field **g** have the same magnitude.

velocity) in the presence of a uniform gravitational field $\mathbf{g} = -\mathbf{a}$, as shown in Figure 34-18*b*. If objects are dropped in the compartment, they will fall to the "floor" with an acceleration $\mathbf{g} = -\mathbf{a}$. If people stand on a spring scale, it will read their "weight" of magnitude ma.

Einstein assumed that the principle of equivalence applies to all physics and not just to mechanics. In effect, he assumed that there is no experiment of any kind that can distinguish uniformly accelerated motion from the presence of a gravitational field. We will look qualitatively at a few of the consequences of this assumption.

The first consequence of the principle of equivalence we will discuss, the deflection of a light beam in a gravitational field, was one of the first to be tested experimentally. Figure 34-19 shows a beam of light entering a compartment that is accelerating. Successive positions of the compartment at equal time intervals are shown in Figure 34-19*a*. Because the compartment is accelerating, the distance it moves in each time interval increases with time. The path of the beam of light as observed from inside the compartment is therefore a parabola, as shown in Figure 34-19*b*. But according to the principle of equivalence, there is no way to distinguish between an accelerating compartment and one moving with uniform velocity in a uniform gravitational field. We conclude, therefore, that a beam of light, like objects having mass, will accelerate in a gravitational field. For example, near the surface of the earth, light will fall with an acceleration of 9.81 m/s^2. This is difficult to observe because of the enormous speed of light. For example, in a distance of 3000 km, which takes about 0.01 s to traverse, a beam of light should fall

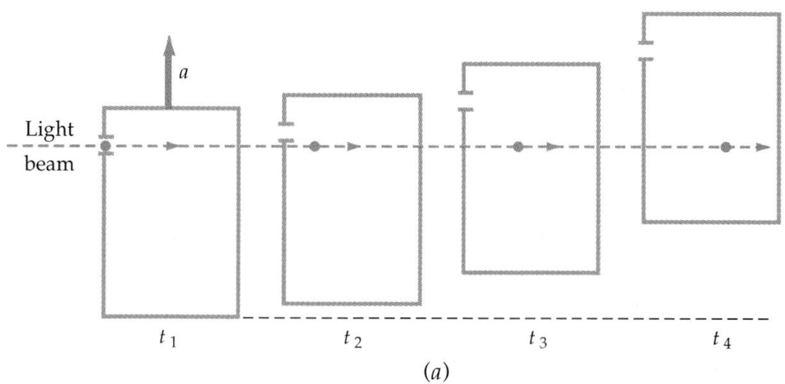

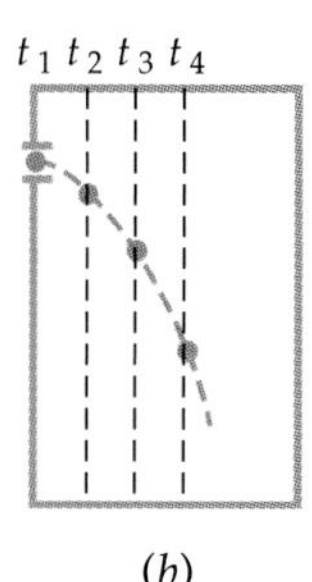

Figure 34-19 (*a*) A light beam moving in a straight line through a compartment that is undergoing uniform acceleration. The position of the beam is shown at equally spaced times t_1, t_2, t_3, and t_4. (*b*) In the reference frame of the compartment, the light travels in a parabolic path as a ball would if it were projected horizontally. The vertical displacements are greatly exaggerated in both (*a*) and (*b*) for emphasis.

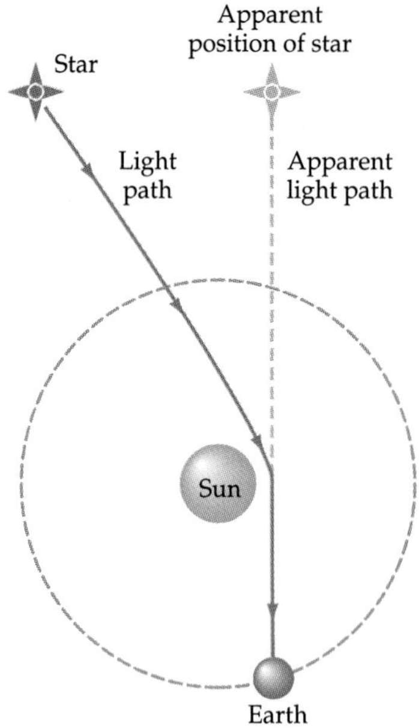

Figure 34-20 The deflection (greatly exaggerated) of a beam of light due to the gravitational attraction of the sun.

about 0.5 mm. Einstein pointed out that the deflection of a light beam in a gravitational field might be observed when light from a distant star passes close to the sun, as illustrated in Figure 34-20. Because of the brightness of the sun, such a star cannot ordinarily be seen. Such a deflection was first observed in 1919 during an eclipse of the sun. This well-publicized observation brought instant worldwide fame to Einstein.

A second prediction from Einstein's theory of general relativity, which we will not discuss in detail, is the excess precession of the perihelion of the orbit of Mercury of about 0.01° per century. This effect had been known and unexplained for some time, so, in a sense, explaining it constituted an immediate success of the theory.

A third prediction of general relativity concerns the change in time intervals and frequencies of light in a gravitational field. In Chapter 10, we found that the gravitational potential energy between two masses M and m a distance r apart is

$$U = -\frac{GMm}{r}$$

where G is the universal gravitational constant, and the point of zero potential energy has been chosen to be when the separation of the masses is infinite. The potential energy per unit mass near a mass M is called the *gravitational potential* ϕ:

$$\phi = -\frac{GM}{r} \qquad 34\text{-}38$$

According to the general theory of relativity, clocks run more slowly in regions of low gravitational potential. (Since the gravitational potential is negative, as can be seen from Equation 34-38, low gravitational potential occurs near the mass where the *magnitude* of the potential is large.) If Δt_1 is a time interval between two events measured by a clock where the gravitational potential is ϕ_1 and Δt_2 is the interval between the same events as measured by a clock where the gravitational potential is ϕ_2, general relativity predicts

(*a*) This quartz sphere in the top part of the container is probably the world's most perfectly round object. It is designed to spin as a gyroscope in a satellite orbiting the earth. General relativity predicts that the rotation of the earth will cause the axis of rotation of the gyroscope to precess in a circle at a rate of about 1 revolution in 100,000 years. (*b*) This extremely accurate hydrogen maser clock was launched in a satellite in 1976, and its time was compared to that of an identical clock on earth. In accordance with the prediction of general relativity, the clock on earth, where the gravitational potential was lower, "lost" about 4.3×10^{-10} s each second compared with the clock orbiting the earth at an altitude of about 10,000 km.

(*a*) (*b*)

that the fractional difference between these times will be approximately

$$\frac{\Delta t_2 - \Delta t_1}{\Delta t} = \frac{1}{c^2}(\phi_2 - \phi_1) \tag{34-39}$$

(Since this shift is usually very small, it does not matter by which interval we divide on the left side of the equation.) A clock in a region of low gravitational potential will therefore run slower than one in a region of high potential. Since a vibrating atom can be considered to be a clock, the frequency of vibration of an atom in a region of low potential, such as near the sun, will be lower than that of the same atom on earth. This shift toward a lower frequency and therefore a longer wavelength is called the **gravitational redshift.**

As our final example of the predictions of general relativity, we mention **black holes,** which were first predicted by Oppenheimer and Snyder in 1939. According to the general theory of relativity, if the density of an object such as a star is great enough, its gravitational attraction will be so great that once inside a critical radius, nothing can escape, not even light or other electromagnetic radiation. (The effect of a black hole on objects outside the critical radius is the same as that of any other mass.) A remarkable property of such an object is that nothing that happens inside it can be communicated to the outside. As sometimes occurs in physics, a simple but incorrect calculation gives the correct results for the relation between the mass and the critical radius of a black hole. In newtonian mechanics, the speed needed for a particle to escape from the surface of a planet or star of mass M and radius R is given by Equation 10-24:

$$v_e = \sqrt{\frac{2GM}{R}}$$

If we set the escape speed equal to the speed of light and solve for the radius, we obtain the critical radius R_S, called the **Schwarzschild radius:**

$$R_S = \frac{2GM}{c^2} \tag{34-40}$$

For an object with a mass equal to that of our sun to be a black hole, its radius would have to be about 3 km. Since no radiation is emitted from a black hole and its radius is expected to be small, the detection of a black hole is not easy. The best chance of detection would occur if a black hole were a companion to a normal star in a binary star system. The black hole would affect a number of properties of its visible companion. Measurements of the doppler shift of the light from the normal star, for example, might allow a computation of the mass of the unseen companion to determine whether it is great enough to be a black hole. At present there are several excellent candidates—one in the constellation Cygnus, one in the Small Magellanic Cloud, and perhaps one in our own galaxy—but the evidence is not conclusive.

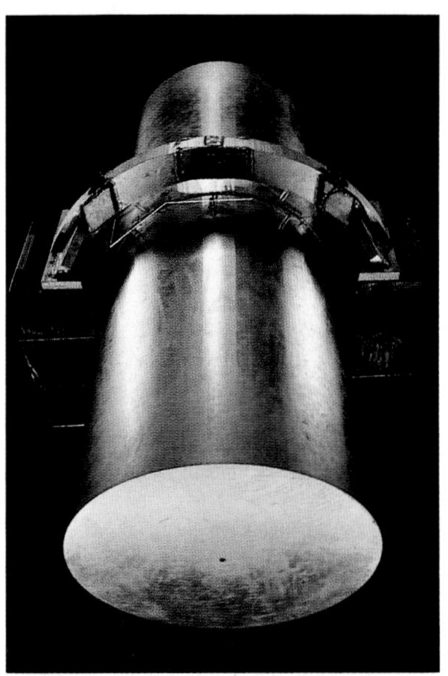

This antenna, consisting of a 1400-kg aluminum cylinder freely suspended by a steel cable, was built by Joseph Weber, David Zippy, and Robert Foward at the University of Maryland to detect gravitational waves. In theory, the antenna should vibrate as the gravity waves pass through it.

Summary

1. The special theory of relativity is based on two postulates of Albert Einstein:
   Postulate 1. Absolute, uniform motion cannot be detected.

Postulate 2. The speed of light is independent of the motion of the source.

An important implication of these postulates is

Postulate 2 (Alternate). Every observer measures the same value for the speed of light independent of the relative motion of the sources and observer.

All of the results of special relativity can be derived from these postulates.

2. The Michelson–Morley experiment was an attempt to measure the absolute velocity of the earth by comparing the speed of light in the direction of motion of the earth with that in a direction perpendicular to that motion. Their null result for the difference in these speeds is consistent with Einstein's postulates.

3. The Lorentz transformation relates the coordinates x, y, and z and the time t of an event seen in frame S to the coordinates x', y', and z' and the time t' of the same event as seen in frame S', which is moving with speed V relative to S:

$$x = \gamma(x' + Vt') \qquad y = y' \qquad z = z'$$

$$t = \gamma\left(t' + \frac{Vx'}{c^2}\right)$$

where

$$\gamma = \frac{1}{\sqrt{1 - V^2/c^2}}$$

The inverse transformation is

$$x' = \gamma(x - Vt) \qquad y' = y \qquad z' = z$$

$$t' = \gamma\left(t - \frac{Vx}{c^2}\right)$$

The transformation equations for velocities are

$$u_x = \frac{u_x' + V}{1 + Vu_x'/c^2}$$

$$u_y = \frac{u_y'}{\gamma(1 + Vu_x'/c^2)}$$

$$u_z = \frac{u_z'}{\gamma(1 + Vu_x'/c^2)}$$

The inverse velocity transformation equations are

$$u_x' = \frac{u_x - V}{1 - Vu_x/c^2}$$

$$u_y' = \frac{u_y}{\gamma(1 - Vu_x/c^2)}$$

$$u_z' = \frac{u_z}{\gamma(1 - Vu_x/c^2)}$$

4. The time interval measured between two events that occur at the same point in space in some reference frame is called the proper time. In another reference frame in which the events occur at different places, the time interval between the events is longer by the factor γ. This result is known as time dilation. A related phenomenon is length contraction. The length of an object measured in a frame in which it is at rest is called its proper length L_p. When measured in another reference frame, the length of the object is L_p/γ.

5. Two events that are simultaneous in one reference frame are not simultaneous in another frame that is moving relative to the first. If two clocks are synchronized in the frame in which they are at rest, they will be out of synchronization in another frame. In the frame in which they are moving, the "chasing" clock leads by an amount $\Delta t_s = L_p V/c^2$, where L_p is the proper distance between the clocks.

6. The relativistic momentum of a particle is related to its mass and velocity by

$$\mathbf{p} = \frac{m_0 \mathbf{u}}{\sqrt{1 - u^2/c^2}}$$

where m_0 is the rest mass of the particle.

7. The kinetic energy of a particle is given by

$$K = \frac{m_0 c^2}{\sqrt{1 - u^2/c^2}} - m_0 c^2 = \frac{m_0 c^2}{\sqrt{1 - u^2/c^2}} - E_0$$

where

$$E_0 = m_0 c^2$$

is the rest energy. The total energy is

$$E = K + E_0 = \frac{m_0 c^2}{\sqrt{1 - u^2/c^2}}$$

The speed of a particle is related to its momentum and its total energy by

$$\frac{u}{c} = \frac{pc}{E}$$

The total energy is related to the momentum and rest energy by

$$E^2 = p^2c^2 + (m_0 c^2)^2$$

For particles with energies much greater than their rest energies, a useful approximation is

$$E \approx pc \qquad \text{for } E \gg m_0 c^2$$

This is an exact equation for particles of zero rest mass such as photons.

8. The total rest mass of bound systems of particles, such as nuclei or atoms, is less than the sum of the rest masses of the particles making up the system. The difference in mass times c^2 equals the binding energy of the system. The binding energy is the energy that must be added to break up the system into its parts. The binding energies of electrons in atoms are of the order of eV or keV, leading to a negligible difference in rest mass. The binding energies of nuclei are of the order of several MeV, and the difference in rest mass is noticeable.

9. The basis of the general theory of relativity is the principle of equivalence: A homogeneous gravitational field is completely equivalent to a uniformly accelerated reference frame. Important consequences of general relativity include the bending of light in a gravitational field, the prediction of the precession of the perihelion of the orbit of Mercury, the gravitational redshift, and probably the existence of black holes.

Suggestions for Further Reading

Bondi, Hermann: *Relativity and Common Sense: A New Approach to Einstein*, Doubleday, Garden City, New York, 1964.

This book uses familiar phenomena to help show how logical and easy it is to understand the ideas of special relativity.

Chaffee, Frederick H., Jr.: "The Discovery of a Gravitational Lens," *Scientific American*, November 1980, p. 70.

General relativity predicts that light should be deflected by concentrations of matter. This article describes how an elliptical galaxy can act as a giant lens in space.

Gamow, George: "Gravity," *Scientific American*, March 1961, p. 94.

Einstein's general theory of relativity is explained in an entertaining and nonmathematical fashion.

Goldberg, Stanley: *Understanding Relativity: Origin and Impact of a Scientific Revolution*, Birkhaeuser, Boston, 1984.

This book examines the intellectual and social context from which Einstein's special theory grew and the theory's early reception by communities of scientists in four countries.

MacKeown, P. K.: "Gravity is Geometry," *The Physics Teacher*, vol. 22, 1984, p. 557.

This article is an excellent, brief exposition of the ideas of general relativity.

Marder, L.: *Time and the Space Traveller*, George Allen & Unwin, Ltd., London, 1971.

This book presents some of the arguments which have been made in the long and colorful debate over the twin paradox. It also examines some practical limitations of space travel, the implications of time dilation for the long-distance space traveller, and the nature of living clocks.

Mook, Delo E., and Thomas Vargish: *Inside Relativity*, Princeton University Press, Princeton, 1987.

This is a book for nonscientists written by two scholars, one working in the physical sciences and the other in the humanities. The book provides a historical and scientific context for Einstein's work, and explains the special and general theories with the aid of drawings and graphs but no mathematics.

Schwinger, Julian: *Einstein's Legacy: The Unity of Space and Time*, Scientific American Books, Inc., New York, 1986.

A modern and well-illustrated exposition of the special and general theories of relativity and some of their consequences.

Shankland, R. S.: "The Michelson–Morley Experiment," *Scientific American*, November 1964, p. 107.

This article sets the experiment in its historical context and considers its influence on the development of the theory of relativity.

Will, Clifford M.: *Was Einstein Right?: Putting General Relativity to the Test*, Basic Books, Inc., New York, 1986.

Starting around 1960, new discoveries in astronomy motivated a renewed interest in experimentally testing predictions of general relativity. This book, written by a physicist who began his career during this relativity "renaissance," describes the tests with great enthusiasm.

Review

A. Objectives: After studying this chapter, you should:

1. Be able to discuss the results and significance of the Michelson–Morley experiment.

2. Be able to state the Einstein postulates of special relativity.

3. Be able to use the Lorentz transformation to derive expressions for time dilation and length contraction and to solve problems in which time and space intervals in different reference frames are compared.

4. Be able to discuss the lack of synchronization of clocks in moving reference frames.

5. Be able to discuss the twin paradox.

6. Be able to state the definition of relativistic momentum and the equations relating to kinetic energy and total energy of a particle to its speed.

7. Be able to discuss the relation between mass and energy in special relativity and compute the binding energy of various systems from the known rest masses of their constituents.

8. Be able to state the principle of equivalence and discuss three predictions derived from it.

B. Define, explain, or otherwise identify:

Reference frame
Inertial reference frame
Newtonian relativity
Ether
Michelson–Morley experiment
Einstein's postulates
Galilean transformation
Lorentz transformation
Proper time

Time dilation
Length contraction
Proper length
Lorentz–FitzGerald contraction
Synchronized clocks
Simultaneity
Redshift
Twin paradox
Relativistic momentum
Relativistic mass
Rest mass
Rest energy
Relativistic energy
Binding energy
Principle of equivalence
Gravitational redshift
Black hole
Schwarzschild radius

C. True or false: If the statement is true, explain why it is true. If it is false, give a counterexample.

1. The speed of light is the same in all reference frames.
2. Proper time is the shortest time interval between two events.
3. Absolute motion can be determined by means of length contraction.
4. The light-year is a unit of distance.
5. Simultaneous events must occur at the same place.
6. If two events are not simultaneous in one frame, they cannot be simultaneous in any other frame.
7. If two particles are tightly bound together by strong attractive forces, the rest mass of the system is less than the sum of the masses of the individual particles when separated.

Problems

Level I

34-1 Newtonian Relativity

There are no problems for this section.

34-2 The Michelson–Morley Experiment

1. In one series of measurements of the speed of light, Michelson used a path length L of 27.4 km (17 mi). (*a*) What is the time needed for light to make the round-trip distance of $2L$? (*b*) What is the classical correction term in seconds in Equation 34-1, assuming earth's speed is $v = 10^{-4}c$? (*c*) From about 1600 measurements, Michelson quoted the result for the speed of light as 299,796 ± 4 km/s. Is this experiment accurate enough to be sensitive to the correction term in Equation 34-1?

2. An airplane flies with speed u relative to still air from point A to point B and returns. Compare the time required for the round trip when the wind blows from A to B with speed v with that when the wind blows perpendicularly to the line AB with speed v.

34-3 Einstein's Postulates

There are no problems for this section.

34-4 The Lorentz Transformation

3. The proper mean lifetime of pions is 2.6×10^{-8} s. If a beam of pions has a speed of $0.85c$, (*a*) what would their mean lifetime be as measured in the laboratory? (*b*) How far would they travel, on the average, before they decay? (*c*) What would your answer be to part (*b*) if you neglect time dilation?

4. (*a*) In the reference frame of the pion in Problem 3, how far does the laboratory travel in a typical lifetime of 2.6×10^{-8} s? (*b*) What is this distance in the laboratory's frame?

5. The proper mean lifetime of a muon is 2 μs. Muons in a beam are traveling at $0.999c$. (*a*) What is their mean lifetime as measured in the laboratory? (*b*) How far do they travel, on the average, before they decay?

6. (*a*) In the reference frame of the muon in Problem 5, how far does the laboratory travel in a typical lifetime of 2 μs? (*b*) What is this distance in the laboratory's frame?

7. A spaceship of proper length 100 m passes you at a high speed. You measure the length of the spaceship to be 85 m. What was the speed of the spaceship?

8. A spaceship departs from earth for the star Alpha Centauri, which is 4 light-years away. The spaceship travels at $0.75c$. How long does it take to get there (*a*) as measured on earth and (*b*) as measured by a passenger on the spaceship?

9. A spaceship travels to a star 95 light-years away at a speed of 2.2×10^8 m/s. How long does it take to get there (*a*) as measured on earth and (*b*) as measured by a passenger on the spaceship?

10. The mean lifetime of a pion traveling at high speed is measured to be 7.5×10^{-8} s. Its lifetime when measured at rest is 2.6×10^{-8} s. How fast is the pion traveling?

11. How fast must a muon travel so that its mean lifetime is 46 μs if its mean lifetime at rest is 2 μs?

12. A meterstick moves with speed $V = 0.8c$ relative to you in the direction parallel to the stick. (*a*) Find the length of the stick as measured by you. (*b*) How long does it take for the stick to pass you?

13. How fast must a meterstick travel relative to you in the direction parallel to the stick so that its length as measured by you is 50 cm?

14. Use the binomial expansion (Equation 34-2) to derive the following results for the case when V is much less than c, and use the results when applicable in the following problems:

(*a*) $\gamma \approx 1 + \frac{1}{2}\frac{V^2}{c^2}$ (*b*) $\frac{1}{\gamma} \approx 1 - \frac{1}{2}\frac{V^2}{c^2}$

(*c*) $\gamma - 1 \approx 1 - \frac{1}{\gamma} \approx \frac{1}{2}\frac{V^2}{c^2}$

15. Supersonic jets achieve maximum speeds of about $(3 \times 10^{-6})c$. (*a*) By what percentage would you see a jet traveling at this speed contracted in length? (*b*) During a time of 1 y = 3.15×10^7 s on your clock, how much time would elapse on the pilot's clock? How many minutes are lost by the pilot's clock in 1 y of your time?

16. How great must the relative speed of two observers be for the time-interval measurements to differ by 1 percent? (See Problem 14.)

34-5 Clock Synchronization and Simultaneity

Problems 17 through 21 refer to the following situation: An observer in S′ lays out a distance L′ = 100c·min between points A′ and B′ and places a flashbulb at the midpoint C′. She arranges for the bulb to flash and for clocks at A′ and B′ to be started at zero when the light from the flash reaches them (see Figure 34-21). Frame S′ is moving to the right with speed 0.6c relative to an observer C in S who is at the midpoint between A′ and B′ when the bulb flashes. At the instant he sees the flash, observer C sets his clock to zero.

Figure 34-21 Problems 17 through 21.

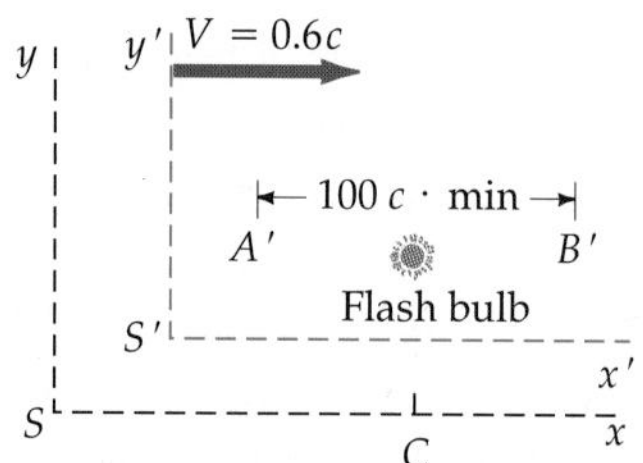

17. What is the separation distance between clocks A' and B' according to the observer in S?

18. As the light pulse from the flashbulb travels toward A' with speed c, A' travels toward C with speed $0.6c$. Show that the clock in S reads 25 min when the flash reaches A'. (*Hint*: In time t, the light travels a distance ct and A' travels $0.6ct$. The sum of these distances must equal the distance between A' and the flashbulb as seen in S.)

19. Show that the clock in S reads 100 min when the light flash reaches B', which is traveling away from C with speed $0.6c$. (See the hint for Problem 18.)

20. The time interval between the reception of the flashes at A' and B' in Problems 18 and 19 is 75 min according to the observer in S. How much time does he expect to have elapsed on the clock at A' during this 75-min interval?

21. The time interval calculated in Problem 20 is the amount that the clock at A' leads that at B' according to the observer in S. Compare this result with $L_p V/c^2$.

34-6 The Doppler Effect

22. How fast must you be moving toward a red light (λ = 650 nm) for it to appear green (λ = 525 nm)?

23. A distant galaxy is moving away from us at a speed of 1.85×10^7 m/s. Calculate the fractional redshift $(\lambda' - \lambda_0)/\lambda_0$ in the light from this galaxy.

24. Show that if V is much less than c, the doppler shift is given approximately by $\Delta f/f_0 \approx \pm V/c$.

25. A distant galaxy is moving away from the earth with a speed that results in each wavelength received on earth being shifted such that $\lambda' = 2\lambda_0$. Find the speed of the galaxy relative to the earth.

26. Sodium light of wavelength 589 nm is emitted by a source that is receding from the earth with speed V. The wavelength measured in the frame of the earth is 620 nm. Find V.

27. A student on earth hears a tune on her radio that seems to be coming from a record that is being played too fast. She has a 33-rev/min record of that tune and determines that the tune sounds the same as when her record is played at 78 rev/min, that is, the frequencies are all too high by a factor of 78/33. If the tune is being played correctly, but is being broadcast by a spaceship that is approaching the earth at speed V, determine V.

34-7 The Twin Paradox

28. A friend of yours who is the same age as you travels at $0.999c$ to a star 15 light-years away. She spends 10 y on one of the star's planets and then returns at $0.999c$. How long has she been away (*a*) as measured by you and (*b*) as measured by her?

34-8 The Velocity Transformation

29. Two spaceships are approaching each other. (*a*) If the speed of each is $0.6c$ relative to the earth, what is the speed of one relative to the other? (*b*) If the speed of each relative to the earth is 30,000 m/s (about 100 times the speed of sound), what is the speed of one relative to the other?

30. A light beam moves along the y' axis with speed c in frame S', which is moving to the right with speed V relative to frame S. (*a*) Find the x and y components of the velocity of the light beam in frame S. (*b*) Show that the magnitude of the velocity of the light beam in S is c.

31. A spaceship is moving east at speed $0.90c$ relative to the earth. A second spaceship is moving west at speed $0.90c$ relative to the earth. What is the speed of one spaceship relative to the other?

32. A particle moves with speed $0.8c$ along the x'' axis of frame S'', which moves with speed $0.8c$ along the x' axis relative to frame S'. Frame S' moves with speed $0.8c$ along the x axis relative to frame S. (*a*) Find the speed of the particle relative to frame S'. (*b*) Find the speed of the particle relative to frame S.

34-9 Relativistic Momentum;
34-10 Relativistic Energy

33. How much rest mass must be converted into energy (*a*) to produce 1 J and (*b*) to keep a 100-W light bulb burning for 10 years?

34. Sketch a graph of the momentum p of a particle versus its speed u.

35. (*a*) Calculate the rest energy in 1 g of dirt. (*b*) If you could convert this energy into electrical energy and sell it for 10 cents per kilowatt-hour, how much money would you get? (*c*) If you could power a 100-W light bulb with this energy, for how long could you keep the bulb lit?

36. Find the ratio of the total energy to the rest energy of a particle of rest mass m_0 moving with speed (*a*) $0.1c$, (*b*) $0.5c$, (*c*) $0.8c$, and (*d*) $0.99c$.

37. An electron with rest energy of 0.511 MeV moves with speed $u = 0.2c$. Find its total energy, kinetic energy, and momentum.

38. A muon has a rest energy of 105.7 MeV. Calculate its rest mass in kilograms.

39. A proton with rest energy of 938 MeV has a total energy of 1400 MeV. (*a*) What is its speed? (*b*) What is its momentum?

40. The total energy of a particle is twice its rest energy. (*a*) Find u/c for the particle. (*b*) Show that its momentum is given by $p = \sqrt{3}m_0c$.

41. For the fusion reaction of Example 34-11, calculate the number of reactions per second that are necessary to generate 1 kW of power.

42. Use Table 34-1 to find how much energy is needed to remove one proton from ^{4}He, leaving ^{3}H plus a proton.

43. A free neutron decays into a proton plus an electron:

$$n \rightarrow p + e$$

Use Table 34-1 to calculate the energy released in this reaction.

44. How much energy would be required to accelerate a particle of mass m_0 from rest to (*a*) $0.5c$, (*b*) $0.9c$, and (*c*) $0.99c$? Express your answers as multiples of the rest energy.

45. If the kinetic energy of a particle equals its rest energy, what error is made by using $p = m_0u$ for its momentum?

46. In another nuclear fusion reaction, ^{2}H nuclei are combined to produce ^{4}He. (*a*) How much energy is released in this reaction? (*b*) How many such reactions must take place per second to produce 1 kW of power?

34-11 General Relativity

There are no problems for this section.

Level II

47. A friend of yours who is the same age as you travels to the star Alpha Centauri, which is 4 light-years away and returns immediately. He claims that the entire trip took just 6 y. How fast did he travel?

48. Use Equations 34-28 and 34-32 to derive the equation $E^2 = p^2c^2 + m_0^2c^4$.

49. If a plane flies at a speed of 2000 km/h, for how long must it fly before its clock loses 1 s because of time dilation?

50. Use the binomial expansion (Equation 34-2) and Equation 34-34 to show that when $pc \ll m_0c^2$, the total energy is given approximately by

$$E \approx m_0c^2 + \frac{p^2}{2m_0}$$

51. A clock is placed in a satellite that orbits the earth with a period of 90 min. By what time interval will this clock differ from an identical clock on earth after 1 y? (Assume that special relativity applies.)

52. *A* and *B* are twins. *A* travels at $0.6c$ to Alpha Centauri (which is $4c$·years from earth as measured in the reference frame of the earth) and returns immediately. Each twin sends the other a light signal every 0.01 year as measured in her own reference frame. (*a*) At what rate does *B* receive signals as *A* is moving away from her? (*b*) How many signals does *B* receive at this rate? (*c*) How many total signals are received by *B* before *A* has returned? (*d*) At what rate does *A* receive signals as *B* is receding from her? (*e*) How many signals does *A* receive at this rate? (*f*) What is the total number of signals received by *A*? (*g*) Which twin is younger at the end of the trip, and by how many years?

53. In frame *S*, event *B* occurs 2 μs after event *A*, which occurs at $\Delta x = 1.5$ km from event *A*. How fast must an observer be moving along the $+x$ axis so that events *A* and *B* occur simultaneously? Is it possible for event *B* to precede event *A* for some observer?

54. Observers in reference frame *S* see an explosion located at $x_1 = 480$ m. A second explosion occurs 5 μs later at $x_2 = 1200$ m. In reference frame S', which is moving along the $+x$ axis at speed *V*, the explosions occur at the same point in space. What is the separation in time between the two explosions as measured in S'?

55. An interstellar spaceship travels from the earth to a distant star system $12c$·years away (as measured in the earth's frame). The trip takes 15 years as measured on the ship. (*a*) What is the speed of the ship relative to the earth? (*b*) When the ship arrives, it sends a signal to the earth. How long after the ship leaves the earth will it be before the earth receives the signal?

56. Show that the speed *u* of a particle of mass m_0 and total energy *E* is given by

$$\frac{u}{c} = \left[1 - \frac{(m_0c^2)^2}{E^2}\right]^{1/2}$$

and that when *E* is much greater than m_0c^2, this can be approximated by

$$\frac{u}{c} \approx 1 - \frac{(m_0c^2)^2}{2E^2}$$

Find the speed of an electron with kinetic energy of (*b*) 0.51 MeV and (*c*) 10 MeV.

57. Two spaceships, each 100 m long when measured at rest, travel toward each other with speeds of $0.85c$ relative to the earth. (*a*) How long is each ship as measured by someone on earth? (*b*) How fast is each ship traveling as measured by an observer on the other? (*c*) How long is one ship when measured by an observer on the other? (*d*) At time $t = 0$ on earth, the fronts of the ships are together as they just begin to pass each other. At what time on earth are their ends together? (*e*) Sketch diagrams in the frame of one of the ships showing the passing of the other ship.

58. In the Stanford linear collider, small bundles of electrons and positrons are fired at each other. In the laboratory's frame of reference, each bundle is about 1 cm long and 10 μm in diameter. In the collision region, each particle has an energy of 50 GeV, and the electrons and positrons are moving in opposite directions. (*a*) How long and how wide is each bundle in its own reference frame? (*b*) What must be the minimum proper length of the accelerator for a bundle to have both its ends simultaneously in the accelerator in its own reference frame? (The actual length of the accelerator is less than 1000 m.) (*c*) What is the length of a positron bundle in the reference frame of the electron bundle?

59. An electron with rest energy of 0.511 MeV has a total energy of 5 MeV. (*a*) Find its momentum in units of MeV/c from Equation 34-34. (*b*) Find the ratio of its speed u to the speed of light.

60. The rest energy of a proton is about 938 MeV. If its kinetic energy is also 938 MeV, find (*a*) its momentum and (*b*) its speed.

61. What percent error is made in using $\frac{1}{2}m_0u^2$ for the kinetic energy of a particle if its speed is (*a*) $0.1c$ and (*b*) $0.9c$?

62. A rocket with a proper length of 1000 m moves in the $+x$ direction at $0.6c$ with respect to an observer on the ground. An astronaut stands at the rear of the rocket and fires a bullet toward the front of the rocket at $0.8c$ relative to the rocket. How long does it take the bullet to reach the front of the rocket (*a*) as measured in the frame of the rocket, (*b*) as measured in the frame of the ground, and (*c*) as measured in the frame of the bullet?

63. A rocket with a proper length of 700 m is moving to the right at a speed of $0.9c$. It has two clocks, one in the nose and one in the tail, that have been synchronized in the frame of the rocket. A clock on the ground and the nose clock on the rocket both read $t = 0$ as they pass. (*a*) At $t = 0$, what does the tail clock on the rocket read as seen by an observer on the ground? When the tail clock on the rocket passes the ground clock, (*b*) what does the tail clock read as seen by an observer on the ground, (*c*) what does the nose clock read as seen by an observer on the ground, and (*d*) what does the nose clock read as seen by an observer on the rocket? (*e*) At $t = 1$ h, as measured on the rocket, a light signal is sent from the nose of the rocket to an observer standing by the ground clock. What does the ground clock read when the observer receives this signal? (*f*) When the observer on the ground receives the signal, he sends a return signal to the nose of the rocket. When is this signal received at the nose of the rocket as seen on the rocket?

64. Derive Equation 34-24*a* for the frequency received by an observer moving with speed V toward a stationary source of electromagnetic waves.

65. Frames S and S' are moving relative to each other along the x and x' axis. They set their clocks to $t = 0$ when their origins coincide. In frame S, event 1 occurs at $x_1 = 1.0c\cdot$year and $t_1 = 1$ y and event 2 occurs at $x_2 = 2.0c\cdot$year and $t_2 = 0.5$ y. These events occur simultaneously in frame S'. (*a*) Find the magnitude and direction of the velocity of S' relative to S. (*b*) At what time do both these events occur as measured in S'?

66. An observer in frame S standing at the origin observes two flashes of colored light separated spatially by $\Delta x = 2400$ m. A blue flash occurs first, followed by a red flash 5 μs later. An observer in S' moving along the x axis at speed V relative to S also observes the flashes 5 μs apart and with a separation of 2400 m, but the red flash is observed first. Find the magnitude and direction of V.

67. The sun radiates energy at the rate of about 4×10^{26} W. Assume that this energy is produced by a reaction whose net result is the fusion of 4 H nuclei to form 1 He nucleus, with the release of 25 MeV for each He nucleus formed. Calculate the sun's loss of rest mass per day.

68. A spaceship of mass 10^6 kg is coasting through space when it suddenly becomes necessary to accelerate. The ship ejects 10^3 kg of fuel in a very short time at a speed of $c/2$ relative to the ship. (*a*) Neglecting any change in the rest mass of the system, calculate the speed of the ship in the frame in which it was initially at rest. (*b*) Calculate the speed of the ship using classical, newtonian mechanics. (*c*) Use your results from (*a*) to estimate the change in the rest mass of the system.

69. Reference frame S' is moving along the x' axis at $0.6c$ relative to frame S. A particle that is originally at $x' = 10$ m at $t_1' = 0$ is suddenly accelerated and then moves at a constant speed of $c/3$ in the $-x'$ direction until time $t_2' = 60$ m/c, when it is suddenly brought to rest. As observed in frame S, find (*a*) the speed of the particle, (*b*) the distance and direction the particle traveled from t_1' to t_2', and (*c*) the time the particle traveled.

70. Show that

$$d\left(\frac{m_0u}{\sqrt{1-u^2/c^2}}\right) = m_0\left(1-\frac{u^2}{c^2}\right)^{-3/2} du$$

71. Two protons approach each other head on at $0.5c$ relative to reference frame S'. (*a*) Calculate the total kinetic energy of the two protons as seen in frame S'. (*b*) Calculate the total kinetic energy of the protons as seen in reference frame S, which is moving with speed $0.5c$ relative to S' such that one of the protons is at rest.

72. A particle of rest mass 1 MeV/c^2 and kinetic energy 2 MeV collides with a stationary particle of rest mass 2 MeV/c^2. After the collision, the particles stick together. Find (*a*) the speed of the first particle before the collision, (*b*) the total energy of the first particle before the collision, (*c*) the initial total momentum of the system, (*d*) the total kinetic energy after the collision, and (*e*) the rest mass of the system after the collision.

73. The radius of the orbit of a charged particle in a magnetic field is related to the momentum of the particle by

$$p = BqR \qquad 34\text{-}41$$

This equation holds classically for $p = mu$ and relativistically for $p = m_0u/\sqrt{1 - u^2/c^2}$. An electron with kinetic energy of 1.50 MeV moves in a circular orbit perpendicular to a uniform magnetic field $B = 5 \times 10^{-3}$ T. (*a*) Find the radius of the orbit. (*b*) What result would you obtain if you used the classical relations $p = mu$ and $K = p^2/2m$?

74. Oblivious to economics and politics, physicists propose building a circular accelerator around the earth's circumference using bending magnets that provide a magnetic field of magnitude 1.5 T. (*a*) What would be the kinetic energy of protons orbiting in this field in a circle of radius R_E? (See Problem 73.) (*b*) What would be the period of rotation of these protons?

75. In a simple thought experiment, Einstein showed that there is mass associated with electromagnetic radiation. Consider a box of length L and mass M resting on a frictionless surface. At the left wall of the box is a light source that emits radiation of energy E, which is absorbed at the right wall of the box. According to classical electromagnetic theory, this radiation carries momentum of magnitude $p = E/c$ (Equation 29-24). (*a*) Find the recoil velocity of the box such that momentum is conserved when the light is emitted. (Since p is small and M is large, you may use classical mechanics.) (*b*) When the light is absorbed at the right wall of the box, the box stops, so the total momentum remains zero. If we neglect the very small velocity of the box, the time it takes for the radiation to travel across the box is $\Delta t = L/c$. Find the distance moved by the box in this time. (*c*) Show that if the center of mass of the system is to remain at the same place, the radiation must carry mass $m = E/c^2$.

76. An antiproton $\bar{p}$ has the same rest energy as a proton. It is created in the reaction $p + p \rightarrow p + p + p + \bar{p}$. In an experiment, protons at rest in the laboratory are bombarded with protons of kinetic energy K_L, which must be great enough so that kinetic energy equal to $2m_0c^2$ can be converted into the rest energy of the two particles. In the frame of the laboratory, the total kinetic energy cannot be converted into rest energy because of conservation of momentum. However, in the zero-momentum reference frame in which the two initial protons are moving toward each other with equal speed u, the total kinetic energy can be converted into rest energy. (*a*) Find the speed of each proton u such that the total kinetic energy in the zero-momentum frame is $2m_0c^2$. (*b*) Transform to the laboratory's frame in which one proton is at rest, and find the speed u' of the other proton. (*c*) Show that the kinetic energy of the moving proton in the laboratory's frame is $K_L = 6m_0c^2$.

Level III

77. A stick of proper length L_p makes an angle θ with the x axis in frame S. Show that the angle θ' made with the x' axis in frame S', which is moving along the $+x$ axis with speed V, is given by $\tan\theta' = \gamma \tan\theta$ and that the length of the stick in S' is

$$L' = L_p\left[\frac{1}{\gamma^2}\cos^2\theta + \sin^2\theta\right]^{1/2}$$

78. Show that if a particle moves at an angle θ with the x axis with speed u in frame S, it moves at an angle θ' with the x' axis in S' given by

$$\tan\theta' = \frac{\sin\theta}{\gamma(\cos\theta - V/u)}$$

where the frame S' is moving with speed V relative to S.

79. For the special case of a particle moving with speed u along the y axis in frame S, show that its momentum and energy in frame S' are related to its momentum and energy in S by the transformation equations

$$p_x' = \gamma\left(p_x - \frac{VE}{c^2}\right) \qquad p_y' = p_y \qquad p_z' = p_z$$

$$\frac{E'}{c} = \gamma\left(\frac{E}{c} - \frac{Vp_x}{c}\right)$$

Compare these equations with the Lorentz transformation for x', y', z', and t'. These equations show that the quantities p_x, p_y, p_z, and E/c transform in the same way as do x, y, z, and ct.

80. The equation for the spherical wavefront of a light pulse that begins at the origin at time $t = 0$ is $x^2 + y^2 + z^2 - (ct)^2 = 0$. Using the Lorentz transformation, show that such a light pulse also has a spherical wavefront in frame S' by showing that $x'^2 + y'^2 + z'^2 - (ct')^2 = 0$ in S'.

81. In Problem 80, you showed that the quantity $x^2 + y^2 + z^2 - (ct)^2$ has the same value (0) in both S and S'. Such a quantity is called an *invariant*. From the results of Problem 79, the quantity $p_x^2 + p_y^2 + p_z^2 - (E/c)^2$ must also be an invariant. Show that this quantity has the value $-m_0c^2$ in both the S and S' reference frames.

82. Two events in S are separated by a distance $D = x_2 - x_1$ and a time $T = t_2 - t_1$. (*a*) Use the Lorentz transformation to show that in frame S', which is moving with speed V relative to S, the time separation is $t_2' - t_1' = \gamma(T - VD/c^2)$. (*b*) Show that the events can be simultaneous in frame S' only if D is greater than cT. (*c*) If one of the events is the *cause* of the other, the separation D must be less than cT since D/c is the smallest time that a signal can take to travel from x_1 to x_2 in frame S. Show that if D is less than cT, t_2' is greater than t_1' in all reference frames. This shows that if the cause precedes the effect in one frame, it must precede it in all reference frames. (*d*) Suppose that a signal could be sent with speed $c' > c$ so that in frame S the cause precedes the effect by the time $T = D/c'$. Show that there is then a reference frame moving with speed V less than c in which the effect precedes the cause.

83. Two identical particles of rest mass m_0 are each moving toward the other with speed u in frame S. The particles collide inelastically with a spring that locks shut (Figure 34-22) and come to rest in S, and their initial kinetic energy is transformed into potential energy. In this problem you are going to show that the conservation of momentum in reference frame S', in which one of the particles is initially at rest, requires that the total rest mass of the system after the collision be $2m_0/\sqrt{1 - u^2/c^2}$. (*a*) Show that the speed of the particle not at rest in frame S' is

$$u' = \frac{2u}{1 + u^2/c^2}$$

and use this result to show that

$$\sqrt{1 - \frac{u'^2}{c^2}} = \frac{1 - u^2/c^2}{1 + u^2/c^2}$$

(*b*) Show that the initial momentum in frame S' is $p' = 2m_0u/(1 - u^2/c^2)$. (*c*) After the collision, the composite particle moves with speed u in S' (since it is at rest in S). Write the total momentum after the collision in terms of the final rest M_0, and show that the conservation of momentum implies that $M_0 = 2m_0/\sqrt{1 - u^2/c^2}$. (*d*) Show that the total energy is conserved in each reference frame.

84. A horizontal turntable rotates with angular speed ω. There is a clock at the center of the turntable and one at a distance r from the center. In an inertial reference frame, the clock at distance r is moving with speed $u = r\omega$. (*a*) Show that from time dilation in special relativity, time intervals Δt_0 for the clock at rest and Δt_r for the moving clock are related by

$$\frac{\Delta t_r - \Delta t_0}{\Delta t_0} \approx -\frac{r^2\omega^2}{2c^2} \qquad \text{if } r\omega \ll c$$

(*b*) In a reference frame rotating with the table, both clocks are at rest. Show that the clock at distance r experiences a pseudoforce $F_r = mr\omega^2$ in this accelerated frame and that this is equivalent to a difference in gravitational potential between r and the origin of $\phi_r - \phi_0 = -\frac{1}{2}r^2\omega^2$. Use this potential difference in Equation 34-39 to show that in this frame the difference in time intervals is the same as in the inertial frame.

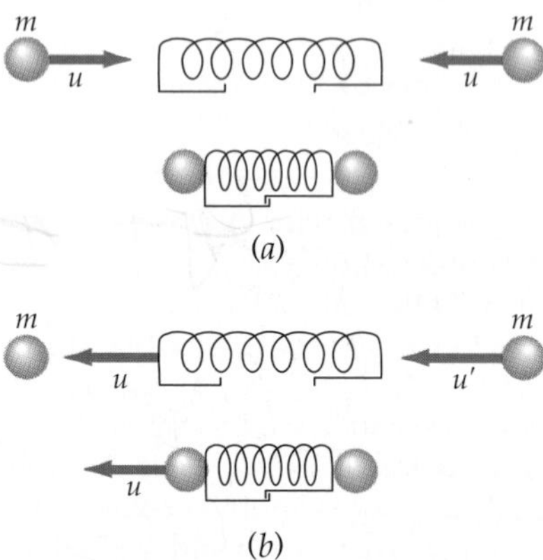

Figure 34-22 Problem 83. An inelastic collision between two identical objects (*a*) in the zero-momentum reference frame S and (*b*) in frame S', which is moving to the right with speed $V = u$ relative to frame S such that one of the particles is initially at rest. The spring, which is assumed to be massless, is merely a device for visualizing the storage of potential energy.

Chapter 35

The Origins of Quantum Theory

The continuous visible spectrum (*top*) along with the characteristic optical line spectrum (*from top to bottom*) emitted by hydrogen, helium, barium, and mercury. Niels Bohr's model of the hydrogen atom, which could be used to predict the wavelengths of the observed spectrum of hydrogen, was one of the great triumphs of twentieth century physics and was an important step in the development of our understanding of the microscopic world.

In Chapter 34, we saw that Newton's laws must be modified when they are applied to objects that move at speeds comparable to the speed of light. In the last 20 years of the nineteenth century and the first 30 years of the twentieth century, many startling discoveries, both experimental and theoretical, demonstrated that the laws of classical physics also break down when they are applied to microscopic systems, such as the particles within an atom. This failure is as drastic as the failure of newtonian mechanics at high speeds. The interior of the atom can be described only in terms of *quantum theory* (sometimes called *quantum mechanics* or *wave mechanics*), which requires the modification of some of our fundamental ideas about the relationships between physical theory and the physical world. Table 35-1 lists the approximate dates of some of the important experiments performed and theories proposed between 1881 and 1932.

The development of quantum theory was very different from that of the theory of relativity. In a sense, special relativity was presented as a complete theory in 1905 (and the general theory, in 1916) by a single scientist, Albert Einstein. Quantum theory, on the other hand, was developed over a long period by many different people. Many of the discoveries initially seemed unrelated, and it wasn't until the late 1920s that any consistent theory emerged. This theory is now the basis of our understanding of the micro-

Table 35-1 **Approximate Dates of Some Important Experiments and Theories, 1881–1932**

1881	Michelson obtains null result for absolute velocity of earth
1884	Balmer finds empirical formula for spectral lines of hydrogen
1887	Hertz produces electromagnetic waves, verifying Maxwell's theory and accidently discovering photoelectric effect
1887	Michelson repeats his experiment with Morley, again obtaining null result
1895	Röntgen discovers x rays
1896	Becquerel discovers nuclear radioactivity
1897	J. J. Thomson measures e/m for cathode rays, showing that electrons are fundamental constituents of atoms
1900	Planck explains blackbody radiation using energy quantization involving new constant h
1900	Lenard investigates photoelectric effect and finds energy of electrons independent of light intensity
1905	Einstein proposes special theory of relativity
1905	Einstein explains photoelectric effect by suggesting quantization of radiation
1907	Einstein applies energy quantization to explain temperature dependence of heat capacities of solids
1908	Rydberg and Ritz generalize Balmer's formula to fit spectra of many elements
1909	Millikan's oil-drop experiment shows quantization of electric charge
1911	Rutherford proposes nuclear model of atom based on alpha-particle scattering experiments of Geiger and Marsden
1912	Friedrich and Knipping and von Laue demonstrate diffraction of x rays by crystals showing that x rays are waves and that crystals are regular arrays
1913	Bohr proposes model of hydrogen atom
1914	Moseley analyzes x-ray spectra using Bohr model to explain periodic table in terms of atomic number
1914	Franck and Hertz demonstrate atomic energy quantization
1915	Duane and Hunt show that the short-wavelength limit of x rays is determined from quantum theory
1916	Wilson and Sommerfeld propose rules for quantization of periodic systems
1916	Millikan verifies Einstein's photoelectric equation
1923	Compton explains x-ray scattering by electrons as collision of photon and electron and verifies results experimentally
1924	De Broglie proposes electron waves of wavelength h/p
1925	Schrödinger develops mathematics of electron wave mechanics
1925	Heisenberg invents matrix mechanics
1925	Pauli states exclusion principle
1927	Heisenberg formulates uncertainty principle
1927	Davisson and Germer observe electron wave diffraction by single crystal
1927	G. P. Thomson observes electron wave diffraction in metal foil
1928	Gamow and Condon and Gurney apply quantum mechanics to explain alpha-decay lifetimes
1928	Dirac develops relativistic quantum mechanics and predicts existence of positron
1932	Chadwick discovers neutron
1932	Anderson discovers positron

scopic world. It is extremely successful, yet there is still debate about its philosophical interpretations. As with special relativity, quantum theory reduces to classical physics when it is applied to macroscopic (large-scale) systems, that is, to objects in our familiar, everyday world.

The origins of quantum theory were not, strangely enough, in the discoveries of radioactivity or x rays or atomic spectra but in thermodynamics. In his study of the radiation spectrum of a blackbody, Max Planck found that he could reconcile theory and experiment if he assumed that radiant energy was emitted and absorbed not continuously but in discrete lumps or quanta. It was Einstein who first recognized that this quantization of radiant energy was not just a calculational device but a general property of radiation. Niels Bohr then applied Einstein's ideas of energy quantization to the energy of an atom and proposed a model of the hydrogen atom that was spectacularly successful in calculations of the wavelengths of the radiation emitted by hydrogen. In this chapter, we will look qualitatively at the origins of the idea of energy quantization.

35-1 The Origin of the Quantum Constant: Blackbody Radiation

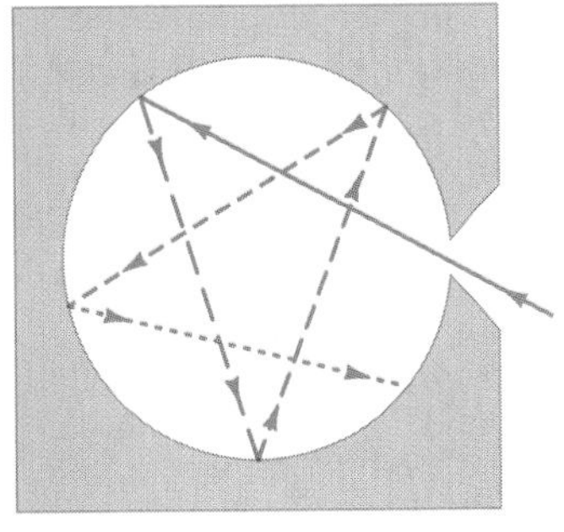

Figure 35-1 Cavity approximating an ideal blackbody. Radiation entering the cavity has little chance of leaving. It is usually completely absorbed.

One of the most puzzling phenomena studied near the end of the nineteenth century was the spectral distribution of blackbody radiation.* A blackbody is an ideal system that absorbs all the radiation incident on it. It can be approximated by a cavity with a very small opening, as illustrated in Figure 35-1. The characteristics of the radiation in such a cavity depend only on the temperature of the walls. At ordinary temperatures (below about 600°C), the thermal radiation emitted by a blackbody is not visible because the energy is concentrated in the infrared region of the electromagnetic spectrum. As the body is heated, the amount of energy radiated increases (according to the Stefan–Boltzmann law, Equation 16-21), and the concentration of energy moves to shorter wavelengths. Between about 600 and 700°C, there is enough energy in the visible spectrum for the body to glow a dull red. At higher temperatures, it becomes bright red or even "white hot."

Figure 35-2 shows the power radiated by a blackbody as a function of wavelength for three different temperatures. These curves are known as spectral distribution curves. The quantity P in this figure is the power radiated per unit wavelength. It is a function of both the wavelength λ and the temperature T and is called the spectral distribution function. This function $P(\lambda,T)$ has a maximum at a wavelength λ_{max} that varies inversely with temperature according to Wien's displacement law (Equation 16-25) discussed in Section 16-3:

$$\lambda_{max} = \frac{2.898 \text{ mm}\cdot\text{K}}{T}$$

Portrait of Max Planck (1858–1947).

The spectral distribution function $P(\lambda,T)$ can be calculated from classical thermodynamics in a straightforward way, and the result can be compared with the experimentally obtained curves of Figure 35-2. The result of this classical calculation, known as the **Rayleigh–Jeans law,** is

$$P(\lambda,T) = 8\pi kT\lambda^{-4} \qquad 35\text{-}1$$

where k is Boltzmann's constant. This result agrees with experimental results in the region of long wavelengths, but it disagrees violently at short wavelengths. As λ approaches zero, the experimentally determined $P(\lambda,T)$

*Blackbodies were discussed briefly in Section 16-3. Some of that discussion is repeated here.

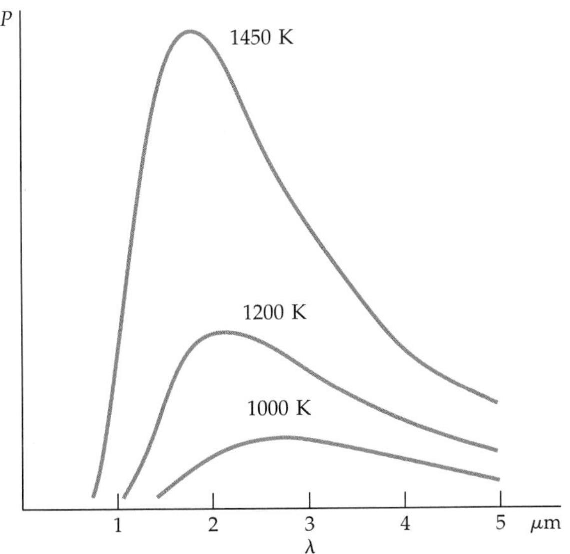

Figure 35-2 Spectral distribution of radiation from a blackbody for three different temperatures.

also approaches zero, but the calculated function approaches infinity because it is proportional to λ^{-4}. Thus, according to the classical calculation, blackbodies radiate an infinite amount of energy concentrated in the very short wavelengths. This result was known as the **ultraviolet catastrophe.**

In 1900, the German physicist Max Planck announced that by making a strange modification in the classical calculation he could derive a function $P(\lambda,T)$ that agreed with the experimental data at all wavelengths. Planck's result is shown in Figure 35-3 along with experimental data and the Rayleigh–Jeans law. Planck first found an empirical function that fit the data and then searched for a way to modify the usual calculation. He found that he could "derive" this function if he made the unusual assumption that the energy emitted and absorbed by the blackbody was not continuous but was instead emitted or absorbed in discrete packets or **quanta.** Planck found that the size of an energy quantum is proportional to the frequency of the radiation:

Quantization of energy of radiation

$$E = hf \tag{35-2}$$

where h is the proportionality constant now known as **Planck's constant.** The value of h was determined by Planck by fitting his function to the experimentally obtained data. The accepted value of this constant is now

$$h = 6.626 \times 10^{-34}\ \text{J·s} = 4.136 \times 10^{-15}\ \text{eV·s} \tag{35-3}$$

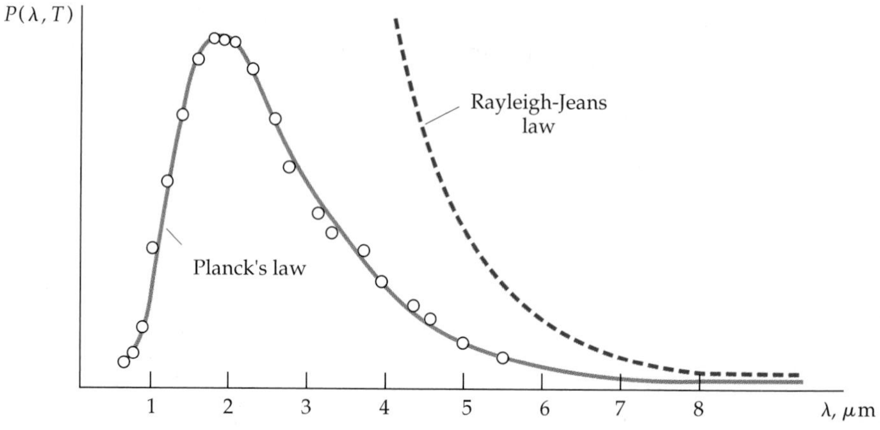

Figure 35-3 Spectral distribution of blackbody radiation versus wavelength at T = 1600 K. The classical theoretical calculation gives the Rayleigh–Jeans law, which agrees with experimental results at very long wavelengths but disagrees with them violently at short wavelengths.

Planck was unable to fit the constant h into the framework of classical physics. The fundamental importance of his assumption of energy quantization, implied by Equation 35-2, was not generally appreciated until Einstein applied similar ideas to explain the photoelectric effect and suggested that quantization is a fundamental property of electromagnetic radiation.

35-2 The Photoelectric Effect

In 1905, Einstein used Planck's idea of energy quantization to explain the photoelectric effect. (His paper on the photoelectric effect appeared in the same journal that contained his special theory of relativity.) Einstein's work marked the beginning of quantum theory, and for it he received the Nobel prize for physics. Whereas Planck looked at energy quantization in his blackbody-radiation theory as a calculational device, Einstein made the bold suggestion that energy quantization is a fundamental property of electromagnetic energy. Three years later, he applied the idea of energy quantization to molecular energies to clear up another puzzle in physics—the discrepancy between the specific heats calculated from the equipartition theorem and those observed experimentally at low temperatures. Later, the ideas of energy quantization were applied to atomic energies by Niels Bohr in the first explanation of atomic spectra.

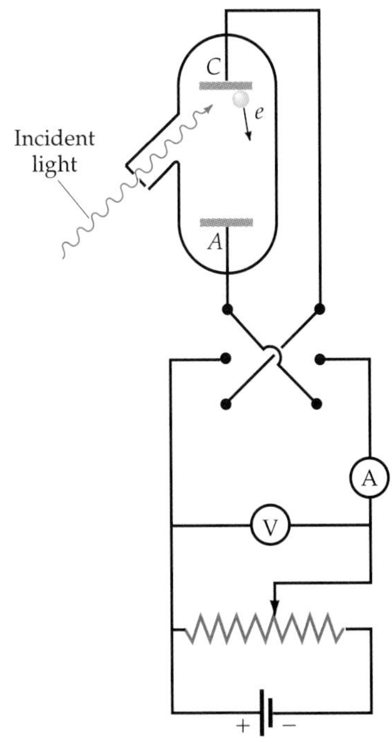

Figure 35-4 Schematic drawing of the apparatus for studying the photoelectric effect. Light strikes the cathode C and ejects electrons. The number of electrons that reach the anode A is measured by the current in the ammeter. The anode can be made positive or negative with respect to the cathode to attract or repel the electrons.

The photoelectric effect was discovered by Hertz in 1887 and was studied by Lenard in 1900. Figure 35-4 shows a schematic diagram of the basic apparatus. When light is incident on the clean metal surface of the cathode C, electrons are emitted. If some of these electrons strike the anode A, there is a current in the external circuit. The number of emitted electrons that reach the anode can be increased or decreased by making the anode positive or negative with respect to the cathode. Let V be the difference in potential between the cathode and the anode. Figure 35-5 shows the current versus V for two values of the intensity of the light incident on the cathode. When V is positive, the electrons are attracted to the anode. At sufficiently large values of V, all the emitted electrons reach the anode, and the current is at its maximum value. A further increase in V does not affect the current. Lenard observed that the maximum current is proportional to the light intensity. When V is negative, the electrons are repelled from the anode. Only electrons with initial kinetic energies $\frac{1}{2}mv^2$ that are greater than $|eV|$ can then reach the anode. From Figure 35-5, we can see that if V is less than $-V_0$, no electrons reach the anode. The potential V_0 is called the **stopping potential.** It is related to the maximum kinetic energy of the emitted electrons by

$$(\tfrac{1}{2}mv^2)_{\text{max}} = eV_0$$

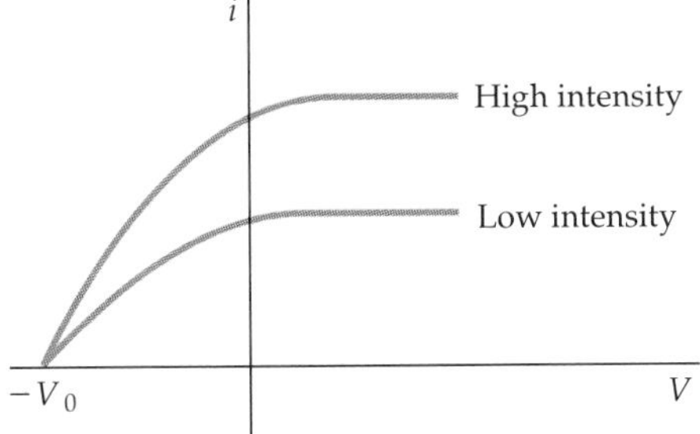

Figure 35-5 Photoelectric current i versus voltage V for two values of light intensity. There is no current when V is less than $-V_0$. The saturation current observed for large values of V is proportional to the intensity of the incident light.

The experimental result that V_0 is independent of the intensity of the incident light was surprising. Classically, increasing the rate of light energy falling on the cathode should increase the energy absorbed by an electron and should therefore increase the maximum kinetic energy of the electrons emitted. Apparently, this is not what happens. In 1905, Einstein demonstrated that this experimental result can be explained if light energy is not distributed continuously in space but rather is quantized in small bundles called **photons.** The energy of each photon is hf, where f is the frequency and h is Planck's constant. An electron emitted from a metal surface exposed to light receives its energy from a single photon. When the intensity of light of a given frequency is increased, more photons fall on the surface in unit time, but the energy absorbed by each electron is unchanged. If ϕ is the energy necessary to remove an electron from a metal surface, the maximum kinetic energy of the electrons emitted will be

Einstein's photoelectric equation

$$(\tfrac{1}{2}mv^2)_{max} = eV_0 = hf - \phi \qquad 35\text{-}4$$

The quantity ϕ, called the **work function,** is a characteristic of the particular metal. Some electrons will have kinetic energies less than $hf - \phi$ because of the loss of energy from traveling through the metal. Equation 35-4 is known as **Einstein's photoelectric equation.** From it, we can see that the slope of V_0 versus f should equal h/e.

Einstein's photoelectric equation was a bold prediction, for at the time it was made there was no evidence that Planck's constant had any applicability outside of blackbody radiation, and there were no experimental data on the stopping potential V_0 as a function of frequency. Experimental verification of Einstein's theory was quite difficult. Careful experiments by R. C. Millikan, reported first in 1914 and then in more detail in 1916, showed that Einstein's equation was correct and that measurements of h agreed with the value found by Planck. Figure 35-6 shows a plot of Millikan's data.

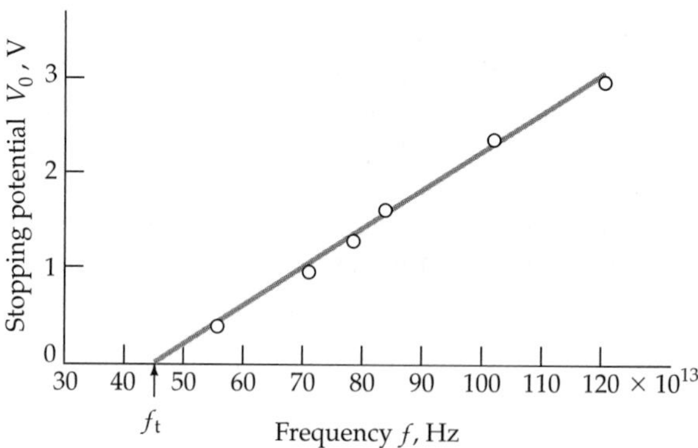

Figure 35-6 Millikan's data for the stopping potential V_0 versus frequency f for the photoelectric effect. The data fall on a straight line that has a slope h/e, as predicted by Einstein a decade before the experiment was performed.

Photons with frequencies less than a **threshold frequency** f_t, and therefore with wavelengths greater than a **threshold wavelength** λ_t, do not have enough energy to eject an electron from a particular metal. The threshold frequency and the corresponding threshold wavelength can be related to the work function ϕ by setting the maximum kinetic energy of the electrons equal to zero in Equation 35-4. Then

$$\phi = hf_t = \frac{hc}{\lambda_t} \qquad 35\text{-}5$$

Work functions for metals are typically a few electron volts. Since wavelengths are usually given in nanometers and energies in electron volts, it is useful to have the value of hc in electron volt–nanometers:

$$hc = (4.14 \times 10^{-15}\ \text{eV·s})(3 \times 10^8\ \text{m/s}) = 1.24 \times 10^{-6}\ \text{eV·m}$$

or

$$hc = 1240\ \text{eV·nm} \qquad 35\text{-}6$$

Example 35-1

Calculate the photon energy for light of wavelengths 400 nm (violet) and 700 nm (red). These are approximately the extreme wavelengths in the visible spectrum.

A collection of photomultiplier tubes used to detect very weak light. The face of each tube is a photosensitive area that emits electrons, via the photoelectric effect, when struck by photons. Each electron is accelerated and strikes a metal electrode, resulting in the emission of several more electrons—which are in turn accelerated and strike other electrodes. The electron beam cascades down the tube until it strikes the anode and produces a measurable electric current.

Using Equation 35-2, we have

$$E = hf = \frac{hc}{\lambda} = \frac{1240 \text{ eV}\cdot\text{nm}}{400 \text{ nm}} = 3.1 \text{ eV}$$

for $\lambda = 400$ nm. For $\lambda = 700$ nm, the photon energy is 4/7 that for $\lambda =$ 400 nm or 1.77 eV. We can see from these calculations that visible light contains photons with energies that range from about 1.8 to 3.0 eV.

Example 35-2

The intensity of sunlight at the earth's surface is approximately 1400 W/m^2. Assuming the average photon energy is 2 eV (corresponding to a wavelength of about 600 nm), calculate the number of photons that strike an area of 1 cm^2 in one second.

Since 1 watt is 1 joule per second, the energy striking the earth's surface in one second is 1400 J/m^2. The energy per second per square centimeter is then

$$\frac{1400 \text{ J}}{\text{m}^2} \times \frac{1 \text{ m}^2}{(100 \text{ cm})^2} = 0.14 \text{ J/cm}^2$$

If N is the number of 2-eV photons that together have a total energy of 0.14 J, we have

$$N(2 \text{ eV}) = 0.14 \text{ J}$$

$$N = \frac{0.14 \text{ J}}{2 \text{ eV}} \times \frac{1 \text{ eV}}{1.6 \times 10^{-19} \text{ J}}$$

$$= 4.38 \times 10^{17} \text{ photons}$$

This is an enormous number. In most everyday situations, the number of photons is so great that a few more or less make no difference. That is, quantization is not noticed.

Example 35-3

The threshold wavelength for potassium is 564 nm. (*a*) What is the work function for potassium? (*b*) What is the stopping potential when light of wavelength 400 nm is incident on potassium?

(*a*) From Equation 35-5, we have for the work function

$$\phi = hf_t = \frac{hc}{\lambda_t} = \frac{1240 \text{ eV·nm}}{564 \text{ nm}} = 2.20 \text{ eV}$$

(*b*) The energy of a photon with a wavelength of 400 nm was calculated in Example 35-1 to be 3.1 eV. The maximum kinetic energy of the emitted electrons is then

$$(\tfrac{1}{2}mv^2)_{max} = eV_0 = hf - \phi = 3.10 \text{ eV} - 2.20 \text{ eV} = 0.90 \text{ eV}$$

The stopping potential is therefore 0.90 V.

Exercise

Find the energy of a photon corresponding to electromagnetic radiation in the FM radio band of wavelength 3 m. (Answer: 4.13×10^{-7} eV)

Exercise

Find the wavelength of a photon whose energy is (*a*) 0.1 eV, (*b*) 1 keV, and (*c*) 1 MeV. [Answers: (*a*) 12.4 μm, (*b*) 1.24 nm, (*c*) 1.24 pm]

Another interesting feature of the photoelectric effect is the absence of any lag between the time the light first strikes the metal and the time the electrons appear. In the classical theory, given the intensity (the power per unit area), the time it takes for enough energy to fall on the area of an atom to eject an electron can be calculated. However, even when the intensity is so small that such a calculation gives a time lag of hours, essentially no time lag is observed. The explanation of this result is simple. When the intensity is low, the number of photons hitting the metal per unit time is very small, but each photon has enough energy to eject an electron. There is therefore a good chance that one photon will be absorbed immediately. The classical calculation gives the correct *average* number of electrons ejected per unit time.

35-3 X Rays

While working with a cathode-ray tube in 1895, W. Röntgen discovered that "rays" from the tube could pass through materials that were opaque to light and activate a fluorescent screen or photographic film. These rays originated from a point where the electrons in the tube hit a target within the tube or the glass tube itself. Röntgen was not able to deflect these rays in a magnetic field, as would be expected if they were charged particles, nor was he able to observe diffraction or interference, as would be expected if they were waves. He therefore gave the rays the somewhat mysterious name **x rays.** Röntgen investigated these rays extensively and found that all materials were transparent to them to some degree and that the degree of transparency decreased with increasing density of the material. This fact led to the medical use of x rays within months after Röntgen's first paper. Röntgen was the first recipient of the Nobel Prize for physics in 1901.

Since classical electromagnetic theory predicts that electric charges will radiate electromagnetic waves when they are accelerated (or decelerated), it

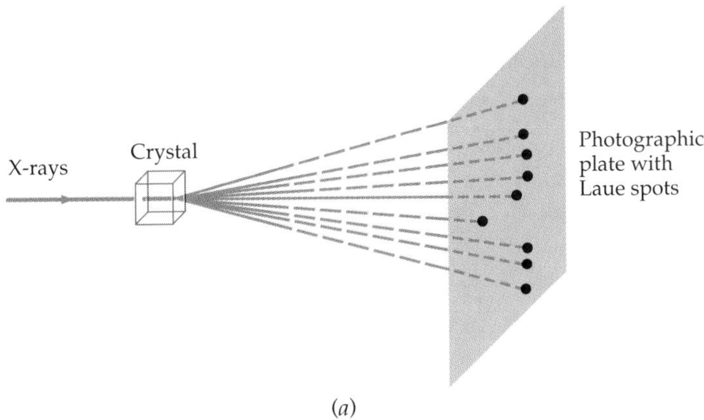

(a)

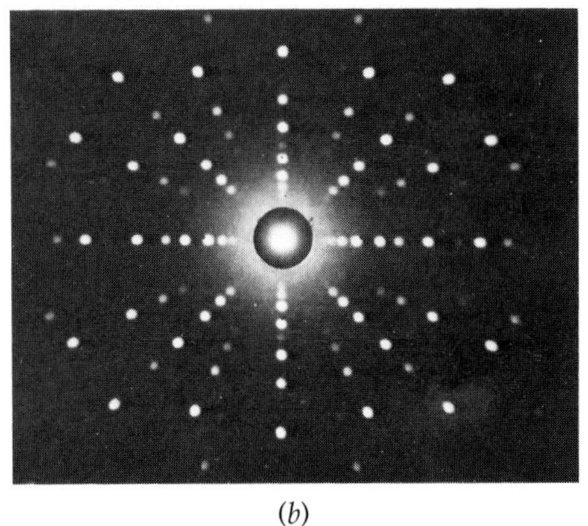

(b)

Figure 35-7 (*a*) Schematic diagram of the Laue experiment. The crystal acts as a three-dimensional grating that diffracts the x-ray beam and produces a regular array of spots called a *Laue pattern* on a photographic plate. (*b*) A modern Laue x-ray diffraction pattern from a niobium diboride crystal using 20-kV x rays from a molybdenum target.

was natural to assume that x rays are electromagnetic waves produced when electrons decelerate as they are stopped by a target. A few years later, a slight broadening of an x-ray beam after it passed through slits a few thousandths of a millimeter wide was observed. This was assumed to be due to diffraction, and the wavelength of x rays was estimated to be about 0.1 nm. In 1912, M. Laue suggested that, since the wavelengths of x rays were of the same order of magnitude as the spacing of the atoms in a crystal, the regular array of atoms in a crystal might act as a three-dimensional grating for the diffraction of x rays. Acting on this suggestion, W. Friedrich and P. Knipping allowed a collimated beam of x rays to pass through a crystal and strike a photographic plate (Figure 35-7*a*). In addition to the central beam, they observed a regular array of spots like those shown in Figure 35-7*b*. From an analysis of the positions of the spots, they were able to calculate that their x-ray beam had wavelengths ranging from about 0.01 to 0.05 nm. This important experiment confirmed two important assumptions: (1) x rays are electromagnetic radiation and (2) the atoms in crystals are arranged in a regular array.

Figure 35-8 shows a plot of intensity versus wavelength for the spectrum emitted from a typical x-ray tube, in which a target (molybdenum in this case) is bombarded with electrons. The spectrum consists of a series of sharp lines called the **characteristic spectrum** superimposed on a continuous spectrum called a **bremsstrahlung spectrum** (from the German for "braking radiation"). The line spectrum is characteristic of the target material and varies from element to element. It is similar to the optical spectrum of the elements except the x-ray spectrum involves transitions of the inner atomic electrons whereas the optical spectrum involves transitions of the outer atomic electrons. We will discuss both the optical spectrum and the characteristic x-ray spectrum in Chapter 37 in the extended version of this book. The continuous bremsstrahlung spectrum is produced by the rapid deceleration of the bombarding electrons when they crash into the target. If the voltage across the x-ray tube is V, the maximum kinetic energy of the electrons is eV when they hit the target. Often several photons are emitted as an electron slows down. However, sometimes just one photon with the maximum energy eV is emitted. Since the wavelength of a photon varies inversely with its energy ($\lambda = hc/hf = hc/E$), the minimum wavelength in the bremsstrahlung spectrum corresponds to a photon with the maximum energy eV. The minimum wavelength is called the **cutoff wavelength** and is labeled λ_m in the figure. The cutoff wavelength is related to the voltage of the x-ray tube by

$$\lambda_m = \frac{hc}{E} = \frac{hc}{eV} \qquad 35\text{-}7$$

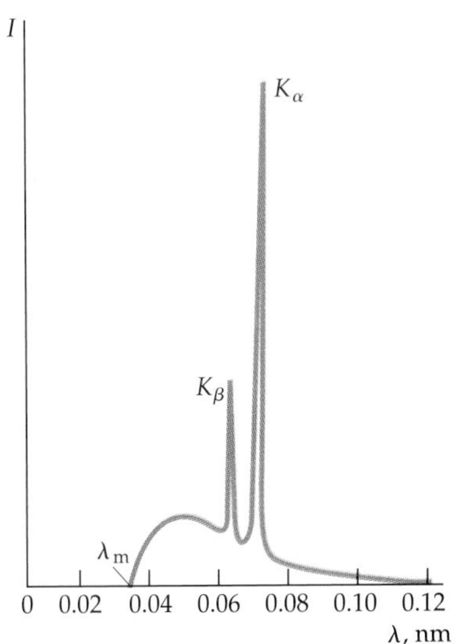

Figure 35-8 X-ray spectrum of molybdenum. The sharp peaks labeled K_α and K_β are characteristic of the target element. The cutoff wavelength λ_m is independent of the target element and is related to the voltage of the x-ray tube by $\lambda_m = hc/eV$.

Example 35-4

What is the minimum wavelength of the x rays emitted by a television picture tube with a voltage of 2000 V?

The maximum kinetic energy of the electrons is 2000 eV, so this will be the maximum energy of the photons in the x-ray spectrum. The wavelength of a photon of this energy is the cutoff wavelength, which from Equation 35-7 is

$$\lambda_m = \frac{hc}{E} = \frac{1240\text{ eV}\cdot\text{nm}}{2000\text{ eV}} = 0.62\text{ nm}$$

Exercise

An x-ray tube operates at a potential of 30 kV. What is the minimum wavelength of the continuous x-ray spectrum from this tube? (Answer: 0.041 nm)

35-4 Compton Scattering

Further evidence of the correctness of the photon concept was furnished by Arthur H. Compton, who measured the scattering of x rays by free electrons. According to classical theory, when an electromagnetic wave of frequency f_1 is incident on material containing charges, the charges will oscillate with this frequency and will reradiate electromagnetic waves of the same frequency. Compton pointed out that if this interaction were described as a scattering process involving a collision between a photon and an electron, the electron would recoil and thus absorb energy. The scattered photon would then have less energy and therefore a lower frequency than the incident photon.

According to classical theory, the energy and momentum of an electromagnetic wave are related by

$$E = pc \tag{35-8}$$

This result is consistent with the relativistic expression relating the energy and momentum of a particle (Equation 34-34),

$$E^2 = p^2c^2 + (mc^2)^2$$

if the mass m of the photon is assumed to be zero. Figure 35-9 shows the geometry of a collision between a photon of wavelength λ_1 and an electron initially at rest. Compton related the scattering angle θ to the incident and scattered wavelengths λ_1 and λ_2 by treating the scattering as a relativistic-mechanics problem and using the conservation of energy and momentum. Let $\mathbf{p}_1$ be the momentum of the incident photon, $\mathbf{p}_2$ be that of the scattered photon, and $\mathbf{p}_e$ be that of the recoiling electron. The conservation of momentum gives

$$\mathbf{p}_1 = \mathbf{p}_2 + \mathbf{p}_e \tag{35-9}$$

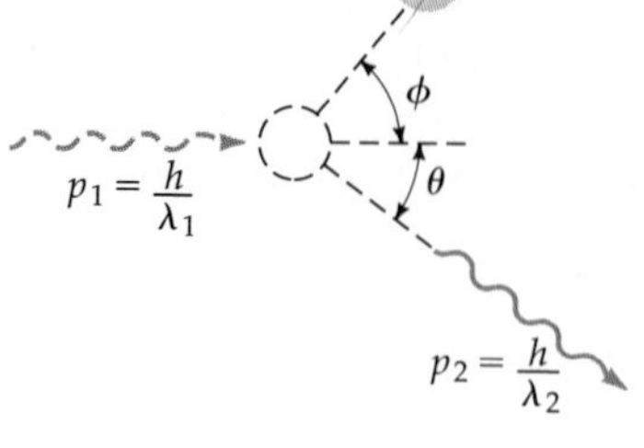

m p_e ϕ θ $p_1 = \frac{h}{\lambda_1}$ $p_2 = \frac{h}{\lambda_2}$

Figure 35-9 Compton scattering of an x ray by an electron. The scattered photon has less energy and therefore a greater wavelength than the incident photon because of the recoil energy of the electron. The change in wavelength is found from conservation of energy and momentum.

or

$$\mathbf{p}_e = \mathbf{p}_1 - \mathbf{p}_2$$

Taking the dot product of each side with itself, we obtain

$$p_e^2 = p_1^2 + p_2^2 - 2\mathbf{p}_1 \cdot \mathbf{p}_2$$

or

$$p_e^2 = p_1^2 + p_2^2 - 2p_1p_2 \cos \theta \qquad 35\text{-}10$$

The energy before the collision is $p_1c + mc^2$, where mc^2 is the rest energy of the electron. After the collision, the electron has energy $\sqrt{(mc^2)^2 + p_e^2c^2}$. Conservation of energy then gives

$$p_1c + mc^2 = p_2c + \sqrt{(mc^2)^2 + p_e^2c^2} \qquad 35\text{-}11$$

Compton eliminated the electron momentum p_e from Equations 35-10 and 35-11 and expressed the photon momenta in terms of the wavelengths to obtain an equation relating the incident and scattered wavelengths λ_1 and λ_2 and the angle θ. The algebraic details are left as a problem (see Problem 60). Compton's result is

$$\lambda_2 - \lambda_1 = \frac{h}{mc}(1 - \cos \theta) \qquad 35\text{-}12$$

The change in wavelengths is independent of the original wavelength. The quantity h/mc depends only on the mass of the electron. It has the dimension of length and is called the **Compton wavelength.** Its value is

$$\lambda_C = \frac{h}{mc} = \frac{hc}{mc^2} = \frac{1240 \text{ eV}\cdot\text{nm}}{5.11 \times 10^5 \text{ eV}} = 2.43 \times 10^{-12} \text{ m} = 2.43 \text{ pm} \qquad 35\text{-}13$$

Because $\lambda_2 - \lambda_1$ is small, it is difficult to observe unless λ_1 is so small that the fractional change $(\lambda_2 - \lambda_1)/\lambda_1$ is appreciable. Compton used x rays of wavelength 71.1 pm. The energy of a photon of this wavelength is $E = hc/\lambda = (1240 \text{ eV}\cdot\text{nm})/(0.0711 \text{ nm}) = 17.4$ keV. Since this is much greater than the binding energy of the valence electrons in atoms (which is of the order of a few electron volts), these electrons can be considered to be essentially free. Compton's experimental results for $\lambda_2 - \lambda_1$ as a function of the scattering angle θ agreed with Equation 35-12, thereby confirming the correctness of the photon concept.

Example 35-5

Calculate the percentage change in wavelength observed in the Compton scattering of 20-keV photons at $\theta = 60°$.

The change in wavelength at $\theta = 60°$ is given by Equation 35-12:

$$\lambda_2 - \lambda_1 = \lambda_C(1 - \cos \theta) = (2.43 \text{ pm})(1 - \cos 60°) = 1.22 \text{ pm}$$

The wavelength of the incident 20-keV photons is

$$\lambda_1 = \frac{1240 \text{ eV}\cdot\text{nm}}{20{,}000 \text{ eV}} = 0.062 \text{ nm} = 62 \text{ pm}$$

The percentage change in wavelength is thus

$$\frac{\lambda_2 - \lambda_1}{\lambda_1} = \frac{1.22 \text{ pm}}{62 \text{ pm}} \times 100\% = 1.97\%$$

35-5 Quantization of Atomic Energies: The Bohr Model

The most famous application of energy quantization to microscopic systems was that of Niels Bohr. In 1913, Bohr proposed a model of the hydrogen atom that had spectacular success in calculations of the wavelengths of the lines in the known hydrogen spectrum and in predicting new lines (later found experimentally) in the infrared and ultraviolet spectra.

Near the turn of the century, much data was collected on the emission of light by atoms in a gas when they are excited by an electric discharge. Viewed through a spectroscope with a narrow-slit aperture, this light appears as a discrete set of lines of different colors or wavelengths; the spacing and intensities of the lines are characteristic of the element. It was possible to determine the wavelengths of these lines accurately, and much effort went into finding regularities in the spectra. In 1884, a Swiss schoolteacher, Johann Balmer, found that the wavelengths of some of the lines in the spectrum of hydrogen can be represented by the formula

$$\lambda = (364.6 \text{ nm}) \frac{m^2}{m^2 - 4} \tag{35-14}$$

where m is an integer that takes on the values $m = 3, 4, 5, \ldots$. Figure 35-10 shows the set of spectral lines of hydrogen, now known as the **Balmer series,** whose wavelengths are given by Equation 35-14.

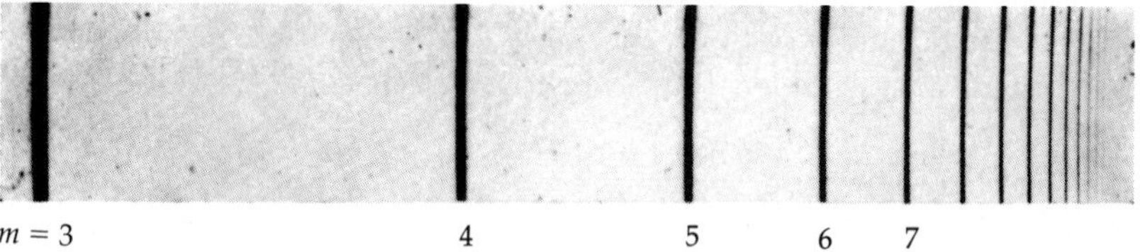

Figure 35-10 The Balmer series for light emitted from the hydrogen atom. The wavelengths of these lines are given by Equation 35-14 for different values of the integer m.

Balmer suggested that his formula might be a special case of a more general expression that would be applicable to the spectra of other elements. Such an expression, found by Johannes R. Rydberg and Walter Ritz and known as the **Rydberg–Ritz formula,** gives the reciprocal wavelength as

$$\frac{1}{\lambda} = RZ^2\left(\frac{1}{n_2^2} - \frac{1}{n_1^2}\right) \qquad n_1 > n_2 \tag{35-15}$$

This formula is valid not only for hydrogen, with atomic number $Z = 1$, but also for heavier atoms of nuclear charge Ze from which all electrons but one have been removed. R, called the **Rydberg constant** or simply the **Rydberg,** is the same for all spectral series of the same element and varies only slightly in a regular way from element to element. For very massive elements, R approaches the value

$$R_\infty = 10.97373\ \mu\text{m}^{-1} \tag{35-16}$$

If we take the reciprocal of Equation 35-14 for the Balmer series, we obtain

$$\frac{1}{\lambda} = \frac{1}{364.6 \text{ nm}}\left(\frac{m^2 - 4}{m^2}\right) = \frac{1}{364.6 \text{ nm}}\left(\frac{1}{1} - \frac{4}{m^2}\right)$$

$$= \frac{4}{364.6 \text{ nm}}\left(\frac{1}{4} - \frac{1}{m^2}\right) = 10.97\ \mu\text{m}^{-1}\left(\frac{1}{2^2} - \frac{1}{m^2}\right)$$

We can thus see that the Balmer formula is indeed a special case of the Rydberg–Ritz formula (Equation 35-15) for hydrogen with $n_2 = 2$ and $n_1 =$

m. The Rydberg–Ritz formula and various modifications of it were very successful in predicting other spectra. For example, hydrogen lines outside the visible spectrum were predicted and found. Setting $n_2 = 1$ in Equation 35-15 leads to a series in the ultraviolet region called the *Lyman series,* whereas setting $n_2 = 3$ leads to the *Paschen series* in the infrared region.

Many attempts were made to construct a model of the atom that would yield these formulas for its radiation spectrum. The most popular early model, developed by J. J. Thomson, considered electrons to be embedded in various arrangements in some kind of fluid that contained most of the mass of the atom and had enough positive charge to make the atom electrically neutral. Thomson's model, called the "plum pudding" model, is illustrated in Figure 35-11. Since classical electromagnetic theory predicted that a charge oscillating with a frequency f would radiate light of the same frequency, Thomson searched for configurations of electrons that were stable and had normal modes of vibration with frequencies equal to those of the spectrum of the atom. A difficulty with this model and all others was that electric forces alone cannot produce stable equilibrium. Thomson was unable to find a configuration of electrons that predicted the observed frequencies for any atom.

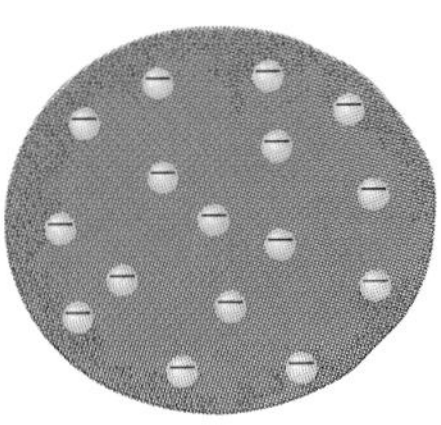

Figure 35-11 J. J. Thomson's "plum pudding" model of the atom. In this model, the negative electrons are embedded in a fluid of positive charge. For a given configuration of electrons in such a system, the resonance frequencies of the oscillations of the electrons can be calculated. According to classical theory, the atom should radiate light with a frequency equal to the frequency of oscillation of the electrons. Thomson could not find any configuration of electrons that would give frequencies that agreed with the measured frequencies of the spectrum of any atom.

The Thomson model was essentially ruled out by a set of experiments performed by H. W. Geiger and E. Marsden under the supervision of E. Rutherford around 1911 in which alpha particles from radioactive radium were scattered by atoms in a gold foil. Rutherford showed that the number of alpha particles scattered at large angles could not be accounted for by an atom in which the positive charge was distributed throughout the atomic volume (known to be about 0.1 nm in diameter). The findings required that the positive charge and most of the mass of the atom be concentrated in a very small region, now called the nucleus, with a diameter of the order of 10^{-6} nm = 1 fm. (Before the establishment of the SI, the femtometer, 1 fm = 10^{-15} m, was called a *fermi* after the Italian physicist Enrico Fermi.)

Niels Bohr, who was working in the Rutherford laboratory at the time, proposed a model of the hydrogen atom that combined the work of Planck, Einstein, and Rutherford and successfully predicted the observed spectra. Bohr assumed that the electron in the hydrogen atom moved under the influence of the Coulomb attraction between it and the positive nucleus according to classical mechanics, which predicts circular or elliptical orbits with the force center at one focus, as in the motion of the planets around the sun. For simplicity he chose a circular orbit as shown in Figure 35-12. Although mechanical stability is achieved because the Coulomb attractive force provides the centripetal force necessary for the electron to remain in orbit, such an atom would be unstable electrically according to classical theory because the electron must accelerate when moving in a circle and must therefore radiate electromagnetic energy of a frequency equal to that of its motion. According to classical electromagnetic theory, such an atom would quickly collapse because the electron would spiral into the nucleus as it radiates away its energy.

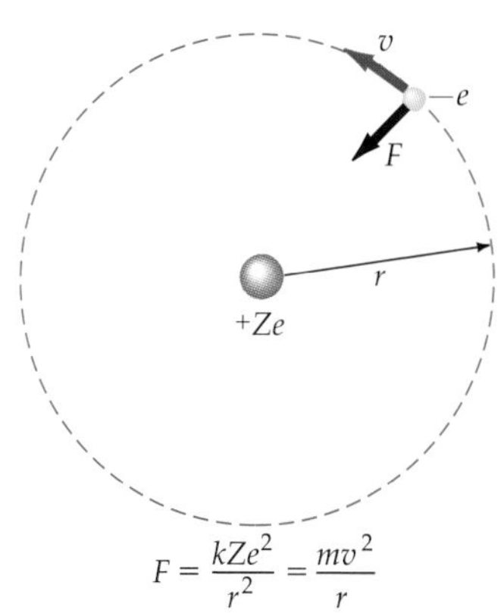

Figure 35-12 An electron of charge $-e$ traveling in a circular orbit of radius r around the nuclear charge $+Ze$. The attractive electrical force kZe^2/r^2 provides the centripetal force to hold the electron in its orbit.

Bohr's first postulate: nonradiating orbits

Bohr "solved" this difficulty by modifying the laws of electromagnetism and *postulating* that the electron could only move in certain nonradiating orbits. This idea is referred to as Bohr's first postulate. He called these stable orbits **stationary states.** The atom radiates only when the electron somehow makes a transition from one stationary state to another. The frequency of the radiation is not the frequency of the electron's motion in either stable orbit. Instead, it is related to the energies of the orbits by

Bohr's second postulate: photon frequency from energy conservation

$$f = \frac{E_i - E_f}{h} \tag{35-17}$$

where h is Planck's constant and E_i and E_f are the total energies in the initial and final orbits, respectively. This assumption, which is equivalent to the assumption of conservation of energy with the emission of a photon, is a key one in the Bohr theory because it deviates from classical theory, which requires the frequency of radiation to be that of the motion of the charged particle.

If the nuclear charge is $+Ze$ and there is only one electron of charge $-e$, the potential energy at a distance r (see Equation 20-8) is

$$U = -\frac{kZe^2}{r}$$

where k is the Coulomb constant. (For hydrogen, $Z = 1$, but it is convenient not to specify Z at this time so that the results can be applied to other hydrogen-like atoms.) The total energy of the electron moving in a circular orbit with a speed v is then

$$E = \frac{1}{2}mv^2 + U = \frac{1}{2}mv^2 - \frac{kZe^2}{r}$$

The kinetic energy can be obtained as a function of r by using Newton's second law $F = ma$. Setting the Coulomb attractive force equal to the mass times the centripetal acceleration, we obtain

$$\frac{kZe^2}{r^2} = m\frac{v^2}{r}$$

or

$$\frac{1}{2}mv^2 = \frac{1}{2}\frac{kZe^2}{r} \tag{35-18}$$

For circular orbits, the kinetic energy equals half the magnitude of the potential energy, a result that holds for circular motion in any inverse-square force field. The total energy is then

$$E = \frac{1}{2}\frac{kZe^2}{r} - \frac{kZe^2}{r} = -\frac{1}{2}\frac{kZe^2}{r} \tag{35-19}$$

Using Equation 35-17 for the frequency of radiation emitted when the electron changes from orbit 1 of radius r_1 to orbit 2 of radius r_2, we obtain

$$f = \frac{E_1 - E_2}{h} = \frac{1}{2}\frac{kZe^2}{h}\left(\frac{1}{r_2} - \frac{1}{r_1}\right) \tag{35-20}$$

To obtain the Rydberg–Ritz formula, $f = c/\lambda = cR(1/n_2^2 - 1/n_1^2)$, it is evident that the radii of stable orbits must be proportional to the squares of integers. Bohr searched for a quantum condition for the radii of the stable orbits that would yield this result. After much trial and error, he found that he could obtain it if he postulated that the angular momentum of the electron in a stable orbit equals an integer times Planck's constant divided by 2π. Since the angular momentum of a circular orbit is just mvr, this postulate is

$$mvr = \frac{nh}{2\pi} = n\hbar \tag{35-21}$$

Bohr's third postulate: quantized angular momentum

where

$$\hbar = \frac{h}{2\pi} = 1.05 \times 10^{-34}\ \text{J·s}$$

(The constant $\hbar = h/2\pi$, read "h bar", is often more convenient to use than h itself, just as the angular frequency $\omega = 2\pi f$ is often more convenient to use than the frequency f.) We can determine r by eliminating v between Equations 35-18 and 35-21. Solving Equation 35-21 for v and squaring, and using Equation 35-18, we obtain

$$v^2 = n^2\frac{\hbar^2}{m^2r^2} = \frac{kZe^2}{mr}$$

Solving for r we obtain

$$r = n^2\frac{\hbar^2}{mkZe^2} = n^2\frac{a_0}{Z} \tag{35-22}$$

where

$$a_0 = \frac{\hbar^2}{mke^2} \approx 0.0529\ \text{nm} \tag{35-23}$$

Bohr radius

is called the first **Bohr radius.** Combining Equations 35-22 and 35-20, we get

$$f = Z^2\frac{mk^2e^4}{4\pi\hbar^3}\left(\frac{1}{n_2^2} - \frac{1}{n_1^2}\right) \tag{35-24}$$

If we compare this expression for $f = c/\lambda$ with the empirical Rydberg–Ritz formula (Equation 35-15), we obtain for the Rydberg constant

$$R = \frac{mk^2e^4}{4\pi c\hbar^3} \tag{35-25}$$

Using the values of m, e, and $\hbar$ known in 1913, Bohr calculated R and found his result to agree (within the limits of the uncertainties of the constants) with the value obtained from spectroscopy. Figure 35-13 illustrates the Bohr model of the hydrogen atom.

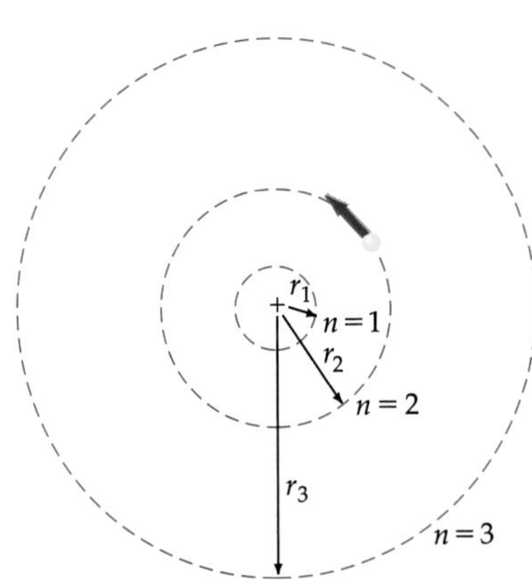

Figure 35-13 Stable orbits in the Bohr model of the hydrogen atom. The radii of the stable orbits are given by $r_n = n^2a_0$, where n is an integer and a_0 is the smallest radius.

The possible values of the energy of the hydrogen atom predicted by the Bohr model and given by Equation 35-19, when r is given by Equation 35-22, are

$$E_n = -\frac{k^2e^4m}{2\hbar^2}\frac{Z^2}{n^2} = -Z^2\frac{E_0}{n^2} \tag{35-26}$$

Energy levels

where

$$E_0 = \frac{k^2e^4m}{2\hbar^2} \approx 13.6\ \text{eV} \tag{35-27}$$

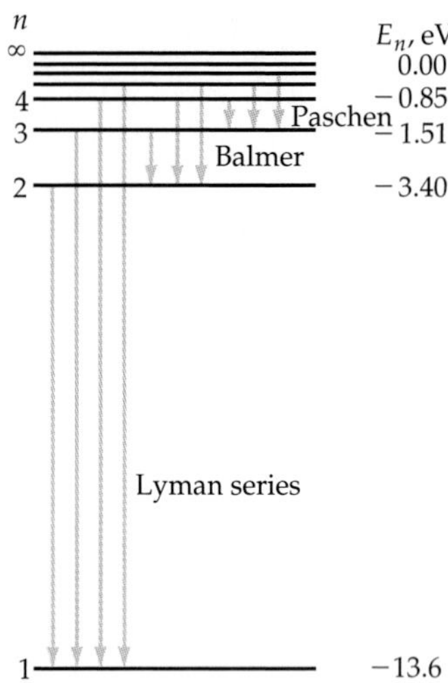

Figure 35-14 An energy-level diagram for hydrogen showing a few of the transitions in the Lyman, Balmer, and Paschen series. The energies of the levels are given by Equation 35-26.

It is sometimes convenient to represent these energies in an **energy-level diagram,** as in Figure 35-14. The lowest energy level is called the ground state. The energy of the hydrogen atom in the ground state is −13.6 eV. The highest energy state E_n is reached as $n \rightarrow \infty$ or $r \rightarrow \infty$, as can be seen from Equations 35-26 and 35-22. This process of removing the electron from an atom is termed **ionization.** The ionization energy of the hydrogen atom has been measured as 13.6 eV. This is thus the energy with which the electron is bound to the atom.

Various series of radiative transitions are indicated in Figure 35-14 by vertical arrows between the energy levels. The frequency of the light emitted in one of these transitions is the energy difference divided by h (Equation 35-17). At the time of Bohr's paper (1913), the Balmer series, corresponding to $n_2 = 2$ and $n_1 = 3, 4, 5, \ldots$, and the Paschen series, corresponding to $n_2 = 3$ and $n_1 = 4, 5, 6, \ldots$, were known. In 1916, T. Lyman found the series corresponding to $n_2 = 1$, and in 1922 and 1924, F. Brackett and H. A. Pfund, respectively, found series corresponding to $n_2 = 4$ and $n_2 = 5$. As can be determined by computing the wavelengths of these series, only the Balmer series lies in the visible portion of the electromagnetic spectrum.

In our derivations, we have assumed that the electron revolves around a stationary nucleus. This is equivalent to assuming that the nucleus has infinite mass. Since the mass of the hydrogen nucleus is not infinite but about 2000 times that of the electron, a correction must be made for the motion of the nucleus. This correction leads to a very slight dependence of the Rydberg constant, as given in Equation 35-25, on the nuclear mass, in precise agreement with the observed variation.

Example 35-6

Find the energy and wavelength of the line with the longest wavelength in the Lyman series.

From Figure 35-14, we can see that the Lyman series corresponds to transitions ending at the ground-state energy, $E_f = E_1 = -13.6$ eV. Since λ varies inversely with energy, the transition with the longest wavelength is the transition with the lowest energy, which is that from the first excited state $n = 2$ to the ground state $n = 1$. The energy of the first excited state is $E_2 = (-13.6 \text{ eV})/4 = -3.40$ eV. Since this is 10.2 eV above the ground-state energy, the energy of the photon emitted is 10.2 eV. The wavelength of this photon is

$$\lambda = \frac{hc}{\Delta E} = \frac{1240 \text{ eV}\cdot\text{nm}}{10.2 \text{ eV}} = 121.6 \text{ nm}$$

This photon is outside the visible spectrum and in the ultraviolet region. Since all the other lines in the Lyman series have even greater energies and shorter wavelengths, the Lyman series is completely in the ultraviolet region.

Exercise

Find the shortest wavelength for a line in the Lyman series. (Answer: 91.2 nm)

Questions

1. If an electron moves to a larger orbit, does its total energy increase or decrease? Does its kinetic energy increase or decrease?
2. How does the spacing of adjacent energy levels change as n increases?
3. What is the energy of the photon with the shortest wavelength emitted by the hydrogen atom?

35-6 Electron Waves and Quantum Theory

In 1924, a French student, L. de Broglie, suggested in his dissertation that electrons may have wave properties. His reasoning was based on the symmetry of nature. Since light was known to have both wave and particle properties, perhaps matter—especially electrons—might also have both wave and particle characteristics. This suggestion was highly speculative since there was no evidence at that time of any wave aspects of electrons. For the frequency and wavelength of electron waves, de Broglie chose the equations

$$f = \frac{E}{h} \qquad 35\text{-}28$$

$$\lambda = \frac{h}{p} \qquad 35\text{-}29$$

where p is the momentum and E is the energy of the electron. Equation 35-28 is the same as the Planck–Einstein equation for the energy of a photon. Equation 35-29 also holds for photons, as can be seen from

$$\lambda = \frac{c}{f} = \frac{hc}{hf} = \frac{hc}{E}$$

Since the momentum of a photon is related to its energy by $E = pc$, we have

$$\lambda = \frac{hc}{pc} = \frac{h}{p}$$

De Broglie's equations are thought to apply to all matter. However, for macroscopic objects, the wavelengths calculated from Equation 35-29 are so small that it is impossible to observe the usual wave properties of interference or diffraction. Even a particle as small as 1 μg is much too massive for any wave characteristics to be noticed, as we will see in the following example.

Example 35-7

Find the de Broglie wavelength of a particle of mass 10^{-6} g moving with a speed of 10^{-6} m/s.

From Equation 35-29, we have

$$\lambda = \frac{h}{p} = \frac{h}{mv} = \frac{6.63 \times 10^{-34}\ \text{J}\cdot\text{s}}{(10^{-9}\ \text{kg})(10^{-6}\ \text{m/s})}$$

$$= 6.63 \times 10^{-19}\ \text{m}$$

Since the wavelength found in Example 35-7 is much smaller than any possible apertures or obstacles (the diameter of the nucleus of an atom is about 10^{-15} m, roughly 10,000 times this wavelength), diffraction or interference of such waves cannot be observed. As we have discussed, the propagation of waves of very small wavelength is indistinguishable from the propagation of particles. Note that the momentum in Example 35-7 is extremely small. A macroscopic particle with a greater momentum would have an even smaller de Broglie wavelength. We therefore do not observe the wave properties of such macroscopic objects as baseballs or billiard balls.

Exercise

Find the de Broglie wavelength of a baseball of mass 0.17 kg moving at 100 km/h. (Answer: 1.4×10^{-34} m)

The situation is different for low-energy electrons. Consider an electron with kinetic energy K. If the electron is nonrelativistic, its momentum is found from

$$K = \frac{p^2}{2m}$$

or

$$p = \sqrt{2mK}$$

Its wavelength is then

$$\lambda = \frac{h}{p} = \frac{h}{\sqrt{2mK}} = \frac{hc}{\sqrt{2mc^2K}}$$

Using $hc = 1240$ eV·nm and $mc^2 = 0.511$ MeV, we obtain

$$\lambda = \frac{1240 \text{ eV·nm}}{\sqrt{2(0.511 \times 10^6 \text{ eV})K}}$$

or

$$\lambda = \frac{1.226}{\sqrt{K}} \text{ nm} \qquad K \text{ in electron volts} \qquad 35\text{-}30$$

From Equation 35-30, we can see that electrons with energies of the order of tens of electron volts have de Broglie wavelengths of the order of nanometers. This is the order of magnitude of the size of the atom and the spacing of atoms in a crystal. Thus, when electrons with energies of the order of 10 eV are incident on a crystal, they are scattered in much the same way as are x rays of the same wavelength.

Exercise

Find the wavelength of an electron whose kinetic energy is 10 eV. (Answer: 0.388 nm)

The crucial test for the existence of wave properties of electrons was the observation of diffraction and interference of electron waves. This was first accomplished accidentally in 1927 by C. J. Davisson and L. H. Germer as they were studying electron scattering from a nickel target at the Bell Telephone Laboratories. After heating the target to remove an oxide coating that had accumulated during an accidental break in the vacuum system, Davisson and Germer found that the intensity of the scattered electrons as a function of the scattering angle showed maxima and minima. Their target had crystallized, and by accident they had observed electron diffraction. They then prepared a target consisting of a single crystal of nickel and investigated this phenomenon extensively. Figure 35-15 illustrates their experiment. Electrons from an electron gun are directed at a crystal and are detected at some ϕ that can be varied. Figure 35-16 shows a typical pattern observed. There is a strong scattering maximum at an angle of 50°. The angle for maximum intensity of scattering of waves from a crystal depends on the wavelength of the waves and the spacing of the atoms in the crystal. Using the known spacing of atoms in their crystal, Davisson and Germer calculated the wavelength that could produce such a maximum and found that it agreed with the de Broglie equation (Equation 35-29) for the electron energy they were using. By varying the energy of the incident electrons, they could vary the electron wavelengths and produce maxima and minima at different locations in the diffraction patterns. In all cases, the measured wavelengths agreed with de Broglie's hypothesis.

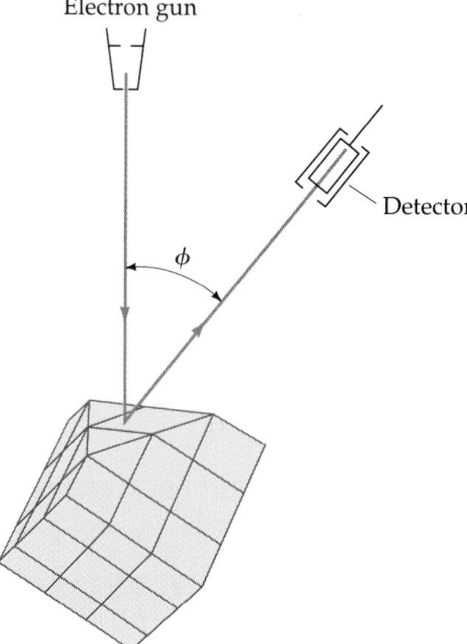

Figure 35-15 The Davisson–Germer experiment. Electrons from the electron gun incident on a crystal are scattered into a detector at some angle ϕ that can be varied.

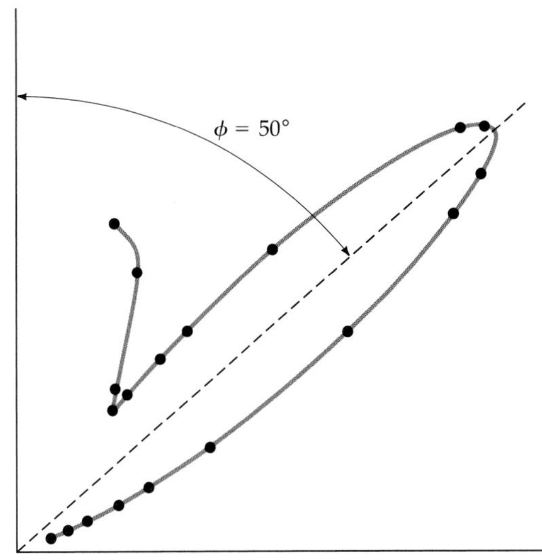

Figure 35-16 Plot of intensity versus angle for the scattered electron in the Davisson–Germer experiment. If this pattern is assumed to be a diffraction–interference pattern, the wavelength of the electrons can be calculated from the known spacing of the atoms in the crystal and the position of the maximum. The result agrees with the de Broglie hypothesis for the wavelength of an electron.

In the same year G. P. Thomson (son of J. J. Thomson) also observed electron diffraction in the transmission of electrons through thin metal foils. A metal foil consists of tiny, randomly oriented crystals. The diffraction pattern resulting from such a foil is a set of concentric circles. Since Thomson performed his experiment, diffraction has been observed for neutrons, protons, and other particles. Figure 35-17*a* to *c* shows the diffraction patterns of x rays, electrons, and neutrons of similar wavelength transmitted through thin metal foils. Figure 35-17*d* shows a diffraction pattern produced by electrons incident on two narrow slits. This experiment is equivalent to Young's famous double-slit diffraction–interference experiment with light. The pattern is identical to that observed with photons of the same wavelength.

Shortly after the wave properties of the electron were demonstrated, it was suggested that electrons rather than light might be used to "see" small objects. Today, the electron microscope is an important research tool. Figure

(*a*)

(*b*)

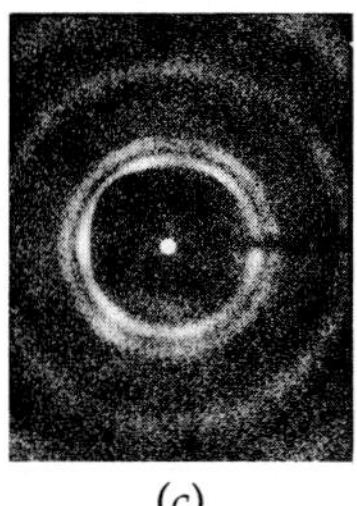
(*c*)

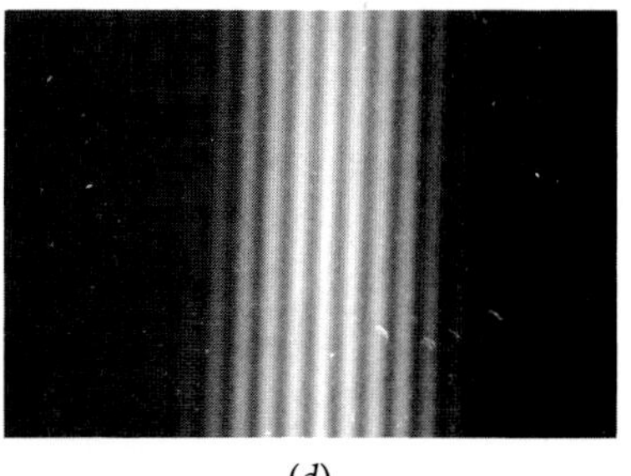
(*d*)

Figure 35-17 Diffraction pattern produced by (*a*) x rays and (*b*) electrons incident on an aluminum foil target and (*c*) neutrons incident on a target of polycrystalline copper. Note the similarity in the patterns produced. (*d*) A two-slit electron diffraction–interference pattern. This pattern is the same as that obtained with photons.

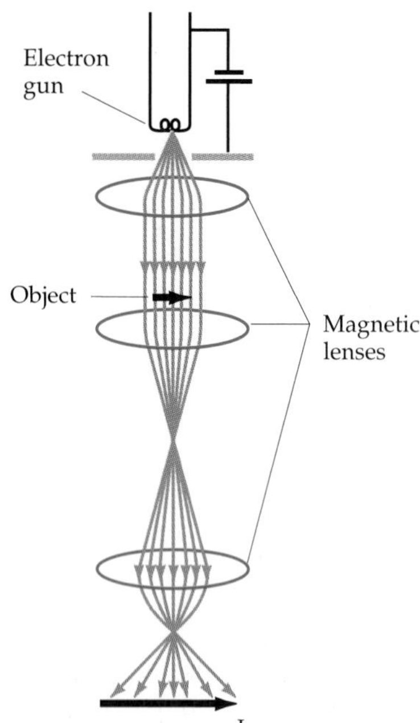

Figure 35-18 Electron microscope. Electrons from a heated filament (the electron gun) are accelerated by a large voltage difference. The electron beam is made parallel by a magnetic focusing lens. The electrons strike a thin target and are then focused by a second magnetic lens that is equivalent to the objective lens in an ordinary microscope. The third magnetic lens takes the place of the eyepiece in a microscope. It projects the electron beam onto a fluorescent screen for viewing the image.

35-18 illustrates the features of an electron microscope. The electron beam is made parallel and focused by specially designed magnets that serve as lenses. The energy of the electrons is typically 100 keV, resulting in a wavelength of about 0.004 nm. The target specimen must be very thin so that the transmitted beam will not be slowed down or scattered too much. The final image is projected onto a fluorescent screen or film. Various distortions resulting from focusing problems with the magnetic lenses limit the resolution to a few tenths of a nanometer, which is about a thousand times better than can be achieved with visible light.

Standing Waves and Energy Quantization

De Broglie pointed out that the Bohr quantum condition (Equation 35-21) for the angular momentum of the electron in a hydrogen atom is equivalent to a standing-wave condition. This condition states that

$$mvr = n\frac{h}{2\pi}$$

Substituting h/λ for the momentum mv gives

$$\frac{h}{\lambda}r = n\frac{h}{2\pi}$$

or

$$n\lambda = 2\pi r = C \qquad 35\text{-}31$$

where C is the circumference of the Bohr orbit. Thus, Bohr's quantum condition is equivalent to saying that an integral number of electron waves must fit into the circumference of the circular orbit as shown in Figure 35-19.

Example 35-8

The kinetic energy of the electron in the ground (lowest energy) state of the hydrogen atom is 13.6 eV. (Its potential energy is −27.2 eV and its total energy is −13.6 eV, leading to a binding energy of 13.6 eV.) Find the de Broglie wavelength for this electron.

Using $K = 13.6$ eV in Equation 35-30, we have

$$\lambda = \frac{1.226}{\sqrt{13.6}}\text{ nm} = 0.332\text{ nm} = 2\pi(0.0529\text{ nm})$$

This is the circumference of the first Bohr orbit in the hydrogen atom.

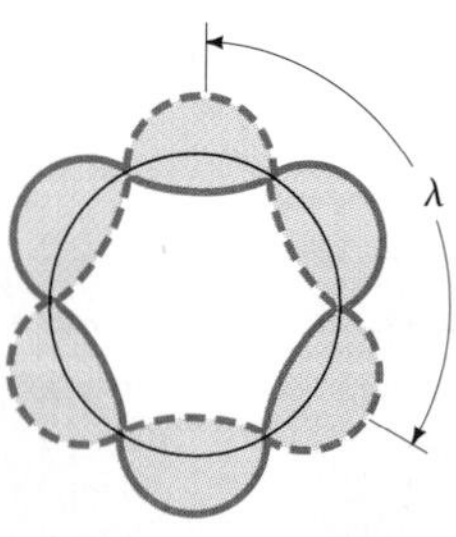

Figure 35-19 Standing waves around the circumference of a circle.

The fitting of an integral number of electron waves into the circumference of a Bohr orbit is similar to the fitting of an integral number of half wavelengths into the length of a string or organ pipe, as in standing waves on strings or standing sound waves. In classical wave theory, standing waves lead to a quantization of frequency. For example, for standing waves on a string of length L fixed at both ends (Figure 35-20), the standing wave condition is

$$n\frac{\lambda}{2} = L$$

For waves traveling with a speed v, the frequency of such standing waves on a string is then given by

$$f = \frac{v}{\lambda} = n\frac{v}{2L}$$

If energy is associated with the frequency of a standing wave, as in Equation 35-28, then standing waves imply quantized energies.

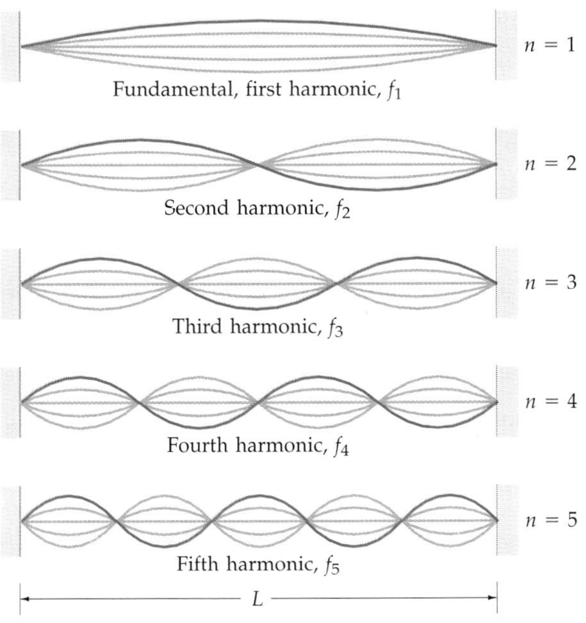

Figure 35-20 Standing waves on a string fixed at both ends. The frequencies of these waves are quantized; that is, they may have only certain values given by $f_n = nf_1$, where f_1 is the fundamental frequency.

The idea of explaining the discrete energy states of matter by standing waves led to the development of a detailed mathematical theory by Erwin Schrödinger and others in 1928. In this theory, known as **quantum theory, quantum mechanics,** or **wave mechanics,** the electron is described by a wave function ψ that obeys a wave equation that is somewhat similar to the classical wave equations for sound and light waves. The frequency and wavelength of electron waves are related to the energy and momentum of the electron just as the frequency and wavelength of light waves are related to the energy and momentum of photons. Schrödinger solved the standing-wave problem for the hydrogen atom, the simple harmonic oscillator, and other systems of interest. He found that the allowed frequencies combined with the de Broglie relation $E = hf$ led to the set of energy levels for the hydrogen atom found by Bohr (Equation 35-26), thereby demonstrating that quantum theory provides a general method of finding the quantized energy levels for a given system. Quantum theory is the basis for our understanding of the modern world, from the inner workings of the atomic nucleus to the radiation spectrum of distant galaxies in cosmology.

Summary

1. The energy in electromagnetic radiation is not continuous but comes in quanta with energies given by

$$E = hf = \frac{hc}{\lambda}$$

where f is the frequency, λ is the wavelength, and h is Planck's constant, which has the value

$$h = 6.626 \times 10^{-34}\ \text{J·s} = 4.136 \times 10^{-15}\ \text{eV·s}$$

The quantity hc occurs often in calculations and has the value

$$hc = 1240\ \text{eV·nm}$$

The quantum nature of light is exhibited in the photoelectric effect, in which a photon is absorbed by an atom with the emission of an electron, and in Compton scattering, in which a photon collides with a free electron and emerges with reduced energy and therefore a greater wavelength.

2. X rays are emitted when electrons are decelerated by crashing into a target in an x-ray tube. An x-ray spectrum consists of a series of sharp lines called the characteristic spectrum superimposed on the continuous bremsstrahlung spectrum. The minimum wavelength in the bremsstrahlung spectrum λ_m corresponds to the maximum energy of the emitted photon, which equals the maximum kinetic energy of the electrons eV, where V is the voltage of the x-ray tube. The minimum wavelength is then given by

$$\lambda_m = \frac{hc}{eV}$$

3. The wavelengths of x rays are typically a few nanometers, which is also approximately equal to the spacing of atoms in a crystal. Diffraction maxima are observed when x rays are scattered from a crystal, indicating that x rays are electromagnetic waves and that the atoms in crystals are arranged in a regular array.

4. In order to derive the Balmer formula for the spectrum of the hydrogen atom, Bohr proposed the following postulates:

Postulate 1: The electron in the hydrogen atom can move only in certain nonradiating circular orbits called stationary states.

Postulate 2: The atom radiates a photon when the electron makes a transition from one stationary orbit to another. The frequency of the photon is given by

$$f = \frac{E_i - E_f}{h}$$

where E_i and E_f are the initial and final energies of the atom.

Postulate 3: The radius (and hence the energy) of a stationary state orbit is determined by classical physics together with the quantum condition that the angular momentum of the electron must equal an integer times Planck's constant divided by 2π:

$$mvr = \frac{nh}{2\pi} = n\hbar$$

where $\hbar = h/2\pi = 1.05 \times 10^{-34}$ J·s.

These postulates lead to allowed energy levels in the hydrogen atom given by

$$E_n = -\frac{k^2e^4m}{2\hbar^2}\frac{Z^2}{n^2} = -Z^2\frac{E_0}{n^2}$$

where n is an integer and

$$E_0 = \frac{k^2e^4m}{2\hbar^2} \approx 13.6 \text{ eV}$$

The radii of the stationary orbits are given by

$$r = n^2\frac{\hbar^2}{mkZe^2} = n^2\frac{a_0}{Z}$$

where

$$a_0 = \frac{\hbar^2}{mke^2} \approx 0.0529 \text{ nm}$$

is the first Bohr radius.

5. The wave nature of electrons was first suggested by de Broglie, who postulated the equations

$$f = \frac{E}{h} \quad \text{and} \quad \lambda = \frac{h}{p}$$

for the frequency and wavelength of electron waves. With these equations, the Bohr quantum condition can be understood as a standing-wave condition. The wave nature of electrons was observed experimentally first by Davisson and Germer and later by G. P. Thomson, who measured the diffraction and interference of electrons.

6. The mathematical theory of the wave nature of matter is known as quantum theory. In this theory, the electron is described by a wave function that obeys a wave equation. Energy quantization arises from standing-wave conditions applied to electrons in various systems. Quantum theory is the basis for our understanding of the physical nature of the modern world.

Suggestions for Further Reading

Feinberg, Gerald: "Light," *Scientific American*, September 1968, p. 50.

This article presents an introduction to our present understanding of light as a phenomenon of both wavelike and particlelike properties, as manifested by diffraction, two-slit interference, the photoelectric effect, and blackbody radiation.

Moran, Paul R., R. Jerome Nickles, and James A. Zagzebski: "The Physics of Medical Imaging," *Physics Today*, vol. 36, no. 7, 1983, p. 36.

This article briefly describes such new medical imaging techniques as digital subtraction angiography, computed tomography (CAT), nmr imaging (MRI), positron-emission tomography (PET), and ultrasound imaging.

Wheeler, John Archibald: "Niels Bohr, the Man," *Physics Today*, vol. 38, no. 10, 1985, p. 66.

Bohr's very personal approach to science is recounted by a former collaborator, who is himself a highly respected physicist. This article appears as part of a special issue commemorating the centennial of Bohr's birth.

Review

A. Objectives: After studying this chapter you should:

1. Be able to sketch the spectral distribution curve for blackbody radiation and the curve predicted by the Rayleigh–Jeans law.

2. Be able to discuss the photoelectric effect and state the Einstein equation describing it.

3. Be able to discuss how the photon concept explains all the features of the photoelectric effect and the Compton scattering of x rays.

4. Be able to sketch a typical x-ray spectrum and relate the minimum wavelength of the spectrum to the voltage of the x-ray tube.

5. Be able to state the Bohr postulates and describe the Bohr model of the hydrogen atom.

6. Be able to draw an energy-level diagram for hydrogen, indicate on it transitions involving the emission of a photon, and use it to calculate the wavelengths of the emitted photons.

7. Be able to state the de Broglie relations for the frequency and wavelength of electron waves and use them and the standing-wave condition to derive the Bohr condition for the quantization of angular momentum in the hydrogen atom.

8. Be able to discuss the experimental evidence for the existence of electron waves.

B. Define, explain, or otherwise identify:

Blackbody radiation
Rayleigh–Jeans law
Ultraviolet catastrophe
Quanta
Planck's constant
Photoelectric effect
Stopping potential
Photons
Work function
Einstein's photoelectric equation
Threshold frequency
Threshold wavelength
X rays
Characteristic spectrum
Bremsstrahlung spectrum
Cutoff wavelength
Compton wavelength
Balmer series
Rydberg–Ritz formula
Rydberg
Stationary states
Bohr radius
Energy-level diagram
Ionization
Quantum theory
Quantum mechanics
Wave mechanics

C. True or false: If the statement is true, explain why it is true. If it is false, give a counterexample.

1. The spectral distribution of radiation in a blackbody depends only on the temperature of the body.

2. In the photoelectric effect, the maximum current is proportional to the intensity of the incident light.

3. The work function of a metal depends on the frequency of the incident light.

4. The maximum kinetic energy of electrons emitted in the photoelectric effect varies linearly with the frequency of the incident light.

5. The energy of a photon is proportional to its frequency.

6. One of Bohr's assumptions is that atoms never radiate light.

7. In the Bohr model, the energy of a hydrogen atom is quantized.

8. In the ground state of the hydrogen atom, the potential energy is −27.2 eV.

9. The de Broglie wavelength of an electron varies inversely with its momentum.

10. Electrons can be diffracted.

11. Neutrons can be diffracted.

12. An electron microscope is used to look at electrons.

Problems

Level I

35-1 The Origin of the Quantum Constant: Blackbody Radiation

There are no problems for this section

35-2 The Photoelectric Effect

1. Find the photon energy in joules and in electron volts for an electromagnetic wave in the FM radio band of frequency 100 MHz.

2. Repeat Problem 1 for an electromagnetic wave in the AM radio band of frequency 900 kHz.

3. What is the frequency of a photon of energy (*a*) 1 eV, (*b*) 1 keV, and (*c*) 1 MeV?

4. Find the photon energy for light of wavelength (*a*) 450 nm, (*b*) 550 nm, and (*c*) 650 nm.

5. Find the range of photon energies in the visible spectrum, which ranges from wavelengths of 400 to 700 nm.

6. Find the photon energy if the wavelength is (*a*) 0.1 nm (about 1 atomic diameter) and (*b*) 1 fm (1 fm = 10^{-15} m, about 1 nuclear diameter).

7. The work function for tungsten is 4.58 eV. (*a*) Find the threshold frequency and wavelength for the photoelectric effect. Find the stopping potential if the wavelength of the incident light is (*b*) 200 nm and (*c*) 250 nm.

8. When light of wavelength 300 nm is incident on potassium, the emitted electrons have maximum kinetic energy of 2.03 eV. (*a*) What is the energy of the incident photon?

(*b*) What is the work function for potassium? (*c*) What would be the stopping potential if the incident light had a wavelength of 430 nm? (*d*) What is the threshold wavelength for the photoelectric effect with potassium?

9. The threshold wavelength for the photoelectric effect for silver is 262 nm. (*a*) Find the work function for silver. (*b*) Find the stopping potential if the incident radiation has a wavelength of 175 nm.

10. The work function for cesium is 1.9 eV. (*a*) Find the threshold frequency and wavelength for the photoelectric effect. Find the stopping potential if the wavelength of the incident light is (*b*) 250 nm and (*c*) 350 nm.

11. A light beam of wavelength 400 nm has an intensity of 100 W/m^2. (*a*) What is the energy of each photon in the beam? (*b*) How much energy strikes an area of 1 cm^2 perpendicular to the beam in 1 s? (*c*) How many photons strike this area in 1 s?

35-3 X Rays

12. An x-ray tube operates at a potential of 460 kV. What is the minimum wavelength of the continuous x-ray spectrum from this tube?

13. The minimum wavelength in the continuous x-ray spectrum from a television tube is 0.134 nm. What is the voltage of the tube?

14. What is the minimum wavelength of the continuous x-ray spectrum from a television tube operating at 2500 V?

35-4 Compton Scattering

15. Find the shift in wavelength of photons scattered at $\theta = 60°$.

16. When photons are scattered by electrons in carbon, the shift in wavelength is 0.33 pm. Find the scattering angle.

17. Find the momentum of a photon in eV/*c* and in kg · m/s if the wavelength is (*a*) 400 nm, (*b*) 2 nm, (*c*) 0.1 nm, and (*d*) 3 cm.

18. The wavelength of Compton-scattered photons is measured at $\theta = 90°$. If $\Delta\lambda/\lambda$ is to be 1.5 percent, what should the wavelength of the incident photons be?

19. Compton used photons of wavelength 0.0711 nm. (*a*) What is the energy of these photons? (*b*) What is the wavelength of the photon scattered at $\theta = 180°$? (*c*) What is the energy of the photon scattered at this angle?

20. For the photons used by Compton, find the momentum of the incident photon and that of the photon scattered at 180°, and use momentum conservation to find the momentum of the recoil electron in this experiment (see Problem 19).

35-5 Quantization of Atomic Energies: The Bohr Model

21. Use the known values of the constants in Equation 35-22 to show that a_0 is approximately 0.0529 nm.

22. The wavelength of the longest wavelength of the Lyman series was calculated in Example 35-6. Find the wavelengths for the transitions (*a*) $n_1 = 3$ to $n_2 = 1$ and (*b*) $n_1 = 4$ to $n_2 = 1$. (*c*) Find the shortest wavelength in the Lyman series.

23. Find the photon energy for the three longest wavelengths in the Balmer series and calculate the wavelengths.

24. (*a*) Find the photon energy and wavelength for the series limit (shortest wavelength) in the Paschen series ($n_2 = 3$). (*b*) Calculate the wavelengths for the three longest wavelengths in this series and indicate their positions on a horizontal linear scale.

25. Repeat Problem 24 for the Brackett series ($n_2 = 4$).

26. A hydrogen atom is in its tenth excited state according to the Bohr model ($n = 11$). (*a*) What is the radius of the Bohr orbit? (*b*) What is the angular momentum of the electron? (*c*) What is the electron's kinetic energy? (*d*) What is the electron's potential energy? (*e*) What is the electron's total energy?

35-6 Electron Waves and Quantum Theory

27. Use Equation 35-30 to calculate the de Broglie wavelength for an electron of kinetic energy (*a*) 2.5 eV, (*b*) 250 eV, (*c*) 2.5 keV, and (*d*) 25 keV.

28. An electron is moving at $v = 2.5 \times 10^5$ m/s. Find its de Broglie wavelength.

29. An electron has a wavelength of 200 nm. Find (*a*) its momentum and (*b*) its kinetic energy.

30. Through what potential must an electron be accelerated so that its de Broglie wavelength is (*a*) 5 nm and (*b*) 0.01 nm?

31. A thermal neutron in a reactor has kinetic energy of about 0.02 eV. Calculate the de Broglie wavelength of this neutron from

$$\lambda = \frac{hc}{\sqrt{2mc^2K}}$$

where $mc^2 = 940$ MeV is the rest energy of the neutron.

32. Find the de Broglie wavelength of a proton (rest energy $mc^2 = 938$ MeV) that has a kinetic energy of 2 MeV. (See Problem 31.)

33. A proton is moving at $v = 0.003c$, where *c* is the speed of light. Find its de Broglie wavelength.

34. What is the kinetic energy of a proton whose de Broglie wavelength is (*a*) 1 nm and (*b*) 1 fm?

35. Find the de Broglie wavelength of a baseball of mass 0.145 kg moving at 30 m/s.

36. The energy of the electron beam in Davisson and Germer's experiment was 54 eV. Calculate the wavelength for these electrons.

37. The distance between Li^+ and Cl^- ions in a LiCl crystal is 0.257 nm. Find the energy of electrons that have wavelengths equal to this spacing.

38. An electron microscope uses electrons of energy 70 keV. Find the wavelength of these electrons.

Level II

39. An x ray undergoes Compton scattering and emerges with a wavelength of 0.20 nm at a scattering angle of 100°. What was the initial energy of the x-ray photon?

40. When the kinetic energy of an electron is much greater than its rest energy, the relativistic approximation $E \approx pc$ is good. (*a*) Show that in this case photons and electrons of the same energy have the same wavelength. (*b*) Find the de Broglie wavelength of an electron of energy 200 MeV.

41. Suppose that a 100-W source radiates light of wavelength 600 nm uniformly in all directions and that the eye can detect this light if only 20 photons per second enter a dark-adapted eye having a 7-mm diameter pupil. How far from the source can the light be detected under these rather extreme conditions?

42. Data for stopping potential versus wavelength for the photoelectric effect using sodium are

λ, nm	200	300	400	500	600
V_0, V	4.20	2.06	1.05	0.41	0.03

Plot these data so as to obtain a straight line and from your plot find (*a*) the work function, (*b*) the threshold frequency, and (*c*) the ratio h/e.

43. The diameter of the pupil of the eye is about 5 mm. (It can vary from about 1 mm to 8 mm). Find the intensity of light of wavelength 600 nm such that 1 photon per second passes through the pupil.

44. Show that the speed of an electron in the *n*th Bohr orbit of hydrogen is given by $v_n = e^2/2\epsilon_0 hn$.

45. A light bulb radiates 90 W uniformly in all directions. (*a*) Find the intensity at a distance of 1.5 m. (*b*) If the wavelength is 650 nm, find the number of photons per second that strike a 1-cm^2 area oriented so that its normal is along the line to the bulb.

46. How many head-on Compton scattering events are necessary to double the wavelength of a photon having initial wavelength 200 pm?

47. An x-ray photon of wavelength 6 pm makes a head-on collision with an electron so that it is scattered by an angle of 180°. (*a*) What is the change in wavelength of the photon? (*b*) What is the energy lost by the photon? (*c*) What is the kinetic energy of the scattered electron?

48. A 0.200-pm photon scatters from a free electron that is initially at rest. For what photon scattering angle will the kinetic energy of the recoiling electron equal the energy of the scattered photon?

49. The binding energy of an electron is the minimum energy required to remove the electron from its ground state to a large distance from the nucleus. (*a*) What is the binding energy for the hydrogen atom? (*b*) What is the binding energy for He^+? (*c*) What is the binding energy for Li^{2+}?

50. A hydrogen atom has its electron in the $n = 2$ state. The electron makes a transition to the ground state. (*a*) What is the energy of the photon according to the Bohr model? (*b*) If angular momentum is conserved, what is the angular momentum of the photon? (*c*) The linear momentum of the emitted photon is E/c. If we assume conservation of linear momentum, what is the recoil velocity of the atom? (*d*) Find the recoil kinetic energy of the atom in electron volts. By what percent must the energy of the photon calculated in part (*a*) be corrected to account for this recoil energy?

51. A particle of mass *m* moves in a one-dimensional box of length *L*. (Take the potential energy of the particle in the box to be zero so that its total energy is its kinetic energy $p^2/2m$). Its energy is quantized by the condition $n(\lambda/2) = L$, where λ is the de Broglie wavelength of the particle and *n* is an integer. (*a*) Show that the allowed energies are given by

$$E_n = n^2 E_1 \qquad \text{where } E_1 = h^2/8mL^2$$

(*b*) Evaluate E_n for an electron in a box of size $L = 0.1$ nm and make an energy-level diagram for the states from $n = 1$ to $n = 5$. Use Bohr's second postulate $f = \Delta E/h$ to calculate the wavelength of electromagnetic radiation emitted when the electron makes a transition from (*c*) $n = 2$ to $n = 1$, (*d*) $n = 3$ to $n = 2$, and (*e*) $n = 5$ to $n = 1$.

52. (*a*) Use the results of Problem 51 to find the energy of the ground state ($n = 1$) and the first two excited states of a proton in a one-dimensional box of length $L = 10^{-15}$ m $= 1$ fm. (These are of the order of magnitude of nuclear energies.) Calculate the wavelength of electromagnetic radiation emitted when the proton makes a transition from (*b*) $n = 2$ to $n = 1$, (*c*) $n = 3$ to $n = 2$, and (*d*) $n = 3$ to $n = 1$.

53. (*a*) Use the results of Problem 51 to find the energy of the ground state ($n = 1$) and the first two excited states of a proton in a one-dimensional box of length 0.2 nm (about the diameter of a H_2 molecule). Calculate the wavelength of electromagnetic radiation emitted when the proton makes a transition from (*b*) $n = 2$ to $n = 1$, (*c*) $n = 3$ to $n = 2$, and (*d*) $n = 3$ to $n = 1$.

54. (*a*) Use the results of Problem 51 to find the energy of the ground state ($n = 1$) and the first two excited states of a small particle of mass 1 μg confined to a one-dimensional box of length 1 cm. (*b*) If the particle moves with a speed of 1 mm/s, calculate its kinetic energy and find the approximate value of the quantum number *n*.

55. In the center-of-mass reference frame of the electron and the nucleus of an atom, the electron and nucleus have equal and opposite momenta of magnitude *p*. (*a*) Show that the total kinetic energy of the electron and nucleus can be written

$$K = \frac{p^2}{2\mu}$$

where

$$\mu = \frac{m_e M}{m_e + M} = \frac{m_e}{1 + m_e/M}$$

is called the reduced mass, m_e is the mass of the electron, and *M* is the mass of the nucleus. It can be shown that the motion of the nucleus can be accounted for by replacing the mass of the electron by the reduced mass. (*b*) Use Equation 35-25 with *m* replaced by μ to calculate the Rydberg for hydrogen ($M = m_p$) and for a very massive nucleus ($M = \infty$). (*c*) Find the percentage correction for the ground-state energy of the hydrogen atom due to the motion of the proton.

56. The kinetic energy of rotation of a diatomic molecule can be written $K = L^2/2I$, where L is its angular momentum and I is the moment of inertia. (*a*) Assuming that the angular momentum is quantized as in the Bohr model of the hydrogen atom, show that the energy is given by $K_n = n^2K_1$, where $K_1 = \hbar^2/2I$. (*b*) Make an energy-level diagram for such a molecule. (*c*) Estimate K_1 for the hydrogen molecule assuming the separation of the atoms to be $r = 0.1$ nm and considering rotation about an axis through the center of mass and perpendicular to the line joining the atoms. Express your answer in electron volts. (*d*) When K_1 is greater than kT (where k is Boltzmann's constant), molecular collisions do not result in rotation and so rotational energy does not contribute to the internal energy of the gas. Use your result of part (*c*) to find the critical temperature $T_c = K_1/k$.

Level III

57. This problem is one of estimating the time lag (expected classically but not observed) in the photoelectric effect. Let the intensity of the incident radiation be 0.01 W/m^2. (*a*) If the area of the atom is 0.01 nm^2, find the energy per second falling on an atom. (*b*) If the work function is 2 eV, how long would it take classically for this much energy to fall on one atom?

58. A photon cannot transfer all of its energy to a single free electron. Prove this by considering the problem of conservation of energy and momentum.

59. An electron and positron are moving towards each other with equal speeds of 3×10^6 m/s. The two particles annihilate each other and produce two photons of equal energy. (*a*) What were the de Broglie wavelengths of the electron and positron? Find the (*b*) energy, (*c*) momentum, and (*d*) wavelength of each photon.

60. (*a*) Solve Equation 35-11 for ${p_e}^2$ to obtain $p_e^2 = p_1^2 + p_2^2 - 2p_1p_2 + 2mc(p_1 - p_2)$. (*b*) Eliminate p_e^2 from your result in part (*a*) and Equation 35-10 to obtain $mc(p_1 - p_2) = p_1p_2(1 - \cos\theta)$. (*c*) Multiply both sides of your result in part (*b*) by h/mcp_1p_2 and use the de Broglie relation $h/p = \lambda$ to obtain the Compton formula (Equation 35-12).

61. The total energy density of radiation in a blackbody is given by

$$\eta = \int f(\lambda, T)\, d\lambda$$

where $f(\lambda, T)$ is given by the Planck formula

$$f(\lambda, T) = \frac{8\pi hc\lambda^{-5}}{e^{hc/\lambda kT} - 1}$$

Change the variable to $x = hc/\lambda kT$ and show that the total energy density can be written

$$\eta = \left(\frac{kT}{hc}\right)^4 8\pi hc \int_0^\infty \frac{x^3}{e^x - 1}\, dx = \alpha T^4$$

where α is some constant independent of T. This shows that the energy density in a blackbody is proportional to T^4.

62. The frequency of revolution of an electron in a circular orbit of radius r is $f_{rev} = v/2\pi r$, where v is the speed. (*a*) Show that in the nth stationary state

$$f_{rev} = \frac{k^2Z^2e^4m}{2\pi\hbar^3}\frac{1}{n^3}$$

(*b*) Show that when $n_1 = n$, $n_2 = n - 1$, and n is much greater than 1,

$$\frac{1}{n_2^2} - \frac{1}{n_1^2} \approx \frac{2}{n^3}$$

(*c*) Use your result in part (*b*) in Equation 35-24 to show that in this case the frequency of radiation emitted equals the frequency of motion. This result is an example of Bohr's correspondence principle: when n is large, so that the energy difference between adjacent states is a small fraction of the total energy, classical and quantum physics must give the same results.

Chapter 36

Quantum Mechanics

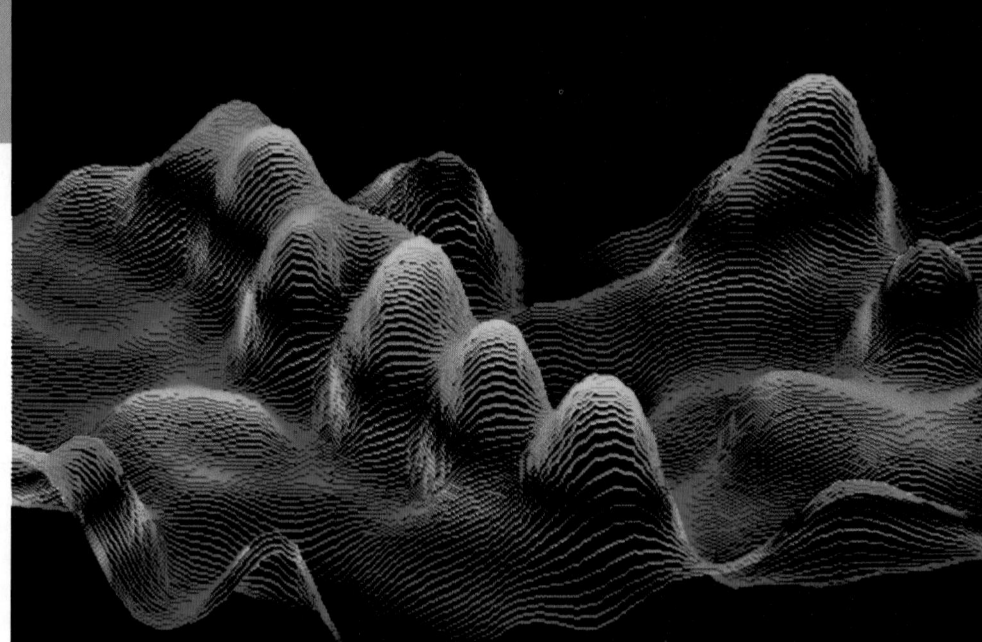

False color scanning tunneling micrograph of a segment of a DNA molecule magnified about one and a half million times. A sample of double-stranded DNA was dissolved in a salt solution and deposited on graphite prior to being imaged in air by the microscope. The row of orange-yellow peaks corresponds to the ridges of the DNA double helix.

In the previous chapter, we saw that light, which was thought to be a wave phenomenon, has particle properties and that electrons and other massive objects have wave properties. De Broglie's ideas about the wave nature of electrons were developed into a detailed mathematical theory by Erwin Schrödinger in 1926. In this theory, the electron is described by a wave function that obeys a wave equation called the Schrödinger equation, which is somewhat similar to the classical wave equations for sound and light waves. The frequency and wavelength of electron waves are related to the energy and momentum of the electron just as the frequency and wavelength of light waves are related to the energy and momentum of a photon. The diffraction and interference of electron waves observed by Davisson and Germer and others is a natural consequence of the propagation of these waves. The quantization of energy in atoms, molecules, and other microscopic systems results from standing-wave patterns of electron waves in these systems.

In this chapter, we will look at some of the properties of electron waves and see how the Schrödinger equation leads to the quantization of energy.

36-1 The Electron Wave Function

In our study of classical waves, such as waves on a string, sound waves, or light waves, we found that the energy density (the energy per unit volume in a wave) is proportional to the square of the wave function of the wave. The intensity, which equals the energy density times the wave speed (Equation 14-20 in Section 14-3), is also proportional to the square of the wave function. For waves on a string, the wave function is the displacement of the string $y(x, t)$. For sound waves in air, the wave function is the displacement of the air molecules from their equilibrium positions or, alternatively, the pressure variations due to the sound wave. For light and other electromagnetic waves, the wave function is the electric field $\mathcal{E}$ associated with the wave.*

The wave function for electron waves (or other matter waves) is designated by the Greek letter psi (Ψ). The wave function is a solution of a wave equation called the Schrödinger equation, which we will discuss later in this chapter, just as the wave function $\mathcal{E}$ for light waves is a solution of the classical wave equation for light. When Schrödinger first published his wave equation for electrons, it was not clear to him or to anyone else just what the wave function Ψ represents. We can get a hint as to how to interpret Ψ by considering the quantization of light waves.

Since the energy per unit volume in a light wave is proportional to $\mathcal{E}^2$ and the energy is quantized in units of hf for each photon, we expect that the number of photons in a unit volume is proportional to $\mathcal{E}^2$. Let us consider Young's famous double-slit experiment (Figure 36-1). The pattern observed on the screen is determined by the interference of the waves from the slits. At a point P_1 on the screen where the wave from one slit is 180° out of phase with that from the other, the resultant electric field $\mathcal{E}$ is zero. There is no light energy at that point and the point is dark. At points such as P_2, where the waves from the slits are in phase, $\mathcal{E}$ is maximum and the points are bright. If the intensity of the light incident on the slits is reduced, we can still observe the interference pattern if we replace the screen by a film and wait a sufficient time to expose the film.

The interaction of light with film is a quantum phenomenon. If we expose the film for a very short time with a low-intensity light source, we do not see merely a weaker version of the high-intensity pattern. Instead, we see "dots" on the film caused by the interactions of the individual photons

*We use a script $\mathcal{E}$ here for the electric field so as not to confuse it with the energy E.

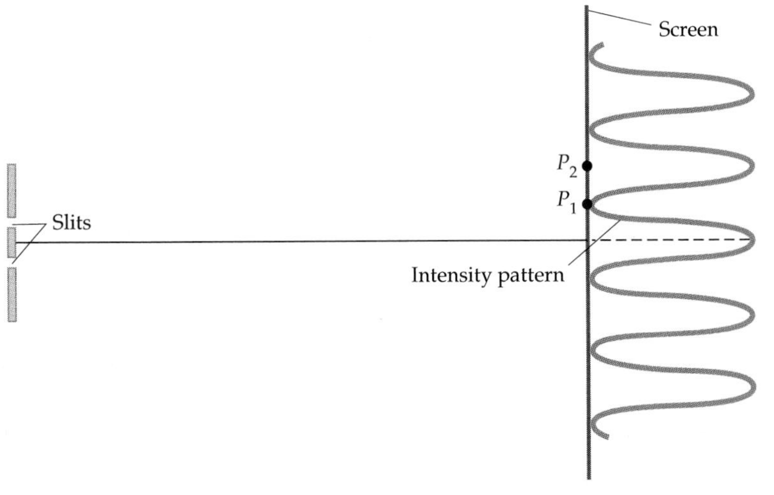

Figure 36-1 Young's double-slit experiment. Many photons go to point P_2 but no photons go to P_1. This experiment can be done with a light intensity that is so low that only one photon at a time arrives at the two slits and at the screen. The intensity pattern must then be interpreted as a measure of the probability of an individual photon arriving at a particular point.

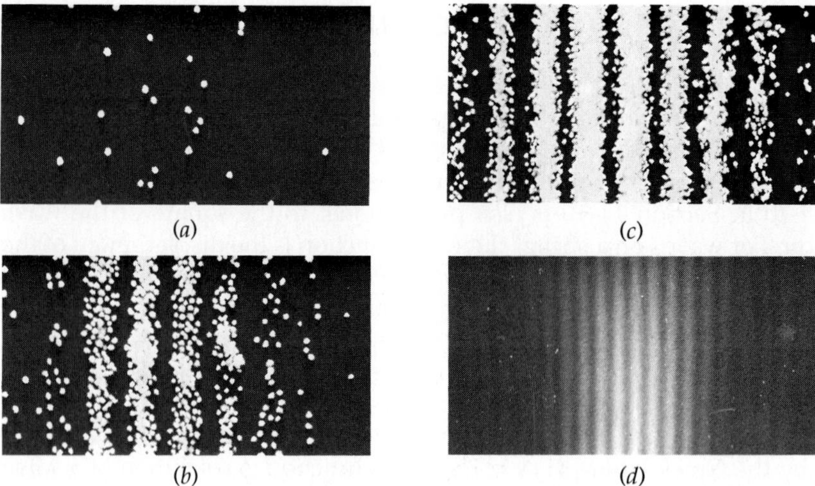

Figure 36-2 Growth of the two-slit interference pattern. Drawing (*a*) shows the expected pattern after the film has been exposed to about 28 photons or electrons. Drawings (*b*) and (*c*) show the expected patterns for 1000 and 10,000 photons or electrons, respectively. Note that there are no dots in the regions of the interference minima. The photograph (*d*) is an actual two-slit electron interference pattern resulting from film being exposed to millions of electrons. The pattern is identical to that usually obtained with photons.

(Figure 36-2). At points where the waves from the slits interfere destructively, there are no dots; that is, no photons arrive at these points. At points where the waves interfere constructively, there are many dots, indicating that many photons arrive at these points. When the exposure time is very short and the light source is weak, random fluctuations from the average locations predicted by the wave theory occur, and the quantum nature of light is clearly evident. If the exposure time is long or the light source is strong so that many photons interact with the film, the fluctuations average out and the quantum nature of light is not noticed. The interference pattern depends on the total number of photons interacting with the film and not on the rate of interaction. *Even when the intensity is so low that only one photon at a time hits the film, the wave theory predicts the correct average pattern.* For low intensities, we therefore interpret $\mathcal{E}^2$ to be proportional to the probability of detecting a photon in a unit volume of space.

At points on the film or screen where $\mathcal{E}^2$ is zero, no photons are observed, whereas at points where $\mathcal{E}^2$ is large they are most likely to be observed. As we have seen, we obtain the same double-slit pattern if we use electrons instead of photons. Figure 36-3 shows the interference patterns with just a few electrons and with many electrons.

In the wave theory of electrons, the motion of a single electron is described by the wave function Ψ, which is a solution of the Schrödinger wave

Figure 36-3 Actual electron interference patterns at increasing electron-beam densities filmed from a television monitor.

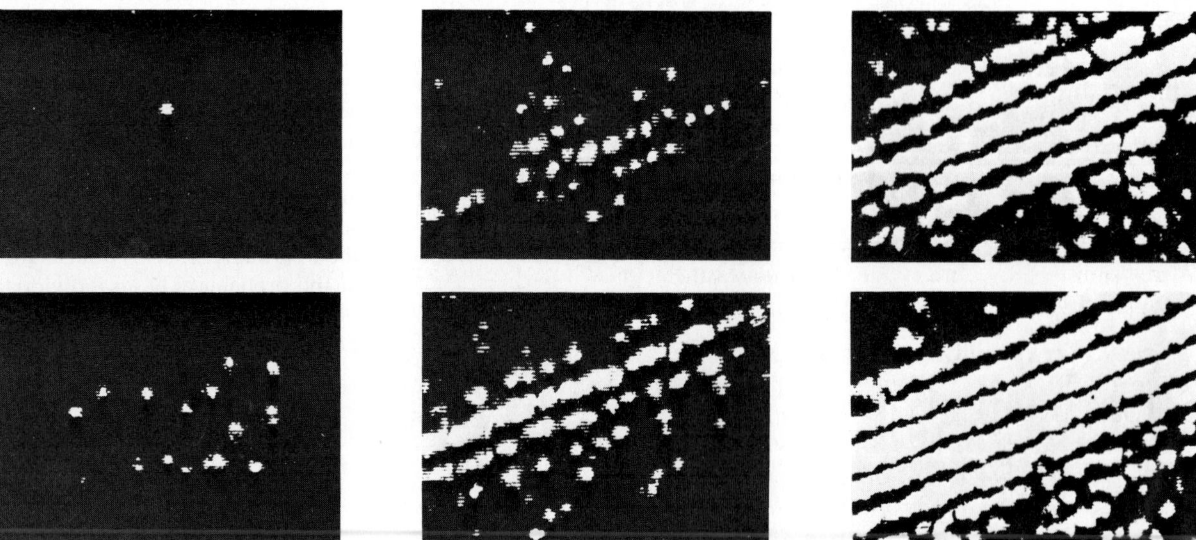

equation. Because the Schrödinger equation contains the imaginary number $i = \sqrt{-1}$ (Section 36-5), the wave functions that describe the motion of an electron are not necessarily real, that is, they may be complex. Since a probability must be a real number, the quantity analogous to $\mathcal{E}^2$ for photons that describes the probability of finding an electron in some region of space is $|\Psi|^2$.

The probability of finding an electron in a unit volume of space is proportional to $|\Psi|^2$.

The probability of finding the electron in some volume element dV must also be proportional to the size of the volume element dV. In one dimension, the probability of finding an electron in some interval dx is $|\Psi|^2\, dx$. If we call this probability $P(x)\, dx$, where $P(x)$ is the **probability distribution function** (also called the **probability density**), we have

$$P(x) = |\Psi|^2 \qquad 36\text{-}1$$

Probability density

Example 36-1

A classical point particle moves back and forth with constant speed between two walls at $x = 0$ and $x = 8$ cm. (*a*) What is the probability density $P(x)$? (*b*) What is the probability of finding the particle at $x =$ 2 cm? (*c*) What is the probability of finding the particle between $x =$ 3.0 cm and $x = 3.4$ cm?

(*a*) We do not know the initial position of the particle. Since the particle moves with constant speed, it is equally likely to be anywhere in the region $0 < x < 8$ cm. The probability density $P(x)$ is therefore constant, independent of x, for $0 < x < 8$ cm, and zero outside of this range:

$$\begin{aligned} P(x) &= P_0 \qquad 0 < x < 8 \text{ cm} \\ &= 0 \qquad x < 0 \text{ or } x > 8 \text{ cm} \end{aligned}$$

The probability of finding the particle in the interval dx at point x_1 *or* at point x_2 is the sum of the separate probabilities $P(x_1)\, dx + P(x_2)\, dx$. Since the particle must certainly be somewhere, the sum of the probabilities over all possible values of x must equal 1:

$$\int_{-\infty}^{+\infty} P(x)\, dx = \int_0^{8 \text{ cm}} P_0\, dx = P_0(8 \text{ cm}) = 1$$

Note that we need only integrate from 0 to 8 cm because $P(x)$ is zero outside this range. From this result we see that $P_0 = 1/(8 \text{ cm})$.

(*b*) The probability of finding the particle in some range dx is proportional to dx. Since $dx = 0$, the probability of finding the particle at the point $x = 2$ cm is 0. Alternatively, since there is an infinite number of points between $x = 0$ and $x = 8$ cm, and the particle is equally likely to be at any point, the chance that the particle will be at one particular point must be zero.

(*c*) Since the probability density is constant, the probability of a particle being in some range Δx in the region $0 < x < 8$ cm is $P_0\, \Delta x$. The probability of the particle being in the region 3.0 cm $< x <$ 3.4 cm is thus

$$P_0\, \Delta x = \left(\frac{1}{8 \text{ cm}}\right) 0.4 \text{ cm} = 0.05$$

36-2 Electron Wave Packets

In Chapter 13, we found that a harmonic wave on a string is represented by the wave function

$$y(x, t) = A \sin (kx - \omega t) \qquad 36\text{-}2$$

where k is the wave number, which is related to the wavelength λ by

$$k = \frac{2\pi}{\lambda} \qquad 36\text{-}3$$

and ω is the angular frequency, which is related to the frequency by

$$\omega = 2\pi f \qquad 36\text{-}4$$

The velocity of the wave is related to the frequency and the wavelength by

$$v = f\lambda = \left(\frac{\omega}{2\pi}\right)\left(\frac{2\pi}{k}\right) = \frac{\omega}{k} \qquad 36\text{-}5$$

Equation 36-2 also describes a harmonic sound wave if the displacement of the string $y(x, t)$ is replaced by the displacement of air molecules $s(x, t)$ or by the pressure variation $p(x, t)$. We can use Equation 36-2 to describe a harmonic electron wave by replacing the displacement $y(x, t)$ with the electron wave function $\Psi(x, t)$.

In Section 14-7 we discussed the fact that a harmonic wave, which has a single frequency ω and wave number k, has no beginning or end in space or time. To represent a pulse that is localized in space, we need a wave packet, that is, a group of harmonic waves containing a continuous distribution of frequencies and wave numbers. An electron that is completely unlocalized, that is, one that can be anywhere in space, can be represented by a single harmonic wave. However, a wave packet is needed to describe an electron that is localized in space. We will illustrate some of the properties of wave packets by considering a very simple group consisting of just two waves of equal amplitude and nearly equal frequencies and wave numbers. Such a group was used to describe the phenomenon of beats in Section 14-4. Let the wave numbers be k_1 and k_2 and the angular frequencies be ω_1 and ω_2. The sum of the two waves is

$$\Psi(x, t) = A_0 \sin (k_1 x - \omega_1 t) + A_0 \sin (k_2 x - \omega_2 t)$$

where A_0 is the amplitude of each wave. Using

$$\sin \theta_1 + \sin \theta_2 = 2 \cos \tfrac{1}{2}(\theta_1 - \theta_2) \sin \tfrac{1}{2}(\theta_1 + \theta_2)$$

for the sum of two sine functions, we obtain for the resultant wave

$$\Psi(x, t) = 2A_0 \cos [\tfrac{1}{2}(k_1 - k_2)x - \tfrac{1}{2}(\omega_1 - \omega_2)t] \sin [\tfrac{1}{2}(k_1 + k_2)x - \tfrac{1}{2}(\omega_1 + \omega_2)t]$$

Using $k_{av} = \frac{1}{2}(k_1 + k_2)$ and $\omega_{av} = \frac{1}{2}(\omega_1 + \omega_2)$ for the average wave number and average angular frequency and $\Delta k = k_1 - k_2$ and $\Delta\omega = \omega_1 - \omega_2$ for the difference in wave numbers and difference in angular frequencies, we have

$$\Psi(x, t) = [2A_0 \cos (\tfrac{1}{2}\, \Delta k\, x - \tfrac{1}{2}\, \Delta\omega\, t)] \sin (k_{av}x - \omega_{av}t) \qquad 36\text{-}6$$

Figure 36-4 shows a sketch of $\Psi(x, t)$ at some particular time as a function of x. The dashed curve is the envelope of the group of two waves given by the factor in brackets in Equation 36-6. The individual waves within the envelope move with the speed $v = \omega_{av}/k_{av}$, which is the phase velocity. If we write the modulating factor in brackets as $\cos \{\frac{1}{2}\, \Delta k\, [x - (\Delta\omega/\Delta k)t]\}$, we can see that the envelope moves with speed $\Delta\omega/\Delta k$. The speed of the envelope is the **group velocity**

$$V_g = \frac{\Delta\omega}{\Delta k} \qquad \text{36-7}$$

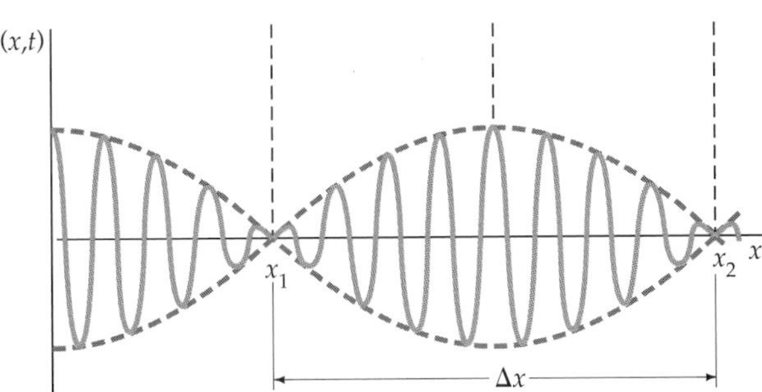

Figure 36-4 The spatial extent of the group Δx is inversely proportional to the difference in wave numbers Δk, where k is related to the wavelength by $k = 2\pi/\lambda$. Identical figures are obtained if $\Psi(x, t)$ is plotted versus time t at a fixed point x. In that case, the extent in time Δt is inversely proportional to the frequency difference $\Delta\omega$.

If x_1 and x_2 are two consecutive values of x for which the envelope is zero, we can take $\Delta x = x_2 - x_1$ to be a measure of the spatial extent of the group. Since the cosine function is zero when its argument is $\frac{1}{2}\pi$, $\frac{3}{2}\pi$, $\frac{5}{2}\pi$, and so on, the values x_1 and x_2 are related by

$$\tfrac{1}{2}\,\Delta k\, x_2 - \tfrac{1}{2}\,\Delta k\, x_1 = \pi$$

or

$$\Delta k\ \Delta x = 2\pi$$

For a particular value of x, the function $\Psi(x, t)$ versus t looks like Figure 36-4 with t replacing x. The extent in time Δt is thus related to $\Delta\omega$ by

$$\Delta\omega\ \Delta t = 2\pi$$

These results are consistent with those discussed in Section 14-7 (Equations 14-34 and 14-35):

$$\Delta k\ \Delta x \sim 1 \qquad \text{36-8}$$

$$\Delta\omega\ \Delta t \sim 1 \qquad \text{36-9}$$

However, the ranges Δx and Δt for our group of just two waves are artificial because the envelope does not remain small outside these ranges.

The wave function for a general wave packet made up of a discrete set of harmonic waves can be written

$$\Psi(x, t) = \Sigma\ A_i \sin\,(k_i x - \omega_i t) \qquad \text{36-10}$$

where A_i is the amplitude of the wave of wave number k_i and angular frequency ω_i. The calculation of the amplitudes A_i needed to construct a wave packet of a given shape is a problem in Fourier series. If we use only a finite number of waves, it is not possible to obtain a wave packet that is small everywhere outside some region of space. To describe an electron that is localized in space, we must construct a wave packet from a continuous distribution of waves. We represent this mathematically by replacing A_i in Equation 36-10 with $A(k)\,dk$ and replacing the sum with an integral. The quantity $A(k)$ is called the distribution function for the wave number k. Either the shape of the wave packet at some fixed time $\Psi(x)$ or the distribution of wave numbers $A(k)$ can be found from the other by the methods of

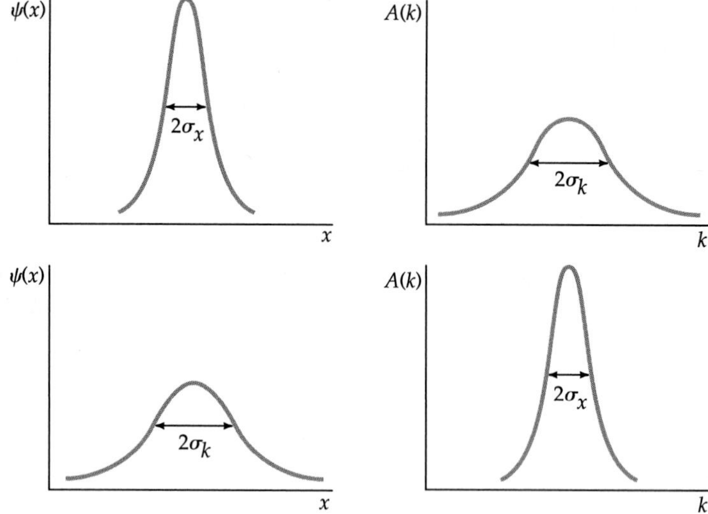

Figure 36-5 A gaussian-shaped wave packet $\psi(x)$ and the corresponding gaussian distribution of wave numbers $A(k)$. The standard deviations of these packets are related by $\sigma_x\sigma_k = \frac{1}{2}$.

Fourier analysis. Figure 36-5 shows a gaussian-shaped wave packet and the corresponding wave-number distribution function for a narrow packet and a wide packet. For this special case, $A(k)$ is also a gaussian function. The standard deviations of these gaussian functions σ_x and σ_k are related by

$$\sigma_k\sigma_x = \tfrac{1}{2} \qquad 36\text{-}11$$

It can be shown that the product of the standard deviations is greater than $\frac{1}{2}$ for a wave packet of any other shape. For a continuous distribution of waves, Equation 36-7 for the group velocity of a wave packet becomes

$$V_g = \frac{d\omega}{dk} \qquad 36\text{-}12$$

The energy and momentum of an electron are related to the frequency and wavelength of the associated electron wave by the de Broglie equations. They are therefore also related to the angular frequency and wave number. Thus, we have

$$p = \frac{h}{\lambda} = \frac{h}{2\pi/k} = \frac{hk}{2\pi}$$

and

$$E = hf = h\frac{\omega}{2\pi}$$

In terms of $\hbar = h/2\pi$, these relations become

$$p = \hbar k \qquad 36\text{-}13$$

and

$$E = \hbar\omega \qquad 36\text{-}14$$

The kinetic energy of an electron moving in free space with no forces acting on it is given by

$$E = \tfrac{1}{2}mv^2 = \frac{p^2}{2m}$$

Substituting $\hbar\omega$ for E and $\hbar k$ for p, we obtain

$$\hbar\omega = \frac{\hbar^2 k^2}{2m} \tag{36-15}$$

Using Equation 36-12 for the group velocity, we obtain

$$V_g = \frac{d\omega}{dk} = \frac{d}{dk}\left(\frac{\hbar k^2}{2m}\right) = \frac{\hbar k}{m} = \frac{p}{m} = v \tag{36-16}$$

Thus, the group velocity equals the velocity of the electron, as we should expect. Note that the phase velocity of the individual waves in the wave packet is not equal to the velocity of the electron:

$$V_p = \frac{\omega}{k} = \frac{\hbar\omega}{\hbar k} = \frac{E}{p} = \frac{p}{2m} = \frac{v}{2}$$

Questions

1. What are Δx and Δk for a purely harmonic wave of a single frequency and wavelength?
2. Which is more important for communication, the phase velocity or the group velocity?

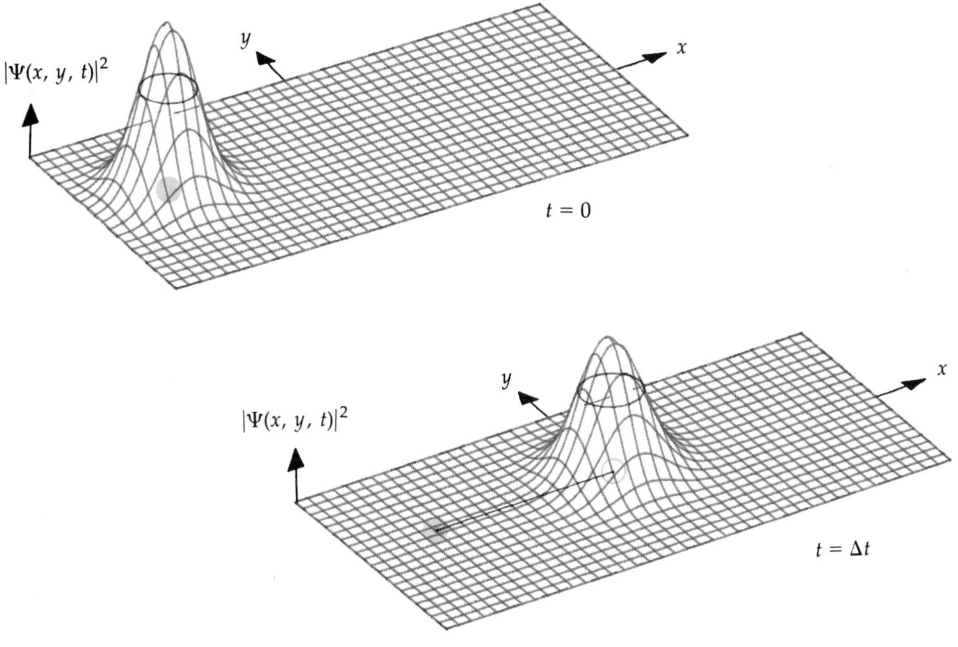

A three-dimensional wave packet representing a particle moving along the x axis. The dot indicates the position of a classical particle. Note that the packet spreads out in the x and y directions. This spreading is due to dispersion, resulting from the fact that the phase velocity of the individual waves making up the packet depends on the wavelength of the waves.

36-3 The Uncertainty Principle

The wave nature of electrons (and other particles) has important consequences. Consider a wave packet $\Psi(x, t)$ representing an electron. The most probable position of the electron is the value of x for which $|\Psi(x, t)|^2$ is a maximum. Since $|\Psi(x, t)|^2$ is proportional to the probability that the electron is at x and $|\Psi(x, t)|^2$ is nonzero for a *range* of values of x, there is an uncertainty in the position of the electron. If we make a number of position measurements on identical electrons, that is, electrons with the same wave function, we will not always obtain the same result. In fact, the distribution function for the results of the measurements will be given by $|\Psi(x, t)|^2$. If the wave packet is very narrow, the uncertainty in position will be small. However, a narrow wave packet must contain a wide range of wave numbers k. Since the momentum is related to the wave number by $p = \hbar k$, a wide range of k values means a wide range of momentum values. If we make a number of momentum measurements on identical electrons, we obtain a distribution of results corresponding to the distribution of wave numbers in the wave packet. Thus, a narrow wave packet that corresponds to a small uncertainty in position also corresponds to a wide distribution of momentum values and therefore to a large uncertainty in momentum. In general, the ranges Δx and Δk are related by Equation 36-8:

$$\Delta k \; \Delta x \sim 1$$

Similarly, a wave packet that is localized in time Δt must contain a range of frequencies $\Delta\omega$, where the ranges are related by Equation 36-9:

$$\Delta\omega \; \Delta t \sim 1$$

These results are inherent properties of waves. If we multiply these equations by $\hbar$ and use $p = \hbar k$ and $E = \hbar\omega$, we obtain

$$\Delta p \; \Delta x \sim \hbar \qquad \text{36-17}$$

and

$$\Delta E \; \Delta t \sim \hbar \qquad \text{36-18}$$

Equations 36-17 and 36-18 provide a statement of the **uncertainty principle** first enunciated by Werner Heisenberg in 1927. Equation 36-17 expresses the fact that the distribution functions for position and momentum cannot both be made arbitrarily narrow; thus, measurements of positions and momentum will have uncertainties whose product is at least as great as $\hbar$. Equation 36-18 expresses the fact that if Δt is the time available for the measurement of the energy of a system, the measurement of the energy will be uncertain by an amount ΔE that is at least as great as $\hbar/\Delta t$. Equation 36-18 has important applications in the determination of the excitation energy of atoms, molecules, and nuclei. For example, if an excited state of an atom has a lifetime τ, its energy can be known only to within about $\hbar/\tau$.

If we define precisely what we mean by the uncertainty in the measurements of position and momentum, we can give a precise statement of the uncertainty principle. If σ_x is the standard deviation for measurements of the wave number k, the product $\sigma_x\sigma_k$ has its minimum value of $\frac{1}{2}$ when the distribution functions are gaussian. If we define Δx and Δp to be the standard deviations, the minimum value of their product is $\frac{1}{2}\hbar$. Thus,

Uncertainty principle for position and momentum

$$\Delta p \; \Delta x \geq \tfrac{1}{2}\hbar \qquad \text{36-19}$$

Similarly,

$$\Delta E \, \Delta t \geq \tfrac{1}{2}\hbar \qquad 36\text{-}20$$

Uncertainty principle for energy and time

Usually, the uncertainty product is much greater than $\hbar/2$. The equality holds only if the measurements of both x and p or both E and t are ideal.

We can get a qualitative understanding of the uncertainty principle by considering the measurement of the position and momentum of a particle. If we know the mass of the particle, we can determine its momentum by measuring its position at two nearby times and computing its velocity. A common way to measure the position of an object is to look at it with light. When we do this, we scatter light from the object and determine the position from the direction in which the light is scattered. If we use light of wavelength λ, we can measure the position only to an uncertainty of the order of λ because of diffraction effects. To reduce the uncertainty in position, we use light of very short wavelength, perhaps even x rays. In principle, there is no limit to the accuracy of such a position measurement because there is no limit on how small a wavelength λ we can use. However, since all electromagnetic radiation carries momentum, the scattering of the radiation by the particle will deflect the radiation and change its original momentum in an uncontrollable way. Since momentum is conserved, the momentum of the particle also changes in an uncontrollable way. According to classical wave theory, this effect on the momentum of the particle could be reduced by reducing the intensity of the radiation. However, the energy and momentum of the radiation are quantized; each photon has momentum h/λ. When the wavelength of the radiation is small, the momentum of each photon will be large and the momentum measurement will have a large uncertainty. This uncertainty cannot be eliminated by reducing the intensity of light; such a reduction merely reduces the number of photons in the beam. To "see" the particle, we must scatter at least one photon. Therefore, the uncertainty in the momentum measurement of the particle will be large if λ is small, and the uncertainty in the position measurement of the particle will be large if λ is large. A detailed analysis shows that the product of these uncertainties will always be at least of the order of Planck's constant h. Of course, we could always "look" at the particles by scattering electrons instead of photons, but we would still have the same difficulty. If we use low-momentum electrons to reduce the uncertainty in the momentum measurement, we have a large uncertainty in the position measurement because of the diffraction of the electrons. The relation between the wavelength and momentum, $\lambda = h/p$, is the same for electrons as for photons.

One consequence of the uncertainty principle is that when a particle is confined in some region of space, it cannot have zero kinetic energy. The minimum energy of a particle is called its **zero-point energy.** Suppose, for example, that a particle is confined to some region of space of length L. The uncertainty in its position is then no greater than L. Consequently, we see from Equation 36-19 that the uncertainty in its momentum Δp is

$$\Delta p \geq \frac{\hbar}{2L} \qquad 36\text{-}21$$

The kinetic energy of the particle is

$$K = \frac{1}{2}mv^2 = \frac{p^2}{2m} \qquad 36\text{-}22$$

The magnitude of the momentum p must be at least as large as its uncertainty Δp ($p = 0 \pm \Delta p$). Therefore,

$$K = \frac{p^2}{2m} \geq \frac{(\hbar^2/4L^2)}{2m} = \frac{\hbar^2}{8mL^2} \qquad \text{36-23}$$

The smaller the region of space L, the greater the minimum kinetic energy.

Example 36-2

A marble of mass 25 g is in a box of length 10 cm. Find the minimum uncertainty in its momentum, its speed v, and its minimum kinetic energy, assuming that $p = \Delta p$.

From Equation 36-19, with $\Delta x = 10$ cm, we have

$$(\Delta p)_{\text{min}} = \frac{\hbar}{2\,\Delta x} = \frac{1.05 \times 10^{-34}\ \text{J·s}}{2(0.1\ \text{m})}$$

$$= 5.3 \times 10^{-34}\ \text{kg·m/s}$$

The speed corresponding to a momentum of this magnitude is

$$v = \frac{p}{m} = \frac{5.3 \times 10^{-34}\ \text{kg·m/s}}{0.025\ \text{kg}} = 2.1 \times 10^{-32}\ \text{m/s}$$

We would be quite safe in saying that the marble is at rest. The minimum kinetic energy is

$$K_{\text{min}} = \frac{(\Delta p_{\text{min}})^2}{2m} = \frac{(5.3 \times 10^{-34}\ \text{kg·m/s})^2}{0.050\ \text{kg}}$$

$$= 5.6 \times 10^{-66}\ \text{J}$$

Because Planck's constant is so small, the uncertainty relation of Equation 36-19 is not significant for macroscopic systems.

Example 36-3

Work Example 36-2 for an electron confined to a region of space of length $L = 0.1$ nm. This distance is of the order of the diameter of an atom.

In this case, the minimum uncertainty in the momentum is

$$(\Delta p)_{\text{min}} = \frac{\hbar}{2\,\Delta x} = \frac{1.05 \times 10^{-34}\ \text{J·s}}{2(10^{-10}\ \text{m})}$$

$$= 5.3 \times 10^{-25}\ \text{kg·m/s}$$

The speed of an electron with momentum of this magnitude is

$$v = \frac{p}{m} = \frac{5.3 \times 10^{-25}\ \text{kg·m/s}}{9.1 \times 10^{-31}\ \text{kg}}$$

$$= 5.8 \times 10^{5}\ \text{m/s}$$

Note that this is a significant speed. The minimum kinetic energy is

$$K_{\text{min}} = \frac{(\Delta p_{\text{min}})^2}{2m} = \frac{(5.3 \times 10^{-25}\ \text{kg·m/s})^2}{2(9.1 \times 10^{-31}\ \text{kg})}$$

$$= 1.5 \times 10^{-19}\ \text{J}$$

This is approximately 1 eV, which is about the order of magnitude of the kinetic energy of an electron in an atom.

Questions

3. Does the uncertainty principle say that the momentum of an electron can never be precisely known?
4. Why is the uncertainty principle not important for macroscopic objects?

36-4 Wave–Particle Duality

We have seen that light, which we ordinarily think of as a wave, exhibits particle properties when it interacts with matter, as in the photoelectric effect or in Compton scattering, and that electrons, which we usually think of as particles, exhibit the wave properties of interference and diffraction. All carriers of momentum and energy, such as electrons, atoms, light, or sound, have both particle and wave characteristics. It might be tempting to say that an electron, for example, is both a wave and a particle, but the meaning of such a statement is not clear. In classical physics, the concepts of waves and particles are mutually exclusive. A **classical particle** behaves like a piece of shot; it can be localized and scattered, it exchanges energy suddenly at a point in space, and it obeys the laws of conservation of energy and momentum in collisions. It does *not* exhibit interference or diffraction. A **classical wave,** on the other hand, behaves like a water wave; it exhibits diffraction and interference, and its energy is spread out continuously in space and time. Nothing can be both a classical particle and a classical wave at the same time.

After Thomas Young observed the two-slit interference pattern with light in 1801, light was thought to be a classical wave. Similarly, after J. J. Thomson's experiment in 1897, in which he deflected electrons in electric and magnetic fields, electrons were thought to be classical particles. We now know that these concepts of classical waves and particles do not adequately describe the complete behavior of any phenomenon.

> Everything propagates like a wave and exchanges energy like a particle.

Often the concepts of the classical particle and the classical wave give the same results. When the wavelength is very small, the propagation of a classical wave cannot be distinguished from that of a classical particle. For waves of very small wavelengths, diffraction effects are negligible, so the waves travel in straight lines. Similarly, interference is not seen for waves of very small wavelength because the interference fringes are too closely spaced to be observed. It then makes no difference which concept we use. When diffraction is negligible, we can think of light as a wave propagating along rays, as in geometrical optics, or as a beam of photon particles. Similarly, we can think of an electron as a wave propagating in straight lines along rays or, more commonly, as a particle.

We can also use either the wave or particle concept to describe exchanges of energy if we have a large number of particles and we are interested only in the average values of energy and momentum exchanges. For example, if we are interested only in the total current (above the threshold) in the photoelectric effect, the wave theory of light correctly predicts that this current is proportional to the intensity of the light.

Question

5. Which is the better model, the classical wave or the classical particle for the description of the propagation of electrons through a crystal? Which is better to describe the interaction of light with a photographic film?

36-5 The Schrödinger Equation

The wave equation governing the motion of electrons (and other particles with mass), which is analogous to the classical wave equation (Equation 29-7), was developed by Schrödinger in 1926 and is now known as the **Schrödinger equation.** Like the classical wave equation, the Schrödinger equation relates the time and space derivatives of the wave function. Schrödinger's reasoning is somewhat difficult to follow and is not important for our purposes. In any case, the Schrödinger equation cannot be derived, just as Newton's laws of motion cannot be derived. The validity of any fundamental equation lies in its agreement with experiment. Although it would be logical merely to postulate the Schrödinger equation, it is helpful to get some idea of what to expect by first considering the wave equation for photons, which is Equation 29-7 with speed $v = c$ and with $y(x, t)$ replaced by the electric field $\mathcal{E}(x, t)$:

$$\frac{\partial^2 \mathcal{E}(x, t)}{\partial x^2} = \frac{1}{c^2}\frac{\partial^2 \mathcal{E}(x, t)}{\partial t^2} \qquad \text{36-24}$$

As we discussed in Chapter 29, a particularly important solution of this equation is the harmonic wave function $\mathcal{E}(x, t) = \mathcal{E}_0 \sin(kx - \omega t)$. Differentiating this function twice with respect to time yields $\partial^2\mathcal{E}/\partial t^2 = -\omega^2\mathcal{E}_0 \sin(kx - \omega t)$, whereas differentiating it twice with respect to x gives $\partial^2\mathcal{E}/\partial x^2 = -k^2\mathcal{E}_0 \sin(kx - \omega t)$. Substitution of $\mathcal{E} = \mathcal{E}_0 \sin(kx - \omega t)$ into Equation 36-24 then gives

$$-k^2 = -\frac{\omega^2}{c^2}$$

or

$$\omega = kc \qquad \text{36-25}$$

Multiplying each side by $\hbar$, we obtain

$$\hbar\omega = \hbar kc$$

or

$$E = pc \qquad \text{36-26}$$

which is the relation between the energy and momentum of a photon.

Let us now use the de Broglie equations for a particle with mass to find the relation between ω and k for an electron that is analogous to Equation 36-25. The energy of a particle of mass m is

$$E = \frac{p^2}{2m} + U \qquad \text{36-27}$$

where U is the potential energy. The de Broglie equations are

$$E = hf = \hbar\omega$$

and

$$p = h/\lambda = \hbar k$$

Substituting these values for E and p into Equation 36-27, we obtain

$$\hbar\omega = \frac{\hbar^2 k^2}{2m} + U \qquad \text{36-28}$$

This differs from Equation 36-25 for photons in that it contains the potential energy U and the angular frequency ω does not vary linearly with k. Note that we get a factor of ω when we differentiate a harmonic wave function with respect to time and a factor of k when we differentiate it with respect to position. We expect, therefore, that the wave equation that applies to electrons will relate the *first* time derivative to the second space derivative and will also involve the potential energy of the electron.

We are now ready to postulate the Schrödinger equation. In one dimension, it has the form

$$-\frac{\hbar^2}{2m}\frac{\partial^2\Psi(x, t)}{\partial x^2} + U(x)\Psi(x, t) = i\hbar\frac{\partial\Psi(x, t)}{\partial t} \qquad \text{36-29}$$

Time-dependent Schrödinger equation

Equation 36-29 is called the **time-dependent Schrödinger equation.** An important difference between the time-dependent Schrödinger equation and the classical wave equation is the explicit appearance of the imaginary number $i = \sqrt{-1}$. The wave functions that satisfy the Schrödinger equation may therefore be complex functions.

Schrödinger's first application of his wave equation was to systems such as the hydrogen atom and the simple harmonic oscillator. He showed that energy quantization for these systems can be explained naturally in terms of standing waves. For such problems we need not consider the time dependence of the wave function. For any wave function that describes a particle in a state of definite energy, the general Schrödinger equation can be simplified by writing the wave function in the form

$$\Psi(x, t) = \psi(x)e^{-i\omega t}$$

where $\psi(x)$ is a function of x only.* The right side of Equation 36-29 is then

$$i\hbar\frac{\partial\Psi(x, t)}{\partial t} = i\hbar(-i\omega)\psi(x)e^{-i\omega t} = \hbar\omega\psi(x)e^{-i\omega t} = E\psi(x)e^{-i\omega t}$$

where we have used $E = \hbar\omega$. Substituting $\Psi(x, t) = \psi(x)e^{-i\omega t}$ into Equation 36-29, we obtain

$$-\frac{\hbar^2}{2m}\frac{\partial^2\psi(x)}{\partial x^2}e^{-i\omega t} + U(x)\psi(x)e^{-i\omega t} = E\psi(x)e^{-i\omega t}$$

Canceling the common factor $e^{-i\omega t}$, we obtain an equation for $\psi(x)$ called the **time-independent Schrödinger equation:**

$$-\frac{\hbar^2}{2m}\frac{d^2\psi(x)}{dx^2} + U(x)\psi(x) = E\psi(x) \qquad \text{36-30}$$

Time-independent Schrödinger equation

The time-independent Schrödinger equation in one dimension is an ordinary differential equation in one variable x and is therefore much easier to handle than Equation 36-29, which is a partial differential equation containing the two variables x and t.

There are important conditions that the wave function $\psi(x)$ must satisfy in addition to being the solution of Equation 36-30. The probability of finding the electron in a region dx at x is $|\psi(x)|^2\,dx$. Since the probability of finding an electron cannot jump discontinuously as we move from one point to a nearby point, the wave function $\psi(x)$ must be continuous in x. Also, for potential-energy functions $U(x)$ that are not infinite, the first derivative $d\psi/dx$ must be continuous. We can see this from Equation 36-30 by multiply-

*It is customary to denote the spatial part of the wave function by a lower case psi (ψ) and the complete wave function with time dependence by an upper case psi (Ψ).

ing each term by $2m/\hbar^2$ and writing $d^2\psi/dx^2$ as $\frac{d}{dx}(d\psi/dx)$. We then have

$$\frac{d}{dx}\left(\frac{d\psi}{dx}\right) = \frac{2m}{\hbar^2}[U(x) - E]\psi(x)$$

or

$$d\left(\frac{d\psi}{dx}\right) = \frac{2m}{\hbar^2}[U(x) - E]\psi(x)\, dx$$

If we let dx approach zero, then $d(d\psi/dx)$, the change in the first derivative, also approaches zero as long as $U(x)$ is not infinite. This is equivalent to saying that the first derivative is continuous in x if $U(x)$ is not infinite.

The probability of finding the electron in dx at point x_1 *or* at point x_2 is the sum of the separate probabilities $P(x_1)\, dx + P(x_2)\, dx$. Since the electron must certainly be somewhere, the sum of the probabilities over all the possible values of x must equal 1. That is,

Normalization condition

$$\int_{-\infty}^{\infty} |\psi|^2\, dx = 1 \tag{36-31}$$

Equation 36-31 is called the **normalization condition.** This condition plays an important role in quantum mechanics. If $\psi(x)$ is to satisfy the normalization condition of Equation 36-31, it must approach zero as x approaches infinity. This places a restriction on the possible solutions of the Schrödinger equation and leads to energy quantization.

36-6 A Particle in a Box

We will now apply the time-independent Schrödinger equation to a traditional though somewhat artificial problem of a particle, such as an electron, confined to a one-dimensional box of length L. Classically, the particle bounces back and forth between the walls of the box, which we assume are at $x = 0$ and $x = L$. The particle is equally likely to be found anywhere in the box, and its energy and momentum can take on any values.

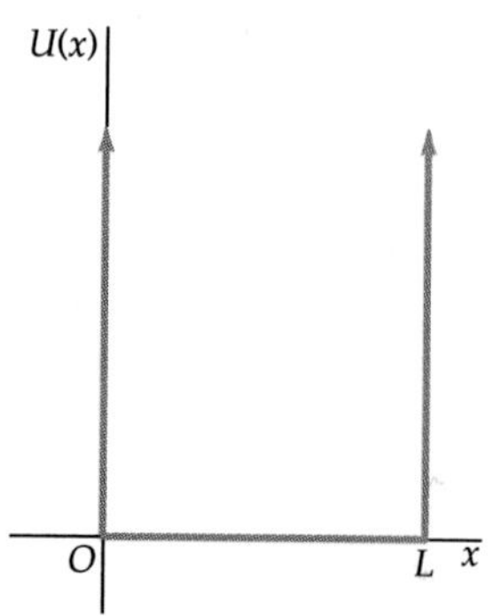

Figure 36-6 The infinite square-well potential energy. For $x < 0$ and $x > L$, the potential energy $U(x)$ is infinite. The particle is confined to the region in the well $0 < x < L$.

According to quantum theory, the particle is described by a wave function ψ that obeys Equation 36-30. The potential energy for this problem is shown in Figure 36-6. It is called an **infinite square-well potential** and is described mathematically by

$$\begin{aligned} U(x) &= 0 \qquad 0 < x < L \\ U(x) &= \infty \qquad x < 0 \text{ or } x > L \end{aligned} \tag{36-32}$$

That is, inside the box, the potential energy is zero, whereas outside the box it is infinite. Since we require the particle to be in the box, we have $\psi(x) = 0$ everywhere outside the box. We then need to solve Equation 36-30 for inside the box subject to the condition that, since the wave function must be continuous, $\psi(x)$ must be zero at $x = 0$ and at $x = L$. Such a condition on the wave function is called a **boundary condition.** (The derivative of the wave function $d\psi/dx$ need not be continuous because the potential energy is infinite for $x < 0$ and $x > L$ in this problem.) Inside the box, Equation 36-30 is

$$-\frac{\hbar^2}{2m}\frac{d^2\psi(x)}{dx^2} = E\psi(x)$$

or

$$\frac{d^2\psi(x)}{dx^2} = -\frac{2mE}{\hbar^2}\psi(x) = -k^2\psi(x) \qquad 36\text{-}33$$

where

$$k^2 = \frac{2mE}{\hbar^2} \qquad 36\text{-}34$$

The general solution of Equation 36-33 can be written as

$$\psi(x) = A \sin kx + B \cos kx \qquad 36\text{-}35$$

where A and B are constants. At $x = 0$, we have

$$\psi(0) = A \sin (k0) + B \cos (k0) = 0 + B$$

The boundary condition $\psi(x) = 0$ at $x = 0$ thus gives $B = 0$, and Equation 36-35 becomes

$$\psi(x) = A \sin kx \qquad 36\text{-}36$$

The boundary condition $\psi(x) = 0$ at $x = L$ gives

$$\psi(L) = A \sin kL = 0 \qquad 36\text{-}37$$

This condition is satisfied if kL is π or any integer times π, that is, if k is restricted to the values k_n given by

$$k_n = n\frac{\pi}{L} \qquad n = 1, 2, 3, \ldots \qquad 36\text{-}38$$

For each value of n there is wave function ψ_n given by

$$\psi_n = A_n \sin \frac{n\pi x}{L} \qquad 36\text{-}39$$

If we write k in terms of the wavelength, $k = 2\pi/\lambda$, Equation 36-38 becomes

$$\frac{2\pi}{\lambda_n} = n\frac{\pi}{L}$$

or

$$L = n\frac{\lambda_n}{2}$$

which is the same as the standing-wave condition for waves on a string fixed at $x = 0$ and at $x = L$. The energy E is related to the wave number k by Equation 36-34:

$$E_n = \frac{\hbar^2 k_n^2}{2m} = n^2\frac{\hbar^2\pi^2}{2mL^2} = n^2\frac{h^2}{8mL^2}$$

or

$$E_n = n^2 E_1 \qquad 36\text{-}40$$

Allowed energies for an infinite square-well potential

where

$$E_1 = \frac{h^2}{8mL^2} \qquad 36\text{-}41$$

Ground-state energy for an infinite square-well potential

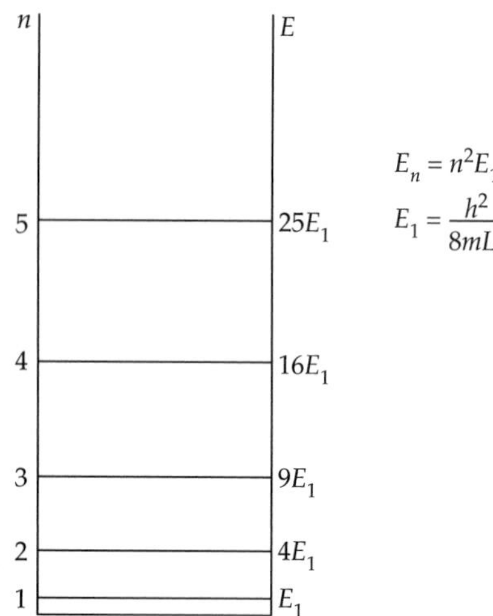

Figure 36-7 Energy-level diagram for the infinite square-well potential. Classically, a particle can have any value of energy. Quantum mechanically, only certain values of energy given by $E_n = n^2E_1 = n^2(h^2/8mL^2)$ are allowed.

is the energy of the lowest state, the ground state. The energy is thus quantized. Figure 36-7 shows the energy-level diagram for the infinite square-well potential. Note that the lowest energy is not zero. According to quantum theory, the particle cannot remain at rest in the box. This result, which is a consequence of the uncertainty principle as discussed in Section 36-3, is a general feature of quantum theory. When a particle is confined to some region of space, it must have a minimum kinetic energy, the zero-point energy. The smaller the region of space, the greater the zero-point energy, as indicated by the fact that E_1 varies as $1/L^2$ in Equation 36-41.

The constant A_n is determined by the normalization condition (Equation 36-31):

$$\int_{-\infty}^{\infty} \psi^2\, dx = \int_0^L A_n^2 \sin^2 \frac{n\pi x}{L}\, dx = 1$$

Note that we need integrate only from $x = 0$ to $x = L$ because $\psi(x)$ is zero everywhere else. Substituting $\theta = n\pi x/L$, we have

$$\int_0^L A_n^2 \sin^2 \frac{n\pi x}{L}\, dx = A_n^2 \frac{L}{n\pi} \int_0^{n\pi} \sin^2 \theta\, d\theta = 1$$

The integral of $\sin^2 \theta$ can be found in tables:

$$\int_0^{n\pi} \sin^2 \theta\, d\theta = \frac{\theta}{2}\bigg|_0^{n\pi} - \frac{\sin 2\theta}{4}\bigg|_0^{n\pi} = \frac{n\pi}{2}$$

The normalization condition thus gives

$$A_n^2 \frac{L}{n\pi} \frac{n\pi}{2} = 1$$

or

$$A_n = \sqrt{\frac{2}{L}}$$

independent of n. The normalized wave functions for the infinite square-well potential are thus

Wave functions for the infinite square-well potential

$$\psi_n = \sqrt{\frac{2}{L}} \sin \frac{n\pi x}{L} \qquad 36\text{-}42$$

The number n is called a **quantum number.** It characterizes the wave function for a particular state and the energy of that state. In our one-dimensional problem, it arises from the boundary condition on the wave function that it be zero at $x = 0$ and $x = L$. In three-dimensional problems, three quantum numbers arise, each associated with a boundary condition in each dimension.

Figure 36-8 shows plots of ψ^2 for the ground state $n = 1$, the first excited state $n = 2$, the second excited state $n = 3$, and the state $n = 10$.* In the ground state, the particle is most likely to be found near the center of the box, as indicated by the maximum value of ψ^2 at $x = L/2$. In the first excited state, the particle is never found exactly in the center of the box because ψ^2 is zero at $x = L/2$. For very large values of n, the maxima and minima of ψ^2 are very close together, as illustrated for $n = 10$. The average value of ψ^2 is

*Since $\psi(x)$ is real in this case, $|\psi|^2 = \psi^2$.

indicated in this figure by the dashed line. For very large values of n, the maxima are so closely spaced that ψ^2 cannot be distinguished from its average value. The fact that $(\psi^2)_{av}$ is constant across the whole box means that the particle is equally likely to be found anywhere in the box—the same as the classical result. This is an example of **Bohr's correspondence principle:**

> In the limit of very large quantum numbers, the classical calculation and the quantum calculation must yield the same results.

Bohr's correspondence principle

The region of very large quantum numbers is also the region of very large energies. It can be shown that for large energies, the percentage change in energy between adjacent quantum states is very small, so energy quantization is not important (see Problem 48).

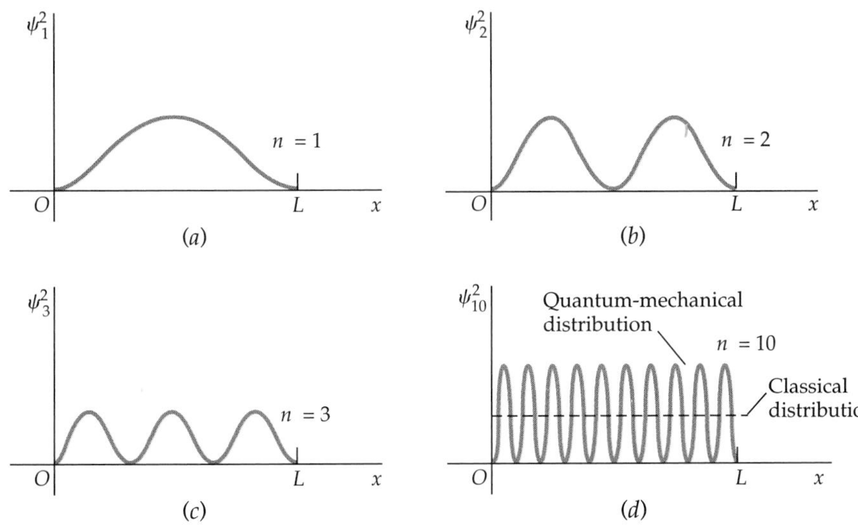

Figure 36-8 ψ^2 versus x for a particle in a box of length L for (*a*) the ground state, $n = 1$; (*b*) the first excited state, $n = 2$; (*c*) the second excited state, $n = 3$; and (*d*) the state $n = 10$. For $n = 10$, the maxima and minima of ψ^2 are so close together that individual maxima may be hard to distinguish. The average value of ψ^2 is indicated by the dashed line. It gives the classical prediction that the particle is equally likely to be found at any point in the box.

We are so accustomed to thinking of the electron as a classical particle that we tend to think of an electron in a box as a particle bounding back and forth between the walls. But the probability distributions shown in Figure 36-8 are stationary; that is, they do not depend on time. (The analogous patterns for the electron in the hydrogen atom discussed in the next chapter are the quantum theoretical counterparts of the stationary orbits of the Bohr model.) An alternative picture of the electron is a cloud of charge with the charge density proportional to ψ^2. Figure 36-8 can then be thought of as plots of the charge density versus x for the various states. In the ground state, $n = 1$, the electron cloud is centered in the middle of the box and is spread out over most of the box, as indicated in Figure 36-8*a*. In the first excited state, $n = 2$, the charge density of the electron cloud has two maxima, as indicated in Figure 36-8*b*. For very large values of n, there are many closely spaced maxima and minima in the charge density resulting in an average charge density that is approximately uniform throughout the box. This electron-cloud picture is very useful in understanding the structure of atoms and molecules, which will be discussed in the following chapters. However, it should be noted that whenever an electron is observed to interact with matter or radiation, it is always observed as a single charge.

Example 36-4

(*a*) Find the energy in the ground state of an electron confined to a one-dimensional box of length $L = 0.1$ nm. (This box is roughly the size of an atom.) (*b*) Make an energy-level diagram and find the wavelengths of the photons emitted for all transitions beginning at state $n = 3$ or less and ending at a lower energy state.

(*a*) The energy in the ground state is given by Equation 36-41. Multiplying the numerator and denominator by c^2, we obtain an expression in terms of hc and the rest energy mc^2:

$$E_1 = \frac{(hc)^2}{8mc^2L^2} \tag{36-43}$$

Substituting $hc = 1240$ eV·nm and $mc^2 = 0.511$ MeV, we obtain

$$E_1 = \frac{(1240 \text{ eV·nm})^2}{8(5.11 \times 10^5 \text{ eV})(0.1 \text{ nm})^2} = 37.6 \text{ eV}$$

This is greater than the minimum energy of about 1 eV that we found from the uncertainty principle in Example 36-3. It is of the same order of magnitude as the kinetic energy of the electron in the ground state of the hydrogen atom, which is 13.6 eV. In that case, the wavelength of the electron equals the circumference of a circle of radius 0.0529 nm, or about 0.33 nm, whereas for the electron in a one-dimensional box of length 0.1 nm, the wavelength in the ground state is $2L = 0.2$ nm.

(*b*) The energies of this system are given by

$$E_n = n^2E_1 = n^2(37.6 \text{ eV})$$

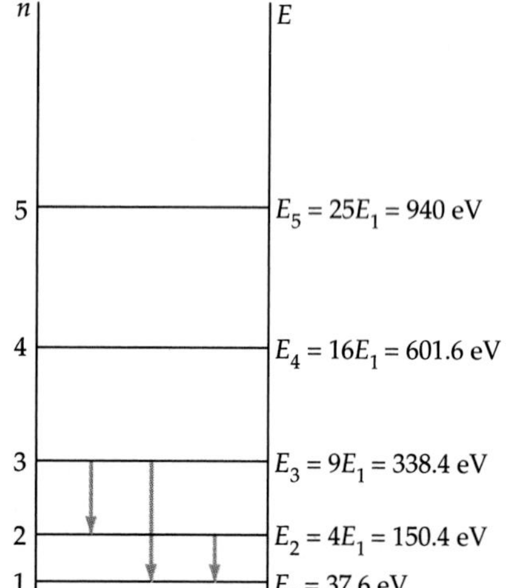

Figure 36-9 Energy-level diagram for Example 36-4. Transitions from the state $n = 3$ to the states $n = 2$ and $n = 1$, and from the state $n = 2$ to $n = 1$, are indicated by the vertical arrows.

Figure 36-9 shows these energies in an energy-level diagram. The energy of the first excited state is $E_2 = 4(37.6 \text{ eV}) = 150.4$ eV, and that of the second excited state is $E_3 = 9(37.6 \text{ eV}) = 338.4$ eV. The possible transitions from level 3 to level 2, from level 3 to level 1, and from level 2 to level 1 are indicated by the vertical arrows on the diagram. The energies of these transitions are

$$\Delta E_{3\to2} = 338.4 \text{ eV} - 150.4 \text{ eV} = 188 \text{ eV}$$

$$\Delta E_{3\to1} = 338.4 \text{ eV} - 37.6 \text{ eV} = 300.8 \text{ eV}$$

and

$$\Delta E_{2\to1} = 150.4 \text{ eV} - 37.6 \text{ eV} = 112.8 \text{ eV}$$

The photon wavelengths for these transitions are

$$\lambda_{3\to2} = \frac{hc}{\Delta E_{3\to2}} = \frac{1240 \text{ eV·nm}}{188 \text{ eV}} = 6.60 \text{ nm}$$

$$\lambda_{3\to1} = \frac{hc}{\Delta E_{3\to1}} = \frac{1240 \text{ eV·nm}}{300.8 \text{ eV}} = 4.12 \text{ nm}$$

$$\lambda_{2\to1} = \frac{hc}{\Delta E_{2\to1}} = \frac{1240 \text{ eV·nm}}{112.8 \text{ eV}} = 11.0 \text{ nm}$$

Exercise

Calculate the wavelength of the photon emitted if the electron in Example 36-4 makes a transition from $n = 4$ to $n = 3$. (Answer: 4.71 nm)

Example 36-5

A particle is in the ground state of an infinite square-well potential. Find the probability of finding the particle (*a*) in the region $0 < x < \frac{1}{4}L$, and (*b*) in $\Delta x = 0.01L$ at $x = \frac{1}{2}L$.

(*a*) The probability of finding the particle in some range dx is

$$P(x)\,dx = \psi^2(x)\,dx = \frac{2}{L}\sin^2\frac{\pi x}{L}\,dx$$

We find the probability of finding the particle in the region $0 < x < \frac{1}{4}L$ by integrating this expression over this range:

$$P = \int_0^{L/4} \frac{2}{L}\sin^2\frac{\pi x}{L}\,dx$$

Substituting $\theta = \pi x/L$, we obtain

$$P = \frac{2}{L}\frac{L}{\pi}\int_0^{\pi/4} \sin^2\theta\,d\theta$$

$$= \frac{2}{\pi}\left[\frac{\theta}{2}\Big|_0^{\pi/4} - \frac{\sin 2\theta}{4}\Big|_0^{\pi/4}\right]$$

$$= \frac{2}{\pi}\left(\frac{\pi}{8} - \frac{1}{4}\right) = 0.091$$

The chance of finding the particle in the region $0 < x < \frac{1}{4}L$ is thus about 9.1 percent. This probability is indicated by the shaded region in Figure 36-10.

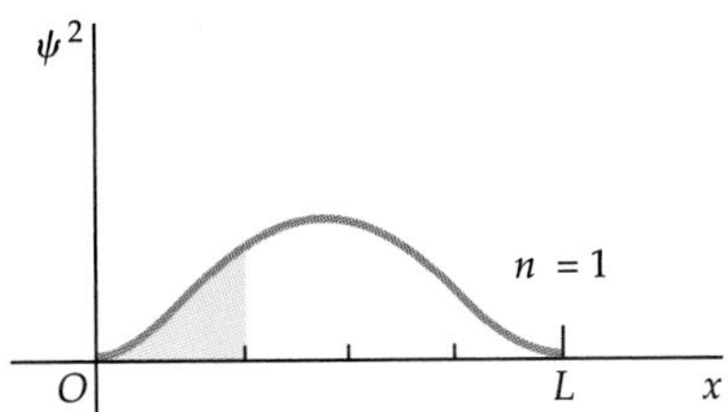

Figure 36-10 The probability density $\psi^2(x)$ versus x for a particle in the ground state of an infinite square-well potential. The probability of finding the particle in the region $0 < x < \frac{1}{4}L$ equals the shaded area.

(*b*) Since the region $\Delta x = 0.01L$ is very small compared with L, we do not need to integrate. The approximate probability is

$$P = \psi^2(x)\,\Delta x = \frac{2}{L}\sin^2\frac{\pi x}{L}\,\Delta x$$

Substituting $x = \frac{1}{2}L$ and $\Delta x = 0.01L$, we obtain

$$P = \frac{2}{L}\left(\sin^2\frac{\pi}{2}\right)(0.01L)$$

$$= \frac{2}{L}(1.0)(0.01L) = 0.02$$

There is thus a 2 percent chance of finding the particle in the region $\Delta x = 0.01L$ at $x = \frac{1}{2}L$.

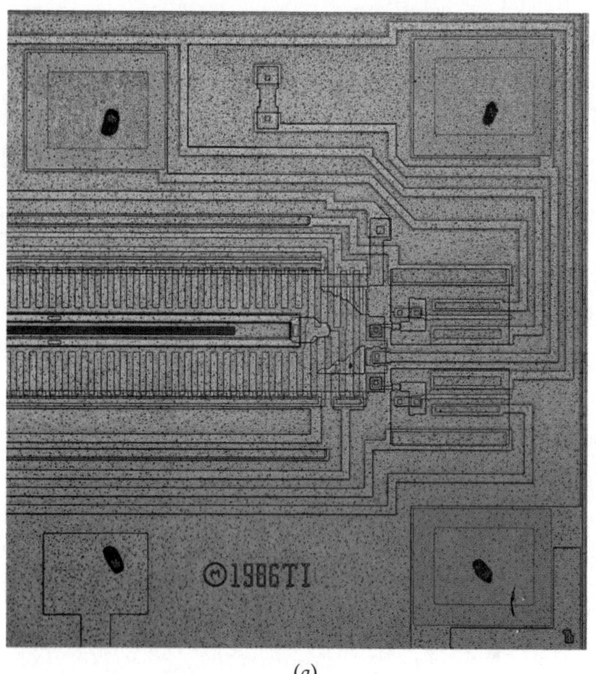

(a)

(b)

Charge-coupled devices (CCDs) are on the forefront of imaging technology. They are efficient and fast, and their output is stored electronically so that it is easily processed and manipulated by computer. Typically 40 to 80 percent of the photons incident on a CCD surface are converted into a stored electrical signal, allowing for short exposure times and a very low detection threshold. This compares to the 2 or 3 percent of incoming photons that react with a film's light-sensitive atoms to produce exposed film grains. Also, unlike a photographic film, the response of a CCD is directly proportional to the amount of incoming light, making possible a much more precise measurement of data.

A CCD is a three-layer semiconductor—the top layer is a series of metallic electrodes, the bottom layer is a silicon crystal, and the middle layer is an insulator separating the two. Light striking silicon in the semiconductor frees electrons, which accumulate in potential wells at the surface of the silicon. Each well in the two-dimensional array on the silicon surface stores an amount of charge that is proportional to the number of photons that strike

Continued

36-7 A Particle in a Finite Square Well

The quantization of energy that we found for a particle in an infinite square well is a general result that follows from the solution of the Schrödinger equation for any particle confined to some region of space. We will illustrate this by considering the qualitative behavior of the wave function for a slightly more general potential-energy function, the finite square well shown in Figure 36-11. This potential-energy function is described mathematically by

$$U(x) = U_0 \qquad x < 0$$
$$U(x) = 0 \qquad 0 < x < L$$
$$U(x) = U_0 \qquad x > L$$

This potential-energy function is discontinuous at $x = 0$ and $x = L$, but it is finite everywhere. The solutions of the Schrödinger equation for this type of potential-energy function depend on whether the total energy E is greater or less than U_0. We will not discuss the case of $E > U_0$, except to remark that in that case the particle is not confined and any value of the energy is allowed; that is, there is no energy quantization. Here we assume that $E < U_0$.

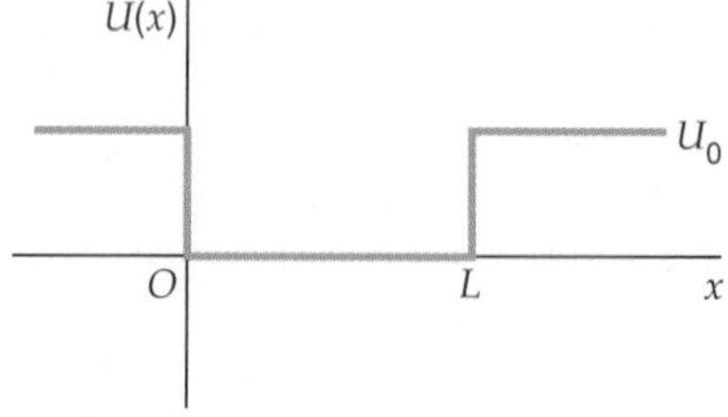

Figure 36-11 The finite square-well potential energy.

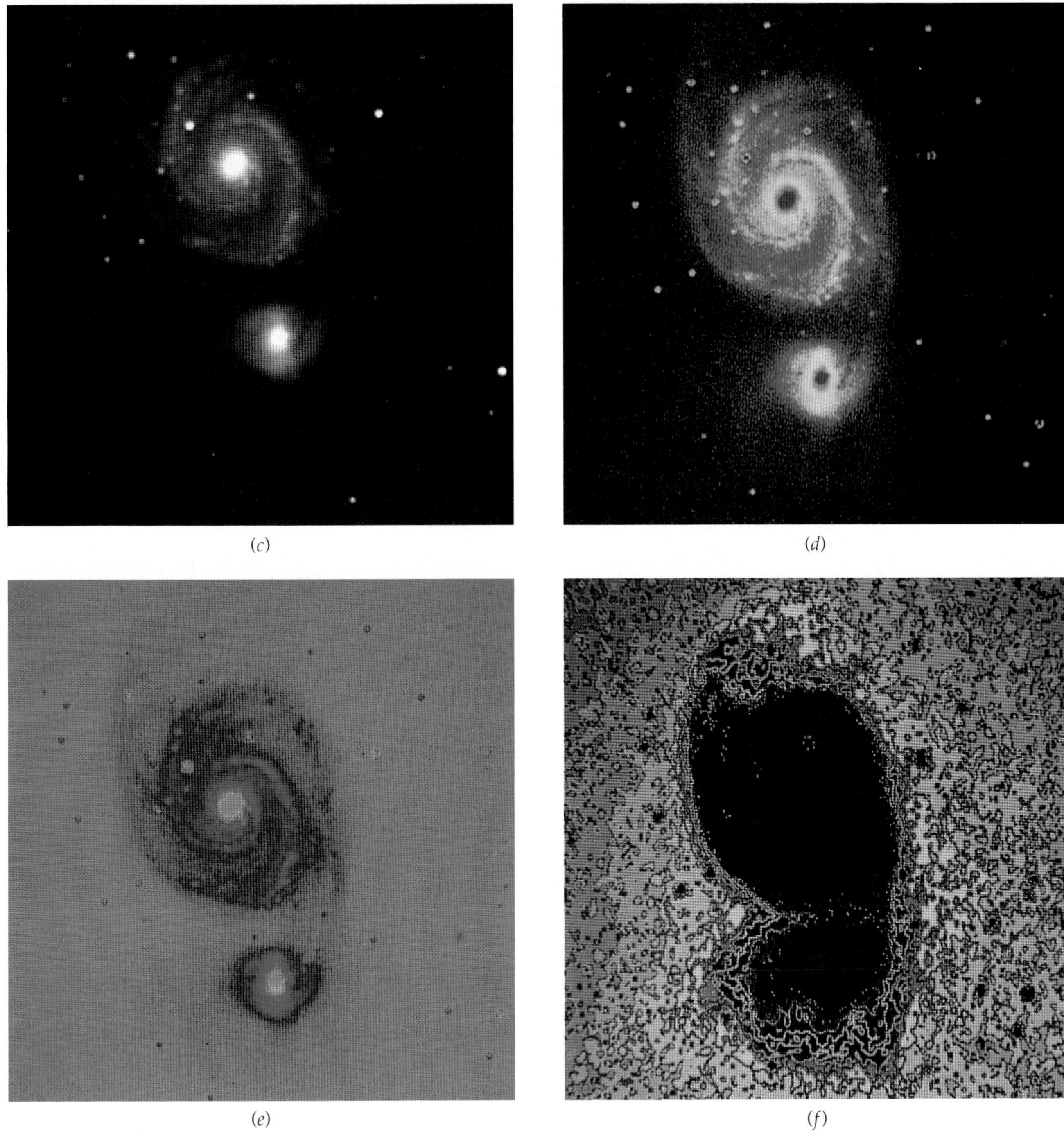

(c) (d) (e) (f)

the surface in the region of the well. The charge is dumped electronically into a computer that records the location and amount of charge in each well. A conventional TV monitor can then be used to reconstruct a digitized version of the original image.

(*a*) A close-up of part of a CCD. The horizontal bar emerging from the left is the photosensitive area (called a "pixel"). The vertical segments above and below it (called a "transfer register") contain the succession of electrodes that transfer accumulated charge packets along a line of potential wells, from left to right, eventually depositing them in an amplifier located in the central right portion of the chip. (*b*) This platinum-silicide CCD chip responds to infrared wavelengths. It contains pixels in a 320 by 244 array. (*c*) An unprocessed CCD image of spiral galaxy Messier 51 and companion galaxy. (*d*) An image generated from the data contained in *c* in which false colors have been assigned, corresponding to different intensity ranges. (*e*) An image that again, like *d*, has been generated from the data in *c* and enhanced and colorized by computer. (*f*) This time the image has been processed for maximum contrast and contoured to show detail in the outer rims of the galaxies.

Inside the well, $U(x) = 0$, and the time-independent Schrödinger equation is the same as for the infinite well (Equation 36-33):

$$\frac{d^2\psi(x)}{dx^2} = -k^2\psi(x)$$

with

$$k^2 = \frac{2mE}{\hbar^2}$$

The general solution is of the form

$$\psi(x) = A \sin kx + B \cos kx$$

In this case, $\psi(x)$ is not zero at $x = 0$, so B is not zero. Outside the well, the time-independent Schrödinger equation is

$$\frac{d^2\psi(x)}{dx^2} = \frac{2m}{\hbar^2}(U_0 - E)\psi(x) = \alpha^2\psi(x) \qquad \text{36-44}$$

where

$$\alpha^2 = \frac{2m}{\hbar^2}(U_0 - E) > 0 \qquad \text{36-45}$$

The wave functions and allowed energies can be found by solving Equation 36-44 for $\psi(x)$ outside the well and then requiring that $\psi(x)$ and $d\psi(x)/dx$ be continuous at the boundaries $x = 0$ and $x = L$. The solution of Equation 36-44 is not difficult [for positive values of x, it is of the form $\psi(x) = Ce^{-\alpha x}$], but applying the boundary conditions involves much tedious algebra and is not important for our purpose. The important feature of Equation 36-44 is that the second derivative of $\psi(x)$, which is related to the curvature of the wave function, has the same sign as the wave function ψ. If ψ is positive, $d^2\psi/dx^2$ is also positive and the wave function curves away from the axis as shown in Figure 36-12a. Similarly, if ψ is negative, $d^2\psi/dx^2$ is negative and ψ again curves away from the axis as shown in Figure 36-12b. This behavior is very different from that inside the well, where ψ and $d^2\psi/dx^2$ have opposite signs so that ψ always curves toward the axis like a sine or cosine function. Because of this behavior outside the well, $\psi(x)$ becomes infinite as x approaches $\pm\infty$ for most values of the energy. That is, $\psi(x)$ is not well behaved outside the well. Though they satisfy the Schrödinger equation, such functions are not proper wave functions because they cannot be normalized. Only for certain values of the energy do the wave functions approach 0 as $|x|$ becomes very large. These energy values are the allowed energies for the finite square well.

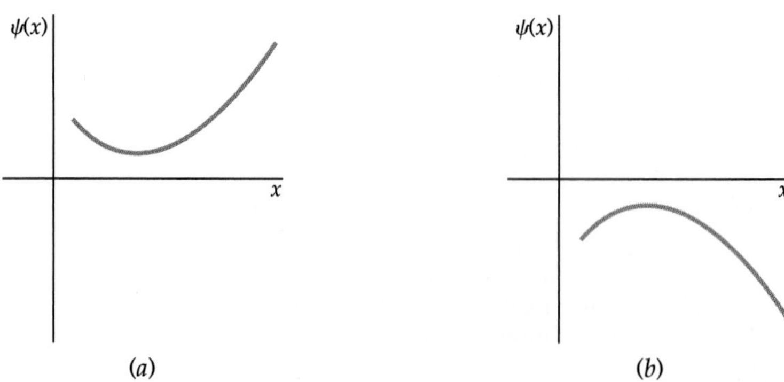

Figure 36-12 (a) A positive function with positive curvature. (b) A negative function with negative curvature.

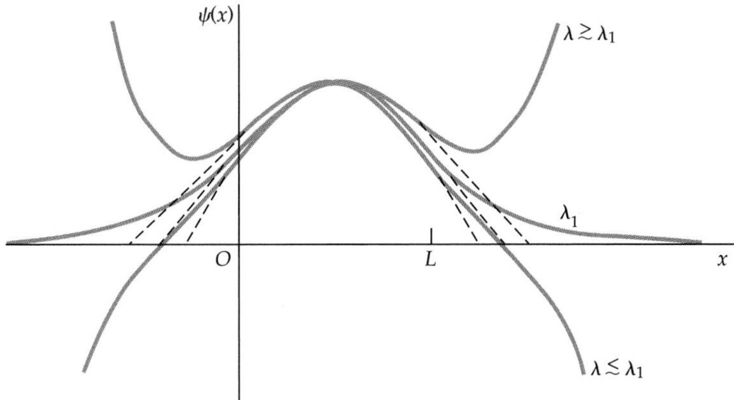

Figure 36-13 Functions satisfying the Schrödinger equation with wavelengths near the wavelength λ_1 corresponding to the ground-state energy $E_1 = h^2/2m\lambda_1^2$ in the finite well. If λ is slightly greater than λ_1, the function approaches infinity as shown in Figure 36-12*a*. At the critical wavelength λ_1, the function and its slope approach zero together. If λ is slightly less than λ_1, the function crosses the x axis while the slope is still negative. The slope then becomes more negative because its rate of change $d^2\psi/dx^2$ is now negative. This function approaches negative infinity as x approaches infinity.

Figure 36-13 shows a well-behaved wave function with a wavelength λ_1 inside the well corresponding to the ground-state energy. The behavior of the wave functions corresponding to nearby wavelengths and energies are also shown. Figure 36-14 shows the wave functions and probability distributions for the ground state and first two excited states. From this figure, we can see that the wavelengths inside the well are slightly longer than the corresponding wavelengths for the infinite well (Figure 36-8), so the corresponding energies are slightly less than those for the infinite well. Another feature of the finite-well problem is that there are only a finite number of allowed energies depending on the size of U_0. For very small values of U_0, there is only one allowed energy.

Note that the wave function penetrates beyond the edges of the well at $x = L$ and $x = 0$, indicating that there is some small probability of finding the particle in the region in which its total energy E is less than its potential energy U_0. This region is called the *classically forbidden region* because the kinetic energy, which is $E - U_0$, would be negative when $U_0 > E$. Since negative kinetic energy has no meaning in classical physics, it is interesting

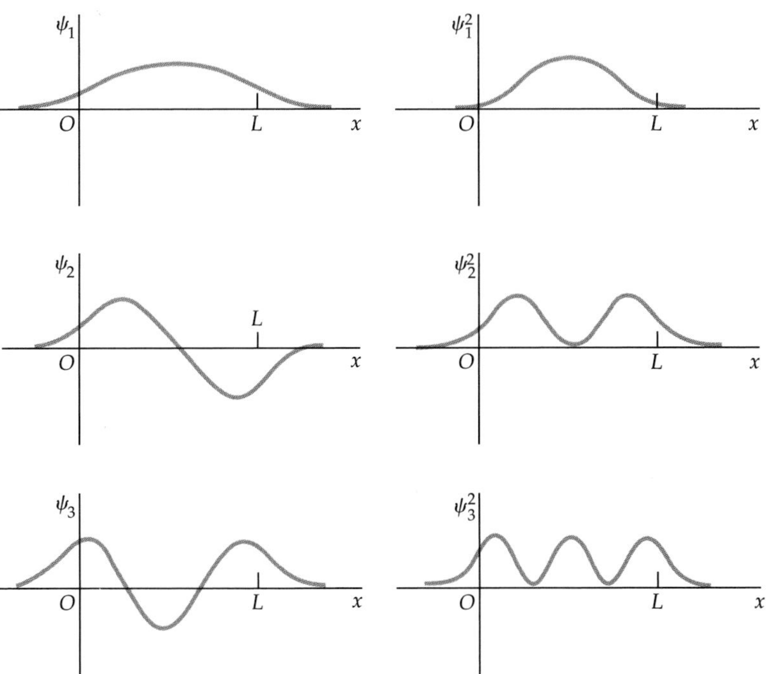

Figure 36-14 Graphs of the wave functions $\psi_n(x)$ and probability distributions $\psi_n^2(x)$ for $n = 1$, 2, and 3 for the finite square well. Compare these graphs with those of Figure 36-8 for the infinite square well, where the wave functions are zero at $x = 0$ and $x = L$. The wavelengths here are slightly longer than the corresponding ones for the infinite well, so the allowed energies are somewhat smaller.

to speculate on the result of an attempt to observe the particle in the classically forbidden region. It can be shown from the uncertainty principle that if an attempt is made to localize the particle in the classically forbidden region, such a measurement introduces an uncertainty in the momentum of the particle corresponding to a minimum kinetic energy that is greater than $U_0 - E$, which is just great enough to prevent us from measuring a negative kinetic energy. The penetration of the wave function into a classically forbidden region does have important consequences in barrier penetration, which will be discussed in Section 36-9.

Much of our discussion of the finite-well problem applies to any problem in which $E > U(x)$ in some region and $E < U(x)$ outside that region. Consider the potential energy $U(x)$ shown in Figure 36-15. Inside the well, the Schrödinger equation is

$$\frac{d^2\psi(x)}{dx^2} = -k^2\psi(x)$$

where $k^2 = 2m[E - U(x)]/\hbar^2$ now depends on x. The solutions of this equation are no longer simple sine or cosine functions because the wave number $k = 2\pi/\lambda$ now varies with x, but since $d^2\psi/dx^2$ and ψ have opposite signs, ψ will always curve toward the axis and the solutions will oscillate. Outside the well, $d^2\psi/dx^2$ and ψ have the same sign, so ψ will curve away from the axis and there will be only certain values of E for which solutions exist that approach zero as x approaches infinity.

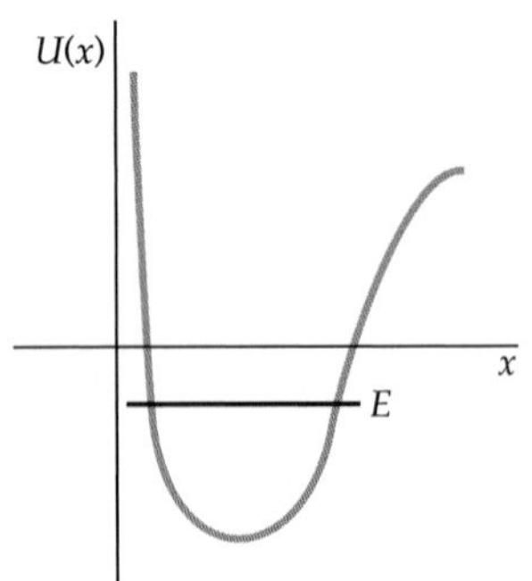

Figure 36-15 Arbitrary well-type potential with possible energy E. Inside the well where E is greater than $U(x)$, $\psi(x)$ and $d^2\psi/dx^2$ have opposite signs, and the wave function will oscillate. Outside the well, $\psi(x)$ and $d^2\psi/dx^2$ have the same sign and, except for certain values of E, the wave functions will not be well behaved.

36-8 Expectation Values

The solution of a classical mechanics problem is typically specified by giving the position of a particle as a function of time. As we have discussed, the wave nature of matter prevents us from doing this for microscopic systems. Instead, we find the wave function $\Psi(x, t)$ and the probability distribution $|\Psi(x, t)|^2$. The most that we can know is the probability of measuring a certain value of position x. If we measure the position for a large number of identical systems, we get a range of values corresponding to the probability distribution. The average value of x obtained from such measurements is called the **expectation value** and written $\langle x \rangle$.

Expectation value

The expectation value of x is the same as the average value of x that we would expect to obtain from a measurement of the positions of a large number of particles with the same wave function $\Psi(x, t)$.

Suppose we have a large number N of identical systems. The number of systems for which the particle is in a particular region dx at some value x is given by $NP(x)\,dx = N|\Psi(x, t)|^2\,dx$. To find the average value of x, we sum all the values of x and then divide by the number of terms in the sum. The values of x in this sum are given by $xNP(x)\,dx = xN|\Psi(x, t)|^2\,dx$, which is the value of x times the number of times it occurs. Since x is continuous, a sum over all the possible values of x is performed by integrating over dx. When we divide by the total number of measurements N, we obtain for the average or expectation value

Expectation value of x defined

$$\langle x \rangle = \int x|\Psi(x, t)|^2\,dx \qquad 36\text{-}46$$

As we have seen, for a particle in a state of definite energy, the probability distribution is independent of time. The expectation value is then given by

$$\langle x\rangle = \int x|\psi(x)|^2\,dx \qquad \text{36-47}$$

Expectation value of x for time-independent state

The expectation value of any function $f(x)$ is given by

$$\langle f(x)\rangle = \int f(x)|\psi(x)|^2\,dx \qquad \text{36-48}$$

Expectation value of f(x) defined

Example 36-6

Find (*a*) $\langle x\rangle$ and (*b*) $\langle x^2\rangle$ for a particle in the ground state of an infinite square-well potential.

(*a*) The wave function for the ground state in an infinite square-well potential is given by Equation 36-42 with $n = 1$:

$$\psi = \sqrt{\frac{2}{L}}\sin\frac{\pi x}{L}$$

The expectation value of x is then

$$\langle x\rangle = \int_{-\infty}^{+\infty} x\psi^2(x)\,dx = \int_0^L x\frac{2}{L}\sin^2\frac{\pi x}{L}\,dx$$

$$= \frac{2}{L}\left(\frac{L}{\pi}\right)^2\int_0^{\pi}\theta\sin^2\theta\,d\theta$$

where we have substituted $\theta = \pi x/L$. This integral can be found in tables. Its value is

$$\int_0^{\pi}\theta\sin^2\theta\,d\theta = \left[\frac{\theta^2}{4} - \frac{\theta\sin 2\theta}{4} - \frac{\cos 2\theta}{8}\right]_0^{\pi} = \frac{\pi^2}{4}$$

The expectation value of x is thus

$$\langle x\rangle = \frac{2L}{\pi^2}\left(\frac{\pi^2}{4}\right) = \frac{L}{2}$$

which is what we would expect because the probability distribution is symmetric about the midpoint of the well.

(*b*) The expectation value of x^2 is

$$\langle x^2\rangle = \int_{-\infty}^{+\infty} x^2\psi^2(x)\,dx = \int_0^L x^2\frac{2}{L}\sin^2\frac{\pi x}{L}\,dx$$

$$= \frac{2}{L}\left(\frac{L}{\pi}\right)^3\int_0^{\pi}\theta^2\sin^2\theta\,d\theta$$

Again, we obtain the integral from tables:

$$\int_0^{\pi}\theta^2\sin^2\theta\,d\theta = \left[\frac{\theta^3}{6} - \left(\frac{\theta^2}{4} - \frac{1}{8}\right)\sin 2\theta - \frac{\theta\cos 2\theta}{4}\right]_0^{\pi} = \frac{\pi^3}{6} - \frac{\pi}{4}$$

Then,

$$\langle x^2\rangle = \frac{2L^2}{\pi^3}\left(\frac{\pi^3}{6} - \frac{\pi}{4}\right) = L^2\left(\frac{1}{3} - \frac{1}{2\pi^2}\right) = 0.283L^2$$

Exercise

A six-sided die has the number 1 painted on three sides and the number 2 painted on the other three sides. (*a*) What is the probability of a 1 coming up when the die is thrown? (*b*) What is the expectation value of the number that comes up when the die is thrown? [Answers: (*a*) 0.50, (*b*) 1.5]

Question

6. Can the expectation value of x ever equal a value that has zero probability of being measured?

36-9 Reflection and Transmission of Electron Waves: Barrier Penetration

In Sections 36-6 and 36-7, we were concerned with bound-state problems in which the potential energy is larger than the total energy for large values of $|x|$. In this section, we consider some simple examples of unbound states for which E is greater than $U(x)$. For these problems, $d^2\psi/dx^2$ and ψ have opposite signs, so $\psi(x)$ curves toward the axis and does not become infinite at large values of $|x|$.

Step Potential

Consider a particle of energy E moving in a region in which the potential energy is the step function

$$U(x) = 0 \qquad x < 0$$
$$U(x) = U_0 \qquad x > 0$$

as shown in Figure 36-16. We are interested in what happens when a particle moving from left to right encounters the step.

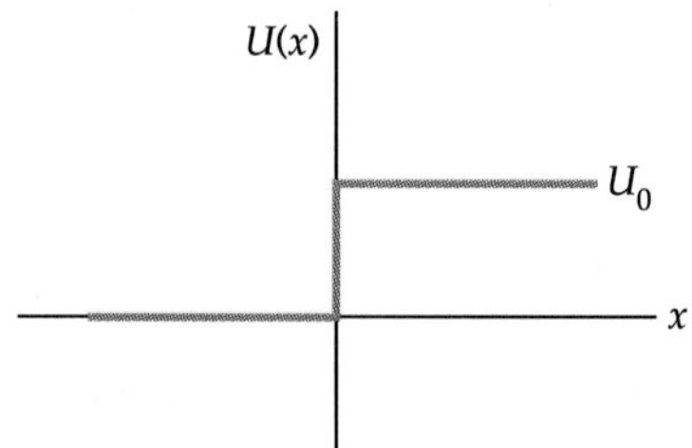

Figure 36-16 Step potential. A classical particle incident from the left, with total energy E greater than U_0, is always transmitted. The change in potential energy at $x = 0$ merely provides an impulsive force that reduces the speed of the particle. A wave incident from the left is partially transmitted and partially reflected because the wavelength changes abruptly at $x = 0$.

The classical answer is simple. To the left of the step, the particle moves with a speed $v = \sqrt{2E/m}$. At $x = 0$, an impulsive force acts on the particle. If the initial energy E is less than U_0, the particle will be turned around and will then move to the left at its original speed; that is, the particle will be reflected by the step. If E is greater than U_0, the particle will continue to move to the right but with reduced speed given by $v = \sqrt{2(E - U_0)/m}$. We can picture this classical problem as a ball rolling along a level surface and coming to a steep hill of height h given by $mgh = U_0$. If the initial kinetic energy of the ball is less than mgh, the ball will roll part way up the hill and then back down and to the left along the lower surface at its original speed. If E is greater than mgh, the ball will roll up the hill and proceed to the right at a lesser speed.

The quantum-mechanical result is similar when E is less than U_0. Figure 36-17 shows the wave function for the case $E < U_0$. The wave function does not go to zero at $x = 0$ but rather decays exponentially, like the wave function for the bound state in a finite square-well problem. The wave penetrates slightly into the classically forbidden region $x > 0$, but it is eventually completely reflected. This problem is somewhat similar to that of total internal reflection in optics.

For $E > U_0$, the quantum-mechanical result differs markedly from the classical result. At $x = 0$, the wavelength changes abruptly from $\lambda_1 = h/p_1 = h/\sqrt{2mE}$ to $\lambda_2 = h/p_2 = h/\sqrt{2m(E - U_0)}$. We know from optics that when the wavelength changes suddenly part of the wave is reflected and part is transmitted. Since the motion of an electron (or other particle) is governed by a wave equation, the electron will likewise be sometimes transmitted and sometimes reflected. The probabilities of reflection and transmission can be calculated by solving the Schrödinger equation in each region of space and comparing the amplitudes of the transmitted and reflected waves with that of the incident wave. This calculation and its result are similar to finding the fraction of light reflected from an air–glass surface. If R is the probability of reflection, called the **reflection coefficient,** this calculation gives

$$R = \frac{(k_1 - k_2)^2}{(k_1 + k_2)^2} \qquad 36\text{-}49$$

where k_1 is the wave number for the incident wave and k_2 is that for the transmitted wave. This result is the same as that in optics for the reflection of light at normal incidence from the boundary between two media having different indexes of refraction n (Equation 30-3). The probability of transmission T, called the **transmission coefficient,** can be calculated from the reflection coefficient since the probability of transmission plus the probability of reflection must equal 1:

$$T + R = 1 \qquad 36\text{-}50$$

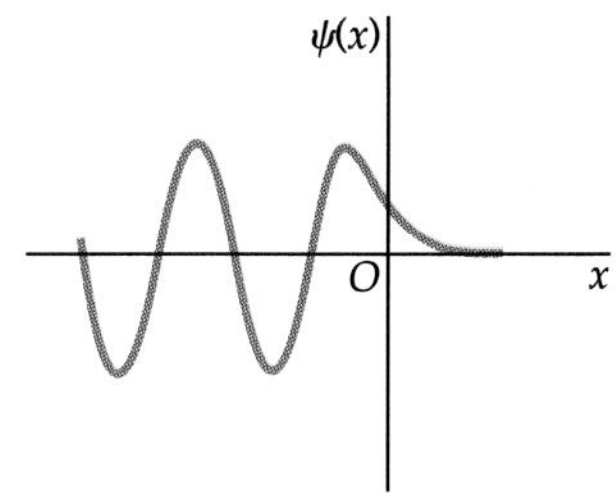

Figure 36-17 When the total energy E is less than U_0, the wave function penetrates slightly into the region $x > 0$. However, the probability of reflection for this case is 1, so no energy is transmitted.

Exercise

Express the index of refraction of light n in terms of the wave number k, and show that Equation 30-3 for the reflection of light at normal incidence is the same as Equation 36-49.

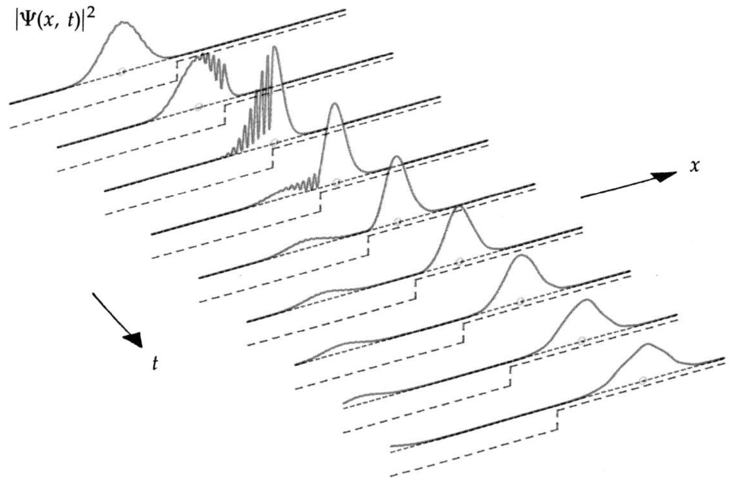

Time development of a one-dimensional wave packet representing a particle incident on a step potential for $E > U_0$. The position of a classical particle is indicated by the dot. Note that part of the packet is transmitted and part is reflected.

Barrier Penetration

Figure 36-18*a* shows a rectangular barrier potential of height U_0 and width a given by

$$U(x) = 0 \qquad x < 0$$
$$U(x) = U_0 \qquad 0 < x < a$$
$$U(x) = 0 \qquad x > a$$

We consider a particle of energy E, which is slightly less than U_0, that is incident on the barrier from the left. Classically, the particle would always be reflected. However, a wave incident from the left does not decrease immediately to zero at the barrier but will instead decay exponentially in the classically forbidden region $0 < x < a$. Upon reaching the far wall of the barrier ($x = a$), the wave function must join smoothly to a sinusoidal wave function to the right of the barrier as shown in Figure 36-18*b*. This implies that there is some probability of the particle (which is represented by the wave function) being found on the far side of the barrier even though, classically, it should never pass through the barrier. For the case in which the quantity $\alpha a = \sqrt{2ma^2(U_0 - E)/\hbar^2}$ is much greater than 1, the transmission coefficient is proportional to $e^{-2\alpha a}$:

Transmission through a barrier

$$T \propto e^{-2\alpha a} \qquad 36\text{-}51$$

with $\alpha = \sqrt{2m(U_0 - E)/\hbar^2}$.

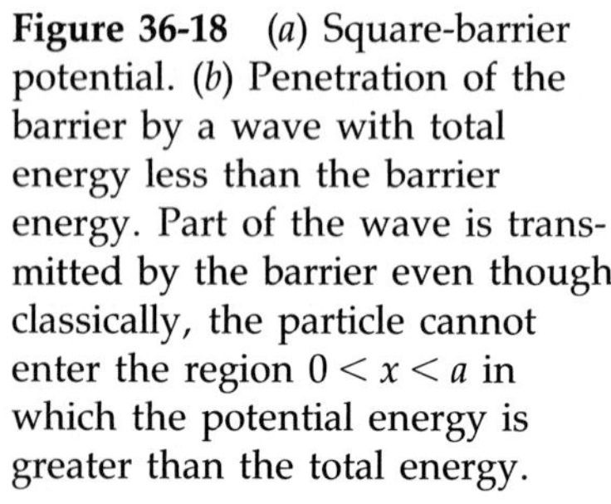

Figure 36-18 (*a*) Square-barrier potential. (*b*) Penetration of the barrier by a wave with total energy less than the barrier energy. Part of the wave is transmitted by the barrier even though, classically, the particle cannot enter the region $0 < x < a$ in which the potential energy is greater than the total energy.

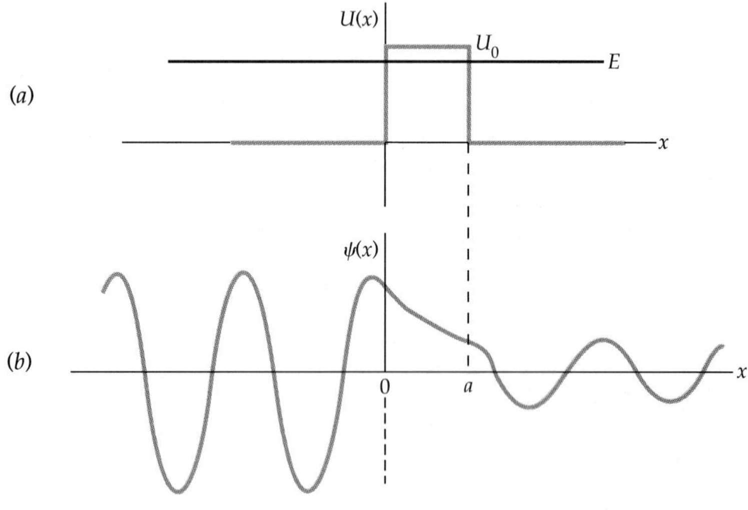

The probability of penetration of the barrier thus decreases exponentially with the barrier thickness a and with the square root of the relative barrier height $(U_0 - E)$.

The penetration of a barrier is not unique to quantum mechanics. In Figure 36-19, a light ray is incident on a glass–air surface. At an angle greater than the critical angle, total reflection occurs. But because of the wave nature of light, the electric field $\mathcal{E}$ does not immediately drop to zero at the surface but rather decreases exponentially and becomes negligible within a few wavelengths of the surface. If another piece of glass is brought near the surface—as shown in Figure 36-19—some of the light is transmitted across the barrier. (This effect can be demonstrated with a laser beam and two 45° prisms.) Similarly, Figure 36-20 shows barrier penetration by water waves in a ripple tank.

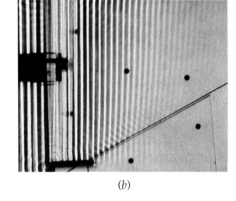

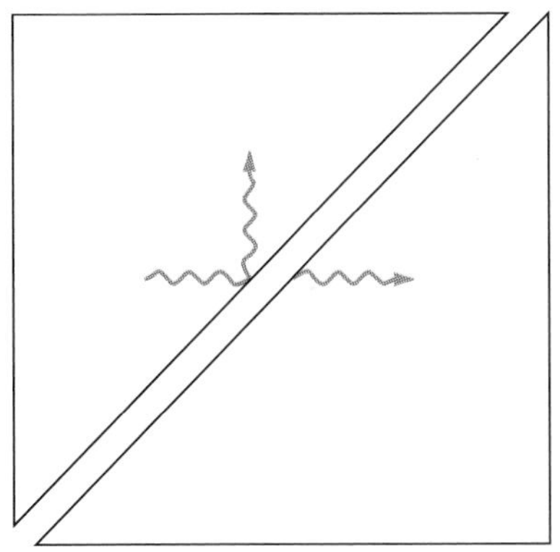

Figure 36-19 Penetration of an optical barrier. If the second prism is close enough to the first, part of the wave penetrates the air barrier even when the angle of incidence in the first prism is greater than the critical angle.

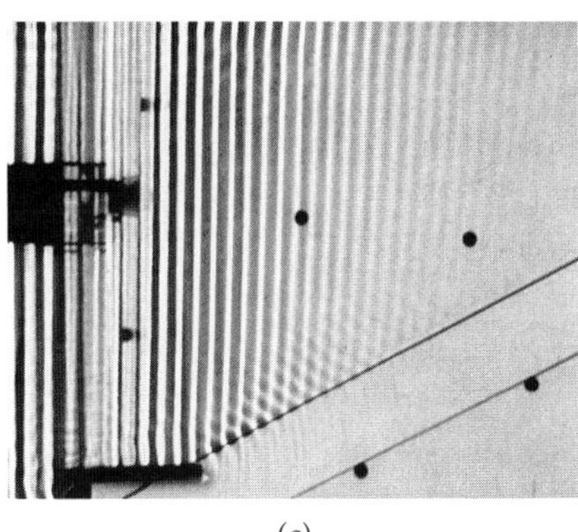

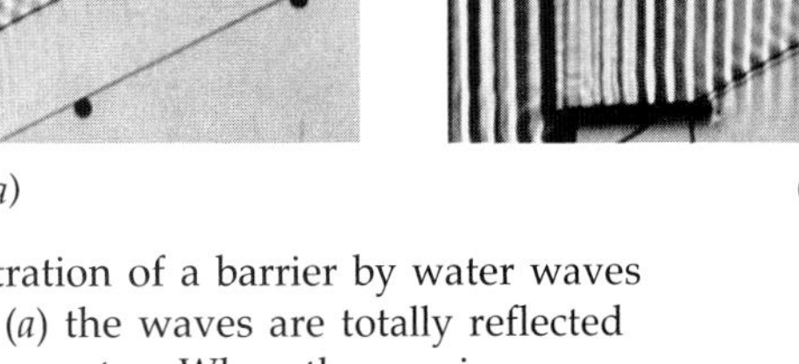

(*a*) (*b*)

Figure 36-20 Penetration of a barrier by water waves in a ripple tank. In (*a*) the waves are totally reflected from a gap of deeper water. When the gap is very narrow, as in (*b*), a transmitted wave appears.

The theory of barrier penetration was used by George Gamow in 1928 to explain the enormous variation in the half-lives for α decay of radioactive nuclei. (Alpha particles are helium nuclei, which consist of two protons and two neutrons tightly bound together.) In general, the smaller the energy of the emitted α particle, the longer the half-life. The energies of α particles from natural radioactive sources range from about 4 to 7 MeV, whereas the half-lives range from about 10^{-5} second to 10^{10} years. Gamow represented a radioactive nucleus by a potential well containing an α particle as shown in Figure 36-21. Without knowing very much about the nuclear force that is exerted on the α particle within the nucleus, he represented it by a square well. Just outside the well, the α particle with its charge of $+2e$ is repelled by the nucleus with its charge $+Ze$, where Ze is the remaining nuclear charge. This force is represented by the Coulomb potential energy $+k(2e)(Ze)/r$. The energy E is the measured kinetic energy of the emitted α particle, because when it is far from the nucleus its potential energy is zero. After the α particle is formed inside the radioactive nucleus, it bounces back and forth inside the nucleus, hitting the barrier at the nuclear radius R. Each time it strikes the barrier, there is some small probability of its penetrating and appearing outside the nucleus. We can see from Figure 36-21 that a small increase in E reduces the relative height of the barrier $U - E$ and also its thickness. Because the probability of penetration is so sensitive to the barrier thickness and relative height, a small increase in E leads to a large increase in the probability of transmission and therefore to a shorter lifetime. Gamow was able to derive an expression for the half-life as a function of E that is in excellent agreement with experimental results.

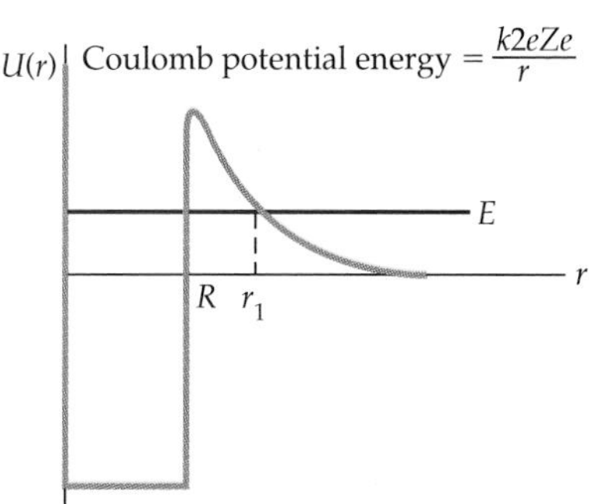

Figure 36-21 Model of a potential-energy function for an α particle in a radioactive nucleus. The strong attractive nuclear force when r is less than the nuclear radius R can be approximately described by the potential well shown. Outside the nucleus the nuclear force is negligible, and the potential is given by Coulomb's law, $U(r) = +k(2e)(Ze)/r$, where Ze is the nuclear charge and $2e$ is the charge of the α particle.

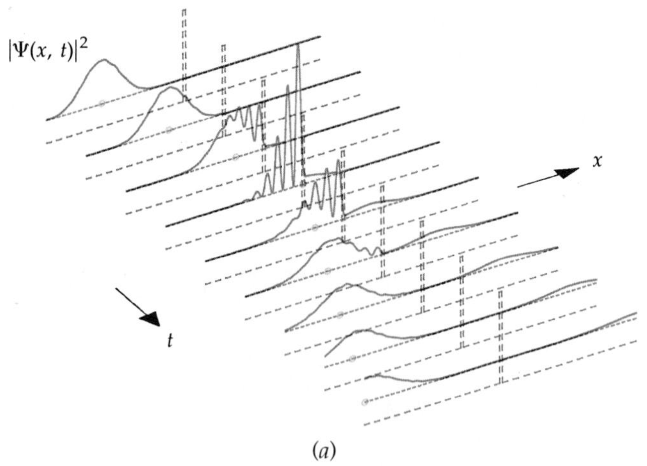

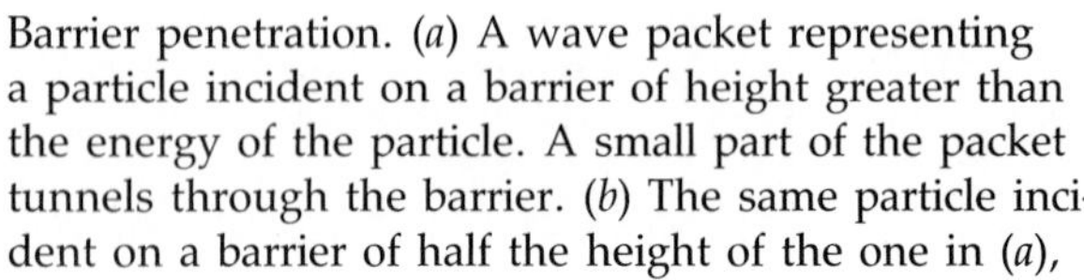

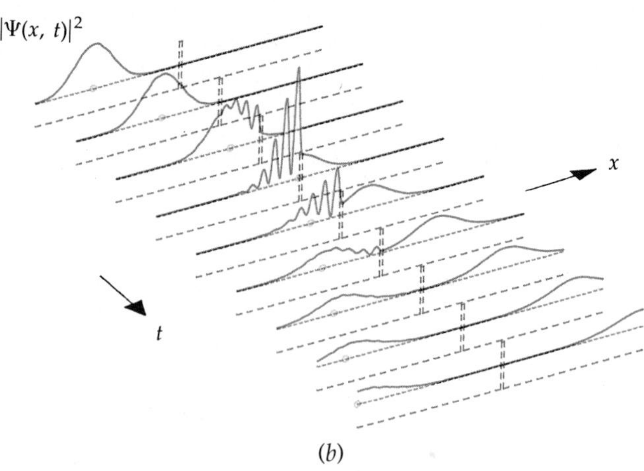

Barrier penetration. (*a*) A wave packet representing a particle incident on a barrier of height greater than the energy of the particle. A small part of the packet tunnels through the barrier. (*b*) The same particle incident on a barrier of half the height of the one in (*a*), but still greater than the energy of the particle. The probability of transmission is much greater as indicated by the size of the transmitted packet. In both drawings, the position of a classical particle is indicated by a dot.

In the **scanning tunneling electron microscope** developed in the 1980s, a thin space between a material specimen and a tiny probe acts as a barrier to electrons bound in the specimen. A small voltage applied between the probe and specimen causes the electrons to tunnel through the vacuum separating the two surfaces if the surfaces are close enough together. The tunneling current is extremely sensitive to the size of the gap between the probe and specimen. If a constant tunneling current is maintained as the probe scans the specimen, the surface of the specimen can be mapped out by the motions of the probe. In this way, the surface features of a specimen can be measured with a resolution of the order of the size of an atom.

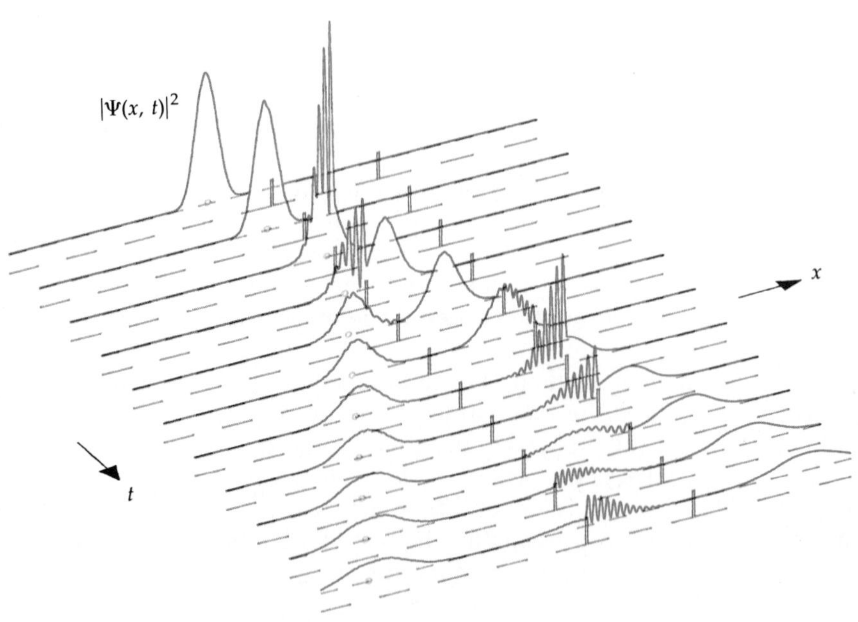

Wave packet representing a particle incident on two barriers. At each encounter, part of the packet is transmitted and part reflected, resulting in part of the packet being trapped between the barriers.

36-10 The Schrödinger Equation in Three Dimensions

The one-dimensional time-independent Schrödinger equation (Equation 36-30) is easily extended to three dimensions. In rectangular coordinates, it is

$$-\frac{\hbar^2}{2m}\left(\frac{\partial^2\psi}{\partial x^2}+\frac{\partial^2\psi}{\partial y^2}+\frac{\partial^2\psi}{\partial z^2}\right)+U\psi=E\psi \qquad 36\text{-}52$$

where the wave function ψ and the potential energy U are generally functions of all three coordinates, x, y, and z. To illustrate some of the features of problems in three dimensions, we consider a particle in a three-dimensional infinite square well given by $U(x, y, z) = 0$ for $0 < x < L$, $0 < y < L$, and $0 < z < L$. Outside this cubical region $U(x, y, z) = \infty$. For this problem, the wave function must be zero at the edges of the well. The solution of Equation 36-52 can be written as

$$\psi(x, y, z) = A \sin k_1x \sin k_2y \sin k_3z \qquad 36\text{-}53$$

where the constant A is determined by normalization. Inserting this solution into Equation 36-52, we obtain for the energy

$$E=\frac{\hbar^2}{2m}(k_1^2+k_2^2+k_3^2)$$

which is equivalent to $E = (p_x^2 + p_y^2 + p_z^2)/2m$, with $p_x = \hbar k_1$ and so on. The wave function will be zero at $x = L$ if $k_1 = n_1\pi/L$, where n_1 is an integer. Similarly, the wave function will be zero at $y = L$ if $k_2 = n_2\pi/L$, and it will be zero at $z = L$ if $k_3 = n_3\pi/L$. The energy is thus quantized to the values

$$E_{n_1,n_2,n_3}=\frac{\hbar^2\pi^2}{2mL^2}(n_1^2+n_2^2+n_3^2) \qquad 36\text{-}54$$

where n_1, n_2, and n_3 are integers. Note that the energy and wave function are characterized by three quantum numbers, each arising from a boundary condition for one of the coordinates.

The lowest energy state (the ground state) for the cubical well occurs when $n_1 = n_2 = n_3 = 1$ and has the value

$$E_{1,1,1}=\frac{3\hbar^2\pi^2}{2mL^2}$$

The first excited energy level can be obtained in three different ways: $n_1 = 2$, $n_2 = n_3 = 1$; $n_2 = 2$, $n_1 = n_3 = 1$; or $n_3 = 2$, $n_1 = n_2 = 1$. Each has a different wave function. For example, the wave function for $n_1 = 2$ and $n_2 = n_3 = 1$ is

$$\psi_{2,1,1}=A\sin\frac{2\pi x}{L}\sin\frac{\pi y}{L}\sin\frac{\pi z}{L} \qquad 36\text{-}55$$

An energy level with which more than one wave function is associated is said to be **degenerate.** In this case, there is threefold degeneracy. Degeneracy is related to the spatial symmetry of the problem. If, for example, we consider a noncubic well where $U = 0$ for $0 < x < L_1$, $0 < y < L_2$, and $0 < z < L_3$, the boundary conditions at the edges would lead to the quantum conditions $k_1L_1 = n_1\pi$, $k_2L_2 = n_2\pi$, and $k_3L_3 = n_3\pi$, and the total energy would be

$$E_{n_1,n_2n_3}=\frac{\hbar^2\pi^2}{2m}\left(\frac{n_1^2}{L_1^2}+\frac{n_2^2}{L_2^2}+\frac{n_3^2}{L_3^2}\right) \qquad 36\text{-}56$$

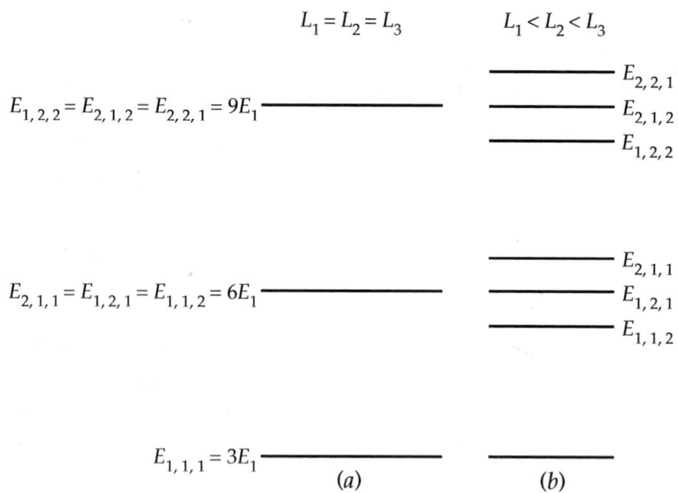

Figure 36-22 Energy-level diagrams for (*a*) a cubic infinite well and (*b*) a noncubic infinite well. In (*a*) the energy levels are degenerate; that is, there are two or more wave functions having the same energy. The degeneracy is removed when the symmetry of the potential is removed, as in (*b*).

These energy levels are not degenerate if L_1, L_2, and L_3 are not equal. Figure 36-22 shows the energy levels for the ground state and first two excited states for an infinite cubic well in which the excited states are degenerate and for a noncubic infinite well in which L_1, L_2, and L_3 are slightly different so that the excited levels are slightly split apart and the degeneracy is removed.

36-11 The Schrödinger Equation for Two Identical Particles

Our discussion of quantum mechanics has thus far been limited to situations in which a single particle moves in some force field characterized by a potential-energy function U. The most important physical problem of this type is the hydrogen atom, in which a single electron moves in the Coulomb potential of the proton nucleus. This problem, which we will consider in some detail in Chapter 37, is actually a two-body problem since the proton also moves in the field of the electron. However, the motion of the proton requires only a very small correction to the energy of the atom that is easily made in both classical and quantum mechanics. When we consider more complicated problems, such as the helium atom, we must apply quantum mechanics to two or more electrons moving in an external field. Such problems are complicated by the interaction of the electrons with each other and also by the fact that the electrons are identical.

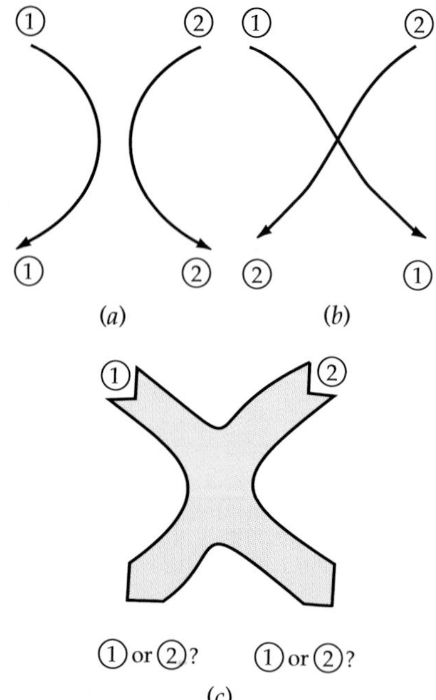

Figure 36-23 Two possible classical electron paths are shown in (*a*) and (*b*). The electrons can be distinguished classically. Because of the quantum-mechanical wave properties of the electrons, the paths are spread out, as indicated by the shaded region in (*c*). It is impossible to distinguish which electron is which after they separate.

The interaction of two electrons with each other is electromagnetic and is essentially the same as that expected classically for two charged particles. The Schrödinger equation for an atom with two or more electrons cannot be solved exactly, so approximation methods must be used. This is not very different from the situation in classical problems with three or more particles. However, the complications arising from the identity of electrons are purely quantum mechanical and have no classical counterpart. They are due to the fact that it is impossible to keep track of which electron is which. Classically, identical particles can be identified by their positions, which can be determined with unlimited accuracy. This is impossible quantum mechanically because of the uncertainty principle. Figure 36-23 offers a schematic illustration of the problem.

The indistinguishability of identical particles has important consequences. For instance, consider the very simple case of two identical, noninteracting particles in a one-dimensional infinite square well. The time-independent Schrödinger equation for two particles, each of mass m, is

$$-\frac{\hbar^2}{2m}\frac{\partial^2\psi(x_1, x_2)}{\partial x_1^2} - \frac{\hbar^2}{2m}\frac{\partial^2\psi(x_1, x_2)}{\partial x_2^2} + U\psi(x_1, x_2) = E\psi(x_1, x_2) \qquad 36\text{-}57$$

where x_1 and x_2 are the coordinates of the two particles. If the particles interact, the potential energy U contains terms with both x_1 and x_2 that cannot be separated into separate terms containing only x_1 or x_2. For example, the electrostatic repulsion of two electrons in one dimension is represented by the potential energy $ke^2/|x_2 - x_1|$. However, if the particles do not interact (as we are assuming here), we can write $U = U_1(x_1) + U_2(x_2)$. For the infinite square well, we need only solve the Schrödinger equation inside the well where $U = 0$, and require that the wave function be zero at the walls of the well. With $U = 0$, Equation 36-57 looks just like that for a particle in a two-dimensional well (Equation 36-52, with no z and with y replaced by x_2).

Solutions of this equation can be written in the form

$$\psi_{n,m} = \psi_n(x_1)\psi_m(x_2) \qquad 36\text{-}58$$

where ψ_n and ψ_m are the single-particle wave functions for a particle in an infinite well, and n and m are the quantum numbers of particles 1 and 2, respectively. For example, for $n = 1$ and $m = 2$, the wave function is

$$\psi_{1,2} = A \sin\frac{\pi x_1}{L} \sin\frac{2\pi x_2}{L} \qquad 36\text{-}59$$

The probability of finding particle 1 in dx_1 *and* particle 2 in dx_2 is $\psi_{n,m}^2(x_1, x_2)\, dx_1\, dx_2$, which is just the product of the separate probabilities $\psi_n^2(x_1)\, dx_1$ and $\psi_m^2(x_2)\, dx_2$. However, even though we have labeled the particles 1 and 2, we cannot distinguish which is in dx_1 and which is in dx_2 if they are identical. The mathematical descriptions of identical particles must be the same if we interchange the labels. The probability density $\psi^2(x_1, x_2)$ must therefore be the same as $\psi^2(x_2, x_1)$:

$$\psi^2(x_2, x_1) = \psi^2(x_1, x_2) \qquad 36\text{-}60$$

Equation 36-60 is satisfied if $\psi(x_2, x_1)$ is either **symmetric** or **antisymmetric** on the exchange of particles. That is, either

$$\psi(x_2, x_1) = \psi(x_1, x_2) \qquad \text{symmetric} \qquad 36\text{-}61$$

or

$$\psi(x_2, x_1) = -\psi(x_1, x_2) \qquad \text{antisymmetric} \qquad 36\text{-}62$$

Note that the wave functions given by Equations 36-58 and 36-59 are neither symmetric or antisymmetric. If we interchange x_1 and x_2 in these wave functions, we get a different wave function, which implies that the particles can be distinguished.

We can find symmetric and antisymmetric wave functions that are solutions of the Schrödinger equation by adding or subtracting $\psi_{n,m}$ and $\psi_{m,n}$. Adding them, we obtain

$$\psi_S = A'[\psi_n(x_1)\psi_m(x_2) + \psi_n(x_2)\psi_m(x_1)] \qquad \text{symmetric}$$

and subtracting them, we obtain

$$\psi_A = A'[\psi_n(x_1)\psi_m(x_2) - \psi_n(x_2)\psi_m(x_1)] \qquad \text{antisymmetric}$$

For example, the symmetric and antisymmetric wave functions for the first excited state of two identical particles in an infinite square well would be

$$\psi_S = A'\left[\sin\frac{\pi x_1}{L}\sin\frac{2\pi x_2}{L} + \sin\frac{\pi x_2}{L}\sin\frac{2\pi x_1}{L}\right]$$

and

$$\psi_A = A'\left[\sin\frac{\pi x_1}{L}\sin\frac{2\pi x_2}{L} - \sin\frac{\pi x_2}{L}\sin\frac{2\pi x_1}{L}\right]$$

There is an important difference between antisymmetric and symmetric wave functions. If $n = m$, the antisymmetric wave function is identically zero for all values of x_1 and x_2 whereas the symmetric wave function is not. Thus, if the wave function describing two identical particles is antisymmetric, the quantum numbers n and m of two particles cannot be the same. This is an example of the **Pauli exclusion principle,** which was first stated by Wolfgang Pauli for electrons in an atom:

Pauli exclusion principle

No two electrons in an atom can have the same quantum numbers.

(As we will see in Chapter 37, the state of an electron in an atom is described by four quantum numbers, one associated with each space coordinate and one associated with electron spin, which will be discussed in that chapter.) It is found that electrons, protons, neutrons, and some other particles have antisymmetric wave functions and obey the Pauli exclusion principle. These particles are called **fermions.** Other particles, such as α particles, deuterons, photons, and mesons, have symmetric wave functions and do not obey the Pauli exclusion principle. These particles are called **bosons.**

Summary

1. The state of a particle, such as an electron, is described by its wave function Ψ, which is the solution of the Schrödinger wave equation. The absolute square of the wave function $|\Psi|^2$ measures the probability of finding the particle in some region of space.
2. A harmonic wave of a single angular frequency ω and wave number k can represent an electron that is completely unlocalized and can be anywhere in space. In terms of the angular frequency and wave number, the de Broglie equations are

$$E = \hbar\omega$$

and

$$p = \hbar k$$

A localized electron can be represented by a wave packet, which is a group of waves of nearly equal frequencies and wavelengths. The wave packet moves with a group velocity

$$V_g = \frac{d\omega}{dk}$$

which equals the velocity of the electron.

3. Wave–particle duality leads to the uncertainty principle, which states that the product of the uncertainty in a measurement of the position of a particle and the uncertainty in a measurement of its momentum must be greater than $h/4\pi$, where h is Planck's constant:

$$\Delta x \ \Delta p \geq \frac{h}{4\pi} = \frac{1}{2}\hbar$$

Similarly, the uncertainty in the energy ΔE is related to the time interval Δt required to measure the energy by

$$\Delta E \ \Delta t \geq \tfrac{1}{2}\hbar$$

An important consequence of the uncertainty principle is that a particle confined in space has a minimum energy called the zero-point energy.

4. Light, electrons, neutrons, and all other carriers of momentum and energy exhibit both wave and particle properties. Everything propagates like a wave, exhibiting diffraction and interference, but exchanges energy in discrete lumps like a particle. Because the wavelengths of macroscopic objects are so small, diffraction and interference are not observed. Also, when a macroscopic amount of energy is exchanged, so many quanta are involved that the particle nature of the energy is not evident.

5. The wave function $\Psi(x, t)$ obeys the time-dependent Schrödinger equation

$$-\frac{\hbar^2}{2m}\frac{\partial^2\Psi(x, t)}{\partial x^2} + U(x)\Psi(x, t) = i\hbar\frac{\partial\Psi(x, t)}{\partial t}$$

For any wave function that describes a particle in a state of definite energy, the time-dependent Schrödinger equation can be simplified by writing the wave function in the form

$$\Psi(x, t) = \psi(x)e^{-i\omega t}$$

where $\psi(x)$ is a function of x only. This leads to the time-independent Schrödinger equation

$$-\frac{\hbar^2}{2m}\frac{d^2\psi(x)}{dx^2} + U(x)\psi(x) = E\psi(x)$$

In addition to satisfying the Schrödinger equation, a wave function $\psi(x)$ must be continuous and must have a continuous first derivative $d\psi/dx$. Since the probability of finding an electron somewhere must be 1, the wave function must obey the normalization condition

$$\int_{-\infty}^{\infty} |\psi|^2 \, dx = 1$$

This condition implies the boundary condition that ψ must approach 0 as x approaches $\pm\infty$. Such boundary conditions lead to the quantization of energy.

6. A one-dimensional box of length L is described by the infinite square-well potential

$$U(x) = 0 \qquad 0 < x < L$$

$$U(x) = \infty \qquad x < 0 \text{ or } x > L$$

The wavelength for a particle confined in such a box obeys the standing-wave condition

$$n\frac{\lambda}{2} = L$$

This results in the energy of the particle being quantized to the values

$$E_n = n^2 E_1$$

where E_1 is the ground-state energy given by

$$E_1 = \frac{h^2}{8mL^2}$$

The number n is called a quantum number.

The wave functions for a particle in a box are given by

$$\psi_n = \sqrt{\frac{2}{L}} \sin \frac{n\pi x}{L}$$

7. An electron in a stationary state can be pictured as a cloud of charge with charge density proportional to $|\psi|^2$.

8. When the quantum numbers of a system are very large, quantum calculations and classical calculations agree—a result known as Bohr's correspondence principle.

9. In a finite well of height U_0, there is a finite number of allowed energies, which are slightly less than the corresponding energies in an infinite well.

10. The expectation value of x is defined by

$$\langle x \rangle = \int x|\Psi(x, t)|^2 \, dx$$

The expectation value of x is the same as the average value of x that we would expect to obtain from a measurement of the positions of a large number of particles with the same wave function $\Psi(x, t)$.

For a particle in a state of definite energy, the expectation value of x is

$$\langle x \rangle = \int x|\psi(x)|^2 \, dx$$

The expectation value of any function $f(x)$ is given by

$$\langle f(x) \rangle = \int f(x)|\psi(x)|^2 \, dx$$

11. When the potential changes abruptly over a small distance, a particle may be reflected even though $E > U(x)$. A particle may also penetrate a region in which $E < U(x)$. Reflection and penetration of de Broglie waves are similar to those for other kinds of waves.

12. When more than one wave function is associated with the same energy level, the energy level is said to be degenerate. Degeneracy arises because of spatial symmetry.

13. A wave function that describes two identical particles must be either symmetric or antisymmetric when the coordinates of the particles are exchanged. Fermions, which include electrons, protons, and neutrons, are described by antisymmetric wave functions and obey the Pauli exclusion principle, which states that no two particles can have the same quantum number. Bosons, which include α particles, deuterons, photons, and mesons, have symmetric wave functions and do not obey the Pauli exclusion principle.

Scanning Tunneling Microscopy

Ellen D. Williams
University of Maryland

All scientists believe that matter consists of atoms. However, the evidence for the existence of atoms is mostly indirect. There are only three known methods for actually imaging individual atoms and thus studying their behavior directly. These are transmission electron microscopy, field ion microscopy, and scanning tunneling microscopy. Of these three, the last is the most recently developed and the most versatile. Gert Binnig and Heinrich Rohrer were awarded the Nobel Prize in Physics in 1986 for the development of this technique, jointly with Ernst Ruska who was honored for his work on the development of electron microscopy.

To understand scanning tunneling microscopy (STM), we first need to understand a little bit about the behavior of electrons in metals (or any other electrically conducting material). A metal consists of a large number of atoms, which are held together by the electrostatic forces acting between the electrons and the nuclei of the atoms. Most of the electrons are bound tightly to the individual nuclei, just as in the case of isolated atoms. However, the electrons that are farthest from the nuclei feel a relatively weak electrostatic attraction, and thus are free to wander about in the space between the nuclei. These are called the conduction electrons because they are the ones that carry (or conduct) electric current. It is a pretty good approximation to treat these electrons as if they are moving in a nearly constant attractive potential, so that they behave very much like particles in a one-dimensional box. We draw an energy-level diagram similar to Figure 36-7 for the conduction electrons as shown in Figure 1.

Continued

Ellen Williams was born in Oshkosh, Wisconsin (so had a connection with Paul Tipler right from the start, see page xv). After completing her BS in chemistry at Michigan State University, she earned her doctorate at the California Institute of Technology, studying the properties of thin layers of molecules on surfaces. Since then, she has been at the Department of Physics and Astronomy at the University of Maryland, where her work in surface physics has continued. This has focused on investigation of the structures and properties of very thin layers of metals on silicon. Enhanced understanding of this is crucial to the development of ultrafast microchip technology and other applications. The use of scanning tunneling microscopy has been of central importance in her research.

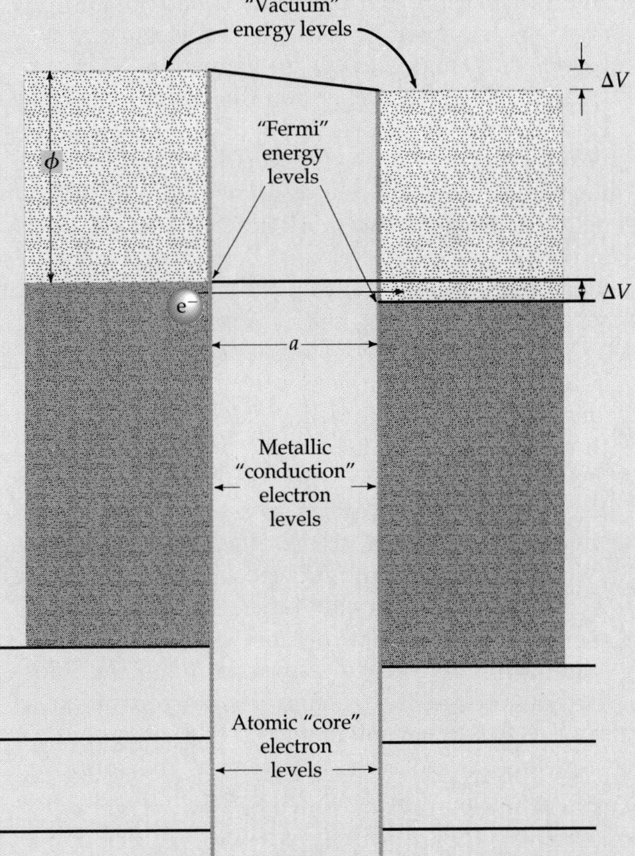

Figure 1 Energy-level diagram for the conduction electrons in two metals separated by a distance a. The electrons move as essentially free particles within the metal. The most weakly bound electrons have the highest energy (the Fermi energy E_F) and are held within the solid by a potential-energy barrier known as the work function ϕ. In order to induce a measurable electron tunneling current, a small voltage difference ΔV is applied between the two metals. This offsets their energy levels as shown, so that electrons can flow from the occupied states near E_F on the more negative metal (on the left in the figure) to the unoccupied states just above E_F on the more positive metal (on the right in the figure).

Because there are large numbers of levels ($\sim 10^{23}$), they overlap to provide a continuous distribution of states available to the conduction electrons. Only the lower energy levels are occupied by electrons. The energy of the most weakly bound electrons is called the Fermi energy (E_F). The electrons at the Fermi energy are held in the metal by an energy barrier of about 5 eV, the work function (ϕ) described in Section 35-2. Classically, these electrons can never leave the metal unless they are given the energy necessary to go over this potential barrier. Quantum mechanically, however, electrons near the Fermi energy can tunnel through the potential barrier, as discussed in Sections 36-4 to 36-9. By placing two pieces of metal close to one another, as shown in Figure 1, a finite square-well barrier like that shown in Figure 36-18 can be created. The probability for the electrons at the Fermi energy to tunnel through the barrier is proportional to $e^{-\alpha a}$, where a is the distance separating the two pieces of metal and α depends on the barrier height (in this case the work function), as discussed in Section 36-9. As explained below, this exponential dependence of the transmission probability on separation is what makes STM possible.

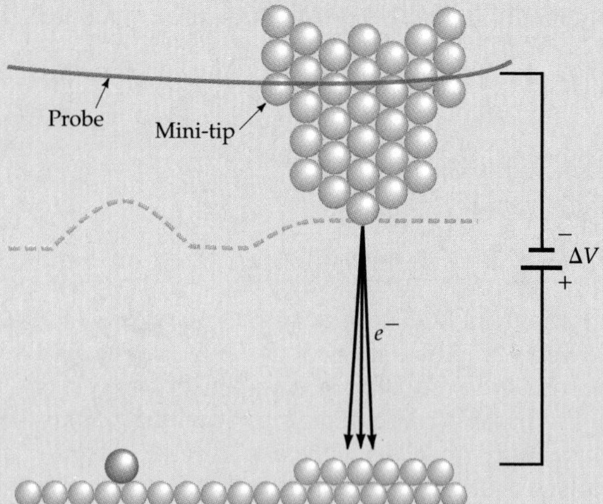

Figure 2 Schematic illustration of the path of a probe (dashed line) scanned across a surface while maintaining constant tunneling current. If the probe is very large (as illustrated by the solid line), tunneling occurs over a large area, and atomic features cannot be resolved. However, if the probe has a mini-tip of atomic dimensions, then tunneling occurs into a small area, allowing very small features (even individual atoms) to be imaged.

The mechanism of STM is illustrated in Figure 2. If a pointed metal probe is placed sufficiently close to a sample and a small voltage (say ΔV about 10 mV) is applied between the probe and sample, then electron tunneling can occur. The net flow of electrons can be measured as a tunneling current, which is proportional to the transmission probability. If we then scan the probe back and forth above the sample, any bumps on the sample surface will change the separation. Because of the exponential relationship between the separation and the transmission probability, changes in the separation as small as 0.01 nm result in measurable changes in the tunneling current. Measurement of the tunneling current while scanning thus generates a topographic map of the surface. Thus, in principle, it is possible to measure the topography of the surface using STM. In practice, formidable experimental problems arise in trying to image individual atoms on the surface. The challenges lie in three general areas: vibrations, tip sharpness, and position control.

Vibrations are important because the separation between the sample and probe must be very small. For a work function of around 5 eV, a separation of only a few nanometers (comparable to the size of atoms) is needed. For such a small separation, a minor perturbation such as vibrations set up by a sneeze can jam the probe right into the sample, ruining the experiment. The most common source of vibrations is motion of the floor, which typically has an amplitude of about 1 μm—a thousand times larger than the tip–sample separation that must be maintained. Thus, very careful engineering is required to make the instrument rigid and to isolate it from these external disturbances.

The second problem—probe sharpness—determines how small a structure can be imaged on a surface. Electrochemical etching can be used to sharpen the end of a metal wire to a radius of about 1 μm (1000 nm). A probe with such a large surface area would allow tunneling to occur over a large region of the sample surface. In order to resolve small features such as atoms, it is necessary to have a probe comparable in size to the features. In trying to fabricate such a probe, we are rewarded for not being able to do a perfect job in making micron-radius probes. The metal wires that we polish electrochemically are rough on an atomic scale: Their surfaces bear many mini-tips like the one illustrated in Figure 2. The end of such a mini-tip will present a single atom (perhaps a few) close to the surface. The exponential dependence of transmission probability on separation then guarantees that tunneling will occur preferentially from the end of the mini-tip.

The third problem in STM is that of position control. How is it possible to move the probe

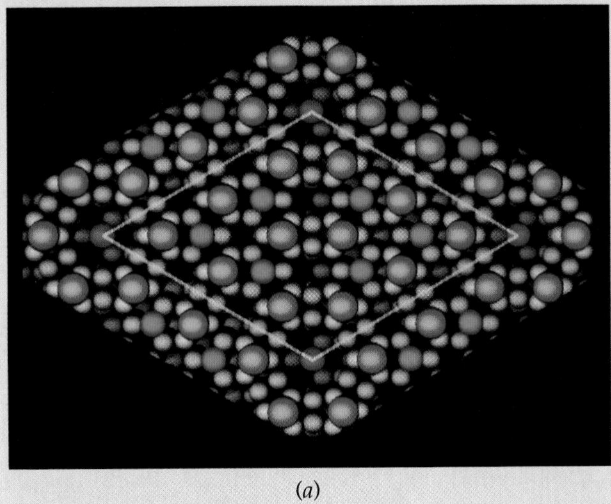

(*a*)

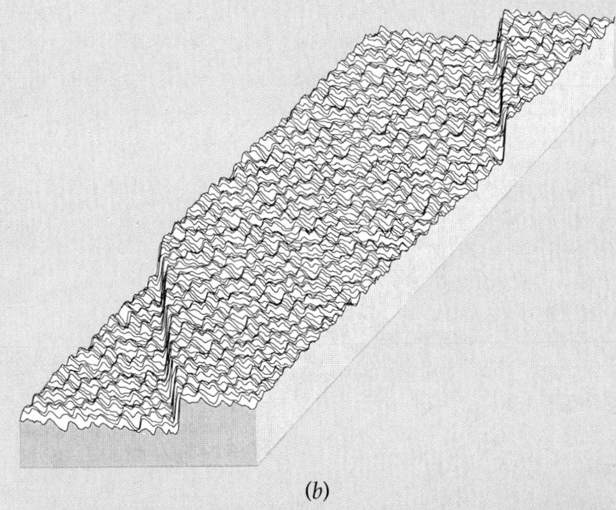

(*b*)

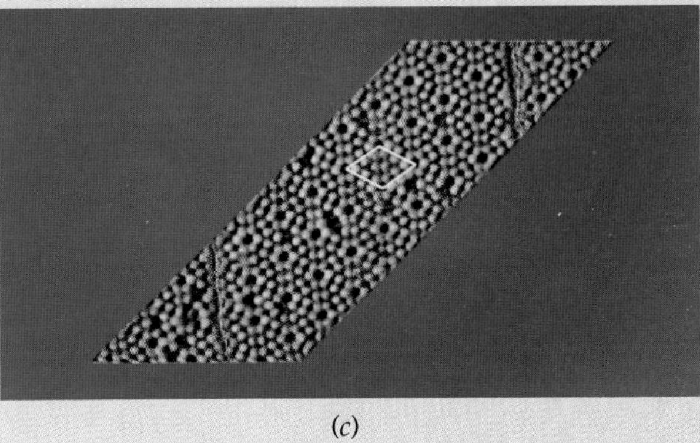

(*c*)

Figure 3 Scanning tunneling microscope used to image atoms and steps on a silicon surface. (*a*) A model of the atoms on a Si surface. The red circles represent atoms that protrude highest above the surface, the blue circles are atoms in a lower layer, and the gray circles are atoms in a still lower layer. The white diamond shows the repeating unit of the structure. The length of each side of this unit cell is 2.7 nm. (*b*) The traces of height versus position (line scans) obtained by an STM over an area of approximately 10 nm by 35 nm. (*c*) The data shown in (*b*) presented in a gray-scale representation, allowing the imaging of the highest layer of atoms [see red circles in the atomic model of (*a*)] to be immediately recognized. The unit cell of the structure is indicated for comparison with (*a*).

around with controllable displacements of less than 0.1 nm? The answer lies in a special type of material known as piezoelectric ceramic. This material expands and contracts when an external voltage is applied to electrodes on opposite faces. Typically, expansions are on the order of a few tenths of a nanometer per applied volt. As a result, a probe attached to a piece of piezoelectric ceramic can be moved with great precision by application of external voltages.

When Binnig and Rohrer first demonstrated that these challenges could be surmounted, it generated tremendous excitement because it opened the possibility of answering fundamental questions about the properties of surfaces, as well as a wealth of potential practical applications. The power of STM is illustrated in Figure 3, where a model of the atomic structure of atoms on the surface of silicon is compared with STM images of the real surface. The data from an STM scan of the surface consist of values of the surface height versus position. This can be immediately presented in the form of a line scan, as shown by the dashed line in Figure 2. The line scan image of Figure 3*b* is easier to visualize if the data are represented by a gray scale as in Figure 3*c*. Here, the height at each point is represented by the intensity of color—ranging from white for the highest points to black for the lowest—showing a striking correspondence to the atomic model in Figure 3*a*. The deep holes correspond to the positions of missing atoms in the model, and the bright spots are due to the atoms that protrude above the average surface plane. There are also two lines where the surface abruptly changes height in this image. These surface steps are important in practical processes such as crystal growth and microfabrication.

In addition to fundamental studies of the physics of atoms at surfaces, STM has a range of potentially practical applications, in part because STM is quite insensitive to its microscopic environment. For the description of tunneling presented above, the material in the gap between the sample and probe is not too important. Tunneling microscopes operate in vacuum, air, liquid helium, oil, water, and even in electrolytic solutions. This makes it possible to

Continued

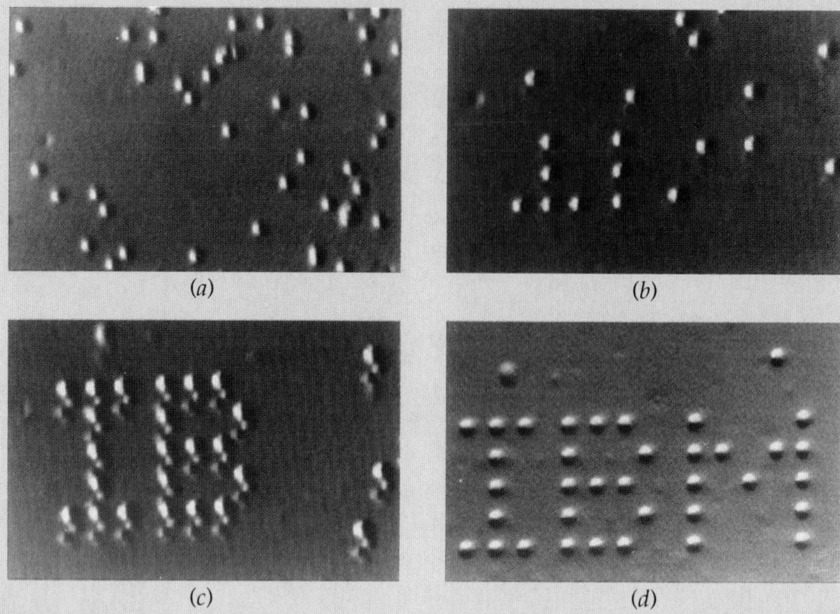

Figure 4 A sequence of STM images taken during the construction of a patterned array of xenon atoms on a nickel surface at a temperature of 4 K. Xenon atoms were allowed to stick randomly on the surface from the gas phase (upper left). The STM tip was then used to "pull" the atoms one by one across the surface to spell out the name of the company that sponsored the development of STM.

apply the STM to such important problems as imaging DNA in a biological environment and observing the surfaces of battery electrodes while they are operating. Variations of STM have also been developed capable of imaging samples that are not conductors (atomic force microscopy) and of imaging the magnetic properties at surfaces. Perhaps the most stunning possibility is that of using STM to write with atomic resolution. Features a few nanometers wide have been written by using the probe to scratch or dent the surface directly or by using the tunneling current to heat the surface. However, the ultimate limit of resolution has been demonstrated by using the probe to pull *individual atoms* of xenon around on a surface to spell out a message, as illustrated in Figure 4.

Scanning tunneling microscopy is a practical demonstration of quantum mechanics and an illustration that understanding of basic concepts of physics can yield tremendous gains in advanced technology. It is also an object lesson in the long-term and often unforeseeable benefits that accrue from developing fundamental ideas. The scientists who first explored the physical possibility of tunneling during the early part of this century would be amazed and delighted to see its application in STM.

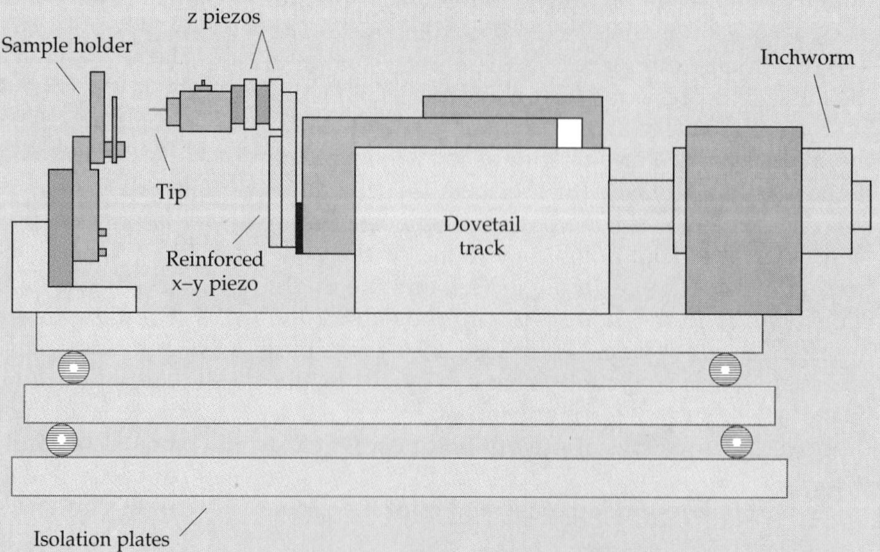

Figure 5 Schematic diagram of an STM. The sample holder is rigidly mounted to the top plate in a stack of isolation plates. The tip is fixed onto the x, y, z scanning piezos. In order to position the tip so tunneling can occur, these piezos are mounted on a heavy block that can slide on a dovetail track. The block is then pushed by an electronically controlled "walker" device called an inchworm, which can move forward and back in steps of 4 nm.

Suggestions for Further Reading

Everhart, Thomas E., and Thomas L. Hayes: "The Scanning Electron Microscope," *Scientific American*, January 1972, p. 54.

This article describes how the interaction between a beam of high-energy electrons and matter is used by the scanning electron microscope to create an image of three-dimensional appearance.

Hey, Tony, and Patrick Walters: *The Quantum Universe*, Cambridge University Press, Cambridge, 1987.

This "coffee-table quantum mechanics" book examines some of the many fields in which quantum theory has had an impact.

Rae, Alastair: *Quantum Physics: Illusion or Reality?*, Cambridge University Press, Cambridge, 1986.

This small book seeks to explain to the general reader the conceptual problems raised by quantum physics.

Shimony, Abner: "The Reality of the Quantum World," *Scientific American*, January 1988, p. 46.

This article explains why "common sense" (including Einstein's!) is unsuccessful in predicting the outcome of some recent experiments.

Review

A. Objectives: After studying this chapter, you should:

1. Know the properties of wave packets and, in particular, the relations between Δx and Δk, and between $\Delta\omega$ and Δt.

2. Be able to discuss the uncertainty principle and some of its consequences.

3. Be able to discuss particle–wave duality.

4. Be able to write the time-dependent and time-independent Schrödinger equations.

5. Be able to solve the time-independent Schrödinger equation for the infinite square-well problem and discuss how energy quantization arises.

6. Be able to sketch $\psi(x)$ and $\psi^2(x)$ for the infinite and finite square-well potentials.

7. Be able to draw an energy-level diagram for the infinite square-well potential.

8. Be able to state Bohr's correspondence principle.

9. Know how expectation values are calculated from the wave function for a particular state.

10. Be able to discuss qualitatively the reflection and transmission of waves at a barrier.

11. Know how quantum numbers arise in the solution of the Schrödinger equation in more than one dimension.

12. Be able to discuss the general features of the solution of the Schrödinger equation for two particles in an infinite square well.

B. Define, explain, or otherwise identify:

Probability distribution function
Group velocity
Uncertainty principle
Zero-point energy
Wave–particle duality
Classical particle
Classical wave
Time-dependent Schrödinger equation
Time-independent Schrödinger equation
Normalization condition
Infinite square-well potential
Boundary condition
Quantum number
Bohr's correspondence principle
Expectation value
Reflection coefficient
Transmission coefficient
Barrier penetration
Scanning tunneling electron microscope
Degeneracy
Symmetric wave function
Antisymmetric wave function
Pauli exclusion principle
Fermion
Boson

C. True or false: If the statement is true, explain why it is true. If it is false, give a counterexample.

1. The velocity of an electron is the same as the phase velocity of the wave describing it.

2. It is impossible in principle to know precisely the position of an electron.

3. A particle that is confined to some region of space cannot have zero energy.

4. All phenomena in nature are adequately described by classical wave theory.

5. The Schrödinger equation follows from Newton's laws of motion.

6. Boundary conditions on the wave function lead to energy quantization.

7. The expectation value of a quantity is the value that you expect to measure.

8. The penetration of a barrier by a wave has no physical significance.

9. Bosons do not obey the Pauli exclusion principle.

Problems

Level I

36-1 The Electron Wave Function

1. A 100-g rigid sphere of radius 1 cm has a kinetic energy of 2 J and is confined to move in a force-free region between two rigid walls separated by 50 cm. (*a*) What is the probability of finding the center of the sphere exactly midway between the two walls? (*b*) What is the probability of finding the center of the sphere between the 24.9- and 25.1-cm marks?

2. For the sphere described in Problem 1, make a graph of the classical probability density $P(x)$ as a function of position between the walls.

36-2 Electron Wave Packets

3. The wave function describing a state of an electron confined to move along the x axis is given at time zero by $\psi(x, 0) = Ae^{-x^2/4\sigma^2}$. Find the probability of finding the electron in a region dx centered at (*a*) $x = 0$, (*b*) $x = \sigma$, and (*c*) $x = 2\sigma$. (*d*) Where is the electron most likely to be found?

4. At time zero, three finite waves have the form cos kx with the wave numbers given by $k_1 - \delta$, k_1, and $k_1 + \delta$. The three waves are chopped so that there respectively are 5, 6, and 7 wavelengths in the region of space $-L < x < L$. The amplitude of the wave for $k = k_1$ is twice the amplitude of the other two. They are in phase at $x = 0$ but interfere destructively at the ends of the wave packet. Make a careful graph of each of the waves and of the resultant of the three chopped wave trains.

5. What is the approximate value of $\Delta x\ \Delta k$ for the wave packet formed in Problem 4?

6. An electron that is not localized in space is described by the wave function $\psi = A \sin(kx - \omega t)$. The kinetic energy of the electron is 1 keV. Find k and ω.

36-3 The Uncertainty Principle

7. Suppose a one-dimensional wave packet represents the state of a 1-ng particle moving along the x axis. The velocity of the wave packet is 10^{-2} m/s with a statistical spread of $\pm 10^{-4}$ m/s. What is the minimum corresponding statistical spread in the position predicted for the particle?

8. If the uncertainty in the position of a wave packet representing the state of a quantum system particle is equal to its de Broglie wavelength, how does the uncertainty in momentum compare with the value of the momentum of the particle?

9. Suppose that the wave function describing the state of an electron predicted a statistical spread in the velocity of 10^{-5} m/s. What is the corresponding statistical spread in the position of the electron?

10. If an excited state of an atom is known to have a lifetime of 10^{-7} s, what is the statistical spread in the energy of photons emitted by such atoms in the spontaneous decay to the ground state?

11. A wave pulse of frequency f_0 has a duration of Δt. It travels with speed v and occupies a region of space $\Delta x = v\ \Delta t$ as shown in Figure 36-24. Let N be the approximate number of waves in Δx. (*a*) How are N, f_0, and Δt related? (*b*) What is the approximate uncertainty in the wave number k?

Figure 36-24 Problem 11.

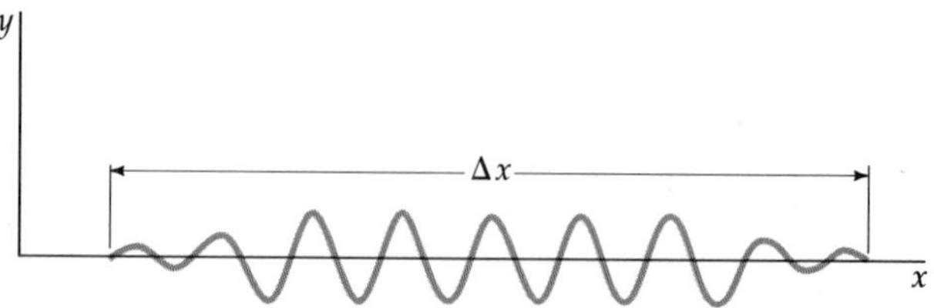

36-4 Wave–Particle Duality

12. What is the de Broglie wavelength of a neutron with speed 10^6 m/s?

13. Suppose you have a spherical object of mass 4 g moving at 100 m/s. What size aperture is necessary for the object to show diffraction? Show that no common objects would be small enough to squeeze through such an aperture.

14. A neutron has a kinetic energy of 10 MeV. What size object is necessary to observe neutron diffraction effects? Is there anything in nature of this size that could serve as a target to demonstrate the wave nature of 10-MeV neutrons?

15. What is the de Broglie wavelength of an electron accelerated from rest through a potential difference of 200 volts? What are some common targets that could demonstrate the wave nature of such an electron?

36-5 The Schrödinger Equation

16. Show that the wave function $\Psi(x, t) = Ae^{kx-\omega t}$ does not satisfy the time-dependent Schrödinger equation.

17. Show that $\Psi(x, t) = Ae^{i(kx-\omega t)}$ satisfies both the time-dependent Schrödinger equation and the classical wave equation (Equation 36-24).

18. In a region of space, a particle has a wave function given by $\psi(x) = Ae^{-x^2/2L^2}$, and energy $\hbar^2/2mL^2$, where L is some length. (*a*) Find the potential energy as a function of x, and sketch U versus x. (*b*) What is the classical potential that has this dependence?

19. (*a*) For Problem 18, find the kinetic energy as a function of x. (*b*) Show that $x = L$ is the classical turning point. (*c*) The potential energy of a simple harmonic oscillator in terms of its angular frequency ω is given by $U = \frac{1}{2}m\omega^2x^2$. Compare this with your answer to part (*a*) of Problem 18, and show that the total energy for this wave function can be written $E = \frac{1}{2}\hbar\omega$.

36-6 A Particle in a Box

20. A particle of mass m is confined to a tube of length L.

(*a*) Use the uncertainty relationship to estimate the smallest possible energy. (*b*) Assume that the inside of the tube is a force-free region and that the particle makes elastic reflections at the tube ends. Use Schrödinger's equation to find the ground-state energy for the particle in the tube. Compare the answer to that of part (*a*).

21. (*a*) What is the wavelength associated with the particle of Problem 20 if the particle is in its ground state? (*b*) What is the wavelength if the particle is in its second excited state (quantum number $n = 3$)? (*c*) Use de Broglie's relationship to find the magnitude for the momentum of the particle in its ground state. (*d*) Show that $p^2/2m$ gives the correct energy for the ground state of this particle in the box.

22. Sketch the wave function $\psi(x)$ and the probability distribution $\psi^2(x)$ for the state $n = 4$ of the infinite square-well potential given by Equation 36-32.

23. A particle is in a box of size L. Calculate the ground-state energy if (*a*) the particle is a proton and $L = 0.1$ nm, a typical size for a molecule; (*b*) the particle is a proton and $L = 1$ fm, a typical size for a nucleus.

24. A particle is in the ground state of an infinite square-well potential given by Equation 36-32. Find the probability of finding the particle in the interval $\Delta x = 0.002\ L$ at (*a*) $x = L/2$, (*b*) $x = 2L/3$, and (*c*) $x = L$. (Since Δx is very small, you need not do any integration.)

25. Do Problem 24 for a particle in the first excited state ($n = 2$) of an infinite square-well potential.

26. Do Problem 24 for a particle in the second excited state ($n = 3$) of an infinite square-well potential.

27. A mass of 10^{-6} g is moving with a speed of about 10^{-1} cm/s in a box of length 1 cm. Treating this as a one-dimensional infinite square-well potential, calculate the approximate value of the quantum number n.

28. (*a*) For the classical particle of Problem 27, find Δx and Δp, assuming that $\Delta x/L = 0.01$ percent and $\Delta p/p = 0.01$ percent. (*b*) What is $(\Delta x\ \Delta p)/\hbar$?

36-7 A Particle in a Finite Square Well

29. Sketch (*a*) the wave function and (*b*) the probability distribution for the $n = 4$ state for the finite square-well potential.

30. Sketch (*a*) the wave function and (*b*) the probability distribution for the $n = 5$ state for the finite square-well potential.

36-8 Expectation Values

31. Find (*a*) $\langle x \rangle$ and (*b*) $\langle x^2 \rangle$ for the first excited state ($n = 2$) in an infinite square-well potential.

32. Find (*a*) $\langle x \rangle$ and (*b*) $\langle x^2 \rangle$ for the second excited state ($n = 3$) in an infinite square-well potential.

33. (*a*) Show that the classical probability distribution function for a particle in a one-dimensional infinite square-well potential of length L is given by $P(x) = 1/L$. (*b*) Use your result in (*a*) to find $\langle x \rangle$ and $\langle x^2 \rangle$ for a classical particle in such a well.

36-9 Reflection and Transmission of Electron Waves: Barrier Penetration

34. A free particle of mass m with wave number k_1 is traveling to the right. At $x = 0$, the potential jumps from zero to U_0 and remains at this value for positive x. (*a*) If the total energy is $E = \hbar^2 k_1^2/2m = 2U_0$, what is the wave number k_2 in the region $x > 0$? Express your answer in terms of k_1 and in terms of U_0. (*b*) Calculate the reflection coefficient R at the potential step. (*c*) What is the transmission coefficient T? (*d*) If one million particles with wave number k_1 are incident upon the potential step, how many particles are expected to continue along in the positive x direction? How does this compare with the classical prediction?

35. Suppose that the potential jumps from zero to $-U_0$ at $x = 0$ so that the free particle speeds up instead of slowing down. The wave number for the incident particle is again k_1, and the total energy is $2U_0$. (*a*) What is the wave number for the particle in the region of positive x? (*b*) Calculate the reflection coefficient R at $x = 0$. (*c*) What is the transmission coefficient T? (*d*) If one million particles with wave number k_1 are incident upon the potential step, how many particles are expected to continue along in the positive x direction? How does this compare with the classical prediction?

36. Work Problem 34 for the case in which the energy of the incident particle is $1.01U_0$ instead of $2U_0$.

Section 36-10 The Schrödinger Equation in Three Dimensions

37. A particle is confined to a three-dimensional box that has sides L_1, $L_2 = 2L_1$, and $L_3 = 3L_1$. Give the quantum numbers n_1, n_2, n_3 that correspond to the lowest ten energy levels of this box.

38. Give the wave functions for the lowest ten energy levels of the particle in Problem 37.

39. (*a*) Repeat Problem 37 for the case $L_2 = 2L_1$ and $L_3 = 4L_1$. (*b*) What quantum numbers correspond to degenerate energy levels?

40. Give the wave functions for the lowest ten energy levels of the particle in Problem 39.

41. A particle moves in a potential well given by $U(x, y, z) = 0$ for $-L/2 < x < L/2$, $0 < y < L$, and $0 < z < L$, and $U = \infty$ outside these ranges. (*a*) Write an expression for the ground-state wave function for this particle. (*b*) How do the allowed energies compare with those for a box having $U = 0$ for $0 < x < L$, rather than for $-L/2 < x < L/2$?

Section 36-11 The Schrödinger Equation for Two Identical Particles

42. Show that Equation 36-59 satisfies Equation 36-57 with $U = 0$, and find the energy of this state.

43. What is the ground-state energy of ten noninteracting bosons in a one-dimensional box of length L?

44. What is the ground-state energy of ten noninteracting fermions, such as neutrons, in a one-dimensional box of length L? (Because of the quantum number associated with spin to be discussed in Chapter 37, each spatial state can hold two neutrons.)

Level II

45. A proton is in an infinite square-well potential given by Equation 36-32 with $L = 1$ fm. (*a*) Find the ground-state energy in MeV. (*b*) Make an energy-level diagram for this system. Calculate the wavelength of the photon emitted for the transitions (*c*) $n = 2$ to $n = 1$, (*d*) $n = 3$ to $n = 2$, and (*e*) $n = 3$ to $n = 1$.

46. Suppose that the bottom of the infinite potential well of Equation 36-32 is changed from zero potential to $+U_0$. Solve the time-independent Schrödinger equation, and find the wave functions and allowed energies.

47. A particle is in the ground state of an infinite square-well potential given by Equation 36-32. Calculate the probability that the particle will be found in the region (*a*) $0 < x < \frac{1}{2}L$, (*b*) $0 < x < \frac{1}{3}L$, and (*c*) $0 < x < \frac{3}{4}L$.

48. (*a*) Show that for large n, the fractional difference in energy between state n and state $n + 1$ for a particle in an infinite square well is given approximately by

$$\frac{E_{n+1} - E_n}{E_n} \approx \frac{2}{n}$$

(*b*) What is the approximate percentage energy difference between the states $n_1 = 1000$ and $n_2 = 1001$? (*c*) Comment on how this result is related to Bohr's correspondence principle.

49. A particle is in the first excited state ($n = 2$) of an infinite square-well potential given by Equation 36-32. (*a*) Sketch $\psi^2(x)$ versus x for this state. (*b*) What is the expectation value $\langle x \rangle$ for this state? (*c*) What is the probability of finding the particle in some small region dx centered at $x = \frac{1}{2}L$? (*d*) Are your answers for (*b*) and (*c*) contradictory? If not, explain.

50. A particle of mass m moves in a region in which the potential energy is constant $U(x) = U_0$. (*a*) Show that neither $\Psi(x, t) = A \sin (kx - \omega t)$ nor $\Psi(x, t) = A \cos (kx - \omega t)$ satisfies the time-dependent Schrödinger equation. (*Hint:* If $C_1 \sin \phi + C_2 \cos \phi = 0$ for all values of ϕ, then C_1 and C_2 must be zero.) (*b*) Show that $\Psi(x, t) = A[\cos (kx - \omega t) + i \sin (kx - \omega t)] = Ae^{i(kx - \omega t)}$ does satisfy the time-independent Schrödinger equation providing that k, U_0, and ω are related by Equation 36-28.

51. Repeat Problem 47 for a particle in the first excited state of the infinite square-well potential.

52. For the square-well wave functions

$$\psi_n(x) = \sqrt{2/L} \sin (n\pi x/L) \qquad n = 1, 2, 3, \ldots$$

corresponding to an infinite square well of length L, show that

$$\langle x^2 \rangle = \frac{L^2}{3} - \frac{L^2}{2n^2\pi^2}$$

53. A 10-eV electron is incident on a potential barrier of height 25 eV and width of 1 nm. (*a*) Use Equation 36-51 to calculate the order of magnitude of the probability that the electron will tunnel through the barrier. (*b*) Repeat your calculation for a width of 0.1 nm.

54. The standard deviation of measurements of the position of a particle is defined as the square root of the average value of the square of $(x - \langle x \rangle)$:

$$\sigma_x = \sqrt{\langle (x - \langle x \rangle)^2 \rangle}$$

Show that this can be written

$$\sigma_x = \sqrt{\langle x^2 \rangle - \langle x \rangle^2}$$

55. Use the result of Problem 54 to calculate σ_x for a classical particle in a box for which the probability distribution is $P(x) = 1/L$ for $0 < x < L$, and $P(x) = 0$ for $x < 0$ and for $x > L$.

56. (*a*) Use the results of Problems 52 and 54 to calculate σ_x for a particle in an infinite square-well potential for a general state n. (*b*) Evaluate your result for $n = 1$. (*c*) Show that for large n, your result approaches that for a classical particle found in Problem 55.

57. Use Equation 36-51 to calculate the order of magnitude of the probability that a proton will tunnel out of a nucleus in one collision with the nuclear barrier if it has energy 6 MeV below the top of the potential barrier and the barrier thickness is 10^{-15} m.

58. The minimum kinetic energy of a particle confined to a region of space of length L satisfies the inequality given in Equation 36-23. The ground-state energy is obtained from Schrödinger's equation applied to a particle in a box, where the potential energy may be taken to be zero inside the box. Show that $p^2/2m$ is the ground-state energy, and show that Equation 36-23 is satisfied.

59. Quantum mechanics predicts that any particle localized in space has a nonzero velocity and consequently can never be at rest. Consider a Ping-Pong ball of diameter 2 cm and mass 2 g that can move back and forth in a box of length 2.001 cm. Hence, the space in which the ball moves is only 0.001 cm in length. (*a*) What is the minimum speed of the Ping-Pong ball according to Schrödinger's equation? (*b*) What is the period of one oscillation?

60. A particle of mass m is in an infinite square-well potential given by

$$U = \infty \qquad x < -\tfrac{1}{2}L$$

$$U = 0 \qquad -\tfrac{1}{2}L < x < +\tfrac{1}{2}L$$

$$U = \infty \qquad +\tfrac{1}{2}L < x$$

Since this potential is symmetric about the origin, the probability density $\psi^2(x)$ must also be symmetric. (*a*) Show that this implies that either $\psi(-x) = \psi(x)$ or $\psi(-x) = -\psi(x)$. (*b*) Show that the proper solutions of the time-independent Schrödinger equation can be written

$$\psi(x) = \sqrt{\frac{2}{L}} \cos \frac{n\pi x}{L} \qquad n = 1, 3, 5, 7, \ldots$$

and

$$\psi(x) = \sqrt{\frac{2}{L}} \sin \frac{n\pi x}{L} \qquad n = 2, 4, 6, 8, \ldots$$

(*c*) Show that the allowed energies are the same as those for the infinite square well given by Equation 36-32. (*d*) Sketch the wave functions and probability distributions for the ground state and the first excited state.

61. Calculate $\langle x \rangle$ and $\langle x^2 \rangle$ for the ground state of the infinite square-well potential in Problem 60.

62. A particle moves in a potential given by $U(x) = A|x|$. Without attempting to solve the Schrödinger equation, sketch the wave function for (*a*) the ground-state energy of a particle inside this potential and (*b*) the first excited state for this potential.

63. Calculate $\langle x \rangle$ and $\langle x^2 \rangle$ for the first excited state of the infinite square-well potential in Problem 60.

64. A particle moves freely in the two-dimensional region defined by $0 \le x \le L$ and $0 \le y \le L$. (*a*) Find the wave function satisfying Schrödinger's equation. (*b*) Find the corresponding energies. (*c*) Find the lowest two states that are degenerate. Give the quantum numbers for this case. (*d*) Find the lowest three states that have the same energy. Give the quantum numbers for the three states having the same energy.

65. What is the next energy level above that found in Problem 64 for a particle in a two-dimensional square box for which the degeneracy is greater than 2?

66. In Problem 18 the wave function $\psi_0(x) = Ae^{-x^2/2L^2}$ represents the ground-state energy of a harmonic oscillator. (*a*) Show that $\psi_1 = L\, d\psi_0(x)/dx$ is also a solution of Schrödinger's equation. (*b*) What is the energy of this new state? (*c*) From a look at the nodes of this wave function, how would you classify this excited state?

Level III

67. For the wave functions in Equation 36-39, show that for any two functions $\psi_n(x)$ and $\psi_m(x)$

$$\int \psi_n(x)\psi_m(x)\, dx = 0$$

unless the integer $n = m$. This is the important orthogonality property of the wave functions of quantum mechanics.

68. A particle of mass m near the earth's surface at $z = 0$ can be described by the potential energy

$$U = mgz \qquad z > 0$$

$$U = \infty \qquad z < 0$$

For some positive value of total energy E, indicate the classically allowed region on a sketch of $U(z)$ versus z. Sketch also the kinetic energy versus z. The Schrödinger equation for this problem is quite difficult to solve. Using arguments similar to those in Section 36-7 about the curvature of the wave function as given by the Schrödinger equation, sketch your "guesses" for the shape of the wave function for the ground state and the first two excited states.

69. Use the Schrödinger equation to show that the expectation value of the kinetic energy of a particle is given by

$$\langle K \rangle = \int_{-\infty}^{+\infty} \psi(x)\left(-\frac{\hbar^2}{2m}\frac{d^2\psi(x)}{dx^2}\right) dx$$

70. Show that Equations 36-49 and 36-50 imply that the transmission coefficient for particles of energy E incident on a step barrier $U_0 < E$ is given by

$$T = \frac{4k_1k_2}{(k_1 + k_2)^2} = \frac{4r}{(1 + r)^2}$$

where $r = k_2/k_1$.

71. (*a*) Show that for the case of a particle of energy E incident on a step barrier $U_0 < E$, the wave numbers k_1 and k_2 are related by

$$\frac{k_2}{k_1} = r = \sqrt{1 - (U_0/E)}$$

Use this and the results of Problem 70 to calculate the transmission coefficient T and the reflection coefficient R for the case (*b*) $E = 1.2U_0$, (*c*) $E = 2.0U_0$, and (*d*) $E = 10.0U_0$.

72. Consider the one-dimensional, classical wave equation, Equation 13-34*b*. (*a*) Substitute the trial function $\Psi(x, t) = \psi(x)f(t)$ into the classical wave equation to obtain

$$\frac{1}{\psi(x)}\frac{d^2\psi(x)}{dx^2} = \frac{1}{v^2 f(t)}\frac{d^2 f(t)}{dt^2} \qquad \text{36-63}$$

In Equation 36-63, the variables x and t are separated. Since the left side is a function of x only, it cannot depend on t. Similarly, the right side cannot depend on x. Therefore, both sides must equal some constant. (*b*) Set each side of Equation 36-63 equal to the constant $-k^2$, and show that the solution for $f(t)$ can be expressed as $e^{\pm i\omega t}$, where $\omega = kv$. (*c*) Show that the time-independent equation for x is given by

$$\frac{d^2\psi(x)}{dx^2} + k^2\psi = 0$$

73. (*a*) Repeat the use of the method of separation of variables, used in Problem 72, on Schrödinger's time-dependent equation to obtain

$$-\frac{\hbar^2}{2m}\frac{d^2\psi(x)/dx^2}{\psi(x)} + U(x) = i\hbar\,\frac{df/dt}{f} \qquad \text{36-64}$$

(*b*) Since the left side of Equation 36-64 does not vary with x, the right side also cannot vary with x. Similarly, neither side can vary with t; thus, they both must equal the same constant E. Show that this implies that $f(t)$ is given by

$$f(t) = e^{-i\omega t}$$

where $\omega = E/\hbar$. Use the de Broglie relation to argue that E must be the total energy. (*c*) Use the left-hand side of Equation 36-64 to obtain Equation 36-30.

Chapter 37

Atoms

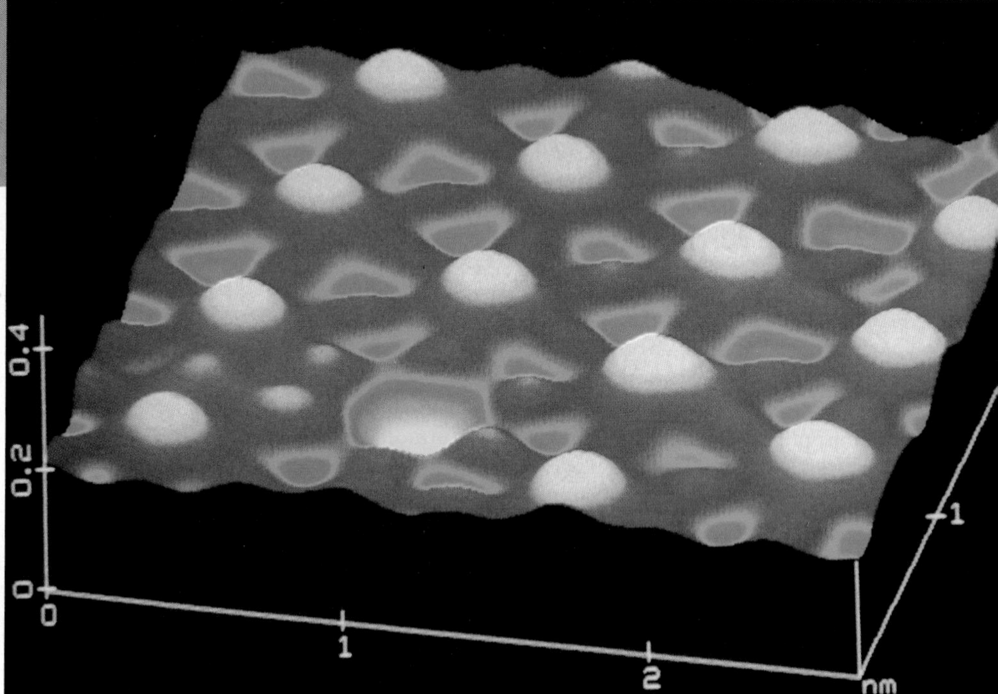

A scanning-tunneling microscope image of iodine atoms (pink) absorbed on platinum. The yellow pocket reflects the gap where an iodine atom has been dislodged.

Slightly more than 100 different elements have been discovered, 92 of which are found in nature. Each is characterized by an atom that contains a number of protons Z, an equal number of electrons, and a number of neutrons N. The number of protons Z is called the **atomic number.** The lightest atom, hydrogen (H), has $Z = 1$; the next lightest, helium (He), has $Z = 2$; the next lightest, lithium (Li), has $Z = 3$; and so forth. Nearly all the mass of the atom is concentrated in a tiny nucleus, which contains the protons and neutrons. The nuclear radius is typically about 1 to 10 fm (1 fm $= 10^{-15}$ m). The distance between the nucleus and the electrons is about 0.1 nm = 100,000 fm. This distance determines the "size" of the atom.

The chemical and physical properties of an element are determined by the number and arrangement of the electrons in the atom. Because each proton has a positive charge $+e$, the nucleus has a total positive charge $+Ze$. The electrons are negatively charged ($-e$), so they are attracted to the nucleus and repelled by each other. The electrons are arranged in shells. The first shell has up to two electrons. The second shell, about four times farther out from the nucleus than the first, can contain up to eight electrons. The third shell, about nine times farther out than the first, can contain up to eighteen electrons. This shell structure accounts for the periodic nature of the properties of the elements shown in the periodic table (Appendix E). Elements with a single electron in an outer shell (hydrogen, lithium, sodium, etc.) or those with a single vacancy in an outer shell (fluorine, chlorine, bromine, etc.) are very active chemically and readily combine to form molecules. Those with completely filled outer shells (helium, neon, argon, etc.) are more or less chemically inert. The calculation of the electron configurations of the atoms and the determination of the resulting chemical properties were great triumphs of quantum mechanics in the 1920s.

Since electrons and protons have equal but opposite charges and there are an equal number of electrons and protons in an atom, atoms are electrically neutral. Atoms that lose or gain one or more electrons are then electrically charged and are called *ions*. Atoms with just one electron in an outer shell (such as sodium, which has eleven electrons) tend to lose it readily and become positive ions; those lacking just one electron for a complete shell tend to gain an electron to become negative ions (for example, chlorine). Atoms bond together to form molecules, such as H_2O, or solids. This bonding involves only the outer electrons called **valence electrons.** In a molecule or solid, the separation of the atomic nuclei is about one atomic diameter, which is of the order of 0.1 nm.

In this chapter, we will apply our knowledge of quantum mechanics from Chapter 36 to give a qualitative description of the simplest atom, the hydrogen atom. We will then discuss qualitatively the structure of other atoms and the periodic table of the elements.

37-1 Quantum Theory of the Hydrogen Atom

Despite its spectacular successes, the Bohr model of the hydrogen atom described in Section 35-5 had many shortcomings. There was no justification for the postulates of stationary states or the quantization of angular momentum other than the fact that these postulates led to energy levels that agreed with spectroscopic data. Furthermore, the Bohr model gave no information about the intensities of the spectral lines, and attempts to apply it to more complicated atoms had little success. The quantum-mechanical theory resolved these difficulties. The stationary states of the Bohr model correspond to the standing-wave solutions of the Schrödinger wave equation. Energy quantization is a direct consequence of the frequency quantization that results from standing waves and the de Broglie relation $E = hf$. The quantized energies resulting from the standing-wave solutions of the Schrödinger equation agree with those obtained from the Bohr model and with experiment. The quantization of angular momentum that had to be postulated in the Bohr model is predicted by the quantum theory.

In quantum theory, the electron is described by its wave function ψ. The absolute square of the electron wave function $|\psi|^2$ gives the probability of finding the electron in some region of space. Boundary conditions on the wave function lead to the quantization of the wavelengths and frequencies and thereby to the quantization of the electron energy.

We can treat the hydrogen atom as a stationary nucleus, the proton, that has a single particle, an electron, moving with kinetic energy $p^2/2m$ and potential energy $U(r)$ due to the electrostatic attraction of the proton:

$$U(r) = -k\frac{Ze^2}{r} \tag{37-1}$$

We include the factor Z, which is 1 for hydrogen, so that we can apply our results to other, one-electron atoms, such as ionized helium He^+, for which $Z = 2$. The time-independent Schrödinger equation for a particle of mass m moving in three dimensions is given by Equation 36-52:

$$-\frac{\hbar^2}{2m}\left(\frac{\partial^2\psi}{\partial x^2} + \frac{\partial^2\psi}{\partial y^2} + \frac{\partial^2\psi}{\partial z^2}\right) + U\psi = E\psi \tag{37-2}$$

Since the potential energy $U(r)$ depends only on the radial distance $r = \sqrt{x^2 + y^2 + z^2}$, the problem is most conveniently treated using the spherical coordinates r, θ, and ϕ, which are related to the rectangular coordinates x, y, and z by

$$z = r \cos \theta$$
$$x = r \sin \theta \cos \phi \qquad 37\text{-}3$$
$$y = r \sin \theta \sin \phi$$

These relations are shown in Figure 37-1. The transformation of the three-dimensional Schrödinger equation into spherical coordinates is straightforward but involves much tedious calculation, which we will omit. The result is

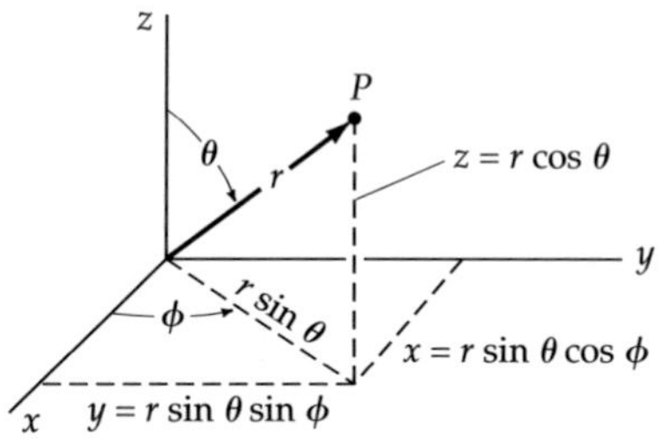

$$-\frac{\hbar^2}{2m}\frac{1}{r^2}\frac{\partial}{\partial r}\left(r^2 \frac{\partial \psi}{\partial r}\right) - \frac{\hbar^2}{2mr^2}\left[\frac{1}{\sin \theta}\frac{\partial}{\partial \theta}\left(\sin \theta \frac{\partial \psi}{\partial \theta}\right) + \frac{1}{\sin^2 \theta}\frac{\partial^2 \psi}{\partial \phi^2}\right] + U\psi = E\psi \qquad 37\text{-}4$$

Figure 37-1 Geometric relations between spherical and rectangular coordinates.

Despite the formidable appearance of this equation, it was not difficult for Schrödinger to solve because it is similar to other partial differential equations that arise in classical physics and such equations had been thoroughly studied. We will not solve this equation but merely discuss qualitatively some of the interesting features of the wave functions that satisfy it.

The first step in the solution of a partial differential equation such as Equation 37-4 is to separate the variables by writing the wave function $\psi(r, \theta, \phi)$ as a product of functions of each single variable:

$$\psi(r, \theta, \phi) = R(r)f(\theta)g(\phi) \qquad 37\text{-}5$$

where R depends only on the radial coordinate r, f depends only on θ, and g depends only on ϕ. When this form of $\psi(r, \theta, \phi)$ is substituted into Equation 37-4, the partial differential equation can be transformed into three ordinary differential equations, one for $R(r)$, one for $f(\theta)$, and one for $g(\phi)$. The potential energy $U(r)$ appears only in the equation for $R(r)$, which is called the **radial equation.** The particular form of $U(r)$ given in Equation 37-1 therefore has no effect on the solutions of the equations for $f(\theta)$ and $g(\phi)$. These solutions are applicable to any central-field problem, that is, to any problem in which the potential energy depends only on r.

As we saw in Section 36-10, the requirement that the wave function be well behaved so that it is continuous and can be normalized introduces three quantum numbers, each associated with one of the three variables. The solution of the Schrödinger equation in spherical coordinates leads to three quantum numbers that are labeled n, ℓ, and m. The quantum numbers n_1, n_2, and n_3 that we found for a particle in a three-dimensional square well were independent of one another, but the quantum numbers associated with wave functions in spherical coordinates are interdependent. The possible values of these quantum numbers are

Quantum numbers for the hydrogen atom

$$n = 1, 2, 3, \ldots$$
$$\ell = 0, 1, 2, \ldots, n-1 \qquad 37\text{-}6$$
$$m = -\ell, -\ell+1, -\ell+2, \ldots, +\ell$$

That is, n can be any integer; ℓ can be 0 or any integer up to $n - 1$; and m can have $2\ell + 1$ possible values, ranging from $-\ell$ to $+\ell$ in integral steps.

The number n is called the **principal quantum number.** It is associated with the dependence of the wave function on the distance r and therefore with the probability of finding the electron at various distances from the nucleus. The quantum numbers ℓ and m are associated with the angular momentum of the electron and with the angular dependence of the electron wave function. The quantum number ℓ is called the **orbital quantum number.** The orbital angular momentum L of the electron is related to ℓ by

$$L = \sqrt{\ell(\ell + 1)}\,\hbar \tag{37-7}$$

The quantum number m is called the **magnetic quantum number.** It is related to the component of the angular momentum in some direction. Ordinarily, all directions are equivalent, but one particular direction can be specified by placing the atom in a magnetic field. If the z direction is chosen for the magnetic field, the z component of the angular momentum of the electron is given by

$$L_z = m\hbar \tag{37-8}$$

If we measure the angular momentum of the electron in units of $\hbar$, we see that the angular momentum is quantized to the value $\sqrt{\ell(\ell + 1)}$ units and that its component along any direction can have only the $2\ell + 1$ values ranging from $-\ell$ to $+\ell$ units. Figure 37-2 shows a vector-model diagram illustrating the possible orientations of the angular-momentum vector for $\ell = 2$. Note that only specific values of θ are allowed; that is, the directions in space are quantized.

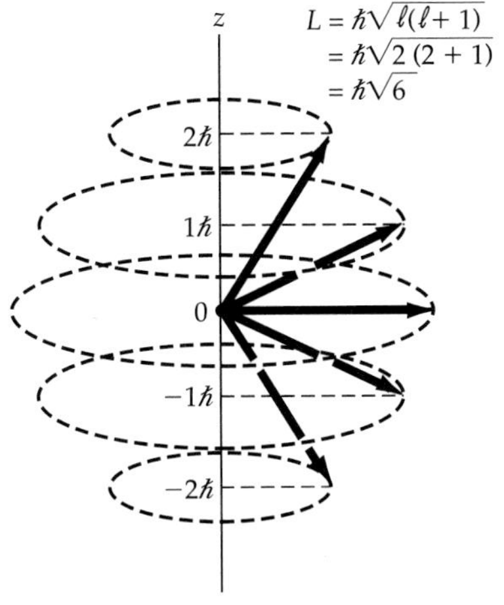

Figure 37-2 Vector-model diagram illustrating the possible values of the z component of the angular-momentum vector for the case $\ell = 2$.

Example 37-1

If an atom has an angular momentum characterized by the quantum number $\ell = 2$, what are the possible values of L_z, and what is the smallest possible angle between **L** and the z axis?

The possible values of L_z are $m\hbar$, where the $2\ell + 1 = 5$ values of m are $-2, -1, 0, +1$, and $+2$. The magnitude of **L** is $L = \sqrt{\ell(\ell + 1)}\,\hbar = \sqrt{6}\,\hbar$. From Figure 37-1, the angle between **L** and the z axis is given by

$$\cos\theta = \frac{L_z}{L} = \frac{m\hbar}{\sqrt{\ell(\ell + 1)}\,\hbar} = \frac{m}{\sqrt{\ell(\ell + 1)}}$$

The smallest angle occurs when $m = +\ell$ or $-\ell$, which for $\ell = 2$ gives $\cos\theta = 2/\sqrt{6} = 0.816$ or $\theta = 35.3°$. We note the somewhat strange result that the angular-momentum vector cannot lie along the z axis. This is related to an uncertainty relation for angular momentum that implies that no two components of the angular momentum can be precisely known except when $\ell = 0$.

Exercise

An atom has an angular momentum characterized by the quantum number $\ell = 4$. What are the possible values of m? (Answer: $-4, -3, -2, -1, 0, 1, 2, 3, 4$)

The allowed energies of the hydrogen atom and other one-electron atoms that result from the solution of the Schrödinger equation with $U(r)$ given by Equation 37-1 are given by

$$E_n = -\frac{Z^2E_0}{n^2} \qquad n = 1, 2, 3, \ldots \tag{37-9}$$

Energy levels for hydrogen

where

$$E_0 = \frac{k^2 e^4 m_e}{2\hbar^2} \approx 13.6 \text{ eV}$$

These energies are the same as in the Bohr model. Note that the energy is negative, indicating that the electron is bound to the nucleus (thus the term *bound state*), and that it depends only on the principal quantum number n. The fact that the energy does not depend on ℓ is a peculiarity of the inverse-square force and holds only for the hydrogen atom. For more complicated atoms having several electrons, the interaction of the electrons leads to a dependence of the energy on the orbital quantum number. In general, the lower the value of ℓ, the lower the energy for such atoms. Since, in general, there is no preferred direction in space, the energy for any atom does not depend on the magnetic quantum number m, which is related to the z component of the angular momentum. The energy does depend on m if the atom is in a magnetic field.

Figure 37-3 shows an energy-level diagram for hydrogen. This diagram is similar to Figure 35-14, except that the states with the same value of n but different values of ℓ are shown separately. These states (called *terms*) are referred to by giving the value of n along with a code letter: S for $\ell = 0$, P for $\ell = 1$, D for $\ell = 2$, and F for $\ell = 3$. (These code letters are remnants of spectroscopists' descriptions of various spectral lines as *sharp, principal, diffuse,* and *fundamental*. For values greater than 3, the letters follow alphabetically; thus, G is used for $\ell = 4$ and so forth.) When an atom makes a transition from one allowed energy state to another, electromagnetic radiation in the form of a photon is emitted or absorbed. Such transitions result in spectral lines that are characteristic of the atom. The transitions obey the **selection rules**

$$\Delta m = 0 \text{ or } \pm 1$$
$$\Delta \ell = \pm 1 \qquad \text{37-10}$$

These selection rules are related to the conservation of angular momentum and to the fact that the photon itself has an intrinsic angular momentum that

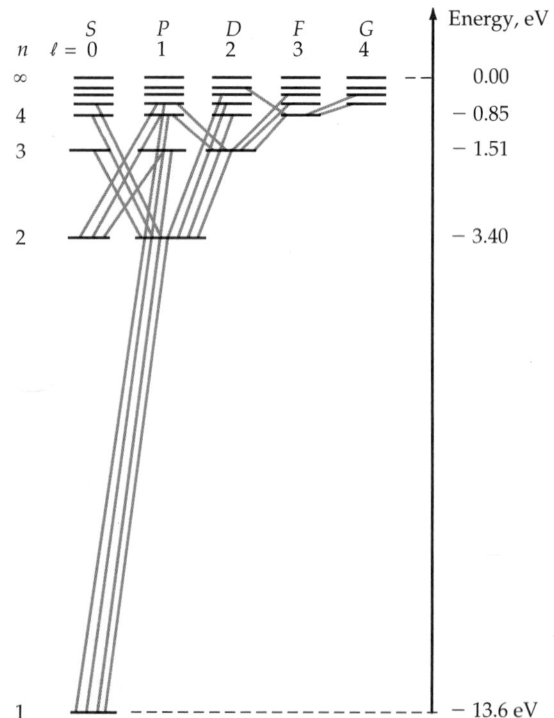

Figure 37-3 Energy-level diagram for hydrogen. The diagonal lines show transitions that involve emission or absorption of radiation and obey the selection rule $\Delta\ell = \pm 1$. States with the same value of n but different values of ℓ have the same energy $-E_0/n^2$, where $E_0 =$ 13.6 eV as in the Bohr model.

has a maximum component along any axis of $1\hbar$. The wavelengths of the spectral lines emitted by hydrogen (and by other atoms) are related to the energy levels by the Bohr formula

$$hf = \frac{hc}{\lambda} = E_i - E_f \qquad 37\text{-}11$$

where E_i and E_f are the energies of the initial and final states.

37-2 The Hydrogen-Atom Wave Functions

The wave functions that are solutions of the Schrödinger equation are characterized by the quantum numbers and are written $\psi_{n\ell m}$. For any given value of n, there are n possible values of ℓ ($\ell = 0, 1, \ldots, n-1$), and for each value of ℓ, there are $2\ell + 1$ possible values of m. Since the energy depends only on n, there are generally many different wave functions that correspond to the same energy, except at the lowest energy level, for which $n = 1$ and therefore ℓ and m must be 0. As discussed previously, the origins of this degeneracy are the $1/r$ dependence of the potential energy and the fact that, in the absence of any external fields, there is no preferred direction in space.

In the lowest energy state, the ground state, the principal quantum number n has the value 1, ℓ is 0, and m is 0. The energy is -13.6 eV, the same as in the Bohr model, but the angular momentum is zero rather than $1\hbar$ as in the Bohr model. The wave function for the ground state is

$$\psi_{100} = C_{100}e^{-Zr/a_0}$$

where

$$a_0 = \frac{\hbar^2}{m_e ke^2} = 0.0529 \text{ nm}$$

is the first Bohr radius and C_{100} is a constant that is determined by normalization. In three dimensions, the normalization condition is

$$\int \psi^2 \, dV = 1$$

where dV is a volume element and the integration is performed over all space. In spherical coordinates, the volume element (Figure 37-4) is

$$dV = (r \sin\theta \, d\phi)(r \, d\theta) \, dr = r^2 \sin\theta \, dr \, d\theta \, d\phi$$

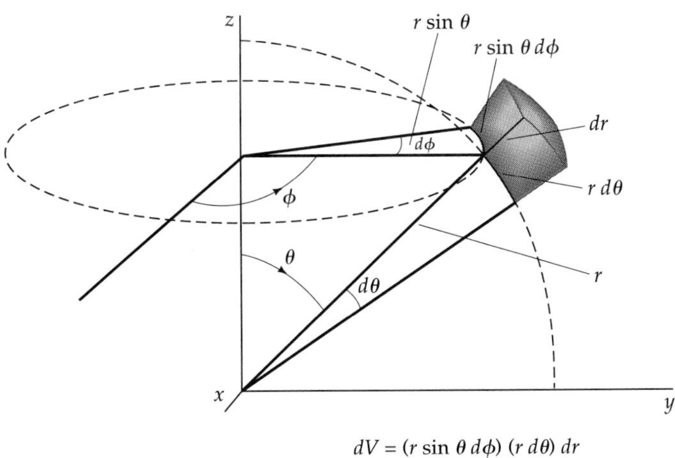

Figure 37-4 Volume element in spherical coordinates.

We integrate over all space by integrating over ϕ from $\phi = 0$ to $\phi = 2\pi$, over θ from $\theta = 0$ to $\theta = \pi$, and over r from $r = 0$ to $r = \infty$. The normalization condition is thus

$$\int \psi^2\,dV = \int_0^\infty \int_0^\pi \int_0^{2\pi} \psi^2 r^2 \sin\theta\,d\phi\,d\theta\,dr$$

$$= \int_0^{2\pi} d\phi \int_0^\pi \sin\theta\,d\theta \int_0^\infty \psi^2 r^2\,dr$$

$$= \int_0^{2\pi} d\phi \int_0^\pi \sin\theta\,d\theta \int_0^\infty C_{100}^2 r^2 e^{-2Zr/a_0}\,dr = 1$$

Since there is no θ or ϕ dependence in ψ_{100}, the integration over the angles gives 4π. From a table of integrals, we obtain

$$\int_0^\infty r^2 e^{-2Zr/a_0}\,dr = \frac{a_0^3}{4Z^3}$$

Then

$$4\pi C_{100}^2\left(\frac{a_0^3}{4Z^3}\right) = 1$$

and

$$C_{100} = \frac{1}{\sqrt{\pi}}\left(\frac{Z}{a_0}\right)^{3/2}$$

The normalized ground-state wave function is thus

$$\psi_{100} = \frac{1}{\sqrt{\pi}}\left(\frac{Z}{a_0}\right)^{3/2} e^{-Zr/a_0} \qquad 37\text{-}12$$

The probability of finding the electron in the volume dV is $\psi^2\,dV$. The probability density ψ^2 is illustrated in Figure 37-5. Note that this probability density is spherically symmetric; that is, it depends only on r and not on θ or ϕ. The probability density is maximum at the origin. We are more often interested in the probability of finding the electron at some radial distance r between r and $r + dr$. This radial probability $P(r)\,dr$ is the probability density ψ^2 times the volume of the spherical shell of thickness dr, which is $dV =$

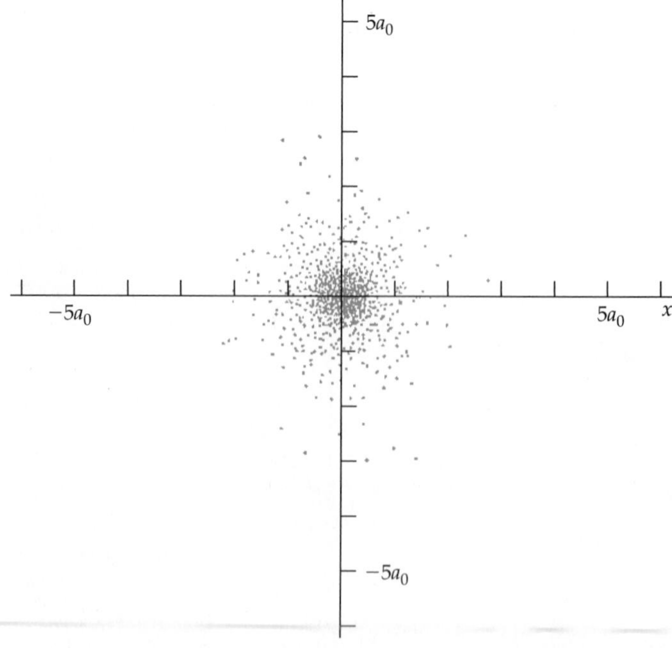

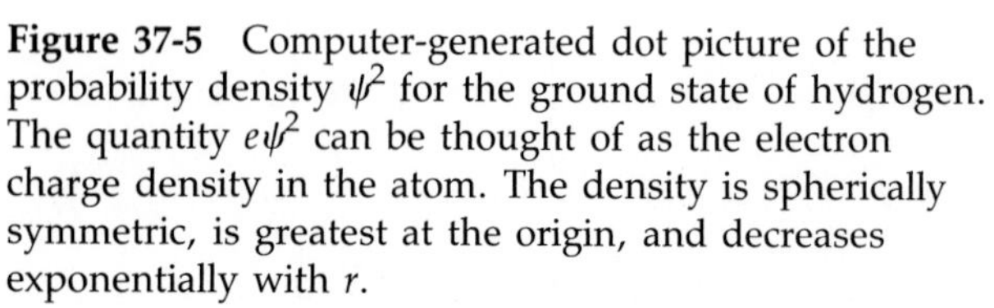
Figure 37-5 Computer-generated dot picture of the probability density ψ^2 for the ground state of hydrogen. The quantity $e\psi^2$ can be thought of as the electron charge density in the atom. The density is spherically symmetric, is greatest at the origin, and decreases exponentially with r.

(a)

(b)

(*a*) The tip of a tungsten probe used in a scanning-tunneling microscope (STM). Some probes have tips only one atom wide. During operation, a small current flows from the surface of the material that is being imaged onto the probe tip. The current results from electrons, on the surface of the material, tunneling through the potential difference between the surface and the probe. Because the tunneling current decreases exponentially with distance, it is very sensitive to changes in that distance. (See essay, *Scanning Tunneling Microscopy*, at the end of Chapter 36.) (*b*) The microcantilever of an atomic-force microscope (ATM). Repulsion between electrons on the surface that is being imaged and electrons on the cantilever causes the cantilever to deflect. The degree of deflection can be closely calibrated to the distance the cantilever is from the surface. Both this image and that in part *a* were taken using a scanning electron microscope (SEM).

$4\pi r^2\,dr$, at a distance r. The probability of finding the electron in the range from r to $r + dr$ is thus $P(r)\,dr = \psi^2 4\pi r^2\,dr$, and the **radial probability density** is

$$P(r) = 4\pi r^2 \psi^2 \qquad \text{37-13}$$

Radial probability density

For the hydrogen atom in the ground state, the radial probability density is

$$P(r) = 4\pi r^2 \psi^2 = 4\pi r^2 C_{100}^2 e^{-2Zr/a_0} = 4\left(\frac{Z}{a_0}\right)^3 r^2 e^{-2Zr/a_0} \qquad \text{37-14}$$

Figure 37-6 shows the radial probability density $P(r)$ as a function of r. The maximum value of $P(r)$ occurs at $r = a_0/Z$, which for $Z = 1$ is the first Bohr radius. In contrast to the Bohr model in which the electron stays in a well-defined orbit at $r = a_0$, we see that it is possible for the electron to be found at any distance from the nucleus. However, the most probable distance is a_0 (assuming $Z = 1$), and the chance of finding the electron at a much different distance is small. It is sometimes useful to think of the electron in an atom as a charged cloud of charge density $\rho = e\psi^2$, but we must remember that an electron is always observed to interact with matter or radiation as a single charge.

In the first excited state, $n = 2$ and ℓ can be either 0 or 1. For $\ell = 0$, $m = 0$, and we again have a spherically symmetric wave function, this time given by

$$\psi_{200} = C_{200}\left(2 - \frac{Zr}{a_0}\right)e^{-Zr/2a_0} \qquad \text{37-15}$$

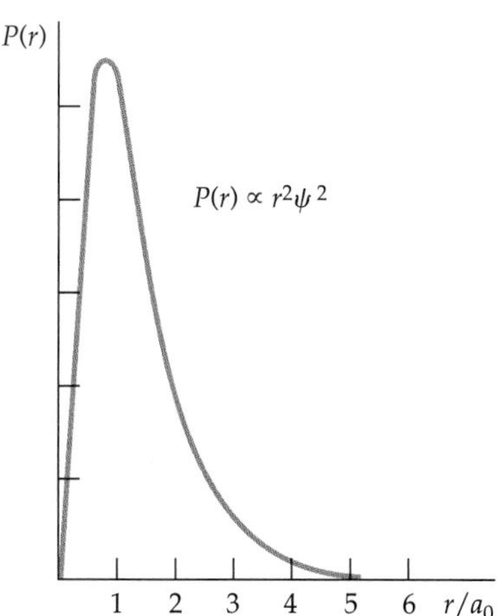

Figure 37-6 Radial probability density $P(r)$ versus r/a_0 for the ground state of the hydrogen atom. $P(r)$ is proportional to $r^2\psi^2$. The value of r for which $P(r)$ is maximum is the most probable distance $r = a_0$.

For $\ell = 1$, m can be $+1$, 0, or -1. The corresponding wave functions are

$$\psi_{210} = C_{210}\frac{Zr}{a_0}e^{-Zr/2a_0}\cos\theta \qquad \text{37-16}$$

$$\psi_{21\pm1} = C_{211}\frac{Zr}{a_0}e^{-Zr/2a_0}\sin\theta\, e^{\pm i\phi} \qquad \text{37-17}$$

In general, the wave functions for $m \neq 0$ are proportional to $e^{im\phi}$ and are therefore not real. As we discussed in Chapter 36, the probability density for complex functions is given by $|\psi|^2$. We may write $|\psi|^2$ as

$$|\psi|^2 = \psi^*\psi \qquad 37\text{-}18$$

where ψ^* is the complex conjugate of ψ, which is obtained from ψ by replacing i by $-i$ wherever it appears.† Since $(e^{im\phi})^* = e^{-im\phi}$ and $(e^{im\phi})^*(e^{im\phi}) = e^{-im\phi}e^{im\phi} = 1$, the probability densities do not depend on m or ϕ even though the wave functions do.

Figure 37-7 shows the probability density $\psi^*\psi$ for $n = 2$, $\ell = 0$, $m = 0$ (Figure 37-7*a*); for $n = 2$, $\ell = 1$, $m = 0$ (Figure 37-7*b*); and for $n = 2$, $\ell = 1$, $m = \pm 1$ (Figure 37-7*c*). An important feature of these plots is that the electron cloud is spherically symmetric for $\ell = 0$ and is not spherically symmetric for $\ell \neq 0$. These angular distributions of the electron charge density depend only on the values of ℓ and m and not on the radial part of the wave function. Similar charge distributions for the valence electrons of more complicated atoms play an important role in the chemistry of molecular bonding, which we will discuss in Chapter 38.

Figure 37-7 Computer-generated dot picture of the probability densities $|\psi|^2$ for the electron in the $n = 2$ states of hydrogen. (*a*) For $\ell = 0$, $|\psi|^2$ is spherically symmetric. (*b*) For $\ell = 1$ and $m = 0$, $|\psi|^2$ is proportional to $\cos^2\theta$. (*c*) For $\ell = 1$ and $m = +1$ or -1, $|\psi|^2$ is proportional to $\sin^2\theta$.

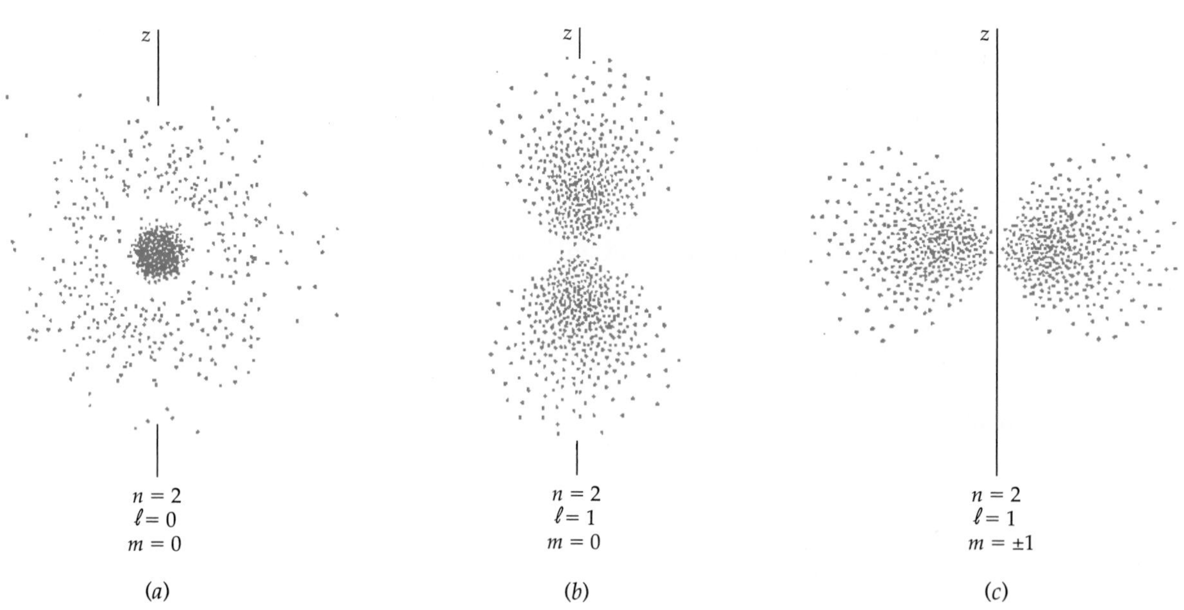

Figure 37-8 shows the probability of finding the electron at a distance r as a function of r for $n = 2$ when $\ell = 1$ and when $\ell = 0$. We can see from the figure that the probability distribution depends on ℓ as well as on n. In particular, we can see that for very small values of r ($r < a_0$), the S state ($\ell = 0$) has a larger probability density than does the P state ($\ell = 1$). This will be important when we consider atoms with more than one electron.

For $n = 1$, we found that the most likely distance between the electron and the nucleus is a_0, the first Bohr radius, whereas for $n = 2$, $\ell = 1$, it is $4a_0$. These are the orbital radii for the first and second Bohr orbits (Equation 35-22). For $n = 3$ (and $\ell = 2$),‡ the most likely distance between the electron and nucleus is $9a_0$, the radius of the third Bohr orbit.

†Every complex number can be written in the form $z = a + bi$, where a and b are real numbers and $i = \sqrt{-1}$. The magnitude or absolute value of z is defined as $\sqrt{a^2 + b^2}$. The complex conjugate of z is $z^* = a - bi$, so $z^*z = (a - bi)(a + bi) = a^2 + b^2 = |z|^2$.

‡The correspondence with the Bohr model is closest for the maximum value of ℓ, which is $n - 1$.

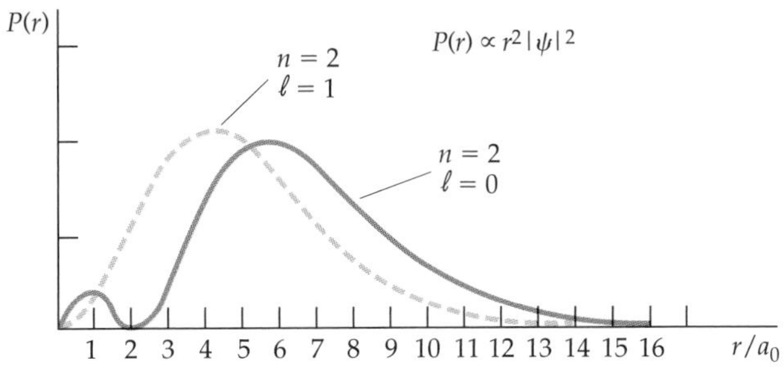

Figure 37-8 Radial probability density $P(r)$ versus r/a_0 for the $n = 2$ states of hydrogen. For $\ell = 1$, $P(r)$ is maximum at the Bohr value $r = 2^2a_0$. For $\ell = 0$, there is a maximum near this value and a smaller submaximum near the origin.

Example 37-2

For the ground state of the hydrogen atom, find the probability of finding the electron in the range $\Delta r = 0.02a_0$ at (*a*) $r = a_0$ and (*b*) $r = 2a_0$.

Because the range Δr is so small, the variation in the radial probability density $P(r)$ can be neglected. The probability of finding the electron in Δr at $r = a_0$ is then

$$P(r)\,dr \approx P(r)\,\Delta r$$

(*a*) Using Equation 37-14 for $P(r)$, with $Z = 1$, we obtain for $r = a_0$

$$P(r)\,\Delta r = 4\left(\frac{1}{a_0}\right)^3 r^2 e^{-2r/a_0}\,\Delta r = 4\left(\frac{1}{a_0}\right)^3 a_0^2 e^{-2}(0.02a_0) = 0.0108$$

Thus, there is about a 1 percent chance of finding the electron in this range at $r = a_0$.

(*b*) For $r = 2a_0$, we obtain

$$P(r)\,\Delta r = 4\left(\frac{1}{a_0}\right)^3 r^2 e^{-2r/a_0}\,\Delta r = 4\left(\frac{1}{a_0}\right)^3 4a_0^2 e^{-4}(0.02a_0) = 0.00586$$

37-3 Magnetic Moments and Electron Spin

When a spectral line of hydrogen is viewed under high resolution, it is found to consist of two closely spaced lines. (The spectral lines for other elements may consist of more than two closely spaced lines.) This splitting of a line into two or more closely spaced lines is called **fine structure.** To explain this fine structure and to clear up a major difficulty with the quantum-mechanical explanation of the periodic table (which will be discussed in the next section), W. Pauli suggested in 1925 that the electron has an additional quantum number that can take on just two values. In the same year, S. Goudsmit and G. Uhlenbeck, graduate students at Leiden, suggested that this fourth quantum number was the z component m_s of an intrinsic angular momentum of the electron called **electron spin.** In the particle model, the electron is pictured as a spinning ball that orbits the nucleus (Figure 37-9), much as the earth rotates about its axis as it revolves around the sun. If this intrinsic spin angular momentum is to be described by a quantum number s, like the orbital quantum number ℓ, we expect the z component to have $2s + 1$ possible values just as there are $2\ell + 1$ possible z components of the orbital angular momentum. If m_s is to have just two possible values, s must be $\frac{1}{2}$. Then m_s can be either $m_s = -s = -\frac{1}{2}$ or $m_s = +s = +\frac{1}{2}$ corresponding to the z components of intrinsic electron spin of $+\frac{1}{2}\hbar$ and $-\frac{1}{2}\hbar$.

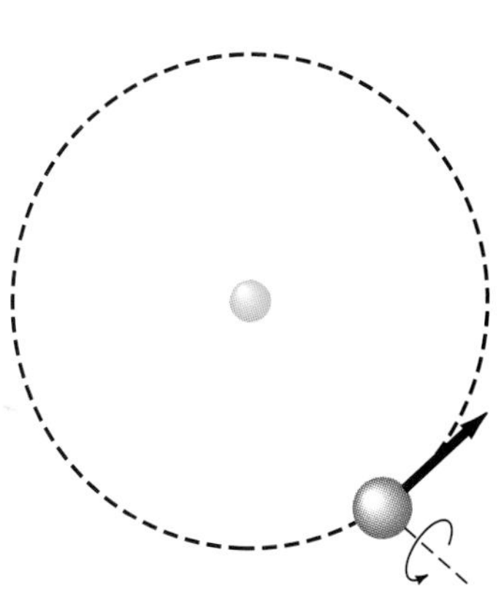

Figure 37-9 The electron can be pictured as a spinning ball that orbits the nucleus somewhat like the spinning earth orbits the sun.

One consequence of electron spin is that the electron possesses an intrinsic magnetic moment—a result to be expected since a spinning charge is equivalent to a set of current loops. In Chapter 27, we saw that the magnetic moment† $\boldsymbol{\mu}$ of a rotating charged system is related to its angular momentum by Equation 27-9:

$$\boldsymbol{\mu} = \frac{q}{2m_q}\mathbf{L} \qquad 37\text{-}19$$

where m_q is the mass of the particle that has a charge q. Substituting $q = -e$ and $m_q = m_e$ for the electron, we have

$$\boldsymbol{\mu} = -\frac{e}{2m_e}\mathbf{L} \qquad 37\text{-}20$$

Applying this equation to the orbital angular momentum of the electron in the hydrogen atom, we have for the magnitude and z component of the magnetic moment

$$\mu = \frac{e}{2m_e}L = \frac{e}{2m_e}\sqrt{\ell(\ell+1)}\,\hbar = \sqrt{\ell(\ell+1)}\,\mu_B \qquad 37\text{-}21$$

and

$$\mu_z = -\frac{e}{2m_e}m\hbar = -\frac{e\hbar}{2m_e}m = -m\mu_B \qquad 37\text{-}22$$

where μ_B is the Bohr magneton, which has the value

$$\mu_B = \frac{e\hbar}{2m_e} = 9.27 \times 10^{-24}\ \text{J/T} = 5.79 \times 10^{-5}\ \text{eV/T}$$

We thus see that the quantization of angular momentum implies the quantization of magnetic moment.

Since the electron has an intrinsic spin angular momentum characterized by the quantum numbers s and m_s, we would expect it to have an intrinsic magnetic moment given by Equation 37-21 and a z component of its magnetic moment given by Equation 37-22, with ℓ replaced by s and m replaced by m_s. In particular, we would expect the z component of the intrinsic magnetic moment of the electron to be $\pm\frac{1}{2}\mu_B$. However, the measured value is twice this expected value; that is, the electron has an intrinsic magnetic moment of one Bohr magneton, not one-half Bohr magneton. It is customary to write the relation between the z component of any type of angular momentum J_z and the z component of the magnetic moment μ_z as

$$\mu_z = -g\mu_B\frac{J_z}{\hbar} \qquad 37\text{-}23$$

where g, called the **gyromagnetic ratio,** has the values $g_\ell = 1$ for orbital angular momentum and $g_s = 2$ for spin. Precise measurements indicate that $g_s = 2.00232$. This result and the phenomenon of electron spin itself was predicted by P. Dirac in 1927. He combined special relativity and quantum mechanics into a relativistic wave equation called the **Dirac equation.** The fact that the intrinsic magnetic moment of the electron is approximately twice what we would expect makes it clear that the simple model of the electron as a spinning ball should not be taken literally. Like the Bohr model

†Because the magnetic quantum number is designated by the symbol m, we will now use μ for the magnetic moment.

of the atom, the simple classical picture is useful for describing quantum-mechanical calculations, and it often provides useful guidelines as to what to expect from an experiment.

The interaction of the magnetic moments associated with the orbital angular momentum and the spin angular momentum of the electron gives rise to a small splitting of the energy levels of hydrogen and other atoms, which is responsible for the fine-structure splitting of the spectral lines. Since the spin angular momentum can have two possible orientations relative to some direction in space, the electron's intrinsic spin magnetic moment can have two possible orientations relative to the magnetic moment associated with the orbital angular momentum. These two orientations are usually described by saying that the two magnetic moments can be parallel or antiparallel to each other. Because of the interaction of the magnetic moments, the energies of these two orientations are slightly different, resulting in the splitting of each energy level in the hydrogen atom (except those with $\ell = 0$) into two nearly equal energy levels.

The addition of the electron-spin quantum number completes the quantum-mechanical description of the hydrogen atom. The wave functions for the electron in the hydrogen atom are characterized by four quantum numbers: n, ℓ, m, and m_s.

Question

1. The Bohr theory and the Schrödinger theory of the hydrogen atom give the same results for energy levels. Discuss the advantages and disadvantages of each model.

37-4 The Stern–Gerlach Experiment

In addition to explaining the fine structure and the periodic table, the proposal of electron spin explained an interesting experiment first performed by O. Stern and W. Gerlach in 1922. In this experiment, atoms from an oven are collimated and sent through a magnet whose poles are shaped so that the magnetic field B_z is not constant in space but increases slightly with z (Figure 37-10). The behavior of a magnetic moment in a magnetic field can be visualized by considering a small bar magnet (Figure 37-11). When the magnet is

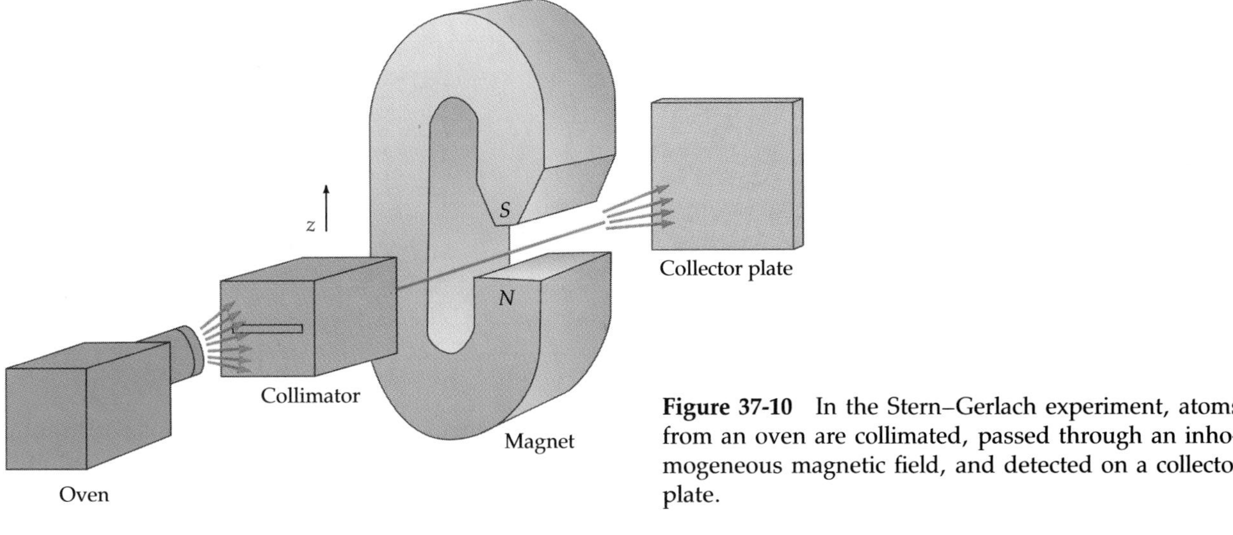

Figure 37-10 In the Stern–Gerlach experiment, atoms from an oven are collimated, passed through an inhomogeneous magnetic field, and detected on a collector plate.

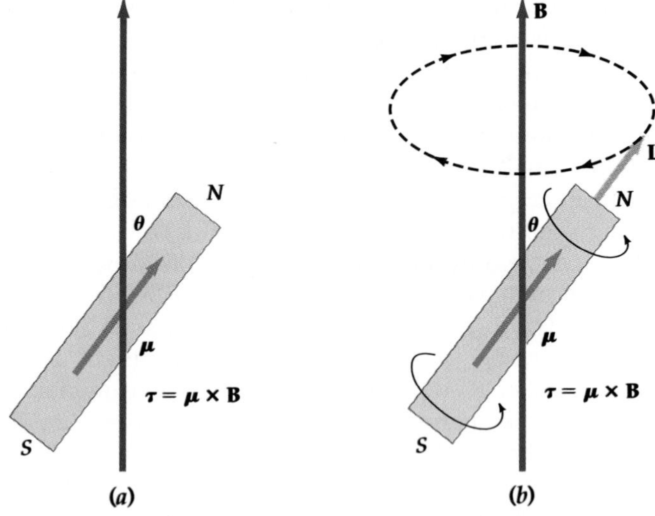

Figure 37-11 Bar-magnet model of a magnetic moment. (*a*) In an external magnetic field, the moment experiences a torque that tends to align it with the field. (*b*) If the magnet is spinning, the torque causes the spin axis to precess about the direction of the external field.

placed in a uniform external magnetic field **B**, there is a torque $\boldsymbol{\tau} = \boldsymbol{\mu} \times \mathbf{B}$ that tends to align the magnet with the direction of the field **B**. If the magnet is spinning about its axis, the effect of the torque is to make the spin axis precess about the direction of the external field. If the magnetic field is not uniform, the force on one pole of the magnet will be greater or less than that on the other pole, depending on the orientation of the magnet. Figure 37-12 illustrates the effects of such a field on three bar magnets of different orientations. In addition to the torque, which merely causes the magnetic moment to precess about the field direction, there is a net force in the positive or negative z direction given by

$$F_z = \mu_z \frac{dB_z}{dz} \tag{37-24}$$

This force deflects the magnet up or down by an amount that depends on the inhomogeneity of the magnetic field and on the z component of the magnetic moment.

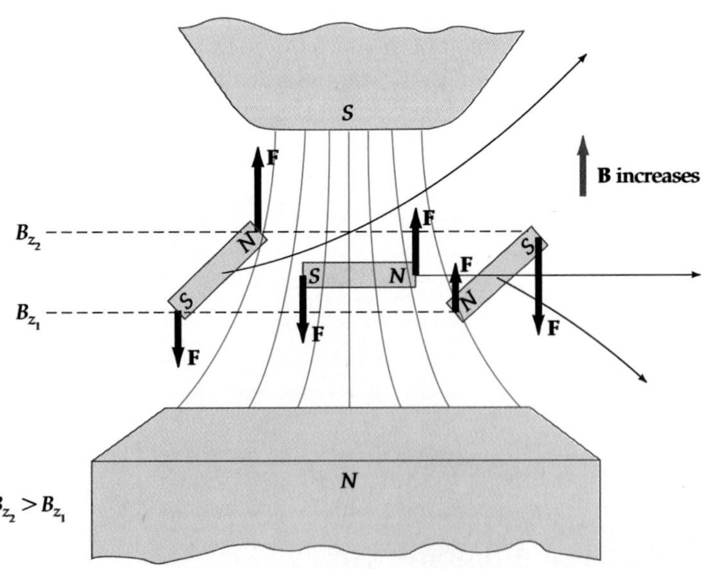

Figure 37-12 In an inhomogeneous magnetic field, a bar magnet experiences a net force that depends on the magnet's orientation. Here the force is stronger near the south pole of the large magnet, so B_z increases in the z direction.

Classically, one would expect a continuum of deflections corresponding to the continuum of possible orientations of the magnetic moments. However, since the magnetic moment of an atom is quantized, quantum mechanics predicts that the z component of the magnetic moment can have

$2j + 1$ values, where j is a quantum number associated with the total angular momentum of the atom that results from the combination of the orbital and spin angular momenta of the electrons in the atom. This experiment thus measures the angular momentum of the atom. If $j = 0$, corresponding to zero angular momentum for the atom, the magnetic moment will be zero and there will be no deflection of the atoms. If $j = 1$, there will be three possible orientations of the z component of the magnetic moment, and the original beam of atoms will be split into three beams. If $j = \frac{1}{2}$, the original beam will be split into two beams corresponding to the two possible orientations of the magnetic moment.

This experiment was performed by Stern and Gerlach in 1922 using silver atoms and by Phipps and Taylor in 1927 using hydrogen atoms. In each case, the beam was split into two beams as shown in Figure 37-13. Since the ground state of hydrogen has $\ell = 0$, we would expect no splitting were it not for electron spin. The net angular momentum of the hydrogen atom in its ground state is just the spin angular momentum of the electron.† The splitting of the beam of hydrogen atoms into two beams confirms the result that the spin angular momentum of the electron can have just two orientations corresponding to the quantum number $s = \frac{1}{2}$. The quantization of the orientation of the magnetic moment of the electron to either of two directions in space is called **space quantization.**

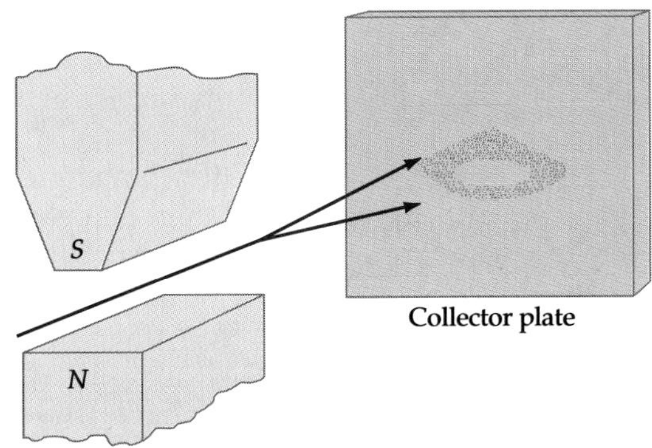

Figure 37-13 Results of the Stern–Gerlach experiment. The atomic beam is split into two beams, indicating that the magnetic moments of the atoms are quantized to two orientations in space. The shape of the upper line is due to the greater inhomogeneity of the magnetic field near the upper pole face.

Exercise

A beam of atoms is split into five lines when passing though an inhomogeneous magnetic field. What is the quantum number associated with the angular momentum of the atoms? (Answer: 2)

37-5 Addition of Angular Momenta and the Spin–Orbit Effect

In general, an electron in an atom has both orbital angular momentum **L** characterized by the quantum number ℓ and spin angular momentum **S** characterized by the quantum number s. Analogous classical systems that have two kinds of angular momentum are the earth, which is spinning about its axis of rotation in addition to revolving about the sun, or a precess-

†The nucleus of an atom also has angular momentum and, therefore, a magnetic moment. However, the mass of the nucleus is about 2000 times that of the electron for hydrogen and is greater still for other atoms. From Equation 37-19, we expect the magnetic moment of the hydrogen nucleus to be on the order of 1/2000 of a Bohr magneton since m_q is now the mass of a proton rather than an electron.

ing gyroscope, which has angular momentum of precession in addition to its spin. Classically, the total angular momentum **J** is given by

$$\mathbf{J} = \mathbf{L} + \mathbf{S} \qquad 37\text{-}25$$

This is an important quantity because the net torque on a system equals the rate of change of the total angular momentum, and in the case of central forces, the total angular momentum is conserved. For a classical system, the magnitude of the total angular momentum **J** can have any value between $L + S$ and $L - S$. We have seen that angular momentum in quantum mechanics is more complicated. Both **L** and **S** are quantized, and their directions are restricted. Quantum mechanics also limits the possible values of the total angular momentum **J**. For an electron with orbital angular momentum characterized by the quantum number ℓ and spin $s = \frac{1}{2}$, the total angular momentum **J** has the magnitude $\sqrt{j(j+1)}\,\hbar$, where the quantum number j can be either

$$j = \ell + \tfrac{1}{2}$$

or

$$j = \ell - \tfrac{1}{2} \qquad \ell \neq 0 \qquad 37\text{-}26$$

(For $\ell = 0$, the total angular momentum is simply the spin, so $j = \frac{1}{2}$.) In Figure 37-14, vector diagrams illustrate the two possible combinations $j = \frac{3}{2}$ and $j = \frac{1}{2}$ for the case of $\ell = 1$. The lengths of the vectors are proportional to $\sqrt{\ell(\ell+1)}\,\hbar$, $\sqrt{s(s+1)}\,\hbar$, and $\sqrt{j(j+1)}\,\hbar$. The spin and orbital angular momenta are said to be parallel when $j = \ell + s$ and antiparallel when $j = \ell - s$.

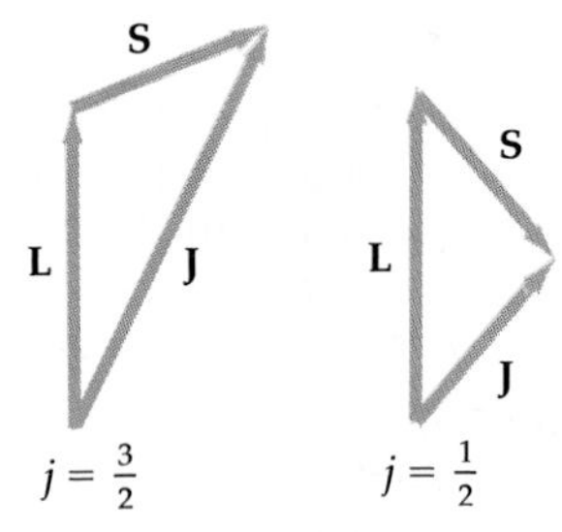

Figure 37-14 Vector diagrams illustrating the addition of orbital and spin angular momentum for the case $\ell = 1$ and $s = \frac{1}{2}$. There are two possible values of the quantum number for the total angular momentum, $j = \ell + s = \frac{3}{2}$ and $j = \ell - s = \frac{1}{2}$.

Equation 37-26 is a special case of a more general rule for combining two angular momenta that is useful when dealing with more than one particle. For example, there are two electrons in helium, each with spin, orbital, and total angular momenta. The general rule is

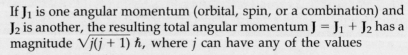

> If $\mathbf{J}_1$ is one angular momentum (orbital, spin, or a combination) and $\mathbf{J}_2$ is another, the resulting total angular momentum $\mathbf{J} = \mathbf{J}_1 + \mathbf{J}_2$ has a magnitude $\sqrt{j(j+1)}\,\hbar$, where j can have any of the values
>
> $$j = j_1 + j_2,\ j_1 + j_2 - 1,\ \ldots,\ |j_1 - j_2| \qquad 37\text{-}27$$

The z component of the total angular momentum can then have any of the $2j + 1$ values

$$m_j = -j,\ -j + 1,\ \ldots,\ +j$$

Example 37-3

Two electrons each have zero orbital angular momentum. What are the possible quantum numbers for the total angular momentum of the two-electron system?

In this case, $j_1 = j_2 = \frac{1}{2}$. Equation 37-27 then gives two possible results, $j = 1$ and $j = 0$. These results are commonly called parallel spins and antiparallel spins. For $j = 1$, the quantum number m_j can be -1, 0, or $+1$; for $j = 0$, m_j must be 0. Thus the $j = 1$ state is commonly referred to as the **triplet state** and the $j = 0$ state is referred to as the **singlet state**.

Example 37-4

An electron in an atom has orbital angular momentum $\mathbf{L}_1$ with quantum number $\ell_1 = 2$ and a second electron has orbital angular momentum $\mathbf{L}_2$ with quantum number $\ell_2 = 3$. What are the possible quantum numbers ℓ for the total orbital angular momentum $\mathbf{L} = \mathbf{L}_1 + \mathbf{L}_2$?

Since $\ell_1 + \ell_2 = 2 + 3 = 5$ and $|\ell_1 - \ell_2| = |2 - 3| = 1$, the possible values of ℓ are 5, 4, 3, 2, and 1.

In spectroscopic notation, the total angular-momentum quantum number of an atomic state is written as a subscript after the code letter describing the orbital angular momentum. For example, the ground state of hydrogen is written $1S_{1/2}$, where the number preceding the code letter indicates the value of n. The $n = 2$ states can have either $\ell = 0$ or $\ell = 1$, and the $\ell = 1$ state can have either $j = \frac{3}{2}$ or $j = \frac{1}{2}$. These states are thus denoted by $2S_{1/2}$, $2P_{3/2}$, and $2P_{1/2}$. Thus the code denotes $n(\ell)_j$.

Atomic states with the same values of n and ℓ but different values of j have slightly different energies because of the interaction between the spin of the electron and its orbital motion. This effect is called the **spin–orbit effect.** The resulting splitting of the spectral line, such as the splitting that results from the transitions $2P_{3/2} \rightarrow 2S_{1/2}$ and $2P_{1/2} \rightarrow 2S_{1/2}$ in hydrogen, is called **fine-structure splitting.** We can understand the spin–orbit effect qualitatively from the simple Bohr-model picture shown in Figure 37-15. In this figure, the electron moves in a circular orbit with speed v around a fixed proton. The orbital angular momentum **L** is up. In the frame of reference of the electron, the proton is moving in a circle around it, thus constituting a circular loop of current that produces a magnetic field **B** at the position of the electron. The direction of **B** is up, parallel to **L**. The potential energy of a magnetic moment in a magnetic field depends on its orientation and is given by Equation 27-13:

$$U = -\boldsymbol{\mu}\cdot\mathbf{B} = -\mu_z B \qquad 37\text{-}28$$

The potential energy is lowest when the magnetic moment is parallel to **B** and highest when it is antiparallel. Since the magnetic moment of the electron is directed opposite to its spin (because the electron has a negative charge), the spin–orbit energy is highest when the spin is parallel to **B** and thus to **L**. The energy of the $2P_{3/2}$ state in hydrogen, in which **L** and **S** are parallel, is therefore slightly higher than the $2P_{1/2}$ state, in which **L** and **S** are antiparallel (Figure 37-16).

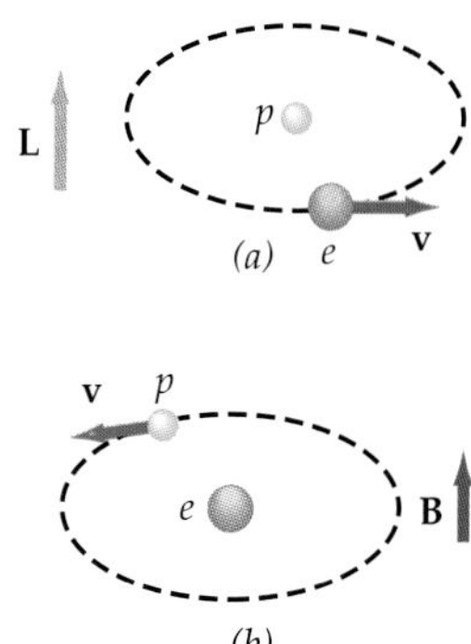

Figure 37-15 (*a*) An electron moving about a proton with angular momentum **L** up. (*b*) The magnetic field **B** seen by the electron due to the apparent (relative) motion of the proton is also up. When the electron spin is parallel to **L**, the magnetic moment is antiparallel to **L** and **B**, so the spin–orbit energy is at its greatest.

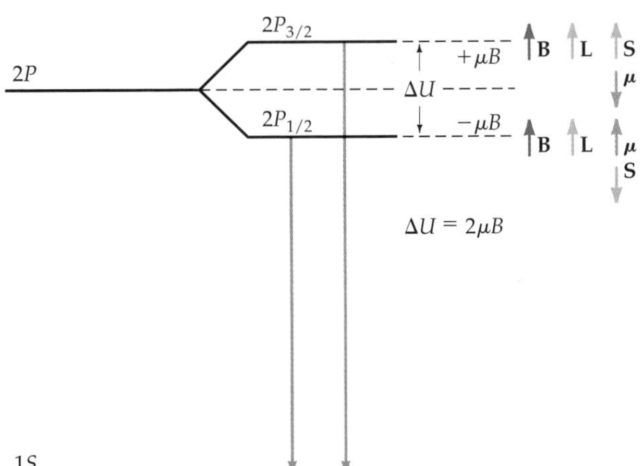

Figure 37-16 Fine-structure energy-level diagram. On the left, the levels in the absence of a magnetic field are shown. The effect of an applied field is shown on the right. Because of the spin–orbit interaction, the magnetic field splits the $2P$ level into two energy levels, with the $j = \frac{3}{2}$ level having slightly greater energy than the $j = \frac{1}{2}$ level. The spectral line due to the transition $2P \rightarrow 1S$ is therefore split into two lines of slightly different wavelengths.

Example 37-5

The fine-structure splitting of the $2P_{3/2}$ and $2P_{1/2}$ levels in hydrogen is 4.5×10^{-5} eV. If the $2p$ electron sees an internal magnetic field **B**, the spin–orbit energy splitting will be of the order of $\Delta E = 2\mu_B B$, where μ_B is the Bohr magneton. From this, estimate the magnetic field that the $2p$ electron in hydrogen experiences.

We have

$$\Delta E = 2\mu_B B = 4.5 \times 10^{-5}\ \text{eV}$$

$$B = \frac{4.5 \times 10^{-5}\ \text{eV}}{2\mu_B} = \frac{4.5 \times 10^{-5}\ \text{eV}}{2(5.79 \times 10^{-5}\ \text{eV/T})} = 0.39\ \text{T}$$

When an atom is placed in an external magnetic field **B**, the energy of the atomic state characterized by the angular-momentum quantum number j is split into $2j + 1$ energy levels corresponding to the $2j + 1$ possible values of the z component of the angular momentum and therefore to the $2j + 1$ possible values of the z component of the magnetic moment. This splitting of the energy levels in the atom gives rise to a splitting of the spectral lines emitted by the atom. The splitting of the spectral lines of an atom placed in an external magnetic field was discovered by P. Zeeman and is known as the **Zeeman effect.**

37-6 The Periodic Table

In the quantum-mechanical theory, the state of each electron in atoms with more than one electron is described by the quantum numbers n, ℓ, m, and m_s. The energy of the electron is determined mainly by the principal quantum number n (which is related to the radial dependence of the wave function) and by the orbital angular-momentum quantum number ℓ. Generally, the lower the values of n and ℓ, the lower the energy. The dependence of the energy on ℓ is due to the interaction of the electrons in the atom with each other. In hydrogen, of course, there is only one electron, and the energy is independent of ℓ. The specification of n and ℓ for each electron in an atom is called the **electron configuration.** Customarily, ℓ is specified according to the same code used to label the states of the hydrogen atom rather than by its numerical value. The code is

Code	s	p	d	f	g	h
ℓ value	0	1	2	3	4	5

Capital letters are used to specify atomic states, and lowercase letters are used for individual electron states. The n values are sometimes referred to as shells, which are identified by another letter code: $n = 1$ denotes the K shell; $n = 2$, the L shell; and so on.

An important principle that governs the electron configuration of atoms is the **Pauli exclusion principle:**

> No two electrons in an atom can be in the same quantum state; that is, no two electrons can have the same set of values for the quantum numbers n, ℓ, m, and m_s.

As we saw in Section 36-11, the exclusion principle is related to the fact that the wave function describing the atom must be antisymmetric in exchange of any two electrons. Using the exclusion principle and the restrictions on the quantum numbers discussed in the previous section (n is an integer, ℓ is an integer that ranges from 0 to $n - 1$, m can have $2\ell + 1$ values from $-\ell$ to ℓ in integral steps, and m_s can be either $+\frac{1}{2}$ or $-\frac{1}{2}$), we can understand much of the structure of the periodic table. We have already discussed the lightest element, hydrogen, which has just one electron. In the ground (lowest energy) state, the electron has $n = 1$ and $\ell = 0$, with $m = 0$ and $m_s = +\frac{1}{2}$ or $-\frac{1}{2}$. We call this a 1s electron. The 1 signifies that $n = 1$, and the s signifies that $\ell = 0$.

As electrons are added to make the heavier atoms, the electrons go into those states that will give the lowest total energy consistent with the Pauli exclusion principle.

Helium ($Z = 2$)

The next element after hydrogen is helium ($Z = 2$), which has two electrons. In the ground state, both electrons are in the K shell with $n = 1$, $\ell = 0$, and $m = 0$; one electron has $m_s = +\frac{1}{2}$ and the other has $m_s = -\frac{1}{2}$. This configuration is lower in energy than any other two-electron configuration. The resultant spin of the two electrons is zero. Since the orbital angular momentum is also zero, the total angular momentum is zero. The electron configuration for helium is written $1s^2$. The 1 signifies that $n = 1$, the s signifies that $\ell = 0$, and the 2 signifies that there are two electrons in this state. Since ℓ can be only 0 for $n = 1$, these two electrons fill the K ($n = 1$) shell. The energy required to remove an electron from an atom is called the **ionization energy.** The ionization energy is the binding energy of the last electron placed in the atom. For helium, the ionization energy is 24.6 eV, which is relatively large. Helium is therefore basically inert.

Example 37-6

Calculate the energy of interaction of the two electrons in the ground state of the helium atom and use it to find the average separation of the two electrons.

If the electrons did not interact, the energy of each electron in helium would be given by Equation 37-9 with $Z = 2$. For the ground state, $n = 1$ and

$$E_1 = -\frac{Z^2E_0}{n^2} = -\frac{(2)^2(13.6\text{ eV})}{(1)^2} = -54.4\text{ eV}$$

The total energy of the two electrons in the ground state would be twice this, $2(-54.4\text{ eV}) = -108.8$ eV. If one electron were removed, the energy of the other electron would be -54.4 eV. Thus, the energy needed to remove one electron would be 54.4 eV, which would be the ionization energy. Since the measured ionization energy is 24.6 eV, the energy of the ground state of helium must be $-(54.4\text{ eV} + 24.6\text{ eV}) = -79.0$ eV. This is 29.8 eV higher than -108.8 eV, so the energy of interaction of the two electrons in the ground state of helium is 29.8 eV.

The energy of interaction of two electrons a distance r apart is the potential energy $U = ke^2/r$. Setting this equal to 29.8 eV, we obtain

$$U = \frac{ke^2}{r} = 29.8\text{ eV}$$

Since we know that $ke^2/a_0 = 13.6$ eV, it is convenient to multiply this equation by a_0/a_0 before solving it for r. We then obtain

$$U = \frac{ke^2}{r}\frac{a_0}{a_0} = \frac{ke^2}{a_0}\frac{a_0}{r} = (13.6 \text{ eV})\frac{a_0}{r} = 29.8 \text{ eV}$$

$$r = \left(\frac{13.6 \text{ eV}}{29.8 \text{ eV}}\right)a_0 = 0.456a_0 = 0.024 \text{ nm}$$

where we have used $a_0 = 0.0529$ nm. This separation is approximately equal to the radius of the first Bohr orbit for an electron in helium, which is $r_1 = a_0/Z = \frac{1}{2}a_0$.

Lithium ($Z = 3$)

The next element, lithium, has three electrons. Since the K shell is completely filled with two electrons, the third electron must go into a higher energy shell. The next lowest energy shell after $n = 1$ is the $n = 2$ or L shell. The outer electron is much farther from the nucleus than are the two inner, $n = 1$ electrons. It is most likely to be found at the radius of the second Bohr orbit, which is four times the radius of the first Bohr orbit. (The radii of the Bohr orbits, as given by Equation 35-22, are proportional to n^2.)

The nuclear charge is partially screened from the outer electron by the two inner electrons. Recall that the electric field outside a spherically symmetric charge density is the same as if all the charge were at the center of the sphere. If the outer electron were completely outside of the charge cloud of the two inner electrons, the electric field it would see would be that of a single charge $+e$ at the center due to the nuclear charge of $+3e$ and the charge $-2e$ of the inner electron cloud. However, the outer electron does not have a well-defined orbit; instead, it is itself a charge cloud that penetrates the charge cloud of the inner electrons to some extent. Because of this penetration, the effective nuclear charge $Z'e$ is somewhat greater than $+1e$. The energy of the outer electron at a distance r from a point charge $+Z'e$ is

$$E = -\frac{1}{2}\frac{kZ'e^2}{r} \qquad 37\text{-}29$$

(a)

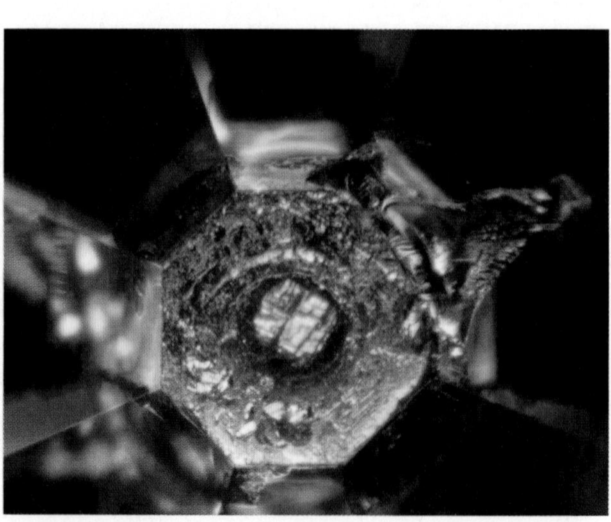

(b)

(a) A diamond anvil cell, in which the facets of two diamonds (about 1 mm^2 each) are used to compress a sample substance, subjecting it to very high pressure. (b) Samarium monosulfide (SmS) is normally a black, dull-looking semiconductor. When it is subjected to pressure above 7000 atm, an electron from the 4f state is dislocated into the 5d state. The resulting compound glitters like gold and behaves like a metal.

(This is Equation 35-19 with the nuclear charge $+Ze$ replaced by $+Z'e$.) The greater the penetration of the inner electron cloud, the greater the effective nuclear charge $Z'e$ and the lower the energy. Because the penetration is greater for lower ℓ values (see Figure 37-8), the energy of the outer electron in lithium is lower for the s state ($\ell = 0$) than for the p state ($\ell = 1$). The electron configuration of lithium in the ground state is therefore $1s^22s$. The ionization energy of lithium is only 5.39 eV. Because its outer electron is so loosely bound to the atom, lithium is very active chemically. It behaves like a "one-electron atom," and its spectrum is similar to that of hydrogen.

Beryllium ($Z = 4$)

The fourth electron has the least energy in the $2s$ state. There can be two electrons with $n = 2$, $\ell = 0$, and $m = 0$ because of the two possible values for the spin quantum number m_s. The configuration of beryllium is thus $1s^22s^2$.

Boron to Neon ($Z = 5$ to $Z = 10$)

Since the $2s$ subshell is filled, the fifth electron must go into the next available (lowest energy) subshell, which is the $2p$ subshell, with $n = 2$ and $\ell = 1$. Since there are three possible values of m ($+1$, 0, and -1) and two values of m_s for each value of m, there can be six electrons in this subshell. The electron configuration for boron is $1s^22s^22p$. The electron configurations for the elements carbon ($Z = 6$) to neon ($Z = 10$) differ from that for boron only in the number of electrons in the $2p$ subshell. The ionization energy increases with Z for these elements, reaching the value of 21.6 eV for the last element in the group, neon. Neon has the maximum number of electrons allowed in the $n = 2$ shell. Its electron configuration is $1s^22s^22p^6$. Because of its very high ionization energy, neon, like helium, is basically chemically inert. The element just before neon, fluorine, has a "hole" in the $2p$ subshell; that is, it has room for one more electron. It readily combines with elements such as lithium that have one outer electron. Lithium, for example, will donate its single outer electron to the fluorine atom to make an F^- ion and a Li^+ ion. These ions then bond together to form a molecule of lithium fluoride.

Sodium to Argon ($Z = 11$ to $Z = 18$)

The eleventh electron must go into the $n = 3$ shell. Since this electron is very far from the nucleus and from the inner electrons, it is weakly bound in the sodium ($Z = 11$) atom. The ionization energy of sodium is only 5.14 eV. Sodium therefore combines readily with atoms such as fluorine. With $n = 3$, the value of ℓ can be 0, 1, or 2. Because of the lowering of the energy due to penetration of the electron shield formed by the other ten electrons (similar to that discussed for lithium) the $3s$ state is lower than the $3p$ or $3d$ states. This energy difference between subshells of the same n value becomes greater as the number of electrons increases. The electron configuration of sodium is $1s^22s^22p^63s^1$. As we move to elements with higher values of Z, the $3s$ subshell and then the $3p$ subshell fill. These two subshells can accommodate $2 + 6 = 8$ electrons. The configuration of argon ($Z = 18$) is $1s^22s^22p^63s^23p^6$. One might expect the nineteenth electron to go into the third subshell (the d subshell with $\ell = 2$), but the penetration effect is now so strong that the energy of the next electron is lower in the $4s$ subshell than in the $3d$ subshell. There is thus another large energy difference between the eighteenth and nineteenth electrons, and so argon, with its full $3p$ subshell, is basically stable and inert.

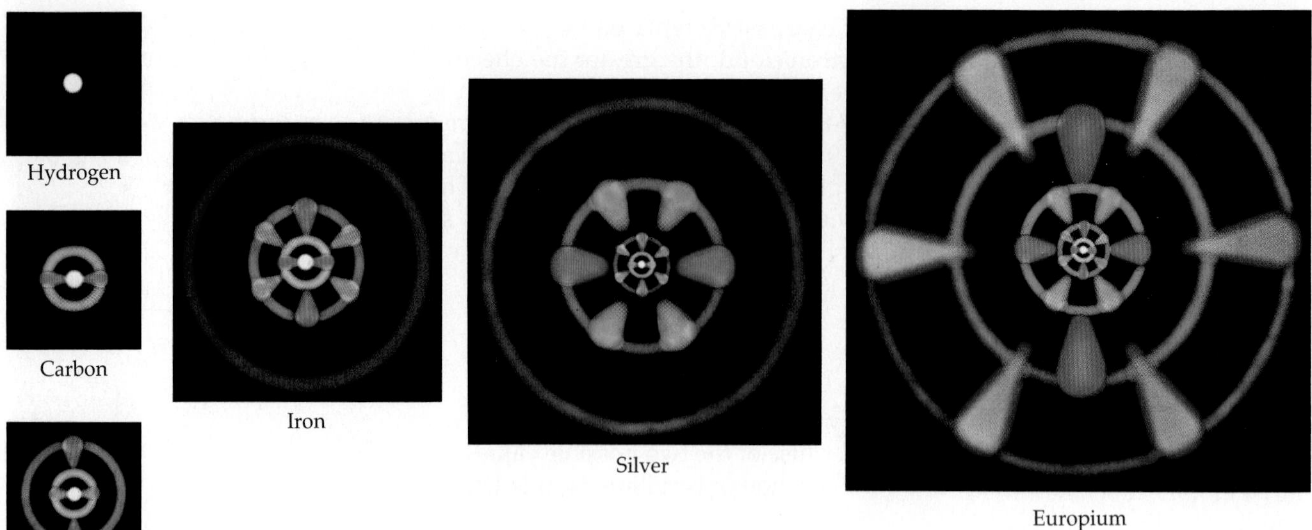

A schematic depiction of the electron configurations in atoms. The spherically symmetric *s* states can contain 2 electrons and are colored white and blue. The dumbbell-shaped *p* states can contain up to 6 electrons and are colored orange. The *d* states can contain up to 10 electrons and are colored yellow-green. The *f* states can contain up to 14 electrons and are colored purple.

Elements with $Z > 18$

The nineteenth electron in potassium (Z = 19) and the twentieth electron in calcium (Z = 20) go into the 4*s* rather than the 3*d* subshell. The electron configurations of the next ten elements, scandium (Z = 21) through zinc (Z = 30), differ only in the number of electrons in the 3*d* shell, except for chromium (Z = 24) and copper (Z = 29), each of which has only one 4*s* electron. These ten elements are called **transition elements.** Since their chemical properties are mainly due to their 4*s* electrons, they are quite similar chemically.

Figure 37-17 shows a plot of the first ionization energy (the energy needed to remove one electron from an atom) versus Z for Z = 1 to Z = 60. The peaks in ionization energy at Z = 2, 10, 18, 36, and 54 mark the closing of a shell or subshell. Table 37-1 (see pp. 1240–1241) gives the electron configurations of all the elements.

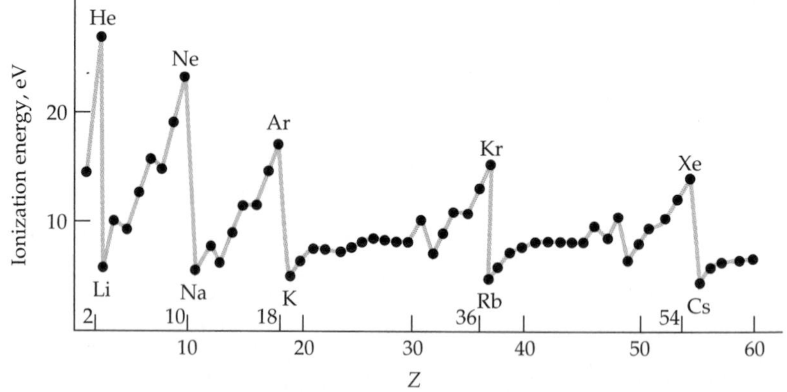

Figure 37-17 First ionization energy versus Z for Z = 1 to Z = 60. This energy is the binding energy of the last electron in the atom. The binding energy increases with Z until a shell is closed at Z = 2, 10, 18, 36, and 54. Elements with a closed shell plus one outer electron, such as sodium (Z = 11), have very low binding energies because the outer electron is very far from the nucleus and is shielded by the inner core electrons.

Questions

2. Why is the energy of the 3*s* state considerably lower than that of the 3*p* state in sodium, whereas in hydrogen these states have essentially the same energy?
3. Discuss the evidence from the periodic table of the need for a fourth quantum number. How would the properties of helium differ if there were only three quantum numbers, *n*, *ℓ*, and *m*?

37-7 Optical and X-Ray Spectra

Optical Spectra

When an atom is in an excited state (that is, when one or more of its electrons is in an energy state above the ground state), the electrons make transitions to lower energy states and, in doing so, emit electromagnetic radiation. The frequency of this radiation is related to the initial and final energy states of the electron by the Bohr formula (Equation 35-17), $f = (E_i - E_f)/h$, where E_i and E_f are the initial and final energies and h is Planck's constant. The wavelength of the radiation is, of course, related to the frequency by $\lambda = c/f$. An atom can be excited to a higher energy state by bombarding it with a beam of electrons in a glass tube with a high voltage across it. Since the excited energy states of an atom are discrete rather than continuous, only certain wavelengths are emitted. The spectral lines at these wavelengths constitute the emission spectrum of the atom.

To understand atomic spectra, we need to understand the excited states of an atom. The situation for an atom with many electrons is, in general, much more complicated than that of hydrogen with its one electron. An excited state of the atom may involve a change in the state of any one of its electrons or even two or more of its electrons. Fortunately, in most cases, an excited state of an atom involves the excitation of just one of the electrons in the atom. The energies of excitation of the outer, valence electrons of an atom are of the order of a few electron volts. Transitions involving these electrons result in photons in or near the visible or **optical spectrum.** (Recall that the energies of visible photons range from about 1.8 eV to 3 eV.) Excitation energies can often be calculated using a simple model in which the atom is pictured as a single electron plus a stable core consisting of the nucleus and the other, inner electrons. This model works particularly well for the alkali metals—lithium, sodium, potassium, rubidium, and cesium—the elements in the first column of the periodic table.

A neon sign outside a Chinatown restaurant in Paris. Neon atoms in the tube are excited by an electron current passing through the tube. The excited neon atoms emit light in the visible range as they decay toward their ground states. The colors of neon signs result from the characteristic red-orange spectrum of neon plus the color of the glass tube itself.

Figure 37-18 shows an energy-level diagram for the optical transitions in sodium, whose electrons consist of a neon core plus one outer electron. Since the spin angular momentum of the core adds up to zero, the spin of each state of sodium is $\frac{1}{2}$. Because of the spin–orbit effect, the states with $j = \ell - \frac{1}{2}$ have a slightly lower energy than those with $j = \ell + \frac{1}{2}$. Each state (except for the S states) is therefore a doublet. The doublet splitting is very small and is not evident with the energy scale used in the figure. The states are labeled using spectroscopic notation in which the superscript 2 before the letter code indicates that the state is a doublet. Thus, $^2P_{3/2}$ (read as "doublet P three halves") denotes a state in which $\ell = 1$ and $j = \frac{3}{2}$. (The S states are customarily labeled as if they were doublet even though they are not.)

In the first excited state, the outer electron is excited from the $3s$ level to the $3p$ level, which is about 2.1 eV above the ground state. The energy difference between the $P_{3/2}$ and $P_{1/2}$ states due to the spin–orbit effect is about 0.002 eV. Transitions from these states to the ground state give the familiar yellow doublet of sodium:

$$3p(^2P_{1/2}) \rightarrow 3s(^2S_{1/2}) \qquad \lambda = 589.59 \text{ nm}$$

$$3p(^2P_{3/2}) \rightarrow 3s(^2S_{1/2}) \qquad \lambda = 588.99 \text{ nm}$$

It is important to distinguish between the doublet energy states and doublet spectral lines. All transitions beginning or ending on an S state give doublet lines because they involve one doublet state and one singlet state. [The selection rule $\Delta\ell = \pm 1$ (Equation 37-10) rules out transitions between two S states.] There are four possible energy differences between two doublet

Table 37-1 **Electron configurations of the atoms in their ground states. For some of the rare-earth elements ($Z = 57$ to 71) and the heavy elements ($Z > 89$) the configurations are not firmly established.**

			Shell: *K*	*L*		*M*			*N*				*O*				*P*			*Q*
			n: 1	2		3			4				5				6			7
Z		Element	ℓ: *s*	*s*	*p*	*s*	*p*	*d*	*s*	*p*	*d*	*f*	*s*	*p*	*d*	*f*	*s*	*p*	*d*	*s*
1	H	hydrogen	1																	
2	He	helium	2																	
3	Li	lithium	2	1																
4	Be	beryllium	2	2																
5	B	boron	2	2	1															
6	C	carbon	2	2	2															
7	N	nitrogen	2	2	3															
8	O	oxygen	2	2	4															
9	F	fluorine	2	2	5															
10	Ne	neon	2	2	6															
11	Na	sodium	2	2	6	1														
12	Mg	magnesium	2	2	6	2														
13	Al	aluminum	2	2	6	2	1													
14	Si	silicon	2	2	6	2	2													
15	P	phosphorus	2	2	6	2	3													
16	S	sulfur	2	2	6	2	4													
17	Cl	chlorine	2	2	6	2	5													
18	Ar	argon	2	2	6	2	6													
19	K	potassium	2	2	6	2	6	.	1											
20	Ca	calcium	2	2	6	2	6	.	2											
21	Sc	scandium	2	2	6	2	6	1	2											
22	Ti	titanium	2	2	6	2	6	2	2											
23	V	vanadium	2	2	6	2	6	3	2											
24	Cr	chromium	2	2	6	2	6	5	1											
25	Mn	manganese	2	2	6	2	6	5	2											
26	Fe	iron	2	2	6	2	6	6	2											
27	Co	cobalt	2	2	6	2	6	7	2											
28	Ni	nickel	2	2	6	2	6	8	2											
29	Cu	copper	2	2	6	2	6	10	1											
30	Zn	zinc	2	2	6	2	6	10	2											
31	Ga	gallium	2	2	6	2	6	10	2	1										
32	Ge	germanium	2	2	6	2	6	10	2	2										
33	As	arsenic	2	2	6	2	6	10	2	3										
34	Se	selenium	2	2	6	2	6	10	2	4										
35	Br	bromine	2	2	6	2	6	10	2	5										
36	Kr	krypton	2	2	6	2	6	10	2	6										
37	Rb	rubidium	2	2	6	2	6	10	2	6	.	.	1							
38	Sr	strontium	2	2	6	2	6	10	2	6	.	.	2							
39	Y	yttrium	2	2	6	2	6	10	2	6	1	.	2							
40	Zr	zirconium	2	2	6	2	6	10	2	6	2	.	2							
41	Nb	niobium	2	2	6	2	6	10	2	6	4	.	1							
42	Mo	molybdenum	2	2	6	2	6	10	2	6	5	.	1							
43	Tc	technetium	2	2	6	2	6	10	2	6	6	.	1							
44	Ru	ruthenium	2	2	6	2	6	10	2	6	7	.	1							
45	Rh	rhodium	2	2	6	2	6	10	2	6	8	.	1							
46	Pd	palladium	2	2	6	2	6	10	2	6	10	.	.							
47	Ag	silver	2	2	6	2	6	10	2	6	10	.	1							
48	Cd	cadmium	2	2	6	2	6	10	2	6	10	.	2							
49	In	indium	2	2	6	2	6	10	2	6	10	.	2	1						
50	Sn	tin	2	2	6	2	6	10	2	6	10	.	2	2						
51	Sb	antimony	2	2	6	2	6	10	2	6	10	.	2	3						
52	Te	tellurium	2	2	6	2	6	10	2	6	10	.	2	4						

Table 37-1 *(Continued)*

		Shell:	K	L		M			N				O				P			Q
		n:	1	2		3			4				5				6			7
Z		Element ℓ:	s	s	p	s	p	d	s	p	d	f	s	p	d	f	s	p	d	s
53	I	iodine	2	2	6	2	6	10	2	6	10	.	2	5						
54	Xe	xenon	2	2	6	2	6	10	2	6	10	.	2	6						
55	Cs	cesium	2	2	6	2	6	10	2	6	10	.	2	6	.	.	1			
56	Ba	barium	2	2	6	2	6	10	2	6	10	.	2	6	.	.	2			
57	La	lanthanum	2	2	6	2	6	10	2	6	10	.	2	6	1	.	2			
58	Ce	cerium	2	2	6	2	6	10	2	6	10	1	2	6	1	.	2			
59	Pr	praseodymium	2	2	6	2	6	10	2	6	10	3	2	6	.	.	2			
60	Nd	neodymium	2	2	6	2	6	10	2	6	10	4	2	6	.	.	2			
61	Pm	promethium	2	2	6	2	6	10	2	6	10	5	2	6	.	.	2			
62	Sm	samarium	2	2	6	2	6	10	2	6	10	6	2	6	.	.	2			
63	Eu	europium	2	2	6	2	6	10	2	6	10	7	2	6	.	.	2			
64	Gd	gadolinium	2	2	6	2	6	10	2	6	10	7	2	6	1	.	2			
65	Tb	terbium	2	2	6	2	6	10	2	6	10	9	2	6	.	.	2			
66	Dy	dysprosium	2	2	6	2	6	10	2	6	10	10	2	6	.	.	2			
67	Ho	holmium	2	2	6	2	6	10	2	6	10	11	2	6	.	.	2			
68	Er	erbium	2	2	6	2	6	10	2	6	10	12	2	6	.	.	2			
69	Tm	thulium	2	2	6	2	6	10	2	6	10	13	2	6	.	.	2			
70	Yb	ytterbium	2	2	6	2	6	10	2	6	10	14	2	6	.	.	2			
71	Lu	lutetium	2	2	6	2	6	10	2	6	10	14	2	6	1	.	2			
72	Hf	hafnium	2	2	6	2	6	10	2	6	10	14	2	6	2	.	2			
73	Ta	tantalum	2	2	6	2	6	10	2	6	10	14	2	6	3	.	2			
74	W	tungsten (wolfram)	2	2	6	2	6	10	2	6	10	14	2	6	4	.	2			
75	Re	rhenium	2	2	6	2	6	10	2	6	10	14	2	6	5	.	2			
76	Os	osmium	2	2	6	2	6	10	2	6	10	14	2	6	6	.	2			
77	Ir	iridium	2	2	6	2	6	10	2	6	10	14	2	6	7	.	2			
78	Pt	platinum	2	2	6	2	6	10	2	6	10	14	2	6	9	.	1			
79	Au	gold	2	2	6	2	6	10	2	6	10	14	2	6	10	.	1			
80	Hg	mercury	2	2	6	2	6	10	2	6	10	14	2	6	10	.	2			
81	Tl	thallium	2	2	6	2	6	10	2	6	10	14	2	6	10	.	2	1		
82	Pb	lead	2	2	6	2	6	10	2	6	10	14	2	6	10	.	2	2		
83	Bi	bismuth	2	2	6	2	6	10	2	6	10	14	2	6	10	.	2	3		
84	Po	polonium	2	2	6	2	6	10	2	6	10	14	2	6	10	.	2	4		
85	At	astatine	2	2	6	2	6	10	2	6	10	14	2	6	10	.	2	5		
86	Rn	radon	2	2	6	2	6	10	2	6	10	14	2	6	10	.	2	6		
87	Fr	francium	2	2	6	2	6	10	2	6	10	14	2	6	10	.	2	6	.	1
88	Ra	radium	2	2	6	2	6	10	2	6	10	14	2	6	10	.	2	6	.	2
89	Ac	actinium	2	2	6	2	6	10	2	6	10	14	2	6	10	.	2	6	1	2
90	Th	thorium	2	2	6	2	6	10	2	6	10	14	2	6	10	.	2	6	2	2
91	Pa	protactinium	2	2	6	2	6	10	2	6	10	14	2	6	10	1	2	6	2	2
92	U	uranium	2	2	6	2	6	10	2	6	10	14	2	6	10	3	2	6	1	2
93	Np	neptunium	2	2	6	2	6	10	2	6	10	14	2	6	10	4	2	6	1	2
94	Pu	plutonium	2	2	6	2	6	10	2	6	10	14	2	6	10	6	2	6	.	2
95	Am	americium	2	2	6	2	6	10	2	6	10	14	2	6	10	7	2	6	.	2
96	Cm	curium	2	2	6	2	6	10	2	6	10	14	2	6	10	7	2	6	1	2
97	Bk	berkelium	2	2	6	2	6	10	2	6	10	14	2	6	10	8	2	6	1	2
98	Cf	californium	2	2	6	2	6	10	2	6	10	14	2	6	10	10	2	6	.	2
99	Es	einsteinium	2	2	6	2	6	10	2	6	10	14	2	6	10	11	2	6	.	2
100	Fm	fermium	2	2	6	2	6	10	2	6	10	14	2	6	10	12	2	6	.	2
101	Md	mendelevium	2	2	6	2	6	10	2	6	10	14	2	6	10	13	2	6	.	2
102	No	nobelium	2	2	6	2	6	10	2	6	10	14	2	6	10	14	2	6	.	2
103	Lw	lawrencium	2	2	6	2	6	10	2	6	10	14	2	6	10	14	2	6	1	2

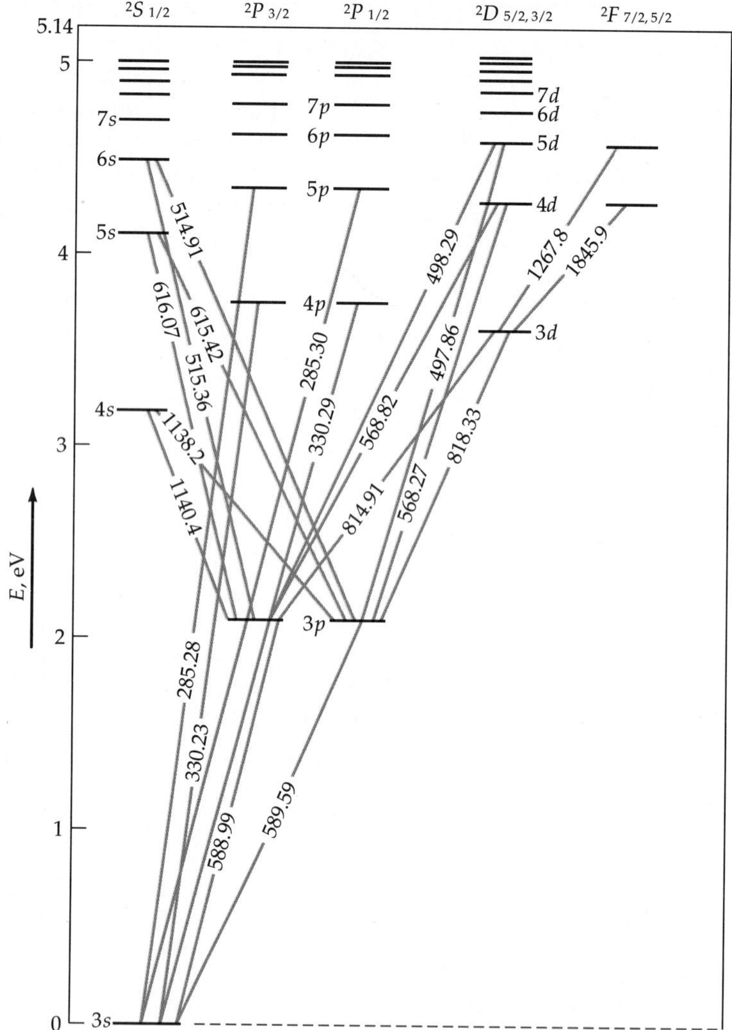

Figure 37-18 Energy-level diagram for sodium. The diagonal lines show observed optical transitions, with wavelengths given in nanometers. The energy of the ground state has been chosen as the zero point for the scale on the left.

states. One of these is ruled out by a selection rule on j, which is

$$\Delta j = \pm 1 \text{ or } 0 \qquad \text{allowed}$$

but

$$j = 0 \rightarrow j = 0 \qquad \text{forbidden}$$

Transitions between doublet states therefore result in triplet spectral lines. The energy levels and optical spectra of the other alkali metals are similar to those for sodium.

The optical spectra for elements that have two outer electrons, such as helium, beryllium, and magnesium, are considerably more complex because of the interaction between the two outer electrons.

X-Ray Spectra

The energy needed to excite an inner, core electron, for example, an electron in the $n = 1$ state (K shell), is much greater than that needed to excite an outer, valence electron. An inner electron cannot be excited to any of the filled states, such as the $n = 2$ states in sodium, because of the Pauli exclusion principle. Therefore, the energy required to excite an inner electron to an unoccupied state is typically of the order of several thousand electron volts (keV). An inner electron can be excited by bombarding the atom with a high-energy electron beam in, for example, an x-ray tube. If an electron is knocked out of the $n = 1$ (K) shell, there is a vacancy left in this shell. This vacancy can be filled by an electron from the L shell or a higher shell that makes a transition to the K shell. The photons emitted by electrons making such transitions have energies of the order of 1 keV. They constitute the **characteristic x-ray spectrum** of an element, appearing as sharp peaks in the continuous x-ray spectrum of the element, as shown for molybdenum in Figure 37-19. Spectral lines arising from transitions that end at the $n = 1$ (K) shell make up the K series of the characteristic x-ray spectrum of an element. For instance, the K_α line in the figure arises from transitions from the $n = 2$ (L) shell to the $n = 1$ (K) shell, and the K_β line arises from transitions from the $n = 3$ shell to the $n = 1$ shell. A second series, the L series, is produced by transitions from higher energy states to a vacated place in the $n = 2$ (L) shell.

We can use the Bohr theory to calculate the approximate frequencies of characteristic x-ray spectra. According to the Bohr model, the energy of a single electron in a state n is given by Equation 37-9:

$$E_n = -Z^2 \frac{13.6 \text{ eV}}{n^2}$$

Since for any atom other than hydrogen there are two electrons in the innermost shell, the K shell, the effective charge seen by one of the electrons is less than Ze because of the shielding due to the other electron. If the effective charge is $(Z - 1)e$, the energy of an electron in the K shell is given by this equation when $n = 1$ and Z is replaced by $Z - 1$:

$$E_1 = -(Z - 1)^2(13.6 \text{ eV})$$

The energy of an electron in state n (provided that the effective charge is the same) is given by

$$E_n = -(Z - 1)^2 \frac{13.6 \text{ eV}}{n^2} \qquad 37\text{-}30$$

When an electron from state n drops into the vacated state in the $n = 1$ shell, a photon of energy $E_n - E_1$ is emitted. The wavelength of this photon is

$$\lambda = \frac{hc}{E_n - E_1} = \frac{hc}{(Z - 1)^2(13.6 \text{ eV})(1 - 1/n^2)} \qquad 37\text{-}31$$

In 1913, the English physicist H. Moseley measured the wavelengths of the characteristic x-ray spectra for about 40 elements. From his data he was able to determine the atomic number Z for each element.

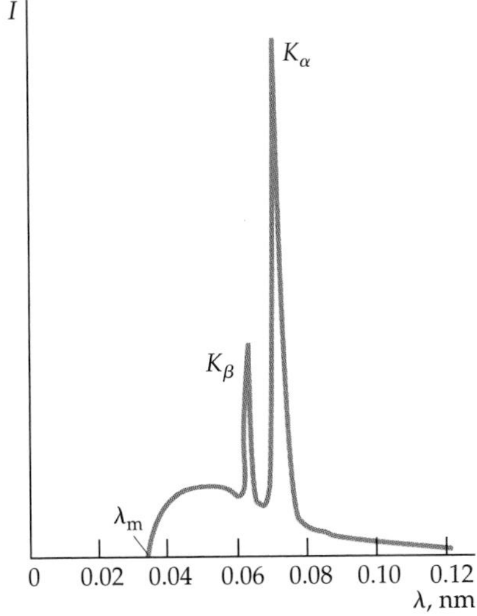

Figure 37-19 X-ray spectrum of molybdenum. The sharp peaks labeled K_α and K_β are characteristic of the element. The cutoff wavelength λ_m is independent of the target element and is related to the voltage V of the x-ray tube by $\lambda_m = hc/eV$.

Example 37-7

Calculate the wavelength of the K_α x-ray line for molybdenum $(Z = 42)$ and compare it with the value $\lambda = 0.0721$ nm measured by Moseley.

The K_α line corresponds to a transition from $n = 2$ to $n = 1$. The wavelength is given by Equation 37-31 with $Z = 42$ and $n = 2$:

$$\lambda = \frac{hc}{(41)^2(13.6 \text{ eV})(1 - \frac{1}{4})} = \frac{1240 \text{ eV}\cdot\text{nm}}{(41)^2(13.6 \text{ eV})(3/4)} = 0.0723 \text{ nm}$$

This result is in good agreement with the measured value.

Questions

4. Would you expect the optical spectrum of potassium to be like that of hydrogen or that of helium?
5. Would you expect the optical spectrum of beryllium to be like that of hydrogen or that of helium?

37-8 Absorption, Scattering, and Stimulated Emission

Information about the energy levels of an atom is usually obtained from the radiation emitted when the atom makes a transition from an excited state to a state of lower energy. We can also obtain information about such energy levels from the absorption spectrum. When atoms are irradiated with a continuous spectrum of radiation, the transmitted radiation shows dark lines corresponding to absorption of light at discrete wavelengths. The absorption spectra of atoms were the first line spectra observed. Since at normal temperatures atoms and molecules are in either their ground states or low-lying excited states, absorption spectra are usually simpler than emission spectra. For example, only those lines corresponding to the Lyman emission series are seen in the absorption spectrum of atomic hydrogen because nearly all the atoms are initially in their ground states.

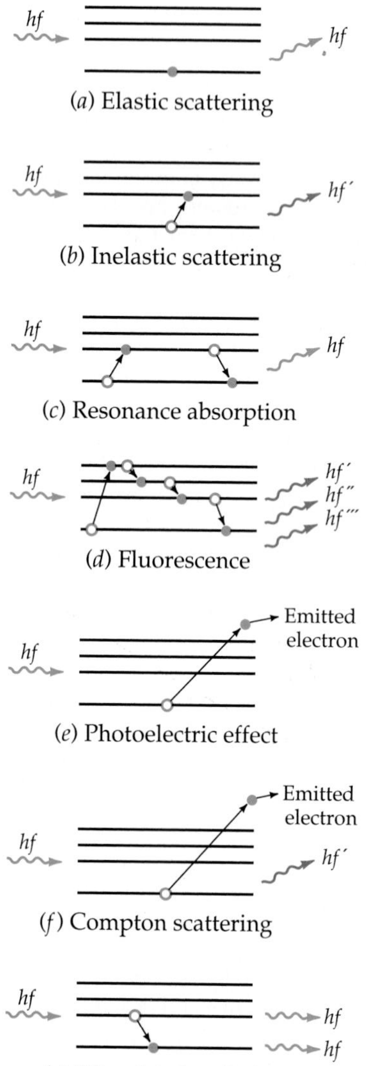

Figure 37-20 illustrates several interesting phenomena in addition to absorption that can occur when a photon is incident on an atom. In Figure 37-20*a*, the energy of the incoming photon is too small to excite the atom to one of its excited states, so the atom remains in its ground state and the photon is said to be scattered. Since the incoming and outgoing or scattered photons have the same energy the scattering is said to be elastic. If the wavelength of the incident light is large compared with the size of the atom, the scattering can be described in terms of classical electromagnetic theory and is called **Rayleigh scattering** after Lord Rayleigh, who worked out the theory in 1871. The probability of Rayleigh scattering varies as $1/\lambda^4$. This means that blue light is scattered much more readily than red light, which accounts for the bluish color of the sky. The removal of blue light by Raleigh scattering also accounts for the reddish color of the transmitted light seen in sunsets.

Figure 37-20*b* shows **inelastic scattering,** which occurs when the incident photon has enough energy to cause the atom to make a transition to an excited state. The energy of the scattered photon hf' is related to the energy of the incident photon hf by

$$hf' = hf - \Delta E$$

where ΔE is the excitation energy, which is the difference between the energy of the ground state and the energy of the excited state. Inelastic scattering of light from molecules was first observed by the Indian physicist C. V. Raman and is therefore often referred to as **Raman scattering.**

In Figure 37-20*c*, the energy of the incident photon is just equal to the difference in energy between the ground state and the first excited state of the atom. The atom makes a transition to its first excited state and then after a short delay makes a transition back to its ground state with the emission of a photon whose energy is equal to that of the incident photon. This multistep process is called **resonance absorption.** The phase of the emitted photon is not correlated with the phase of the incident photon. The emission of a photon as an atom spontaneously makes a transition to a lower state is called **spontaneous emission.**

In Figure 37-20*d*, the energy of the incident photon is great enough to excite the atom to one of its higher excited states. The atom then loses its energy by spontaneous emission as it makes one or more transitions to lower energy states. A common example occurs when the atom is excited by ultraviolet light and emits visible light as it returns to its ground state. This process is called **fluorescence.** Since the lifetime of a typical excited atomic energy state is of the order of 10^{-8} s, this process appears to occur instantaneously. However, some excited states have much longer lifetimes—of the

hf hf
(*a*) Elastic scattering

hf hf′
(*b*) Inelastic scattering

hf hf
(*c*) Resonance absorption

hf hf′ hf″ hf‴
(*d*) Fluorescence

hf Emitted electron
(*e*) Photoelectric effect

hf Emitted electron hf′
(*f*) Compton scattering

hf hf hf
(*g*) Stimulated emission

Figure 37-20 Phenomena that can occur when a photon is incident on an atom.

(a)

(b)

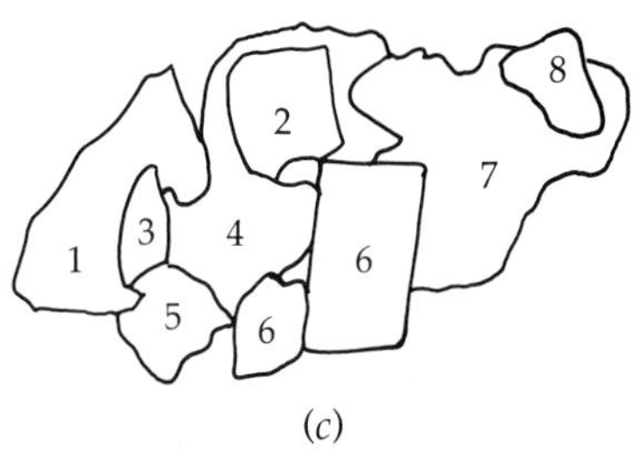

(c)

A collection of minerals in (a) daylight and (b) ultraviolet light (sometimes called "black light"). Identified by number in the schematic (c), they are 1 powellite, 2 willemite, 3 scheelite, 4 calcite, 5 calcite and willemite composite, 6 optical calcite, 7 willemite, 8 opal. The change in color is due to the minerals fluorescing under the ultraviolet light. In the case of optical calcite, both fluorescence and phosphorescence are occurring.

order of milliseconds or occasionally seconds or even minutes. Such a state is called a **metastable state. Phosphorescent materials** have metastable states, and so emit light long after the original excitation.

Figure 37-20*e* illustrates the photoelectric effect, in which the absorption of the photon ionizes the atom by causing the emission of an electron. Figure 37-20*f* illustrates Compton scattering, which occurs if the energy of the incident photon is much greater than the ionization energy. Note that in Compton scattering, a photon is emitted, whereas in the photoelectric effect, the photon is absorbed with none emitted.

Figure 37-20*g* illustrates **stimulated emission.** This process occurs if the atom or molecule is initially in an excited state of energy E_2, and the energy of the incident photon is equal to $E_2 - E_1$, where E_1 is the energy of a lower state or the ground state. In this case, the oscillating electromagnetic field associated with the incident photon stimulates the excited atom or molecule which then emits a photon in the same direction as the incident photon and in phase with it. In spontaneous emission, the phase of the light from one atom is unrelated to that from another atom, so the resulting light is incoherent. However, in stimulated emission, the phase of the light emitted from one atom is related to that emitted by every other atom, so the resulting light is coherent. As a result, interference of the light from different atoms can be observed.

In addition to being excited by incident radiation, atoms may be excited to higher energy states by collisions with electrons or with other atoms.

37-9 The Laser

The **laser** (*l*ight *a*mplification by *s*timulated *e*mission of *r*adiation) is a device that produces a strong beam of coherent photons by stimulated emission. Consider a system of atoms that have a ground-state energy E_1 and an excited-state energy E_2. If these atoms are irradiated by photons of energy $E_2 - E_1$, those atoms in the ground state can absorb a photon and make the transition to state E_2, whereas those atoms already in the excited state may be stimulated to decay back to the ground state. The relative probabilities of absorption and stimulated emission were first worked out by Einstein, who showed them to be equal. Ordinarily, at normal temperatures, nearly all the atoms will initially be in the ground state, so absorption will be the main effect. To produce more stimulated-emission transitions than absorption transitions, we must arrange to have more atoms in the excited state than in the ground state. This condition is called **population inversion.** It can be achieved if the excited state is a metastable state. Population inversion is often obtained by a method called **optical pumping** in which atoms are "pumped" up to energy levels of energy greater than E_2 by the absorption of an intense auxiliary radiation. The atoms then decay down to state E_2 by either spontaneous emission or by nonradiative transitions such as those due to collisions.

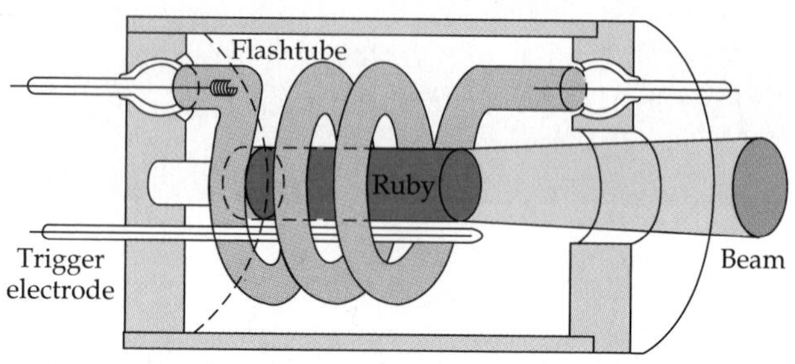

Figure 37-21 Schematic diagram of the first ruby laser.

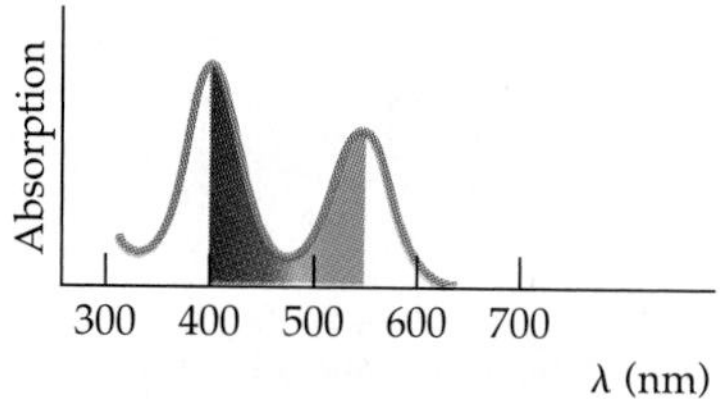

Figure 37-22 Absorption versus wavelength for Cr^{3+} in ruby. Ruby appears red because of the strong absorption of green and blue light by the chromium ions.

Figure 37-21 shows a schematic diagram of the first laser, a ruby laser built by Theodore Maiman in 1960. It consists of a small rod of ruby (a few centimeters long) surrounded by a helical gaseous flashtube. The ends of the ruby rod are flat and perpendicular to the axis of the rod. Ruby is a transparent crystal of Al_2O_3 with a small amount (about 0.05 percent) of chromium. It appears red because the chromium ions (Cr^{3+}) have strong absorption bands in the blue and green regions of the visible spectrum as shown in Figure 37-22. The energy levels of chromium that are important for the operation of a ruby laser are shown in Figure 37-23.

When the flashtube is fired, there is an intense burst of light lasting a few milliseconds. Absorption excites many of the chromium ions to the bands of energy levels indicated by the shading in Figure 37-23. The chromium ions then relax, giving up their energy to the crystal in nonradiative transitions and drop down to a pair of metastable states labeled E_2 in the figure. These metastable states are about 1.79 eV above the ground state. If

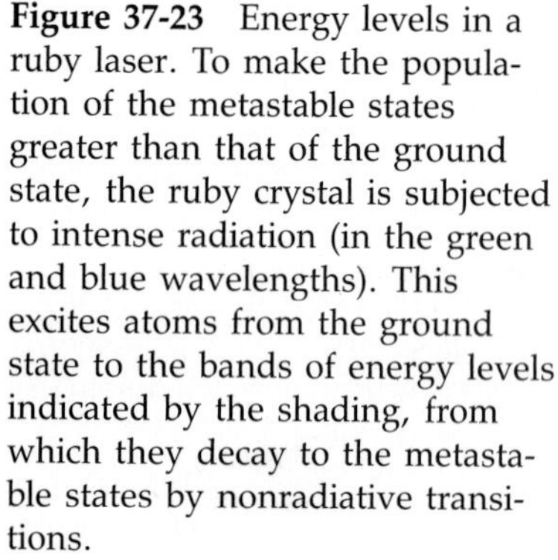

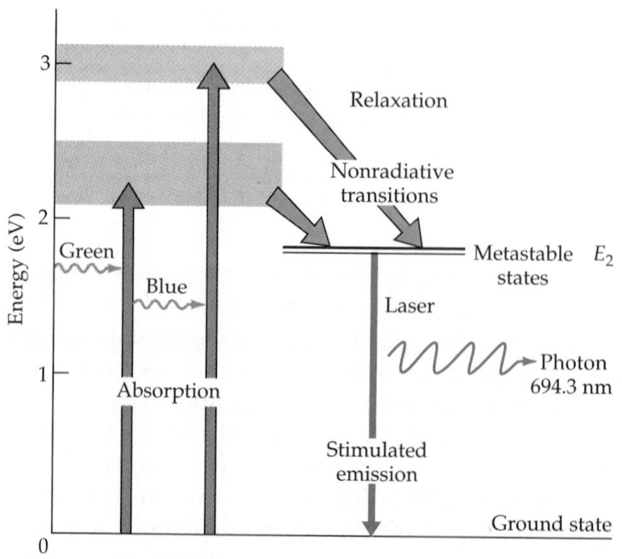

Figure 37-23 Energy levels in a ruby laser. To make the population of the metastable states greater than that of the ground state, the ruby crystal is subjected to intense radiation (in the green and blue wavelengths). This excites atoms from the ground state to the bands of energy levels indicated by the shading, from which they decay to the metastable states by nonradiative transitions.

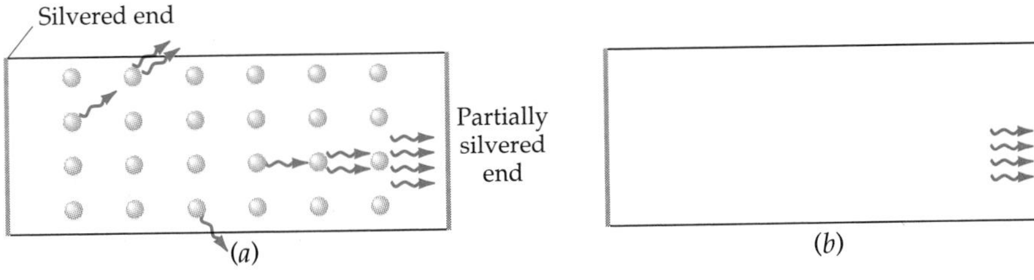

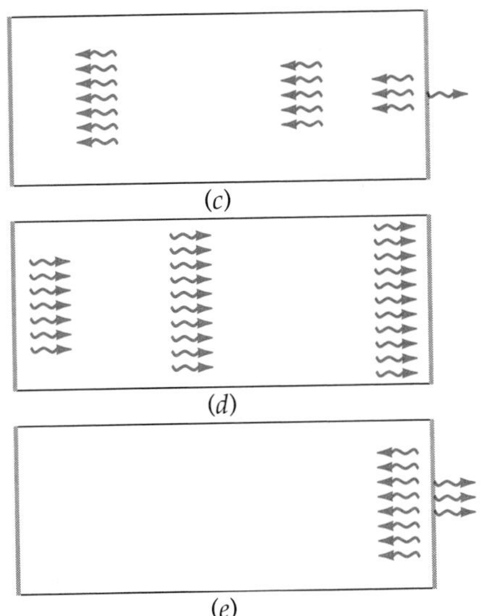

Figure 37-24 Buildup of photon beam in a laser. In (a), some of the atoms spontaneously emit photons, some of which travel to the right and stimulate other atoms to emit photons parallel to the axis of the crystal. In (b), four photons strike the partially silvered right face of the laser. In (c), one photon has been transmitted and the others have been reflected. As these photons traverse the laser crystal, they stimulate other atoms to emit photons and the beam builds up. By the time the beam reaches the right face again in (d), it comprises many photons. In (e), some of these photons are transmitted and the rest are reflected.

the flash is intense enough, more atoms will make the transition to the states E_2 than remain in the ground state. As a result, the populations of the ground state and the metastable states become inverted. When some of the atoms in the states E_2 decay to the ground state by spontaneous emission, they emit photons of energy 1.79 eV and wavelength 694.3 nm. Some of these photons stimulate other excited atoms to emit photons of the same energy and wavelength.

In the ruby laser, both ends of the crystal are silvered such that one end is almost totally reflecting (about 99.9 percent) and the other end is only partially reflecting (about 99 percent) so that some of the beam is transmitted. If the ends are parallel, standing waves are set up, and an intense beam of coherent light emerges through the partially silvered end. Figure 37-24 illustrates the buildup of the beam inside the laser. When photons traveling parallel to the axis of the crystal strike the silvered ends, all are reflected from the back face and most are reflected from the front face, with a few escaping through the partially silvered front face. During each pass through the crystal, the photons stimulate more and more atoms so that the photon beam builds up and an intense beam is emitted.

Modern ruby lasers generate intense light beams with energies ranging from 50 J to 100 J in pulses lasting a few milliseconds. The beam can have a diameter as small as 1 mm and an angular divergence as small as 0.25 milliradian to about 7 milliradians.

In 1961, the first successful operation of a continuous helium–neon gas laser was announced by Ali Javan, W. R. Bennet, Jr., and D. R. Herriott. Figure 37-25 shows a schematic diagram of the type of helium–neon laser commonly used for physics demonstrations. It consists of a gas tube containing 15 percent helium gas and 85 percent neon gas. A totally reflecting flat mirror is mounted at one end of the gas tube and a partially reflecting concave mirror is placed at the other end. The concave mirror focuses parallel light at the flat mirror and also acts as a lens that transmits part of the light so that it emerges as a parallel beam.

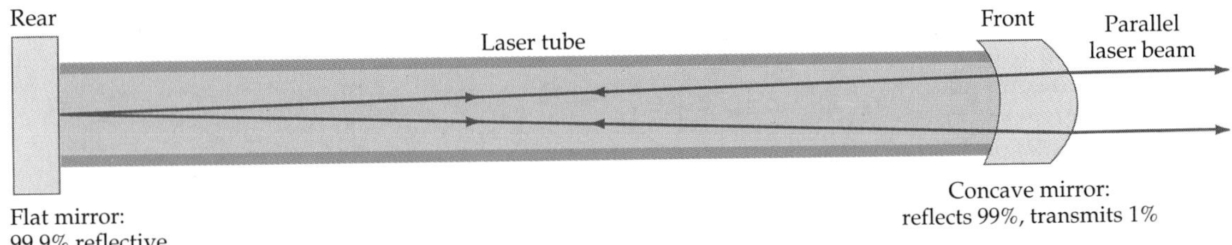

Figure 37-25 Schematic drawing of a helium–neon laser. The use of a concave mirror rather than a second plane mirror makes the alignment of the mirrors less critical than it is for the ruby laser. The concave mirror on the right also serves as a lens that focuses the emitted light into a parallel beam.

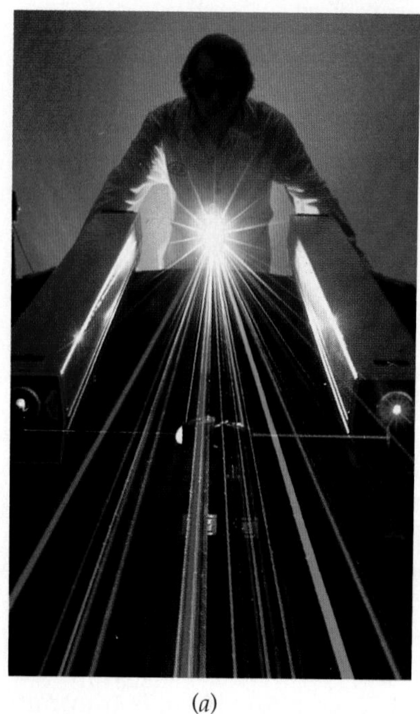

(*a*) Beams from a krypton and an argon laser, split into their component wavelengths. In these gas lasers, krypton and argon atoms have been stripped of multiple electrons, forming positive ions. The light-emitting energy transitions occur when excited electrons in the ions decay from one upper energy level to another. Here, several energy transitions are occuring at once, each corresponding to emitted light of a different wavelength. (*b*) Fluorescent organic compounds called dyes, dissolved in a solvent, are used in tuneable dye lasers. An external light source excites the dye molecules, which decay to their ground state in a series of steps, emitting light during some of the transitions. The dye molecules (which are large molecules, with multiple ring structures) have a range of vibrational and rotational, as well as electronic, energy states. The energy levels are closely enough spaced to allow a near continuum of possible transitions. In the laser, the dyes are contained in adjustable resonant cavities. Light emitted by the dye oscillates in the cavity, which is tuned to amplify only certain wavelengths.

Population inversion is achieved somewhat differently in the helium–neon laser than in the ruby laser. Figure 37-26 shows the energy levels of helium and neon that are important for operation of the laser. (The complete energy-level diagrams for helium and neon are considerably more complicated.) Helium has an excited energy state $E_{2,\text{He}}$ that is 20.61 eV above its ground state. Helium atoms are excited to state $E_{2,\text{He}}$ by an electric discharge. Neon has an excited state $E_{3,\text{Ne}}$ that is 20.66 eV above its ground state. This is just 0.05 eV above the first excited state of helium. The neon atoms are excited to state $E_{3,\text{Ne}}$ by collisions with excited helium atoms. The kinetic energy of the helium atoms provides the extra 0.05-eV energy needed to excite the neon atoms. There is another excited state of neon $E_{2,\text{Ne}}$ that is 18.70 eV above its ground state and 1.96 eV below state $E_{3,\text{Ne}}$. Since state $E_{2,\text{Ne}}$ is normally unoccupied, population inversion between states $E_{3,\text{Ne}}$ and $E_{2,\text{Ne}}$ is obtained immediately. The stimulated emission that occurs between these states results in photons of energy 1.96 eV and wavelength 632.8 nm, which produces a bright red light. After stimulated emission, the atoms in state $E_{2,\text{Ne}}$ decay to the ground state by spontaneous emission.

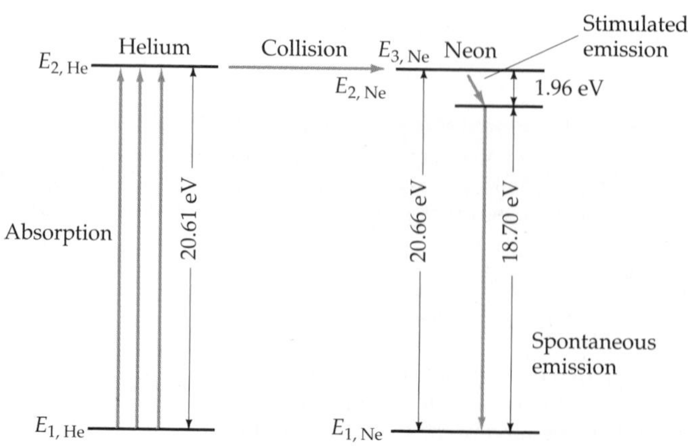

Figure 37-26 Energy levels of helium and neon that are important for the helium–neon laser. The helium atoms are excited by electrical discharge to an energy state 20.61 eV above the ground state. They collide with neon atoms, exciting some neon atoms to an energy state 20.66 eV above the ground state. Population inversion is thus achieved between this level and one 1.96 eV below it. The spontaneous emission of photons of energy 1.96 eV stimulates other atoms in the upper state to emit photons of energy 1.96 eV.

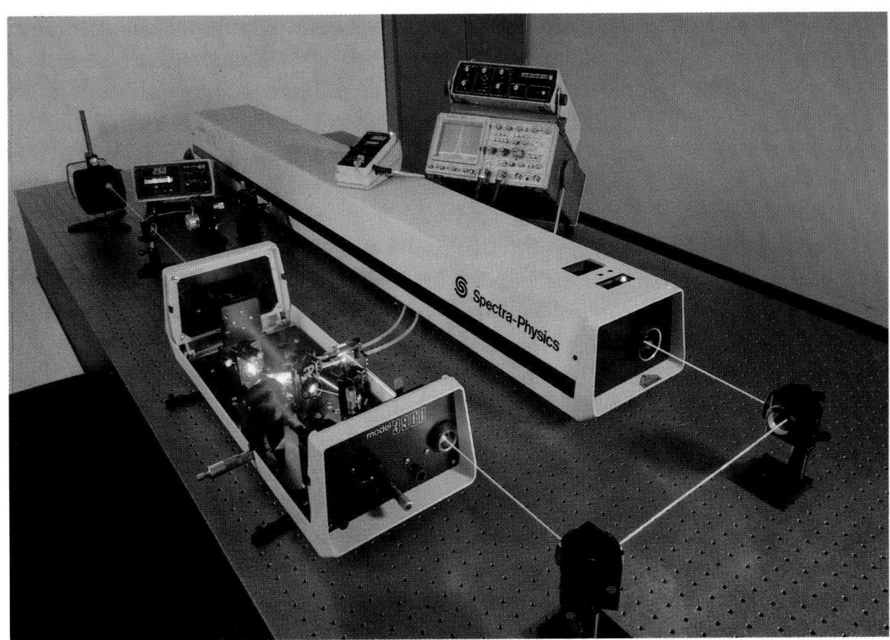

(c)

(d)

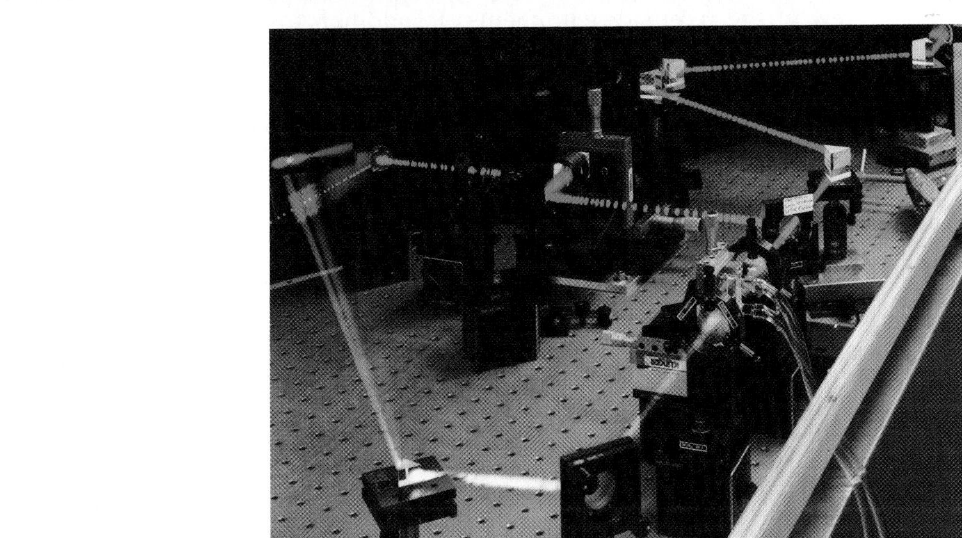

(e)

(c) The titanium-sapphire laser is presently state-of-the-art in tuneable lasers. The blue beam is laser light acting as an optical pump for the titanium-sapphire laser, whose cover is removed. Vibrational energy levels in the titanium atoms embedded in the sapphire crystal are superimposed on the normal energy levels of the aluminum and oxygen atoms that compose the sapphire. Sapphire atoms are excited to higher energy states by absorbing the light from the optical pump. They emit red light and return to a lower energy state corresponding to the lowest vibrational energy mode of the crystal. Because the electronic energy transitions can terminate anywhere in the band of the finely spaced vibrational states, the light is emitted in a broad range of wavelengths. This light is diverted to a tuning cavity where a particular wavelength can be extracted (for instance, by adjusting the angle of a diffraction grating). (d) A semiconductor laser, shown for scale in the eye of a needle. *pn* Junction semiconductors (Chapter 39) are constructed by doping a crystal with small amounts of an impurity element. When an external voltage is applied across the semiconductor, free electrons from the impurity element combine with holes in the crystal lattice and light energy is released. In the cleaved-coupled cavity design shown here, opposite ends of the semiconductor crystal are cleaved to form reflective facets. The reflected light itself is partially absorbed by electron-hole pairs, causing additional electron-hole combinings and thereby amplifying the net emitted light. (e) A femtosecond pulsed laser. By a technique known as "modelocking," different excited modes within a laser's cavity can be made to interfere with one another and create a series of ultrashort pulses, picoseconds long, that correspond to the time it takes light to bounce back and forth once within the cavity. It has been possible to compress such pulses even further, once they have left the laser. Ultrashort pulses have been used as probes to study the behavior of molecules during chemical reactions.

Note that there are four energy levels involved in the helium–neon laser, whereas the ruby laser involved only three levels. In a three-level laser, population inversion is difficult to achieve because more than half the atoms in the ground state must be excited. In a four-level laser, population inversion is easily achieved because the state after stimulated emission is not the ground state but an excited state that is normally unpopulated.

A laser beam is coherent, very narrow, and intense. Its coherence makes the laser beam useful in the production of holograms, which were discussed in Section 33-11. The precise direction and small angular spread of the beam make it useful as a surgical tool for destroying cancer cells or reattaching a detached retina. Lasers are also used by surveyors for precise alignment over large distances. Distances can be accurately measured by reflecting a laser pulse from a mirror and measuring the time the pulse takes to travel to the mirror and back. The distance to the moon has been measured to within a few centimeters using a mirror placed on the moon for that purpose. Laser beams are also used in fusion research. An intense laser pulse is focused on tiny pellets of deuterium–tritium in a combustion chamber. The beam heats the pellets to temperatures of the order of 10^8 K in a very short time, causing the deuterium and tritium to fuse and release energy.

Laser technology is advancing so fast that it is possible to mention only a few of the recent developments. In addition to the ruby laser, there are many other solid-state lasers with output wavelengths ranging from about 170 nm to about 3900 nm. Lasers that generate more than 1 kW of continuous power have been constructed. Pulsed lasers can now deliver nanosecond pulses of power exceeding 10^9 W. Various gas lasers can now produce beams of wavelengths ranging from the far infrared to the ultraviolet. Semiconductor lasers (also known as diode lasers or junction lasers) the size of a pinhead can develop 200 mW of power. Liquid lasers using chemical dyes can be tuned over a range of wavelengths (about 70 nm for continuous lasers and more than 170 nm for pulsed lasers). A relatively new laser, the free-electron laser, extracts light energy from a beam of free electrons moving through a spatially varying magnetic field. The free-electron laser has the potential for very high power and high efficiency and can be tuned over a large range of wavelengths. There appears to be no limit to the variety and uses of modern lasers.

Question

6. Why is helium needed in a helium–neon laser? Why not just use neon?

Summary

1. In quantum theory, the hydrogen atom is described by a wave function, the square of which gives the probability of finding the electron in a given region of space. The wave function is characterized by four quantum numbers:

$$n = 1, 2, 3, \ldots$$

$$\ell = 0, 1, \ldots, n - 1$$

$$m = -\ell, -\ell + 1, \ldots, +\ell$$

$$m_s = +\tfrac{1}{2} \text{ or } -\tfrac{1}{2}$$

The energy of the hydrogen atom depends only on the principal quantum number n and is the same as in the Bohr model. In the ground state, $n = 1$, $\ell = 0$, and $m = 0$, and the probability distribution is spherically symmetric, with the electron most likely to be found near the first Bohr radius. It is convenient to think of the electron as a charged cloud with a charge density proportional to the probability distribution.

2. In multielectron atoms, the energy of the electron is determined mainly by the principal quantum number n (which is related to the radial dependence of the wave function) and by the orbital angular-momentum quantum number ℓ. Generally, the lower the values of n and ℓ, the lower the energy. The specification of n and ℓ for all the electrons in an atom is called the electron configuration. Customarily, ℓ is specified by a letter code rather than by its numerical value. The code is

Code	*s*	*p*	*d*	*f*	*g*	*h*
ℓ value	0	1	2	3	4	5

3. An important principle that governs the electron configurations of atoms is the Pauli exclusion principle. It states that no two electrons in an atom can be in the same quantum state; that is, no two electrons can have the same set of values for the quantum numbers n, ℓ, m, and m_s. Using the exclusion principle, we can understand much of the structure of the periodic table of the elements.

4. Atomic spectra consist of optical spectra and x-ray spectra. Optical spectra can be understood in terms of transitions between energy levels of a single outer electron moving in the field of the nucleus and core electrons of the atom. Characteristic x-ray spectra result from the excitation of an inner core electron and the subsequent filling of the resulting vacancy by another electron in the atom.

5. Stimulated emission occurs if an atom is initially in an excited state and a photon of energy equal to the excitation energy is incident on the atom. The oscillating electromagnetic field of the incident photon stimulates the excited atom to emit another photon in the same direction and in phase with the incident photon. The operation of a laser depends on population inversion, in which there are more atoms in an excited state than in the ground state or a lower state. A laser produces an intense, coherent, and narrow beam of photons.

Trapped Atoms and Laser Cooling

D. J. Wineland
National Institute of Standards and Technology, Boulder, Colorado

Physicists are often interested in investigating the energy-level structure of atoms with high accuracy. Perhaps the most important reason for this is the desire to test with precision the theories that predict this energy structure. If a deviation, no matter how small, occurs between the theoretically predicted structure and the experimental measurements, the theory must be modified. What is most interesting is to find the difference to be explained by some new physical effect which must then be incorporated into the theory.

Spectroscopy is an experimental procedure for measuring the energy-level structure of atoms. It involves measuring the energy differences between atomic states by observing the frequency f (or equivalently, the wavelength) of the radiation that is emitted or absorbed in transitions between atomic states. The energy difference between two particular states is $E_2 - E_1 = hf$, where h is Planck's constant (Equation 35-17). The energy levels of the atom can then be constructed from series of such energy differences.

David Wineland has been a staff physicist at the National Institute of Standards and Technology in Boulder, Colorado since 1975. After receiving his bachelor's degree from Berkeley he attended Harvard University where he received his PhD in 1970. From 1970 to 1975, he worked at the University of Washington. There, he learned about trapping techniques from Hans Dehmelt who shared the 1989 Physics Nobel Prize for the development of ion traps. When not in the lab, he enjoys bicycling and skiing.

Measurements of energy differences $(E_2 - E_1)$ are always accompanied by some measurement uncertainty ΔE. In the case of absorption or emission spectra, the uncertainty in energy corresponds to the width of spectral lines: the smaller the uncertainty ΔE, the smaller the range of frequencies spanned, that is, the narrower the lines. Furthermore, the highest resolution (smallest ΔE) can be obtained when the time for transition between states Δt is as long as possible. This is a consequence of the time-energy uncertainty relation, which states that the time over which an energy is measured Δt and the uncertainty of that energy ΔE are related by $\Delta E \Delta t \sim \hbar$ (Section 36-3). In practice, Δt cannot be made arbitrarily long because processes other than emission or absorption—such as collisions with other atoms—can intervene and significantly alter the state of the atom. Such an intervention cuts short the time the atom has to emit or absorb in its undisturbed state, and therefore cuts short the time available for measuring the energy of such an emission or absorption.

The atoms in a sample being analyzed (for example, helium atoms emerging from a small hole in a helium reservoir) are normally moving relative to the radiation source (which is fixed in the laboratory), so the apparent frequencies of light emitted by the atoms are shifted by the doppler effect (Section 34-6). ^{4}He atoms at room temperature move with speeds of about 1.5×10^5 cm/s; therefore the apparent emission frequency of an atom moving toward the radiation source would be shifted to a higher value by the fraction $v/c = 5 \times 10^{-6}$. (The frequency of absorption would be shifted to a lower value by the same amount.) This is a significant amount in very high resolution studies.

The energy resolution can be increased (ΔE made small) and the doppler shift reduced if the atoms are confined to a region of space. By holding the atoms, we can make the time for transition Δt large and ΔE small. As the atoms are held in a localized region, their velocity and associated doppler shift average to zero. The trick in this procedure is to hold the atoms in such a way that when they strike the walls of the container, the atomic states are not distorted to such a degree that the measurement is rendered inaccurate. Various methods are employed to do this; for atomic or molecular ions we can use special configurations of electric and magnetic fields that act on the charge of the ion to pro-

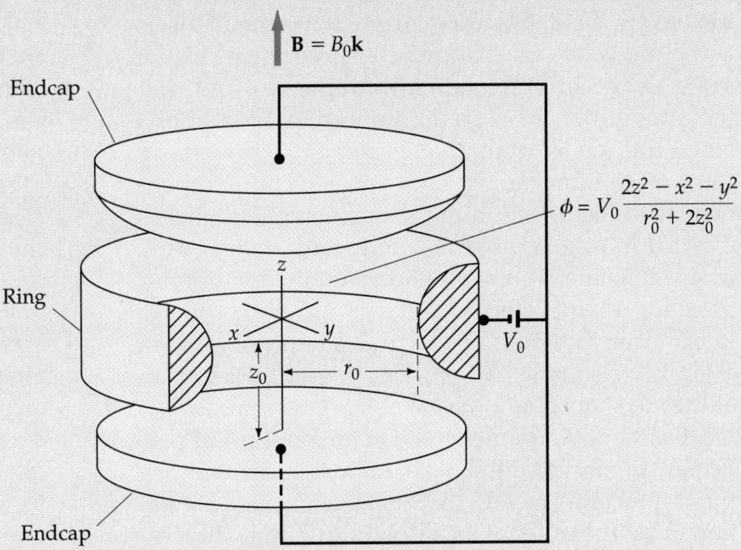

Figure 1 Schematic representation of the electrode configuration for the Penning or Paul trap. Electrode surfaces are figures of revolution about the z axis and are made as close as possible to be equipotentials of the quadratic potential ϕ discussed in the text.

vide confinement. It turns out that these electric and magnetic fields perturb the ion's internal structure very little, so that they do not limit the accuracy of the observation. The "walls" of the container provided by the fields are, in effect, very soft.

Ion Trapping

Several kinds of electromagnetic "traps" are used to confine ions. Cyclotrons (pp. 795–797) and synchrotrons are used for accelerating and storing high-energy elementary particles. Here we'll briefly describe one kind of trap for atomic spectroscopy. This trap, shown in Figure 1, uses static electric and magnetic fields and is usually called a Penning trap, after F. M. Penning who reported the basic device in 1936.

Assume we are interested in confining positively charged ions, for example Be^+ or Hg^+, where one electron has been removed from the neutral atom. If we put a positive voltage V_0 on the two endcap electrodes and leave the central ring at ground potential, the ions experience a force in a direction toward the xy plane; in other words, they are confined as to how far they can travel up or down in the z direction. Unfortunately, this configuration of electrodes causes the ions to feel a radial electric force away from the z axis of the trap, so they accelerate rapidly into the ring electrode. This difficulty can be overcome if we superpose a static magnetic field $\mathbf{B}_0$ whose field lines are parallel to the z axis. This axial magnetic field acting with the radial electric force causes the ions to orbit about the z axis of the trap and confines the ions in all three dimensions.

Let us analyze quantitatively the motion of an ion in a common type of Penning trap. One that is often used has electrode surfaces that conform to equipotentials of the potential (in cartesian coordinates)

$$\phi = V_0(2z^2 - x^2 - y^2)/(r_0^2 + 2z_0^2)$$

This is provided, to a good approximation, by the electrode shapes shown in Figure 2. The particular shape is that generated by rotating a hyperbola around the z axis. From this potential, we can find

Continued

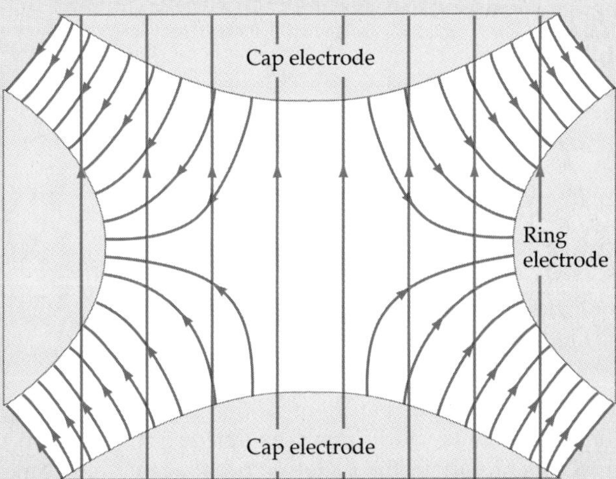

Figure 2 Electric (red) and magnetic (blue) field lines of trap shown in Figure 1.

the electric field (Equation 20-21) along the z axis as $E_z = -d\phi/dz \propto -z$. Thus the electric force on the ion is always in a direction toward the xy plane, and the strength of the force is proportional to the distance the ion is from the xy plane. The magnetic force on the ion can act only in a direction perpendicular to the direction of the magnetic field, meaning that in this case the magnetic force will act only in the xy plane. So the only force on the ion in the z direction is the electric force. Since $F_z \propto -z$, we know from Chapter 12 that the motion of the ion must be harmonic. It is in fact this property of an electrode shaped as a hyperboloid of revolution that makes it a popular one to use in an ion trap. In particular, the fact that the ion's motion is harmonic (in the z direction) means that its frequency of oscillation is independent of its amplitude of oscillation (p. 374). From the results of Chapter 12, we can work out the frequency of oscillation in the z direction, f_z, and find

$$f_z^2 = qV_0/\pi^2 m(r_0^2 + 2z_0^2)$$

where q and m are the ion's charge and mass respectively. To get an idea of the magnitude of this frequency, assume some typical values of parameters encountered in the laboratory. For $V_0 = 1$ V, $m \approx 9$ atomic mass units (e.g., $^9Be^+$ ions), $r_0 = z_0\sqrt{2} = 1$ cm, we find $f_z \approx 74$ kHz.

A charged particle in a magnetic field executes circular motion around the field lines at the cyclotron frequency $f_c = qB/2\pi m$ (Equation 24-8). However the addition of the potential ϕ gives rise to a radial electric field whose strength is proportional to the distance of the ion from the z axis. As this radial electric field is perpendicular to the magnetic field, the center of the ions' cyclotron orbits drifts in a direction perpendicular to the magnetic and electric field lines (Equation 24-9). For the cylindrical geometry of the trap, this drift motion is a circular orbit about the z axis. Therefore the overall motion in the direction normal to the trap axis is a composite of circular motion at approximately the cyclotron frequency (which is actually shifted slightly by the presence of the radial electric field) and a circular motion around the z axis due to the crossed electric and magnetic fields. The latter is usually called the magnetron motion. The energy extracted from this motion in a "magnetron" tube is what provides heat in a microwave oven (see photo on p. 962). The complete motion of the ion in the Penning trap is summarized in Figure 3*a*, *b*.

If the ions collide with background gas atoms, the radius of the magnetron motion gradually increases until the ions strike the ring electrode and stick to it. To prevent this, the trap apparatus is nor-

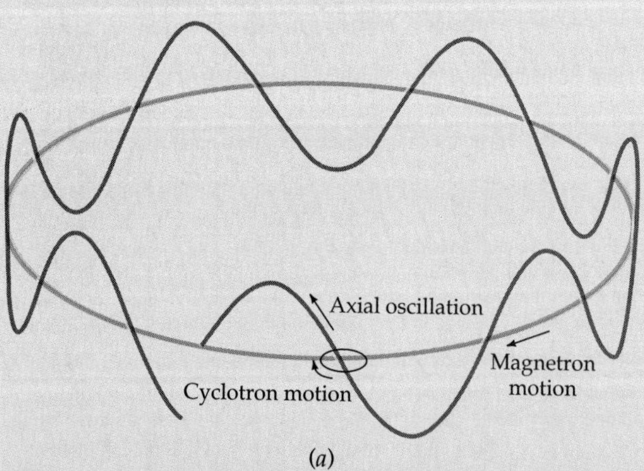

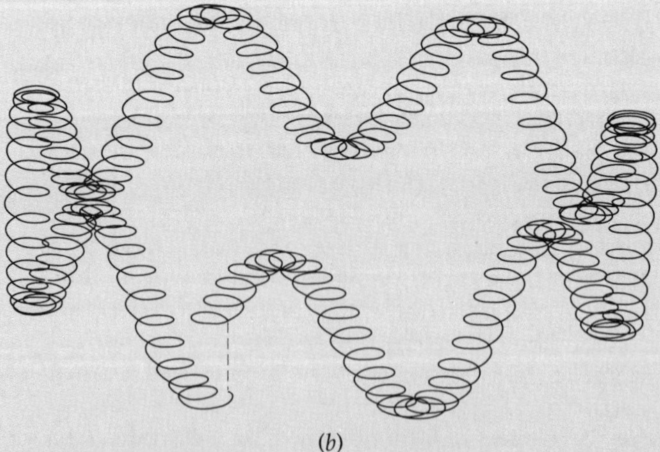

Figure 3 (*a*) Three modes of motion of the ion in the trap combine to yield a complex trajectory (*b*). The ion revolves rapidly in the cyclotron orbit; at the same time the center of the cyclotron orbit follows the much larger circular path of the magnetron motion; all the while the ion is oscillating along the axis of the trap.

Figure 4 Electrodes for a small Paul trap mounted on a penny to give the size. In the experiment, these electrodes are mounted inside a quartz vacuum enclosure where the vacuum is about 10^{-8} Pa or about 10^{-13} atm. The electrodes on either end are held at ground potential and a potential of about 500 V which oscillates sinusoidally at a frequency around 20 MHz is applied to the center ring electrode. The quartz vacuum enclosure allows the ultraviolet light scattered from the ion to pass and be photographed.

mally installed in an evacuated envelope (for example, a sealed glass tube) where the pressure inside the envelope is on the order of 10^{-8} Pa (about 10^{-13} atm). Under these conditions, the ions can be stored in the trap for many days.

Another kind of ion trap uses the same electrode shapes as the Penning trap but confines ions by the action of an oscillating electric potential $V_0 \cos \omega t$ applied between the endcaps and ring electrode. This trap is called the Paul trap after Wolfgang Paul who proposed it in the early 1950s (Figure 4). Figure 5 shows an ultraviolet photograph of a single Hg^+ ion confined in a miniature Paul trap. The high degree of localization obtained with the trap allows this kind of photograph to be made.

With both kinds of traps, ions, electrons, and even more exotic particles such as positrons or antiprotons can be confined for such long times that the resolution of energy measurements is no longer limited by the residence time of the ions in the trap.

Various types of traps can be used for neutral atoms. These provide trapping by electric and magnetic fields but (because of the overall charge neutrality of the atom) the fields must act on the electric or magnetic dipole moment of the atom.

Continued

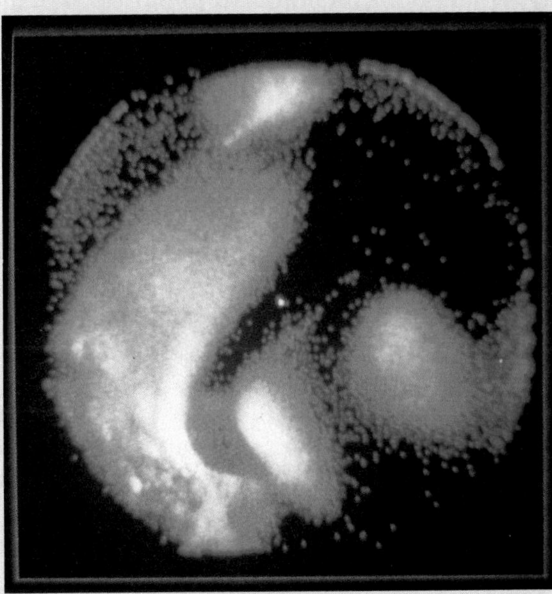

Figure 5 False-color image of a *single* Hg^+ ion (small isolated dot near the center) stored in the Paul trap shown in Figure 4. The other shapes are due to light reflected from the trap electrodes. The inner diameter of the ring electrode in this trap was about 0.9 mm. The ion is actually confined to a region of space much smaller than indicated by the size of the dot but the lens used in this experiment limited the resolution obtained.

High Accuracy Atomic Spectroscopy

Accuracies of energy measurements in certain atomic spectroscopy experiments are now better than 1 part in 10^{13}. This means that when we measure the frequency of an atomic transition, we know all of the environmental perturbations on the atom to such an extent that we can tell you the frequency at which the atom would absorb the radiation if it were isolated and at rest in space. If the accuracy is 1 part in 10^{13}, the inaccuracy of our prediction of the frequency is only 10^{-13} of the value of the frequency. As a comparison, if we could measure the distance between two points on the east and west coasts of the United States to an accuracy of 1 part in 10^{13}, the error in our measurement would be about 5×10^{-5} cm (the distance of one wavelength of visible light).

With such high accuracies in atomic spectroscopy, we must consider a number of effects that are typically small enough that we do not ordinarily worry about them. One such effect is called the second-order doppler shift. It can be derived by expressing Equation 34-24*a* or 34-24*b* in a power series in V/c. The term proportional to V/c is the same expression as was derived in Section 14-9. The second-order doppler shift is the term proportional to $(V/c)^2$. This is the effect caused by relativistic time dilation; because the ions (or atoms) are moving with respect to our radiation source, which is stationary in the lab, time moves more slowly for them. Hence, when we measure the frequencies of their transitions, we measure a value which is slightly lower than the frequency we would measure if they were at rest. This is not a particularly big effect—for $^9Be^+$ ions stored in a trap where the kinetic energies are near room temperature (about 300 K), the magnitude of this shift is fractionally $V^2/2c^2 \approx 5 \times 10^{-12}$. However, it is a difficult shift to measure accurately as it is hard to measure the ions' velocity distribution precisely. One approach to this problem is to reduce the temperature of the ions; an effective method is laser cooling.

Laser Cooling

We are familiar with the use of lasers to provide heat (e.g., laser surgery, welding, and inertial confinement fusion). However, as explained below, laser light has now also been used to cool small samples of trapped ions and atoms to very low temperatures—in some cases to about 1 μK. This cooling results from the mechanical momentum imparted to the atoms when they scatter light; by suitable arrangement of the laser beam's frequency and position, the atoms can be made to scatter light only when this scattering causes their momentum to be reduced.

That electromagnetic radiation can impart momentum to matter was known to James Clerk Maxwell in the late nineteenth century. Albert Einstein used the discrete momentum changes imparted to atoms by electromagnetic radiation in his theoretical studies of thermal equilibrium between radiation and matter. In 1933, Otto Frisch demonstrated experimentally the transfer of momentum from photons to atoms by deflecting a beam of sodium atoms with resonance radiation from a lamp. With the development of tunable lasers in the 1970s, such effects could be much more pronounced. Recently, the narrow spectral width of lasers has also allowed cooling of atoms by these mechanical forces. The simplest form of laser cooling, and the one most commonly used, is called doppler cooling. It relies on the high spectral purity of lasers, the fact that atoms tend to absorb light only at particular frequencies, and the frequency shift of the light (as viewed by the atom) due to the doppler effect.

First, suppose we use a tunable laser to measure the absorption spectrum of an atom near one of its optical transitions. If we could hold the atom stationary, the absorption would be strongest at a particular frequency f_0 and have a narrow range Δf over which it absorbs most strongly. For the transition of interest in $^9Be^+$ ions, $f_0 \approx 10^{15}$ Hz and $\Delta f \approx$ 20 MHz.

Suppose we now release the atom and subject it to a laser beam coming from the left. Assume this laser has frequency f_L where $f_L < f_0$. If the atom moves to the left with velocity V, then in the atom's reference frame, the laser light appears to have a frequency approximately equal to $f_L(1 + V/c)$ due to the doppler frequency shift (Section 34-7). For a particular value of V, $f_L(1 + V/c) \approx f_0$, and the atom absorbs and re-emits photons at a high rate. On absorption, the photon's momentum is transferred to the atom and thereby reduces the atom's momentum by approximately h/λ where λ is the wavelength of the laser radiation (see Equation 35-29). However, the photon re-emission is spatially symmetric, so on the average, there is no net momentum imparted to the atom by re-emission. Hence, on the average, the atom's momentum is reduced by h/λ for each scattering event. This is really no different than the collisions described for macroscopic bodies in Section 7-7, since all we need to take care of is conservation of energy and momentum.

If, instead, the atom moves to the right, each scattering event increases the atom's momentum by h/λ. However, the scattering rate is much less for atoms moving to the right because the frequency of the radiation (in the atom's frame) is now $f_L(1 - V/c) < f_L < f_0$ and the laser appears to be tuned away from the atom's resonance. This asymmetry in the scattering rate, and in the accompanying transferred momentum, for atoms moving left or right, gives rise to a net cooling effect. If an atom is subjected to three mutually orthogonal, intersecting pairs of counterpropagating laser beams tuned to $f_L < f_0$, the atom feels a damping or cooling force independent of which direction it moves. Such a configuration has been called "optical molasses."

The randomness in the times of absorption and the randomness in the direction of photon re-emission act like random impulses on the atom, which counteract the cooling effect. These random impulses, which cause heating, reach a balance with the cooling when the effective temperature of the atoms reaches a minimum value equal to $h\,\Delta f/2k$ (k = Boltzmann's constant). For many atoms, this temperature is around 1 mK or less. At 1 mK, the second-order doppler shift of $^9Be^+$ ions is about 1.5×10^{-17}, so by the use of laser cooling we significantly reduce the perturbation to the measured frequency caused by the second-order doppler shift.

Other Applications of Trapping and Cooling

In the above, we have discussed how the techniques of atom trapping and cooling can be used for precision atomic spectroscopy. It appears that these techniques can also be used to advantage for other purposes. Some examples are briefly discussed here.

Atomic Clocks The regular oscillations or vibrations of atoms and molecules can be likened to the oscillations of the pendulum in a grandfather clock. To make a clock based on atoms, we can tune the frequency of a radiation source until we drive a particular transition in an atom with maximum probability. If we then count the oscillations of the radiation source and wait until a certain number of cycles have elapsed, we define a unit of time. The nice thing about atoms is that, as far as we know, all atoms of a particular kind (such as $^9Be^+$ ions) are the same. No matter where two people are in the universe, if they agree to synchronize their radiation sources to a particular transition in a particular atom, when they count a given number of cycles, the unit of time they measure will be the same independent of a direct comparison. This is to be contrasted to pendulum clocks where no matter how much care is taken in their construction, they will oscillate at slightly different frequencies as it is hard to make the length of the pendula the same. Eventually we hope that, with the aid of trapping and cooling techniques, we will be able to make a clock that will be accurate to 1 second over the age of the universe. Accurate clocks are very useful in satellite and deep-space navigation systems.

Collision Studies Atomic collision studies at extremely low energies are now possible. At very low temperatures, the atom's de Broglie wavelength is long and quantum-mechanical effects are very important in describing the collisions. If the atom's de Broglie wavelength is large compared to the attractive region near a material surface, the atom may experience only the repulsive part of the surface and elastically bounce rather than stick. This may help provide nearly ideal atom "boxes."

Atom Manipulation Optical forces, such as those used in laser cooling, have been used to slow neutral atoms, steer atomic beams, and make atom "traps." The traps provided by optical forces are typically shallow (trap depths corresponding to a few kelvins). Therefore atoms from an atomic beam can be first slowed by overlapping them with a counterpropagating laser beam whose cooling force can stop the atoms at the position of the trap. In one trap called "optical tweezers," which uses the forces of a focused laser beam, the trapped atoms can be moved to different spatial locations by simply moving the laser beam. Applications of atom traps may include storage and manipulation of atomic antimatter, which must avoid contact with ordinary matter to prevent annihilation.

Condensed Matter Collections of atomic ions contained in ion traps (for instance, Figure 6) can be viewed as plasmas. At the very low temperatures provided by laser cooling, the Coulomb potential energy between adjacent ions exceeds their kinetic energy and the ions show regular spatial structure. In Figure 7, this regular spatial structure takes the form of shells of

Continued

Figure 6 Photograph of the Penning trap used to confine the $^9Be^+$ ions shown in Figure 7. In this trap, which has cylindrical geometry (inner diameter of the cylinders ≈ 2.5 cm), the sections at either end have the function of the endcaps of the trap shown in Figure 1. They are at a positive potential with respect to the central electrodes. A uniform magnetic field from a superconducting magnet (not shown) is parallel to the axis of the cylinders. This trap is also mounted in a quartz vacuum enclosure. Photographs of the ions are made by viewing along the axis of the cylinders.

ions. If a sample of weakly interacting atoms (e.g., atomic hydrogen) is trapped and sufficiently cooled, it may be possible to observe a transition to a state where the wave functions of the atoms are all the same and occupy the same region of space. This phenomenon, which has yet to be observed, is called Bose–Einstein condensation.

These topics have been developed in more detail by P. Ekstrom and D. Wineland, "The Isolated Electron," *Scientific American,* August 1980, pp. 104–121; D. J. Wineland and W. M. Itano, "Laser Cooling," *Physics Today* vol. 40, no. 6, June 1987, pp. 34–40; W. D. Phillips, and H. J. Metcalf, "Cooling and Trapping Atoms," *Scientific American,* March 1987, pp. 50–56. J. J. Bollinger and D. J. Wineland, "Microplasmas," *Scientific American,* January 1990, pp. 124–130. C. Cohen-Tannoudji and W. D. Phillips, *Physics Today,* vol. 43, no. 10, October 1990, pp. 33–40.

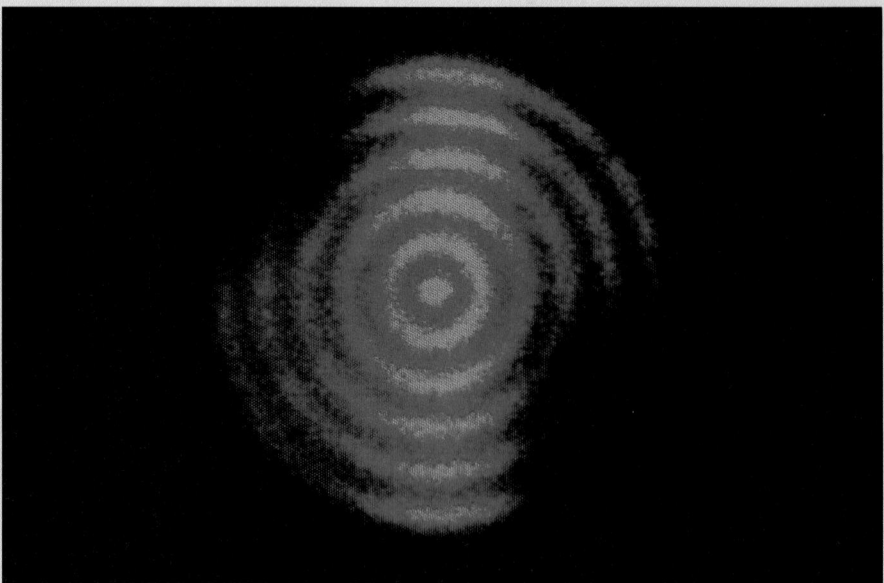

Figure 7 Ultraviolet photograph of a small Be^+ ion plasma which has been stored in the Penning trap shown in Figure 6 and laser cooled to about 10 mK. This picture was taken by viewing the ion plasma along the z axis through one of the endcaps. At low temperatures, plasmas become "strongly coupled" and show spatial structure. Here, this structure takes the form of cylindrical shells, which have been partially illuminated by a laser beam. The diameter of the outer shell in this picture is about 150 μm.

Suggestions for Further Reading

Leith, Emmett N., and Juris Upatnicks: "Photography by Laser," *Scientific American*, June 1965, p. 24.

The interference of coherent light produced by a laser is employed in wavefront reconstruction photography, more commonly known as holography.

Schawlow, Arthur L.: "Laser Light," *Scientific American*, September 1968, p. 120.

How lasers work and how laser light differs from ordinary light are discussed in this article.

Schewe, Phillip F.: "Lasers," *The Physics Teacher*, vol. 19, no. 8, 1981, p. 534.

This is an excellent and comprehensive exposition of the principles of laser operation, the types of lasers in use today, and application of laser light.

Vali, Victor: "Measuring Earth Strains by Laser," *Scientific American*, December 1969, p. 88.

One important modern use for laser interferometers is to measure small changes in the compression of the earth's crust, both steady and sudden, as in earthquakes.

Walker, Jearl: "The Amateur Scientist: The Colors Seen in the Sky Offer Lessons in Optical Scattering," *Scientific American*, January 1988, p. 102.

The scattering of light by molecules and dust particles explains blue skies, red sunsets, and other lesser-known atmospheric color phenomena.

Walker, Jearl: "The Amateur Scientist: The Spectra of Streetlights Illuminate Basic Principles of Quantum Mechanics," *Scientific American*, January 1984, p. 138.

This straightforward account of the development and application of quantum mechanics to the explanation of atomic spectra is illustrated with novel photographs.

Zare, Richard N.: "Laser Separation of Isotopes," *Scientific American*, February 1977, p. 86.

This article explains how laser irradiation of atoms or molecules in a beam can be useful in separating the isotopes of an element.

Review

A. Objectives: After studying this chapter, you should:

1. Know the origins of the quantum numbers n, ℓ, and m and the possible values of these numbers.

2. Know the relation between the quantum numbers n, ℓ, and m and the quantization of energy and angular momentum in the hydrogen atom.

3. Be able to compare the Schrödinger and Bohr models of the hydrogen atom.

4. Be able to sketch the wave function and probability distribution functions for the ground state of hydrogen.

5. Know the connection between magnetic moment and angular momentum, and be able to describe the Stern–Gerlach experiment.

6. Know the rules for the combination of angular momenta, and be able to discuss qualitatively the spin–orbit effect.

7. Be able to discuss the shell structure of atoms and the periodic table.

8. Be able to compare the optical and x-ray spectra of an atom.

9. Be able to describe the operation of a ruby laser and a helium–neon laser.

B. Define, explain, or otherwise identify:

Atomic number
Valence electron
Radial equation for hydrogen
Principal quantum number
Orbital quantum number
Magnetic quantum number
Selection rules
Radial probability density
Fine structure
Electron spin
Gyromagnetic ratio
Dirac equation
Stern–Gerlach experiment
Space quantization
Triplet state
Singlet state
Spin–orbit effect
Fine-structure splitting
Zeeman effect
Electron configuration
Pauli exclusion principle
Ionization energy
Transition elements
Optical spectrum
Characteristic x-ray spectrum
Rayleigh scattering
Inelastic scattering
Raman scattering
Resonance absorption
Spontaneous emission
Fluorescence
Metastable state
Phosphorescent materials
Stimulated emission
Laser
Population inversion
Optical pumping

C. True or false: If the statement is true, explain why it is true. If it is false, give a counterexample.

1. No two electrons can be in the same quantum state.

2. Elements with one electron outside a closed shell have small ionization energies and are chemically active.

3. Visible light results from transitions involving only the outermost electrons in an atom.

4. Characteristic x rays result from transitions made by inner electrons.

Problems

Level I

37-1 Quantum Theory of the Hydrogen Atom

1. For $\ell = 1$, find (*a*) the magnitude of the angular momentum L and (*b*) the possible values of m. (*c*) Draw to scale a vector diagram showing the possible orientations of **L** with the z axis.

2. Work Problem 1 for $\ell = 3$.

3. If $n = 3$, (*a*) what are the possible values of ℓ? (*b*) For each value of ℓ in (*a*), list the possible values of m. (*c*) Using the fact that there are two quantum states for each value of ℓ and m because of electron spin, find the total number of electron states with $n = 3$.

4. Find the total number of electron states with (*a*) $n = 2$ and (*b*) $n = 4$. (See Problem 3.)

5. The moment of inertia of a phonograph record is about 10^{-3} kg·m^2. (*a*) Find the angular momentum $L = I\omega$ when it rotates at $\omega/2\pi = 33.3$ rev/min and (*b*) find the approximate value of the quantum number ℓ.

6. Find the minimum value of the angle θ between **L** and the z axis for (*a*) $\ell = 1$, (*b*) $\ell = 4$, and (*c*) $\ell = 50$.

7. What are the possible values of n and m if (*a*) $\ell = 3$, (*b*) $\ell = 4$, and (*c*) $\ell = 0$?

8. What are the possible values of n and ℓ if (*a*) $m = 0$, (*b*) $m = -1$, and (*c*) $m = 2$?

37-2 The Hydrogen-Atom Wave Functions

9. For the ground state of the hydrogen atom, find the values of (*a*) ψ, (*b*) ψ^2, and (*c*) the radial probability density $P(r)$ at $r = a_0$. Give your answers in terms of a_0.

10. (*a*) If spin is not included, how many different wave functions are there corresponding to the first excited energy level $n = 2$ for hydrogen? (*b*) List these functions by giving the quantum numbers for each state.

11. For the ground state of the hydrogen atom, find the probability of finding the electron in the range $\Delta r = 0.03a_0$ at (*a*) $r = a_0$ and (*b*) $r = 2a_0$.

12. Show that the radial probability density for the $n = 2$, $\ell = 1$, $m = 0$ state of a one-electron atom can be written as

$$P(r) = A \cos^2\theta \; r^4 e^{-Zr/a_0}$$

where A is a constant.

13. The value of the constant C_{200} in Equation 37-15 is

$$C_{200} = \frac{1}{4\sqrt{2\pi}}\left(\frac{Z}{a_0}\right)^{3/2}$$

Find the values of (*a*) ψ, (*b*) ψ^2, and (*c*) the radial probability density $P(r)$ at $r = a_0$ for the state $n = 2$, $\ell = 0$, $m = 0$ in hydrogen. Give your answers in terms of a_0.

14. Find the probability of finding the electron in the range $\Delta r = 0.02a_0$ at (*a*) $r = a_0$ and (*b*) $r = 2a_0$ for the state $n = 2$, $\ell = 0$, $m = 0$ in hydrogen. (See Problem 13 for the value of C_{200}.)

37-3 Magnetic Moments and Electron Spin; 37-4 The Stern–Gerlach Experiment

15. The potential energy of a magnetic moment in an external magnetic field is given by $U = -\boldsymbol{\mu}\cdot\mathbf{B}$. (*a*) Calculate the difference in energy between the two possible orientations of an electron in a magnetic field $\mathbf{B} = 0.600$ T **k**. (*b*) If these electrons are bombarded with photons of energy equal to this energy difference, "spin flip" transitions can be induced. Find the wavelength of the photons needed for such transitions. This phenomenon is called **electron-spin resonance**.

16. Calculate the force on an electron in an inhomogeneous magnetic field for which $dB_z/dz = 850$ T/m.

17. How many beams would be expected in a Stern–Gerlach experiment done with atoms for which the spin is zero but the orbital quantum number $\ell = 1$?

18. Consider the magnet orientations shown in Figure 37-12. Could the magnet shown in the middle of the figure be an electron? Explain.

19. A convenient unit for the magnetic moment of nuclei is the nuclear magneton $\mu_N = e\hbar/2m_p$, where m_p is the mass of the proton. Calculate the magnitude of the nuclear magneton in (*a*) joules per tesla and (*b*) electron volts per gauss.

37-5 Addition of Angular Momenta and the Spin–Orbit Effect

20. The total angular momentum of a hydrogen atom in a certain excited state has the quantum number $j = \frac{1}{2}$. What can you say about the orbital angular-momentum quantum number ℓ?

21. The total angular momentum of a hydrogen atom in a certain excited state has the quantum number $j = 1\frac{1}{2}$. What can you say about the orbital angular-momentum quantum number ℓ?

22. A hydrogen atom is in the state $n = 3$, $\ell = 2$. (*a*) What are the possible values of j? (*b*) What are the possible values of the magnitude of the total angular momentum including spin? (*c*) What are the possible z components of the total angular momentum?

23. A deuteron is a nucleus with one proton and one neutron, each having spin $\frac{1}{2}$. (*a*) What are the possible values of the total spin quantum number of the deuteron ($\ell = 0$)? (*b*) In the ground state, the deuteron has $\ell = 0$ and $s = 1$. What is the magnitude of the angular momentum of the deuteron? (*c*) Draw a vector diagram illustrating the spins of the proton, neutron, and deuteron, and find the angle between the spins of the neutron and proton.

24. List all the spectroscopic state designations in atomic hydrogen for $n = 2$ and $n = 4$, including the label for total angular momentum.

37-6 The Periodic table

25. Write the electron configuration of (*a*) carbon and (*b*) oxygen.

26. Write the electron configuration of (*a*) aluminum and (*b*) chromium.

27. What element has the electron configuration (*a*) $1s^22s^22p^63s^23p^2$ and (*b*) $1s^22s^22p^63s^23p^64s^2$?

28. The properties of iron ($Z = 26$) and cobalt ($Z = 27$), which have adjacent atomic numbers, are similar, whereas the properties of neon ($Z = 10$) and sodium ($Z = 11$), which also have adjacent atomic numbers, are very different. Explain why.

29. In Figure 37-17, there are small dips in the ionization-energy curve at $Z = 31$ (gallium) and $Z = 49$ (indium) that are not labeled. Explain these dips using the electron configurations of these atoms given in Table 37-1.

30. Which of the following elements would you expect to have a ground state split by the spin–orbit interaction: Li, B, Na, Al, K, Cu, Ga, Ag? *Hint:* Use Table 37-1 to see which elements have $\ell = 0$ in their ground state and which do not.

31. If the outer electron in lithium moves in the $n = 2$ Bohr orbit, the effective nuclear charge would be $Z'e = 1e$, and the energy of the electron would be $-13.6\text{ eV}/2^2 = -3.4$ eV. However, the ionization energy of lithium is 5.39 eV, not 3.4 eV. Use this fact and Equation 37-29 to calculate the effective nuclear charge Z' seen by the outer electron in lithium. Assume that $r = 4a_0$ for the outer electron.

32. Give the possible values of the z component of the orbital angular momentum of (*a*) a d electron and (*b*) an f electron.

33. Separate the following six elements—potassium, calcium, titanium, chromium, manganese, and copper—into two groups of three each such that those in a group have similar properties.

37-7 Optical and X-Ray Spectra

34. The optical spectra of atoms with two electrons in the same outer shell are similar, but they are quite different from the spectra of atoms with just one outer electron because of the interaction of the two electrons. Separate the following elements into two groups such that those in each group have similar spectra: lithium, beryllium, sodium, magnesium, potassium, calcium, chromium, nickel, cesium, and barium.

35. Give the possible electron configurations for the first excited state of (*a*) hydrogen, (*b*) sodium, and (*c*) helium.

36. Which of the following elements should have optical spectra similar to that of hydrogen and which should have optical spectra similar to that of helium: Li, Ca, Ti, Rb, Ag, Cd, Ba, Hg, Fr, and Ra?

37. (*a*) Calculate the next two longest wavelengths after the K_α line in the K series of molybdenum. (*b*) What is the shortest wavelength in this series?

38. The wavelength of the K_α line for a certain element is 0.3368 nm. What is the element?

39. The wavelength of the K_α line for a certain element is 0.0794 nm. What is the element?

40. Calculate the wavelength of the K_α line of rhodium.

41. Calculate the wavelength of the K_α line of (*a*) magnesium ($Z = 12$) and (*b*) copper ($Z = 29$).

42. (*a*) Calculate the energy of the electron in the K shell for tungsten using $Z - 1$ for the effective nuclear charge. (*b*) The experimental result for this energy is 69.5 keV. Assume that the effective nuclear charge is $(Z - \sigma)$, where σ is the screening constant, and calculate σ from the experimental result for the energy.

37-8 Absorption, Scattering, and Stimulated Emission

There are no problems for this section.

37-9 The Laser

43. A pulse from a ruby laser has an average power of 10 MW and lasts 1.5 ns. (*a*) What is the total energy of the pulse? (*b*) How many photons are emitted in this pulse?

44. A helium–neon laser emits light of wavelength 632.8 nm and has a power output of 4 mW. How many photons are emitted per second by this laser?

45. A laser beam is aimed at the moon a distance 3.84×10^8 m away. The angular spread of the beam is given by the diffraction formula, $\sin\theta = 1.22\lambda/D$, where D is the diameter of the laser tube. Calculate the size of the beam on the moon for $D = 10$ cm and $\lambda = 600$ nm.

Level II

46. The angular momentum of the yttrium atom in the ground state is characterized by the quantum number $j = 1\frac{1}{2}$. How many lines would you expect to see if you could perform the Stern–Gerlach experiment with yttrium atoms?

47. Find the final state (or the final kinetic energy) of the hydrogen-atom electron if a photon of (*a*) 12.09 eV and (*b*) 20 eV is absorbed by the hydrogen atom when it is in the ground state.

48. The total energy of the electron of momentum p and mass m a distance r from the proton in the hydrogen atom is given by

$$E = \frac{p^2}{2m} - \frac{ke^2}{r}$$

where k is the Coulomb constant. If we assume that the minimum value of p^2 is $p^2 \approx (\Delta p)^2 = \hbar^2/r^2$, where Δp is the uncertainty in p and we have taken $\Delta r \sim r$ for the order of magnitude of the uncertainty in position, the energy is

$$E = \frac{\hbar^2}{2mr^2} - \frac{ke^2}{r}$$

Find the radius r_m for which this energy is a minimum, and calculate the minimum value of E in electron volts.

49. The L shell in an atom has $n = 2$ and $\ell = 1$. For atoms with eight or more electrons, there are two electrons in the K shell and six in the L shell. An electron in the L shell is thus shielded from the nuclear charge by the two electrons in the K shell and is partially shielded by the five other electrons in the L shell. There may also be some shielding due to the penetration of the wave functions of the electrons in outer shells. The frequencies for x rays in the L series involve transitions from some state $n_1 > 2$ to the state $n_2 = 2$. Equation 37-30 gives a good approximation for the initial and final electron energies of L-series x rays if $(Z - 1)$ in that equation is replaced by $(Z - 7.4)$. Use this result to calculate the shortest wavelength in the L series for (*a*) molybdenum ($Z = 42$) and (*b*) zinc ($Z = 30$).

50. For $\ell = 2$, (*a*) what is the minimum value of $L_x^2 + L_y^2$? (*b*) What is the maximum value of $L_x^2 + L_y^2$? (*c*) What is the value of $L_x^2 + L_y^2$ for $\ell = 2$ and $m = 1$?

51. The L_α x-ray line arising from the transition from $n_1 = 3$ to $n_2 = 2$ for a certain element has a wavelength of 0.3617 nm. What is the element? (See Problem 49.)

52. A particle of mass m moves in a circle of radius r with speed v in the xy plane around the z axis. Consider the time when the particle is on the y axis. Multiply the uncertainty product $\Delta x\, \Delta p$ by r/r to obtain an uncertainty relation between the angle ϕ and the z component of the angular momentum.

53. (*a*) Calculate the energies in electron volts of the photons corresponding to the two yellow lines in the optical spectrum of sodium whose wavelengths are 589.0 nm and 589.6 nm. (*b*) The difference in energies of these photons equals the difference in energy ΔE between the $3P_{3/2}$ and $3P_{1/2}$ states in sodium. Calculate ΔE. (*c*) Estimate the magnetic field that the $3p$ electron in sodium experiences.

54. Find the minimum angle between the angular momentum $\mathbf{L}$ and the z axis for a general value of ℓ, and show that for large values of ℓ, $\theta_{\text{min}} \approx 1/\sqrt{\ell}$.

55. The wavelengths of the photons emitted by potassium corresponding to transitions from the $4P_{3/2}$ and $4P_{1/2}$ states to the ground state are 766.41 nm and 769.90 nm. (*a*) Calculate the energies of these photons in electron volts. (*b*) The difference in the energies of these photons equals the difference in energy ΔE between the $4P_{3/2}$ and $4P_{1/2}$ states in potassium. Calculate ΔE. (*c*) Estimate the magnetic field that the $4p$ electron in potassium experiences.

56. If a classical system does not have a constant charge-to-mass ratio throughout the system, the magnetic moment can be written

$$\mu = g\frac{Q}{2M}L$$

where Q is the total charge, M is the total mass, and $g \neq 1$. (*a*) Show that $g = 2$ for a solid cylinder that spins about its axis and has a uniform charge on its cylindrical surface. (*b*) Show that $g = 2\frac{1}{2}$ for a solid sphere that has a ring of charge on its surface at the equator.

57. In a Stern–Gerlach experiment hydrogen atoms in their ground state move with speed $v_x = 14.5$ km/s. The magnetic field is in the z direction and its maximum gradient is given by $dB_z/dz = 600$ T/m. (*a*) Find the maximum acceleration of the hydrogen atoms. (*b*) If the region of the magnetic field extends over a distance $\Delta x = 75$ cm, and there is an additional 1.25 m from the edge of the field to the detector, find the maximum distance between the two lines on the detector.

58. The radial probability distribution function for a one-electron atom in its ground state can be written $P(r) = Cr^2e^{-2Zr/a_0}$, where C is a constant. Show that $P(r)$ has its maximum value at $r = a_0/Z$.

59. Consider a system of two electrons, each with $\ell = 1$ and $s = \frac{1}{2}$. (*a*) Neglecting spin, find the possible values of the quantum number for the total orbital angular momentum $\mathbf{L} = \mathbf{L}_1 + \mathbf{L}_2$. (*b*) What are the possible values of the quantum number S for the total spin $\mathbf{S} = \mathbf{S}_1 + \mathbf{S}_2$? (*c*) Using the results of parts (*a*) and (*b*), find the possible quantum numbers j for the combination $\mathbf{J} = \mathbf{L} + \mathbf{S}$. (*d*) What are the possible quantum numbers j_1 and j_2 for the total angular momentum of each particle? (*e*) Use the results of part (*d*) to calculate the possible values of j from the combination of j_1 and j_2. Are these the same as in part (*c*)?

60. (*a*) Show that the radial probability distribution for the $n = 2$, $\ell = 1$ energy levels of a one-electron atom can be written $P(r) = Ar^4e^{-Zr/a_0}$, where A depends on θ but not on r. (*b*) Show that $P(r)$ is maximum at $r = 4a_0/Z$.

61. The radius of a proton is about $R_0 = 10^{-15}$ m. The probability that the hydrogen-atom electron is inside the proton is

$$P = \int_0^{R_0} P(r)\, dr$$

where $P(r)$ is the radial probability density. Calculate this probability for the ground-state of hydrogen. *Hint:* Show that $e^{-2r/a_0} \approx 1$ for $r \leq R_0$ is valid for this calculation.

62. Show that the expectation value of r for the electron in the ground state of a one-electron atom is $\langle r\rangle = \frac{3}{2}a_0/Z$.

63. Show that the expectation value of the potential energy $U(r) = -kZe^2/r$ for the ground state of a one-electron atom is given by $\langle U(r)\rangle = -2Z^2E_0$.

64. Show by direct substitution that the ground-state wave function for hydrogen given by Equation 37-12 satisfies the Schrödinger equation in spherical coordinates (Equation 37-4).

Level III

65. Show by direct substitution that the wave function ψ_{210} given by Equation 37-16 satisfies the Schrödinger equation in spherical coordinates (Equation 37-4).

66. Show that the number of states in the hydrogen atom for a given n is $2n^2$.

67. Find the two values of r for which $P(r)$ is a maximum for the $2s$ wave function of hydrogen.

68. Derive the expression for C_{200} given in Problem 13.

69. Calculate the probability that the electron in the ground state of a hydrogen atom is in the region $0 < r < a_0$.

Chapter 38

Molecules

Hemoglobin is an iron-containing protein found in red-blood cells that transports oxygen throughout the body. Shown here is a model of part of a molecule synthesized in the laboratory and designed to function, potentially, as artificial human hemoglobin. The notch in the center is the binding site for oxygen. The two hemispheres are compounds known as cyclodextrins made up of carbon, hydrogen, and oxygen. They encage a molecule formed of iron and a porphyrin, a nitrogen-containing organic compound found universally in protoplasm. The artificial hemoglobin has the interesting property that its subunits will spontaneously self-assemble themselves. Each individual subunit is held together by covalent chemical bonds, whereas the subunits themselves bind to one another by electrostatic and "water-avoiding" interactions. Though complicated molecules routinely assemble themselves at high speed in living organisms, chemists are only now beginning to duplicate this process in the test tube.

Single atoms rarely occur in nature. Instead, atoms usually bond together to form molecules or solids. Molecules may exist as separate entities, as in gaseous oxygen (O_2) or nitrogen (N_2), or they may bond together to form liquids or solids. A molecule is the smallest constituent of a substance that retains the chemical properties of the substance. The study of the properties of molecules thus forms the basis for theoretical chemistry.

The application of quantum mechanics to molecular physics has been spectacularly successful in explaining the structure of molecules and the complexity of their spectra. It has also provided answers to such puzzling questions as why two hydrogen atoms join together to form a molecule but three hydrogen atoms do not. As in atomic physics, the quantum-mechanical calculations in molecular physics are very difficult, so much of our discussion will be qualitative.

There are two extreme views that we can take of a molecule. Consider, for example, gaseous hydrogen (H_2). We can think of it either as two hydrogen atoms somehow joined together or as a quantum-mechanical system consisting of two protons and two electrons. The latter view is more fruitful in this case because neither of the electrons in the H_2 molecule can be identified as belonging to either proton. Instead, the wave function for each electron is spread out in space throughout the whole molecule. For more complicated molecules, however, an intermediate view is useful. For example, the nitrogen molecule N_2 has 14 protons and 14 electrons, but only two of the electrons take part in the bonding. The electron configuration of a nitrogen atom in the ground state is $1s^22s^22p^3$. Of the three electrons in the $2p$ state, two have their spins paired; that is, their spins are antiparallel, so their resultant spin is zero. The single electron with the unpaired spin is the most free to take part in the bonding of one nitrogen atom to another. We can therefore consider the N_2 molecule as consisting of two N^+ ions and two electrons that belong to the molecule as a whole. The molecular wave functions for these bonding electrons are called **molecular orbitals.** In many cases, molecular wave functions can be constructed by combining the atomic wave functions with which we are familiar.

38-1 Molecular Bonding

The two principal types of bonds responsible for the formation of molecules are the ionic bond and the covalent bond. Other types of bonds that are important in the bonding of liquids and solids are van der Waals bonds, metallic bonds, and hydrogen bonds. In many cases, bonding is a mixture of these mechanisms.

The Ionic Bond

The simplest type of bond is the **ionic bond,** which is found in most salts. Consider sodium chloride (NaCl) as an example. The sodium atom has one $3s$ electron outside a stable core. It takes just 5.14 eV to remove this electron from sodium (see Figure 37-17). The ionization energies for the other alkali metals are also low. The removal of one electron from sodium leaves a positive ion with a spherically symmetric, closed-shell electron core. Chlorine, on the other hand, is only one electron short of having a closed shell. The energy released by an atom's acquisition of one electron is called its **electron affinity,** which in the case of chlorine is 3.61 eV. The acquisition of one electron by chlorine results in a negative ion with a spherically symmetric, closed-shell electron core. Thus, the formation of a Na^+ ion and a Cl^- ion by the donation of one electron of sodium to chlorine requires only 5.14 eV − 3.61 eV = 1.53 eV at infinite separation. The electrostatic potential energy of the two ions when they are a distance r apart is $-ke^2/r$. When the separation of the ions is less than about 0.94 nm, the negative potential energy of attraction is of greater magnitude than the 1.53 eV of energy needed to create the ions. Thus, at separation distances less than 0.94 nm it is energetically favorable (that is, the total energy of the system is reduced) for the sodium atom to donate an electron to the chlorine atom to form NaCl.

Since the electrostatic attraction increases as the ions get closer together, it might seem that equilibrium could not exist. However, when the separation of the ions is very small, there is a strong repulsion that is quantum mechanical in nature and is related to the exclusion principle. This **exclusion-principle repulsion** is responsible for the repulsion of the atoms in all molecules (except H_2) no matter what the bonding mechanism is. (In H_2, the repulsion is simply that of the two positively charged protons.)

We can understand the exclusion-principle repulsion qualitatively as follows. When the ions are very far apart, the wave function for a core electron in one of the ions does not overlap that of any electron in the other ion. We can distinguish the electrons by the ion to which they belong. This means that electrons in the two ions can have the same quantum numbers because they occupy different regions of space. (Recall from our discussion in Section 36-11 that the exclusion principle is related to the fact that the wave function for two identical electrons is antisymmetric on the exchange of the electrons and that an antisymmetric wave function for two electrons with the same quantum numbers is zero if the space coordinates of the electrons are the same.) However, as the distance between the ions decreases, the wave functions of the core electrons begin to overlap; that is, the electrons in the two ions begin to occupy the same region of space. Because of the exclusion principle, some of these electrons must go into higher-energy quantum states. But energy is required to shift the electrons into higher-energy quantum states. This increase in energy when the ions are pushed closer together is equivalent to a repulsion of the ions. This is not a sudden process. The energy states of the electrons change gradually as the ions are brought together. A sketch of the potential energy of the Na^+ and Cl^- ions versus separation is shown in Figure 38-1. The energy is lowest at an equilibrium separation of about 0.236 nm. At smaller separations, the energy rises steeply as a result of the exclusion principle. The energy required to separate the ions and form neutral sodium and chlorine atoms is called the **dissociation energy,** which is about 4.26 eV for NaCl.

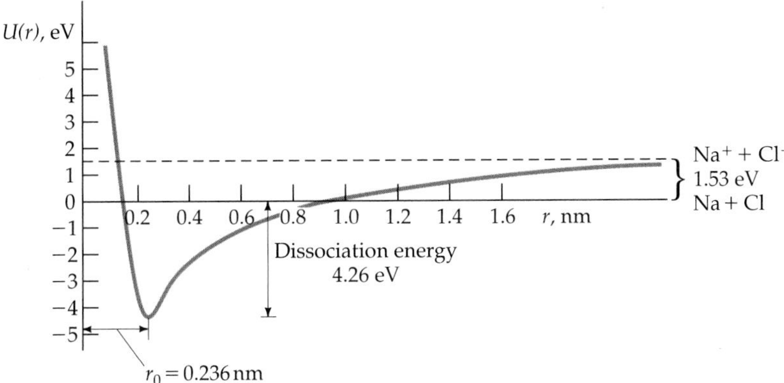

Figure 38-1 Potential energy for Na^+ and Cl^- ions as a function of separation distance r. The energy at infinite separation was chosen to be 1.53 eV, corresponding to the energy needed to form the ions from neutral atoms. The minimum energy for this curve is at the equilibrium separation $r_0 = 0.236$ nm for the ions in the molecule.

The equilibrium separation distance of 0.236 nm is for gaseous, diatomic NaCl, which can be obtained by evaporating solid NaCl. Normally, NaCl exists in a cubic crystal structure, with the Na^+ and Cl^- ions at the alternate corners of a cube. The separation of the ions in a crystal is somewhat larger, about 0.28 nm. Because of the presence of neighboring ions of opposite charge, the Coulomb energy per ion pair is lower when the ions are in a crystal.

Example 38-1

The electron affinity of fluorine is 3.45 eV, and the equilibrium separation of sodium fluoride (NaF) is 0.193 nm. (*a*) How much energy is needed to form Na^+ and F^- ions from neutral sodium and fluorine atoms? (*b*) What is the electrostatic potential energy of the Na^+ and F^- ions at their equilibrium separation? (*c*) The dissociation energy of NaF is 4.99 eV. What is the energy due to repulsion of the ions at the equilibrium separation?

(*a*) Since the energy needed to ionize sodium is 5.14 eV, the energy needed to form Na^+ and F^- ions from the neutral sodium and fluorine atoms is 5.14 eV − 3.45 eV = 1.69 eV.

(*b*) The electrostatic potential energy of Na^+ and F^- ions at their equilibrium separation (with $U = 0$ at infinite separation) is

$$U = -\frac{ke^2}{r} = -\frac{(8.99 \times 10^9\ \text{N}\cdot\text{m}^2/\text{C}^2)(1.60 \times 10^{-19}\ \text{C})^2}{1.93 \times 10^{-10}\ \text{m}}$$

$$= -1.19 \times 10^{-18}\ \text{J}$$

Converting this to electron volts, we obtain

$$U = -1.19 \times 10^{-18}\ \text{J}\ \frac{1\ \text{eV}}{1.60 \times 10^{-19}\ \text{J}} = -7.45\ \text{eV}$$

(*c*) If we choose the potential energy at infinity to be 1.69 eV (the energy needed to form the Na^+ and F^- ions from neutral sodium and fluorine atoms), the electrostatic potential energy is

$$U = -\frac{ke^2}{r} + 1.69\ \text{eV}$$

At the equilibrium separation, this energy is $U = -7.45\ \text{eV} + 1.69\ \text{eV} = -5.76\ \text{eV}$. Since the dissociation energy is 4.99 eV, the energy due to repulsion of the Na^+ and F^- ions at the equilibrium separation must be $5.76\ \text{eV} - 4.99\ \text{eV} = 0.77\ \text{eV}$.

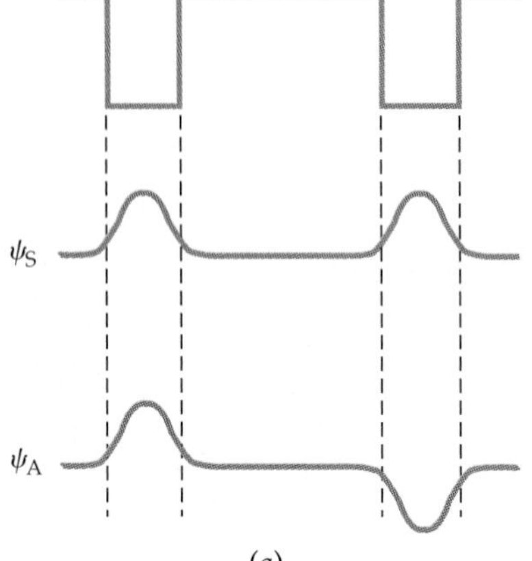

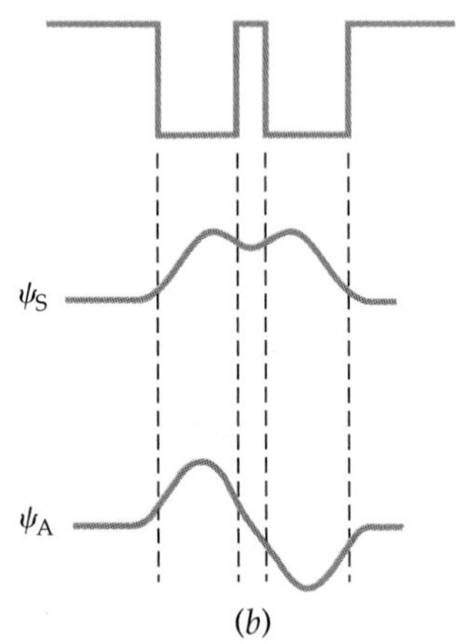

Figure 38-2 (*a*) Two square wells that are far apart. The electron wave function can be either symmetric (ψ_S) or antisymmetric (ψ_A) in space. The probability distributions and energies are the same for the two wave functions. (*b*) Two square wells that are close together. The symmetric wave function is larger between the wells than the antisymmetric wave function.

The Covalent Bond

A completely different mechanism, the **covalent bond,** is responsible for the bonding of identical or similar atoms to form such molecules as gaseous hydrogen (H_2), nitrogen (N_2), and carbon monoxide (CO). If we calculate the energy needed to form H^+ and H^- ions by the transfer of an electron from one atom to the other and then add this energy to the electrostatic potential energy, we find that there is no separation distance for which the total energy is negative. The bond thus cannot be ionic. Instead, the attraction of two hydrogen atoms is an entirely quantum-mechanical effect. The decrease in energy when two hydrogen atoms approach each other is due to the sharing of the two electrons by both atoms. It is intimately connected with the symmetry properties of the wave functions of the electrons.

We can gain some insight into covalent bonding by considering a simple, one-dimensional, quantum-mechanics problem of two finite square wells. We first consider a single electron that is equally likely to be in either well. Since the wells are identical, symmetry requires that ψ^2 be symmetric about the midpoint between the wells. Then ψ must be either symmetric or antisymmetric. The two possibilities for the ground state are shown in Figure 38-2*a* for the case in which the wells are far apart and in Figure 38-2*b* for the case in which the wells are close together. An important feature of Figure 38-2*b* is that in the region between the wells the symmetric wave function is large and the antisymmetric wave function is small.

Now consider adding a second electron to the two wells. The total wave function for the two electrons must be antisymmetric on exchange of the electrons. (Note that exchanging the electrons in the wells is the same as exchanging the wells.) The total wave function for two electrons can be written as a product of a space part and a spin part.

To understand the symmetry of the total wave function, we must first digress and consider the symmetry of the spin part of the wave function. When the two electrons have parallel spins ($S = 1$), the spin state $\phi_{1,+1}$, corresponding to $m_S = +1$, can be written as

$$\phi_{1,+1} = \uparrow_1 \uparrow_2 \qquad S = 1,\ m_S = +1 \tag{38-1}$$

where $\uparrow_1$ means that electron 1 is "up" ($m_1 = +\frac{1}{2}$) and $\uparrow_2$ means the same for electron 2. Similarly, the spin state for $m_S = -1$ can be written as

$$\phi_{1,-1} = \downarrow_1 \downarrow_2 \qquad S = 1,\ m_S = -1 \tag{38-2}$$

where $\downarrow_1$ means that electron 1 is "down" ($m_1 = -\frac{1}{2}$). Note that both of these states are symmetric upon exchange of the electrons. The spin state corresponding to $S = 1$ and $m_S = 0$ is not quite so obvious. It turns out to be proportional to

$$\phi_{1,0} = \uparrow_1 \downarrow_2 + \uparrow_2 \downarrow_1 \qquad S = 1,\ m_S = 0 \tag{38-3}$$

This spin state is also symmetric upon exchange of the electrons. The spin state for two electrons with antiparallel spins ($S = 0$) is

$$\phi_{0,0} = \uparrow_1 \downarrow_2 - \uparrow_2 \downarrow_1 \qquad S = 0,\ m_S = 0 \tag{38-4}$$

This spin state is antisymmetric upon exchange of electrons.

We thus have the important result that the spin part of the wave function is symmetric for parallel spins and antisymmetric for antiparallel spins. To obtain a total wave function that is antisymmetric, we form the product of a symmetric space state and an antisymmetric spin state, or of an antisymmetric space state and a symmetric spin state. This gives us the following result:

For the total wave function of two electrons to be antisymmetric, the space part of the wave function must be antisymmetric for parallel spins ($S = 1$) and symmetric for antiparallel spins ($S = 0$).

We can now consider the problem of two hydrogen atoms. Figure 38-3*a* shows a symmetric wave function ψ_S and an antisymmetric wave function ψ_A for two hydrogen atoms that are far apart, and Figure 38-3*b* shows the same two wave functions for two hydrogen atoms that are close together. The squares of these two wave functions are shown in Figure 38-3*c*. Note that the probability distribution $|\psi|^2$ in the region between the protons is large for the symmetric wave function and small for the antisymmetric wave function. Thus, when the spins of the two electrons are antiparallel, which means that the space part of the wave function is symmetric, the electrons are often found in the region between the protons, and the protons are bound together by the negatively charged electrons between them. The negatively charged electron cloud representing these electrons is concentrated in the space between the protons, as shown in the upper part of Figure 38-3*c*. Conversely, when the electron spins are parallel, which means that the space part of the wave function is antisymmetric, the electrons spend little time between the protons, and the atoms do not bind together to form a molecule. In this case, the electron cloud is not concentrated in the space between the protons, as shown in the lower part of Figure 38-3*c*.

The total electrostatic potential energy for the H_2 molecule consists of the positive energy of repulsion of the two electrons and the negative poten-

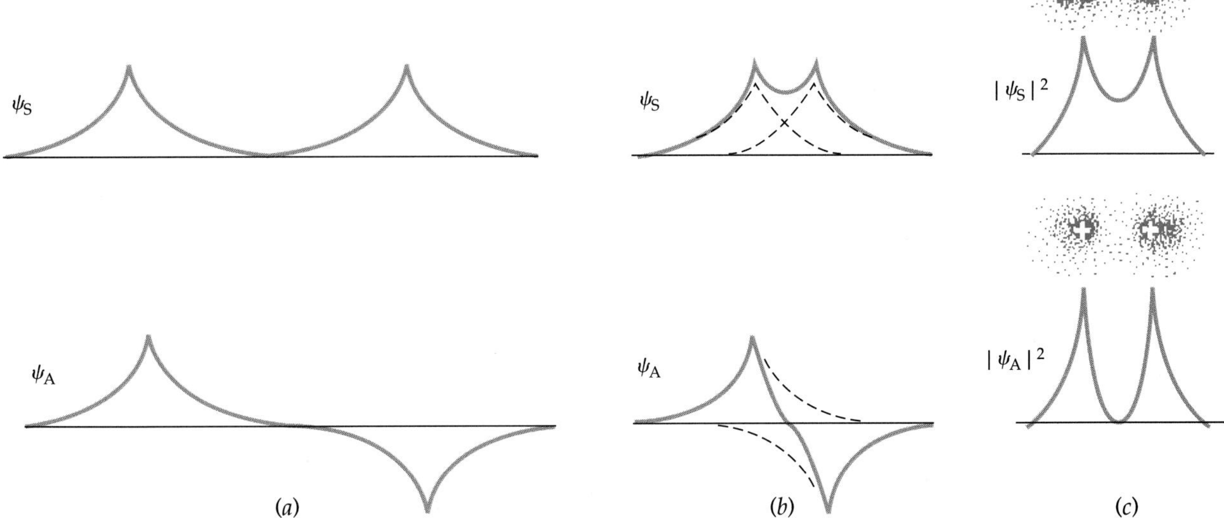

Figure 38-3 One-dimensional symmetric and antisymmetric wave functions for two hydrogen atoms (*a*) far apart and (*b*) close together. (*c*) Electron probability distributions $|\psi^2|$ for the wave functions in (*b*). For the symmetric wave function on the left, the electron charge density is large between the protons. This negative charge density holds the protons together in the hydrogen molecule H_2. For the antisymmetric wave functions on the right, the electron charge density is not large between the protons, and the atoms do not bond together to form a molecule.

tial energy of attraction of each electron for each proton. Figure 38-4 shows the electrostatic potential energy for two hydrogen atoms versus separation for the case in which the space part of the electron wave function is symmetric (U_S) and for the case in which it is antisymmetric (U_A). We can see that the potential energy for the symmetric state is the lower of the two and that the shape of this potential-energy curve is similar to that for ionic bonding. The equilibrium separation for H_2 is $r_0 = 0.074$ nm, and the binding energy is 4.52 eV. For the antisymmetric state, the potential energy is never negative, and there is no bonding.

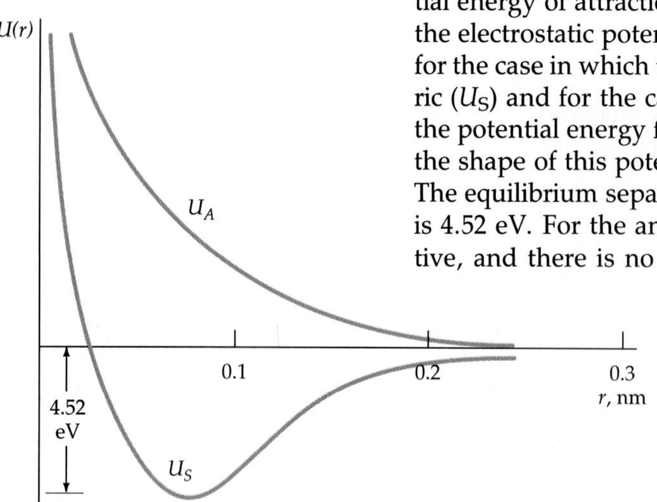

Figure 38-4 Potential energy versus separation for two hydrogen atoms. The curve labeled U_S is for a wave function with a symmetric space part, and the curve labeled U_A is for a wave function with an antisymmetric space part.

We can now see why three hydrogen atoms do not bond to form H_3. If a third hydrogen atom is brought near an H_2 molecule, the third electron cannot be in a 1*s* state and have its spin antiparallel to the spin of both of the other electrons. If this electron is in an antisymmetric space state with respect to exchange with one of the electrons, the repulsion of this atom is greater than the attraction of the other. As the three atoms are pushed together, the third electron is, in effect, forced into a higher quantum-energy state by the exclusion principle. The bond between two hydrogen atoms is called a **saturated bond** because there is no room for another electron. The two shared electrons essentially fill the 1*s* states of both atoms.

We can also see why two helium atoms do not normally bond together to form the He_2 molecule. There are no valence electrons that can be shared. The electrons in the closed shells are forced into higher energy states when the two atoms are brought together. At low temperatures or high pressures, helium atoms do bond together due to van der Waals forces, which we will discuss next. This bonding is so weak that at atmospheric pressure helium boils at 4 K, and it does not form a solid at any temperature unless the pressure is greater than about 20 atm.

When two identical atoms bond, as in O_2 or N_2, the bonding is purely covalent. However, the bonding of two dissimilar atoms is often a mixture of covalent and ionic bonding. Even in NaCl, the electron donated by sodium to chlorine has some probability of being at the sodium atom because its wave function does not suddenly fall to zero. Thus, this electron is partially shared in a covalent bond, although this bonding is only a small part of the total bond, which is mainly ionic.

A measure of the degree to which a bond is ionic or covalent can be obtained from the electric dipole moment of the molecule. For example, if the bonding in NaCl were purely ionic, the center of positive charge would be at the Na^+ ion and the center of negative charge would be at the Cl^- ion. The electric dipole moment would have the magnitude

$$p_{\text{ionic}} = er_0 \qquad 38\text{-}5$$

where r_0 is the equilibrium separation of the ions. Thus, the dipole moment of NaCl would be

$$p_{\text{ionic}} = er_0$$
$$= (1.60 \times 10^{-19}\ \text{C})(2.36 \times 10^{-10}\ \text{m}) = 3.78 \times 10^{-29}\ \text{C·m}$$

The actual measured electric dipole moment of NaCl is

$$p_{\text{measured}} = 3.00 \times 10^{-29}\ \text{C·m}$$

We can define the ratio of p_{measured} to p_{ionic} as the fractional amount of ionic bonding. For NaCl, this ratio is $3.00/3.78 = 0.79$. Thus, the bonding in NaCl is about 79 percent ionic.

Exercise

The equilibrium separation of HCl is 0.128 nm and its measured electric dipole moment is 3.60×10^{-30} C·m. What is the percentage of ionic bonding in HCl? (Answer: 18 percent)

The van der Waals Bond

Any two separated molecules will be attracted to one another by electrostatic forces called van der Waals forces. So will any two atoms that do not form ionic or covalent bonds. The **van der Waals bonds** formed because of these forces are much weaker than those already discussed. At high enough temperatures, these forces are not strong enough to overcome the ordinary thermal agitation of the atoms or molecules, but at sufficiently low temperatures, thermal agitation becomes negligible, and the van der Waals forces will cause virtually all substances to condense into a liquid and then a solid form. (Helium is the only element that does not solidify at any temperature at atmospheric pressure.)

The van der Waals forces arise from the interaction of the instantaneous electric dipole moments of the molecules. Figure 38-5 shows how two polar molecules—molecules with permanent electric dipole moments, such as H_2O—can bond. The electric field due to the dipole moment of one molecule orients the other molecule such that the two dipole moments attract.

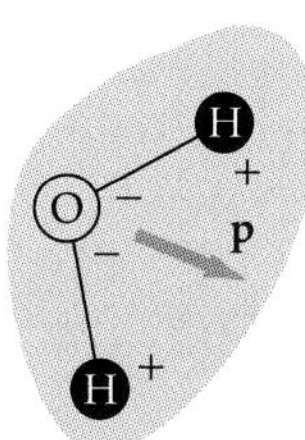

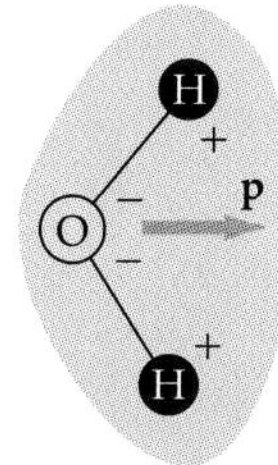

Figure 38-5 The bonding of H_2O molecules due to the attraction of the electric dipoles. The dipole moment of each molecule is indicated by **p**.

Nonpolar molecules also attract other nonpolar molecules via the van der Waals forces. Although nonpolar molecules have zero electric dipole moments on the average, they have instantaneous dipole moments that are generally not zero because of fluctuations in the positions of the charges. When two nonpolar molecules are near each other, the fluctuations in the instantaneous dipole moments tend to become correlated so as to produce attraction. This is illustrated in Figure 38-6.

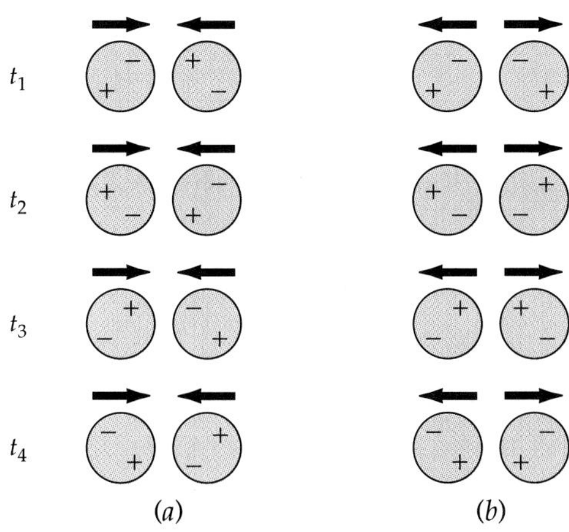

Figure 38-6 The van der Waals attraction of molecules with zero average dipole moments. (*a*) Possible orientations of the instantaneous dipole moments at different times leading to attraction. (*b*) Possible orientations leading to repulsion. The electric field of the instantaneous dipole moment of one molecule tends to polarize the other molecule. Thus, the orientations in (*a*) leading to attraction are much more likely than those in (*b*) leading to repulsion.

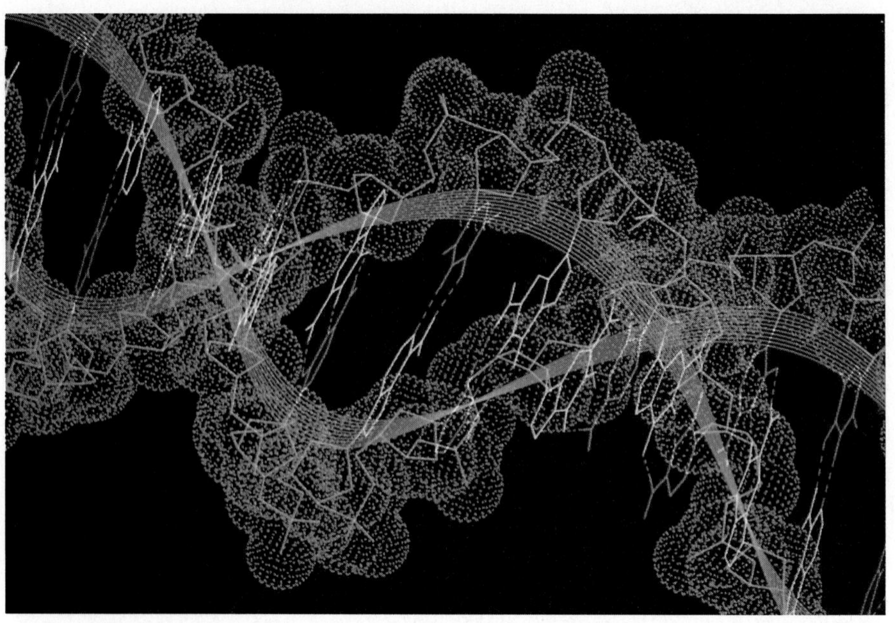

Figure 38-7 The DNA molecule.

The Hydrogen Bond

Another bonding mechanism of great importance is the hydrogen bond. It often holds groups of molecules together and is responsible for the cross-linking that allows giant biological molecules and polymers to hold their fixed shapes. The well-known helical structure of DNA is due to hydrogen-bond linkages across turns of the helix (Figure 38-7). The hydrogen bond is formed by the sharing of a proton (the nucleus of the hydrogen atom) between two atoms, frequently two oxygen atoms. This sharing of a proton is similar to the sharing of electrons responsible for the covalent bond already discussed. It is facilitated by the small mass of the proton and by the absence of inner-core electrons in hydrogen.

The Metallic Bond

The bonding of atoms in a metal is different from the bonding of atoms in a molecule. In a metal, two atoms do not bond together by exchanging or sharing an electron to form a molecule. Instead, each valence electron is shared by many atoms. The bonding is thus distributed throughout the entire metal. A metal can be thought of as a lattice of positive ions held together by a "gas" of essentially free electrons that roam throughout the solid. In the quantum-mechanical picture, these free electrons form a cloud of negative charge density between the positively charged lattice ions that holds the ions together. In this respect, the metallic bond is somewhat similar to the covalent bond. However, with the metallic bond, there are far more than just two atoms involved, and the negative charge is distributed uniformly throughout the volume of the metal. The number of free electrons varies from metal to metal but is of the order of one per atom.

Questions

1. Why would you expect the separation distance between the two protons to be smaller in the H_2^+ ion than in the H_2 molecule?
2. Would you expect the NaCl molecule to be polar or nonpolar?
3. Would you expect the N_2 molecule to be polar or nonpolar?
4. Does neon occur naturally as Ne or Ne_2? Why?

38-2 Polyatomic Molecules

Molecules with more than two atoms range from such relatively simple molecules as water, which has a molecular mass of 18, to such giants as proteins, which can have molecular masses of hundreds of thousands or even a million. As with diatomic molecules, the structure of polyatomic molecules can be understood by applying basic quantum mechanics to the bonding of individual atoms. The bonding mechanisms for most polyatomic molecules are the covalent bond and the hydrogen bond. We will discuss only some of the simplest polyatomic molecules—H_2O, NH_3, and CH_4—to illustrate both the simplicity and complexity of the application of quantum mechanics to molecular bonding.

The basic requirement for the sharing of electrons in a covalent bond is that the wave functions of the valence electrons in the individual atoms must overlap as much as possible. As our first example, we will consider the water molecule. The ground-state configuration of the oxygen atom is $1s^2 2s^2 2p^4$. The 1s and 2s electrons are in closed-shell states and do not contribute to the bonding. The 2p shell has room for six electrons, two in each of the three space states corresponding to $\ell = 1$. In an isolated atom, we describe these space states by the hydrogen-like wave functions corresponding to $\ell = 1$ and $m = +1$, 0, and -1. Since the energy is the same for these three space states, we could equally well use any linear combination of these wave functions. When an atom participates in molecular bonding, certain combinations of these atomic wave functions are important. These combinations are called the p_x, p_y, and p_z **atomic orbitals.** The angular dependence of these orbitals is

$$p_x \propto \sin\theta \cos\phi \tag{38-6}$$

$$p_y \propto \sin\theta \sin\phi \tag{38-7}$$

$$p_z \propto \cos\theta \tag{38-8}$$

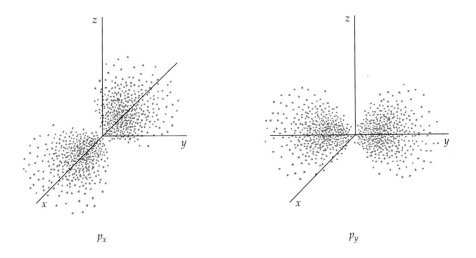

The electron charge distribution is maximum along the x, y, or z axis for these orbitals as shown in Figure 38-8. When oxygen is in an H_2O molecule, maximum overlap of the electron wave functions occurs when two of the four 2p electrons are paired with their spins antiparallel in one of the orbitals, for example, p_z, one of the other electrons is in the p_x orbital, and the other electron is in the p_y orbital. Each of the unpaired electrons in the p_x and p_y orbitals forms a bond with the electron of a hydrogen atom as shown in Figure 38-9. Because of the repulsion of the two hydrogen atoms, the angle between the O–H bonds is actually greater than 90°. The effect of this repulsion can be calculated, and the result is in agreement with the measured angle of 104.5°.

Figure 38-8 Computer-generated dot plot illustrating the spatial dependence of the electron charge distribution in the p_x, p_y, and p_z atomic orbitals.

p_x

p_y

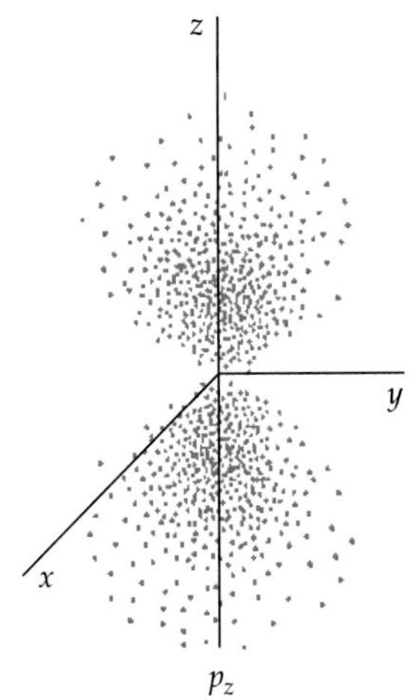

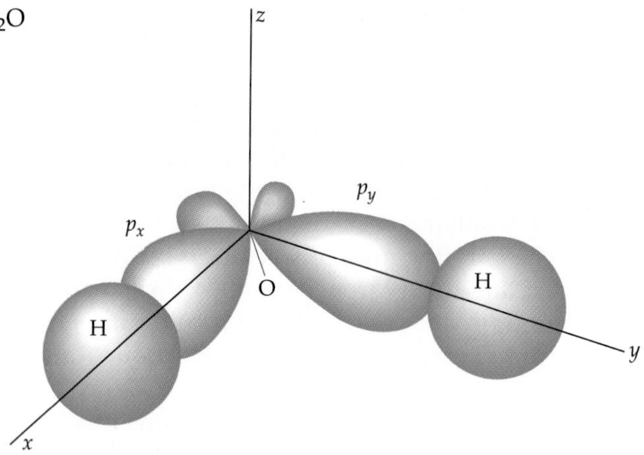

Figure 38-9 Electron charge distribution in the H_2O molecule.

Similar reasoning leads to an understanding of the bonding in NH_3. In the ground state, nitrogen has three electrons in the $2p$ state. When these three electrons are in the p_x, p_y, and p_z atomic orbitals, they bond to the electrons of hydrogen atoms. Again, because of the repulsion of the hydrogen atoms, the angles between the bonds are somewhat larger than 90°.

The bonding of carbon atoms is somewhat more complicated. Carbon forms a wide variety of different types of molecular bonds, leading to a great diversity in the kinds of organic molecules that contain carbon. The ground-state configuration of carbon is $1s^22s^22p^2$. From our previous discussion, we might expect carbon to be divalent, with the two $2p$ electrons forming bonds at approximately 90°. However, one of the most important features of the chemistry of carbon is that tetravalent carbon compounds, such as CH_4, are overwhelmingly favored.

The observed valence of 4 for carbon comes about in an interesting way. One of the first excited states of carbon occurs when a $2s$ electron is excited to a $2p$ state, giving a configuration of $1s^22s^12p^3$. In this excited state, we can have four unpaired electrons, one each in the $2s$, $2p_x$, $2p_y$, and $2p_z$ atomic orbitals. We might expect there to be three similar bonds corresponding to the three p orbitals and one different bond corresponding to the s orbital. However, when carbon forms tetravalent bonds, these four atomic orbitals become mixed and form four new *equivalent* molecular orbitals called **hybrid orbitals.** This mixing of atomic orbitals, called **hybridization,** is probably the most important new feature involved in the physics of complex molecular bonds. Figure 38-10 shows the tetrahedral structure of the methane molecule (CH_4), and Figure 38-11 shows the structure of the ethane molecule (CH_3–CH_3), which is similar to two joined methane molecules in which one of the C–H bonds is replaced with a C–C bond.

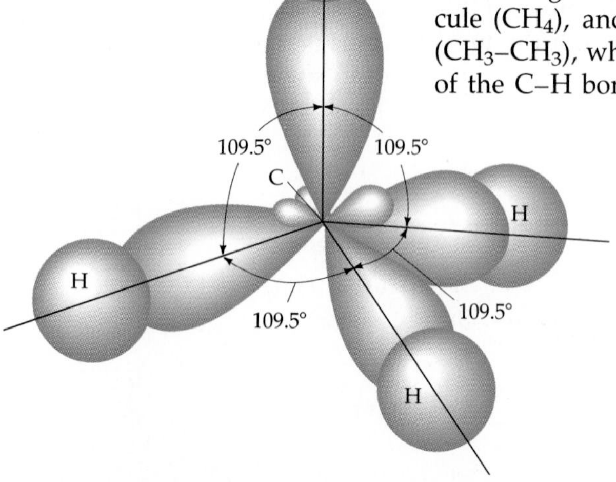

Figure 38-10 Electron charge distribution in the CH_4 molecule (methane).

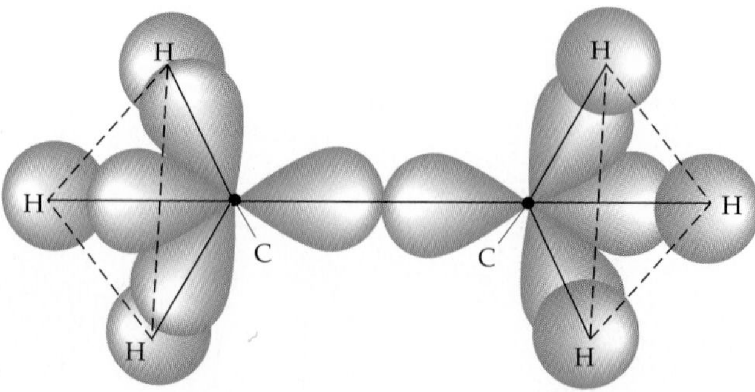

Figure 38-11 Electron charge distribution in the CH_3–CH_3 molecule (ethane).

Carbon orbitals can also hybridize with the s, p_x, and p_y orbitals combining to form three hybrid orbitals in the xy plane with 120° bonds and the p_z orbital remaining unmixed. An example of this configuration is graphite, in which the bonds in the xy plane provide the strongly layered structure characteristic of the material. Structures of some common molecules are shown in Table 38-1 on pages 1274–1275.

38-3 Energy Levels and Spectra of Diatomic Molecules

As is the case with an atom, a molecule often emits electromagnetic radiation when it makes a transition from an excited energy state to a state of lower energy. Conversely, a molecule can absorb radiation and make a transition from a lower energy state to a higher energy state. The study of molecular emission and absorption spectra thus provides us with information about the energy states of molecules. For simplicity, we will consider only diatomic molecules here.

The energy of a molecule can be conveniently separated into three parts: electronic due to the excitation of the electrons of the molecule, vibrational due to the oscillations of the atoms of the molecule, and rotational due to the rotation of the molecule about its center of mass. The magnitudes of these energies are sufficiently different that they can be treated separately. The energies due to the electronic excitations of a molecule are of the order of magnitude of 1 eV, the same as for the excitation of an atom. The energies of vibration and rotation are much smaller than this.

Rotational Energy Levels

Figure 38-12 shows a simple schematic model of a diatomic molecule consisting of a mass m_1 and a mass m_2 separated by a distance r and rotating about its center of mass. Classically, the kinetic energy of rotation (see Section 8-3) is

$$E = \tfrac{1}{2}I\omega^2 \qquad 38\text{-}9$$

where I is the moment of inertia and ω is the angular frequency of rotation. If we write this in terms of the angular momentum $L = I\omega$, we have

$$E = \frac{(I\omega)^2}{2I} = \frac{L^2}{2I} \qquad 38\text{-}10$$

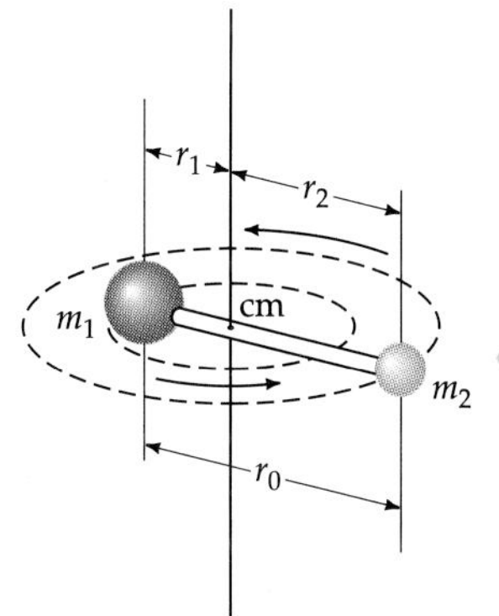

Figure 38-12 Diatomic molecule rotating about an axis through its center of mass.

The solution of the Schrödinger equation for rotation leads to quantization of the angular momentum with values given by

$$L^2 = \ell(\ell + 1)\hbar^2 \qquad \ell = 0, 1, 2, \ldots \qquad 38\text{-}11$$

where ℓ is the **rotational quantum number.** This is the same quantum condition on angular momentum that holds for the orbital angular momentum of an electron in an atom. Note, however, that L in Equation 38-10 refers to the angular momentum of the entire molecule rotating about its center of mass. The energy levels of a rotating molecule are therefore given by

$$E = \frac{\ell(\ell + 1)\hbar^2}{2I} = \ell(\ell + 1)E_{0r} \qquad \ell = 0, 1, 2, \ldots \qquad 38\text{-}12$$

Rotational energy levels

Table 38-1 **Some Common Molecules**

Oxygen atoms **Hydrogen atoms** **Carbon atoms**  **Nitrogen atoms**  **Sulfur atoms**

ESTRADIOL $C_{18}H_{24}O_2$

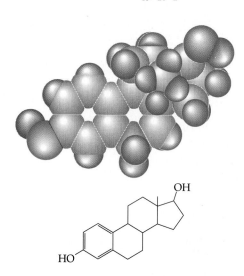

Estradiol is one of the principal female sex hormones. It is released at puberty and decreases in abundance at menopause.

CHOLESTEROL $C_{27}H_{46}O$

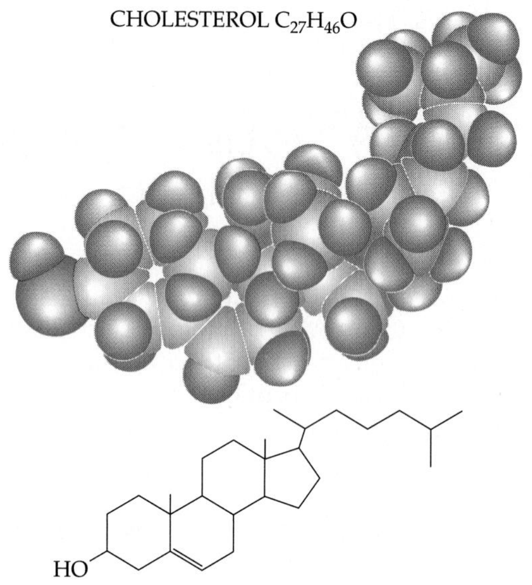

Cholesterol is produced in the liver and plays an essential role in metabolism in the body.

2-FURYLMETHANETHIOL C_5H_6OS

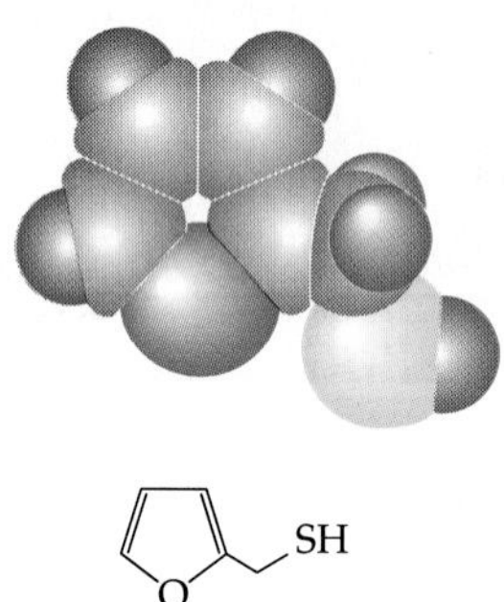

2-Furylmethanethiol is one of the molecules responsible for the aroma of coffee.

CAFFEINE $C_8H_{10}O_2N_4$

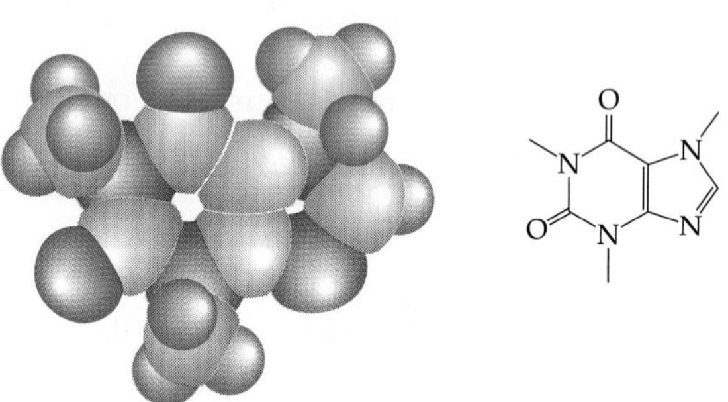

Caffeine is a stimulant found in coffee, tea, and soft drinks.

CITRIC ACID $C_6H_8O_7$

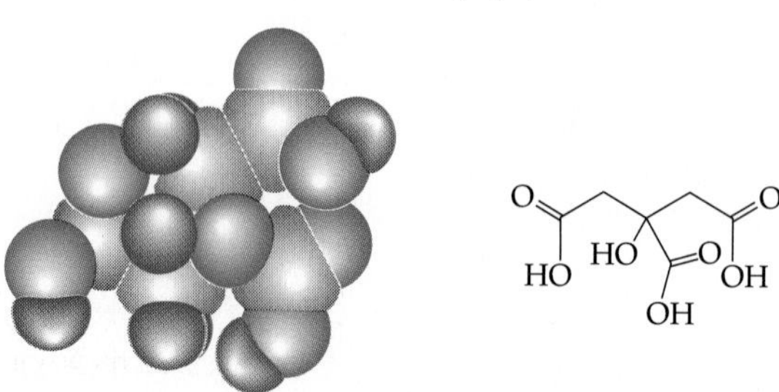

Citric acid is found in fruits such as lemons, grapefruit, and oranges.

FRUCTOSE $C_6H_{12}O_6$

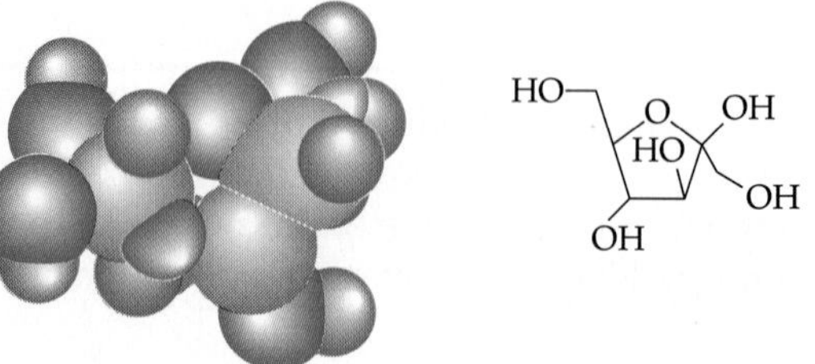

Fructose is a sugar found in honey and many fruits. It is also the sugar that powers the motion of sperm.

MORPHINE $C_{17}H_{19}O_3N$

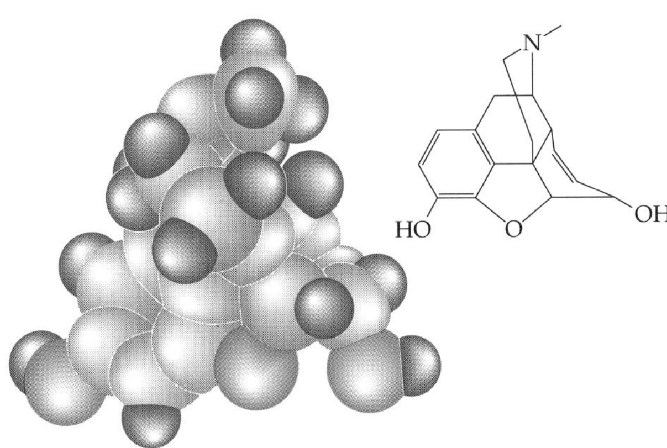

Morphine is the principal component of opium, which acts on the central nervous system and can induce addiction.

VANILLIN $C_8H_8O_3$

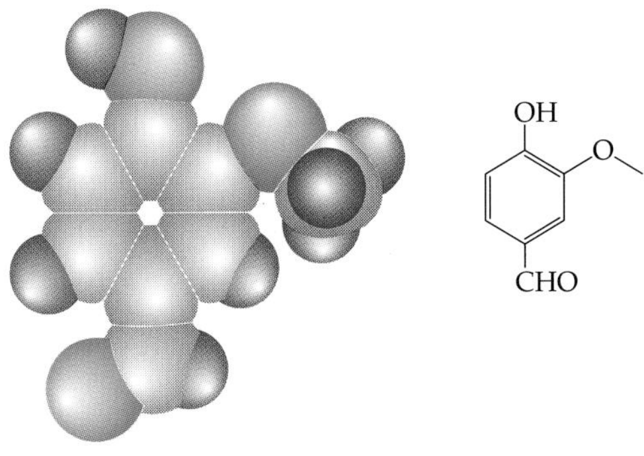

Vanillin is the essential component of oil of vanilla used to flavor many foods.

PELARGONIDIN $C_{15}H_{11}O_5$

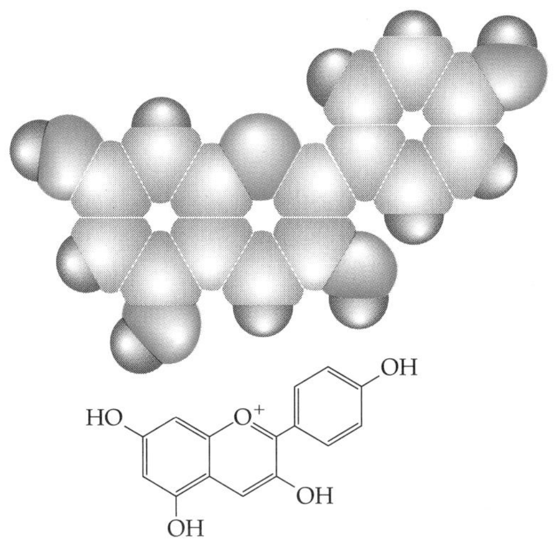

Pelargonidin is one of the molecules responsible for the red color of raspberries, strawberries, and apples.

para-HYDROXYPHENOL-2-BUTANONE $C_{10}H_{12}O_2$

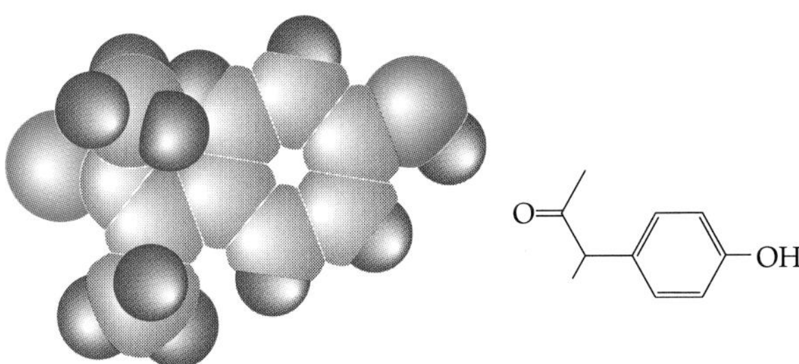

Para-hydroxyphenol-2-butanone is the molecule chiefly responsible for the aroma of raspberries.

METHYL CYANOACRYLATE $C_5H_5O_2N$

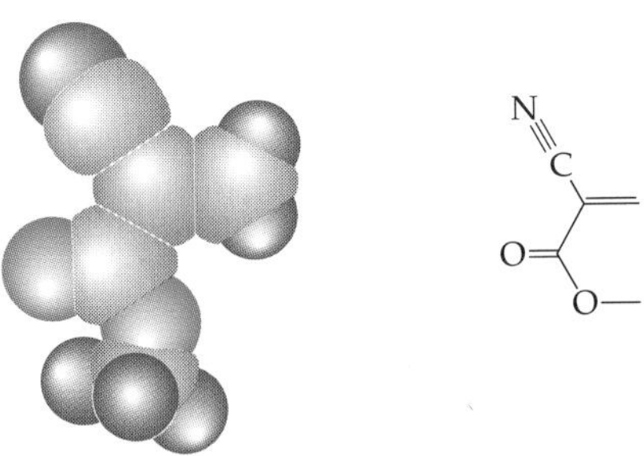

Methyl cyanoacrylate is the substance used in *Super Glue.*

TRINITROTOLUENE $C_7H_5O_6N_3$

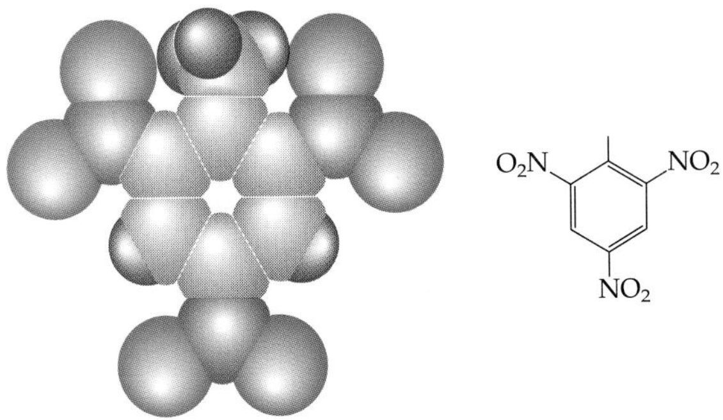

When the oxygen atoms of trinitrotoluene are nudged slightly, they combine with the carbon and hydrogen atoms to produce CO_2, H_2O, and gaseous N_2—thus converting the compact molecule into a suddenly expanding gas and giving TNT its explosiveness.

where E_{0r} is the characteristic rotational energy of a particular molecule, which is inversely proportional to its moment of inertia:

Characteristic rotational energy

$$E_{0r} = \frac{\hbar^2}{2I} \quad \text{38-13}$$

A measurement of the rotational energy of a molecule from its rotational spectrum can be used to determine the moment of inertia of the molecule, which can then be used to find the separation of the atoms in the molecule. The moment of inertia about an axis through the center of mass of a diatomic molecule (see Figure 38-12) is

$$I = m_1 r_1^2 + m_2 r_2^2$$

Using $m_1 r_1 = m_2 r_2$, which relates the distances r_1 and r_2 from the atoms to the center of mass, and $r_0 = r_1 + r_2$ for the separation of the atoms, we can write the moment of inertia as (see Problem 22)

$$I = \mu r_0^2 \quad \text{38-14}$$

where μ, called the **reduced mass,** is

Reduced mass

$$\mu = \frac{m_1 m_2}{m_1 + m_2} \quad \text{38-15}$$

If the masses are equal ($m_1 = m_2 = m$), as in H_2 and O_2, the reduced mass $\mu = \frac{1}{2}m$ and

$$I = \tfrac{1}{2}mr_0^2 \quad \text{38-16}$$

A unit of mass convenient for discussing atomic and molecular masses is the **unified mass unit** u, which is defined as one-twelfth the mass of the neutral carbon-12 (^{12}C) atom. The mass of one ^{12}C atom is thus 12 u. The mass of an atom in unified mass units is therefore numerically equal to the molar mass of the atom in grams. The unified mass unit is related to the gram and kilogram by

$$1 \text{ u} = \frac{1 \text{ g}}{N_A} = \frac{10^{-3} \text{ kg}}{6.0221 \times 10^{23}} = 1.6606 \times 10^{-27} \text{ kg} \quad \text{38-17}$$

where N_A is Avogadro's number.

Example 38-2

Find the reduced mass of the HCl molecule.

From the periodic table in Appendix E, the mass of the hydrogen atom is 1.01 u and that of the chlorine atom is 35.5 u. The reduced mass of HCl is therefore

$$\mu = \frac{m_1 m_2}{m_1 + m_2} = \frac{(1.01 \text{ u})(35.5 \text{ u})}{1.01 \text{ u} + 35.5 \text{ u}} = 0.982 \text{ u}$$

Note that the reduced mass is less than the mass of either atom in the molecule and that it is approximately equal to the mass of the hydrogen atom. When one atom of a diatomic molecule is much more massive than the other, the center of mass of the molecule is approximately at the center of the more massive atom, and the reduced mass is approximately equal to the mass of the lighter atom.

Example 38-3

Estimate the characteristic rotational energy of an O_2 molecule, assuming that the separation of the atoms is 0.1 nm.

If r_0 is the separation of the atoms in the oxygen molecule, the moment of inertia is

$$I = \tfrac{1}{2}mr_0^2$$

where m is the mass of an oxygen atom. Substituting this into Equation 38-13, we obtain

$$E_{0r} = \frac{\hbar^2}{mr_0^2}$$

Using $m = 16\text{ u} = 16(1.66 \times 10^{-27}\text{ kg}) = 2.66 \times 10^{-26}\text{ kg}$ and $r_0 = 10^{-10}\text{ m}$, we obtain

$$E_{0r} = \frac{\hbar^2}{mr_0^2} = \frac{(1.05 \times 10^{-34}\text{ J·s})^2}{(2.66 \times 10^{-26}\text{ kg})(10^{-10}\text{ m})^2}$$
$$= 4.14 \times 10^{-23}\text{ J} = 2.59 \times 10^{-4}\text{ eV}$$

We can see from Example 38-3 that the rotational energy levels are several orders of magnitude smaller than those due to electron excitation, which have energies of the order of 1 eV or higher. Transitions within a given set of rotational energy levels yield photons in the far infrared region of the electromagnetic spectrum. Note that the rotational energies are small compared with the typical thermal energy kT at normal temperatures. For $T = 300$ K, for example, kT is about 2.6×10^{-2} eV. Thus, at ordinary temperatures, a molecule can be easily excited to the lower rotational energy levels by collisions with other molecules. But such collisions cannot excite the molecule to its electronic energy levels above the ground state.

Vibrational Energy Levels

The quantization of energy in a simple harmonic oscillator was one of the first problems solved by Schrödinger in his paper proposing his wave equation. Solving the Schrödinger equation for a simple harmonic oscillator gives

$$E_\nu = (\nu + \tfrac{1}{2})hf \qquad \nu = 0, 1, 2, 3, \ldots \tag{38-18}$$

Vibrational energy levels

where f is the frequency of the oscillator and ν is the **vibrational quantum number.*** An interesting feature of this result is that the energy levels are equally spaced with intervals equal to hf. The frequency of vibration of a diatomic molecule can be related to the force exerted by one atom on the other. Consider two objects of mass m_1 and m_2 connected by a spring of force constant K. The frequency of oscillation of this system can be shown to be (see Problem 34)

$$f = \frac{1}{2\pi}\sqrt{\frac{K}{\mu}} \tag{38-19}$$

where μ is the reduced mass given by Equation 38-15. The effective force constant of a diatomic molecule can thus be determined from a measurement of the frequency of oscillation of the molecule.

*We use ν (the Greek letter nu) here rather than n so as not to confuse the vibrational quantum number with the principal quantum number n for electronic energy levels.

A selection rule on transitions between vibrational states (of the same electronic state) requires that ν can change only by ± 1, so the energy of a photon emitted by such a transition is hf and the frequency is f, the same as the frequency of vibration. There is a similar selection rule that ℓ must change by ± 1 for transitions between rotational states. A typical measured frequency of a transition between vibrational states is 5×10^{13} Hz, which gives for the order of magnitude of vibrational energies

$$E \sim hf = (4.14 \times 10^{-15}\ \text{eV}\cdot\text{s})(5 \times 10^{13}\ \text{s}^{-1}) = 0.2\ \text{eV}$$

Note that a typical vibrational energy is about 1000 times greater than the typical rotational energy E_{0r} of the O_2 molecule we found in Example 38-3 and about 8 times greater than the typical thermal energy $kT = 0.026$ eV at $T = 300$ K.

Example 38-4

The frequency of vibration of the CO molecule is 6.42×10^{13} Hz. What is the effective force constant for this molecule?

Using 12 u for the mass of the carbon atom and 16 u for the mass of the oxygen atom, we obtain for the reduced mass

$$\mu = \frac{m_1 m_2}{m_1 + m_2} = \frac{(12\ \text{u})(16\ \text{u})}{12\ \text{u} + 16\ \text{u}} = 6.86\ \text{u}$$

Multiplying by 1.66×10^{-27} kg/u to convert to SI units, we obtain

$$\mu = (6.86\ \text{u})(1.66 \times 10^{-27}\ \text{kg/u}) = 1.14 \times 10^{-26}\ \text{kg}$$

Solving Equation 38-19 for the effective force constant K, we obtain

$$K = (2\pi f)^2 \mu = 4\pi^2 (6.42 \times 10^{13}\ \text{Hz})^2 (1.14 \times 10^{-26}\ \text{kg})$$
$$= 1.86 \times 10^3\ \text{N/m}$$

Emission Spectra

Figure 38-13 shows schematically some electronic, vibrational, and rotational energy levels of a diatomic molecule. The vibrational levels are labeled with the quantum number ν and the rotational levels are labeled with ℓ. The lower vibrational levels are evenly spaced, with $\Delta E = hf$. For higher vibra-

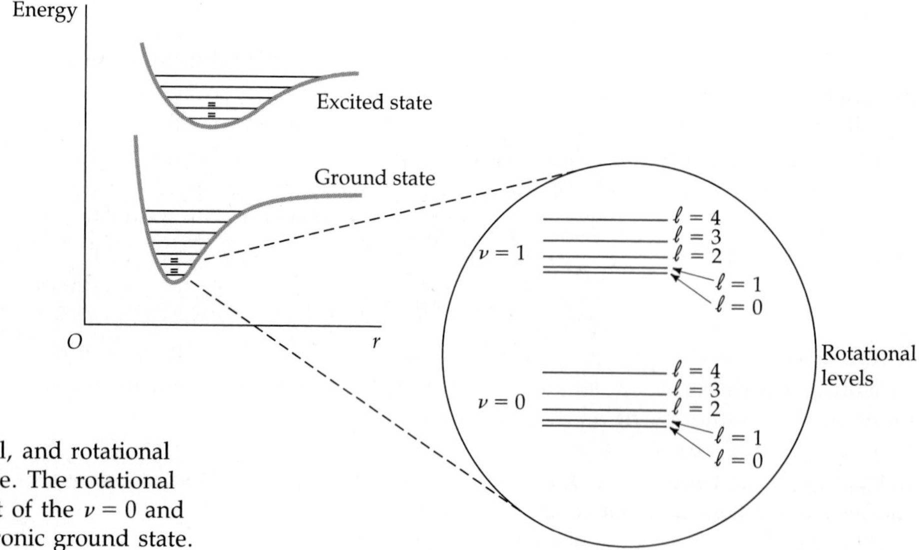

Figure 38-13 Electronic, vibrational, and rotational energy levels of a diatomic molecule. The rotational levels are shown in an enlargement of the $\nu = 0$ and $\nu = 1$ vibrational levels of the electronic ground state.

tional levels, the approximation that the vibration is simple harmonic is not valid and the levels are not quite evenly spaced. Note that the potential-energy curves representing the force between the two atoms in the molecule do not have exactly the same shape for the electronic ground and excited states. This implies that the fundamental frequency of vibration f is different for different electronic states. For transitions between vibrational states of different electronic states, the selection rule $\Delta\nu = \pm 1$ does not hold. Such transitions result in the emission of photons of wavelengths in or near the visible spectrum, so the emission spectrum of a molecule for electronic transitions is also sometimes called the optical spectrum.

The spacing of the rotational levels increases with increasing values of ℓ. Since the energies of rotation are so much smaller than those of vibrational or electronic excitation of a molecule, molecular rotation shows up in optical spectra as a fine splitting of the spectral lines. When the fine structure is not resolved, the spectrum appears as bands as shown in Figure 38-14*a*. Close inspection of these bands reveals that they have a fine structure due to the rotational energy levels, as shown in the enlargement in Figure 38-14*b*.

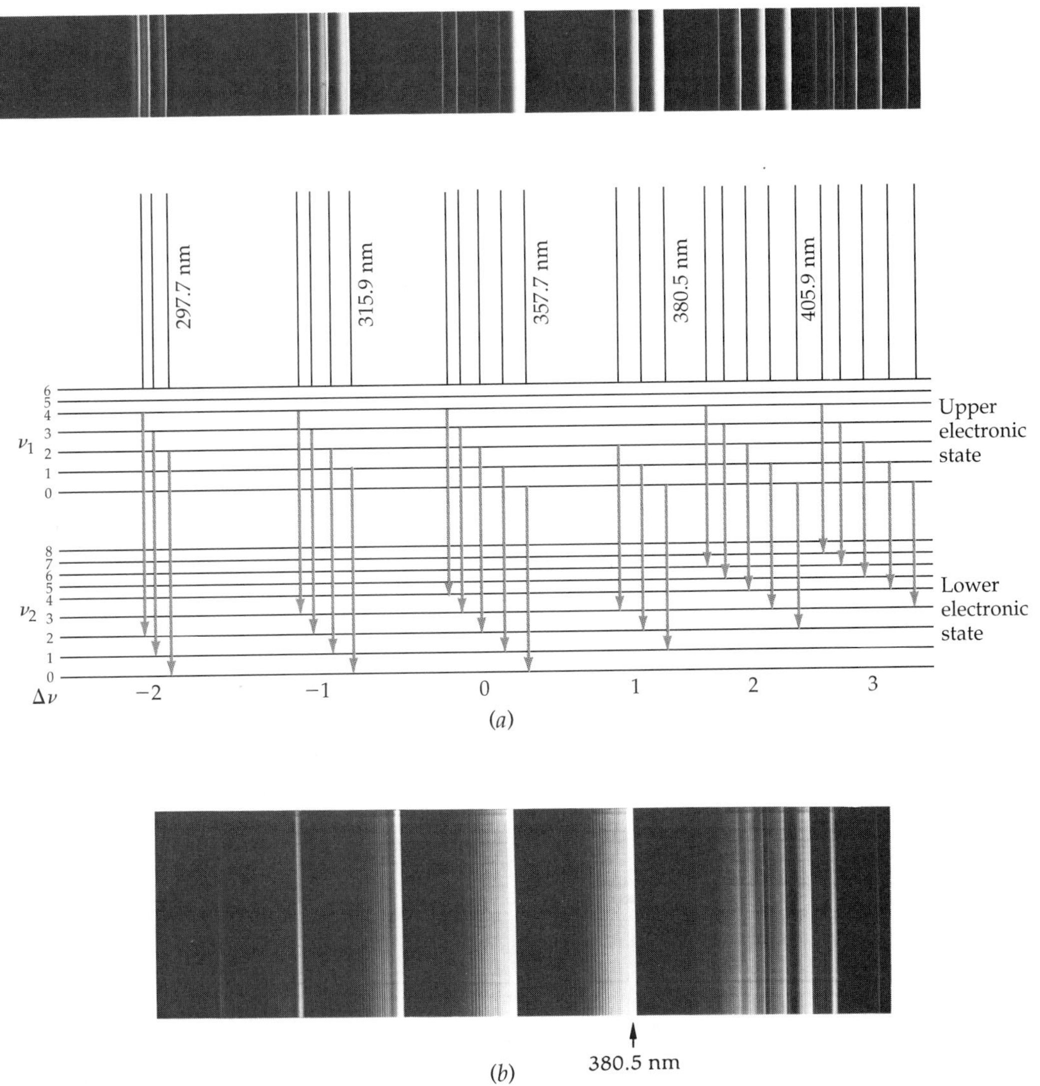

Figure 38-14 Part of the emission spectrum of N_2. (*a*) These components of the band are due to transitions between the vibrational levels of two electronic states, as indicated in the diagram. (*b*) An enlargement of part of *a* shows that the apparent lines in *a* are in fact band heads with structure caused by rotational levels.

Absorption Spectra

Much molecular spectroscopy is done using infrared absorption techniques in which only the vibrational and rotational energy levels of the ground-state electronic level are excited. For ordinary temperatures, the vibrational energies are sufficiently large in comparison with the thermal energy kT that most of the molecules are in the lowest vibrational state $\nu = 0$, for which the energy is $E_0 = \frac{1}{2}hf$. The transition from $\nu = 0$ to $\nu = 1$ is the predominant transition in absorption. The rotational energies, however, are sufficiently less than kT that the molecules are distributed among several rotational energy states. If the molecule is originally in a vibrational state characterized by $\nu = 0$ and a rotational state characterized by the quantum number ℓ, its initial energy is

$$E_\ell = \tfrac{1}{2}hf + \ell(\ell + 1)E_{0r} \qquad 38\text{-}20$$

where E_{0r} is given by Equation 38-13. From this state, two transitions are permitted by the selection rules. For a transition to the next highest vibrational state $\nu = 1$ and a rotational state characterized by $\ell + 1$, the final energy is

$$E_{\ell+1} = \tfrac{3}{2}hf + (\ell + 1)(\ell + 2)E_{0r} \qquad 38\text{-}21$$

For a transition to the next highest vibrational state and to a rotational state characterized by $\ell - 1$, the final energy is

$$E_{\ell-1} = \tfrac{3}{2}hf + (\ell - 1)\ell E_{0r} \qquad 38\text{-}22$$

The energy differences are

$$\Delta E_{\ell\to\ell+1} = E_{\ell+1} - E_\ell = hf + 2(\ell + 1)E_{0r} \qquad 38\text{-}23$$

where $\ell = 0, 1, 2, \ldots$, and

$$\Delta E_{\ell\to\ell-1} = E_{\ell-1} - E_\ell = hf - 2\ell E_{0r} \qquad 38\text{-}24$$

where $\ell = 1, 2, 3, \ldots$. (In Equation 38-24, ℓ begins at $\ell = 1$ because from $\ell = 0$ only the transition $\ell \to \ell + 1$ is possible.) Figure 38-15 illustrates these transitions. The frequencies of these transitions are given by

$$f_{\ell\to\ell+1} = \frac{\Delta E_{\ell\to\ell+1}}{h} = f + \frac{2(\ell + 1)E_{0r}}{h} \qquad \ell = 0, 1, 2, \ldots \qquad 38\text{-}25$$

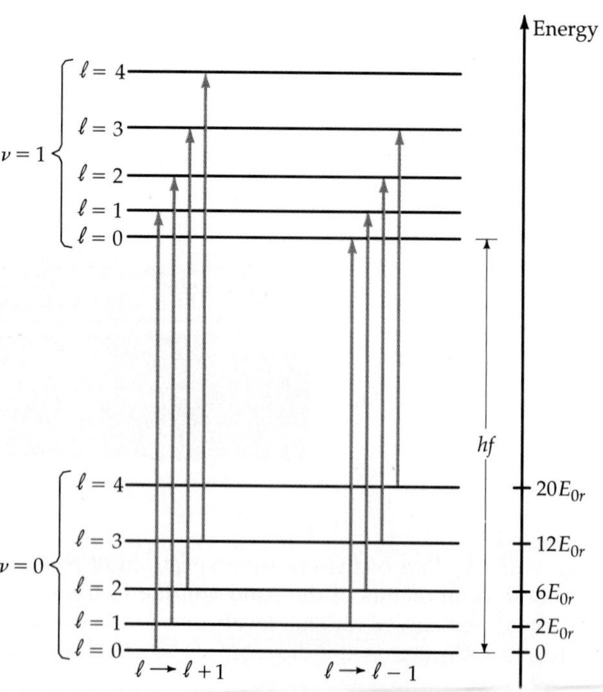

Figure 38-15 Absorptive transitions between the lowest vibrational states $\nu = 0$ and $\nu = 1$ in a diatomic molecule. These transitions obey the selection rule $\Delta\ell = \pm 1$ and fall into two bands. The energies of the $\ell \to \ell + 1$ band are $hf + 2E_{0r}$, $hf + 4E_{0r}$, $hf + 6E_{0r}$, and so forth, whereas the energies of the $\ell \to \ell - 1$ band are $hf - 2E_{0r}$, $hf - 4E_{0r}$, $hf - 6E_{0r}$, and so forth.

and

$$f_{\ell\to\ell-1} = \frac{\Delta E_{\ell\to\ell-1}}{h} = f - \frac{2\ell E_{0r}}{h} \qquad \ell = 1, 2, 3, \ldots \qquad 38\text{-}26$$

The frequencies for the transitions $\ell \to \ell + 1$ are thus $f + 2(E_{0r}/h)$, $f + 4(E_{0r}/h)$, $f + 6(E_{0r}/h)$, and so forth; those corresponding to the transition $\ell \to \ell - 1$ are $f - 2(E_{0r}/h)$, $f - 4(E_{0r}/h)$, $f - 6(E_{0r}/h)$, and so forth. We thus expect the absorption spectrum to contain frequencies equally spaced by $2E_{0r}/h$ except for a gap of $4E_{0r}/h$ at the vibrational frequency f as shown in Figure 38-16. A measurement of the position of the gap gives f, and a measurement of the spacing of the absorption peaks gives E_{0r}, which is inversely proportional to the moment of inertia of the molecule.

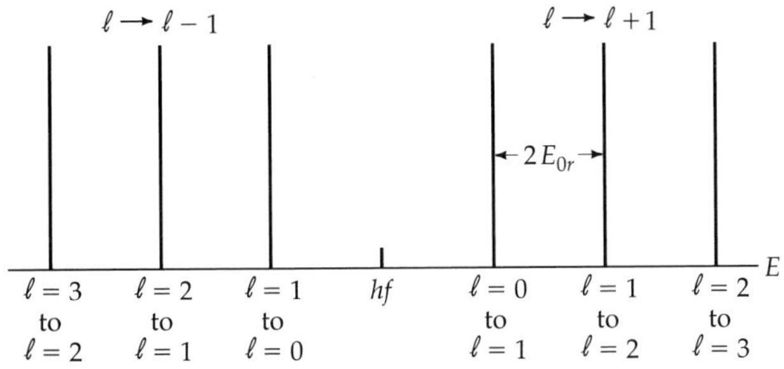

Figure 38-16 Expected absorption spectrum of a diatomic molecule. The right branch corresponds to the transitions $\ell \to \ell + 1$ and the left branch to the transitions $\ell \to \ell - 1$. The lines are equally spaced by $2E_{0r}$. The energy midway between the branches is hf, where f is the frequency of vibration of the molecule.

Figure 38-17 shows the absorption spectrum of HCl. The double-peak structure results from the fact that chlorine occurs naturally in two isotopes. ^{35}Cl and ^{37}Cl, which have different moments of inertia. If all the rotational levels were equally populated initially, we would expect the intensities of each absorption line to be equal. However, the population of a rotational level ℓ is proportional to the degeneracy of the level, that is, to the number of states with the same value of ℓ, which is $2\ell + 1$, and to the Boltzmann factor $e^{-E/kT}$, where E is the energy of the state. For low values of ℓ, the population increases slightly because of the degeneracy factor, whereas for higher values of ℓ, the population decreases because of the Boltzmann factor. The intensities of the absorption lines therefore increase with ℓ for low values of ℓ and then decrease with ℓ for high values of ℓ, as can be seen from the figure.

Figure 38-17 Absorption spectrum of the diatomic molecule HCl. The double-peak structure results from the two isotopes of chlorine, ^{35}Cl (abundance 75.5 percent) and ^{37}Cl (abundance 24.5 percent). The intensities of the peaks vary because the population of the initial state depends on ℓ.

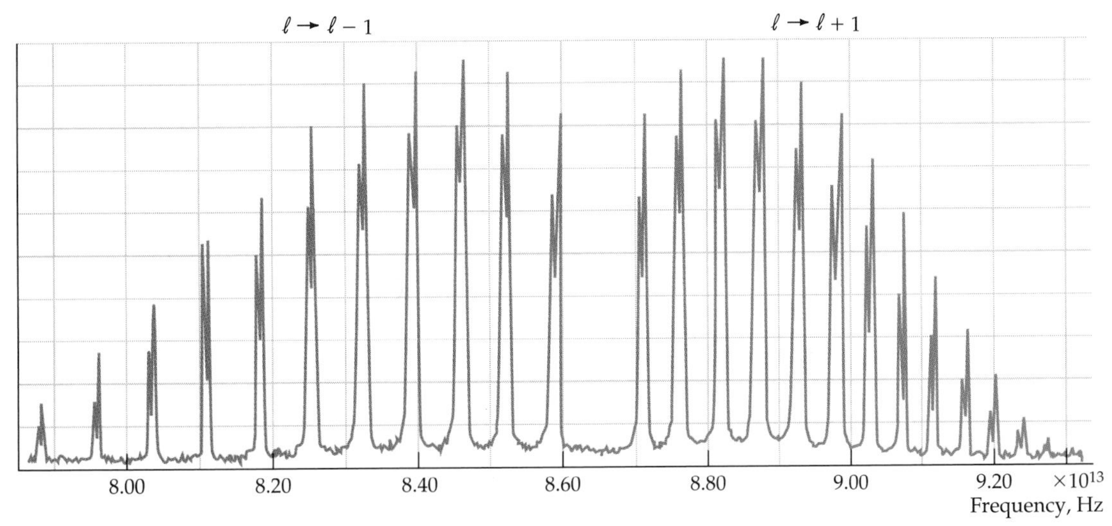

Questions

5. How does the effective force constant calculated for the CO molecule in Example 38-4 compare with the force constant of an ordinary spring?
6. Why can an atom absorb radiation only from the ground state whereas a diatomic molecule can absorb radiation from many different rotational states?

Summary

1. Bonding mechanisms for atoms and molecules include ionic, covalent, van der Waals, hydrogen, and metallic bonds. Ionic bonds result when an electron is transferred from one atom to another, resulting in a positive ion and a negative ion that bond together. The covalent bond is a quantum-mechanical effect that arises from the sharing of one or more electrons by identical or similar atoms. The van der Waals bonds are weak bonds that result from the interaction of the instantaneous electric dipole moments of molecules. The hydrogen bond results from the sharing of a hydrogen atom by other atoms. In the metallic bond, the positive lattice ions of the metal are held together by a cloud of negative charge comprised of free electrons.
2. A diatomic molecule formed from two identical atoms, such as O_2, must bond by covalent bonding. The bonding of two nonidentical atoms is often a mixture of covalent and ionic bonding. The percentage of ionic bonding can be found from the ratio of the measured electric dipole moment to the ionic electric dipole moment defined by

$$p_{\text{ionic}} = er_0$$

where r_0 is the equilibrium separation of the ions.
3. The shapes of such polyatomic molecules as H_2O and NH_3 can be understood from the spatial distribution of the atomic-orbital or molecular-orbital wave functions. The tetravalent nature of the carbon atom is a result of the hybridization of the $2s$ and $2p$ atomic orbitals.
4. The rotational energies of a diatomic molecule are quantized to the values

$$E_\ell = \frac{\ell(\ell+1)\hbar^2}{2I} = \ell(\ell+1)E_{0r} \qquad \ell = 0, 1, 2, \ldots$$

where

$$E_{0r} = \frac{\hbar^2}{2I}$$

and I is the moment of inertia of the molecule. The moment of inertia of a diatomic molecule is related to the equilibrium separation r_0 by

$$I = \mu r_0^2$$

where

$$\mu = \frac{m_1 m_2}{m_1 + m_2}$$

is the reduced mass.

5. The vibrational energies of a diatomic molecule are quantized to the values

$$E_\nu = (\nu + \tfrac{1}{2})hf \qquad \nu = 0, 1, 2, 3, \ldots$$

where f is the frequency of vibration of the molecule, which is related to the effective force constant K by

$$f = \frac{1}{2\pi}\sqrt{\frac{K}{\mu}}$$

6. The optical spectra of molecules have a band structure due to transitions between rotational levels. Information about the structure and bonding of a molecule can be found from its rotational and vibrational absorption spectrum involving transitions from one vibrational–rotational level to another. These transitions obey the selection rules

$$\Delta\nu = \pm 1$$

$$\Delta\ell = \pm 1$$

Suggestions for Further Reading

Atkins, P. W.: *Molecules*, Scientific American Library, A Division of HPHLP, New York, 1987.

This book illustrates the structure and describes some properties of a selection of molecules.

Nassau, Kurt: *The Physics and Chemistry of Color: The Fifteen Causes of Color*, John Wiley & Sons, New York, 1983.
Nassau, Kurt: "The Causes of Color," *Scientific American*, October 1980, p. 124.

This fascinating book and article discuss mechanisms of color production in such objects as hot bodies, neon signs, organic dyes, gemstones, metals, and semiconductors.

Pauling, Linus, and Roger Hayward: *The Architecture of Molecules*, W. H. Freeman and Company, San Francisco and London, 1964.

This is a picture book illustrating the relative positions of the various atoms in some common molecules and solids, with commentary by one of the men most responsible for our modern understanding of the chemical bond.

Ronn, Avigdor M.: "Laser Chemistry," *Scientific American*, May 1979, p. 114.

This article explains how laser light of the correct frequency can efficiently induce the electronic or molecular transitions required for a desired chemical reaction to occur.

Review

A. Objectives: After studying this chapter, you should:

1. Be able to list the ways atoms can bond together and describe each briefly.

2. Be able to compare ionic and covalent bonding.

3. Know the general shapes of the H_2O and NH_3 molecules and understand the origin of these shapes.

4. Be able to describe the general features of the energy-level diagram for a diatomic molecule and discuss the vibrational–rotational spectrum.

5. Be able to discuss why only certain lines in the emission spectrum of a molecule are seen in the absorption spectrum.

B. Define, explain, or otherwise identify:

Molecular orbitals
Ionic bond
Electron affinity
Exclusion-principle repulsion
Dissociation energy
Covalent bond
Saturated bond
The van der Waals bond
Hydrogen bond
Metallic bond
Atomic orbital
Hybrid orbital
Hybridization
Rotational quantum number
Reduced mass
Unified mass unit
Vibrational quantum number

C. True or false: If the statement is true, explain why it is true. If it is false, give a counterexample.

1. Ionic bonds are formed by atoms whose outer electron shells are filled.

2. The repulsive force between the two oxygen atoms in the O_2 molecule is due to the Coulomb repulsion of the nuclei of the atoms.

3. The dissociation energy of NaCl is the energy needed to move the Na^+ and Cl^- ions to an infinite separation.

4. Covalent bonding is a rare occurrence in nature.

5. The energies of vibration and rotation of a molecule are usually much greater than the energy of electronic excitation.

Problems

Level I

38-1 Molecular Bonding

1. Calculate the separation of Na^+ and Cl^- ions for which the potential energy is -1.53 eV.

2. The dissociation energy of Cl_2 is 2.48 eV. Consider the formation of an NaCl molecule by the reaction

$$\text{Na} + \tfrac{1}{2}\text{Cl}_2 \rightarrow \text{NaCl}$$

Is this reaction endothermic (requiring energy) or exothermic (giving off energy)? How much energy per molecule is required or given off?

3. What kind of bonding mechanism would you expect for (*a*) the KCl molecule, (*b*) the O_2 molecule, and (*c*) copper atoms in a solid?

4. What type of bonding mechanism would you expect for (*a*) NaF, (*b*) KBr, and (*c*) N_2?

5. The dissociation energy is sometimes expressed in kilocalories per mole. (*a*) Find the relation between electron volts per molecule and kilocalories per mole. (*b*) Find the dissociation energy of molecular NaCl in kilocalories per mole.

6. What kind of bonding mechanism would you expect for (*a*) KF, (*b*) NO, and (*c*) silver atoms in a solid?

7. The equilibrium separation of the atoms in the HF molecule is 0.0917 nm, and its measured electric dipole moment is 6.40×10^{-30} C·m. What percentage of the bonding is ionic?

8. Repeat Problem 7 for CsCl, for which the equilibrium separation is 0.291 nm and the measured electric dipole moment is 3.48×10^{-29} C·m.

9. The equilibrium separation of CsF is 0.2345 nm. If its bonding is 70 percent ionic, what should its measured electric dipole moment be?

38-2 Polyatomic Molecules

10. Using Table 37-1, find other elements with the same subshell electron configuration in the two outermost orbitals as carbon. Would you expect the same type of hybridization for these elements as occurs for carbon?

38-3 Energy Levels and Spectra of Diatomic Molecules

11. The characteristic rotational energy E_{0r} for the N_2 molecule is 2.48×10^{-4} eV. From this, find the separation distance of the nitrogen atoms in N_2.

12. Explain why the moment of inertia of a diatomic molecule increases slightly with increasing angular momentum.

13. Show that when one atom in a diatomic molecule is much more massive than the other, the reduced mass is approximately equal to the mass of the lighter atom.

14. Calculate the reduced mass in unified mass units for (*a*) H_2, (*b*) N_2, (*c*) CO, and (*d*) HCl.

15. The separation of the oxygen atoms in O_2 is actually slightly greater than the 0.1 nm assumed in Example 38-3, and the characteristic energy of rotation E_{0r} is 1.78×10^{-4} eV rather than the result obtained in that example. Use this value to calculate the separation distance of the oxygen atoms in O_2.

Level II

16. We are often interested in finding the quantity ke^2/r in electron volts when r is given in nanometers. Show that $ke^2 = 1.44$ eV·nm.

17. The equilibrium separation of the K^+ and Cl^- ions in KCl is about 0.267 nm. (*a*) Calculate the potential energy of attraction of the ions, assuming them to be point charges at this separation. (*b*) The ionization energy of potassium is 4.34 eV and the electron affinity of chlorine is 3.61 eV. Find the dissociation energy for KCl, neglecting any energy of repulsion (see Figure 38-1). (*c*) The measured dissociation energy is 4.43 eV. What is the energy due to repulsion of the ions at the equilibrium separation?

18. Indicate the mean value of r for two vibration levels on the potential-energy curve for a diatomic molecule, and show that because of the asymmetry in the curve, r_{av} increases with increasing vibrational energy, which is why solids expand when heated.

19. (*a*) Calculate the potential energy of attraction between the Na^+ and Cl^- ions of NaCl at the equilibrium separation $r_0 = 0.236$ nm, and compare this result with the dissociation energy given in Figure 38-1. (*b*) What is the energy due to repulsion of the ions at the equilibrium separation?

20. Show that the reduced mass is smaller than either mass in a diatomic molecule.

21. The equilibrium separation of the K^+ and F^- ions in KF is about 0.217 nm. (*a*) Calculate the potential energy of attraction of the ions, assuming them to be point charges

at this separation. (*b*) The ionization energy of potassium is 4.34 eV and the electron affinity of fluorine is 3.45 eV. Find the dissociation energy of KF, neglecting any energy of repulsion. (*c*) The measured dissociation energy is 5.07 eV. Calculate the energy due to repulsion of the ions at the equilibrium separation.

22. Derive Equations 38-14 and 38-15 for the moment of inertia in terms of the reduced mass of a diatomic molecule.

23. The effective force constant for the HF molecule is 970 N/m. Find the frequency of vibration for this molecule.

24. The frequency of vibration of the NO molecule is 5.63×10^{13} Hz. Find the effective force constant for NO.

25. Use the equilibrium separation of the K^+ and Cl^- ions given in Problem 17 and the reduced mass of KCl to calculate the characteristic rotational energy E_{0r} of KCl.

26. The central frequency for the absorption band of HCl shown in Figure 38-17 is at $f = 8.66 \times 10^{13}$ Hz, and the absorption peaks are separated by about $\Delta f = 6 \times 10^{11}$ Hz. Use this information to find (*a*) the lowest vibrational energy for HCl, (*b*) the moment of inertia of HCl, (*c*) the equilibrium separation of the atoms.

27. Calculate the effective force constant for HCl from its reduced mass and the fundamental vibrational frequency obtained from Figure 38-17.

28. The potential energy between two atoms in a molecule can often be described rather well by the Lenard–Jones potential, which can be written

$$U = U_0\left[\left(\frac{a}{r}\right)^{12} - 2\left(\frac{a}{r}\right)^{6}\right]$$

where U_0 and a are constants. (*a*) Find the interatomic separation r_0 in terms of a for which the potential energy is minimum. (*b*) Find the corresponding value of U_{min}. (*c*) Use Figure 38-4 to obtain numerical values for r_0 and U_0 for the H_2 molecule. Express your answers in nanometers and electron volts.

29. Make a plot of the potential energy $U(r)$ versus the internuclear separation r for the H_2 molecule. Use the Lenard–Jones potential determined in Problem 28 and plot each term separately, together with the total $U(r)$.

30. In this problem, you are to find how the van der Waals force between a polar and a nonpolar molecule depends on the distance between the molecules. Let the dipole moment of the polar molecule be in the x direction and the nonpolar molecule be a distance x away. (*a*) How does the electric field due to an electric dipole depend on the distance x? (*b*) Use the facts that the potential energy of an electric dipole of moment p in an electric field $\mathbf{E}$ is $U = -\mathbf{p}\cdot\mathbf{E}$ and that the induced dipole moment of the nonpolar molecule is proportional to $\mathbf{E}$ to find how the potential energy of interaction of the two molecules depends on separation distance. (*c*) Using $F_x = -dU/dx$, find the x dependence of the force between the two molecules.

31. Find the dependence of the force between two polar molecules on separation distance. (See Problem 30.)

32. Use the infrared absorption spectrum of HCl given in Figure 38-17 to obtain (*a*) the characteristic rotational energy E_{0r} in electron volts and (*b*) the vibrational frequency f and the vibrational energy hf in electron volts.

33. For a molecule such as CO, which has a permanent electric dipole moment, radiative transitions obeying the selection rule $\Delta\ell = \pm 1$ between two rotational energy levels of the same vibrational energy state are allowed. (That is, the selection rule $\Delta\nu = \pm 1$ does not hold.) (*a*) Find the moment of inertia of CO for which $r_0 =$ 0.113 nm, and calculate the characteristic rotational energy E_{0r} in electron volts. (*b*) Make an energy-level diagram for the rotational levels for $\ell = 0$ to $\ell = 5$ for some vibrational level. Label the energies in electron volts, starting with $E = 0$ for $\ell = 0$. (*c*) Indicate on your diagram transitions that obey $\Delta\ell = -1$ and calculate the energy of the photons emitted. (*d*) Find the wavelength of the photon emitted for each transition in (*c*). In what region of the electromagnetic spectrum are these photons?

Level III

34. Two objects of mass m_1 and m_2 are attached to a spring of force constant K and equilibrium length r_0. (*a*) Show that when m_1 is moved a distance Δr_1 from its equilibrium position, the force exerted by the spring is

$$F = -K\,\frac{m_1 + m_2}{m_2}\,\Delta r_1$$

(*b*) Show that the frequency of vibration is $f = (1/2\pi)\sqrt{K/\mu}$, where μ is the reduced mass.

35. (*a*) Calculate the reduced mass for the $H^{35}Cl$ and $H^{37}Cl$ molecules and the fractional difference $\Delta\mu/\mu$. (*b*) Show that the mixture of isotopes in HCl leads to a fractional difference in the frequency of a transition from one rotational state to another given by $\Delta f/f = -\Delta\mu/\mu$. (*c*) Compute $\Delta f/f$ and compare your result with Figure 38-17.

36. In calculating the rotational energy levels of a diatomic molecule, we did not consider rotation of the molecule about the line joining the atoms. (*a*) Estimate the moment of inertia of the H_2 molecule about this line. (*b*) Use your results for (*a*) to estimate the typical rotational energy E_{0r} for rotation about the line joining the atoms. (*c*) Compare your answer in (*b*) with the typical thermal energy kT at $T = 300$ K.

Chapter 39

Solids

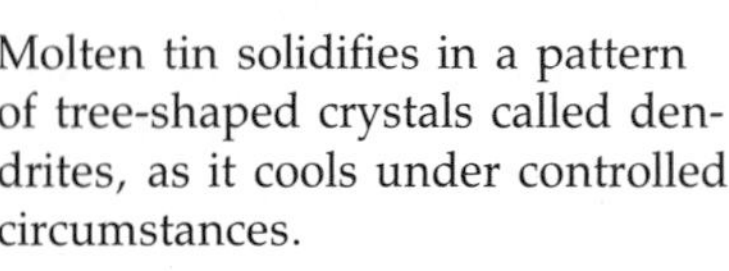

Molten tin solidifies in a pattern of tree-shaped crystals called dendrites, as it cools under controlled circumstances.

The many and varied properties of solids have intrigued us for centuries. Technological developments involving metals and alloys have shaped the courses of civilizations, and the symmetry and beauty of naturally occurring, large single crystals have consistently captured our imaginations. However, the origins of the physical properties of solids were not understood at all until the development of quantum mechanics. The application of quantum mechanics to solids has provided the basis for much of the technological progress of modern times. We will briefly study some aspects of the structure of solids and then concentrate on their electrical properties.

39-1 The Structure of Solids

In our everyday world, we see matter in three phases: gases, liquids, and solids. In a gas, the average distance between two molecules is large compared with the size of a molecule. The molecules have little influence on one another except during their frequent but brief collisions. In a liquid or solid, the molecules are close together and exert forces on one another that are comparable to the forces that bind atoms into molecules. In a liquid, the molecules form temporary short-range bonds that are continually broken and reformed due to the thermal kinetic energy of the molecules. The strength of these bonds depends on the type of molecule. For example, the bonds between helium atoms are the very weak van der Waals bonds, and helium does not liquefy at atmospheric pressure unless the temperature is 4.2 K or lower.

If a liquid is cooled slowly so that the kinetic energy of its molecules is reduced slowly, the molecules may arrange themselves in a regular crystalline array, producing the maximum number of bonds and leading to a minimum potential energy. However, if the liquid is cooled rapidly so that its internal energy is removed before the molecules have a chance to arrange themselves, the solid formed is often not crystalline but instead resembles a snapshot of the liquid. Such a solid is called an **amorphous solid.** It displays short-range order but not the long-range order (over many molecular diameters) that is characteristic of a crystal. Glass is a typical amorphous solid. A characteristic of the long-range ordering of a crystal is that it has a well-defined melting point, whereas an amorphous solid merely softens as its temperature is increased. Many materials may solidify into either an amorphous or a crystalline state depending on how they are prepared. Others exist only in one form or the other. Most common solids are polycrystalline; that is, they consist of many single crystals that meet at grain boundaries. The size of a single crystal is typically a fraction of a millimeter. However, large single crystals do occur naturally and can be produced artificially. We will discuss only simple crystalline solids in this chapter.

The most important property of a single crystal is the symmetry and regularity of its structure. It can be thought of as having a single unit structure that is repeated throughout the crystal. This smallest unit of a crystal is called the **unit cell.** The structure of the unit cell depends on the type of bonding between the atoms, ions, or molecules in the crystal. If more than one kind of atom is present, the structure will also depend on the relative sizes of the atoms. The bonding mechanisms are those discussed in Chapter 38—ionic, covalent, metallic, hydrogen, and van der Waals. Figure 39-1 shows the structure of the ionic crystal sodium chloride (NaCl). The Na^+ and Cl^- ions are spherically symmetric, and the Cl^- ion is approximately twice as large as the Na^+ ion. The minimum potential energy for this crystal occurs when an ion of either kind has six nearest neighbors of the other kind. This structure is called *face-centered-cubic (fcc).* Note that the Na^+ and Cl^- ions in solid NaCl are *not* paired into NaCl molecules.

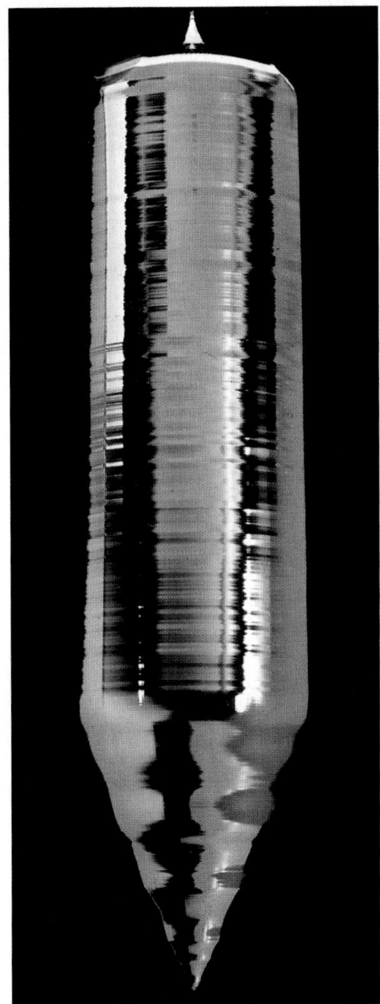

Synthetic crystal silicon is created beginning with a raw material containing silicon (for instance, common beach sand), purifying out the silicon, and melting it. From a seed crystal, the molten silicon grows into a cylindrical crystal, such as the one shown here. The crystals (typically about 1.3 m long) are formed under highly controlled conditions to ensure that they are flawless, and sliced into thousands of thin wafers, onto which the layers of an integrated circuit are etched.

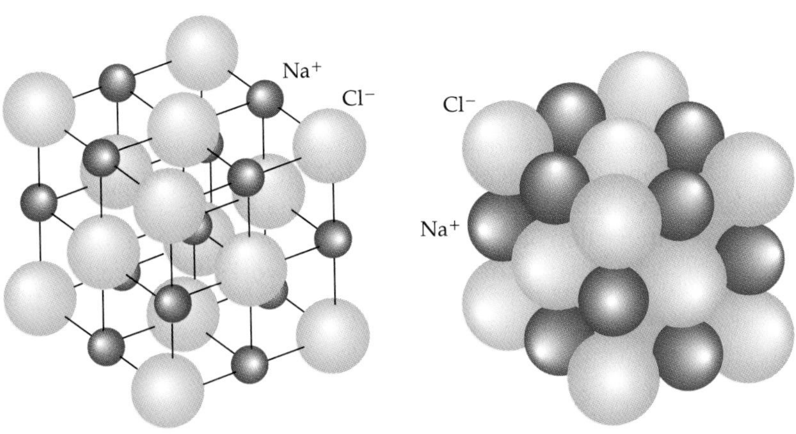

Figure 39-1 Face-centered-cubic structure of the NaCl crystal.

(a) (b)

(a) The hexagonal symmetry of a snowflake arises from a hexagonal symmetry in its lattice of hydrogen and oxygen atoms. (b) NaCl (salt) crystals, magnified about 30 times. The crystals are built up from a cubic lattice of sodium and chloride ions. In the absence of impurities, an exact cubic crystal is formed. This (false-color) scanning electron micrograph shows that in practice the basic cube is often disrupted by dislocations, giving rise to crystals with a wide variety of shapes. The underlying cubic symmetry, though, remains evident. (c) A crystal of quartz (SiO_2, silicon dioxide), the most abundant and widespread mineral on earth. If molten quartz is allowed to solidify without crystallizing, it will form glass. (d) A soldering iron tip, ground down to reveal the copper core within its iron sheath. Visible in the iron is its underlying microcrystalline structure.

The net attractive part of the potential energy of an ion in a crystal can be written

$$U_{att} = -\alpha \frac{ke^2}{r} \tag{39-1}$$

where r is the separation distance between neighboring ions (which is 0.281 nm for the Na^+ and Cl^- ions in crystalline NaCl); and α, called the **Madelung constant,** depends on the geometry of the crystal. If only the 6 nearest neighbors of each ion were important, α would be 6. However, in addition to the 6 neighbors of the opposite charge at a distance r, there are 12 ions of the same charge at a distance $\sqrt{2}r$, 8 ions of opposite charge at a distance $\sqrt{3}r$, and so on. The Madelung constant is thus an infinite sum:

$$\alpha = 6 - \frac{12}{\sqrt{2}} + \frac{8}{\sqrt{3}} - \cdots \tag{39-2}$$

The result* for face-centered-cubic structures is $\alpha = 1.7476$.

When Na^+ and Cl^- ions are very close together, they repel each other because of the overlap of their electrons and the exclusion-principle repulsion discussed in Section 38-1. A simple empirical expression for the potential energy associated with this repulsion that works fairly well is

$$U_{rep} = \frac{A}{r^n}$$

*A large number of terms are needed to calculate the Madelung constant accurately because the sum converges very slowly.

(c)

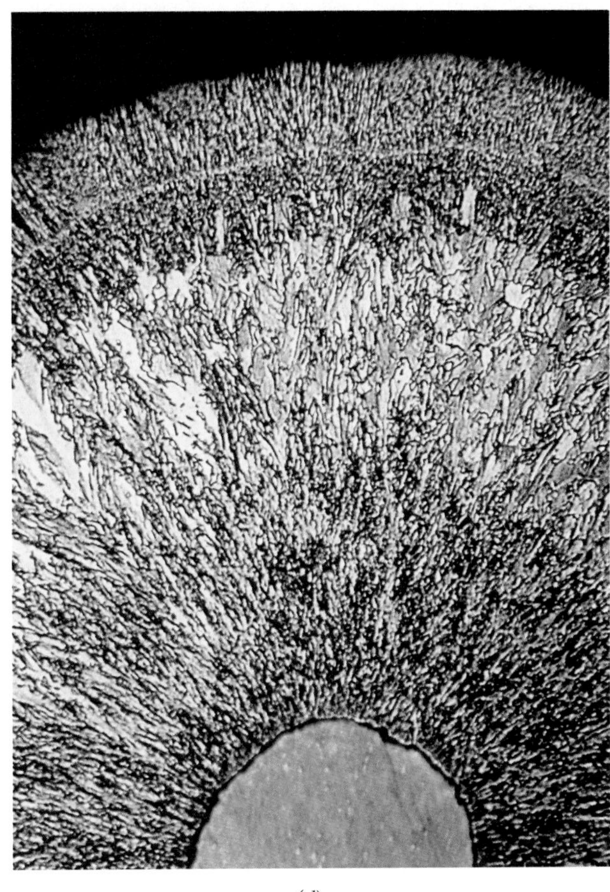

(d)

where A and n are constants. The total potential energy of an ion is then

$$U = -\alpha \frac{ke^2}{r} + \frac{A}{r^n} \tag{39-3}$$

The equilibrium separation $r = r_0$ is that at which the force $F = -dU/dr$ is zero. Differentiating and setting $dU/dr = 0$ at $r = r_0$, we obtain

$$A = \frac{\alpha k e^2 {r_0}^{n-1}}{n} \tag{39-4}$$

Exercise
Derive Equation 39-4.

The total potential energy can thus be written

$$U = -\alpha \frac{ke^2}{r_0}\left[\frac{r_0}{r} - \frac{1}{n}\left(\frac{r_0}{r}\right)^n\right] \tag{39-5}$$

At $r = r_0$, we have

$$U(r_0) = -\alpha \frac{ke^2}{r_0}\left(1 - \frac{1}{n}\right) \tag{39-6}$$

If we know the equilibrium separation r_0, the value of n can be found approximately from the **dissociation energy** of the crystal, which is the energy needed to break up the crystal into atoms.

Example 39-1

Calculate the equilibrium spacing r_0 for NaCl from the measured density of NaCl, which is $\rho = 2.16\ \text{g/cm}^3$.

We consider each ion to occupy a cubic volume of side r_0. The mass of 1 mol of NaCl is 58.4 g, which is the sum of the atomic masses of sodium and chlorine. The ions occupy a volume of $2N_A r_0^3$, where $N_A = 6.02 \times 10^{23}$ is Avogadro's number. The density is thus related to r_0 by

$$\rho = \frac{m}{V} = \frac{m}{2N_A r_0^3}$$

Then

$$r_0^3 = \frac{m}{2N_A \rho} = \frac{58.4\ \text{g}}{2(6.02 \times 10^{23})(2.16\ \text{g/cm}^3)} = 2.24 \times 10^{-23}\ \text{cm}^3$$

$$r_0 = 2.82 \times 10^{-8}\ \text{cm} = 0.282\ \text{nm}$$

The measured dissociation energy of NaCl is 770 kJ/mol. Using 1 eV = 1.602×10^{-19} J, and the fact that 1 mol of NaCl contains N_A pairs of ions, we can express the dissociation energy in electron volts per ion pair. The conversion between electron volts per ion pair and kilojoules per mole is

$$1\ \frac{\text{eV}}{\text{ion pair}} \times \frac{6.022 \times 10^{23}\ \text{ion pairs}}{\text{mol}} \times \frac{1.602 \times 10^{-19}\ \text{J}}{1\ \text{eV}}$$

The result is

$$1\ \frac{\text{eV}}{\text{ion pair}} = 96.47\ \frac{\text{kJ}}{\text{mol}} \qquad 39\text{-}7$$

Thus 770 kJ/mol = 7.98 eV per ion pair. Substituting −7.98 eV for $U(r_0)$, 0.282 nm for r_0, and 1.75 for α in Equation 39-6, we can solve for n. The result is $n = 9.35 \approx 9$.

Most ionic crystals, such as LiF, KF, KCl, KI, and AgCl, have a face-centered-cubic structure. Some elemental solids that have this structure are silver, aluminum, gold, calcium, copper, nickel, and lead.

Figure 39-2 shows the structure of CsCl, which is called the *body-centered-cubic (bcc) structure.* In this structure, each ion has eight nearest neighbor ions of the opposite charge. The Madelung constant for these crystals is 1.7627. Elemental solids with this structure include barium, cesium, iron, potassium, lithium, molybdenum, and sodium.

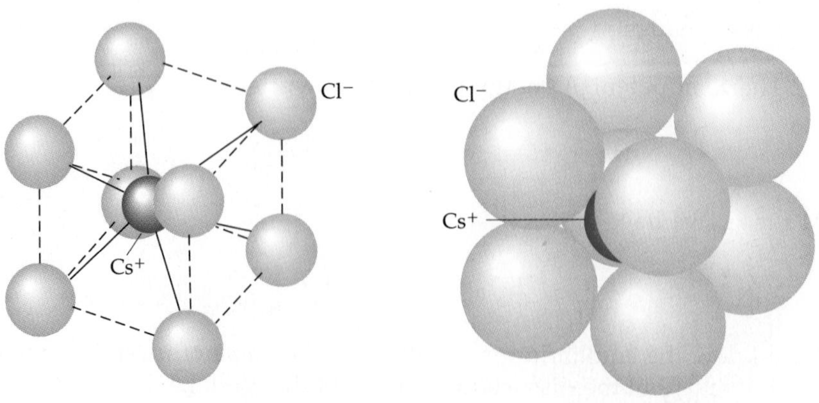

Figure 39-2 Body-centered-cubic structure of the CsCl crystal.

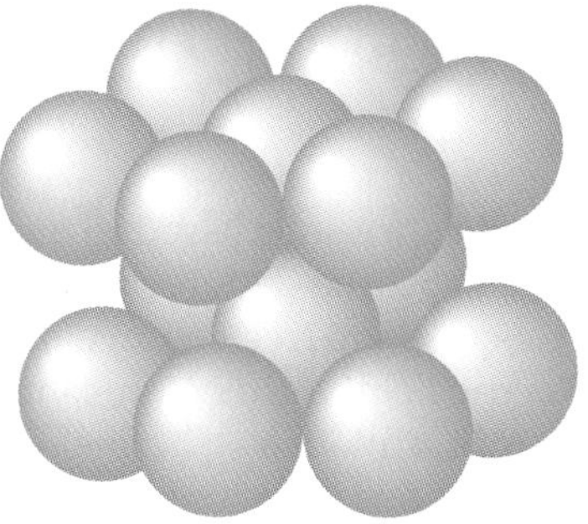

Figure 39-3 Hexagonal close-packed crystal structure.

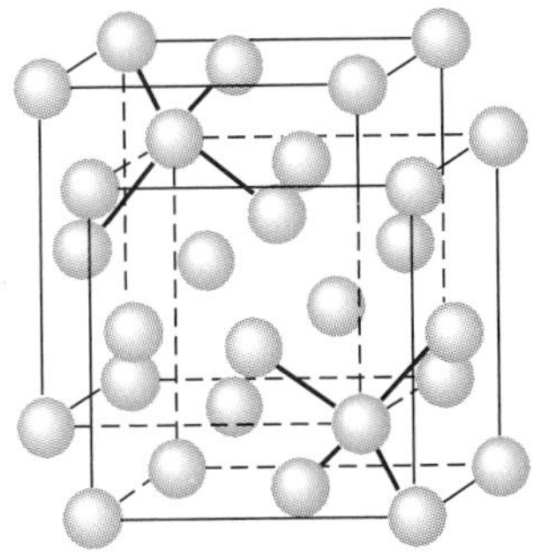

Figure 39-4 Diamond crystal structure. This structure can be considered to be a combination of two interpenetrating face-centered-cubic structures.

Figure 39-3 shows another important crystal structure: the *hexagonal close-packed (hcp) structure.* It is obtained by stacking identical spheres, such as bowling balls. In the first layer, each ball touches six others; thus, the name *hexagonal.* In the next layer, each ball fits into a triangular depression of the first layer. In the third layer, each ball fits into a triangular depression of the second layer, so it lies directly over a ball in the first layer. Elemental solids with *hcp* structure include beryllium, cadmium, cerium, magnesium, osmium, and zinc.

In some solids with covalent bonding, the crystal structure is determined by the directional nature of the bonds. Figure 39-4 illustrates the diamond structure of carbon, in which each atom is bonded to four others as a result of hybridization, which we discussed in Section 38-2. This is also the structure of germanium and silicon.

Questions

1. Why is r_0 different for solid NaCl than for the diatomic molecule?
2. Why would you not expect NaCl to have a *hcp* structure?

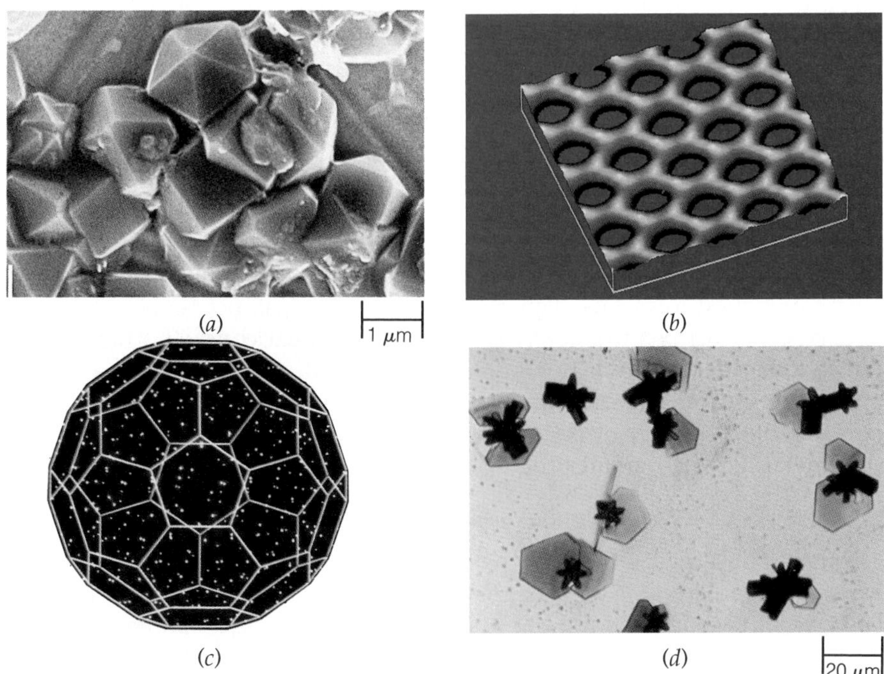

Carbon exists in three well-defined crystalline forms: diamond, graphite, and fullerenes (short for "buckminsterfullerenes") —the third of which was predicted and discovered only a few years ago. The forms differ in how the carbon atoms are packed together in a lattice. A fourth form of carbon, in which no well-defined crystalline form exists, is common charcoal. (*a*) Synthetic diamonds, magnified about 50,000 times. In diamond, each carbon atom is centered in a tetrahedron of four other carbon atoms. The strength of these bonds accounts for the hardness of a diamond. (*b*) An atomic-force micrograph of graphite. In graphite, carbon atoms are arranged in sheets, each sheet made up of atoms in hexagonal rings. The sheets slide easily across one another, a property that allows graphite to function as a lubricant. (*c*) A single sheet of carbon rings can be closed on itself if certain rings are allowed to be pentagonal, instead of hexagonal. A computer-generated image of the smallest such structure, C_{60}, is shown here. Each of the 60 vertices corresponds to a carbon atom; 20 of the faces are hexagons and 12 are pentagons. The same geometric pattern is encountered in a soccer ball. (*d*) Fullerene crystals, in which C_{60} molecules are close-packed. The smaller crystals tend to form thin brownish platelets; larger crystals are usually rod-like in shape. Fullerenes exist in which more than 60 carbon atoms appear. In the crystals shown here, about one-sixth of the molecules are C_{70}.

39-2 The Classical Free-Electron Theory of Metals

The classical free-electron model of metals was discussed in Section 22-5. In this model, a metal is pictured as a regular three-dimensional array of atoms or ions with a large number of electrons that are free to move throughout the whole metal. In the absence of an electric field, the electrons move about the metal much as gas molecules move in a container. Thermal equilibrium is maintained by collisions of electrons with the lattice ions. In Section 22-5, we found an expression (Equation 22-29) for the resistivity ρ in terms of the number of electrons per unit volume n and the collision time τ, which is the average time between collisions:

$$\rho = \frac{m_e}{ne^2\tau} \qquad 39\text{-}8$$

where m_e is the electron mass and e is its charge. The resistivity can also be expressed in terms of the mean free path $\lambda = v_{av}\tau$, where v_{av} is the mean speed due to random thermal motion. Substituting $\tau = \lambda/v_{av}$ into Equation 39-8, we obtain

$$\rho = \frac{m_e v_{av}}{ne^2\lambda} \qquad 39\text{-}9$$

In the classical theory, the mean speed v_{av} is calculated from the equipartition theorem and is of the order of $\sqrt{kT/m_e}$. The mean free path is related to the radius of the lattice ions r and the number of electrons per unit volume n by Equation 22-32:

$$\lambda = \frac{1}{n\pi r^2} \qquad 39\text{-}10$$

As we discussed in Chapter 22, this model successfully predicts Ohm's law, but it has several defects. When classical methods are used to find the mean free path and the mean speed, the magnitude of the resistivity calculated from Equation 39-9 is about six times the measured value at $T = 300$ K. In addition, the temperature dependence is not correct. The temperature dependence of the resistivity in Equation 39-9 is given completely by the mean speed v_{av}, which is proportional to $\sqrt{T}$. Thus, this equation does not reflect the observed linear dependence on temperature. Finally, the classical model says nothing about why some materials are conductors, others are insulators, and still others are semiconductors.

Heat Conduction and Heat Capacity

Good conductors of electricity are also good conductors of heat. The classical theory assumes that this is because the electron gas is mainly responsible for heat conduction in metals. This theory, which we will not discuss in detail, adequately describes the qualitative features of heat conduction in metals, but its quantitative predictions do not agree with experiment.

If the electron gas has a Maxwell–Boltzmann distribution, it should have a mean kinetic energy of $\frac{3}{2}kT$, and we would expect the molar heat capacity of a metal to be $\frac{3}{2}R$ greater than that of an insulator. That is,

$$C_{mv} = (3R)_{\text{lattice vibrations}} + (\tfrac{3}{2}R)_{\text{electron gas}} = \tfrac{9}{2}R \qquad 39\text{-}11$$

This is not observed. The molar heat capacity of metals is very nearly $3R$. At high temperatures, it is slightly greater, but the increase is nowhere near the value of $\frac{3}{2}R$ predicted by the classical theory. The increase is, in fact, proportional to the temperature, and at $T = 300$ K it is only about $0.02R$.

39-3 The Fermi Electron Gas

One of the difficulties of the classical free-electron theory of metals is related to the assumption that the average energy of the electrons is $\frac{3}{2}kT$. This result follows from the equipartition theorem, which applies to any system of particles that obey the Maxwell–Boltzmann energy distribution (Equation 15-33). However, because of the exclusion principle, the energy distribution of the free electrons in a metal is not even approximately given by the Maxwell–Boltzmann distribution, and the average energy is not $\frac{3}{2}kT$. We will consider first the energy distribution of electrons at temperature $T = 0$, which can be calculated rather easily and is a very good approximation to the distribution at other temperatures, even those as high as several thousand kelvins.

The Energy Distribution at $T = 0$

Classically, all the electrons in a conductor should have zero kinetic energy at $T = 0$. As the conductor is heated, the lattice ions acquire an average kinetic energy of $\frac{3}{2}kT$, which is imparted to the electron gas by the collisions between the electrons and the ions. In equilibrium, the electrons would be expected to have a mean kinetic energy of $\frac{3}{2}kT$.

Since the electrons are confined to the space occupied by the metal, it is clear from the uncertainty principle that, even at $T = 0$, an electron cannot have zero kinetic energy. Furthermore, the exclusion principle prevents more than two electrons (with opposite spins) from being in the lowest energy level. At $T = 0$, we expect the electrons to have the lowest energies consistent with the exclusion principle. It is instructive to consider first a one-dimensional model.

Consider N electrons in a one-dimensional infinite square-well potential of length L. The allowed energies were calculated in Chapter 36 and are given by Equation 36-39:

$$E_n = n^2 E_1 \qquad n = 1, 2, 3, \ldots \tag{39-12}$$

where the energy of the lowest level E_1 (Equation 36-40) is

$$E_1 = \frac{h^2}{8m_eL^2} \tag{39-13}$$

We can put two electrons with opposite spins in level $n = 1$, two in level $n = 2$, and so on. The N electrons will thus fill up $N/2$ levels from $n = 1$ to $n = N/2$. The energy of the last filled (or half-filled) level is called the **Fermi energy at $T = 0$.** We can calculate this Fermi energy E_F for N electrons by setting $n = N/2$ in Equation 39-12:

$$E_F = E_{N/2} = \left(\frac{N}{2}\right)^2 E_1 = \frac{h^2}{32m_e}\left(\frac{N}{L}\right)^2 = \frac{(hc)^2}{32m_ec^2}\left(\frac{N}{L}\right)^2 \tag{39-14}$$

where we have multiplied the numerator and denominator by c^2 to simplify numerical calculations. We can see that the Fermi energy is a function of the number of electrons per unit length, which is the number density in one dimension. The number density of free electrons in copper, if we assume one free electron per atom, is $8.47 \times 10^{22}/\text{cm}^3$. In one dimension, this corresponds to

$$\frac{N}{L} = (8.47 \times 10^{22}/\text{cm}^3)^{1/3} = 4.39 \times 10^7/\text{cm} = 4.39/\text{nm}$$

The Fermi energy is then

$$E_F = \frac{(hc)^2}{32m_ec^2}\left(\frac{N}{L}\right)^2 = \frac{(1240\ \text{eV}\cdot\text{nm})^2(4.39/\text{nm})^2}{32(5.11 \times 10^5\ \text{eV})} = 1.82\ \text{eV}$$

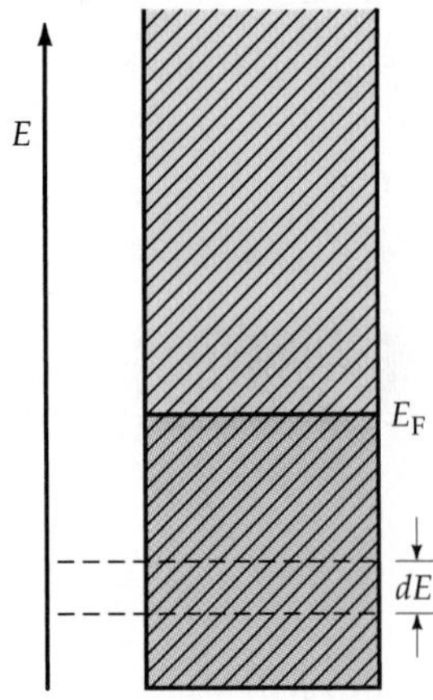

Figure 39-5 Energy levels in a one-dimensional square well. The Fermi energy E_F at $T = 0$ is the energy of the highest occupied level. The levels are so closely spaced they can be assumed to be continuous. The number of states between E and $E + dE$ is $g(E)\,dE$, where $g(E)$ is the density of states.

The average energy is the total energy divided by the number of particles:

$$E_{av} = \frac{1}{N}\sum_{n=1}^{N/2} 2n^2E_1$$

Since $N/2 \gg 1$, we can approximate the sum using an integral:

$$\sum_{n=1}^{N/2} n^2 \approx \int_0^{N/2} n^2\,dn = \frac{1}{3}\left(\frac{N}{2}\right)^3$$

Thus

$$E_{av} = \frac{2E_1}{N}\,\frac{1}{3}\left(\frac{N}{2}\right)^3$$

$$= \frac{1}{3}\left(\frac{N}{2}\right)^2 E_1 = \frac{1}{3}E_F \qquad 39\text{-}15$$

Our one-dimensional calculation thus gives an average energy of about $\frac{1}{3}(1.82 \text{ eV}) \approx 0.6$ eV. The temperature at which the average energy would be 0.6 eV for a one-dimensional Maxwell–Boltzmann distribution, obtained from $\frac{1}{2}kT = 0.6$ eV, is about 14,000 K.

Since the energy states are so close together, we can assume that they are continuous (Figure 39-5). Let $n(E)\,dE$ be the number of particles with energies between E and $E + dE$. We can write this distribution function as

Energy distribution function

$$n(E)\,dE = g(E)\,dE\,F \qquad 39\text{-}16$$

where $g(E)\,dE$ is the number of states in dE, and F is the **Fermi factor,** which is the probability that a state will be occupied. At $T = 0$, the Fermi factor is 1 for states of energy less than E_F and 0 for states of energy greater than E_F:

Fermi factor at $T = 0$

$$\begin{aligned} F &= 1 \qquad E < E_F \\ F &= 0 \qquad E > E_F \end{aligned} \qquad 39\text{-}17$$

The quantity $g(E)$ in Equation 39-16 is called the **density of states.** In one dimension, it is given by

$$g(E) = 2\,\frac{dn}{dE}$$

where $E = n^2E_1$ and the 2 is for the two spin states per space state. Then

$$dE = 2nE_1\,dn = 2\left(\frac{E}{E_1}\right)^{1/2} E_1\,dn = 2(EE_1)^{1/2}\,dn$$

and

$$g(E) = E_1^{-1/2}E^{-1/2} \qquad 39\text{-}18$$

The energy distribution function is then

$$n(E) = E_1^{-1/2}E^{-1/2}F$$

In three dimensions, it is more difficult to count the number of states, so we will just give the results. The Fermi energy in three dimensions at $T = 0$ is given by

$$E_F = \frac{h^2}{8m_e}\left(\frac{3N}{\pi V}\right)^{2/3} = \frac{(hc)^2}{8m_ec^2}\left(\frac{3N}{\pi V}\right)^{2/3} \qquad \text{39-19}a$$

Fermi energy in three dimensions at $T = 0$

where V is the volume of the metal. As in one dimension, the Fermi energy depends on the number density N/V. Table 39-1 lists the free-electron number densities and the calculated Fermi energies at $T = 0$ for several metals.

Table 39-1 **Free-Electron Number Densities and Fermi Energies at $T = 0$ for Selected Elements**

Element	N/V, electrons per cm^3	E_F, eV
Al aluminum	18.1×10^{22}	11.7
Ag silver	5.86×10^{22}	5.50
Au gold	5.90×10^{22}	5.53
Cu copper	8.47×10^{22}	7.04
Fe iron	17.0×10^{22}	11.2
K potassium	1.4×10^{22}	2.11
Li lithium	4.70×10^{22}	4.75
Mg magnesium	8.60×10^{22}	7.11
Mn manganese	16.5×10^{22}	11.0
Na sodium	2.65×10^{22}	3.24
Sn tin	14.8×10^{22}	10.2
Zn zinc	13.2×10^{22}	9.46

Example 39-2

Calculate the Fermi energy at $T = 0$ for copper.

Using Equation 39-19*a*, we obtain for the Fermi energy

$$E_F = \frac{(1240\ \text{eV}\cdot\text{nm})^2}{8(5.11 \times 10^5\ \text{eV})}\left(\frac{3N}{\pi V}\right)^{2/3} = (0.365\ \text{eV}\cdot\text{nm}^2)\left(\frac{N}{V}\right)^{2/3} \qquad \text{39-19}b$$

From Table 39-1, the number density of electrons in copper is $8.47 \times 10^{22}/\text{cm}^3 = 84.7/\text{nm}^3$. Then

$$E_F = (0.365\ \text{eV}\cdot\text{nm}^2)\ (84.7/\text{nm}^3)^{2/3} = 7.04\ \text{eV}$$

Note that the Fermi energy is much greater than kT at ordinary temperatures.

Exercise

Use Equation 39-19*b* to calculate the Fermi energy at $T = 0$ for gold. (Answer: 5.53 eV)

For the three-dimensional case, we again write the energy distribution as $g(E)F$, where F is the Fermi factor given by Equation 39-17, and $g(E)$ is the density of states in three dimensions, which is given by

$$g(E) = \frac{3N}{2}\ E_F^{-3/2}\ E^{1/2} \qquad \text{39-20}$$

Density of states in three dimensions

The average energy at $T = 0$ is calculated from

$$E_{av} = \frac{\int_0^{E_F} Eg(E)\,dE}{\int_0^{E_F} g(E)\,dE} = \frac{1}{N}\int_0^{E_F} Eg(E)\,dE$$

The result is

Average energy at T = 0

$$E_{av} = \tfrac{3}{5}E_F \qquad 39\text{-}21$$

Exercise

Using Equation 39-20, show that $\int_0^{E_F} g(E)\,dE = N$.

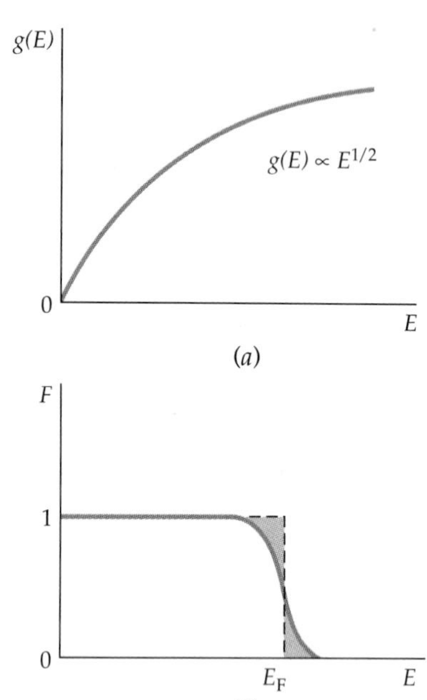

Figure 39-6 The Fermi energy distribution function $n(E)$ shown in (*c*) is the product of the density of states $g(E)$ shown in (*a*) and the Fermi factor F shown in (*b*). The dashed curves in (*b*) and (*c*) show the Fermi factor and energy distribution at $T = 0$. At higher temperatures, some electrons with energies near the Fermi energy are excited, as indicated by the shaded regions in (*b*) and (*c*).

The Energy Distribution at Temperature T

At temperatures greater than $T = 0$, some electrons will gain energy and occupy higher-energy states. The Fermi factor will then be slightly different from that given by Equation 39-17. However, an electron cannot move to a higher or lower state unless it is unoccupied. Since the kinetic energy of the lattice ions is of the order of kT, electrons cannot gain much more energy than kT in collisions with the lattice. Therefore, only those electrons with energies within about kT of the Fermi energy can gain energy as the temperature is increased. At 300 K, kT is only 0.026 eV, so the exclusion principle prevents all but a very few electrons near the top of the energy distribution from gaining energy through random collisions with the lattice ions. Figure 39-6 shows the density of states, the Fermi factor, and the product of these, which gives the energy distribution.

The **Fermi temperature** T_F is defined by

$$kT_F = E_F \qquad 39\text{-}22$$

For temperatures much lower than the Fermi temperature, the average energy of the lattice ions will be much less than the Fermi energy, and the electron energy distribution will not differ greatly from that at $T = 0$.

Example 39-3

Find the Fermi temperature for copper.

Using $E_F = 7.04$ eV and $k = 1.38 \times 10^{-23}$ J/K $= 8.62 \times 10^{-5}$ eV/K in Equation 39-22, we obtain

$$T_F = \frac{E_F}{k} = \frac{7.04\text{ eV}}{8.62 \times 10^{-5}\text{ eV/K}} = 81{,}700\text{ K}$$

We can see from this example that the Fermi temperature of copper is much greater than any temperature T for which copper remains a solid.

The complete quantum-mechanical distribution function for electrons at any temperature is called the **Fermi–Dirac distribution.** The energy distribution function $n(E)$ is given by Equation 39-16. The Fermi factor for temperatures much less than the Fermi temperature is sketched in Figure 39-6*b*. At temperatures above $T = 0$, there is no energy below which all states are full and above which all states are empty, so we must alter our definition of the Fermi energy:

The Fermi energy E_F at any temperature T is defined as that energy at which the probability of a state being occupied is $\frac{1}{2}$; that is, it is the energy at which the Fermi factor F has the value $\frac{1}{2}$.

Fermi energy at temperature T

For all but extremely high temperatures, the difference between the Fermi energy at temperature T and that at $T = 0$ is very small because only those electrons within about kT of the Fermi energy can be excited to higher energy states. The Fermi factor F at temperature T is given by

$$F = \frac{1}{e^{(E-E_F)/kT} + 1} \qquad 39\text{-}23$$

Fermi factor at temperature T

We can see from Equation 39-23 that as $T \to 0$, $e^{(E-E_F)/kT} \to 0$ for $E < E_F$ and $e^{(E-E_F)/kT} \to \infty$ for $E > E_F$, which is consistent with Equation 39-17. We can also see that for those few electrons with energies much greater than the Fermi energy, the Fermi factor approaches $e^{(E_F-E)/kT}$, which is proportional to $e^{-E/kT}$. Thus, the high-energy tail of the Fermi–Dirac energy distribution decreases as $e^{-E/kT}$, just like the classical Maxwell–Boltzmann energy distribution. In this energy region, there are many unoccupied energy states and few electrons, so the Pauli exclusion principle is not important and the distribution approaches the classical distribution. This result is important because it applies to the conduction electrons in semiconductors, as will be discussed in Sections 39-5 and 39-6.

Contact Potential

When two different metals are placed in contact, a potential difference develops between them. Figure 39-7*a* shows the energy levels for two metals with different Fermi energies E_{F1} and E_{F2} and different work functions ϕ_1 and ϕ_2. When the metals are in contact, the total energy of the system is lowered if electrons near the boundary move from the metal with the higher Fermi energy into the metal with the lower Fermi energy until the Fermi energies of the two metals are the same, as shown in Figure 39-7*b*. When equilibrium is established, the metal with the lower initial Fermi energy is negatively charged and that with the higher initial Fermi energy is positively charged, so there is a potential difference between them. This potential difference, called the **contact potential,** equals the difference in the work functions of the two metals divided by the electronic charge e:

$$V_{\text{contact}} = \frac{\phi_1 - \phi_2}{e} \qquad 39\text{-}24$$

Table 39-2 lists the work functions for several metals.

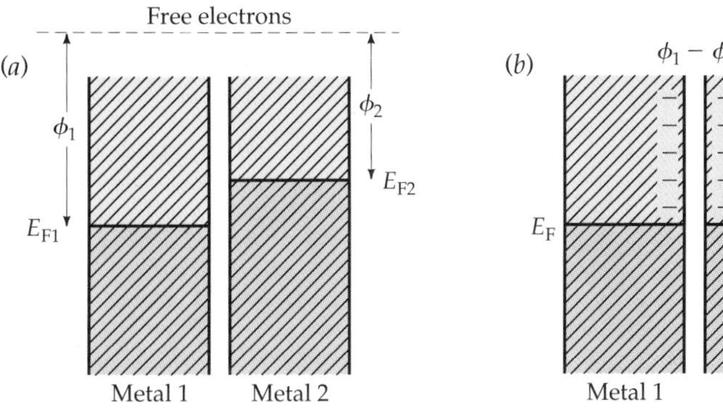

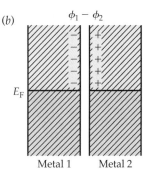

Figure 39-7 (*a*) Energy levels for two different metals with different Fermi energies and work functions. (*b*) When the metals are in contact, electrons flow from the metal that initially has the higher Fermi energy to the metal that initially has the lower Fermi energy until the Fermi energies are equal.

Table 39-2 **Work Functions for Some Metals**

Metal	ϕ, eV
Ag silver	4.7
Au gold	4.8
Ca calcium	3.2
Cu copper	4.1
K potassium	2.1
Mn manganese	3.8
Na sodium	2.3
Ni nickel	5.2

Example 39-4

The threshold wavelength for the photoelectric effect is 271 nm for tungsten and 262 nm for silver. What is the contact potential developed when silver and tungsten are placed in contact?

From Equation 35-5, the work function ϕ is related to the threshold wavelength λ_t by

$$\phi = \frac{hc}{\lambda_t}$$

The work function for tungsten ϕ_W is thus

$$\phi_W = \frac{hc}{\lambda_t} = \frac{1240\ \text{eV·nm}}{271\ \text{nm}} = 4.58\ \text{eV}$$

and that for silver is

$$\phi_{Ag} = \frac{1240\ \text{eV·nm}}{262\ \text{nm}} = 4.73\ \text{eV}$$

The contact potential is thus

$$V_{\text{contact}} = \frac{\phi_{Ag} - \phi_W}{e} = 4.73\ \text{V} - 4.58\ \text{V} = 0.15\ \text{V}$$

39-4 Quantum Theory of Electrical Conduction

With two relatively simple but important quantum-mechanical modifications of the classical free-electron theory, we can understand the electrical conduction of metals. First, we must replace the classical Maxwell–Boltzmann energy distribution with the Fermi–Dirac distribution of energies in the electron gas as discussed in the previous section. Second, we must consider the effect of the wave properties of electrons on the scattering of electrons by the lattice ions. We will discuss this modification qualitatively.

We might expect that most of the electrons would not participate in the conduction of electricity because of the exclusion principle, but this is not the case because the electric field accelerates all the electrons together. Figure 39-8 shows the Fermi factor in one dimension versus velocity for an ordinary temperature (such as $T = 300$ K) that is small compared with T_F. The factor is approximately 1 for speeds v_x in the range $-u_F < v_x < u_F$, where the Fermi speed u_F is the speed corresponding to the Fermi energy:

$$u_F = \sqrt{\frac{2E_F}{m_e}} \qquad 39\text{-}25$$

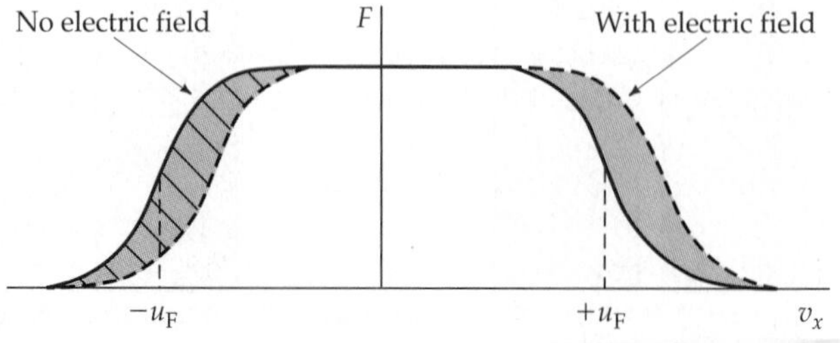

Figure 39-8 Fermi factor versus velocity in one dimension with no electric field (solid) and with an electric field in the $+x$ direction (dashed). The difference is greatly exaggerated.

Example 39-5

Calculate the Fermi speed for copper.

Using $E_F = 7.04$ eV, we obtain

$$u_F = \sqrt{\frac{2(7.04 \text{ eV})}{9.11 \times 10^{-31} \text{ kg}}\left(\frac{1.6 \times 10^{-19} \text{ J}}{1 \text{ eV}}\right)} = 1.57 \times 10^6 \text{ m/s}$$

The dashed curve in Figure 39-8 shows the Fermi factor after the electric field has been acting for some time t. Although all of the electrons have been shifted to higher velocities, the net effect is equivalent to shifting only the electrons near the Fermi level. We can use the classical equation for the resistivity (Equation 39-9) if we use the Fermi speed u_F in place of v_{av}:

$$\rho = \frac{m_e u_F}{ne^2\lambda} \qquad 39\text{-}26$$

We now have two problems. First, since the Fermi speed u_F is approximately independent of temperature, the resistivity given by Equation 39-26 is independent of temperature unless the mean free path depends on it. The second problem concerns magnitudes. As mentioned earlier, the classical expression for resistivity using v_{av} calculated from the Maxwell–Boltzmann distribution gives values about 6 times too large at $T = 300$ K. Since the Fermi speed u_F is about 16 times the Maxwell–Boltzmann value of v_{av}, the magnitude of ρ predicted by Equation 39-26 will be more than 6 times greater than the experimentally determined value.

The resolution of both of these problems lies in the way the value of the mean free path is calculated. If we use u_F from Equation 39-25 and the experimental value for the resistivity, $\rho \approx 1.7 \times 10^{-8}$ Ω·m, we can solve Equation 39-26 for the mean free path. The result is $\lambda \approx 40$ nm, which is about 100 times the value of 0.4 nm calculated from Equation 39-10 using $r \approx 0.1$ nm as the radius of the copper ion (see Example 22-13). The reason for this large discrepancy between the classical calculation of the mean free path and the "experimental" result calculated from Equation 39-26 is that the wave nature of the electron must be taken into account. The collision of an electron with a lattice ion is not similar to the collision of a baseball and a tree. Instead, it involves the scattering of the electron wave by the regularly spaced lattice. Detailed calculations show that there is no scattering of the electron waves by a *perfectly* ordered crystal, and that the mean free path is infinite. The scattering of electron waves arises because of imperfections in the crystal lattice, most commonly due to impurities, or thermal vibrations of the lattice ions.

In Equation 39-10 for the classical mean free path, the quantity πr^2 can be thought of as the area A of the lattice ion as seen by an electron. The mean free path is then given by

$$\lambda = \frac{1}{nA} \qquad 39\text{-}27$$

Figure 39-9*a* depicts the classical picture in which the lattice ions have an area $A = \pi r^2$. According to the quantum-mechanical theory of electron scattering, however, the area A has nothing to do with the size of the lattice ions. Instead, it depends on deviations of the lattice ions from a perfectly ordered array. Figure 39-9*b* depicts the quantum picture in which the lattice ions are points that have no size but present an area $A = \pi r_0^2$, where r_0 is the amplitude of thermal vibrations, to the electrons. This area is proportional to the square of the amplitude of the vibrations. From our knowledge of simple

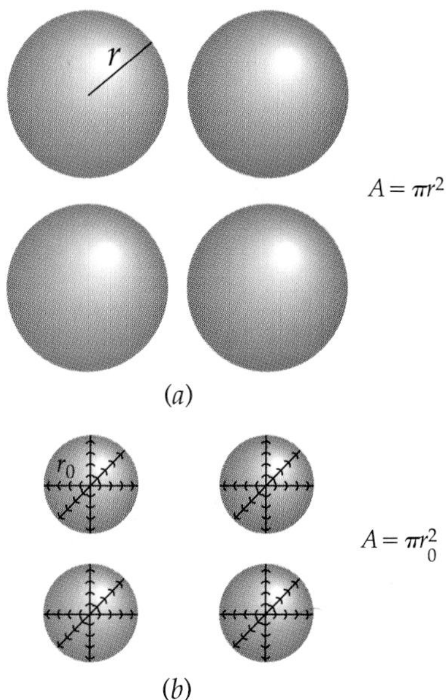

Figure 39-9 (*a*) Classical picture of the lattice ions as spherical balls of radius r that present an area πr^2 to the electrons. (*b*) Quantum-mechanical picture of the lattice ions as points that are vibrating in three dimensions. The area presented to the electrons is πr_0^2, where r_0 is the amplitude of oscillation of the ions.

harmonic motion, we know that the energy of vibration is also proportional to the square of the amplitude. Thus, the effective area A is proportional to the energy of vibration of the lattice ions. From the equipartition theorem, we know that the energy of vibration is proportional to kT, where k is Boltzmann's constant. Thus, the mean free path λ is proportional to $1/T$, so the resistivity is proportional to T in agreement with experiment.

The effective area A due to thermal vibrations can be calculated, and it turns out to be about 100 times smaller at $T = 300$ K than the actual area πr^2 of a lattice ion, thus giving values for the resistivity that are in agreement with experiment. We see, therefore, that the free-electron model of metals gives a good account of electrical conduction if the classical mean speed v_{av} is replaced by the Fermi speed u_{F} and if the collisions between electrons and the lattice ions are interpreted in terms of the scattering of electron waves, for which only deviations from a perfectly ordered lattice are important.

The presence of impurities in a metal also causes deviations from perfect regularity in the crystal lattice. The effects of impurities on resistivity are approximately independent of temperature. The resistivity of a metal containing impurities can be written $\rho = \rho_{\text{t}} + \rho_{\text{i}}$, where ρ_{t} is the resistivity due to the thermal motion of the lattice ions and ρ_{i} is the resistivity due to impurities. Figure 39-10 shows typical resistance-versus-temperature curves for metals with impurities. As the temperature approaches zero, ρ_{t} approaches zero and the resistivity approaches the constant ρ_{i} due to impurities.

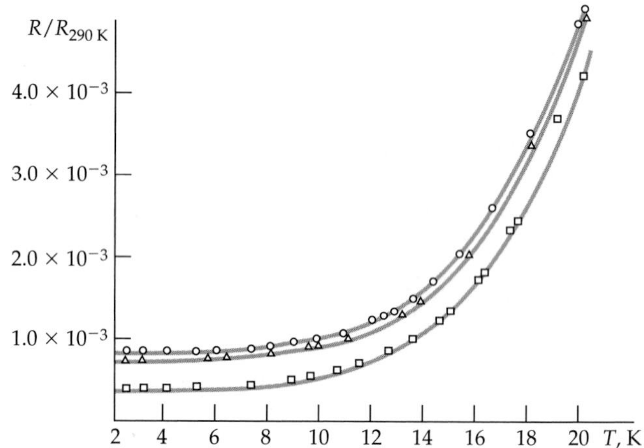

Figure 39-10 Relative resistance versus temperature for three samples of sodium. The three curves have the same temperature dependence but different magnitudes because of differing amounts of impurities in the samples.

Heat Conduction and Heat Capacity

Quantum-mechanical modifications of the theory of heat conduction in metals also bring that theory into good agreement with experimental measurements. By replacing the Maxwell–Boltzmann energy distribution of the electrons with the Fermi–Dirac distribution, we can understand why the contribution of the electron gas to the heat capacity of a metal is much less than the value $\frac{3}{2}R$ predicted by classical theory. At $T = 0$, the average energy of the electrons is $\frac{3}{5}E_{\text{F}}$, so the total energy is $E = \frac{3}{5}NE_{\text{F}}$. At some temperature T, only those electrons with energies near the Fermi energy can be excited by random collisions with the lattice ions, which have an average energy of the order of kT. The fraction of the electrons that are excited is of the order of kT/E_{F}, and their energy is increased from that at $T = 0$ by an amount that is of the order of kT. We can thus express the energy of N electrons at temperature T as

$$E = \frac{3}{5}NE_F + \alpha N \frac{kT}{E_F} kT \tag{39-28}$$

where α is some constant that we expect to be of the order of 1 if our reasoning is correct. The calculation of α involves the complete Fermi–Dirac distribution at an arbitrary temperature T and is quite difficult. The result is $\alpha = \pi^2/4$. Using this result and $E_F/k = T_F$, the Fermi temperature, we obtain for the contribution of the electron gas to the heat capacity at constant volume

$$C_v = \frac{dU}{dT} = 2\alpha Nk \frac{kT}{E_F} = \frac{\pi^2}{2} nR \frac{T}{T_F}$$

For one mole, $Nk = R$, the gas constant. The molar heat capacity at constant volume is then

$$C_{mv} = \frac{\pi^2}{2} R \frac{T}{T_F} \tag{39-29}$$

We can see that because of the large value of T_F, the contribution of the electron gas is a small fraction of R at ordinary temperatures. Because T_F = 81,700 K for copper, the molar heat capacity of the electron gas at T = 300 K is

$$C_{mv} = \frac{\pi^2}{2}\left(\frac{300\text{ K}}{81{,}700}\right) R \approx 0.02R$$

which is in good agreement with experiment.

39-5 Band Theory of Solids

We have seen that, if the electron gas is treated as a Fermi gas and the electron–lattice collisions are treated as the scattering of electron waves, the free-electron model gives a good account of the thermal and electrical properties of conductors. However, this simple model gives no indication of why one material is a conductor and another is an insulator. Resistivities vary enormously between insulators and conductors. For example, the resistivity of a typical insulator, such as quartz, is of the order of 10^{16} Ω·m, whereas that of a typical conductor is of the order of 10^{-8} Ω·m. To understand why some materials conduct and others do not, we must refine the free-electron model and consider the effect of the lattice on the electron energy levels.

We begin by considering the energy levels of the individual atoms as they are brought together. As we have seen, the allowed energy levels in an isolated atom are often far apart. For example, in hydrogen, the energy for $n = 1$ is -13.6 eV and that for $n = 2$ is $(-13.6\text{ eV})/4 = -3.4$ eV. Let us consider two identical hydrogen atoms, and focus our attention on one particular energy level, such as $n = 2$. When the atoms are far apart, the energy of this level is the same for each atom. As the atoms are brought close together, the energy of this level for each atom changes because of the influence of the other atom. As a result, the $n = 2$ level splits into two levels of slightly different energies for the two-atom system.

If we have N identical atoms, a particular energy level in the isolated atom splits into N different, nearly equal energy levels when the atoms are close together. Figure 39-11 shows the energy splitting of the 1s and 2s

energy levels for six atoms as a function of the separation of the atoms. In a macroscopic solid, N is very large—of the order of Avogadro's number—so each energy level splits into a very large number of levels called a **band.** Because the number of levels in the band is so large, the levels are spaced almost continuously within the band. There is a separate band of levels for each particular energy level of the isolated atom. The bands may be widely separated in energy, they may be close together, or they may even overlap, depending on the kind of atom and the type of bonding in the solid.

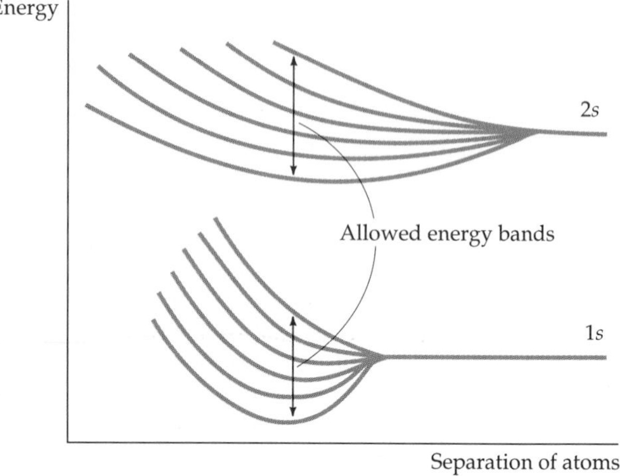

Figure 39-11 Energy splitting of the 1s and 2s energy levels for six atoms as a function of the separation of the atoms.

We can now understand why some solids are conductors and others are insulators. Consider sodium. There is room for two electrons in the 3s state of each atom, but each separate sodium atom has only one 3s electron. Therefore, when N sodium atoms are bound in a solid, the 3s energy band is only half filled. In addition, the empty 3p band overlaps the 3s band. The allowed energy bands of sodium are shown schematically in Figure 39-12. The occupied levels are shaded. We can see that many allowed energy states are available just above the filled ones, so the valence electrons can easily be raised to a higher-energy state by an electric field. Accordingly, sodium is a good conductor. Magnesium, on the other hand, has two 3s electrons, so the 3s band is filled. However, like sodium, the empty 3p band overlaps the 3s band, so magnesium is also a conductor.

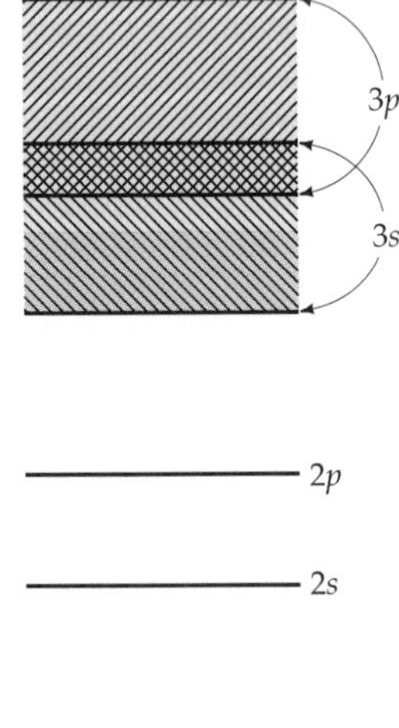

Figure 39-12 Energy band structure of sodium. The empty 3p band overlaps the half-filled 3s band. Just above the filled states are many empty states into which electrons can be excited by an electric field, so sodium is a conductor.

The band structure of an ionic crystal, such as NaCl, is quite different. The energy bands arise from the energy levels of the Na^+ and Cl^- ions. Both of these ions have a closed-shell configuration, so the highest occupied band in NaCl is completely full. The next allowed band, which is empty, arises from the excited states of Na^+ and Cl^-. There is a large energy gap between the filled band and this empty band. A typical electric field applied to NaCl will be too weak to excite an electron from the upper energy levels of the filled band across the large gap into the lower energy levels of the empty band, so NaCl is an insulator. When an applied electric field is sufficiently strong to cause an electron to be excited to the empty band, the phenomenon is called dielectric breakdown.

Figure 39-13 shows four possible kinds of band structures for a solid. The band occupied by the outermost, valence electrons is called the **valence band.** The lowest band in which there are unoccupied states is called the **conduction band.** In sodium, the valence band is only half filled, so the valence band is also the conduction band. In magnesium, the filled 3s band and the empty 3p bands overlap, forming a combined valence–conduction band that is only partially filled, so magnesium is also a conductor.

The band structure for a conductor such as copper is shown in Figure 39-13*a*. The lower bands are filled with the inner electrons of the atoms. According to the Pauli exclusion principle, no more electrons can occupy levels in these bands. The uppermost band that contains electrons is only about half full. Because these electrons are the electrons that form the metallic bond, this band is the valence band. In the normal state, at low temperatures, the lower half of the valence band is filled and the upper half is empty, so this band is also the conduction band. At higher temperatures, a few of the electrons are in the higher energy states of this band because of thermal excitation, but there are still many unfilled energy states above the filled ones. When an electric field is established in the conductor, the electrons in the conduction band are accelerated, which means that their energy is increased. This is consistent with the Pauli exclusion principle because there are many empty energy states just above those occupied by electrons in this band. These electrons are thus the conduction electrons.

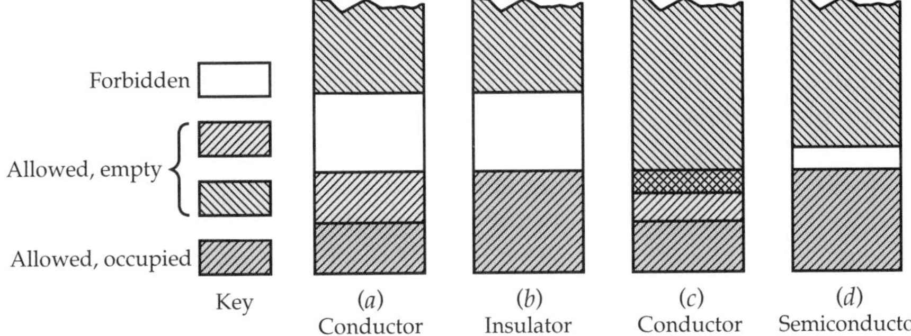

Figure 39-13 Four possible band structures for a solid. (*a*) A typical conductor. The valence band is only partially full, so electrons can be easily excited to nearby energy states. (*b*) A typical insulator. There is a forbidden band with a large energy gap between the filled valence band and the conduction band. (*c*) A conductor in which the allowed energy bands overlap. (*d*) A semiconductor. The energy gap between the filled valence band and the conduction band is very small, so some electrons are excited to the conduction band at normal temperatures, leaving holes in the valence band.

Figure 39-13*b* shows the band structure for a typical insulator. At $T = 0$ K, the highest energy band that contains electrons is completely full. The next energy band containing empty energy states, the conduction band, is separated from the last filled band by an energy gap. At $T = 0$, the conduction band is empty. At ordinary temperatures, a few electrons can be excited to states in this band, but most cannot be because the energy gap is large compared with the energy an electron might obtain by thermal excitation, which on the average is of the order of $kT \approx 0.026$ eV at $T = 300$ K. This energy gap is sometimes referred to as the **forbidden energy band.** Very few electrons can be thermally excited to the nearly empty conduction band, even at fairly high temperatures. When an electric field is established in the solid, electrons cannot be accelerated because there are no empty energy states at nearby energies. We describe this by saying that there are no free electrons. The small conductivity that is observed is due to the very few electrons that are thermally excited into the nearly empty conduction band.

In Figure 39-13*c* the valence and conduction bands overlap. A material such as magnesium with this type of band structure is a conductor.

In some materials the energy gap between the top filled band and the empty conduction band is very small, as shown in Figure 39-13*d*. At $T = 0$, there are no electrons in the conduction band and the material is an insulator. At ordinary temperatures, there are an appreciable number of electrons in the conduction band due to thermal excitation. Such a material is called an **intrinsic semiconductor.** In the presence of an electric field, the electrons in the conduction band can be accelerated because there are empty states nearby. Also, for each electron in the conduction band there is a vacancy, or hole, in the nearly filled valence band. In the presence of an electric field, electrons in this band can be excited to a vacant energy level. This contributes to the electric current and is most easily described as the motion of a hole in the direction of the field and opposite the motion of the electrons. The hole thus acts like a positive charge. An analogy of a two-lane, one-way road with one lane full of parked cars and the other empty may help to visualize the conduction of holes. If a car moves out of the filled lane into the empty lane, it can move ahead freely. As the other cars move up to occupy the space left, the empty space propagates backwards in the direction opposite the motion of the cars. Both the forward motion of the car in the nearly empty lane and the backward propagation of the empty space contribute to a net forward propagation of the cars.

An interesting characteristic of semiconductors is that the conductivity increases (and the resistivity decreases) as the temperature increases, which is contrary to the case for normal conductors. The reason is that as the temperature is increased, the number of free electrons is increased because there are more electrons in the conduction band. The number of holes in the valence band is also increased of course. In semiconductors, the effect of the increase in the number of charge carriers, both electrons and holes, exceeds the effect of the increase in resistivity due to the increased scattering of the electrons by the lattice ions due to thermal vibrations. Semiconductors therefore have a negative temperature coefficient of resistivity.

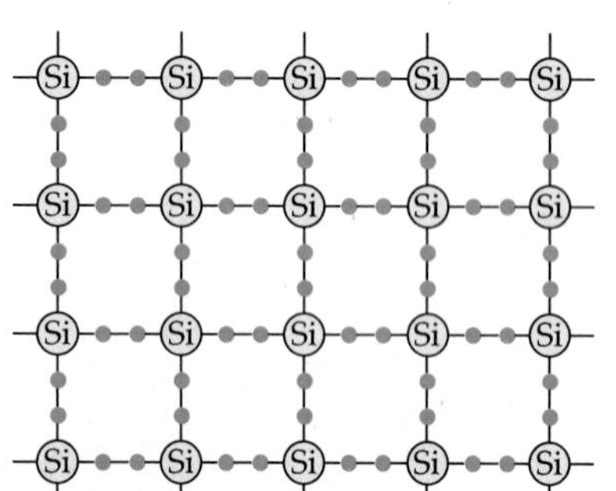

Figure 39-14 A two-dimensional schematic illustration of solid silicon. Each atom forms a covalent bond with four neighbors, sharing one of its four valence electrons with each neighbor.

A typical intrinsic semiconductor is silicon, which has four valence electrons in the $n = 3$ shell. In a solid crystal of silicon, each atom forms a covalent bond with four neighboring atoms and shares one of its valence electrons with each neighbor, as illustrated schematically in Figure 39-14. Figure 39-15 shows the energy level structures of carbon, silicon, and germanium. The 2*s* and 2*p* states of carbon, the 3*s* and 3*p* states of silicon, and 4*s* and 4*p* states of germanium have room for eight electrons. These states split into two hybrid states, each with room for four electrons. The lower hybrid state is filled with the valence electrons and the upper one is empty. At the separation of about 0.15 nm for carbon, the energy gap between these states is about 7 eV, so carbon is an insulator. For silicon and germanium, the separation is about 0.24 nm and the energy gap is about 1 eV, so these elements are intrinsic semiconductors.

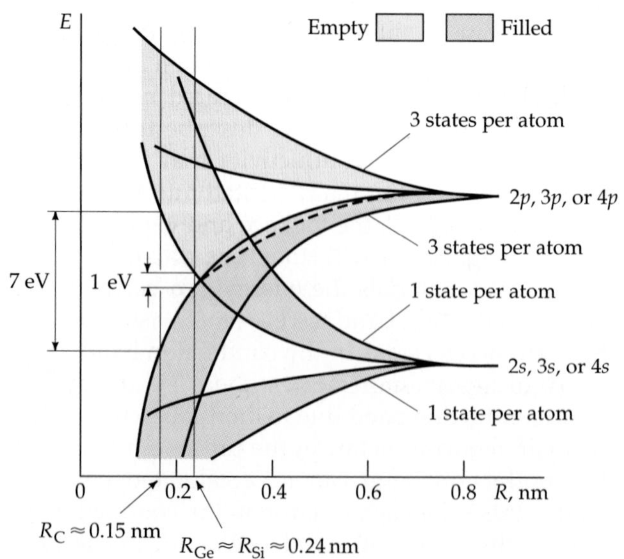

Figure 39-15 Splitting of the 2*s* and 2*p* states of carbon, the 3*s* and 3*p* states of silicon, or the 4*s* and 4*p* states of germanium versus the separation of the atoms. The energy gap between the filled states and empty states is about 7 eV for carbon but only about 1 eV for silicon and germanium.

39-6 Impurity Semiconductors

Most semiconductor devices, such as the semiconductor diode and the transistor, make use of **impurity semiconductors,** which are created through the controlled addition of certain impurities to intrinsic semiconductors. This process is called **doping.** Figure 39-16*a* is a schematic illustration of silicon doped with a small amount of arsenic such that arsenic atoms replace a few of the silicon atoms in the crystal lattice. Arsenic has five electrons in its valence shell rather than the four of silicon. Four of these electrons take part in covalent bonds with the four neighboring silicon atoms, and the fifth electron is very loosely bound to the atom. This extra electron occupies an energy level that is just slightly below the conduction band in the solid, and it is easily excited into the conduction band, where it can contribute to electrical conduction.

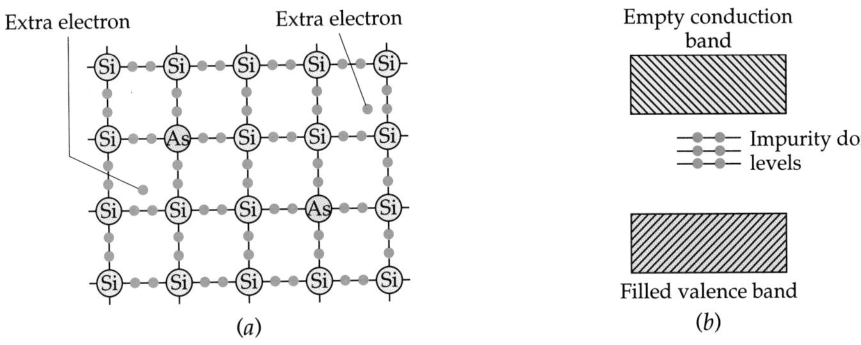

Figure 39-16 (*a*) A two-dimensional schematic illustration of silicon doped with arsenic. Because arsenic has five valence electrons, there is an extra, weakly bound electron that is easily excited to the conduction band, where it can contribute to electrical conduction. (*b*) Band structure of an *n*-type semiconductor such as silicon doped with arsenic. The impurity atoms provide filled energy levels that are just below the conduction band. These levels donate electrons to the conduction band.

The effect on the band structure of a silicon crystal achieved by doping it with arsenic is shown in Figure 39-16*b*. The levels shown just below the conduction band are due to the extra electrons of the arsenic atoms. These levels are called **donor levels** because they donate electrons to the conduction band without leaving holes in the valence band. Such a semiconductor is called an ***n*-type semiconductor** because the major charge carriers are *negative* electrons. The conductivity of a doped semiconductor can be controlled by controlling the amount of impurity added. The addition of just one part per million can increase the conductivity by several orders of magnitude.

Another type of impurity semiconductor can be made by replacing a silicon atom with a gallium atom that has 3 electrons in its valence level (Figure 39-17*a*). The gallium atom accepts electrons from the valence band to complete its four covalent bonds, thus creating a hole in the valence band. The effect on the band structure of silicon achieved by doping it with gallium is shown in Figure 39-17*b*. The empty levels shown just above the valence band are due to the holes from the ionized gallium atoms. These levels are called **acceptor levels** because they accept electrons from the filled valence band when these electrons are thermally excited to a higher energy state.

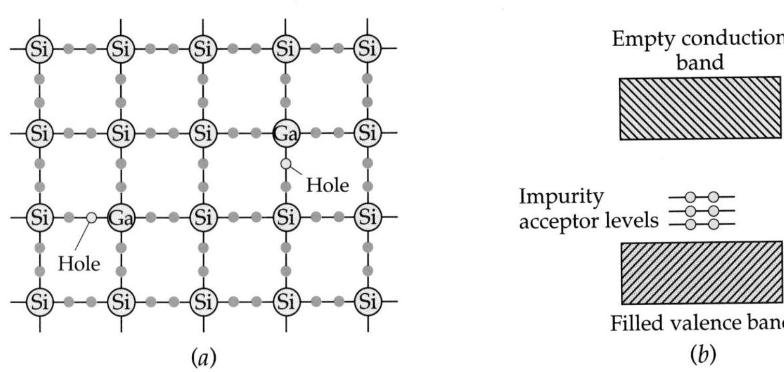

Figure 39-17 (*a*) A two-dimensional schematic illustration of silicon doped with gallium. Because gallium has only three valence electrons, there is a hole in one of its bonds. As electrons move into the hole, the hole moves about, contributing to the conduction of electrical current. (*b*) Band structure of a *p*-type semiconductor such as silicon doped with gallium. The impurity atoms provide empty energy levels just above the filled valence band that accept electrons from the valence band.

This creates holes in the valence band that are free to propagate in the direction of an electric field. Such a semiconductor is called a ***p*-type semiconductor** because the charge carriers are *positive* holes. The fact that conduction is due to the motion of holes can be verified by the Hall effect, which we discussed in Section 24-4.

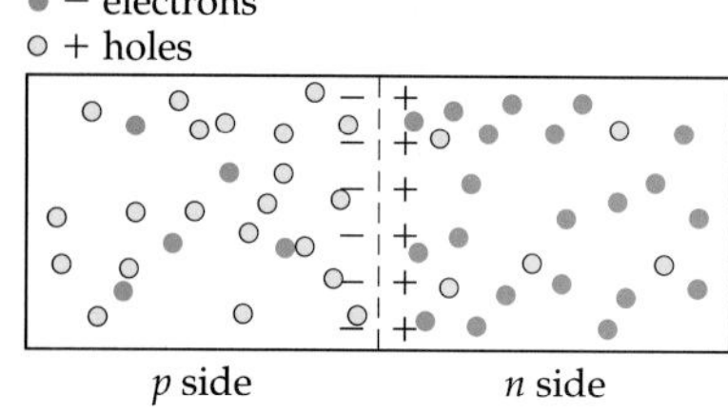

Figure 39-18 A *pn* junction. Because of the difference in their concentrations, holes diffuse from the *p* side to the *n* side and electrons diffuse from the *n* side to the *p* side. As a result, there is a double layer of charge at the junction with the *p* side being negative and the *n* side being positive.

39-7 Semiconductor Junctions and Devices

Semiconductor devices such as diodes and transistors make use of *n*-type and *p*-type semiconductors joined together as shown in Figure 39-18. In practice, the two types of semiconductors are often a single silicon crystal doped with donor impurities on one side and acceptor impurities on the other. The region in which the semiconductor changes from a *p*-type to an *n*-type is called a **junction.**

When an *n*-type and a *p*-type semiconductor are placed in contact, the initially unequal concentrations of electrons and holes result in the diffusion of electrons across the junction from the *n* side to the *p* side and holes from the *p* side to the *n* side until equilibrium is established. The result of this diffusion is a net transport of positive charge from the *p* side to the *n* side. Unlike the case when two different metals are in contact, the electrons cannot travel very far from the junction region because the semiconductor is not a particularly good conductor. The diffusion of electrons and holes therefore creates a double layer of charge at the junction similar to that on a parallel-plate capacitor. There is thus a potential difference V across the junction, which tends to inhibit further diffusion. In equilibrium, the *n* side with its net positive charge will be at a higher potential than the *p* side with its net negative charge. In the junction region, there will be very few charge carriers of either type, so the junction region has a high resistance. Figure 39-19 shows the energy level diagram for a *pn* junction. The junction region is also called the **depletion region** because it has been depleted of charge carriers.

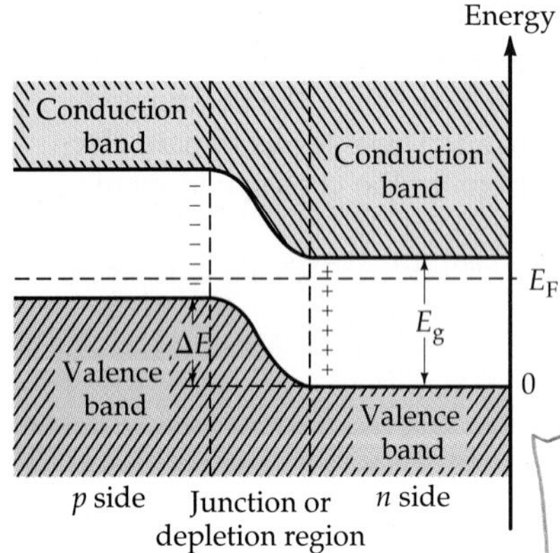

Figure 39-19 Electron energy levels for an unbiased *pn* junction.

A semiconductor with a *pn* junction can be used as a simple diode rectifier. In Figure 39-20, an external potential difference has been applied across the junction by connecting a battery and resistor to the semiconductor. When the positive terminal of the battery is connected to the *p* side of the junction, as shown in Figure 39-20*a*, the diode is said to be **forward biased.** Forward biasing lowers the potential across the junction. The diffusion of electrons and holes is thereby increased as they attempt to reestablish equilibrium, resulting in a current in the circuit. If the positive terminal of the battery is connected to the *n* side of the junction as shown in Figure 39-20*b*, the diode is said to be **reverse biased.** Reverse biasing tends to increase the potential difference across the junction, thereby further inhibiting diffusion. Figure 39-21 shows a plot of current versus voltage for a typical semiconductor junction. Essentially, the junction conducts only in one direction, the same as the vacuum-tube diode discussed in Section 28-7. Junction diodes have replaced vacuum diodes in nearly all applications except when a very high current is required.

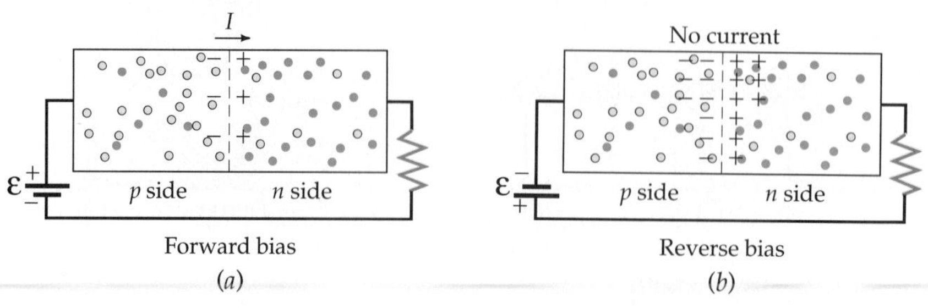

Figure 39-20 A *pn*-junction diode. (*a*) Forward-biased *pn* junction. The applied potential difference enhances the diffusion of holes from the *p* side to the *n* side and electrons from the *n* side to the *p* side, resulting in a current *I*. (*b*) Reverse-biased *pn* junction. The applied potential difference inhibits the further diffusion of holes and electrons, so there is no current.

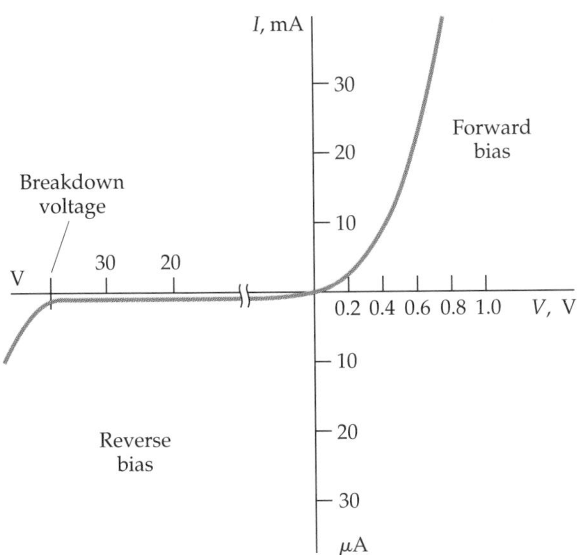

Figure 39-21 Current versus applied voltage across a *pn* junction. Note the different scales for the forward and reverse bias conditions.

Note that the current in Figure 39-21 suddenly increases in magnitude at extreme values of reverse bias. In such large electric fields, electrons are stripped from their atomic bonds and accelerated across the junction. These electrons, in turn, cause others to break loose. This effect is called **avalanche breakdown.** Although such a breakdown can be disastrous in a circuit where it is not intended, the fact that it occurs at a sharp voltage value makes it of use in a special voltage reference standard known as a **Zener diode.**

An interesting effect that we can discuss only qualitatively occurs if both the *n* side and *p* side of a *pn*-junction diode are so heavily doped that the donors on the *n* side provide so many electrons that the lower part of the conduction band is practically filled and the acceptors on the *p* side accept so many electrons that the upper part of the valence band is nearly empty. Figure 39-22*a* shows the energy-level diagram for this situation. Because the depletion region is now so narrow, electrons can easily penetrate the potential barrier across the junction. This flow of electrons is called a **tunneling current,** and such a heavily doped diode is called a **tunnel diode.**

At equilibrium with no bias, there is an equal tunneling current in each direction. When a small bias voltage is applied across the junction, the energy-level diagram is as shown in Figure 39-22*b*, and the tunneling of electrons from the *n* side to the *p* side is increased whereas that in the opposite direction is decreased. This tunneling current in addition to the usual current due to diffusion results in a considerable net current. When the bias voltage is increased slightly, the energy-level diagram is as shown in Figure 39-22*c*, and the tunneling current is decreased. Although the diffusion current is increased, the net current is decreased. At large bias voltages, the

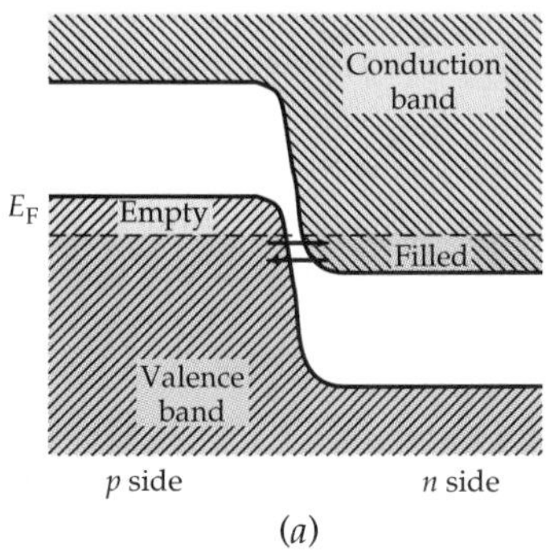

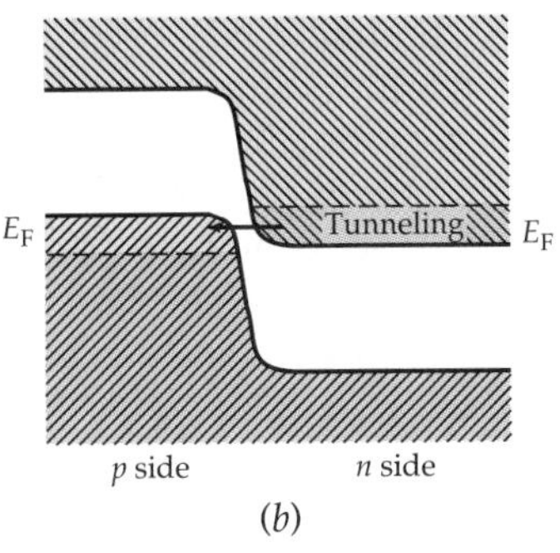

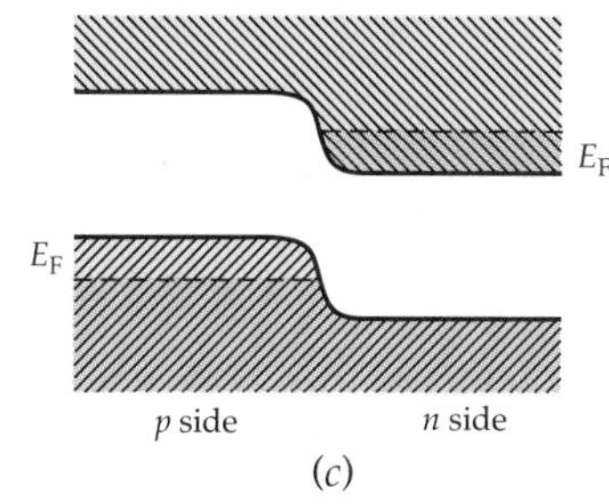

Figure 39-22 Electron energy levels for a heavily doped *pn*-junction tunnel diode. (*a*) With no bias voltage, some electrons tunnel in each direction. (*b*) With a small bias voltage, the tunneling current is enhanced in one direction, making a sizable contribution to the net current. (*c*) With further increases in the bias voltage, the tunneling current decreases dramatically.

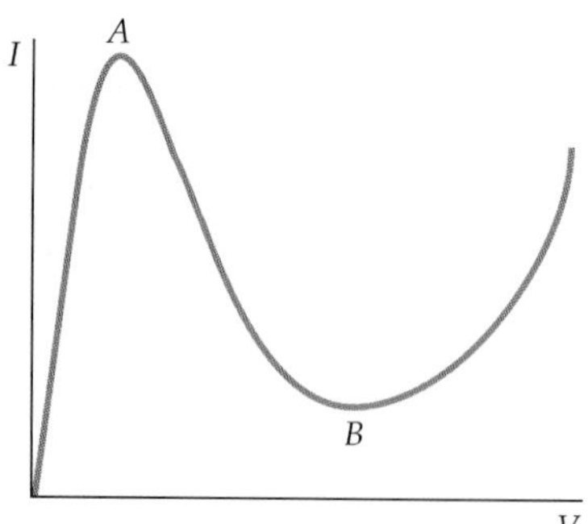

Figure 39-23 Current versus applied voltage for a tunnel diode. Up to point *A*, an increase in the bias voltage enhances tunneling. Between points *A* and *B*, an increase in the bias voltage inhibits tunneling. After point *B*, the tunneling is negligible, and the diode behaves like an ordinary *pn*-junction diode.

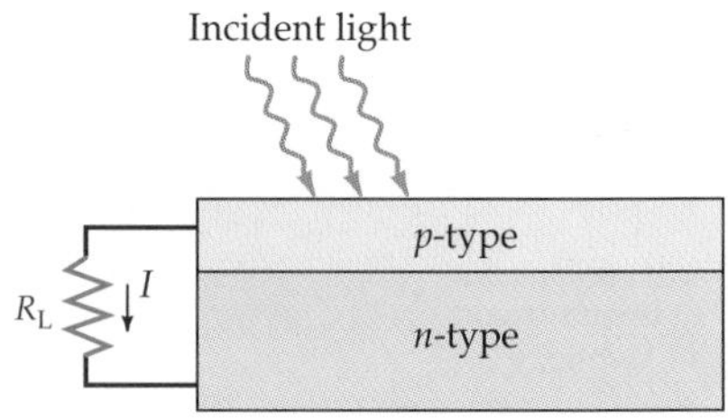

Figure 39-24 A *pn*-junction semiconductor as a solar cell. When light strikes the *p*-type region, electron–hole pairs are created, resulting in a current through the load resistance R_L.

tunneling current is completely negligible, and the total current increases with increasing bias voltage due to diffusion as in an ordinary *pn*-junction diode. Figure 39-23 shows the current-versus-voltage curve for a tunnel diode. Such diodes are used in electric circuits because of their very fast response time. When operated near the peak in the current-versus-voltage curve, a small change in bias voltage results in a large change in the current.

Another use for the *pn*-junction semiconductor is the **solar cell,** which is illustrated schematically in Figure 39-24. When a photon of energy greater than the gap energy (1.1 eV in silicon) strikes the *p*-type region, it can excite an electron from the valence band into the conduction band, leaving a hole in the valence band. This region is already rich in holes. Some of the electrons created by the photons will recombine with holes, but some will migrate to the junction. From there they are accelerated into the *n*-type region by the electric field between the double layer of charge. This creates an excess negative charge in the *n*-type region and excess positive charge in the *p*-type region. The result is a potential difference between the two regions, which in practice is about 0.6 V. If a load resistance is connected across the two regions, a charge flows through the resistance. Some of the incident light energy is thus converted into electrical energy. The current in the resistor is proportional to the number of incident photons, which is in turn proportional to the intensity of the incident light.

There are many other applications of semiconductors with *pn* junctions Particle detectors called **surface-barrier detectors** consist of a *pn*-junction semiconductor with a large reverse bias so that there is ordinarily no current. When a high-energy particle, such as an electron, passes through the semiconductor, it creates many electron–hole pairs as it loses energy. The resulting current pulse signals the passage of the particle. **Light-emitting diodes (LEDs)** are *pn*-junction semiconductors with a large forward bias that produces a large excess concentration of electrons on the *p* side and holes on the *n* side of the junction. Under these conditions, the diode emits light as the electrons and holes recombine. This is essentially the reverse of the process that occurs in a solar cell, in which electron–hole pairs are created by the absorption of light. LEDs are commonly used as displays for digital watches and calculators.

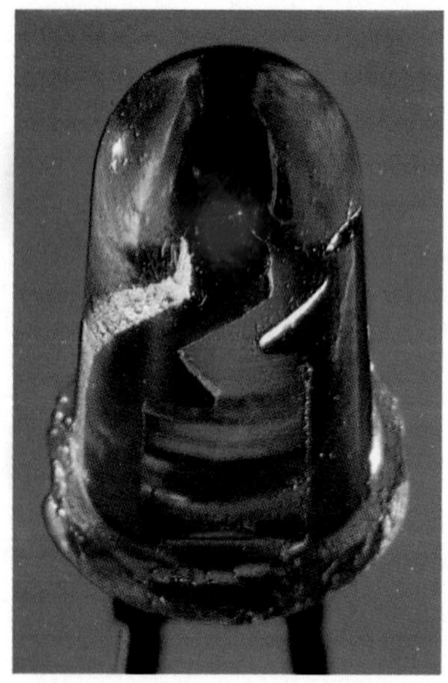

A light-emitting diode (LED).

Transistors

The transistor, invented in 1948 by William Shockley, John Bardeen, and Walter H. Brattain, has revolutionized the electronics industry and our everyday world. A simple junction transistor consists of three distinct semiconductor regions called the **emitter,** the **base,** and the **collector.** The base is a very thin region of one type of semiconductor sandwiched between two regions of the opposite type. The emitter semiconductor is much more heavily doped than either the base or the collector. In an *npn* transistor, the emitter and collector are *n*-type semiconductors and the base is a *p*-type semiconductor; in a *pnp* transistor, the base is an *n*-type semiconductor and the emitter and collector are *p*-type semiconductors. The emitter, base, and

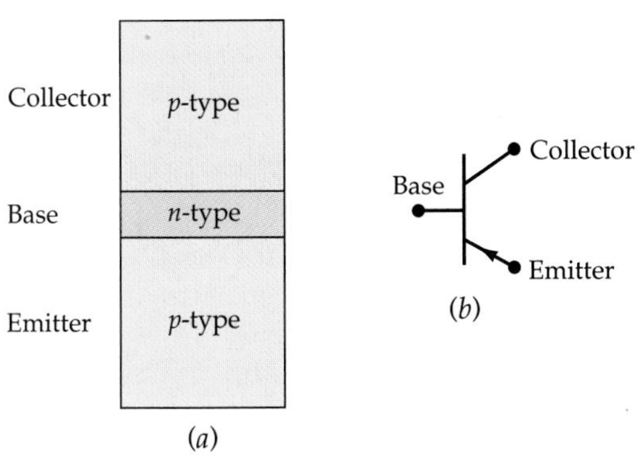

Figure 39-25 A *pnp* transistor. The heavily doped emitter emits holes that pass through the thin base to the collector. (*b*) Symbol for a *pnp* transistor in a circuit. The arrow points in the direction of the conventional current, which is the same as that of the emitted holes.

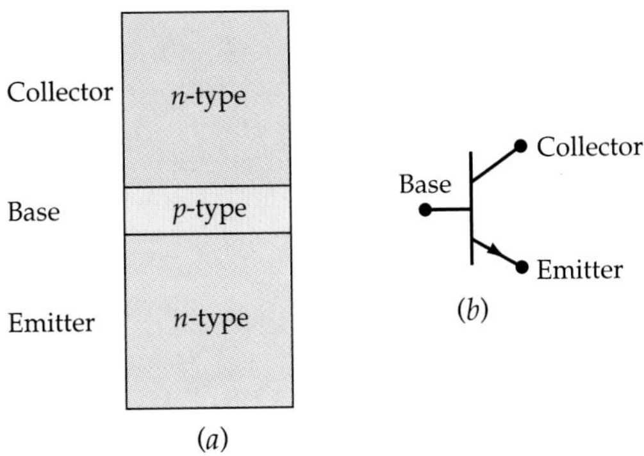

Figure 39-26 An *npn* transistor. The heavily doped emitter emits electrons that pass through the thin base to the collector. (*b*) Symbol for an *npn* transistor. The arrow points in the direction of the conventional current, which is opposite the direction of the emitted electrons.

collector behave somewhat similarly to the cathode, grid, and plate in a vacuum-tube triode (Section 28-7), except that in a *pnp* transistor it is holes that are emitted rather than electrons.

Figures 39-25 and 39-26 show, respectively, a *pnp* transistor and an *npn* transistor with the symbols used to represent each transistor in circuit diagrams. We see that a transistor consists of two *pn* junctions. We will discuss the operation of a *pnp* transistor. The operation of an *npn* transistor is similar.

In normal operation, the emitter–base junction is forward biased, and the base–collector junction is reverse biased, as shown in Figure 39-27. The heavily doped *p*-type emitter emits holes that flow across the emitter–base junction into the base. Because the base is very thin, most of these holes flow across the base into the collector. This flow constitutes a current I_c from the emitter to the collector. However, some of the holes recombine in the base producing a positive charge that inhibits the further flow of current. To prevent this, some of the holes that do not reach the collector are drawn off the base as a base current I_b in a circuit connected to the base. In Figure 39-27, therefore, I_c is almost but not quite equal to I_e, and I_b is much smaller than either I_c or I_e. It is customary to express I_c as

$$I_c = \beta I_b \qquad 39\text{-}30$$

where β is called the **current gain** of the transistor. Transistors can be designed to have values of β as low as 10 or as high as several hundred.

Figure 39-28 shows a simple *pnp* transistor used as an amplifier. A small time-varying input voltage v_s is connected in series with a bias voltage V_{eb}.

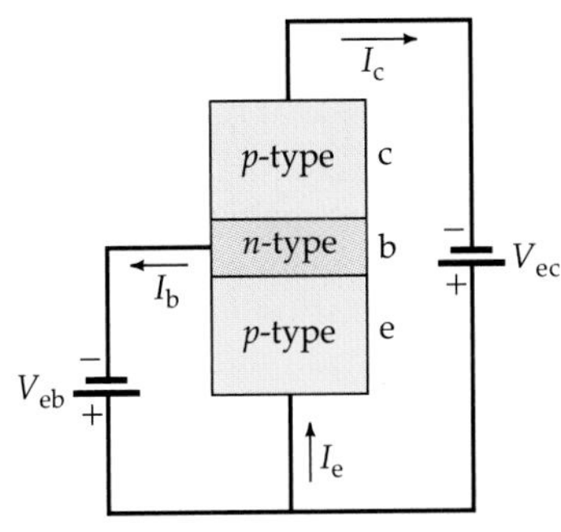

Figure 39-27 A *pnp* transistor biased for normal operation. Holes from the emitter can easily diffuse across the base, which is only tens of nanometers thick. Most of the holes flow to the collector, producing the current I_c.

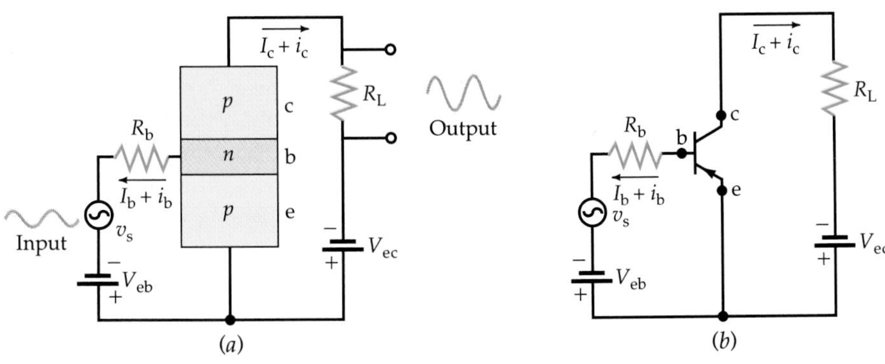

Figure 39-28 (*a*) A *pnp* transistor used as an amplifier. A small change i_b in the base current results in a large change i_c in the collector current. Thus a small signal in the base circuit results in a large signal across the load resistor R_L in the collector circuit. (*b*) The same circuit as (*a*) with the conventional symbol for the transistor.

(a)

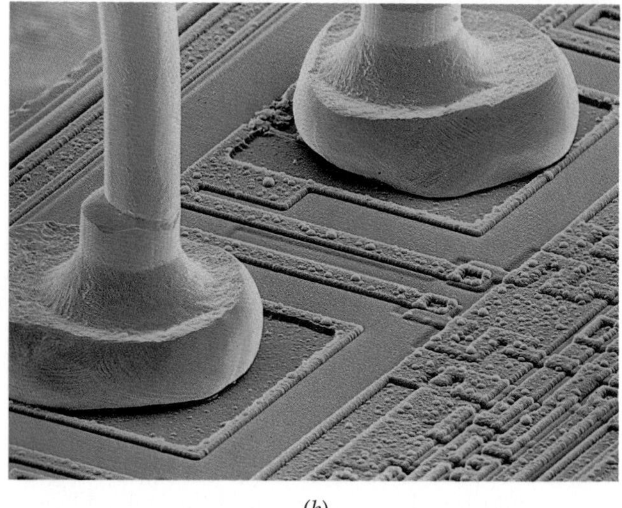

(b)

Integrated circuits (often called ICs or chips) combine "active" electronic devices (transistors and diodes) with "passive" ones (capacitors and resistors) on a single semiconductor crystal. (a) This particular chip, shown connected to 44 conductor leads, has an actual size of 6.4 mm square. It is used to format digitized voice and data signals, so they can share a single transmission line. (b) A scanning electron micrograph showing two conductor leads precision bonded to the edge of a chip (magnification: ×163).

The base current is then the sum of a steady current I_b produced by the bias voltage V_{eb} and a varying current i_b due to the signal voltage v_s. Because v_s may at any instant be either positive or negative, the bias voltage V_{eb} must be large enough to ensure that there is always a forward bias on the emitter–base junction. The collector current will consist of two parts: a direct current $I_c = \beta I_b$ and an alternating current $i_c = \beta i_b$. We thus have a current amplifier in which the time-varying output current i_c is β times the input current i_b. In such an amplifier, the steady currents I_c and I_b, although essential to the operation of the transistor, are usually not of interest. The input signal voltage v_s is related to the base current by Ohm's law:

$$i_b = \frac{v_s}{R_b + r_b} \qquad 39\text{-}31$$

where r_b is the internal resistance of the transistor between the base and emitter. Similarly, the collector current i_c produces a voltage v_L across the output or load resistance R_L given by

$$v_L = i_c R_L \qquad 39\text{-}32$$

Using Equation 39-30, we have

$$i_c = \beta i_b = \beta \frac{v_s}{R_b + r_b}$$

The output voltage is thus related to the input voltage by

$$v_L = \beta \frac{R_L}{R_b + r_b} v_s \qquad 39\text{-}33$$

The ratio of the output voltage to the input voltage is the **voltage gain** of the amplifier:

$$\text{Voltage gain} = \frac{v_L}{v_s} = \beta \frac{R_L}{R_b + r_b} \qquad 39\text{-}34$$

A typical amplifier, such as that in a tape player, has several transistors similar to the one in Figure 39-28 connected in series so that the output of one transistor serves as the input for the next. Thus, the very small voltage produced by the passage of the magnetized tape past the pickup heads controls the large amounts of power required to drive the loudspeakers. The power delivered to the speakers is supplied by the sources of direct voltage connected to each transistor.

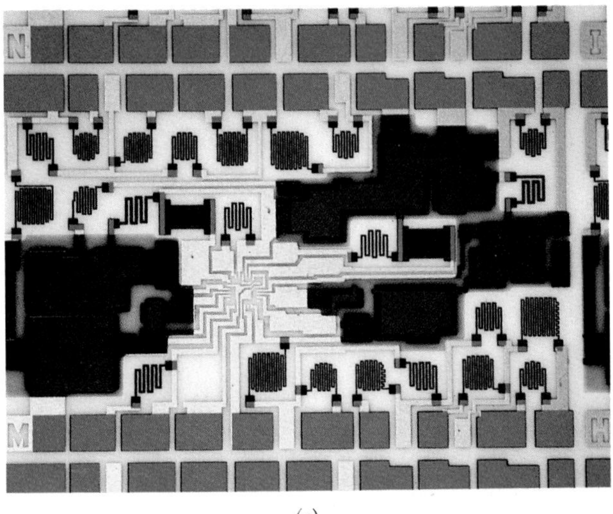

(c)

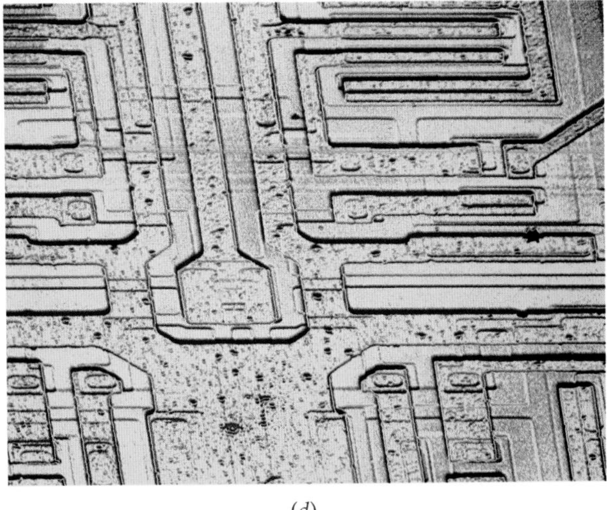

(d)

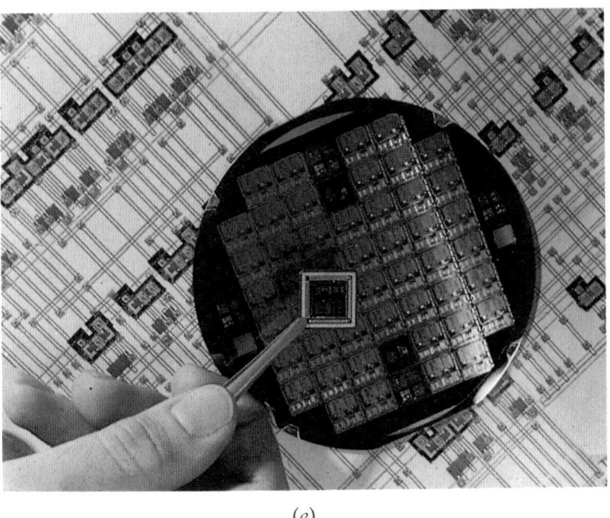

(e)

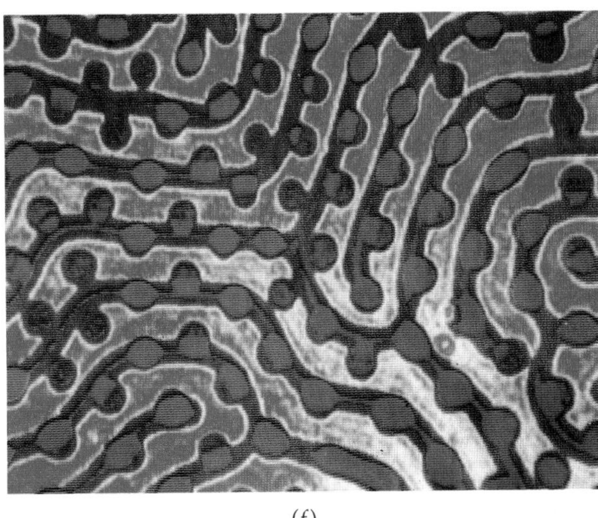

(f)

(c) Capacitors (orange blocks), resistors (brown blocks and meandering black lines), and conductors (gold lines) on a ceramic base, formed here by metal films only a few tenths of a micrometer thick. No means have been found to directly fabricate inductors (the other passive circuit component) on ICs; they are simulated with other circuitry or appended to a chip as discrete components. (d) A scanning electron micrograph of metal oxide silicon (MOS) transistors in patterned layers (magnification ×106). MOS transistors are manufactured by heating an original silicon wafer to about 1000°C, causing a layer of silicon dioxide (SiO_2) to form on its surface. This is coated with a photoresist and exposed to light through a mask. Unexposed (masked) windows of photoresist are etched away with a developer, exposing the silicon dioxide, which is etched away with acid. The exposed (unmasked) areas are resistant to the developer and are not affected. The wafer is again heated and this time doped, via a diffusion process, with a *p*-type impurity, forming *pn* junctions in the *n*-type silicon. The chip is covered with a contact metal (typically aluminum), which bonds to the SiO_2 that has reformed in windows while the chip was heated and doped. The contact metal itself is patterned in a final photo-etching process. Entire microchips are fabricated by an elaboration, using many masks, of this process. (e) The chip in the tweezers holds 150,000 transistors. Beneath it is a 4-inch wide silicon wafer, awaiting dicing, on which a group of chips have been fabricated simultaneously. In the background is a detail of the "stare plot," the layout of the chip's circuits. (f) Magnetized domains ("bubbles"), blue in this video micrograph, flow along channels in a thin-film garnet memory crystal. Magnetic bubble memory chips are the integrated-circuit analog to magnetic recording tape and disks. The bubbles (actually, cylinders seen in cross section) are created when the garnet is placed between two permanent magnets. They represent regions whose magnetic polarity points in a direction opposite to that of the surrounding crystal. An additional external magnetic field manipulates the position of the bubbles. (Garnet is easy to magnetize, up or down, along a particular axis—but hard to magnetize perpendicular to that axis. This property is necessary for the formation and movement of bubbles.) Storage sites for bubbles are established using a layer of ferromagnetic material deposited on the surface of the crystal; the presence or absence of a bubble at a site can be used to represent a bit of data.

A fixed pattern of light and dark is established in a photorefractive crystal, for instance by letting two laser beams of the same frequency interfere within the crystal.

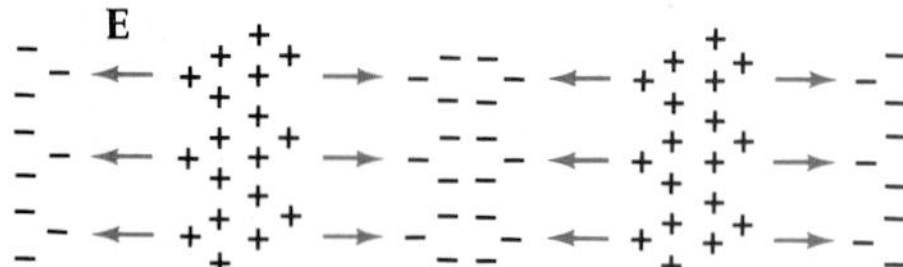

Responding to the electric fields of the light, free electrons or holes move away from bright regions of the crystal. The resulting build-up of charge creates a static electric field within the crystal.

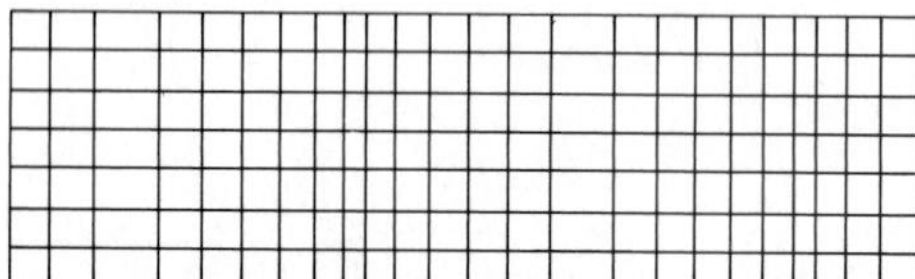

The static electric field acts on the atoms of the crystal lattice. The positions of the atoms shift and the lattice becomes distorted. Distortions in the lattice cause light to propagate faster in some regions of the crystal and slower in others: that is, the refractive index of the crystal is altered.

(*a*)

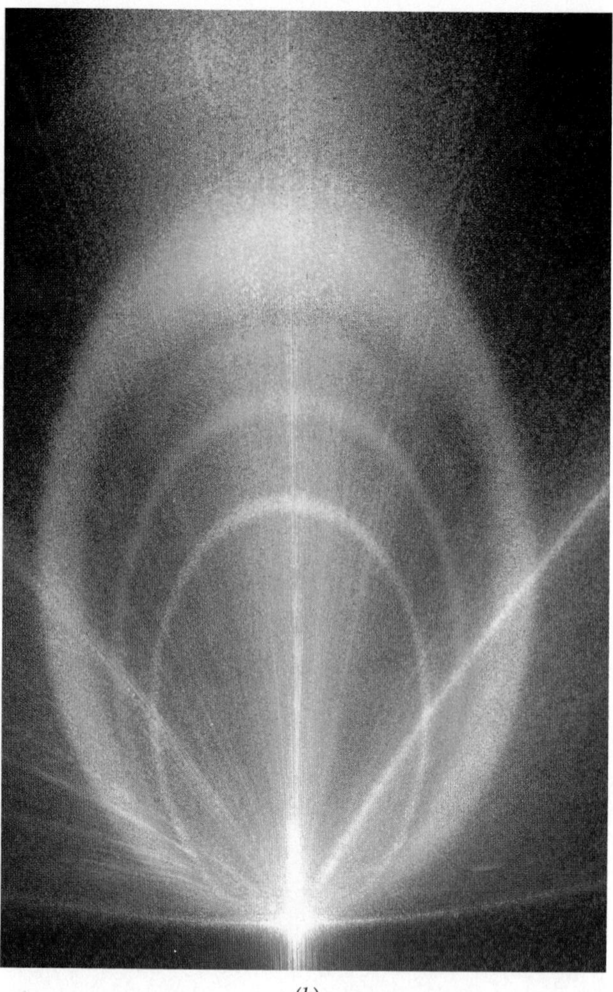

(*b*)

Photorefractivity. Light incident on a certain class of materials called "photorefractive crystals" causes a rearrangement of their crystal lattice, which in turn alters the optical properties of that lattice. This behavior is due to impurities or defects in the photorefractive crystal. These defects are a source of free electrons or holes that migrate away from illuminated areas in the crystal. The resulting rearrangement of free charge gives rise to electric fields that act on the atoms in the crystal lattice, distorting its shape. This process is summarized in schematic (*a*). Here, the incident light is from two interfering laser beams. The interference pattern of these beams is translated into a periodic pattern in the crystal lattice itself. Such a pattern is sometimes referred to as a "refractive-index grating." Typically, the grating pattern will be phase shifted with respect to the interference pattern of the light beams. One of the beams will tend to interfere constructively with light scattered by the grating, the other destructively. As a result, the first beam will be intensified passing through the crystal, and the second beam will be reduced. Figure (*b*) shows the result of the "beam-fanning effect" occurring in the photorefractive crystal barium titanate ($BaTiO_3$). Multicolored incident laser beams are partially scattered by defects in the crystal. An interference pattern arises between the incident and scattered beams, and the resulting deformation of the crystal lattice changes the lattice's index of refraction. The incident beams are refracted to a different angle and a new interference pattern arises, causing a further change in the index of refraction.

Beam fanning lies at the heart of new devices called "phase-conjugate mirrors." Such mirrors produce beams that retrace the path of any incident beam exactly, with the leading edge of the incident wavefront appearing last in the phase-conjugate beam. (In reflection from an ordinary mirror the leading edge of the incident wavefront emerges first.) Incident light undergoing beam fanning in a photorefractive material may be preferentially swept into a corner of the crystal where it is twice internally reflected. The path of such light forms a loop, in which the exiting beam is phase conjugate to the incoming beam. Such a configuration is shown in (*c*) inside a crystal of barium titanate. The volume refractive-index grating generated in the crystal by these loops produces the phase-conjugate beam. A phase-conjugate beam can be used to restore its incident beam if the incident beam has been overlaid with noise, or in other ways distorted.

Mutually reinforcing beams can give rise to spectacular effects. For instance, if a laser beam is sent

Continued

(c)

(d)

separately into each of two crystals positioned, say, a meter apart, a third beam of light will, after a few seconds, spontaneously arise between the two crystals—connecting them. Light scattered by each crystal is phase conjugated by the other crystal. Such a phase-conjugate beam is aimed directly at the other crystal and so contributes to a growing beam connecting the two.

Refractive index gratings, if formed from a reference beam interfering with a beam scattered from an object, are effectively holograms stored in the photorefractive material. Such holograms are often referred to as being "real-time," since they can be generated and erased continuously in a crystal. The use of applied light to control the transmission of light in crystals with impurities is generally reminiscent of the use of applied voltages to control electrical conduction and electrical properties of semiconductors. The hope thus arises that an "integrated optics" technology can be developed that might supersede the microelectronics of integrated circuits. Early prototypes along these lines have, in fact, been developed—one of which is shown in (*d*).

39-8 Superconductivity

In Section 22-2, we learned that there are some materials called superconductors that have resistivities of zero below a critical temperature I_c, which varies from material to material. Table 39-3 lists some superconducting materials and their critical temperatures. In the presence of a magnetic field B, the critical temperature is lower than it is when there is no field. As the magnetic field increases, the critical temperature decreases. If the magnetic field is greater than some critical field B_c, superconductivity does not exist at any temperature.

Consider a superconducting material that is originally at a temperature greater than the critical temperature and is in the presence of a small external magnetic field $B < B_c$. We now cool the material below the critical temperature so that it becomes superconducting. Since the resistance is now zero, there can be no emf in the superconductor. Thus, from Faraday's law, the magnetic field in the superconductor cannot change. We therefore expect from classical physics that the magnetic field in the superconductor will remain constant. However, it is observed experimentally that when a superconductor is cooled below the critical temperature in an external magnetic field, the magnetic-field lines are expelled from the superconductor and thus the magnetic field inside the superconductor is zero (Figure 39-29). This effect was discovered by Meissner and Ochsenfeld in 1933 and is now known as the **Meissner effect.** The mechanism by which the magnetic-field lines are expelled is an induced superconducting current on the surface of

Table 39-3 **Critical Temperatures for Some Superconducting Materials**

Material	T_c, K
Elements	
Al aluminum	1.14
Hg mercury	4.15
In indium	3.40
Pb lead	7.19
Sn tin	3.72
Ta tantalum	4.48
Compounds	
Nb_3Sn	18.05
Nb_3Ge	23.2
NbN	16.0
V_3Ga	16.5
V_3Si	17.1
La_3In	10.4

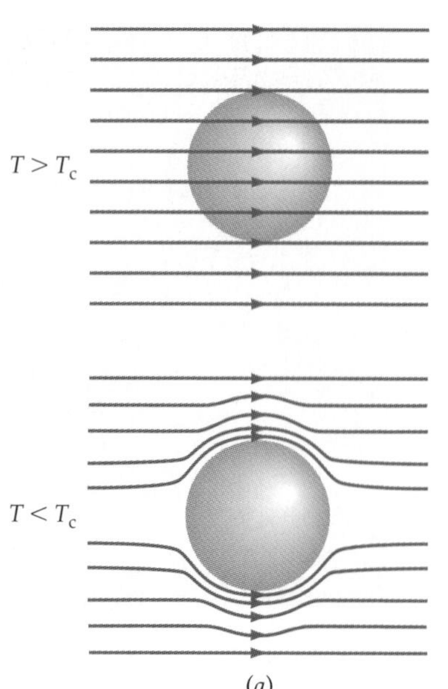

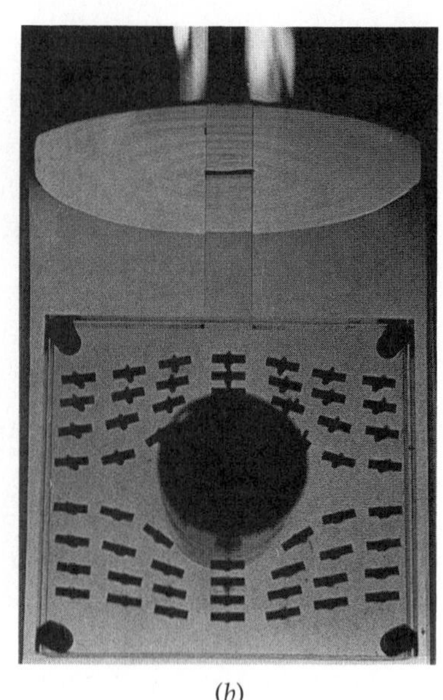

Figure 39-29 (*a*) The Meissner effect in a superconducting sphere cooled in a constant applied magnetic field. As the temperature drops below the critical temperature T_c, the magnetic-field lines are expelled from the sphere. (*b*) Demonstration of the Meissner effect. A superconducting tin cylinder is situated with its axis perpendicular to a horizontal magnetic field. The directions of the field lines near the cylinder are indicated by weakly magnetized compass needles mounted in a Lucite sandwich so that they are free to turn.

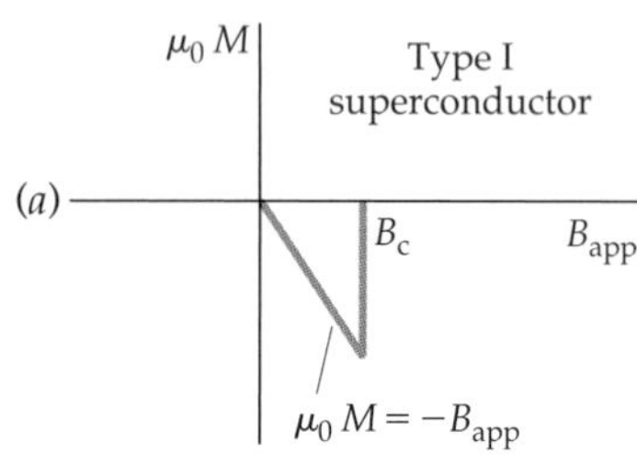

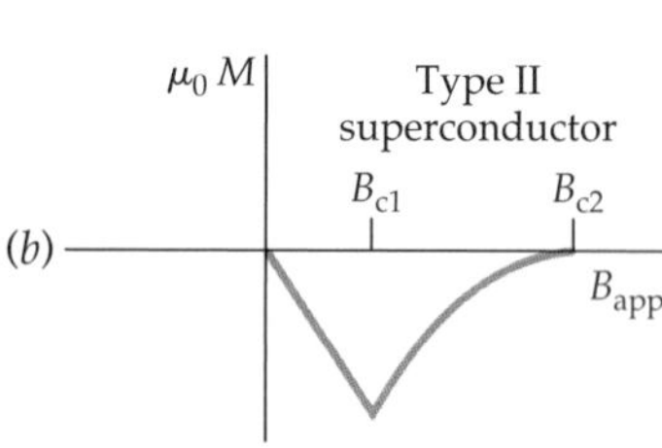

Figure 39-30 Plots of μ_0 times the magnetization M versus applied magnetic field for type I and type II superconductors. (*a*) In a type I superconductor, the resultant magnetic field is zero below a critical applied field B_c because the field due to induced currents on the surface of the superconductor exactly cancels the applied field. Above the critical field, the material is a normal conductor and the magnetization is too small to be seen on this scale. (*b*) In a type II superconductor, the magnetic field starts to penetrate the superconductor at a field B_{c1}, but the material remains superconducting up to the field B_{c2} after which it becomes a normal conductor.

the superconductor. The magnetic levitation shown on page 596 results from the repulsion between the permanent magnet producing the external field and the magnetic field produced by the currents induced in the superconductor. Only certain superconductors called **type I superconductors** exhibit the complete Meissner effect. Figure 39-30*a* shows a plot of the magnetization M times μ_0 versus the applied magnetic field B_{app} for a type I superconductor. For a magnetic field less than the critical field B_c, the magnetic field $\mu_0 M$ induced in the superconductor is equal and opposite to the external magnetic field; that is, the superconductor is a perfect diamagnet. The values of B_c for type I superconductors are always too small for such materials to be useful in the coils of a superconducting magnet.

Other materials, known as **type II superconductors,** have a magnetization curve similar to that in Figure 39-30*b*. Such materials are usually alloys or metals that have large resistivities in the normal state. Type II superconductors exhibit the electrical properties of superconductors except for the Meissner effect up to the critical field B_{c2}, which may be several hundred times the typical values of critical fields for type I superconductors. For example, the alloy Nb_3Ge has a critical field $B_{c2} = 34$ T. Such materials can be used for high-field superconducting magnets. Below the critical field B_{c1}, the behavior of a type II superconductor is the same as that of a type I superconductor. In the region between fields B_{c1} and B_{c2}, the superconductor is said to be in a vortex state.

The BCS Theory

It had been recognized for some time that superconductivity is due to a collective action of the conducting electrons. In 1957, John Bardeen, Leon Cooper, and Bob Schrieffer published a successful theory of superconductivity now known as the **BCS theory.** According to this theory, the electrons in a superconductor are coupled in pairs at low temperatures. The coupling comes about because of the interaction between electrons and the crystal lattice. One electron interacts with the lattice and perturbs it. The perturbed lattice interacts with another electron in such a way that there is an attraction between the two electrons that at low temperatures can exceed the Coulomb repulsion between them. The electrons form a bound state called a **Cooper pair.** The electrons in a Cooper pair have opposite spins and equal and opposite linear momenta. They thus form a system with zero spin and zero momentum. Each Cooper pair may be considered as a single particle with zero spin. Such a particle does not obey the Pauli exclusion principle, so any number of Cooper pairs may be in the same quantum state with the same energy. In the ground state of a superconductor (at $T = 0$), all the electrons are in Cooper pairs and all the Cooper pairs are in the same energy state. In the superconducting state, the Cooper pairs are correlated so that they all act together. In order for the electrons in a superconducting state to absorb or emit energy, the binding of the Cooper pairs must be broken. The energy needed to break up a Cooper pair is similar to that needed to break up a molecule into its constituent atoms. This energy is called the **superconducting energy gap** E_g. In the BCS theory, this energy at absolute zero is predicted to be

$$E_g = 3.5kT_c \qquad 39\text{-}35$$

Example 39-6

Calculate the superconducting energy gap for mercury predicted by the BCS theory, and compare your result with the measured value of 1.65×10^{-3} eV.

From Table 39-3, we have $T_c = 4.15$ K for mercury. The BCS prediction for the energy gap is then

$$E_g = 3.5kT_c = 3.5(1.38 \times 10^{-23}\ \text{J/K})(4.15\ \text{K})\ \frac{1\ \text{eV}}{1.6 \times 10^{-19}\ \text{J}}$$

$$= 1.25 \times 10^{-3}\ \text{eV}$$

This differs from the measured value of 1.65×10^{-3} eV by 24 percent.

Note that the energy gap for a typical superconductor is much smaller than the energy gap for a typical semiconductor, which is of the order of 1 eV. As the temperature is increased from $T = 0$, some of the Cooper pairs are broken. The resulting individual (unpaired) electrons interact with the Cooper pairs, reducing the energy gap until at $T = T_c$ the energy gap is zero (Figure 39-31).

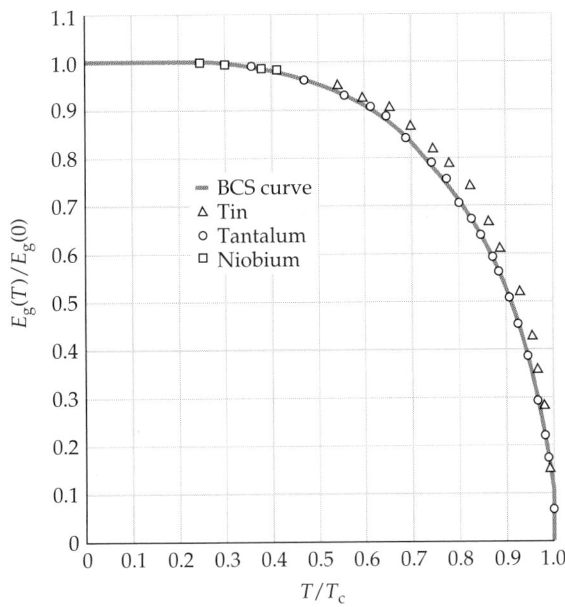

Figure 39-31 Ratio of the energy gap at temperature T to that at $T = 0$ as a function of the relative temperature T/T_c. The solid curve is that predicted by the BCS theory.

The Cooper pairs that we have discussed so far have zero momentum, so there are as many electrons traveling in one direction as the other and there is no current. Cooper pairs can also be formed with a net momentum P rather than zero momentum, but all the pairs have the same momentum. In this state, current is carried by the Cooper pairs. In ordinary conductors, resistance is present because the current carriers can be scattered with a change in momentum. As we have discussed, this scattering may be due to impurity atoms or thermal vibrations of the lattice ions. In a superconductor, the Cooper pairs are constantly scattering each other, but since the total momentum remains constant in this process, there is no change in the current. A Cooper pair cannot be scattered by a lattice ion because all the pairs act together. The only way that the current can be decreased by scattering is if a pair is broken up, which requires energy greater or equal to the energy gap E_g. At reasonably low currents, scattering events in which the total momentum of a Cooper pair is changed are completely prohibited, so there is no resistance.

Flux Quantization

Consider a superconducting ring of area A carrying a current. There can be a magnetic flux $\phi_m = B_n A$ through the ring due to the current in the ring and perhaps also due to other currents external to the ring. According to Faraday's law, if the flux changes, an emf will be induced in the ring that is proportional to the rate of change of the flux. But there can be no emf in the ring because it is resistanceless. The flux through the ring is thus frozen and cannot change. Another effect that results from the quantum-mechanical treatment of superconductivity is that the total flux through the loop is quantized and is given by

$$\phi_m = n\frac{h}{2e} \qquad n = 1, 2, 3, \ldots \tag{39-36}$$

The smallest unit of flux, called a **fluxon,** is

$$\phi_0 = \frac{h}{2e} = 2.0678 \times 10^{-15}\ \text{T}\cdot\text{m}^2 \tag{39-37}$$

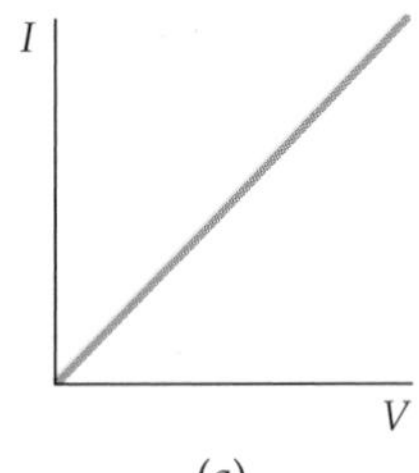

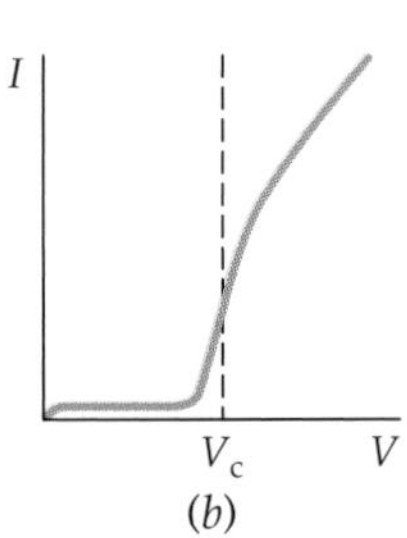

Figure 39-32 Tunneling current versus voltage for a junction of two metals separated by a thin oxide layer. (*a*) When both metals are normal metals, the current is proportional to the voltage as predicted by Ohm's law. (*b*) When one metal is a normal metal and one is a superconductor, the current is approximately zero until the applied voltage exceeds the critical voltage $V_c = E_g/2e$.

Tunneling

In Section 36-8, we discussed barrier penetration—the tunneling of a single particle through a potential barrier. The tunneling of electrons from one metal to another can be observed by separating the two metals with a thin layer only a few nanometers thick of an insulating material such as aluminum oxide. When both metals are normal metals (not superconductors), the current resulting from the tunneling of electrons through the insulating layer obeys Ohm's law for low applied voltages (Figure 39-32*a*). When one of the metals is a normal metal and the other is a superconductor, there is no current (at absolute zero) unless the applied voltage V is greater than a critical voltage $V_c = E_g/2e$, where E_g is the superconductor energy gap. Figure 39-32*b* shows the plot of current versus voltage for this situation. The current jumps abruptly when the energy $2eV$ absorbed by a Cooper pair is great enough to break up the pair. (At temperatures above absolute zero, there is a small current because some of the electrons in the superconductor are thermally excited above the energy gap and are therefore not paired.) The superconducting energy gap can thus be accurately measured by measuring the critical voltage V_c.

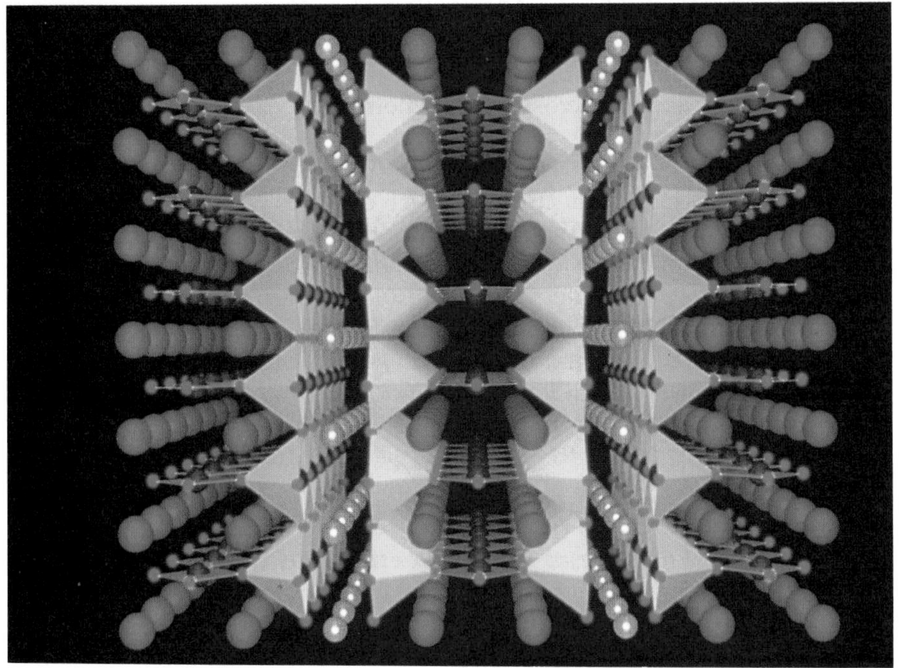

(*a*)

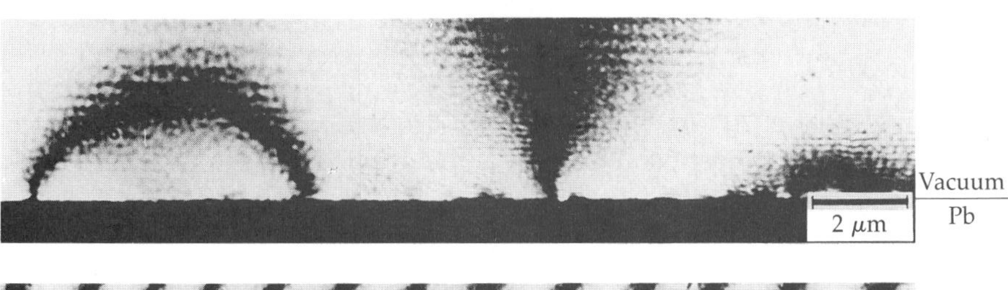

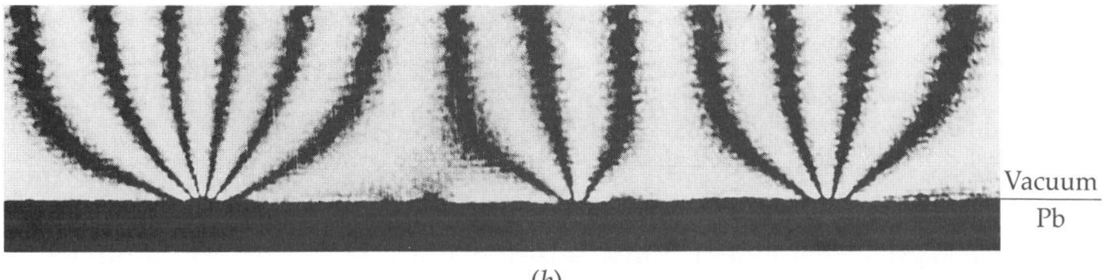

(*b*)

(*c*)

(*a*) A computer-generated image of the crystal structure of the high-temperature superconducting material yttrium-barium-copper oxide. (*b*) Fluxons penetrating a superconducting film. The image has been formed by a new technique—electron holography—in which coherent electron beams are used in place of coherent light beams to create a hologram. Electrons passing by a magnetic field are phase shifted; that is, the phase term in their wave function changes. (The shift arises from a phenomenon known as the Aharonov–Bohm effect.) By superposing such a phase-shifted beam with an unshifted reference beam, an interference pattern is created that can be interpreted as an image of the magnetic field. For the upper image, a magnetic field was applied perpendicular to a thin superconducting lead film. When the field was weak it was expelled by the Meissner effect. A stronger field, however, penetrated the film. The fluxons shown arose from vortexes of current set up in the superconductor—not from the applied field directly. In the upper right is an isolated fluxon; in the upper left is an antiparallel pair of fluxons. The lower micrograph, in which the lead film is thicker, shows penetration by bundles of fluxons. (*c*) A magnetohydrodynamic-powered ship, whose silent engine has no moving parts, is an application of superconductors. The propulsion system consists of a duct encased by an electromagnet. The windings around the magnet are superconductors. An electric current is passed through seawater in the duct and a resulting $I\boldsymbol{\ell} \times \mathbf{B}$ force drives water out the back, propelling the ship forward. Reversing the current reverses the thrust.

In 1962, Brian Josephson proposed that when two superconductors form a junction, now called a **Josephson junction,** Cooper pairs could tunnel from one superconductor to the other with no resistance. The current is observed with no voltage applied across the junction and is given by

$$I = I_{\text{max}} \sin (\phi_2 - \phi_1) \qquad 39\text{-}38$$

where I_{max} is the maximum current, which depends on the thickness of the barrier, ϕ_1 is the phase of the wave function for the Cooper pairs in one of the superconductors, and ϕ_2 is the phase of the corresponding wave function in the other superconductor. This result has been observed experimentally and is known as the **dc Josephson effect.**

Two Josephson junctions.

Josephson also predicted that if a dc voltage were applied across a Josephson junction, there would be a current that alternates with frequency f given by

$$f = \frac{2eV}{h} \qquad 39\text{-}39$$

This result, known as the **ac Josephson effect,** has been observed experimentally, and careful measurement of the frequency allows a precise determination of the ratio e/h. Because frequency can be measured so accurately, the ac Josephson effect is also used to establish precise voltage standards. The inverse effect, in which the application of an alternating voltage across a Josephson junction results in a dc current, has also been observed.

Example 39-7

Using $e = 1.602 \times 10^{-19}$ C and $h = 6.626 \times 10^{-34}$ J·s, calculate the frequency of the Josephson current if the applied voltage is 1 μV.

From Equation 39-39, we obtain

$$f = \frac{2eV}{h} = \frac{2(1.602 \times 10^{-19}\text{ C})(10^{-6}\text{ V})}{6.626 \times 10^{-34}\text{ J·s}} = 4.836 \times 10^{8}\text{ Hz}$$

$$= 483.6\text{ MHz}$$

There is a third effect observed with Josephson junctions. When a dc magnetic field is applied through a superconducting ring containing two Josephson junctions, the maximum supercurrent shows interference effects that depend on the intensity of the magnetic field (Figure 39-33). This effect can be used to measure very weak magnetic fields. It is the basis for a device called a **SQUID** (for Superconducting Quantum Interference Device) that can detect magnetic fields as low as 10^{-14} T. Such a device can detect the magnetic fields produced by the tiny currents flowing in the heart and brain. (See the essay at the end of this chapter.)

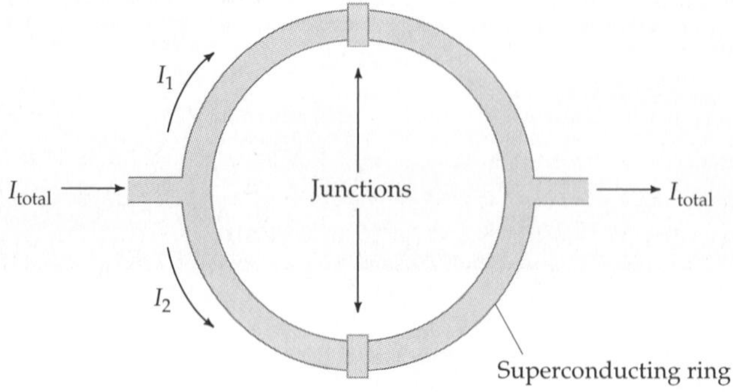

Figure 39-33 A superconducting ring with two Josephson junctions. When there is no applied magnetic field through the ring, the currents I_1 and I_2 are in phase. A very small applied magnetic field produces a phase difference in the two currents that produces interference in the total current exiting the ring.

High-Temperature Superconductivity

Until recently, the highest known critical temperature for a superconductor was 23.2 K for the alloy Nb_3Ge. In 1986, Bednorz and Muller found that an oxide of lanthanum, barium, and copper became superconducting at 30 K. Soon afterwards, in 1987, superconductivity with a critical temperature of 92 K was found in a copper oxide containing yttrium and barium ($YBa_2Cu_3O_7$). Since then, several copper oxides have been found with critical temperatures as high as 125 K. Table 39-4 lists some of the new, high-temperature superconductors along with their critical temperatures. These discoveries have revolutionized the study of superconductivity because relatively inexpensive liquid nitrogen, which boils at 77 K, can be used for a coolant. However, there are many problems, such as the brittleness of ceramics, that make these new superconductors difficult to use.

The new, high-temperature superconductors are all type II superconductors with very high upper critical fields. For some, B_{c2} is estimated to be as high as 100 T. Although the BCS theory appears to be the correct starting place for understanding these new superconductors, they have many features that are not clearly understood. Thus, there is much work, both experimental and theoretical, to be done.

Table 39-4 **Critical Temperature for Some New, High-Temperature Superconducting Materials**

Material	T_c, K
LaBaCuO	30
La_2CuO_4	40
$YBa_2Cu_3O_7$	92
$DyBa_2Cu_3O_7$	92.5
BiSrCaCuO	120
TlBaCaCuO	125

Summary

1. Solids are often found in crystalline form in which a small structure called the unit cell is repeated over and over. A crystal may have a face-centered-cubic, body-centered-cubic, hexagonal close-packed, or other structure depending on the type of bonding between the atoms, ions, or molecules in the crystal and on the relative sizes of the atoms if there are more than one kind as in NaCl.

2. The attractive part of the potential energy of an ion in an ionic crystal can be written

$$U_{\text{att}} = -\frac{\alpha k e^2}{r}$$

where r is the separation distance between neighboring ions and α is the Madelung constant, which depends on the geometry of the crystal and is of the order of 1.8. The repulsive part of the potential energy is due to the exclusion principle. An empirical expression that works fairly well is

$$U_{\text{rep}} = \frac{A}{r^n}$$

where n is about 9.

3. In the classical free-electron theory of metals, the electrical resistivity is given by

$$\rho = \frac{m_e v_{\text{av}}}{n e^2 \lambda}$$

where v_{av} is the mean speed of the electrons in the electron gas and λ is the mean free path between collisions. In the classical theory, v_{av} is given by the Maxwell–Boltzmann distribution and λ is related to the radius of a lattice ion r and the number density of ions n by

$$\lambda = \frac{1}{n\pi r^2}$$

The classical free-electron theory accounts for Ohm's law, but it gives the wrong temperature dependence of the resistivity and it gives magnitudes for the resistivity that do not agree with experiment. Also, the classical theory predicts that the heat capacity for metals should be $\frac{3}{2}R$ higher than that for other solids, which is not observed.

4. In the quantum-mechanical free-electron theory of metals, the Maxwell–Boltzmann distribution is replaced by the Fermi–Dirac distribution and the wave nature of electron scattering is taken into account. In the Fermi–Dirac distribution at $T = 0$, all the energy states below a certain energy called the Fermi energy E_F are filled and all those above this energy are empty. At higher temperatures, some electrons with energies of the order of kT below the Fermi energy are excited to energy states of the order of kT above that level. The Fermi energy at $T = 0$ depends on the electron number density $n = N/V$ and is given by

$$E_F = \frac{(hc)^2}{8m_ec^2}\left(\frac{3N}{\pi V}\right)^{2/3} = (0.365\ \text{eV}\cdot\text{nm}^2)\left(\frac{N}{V}\right)^{2/3}$$

The Fermi energy for copper is about 7 eV, which is much greater than kT at ordinary temperatures. At a temperature T, the Fermi energy is defined to be that energy for which the probability of being occupied by an electron is $\frac{1}{2}$. The difference between the Fermi energy at a temperature T and that at $T = 0$ is usually negligible.

5. When two different metals are placed in contact, electrons flow from the metal with the higher Fermi energy to the metal with a lower Fermi energy until the Fermi energies of the two metals are equal. In equilibrium, there is a potential difference between the metals called the contact potential, which is equal to the difference in the work functions of the two metals divided by the electronic charge e:

$$V_{\text{contact}} = \frac{\phi_1 - \phi_2}{e}$$

6. In the quantum-mechanical theory of conduction, the quantity v_{av} in the expression for ρ is replaced by the Fermi speed u_F, which is essentially independent of temperature, and λ is interpreted as the mean free path for electrons in a lattice in which the ions are vibrating points. In the quantum theory, the contribution of the electron gas to the heat capacity is small because only those electrons within kT of the Fermi energy can be randomly excited.

7. When many atoms are brought together to form a solid, the individual energy levels are split into bands of allowed energies. The splitting depends on the type of bonding and the lattice separation. In a conductor, the uppermost band containing electrons is only partially full, so there are many available states for excited electrons. In an insulator, the uppermost band containing electrons, the valence band, is completely full and there is a large energy gap between it and the next allowed band, the conduction band. In a semiconductor, the energy gap between the filled valence band and the empty conduction band is small; so at ordinary temperatures, an appreciable number of electrons are thermally excited into the conduction band.

8. The conductivity of a semiconductor can be greatly increased by doping.

In an n-type semiconductor, the doping adds electrons just below the conduction band. In a p-type semiconductor, holes are added just above the valence band. A junction between an n-type and p-type semiconductor has applications in many devices, such as diodes, solar cells, and light emitting diodes.

9. A transistor consists of a very thin semiconductor of one type sandwiched between two semiconductors of the opposite type. Transistors are used in amplifiers because a small variation in the base current results in a large variation in the collector current.

10. In a superconductor, the resistance drops suddenly to zero below a critical temperature T_c. In the presence of an external magnetic field, the critical temperature is lowered, and above a critical field B_c, there is no superconductivity. Magnetic-field lines are completely expelled and $\mathbf{B} = 0$ in type I superconductors. This is known as the Meissner effect. Type II superconductors have two critical fields. Below the first, B_{c1}, type II superconductors behave like type I superconductors. Above the second critical field, B_{c2}, superconductivity does not exist. The region between the two critical fields is called the vortex state. Superconductivity exists in this region but the Meissner effect is incomplete.

11. Superconductivity is described by a theory of quantum mechanics called the BCS theory in which the free electrons form Cooper pairs. The energy needed to break up a Cooper pair is called the energy gap E_g. When all the electrons are paired, individual electrons cannot be scattered by a lattice ion, so the resistance is zero.

12. The magnetic flux through a superconducting ring is quantized and takes on the values

$$\phi_m = n\frac{h}{2e} \qquad n = 1, 2, 3, \ldots$$

The smallest unit of flux $\phi_0 = h/2e = 2.0678 \times 10^{-15}\ \text{T·m}^2$ is called a fluxon.

13. When a normal conductor is separated from a superconductor by a thin layer of oxide, electrons can tunnel through the energy barrier if the bias voltage across the layer is $E_g/2e$, where E_g is the energy needed to break up a Cooper pair. The energy gap E_g can be determined by a measurement of the tunneling current versus the bias voltage. A system of two superconductors separated by a thin oxide layer is called a Josephson junction. A dc current is observed to tunnel through such a junction where there is no bias voltage. This is called the dc Josephson effect. When a dc voltage V is applied across a Josephson junction, an ac current is observed with a frequency $f = 2eV/h$. This is called the ac Josephson effect. Measurement of the frequency of this current allows a precise determination of the ratio e/h.

14. Recently a new type of superconductor called a high-temperature superconductor has been discovered. Such superconductors have critical temperatures as high as 125 K and can support magnetic fields as large as 100 T. These new superconductors are only partially understood in terms of the BCS theory.

SQUIDs

Samuel J. Williamson
New York University

Who would predict that the wave nature of an electron lets us study the extremely weak magnetic field of the human brain! This curious relationship is possible because of a device that relies on the quantum mechanical properties of electron waves for such exquisite sensitivity. It is known as the "SQUID", for Superconducting QUantum Interference Device.

Foundations

The SQUID's birth took place in the laboratories of low temperature physicists who were fascinated by superconducting phenomena. But the understanding of its principles rests on the theory developed by Brian Josephson in 1962 of how a pair of wave functions are coupled when two superconductors are separated by only a thin barrier of nonsuperconducting material. A practical example would be two films of niobium, one deposited on top of the other but separated by a thin layer of highly resistive niobium oxide (a "Nb-NbO-Nb" junction). If a source feeds electrical current into one film and takes it from the other at a temperature exceeding T_c for niobium, the corresponding electric field causes a multitude of individual electrons to diffuse across the barrier. The resistance of a typical-size example above the superconducting critical temperature T_c—which is 9.2 K for niobium—would be about 1 Ω.

This junction produces intriguing effects when cooled below T_c. Cooper pairs in each Nb film are described by a wave function, and these two states penetrate the barrier from opposite sides, with amplitudes decreasing exponentially with depth. If the barrier is sufficiently thin, the overlap of the two states near the center couples one to the other. The thinner the insulator, the stronger the coupling. This is called a tunnel junction, because Cooper pairs "tunnel" through the potential barrier where classically they would be prohibited.

The equation for the dc Josephson effect (39-38) shows that the current through a tunnel junction depends on the maximum superconducting current I_{max} and the phase difference between the wave functions on opposite sides. In practical devices such as the Nb-NbO-Nb tunnel junction, the value of I_{max} lies between 1 μA and 1 mA. When current provided to a junction just exceeds I_{max}, the voltage appearing across the barrier is much less than expected for a normal resistive barrier. The resistance is associated with the ac Josephson effect: Cooper pairs radiate energy at the Josephson frequency (Equation 39-39) as their potential energy decreases on crossing the junction.

Samuel J. Williamson is University Professor at New York University with appointments as Professor of Physics and Neural Science in the School of Arts and Science, and Adjunct Professor of Physiology and Biophysics in the School of Medicine at New York University. He received his SB and ScD from Massachusetts Institute of Technology. With a background in low-temperature physics, during the past 15 years he has developed techniques for applying SQUID magnetic-field sensors to studies of neuronal activity in the human brain. He is co-director of the Neuromagnetism Laboratory of NYU, where research focuses on the relationships between human physiology, performance, and cognition. He has written and edited several books in the physical sciences, and been actively involved in organizing international conferences on biomagnetism.

Basic Principles

How a SQUID works can be seen by considering a common type comprised of two junctions connected in parallel, as illustrated in Figure 39-33. (In practice, a resistor is also connected across each junction to minimize hysteresis effects associated with the small capacitance of the junction, but we shall disregard this technical point.) This is known as a dc SQUID, because dc current is provided at one end and taken from the other by attaching a current source. The strength of this "bias" current is chosen to exceed the maximum current I_{max} in both junctions, so that a voltage V appears across them.

When a magnetic field from an external source is applied to the area within the SQUID, the wavelength of the Cooper pairs taking one path around the area is increased, while the wavelength of those taking the other path is decreased. This is a remarkable quantum-mechanical property of a magnetic field—that its presence within the loop has a different effect on moving charges that pass around it in a clockwise sense from those that pass around it in a counterclockwise sense. The net result is that when the two currents rejoin they will generally have different phases. An interference phenomenon takes place that is very much like that of two-slit interference in optics (Section 33-4). Constructive and destructive interference take place in succession if the strength of the external magnetic field is increased steadily. Correspondingly, the voltage appearing across the junctions increases and decreases periodically with increasing field.

Sensitivity

It should be no surprise that the period for one complete cycle in the SQUID's voltage corresponds to a change of flux within the SQUID equal to the fluxon $\phi_0 = h/2e$. Therefore, to go from one peak in the voltage to the next requires a flux change of only 2.07×10^{-15} T·m^2. If the area within the SQUID is 1 mm $\times$ 1 mm, this corresponds to a change in the external field of about 2×10^{-9} T. The value is impressively small compared with the earth's steady field of 60×10^{-6} T. Nevertheless, many orders of magnitude of greater sensitivity can be achieved by adding one more feature to the circuit.

A coil consisting of a few turns of wire placed near the SQUID is provided a current by a separate electronic circuit that can monitor the SQUID's voltage. If the SQUID's voltage begins to depart from, say, a minimum value because of a change in the external field, the circuit senses this and adjusts the current so that the coil's magnetic field counters the change in the external field. In this way, the total flux within the SQUID is held nearly constant. If a resistor is connected in series with the coil, the voltage across the resistor provides a continuous measurement of the external field. Such a circuit is so sensitive that a modern SQUID system detects changes of field amounting to only $10^{-6}\phi_0$. An example is shown in Figure 1. These SQUID sensors are being applied to a variety of magnetic studies where high sensitivity is essential.

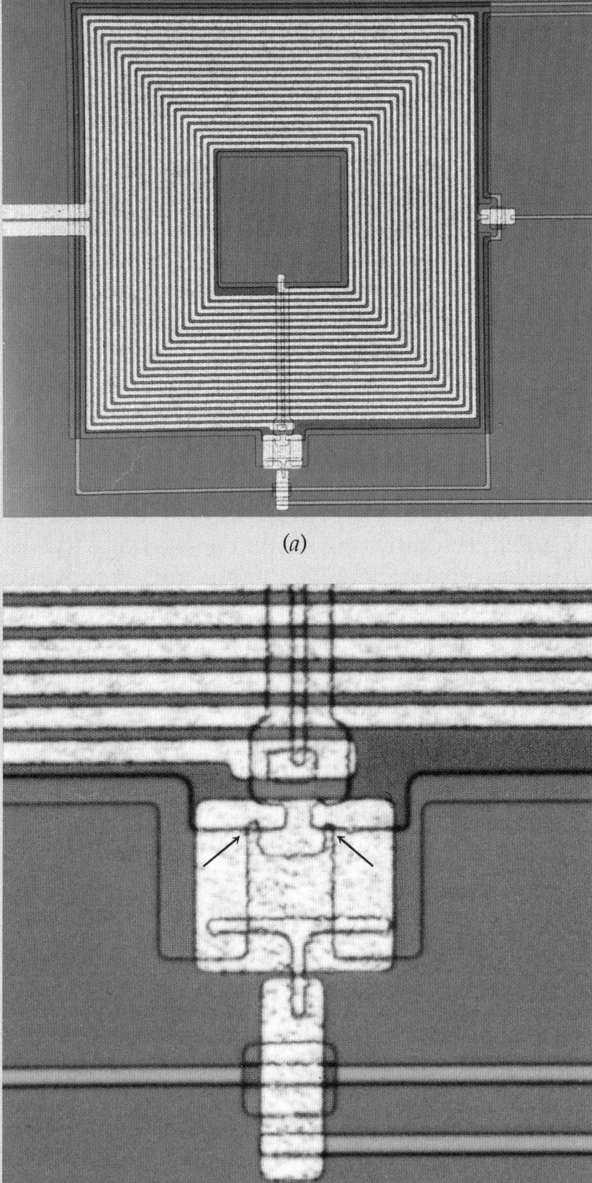

Figure 1 (*a*) Photomicrograph of a SQUID fabricated by depositing thin layers of appropriate materials on an insulating base. Because the layers are so thin, it is possible to see through them to layers beneath. The dominant feature is a 25-turn spiral input coil, only 0.3 mm on a side. Connections to the detection coil that responds to the field of interest will be attached to the two broad pads halfway up on the left side. The dc bias current for the two Josephson junctions comes in via the lead at the bottom right and leaves by the lead halfway up at the right. The bottom lead, shown magnified in (*b*), runs to the junctions, each only 2 μm long, which are the small tips of the U-shaped pad (indicated by arrows) just below the bottom of the input coil.

Biomagnetic Research

Now we arrive at the point where we can relate the SQUID's capabilities to intriguing applications in brain research. Nerve cells ("neurons") in the surface layer, or cortex, of the brain respond to excitations from their neighbors by admitting ions through their membranes, a process that over a period of 10 to 100 ms produces an electrical current flowing along the length of the cell. There is much interest in studying this activity because that is how information from the sense organs such as the eyes, ears, and fingers is processed. The magnetic field of a single neuron is too weak to be detected outside the scalp, but coherent activity of populations of

Continued

cells can be detected with a SQUID sensor, as shown in Figure 2. The sensing elements may be immersed in a bath of liquid helium (at 4.2 K) kept in a vacuum-insulated container called a dewar.

The principal challenge in biomagnetic measurements is not sensitivity for detecting the very weak signals, for field sensitivities as impressive as 3×10^{-15} T have been achieved (in comparison with neuronal signals that range up to 500×10^{-15} T). The major problem is environmental noise that can be 10^5 times greater than the fields of interest. One way to overcome much of this is by doing measurements within a magnetically shielded room. Another is by employing a "detection coil" having a geometry that rejects much of the environmental noise. The detection coil conveys field information to the SQUID by a "flux transporter", a closed superconducting circuit consisting of the detection coil and a smaller "input coil" positioned near the SQUID. Because of flux quantization, when the brain applies magnetic flux to the detection coil, current flows around the entire transporter to cancel the applied flux and maintain the total flux within it invariant. The SQUID experiences the correspondingly proportional field from the input coil. A detection coil wound as a gradiometer discriminates against distant noise sources, which produce relatively uniform fields, while retaining sensitivity to sources close by. A "first-order" gradiometer (one clockwise loop placed near a counter-clockwise loop) will not respond to a uniform field, for the positive flux imposed on one loop is cancelled by the negative flux imposed on the other. Yet it retains sensitivity to a field source that is placed much closer to one loop than the other. Figure 2 shows a second-order gradiometer, which provides even better discrimination against distant sources, for it may be viewed as two first-order gradiometers mounted back to back. Neither first-order gradiometer responds to a uniform field, and the combination of two first-order gradiometers will not respond to a uniform field gradient.

Measurements of magnetic field patterns across the scalp provide information from which the locations and strengths of underlying brain activity may be deduced (Figure 3). This is possible because the head is transparent to magnetic fields, so they emerge without distortion. As a result, the site of an active area in the cortex can be located with an accuracy of better than 3 mm. For instance, it was dis-

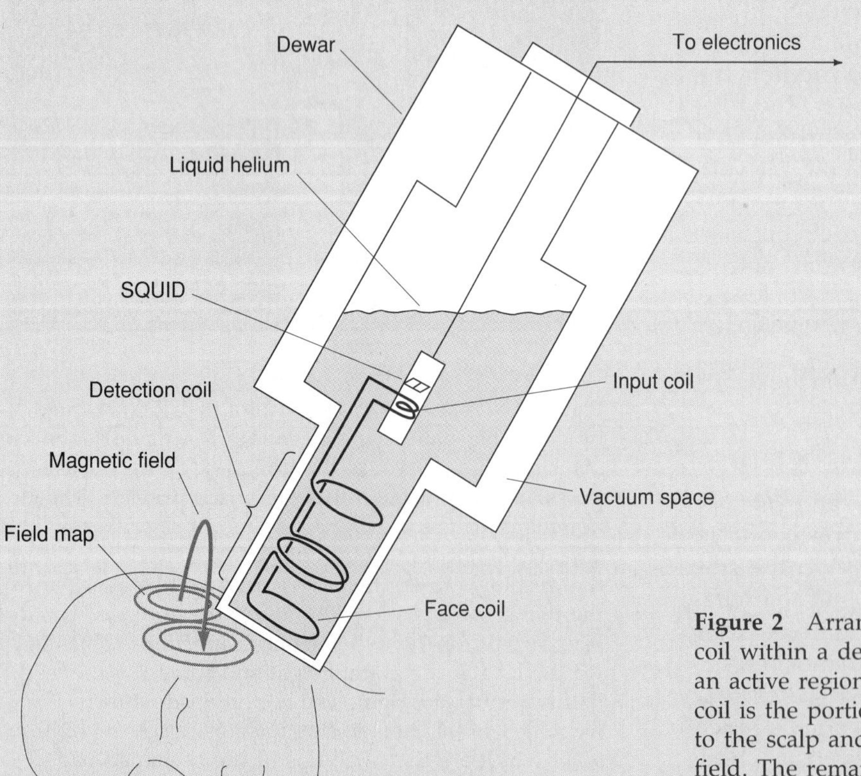

Figure 2 Arrangement of a SQUID and detection coil within a dewar, by which the magnetic field of an active region of the brain can be detected. The face coil is the portion of the detection coil that is closest to the scalp and receives the strongest biomagnetic field. The remaining turns of wire of the detection coil are arranged so that the net magnetic flux of a uniform field, or uniform field gradient, imposed by environmental field noise is made zero.

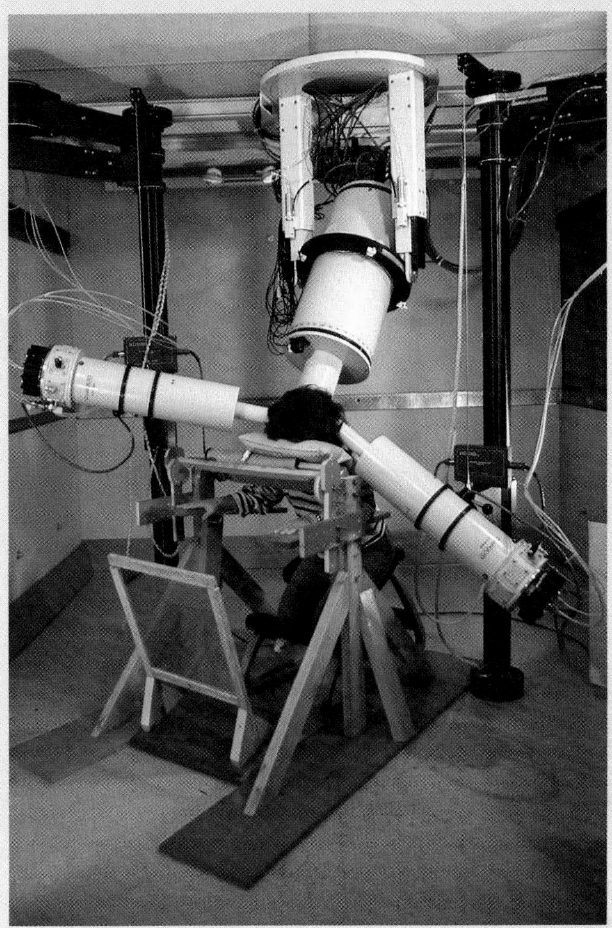

Figure 3 Three SQUID systems positioned to monitor the sequence of brain activity of a volunteer subject who responds by pressing an electrical switch when a visual display is projected onto the screen in front. The dewar suspended from the ceiling contains five SQUID sensors to monitor the response of visual cortex at five closely-spaced positions on the scalp. Each of the two dewars directed from the sides contains a single sensor to monitor the motor cortex controlling left and right finger motions. The SQUIDs in these smaller dewars are cooled by refrigerators that depend on the principle that high-pressure helium gas cools when it is allowed to expand rapidly at these low temperatures.

covered magnetically that humans have a "tone map" across the cortex concerned with auditory functions, located a few centimeters above the ear. Tones of different frequencies evoke activity at different locations. The cumulative distance across the cortex from one site of activity to another increases with the logarithm of the frequency (just as the piano keyboard is arranged). It was also discovered that tones of different intensity have relatively little effect on the strength of the neuronal response. Instead, the site of activity shifts across the cortex with increasing intensity, following a track that is almost at a right angle to the tone sequence. Again this is a logarithmic map, where the distance increases as the logarithm of the intensity of the tone. The strength of neuronal response is always the same, for easily heard tones. What affects the strength of activity is attention: neuronal response is more intense when a person pays attention to a sound than when it is disregarded. Many research groups are using noninvasive SQUID detectors to elucidate sensory and higher-level functions of the human brain.

Additional Applications

SQUIDs have a variety of applications in which extremely weak magnetic fields or currents must be measured. For instance, they are used in geophysical studies to monitor the low-frequency magnetic fields of electromagnetic waves that are reflected between the earth and ionosphere when solar activity creates disturbances in the upper atmosphere. Because these waves penetrate to considerable depth in the ground, measurements of the magnetic field and electric field (the latter obtained with electrodes placed in the soil) provide information that is useful in locating deposits of oil. It is also possible to locate bodies of hot water that may provide a geothermal source of energy.

Another application helps in the search for gravity waves. Here the goal is to detect minute changes in the shape of a massive bar of aluminum that would indicate the fleeting presence of such a wave. One type of detector would depend on a SQUID to indicate the movement of a nearby source of magnetic field, such as a small electromagnet mounted on the bar. Other uses in physics include measurements of extremely small changes in the resistance of a conductor, where a SQUID monitors the imbalance of current in an electrical bridge circuit that is produced by a change in resistance when some property such as temperature is altered.

Thus we see how the wave nature of the electron, exploited in the design of the SQUID sensor, provides opportunities for intriguing studies of very weak magnetic fields. One of the least expected when SQUIDs were first developed was the activity of neurons of the human brain!

Suggestions for Further Reading

Bentley, W. A., and W. J. Humphreys: *Snow Crystals*, Dover, New York, 1962.

A compilation of over 2000 photographs of snowflakes.

Feynman, Richard: *Lectures on Physics*, Vol. III, Addison-Wesley, Reading, Massachusetts, 1964.

The final volume of the three-volume set is devoted to quantum mechanics and its applications to matter. Based on lectures for freshmen and sophomores, this classic work has been enjoyed by students and faculty at all levels.

Hecht, Jeff: *The Laser Guidebook*, McGraw-Hill, New York, 1986.

A nonmathematical account of state-of-the-art laser technology.

Horowitz, Paul, and Winfield Hill: *The Art of Electronics*, 2nd ed., Cambridge University Press, Cambridge, 1989.

This text on electronic circuit design emphasizes intuitive and nonmathematical techniques. No previous exposure to electronics is assumed.

Kittel, Charles: *Introduction to Solid State Physics*, 6th ed., J. Wiley & Sons, New York, 1986.

The standard senior or beginning graduate text in solid state physics. Sections of the book are, nonetheless, accessible to less-advanced readers.

Rose-Innes, A. C., and E. H. Rhoderick: *Introduction to Superconductivity*, 2nd ed., Pergamon Press, Oxford, 1978.

Written for undergraduates or beginning graduate students. No previous knowledge of superconductivity is assumed.

Review

A. Objectives: After studying this chapter, you should:

1. Know what the Madelung constant is and why it differs from 1.

2. Be able to discuss the successes and failures of the classical free-electron theory of metals.

3. Be able to derive an expression for the Fermi energy at $T = 0$ in one dimension.

4. Be able to discuss qualitatively the band theory of solids.

5. Be able to discuss the effects of adding impurities to intrinsic semiconductors.

6. Know the general features of the current-versus-voltage curve for a typical *pn*-semiconductor junction.

7. Be able to discuss the characteristics and operation of a transistor.

8. Be able to discuss qualitatively the BCS theory of superconductivity.

B. Define, explain, or otherwise identify:

Amorphous solid
Unit cell
Madelung constant
Dissociation energy
Fermi energy at $T = 0$
Fermi factor
Density of states
Fermi temperature
Fermi–Dirac distribution
Contact potential
Valence band
Conduction band
Forbidden energy band
Intrinsic semiconductor
Impurity semiconductor
Doping
Donor levels
n-type semiconductor
Acceptor levels
p-type semiconductor
Junction
Depletion region
Forward bias
Reverse bias
Avalanche breakdown
Zener diode
Tunnel diode
Solar cell
Surface-barrier detector
Light-emitting diode
Emitter
Base
Collector
Current gain
Voltage gain
Meissner effect
Type I superconductors
Type II superconductors
BCS Theory
Cooper pair
Superconducting energy gap
Fluxon
Josephson junction
dc Josephson effect
ac Josephson effect
SQUID

C. True or false: If the statement is true, explain why it is true. If it is false, give a counterexample.

1. Solids that are good electrical conductors are usually good heat conductors.

2. The classical free-electron theory adequately explains the heat capacity of metals.

3. At $T = 0$, the Fermi factor is either 1 or 0.

4. The Fermi energy is the average energy of an electron in a solid.

5. The contact potential between two metals is proportional to the difference in the work functions of the two metals.

6. At $T = 0$, an intrinsic semiconductor is an insulator.

7. Semiconductors conduct current in one direction only.

Problems

Level 1

39-1 The Structure of Solids

1. Suppose that hard spheres of radius R are located at the corners of a unit cell with a simple cubic structure. (*a*) If the hard spheres touch so as to take up the minimum volume possible, what is the size of the unit cell? (*b*) What fraction of the volume of the cubic structure is occupied by the hard spheres?

2. Calculate the distance r_0 between the K^+ and the Cl^- ions in KCl, assuming that each ion occupies a cubic volume of side r_0. The molar mass of KCl is 74.55 g/mol and its density is 1.984 g/cm^3.

3. The distance between the Li^+ and Cl^- ions in LiCl is 0.257 nm. Use this and the molecular mass of LiCl (42.4 g/mol) to compute the density of LiCl.

4. Find the value of n in Equation 39-6 that gives the measured dissociation energy of 741 kJ/mol for LiCl, which has the same structure as NaCl and for which $r_0 =$ 0.257 nm.

39-2 The Classical Free-Electron Theory of Metals

5. A measure of the density of the free-electron gas in a metal is the distance r_s, which is defined as the radius of the sphere whose volume equals the volume per conduction electron. (*a*) Show that $r_s = (3/4\pi n)^{1/3}$, where n is the free-electron number density. (*b*) Calculate r_s for copper in nanometers.

6. (*a*) Given a mean free path $\lambda = 0.4$ nm and a mean speed $v_{av} = 1.17 \times 10^5$ m/s for the current flow in copper at a temperature of 300 K, calculate the classical value for the resistivity ρ of copper. (*b*) The classical model suggests that the mean free path is temperature independent and that v_{av} depends on temperature. From this model, what would ρ be at 100 K?

39-3 The Fermi Electron Gas

7. Calculate the number density of free electrons in (*a*) Ag ($\rho = 10.5$ g/cm^3) and (*b*) Au ($\rho = 19.3$ g/cm^3), assuming one free electron per atom, and compare your results with the values listed in Table 39-1.

8. Calculate the number density of free electrons for (*a*) Mg ($\rho = 1.74$ g/cm^3) and (*b*) Zn ($\rho = 7.1$ g/cm^3), assuming two free electrons per atom, and compare your results with the values listed in Table 39-1.

9. The density of aluminum is 2.7 g/cm^3. How many free electrons are there per aluminum atom? (Use Table 39-1 for the number density.)

10. The density of potassium is 0.851 g/cm^3. How many free electrons are there per potassium atom? (Use Table 39-1 for the number density.)

11. The density of tin is 7.3 g/cm^3. How many free electrons are there per tin atom? (Use Table 39-1 for the number density.)

12. Calculate the Fermi energy for (*a*) Al, (*b*) K, and (*c*) Sn using the number densities given in Table 39-1.

13. Calculate the Fermi temperature for (*a*) Al, (*b*) K, and (*c*) Sn.

14. Find the average energy of the conduction electrons at $T = 0$ in (*a*) copper and (*b*) lithium.

15. Calculate the (*a*) Fermi energy and (*b*) Fermi temperature for iron at $T = 0$.

16. (*a*) Which two metals in Table 39-2 develop the greatest potential when placed in contact? (*b*) What is the value of that contact potential?

17. (*a*) Which two metals in Table 39-2 develop the least potential when placed in contact? (*b*) What is the value of that contact potential?

18. Calculate the contact potential between (*a*) Ag and Cu, (*b*) Ag and Ni, and (*c*) Ca and Cu.

39-4 Quantum Theory of Electrical Conduction

19. Use Equation 39-28 with $\alpha = \pi^2/4$ to calculate the average energy of an electron in copper at $T = 300$ K. Compare your result with the average energy at $T = 0$ and the classical result of $\frac{3}{2}kT$.

20. What is the speed of a conduction electron whose energy is equal to the Fermi energy E_F for (*a*) Na, (*b*) Au, and (*c*) Sn?

21. The resistivities of Na, Au, and Sn at $T = 273$ K are 4.2 $\mu\Omega\cdot$cm, 2.04 $\mu\Omega\cdot$cm, and 10.6 $\mu\Omega\cdot$cm, respectively. Use these values and the Fermi speeds calculated in Problem 20 to find the mean free paths λ for the conduction electrons in these elements.

22. At what temperature is the heat capacity due to the electron gas in copper equal to 10 percent of that due to lattice vibrations?

39-5 Band Theory of Solids

23. The energy gap between the valence band and the conduction band in silicon is 1.14 eV at room temperature. What is the maximum wavelength of a photon that will excite an electron from the top of the valence band to the bottom of the conduction band?

24. Work Problem 23 for germanium, for which the energy gap is 0.74 eV.

25. Work Problem 23 for diamond, for which the energy gap is 7.0 eV.

26. A photon of wavelength 3.35 μm has just enough energy to raise an electron from the valence band to the conduction band in a lead sulfide crystal. (*a*) Find the energy gap between these bands in lead sulfide. (*b*) Find the temperature T for which kT equals this energy gap.

39-6 Impurity Semiconductors

27. What type of semiconductor is obtained if silicon is doped with (*a*) aluminum and (*b*) phosphorus? (See Table 37-1 for the electron configurations of these elements.)

28. What type of semiconductor is obtained if silicon is doped with (*a*) indium and (*b*) antimony? (See Table 37-1 for the electron configurations of these elements.)

29. The donor energy levels in an *n*-type semiconductor are 0.01 eV below the conduction band. Find the temperature for which $kT = 0.01$ eV.

39-7 Semiconductor Junctions and Devices

30. Simple theory for the current versus the bias voltage across a *pn* junction yields the equation

$$I = I_0(e^{eV_b/kT} - 1)$$

Sketch I versus V_b for both positive and negative values of V_b using this equation.

31. For a temperature of 300 K, use the equation in Problem 30 to find the bias voltage V_b for which the exponential term has the value (*a*) 10 and (*b*) 0.1.

39-8 Superconductivity

32. (*a*) Use Equation 39-35 to calculate the superconducting energy gap for tin and compare your result with the measured value of 6×10^{-4} eV. (*b*) Use the measured value to calculate the wavelength of a photon having sufficient energy to break up Cooper pairs in tin at $T = 0$.

33. Repeat Problem 32 for lead, which has a measured energy gap of 2.73×10^{-3} eV.

Level II

34. Suppose identical bowling balls of radius R are packed into a hexagonal close-packed structure. What fraction of the available volume of the unit cell is filled by the bowling balls?

35. Estimate the fraction of free electrons in copper that are in excited states above the Fermi energy at (*a*) room temperature of 300 K and (*b*) 1000 K.

36. A one-dimensional model of an ionic crystal consists of a line of alternating positive and negative ions with distance r_0 between each ion. (*a*) Show that the potential energy of attraction of one ion in the line is

$$V = -\frac{2ke^2}{r_0}\left(1 - \frac{1}{2} + \frac{1}{3} - \frac{1}{4} + \frac{1}{5} - \cdots\right)$$

(*b*) Using the result that

$$\ln(1 + x) = x - \frac{x^2}{2} + \frac{x^3}{3} - \frac{x^4}{4} + \cdots$$

show that the Madelung constant for this one-dimensional model is $\alpha = 2 \ln 2 = 1.386$.

37. Estimate the Fermi energy of zinc from its electronic molar heat capacity of $(3.74 \times 10^{-4}\ \text{J/mol·K}^2)T$.

38. Derive Equation 39-21 for the average energy of a Fermi electron gas at $T = 0$.

39. (*a*) How many atoms are in a 10-kg block of copper? (*b*) If there are eight conduction states per atom, find the number of states. (*c*) If these states are filled to the Fermi energy level, find the average separation in energy between the states. (*d*) Compare this average spacing to thermal energy kT at $T = 300$ K.

40. The density of the electron states in a metal can be written $g(E) = AE^{1/2}$, where A is a constant and E is measured from the bottom of the conduction band. (*a*) Show that the total number of states is $\frac{2}{3}AE_F^{3/2}$. (*b*) About what fraction of the conduction electrons are within kT of the Fermi energy? (*c*) Evaluate this fraction for copper at $T = 300$ K.

41. What energy is required to remove one ion pair from NaCl and convert the ions into an NaCl molecule? The energy is called the *cohesive energy per ion pair* for NaCl.

42. (*a*) Calculate the number of atoms in 10 g of KBr and from this the number of energy states in the conduction band, assuming eight states per atom. (*b*) The width of the KBr valence band is measured using x-ray spectroscopy to be about 1.5 eV. Estimate the average density of electron states in the valence band for a 10-g crystal. (*c*) How does the number of electron states per unit of energy change if the mass of the crystal is increased to 1 kg?

43. In Figure 39-28 for the *pnp*-transistor amplifier, suppose $R_b = 2$ kΩ and $R_L = 10$ kΩ. Suppose further that 10-μA ac base current generates a 0.5-mA ac collector current. What is the voltage gain of the amplifier?

44. Germanium can be used to measure the energy of incident particles. Consider a 660-keV gamma ray emitted from ^{137}Cs. (*a*) Given that the band gap in germanium is 0.72 eV, how many electron–hole pairs can be generated by this incident gamma ray? (*b*) The number of pairs N in part (*a*) will have statistical fluctuations given by $\pm\sqrt{N}$. What then is the energy resolution of this detector in this photon energy region?

45. A "good" silicon diode has a current–voltage characteristic given by

$$I = I_0(e^{eV_b/kT} - 1)$$

Let $kT = 0.025$ eV (room temperature) and the saturation current $I_0 = 1$ nA. (*a*) Show that for small reverse-bias voltages, the resistance is 25 MΩ. *Hint:* Do a Taylor expansion of the exponential function, or use your calculator and enter small values for V_b. (*b*) Find the dc resistance for a reverse bias of 0.5 V. (*c*) Find the dc resistance for a 0.5-V forward bias. What is the current in this case? (*d*) Calculate the ac resistance dV/dI for a 0.5-V forward bias.

46. The relative binding of the extra electron in the arsenic atom that replaces an atom in silicon or germanium can be understood from a calculation of the first Bohr orbit of this electron in these materials. Four of arsenic's outer electrons form covalent bonds, so the fifth electron sees a singularly charged center of attraction. This model is a modified hydrogen atom. In the Bohr model of the hydrogen atom, the electron moves in free space at a radius a_0 given by

$$a_0 = \frac{\epsilon_0 h^2}{\pi m_e e^2}$$

When an electron moves in a crystal, we can approximate the effect of the other atoms by replacing ϵ_0 with $\kappa\epsilon_0$ and m_e with an effective mass for the electron. For silicon κ is 12 and the effective mass is about $0.2m_e$, and for germanium κ is 16 and the effective mass is about $0.1m_e$. Estimate the Bohr radii for the outer electron as it orbits the impurity arsenic atom in silicon and germanium.

47. Modify Equation 37-9 for the ground-state energy of

the hydrogen atom in the spirit of the preceding problem by replacing ϵ_0 by $\kappa\epsilon_0$ and m_e by an effective mass for the electron to estimate the binding energy of the extra electron of an impurity arsenic atom in (*a*) silicon and (*b*) germanium. (*c*) Suppose you dope silicon with one part per million of arsenic. What fraction of these arsenic atoms would supply an electron to the conduction band at room temperature?

48. The number of electrons in the conduction band of an insulator or intrinsic semiconductor is governed chiefly by the Fermi factor, which in these cases is $e^{-(E-E_F)/kT}$, where E_F is the Fermi energy, which is approximately midway between the nearly filled valence band the nearly empty conduction band. If E_g is the energy gap between the valence and conduction bands, and E is measured from the top of the valence band, $E_F = \frac{1}{2}E_g$. The Fermi factor at the bottom of the conduction band $E = E_g$ is then $e^{-E_g/2kT}$. Calculate this Fermi factor at $T = 300$ K for a typical gap energy of (*a*) 6.0 eV for an insulator and (*b*) 1.0 eV for a semiconductor. Discuss the significance of these results if there are 10^{22} valence electrons per cubic centimeter and $e^{-(E-E_F)/kT}$ is the probability of one electron having an energy E in the conduction band.

49. The pressure of an ideal gas is related to the average energy of the gas particles by $PV = \frac{2}{3}NE_{av}$, where N is the number of particles and E_{av} is the average energy. Use this to calculate the pressure of the Fermi electron gas in copper in newtons per square meter, and compare your result with atmospheric pressure, which is about 10^5 N/m^2. (*Note:* The units are most easily handled by using the conversion factors 1 N/m^2 = 1 J/m^3 and 1 eV = 1.6×10^{-19} J.)

50. The bulk modulus B of a material can be defined by

$$B = -V\frac{\partial P}{\partial V}$$

(*a*) Use the ideal-gas relation $PV = \frac{2}{3}NE$ and Equations 39-19 and 39-21 to show that

$$P = \frac{2NE_F}{5V} = CV^{-5/3}$$

where C is a constant independent of V. (*b*) Show that the bulk modulus of the Fermi electron gas is therefore

$$B = \frac{5}{3}P = \frac{2NE_F}{3V}$$

(*c*) Compute the bulk modulus in newtons per square meter for the Fermi electron gas in copper and compare your result with the measured value of 134×10^9 N/m^2.

51. The resistivity of pure copper is increased by about 1×10^{-8} Ω·m by the addition of 1 percent (by number of atoms) of an impurity throughout the metal. The mean free path depends on both the impurity and the oscillations of the lattice ions according to the equation

$$\frac{1}{\lambda} = \frac{1}{\lambda_t} + \frac{1}{\lambda_i}$$

(*a*) Estimate λ_i from the given data. (*b*) If d is the effective diameter of an impurity lattice ion seen by an electron, the scattering cross section is d^2. Estimate d^2 from $d = 2r$, where r is related to λ_i by Equation 39-10.

Level III

52. (*a*) Calculate the force $F = -dU/dr$ from Equation 39-5 and show that

$$F = \alpha\frac{ke^2}{r_0^2}\left(\frac{r_0^{n+1}}{r^{n+1}} - \frac{r_0^2}{r^2}\right)$$

(*b*) Note that $F = 0$ when $r = r_0$. Write $r = r_0 + \Delta r = r_0(1 + \epsilon)$, where $\epsilon = \Delta r/r_0$, and use the binomial expansion $(1 + \epsilon)^n = 1 + n\epsilon + n(n-1)\epsilon^2/2$ to write F as a power series in Δr and show that, when $r = r_0 + \Delta r$, F is given by

$$F = -C\,\Delta r + B(\Delta r)^2 + \cdots$$

where

$$C = \alpha\frac{(n-1)ke^2}{r_0^3}$$

and

$$B = \alpha\frac{(n^2 + 3n - 4)ke^2}{2r_0^4}$$

(Note that the k in ke^2 in the expressions for C and B is the Coulomb constant.)

53. The quantity C in Problem 52 is the force constant for a "spring" consisting of a line of alternating positive and negative ions. If these ions are displaced slightly from their equilibrium separation r_0, they will vibrate with a frequency

$$f = \frac{1}{2\pi}\sqrt{\frac{C}{m}}$$

(*a*) Use the values of α, n, and r_0 for NaCl and the reduced mass for the NaCl molecule to calculate this frequency. (*b*) Calculate the wavelength of electromagnetic radiation corresponding to this frequency, and compare your result with the characteristic strong infrared absorption bands in the region of about $\lambda = 61$ μm that are observed for NaCl.

54. The term $B(\Delta r)^2$ in Problem 52 is related to the coefficient of thermal expansion of a solid. The time average of the force must be zero since there is no net acceleration averaged over time. Therefore, we have

$$\overline{\Delta r} = \frac{B}{C}\,\overline{(\Delta r)^2}$$

(*a*) Use the equipartition result that

$$\tfrac{1}{2}C\,\overline{(\Delta r)^2} = \tfrac{1}{2}kT$$

to show that $\overline{\Delta r} = (kB/C^2)T$, where k is Boltzmann's constant. (*b*) Evaluate the coefficient of thermal expansion kB/C^2r_0 for NaCl, and compare your result with the measured value of about 4×10^{-6}/K.

55. Consider a model for a metal in which the lattice of positive ions forms a container for a classical electron gas with n electrons per unit volume. In equilibrium, the average electron velocity is zero, but the application of an electric field produces an acceleration of the electrons. If we use a relaxation time τ to account for the electron–lattice collisions, then we have the equation

$$m\frac{dv}{dt} + \frac{m}{\tau}v = -eE$$

(*a*) Solve the equation for the drift velocity in the direction of the applied electric field. (*b*) Verify that Ohm's law is valid, and find the resistivity as a function of n, e, m, and the relaxation time τ.

Chapter 40

Nuclei

A nuclear power plant in West Germany. The fission reactor core is housed in the hemispherical containment structure at the center of the picture. To its left are two large cooling towers. A schematic of such a facility is shown on page 1350.

The first information about the atomic nucleus came from the discovery of radioactivity by A. H. Becquerel in 1896. The rays emitted by radioactive nuclei were studied by many physicists in the early decades of the twentieth century. They were first classified by E. Rutherford as alpha, beta, and gamma rays according to their ability to penetrate matter and ionize air: Alpha radiation penetrates the least and produces the most ionization and gamma radiation penetrates the most and produces the least ionization. It was later found that alpha rays are helium nuclei, beta rays are electrons or positrons, and gamma rays are high-energy photons, that is, electromagnetic radiation of very short wavelength. (A positron is an antielectron, a particle identical to the electron except that its charge is positive.) The alpha-particle scattering experiments of H. W. Geiger and E. Marsden in 1911 and the successes of the Bohr model of the atom led to the modern picture of an atom as consisting of a tiny, massive nucleus with a radius of the order of 1 to 10 fm (1 fm = 10^{-15} m) surrounded by a cloud of electrons at a relatively great distance, of the order of 0.1 nm = 100,000 fm, from the nucleus.

Artificial nuclear disintegration was first observed in 1919 when Rutherford detected protons that were emitted when he bombarded nitrogen with alpha particles. Such experiments were extended to many other elements in the next few years.

In 1932, the neutron was discovered by J. Chadwick, the positron was discovered by C. Anderson, and the first nuclear reaction using artificially accelerated particles was observed by J. D. Cockcroft and E. T. S. Walton. It is therefore reasonable to mark that year as the beginning of modern nuclear physics. With the discovery of the neutron, it became possible to understand some of the properties of nuclear structure. The advent of nuclear accelerators removed the severe limitations on particle type and energy imposed by the need to use naturally occurring radioactive sources, and made many additional experimental studies possible.

In this chapter, we will first discuss some of the general properties of the atomic nucleus and the important features of radioactivity. We will then look at some nuclear reactions, including the important reactions of fission and fusion. Finally, we will look at the interactions of nuclear particles with matter, a subject important for understanding the detection of nuclear particles, the shielding of reactors, and the effects of radiation on the human body. Our discussions will be descriptive and phenomenological, with the aim of presenting general information rather than a theoretical understanding of nuclear physics.

40-1 Properties of Nuclei

The nucleus of an atom contains just two kinds of particles: protons and neutrons. (The normal hydrogen nucleus consists of a single proton.) These particles have approximately the same mass. (The neutron is about 0.2 percent more massive.) The proton has a charge of $+e$ and the neutron is uncharged. The number of protons Z is the **atomic number** of the atom, which also equals the number of electrons in the atom. The number of neutrons N is approximately equal to Z for light nuclei and is somewhat greater than Z for heavier nuclei. The total number of nucleons $A = N + Z$ is called the **mass number** of the nucleus. (The term **nucleon** refers to either a neutron or a proton.) A particular nuclear species, called a **nuclide,** is designated by its atomic symbol (H for hydrogen, He for helium, and so forth) with the mass number A as a presuperscript. Two or more nuclides with the same atomic number Z but different N and A numbers are called **isotopes.** The lightest element, hydrogen, has three isotopes: ordinary hydrogen (^{1}H), whose nucleus is just a single proton; deuterium (^{2}H), whose nucleus contains one proton and one neutron; and tritium (^{3}H), whose nucleus contains one proton and two neutrons. Although the mass of the deuterium atom is about twice that of the ordinary hydrogen atom and that of the tritium atom is three times as great, these three atoms have nearly identical chemical properties because they each have one electron. On the average, there are about 2.6 stable isotopes for each element, although some have only one and others have five or six. The most common isotope of the second lightest atom, helium, is ^{4}He. The nucleus of the ^{4}He atom is also known as an alpha particle. Another isotope of helium is ^{3}He.

Inside the nucleus, the nucleons exert strong attractive forces on their nearby neighbors. This force, called the **strong nuclear force** or the **hadronic force,** is much stronger than the electrostatic force of repulsion between the protons and is very much stronger than the gravitational forces between the nucleons. (Gravity is so weak that it can always be neglected in nuclear physics.) The strong nuclear force is roughly the same between two neutrons, two protons, or a neutron and a proton. Two protons, of course, also exert a repulsive electrostatic force on each other due to their charges, which tends to weaken the attraction between them somewhat. The strong nuclear force decreases rapidly with distance, and it is negligible when two nucleons are more than a few femtometers apart.

Nuclear Size and Shape

The size and shape of the nucleus can be determined by bombarding it with high-energy particles and observing the scattering (similar to the experiments of Geiger and Marsden) or, in some cases, from measurements of its radioactivity. The results depend somewhat on the kind of experiment. For example, a scattering experiment using electrons measures the charge distribution of the nucleus, whereas one using neutrons determines the region of influence of the strong nuclear force. Despite these differences, a wide variety of experiments suggest that most nuclei are approximately spherical, with radii given approximately by

Nuclear radius

$$R = R_0 A^{1/3} \tag{40-1}$$

where R_0 is about 1.5 fm. The fact that the radius of a spherical nucleus is proportional to $A^{1/3}$ implies that the volume of the nucleus is proportional to A. Since the mass of the nucleus is also approximately proportional to A, the densities of all nuclei are approximately the same. The fact that a drop of liquid also has a constant density independent of its size has led to a model in which the nucleus is viewed as analogous to a drop of liquid. This model has proved quite successful in understanding nuclear behavior, especially the fission of heavy nuclei.

N and Z Numbers

For light nuclei, the greatest stability is achieved when the numbers of protons and neutrons are approximately equal, that is, when $N \approx Z$. For heavier nuclei, electrostatic repulsion between the protons leads to greater stability when there are more neutrons than protons. We can see this by looking at the N and Z numbers for the most abundant isotopes of some representative elements: for ${}^{16}_{8}\mathrm{O}$, $N = 8$ and $Z = 8$; for ${}^{40}_{20}\mathrm{Ca}$, $N = 20$ and $Z = 20$; for ${}^{56}_{26}\mathrm{Fe}$, $N = 30$ and $Z = 26$; for ${}^{207}_{82}\mathrm{Pb}$, $N = 125$ and $Z = 82$; and for ${}^{238}_{92}\mathrm{U}$, $N = 146$ and $Z = 92$. (The atomic number Z has been included here as a presubscript of the atomic symbol for emphasis. It is not actually needed because the atomic number is implied by the atomic symbol.)

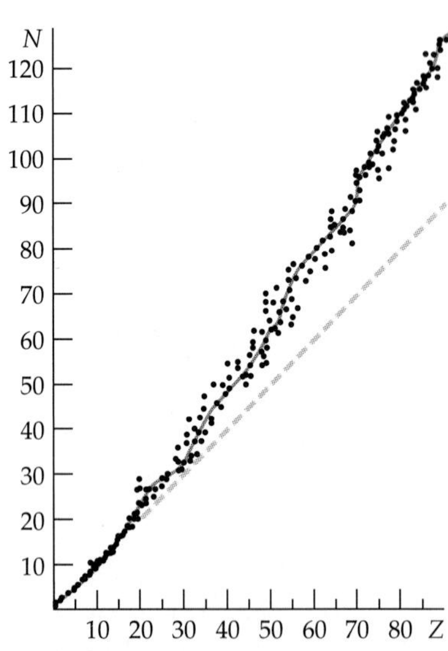

Figure 40-1 Plot of number of neutrons N versus number of protons Z for the stable nuclides. The dashed line is $N = Z$.

Figure 40-1 shows a plot of N versus Z for the known stable nuclei. The straight line $N = Z$ is followed for small values of N and Z. We can understand this tendency for N and Z to be equal by considering the total energy of A particles in a one-dimensional box (see Section 36-6). Figure 40-2 shows the energy levels for eight neutrons and for four neutrons and four protons. Because of the exclusion principle, only two identical particles (with opposite spins) can be in the same space state. Since protons and neutrons are not identical, we can put two each in a state as in Figure 40-2b. Thus, the total energy for four protons and four neutrons is less than that for eight neutrons (or eight protons) as in Figure 40-2a. When the Coulomb energy of repulsion is included, this result changes somewhat. This potential energy is proportional to Z^2. For large values of A (and therefore for large values of Z), the

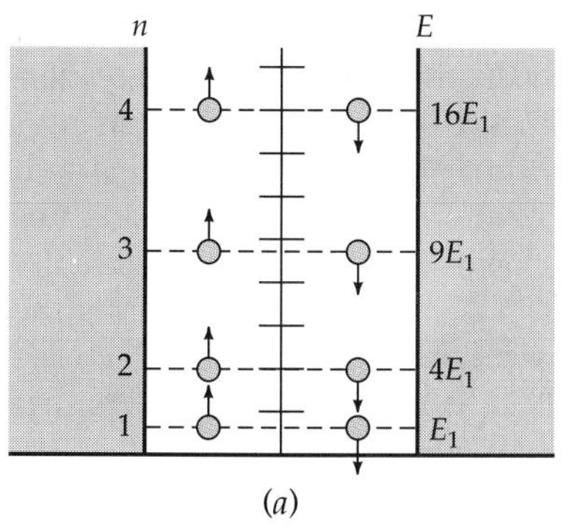

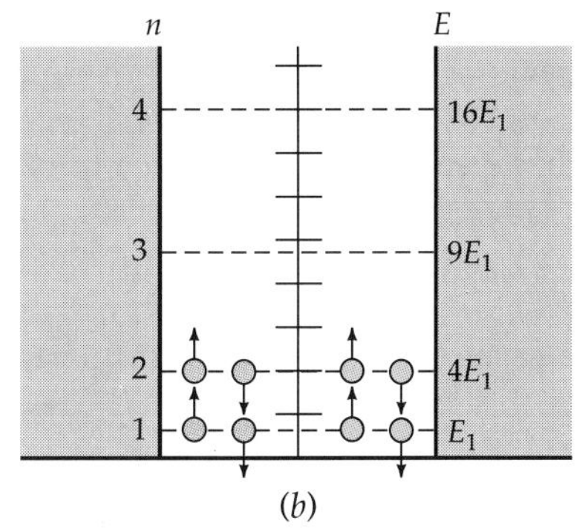

Figure 40-2 (*a*) Eight neutrons in a one-dimensional box. In accordance with the exclusion principle, only two neutrons (with opposite spins) can be in a given energy level. (*b*) Four neutrons and four protons in a one-dimensional box. Because protons and neutrons are not identical particles, two of each can be in each energy level. The total energy is much less for this case than for that in (*a*).

total energy may be increased less by adding two neutrons than by adding one neutron and one proton because of the electrostatic repulsion. This explains why $N > Z$ for the heavier nuclei.

Mass and Binding Energy

In our study of relativistic energy (Section 34-10), we saw that the mass of a nucleus is not equal to the sum of the masses of the individual nucleons that make up the nucleus. When two or more nucleons fuse together to form a nucleus, the total rest mass decreases and energy is given off. Conversely, to break up a nucleus into its parts, energy must be put into the system to increase the rest mass. The energy involved is c^2 times the change in mass, where c is the speed of light in vacuum. The difference between the rest energy of the parts of a nucleus and the rest energy of the nucleus is the total **binding energy** of the nucleus.

As we saw in Section 38-3, atomic and nuclear masses are often given in unified mass units (u), which are defined as one-twelfth of the mass of the neutral carbon-12 atom. The rest energy of one unified mass unit is

$$(1\ \text{u})c^2 = 931.5\ \text{MeV} \qquad 40\text{-}2$$

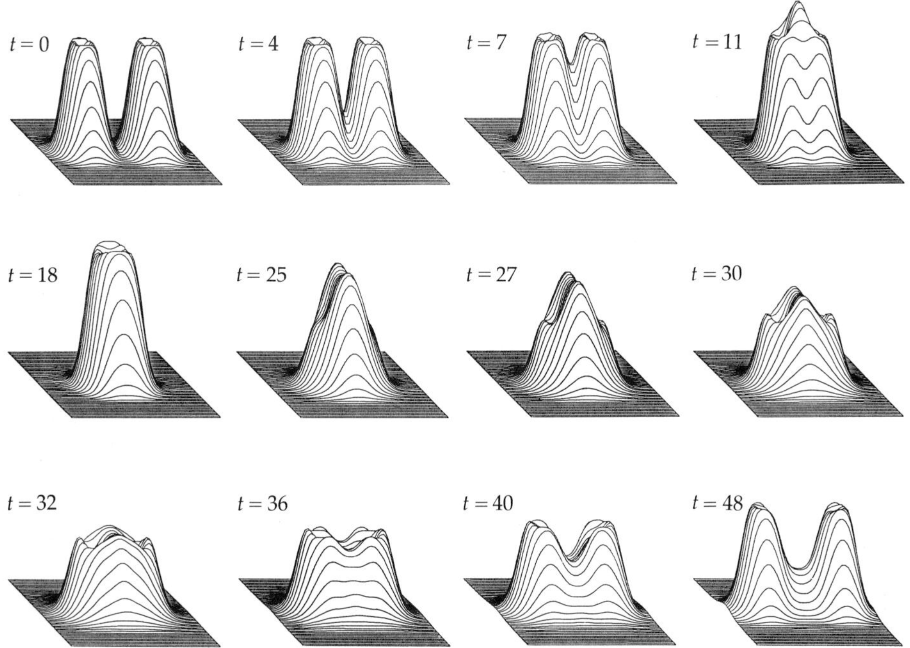

A computer simulation of a collision between two carbon-12 nuclei. The incoming carbon nucleus is considered to have an energy of 768 MeV, distributed equally among the 12 nucleons. The base of the plot corresponds to an area 18 fm on a side. The vertical dimension represents the density of protons and neutrons on a scale where the original height of each nucleus corresponds to 1.5×10^{11} nucleons/fm^3. Increments of time t are in units of 3.3×10^{-24} s. At certain energies, colliding nuclei bind together for a brief period before separating. Afterward, some of the initial kinetic energy appears in the form of internal excitations of the nuclei.

Consider ^{4}He, for example, which consists of two protons and two neutrons. The mass of an atom can be accurately measured using a mass spectrometer. The mass of the ^{4}He atom is 4.002603 u. This includes the masses of the two electrons in the atom. The mass of the ^{1}H atom is 1.007825 u, and that of the neutron is 1.008665 u. The sum of the masses of two ^{1}H atoms plus two neutrons is 2(1.007825 u) + 2(1.008665 u) = 4.032980 u, which is greater than the mass of the ^{4}He atom by 0.030377 u. Note that by using the masses of two ^{1}H atoms rather than two protons, the masses of the electrons in the atoms are accounted for. We do this because it is atomic masses that are measured directly and listed in mass tables.

We can find the binding energy of the ^{4}He nucleus from the mass difference of 0.030377 u by using the conversion factor 1 uc^2 = 931.5 MeV from Equation 40-2:

$$(0.030377\ \text{u})c^2 = 0.030377\ \text{u}c^2 \times \frac{931.5\ \text{MeV}}{1\ \text{u}c^2} = 28.30\ \text{MeV}$$

The total binding energy of ^{4}He is thus 28.3 MeV. In general, the binding energy E_b of the nucleus of an atom of atomic mass M_A containing Z protons and N neutrons is found by calculating the difference between the mass of the parts and the mass of the nucleus and then multiplying by c^2:

Total nuclear binding energy

$$E_b = (ZM_H + Nm_n - M_A)c^2 \qquad 40\text{-}3$$

where M_H is the mass of the ^{1}H atom and m_n is the mass of the neutron. Note again that this formula is written in terms of atomic masses rather than nuclear masses. The mass of the Z electrons in the term ZM_H is canceled by the mass of the Z electrons in the term M_A. The atomic masses of the neutron and some selected isotopes are listed in Table 40-1.

Table 40-1 **Atomic Masses of the Neutron and Selected Isotopes**

Element	Symbol	Z	Atomic mass, u
Neutron	n	0	1.008 665
Hydrogen	^{1}H	1	1.007 825
(deuterium)	^{2}H or D	1	2.014 102
(tritium)	^{3}H or T	1	3.016 050
Helium	^{3}He	2	3.016 030
	^{4}He	2	4.002 603
Lithium	^{6}Li	3	6.015 125
Boron	^{10}B	5	10.012 939
Carbon	^{12}C	6	12.000 000
	^{14}C	6	14.003 242
Oxygen	^{16}O	8	15.994 915
Sodium	^{23}Na	11	22.989 771
Potassium	^{39}K	19	38.963 710
Iron	^{56}Fe	26	55.939 395
Copper	^{63}Cu	29	62.929 592
Silver	^{107}Ag	47	106.905 094
Gold	^{197}Au	79	196.966 541
Lead	^{208}Pb	82	207.976 650
Polonium	^{212}Po	84	211.989 629
Radon	^{222}Rn	86	222.017 531
Radium	^{226}Ra	88	226.025 360
Uranium	^{238}U	92	238.048 608
Plutonium	^{242}Pu	94	242.058 725

Example 40-1

Find the binding energy of the last neutron in ^{4}He.

From Table 40-1, the rest mass of ^{4}He is 4.00260 u and that of ^{3}He is 3.01603 u. The rest mass of ^{3}He plus that of the neutron is 3.01603 u + 1.00866 u = 4.02469 u. This is greater than the rest mass of ^{4}He by 4.02469 u − 4.00260 u = 0.02209 u. The binding energy of the last neutron is thus

$$(\Delta m)c^2 = (0.02209 \text{ u})c^2 \times \frac{931.5 \text{ MeV}}{1 \text{ u}c^2} = 20.58 \text{ MeV}$$

Once the atomic mass of a nuclide has been determined, its binding energy can be computed from Equation 40-3. Figure 40-3 shows the binding energy per nucleon E_b/A versus A. The mean value of E_b/A is about 8.3 MeV. The flatness of this curve for $A > 50$ shows that E_b is approximately proportional to the number of nucleons A in the nucleus. This indicates that there is saturation of nuclear bonds in the nucleus. That is, each nucleon bonds to only a certain maximum number of other nucleons independent of the total number of nucleons in the nucleus. If there were no saturation and each nucleon bonded to every other nucleon, there would be $A - 1$ bonds for each nucleon and a total of $A(A - 1)$ bonds altogether. The total binding energy, which is a measure of the energy needed to break all the bonds between nucleons, would then be proportional to $A(A - 1)$, and E_b/A would not be approximately constant. Figure 40-3 indicates that, above a certain value of A, there is a fixed number of bonds per nucleon, as would be the case if each nucleon were attracted only to its nearest neighbors. Such a situation also leads to a constant nuclear density, which is consistent with measurements of nuclear radii. The steep rise in the curve for low values of A is due to the increase in the number of nearest neighbors and therefore to the increase in the number of bonds per nucleon. The gradual decrease in E_b/A at high values of A is due to the Coulomb repulsion between the protons, which increases as Z^2 and decreases the binding energy. For values of A greater than about 260, the Coulomb repulsion becomes so great that the nucleus is unstable and undergoes spontaneous fission.

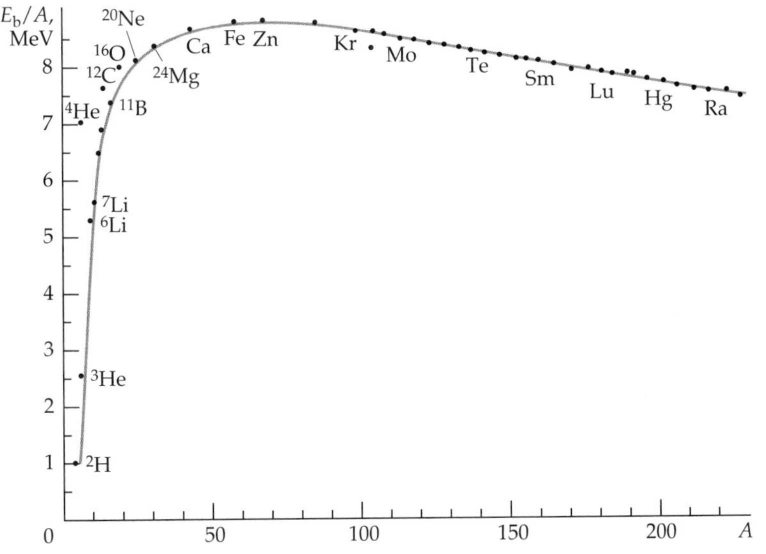

Figure 40-3 The binding energy per nucleon versus the mass number A. For nuclei with values of A greater than about 50, the curve is approximately flat, indicating that the total binding energy is approximately proportional to A.

Spin and Magnetic Moment

The spin quantum number of the neutron and the proton is $\frac{1}{2}$. The angular momentum of the nucleus is a combination of the spin angular momenta of the nucleons plus any orbital angular momenta due to the motion of the nucleons. The resultant angular momentum is called the **nuclear spin I**. The spin of all nuclei with an even number of protons and an even number of neutrons is zero. Evidently, the nucleons couple together in pairs in such a way that their angular momenta add to zero, as is often the case for electrons in atoms.

In Section 37-3, we found that the magnetic moment associated with the angular momentum of an electron in an atom was of the order of a Bohr magneton $\mu_B = e\hbar/2m_e$ and that the magnetic moment associated with electron spin was 1 μ_B in agreement with the Dirac relativistic wave equation rather than the $\frac{1}{2}$ μ_B predicted by classical physics. The magnetic moment of the nucleus is of the order of the **nuclear magneton**

$$\mu_N = \frac{e\hbar}{2m_p} = 5.05 \times 10^{-27} \text{ J/T} = 3.15 \times 10^{-8} \text{ eV/T} \qquad 40\text{-}4$$

where m_p is the mass of the proton. If the proton and neutron obeyed the Dirac relativistic wave equation as the electron does, the (z components of the) magnetic moments associated with their spins would be 1 μ_N for the proton and 0 for the neutron because it has no charge. The experimentally determined moment of the proton is

$$(\mu_z)_{\text{proton}} = +2.7928\ \mu_N \qquad 40\text{-}5$$

and that of the neutron is

$$(\mu_z)_{\text{neutron}} = -1.9135\ \mu_N \qquad 40\text{-}6$$

The negative sign for the z component of the magnetic moment of the neutron indicates that its direction is opposite that of the neutron's spin. It is interesting to note that the deviations of these moments from those predicted by the Dirac equation are of about the same magnitude—1.91 μ_N for the neutron and 1.79 μ_N for the proton. These deviations are due to the fact that the proton and neutron are not point particles like the electron but are instead composite particles made up of elementary particles called quarks, which we will discuss in Chapter 41.

Questions

1. How does the strong nuclear force differ from the electromagnetic force?
2. What property of the strong nuclear force is indicated by the fact that all nuclei have about the same density?
3. The mass of ^{12}C, which contains six protons and six neutrons, is exactly 12.000 u by the definition of the unified mass unit. Why isn't the mass of ^{16}O, which contains eight protons and eight neutrons, exactly 16.000 u?

40-2 Nuclear Magnetic Resonance

In Section 37-5, we saw that the energy levels of the atom were split in the presence of an external magnetic field (the Zeeman effect) because of the interaction of the atomic magnetic moment and the field. Since nuclei also have magnetic moments, the energy of a nucleus is also split in the presence of a magnetic field. We will consider only the simplest case, the hydrogen atom for which the nucleus is a single proton.

The potential energy of a magnetic moment $\boldsymbol{\mu}$ in an external magnetic field $\mathbf{B}$ is given by (Equation 27-13)

$$U = -\boldsymbol{\mu} \cdot \mathbf{B} \qquad \text{40-7}$$

The potential energy is lowest when the magnetic moment is aligned with the field and highest when it is in the opposite direction. Since the spin quantum number of the proton is $\frac{1}{2}$, the proton's magnetic moment has two possible orientations in an external magnetic field: parallel to the field (spin up) or antiparallel to the field (spin down). The difference in energy of these two orientations (Figure 40-4) is

$$\Delta E = 2(\mu_z)_p B \qquad \text{40-8}$$

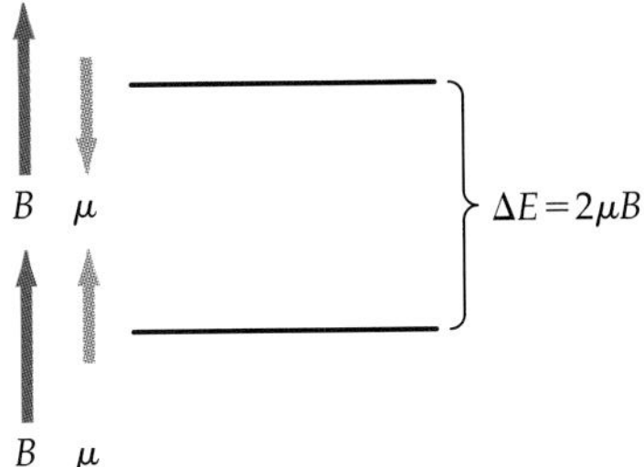

Figure 40-4 A proton has two energy states in the presence of a magnetic field corresponding to whether the magnetic moment of the proton is aligned parallel or antiparallel to the field.

When hydrogen atoms are irradiated with photons of energy ΔE, some of the nuclei are induced to make transitions from the lower state to the upper state by resonance absorption. These nuclei then decay back to the lower state, emitting photons of energy ΔE. The frequency of the photons absorbed and emitted is found from

$$hf = \Delta E = 2(\mu_z)_p B$$

In a magnetic field of 1 T, this energy is

$$\Delta E = 2(\mu_z)_p B$$

$$= 2(2.79\ \mu_N)\left(\frac{3.15 \times 10^{-8}\ \text{eV/T}}{1\ \mu_N}\right)(1\ \text{T})$$

$$= 1.76 \times 10^{-7}\ \text{eV}$$

and the frequency of the photons is

$$f = \frac{\Delta E}{h} = \frac{1.76 \times 10^{-7}\ \text{eV}}{4.14 \times 10^{-15}\ \text{eV·s}}$$

$$= 4.25 \times 10^{7}\ \text{Hz} = 42.5\ \text{MHz}$$

This frequency is in the radio band of the electromagnetic spectrum, and the radiation is called **RF** (radiofrequency) **radiation.** The measurement of this resonance frequency for free protons can be used to determine the magnetic moment of the proton.

When a hydrogen atom is in a molecule, the magnetic field at the nucleus is the sum of the external magnetic field and the local magnetic field due to the atoms and nuclei of the surrounding material. Since the resonance frequency is proportional to the total magnetic field seen by the proton, a measurement of this frequency can give information about the internal magnetic field seen in the molecule. This is called **nuclear magnetic resonance.**

Nuclear magnetic resonance is used as an alternative to x rays or ultrasound for imaging. A patient can be placed in a magnetic field that is constant in time but not in space. When the patient is irradiated by a broadband RF source, the resonance frequency of the absorbed and emitted RF photons is then dependent on the value of the magnetic field, which can be related to specific positions in the body of the patient. Since the energy of the photons is much less than the energy of molecular bonds and the intensity used is low enough so that it produces negligible heating, the RF photons produce little (if any) biological damage.

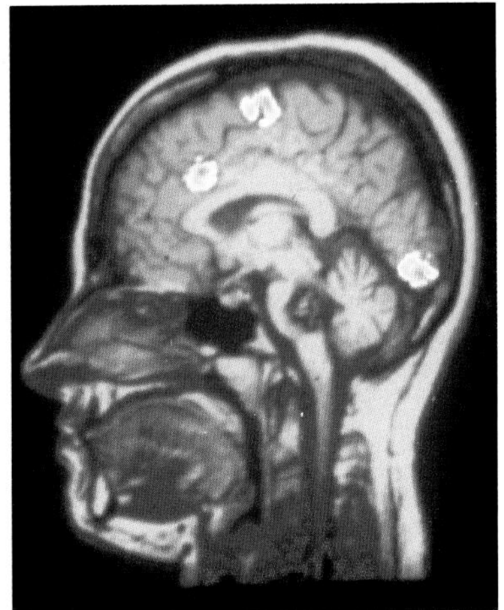

Positron emission tomography images three areas (small pink dots surrounded by white) in the left brain that are active during language tasks. The area in the back is active during reading, the middle area during speech, and the front area while thinking about the meaning of a word. A PET image, which is here superposed over an NMR scan (blue and purple), is obtained by injecting a radioactive substance, for instance an isotope of oxygen, into the body. As the tracer circulates, it emits positrons. The annihilation of the positrons with electrons produces pairs of γ rays from which the location of the reaction can be inferred. An increase in the reaction rate can be interpreted as an increase in blood flow, corresponding to an increase in physiological activity.

40-3 Radioactivity

Nuclei that are not stable are radioactive; that is, they decay into other nuclei by the emission of radiation. The term radiation refers to the emission of particles, such as electrons, neutrons, or alpha particles, as well as to electromagnetic radiation. The three kinds of radioactivity—**alpha** (α) **decay, beta** (β) **decay,** and **gamma** (γ) **decay**—were named before it was known that alpha particles are ^{4}He nuclei, beta particles are electrons (e^-) or positrons (e^+), and gamma rays are photons.

In 1900, Rutherford discovered that the rate of emission of radioactive particles from a substance is not constant over time but decreases exponentially. *This exponential time dependence is characteristic of all radioactivity and indicates that radioactive decay is a statistical process.* Because each nucleus is well shielded from others by the atomic electrons, pressure and temperature changes have little or no effect on the rate of radioactive decay or other nuclear properties.

Let N be the number of radioactive nuclei at some time t. For a statistical decay process in which the decay of any individual nucleus is a random event, we expect the number of nuclei that decay in some time interval dt to be proportional to N and to dt. Because of these decays, the number N will decrease. The change in N is given by

$$dN = -\lambda N\, dt \tag{40-9}$$

where λ is a constant of proportionality called the **decay constant.** The rate of change of N, dN/dt, is proportional to N. This is characteristic of exponential decay. To solve Equation 40-9 for N, we first divide each side by N, thus separating the variables N and t:

$$\frac{dN}{N} = -\lambda\, dt$$

Integrating, we obtain

$$\ln N = -\lambda t + C \tag{40-10}$$

where C is some constant of integration. Taking the exponential of each side of Equation 40-10, we obtain

$$N = e^{-\lambda t + C} = e^C e^{-\lambda t}$$

or

$$N = N_0 e^{-\lambda t} \tag{40-11}$$

where $N_0 = e^C$ is the number of nuclei at $t = 0$. The number of radioactive decays per second is called the decay rate R:

Decay rate

$$R = -\frac{dN}{dt} = \lambda N = \lambda N_0 e^{-\lambda t} = R_0 e^{-\lambda t} \tag{40-12}$$

where

$$R_0 = \lambda N_0 \tag{40-13}$$

is the rate of decay at time $t = 0$. The decrease in the decay rate R is the quantity that is determined experimentally.

The average or **mean lifetime** τ is the reciprocal of the decay constant:

$$\tau = \frac{1}{\lambda} \qquad 40\text{-}14$$

(See Problem 54.) The mean lifetime is analogous to the time constant in the exponential decrease in the charge on a capacitor in an RC circuit that we discussed in Section 23-2. After a time equal to the mean lifetime the number of radioactive nuclei and the decay rate have each decreased to 37 percent of their original values. The **half-life** $t_{1/2}$ is defined as the time it takes for the number of nuclei and the decay rate to decrease by half. Setting $t = t_{1/2}$ and $N = N_0/2$ in Equation 40-11, we obtain

$$\frac{N_0}{2} = N_0 e^{-\lambda t_{1/2}} \qquad 40\text{-}15$$

Solving for $t_{1/2}$, we obtain

$$t_{1/2} = \frac{\ln 2}{\lambda} = \frac{0.693}{\lambda} = 0.693\tau \qquad 40\text{-}16$$

Exercise

Show the mathematical steps needed to obtain Equation 40-16 from Equation 40-15.

Figure 40-5 shows a plot of N versus t. If we multiply the numbers on the N axis by λ, this graph becomes a plot of R versus t. After each time interval of one half-life, the number of nuclei left and the decay rate have decreased to half of their previous values. For example, if the decay rate is R_0 initially, it will be $\frac{1}{2}R_0$ after one half-life, $(\frac{1}{2})(\frac{1}{2})R_0$ after two half-lives, and so forth. After n half-lives, the decay rate will be

$$R = \left(\frac{1}{2}\right)^n R_0 \qquad 40\text{-}17$$

The half-lives of radioactive nuclei vary from very small times (less than 1 μs) to very large times (up to 10^{16} y).

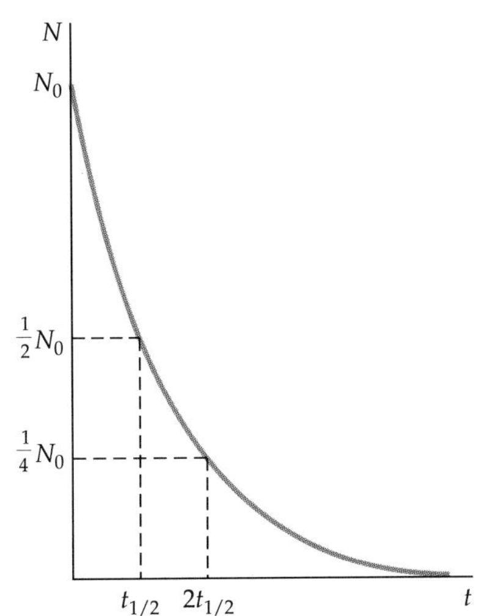

Figure 40-5 Exponential radioactive decay. After each half-life $t_{1/2}$, the number of nuclei remaining has decreased by one-half. The decay rate $R = \lambda N$ has the same time dependence.

Example 40-2

A radioactive source has a half-life of 1 min. At time $t = 0$, it is placed near a detector, and the counting rate (the number of decay particles detected per unit time) is observed to be 2000 counts/s. Find the counting rate at times $t = 1$, 2, 3, and 10 min.

Since the half-life is 1 min, the counting rate will be half as great at $t = 1$ min as at $t = 0$, so $r_1 = 1000$ counts/s at 1 min, $r_2 = 500$ counts/s at $t = 2$ min, and $r_3 = 250$ counts/s at 3 min. At $t = 10$ min $= 10t_{1/2}$, the rate will be $r_{10} = (\frac{1}{2})^{10}(2000) = 1.95$ counts/s ≈ 2 counts/s.

Example 40-3

If the detection efficiency in Example 40-2 is 20 percent, how many radioactive nuclei are there at time $t = 0$? At time $t = 1$ min? How many nuclei decay in the first minute?

The detection efficiency depends on the distance from the source to the detector and the probability that a radioactive decay particle entering the detector will produce a count. If the source is smaller than the detector and is placed very close to the detector, about half the emitted particles will enter the detector. If the counting rate at $t = 0$ is 2000 counts/s and the efficiency is 20 percent, the decay rate at $t = 0$ must be 10,000 s^{-1}. The number of radioactive nuclei can be found from Equation 40-12:

$$R = \lambda N$$

The decay constant is related to the half-life by Equation 40-16:

$$\lambda = \frac{0.693}{t_{1/2}} = \frac{0.693}{1 \text{ min}}$$

The number of nuclei at $t = 0$ is therefore

$$N = \frac{R}{\lambda} = \frac{10{,}000 \text{ s}^{-1}}{0.693 \text{ min}^{-1}} \times \frac{60 \text{ s}}{1 \text{ min}} = 8.66 \times 10^5$$

At time $t = 1$ min $= t_{1/2}$, there are half as many nuclei as at $t = 0$, so $N_1 = \frac{1}{2}(8.66 \times 10^5) = 4.33 \times 10^5$. The number of nuclei that decay in the first minute is therefore 4.33×10^5.

The SI unit of radioactive decay is the **becquerel** (Bq), which is defined as one decay per second:

$$1 \text{ Bq} = 1 \text{ decay/s} \qquad 40\text{-}18$$

A historical unit that applies to all types of radioactivity is the **curie** (Ci), which is defined as

$$1 \text{ Ci} = 3.7 \times 10^{10} \text{ decays/s} = 3.7 \times 10^{10} \text{ Bq} \qquad 40\text{-}19$$

The curie is the rate at which radiation is emitted by 1 g of radium. Since this is a very large unit, the millicurie (mCi) or microcurie (μCi) is often used.

Beta Decay

Beta decay occurs in nuclei that have too many or too few neutrons for stability. The energy released in β decay can be determined by computing the difference between the rest mass of the original nucleus and that of the decay products. In β decay, A remains the same while Z either increases by 1 (β^- decay) or decreases by 1 (β^+ decay).*

The simplest example of β decay is the decay of the free neutron into a proton plus an electron. (The half-life of a free neutron is about 10.8 min.) The energy of decay is 0.782 MeV, which is the difference between the rest energy of the neutron and that of the proton plus electron. More generally, in β^- decay, a nucleus of mass number A and atomic number Z decays into a nucleus, referred to as the **daughter nucleus,** of mass number A and atomic number $Z' = Z + 1$ with the emission of an electron. If the decay energy is shared by the daughter nucleus and the emitted electron, the energy of the electron is uniquely determined by the conservation of energy and momentum. Experimentally, however, the energies of the electrons emitted in β^- decay are observed to vary from zero to the maximum energy available. A typical energy spectrum for these electrons is shown in Figure 40-6.

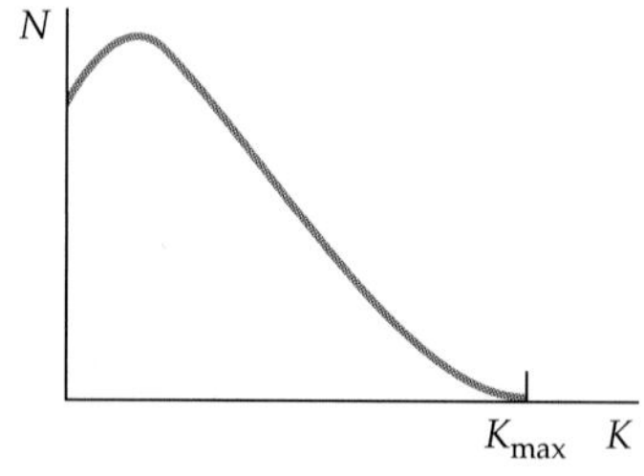

Figure 40-6 Number of electrons emitted in β^- decay versus kinetic energy. The fact that all the electrons do not have the same energy K_{max} suggests that another particle that shares the energy available for decay is emitted.

To explain the apparent nonconservation of energy in beta decay, W. Pauli in 1930 suggested that a third particle, which he called the **neutrino,** is also emitted. The mass of the neutrino was originally assumed to be

*The notations β^- and e^- are used interchangeably for the electron. Similarly, β^+ and e^+ are used interchangeably for the positron.

zero because the maximum energy of the emitted electrons is equal to the total available for the decay. In 1948, measurements of the momenta of the emitted electron and the recoiling nucleus showed that the neutrino was also needed for the conservation of linear momentum in β decay. The neutrino was first observed experimentally in 1957. It is now known that there are at least three kinds of neutrinos, one (ν_e) associated with electrons, one (ν_μ) associated with muons, and one (ν_τ) not yet observed experimentally, associated with the newly discovered tau particle τ. Moreover, each neutrino has an antiparticle, written $\overline{\nu}_e$, $\overline{\nu}_\mu$, and $\overline{\nu}_\tau$. It is the electron antineutrino that is emitted in the decay of a neutron, which is written

$$n \rightarrow p + \beta^- + \overline{\nu}_e \qquad 40\text{-}20$$

The mass of the electron neutrino or electron antineutrino is known to be less than 4×10^{-5} times the mass of the electron.

In β^+ decay, a proton changes into a neutron with the emission of a positron (and a neutrino). A free proton cannot decay by positron emission because of conservation of energy (the rest mass of the neutron plus the positron is greater than that of the proton), but because of binding-energy effects, a proton inside a nucleus can decay. A typical β^+ decay is

$$^{13}_{7}\text{N} \rightarrow {}^{13}_{6}\text{C} + \beta^+ + \nu_e \qquad 40\text{-}21$$

The electrons or positrons emitted in β decay do not exist inside the nucleus. They are created in the process of decay, just as photons are created when an atom makes a transition from a higher to a lower energy state.

An important example of β decay is that of ^{14}C, which is used in **radioactive carbon dating:**

$$^{14}\text{C} \rightarrow {}^{14}\text{N} + \beta^- + \overline{\nu}_e \qquad 40\text{-}22$$

The half-life for this decay is 5730 y. The radioactive isotope ^{14}C is produced in the upper atmosphere in nuclear reactions caused by cosmic rays. The chemical behavior of carbon atoms with ^{14}C nuclei is the same as those with ordinary ^{12}C nuclei. For example, atoms with these nuclei combine with oxygen to form CO_2 molecules. Since living organisms continually exchange CO_2 with the atmosphere, the ratio of ^{14}C to ^{12}C in a living organism is the same as the equilibrium ratio in the atmosphere, which is about 1.3×10^{-12}. After an organism dies, it no longer absorbs ^{14}C from the atmosphere, so the ratio of ^{14}C to ^{12}C continually decreases due to the radioactive decay of ^{14}C. The number of ^{14}C decays per minute per gram of carbon in a living organism can be calculated from the known half-life of ^{14}C and the number of ^{14}C nuclei in a gram of carbon. The result is that there are about 15.0 decays per minute per gram of carbon in a living organism. Using this result and the measured number of decays per minute per gram of carbon in a nonliving sample of bone, wood, or other object containing carbon, we can determine the age of the sample. For example, if the measured rate were 7.5 decays per minute per gram, the sample would be one half-life = 5730 years old.

Example 40-4

A bone containing 200 g of carbon has a β-decay rate of 400 decays/min. How old is the bone?

We first obtain a rough estimate of the age of the bone. If the bone were from a living organism, we would expect the decay rate to be (15 decays/min·g)(200 g) = 3000 decays/min. Since 400/3000 is roughly 1/8 (actually 1/7.5), the sample must be about three half-lives old, which is about 3(5730) y = 17,190 y.

To find the age of the bone more accurately, we note that after n half-lives, the decay rate will have decreased by a factor of $(\frac{1}{2})^n$. We can find n from

$$\left(\frac{1}{2}\right)^n = \frac{400}{3000}$$

or

$$2^n = \frac{3000}{400} = 7.5$$

We solve for n by taking the logarithm of each side:

$$n \ln 2 = \ln 7.5$$

$$n = \frac{\ln 7.5}{\ln 2} = 2.91$$

The age of the bone is therefore

$$t = nt_{1/2} = 2.91(5730 \text{ y}) = 16{,}700 \text{ y}$$

Gamma Decay

In gamma decay a nucleus in an excited state decays to a lower-energy state by the emission of a photon. This is the nuclear counterpart of spontaneous emission of photons by atoms and molecules. Unlike β or α decay, the radioactive nucleus remains the same nucleus as it decays via γ decay. Since the spacing of the nuclear energy levels is of the order of 1 MeV (as compared with spacing of the order of 1 eV in atoms), the wavelengths of the emitted photons are of the order of 1 pm:

$$\lambda = \frac{hc}{E} \approx \frac{1240 \text{ eV}\cdot\text{nm}}{1 \text{ MeV}} = 0.00124 \text{ nm} = 1.24 \text{ pm}$$

The emission of gamma rays normally happens very quickly and is observed only because it usually follows either α or β decay. For example, if a radioactive parent nucleus decays by β decay to an excited state of the daughter nucleus, the daughter nucleus then decays to its ground state by γ emission. The mean lifetime for γ decay is often very short. Direct measurements of mean lifetimes as short as about 10^{-11} s are possible. Measurements of mean lifetimes shorter than 10^{-11} s are difficult, but they can sometimes be made by indirect methods.

A few γ emitters have very long lifetimes, of the order of hours. Nuclear energy states that have such long lifetimes are called **metastable states.**

Alpha Decay

All very heavy nuclei ($Z > 83$) are theoretically unstable to alpha decay because the mass of the original radioactive nucleus is greater than the sum of the masses of the decay products—an α particle and the daughter nucleus. Consider the decay of ^{232}Th ($Z = 90$) into ^{228}Ra ($Z = 88$) plus an α particle. This is written as

$$^{232}\text{Th} \rightarrow {}^{228}\text{Ra} + \alpha = {}^{228}\text{Ra} + {}^{4}\text{He} \qquad 40\text{-}23$$

The mass of the ^{232}Th atom is 232.038124 u. The mass of the daughter atom ^{228}Ra is 228.031139 u. Adding to this 4.002603 u for the mass of ^{4}He, we get 232.033742 u for the total mass of the decay products. This is less than the mass of ^{232}Th by 0.004382 u, which multiplied by 931.5 MeV/c^2 gives

4.08 MeV/c^2 for the excess rest mass of ^{232}Th over that of the decay products. The isotope ^{232}Th is therefore theoretically unstable to α decay. This decay does in fact occur in nature with the emission of an α particle of kinetic energy 4.08 MeV. (The kinetic energy of the α particle is actually somewhat less than 4.08 MeV because some of the decay energy is shared by the recoiling ^{228}Ra nucleus.)

In general, when a nucleus emits an α particle, both N and Z decrease by 2 and A decreases by 4. The daughter of a radioactive nucleus is often itself radioactive and decays by either α or β decay or both. If the original nucleus has a mass number A that is 4 times an integer, the daughter nucleus and all those in the chain will also have mass numbers equal to 4 times an integer. Similarly, if the mass number of the original nucleus is $4n + 1$, where n is an integer, all the nuclei in the decay chain will have mass numbers given by $4n + 1$, with n decreasing by one at each decay. We can see, therefore, that there are four possible α-decay chains, depending on whether A equals $4n$, $4n + 1$, $4n + 2$, or $4n + 3$, where n is an integer. All but one of these decay chains are found in nature. The $4n + 1$ series is not found because its longest-lived member (other than the stable end product ^{209}Bi) is ^{237}Np, which has a half-life of only 2×10^6 y. As this is much less than the age of the earth, this series has disappeared.

Figure 40-7 shows the thorium series, for which $A = 4n$. It begins with an α decay from ^{232}Th to ^{228}Ra. The daughter nuclide of an α decay is on the left or neutron-rich side of the stability curve (the dashed line in the figure), so it often decays by β^- decay. In the thorium series, ^{228}Ra decays by β^- decay to ^{228}Ac, which in turn decays by β^- decay to ^{228}Th. There are then four α decays to ^{212}Pb, which decays by β^- decay to ^{212}Bi. The series branches at ^{212}Bi, which decays either by α decay to ^{208}Tl or by β^- decay to ^{212}Po. The branches meet at the stable lead isotope ^{208}Pb.

The energies of α particles from natural radioactive sources range from about 4 to 7 MeV, and the half-lives range from about 10^{-5} s to 10^{10} y. In general, the smaller the energy of the emitted α particle, the longer the half-life. As we discussed in Section 36-9, the enormous variation in half-lives was explained by George Gamow in 1928. He considered α decay to be

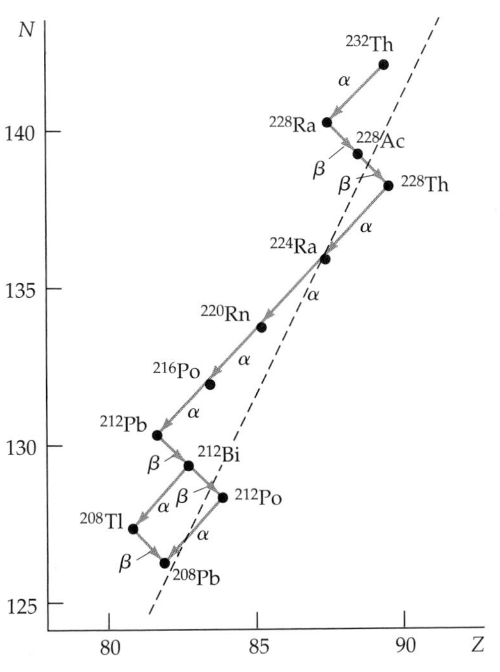

Figure 40-7 The thorium ($4n$) α-decay series. The dashed line is the curve of stability.

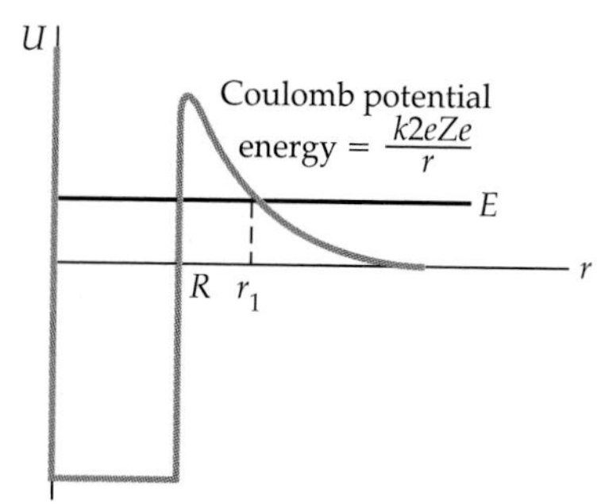

Figure 40-8 A model of the potential energy for an α particle and a nucleus. The strong attractive nuclear force that exists for values of r less than the nuclear radius R is indicated by the potential well. Outside the nucleus, the nuclear force is negligible, and the potential energy is given by Coulomb's law $U = +k2eZe/r$, where Ze is the nuclear charge and $2e$ is the charge of the α particle. The energy E is the kinetic energy of the α particle when it is far away from the nucleus. A small increase in E reduces the relative height of the barrier and also its thickness, leading to a much greater chance of penetration. An increase in the energy of the emitted α particles by a factor of 2 results in a reduction of the half-life by a factor of more than 10^{20}.

a process in which an α particle is first formed inside a nucleus and then tunnels through the Coulomb barrier (Figure 40-8). A slight increase in the energy of the α particle reduces the relative height $U - E$ of the barrier and also the thickness. Because the probability of penetration is so sensitive to the relative height and thickness of the barrier, a small increase in E leads to a large increase in the probability of barrier penetration and therefore to a shorter lifetime. Gamow was able to derive an expression for the half-life as a function of E that is in excellent agreement with experimental results.

Questions

4. Why do extreme changes in the temperature or pressure of a radioactive sample have little or no effect on the radioactivity?
5. Why is the decay series $A = 4n + 1$ not found in nature?
6. A decay by α emission is often followed by β decay. When this occurs, it is by β^- and not β^+ decay. Why?
7. The half-life of ^{14}C is much less than the age of the universe, yet ^{14}C is found in nature. Why?
8. What effect would a long-term variation in cosmic-ray activity have on the accuracy of ^{14}C dating?

40-4 Nuclear Reactions

Information about nuclei is typically obtained by bombarding them with various particles and observing the results. Although the first experiments of this type were limited by the need to use naturally occurring radiation, they produced many important discoveries. In 1932 J. D. Cockcroft and E. T. S. Walton succeeded in producing the reaction

$$p + {}^7\text{Li} \rightarrow {}^8\text{Be} \rightarrow {}^4\text{He} + {}^4\text{He}$$

using artificially accelerated protons. At about the same time, the Van de Graaff electrostatic generator was built (by R. Van de Graaff in 1931), as was the first cyclotron (by E. O. Lawrence and M. S. Livingston in 1932). Since then, enormous advances in the technology for accelerating and detecting particles have been made, and many nuclear reactions have been studied.

When a particle is incident on a nucleus, several different things can happen. The incident particle may be scattered elastically or inelastically, in which case the nucleus is left in an excited state and decays by emitting photons (or other particles); or the incident particle may be absorbed by the nucleus, and another particle or particles may be emitted.

When energy is released by a nuclear reaction, the reaction is said to be an **exothermic reaction.** The amount of energy released is called the **Q value** of the reaction. In an exothermic reaction, the total mass of the incoming particles is greater than that of the outgoing particles. The Q value equals c^2 times this mass difference. If the total mass of the incoming particles is less than that of the outgoing particles, energy is required for the reaction to take place, and the reaction is said to be an **endothermic reaction.** The Q value of an endothermic reaction is also equal to c^2 times the initial mass minus the final mass, but in this case the Q value is negative. An endothermic reaction cannot take place below a certain threshold energy. The threshold energy is usually somewhat greater than $|Q|$ because the outgoing particles must have some kinetic energy to conserve momentum.

A measure of the effective size of a nucleus for a particular nuclear reaction is the **cross section** σ. If I is the number of incident particles per unit time per unit area (the incident intensity) and R is the number of reactions per unit time per nucleus, the cross section is

$$\sigma = \frac{R}{I} \qquad 40\text{-}24$$

The cross section σ has the dimensions of area. Since nuclear cross sections are of the order of the square of the nuclear radius, a convenient unit for them is the **barn,** which is defined as

$$1 \text{ barn} = 10^{-28} \text{ m}^2 \qquad 40\text{-}25$$

The cross section for a particular reaction is a function of energy. For an endothermic reaction, it is zero for energies below the threshold energy.

Example 40-5

Find the Q value of the reaction

$$p + {}^7\text{Li} \rightarrow {}^4\text{He} + {}^4\text{He}$$

and state whether the reaction is exothermic or endothermic. The atomic mass of ^{7}Li is 7.016004 u.

Using 1.007825 u for the mass of ^{1}H and 4.002603 u for the mass of ^{4}He from Table 40-1, we have for the total mass of the original particles

$$m_i = 1.007825 \text{ u} + 7.016004 \text{ u} = 8.023829 \text{ u}$$

and for the total mass of the final particles

$$m_f = 2(4.002603 \text{ u}) = 8.005206 \text{ u}$$

Since the initial mass is greater than the final mass by

$$\Delta m = m_i - m_f = 8.023829 \text{ u} - 8.005206 \text{ u} = 0.018623 \text{ u}$$

mass is converted into energy and the reaction is exothermic. The Q value is positive and is given by

$$Q = (\Delta m)c^2 = (0.018623 \text{ u})c^2(931.5 \text{ MeV}/\text{u}c^2) = 17.35 \text{ MeV}$$

Note that we used the mass of atomic hydrogen rather than that of the proton and the atomic masses of the ^{7}Li and ^{4}He atoms rather than the masses of the individual nuclei so that the masses of the eight electrons on each side of the reaction cancel.

Reactions with Neutrons

Nuclear reactions involving neutrons are important for understanding nuclear reactors. The most likely reaction with a nucleus for a neutron of energy of more than about 1 MeV is scattering. However, even if the scattering is elastic, the neutron loses some energy to the nucleus because the nucleus recoils. If a neutron is scattered many times in a material, its energy decreases until it is of the order of the energy of thermal motion kT, where k is Boltzmann's constant and T is the absolute temperature. (At ordinary room temperatures, kT is about 0.025 eV.) The neutron is then equally likely to gain or lose energy from a nucleus when it is elastically scattered. A neutron with energy of the order of kT is called a **thermal neutron.**

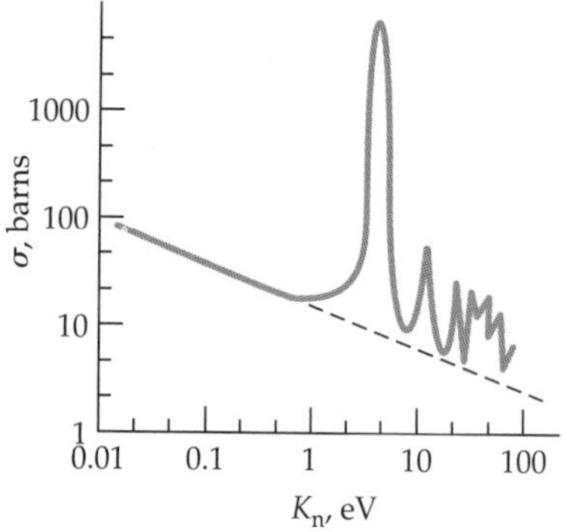

Figure 40-9 Neutron-capture cross section for silver versus energy of the neutron. The straight line indicates the $1/v$ dependence of the cross section, which is proportional to the time spent by the neutron near the silver nucleus. Superimposed on this dependence are a large resonance and several smaller resonances.

At low energies, a neutron is likely to be captured, with the emission of a γ ray from the excited nucleus. Figure 40-9 shows the neutron-capture cross section for silver as a function of the energy of the neutron. The large peak in this curve is called a **resonance.** Except for the resonance, the cross section varies fairly smoothly with energy, decreasing with increasing energy roughly as $1/v$, where v is the speed of the neutron. We can understand this energy dependence as follows: Consider a neutron moving with speed v near a nucleus of diameter $2R$. The time it takes the neutron to pass the nucleus is $2R/v$. Thus, the neutron-capture cross section is proportional to the time spent by the neutron in the vicinity of the nucleus. The dashed line in Figure 40-9 indicates this $1/v$ dependence. At the maximum of the resonance, the value of the cross section is very large ($\sigma > 5000$ barns) compared with a value of only about 10 barns just past the resonance. Many elements show similar resonances in their neutron-capture cross sections. For example, the maximum cross section for ^{113}Cd is about 57,000 barns. This material is thus very useful for shielding against low-energy neutrons.

An important nuclear reaction that involves neutrons is fission, which will be discussed in the next section.

Questions

9. What is meant by the cross section for a nuclear reaction?
10. Why is the neutron-capture cross section (excluding resonances) proportional to $1/v$?
11. What is meant by the Q value of a reaction?

(*a*)

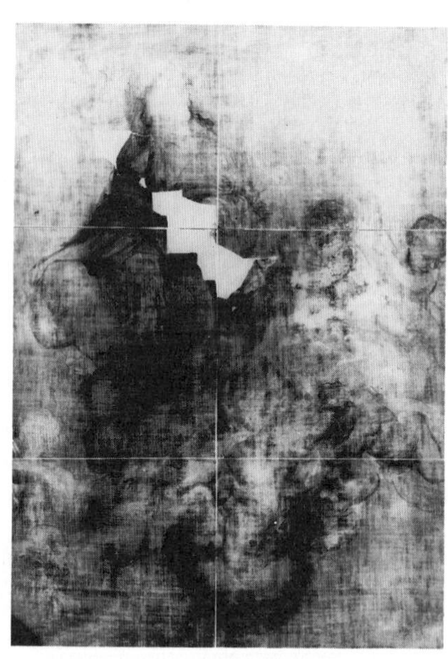

(*b*)

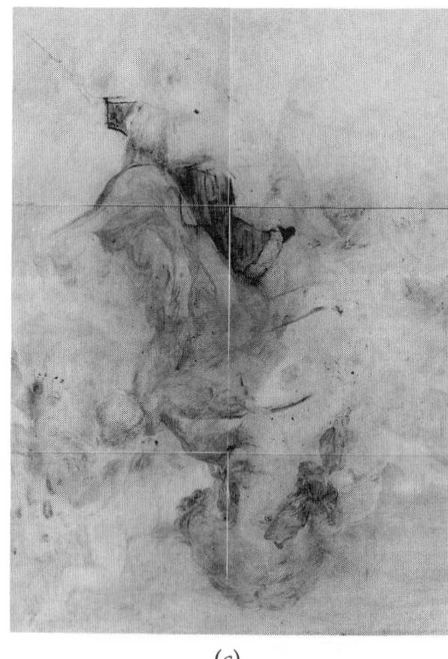

(*c*)

Hidden layers in paintings are analyzed by bombarding the painting with neutrons and observing the radiative emissions from nuclei that have captured a neutron. Different elements used in the painting have different half-lives. (*a*) Van Dyck's painting "Saint Rosalie Interceding for the Plague-Stricken of Palermo." The black and white images in (*b*) and (*c*) were formed using a special film sensitive to electrons emitted by the radioactively decaying elements. Image (*b*), taken a few hours after the neutron irradiation, reveals the presence of manganese, found in umber, a dark earth-pigment used for the painting's base layer. (Blank areas show where modern repairs, free of manganese, have been made.) The image in (*c*) was taken four days later, after the umber emissions had died away and when phosphorus, found in charcoal and boneblack, was the main radiating element. Upside down, is revealed a sketch of Van Dyck himself. The self-portrait, executed in charcoal, had been overpainted by the artist.

40-5 Fission, Fusion, and Nuclear Reactors

Figure 40-10 shows a plot of the nuclear mass difference per nucleon $(M - Zm_p - Nm_n)/A$ in units of MeV/c^2 versus A. This is just the negative of the binding-energy curve shown in Figure 40-3. From Figure 40-10, we can see that the rest mass per nucleon for both very heavy ($A \approx 200$) and very light ($A \lesssim 20$) nuclides is more than that for nuclides of intermediate mass. Thus, energy is released when a very heavy nucleus, such as ^{235}U, breaks up into two lighter nuclei—a process called **fission**—or when two very light nuclei, such as ^{2}H and ^{3}H, fuse together to form a nucleus of greater mass—a process called **fusion.**

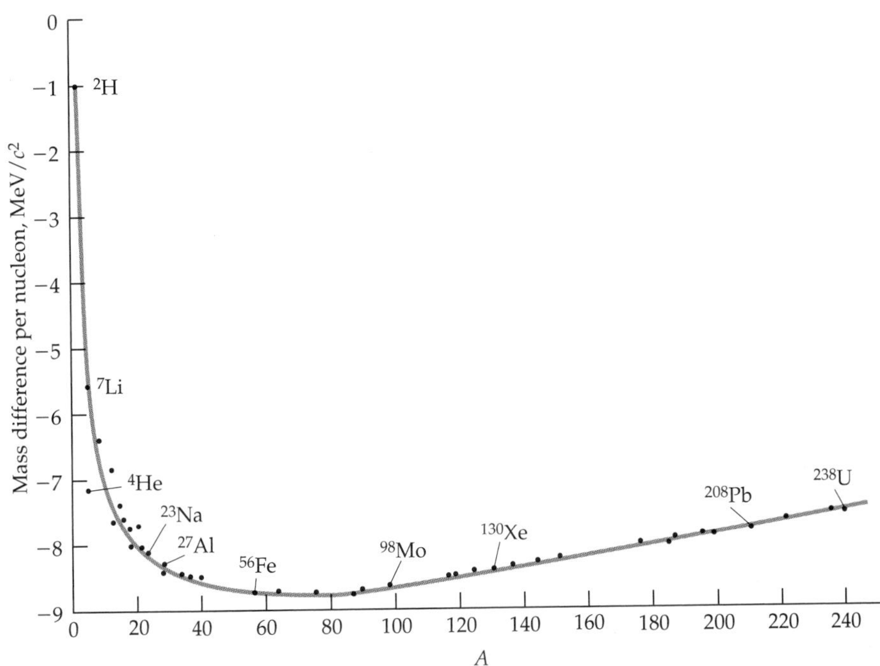

Figure 40-10 Plot of mass difference per nucleon $(M - Zm_p - Nm_n)/A$ in units of MeV/c^2 versus A. The rest mass per nucleon is less for intermediate mass nuclei than for either very light or very heavy nuclei.

The application of both fission and fusion to the development of nuclear weapons has had a profound effect on our lives during the past 45 years. The peaceful application of these reactions to the development of energy resources may have an even greater effect in the future. In this section, we will look at some of the features of fission and fusion that are important for their application in reactors to generate power.

Fission

Very heavy nuclei ($Z > 92$) are subject to spontaneous fission. They break apart into two nuclei even if left to themselves with no outside disturbance. We can understand this by considering the analogy of a charged liquid drop. If the drop is not too large, surface tension can overcome the repulsive forces of the charges and hold the drop together. There is, however, a certain maximum size beyond which the drop will be unstable and will spontaneously break apart. Spontaneous fission puts an upper limit on the size of a nucleus and therefore on the number of elements that are possible.

Some heavy nuclei, uranium and plutonium in particular, can be induced to fission by the capture of a neutron. In the fission of ^{235}U, for example, the uranium nucleus is excited by the capture of a neutron, causing it to split into two nuclei and emit several neutrons. The Coulomb force of repulsion drives the fission fragments apart, with the energy eventually showing up as thermal energy. Consider, for example, the fission of a nucleus of mass number $A = 200$ into two nuclei of mass number $A = 100$. Since the rest energy for $A = 200$ is about 1 MeV per nucleon greater than that for $A = 100$, about 200 MeV per nucleus is released in such a fission. This is a large amount of energy. By contrast, in the chemical reaction of combustion, only about 4 eV of energy is released per molecule of oxygen consumed.

Example 40-6

Calculate the total energy in kilowatt-hours released in the fission of 1 g of ^{235}U, assuming that 200 MeV is released per fission.

Since 1 mol of ^{235}U has a mass of 235 g and contains $N_A = 6.02 \times 10^{23}$ nuclei, the number of ^{235}U nuclei in 1 g is

$$N = \frac{6.02 \times 10^{23}\ \text{nuclei/mol}}{235\ \text{g/mol}} = 2.56 \times 10^{21}\ \text{nuclei/g}$$

The energy released per gram is then

$$\frac{200\ \text{MeV}}{\text{nucleus}} \times \frac{2.56 \times 10^{21}\ \text{nuclei}}{1\ \text{g}} \times \frac{1.6 \times 10^{-19}\ \text{J}}{1\ \text{eV}} \times \frac{1\ \text{h}}{3600\ \text{s}} \times \frac{1\ \text{kW}}{1000\ \text{J/s}}$$

$$= 2.28 \times 10^4\ \text{kW·h/g}$$

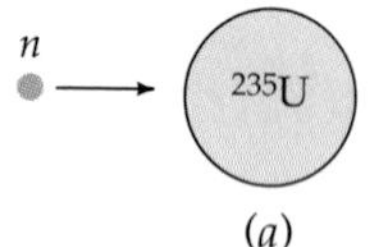

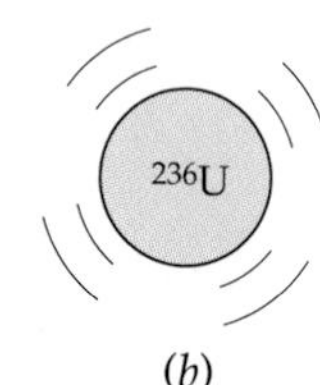

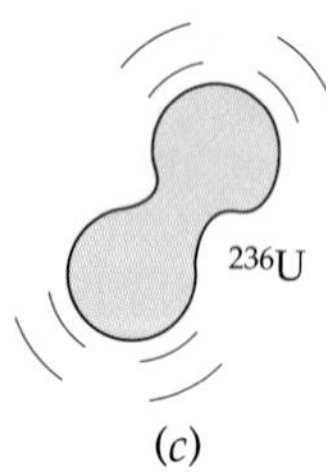

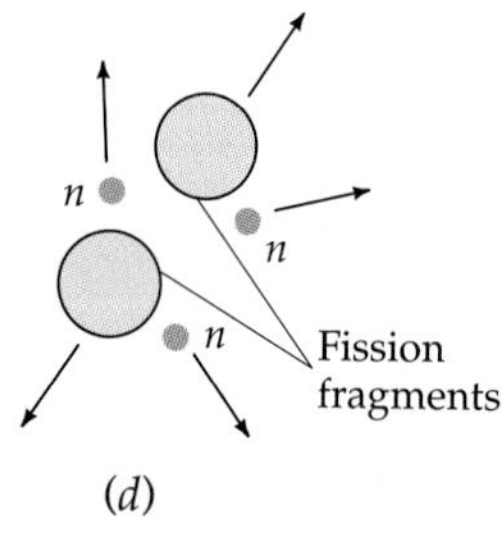

Figure 40-11 Schematic illustration of nuclear fission. (*a*) The absorption of a neutron by ^{235}U leads to (*b*) ^{236}U in an excited state. In (*c*), the oscillation of ^{236}U has become unstable. (*d*) The nucleus splits apart into two nuclei of medium mass and emits several neutrons that can produce fission in other nuclei.

The fission of uranium was discovered in 1939 by Hahn and Strassmann, who found, by careful chemical analysis, that medium-mass elements (such as barium and lanthanum) were produced in the bombardment of uranium with neutrons. The discovery that several neutrons are emitted in the fission process led to speculation concerning the possibility of using these neutrons to cause further fissions, thereby producing a chain reaction. When ^{235}U captures a neutron, the resulting ^{236}U nucleus emits gamma rays as it deexcites to the ground state about 15 percent of the time and undergoes fission about 85 percent of the time. The fission process is somewhat analogous to the oscillation of a liquid drop, as shown in Figure 40-11. If the oscillations are violent enough, the drop splits in two. Using the liquid-drop model, Bohr and Wheeler calculated the critical energy E_c needed by the ^{236}U nucleus to undergo fission. (^{236}U is the nucleus formed momentarily by the capture of a neutron by ^{235}U.) For this nucleus, the critical energy is 5.3 MeV, which is less than the 6.4 MeV of excitation energy produced when ^{235}U captures a neutron. The capture of a neutron by ^{235}U therefore produces an excited state of the ^{236}U nucleus that has more than enough energy to break apart. On the other hand, the critical energy for fission of the ^{239}U nucleus is 5.9 MeV. The capture of a neutron by a ^{238}U nucleus produces an excitation energy of only 5.2 MeV. Therefore, when a neutron is captured by ^{238}U to form ^{239}U, the excitation energy is not great enough for fission to occur. In this case, the excited ^{239}U nucleus deexcites by γ or α emission.

A fissioning nucleus can break into two medium-mass fragments in many different ways, as shown in Figure 40-12. Depending on the particular

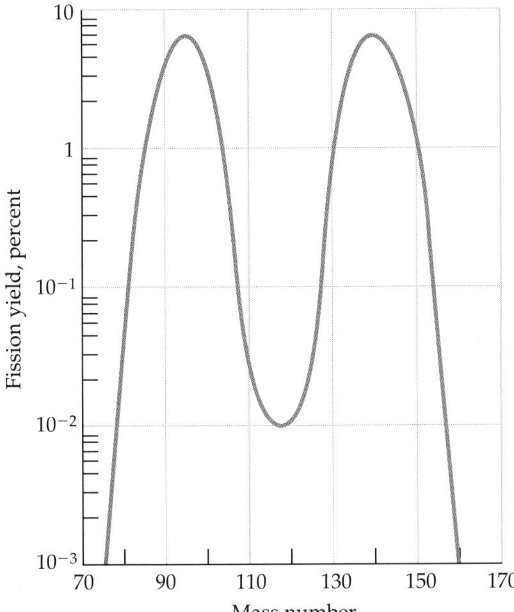

Figure 40-12 Distribution of the possible fission fragments of ^{235}U. The splitting of ^{235}U into two fragments of unequal mass is more likely than its splitting into fragments of equal mass.

reaction, 1, 2, or 3 neutrons may be emitted. The average number of neutrons emitted in the fission of ^{235}U is about 2.5. A typical fission reaction is

$$n + {}^{235}U \rightarrow {}^{141}Ba + {}^{92}Kr + 3n$$

Nuclear Fission Reactors

To sustain a chain reaction in a fission reactor, one of the neutrons (on the average) emitted in the fission of ^{235}U must be captured by another ^{235}U nucleus and cause it to fission. The **reproduction constant** k of a reactor is defined as the average number of neutrons from each fission that cause a subsequent fission. The maximum possible value of k is 2.5, but it is normally less than this for two important reasons: (1) Some of the neutrons may escape from the region containing fissionable nuclei, and (2) some of the neutrons may be captured by nonfissioning nuclei in the reactor. If k is exactly 1, the reaction will be self-sustaining. If it is less than 1, the reaction will die out. If k is significantly greater than 1, the reaction rate will increase rapidly and "run away." In the design of nuclear bombs, such a runaway reaction is desired. In power reactors, the value of k must be kept very nearly equal to 1.

Since the neutrons emitted in fission have energies of the order of 1 MeV, whereas the cross section for neutron capture leading to fission in ^{235}U is largest at small energies, the chain reaction can be sustained only if the neutrons are slowed down before they escape from the reactor. At high energies (1 to 2 MeV), neutrons lose energy rapidly by inelastic scattering from ^{238}U, the principal constituent of natural uranium. (Natural uranium contains 99.3 percent ^{238}U and only 0.7 percent fissionable ^{235}U.) Once the neutron energy is below the excitation energies of the nuclei in the reactor (about 1 MeV), the main process of energy loss is by elastic scattering, in which a fast neutron collides with a nucleus at rest and transfers some of its kinetic energy to that nucleus. Such energy transfers are efficient only if the masses of the two bodies are comparable. A neutron will not transfer much energy in an elastic collision with a heavy uranium nucleus. Such a collision is like one between a marble and a billiard ball. The marble will be deflected by the much more massive billiard ball, but very little of its kinetic energy will be transferred to the billiard ball. A **moderator** consisting of material such as water or carbon that contains light nuclei is therefore placed around the fissionable material in the core of the reactor to slow down the neutrons. The neutrons are slowed down by elastic collisions with the nuclei of the moderator until they are in thermal equilibrium with the moderator. Because of the relatively large neutron-capture cross section of the hydrogen nucleus in water, reactors using ordinary water as a moderator cannot easily achieve $k \approx 1$ unless they use enriched uranium, in which the ^{235}U content has been increased from 0.7 to between 1 and 4 percent. Natural uranium can be used if heavy water (D_2O) is used instead of ordinary (light) water (H_2O) as the moderator. Although heavy water is expensive, most Canadian reactors use it for a moderator to avoid the cost of constructing uranium-enrichment facilities.

Figure 40-13 shows some of the features of a pressurized-water reactor commonly used in the United States to generate electricity. Fission in the core heats the water in the primary loop, which is closed, to a high temperature. This water, which also serves as the moderator, is under high pressure to prevent it from boiling. The hot water is pumped to a heat exchanger, where it heats the water in the secondary loop and converts it to steam, which is then used to drive the turbines that produce electrical power. Note that the water in the secondary loop is isolated from that in the primary loop to prevent its contamination by the radioactive nuclei in the reactor core.

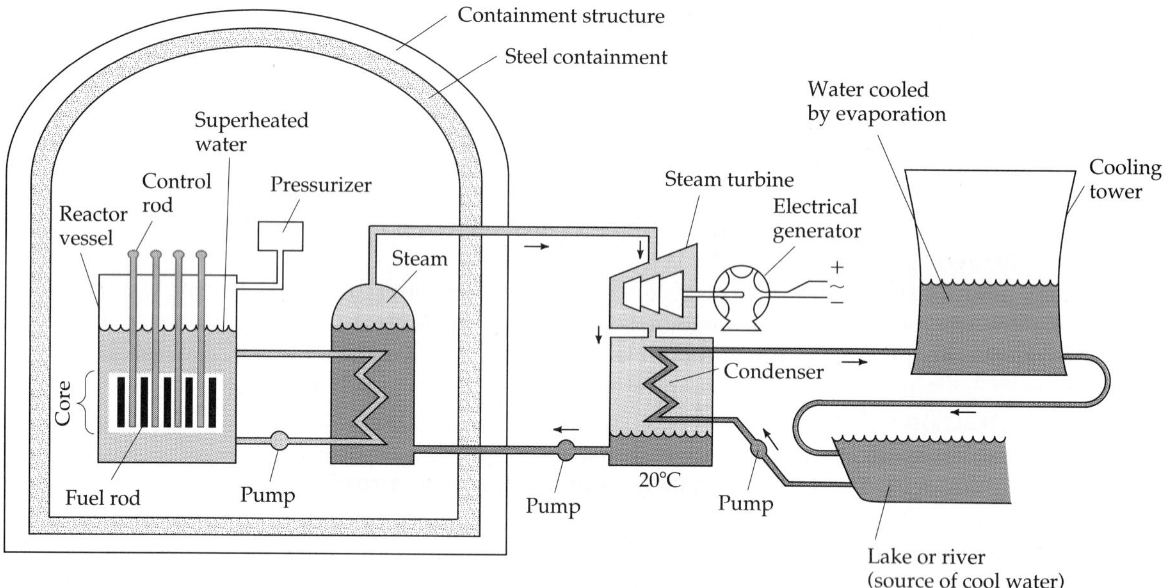

Figure 40-13 Simplified drawing of a pressurized-water reactor. The water in contact with the reactor core serves as both the moderator and the heat-transfer material. It is isolated from the water used to produce the steam that drives the turbines. Many features, such as the back-up cooling mechanisms, are not shown here.

The ability to control the reproduction factor k precisely is important if a power reactor is to be operated safely. Both natural negative-feedback mechanisms and mechanical methods of control are used. If k is greater than 1 and the reaction rate increases, the temperature of the reactor increases. If water is used as a moderator, its density decreases with increasing temperature and it becomes a less effective moderator. A second important control method is the use of control rods made of a material, such as cadmium, that has a very large neutron-capture cross section. When the reactor is started up, the control rods are inserted so that k is less than 1. As the rods are gradually withdrawn from the reactor, fewer neutrons are captured by them and k increases to 1. If k becomes greater than 1, the rods are inserted again.

Mechanical control of the reaction rate of a nuclear reactor using control rods is possible only because some of the neutrons emitted in the fission process are **delayed neutrons.** The time needed to slow down a neutron from 1 or 2 MeV to the thermal-energy level is only of the order of a millisecond. If all the neutrons emitted in fission were prompt neutrons, that is, emitted immediately in the fission process, mechanical control would not be possible because the reactor would run away before the rods could be inserted. However, about 0.65 percent of the neutrons emitted are delayed by an average time of about 14 s. These neutrons are emitted not in the fission process itself but in the decay of the fission fragments. The effect of the delayed neutrons can be seen in the following examples.

Example 40-7

If the average time between fission generations (the time it takes for a neutron emitted in one fission to cause another) is 1 ms = 0.001 s and the reproduction constant is 1.001, how long will it take for the reaction rate to double?

If $k = 1.001$, the reaction rate after N generations is 1.001^N. Setting this rate equal to 2 and solving for N, we obtain

$$(1.001)^N = 2$$

$$N \ln 1.001 = \ln 2$$

$$N = \frac{\ln 2}{\ln 1.001} = 693 \approx 700$$

It thus takes about 700 generations for the reaction rate to double. The time for 700 generations is 700(0.001 s) = 0.70 s. This is not enough time for mechanical control by the insertion of control rods.

Example 40-8

Assuming that 0.65 percent of the neutrons emitted are delayed by 14 s, find the average generation time and the doubling time if $k = 1.001$.

Since 99.35 percent of the generation times are 0.001 s and 0.65 percent are 14 s, the average generation time is

$$t_{av} = 0.9935(0.001 \text{ s}) + 0.0065(14 \text{ s}) = 0.092 \text{ s}$$

Note that these few delayed neutrons increase the generation time by nearly a hundredfold. The time for 700 generations is

$$700(0.092 \text{ s}) = 64.4 \text{ s}$$

This is plenty of time for the mechanical insertion of control rods.

Because of the limited supply of natural uranium, the small fraction of ^{235}U in natural uranium, and the limited capacity of enrichment facilities, reactors based on the fission of ^{235}U cannot meet our energy needs for very long. A promising alternative is the **breeder reactor.** When the relatively plentiful but nonfissionable ^{238}U nucleus captures a neutron, it decays by β decay (with a half-life of 20 min) to ^{239}Np, which in turn decays by β decay (with a half-life of 2 days) to the fissionable nuclide ^{239}Pu. Since ^{239}Pu fissions with fast neutrons, no moderator is needed. A reactor initially fueled with a mixture of ^{238}U and ^{239}Pu will breed as much fuel as it uses or more if one or more of the neutrons emitted in the fission of ^{239}Pu is captured by ^{238}U. Practical studies indicate that a typical breeder reactor can be expected to double its fuel supply in 7 to 10 years.

The inside of a nuclear power plant in Kent, England. A technician is standing on the reactor charge transfer plate, into which uranium fuel rods fit.

There are two major safety problems inherent with breeder reactors. The fraction of delayed neutrons is only 0.3 percent for the fission of ^{239}Pu, so the time between generations is much less than that for ordinary reactors. Mechanical control is therefore much more difficult. Also, since the operating temperature of a breeder reactor is relatively high and a moderator is not desired, a heat-transfer material such as liquid sodium metal is used rather than water (which is the moderator as well as the heat-transfer material in an ordinary reactor). If the temperature of the reactor increases, the resulting decrease in the density of the heat-transfer material leads to positive feedback since it will absorb fewer neutrons than before. Because of these safety considerations, breeder reactors are not yet in use in the United States. There are, however, several in operation in France, Great Britain, and the Soviet Union.

Safety of Fission Reactors

There has been heated debate about the safety of fission reactors, particularly since the disastrous accident at Chernobyl in the U.S.S.R. in 1986 and the much less serious accident at the Three Mile Island plant in the United States in 1979. A common fear is that a reactor might blow up as a uranium bomb. This is virtually impossible. Even in light-water reactors, the enriched uranium used contains only 1 to 4 percent ^{235}U, whereas a uranium bomb typically requires enriched uranium with 90 percent ^{235}U. Another fear is **meltdown,** the melting of the fuel core because of the heat produced by the radioactive decay of the fission fragments that occurs even after the reactor is shut down. If the cooling system fails, it is possible that the core would melt and, in a worst-case scenario, plunge through the containment building into the ground. Meltdown did not occur at either Chernobyl or Three Mile Island.

A much more serious fear is that radioactive material may be released into the atmosphere, as did occur at Chernobyl. The reactor at Chernobyl was a graphite-moderated reactor designed to produce plutonium for weapons as well as electrical power. It was run at a very high power sufficient to ignite the graphite if the cooling system failed. There are no comparable dual-purpose reactors outside the Soviet Union that are operated in this way. A similar accident with water-cooled reactors is probably not possible. Furthermore, a common safety feature not used at Chernobyl is a containment building with walls of concrete and steel at least 1 m thick.

With any type of fission reactor, there is the problem of the storage of the long-lived radioactive waste products produced. Despite the fact that elaborate storage methods are used, their long-term efficacy is always open to question. In assessing the danger of nuclear reactors, however, we should compare them with the dangers of ordinary power plants, such as coal-burning plants. The deaths in the United States due to lung disease caused by coal-burning plants number in the tens of thousands in a single year.

Fusion

In fusion, two light nuclei such as deuterium (^{2}H) and tritium (^{3}H) fuse together to form a heavier nucleus. A typical fusion reaction is

$$^2\text{H} + {}^3\text{H} \rightarrow {}^4\text{He} + n + 17.6 \text{ MeV} \qquad 40\text{-}26$$

The energy released in fusion depends on the particular reaction. For the ^{2}H + ^{3}H reaction, it is 17.6 MeV. Although this is less than the energy released in a fission reaction, it is a greater amount of energy per unit mass. The energy released in this fusion reaction is (17.6 MeV)/(5 nucleons) = 3.52 MeV per nucleon. This is about 3.5 times as great as the 1 MeV per nucleon released in fission.

The production of power from the fusion of light nuclei holds great promise because of the relative abundance of the fuel and the absence of some of the dangers inherent in fission reactors. Unfortunately, the technology necessary to make fusion a practical source of energy has not yet been developed. We will consider the ^{2}H + ^{3}H reaction; other reactions present similar problems.

Because of the Coulomb repulsion between the ^{2}H and ^{3}H nuclei, very large kinetic energies, of the order of 1 MeV, are needed to get the nuclei close enough together for the attractive nuclear forces to become effective and cause fusion. Such energies can be obtained in an accelerator, but since the scattering of one nucleus by the other is much more probable than fusion, the bombardment of one nucleus by another in an accelerator requires the input of more energy than is recovered. To obtain energy from fusion, the particles must be heated to a temperature great enough for the fusion reaction to occur as the result of random thermal collisions. Because a significant number of particles have kinetic energies greater than the mean kinetic energy $\frac{3}{2}kT$ and because some particles can tunnel through the Coulomb barrier, a temperature T corresponding to $kT \approx 10$ keV is adequate to ensure that a reasonable number of fusion reactions will occur if the density of particles is sufficiently high. The temperature corresponding to $kT = 10$ keV is of the order of 10^8 K. Such temperatures occur in the interiors of stars, where such reactions are common. At these temperatures, a gas consists of positive ions and negative electrons and is called a **plasma.** One of the problems arising in attempts to produce controlled fusion reactions is that of confining the plasma long enough for the reactions to take place. In the interior of the sun the plasma is confined by the enormous gravitational field of the sun. In a laboratory on earth, confinement is a difficult problem.

The energy required to heat a plasma is proportional to the density of its ions n, whereas the collision rate is proportional to n^2, the square of the density. If τ is the confinement time, the output energy is proportional to $n^2\tau$. If the output energy is to exceed the input energy, we must have

$$C_1 n^2 \tau > C_2 n$$

where C_1 and C_2 are constants. In 1957, the British physicist J. D. Lawson evaluated these constants from estimates of the efficiencies of various hypothetical fusion reactors and derived the following relation between density and confinement time, known as **Lawson's criterion:**

Lawson's criterion

$$n\tau > 10^{20}\ \text{s·particles/m}^3 \qquad \text{40-27}$$

If Lawson's criterion is met and the thermal energy of the ions is great enough ($kT \sim 10$ keV), the energy released by a fusion reactor will just equal the energy input; that is, the reactor will just break even. For the reactor to be practical, much more energy must be released.

Two schemes for achieving Lawson's criterion are currently under investigation. In one scheme, **magnetic confinement,** a magnetic field is used to confine the plasma (see Section 24-2). In the most common arrangement, first developed in the U.S.S.R. and called the Tokamak, the plasma is confined in a large toroid. The magnetic field is a combination of the doughnut-shaped magnetic field due to the windings of the toroid and the self-field due to the current of the circulating plasma. The break-even point has been achieved recently using magnetic confinement, but we are still a long way from building a practical fusion reactor.

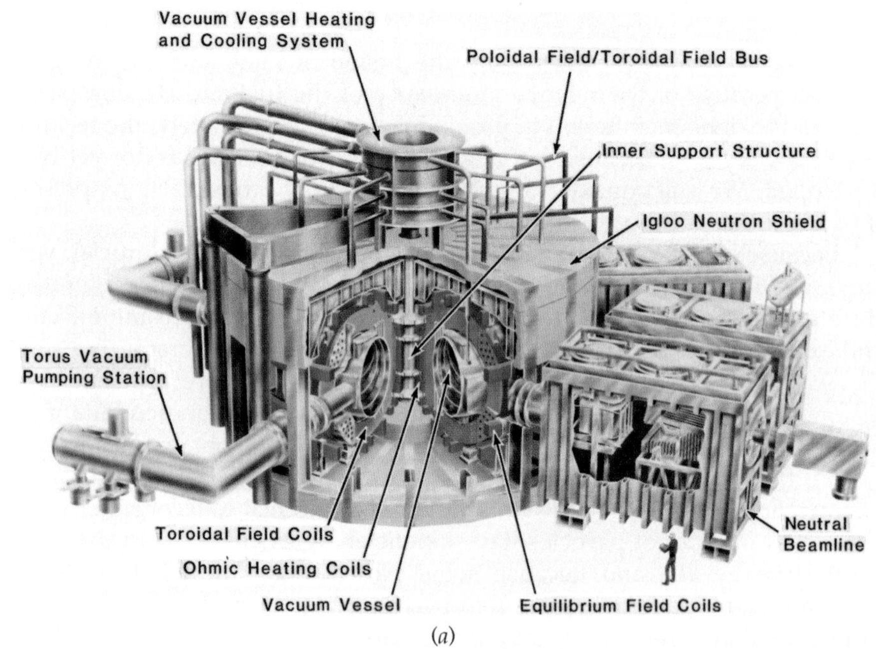

(*a*)

(*b*)

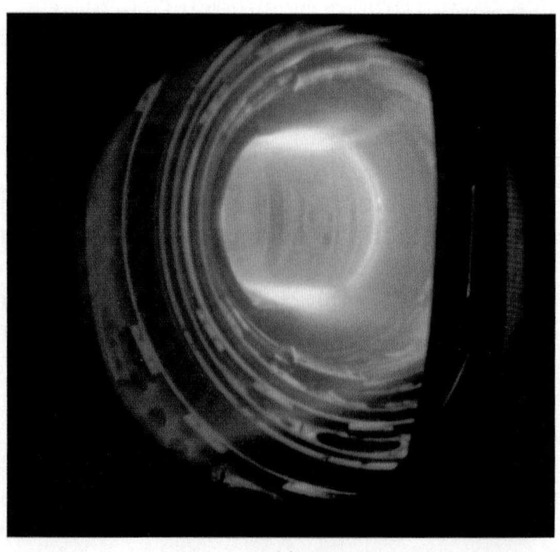

(*c*)

(*a*) Schematic of the tokamak fusion test reactor. The toroidal coils, encircling the doughnut-shaped vacuum vessel, are designed to conduct current for 3-s pulses, separated by waiting times of 5 min. Pulses peak at 73,000 A, producing a magnetic field of 5.2 T. This field is the principal means of confining the deuterium–tritium plasma that circulates within the vacuum vessel. Current for the pulses is delivered by converting the rotational energy of two 600-ton flywheels. Sets of poloidal coils, perpendicular to the toroidal coils, carry an oscillating current that generates a current through the confined plasma itself, heating it ohmically. Additional poloidal fields help stabilize the confined plasma. Between four and six neutral-beam injection systems (only one of which is shown in the schematic) are used to inject high-energy deuterium atoms into the deuterium–tritium plasma, heating it beyond what could be obtained ohmically—ultimately to the point of fusion. (*b*) The TFTR itself. The diameter of the vacuum vessel is 7.7 m (see also photo on p. 830). (*c*) An 800-kA plasma, lasting 1.6 s, as it discharges within the vacuum vessel.

(a)

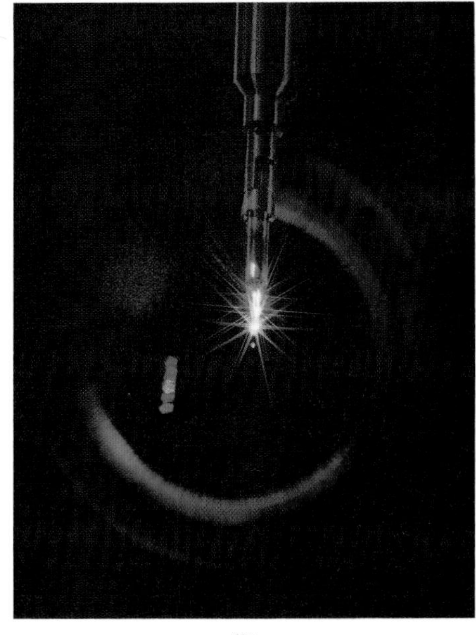

(b)

(*a*) The Nova target chamber, an aluminum sphere approximately 5 m in diameter, inside which ten beams from the world's most powerful laser converge onto a hydrogen-containing pellet 0.5 mm in diameter. The resulting fusion reaction is visible as a tiny star (*b*), lasting 10^{-10} s, releasing 10^{13} neutrons.

In a second scheme, called **inertial confinement,** a pellet of frozen solid deuterium and tritium is bombarded from all sides by intense pulsed laser beams of energies of the order of 10^6 J lasting about 10^{-8} s. (Intense ion beams are also used.) Computer simulation studies indicate the pellet should be compressed to about 10^4 times its normal density and heated to a temperature greater than 10^8 K. This should produce about 10^6 J of fusion energy in 10^{-10} s, which is so brief that confinement is achieved by inertia alone.

Because the break-even point is just barely being achieved in magnetic-confinement fusion, and because the building of a fusion reactor involves many practical problems that have not yet been solved, the availability of fusion to meet our energy needs is not expected for at least several decades. However, fusion holds great promise as an energy source for the future.

Questions

12. Why isn't there an element with $Z = 130$?

13. Why is a moderator needed in an ordinary nuclear-fission reactor?

14. Explain why water is more effective than lead in slowing down fast neutrons.

15. What happens to the neutrons produced in fission that do not produce another fission?

16. What is the advantage of a breeder reactor over an ordinary one? What are the disadvantages?

17. Why does fusion occur spontaneously in the sun but not on earth?

40-6 The Interaction of Particles with Matter

In this section, we will discuss briefly the main interactions of charged particles, neutrons, and photons with matter. Understanding these interactions is important for study of nuclear detectors, shielding, and the effects of radiation on living organisms. In each case, we will examine the principal factors involved in stopping or attenuating a beam of particles.

Charged Particles

When a charged particle traverses matter, it loses energy mainly through collisions with electrons. This often leads to the ionization of the atoms in the matter, in which case the particle leaves a trail of ionized atoms in its path. If the energy of the particle is large compared with the ionization energies of the atoms, the energy loss in each encounter with an electron will be only a small fraction of the particle's energy. (A heavy particle cannot lose a large fraction of its energy to a free electron because of conservation of momentum. For example, when a billiard ball collides with a marble, only a very small fraction of the energy of the billiard ball can be lost.) Since the number of electrons in matter is so large, we can treat the loss of energy as continuous. After a fairly well-defined distance, called the **range,** the particle will have lost all its kinetic energy and will come to a stop. Near the end of the range, the view of energy loss as continuous is not valid because individual encounters are important. For electrons, this can lead to a significant statistical variation in path length, but for protons and other heavy particles with energies of several MeV or more, the path lengths vary by only a few percent or less.

Figure 40-14 shows the rate of energy loss per unit path length $-dK/dx$ versus the energy of the ionizing particle. We can see from this figure that the rate of energy loss $-dK/dx$ is maximum at low energies and that at high energies it is approximately independent of the energy. Particles with kinetic energies greater than their rest energies mc^2 are called **minimum ionizing particles.** Their energy loss per unit path length is approximately constant, and their range is roughly proportional to their energy. Figure 40-15 shows the range-versus-energy curve for protons in air.

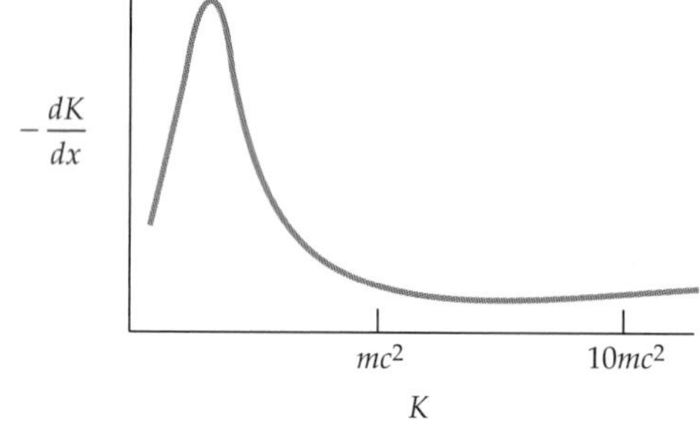

Figure 40-14 Energy loss per unit path length $-dK/dx$ versus kinetic energy for a charged particle. For particles with kinetic energies greater than their rest energies, the energy loss per unit path length is approximately constant. The distance traveled by such a particle is therefore approximately proportional to its energy.

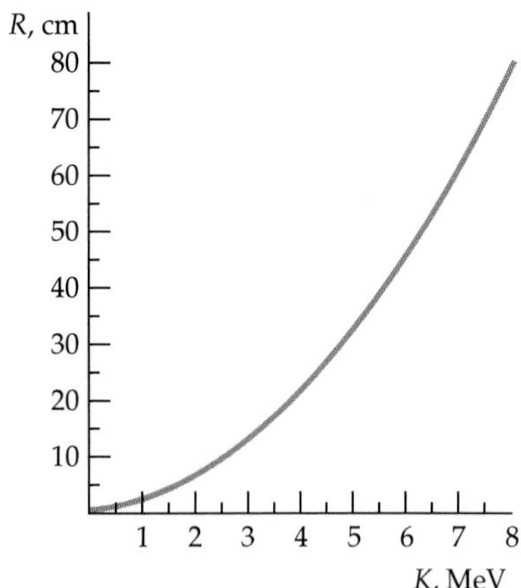

Figure 40-15 Range versus kinetic energy for protons in dry air. Except at low energies, the relationship between range and energy is approximately linear.

Since a charged particle loses energy through collisions with the electrons in a material, the greater the number of electrons, the greater the rate of energy loss. The energy loss rate $-dK/dx$ is approximately proportional to the density of the material. For example, the range of a 6-MeV proton is about 40 cm in air; but in water, which is about 800 times more dense than air, its range is only 0.5 mm.

If the energy of the charged particle is large compared with its rest energy, the particle also radiates some energy away in the form of bremsstrahlung.

The fact that the rate of energy loss for heavy charged particles is large at very low energies (as shown by the low-energy peak in Figure 40-14) has important applications in nuclear radiation therapy. Figure 40-16 shows a plot of energy loss versus penetration distance for charged particles in water. Most of the energy is lost near the end of the range. The peak in this curve is called the **Bragg peak.** A beam of heavy charged particles can be used to destroy cancer cells at a given depth in the body without destroying other, healthy cells if the energy is carefully chosen so that most of the energy loss occurs at the proper depth.

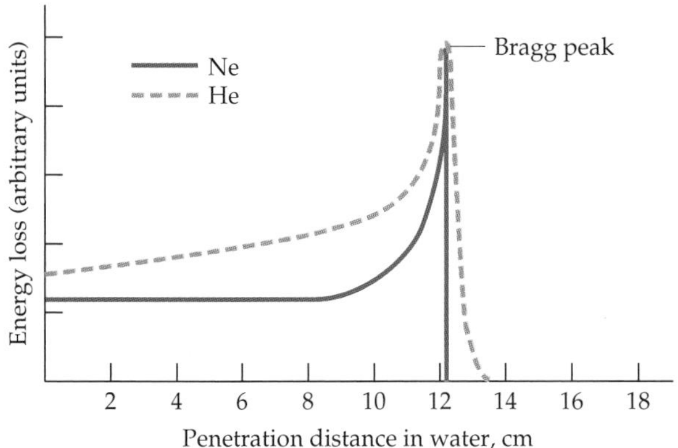

Figure 40-16 Energy loss of helium ions and neon ions in water versus penetration distance. Most of the energy loss occurs near the end of the path, as shown by the Bragg peak. The heavier the ion, the narrower the peak.

Neutrons

Since neutrons are uncharged, they do not interact with electrons in matter. Neutrons are removed from a beam by nuclear scattering or by capture. For energies that are large compared with thermal energies (kT), the most important processes are elastic and inelastic scattering. If we have a collimated neutron beam, any scattering or absorption will remove neutrons from the beam. This is very different from the case of a charged particle, which undergoes many collisions that decrease the energy of the particle but do not remove it from the beam until its energy is essentially zero. A neutron is removed from the beam at its first collision.

The chance of a neutron being removed from a beam within a given path distance is proportional to the number of neutrons in the beam and to the path distance. Let σ be the total cross section for the scattering plus the absorption of a neutron. If I is the incident intensity of the neutron beam (the number of particles per unit time per unit area), the number of neutrons removed from the beam per unit time will be $R = \sigma I$ per nucleus (Equation 40-24). If n is the number density of the nuclei (the nuclei per unit volume) and A is the area of the incident beam, the number of nuclei encountered in a distance dx is $nA\,dx$. The number of neutrons removed from the beam in a distance dx is thus

$$-dN = \sigma I(nA\,dx) = \sigma nN\,dx \qquad 40\text{-}28$$

where $N = IA$ is the total number of neutrons per unit time in the beam. Solving Equation 40-28 for N, we obtain

$$N = N_0 e^{-\sigma n x} \qquad 40\text{-}29$$

If we divide each side of Equation 40-29 by the area of the beam, we obtain a similar equation for the intensity of the beam:

$$I = I_0 e^{-\sigma n x} \qquad 40\text{-}30$$

We thus have an exponential decrease in the neutron intensity with penetration. After a certain characteristic distance, half the neutrons in a beam are removed. Then after an equal distance, half of the remaining neutrons are removed, and so on. Thus, there is no well-defined range.

At the half-penetration distance $x_{1/2}$, the number of neutrons will be $\frac{1}{2}N_0$. From Equation 40-29,

$$\frac{1}{2}N_0 = N_0 e^{-\sigma n x_{1/2}}$$

$$e^{\sigma n x_{1/2}} = 2$$

$$x_{1/2} = \frac{\ln 2}{\sigma n} \qquad 40\text{-}31$$

Example 40-9

The total cross section for the scattering and absorption of neutrons of a certain energy is 0.3 barn for copper. (*a*) Find the fraction of neutrons of that energy that penetrates 10 cm in copper. (*b*) To what distance will half the neutrons penetrate?

(*a*) Using $n = 8.45 \times 10^{28}$ nuclei/m^3 for copper, we have

$$\sigma n x = (0.3 \times 10^{-28}\ \text{m}^2)(8.45 \times 10^{28}/\text{m}^3)(0.10\ \text{m}) = 0.254$$

According to Equation 40-29, if we have N_0 neutrons at $x = 0$, the number at $x = 0.10$ m is

$$N = N_0 e^{-\sigma n x} = N_0 e^{-0.254} = 0.776 N_0$$

The fraction that penetrates 10 cm is thus 0.776 or 77.6 percent.

(*b*) For $n = 8.45 \times 10^{28}$ nuclei/m^3 and $\sigma = 0.3 \times 10^{-28}$ m^2, we have from Equation 40-31

$$x_{1/2} = \frac{\ln 2}{(0.3 \times 10^{-28}\ \text{m}^2)(8.45 \times 10^{28}/\text{m}^3)} = \frac{0.693}{2.54}\ \text{m}$$
$$= 0.273\ \text{m} = 27.3\ \text{cm}$$

Photons

The intensity of a photon beam, like that of a neutron beam, decreases exponentially with distance in an absorbing material. The intensity versus penetration is given by Equation 40-30, where σ is the total cross section for absorption and scattering. The important processes that remove photons from a beam are the photoelectric effect, Compton scattering, and pair production. The total cross section for absorption and scattering is the sum of the partial cross sections for these three processes: σ_{pe}, σ_{cs}, and σ_{pp}. These partial cross sections and the total cross section are shown as functions of energy in Figure 40-17. The cross section for the photoelectric effect dominates at very low energies (less than 1 MeV, which are not shown in the figure), but it decreases rapidly with increasing energy. If the photon energy is large compared with the binding energy of the electrons (a few keV), the electrons can be considered to be free and Compton scattering is the principal mechanism for the removal of photons from the beam. If the photon energy is greater than $2m_ec^2 = 1.02$ MeV, the photon can disappear, with the creation of an electron–positron pair, a process called **pair production.** The cross section for pair production increases rapidly with the photon energy and is the dominant component of the total cross section at high ener-

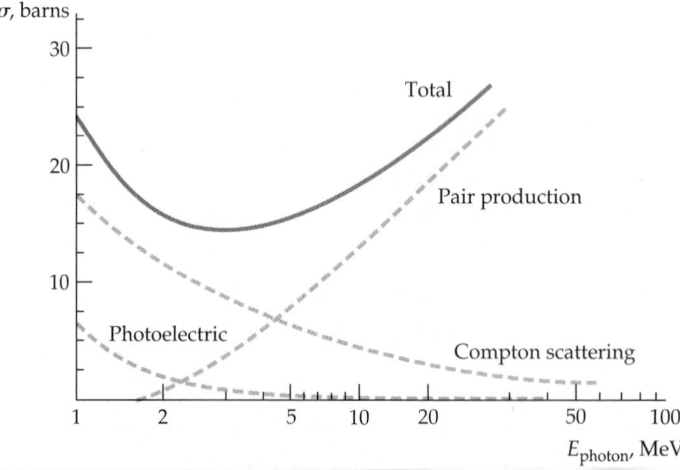

Figure 40-17 Photon interaction cross sections versus energy for lead. The total cross section is the sum of the cross sections for the photoelectric effect, Compton scattering, and pair production.

gies. Pair production cannot occur in free space. Momentum conservation requires that a nucleus be nearby to absorb momentum by recoil. The cross section for pair production is proportional to Z^2 of the absorbing material.

Dosage

The biological effects of radiation are principally due to the ionization it produces. Even a small amount of ionization can seriously disrupt the functioning of sensitive living cells or even kill them. Three different units are used to measure these effects: the roentgen, the rad, and the rem.

The **roentgen** (R) is defined as the amount of radiation that produces $\frac{1}{3} \times 10^{-9}$ C of electric charge (either positive ions or electrons) in 1 cm^3 of dry air at standard conditions. It is a measure of exposure to radiation. The roentgen has been largely replaced by the **rad** (*r*adiation *a*bsorbed *d*ose), a measure of the energy absorbed, which is defined as the amount of radiation that deposits 10^{-2} J/kg of energy in any material. The SI unit, joules per kilogram, is called a **gray** (Gy). Thus

$$1 \text{ rad} = 10^{-2} \text{ Gy} \qquad 40\text{-}32$$

Since 1 R is equivalent to the deposit of about 8.7×10^{-3} J/kg of energy, the rad and the roentgen are roughly equal.

The amount of biological damage depends not only on the energy absorbed, which is equivalent to the number of ion pairs formed, but also on the spacing of the ion pairs. If the ion pairs are closely spaced, as in the ionization caused by α particles, the biological effect is increased. The unit **rem** (*r*oentgen *e*quivalent in *m*an) is the dose that has the same biological effect as 1 rad of β or γ radiation:

$$1 \text{ rem} = 1 \text{ rad} \times \text{RBE} \qquad 40\text{-}33$$

where RBE is the *r*elative *b*iological *e*ffectiveness factor. Table 40-2 gives approximate values of the RBE factors for different types of radiation. The SI unit for dose equivalent is the seivert (Sv), which is defined as the product of the gray and the RBE:

$$1 \text{ Sv} = 1 \text{ Gy} \times \text{RBE} = 100 \text{ rem} \qquad 40\text{-}34$$

Table 40-3 compares the various radiation units we have discussed.

Our knowledge of the effects of large radiation doses comes mainly from studies of victims of atomic bomb explosions. Doses under 25 rem over the entire body seem to have no immediate effects. Doses over 100 rem damage the blood-forming tissues, and those over 500 rem usually lead to death in a short time.

The long-term effects of sublethal doses acquired over a period of time are more difficult to measure. The chances of dying of cancer are doubled by

Table 40-2 **Approximate RBE Factors**

Type of radiation	RBE factor
Photons < 4 MeV	1
Photons > 4 MeV	0.7
β particles < 30 keV	1.7
β particles > 30 keV	1
Slow neutrons	4 or 5
Fast neutrons	10
Protons	10
α particles	10
Heavy ions	20

Table 40-3 **Radiation and Dose Units***

	Customary unit		SI unit		
Quantity	Name	Symbol	Name	Symbol	Conversion
Energy	electron volt	eV	joule	J	1 MeV = 1.602×10^{-13} J
Exposure	roentgen	R	coulomb per kilogram	C/kg	1 R = 2.58×10^{-4} C/kg
Absorbed dose	rad	rad or rd	gray	Gy = J/kg	1 rad = 10^{-2} J/kg = 10^{-2} Gy
Dose equivalent	rem	rem	seivert	Sv	1 rem = 10^{-2} Sv
Activity	curie	Ci	becquerel	Bq = 1/s	1 Ci = 3.7×10^{10} decays/s = 3.7×10^{10} Bq

*Adapted from S. C. Bushong, *The Physics Teacher*, vol. 15, no. 3, p. 135, 1977.

a dose of somewhere between 100 and 500 rem. Not much is known about the effects of very low-level doses. It is possible that there is some threshold dose below which the damage done is repaired so that there is no resulting increase in the chance of cancer. But it is also possible (and is generally believed) that there is no such threshold and that the cancer-causing effects of radiation are proportional to the cumulative dose even at low levels.

Some typical human radiation exposures are listed in Table 40-4. The internal dose listed in this table is from radioactive nuclei, such as potassium-40 and uranium and its decay products, inside our bodies. Most of the radioactive fallout due to nuclear weapons testing is strontium-90 and cesium-137, both of which have half-lives of about 30 y. If there is no further testing, this source of radiation will eventually become negligible. We are shielded from most cosmic rays by the atmosphere. The dose we receive is about 40 mrem/y at sea level; it increases by about 1 mrem/y for every 30 m of altitude.

Table 40-4 **Average Radiation Doses Received by the United States Population**

Radiation source	Average effective dose (mrem/y)
Cosmic rays	45
Internal radioactive nuclides	35
Building materials	40
Ground	11
Air	5
Diagnostic x rays	70
Global fallout	4
Television	1
Nuclear power	0.003

One of the decay products of uranium is radon-222 (^{222}Rn), which decays by α emission with a half-life of 3.82 days. This decay is followed by other α and β decays that result in the relatively stable lead-210, which has a half-life of 22 y. Since radon is an inert gas, it diffuses through materials without interacting with them chemically. It has recently been recognized as a health hazard because it seeps into homes from the ground, where it accumulates and can be breathed into our lungs.

The largest source of artificial exposure to radiation is currently medical diagnostic x rays. The dose received varies enormously, depending on the type of machine used, the sensitivity of the film, and so forth. For a chest x ray, some mobile units give doses of 1000 millirems, and the average dose is around 200 millirems. If the best procedures are used, however, the dose from a chest x ray can be limited to 6 millirems.

Because of our lack of knowledge of the risks of radiation, we should clearly limit our exposure to it as much as possible. Table 40-5 lists some of the dose limits recommended by the National Council on Radiation Protection.

Table 40-5 **Recommended Dose Limits***

	Maximum permissible dose equivalent for occupational exposure
Combined whole-body occupational exposure	
Prospective annual limit	5 rems in any one year
Retrospective annual limit	10–15 rems in any one year
Long-term accumulation to age N years	$(N - 18) \times 5$ rems
Skin	15 rems in any one year
Hands	75 rems in any one year (25 per quarter)
Forearms	30 rems in any one year (10 per quarter)
Other organs, tissues, and organ systems	15 rems in any one year (5 per quarter)
Fertile women (with respect to fetus)	0.5 rem in gestation period
Dose limits for nonoccupationally exposed	
Population average	0.17 rem in any one year
An individual in the population	0.5 rem in any one year
Students	0.1 rem in any one year

*Adapted from *NCRP Rep.* 39, 1971, as given in S. C. Bushong, *The Physics Teacher*, vol. 15, no. 3, p. 135, 1977.

Summary

1. Nuclei have N neutrons, Z protons, and a mass number $A = N + Z$. Two or more nuclei having the same value of Z but different values of N and A are called isotopes.

2. For light nuclei, N and Z are approximately equal, whereas for heavy nuclei, N is greater than Z.

3. Most nuclei are approximately spherical in shape and have a volume that is proportional to A, implying that nuclear density is independent of A. The radius of a nucleus is given approximately by

$$R = R_0 A^{1/3}$$

where R_0 is about 1.5 fm.

4. The mass of a stable nucleus is less than the sum of the masses of its nucleons. The mass difference times c^2 equals the binding energy of the nucleus. The binding energy is approximately proportional to the mass number A.

5. Nuclei have magnetic moments associated with their spin. The magnetic moment of the proton is $+2.7928\ \mu_N$ and that of the neutron is $-1.9135\ \mu_N$, where

$$\mu_N = \frac{e\hbar}{2m_p} = 5.05 \times 10^{-27}\ \text{J/T} = 3.15 \times 10^{-8}\ \text{eV/T}$$

is the nuclear magneton.

6. A proton may have its magnetic moment aligned parallel or antiparallel to an external magnetic field. The difference in energy between these two states is given by

$$\Delta E = 2(\mu_z)_p B$$

Transitions between these two states can be induced by RF photons. The resonance absorption of these RF photons is called nuclear magnetic resonance.

7. Unstable nuclei are radioactive and decay by emitting α particles (^{4}He nuclei), β particles (electrons or positrons), or γ rays (photons). All radioactivity is statistical in nature and follows an exponential decay law:

$$N = N_0 e^{-\lambda t}$$

where λ is the decay constant. The rate of decay is given by

$$R = \lambda N = R_0 e^{-\lambda t}$$

The time it takes for the number of nuclei or the decay rate to decrease by half is called the half-life:

$$t_{1/2} = \frac{0.693}{\lambda}$$

Half-lives of alpha decay range from a fraction of a second to millions of years. For β decay they range up to hours or days and for γ decay half-lives are usually less than a microsecond. The number of decays per second of 1 g of radium is the curie, which equals 3.7×10^{10} decays/s = 3.7×10^{10} Bq.

8. The cross section σ is a measure of the effective size of a nucleus for a particular nuclear reaction. Cross sections are measured in barns, where 1 barn = 10^{-28} m^2. An important nuclear reaction is neutron capture, in which a neutron is captured by a nucleus with the emission of a photon. The neutron-capture cross section exhibits strong resonances superimposed on a $1/v$ dependence.

9. Fission occurs when some heavy elements, such as ^{235}U or ^{239}Pu, capture a neutron and split apart into two medium-mass nuclei. The two nuclei then fly apart because of electrostatic repulsion, releasing a large amount of energy. A chain reaction is possible because several neutrons are emitted by a nucleus when it undergoes fission. A chain reaction can be sustained if, on the average, one of the emitted neutrons is slowed down by scattering in the reactor and is then captured by another fissionable nucleus. Very heavy nuclei ($Z > 92$) are subject to spontaneous fission.

10. A large amount of energy is also released when two light nuclei, such as ^{2}H and ^{3}H, fuse together. Fusion takes place spontaneously inside the sun and other stars, where the temperature is great enough (about 10^8 K) for thermal motion to bring the charged hydrogen ions close enough together to fuse. Although controlled fusion holds great promise as a future energy source, practical difficulties have thus far prevented its development.

11. Charged particles lose energy nearly continuously when traversing matter because they interact with the electrons in the matter. They have fairly well-defined ranges in a particular material that are roughly proportional to the energy of the particle and vary inversely with the density of the material. Neutrons and photons do not have well-defined ranges. Instead, the intensity of a neutron or photon beam decreases exponentially with the penetration distance. Neutrons are removed from a beam by scattering or by capture. Photons are absorbed by the photoelectric effect at very low energies and by pair production at high energies. At intermediate energies they undergo Compton scattering from electrons in the material.

12. Radiation dosages are measured in rads, where 1 rad is the amount of radiation that deposits 10^{-2} J/kg of energy in any material. A rem is the dose that has the same biological effect as 1 rad of β or γ radiation. It equals the rad multiplied by the relative biological effectiveness factor RBE, which is approximately 1 for gamma and beta rays, about 4 or 5 for slow neutrons, about 10 for fast neutrons, and between 10 and 20 for alpha particles of energies of the order of 5 to 10 MeV.

Suggestions for Further Reading

Bulletin of the Atomic Scientists, published monthly except for July and August by the Educational Foundation for Nuclear Science, Chicago, Illinois.

Founded in 1945 by Western scientists concerned about the consequences of our new-found ability to induce nuclear fission, this magazine has numbered among its sponsors many eminent physicists including Albert Einstein. Its articles deal with current issues relating mainly to nuclear weapons and the policies of countries possessing them but also to nuclear reactors.

Černy, Joseph, and Arthur M. Poskanzer: "Exotic Light Nuclei," *Scientific American,* June 1978, p. 60.

Many isotopes of most of the light elements can be produced, but some decay so quickly that there is only just sufficient time to detect them before they are gone.

Engelman, Donald M., and Peter B. Moore: "Neutron-Scattering Studies of the Ribosome," *Scientific American,* October 1976, p. 44.

This article describes how a beam of neutrons can be used to obtain information about the relative positions of large molecules in biological structures such as the ribosome.

Furth, Harold P.: "Progress Toward a Tokamak Fusion Reactor," *Scientific American*, August 1979, p. 50.

Nuckolls, John H.: "The Feasibility of Inertial-Confinement Fusion," *Physics Today*, vol. 35, no. 9, 1982, p. 24.

These articles describe two possible methods of achieving the fusion of light nuclei under controlled conditions suitable for power generation.

Levi, Barbara G.: "Radionuclide Releases from Severe Accidents at Nuclear Power Plants," *Physics Today*, vol. 38, no. 5, 1985, p. 67.

Lewis, Harold W.: "The Safety of Fission Reactors," *Scientific American*, March 1980, p. 53.

Both of these articles deal with risk-assessment studies of American fission reactor technology undertaken after the Three Mile Island reactor accident.

Perrow, Charles: *Normal Accidents: Living with High-Risk Technologies*, Basic Books, New York, 1984.

This book is also the result of a study undertaken after the Three Mile Island reactor accident. It provides an alternative view to that of risk assessment in making decisions about such complex, high-risk systems as fission reactors.

Schramm, David N.: "The Age of the Elements," *Scientific American*, January 1974, p. 69.

The age of the universe and structures in it can be determined from abundance ratios of various radioactive elements and their daughter nuclei, based on certain assumptions about the processes by which they were formed.

Schroeer, Dietrich: *Science, Technology, and the Nuclear Arms Race*, John Wiley & Sons, New York, 1984.

This book describes in a clear and understandable way the principles of operation of modern fission and fusion bombs. The technology of bomb delivery systems, nuclear deterrence, alternatives to deterrence, and arms control and disarmament are discussed with reference to the specific technologies involved.

Review

A. Objectives: After studying this chapter, you should:

1. Be able to give the order of magnitude of the radius of an atom and of a nucleus.

2. Be able to sketch the N-versus-Z curve for stable nuclei.

3. Be able to sketch the binding energy per nucleon versus A and discuss the significance of this curve for fission and fusion.

4. Know the exponential law of radioactive decay and be able to work problems using it.

5. Be able to describe the nuclear-fission chain reaction and discuss the advantages and disadvantages of fission reactors.

6. Be able to state Lawson's criterion for nuclear-fusion reactors.

7. Be able to discuss the chief mechanisms by which particles lose energy in matter and explain why some particles have well-defined ranges and others do not.

8. Be able to discuss the three units of radiation dosage: the roentgen, the rad, and the rem.

B. Define, explain, or otherwise identify:

Atomic number
Mass number
Nucleon
Nuclide
Isotopes
Strong nuclear force
Hadronic force
Binding energy
Nuclear spin
Nuclear magneton
RF radiation
Nuclear magnetic resonance
Alpha decay
Beta decay
Gamma decay
Decay constant
Mean lifetime
Half-life
Becquerel
Curie
Daughter nucleus
Neutrino
Radioactive carbon dating
Metastable state
Exothermic reaction
Q value of reaction
Endothermic reaction
Cross section
Barn
Thermal neutron
Resonance
Fission
Fusion
Reproduction constant
Moderator
Delayed neutron
Breeder reactor
Meltdown
Plasma
Lawson's criterion
Magnetic confinement
Inertial confinement
Range
Minimum ionizing particles
Bragg peak
Pair production
Roentgen
Rad
Gray
Rem

C. True or false: If the statement is true, explain why it is true. If it is false, give a counterexample.

1. The atomic nucleus contains protons, neutrons, and electrons.

2. The mass of ^{2}H is less than the mass of a proton plus a neutron.

3. After two half-lives, all the radioactive nuclei in a given sample have decayed.

4. Exothermic reactions have no threshold energy.

5. In a breeder reactor, fuel can be produced as fast as it is consumed.

6. The rad is a measure of the energy deposited per unit volume in matter.

Problems

Level I

40-1 Properties of Nuclei

1. Give the symbols for two other isotopes of (*a*) ^{14}N, (*b*) ^{56}Fe, and (*c*) ^{118}Sn.

2. Calculate the binding energy and the binding energy per nucleon from the masses given in Table 40-1 for (*a*) ^{12}C, (*b*) ^{56}Fe, and (*c*) ^{238}U.

3. Repeat Problem 2 for (*a*) ^{6}Li, (*b*) ^{39}K, and (*c*) ^{208}Pb.

4. Use Equation 40-1 to compute the radii of the following nuclei: (*a*) ^{16}O, (*b*) ^{56}Fe, and (*c*) ^{197}Au.

5 Find the energy needed to remove a neutron from (*a*) ^{4}He and (*b*) ^{7}Li (mass, 7.016004 u).

6. (*a*) Given that the mass of a nucleus of mass number A is approximately $m = CA$, where C is a constant, find an expression for the nuclear density in terms of C and the constant R_0 in Equation 40-1. (*b*) Compute the value of this nuclear density in grams per cubic centimeter using the fact that C has the approximate value of 1 g per Avogadro's number of nucleons.

7. Derive Equation 40-2; that is, show that the rest energy of one unified mass unit is 931.5 MeV.

8. Use Equation 40-1 for the radius of a spherical nucleus and the approximation that the mass of a nucleus of mass number A is A u to calculate the density of nuclear matter in grams per cubic centimeter.

40-2 Nuclear Magnetic Resonance

There are no problems for this section.

40-3 Radioactivity

9. The counting rate from a radioactive source is 4000 counts/s at time $t = 0$. After 10 s, the counting rate is 1000 counts/s. (*a*) What is the half-life? (*b*) What is the counting rate after 20 s?

10. A certain source gives 2000 counts/s at time $t = 0$. Its half-life is 2 min. (*a*) What is the counting rate after 4 min? (*b*) After 6 min? (*c*) After 8 min?

11. The counting rate from a radioactive source is 6400 counts/s. The half-life of the source is 10 s. Make a plot of the counting rate as a function of time for times up to 1 min. What is the decay constant for this source?

12. The counting rate from a radioactive source is 8000 counts/s at time $t = 0$, and 10 min later the rate is 1000 counts/s. (*a*) What is the half-life? (*b*) What is the decay constant? (*c*) What is the counting rate after 20 min?

13. The half-life of radium is 1620 y. Calculate the number of disintegrations per second of 1 g of radium, and show that the disintegration rate is approximately 1 Ci.

14 A radioactive silver foil ($t_{1/2} = 2.4$ min) is placed near a Geiger counter and 1000 counts/s are observed at time $t = 0$. (*a*) What is the counting rate at $t = 2.4$ min and at $t = 4.8$ min? (*b*) If the counting efficiency is 20 percent, how many radioactive nuclei are there at time $t = 0$? At time $t = 2.4$ min? (*c*) At what time will the counting rate be about 30 counts/s?

15 The stable isotope of sodium is ^{23}Na. What kind of radioactivity would you expect of (*a*) ^{22}Na and (*b*) ^{24}Na?

16. Use Table 40-1 to calculate the energy in MeV for the α decay of (*a*) ^{226}Ra and (*b*) ^{242}Pu.

17. A sample of wood contains 10 g of carbon and shows a ^{14}C decay rate of 100 counts/min. What is the age of the sample?

18. A bone claimed to be 10,000 years old contains 15 g of carbon. What should the decay rate of ^{14}C be for this bone?

19. A sample of animal bone unearthed at an archaeological site is found to contain 175 g of carbon, and the decay rate of ^{14}C in the sample is measured to be 8.1 Bq. How old is the bone?

20. A sample of a radioactive isotope is found to have an activity of 115.0 Bq immediately after it is pulled from the reactor that formed it. Its activity 2 h 15 min later is measured to be 85.2 Bq. (*a*) Calculate the decay constant and the half-life of the sample. (*b*) How many radioactive nuclei were there in the sample initially?

40-4 Nuclear Reactions

21 Using Table 40-1, find the Q values for the following reactions: (*a*) $^{1}\text{H} + {}^{3}\text{H} \rightarrow {}^{3}\text{He} + n + Q$ and (*b*) $^{2}\text{H} + {}^{2}\text{H} \rightarrow {}^{3}\text{He} + \text{n} + Q$.

22. Using Table 40-1, find the Q values for the following reactions: (*a*) $^{2}\text{H} + {}^{2}\text{H} \rightarrow {}^{3}\text{H} + {}^{1}\text{H} + Q$, (*b*) $^{2}\text{H} + {}^{3}\text{He} \rightarrow {}^{4}\text{He} + {}^{1}\text{H} + Q$, and (*c*) $^{6}\text{Li} + n \rightarrow {}^{3}\text{H} + {}^{4}\text{He} + Q$.

40-5 Fission, Fusion, and Nuclear Reactors

23. Assuming an average energy of 200 MeV per fission, calculate the number of fissions per second needed for a 500-MW reactor.

24. If the reproduction factor in a reactor is $k = 1.1$, find the number of generations needed for the power level to (*a*) double, (*b*) increase by a factor of 10, and (*c*) increase by a factor of 100. Find the time needed in each case if (*d*) there are no delayed neutrons, so the time between generations is 1 ms, and (*e*) there are delayed neutrons that make the average time between generations 100 ms.

25. Compute the temperature T for which $kT = 10$ keV, where k is Boltzmann's constant.

40-6 The Interaction of Particles with Matter

26 The range of 4-MeV α particles in air ($\rho = 1.29\ \text{mg/cm}^3$) is 2.5 cm. Assuming the range to be inversely proportional to the density of matter, find the range of 4-MeV α particles in (*a*) water and (*b*) lead ($\rho = 11.2\ \text{g/cm}^3$).

27. The range of 6-MeV protons in air is approximately 45 cm. Find the approximate range of 6-MeV protons in (*a*) water and (*b*) lead (see Problem 26).

28. A neutron beam has its intensity reduced by a factor of 2 at a penetration distance of 3 cm in iron. How great a thickness of iron is needed to reduce the intensity by a factor of (*a*) 8 and (*b*) 128?

29. The number density of iron nuclei is $n = 8.50 \times 10^{28}$ nuclei/m^3. Find the total scattering and absorption cross section for the neutron beam of Problem 28.

30. A 1.0-cm-thick piece of lead shielding reduces the intensity of a beam of 15-MeV γ rays by a factor of 2. (*a*) By how much will 5 cm of lead reduce the intensity of this beam? (*b*) Approximately what thickness of lead is needed to reduce the intensity by a factor of 1000?

31. Use the data of Problem 30 and the number density of lead $n = 3.30 \times 10^{28}$ nuclei/m^3 to find the total cross section for the removal of 15-MeV photons in lead.

Level II

32. (*a*) Show that $ke^2 = 1.44$ MeV · fm, where k is the Coulomb constant and e is the electron charge. (*b*) Show that $hc = 1240$ MeV · fm.

33. Twelve nucleons are in a one-dimensional infinite square well of length $L = 3$ fm. (*a*) Using the approximation that the mass of a nucleon is 1 u, find the lowest energy of a nucleon in the well. Express your answer in MeV. What is the ground-state energy of the system of 12 nucleons in the well if (*b*) all the nucleons are neutrons so that there can be only 2 in each state and (*c*) 6 of the nucleons are neutrons and 6 are protons so that there can be 4 nucleons in each state? (Neglect the energy of Coulomb repulsion of the protons.)

34. Derive the result that the activity of 1 g of natural carbon due to the β decay of ^{14}C is 15 decays/min = 0.25 Bq.

35. A 0.05394-kg sample of ^{144}Nd (atomic mass 143.91 u) emits an average of 2.36 α particles each second. Find the decay constant in s^{-1} and the half-life in years.

36. (*a*) Show that if the decay rate is R_0 at time $t = 0$ and R_1 at some later time t_1, the decay constant is given by

$$\lambda = t_1^{-1} \ln \frac{R_0}{R_1}$$

and the half-life is given by

$$t_{1/2} = \frac{0.693\, t_1}{\ln (R_0/R_1)}$$

(*b*) Use these results to find the decay constant and the half-life if the decay rate is 1200 Bq at $t = 0$ and 800 Bq at $t_1 = 60$ s.

37. A 1.00-mg sample of substance of atomic mass 59.934 u emits β particles with an activity of 1.131 Ci. Find the decay constant for this substance in s^{-1} and its half-life in years.

38. The counting rate from a radioactive source is measured every minute. The resulting counts per second are 1000, 820, 673, 552, 453, 371, 305, 250. Plot the counting rate versus time on semilog graph paper, and use your graph to find the half-life of the source.

39. A sample of radioactive material is initially found to have an activity of 115.0 decays/min. After 4 d 5 h, its activity is measured to be 73.5 decays/min. (*a*) Calculate the half-life of the material. (*b*) How long (from the initial time) will it take for the sample to reach an activity level of 10.0 decays/min?

40. (*a*) Use the atomic masses $m = 14.00324$ u for $^{14}_{6}$C and $m = 14.00307$ u for $^{14}_{7}$N to calculate the Q value (in MeV) for the beta decay

$$^{14}_{6}\text{C} \rightarrow {}^{14}_{7}\text{N} + \beta^- + \bar{\nu}_e$$

(*b*) Explain why you do not need to add the mass of the β^- to that of atomic $^{14}_{7}$N for this calculation.

41. (*a*) Use the atomic masses $m = 13.00574$ u for $^{13}_{7}$N and $m = 13.003354$ u for $^{13}_{6}$C to calculate the Q value (in MeV) for the beta decay

$$^{13}_{7}\text{N} \rightarrow {}^{13}_{6}\text{C} + \beta^+ + \nu_e$$

(*b*) Explain why you need to add two electron masses to the mass of $^{13}_{6}$C in the calculation of the Q value for this reaction.

42. The electrostatic potential energy of two charges q_1 and q_2 separated by a distance r is $U = kq_1q_2/r$, where k is the Coulomb constant. (*a*) Use Equation 40-1 to calculate the radii of ^{2}H and ^{3}H. (*b*) Find the electrostatic potential energy when these two nuclei are just touching, that is, when their centers are separated by the sum of their radii.

43. (*a*) Calculate the radii of $^{141}_{56}$Ba and $^{92}_{36}$Kr from Equation 40-1. (*b*) Assume that after the fission of ^{235}U into ^{141}Ba and ^{92}Kr, the two nuclei are momentarily separated by a distance r equal to the sum of the radii found in (*a*), and calculate the electrostatic potential energy for these two nuclei at this separation. (See Problem 42.) Compare your result with the measured fission energy of 175 MeV.

44. (*a*) Find the wavelength of a particle in the ground state of a one-dimensional infinite square well of length $L = 2$ fm. (*b*) Find the momentum in units of MeV/c for a particle with this wavelength. (*c*) Show that the total energy of an electron with this wavelength is approximately $E \approx pc$. (*d*) What is the kinetic energy of an electron in the ground state of this well? This calculation shows that if an electron were confined in a region of space as small as a nucleus, it would have a very large kinetic energy.

45. The intensity of a neutrino beam decreases exponentially with distance according to Equation 40-30b, just like that of a beam of neutrons or photons. The absorption cross section for neutrinos is of the order of 10^{-20} barn. Find the thickness of iron ($n = 8.50 \times 10^{28}$ nuclei/m^3) needed to reduce the intensity of a neutrino beam by a factor of e. Compare this thickness with the distance from the earth to the sun (about 150 Gm).

46. A shielded γ source yields a dose rate of 0.0500 rad/h at a distance of 1.00 m for an average-sized person. If workers are allowed a maximum dose rate of 5.00 rem/y, how close to the source may they work, assuming that they work 2000 hours per year? Assume that the radiation intensity follows the inverse-square law. (It will actually fall off more rapidly than $1/r^2$ because of the absorption of photons by the air, so the result for this problem will give a better-than-permissible value.)

47. In 1989, researchers claimed to have achieved fusion in an electrochemical cell at room temperature. They claimed a power output of 4 W from deuterium fusion reactions in the palladium electrode of their apparatus. (*a*) If the two most likely reactions are

$$^2\text{H} + {}^2\text{H} \rightarrow {}^3\text{He} + n + 3.27 \text{ MeV}$$

and

$$^2\text{H} + {}^2\text{H} \rightarrow {}^3\text{H} + {}^1\text{H} + 4.03 \text{ MeV}$$

with 50 percent of the reactions going by each branch, how many neutrons per second would we expect to be emitted in the generation of 4 W of power? (*b*) If one-tenth of these neutrons were absorbed by the body of an 80.0-kg worker near the device, and if each absorbed neutron carries an average energy of 0.5 MeV with an RBE of 4, to what radiation dose rate in rems per hour would this correspond? (*c*) How long would it take for a person to receive a total dose of 500 rems? (This is the dose that is usually lethal to half of those receiving it.)

48. The total energy consumed in the United States in 1 y is about 7.0×10^{19} J. How many kilograms of ^{235}U would be needed to provide this amount of energy if we assume that 200 MeV of energy is released by each fissioning uranium nucleus, that all of the uranium atoms undergo fission, and that all of the energy-conversion mechanisms used are 100 percent efficient?

49. The rubidium isotope ^{87}Rb is a β emitter with a half-life of 4.9×10^{10} y that decays into ^{87}Sr. It is used to determine the age of rocks and fossils. Rocks containing the fossils of early animals contain a ratio of ^{87}Sr to ^{87}Rb of 0.0100. Assuming that there were no ^{87}Sr present when the rocks were formed, calculate the age of these fossils.

50. (*a*) How many α decays and how many β decays must a ^{222}Rn nucleus undergo before it becomes a ^{210}Pb nucleus? (*b*) Calculate the total energy released in the decay of one ^{222}Rn nucleus to ^{210}Pb. (The mass of ^{210}Pb is 209.984187 u.)

51. The EPA standard for the maximum indoor radon exposure is 4 pCi per liter of air. (*a*) If the lung capacity of a person is 3.5 L, how many atoms of ^{222}Rn are in the lungs of a person in a room that has the maximum allowed amount of radon? (*b*) If the total energy absorbed in the lungs from the decay of one ^{222}Rn nucleus to ^{210}Pb is 20.3 MeV and the lungs have a mass of 2.0 kg, what dose in rems does a person who breathes nothing but contaminated air for 1 y receive? (*c*) Assuming that the risk of lung cancer in nonsmokers is proportional to the dose in rems of the radiation received by the lungs, by what factor is the probability of lung cancer increased for this person, assuming a background dose of 150 mrem per year?

52. A fusion reactor using only deuterium for fuel would have the following two reactions taking place in it:

$$^2\text{H} + {}^2\text{H} \rightarrow {}^3\text{He} + n + 3.27 \text{ MeV}$$

and

$$^2\text{H} + {}^2\text{H} \rightarrow {}^3\text{H} + {}^1\text{H} + 4.03 \text{ MeV}$$

The ^{3}H produced in the second reaction reacts immediately with another ^{2}H to produce

$$^3\text{H} + {}^2\text{H} \rightarrow {}^4\text{He} + n + 17.7 \text{ MeV}$$

The ratio of ^{2}H to ^{1}H atoms in naturally occurring hydrogen is 1.5×10^{-4}. How much energy would be produced from 4 L of water if all of the ^{2}H nuclei undergo fusion?

Level III

53. The fusion reaction between ^{2}H and ^{3}H is

$$^3\text{H} + {}^2\text{H} \rightarrow {}^4\text{He} + n + 17.7 \text{ MeV}$$

Using the conservation of momentum and the given Q value, find the final energies of both the ^{4}He nucleus and the neutron, assuming that the initial momentum of the system is zero.

54. If there are N_0 radioactive nuclei at time $t = 0$, the number that decay in some time interval dt at time t is $-dN = \lambda N_0 e^{-\lambda t}\, dt$. If we multiply this number by the lifetime t of these nuclei, sum over all the possible lifetimes from $t = 0$ to $t = \infty$, and divide by the total number of nuclei, we get the mean lifetime τ:

$$\tau = \frac{1}{N_0}\int_0^\infty t\,|dN| = \int_0^\infty t\lambda e^{-\lambda t}\, dt$$

Show that $\tau = 1/\lambda$.

55. Assume that a neutron decays into a proton plus an electron without the emission of a neutrino. The energy shared by the proton and electron is then 0.782 MeV. In the rest frame of the neutron, the total momentum is zero, so the momentum of the proton must be equal and opposite that of the electron. This determines the relative energies of the two particles, but because the electron is relativistic, the exact calculation of these relative energies is somewhat difficult. (*a*) Assume that the kinetic energy of the electron is 0.782 MeV and calculate the momentum p of the electron in units of MeV/c. (*Hint:* Use Equation 34-34.) (*b*) From your result for (*a*), calculate the kinetic energy $p^2/2m_p$ of the proton. (*c*) Since the total energy of the electron plus proton is 0.782 MeV, the calculation in (*b*) gives a correction to the assumption that the energy of the electron is 0.782 MeV. What percentage of 0.782 MeV is this correction?

56. Consider a neutron of mass m moving with speed v_L and making an elastic head-on collision with a nucleus of mass M that is at rest in the laboratory frame of reference. (*a*) Show that the speed of the center of mass in the lab frame is $V = mv_L/(m + M)$. (*b*) What is the speed of the nucleus in the center-of-mass frame before the collision? After the collision? (*c*) What is the speed of the nucleus in the lab frame after the collision? (*d*) Show that the energy of the nucleus after the collision in the lab frame is

$$\frac{1}{2}M(2V)^2 = \frac{4mM}{(m+M)^2}\left(\frac{1}{2}mv_L^2\right)$$

(*e*) Show that the fraction of the energy lost by the neutron in this elastic collision is

$$\frac{-\Delta E}{E} = \frac{4mM}{(m+M)^2} = \frac{4(m/M)}{(1+m/M)^2}$$

57. (*a*) Use the result of part (*e*) of Problem 56 (Equation 40-35) to show that after N head-on collisions of a neutron with carbon nuclei at rest, the energy of the neutron is approximately $0.714^N E_0$, where E_0 is its original energy. (*b*) How many head-on collisions are required to reduce the energy of the neutron from 2 MeV to 0.02 eV, assuming stationary carbon nuclei?

58. On the average, a neutron loses 63 percent of its energy in a collision with a hydrogen atom and 11 percent of its energy in a collision with a carbon atom. Calculate the number of collisions needed to reduce the energy of a neutron from 2 MeV to 0.02 eV if the neutron collides with (*a*) hydrogen atoms and (*b*) carbon atoms. (See Problem 57.)

59. Energy is generated in the sun and other stars by fusion. One of the fusion cycles, the proton–proton cycle, consists of the following reactions:

$$^1\text{H} + {}^1\text{H} \rightarrow {}^2\text{H} + \beta^+ + \nu_e$$

$$^1\text{H} + {}^2\text{H} \rightarrow {}^3\text{He} + \gamma$$

followed by

$$^1\text{H} + {}^3\text{He} \rightarrow {}^4\text{He} + \beta^+ + \nu_e$$

(*a*) Show that the net effect of these reactions is

$$4\,{}^1\text{H} \rightarrow {}^4\text{He} + 2\beta^+ + 2\nu_e + \gamma$$

(*b*) Show that rest energy of 24.7 MeV is released in this cycle (not counting the energy of 1.02 MeV released when each positron meets an electron and the two annihilate). (*c*) The sun radiates energy at the rate of about 4×10^{26} W. Assuming this is due to the conversion of four protons into helium plus γ rays and neutrinos, which releases 26.7 MeV, what is the rate of proton consumption in the sun? How long will the sun last if it continues to radiate at its present level? [Assume that protons constitute about half of the total mass (2×10^{30} kg) of the sun.]

60. Radioactive nuclei with a decay constant of λ are produced in an accelerator at a constant rate R_p. The number of radioactive nuclei N then obeys the equation $dN/dt = R_p - \lambda N$. (*a*) If N is zero at $t = 0$, sketch N versus t for this situation. (*b*) The isotope ^{62}Cu is produced at a rate of 100 per second by placing ordinary copper (^{63}Cu) in a beam of high-energy photons. The reaction is

$$\gamma + {}^{63}\text{Cu} \rightarrow {}^{62}\text{Cu} + n$$

^{62}Cu decays by β decay with a half-life of 10 min. After a time long enough so that $dN/dt \approx 0$, how many ^{62}Cu nuclei are there?

61. A common medical imaging procedure requires the injection of 0.3 mCi of technetium-99, which is in an excited metastable state and emits γ rays with half-life of 6.0 h. The energy of the emitted photons is 0.143 MeV. If the body is able to excrete 60 percent of the technetium in it every hour, what is the total radiation dose that would be given to a 50-kg woman?

Chapter 41

Elementary Particles

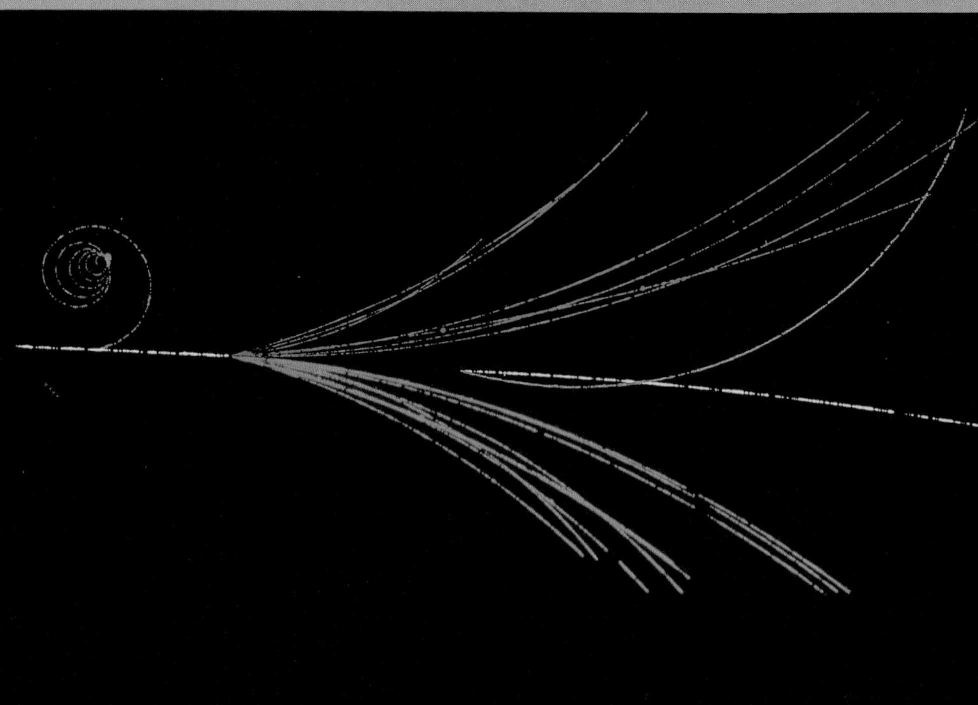

Tracks in a bubble chamber produced by an incoming high-energy proton (yellow) incident from the left, colliding with a proton at rest. The small green spiral is an electron knocked out of an atom. It curves to the left because of an external magnetic field in the chamber. The collision produces seven negative particles (π^-)(blue), a neutral particle Λ^0 that leaves no track, nine positive particles (red) that include seven π^+, a K^+, and a proton. The Λ^0 travels in the original direction of the incoming proton before decaying into a proton (yellow) and a π^- (purple).

In Dalton's atomic theory of matter (1808), the atom was considered to be the smallest indivisible constituent of matter, that is, an elementary particle. Then, with the discovery of the electron by Thomson (1897), the Bohr theory of the nuclear atom (1913), and the discovery of the neutron (1932), it became clear that atoms and even nuclei have considerable structure. For a time, it was thought there were just four "elementary" particles: the proton, neutron, electron, and photon. However, the positron or antielectron was discovered in 1932, and shortly thereafter, the muon, the pion, and many other particles were predicted and discovered.

Since the 1950s, enormous sums of money have been spent constructing particle accelerators of greater and greater energies in hopes of finding particles predicted by various theories. At present, we know of several hundred particles that at one time or another have been considered to be elementary, and research teams at the giant accelerator laboratories around the world are searching for and finding new particles. Some of these have such short lifetimes (of the order of 10^{-23} s) that they can be detected only indirectly. Many are observed only in nuclear reactions with high-energy accelerators. In addition to the usual particle properties of mass, charge, and spin, new properties have been found and given whimsical names such as strangeness, charm, color, topness, and bottomness.

In this chapter, we will first look at the various ways of classifying the multitude of particles that have been found. We will then describe the current theory of elementary particles, called the *standard model,* in which all matter in nature—from the exotic particles produced in the giant accelerator laboratories to ordinary grains of sand—is considered to be constructed from just two families of elementary particles, leptons and quarks.

41-1 Hadrons and Leptons

All the different forces observed in nature, from ordinary friction to the tremendous forces involved in supernova explosions, can be understood in terms of the four basic interactions. In order of decreasing strength, these are

1. The strong nuclear interaction (also called the hadronic interaction)
2. The electromagnetic interaction
3. The weak (nuclear) interaction
4. The gravitational interaction

Molecular forces and most of the everyday forces we observe between macroscopic objects (for example, friction, contact forces, and forces exerted by springs and strings) are complex manifestations of the electromagnetic interaction. Although gravity plays an important role in our lives, it is so weak compared with the other forces that its role in the interactions between elementary particles is essentially negligible. The weak interaction describes the interaction between electrons or positrons and nucleons that results in beta decay, which we discussed in Chapter 40. The strong interaction describes the forces between nucleons (neutrons and protons) that hold nuclei together. The four basic interactions provide a convenient structure for the classification of particles. Some particles participate in all four interactions, whereas others participate in only some of them.

Particles that interact via the strong interaction are called **hadrons.** There are two kinds of hadrons: **baryons,** which have spin $\frac{1}{2}$ (or $\frac{3}{2}$, $\frac{5}{2}$, and so on); and **mesons,** which have zero or integral spin. Baryons, which include nucleons, are the most massive of the elementary particles. Mesons have intermediate masses between the mass of the electron and the mass of the proton.

The existence of mesons was predicted in 1935 by the Japanese physicist H. Yukawa in a theory of nuclear forces that involved the exchange of a particle whose mass is related to the range of the strong nuclear force. Yukawa's theory was analogous to the quantum theory of the electrodynamic interaction. In classical electrodynamics, we avoid action at a distance by describing the force between two electric charges as being carried by the electromagnetic field. In an alternative description, called **quantum electrodynamics (QED),** the electromagnetic field is described in terms of photons, or field quanta, that are emitted by one charged particle and absorbed by another. Because these photons are not observed directly, they are called **virtual photons.** The emission of a virtual photon by a particle such as an electron violates the conservation of energy. However, in quantum mechanics, energy conservation can be violated if it occurs for such a short time that it cannot be observed in accordance with the uncertainty principle. If ΔE is the energy needed to create a virtual photon, such a process is allowed if the photon is absorbed by another particle (or is reabsorbed by the emitting particle) in time Δt given by $\Delta t \approx \hbar/\Delta E$.

Yukawa described the strong force between two nucleons in terms of the emission and absorption of a virtual particle of mass m_π. To avoid violating the conservation of energy, this particle must be absorbed in time Δt given by the uncertainty relation $\Delta t \approx \hbar/\Delta E$ with $\Delta E = m_\pi c^2$:

$$\Delta t \approx \frac{\hbar}{\Delta E} \approx \frac{\hbar}{m_\pi c^2} \qquad 41\text{-}1$$

Since no particle can travel faster than the speed of light c, the maximum distance this particle can travel is

$$d = c\,\Delta t = \frac{\hbar}{m_\pi c} \qquad 41\text{-}2$$

Setting this distance equal to the range of the nuclear force, which is about 1.5×10^{-15} m, Yukawa obtained an estimate of the mass m_π of his new particle. Solving Equation 41-2 for m_π and multiplying by c^2, we obtain for the rest energy of the particle

$$m_\pi c^2 \approx \frac{\hbar c}{d} = \frac{(6.58 \times 10^{-16}\ \text{eV·s})(3 \times 10^8\ \text{m/s})}{1.5 \times 10^{-15}\ \text{m}}$$

$$\approx 1.30 \times 10^8\ \text{eV} = 130\ \text{MeV}$$

The π meson or **pion** discovered in 1947 has just the properties described by Yukawa.

Particles that decay via the strong interaction have very short lifetimes of the order of 10^{-23} s, which is about the time it takes light to travel a distance equal to the diameter of a nucleus. On the other hand, particles that decay via the weak interaction have much longer lifetimes of the order of 10^{-10} s.

Hadrons are rather complicated entities with complex structures. If we use the term "elementary particle" to mean a point particle without structure that is not constructed from some more elementary entities, hadrons do not fit the bill. It is now believed that all hadrons are composed of more fundamental entities called *quarks,* which are truly elementary particles. We will discuss quarks in Section 41.4. Table 41-1 lists some of the properties of the hadrons that are stable against decay via the strong interaction.

Table 41-1 **Hadrons That Are Stable Against Decay Via the Strong Nuclear Interaction**

Name	Symbol	Mass, MeV/c^2	Spin, $\hbar$	Charge, e	Antiparticle	Mean lifetime, s	Typical decay products†
Baryons							
Nucleon	p (proton)	938.3	$\frac{1}{2}$	+1	p^-	infinite	
	n (neutron)	939.6	$\frac{1}{2}$	0	$\overline{n}$	930	$p + e^- + \overline{\nu}_e$
Lambda	Λ^0	1116	$\frac{1}{2}$	0	$\overline{\Lambda^0}$	2.5×10^{-10}	$p + \pi^-$
Sigma	Σ^+	1189	$\frac{1}{2}$	+1	$\overline{\Sigma^-}$	0.8×10^{-10}	$n + \pi^+$
	Σ^0	1193	$\frac{1}{2}$	0	$\overline{\Sigma^0}$	10^{-20}	$\Lambda^0 + \gamma$
	Σ^-	1197	$\frac{1}{2}$	−1	$\overline{\Sigma^+}$	1.7×10^{-10}	$n + \pi^-$
Xi	Ξ^0	1315	$\frac{1}{2}$	0	$\overline{\Xi^0}$	3.0×10^{-10}	$\Lambda^0 + \pi^0$
	Ξ^-	1321	$\frac{1}{2}$	−1	$\overline{\Xi^+}$	1.7×10^{-10}	$\Lambda^0 + \pi^-$
Omega	Ω^-	1672	$\frac{3}{2}$	−1	$\overline{\Omega^+}$	1.3×10^{-10}	$\Xi^0 + \pi^-$
Mesons							
Pion	π^+	139.6	0	+1	π^-	2.6×10^{-8}	$\mu^+ + \nu_\mu$
	π^0	135	0	0	π^0	0.8×10^{-16}	$\gamma + \gamma$
	π^-	139.6	0	−1	π^+	2.6×10^{-8}	$\mu^- + \overline{\nu}_\mu$
Kaon	K^+	493.7	0	+1	K^-	1.24×10^{-8}	$\pi^+ + \pi^0$
	K^0	497.7	0	0	$\overline{K^0}$	0.88×10^{-10} and 5.2×10^{-8}‡	$\pi^+ + \pi^-$; $\pi^+ + e^- + \overline{\nu}_e$
Eta	η^0	549	0	0		2×10^{-19}	$\gamma + \gamma$

†Other decay modes also occur for most particles.
‡The K^0 has two distinct lifetimes, sometimes referred to as K^0_{short} and K^0_{long}. All other particles have a unique lifetime.

Particles that participate in the weak interaction but not in the strong interaction are called **leptons.** These include electrons, muons, and neutrinos, which are all less massive than the lightest hadron. The word *lepton,* meaning "light particle," was chosen to reflect the relatively small mass of these particles. However, the most recently discovered lepton, the *tau,* found by Perl in 1975, has a mass of about 1780 MeV/c^2, nearly twice that of the proton (938 MeV/c^2), so we now have a "heavy lepton." As far as we know, leptons are point particles with no structure and can be considered to be truly elementary in the sense that they are not composed of other particles.

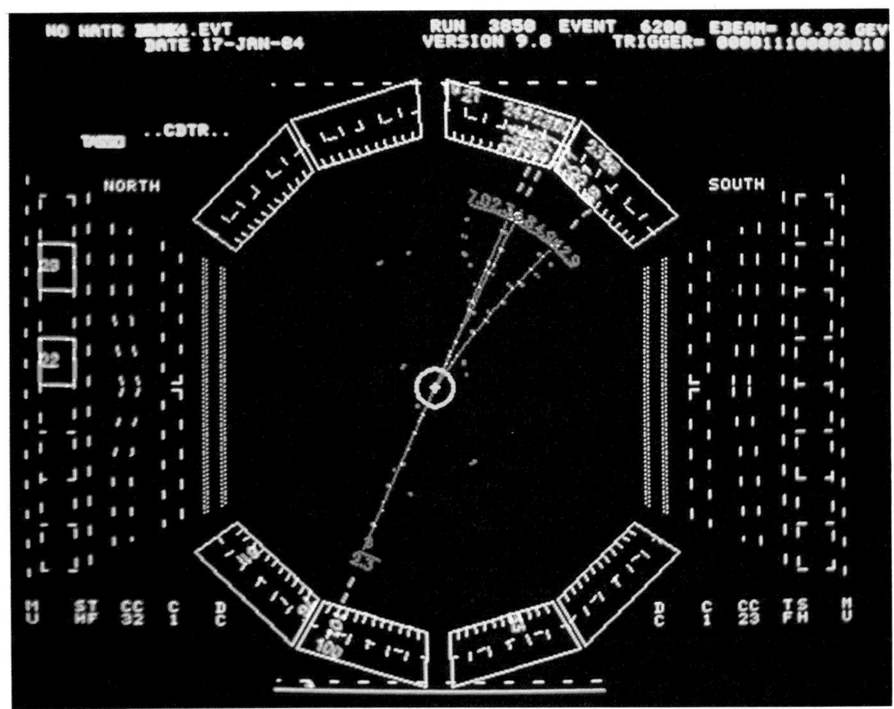

(*a*)

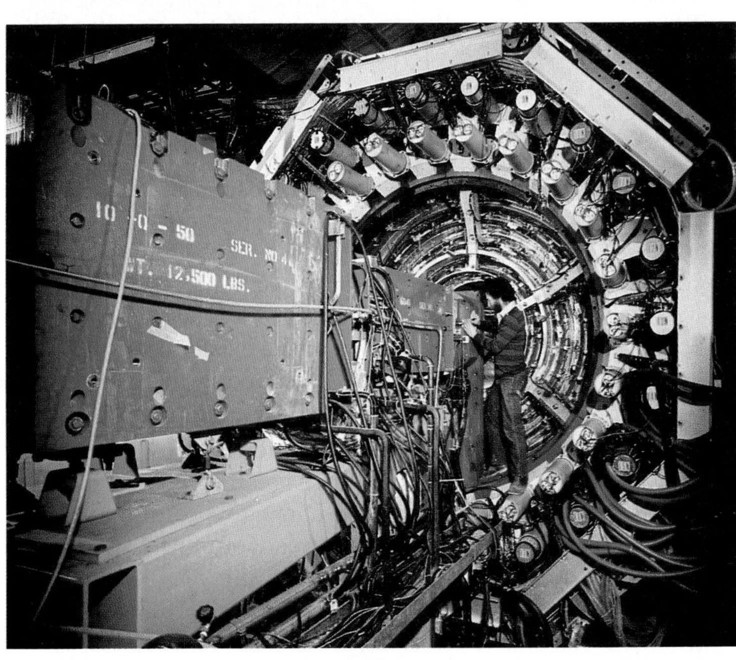

(*b*)

(*a*) A computer display of the production and decay of a τ^+ and τ^- pair. An electron and positron annihilate at the center marked by the yellow cross—producing a τ^+ and τ^- pair, which travel in opposite directions, but quickly decay while still inside the beam pipe (yellow circle). The τ^+ decays into two invisible neutrinos and a μ^+, which travels toward the bottom left. Its track in the drift chamber is calculated by a computer and indicated in red. It penetrates the lead-argon counters outlined in purple and is detected at the blue dot near the bottom blue line that marks the end of a muon detector. The τ^- decays into three charged pions (red tracks moving upward) plus invisible neutrinos. (*b*) The Mark I detector, built by a team from the Stanford Linear Accelerator Center (SLAC) and the Lawrence Berkeley Laboratory, became famous for many discoveries, including the ψ/J meson and the τ lepton. Tracks of particles are recorded by wire spark chambers wrapped in concentric cylinders around the beam pipe extending out to the ring where physicist Carl Friedberg has his right foot. Beyond this are two rings of protruding tubes, housing photomultipliers that view various scintillation counters. The rectangular magnets at the left guide the counterrotating beams that collide in the center of the detector.

It is currently thought that there are six leptons, each of which has an antiparticle. They are the electron, the muon, and the tau and a distinct neutrino associated with each of these three particles. (The neutrino associated with the tau has not yet been observed experimentally.) The masses of these particles are quite different. The mass of the electron is 0.511 MeV/c^2, the mass of the muon is 105 MeV/c^2, and that of the tau is 1780 MeV/c^2. The neutrinos are thought to be massless, but there is considerable debate as to the possibility that they may have a very small but nonzero mass, perhaps of the order of a few eV/c^2. Experiments designed to detect neutrinos emitted from the sun have found a much smaller number than expected, which could be explained if the mass of the neutrino were not zero. In addition, a mass as small as 40 eV/c^2 for the neutrino would have great cosmological significance. The answer to the question of whether the universe will continue to expand indefinitely or will reach a maximum size and begin to contract depends on the total mass in the universe. Thus, the answer could depend on whether the rest mass of the neutrino is merely small rather than zero. The observation of electron neutrinos from the supernova 1987A puts an upper limit on the mass of these neutrinos. Since the velocity of a particle with mass depends on its energy, the arrival time of a burst of neutrinos with mass from a supernova would be spread out in time. The fact that the electron neutrinos from the 1987 supernova all arrived at the earth within 13 seconds of one another results in an upper limit of about 16 eV/c^2 for their mass. This upper limit does not rule out the possibility of zero mass for these neutrinos.

Questions

1. How are baryons and mesons similar? How are they different?
2. The muon and the pion have nearly the same mass. How do these particles differ?

41-2 Spin and Antiparticles

One important characteristic of a particle is its intrinsic spin angular momentum. We have already discussed the fact that the electron has a quantum number m_s that corresponds to the z component of its intrinsic spin characterized by the quantum number $s = \frac{1}{2}$. Protons, neutrons, neutrinos, and the various other particles that also have an intrinsic spin characterized by the quantum number $s = \frac{1}{2}$ are called **spin-$\frac{1}{2}$ particles.** Particles that have spin $\frac{1}{2}$ (or $\frac{3}{2}$, $\frac{5}{2}$, . . .) are called **fermions** and obey the Pauli exclusion principle. Particles such as pions and other mesons have zero spin or integral spin ($s = 0, 1, 2, \ldots$). These particles are called **bosons** and do not obey the Pauli exclusion principle. Any number of these particles can be in the same quantum state.

Spin-$\frac{1}{2}$ particles are described by the Dirac equation, an extension of the Schrödinger equation to include special relativity. One feature of Dirac's theory proposed in 1927 is the prediction of the existence of antiparticles. In special relativity, the energy of a particle is related to the mass and momentum of the particle by $E = \pm\sqrt{p^2c^2 + m^2c^4}$. We usually choose the positive solution and dismiss the negative-energy solution with a physical argument. However, the Dirac equation requires the existence of wave functions that correspond to the negative-energy states. Dirac got around this difficulty by postulating that all the negative-energy states were filled and would therefore not be observable. Only holes in the "infinite sea" of negative-energy states would be observed. This interpretation received little attention until the positron was discovered in 1932 by Carl Anderson.

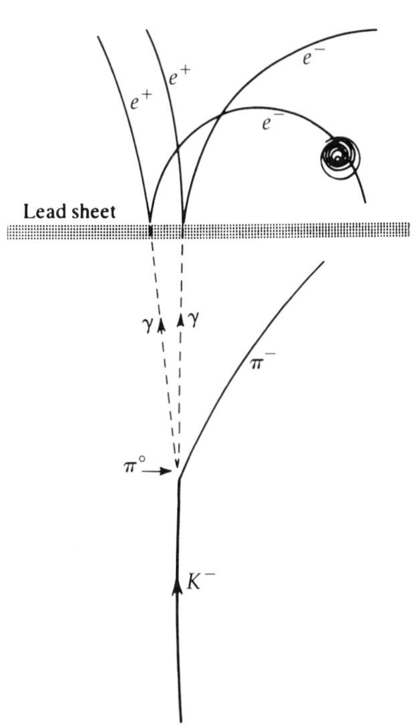

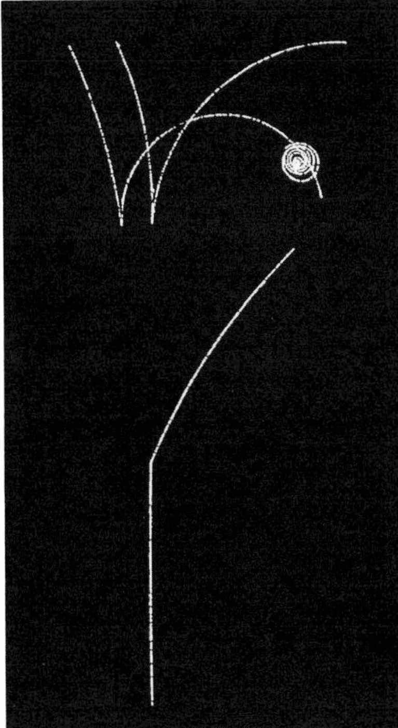

A negative kaon (K^-) enters a bubble chamber from the bottom and decays into a π^-, which moves off to the right, and a π^0, which immediately decays into two photons whose paths are indicated by the dashed lines in the drawing. Each photon interacts in the lead sheet, producing an electron-positron pair. The spiral at the right is another electron that has been knocked out of an atom in the chamber. (Other extraneous tracks have been removed from the photograph.)

Antiparticles are never created alone but always in particle–antiparticle pairs. In the creation of an electron–positron pair by a photon, the energy of the photon must be greater than the rest energy of the electron plus the positron, which is $2m_ec^2 \approx 1.02$ MeV, where m_e is the mass of the electron. Although the positron is stable, it has only a short-term existence in our universe because of the large supply of electrons in matter. The fate of a positron is annihilation according to the reaction

$$e^+ + e^- \rightarrow \gamma + \gamma \qquad 41\text{-}3$$

The probability of this reaction is large only if the positron is at rest or nearly at rest. Two photons moving in opposite directions are needed to conserve linear momentum.

The fact that we call electrons *particles* and positrons *antiparticles* does not imply that positrons are less fundamental than electrons. It merely reflects the nature of our part of the universe. If our matter were made up of negative protons and positive electrons, then positive protons and negative electrons would suffer quick annihilation and would be called antiparticles.

The antiproton (p^-) was discovered in 1955 by E. Segré and O. Chamberlain using a beam of protons in the Bevatron at Berkeley to produce the reaction*

$$p^+ + p^+ \rightarrow p^+ + p^+ + p^+ + p^- \qquad 41\text{-}4$$

The creation of a proton–antiproton pair (Figure 41-1) requires kinetic energy of at least $2m_pc^2 = 1877$ MeV $= 1.877$ GeV in the zero-momentum

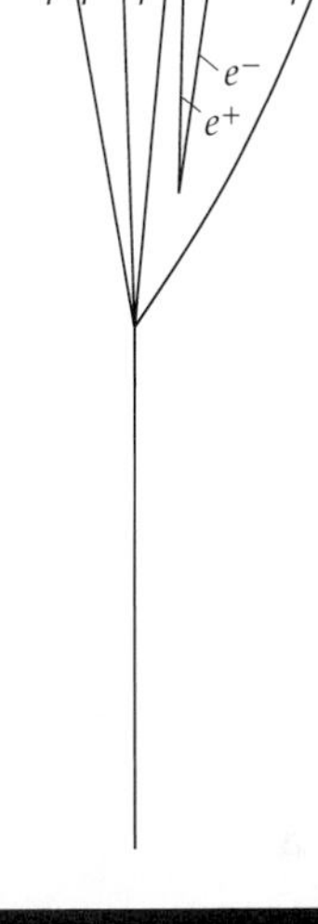

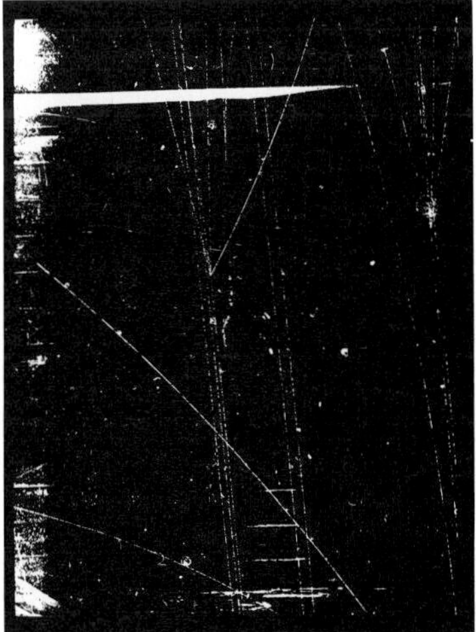

Figure 41-1 Bubble-chamber tracks showing the creation of a proton–antiproton pair in the collision of an incident 25-GeV proton with a stationary proton in liquid hydrogen.

*The antiproton is sometimes denoted by $\bar{p}$ rather than p^-. For neutral particles, such as the neutron, the bar must be used to denote the antiparticle. Thus the antineutron is denoted by $\bar{n}$. The normal electron and proton are often denoted by e and p without the minus or plus superscripts.

reference frame in which the two protons approach each other with equal and opposite momenta. In the laboratory frame in which one of the protons is initially at rest, the kinetic energy of the incoming proton must be at least $6m_pc^2 = 5.63$ GeV (see Problem 76 of Chapter 34). This energy was not available in laboratories before the development of high-energy accelerators in the 1950s. Antiprotons annihilate with protons to produce two gamma rays in a reaction similar to that in Equation 41-3.

Example 41-1

A proton and an antiproton at rest annihilate according to the reaction

$$p^+ + p^- \rightarrow \gamma + \gamma$$

Find the energies and wavelengths of the photons.

Since the proton and the antiproton are at rest, conservation of momentum requires that the two photons created in their annihilation have equal and opposite momenta and therefore equal energies. Since the total energy on the left side of the reaction is $2m_pc^2$, the energy of each photon is

$$E_\gamma = m_pc^2 = 938 \text{ MeV}$$

The wavelength is

$$\lambda = \frac{c}{f} = \frac{hc}{hf} = \frac{hc}{E_\gamma} = \frac{1240 \text{ eV}\cdot\text{nm}}{9.38 \times 10^8 \text{ eV}} = 1.32 \times 10^{-15} \text{ m} = 1.32 \text{ fm}$$

Air view of the European Laboratory for Particle Physics (CERN) just outside of Geneva, Switzerland. The large circle shows the Large Electron-Positron collider (LEP) tunnel that is 27 kilometers in circumference. The irregular dashed line is the border between France and Switzerland.

The tunnel of the proton–antiproton collider at CERN. The same bending magnets and focusing magnets can be used for protons or antiprotons moving in opposite directions. The rectangular box in the foreground is a focusing magnet; the next four boxes are bending magnets.

41-3 The Conservation Laws

One of the maxims of nature is "anything that can happen does." If a conceivable decay or reaction does not occur, there must be a reason. The reason is usually expressed in terms of a conservation law. The conservation of energy rules out the decay of any particle for which the total rest mass of the decay products would be greater than the initial rest mass of the particle before decay. The conservation of linear momentum requires that when an electron and positron at rest annihilate, two photons must be emitted. Angular momentum must also be conserved in a reaction or decay. A fourth conservation law that restricts the possible particle decays and reactions is that of electric charge. The net electric charge before a decay or reaction must equal the net charge after the decay or reaction.

There are two additional conservation laws that are important in the reactions and decays of elementary particles: the conservation of baryon number and the conservation of lepton number. Consider the possible decay

$$p \rightarrow \pi^0 + e^+$$

This decay would conserve charge, energy, angular momentum, and linear momentum, but it does not occur. It does not conserve either lepton number or baryon number. The conservation of lepton number and baryon number implies that whenever a lepton or baryon particle is created, an antiparticle of the same type is also created. We assign the **lepton number** $L = +1$ to all leptons, $L = -1$ to all antileptons, and $L = 0$ to all other particles. Similarly, the **baryon number** $B = +1$ is assigned to all baryons, $B = -1$ to all antibaryons, and $B = 0$ to all other particles. The baryon and lepton numbers cannot change in a reaction or decay. The conservation of baryon number along with the conservation of energy implies that the least massive baryon, the proton, must be stable.

The conservation of lepton number implies that the neutrino emitted in the beta decay of the free neutron is an antineutrino:

$$n \rightarrow p^+ + e^- + \bar{\nu}_e \qquad \text{41-5}$$

The fact that neutrinos and antineutrinos are different is illustrated by an experiment in which ^{37}Cl is bombarded with an intense antineutrino beam from the decay of reactor neutrons. If neutrinos and antineutrinos were the same, we would expect the following reaction:

$$^{37}\text{Cl} + \bar{\nu}_e \rightarrow {}^{37}\text{Ar} + e^- \qquad \text{41-6}$$

This reaction is not observed. However, if protons are bombarded with antineutrinos, the reaction

$$p + \bar{\nu}_e \rightarrow n + e^+ \qquad 41\text{-}7$$

is observed. Note that the lepton number is -1 on the left side of reaction 41-6 and $+1$ on the right side. But the lepton number is -1 on both sides of reaction 41-7.

Not only are neutrinos and antineutrinos distinct particles, but the neutrinos associated with electrons are distinct from the neutrinos associated with muons. It is also believed that the recently discovered heavy lepton, the tau, has a neutrino associated with it. Electronlike leptons (e and ν_e), muonlike leptons (μ and ν_μ), and presumably taulike leptons (τ and ν_τ) are each separately conserved. This is easily handled by assigning separate lepton numbers L_e, L_μ, and L_τ to the particles. For e and ν_e, $L_e = +1$; for their antiparticles, $L_e = -1$; and for all other particles, $L_e = 0$. The lepton numbers L_μ and L_τ are similarly assigned.

Example 41-2

What conservation laws (if any) are violated by the following decays?
(*a*) $n \rightarrow p + \pi^-$ (*b*) $\Lambda^0 \rightarrow p^- + \pi^+$ (*c*) $\mu^- \rightarrow e^- + \gamma$

(*a*) There are no leptons in this decay, so there is no problem with the conservation of lepton number. The net charge is zero before and after the decay, so charge is conserved. Also, the baryon number is $+1$ before and after the decay. However, the rest energy of the proton (938.3 MeV) plus that of the pion (139.6 MeV) is greater than the rest energy of the neutron (939.6 MeV). Thus, this decay violates the conservation of energy.

(*b*) Again, there are no leptons involved, and the net charge is zero before and after the decay. Also, the rest energy of the Λ^0 (1116 MeV) is greater than the rest energy of the antiproton (938.3 MeV) plus that of the pion (139.6 MeV), so energy is conserved with the loss in rest energy equaling the gain in kinetic energy of the decay products. However, this decay does not conserve baryon number, which is $+1$ for the Λ^0 and -1 for the antiproton.

(*c*) This reaction does not conserve muon lepton number or electron lepton number. The muon does decay via

$$\mu^- \rightarrow e^- + \bar{\nu}_e + \nu_\mu$$

which does conserve both muon and electron lepton numbers.

There are some conservation laws that are not universal but apply only to certain kinds of interactions. In particular, there are quantities that are conserved in decays and reactions that occur via the strong interaction but not in decays or reactions that occur via the weak interaction. One of these quantities that is particularly important is **strangeness** introduced by M. Gell-Mann and K. Nishijima in 1952 to explain the strange behavior of the heavy baryons and mesons. Consider the reaction

$$p + \pi^- \rightarrow \Lambda^0 + K^0 \qquad 41\text{-}8$$

The cross section for this reaction is large, as would be expected if it takes place via the strong interaction. However, the decay times for both the Λ^0 and K^0 are of the order of 10^{-10} s, which is characteristic of the weak interaction, rather than 10^{-23} s, which would be expected for the strong interaction. Other particles showing similar behavior were called **strange particles.**

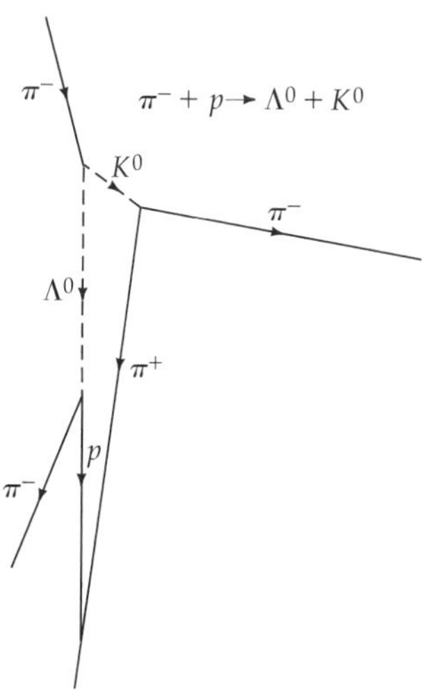

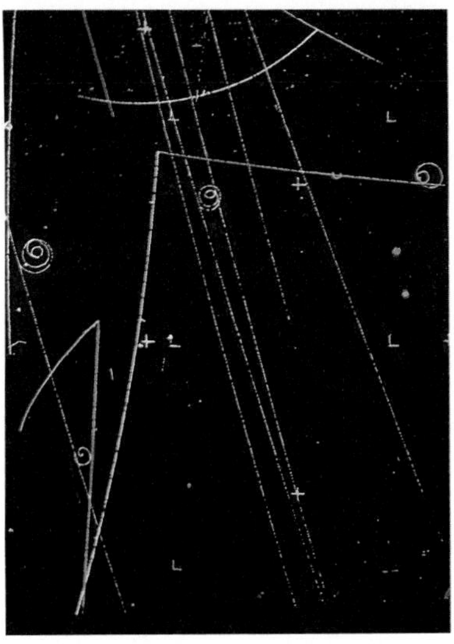

An early photograph of bubble-chamber tracks at the Lawrence Berkeley Laboratory, showing the production and decay of two strange particles, the K^0 and the Λ^0. These neutral particles are identified by the tracks of their decay particles. The lambda particle was named because of the similarity of the tracks of its decay particles and the Greek letter Λ. (The blue tracks are particles not involved in the reaction of Equation 41-8.)

These particles are always produced in pairs and never singly, even when all other conservation laws are met. This behavior is described by assigning a new property called strangeness to these particles. The strangeness of the ordinary hadrons—the nucleons and pions—was arbitrarily taken to be zero. The strangeness of the K^0 was arbitrarily chosen to be $+1$. Therefore, the strangeness of the Λ^0 particle must be -1 so that strangeness is conserved in the reaction of Equation 41-8. The strangeness of other particles could then be assigned by looking at their various reactions and decays. In reactions and decays that occur via the strong interaction, strangeness is conserved. In those that occur via the weak interaction, the strangeness can change by ± 1.

Figure 41-2 shows the masses of the baryons and mesons that are stable against decay via the strong interaction versus strangeness. We can see from this figure that these particles cluster in multiplets of one, two, or three particles of approximately equal mass, and that the strangeness of a multiplet of particles is related to the "center of charge" of the multiplet.

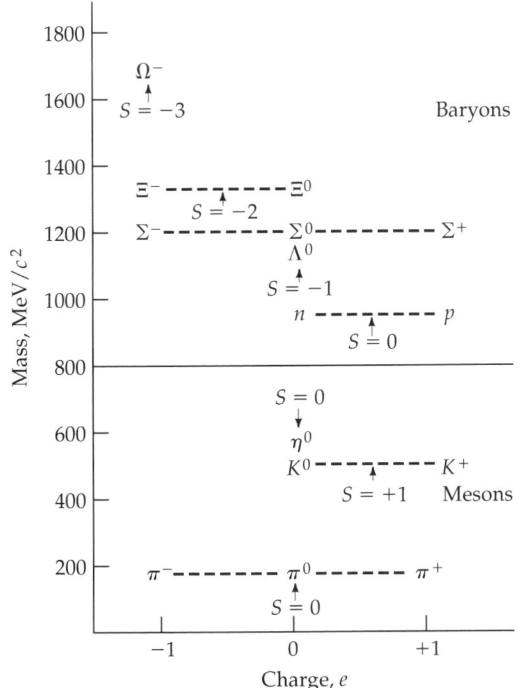

Figure 41-2 The strangeness of hadrons shown on a plot of rest mass versus charge. The strangeness of a baryon-charge multiplet is related to the number of places the center of charge of the multiplet is displaced from that of the nucleon doublet. For each displacement of $\frac{1}{2}e$, the strangeness changes by ± 1. For mesons, the strangeness is related to the number of places the center of charge is displaced from that of the pion triplet. Because of the unfortunate original assignment of $+1$ for the strangeness of kaons, all of the baryons that are stable against decay via the strong interaction have negative or zero strangeness.

Example 41-3

State whether the following decays can occur via the strong interaction, via the weak interaction, or not at all:

(*a*) $\Sigma^+ \rightarrow p + \pi^0$ (*b*) $\Sigma^0 \rightarrow \Lambda^0 + \gamma$ (*c*) $\Xi^0 \rightarrow n + \pi^0$

We first note the mass of each decaying particle is greater than that of the decay products, so there is no problem with energy conservation in any of the decays. In addition, there are no leptons involved in any of the decays, and charge and baryon number are both conserved in all the decays.

(*a*) From Figure 41-2, we can see that the strangeness of the Σ^+ is -1 whereas the strangeness of both the proton and the pion is zero. This decay is possible via the weak interaction but not the strong interaction. It is, in fact, one of the decay modes of the Σ^+ particle with a lifetime of the order of 10^{-10} s.

(*b*) Since the strangeness of both the Σ^0 and Λ^0 is -1, this decay can proceed via the strong interaction. It is, in fact, the dominant mode of decay of the Σ^0 particle with a lifetime of about 10^{-20} s.

(*c*) The strangeness of the Ξ^0 is -2 whereas the strangeness of both the neutron and pion is zero Since strangeness cannot change by 2 in a decay or reaction, this decay cannot occur.

Question

3. How can you tell whether a decay proceeds via the strong interaction or the weak interaction?

41-4 The Quark Model

We have seen that leptons appear to be truly elementary particles in that they do not break down into smaller entities and they seem to have no measurable size or structure. Hadrons, on the other hand, are complex particles with size and structure, and they decay into other hadrons. Furthermore, at the present time, there are only six known leptons, whereas there are many more hadrons. Table 41-1 includes only hadrons that are stable against decay via the strong interaction. Hundreds of other hadrons have been discovered, and their properties, such as charge, spin, mass, strangeness, and decay schemes, have been measured.

The most important advance in our understanding of elementary particles was the quark model proposed by M. Gell-Mann and G. Zweig in 1963. According to this model, all hadrons are thought to consist of combinations of two or three truly elementary particles called **quarks.** (The name *quark* was chosen by Gell-Mann from a quotation from *Finnegans Wake* by James Joyce.) In the original model, quarks came in three types, called **flavors,** labeled u, d, and s (for *up*, *down*, and *strange*). An unusual property of quarks is that they carry fractional electron charges. The charge of the u quark is $+\frac{2}{3}e$ and that of the d and s quarks is $-\frac{1}{3}e$. Each quark has spin $\frac{1}{2}\hbar$ and a baryon number of $\frac{1}{3}$. The strangeness of the u and d quark is 0 and that of the s quark is -1. Each quark has an antiquark with the opposite electric charge, baryon number, and strangeness. These properties are listed in Table 41-2. Baryons consist of three quarks (or three antiquarks for antiparticles), whereas mesons consist of a quark and an antiquark, giving them a baryon number $B = 0$, as required. The proton consists of the combination *uud* and the neutron, *udd*. Baryons with a strangeness $S = -1$ contain one s quark. All the particles listed in Table 41-1 can be constructed from these

Table 41-2 **Properties of Quarks and Antiquarks**

Flavor	Spin	Charge	Baryon number	Strangeness	Charm	Topness	Bottomness
Quarks							
u (up)	$\frac{1}{2}\hbar$	$+\frac{2}{3}e$	$+\frac{1}{3}$	0	0	0	0
d (down)	$\frac{1}{2}\hbar$	$-\frac{1}{3}e$	$+\frac{1}{3}$	0	0	0	0
s (strange)	$\frac{1}{2}\hbar$	$-\frac{1}{3}e$	$+\frac{1}{3}$	−1	0	0	0
c (charmed)	$\frac{1}{2}\hbar$	$+\frac{2}{3}e$	$+\frac{1}{3}$	0	+1	0	0
t (top)	$\frac{1}{2}\hbar$	$+\frac{2}{3}e$	$+\frac{1}{3}$	0	0	+1	0
b (bottom)	$\frac{1}{2}\hbar$	$-\frac{1}{3}e$	$+\frac{1}{3}$	0	0	0	+1
Antiquarks							
$\bar{u}$	$\frac{1}{2}\hbar$	$-\frac{2}{3}e$	$-\frac{1}{3}$	0	0	0	0
$\bar{d}$	$\frac{1}{2}\hbar$	$+\frac{1}{3}e$	$-\frac{1}{3}$	0	0	0	0
$\bar{s}$	$\frac{1}{2}\hbar$	$+\frac{1}{3}e$	$-\frac{1}{3}$	+1	0	0	0
$\bar{c}$	$\frac{1}{2}\hbar$	$-\frac{2}{3}e$	$-\frac{1}{3}$	0	−1	0	0
$\bar{t}$	$\frac{1}{2}\hbar$	$-\frac{2}{3}e$	$-\frac{1}{3}$	0	0	−1	0
$\bar{b}$	$\frac{1}{2}\hbar$	$+\frac{1}{3}e$	$-\frac{1}{3}$	0	0	0	−1

three quarks and three antiquarks.* The great strength of the quark model is that all the allowed combinations of three quarks or quark–antiquark pairs result in known hadrons. Strong evidence for the existence of quarks inside a nucleon is provided by high-energy scattering experiments called *deep inelastic scattering*. In these experiments, a nucleon is bombarded with electrons, muons, or neutrinos of energies from 15 to 200 GeV. Analyses of particles scattered at large angles indicate the presence within the nucleon of spin-$\frac{1}{2}$ particles of sizes much smaller than that of the nucleon. These experiments are analogous to Rutherford's scattering of α particles by atoms in which the presence of a tiny nucleus in the atom was inferred from the large-angle scattering of the α particles.

Example 41-4

What are the properties of the particles made up of the following quarks: (*a*) $u\bar{d}$, (*b*) $\bar{u}d$, (*c*) dds, and (*d*) uss?

(*a*) Since $u\bar{d}$ is a quark–antiquark combination, it has baryon number 0 and is therefore a meson. There is no strange quark here, so the strangeness of the meson is zero. The charge of the up quark is $+\frac{2}{3}e$ and that of the anti-down quark is $+\frac{1}{3}e$, so the charge of the meson is $+1e$. This is the quark combination of the π^+ meson.

(*b*) The particle $\bar{u}d$ is also a meson with zero strangeness. Its electric charge is $-\frac{2}{3}e + (-\frac{1}{3}e) = -1e$. This is the quark combination of the π^- meson.

(*c*) The particle dds is a baryon with strangeness -1 since it contains one strange quark. Its electric charge is $-\frac{1}{3}e - \frac{1}{3}e - \frac{1}{3}e = -1e$. This is the quark combination for the Σ^- particle.

(*d*) The particle uss is a baryon with strangeness -2. Its electric charge is $+\frac{2}{3}e - \frac{1}{3}e - \frac{1}{3}e = 0$. This is the quark combination for the Ξ^0 particle.

*The correct quark combinations of hadrons are not always obvious because of the symmetry requirements on the total wave function. For example, the π^0 meson is represented by a linear combination of $u\bar{u}$ and $d\bar{d}$.

In 1967, a fourth quark was proposed to explain some discrepancies between experimental determinations of certain decay rates and calculations based on the quark model. The fourth quark is labeled c for a new property called **charm.** Like strangeness, charm is conserved in strong interactions but changes by ± 1 in weak interactions. In 1975, a new heavy meson called the **ψ/J particle** (or simply the **ψ particle**) was discovered that has the properties expected of a $c\bar{c}$ combination. Since then other mesons with combinations such as $c\bar{d}$ and $\bar{c}d$, as well as baryons containing the charmed quark, have been discovered. Two more quarks labeled t and b (for *t*op and *b*ottom or, as some prefer, *t*ruth and *b*eauty) have been proposed. In 1977, a massive new meson called the **Υ meson** or **bottomonium,** which is considered to have the quark combination $b\bar{b}$, was discovered. At the present time there is no direct evidence of the existence of the top quark.

The six quarks and six leptons (and their antiparticles) are thought to be the fundamental, elementary particles of which all matter is composed. Table 41-3 lists some of the properties of quarks and leptons. In this table, the masses given for neutrinos are upper limits, and those given for quarks are educated guesses. There is experimental evidence for the existence of each of these particles except for the top quark. The existence of the top quark is strongly suspected because of symmetry between the number of quarks and the number of leptons.

Table 41-3 **Fundamental Particles and Their Approximate Masses**[†]

	Light	Medium	Heavy	Charge
Quarks	$u(\sim 400\ \text{MeV}/c^2)$	$c(\sim 1.5\ \text{GeV}/c^2)$	$t(>89\ \text{GeV}/c^2)$	$+\frac{2}{3}e$
	$d(\sim 700\ \text{MeV}/c^2)$	$s(\sim 0.15\ \text{GeV}/c^2)$	$b(\sim 4.7\ \text{GeV}/c^2)$	$-\frac{1}{3}e$
Leptons	$e(0.511\ \text{MeV}/c^2)$	$\mu(106\ \text{MeV}/c^2)$	$\tau(1.78\ \text{GeV}/c^2)$	$-1e$
	$\nu_e(<16\ \text{eV}/c^2)$[‡]	$\nu_\mu(<300\ \text{keV}/c^2)$[‡]	$\nu_\tau(<40\ \text{MeV}/c^2)$[‡]	0

[†]Because quarks are always bound in mesons or baryons, their masses are not well understood. The values given here are merely educated guesses.

[‡]These masses are upper limits. Neutrinos may be massless particles.

Quark Confinement

Despite considerable experimental effort, no isolated quark has ever been observed. It is now believed that it is impossible to obtain an isolated quark. This would be true, for example, if the force between two quarks remains constant regardless of separation distance, rather than decreasing with increasing separation distance as is the case for other fundamental forces such as the electric force between two charges, the gravitational force between two masses, and the strong nuclear force between two hadrons. As a result, the potential energy of two quarks increases with increasing separation distance so that an infinite amount of energy would be needed to separate the quarks completely.

When a large amount of energy is added to a quark system such as a nucleon, a quark–antiquark pair is created and the original quarks remain confined within the original system. Because quarks cannot be isolated, but are always bound in a baryon or meson, the mass of a quark cannot be accurately known, which is why the masses listed in Table 41-3 are merely educated guesses.

Questions

4. How can you tell whether a particle is a meson or a baryon by looking at its quark content?
5. Are there any quark–antiquark combinations that result in a nonintegral electric charge?

41-5 Field Particles

In addition to the six fundamental leptons and six fundamental quarks, there are other particles called *field particles* or *field quanta* that are associated with the forces exerted by one elementary particle on another. As we have seen, in quantum electrodynamics the electromagnetic field of a single charged particle is described by virtual photons that are continuously being emitted and reabsorbed by the particle. If we put energy into the system by accelerating the charge, some of these virtual photons can be "shaken off" and become real, observable photons. The photon is said to mediate the electromagnetic interaction. Each of the four basic interactions can be described in this way.

The field quantum associated with the gravitational interaction, called the **graviton,** has not yet been observed. The gravitational "charge" analogous to electric charge is mass.

The weak interaction is thought to be mediated by three field quanta called **vector bosons:** W^+, W^-, and Z^0. These particles were predicted by S. Glashow, A. Salam, and S. Weinberg in theory called the *electroweak theory,* which we will discuss in the next section. The W and Z particles were first observed in 1983 by a group of over a hundred scientists lead by C. Rubbia using the high-energy accelerator at CERN in Geneva, Switzerland. The masses of the $W^\pm$ particles (about 80 GeV/c^2) and the Z particle (about 91 GeV/c^2) measured in this experiment were in excellent agreement with those predicted by the electroweak theory. (The W^- particle is the antiparticle of the W^+ particle, so they must have identical masses.)

The field quanta associated with the strong force between quarks are called **gluons.** Isolated gluons have not been observed experimentally. The "charge" responsible for the strong interactions comes in three varieties labeled *red, green,* and *blue* (analogous with the three primary colors), and the strong charge is called the **color charge.** The field theory for strong interactions, analogous to quantum electrodynamics for electromagnetic interactions, is called **quantum chromodynamics (QCD).**

Table 41-4 lists the bosons responsible for mediating the basic interactions.

Table 41-4 **Bosons that Mediate the Basic Interactions**

Interaction	Boson	Spin	Mass	Electric charge
Strong	g (gluon)	1	0	0
Weak	$W^\pm$	1	79.8 GeV/c^2	$\pm 1e$
	Z^0	1	91.2 GeV/c^2	0
Electromagnetic	γ (photon)	1	0	0
Gravitational	Graviton†	2	0	0

†Not yet observed

Example 41-5

If a particular weak interaction is mediated by emission of a virtual Z particle, calculate the order of magnitude of the range of the interaction.

From the uncertainty principle, the time that a virtual Z particle can exist without violating the conservation of energy is

$$\Delta t \approx \frac{\hbar}{\Delta E} \approx \frac{\hbar}{m_Z c^2}$$

The range is then given by Equation 41-2:

$$R = c\ \Delta t \approx \frac{\hbar c}{m_Z c^2} = \frac{(6.58 \times 10^{-16}\ \text{eV}\cdot\text{s})(3 \times 10^{8}\ \text{m/s})}{91 \times 10^{9}\ \text{eV}} \approx 2 \times 10^{-18}\ \text{m}$$

41-6 The Electroweak Theory

In the **electroweak theory,** the electromagnetic and weak interactions are considered to be two different manifestations of a more fundamental electroweak interaction. At very high energies ($\gg$100 GeV), the electroweak interaction would be mediated by four bosons. From symmetry considerations, these would be a triplet consisting of W^+, W^0, and W^-, all of equal mass, and a singlet B^0 of some other mass. Neither the W^0 nor the B^0 would be observed directly, but one linear combination of the W^0 and the B^0 would be the Z^0 and another would be the photon. At ordinary energies, the symmetry is broken. This leads to the separation of the electromagnetic interaction mediated by the massless photon and the weak interaction mediated by the W^+, W^-, and Z^0 particles. The fact that the photon is massless and that the W and Z particles have masses of the order of 100 GeV/c^2 shows that the symmetry assumed in the electroweak theory does not exist at lower energies.

The symmetry-breaking mechanism is called a **Higgs field,** which requires a new boson, the **Higgs boson,** whose rest energy is expected to be of the order of 1 TeV (1 TeV = 10^{12} eV). The Higgs boson has not yet been observed. Calculations show that Higgs bosons (if they exist) should be produced in head-on collisions between protons of energies of the order of 20 TeV. Such energies are not presently available but will be in a proposed new superconducting super collider (SSC) scheduled to be completed in the mid 1990s.

41-7 The Standard Model

The combination of the quark model, electroweak theory, and quantum chromodynamics is called the **standard model.** In this model, the fundamental particles are the leptons and quarks, each of which comes in six flavors as shown in Table 41-3, and the force carriers are the photon, the $W^{\pm}$ and Z particles, and the gluons (of which there are eight types). The leptons and quarks are all spin-$\frac{1}{2}$ fermions, which obey the Pauli exclusion principle, and the force carriers are integral-spin bosons, which do not obey the Pauli exclusion principle. Every force in nature is due to one of the four basic interactions: strong, electromagnetic, weak, and gravitational. A particle experiences one of the basic interactions if it carries a charge associated with that interaction. Electric charge is the familiar charge that we have studied previously. Weak charge, also called flavor charge, is carried by leptons and quarks. The charge associated with the strong interaction is called color charge and is carried by quarks and gluons but not by leptons. The charge associated with the gravitational force is mass. It is important to note that the photon, which mediates the electromagnetic interaction, does not carry electric charge. Similarly, the $W^{\pm}$ and Z particles, which mediate the weak interaction, do not carry weak charge. However, the gluons, which mediate the strong interaction, do carry color charge. This fact is related to the confinement of quarks as discussed in Section 41-4.

All matter is made up of leptons or quarks. There are no known composite particles consisting of leptons bound together by the weak force. Leptons exist only as isolated particles. Hadrons (baryons and mesons) are composite particles consisting of quarks bound together by the color charge. A result of QCD theory is that only color-neutral combinations of quarks are allowed. Three quarks of different colors can combine to form color-neutral baryons, such as the neutron and proton. Mesons contain a quark and an antiquark and are also color-neutral. Excited states of hadrons are considered to be different particles. For example, the Δ^+ particle is an excited state of the proton. Both are made up of the *uud* quarks, but the proton is in the ground state with spin $\frac{1}{2}$ and a rest energy of 938 MeV, whereas the Δ^+ particle is the first excited state with spin $\frac{3}{2}$ and a rest energy of 1232 MeV. The two u quarks can be in the same spin state in the Δ^+ without violating the exclusion principle because they have different color. All baryons eventually decay to the lightest baryon, the proton. The proton cannot decay because the conservation of energy and baryon number.

The strong interaction has two parts, the fundamental or color interaction and what is called the *residual strong interaction*. The fundamental interaction is responsible for the force exerted by one quark on another and is mediated by gluons. The residual strong interaction is responsible for the force between color-neutral nucleons, such as the neutron and proton. This force is due to the residual strong interactions between the color-charged quarks that make up the nucleons and can be viewed as being mediated by the exchange of mesons. The residual strong interaction between color-neutral nucleons can be thought of as analogous to the residual electromagnetic interaction between neutral atoms that bind them together to form molecules. Table 41-5 lists some of the properties of the basic interactions.

Table 41-5 **Properties of the Basic Interactions**

	Interaction				
				Strong	
	Gravitational	Weak	Electromagnetic	Fundamental	Residual
Acts on	Mass	Flavor	Electric charge	Color charge	
Particles experiencing	All	Quarks, leptons	Electrically charged	Quarks, gluons	Hadrons
Particles mediating	Graviton	$W^{\pm}$, Z	γ	Gluons	Mesons
Strength for two quarks at 10^{-18} m†	10^{-41}	0.8	1	25	(Not applicable)
Strength for two protons in nucleus†	10^{-36}	10^{-7}	1	(Not applicable)	20

†Strengths are relative to electromagnetic strength

For each particle there is an antiparticle. A particle and its antiparticle have identical mass and spin but opposite electric charge. For leptons, the lepton numbers L_e, L_μ, and L_τ of the antiparticles are the negatives of the corresponding numbers for the particles. For example, the lepton number for the electron is $L_e = +1$ and that for the positron is $L_e = -1$. For hadrons, the baryon number, strangeness, charm, topness, and bottomness are the sums of those quantities for the quarks that make up the hadron. The number of each antiparticle is the negative of the number for the corresponding particle. For example, the lambda particle Λ^0, which is made up of the *uds* quarks, has $B = 1$ and $S = -1$, whereas its antiparticle $\overline{\Lambda}^0$, which is made up of the $\overline{u}\overline{d}\overline{s}$ quarks, has $B = -1$ and $S = +1$. A particle such as the photon γ or the Z^0 particle that has zero electric charge, $B = 0$, $L = 0$, $S = 0$, and zero charm, topness, and bottomness, is its own antiparticle. Note that the K^0 meson ($d\overline{s}$) has a zero value for all of these quantities except strangeness, which is +1. Its antiparticle, the $\overline{K}^0$ meson ($\overline{d}s$) has strangeness −1, which makes it distinct from the K^0. The π^+ ($u\overline{d}$) and π^- ($\overline{u}d$) are somewhat special in that they have electric charge but zero values for L, B, and S. They are antiparticles of each other, but since there is no conservation law for mesons, it is impossible to say which is the particle and which is the antiparticle. Similarly, the W^+ and W^- are antiparticles of each other.

41-8 Grand Unification Theories

With the success of the electroweak theory, attempts have been made to combine the strong, electromagnetic, and weak interactions in various **grand unification theories** known as **GUTs**. In one of these theories, leptons and quarks are considered to be two aspects of a single class of particles. Under certain conditions, a quark could change into a lepton and vice versa, even though this would appear to violate the conservation of lepton number and baryon number. One of the exciting predictions of this theory is that the proton is not stable but merely has a very long lifetime of the order of 10^{31} y. Such a long lifetime makes proton decay difficult to observe.

It was Einstein's dream to be able to describe all the forces in nature through one unified theory. Whether this will ever be accomplished is an open question. There is considerable experimental effort underway to observe the decay of the proton and to test other predictions of the grand unification theories. At the same time, much theoretical work is being done to refine these theories and construct others to further our understanding of our universe.

Summary

1. There are four basic interactions: strong, electromagnetic, weak, and gravitational. All particles with mass experience the force due to the gravitational interaction. All particles with electric charge experience the force due to the electromagnetic interaction. The "charge" associated with the weak interaction is called flavor. Quarks and leptons have flavor and experience the weak interaction. The "charge" associated with the strong interaction is called color. Quarks and gluons have color and experience the strong interaction. Hadrons (baryons and mesons) experience a residual strong interaction resulting from the fundamental strong interaction between the quarks that make up the hadrons.

2. In quantum field theory, each interaction is mediated by the exchange of one or more field particles. The field particle associated with the electromagnetic interaction is the photon, and that associated with the gravitational interaction is the graviton, which has not yet been observed. The field particles associated with the weak interaction are the W^+, W^-, and Z^0. The field particles associated with the strong interaction between quarks are called gluons. All the field particles are bosons. The graviton has spin 2, whereas all of the other field particles have spin 1. The residual strong interaction between hadrons is mediated by mesons, which are also bosons with spin 0 (or 1).
3. There are two families of fundamental particles, leptons and quarks, each containing six members. It is thought that these particles have no size and no internal structure. The leptons are spin-$\frac{1}{2}$ fermions: the electron e and its neutrino ν_e, the muon μ and its neutrino ν_μ, and the tau τ and its neutrino ν_τ. The electron, muon, and tau have mass, electric charge, and flavor but not color, so they participate in the gravitational, electromagnetic, and weak interactions but not the strong interaction. The neutrinos have flavor but no electric charge and no color. They may be massless, but whether or not this is so is still an open question. The quarks are also spin-$\frac{1}{2}$ fermions. The six quarks are called up u, down d, strange s, charmed c, top t, and bottom b. The top quark has not yet been observed. The quarks participate in all of the basic interactions. Because they are always confined in mesons or baryons, their masses can only be estimated.
4. Hadrons are composite particles that are made up of quarks. There are two types, baryons and mesons. Baryons, which include the neutron and proton, are fermions of half-integral spin consisting of three quarks. Mesons, which include pions and kaons, have zero or integral spin. Hadrons interact with each other via the residual strong interaction.
5. Some quantities, such as energy, momentum, electric charge, angular momentum, baryon number, and each of the three lepton numbers, are strictly conserved in all reactions and decays. Others, such as strangeness and charm, are conserved in reactions and decays that proceed via the strong interaction but not in those that proceed via the weak interaction.
6. Particles and their antiparticles have identical masses but opposite values for their other properties, such as charge, lepton number, baryon number, and strangeness. Particle–antiparticle pairs can be produced in various nuclear reactions if the energy available is greater than $2mc^2$, where m is the mass of the particle.

Suggestions for Further Reading

Bloom, Elliott D., and Gary J. Feldman: "Quarkonium," *Scientific American*, May 1982, p. 66.

How quark–antiquark pairs have been observed and are being investigated in order to learn more about the color force binding them together is discussed in this article.

Close, Frank, Michael Martin, and Christine Sutton: *The Particle Explosion*, Oxford University Press, Oxford, England, 1987.

This colorful book describes the discovery of the subatomic particles: the people involved and the instruments they used.

Glashow, Sheldon: "Tangled in Superstring: Some Thoughts on the Predicament Physics Is In," *The Sciences*, May/June 1988, p. 22.

A Nobel Prize–winning physicist raises doubts about a recent theoretical approach to particle physics.

Jackson, J. David, Maury Tigner, and Stanley Wojcicki: "The Superconducting Supercollider," *Scientific American*, March 1986, p. 66.

This article explains why a giant, new colliding-beam particle accelerator 84 km in circumference is needed to test possible extensions of the electroweak theory and new theories such as supersymmetry.

LoSecco, J. M., Frederick Reines, and Daniel Sinclair: "The Search for Proton Decay," *Scientific American*, June 1985, p. 54.

Experimenters who have taken part in the design and operation of the largest of the water Cerenkov proton-decay detectors describe the history of the search and the contribution of their experiment to it.

Quigg, Chris: "Elementary Particles and Forces," *Scientific American*, April 1985, p. 84.

This article provides an overview of leptons, hadrons, and quarks; the fundamental interactions; the unified theory of the electromagnetic and weak interactions; and possible further unification encompassing the strong force. New experimental directions such as the search for proton decay and the Superconducting Supercollider are also discussed.

Sutton, Christine: *The Particle Connection: The Most Exciting Scientific Chase Since DNA and the Double Helix*, Simon and Schuster, New York, 1984.

An excellent 175-page account of the search for the particles that mediate the electroweak interaction: W^+, W^-, *and* Z^0. *The book requires as background only an introductory physics course.*

von Baeyer, Hans Christian: "The Voltage Makers," *The Sciences*, January/February 1988, p. 6.

An alternative to the Superconducting Supercollider (SSC), the "beam transformer," may allow further progress in particle physics.

von Baeyer, Hans Christian: "The Atomic Cathedral," *The Sciences*, January/February 1987, p. 8.

Some reflections on the medieval and the modern quest for knowledge, inspired by similarities in the principles of design of cathedrals and particle accelerators.

Weinberg, Steven: "The Decay of the Proton," *Scientific American*, June 1981, p. 64.

One of the three who shared the 1979 Nobel Prize in physics for the theory unifying the electromagnetic and weak interactions describes why attempts at unification of these two forces with the strong force lead to predictions of an extremely small but finite rate of decay for the proton.

Review

A. Objectives: After studying this chapter, you should:

1. Be able to list the four basic interactions and name some of the particles that participate in each interaction.

2. Be able to name the particle or particles that mediate each of the basic interactions.

3. Be able to discuss the difference between hadrons and leptons and between baryons and mesons.

4. Be able to discuss the quark model of hadrons.

5. Be able to apply the various conservation laws to tell whether a given decay or reaction can proceed via the strong interaction, the weak interaction, or not at all.

B. Define, explain, or otherwise identify:

Strong interaction
Electromagnetic interaction
Gravitational interaction
Weak interaction
Hadron
Baryon
Meson
Quantum electrodynamics (QED)
Virtual photon
Pion
Lepton
Spin-$\frac{1}{2}$ particles
Fermion
Boson
Baryon number
Lepton number
Strangeness
Strange particle
Quark
Flavor
Charm
ψ/J particle
ψ particle
Υ meson
Bottomonium
Graviton
W particle
Z particle
Gluon
Color charge
Quantum chromo dynamics (QCD)
Electroweak theory
Higgs field
Higgs boson
Standard model
Grand Unification theories (GUTs)

C. True or false: If the statement is true, explain why it is true. If it is false, give a counterexample.

1. Leptons are fermions.

2. All baryons are hadrons.

3. All hadrons are baryons.

4. Mesons are spin-$\frac{1}{2}$ particles.

5. Leptons consist of three quarks.

6. The times for decays via the weak interaction are typically much longer than those for decays via the strong interaction.

7. The electron interacts with the proton via the strong interaction.

8. Strangeness is not conserved in weak interactions.

9. Neutrons have no charm.

Problems

In the following problems, use Figure 41-2 to find the strangeness of hadrons.

Level I

41-1 Hadrons and Leptons

1. Suppose the force between two nucleons were mediated by a kaon. What would be the approximate range of the force?

41-2 Spin and Antiparticles

2. Two pions at rest annihilate according to the reaction $\pi^+ + \pi^- \rightarrow \gamma + \gamma$. (*a*) Why must the energies of the two gamma rays be equal? (*b*) Find the energy of each gamma ray. (*c*) Find the wavelength of each gamma ray.

3. Find the minimum energy of the photon needed for the following pair-production reactions: (*a*) $\gamma \rightarrow \pi^+ + \pi^-$, (*b*) $\gamma \rightarrow p + p^-$, and (*c*) $\gamma \rightarrow \mu^- + \mu^+$.

41-3 The Conservation Laws

4. State which of the decays or reactions that follow violate one or more of the conservation laws, and give the law or laws violated in each case: (*a*) $p^+ \rightarrow n + e^+ + \bar{\nu}_e$, (*b*) $n \rightarrow p^+ + \pi^-$, (*c*) $e^+ + e^- \rightarrow \gamma$, (*d*) $p + p^- \rightarrow \gamma + \gamma$, and (*e*) $\nu_e + p \rightarrow n + e^+$.

5. Determine the change in strangeness in each reaction that follows, and state whether the reaction can proceed via the strong interaction, the weak interaction, or not at all: (*a*) $\Omega^- \rightarrow \Xi^0 + \pi^-$, (*b*) $\Xi^0 \rightarrow p + \pi^- + \pi^0$, and (*c*) $\Lambda^0 \rightarrow p^+ + \pi^-$.

6. Determine the change in strangeness for each decay, and state whether the decay can proceed via the strong interaction, the weak interaction, or not at all: (*a*) $\Omega^- \rightarrow \Lambda^0 + K^-$ and (*b*) $\Xi^0 \rightarrow p + \pi^-$.

7. Determine the change in strangeness for each decay, and state whether the decay can proceed via the strong interaction, the weak interaction, or not at all: (*a*) $\Omega^- \rightarrow \Lambda^0 + \bar{\nu}_e + e^-$ and (*b*) $\Sigma^+ \rightarrow p + \pi^0$.

8. (*a*) Which of the following decays of the tau particle is possible?

$$\tau \rightarrow \mu^- + \bar{\nu}_\mu + \nu_\tau$$

$$\tau \rightarrow \mu^- + \nu_\mu + \bar{\nu}_\tau$$

(*b*) Calculate the kinetic energy of the decay products for the decay that is possible.

41-4 The Quark Model

9. Find the baryon number, charge, and strangeness for the following quark combinations and identify the hadron: (*a*) *uud*, (*b*) *udd*, (*c*) *uus*, (*d*) *dds*, (*e*) *uss*, and (*f*) *dss*.

10. Repeat Problem 9 for the following quark combinations: (*a*) $u\bar{d}$, (*b*) $\bar{u}d$, (*c*) $u\bar{s}$, and (*d*) $\bar{u}s$.

11. The Δ^{++} particle is a baryon that decays via the strong interaction. Its strangeness, charm, topness, and bottomness are all zero. What combination of quarks gives a particle with these properties?

12. Find a possible combination of quarks that gives the correct values for electric charge, baryon number, strangeness for (*a*) K^+ and (*b*) K^0.

13. The D^+ meson has no strangeness, but it has charm of +1. (*a*) What is a possible quark combination that will give the correct properties for this particle? (*b*) Repeat (*a*) for the D^- meson, which is the antiparticle of the D^+.

14. Find a possible combination of quarks that gives the correct values for electric charge, baryon number, strangeness for (*a*) K^- (the K^- is the antiparticle of the K^+) and (*b*) $\bar{K}^0$.

41-5 Field Particles

15. What is the approximate range of a weak interaction that is mediated by the W^+ particle?

16. The grand unification theory predicts an *X* particle that would change a quark into a lepton and vice versa. What would be the approximate range of a virtual *X* particle if its mass were 10^{15} GeV/c^2?

41-6 The Electroweak Theory

There are no problems for this section.

41-7 The Standard Model

17. (*a*) What conditions are necessary for a particle and its antiparticle to be the same? Find the antiparticle for (*b*) π^0 and (*c*) Ξ^0.

41-8 Grand Unification Theories

There are no problems for this section.

Level II

18. Find a possible quark combination for the following particles: (*a*) Λ^0, (*b*) p^-, and (*c*) Σ^-.

19. Find a possible quark combination for the following particles: (*a*) $\bar{n}$, (*b*) Ξ^0, and (*c*) Σ^+.

20. Find a possible quark combination for the following particles: (*a*) Ω^- and (*b*) Ξ^-.

21. State the properties of the particles made up of the following quarks: (*a*) *ddd*, (*b*) $u\bar{c}$, (*c*) $u\bar{b}$, and (*d*) $\bar{s}\bar{s}\bar{s}$.

22. Consider the following decay chain:

$$\Xi^0 \rightarrow \Lambda^0 + \pi^0$$

$$\Lambda^0 \rightarrow p + \pi^-$$

$$\pi^0 \rightarrow \gamma + \gamma$$

$$\pi^- \rightarrow \mu^- + \bar{\nu}_\mu$$

$$\mu^- \rightarrow e^- + \bar{\nu}_e + \nu_\mu$$

(*a*) Are all the final products shown stable? If not, finish the decay chain. (*b*) Write the overall decay reaction for Ξ^0 to the final products. (*c*) Check the overall decay reaction for the conservation of electric charge, baryon number, lepton number, and strangeness. (*d*) In the first step of the chain, could the Λ^0 have been a Σ^0?

23. Consider the following decay chain:

$$\Omega^- \rightarrow \Xi^0 + \pi^-$$
$$\Xi^0 \rightarrow \Sigma^+ + e^- + \bar{\nu}_e$$
$$\pi^- \rightarrow \mu^- + \bar{\nu}_\mu$$
$$\Sigma^+ \rightarrow n + \pi^+$$
$$\pi^+ \rightarrow \mu^+ + \nu_\mu$$
$$\mu^+ \rightarrow e^+ + \bar{\nu}_\mu + \nu_e$$
$$\mu^- \rightarrow e^- + \bar{\nu}_e + \nu_\mu$$

(*a*) Are all the final products shown stable? If not, finish the decay chain. (*b*) Write the overall decay reaction for Ω^- to the final products. (*c*) Check the overall decay reaction for the conservation of electric charge, baryon number, lepton number, and strangeness.

24. Test the following decays for violation of the conservation of energy, electric charge, baryon number, and lepton number:
(*a*) $n \rightarrow \pi^+ + \pi^- + \mu^+ + \mu^-$
(*b*) $\pi^0 \rightarrow e^+ + e^- + \gamma$
Assume that linear and angular momentum are conserved. State which conservation laws (if any) are violated in each decay.

25. Test the following decays for violation of the conservation of energy, electric charge, baryon number, and lepton number:
(*a*) $\Lambda^0 \rightarrow p + \pi^-$
(*b*) $\Sigma^- \rightarrow n + p^-$
(*c*) $\mu^- \rightarrow e^- + \bar{\nu}_e + \nu_\mu$
Assume that linear and angular momentum are conserved. State which conservation laws (if any) are violated in each decay.

Level III

26. (*a*) Calculate the total kinetic energy of the decay products for the decay

$$\Lambda^0 \rightarrow p + \pi^-$$

Assume the Λ^0 is initially at rest. (*b*) Find the ratio of the kinetic energy of the pion to the kinetic energy of the proton. (*c*) Find the kinetic energies of the proton and the pion for this decay.

27. A Σ^0 particle at rest decays into a Λ^0 plus a photon. (*a*) What is the total energy of the decay products? (*b*) Assuming that the kinetic energy of the Λ^0 is negligible compared with the energy of the photon, calculate the approximate momentum of the photon. (*c*) Use your result for (*b*) to calculate the kinetic energy of the Λ^0. (*d*) Use your result for (*c*) to obtain a better estimate of the momentum and the energy of the photon.

28. In this problem, you will calculate the difference in the time of arrival of two neutrinos of different energy from a supernova that is 170,000 light-years away. Let the energies of the neutrinos be $E_1 = 20$ MeV and $E_2 = 5$ MeV, and assume that the rest mass of a neutrino is 20 eV/c^2. Because their total energy is so much greater than their rest energy, the neutrinos have speeds that are very nearly equal to c and energies that are approximately $E \approx pc$. (*a*) If t_1 and t_2 are the times it takes for neutrinos of speeds u_1 and u_2 to travel a distance x, show that

$$\Delta t = t_2 - t_1 = x\frac{u_1 - u_2}{u_1 u_2} \approx \frac{x\,\Delta u}{c^2}$$

(*b*) The speed of a neutrino of rest mass m_0 and total energy E can be found from Equation 34-32. Show that when $E >> m_0c^2$, the speed u is given approximately by

$$\frac{u}{c} \approx 1 - \frac{1}{2}\left(\frac{m_0c^2}{E}\right)^2$$

(*c*) Use the results for (*b*) to calculate $u_1 - u_2$ for the energies and rest mass given, and calculate Δt from the result for (*a*) for $x = 170{,}000$ $c{\cdot}$y. (*d*) Repeat the calculation in (*c*) using $m_0c^2 = 40$ eV for the rest energy of a neutrino.

Chapter 42

Astrophysics and Cosmology

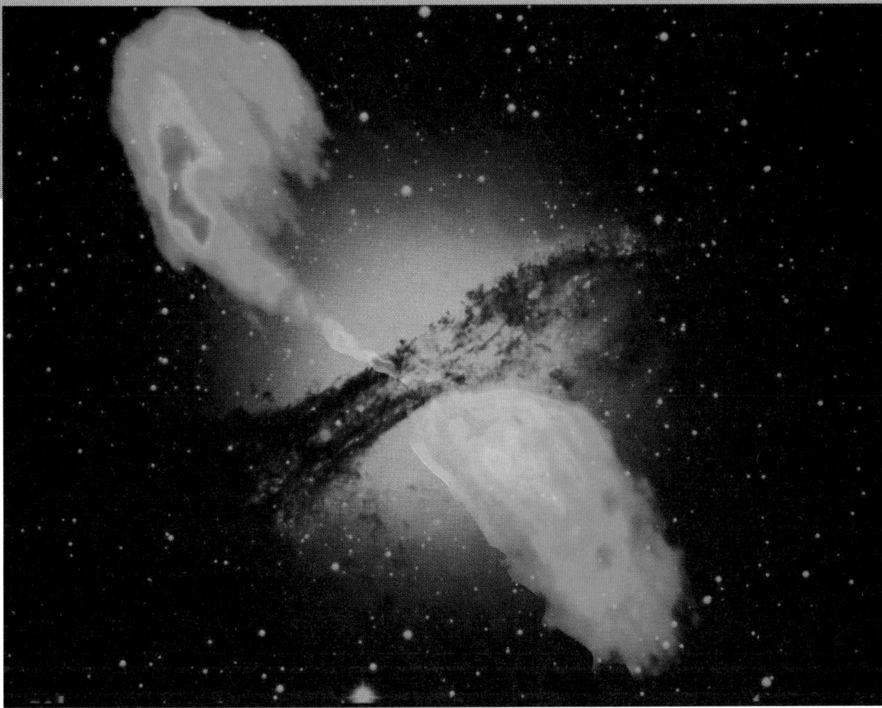

An optical image of galaxy NGC 5128, over which has been superposed a radio image of Centaurus A, a radio source within the galaxy. The optical appearance of the galaxy is very peculiar in that it shows a heavy dust zone surrounding the system in a plane perpendicular to the galaxy's rotation. The intense double radio lobes of Centaurus A extend across 0.12° of the sky, corresponding to a distance of 54,800 $c \cdot y$. The core, or nucleus, of the galaxy is also a source of radio emissions. Not shown are much fainter and larger radio lobes that surround the intense lobes and extend across 10° of the sky. (The full moon covers only 0.5° of the sky.) The thin cone-shaped blue and red emission extending from the nucleus to the northern lobe is a jet of high-velocity gas 15,000 $c \cdot y$ in length. The faint stars surrounding the galaxy lie in shells that computer analyses associate with collisions of galaxies and suggest that Centaurus A is the result of such a collision that occurred about one billion years ago (see also page 209).

Physics is an experimental science. The formulation and acceptance of our current understanding of the physical world, from Newton's laws through Maxwell's equations to relativity theory and quantum mechanics, are based on countless experimental observations. In this chapter, we look outward from the earth into the cosmos and apply the principles and techniques of physics first to the composition and evolution of stars, a branch of physics called **astrophysics,** and then to the large-scale structure of the universe, a field called **cosmology.**

When observing stars and galaxies, astrophysicists and cosmologists are limited to examining the electromagnetic radiation and occasional particles emitted at times past that happen to have traveled to the vicinity of the earth, arriving at the moment of observation. The information thus gained, together with the fundamental assumption that the laws of physics discovered here on the earth are also valid throughout the universe, forms the basis for their work. During most of history, the instrument used for studying the cosmos has been the human eye. Though well adapted to life on earth, the eye is a relatively poor instrument for the scientific examination of the sky because it stores information for only a small fraction of a second before that information is transmitted to the brain for analysis. Today, most of our information about the distant universe is received through telescopes (see Chapter 32).

42-1 Our Star, the Sun

As we look outward from the earth and beyond the moon, the most obvious object in the sky is, of course, the sun. It is important to us for several reasons: The light that reaches us from the sun is responsible for life on the earth. It sustains a comfortable average temperature on the earth's surface and is the ultimate source of virtually all of our energy. Since the sun contains nearly all of the mass of the solar system, it also provides the gravitational force that binds our planet to the system. But most important for our purposes in this section, the sun is the only star of the 100 billion or so in the Milky Way galaxy that is close enough for us to examine its surface. All of the others are so far away that they appear only as point sources when viewed by even the largest telescopes. What we learn from studies of our star not only provides us with a more complete understanding of the processes taking place on it but surely applies to other stars as well.

The Surface and Atmosphere of the Sun

We can see only the thin layer of the sun, the **photosphere,** which emits the light that makes the sun visible. The photosphere is generally considered to be the surface of the sun. The emitted energy per second per square meter that arrives from the sun at the top of the earth's atmosphere is called the **solar constant** f. It has been measured to be

$$f = 1.36 \times 10^3 \text{ W/m}^2 \tag{42-1}$$

This quantity for stars other than the sun is called the apparent brightness, as we will see in Section 42-3. Using the solar constant, the earth–sun distance of 1 AU = 1.5×10^8 km, and the conservation of energy, we can calculate the **luminosity** L, which is the total power radiated by the sun or by any star. The area A of a sphere with a radius of 1 AU is

$$A = 4\pi r^2 = 4\pi(1.50 \times 10^{11} \text{ m})^2$$

At that radius, each square meter receives energy from the sun at the rate given by the solar constant. Therefore, the sun's luminosity $L_\odot$ is given by

$$L_\odot = Af = 4\pi(1.50 \times 10^{11} \text{ m})^2(1.36 \times 10^3 \text{ W/m}^2) = 3.85 \times 10^{26} \text{ W} \tag{42-2}$$

This is truly enormous power. If we could put a 1000-MW electricity-generating plant on each square meter of the earth's surface, all of them combined would produce only 0.1 percent of the power produced by the sun.

If we assume that the sun radiates as a blackbody, we can use the luminosity of the sun along with its radius (6.96×10^8 m) to calculate the effective temperature at the surface of the sun from the Stefan–Boltzmann law (Equation 16-21). It states that the intensity I (the power per unit area) radiated by a blackbody in thermal equilibrium is proportional to the fourth power of its surface temperature:

$$I = \sigma T^4 \tag{42-3}$$

where $\sigma = 5.67 \times 10^{-8}$ W/m²·K⁴ and T is the absolute temperature. If the radius of the sun is $R_\odot$, the intensity radiated at the surface of the sun is

$$I = \frac{L_\odot}{4\pi R_\odot^2} \tag{42-4}$$

The effective temperature T_e of the surface of the sun is defined as the temperature for which the intensity radiated satisfies the Stefan–Boltzmann law for a blackbody:

$$I = \frac{L_\odot}{4\pi R_\odot^2} = \sigma T_e^4$$

Solving for T_e, we obtain

$$T_e = \left(\frac{I}{\sigma}\right)^{1/4} = \left(\frac{L_\odot}{4\pi R_\odot^2 \sigma}\right)^{1/4} \qquad 42\text{-}5$$

Example 42-1

Use the Stefan–Boltzmann law to calculate the effective temperature of the photosphere.

Using $L_\odot = 3.85 \times 10^{26}$ W in Equation 42-5, we have

$$T_e = \left(\frac{L_\odot}{4\pi R_\odot^2 \sigma}\right)^{1/4} = \left[\frac{3.85 \times 10^{26}\ \text{W}}{4\pi(6.96 \times 10^8\ \text{m})^2(5.67 \times 10^{-8}\ \text{W/m}^2\cdot\text{K}^4)}\right]^{1/4}$$

$$= 5800\ \text{K}$$

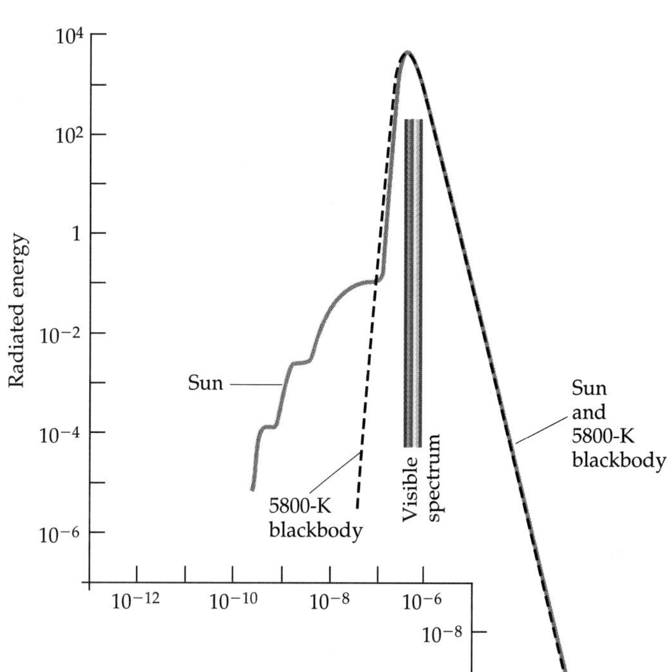

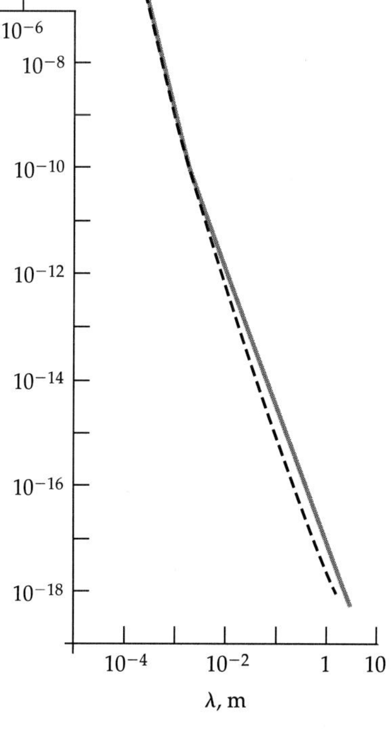

Figure 42-1 The spectral distribution of the energy emitted by the sun closely matches that of a blackbody at 5800 K. The discrepancies are mainly due to the fact that the photosphere is not in thermal equilibrium. The hump at short wavelengths is due to x rays emitted by the corona, which is at a much higher temperature.

The intensity of solar radiation has been measured at wavelengths ranging from about 10^{-13} m in the gamma-ray region to nearly 10 m in the radio region, a range accounting for over 99 percent of the sun's emitted power. Over much of this span, the solar spectrum is quite well predicted by Planck's law of blackbody radiation (see Chapter 35) with $T = 5800$ K as shown in Figure 42-1. The distribution peaks in the yellow region of the visible wavelengths, which is of course why the sun looks yellow. This agreement between the measured and theoretical spectra is very constant and is one of the characteristics of the **quiet sun.**

If we examine the edge of the solar disc, called the **limb,** we see that it is sharply demarcated and darker than the rest of the sun. From the sharpness of the limb, we conclude that the photosphere is very thin from the following reasoning: Atmospheric turbulence during daylight limits the angular resolution of optical telescopes to about 1 arc second (1/3600 of a degree) or more. At the distance of the sun, this corresponds to about 700 km. As we look at the sun, the angle over which the gas of the photosphere changes from rarefied and transparent to optically dense and opaque is smaller than we can resolve. Therefore, the photosphere must be less than 700 km thick, which is only about 0.1 percent of the solar radius.

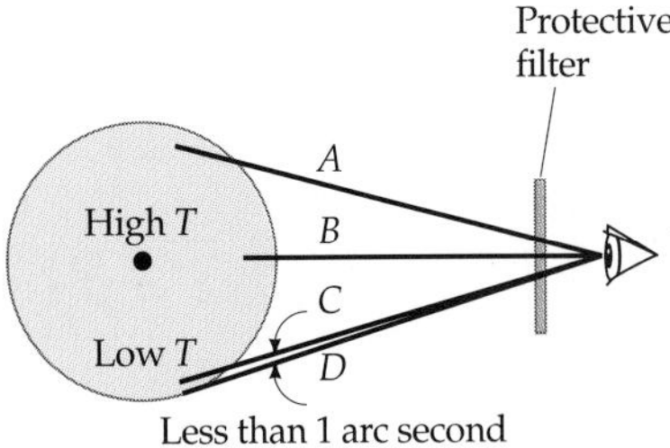

Figure 42-2 Because the photosphere is more transparent when viewed at normal incidence than when viewed at a grazing angle, the light traveling along path *B* originates deeper in the sun than the light traveling along path *A* and therefore looks brighter. The limb looks darker. The change in brightness from path *C* to path *D* is smaller than we can resolve, so the limb looks sharp.

The relatively dark appearance of the limb tells us about the temperature gradient in the sun's atmosphere. Figure 42-2 shows two paths *A* and *B* for viewing the sun. Because the photosphere is more transparent when viewed at normal incidence than when viewed at a grazing angle, the light traveling along path *B* originates deeper in the sun than light traveling along path *A*. Since the interior is hotter than the outer layers, the light traveling along path *B* originates in a hotter (brighter) part of the sun than the light traveling path *A*. Thus, the light from the limb appears darker (cooler). By measuring the change in brightness from path *A* to path *B*, we can determine the temperature gradient in the photosphere. It is shown in the left portion of Figure 42-3. Notice in the right portion of Figure 42-3 that the temperature begins to rise sharply, accompanying the transition from the sun's surface, the photosphere, into the solar atmosphere.

Outside the photosphere are two layers of the sun's atmosphere that are not generally seen because of the brightness of the photosphere. The innermost of the two layers of the solar atmosphere, the **chromosphere,** is visible for the first few seconds of totality during a solar eclipse. Under high resolution, the chromosphere resembles a field of burning grass, although each burning "blade" is about 700 km thick and 7000 to 10,000 km high and lasts for only 5 to 15 minutes. Spectral examination indicates that the temperature of the chromosphere *increases* with distance above the photosphere, averaging about 15,000 K.

When the totality of the eclipse blocks out the chromosphere, the outer layer of the sun's atmosphere, the **corona,** becomes visible. It is decidedly nonuniform in thickness, consisting of faint white streamers that extend two to three solar diameters into space, as shown in Figure 42-4. The temperature of the corona is approximately 2,000,000 K. Radiation from the corona would overpower that from the 5800-K photosphere, except for the fact that the gas of the corona is so rarefied that the total energy it emits is miniscule compared to that emitted by the photosphere. It does, however, account for the relatively high intensity of x rays emitted by the sun, which show up in Figure 42-1 as a deviation from the spectral distribution of the blackbody at short wavelengths. It is thought that the extreme temperatures in the corona are produced by acoustic waves generated in the sun's interior that build into shock waves in the corona. These shock waves heat the gases of the outer atmosphere and give the particles so much energy that even the sun's intense gravity cannot confine them. These high-energy particles, mostly electrons and protons, stream outward from the corona continuously. They form the **solar wind** that pervades the entire solar system.

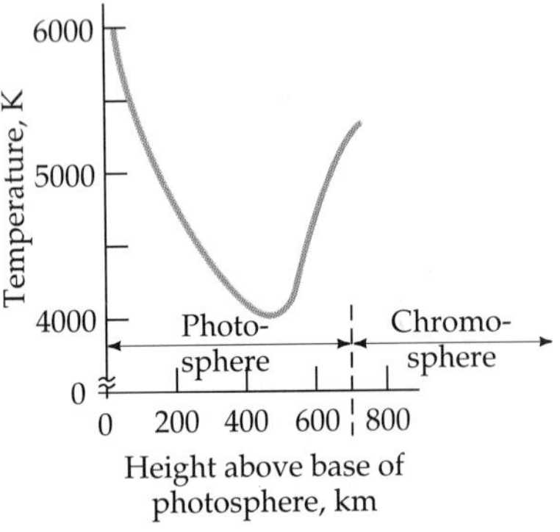

Figure 42-3 The temperature of the sun decreases from the base of the photosphere outward to a minimum at about 500 km. It then increases sharply to an average of about 15,000 K in the chromosphere.

Figure 42-4 The hot but rarefied corona becomes visible during a total solar eclipse.

The Sun's Interior

We cannot see through the photosphere into the interior of the sun. Consequently, our understanding of the processes that occur there is purely theoretical. With the single exception of solar neutrinos, no radiation or particles originating in the interior reach us directly. (As we will see, the single bit of direct observational evidence we do receive is at odds with the current theory.)

For simplicity, theoretical models usually consider the sun to be a nonrotating star in hydrostatic equilibrium. This means that the outward pressure at any point, which is presumed to be due to the energy-conversion processes occurring within the sun, is exactly balanced by the inward pressure of gravity. Although the mean density of the sun (1.4 g/cm^3) is not much different from that of the earth (5.5 g/cm^3), the enormous pressures that exist in the solar interior substantially exceed those that correspond to the electrodynamic forces that bind electrons to nuclei. Thus, the matter in the interior of the sun—and certainly that within the **core,** the central region in which temperatures are high enough to allow hydrogen fusion—must surely be in the plasma (ionized) state.

Example 42-2

Show that neutral hydrogen is unlikely to exist in the sun's interior.

The pressure at the center of the sun P_c is of the order $P_c = \mu g$, where μ = mass/unit surface area $\approx M_\odot/R_\odot^2$ and $g = \frac{1}{2}GM_\odot/R_\odot^2$ is the average acceleration of gravity in the sun (see Chapter 10). The pressure turns out to be about 10^{16} N/m^2. This is the pressure pushing on the surface of a hydrogen atom near the sun's center. The resistance to this gravitational pressure would come from the Coulomb force tending to hold the atom together. That pressure is given by the Coulomb attraction between the proton and electron per unit surface area of the atom. Using the Bohr radius a_0 for hydrogen, we have

$$\frac{F}{A} = \frac{ke^2/a_0^2}{4\pi a_0^2} = \frac{ke^2}{4\pi a_0^4} = \frac{(9 \times 10^9)(1.6 \times 10^{-19})^2}{4\pi(0.5 \times 10^{-10})^4}$$

$$= 2.9 \times 10^{12}\ \text{N/m}^2$$

Thus, the gravitational pressure in the sun's interior, at least near the center, exceeds that tending to hold the hydrogen atoms together by about a factor of 10,000—making it unlikely that neutral hydrogen atoms could exist there.

However, given the sun's density, the particles even in the depths of the core are still relatively far apart, so the plasma behaves much like an ideal gas. This allows us to calculate the core temperature from the ideal-gas law. It is found to be 1.5×10^7 K.

The Source of the Sun's Energy

Using the value for the luminosity of the sun that we computed earlier, the present energy content of the sun, as calculated from thermodynamics, would be radiated away in about 3×10^7 y. Since life has existed on earth for approximately a hundred times that long, we can conclude that the sun has been radiating at close to its present luminosity for at least 3×10^9 y. Therefore, the sun must have a supply of energy far larger than that represented by the hot plasma and the observed radiation field.

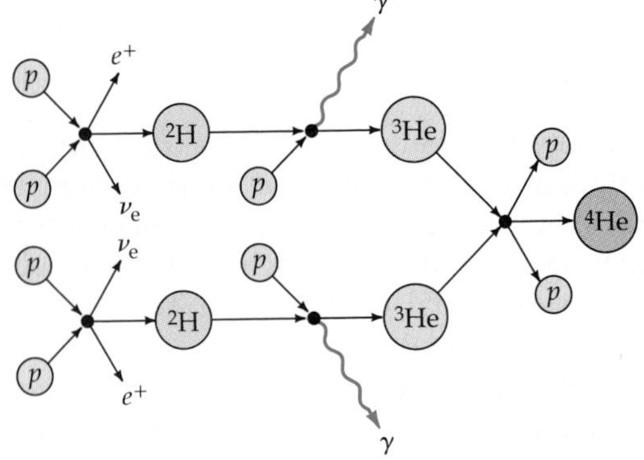

Figure 42-5 The proton–proton cycle is the primary source of the sun's energy. The neutrino created in the initial reaction escapes from the core. The net energy produced in the cycle is about 26.7 MeV.

The source of the sun's energy is nuclear fusion. Current theory proposes that as the young sun contracted its temperature rose. Eventually, the temperature of the core reached about 1.5×10^7 K, which is high enough for the hydrogen nuclei (protons) in the plasma to have sufficient energy on the average (about 1 keV) to fuse into helium nuclei. This reaction, actually a chain of reactions, was first proposed by H. A. Bethe in 1938 and is referred to as the **proton–proton cycle** (see Figure 42-5). The first reaction in this chain is

$$^1\text{H} + {}^1\text{H} \longrightarrow {}^2\text{H} + e^+ + \nu_e + 1.44 \text{ MeV} \qquad 42\text{-}6$$

The probability of this reaction occurring is very low except for those protons in the high-energy tail of the Maxwell–Boltzmann distribution. This sets a limit on the rate at which the sun can produce energy and thus ensures a long lifetime for the sun and similar stars. This limit is sometimes called the "bottleneck" in the solar-fusion cycle. Once ^{2}H (deuterium) is formed via Reaction 42-6, the following reaction becomes probable:

$$^2\text{H} + {}^1\text{H} \longrightarrow {}^3\text{He} + \gamma + 5.49 \text{ MeV} \qquad 42\text{-}7$$

It is followed by

$$^3\text{He} + {}^3\text{He} \longrightarrow {}^4\text{He} + 2\,{}^1\text{H} + \gamma + 12.86 \text{ MeV} \qquad 42\text{-}8$$

This process by which hydrogen nuclei are "burned" to helium nuclei is shown schematically in Figure 42-5. There are other possible reactions for converting ^{3}He to ^{4}He, all of which have the same net Q-value. Their rates, however, differ depending on the composition and temperature of the interior.

The neutrinos produced in the proton–proton cycle escape from the core, providing our only means for direct observation of the sun's interior. The measured luminosity $L_\odot$ and the known total Q-value of the proton–proton cycle enable a calculation of the total reaction rate. In addition, the alternative reactions for ^{3}He have different neutrino energy spectra, thus providing a way of determining the relative contributions of each reaction and gaining information about the core's composition and temperature. However, the measured rate at which solar neutrinos arrive at the earth is less than half that predicted by theoretical calculations based on the standard solar model. This discrepancy is referred to as the **solar-neutrino problem.**

The as-yet-unresolved solar-neutrino problem has several implications, two of which are particularly important for our purposes. First, there may be a serious gap in our understanding of the properties and behavior of neutrinos. Second, if our theoretical understanding of neutrinos is essentially accurate, then there is a serious error in the current standard solar model. Such an error would have far-reaching ramifications for theories of stellar evolution. For example, Stephen Hawking suggests the possibility that part of the sun's emission of energy may arise from gravitational energy that is released when mass falls into a small black hole at the sun's center. This means that there would be less fusion occurring than current theory suggests and, hence, fewer neutrinos.

The Active Sun

In addition to the relatively stable phenomena that we have discussed, the sun exhibits a number of transient phenomena, most of them associated with its magnetism. We noted earlier that the solar interior must be primarily a plasma composed of protons and electrons. The sun rotates with different angular velocities at different latitudes. At any given latitude, it probably has different angular velocities at different distances from the spin axis as well. The complex motions resulting from this differential rotation and from the rise and fall of charged particles in the convection zone between the core and the photosphere are probably the source of the sun's chaotic magnetic-field structure (see Figure 42-6). This transient structure may have localized magnetic-field strengths exceeding 1 T on occasion.

The transient structure is superimposed onto a general average magnetic field of about 10^{-4} T. The origin of this general field is not known, except that it is not a remnant from the sun's formation since any primeval field would have decayed away by now. Its presence poses formidable problems for any theoretical solar model. Not only must the model explain the origin of the general field, but it must also account for the fact that its polarity reverses every 11 years, in step with the **sunspot cycle.**

Sunspots, dark blemishes on the solar disc, were first reported in pretelescope times and were observed with a telescope by Galileo in 1610. They originate in the following way, according to one of the current models: As shown in Figure 42-7, the sun's magnetic-field lines are distorted into bundles or tubes by the sun's differential rotation. Occasionally, vertical movements in the convection zone may push a bundle through the surface. The area where it leaves the surface and the area where it returns to the surface become the sunspots. They appear darker than the adjacent photosphere, which means that they are cooler, typically around 3800 K. One of the pair of spots will have a magnetic north pole and the other will have a south pole. If the bundle of field lines doesn't protrude completely through the photosphere, only a single sunspot is formed.

The number of sunspots per year varies regularly from about 50 to about 150 in a cycle of 11 years, as can be seen in Figure 42-8. Early in each new cycle, the sunspots form at a latitude of about 30°. As the sun progresses through its 11-year cycle, the spots form progressively closer to the equator. There is an additional cyclical variation in the annual number of sunspots with a period of about 100 years that is also apparent in Figure 42-8. Currently, there is no theoretical explanation for these regularities.

Solar flares are violent, storm-like phenomena that appear to be associated with the large magnetic fields in the vicinity of sunspots. A spectacular flare is shown in photo (*a*) on page 790. There is, however, no generally

Figure 42-6 The magnetic-field lines of the sun show a chaotic structure.

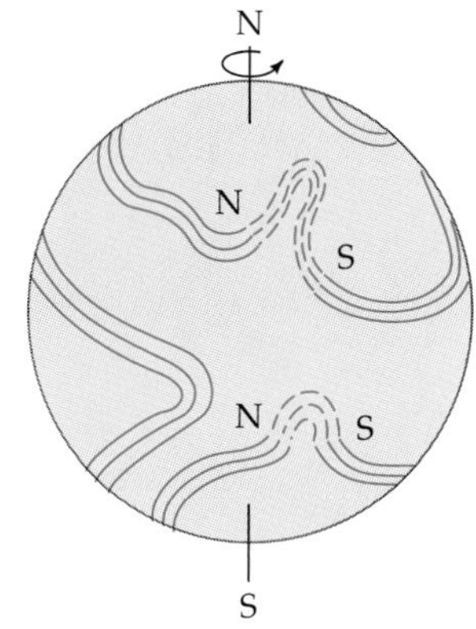

Figure 42-7 The magnetic-field lines of the sun are distorted by the differential rotation of the sun. Sunspots occur where a bundle of field lines leaves and reenters the surface.

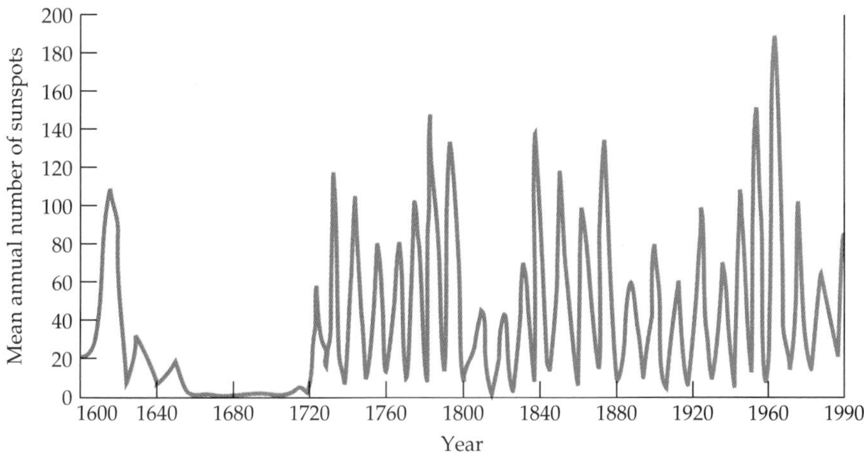

Figure 42-8 Number of sunspots versus year. This number has varied regularly on an 11-year cycle for more than 270 years. The unexplained absence of sunspots between about 1650 and 1700, referred to as the Maunder minimum, coincides approximately with a period of unusually low temperatures in Europe known as the Little Ice Age.

accepted model to explain them. Solar flares erupt explosively, ejecting particles and emitting radiation ranging from the x-ray through the radio wavelengths of the spectrum. They last anywhere from a few minutes to a few hours and can have temperatures as high as 5×10^6 K. The particles ejected by solar flares reach the earth within a day or so and often produce auroras as they interact with the earth's magnetic field. (See the essay following Chapter 26.) Solar flares can disrupt some types of radio transmissions and, on rare occasions, can generate surges in high-voltage transmission lines.

Two other transient solar phenomena are plages and filaments. **Plages** are bright (hotter) areas adjacent to the dark sunspots. The evolution of plages suggest that they are areas of increased mass density, resulting perhaps from the movement of the magnetic field bundles generating from the sunspots. **Filaments** are dark, thin lines that thread their way across the disc, sometimes for thousands of kilometers. They do not lie on the surface but extend out into space, sometimes more than 100,000 kilometers, in graceful loops and swirls. Filaments that are seen projecting into space at the sun's edge are called **prominences.** They may erupt and disappear quickly or persist for several weeks. Although prominences appear to be closely related to the shape of the magnetic field, as with other transient features, there is no model that fully accounts for them.

42-2 The Stars

On clear, dark nights, we can see about 6000 stars without the aid of a telescope. The sight is incredibly beautiful and must surely have been just as awesome to our forbearers as it is to us. A cursory glance at the night sky reveals the following features: The distribution of stars is not uniform, the stars do not all have the same brightness, and there is a dim, irregular band of light bisecting the sky. In this section, we will investigate these features.

The hazy band of light that stretches across the entire sky is the Milky Way. With the aid of a small telescope or even binoculars, the band is resolved into a mass of individual stars. It is part of a huge galaxy containing an estimated 10^{11} stars that are bound together gravitationally in our region of the universe. (The term *galaxy* is derived from the Greek word for "milk.") Most of the stars visible to the unaided eye seen in any direction are simply those members of the Milky Way galaxy that are close enough to the earth to be individually resolved by the eye.

Constellations

Chance groupings in the celestial pattern, usually among the brighter individual stars, are called **constellations.** They were associated by ancient peoples with persons, gods, and objects from their histories, religions, and myths, probably as mnemonic devices. The constellations, as well as several prominent stars, have always had practical uses. For centuries, seafarers have used the Pole Star (in the northern hemisphere) and the Southern Cross (in the southern hemisphere) as aids in navigation. In ancient Egypt, the pharaoh's advisors learned to predict the life-sustaining annual flooding of the Nile by watching for the first appearance of the bright star of Sirius above the horizon in the early spring. Today, eighty-eight constellations (see Figure 42-9 for some of them) are used by astronomers to identify sections of the sky. For example, the center of the Milky Way galaxy is said to be "in Sagittarius," meaning that it is in the direction of the constellation Sagittarius. (The center of the Galaxy is actually more than ten times farther from the sun than are the stars that form the constellation.)

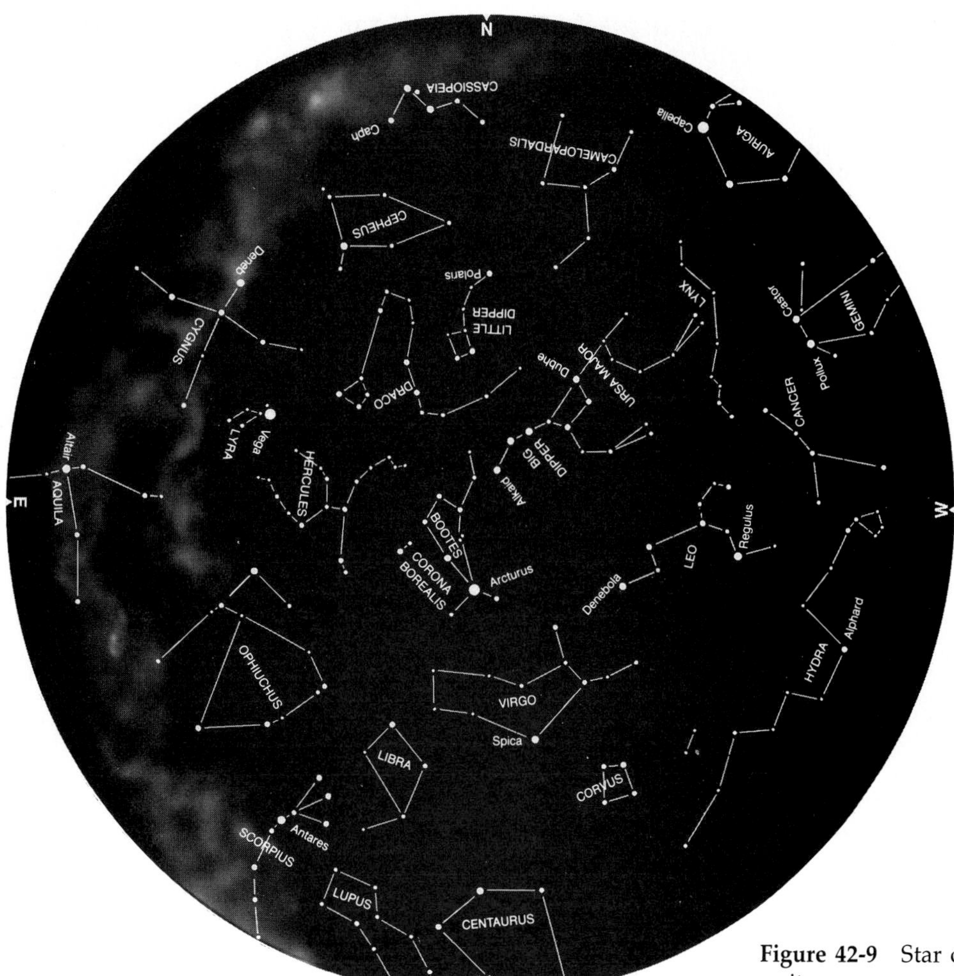

Figure 42-9 Star chart of the sky as it appears on a spring evening at latitude 40° north, showing many of the constellations visible. During the night, the entire pattern revolves about 120° about the Pole Star. To use the chart, hold it (or a copy) in front of you with the S (south) at the bottom while you face south. Match the lower half to the stars that you see. Then rotate the chart, putting the W at the bottom, face west, and again match the lower half to the stars you see, and so on.

Stellar Populations

One characteristic of our Galaxy is that certain regions of it have many more stars than other nearby regions. Such concentrations are called **star clusters.** Clusters are groups of stars that collectively have a small angular diameter, implying that they are all at about the same distance from us. There are two types of star clusters: galactic clusters and globular clusters. **Galactic clusters,** also called **open clusters,** may contain from about 20 to several hundred stars. One such cluster, photographed by the Hubble space telescope, is shown on page 1057. All of them appear to have very similar compositions, as inferred from studies of their optical spectra. About 70 percent of their mass is hydrogen, another 28 percent or so is helium, and 2 to 3 percent consists of elements heavier than helium. Stars with this characteristic composition, like our sun, are referred to as **population I stars. Globular clusters** may consist of 10^3 to 10^6 stars in a compact, roughly spherical group. Their concentrations of elements heavier than helium are all very similar and are much lower than those of population I stars, typically 0.1 to 0.01 percent. These are called **population II stars.**

Population I stars are thought to be current-generation stars that formed after the gas and dust that exist between these stars had been enriched by the products of ancient fusion reactions in the early universe. The lower concentrations of heavier elements in the population II stars suggest that they are of a previous generation, hence older than those of population I. The fact that they are found in regions of space where there is little dust or gas tends to support this interpretation.

Figure 42-10 A hand drawing of the Milky Way from the perspective of a viewer at the sun.

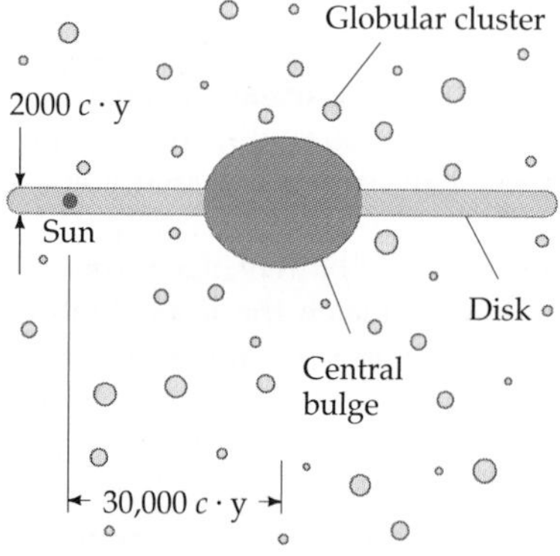

Figure 42-11 A diagram of the presently accepted structure of the Milky Way galaxy based on the work of Harlow Shapley.

The Structure of the Milky Way Galaxy

Figure 42-10 is a map of the Milky Way viewed from the location of the sun. The size and shape of the Galaxy are not at all obvious—hardly surprising from the perspective of an observer inside the Galaxy itself. However, painstaking counts of the number of stars per unit volume in various directions have revealed that the Milky Way galaxy is basically a huge disk. Up until the early 1900s, astronomers thought the sun was located at the disk's center. The true size and shape of the Galaxy (see Figure 42-11) were deduced by Harlow Shapley in 1917 through a brilliant analysis of the distribution of globular clusters. He discovered that 200 or so globular clusters are distributed approximately spherically in space and proposed that the center of that distribution coincided with the center of our Galaxy. The center lies about 30,000 light-years from the sun. It has been said that Shapley dethroned the sun from the center of the Galaxy much as Copernicus had dethroned the earth from the center of the universe.

Following Shapley's work, astronomers studying nearby galaxies with the aid of new, high-resolution telescopes found that the distribution of stars within those systems, many of which have open-spiral structures like Andromeda shown on page 9 and Centaurus shown in Figure 42-19*b*, depends in part upon the ages and compositions of the stars, with open clusters being found mainly in the arms of the spirals. Making the reasonable assumption that such distribution patterns would also hold for the Milky Way galaxy, meticulous measurements of the distances to about 200 open clusters enabled the identification of three spiral arms for the Milky Way galaxy. Thus, if we could look down on the Milky Way galaxy from the galactic north pole, it would look much like Figure 42-12*a*.

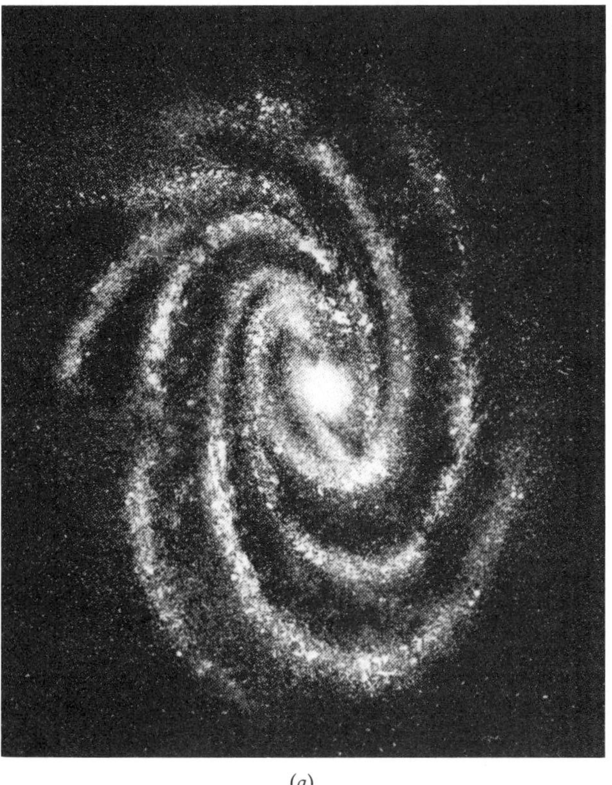

(a)

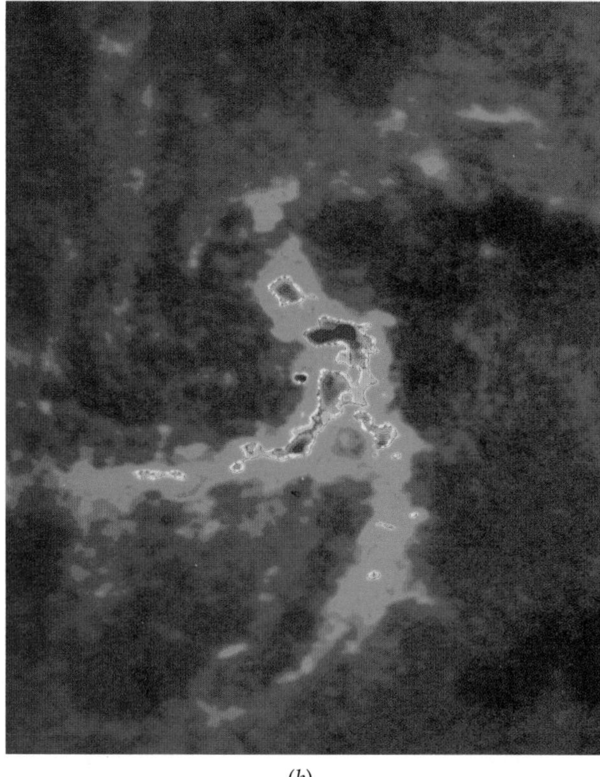

(b)

Figure 42-12 (a) The combination of observations in the visible and radio regions of the spectrum reveal a spiral structure for the Milky Way. To an observer looking down on the Galaxy from about a million parsecs, the Milky Way might look like this. The cross marks the position of the sun. (b) Viewed from the earth, the center of the Galaxy is obscured by clouds of dust and gas that prevent visible light from reaching us; however, it contains several areas of strong radio emission, the strongest of which is Sagittarius A, a compact radio source that appears to dominate the large-scale motion of the galactic center. This is a radio image (taken at 6-cm wavelength) of the inner 8 c·y of the Milky Way. The dark red spot at the very center is Sagittarius A, which some astronomers think may contain a huge black hole. This image was made using the Very Large Array, a radio frequency interferometer made of 27 synchronized antennae with an effective diameter of about 20 km, located in New Mexico. Its resolution is better than that of the best ground-based optical telescopes by about a factor of five.

The Mass (and Missing Mass) of the Milky Way Galaxy

Using the doppler effect, J. Oort and B. Lindblad first demonstrated in 1926 that the Galaxy is rotating. The sun is apparently moving in a circular orbit at a speed of about 2.5×10^5 m/s toward the constellation Cygnus. Assuming that the sun's speed is constant, we can compute the length of the galactic year, that is, the time for the Galaxy to complete one revolution, and the mass of the Galaxy. Since the sun is 30,000 c·y from the galactic center, the galactic year is 2.3×10^8 years (see Problem 3).

Example 42-3

Estimate the mass of that part of the Galaxy that lies inside the sun's orbit from Newton's law of gravity and the sun's orbital speed.

Setting the gravitational force on the sun equal to the mass of the sun $M_\odot$ times its centripetal acceleration, we obtain

$$\frac{GM_\odot M_G}{R^2} = \frac{M_\odot v^2}{R} \qquad 42\text{-}9$$

where M_G is the mass of the Galaxy, v is the sun's orbital speed, R is the distance from the sun to the galactic center, and G is the gravitational constant. Solving for M_G, we obtain

$$M_G = \frac{v^2 R}{G} = \frac{(2.5 \times 10^5 \text{ m/s})^2(30{,}000\ c\cdot\text{y})}{6.67 \times 10^{-11} \text{ N}\cdot\text{m}^2/\text{kg}^2}$$

$$= 2.66 \times 10^{41} \text{ kg} = (1.3 \times 10^{11})\ M_\odot$$

Thus, if the sun's mass is a representative average for the stars of the Milky Way, the Galaxy contains some 1.3×10^{11} stars.

If we add together the masses of all of the visible stars in the Galaxy, including those beyond the sun's orbit, plus all of the dust and gas clouds, even if we confine the examination just to the solar neighborhood where seeing through the interstellar dust and gas is not a serious problem, we can account for only about 10 percent of the gravitational mass necessary to hold the Galaxy together. This discrepancy is referred to as the **missing-mass problem.** It exists for all galaxies and, indeed, for the universe itself. Various solutions to the problem, such as black holes, the possibility that neutrinos have mass, and as yet undiscovered, weakly interacting massive particles called WIMPs, are under intense investigation and debate but as yet there is no clear experimental support for any of them.

42-3 The Evolution of Stars

While no universally accepted theory of stellar formation exists, it is generally agreed that stars are formed from the massive clouds of dust and gas that exist throughout space. At some point in the swirling cloud, gravitational attraction begins to cause aggregations of matter to collect. These contract further due to gravity, attracting still more matter to them and eventually, if the cloud has sufficient mass, increasing the temperature to that necessary to initiate fusion; and a star is born.

In this section, we discuss how stars evolve once they have been formed. Two characteristics of stars are important for this discussion, the luminosity L and the effective temperature T_e. The effective temperature of a star is difficult to measure. It is usually inferred from a comparison of the spectral distribution of its radiation with that of a blackbody or from measurements of the absorption lines of hydrogen and helium in the atmosphere of the star.

The luminosity is the total power radiated by the star. It is determined from the apparent brightness of the star at the earth f (called the solar constant for the sun) and the distance r from the earth to the star (Equation 42-2):

$$L = 4\pi r^2 f \qquad \text{42-10}$$

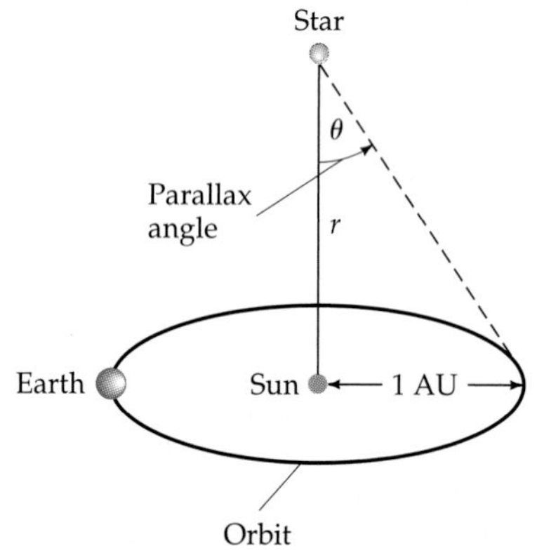

Figure 42-13 The parallax method of finding distances to nearby stars. A parsec is the distance r for which the parallax angle subtended by 1 AU is 1″.

Determining the distance to a star is generally a very difficult task. For stars that are relatively close, the distance can be determined from the apparent motion of the star due to the motion of the earth around the sun. During one complete revolution of the earth, a star appears to move in a circle of angular radius θ called the **parallax angle** as shown in Figure 42-13. The parallax angle is given by

$$\theta = \frac{1\ \text{AU}}{r} \qquad \text{42-11}$$

Astronomical distances are often measured in parsecs or light-years. One parsec is that distance at which 1 AU subtends an angle of 1 arc second (1″), which equals 1/3600 of a degree. Setting $\theta = 1''$ in Equation 42-11, we obtain

$$1\ \text{parsec} = \frac{1\ \text{AU}}{1''} \times \frac{3600''}{1^\circ} \times \frac{180^\circ}{\pi\ \text{rad}} = 2.0626 \times 10^5\ \text{AU} \qquad \text{42-12}$$

Using $1\ \text{AU} = 1.496 \times 10^{11}$ m and $1\ c{\cdot}y = 9.461 \times 10^{15}$ m, we can express the parsec in terms of meters or light-years:

$$1\ \text{parsec} = 3.086 \times 10^{16}\ \text{m} = 3.26\ c{\cdot}y \qquad \text{42-13}$$

Example 42-4

Proxima Centauri is the star closest to the sun. By measuring the apparent change in the direction of Proxima Centauri between two observations made six months apart, the parallax angle θ is found to be 0.765″. How far is it to Proxima Centauri?

Since 1 AU/1″ equals 1 parsec, we have for $\theta = 0.765''$

$$r = \frac{1\text{ AU}}{\theta} = \frac{1\text{ AU}}{0.765''} = \frac{1\text{ AU}}{1''}\,\frac{1''}{0.765''} = 1.31\text{ parsecs} = 4.27\ c\cdot\text{y}$$

The parallax-angle method of Example 42-4 can be used for only about 8000 stars that are relatively close to the sun. For the rest, the parallax angle is immeasurably small. In other situations, more indirect measurements of distance are necessary. These involve complex analyses of intensity variations over time for particular types of pulsating stars found in clusters. Thus, the distances to many clusters can be found. There is as yet no accurate method for determining the distances to individual, nonpulsating stars that are far away.

The various states of stars can be conveniently displayed by plotting the luminosity L versus the effective temperature T_e. The result is called a **Hertzprung–Russell (H–R) diagram.** Figure 42-14 shows an H–R diagram for some stars of representative masses. Each point represents a single star.

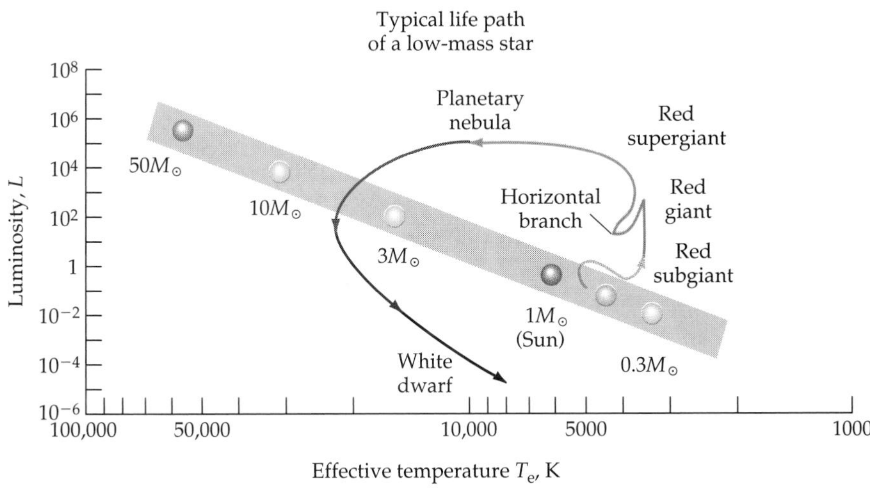

Figure 42-14 A Hertzprung–Russell (H–R) diagram. The points shown are the locations of stars of the indicated masses that have just formed. The shaded band is called the main sequence.

The large majority of stars on an H–R diagram fall in the shaded band called the **main sequence.** Main-sequence stars are normal in that they are homogeneous mixtures except in the core, they have essentially the same chemical composition, and they are fusing hydrogen into helium via one or another of the nuclear reactions discussed earlier. When stars leave the main sequence, they do so by expanding. Thus, stars in the main sequence are often called **main-sequence dwarfs.**

The location of a star along the main sequence in the H–R diagram depends on its luminosity, which is primarily dependent on the mass of the star. The masses of stars range from about $0.08M_\odot$ to about $60M_\odot$, where $M_\odot$ is the mass of our sun. Gaseous objects with masses less than about $0.08M_\odot$ do not have enough gravity for their central cores to be compressed sufficiently to generate the high temperature necessary for sustaining the fusion reactions needed for energy emission. Objects with masses greater than $60M_\odot$ would generate such enormous internal temperatures that the outward radiation pressure would exceed the gravity-generated inward pressure. Such a system would be very unstable, if indeed it could form at all.

The luminosity of a star is approximately proportional to the fourth power of its mass:

$$L \propto M^4 \tag{42-14}$$

The lifetime of a star t_L is proportional to the total available energy, which is proportional to the star's mass ($E = Mc^2$), and inversely proportional to the rate of energy emission, which is the luminosity:

$$t_L = \frac{E}{L} \propto \frac{Mc^2}{M^4} \propto M^{-3} \tag{42-15}$$

Thus, more massive stars burn their hydrogen more quickly than do less massive stars. (Equation 42-15 doesn't work for very small or very large stars because the luminosity–mass relationship of Equation 42-14 is only an average result. The exponent in Equation 42-15 is larger in magnitude for very small stars and smaller for very large stars.)

Considerations of energy balance for stars on the main sequence lead to the approximate proportionality of the radius and the mass:

$$R \propto M \tag{42-16}$$

Combining this with Equation 42-5, which relates the effective temperature to the luminosity per unit area, we can relate the effective temperature to the mass of the star:

$$T_e = \left(\frac{L}{4\pi R^2 \sigma}\right)^{1/4} \propto \left(\frac{M^4}{M^2}\right)^{1/4} \propto M^{1/2} \tag{42-17}$$

Thus, stars with larger masses have higher effective temperatures and, hence, higher luminosities than do those with lower masses. It is on the basis of Equations 42-5 and 42-17 that the stellar masses were plotted on the H–R diagram in Figure 42-14.

As a star ages, it consumes its primary fuel, hydrogen. What happens to it as the hydrogen supply in the core becomes exhausted depends on its initial mass. Low-mass and high-mass stars follow somewhat different evolutionary paths. In either case, however, the fundamental processes involved are successive nuclear reactions fueled by the product of the previous cycle. Thus, after the hydrogen in the core has fused to helium, the star must begin fusing helium in a cycle that eventually forms carbon. Before this can occur, the core must heat up still further to the 10^8 K necessary to initiate helium fusion. The chain of events involved in this process is complex and beyond the scope of this book. However, its result for low-mass stars is that the radius (and therefore the surface area) increases while the luminosity remains nearly constant. Thus, the intensity (the luminosity per unit area) and, consequently, the effective temperature decrease and the radiation emitted shifts to longer wavelengths as the star expands to become a **red subgiant.** The photosphere rapidly becomes more transparent as T_e decreases, thus increasing the luminosity and effectively limiting the decrease in temperature. The star is then a **red giant.** The track of a typical evolving low-mass star is shown on the H–R diagram in Figure 42-14.

Helium ignition results in the star again increasing its effective temperature and moving to the **horizontal branch.** When the helium in the core is exhausted, the star begins fusing carbon and ascends the red-giant branch again becoming a red supergiant. What happens after this is not clear.

Figure 42-15 Nebula 30 Doradus, also known as the Tarantula nebula. The shape of the rapidly expanding gas cloud changes over time. (Compare this photo with that of the Helix nebula on page 538.) Located in the Large Magellanic Cloud, the Tarantula contains one of the most massive stars known, as well as supernova SN1987A, the very bright star at the lower right.

Through a combination of events that includes the loss of considerable mass and perhaps passing once (or more) through a **planetary-nebula** stage, such as the nebula shown in Figure 42-15 and on page 538, the star becomes a white dwarf, slowly cooling toward thermal equilibrium with the universe. We will discuss white dwarfs further in Section 42-5.

High-mass stars—those with masses greater than about $6M_\odot$—evolve much more quickly than low-mass ones, as predicted by Equation 42-15. In addition, they have sufficient initial mass to generate gravitationally the high pressures and temperatures necessary to ignite the fusion reactions with oxygen, neon, and then silicon to produce, ultimately, iron. These reactions occur with phenomenal speed and lead to catastrophic events that will be discussed in the next section.

42-4 Cataclysmic Events

Huge explosions and other sorts of cataclysmic events are a natural part of the life cycle of stars. Stars formed in swirling clouds of gas move along the H–R diagram, incorporating such occurrences into their evolution and forming in the process the elements needed to form new stars. Why these events occur is the subject of this section.

Novae

More than half of all stars are estimated to be members of **binary pairs** or even larger associations. These stars orbit their common center of mass as the group moves with the rotation of the galaxy. The periods of binaries vary from a few hours for those with the companions very close to each other to millions of years for those with the companions separated by thousands of astronomical units. Here, we are interested in close binaries.

A complete analysis of the interactions between the two stars forming a close binary is beyond the scope of this book, but a qualitative explanation will suffice. Consider a binary whose stars of masses M_1 and M_2 rotate about their common center of mass in circular orbits. An observer at rest in the rotating system experiences a net force that is the sum of the gravitational forces due to the two stars and the pseudoforces due to rotation. Figure

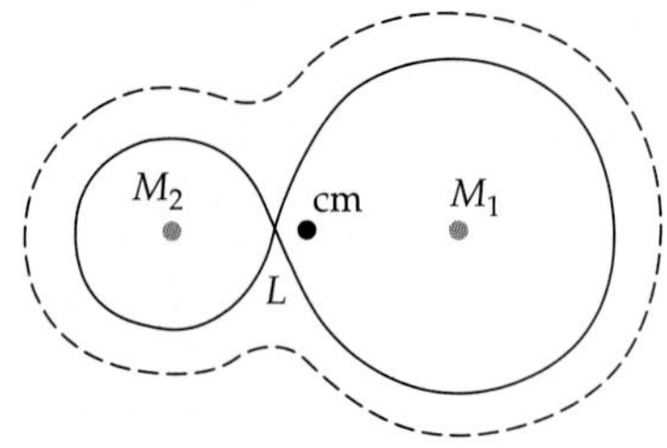

Figure 42-16 Cross sections of two gravitational equipotential surfaces for binary stars of mass M_1 and M_2. The point labeled L is one of five Lagrangian points where the gravitational potential is a minimum and the net force is zero.

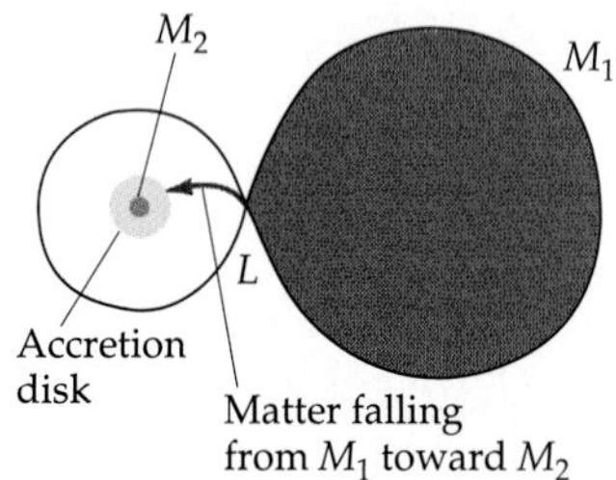

Figure 42-17 Material from M_1 pouring through the Lagrangian point into the Roche lobe of M_2 forms an accretion disk in the equatorial plane of M_2. Material arriving later hits the disk, generating a high-temperature impact area. This causes novae to flicker irregularly.

42-16 shows an equipotential surface about a binary pair. It is easy to visualize that there is a point along the line joining the centers of the two stars where the net potential is a minimum. At this point, the net force due to the combined effects of the rotation and the gravitational attraction of the masses M_1 and M_2 is zero. This point is a **Lagrangian point.** The three-dimensional equipotential surface that includes the Lagrangian point forms an envelope around each star called the **Roche lobe.**

Now consider what happens when, through natural evolution, one of the stars, say M_1, begins expanding and fills its Roche lobe. The photosphere of any star "sees" a vacuum outside the surface, the outward pressure at any point being balanced by gravity. But at the Lagrangian point there is no gravity. Thus, material from M_1 pours through the Lagrangian point into the Roche lobe of M_2. Once inside it is gravitationally attracted toward M_2. Since the system is rotating, the material from M_1 doesn't simply move directly toward M_2 because of the Coriolis force, but instead forms a spiralling **accretion disk** (see Figure 42-17).

If M_2 is a normal star, nothing of great consequence occurs, but if it is a white dwarf, then cataclysmic events called **novae** can occur. We will mention two possibilities. Material flowing through the Lagrangian point into the accretion disk is stored there until some instability occurs in the disk that results in the dumping of material onto the surface of the white dwarf. The impact heats the surface, causing a sudden brief increase in intensity by a factor of from 10 to 100. These events recur at intervals of from a few weeks for **dwarf novae** to hundreds or thousands of years for **recurrent novae.** Between these sudden bursts in intensity, the novae flicker as described in the caption to Figure 42-17.

For **classical novae,** which eject substantial amounts of material into space and can brighten by a factor of a million within a few days, astrophysicists suggest that the sudden dumping of material from the accretion disk onto the hot surface of the white dwarf may result in the buildup of sufficient hydrogen to initiate a thermonuclear explosion. After the blast, the system returns to a more quiescent state, pending the accumulation of more hydrogen in the disk. The theoretical problems involved in explaining such an event are formidable, however, and no general agreement on the mechanism exists.

Supernovae

The **supernova**—the catastrophic explosion of an entire star—is, perhaps surprisingly, somewhat more clearly understood than the nova. First, note that supernovae are not just big novae. Their origin is completely different. In Section 42-3, we saw what occurs in a star as it uses up the hydrogen in the core and begins moving off the main sequence of the H–R diagram. The star begins to fuse helium and then carbon. If it is a low-mass star, it has insufficient gravitational energy to ignite the fusion of heavier nuclei in quantity.

For massive stars, however, the situation is different. If the mass is greater than about $8M_\odot$, gravity is strong enough to continue to draw mass from the middle layers into the core as the core uses up fuel. The increasing temperatures, exceeding 10^8 K, are sufficient to ignite fusion in neon and silicon ultimately producing iron. As we saw in Chapter 40, the specific binding energy of iron is the highest in the periodic table. Fusing elements above iron doesn't produce energy; it requires energy. Thus, when the core has been fused to iron, there is nowhere else to go via thermonuclear reactions. With no counteracting outward pressure from nuclear reactions, gravitational contraction continues even more rapidly and the core continues to heat up until it exceeds 10^9 K. At that point, the radiation within the star is intense and the iron nuclei undergo photodisintegration into helium and neutrons, sucking energy from the core and accelerating the gravitational collapse:

$$^{56}_{26}\text{Fe} \longrightarrow 13\,^{4}_{2}\text{He} + 4n \qquad \text{42-18}$$

The helium nuclei then begin to photodisintegrate, extracting enormous amounts of energy to overcome the binding energy of helium:

$$^{4}_{2}\text{He} \longrightarrow 2p + 2n \qquad \text{42-19}$$

The core is now in gravitational free fall, compressing the electrons and protons into neutrons via inverse beta decay:

$$p + e^- \longrightarrow n + \nu_e \qquad \text{42-20}$$

What happens next is a matter of intense theoretical conjecture that we will explore in Section 42-5.

What happens to the envelope of the star—the material outside of the core—although also unclear theoretically, is certainly apparent visually. The entire envelope is blown away in an incredibly massive explosion. This is a supernova. Supernovae are extremely rare, but scientists were fortunate enough to observe one in 1987 only 170,000 $c{\cdot}$y away in the Large Magellanic Cloud, a small, irregular galaxy that is a companion to the Milky Way galaxy. Called SN1987A, it was the first supernova to occur close enough to be visible to the unaided eye since 1604, when both Kepler and Galileo saw one. Two others were recorded earlier, in 1006 and 1054, the latter documented by Chinese astronomers and still visible as the Crab Nebula. Several others have been observed with telescopes.

At its peak light output, a supernova typically shines more brightly than the entire galaxy in which it is located. The spectra of supernovae reveal the presence of elements throughout the entire periodic table. This indicates that some of the energy removed from the core following the production of iron is used to produce elements of even higher atomic numbers. The supernova ejects some of this material into space, where it eventually contributes to the formation of a new generation of stars and their planets via condensation. Such events undoubtedly preceded the birth of the sun and the formation of the earth. We are, as has been said before, "made of the stuff of stars."

There are two kinds of supernovae: *type I* and *type II*. The preceding description applies to type II supernovae. Type I supernovae occur among population II stars, which tend to be of low mass. (Don't be confused by the apparent inconsistency in nomenclature.) This contradiction with our earlier discussion of the evolution of low-mass stars currently has no theoretical explanation.

42-5 Final States of Stars

The cataclysmic events that occur near the end of the life of a star lead to one of only three possible final states: a degenerate dwarf, a neutron star, or a black hole. The mass of the star, particular that of the core, appears to be the primary factor in determining the final state.

Degenerate Dwarfs

Stars whose masses are less than about $6M_\odot$ follow an evolutionary track on the H–R diagram that takes them through one or more periods of substantial mass loss from the envelope. How this occurs is not clear, but the ejected mass, which is heated to a glowing planetary nebula by the hot core, leaves behind a **degenerate dwarf,** also called a **white dwarf** at this stage because it is literally white hot. Its mass is typically about $1M_\odot$ and its radius of the order of 10^7 m, which is about the radius of the earth. Thus, the density of a typical white dwarf is about 5×10^5 g/cm^3. A coin the size of a penny made from the material of a white dwarf would have a mass of over 200 kg.

Thermonuclear reactions have ceased in the white dwarf, so there is no outward pressure due to them from within the star. The star therefore collapses because of the inward gravitational pressure until the exclusion principle prevents the atomic electrons from coming any closer together. This effect is similar to the exclusion-principle repulsion between atoms in a molecule that we discussed in Chapter 38. It results in an outward pressure that is larger even than the thermal pressure of the hot core. It is this **electron-degeneracy pressure** that supports the white dwarf. When the outward electron-degeneracy pressure equals the inward pressure due to gravity, the star stops contracting.

Explicit derivation of the expression for the electron-degeneracy pressure leads to a nonrelativistic relation between the dwarf's radius R and mass M:

$$R = (3.1 \times 10^{17}\ \text{m}\cdot\text{kg}^{1/3})\left(\frac{Z}{A}\right)^{5/3} M^{-1/3} \qquad \text{42-21}$$

where Z is the atomic number and A is the atomic mass number of the material of the star. Note the interesting result that the larger the mass, the smaller the radius. For example, a white dwarf with a mass of $1M_\odot$ will have a radius smaller than one with a mass of $0.5M_\odot$. This raises the interesting question of whether, when the electrons become relativistic, the mass might become large enough for the radius of the dwarf to shrink to zero. Although Equation 42-21 does not formally allow that possibility until M approaches infinity, S. Chandrasekhar derived the corresponding relativistic relation and found that the radius would go to zero when the mass reaches about $1.4M_\odot$. This mass is called the **Chandrasekhar limit.** Its validity is strongly supported by the fact that the masses of all the white dwarfs that have been measured are less than that value.

The white dwarf continually loses heat to space and, without a nuclear furnace, it slowly cools off and dims. When it is no longer visible, it has become a **black dwarf** and continues to cool toward thermal equilibrium with the universe. It is likely that no white dwarfs have yet reached this final stage.

Neutron Stars

In the discussion of supernovae, we saw that the enormous pressures in the core forced inverse beta decay to occur, converting the core into neutrons. If the mass of the core after the explosion is greater than the Chandrasekhar limit, what happens? We can get an idea by considering the neutrons to be an ideal gas of fermions and deriving a nonrelativistic expression for the mass–radius relation analogous to Equation 42-21. The result is

$$R = (1.6 \times 10^{14}\ \text{m}\cdot\text{kg}^{1/3})\, M^{-1/3} \qquad \text{42-22}$$

where M is the mass of the core in kilograms and R is the radius of the core in meters. Such a star is called a **neutron star,** since the envelope was blown away in the supernova and all that is left is the core consisting of neutrons. For $M = 1M_\odot$, Equation 42-22 yields $R = 1.27 \times 10^4$ m $= 12.7$ km.

The density of the neutron star is about 1.2×10^{14} g/cm^3. This is only slightly less than the density of the neutron itself, which is about 4×10^{14} g/cm^3. Thus, we can conclude that the gravitational pressure of the neutron star is balanced by the repulsive component (due to the exclusion principle) of the strong nuclear force between the neutrons. As you might guess from our earlier discussion, gravity can overcome even this resisting pressure. The mass corresponding to the gravity at which that occurs is the maximum mass possible for a neutron star, a mass analogous to the Chandrasekhar limit for white dwarfs. Current theory puts the maximum mass of a neutron star at between $1.7M_\odot$ and $3M_\odot$. The few neutron stars that have been tentatively identified and measured all have masses below this limit.

Regularly pulsing radio sources, called **pulsars,** discovered in 1967 in nebulae such as the Crab Nebula that are remnants of supernovae, are thought to be neutron stars. Current theory suggests that the radiation is emitted as the result of the charged particles emitted by the neutron star that are accelerated along the star's magnetic-field lines as a consequence of the star's rapid rotation as illustrated in Figure 42-18. The Crab pulsar also corresponds to an optical variable as illustrated on page 237. It emits energy at an incredible 3×10^{31} W. Its period is equally incredible, only 0.033 s, one of the shortest known.

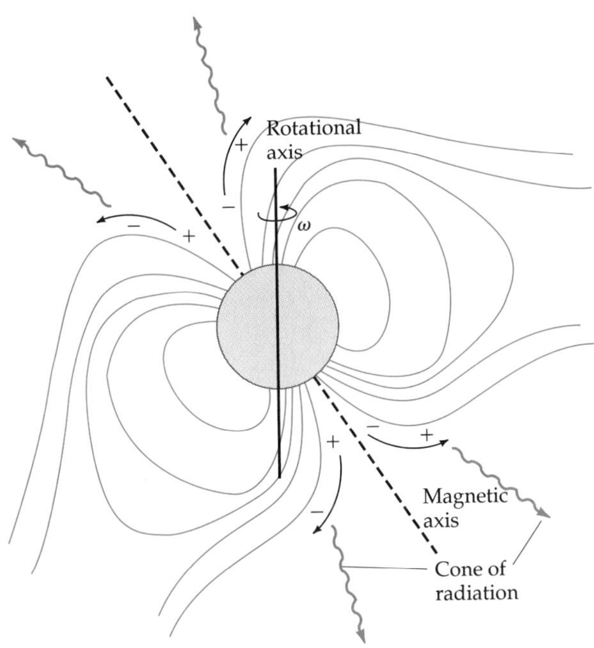

Figure 42-18 The neutron star acquires much of the original star's angular momentum and magnetic field, causing it to rotate rapidly while dragging along a distorted magnetosphere. Accelerated charged particles radiate in a cone about the rotating magnetic axis like a cosmic lighthouse.

As it emits energy into space, the neutron star also slowly cools, approaching thermal equilibrium with the universe.

Black Holes

What happens when the mass of the core remaining after a supernova exceeds the $1.7M_\odot$ to $3M_\odot$-upper limit for the formation of a neutron star? In Chapter 10, we saw that the velocity necessary for an object of mass m to escape from a large object of mass M is found by equating the gravitational potential energy at the surface of M to the kinetic energy necessary to escape. This results in

$$v_e = \left(\frac{2GM}{R}\right)^{1/2} \qquad 42\text{-}23$$

For a neutron star with $M = 1.0M_\odot$, $v_e = 1.3 \times 10^8$ m/s, which is more than 40 percent of the speed of light. If there were no relativistic and quantum-mechanical effects, the escape velocity would equal c when

$$R_S = \frac{2GM}{c^2} \qquad 42\text{-}24$$

where R_S is called the **Schwarzschild radius.** Thus, if an incipient neutron star is so massive that its radius is less than R_S, no object with mass can escape from its surface. In addition, radiation of wavelength λ emitted at some distance R from mass M is shifted to a longer wavelength λ' according to the **gravitational redshift** described in Section 34-11, which is given by

$$\frac{\lambda'}{\lambda} = \left(1 - \frac{v_e^2}{c^2}\right)^{-1/2} = \left(1 - \frac{2GM}{c^2R}\right)^{-1/2} = \left(1 - \frac{R_S}{R}\right)^{-1/2} \tag{42-25}$$

If R shrinks to the Schwarzschild radius, then λ' approaches infinity and the energy ($E = hf = hc/\lambda$) approaches zero. Thus, if R is less than R_S, no energy can escape the surface as radiation, either. Such an object is called a **black hole** because it neither emits nor reflects radiation or mass and, hence, appears absolutely black.

The radius of a black hole with a mass of $1M_\odot$, if there is such an object, would be only about 3 km. Thus far, there have been no confirmed observations of black holes, although a number of possible ones are the subject of intense research. Many astrophysicists currently believe that a massive black hole is located at the center of the Milky Way galaxy and may account for part of the "missing mass" of the Galaxy. Unlike degenerate dwarfs and neutron stars, black holes are not cooling toward thermal equilibrium with the universe.

42-6 Galaxies

In Section 42-2, we saw that the Milky Way galaxy is shaped like a spiralled disk with a central bulge located about 30,000 $c \cdot y$ from the sun. The disk is surrounded by a roughly spherical "halo" of globular clusters comprised mostly of population II stars, which are also part of our Galaxy. We will now look at some of the characteristics of galaxies in general.

Material Between the Stars

"Holes in the sky"—regions where no stars are seen—have been observed since the early days of astronomy and were assumed to be empty space. However, studies of open clusters led to the discovery about 60 years ago of a more or less continuous distribution of tiny dust particles called **interstellar dust** between the stars. Consisting of solid specks of silicates and carbides averaging only a few hundred nanometers in diameter (approximately matching the wavelengths of visible light), the interstellar dust both absorbs and scatters some of the starlight striking it. Since blue light scatters more efficiently than red light, starlight is reddened on its trip to us, just as sunlight is reddened at sunset. Although the dust seems to pervade the entire Galaxy, its concentrations are very low. The vacuum in interstellar space is far better than the best vacuum obtainable in the laboratory.

Spectroscopic studies of binaries reveal some absorption lines that are not doppler shifted. In 1904, J. F. Hartmann reasoned correctly, although not to universal acceptance, that the unshifted lines result from the absorption of light from the binary by an intervening gas cloud rather than by gas in the atmosphere of the star. Though still difficult to demonstrate conclusively in all cases, the existence of interstellar gas clouds is now generally accepted. The gas clouds are composed mainly (some exclusively) of hydrogen.

Together, the interstellar dust and the clouds of gas account for an estimated 2 to 3 percent of the mass of our Galaxy. It is nearly certain that there is not enough unseen gas and dust to account for the Galaxy's "missing mass."

Gaseous Nebulae

Though most gas clouds, or nebulae, in interstellar space are irregular in shape, a few are circular, leading to speculation that they are self-gravitating and represent the very early stages of the formation of new stars. Some large hydrogen clouds have spherical regions of ionized hydrogen, with the demarcation between the H and H^+ regions being quite sharp. Astrophysicists believe that the ionized region is maintained by ultraviolet photons with frequencies above the Lyman limit that are emitted by a hot, newly formed star at the center of the region. The view that new stars form in nebulae in an ongoing process is strongly supported by the observation that, although it is of the order of 10^{10} y old, our Galaxy contains main-sequence stars that are no more than 2 to 3×10^6 y old. Furthermore, high-resolution radioastronomy has in recent years located numerous newly forming stars embedded in clouds of dust and gas that are completely opaque to optical wavelengths.

Classification of Galaxies

Although fuzzy, extended objects, at one time called "nebulae," that were obviously not stars have been observed in the night sky since the 1700s, what and where they were was a matter of active scientific debate until well into the twentieth century. The answer had to await the development of telescopes with sufficient resolution and light-gathering power and a theoretical means of computing distances from observations made with them. These came together in the mid-1920s when Edwin Hubble used the 2.5-m telescope on Mount Wilson, the largest in existence at the time, to measure the intensities of rare stars, called Cepheid variables,* that he discovered in three "nebulae." One of those nebulae, the great spiral Andromeda, he measured to be 2×10^6 $c{\cdot}$y away. In one stroke, he was able to demonstrate that the "nebulae" were in fact galaxies much like our own, as had first been suggested by the philosopher Emmanuel Kant 150 years earlier, and that they were far outside the Milky Way galaxy. Exploring Hubble's discovery will take us into the realm of cosmology, the study of the universe.

Following his discovery that "nebulae" were in reality distant galaxies, Hubble conducted a systematic study of the enormous number that were visible. He found that all but a very few fit into four general categories. Most have regular geometrical shapes and occur in two varieties: **ellipticals,** which are roundish, rather like a football; and **disks.** The disks, in turn, have two subgroups: **ordinary spirals** and **barred spirals.** The small percentage that did not have regular shapes he called **irregular galaxies.** Figure 42-19 shows an example of each type of galaxy.

In addition to their geometrical differences, the four types of galaxies have other dissimilarities. A large fraction of the motion of the stars in spirals is rotational about the galactic center, whereas the motion of the stars in ellipticals is generally random with only a relatively small rotational component. Ellipticals also seem to have very little interstellar gas and dust, whereas spirals and many irregular galaxies have substantial amounts. The fact that most ellipticals have no young stars is probably a consequence of that lack. With a few exceptions, ellipticals are much smaller than spirals, typically having only about 20 percent of the diameter of an average spiral like the Milky Way galaxy and only a thousandth of the mass.

*Cepheid variables are rare stars for which there exists a relation connecting the period of intensity variation to the brightness and, hence, to the distance from the sun. They are the primary key to our knowledge of astronomical distances. Polaris, the current Pole Star, is a Cepheid variable.

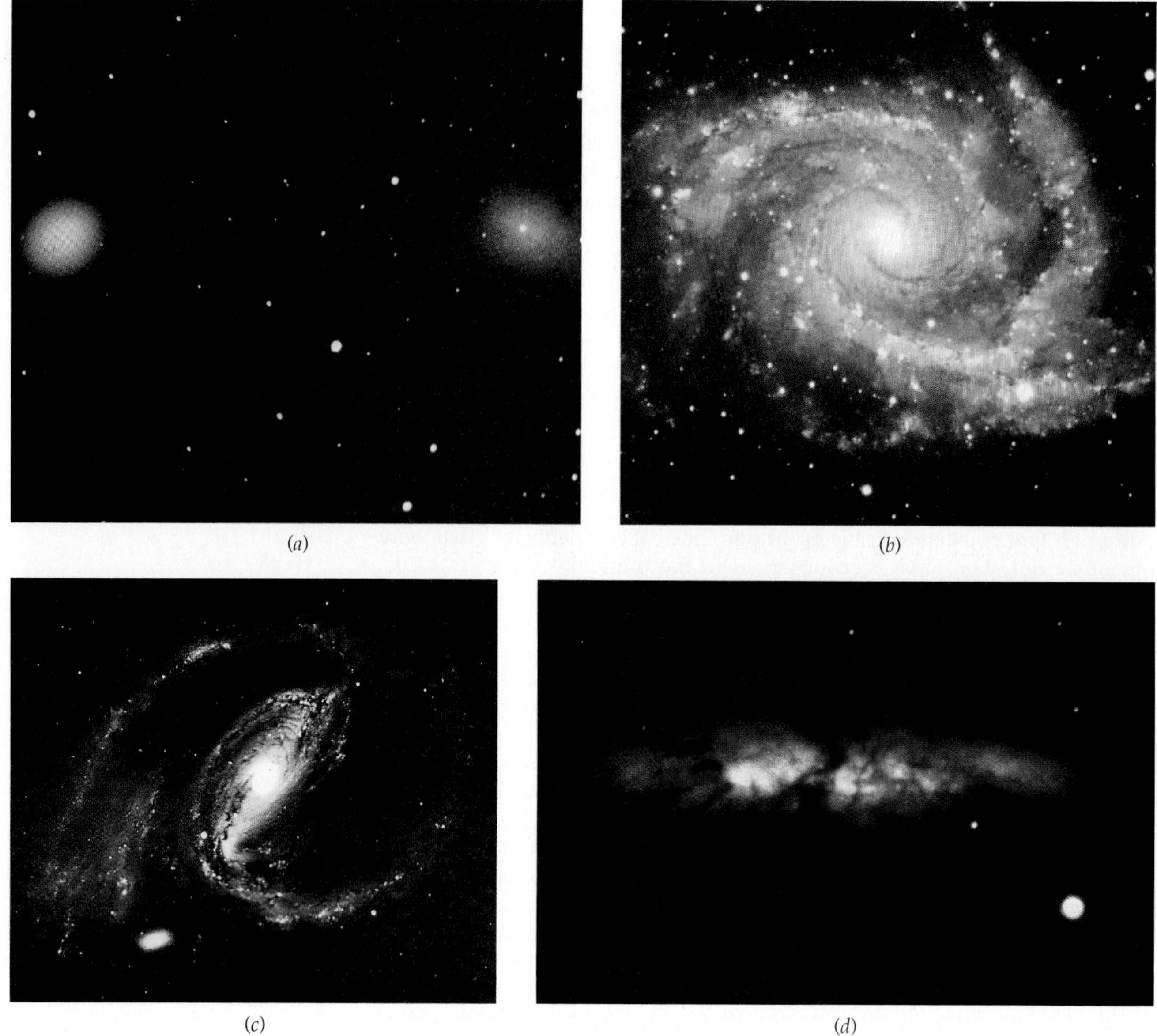

(a) (b) (c) (d)

Figure 42-19 Examples of the four types of galaxies in Hubble's classification scheme: (*a*) elliptical, (*b*) ordinary spiral, (*c*) barred spiral, and (*d*) irregular. The Milky Way galaxy is thought to be an ordinary spiral.

Quiet and Active Galaxies

Most of the approximately 10^{10} galaxies in the observable universe appear to be **quiet galaxies;** that is, there is very little activity other than what might be expected for such dynamic systems. The vast majority of these galaxies are so distant that our instruments cannot resolve internal details. Therefore, only the composite spectra and apparent brightness f for the entire galaxy can be observed. The range of velocities Δv among the stars of the regular galaxies, measured by the doppler broadening of the spectral lines, turns out to be related to the total luminosity L by

$$L \propto (\Delta v)^4 \qquad 42\text{-}26$$

Since L is related to f and r, the distance to the galaxy, by Equation 42-10, the distance r can be found from measurements of the doppler shifts and the brightness of the galaxy.

In a very small percentage of galaxies, something extremely violent, even by comparison with stellar supernovae, is occurring. They are called **active galaxies.** There are several distinct types, some of which may not even be galaxies at all. The first discovered were **Seyfert galaxies,** named after Carl Seyfert, their discoverer. They are spirals with extremely bright central cores, or nuclei. In many of them, the light coming from the core exceeds that from all of the stars in the galaxy and, incredibly, may vary in intensity by a factor of two or more in less than a year. Such a rapid variation in the total intensity means that the source must be less than 1 light-year in extent even though it produces as much energy as 10^{11} stars. Even more incredible is the fact that the light emitted by a Seyfert galaxy is an emission line spectrum, not a continuum with absorption lines typical of stars. This suggests that its enormous energy is not coming from thermonuclear reactions. The source is still a mystery.

A similar sort of extreme activity occurs in a few ellipticals called **N galaxies** and **BL Lac objects.** N galaxies are elliptical counterparts of Seyfert galaxies, that is, they have very bright centers. BL Lac objects seem to be like N galaxies, but exhibit substantial short-term intensity variations. In these, an intensity variation of a factor of two can occur within a week and a complete reversal of the polarization of the emitted light can occur within one day, suggesting that the energy source is only 1 light-day in diameter. BL Lac objects are now thought to be giant ellipticals about 10^9 $c{\cdot}y$ from earth.

Some of the giant ellipticals are also strong emitters in the radio region of the spectrum. Study of these **radio galaxies** has been intense, and the results have been astonishing. For example, the radio source Centaurus A is double-lobed with a small, radio-emitting nucleus midway between the lobes (see page 1389). It is one of the largest radio-emitting objects in the universe. Analyses of its spectra indicate that the energy release represented by the radiation we see amounted to 10^{56} J, which is about the equivalent of all the stars in the Milky Way galaxy undergoing supernova explosions simultaneously. The nature of such a colossal event is a mystery.

In a universe of strange phenomena, **quasars,** short for *quasistellar radio sources,* are among the strangest. Their optical images look like stars; that is, they have no resolved structure. Their spectra, however, resemble that of a Seyfert galaxy. Resolved radio images of some quasars show that a few of them are double-lobed, like the radio galaxies, which makes their identification ambiguous. In addition, there is a group of objects about 20 times more numerous than quasars called **quasistellar objects,** or **QSOs.** These are like quasars in every major way, except that they are not radio emitters.

Perhaps the strangest thing about the quasars is the magnitude of the redshift of their spectra, which is very large. *If* it is due to the doppler effect alone, it implies that some quasars are receding directly away from us at speeds greater than $0.9c$, which is larger than the speed of the general expansion of the universe. This would make them the most distant massive objects, of the order of 10^{10} $c{\cdot}y$ from earth. Their apparent optical brightness f and their great distance imply power outputs of 10^{40} W, greater than that of 10^{12} suns. Not only that, but the intensities of some of them vary over only a few hours, suggesting dimensions of only a few light-hours.

The nature of quasars and QSOs is an unresolved scientific issue. There are two schools of thought. Many astrophysicists think that they are not at any great distance from us but rather are at distances consistent with their brightness. Why they should exhibit such large redshifts is then perplexing because the quasar would violate Hubble's law (discussed next), which is a very simple and very general relation between the velocities of extragalactic systems relative to us and their distances from us. Other astrophysicists feel that the large redshifts are good indicators of the objects' distances from us, which makes their prodigious energy output the issue. The suggestion has been made that the source of the colossal implied power is matter from colliding galaxies falling onto an enormous black hole at the center of one (or both) of the galaxies.

Hubble's Law

E. P. Hubble was the first astronomer to recognize that there is a relation between the redshifts in the spectra of galaxies and their distances from us. This relation is illustrated in Figure 42-20 for a group of spiral galaxies used by astronomers for calibrating distances. Provided that the redshift is due to the doppler effect, the recession velocity v of a galaxy is related to its distance r from us by **Hubble's law:**

$$v = Hr \tag{42-27}$$

where H is the **Hubble constant.** Figure 42-21 shows the redshifted spectra of five galaxies whose distances from us range from 2.6 megaparsecs to 287.5 megaparsecs.

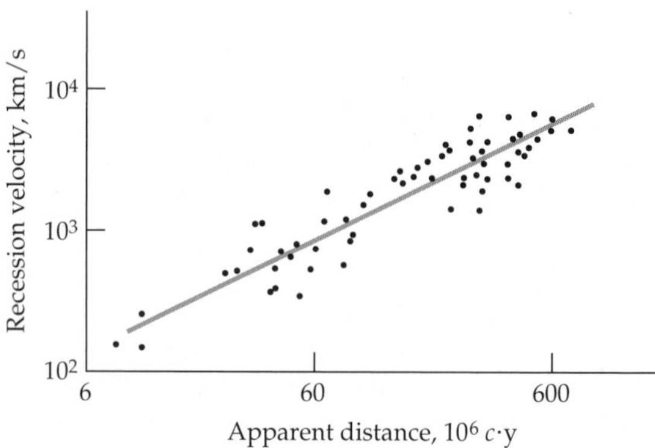

Figure 42-20 A plot of the recession velocities of individual galaxies versus apparent distance illustrates Hubble's law.

In principle, the value of H is easy to obtain since it relies on the direct calculation of v from redshift measurements. However, recall that astronomical distances are very difficult to obtain and that they have been computed for only a fraction of the 10^{10} or so galaxies in the observable universe. Thus, the value of H changes as distance calibration data is refined. The currently accepted value of the Hubble constant is

$$H = \frac{23 \text{ km/s}}{10^6 \; c\cdot\text{y}} \tag{42-28}$$

Notice that the basic dimension of H is reciprocal time. The quantity $1/H$ is called the **Hubble age** and equals about 1.3×10^{10} y. This would correspond to the age of the universe if the gravitational pull on the receding galaxies were ignored.

Exercise

Show that $1/H = 1.3 \times 10^{10}$ y.

Example 42-5

Redshift measurements of a galaxy in the constellation Virgo yields a recession velocity of 1200 km/s. How far is it to that galaxy?

From Hubble's law, we obtain

$$r = \frac{v}{H} = (1200 \text{ km/s})\frac{10^6 \; c\cdot\text{y}}{23 \text{ km/s}} = 52 \times 10^6 \; c\cdot\text{y}$$

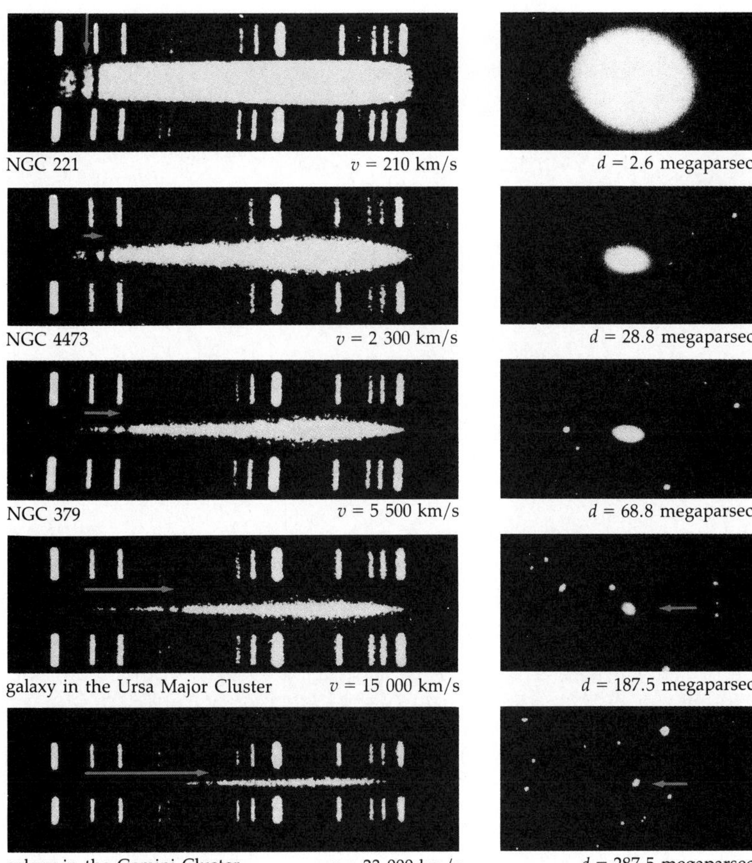

Figure 42-21 The redshift of the Ca, H, and K absorption spectral lines are shown for five galaxies at different distances from us. The line spectra above and below the absorption spectrum are standards used for determining the amount of shift accurately.

Hubble's law tells us that the galaxies are all rushing away from us, with those the farthest away moving the fastest. However, there is no reason why our location should be special. An observer in any galaxy would make the same observations and compute the same Hubble constant (see Problem 20). Thus, Hubble's law suggests that all of the galaxies are receding from each other at an average speed of 23 km/s per 10^6 c·y of separation. In other words, the universe is expanding. This is a profound discovery with enormous theoretical implications.

An obvious question is whether there are other observational results that support Hubble's law. For example, is the observed expansion general, or could it be a statistical accident—a consequence of our having to date measured the redshifts of "only" about 30,000 of the 10^{10} galaxies in the observable universe? Thus, redshift surveys of the universe are an important first step in studying Hubble's expansion. Such surveys have been underway for several years and about 10^{-5} of the volume of the visible universe has now been mapped. These surveys have yielded several unexpected discoveries, but have not yet answered the question conclusively. There are huge voids in space—regions where the density of galaxies is only 20 percent or so of the average for the universe. The galaxies themselves tend to be grouped into local clusters of a dozen or so and the local clusters into superclusters of a few thousand. In addition, the galaxies tend to lie on thin, sheet-like structures. The largest detected thus far is called the "Great Wall" by its discoverers, Margaret J. Geller and John P. Huchra (see Figure 42-22). How such structures might have evolved in the general expansion described by Hubble's law presents a serious challenge to existing theoretical descriptions of the development of the large-scale structure of the universe.

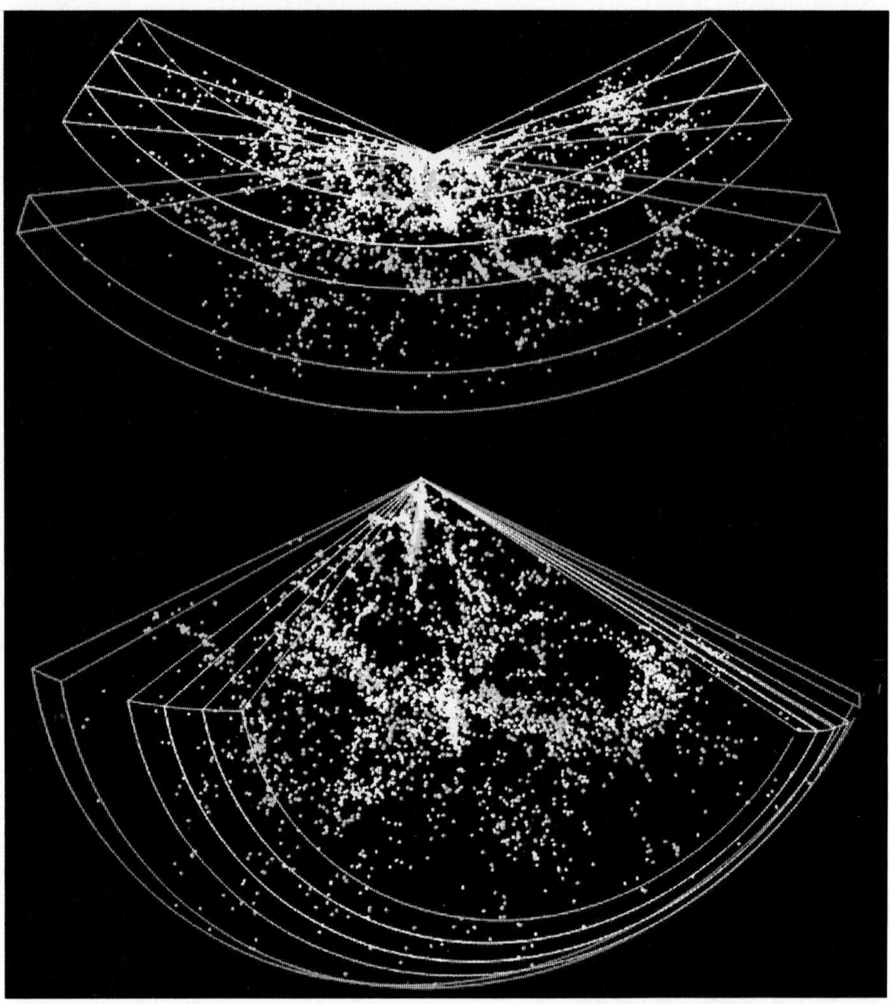

Figure 42-22 These are two views of approximately 4000 galaxies included in the redshift survey of Margaret Geller and John Huchra. The Milky Way is at the vertex. The three layers outlined in white cover latitudes from 26.5° to 44.5° above the celestial equator (called declination by astronomers); the layer outlined in orange covers declinations from 8.5° to 14.5°. Each layer covers the celestial longitude (called right ascension by astronomers) range from 8^h to 17^h, or about 37 percent of the celestial equatorial circle. Galaxies shown in white are in the white-outlined layers; those shown in yellow are in the orange layer. The Great Wall stands out clearly in the lower view, running approximately parallel to the outer boundary of the survey—about halfway from the vertex. Note the huge voids in space, some as large as 10^8 c·y.

42-7 Gravitation and Cosmology

We have seen that Hubble's law leads inescapably to the conclusion that the universe is expanding, and it provides us with a measure, $1/H$, of how long ago that expansion began. In this section, we will examine the basic theoretical framework that suggests possible tests of that conclusion, in addition to the redshift survey.

The basis for this discussion is the philosophical view that the universe is homogeneous and isotropic at any instant in time. That is, at any given instant, the universe has the same physical properties everywhere and looks the same in all directions from every location. This point of view is called the **cosmological principle.** Note that Hubble's law is consistent with the cosmological principle.

We have already seen that the cosmological principle clearly does *not* hold on a local scale. Galaxies are clustered into local groups. Even on a scale of 10^8 c·y, the dimension typical of galactic superclusters, the universe is neither homogeneous nor isotropic. However, when maps of very distant space are examined (see Figure 42-23), the distribution of galaxies does appear to be homogeneous and isotropic. Whether redshift maps like that shown in Figure 42-22, which extends to about 4×10^8 c·y, will begin to show homogeneity and isotropy as they are extended into deep space remains to be seen.

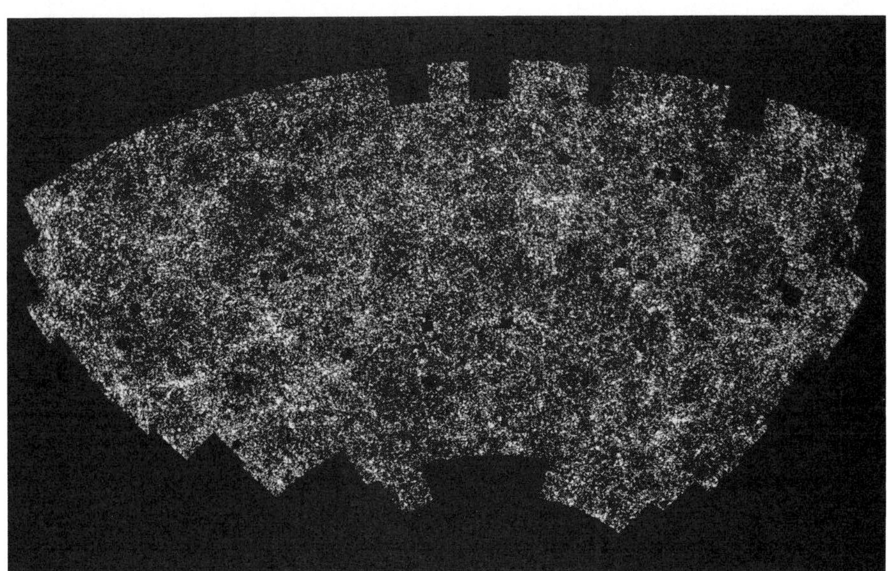

Figure 42-23 A map showing approximately two million galaxies ranging up to 2×10^9 $c{\cdot}y$ away. The distribution of the galaxies looks essentially homogeneous and isotropic. This is a composite of 185 contiguous photos taken by the Schmidt telescope at the European Southern Observatory. The south galactic pole is at the bottom center. The four blank squares at the top are covered by photos not yet analyzed.

The Critical Mass Density of the Universe

We noted earlier that the Hubble age $1/H = 1.3 \times 10^{10}$ y ignores the effect of gravity. Our expectation is that gravity tends to slow the expansion over time. Is the gravity in the universe strong enough eventually to reverse the expansion and cause the universe to collapse? Or will the expansion continue forever? The answer depends on the mass density of the universe. We can understand this by considering the motion of a single galaxy of mass m at a very large distance R from the earth. Let M be the total mass of all the galaxies within the spherical volume of radius R. The gravitational potential energy of the galaxy is $-GMm/R$. (This is analogous to Equation 10-25 for the potential energy of a particle at a distance r from the earth.) The total energy of the galaxy is

$$E = K + U = \tfrac{1}{2}mv^2 - \frac{GMm}{R} \qquad \text{42-29}$$

In Chapter 10, we saw that if we project an object with some speed v from the earth, the object will escape if its total energy is greater than or equal to zero, but if the total energy is negative, the particle will eventually stop and fall back to the earth. Similarly, if the total energy of the galaxy is greater than or equal to zero, it will continue to move away from the earth forever, but if the total energy is negative, the galaxy will eventually stop moving away from the earth and start moving back toward the earth. We can see from Equation 42-29 that the total energy of the galaxy depends on the total mass M within the spherical volume R; that is, it depends on the mass density $\rho = M/(\frac{4}{3}\pi R^3)$. We can find the critical mass density of the universe ρ_c by setting the total energy in Equation 42-29 equal to zero:

$$\tfrac{1}{2}mv^2 = \frac{GMm}{R}$$

Substituting $v = HR$ from Hubble's law (Equation 42-27), we obtain

$$\tfrac{1}{2}m(HR)^2 = \frac{GMm}{R}$$

$$\tfrac{1}{2}H^2 = \frac{GM}{R^3}$$

Then

$$\rho_c = \frac{M}{\frac{4}{3}\pi R^3} = \frac{3H^2}{8\pi G} \qquad \text{42-30}$$

Using present values for H and G, we obtain for the critical mass density of the universe

$$\rho_c \sim 10^{-26}\ \text{kg/m}^3$$

This corresponds to about five hydrogen atoms per cubic meter of space.

Determining the present mass density of the universe ρ_0 is thus an important goal. If it is larger than ρ_c, the expansion will reverse and the universe will collapse. If it is smaller, the expansion will continue forever. If it should happen that $\rho_0 = \rho_c$, the universe will coast to a stop but will not begin to contract. It should also be clear that if ρ_0 is greater than ρ_c now, it will always be so because it is actually the conservation of energy that determines whether contraction or continued expansion will occur. Since ρ_0 must decline over time as expansion progresses, the Hubble constant must also decline over time to ensure that ρ_0 remains larger than ρ_c. In other words, the Hubble constant must be a function of time. The value of ρ_0 based on the *visible* universe is only about 4 percent of ρ_c, suggesting that the universe will expand forever. However, the missing mass of the universe discussed earlier affects the value of ρ_0. No generally accepted value of ρ_0 has yet been found, although current estimates put it close to ρ_c.

At this point in time, the emphasis of investigations in cosmology is still centered on developing a basic cosmological model against which to compare the many observational discoveries made by astronomers and astrophysicists, only a few of which we have discussed briefly. To be sure, there is a good candidate for a comprehensive model, the Big Bang. We will review some of its successes and some of the questions it hasn't answered in the final section of the chapter.

42-8 Cosmogenesis

Following his completion of the general relativity theory in 1915, Einstein turned to cosmology. He based his early work on the assumption that the universe was not only homogeneous and isotropic but also constant in time. This is sometimes called the **perfect cosmological principle.** He quickly discovered that such a static universe, like that described by Newton's gravitational theory, is empty; that is, it contains no mass. He accounted for mass by adding a new force of unknown origin via the **cosmological constant,** thereby committing what he later described as the biggest blunder of his life. On learning of Hubble's discovery of the expansion of the universe, he abandoned the cosmological constant. Others, however, were not so quick to give up on the philosophically attractive static model. They argued that the observed expansion would not result in a decrease in the mass density of the universe if new matter were being created in space at a rate sufficient to maintain the density of a steady-state universe.

One difficulty with the steady-state model of the universe is a problem known as Olber's paradox. If there is a uniform distribution of stars throughout an infinite space, then no matter in which direction you look, you will eventually see a star. Therefore, the night sky should look as bright as the surface of the average star. (This is analogous to standing in an infinitely large forest in which all the trees are painted white. Along any line of sight, you will eventually see a white tree, so you should see white in all directions.) Why then is the night sky dark? This dilemma is called Olber's paradox after the nineteenth-century physician–astronomer who publicized it widely. The solution offered by Olber himself was that interstellar dust absorbs the light from distant stars. This is no help since the dust would eventually be heated to glowing and the night sky should still be bright.

The solution to this problem came with Hubble's discovery of the expansion of the universe. Since the velocity of light is finite, looking into space means looking back in time. As we look deeper and deeper into space, we are eventually looking at a time before the stars began to form, that is, at a time greater than the Hubble age. (In terms of our forest analogy, the distant trees have not yet been painted white. Therefore, if the separation of the trees is great enough, many lines of sight will end on dark trees.)

The Big Bang

Two major astrophysical discoveries made in the 1960s were the first of several that have convinced most scientists that the universe is not constant in time but was initiated by a single event at a particular time in the past, the **Big Bang,** and is evolving over time. The first of the two major discoveries that supported the evolving-universe model was Martin Ryle's discovery that there are more distant radio galaxies than nearby ones. Since distant observations correspond to earlier times, this meant that the universe had looked different at earlier times than it does now; that is, it has evolved.

The second discovery was monumental, as important as Hubble's discovery of the expansion of the universe itself. In investigating ways of accounting for the cosmic abundance of elements heavier than hydrogen, cosmologists recognized that nucleosynthesis in stars could explain the abundance of elements heavier than helium but not that of helium. Helium must therefore have been formed during the Big Bang. Synthesizing the amount of helium to account for its present abundance requires that the Big Bang would occur at an extremely high initial temperature to provide the necessary reaction rate before fusion was shut down by the decreasing density of the very rapid initial expansion. The high temperature implies a corresponding thermal (blackbody) radiation field that would cool as the expansion progressed. Theoretical analysis predicted that from the estimated time of the Big Bang to the present, the remnants of the radiation field should have cooled to a temperature of about 3 K, corresponding to a blackbody spectrum with peak wavelength λ_{max} in the microwave region. In 1965, the predicted cosmic background radiation was discovered by Arno Penzias and Robert Wilson at the Bell Labs. Since this landmark discovery, careful analysis has established that the temperature of the background field is 2.7 ± 0.1 K and has shown that it has the isotropic distribution in space that is absolutely essential for a universe that satisfies the cosmological principle.

The Very Early History of the Universe

What was the Big Bang like? The singular event that initiated the expansion of the universe must have been a huge explosion that occurred throughout the entire universe. We don't know whether the universe at the instant of the explosion occupied a small volume (nearly a point) or an infinite one since we don't yet know if the present average density is greater or less than the critical density, that is, whether the universe is infinite or finite. In either case, it is the size of space itself that has been expanding ever since.

Most cosmologists currently favor a theoretical description of the evolution of the universe following the Big Bang called the **standard model.** It relies heavily on recent experimental discoveries and theoretical advances in particle physics and reflects the increasing overlap of frontier research in those areas of physics over the past several years. The standard model's account of how the universe evolved from $t = 0$ to the present, when $t = 10^{10}$ years, is outlined in the following discussion and illustrated in Figure 42-24.

Initially, the four forces of nature (strong, electromagnetic, weak, and gravity) were unified into a single force. Physicists have been successful in developing theoretical descriptions that unify the first three, but a theory of quantum gravity, needed for the extreme densities of the single-force period, does not yet exist. Consequently, until the cooling universe "froze" or "condensed out" the gravitational force at about 10^{-43} s after the Big Bang when the temperature was still 10^{32} K, we have no means of describing what was occurring. At this point, the average energy of the particles created would have been about 10^{19} GeV. As the universe continued to cool below 10^{32} K, the three forces other than gravity remained unified and are described by the grand unification theories (GUTs). Quarks and leptons were indistinguishable and particle quantum numbers were not conserved. It was during this period that a slight excess of quarks over antiquarks occurred, roughly 1 in 10^9, that ultimately resulted in the matter that we now observe in the universe.

At 10^{-35} s, the universe had expanded sufficiently to cool to about 10^{27} K, at which point another phase transition occurred as the strong force condensed out of the GUTs group, leaving only the electromagnetic and weak forces still unified as the **electroweak force.** During this period, the previously free quarks in the dense mixture of roughly equal numbers of quarks, leptons, their antiparticles, and photons began to combine into hadrons and their antiparticles, including the nucleons. By the time the universe had cooled to about 10^{13} K, at about $t = 10^{-6}$ s, the hadrons had mostly disappeared. This is because 10^{13} K corresponds to $kT \sim 1$ GeV, which is the minimum energy needed to create nucleons and antinucleons from the photons present via the reactions

$$\gamma \longrightarrow p^+ + p^- \qquad \text{42-31}a$$

and

$$\gamma \longrightarrow n + \overline{n} \qquad \text{42-31}b$$

The particle–antiparticle pairs annihilated and there was no new production to replace them. Only the slight earlier excess of quarks led to a slight excess of protons and neutrons over their antiparticles. The annihilations resulted in photons and leptons, and after about $t = 10^{-4}$ s, those particles in roughly equal numbers dominated the universe. This was the **lepton era.** At about $t = 10$ s, the temperature had fallen to 10^{10} K ($kT \sim 1$ MeV). Further expansion and cooling dropped the average photon energy below that needed to form an electron–positron pair. Annihilation then removed all of the positrons as it had the antiprotons and antineutrons earlier, leaving only the small excess of electrons arising from charge conservation, and the **radiation era** began. The particles present were primarily photons and neutrinos.

Within a few more minutes, the temperature dropped sufficiently to enable fusing protons and neutrons to form nuclei that were not immediately photodisintegrated. Deuterium, helium, and a bit of lithium were produced in this **nucleosynthesis period,** but the rapid expansion soon dropped the temperature too low for the fusion to continue and the formation of heavier elements had to await the birth of stars.

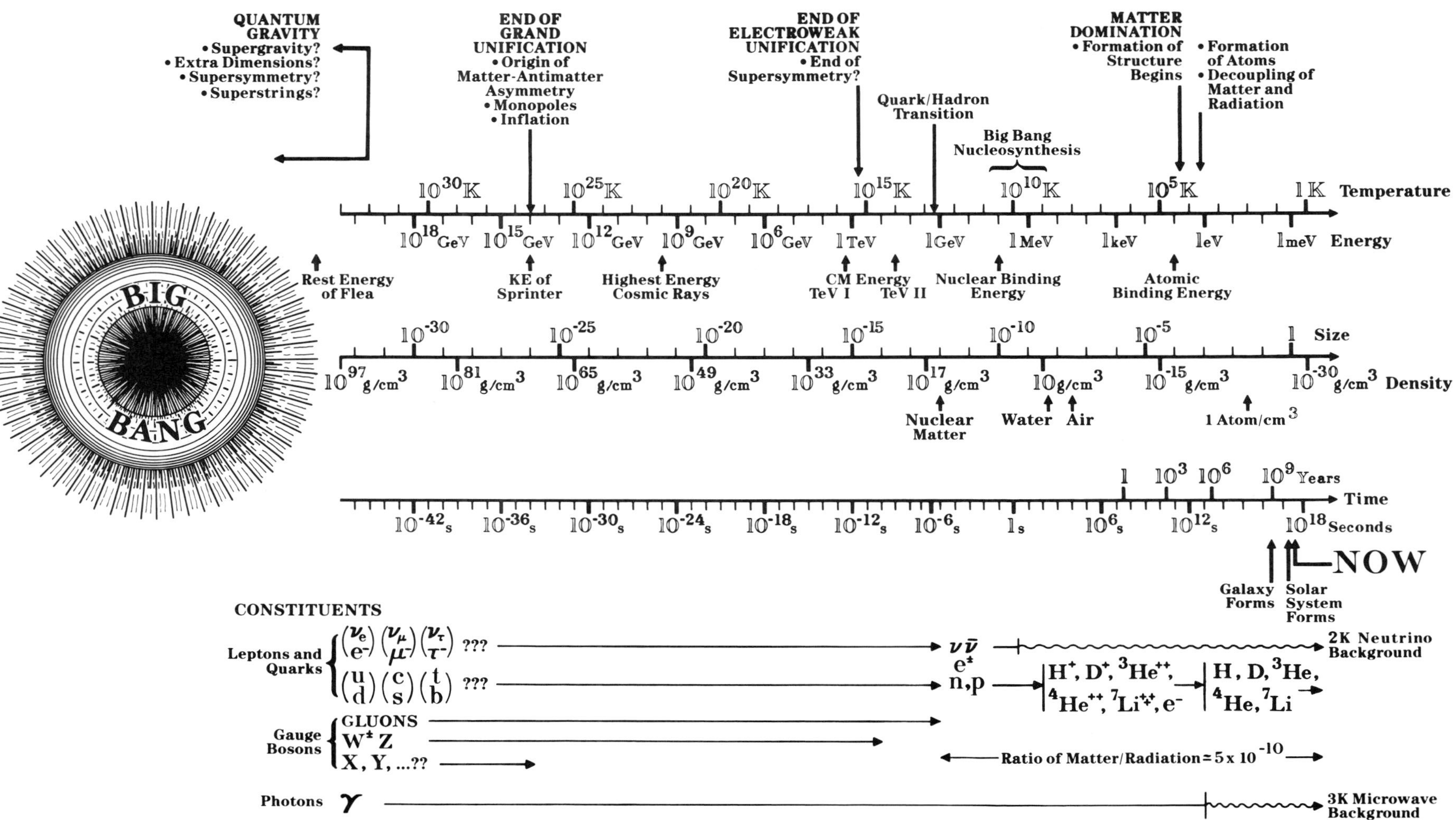

Figure 42-24 The evolution of the universe from time $t = 0$ to the present according to the standard model.

A long time later, when the temperature had dropped to about 3000 K as the universe grew to about 1/1000 of its present size, kT dropped below typical atomic ionization energies and atoms were formed. By then the expansion had redshifted the radiation field so that the total radiation energy was about equal to the energy represented by the remaining mass. As expansion and cooling continued, the energy of the steadily redshifting radiation steadily declined until, at $t = 10^{10}$ y (now), matter came to dominate the universe, with its energy density exceeding that of the 2.7-K radiation remaining from the Big Bang by a factor of about 1000.

Unanswered Questions and the Limits of Knowledge

The standard model of the evolution of the universe and the current theories of stellar and galactic genesis and evolution have been amazingly successful. Still, some fundamental questions that have arisen during our discussions are yet unanswered. Will the universe expand forever or rebound to a point and repeat the Big Bang? The answer depends on whether the present average mass density is greater or less than the critical density of about 10^{-26} kg/m^3. The uncertainty in the current measurement would allow either possibility, but the value is tantalizingly close to the critical value. If it does equal the critical value, an intriguing additional question is, "Why?" We have noted the serious problem of the missing mass of the universe and how it might be explained. The possible explanations—black holes and massive neutrinos—present other major questions. Answering some of them requires that we probe at the current limits of physical knowledge. For example, near a mass m, general relativity prevents our seeing events occurring at dimensions less than L, the **event horizon:**

$$L = \frac{Gm}{c^2} \tag{42-32}$$

On the other hand, the uncertainty principle in quantum theory places this limit at the Compton wavelength λ_C:

$$\lambda_C = \frac{h}{mc} \tag{42-33}$$

Equating these yields the **Planck mass** $m = 5.5 \times 10^{-8}$ kg. The length $L = \lambda_C \approx 10^{-35}$ m is called the **Planck length** and the time for light to travel across that length,

$$t = \left(\frac{Gh}{c^5}\right)^{1/2} = 1.35 \times 10^{-43}\ \text{s} \tag{42-34}$$

is called the **Planck time.** Thus, for Planck time the mass density of the universe is such that the mass m is contained within a volume of dimensions $L^3 \sim (10^{-35}\ \text{m})^3$. Relativistic space-time is no longer a continuum and even a new theory of gravity—quantum gravity or supergravity—is needed.

Perhaps it really is as some cosmologists have suggested: If the universe had evolved even slightly differently than it has, perhaps due to a slightly different value for h or e or some other fundamental constant, life on earth and maybe the earth itself would be impossible. This is the **anthropic principle,** which holds that the universe looks as it does because we are here to see it.

Summary

1. The luminosity L of a star is the power that it emits. It is related to the power per unit area received at the earth, called the apparent brightness f, and the distance r to the star by

$$L = 4\pi r f$$

In the case of the sun, where r is accurately known, f is called the solar constant.

2. Computations of stellar surface temperatures are based on Planck's law of blackbody radiation, an assumption well supported by detailed measurements of the sun's electromagnetic radiation spectrum.

3. The source of the sun's energy is the proton–proton fusion cycle which starts with the nuclear reaction

$$^1\mathrm{H} + {}^1\mathrm{H} \rightarrow {}^2\mathrm{H} + e^+ + \nu_e + 1.44\ \mathrm{MeV}$$

4. Stars are classed as population I or population II, based on their compositions. Population I stars have 2 to 3 percent of their mass composed of elements heavier than He and are considered to be the younger of the two classes. Population II stars are nearly devoid of elements heavier than He.

5. The Milky Way is a spiral galaxy consisting of about 10^{10} stars. The sun is located about 30,000 $c{\cdot}\mathrm{y}$ from the center of the Galaxy, which is in the direction of the constellation Sagittarius from us. Approximately 90 percent of the gravitational mass of the Galaxy consists of dark, that is, nonluminous matter.

6. The Hertzprung–Russell diagram relates the luminosity of stars to their effective temperatures. Both quantities are related to stellar masses:

$$L \propto M^4 \qquad T \propto M^{1/2}$$

Stars that are "burning" hydrogen into helium fall on the main sequence of the diagram.

7. Following the exhaustion of their hydrogen fuel supply, stars evolve along different paths in the H–R diagram that depend primarily on their initial masses, eventually reaching one of three possible final states: white dwarf, neutron star, or black hole. It is in cataclysmic events that precede the latter two final states that elements heavier than Fe are formed.

8. If the mass of an incipient neutron star is so large that its radius is less than the Schwarzchild radius

$$R_S = 2GM/c^2$$

then no radiation or object with mass can escape its surface. It is a black hole.

9. Galaxies outside the Milky Way were first identified by Edwin Hubble who also showed that, with rare exceptions, they could be grouped into four general classes: spirals, barred spirals, ellipticals, and irregulars.

10. Hubble's law relates the recession velocity of a galaxy, determined from the redshift of its spectrum, to the distance of the galaxy from us:

$$v = Hr$$

where the Hubble constant $H = 23$ km/s per million light-years. From Hubble's law, we conclude that the universe is expanding and that the expansion began approximately $1/H$ years ago.

11. The critical density of the universe is $\rho_c \approx 10^{-26}$ kg/m^3, which is the density at which the kinetic energy of a galaxy equals its gravitational potential energy. If the density of the universe happens to equal this value, the expansion will ultimately coast to a halt. Lower values will result in expansion forever. Higher values will cause an eventual reversal of the present expansion.

12. The model currently used to describe the evolution of the universe is the standard model, in which the universe began with the Big Bang approximately 10^{10} years ago.

13. The standard model is supported by substantial experimental observations, including the isotropic, 2.7-K, background blackbody radiation spectrum. There are also many fundamental questions for which it has not yet provided answers.

Suggestions for Further Reading

Pipkin, Francis M.: "Gravity Up in the Air: Are the Laws of the Universe in Doubt?" *The Sciences*, May/June 1984, p. 42.

This article examines cosmological theories in which the gravitational "constant," G, varies over time.

Schramm, David N. and Gary Steigman: "Particle Accelerators Test Cosmological Theory," *Scientific American*, June 1988, p. 66.

This article describes how cosmological theories have set limits on the possible number of families of elementary particles and how this prediction is being tested.

von Baeyer, Hans Christian: "Creatures of the Deep," *The Sciences*, March/April 1989, p. 2.

The problem of "dark matter," which may make up most of the mass of the universe, puzzles cosmologists, astronomers, and particle physicists.

Review

A. Objectives: After studying this chapter, you should:

1. Know the source of the sun's energy, the general structure of the sun's outer layers, and the approximate value of its surface temperature; and understand the origin of sunspots.

2. Know the approximate size of the Milky Way galaxy in light-years and the sun's approximate location in the Galaxy; and know how the luminosity of a star, its effective temperature, and its mass are related by the Hertzprung–Russell diagram.

3. Know about the missing-mass problem in the Galaxy and in the universe and how it relates to the future expansion.

4. Be able to describe the evolutionary path of a medium size star, like the sun; and know how the mass of a star determines which of the possible final states—white dwarf, neutron star, or black hole—it will reach.

5. Know how galaxies are classified and how they are distributed throughout the universe; and know about the prodigious power emitted by active galaxies.

6. Be able to compute the distances to galaxies with Hubble's law, using the recession velocity determined from the doppler effect.

7. Know about the cosmological principle and the general expansion of the universe.

8. Know how the current theory of cosmogenesis, the standard model, accounts for the evolution of the universe to the present time and suggests what its future may be.

B. Define, explain, or otherwise identify:

Astrophysics	Proton–proton cycle
Cosmology	Solar-neutrino problem
Photosphere	Sunspot cycle
Solar constant	Sunspots
Luminosity	Solar flares
Quiet sun	Plages
Limb	Filaments
Chromosphere	Prominences
Corona	Constellations
Solar wind	Star clusters
Core	Galactic clusters

Open clusters
Population I stars
Globular clusters
Population II stars
Missing-mass problem
Parallax angle
Hertzprung–Russell (H–R) diagram
Main sequence
Main-sequence dwarfs
Red subgiant
Red giant
Horizontal branch
Planetary nebula
Binary pairs
Lagrangian point
Roche lobe
Accretion disk
Nova
Dwarf nova
Recurrent nova
Classical nova
Supernova
Degenerate dwarf
White dwarf
Electron-degeneracy pressure
Chandrasekhar limit
Black dwarf
Neutron star
Pulsar
Schwarzschild radius
Gravitational redshift
Black hole
Interstellar dust
Ellipticals
Disks
Ordinary spirals
Barred spirals
Irregular galaxies
Quiet galaxies
Active galaxies
Seyfert galaxies
N galaxies
BL Lac objects
Radio galaxies
Quasars
Quasistellar objects (QSOs)
Hubble's law
Hubble constant
Hubble age
Cosmological principle
Perfect cosmological principle
Cosmological constant
Big Bang
Standard model
Electroweak force
Lepton era
Radiation era
Nucleosynthesis period
Event horizon
Planck mass
Planck length
Planck time
Anthropic principle

C. True or false: If a statement is true, explain why it is true. If it is false, give a counterexample.

1. The part of the sun that we see, the surface or photosphere, cannot be more than about 700 km thick.

2. A significant portion of the solar energy absorbed by the earth is furnished by neutrinos produced in the sun's proton–proton fusion cycle.

3. The lifetime of a star whose mass is five times that of the sun will be only one-fifth of the sun's lifetime.

4. Supernova SN1987A didn't really occur in 1987.

5. Unlike white dwarfs and neutron stars, black holes are not cooling toward thermal equilibrium with the universe.

6. There is no physical evidence to support the theory that the universe is expanding.

7. The sun is located at the center of the Milky Way.

8. According to the standard model, when the temperature of the expanding universe fell below about ten billion K, the formation of electron–positron pairs ceased.

Problems

Level I

42-1 Our Star, the Sun

1. Measurement of the doppler shift of spectral lines in light from the east and west limbs of the sun at the solar equator reveals that the tangential velocities of the limbs differ by 4 km/s. Use this result to compute the approximate period of the sun's rotation. ($R_{\odot} = 6.96 \times 10^5$ km)

2. The gravitational potential energy U of a self-gravitating spherical body of mass M and radius R is a function of the details of the mass distribution. For the sun, $U_{\odot} = -2GM_{\odot}^2/R_{\odot}$. What would be the approximate lifetime of the sun, if the source of its emitted energy were entirely derived from gravitational contraction? ($M_{\odot} = 1.99 \times 10^{30}$ kg)

42-2 The Stars

3. The sun is moving with speed 2.5×10^5 m/s in a circular orbit about the center of the Galaxy. How long (in earth years) does it take to complete one orbit? How many orbits has it completed since it was formed?

4. The reason massive neutrinos are considered a candidate for solving the missing-mass problem is that, at the conclusion of the lepton era, the universe contained about equal numbers of photons and neutrinos. They are still here, for the most part. The photons can be observed and their density is measured to be about 500 photons/cm^3; thus, there must be about that number density of neutrinos in the universe, too. If neutrinos have a mass m_ν and if the cosmological expansion has reduced their average speed so that their energy is now primarily rest mass, what would be the individual neutrino mass (in eV/c^2) necessary to account for the missing mass of the universe? Recall that the observed mass accounts for only about 10 percent of that needed to bond the universe.

42-3 The Evolution of Stars

5. A unit of length often used by astronomers to measure distances in nearby space is the parsec, defined as the distance at which a star subtends a parallax angle of one arc second. (See Equation 42-11 and Example 42-4.) The practical limit of such measurements is 0.01 arc second. (*a*) How many light-years is 1 parsec? (*b*) If the density of stars in the sun's region of the Milky Way galaxy is 0.08 stars/(parsec)^3, how many stars could, in principle, have their distances from us measured by the trigonometric parallax method?

6. Astronomers often use the *apparent magnitude m* as a means to compare the visual brightness of stars and then relate the comparison to the luminosity and distance to standard stars, such as the sun. (See Equation 42-10.) The difference in the apparent magnitudes of two stars m_1 and m_2 is defined as $m_1 - m_2 = 2.5 \log(f_1/f_2)$, a relation based on the logarithmic response of the human eye to the brightness of objects. Pollux, one of the "twins" in the constellation Gemini, has apparent magnitude 1.16 and is 12 parsecs away. Betelgeuse, the star at Orion's right shoulder, has apparent magnitude 0.41. How far away is Betelgeuse, if they have the same luminosity?

7. Using the H–R diagram (Figure 42-14), determine the effective temperature and the luminosity of a star whose mass is (*a*) $0.3M_\odot$ and (*b*) $3M_\odot$. (*c*) Find the radius of each star. (*d*) Determine their expected lifetimes, relative to that of the sun.

42-4 Cataclysmic Events

8. Compute the energy required (in MeV) to produce each of the photodisintegration reactions in Equations 42-18 and 42-19.

9. The gas shell of a planetary nebula shown in Figure 42-15 is expanding at 24 km/s. Its diameter is 1.5 *c*·y. (*a*) How old is the gas shell? (*b*) If the central star of the planetary nebula is 12 times as luminous as the sun and 1.36 times hotter, what is the radius of the central star in units of $R_\odot$?

42-5 Final States of Stars

10. Calculate the Schwarzschild radius of a star whose mass is equal to that of (*a*) the sun, (*b*) Jupiter, (*c*) the earth. (The mass of Jupiter is approximately 318 times that of the earth.)

11. Consider a neutron star whose mass equals $2M_\odot$. (*a*) Compute the star's radius. (*b*) If the neutron star is rotating at 0.5 rev/s and its density is uniform, what is its rotational kinetic energy? (*c*) If its rotation slows by 1 part in 10^8 per day and the lost kinetic energy is all radiated, what is the star's luminosity?

12. If the 90 percent of the Milky Way's mass that is "missing" resides entirely in a large black hole at the center of the Galaxy, what would be the black hole's (*a*) mass and (*b*) radius?

42-6 Galaxies

13. A particular galaxy is observed to have a recession velocity of 72,000 km/s. (*a*) Find the distance to the galaxy. (*b*) What is the upper bound to the present age of the universe according to the Big Bang expansion theory. (*c*) The value of Hubble's constant depends critically on calibration distance measurements, which are difficult to make. If the calibration distance measurements are in error by 10 percent, by how much is the age calculation in (*b*) in error?

14. The bright core of a certain Seyfert galaxy had a luminosity of $10^{10}\ L_\odot$. The luminosity increased by 100 percent in a period of 18 months. Show that this means the energy source of the core is about 9.5×10^4 AU in diameter. How does this compare to the diameter of the Milky Way galaxy?

42-7 Gravitation and Cosmology

15. Evaluate Equation 42-30 for the critical density of the universe.

42-8 Cosmogenesis

16. Cosmological theory suggests that the average distance between galaxies and, hence, the size of the universe is inversely proportional to the absolute temperature. If this is true, how large was the universe relative to its present size (*a*) 2000 years ago, (*b*) 10^6 years ago, (*c*) when $t = 1$ s after the Big Bang, (*d*) when $t = 10^{-6}$ s, and (*e*) $t = 10$ s. (See Figure 42-24.)

17. Determine the mass density of the universe for $t =$ Planck time. How does this value compare to the mass density of the proton? Of osmium (density, 2.25×10^4 kg/m^3)?

18. At what wavelength is the blackbody radiation distribution of the cosmic microwave background at a maximum?

19. How long after the Big Bang did it take the universe to cool to the threshold temperature for the formation of muons? What would be the mass of a particle that could be formed by the average energy of the current 2.7-K background radiation?

Level II

20. If Hubble's law is true for an observer in the Milky Way galaxy, prove that it must also be true for observers in other galaxies. (*Hint:* Use the vector property of the velocity.)

21. Supernova SN1987A was first visible at the earth in 1987. (*a*) How many years before 1987 did the explosion occur? (*b*) If protons with 100 GeV of kinetic energy were produced in the event, when should they arrive at earth? (See Chapter 41, Problem 28*b*.)

22. Assume that the sun was composed of 70 percent hydrogen when it first formed. (*a*) How many hydrogen nuclei were there in the sun at that time? (*b*) How much energy would ultimately be released if all of the hydrogen nuclei fused into helium? (*c*) Astrophysicists have predicted that the sun can radiate energy at its current rate until about 23 percent of the hydrogen has been burned. What total lifetime for the sun does that prediction imply?

23. Consider an eclipsing binary, consisting of two stars of mass m_1 and m_2 separated by a distance r, whose orbital plane is parallel to our line of sight. Doppler measurements of the radial velocity of each component of the binary are shown below. Assume that the orbit of each star about the center of mass is circular. (*a*) What is the period T and angular frequency ω of the binary? (*b*) Show that $m_1 + m_2 = \omega^2 r^3/G$. (*c*) Compute the values of m_1, m_2, and r from the data in the v versus t graph.

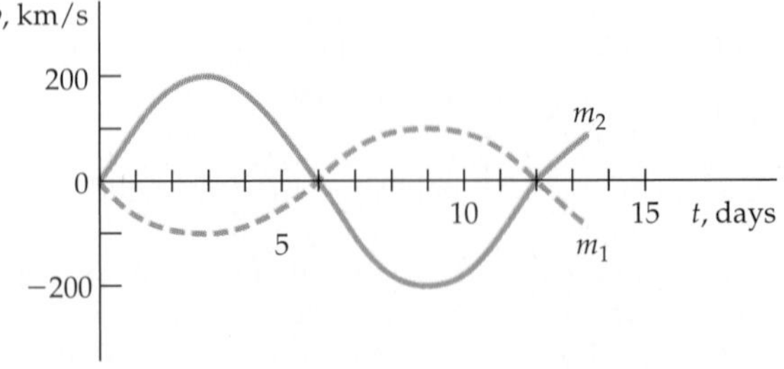

24. Prove that the total energy of the earth's orbital motion $E = (mv^2/2) + (-GM_\odot m/r)$ is equal to one-half of its gravitational potential energy $(-GM_\odot m/r)$, where r is the earth's orbital radius.

25. Given the currently accepted value of the Hubble constant and the fact that the average matter density of the universe is one H-atom/m^3, what creation rate of new H-atoms would be necessary in a steady-state model to maintain the present mass density, even though the universe is expanding? [Give your answer in (H-atoms/m^3)/10^6 y.] Would you expect such a spontaneous creation rate to be readily observable?

Level III

26. The ability of a planet to retain particular gases in an atmosphere depends on the temperature that its atmosphere has (or would have) and the escape velocity for that body. In general, if the average speed of a particular gas molecule exceeds $\frac{1}{6}$ of the escape velocity, that gas will disappear from the atmosphere in about 10^8 years. (*a*) Graph the average speed of H_2O, CO_2, O_2, CH_4, H_2, and He from 50 K to 1000 K. On the same graph, show the points representing $\frac{1}{6}$ of the escape velocity versus average temperature of the atmosphere for the bodies in the table below. (*b*) Show that the escape speed from a planet is $v = v_{\text{Earth}}\sqrt{\alpha/\beta}$. (*c*) Of the six gases plotted, above, which would probably be found in the current atmospheres of the solar system bodies in the table, and which would not? Explain *each* answer briefly.

Atmospheric Temperatures

Average T_{Atmo}(K)	Body	α Mass (Earth = 1.0)	β Diameter (Earth = 1.0)
300	Earth	1.00	1.00
390	Venus	0.81	0.95
600	Mercury	0.06	0.38
150	Jupiter	318.00	11.00
60	Neptune	17.00	3.90
290	Mars	0.11	0.53

27. Suppose that the sun's luminosity increases by a factor of 100 as it evolves into a red giant star. Show that the earth's oceans will evaporate, but the water vapor will not escape from the atmosphere. (See Problem 26, Equation 10-24, and Chapter 15.)

28. The approximate mass of dust in the Galaxy can be computed from the observed extinction of starlight. Assuming the mean radius of dust grains to be R with a uniform number density of n grains/cm^3, (*a*) show that the mean free path d_0 of a photon in interstellar dust is given by $d_0 = 1/n\pi R^2$. (*b*) Starlight traveling toward an earth observer a distance d from the star has intensity $I = I_0\, e^{-d/d_0}$. In the vicinity of the sun, measurement of I yields $d_0 = 3000$ c·y. If $R = 10^{-5}$ cm, calculate n. (*c*) The average mass density of solid material in the Galaxy is 2 g/cm^3; and in the disk, the density of stars is about $1M_\odot/300$. Compute the ratio of the mass of dust in 300 $(c\cdot\text{y})^3$ to $1M_\odot$.

29. Supernova SN1987A certainly produced some heavy elements. To illustrate this, calculate the energy required to fuse two ^{56}Fe-atoms into one ^{112}Cd-atom. Compare this with the energy released in fusing 56 ^{1}H-atoms into one ^{56}Fe-atom, using the proton–proton cycle, and show that more than enough energy is available to produce a ^{112}Cd atom (mass, 111.902762 u).

Appendix A

Review of Mathematics

In this appendix, we will review some of the basic results of algebra, geometry, trigonometry, and calculus. In many cases, we will merely state results without proof. Table A-1 lists some mathematical symbols.

Equations

The following operations can be performed on mathematical equations to facilitate their solution:

1. The same quantity can be added to or subtracted from each side of the equation.
2. Each side of the equation can be multiplied or divided by the same quantity.
3. Each side of the equation can be raised to the same power.

It is important to understand that the preceding rules apply to each *side* of the equation and not to each *term* in the equation.

Table A-1 **Mathematical Symbols**

$=$	is equal to
$\neq$	is not equal to
$\approx$	is approximately equal to
$\sim$	is of the order of
$\propto$	is proportional to
$>$	is greater than
$\geq$	is greater than or equal to
$\gg$	is much greater than
$<$	is less than
$\leq$	is less than or equal to
$\ll$	is much less than
Δx	change in x
$\lvert x \rvert$	absolute value of x
$n!$	$n(n-1)(n-2)\cdots 1$
Σ	sum
lim	limit
$\Delta t \to 0$	Δt approaches zero
$\frac{dx}{dt}$	derivative of x with respect to t
$\frac{\partial x}{\partial t}$	partial derivative of x with respect to t
$\int$	integral

Example A-1

Solve the following equation for x:

$$(x - 3)^2 + 7 = 23$$

We first subtract 7 from each side of the equation to obtain $(x - 3)^2 = 16$. Taking the square root of each side, which is the same as raising each side to the $\frac{1}{2}$ power, we obtain $\pm(x - 3) = \pm 4$. We have included the plus-or-minus signs because either $(+4)^2 = 16$ or $(-4)^2 = 16$. We do not need to write $\pm$ on both sides of the equation as all the possibilities are included in $x - 3 = \pm 4$. We can now solve for x by adding 3 to each side. There are two solutions: $x = 4 + 3 = 7$ and $x = -4 + 3 = -1$. These values can be checked by substituting them into the original equation.

Example A-2

Solve the following equation for x:

$$\frac{1}{x} + \frac{1}{4} = \frac{1}{3}$$

This type of equation occurs both in geometric optics and in analyses of electric circuits. Although it is easy to solve, errors are often made. We solve it by first subtracting $\frac{1}{4}$ from each side to obtain

$$\frac{1}{x} = \frac{1}{3} - \frac{1}{4} = \frac{4}{12} - \frac{3}{12} = \frac{1}{12}$$

We then multiply each side by $12x$ to obtain $x = 12$. Note that this is equivalent to taking the reciprocal of each side of the equation. A typical mistake in handling this type of equation is to take the reciprocal of each term first to obtain $x + 4 = 3$. This operation is not allowed; it changes the relative values of each side of the equation and leads to incorrect results.

Direct and Inverse Proportion

The relationships of direct proportion and inverse proportion are so important in physics that they deserve special consideration. Often much algebraic manipulation can be avoided through a simple knowledge of these relationships. Suppose, for example, that you work for 5 days at a certain pay rate and earn \$400. How much would you earn at the same pay rate if you worked 8 days? In this problem, the money earned is *directly proportional* to the time worked. We can write an equation relating the money earned M to the time worked t using a constant of proportionality R:

$$M = Rt$$

The constant of proportionality in this case is the pay rate. We can express R in dollars per day. Since \$400 was earned in 5 d, the value of R is \$400/(5 d) = \$80/d. In 8 d, the amount earned is therefore

$$M = (\$80/\text{d})(8\text{ d}) = \$640$$

However, we do not have to find the pay rate explicitly to work the problem. Since the amount earned in 8 d is $\frac{8}{5}$ times that earned in 5 d, this amount is

$$M = \tfrac{8}{5}(\$400) = \$640$$

We can use a similar example to illustrate inverse proportion. If you get a 25 percent raise, how long would you need to work to earn \$400? Here we consider R to be a variable and we wish to solve for t:

$$t = \frac{M}{R}$$

In this equation, the time t is *inversely proportional* to the pay rate R. Thus, if the new rate is $\frac{5}{4}$ times the old rate, the new time will be $\frac{4}{5}$ times the old time or 4 d.

There are some situations in which one quantity varies as the square or some other power of another quantity where the ideas of proportionality are also very useful. Suppose, for example, that a 10-in-diameter pizza costs \$8.50. How much would you expect a 12-in-diameter pizza to cost? We expect the cost of a pizza to be approximately proportional to the amount of its contents, which is proportional to the area of the pizza. Since the area is in turn proportional to the square of the diameter, the cost should be proportional to the square of the diameter. If we increase the diameter by a factor of 12/10, the area increases by a factor of $(12/10)^2 = 1.44$, so we should expect the cost to be (1.44)(\$8.50) = \$12.24.

Example A-3

The intensity of light from a point source varies inversely with the square of the distance from the source. If the intensity is 3.20 W/m^2 at 5 m from a source, what is it at 6 m from the source?

The equation expressing the fact that the intensity varies inversely with the square of the distance can be written

$$I = \frac{C}{r^2}$$

where C is some constant. Then, if $I_1 = 3.20$ W/m^2 at $r_1 = 5$ m and I_2 is the unknown intensity at $r_2 = 6$ m, we have

$$\frac{I_2}{I_1} = \frac{C/r_2^2}{C/r_1^2} = \frac{r_1^2}{r_2^2} = \left(\frac{5}{6}\right)^2 = 0.694$$

The intensity at 6 m from the source is thus

$$I_2 = 0.694(3.20 \text{ W/m}^2) = 2.22 \text{ W/m}^2$$

Linear Equations

An equation in which the variables occur only to the first power is said to be linear. A linear equation relating y and x can always be put into the standard form

$$y = mx + b \qquad \text{A-1}$$

where m and b are constants that may be either positive or negative. Figure A-1 shows a graph of the values of x and y that satisfy Equation A-1. The constant b, called the **intercept,** is the value of y at $x = 0$. The constant m is the **slope** of the line, which equals the ratio of the change in y to the corresponding change in x. In the figure, we have indicated two points on the line, x_1, y_1 and x_2, y_2, and the changes $\Delta x = x_2 - x_1$ and $\Delta y = y_2 - y_1$. The slope m is then

$$m = \frac{y_2 - y_1}{x_2 - x_1} = \frac{\Delta y}{\Delta x}$$

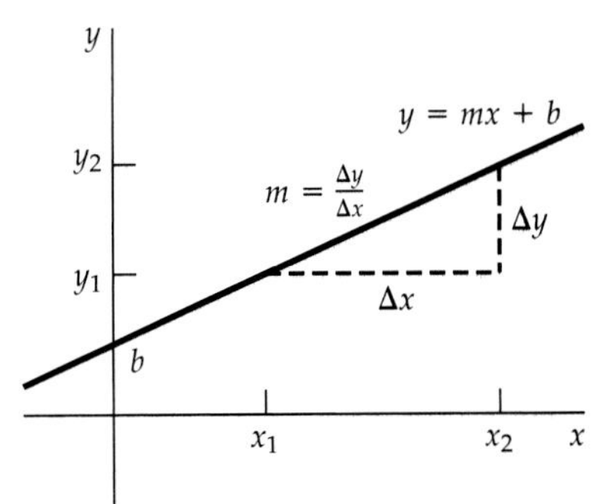

Figure A-1 Graph of the linear equation $y = mx + b$, where b is the intercept and $m = \Delta y/\Delta x$ is the slope.

If x and y are both unknown, there is no unique solution for their values. Any pair of values x_1, y_1 on the line in Figure A-1 will satisfy the equation. If we have two equations, each with the same two unknowns x and y, the equations can be solved simultaneously for the unknowns.

Example A-4

Find the values of x and y that satisfy

$$3x - 2y = 8 \qquad \text{A-2}$$

and

$$y - x = 2 \qquad \text{A-3}$$

Figure A-2 shows a graph of each of these equations. At the point where the lines intersect, the values of x and y satisfy both equations. We can solve two simultaneous equations by first solving either equation for one variable in terms of the other variable and then substituting the result into the other equation. From Equation A-3, we have

$$y = x + 2$$

Substituting this value for y in Equation A-2, we obtain

$$3x - 2(x + 2) = 8$$

$$3x - 2x - 4 = 8$$

$$x = 12$$

Then

$$y = x + 2 = 14$$

An alternative method that is sometimes easier is to multiply one equation by a constant such that one of the unknown terms is eliminated

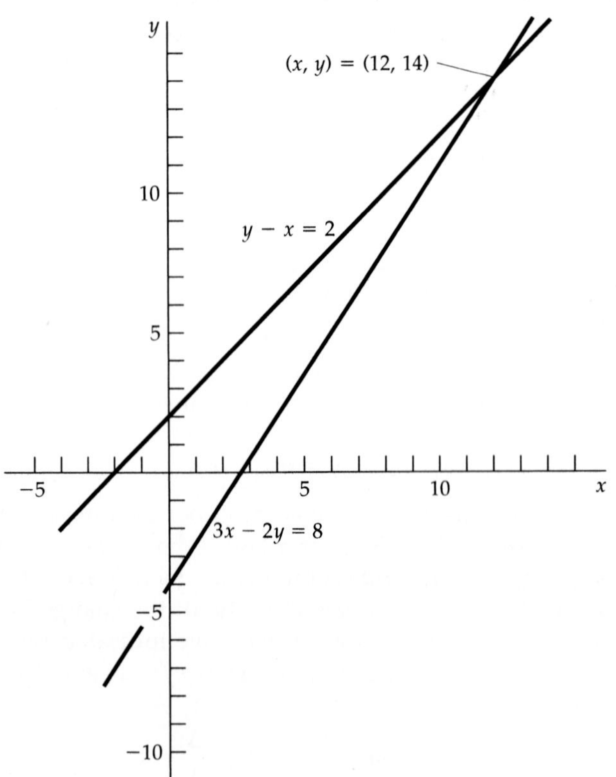

Figure A-2 Graph of Equations A-2 and A-3. At the point where the lines intersect, the values of x and y satisfy both equations.

when the equations are added or subtracted. If we multiply Equation A-3 by 2, we can add the resulting equation to Equation A-2 and eliminate y:

$$3x - 2y = 8$$

$$2y - 2x = 4$$

Adding, we obtain $3x - 2x = 12$ or $x = 12$, as before.

Factoring

Equations can often be simplified by factoring. Three important examples are

1. Common factor: $2ax + 3ay = a(2x + 3y)$
2. Perfect square: $x^2 \pm 2xy + y^2 = (x \pm y)^2$
3. Difference of squares: $x^2 - y^2 = (x + y)(x - y)$

The Quadratic Formula

An equation that contains a variable to the second power is called a *quadratic equation.* The standard form for a quadratic equation is

$$ax^2 + bx + c = 0 \qquad \text{A-4}$$

where a, b, and c are constants. The general solution of this equation is

$$x = -\frac{b}{2a} \pm \frac{1}{2a}\sqrt{b^2 - 4ac} \qquad \text{A-5}$$

When b^2 is greater than $4ac$, there are two solutions corresponding to the + and − signs. Figure A-3 shows a graph of y versus x where $y = ax^2 + bx + c$. The curve, called a **parabola,** crosses the x axis twice. The values of x for which $y = 0$ are the solutions to Equation A-4. When $b^2 < 4ac$, the graph of y versus x does not intersect the x axis, as is shown in Figure A-4, and there are no real solutions to Equation A-4. When $b^2 = 4ac$, the graph of y versus x is tangent to the x axis at the point $x = -b/2a$.

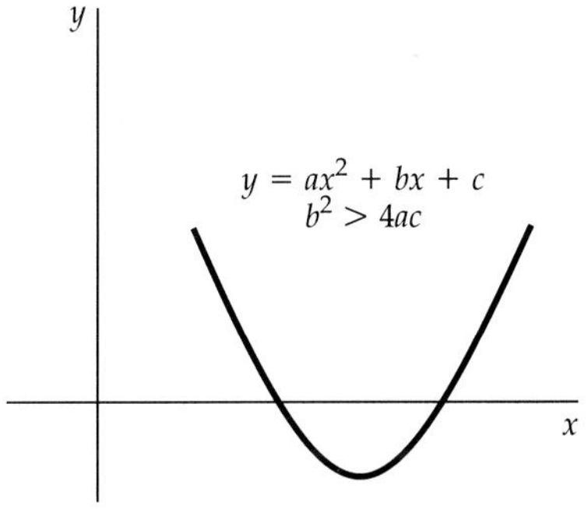

$y = ax^2 + bx + c$, $b^2 > 4ac$

Figure A-3 Graph of y versus x when $y = ax^2 + bx + c$ for the case $b^2 > 4ac$. The two values of x for which $y = 0$ satisfy the quadratic equation (Equation A-4).

Exponents

The notation x^n stands for the quantity obtained by multiplying x times itself n times. For example, $x^2 = x \cdot x$ and $x^3 = x \cdot x \cdot x$. The quantity n is called the **power,** or the **exponent,** of x. When two powers of x are multiplied, the exponents are added:

$$(x^m)(x^n) = x^{m+n} \qquad \text{A-6}$$

This can be readily seen from an example:

$$x^2x^3 = (x \cdot x)(x \cdot x \cdot x) = x^5$$

Any number raised to the 0 power is defined to be 1:

$$x^0 = 1 \qquad \text{A-7}$$

Then

$$x^nx^{-n} = x^0 = 1$$

$$x^{-n} = \frac{1}{x^n} \qquad \text{A-8}$$

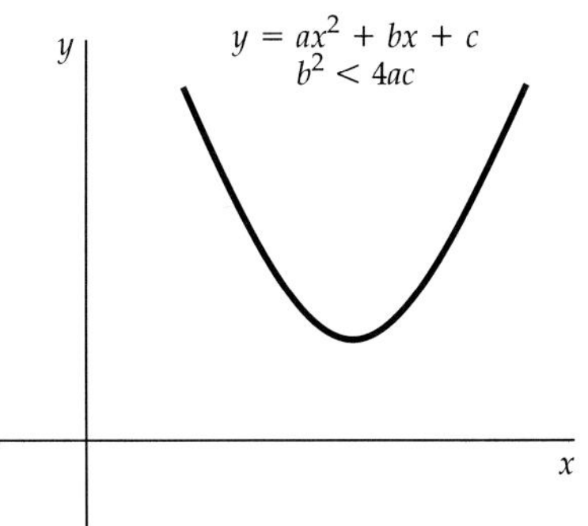

Figure A-4 Graph of y versus x when $y = ax^2 + bx + c$ for the case $b^2 < 4ac$. In this case, there are no (real) values of x for which $y = 0$.

When two powers are divided, the exponents are subtracted:

$$\frac{x^n}{x^m} = x^n x^{-m} = x^{n-m} \qquad \text{A-9}$$

Using these rules, we have

$$x^{1/2} \cdot x^{1/2} = x$$

so

$$x^{1/2} = \sqrt{x}$$

When a power is raised to another power, the exponents are multiplied:

$$(x^n)^m = x^{nm} \qquad \text{A-10}$$

Logarithms

When y is related to x by $y = a^x$, the number x is said to be the logarithm of y to the base a and is written

$$x = \log_a y$$

If $y_1 = a^n$ and $y_2 = a^m$, then

$$y_1 y_2 = a^n a^m = a^{n+m}$$

and

$$\log_a y_1 y_2 = n + m = \log_a y_2 + \log_a y_1 \qquad \text{A-11}$$

It then follows that

$$\log_a y^n = n \log_a y \qquad \text{A-12}$$

Since $a^1 = a$ and $a^0 = 1$,

$$\log_a a = 1 \qquad \text{A-13}$$

and

$$\log_a 1 = 0 \qquad \text{A-14}$$

There are two bases in common use: base 10, called **common logarithms,** and base e ($e = 2.728 \ldots$), called **natural logarithms.** When no base is specified, the base is understood to be 10. Thus, $\log 100 = \log_{10} 100 = 2$ since $100 = 10^2$.

The symbol ln is used for natural logarithms. Thus,

$$y = \ln x \qquad \text{A-15}$$

implies

$$x = e^y \qquad \text{A-16}$$

Logarithms can be changed from one base to another. Suppose that

$$z = \log x \qquad \text{A-17}$$

Then

$$10^z = x \qquad \text{A-18}$$

Taking the natural logarithm of both sides of Equation A-18, we obtain

$$z \ln 10 = \ln x$$

or

$$\ln x = (\ln 10) \log x \qquad \text{A-19}$$

The Exponential Function

When the rate of change of a quantity is proportional to the quantity itself, the quantity increases or decreases exponentially. An example of *exponential decrease* is nuclear decay. If N is the number of radioactive nuclei at some time, then the change dN in some very small time interval dt will be proportional to N and to dt:

$$dN = -\lambda N\, dt$$

where the constant of proportionality λ is the decay rate. The function N satisfying this equation is

$$N = N_0\, e^{-\lambda t} \qquad \text{A-20}$$

where N_0 is the number at time $t = 0$. Figure A-5 shows N versus t. A characteristic of exponential decay is that N decreases by a constant factor in a given time interval. The time interval for N to decrease to half its original value is its half-life $t_{1/2}$, which is related to the decay rate by

$$t_{1/2} = \frac{\ln 2}{\lambda} = \frac{0.693}{\lambda} \qquad \text{A-21}$$

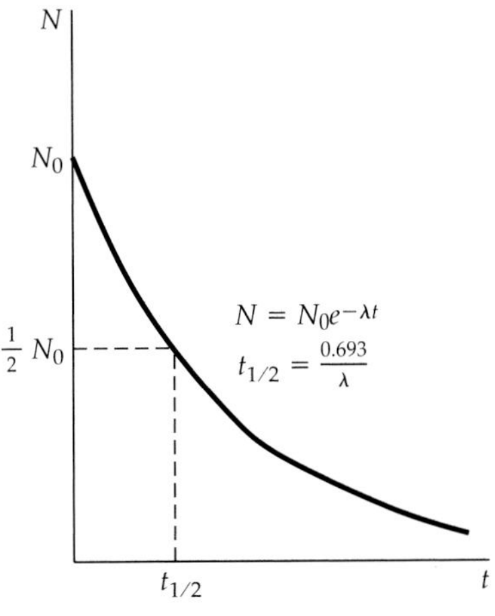

Figure A-5 Graph of N versus t when N decreases exponentially. The time $t_{1/2}$ is the time it takes for N to decrease by one-half.

An example of *exponential increase* is population growth. If the number of organisms is N, the change in N after a small time interval dt is given by

$$dN = +\lambda N\, dt$$

where λ is a constant that characterizes the rate of increase. The function N satisfying this equation is

$$N = N_0\, e^{\lambda t} \qquad \text{A-22}$$

A graph of this function is shown in Figure A-6. An exponential increase is characterized by a doubling time T_2, which is related to λ by

$$T_2 = \frac{\ln 2}{\lambda} = \frac{0.693}{\lambda} \qquad \text{A-23}$$

If the rate of increase λ is expressed as a percentage, $r = \lambda/100\%$, the doubling time is

$$T_2 = \frac{69.3}{r} \qquad \text{A-24}$$

For example, if the population increases by 2 percent per year, the population will double every $69.3/2 \approx 35$ years. Table A-2 lists some useful relations for exponential and logarithmic functions.

Table A-2 Exponential and Logarithmic Functions

$e = 2.71828 \qquad e^0 = 1$

If $y = e^x$, then $x = \ln y$.

$e^{\ln x} = x$

$e^x e^y = e^{(x+y)}$

$(e^x)^y = e^{xy} = (e^y)^x$

$\ln e = 1 \qquad \ln 1 = 0$

$\ln xy = \ln x + \ln y$

$\ln \frac{x}{y} = \ln x - \ln y$

$\ln e^x = x \qquad \ln a^x = x \ln a$

$\ln x = (\ln 10) \log x$
$= 2.3026 \log x$

$\log x = \log e \ln x = 0.43429 \ln x$

$e^x = 1 + x + \frac{x^2}{2!} + \frac{x^3}{3!} + \cdots$

$\ln (1 + x) = x - \frac{x^2}{2} + \frac{x^3}{3} - \frac{x^4}{4} + \cdots$

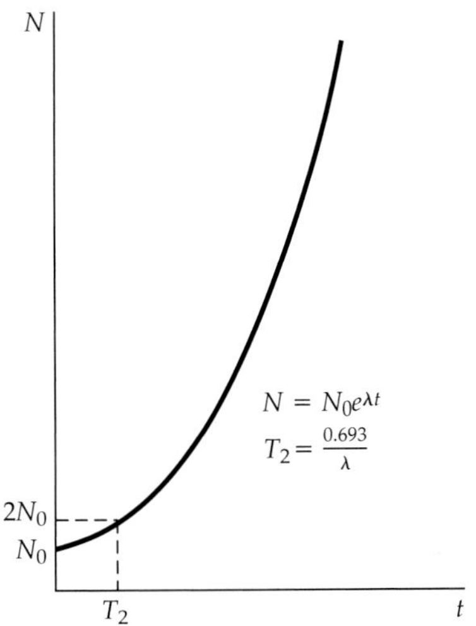

Figure A-6 Graph of N versus t when N increases exponentially. The time T_2 is the time it takes for N to double.

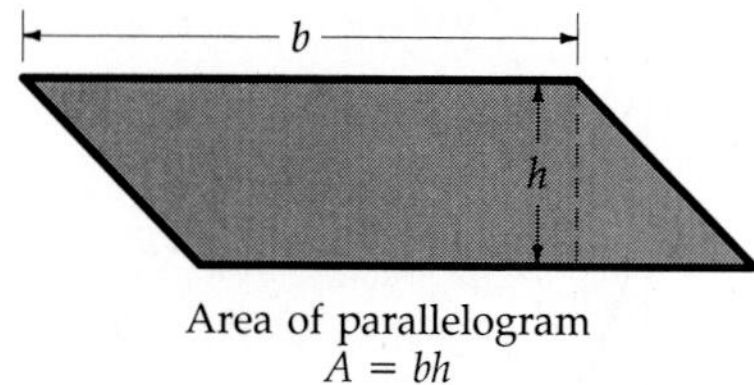

Figure A-7 Area of a parallelogram.

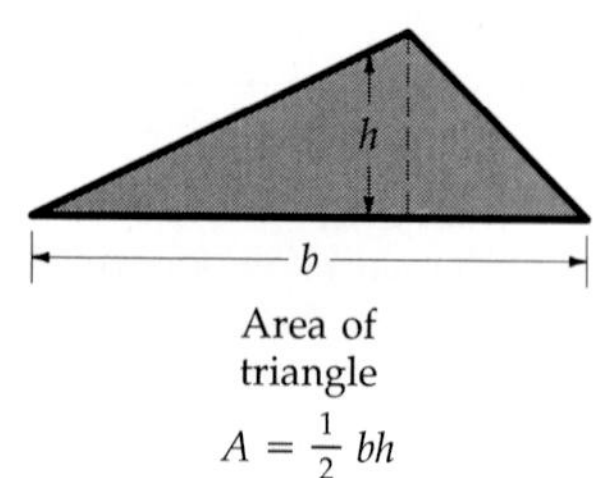

Figure A-8 Area of a triangle.

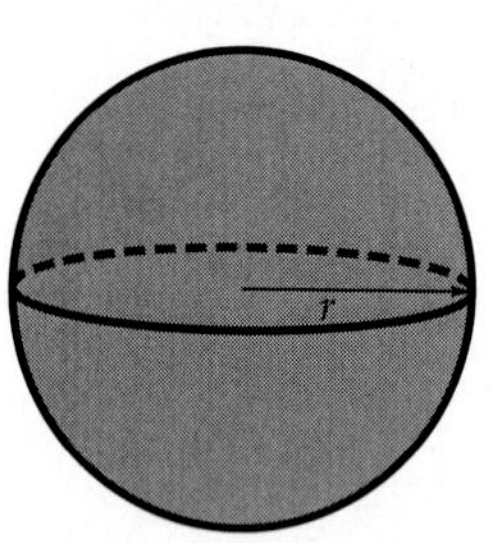

Figure A-9 Surface area and volume of a sphere.

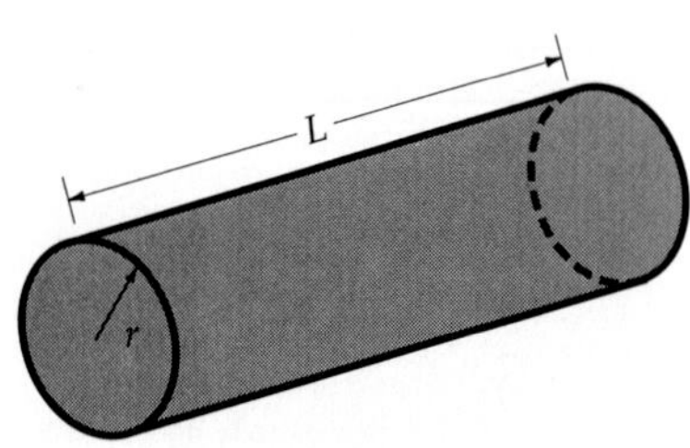

Figure A-10 Surface area (not including the end faces) and volume of a cylinder.

Geometry

The ratio of the circumference of a circle to its diameter is a natural number π, which has the approximate value

$$\pi = 3.141592$$

The circumference of a circle C is thus related to its diameter d and its radius r by

$$C = \pi d = 2\pi r \qquad \text{circumference of circle} \qquad \text{A-25}$$

The area of a circle is

$$A = \pi r^2 \qquad \text{area of circle} \qquad \text{A-26}$$

The area of a parallelogram is the base b times the height h (Figure A-7) and that of a triangle is one-half the base times the height (Figure A-8). A sphere of radius r (Figure A-9) has a surface area given by

$$A = 4\pi r^2 \qquad \text{spherical surface area} \qquad \text{A-27}$$

and a volume given by

$$V = \tfrac{4}{3}\pi r^3 \qquad \text{spherical volume} \qquad \text{A-28}$$

A cylinder of radius r and length L (Figure A-10) has surface area (not including the end faces) of

$$A = 2\pi rL \qquad \text{cylindrical surface} \qquad \text{A-29}$$

and volume of

$$V = \pi r^2 L \qquad \text{cylindrical volume} \qquad \text{A-30}$$

Trigonometry

The angle between two intersecting straight lines is measured as follows. A circle is drawn with its center at the intersection of the lines, and the circular arc is divided into 360 parts called **degrees.** The number of degrees in the arc between the lines is the measure of angle between the lines. For very small angles, the degree is divided into minutes (′) and seconds (″) with $1' = 1°/60$ and $1'' = 1'/60 = 1°/3600$. For scientific work, a more useful measure of an angle is the radian (rad), which is defined as the length of the circular arc between the lines divided by the radius of the circle (Figure A-11). If s is the arc length and r is the radius of the circle, the angle θ measured in radians is

$$\theta = \frac{s}{r} \qquad \text{A-31}$$

Since the angle measured in radians is the ratio of two lengths, it is dimensionless. The relation between radians and degrees is

$$360° = 2\pi \text{ rad}$$

or

$$1 \text{ rad} = \frac{360°}{2\pi} = 57.3° \qquad \text{A-32}$$

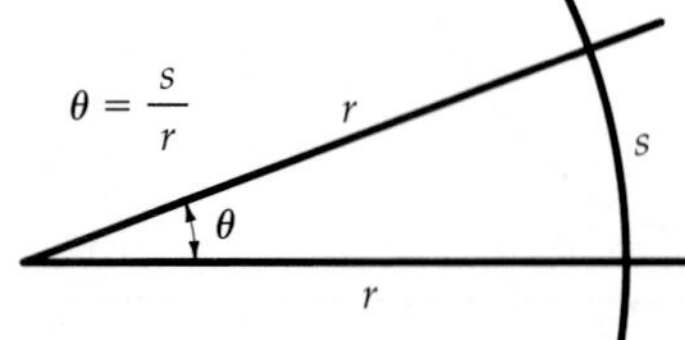

Figure A-11 The angle θ in radians is defined to be the ratio s/r, where s is the arc length intercepted on a circle of radius r.

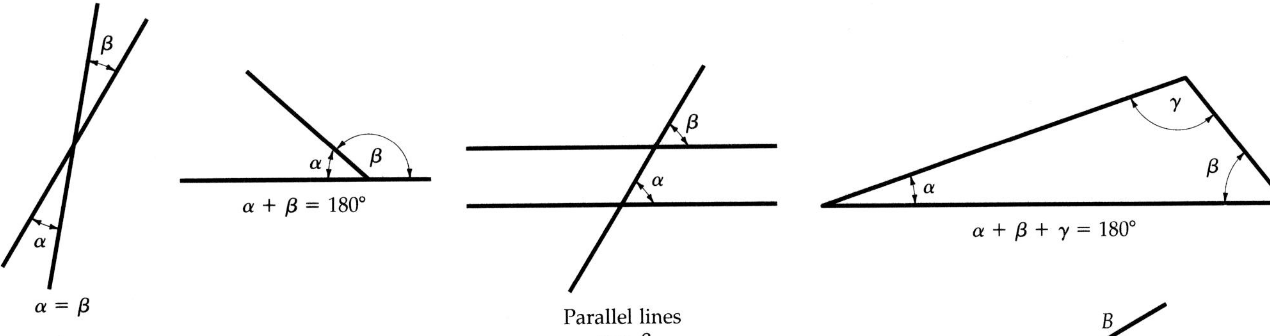

Figure A-12 shows some useful relations for angles.

Figure A-13 shows a right triangle formed by drawing the line BC perpendicular to AC. The lengths of the sides are labeled a, b, and c. The trigonometric functions $\sin\theta$, $\cos\theta$, and $\tan\theta$ for an acute angle θ are defined as

$$\sin\theta = \frac{a}{c} = \frac{\text{opposite side}}{\text{hypotenuse}} \qquad \text{A-33}$$

$$\cos\theta = \frac{b}{c} = \frac{\text{adjacent side}}{\text{hypotenuse}} \qquad \text{A-34}$$

$$\tan\theta = \frac{a}{b} = \frac{\text{opposite side}}{\text{adjacent side}} = \frac{\sin\theta}{\cos\theta} \qquad \text{A-35}$$

Three other trigonometric functions are the reciprocals of these functions and are defined as

$$\sec\theta = \frac{c}{b} = \frac{1}{\cos\theta} \qquad \text{A-36}$$

$$\csc\theta = \frac{c}{a} = \frac{1}{\sin\theta} \qquad \text{A-37}$$

$$\cot\theta = \frac{b}{a} = \frac{1}{\tan\theta} = \frac{\cos\theta}{\sin\theta} \qquad \text{A-38}$$

The angle θ whose sine is x is called the arcsine of x, and is written $\sin^{-1}x$. That is, if

$$\sin\theta = x$$

then

$$\theta = \arcsin x = \sin^{-1} x \qquad \text{A-39}$$

The arcsine is the inverse of the sine. The inverse of the cosine and tangent are defined similarly. The angle whose cosine is y is the arccosine of y. That is, if

$$\cos\theta = y$$

then

$$\theta = \arccos y = \cos^{-1} y \qquad \text{A-40}$$

The angle whose tangent is z is the arctangent of z. That is, if

$$\tan\theta = z$$

$$\theta = \arctan z = \tan^{-1} z \qquad \text{A-41}$$

The pythagorean theorem

$$a^2 + b^2 = c^2 \qquad \text{A-42}$$

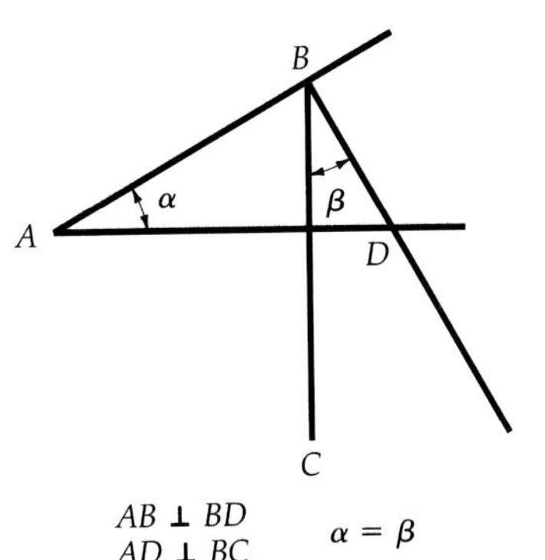

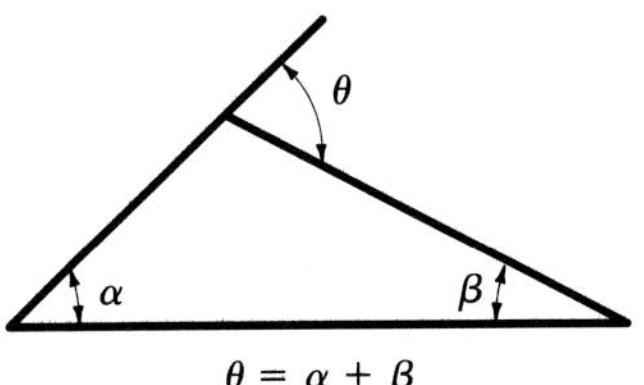

Figure A-12 Some useful relations for angles.

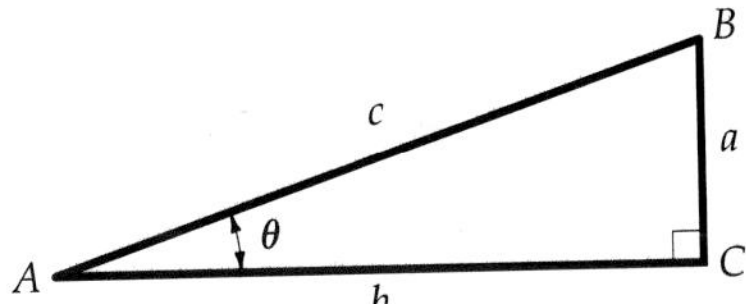

Figure A-13 A right triangle with sides of length a and b and a hypotenuse of length c.

gives some useful identities. If we divide each term in this equation by c^2, we obtain

$$\frac{a^2}{c^2} + \frac{b^2}{c^2} = 1$$

or, from the definitions of sin θ and cos θ,

$$\sin^2 \theta + \cos^2 \theta = 1 \qquad \text{A-43}$$

Similarly, we can divide each term in Equation A-42 by a^2 or b^2 and obtain

$$1 + \cot^2 \theta = \csc^2 \theta \qquad \text{A-44}$$

and

$$1 + \tan^2 \theta = \sec^2 \theta \qquad \text{A-45}$$

These and other useful trigonometric formulas are listed in Table A-3.

Table A-3 **Trigonometric Formulas**

$\sin^2 \theta + \cos^2 \theta = 1 \qquad \sec^2 \theta - \tan^2 \theta = 1 \qquad \csc^2 \theta - \cot^2 \theta = 1$

$\sin 2\theta = 2 \sin \theta \cos \theta$

$\cos 2\theta = \cos^2 \theta - \sin^2 \theta = 2 \cos^2 \theta - 1 = 1 - 2 \sin^2 \theta$

$$\tan 2\theta = \frac{2 \tan \theta}{1 - \tan^2 \theta}$$

$$\sin \frac{1}{2}\theta = \sqrt{\frac{1 - \cos \theta}{2}} \qquad \cos \frac{1}{2}\theta = \sqrt{\frac{1 + \cos \theta}{2}} \qquad \tan \frac{1}{2}\theta = \sqrt{\frac{1 - \cos \theta}{1 + \cos \theta}}$$

$\sin (A \pm B) = \sin A \cos B \pm \cos A \sin B$

$\cos (A \pm B) = \cos A \cos B \mp \sin A \sin B$

$$\tan (A \pm B) = \frac{\tan A \pm \tan B}{1 \mp \tan A \tan B}$$

$\sin A \pm \sin B = 2 \sin [\tfrac{1}{2}(A \pm B)] \cos [\tfrac{1}{2}(A \mp B)]$

$\cos A + \cos B = 2 \cos [\tfrac{1}{2}(A + B)] \cos [\tfrac{1}{2}(A - B)]$

$\cos A - \cos B = 2 \sin [\tfrac{1}{2}(A + B)] \sin [\tfrac{1}{2}(B - A)]$

$$\tan A \pm \tan B = \frac{\sin (A \pm B)}{\cos A \cos B}$$

Example A-5

Use the isosceles right triangle shown in Figure A-14 to find the sine, cosine, and tangent of 45°.

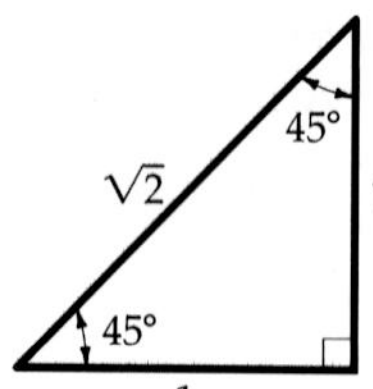

Figure A-14 An isosceles right triangle for Example A-5.

It is clear from the figure that the two acute angles of this triangle are equal. Since the sum of the three angles in a triangle must equal 180° and the right angle is 90°, each acute angle must be 45°. If we multiply each side of any triangle by a common factor, we obtain a similar triangle with the same angles as the first. Since the trigonometric functions involve the ratios of only two sides of a triangle, we can choose any convenient length for one side. Let the equal sides of this triangle have a length of 1 unit. The length of the hypotenuse can then be found from the pythagorean theorem:

$$c = \sqrt{a^2 + b^2} = \sqrt{1^2 + 1^2} = \sqrt{2} \text{ units}$$

The sine, cosine, and tangent for the angle 45° are then given by Equations A-33, A-34, and A-35, respectively:

$$\sin 45° = \frac{1}{\sqrt{2}} = 0.707 \qquad \cos 45° = \frac{1}{\sqrt{2}} = 0.707 \qquad \tan 45° = \frac{1}{1} = 1$$

Example A-6

The sine of 30° is exactly $\frac{1}{2}$. Find the ratios of the sides of a 30–60° right triangle.

This common triangle is shown in Figure A-15. We choose a length of 1 unit for the side opposite the 30° angle. The hypotenuse is then obtained from the fact that sin 30° = 0.5:

$$c = \frac{a}{\sin 30°} = \frac{1}{0.5} = 2$$

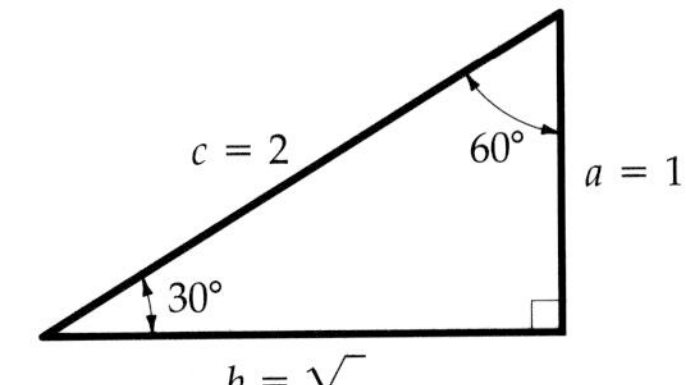

Figure A-15 A 30–60° right triangle for Example A-6.

The length of the side opposite the 60° angle is found from the pythagorean theorem:

$$b = \sqrt{c^2 - a^2} = \sqrt{2^2 - 1^2} = \sqrt{3}$$

From these results, we can obtain the following trigonometric functions for the angles 30° and 60°:

$$\cos 30° = \frac{b}{c} = \frac{\sqrt{3}}{2} = 0.866$$

$$\tan 30° = \frac{a}{b} = \frac{1}{\sqrt{3}} = 0.577$$

$$\sin 60° = \frac{b}{c} = \cos 30° = \frac{\sqrt{3}}{2} = 0.866$$

$$\cos 60° = \frac{a}{c} = \sin 30° = 0.500$$

$$\tan 60° = \frac{b}{a} = \frac{\sqrt{3}}{1} = 1.732$$

For small angles, the length a is nearly equal to the arc length s, as can be seen in Figure A-16. The angle $\theta = s/c$ is therefore nearly equal to $\sin\theta = a/c$:

$$\sin\theta \approx \theta \qquad \text{for small values of } \theta \qquad \text{A-46}$$

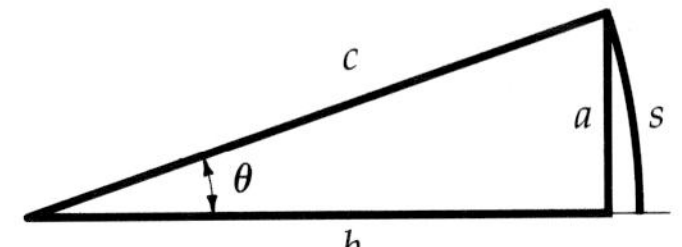

Figure A-16 For small angles, $\sin\theta = a/c$, $\tan\theta = a/b$, and the angle $\theta = s/c$ are all approximately equal.

Similarly, the lengths c and b are nearly equal, so $\tan\theta = a/b$ is nearly equal to both θ and $\sin\theta$ for small values of θ:

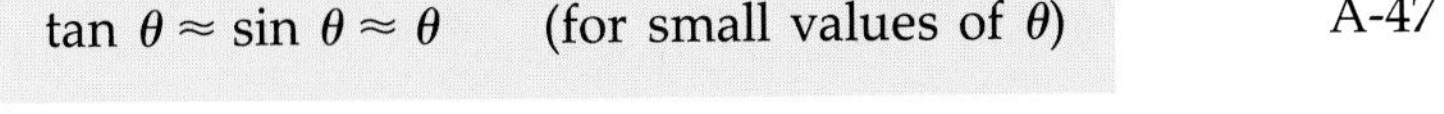

$$\tan\theta \approx \sin\theta \approx \theta \qquad \text{(for small values of } \theta) \qquad \text{A-47}$$

Equations A-46 and A-47 hold only if θ is measured in radians. Since $\cos\theta = b/c$ and these lengths are nearly equal for small values of θ, we have

$$\cos\theta \approx 1 \qquad \text{for small values of } \theta \qquad \text{A-48}$$

Example A-7

By how much do $\sin\theta$, $\tan\theta$, and θ differ when $\theta = 15°$?

This angle in radians is

$$\theta = 15° \frac{2\pi \text{ rad}}{360°} = 0.262 \text{ rad}$$

Using a calculator or a table of trigonometric functions, we obtain

$$\sin 15° = 0.259$$

and

$$\tan 15° = 0.268$$

Thus, $\sin\theta$ and θ (in radians) differ by 0.003 or about 1 percent, and $\tan\theta$ and θ differ by 0.006 or about 2 percent. For small angles, the approximation $\theta \approx \sin\theta \approx \tan\theta$ is even more accurate.

Example A-7 shows that if accuracy of a few percent is needed, small angle approximations can be used only for angles of about 15° or less. Figure A-17 shows graphs of θ, $\sin\theta$, and $\tan\theta$ versus θ, for small values of θ.

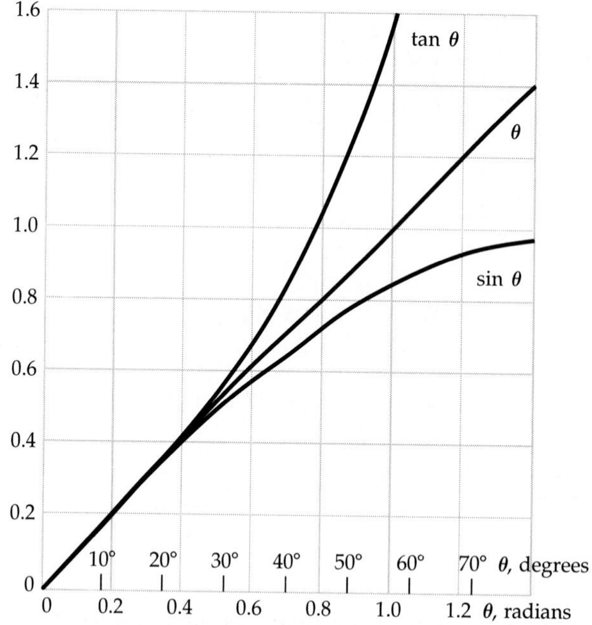

Figure A-17 Graphs of $\tan\theta$, θ, and $\sin\theta$ versus θ for small values of θ.

Figure A-18 shows an obtuse angle with its vertex at the origin and one side along the x axis. The trigonometric functions for a general angle such as this are defined by

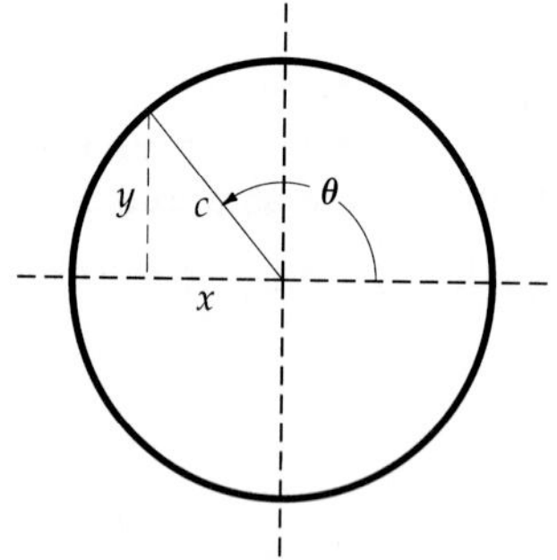

Figure A-18 Diagram for defining the trigonometric functions for an obtuse angle.

$$\sin\theta = \frac{y}{c} \quad \text{A-49}$$

$$\cos\theta = \frac{x}{c} \quad \text{A-50}$$

$$\tan\theta = \frac{y}{x} \quad \text{A-51}$$

Figure A-19 shows plots of these functions versus θ. All trigonometric functions have a period of 2π. That is, when an angle changes by 2π rad, the function returns to its original value. Thus, $\sin(\theta + 2\pi) = \sin\theta$ and so forth. Some other useful relations are

$$\sin(\pi - \theta) = \sin\theta \quad \text{A-52}$$

$$\cos(\pi - \theta) = -\cos\theta \quad \text{A-53}$$

$$\sin(\pi/2 - \theta) = \cos\theta \quad \text{A-54}$$

$$\cos(\pi/2 - \theta) = \sin\theta \quad \text{A-55}$$

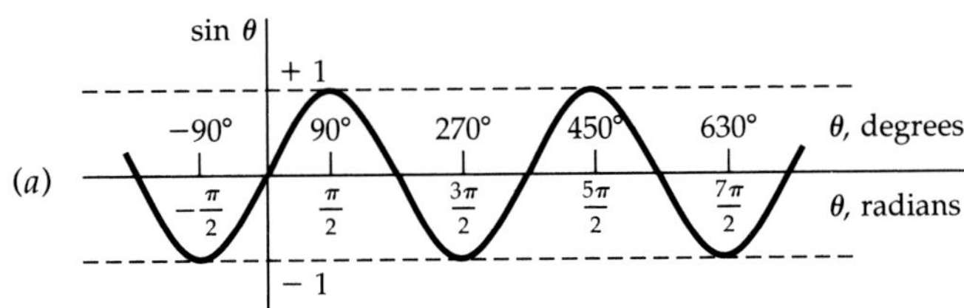

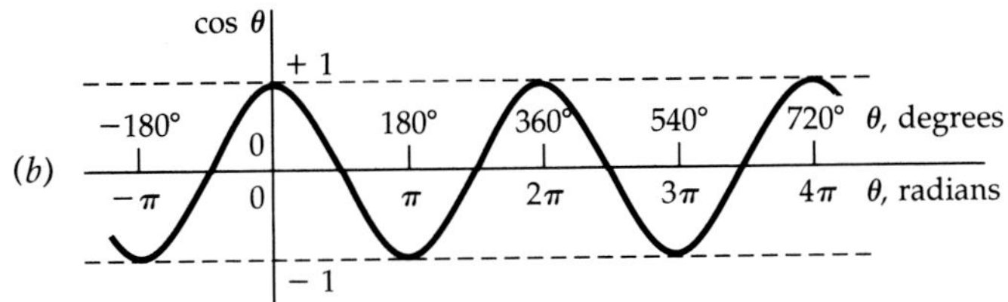

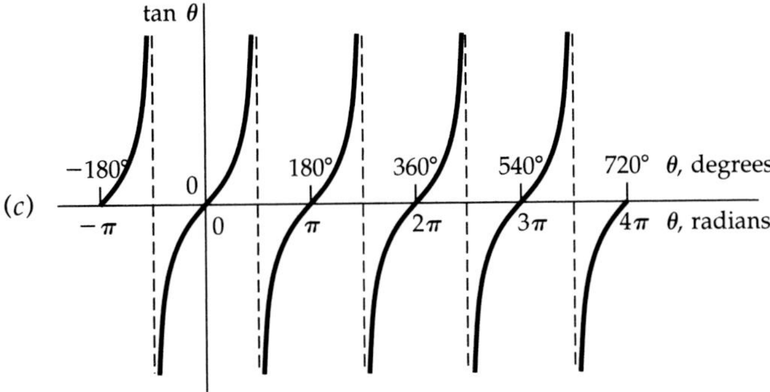

Figure A-19 The trigonometric functions sin θ, cos θ, and tan θ versus θ.

The trigonometric functions can be expressed as power series in θ. The series for sin θ and cos θ are

$$\sin\theta = \theta - \frac{\theta^3}{3!} + \frac{\theta^5}{5!} - \frac{\theta^7}{7!} + \cdots \qquad \text{A-56}$$

$$\cos\theta = 1 - \frac{\theta^2}{2!} + \frac{\theta^4}{4!} - \frac{\theta^6}{6!} + \cdots \qquad \text{A-57}$$

When θ is small, good approximations are obtained using only the first few terms in the series.

The Binomial Expansion

The binomial theorem is very useful for making approximations. One form of this theorem is

$$(1+x)^n = 1 + nx + \frac{n(n-1)}{2!}x^2 + \frac{n(n-1)(n-2)}{3!}x^3 + \frac{n(n-1)(n-2)(n-3)}{4!}x^4 + \cdots \qquad \text{A-58}$$

If n is a positive integer, there are $n + 1$ terms in this series. If n is a real number other than a positive integer, there are an infinite number of terms. The series is valid for any value of n if x^2 is less than 1. It is also valid for $x^2 = 1$ if n is positive. The series is particularly useful if $|x|$ is much less than 1. Then each term is much smaller than the previous term and we can drop all but the first two or three terms in the equation. If $|x|$ is much less than 1, we have

$$(1+x)^n \approx 1 + nx \qquad |x| \ll 1 \qquad \text{A-59}$$

Example A-8

Use Equation A-59 to find an approximate value for the square root of 101.

We first restate the problem in the form $(1 + x)^n$ with x much less than 1:

$$(101)^{1/2} = (100 + 1)^{1/2} = (100)^{1/2}\,(1 + 0.01)^{1/2} = 10(1 + 0.01)^{1/2}$$

We can now use Equation A-59 with $n = \frac{1}{2}$ and $x = 0.01$:

$$(1 + 0.01)^{1/2} \approx 1 + \tfrac{1}{2}(0.01) = 1.005$$

Then

$$(101)^{1/2} \approx 10(1.005) = 10.05$$

We can get an idea of the accuracy of this approximation by looking at the first term in Equation A-58 that has been neglected. This term is

$$\frac{n(n-1)}{2}\,x^2$$

Since x is 0.01, x^2 is 0.0001 and

$$\frac{n(n-1)}{2}\,x^2 = \frac{\frac{1}{2}(-\frac{1}{2})}{2}\,(0.0001) = -\frac{0.0001}{8} \approx -0.00001 = -0.001\%$$

We therefore expect our answer to be correct to within about 0.001 percent. The value of $(101)^{1/2}$ to eight significant figures is 10.049876, which differs from our approximate value of 10.05 by 0.000124 or about 0.001% of 10.05.

Complex Numbers

A general complex number z can be written

$$z = a + bi \qquad \text{A-60}$$

where a and b are real numbers and $i = \sqrt{-1}$. The quantity a is called the real part and the quantity ib is called the imaginary part of z. We can represent a complex number in a plane as shown in Figure A-20, where the x axis is the real axis and the y axis is the imaginary axis. We can use the relations $a = r\cos\theta$ and $b = r\sin\theta$ from Figure A-20 to write the complex number z in polar coordinates:

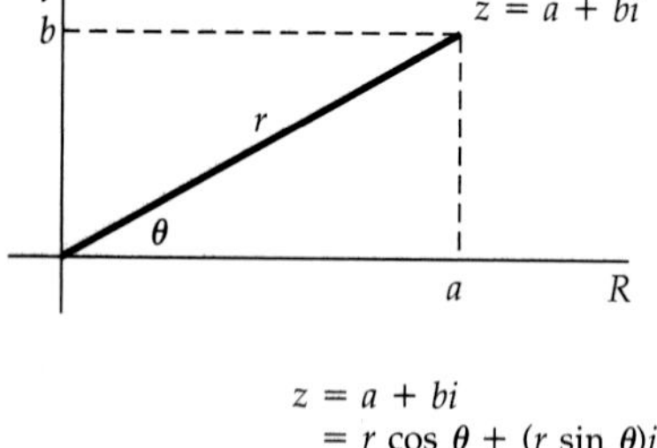

$$z = a + bi = r\cos\theta + (r\sin\theta)i = r(\cos\theta + i\sin\theta)$$

Figure A-20 Representation of a complex number in a plane. The real part of the complex number is plotted along the horizontal axis, and the imaginary part is plotted along the vertical axis.

$$z = r\cos\theta + (r\sin\theta)\,i \qquad \text{A-61}$$

where $r = \sqrt{a^2 + b^2}$ is called the magnitude of z.

When complex numbers are added or subtracted, the real and imaginary parts are added or subtracted separately:

$$z_1 + z_2 = (a_1 + b_1 i) + (a_2 + b_2 i) = (a_1 + a_2) + (b_1 + b_2)i \qquad \text{A-62}$$

However, when two complex numbers are multiplied, each part of one number is multiplied by each part of the other number:

$$\begin{aligned} z_1 z_2 &= (a_1 + b_1 i)(a_2 + b_2 i) = a_1 a_2 + b_1 b_2 i^2 + (a_1 b_2 + a_2 b_1)i \qquad \text{A-63} \\ &= a_1 a_2 - b_1 b_2 + (a_1 b_2 + a_2 b_1)i \end{aligned}$$

where we have used $i^2 = -1$.

The complex conjugate of a complex number z^* is that number obtained by replacing i with $-i$:

$$z^* = (a + bi)^* = a - bi \qquad \text{A-64}$$

The product of a complex number and its complex conjugate equals the square of the magnitude of the number:

$$zz^* = (a + bi)(a - bi) = a^2 + b^2 \qquad \text{A-65}$$

A particularly useful function of a complex number is the exponential $e^{i\theta}$. Using the expansion for e^x given in Table A-2, we have

$$e^{i\theta} = 1 + i\theta + \frac{(i\theta)^2}{2!} + \frac{(i\theta)^3}{3!} + \frac{(i\theta)^4}{4!} + \cdots$$

Using $i^2 = -1$, $i^3 = -i$, $i^4 = +1$, and so forth and separating the real parts from the imaginary parts, this expansion can be written

$$e^{i\theta} = 1 - \frac{\theta^2}{2!} + \frac{\theta^4}{4!} + \cdots + i\left(\theta - \frac{\theta^3}{3!} + \cdots\right)$$

Comparing this result with Equations A-56 and A-57, we can see that

$$e^{i\theta} = \cos\theta + i\sin\theta \qquad \text{A-66}$$

Using this result, we can express a general complex number as an exponential:

$$z = a + bi = r\cos\theta + (r\sin\theta)i = re^{i\theta} \qquad \text{A-67}$$

where $r = \sqrt{a^2 + b^2}$.

Differential Calculus

When we say that x is a function of t, we mean that for each value of t there is a corresponding value of x. An example is $x = At^2$, where A is a constant. To indicate that x is a function of t, we sometimes write $x(t)$ for x. Figure A-21 is a graph of x versus t for a typical function $x(t)$. At a particular value $t = t_1$, x has the value of x_1 as indicated. At another value t_2, x has the value x_2. The change in t, $t_2 - t_1$, is written $\Delta t = t_2 - t_1$ and the corresponding change in x is written $\Delta x = x_2 - x_1$. The ratio $\Delta x/\Delta t$ is the slope of the straight line connecting (x_1, t_1) and (x_2, t_2). If we make Δt smaller and smaller, the line connecting (x_1, t_1) and (x_2, t_2) approaches the line that is tangent to the curve at the point (x_1, t_1). The slope of this tangent line is called the derivative of x with respect to t and is written dx/dt:

$$\frac{dx}{dt} = \lim_{\Delta t \to 0} \frac{\Delta x}{\Delta t} \qquad \text{A-68}$$

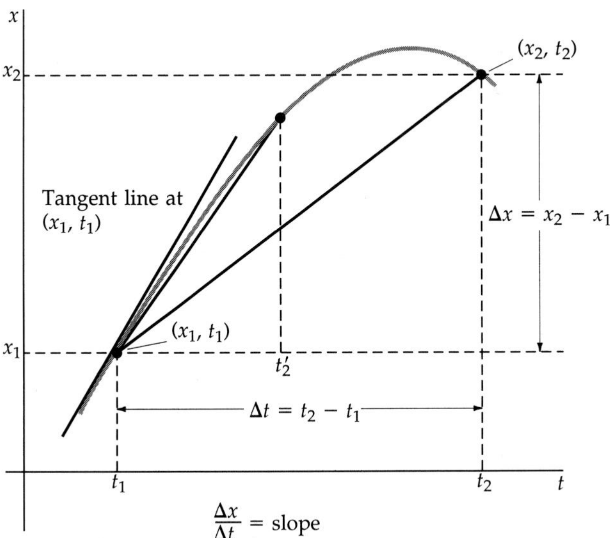

Figure A-21 Graph of a typical function $x(t)$. The points (x_1, t_1) and (x_2, t_2) are connected by a straight line. The slope of this line is $\Delta x/\Delta t$. As the time interval beginning at t_1 is decreased, the slope for that interval approaches the slope of the line tangent to the curve at time t_1, which is the derivative of x with respect to t.

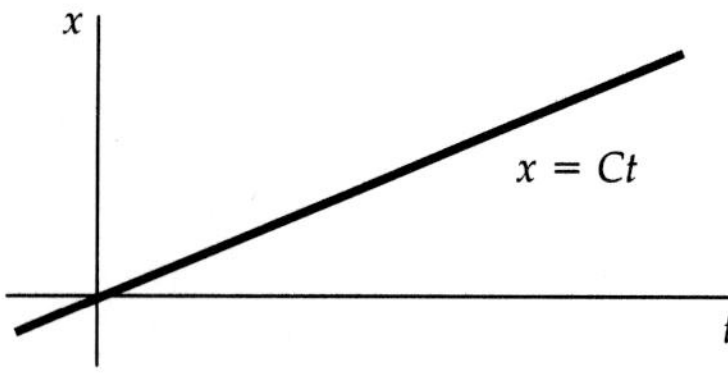

Figure A-22 Graph of the linear function $x = Ct$. This function has a constant slope C.

The derivative of a function of t is another function of t. If x is a constant, the graph of x versus t is a horizontal line with zero slope. The derivative of a constant is thus zero. In Figure A-22, x is proportional to t:

$$x = Ct$$

This function has a constant slope equal to C. Thus the derivative of Ct is C. Table A-4 lists some properties of derivatives and the derivatives of some particular functions that occur often in physics. It is followed by comments aimed at making these properties and rules clearer. More detailed discussion can be found in most calculus books.

Table A-4 **Properties of Derivatives and Derivatives of Particular Functions**

Linearity

1. The derivative of a constant times a function equals the constant times the derivative of the function:

$$\frac{d}{dt}[Cf(t)] = C\frac{df(t)}{dt}$$

2. The derivative of a sum of functions equals the sum of the derivatives of the functions:

$$\frac{d}{dt}[f(t) + g(t)] = \frac{df(t)}{dt} + \frac{dg(t)}{dt}$$

Chain rule

3. If f is a function of x and x is in turn a function of t, the derivative of f with respect to t equals the product of the derivative of f with respect to x and the derivative of x with respect to t:

$$\frac{d}{dt}f(x) = \frac{df}{dx}\frac{dx}{dt}$$

Derivative of a product

4. The derivative of a product of functions $f(t)g(t)$ equals the first function times the derivative of the second plus the second function times the derivative of the first:

$$\frac{d}{dt}[f(t)g(t)] = f(t)\frac{dg(t)}{dt} + \frac{df(t)}{dt}g(t)$$

Reciprocal derivative

5. The derivative of t with respect to x is the reciprocal of the derivative of x with respect to t, assuming that neither derivative is zero:

$$\frac{dx}{dt} = \left(\frac{dt}{dx}\right)^{-1} \quad \text{if} \quad \frac{dt}{dx} \neq 0$$

Derivatives of particular functions

6. $\dfrac{dC}{dt} = 0$ where C is a constant

7. $\dfrac{d(t^n)}{dt} = nt^{n-1}$

8. $\dfrac{d}{dt}\sin\omega t = \omega\cos\omega t$

9. $\dfrac{d}{dt}\cos\omega t = -\omega\sin\omega t$

10. $\dfrac{d}{dt}\tan\omega t = \omega\sec^2\omega t$

11. $\dfrac{d}{dt}e^{bt} = be^{bt}$

12. $\dfrac{d}{dt}\ln bt = \dfrac{1}{t}$

Example A-9

Find the derivative of $x = at^2 + bt + c$, where a, b, and c are constants.

From rule 2, we can differentiate each term separately and add the results. Using rules 1 and 7, we have

$$\frac{d(at^2)}{dt} = 2at^1 = 2at$$

Similarly, $d(bt)/dt = b$ and $dc/dt = 0$. Adding these results, we obtain

$$\frac{dx}{dt} = 2at + b$$

Comments on Rules 1 through 5

Rules 1 and 2 follow from the fact that the limiting process is linear. We can understand rule 3, the chain rule, by multiplying $\Delta f/\Delta t$ by $\Delta x/\Delta x$ and noting that, since x is a function of t, both Δx and Δf approach zero as Δt approaches zero. Since the limit of a product of two functions equals the product of their limits, we have

$$\lim_{\Delta t \to 0} \frac{\Delta f}{\Delta t} = \lim_{\Delta t \to 0} \frac{\Delta f}{\Delta x}\frac{\Delta x}{\Delta t} = \left(\lim_{\Delta x \to 0} \frac{\Delta f}{\Delta x}\right)\left(\lim_{\Delta t \to 0} \frac{\Delta x}{\Delta t}\right) = \frac{df}{dx}\frac{dx}{dt}$$

Rule 4 is not immediately apparent. The derivative of a product of functions is the limit of the ratio

$$\frac{f(t + \Delta t)g(t + \Delta t) - f(t)g(t)}{\Delta t}$$

If we add and subtract the quantity $f(t + \Delta t)g(t)$ in the numerator, we can write this ratio as

$$\frac{f(t + \Delta t)g(t + \Delta t) - f(t + \Delta t)g(t) + f(t + \Delta t)g(t) - f(t)g(t)}{\Delta t}$$

$$= f(t + \Delta t)\left[\frac{g(t + \Delta t) - g(t)}{\Delta t}\right] + g(t)\left[\frac{f(t + \Delta t) - f(t)}{\Delta t}\right]$$

As Δt approaches zero, the terms in brackets become $dg(t)/dt$ and $df(t)/dt$, respectively, and the limit of the expression is

$$f(t)\frac{dg(t)}{dt} + g(t)\frac{df(t)}{dt}$$

Rule 5 follows directly from the definition:

$$\frac{dx}{dt} = \lim_{\Delta t \to 0} \frac{\Delta x}{\Delta t} = \lim_{\Delta t \to 0} \left(\frac{\Delta t}{\Delta x}\right)^{-1} = \lim_{\Delta x \to 0} \left(\frac{\Delta t}{\Delta x}\right)^{-1} = \left(\lim_{\Delta x \to 0} \frac{\Delta t}{\Delta x}\right)^{-1}$$

Comments on Rule 7

We can obtain this important result using the binomial expansion. We have

$$f(t) = t^n$$

$$f(t + \Delta t) = (t + \Delta t)^n = t^n\left(1 + \frac{\Delta t}{t}\right)^n$$

$$= t^n\left[1 + n\frac{\Delta t}{t} + \frac{n(n-1)}{2!}\left(\frac{\Delta t}{t}\right)^2 + \frac{n(n-1)(n-2)}{3!}\left(\frac{\Delta t}{t}\right)^3 + \cdots\right]$$

Then

$$f(t - \Delta t) - f(t) = t^n \left[n \frac{\Delta t}{t} + \frac{n(n-1)}{2!} \left(\frac{\Delta t}{t} \right)^2 + \cdots \right]$$

and

$$\frac{f(t - \Delta t) - f(t)}{\Delta t} = nt^{n-1} + \frac{n(n-1)}{2!} t^{n-2} \Delta t + \cdots$$

The next term omitted from the last sum is proportional to $(\Delta t)^2$, the following to $(\Delta t)^3$, and so on. Each term except the first approaches zero as Δt approaches zero. Thus

$$\frac{df}{dt} = \lim_{\Delta x \to 0} \frac{f(t + \Delta t) + f(t)}{\Delta t} = nt^{n-1}$$

Comments on Rules 8 to 10

We first write $\sin \omega t = \sin \theta$ with $\theta = \omega t$ and use the chain rule,

$$\frac{d \sin \theta}{dt} = \frac{d \sin \theta}{d\theta} \frac{d\theta}{dt} = \omega \frac{d \sin \theta}{d\theta}$$

We then use the trigonometric formula for the sine of the sum of two angles θ and $\Delta\theta$:

$$\sin (\theta + \Delta\theta) = \sin \Delta\theta \cos \theta + \cos \Delta\theta \sin \theta$$

Since $\Delta\theta$ is to approach zero, we can use the small-angle approximations

$$\sin \Delta\theta \approx \Delta\theta \qquad \text{and} \qquad \cos \Delta\theta \approx 1$$

Then

$$\sin (\theta + \Delta\theta) \approx \Delta\theta \cos \theta + \sin \theta$$

and

$$\frac{\sin (\theta + \Delta\theta) - \sin \theta}{\Delta\theta} \approx \cos \theta$$

Similar reasoning can be applied to the cosine function to obtain rule 9.

Rule 10 is obtained by writing $\tan \theta = \sin \theta / \cos \theta$ and applying rule 4 along with rules 8 and 9.

$$\frac{d}{dt} (\tan \theta) = \frac{d}{dt} (\sin \theta)(\cos \theta)^{-1} = \sin \theta \frac{d}{dt} (\cos \theta)^{-1} + \frac{d (\sin \theta)}{dt} (\cos \theta)^{-1}$$

$$= \sin \theta \, (-1)(\cos \theta)^{-2}(-\sin \theta) + (\cos \theta)(\cos \theta)^{-1}$$

$$= \frac{\sin^2 \theta}{\cos^2 \theta} + 1 = \tan^2 \theta + 1 = \sec^2 \theta$$

Comments on Rule 11

Again we use the chain rule

$$\frac{de^\theta}{dt} = b \frac{de^\theta}{d\theta} \qquad \text{with} \qquad \theta = bt$$

and the series expansion for the exponential function:

$$e^{\theta + \Delta\theta} = e^\theta e^{\Delta\theta} = e^\theta \left[1 + \Delta\theta + \frac{(\Delta\theta)^2}{2!} + \frac{(\Delta\theta)^3}{3!} + \cdots \right]$$

Then

$$\frac{e^{\theta + \Delta\theta} - e^\theta}{\Delta\theta} = e^\theta + e^\theta \frac{\Delta\theta}{2!} + e^\theta \frac{(\Delta\theta)^2}{3!} + \cdots$$

As $\Delta\theta$ approaches zero, the right side of the equation above approaches e^θ.

Comments on Rule 12

Let

$$y = \ln bt$$

Then

$$e^y = bt \qquad \text{and} \qquad \frac{dt}{dy} = \frac{1}{b}\, e^y = t$$

Then using rule 5, we obtain

$$\frac{dy}{dt} = \left(\frac{dt}{dy}\right)^{-1} = \frac{1}{t}$$

Integral Calculus

Integration is related to the problem of finding the area under a curve. It is also the inverse of differentiation. Figure A-23 shows a function $f(t)$. The area of the shaded element is approximately $f_i\,\Delta t_i$, where f_i is evaluated anywhere in the interval Δt_i. This approximation improves if Δt_i is very small. The total area from t_1 to t_2 is found by summing all the area elements from t_1 to t_2 and taking the limit as each Δt_i approaches zero. This limit is called the integral of f over t and is written

$$\int_{t_1}^{t_2} f\,dt = \text{area} = \lim_{\Delta t_i \to 0} \sum_i f_i\,\Delta t_i$$

If we integrate some function $f(t)$ from t_1 to some general value of t, we obtain another function of t. Let us call this function y:

$$y = \int_{t_1}^{t} f\,dt$$

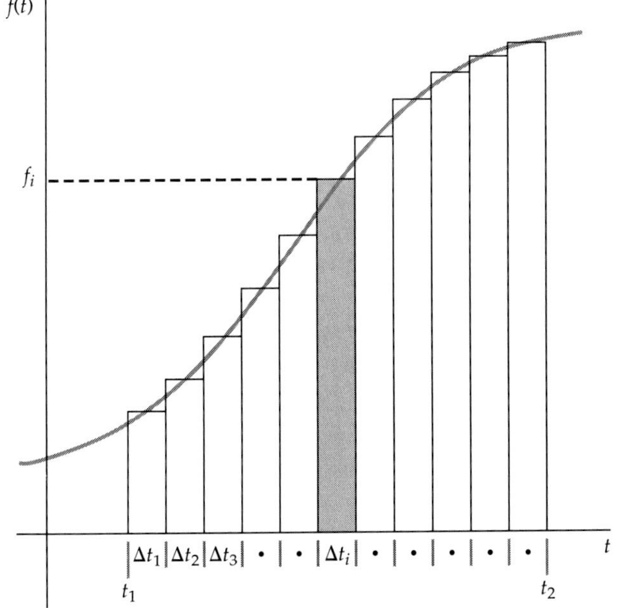

Figure A-23 A general function $f(t)$. The area of the shaded element is approximately $f_i\,\Delta t_i$, where f_i is evaluated anywhere in the interval.

The function y is the area under the f-versus-t curve from t_1 to a general value t. For a small interval Δt, the change in the area Δy is approximately $f\,\Delta t$.

$$\Delta y \approx f\,\Delta t$$

$$f \approx \frac{\Delta y}{\Delta t}$$

If we take the limit as Δt approaches 0, we can see that f is the derivative of y:

$$f = \frac{dy}{dt}$$

The relation between y and f is often written

$$y = \int f\,dt$$

where $\int f\,dt$ is called an **indefinite integral.** To evaluate an indefinite integral, we find the function y whose derivative is f. The definite integral of f from t_1 to t_2 is $y(t_1) - y(t_2)$, where $df/dt = y$:

$$\int_{t_1}^{t_2} f\,dt = y(t_2) - y(t_1)$$

Example A-10

Find the indefinite integral of $f(t) = t$.

The function whose derivative is t is $\frac{1}{2}t^2$ plus any constant. Thus,

$$\int t\,dt = \tfrac{1}{2}t^2 + C$$

where C is any constant.

Table A-5 lists some important integration formulas. More extensive lists of differentiation and integration formulas can be found in handbooks such as Herbert Dwight's *Tables of Integrals and Other Mathematical Data*, fourth edition, Macmillan Publishing Company, Inc., New York, 1961.

Table A-5 **Integration Formulas†**

1. $\int A\,dt = At$		5. $\int e^{bt}\,dt = \frac{1}{b}e^{bt}$
2. $\int At\,dt = \frac{1}{2}At^2$		6. $\int \cos\omega t\,dt = \frac{1}{\omega}\sin\omega t$
3. $\int At^n\,dt = A\frac{t^{n+1}}{n+1}$	$n \neq -1$	7. $\int \sin\omega t\,dt = -\frac{1}{\omega}\cos\omega t$
4. $\int At^{-1}\,dt = A\ln t$		

†In these formulas, A, b, and ω are constants. An arbitrary constant C can be added to the right side of each equation.

Appendix B

SI Units

Basic Units	
Length	The *meter* (m) is the distance traveled by light in a vacuum in 1/299,792,458 s
Time	The *second* (s) is the duration of 9,192,631,770 periods of the radiation corresponding to the transition between the two hyperfine levels of the ground state of the ^{133}Cs atom
Mass	The *kilogram* (kg) is the mass of the international standard body preserved at Sèvres, France
Current	The *ampere* (A) is that current in two very long parallel wires 1 m apart that gives rise to a magnetic force per unit length of 2×10^{-7} N/m
Temperature	The *kelvin* (K) is 1/273.16 of the thermodynamic temperature of the triple point of water
Luminous intensity	The *candela* (cd) is the luminous intensity, in the perpendicular direction, of a surface of area 1/600,000 m^2 of a blackbody at the temperature of freezing platinum at a pressure of 1 atm

Derived Units		
Force	newton (N)	$1\ N = 1\ kg{\cdot}m/s^2$
Work, energy	joule (J)	$1\ J = 1\ N{\cdot}m$
Power	watt (W)	$1\ W = 1\ J/s$
Frequency	hertz (Hz)	$1\ Hz = s^{-1}$
Charge	coulomb (C)	$1\ C = 1\ A{\cdot}s$
Potential	volt (V)	$1\ V = 1\ J/C$
Resistance	ohm (Ω)	$1\ \Omega = 1\ V/A$
Capacitance	farad (F)	$1\ F = 1\ C/V$
Magnetic field	tesla (T)	$1\ T = 1\ N/A{\cdot}m$
Magnetic flux	weber (Wb)	$1\ Wb = 1\ T{\cdot}m^2$
Inductance	henry (H)	$1\ H = 1\ J/A^2$

Appendix C

Numerical Data

Terrestrial Data

Acceleration of gravity g	
Standard value	9.80665 m/s^2
	32.1740 ft/s^2
At sea level, at equator†	9.7804 m/s^2
At sea level, at poles†	9.8322 m/s^2
Mass of earth M_E	5.98×10^{24} kg
Radius of earth R_E, mean	6.37×10^6 m
	3960 mi
Escape speed $\sqrt{2R_E g}$	1.12×10^4 m/s
	6.95 mi/s
Solar constant‡	1.35 kW/m^2
Standard temperature and pressure (STP):	
Temperature	273.15 K
Pressure	101.325 kPa
	1.00 atm
Molar mass of air	28.97 g/mol
Density of air (STP), ρ_{air}	1.293 kg/m^3
Speed of sound (STP)	331 m/s
Heat of fusion of H_2O (0°C, 1 atm)	333.5 kJ/kg
Heat of vaporization of H_2O (100°C, 1 atm)	2.257 MJ/kg

†Measured relative to the earth's surface.
‡Average power incident normally on 1 m^2 outside the earth's atmosphere at the mean distance from the earth to the sun.

Astronomical Data

Earth	
Distance to moon†	3.844×10^8 m
	2.389×10^5 mi
Distance to sun, mean†	1.496×10^{11} m
	9.30×10^7 mi
	1.00 AU
Orbital speed, mean	2.98×10^4 m/s
Moon	
Mass	7.35×10^{22} kg
Radius	1.738×10^6 m
Period	27.32 d
Acceleration of gravity at surface	1.62 m/s^2
Sun	
Mass	1.99×10^{30} kg
Radius	6.96×10^8 m

†Center to center.

Physical Constants		
Gravitational constant	G	$6.672\,6 \times 10^{-11}$ N·m²/kg²
Speed of light	c	$2.997\,924\,58 \times 10^{8}$ m/s
Electron charge	e	$1.602\,177 \times 10^{-19}$ C
Avogadro's number	N_A	$6.022\,137 \times 10^{23}$ particles/mol
Gas constant	R	$8.314\,51$ J/mol·K
		$1.987\,22$ cal/mol·K
		$8.205\,78 \times 10^{-2}$ L·atm/mol·K
Boltzmann's constant	$k = R/N_A$	$1.380\,658 \times 10^{-23}$ J/K
		$8.617\,385 \times 10^{-5}$ eV/K
Unified mass unit	$u = (1/N_A)$ g	$1.660\,540 \times 10^{-24}$ g
Coulomb constant	$k = 1/4\pi\epsilon_0$	$8.987\,551\,788 \times 10^{9}$ N·m²/C²
Permittivity of free space	ϵ_0	$8.854\,187\,817 \times 10^{-12}$ C²/N·m²
Permeability of free space	μ_0	$4\pi \times 10^{-7}$ N/A²
		$1.256\,637 \times 10^{-6}$ N/A²
Planck's constant	h	$6.626\,076 \times 10^{-34}$ J·s
		$4.135\,669 \times 10^{-15}$ eV·s
	$\hbar = h/2\pi$	$1.054\,573 \times 10^{-34}$ J·s
		$6.582\,122 \times 10^{-16}$ eV·s
Mass of electron	m_e	$9.109\,390 \times 10^{-31}$ kg
		$510.999\,1$ keV/c^2
Mass of proton	m_p	$1.672\,623 \times 10^{-27}$ kg
		$938.272\,3$ MeV/c^2
Mass of neutron	m_n	$1.674\,929 \times 10^{-27}$ kg
		$939.565\,6$ MeV/c^2
Bohr magneton	$m_B = e\hbar/2m_e$	$9.274\,015\,4 \times 10^{-24}$ J/T
		$5.788\,382\,63 \times 10^{-5}$ eV/T
Nuclear magneton	$m_n = e\hbar/2m_p$	$5.050\,786\,6 \times 10^{-27}$ J/T
		$3.152\,451\,66 \times 10^{-8}$ eV/T
Magnetic flux quantum	$\phi_0 = h/2e$	$2.067\,834\,6 \times 10^{-15}$ T·m²
Quantized Hall resistance	$R_K = h/e^2$	$2.581\,280\,7 \times 10^{4}$ Ω
Rydberg constant	R_H	$1.097\,373\,153\,4 \times 10^{7}$ m⁻¹
Josephson frequency–voltage quotient	$2e/h$	$4.835\,979 \times 10^{14}$ Hz/V
Compton wavelength	$\lambda_C = h/m_e c$	$2.426\,310\,58 \times 10^{-12}$ m

For additional data, see the front and back endpapers and the following tables in the text

Appendix D

Conversion Factors

Conversion factors are written as equations for simplicity; relations marked with an asterisk are exact.

Length

1 km = 0.6215 mi

1 mi = 1.609 km

1 m = 1.0936 yd = 3.281 ft = 39.37 in

*1 in = 2.54 cm

*1 ft = 12 in = 30.48 cm

*1 yd = 3 ft = 91.44 cm

1 lightyear = 1 c·y = 9.461×10^{15} m

*1 Å=0.1 nm

Area

*1 $m^2 = 10^4\ cm^2$

1 $km^2 = 0.3861\ mi^2$ = 247.1 acres

*1 $in^2 = 6.4516\ cm^2$

1 $ft^2 = 9.29 \times 10^{-2}\ m^2$

1 $m^2 = 10.76\ ft^2$

*1 acre = 43,560 ft^2

1 mi^2 = 640 acres = 2.590 km^2

Volume

*1 $m^3 = 10^6\ cm^3$

*1 L = 1000 $cm^3 = 10^{-3}\ m^3$

1 gal = 3.786 L

1 gal = 4 qt = 8 pt = 128 oz = 231 in^3

1 $in^3 = 16.39\ cm^3$

1 $ft^3 = 1728\ in^3$ = 28.32 L = $2.832 \times 10^4\ cm^3$

Time

*1 h = 60 min = 3.6 ks

*1 d = 24 h = 1440 min = 86.4 ks

1 y = 365.24 d = 31.56 Ms

Speed

1 km/h = 0.2778 m/s = 0.6215 mi/h

1 mi/h = 0.4470 m/s = 1.609 km/h

1 mi/h = 1.467 ft/s

Angle and Angular Speed

*π rad = 180°

1 rad = 57.30°

1° = 1.745×10^{-2} rad

1 rev/min = 0.1047 rad/s

1 rad/s = 9.549 rev/min

Mass

*1 kg = 1000 g

*1 tonne = 1000 kg = 1 Mg

1 u = 1.6606×10^{-27} kg

1 kg = 6.022×10^{23} u

1 slug = 14.59 kg

1 kg = 6.852×10^{-2} slug

1 u = 931.50 MeV/c^2

Density

*1 g/cm^3 = 1000 kg/m^3 = 1 kg/L

(1 g/cm^3)g = 62.4 lb/ft^3

Force

1 N = 0.2248 lb = 10^5 dyn

1 lb = 4.4482 N

(1 kg)g = 2.2046 lb

Pressure

*1 Pa = 1 N/m^2

*1 atm = 101.325 kPa = 1.01325 bars

1 atm = 14.7 lb/in^2 = 760 mmHg

= 29.9 inHg = 33.8 ftH_2O

1 lb/in^2 = 6.895 kPa

1 torr = 1 mmHg = 133.32 Pa

1 bar = 100 kPa

Energy

*1 kW·h = 3.6 MJ

*1 cal = 4.1840 J

1 ft·lb = 1.356 J = 1.286 × 10^{-3} Btu

*1 L·atm = 101.325 J

1 L·atm = 24.217 cal

1 Btu = 778 ft·lb = 252 cal = 1054.35 J

1 eV = 1.602 × 10^{-19} J

1 u·c^2 = 931.50 MeV

*1 erg = 10^{-7} J

Power

1 horsepower = 550 ft·lb/s = 745.7 W

1 Btu/min = 17.58 W

1 W = 1.341 × 10^{-3} horsepower

= 0.7376 ft·lb/s

Magnetic Field

*1 G = 10^{-4} T

*1 T = 10^4 G

Thermal Conductivity

1 W/m·K = 6.938 Btu·in/h·ft^2·F°

1 Btu·in/h·ft^2·F° = 0.1441 W/m·K

Appendix E Periodic Table

1	2	3	4	5	6	7	8	9	10	11	12	13	14	15	16	17	18
1 **H** 1.00797																	2 **He** 4.003
3 **Li** 6.941	4 **Be** 9.012											5 **B** 10.81	6 **C** 12.011	7 **N** 14.007	8 **O** 15.9994	9 **F** 19.00	10 **Ne** 20.179
11 **Na** 22.990	12 **Mg** 24.31											13 **Al** 26.98	14 **Si** 28.09	15 **P** 30.974	16 **S** 32.064	17 **Cl** 35.453	18 **Ar** 39.948
19 **K** 39.102	20 **Ca** 40.08	21 **Sc** 44.96	22 **Ti** 47.88	23 **V** 50.94	24 **Cr** 52.00	25 **Mn** 54.94	26 **Fe** 55.85	27 **Co** 58.93	28 **Ni** 58.69	29 **Cu** 63.55	30 **Zn** 65.38	31 **Ga** 69.72	32 **Ge** 72.59	33 **As** 74.92	34 **Se** 78.96	35 **Br** 79.90	36 **Kr** 83.80
37 **Rb** 85.47	38 **Sr** 87.62	39 **Y** 88.906	40 **Zr** 91.22	41 **Nb** 92.91	42 **Mo** 95.94	43 **Tc** (98)	44 **Ru** 101.1	45 **Rh** 102.905	46 **Pd** 106.4	47 **Ag** 107.870	48 **Cd** 112.41	49 **In** 114.82	50 **Sn** 118.69	51 **Sb** 121.75	52 **Te** 127.60	53 **I** 126.90	54 **Xe** 131.29
55 **Cs** 132.905	56 **Ba** 137.33	57–71 **Rare** **Earths**	72 **Hf** 178.49	73 **Ta** 180.95	74 **W** 183.85	75 **Re** 186.2	76 **Os** 190.2	77 **Ir** 192.2	78 **Pt** 195.09	79 **Au** 196.97	80 **Hg** 200.59	81 **Tl** 204.37	82 **Pb** 207.19	83 **Bi** 208.98	84 **Po** (210)	85 **At** (210)	86 **Rn** (222)
87 **Fr** (223)	88 **Ra** (226)	89–103 **Actinides**	104 **Rf** (261)	105 **Ha** (260)	106 (263)	107 (262)	108 (265)	109 (266)									

Rare Earths (Lanthanides)	57 **La** 138.91	58 **Ce** 140.12	59 **Pr** 140.91	60 **Nd** 144.24	61 **Pm** (147)	62 **Sm** 150.36	63 **Eu** 152.0	64 **Gd** 157.25	65 **Tb** 158.92	66 **Dy** 162.50	67 **Ho** 164.93	68 **Er** 167.26	69 **Tm** 168.93	70 **Yb** 173.04	71 **Lu** 174.97
Actinides	89 **Ac** 227.03	90 **Th** 232.04	91 **Pa** 231.04	92 **U** 238.03	93 **Np** 237.05	94 **Pu** (244)	95 **Am** (243)	96 **Cm** (247)	97 **Bk** (247)	98 **Cf** (251)	99 **Es** (252)	100 **Fm** (257)	101 **Md** (258)	102 **No** (259)	103 **Lr** (260)

The 1–18 group designation has been recommended by the International Union of Pure and Applied Chemistry (IUPAC).

Atomic Numbers and Atomic Masses

Element	Symbol	Atomic number	Atomic mass	Element	Symbol	Atomic number	Atomic mass
Actinium	Ac	89	227.03	Mercury	Hg	80	200.59
Aluminum	Al	13	26.98	Molybdenum	Mo	42	95.94
Americium	Am	95	(243)	Neodymium	Nd	60	144.24
Antimony	Sb	51	121.75	Neon	Ne	10	20.179
Argon	Ar	18	39.948	Neptunium	Np	93	237.05
Arsenic	As	33	74.92	Nickel	Ni	28	58.69
Astatine	At	85	(210)	Niobium	Nb	41	92.91
Barium	Ba	56	137.3	Nitrogen	N	7	14.007
Berkelium	Bk	97	(247)	Nobelium	No	102	(259)
Beryllium	Be	4	9.012	Osmium	Os	76	190.2
Bismuth	Bi	83	208.98	Oxygen	O	8	15.9994
Boron	B	5	10.81	Palladium	Pd	46	106.4
Bromine	Br	35	79.90	Phosphorus	P	15	30.974
Cadmium	Cd	48	112.41	Platinum	Pt	78	195.09
Calcium	Ca	20	40.08	Plutonium	Pu	94	(244)
Californium	Cf	98	(251)	Polonium	Po	84	(210)
Carbon	C	6	12.011	Potassium	K	19	39.102
Cerium	Ce	58	140.12	Praseodymium	Pr	59	140.91
Cesium	Cs	55	132.905	Promethium	Pm	61	(147)
Chlorine	Cl	17	35.453	Protactinium	Pa	91	231.04
Chromium	Cr	24	52.00	Radium	Ra	88	(226)
Cobalt	Co	27	58.93	Radon	Rn	86	(222)
Copper	Cu	29	63.55	Rhenium	Re	75	186.2
Curium	Cm	96	(247)	Rhodium	Rh	45	102.905
Dysprosium	Dy	66	162.50	Rubidium	Rb	37	85.47
Einsteinium	Es	99	(252)	Ruthenium	Ru	44	101.1
Erbium	Er	68	167.26	Rutherfordium	Rf	104	(261)
Europium	Eu	63	152.0	Samarium	Sm	62	150.36
Fermium	Fm	100	(257)	Scandium	Sc	21	44.96
Fluorine	F	9	19.00	Selenium	Se	34	78.96
Francium	Fr	87	(223)	Silicon	Si	14	28.09
Gadolinium	Gd	64	157.25	Silver	Ag	47	107.870
Gallium	Ga	31	69.72	Sodium	Na	11	22.990
Germanium	Ge	32	72.59	Strontium	Sr	38	87.62
Gold	Au	79	196.97	Sulfur	S	16	32.064
Hafnium	Hf	72	178.49	Tantalum	Ta	73	180.95
Hahnium	Ha	105	(260)	Technetium	Tc	43	(98)
Helium	He	2	4.003	Tellurium	Te	52	127.60
Holmium	Ho	67	164.93	Terbium	Tb	65	158.92
Hydrogen	H	1	1.00797	Thallium	Tl	81	204.37
Indium	In	49	114.82	Thorium	Th	90	232.04
Iodine	I	53	126.90	Thulium	Tm	69	168.93
Iridium	Ir	77	192.2	Tin	Sn	50	118.69
Iron	Fe	26	55.85	Titanium	Ti	22	47.88
Krypton	Kr	36	83.80	Tungsten	W	74	183.85
Lanthanum	La	57	138.91	Uranium	U	92	238.03
Lawrencium	Lr	103	(260)	Vanadium	V	23	50.94
Lead	Pb	82	207.19	Xenon	Xe	54	131.29
Lithium	Li	3	6.941	Ytterbium	Yb	70	173.04
Lutetium	Lu	71	174.97	Yttrium	Y	39	88.906
Magnesium	Mg	12	24.31	Zinc	Zn	30	65.38
Manganese	Mn	25	54.94	Zirconium	Zr	40	91.22
Mendelevium	Md	101	(258)				

Illustration Credits

Part Openers
Part 1 p. xix, 18 © James Sugar/Black Star; **p. 19** © D. Cavagnaro/Peter Arnold, Inc.
Part 2 p. xxii, 366–367 © Michael Freeman.
Part 3 p. xxiii, 484–485 Sandia National Laboratories.
Part 4 p. xxiv, 596–597 © 1988 Richard Megna/Fundamental Photos.
Part 5 p. xxvii, 972–973 Courtesy AT&T Archives.
Part 6 p. xxviii, 1098–1099 Courtesy AT&T Archives.

Chapter 1
Opener p. 1 The Granger Collection; **p. 3 (a)** The Granger Collection; **(b)** Frequency Electronics, Inc.; **p. 4 (a)** McDonald Observatory; **(b)** Bruce Coleman; **p. 5** Owen Franken/Stock, Boston; **p. 9 (a)** IBM Almaden Research Center; **(b)** © Lennart Nilsson, from *The Body Victorious,* Delacourt Press, 1985; **(c)** Kent and Donnan Dannen/Photo Researchers; **(d)** N.A.S.A. (68-HC-74); **(e)** Smithsonian Institution; **p. 11** B. A. Watkinson; **p. 12** Fermi Institute, University of Chicago; **p. 14** AIP Niels Bohr Library.

Chapter 2
Opener p. 20 © George D. Lepp/Comstock; **p. 24** © Gunter Ziesler/Peter Arnold, Inc.; **p. 29** Estate of Harold E. Edgerton/Courtesy Palm Press, Inc.; **p. 30** © Sidney Harris; **p. 34 (a)** Courtesy Stanford Linear Accelerator Center, U.S. Department of Energy; **(b)** John Seeman, SLAC; **p. 35** Courtesy General Motors Corp.

Chapter 3
Opener p. 47 © Lewis Portnoy/Spectra-Action, Inc.; **p. 60** © 1968 Fundamental Photographs; **p. 63** Scala/Art Resource; **p. 64** *PSSC Physics,* 2nd ed., 1965. D. C. Heath & Co. and Education Development Center, Inc., Newton, Massachusetts; **p. 65** N.A.S.A. (82-HC-78); **Figure 3-27** AIP Niels Bohr Library; **p. 68** N.A.S.A. (81-HC-408); **p. 75** Courtesy New York Public Library.

Chapter 4
Opener p. 77 The Granger Collection; **p. 79** (*left*) © George Hall/Woodfin Camp and Assoc.; (*right*) © Arthur Tilley/FPG International; **p. 83** © Berenice Abbott/Commerce Graphics, Ltd., Inc.; **p. 84** N.A.S.A. (76-HC-6); **p. 87** Culver Pictures; **p. 88 (a)** © Cotton Coulson/Woodfin Camp and Assoc.; **(b)** © 1972 Gary Ladd; **p. 89 (c)** Los Alamos Scientific Laboratory; **(d)** © Science Photo Library/Photo Researchers; **p. 90** © Fundamental Photographs; **Figure 4-7 (b)** © Michael Abbey/Photo Researchers; **p. 95** Museum of Modern Art, New York; **p. 105** © David de Lossy/The Image Bank.

Chapter 5
Opener p. 107 Courtesy Center for Engineering Design, University of Utah, Salt Lake City; **p. 108 (a)** F. P. Bowden and D. Tabor, *Friction and Lubrication of Solids,* Oxford University Press, 1950. Reprinted by permission of the publisher; **(b)** Uzi Landman, W. David Luedtke, Georgia Institute of Technology; **Figure 5-5** © Jean-Claude Lejeune/Stock, Boston; **Figure 5-10 (b)** © Malraux Photography/The Image Bank; **p. 115** © Robert Alexander/Photo Researchers; **p. 116** (*top*) © Fundamental Photographs; **Figure 5-12 (b)** © Tom Walker/Stock, Boston; **p. 117 (a)** © N.A.S.A. (81-HC-945); **(b)** © Jonathan Blair/Woodfin Camp and Assoc.; **p. 123** Sandia National Laboratories; **p. 124** N.A.S.A. (85-HC-345); **p. 131** © P. Thomann/The Image Bank.

Chapter 6
Opener p. 135 © 1990 Estate of Harold E. Edgerton/Courtesy Palm Press, Inc.; **p. 139** Naples National Museum/Scala/Art Resource; **p. 142** © L. Grant/FPG International; **p. 148** © Steve Leonard/Black Star; **Figure 6-12 (b)** © T. Alvez/FPG International; **p. 156** © 1989 M. C. Escher Heirs/Cordon Art, Baarn, Holland; **p. 157** © 1973 Berenice Abbott/Photo Researchers; **p. 167** (*left*) USGS; (*right*) New York State Commerce Department; **p. 168 (a)** © Stan Sholik/FPG International; **(b)** © Leicester University/Science Photo Library/Photo Researchers; **p. 169** © Visual Horizons/FPG International.

Chapter 7
Opener p. 182 © David Parker/Science Photo Library/Photo Researchers; **Figure 7-1** © Estate of Harold E. Edgerton/Courtesy Palm Press, Inc.; **p. 198** © Estate of Harold E. Edgerton/Courtesy Palm Press, Inc.; **Figure 7-23** © Estate of Harold E. Edgerton/Courtesy Palm Press, Inc.; **Figure 7-24 (b)** © Estate of Harold E. Edgerton/Courtesy Palm Press, Inc.; **p. 205** Courtesy of Mercedes-Benz of North America, Montvale, New Jersey; **p. 209 (a)** © 1973 Berenice Abbott/Photo Researchers; **(b)** Brookhaven National Laboratory; **(c) (1–4)** Alar and Juri Toomre, from *Cambridge*

Atlas of Astronomy, Cambridge University Press, 1985; **(d)** Kitt Peak National Observatory, from *Cambridge Atlas of Astronomy,* Cambridge University Press, 1985; **p. 211** © Dr. David Jones/Science Photo Library/Photo Researchers; **p. 215 (a)** N.A.S.A. (393-479)/Shostal Superstock; **(b)** © Bill Curtsinger/Photo Researchers; **p. 220** Courtesy Central Scientific Company.

Chapter 8
Opener p. 227 © David F. Malin/Anglo-Australian Telescope Board; **p. 229** Courtesy Center for Engineering Design, University of Utah, Salt Lake City; **p. 232** Culver Pictures; **p. 236** PAR/NYC, Inc. Archives; **p. 237 (a)** and **(b)** © David Malin, Anglo-Australian Telescope Board; **Figure 8-21** Focus on Sports; **Figure 8-22** © Estate of Harold E. Edgerton/Courtesy Palm Press, Inc.; **p. 247** © Dick Luria/FPG International; **p. 248 (a)** and **(b)** N.A.S.A. (84-HC-582); **p. 254** National Oceanographic and Atmospheric Administration; **p. 257** © Dick Luria/Photo Researchers; **p. 261** (*left*) Richard Minnix; (*right*) Courtesy Central Scientific Company.

Chapter 9
Opener p. 275 © Tom McHugh/Photo Researchers; **p. 276** Manitowoc Engineering Co., Manitowoc, Wisconsin; **p. 277** © Mike and Carol Werner/Comstock; **p. 278** © Malcolm Hanes/Bruce Coleman; **p. 281** © Marc Romanelli/The Image Bank; **p. 286** © Rick Rickman/Black Star; **p. 287** (*left*) © Yulsman/The Image Bank; (*right*) Demetrios Zangos.

Chapter 10
Opener p. 295 Collection of Historical Scientific Instruments, Harvard University; **p. 296** © Erich Lessing/Magnum; **p. 298 (a)** N.A.S.A.; **(b)** © Dennis di Cicco/Peter Arnold, Inc.; **p. 302** © Erich Lessing, Culture and Fine Arts Archive, Vienna; **Figure 10-8 (a)** Courtesy Deutsches Museum, Munich; **p. 306** Courtesy Central Scientific Company; **p. 308 (a)** *Physics Today,* May 1988; **(b)** N.A.S.A.; **p. 313** N.A.S.A. (69-HC-905); **p. 317 (b)** N.A.S.A. (80-HC-627); **Figure 1 p. 322** By permission of the Syndics of Cambridge University Library, Additional MS 3975, p. 15; **Figure 2 p. 323** From Voltaire, *The Elements of Sir Isaac Newton's Philosophy,* 1738, New York: Frank Cass, 1967, p. 97; **Figure 3 p. 323** From Newton, *Opticks,* 1730; **p. 327** Art Resource.

Chapter 11
Opener p. 331 Dr. Tai Il Mah, from *Popular Science,* June 1990; **p. 332** © Daemmrich/Stock, Boston; **p. 335 (a)** Scala/Art Resource; **(b)** © Jonathan Watts/Science Photo Library/Photo Researchers; **p. 341 (a)** © David Burnett/Woodfin Camp and Assoc.; **(b)** PAR/NYC, Inc. Archives; **p. 342** © Chuck O'Rear/Woodfin Camp and Assoc.; **p. 344 (a)** and **(b)** © Estate of Harold E. Edgerton/Courtesy Palm Press, Inc.; **p. 346 (a)** © Jerry Howard/Stock, Boston; **(b)** © Alan Oddie/Photo Edit; **Figure 11-16 (a)** © Richard Megna/Fundamental Photographs; **(b)** © Estate of Harold E. Edgerton/Courtesy Palm Press, Inc.; **p. 348** Picker International; **p. 350 (a)** and **(b)** Office National d'Etudes et de Rechèrches Aérospatiales; **p. 353** Dr. Owen M. Griffin, Naval Research Laboratory; **Figure 3 p. 359** Jack Zohrt/FPG International; **Figure 4 p. 359** Dave Custer, 1986, photo courtesy of Milliken Research Associates, Inc.

Chapter 12
Opener p. 368 © Fundamental Photographs; **p. 371** N.A.S.A. (73-HC-787); **p. 376** Institute for Marine Dynamics; **p. 383** © Berenice Abbott/Commerce Graphics Ltd., Inc.; **p. 388** © Estate of Harold E. Edgerton/Courtesy Palm Press, Inc.; **p. 390** Courtesy of Monroe Auto Equipment, Monroe, Michigan; **p. 392** Courtesy Citicorp Building, New York City; **p. 394** Royal Swedish Academy of Music, courtesy Prof. Thomas D. Rossing, Northern Illinois University, DeKalb; **p. 395** Courtesy Central Scientific Company **p. 402** (*top*) From H.-O. Peitgen, P. H. Richter, *The Beauty of Fractals,* Springer-Verlag, New York, 1986; (*bottom*) Prof. Julio M. Ottino, University of Massachusetts, Amherst.

Chapter 13
Opener p. 409 (c) © Hashi; **Figure 13-1 (b)** *PSSC Physics,* 2nd ed., 1965. D. C. Heath & Co. and Education Development Center, Inc., Newton, Massachusetts; **Figure 13-3 (b)** and **(d)** *PSSC Physics* 2nd ed., 1965. D. C. Heath & Co. and Education Development Center Inc., Newton, Massachusetts; **Figure 13-5 (a)** and **(b)** © Berenice Abbott/Photo Researchers; **p. 414 (a)** and **(b)** © Four By Five, Inc.; **p. 421 (a,b,c,d)** *PSSC Physics,* 2nd ed., 1965. D. C. Heath & Co. and Education Development Center, Inc., Newton, Massachusetts; **p. 422** © David Yost/Courtesy Steinway & Sons; **p. 424 (a,b,c)** University of Washington; **p. 428** Prof. Thomas D. Rossing, Northern Illinois University, DeKalb.

Chapter 14
Opener p. 439 From Winston E. Kock, *Lasers and Holography,* Dover Publications, New York, 1981; **Figure 14-4** *PSSC Physics,* 2nd ed., 1965. D. C. Heath & Co. and Education Development Center, Inc., Newton, Massachusetts; **Figure 14-10 (a)** © 1973 Berenice Abbott/Photo Researchers; **p. 459 (a)** From Adolphe Ganot, *Elementary Treatise on Physics,* William Wood & Publishers Ltd., 1910; **(b)** © 1989 Mike Chesser, Yamaha BBDO Worldwide Network, Los Angeles; **p. 460** Davis Hathaway/N.A.S.A./Courtesy of National Optical Astronomy Observatories; **p. 462 (a)** Courtesy Naval Research Laboratory; **(b)** Courtesy San Francisco Symphony; **Figure 14-28** and **Figure 14-30** *PSSC Physics,* 2nd ed., 1965. D. C. Heath & Co. and Education Development Center, Inc., Newton, Massachusetts; **p. 463** Dr. John S. Shelton; **p. 464 (a)** © 1986 Howard Sochurek/Medichrome; **(b)** N.A.S.A. (81-HC-670); **Figure 14-31 (a)** *PSSC Physics,* 2nd ed., 1965. D. C. Heath & Co. and Education Development Center, Inc., Newton, Massachusetts; **Figure 14-33 (b)** *PSSC Physics,*

2nd ed., 1965. D. C. Heath & Co. and Education Development Center, Inc., Newton, Massachusetts; **p. 469 (a)** Sandia National Laboratories; **(b)** © Robert de Gast/Photo Researchers; **(c)** © Estate of Harold E. Edgerton/Courtesy Palm Press, Inc.; **(d)** Sandia National Laboratories; **Figure 7 p. 476** New Scientist/IPC Magazines Limited/World Press Network 1990.

Chapter 15
Opener p. 487 Lockheed Corporation; **Figure 15-2** © Damien Lovegrove/Science Photo Library/Photo Researchers; **p. 488 (a)** © Michael Melford/The Image Bank; **(b)** Courtesy Central Scientific Company; **(c)** Instituto e Museo della Storia Scienza; **Figure 15-5 (b)** © Demetrios Zangos; **p. 491** Dr. William Mangum/National Bureau of Standards; **p. 494** © Richard Choy/Peter Arnold, Inc.; **p. 505 (a)** N.A.S.A. (78-HC-575); **(b)** N.A.S.A./Jet Propulsion Laboratory; **p. 507 (a,b,c)** Courtesy Central Scientific Company; **p. 510 (a)** © Alison Wright/Image Works; **(b)** © Eric Neurath/Stock, Boston.

Chapter 16
Opener p. 517 Lana Berkovich. Photo caption by kind cooperation of Robert Serlin, Phoenix Pipe and Tube Co., Phoenixville, Pennsylvania; **p. 522 (a)** Micrograph produced on ESEM/Courtesy of Electro Scan Corporation, Wilmington, Massachusetts; **(b)** © Manfred Kage/Peter Arnold, Inc.; **p. 523 (c)** Dr. Mary Neubert, Liquid Crystal Institute, Kent State University; **(d)** © Don Gray/f/Stop Pictures; **(e)** © Ken Karp; **p. 529** Dr. Al Bartlett, University of Colorado, Boulder; **p. 530** From *Proceedings of Third International Heat Transfer Conference*, The Science Press, Ephrata, Pennsylvania, 1966; **p. 531 (a)** © Dr. R. P. Clark and M. Goff/Science Photo Library/Photo Researchers; **(b)** N.A.S.A. (75-HC-620); **p. 533** © 1990 John W. Warden. Photo caption by kind cooperation of David A. Jensen, Alyeska Pipeline Service Co.; **p. 534** The Royal Society; **Figure 16-7 (b)** Science Museum, London; **Figure 16-10 (b)** © David F. Malin/Anglo-Australian Telescope Board; **p. 549** © Will and Deni McIntyre/Photo Researchers; **p. 556** N.A.S.A. (83-HC-227).

Chapter 17
Opener p. 563 Sandia National Laboratories; **pp. 566–567** R. F. Sawyer, University of California, Berkeley; **p. 571** © J. M. Mejuto/FPG International; **p. 574 (a)** © Michael Collier/Stock, Boston; **(b)** © Jean-Pierre Horlin/The Image Bank; **p. 575 (c)** Sandia National Laboratories; **(d)** © Peter Miller/The Image Bank; **(e)** Sandia National Laboratories; **p. 583** © Sidney Harris.

Chapter 18
Opener p. 598 Lawrence Berkeley Laboratory/Science Photo Library/Photo Researchers; **Figure 18-1** *PSSC Physics*, 2nd ed., 1965. D. C. Heath & Co. and Education Development Center, Inc., Newton, Massachusetts; **p. 600** (*top*) Courtesy National Institute of Standards and Technology; (*margin*) Bruce Terris/IBM Almaden Research Center; **p. 603** (*top left*) © Grant Heilman; (*top right*) Ann Ronan Picture Library; (*margin*) Burndy Library, Norwalk, Connecticut; **Figure 18-16 (b)** Harold M. Waage; **Figure 18-17 (b)** Harold M. Waage; **Figure 18-18 (b)** Harold M. Waage; **p. 615** Courtesy of Hulon Forrester/Video Display Corporation, Tucker, Georgia; **Figure 18-25** Courtesy Tripos-Evans and Sutherland Corporation.

Chapter 19
Opener p. 624 © Dagmar Hailer-Hamann/Peter Arnold, Inc.; **Figure 19-4 (b)** Ben Damsky Electric Power Research Institute; **Figure 19-22 (b)** Runk/Schoenberger from Grant Heilman; **p. 645** Harold M. Waage.

Chapter 20
Opener p. 656 © 1990 Richard Megna/Fundamental Photographs; **p. 659** © Mark Antman/The Image Works; **Figure 20-19 (b)** © Karen R. Preuss; **(c)** © Hank Morgan/Rainbow; **p. 678 (a)** Stanford Linear Accelerator Center/Science Photo Library/Photo Researchers; **(b)** © 1986 Wetmore/Photo Researchers; **Figure 20-20 (a)** Harold M. Waage; **p. 683** Courtesy Ohio Edison Company; **p. 684** Courtesy Xerox Corporation.

Chapter 21
Opener p. 690 Courtesy Tusonix, Tucson, Arizona; **Figure 21-2 (b)** Harold M. Waage; **p. 693** © Mark Antman/The Image Works; **p. 697 (a,b)** © Bruce Iverson; **(c)** © Manfred Kage/Peter Arnold, Inc.; **p. 700** © Lawrence Livermore National Laboratory; **p. 703 (a)** © Picturepoint Ltd., London; **(b)** © Paul Brierly; **(c)** © Lenman Connell/The Stock Market; **(d)** © Robert Essel/The Stock Market; **p. 707** Courtesy Lockheed Corporation.

Chapter 22
Opener p. 716 © Rabbit System, Santa Monica, California. Photo caption by kind cooperation of Larry Albright; **p. 719** Courtesy Sandia National Laboratories; **p. 723** © Chris Rogers/The Stock Market; **p. 724** Courtesy AT&T Archives; **p. 726 (a)** © 1985 Bob O'Shaughnessy/The Stock Market; **(b)** Courtesy EG&G Wakefield; **(c)** Courtesy Liquid Crystal Applications, Inc.; **Figure 22-11** © Paul Silverman/Fundamental Photographs; **p. 730 (a)** © *Popular Science*, June 1990; **(b)** © Paul Brierly; **(c)** © Coco McCoy/Rainbow; **(f)** Ann Ronan Picture Library; **p. 735** Donald Calabrese/Rome Air Development Center, Grittiss AFB, New York; **p. 740 Figure 1 (b)** Carolina Biological Supply Company; **p. 743 Figure 6 (d)** and (*right*) © Lennart Nilsson.

Chapter 23
Opener p. 749 Courtesy AT&T Archives; **p. 766** UPI/The Bettmann Archive; **p. 767 (a,b)** Courtesy Central Scientific Company; **p. 770 (a)** © Lennart Nilsson; **(b)** © Bruce Iverson; **p. 771 (c)** Courtesy Sandia National Laboratories; **(d)** © Joel E. Arem;

(e) © Lennart Nilsson; **(f)** Courtesy of Omega Engineering, Inc.

Chapter 24

Opener p. 781 © Stadler GmbH, Courtesy Transrapid International, Munich; **Figure 24-1** Fred Weiss; **Figure 24-8 (b)** © 1990 Richard Megna/Fundamental Photographs; **p. 788 (a)** Larry Langrill; **(b)** © Lawrence Berkeley Laboratory/Science Photo Library; **Figure 24-10 (b)** Carl E. Nielsen; **p. 790 (a)** © N.A.S.A. (74-HC-260); **(b)** Courtesy NRAO/AUI; **Figure 24-14 (b)** Courtesy Cavendish Laboratory, University of Cambridge; **p. 793** Courtesy Central Scientific Company; **Figure 24-16 (b)** © Steven Cohen, Rockefeller University; **p. 796 (a)** © 1980 Smithsonian Institution; **(b)** Courtesy Scanditronix Inc., Essex, Massachusetts.

Chapter 25

Opener p. 811 Bob Williamson, Oakland University, Rochester, Michigan; **p. 815 (a,b)** © 1990 Richard Megna/Fundamental Photographs; **Figure 25-8** © 1990 Richard Megna/Fundamental Photographs; **Figure 25-11 (b)** © 1990 Richard Megna/Fundamental Photographs; **p. 820 (a,b)** © Bruce Iverson; **p. 821 (a)** Courtesy CERN; **(b)** Courtesy Brookhaven National Laboratory; **Figure 25-15 (b)** © 1990 Richard Megna/Fundamental Photographs; **p. 825** Courtesy F. W. Bell Co.; **Figure 25-20** Clarence Bennett, Oakland University, Rochester, Michigan; **p. 830 (a,b)** Courtesy Princeton University Plasma Physics Laboratory.

Chapter 26

Opener p. 840 © Bruce Iverson; **Figure 26-1** © 1990 Richard Megna/Fundamental Photographs; **p. 843** Courtesy National Portrait Gallery, London; **p. 855** Sun Yihons, *China Reconstructs*, October, 1985; **p. 856 (a,c)** Courtesy U.S. Dept. of the Interior, Dept. of Reclamation; **(b)** © Lee Langum/Photo Researchers; **Figure 26-25 (b)** © Michael Holford, Collection of the Science Museum, London; **p. 859** *(bottom)* Clarence Bennett, Oakland University, Rochester, Michigan; **p. 866** *(bottom left)* Y. Yoshida; **Figure 1** Prof. Neal Brown, Geophysical Institute, University of Alaska, Fairbanks; **Figure 2 (a)** N.A.S.A. (85-HC-148); **Figure 3 p. 867** Courtesy Prof. Syun-Ichi Akasofu; **Figure 2 (b)** Courtesy Prof. Lou Frank, University of Iowa.

Chapter 27

Opener p. 878 Robert J. Celotta, National Institute of Standards and Technology; **p. 885** J. F. Allen, St. Andrews University, Scotland; **p. 886** © Paul Silverman/Fundamental Photographs; **p. 887 (a)** © Akira Tonomura, Hitachi Advanced Research Laboratory, Hatoyama, Japan; **(b,c)** © Bruce Iverson; **p. 889 (a)** © Seagate Technologies; **(b)** John Mamin, IBM Almaden Research Center; **p. 891** © Bill Pierce/Time Magazines, Inc.

Chapter 28

Opener p. 898 © 1990 C. J. Allen/New England Hydro Transmission Corporation; **p. 906 (a)** Courtesy Intel Corporation; **(b)** Courtesy Sperry Corporation; **p. 916** © George H. Clark Radioana Collection-Archive Center, National Museum of American History; **p. 920 (a)** Courtesy Champion Spark Plug Co; **(b)** © Bruce Iverson; **p. 921 (a)** © Yoav/Phototake; **(b)** © Daniel S. Brody/Stock, Boston; **p. 923** Lana Berkovich; **p. 925 (a)** Photo by John Petit/Courtesy J. W. Stokes, The Vestal Press Ltd.; **(b)** © Bruce Iverson; **p. 926** Courtesy Dr. H. H. Busta, Amoco Research Center, Naperville, Illinois; **Figure 6, p. 932 (a-d)** © Lloyd Wolf; **Figure 7, p. 933** © Bruce Iverson.

Chapter 29

Opener p. 943 N.A.S.A. (70-HC-220); **p. 948** © Doug Johnson/SPL/Science Source; **p. 952** NOAA-N.A.S.A./JPL; **p. 955** Dave Cooke, University of Chicago; **p. 956 (a)** Courtesy AT&T Archives; **(b)** California Institute of Technology; **p. 959 (a)** © 1984 Frank Zullo; **(b)** © Galen Rowell/Peter Arnold, Inc.; **p. 962** *(left)* Alistair Steyn-Ross, University of Waikato, New Zealand/Courtesy *The Physics Teacher;* *(right)* Courtesy Brookhaven National Laboratory; **p. 967** Courtesy Cavendish Laboratory, University of Cambridge; **Figure 29-13** Courtesy Central Scientific Company.

Chapter 30

Opener p. 974 Robert Greenler; **p. 976 (a)** Courtesy Optical Coating Laboratory, Inc. (OCLI); **(b)** © Chuck O'Rear/West Light; **p. 977 (c)** © William James Warren/West Light; **(d,e)** © Chuck O'Rear/West Light; **p. 980 (a)** M. A. Duguay and A. T. Mattick, Bell Telephone Laboratories, Murray Hill, New Jersey; **(b)** © 1989 Anglo-Australian Telescope Board; **Figure 30-8 (a)** Courtesy Battelle-Northwest Laboratories; **(b)** © 1987 Ken Kay/Fundamental Photographs; **Figure 30-10 (b)** © 1987 Pete Saloutos/The Stock Market; **p. 984 (a)** Ray Curletti/Courtesy of *Audio* Magazine; **(b)** Demetrios Zangos, with kind cooperation of Photographic Sciences Corporation, Webster, New York; **(c)** © Leonard Lessin/Peter Arnold, Inc.; **(d)** Courtesy University of Miami, Music Engineering; **(e)** Courtesy Digital Instruments, Santa Barbara, California; **Figure 30-13 (b)** © 1990 Richard Megna/Fundamental Photographs; **Figure 30-16** *PSSC Physics,* 2nd ed., 1965. D. C. Heath & Co. and Education Development Center, Inc., Newton, Massachusetts; **Figure 30-17 (b)** © 1987 Ken Kay/Fundamental Photographs; **Figure 30-19 (b)** © 1987 Ken Kay/Fundamental Photographs; **p. 989 (a)** © Lawrence Manning/West Light; **(b)** © Dan Boyd/Courtesy Naval Research Laboratory; **Figure 30-20 (b)** Courtesy AT&T Archives; **(c)** © 1983 C. Falco/Photo Researchers; **Figure 30-21 (c)** © Robert Greenler; **Figure 30-23 (b)** © Paul Silverman/Fundamental Photographs; **p. 994 (a)** © Robert Greenler; **(b)** Giovanni DeAmici, NSF,

Lawrence Berkeley Laboratory; **Figure 30-35 (a,b)** Larry Langrill; **p. 999 (a,b)** Glen A. Izett, U.S. Geological Survey, Denver, Colorado; **(c,d)** Dr. Anthony J. Gow, Cold Regions Research and Engineering Laboratory, Hanover, New Hampshire; **(e)** © Sepp Seitz/Woodfin Camp and Associates, Inc.; **p. 1001 (a)** © 1970 Fundamental Photographs; **(b)** © 1990 PAR/NYC, Inc./Photo by Elizabeth Algieri; **p. 1003** © 1987 Paul Silverman/ Fundamental Photographs; **Figure 2 p. 1005** Robert Greenler; **Figures 3, 4 p. 1006** Robert Greenler.

Chapter 31
Opener p. 1012 © Dagmar Hailer-Hamann/Peter Arnold, Inc.; **Figure 31-2** Demetrios Zangos; **p. 1015** © 1982 Jon Brenneis Photo; **Figure 31-12 (c)** © 1990 Richard Megna/Fundamental Photographs; **p. 1020 (a,b)** A.G.E. FotoStock/© Peter Arnold, Inc.; **p. 1021 (a,b)** © 1990 Richard Megna/Fundamental Photographs; **p. 1025** © 1990 Richard Megna/Fundamental Photographs; **Figure 31-26 (b)** Nils Abramson; **(d)** © 1974 Fundamental Photographs; **Figure 31-27 (c)** © Fundamental Photographs; **p. 1032** © 1990 Richard Megna/Fundamental Photographs; **p. 1033** © Bohdan Hrynewych/Stock, Boston.

Chapter 32
Opener p. 1041 Robert Parshall, Biomedical Visualization Laboratories, University of Illinois at Chicago; **p. 1044 (a,b)** © Lennart Nilsson; **p. 1045 (c)** © Lennart Nilsson; **(d)** Courtesy IMEC and University of Pennsylvania, Dept. of Electrical Engineering; **p. 1049** (*top*) Courtesy Nikon, Inc.; (*margin*) © Chuck O'Rear/West Light; **Figure 32-8 (a)** © 1983 Jonathan Levine; **p. 1051 (a)** Cornell University; **(b)** M. Issacson, M. Barshatsky, J. Cline/Cornell University; **p. 1053 (a)** Scala/Art Resource; **(b)** © Royal Astronomical Society Library; **(c)** © Lick Observatory, University of California Regents; **(d)** California Institute of Technology; **(e)** © 1980 Gary Ladd; **p. 1054 (a,b,d)** © California Association for Research in Astronomy; **(c)** © Ray Ellis/Photo Researchers, Inc.; **p. 1056 (a)** N.A.S.A. (531-76-026); **(b)** N.A.S.A. (531-76-0390); **p. 1057 (c,d,e)** N.A.S.A. (90-HC-508).

Chapter 33
Opener p. 1061 © Ken Kay/Fundamental Photographs; **p. 1063** © 1990 Richard Megna/Fundamental Photographs; **Figure 33-3 (a)** Courtesy Bausch & Lomb; **Figure 33-5 (a,b)** Courtesy T. A. Wiggins; **p. 1066 (a,b)** Optical Coating Laboratory, Inc. (OCLI); **Figure 33-7** *PSSC Physics,* 2nd ed., 1965. D. C. Heath & Co. and Education Development Center, Newton, Massachusetts; **Figure 33-9 (a)** Courtesy Michel Cagnet; **Figure 33-18 (a,b,c)** Courtesy Michel Cagnet; **Figure 33-19 (a)** Courtesy Michel Cagnet; **Figure 33-27 (a)** Courtesy Michel Cagnet; **Figure 33-29 (a,b)** M. Cagnet, M. Françon, J. C. Thrierr, *Atlas of Optical Phenomena;* **Figure 33-30 (a)** Courtesy Battelle-Northwest Laboratories; **Figure 33-31** Courtesy Michel Cagnet; **Figure 33-32** Courtesy Michel Cagnet; **Figure 33-34 (a,b)** Courtesy Michel Cagnet; **Figure 33-36 (a)** The Bettmann Archive; **(b)** Clarence Bennett, Oakland University, Rochester, Michigan; **(c)** Courtesy Holotek Ltd., Rochester, New York; **(d)** NRAO/AUI/ Science Photo Library; **p. 1089 (a)** © Philippe Plailly/ Science Photo Library; **(b,c)** © Ronald R. Erickson, 1981. Hologram by Nicklaus Phillips, 1978, for Digital Equipment Corporation; **(d)** © 1983 Ronald R. Erickson; **(e)** © Chuck O'Rear/West Light.

Chapter 34
Opener p. 1100 Albert Einstein Archives/AIP Niels Bohr Library; **p. 1101** Courtesy NRAO/AUI; **p. 1126** C. Powell, P. Fowler & D. Perkins/Science Photo Library/Photo Researchers; **p. 1134 (a)** © Michael Freeman; **(b)** N.A.S.A. (76-HC-612); **p. 1135** © Michael Freeman.

Chapter 35
Opener p. 1145 Adapted from Eastman Kodak and Wabash Instrument Corporation; **p. 1147** Max Planck Institute, Berlin; **p. 1151** Courtesy Thorn EMI Electron Tubes Ltd.; **Figure 35-7 (b)** Courtesy General Electric Company; **Figure 35-10** From G. Herzberg, *Annalen de Physick,* Vol. 84, p. 565, 1927; **Figure 35-17 (a,b)** *PSSC Physics,* 2nd ed., 1965. D. C. Heath & Co. and Education Development Center, Newton, Massachusetts; **(c)** © C. G. Shull; **(d)** © Claus Jönsson.

Chapter 36
Opener p. 1172 Lawrence Livermore Laboratory/ Science Photo Library/Photo Researchers; **Figure 36-2 (a,b,c)** E. R. Huggins; **(d)** Claus Jönsson; **Figure 36-3** From G. F. Missiroli and G. Pozzi, *American Journal of Physics,* Vol. 44, No. 3, 1976, p. 306; **p. 1179** S. Brandt and H. D. Dahmen, from *The Picture Book of Quantum Mechanics.* The color graphics reproduced here were made using the program contained in the book *Quantum Mechanics on the Personal Computer* by S. Brandt and H. D. Dahmen (Springer-Verlag, 1989); **p. 1192 (a)** Courtesy of Texas Instruments; **(b)** Courtesy of David Sarnoff Research Center; **p. 1193 (c,d,e,f)** Images by Patterson Electronics and computer processing by John Sanford; **p. 1199** S. Brandt and H. D. Dahmen, op. cit.; **Figure 36-20 (a,b)** *PSSC Physics,* 2nd ed., 1965. D. C. Heath & Co. and Education Development Center, Newton, Massachusetts; **p. 1202 (a,b)** and (*bottom*) S. Brandt and H. D. Dahmen, op. cit.; **Figure 3, p. 1211 (a)** R. Trump/IBM Research; **(b,c)** Courtesy of E. D. Williams; **Figure 4 p. 1212** D. M. Eigler, and E. K. Schweizer, IBM Almaden Research Center.

Chapter 37
Opener p. 1218 Dr. Bruce Schardt/Courtesy Digital Instruments, Inc.; **p. 1225 (a)** Courtesy of IBM

Research; **(b)** Courtesy of Park Scientific Instruments, Mountain View, California; **p. 1236 (a,b)** A. Jayaraman, AT&T Bell Labs; **p. 1238** David Parker/Science Photo Library/Photo Researchers; **p. 1239** © Robert Landau/West Light; **p. 1245 (a,b)** © 1991 Paul Silverman/Fundamental Photographs; **p. 1248 (a,b)** Chuck O'Rear/West Light; **p. 1249 (c)** Courtesy of Spectra-Physics Lasers; **(d)** © Bell Labs/Science Source/Photo Researchers; **(e)** Courtesy of Ahmed H. Zewail, California Institute of Technology; **Figures 4, 5 p. 1255** Courtesy of David Wineland; **Figures 6,7 p. 1258** Courtesy of David Wineland.

Chapter 38

Opener p. 1263 Courtesy David S. Lawrence, SUNY Buffalo; **Figure 38-7** © Will and Deni McIntyre/Photo Researchers; **Figure 38-14 (a,b)** Courtesy Dr. J. A. Marquisee.

Chapter 39

Opener p. 1286 Courtesy of AT&T Archives; **p. 1287** Courtesy of Museum of Modern Art, New York; **p. 1288 (a)** Richard Walters/*Discover;* **(b)** Dr. Jeremy Burgess/Science Photo Library/Photo Researchers; **p. 1289 (c)** © 1976 Thomas R. Taylor/Photo Researchers; **(d)** Courtesy of AT&T Archives; **p. 1291 (a)** Chris Kovach/*Discover;* **(b)** Srinivas Manne, University of California-Santa Barbara; **(c)** Dr. F. A. Quiocho and J. C. Spurlino/Howard Hughes Medical Institute, Baylor College of Medicine; **(d)** W. Krätschmer/Max-Planck Institute for Nuclear Physics; **p. 1308** © 1983 C. Falco/Photo Researchers; **p. 1310 (a)** Courtesy of AT&T Archives; **(b)** Dr. Jeremy Burgess/Science Photo Library/Photo Researchers; **p. 1311 (c)** Courtesy of AT&T Archives; **(d)** Dr. Jeremy Burgess/Science Photo Library/Photo Researchers; **(e,f)** Courtesy of AT&T Archives; **p. 1312 (a)** Adapted from "The Photorefractive Effect," by David Pepper, Jack Feinberg, and Nicolai Kukhtarev. © by Scientific American, Inc. 1990; **(b)** Roger S. Cudney/Courtesy of Jack Feinberg, University of Southern California; **p. 1313 (c)** Jack Feinberg, "Self-Pumped, Continuous-Wave Phase Conjugator Using Internal Reflection," *Optics Letters,* Vol. 7, 1982, pp. 486–488; **(d)** Courtesy of AT&T Archives; **Figure 39-29 (b)** A. Leitner/Rensselaer Polytechnic Institute; **p. 1317 (a)** Courtesy of IBM Research; **(b)** Akira Tonomura, Hitachi Ltd., Saitama, Japan; **(c)** © Fujifotos/The Image Works; **p. 1318** © 1983 C. Falco/Photo Researchers; **p. 1322** © Peter Freed/New York University; **Figure 1 (a,b) p. 1323** Courtesy of IBM Thomas J. Watson Research Center; **Figure 3 p. 1325** Courtesy of Samuel J. Williamson.

Chapter 40

Opener p. 1330 © Hans Wolf/The Image Bank; **p. 1333** Courtesy of Ronald Y. Cusson; **p. 1337** Courtesy of Malinckrodt Institute of Radiology; **p. 1346 (a)** © 1991 by the Metropolitan Museum of Art, New York; **(b,c)** Courtesy of Paintings Conservation Dept., Metropolitan Museum of Art, New York; **p. 1352** © Jerry Mason/Science Photo Library/Photo Researchers; **p. 1354 (a,b,c)** Courtesy of Princeton Plasma Physics Laboratory; **p. 1355 (a,b)** Courtesy of Lawrence Livermore National Laboratory, U.S. Department of Energy.

Chapter 41

Opener p. 1368 Lawrence Berkeley Laboratory/Science Photo Library/Photo Researchers; **p. 1371 (a)** Science Photo Library/Photo Researchers; **(b)** Lawrence Berkeley Laboratory/Science Photo Library/Photo Researchers; **p. 1373 (a)** From *The Particle Explosion,* by Frank Close, Michael Marten, and Christine Sutton, Oxford University Press, 1987, p. 130; **(b)** Lawrence Berkeley Laboratory/Science Photo Library/Photo Researchers; **Figure 41-1** Richard Ehrlich; **p. 1374** Courtesy of CERN; **p. 1375** Courtesy of CERN; **p. 1377** Lawrence Berkeley Laboratory/Science Photo Library/Photo Researchers.

Chapter 42

Opener p. 1389 Jack Burns/University of Southern California and the Anglo-Australian Telescope Board; **Figure 42-4** Patrick Wiggins/ © Hansen Planetarium; **Figure 42-9** Helmut Wimmer/Courtesy of *Natural History;* **Figure 42-10** Lund Observatory, Sweden; **Figure 42-12 (a)** From *Foundations of Astronomy,* 1986 ed., by Michael A. Seeds, © 1986 by Wadsworth, Inc., reprinted by permission of the publisher. Adapted from a study by G. De Vaucouleurs and W. D. Pence, *The Astronomical Journal 83,* No. 10, October 1978, p. 1163; **(b)** NRAO/AUI and Farhad Yusef-Zadeh/Northwestern University; **Figure 42-15** © Anglo-Australian Telescope Board; **Figure 42-19 (a)** U.S. Naval Observatories; **(b)** © 1980 Anglo-Australian Telescope Board; **(c)** David Malin/ © Anglo-Australian Telescope Board; **(d)** California Institute of Technology; **Figure 42-21** California Institute of Technology; **Figure 42-22** Margaret J. Geller/ © 1989 by the American Association for the Advancement of Science. From "Mapping the Universe," by Margaret J. Geller and John P. Huchra, *Science,* Vol. 246, November 17, 1989, pp. 897–903; **Figure 42-23** S. Maddox, W. Sutherland, G. Efstathiou, and J. Loveday/University of Oxford; **Figure 42-24** © 1984 Fermilab.

Answers

These answers are calculated using $g = 9.81\ \text{m/s}^2$ unless otherwise specified in the exercise or problem. The results are usually rounded to three significant figures. Differences in the last figure can easily result from differences in rounding the input data and are not important.

Chapter 1

True or False

1. True
2. False; for example, $x = vt$ where the speed v and the time t have different dimensions
3. True

Problems

1. (*a*) 1 MW (*b*) 2 mg (*c*) 3 μm (*d*) 30 ks
3. (*a*) 1 picoboo (*b*) 1 gigalow (*c*) 1 microscope (*d*) 1 attoboy (*e*) 1 megaphone (*f*) 1 nanogoat (*g*) 1 terabull
5. (*a*) C_1, ft; C_2, ft/s (*b*) C_1, ft/s^2 (*c*) C_1, ft/s^2 (*d*) C_1, ft; C_2, s^{-1} (*e*) C_1, ft/s; C_2, s^{-1}
7. (*a*) 62.1 mi/h (*b*) 23.6 in (*c*) 91.4 m
9. (1.61 km/h)/(1 mi/h)
11. (*a*) 36.00 km/h·s (*b*) 10.00 m/s^2 (*c*) 88 ft/s (*d*) 27 m/s
13. 4046.9 m^2
15. (*a*) C_1, L; C_2, L/T (*b*) C_1, L/T^2 (*c*) C_1, L/T^2 (*d*) C_1, L; C_2, T^{-1} (*e*) C_1, L/T; C_2, T^{-1}
17. $F = mv^2/r$
19. (*a*) 30,000 (*b*) 0.0062 (*c*) 0.000004 (*d*) 217,000
21. (*a*) 1.14×10^5 (*b*) 2.25×10^{-8} (*c*) 8.27×10^3 (*d*) 6.27×10^2
23. 1 in = 3.63×10^6 cell membranes
25. (*a*) 1690 (*b*) 5 (*c*) 5.6 (*d*) 10
27. 3.51 Mm
29. (*a*) 6000 km (*b*) 8760 tankers/y (*c*) 43.8 billion dollars
31. (*a*) 1.41×10^{17} kg/m^3 (*b*) 216 m
33. (*a*) 4.92×10^9 m (*b*) 12.8 $r_{\text{e-m}}$
35. (*a*) $T = Cm^n$, with $n = 0.5$ and $C = 1.81$ s/kg$^{1/2}$ (*b*) Those points with the greatest deviation are at $T =$ 1.05 and 1.75 s.

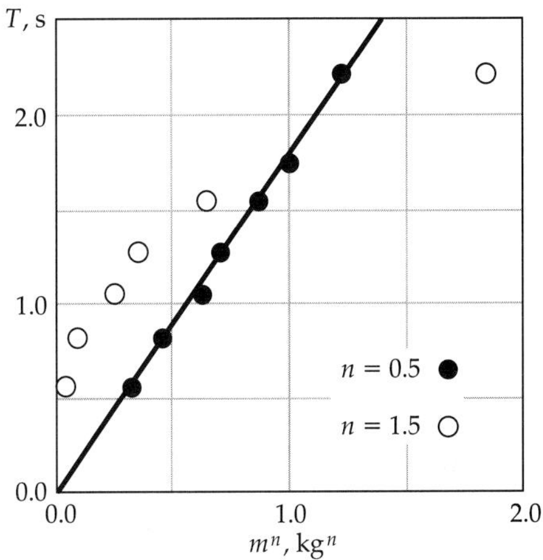

37. (*a*) $T = C\sqrt{L/g}$ (**b**) example results: $L = 1$ m, $T =$ 2 s; $L = 0.5$ m, $T = 1.4$ s (*c*) $C = 2\pi$
39. (*a*) R increases with increasing H and with increasing v (*b*) $R = v\sqrt{2H/g}$
41. (*a*) 8×10^8 bits = 1×10^8 bytes (*b*) $\approx$ 500 books

Chapter 2

True or False **1.** False; this equation holds only for constant acceleration **2.** False; a ball at the top of its flight is momentarily at rest but has acceleration g **3.** False; for example, motion with constant velocity **4.** True **5.** True; by definition of average velocity

Problems

1. (*a*) 24 km/h (*b*) −12 km/h (*c*) 0 km/h (*d*) 16 km/h
3. (*a*) −2 m/s (*b*) 2.25 m/s (*c*) −0.3 m/s
5. (*a*) 260 km (*b*) 65 km/h
7. (*a*) 1.82 h (*b*) 4.85 h (*c*) 945 km/h (*d*) 621 km/h
9. (*a*) 8.33 min (*b*) 1.28 s (*c*) 1 lightyear = 9.47×10^{12} km = 5.88×10^{12} mi
11. $66\frac{2}{3}$ km/h
13. (*a*) 1 m/s (*b*) 2 m/s
15. (*a*) 1 m/s (*b*) 0.7 m/s (*c*) 8 s
17. −2 m/s^2
19. (*a*) 80 m/s (*b*) 400 m (*c*) 40 m/s
21. 15.6 m/s^2
23. $t = 4.47$ s, $v = 44.7$ m/s, $v_{\text{av}} = 22.4$ m/s
25. 4.59 km
27. (*a*) 25.1 s (*b*) 3.08 km
29. (*a*) The area under the curve in the indicated time interval is −36 m.

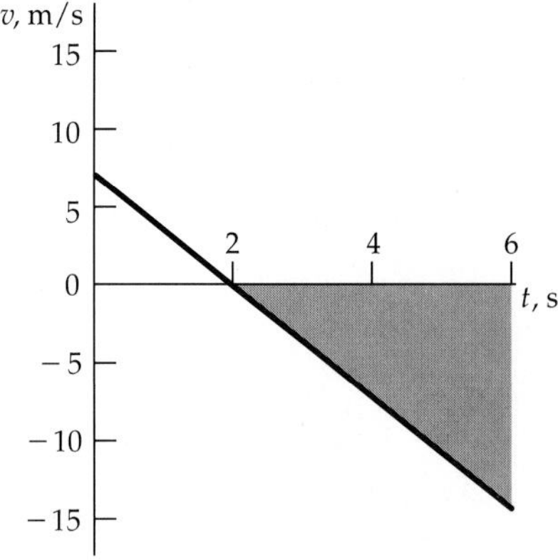

(b) $x(t) = 7t - 2t^2 + c$, $\Delta x = -36$ m (c) -9 m/s

31. (a) 0 m/s (b) $\frac{1}{3}$ m/s (c) -2 m/s (d) 1 m/s

33. (a)

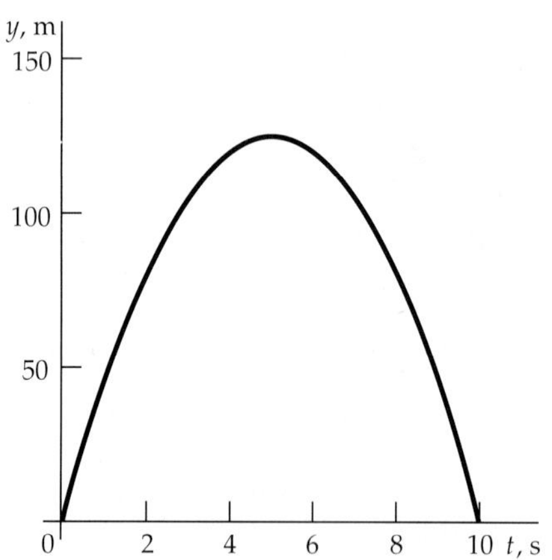

(b) $t = 0$ to 1 s, $v_{av} = 45$ m/s;
$t = 1$ to 2 s, $v_{av} = 35$ m/s;
$t = 2$ to 3 s, $v_{av} = 25$ m/s; $t = 3$ to 4 s, $v_{av} = 15$ m/s;
$t = 4$ to 5 s, $v_{av} = 5$ m/s; $t = 5$ to 6 s, $v_{av} = -5$ m/s;
$t = 6$ to 7 s, $v_{av} = -15$ m/s;
$t = 7$ to 8 s, $v_{av} = -25$ m/s;
$t = 8$ to 9 s, $v_{av} = -35$ m/s;
$t = 9$ to 10 s, $v_{av} = -45$ m/s.

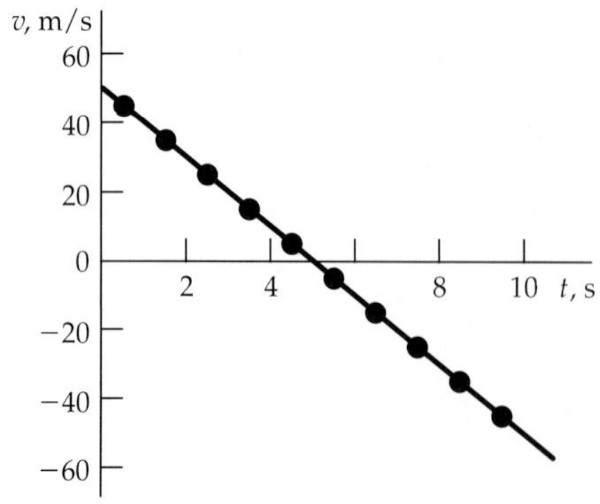

(c) $v(t) = -10t + 50$

35. (a) Velocity is greatest for $t \approx 2$ to 3 s

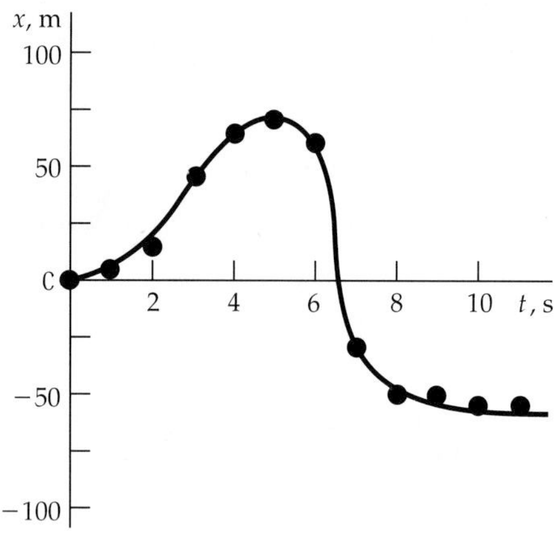

(b) Speed is least near $t = 5$ s, and for $t \approx$ 8 s and $t \approx 11$ s (c) Same as part (b). (d) Speed is constant for $t > 10$ s (e) Acceleration is positive for $t \approx 0$ s to 3 s and for $t \approx 7$ s to 10 s. (f) Acceleration is negative for $t \approx 3$ s to 7 s.

37. (a) -7.67 m/s (b) 6.26 m/s (c) 697 m/s^2 upward

39. (a) $x(t) = (C/6)t^3 + Dt + E$, where E is a constant in meters. (b) At $t = 5$ s, we find $v = 37.5$ m/s and $x =$ 62.5 m.

41. $\Delta x = 4\frac{1}{3}$ m. The average velocity is not equal to the mean of the initial and final values when the acceleration is not constant, as is the case here.

43.

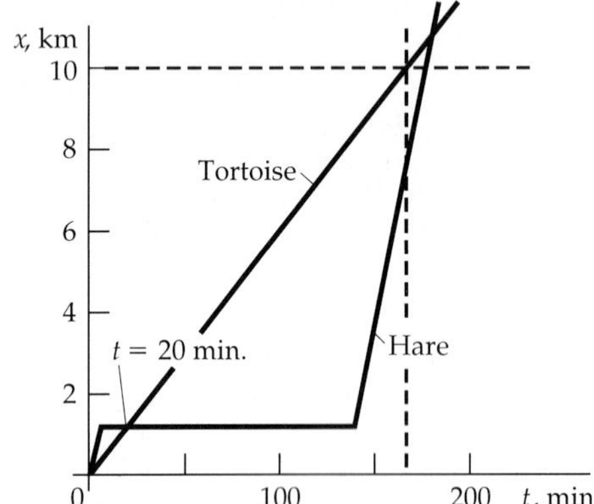

Tortoise passes Hare at $t = 20$ min. When Tortoise finishes, Hare is 2.4 km behind. Hare could nap up to 125 min without losing the race.

45. (a) t_0, t_1 (b) t_3, t_4, t_6, t_7 (c) t_2, t_5 (d) t_4 (e) t_2, t_6 (f) t_0, t_1, t_3, t_5, t_7

47. For Figure 2–22a, (a) $a > 0$ for $t \approx 3$ to 6 s; $a < 0$ for $t \approx 1$ to 3 s; $a = 0$ near $t = 3$ and for $t = 6$ s to 7 s. (b) $a =$ constant where the curve is straight. (c) $v = 0$ at $t \approx 8.75$ s. For Figure 2–22b, (a) $a > 0$ for $t \approx 5$ to $t \approx 7$ s; $a < 0$ for $t \approx 0$ to 3 s and $t > 7$ s; $a = 0$ for $t \approx 3$ to 5 s (b) $a =$ constant on the parabolic and straight-line segments of the curve. (c) $v = 0$ near $t = 2$, 6, and 8 s.

49. (*a*) 34.7 s (*b*) 1.21 km

(*c*)

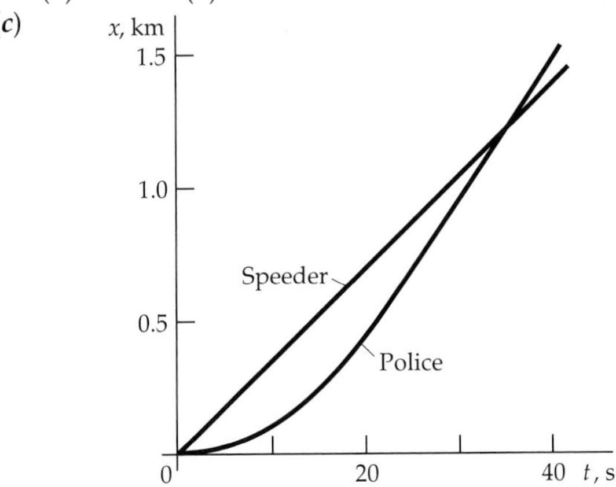

51. $v_0 = 40$ cm/s; $a = -6.88$ cm/s^2

53. (*a*) $6\frac{2}{3}$ s (*b*) $6\frac{2}{3}$ s

55. 18.4 m

57. (*a*) 26.1 m (*b*) 23.4 m/s

59.

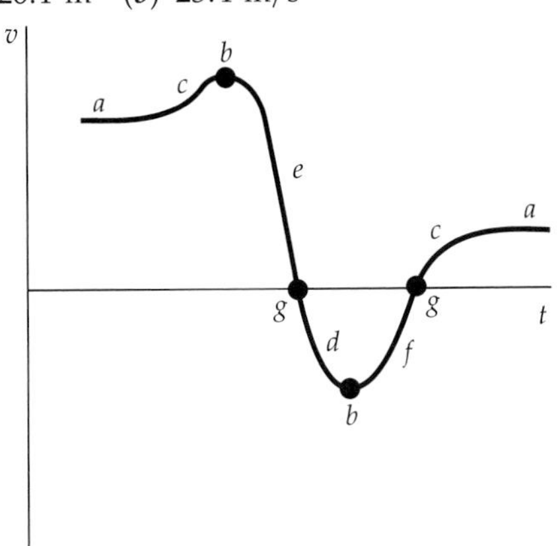

61. (*a*) Yes

(*b*)

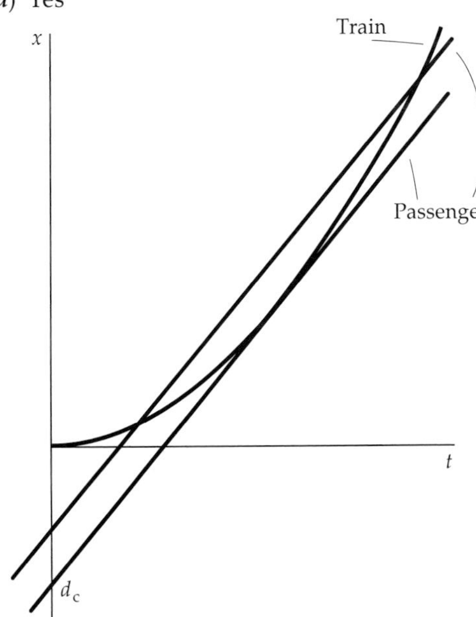

(*c*) The train's speed is 8 m/s, its average speed is 4 m/s and $d_c = 32$ m.

63. The balls collide two-thirds of the way to the top of the building.

65. (*a*)

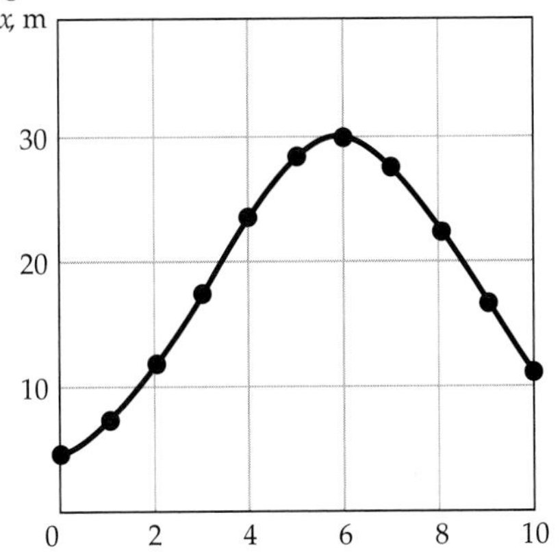

(*b*)

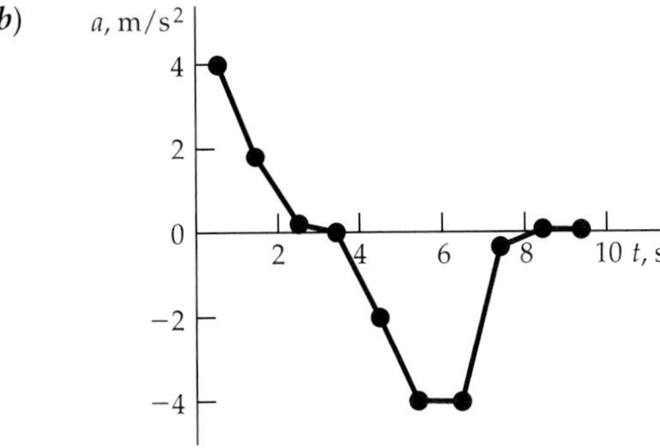

67. (*a*) 0.875 cm/s

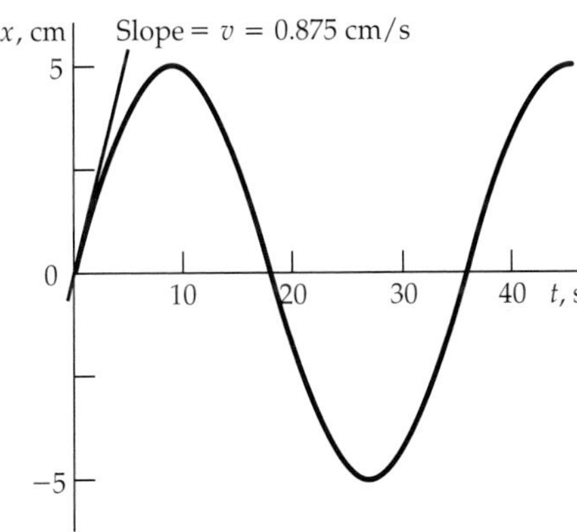

(*b*) The average velocities are 0.723, 0.835, 0.857, 0.871, 0.874, and 0.875 cm/s for the time intervals ending at 6, 3, 2, 1, 0.5, and 0.25 s, respectively. (*c*) $dx/dt = A\omega \cos \omega t = 0.875$ cm/s at $t = 0$ s.

69. (*a*)

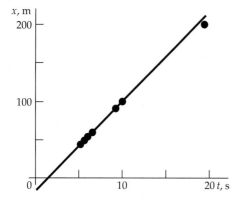

(*b*) $x = \frac{1}{2}v_0T + aT(t - T)$, or $x = v_0(t - \frac{1}{2}T)$ (*c*) The slope is ≈ 11.6 m/s $= v_0$, and the intercept with the time axis is ~1.3 s. Thus, $T = 2.6$ s and $a = 4.46$ m/s^2 (*d*) The sprint is run at top speed. For longer races, top speed cannot be maintained and a runner must run at a slower, but sustainable, pace
71. (*a*) Initially the particle accelerates with the acceleration of gravity, but as its velocity increases, its acceleration diminishes. When $v = g/b$, the acceleration becomes zero. (*b*) The figure shows a rough sketch of v versus t

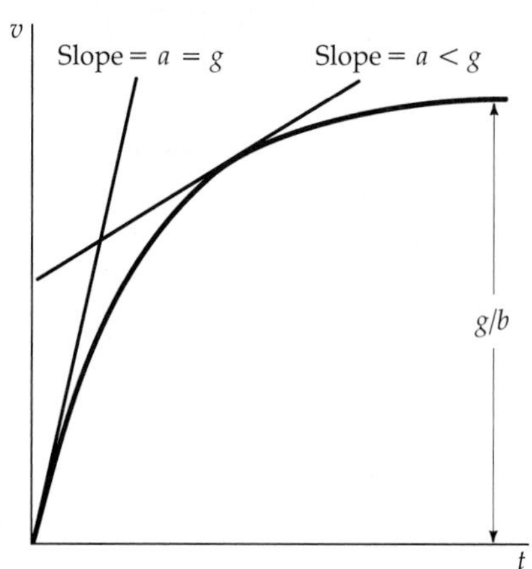

73. $x(t) = x_0e^{A(t-t_0)}$, where $A = 1\ \mathrm{s}^{-1}$
75. (*a*)

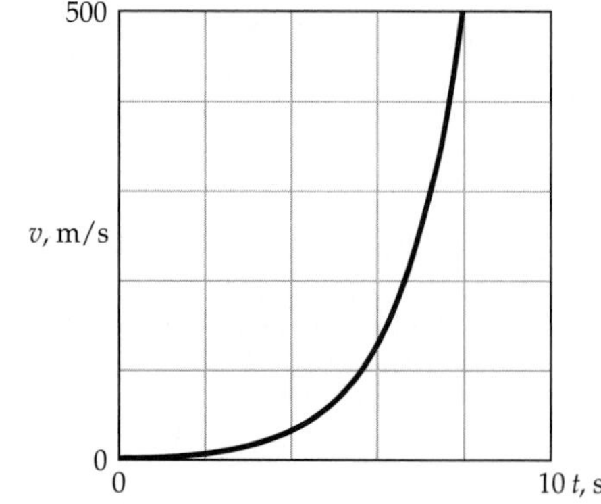

(*b*) 295 m/s

Chapter 3

True or False **1.** False; for example, equal and opposite vectors **2.** True **3.** False; for example, projectile motion **4.** False; for example, circular motion **5.** True **6.** True **7.** True

Problems

1.

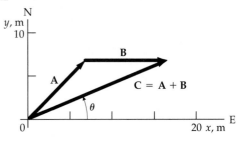

18.5 m, 22.5° north of east.
3. (*a*) $t = 15$ s, $\Delta r = 14.1$ m, and $\theta = 45°$; $t = 30$ s, $\Delta r = 20$ m, and $\theta = 90°$; $t = 45$ s, $\Delta r = 14.1$ m, and $\theta = 135°$; $t = 60$ s, $\Delta r = 0$ m (*b*) $\Delta r = 14.1$ m, $\theta = 45°$; $\Delta r = 14.1$ m, $\theta = 135°$; $\Delta r = 14.1$ m, $\theta = 225°$; $\Delta r = 14.1$ m, $\theta = 315°$ (*c*) same magnitude, but rotated by 90° (*d*) same magnitude but in the opposite direction
5. (*a*) 1 km (*b*) 57.3° north of east (*c*) 1/5
7. (*a*) $A_x = 8.66$ m, $A_y = 5$ m (*b*) $A_x = 3.54$ m, $A_y = 3.54$ m (*c*) $A_x = 3.5$ km, $A_y = 6.06$ km (*d*) $A_x = 0$ km, $A_y = 5$ km (*e*) $A_x = -13$ km/s, $A_y = 7.5$ km/s (*f*) $A_x = -5$ m/s, $A_y = -8.66$ m/s (*g*) $A_x = 0$ m/s^2, $A_y = -8$ m/s^2
9. (*a*) $A_x = 1.41$ m $= A_y$, $B_x = 1.73$ m, $B_y = -1$ m (*b*) $C_x = 3.15$ m, $C_y = 0.414$ m, $C = 3.17$ m, $\theta = 7.5°$ (*c*) $C_x = -0.318$ m, $C_y = 2.41$ m, $C = 2.44$ m, $\theta = 97.5°$
11. (*a*) $A = 8.06$, $\theta = 240°$; $B = 3.61$, $\theta = -33.7°$; $C = 9.06$, $\theta = 264°$ (*b*) $A = 4.12$, $\theta = -76°$; $B = 6.32$, $\theta = 71.6°$; $C = 3.61$, $\theta = 33.7°$
13. (*a*) $\mathbf{v} = 5$ m/s $\mathbf{i} + 8.66$ m/s $\mathbf{j}$ (*b*) $\mathbf{A} = (-3.54\ \mathrm{m})(\mathbf{i} + \mathbf{j})$ (*c*) $\Delta\mathbf{r} = 14$ m $\mathbf{i} - 6$ m $\mathbf{j}$
15. $A_y/A_x = \pm B_y/B_x$
17.

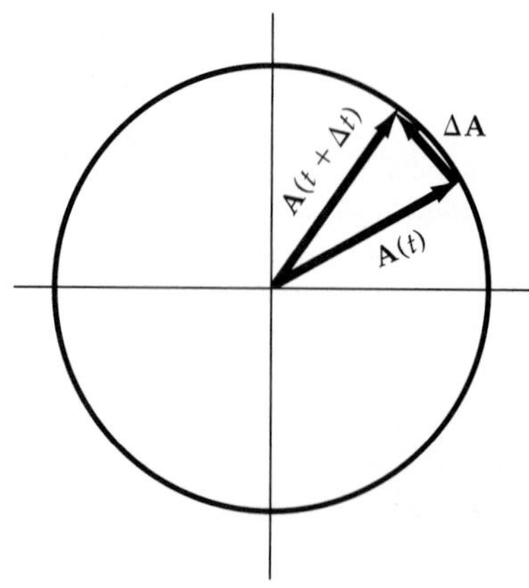

$\Delta\mathbf{A}$ becomes perpendicular to $\mathbf{A}(t)$.
19. $v_{av} = 14.7$ km/h, $\theta = -16.3°$

21. (*a*)

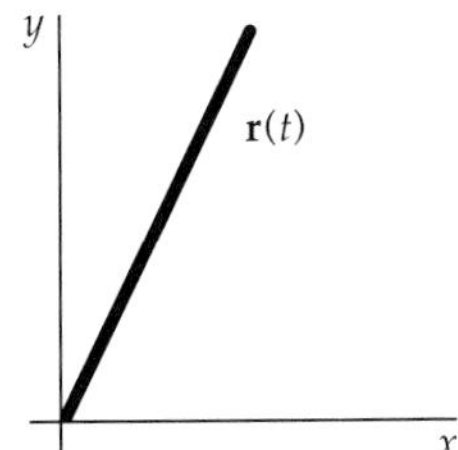

(*b*) $\mathbf{v} = d\mathbf{r}/dt = 5\ \text{m/s}\ \mathbf{i} + 10\ \text{m/s}\ \mathbf{j}$; $v = 11.2$ m/s
23. (*a*) *AB*, north; *BC*, northeast; *CD*, east; *DE*, southeast; *EF*, south (*b*) *AB*, north; *BC*, southeast; *CD*, acceleration = 0; *DE*, southwest; *EF*, north (*c*) The acceleration is greater on segment *DE*
25. (*a*) $\mathbf{v}_{av} = 33\frac{1}{3}\ \text{m/s}\ \mathbf{i} + 26\frac{2}{3}\ \text{m/s}\ \mathbf{j}$ (*b*) $\mathbf{a}_{av} = -3\ \text{m/s}^2\ \mathbf{i} - 1.77\ \text{m/s}^2\ \mathbf{j}$
27. (*a*) west of north by $\theta = 13.1°$ (*b*) 300 km/h
29. 0.533 s
31. (*a*) 2294 m (*b*) 43.2 s (*c*) 9174 m
33. $\mathbf{v} = 15$ m/s $\mathbf{i}$, where $\mathbf{i}$ is taken to be horizontal; $a = 9.81\ \text{m/s}^2$ downward

37. (*a*) $a_r \rightarrow 4a_r$ (*b*) $a_r \rightarrow \frac{1}{2}a_r$ (*c*) A perfectly sharp turn implies $r = 0$, which would require an infinitely large acceleration for any finite speed.
39. 29.9 rev/min
41. (*a*) $\mathbf{a} = 2\ \text{m/s}^2\ \mathbf{i} - 3.5\ \text{m/s}^2\ \mathbf{j}$ (*b*) $\mathbf{v} = (1 + 2t)$ m/s $\mathbf{i} + (1 - 3.5t)$ m/s $\mathbf{j}$ (*c*) $\mathbf{r} = (4 + t + t^2)$ m $\mathbf{i} + (3 + t - 1.75t^2)$ m $\mathbf{j}$
43. (*a*) 31.6 m/s^2 (*b*) 26.1 m/s^2 (*c*) 23.9 m/s^2
45. $\mathbf{v} = 30\mathbf{i} + (40 - 10t)\mathbf{j}$, $\mathbf{a} = d\mathbf{v}/dt = -10\mathbf{j}$
47. (*a*) 8.14 m/s (*b*) 23.2 m/s
49. The projectile covers a horizontal distance $x = 408$ m
51. (*a*) -120 m $\mathbf{i}$ + 4 m $\mathbf{j}$ (*b*) -20 m/s $\mathbf{i}$ − 12 m/s $\mathbf{j}$ (*c*) $-2\ \text{m/s}^2\ \mathbf{j}$
53. 3.20 m/s in the direction of the thrower
55. 14.2 m/s = 31.7 mi/h
57. 4.29 m
59. 0.785 m
61. (*a*) 2 s (*b*) π m/s (*c*) $-\mathbf{j}$ (*d*) $a_r = \pi^2\ \text{m/s}^2$, $a_t = \pi/2\ \text{m/s}^2$ (*e*) $a = 9.99\ \text{m/s}^2$, $\theta = 189°$
63. 34.6 m/s
65. (*a*) 20.0 m/s (*b*) 2.87 m
67. (*a*) Note that $r = 4$ m for all time; thus, the path is a circle of radius 4 m centered on the origin. (*b*) $\mathbf{v} = (8\pi\ \text{m/s})(\cos 2\pi t\ \mathbf{i} - \sin 2\pi t\ \mathbf{j})$ (*c*) $\mathbf{a} = (-16\pi^2\ \text{m/s}^2)(\sin 2\pi t\ \mathbf{i} + \cos 2\pi t\ \mathbf{j}) = (-4\pi^2\ \text{s}^{-2})\mathbf{r}$; hence $\mathbf{a}$ is radial, and in particular, is radially inward as indicated by the minus sign.
69. (*a*) $H = W \tan\theta$ (*b*)

$$v_0 = \frac{W}{\cos\theta}\sqrt{\frac{g}{2(W\tan\theta - H)}}$$

71. (*a*) 54.3 m/s (*b*) 269 m (*c*) 6.56 s
73. (*a*) $\mathbf{v} = (10 - 10 \sin 2t)\mathbf{i} - (10 \cos 2t)\mathbf{j}$ (*b*) The particle is in contact with the surface at $5\pi/2$, $25\pi/2$, . . . = 7.85, 39.2, ...

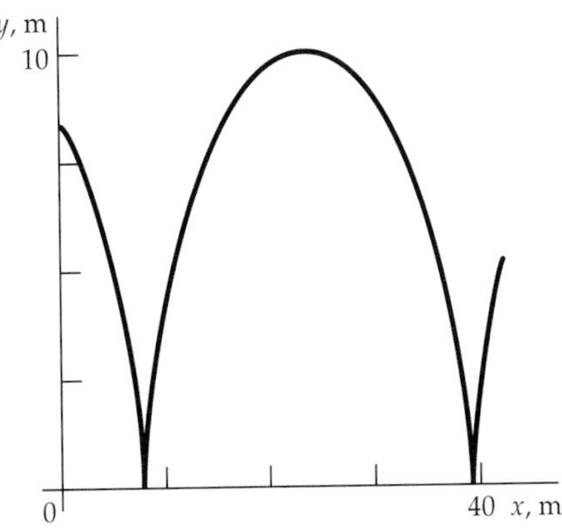

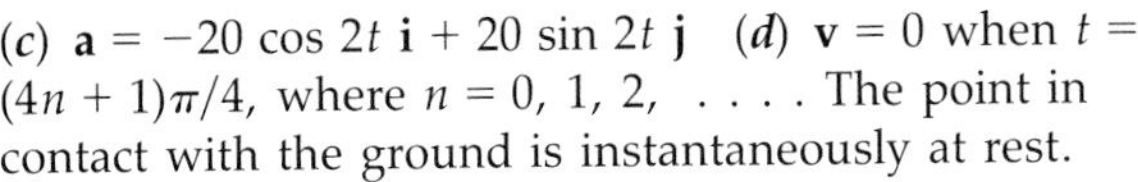
(*c*) $\mathbf{a} = -20 \cos 2t\ \mathbf{i} + 20 \sin 2t\ \mathbf{j}$ (*d*) $\mathbf{v} = 0$ when $t = (4n + 1)\pi/4$, where $n = 0, 1, 2, \ldots$. The point in contact with the ground is instantaneously at rest.

Chapter 4

True or False **1.** True **2.** False; the forces acting sum to zero **3.** False; the *acceleration* is in the direction of the unbalanced force **4.** True **5.** False **6.** False

Problems
1. (*a*) 8 m/s^2 (*b*) $m_2 = \frac{1}{2}m_1$ (*c*) 8/3 m/s^2
3. (*a*) $a = 7.07\ \text{m/s}^2$ in a direction bisecting the angle between the two forces (*b*) $a = 14.0\ \text{m/s}^2$ in a direction 14.6° away from $2\mathbf{F}_0$ toward $\mathbf{F}_0$
5. 12 m/s^2
7. (*a*) 6 m/s (*b*) 9 m
9. 10 kg
11.

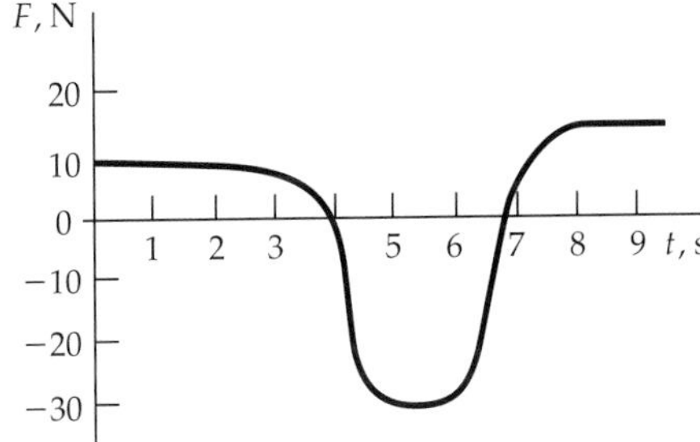

13. *AB*, force vertically upward; *BC*, force radial toward center of circular arc; *CD*, no force; *DE*, force is toward center of circular arc; *EF*, force acts vertically upward
15. (*a*) 79.4 kg (*b*) 7.94×10^4 g
17. (*a*) $W = 784.8$ N = 176 lb (*b*) $W = 716$ N = 161 lb (*c*) $m = 80$ kg = 5.48 slugs
19.

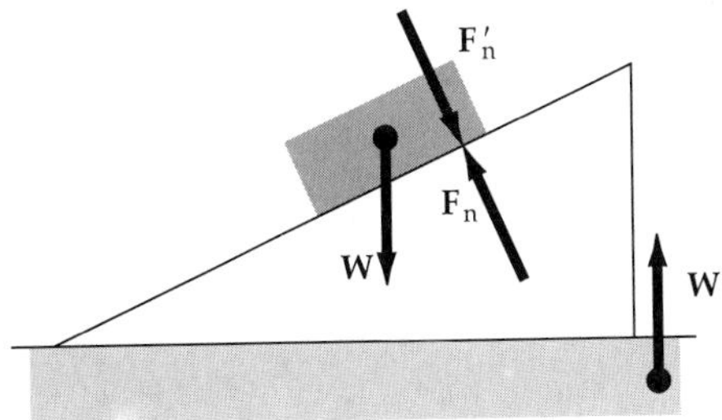

21. (*a*) 1.28 s (*b*) 8.7 y
23. $5\frac{1}{3}$ m/s^2
25. (*a*) 98.1 N (*b*) 98.1 N (*c*) 49.05 N (*d*) 49.05 N
27. (*a*) 8.77 m/s^2 (*b*) 26.3 N
29. (*a*) 132 lb (*b*) 105 lb (*c*) Same as part (*b*)
31. (*a*) $T = w/(2 \sin \theta)$, $T_{min} = \frac{1}{2}w$, $T_{max} \to \infty$ (*b*) 19.6 N
33. (*a*) 19.6 N (*b*) 19.6 N (*c*) 39.6 N (*d*) For $t = 0$ to 2 s, $T = 19.6$ N; for $t = 2$ to 4 s, $T = 9.62$ N
35. (*a*) 11,810 N (*b*) 9810 N (*c*) T = 7810 N
37. (*a*) 3821 N (*b*) 5732 N
39. 500 N
41. (*a*) 149 N (*b*) 3.41 m/s^2
43. **F** = 329 N = 73.9 lb directed toward the man.
45. (*a*) 17 m/s^2 (*b*) If a is made greater, the block will accelerate up the wedge.
47. (*a*) 398 N (*b*) 368 N
49. (*a*) The direction of deflection is opposite to the direction of acceleration. (*c*) $\theta = 9.31°$ in the forward direction.
51. $\theta = 51.6°$

Chapter 5

True or False **1.** False **2.** False; it opposes the relative motion of the surfaces **3.** True **4.** True **5.** False; round-off errors increase and may be important

Problems

1. 0.417
3. (*a*) Since $\mu_s > \mu_k$ in general, the car's acceleration is greater when the wheels do not spin. (*b*) 0.212
5. (*a*) 0.289 (*b*) 600 N
7. 4.57°
9. (*a*) M/T, kg/s (*b*) M/L, kg/m
11. 2.79×10^{-4} kg/m
13. 205 km/h
15. (*a*) $a = 2.45$ m/s^2, $T = 36.8$ N (*b*) $T = m_1 m_2 g(1 + \sin \theta)/(m_1 + m_2)$
17. Consider first the mass m_2. If we take radially inward as positive, then $T_1 = 4\pi^2[m_1L_1 + m_2(L_1 + L_2)]/T^2$ and $T_2 = m_2v_2^2/(L_1 + L_2)$.
19. (*a*) In the inertial frame, the object moves with constant speed. In the boxcar frame, the object has an acceleration $a = -5$ m/s^2. (*b*) 4 s
21. In the inertial frame, the object slows with time, then sticks to the floor and accelerates with the boxcar. In the boxcar frame, the object slows, then comes to rest and remains at rest.
23. 6.61 cm
25. 24.5 m/s
27. 83.9 m
29. (*a*) 10 m/s^2 (*b*) $T = (100 - 8x)$ N
31. (*a*) 400 N (*b*) 625 N
33. 23.6 rev/min
35. (*a*) 16.4 m/s^2 (*b*) 19.6 N (*c*) The frictional force will not increase if the acceleration is increased.
37. (*a*) The force initially decreases with θ since the vertical component of **F** will decrease the normal force $\mathbf{F}_n$ exerted by the table. As θ increases further, however, the horizontal component of **F** decreases so much that **F** must become larger again in order to move the box (*b*) $F = \mu_s mg/(\cos \theta + \mu_s \sin \theta)$; F is a minimum near 31°.

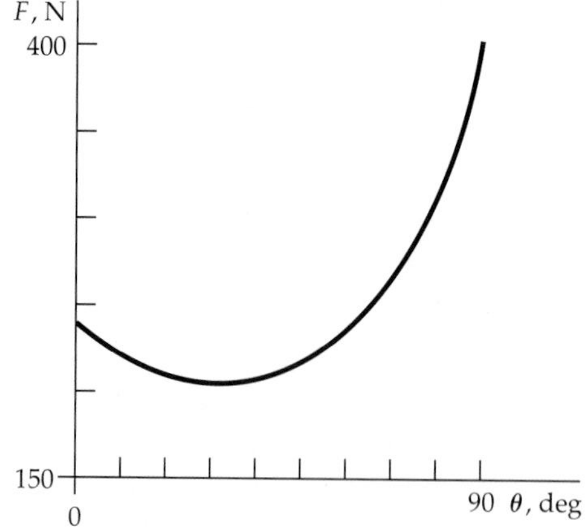

39. (*a*) 51.7 km/h (*b*) 27.6 km/h
41. (*a*) 80 N (*b*) $F_{net} = 600$ N, $F = 680$ N (*c*) 6.8 m/s^2
43. (*a*) $a = \frac{1}{2}$ m/s^2, $F_c = 2$ N (*b*) $F_c = m_2F/(m_1 + m_2)$
45. (*a*) The maximum m is 3.39 kg; the minimum m is 0.614 kg. (*b*) 9.81 N
47. (*a*) 17.7 N (*b*) $a = 1.47$ m/s^2, $f = 2.94$ N (*c*) $a_2 = 1.96$ m/s^2, $a_4 = 7.85$ m/s^2, $f = 3.92$ N
49. (*a*) $F(\theta) = \mu_s mg/(\cos \theta + \mu_s \sin \theta)$ (*b*) $dF/d\theta = \mu_s mg(\sin \theta - \mu_s \cos \theta)/(\cos \theta + \mu_s \sin \theta)^2$, $\theta = \tan^{-1}(0.6) = 31°$
51. (*a*) 7.25 m/s (*b*) 0.536
53. $v(t) = v_0/[1 + (\mu_k/r)v_0t]$

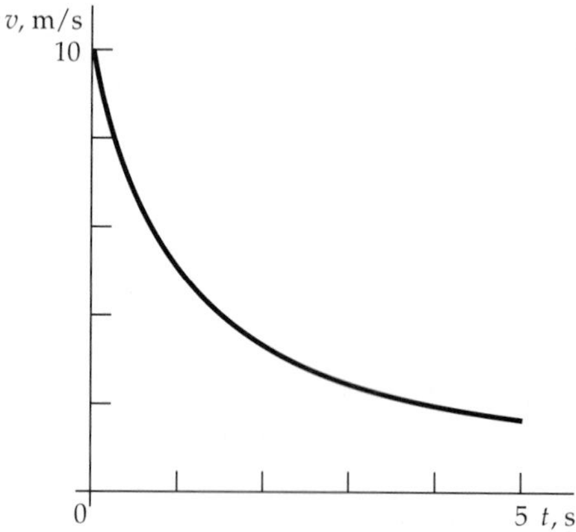

55. $v_{max} = 56.0$ km/h, $v_{min} = 20.1$ km/h
57. $a = -1.12$ m/s^2, $T = 44.7$ N
59. (*a*) $F_n = 4\pi^2mrB^2/3600$ (*b*) Relative to the compartment, the mass falls to the floor. From the point of view of the rotating frame of reference, this is caused by a pseudoforce of magnitude mv^2/r acting radially outward. (*c*) In an inertial frame, no forces act on the mass; therefore, it travels in a straight line with the same velocity it had when dropped. This straight-line path intercepts the circular path of the compartment floor; thus, the mass is seen to drop to the floor.

Chapter 6

True or False **1.** False **2.** True **3.** False **4.** True **5.** False; it is a unit of energy **6.** False **7.** False **8.** True

Problems

1. (*a*) 7200 J (*b*) 1800 J (*c*) 28,800 J
3. 1 J = 0.738 ft·lb
5. (*a*) 9 J (*b*) 12 J (*c*) 4.58 m/s
7. $\frac{81}{4}$ C
9. (*a*) F = 327, 245, and 196 N (*b*) 981 J, independent of L (*c*) The work is the same in each case, but the force required is less as L becomes greater
11. 18 m^2
13. (*a*) 142° (*b*) 101° (*c*) 90°
15. (*a*) $\mathbf{A}\cdot\mathbf{i} = A_x(\mathbf{i}\cdot\mathbf{i}) + A_y(\mathbf{j}\cdot\mathbf{i}) + A_z(\mathbf{k}\cdot\mathbf{i}) = A_x$ (*b*) $\mathbf{A}/\sqrt{A_x^2 + A_y^2 + A_z^2}$ (*c*) 2
17. (*a*) nonconservative (*b*) conservative (*c*) conservative (*d*) nonconservative
19. (*a*) 392 J (*b*) $x = 2.45$ m, $v = 4.91$ m/s (*c*) $K = 24.1$ J, $U = 368$ J (*d*) $K = 392$ J, $v = 19.8$ m/s
21. (*a*) $U(x) - U(x_0) = -4(x - x_0)$ (*b*) $U(x) = -4x + 24$ (*c*) $U(x) = -4x + 36$
23. (*a*) A, negative; B, zero; C, positive; D, zero; E, negative; F, zero (*b*) Point C would appear to have the force with greatest magnitude, followed closely by point A. (*c*) Points B, D, and F are equilibrium points. B is unstable, D is stable, and F is neutral.
25.

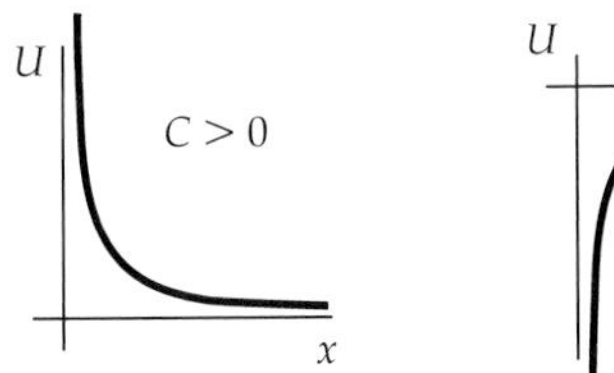

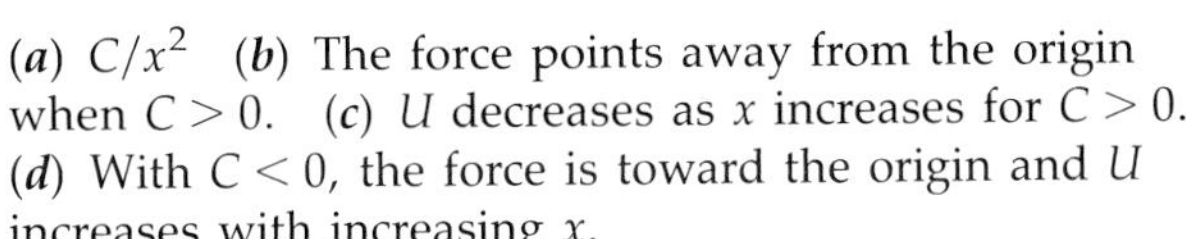

(*a*) C/x^2 (*b*) The force points away from the origin when $C > 0$. (*c*) U decreases as x increases for $C > 0$. (*d*) With $C < 0$, the force is toward the origin and U increases with increasing x.
27. (*a*) The work done by the woman is 1.22 J. The work done by the spring is −1.22 J. (*b*) 1.10 m/s
29. (*a*) 0.858 m. (*b*) After coming to rest, the object is accelerated by the spring. The kinetic energy gained by the object is just enough to allow it to rise again to the original height of 5 m. Assuming no nonconservative forces act on the object, it will continue to oscillate back and forth indefinitely.
31. (*a*) The object stops 2.5 m below the point Q, slides back to the left, rising 2.5 m above the point P, and continues this motion. (*b*) The object passes the point Q with a speed of 2.10 m/s. (*c*) 9.90 m/s
33. 5.05 m
35. (*a*) 104 J (*b*) −70.2 J (*c*) 33.7 J (*d*) 2.90 m/s
37. (*a*) 7.67 m/s (*b*) −58.9 J (*c*) 1/3
39. (*a*) 3.14×10^{16} J (*b*) 7.47 megatons TNT
41. (*a*) No work is done on the skater (*b*) 560 J (*c*) The kinetic energy of the skater results from the conversion of chemical energy in the muscles
43. (*a*) 1.67 m/s (*b*) 15 J
45. 1.37×10^6 kW
47. (*a*) 8.64×10^6 J (*b*) 2065 kcal
49. (*a*) 60° (*b*) $v^2 = \frac{2}{3}gL(2 - \cos\theta_0)$
51. (*a*) −11 J, −10 J, −7 J, −3 J, 0 J, 1 J, 0 J, −2 J, and −3 J
(*b*)

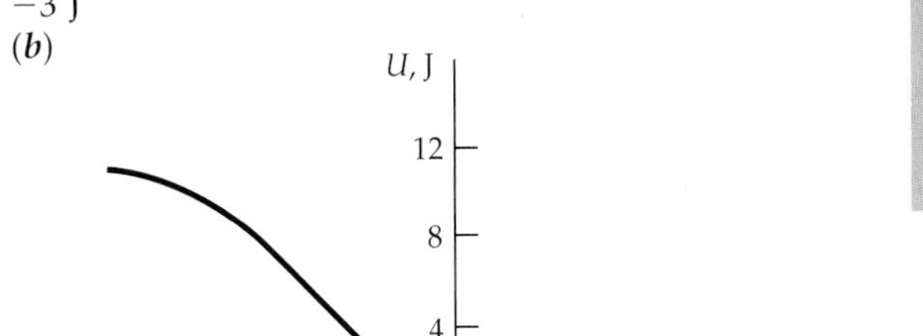

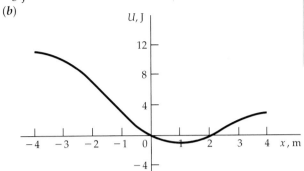

53. (*a*) 11/4 J (*b*) 6.13 J (*c*) 2.02 m/s (*d*) 14.25/4 J (*e*) 2.15 m/s
55. (*a*) 57.6 J (*b*) $U = 14.4$ J, $K = 43.2$ J (*c*) 175 J (*d*) 41.9 m/s
57. $U(x) = \frac{1}{3}ax^3$

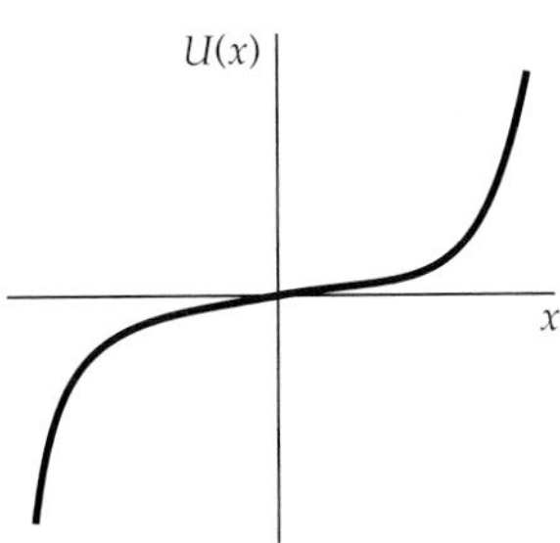

59 (*a*) 0.989 m (*b*) 0.783 m (*c*) 1.54 m
61. (*a*) $M(y) = 40 - \frac{1}{2}y$ (*b*) 1.18×10^4 J
63. (*a*) The forces and the work done by them are: gravity, $-mgL \sin 60°$; friction, $-\mu_k(mg \cos 60°)L$; the normal force, no work. (*b*) 0.451 m (*c*) The work done by gravity, friction, and the normal force is $mgL \sin 60°$, $-\mu_k mgL \cos 60°$, and zero, respectively. (*d*) 2.52 m/s
65. (*a*) $W_{nc} = -\frac{3}{4}(\frac{1}{2}mv_0^2)$ (*b*) $\mu_k = -W_{nc}/(2\pi rmg)$ (*c*) $\frac{1}{3}$
67. y_0
69. 16.0 hp
71. (*a*) 0 (*b*) 78 J
73. (*a*) $U(y) = 2Mg[(d^2 + y^2)^{1/2} - d] - mgy$
(*b*)

$$y_0 = d\frac{\left(\dfrac{m}{2M}\right)}{\sqrt{1 - \dfrac{m^2}{4M^2}}}$$

75. 1.45 hp
77. (*a*) $mg(h - 2R)$ (*b*) $2g(h - 2R)/R$ (*c*) $2.5R$
79. $T_{rest} = mg - K/L$, $T_{bottom} = mg + 2K/L = T_{rest} + 3K/L$
81. (*a*)

$$v(\theta) = \sqrt{v_0^2 + 2gR(1 - \cos\theta)}$$

(*b*) $\cos\theta = \frac{2}{3} + v_0^2/(3gR)$

83. (*a*) $x = 0$ m, $x = 2$ m, and $x > 3$ m (*b*) A plot of U versus x is presented below

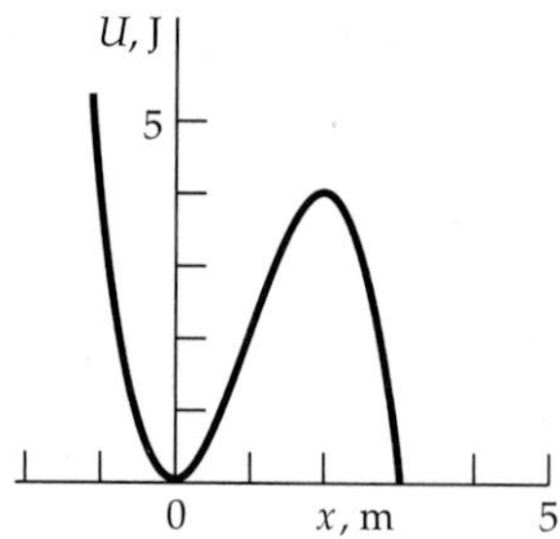

(*c*) The equilibria at $x = 0$ m, $x = 2$ m, and $x > 3$ m are stable, unstable, and neutral, respectively (*d*) 2 m/s
85. For the path from $x = 4$ m, $y = 1$ m to $x = 4$ m, $y = 4$ m, $W = 36A$. For a path consisting of three parts—horizontal from $x = 4$ m to $x = 5$ m with $y = 1$ m; vertical from $y = 1$ m to $y = 4$ m with $x = 5$ m; horizontal from $x = 5$ m to $x = 4$ m with $y = 4$ m—$W = 45A$
87. (*a*)

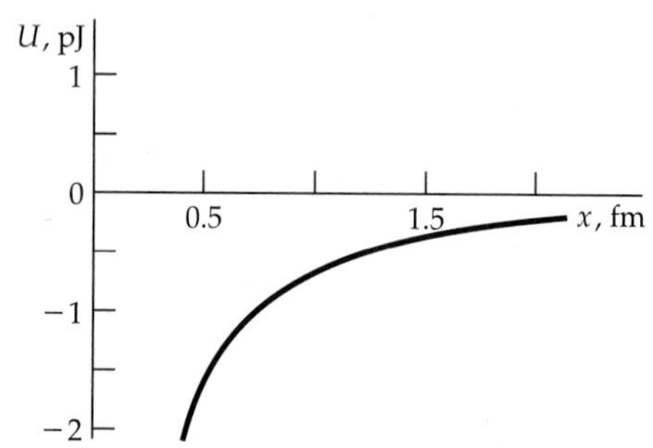

(*b*)

$$F_x = -\frac{dU}{dx} = -U_0 e^{-x/a}\left[\frac{a}{x^2} + \frac{1}{x}\right]$$

(*c*) $F_x(x = 2a)/F_x(x = a) = 0.138$
(*d*) $F_x(x = 5a)/F_x(x = a) = 0.00220$
89. (*a*)

$$\frac{dx}{\sqrt{E - U}} = \sqrt{\frac{2}{m}}\,dt$$

(*b*)

$$\frac{dx}{\sqrt{(A^2 - x^2)}} = \sqrt{\frac{k}{m}}\,dt$$

(*c*)

$$x = A\cos\left(\sqrt{\frac{k}{m}}\,t\right)$$

Chapter 7

True or False **1.** True **2.** False; it is true only in the center-of-mass reference frame **3.** True **4.** True **6.** True

Problems

1. 0.233 m
3. $X_{cm} = 6\frac{2}{3}$ m, $Y_{cm} = 6\frac{2}{3}$ m
5. $X_{cm} = 52.5$ cm, $Y_{cm} = 22.5$ cm
7. $X_{cm} = 2$ cm, $Y_{cm} = 1.4$ cm
9. 4 m/s to the east
11. (*a*) 11.8 m/s **i** (*b*) 32.9 m
13. 1.5 m/s, in the direction opposite the girl's motion
15. 0.192 m/s
17. (*a*) The ball moves in pure translational motion (*b*) The ball rotates about the center of mass
19. (*a*) 43.5 J (*b*) 1.5 m/s (*c*) $u_1 = 3.5$ m/s, $u_2 = -3.5$ m/s, taking motion to the right to be positive. (*d*) 36.75 J (*e*) $K_{tot} - K_{rel} = 6.75$ J. For the center of mass, $K_{cm} = \frac{1}{2}MV_{cm}^2 = 6.75$ J.
21. 0.652 m/s
23. 3.13 m/s
25. (*a*) $v_1 = v_2 = 3.75$ m/s (*b*) $v_{1f} = 2.5$ m/s, $v_{2f} = 4.5$ m/s
27. (*a*) 0.588 m/s (*b*) The collision is inelastic
29. (*a*) One-fifth of the mechanical energy is lost with each bounce. (*b*) 0.894
31. (*a*) The cue ball moves at an angle of 60° on the opposite side of the original direction from the 8 ball. (*b*) $v_c = 2.5$ m/s, $v_8 = 4.33$ m/s
33. (*a*) 10.75 kg·m/s (*b*) 1.34×10^3 N
35. (*a*) 6 kg·m/s (*b*) 4.62×10^3 N
37. 3290 N
39. 1.2×10^6 N
41. (*a*) 4.83 km/s (*b*) 6.91 km/s (*c*) 13.8 km/s
43. 450 m/s
45. (*a*) 212 m from the launch site (*b*) $K_i = 3600$ J, $K_f = \frac{1}{2}(2\text{ kg})(34.6\text{ m/s})^2 + \frac{1}{2}(4\text{ kg})(69.2\text{ m/s})^2 = 10{,}800$ J. The explosion released at least $K_f - K_i = 7200$ J of energy
47. $v_{1f} = -1.2$ m/s; $v_{2f} = 0.8$ m/s
49. (*a*) 1700 km below the earth's surface (*b*) The predominant external force is that exerted by the sun, though the other planets and every other body in the universe also act on the system. (*c*) To a very good approximation, the center of mass accelerates directly toward the sun. (*d*) 9340 km radially every 14 days
51. $r_{cm} = \frac{r}{14}$, where positive is taken as the direction away from the hole
53. (*a*) 1.22 m/s (*b*) 1.17 m/s
55. (*a*) 58.8 N (*b*) Yes. The weight of the ball is $w = 1.47$ N, only about 2.5 percent of the force exerted during the throw
57.

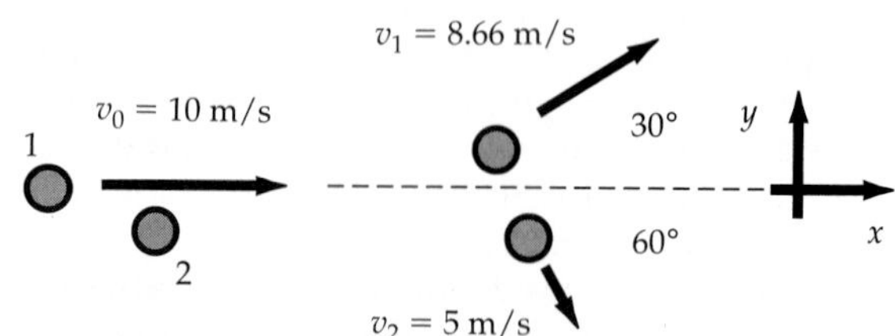

$v_1 = 8.66$ m/s; $v_2 = 5$ m/s
59. (*a*) 4.8 kg·m/s (*b*) 1600 N (*c*) 2.4 kg·m/s (*d*) 19.2 N
61. 36.4 m/s
63. $m_1^2 v^2/(8gm_2^2)$
65. (*a*) 3.6×10^5 N (*b*) 120 s (*c*) 1720 m/s
67. (*a*) 0.0165 N (*b*) The weight of a single water droplet is $mg = 0.0981$ N
69. 2252 m/s

71. $v_{1f} = v_{2f} = 0.693v_{ci}$, $v_{cf} = -0.2v_{ci}$. The cue ball returns along its original direction of motion; the other two balls move at 30° above and below the original motion of the cue ball.

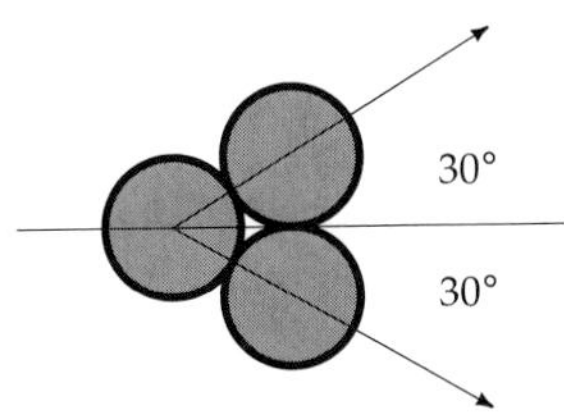

73.

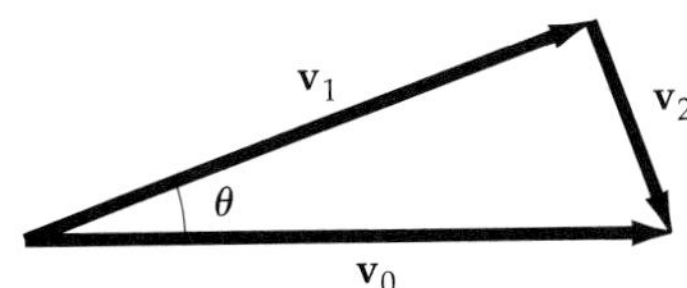

$E_2 = (\sin^2 \theta)\, E_0$

75. $(X_{cm}, Y_{cm}) = (6.67 \text{ m}, 3.33 \text{ m})$

77. (*a*) 9.66 km/s (*b*) 8.68 km/s (*c*) The rocket rises $\approx$ 400 km. Since this is only $\approx$ 7 percent of the earth's radius, the variation of g can be safely neglected.

Chapter 8

True or False **1.** False **2.** True **3.** True **4.** True **5.** False **6.** False; if the net torque is zero, the *rate of change* of the angular momentum is zero **7.** False; it is true for a rigid body.

Problems

1. (*a*) 0.2 rad/s (*b*) 0.955 rev
3. (*a*) -0.0291 rad/s^2 (*b*) 1.75 rad/s (*c*) $33\frac{1}{3}$ rev
5. 0.625 s
7. (*a*) 0.628 rad/s^2 (*b*) 6.28 rad/s (*c*) 11.3 m
9. (*a*) -7.33 rad/s^2 (*b*) -0.0359 N·m
11. (*a*) $\tau = FR = 1.5$ N·m, $\alpha = 100$ rad/s^2 (*b*) $\omega = 400$ rad/s
13. (*a*) 56 kg·m^2 (*b*) 112 J
15. (*a*) 0.136 J (*b*) 356 rev/min
17. 28 kg·m^2
19. (*a*) $I_x = 28$ kg·m^2 (*b*) $I_y = 28$ kg·m^2 (*c*) $I_z = 56$ kg·m^2
21. $\frac{1}{4}MR^2$
23. (*a*) 60 kg·m^2/s (*b*) 75 kg·m^2 (*c*) 0.8 rad/s
25. (*a*) L is doubled (*b*) L is doubled
27. (*a*) zero (*b*) $v_1/v_2 = r_2/r_1$
29. (*a*) K_{rot} is 28.6 percent of the total kinetic energy (*b*) K_{rot} is 33.3 percent of the total kinetic energy (*c*) K_{rot} is 50 percent of the total kinetic energy
31. (*a*) $\frac{5}{7}g \sin\theta$ (*b*) $(2/7)\, mg \sin\theta$ (*c*) $\tan\theta = (7/2)\mu_s$
33. $\mathbf{F} = -F\mathbf{i}$; $\mathbf{r} = R\mathbf{j}$; $\boldsymbol{\tau} = FR\mathbf{k}$
35. (*a*) 36**k** (*b*) -36**j** (*c*) 12**k**
37. (*a*) -25.8 kg·m^2/s **k** (*b*) 12.9 N·m **k**
39. (*a*) 20.8 kg·m^2/s (*b*) 0.520 rad/s (*c*) 12.1 s (*d*) 0.172 kg·m^2/s, upward (downward) when the angular momentum due to the spin of the wheel is away (toward) the pivot
41. (*a*) 2.5 N (*b*) 25 cm from the axis in a direction away from the center of the disk
43. (*a*) 2.4 N·m (*b*) 66.7 rad/s^2 (*c*) 200 rad/s (*d*) 720 J (*e*) 7.2 kg·m^2/s (*f*) 300 rad (*g*) The work done by the torque is $W = 720$ J
45. (*a*) 283 kg·m^2/s (*b*) 9.43 N·m (*c*) 9.43 N·m (*d*) 31.4 N
47. (*a*) 1.33×10^{-5} kg·m^2/s (*b*) 1.33×10^{-5} kg·m^2/s (*c*) 1.33×10^{-5} kg·m^2/s (*d*) $L_{cm} = 8.83 \times 10^{-5}$ kg·m^2/s or -6.18×10^{-5} kg·m^2/s, depending on the direction of $\mathbf{v}_{cm}$.
49. $K_{rot} = 2.60 \times 10^{29}$ J; $K_{orb} = 2.67 \times 10^{33}$ J
51. (*a*) 13.6 kg·m^2 (*b*) 7.14 N·m
53. 1.95 m
55. (*a*) 5.42 rad/s (*b*) $(5/2)mg$
57. $a = 0.2$ m/s^2; $v = at$; $x = x_0 + \frac{1}{2}at^2$; $\alpha = 3.33$ rad/s^2; $\omega = \alpha t$
59. (*a*)
$$a = \frac{g}{\left(1 + \dfrac{2}{5}\dfrac{M}{m}\right)}$$
(*b*) $T = \frac{2}{5}Ma$
61. (*a*) 2.7R (*b*) 2.5R
63. (*b*) $\mathbf{T} = \frac{1}{3}M\mathbf{g}$
65. (*a*) 5 rev/s (*b*) 592 J (*c*) The additional energy is supplied by the man's muscles
67. (*a*) 5.36 rad/s (*b*) 9.99 J
69. (*a*) 552 N (*b*) 1.77 rad/s^2 (*c*) 2.17 rad/s
71. 1.13×10^{-4} s
73. (*a*) 0.0948 m/s^2 (*b*) $T_1 = 4.952$ N; $T_2 = 4.955$ N; $T_2 - T_1 = 2.37 \times 10^{-3}$ N (*c*) If you neglect the motion of the pulley, that is, if you set $a = -0.0971$ m/s^2, $T_1 = T_2 = 4.95$ N
75. (*b*)
$$W = -\frac{L_0^2}{m}\int_{r_0}^{r_f} \frac{1}{r^3}\,dr = \frac{L_0^2}{2m}(r_f^{-2} - r_0^{-2})$$
(*c*) $v_f = L_0/mr_f$, $\Delta K = L_0^2(r_f^{-2} - r_0^{-2})/2m = W$ (*d*) 0.282 m
77. The day would lengthen by 0.554 s
79. (*a*) $I = 0.04$ kg·m^2 (*b*) $I = 0.0415$ kg·m^2, about 3.5 percent greater than the approximate result
81. (*a*) $s_1 = (12/49)v_0^2/\mu_k g$; $t_1 = (2/7)v_0/\mu_k g$; $v_1 = (5/7)v_0$ (*b*) $s_1 = 3.99$ m; $t_1 = 0.583$ s; $v_1 = 5.71$ m/s
83. (*a*) $v_{cm} = (L/2)\omega = (3/2)\, P_0x/ML$ (*b*) $P = P_0(3x/2L - 1)$
85. (*a*) 0.0443 m/s^2 (*b*) 2.21 rad/s^2 (*c*) 3.63 J (*d*) 399 J
87. (*a*) 200 m/s (*b*) 8000 rad/s (*c*) 257 m/s (*d*) 11.7 s
89.

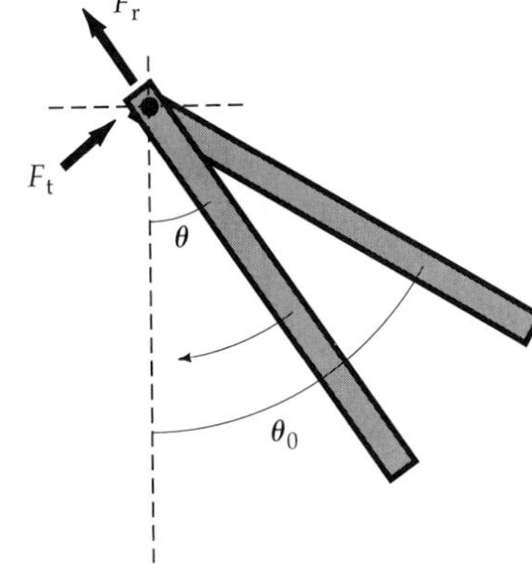

$F_r = \frac{1}{2}Mg(5 \cos \theta - 3 \cos \theta_0)$; $F_t = \frac{1}{4}Mg \sin \theta$
91. (*a*) $(11/7)v_0$ (*b*) $4v_0/(7\mu_k g)$ (*c*) $(36/49)v_0^2/\mu_k g$
93. 54.0°

Chapter 9

True or False **1.** False; the resultant torque must also be zero **2.** True **3.** True **4.** False **5.** False

Problems
1. 1.4 m away from the center
3. $F_B = 417$ N; $F_J = 270$ N
5. $F_1 = 117$ N; $F_2 = 333$ N
7. (*b*) Part (*a*) gives the maximum angle the leg may make with the vertical before the foot slips. (*c*) On ice μ_s is very small; thus the maximum θ is also small. This implies that short steps must be taken
9. The U-shaped figure has its center of gravity on the vertical line about which the shape is symmetric. The height of the center of gravity above the base of the "U" is $Y_{cg} = 4L/5$. For the L-shaped wire, choosing the origin at the bend in the "L" gives $X_{cg} = L/6$, $Y_{cg} = 2L/3$. For the triangular shape, the center of gravity is on the vertical line of symmetry, and $Y_{cg} = \sqrt{15}L/5$
11. 99.0 cm from the man's feet
13. Let F_1 and F_2 denote the forces exerted by the left and right supports, respectively. $F_1 = 245$ N; $F_2 = 736$ N
15. (*a*) 5400 N (*b*) 9
17. $\frac{1}{2}F(\sqrt{3}b - a)$
19. (*a*) $\tau = Fa$ (*b*) The effective point of application of the normal force is a distance $a/6$ in from the corner of the cube. (*c*) $\frac{1}{2}Mg$
21. $x_{cg} = 0.944a$, $y_{cg} = a$
23. Let F_1 and F_2 be the forces exerted on the cylinder by the left and right planes, respectively. $F_1 = 0.866W$; $F_2 = 0.5W$
25. (*a*) 42.4 N (*b*) 57.0 N, along a direction 7.13° above the strut
27. 26.6°
29. (*a*) 450 N (*b*) 241 N (*c*) The tension decreases as the brace is moved downward
31.

$$\mathbf{F} = \frac{M\mathbf{g}\sqrt{h(2R - h)}}{(R - h)}$$

33. 9 m
35. (*a*)

$$\mathbf{N} = M\mathbf{g} - \mathbf{F}\sqrt{\frac{(2R - h)}{h}}$$

(*b*) $-\mathbf{F}$
(*c*)

$$F_{cy} = F\sqrt{\frac{(2R - h)}{h}}$$

37. (*a*) 49.1 N (*b*) 73.6 N·m (*c*) Let F_1 and F_2 refer to the forces exerted on the end of the rod and 0.1 m in from the end, respectively. With upward taken as positive, $F_2 = 736$ N and $F_1 = -687$ N
39. F_1 and F_2 correspond to the forces exerted by the left and right supports, respectively. $F_1 = -1938$ N, a tension; $F_2 = 2919$ N, a compression
41. $a = g/3 = 3.27$ m/s^2
43. $x_{cg} = 0$, $y_{cg} = 2R/\pi$
45. 566 N
47. 61.6°
49. (*a*) $0.361mg$ (*b*) $0.313mg$ (*c*) $0.820mg$

Chapter 10

True or False **1.** False **2.** True **3.** True

Problems
1. 2.92 AU
3. 10^4 AU
5. 248 y
7. 2.63×10^6 m/s
9. (*a*) 1.16×10^6 s (*b*) 8.79×10^{25} kg
11. 6.02×10^{24} kg
13. 5.97×10^{24} kg
15. 10 times as much
17. (*a*) 4.80×10^{-9} N (*b*) 8.64×10^{-10} N·m
19. (*a*) 5.77 kg (*b*) gravitational
21. 5330 mi/h
23. (*a*) -6.25×10^9 J (*b*) -3.13×10^9 J (*c*) 7.92×10^3 m/s
25. (*a*) 0 (*b*) 0 (*c*) 3.20×10^{-9} m/s^2
27.

$$W = \frac{GMm_0}{R}$$

29. (*a*) $GM_1m/(3a)^2 + GM_2m/(2.2a)^2$ (*b*) $GM_1m/(1.1a)^2$ (*c*) 0
31. $\sqrt{3}v_e = 1.94 \times 10^4$ m/s
33. 51.2 km/s
35. $h = 0.147\ R_E = 935$ km
37. (*a*) 3.31 y (*b*) 1.49×10^{29} kg (*c*) Planet 2. Planet 2 also has the greater total energy. (*d*) $v_A = v_P/1.8$
39. (*b*) $\mathbf{g} = \mathbf{F}/m_0 = [-2Gmx/(x^2 + a^2)^{3/2}]\mathbf{i}$
41. 9.66×10^{-8} N
43. (*a*) 7.38 m (*b*) 0.0319 mm
45. (*a*) 2.74×10^{-3} N (*b*) 0.480 N, a value 175 times that due to the moon (*c*) -0.06 percent
47. 1.88 N
49.

$$\frac{GMm}{a(a + L)}$$

51. (*a*) $F_r = GMm_0/r$ for $r > R$ and $F_r = 0$ for $r < R$. (*b*) $U(r) = -GMm_0/r$ (*c*) Since $dU = -F_r\,dr = 0$ everywhere inside the shell, it follows that U is constant in value there. From the results of part (*c*), we conclude that $U(r) = -GMm_0/R$ for $r \le R$
(*d*)

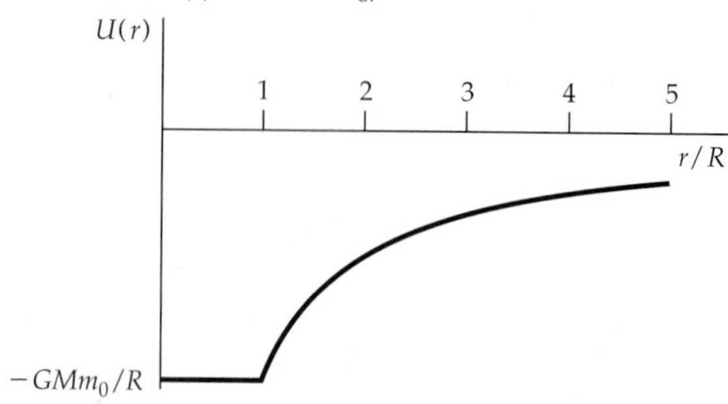

53. (*b*) $U(x) = -(GMm_0/L) \ln[(x + L/2)/(x - L/2)]$ for $x > \frac{1}{2}L$. (*c*) $F_x = -GMm_0/(x^2 - L^2/4)$
55.

$$g = F/m_0 = GM\left[\frac{1}{x^2} - \frac{1}{8(x - R/2)^2}\right]$$

57. $\omega = (GM/R^3)^{1/2} = (\frac{4}{3}G\pi\rho_0)^{1/2}$
59. (*a*) $F_s/F_m = (M_s/M_m)(r_m/r_s)^2$. Substituting the appropriate numerical values yields 169 for this ratio, showing that the sun exerts by far the greater force.
(*c*) Numerically, this is equal to 0.422, showing that the moon has the greater effect in causing the tides in earth's oceans
61. (*a*) $\mathbf{F} = -(GMm/d^2)[1 - \frac{1}{4}d^3/(d^2 + R^2/4)^{3/2}]\mathbf{i}$
(*b*) $\mathbf{F} = -0.821(GMm/R^2)\mathbf{i}$

Chapter 11

True or False 1. False 2. True 3. False; it can float because of surface tension 4. True

Problems
1. 0.676 kg
3. 103 kg
5. 54 kg
7. 0.976 mm
9. 5.01°
11. (*a*) 10.3 m (*b*) 0.757 m
13. $F = 6.46 \times 10^4$ N The table doesn't collapse because air pressure also exerts an upward force on the bottom surface of the card table
15. 230 N
17. (*a*) $F = 8.97 \times 10^4$ N (*b*) $F = 5.02 \times 10^4$ N
(*c*) If a window is rolled down the pressure on the two sides of the door will equalize, and it may then be opened
19. 1060 kg/m^3
21. 15.8 cm
23. 4.36 N
25. (*a*) 11.1×10^3 kg/m^3 (*b*) lead
27. 0.0233 N/m
29. (*a*) 4.58 L/min (*b*) 763 cm^2
31. (*a*) 12 m/s (*b*) 133 kPa (*c*) The flow rates are identical in the two sections
33. $F = 1.31 \times 10^5$ N = 29,400 lb
35. 1.43 mm
37. 1.49 ft
39. 183 m^3
41. $2[h(H - h)]^{1/2}$
43. 0.441 Pa
45. 3.89 kg
47. The readings on the upper and lower scales are 12.4 N and 36.7 N, respectively
49. 0.25×10^3 kg/m^3
51. (*a*) 21.2 kg (*b*) 636 N·s
(*c*)

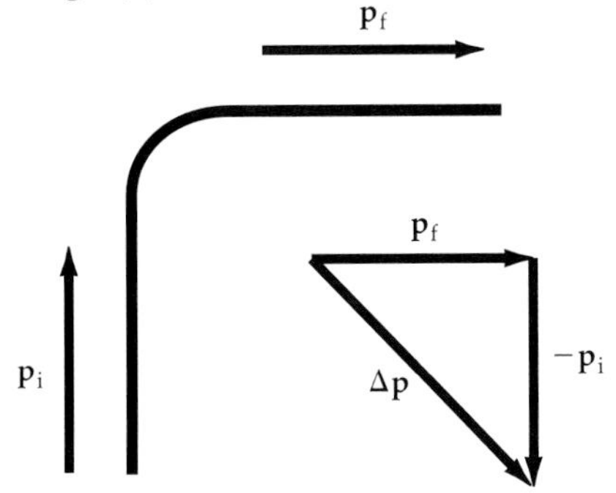

$\Delta p = 900$ N·s, $F = 900$ N

53. (*a*) $dF = (\rho g y)(w\,dy)$
(*b*)
$$F = \tfrac{1}{2}\rho g w d^2 = 9.20 \times 10^7 \text{ N}$$
(*c*) The reason atmospheric pressure may be neglected is that it acts equally on both sides of the dam and thus gives rise to no net force
55. (*a*) 34.7 cm (*b*) 7.76×10^{-2} J
57. (*a*) 14.7 g (*b*) 1.03×10^3 kg/m^3
59. (*b*) 1.12 mm
61. 5.02 h
63. (*c*)
$$t = \frac{1}{\sqrt{2g}}\frac{2A_1}{A_2}(\sqrt{H} - \sqrt{h})$$
or
$$h = \left(\frac{\sqrt{H} - \sqrt{g}\,A_2 t}{\sqrt{2}\,A_1}\right)^2$$
(*d*) $t = (2H/g)^{1/2}(A_1/A_2) = 1.77$ h

Chapter 12

True or False 1. False; it is independent of the amplitude 2. True 3. True 4. False; it is simple harmonic only for small angular displacements 5. True 6. True 7. True 8. True 9. True

Problems
1. (*a*) 7.96 Hz (*b*) 0.126 s (*c*) 0.1 m
(*d*) 5 m/s (*e*) 250 m/s^2 (*f*) $t = 0.0314$ s; $a = 0$
3. (*a*) 474 N/m (*b*) 0.5 s (*c*) 1.26 m/s
(*d*) 15.8 m/s^2
5. 0.801 Hz
7. (*a*) 1.49 kg (*b*) 5.84 Hz (*c*) 0.171 s
9. (*a*) 62.8 cm/s (*b*) 790 cm/s^2
11. (*a*) $f = 0.477$ Hz, $T = 2.09$ s (*b*) 0.283 m
(*c*) −0.262 m
13. (*a*) $x(t) = (0.25\text{ m}) \cos (4\pi/3)t$
(*b*) $v = -(\pi/3\text{ m/s}) \sin (4\pi/3)t$
(*c*) $a = -(4\pi^2/9\text{ m/s}^2) \cos (4\pi/3)t$
15. (*a*) 1.05 m/s (*b*) 4.39 m/s^2 (*c*) 1.13 s
17. (*a*) 0.314 m/s (*b*) $2\pi/3$ rad/s (*c*) $x = (0.15\text{ m}) \cos 2\pi t/3$
19. 1.92 J
21. (*a*) 3 cm (*b*) 77.5 cm/s
23. (*a*) 0.270 J (*b*) −0.736 J (*c*) 1.01 J
(*d*) 0.270 J
25. (*a*) 1.90 cm (*b*) 0.054 J (*c*) 0.223 J
(*d*) 0.277 J
27. 12.3 s
29. 11.7 s
31. 1.10 s
33. 0.504 kg·m^2
35. (*a*) 0.444 s (*b*) 0.180 J (*c*) $Q = 628$; $b = 0.045$ kg/s
37. (*a*) The energy decreases by 10 percent each cycle
(*b*) 30 s (*c*) 62.8
39. (*a*) 314 (*b*) 6 rad/s
41. 0.420 J

43. (*a*)

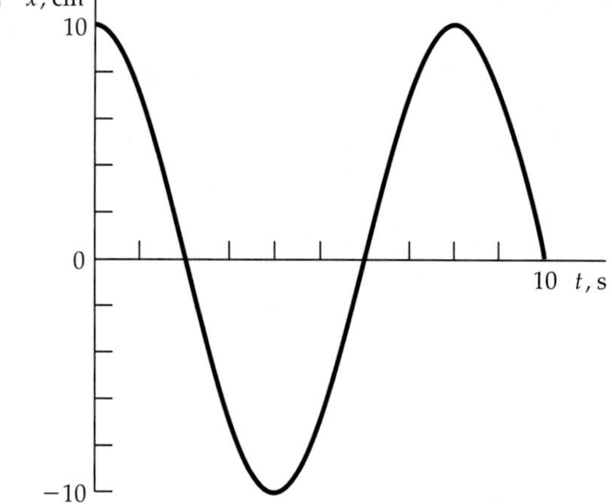

(*b*) 2.93 cm, 7.07 cm, 7.07 cm, 2.93 cm
45. 2.62 s
47. 12.9 Hz
49. (*a*) 70 J (*b*) 22 J (*c*) 7.33 W
51. (*a*) 0.314 (*b*) 0.000313 percent
53. (*a*) 1.57 percent (*b*) $E_n = (0.984)^n E_0$
(*c*) $(0.430)E_0$
55. (*a*) 7.85 cm (*b*) 3.92 cm (*c*) 0.397 s
(*d*) 62.0 cm/s; $t = 0.0993$ s
57. 9.10 Hz
59. (*a*) 5.23×10^4 N/m (*c*) 2×10^{11} N/m
61. (*a*) $A = 0.141$ m; $T = 0.444$ s (*b*) $A = 0.231$ m; $T =$ 0.363 s (*c*) For the inelastic collision, we have $x(t) = (0.141\text{ m}) \cos [(14.1\text{ rad/s})t - \pi/2]$; $I_2 = 4$ kg·m/s. For the elastic collision, $x(t) = (0.231\text{ m}) \cos [(17.3\text{ rad/s})t - \pi/2]$; $I_2 = 8$ kg·m/s.
63. (*a*) 1.55 cm (*b*) 1.72 m/s (*c*) 16.6 cm
65. (*a*) $F_r = -(GM_Em/R_E^3)r = -kr$, so $F_r \sin\theta = -kr \sin\theta = -kx$. The motion is simple harmonic, with a force constant $k = GM_Em/R_E^3$.
(*b*) $T = 2\pi(m/k)^{1/2} = 2\pi(R_E/g)^{1/2} = 84.4$ min
67. (*a*) 29.4 (*b*) 3.53 s (*c*) 6 J
69. (*c*) The error is 0.008 percent. For the error to be 1 percent the requirement is $r = 22.4$ cm
71. answer given in problem
73. answer given in problem

Chapter 13

True or False **1.** True **2.** False **3.** False; even though $v = f\lambda$, the speed depends only on the tension and mass density. When λ changes, f also changes so that v remains the same. **4.** True **5.** True

Problems

1.

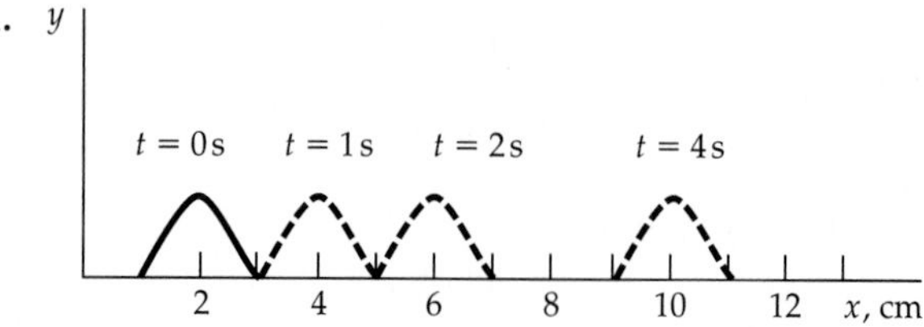

3.

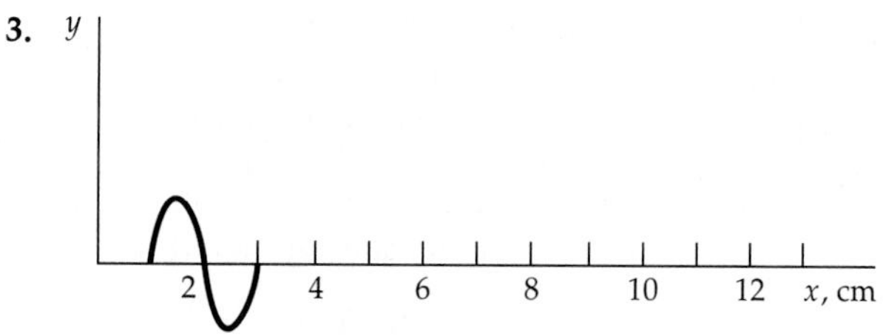

5.

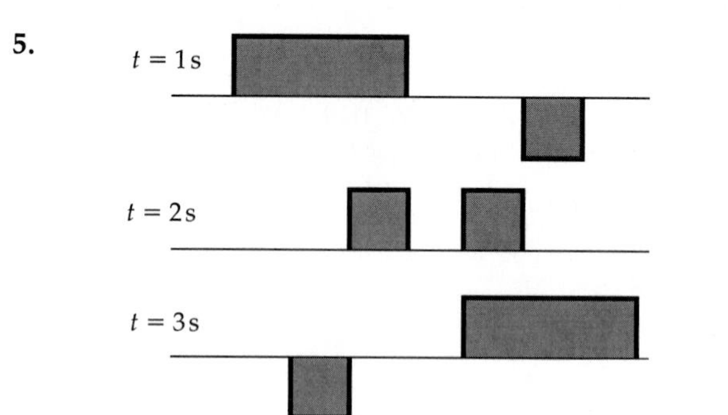

7. (*a*)

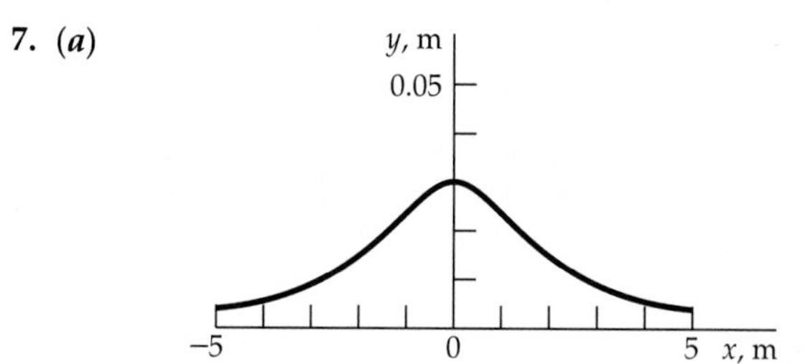

(*b*) $y(x, t) = (0.12\text{ m}^3)/[(2.00)^2 + (x - 10t)^2]\text{ m}^2$
(*c*) $y(x, t) = (0.12\text{ m}^3)/[(2.00)^2 + (x + 10t)^2]\text{ m}^2$
9. 251 m/s
11. (*a*) 20 m/s (*b*) 28.3 m/s (*c*) 14.1 m/s
13. 0.252 s
15. (*a*) 66.7 m/s (*b*) 889 N
17. (*a*) $y = y_0 \sin k(x - vt)$ (*b*) $y = y_0 \sin 2\pi(x/\lambda - ft)$
(*c*) $y = y_0 \sin 2\pi(x/\lambda - t/T)$ (*d*) $y = y_0 \sin (2\pi/\lambda)(x - vt)$ (*e*) $y = y_0 \sin 2\pi f(x/v - t)$
19. $y = (0.02\text{ m}) \sin 12\pi(x - 10t)$
21. $y = (0.5\text{ m}) \cos (2\pi/50)(x - 10t)$
23. 9.87 W
25. (*a*) 3.86 cm (*b*) 3.46 cm
27. $A_{\text{resul}} = 2A \cos \frac{1}{2}\delta = 2(0.05\text{ m}) \cos \pi/4 = 7.07$ cm
29. (*a*) $\lambda = 2$ m, $f = 25$ Hz (*b*) $y = (4 \times 10^{-3}\text{ m}) \cos 50\pi t \sin \pi x$
31. (*a*) 1.25 Hz (*b*) 3.75 Hz (*c*) 6.25 Hz
33. 141 Hz
35. (*a*) $\lambda = 31.4$ cm, $f = 47.7$ Hz (*b*) 1500 cm/s
(*c*) 62.8 cm

37. (*a*)

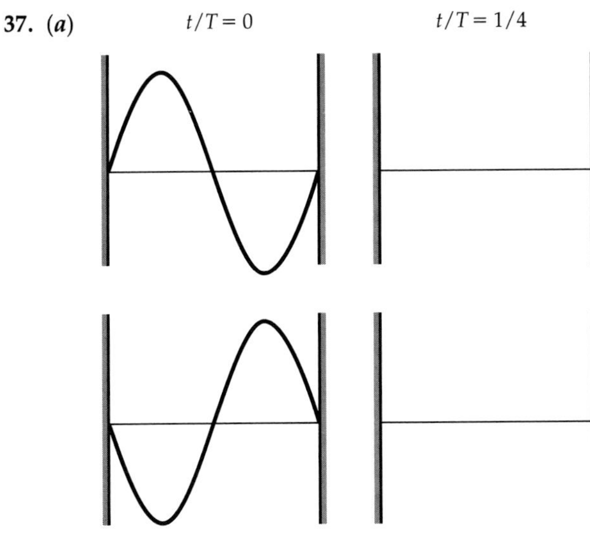

(*b*) 0.0126 s (*c*) The energy of the wave is kinetic
39. (*a*) 2.66 m (*b*) 0.666 m (*c*) 160 m/s
41. (*a*) 5/3, 7/5 (*b*) Since these are successive frequencies, it is clear that the even harmonics are missing, as is the case for a string fixed on one end only. (*c*) 25 Hz (*d*) The frequencies 75, 125, and 175 Hz are the third, fifth, and seventh harmonics, respectively. (*e*) 4 m
43. For $y(x, t) = A \sin kx \cos \omega t$, it is clear that $\partial^2 y/\partial t^2 = -\omega^2 y$ and $\partial^2 y/\partial x^2 = -k^2 y$. Therefore, $\partial^2 y/\partial x^2 = (k^2/\omega^2)\partial^2 y/\partial t^2 = (1/\omega^2)\partial^2 y/\partial t^2$, and thus $y(x,t)$ satisfies the wave equation
45. (*a*) 75 Hz (*b*) 375 Hz is the fifth harmonic, and 450 Hz is the sixth harmonic. (*c*) 2 m
47. (*a*) $f = 400$ Hz, $T = 2.5 \times 10^{-3}$ s (*b*) 316 m/s (*c*) $\lambda = 0.791$ m, $k = 7.95\ \text{m}^{-1}$ (*d*) $(0.5 \times 10^{-3}\ \text{m}) \sin 2\pi(x/\lambda - ft)$ (*e*) $v_{max} = 1.26$ m/s, $a_{max} = 3.16 \times 10^3\ \text{m/s}^2$ (*f*) 2.50 W
49. (*a*) $v = -(7.54\ \text{m/s}) \sin 2.36x \sin 377t$
(*b*) The maximum speed occurs at the point where $\sin 2.36x = 1$ or $x = 0.666$ m; $v_{max} = 7.54$ m/s
(*c*) $a = -(2.84 \times 10^3\ \text{m/s}^2) \sin 2.36x \cos 377t$ (*d*) The greatest acceleration also occurs at $x = 0.666$ m; $a_{max} = 2.84 \times 10^3\ \text{m/s}^2$
51. (*a*) 1.11 W (*b*) The amplitude must be increased by a factor of 10 or the frequency must be increased by a factor of 10, which implies that the tension must be increased by 10,000 times its original value. (*c*) The frequency of the power source is probably easiest to change
53. (*a*)

$$K = \frac{1}{2}\mu n^2 \omega_1^2 A^2 \sin^2 n\omega_1 t \int_0^L \sin^2 \frac{n\pi x}{L}\, dx$$

(*b*) $K = \frac{1}{4}\mu n^2 \omega_1^2 A_n^2 L$ (*c*) $y(x, t) = 0$ for all x (*d*) $K = (\frac{1}{2}\mu\omega_1^2 L)(n^2 A_n^2)$
55. (*a*) $df_n/f_n = \frac{1}{2}dF/F$ (*b*) A 1.54 percent increase in the tension is required
57. (*a*) $y(x, t) = (0.01\ \text{m}) \sin (2\pi x/3 - 160\pi t)$
(*b*) 80 Hz (*c*) 0.0125 s (*d*) 5.03 m/s
59. (*a*) 0.8 m (*b*) 480 N (*c*) The finger should be placed 9.23 cm from the end
61. (*a*) 60 cm (*b*) $2\pi/5$ (*c*) 24 m/s
63. (*a*)

$$v = -A_1\omega_1 \sin \omega_1 t \sin k_1 x - A_2\omega_2 \sin \omega_2 t \sin k_2 x$$

(*b*)

$$dK = \tfrac{1}{2}\mu v^2\, dx = \tfrac{1}{2}\mu[A_1^2\omega_1^2 \sin^2 \omega_1 t \sin^2 k_1 x + A_2\omega_2^2 \sin^2 \omega_2 t \sin^2 k_2 x + 2A_1A_2\omega_1\omega_2 \sin \omega_1 t \sin \omega_2 t \sin k_1 x \sin k_2 x]\, dx$$

(*c*) $\frac{1}{4}\mu L\omega_1[(n_1A_1)^2 + (n_2A_2)^2]$
65. (*a*)

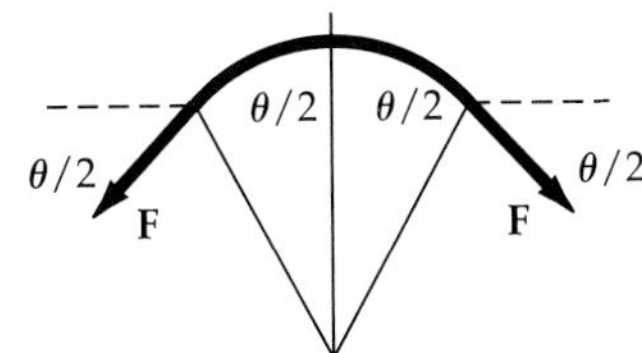

$2F \sin \dfrac{\theta}{2} \approx F\theta = mv_0^2/R = \mu R\theta v_0^2/R$, so $F = \mu v_0^2$. (*b*) v_0
(*c*) A forward-moving pulse always remains in contact with the ground (where it was generated), whereas a backward-moving pulse appears to make two revolutions for one rotation of the loop
67. (*b*) 2.21 s
69. (*a*) 4 m/s (*b*) 0.00424 m (*c*) $y(x, t) = (0.00424\ \text{m}) \cos (12.5x - 50t)$

Chapter 14

True or False **1.** False; sound waves are longitudinal **2.** False **3.** False; it has 1000 times the intensity **4.** False **5.** False **6.** True **7.** False **8.** True **9.** True **10.** True **11.** False

Problems
1. $[(B/\rho)^{1/2}] = [(\text{kg/m}\cdot\text{s}^2)/(\text{kg/m}^3)]^{1/2} = (\text{m/s})$
3. 5.09×10^3 m/s
5. $2.70 \times 10^{10}\ \text{N/m}^2$
7. (*a*) 1.30 m (*b*) 0.649 m
9. 8.27×10^{-2} Pa
11. (*a*) zero (*b*) 3.67×10^{-6} m
13. (*a*) 138 Pa (*b*) 21.6 W/m² (*c*) 0.216 W
15. (*a*) $10^{-11}\ \text{W/m}^2$ (*b*) $2 \times 10^{-12}\ \text{W/m}^2$
(*c*) For $\beta = 3$ dB, $p_0 = 4.18 \times 10^{-5}$ Pa; for $\beta = 10$ dB, $p_0 = 9.37 \times 10^{-5}$ Pa
17. 99/100 of the acoustic power must be eliminated to reduce β by 20 dB
19. (*a*) 90° (*b*) $\sqrt{2}A$
21. (*a*) 85 Hz; 255 Hz (*b*) Some sound will reflect off walls and other surfaces, thus giving a variety of path differences
23. (*a*) zero (*b*) $2I_0$ (*c*) $4I_0$
25. (*a*) $\lambda/4$ (*b*) The smallest magnitude of $r_2 - r_1$ occurs for $r_2 - r_1 = -\lambda/4$, and the smallest positive value is $r_2 - r_1 = 3\lambda/4$
27. 437 Hz
29. (*a*) 17 Hz (*b*) 8.5 Hz
31. (*a*) 2267 Hz (*b*) The eighth harmonic
33. (*a*) 3400 Hz, 10,200 Hz and 17,000 Hz
(*b*) We might expect the ear to be more sensitive near the above resonant frequencies, since these special

frequencies produce standing waves within the ear canal

35. (*a*) $N \approx \Delta t/T = f_0\,\Delta t$ (*b*) $\lambda \approx \Delta x/N$ (*c*) $k = 2\pi/\lambda \approx 2\pi N/\Delta x$ (*d*) N is uncertain because the waveform dies out gradually, rather than stopping abruptly at some time. (*e*) $\Delta k = (2\pi/\Delta x)\,\Delta N$

37. (*a*) $\lambda = 15$ m and $f = 20$ pies/min (*b*) $\lambda = 13.5$ m; $f = 22.2$ pies/min (*c*) $\lambda = 15$ m; $f = 22$ pies/min

39. (*a*) 1.3 m (*b*) 262 Hz

41. (*a*) 2.1 m (*b*) 162 Hz

43. (*a*) 80 m/s (*b*) 420 m/s (*c*) 1.7 m (*d*) 247 Hz

45. 529 Hz; 474 Hz

47. 3.42 m/s

49. (*a*) $I_1/I_2 = (M_1/M_2)^{1/2} = 4$ (*b*) $I_1/I_2 = s_{01}/s_{02} = \frac{1}{4}$ (*c*) $s_{01}/s_{02} = [\rho_2 v_2/\rho_1 v_1]^{1/2} = \frac{1}{2}$; $p_{01}/p_{02} = \rho_1 v_1 s_{01}/\rho_2 v_2 s_{02} = 2$

51. A reasonable time for Δt might be $\frac{1}{4}$ to $\frac{1}{2}$ s, giving a distance to the wall of 20 to 40 m

53. (*a*) At a node, the powder is not vibrated; hence, it collects there. (*b*) $v = \lambda f = 2Sf$ (*c*) $S = 5$ cm, $f = v/\lambda = 3400$ Hz (*d*) Suppose the tube is 1 m long. The lowest frequency in air is $f = 85$ Hz, the highest frequency is 8500 Hz. For the same tube, the appropriate frequencies in helium would be 35 to 3500 Hz

55. $v = 338$ m/s. This method lacks accuracy since the antinode generally does not occur exactly at the end of an open pipe

57. (*a*) $dv/dT = \frac{1}{2}(\gamma R/MT)^{1/2} = \frac{1}{2}(v/T)$ (*b*) 4.95 percent (*c*) $v = 347.4$ m/s. An exact calculation gives $v = (331 \text{ m/s})[(273 + 27)/273]^{1/2} = 347$ m/s, showing the approximate result to be quite accurate

59. (*a*) 0 dB (*b*) 66 dB (*c*) 63 dB

61. (*a*) $I_1 = 1.99 \times 10^{-5}$ W/m^2; $I_2 = 0.884 \times 10^{-5}$ W/m^2 (*b*) $[(I_1)^{1/2} + (I_2)^{1/2}]^2 = 5.53 \times 10^{-5}$ W/m^2 (*c*) $[(I_1)^{1/2} - (I_2)^{1/2}] = 0.221 \times 10^{-5}$ W/m^2 (*d*) $I_1 + I_2 = 2.87 \times 10^{-5}$ W/m^2

63. (*a*) 4.5 m/s (*b*) 22.1 beats/s (*c*) 22 beats/s

65. (*a*) 55.1 N/m^2 (*b*) 3.46 W/m^2 (*c*) 0.245 W

67. 0.773 m/s

69. (*a*) $r = 100$ m (*b*) 0.126 W

71. (*a*) 100 dB (*b*) 25.1 W (*c*) 2 m (*d*) 96.5 dB

73. 3.97×10^{-5} W

75. (*a*) 26 percent increase per year. This seems to be an unreasonably large increase. (*b*) 3.01 years

77. 87.8 dB

79. answer given in problem

81. (*a*) 0.279 m (*b*) 1.22×10^3 Hz (*c*) $\theta_3 = 24.7°$; $\theta_4 = 33.9°$; $\theta_5 = 44.2°$; $\theta_6 = 56.9°$; $\theta_7 = 77.6°$ (*d*) 4°

83. answer given in problem

85. (*a*) (80, 0) m; (75.2, ±27.2) m; (58.7, ±54.4) m (*b*) (78.8, ±13.6) m; (68.8, ±40.8) m

87. answer given in problem

Chapter 15

True or False **1.** False **2.** False **3.** False; the degree size is also different **4.** True **5.** False **6.** False; it could remain the same or even decrease depending on the change in the volume **7.** True **8.** True **9.** True

Problems

1. 10.4°F to 19.4°F

3. 56.7°C and −62.2°C

5. (*a*) 8.4 cm (*b*) 107°C

7. −320.44°F

9. (*a*) 54.9 torr (*b*) 3704 K

11. 30.0264 cm

13. 2.10 km

15. $V_2 = 1.16V_1$

17. 1.79 mol; 1.07×10^{24} molecules

19. (*a*) 3.66×10^3 moles (*b*) 59.9 moles

21. 152 J

23. $v_{rms} = 4.99 \times 10^5$ m/s; $K_{av} = \frac{3}{2}kT = 2.07 \times 10^{-16}$ J

25.

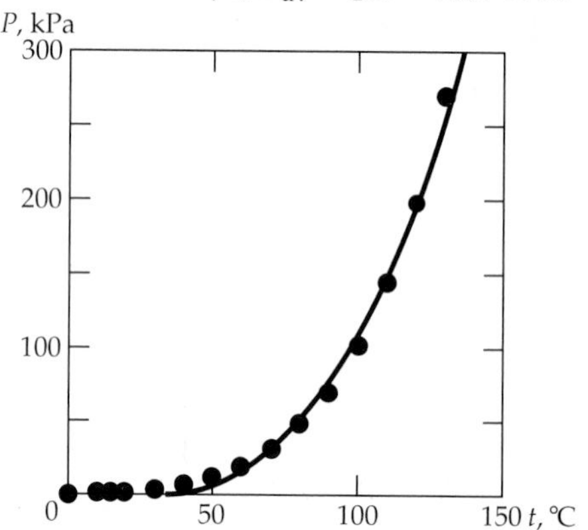

(*a*) 88.2°C (*b*) 80.4°C (*c*) 173 kPa

27. 3.39 kPa

29. 70.8

31. (*a*) 1846 m/s (*b*) 461 m/s (*c*) 393 m/s (*d*) One-sixth of the escape speed on Mars is 833 m/s; hence, O_2 and CO_2 should still be found on Mars, but not H_2

33. If $t_F = -40°$F, then $t_C = -40°$C.

35. (*a*) 74.98 cm (*b*) 75.02 cm

37. (*a*) $L_B' - L_A' = L_B - L_A$ if $\alpha_B L_B = \alpha_A L_A$ or $L_A/L_B = \alpha_B/\alpha_A$ (*b*) 182 cm

39. $V = 3.94 \times 10^{-23}$ cm^3; $r = 2.11$ Å

41. (*a*) 231 kPa (*b*) 201 kPa

43. 0.78 L

45. 375 K

47. $K_{av} = 6.21 \times 10^{-21}$ J. The change in gravitational potential energy is $mgh = 7.82 \times 10^{-26}$ J

49. (*a*) 4.7 m/s (*b*) 5.09 m/s

51.

$$\int_0^\infty f(v)\,dv = \frac{4}{\sqrt{\pi}}\left(\frac{m}{2kT}\right)^{3/2}\frac{\sqrt{\pi}}{4}\left(\frac{m}{2kT}\right)^{-3/2} = 1$$

53. $dP/dV = -nRT/(V - nb)^2 + 2an^2/V^3$ and $d^2P/dV^2 = 2nRT/(V - nb)^3 - 6an^2/V^4$. Setting these two derivatives to zero yields $V_c = 3nb$, which is $3b$ per mole

55. (*a*) $B = 3.95 \times 10^3$ K, $R_0 = 3.89 \times 10^{-3}$ ohms (*b*) 1320 ohms (*c*) For the ice point and the steam point, dR/dT is −393 ohms/K and −4.37 ohms/K, respectively. The thermistor is more sensitive at the ice point since the resistance changes more there for a given change in temperature

57. 1.34×10^5 N

59. 400.5 K

61. (*a*) 1.61×10^5 K (*b*) 1.01×10^4 K (*c*) For H_2, $f(v) \approx 10^{-9}$; for O_2, $f(v) \approx 10^{-106}$ (*d*) On the moon, the escape speed is $v_m = 2.38 \times 10^3$ m/s. O_2 and H_2 have an rms speed equal to v_m when the temperature is 7.28×10^3 K and 455 K, respectively

Chapter 16

True or False **1.** True **2.** False **3.** True **4.** False; it is proportional to the fourth power of T **5.** False **6.** False **7.** True **8.** True **9.** False **10.** True

Problems

1. (*a*) 1.046×10^7 J (*b*) 121 W
3. 0.0924 kcal/kg·K
5. (*a*) 0°C (*b*) 125 g
7. 99.8 g
9. 2073 Btu/h
11. (*a*) $I_{Al} = 569$ W, $I_{Cu} = 962$ W (*b*) $I_{total} = I_{Al} + I_{Cu} = 1531$ W (*c*) 0.0522 K/W
13. 9470 nm
15. 874 kJ
17. (*a*) 0.117 K (*b*) 1.74 K
19. (*a*)

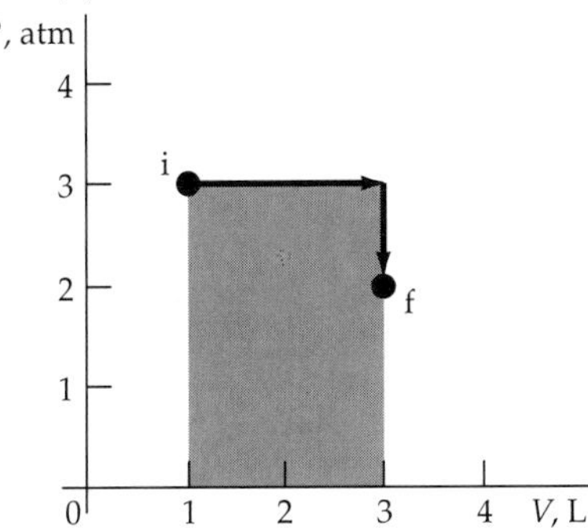

$W = 608$ J (*b*) 254 cal
21. (*a*)

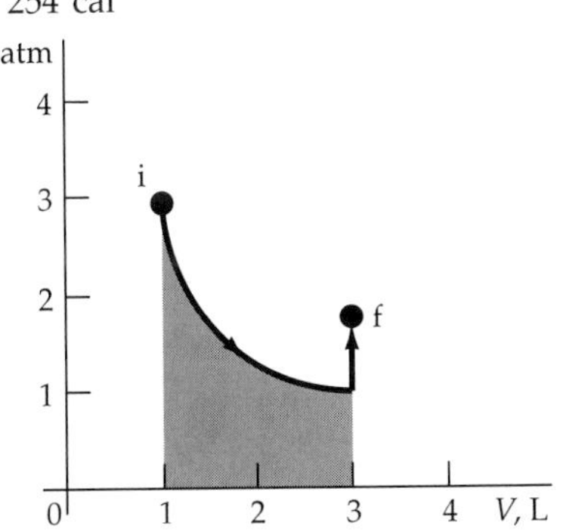

$W = 3.30$ L·atm (*b*) 189 cal
23. (*a*) 1.57 kJ (*b*) 1.57 kJ
25. (*a*) 0.713 J/g·K (*b*) 252 J/L
27. (*a*) $\Delta U = 6.24$ kJ; $W = 0$; $Q = 6.24$ kJ (*b*) $\Delta U = 6.24$ kJ; $Q = 8.73$ kJ; $W = 2.49$ kJ (*c*) $W = 2.49$ kJ
29. (*a*) $V_i = 2.24$ L; $V_f = 5.89$ L (*b*) $T_f = 143$ K (*c*) 1.62 kJ
31. (*a*) 28.5°C (*b*) 15.5°C
33. (*a*) 0.0920 J/kg·K (*b*) 0.0584 J/kg
35. (*a*) 34.0 km (*b*) yes (*c*) The potential energy released is the same with or without air resistance, so there should be little net effect
37. (*a*) 2.99°C (*b*) 199.85 (*c*) no
39. 18
41. 101.3°C
43. 47.8 cm
45. 96.6 g
47. 0.127 g
49. (*a*) For an ideal gas, U depends only on T; hence, for a given ΔT there is a unique ΔU. (*b*) $\Delta U = Q - W = (C_p - nR)\,\Delta T = C_v\,\Delta T$
51. (*a*) 3.40 kJ (*b*) $U_f = 3.70$ kJ, $W = 200$ J (*c*) $U_f = 3.90$ kJ, $W = 0$
53. (*a*) 263 K (*b*) 10.8 L (*c*) 1.48 kJ (*d*) −1.48 kJ
55. (*a*)

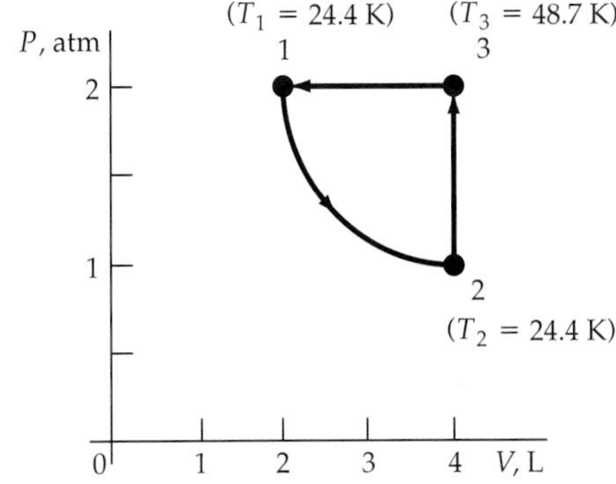

(*b*) For 1 to 2, $W = 2.77$ L·atm; $Q = 2.77$ L·atm. For 2 to 3, $W = 0$; $Q = 6.0$ L·atm. For 3 to 1, $W = -4$ L·atm; $Q = -10.0$ L·atm. (*c*) $T_1 = 24.4$ K; $T_2 = T_1$; $T_3 = 48.7$ K
57. 626°C = 1159°F
59. 577 J
61. (*a*) $\frac{1}{2}P_0$ (*b*) diatomic (*c*) No change occurs during the isothermal expansion; during the adiabatic compression, the kinetic energy increases
63. answer given in problem
65. answer given in problem
67. (*a*)

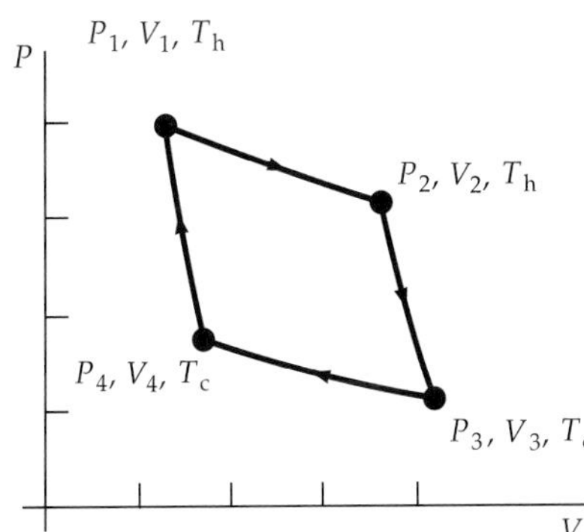

Chapter 17

True or False **1.** False; it can be in a noncyclic process **2.** False **3.** False **4.** False **5.** False **6.** True **7.** False **8.** True

Problems

1. (*a*) 500 J (*b*) 400 J
3. (*a*) 0.4 (*b*) 80 W

5. (*a*) 1.67 (*b*) 0.375
7. answer given in problem
9. answer given in problem
11. (*a*) $\frac{1}{3}$ (*b*) 33.3 J (*c*) 66.7 J (*d*) 2
13. (*a*) 13.7 (*b*) 8.77
15. (*a*) 0.51 (*b*) 102 kJ (*c*) 98 kJ
17. (*a*) 5.26 (*b*) 3.19 kW (*c*) 5.32 kW
19. (*a*) 303 kJ (*b*) 212 kJ
21. (*a*) 11.5 J/K (*b*) zero
23. (*a*) 50 J (*b*) 0.167 J/K (*c*) zero (*d*) The answers to parts (*a*) and (*b*) stay the same; the answer to part (*c*) is $\Delta S_u > 0$
25. (*a*) 11.5 J/K (*b*) 11.5 J/K
27. 0.417 J/K
29. 6.05 kJ/K
31. (*a*) Process 1 is more wasteful. (*b*) $\Delta S_1 =$ 1.67 J/K, $\Delta S_2 = 0.833$ J/K
33.

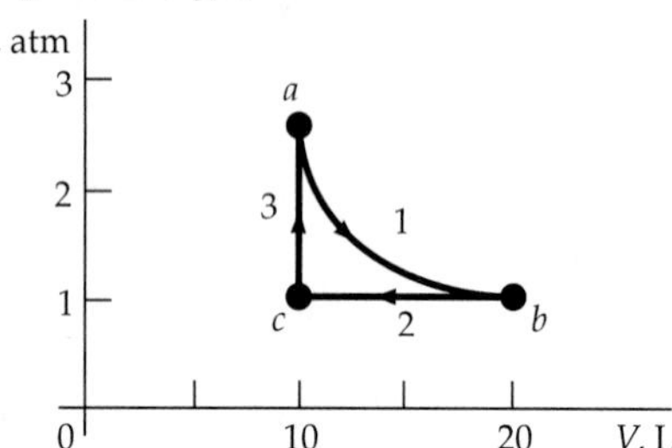

$\epsilon = W/Q_3 = 0.146$
35. (*a*) The maximum possible efficiency is 40.5 percent. (*b*) 1.68×10^9 J
37. (*a*)

(*b*) $V_3 = 2.30$ L; $T_3 = 345$ K (*c*) 5.15 L·atm (*d*) 0.232
39. (*a*) 10.1°C (*b*) 22 J/K
41. 10.7 J/K
43. 1.97×10^3 J/K
45. (*a*) 0.2 (*b*) $\Delta S_{\text{engine}} = 0$; $\Delta S_h = -2.5$ J/K; $\Delta S_c =$ 4.0 J/K; $\Delta S_u = 1.5$ J/K (*c*) 0.5 (*d*) 500 J
47. 93.9 W
49. (*a*) The heat input is $Q_h = C_v\,(T_c - T_b)$, and the magnitude of the heat output is $|Q_c| = C_v(T_d - T_a)/(T_c - T_b)$. (*c*) 0.565 (*d*) In a real engine, the processes are not truly adiabatic, nor are they quasi-static
51. $\epsilon = (T_h - T_c)/\{T_h + c_v(T_h - T_c)/[R \ln (V_a/V_b)]\}$
53. answer given in problem
55. (*a*) $T_1 = 301$ K; $T_2 = T_3 = 601$ K (*b*) $Q_{12} = 3.75$ kJ; $Q_{23} = 3.47$ kJ; $Q_{31} = -6.25$ kJ (*c*) 0.134
57. answer given in problem
59. answer given in problem

Chapter 18

True or False **1.** False; it points toward a negative charge **2.** True (except for the charges on quarks which are $e/3$ or $2e/3$, but isolated quarks have not been found) **3.** False; they diverge from positive point charges **4.** True **5.** True

Problems

1. 5×10^{12} electrons
3. 4.82×10^7 C
5. (*a*)

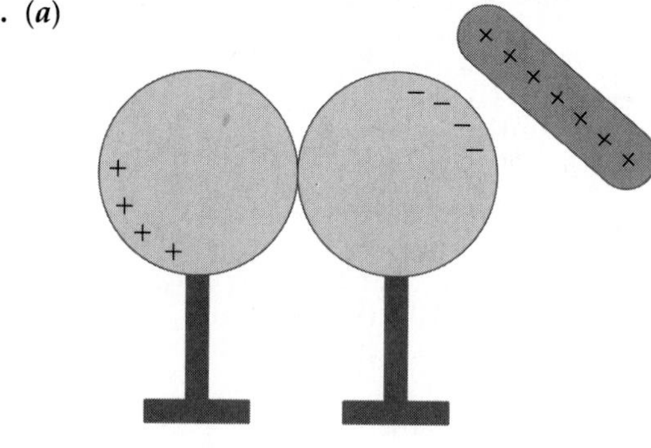

(*b*)

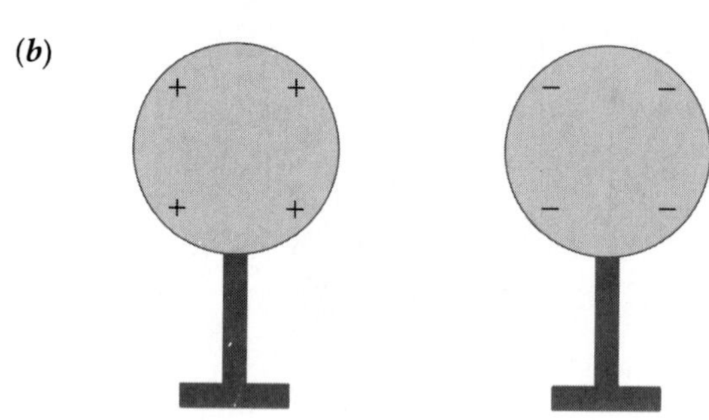

7. 1.50×10^{-2} N **i**
9. 2.96×10^{-5} N along the diagonal, away from the −3-nC charge
11. (*a*) 999 N/C **i** (*b*) −360 N/C **i**

(*c*)

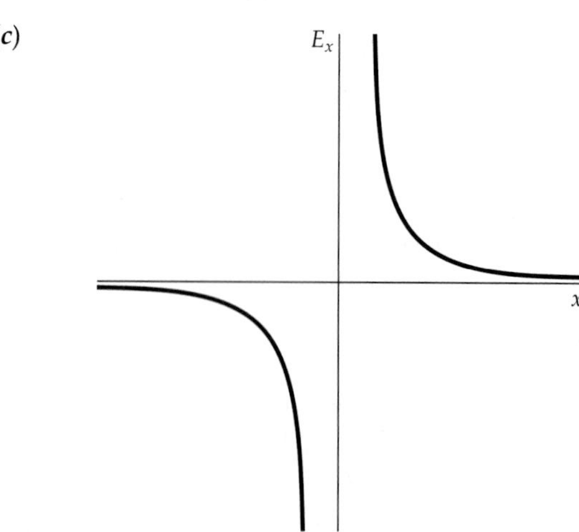

13. (*a*) 3.45×10^4 N/C **i** (*b*) 6.90×10^{-5} N **i**
15. 8.18×10^5 N/C, upward

17. (*a*) The particle on the left has the greater charge by a factor of 4 (*b*) The particles on the left and right are positive and negative, respectively (*c*) The field is strong above and below the particle on the left; the field is weak to the right and to the left of the two particles

19.

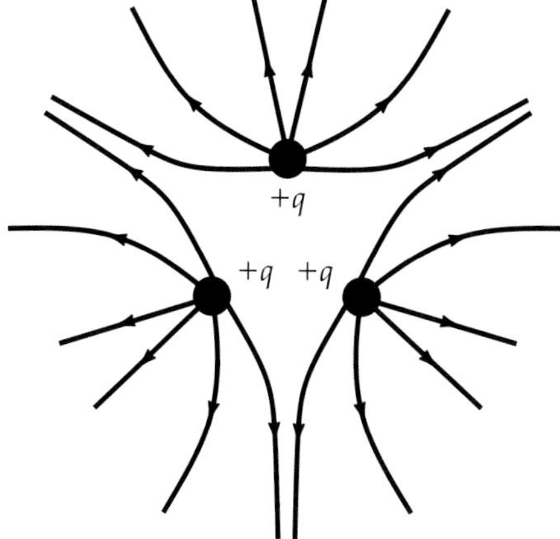

21. (*a*) 1.76×10^{11} C/kg (*b*) 1.76×10^{13} m/s^2, in the direction opposite **E** (*c*) 0.171 μs (*d*) 25.6 cm
23. (*a*) 7.03×10^{13} m/s^2 (*b*) 5×10^{-8} s (*c*) 8.78 cm in the negative y direction
25. (*a*) 8×10^{-18} C·m

(*b*)

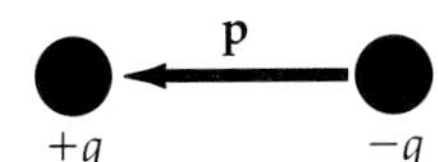

27. (*a*) 3.3×10^{-7} percent (*b*) 32.4 N
29. (*a*) $E = 1.90 \times 10^3$ N/C, $\theta = 235°$ (*b*) $F = 3.04 \times 10^{-16}$ N, $\theta = 235°$
31. (*a*) 3.21×10^3 N/C (*b*) -5.88×10^6 N/C
33. (*a*) 6.4 mm below the tube axis (*b*) 17.7° below the tube axis (*c*) 4.48 cm below the tube axis
35. (*a*) 4 μC and 2 μC (*b*) +7.12 μC and −1.12 μC
37. (*c*) For large values of x, the system is essentially the same as one with a charge $2q$ located at the origin
39. $\mathbf{E} = -\dfrac{2kqa}{y(y^2 + a^2)}\mathbf{i} \rightarrow -\dfrac{kp}{y^3}\mathbf{i}$
41. (*b*) 0.241 μC
43. (*a*) The equilibrium is unstable to displacements along the x axis and stable to displacements along the y axis (*b*) The equilibrium is unstable to displacements along the y axis and stable to displacements along the x axis (*c*) $-q/4$ (*d*) If the $+q$ charges are fixed in place, the system is stable to displacements along the y axis, as in part (*b*). If all three charges are free to move, the system is unstable to any displacement
45. (*a*) For $+q$, $\mathbf{F} = (q)C(x_1 + a)\mathbf{i}$, for $-q$, $\mathbf{F} = (-q)C(x_1 - a)\mathbf{i}$
47. (*a*) $\mathbf{E} = (-3kqa^2/x^4)\mathbf{i}$ (*b*) $\mathbf{E} = (6kq/y^4)\mathbf{j}$

Chapter 19

True or False 1. False; the flux through the surface must be zero, but **E** need not be zero everywhere 2. False; it holds for any charge distribution, but is useful for finding **E** only for symmetric distributions 3. True 4. True 5. False 6. False; it can be positive in some regions and negative in others 7. False; for example, **E** is continuous at the edge of a spherical volume charge. **E** is discontinuous at points where there is a surface charge density σ 8. True

Problems
1. (*a*) 17.5 nC (*b*) 26.2 N/C (*c*) 4.37 N/C (*d*) 2.57×10^{-3} N/C (*e*) For a point charge $E_x = 2.52 \times 10^{-3}$ N/C, about 2 percent lower than the correct result for a line charge
3. (*a*) 4.69×10^5 N/C (*b*) 1.13×10^6 N/C (*c*) 1.54×10^3 N/C (*d*) 1.55×10^3 N/C, about 0.07 percent greater than for the ring of charge
5. (*a*) 2.00×10^5 N/C (*b*) 2.54 N/C
7. $a/(3)^{1/2}$
9. (*a*) (0.804)(2πkσ) (*b*) (0.553)(2πkσ) (*c*) (0.427)(2πkσ) (*d*) (0.293)(2πkσ) (*e*) (0.106)(2πkσ) (*f*)

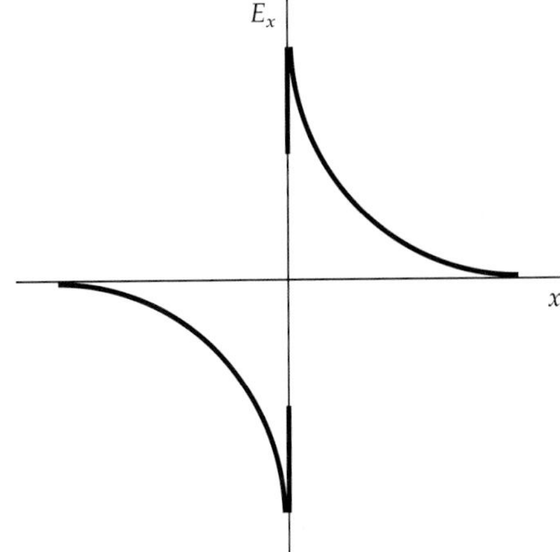

11. (*a*)

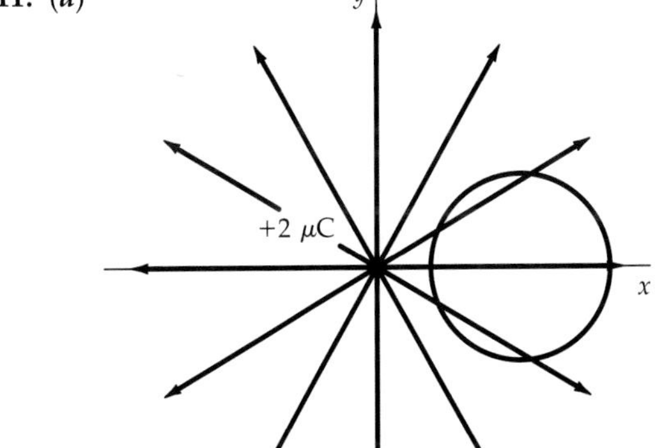

There are lines entering and leaving the surface. (*b*) zero (*c*) zero
13. (*a*) N (*b*) $N/6$ (*c*) q/ϵ_0 (*d*) $q/6\epsilon_0$ (*e*) Parts (*b*) and (*d*) would change
15. (*a*) 3.14 m^2 (*b*) 7.19×10^4 N/C (*c*) 2.26×10^5 N·m^2/C (*d*) No (*e*) 2.26×10^5 N·m^2/C
17. (*a*) 0.407 nC (*b*) 0 (*c*) 0 (*d*) 984 N/C (*e*) 366 N/C

19. (*a*) $Q = 40.7$ nC (*b*) $E_r = 0$ (*c*) $E_r = 0$ (*d*) $E_r =$ 999 N/C (*e*) $E_r = 610$ N/C
21. (*a*) $E = 0$ for $r < R_1$, $E = kq_1/r^2$ for $R_1 < r < R_2$, $E = k(q_1 + q_2)/r^2$ for $r > R_2$ (*b*) $|q_1/q_2| = 1$, and the signs of q_1 and q_2 are opposite (*c*) The electric-field lines corresponding to part (*b*) for $q_1 > 0$ are

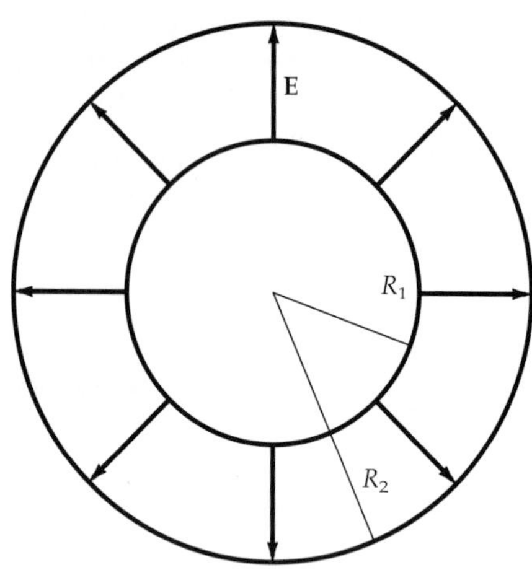

23. 1.15×10^5 N/C
25. 9.41×10^3 N/C
27. (*a*) For $r < a$, $E = kq/r^2$; for $a < r < b$, $E = 0$; for $r > b$, $E = kq/r^2$ (*b*) The electric-field lines are

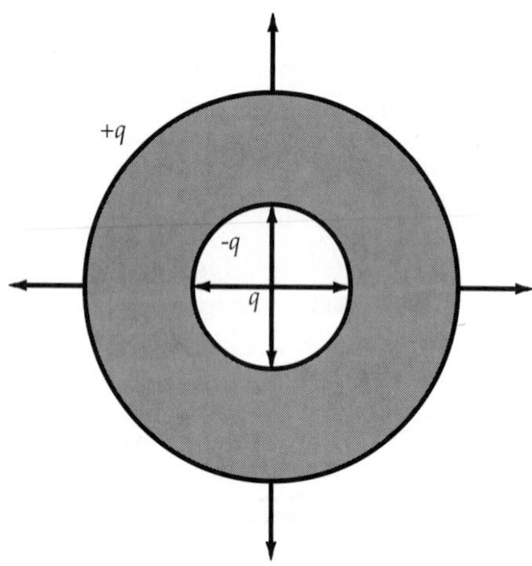

(*c*) On the inner surface, $\sigma = -q/4\pi a^2$; on the outer surface, $\sigma = q/4\pi b^2$
29. -1.18×10^{-12} C/m^3
31. (*a*) $\mathbf{E} = 9.41 \times 10^4$ N/C $\mathbf{i}$ (*b*) $\mathbf{E} = 3.36 \times 10^4$ N/C $\mathbf{j}$ (*c*) $\mathbf{E} = (1.56 \times 10^4 \text{ N/C})(2\mathbf{i} - 3\mathbf{j})/\sqrt{13}$
33. (*a*) At the center of the electron sphere of charge (*b*) The equilibrium position is a distance $d = E_0R^3/kZe$ from the center of the electron sphere (*c*) E_0R^3/k
35. (*a*) For the inner surface the induced charge is -2.5 μC, and the surface charge density is -0.553 μC/m^2; for the outer surface, the induced charge is 2.5 μC, and the surface charge density is 0.246 μC/m^2 (*b*) For $r < r_1$, $E = kq_1/r^2$; for $r_1 < r < r_2$, $E = 0$; for $r > r_2$, $E = kq_1/r^2$ (*c*) The results for the inner surface are unchanged. For the outer surface, the total charge is 6 μC, and the surface charge density is 0.589 μC/m^2; for $r < r_1$, $E = kq_1/r^2$; for $r_1 < r < r_2$, $E = 0$; for $r > r_2$, $E = k(q_1 + q_2)/r^2$
37. answer given in problem
39. (*a*) $E = 2.04 \times 10^5$ N/C, $\theta = 56.3°$ counterclockwise from the positive x axis (*b*) $E = 2.63 \times 10^5$ N/C, $\theta =$ 153° counterclockwise from the positive x axis
41. Total charge $q = \rho[\frac{4}{3}\pi(b^3 - a^3)]$; for $r < a$, $E = 0$; for $a < r < b$, $E = k(\frac{4}{3}\pi\rho)(r^3 - a^3)/r^2$; for $r > b$, $E = kq/r^2 = k(\frac{4}{3}\pi\rho)(b^3 - a^3)/r^2$
43. (*a*) $q_2/q_1 = r_2/r_1$; the element s_1 gives the greater field (*b*) Each element produces a field pointing away from it, along a line from its center to the point P; the total field points away from s_1 (*c*) 0 (*d*) $q_2/q_1 = r_2^2/r_1^2$; each element produces a field of the same magnitude and pointing away; the total field is zero; for $E \propto 1/r$, the total field would point away from s_2
45. $E_x = -k\lambda/y$, $E_y = k\lambda/y$
47. For $r < a$, $E = 0$; for $a < r < b$, $E = 2\pi\rho k(r^2 - a^2)/r$; for $r > b$, $E = 2\pi\rho k(b^2 - a^2)/r$
49. (*a*) $Q = 2\pi BR^2$ (*b*) for $r < R$, $E_r = 2\pi kB$; for $r \geq R$, $E_r = 2\pi kBR^2/r^2 = kQ/r^2$

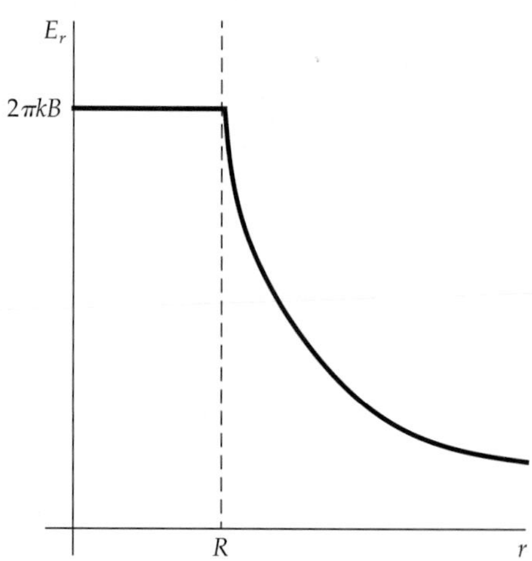

51. $F = kQq/[R(R + d)]$
53. $E = kQx/[(x^2 + L^2/4)(x^2 + L^2/2)^{1/2}]$, where $Q = 4L\lambda$ is the total charge on the square; for a ring of radius $r = L/2$, $E = kQx/(x^2 + L^2/4)^{3/2}$
55. (*b*) Half of the field just outside a conductor is due to the charge in the area ΔA, and the other half is due to all other charges; it is only this latter half that contributes to the force (*c*) 14.3 N/m^2

Chapter 20

True or False **1.** False; if $\mathbf{E} = 0$ in some region, V is constant in that region, but not necessarily zero. **2.** True; if $V =$ constant, $-dV/dx = 0$ **3.** False; $\mathbf{E}$ depends on the spatial rate of change of V, not on its value at any point **4.** True **5.** True **6.** True **7.** False; dielectric breakdown depends on the magnitude of the electric field E, not on the magnitude of the potential. It occurs in air when $E = 3$ MV/m

Problems

1. (*a*) 2.4×10^{-2} J (*b*) -2.4×10^{-2} J (*c*) -8000 V (*d*) $(-2$ kV/m$)x$ (*e*) 4000 V $-$ (2 kV/m)x (*f*) 2000 V $-$ (2 kV/m)x
3. (*a*) Positive (*b*) 25,000 V/m
5. (*a*) N/C·m = kg/C·s^2 = V/m^2 (*b*) $q_0ax^2/2$ (*c*) $V(x) = -ax^2/2$
7. (*a*) 1.29×10^4 V (*b*) 7.55×10^3 V (*c*) 4.44×10^3 V
9. (*a*) 2.68×10^5 V (*b*) 1.91×10^5 V
11. (*a*) 0.0487 J (*b*) 0 J (*c*) -0.0232 J (*d*) -0.0127 J
13. (*a*) 0.190 J (*b*) -0.0634 J (*c*) -0.0634 J
15. (*a*) Just outside the shell, $E = 6.24 \times 10^3$ V/m; just inside the shell, $E = 0$ (*b*) $V = 749$ V, both just inside and just outside the shell (*c*) $V = 749$ V, $E = 0$
17. (*a*) 6.02×10^3 V (*b*) -1.27×10^4 V (*c*) -4.23×10^4 V
19. (*a*) At $x = 3$ m, $V(x) = 8.99 \times 10^3$ V; at $x = 3.01$ m, $V(x) = 8.96 \times 10^3$ V (*b*) The potential decreases as x increases; $-\Delta V/\Delta x = 2.97 \times 10^3$ V/m (*c*) $E = 2.997 \times 10^3$ V/m (*d*) At $x = 3$ m, $y = 0.01$ m, $V = 8.99 \times 10^3$ V; V is nearly the same for the two points in question because they are approximately on an equipotential surface
21. (*a*) -3000 V/m (*b*) -3000 V/m (*c*) 3000 V/m (*d*) Zero
23. 0.506 mm
25. (*a*) ± 8.54 μC (*b*) $\pm 4.80 \times 10^5$ V
27. 26.6 μC/m^2
29. 250 W
31. (*a*) $E_x = 2\sqrt{2}kq/a^2$, $E_y = 0$ (*b*) $3\sqrt{2}kq/a$ (*c*) $v = q(6\sqrt{2}k/ma)^{1/2}$
33. $V(x = 2\text{ m}) - V(x = 1\text{ m}) = -7500$ V
35. (*a*) 3.10×10^7 m/s (*b*) 2.5×10^6 V/m
37. (*a*) 234 MeV (*b*) 2.67×10^{16} fissions per second
39. (*a*) 30,000 eV (*b*) 4.8×10^{-15} J (*c*) 1.03×10^8 m/s
41. $kq(1/a - 1/b)$
43. $V_a - V_b = (2kq/L) \ln (b/a)$
45. (*a*)

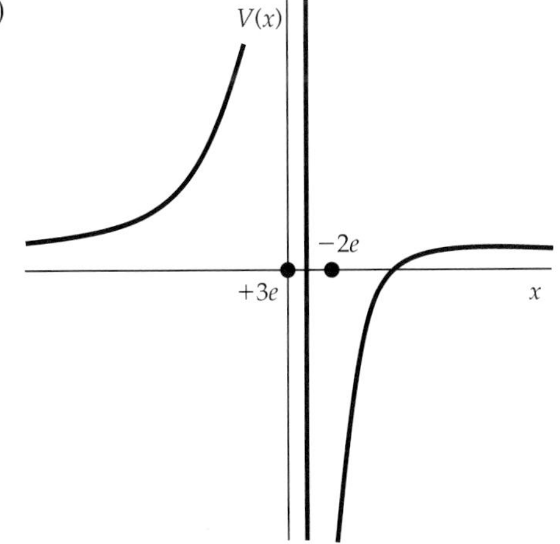

(*b*) $x = 3a/5$, $x = 3a$, $x = +\infty$, $x = -\infty$ (*c*) $2ke^2/a$
47. (*a*) $V(x) = kq(1/|x| - 3/|x - 1|)$ (*b*) $x = -0.5$ m, $x = 0.25$ m, $x = +\infty$, $x = -\infty$ (*c*) At $x = -0.5$ m, $E_x = -8kq/3$; at $x = 0.25$ m, $E_x = 64kq/3$; $E_x = 0$ at $x = \pm\infty$

(*d*)

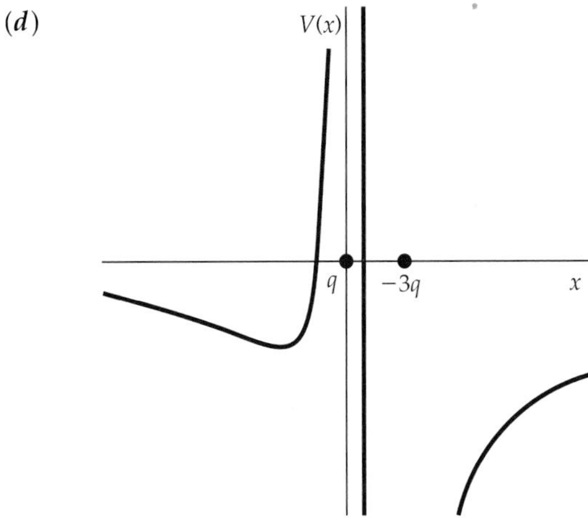

49. 1.45×10^{-7} J $= 9.03 \times 10^{11}$ eV
51. (*a*) $kQ^2(4 + \sqrt{2})/2L$ (*b*) $kQ^2(2 + \sqrt{2})/2L$ (*c*) kQ^2/L (*d*) 0
53. $\sigma_t = 9$ μC/m^2, $\sigma_b = 3$ μC/m^2
55. (*a*) $V(x) = kQ/(x^2 + a^2)^{1/2} + kQ'/|x - 2a|$ (*b*) For $x < 2a$, $E_x = kQx/(x^2 + a^2)^{3/2} - kQ'/(x - 2a)^2$; for $x > 2a$, $E_x = kQx/(x^2 + a^2)^{3/2} + kQ'/(x - 2a)^2$
57. $E_x = -8$ V/m, $E_y = -2$ V/m, $E_z = -1$ V/m
59. (*a*) $V(r) = kQ/r$ (*b*) $V(r) = (kQ/2R)(3 - r^2/R^2)$ (*c*) $3kQ/2R$ (*d*)

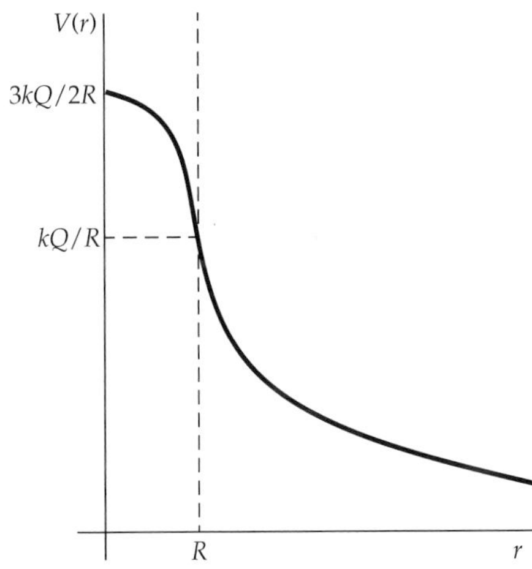

61. (*b*) $E_x = 3kpzx/r^5$, $E_y = 3kpzy/r^5$, $E_z = -kp/r^3 + 3kpz^2/r^5$
63. (*a*) $V(a) = kQ(1/b - 1/c) = V(b)$, $V(c) = 0$ (*b*) $V(a) = V(c) = 0$, $V(b) = -kQ_a(b - a)/ba = kQ(c - b)(b - a)/[(c - a)b^2]$; $Q_a = -Q(a/b)(c - b)/(c - a)$, $Q_c = -Q(c/b)(b - a)/(c - a)$; $Q_b = Q$
65. (*a*) $\mathbf{v}(x) = -[(kQ^2/2m)(1/x - 1/a)]^{1/2}\,\mathbf{i}$ (*b*) $t = (\pi/2)(2ma^3/kQ^2)^{1/2}$

Chapter 21

True or False **1.** False; C is the ratio of the charge to the voltage **2.** False; it depends only on the area and separation of the plates **3.** False; although $C = Q/V$, V is proportional to Q, so the ratio does not depend on Q **4.** True **5.** True **6.** True **7.** True

Problems

1. (*a*) 1.69×10^7 m^2 (*b*) 4117 m or 2.56 mi
3. 8×10^{-8} F
5. 22.1 μF
7. 2.71 nF
9. (*a*) 2.08 (*b*) 45.2 cm^2 (*c*) 5.2 nC
11. 2.22×10^{-5} J
13. (*a*) 0.625 J (*b*) 1.875 J
15. (*a*) 10^5 V/m (*b*) 0.0443 J/m^3 (*c*) 8.85×10^{-5} J (*d*) 1.77×10^{-8} F (*e*) 8.85×10^{-5} J
17. (*a*) 30 μF (*b*) 6 V (*c*) The charge on the 10-μF capacitor is 60 μC, and the charge on the 20-μF capacitor is 120 μC
19. (*a*) 24 μC (*b*) 4 μF
21. 2 μF
23. $C_{eq} = (C_1C_2 + C_2C_3 + C_1C_3)/(C_1 + C_3)$
25. (*a*) 0.05 mm (*b*) 235 cm^2
27. (*a*) 7.91 m^2 (*b*) 22.9 V (*c*) 3.66×10^{-5} J (*d*) 210 μC
29. (*a*) 15.2 μF (*b*) The 12-μF capacitor has a charge of 2400 μC; both the 4-μF and the 15-μF capacitors have charges of 632 μC (*c*) 0.303 J
31. (*a*) The charge on the 20-pF capacitor is 1.71×10^{-8} C; the charge on the 50-pF capacitor is 4.29×10^{-8} C (*b*) The initial energy is 9×10^{-5} J; the final energy is 2.57×10^{-5} J, so energy is lost when the capacitors are connected
33. (*a*) The maximum equivalent capacitance occurs when the capacitors are connected in parallel, so the appropriate combination is three 5-μF capacitors in parallel (*b*) The other possible equivalent capacitances are 10/3 μF, 7.5 μF, and 5/3 μF
35. (*a*) 6 V (*b*) The initial and final energies are 1.15×10^{-3} J and 2.88×10^{-4} J, respectively
37. (*a*) 1200 V (*b*) 6.4×10^{-4} J
39. (*a*) 1.67×10^{-8} F (*b*) 1.17×10^{-9} C (*c*) 7×10^6 V/m
41. (*a*) 5 μF (*b*) 133 V
43. (*a*) 2.28×10^{-9} F (*b*) 6.67×10^{-5} C
45. Connect four of the capacitors in series for an equivalent capacitance of 0.5 μF and a breakdown voltage of 400 V; now connect four such groups of four capacitors in parallel for a final capacitance of 2 μF
47. (*a*) $C_{eq} = \epsilon_0 b[(\kappa - 1)x + a]/d$ (*b*) For $x = 0$, $C_{eq} = \epsilon_0 ba/d$; for $x = a$, $C_{eq} = \kappa\epsilon_0 ba/d$
49. (*a*) $2C_0$ (*b*) $11C_0$
51. (*a*) 3.33×10^{-4} m (*b*) 3.77 m^2
53. (*a*) 40 V (*b*) 1.49×10^{-5} m^2 (*c*) 6
55. (*a*) 2.51×10^3 m^3 (*b*) 5.02×10^{-2} m^3
57. (*a*) $Q_1 = C_1V$, $Q_2 = \kappa C_1V$ (*b*) $U_i = \frac{1}{2}(1 + \kappa)C_1V^2$ (*c*) $U_f = \frac{1}{4}(1 + \kappa)^2C_1V^2 = \frac{1}{2}(1 + \kappa)U_i$ (*d*) $\frac{1}{2}(1 + \kappa)V$
59. (*a*) 0.255 J (*b*) The capacitor without a dielectric has a charge of 10^{-3} C, and the capacitor with the dielectric has a charge of 3.5×10^{-3} C (*c*) Both capacitors now have the same charge, 2.25×10^{-3} C (*d*) 0.506 J
61. (*a*) 0.001 J (*b*) The capacitor without a dielectric has the charge 47.6 μC; the capacitor with the dielectric has the charge 152 μC (*c*) 4.76×10^{-4} J
63. 2.55×10^{-6} J
65. (*a*) $E = 2kQ/rL$, $\eta = 2\epsilon_0 k^2Q^2/r^2L^2$ (*b*) $(kQ^2/rL)\,dr$ (*c*) $(kQ^2/L)\ln(R_2/R_1)$
67. (*a*) $(Q/2\pi\epsilon_0\kappa L)\ln(b/a)$ (*b*) $Q/2\pi aL$ (*c*) $-Q/2\pi aL$ (*d*) $-Q(\kappa - 1)/2\pi aL\kappa$ (*e*) $Q(\kappa - 1)/2\pi aL\kappa$ (*f*) $(kQ^2/L\kappa)\ln(b/a)$ (*g*) $(kQ^2/L\kappa)(\kappa - 1)\ln(b/a)$
69. (*a*) $3\epsilon_0 A/(y_0 \ln 4)$ (*b*) $\sigma_{top} = -3\sigma/4$, $\sigma_{bot} = 0$ (*c*) $(3\sigma/y_0)(1 + 3y/y_0)^{-2}$

Chapter 22

True or False **1.** False; $R = V/I$ is the definition of resistance. Ohm's Law states that R is independent of I or V **2.** False; they drift in the opposite direction **3.** True **4.** True **5.** True **6.** False

Problems

1. (*a*) 600 C (*b*) 3.75×10^{21}
3. 0.4 A
5. (*a*) $v/2\pi r$ (*b*) $vq/2\pi r$
7. (*a*) 3.21×10^{13} protons/m^3 (*b*) 3.75×10^{17} (*c*) $q = It = (10^{-3}\text{ C/s})t$
9. (*a*) 1 V (*b*) 0.1 V/m
11. (*a*) $V_{Cu} = IL\rho_{Cu}/A$, $V_{Fe} = IL\rho_{Fe}/A$, $V_{Fe}/V_{Cu} = 5.88$ (*b*) **E** is greater in the iron wire
13. (*a*) 0.0275 Ω (*b*) 0.030 Ω
15. 0.182 Ω
17. 45.6°C
19. 250 W
21. (*a*) 5 mA (*b*) 50 V
23. (*a*) 0.707 A (*b*) 7.07 V
25. 180 J
27. (*a*) 240 W (*b*) 228 W (*c*) 4.32×10^4 J (*d*) 2160 J
29. \$77.76
31. (*a*) 4.5 Ω (*b*) $I_3 = 2.67$ A, $I_2 = 2$ A, $I_6 = 2/3$ A
33. (*a*) 6 Ω (*b*) The top two resistors both have a current of 2/3 A; the two 6-Ω resistors in parallel each have a current of 2/3 A, and the single 6-Ω resistor on the bottom has a current of 4/3 A
35. (*b*) No effect
37. (*a*) $I_3 = 30/19$ A, $I_2 = 12/19$ A, $I_4 = 6/19$ A (*b*) 9.47 W
39. 0.0314 Ω
41. (*b*)

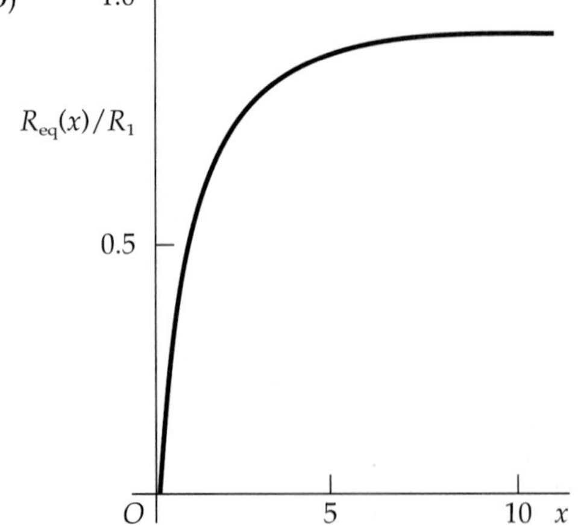

43. (*a*) 0.05 A (*b*) 5000 W
45. (*a*) $E_{Cu} = 0.0433$ V/m, $E_{Fe} = 0.255$ V/m (*b*) $V_{Cu} = 3.46$ V, $V_{Fe} = 12.5$ V (*c*) $R_{eq} = 7.97$ Ω, $R_{Cu} = 1.73$ Ω, $R_{Fe} = 6.24$ Ω

47. 382°C
49. (*a*) \$150.77 (*b*) 3.79 cents/kW·h
51. (*a*) 6.91×10^6 J (*b*) 12.8 h
53. (*a*) 3×10^{-3} kW·h (*b*) \$2663 per kW·h (*c*) 0.027 cents
55. (*a*) 0.03 Ω (*b*) 0.3 percent (*c*) 7.5C°
57. (*a*) 79.6 Ω (*b*) 318 Ω
59. (*a*) 15 A (*b*) 11.2 Ω (*c*) 1.28 kW
61. (*a*) 5.14×10^{-3} Ω (*b*) 0.462 V (*c*) 41.6 W
63. (*b*) $L_{Cu} = 264L_C$
65. $R = \rho\pi/[t \ln (b/a)]$
67. (*a*) $R = (\rho/2\pi L) \ln (b/a)$ (*b*) 2.05 A
69. $R = \rho L/\pi ab$

Chapter 23

True or False 1. True 2. False; it is the time for the charge to be reduced by a factor of e^{-1} 3. False; the voltmeter is placed in parallel with the resistor

Problems

1. (*a*) $I = 1.13$ A, $P = 6.79$ W (*b*) $I = 0.583$ A, $P =$ 3.50 W
3. (*a*) 3 V (*b*) 1 Ω
5. (*a*) 1 A (*b*) 12 W delivered by the emf on the left; 6 W absorbed by the emf on the right (*c*) 2 W for the 2-Ω resistor; 4 W for the 4-Ω resistor
7. (*a*) $I_4 = 2/3$ A, $I_3 = 8/9$ A, $I_6 = 14/9$ A (*b*) $V_b - V_a = -28/3$ V (*c*) 8 W supplied by the left-side emf; 32/3 W by the right-side emf
9. (*a*) 6×10^{-4} C (*b*) 0.2 A (*c*) 3×10^{-3} s (*d*) 8.12×10^{-5} C
11. 4.81×10^7Ω
13. (*a*) 5.69 μC (*b*) 1.10 μC/s (*c*) 1.10 μA (*d*) 6.62×10^{-6} W (*e*) 2.44×10^{-6} W (*f*) 4.19×10^{-6} J/s
15. (*a*) 0.0841 Ω (*b*) 4027 Ω
17. (*a*) 0.168 Ω (*b*) 0.168 Ω (*c*) 2.14×10^6Ω
19. (*a*) 910 Ω (*b*) 1000 Ω (*c*) 9000 Ω
21. $R_1 = 7582$ Ω, $R_2 = 69{,}231$ Ω, $R_3 = 692{,}308$ Ω
23. (*a*) Parallel connection supplies greater power (*b*) Series connection supplies greater power
25. (*a*) The currents in the 1-Ω and 6-Ω resistors are 2 A and 1 A, respectively; the current in the horizontal 2-Ω resistor is 2 A; the current in the vertical 2-Ω resistor is 1 A (*b*) The 8-V emf supplies 16 W, the upper 4-V emf supplies 8 W, and the lower 4-V emf absorbs 4 W (*c*) The power dissipated in the resistors is 4 W, 6 W, 8 W, and 2 W, where the order is the same as in part (*a*)
27. (*b*) 9.65 h
29. (*a*) Connecting the batteries in parallel gives the largest current for $R < 0.4$ Ω; connecting them in series works best for $R > 0.4$ Ω (*b*) 10.7 A parallel (*c*) 6.67 A series (*d*) 5.45 A series (*e*) 4.44 A series
31. (*a*) Connect the galvanometer in series with a resistor $R = 999{,}800$ Ω (*b*) Connect the galvanometer in parallel with a resistor $R = 10^{-3}$ Ω
33. (*a*) $I_{a\to b} = 2$ A; $I_{b\to c} = 1.5$ A; $I_{b\to h} = 0.5$ A; $I_{d\to h} =$ 1.5 A; $I_{c\to f} = 0.5$ A; $I_{h\to g} = 1.5$ A (*b*) $V_a = 0$ V, $V_b =$ 24 V, $V_c = 21$ V, $V_d = 15$ V, $V_e = 15$ V, $V_f = 5$ V, $V_g = 0$ V, $V_h = 12$ V
35. (*a*) 50/3 Ω (*b*) For the top three and lower three resistors, each has a current of 2/5 A; for the middle three resistors, the first and last have a current of 4/5 A, and the current in the middle resistor is 2/5 A
37. $\mathcal{E}=7$ V, $R = 14$ Ω
39. (*a*) 3.42 A (*b*) 0.962 A (*c*) $Q_{10} = 260$ μC, $Q_5 =$ 130 μC
41. (*a*) 10^{-4} A (*b*) 6.67×10^{-5} A (*c*) 40 V
43. (*a*) 43.9 Ω (*b*) 300 Ω (*c*) 3800 Ω
45. (*a*) The current in the car battery is −57.0 A + (10 A/h)*t*, where the minus sign means that current is entering the battery; the current in the second battery is 63.0 A − (10 A/h)*t* (*b*)

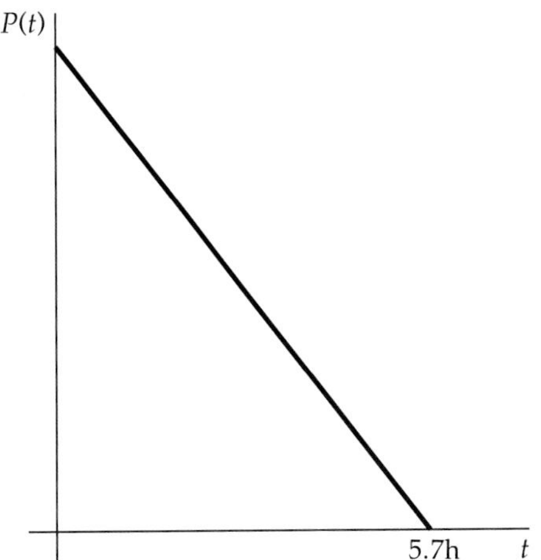

47. (*a*) Case *a* is preferred for small *R*, case *b* for large *R*; case *a* is the correct configuration for an ideal voltmeter with R_v infinitely large; if *R* is comparable to R_v then case *b* compensates for the fact that a finite current flows through the voltmeter (*b*) Case *a*, $R =$ 0.498 Ω; case *b*, $R = 0.6$ Ω (*c*) Case *a*, $R = 2.91$ Ω; case *b*, $R = 3.10$ Ω (*d*) Case *a*, $R = 44.4$ Ω; case *b*, $R = 80.1$ Ω
49. answers given in problem
51. answer given in problem
53. $R_{eq} = \frac{1}{3}R$
55. $\frac{1}{3}R$
57. $R_{eq} = (1 + \sqrt{3})R$
59. $I_{10} = 104.4/141$ A, $I_{40} = 66.6/141$ A, $I_{30} = 54/141$ A, $I_{80} = 50.4/141$ A, $I_{20} = 120.6/141$ A
61. (*a*) 4.17×10^{-5} A (*b*) 2.78×10^{-5} A (*c*) $I(t) =$ $(2.78 \times 10^{-5}\text{ A})\, e^{-t/(1.5\text{ s})}$
63. (*a*) $I(t) = (V_0/R)\, e^{-t/RC}$, where $C = C_{eq} =$ $C_1C_2/(C_1 + C_2)$ (*b*) $P(t) = (V_0^2/R)\, e^{-2t/RC}$ (*c*) $U = \frac{1}{2}C_{eq}V_0^2$

65. (*a*) $\mathcal{E}I(t) = (\mathcal{E}^2/R)\, e^{-t/RC}$ (*b*) $I(t)^2R = (\mathcal{E}^2/R)\, e^{-2t/RC}$ (*c*) $dU/dt = (\mathcal{E}^2/R)\, e^{-t/RC} - (\mathcal{E}^2/R)\, e^{-2t/RC}$

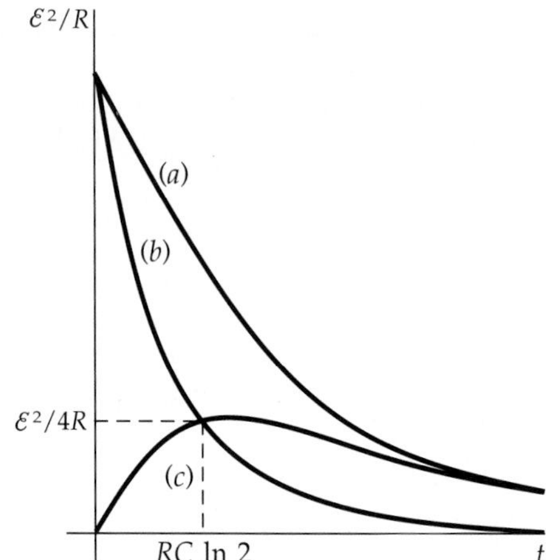

(*d*) $(dU/dt)_{max} = \mathcal{E}^2/4R$, $t = RC \ln 2$

Chapter 24

True or False **1.** True **2.** True **3.** True **4.** False; it is independent of the radius **5.** True

Problems

1. -1.25×10^{-12} N **j**
3. (*a*)-7.17×10^{-13} N **j** (*b*) 5.12×10^{-13} N **i** (*c*) 0 (*d*) 8.19×10^{-13} N **i** $- 6.14 \times 10^{-13}$ N **j**
5. 1 N
7. 14.0 N/m **k**
9. (*a*) 2.20 mm (*b*) $f = 9.08 \times 10^9$ Hz, $T = 1.10 \times 10^{-10}$ s
11. (*a*)-1.05×10^4 N/C **k** (*b*) No
13. (*a*) 1.42 km (*b*) 28.5 m
15. (*a*) 2.13×10^{-7} Hz (*b*) 46.0 MeV (*c*) Both the frequency and the kinetic energy will be reduced by a factor of 2
17. (*a*) 0.302 A·m^2 (*b*) 0.131 N·m
19. 2.83×10^{-5} N·m
21. (*a*) 0 (*b*) 2.7×10^{-3} N·m
23. (*a*) 2.125 N·m/T **i** (*b*)-3.40 N·m **j** + 5.31 N·m **k**
25. (*a*) 1.07×10^{-4} m/s (*b*) 5.85×10^{28} electrons/m^3
27. (*a*) 3.69×10^{-5} m/s (*b*) 1.48 μV
29. 1.02×10^{-3} V
31. 0.0864 N **i** $-$ 0.0648 N **j**
33. (*a*) 7.35 mm (*b*) 6.64×10^{-5} T
35. (*a*) The normal points 37° below the x axis (*b*) $\hat{\mathbf{n}} = 0.799\,\mathbf{i} - 0.602\,\mathbf{j}$ (*c*) $\mathbf{m} = 0.335$ A·m^2 **i** $-$ 0.253 A·m^2 **j** (*d*) 0.503 N·m **k**
37. answers given in problem
39. $r_d/r_p = \sqrt{2}$, $r_\alpha/r_p = 1$
41. (*a*) $v_p/v_\alpha = 2$ (*b*) $K_p/K_\alpha = 1$ (*c*) $L_p/L_\alpha = \frac{1}{2}$
43. $I = Mg/\pi RB$
45. answer given in problem
47. answers given in problem
49. (*a*) 1.6×10^{-18} N **j** (*b*) 10 V/m **j** (*c*) 20 V
51. (*a*) $B = (Mg/IL) \tan\theta$ (*b*) $a = g \sin\theta$, uphill
53. $T = 2\pi(M/\pi IB)^{1/2}$
55. answer given in problem
57. answer given in problem
59. answers given in problem
61. answer given in problem

Chapter 25

True or False **1.** False **2.** True **3.** False; it varies inversely with distance **4.** False; it is useful in finding **B** only if there is symmetry, but it is valid for any continuous current **5.** True

Problems

1. (*a*) $\mathbf{B} = -9 \times 10^{-12}$ T **k** (*b*) $\mathbf{B} = -3.6 \times 10^{-11}$ T **k** (*c*) $\mathbf{B} = 3.6 \times 10^{-11}$ T **k** (*d*) $\mathbf{B} = 9 \times 10^{-12}$ T **k**
3. (*a*) 0 (*b*) -3.56×10^{-23} T **k** (*c*) 4×10^{-23} T **k**
5. 12.5 T
7. -9.6×10^{-12} T **i**
9. 11.1 A

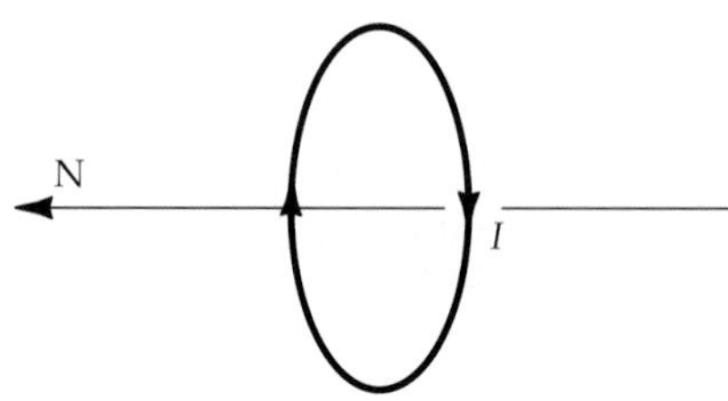

11. 6.98×10^{-4} T
13. (*a*) $x = \pm 5.72$ cm (*b*) $x = \pm 13.6$ cm (*c*) $x = \pm 29.8$ cm
15. (*a*) -8.89×10^{-5} T **k** (*b*) 0 (*c*) 8.89×10^{-5} T **k** (*d*) -1.6×10^{-4} T **k**
17. (*a*) -1.78×10^{-4} T **k** (*b*) -1.33×10^{-4} T **k** (*c*) -1.78×10^{-4} T **k** (*d*) 1.07×10^{-4} T **k**
19. (*a*) 6.4×10^{-5} T **j** (*b*) -4.8×10^{-5} T **k**
21. The fields caused by the wire segments, going from left to right, are 0, 56.6 μT, 113 μT, 56.6 μT, and 0; all of the fields are into the page; the total field is 226 μT into the page
23. 9.47 A

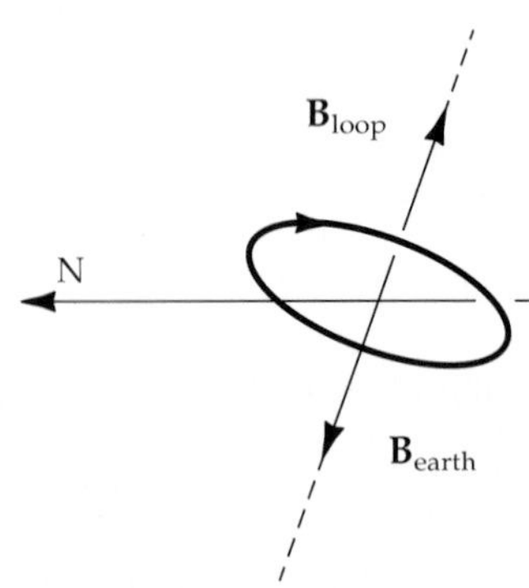

25. (*a*) Antiparallel (*b*) 39.3 mA
27. 28 A
29. (*a*) 4.5×10^{-4} N/m to the right (*b*) 30 μT down
31. (*a*) C_1, (8 A)μ_0; C_2, 0; C_3, (-8 A)μ_0 (*b*) None of them

33. (*a*) 8×10^{-4} T (*b*) 4×10^{-3} T (*c*) 2.86×10^{-3} T
(*d*)

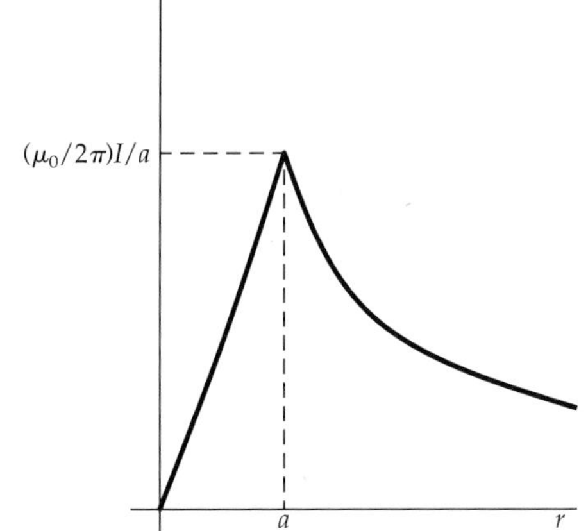

35. (*a*) 0.0273 T (*b*) 0.0200 T
37. (*a*) 3.2×10^{-16} N, in the direction opposite to the current (*b*) 3.2×10^{-16} N, away from the wire (*c*) 0
39. (*a*) $\pi(\mu_0 I/L) = 3.14(\mu_0 I/L)$ (*b*) $(8\sqrt{2}/\pi)(\mu_0 I/L) = 3.60(\mu_0 I/L)$ (*c*) $(27/2\pi)(\mu_0 I/L) = 4.30(\mu_0 I/L)$
41. (*a*) $3\sqrt{2}\mu_0 I^2/4\pi a$, along the diagonal toward the opposite corner (*b*) $\sqrt{2}\mu_0 I^2/4\pi a$, along the diagonal away from the opposite corner
43. (*a*) 2.26×10^{-5} T·m (*b*) 1.13×10^{-5} T·m (*c*) 0
45. $\tau = \pi r_2^2 \mu_0 N_1 N_2 I_1 I_2 / 2r_1 = 1.97 \times 10^{-6}$ N·m, where the subscripts 1 and 2 refer to the large and small coil, respectively
47. 3.18 cm
49. (*a*) 80 A, in the positive z direction.
(*b*) -2.4×10^{-4} T **j**
51. (*a*)

$$B = \frac{\mu_0 I}{2\pi R}\sin\theta,$$

where R is the perpendicular distance from the point P to the wire (*b*) For a polygon of N sides,

$$B = \frac{N\mu_0 I}{2\pi R}\sin(\pi/N);$$

for large values of N, the field approaches $\mu_0 I/2R$
53. (*a*) $x = 5$ cm, $B = 0.0540$ T; $x = 7$ cm, $B = 0.0539$; $x = 9$ cm, $B = 0.0526$ T; $x = 11$ cm, $B = 0.0486$ T (*b*)

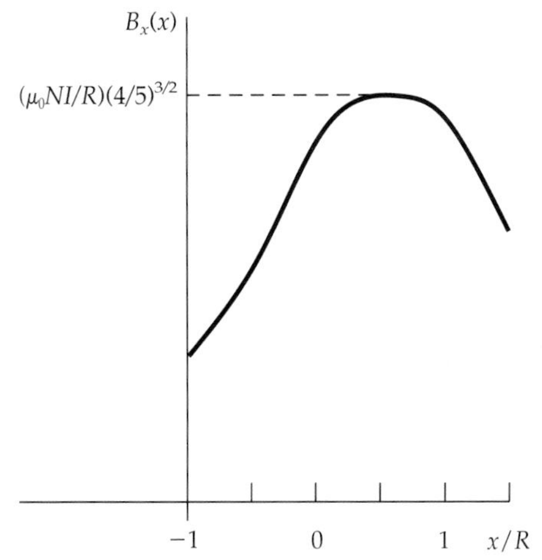

55. (*a*) The force on each of the horizontal segments is 0.251×10^{-4} N, down on the upper segment and up on the lower segment; the force on the left vertical segment is 1×10^{-4} N to the right, and the force on the right vertical segment is 0.286×10^{-4} N to the left (*b*) 0.714×10^{-4} N to the right
57. $B_x = (\mu_0 I L^2/2\pi x^3)(1 + L^2/4x^2)^{-1}(1 + 2L^2/4x^2)^{-1/2}$
59. (*a*) $B_y = \dfrac{\mu_0 I}{2\pi(R^2 - a^2)}\left[\dfrac{R}{2} - \dfrac{a^2}{2R - b}\right]$
(*b*) $B_x = \dfrac{\mu_0 I}{\pi(R^2 - a^2)}\left[\dfrac{a^2 R}{4R^2 + b^2} - \dfrac{R}{4}\right]$
$B_y = \dfrac{\mu_0 I a^2 b}{2\pi(R^2 - a^2)(4R^2 + b^2)}$
61. (*c*) $B_x = \frac{1}{2}\mu_0\sigma\omega[(R^2 + 2x^2)/(x^2 + R^2)^{1/2} - 2x]$
63. answer given in problem
65. (*a*) $dm = (N/L)I\pi R^2\,dx = nIA\,dx$

Chapter 26

True or False **1.** False; it depends only on the rate of change of the flux **2.** True **3.** True **4.** False **5.** True

Problems

1. (*a*) 5×10^{-4} Wb (*b*) 4.33×10^{-4} Wb (*c*) 2.5×10^{-4} Wb (*d*) 0
3. 7.58×10^{-4} Wb
5. (*a*) 8.48×10^{-3} Wb (*b*) 7.97×10^{-3} Wb
7. (*a*) 8.48×10^{-3} Wb (*b*) 133 turns
9. 199 T/s
11. (*a*)

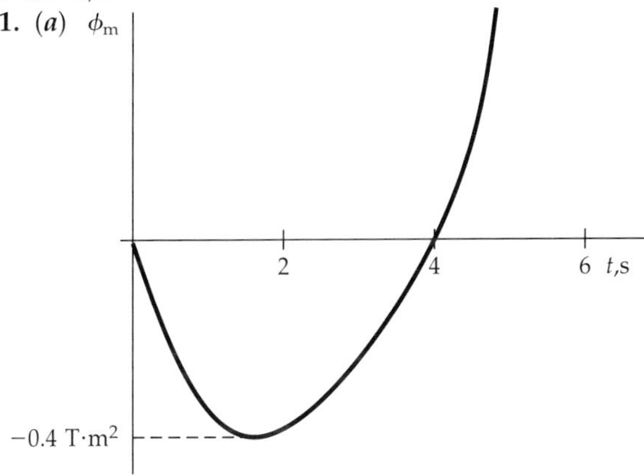

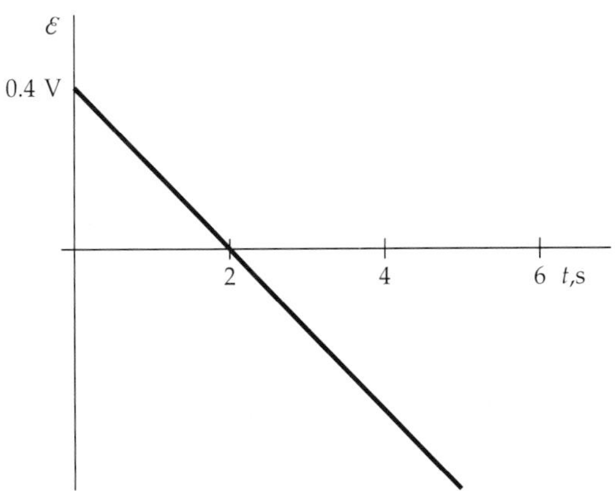

(*b*) At $t = 2$ s, ϕ_m has its maximum negative value; ϕ_m increases indefinitely as t goes to infinity. At $t = 2$ s the emf is zero. (*c*) $\phi_m = 0$ at $t = 0$ and $t = 4$ s; at $t = 0$ s, $\mathcal{E} = 0.4$ V, and at $t = 4$ s, $\mathcal{E} = -0.4$ V

13. 2.8×10^{-4} C
15. (*a*) 7.07×10^{-3} V (*b*) 6.64×10^{-3} V
17. (*a*) 3.1×10^{-3} Wb (*b*) 2.21×10^{-3} V
19. (*a*)

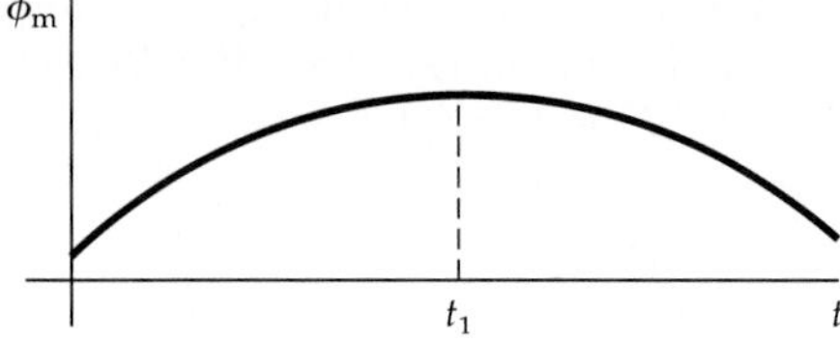

(*b*)

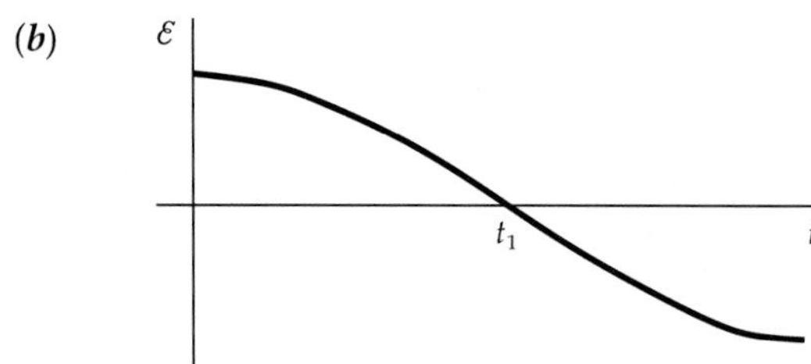

21. (*a*)

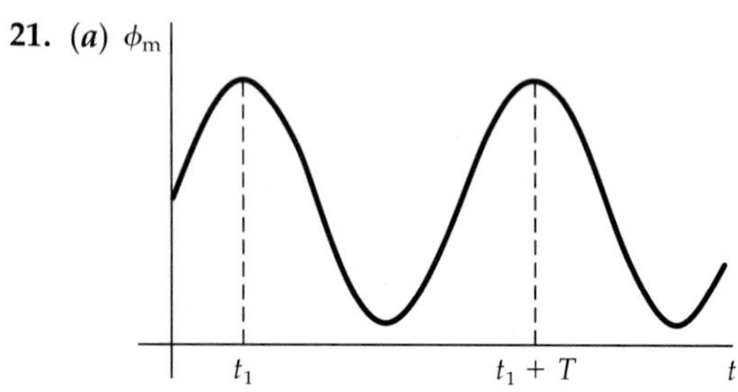

(*b*)

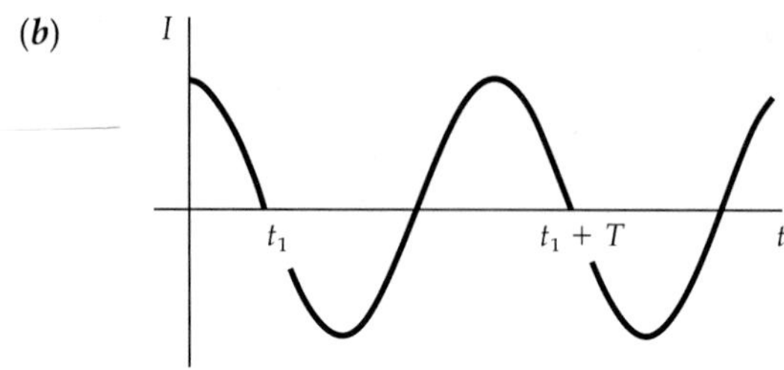

23. 400 m/s
25. (*a*) 3.6 V (*b*) 3 A (*c*) 1.8 N (*d*) 10.8 W (*e*) 10.8 W
27. 0.332 T
29. 0.707 T
31. $\phi_m = LI_0 \sin 2\pi ft$

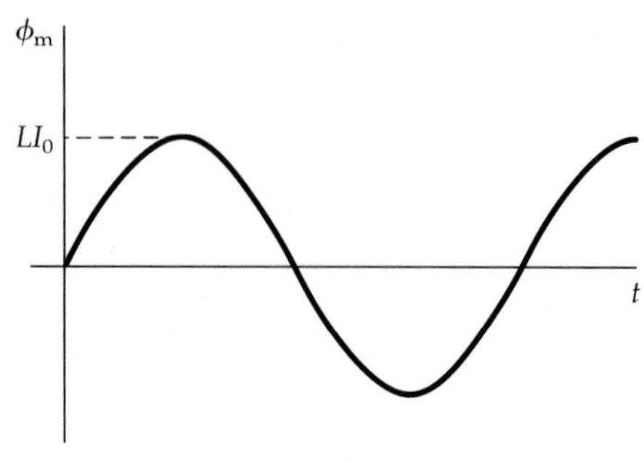

$\mathcal{E} = -LI_0 2\pi f \cos 2\pi ft$

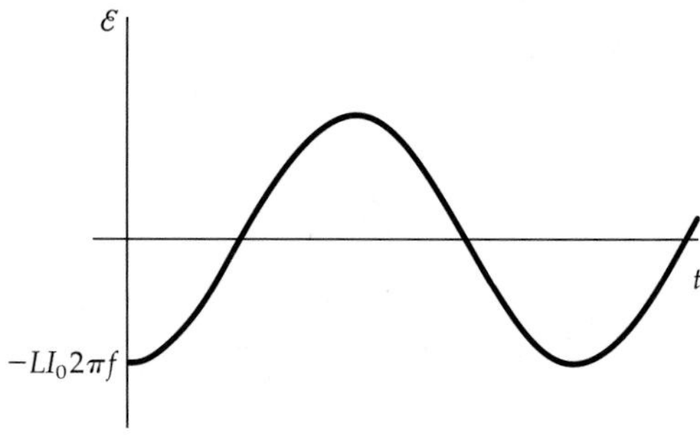

33. 1.89 mH
35. (*a*) $I = 0$ A, $dI/dt = 25$ A/s (*b*) $I = 2.27$ A, $dI/dt = 20.5$ A/s (*c*) $I = 7.90$ A, $dI/dt = 9.20$ A/s (*d*) $I = 10.8$ A, $dI/dt = 3.38$ A/s
37. (*a*) 13.5 mA (*b*) 7.44×10^{-44} A
39. (*a*) $t = 0$ s, $P = 47.7$ W; $t = 100$ s, $P = 48.0$ W (*b*) $t = 0$ s, $I^2R = 47.4$ W; $t = 100$ s, $I^2R = 48.0$ W (*c*) $t = 0$ s, $dU_m/dt = 0.321$ W; $t = 100$ s, $dU_m/dt = 0$ W
41. (*a*) 3.98×10^5 J (*b*) 4.43×10^{-4} J (*c*) 3.98×10^5 J
43. $B\pi R^2$
45. (*a*) $\mu_0 nIN\pi R_1^2$ (*b*) $\mu_0 nIN\pi R_3^2$
47. (*b*) 275 rad/s
49. (*a*) 13.9 Ω (*b*) 214 V
51. (*a*) 88.1 ms (*b*) 35.2 mH
53. (*a*) 2.41 s (*b*) 20.1 s
55. answer given in problem
57. (*a*) $I_B = I_{10} = I_L = 1$ A, $I_{100} = 0$ A (*b*) 100 V (*c*) $I(t) = (1\text{ A})\, e^{-50t}$
59. (*a*) For the resistor, $dI/dt = 9000$ A/s; for the inductors, $dI_1/dt = 3000$ A/s and $dI_2/dt = 6000$ A/s, where 1 and 2 refer to the 8-mH and 4-mH inductors, respectively (*b*) 1.6 A
61. (*a*) 0.0536 J (*b*) 447 J/m^3 (*c*) 0.0335 T (*d*) 447 J/m^3
63. (*a*) $E = \frac{1}{2}\mu_0 nrI_0\omega \cos \omega t$ (*b*) $E = \frac{1}{2}\mu_0 n(R^2/r)I_0\omega \cos \omega t$
65. (*a*) $0 \text{ s} \le t \le 4.17$ s, $\phi_m = (2.04 \times 10^{-3}\text{ Wb/s})t$; $4.17 \text{ s} \le t \le 8.33$ s, $\phi_m = 8.5 \times 10^{-3}$ Wb; $8.33 \text{ s} \le t \le 12.5$ s, $\phi_m = 8.5 \times 10^{-3}\text{ Wb} - (2.04 \times 10^{-3}\text{ Wb/s})(t - 8.33)$; $t > 12.5$ s, $\phi_m = 0$

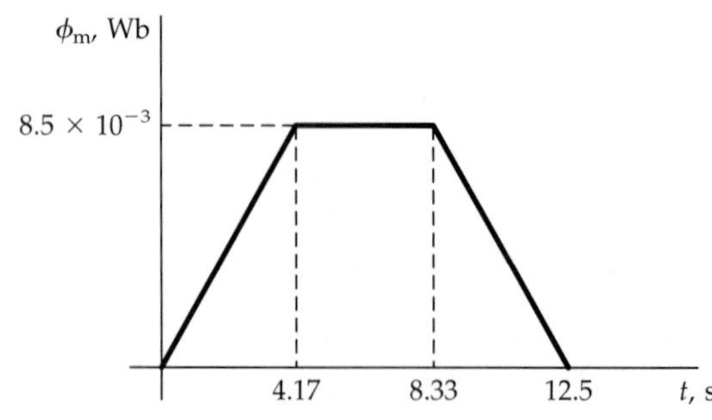

(*b*) 0 s ≤ t ≤ 4.17 s, $\mathcal{E} = -2.04 \times 10^{-3}$ V; 4.17 s ≤ t ≤ 8.33 s, $\mathcal{E} = 0$ V; 8.33 s ≤ t ≤ 12.5 s, $\mathcal{E} = 2.04 \times 10^{-3}$ V; $t > 12.5$ s, $\mathcal{E} = 0$

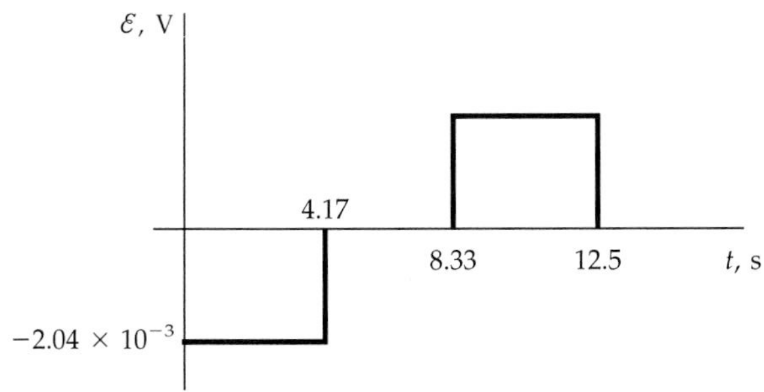

67. (*a*) $F = (\mathcal{E} - B\ell v)B\ell/R = m\,dv/dt$ (*b*) $v_t = \mathcal{E}/B\ell$ (*c*) 0
69. answers given in problem
71. (*a*) 1.38×10^{-4} V/m (*b*) 5.51×10^{-6} V
73. answers given in problem
75. (*a*) $(\mu_0 I/\pi)\ln[(d-a)/a]$ (*b*) $(\mu_0/\pi)\ln[(d-a)/a]$
77. (*c*) $\frac{1}{2}\ell^2 B\theta$
79. 12 mH
81. $(\mu_0 I\ell/2\pi)\ln(r_2/r_1)$

Chapter 27

True or False **1.** True theoretically, but it is masked by paramagnetism or ferromagnetism in materials whose molecules have permanent dipole moments **2.** True **3.** False **4.** True **5.** False

Problems
1. (*a*) $B_{app} = 0.0101$ T, $B = 0.0101$ T (*b*) $B_{app} = 0.0101$ T, $B = 1.52$ T
3. $B_{app} = 0.0101$ T, $M = 0.183$ A/m, $B = 0.0101$ T
5. (*a*) The field decreases by 6.8×10^{-3} percent (*b*) The self-inductance also decreases by 6.8×10^{-3} percent
7. (*a*) 0.0628 T (*b*) 0.0628 T to three places, though greater than (*a*) by 2.3×10^{-3} percent (*c*) 0.0628 T to three places, though less than (*a*) by 2.6×10^{-3} percent
9. $m = 1.69\, m_B$
11. (*a*) $M_s = 5.58 \times 10^5$ A/m, $\mu_0 M_s = 0.701$ T (*b*) 5.23×10^{-4} (*c*) Diamagnetic effects have been neglected, and these effects tend to reduce the susceptibility
13. $M = 6.87 \times 10^5$ A/m, $B = 0.864$ T
15. (*a*) 0.0126 T (*b*) 1.36×10^6 A/m (*c*) 137
17. (*a*) 0.0603 T (*b*) 24 A
19. (*a*) 8.12×10^3 A/m (*b*) 1.62×10^{21} electrons (*c*) 24.4 A
21. (*a*) 6×10^{13} m^3 (*b*) 24.3 km
23. $n_N = 418 n_O$
25. $B_{app} = \mu_0 NI/2\pi R$, $B = \mu_0(NI/2\pi R + M)$
27. (*a*) 1.42×10^6 A/m (*b*) $K_m = 90$, $\mu = 90\mu_0 = 1.13 \times 10^{-4}$ T·m/A, $\chi_m = 89$
29. (*a*) 15.1 T (*b*) 1.2×10^7 A/m (*c*) 0.0302 T
31. answer given in problem
33. (*a*) 0.0524 A·m^2 (*b*) 7.70×10^5 A/m (*c*) 2.31×10^4 A
35. answer given in problem
37. (*b*) 1.25 N
39. (*a*) 3.02×10^{-4} T (*b*) 0.121 T (*c*) 14.5 J/m^3
41. -2.21×10^{-5}

Chapter 28

True or False **1.** False; the power dissipated is proportional to I^2 **2.** True **3.** True **4.** True **5.** True **6.** True

Problems
1. (*a*) 0.833 A (*b*) 1.18 A (*c*) 200 W
3. (*a*) 20.8 A (*b*) 29.5 A (*c*) $I_{rms} = 41.7$ A, $I_{max} = 58.9$ A
5. (*a*) 0.377 Ω (*b*) 3.77 Ω (*c*) 37.7 Ω
7. 1.59 kHz
9. (*a*) 2.65 MΩ (*b*) 26.5 kΩ (*c*) 26.5 Ω
11.

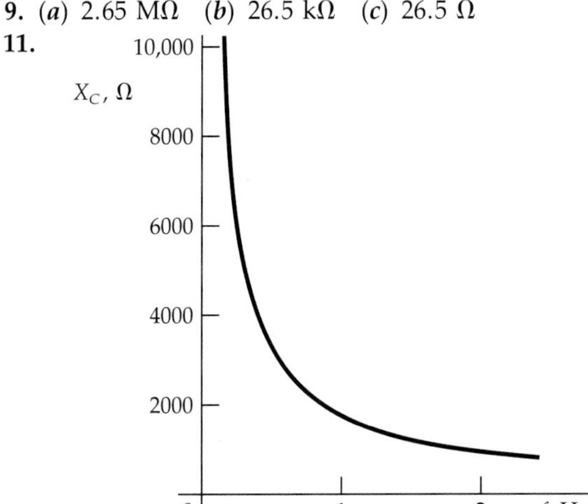

13. (*a*) 15.9 kHz (*b*) 159 Hz (*c*) 1.59 MHz
15. answer given in problem
17. 88 mH
19. (*a*) 2.25 mJ (*b*) 712 Hz (*c*) 0.671 A
21. (*a*) 1.13 kHz (*b*) $X_C = 79.6$ Ω, $X_L = 62.8$ Ω (*c*) $Z = 17.5$ Ω, $I_{rms} = 4.04$ A (*d*) −73.4°
23. 2002
25. (*a*) 14.1 (*b*) 79.6 Hz (*c*) 0.275
27. (*a*) 0.553 (*b*) 0.663 A (*c*) 44 W
29. (*a*) A step-down transformer (*b*) 2.4 V rms (*c*) 5 A
31. 10.4 turns for 2.5 V; 31.3 turns for 7.5 V; 37.5 turns for 9 V
33. (*a*) 1.75 A (*b*) 2.47 A
35. (*a*) 12 V (*b*) 8.49 V
37. 60 V
39. (*a*)

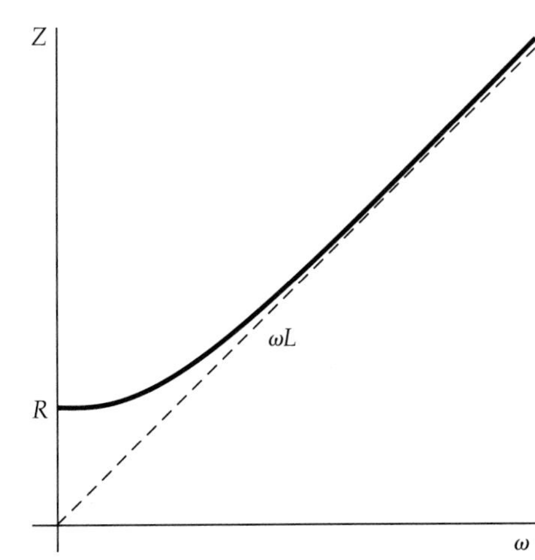

(*b*)

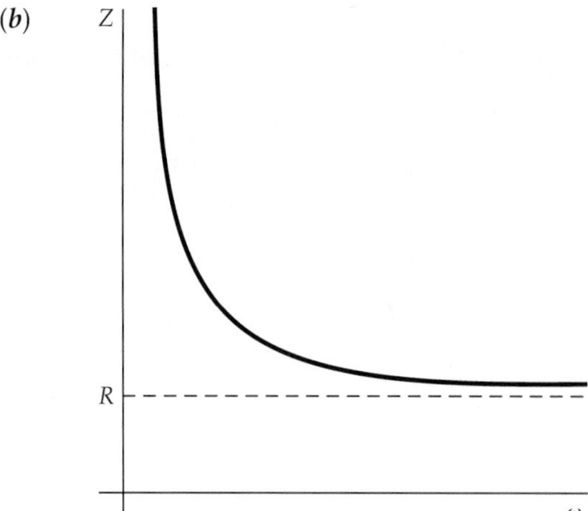

(*c*)

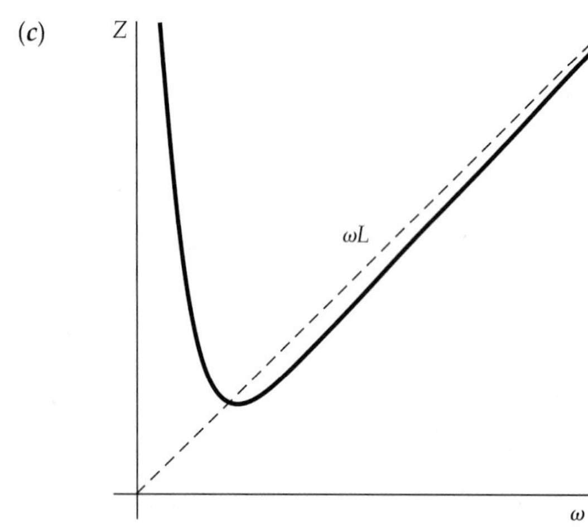

41. (*a*) 396 Ω (*b*) 50 V
43. 29.2 mH
45. (*a*) 15 W (*b*) 15 Ω (*c*) 0.235 H
47. (*a*) 6 Ω (*b*) 35.5 mH
49. (*a*) $C = 18.8\ \mu F$, $I_{rms} = 0.531$ A (*b*) 25 V
51. (*b*) $\delta = -\pi/2 + \omega RC$ (*c*) $\delta = \pi/2 - R/L\omega$
53. (*a*) 80.3 V (*b*) 77.8 V (*c*) 165 V (*d*) 112 V (*e*) 182 V
55. 0.935 μF
57. $R = 933\ \Omega$, $C = 0.517$ pF
59. (*b*) $\omega = 1/\sqrt{3}RC$ (*c*)

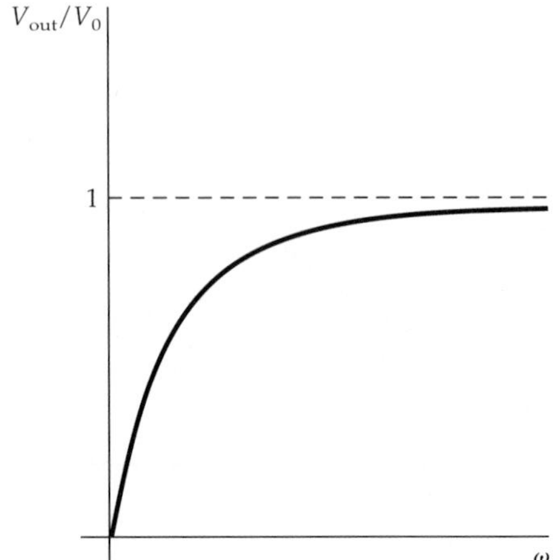

61. (*a*) $V_{out} = (9.95\text{ V}) \cos 100t + (0.995\text{ V}) \cos 10{,}000t$ (*b*) 10
63. (*a*) $L = 0.8$ mH, $C = 12.5\ \mu F$ (*b*) 1.6 (*c*) 2 A
65. (*a*) 933 W (*b*) 7.71 Ω (*c*) 99.8 μF (*d*) Add a capacitance of 40.9 μF
67. (*c*)

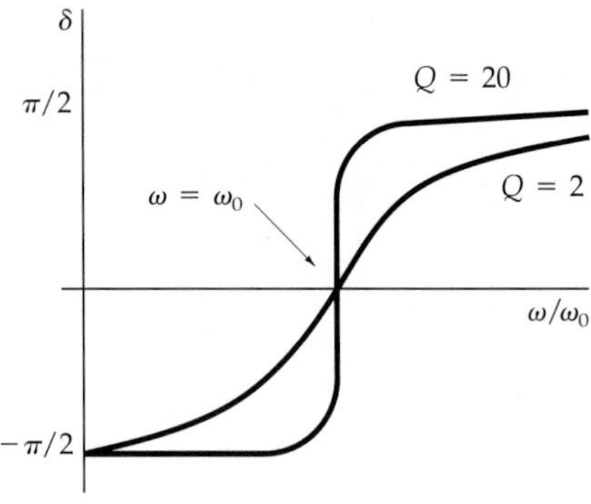

69. (*a*) 12 Ω (*b*) $R = 7.2\ \Omega$, $X = 9.6\ \Omega$ (*c*) Capacitive
71. (*a*) 4 mH (*b*) 0.1 A
73. answers given in problem
75. answer given in problem
77. answers given in problem

Chapter 29

True or False **1.** False **2.** True **3.** True **4.** True **5.** False **6.** True

Problems

1. (*a*) 3.4×10^{14} V/m·s (*b*) 5 A
3. answer given in problem
5. (*a*) 3.33×10^{-7} N/m² (*b*) 194 V/m (*c*) 647 nT
7. answers given in problem
9. 6.93×10^{-8} N
11. answer given in problem
13. (*a*) 300 m (*b*) 3 m
15. 3×10^{18} Hz
17. (*b*) $(-2.36 \times 10^{-5}\text{ A}) \sin 500\pi t$
19. (*a*) $I/9$ (*b*) $I/2$ (*c*) $I/16$
21. (*a*) $E_{max} = 12$ V/m, $B_{max} = 4 \times 10^{-8}$ T (*b*) $E_{max} = 0.12$ V/m, $B_{max} = 4 \times 10^{-10}$ T (*c*) $E_{max} = 1.2 \times 10^{-3}$ V/m, $B_{max} = 4 \times 10^{-12}$ T
23. 111 m² or 10.5 by 10.5 m
25. (*a*) 2×10^{-12} N (*b*) 6×10^{-12} N
27. (*a*) 3 m (*b*) 5.31×10^5 J/m³ (*c*) $E_{max} = 3.46 \times 10^8$ V/m, $B_{max} = 1.15$ T
29. (*a*) Positive *x* direction (*b*) $\lambda = 0.628$ m, $f = 4.77 \times 10^8$ Hz (*c*) $\mathbf{E} = (194\text{ V/m}) \cos [10x - (3 \times 10^9)t]\mathbf{j}$, $\mathbf{B} = (0.647 \times 10^{-6}\text{ T}) \cos [10x - (3 \times 10^9)t]\mathbf{k}$
31. (*a*) 1417 W/m (*b*) 902 W/m² (*c*) $E_{rms} = 583$ V/m (*d*) 1.94×10^{-6} T
33. (*a*) 279 K (*b*) 245 K
35. 3.42×10^6 W/m²
37. 7.25×10^{-3} V
39. (*a*) $(5 \times 10^{-5}\text{ V}) \cos 10^6 t$ (*b*) $(4.19 \times 10^{-8}\text{ V}) \sin 10^6 t$
41. (*a*) $At/100\rho d$, where *d* is the separation between the plates (*b*) $\kappa\epsilon_0 A/100d$ (*c*) $\kappa\epsilon_0\rho$
43. (*a*) $V_0[(1/R) \sin \omega t + (\epsilon_0 A\omega/d) \cos \omega t]$ (*b*) $(\mu_0/2\pi)[(V_0/rR) \sin \omega t + (\epsilon_0\omega\pi V_0 r/d) \cos \omega t]$ (*c*) $\tan \delta = \epsilon_0 A\omega R/d$

45. (*a*) $\rho I/\pi a^2$ (*b*) $\mu_0 I/2\pi a$ (*c*) $\mathbf{S} = \rho I^2/2\pi^2 a^3$, radially inward (*d*) $S(2\pi aL) = I^2(\rho L/A) = I^2 R$
47. 5.74×10^{-7} m
49. answer given in problem
51. (*a*) 3.12×10^{-6} rad/s^2 (*b*) 3.88 days (*c*) no

Chapter 30

True or False **1.** True **2.** False **3.** False; it is greater when the second medium has a lower index of refraction—for example, for refraction from water to air **4.** False; if it were, there would be no rainbow **5.** True

Problems

1. 2.11×10^6 *c*-year
3. ± 0.3 m
5. 92 percent
7. (*a*) 27.1° (*b*) 41.7° (*c*) 70.1° (*d*) Totally reflected
9. $v_{water} = 2.26 \times 10^8$ m/s, $v_{glass} = 2 \times 10^8$ m/s
11. (*a*) 50.2° (*b*) 38.8° (*c*) 26.3°
13. 62.5°
15. (*a*) Yes (*b*) Since the ball moves with constant speed, least distance implies least time
17. (*a*) $I_0/8$ (*b*) $3I_0/32$
19. 35.3°
21.

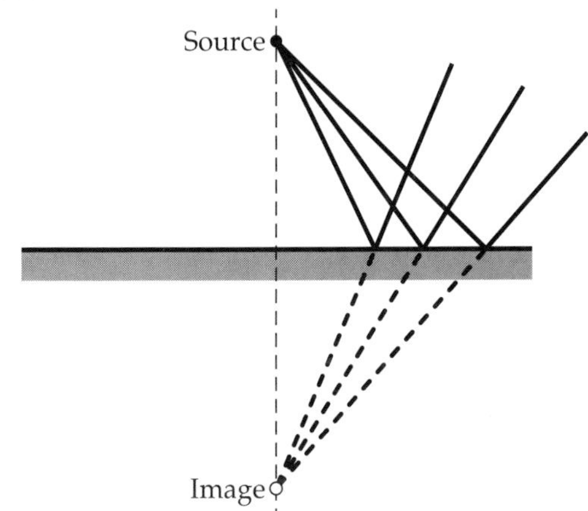

23. $\Delta t = 0.505$ s
25. 102 m^2
27. answer given in problem
29. answer given in problem
31. (*b*) 3.47°
33. $\theta_{violet} = 27.0°$, $\theta_{red} = 27.3°$
35. (*b*) The critical angle is greater
37. answer given in problem
39. (*a*) $I_0(\cos \pi/2N)^{2N}$ (*b*) $I_0/4$ (*c*) $0.976I_0$ (*d*) Perpendicular to initial polarization
41. (*a*) 1.33 (*b*) 37.2° (*c*) $\theta_r = 48.7°$; no, the refraction into the liquid doesn't change the outcome; the final beam is parallel to the liquid surface
43. answer given in problem
45. answer given in problem
47. answer given in problem
49. $\theta_r = 14.5°$; the refracted ray is bent toward the perpendicular

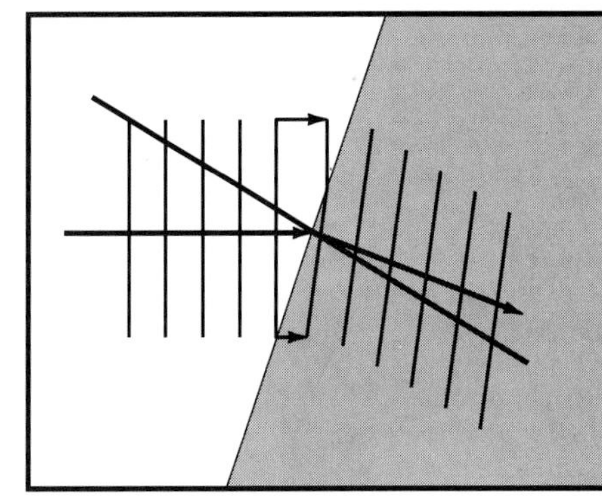

51. (*a*) $d\phi_d/d\theta_1 = 2 - (4 \cos \theta_1)/(n^2 - \sin^2 \theta_1)^{1/2}$
53. 2.18 cm
55. $\frac{1}{2}[1 - (1 - 1/n^2)^{1/2}]$

Chapter 31

True or False **1.** True **2.** False **3.** True **4.** False; spherical aberration occurs for rays far from the axis of the mirror **5.** True **6.** False; for example, the image distance is negative for a simple magnifier **7.** True

Problems

1. The eye can see the image from anywhere between rays 1 and 2

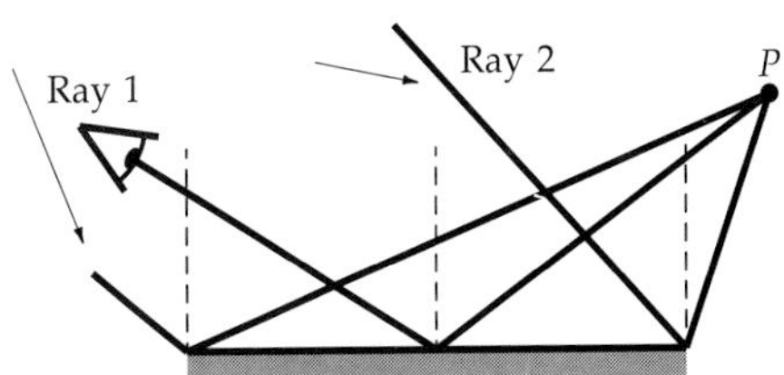

3. (*a*) 0.81 m (*b*) The bottom of the mirror should be 0.735 m above the floor

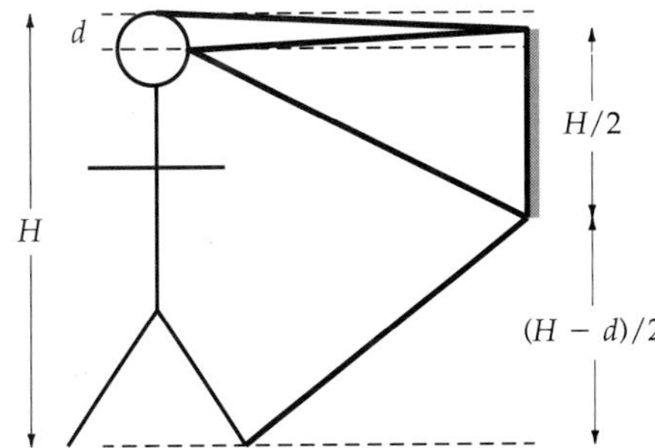

5. (*a*)

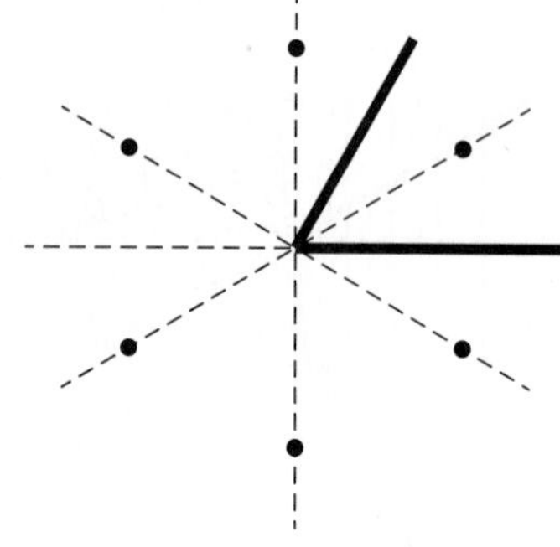

(*b*)

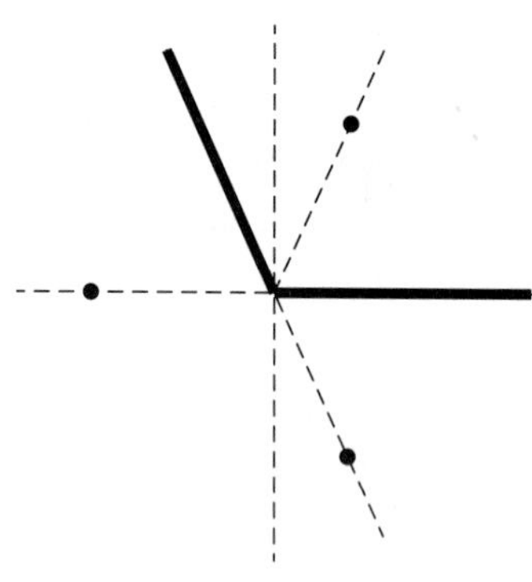

7. (*a*) $s' = 25$ cm, $m = -0.25$, real, inverted, reduced (*b*) $s' = 40$ cm, $m = -1$, real, inverted, same size (*c*) $s' = \infty$, $m = -\infty$, real, inverted, enlarged (*d*) $s' = -20$ cm, $m = 2$, virtual, erect, enlarged
9. (*a*) $s' = -16.7$ cm, $m = 0.167$, virtual, erect, reduced (*b*) $s' = -13.3$ cm, $m = 0.333$, virtual, erect, reduced (*c*) $s' = 10$ cm, $m = 0.5$, virtual, erect, reduced (*d*) $s' = -6.67$ cm, $m = 0.667$, virtual, erect, reduced
11. (*a*) 0.566 m (*b*) Behind (*c*) 0.113 m
13. (*a*) 5.13 cm (*b*) Concave
15. (*a*) $s' = 30$ cm, real

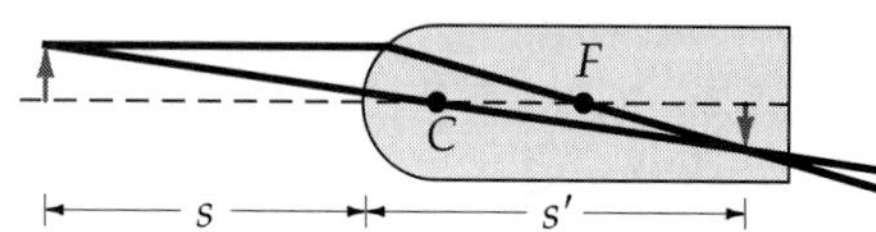

(*b*) $s' = -15$ cm, virtual

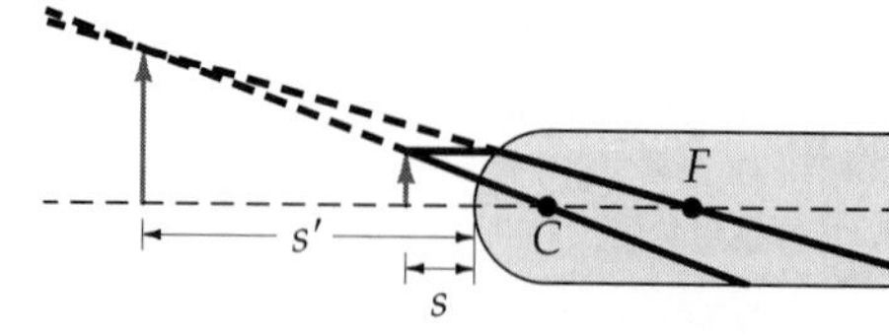

(*c*) $s' = 15$ cm, real; image is of zero size and located at F

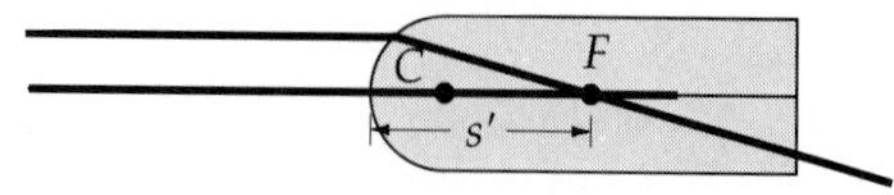

17. (*a*) $s' = -10$ cm, virtual

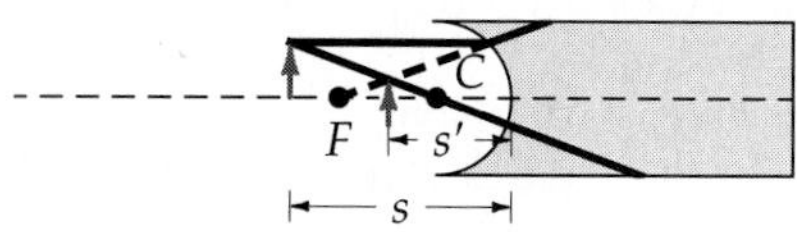

(*b*) $s' = -5$ cm, virtual; paraxial rays starting at C are undeflected, thus the image and object are identical

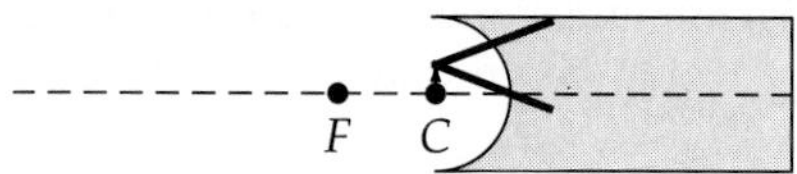

(*c*) $s' = -15$ cm, virtual; image is of zero size and located at F

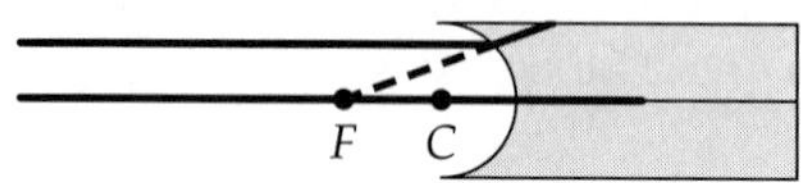

19. (*a*) $s' = -14.9$ cm, virtual

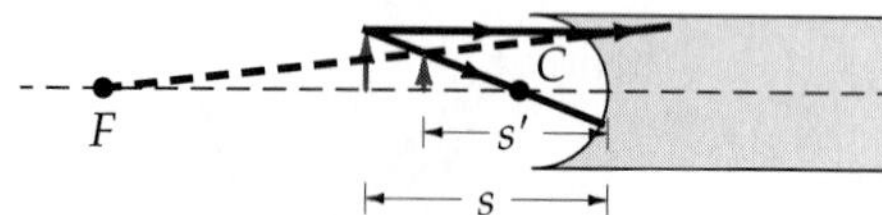

(*b*) $s' = -5$ cm, virtual; paraxial rays starting at C are undeflected, thus the image and object are identical

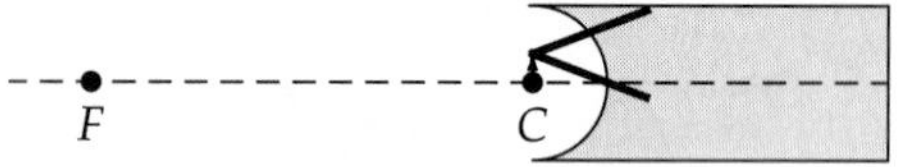

(*c*) $s' = -44.1$ cm, virtual; image is of zero size and located at F

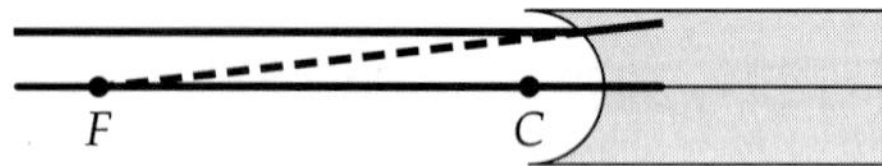

21. (*a*) −0.839 m (*b*) 0.446
23. (*a*) $r_1 = 6$ cm, $r_2 = -6$ cm

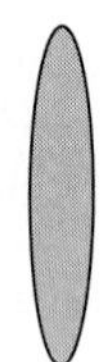

(*b*) $r_1 = -6$ cm, $r_2 = 6$ cm

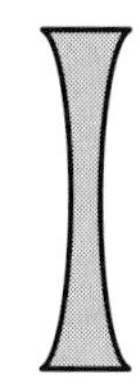

25. (*a*) −30.3 cm (*b*) −22.0 cm (*c*) 0.275 (*d*) Virtual, upright
27. (*a*) −33.3 cm

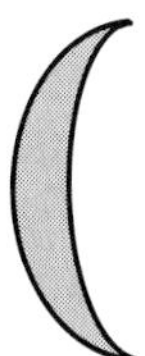

(*b*) 33.3 cm

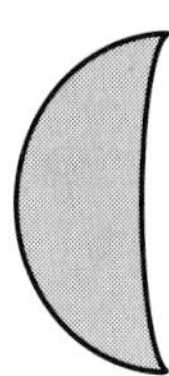

(*c*) −33.3 cm

29. (*a*) $s' = 40$ cm, $m = -1$, real, inverted (*b*) $s' =$ 20 cm, $m = 2$, real, erect (*c*) $s' = -17.1$ cm, $m = 0.429$, virtual, erect (*d*) $s' = -7.5$ cm, $m = 0.75$, virtual, erect
31. $s' = 10$ cm, $m = -1$

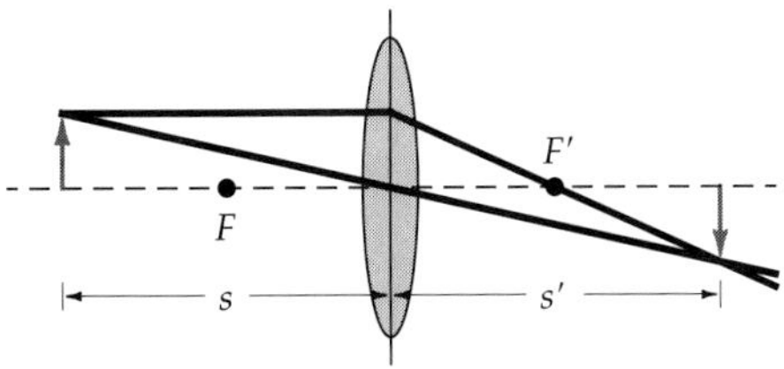

33. (*a*) $s = 5$ cm, $s' = -10$ cm

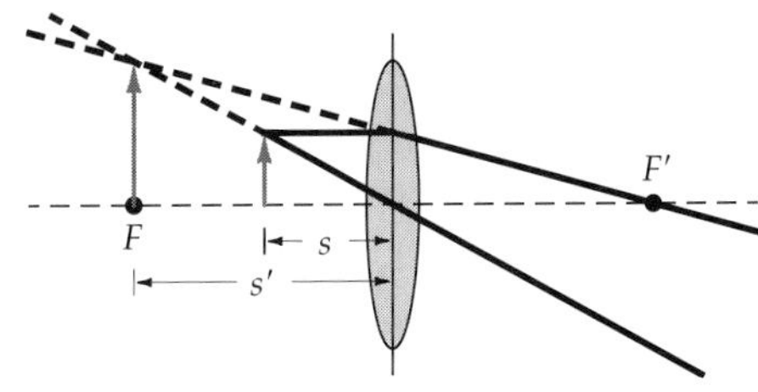

(*b*) $s = 15$ cm, $s' = 30$ cm

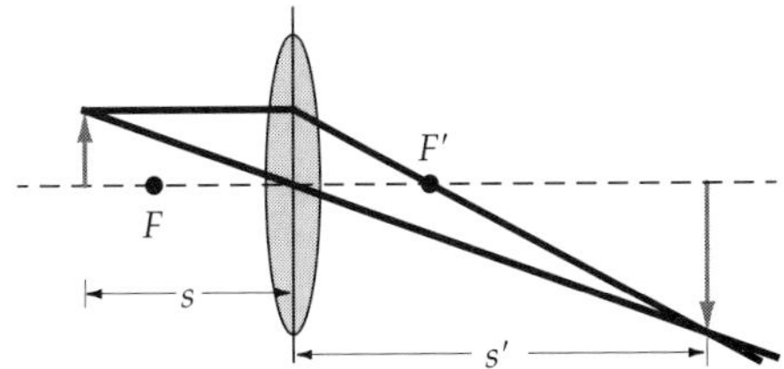

35. (*a*) 30 cm to the far side of the second lens

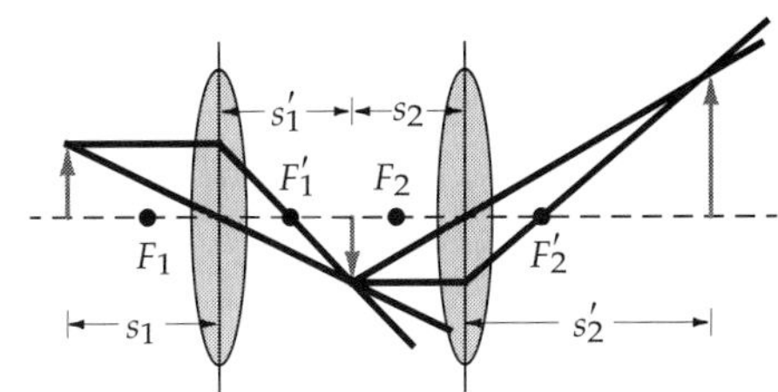

(*b*) Real, erect (*c*) 2
37. (*a*) 10.6 cm (*b*) 9.43 cm
39. (*a*) −66.7 cm (*b*) Virtual
41. The mirror should be moved 91 cm away from the object
43. Concave, $f = 90$ cm
45. (*a*) −128 cm (*b*) 14.7 cm (*c*) Real
47. (*a*) $r_2 = 35.0$ cm, concave

(*b*)

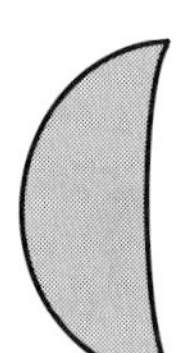

49. 4.24 cm
51. (*a*) −1.33 m (*b*) Convex
53. (*a*) 9.52 cm (*b*) −1.19

(*c*)

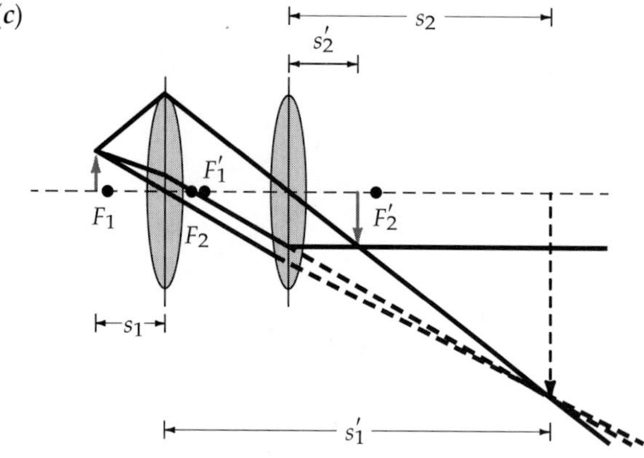

55. The final image is at the left focal point of the second lens; the image is erect and the same size as the object.

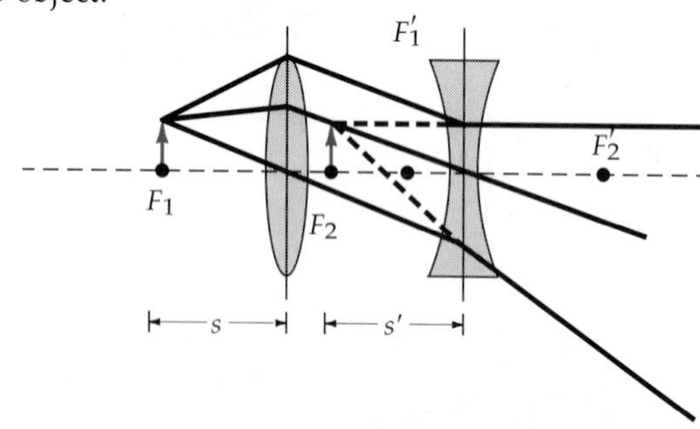

57. (*a*) 18 cm to left of lens (*b*) Real, upright

(*c*)

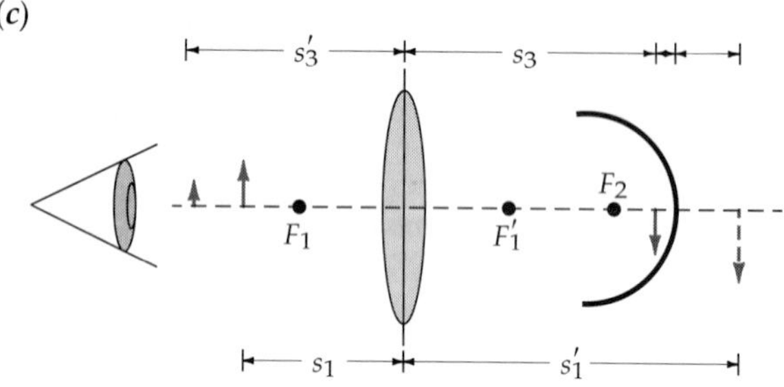

59. 200 cm or 30 cm, with 30 cm preferred
61. 36.8 cm
63. (*a*) The final image is 0.9 cm behind the back surface. (*b*) The final image is on the back surface.
65. (*a*) 1.8 m from screen (*b*) 45 cm
67. (*b*) 17.5 cm
69. answer given in problem

Chapter 32

True or False **1.** True **2.** True **3.** True **4.** True **5.** False; it varies inversely with the square of the *f*-number **6.** True **7.** True **8.** False; it is inverted and smaller than the object **9.** False; it uses a mirror for its objective

Problems

1. 0.278 cm
3. (*a*) 103 cm (*b*) 0.972 diopters
5. 44.4 cm
7. 0.714 cm; the actual radius would probably be greater
9. 6
11. 5
13. 35.7 mm
15. 1.3 mm
17. (*a*) $\approx\frac{1}{64}$ s (*b*) $\approx\frac{1}{120}$ s (*c*) $\approx\frac{1}{250}$ s (*d*) $\approx\frac{1}{500}$ s (*e*) $\approx\frac{1}{1000}$ s
19. −267
21. (*a*) 20 cm (*b*) −4 (*c*) −20 (*d*) 6.25 cm
23. (*a*) 0.9 cm (*b*) 0.18 rad (*c*) −20
25. (*a*) 25 (*b*) −134
27. (*a*) 3 (*b*) 4
29. 3.7 m
31. $f_e = 4$ cm, $f_o = 28$ cm
33. (*b*)

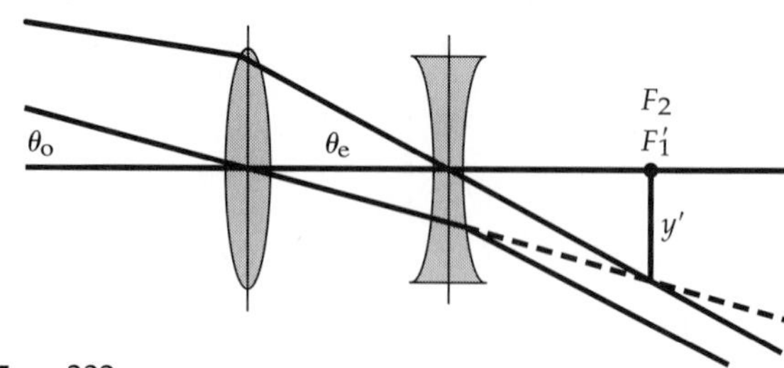

35. −232
37. 0.00667
39. (*a*) 1.67 cm (*b*) 0.508 cm in front of the objective (*c*) 0.496 cm

Chapter 33

True or False **1.** False **2.** True **3.** True **4.** True **5.** True

Problems

1. (*a*) Incoherent (*b*) Coherent (*c*) Coherent (*d*) Incoherent (*e*) Coherent
3. 164°
5. (*a*) The top of the film approaches zero thickness, so the phase difference approaches 180°. (*b*) Violet (*c*) The top of the film is white, the color of the first band is red.
7. 115 nm
9. (*a*) 7.2 μm (*b*) 1.44
11. 8.33 fringes/cm
13. (*a*) 50 μm (*b*) No (*c*) 0.5 mm
15. 695 nm
17. $E = 3.61 \sin(\omega t - 56.3°)$
19. (*a*) $A' = 2.73\ E_0$, $\delta' = 30°$

30°
A'
δ
30°

(*b*) $A' = 2E_0$, $\delta' = 60°$

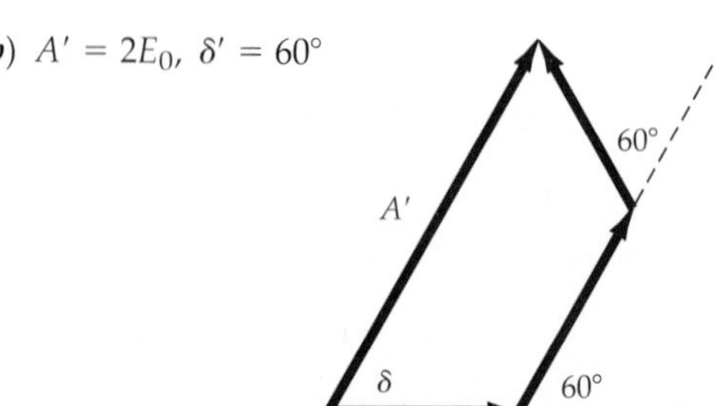

(*c*) $A' = E_0$, $\delta' = 90°$

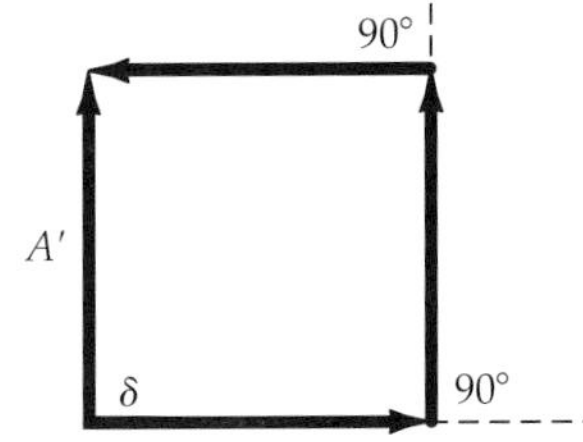

(*d*) $A' = 0$

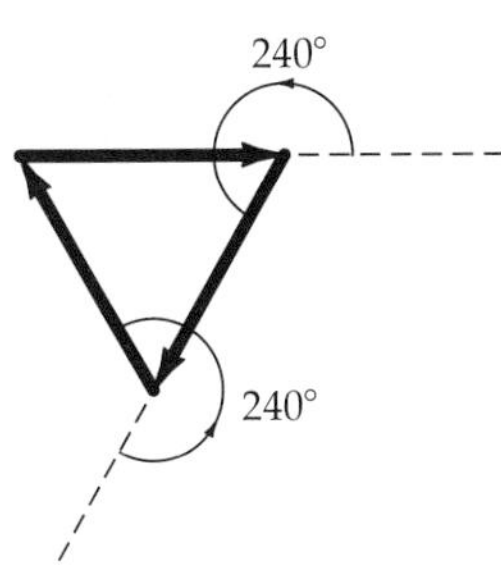

21. (*a*) $\theta_1 = \lambda/d$, $\theta_{\text{min}} = \lambda/5d$

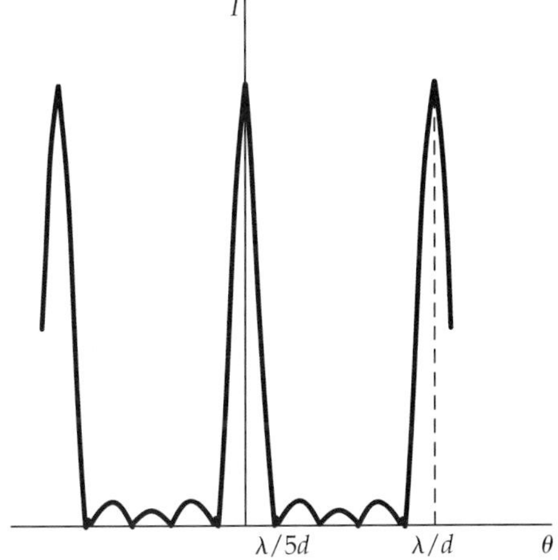

23. The separation between the slits is d, and the condition for an interference maximum is $d \sin \theta = m\lambda$; the width of an individual slit is a, and the condition for a diffraction minimum is $a \sin \theta = m\lambda$.
25. (*a*) 2 cm (*b*) 20 cm (*c*) 2.31 m
27. 3.01 cm
29. 39 fringes
31. (*a*) 8.54×10^{-3} rad (*b*) 6.83 cm
33. (*a*) 55.6 km (*b*) 55.6 m
35. 33.6 mm
37. 484 m
39. $\lambda_1 = 486$ nm, $\lambda_2 = 660$ nm
41. (*a*) 0.0231° (*b*) 0.145 cm
43. (*a*) 0.30° (*b*) 8
45. 4.54×10^6 km
47. (*a*) 0.6 μm (*b*) 400 nm, 514 nm, 720 nm (*c*) 400 nm, 514 nm, 720 nm
49. (*a*) 0.353 m, 0.707 m (*b*) 88.4 μm (*c*) 8000
51. (*a*) 97.8 nm (*b*) No (*c*) $I_{400} = 0.273 I_{\text{max}}$, $I_{700} = 0.124 I_{\text{max}}$
53. (*b*) The width of the principal interference maximum is 6 mm for four sources, 12 mm for two sources.
55. (*a*) 0.242 rad (*b*) 0.08 rad, 0.161 rad (*c*) 0.02 rad (*d*) Intensity as a function of angle is shown in the figure.

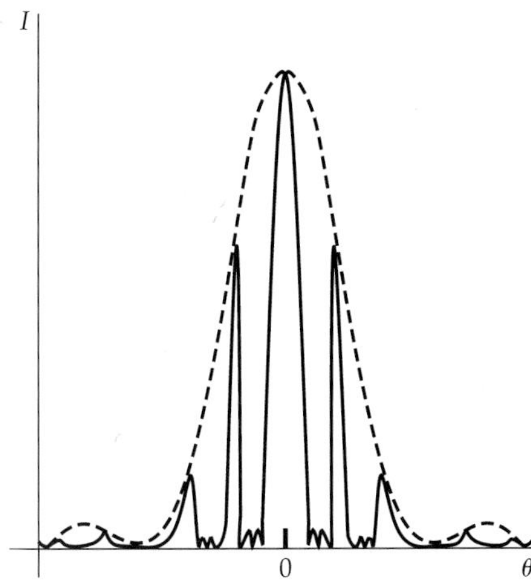

The dotted line shows the envelope produced by the single-slit diffraction pattern.

57. 20.5 m
59. 13.0°
61. answer given in problem
63. (*c*) Reversed (*d*) 68 fringes (*e*) 1.14 cm (*f*) The fringes will be closer together.
65. (*b*) 491 (*c*) 0.988 mm
67. (*a*) 3.84 μm (*b*) 1.91
69. (*a*) $I = I_{\text{max}} \cos^2 [(\pi/2) \sin \theta]$

(*b*) $I(\theta)$

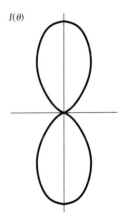

71. (*a*) $\phi = \pm 2.86\pi, \pm 4.92\pi, \pm 6.94\pi$ (*b*) Same as part (*a*)
73. answer given in problem

Chapter 34

True or False **1.** True **2.** True **3.** False **4.** True **5.** False **6.** False **7.** True

Problems

1. (*a*) 0.183 ms (*b*) 1.83×10^{-12} s (*c*) No
3. (*a*) 4.94×10^{-8} s (*b*) 12.6 m (*c*) 6.63 m
5. (*a*) 44.7 μs (*b*) 13.4 km
7. 0.527*c*
9. (*a*) 130 y (*b*) 88.1 y

11. $0.9991c$
13. 2.60×10^8 m/s
15. (*a*) 4.5×10^{-10} percent (*b*) The time elapsed on the pilot's clock is 3.15×10^7 s $- 1.42 \times 10^{-4}$ s; the time lost in minutes is 2.37×10^{-6} min.
17. 80 c·min
19. answer given in problem
21. $L_p V/c^2 = 60$ min
23. 0.0637
25. $0.6c$
27. $0.696c$
29. (*a*) $-0.882c$ (*b*) $-60{,}000$ m/s $+ 6 \times 10^{-4}$ m/s
31. $-0.994c$
33. (*a*) 1.11×10^{-17} kg (*b*) 0.351 mg
35. (*a*) 9×10^{13} J (*b*) \$2.5 million (*c*) 28,571 y
37. $E = 0.522$ MeV, $K = 1.05 \times 10^{-2}$ MeV, $p = 0.104$ MeV/c
39. (*a*) 2.23×10^8 m/s (*b*) 1039 MeV/c
41. 3.55×10^{14} reactions/s
43. 0.782 MeV
45. 50 percent
47. $0.8c$
49. 1.85×10^4 y
51. 9.61 ms
53. The required speed is $0.4c$; event B precedes event A for an observer moving with a speed $0.4c < v < c$.
55. (*a*) $0.625c$ (*b*) 31.2 y
57. (*a*) 52.7 m (*b*) $-0.987c$ (*c*) 16.1 m (*d*) 2.07×10^{-7} s (*e*)

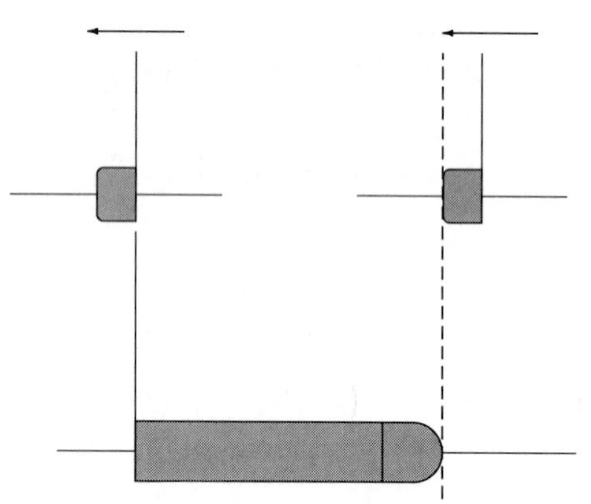

59. (*a*) 4.97 MeV/c (*b*) 0.995
61. (*a*) 0.75 percent (*b*) 68.7 percent
63. (*a*) 630 m/c (*b*) 777 m/c (*c*) 148 m/c (*d*) 778 m/c (*e*) 4.36 h (*f*) 19 h
65. (*a*) To the left with speed $0.5c$ (*b*) 1.73 y
67. 3.84×10^{14} kg/day
69. (*a*) $0.333c$ (*b*) 20 m in the $+x$ direction (*c*) 60 m/c
71. (*a*) 290 MeV (*b*) 625 MeV
73. (*a*) 1.30 m (*b*) 0.825 m
75. (*a*) E/Mc (*b*) EL/Mc^2
77. answer given in problem
79. answer given in problem
81. answer given in problem
83. answers given in problem

Chapter 35

True or False **1.** True **2.** True **3.** False **4.** True **5.** True **6.** False **7.** True **8.** True **9.** True **10.** True **11.** True **12.** False

Problems

1. $E = 6.626 \times 10^{-26}$ J $= 4.14 \times 10^{-7}$ eV
3. (*a*) 2.41×10^{14} Hz (*b*) 2.41×10^{17} Hz (*c*) 2.41×10^{20} Hz
5. $E_{400} = 3.11$ eV, $E_{700} = 1.77$ eV
7. (*a*) $f_t = 1.11 \times 10^{15}$ Hz, $\lambda_t = 271$ nm (*b*) 1.63 V (*c*) 0.39 V
9. (*a*) 4.74 eV (*b*) 2.36 V
11. (*a*) 4.97×10^{-19} J (*b*) 0.01 J (*c*) 2.01×10^{16} photons/s
13. 9.27×10^3 V
15. 1.215 pm
17. (*a*) $p = 1.66 \times 10^{-27}$ kg·m/s $= 3.11$ eV/c (*b*) $p = 3.32 \times 10^{-25}$ kg·m/s $= 621$ eV/c (*c*) $p = 6.63 \times 10^{-24}$ kg·m/s $= 12.4$ keV/c (*d*) $p = 2.21 \times 10^{-32}$ kg·m/s $= 4.14 \times 10^{-5}$ eV/c
19. (*a*) 17.5 keV (*b*) 76.0 pm (*c*) 16.3 keV
21. answer given in problem
23. $\Delta E_{3\to2} = 1.89$ eV, $\lambda_{3\to2} = 656$ nm; $\Delta E_{4\to2} = 2.55$ eV, $\lambda_{4\to2} = 486$ nm; $\Delta E_{5\to2} = 2.86$ eV, $\lambda_{5\to2} = 434$ nm
25. (*a*) $\Delta E_{\infty\to4} = 0.850$ eV, $\lambda_{\infty\to4} = 1459$ nm (*b*) $\lambda_{5\to4} = 4052$ nm, $\lambda_{6\to4} = 2627$ nm, $\lambda_{7\to4} = 2168$ nm

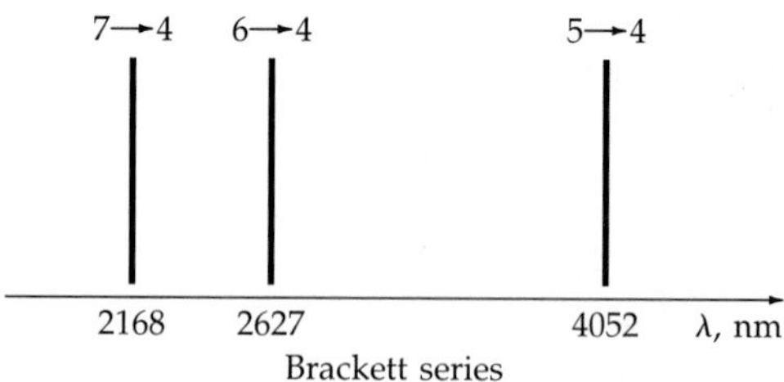

27. (*a*) 0.775 nm (*b*) 0.0775 nm (*c*) 0.0245 nm (*d*) 0.00775 nm
29. (*a*) 3.313×10^{-27} kg·m/s (*b*) 6.024×10^{-24} J
31. 0.203 nm
33. 4.40×10^{-13} m
35. 1.52×10^{-34} m
37. 22.8 eV
39. 6.31 keV
41. 6.80×10^6 m $= 4225$ mi
43. 1.69×10^{-14} W/m^2
45. (*a*) 3.18 W/m^2 (*b*) 1.04×10^{15} photons/s
47. (*a*) 4.86 pm (*b*) 92.7 keV (*c*) 92.7 keV
49. (*a*) 13.6 eV (*b*) 54.4 eV (*c*) 122.4 eV

51. (*b*) $E_n = n^2(37.6 \text{ eV})$

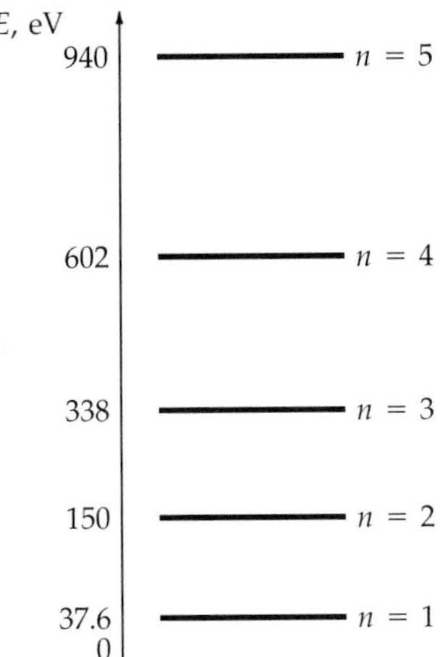

(*c*) 11.0 nm (*d*) 6.60 nm (*e*) 1.37 nm
53. (*a*) $E_1 = 5.13 \times 10^{-3}$ eV, $E_2 = 2.05 \times 10^{-2}$ eV, $E_3 = 4.61 \times 10^{-2}$ eV (*b*) 80.8 μm (*c*) 48.5 μm (*d*) 30.3 μm
55. (*b*) $R_\infty = 10.97373\ \mu\text{m}^{-1}$; $R_H = 10.96776\ \mu\text{m}^{-1}$ (*c*) 0.0545 percent
57. (*a*) 10^{-22} W (*b*) 53 min
59. (*a*) 2.42×10^{-10} m (*b*) 0.512 MeV (*c*) 2.73×10^{-22} kg·m/s (*d*) 2.42 pm
61. answer given in problem

Chapter 36

True or False **1.** False; the electron velocity equals the group velocity of the wave packet describing it **2.** False; the position can be known precisely if the momentum is completely undetermined **3.** True **4.** False; energy exchanges, such as in the photoelectric effect, require a particle model **5.** False **6.** True **7.** False; the expectation value is the *average* of the measurement for a large number of identical systems. It may, or may not, be one of the possible values measured **8.** False; α decay is a physical example of the penetration of a barrier by a wave **9.** True

Problems

1. (*a*) 0 (*b*) 0.00417
3. (*a*) $|A|^2\,dx$ (*b*) $|A|^2e^{-1/2}\,dx$ (*c*) $|A|^2e^{-2}\,dx$ (*d*) The electron is most likely to be found in a region centered at $x = 0$.
5. $\Delta x\,\Delta k \approx 4\pi$
7. 2.64×10^{-19} m
9. 5.79 m
11. (*a*) $f_0 = N/\Delta t$ (*b*) $\Delta k \approx 1/\Delta x = (v\,\Delta t)^{-1}$
13. 1.66×10^{-33} m
15. $\lambda = 8.68 \times 10^{-11}$ m, which is roughly the size of an atom
17. answer given in problem
19. (*a*) $K(x) = (\hbar^2/2mL^2)(1 - x^2/L^2)$
21. (*a*) $2L$ (*b*) $2L/3$ (*c*) $h/2L$ (*d*) $p^2/2m = h^2/8mL^2$
23. (*a*) 0.0205 eV (*b*) 205 MeV
25. (*a*) 0 (*b*) 0.003 (*c*) 0
27. 3.02×10^{19}
29. (*a*)

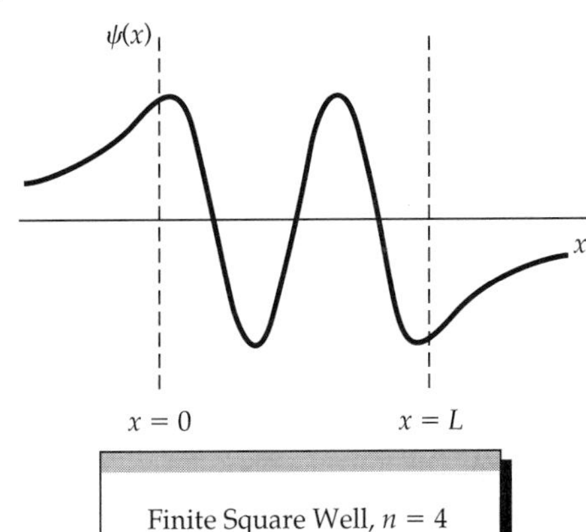

(*b*)

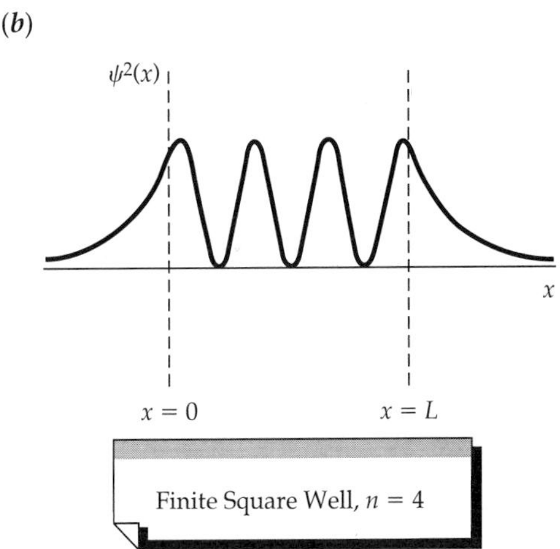

31. (*a*) $L/2$ (*b*) $0.321L^2$
33. (*a*) answer given in problem (*b*) $< x > = L/2$, $< x^2 > = L^2/3$
35. (*a*) $k_2 = (6mU_0)^{1/2}/\hbar = \sqrt{3/2}k_1$ (*b*) 0.0102 (*c*) 0.990 (*d*) 9.90×10^5 particles are expected to continue past the potential step. Classically, 100 percent of the particles are transmitted.
37.

n_1	n_2	n_3	$E/(h^2/288mL_1^2)$
1	1	1	49
1	1	2	61
1	2	1	76
1	1	3	81
1	2	2	88
1	2	3	108
1	1	4	109
1	3	1	121
1	3	2	133
1	2	4	136

39. *(a)*

n_1	n_2	n_3	$E/(h^2/128mL_1^2)$
1	1	1	21
1	1	2	24
1	1	3	29
1	2	1	33
1	1	4	36
1	2	2	36
1	2	3	41
1	1	5	45
1	2	4	48
1	3	1	53
1	1	6	56
1	3	2	56

(b) $n_1 = 1$, $n_2 = 1$, $n_3 = 4$ and $n_1 = 1$, $n_2 = 2$, $n_3 = 2$ both give 36; $n_1 = 1$, $n_2 = 1$, $n_3 = 6$ and $n_1 = 1$, $n_2 = 3$, $n_3 = 2$ both give 56
41. (a) $\psi(x, y, z) = A \cos(\pi x/L) \sin(\pi y/L) \sin(\pi z/L)$
(b) The energy levels are the same.
43. $10E_1$
45. *(a)* 205 MeV *(b)*

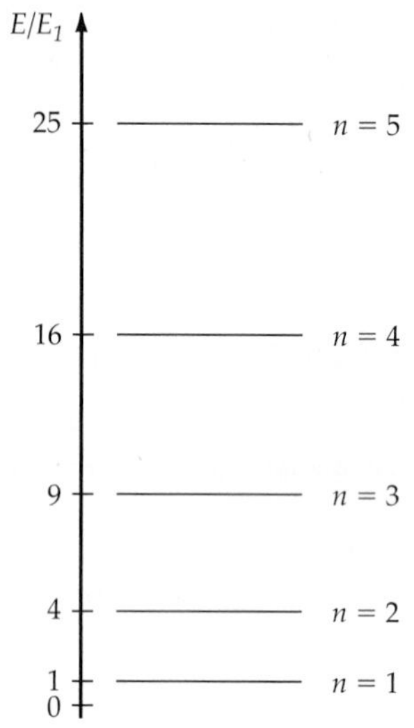

(c) 2.02×10^{-15} m *(d)* 1.21×10^{-15} m *(e)* 7.57×10^{-16} m
47. *(a)* 0.5 *(b)* 0.196 *(c)* 0.909
49. *(a)*

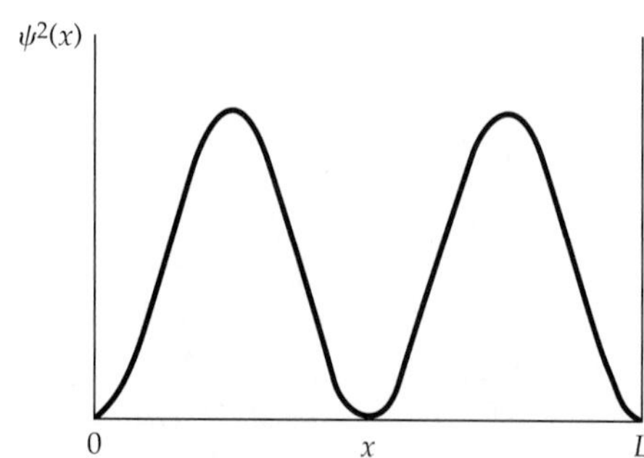

(b) $L/2$ *(c)* 0 *(d)* There is no contradiction.
51. *(a)* 0.5 *(b)* 0.402 *(c)* 0.75
53. (a) $T \approx 10^{-18}$ *(b)* $T \approx 10^{-2}$
55. $0.289L$
57. $T \approx 10^{-1}$
59. *(a)* 1.66×10^{-26} m/s *(b)* $T = 1.21 \times 10^{21}$ s = 3.82×10^{13} y
61. $< x > = 0$, $< x^2 > = L^2(1/12 - 1/2\pi^2)$
63. $< x > = 0$, $< x^2 > = L^2(1/12 - 1/8\pi^2)$
65. $E = 65h^2/8mL^2$ is four-fold degenerate, with quantum numbers $n = 7$, $m = 4$; $n = 4$, $m = 7$; $n = 8$, $m = 1$; $n = 1$, $m = 8$.
67. answer given in problem
69. answer given in problem
71. *(b)* $T = 0.823$, $R = 0.177$ *(c)* $T = 0.971$, $R = 0.0294$ *(d)* $T = 0.9993$, $R = 6.93 \times 10^{-4}$
73. answers given in problem

Chapter 37

True or False **1.** True **2.** True **3.** True **4.** True

Problems

1. *(a)* $\sqrt{2}\hbar$ *(b)* $-1, 0, +1$ *(c)*

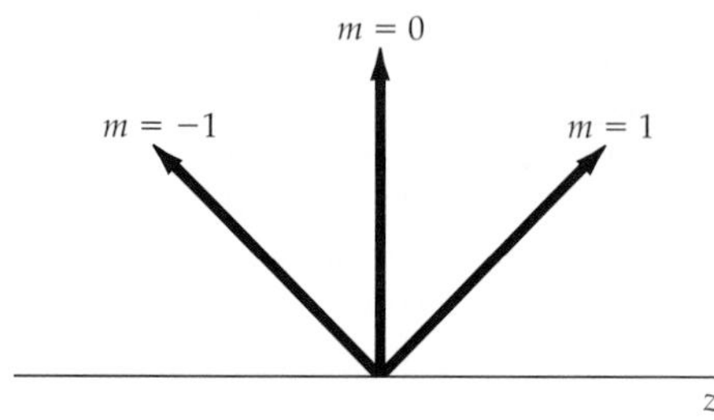

3. *(a)* $\ell = 0, 1, 2$ *(b)* For $\ell = 0$, $m = 0$; for $\ell = 1$, $m = -1, 0, 1$; for $\ell = 2$, $m = -2, -1, 0, 1, 2$ *(c)* 18
5. *(a)* 3.49×10^{-3} kg · m²/s *(b)* 3.31×10^{31}
7. *(a)* $n \geq 4$, $m = -3, -2, -1, 0, 1, 2, 3$ *(b)* $n \geq 5$, $m = -4, -3, -2, -1, 0, 1, 2, 3, 4$ *(c)* $n \geq 1$, $m = 0$
9. *(a)* $\psi = (1/\sqrt{\pi})(1/a_0)^{3/2}e^{-1}$ *(b)* $\psi^2 = (1/\pi a_0^3)e^{-2}$ *(c)* $P(r) = (4/a_0)e^{-2}$
11. *(a)* 0.0162 *(b)* 0.00879
13. *(a)* $\psi = (1/4\sqrt{2\pi})(1/a_0)^{3/2}(2 - Z)e^{-Z/2}$ *(b)* $\psi^2 = (1/32\pi a_0^3)(2 - Z)^2e^{-Z}$ *(c)* $P(r) = (1/8a_0)(2 - Z)^2e^{-Z}$
15. *(a)* 6.95×10^{-5} eV *(b)* 1.79 cm
17. 3
19. *(a)* 5.05×10^{-27} J/T *(b)* 3.15×10^{-12} eV/G
21. $\ell = 1$ or $\ell = 2$
23. *(a)* $j = 1$ or $j = 0$ *(b)* $\sqrt{2}\hbar$
(c)

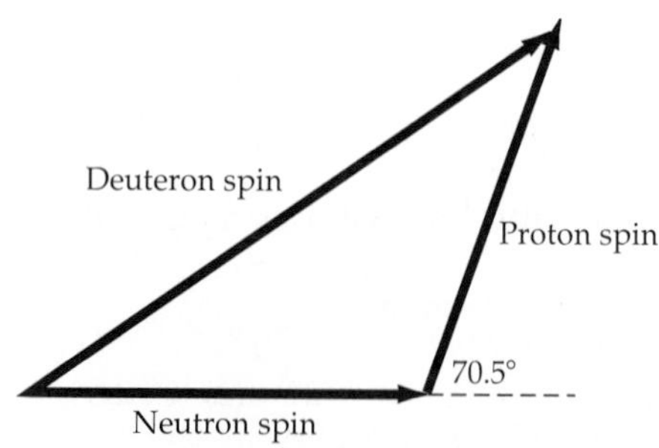

25. *(a)* $1s^22s^22p^2$ *(b)* $1s^22s^22p^4$
27. *(a)* Silicon *(b)* Calcium

29. Both gallium and indium have a single p-shell electron shielded from the nucleus by the inner closed shells, making it relatively easy to remove.
31. 1.59
33. Potassium, chromium and copper each have a single 4s electron as the outer electron; calcium, titanium, and manganese each have two 4s electrons as the outer electrons.
35. (*a*) 2s or 2p (*b*) $1s^2 2s^2 2p^6 3p$ (*c*) 1s2s
37. (*a*) 0.0611 nm, 0.0580 nm (*b*) 0.0543 nm
39. Zirconium
41. (*a*) 1.01 nm (*b*) 0.155 nm
43. (*a*) 15 mJ (*b*) 5.24×10^{16}
45. 2.81 km
47. (*a*) 3p (*b*) $K = 6.4$ eV
49. (*a*) 0.305 nm (*b*) 0.715 nm
51. Tin
53. (*a*) $E = 2.10505$ eV and $E = 2.10291$ eV (*b*) $\Delta E = 2.14 \times 10^{-3}$ eV (*c*) $B = 18.5$ T
55. (*a*) $E = 1.61774$ eV and $E = 1.61041$ eV (*b*) $\Delta E = 7.33 \times 10^{-3}$ eV (*c*) $B = 63.3$ T
57. (*a*) 3.32×10^6 m/s^2 (*b*) $\Delta z = (a\, \Delta x_1/v_0^2)(\Delta x_1 + 2\, \Delta x_2)$, with $\Delta x_1 = 0.75$ m and $\Delta x_2 = 1.25$ m
59. (*a*) 2, 1, 0 (*b*) 1, 0 (*c*)

ℓ	s	j
2	1	3, 2, 1
2	0	2
1	1	2, 1, 0
1	0	1
0	1	1
0	0	0

(*d*) 3/2, 1/2 (*e*)

j_1	j_2	j
3/2	3/2	3, 2, 1, 0
3/2	1/2	2, 1
1/2	3/2	2, 1
1/2	1/2	1, 0

The same values of j occur here as in part (*c*).
61. $P = 9.01 \times 10^{-15}$
63. answer given in problem
65. answer given in problem
67. $r = (3 \pm \sqrt{5})a_0/Z$
69. 0.323

Chapter 38

True or False 1. False **2.** False; it is due to exclusion-principle repulsion **3.** False; the dissociation energy is the energy to separate the ions *and* form neutral atoms from the ions. For NaCl, the dissociation energy is 1.53 eV less than the energy needed to remove the ions to an infinite separation **4.** False; it is very common **5.** False; they are much less than the energy of electronic excitation

Problems
1. 0.940 nm
3. (*a*) Ionic bonding (*b*) Covalent bonding (*c*) Metallic bonding
5. (*a*) 1 eV/molecule = 23.0 kcal/mole (*b*) 98.1 kcal/mole
7. 43.6 percent
9. 2.63×10^{-29} Cm
11. 0.110 nm
13. answer given in problem
15. 0.121 nm
17. (*a*) −5.39 eV (*b*) 4.66 eV (*c*) 0.230 eV
19. (*a*) $U = -6.09$ eV. From the figure the dissociation energy is 5.79 eV. (*b*) 0.305 eV
21. (*a*) −6.64 eV (*b*) 5.75 eV (*c*) 0.680 eV
23. 1.25×10^{14} Hz
25. 1.58×10^{-5} eV
27. 477 N/m
29. The Lenard-Jones potential appropriate for an H_2 molecule is shown in the figure as curve (1).

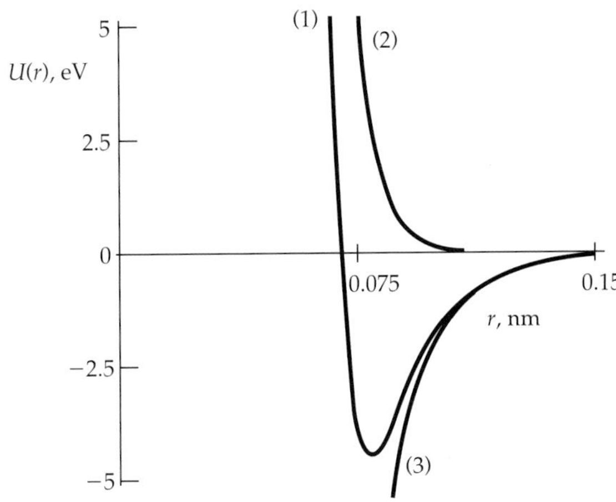

Curve (2) shows the $U_0(a/r)^{12}$ term, and curve (3) shows the $-2U_0(a/r)^6$ term.
31. $F \sim x^{-4}$
33. (*a*) $I = 1.45 \times 10^{-46}$ kg · m^2. $E_{0r} = 0.000239$ eV (*b*) Energy levels for $\ell = 0$ to $\ell = 5$ are shown in the figure. Choosing $E = 0$ for the $\ell = 0$ state yields $E = \ell(\ell + 1)E_{0r}$ for general ℓ.

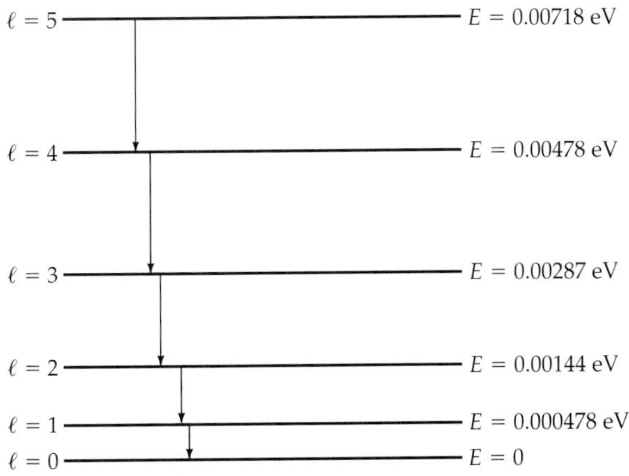

(*c*) $\ell = 1 \rightarrow \ell = 0$, $E = 0.000478$ eV; $\ell = 2 \rightarrow \ell = 1$, $E = 0.000956$ eV; $\ell = 3 \rightarrow \ell = 2$, $E = 0.00144$ eV; $\ell = 4 \rightarrow \ell = 3$, $E = 0.00191$ eV; $\ell = 5 \rightarrow \ell = 4$, $E = 0.00239$ eV. (*d*) $\ell = 1 \rightarrow \ell = 0$, $\lambda = 2.60$ mm; $\ell = 2 \rightarrow \ell = 1$, $\lambda = 1.30$ mm; $\ell = 3 \rightarrow \ell = 2$, $\lambda = 0.865$ mm; $\ell = 4 \rightarrow \ell = 3$, $\lambda = 0.649$ mm; $\ell = 5 \rightarrow \ell = 4$, $\lambda = 0.519$ mm. These photons fall in the microwave and short radio wave portion of the electromagnetic spectrum.

35. (*a*) $\mu = 0.9722$ u for $H^{35}Cl$, and $\mu = 0.9737$ for $H^{37}Cl$; $\Delta\mu/\mu = 0.00153$
(*b*) Answer given in problem
(*c*) $\Delta f/f = -0.00153$; According to Figure 38-17, the major peak is at approximately $f_{major} = 9.21 \times 10^{15}$ Hz. The associated minor peak should therefore be at $f_{minor} = f_{major} + \Delta f = 9.21 \times 10^{15}$ Hz $- 0.00153(9.21 \times 10^{15}$ Hz$) = 9.196 \times 10^{15}$ Hz, which is in agreement with the location of the minor peak in the figure.

Chapter 39

True or False **1.** True **2.** False; the heat capacity due to the free electrons is much less than that predicted by the classical theory **3.** True **4.** False **5.** True **6.** True **7.** False

Problems

1. (*a*) The unit cell is a cube with sides of length $2R$. (*b*) 0.524
3. 2.07 g/cm^3
5. (*b*) 0.141 nm
7. (*a*) 5.86×10^{22} electrons/cm^3 (*b*) 5.90×10^{22} electrons/cm^3
9. 3.00
11. 4.00
13. (*a*) 1.36×10^5 K (*b*) 2.46×10^4 K (*c*) 1.18×10^5 K
15. (*a*) 11.2 eV (*b*) 1.30×10^5 K
17. (*a*) Silver and gold (*b*) 0.1 V
19. At $T = 0$, $U/N = 4.2240$ eV; at $T = 300$ K, $U/N = 4.2242$ eV. The classical prediction is $U/N = 0.0388$ eV
21. (*a*) 3.42×10^{-8} m (*b*) 4.11×10^{-8} m (*c*) 4.29×10^{-9} m
23. 1.09×10^{-6} m
25. 1.77×10^{-7} m
27. (*a*) *p*-type semiconductor (*b*) *n*-type semiconductor
29. 116 K
31. (*a*) 0.0596 V (*b*) −0.0596 V
33. (*a*) 2.17×10^{-3} eV (*b*) 0.455 mm
35. (*a*) 3.67×10^{-3} (*b*) 1.22×10^{-2}
37. 9.46 eV
39. (*a*) 9.48×10^{25} atoms (*b*) 7.58×10^{26} states (*c*) 9.29×10^{-27} eV/state (*d*) The thermal energy is 0.0259 eV, which is 2.79×10^{24} greater than the average spacing between conduction states.
41. 2.19 eV
43. 250
45. (*b*) 500 MΩ (*c*) $R = 1.03\ \Omega$, $I = 0.485$ A (*d*) 0.0515 Ω
47. (*a*) 0.0189 eV (*b*) 0.00531 eV (*c*) 48.2 percent
49. $P = 3.82 \times 10^{10}$ N/m^2 $= 3.78 \times 10^5$ atm
51. (*a*) 66.1 nm (*b*) 2.27×10^{-20} m^2
53. (*a*) 1.28×10^{13} Hz (*b*) $\lambda = 23.4\ \mu$m, which is of the same order of magnitude as the absorption bands at 61 μm
55. (*a*) $v_d = -eE\tau/m$ (*b*) $\rho = m/ne^2\tau$

Chapter 40

True or False **1.** False; it does not contain electrons **2.** True **3.** False; after 2 half-lives, $\frac{3}{4}$ of the nuclei in a sample have decayed and $\frac{1}{4}$ remain **4.** True **5.** True **6.** False; it is a measure of the energy deposited per unit mass

Problems

1. (*a*) ^{15}N, ^{13}N (*b*) ^{57}Fe, ^{58}Fe (*c*) ^{117}Sn, ^{119}Sn
3. (*a*) 31.99 MeV, 5.332 MeV/nucleon (*b*) 333.7 MeV, 8.557 MeV/nucleon (*c*) 1636 MeV, 7.868 MeV/nucleon
5. (*a*) 20.58 MeV (*b*) 7.253 MeV
7. answer given in problem
9. (*a*) 5 s (*b*) 250 counts/s
11.

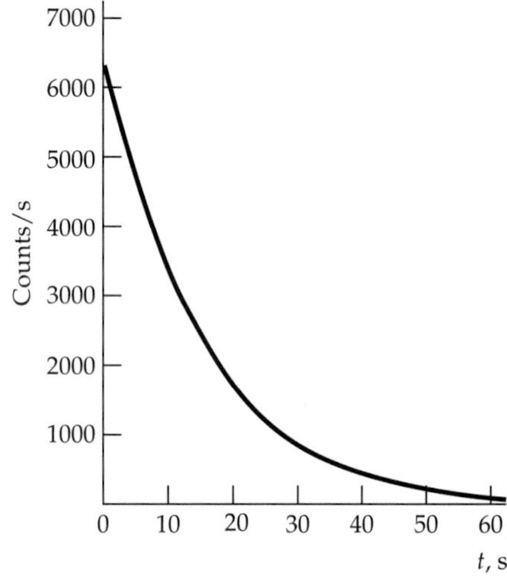

$\lambda = 0.0693$ s^{-1}
13. 3.61×10^{10} decays/s, which is approximately 1 Ci $= 3.7 \times 10^{10}$ decays/s
15. (*a*) ^{22}Na can decay by β^+ emission to become ^{22}Ne. (*b*) ^{24}Na can decay by emission of a β^- to become ^{24}Mg.
17. 3352 y
19. 13,940 y
21. (*a*) −0.7638 MeV (*b*) 3.269 MeV
23. 1.56×10^{19} fissions/s
25. 1.16×10^8 K
27. (*a*) 0.0581 cm (*b*) 0.00518 cm
29. 2.72 barns
31. 21.0 barns
33. (*a*) 22.96 MeV (*b*) 4.178 MeV (*c*) 1.286 MeV
35. $\lambda = 1.05 \times 10^{-23}$ s^{-1}, $\tau_{1/2} = 2.10 \times 10^{15}$ y
37. $\lambda = 4.16 \times 10^{-9}$ s^{-1}, $\tau_{1/2} = 5.27$ y
39. (*a*) 156 h (*b*) 551 h
41. (*a*) 1.20 MeV (*b*) The mass of ^{13}C does not include the mass of the seventh electron on the atom and the mass of the emitted β^+.
43. (*a*) For ^{141}Ba, $R = 7.81$ fm; for ^{92}Kr, $R = 6.77$ fm. (*b*) 200 MeV
45. 7.84×10^7 AU
47. (*a*) 3.42×10^{12} neutrons/s (*b*) 493 rem/h (*c*) 1.01 h
49. 7.03×10^8 y
51. (*a*) 2.47×10^5 atoms (*b*) 26.6 mrem (*c*) 1.177
53. $E_n = 14.1$ MeV, $E_{He} = 3.56$ MeV
55. (*a*) 1.19 MeV/*c* (*b*) 0.752 keV (*c*) 0.0962 percent
57. (*b*) 55

59. (*c*) The rate of proton consumption is 3.74×10^{38} protons/s. At this rate the sun will last 50.5×10^9 y.
61. 1.77×10^{-3} rem

Chapter 41

True or False **1.** True **2.** True **3.** False; hadrons include both baryons and mesons **4.** False **5.** False **6.** True **7.** False; the electron interacts with the proton via the weak interaction (as well as the electromagnetic interaction and the gravitational interaction) **8.** True; strangeness can change by ±1 in weak interactions **9.** True

Problems

1. 0.397 fm
3. (*a*) 279.2 MeV (*b*) 1878 MeV (*c*) 212 MeV
5. (*a*) $\Delta S = +1$, the reaction can proceed via the weak interaction (*b*) $\Delta S = +2$, the reaction is not allowed (*c*) $\Delta S = +1$, the reaction can proceed via the weak interaction
7. (*a*) $\Delta S = +2$, the reaction is not allowed (*b*) $\Delta S = +1$, the reaction can proceed via the weak interaction
9.

	Quark structure	Baryon number	Charge, *e*	Strangeness	Hadron identity
(*a*)	*uud*	+1	+1	0	p^+
(*b*)	*udd*	+1	0	0	n
(*c*)	*uus*	+1	+1	−1	Σ^+
(*d*)	*dds*	+1	−1	−1	Σ^-
(*e*)	*uss*	+1	0	−2	Ξ^0
(*f*)	*dss*	+1	−1	−2	Ξ^-

11. *uuu*
13. (*a*) $c\bar{d}$ (*b*) $\bar{c}d$
15. 2.44×10^{-18} m
17. (*a*) It is necessary that its charge, lepton number, baryon number, strangeness, charm, topness, and bottomness all be zero. (*b*) π^0 (*c*) Ξ^0
19. (*a*) $\bar{u}\bar{d}\bar{d}$ (*b*) *uss* (*c*) *uus*
21. (*a*) Charge −1*e*, baryon number 1, and strangeness 0 (*b*) Charge 0, baryon number 0, strangeness 0, and charm −1 (*c*) Charge +1*e*, baryon number 0, strangeness 0, and bottomness −1 (*d*) charge +1*e*, baryon number−1, and strangeness 3
23. (*a*) The final products shown are not all stable. The neutron decays to yield $p + e + \bar{\nu}_e$. (*b*) $\Omega^- \rightarrow p + 3e + e^+ + 3\bar{\nu}_e + \nu_e + 2\bar{\nu}_\mu + 2\nu_\mu$ (*c*) All quantities are conserved except for strangeness, which changes by + 3
25. (*a*) All conservation laws satisfied (*b*) Energy and baryon number conservation are violated (*c*) All conservation laws satisfied
27. (*a*) 1193 MeV (*b*) 77 MeV/*c* (*c*) 2.66 MeV (*d*) $E = 74.3$ MeV, $p = 74.3$ MeV/*c*

Chapter 42

True or False **1.** True **2.** False **3.** False; since the lifetime varies as $M_\odot^{-3}$, it will be 1/125 the sun's lifetime **4.** True; it occurred about 170,000 years ago **5.** True **6.** False **7.** False **8.** True

Problems

1. 25.3 d
3. The time for one orbit is 2.26×10^8 y. The sun has completed about 44 orbits.
5. (*a*) 3.26 $c \cdot$ y (*b*) 3.35×10^5 stars
7. (*a*) $T_e \approx 3300$ K, $L \approx 5 \times 10^{-2}\ L_\odot$ (*b*) $T_e \approx 13{,}500$ K, $L \approx 10^2\ L_\odot$ (*c*) $R_{0.3} = 0.3R_\odot$, $R_3 = 3R_\odot$ (*d*) For $M = 0.3M_\odot$, $t_L = 37t_{L\odot}$; for $M = 3M_\odot$; $t_L = 0.037t_{L\odot}$
9. (*a*) 9360 y (*b*) $1.86R_\odot$
11. (*a*) 10.1 km (*b*) 8.01×10^{38} J (*c*) 1.85×10^{26} W
13. (*a*) 3.13×10^9 $c \cdot$ y (*b*) 1.30×10^{10} y (*c*) 10 percent
15. 1.06×10^{-26} kg/m^3
17. Planck density = 5.5×10^{97} kg/m^3, proton density = 1.67×10^{18} kg/m^3, and osmium density = 2.45×10^4 kg/m^3.
19. The muon threshold was at roughly 10^{-3} s after the Big Bang. At the current temperature, the maximum rest mass that could be created is of order of magnitude 10^{-40} kg.
21. (*a*) 170,000 y (*b*) 7.34 y after the light, in 1994
23. (*a*) $T = 12$ d, $\omega = 6.06 \times 10^{-6}$ rad/s (*c*) $m_1 = 4.45 \times 10^{31}$ kg, $m_2 = 2.23 \times 10^{31}$ kg, $r = 4.95 \times 10^{10}$ m
25. 9.68×10^{-4} H atoms/m^3. It is unlikely this rate would be observable.
27. The earth's new temperature would be 949 K, enough to boil water but not to remove it from the atmosphere.
29. 26.42 MeV needed to form ^{112}Cd, 462 MeV released in forming ^{56}Fe

Index

Some Conversion Factors

1 m = 39.37 in = 3.281 ft = 1.094 yd

1 m = 10^{15} fm = 10^{10} Å = 10^9 nm

1 km = 0.6215 mi

1 mi = 5280 ft = 1.609 km

1 lightyear = 1 c·y = 9.461×10^{15} m

1 in = 2.540 cm

1 L = 10^3 cm^3 = 10^{-3} m^3 = 1.057 qt

1 h = 3.6 ks

1 y = 365.24 d = 3.156×10^7 s

1 km/h = 0.278 m/s = 0.6215 mi/h

1 ft/s = 0.3048 m/s = 0.6818 mi/h

1 rev = 2π rad = 360°

1 rad = 57.30°

1 rev/min = 0.1047 rad/s

1 slug = 14.59 kg

1 tonne = 10^3 kg = 1 Mg

1 atm = 101.3 kPa = 1.013 bar = 76.00 cmHg = 14.70 lb/in^2

1 N = 10^5 dyn = 0.2248 lb

1 lb = 4.448 N

1 Pa·s = 10 poise

1 J = 10^7 erg = 0.7373 ft·lb = 9.869×10^{-3} L·atm

1 kW·h = 3.6 MJ

1 cal = 4.184 J = 4.129×10^{-2} L·atm

1 L·atm = 101.3 J = 24.22 cal

1 eV = 1.602×10^{-19} J

1 Btu = 778 ft·lb = 252 cal = 1054 J

1 horsepower = 550 ft·lb/s = 746 W

1 W/m·K = 6.938 Btu·in/h·ft^2·°F

1 T = 10^4 G

1 kg weighs about 2.205 lb